HANDBOOK OF PHYSIOLOGY

Section 14: Cell Physiology

HANDBOOK OF PHYSIOLOGY

A critical, comprehensive presentation of
physiological knowledge and concepts

Section 14: Cell Physiology

Edited by

JOSEPH F. HOFFMAN

Department of Cellular and Molecular Physiology
Yale University School of Medicine

James D. Jamieson

Department of Cell Biology
Yale University School of Medicine

New York   Oxford
Published for the American Physiological Society
by Oxford University Press
1997

Oxford University Press

Oxford   New York
Athens   Auckland   Bangkok   Bogota
Bombay   Buenos Aires   Calcutta   Cape Town
Dar es Salaam   Delhi   Florence   Hong Kong   Istanbul
Karachi   Kuala Lumpur   Madras   Madrid
Melbourne   Mexico City   Nairobi   Paris
Singapore   Taipei   Tokyo   Toronto

and associated companies in
Berlin   Ibadan

Copyright © 1997 by Oxford University Press, Inc.

Published by Oxford University Press, Inc.
198 Madison Avenue, New York, New York 10016

Oxford is a registered trademark of Oxford University Press

Library of Congress Cataloging-in-Publication Data

Cell physiology / edited by Joseph F. Hoffman, James D. Jamieson.
      p.   cm. — (Handbook of physiology ; section 14)
      Includes bibliographical references and index.
      ISBN 0-19-507172-7
      1. Cell physiology. I. Hoffman, Joseph F. II. Jamieson, James
D. III. American Physiological Society (1887–   ) IV. Series:
Handbook of physiology (Bethesda, Md.) ; section 14.
      [DNLM: 1. Cells—physiology.   QT 104 H2361 1977 sect.14]
QP6.H25   1977 sect. 14
[QH631]
571 s—dc21
[571.6]
DNLM/DLC
for Library of Congress   97-4219
                                                                 CIP

9 8 7 6 5 4 3 2 1

Printed in the United States of America
on acid-free paper

Preface

The Handbook of Cell Physiology represents a new undertaking of the American Physiological Society (APS) in recognition of the continuity of biomedical science that encompasses the reductive as well as the integrative aspects of our science. In focusing broadly on cellular and molecular areas of research, the book provides a framework for applying this information to the physiology of organs and organisms.

It may interest readers to know how this *Handbook* came about. The idea was first suggested to the APS by members of the Society of General Physiologists who happened also to be APS members. Because the APS has always considered cellular and general physiology part of its just domain, their response was positive and enthusiastic. As a result, we hope that the Handbook series is more complete by the inclusion of the present volume.

Inevitably, some areas of cellular physiology are not covered here. This particularly applies to excitable tissues, such as nerve and muscle. Most of those areas have been or will be covered in other Handbook of Physiology sections. Because of the blurring of boundaries between different fields of physiology, topics that might have been separate are being folded into broader frames of reference. In general, more cellular and molecular information is now included in integrative physiology. Nevertheless, it should be clear that the topics covered in this volume represent most of the key areas of the field.

In line with the general mission of the Handbook of Physiology, the chapters provide authoritative reviews of their topics. They begin with basic membrane processes that include membrane structure as it relates to function, the biophysics of membrane transport, and cell volume regulatory mechanisms. Then a group of chapters reviews in depth the organization and function of the plasma membrane and the intracellular organelles involved in membrane trafficking and the biogenesis of cell polarity. Energy generation and trandsduction that subserve cellular function are covered in another series of chapters. Critical reviews of the structure and function of the cytoskeleton and the relationship of the cytoskeleton to events regulated by interaction with the extracellular matrix are presented. Chapters on integrative aspects of cellular physiology including immunobiology, cell–cell interactions, fertilization, and interactions of cells with the extracellular matrix round out our survey of contemporary cellular physiology.

We hope that this volume will provide insights for any reader interested in understanding how cells work and in pursuing the elusive goal, expounded by Claude Bernard, of searching for the "elemental condition for the phenomenon of life."

It is a genuine pleasure to acknowledge the unfailing assistance of Ms. Leah Sanders in all phases of bringing this volume to fruition. We are also grateful to the Publication Committee of the American Physiological Society, as well as Oxford University Press, for their encouragement, patience, and support.

New Haven, Conn.   J.F.H.
December 1996   J.D.J.

Contents

Contributors

Merton Bernfield, M.D.
Department of Pediatrics
Harvard Medical School
Boston, Massachusetts

R. Blumenthal, Ph.D.
Section on Membrane Structure and Function
National Cancer Institute
National Institutes of Health
Bethesda, Maryland

Michael J. Caplan, M.D., Ph.D.
Department of Physiology
Yale University School of Medicine
New Haven, Connecticut

Carol D. Cianci, M.S.
Department of Pathology
Yale University School of Medicine
New Haven, Connecticut

David E. Clapham, M.D., Ph.D.
Department of Pharmacology
Mayo Foundation
Rochester, Minnesota

B. Crabtree, D. Phil.†
Rowelt Research Institute
Aberdeen, Scotland

D. S. Dimitrov, Ph.D.
Section on Membrane Structure and Function
National Cancer Institute
National Institutes of Health
Bethesda, Maryland

David Epel, Ph.D.
Department of Biological Sciences
Hopkins Marine Station
Stanford University
Pacific Grove, California

Fredric S. Fay, Ph.D.
Biomedical Imaging Group
Department of Physiology
University of Massachusetts Medical School
Worcester, Massachusetts

Harvey M. Florman, Ph.D.
Department of Anatomy and Cellular Biology
Tufts University School of Medicine
Boston, Massachusetts

Ephraim Fuchs, M.D.
Johns Hopkins Oncology Center
Baltimore, Maryland

Hans-Peter Hauri, Ph.D.
Department of Pharmacology
Biozentrum, University of Basel
Basel, Switzerland

Karen E. Hedin, Ph.D.
Department of Pharmacology
Mayo Foundation
Rochester, Minnesota

C. Huang, Ph.D.
Department of Biochemistry
University of Virginia
Health Sciences Center
Charlottesville, Virginia

Donald E. Ingber, M.D., Ph.D.
Departments of Pathology and Surgery
Children's Hospital
and Harvard Medical School
Boston, Massachusetts

James D. Jamieson, M. D., Ph.D. (Editor)
Department of Cell Biology
Yale University School of Medicine
New Haven, Connecticut

Michael L. Jennings, Ph.D.
Department of Physiology and Biophysics
University of Arkansas for Medical Sciences
Little Rock, Arkansas

Scott P. Kennedy, Ph.D.
Department of Pathology
Yale University School of Medicine
New Haven, Connecticut

Hynda K. Kleinman, Ph.D.
National Institute of Dental Research
National Institutes of Health
Bethesda, Maryland

Grigory P. Krapivinsky, Ph.D.
Department of Pharmacology
Mayo Foundation
Rochester, Minnesota

Robert I. Macey, Ph.D.
Department of Molecular and Cell Biology
University of California
Berkeley, California

Mark A. Milanick, Ph.D.
Department of Physiology
University of Missouri School of Medicine
and Dalton Cardiovascular Research Center
Columbia, Missouri

† Deceased.

Jon S. Morrow, M.D., Ph.D.
Department of Pathology
Yale University School of Medicine
New Haven, Connecticut

Teresa F. Moura, Ph.D.
Faculdade de Ciências e Tecnologia UNL and
    Instituto de Tecnologia Química e Biológica
Universidade Nova de Lisboa
Lisbon, Portugal

E. A. Newsholme, D.Sc.
Merton College
University of Oxford
Oxford, England

Chris Norbury, Ph.D.
Imperial Cancer Research Fund
    Molecular Oncology Laboratory
University of Oxford Institute of Molecular Medicine
John Radcliffe Hospital
Oxford, England

Carmen M. Pérez-Terzic, M.D.
Department of Pharmacology
Mayo Foundation
Rochester, Minnesota

Robert W. Putnam, Ph.D.
Department of Physiology and Biophysics
Wright State University School of Medicine
Dayton, Ohio

N. B. Reppas, B.A.
Balliol College
University of Oxford
Oxford, England

Luis Reuss, M.D.
Department of Physiology and Biophysics
The University of Texas Medical Branch
Galveston, Texas

Jacqueline A. Reynolds, Ph.D.
Easingwold
York, North Yorkshire
England

David L. Rimm, M.D., Ph.D.
Department of Pathology
Yale University School of Medicine
New Haven, Connecticut

Enrique Rodriguez-Boulan, M.D.
Department of Cell Biology and Anatomy
Cornell University Medical College
New York, New York

Albert Roos, M.D.
Department of Cell Biology and Physiology
Washington University School of Medicine
St. Louis, Missouri

M. B. Sankaram, Ph.D.
Department of Biochemistry
University of Virginia
Health Sciences Center
Charlottesville, Virginia

Peter Satir, Ph.D.
Department of Anatomy and Structural Biology
Albert Einstein College of Medicine
Bronx, New York

Anja Schweizer, M.D.
Department of Medicine
Washington University
School of Medicine
St. Louis, Missouri

Dola Sengupta, Ph.D.
Department of Cell Biology
Yale University School of Medicine
New Haven, Connecticut

Michael P. Sheetz, Ph.D.
Department of Physiology
Duke University Medical Center
Durham, North Carolina

John H. Sinard, M.D., Ph.D.
Department of Pathology
Yale University School of Medicine
New Haven, Connecticut

Vladimir Skulachev, D.Sc.
A. N. Belozersky Institute of Physico-Chemical Biology
Moscow State University
Moscow, Russia

George N. Somero, Ph.D.
Hopkins Marine Station
Stanford University
Pacific Grove, California

Lisa Stehno-Bittel, Ph.D.
Department of Pharmacology
Mayo Foundation
Rochester, Minnesota

Charles Tanford, Ph.D.
Easingwold
York, North Yorkshire
England

T. E. Thompson, Ph.D.
Department of Biochemistry
University of Virginia
Health Sciences Center
Charlottesville, Virginia

Jack H. Valentijn, Ph.D.
Department of Cell Biology
Yale University
School of Medicine
New Haven, Connecticut

Bratislav Velimirovic, M.D.
Department of Pharmacology
Mayo Foundation
Rochester, Minnesota

Paul M. Wassarman, Ph.D.
Department of Cell Biology and Anatomy
Mount Sinai School of Medicine
New York, New York

Scott A. Weed, Ph.D.
Department of Pathology
Yale University
School of Medicine
New Haven, Connecticut

Kevin Wickman, Ph.D.
Department of Pharmacology
Mayo Foundation
Rochester, Minnesota

Paul H. Yancey, Ph.D.
Department of Biology
Whitman College
Walla Walla, Washington

HANDBOOK OF PHYSIOLOGY

Section 14: Cell Physiology

1. Optical methods in cell physiology

FREDRIC S. FAY | *Biomedical Imaging Group, Department of Physiology, University of Massachusetts Medical School, Worcester, Massachusetts*

CHAPTER CONTENTS

THE FUNCTIONS CARRIED OUT BY ALL CELLS reflect the chemical reactions going on within them. It has become increasing clear that those reactions occur in a highly organized fashion. Thus, one cannot hope to understand the function of cells simply by knowing what enzymes, substrates, and products exist inside the cell using information about reactions in a well-mixed solution. This becomes apparent simply by considering the fact that many of the reactions carried out by a cell require chemical conditions (for example pH, pCa, [ATP]) that are mutually incompatible. Evolution has apparently dealt with this dilemma by carrying out specific reaction sequences or pathways inside discrete compartments that have a unique chemical composition. The segregation of molecules involved in different pathways to specific compartments or domains also confers kinetic advantages to such compartmentalized pathways, as intermediates have diminished distance, and therefore time, between sequential steps. Such reac-

tion cascades are often controlled by the docking of a key enzyme in that pathway, hence understanding of when and where within a cell a reaction is triggered often requires knowledge of where specific molecules are within the cell. Bulk measurements of levels of specific molecules cannot tell us how and why cell function suddenly changes in response to a specific stimulus. While cells carry out reactions that seemingly counteract each other (for example, active transport of Na^+ and Na^+:H^+ exchange), by localizing such reactions to different regions of the cell, transcellular transport of important solutes can be effected. Again, understanding of how cells accomplish such processes requires knowledge about where specific molecules, and the reactions that they control, are distributed within a cell.

Recent advances in optical methods provide a means for cellular physiologists to obtain information about the local concentrations of important bioactive molecules or ions inside cells, often in a manner that is compatible with the continued normal functioning of those cells (41, 101). As a consequence, physiologists can begin to look for the chemical or ionic events that are likely to trigger a specific cell function. Hypotheses regarding the role of a specific ion or molecule in mediating a particular cellular response can also be tested using other optical methods that allow physiologists to change the concentration of a specific ion or molecule inside a cell using light to photorelease a substance of interest from a caged photolabile precursor (71). Information about the extent to which a molecule of interest is bound to sites within a cell, presumably causing an effect on the activity of the molecule to which it is bound, can also be obtained by modern optical techniques (10). Optical techniques can also provide information about which molecule(s) are closely apposed to a molecule of interest (97).

While these new approaches are based for the most part on well-understood interactions of light with molecules, the revolution in the application of these approaches to problems of interest to cellular physiologists arises because of the development of new, highly specific fluorescent probes for a wide range of bioactive molecules, coupled with the development of methods for obtaining and analyzing images of the faint signals arising from such probes in single cells. The goal of this chapter is, first, to describe the general physiological questions for which these modern methods have been or can be used. Second, the optical approaches to these problems and the principles underlying them will be discussed. The final section will focus on strategies used to acquire and analyze images that arise from the use of these methods. The intent is to provide an overview of available methods, how they work, and strategic concerns involved in their use. Numerous excellent monographs (68, 70, 80, 82, 97, 107) contain detailed information on many of the methods that will be discussed.

This chapter will focus almost exclusively on optical methods based on fluorescent probes, as they are most relevant to questions of cellular physiology. Fluorescence involves the absorption of a photon by a molecule and the subsequent reemission of a photon of longer wavelength and thus lower energy after a brief time (the excited state lifetime, typically about 10 nss). There are three major reasons for the focus on methods based on fluorescence. First, measurements of fluorescence can be made on living cells without perturbing their normal function. Second, modern fluorescent probes provide strong signals against a dark background, thereby allowing detection of signals with good signal:noise ratio at the level of a single cell or subcellular region. Finally, as described in detail later (see APPROACHES), the fluorescence phenomenon can be used to provide information about many different types of molecular changes that take place in cells.

PHYSIOLOGICAL QUESTIONS

Imaging methods have proven to be valuable tools in understanding the molecular events underlying a wide range of cell functions. For example, to understand how a given stimulus causes a cell to carry out a specific function, imaging methods are providing vital insights into the sequence of events linking the stimulus to the response of the cell. The techniques have been used to determine if the concentration of a suspected second messenger actually changes in response to a given stimulus, and if the change in the concentration of that second messenger is large enough and occurs rapidly enough to act as a mediator of the cell's response. Imaging methods not only provide information about when and how much the concentration of a putative second messenger changes but also where within the cell that change is initiated. If changes are initially localized one can determine if they spread. Such information is important in tracking the site

within cells where the second messenger may be generated and in searching for the molecules involved in upstream signaling that give rise to changes in that second messenger. Clearly, images are also important in assessing where downstream targets believed to underlie a particular response may be located.

If the distribution of the putative targets of a second messenger are known, the observed pattern of the second messenger relative to those targets provides an important test of understanding of the assumed involvement of both second messengers and targets in a physiological response. Clearly if the second messenger does not reach the assumed targets on which it acts, or does so after the overall cellular response takes place, then understanding of the pathway underlying a particular response must be incomplete at best.

Hypotheses regarding the role of a particular molecule in mediating a physiological response can be tested using photolabile caged precursors of that molecule to determine if changing the concentration of the putative second messenger can alone trigger the physiological response. If that second messenger is believed to act in a highly localized manner within the cell, then the light to trigger the photorelease of the molecule can be focused to a discrete region of the cell to specifically investigate that possibility. The ability to locally change the concentration of a bioactive molecule inside a cell can be used to investigate the distance that a given messenger spreads before it is inactivated or removed from the cytoplasm by the cell's own mechanisms.

Optical methods can also be applied to follow the interactions of molecules involved in signaling. Measurements of fluorescence anisotropy can be used to discern whether a molecule of interest is free in the cytoplasm or bound to a target. Images of the distribution of that molecule relative to others can be used to investigate the likely target to which it binds. The technique of fluorescence energy transfer can be used to probe for those interactions with spatial sensitivity in the nanometer range.

APPROACHES

Fixed Cells

Distribution of Molecules. It has long been known that by using probes that specifically bind to molecules of interest inside cells, one can obtain information regarding their distribution. This general approach was first applied to analyze the distribution of proteins (97) and more recently has been extended to map the distribution of RNA and DNA (54). In the case of proteins, antibodies raised against a specific protein have been used to detect the distribution of that protein. For RNA and DNA, complementary oligonucleotides have been employed. While in some studies these protein-specific or nucleotide-specific probes have been directly labeled with a fluorophore, in most studies the fluorophore is contained in a second probe that specifically binds to the primary detector molecule, often at multiple epitopes or sites. The strategy of attaching the fluorophore to a secondary detector has advantages and possible disadvantages. If the secondary detector can bind to multiple sites on the primary detector then the primary signal is amplified, allowing detection of a signal that might otherwise be too weak to be detected if the number of copies of a molecule of interest in a cell is rather low. However, this can be a disadvantage if there is considerable nonspecific binding by the secondary probe, since the signal:noise in the resulting images will be degraded. The chemical conjugation of a fluorophore to a probe and the necessary postconjugation purification of the reaction mixture is fairly straightforward but somewhat time consuming. Therefore a further advantage of the secondary labeling strategy is that a single labeled secondary probe can often be used for many different studies.

Regardless of whether the probe for a given protein, DNA, or RNA is labeled directly or indirectly using a secondary probe containing the fluorophore, none of these probes can pass readily across the membranes of an intact living cell. As a consequence, these methods generally require that cells be chemically treated to make their membranes permeable to large molecules to facilitate both the access of the probe to molecules it specifically binds to in the cell, as well as the subsequent washout of excess probe. To prevent molecules from leaving the cell or rearranging during the permeabilization and subsequent incubations with probes, the cell must first be chemically treated to immobilize the molecules within it. This is usually accomplished with a chemical treatment that cross-links or denatures molecules. While this fixation must be sufficient to prevent movement of molecules when the cell is exposed to various environments, care must be taken not to use fixatives that so highly cross-link the molecules within the cell that the molecule one wants to detect is altered to the degree that it is no longer recognized by the probe.

Cells that have been labeled in this manner can then be imaged using techniques that provide high-resolution quantitative maps of the distribution of a specific molecule. When such images are obtained for cells fixed in different physiological states, those images can be used to determine if the gross intracellular density or distribution of a specific molecule changes as a cell responds to a stimulus. Images at various times after application of a stimulus can be used to assess the timecourse of changes in the distribution of a molecular species relative to the physiological response, thereby providing information relevant to the nature of the linkage between a change in the distribution or concentration of a molecular species and the response.

These general strategies can also be employed to analyze how different molecules are distributed with respect to one another. All that is required is that the probes for the different molecules be labeled with fluorophores whose spectral properties are sufficiently different that their signals can be separated from one another. To date, signals from three different fluorophores have been effectively separated when used on a single sample. By using three different fluorophores in various combinations for different molecular probes, the relative distributions of seven different molecules have been studied in a single cell (86). It is likely that the number of different molecules whose distribution can be assessed in a single cell will increase with the development of new fluorophores covering a broader range of the optical spectrum and with a wider range of shifts between their excitation and emission spectra, the Stokes shift.

Because the assessment of molecular distribution with antibodies or oligonucleotides requires fixation and cell permeabilization, it provides only information on the distribution of molecules at a single point in the life of a cell. If there is considerable variability in that distribution between cells, then many cells will have to be sampled during the response to a physiological stimulus before information about the molecular changes underlying that response can be obtained. Furthermore, small molecules or ions are not amenable to analysis by this general approach, as fixation and permeabilization strategies currently employed do not immobilize such species.

Living Cells

Distribution of Molecules

General considerations. The largest strides in our understanding of how changes in the concentration or distribution of molecules cause physiological responses have come about as a result of our ability to detect and image molecules inside intact living cells. As most bioactive molecules are not intrinsically detectable by optical techniques, the quantitative analysis of such molecules has required the development of chemical indicators. There are now many commercially available compounds that can be used to monitor the concentration of numerous biologically important analytes, almost all of them in intact living cells. In addition, the general strategies for making fluorescent indicators that provide information on the concentration of molecules of biological importance have now been demonstrated (1, 46). Hence it should be possible in the years to come to make continuous measurements of the concentration of virtually all bioactive molecules inside living cells as they respond to stimulation. Recent work on the development of indicators to measure the concentration of Ca^{2+} is illustrative of the issues of concern in developing and using such indicators. Clearly, the creation and use of such compounds has had an enormous impact on biology, since to date over a thousand papers have been published reporting experiments using indicators to probe the role of Ca^{2+} in biological signaling.

In designing and using a fluorescent indicator, two issues of immediate concern are the selectivity and sensitivity of the indicator. An effective indicator must be selective for the analyte of interest in the presence of other molecules or ions that have similar chemical characteristics and that may be present within the same biological compartment. The sensitivity of the indicator must be well matched to the range of concentrations over which the bioactive analyte is expected to vary. While these characteristics can occasionally be found in existing synthetic compounds, they can always be found in the molecules that sense and respond to changes in the concentration of the analyte inside the cell. In the case of Ca^{2+}, the indicators that have been most widely used are based on the structure of the highly specific Ca^{2+} chelator, EGTA [ethylene glycol-bis(β-aminoethyl ether) N, N, N′, N′-tetracetic acid]. However, for more complex analytes such as cyclic adenosine monophosphate (c-AMP) (1) or fatty acids (93), endogenous biological molecules (in these two examples, a plasma fatty acid binding protein and c-AMP-dependent protein kinase) have proven to be most effective as specific sensors. When designing indicators based on proteins that normally bind to a particular analyte in the organism, two general strategies have been employed to obtain a change in a fluorescence signal that is sensitive to analyte binding: In a number of cases, a fluorophore has been inserted at a position in the molecule where changes in the local chemical environment (hydrophobicity, charge density, among others) are known to occur upon binding of the analyte. By choosing a fluorophore whose fluorescence is sensitive to these changes, the binding of the analyte to the protein can be detected. The other successful strategy is to use fluorescence energy transfer to monitor a change in a molecule that occurs upon specific binding of an analyte. Fluorescence energy transfer is the radiationless transfer of energy between one fluorophore (the "donor") whose emission spectrum overlaps significantly with the excitation spec-

trum of another fluorophore (the "acceptor"). This process of energy transfer is exquisitely sensitive to the distance between donor and acceptor fluorophores, is sensitive to (the distance between fluorophores)$^{-6}$, and is effective in the 3–7 nm range. By inserting the donor and acceptor into sites whose separation changes upon binding of an analyte, a signal sensitive to the presence of an analyte has been obtained.

Further concerns about the design and use of a chemical indicator are the dynamic range, strength, and stability of the signal produced by that indicator. Clearly one wants an indicator whose signal characteristics (for example, intensity, spectral characteristics) change significantly as the concentration of the analyte changes. The brightness of an indicator molecule is also an important concern. One wants an indicator that gives a large signal so that the concentration of the probe to be introduced into the cell or subcellular compartment can be kept as low as possible so that it does not significantly buffer changes that take place inside the cell and in that way alter the relationship between the stimulus and the cellular response. Measurement of the relationship between stimulus and response in cells, in the presence and absence of the indicator, should be carried out to find optimum conditions for recording changes in the concentration of an analyte without perturbing the physiology of the cell (108). Increasing the brightness of a fluorescent indicator will clearly improve signal:noise, because noise introduced in the detection system becomes less important, and because the inescapable "shot noise" due to the random nature of the emission process is reduced, being proportional to the square root of the number of detected photons (22). However, the detectability of a given change in analyte concentration will depend on the dynamic range of the fluorescent signal. Thus if two dyes produce the same absolute change in intensity for a given change in the concentration of an analyte of interest but one has a lower starting signal, that indicator will produce a more reliable (that is, higher) signal:noise, measurement of a change in concentration. This is one major reason why indicators that change their fluorescence are preferable to indicators that change their absorbance upon binding to an analyte of interest.

An additional concern in choosing an indicator for use in a living cell is the stability of its signal. An ideal indicator would be capable of reproducibly reporting a change in the concentration of some ion or molecule for an unlimited time. Virtually no fluorescent or luminescent indicator is strictly capable of doing this. Luminescent indicators such as the Ca^{2+} sensitive luminescent protein aequorin typically consume their luminescent precursor when they bind the ion or molecule that activates the light-emitting event (47). This behavior can be taken into account in the calibration procedure, but this assumes that the luminescent molecule is either consumed evenly or diffuses rapidly throughout the cell or tissue, an assumption unlikely to be valid for the most widely used luminescent indicator, the 40 kD Ca^{2+} sensitive luminescent protein, aequorin (16). Unlike the situation with luminescent indicators, fluorescent indicators typically are capable of emitting several thousand photons, but the number of cycles that a typical fluorophore can undergo is not unlimited (13, 69). Ultimately the process of repeated excitation and emission causes the fluorophore to be altered chemically, usually resulting in loss of fluorescence. This process of "photobleaching" is not well understood and may occur by different mechanisms for different fluorophores. In many cases it is thought that photobleaching results from the reaction of a reactive oxygen molecule known as singlet oxygen with the fluorophore (69). The singlet oxygen is itself produced by transfer of energy from the excited fluorophore to an oxygen molecule. It is clearly desirable to select fluorescent indicators that are not rapidly photobleached and to keep the oxygen concentration surrounding the cells only high enough to maintain the cell or tissue under study in a viable state.

Intracellular targeting and calibration of indicators. Once an indicator with appropriate properties is chosen, it must be introduced into the cytoplasm or other cellular compartment from which information is to be obtained. Indicators may be loaded into the cytoplasm of a single cell through a micropipette or, alternatively, into large numbers of cells by chemical or physical methods (72).

Perhaps the most widely used method for trapping indicators inside populations of cells involves incubating cells with a chemically modified form of the indicator where charged groups are masked by coupling them to hydrophobic residues through an ester linkage which can be cleaved by intracellular esterases. Such modified dye molecules pass readily across cell membranes, but, after cleavage of the ester-linked hydrophobic masking groups, the dyes become trapped in that compartment. This strategy has been used to introduce indicator dyes into the cytoplasm of many cell types, and while requiring no special equipment or expertise, more often than not results in some dye being deesterified and trapped in intracellular compartments other than the cytosol (110). This may be either a drawback or an advantage, depending on the goals of the experiment. If dye resides in multiple compartments, the signal from each compartment can be analyzed separately using high-resolution imaging methods or, alternatively, the signal from all but the compartment of interest can be quenched or otherwise extinguished (100). When loading dyes using the deesterification method, it is important to keep in mind the likelihood that dye, and hence the signal recorded from a single cell, may originate from compartments in addition to the cytoplasm. Because the spectral properties and sensitivity to the analyte of interest of the esterified dye are usually quite different from those of the parent compound, it is necessary to establish conditions where the esterified masking groups are well cleaved, often not a trivial task. Incomplete deesterification of dye can be identified, and often compensated for, by performing an in situ calibration of the dye (91, 110).

Dye can also be loaded into populations of cells by both chemical and physical treatments that transiently create small holes in the cell membrane (72). Some fine tuning is usually required to find conditions in which the cells are not permanently damaged by the permeabilization and yet sufficient dye is introduced into the cells. These methods typically transfer a small proportion of the dye from the media into the cells, and they therefore require use of relatively large amounts of indicator, making this approach unsuitable if the indicator is very precious or in short supply.

As an alternative to bulk loading, an indicator can be introduced directly into single cells from a micropipette, either by pressure injection of iontophoresis, or by passive diffusion from the larger orifice of a whole-cell patch pipette. The process can be monitored continuously, and usually represents the most effective way to introduce an indicator into the cytoplasm, though in the whole-cell configuration, careful controls are needed against loss of physiologically important constituents from the cell into the patch pipette.

As discussed briefly earlier, the fact that dye loaded into cells as a membrane-permeant esterified derivative ends up in compartments other than the cytoplasm can be exploited by imaging techniques to monitor chemical changes in other intracellular compartments. While such techniques will significantly improve the separation of signals originating from indicator in different compartments, there will always be some residual blurring and thus spill-over of signals from one compartment to another. This can cause significant problems in the interpretation of results (27). Two strategies have been described to deal further with problems arising from the distribution of indicator in cells. One of these, already mentioned, involves application of physical or chemical methods to remove or quench indicator dye in compartments (typically the cytoplasm) where it may not be wanted (100). Such approaches depend critically upon finding a quenching agent or dialysis strategy that acts only on a single compartment. An alternative strategy is to target an indicator to a specific intracellular compartment by modifying the indicator chemically. To date the most successful targeting approaches use chimeric molecules, in which a peptide corresponding to a sequence that codes for the import of other molecules into an intracellular compartment has been coupled to an indicator. For example, chimeric molecules have been produced by chemically coupling an indicator dye to a peptide (6) or by using molecular biological methods to couple the sequences encoding a biological indicator such as aequorin with a sequence encoding a molecule or region of a biological molecule that is taken up or retained in a specific intracellular compartment (89). The advantage of the molecular biological strategy is that once an appropriate construct has been made, cells can be transfected with the construct so that the indicator is made by the cell itself and targeted to a specific site. This obviates the need to synthesize and then purify an indicator-peptide construct as well as the problem of then introducing it into a cell. Molecular biological approaches have been successfully employed to target the calcium-sensitive bioluminescent protein aequorin, as well as the green fluorescent protein, into specific intracellular compartments (87–89). The aequorin signal is not strong enough to provide measurements of changes in calcium concentration in a specific intracellular compartment at the single cell level, so this strategy has thus far been employed only with populations of cells. In contrast, the green fluorescent protein provides a signal sufficiently robust to be used at the single cell level. However, it has thus far been used principally as a tracer to follow a specific protein within a cell; no constructs have yet been made whose fluorescence is sensitive to changes in a bioactive molecule or ion. This is a very powerful approach whose full potential remains to be developed. Increasing knowledge of the molecular requirements for import or retention of molecules in specific intracellular compartments promises to enhance our capability to target indicators to specific sites within the cell and to measure the chemical environment there.

Dye indicators that respond to changes in the concentration of an analyte do so by changes in their brightness, their spectral properties, or both. The choice of optical properties to be measured will depend on the properties of the indicator. For an indicator that simply changes its brightness upon binding an analyte, a single measurement of the fluorescence intensity usually suffices (101). Ideally the cell containing the indicator is illuminated at the wavelength where the indicator absorbs maximally, and the emitted light is collected with as broad a spectral band pass as possible without contamination from the excitation beam (77). This strategy optimizes use of available photons from the indicator and thus minimizes the need to overilluminate the cell, with attendant problems of photobleaching and possible additional deleterious effects on cell function. While indicators that simply change brightness upon binding an analyte require the simplest of measurements, changes in brightness are among the most difficult to interpret. The difficulty arises because brightness may be influenced by movement of the cell in the field of the microscope, by a change in light scattering by the cell, or by changes in the light source or detector. Even when such artifacts can be eliminated, the conversion of fluorescence intensity to analyte concentration requires some additional information. As indicated in equation 1–1, conversion of fluorescence intensity (F) to analyte concentration $([X])$ requires knowledge of the dissociation constant (K_D) of the indicator, as well as values for the fluorescence intensity of the indicator in that cell in the absence (F_{min}) and presence (F_{max}) of saturating levels of the analyte.

$$[X] = [(F - F_{min})/(F_{max} - F)] \times K_D \qquad (1\text{--}1)$$

By permeabilizing the cell so that the intracellular analyte concentration can be set by the medium, all the necessary constants can be obtained (109). As the properties of many indicator dyes are influenced by the intracellular milieu, it is very important to check the properties of the indicator inside the cell.

Ratiometric strategies. It is desirable to use indicators that change their spectral properties, rather than their brightness, upon binding to an analyte. By measuring the fluorescence intensity at two different excitation or emission wavelengths rapidly relative to rates of cellular motion and then ratioing them, a measure of the extent of binding of the indicator is obtained that is not influenced either by cell movement or by most fluctuations in the fluorescence detection system (44). As shown in equation 1–2, much as was the case for dyes that simply change brightness, it is possible to calculate the concentration of an analyte corresponding to a given value for the fluorescence ratio with knowledge of the dissociation constant of the indicator, limiting values for the fluorescence ratio in the absence (R_{min}) and presence (R_{max}) of saturating levels of analyte; one additional optical constant is required: β, which is the ratio of fluorescence intensities at one of the wavelengths for dye in the absence and presence of saturating analyte. These constants can be obtained using the permeabilization strategies discussed in the preceding section. It is important to keep in mind that the use of this equation to calculate the concentration of analyte from measurements with a ratiometric indicator, or use of equation 1–1 with a nonratiometric indicator, assumes that the analyte and dye are at equilibrium. If the concentration of the analyte is changing so rapidly that it is not in equilibrium with the indicator, errors will be introduced by the use of this equation (53). More complicated analytical strategies must then be used to interpret changes in the fluorescent signal in terms of changes in analyte concentration.

$$[X] = [(R - R_{min})/(R_{max} - R)] \times K_D \times \beta \quad (1-2)$$

When ratiometric indicators are not available for an analyte, one can nonetheless make measurements in a ratiometric manner by coinjecting with the indicator another molecule that acts as a normalizing volume marker (97). The normalizing volume marker is chosen to have different spectral properties from the indicator and to be insensitive to the analyte and we hope, all other intracellular species. A key assumption in using this strategy for imaging studies is that the distribution of the indicator and volume marker are identical throughout the cell. Thus the volume marker should be of similar size and charge to the indicator; but even then their codistribution is not certain. Equivalent distributions can be guaranteed by coupling both an indicator and an analyte-insensitive fluorophore to a single molecule such as dextran. A few such co-anchored fluorophore pairs are commercially available.

Another strategy for obtaining ratiometric measurements of the concentration of an analyte using an indicator that simply changes its brightness upon binding is based upon measurements of fluorescence lifetime (58, 106). Many of the indicators that change their brightness upon binding an analyte exhibit a change in their fluorescence lifetime, that is the time between absorption and reemission of a photon. Changes of almost tenfold in fluorescent lifetime have been reported following Ca^{2+} binding for indicators such as Quin 2 (58). Fluorescence lifetime of dye inside a specimen can be measured in one of two ways (57). The specimen can be illuminated with a brief flash of laser light, and fluorescence emission at successive intervals determined. Alternatively, the specimen can be illuminated with a beam of light at the excitation maximum of the indicator dye whose intensity varies sinusoidally and from the phase delay in the fluorescence emission, the fluorescent lifetimes of the indicator determined. By determining the fraction of the dye population having fluorescent lifetimes corresponding to the free and bound form of the dye, the free concentration of the analyte can be simply calculated. Measurements of fluorescent lifetime have been made in cells using imaging or photometric detection methods. Such measurements have in turn provided information on the concentrations of various analytes. While this clever approach circumvents numerous problems in interpreting data from an indicator that simply changes

brightness upon binding an analyte, it typically takes several seconds to acquire sufficient information to estimate fluorescent lifetimes in a sample, making the technique unsuitable for systems where concentration is changing rapidly. At present the special instrumentation required for this approach is available only in a limited number of laboratories.

Measurements of Molecular Activation. Signals from the indicators described earlier provide information about changes in the concentration of various ions or molecules during the response to a physiological stimulus. While many of the initial signaling events may involve a change in the concentration of a particular ion or molecule through the activation of an ion channel or enzyme respectively, many signaling cascades also involve the covalent modification of some molecule inside the cell. Work on the development of indicators to follow the extent of phosphorylation of specific proteins inside living cells is in its infancy, but the feasibility of monitoring such processes in living cells has been recently demonstrated (83, 84). By incorporating a fluorophore into the myosin regulatory light chain at amino acid position 18, only one amino acid away from the site normally phosphorylated when smooth muscle or non-muscle myosin is activated; the fluorescence of the fluorophore is changed upon activation. By genetically engineering the myosin light chain so that a reactive cysteine was substituted at a site near the site of phosphorylation, a functionally competent, but phosphorylation sensitive, labeled fluorescent myosin light chain was constructed (84). Similar general strategies should allow the extent of phosphorylation or covalent modification (for example, prenylation or methylation) of other molecules involved in regulating specific physiological processes to be monitored. These general strategies have also proved successful in monitoring the activation of proteins such as calmodulin (46), which upon activation (by Ca^{2+} binding) undergo a conformational change. Insertion of an environmentally sensitive fluorophore at a site near this conformational change, provides a measure of the extent of activation of calmodulin in living cells. An additional strategy based on the use of fluorescence energy transfer has also proved successful in monitoring the activation of a molecule resulting from phosphorylation. By attaching a fluorescein and a rhodamine on myosin light and heavy chains, respectively, changes in the extent of phosphorylation have been monitored in tissue cultured cells following serum stimulation (83).

Tracking Movements of Molecules. The responses of a cell, for example those involving the cytoskeleton, often involve changes in the pattern of organization of molecules rather than a change in the concentration or activation of a particular molecule. Strategies have thus been developed specifically to follow changes in the distribution of molecules inside living cells. The approach centers on creating a functionally competent, fluorescently labeled form of a protein (29, 67) or oligonucleotide (105) of interest, and then introducing it into a living cell. One employs the fluorescent analogs, as they have been called, to follow the reorganization of a particular molecular species during the response to a particular stimulus, much as radioactive tracers are used to follow the movement of ions from one compartment to another. Unlike the labeled molecules described in the preceding section, there is no intent to place the fluorophore at a position where it is sensitive to the extent of activation of the molecule. One merely wants to have a labeled form of a particular molecule, which when introduced into a cell and after mixing with the endogenous pool, provides information on the distribution of a particular molecular species during a physiological response. This technique has been widely used to follow changes in the organization of cytoskeletal proteins during many different processes (74, 96).

At the time that this approach was first introduced, proteins, and more recently oligonucleotides, were labeled by direct chemical coupling of reactive fluorophores with free amino or sulfhydryl groups on purified molecules. The reaction mixture was then subjected to purification to isolate labeled, but functionally competent, molecules from both unlabeled molecules and from molecules labeled in such a way as to render them nonfunctional. Recently it has become possible to create fluorescent analogs using molecular biological techniques. A chimeric molecule is created by connecting the sequence coding for a protein of interest with the sequence coding for a naturally fluorescent protein, the green fluorescent protein (GFP) from *Aequorea victoria* (23, 83). GFP is a highly fluorescent molecule consisting of 238 amino acids. Fluorescence results from the auto-cyclization of a serine-tyrosine-glycine sequence and the subsequent dehydrogenation of tyrosine by reaction with molecular oxygen. The resulting fluorophore is buried deep within this globular protein and is thus not susceptible to bleaching by oxygen, resulting in what appears to be a very robust signal. These chimeras may be introduced by simply transfecting a cell with the appropriate construct, or alternatively the fluorescent chimera can be synthesized in some other cell type and then microinjected into a cell of interest. This method has been used to produce and then follow labeled proteins in *C. elegans* (23), yeast (55), and numerous tissue cultured cells and developing embryos (7, 40, 81). As GFP (mol. wt. 27kD) is considerably larger than most typical fluorophores (for example, the molecular weight of fluorescein is about 500), the challenge in applying this technique is to place the sequence coding for GFP at a position in the protein of interest where it does not disrupt its function. Mutants of the native GFP have been created with altered spectral properties, making it possible to examine simultaneously the distribution of two or more different proteins in the same cell by this general approach (48). Forms of GFP that are sensitive to the local chemical environment have not been identified, nor is it likely that, given the highly protected position of the fluorophore, such forms can be made. Thus it is unlikely that the strategy used in the past to make fluorescent analogs that report the extent of activation of a molecule will result from inserting the GFP sequence into a region of a molecule that experiences a change in conformation, hydrophobicity, or charge upon activation. However, it is certainly conceivable that, by employing a pair of GFP mutants with spectral properties appropriate for fluorescence energy transfer, a conformational change associated with activation of a molecule could be detected as a change in the fluorescence energy transfer signal (49). This will require that two different GFP mutants be placed at sites on the molecule where activation causes a significant change in the distance between those sites.

Images of fluorescent analogs have provided much useful information on the distribution of the overall population of molecules in living cells. To understand the processes that produce a given distribution it is important to obtain information about the flow of molecules into or out of a region of interest. For example, if a microtubule or actin filament bundle is growing or shrinking, it is extremely important to know if length changes occur by addition or loss of molecules at a specific point along that structure. To obtain information about the rate and direction of flow of molecules into or out of a region, methods have been developed to use focused light beams to "mark" molecules in a specific region, and to follow the movement of molecules into or out of that area.

The earliest approaches used a powerful laser beam to bleach fluorescent analogs in a cellular region and then followed the recovery of fluorescence in the bleached region to provide information on the extent to which a specific molecule was free in the cytoplasm or bound to immobile sites (11, 33, 112). Information on the rates of exchange between free and bound states can be obtained from such data. By imaging the bleached spot and following it over time, movement of the slowly exchanging bound form of the protein can be detected. A drawback to this approach is that the bleached spot appears as a local decrease in fluorescence against a bright background. As discussed earlier, this is an inherently poor signal:noise situation. The recent development of caged fluorophores (90, 98) has greatly enhanced the information that can be obtained by these kinds of strategies. Caged fluorophores are fluorescent molecules that have been modified with a photolabile group (typically a nitrobenzene derivative) that distorts the electronic structure of the parent compound, thereby greatly diminishing its fluorescence. Upon irradiation with ultraviolet light the photolabile group is split off from the parent fluorophore, which then regains its full fluorescence. As a consequence the fluorescent analogs that have been marked by the light beam are more easily followed against a dark background.

Analysis of Molecular Binding. Many cellular or biochemical processes are activated by the binding or docking of two molecules. Thus it is of interest to be able to ascertain if a molecule of interest is bound within a cell, in what region the binding takes place, and to what molecular species it is binding. Measurements of fluorescence anisotropy provide a means of assessing whether a given molecule is bound, especially to a larger relatively immobile target within a cell (10). As already discussed, fluorescence involves the absorption of a photon by a molecule, which distorts the electronic orbitals of that molecule, raising an electron to a higher energy state. Some of the absorbed energy is reemitted as heat, and the rest emerges as a photon of longer wavelength relative to that absorbed. The electronic orbitals that can absorb the energy of a photon are typically delocalized electrons, often those in aromatic ring structures. As those rings and associated electrons have a defined structure, a photon is preferentially absorbed when its electric vector is oriented in the same direction as the electron orbitals capable of absorbing the photon's energy. The photon that is reemitted typically has its electric vector oriented in a similar direction as the orbital into which it was absorbed, and hence if the fluorophore is stationary the orientation of the electric vector of the exciting and emitted light should be similar. As most fluorophores exhibit some internal structural fluctuations, and in addition tumble in solution, the orientation of the electric vector of the emitted photon is usually rotated slightly relative to that of the absorbed exciting photon. The extent of this change in the orientation of the electric vector will depend on the rate and the extent of motion of the fluorophore, as well as the time that intervenes between absorption and reemission (57). By measuring the intensity of light emitted, whose electric vector (polarization) is parallel and perpendicular to the axis of polar-

ization of the incident exciting light, one can obtain an optical measure of the extent of rotational freedom of a given molecule inside a cell. The measurement of emission intensity, collected through a polarizer oriented parallel ($F_{parallel}$) and perpendicular ($F_{perpendicular}$) to the axis of polarization of the excitation beam, is used to calculate the fluorescence anisotropy according to equation 1–3.

$$\text{Fluorescence Anisotropy} = [(F_{parallel} - F_{perpendicular})/ (F_{parallel} + 2F_{perpendicular})] \quad (1\text{--}3)$$

These measurements require the preparation of a functionally competent fluorescent analog of the molecule of interest. It is important in designing a fluorescent analog for this purpose to select a fluorophore whose lifetime is sufficiently long relative to the time that it takes the molecule to rotate significantly so that binding will cause significant changes in fluorescence anisotropy. In practice one can image the intensity of fluorescence emission through polarized filters and use the intensities measured throughout the cell to ascertain if there are regional differences in the extent of binding of a molecular species. Similar measurements can be made during the response to a physiological stimulus to determine if a molecule changes its extent of binding during the response and where within the cell that change takes place. To date, these techniques have been principally exploited to follow the binding of calmodulin in tissue cultured cells during a wound healing response (43), but they hold great promise for understanding many other signaling mechanisms. There is increasing evidence that a fundamental feature of G-protein–mediated signaling involves the selective binding and unbinding of specific subunits (79), a process that has yet to be studied in a living cell. This is one of many areas where these techniques are likely to prove important.

While fluorescence anisotropy can be used to determine where and when the binding of a molecule of interest changes during the response to a stimulus, it cannot provide information about what it is binding to. Again, optical techniques can be used to address this question as well. There are two general strategies that can be applied. If one has some information about likely sites to which the molecule of interest is bound, then one can use fluorescence energy transfer to test those ideas inside the living cell. Insights into likely sites where a molecule binds might come from in vitro binding studies, two hybrid screens, or from the pattern of distribution of binding evident from fluorescence anisotropy images. The search for a likely site for binding of a molecular species during a physiological response can be narrowed by using antibodies to suspected targets for binding. By examining cells fixed, for example, at the time of peak binding one can investigate whether there is significant colocalization of an antibody directed toward a suspected target and the molecule whose binding changes during a physiological response (67). Once a likely binding target has been identified, studies can be carried out using fluorescence energy-transfer measurements to determine if the suspected target is the site to which the molecule of interest binds during the physiological response. Because of the extreme sensitivity of the fluorescence energy-transfer process to the distance between fluorophores, small changes in the position of one fluorescent analog with respect to the other can be sensed by this technique. Hence subtle changes in binding may become apparent from fluorescence energy-transfer measurements that are not detectable by fluorescence polarization studies (84). It is quite likely that many signal transduction cascades involve steps where predocked molecules change their conformation upon activation and, in turn, alter their binding to other nearby molecules that are part of a preformed multimolecular signaling complex. This kind of signaling could be investigated with considerable advantage using fluorescence energy-transfer methods.

Manipulating the Chemistry of Living Cells Using Photoactivatable Molecules. The use of photoactivatable molecules in optical studies has already been introduced as part of the strategy for monitoring the flow of molecules into and out of a region of interest. In these studies a photolabile group is used to inhibit the inherent fluorescence of a fluorophore to which it is attached. Photolabile groups have also been attached to many other molecules, principally to mask their inherent biological reactivity, thereby providing physiologists with the opportunity to change the chemistry of a cell or a subcellular region with a beam of light (45, 56). These caged molecules can then be used to examine the effect of suddenly changing the concentration of a second messenger or the activity of an enzyme, and thus to test the role of that molecule in a particular physiological response. Photolabile caged second messengers (104), ions such as phosphate or calcium (32), nucleotides (42), neurotransmitters (59), and proteins (31) have been synthesized and used to address a wide range of biological questions to date.

There are a number of distinct advantages of using caged molecules to manipulate the chemistry of a cell. For one, the chemical change can be accomplished far more rapidly by this technique than by conventional perfusion. There are no delays introduced by mixing or diffusion since the precursor molecules can be introduced and time allowed for them to equilibrate in the specimen before illumination with a beam of ultraviolet

light for a brief interval. Adequate photoactivation of a caged molecule may be accomplished in fractions of a millisecond, and the subsequent photodecomposition of the caging group (typically a nitro benzene derivative) may occur in a comparable time (2). Hence the kinetics of steps linking a molecule to some downstream cellular response can be clearly studied. Because the chemistry of a cell can be so suddenly changed using caged precursors of a putative second messenger, desensitization phenomena that often complicate the interpretation of responses to a second messenger are obviated (103). When using a caged molecule that is an inhibitor of a particular enzyme or pathway, the normal response to a stimulus can be obtained in a given cell and then within milliseconds the inhibitor photoreleased and the cell stimulated again to determine if that enzyme or pathway is involved in mediating the cellular response. The same cell serves as its own control and, because the cellular responses can be determined with and without inhibitor in rapid succession, the opportunity for secondary compensations by the cell, for example by upregulation of another redundant transduction pathway, is minimized. The existence of redundant control mechanisms in cells has become clear from the inability to see effects on cellular responses using molecular biological knockout strategies (40).

Finally, an as-of-yet underexploited capability of this optical method is that chemical changes that are highly localized within the cell can be accomplished simply by focusing the ultraviolet uncaging beam on a particular region of a cell. As a consequence one can explore, the possibility that a given second messenger spreads significantly within a cell. If it appears to be acting locally, one can determine if there are regional differences in the responsiveness to that molecule (36). In designing and then using caged molecules, it is important to verify that effects seen after photolysis are due to release of the caged molecule and not the effect of either exposure to the ultraviolet uncaging beam or the photocleaved cage and its decomposition products. Control strategies are well described in the literature (45). While the number of caged molecules is not as extensive as the number of indicator dyes, there continue to be new caged molecules appearing commercially as well as new strategies for caging molecules appearing regularly in the literature. The recent development of caged amino acids that can be incorporated into proteins during standard Merrifield synthesis opens the possibility that a wide range of enzyme activities will be controllable through the use of caged peptides (31). The actual apparatus required to employ caged compounds is relatively simple, involving addition of a flash lamp or ultraviolet laser to a standard microscope; if photolysis is to be accomplished through the microscope optics, an excitation path that transmits in the ultraviolet must be used.

Manipulating Molecules and Organelles with Optical Tweezers.

While the optical tweezer technique is not based on fluorescence per se, it must be part of any discussion of modern optical methods used to investigate physiological problems (94). When light is focused by a high numerical aperture objective on a structure having a higher refractive index than its surroundings, a gradient of force acts on that structure. This force tends to hold such a structure at the point of focus of the light, hence, acting like a tweezer. If the position of the focused light beam is moved, a structure that is "trapped" at the point of focus of the light beam will move with the focused light. While the force exerted by light on a refractile particle is small (typically less than 10 piconewtons), those forces are sufficient to trap and move a single cell in solution or even to manipulate a single organelle inside a living cell (9). The magnitude of the force tending to hold a particle at the center of the focused beam is proportional to the intensity of the light. Hence if an optical tweezer is used as part of a system that can also sense the position of a particle that is trapped in the light beam, optical tweezers can be used to make measurements of force being exerted on that particle by determining the light intensity and thus the force required to just hold the position of the particle constant. By adsorbing single molecules that have "motor activity" onto refractive small plastic beads, remarkable insights have been obtained into the events underlying force generation by single molecules of myosin (37, 75), kinesin (95), and other molecules such as RNA polymerase (113) that move relative to a series of binding sites as part of their action. This technique promises to allow molecular physiologists to study directly the forces, as well as the size and speed, of movements underlying the action of many bioactive molecules.

OPTICAL SIGNAL DETECTION AND ANALYSIS

Overview

Regardless of the indicator dye, fluorescent analog, or other optical probe being used to study a physiological response, the signal that emerges from the specimen is a stream of photons with a characteristic spatial and temporal distribution. There are a number of important issues that must be kept in mind in deciding how to acquire and analyze the optical information emerging from the specimen. First, one clearly wants sufficient signal:noise to measure changes of physiological interest with appropriate spatial and temporal resolution. Second, one wants to make the measurements in a manner that does not perturb the process one is investigating.

A number of issues in turn impact on these requirements. As previously noted there are a finite number of excitation/emission cycles that a fluorophore can undergo before it photobleaches; hence the strategies employed must be efficient in their detection of photons emitted from fluorophores within the specimen. The signal:noise can never be better than the limit imposed by the Poisson uncertainty of photon detection, and therefore optimum strategies are ones that are very efficient in detecting all photons that arrive at the detector and involve little further introduction of noise by the photon detection system. There is a limit to the rate at which photons are emitted from the specimen and thus a limit to the temporal resolution of the measurements that is set by the delay between absorption and reemission of a photon, an inherent property of the fluorophore in the indicator. In practice one should never illuminate so strongly that following emission of a photon there is a very high probability of immediate reexcitation, as complex and destructive photochemical process are likely to occur under these conditions, processes thought to result from the absorption of a photon by an already excited molecule (102). This limit must be kept in mind when planning experiments to follow rapid chemical changes in cells with fluorescent indicators. Where temporal resolution is of paramount concern, strategies must be employed that themselves introduce minimal delays in detection of the pattern of photon emission from the specimen, as one cannot compensate for delays introduced by a particular imaging strategy simply by increasing the illumination intensity without limit. Finally and perhaps most importantly, the poor depth discrimination of the fluorescent microscope seriously limits the contrast and spatial resolution of images of most biological specimens, which are complex three-dimensional structures. This results from the fact that photons emitted from fluorophores at all depths within a cell are collected with almost equal efficiencies and it is only the extent of blurring of signal that varies with the distance away from the plane of focus (34). As a consequence, a single image of a cell containing a fluorescent indicator obtained with a standard fluorescence microscope reflects signal from all indicator molecules within that cell; the signal from fluorophores above or below the plane of focus is, however, blurred in a complex fashion determined by the distance of each fluorophore relative to the plane of focus of the microscope objective. At the very least, one needs to be aware of this property of the standard fluorescent microscope in interpreting data obtained with it. In addition, however, one can employ strategies to reduce this problem.

Fluorescent Microscope Imaging Methods

Two general strategies have been developed in recent years to deal with the problems resulting from the poor depth discrimination of the fluorescence microscope. One uses various modifications to the optics of the fluorescence microscope to obtain an image that originates from a narrower focal plane within the cell. The other employs computational methods to reassign light in a series of images at several focal planes more closely to the points within the cell where it originated. The basic principles behind these approaches and their strengths and weaknesses are discussed in the section that follows.

Confocal Microscopy. Perhaps the most widely employed new microscopic approach to enhance the axial depth discrimination and contrast in biological specimens is confocal microscopy. The concept originated almost 40 years ago with Minsky (73), but versions of the confocal microscope that were useful for physiology have become available only during the past ten years (17, 18). Enhanced depth discrimination is achieved in the confocal microscope by illuminating the specimen with a highly focused beam of light, while fluorescence excited in planes above and below the point of optimum focus is suppressed by placing a small aperture in the emission path at an intermediate image plane. An image of the distribution of a fluorescent indicator in a specimen is built up either by moving the specimen through a stationary beam or by moving the excitation beam and its associated aperture over the specimen. In either case, because fluorescence is measured point by point within a specimen, the method produces complete images of a specimen rather slowly (typical images require exposures of 1 second or more). Increased temporal resolution may be obtained by confining the beam to a limited area of a cell (24) or alternatively, multiple beams and pinhole apertures may be moved in tandem over the field to achieve a degree of parallelism in image acquisition in the confocal mode (17). The density of the beam/aperture arrays is limited by the need to keep the spacing between pinhole apertures sufficiently far apart that light from each point that is illuminated is not able to pass through adjacent apertures (111; note added in proof). The confocal approach, therefore, effectively reduces the contribution of light from out-of-focus planes to an image, but it does so at the price of diminished temporal resolution and is inherently inefficient in its use of photons emitted

from the specimen, as photons emitted by fluorophores above and below the plane of focus are largely thrown away by this method. For a specimen with limited numbers of fluorophores, each capable of undergoing a limited number of excitation and emission cycles, this inefficiency may preclude the use of confocal techniques for indicators or cells with weak signals and will certainly diminish the number of observations that can be made sequentially on the same cell. In such instances, a confocal microscope can still be useful, however, if the confocal aperture(s) is opened up a bit, thereby increasing the efficiency of photon detection yet still rejecting photons that emerge from regions of a cell far away from the plane of focus; in this partial confocal mode the time to image a specimen is also reduced due both to the enhanced signal strength as well as to the larger area from which photons are collected at each point in the specimen. This strategy may be employed with a standard scanned beam system or using a slit aperture that is scanned in only one dimension to build up an image.

Two-Photon, Standing Wave, and Near-Field Fluorescence Microscopy. Two-photon excitation fluorescence microscopy, like confocal microscopy, enhances depth discrimination and contrast in biological specimens (30). The improved depth discrimination is achieved not by enhanced selectivity in the detection of in-focus emitted photons, but rather by increasing the probability that fluorophores in the plane of focus are excited relative to those above and below that plane. This is achieved by excitation with light about twice the wavelength (half the energy) for traditional single-photon excitation of a particular fluorophore. Excitation only occurs when two of these lower energy photons are simultaneously absorbed. Unlike traditional single-photon excitation, where the probability of excitation of a fluorophore varies linearly with the flux density, with two-photon excitation the probability of excitation varies with the square of flux density. Hence the probability of excitation falls off more steeply along the optical axis as the illumination beam becomes defocused. Because the probability of simultaneous absorption of two photons is considerably less than for single-photon excitation, very powerful, and to date expensive, pulsed lasers are required for two-photon excitation. Images are typically generated by moving a highly focused excitation beam over the specimen and digitally producing a composite image of the measured fluorescence intensity distribution. As a consequence the time to generate an image is relatively long, taking several seconds even for well-stained specimens, much as for the confocal microscope. Unlike the confocal microscope, the two-photon microscope does not reject any photons that are emitted from the fluorophores; hence it does not inherently compromise one's ability to make repeated measurements on a cell or image of weakly fluorescent specimens. This is very much an evolving microscopic mode, with new lasers likely to emerge that will diminish the cost of applying this approach, as well as expanding the fluorescent compounds to which it can be applied. At present it seems particularly well suited for focally activating caged compounds which require near-ultraviolet wavelengths (320nm–350 nm) for activation, thus being well matched to two-photon activation by titanium-sapphire lasers (670nm–1100 nm), which have the required power to produce two-photon absorption.

Standing-wave excitation fluorescence microscopy (12), like two-photon excitation microscopy, also enhances depth discrimination and contrast by more selectively exciting fluorophores in the plane of focus of the microscope. This technique sets up an interference pattern between two coherent laser beams along the optical axis of the microscope, causing the excitation intensity to fall off sharply, away from the plane where maximal constructive interference takes place. The constructive interference patterns are periodic, and thus there may be multiple planes in the specimen separated by the wavelength of the excitation beam that are strongly excited. Thus this method has proven most effective in obtaining high-contrast images of cellular regions whose thickness is less than the wavelength of light used to excite the indicator. Again, this is an approach that is evolving, and anticipated progress in synthesizing interference patterns that have a single interference node of maximum brightness may make this approach useful for thicker biological structures. The combination of this or the other optical methods discussed with computational methods that reassign light in a series of images more accurately to the point at which it originates is most likely to yield the highest resolution, contrast, and depth discrimination (34).

Before leaving this discussion of optical methods that enhance resolution and depth discrimination of the fluorescence microscope, near-field optical methods need to be mentioned (14, 60). This clever approach results in a highly restricted excitation of fluorophores in a specimen by delivering the excitation beam through an aperture that is smaller than the wavelength of light and that is placed extremely close to the specimen. In the region very close to the aperture, light exists as a highly collimated beam whose size and shape is determined solely by the aperture from which it emerges. A short distance beyond the aperture, classical diffraction effects take over and the light spreads out; thus, to achieve resolution beyond the limits set by classical optics, the aperture must be kept very close

(within 50 nm, typically) to the specimen. By scanning a light guide (a metal-coated micropipette) over the specimen, a map of fluorescence can be generated with the resolution of the light guide. To date, this approach has been applied to investigate the properties of single molecules on a flat substrate (14, 52, 99). Investigators in this field are currently developing methods to sense the position of the light guide relative to a more irregular surface opening the possibility that in the years to come this method may be used to locate and then study the properties of single molecules on membrane surfaces and possibly at other sites within the cell.

Computational Optical Sectioning Microscopy. All the computational methods for increasing the depth discrimination, resolution, and numerical accuracy of fluorescent images are based on the starting premise that the fluorescence microscope is a linear system (22). What this means is that signal from each fluorescent probe within the cell gives rise to a signal within the image independent of any other, and that the conversion of the signal emanating from each fluorescent molecule to the pattern of light in the image can be described by a single process or function. Furthermore, most of the computational methods assume that the microscope is a spatially invariant system, meaning that the effect of moving an object in the microscope field is simply to change its position in the image in a proportional manner.

The function responsible for collecting light from each fluorophore in a cell and forming a pattern of light in the image can be determined simply by obtaining an image of a infinitesimally small fluorescent particle. A fluorescent bead whose diameter is smaller than the resolution limit of the light microscope is effectively such a fluorescent particle. A complete description of the process or function whereby light from such a small particle at any depth within a cell is formed into an image by the microscope is obtained by acquiring images of the bead as the focus of the microscope is varied over a range of focal positions at regular intervals. The resulting set of images is known as the point-spread function of the microscope. Light from every fluorophore in the cell is represented in the microscope in a manner described exactly by this point-spread function. For example, light from a fluorophore that is in the plane of focus is represented in the microscope image like the image of the fluorescent bead when it is perfectly in focus; light from a fluorophore that is some distance from the plane of focus contributes its light to the image in a pattern identical to the image of the sub-resolution diameter fluorescent bead the same distance from focus. This transformation of the pattern of light-emitting fluorophores in the cell into the distribution of light in the microscope image can be expressed mathematically as shown in equation 1–4, which states that the microscope image (image) results from the convolution (denoted by *) of the distribution of fluorophores within the cell (object) with the microscope's point-spread function; in addition, noise from various sources is also added.

$$Image = (Object*PSF) + Noise \qquad (1-4)$$

As physiologists we want to know the molecular distribution that gave rise to a particular image, and so what we want to do is somehow to reverse what the microscope has done in forming an image. That is, we would like to reverse the convolution with the point-spread function to be able to see more accurately how molecules are distributed within the cell. While there are one-step mathematical procedures that can effect this deconvolution, simple deconvolution amplifies noise present in even the best images and thus does not produce useful results. More complicated procedures are thus required to reverse the blurring introduced by the microscope that utilize additional information in obtaining a more accurate estimate of the molecular distribution in a cell. The principal additional constraint used in these procedures is that estimates of the distribution of fluorescent probes must be non-negative, a powerful but simple statement about the physical nature of the measurement; in addition, the estimate of fluorophore distribution is constrained to be smoothed slightly. An image of fluorophore distribution in a cell is determined by computational methods that search for the best estimate of fluorophore distribution which when convolved with the point-spread function of the microscope gives rise to a series of images at various planes that best match the actual images of the cell (3, 4, 19). The algorithms used to find the best estimates of molecular distribution that incorporate the non-negativity constraint are iterative and, when first introduced, required large and expensive computers. Today they can be run in a few hours on a personal computer and in a few minutes on relatively inexpensive array processors. Other computational procedures have been described that make a number of simplifying assumptions in order to reduce the computational task of obtaining a more accurate view of molecular distribution in a cell (22, 76). While these methods produce images that appear sharper, they cannot guarantee either numerical accuracy or that the pattern of light in the images more accurately reflects the actual distribution of fluorescent indicators in the cell. Such accuracy can only be proven for methods based on a complete description of the process for the formation of images by the fluorescence microscope.

The computational approach to enhancing resolution, contrast, and depth discrimination, unlike the confocal approach, uses all the available photons that are collected by the objective, thus performing well with indicators giving both strong and weak signals. Because it is so efficient, adequate images may be obtained with minimal exposure of the cell to the excitation light, thereby diminishing phototoxicity and photobleaching (66). The resolution obtained by this approach has exceeded that which is theoretically possible with the confocal or two-photon excitation fluorescence microscope (20). In addition, because it employs images obtained by a standard fluorescence microscope where the specimen is uniformly illuminated and an image of fluorescence emission from all points in the cell simultaneously acquired, data collection using this approach is inherently faster, making it well suited for following rapid changes in cells (53, 65); seven optical sections required to generate a high-resolution image of a limited depth of a cell may be obtained in under 20 ms. These computational methods may be combined with any of the new microscopic methods described above to achieve the highest possible resolution, contrast, and depth discrimination. Computational optical sectioning microscopy is not particularly effective on very thick specimens where there is a large amount of out-of-focus light relative to that which is coming from planes at, or near, focus; similarly, it does not work particularly well with indicator signals that are uniform throughout a cell. When studying such specimens it is best to collect images at multiple planes using one of the techniques that select for signals from a limited depth within the cell, and then apply the computational methods to that data.

Finally, it should be stressed that the resolution that can be obtained by any of the aforementioned techniques depends critically on the quality of the optics used to collect the fluorescent images. The efficiency of light collection, the resolution, and the depth discrimination are all directly dependent on the numerical aperture of the objective. In working with high numerical aperture objectives that are corrected for various sources of aberration, it is important that the specimen is mounted in medium with a refractive index and with a coverslip of thickness and refractive index specified by the manufacturer of the microscope objective. Lack of attention to these requirements will introduce spherical aberration into the images, thereby diminishing resolution (51). Furthermore, the point-spread function will vary with distance above the coverslip and hence, deconvolution methods based on the assumption that there is a single point-spread function describing the imaging process at all points in the specimen will not perform optimally. Water immersion objectives should be used for work on living cells or other specimens in aqueous media; although these objectives have an upper limit numerical aperture that is slightly lower than oil immersion objectives, they are the only objective that is properly corrected for work on living cells (38).

Detectors

Unless one is building an imaging system from basic components, commercially available instruments usually come with a light detector as an integral component. In choosing an instrument, the properties of the light detector are extremely important as the signal:noise, and consequently the reliability of one's measurements, will be critically determined by the light detector used. Several fundamental concerns should govern the choice. First, one wants a detector that is very efficient in detecting the photons that impinge on it but adds little or no noise to its output. As the efficiency of photon detection (typically involving conversion of a photon to an electron) varies depending on wavelength, it is important to compare devices at the wavelength of the indicator that will be principally used. Second, one needs to consider the speed with which the photon-detection device can pass its information on to a storage system, usually a computer, if one wants to follow rapidly changing signals from a cell.

There are basically two kinds of devices available for non-imaging measurements of light emission from a cell or cellular region, photomultipliers and photodiodes. Photomultipliers have traditionally been used for this purpose and can have extremely low inherent noise levels (as low as 100 electrons/s). They can adequately detect photons at quite high rates (1×10^8 photons/s) and thus are well suited for following rapidly changing signals. The principal disadvantage of photomultipliers is that they are not particularly efficient at detecting photons impinging on them, which diminishes the reliability of one's measurements. The efficiency of conversion of an incoming photon to an electron is typically 10%–15%, although efficiencies as high as 40% are available with some new photomultipliers whose photocathodes are made of special high-quantum-efficiency materials. This relatively low detection efficiency is a characteristic of the materials used on the photomultiplier's front surface, which is responsible for the conversion of a photon to an electron that is then amplified in the device. Higher efficiencies of conversion of a photon to an electron (as high as 90%) are attainable using photodiodes that employ silicon at the primary detection surface. Avalanche photodiodes have a high-quantum-efficiency silicon front surface and sufficiently high gains to allow photon counting in the face of their

inherent noise levels. Hence such devices might be expected ultimately to replace photomultipliers for photometric applications (8). Their principal limitation is that they cannot be used at relatively high photon fluxes, which affects their usefulness for applications requiring the highest time resolution.

Images of the faint fluorescence in single cells are currently detected with either an image intensifier coupled to a camera whose output is generated usually in video format or, alternatively, with a charge-coupled-device (CCD) camera whose output is digital and does not contain an image intensifier. An image intensifier is a device that converts each photon into an electron, amplifies the charge in a more or less spatially coherent manner, and then re-forms an image so that the intensity distribution in the original image is thus amplified. Image-intensifier systems may have gains of over a million times and thus a single detected photon will be clearly visible (92). These high gains are required so that even faint fluorescent signals can be detected in the face of the considerable noise associated with video output electronics. The efficiency of conversion of an incident photon to an electron by most image intensifiers is typically between 10% and 15%, and while a single photon produces a clearly detectable signal the reliability of the signal at any point in the image is reduced, as five of every six photons incident on the camera are not detected. As discussed earlier, the best signal:noise that can be achieved is that determined by the uncertainty in photon detection, which is equal to the square root of the number of detected photons. Thus image intensifiers cause irretrievable losses in the information available from a cell. Modern CCD cameras have a considerably higher efficiency of photon detection and numerous other characteristics : linear response over a wide dynamic range ($1:10^6$ photons), absence of geometric distortion, and low noise (as low as 1–2 photon equivalents) that make them the detector of choice for imaging fluorescence signals from cells (5, 50). Until recently, however, low noise levels could be achieved only by digitizing the charge at each pixel on the CCD camera at relatively slow rates (200,000 pixels/s), and this limited the speed with which images could be acquired (a 512×512 image took slightly more than a second to be digitized and stored). Recent improvements in electronics and the use of multiple parallel digitization and readout strategies have allowed the full acquisition of a 128×128 pixel image in under 2 ms while still preserving ultra-low noise and high quantum efficiency in some experimental devices. Larger format versions of this technology can be expected in the next few years, opening the possibility of imaging faint fluorescent signals from cells with both high speed and sensitivity.

Image Visualization and Analysis

Implicit in the discussion of the new optical methods for addressing questions of interest to cell physiologists is that the fluorescence due to some probe or indicator in a cell is ultimately stored in digital form. Were it not for this, very few of the strategies described would be possible. The numerical nature of the data generated by modern digital fluorescence microscope systems clearly makes possible the conversion of intensity patterns in an image formed by the microscope into estimates of the concentration of a particular molecule or ion or the degree of binding of some molecule within the cell. The numerical nature of the data also opens up new possibilities for the visualization and analysis of information generated by these new techniques.

The array of numbers (often as large as 10^7 values) describing the distribution of a particular molecule or ion can be displayed using a number of powerful computer graphics software packages (63) that allow a physiologist to "see" the entire data set at once. New display devices allow multiple observers to simultaneously view in stereo a 3D image of molecular distribution (63). Data describing the 3D distribution of molecules inside a cell can be viewed from any desired vantage point so as to search for patterns of organization within the cell. For example, the cell may be viewed from the side to determine if molecules are organized within the cell in a unique manner relative to the cell–substrate interface (21). The data can be displayed so that molecular distribution is viewed from inside the cell looking in any direction, for example to follow the path of elements of the cytoskeleton within the cell or to search for an axis of symmetry about which molecules may be organized (35). One can simply zoom in or out of a region of interest to relate fine details of molecular organization to the overall cell structure. If images are taken over time, one can visualize changes in molecular distribution in time or use the time series to support hypotheses about patterns of organization in regions where the signal:noise in a single image may be poor. Patterns of distribution of two or more different molecular species in the same cell may be superimposed using different colors to convey simultaneously information about each molecular species (21, 78). Various interactive pointing devices can be used to point to specific pixels, regions, or directions in a 3D image and thereby change the display, or in some manner mark the data set as part of an analysis strategy (61, 63, 62, 78).

While there are powerful tools currently available for visualizing data obtained using these new optical methods, considerable work remains to be done. A significant challenge for computer scientists is to develop meaningful visualization metaphors for the display of increasing amounts of information about different molecules from the same cell. Perhaps even more daunting is the task of developing intuitive visualization schemes for the 3D distribution of molecules or ions like Ca^{2+}, which may not be organized into discrete bodies but instead exist throughout the cell, albeit at different densities. Finally, a considerable challenge for the physiologist is to communicate to computer scientists what aspect of a data set needs to be visualized, and in viewing images generated by a particular algorithm, to remember constantly what is, and is not, being displayed to avoid either missing some important principle to be gained from the data or misinterpreting what the image may be saying about one's data.

Visualization of data obtained by these optical methods is intimately linked to the analysis of that data. Visual inspection of images of molecular distribution will often lead to the formation of a hypothesis regarding the linkage between the distribution of a particular ion or molecule and a specific cell function. More rigorous analysis of one's data can then be carried out to test that particular hypothesis. For example, if one suspects that a change in the concentration of a particular ion or molecule occurred in response to a stimulus one can determine how large a change occurred, what the kinetics of that change was, and what the likelihood was (given estimates of the noise associated with the measurements) that the observed change was significant. One can apply statistical tests to patterns of distribution of a particular molecule to determine the probability that the observed pattern is simply due to chance. If the distributions of two different molecular species have been imaged in the same cell, one can utilize statistical methods to determine the likelihood that the observed pattern of distribution of one species relative to another is simply due to chance or instead reflects biological/chemical constraints (78). The development of algorithms and software to analyze patterns of organization of two molecular species in a cell is in its infancy. Clearly it would be of great interest to have techniques that utilize images of molecular distribution to extract underlying basic rules of organization of specific molecules with respect both to other structures and to molecules within the cell.

Many types of molecules are organized into discrete structures inside a cell (for example, membranes, filaments). It is often important to simplify images of such molecules so one can see how these basic structures are organized within a cell or how other molecules are distributed relative to those structures. This requires that these basic structures be identified in images of molecular distribution that are produced by the microscope. There are three types of strategies that have been employed for identifying basic structural elements into which specific molecules are clustered within a cell.

The simplest involves visualizing the images of molecular distribution and then utilizing some computer interface device to interactively mark on the display where filaments or membranes are located. This can be a somewhat subjective and tedious process especially for molecules like tubulin or actin, which are organized into several hundred filaments in a cell. However, once all filaments or membranes within a cell have been identified, that information represents a simpler description of the distribution of a particular molecule. This sparser representation of the distribution of molecules facilitates the storage, display, and understanding of the overall pattern of organization of the molecule within the cell.

To obviate the subjectivity and tedium associated with interactive identification of structures into which molecules may be clustered, methods have also been developed to locate automatically filaments, membranes, or other identified and relatively easily modeled structures into which a given molecular species may be organized. While algorithms for automated feature extraction have been extensively developed for the relatively regular world of machine parts on conveyor belts, or airplanes, or tanks viewed from aerial photographs, new algorithms are required to deal with patterns of organization of molecules inside cells. The structures into which molecules are organized appear in images to have poorly defined boundaries and are often quite variable in size, shape, and orientation. Furthermore, the images of fluorescent probes inside cells are often noisy due both to uncertainties in the labeling by the fluorescent probes, and to noise introduced by the processes involved in photon detection. Algorithms based on mathematical filters whose size, shape, and orientation give a strong response when centered over a short stretch of a filament or a patch of membrane have yielded promising results (25, 26, 35). The information extracted by these filters about parts of filaments or membranes is then combined to complete the description of that structure. In general these methods work best when the algorithms incorporate as much a priori knowledge as possible about the structures into which a molecule is organized and make final decisions about the size, shape, and orientation of those structures using multiple tests of the validity of that assignment. Even then it is desirable always to check the accuracy of the assignment of specific structural elements to specific regions of an image by the automated feature extraction methods. This is best accomplished using visualization tools that allow the simultaneous visualization of molecular distribution and the structural elements that have been created by the automated feature extraction algorithm (62).

While automated feature extraction algorithms have proven useful in a limited number of studies, the current trend is to merge man and machine in this most difficult task. The ability of the trained human observer to analyze complex visual information is often far superior to any algorithms that have been developed and thus the most successful strategies for image feature extraction involve human interaction to initialize the search process or to check and possibly fine tune the results obtained by automated feature extraction algorithms. For example, algorithms based on deformable surfaces that are attracted to bright pixels have been used to find compartment boundaries in cells when a molecule present at high density at that interface has been labeled (78). To avoid problems introduced by noise in the image, we have found that placement of that surface near the compartment edge using interactive computer graphics software helps to ensure best results.

Much effort will have to be put into this area in the years to come. Modern digital imaging microscopes are producing vast amounts of new information about the spatiotemporal characteristics of signals within and between cells. While new phenomena are being uncovered by the new imaging methods, insight into the meaning of new and complex signaling patterns remains to be extracted from much of the data that is being generated. For example, recordings of $[Ca^{2+}]$ in hippocampal slices have revealed that both neurons and glial cells exhibit $[Ca^{2+}]$ waves that spread in complex patterns in response to neurotransmitters (28). How those signal patterns cause information to be stored or processed remains unknown and will require methods to extract, store, and process information about the spatial and temporal nature of the sequence of Ca^{2+} signals.

NOTE ADDED IN PROOF

A recent report by R. Juskaitis, T. Wilson, M. A. A. Neil, and M. Kozubek (Nature 383: 804–806, 1996) describes a novel multiaperture excitation–emission illumination system with much higher aperture densities than previous multiaperture confocal microscopes that yields a confocal image after single algebraic processing.

I thank my colleagues in the Biomedical Imaging Group (Walter Carrington, Kevin Fogarty, Larry Lifshitz, and Richard Tuft) for their helpful discussions throughout my work in this area and specifically during the preparation of this manuscript. Thanks also to Debbie Bennett, Michael Berridge, Martin Bootman, Sir Andrew Huxley, Roger Moreton, and Peter Thorn for their critical reading of this manuscript, and Debbie Roy for help with the references. I gratefully acknowledge grant support from the NSF (BIR 9200027) and the NIH (HL14523 and HL47530) during the preparation of this chapter.

REFERENCES

1. Adams, S. R., A. T. Harootunian, Y. J. Buechler, S. S. Taylor, and R. Y. Tsien. Fluorescence ratio imaging of cyclic-AMP in single cells. *Nature* 349: 694–697, 1991.

2. Adams, S. R., J.P.Y. Kao, G. Grynkiewicz, A. Minta, and R. Y. Tsien. Biologically useful chelators that release Ca^{2+} upon illumination. *J. Am. Chem. Soc.* 110: 3212–3220, 1988.

3. Agard, D. A., Y. Hiraoka, P. Shaw, and J. W. Sedat. Fluorescence microscopy in 3 dimensions. *Methods Cell Biol.* 30: 353–377, 1989.

4. Agard, D. A., and J. W. Sedat. 3-dimensional architecture of a polytene nucleus. *Nature* 302: 676–681, 1983.

5. Aikens, R. Properties of low-light-level and slow-scan detectors. In: *Fluorescent and Luminescent Probes for Biological Activity*, edited by W. T. Mason. San Diego: Academic Press, 1993, p. 277–286.

6. Allbritton, N. L., E. Oancea, M. A. Kuhn, and T. Meyer. Source of nuclear calcium signals. *Proc. Natl. Acad. Sci. U.S.A.* 91: 12458–12462, 1994.

7. Amsterdam, A., S. Lin., and N. Hopkins. The aequorea-Victoria green fluorescent protein can be used as a reporter in live zebrafish embryos. *Dev. Biol.* 171: 123–129, 1995.

8. Art, J. Photon detectors for confocal microscopy. In: *Handbook of Biological and Confocal Microscopy*, edited by J. B. Pawley. New York/London: Plenum Press, 1990, p. 127–139.

9. Ashkin, A., K. Schutze, J. M. Dziedzic, D. Euteneuer, and M. Schliwa. Force generation of organelle transport measured in vivo by an infrared-laser trap. *Nature* 348: 346–348, 1990.

10. Axelrod, D. Fluorescence polarization microscopy. *Methods Cell. Biol.* 29: 333–352, 1989.

11. Axelrod, D., D. E. Koppel, J. Schelessinger, E. Elson, and W. W. Webb. Mobility measurement by analysis of fluorescence photobleaching recovery kinetics. *Biophys. J.* 16: 1055–1069, 1976.

12. Bailey, B., D. L. Farkas, D. L. Taylor, and F. Lanni. Enhancement of axial resolution in fluorescence microscopy by standing-wave excitation. *Nature* 366: 44–48, 1993.

13. Becker, P. L., and F. S. Fay. Photobleaching of fura-2 and its effect on determination of calcium concentrations. *Am. J. Physiol.* 253(Cell Physiol. 22): C613–C618, 1987.

14. Betzig, E., and R. J. Chichester. Single molecules observed by near-field scanning optical microscopy. *Science* 262: 1422–1425, 1993.

15. Betzig, E., and J. K. Trautman. Near-field optics—microscopy, spectroscopy, and surface modification beyond the diffraction limit. *Science* 257: 189–195, 1992.

16. Blinks, J. R., W. G. Wier, P. Hess, and F. G. Prendergast. Measurement of Ca^{2+} concentrations in living cells. *Prog. Biophys. Mol. Biol.* 40: 1–114, 1982.

17. Boyde, A. Stereoscopic images in confocal (tandem scanning) microscopy. *Science* 230: 1270–1272, 1985.

18. Brakenhoff, G. J., H.T.M. Vandervoort, E. A. Vanspronsen, W.A.M. Linnemans, and N. Nanninga. 3-dimensional chromatin distribution in neuro-blastoma nuclei shown by confocal scanning laser microscopy. *Nature* 317: 748–749, 1985.

19. Carrington, W. A., and K. E. Fogarty. 3-D molecular distribution in living cells by deconvolution of optical sections using light microscopy. *Proceedings of the 13th Annual Northeast Bioengineering Conference*, Philadelphia, March, 1987.

20. Carrington, W. A., R. M. Lynch, E.D.W. Moore, G. Isenberg, K. E. Fogarty, and F. S. Fay. Superresolution 3-dimensional images of fluorescence in cells with minimal light exposure. *Science* 268: 1483–1487, 1995.

21. Carter, K. C., D. Bowman, W. Carrington, K. Fogarty, J. A. McNeil, F. S. Fay, and J. B. Lawrence. A three-dimensional view of precursor messenger RNA metabolism within the nucleus. *Science* 259: 1330–1335, 1993.

22. Castleman, K. R. *Digital Image Processing*. New Jersey: Prentice-Hall, Inc., 1979.

23. Chalfie, M., Y. Tu, G. Euskirchen, W. W. Ward, and D. C. Prasher. Green fluorescent protein as a marker for gene-expression. *Science* 263: 802–805, 1994.

24. Cheng, H., W. J. Lederer, and M. B. Cannell. Calcium sparks—elementary events underlying excitation-contraction coupling in heart muscle. *Science* 262: 740–744, 1993.

25. Coggins, J. M., F. S. Fay, and K. E. Fogarty. Development and application of a three-dimensional artificial visual system. In: *Computer Methods and Programs in Biomedicine*, Elsevier Science Publishers, 22: 69–77, 1986.

26. Coggins, J. M., K, E. Fogarty, and F. S. Fay. Interfacing image processing and computer graphics systems using an artificial visual system. *Proceedings of IEEE Conferences on Graphics and Vision Interface*, 1986 p. 44–52.

27. Connor, J. A. Intracellular calcium mobilization by inositol 1,4,5-trisphosphate—intracellular movements and compartmentalization. *Cell Calcium* 14: 185–200, 1993.

28. Cornellbell, A. H., S. M. Finkbeiner, M. S. Cooper, and S. J. Smith. Glutamate induces calcium waves in cultured astrocytes—long-range glial signaling. *Science* 247: 470–473, 1990.

29. Debiasio, R. L., L. L. Wang, G. W. Fisher, and D. L. Taylor. The dynamic distribution of fluorescent analogs of actin and myosin in protrusions at the leading-edge of migrating Swiss 3T3 fibroblasts. *J. Cell. Biol.* 107: 2631–2645, 1988.

30. Denk, W., J. H. Strickler, and W. W. Webb. 2-photon laser scanning fluorescence microscopy. *Science* 248: 73–76, 1990.

31. Drummond, R. M., R. Sreekumar, J. W. Walker, R. E. Carraway, M. Ikebe, and F. S. Fay. A caged peptide inhibitor of calmodulin prevents Ca^{2+}-dependent enhancement of Ca^{2+} current in smooth muscle cells (abstract). *Biophys. J.* 70: A386, 1996.

32. Ellis-Davies, G.C.R., and J. H. Kaplan. Nitrophenyl-EGTA, a photolabile chelator that selectively binds Ca^{2+} with high-affinity and releases it rapidly upon photolysis. *Proc. Nat. Acad. Sci. U.S.A.* 91: 187–191, 1994.

33. Elson, E. L., and H. Qian. Interpretation of fluorescence correlation spectroscopy and photobleaching recovery in terms of molecular interactions. *Methods Cell Biol.* 30: 307–332, 1989.

34. Fay, F. S., W. Carrington, and K. E. Fogarty. 3-dimensional molecular distribution in single cells analyzed using the digital imaging microscope. *J. Microscopy-Oxford* 153: 133–149, 1989.

35. Fay, F. S., K. E. Fogarty, and J. M. Coggins. Analysis of molecular distribution in single cells using a digital imaging microscope. In: *Optical Methods in Cell Physiology*, edited by P. De Weer and B. Salzberg. John Wiley & Sons, 1986, 51–62.

36. Fay, F. S., S. H. Gilbert, and R. A. Brundage. Calcium signaling during chemotaxis. *Calcium Waves, Gradients Oscillations*. Ciba Foundation Symposium 188: 121–135, 1995.

37. Finer, J. T., R. M. Simmons, and J. A. Spudich. Single myosin molecule mechanics—piconewton forces and nanometer steps. *Nature* 368: 113–119, 1994.

38. Gasberg, P. K., A. Horowitz, R. A. Tuft, W. A. Carrington, F. S. Fay, and K. E. Fogarty. Analysis of the true 3-dimensional point spread function and its effects on quantitative fluorescence microscopy [Abstract]. *Biophys. J.* 66: A274, 1994.

39. Gerisch, G., R. Albrecht, C. Heizer, S. Hodgkinson, and M. Maniak. Chemoattractant-controlled accumulation of coronin at the leading edge of dictyostelium cells monitored using green fluorescent protein-coronin fusion protein. *Curr. Biol.* 5: 1280–1285, 1995.

40. Gerisch, G., A. A. Noegel, and M. Schleicher. Genetic alteration of proteins in actin-based motility systems. *Annu. Rev. Physiol.* 53: 607–628, 1991.

41. Giuliano, K. A., P. L. Post, K. M. Hahn, and D. L. Taylor. Fluorescent protein biosensors—measurement of molecular-dynamics in living cells. *Annu. Rev. Biophys. Biomol. Struct.* 24: 405–434, 1995.

42. Goldman, Y. E., M. G. Hibberd, J. A. McCray, and D. R. Trentham. Relaxation of muscle-fibers by photolysis of caged ATP. *Nature* 300: 701–705, 1982.

43. Gough, A. H., and D. L. Taylor. Fluorescence anisotropy imaging microscopy maps calmodulin-binding during cellular contraction and locomotion. *J. Cell Biol.* 121: 1095–1107, 1993.

44. Grynkiewicz, G., M. Poenie, and R. Y. Tsien. A new generation of Ca^{2+} indicators with greatly improved fluorescence properties. *J. Biol. Chem.* 260: 3440–3450, 1985.

45. Gurney, A. M., and H. A. Lester. Light-flash physiology with synthetic photosensitive compounds. *Physiol. Rev.* 67: 583–617, 1987.

46. Hahn, K., R. Debiasio, D. L. Taylor. Patterns of elevated free calcium and calmodulin activation in living cells. *Nature* 359: 736–738, 1992.

47. Hastings, J. W., C. J. Potrikus, S. C. Gupta, M. Kurfurst, and J. C. Makemson. Biochemistry and physiology of bioluminescent bacteria. *Adv. Microb. Physiol.* 26: 235–291, 1985.

48. Heim, R., D. C. Prasher, and R. Y. Tsien. Wavelength mutations and posttranslational autoxidation of green fluorescent protein. *Proc. Natl. Acad. Sci. U.S.A.* 91: 12501–12504, 1994.

49. Heim R., and R. Y. Tsien. Engineering green fluorescent protein for improved brightness, longer wavelengths and fluorescence resonance energy transfer. *Curr. Biol.* 6: 178–192, 1996.

50. Hiraoka, Y., J. W. Sedat, and D. A. Agard. The use of a charge-coupled device for quantitative optical microscopy of biological structures. *Science* 238: 36–41, 1987.

51. Hiraoka, Y., J. W. Sedat, and D. A. Agard. Determination of three-dimensional properties of an optical microscope system. Partial confocal behavior in epifluorescence microscopy. *Biophys. J.* 57: 325–333, 1990.

52. Hwang, J., L. K. Tamm, C. Bohm, T. S. Ramalingam, E. Betzig, and M. Edidin. Nanoscale complexity of phospholipid monolayers investigated by near-field scanning optical microscopy. *Science* 270: 610–614, 1995.

53. Isenberg, G., E. F. Etter, M.-F. Wendt-Gallitelli, A. Schiefer, W. A. Carrington, R. A. Tuft, and F. S. Fay. Intrasarcomere $[Ca^{2+}]$ gradients in ventricular myocytes revealed by high speed digital imaging microscopy. *Proc. Natl. Acad. Sci. U.S.A.* (in press). 93: 5413–5418, 1996.

54. Johnson, C. V., R. H. Singer, and J. B. Lawrence. Fluorescent detection of nuclear-RNA and DNA-implications for genome organization. *Methods Cell Biol.* 35: 73–99, 1991.

55. Kahana, J. A., B. J. Schnapp, and P. A. Silver. Kinetics of spindle pole body separation in budding yeast. *Proc. Natl. Acad. Sci. U.S.A.* 92: 9707–9711, 1995.

56. Kaplan H., B. Forbush, III., and J. F. Hoffman. Rapid photolytic release of adenosine 5′-triphosphate from a protected analogue: utilization by the Na: K pump of human red blood cell ghosts. *Biochem. J.* 17: 1929–1935, 1978.

57. Lakowicz, J. R. *Principles of Fluorescence Spectroscopy.* New York: Plenum Press, 1983.

58. Lakowicz, J. R., H. Szmacinski, K. Nowaczyk, and M. L. Johnson. Fluorescence lifetime imaging of calcium using Quin-2. *Cell Calcium* 3: 131–147, 1992.

59. Lester, H. A., and J. M. Nerbonne. Physiological and pharmacological manipulations with light-flashes. *Annu. Rev. Biophys. Bioeng.* 11: 151–175, 1982.

60. Lewis, A., and K. Lieberman. Near-field optical imaging with a non-evanescently excited high-brightness light source of subwavelength dimensions. *Nature* 354: 214–216, 1991.

61. Lifshitz, L. M. Tracking cells and subcellular features. In *Advances in Image Analysis,* edited by Y. Mahdavieh and R. C. Gonzalez. Bellingram, WA: SPIE Press, 1992, p. 218–243.

62. Lifshitz, L., J. Collins, E. Moore, and J. Gauch. Computer vision and graphics in fluorescence microscopy. *IEEE Proceedings of the Biomedical Imaging Workshop,* Los Alamitos, CA: IEEE Computer Society Press, 1994, p. 166–175.

63. Lifshitz, L. M., K. Fogarty, J. M. Gauch, and E. Moore. Computer vision and graphics in fluorescence microscopy. In: *Visualization in Biomedical Computing,* edited by R. A. Robb. SPIE Proceedings, vol. 1808, Bellingham, WA: SPIE Press, 1992, p. 521–534.

64. Lifshitz, L. M., and S. M. Pizer. A multiresolution hierarchical approach to image segmentation based on intensity extreme. In: *Computer Vision: Advances and Applications,* edited by R. Kasturi and R. Jain. Los Alamitos, CA: IEEE Computer Society Press, 1991, p. 606–617.

65. Loew, L. M., W. Carrington, R. A. Tuft, and F. S. Fay. Physiological cytosolic Ca^{2+} transients evoke concurrent mitochondrial depolarizations. *Proc. Natl. Acad. Sci. U.S.A.* 91: 12579–12583, 1994.

66. Loew, L. M., R. A. Tuft, W. Carrington, and F. S. Fay. Imaging in 5 dimensions—time-dependent membrane potentials in individual mitochondria. *Biophys. J.* 65: 2396–2407, 1993.

67. Lynch, R. M., W. Carrington, K. E. Fogarty, and F. S. Fay. Metabolic modulation of hexokinase association with mitochondria in living cells. *Am. J. Physiol. (Cell. Physiol. 39),* 270: C488–C499, 1996.

68. Mason, W. T., Editor. *Fluorescent and Luminescent Probes for Biological Activity.* San Diego: Academic Press, 1993.

69. Mathies, R. A., and L. Stryer. Single-molecule fluorescence detection: a feasibility study using phycoerythrin. In: *Applications of Fluorescence in the Biomedical Sciences,* edited by D. L. Taylor, A. S. Waggoner, R. F. Murphy, F. Lanni, and R. R. Birge. New York: Liss, 1986, p. 129–140.

70. Matsumato, B., Editor. *Methods in Cell Biology 38: Cell biological applications of confocal microscopy.* San Diego: Academic Press, 1993.

71. McCray, J. A., and D. R. Trentham. Properties and uses of photoreactive caged compounds. *Annu. Rev. Biophys. Chem.* 18: 239–270, 1989.

72. McNeil, P. L. Incorporation of macromolecules into living cells. *Methods in Cell Biol.* 29: 153–173, 1989.

73. Minsky, M. *Microscopy apparatus,* U.S. Patent #3013467, 1957.

74. Mitchison, T. J. Microtubule dynamics and kinetochore function in mitosis. *Annu. Rev. Cell Biol.* 4: 527–549, 1988.

75. Molloy, J. E., J. E. Burns, J. Kendrick-Jones, R. T. Tregear, and D.C.S. White. Movement and force produced by a single myosin head. *Nature.* 378: 209–212, 1995.

76. Monck, J. R., A. F. Oberhauser, T. J. Keating, and J. M. Fernandez. Thin-section ratiometric Ca^{2+} images obtained by optical sectioning of fura-2 loaded mast-cells. *J. Cell Biol.* 116: 745–759, 1992.

77. Moore, E.D.W., P. L. Becker, K. E. Fogarty, D. A. Williams, and F. S. Fay. Ca^{2+} imaging in single living cells—theoretical and practical issues. *Cell Calcium* 11: 2–3, 1990.

78. Moore, E.D.W., K. D. Philipson, W. A. Carrington, K. E. Fogarty, L. M. Lifshitz and F. S. Fay. Coupling of the Na^+/Ca^{2+} Exchanger, Na^+/K^+ Pump and Sarcoplasmic Reticulum in Smooth Muscle. *Nature* 365: 657–660, 1993.

79. Neubig, R. R. Membrane organization in G-protein mechanisms. *FASEB J.* 8: 939–946, 1994.

80. Nuccitelli, R., Editor. *Methods in Cell Biology.* A practical guide to the study of calcium in living cells. San Diego: Academic Press, 1994.

81. Olson, K. R., J. R. McIntosh, and J. B. Olmsted. Analysis of map-4 function in living cells using green fluorescent protein (GFP) chimeras. *J. Cell Biol.* 130: 639–650, 1995.

82. Pawley, J. B., Editor. *Handbook of Biological Confocal Microscopy.* New York/London: Plenum Press, 1990.

83. Post, P. L., R. L. Debiasio, and D. L. Taylor. A fluorescent protein biosensor of myosin-II regulatory light chain phosphorylation reports a gradient of phosphorylated myosin-II in migrating cells. *Mol. Biol. Cell* 6: 1755–1768, 1995.

84. Post, P. L., K. M. Trybus, and D. L. Taylor. A genetically engineered protein-based optical biosensor of myosin II regulatory light chain phosphorylation. *J. Biol. Chem.* 269: 12880–12887, 1994.

85. Prasher, D. C., V. K. Eckenrode, W. W. Ward, F. G. Prendergast, and M. J. Cormier. Primary structure of the aequorea-victoria green-fluorescent protein. *Gene* 111: 229–233, 1992.

86. Ried, T., A. Baldini, T. C. Rand, and D. C. Ward. Simultaneous visualization of 7 different DNA probes by in situ hybridization using combinatorial fluorescence and digital imaging microscopy. *Proc. Natl. Acad. Sci. U.S.A.* 89: 1388–1392, 1992.

87. Rizzuto, R., M. Brini, P. Pizzo, M. Murgia, and T. Pozzan. Chimeric green fluorescent protein as a tool for visualizing subcellular organelles in living cells. *Curr. Biol.* 5: 635–642, 1995.

88. Rizzuto, R., M. Brini, and T. Pozzan. Targeting recombinant aequorin to specific intracellular organelles. *Methods Cell Biol.* 40: 339–358, 1994.

89. Rizzuto, R., A.W.M. Simpson, M. Brini, and T. Pozzan. Rapid changes of mitochondrial Ca^{2+} revealed by specifically targeted recombinant aequorin. *Nature* 360: 768, 1992.

90. Sawin, K. E., J. A. Theriot, and T. J. Mitchison. Photoactivation of fluorescence as a probe for cytoskeletal dynamic in mitosis and cell motility. In: *Fluorescent and Luminescent Probes for Biological Activity,* edited by W. T. Mason. San Diego: Academic Press, 1993, p. 405–419.

91. Scanlon, M., D. A. Williams, and F. S. Fay. A Ca^{2+} insensitive form of fura-2 associated with polymorphonuclear leukocytes. *J. Biol. Chem.* 262: 6308–6312, 1987.

92. Spring, K. R. Detectors for fluorescence microscopy. *Scanning Microscopy* 5: 63–69, 1991.

93. Storch, J., C. Lechene, and A. M. Kleinfeld. Direct determination of free fatty-acid transport across the adipocyte plasma membrane using quantitative fluorescence microscopy. *J. Biol. Chem.* 21: 13473–13476, 1991.

94. Svoboda, K., and S. M. Block. Biological applications of optical forces. *Annu Rev. Biophys. Biomol. Structure* 23: 247–285, 1994.

95. Svoboda, K., C. F. Schmidt, B. J. Schnapp, and S. M. Block. Direct observation of kinesin stepping by optical trapping interferometry. *Nature* 365: 721–727, 1993.

96. Symons, M. H., and T. J. Mitchison. Control of actin polymerization in live and permeabilized fibroblasts. *J. Cell Biol.* 114: 503–513, 1991.

97. Taylor, D. L., and Y. Wang, Editors. *Methods in Cell Biology 29 and 30:* Fluorescence microscopy of living cells in culture. San Diego: Academic Press, 1989.

98. Theriot, J. A., T. J. Mitchison, L. G. Tilney, and D. A. Portnoy. The rate of actin-based motility of intracellular listeria-monocytogenes equals the rate of actin polymerization. *Nature* 357: 257–260, 1992.

99. Trautman, J. K., J. J. Macklin, L. E. Brus, and E. Betzig. Near-field spectroscopy of single molecules at room temperature. *Nature* 369: 40–42, 1994.

100. Tse, F. W., A. Tse, and B. Hille. Cyclic Ca^{2+} changes in intracellular stores of gonadotropes during gonadotropin-releasing hormone-stimulated Ca^{2+} oscillations. *Proc. Natl. Acad. Sci. USA.* 91: 9750–9754, 1994.

101. Tsien, R. Y. Fluorescent probes of cell signaling. *Annu Rev. Neurosci.* 12: 227–253, 1989.

102. Tsien, R. Y., and A. Waggoner. Fluorophores for confocal microscopy: photophysics and photochemistry. In: *Handbook of Biological Confocal Microscopy,* edited by J. B. Pawley. New York/London: Plenum Press, 1990, p. 169–178.

103. Valdivia, H. H., J. H. Kaplan, G.C.R. Ellis-Davies, W. J. Lederer. Rapid adaptation of cardiac ryanodine receptors—modulation by Mg^{2+} and phosphorylation. *Science* 267: 1997–2000, 1995.

104. Walker, J. W., A. V. Somlyo, Y. E. Goldman, A. P. Somlyo, and D. R. Trentham. Kinetics of smooth and skeletal muscle activation by laser-pulse photolysis of caged inositol 1,4,5-trisphosphate. *Nature* 327: 249–252, 1987.

105. Wang, J., L. G. Cao, Y. L. Wang, and T. Pederson. Localization of pre-messenger-RNA at discrete nuclear sites. *Proc. Natl. Acad. Sci. USA.* 88: 7391–7395, 1991.

106. Wang, X. F., A. Periasamy, B. Herman, and D. M. Coleman. Fluorescence lifetime imaging microscopy (flim)-instrumentation and applications. *Crit. Rev. Anal. Chem.* 23: 369–395, 1992.

107. Williams, D. A., and F. S. Fay, Editors. Imaging of cell calcium. *Cell Calcium* 11: 55–250, 1990.

108. Williams, D. A., and F. S. Fay. Calcium transients and resting levels in isolated smooth muscle cells as monitored with Quin-2. *Am. J. Physiol. (Cell Physiol. 19)* 250: C799–C791, 1986.

109. Williams, D. A., and F. S. Fay. Intracellular calibration of the fluorescent calcium indicator fura-2. *Cell Calcium* 11: 75–83, 1990.

110. Williams, D. A., K. E. Fogarty, R. Y. Tsien, and F. S. Fay. Calcium gradients in single smooth muscle cells revealed by the digital imaging microscope. *Nature,* 318: 558–561, 1985.

111. Wilson, T., and C.J.R. Sheppard. *Theory and Practice of Scanning Optical Microscopy.* London: Academic Press, 1984.

112. Wolf, D. E. Designing, building, and using a fluorescence recovery after photobleaching instrument. *Methods Cell Biol.* 30: 271–306, 1989.

113. Yin, H., M. D. Wang, K. Svoboda, R. Landick, S. M. Block, and J. Gelles. Transcription against an applied force. *Science* 270: 1653–1657, 1995.

2. Organization and dynamics of the lipid components of biological membranes

T. E. THOMPSON

M. B. SANKARAM

C. HUANG

Department of Biochemistry, University of Virginia, Health Sciences Center, Charlottesville, Virginia

THE LIPID BILAYER AS A STRUCTURAL AND FUNCTIONAL COMPONENT OF BIOLOGICAL MEMBRANES

THE IDEA THAT IN ALL BIOLOGICAL MEMBRANES THE BASIC PERMEABILITY BARRIER is a lipid bilayer has gained universal acceptance. This construct, first presented in a simple form by Danielli and Davison (48) about sixty years ago, was modified in an important way by Singer and Nicholson (170) in the early seventies. These later workers introduced the idea that all of the individual lipid molecules of the bilayer and the integral proteins of the membrane are free to exhibit a variety of motional modes such as translation, vibration, and rotation. The biological membrane thus became a dynamic structure at the molecular level.

The Singer and Nicholson view of the membrane, known as the fluid mosaic hypothesis (170), has received overwhelming experimental support over the past twenty years. There have, however, been a number of important modifications during this period. Several are of considerable significance. One recognizes the interactions of membrane components with cytoskeletal elements (16). Such interactions alter the time scale of the motional dynamics of many membrane proteins. For example, in the red blood cell the attachment to the cytoskeleton of the anion transporter creates a class of these molecules that are essentially immobile on the time-scale of free translational and rotational diffusion (141, 214). A second important modification of the fluid mosaic hypothesis was the recognition that compositional domain structure exists both in the plane of the membrane and in a transverse plane (54, 187).

The fact that the lipid and protein composition on the two faces is very different has been known for some time. In the case of the lipid bilayer, the molecular composition of the two monolayers comprising the bilayer has proved to be markedly different in almost every system examined (220). It has been shown conclusively that membrane proteins do not rotate about a molecular axis located in the membrane plane; hence, each transmembrane membrane protein presents a unique face to each side of the membrane. In general, there is substantially more peptide mass on the cytoplasmic surface of the membrane than on the opposite surface (71).

Interest in the existence of in-plane domain structure in biological membranes has developed more recently. A variety of studies involving different membranes demonstrate both directly and indirectly the existence of in-plain domains (54). Two important questions are posed by this structure. The first concerns the physiological functions of these domains, the second their origins. At present there are no clear answers to the first question; recently, however, some plausible answers have been proposed (134, 187). As far as the second question is concerned, there are clearly several possibilities. In-plane domains could be the result of strong lateral interactions between membrane proteins, interactions between membrane proteins and cytoskeletal elements, phase structure of the lipid bilayer of the membrane, focal addition or removal of membrane components by vesicular transport systems, or some combination of these causes. The possibility that the basis of domains may be phase structure in the bilayer is of particular interest (187). The phase structure of bilayer systems is discussed later (see under Lamellar Phases).

Let us now turn to a brief survey of the molecular structure of membrane lipids and the lipid composition of cell membranes.

MOLECULAR STRUCTURES OF MEMBRANE LIPIDS

Biological membranes are composed of three general types of lipids: glycerophospholipids, sphingolipids, and sterols. Within these groups there are a number of subgroups within which there are many different individual molecules. The outstanding characteristic of the lipids that form biological membrane bilayers is their amphipathic structure. Molecules with this characteristic have a marked geographical segregation of their polar and non-polar moieties. They are generally elongated with a hydrophilic head connected by a belt region of intermediate polarity to a hydrophobic moiety. Because of this segregation of polarity, amphipathic molecules usually form aggregated structures in all solvents. The aggregates are organized to minimize energetically unfavorable, and maximize favorable, interactions with solvent molecules. The structural details of the aggregate depend upon the characteristics of the amphipathic molecule and the solvent. In excess water, the milieu of biology, membrane lipids with the exception of sterols, usually form bilayer structures spontaneously. These molecular lamellae are two molecules in thickness with the component molecules arranged

so that their polar heads form the two faces and the nonpolar tails, the interior of the lamella. From a structural viewpoint, a bilayer consists of two monolayers with their nonpolar faces in contact. The physical properties of bilayers are not, however, simply related to those of the individual monolayers. The bilayer arrangement eliminates the energetically unfavorable contact between the non-polar tails and water, leaving only the favorable contacts between water and the polar portions of the phospholipids forming the two bilayer faces. Under a variety of circumstances, most of which are nonbiological, some membrane lipids will aggregate and give nonbilayer structures. These forms are discussed later (see under Nonlamellar Phases).

Glycerophospholipids

Glycerophospholipids (phosphoglycerides), the most abundant class of membrane lipids, are esters of glycerol and have the generic structure shown in Figure 2.1. The hydroxyl groups on carbons 1 and 2 of the glycerol backbone are each esterified to a long-chain fatty acid. In biological membranes these two fatty acids are generally between 14 and 24 carbons long and made up of an even number of carbons. In addition, the two fatty acyl groups in any given molecule are different. Generally R_1 is saturated and R_2 is unsaturated with the cis configuration at each double bond. The very polar phosphate moiety linked to the third carbon of glycerol is esterified to R_3 which may be one of several groups. Three of the most common are choline $HOCH_2CH_2N(CH_3)_3{}^+$, ethanolamine $HOCH_2CH_2N^+H_3$, and the amino acid serine $HOCH_2CH(COO^-)N^+H_3$. The three glycerophospholipids thus formed are phosphatidylcholine, phosphatidylethanolamine, and phosphatidylserine, respectively. Less common substituents at R_3 are glycerol, inositol, and phosphatidylglycerol. Within these generic subclasses a variety of lipids are defined by the acyl chains at positions 1 and 2. Table 2.1 illustrates the wide diversity of fatty acyl groups that are found in several different phospholipids in a single-membrane system, the human erythrocyte membrane. A detailed discussion of the structure of this class of molecules is presented later (see under Lamellar Phases) and can be found in Cevc and Marsh (37).

Sphingolipids

The second important class of membrane lipids, sphingolipids, are based on the eighteen carbon amine diol, sphingosine. The most common of these and the only

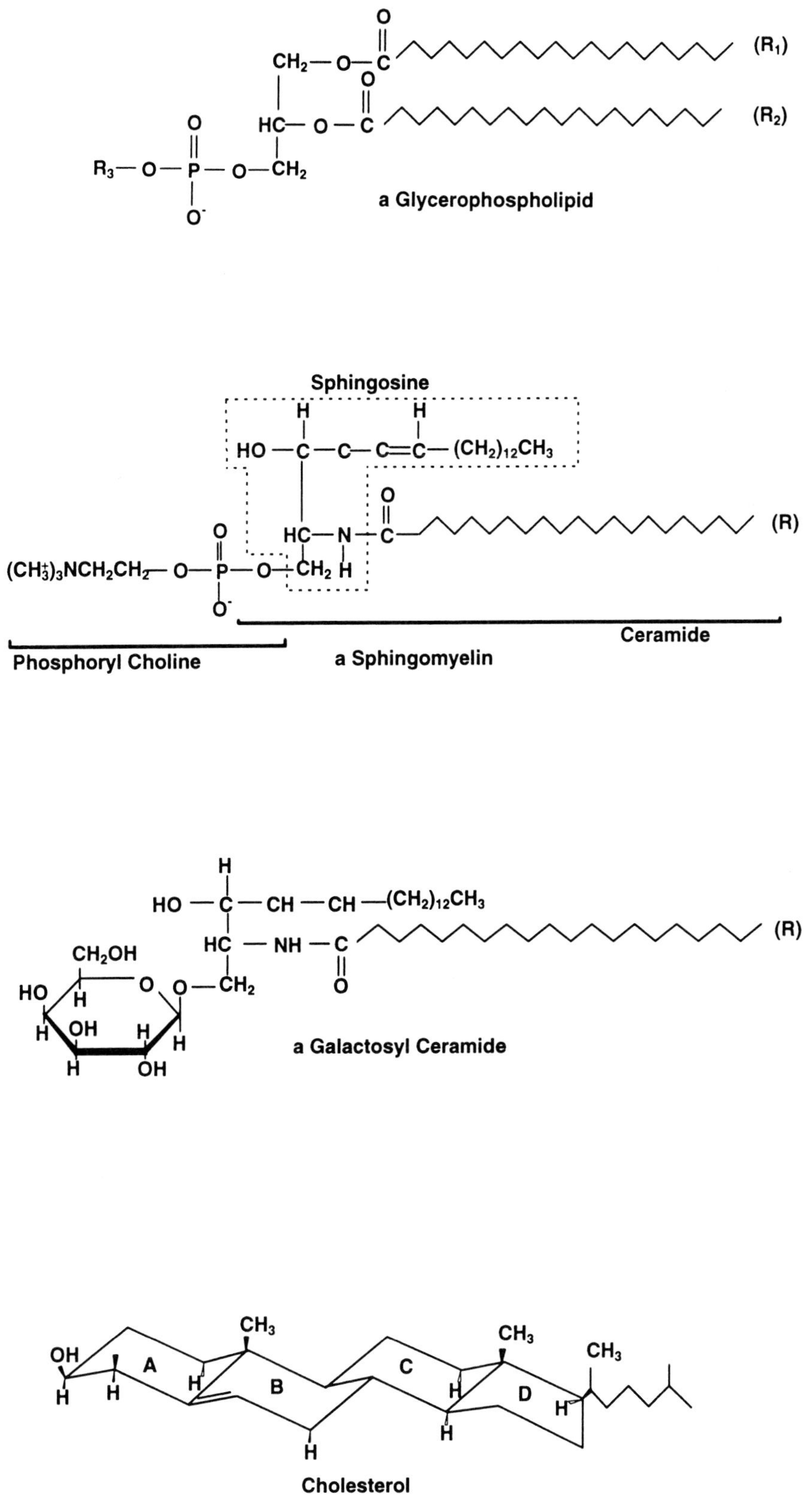

a Glycerophospholipid

Sphingosine

a Sphingomyelin

Phosphoryl Choline

Ceramide

a Galactosyl Ceramide

Cholesterol

FIG. 2.1. Structures of common membrane lipids.

25

one that is a phospholipid is sphingomyelin, shown in Figure 2.1 with the sphingosine backbone contained in the box. The primary alcohol on carbon 1 is in ester linkage to phosphorylcholine. The amino group on carbon 2 is in amide linkage to a long chain acyl group usually from 18 to 26 carbons long, which is saturated or monounsaturated. The variety of acyl chains found in this position in the sphingomyelin of the human erythrocyte are shown in Table 2.1. The amide group and the secondary alcohol on carbon 3 confer hydrogen bond donor capabilities on sphingomyelin in addition to the hydrogen bond acceptor capabilities inherent in the phosphate group. This is in contrast to phosphatidylcholine which has only hydrogen bond acceptor capability. The amide link and the *trans* double bond between carbons 4 and 5 are both coplanar structures. These are the analogs of the two planar ester bond systems in glycerol phosphatides (78). Important motional restrictions are imposed on both types of molecules by these planar structures, as is discussed later (see under Intramolecular Motions). Sphingosine with a long-chain fatty acyl group linked by amide bond to carbon 2 is called a ceramide.

The remaining sphingolipids, of which a large number are known, are glycosphingolipids. The characteristic feature of these molecules is a polysaccharide chain glycosidically linked to the ceramide portion of the molecule. Glucose is the most common linkage hexose, but galactose also occurs in this position. The glycosidic bond linking the polar and nonpolar moieties of these molecules is the β configuration. The simplest glycosphingolipid subclass is a galactosylceramide shown in Figure 2.1. Molecules in this subclass have different acyl chains in amide linkage to carbon 2. In general, glycosphingolipids may have a wide variety of polysaccharides linked to carbon 1 of sphingosine. The most common components of the sugar moieties of these molecules are glucose, galactose, N-acetyl galactosamine, and sialic acid. The class of molecules containing one or more sialic acid residues are referred to as gangliosides. The non-sialic acid–containing glycosphingolipids are called generically neutral glycosphingolipids. Glycosphingolipids are found as minor components localized exclusively on the external surface of the plasma membrane bilayers of most eucaryotic cells. They have been implicated in a wide variety of cell–cell and cell–ligand interactions (188). An excellent review of the structures can be found in Kanfer and Hakomori (99).

Cholesterol

By far the most abundant sterol, the third major type of membrane lipids, is cholesterol (shown in Figure 2.1). The polar moiety of this amphipathic molecule, an equatorial OH group on carbon 3, is quite small relative to the remaining hydrophobic portion. The overall shape of this molecule is that of a rigid, flat system comprised of three fused six-carbon rings and one five-carbon ring. A branched acyl chain of eight

TABLE 2.1. *Gas Chromatographic Analyses of the Fatty Acid Chains in Human Red Cell Phospholipid* *

Chain Length: Unsaturation	Total Phospholipids	Sphingomyelin	Phosphatidylcholine (Lecithin)	Phosphatidylethanolamine	Phosphatidylserine
16:0†	20.1	23.6	31.2	12.9	2.7
18:0	17.0	5.7	11.8	11.5	37.5
18:1	13.3	+	18.9	18.1	8.1
18:2	8.6	+	22.8	7.1	3.1
20:0	+‡	1.9	+	+	+
20:3	1.3	−	1.9	1.5	2.6
22:0	1.9	9.5	1.9	1.5	2.6
20:4	12.6	1.4	6.7	23.7	24.2
23:0	+	2.0	+	+	+
24:0	4.7	22.8	+	+	+
22:4	3.1	−	+	7.5	4.0
24:1	4.8	24.0	+	+	+
22:5	2.0	−	+	4.3	3.4
22:6	4.2	−	2.1	8.2	10.1

*The data are expressed as weight percent of the total.   †This code indicates the number of carbon atoms in the chain and the number of double bonds.   ‡Denotes that the concentration did not exceed 1% of the total.
[Taken with kind permission from p. 30 of Vance and Vance, 1985 (198).]

carbons is attached to carbon 17 of ring D. Two axial methyl groups are located at carbons 10 and 13. The only flexibility exhibited by cholesterol is in the terminal alkyl chain. Unlike the phospholipids and gangliosides, but similar to the neutral glycosphingolipids, there is no formal charge on cholesterol.

LIPID COMPOSITION OF SOME MAMMALIAN CELL MEMBRANES

The lipid compositions of the various membrane systems of the cell differ widely in the same cell and cells of different tissues and organs in the same organism. There are also marked differences in composition between the same membrane system in the same cell type in different organisms. This exceedingly wide diversity of lipid membrane compositions is illustrated in Tables 2.2 and 2.3. Table 2.2 compares the lipid composition of the various subcellular membrane systems of the rat hepatocyte. Generally speaking, the cholesterol concentration is much higher in the plasma membrane than in the subcellular membranes. The same is true of sphingomyelin, as illustrated in Table 2.2. The lipid composition of the plasma membrane of mammalian erythrocytes is compared in Table 2.3. About 25% of the total lipid in these membranes is cholesterol with from 5% to 25% being glycosphingolipid. The remainder is phospholipid. The values given in this table are roughly representative of many mammalian cell types. A review of the phospholipid composition of many cells and subcellular organelles can be found in Ansell et al. (5). Some information about membrane glycolipid composition can be found in Sweeley and Siddiqui (178).

MOLECULAR ORGANIZATION OF MEMBRANE LIPIDS

Lamellar Phases

A fundamental and important concept of biochemistry is that the structural characteristics of biomolecules existing as aggregated assemblies such as the lipid bilayer are ultimately derived from the basic structural and chemical properties of the monomer. It will be shown that a specific packing motif exhibited by phospholipids in the gel–state bilayer depends on the chainlength asymmetry of the lipid molecule in a systematic way. Other properties such as the chain melting temperatures of one-component lipid bilayers and the mixing behavior of two-component phospholipids in the bilayer plane at various temperatures are also related to the structural characteristics of the lipid species

under study. It is thus important to discuss, at least briefly, the fundamental structural characteristics of a typical lipid molecule before turning attention to the bilayer structure of phospholipids. Although the molecular structures of many lipid species are now known (146), the saturated identical-chain C(14):C(14)PC (saturated phosphatidylcholine with 14 carbons in the *sn*-1 acyl chain and 14 carbons in the *sn*-2 acyl chain), or dimyristoylphosphatidylcholine, can be considered a representative species for discussion. Consequently, many structural properties of the various lipid bilayers discussed here are related to those of saturated phosphatidylcholines.

Two x-ray crystallographic structures of C(14): C(14)PC were first described by Pearson and Pascher (148). These structures, called A and B, are different primarily in the orientation of the polar headgroup with respect to the hydrophobic part. The P—O—C—C—N dipoles are 4.3 Å and 4.5 Å long and inclined to the bilayer surface by 17° and 27° for molecules A and B, respectively. In other words, the polar headgroup of this phosphatidylcholine molecule adopts a bent-down conformation towards the bilayer surface. This difference between molecules A and B can be attributed to the rotation of the polar headgroup about the glycerol C(2)—C(3) bond (148). The hydrophobic or diglyceride moieties of the two crystallographic structures, however, are qualitatively similar. The diglyceride moieties of molecules A and B exhibit a common "h" shape geometry in which the glycerol carbons [C(1), C(2), and C(3)], the primary ester oxygen, and all carbons in the *sn*-1 acyl chains are topologically arranged in a virtually straight zigzag plane, while the *sn*-2 acyl chain is bent at C_2 position (Fig. 2.2A). Specifically, four structural features can be associated with the diglyceride moiety of the C(14): C(14)PC molecule in the crystals: *(1)* The hydrocarbon chain is basically a zigzag plane of conformationally regular array of methylene (—CH_2—) units. The C—C bonds in the *sn*-1 acyl chain and the C—C bonds beyond the C_2-atom in the *sn*-2 acyl chain are thus all *trans*. *(2)* The bond that links the glycerol oxygen and the carboxyl carbon is the O—C_1 ester bond, where C_1 is the first carbon atom of the fatty acyl chain (Figure 2.2A). This ester bond has a partial double-bond character owing to the resonance hybrid effect (79); consequently, the primary and secondary ester bonds, which link the *sn*-1 and *sn*-2 acyl chains, respectively, to the glycerol backbone, have *trans* configurations. The energetically preferred *trans* configuration renders the five atoms [C(1 or 2)—O—C_1 (=O)—C_2] in each of the two ester groups coplanar at physiological temperatures. Here, the C(1) and C(2) denote the carbon atoms 1 and 2 of the glycerol backbone (Fig.

TABLE 2.2. *Lipid Composition of Subcellular Organelle Membranes of Rat Hepatocytes* *

Lipid	Plasma Membrane	Microsome	Mitochondrial Inner	Mitochondrial Outer	Golgi	Nuclear
Cholesterol	28.0	6.0	<1.0	6.0	7.6	5.1
Phosphatidylcholine	31.0	55.2	37.9	42.7	24.4	58.3
Sphingomyelin	16.6	3.7	0.8	4.1	6.6	3.0
Phosphatidylethanolamine	14.3	(24.0)	38.3	28.6	9.6	21.5
Phosphatidylserine	2.7		<1.0	<1.0	2.3	3.4
Phosphatidylinositol	4.7	7.7	2.0	7.9	4.7	8.2
Phosphatidic acid and Cardiolipin	1.4	1.5	20.4	8.9	—	<1.0
Lysophosphatidylcholine	1.3	1.9	0.6	1.7	3.2	1.4
Other	—	—	—	—	3.4	—

* Data are weight percent of total lipid.
[Taken with kind permission from p. 36 of Thompson and Huang, 1986 (184)].

2.2A), respectively, and C_1 and C_2 are the first and second carbon atoms of the fatty acyl chain, respectively (Fig. 2.2A). (3) The plane occupied by the five atoms of the primary ester group is approximately perpendicular to the plane of the secondary ester which runs parallel to the bilayer surface. The relative position of the two ester planes is dictated by a unique torsional angle (θ_4) of approximately 60° for the O—C(1)—C(2)—O bond (Fig. 2.2A). (4) The *sn*-2 acyl chain is bent 90° at the C_2 position (Fig. 2.2A). Consequently, the *sn*-1 acyl chain is effectively longer than the *sn*-2 acyl chain despite the same total carbon number in each chain. The two terminal methyl groups of the *sn*-1 and *sn*-2 acyl chains are separated from each other along the chain-axis by 4.67 Å on 3.68 C—C bond lengths along the chain.

Although the detailed structure of a lipid bilayer depends profoundly on the phase in which the lipid molecules are present, the basic feature of a bilayer or lamella can be considered a two-dimensional, bimolecular sheet consisting of two opposing leaflets. In this structure, the hydrocarbon chains of diacyl phospholipids from the opposing leaflets are aggregated to form a two-dimensional hydrophobic core. Immediately next to the hydrophobic core are two layers of the interfacial region in which the secondary ester groups of phospholipids are layered on top of the hydrophobic core in a sandwich manner (Figs. 2.3, 2.4). The oxygens of the phosphate ester group, the phosphate group, and other groups of the polar headgroup form two boundary layers separating the bulk water from the interfacial region of the lipid bilayer (Figs. 2.3, 2.4). The average acyl chain conformation of amphipathic lipids in the two-dimensional hydrocarbon core of the lipid bilayer depends on temperature. In single crystals, the *sn*-1 acyl chain and the *sn*-2 acyl chain beyond C_2 of a

TABLE 2.3. *Lipid Composition of Plasma Membranes of Mammalian Erythrocytes* *

Lipid	Pig	Human	Cat	Rabbit	Horse	Rat
Cholesterol	26.8	26.0	26.8	28.9	24.5	24.7
Phosphatidylcholine	13.9	17.5	18.7	22.3	22.0	31.8
Sphingomyelin	15.8	16.0	16.0	12.5	7.0	8.6
Phosphatidylethanolamine	17.7	16.6	13.6	21.0	12.6	14.4
Phosphatidylserine	10.6	7.9	8.1	8.0	9.4	7.2
Phosphatidylinositol	1.1	1.2	4.5	1.0	<0.2	2.3
Phosphatidic acid	<0.2	0.6	0.5	1.0	<0.2	<0.2
Lysophosphatidylcholine	0.5	0.9	<0.2	<0.2	0.9	2.6
Glycosphingolipids	13.4	11.0	11.9	5.3	23.5	8.3

* Data are weight percent of total lipid.
[Taken with kind permission from p. 27 of Thompson and Huang, 1986 (184)].

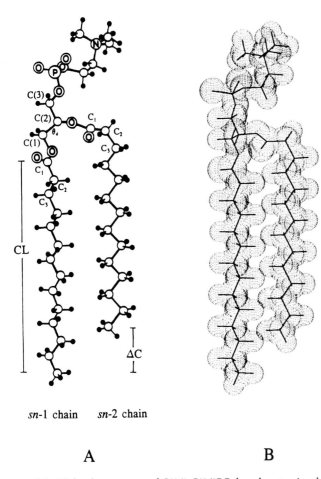

sn-1 chain   *sn*-2 chain

A   B

FIG. 2.2. Molecular structure of C(14):C(14)PC, based on torsional angles of x-ray crystallographic structure B of C(14):C(14)PC dihydrate determined by Pearson and Pascher (148). *A:* Ball-and-stick model. *B:* Computer-generated diagram with van der Waals spheres of atoms.

C(14):C(14)PC molecule adopt an all-*trans* conformation as shown in Figure 2.2. In the lipid bilayer, the two acyl chains of C(14):C(14)PC at temperatures below O°C are also highly ordered mostly with all-*trans* conformations. Upon heating, the lipid bilayer of C(14):C(14)PC, in excess water, undergoes multiple phase transitions. At a temperature slightly above the room temperature (24°C), however, many C—C bonds in the two acyl chains begin to undergo highly cooperative *trans → gauche* isomerizations, resulting in a disordered hydrocarbon core. This sharp order/disorder transition involving the rotational isomerization of the *trans → gauche* bonds along the lipid acyl chain is termed the main phase transition, and the characteristic temperature associated with the sharp transition is called the main phase transition temperature (T_m). Each saturated diacyl phospholipid with long acyl chains has its own T_m value and will be discussed later in this section.

Mostly from physical studies on mixed-chain phosphatidylcholines in excess water, several packing motifs of gel-phase phospholipids in bilayers, at $T < T_m$, have been identified in recent years (for a recent review, see ref. 172). Using molecular mechanics calculations, it has been shown that the specific packing motif depends profoundly on the lateral chain–chain interactions which, in turn, are dictated by the chain-length asymmetry between the *sn*-1 and *sn*-2 acyl chains of the lipid molecule in the lipid bilayer (112). Phospholipids exhibiting the highest degree of chain-length asymmetry such as lysophospholipids form fully interdigitated bilayers at temperatures below the T_m. Phospholipids whose two acyl chains are nearly identical in effective chain lengths such as C(16):C(18)PC form bilayers at $T < T_m$, with a non-interdigitated packing motif. Between these two extremes lipids tend to assemble in excess water into gel-state bilayers with the partially interdigitated or mixed interdigitated motif (89, 172).

Before discussing the various motifs of the interdigitated bilayers it is appropriate to introduce two structural parameters (ΔC and CL) associated with the acyl chain-length asymmetry in lipids at $T < T_m$. Based on the crystal structure of C(14):C(14)PC shown in Figure 2.2A, the *sn*-1 acyl chain is observed to be ester-linked to C(1) of the glycerol backbone. Here, the length of the zigzag plane extending from the glycerol backbone to the chain terminal methyl end is longer than the full length of the *sn*-1 acyl chain by an initial segment comprising the $C(2)—C(1)—O—C_1$ moiety. The *sn*-2 acyl chain, however, is shortened by 1.2 Å due to the sharp bend at C_2. The difference between the extended *sn*-1 acyl chain and the shortened *sn*-2 acyl chain along the long molecular axis is defined as the effective chain-length difference (ΔC) between the two acyl chains. For C(14):C(14)PC in the crystalline state, the value of ΔC is 3.68 C—C bond lengths along the chain (Figure 2.2A); however, this value is reduced to 1.5 for identical-chain phospholipids packed in the gel-state bilayer (219). This value of 1.5 C—C bond lengths is taken as the effective chain-length difference of the reference state (ΔC_{ref}). For a diacyl-saturated–mixed-chain C(X):C(Y)PC (saturated phosphatidylcholine with X carbons in the *sn*-1 acyl chain and Y carbons in the *sn*-2 acyl chain) packed in the gel-state bilayer, ΔC is related to X and Y as follows (127): $\Delta C = |X - Y + \Delta C_{ref}| = |X - Y + 1.5|$, where X and Y are the number of carbons in the *sn*-1 and *sn*-2 acyl chains, respectively. Another structural parameter, CL, of C(X):C(Y)PC refers to the effective chain-length of the longer of the two acyl chains in C—C bond lengths, and is defined as CL = X - 1, if *sn*-1 is the longer chain or CL = Y - 2.5, if *sn*-2 is the longer chain (127). This structural parameter is also illustrated in

A   B

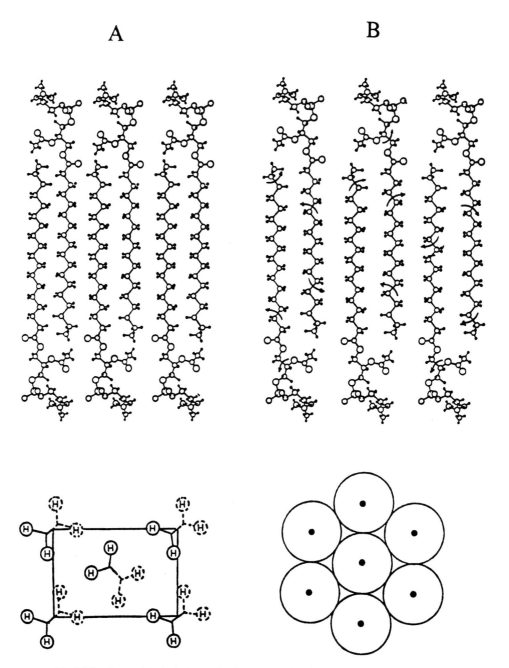

FIG. 2.3. Molecular models depicting molecular *(top)* and acyl chain *(bottom)* packing arrangement of fully hydrated C(18):C(2)PC in crystalline *(A)* and gel *(B)* phases. [Taken with kind permission from Huang et al., 1984 (83).]

Figure 2.2A. The normalized chain-length asymmetry or difference is thus given by the ratio of $\Delta C/CL$. For instance, C(18):C(10)PC has a $\Delta C/CL$ value of $[(|18 - 10 + 1.5|)/(18 - 1)] = 0.56$ and C(10):C(18)PC has a $\Delta C/CL$ value of $[(|10 - 18 + 1.5|)/(18 - 2.5)] = 0.42$. A comparison of these two calculated values of $\Delta C/CL$ indicates that the molecular species of C(18):C(10)PC is more asymmetrical than its positional isomer at $T < T_m$ in the lipid bilayer.

When the normalized chain-length difference of a phospholipid molecule is near unity, this lipid tends to self assemble at $T < T_m$ into the bilayer with a fully interdigitated packing motif. In this packing motif (Figure 2.3) the long acyl chain spans the entire width of the bilayer, resulting in the equivalent of two acyl chain cross-sectional areas per unit headgroup cross-sectional area. Taking C(18):C(2)PC as an example, its ΔC and CL values are 17.5 and 17 C—C bond lengths, respectively, and the ratio of $\Delta C/CL = 1.03$. This highly asymmetric C(18):C(2)PC molecule aggregates in excess water to form micelles at room temperature. Prolonged incubation of the micellar solution at 0°C results in the conversion of micelles into lamellar structures as demonstrated by the line shapes of their ^{31}P nuclear magnetic resonance (NMR) spectra (83). Upon reheating, two phase transitions can be detected at 6.5°C and 18.5°C, respectively. The smaller low-temperature transition corresponds to the interdigitated crystalline to interdigitated gel phase transition, and the larger high-temperature transition corresponds to the interdigitated gel to micelle transition (83). At very low temperatures (< 6.5°C), C(18):C(2)PC molecules with the sn-1 acyl chains in a nearly all-trans interdigitated conformation undergo very small amplitude thermal oscillations or orientational fluctuations about their long molecular axes with rates of about 10^4 s^{-1}. As the C(18):C(2)PC dispersion is heated above 6.5°C, the interdigitated crystalline phase is transformed into an interdigitated gel phase. This thermally induced transition results in rapid rotational motions of C(18):C(2)PC molecules about their long molecular axes; the effect is to transform the relatively immobilized sn-1 acyl chain in the orthorhombic lattice into a randomly oriented sn-1 acyl chain in a hexagonal lattice (Fig. 2.3). When the temperature is increased further to above 18.5°C, the assembly of this highly asymmetric phospholipid undergoes the lamellar → micellar transition. In the micellar form, acyl chains of the C(18):C(2)PC dispersion are in a highly fluctuating disordered state due to the presence of a large number of gauche rotamers in the acyl chain. Moreover, the entire micellar unit now undergoes overall tumbling motions in the aqueous medium.

Lysophosphatidylcholines and platelet activating factors with ΔC/CL values greater than unity have also been demonstrated to form fully interdigitated bilayers at T < T_m (84, 89, 173).

When the longer acyl chain of C(X):C(Y)PC has an effective chain-length which is about twice that of the shorter one, these C(X):C(Y)PC tend to aggregate in excess water into the mixed interdigitated bilayer at T < T_m. This second type of interdigitated packing motif, shown in Figure 2.4, takes into consideration the following experimental observations originally obtained with fully hydrated C(18):C(10)PC using the x-ray diffraction technique (90, 131): (1) The observed bilayer hydrophobic thickness is consistent with the chain length of an all-trans C(18)–acyl chain; (2) the average headgroup area of C(18):C(10)PC in the gel-state bilayer is about three times the cross-sectional area of an all-trans acyl chain; (3) the acyl chains are not tilted in the gel state, since a sharp and symmetric x-ray reflection centered at 4.2 Å is observed in the wide-angle region for fully hydrated C(18):C(10)PC at T < T_m. This packing motif, shown in Figure 2.4, is thus characterized by the longer acyl chain extending completely across the whole width of the hydrocarbon core of the bilayer, whereas the two shorter acyl chains, each from a lipid in the opposing leaflet, meet end-to-end in the bilayer midplane. Recently, asymmetric phospholipids with ΔC/CL = 0.55 such as C(8):C(18)PC and C(22):C(12)PC have been shown from x-ray diffraction experiments to pack into the mixed interdigitated motif in excess water at T < T_m (166, 222).

Among all the asymmetric phospholipids that can form mixed interdigitated bilayers at T < T_m C(18):C(10)PC is perhaps the most extensively studied molecule. Based on high-pressure–Fourier-transform infrared spectroscopic data, the zigzag planes of neighboring acyl chains of C(18):C(10)PC in the mixed interdigitated bilayer are found to be intermolecularly and intramolecularly perpendicular to each other

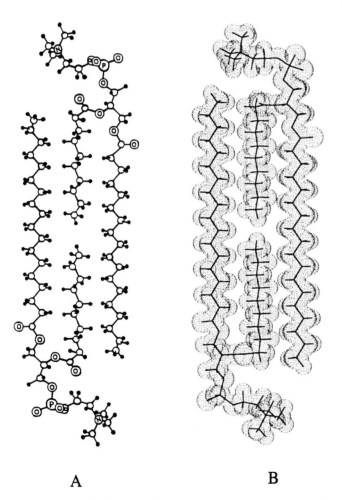

A   B

FIG. 2.4. Molecular models depicting bilayer structure of C(18):C(10)PC with mixed interdigitated packing motif. A: Ball-and-stick model. B: Computer-generated diagram with van der Waals spheres of atoms.

(211). This packing characteristic is illustrated in Figure 2.4. At temperatures above the T_m, the mixed interdigitated motif exhibited by fully hydrated C(18):C(10)PC transforms into the partially interdigitated packing motif with the area per headgroup corresponding to only two acyl chains (129, 131). This disordered state of the lipid bilayer at $T > T_m$ is called the liquid-crystalline or L_α phase. In the L_α phase, the short acyl chain of C(18):C(10)PC in one leaflet meets end-to-end with the longer acyl chain counterpart contributed by the opposing leaflet and vice versa. C(18):C(10)PC is thus the first example in which the acyl chain interdigitation is known for bilayers at temperatures both above and below the T_m. Because the phospholipid packing motif changes from a mixed interdigitated to a partially interdigitated type at the T_m, the bilayer thickness of C(18):C(10)PC remains relatively constant as the bilayer of C(18):C(10)PC undergoes the phase transition at 18.8°C (90, 132).

Since C(18):C(10)PC, C(8):C(18)PC, and C(22):C(12)PC have been well established by x-ray diffraction techniques to form mixed interdigitated bilayers in excess water at $T < T_m$, their transition characteristics can thus be compared with other C(X):C(Y)PC. Based on these comparative studies, it has been inferred that mixed-chain PC with ΔC/CL values in the range of 0.43–0.63 can self assemble in excess water into the mixed interdigitated bilayer at $T < T_m$ (80). Recently, it has been suggested that some 81 molecular species of C(X):C(Y)PC adopt the mixed interdigitated packing motif with predictable T_m values (87).

The third kind of interdigitated packing model observed for phospholipids at $T < T_m$ is the partially interdigitated motif. C(X):C(Y)PC with ΔC/CL values greater than 0.05 but smaller than 0.41, including identical-chain phosphatidylcholines, belong to this class. The general characteristics of this packing motif can be described using the example of C(14):C(14)PC. In single crystals of C(14):C(14)PC dihydrate, two types of C(14):C(14)PC assembly, shown in Figure 2.5, were observed (146). In one assembly, the two all-trans acyl chains of one lipid molecule in one leaflet are aligned along the chain axes with two fully extended acyl chains contributed separately by two lipid molecules in the opposing leaflet (Figure 2.5A). The second assembly shows, however, that the two fully extended acyl chains of one lipid molecule are aligned with the two fully extended acyl chains of a second lipid molecule in the opposing leaflet (Fig. 2.5B). The common features shared by both assemblies are three-fold: (1) The longer chain of one lipid on one side of the bilayer is packed end-to-end with the shorter chain of another lipid molecule in the opposing bilayer leaflet.

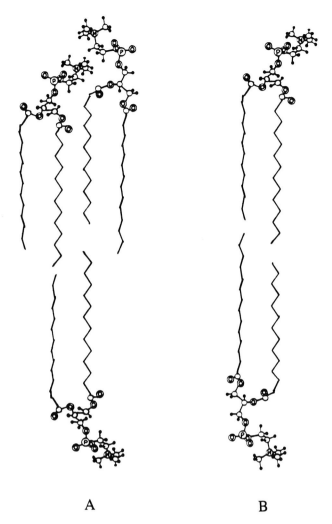

A   B

FIG. 2.5. Two types of partially interdigitated packing motif. A: One C(14):C(14)PC molecule in one leaflet pairs with two C(14):C(14)PC molecules in the opposing leaflet. B: One C(14):C(14)PC molecule pairs with another single C(14):C(14)PC molecule in the opposing leaflet.

(2) The headgroup area per lipid molecule at the lipid/water interface is twice the cross-sectional area of the fully extended acyl chain. (3) The thickness of the hydrophobic core (N) or the distance separating the two nearest carbonyl oxygens along the chain axis in C(X):C(Y)PC bilayers is $[X + Y + (1 - \Delta C_{ref})]$ by assuming the van der Waal contact distance between the two opposing methyl termini to be 3 C—C bond lengths along the chain (81). In gel-state bilayers, the effective chain length difference between two acyl chains in an identical-chain PC, ΔC_{ref}, is approximately 1.5 C—C bond lengths (219); hence, the value of N may be quantitated, in C—C bond lengths, by the expression: $N = X + Y - 0.5$ for all gel-state C(X):C(Y)PC bilayers with the partially interdigitated motif. It should also be mentioned that the major

difference between the two types of partially interdigitated motif, shown in Figure 2.5, lies in the intermolecular chain–chain distance. In Figure 2.5A, the average intermolecular chain–chain distance is indistinguishable from the average intramolecular chain–chain distance. This may not necessarily be true for phospholipids packed in the alternative motif shown in Figure 2.5B.

Unlike the mixed interdigitated bilayer which exhibits a single phase transition upon heating, the partially interdigitated bilayer may undergo multiple phase transitions upon heating. Figure 2.6A shows the first and second thermograms exhibited by C(18):C(10)PC when the aqueous dispersion of C(18):C(10)PC was subjected to repeated heating scans in the high-resolution differential scanning calorimeter (DSC). Clearly the sharp, single transition at 18.8°C with a transition enthalpy, ΔH, of 9.0 kcal/mol is reproduced in the second heating scan. The value of ΔH, determined from the integrated area under the transition peak, gives a measure of the heat required to overcome the energy barrier for the bilayer to convert from the mixed interdigitated gel phase to the L_α phase.

The first DSC heating scan for the aqueous dispersion of C(16):C(16)PC ($\Delta C/CL = 0.100$) is shown in Figure 2.6B. There are three discernible transitions: the sub-, the pre-, and the main transitions. The T_m and ΔH values for these transitions are 19.5°, 33.3°, 41.5°C, and 5.7. 1.0, 9.0 kcal/mol, respectively. These transitions are ascribed to the $L_c \rightarrow L_{\beta'}$, $L_{\beta'} \rightarrow P_{\beta'}$, and $P_{\beta'} \rightarrow L_\alpha$ phase transitions, respectively. Upon reheating,

the $L_c \rightarrow L_{\beta'}$ sub-transition is absent; the $L_{\beta'} \rightarrow P_{\beta'}$ pre-transition is downshifted by about 1°C; the $P_{\beta'} \rightarrow L_\alpha$ main transition is the only reversible transition as shown in Figure 2.6B.

The L_c phase of C(16):C(16)PC has been extensively studied by Raman, Fourier Transform Infrared, and x-ray diffraction techniques. However, conclusions regarding the hydrocarbon chain packing and the relative chain orientation are still controversial. For instance, vibrational spectroscopic studies indicate that the two zigzag planes are parallel to each other (110, 212), whereas x-ray data indicate that the two zigzag planes of C(16):C(16)PC are nearly perpendicular to each other (156). Despite the difference, the acyl chains of C(16):C(16)PC can be regarded as tightly packed laterally in the L_c phase. In the $L_{\beta'}$ phase, the acyl chains of C(16):C(16)PC are tilted with respect to the bilayer normal, and these acyl chains are packed laterally in a disordered orthorhombic hybrid subcell (156). In the $P_{\beta'}$ phase, the lipid molecule as a whole undergoes the rotational motion about its long molecular axis (124). The acyl chains, however, are still mainly in the nearly all-*trans* conformation and remain tilted (~30°) with respect to the bilayer normal. In addition, the $P_{\beta'}$ phase is also characterized by a wave-like surface pattern called the periodic ripple. At $T < T_m$, the lipid bilayer is in the L_α phase. In the L_α phase, the lower segments of the acyl chains undergo the rapid *trans–guache* isomerizations about the C—C bonds; the acyl chain as a whole undergoes some large amplitude thermal oscillations; the lipid molecules them-

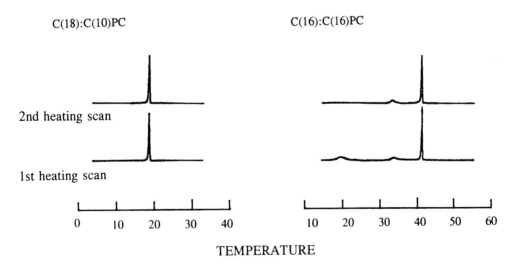

C(18):C(10)PC   C(16):C(16)PC

2nd heating scan

1st heating scan

0   10   20   30   40   10   20   30   40   50   60

TEMPERATURE

FIG. 2.6. Representative DSC heating thermograms for aqueous dispersions of C(18):C(10)PC *(left)* and C(16):C(16)PC *(right)*. The phase transition curves observed at left for C(18):C(10)PC are virtually identical in the first and second heating scans. The first DSC heating scan for C(16):C(16)PC, shown at right, has three discernible transitions; however, only two (the $L_{\beta'} \rightarrow P_{\beta'}$ and $P_{\beta'} \rightarrow L_\alpha$) transitions are detectable in the second heating scan. The lowest-temperature (the $L_c \rightarrow L_{\beta'}$) transition observed in the first heating scan is abolished in the second heating scan.

selves also move laterally in a random manner in each of the two-dimensional planes of the two monolayers in the bilayer. Despite the highly dynamic nature of lipid molecules in the liquid-crystalline phase, it is interesting to point out that the averaged orientation of the acyl chain for C(16):C(16)PC in the liquid-crystalline phase is roughly perpendicular to the bilayer surface, and that the acyl chains are, nevertheless, packed in a one-dimensional, ordered hexagonal lattice. The bilayer thickness, however, is reduced by about 39% as it undergoes the $P_{\beta'} \rightarrow L_{\alpha}$ phase transition (94) due to the presence of appreciable *gauche* rotamers along the acyl chain in the L_{α} phase.

The T_m values of the main phase transitions for C(X):C(Y)PC can be determined calorimetrically by repeated heating scans. Some representative T_m values are given in Table 2.4. For saturated identical-chain and mixed-chain PC, it can be seen from Table 2.4 that the larger the X + Y value, the higher is the T_m value. For lipids with the same X + Y value such as C(16):C(16)PC($\Delta C = 1.5$), C(18):C(14)PC($\Delta C = 5.5$), and C(14):C(18)PC($\Delta C = 2.5$), the T_m value is observed to increase with decreasing value of ΔC, where $\Delta C = |X - Y + 1.5|$. Since the sum of X and Y relates directly to the hydrocarbon thickness (N = X + Y − 0.5), and since the term ΔC represents the chain-length asymmetry, the T_m values presented in Table 2.4 for saturated PC can thus be adequately described by a model in which the T_m value is dominated by the

thickness of the hydrocarbon core of the lipid bilayer in the gel state (N) and it is perturbed by the chain-length asymmetry of the monomeric lipid species (ΔC). To a first approximation, the T_m values can be related to N and ΔC for all saturated phosphatidylcholines which undergo the partially interdigitated gel $\rightarrow L_{\alpha}$ phase transitions as follows (81): $T_m = 158.67 - 3532.18 \, (1/N) - 94.99 \, (\Delta C/N)$. It is therefore interesting to point out that the T_m value of a C(X):C(Y)PC bilayer can be predicted based on the values of X and Y since the structural parameters N and ΔC are simple functions of X and Y only.

When an unsaturated *cis*-double bond is introduced into the acyl chain, the T_m value is seen to decrease appreciably (Table 2.4). Based on calorimetric studies of a series of identical-chain phosphatidylcholines with a *cis*-double bond at different positions along each acyl chain, Barton and Gunstone (15) have shown that the T_m value is influenced by the location of the *cis*-double bond. The value of T_m is minimal, that is, the downshift relative to the corresponding saturated symmetric phosphatidylcholine is maximal, when each *cis*-double bond is located near the center of the acyl chain. Moreover, the T_m value increases progressively as each *cis*-double bond moves toward either end of the fatty acyl chain. It should be mentioned that the introduction of a *cis*-double bond also eliminates the sub- and pretransitions observed for the saturated phosphatidylcholines in excess water.

TABLE 2.4. *Calorimetric T_m Values of the Main Phase Transition (Gel → L_a) of Aqueous Dispersions for Some Selective Phospholipids*

Identical-Chain PC		Identical-Chain PE	
Phospholipid	$T_m(°C)$	Phospholipid	$T_m(°C)$
C(14):C(14)PC	24.1	C(14):C(14)PE	49.6
C(15):C(15)PC	34.0	C(15):C(15)PE	57.3
C(16):C(16)PC	41.5	C(16):C(16)PE	63.0
C(18):C(18)PC	55.3	C(18):C(18)PE	74.0
C(20):C(20)PC	66.4	C(20):C(20)PE	82.5
Mixed-Chain PC		Mixed-Chain PE	
Phospholipid	$T_m(°C)$	Phospholipid	$T_m(°C)$
C(16):C(12)PC	11.3	C(16):C(12)PE	37.4
C(12):C(16)PC	21.7	C(12):C(16)PE	48.0
C(18):C(14)PC	31.2	C(18):C(14)PE	54.9
C(14):C(18)PC	39.2	C(14):C(18)PE	61.6
Monounsaturated PC		Monounsaturated PE	
Phospholipid	$T_m(°C)$	Phospholipid	$T_m(°C)$
C(16):C(18:1Δ^9)PC	−2.6	C(16):C(18:1Δ^9)PE	26.1
C(18):C(18:1Δ^9)PC	5.6	C(18):C(18:1Δ^9)PE	31.5
C(20):C(18:1Δ^9)PC	11.0	C(20):C(18:1Δ^9)PE	33.9

Phosphatidylethanolamines (PE) make up a significant fraction of membrane lipids in many procaryotic organisms and virtually all higher organisms. Structurally, phosphatidylethanolamine differs from phosphatidylcholine in the polar headgroup region. The choline moiety of a phosphatidylcholine molecule is replaced by the smaller ethanolamine moiety in phosphatidylethanolamine. The structural as well as thermodynamic properties of fully hydrated saturated identical-chain phosphatidylethanolamine, C(X):C(X)PE, have been studied extensively over the years (44, 103, 111, 131, 139, 161, 182, 209, 216), after the initial reports from several laboratories that dilauroylphosphatidylethanolamine or C(12):C(12) PE, in excess water, exhibit characteristic phase behavior (41, 121, 162).

As shown in Figure 2.7, the aqueous dispersions of C(12):C(12) PE exhibit a sharp endothermic transition with T_m = 43.0°C and ΔH = 13.6 kcal/mol in the initial DSC heating scan. This transition has been assigned as the $L_c \rightarrow L_\alpha$ phase transition, where L_c is the crystalline or sub-gel phase and L_α is the liquid-crystalline phase. The same sample, however, displays a smaller and down-shifted transition upon immediate reheating with T_m = 30.6°C and ΔH = 3.7 kcal/mol as shown in Figure 2.7. This low-temperature transition is reproducible upon subsequent repeated reheatings, and is assigned as the $L_\beta \rightarrow L_\alpha$ phase transition. The L_β phase is similar to the $L_{\beta'}$ phase observed for identical-chain phosphatidylcholines except that the acyl chains of lipid molecules are oriented perpendicularly to the bilayer surface. The T_m values of the low-temperature transitions for some identical-chain phosphatidylethanolamines are presented in Table 2.4.

The structural and phase characteristics of highly asymmetric mixed-chain phosphatidylethanolamines are less well understood. At $T < T_m$, C(18):C(10)PE (ΔC = 0.55) has been characterized by DSC and ^{31}P NMR spectroscopy, demonstrating that C(18):C(10)PE can self-assemble into the gel-state bilayer with a mixed interdigitated packing motif (128).

Table 2.4 also shows the T_m values for aqueous dispersions of a number of mixed-chain phosphatidylethanolamines. These values are obtained calorimetrically from the second heating scans. Clearly, the T_m values for mixed-chain phosphatidylethanolamines with $\Delta C/CL$ values in the range of 0.08 to 0.37 are consistently higher than the corresponding T_m values obtained with the equivalent mixed-chain phosphatidylcholines, indicating the overall stronger lateral chain–chain interaction in the phosphatidylethanolamine system. The stronger lateral chain–chain interaction is most likely due to a favorable headgroup–headgroup interaction between phosphatidylethanolamines in the

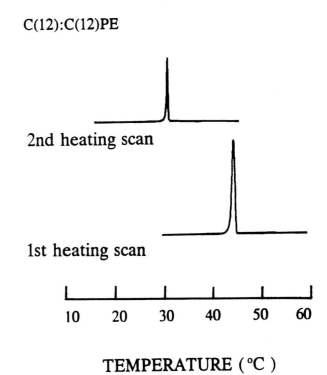

C(12):C(12)PE

2nd heating scan

1st heating scan

10   20   30   40   50   60

TEMPERATURE (°C)

FIG. 2.7. Representative DSC heating thermograms for aqueous dispersions of C(12):C(12)PE. *Bottom:* First heating scan showing $L_c \rightarrow L_\alpha$ phase transition with T_m = 43.0 °C and ΔH = 13.6 kcal/mol, and *top:* second heating scan showing the $L_\beta \rightarrow L_\alpha$ phase transition with lower transition temperature and smaller peak area (T_m = 30.6 °C and ΔH = 3.7 kcal/mol).

gel-state bilayer, which can be attributed to the smaller headgroup of phosphatidylethanolamine and the ability of the amine moiety to undergo H—bond interactions with neighboring phosphatidylethanolamine headgroups.

Binary Phospholipid Mixtures

Thus far the structure and phase behavior of a limited number of one-component phospholipid systems has been discussed. Attention is now directed to the mixing behavior of two-component phospholipids in the bilayer at ambient pressure. The basic information is contained in the temperature-composition phase diagram for the binary lipid system. It has been well-documented that many types of phase diagrams can be exhibited by binary lipid mixtures (125). Some of the most commonly observed phase diagrams are shown in Figure 2.8. These various phase diagrams reflect the miscibility and/or immiscibility of the component lipids in the gel, the liquid-crystalline, and the two coexisting phases over a certain range of the lipid composition.

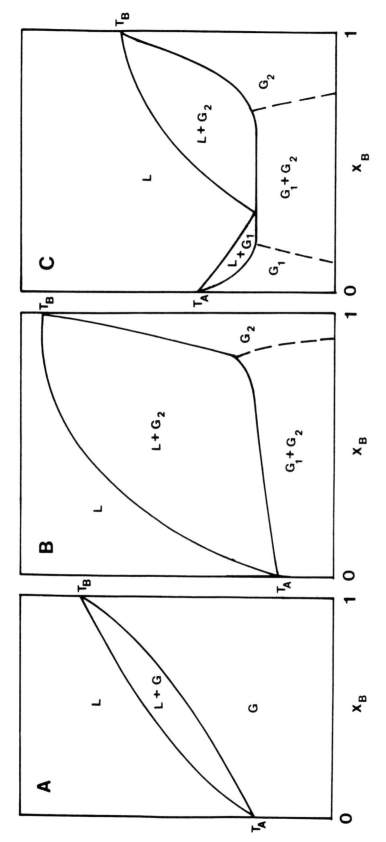

FIG. 2.8. Phase diagrams for (A) an isomorphous system with complete miscibility in both gel (G) and liquid-crystalline (L) phases, (B) a peritectic system showing extensive gel–gel immiscibility region ($G_1 + G_2$), and (C) a eutectic system showing partial gel–gel phase separation ($G_1 + G_2$) combined with liquid-crystalline (L) phase miscibility.

Gel-Phase Miscibility and Fluid-Phase Miscibility—The Isomorphous Phase Diagram. The miscibility of the two molecular species of phosphatidylcholine in the two-dimensional plane of the lipid bilayer must depend on the lateral lipid–lipid interactions which, in turn, can be expected to depend on the structural similarity between the component lipids. Furthermore, the structural characteristics of phospholipid molecules in the bilayer may change drastically as the bilayer undergoes the phase transition.

The simplest binary phospholipid mixture is the isomorphous system in which the component lipid A and component lipid B are completely miscible in both the gel and liquid-crystalline states over the entire composition range (Fig. 2.8A). As an example, the binary mixture of two structurally similar identical-chain phosphatidylcholines such as C(14):C(14)PC and C(16):C(16)PC gives rise to an isomorphous phase diagram characterized by a lens-shaped, two-phase region. The phase diagram of this binary mixture can be constructed readily on the basis of excess heat capacity curves obtained by DSC at different molar ratios of the mixture (120). The curves of the mixtures are slightly broadened in comparison with those of the pure component lipids. Each DSC curve has a characteristic onset and completion temperature positioned at the two sides of the transition curve. The phase diagram is constructed by plotting the onset and completion temperatures as a function of the relative concentration of the higher melting component: these onset and completion temperature points define the solidus and liquidus, respectively, in the temperature–composition binary phase diagram. In between the solidus and liquidus is a lens-shaped enclosed region where the gel (G) and liquid-crystalline (L) phases coexist. The composition and relative amount of each coexisting phase in the (G + L) region can be determined by the tie line and the lever rule (218). Below the solidus is the single phase G region; above the liquidus is the single phase L region.

Another example of isomorphous system involving a binary mixture of two highly asymmetric mixed-chain phosphatidylcholines is C(10):C(22)PC/C(22):C(12)PC (215). Each component lipid in this mixture has a $\Delta C/CL$ value close to 0.54, except that the positions of the effective long and short chains are reversed with respect to the glycerol backbone. The phase diagram of this mixture also exhibits the characteristics shown in Figure 2.8A. The solidus and the liquidus do not display either a flat region or a point of sharp inflection, indicating that C(10):C(22)PC and C(22):C(12)PC are mixed nearly ideally in both the gel and liquid-crystalline states over the entire composition range in the bilayer.

Extensive Gel–Gel Separations with Liquid-Crystalline Phase Miscibility—The Peritectic Phase Diagram. As the chain-length difference between two identical-chain phosphatidylcholines is increased from 2 to 4 methylene units, the phase diagram constructed from the DSC curves changes from lens shape as observed for C(14):C(14)PC/C(16):C(16)PC binary mixtures (Fig. 2.8A) to the peritectic type (Fig. 2.8B) exhibited by C(14):C(14)PC/C(18):C(18)PC mixtures (120). Certain features of the mixing of C(14):C(14)PC and C(18):C(18)PC are obvious from the phase diagram. Over a wide composition range the solidus is flat, indicating the coexistence of two immiscible gel phases (G_1 and G_2). The liquidus in Figure 2.8B has no flat regions, indicating that no liquid-crystalline immiscibility of C(14):C(14)PC and C(18):C(18)PC occurs. However, the shape of the liquidus is indicative of some non-ideality of mixing in the liquid-crystalline state (29, 108).

Similar phase diagrams have also been detected for binary mixtures of C(16):C(18:1Δ^9)PC/C(16):C(16)PC and C(16):C(18:1Δ^9)PC/C(18):C(18)PC (47).

Partial Gel–Gel Phase Separation Combined with Liquid-Crystalline Phase Miscibility—The Eutectic Phase Diagram. When a mixed-chain phosphatidylcholine with a Δ/CL value of about 0.55 is mixed with an identical-chain phosphatidylcholine with the same molecular weight, the phase diagram of this binary mixture usually exhibits the shape which is the hallmark of a eutectic system (Fig. 2.8C). These binary mixtures include C(18):C(10)PC/C(14):C(14)PC, C(10):C(22)PC/C(16):C(16)PC, C(22):C(12)PC/C(17):C(17)PC (31, 115, 171).

As shown in Figure 2.8C, there are three one-phase regions (G_1, G_2, and L), three two-phase regions (G_1 + L, G_2 + L, and G_1 + G_2) and one degenerate three-phase region, the eutectic point. All three systems [C(18):C(10)PC/C(14):C(14)PC, C(10):C(22)PC/C(16):C(16)PC, C(22):C(12)PC/C(17):C(17)PC] have eutectic points located at a composition of ~40 mol% of the higher melting lipid component. The temperature component of the eutectic horizontal, however, varies with the combination of phospholipids.

The mixing behavior can be interpreted based on the physical properties of the component lipids. Mixed-chain C(18):C(10)PC are known, for instance, to form mixed interdigitated bilayers at T < T_m. The mixed interdigitated bilayer of C(18):C(10)PC undergoes a sharp transition at 18.8°C and the bilayer transforms into the L_α phase with a partially interdigitated motif. The bilayer thickness increases slightly, changing from 33 Å at 10°C to 35 Å at 30°C (90). The C(14):C(14)PC, however, undergoes a partially interdigitated gel → L_α phase transition at 24°C, and the thickness decreases appreciably with increasing temperature, changing from 43 Å at 10°C to 35 Å at 30°C (94). Based on the thickness information, one can expect that C(18):C(10)PC and C(14):C(14)PC are immiscible in the bilayer plane at 10°C due to the large mismatch (10 Å) in the bilayer thickness between the two lipid systems. However, C(18):C(10)PC and C(14):C(14)PC are expected to be miscible at 30°C owing to their identical thicknesses. These expectations are indeed borne out by the eutectic phase diagram observed for C(18):C(10)PC/C(14):C(14)PC binary mixtures. Similar explanations can be applied to the binary mixtures of C(10):C(22)PC/C(16):C(16)PC and C(22):C(12)PC/C(17):C(17)PC.

Nonlamellar Phases

Most of the lipids extracted from biological membranes when dispersed by themselves in aqueous media form a thermodynamically stable lamellar liquid-crystalline (L_α) phase. However, it is possible to induce a phase transition from the lamellar phase to a nonlamellar phase by changing the chemical composition, temperature, pressure, pH, degree of hydration, salt concentration, etc. (46, 116, 122, 123, 160, 181). Examples of conditions under which lipids, lipid mixtures, and lipid extracts from biological membranes form three widely studied nonlamellar phases are given in Table 2.5. The three categories include the hexagonal phases of type I and II, H_I and H_{II} respectively, and a class of cubic phases.

The topological features of the H_I, H_{II} and cubic phases are shown in Figure 2.9. The H_I phase is a lyotropic mesophase with hexagonal symmetry consisting of lipid aggregates in a continuous aqueous medium. The core of the aggregate is formed by the hydrophobic chains of the amphipathic lipid molecules with the polar headgroups pointing outward from the center of the cylindrical repeating unit. The H_{II} phase is an inverted hexagonal phase in which the hydrophobic chains point outward from the center of the cylinder and the core is occupied by the aqueous medium. The H_I phase can swell essentially indefinitely with no change in the dimensions of the cylindrical aggregate, whereas the aqueous core of cylinders in the H_{II} phase can be swollen only to a limited extent. The radius of the water cylinder in a typical H_{II} phase is in the range 20–30 Å.

Nonlamellar phases of cubic symmetry can also be classified as the normal, type I, and the inverted, type II, cubic phases. However, since many different cubic space groups are possible, a different nomenclature based on the differential compartmentation of the hydrocarbon and polar components of the system is used (116, 123). In this scheme, the cubic structures are

called bicontinuous, when the structure has regions that are continuous with respect to both water and the hydrocarbon components. An interesting feature of bicontinuous cubic phases is that molecular species soluble in the aqueous phase can sample the entire aqueous medium by diffusion while those confined to the hydrocarbon region can diffuse, in the absence of any obstructions, through the entire hydrocarbon region. The topology of two bicontinuous cubic phases of lipids with space groups $Im\overline{3}m$ or $Pn\overline{3}m$ are shown in Figure 2.9. Both structures are built from bilayer units. In the former case, also known as the plumber's nightmare, the networks are connected by joining bilayer units 6 by 6 along orthogonal axes. In the latter case, also known as the double diamond phase, the bilayer units are joined tetrahedrally to form the water labyrinths. The second class of cubic phases is the discontinuous cubic phase, which is built up either of discontinuous hydrocarbon regions but with continuous water regions or vice versa. When the hydrocarbon region is disconnected, the compartmentation of the two components is similar to that in normal type I phase (e.g., H_I phase). When the aqueous medium is disconnected, one has a structure similar to the inverted, type II phase (e.g., H_{II} phase). The discontinuous cubic phases are, therefore, obtained from monolayer units either in the normal or in the inverted micellar configuration.

As must be clear from the characteristic symmetries of the lamellar and nonlamellar (hexagonal and cubic) phases, they can be described by periodic minimal surfaces using mathematical principles of differential geometry. A minimal surface is defined as one having at each point on the surface a mean curvature of zero. Descriptions of cubic phases as area-minimizing surfaces can be found in the reviews by Lindbloom and Rilfors (116) and by Seddon (160).

Several mechanisms have been proposed to account for the various features associated with nonlamellar phases. One of the early phenomenological theories correlated the molecular shape of a given lipid molecule to its phase preference (91). Thus, cylindrical molecular shapes were suggested to prefer lamellar phases, while cone shaped molecules with larger interfacial areas compared to the hydrocarbon region were predicted to form micellar structures or normal H_I phases. Inverted cone-shaped molecules, with smaller areas at the hydrocarbon/water interface relative to the hydrocarbon region, were suggested to favor the inverted micellar or inverted H_{II} phases. While many lipids that favor either the lamellar or one of the two hexagonal phases conform to this correlation, the model does not correctly predict the formation of intermediate bicontinuous cubic phases. It must also be recognized that the

TABLE 2.5. *Examples of Conditions under Which Lipids, Lipid Mixtures, and Lipid Extracts from Biological Membranes Form Nonlamellar Phases*

Lipid-Containing System	Phase
Dodecylsulfonic acid in 23%–70% water	H_I
Egg lysophosphatidylcholine at 37°C in 22 wt%–52 wt% water	H_I
Diheptanoylphosphatidylcholine in water	H_{II}
Lipid extract from human brain containing phosphatidylethanolamine, phosphatidylcholine and phosphatidylinositol at 37°C and <22 wt% water	H_{II}
Sodium caprylate/decanol/water	H_{II}
Monoglucosyldiacylglycerols from *Acholeplasma laidlawii*	H_{II}
Phosphatidylethanolamines at high temperature or low pH	H_{II}
Cardiolipins at low hydration or high calcium	H_{II}
Bovine brain gangliosides at low hydration	H_{II}
Certain mixtures of phospholipids with diacylglycerols	H_{II}
Dioleoylphosphatidylcholine—1,3-dioleoylglycerol in excess water	Cubic
Dioleoylphosphatidylcholine/monooleoylglycerol/water, 44%/44%/12% water	Cubic
Monooleoylglycerol-cytochrome *c* at low hydration	Cubic
Lipid extract from *Sulfolobus solfataricus*	Cubic

[Compiled from (46, 116, 122, 123, 160, 191).]

model neglects molecular interactions, specifically the geometry dependence of the components of the free energy of interaction. In another theory it has been suggested that the phase behavior is a result of the competition between the tendency of lipid monolayers to (or not to) curl and the hydrocarbon packing strains that result as a consequence of the curling (70). The tendency of the lipid monolayer to curl is given by the intrinsic radius of curvature which minimizes the bending energy of the monolayer. All local interactions, including the hydrophobic effect, headgroup–headgroup and chain–chain interactions, packing interactions arising from the filling of the interstices between the lipid cylinders, surface electrostatics for charged lipids, and hydration interactions arising from the polarization of the water molecules by the lipid headgroups are explicitly included in the model. Using this theory it is possible to isolate effects due to relieving the packing stress from those due to changes in the preferred spontaneous curvature of lipid layers. In a third model the lamellar to H_{II} phase transition has been suggested to take place in a step-wise manner (168). The primary initial event was taken to be the formation of inverted micellar intermediates between apposed bilayers, but not within a single bilayer. These intermediates then either fuse into rod-shaped micellar intermediates, or form line defects, both of which can assemble into the H_{II} phase. Alternatively, the inverted micellar intermediates can fuse with the outer monolayers to form an interlamellar attachment, which is a channel through the pair of bilayers between which the initial intermediate was formed. A progressive formation of such interlamellar attachments eventually results in the formation of bicontinuous cubic phases. By invoking more than one way in which the intermicellar intermediates could further aggregate, this model predicts metastability for the various phases predicted. This feature may well turn out to be one of the most useful ones of the various models, since in many instances the bicontinuous inverted cubic phases are formed during the L_α–H_{II} phase transition. Despite the large collection of data on nonlamellar phases, it is not clear what role, if any, the H_I, H_{II} and the cubic phases might play in physiology.

PHYSICAL PROPERTIES OF BILAYERS

Experimental Bilayer Systems

In the late sixties and early seventies the physical properties of phospholipid bilayers in excess water were studied intensely in a number of laboratories. The ranges of values for bilayer thickness, electrical and permeability parameters, as well as mechanical parameters became of considerable interest in establishing the bilayer as the structural matrix of biological membranes (75, 183). Studies of this type were carried out principally on two types of experimental systems, the Rudin-Mueller (soap film analog) bilayer (138), and

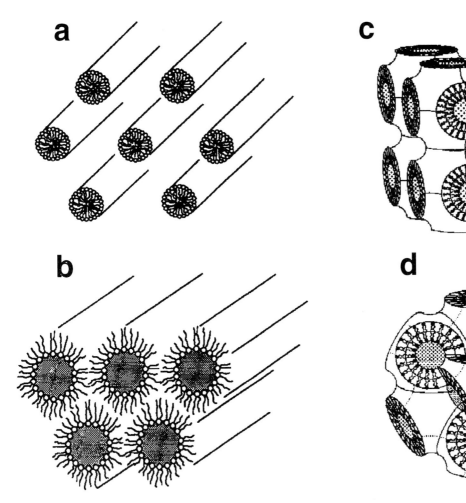

FIG. 2.9. Topology of the normal, type I, and inverse, type II, hexagonal, and cubic phases. *a:* Normal hexagonal phase, H_I, *b:* inverted hexagonal phase, H_{II}, *c:* "plumber's nightmare" bicontinu- ous cubic phase, and *d:* "double diamond" bicontinuous cubic phase. [Adapted from Tate et al., 1991 (181).]

liposome dispersions introduced by Bangham and co-workers (11, 12).

The planar Rudin-Mueller bilayer was the first experimental system to be described. The structure forms when a microdroplet of a solution of a few weight-percent phospholipid in a neutral hydrocarbon, such as decane or tetradecane, is applied to a small hole (diameter < 2 mm) in a plastic septum immersed in an aqueous buffer. The initially thick droplet spontaneously thins to a bilayer, roughly the diameter of the hole, surrounded by and connected to, a torus of bulk lipid solution adhering to the circumference of the hole in the supporting septum. The thinning process is analogous in all respects to the spontaneous thinning of a soap film supported on a wire ring in water-saturated air. The bilayer thus formed in the hole in the plastic septum can be arranged in a cell of suitable design to separate two aqueous compartments. Al-though the area of the bilayer formed in this way for mechanical reasons cannot be larger than several square millimeters, its high electrical resistance and capacitance can be measured easily using electrodes inserted into the two aqueous chambers separated by the bilayer. This experimental setup has been used to determine the specific conductance and capacitance of bilayers made from a wide variety of amphipathic molecules. The transference numbers of many biologically interesting ions as well as ion permeabilities induced by a number of carriers and channels have also been determined using this experimental configuration (183). Two successful electron microscopy studies using OsO_4 fixed, imbedded, and sectioned bilayers of this type, have been reported (73, 202). The thickness was found to be 69–73 Å. Spherical bilayers and semi-spherical bilayers of this type attached to pipettes have been used to measure mass fluxes of many different ions and other molecules using radioactive tracers. Details of the techniques for forming these types of bilayers as well as planar systems are available in comprehensive reviews (63, 92, 183, 189). A useful method devised by Mueller and Montal has been described for the formation of a bilayer in a very small aperture by bringing the hydrophobic faces of two monolayers together (137). This procedure has many advantages including the formation of bilayers from two monolayers of different composition and the inclusion of membrane proteins in one or both initial monolayers.

Liposome dispersions are formed by mechanically dispersing amphipathic lipids, usually phospholipids, in an aqueous buffer (10). The liposomes thus formed consist of concentric topologically closed phospholipid bilayers each separated by a shell of aqueous buffer of about the same thickness as a bilayer. These onion-like spheroidal or cylindroidal objects range in size from a few hundred Angstroms to objects just visible to the naked eye. A liposome consists of a series of concentric water compartments each separated by a bilayer, and of course conversely, a series of bilayers each separated by an aqueous lamella. Information about the structure and dynamics of phospholipids in bilayer array has been obtained for many substances using liposome dispersions (38). The favored techniques are NMR, electron spin resonance (ESR), and fluorescence spectroscopy which employ suitably labeled reporter molecules. Liposome structure can be visualized by several electron microscopic techniques. The bilayer and water layer thicknesses and the orientation of lipid molecules have been determined by low-angle x-ray diffraction and neutron diffraction (38). Methods for making stacks of several hundred bilayers each separated by a water layer on a glass or quartz support have also been described (200). These multilayers can be easily oriented in a magnetic field or optical beam. Single bilayers on a quartz substrate that are suitable for study by several techniques can also be formed (23, 98).

Water soluble molecules present in the aqueous medium during the formation of liposomes are readily trapped inside each internal water compartment. After removal of untrapped molecules in the ambient aqueous phase by molecular sieve chromatography or other means, the time rate of efflux of trapped molecules can be measured. The permeability thus determined for a given molecular species can be compared to that measured for a second type of permeant molecule. Absolute permeability coefficients, however, cannot be determined because of the serial arrangement of the water compartments in liposomes and the unknowable bilayer area involved (10). This difficulty can be overcome by carrying out permeability measurement on

unilamellar vesicles of uniform size. Vesicles of this type can be made by sonication of liposome dispersions (77), or by other disruption methods (114, 179) followed by molecular sieve chromatography to establish a degree of size homogeneity. Minimum size vesicles from 100–200 Å in radius are usually produced by these methods. Because of the small size of these vesicles, many of the physical properties of their bilayers have been shown to differ from those measured in planar bilayers when comparison is possible, as described later (see under BILAYER CURVATURE). The source of these differences lies in the small radius of the vesicle. A bilayer in a formal sense is comprised of two monolayers. As a result, when it is sharply curved as in a small vesicle, the outer monolayer has a significantly larger radius of curvature than does the inner monolayer. In addition, the two radii are of opposite sign. As a result, the molecular packing in the two monolayers is different when viewed along the bilayer normal. In minimum size vesicles of 100–200 Å radius, the packing anomalies are largest. They disappear when the radius is two or three times larger (186). Dispersions of unilamellar vesicles in the diameter range of 1000–3000 Å and relatively uniform in size can be made by extrusion methods (130). Single large unilamellar vesicles up to 50 μm in diameter have been prepared (151) and used to determine the bending and area compression moduli (104). Vesicles of this size are large enough to be studied individually under phase contrast or fluorescence microscopy either visually or using video recording techniques (104).

Bilayer Thickness

The view of the bilayer as a structure formed by apposing the non-polar faces of two monolayers requires that the physical thickness of the structure be equal to or, if the molecular axis is tilted with respect to the bilayer plane, less than twice the extended length of these molecules (109). Although the acyl chains and polar headgroups of biological membrane phospholipids vary greatly as described in section I, twice the extended lengths of these molecules is remarkably similar and is in the range of 50–65 Å. The hydrophobic region of a bilayer, which is composed of the acyl chains of the component molecules, is about 40 Å in thickness. The actual dimensions of specific systems as well as the dimension along the normal to the bilayer surface have been determined by several different means. These include electron microscopy, x-ray and neutron diffraction, optical interference methods, deuterium NMR spectroscopy, and bilayer capacitance measurements. Many of these methods have also been used to examine the thickness of specific biological membranes. Those methods that measure the total material thickness generally show biological membranes to be of the order of 100 Å in thickness. This thickness, which is about 50% greater than that of the bilayer itself is caused by extrinsic membrane proteins and the *trans* and *cis* cytoplasmic domains of intrinsic membrane proteins. It is important to understand that agreement among the values obtained by applying different methods to the same system is not perfect. This is due to the fact that reduction of the primary data to a thickness requires in all cases assumptions specific to each method (159).

Early efforts to stain multilamellar liposomes with OsO_4 followed by embedding in epoxy resin and sectioning for electron microscope examination, a method that proved successful with biological membranes, gave generally unsatisfactory results. However, two successful studies using OsO_4, have been reported (73, 202) which gave a thickness of 69–73 Å. Electron microscope examination of phospholipid liposomes negatively stained with electron-dense materials such as uranyl acetate or phosphotungstate were found to give images in which the alternating bilayer–water space repeat was clearly discernable. Bilayer thickness measurements of these images supported the bimolecular nature of the lipid organization (9). Freeze-fracture technologies for the preparation of specimens for electron microscopy also give images with a similar bilayer thickness (35). Quite recently cryoelectron microscopy has been used with considerable success to examine unilamellar bilayer vesicles, again giving results in accord with the established dimensions for bilayer thickness (203).

A second approach to the determination of thickness is through the measurement of the bilayer specific capacitance. This measurement is easily carried out on planar soap film analog bilayers of the Rudin-Mueller type. Based on the reasonable assumption that the bilayer separating the two aqueous phases is in fact a parallel plate condenser, the thickness, d, can be calculated from the measured capacitance, C and bilayer are A:

$$d = \frac{\epsilon A}{4 \pi C} \qquad (2.1)$$

It is of course necessary to know the value of the dielectric constant, ϵ, of the hydrophobic core of the bilayer. Assuming a reasonable value of 2 for ϵ, the hydrophobic core thicknesses is found to be between 40 and 50 Å (183). This range is in good agreement with other measurements. It is also in excellent agreement with the thickness of the dielectric element of the red blood cell membrane deduced early on from capacitance measurements by Fricke (66).

An alternative and totally independent way of determining bilayer thickness in planar Rudin-Mueller bilayers is based on quantitative measurements of the intensity of light reflected from the bilayer combined with refractive index data obtained from measurements of the Brewster angle (85, 86). These measurements are difficult from both experimental and theoretical standpoints. The most refined study obtained a thickness of 62 ± 2 Å for bilayers prepared from hen egg phosphatidylcholine (42, 43).

Perhaps the most widely applicable and generally useful method for examining the thickness and molecular structure of bilayers is that of low-angle x-ray diffraction. This method was first used to study phospholipid bilayers by Luzzati and coworkers (118, 119). It has been widely applied to systems of many different lipids. The results of these studies are summarized in several reviews (39, 126, 174). Although the liquid crystalline phase is of primary importance in biological membranes, there is increasing evidence that gel phases may be present in membranes. Low-angle x-ray diffraction studies give rather detailed information about the thickness and molecular organization in crystalline, gel, and liquid-crystalline phases.

Figure 2.10 illustrates for dipalmitoylphosphatidylcholine the type of information that can be obtained (174). Three gel phases and one liquid-crystalline phase have been described for this lipid in excess water (94, 156, 157). The most highly ordered is the crystalline L_c phase. The $L_{\beta'}$ gel phase is characterized by having the long molecular axis of the component molecules tilted with respect to the bilayer normal by an angle of 30°–32° (158). Some lipids exhibit an L_β gel phase for which the tilt angle is 0°. The $P_{\beta'}$, seen only in diacyl saturated phosphatidylcholin, is characterized by a ripple structure of regular period from 100–300 Å, depending upon conditions. The molecular organization of this structure is unknown, although several models have been advanced (36, 94, 158). The L_α or liquid crystalline phase is the most disordered with the acyl chains containing many *gauche* conformers and the hexagonal packing order of the molecules extending over much shorter distances than in gel phases. Similar information can be obtained using neutron diffraction on specifically deuterated phospholipids in bilayer array (32). Quite recently Wiener and White have shown how considerably more detailed information can be obtained by a joint refinement of x-ray and neutron diffraction data (206–208). Using this powerful approach with data obtained for a dioleoylphosphatidylcholine bilayer at 23°C, these authors have determined the mean distance from the bilayer center of seven sub-parts of this molecule together with the distribution of 5.36 molecules of bound water per lipid

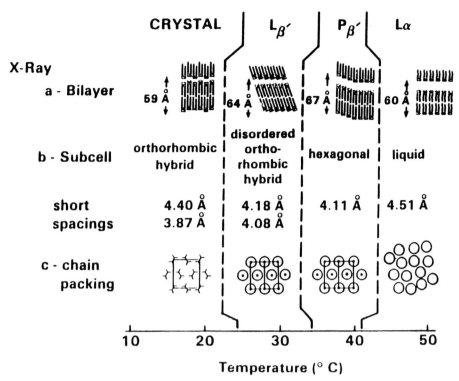

FIG. 2.10. Molecular organization of the four phases of dipalmitoylphosphatidylcholine as determined from low-angle x-ray data. [Taken with kind permission from Figure 12-6 on p. 487 of Small, 1986 (174).]

molecule. This is the most detailed data set for a liquid crystalline phospholipid bilayer that has been obtained to date. These authors have introduced the idea of a dynamic thickness for the bilayer and set limits on this thickness about a time–average value of 50 Å (208). The thickness fluctuations are thermally driven and reflect the existence of an intramolecular gradient of thermal motion that increases in either direction away from the glycerol backbone, the most highly constrained part of the molecule (208).

Electrical Properties

As an electric circuit-element, simple bilayers, unmodified by adventitious impurities or deliberately added dopants, display a resistance and a capacitance but no inductance. The parameters are most easily measured in Rudin-Mueller soap–film analog and Mueller-Montal bilayers. Capacitance measurements are generally in the range of 0.3–0.8 $\mu F/cm^2$. The conductances are very low and usually lie in the range of 10^{-9}–10^{-8} S/cm^2. The actual values of these parameters depend upon the specific phospholipid, the neutral hydrocarbon used in the membrane-forming solution, the temperature, and in some instances, the composition of the aqueous phase. Mueller-Montal hydrocarbon-free bilayers usually have a somewhat higher capacitance than Rudin-Mueller bilayers, (206) which contain some neutral hydrocarbon solvent (74, 152). The thickness of Mueller-Montal bilayers calculated from capacitance measurements is very close to that found by low-angle x-ray diffraction (152). These electrical parameters, determined by dielectric dispersion measurements, have also been obtained for bilayers in vesicle dispersions formed from phosphatidylcholines (150).

It is of considerable interest to note that bilayers can withstand transmembrane voltages up to 300 mV. Bilayers of very small area have been reported to be stable at 500 mV (4). Because of the extreme thinness of the structure, these voltages impose an electric-field strength in excess of 10^5 V/cm across the bilayer. This is a most unusual property for dielectric films but one that is absolutely essential from the biological standpoint because transmembrane voltages in excess of 100 mV are not uncommon in cells.

The ionic conductances for both anions and cations of biological relevance as well as a number of other small ions have been studied in some detail. Table 2.6 lists the bilayer permeabilities for some ions of biological interest. The fact that the very low electrical conductance of the bilayer is due to ions has been established by isotopic flux measurements. The first

TABLE 2.6. *Diffusive Permeability for Transport of Solutes across Phosphatidylcholine Bilayers*

Solute	System	T (°C)	P_d (cm/s)	Reference
H_2O	PC-cholesterol (9/1) SUV	23	6.3×10^{-4}	(217)
H^+	Egg PC LUV	RT	3×10^{-4}	(14)
Na^+	Egg PC SUV	4	$1-2 \times 10^{-14}$	(72)
		4	1.8×10^{-14}	(30)
		25	2.1×10^{-13}	
		37	7.4×10^{-13}	
	Egg PC LUV	RT	1×10^{-12}	(143)
		24	9.5×10^{-13}	(136)
		RT	1.4×10^{-12}	(195)
	DPPC LUV	42 (T_m)	60×10^{-10}	(56)
K^+	Egg PC LUV	RT	7×10^{-11}	(14)
$Rb+$	Egg PC LUV	24	3.3×10^{-12}	(136)
	DPPC LUV	22	1×10^{-10}	(56)
		42 (T_m)	1×10^{-9}	
		56	1×10^{-8}	
Cl^-	Egg PC SUV	4	5×10^{-12}	(72)
		20	1.1×10^{-10}	(193)
	Egg PC LUV	24	7.6×10^{-11}	(136)
		RT	8×10^{-11}	(196)

RT, room temperature; T_m, transition tempertature; PC, phosphatidylcholine; LUV, large uni-lamellar vesicles; SUV, small unilamellar vesicles; DPPC, dipalmitoylphosphatidylcholine.

such study carried out by Pagano and Thompson (145) on spherical bilayers found the Na^+ permeability of 2×10^{-8} cm/s at 30°C to agree well with the value calculated from electrical measurements on the same system. The flux of Cl^-, however, proved to be 2–3 orders of magnitude larger than that calculated from electrical parameters. The origin of this large electrically silent chloride flux was investigated in subsequent studies (153, 192, 193). In unilamellar vesicle dispersions the permeabilities for univalent cations are about 10^{-12} cm/s (40). Unmodified bilayers show very little preference for either cations or anions of biological interest provided the bilayers do not contain an amphipathic lipid with a net nonzero charge. A discussion of the electrical parameters from both the experimental and theoretical viewpoints can be found in several reviews (3, 40, 190).

Permeability to Non-Ionic Solutes

Bilayers exhibit measurable permeabilities to a wide variety on non-ionic solutes. If the solutes are small lipid soluble molecules, the permeabilities are usually large and correlate reasonably well with the partition coefficients of the solutes between hydrophobic and aqueous phases. The correlation is improved if the product of the permeability coefficient and the square root of the molecular weight of the solutes is plotted against the partition coefficients. This correlation leads to a simple model for the permeation process in which the solute partitioning between the aqueous phase and the bilayer is at equilibrium while the rate-limiting step in permeation is the transbilayer diffusion of the solute (93, 204). The permeability properties of bilayers for small lipid-soluble molecules mimic the permeability properties of biological membranes for this class of molecules (93, 204).

In contrast to the larger permeabilities of lipid-soluble molecules, small water-soluble non-electrolytes have, for the most part, very small permeability coefficients reminiscent of those displayed by ions. For example, the phosphatidylcholine bilayer permeabilities for fructose, glucose, and sucrose at 25°C are 4.0×10^{-10}, 0.30×10^{-10}, and 8.0×10^{-14} cm/s, respectively (30). These values, measured for single bilayer vesicles, are one-half to one-third smaller than the values measured for Rudin-Mueller planar bilayers at the same temperature. These differences in the permeability coefficient for non-electrolytes measured in these two types of bilayer systems are probably due to the presence of neutral lipid solvent in the Rudin-Mueller bilayers, which lowers the average viscosity of the bilayer, resulting in a larger transmembrane diffusion coefficient. The bilayer permeabilities of the sugars cited earlier are representative of the permeabilities of the majority of the intermediary metabolites present in cells. The bilayers of cell membranes are thus effectively impermeable to these molecules. As a result, molecules of this type are confined to the cellular compartment where they are reactants and products or their passage across boundary membranes requires specific transport systems.

Since the bilayer permeability increases with increases in the bilayer/water partition coefficient, as the hydrophobic character of the nonelectrolyte increases relative to its hydrophilic character, the bilayer permeability increases. For example the permeability of glycerol and butyramide under the same conditions are 5×10^{-6} and 7.3×10^{-5} cm/s, respectively (40). Tables of permeability coefficients for a wide variety of molecules through bilayers of various types and compositions can be found in several reviews (3, 40, 93, 190, 204). While these general trends are not unexpected, it is surprising that the permeability of phosphatidylcholine bilayers to H_2O is in the range of $(1.7-7.1) \times 10^{-3}$ cm/s at 36°C (3). The lower values are determined for phosphatidylcholine vesicles and the higher values for Rudin-Mueller bilayers. If the classical view of membrane permeation is adopted in which the process is modeled to be equilibrium distribution of the permeant solute between aqueous phases coupled to a diffusion step in crossing the membrane, then the water permeability of the bilayer can be calculated from knowledge of the solubility and diffusion coefficient of water in bulk hydrocarbons (93, 204). In this model, the bilayer is replaced by a thin sheet of isomorphous hydrocarbon. This calculation, first carried out by Finkelstein in 1976, gave quite good agreement with the water permeability coefficients of bilayers of several different compositions (64). It has been argued subsequently that this agreement using the diffusion and partition coefficients obtained for bulk systems is fortuitous and is the result of compensating errors (149, 169). The high measured bilayer permeability coefficients for water are large enough to account for the water permeability of many mammalian cell membranes (183). The interesting problem is the low and regulated water permeability in specialized membranes such as those of the collecting ducts of the kidney. Recently considerable attention has been focussed on proteins that appear to act as water channels in certain biological membranes (201).

In general, bilayers in the liquid-crystalline or fluid state display increasing permeabilities to most solutes with increasing temperatures. The activation energies do not differ very much for different solutes and are usually in the range of 12–25 kcal/mole. It has been suggested that even though the permeabilities for different solutes differ greatly, the similarity in activation energies means that the differences in permeabilities are largely entropically controlled (3). Permeabilities in gel phases are generally several orders of magnitude lower than in fluid phases (22). In some bilayer systems a maximum in ion permeability near the gel/liquid crystalline phase transition has been reported (18). Recently large increases in glucose permeability in two-component, two-phase bilayers have been observed (45a).

Mechanical Properties of Bilayers

Single and multiple bilayer vesicles, as described earlier (see under Experimental Bilayer Systems) are mechanically stable. Disruption of multiple bilayer vesicles or liposomes generally requires the input of considerable energy (77, 114, 179). Soap film analog bilayers of the Rudin-Mueller type are, in contrast, easily disrupted by low frequency vibrations. Because of this fragility, planar bilayers larger than 4–5 mm (75), of the type that span the aperture in the septum separating two aqueous compartments, are unsatisfactory as study objects. It is, however, possible to generate spherical or semi-spherical bilayers of this type with areas of up to 100 mm^2 in either free-floating form (144) or tethered to a 1 ml glass pipette tip (192).

In general, mechanically induced membrane deformations can be described by three first-order conservative relations that relate the stress to the strain through a modulus that is characteristic of the material. These three moduli described the resistance of the membrane to changes in area driven by tension tangents to the membrane plane, to shear focus, and to bending moments. Thus for area dilation or condensation:

$$\frac{(\tau_1 + \tau_2)}{2} = K \frac{\Delta A}{A_0} \qquad (2.2)$$

here τ_1 and τ_2 are the orthogonal tensions in the membrane surface, $\Delta A/A_0$ is fractional area change, and K is the modulus. For surface shear:

$$\frac{(\tau_1 + \tau_2)}{2} = \mu \left[\left(\frac{L}{L_0} \right)^2 - \left(\frac{L_0}{L} \right)^2 \right] \qquad (2.3)$$

here L/L_0 is the fractional in-plane extension at constant area and μ is the shear modulus. Finally for bending:

$$M = B \Delta \left(\frac{1}{R_1} - \frac{1}{R_2} \right) \qquad (2.4)$$

here $1/R_1$ and $1/R_2$ are the orthogonal membrane curvatures, B is the bending modulus and M is the bending moment (58).

Since liquid crystalline bilayers are fluid structures, the shear modulus μ is equal to zero. The dilation and bending moduli have been measured for a number of systems. The general result is that fluid bilayers are "soft" systems when compared to conventional thin polymer films. The most extensive work has been done by Evans and coworkers using large single-bilayer vesicles 10–50 μm in diameter. The experimental procedure depends upon the capture of a vesicle on the tip of a micro-pipette by application of controlled negative pressure. Under these circumstances a short cylinder of bilayer enters the bore of the pipette tip. The pipette tip with tethered vesicle in an aqueous medium is viewed and recorded by video phase contrast microscopy. These workers have shown that small changes in the length of the bilayer cylinder trapped in the pipette tip are proportional to small changes in surface area. The mean tension $(\tau_1 + \tau_2)/2$ is simply related to the suction pressure in the pipette (104). This apparatus can be used to determine the area dilation modulus, K, and the thermal expansivity coefficient α^0, and a rough value for the bending modulus, B. A combination of K and α^0 can be used to calculate a value for the heat of expansion per unit area, ΔH. Representative values are shown in Table 2.7 together with values for the same moduli determined by similar methods for the human erythrocyte. Although the bilayer of the erythrocyte is believed to be fluid, the value of K obtained at 37°C is substantially larger than the simple fluid bilayer value at 15°C. This difference may be explained by the presence of cholesterol at 44 mol% in the human erythrocyte membrane. Cholesterol in simple bilayers at this mole ratio can increase K to comparable values at 15°C (58). A striking difference is noted in the values for the shear modulus, μ. The resistance to shear is quite large in the erythrocyte membrane and comparable to that in solids. There can be little doubt that the spectrin skeleton in contact with the erythrocyte membrane contributes substantially to membrane shape and deformability characteristics. It is interesting to note that the values of K for both bilayers and erythrocyte membranes suggest that these structures are 10–100 times more expandable than are ordinary liquids in bulk (58).

LIPID DYNAMICS IN BILAYER SYSTEMS

In discussing the physical properties of biological membranes, both the structure and the dynamics of the lipid and protein molecules arranged in a two-dimensional bilayer have to be addressed. Structure and dynamics are inseparable since the experimental observables in any experiment are averaged over temporal and spatial ensembles. By choosing a given technique that is most sensitive to molecular motions in a characteristic time-window, one can either obtain a single time-averaged structure for the system, or the correlation times for those motional modes that are neither too slow nor too fast compared to the time scale of the technique. A selection of experimental methods, the range for the time scale of the method (the same as the range of the correlation times to which that technique is most sensitive), and the kinds of intramolecular and whole molecule motions that can be studied are given in Table 2.8.

The correlation times for the various intramolecular and whole molecule motions are usually obtained by NMR, and by electron spin resonance (ESR), and fluorescence spectroscopic techniques (69). In these methods, the time course of recovery of a perturbed magnetization or of a quenched fluorescence signal intensity is monitored. Since this recovery process is influenced by the dynamics of lipids, the correlation times can be

TABLE 2.7. *Static Elastic Parameters*

Lipid*	K (dyn/cm)	μ (dyn/cm)	B (dyn/cm)	α^0 (/°C)	ΔH (erg/cm^2)	T (°C)
Phospholipid Bilayer						
DMPC (L$_\alpha$)	1.5×10^2	0	—	6.8×10^{-3}	—	29
DMPC (L$_{\beta'}$)	8.5×10^2	—	—	1.0×10^{-3}	—	8
SOPC (L$_\alpha$)	2.0×10^2	0	—	3.3×10^{-3}	—	15
Egg PC (L$_\alpha$)	—	0	2.3×10^{-12}	—	100–135	20
Human Red Blood Cell						
	4.0×10^2	5.7×10^{-3}	—	1.2×10^{-3}	175	37

* DMPC, dimyristoylphosphatidylcholine; SOPC, 1-stearoyl-2-oleoylphosphatidylcholine; PC, phosphatidylcholine; L$_\alpha$, fluid phase; L$_{\beta'}$, gel phase.
[All values taken from (60) except the B value, which comes from (165).]

TABLE 2.8. *Approximate Ranges for the Time Scales of a Selection of Common Methods Used for Studying the Dynamics of Lipids in Membranes and Examples of the Kinds of Molecular Processes Studied Using Those Methods*

Technique	Frequency (s^{-1})	Molecular Processes
Infrared and Raman spectroscopy	$10^{13}-10^{15}$	Vibrational motions
Dynamic neutron scattering	$10^{7}-10^{10}$	Configurational isomerism, rotational motion
Dielectric relaxation	$10^{5}-10^{12}$	
Electron spin resonance spectroscopy	$10^{3}-10^{10}$	Configurational isomerism, lateral diffusion, rotational motion, flip–flop
Fluorescence and phosphorescence spectroscopy	$10^{3}-10^{7}$	Configurational isomerism, rotational motion, lateral diffusion, intervesicular lipid exchange
Multinuclear nuclear magnetic resonance	$10^{-1}-10^{11}$	Configurational isomerism, rotational motion, lateral diffusion, wobbling motion

extracted from the time course using standard methods. This approach necessarily gives rise to a time- and space-averaged structure, departures from which are the time-dependent conformations of the molecule. Information regarding the possible structures that the molecule might adopt while undergoing the various motions is not available from such studies. Molecular dynamics calculations have been shown recently to be of great value in this regard (17, 51, 100, 140, 147). In these calculations, an initial structure is assumed for the molecule. Characteristic time scales are assigned for the various motions the molecule can undergo e.g., configurational isomerization, whole molecule rotation, chain wobble, lateral diffusion, etc. Either the molecule as a whole or segments of it are then allowed to move spatially with characteristic time constants. At desired time intervals, the structure can be visually examined. Figure 2.11 shows an example of a computer simulation, in which the headgroups and the acyl chain configurational excursions were the only motions considered (51). The disordering effect on molecular conformations as a function of time is clearly demonstrated in this simulation. The ensemble average properties are computed to compare with experimental spectroscopic data and judge the goodness of the calculation. A wide range of motions, including overall reorientation with nanosecond decay time, are observed in molecular dynamics simulations.

Intramolecular Motions

The polymethylene chains of lipids in bilayers exhibit configurational isomerism by rotations around the individual C—C bonds. As a result, the acyl chains consist of *trans* and *gauche* configurations (135, 163). Increase in temperature, increase in the degree of unsaturation, and decrease in the number of carbons per acyl chain lead to an increase in the population of the *gauche* conformers. The consequence of the shift towards *gauche* conformers is a decrease in the thickness of the lipid bilayer (163). This motion occurs at frequencies up to $10^{10}-10^{11}$ s^{-1} (27). In addition to the configurational isomerization of the acyl chains, the other types of intramolecular motions include a wobbling motion exhibited by the acyl chains and by the headgroup segments. The frequencies of these motions are generally in the range 10^6-10^{10} s^{-1} (6–8, 19, 20, 24–27, 34, 49, 50, 65, 95, 175, 196, 205).

Whole Molecule Motions

The motions of the lipid molecule as a whole include lateral diffusion of lipids in the plane of the membrane (52, 97, 199), rotation about an axis perpendicular to the plane of the bilayer (62), movement of lipids from one monolayer to the other, also known as flip–flop, (102, 210), and spontaneous (96, 155) and mediated exchange (13, 28, 53) of lipids between vesicles. The time scales for these motions vary from several hours for intervesicular lipid exchange to the submicrosecond range for lipid lateral diffusion.

Translation. The translational diffusion coefficient, D, has been measured by a number of methods to be in the range of $10^{-7}-10^{-9}$ cm²/s for phospholipids in the fluid phase (57) and about 10^2-10^3-fold smaller in the gel phase. Integral membrane proteins in fluid phase

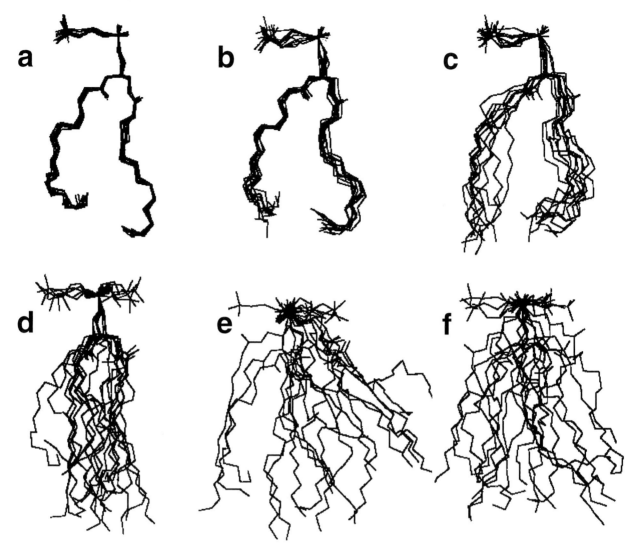

FIG. 2.11. Superposition of ten consecutive time frames of single-molecule simulation. Time interval between superimposed coordinate sets varies from 100 fs (a) to 10 ns (f), increasing by a factor of 10 between each part of figure. [Taken with kind permission from de Loof et al., 1991 (51).]

membranes have D values similar to phospholipids in gel phase membranes (45, 97).

From a theoretical standpoint, lateral diffusion of lipids can be understood by a free-volume model of diffusion (45). The free-volume model involves a microscopic process that describes the diffusion of a molecule in a solvent as consisting of the creation of a hole of a minimum size due to density fluctuations, the movement of the molecule into this hole, and the filling of the void left by the diffusing molecule by solvent molecules.

It is interesting to examine the consequence of translational diffusion on the compositional redistribution of a newly introduced lipid or protein molecule into the membrane. Lipid biosynthesis, lipid hydrolysis by phospholipases, and signal–peptide directed protein translocation are a few examples of this phenomenon. Einstein's diffusion equation, when applied to a cell of 1 μm diameter, shows that it takes about 1s to 4s for the new lipid molecule added to the cell membrane to diffuse laterally in the membrane and return to the original point of its addition. A protein may require several hours to accomplish this task. This calculation assumes that there exist no obstructions to translational diffusion for the lipid and protein molecules. However, microscopic and spectroscopic studies on the diffusion of lipids and proteins in membranes provide evidence to the contrary (54, 55, 154, 199). Many membranes are known to contain a characteristic domain structure. When the membrane substructure involves domains, percolation theory has been found to be of great value in understanding lipid lateral diffusion (1, 2). Percolation theory assumes that the diffusing molecule can only diffuse through one of two types of region in the membrane but not through the other (176).

Flip–Flop. Phospholipid flip–flop refers to the transbilayer movement of lipids from one monolayer to the other of a single bilayer leaflet (102). In model membranes, this is a slow process with half-times ranging from seconds for cholesterol to several hours for phosphatidylcholine. Flip–flop requires the transfer of the polar headgroup of the phospholipid through the hydrophobic interior of the bilayer. Thus, the flip–flop rates are dependent on factors that affect headgroup polarity such as pH and those that affect the hydrophobic character of the membrane such as cholesterol content, temperature, and chain length.

In biological membranes, flip–flop of certain phospholipids may be accelerated by enzymes, known as flippases (221). Interestingly, the phospholipid composition of the outer and the inner leaflets of biological membranes are not the same (210). For instance, the percentage of phosphatidylcholine, phosphatidylethanolamine, and phosphatidylserine in human erythrocytes associated with the inner leaflet is 24, 80, and 100, respectively. This may be the consequence of the slow transbilayer movement of phospholipids. This may also be due to specific lipid-protein interactions that enrich one of the monolayers in phospholipids that interact specifically with the extracellular domains of proteins on that side of the membrane (88).

Rotation. In the fluid phase, the correlation time for rotational motion of phospholipid molecule as a whole about the long axis is about 10^{-7} s (194). The correlation times for oscillatory motions of the acyl chains about the long axis, and rotational motions of the headgroups around the P—O bond in the fluid phase have correlation times of about 10^{-9} s (167, 197). In the gel phase, these correlation times are reduced to the ms range (68, 124, 167, 194, 213).

Spontaneous Inter-Bilayer Lipid Transfer. In model membranes at low lipid concentrations, the spontaneous inter-bilayer lipid transfer has half-times in excess of 50 hours. This process is limited by the low water-solubility of lipids (101, 133, 155). This limitation can be circumvented by increasing the lipid and hence the vesicle concentration, which results in faster lipid transfer rates by a collisional process (96).

In biological membranes, inter-bilayer lipid transfer may be effected not only spontaneously but also by membrane fusion and by phospholipid-specific transfer proteins (13, 28, 53).

BILAYER CURVATURE

As noted earlier (see under Mechanical Properties of Bilayers), phospholipid bilayers are "soft" structures that are easily bent. The minimum radius of curvature that these structures tolerate is about 100 Å. This conclusion is based on the observation that the minimum size vesicle that can be formed by sonication or extrusion is of this radius (77, 114). It is interesting to note that frequently biological membranes exhibit curved shapes, however in most cases the two orthogonal radii of curvature that define the shape are unequal. In many instances at least one of the radii of curvature approaches close to the 100 Å limit (185). In some biological membranes small radii are associated with time-invariant structures; in others, membrane shape changes are dynamic in the time frame of minutes to hours.

The bilayers of small minimum radii curvature vesicles display physical properties that are in general different from the corresponding properties of bilayers with larger radii. This difference in properties vanishes when the vesicle radius is larger than about 400 Å–500 Å (113, 142). For example, the partial specific volume of the lipid in minimum radius vesicle is 4% larger than in planar bilayers (185). The main phase transition temperature, entropy and enthalpy are 37°C, 20 cal deg^{-1}·mol^{-1}, and 6.3 kcal/mol, respectively, in small vesicles in contrast to planar bilayers which have as corresponding values 41.2°C, 27 cal deg^{-1}·mol^{-1} and 8.2 kcal/mol (177). Recently it has been suggested on the basis of ^{13}C NMR relaxation measurements that the lateral diffusion coefficient of phospholipids in sonicated vesicles may be substantially larger than it is in uncurved bilayers (57).

The most interesting effect produced by bilayer curvature is found in systems containing more than one lipid component. In these systems the relative compositions of the inner and outer surfaces of the vesicle bilayer are usually different. Table 2.9 shows representative data for three different two-component bilayers with the total concentration of the two components equimolar in each system. It is clear that when the vesicle radius is about 100 Å, the relative compositions of the two bilayer surfaces can be markedly different. This transmembrane compositional asymmetry is not seen, however, in vesicles with a diameter of 1000 Å (142).

The altered physical properties of sharply curved bilayers must be the result of differential packing con-

TABLE 2.9. *Relative Compositions of Inner and Outer Bilayer Surfaces in Minimum-Sized Vesicles* *

Total Vesicle (mole ratio)	Inner Surface (mole ratio)	Outer Surface (mole ratio)
PG/PC = 1	PG/PC = 2.0	PG/PC = 0.33
PE/PC = 1	PE/PC = 0.65	PE/PC = 3.0
Chol/PC = 1	Chol/PC = 0.89	Chol/PC = 1.3

* PG, phosphatidylglycerol; PC, phosphatidylcholine; PE, phosphatidylethanolamine; Chol, cholesterol.
[Taken with kind permission from p. 5 of Thompson et al., 1974 (185).]

straints on the molecules of the inner and outer mono-layers comprising the bilayer, Figure 2.12 shows a cross section through a 105 Å radius vesicle. If the thickness of the bilayer is 40 Å, and the outer radius is 105 Å, the inner radius is 65 Å. More important than this difference in curvature radii is the fact that the mirror plane symmetry of the bilayer causes the hydrophobic tails of the molecules of inner monolayer to splay outward, while it is the polar heads of the molecules on the outer monolayer which show this splay. It is quite clear that the shapes of the volumes occupied by molecules in the inner monolayer are quite different than those of the molecules in the outer monolayer. Huang and Mason have shown how the detailed geometry of the packing constraints can be determined from the experimentally determined vesicles parameters of shell weight, partial specific volume of the lipid, and the outer/inner surface area ratio (82).

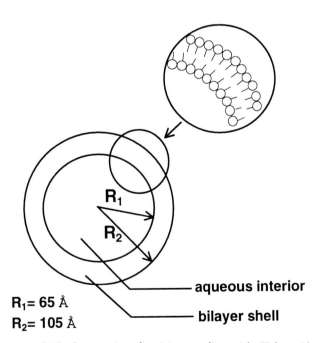

FIG. 2.12. Cross-section of a minimum radius vesicle. [Taken with kind permission from p. 5 of Thompson et al., 1974 (185).]

The difference in packing geometry in the inner monolayer as compared to the outer monolayer must also be the basis for the compositional asymmetry in minimum-sized vesicles. The packing of the headgroups on the inner monolayer is much tighter than in the outer monolayer. In fact, the cross sectional headgroup area is close to that of phospholipids in the gel phase. In contrast the splay of the outer monolayer gives rise to a greater cross sectional headgroup area than is found in uncurved fluid bilayers (82). Thus both phosphatidylethanolamine and cholesterol with smaller headgroup cross sections than phosphatidylcholine would be expected to prefer the inner monolayer, while the negatively charged phosphatidylglycerol would be expected to prefer the outer monolayer because of the unfavorable electrostatic interactions between negatively charged headgroups. These expectations are in qualitative accord with the data in Figure 2.12. The actual values of the relative concentrations in the inner and outer monolayers for sharply curved two-component bilayers can be predicted on theoretical grounds for some systems (33).

From the viewpoint of biological membranes the causal relation between bilayer curvature and trans-membrane compositional asymmetry has interesting implications. It is apparent that there is a connection between membrane curvature and composition, and thus between curvature and function. The causal relationship between curvature and composition can operate in either direction. Thus changes in curvature caused by mechanical means such contractile proteins or microtubules can in principal drive changes in trans-membrane composition, or changes in transmembrane composition can drive changes in curvature that may be of physiological importance (185). A formal phenomenological treatment of the relation between membrane curvature and compositional asymmetry has been presented by Evans (59, 61).

Bilayers as "soft" systems undergo spontaneous curvature (117). This process plays a central role in vesicle formation and the shape fluctuations seen in micron-sized, single lamellar vesicles. The formation of vesicles, usually single or oligolamellar, can be accomplished by several means as described earlier (see under Intramolecular Motions), most commonly by mechanical disruption of multilamellar liposomes, dilution into aqueous buffer of lipids in an organic solvent or in a detergent-solubilized solution. Extrusion of aqueous dispersions of liposomes through very small orifices under high pressure has recently become a method of choice for the preparation of single-lamellar vesicles in the 1000 Å diameter size range. Under many circumstances vesicles from spontaneously from larger multilamellar structures.

When vesicles are prepared by solvent or detergent dilution or by mechanical disruption, it seems certain that they form from small bilayer fragments known as disc micelles (67, 105–107). Since the edge of a disc micelle has an excess-free energy due to the unfavorable interactions of hydrophobic interior of the bilayer at the disc periphery with the aqueous phase, the disc will spontaneously reduce this excess free energy by minimizing the length of the edge per unit mass of lipid. This is accomplished by forming circular discs, by fusion of small discs to form larger discs, and by bending the discs into bell-like shapes. The thermally-driven bending of discs to reduce the edge length is opposed by the inherent stiffness of the bilayer reflected in a bending modulus (see earlier under Mechanical Properties of Bilayers). As the discs grow in size, the energy required for bending becomes less than the energy gain obtained to close the disc to form a vesicle, a structure with zero edge energy. At this point, spontaneous closure of the mouth of the bell occurs. A quantitative treatment of this model leads to estimates of vesicular sizes that are in good agreement with experiment (105–107). Disc micelles can be stabilized by detergent molecules of appropriate amphipathic geometry that associate with the disc edge. As the detergent is removed by dilution or dialysis, discs undergo the same processes to reduce edge energy that ultimately end in disc closure to form vesicles (67).

In all probability the formation of vesicles by extrusion of multilamellar liposomes through very small diameter cylindrical orifices also involves spontaneous curvature of bilayers. The most commonly used procedure is extrusion through polycarbonate filters under high hydrostatic pressure (76). A reasonable mechanism for vesicle formation is as follows (45b): The pores in these filters are long cylindrical channels, typically about 6 μm long and 0.1 μm in diameter. Because of the high flow velocity through the pore, the shear gradient across the pore diameter is very large. As a result, single lamellar tubes of about the length of the pore are formed in it and extruded from the

trans side of the filter. Cylindrical structure of this type have been known for many years in soap films and emulsions. They are mechanically unstable and spontaneously break-up into small vesicles. In the simplest case when the length of the cylinder is equal to, or greater than, the cylinder diameter, a critical condition obtains that causes the cylinder to break spontaneously into two vesicles of unequal size (21). The theoretical understanding of this type of system, first studied by Rayleigh, is well-developed (180).

It has been known for some time that large, spherical, single, or oligolamellar vesicles in the 0.1–10 μm size range spontaneously undergo time-dependent changes in shape that in some instances lead to the pinching off of smaller vesicles processes that can be observed under phase contrast microscopy. A detailed discussion of experimental observations and a theoretical treatment of spontaneous shape changes in vesicular system can be found in two recent reviews (117, 164). It has been suggested that this phenomenon, in all probability, underlies shape changes, fusion, and fission of cell and organelle membranes (117).

REFERENCES

1. Almeida, P.F.F., W.L.C. Vaz, and T. E. Thompson. Lateral diffusion and percolation in two-phase, two-component lipid bilayers. Topology of the solid-phase domains in-plane and across the lipid bilayer. *Biochemistry* 31: 7198–7210, 1992.
2. Almeida, P.F.F., W.L.C. Vaz, and T. E. Thompson. Lateral diffusion in the liquid phases of dimyristoylphosphatidylcholine/cholesterol lipid bilayers: a free volume analysis. *Biochemistry* 31: 6739–6747, 1992.
3. Andersen, O. Permeability properties of unmodified lipid bilayer membranes. In: *Membrane Transport in Biology*, edited by G. Giebisch, D. C. Tosteson, and H. H. Ussing, Berlin: Springer-Verlag, 1978, p. 370–372.
4. Andersen, O. Ion movement through gramicidin A channels. Single-channel measurements at very high potentials. *Biophys. J.* 41: 119–133, 1983.
5. McMurray, W. C. Phospholipids of subcellular organelles and membranes. In: *Form and Function of Phospholipids*, edited by G. B. Ansell, J. N. Hawthorne, and R. M. C. Dawson, Amsterdam: Elsevier, 1973, p. 205–223.
6. Auger, M., D. Carrier, I. C. P. Smith, and H. C. Jarrell. Elucidation of motional modes in glycoglycerolipid bilayers. A ^{2}H NMR relaxation and line-shape study. *J. Am. Chem. Soc.* 112: 1373–1381, 1990.
7. Auger, M., I.C.P. Smith, and H. C. Jarrell. Slow motions in lipid bilayers. Direct detection by two-dimensional solid-state deuterium nuclear magnetic resonance. *Biophys. J.* 59: 31–38, 1991.
8. Auger, M., M.-R. Van Calsteren, I.C.P. Smith, and H. C. Jarrell. Glycerolipids: common features of molecular motion in bilayers. *Biochemistry* 29: 5815–5821, 1990.
9. Bangham, A. D. and D. A. Haydon. Ultrastructure of membranes: biomolecular organization. *Br. Med. Bull.* 24: 124–126, 1968.
10. Bangham, A. D., M. W. Hill, and N.G.A. Miller. Preparation and use of liposomes as models of biological membranes. In:

Methods in Membrane Biology, Volume 1, edited by E. D. Korn, New York: Plenum Press, 1974, p. 1–68.

11. Bangham, A. D., M. M. Standish, and J. C. Watkins. Diffusion of univalent ions across the lamellae of swollen phospholipids. *J. Mol. Biol.* 13: 238–252, 1965.

12. Bangham, A. D., M. M. Standish, and G. Weissman. The action of steroids and streptolysin S on the permeability of phospholipid structures to cations. *J. Mol. Biol.* 13: 253–259, 1965.

13. Bankaitis, V. A., J. R. Aitken, A. E. Cleves, and W. Dowhan. An essential role for a phospholipid transfer protein in yeast golgi function. *Nature* 347: 561–562, 1990.

14. Barchfeld, G. L. and D. W. Deamer. The effect of general anesthetics on the proton and potassium permeability of liposomes. *Biochim. Biophys. Acta* 819: 161–169, 1985.

15. Barton, P. G. and F. D. Gunstone. Hydrocarbon chain packing and molecular motion in phospholipid bilayers formed from unsaturated lecithins. *J. Biol. Chem.* 250: 4470–4476, 1975.

16. Bennet, V. Spectrin, a structural mediator between diverse plasma membrane proteins and the cytoplasm. *Curr. Opin. Cell Biol.* 2: 51–56, 1990.

17. Berkowitz, M. L. and K. Raghavan. Computer simulation of a water/membrane interface. *Langmuir* 7: 1042–1044, 1991.

18. Blok, M. C., E.C.M. Van Der Neut-Kok, L.L.M. Van Deenen, and J. De Gier. The effect of chain length and lipid phase transitions on the selective permeability properties of liposomes. *Biochim. Biophys. Acta* 406: 187–196, 1975.

19. Bloom, M. and E. Sternin. Transverse nuclear spin relaxation in phospholipid bilayer membranes. *Biochemistry* 26: 2101–2105, 1987.

20. Bonmatin, J.-M., I.C.P. Smith, H. C. Jarrell, and D. J. Siminovitch. Use of a comprehensive approach to molecular dynamics in ordered lipid systems: cholesterol reorientation in oriented lipid bilayers. A ^{2}H NMR relaxation case study. *J. Am. Chem. Soc.* 112: 1697–1704, 1990.

21. Boys, C. V. *Soap Bubbles. Their Colors and The Forces That Mold Them,* New York: Dover Publications, 1959, p. 60–61.

22. Bresselers, G.J.M., H. L. Goderis, and P. P. Tobback. Measurement of the glucose permeation rate across phospholipid bilayers using small unilamellar vesicles. Effect of membrane composition and temperature. *Biochim. Biophys. Acta* 772: 374–382, 1984.

23. Brian, A. A. and H. M. McConnell. Allogeneic stimulation of cytotoxic T cells by supported planar membranes. *Proc. Natl. Acad. Sci. U.S.A.* 81: 6159–6163, 1984.

24. Brown, M. F. Theory of spin-lattice relaxation in lipid bilayers and biological membranes. ^{2}H and ^{14}N quadrupolar relaxation. *J. Chem. Phys.* 77: 1576–1599, 1982.

25. Brown, M. F. Theory of spin-lattice relaxation in lipid bilayers and biological membranes. Dipolar relaxation. *J. Chem. Phys.* 80: 2808–2836, 1984.

26. Brown, M. F., A. A. Ribeiro, and G. D. Williams. New view of lipid bilayer dynamics from ^{2}H and ^{13}C NMR relaxation time measurements. *Proc. Natl. Acad. Sci. U.S.A.* 80: 4325–4329, 1983.

27. Brown, M. F., J. Seelig, and U. Haeberlen. Structural dynamics in phospholipid bilayers from deuterium spin-lattice relaxation time measurements. *J. Chem. Phys.* 70: 5045–5053, 1979.

28. Brown, R. E., F. A. Stephenson, T. Markello, Y. Barenholz, and T. E. Thompson. Properties of a specific glycolipid transfer protein from bovine brain. *Chem. Phys. Lipids* 38: 79–93, 1985.

29. Brumbaugh, E. E. and C. Huang. Parameter estimation in binary mixtures of phospholipids. *Methods Enzymol.* 210: 521–539, 1992.

30. Brunner, J., D. E. Graham, H. Hauser, and G. Semenza. Ion and sugar permeabilities of lecithin bilayers: comparison of curved and planar bilayers. *J. Membr. Biol.* 57: 133–141, 1980.

31. Bultmann, T., H-N. Lin, Z-Q. Wang, and C. Huang. Thermotropic and mixing behavior of mixed-chain phosphatidylcholines with molecular weights identical with L-α-dipalmitoylphosphatidylcholine. *Biochemistry* 30: 7194–7202, 1991.

32. Büldt, G. H., J. Gally, J. Seelig, and G. Zaccai. Neutron diffraction studies on phosphatidylcholine model membranes. I. Head group conformation. *J. Mol. Biol.* 134: 673–691, 1979.

33. Carnie, S., J. N. Israelachvili, and B. A. Pailthorpe. Lipid packing and transbilayer asymmetries of mixed lipid vesicles. *Biochim. Biophys. Acta* 554: 340–357, 1979.

34. Carrier, D., J. B. Giziewicz, D. Moir, I.C.P. Smith, and H. C. Jarrell. Dynamics and orientation of glycolipid headgroups by ^{2}H-NMR: gentiobiose. *Biochim. Biophys. Acta* 983: 100–108, 1989.

35. Castello, M. J. and T. Gulik-Krzywicki. Correlated x-ray diffraction and freeze-fracture studies on model membrane systems. Perturbations induced by freeze-fracture preparation procedures. *Biochim. Biophys. Acta* 455: 412–432, 1976.

36. Cevc, G. Polymorphism of the bilayer membranes in the ordered phase and the molecular origin of the lipid pretransition and rippled lamellae. *Biochim. Biophys. Acta* 1062: 59–69, 1991.

37. Cevc, G. and D. Marsh. *Phospholipid Bilayers: Physical Principles and Models,* New York: John Wiley and Sons, 1987, p. 3–12.

38. Cevc, G. and D. Marsh. *Phospholipid Bilayers: Physical Principles and Models,* New York: John Wiley and Sons, 1987, p. 29–46.

39. Cevc, G. and D. Marsh. *Phospholipid Bilayers: Physical Principles and Models,* New York: John Wiley and Sons, 1987, p. 19–24.

40. Cevc, G. and D. Marsh. *Phospholipid Bilayers: Physical Principles and Models,* New York: John Wiley and Sons, 1987, p. 157–195.

41. Chang, H. and R. M. Epand. The existence of a highly ordered phase in fully hydrated dilauroylphosphatidylethanolamine. *Biochim. Biophys. Acta* 728: 319–324, 1983.

42. Cherry, R. J. and D. Chapman. Refractive index determination of lecithin black films. *J. Mol. Biol.* 30: 551–553, 1967.

43. Cherry, R. J. and D. Chapman. Optical properties of black lecithin films. *J. Mol. Biol.* 40: 19–32, 1969.

44. Chowdhry, B. Z., G. Lipka, A. W. Dalziel, and J. M. Sturtevant. Multicomponent phase transitions of diacyl phosphatidylethanolamine dispersions. *Biophys. J.* 45: 901–904, 1984.

45. Clegg, R. M. and W.L.C. Vaz. Translational diffusion of proteins and lipids in artificial lipid bilayer membranes. A comparison of experiment with theory. In: *Progress in Protein-Lipid Interactions,* edited by A. Watts, Amsterdam: Elsevier, 1985, p. 173–229.

45a. Clerq, S. G., and T. E. Thompson. Permeability of dimyristoyl phosphatidylcholine/dipalmitoyl phosphatidylcholine membranes with coexisting gel and liquid-crystalline phases. *Biophys. J.* 68: 2333–2341, 1995.

45b. Clerq, S. G., and T. E. Thompson. A possible mechanism for vesicle formation by extensions. *Biophys. J.* 67: 475–477, 1995.

46. Cullis, P. R. and B. De Kruijff. Lipid polymorphism and the functional roles of lipids in biological membranes. *Biochim. Biophys. Acta* 559: 399–420, 1979.

47. Curatolo, W., B. Sears, and L. J. Neuringer. A calorimetry and deuterium NMR study of mixed model membranes of 1-

palmitoyl-2-oleylphosphatidylcholine and saturated phosphatidylcholines. *Biochim. Biophys. Acta* 817: 261–270, 1985.

48. Danielli, J. F. and H. Davson. A contribution to the theory of the permeability of thin films. *J. Cell. Comp. Physiol.* 5: 495–508, 1935.

49. Davis, J. H. Deuterium magnetic resonance study of the gel and liquid crystalline phases of dipalmitoyl phosphatidylcholine. *Biophys. J.* 27: 339–358, 1979.

50. Davis, J. H. Deuterium nuclear magnetic resonance and relaxation in partially ordered systems. *Adv. Magn. Res.* 13: 195–222, 1989.

51. De Loof, H., S. C. Harvey, J. P. Segrest, and R. W. Pastor. Mean field stochastic boundary molecular dynamics simulation of a phospholipid in a membrane. *Biochemistry* 30: 2099–2113, 1991.

52. Devaux, P., C. J. Scandella, and H. M. McConnell. Spin-spin interactions between spin-labelled phospholipids. *J. Magn. Res.* 9: 474–485, 1973.

53. Dowhan, W. Phospholipid-transfer proteins. *Curr. Opin. Cell Biol.* 3: 621–625, 1991.

54. Edidin, M. Molecular associations in membrane domains. *Curr. Top. Membr. Transplant.* 36: 81–96, 1990.

55. Edidin, M. The variety of cell surface membrane domains. *Comments Mol. Cell. Biophys.* 8: 73–82, 1992.

56. El-Mashak, E. M. and T. Y. Tsong. Ion selectivity of temperature-induced and electric field induced pores in dipalmitoylphosphatidylcholine vesicles. *Biochemistry* 24: 2884–2888, 1985.

57. Ellena, J. F., L. S. Lepore, and D. S. Cafiso. Estimating lipid lateral diffusion in phospholipid vesicles from ^{13}C spin-spin relaxation. *J. Phys. Chem.* 97: 2952–2957, 1993.

58. Evans, E. and D. Needham. Physical properties of surfactant bilayer membranes. *J. Phys. Chem.* 91: 4219–4288, 1987.

59. Evans, E. A. Bending resistance and chemically induced moments in membrane bilayers. *Biophys. J.* 14: 923–931, 1974.

60. Evans, E. A. and R. M. Hochmuth. Mechanochemical properties of membranes. In: *Current Topics in Membranes and Transport,* edited by F. Bonner and A. Kleinzeller, New York: Academic Press, 1978, p. 1–64.

61. Evans, E. A. and R. Skalak. Mechanics and thermodynamics of biomembranes. *Crit. Rev. Bio Eng.* 3: 294–299, 1979.

62. Fajer, P., A. Watts, and D. Marsh. Saturation transfer, continuous wave saturation, and saturation recovery electron spin resonance studies of chain-spin labeled phosphatidylcholines in the low temperature phases of dipalmitoyl phosphatidylcholine bilayers. Effects of rotational dynamics and spin-spin interactions. *Biophys. J.* 61: 879–891, 1992.

63. Fettiplace, R., L.M.G. Gordon, S. B. Hladky, J. Requena, H. P. Zingsheim, and D. A. Haydon. Techniques in the formation and examination of black lipid bilayer membranes. In: *Methods in Membrane Biology, Volume 4,* edited by E. D. Korn, New York: Plenum Press, 1975, p. 1–75.

64. Finkelstein, A. Water and nonelectrolyte permeability of lipid bilayer membranes. *J. Gen. Physiol.* 68: 127–135, 1976.

65. Florio, E., H. C. Jarrell, D. B. Fenske, K. R. Barber, and C.W.M. Grant. Glycosphingolipid interdigitation in phospholipid bilayers examined by deuterium NMR and EPR. *Biochim. Biophys. Acta* 1025: 157–163, 1990.

66. Fricke, H. The electric capacity of suspensions with special reference to blood. *J. Gen. Physiol.* 9: 137–152, 1925.

67. Fromherz, P. and D. Ruppel. Lipid vesicle formation: the transition from open discs to closed shells. *FEBS Lett.* 179: 155–159, 1985.

68. Füldner, H. H. Characterization of a third phase transition in multilamellar dipalmitoyllecithin liposomes. *Biochemistry* 20: 5707–5710, 1981.

69. Grell, E. *Membrane Spectroscopy,* Berlin: Springer-Verlag, 1981, 498 p.

70. Gruner, S. M. Intrinsic curvature hypothesis for biomembrane lipid composition. *Proc. Natl. Acad. Sci. U.S.A.* 82: 3665–3669, 1985.

71. Guidotti, G. Membrane proteins: structure, arrangement and disposition in membranes. In: *Physiology of Membrane Disorders,* edited by T. E. Andreoli, J. F. Hoffman, D. D. Fanestil, and S. G. Schultz, New York: Plenum Press, 1986, p. 45–55.

72. Hauser, H., O. Oldani, and M. C. Philips. Mechanism of ion escape from phosphatidylcholine and phosphatidylserine single bilayer vesicle. *Biochemistry* 12: 4507–4517, 1973.

73. Henn, F. A., G. L. Decker, J. W. Greenawalt, and T. E. Thompson. The properties of lipid bilayer membranes separating two aqueous phases: Electron microscope studies. *J. Mol. Biol.* 24: 51–58, 1967.

74. Henn, F. A. and T. E. Thompson. Properties of lipid bilayer membranes separating two aqueous phases. *J. Mol. Biol.* 31: 227–235, 1968.

75. Henn, F. A. and T. E. Thompson. Synthetic lipid bilayer membranes. *Annu. Rev. Biochem.* 38: 241–262, 1969.

76. Hope, M. J., M. B. Baley, G. Webb, and P. R. Cullis. Production of large unilamellar vesicles by a rapid extrusion procedure. Characterization of size distribution, trapped volume and ability to maintain a membrane potential. *Biochim. Biophys. Acta* 812: 55–65, 1985.

77. Huang, C. Studies on phosphatidylcholine vesicles: formation and physical characteristics. *Biochemistry* 8: 344–352, 1969.

78. Huang, C. Roles of carbonyl oxygens at the bilayer interface in phospholipid-sterol interaction. *Nature* 259: 242–244, 1976.

79. Huang, C. A structural model for the cholesterol-phosphatidylcholine complexes in bilayer membranes. *Lipids* 12: 348–356, 1977.

80. Huang, C. Mixed-chain phospholipids and interdigitated bilayer systems. *Klin. Wochenschrift* 68: 149–165, 1990.

81. Huang, C., S. Li, Z-Q. Wang, and H-N. Lin. Dependence of the bilayer transition temperatures on the structural parameters of phosphatidylcholines. *Lipids* 28: 365–370, 1993.

82. Huang, C. and J. T. Mason. Geometric packing constraints in egg phosphatidylcholine vesicles. *Proc. Natl. Acad. Sci. U.S.A.* 75: 308–310, 1978.

83. Huang, C., J. T. Mason, F. A. Stephenson, and I. W. Levin. Raman and ^{31}P NMR spectroscopic identification of a highly ordered lamellar phase in aqueous dispersions of 1-stearoyl-2-acetyl-*sn*-glycero-3-phosphorylcholine. *J. Phys. Chem.* 88: 6454–6458, 1984.

84. Huang, C., J. T. Mason, F. A. Stephenson, and I. W. Levin. Polymorphic phase behavior of platelet-activating factor. *Biophys. J.* 49: 587–595, 1986.

85. Huang, C. and T. E. Thompson. Properties of lipid bilayer membranes separating two aqueous phases: determination of membrane thickness. *J. Mol. Biol.* 13: 183–193, 1965.

86. Huang, C. and T. E. Thompson. Thickness of bilayer membranes. *J. Mol. Biol.* 16: 576, 1966.

87. Huang, C., Z-Q. Wang, H-N. Lin, and E. E. Brumbaugh. Calorimetric studies of fully hydrated phosphatidylcholines with highly asymmetric acyl chains. *Biochim. Biophys. Acta* 1145: 298–310, 1993.

88. Hubbell, W. L. Transbilayer coupling mechanism for the formation of lipid asymmetry in biological membranes. Application to the photoreceptor disc membranes. *Biophys. J.* 57: 99–108, 1990.

89. Hui, S. W. and C. Huang. X-ray diffraction evidence for fully interdigitated bilayers of 1-stearoyllysophosphatidylcholine. *Biochemistry* 25: 1330–1335, 1986.

90. Hui, S. W., J. T. Mason, and C. Huang. Acyl chain interdigitation in saturated mixed-chain phosphatidylcholine bilayer dispersions. *Biochemistry* 23: 5570–5577, 1984.

91. Israelachvili, J. N., D. J. Mitchell, and B. W. Ninham. Theory of self-assembly of lipid bilayers and vesicles. *Biochim. Biophys. Acta* 470: 185–201, 1977.

92. Jain, M. K. *The Bimolecular Lipid Membrane—A System,* New York: Van Nostrad Reinhold, 1972, p. 52–84.

93. Jain, M. K. *The Bimolecular Lipid Membrane—A System.* New York: Van Nostrad Reinhold, 1972, p. 112–132.

94. Janiak, M. J., D. M. Small, and G. G. Shipley. Nature of the pretransition of synthetic phospholipids: dimyristoyl and dipalmitoyl phosphatidylcholine. *Biochemistry* 15: 4575–4580, 1976.

95. Jarrell, H. C., J. B. Giziewicz, and I.C.P. Smith. Structure and dynamics of a glyceroglycolipid: A ^{2}H NMR study of head group orientation, ordering, and effect on lipid aggregate structure. *Biochemistry* 25: 3950–3957, 1986.

96. Jones, J. D. and T. E. Thompson. Spontaneous phosphatidylcholine transfer by collision between vesicles at high lipid concentration. *Biochemistry* 28: 129–134, 1989.

97. Jovin, T. M. and W.L.C. Vaz. Rotational and translational diffusion in membranes measured by fluorescence and phosphorescence methods. *Methods Enzymol.* 172: 471–512, 1989.

98. Kalb, E., S. Frey, and L. K. Tamm. Formation of supported planar bilayers by fusion of vesicles to supported phospholipid monolayers. *Biochim. Biophys. Acta* 1103: 307–316, 1992.

99. Kanfer, J. N. and S. Hakomori. *Sphingolipid Biochemistry.* New York: Plenum Press, 1983, p. 89–135.

100. Karplus, M. and G. A. Petsko. Molecular dynamics simulations in biology. *Nature* 347: 631–639, 1990.

101. Katz, S. L., H. M. Laboda, L. R. McLean, and M. C. Phillips. Influence of molecular packing and phospholipid types on rates of cholesterol exchange. *Biochemistry* 27: 3416–3423, 1988.

102. Kornberg, R. D. and H. M. McConnell. Inside-outside transitions of phospholipids in vesicle membranes. *Biochemistry* 10: 1111–1120, 1971.

103. Koynova, R. and H-J. Hinz. Metastable behavior of saturated phosphatidylethanolamines: a densitometric study. *Chem. Phys. Lipids* 54: 67–72, 1990.

104. Kwok, R. and E. Evans. Thermoelasticity of large lecithin bilayer vesicles. *Biophys. J.* 35: 637–652, 1981.

105. Lasic, D. D. A molecular model for vesicle formation. *Biochim. Biophys. Acta* 692: 501–502, 1982.

106. Lasic, D. D. A general model for vesicle formation. *J. Theor. Biol.* 124: 35–41, 1987.

107. Lasic, D. D. The mechanism of vesicle formation. *Biochem. J.* 256: 1–11, 1988.

108. Lee, A. G. Lipid phase transitions and phase diagrams. II. Mixtures involving lipids. *Biochim. Biophys. Acta* 472: 285–344, 1977.

109. Lewis, B. A. and D. M. Engelman. Lipid bilayer thickness varies linearly with acyl chain length in fluid phosphatidylcholine vesicles. *J. Mol. Biol.* 166: 211–217, 1983.

110. Lewis, R.N.A.H. and R. N. McElhaney. Structures of the subgel phases of n-saturated diacyl phosphatidylcholine bilayers: FTIR spectroscopic studies of ^{13}C=O and ^{2}H labelled lipids. *Biophys. J.* 67: 63–77, 1992.

111. Lewis, R.N.A.H. and R. N. McElhaney. Calorimetric and spectroscopic studies of the polymorphic phase behavior of a homologous series of n-saturated 1, 2-diacylphosphatidylethanolamines. *Biophys. J.* 64: 1081–1096, 1993.

112. Li, S., Z-Q. Wang, H-N. Lin, and C. Huang. Energy-minimized structures and packing states of a homologous series of mixed-chain phosphatidylcholines: a molecular mechanics study on the diglyceride moieties. *Biophys. J.* 65: 1415–1428, 1993.

113. Lichtenberg, D., E. Friere, C. F. Schmidt, Y. Barenholz, P. L. Felgner, and T. E. Thompson. The effect of surface curvature on the stability, thermodynamic behavior and osmotic activity of single lamellar dipalmitoylphosphatidylcholine vesicles. *Biochemistry* 20: 3462–3467, 1981.

114. Lichtenberg, L. and Y. Barenholz. Liposomes: preparation, characterization and preservation. In: *Methods of Biochemical Analysis,* edited by D. Glick, 1988, p. 337–462.

115. Lin, H-N. and C. Huang. Eutectic phase behavior of 1-stearoyl-2-caprylphosphatidylcholine and dimyristoylphosphatidylcholine mixtures. *Biochim. Biophys. Acta* 946: 178–184, 1988.

116. Lindblom, G. and L. Rilfors. Cubic phases and isotropic structures formed by membrane lipids—possible biological relevance. *Biochim. Biophys. Acta* 988: 221–256, 1989.

117. Lipowsky, R. The conformation of membranes. *Nature* 349: 475–481, 1991.

118. Luzzati, V. X-ray diffraction studies of lipid-water systems. In: *Biological Membranes,* edited by D. Chapman, New York: Academic Press, 1968, p. 71–123.

119. Luzzati, V. and F. Husson. The structure of the liquid crystalline phases of lipid-water systems. *J. Cell Biol.* 12: 207–219, 1962.

120. Mabrey, S. and J. M. Sturtevant. Investigation of phase transitions of lipids and lipid mixtures by high sensitivity differential scanning calorimetry. *Proc. Natl. Acad. Sci. U.S.A.* 73: 3862–3866, 1976.

121. Mantsch, H. H., S. C. Hsi, K. W. Butler, and D. G. Cameron. Studies on the thermotropic behavior of aqueous phosphatidylethanolamines. *Biochim. Biophys. Acta* 728: 325–330, 1983.

122. Mariani, P. The cubic phases. *Curr. Opin. Struct. Biol.* 1: 501–505, 1991.

123. Mariani, P., V. Luzzati, and H. Delacroix. Cubic phases of lipid-containing systems. Structure analysis and biological implications. *J. Mol. Biol.* 204: 165–189, 1988.

124. Marsh, D. Molecular motion in phospholipid bilayers in the gel phase: long axis rotation. *Biochemistry* 19: 1632–1637, 1980.

125. Marsh, D. *CRC Handbook of Lipid Bilayers.* Boca Raton: CRC Press, 1990, 387 p.

126. Marsh, D. *Handbook of Lipid Bilayers.* Boca Raton: CRC Press, 1990, p. 87–120.

127. Mason, J. T., C. Huang, and R. L. Biltonen. Calorimetric investigations of saturated mixed-chain phosphatidylcholine bilayer dispersions. *Biochemistry* 20: 6086–6092, 1981.

128. Mason, J. T. and F. A. Stephenson. Thermotropic properties of saturated mixed acyl phosphatidylethanolamines. *Biochemistry* 29: 590–598, 1990.

129. Mattai, J., P. K. Sripada, and G. G. Shipley. Mixed-chain phosphatidylcholine bilayers: structure and properties. *Biochemistry* 26: 3287–3297, 1987.

130. Mayer, L. D., M. J. Hope, and P. Cullis. Vesicles of variable size produced by a rapid extrusion procedure. *Biochim. Biophys. Acta* 858: 161–168, 1986.

131. McIntosh, T. J. and S. A. Simon. Area per molecule and distribution of water in fully hydrated dilauroylphosphatidylethanolamine bilayers. *Biochemistry* 25: 4948–4952, 1986.

132. McIntosh, T. J., S. A. Simon, J. Ellington, and N. A. Porter. New structural model for mixed-chain phosphatidylcholine bilayers. *Biochemistry* 23: 4038–4044, 1984.

133. McLean, L. R. and M. C. Phillips. Mechanism of cholesterol and phosphatidylcholine exchange or transfer between unilamellar vesicles. *Biochemistry* 20: 2893–2900, 1981.

134. Melo, E.C.C., I.M.G. Lourtie, M. B. Sankaram, T. E. Thompson, and W. L. C. Vaz. Effects of domain connection and disconnection on the yields of in-plane bimolecular reactions in membranes. *Biophys. J.* 63: 1506–1512, 1992.

135. Meraldi, J.-P. and J. Schlitter. A statistical mechanical treatment of fatty acyl chain order in phospholipid bilayers and correlation with experimental data. A. theory. *Biochim. Biophys. Acta* 645: 183–192, 1981.

136. Mimms, L. T., G. Zampighi, Y. Nozaki, C. Tanford, and J. A. Reynolds. Phospholipid vesicle formation and transmembrane protein incorporation using octyl glucoside. *Biochemistry* 20: 833–840, 1981.

137. Montal, M. Formation of bimolecular membranes from lipid monolayers. *Methods Enzymol.* 32: 545–556, 1974.

138. Mueller, P., D. O. Rudin, H. T. Tien, and W. C. Wescott. Reconstitution of cell membrane structure in vitro and its transformation into an excitable system. *Nature* 194: 979–980, 1962.

139. Mulukutla, S. and G. G. Shipley. Structural and thermotropic properties of phosphatidylethanolamine and its N-methyl derivatives. *Biochemistry* 23: 2514–2519, 1984.

140. Nicklas, K., J. Bocker, M. Schlenkrich, J. Brickmann, and P. Bopp. Molecular dynamics studies of the interface between a model membrane and an aqueous solution. *Biophys. J.* 60: 261–272, 1991.

141. Nigg, E. and R. Cherry. Anchorage of a band 3 population at the erythrocyte cytoplasmic membrane surface, protein rotational diffusion measurements. *Proc. Natl. Acad. Sci. U.S.A.* 77: 4702–4706, 1980.

142. Nordlund, J. R., C. F. Schmidt, S. N. Dicken, and T. E. Thompson. Transbilayer distribution of phosphatidylethanolamine in large and small unilamellar vesicles. *Biochemistry* 20: 3237–3241, 1981.

143. Nozaki, Y. and C. Tanford. Proton and hydroxide ion permeability of phospholipid vesicles. *Proc. Natl. Acad. Sci. U.S.A.* 78: 4324–4328, 1981.

144. Pagano, R. and T. E. Thompson. Spherical bilayer membranes. *Biochim. Biophys. Acta* 144: 666–669, 1967.

145. Pagano, R. and T. E. Thompson. Spherical lipid bilayer membranes: Electrical and isotopic studies of ion permeability. *J. Mol. Biol.* 38: 41–57, 1968.

146. Pascher, I., M. Lundmark, P.-G. Nyholm, and S. Sundell. Crystal structures of membrane lipids. *Biochim. Biophys. Acta* 1113: 339–373, 1992.

147. Pastor, R. W., R. M. Venable, and M. Karplus. Model for the structure of the lipid bilayer. *Proc. Natl. Acad. Sci. U.S.A.* 88: 892–896, 1991.

148. Pearson, R. H. and I. Pascher. The molecular structure of lecithin hydrate. *Nature* 281: 499–501, 1979.

149. Peterson, D. C. Water permeation through the lipid bilayer membrane. Test of the liquid hydrocarbon model. *Biochim. Biophys. Acta* 600: 666–677, 1980.

150. Redwood, W. R., W. Takashima, H. P. Schwan, and T. E. Thompson. Dielectric studies on homogeneous phosphatidylcholine vesicles. *Biochim. Biophys. Acta* 255: 557–566, 1992.

151. Reeves, J. P. and R. M. Dowben. Formation and properties of thin-walled phospholipid vesicles. *J. Cell Physiol.* 73: 49–60, 1969.

152. Requena, J. and D. A. Haydon. Lenses and the compression of black lipid membranes by an electric field. *Biophys. J.* 15: 77–81, 1975.

153. Robertson, R. N. and T. E. Thompson. The function of phospholipid polar groups in membranes. *FEBS Lett.* 74: 16–19, 1977.

154. Rodgers, W. and M. Glaser. Characterization of lipid domains in erythrocyte membranes. *Proc. Natl. Acad. Sci. U.S.A.* 88: 1364–1368, 1991.

155. Roseman, M., B. J. Litman, and T. E. Thompson. Transbilayer exchange of phosphatidylethanolamine for phosphatidylcholine and N-acetamidoylphosphatidylethanolamine in single-walled bilayer vesicles. *Biochemistry* 14: 4826–4830, 1975.

156. Ruocco, M. J. and G. G. Shipley. Characterization of the subtransition of hydrated dipalmitoyl phosphatidylcholine bilayers: kinetic hydration and structural study. *Biochim. Biophys. Acta* 691: 309–320, 1982.

157. Ruocco, M. J. and G. G. Shipley. Characterization of the subtransition of hydrated dipalmitoyl phosphatidylcholine bilayers: X-ray diffraction study. *Biochim. Biophys. Acta* 684: 59–66, 1982.

158. Sackmann, E. Physical foundation of the molecular organization and dynamics of membranes. In: *Biophysics*, edited by W. Hoppe, W. Lohman, H. Marlet, and H. Ziegler, Berlin: Springer-Verlag, 1983, p. 425–433.

159. Sankaram, M. B. and T. E. Thompson. Deuterium magnetic resonance study of phase equilibria and membrane thickness in binary phospholipid mixed bilayers. *Biochemistry* 31: 8258–8268, 1992.

160. Seddon, J. M. Structure of the inverted hexagonal (H_{II}) phase, and non-lamellar phase transitions of lipids. *Biochim. Biophys. Acta* 1031: 1–69, 1990.

161. Seddon, J. M., G. Cevc, R. D. Kaye, and D. Marsh. X-ray diffraction study of the polymorphism of hydrated diacyl- and dialkylphosphatidylethanolamines. *Biochemistry* 23: 2634–2644, 1984.

162. Seddon, J. M., K. Harlos, and D. Marsh. Metastability and polymorphism in the gel and fluid bilayer phases of dilauroylphosphatidylethanolamine. Two crystalline forms in excess water. *J. Biol. Chem.* 258: 3850–3854, 1983.

163. Seelig, A. and J. Seelig. The dynamic structure of fatty acyl chains in a phospholipid bilayer measured by deuterium magnetic resonance. *Biochemistry* 13: 4839–4845, 1974.

164. Seifert, U., K. Berndl, and R. Lipowsky. Shape transformations in vesicles: phase diagram for spontaneous curvature and bilayer-coupling models. *Phys. Rev.* 44: 1182–1202, 1991.

165. Servas, R. M., W. Harbich, and W. Helfrich. Measurement of the curvature-elastic modules of egg lecithin bilayers. *Biochim. Biophys. Acta* 436: 900–903, 1976.

166. Shah, J., P. K. Sripada, and G. G. Shipley. Structure and properties of mixed-chain phosphatidylcholine bilayers. *Biochemistry* 29: 4254–4262, 1990.

167. Shepherd, J.C.W. and G. Büldt. Zwitterionic dipoles as a dielectric probe for investigating head group mobility in phospholipid membranes. *Biochim. Biophys. Acta* 514: 83–94, 1978.

168. Siegel, D. P. Inverted micellar intermediates and the transitions between lamellar, cubic and inverted hexagonal lipid phases. I. Mechanism of the L_α-H_{II} phase transitions. *Biophys. J.* 49: 1155–1170, 1986.

169. Simon, S. A. A comment on the water permeability through planar lipid bilayers. *J. Gen. Physiol.* 70: 123–125, 1977.

170. Singer, S. J. and G. L. Nicholson. The fluid mosaic model of the structure of cell membranes. *Science* 175: 720–731, 1972.

171. Sisk, R. B., Z-Q. Wang, H-N. Lin, and C. Huang. Mixing

behavior of identical molecular weight phosphatidylcholines with various chain-length differences in two-component lamellae. *Biophys. J.* 58: 777–784, 1991.

172. Slater, J. L. and C. Huang. Lipid bilayer interdigitation. In: *The Structure of Biological Membranes,* edited by P. L. Yeagle, Boca Raton: CRC Press, 1992, p. 175–210.

173. Slater, J. L., C. Huang, R. G. Adams, and I. W. Levin. Polymorphic phase behavior of lysophosphatidylethanolamine dispersions. A thermodynamic and spectroscopic characterization. *Biophys. J.* 56: 243–252, 1989.

174. Small, D. M. *The Physical Chemistry of Lipids. From Alkanes to Phospholipids.* New York: Plenum Press, 1986, 672 p.

175. Speyer, J. B., R. T Weber, S. K. Das Gupta, and R. G. Griffin. Anisotropic ^{2}H NMR spin-lattice relaxation in L_α-phase cerebroside bilayers. *Biochemistry* 28: 9569–9574, 1989.

176. Stauffer, D. *Introduction to Percolation Theory.* London: Taylor and Francis, 1985, 124 p.

177. Suurkuusk, J., B. R. Lentz, Y. Barenholz, R. L. Biltonen, and T. E. Thompson. A calorimetric and fluorescent probe study of the gel-liquid crystalline phase transition in small, single-walled dipalmitoylphosphatidylcholine vesicles. *Biochemistry* 15: 1393–1401, 1976.

178. Sweeley, C. C. and B. Siddiqui. Chemistry of mammalian glycolipids. In: *The Glycoconjugates, Volume I,* edited by M. I. Horowitz and W. Pigman, New York: American Press, 1977, p. 459–540.

179. Szoka, F. and D. Papahadjopoulos. Comparative properties and methods of preparation of lipid vesicle (liposomes). *Annu. Rev. Biophys. Biomol. Chem.* 9: 467–508, 1980.

180. Tadros, T. F. and B. Vincent. Emulsion stability. In: *Encyclopedia of Emulsion Technology, Volume 1,* edited by P. Becher, New York: Marcel Dekker, 1983, p. 129–285.

181. Tate, M. W., E. F. Eikenberry, D. C. Turner, E. Shyamsunder, and S. M. Gruner. Nonbilayer phases of membrane lipids. *Chem. Phys. Lipids* 57: 147–164, 1991.

182. Tenchov, B. G., A. I. Boyanov, and R. D. Koynova. Lyotropic polymorphism of racemic dipalmitoylphosphatidylethanolamine. A differential scanning calorimetry study. *Biochemistry* 23: 3553–3558, 1984.

183. Thompson, T. E. and F. A. Henn. Experimental phospholipid model membranes. In: *Structure and Function of Membranes of Mitochondria and Chloroplasts,* edited by E. Racker, New York: Van Nostrad Reinhold, 1969, p. 1–52.

184. Thompson, T. E. and C. Huang. Composition and dynamics of lipids in biomembranes. In: *Physiology of Membrane Disorders,* edited by T. Andreoli, R. D. Fannestil, J. F. Hoffman, and S. G. Schulz, New York: Plenum, 1985, p. 25–44.

185. Thompson, T. E., C. Huang, and B. J. Litman. Bilayers and biomembranes: compositional asymmetries induced by surface curvature. In: *The Cell Surface in Development,* edited by A. A. Moscona, New York: John Wiley and Sons, 1974, p. 1–16.

186. Thompson, T. E., B. Lentz, and Y. Barenholz. A calorimetric and fluorescent probe study of phase transitions in phosphatidylcholine liposomes. In: *Biochemistry of Membrane Transport,* edited by G. Semenza and E. Carafoli, Berlin: Springer-Verlag, 1977, p. 47–71.

187. Thompson, T. E., M. B. Sankaram, and R. L. Biltonen. Biological membrane domains: functional significance. *Comments Mol. Cell. Biophys.* 8: 1–15, 1992.

188. Thompson, T. E. and T. W. Tillack. Organization of glycosphingolipids in bilayers and plasma membranes of mammalian cells. *Annu. Rev. Biophys. Biophys. Chem.* 14: 361–386, 1985.

189. Tien, H. T. *Bilayer Lipid Membranes (BLM),* New York: Marcel Dekker, 1974, p. 11–28.

190. Tien, H. T. *Black Lipid Membranes (BLM),* New York: Marcel Dekker, 1974, p. 117–163.

191. Tilcock, C.P.S. Lipid polymorphism. *Chem. Phys. Lipids* 40: 109–125, 1986.

192. Toyoshima, Y. and T. E. Thompson. Chloride flux in bilayer membranes: The electrically silent chloride flux in semispherical bilayers. *Biochemistry* 14: 1518–1524, 1975.

193. Toyoshima, Y. and T. E. Thompson. Chloride flux in bilayer membranes: chloride permeability in aqueous dispersions of single-walled, bilayer vesicles. *Biochemistry* 14: 1525–1531, 1975.

194. Trahms, L., W. D. Klabe, and L. Boroske. ^{1}H NMR study of the three low temperature phases of DPPC-water systems. *Biophys. J.* 42: 285–293, 1983.

195. Ueno, M., C. Tanford, and J. Reynolds. Phospholipid vesicle formation using nonionic detergents with low monomer solubility. Kinetic factors determine vesicle size and permeability. *Biochemistry* 23: 3070–3076, 1984.

196. Ulrich, A. S., F. Volke, and A. Watts. The dependence of phospholipid and head-group mobility on hydration as studied by deuterium-NMR spin-lattice relaxation time measurements. *Chem. Phys. Lipids* 55: 61–66, 1990.

197. Van Der Leeuw, Y.C.W. and G. Stulen. Proton relaxation measurements on lipid membranes oriented at the magic angle. *J. Magn. Reson. Imaging,* 42: 434–445, 1981.

198. Vance, D. F. and J. E. Vance. *Biochemistry of Lipids and Membranes,* San Francisco: Benjamin/Cummings, 1985, p. 30.

199. Vaz, W.L.C. Translational diffusion in phase-separated lipid bilayer membranes. *Comments Mol. Cell. Biophys.* 8: 17–36, 1992.

200. Vaz, W.L.C., E.C.C. Melo, and T. E. Thompson. Translational diffusion and fluid domain connectivity in a two-component, two-phase phospholipid bilayer. *Biophys. J.* 56: 869–876, 1989.

201. Verkman, A. S. Water channels in cell membranes. *Annu. Rev. Physiol.* 54: 97–108, 1992.

202. Ververgaert, P. H. and P. E. Elbers. Ultrastructural analysis of black lipid membranes. *J. Mol. Biol.* 58: 431–437, 1971.

203. Vinson, P. K., Y. Talmon, and A. Walter. Vesicle-micelle transition of phosphatidylcholine and octyl glucoside elucidated by cryo-transmission electron microscopy. *Biophys. J.* 56: 669–681, 1989.

204. Walter, A. and J. W. Gutknecht. Permeability of small nonelectrolytes through lipid bilayer membranes. *J. Membr. Biol.* 90: 207–217, 1986.

205. Weisz, K., G. Grobner, C. Mayer, J. Stohrer, and G. Kothe. Deuteron nuclear magnetic resonance study of the dynamic organization of phospholipid/cholesterol bilayer membranes: molecular properties and viscoelastic properties. *Biochemistry* 31: 1100–1112, 1992.

206. Wiener, M. C. and S. H. White. Fluid bilayer structure determination by the combined use of x-ray and neutron diffraction. I. Fluid bilayer models and the limits of resolution. *Biophys. J.* 59: 162–173, 1991.

207. Wiener, M. C. and S. H. White. Fluid bilayer structure determination by the combined use of x-ray and neutron diffraction. II. The composition space refinement method. *Biophys. J.* 59: 174–185, 1991.

208. Wiener, M. C. and S. H. White. Fluid bilayer structure determination by combined use of x-ray and neutron diffraction. III. The complete structure. *Biophys. J.* 61: 434–447, 1992.

209. Wilkinson, D. A. and J. F. Nagle. Metastability in the phase behavior of dimyristoylphophatidylethanolamine bilayers. *Biochemistry* 23: 1538–1541, 1984.

210. Williamson, P., A. Kulick, A. Zachowski, R. A. Schlegel, and P. F. Devaux. Ca^{2+} induces transbilayer redistribution of all major phospholipids in human erythrocytes. *Biochemistry* 31: 6355–6360, 1992.

211. Wong, P.T.T. and C. Huang. Structural aspects of pressure effects on infrared spectra of mixed-chain phosphatidylcholine assemblies in D_2O. *Biochemistry* 28: 1259–1263, 1989.

212. Wong, P.T.T., D. J. Siminovitch, and H. H. Mantsch. Structure and properties of model membranes: new knowledge from high-pressure vibrational spectroscopy. *Biochim. Biophys. Acta* 947: 139–171, 1988.

213. Woolley, G. A. and B. A. Wallace. Model ion channels: gramicidin and alamethicin. *J. Membr. Biol.* 129: 109–136, 1992.

214. Wyatt, K. and R. Cherry. Both ankyrin and band 4.1 are required to restrict the rotational mobility of band 3 in the human erythrocyte membrane. *Biochim. Biophys. Acta* 1103: 327–330, 1992.

215. Xu, H. and C. Huang. Scanning calorimetric study of fully hydrated asymmetric phosphatidylcholines with one acyl chain twice as long as the other. *Biochemistry* 26: 1036–1043, 1987.

216. Xu, H., F. A. Stephenson, H-N. Lin, and C. Huang. Phase metastability and supercooled metastable state of diundecanoyl-phosphatidylethanolamine bilayers. *Biochim. Biophys. Acta* 943: 63–75, 1988.

217. Ye, R. and A. S. Verkman. Simultaneous optical measurement of osmotic and diffusional water permeability in cells and liposomes. *Biochemistry* 28: 824–829, 1989.

218. Yeh, H. C. Interpretation of phase diagrams. In: *Phase Diagrams: Materials Science and Technology*, edited by A. M. Alper, New York: Academic Press, 1970, p. 167–197.

219. Zaccai, G., G. Büldt, A. Seelig, and J. Seelig. Neutron diffraction studies on phosphatidylcholine model membranes. II. Chain conformation and segmental disorder. *J. Mol. Biol.* 134: 693–706, 1979.

220. Zachowski, A. and P. F. Devaux. Bilayer asymmetry and lipid transport across biomembranes. *Comments Mol. Cell. Biophys.* 6: 63–90, 1989.

221. Zachowski, A., E. Favre, S. Cribier, P. Herve, and P. F. Devaux. Outside-inside translocation of aminophospholipids in the human erythrocyte membrane is maintained by a specific enzyme. *Biochemistry* 25: 2585–2590, 1986.

222. Zhu, T. and M. Caffrey. Thermodynamic, thermomechanical and structural properties of a hydrated asymmetric phosphatidylcholine. *Biophys. J.* 65: 939–954, 1993.

3. Membrane structure/proteins

CHARLES TANFORD

JACQUELINE A. REYNOLDS

Easingwold, York, North Yorkshire, England

PROTEINS ARE THE WORKING PARTS OF THE CELL: the wheels and gears, the fuel injectors, the catalytic convertors. This chapter is a summary of general principles and not a review of the current status of the field of protein chemistry. We attempt to help the reader understand and appreciate the unity in this field and to develop a philosophy of thinking about protein structure as it relates to function. We emphasize the latter because it is surely the most vital aspect of protein chemistry for physiologists—we include some speculative examples, because there is as yet a dearth of absolutely certain structures.

As long ago as 1899 Ernest Overton recognized the functional need for a component other than lipid in cell membranes—a component that could be responsible for phenomena like active transport of inorganic ions, for example. Overton, decades ahead of his time, had demonstrated that passive permeability of cell membranes to a long list of organic substances varies roughly as the solubilities of the same substances in olive oil, and he concluded that the bulk of the membrane must be chemically similar to olive oil, consisting of fats or lipids. But he was astute enough to appreciate that much physiological traffic across membranes is specifically selective, active not passive (in the sense of being thermodynamically uphill), and therefore requires the presence of uniquely designed engines to catalyze the use of metabolic energy for the transport. Neither he nor anyone else could at the time identify proteins as the agents for this function, but that of course is what we now know them to be. (There have been people behind as well as ahead of the times. As recently as 1970, J. D. Robertson denied the presence of proteins within membranes because he could not see them in his electron micrographs; other people tried futilely to address the question of how lipids alone might mediate selective transport.)

Today, of course, proteins are accepted for what they are and what they do, and now we want to understand how they work. Much as a curious child examines a mechanical toy or takes apart a ticking clock, we want to examine proteins in detail, describe all their parts and how they are related to each other. How does an alteration or movement in one part affect the whole and govern the function?

In this context of curious seekers, we know a great deal about proteins already, both generally and specifically. The enormous variety of structures and functions generated by the known exemplars of these macromolecules is a continuing source of fascination. Even the excitement of the new era of molecular genetics has not diminished the enthusiasm for discovering just how a particular protein is put together and how it "does its thing." A gene, a piece of DNA, identified as related to or responsible for some function, assures us of the existence of a protein, likewise related or responsible—the DNA in fact dictates its primary structure, the sequence of amino acids within it. But three-dimensional structure and the performance of function are a quite separate matter. This is the subject we address in this chapter.

ENVIRONMENT

Proteins do not exist or work in a vacuum but in a heterogeneous community of other molecules. Since each individual protein has evolved to fit a particular environment, we must take this into account when considering protein structure and function. At the simplest level (in the context of physiological function, where insoluble structural proteins are of lesser interest) this requires recognition of a minimum of two distinct classes of protein.

Water-soluble proteins. These are the proteins of the cytoplasm or extracellular fluids. They carry out their function in a medium that is essentially 55 *M* water, seasoned with inorganic salts and small organic molecules.

Membrane proteins. These are the proteins designed to reside in membranes and are subject to a mixed solvent system ranging from the cytoplasmic or extracellular aqueous salt solution at some distance from the membrane's external surfaces to essentially pure hydrocarbon in the center of the lipid bilayer. Figure 3.1 is a schematic diagram of this system drawn to scale.

Lipid head groups and lipid hydrocarbon chains are in a constant state of motion, filling all available space. The two-dimensional drawing cannot readily display this aspect but rather is designed to give an instant snapshot of some of the conformations adopted by both the head groups and hydrocarbon chains. Similarly, the outer surfaces of the helices (shown here as smooth) are actually knobby in appearance, as amino acid side chains extending from the helix core differ in size and shape. Note also that water has a significant "solubility" in hydrocarbon media (approximately 1 water molecule per 3300 phospholipid molecules), but it is essential to understand that this is a phenomenon similar to the normal presence of water molecules in the vapor above an aqueous medium; it involves free and isolated single water molecules. These scattered intruders do not affect the distinction between liquid water and a hydrocarbon medium: the unique properties we normally ascribe to an aqueous medium manifest themselves only in the bulk liquid state, for they arise from three-dimensional interactions among a multitude of water molecules.

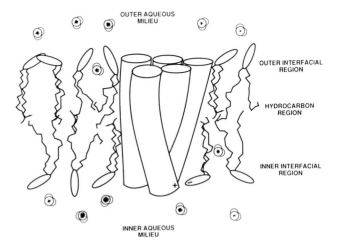

FIG. 3.1. Schematic diagram (drawn to scale) of a phosphatidyl-choline bilayer and segments of a hypothetical intrinsic membrane protein. The bilayer is *fluid* and what is shown is a two-dimensional cross section of it. The apparently empty spaces within it must be imagined as filled with portions of lipid chains that protrude into the pictured cross-section from adjacent cross-sections. The hypothetical protein segments are drawn as a bundle of helices, imagined as parts of a single protein molecule, joined together outside the bilayer domain by other parts of the protein polypeptide chain. The small shaded particles are H_2O molecules. [From Reynolds and McCaslin (21) with permission.]

In Figure 3.1 recognize also that the bulk aqueous media on the two sides of a membrane are different. The inside and outside media of any cell typically bear no resemblance to each other even in the concentrations of simple salts and hydrogen ions, and these differences usually give rise to a membrane potential which in itself can affect the structure and function of membrane bound proteins. Changes in membrane potential, for example, are involved in opening and closing ion channels.

The final structure of a protein is dependent on its environment with the hydrophobic/hydrophilic preference being the principal driving force. This dictates that charged entities of the protein (ionic side chains) must normally be in contact with bulk liquid water. Contrariwise, nonpolar portions (hydrocarbon amino acid side chains) should not be enveloped by bulk water. These preferences are broadly attained in the quest for an optimal stable structure in the course of protein folding—whether it be a water-soluble or membrane-bound protein.

With this principle in mind, we must view a membrane as a continuum of environments: outer and inner interfacial regions intervening between the purely aqueous outside and the purely hydrocarbon inside, as Figure 3.1 shows. This provides a unique dimension for the study of membrane proteins and their structures.

Identical physico-chemical principles are involved, but the membrane represents a graded environment in contrast to the homogeneous environment of water-soluble proteins. A protein that passes all the way through a membrane, with functional sites on both sides, encounters a remarkably kaleidoscopic sequence of environments!

WHAT DO PROTEINS LOOK LIKE? INFORMATION DERIVED FROM WATER-SOLUBLE PROTEINS

All the available evidence suggests that the primary sequence of amino acids is a sufficient basis for folding of a polypeptide chain to its ultimate structure, taking into account the solvent system in which the protein finds itself. The process is largely thermodynamic, a quest for a state of low free energy by striving to satisfy the hydrophobic/hydrophilic preference, with due respect for more general structural requirements—the need to keep bond angles within acceptable limits, the avoidance of squeezing unbonded groups too close together, etc. Compromises must obviously be made in the process. For instance, some hydrophobic residues are nearly always partially exposed to water in the final structure, the unfavorable free energy being balanced by a more favorable placement of charged groups or by solvent interactions.

Nonthermodynamic factors enter into the folding process to a lesser extent, so that the ultimate structure does not necessarily represent the absolute minimal free energy state. We even have examples of temporary helper proteins (chaperonins) that may combine with a polypeptide chain during synthesis or folding to protect against proteolytic degradation or to escort the protein safely from its place of synthesis to its final location. These processes are outside the scope of this chapter, which is focused on structure per se and its relation to function. (One might note in passing, however, that the problem of getting safely to the final destination is exacerbated for membrane proteins because of the environmental problem.)

Hierarchy of Structures

The amino acid sequence of a polypeptide chain is often called its primary structure. Specific regular features, such as the α-helix, the β-sheet, or the β-turn, occur universally in the folding of individual chains and are referred to as secondary structure. The arrangement of these regular segments with respect to each other in adoption of a three-dimensional whole is called tertiary structure. And finally, two or more folded polypeptides may associate to form a larger complex—quaternary structure. The individual polypeptides of the latter complex are customarily called *subunits*.

Subunit Molecular Weights

Individual polypeptides of a complex protein are often referred to—especially in the case of membrane proteins—in the terminology of molecular weights, such as the "35 kilodalton subunit." It needs to be pointed out that this can be a misleading nomenclature. "Molecular weight" is a precisely defined quantity and can nearly always be measured with reasonable accuracy for a purified polypeptide (see the Appendix to this chapter). However, the requisite measurement has normally not been carried out in the naming of subunits—the name is usually based on SDS polyacrylamide gel electrophoresis, which has the advantage of being measurable before purification, as soon as a particular protein is first identified. But electrophoretic mobility is only partially determined by the actual molecular mass and molecular weights implied by subunit names may therefore differ by large amounts from true values based on the primary sequence of a polypeptide.

Conformational Flexibility

All protein molecules display conformational flexibility; they are not ping-pong balls or rigid rods. In some cases conformational flexibility is an obvious requirement for functionality, but proteins commonly undergo a variety of motions, ranging from fractions of Angstroms to several Angstroms, that have no relation to function. Sometimes amino acid side chains are seen to oscillate between different positions on the molecular surface, and lesser movements of atoms occur within the macromolecular interior. Of greatest interest in the context of physiological function, however, are the larger scale movements that occur when a particular protein can exist in two or more readily identifiable, accessible conformational states, such that structural alterations simultaneously affecting different parts of the molecule are involved in transition from one state to another. (For obvious reasons, we will not be concerned here with the most massive conformational changes of this kind, which are encountered in protein denaturation: transitions to random coils, for example, or major changes in helical content induced by detergent binding.)

Two kinds of well-documented conformational transitions from the field of water-soluble proteins may be singled out as relevant to our discussion of specific membrane proteins later in this chapter. *(1)* Proteins containing multiple subunits often exhibit alterations in the relative positions of the subunits as the result of ligand binding. Classical examples of this phenomenon are hemoglobin in the oxygenated and de-oxygenated states, and glycogen phosphorylase in the so-called "T" and "R" states. In both cases, the binding of a small ligand to one portion of the protein leads to a 10°–15° rotation of one subunit with respect to another with only small changes in the bulk of the protein. *(2)* Single polypeptide chain proteins often consist of domains connected by relatively flexible hinge regions which can undergo bending motions. The enzymes hexokinase and lysozyme, for example, contain clefts between two domains of the protein which are either open or closed depending upon substrate occupancy. Closure of the cleft in both cases requires movements of the individual domains over distances of 6 Å–10 Å, accompanied by a sizeable twisting motion.

Structural Families

The explosion of protein structural information over the past few years has led to an equal explosion in the literature correlating all these data. One happy outcome of all this activity is the recognition that many similarities among existing proteins allow us to group them into easily definable "families." Sometimes the structural similarities within a family are not surprising since the physiological functions are similar as, for example, in nucleotide binding proteins. Figures 3.2 and 3.3 illustrate the structural homologies around the ligand binding site of some of these proteins. The common structural feature is a set of alternating α helices and β strands. In Figure 3.2 the set labeled *(a)* includes clathrin uncoating ATPase, actin, and a heat shock protein—all ATP binding proteins. The set labeled *(b)* is called a G-protein fold creating the site for the binding of GTP, and the set labeled *(c)* is generally referred to as the Rossman fold, identified many years ago by M. Rossman in several NAD binding dehydrogenases.

In other cases, proteins that at first sight would surely be considered as unrelated are actually the evolutionary products of a common progenitor and show structural similarity for this reason, with no link to any functional property, as is illustrated in Figure 3.4. Here we have two apparently unrelated proteins, superoxide dismutase and an immunoglobulin, with nearly identical three-dimensional structures. Cyrus Chothia (7) has examined a large volume of DNA sequence data and has pinpointed the clustering into families of this kind. He makes the encouraging estimate that the total number of existing protein structural families (common ancestors) may be less than one thousand. If future work corroborates this conclusion, we can look for-

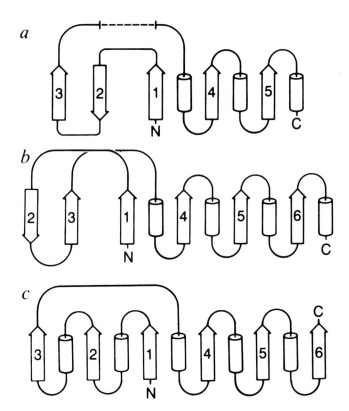

FIG. 3.2. Schematic diagrams showing the topologies of three families of nucleotide binding proteins: *(a)* for ATP, *(b)* for GTP, *(c)* for NAD. All of them contain alternating α-helices and β-strands, shown respectively as cylinders and arrows, but the specified arrangements are different. All six β-strands in *(c)* are parallel, for example, whereas five are parallel and one is antiparallel in *(b)*. [From Brändén (4) with permission.]

ward to a time when new projects in structure/function correlation can often begin in already familiar territory, that is the known database for proteins in the same family.

MEMBRANE PROTEINS

Do all the foregoing principles apply regardless of the solvent system? Yes, but we need to recognize that membrane proteins must have evolved so as to fold into distinct domains, to adapt to the different environments found in and around a membrane. Protein segments running through the core of a lipid bilayer must by and large have a surface devoid of charges or exposed highly polar groups, whereas domains in the aqueous media on either side will be more like ordinary soluble proteins with hydrophilic surfaces. In spite of these complications, the principle that sequence dictates structure continues to apply. It could hardly be otherwise, for then our conceptual edifice of the relation between genes and function would become very tenuous.

A convincing demonstration of this last point is provided by experiments from our own laboratories involving comparison between polypeptide chains destined to become membrane proteins and a lipophilic polypeptide derived from a water-soluble lipoprotein—the 250,000 dalton beta polypeptide of human serum low density lipoprotein (LDL). LDL exists in nature as a soluble complex with a lipid/protein ration of about 4:1, similar to the lipid/protein ratio in cell membranes. The beta polypeptide, like most membrane polypeptides, becomes insoluble in aqueous media when divorced from its associated lipid, but, again like most membrane polypeptides, can be solubilized by detergents.

The crucial experiment to demonstrate essential inbred differences between the two kinds of polypeptide is what is known as reconstition. Detergent is slowly removed from detergent-solubilized protein and replaced by phospholipid under conditions where unilamellar vesicles are formed when all protein is absent. Similar vesicles are formed in the presence of membrane-derived polypeptides, the latter becoming inserted into the membrane. The experiment with the

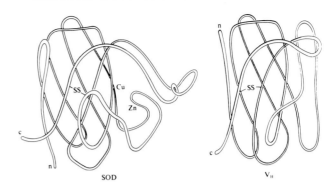

FIG. 3.4. Two functionally unrelated proteins which nevertheless have identical structural topology (and even a similar disulfide bond) over a major part of their three-dimensional structures. The enzyme superoxide dismutase with Cu and Zn atoms at its active center is on the *left*. The variable domain of the heavy chain of immunoglobulin G is on the *right*. The principal structure in both cases is so-called β-barrel, composed of seven strands of β-sheet. (Based on a drawing by D. Richardson.)

beta polypeptide of LDL (25) is essentially a control experiment. It is found that no vesicles are formed at all in this case; the primary structure of the protein dictates spontaneous formation of a water-soluble complex. (Without this control experiment, it might be thought that the inherent drive of phospholipids for vesiculation is the overriding factor in all reconstitutions, subsequent association with lipophilic peptides being a secondary process, not necessarily related to the native structures which a reconstitution experiment is designed to reproduce.)

Reconstitution into vesicles, with recovery of functional properties that had been destroyed in the detergent-solubilized state, has been successfully achieved with most membrane proteins. This gives us the confidence that membrane proteins can be handled in the laboratory with the same facility as water-soluble proteins. In spite of this, however, detailed three-dimensional structures are known for only a handful of membrane proteins, in contrast to the hundreds of known structures of water-soluble proteins. There is no problem in understanding the reason for this. Three-dimensional structures are usually obtained by means of x-ray studies of large and regular protein crystals—the compact and globular molecules of most soluble proteins readily form appropriate crystals; membrane proteins normally have more complicated shapes and do not retain their native structures in the absence of lipids, so that formation of three-dimensional crystals becomes a very difficult problem. Some of the techniques being used to overcome these difficulties are discussed in the appendix to this chapter.

In the present section we shall discuss in some detail the few structures that have actually been determined,

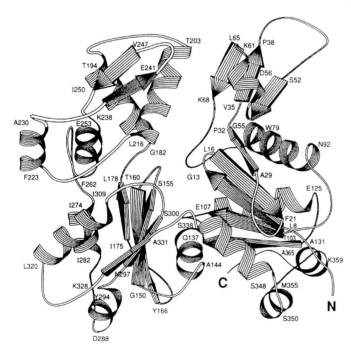

FIG. 3.3. Schematic ribbon diagram for the three-dimensional structure of actin, which contains an ATP binding site similar to *(a)* in Figure 3.2. The letters and numbers in this diagram identify the amino acids in the sequence at which α-helices begin and terminate. The ATP binding site is in the cleft between the domains in the lower center of the picture, each domain containing an arrangement of α-helices and β-sheets such as those in *(a)* of Figure 3.2. [From Kabsch et al. (15) with permission.]

with emphasis on the similarities to and differences from water-soluble proteins.

Bacteriorhodopsin

Nature has provided a membrane-bound protein ideal for structural analysis. The light-driven proton pump, bacteriorhodopsin (from *Halobacterium halobium*), is present in such high concentrations within the lipid bilayer that two-dimensional crystals in the plane of the bilayer are formed spontaneously. Two-dimensional crystals, just one molecule thick in the third dimension, are of course unsuitable for the kind of structural analysis we are accustomed to from three-dimensional crystals. Nevertheless, new techniques have been patiently developed, primarily by Henderson and coworkers (e.g., 12), to extract structural information from the two-dimensional ordered arrays by means of electron microscopy.

The earliest results (13) were at a relatively low resolution of 7 Å and strongly suggested the presence of seven transmembrane helices. More recently, (12) improved instrumentation has permitted elucidation of the structure to 3.5 Å resolution within the plane of the membrane. The seven-helix motif has been confirmed and the binding site for the retinal has been located near the center of the bilayer core. Figure 3.5 shows the overall structure and demonstrates the participation of five of the seven helices in the formation of the retinal binding site. Site-directed mutagenesis had previously identified a number of residues involved in proton transport, and this information together with the three-dimensional structure has led to the proposal that the proton pathway is located in the center of the helices. Charged residues Lys 216, Asp 85, 96, and 212, and Arg 82 are in the channel, well protected from the hydrocarbon milieu of the phospholipid bilayer (see Figs. 3.9 and 3.10 for a discussion of the function). The upper channel is narrow and highly hydrophobic while the lower (proton discharge) channel is considerably wider and hydrophilic in nature, with sufficient space for water molecules. The retinal binding pocket, on the other hand, is highly hydrophobic as would be expected from the hydrophobic nature of the ligand.

Porin

Porins are small proteins (approximately 300 residues long) found in the outer membrane of gram-negative bacteria. They contain channels for rapid diffusion of small polar molecules across the membrane. Unlike bacteriorhodopsin, where the transmembrane domain consists of α-helices, the secondary structure of porin is

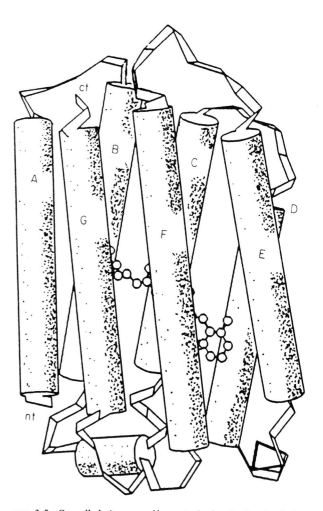

FIG. 3.5. Overall chain trace of bacteriorhodopsin showing helices as solid rods. The N terminus (*nt*) is at the bottom at the extracellular surface; the C terminus (*ct*) is at the top at the cytoplasmic surface. The helices range from 30 to 40 Å in length, sufficient to span the hydrophobic core of the lipid bilayer. The retinal is attached by a Schiff base linkage to Lys 216 (helix G) and is tilted roughly 20° out of the plane of the membrane. [From Henderson et al. (12) with permission.]

nearly 100% β-pleated sheet. Also unlike bacteriorhodopsin, one of the porins (matrix porin of *Escherichia coli*) has been obtained in the form of three-dimensional crystals, first reported in 1980 by Garavito, Rosenbusch, and colleagues (11). The crystals were grown from a detergent solution, with many detergent molecules retained in the crystal to cover the hydrophobic part of the protein surface that would in the native state have been in contact with the hydrocarbon chains of phospholipids. Initially, the quality of the crystals was insufficient for high-resolution x-ray diffraction because (presumably as a result of the presence of detergent) it was difficult to maintain long-range order. We have had to wait for more than a decade for the gradual development of better techniques to obtain a picture of the protein at atomic resolution.

Figure 3.6 is a ribbon diagram of matrix porin from *E. coli*, a relatively nonspecific channel-forming protein for small hydrophilic molecules. The structure is a 16-stranded anti-parallel β-barrel with the amino and carboxy terminals linked by a salt bridge. (Note that the closed topology of this structure ensures that all of the polar main-chain atoms are engaged in interstrand hydrogen bonds and not exposed to the membrane core.) The outer surface has a hydrophobic band approximately 25 Å wide that presumably is in contact with the interior of the lipid bilayer in the membrane. The boundaries of this band are largely phenylalanine and tyrosine groups, suggesting that polarizable aromatic rings are energetically favored at the interface between two quite disparate solvent systems—that is, the hydrophobic interior and the polar head group region. The aqueous pore is in the center of the β-barrel and has the dimensions 7 × 11 Å at one end,

increasing to 15 × 22 Å at the other. The interior of the pore is, as expected, hydrophilic—a cluster of negatively charged amino acid residues at one end and a second cluster of arginine and lysine residues at the other.

Similar structures are observed for phosphoporin from *E. coli* and even a porin obtained from *Rhodobacter capsulatus*, which has a totally nonhomologous sequence with the *E. coli* porins—yet another example of families of protein structures that are related to function and not necessarily to primary amino acid sequence.

Photosynthetic Reaction Centers

The membrane of photosynthetic bacteria contains a large, multisubunit protein, known as the photosynthetic reaction center, which catalyzes photo-activated electron transport. This is another case where it has been possible to obtain three-dimensional crystals in the presence of bound detergent and the three-dimensional structure of this protein from two different sources has been determined to high resolution by x-ray diffraction of such crystals. (1, 2, 9, 10). The complex contains four subunits, H, 258 residues; M, 323 residues; L, 273 residues; plus the already well-known protein cytochrome C, 333 residues. (It is an amusing aside to note that H, L, and M stand for "heavy," "light," and "medium," based on molecular weights inferred from SDS gel electrophoresis before the sequence had been established, a problem we discussed earlier under Subunit Molecular Weights.)

The significant characteristics of the reaction center are ten membrane spanning helices, one from the H subunit and five each from the L and M subunits. These helices are approximately 40 Å long, tilted with respect to the plane of the membrane, and connected by segments in contact with the bulk of the H subunit (cytoplasmic face) and cytochrome C. There is a 25–30 Å band around the complex that is hydrophobic and binds detergent in the crystallized form. There is an exceptionally large number of non-protein prosthetic groups bound to the reaction center—covalently bound heme groups, bacteriochlorophyll b, bacteriopheophytin b, quinones, a single non-heme ferrous iron, and several carotenoids. These prosthetic groups are located in a protective cage made up of the protein subunits.

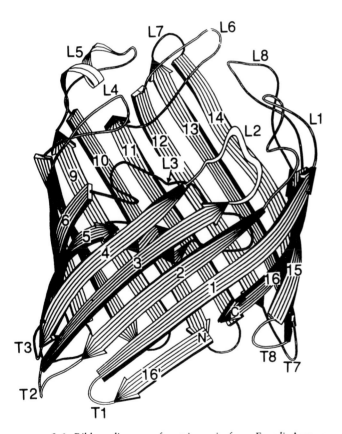

FIG. 3.6. Ribbon diagram of matrix porin from *E. coli*. Arrows represent β-strands and are labelled 1–16 starting from the strand after the first short turn. The short β-strand at the N terminus continues the C-terminal strand β 16. The long loops are denoted L1–L8, the short turns at the other end T1–T8. Loop L2 protrudes towards the viewer. Loop L3 folds inside the barrel. [From Cowan et al. (8) with permission.]

Light-Harvesting Complex

Even low-resolution structural information can be highly useful, as illustrated by the example of the light-harvesting chlorophyll a/b-protein complex, a membrane protein serving as the major antenna of solar energy in plant photosynthesis. This protein is actually a trimer, and structural data were obtained from two-dimensional crystals with a resolution of only 6 Å, which proved to be sufficient to resolve the arrangement of the three subunits.

The resolution is sufficient to identify three membrane-spanning helices in each subunit and to locate the fifteen chlorophyll porphyrins arranged on two levels, corresponding roughly to the two bilayer leaflets. This information alone provides a speculative model of the nature of energy transfer in the operational mechanism of the protein (16).

Concluding Comment

In these admittedly limited examples of the three-dimensional structures of several unrelated membrane proteins we see all the same general principles as for water-soluble proteins. Hydrophobic bands are observed around the outer surfaces that are in contact with the interior of the lipid membrane in the native state. Charged groups appear in aqueous channels. Polarizable aromatic amino acids are frequently seen at the interface between hydrophobic and hydrophilic milieus—just as they are not infrequently observed on the surfaces of water-soluble proteins. No great surprises have appeared and no previously established rules have been broken.

SPECULATIVE STRUCTURES

Given the scarcity of actual three-dimensional structures for membrane proteins, it is not surprising that numerous examples of intelligent, but nevertheless speculative, surmises are found in the literature. In fact, if we seek structural information about the membrane proteins of most burning interest to physiologists—receptors, pumps, ion channels—most fall into this category at the time we are writing.

In all cases we begin with a knowledge of the number of subunits and the primary structure (amino acid sequence) of each one. Additionally, we often have chemical information that can identify segments of this primary structure as the locations of specific binding sites for ATP, cofactors, transported species, etc., and it is often possible to locate sites that are exceptionally susceptible to proteolytic scission. Judicious experiments with intact cells or vesicles can additionally tell us about the location of such sites (and therefore of the particular amino acids identified with them) with respect to the bilayer, that is, the cytoplasmic side versus the extracellular side. Immunological methods can sometimes serve the same purpose. Photoactivata-

ble lipid-soluble probes can be used to covalently label residues in the interior of the membrane.

A very common device is to attempt to identify transmembrane segments of the protein subunits by searching for regions of relatively high hydrophobicity in the amino acid sequence, usually by means of one of a number of computer-based algorithms that have been proposed. Such identification is usually accompanied by the assumption that the hydrophobic segments will exist as α-helices within the membrane. The usual next step is to arrange the putative hydrophobic helices in the membrane in a manner consistent with the location of known binding sites, sites of proteolytic cleavage, etc., on the proper sides of the membrane bilayer. It should be emphasized that it is not a safe assumption to suppose that polypeptides invariably cross a bilayer in a helical conformation, for we have already seen that the transmembrane structure of porin is nearly 100% β-sheet.

An example of one such surmise is shown in Figure 3.7 where speculative polypeptide chain arrangements for three specific ion channels are compared (6). Note the clusters of six putative transmembrane helices in each structure. Note that the proposed structures emphasize the principle that proteins with similar function often have similar structures.

We have deliberately not provided a long list of speculative structures of this kind because they must all be conceived as transient images, expected to change from year to year as more information becomes available. It should be emphasized, however, that the continuing expansion of the database is by no means limited to data that relate directly to aspects of protein structure. Refinement of our understanding of functional details and related theoretical necessities are equally important. In fact, one can conceive of "progress" in this field as a process of convergence between preconceived structural notions (based on physics and chemistry and functional studies) and *actual* structural or structure-related measurements. This mode of investigation of the problem may be disparaged by purists, but is more or less forced on us by the fact that we have as yet little assurance that actual precise structural data will soon become available for more than a handful of physiologically important proteins. In the interim, few of us want to sit on our hands and do nothing to further our understanding.

STRUCTURE–FUNCTION RELATIONSHIP—MINIMAL REQUIREMENTS

The ability to "see" structures that enable us to understand the molecular basis for important physiological functions is, of course, our ultimate goal. In the absence of high-resolution structures for the most interesting membrane proteins from the physiological point of view, published work on structure/function relationships is necessarily mostly speculative. Even if a de-

tailed structure is available, as in the case of bacterio-rhodopsin, the structural interpretation of chemical and physiological data does not necessarily follow automatically and may retain a considerable level of speculation. The classical triumph of structural knowledge, the double helix of DNA, where elucidations of the structure *immediately* suggested a possible copying mechanism for the genetic information, has no parallel in protein chemistry—proteins are more reluctant to yield the secrets of their working mechanisms.

Investigations of structure/function relationships fall broadly into three categories. *(1)* Assertion of minimal requirements, without which the physiological process could not work. These may be taken as *predictions* of what one must look for in the structure of particular catalytic proteins. *(2)* Interplay between precise chemical, kinetic, and (sometimes) immunological data and speculative surmises of the structure. The functional data are in effect tools that enter into the generation of speculative structures; mechanistic and structural insights may emerge together from the interplay. *(3)* Only where the three-dimensional structure is precisely known can we speak in terms of the conceptually straightforward (but not necessarily easy) process of reconciliation between function and structure where all the primary data are unarguably true. We give examples of all three categories: in this section we provide illustrative examples of minimal requirements—imperatives for any plausible model.

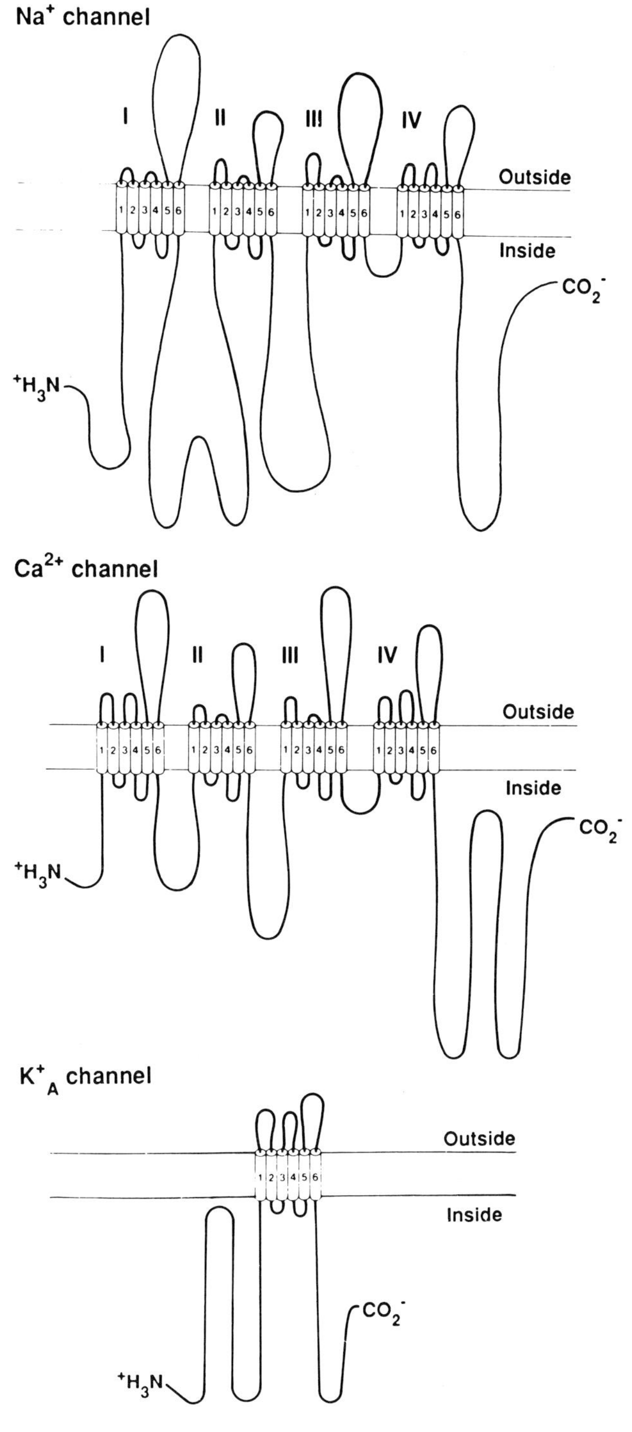

Conformational Change

It is difficult to conceive the action of any protein as a physiological engine without postulating that the protein can exist in at least two distinct conformational states. The situation here is quite different from what applies to many common biochemical enzymes. The mechanism of the latter often involves a single fixed "active site." In the majority of cases small local rearrangement of atomic positions may occur when substrate is bound or dissociated, but the long-range structure is unaffected. "Allosteric" enzymes may in addition be able to exist in two conformational states,

FIG. 3.7. Proposed arrangements of the polypeptide chains of the principal subunits of three transmembrane ion channels: Na^+ channel from rat brain, Ca^{2+} channel from rabbit skeletal muscle, and A-current K^+ channel from *Drosophila*. Na^+ channel polypeptides from various sources range in size from 1832 to 2012 amino acid residues. Similarly, the Ca^+ channel polypeptide consists of 2005 residues. The K^+ channel protein, however, is only 616 residues long. Putative trans-membrane helices are shown as numbered cylinders. [From Catterall (6) with permission.]

in one of which the enzyme has a much higher rate of catalytic activity, but change from one conformation to another need not be part of the intrinsic catalytic mechanism per se. For physiologically interesting proteins, however, long-range major structural changes are usually at the core of physiological function itself, occurring repeatedly each time the protein goes through its sequence of chemical steps.

Ion channels, for example, cannot be conceived without invoking at least two conformations for the protein—channel open and channel closed. Extensive rearrangement of long polypeptide segments is necessary to achieve the transition. This is a factor that complicates the correlation of structure with function because the most rigorous structural information is by its very nature a static picture of a protein. Given the difficulty of obtaining a structure at all for most membrane proteins, we can expect that it will usually be available only for one conformational state. Much imagination will be necessary to postulate a structure or structures for other states, far more than is ever necessary when thinking about "induced fit" of substrate binding sites for enzymes.

Transport Pathway Across the Membrane

One subject amenable to confident theoretical prediction deals with the pathway for translocation of ions across a membrane. Although the proteins responsible for ion movement must themselves be anchored in the membrane by means of an almost entirely nonpolar surface—a belt of sufficient thickness to extend between the opposite headgroup regions—any conceivable pathway for a moving ion must be highly polar or must contain charged groups (*negative* for cations or vice versa). We can be even more specific than this, depending on whether we are dealing with passive or active transport.

In so-called ion channels the physiological process involves an unimpeded rush (thermodynamically *downhill*) of ions across the membrane at an impressively high rate. We must there look for a relatively wide continuous passage across the membrane, which can exist in open and closed states—all ion channels are known to be regulated, and no ion concentration gradient could ever be created or maintained if a channel remained open for an appreciable length of time.

The situation is quite different in the case of ion pumps, however. Here the transport process is thermodynamically *uphill*. Although structural chemists may use the term "channel" to describe a perceived transmembrane pathway, pump proteins could never work if they had in any conformation, at any stage of their reaction cycle, an actual passage freely connecting one side of the membrane to the other. A time much less than the kinetic cycle of a pump protein (typically of the order of milliseconds) would be sufficient for a connecting passage to abolish whatever gradient the pump was trying to establish.

How then can the pump function be carried out? This is clearly an example of the obligatory existence of at least two conformational states, with a transition from one to the other (and back again!) as an essential feature of each individual reaction cycle. The cycle of an ion pump will generally involve one state in which the transport pathway is open exclusively to the uptake side of the membrane, and, later in the cycle, another conformational state for discharge on the opposite side of the membrane. In between, the transported ion must be held bound to the protein. It is easy to see the severe structural constraints that these requirements impose.

Something else that must happen between uptake and discharge is that the transported ion must somehow acquire the free energy (excess chemical potential) that is needed for discharge on the high concentration side of the membrane. This is less easily translated into a unique structural requirement.

Absorption of Light

Several membrane proteins absorb light and use it for a physiological purpose. This obviously requires the presence of light-absorbing chromophores, such as chlorophyll or retinal, which must be intimately involved in the structure/function relation. As for other processes, utilization of light absorption for energy or information requires multiple conformational states in the reaction cycle. The photochemical processes that this involves may be complex and difficult to resolve, but bound chromophores may actually be helpful in early stages of structure determination: the chromophores are large enough to be seen even at low resolution and may provide a useful starting point for tracing the folding of a polypeptide chain, for example. It is probably no coincidence that the list of membrane proteins about which we have useful structural data includes several light-utilizing proteins.

STRUCTURE–FUNCTION RELATIONSHIP—SPECULATIVE MODELS

Voltage-Sensitive Ion Channels

On the basis of the fourfold pattern repetition shown in Figure 3.7 for the Na^+ channel and additional experimental data, Catterall has suggested that the crucial entity for channel function is a transmembrane structure of four homologous domains in a square array. The transport channel is presumed to be formed in the center of the square array by α-helical segments from each of the domains.

The difficult problem of how a conformational change (closed → open) might be triggered by a change in potential across the membrane is addressed by the rather complicated hypothetical sliding helix model illustrated in Figure 3.8. The key experimental datum is that the amino acid sequence of one of the putative transmembrane helices (S4) leads to a spiral ribbon of positively charged arginine residues perpendicular to the membrane. In the model for the closed conformation these charges are proposed to neutralize negative charges from other parts of the channel and, in addition, the resting membrane potential (inside negative) is proposed as a stabilizing force. Depolarization removes this stabilizing force, and it is proposed that this leads to a spiral rotation of the S4 helix, accompanied by an

outward displacement, leaving an unpaired negative charge at the outward surface. The sum of overall charge movements of this kind in all four of the supposedly involved helices is consistent with experimentally measured gating currents that accompany channel opening.

It should be noted that the primary structures shown in Figure 3.7 for the *Drosophila* K^+ channel has only a single transmembrane domain, which does, however, have the requisite sequence homologies with the multiple domains of the two other channel proteins shown in the diagrams. If the above mechanism applies, it would require that this protein exist in the membrane as a tetramer, and there is experimental evidence to support this possibility.

Light-Driven Proton Pump

In bacteriorhodopsin we have a reasonably precise picture of the molecule at essentially atomic resolution, and the knowledge that the initial energizing step is the absorption of light to convert the retinal cofactor from all-*trans* to a 13-*cis* conformation. Figure 3.9 is a schematic diagram of the retinal binding site showing the amino acid residues that are believed to be relevant to the transport mechanism. Based on this information, Henderson and colleagues have proposed a model for proton movement through the membrane (Figure 3.10). As expected, there is no continuous open passage from one side of the membrane to the other in any of the conformational states; in fact, the proton gets across by sequential jumps between proton binding sites. A minimum of three distinct conformational states for the protein is required, and it is important to remember that only one of the proposed structures is experimentally determined (*a* in the figure); the remainder are educated "guesses" that currently fit all available experimental data.

Protons are ubiquitous in protein molecules, and many of them are always freely exchangeable with solvent protons. This precludes an independent analysis for the number of bound protons and allows greater freedom in the formulation of hypotheses—for example, in the present case a mechanism in which it takes several complete reaction cycles for a given proton to cross the membrane. In other words, the proton ejected in a single cycle is not the same one that was taken up in that cycle or even in the preceding one. Such a mechanism would be testable for active transport of most other ions and would be automatically excluded (for example) for the Ca^{+2} pump discussed below, because direct analysis in that case shows that the number of bound ions at any stage of the reaction is equal to the number transported in a single cycle. (This

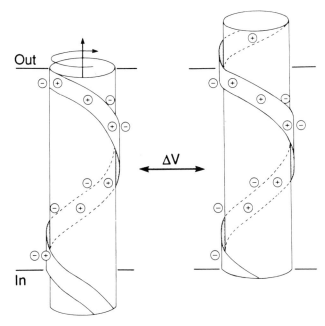

FIG. 3.8. Sliding helix model of voltage-dependent gating. Movement of the S4 helix of domain IV of the Na^+ channel in response to membrane depolarization. The proposed helix is illustrated as a cylinder with a spiral ribbon of positive charge. At the resting membrane potential (*left*) all positively charged residues are paired with fixed negative charges on other transmembrane segments of the channel and the transmembrane segment is held in that position by the negative internal membrane potential. Depolarization reduces the force holding the positive charges in their inward position. The S4 helix is then proposed to undergo a spiral motion through a rotation of approximately 60° and an outward displacement of approximately 5 Å. This movement leaves an unpaired negative charge on the inward surface of the membrane and reveals an unpaired positive charge on the outward surface to give a net charge transfer of +1. [From Catterall (6) with permission.]

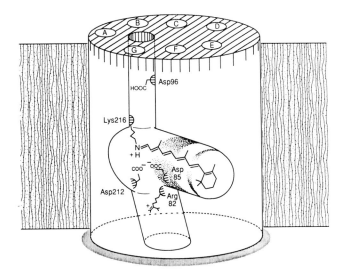

FIG. 3.9. Artistic impression showing the relationship between the key residues Asp 85, Asp 96, Asp 212, Lys 216, and Arg 82 and the retinal binding site, the proton channel and overall molecular boundary for the ground state of bacteriorhodopsin. The cytoplasm is at the top of the diagram. [From Henderson et al. (12) with permission.]

is not intended as an argument against the proposed mechanism for proton transport. It only stresses the greater difficulty of working with proton pumps as compared to many other systems.)

ATP-Driven Ca²⁺ Pump

Hypothetical models for this membrane-bound pump protein from sarcoplasmic reticulum have been constructed from the primary sequence (similar to the models shown in Figure 3.7 for the channel proteins). Four different proposals suggest seven, eight, or ten transmembrane helices, although immunological data would seem to rule out the seven-helix model. (see ref. 19 for a summary). Electron microscopic views of frozen hydrated crystals are of insufficient resolution to be sure of the exact number of transmembrane helices, but the technique does show an extramembranous domain extending 60 Å above the membrane surface with a large pear-shaped head (65 × 40 × 30Å) connected to the surface by a 20 Å diameter stalk. (23).

Figure 3.11 presents a possible model for the translocation pathway of Ca^{2+} ions in this pump. It is known that two ions are transported all the way through in each reaction cycle, unlike the model for proton transport shown in Figure 3.12. Note that the ions are tightly bound in Ca_2E_z and that free energy from ATP hydrolysis is used between that point and the

discharge state to break (or greatly weaken) the bonds that bind the ions to the protein. Note also that we do not know the exact location of the binding sites in either the polypeptide chain sequence or the model based on it.

This protein is one of a class of ATPases that operate via a phosphorylated intermediate in which the terminal phosphate group of ATP is covalently attached to the protein. Although we don't know exactly where the Ca^{2+} binding sites are, both the binding site for ATP and the phosphorylation site have been identified. Both are on the pear-shaped head of the molecule (sticking into the cytoplasm), but the most intriguing fact is that the two sites reside on separate domains of this part of the protein—the phosphorylation site is at a considerable distance from the initial ATP binding site. This suggests the mechanism for energy coupling shown in Figure 3.12 in which the two domains move into close proximity as the protein goes from state E_x to Ca_2E_z.

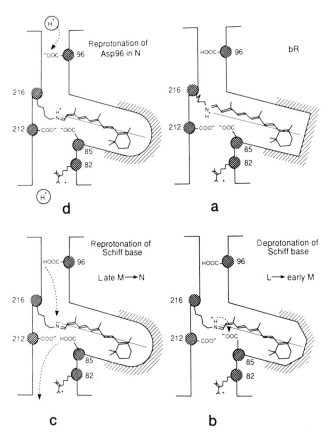

FIG. 3.10. A diagram to describe the 4 proton movements in the photocycle of bacteriorhodopsin. (a) Structure before isomerization of the retinal. (b) Proton transfer from Schiff base to Asp 85. (c) Proton transfers from Asp 96 to Schiff base and from Asp 85 to the extracellular space. (d) Proton transfer from the cytoplasmic surface to Asp 96. [From Henderson et al. (12) with permission.]

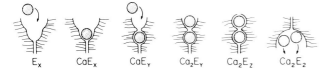

FIG. 3.11. Pictorial representation of the successive stages of the Ca^{+2} binding domain of the sarcoplasmic reticulum calcium pump. The binding cavity is assumed to extend physically into the cytoplasm and its inner surface is assumed to possess a net negative charge. E_1 and E_2 are the names given to the two principal conformational states with access to opposite sides of the membrane. E_x, E_y, and E_z are variants of E_1 which had to be assumed to account for details of the experimental data for Ca^{2+} binding from the cytoplasmic side of the membrane. [From Tanford et al. (24) with permission.]

G-Protein Coupled Receptors

Hormone receptors are functionally among the most fascinating of all membrane proteins. Here no ions or molecules cross the membrane at all, yet binding of a ligand on the outside drastically alters chemical events on the inside. There is as yet little information about the three-dimensional structure of these proteins and even speculative models are at this time probably more tenuous than the models we have discussed for ion transport.

A large family of membrane-bound receptor proteins have been identified that exert their physiological effect through two proteins on the internal side of the membrane, following binding to the receptor from the outside—a guanine nucleotide binding protein and the enzyme adenylate cyclase. In the search for similar structure dictated by the similar function investigators have proposed (based on primary sequence and other indirect data) that these receptors all contain seven transmembrane helices (*à la* bacteriorhodopsin) with considerable sequence homology among themselves (though not with bacteriorhodopsin). Binding of a ligand to the external domain of the receptor is assumed to cause a protein conformational change that allows the binding of a G-protein on the intracellular surface. G-protein containing bound GTP in turn interacts with adenylate cyclase converting it into an active form to produce cyclic AMP, an example of what is called a "second messenger," the hormone being the first. Hydrolysis of the GTP to GDP then converts the G-protein to a form that can no longer activate the adenylate cyclase.

Confirmation of the seven-helix theme for this group of receptors awaits a high-resolution structure determination of at least one member of this family, as does the identification of the ligand-induced conformational change.

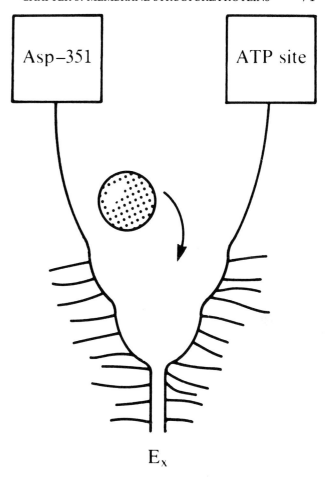

FIG. 3.12. Speculative model for free energy coupling in the sarcoplasmic reticulum calcium pump. It is suggested that the binding of Ca^{+2} (Fig. 3.11) brings the ATP binding region and the phosphorylation site into close proximity, permitting the transfer of the terminal phosphate group of ATP from one site to the other. [From Tanford et al. (24) with permission.]

APPENDIX: METHODS FOR STRUCTURE DETERMINATION

Primary Structure

Primary structure (amino acid sequence) can be determined by direct sequencing methods, which involve chemical or proteolytic cleavage into smaller peptides and subsequent step wise degradation of these peptides into their constituent amino acids. Alternatively, if the gene encoding the protein of interest can be identified and isolated, the DNA sequence will provide the same information. With the advent of new techniques and automated equipment, both these procedures have become nearly routine in many protein chemistry laboratories, and large catalogues of primary sequences are available from the Brookhaven Protein Databank and the EMBL/Swiss Protein Databank.

Molecular Weight

The molecular weight of a polypeptide emerges from knowledge of the amino acid sequence. Before that knowledge is available, the same information can be obtained using rigorous thermodynamic methods—in practice only equilibrium ultracentrifugation is usable for membrane proteins, and even then, of course, the method is valid only if one can be certain that the protein entity subject to measurement is actually in the form of a completely dissociated polypeptide. The advantage of the method is in fact that it will rigorously measure the mass of the molecular species in the solution. At least in principle, it can give independent values for protein and lipid in a particle containing both (22; see also later under Tertiary and Quaternary Structure).

Rate of transport measurements of various types, including electrophoretic mobility, sedimentation velocity, among others, cannot be used to determine molecular weight because the speed of movement is dependent not only upon the mass of the protein but also on its shape and, in the case of electrophoresis, on its total electric charge (see ref. 5 for a treatment of general principles; ref. 20 for a discussion of sodium dodecyl sulfate [SDS] gel electrophoresis).

Secondary Structure

The secondary structure of proteins emerges from the three-dimensional structure once it has been obtained by some suitable means. As we have pointed out in the main text, complete structures are so far few in number. Therefore circular dichroism is frequently used to provide some estimate of α-helix and β-sheet structure. This optical tool provides only approximate information since the optical properties of segments of these types of structure depend to some extent on their length and their nearest neighbors in the polypeptide chain. Another popular procedure is the attempt to predict the presence of α-helix and β-sheet from the known primary structure. This involves studying the preferred positions of the amino acid residues in known protein conformations and calculating a probability factor for the appearance of each residue in either an α-helix or a β-sheet. Statistical methods are then applied to the sequence of a protein of unknown conformation, and the probability of certain regions of the polypeptide chain folding into one of the regular secondary forms is calculated. The database is almost entirely from water-soluble protein structures and may or may not be appropriate for membrane-bound proteins (see ref. 17 for a detailed discussion of molecular modeling).

Tertiary and Quaternary Structure

The term "tertiary structure," referring to the spatial arrangement of a single polypetide beyond the secondary level, has lost most of its practical utility because it is automatically and unambiguously obtained from a complete three-dimensional structure, and because no methods for its separate determination exist. The same is not true for the term "quaternary structure," referring to association of more than one polypetide into a more complex molecular entity. It is often functionally critical and, more important, it is measurable. As previously noted, equilibrium ultracentrifugation can determine the mass of any particle in solution, including for example (in the present context) detergent-solubilized particles that retain functional activity, which may sometimes be usefully compared with similar particles that have lost activity. Quaternary structure can sometimes also be deduced directly from electron microscope pictures.

X-ray Diffraction

X-ray diffraction of three-dimensional crystals can provide the true position of each atom within a protein molecule to a resolution of 2 Å–3 Å. In fact, this is more easily said than done. First, one must obtain a crystal of sufficient size and regularity to acquire interpretable diffraction data—a process that is perhaps more an art than a science. Protein crystals are typically grown from solutions that are not infrequently as high in concentration as 50–100 mg protein per milliliter solvent. The growth process takes weeks, risking the constant danger of contamination. Because the protein structure is governed by interactions not only between the amino acids but more importantly by interactions with the solvent, solvent molecules are an integral and important part of any protein crystal. If the protein is water-soluble, dehydration must not occur during exposure to the x-ray beam. If all or part of the protein is normally in a lipid environment, those regions in contact with the hydrocarbon milieu must be protected by a similar solvent (for example, detergent molecules). In the latter case it has been found experimentally that replacement of lipid by detergents with small polar head groups is the most successful means of growing good three-dimensional crystals (see ref. 11 for a discussion of x-ray crystallography of membrane bound proteins, and ref. 5, for details of x-ray diffraction).

Electron Microscopy

Electron microscopy and electron diffraction techniques have become powerful tools for the examination of membrane-bound proteins. A purified membrane protein embedded in a lipid bilayer can frequently be induced to form a two-dimensional crystalline array. (One naturally occurring system is bacteriorhodopsin described earlier under Bacterhodopsin.) Provided the array is large enough and sufficiently regular, structural data can be obtained from a combination of electron microscopy, optical diffraction of the image, and electron diffraction. By stacking multiple two-dimensional layers above each other, or by tilting a single layer and measuring diffraction at different angles, atomic resolution, approaching x-ray results in quality, is at least in principle attainable.

Because of beam damage to the specimen from the high-energy electrons, the sample must be maintained at very low temperatures ($-120°$ to $-268°C$) and total electron dosage kept as low as possible. Glycerol or glucose is often used to prevent excessive dehydration. An additional experimental problem is presented by the carbon surface of the grid used to support the membrane film—again, attainment of an appropriate surface is more an art than a science. The degree of hydrophobicity of the carbon surface is a function of its age, and much trial and error is necessary to determine which grids are most suitable for maintaining the regular structure of the membrane sample. An excellent discussion of sample preparation and electron microscopic techniques is given by Henderson et al. (12).

Nuclear Magnetic Resonance Spectroscopy

Nuclear magnetic resonance spectroscopy is capable of serving as a structural tool because it can act as a spectroscopic "ruler," measuring distances between pairs of atoms that have nuclear magnetic moments (C^{13}, for example, but not C^{12})—knowledge of the primary sequence is, of course, a prerequisite. The method has been successfully applied in determining protein structures with molecular weights less than 12 kD, and current efforts suggest this size limitation can be extended to perhaps as large as 30 kD. High concentrations of protein are needed (approximately 1–5 mM). The technique is primarily applicable to water-soluble proteins since the presence of large amounts of bound lipid or detergent molecules presents serious problems in interpretation (see ref. 26 for a summary of the current state of the art in this field).

Compilations of Protein Structures

The annual publication, *Macromolecular Structures* provides an up-to-date compilation of all structures published in the previous year. Two general sources include *Protein Architecture: A Practical Approach* (18) and *Introduction to Protein Structure* (3).

REFERENCES

1. Allen, J. P., G. Feher, T. O. Yeates, H. Komiya, and D. C. Rees. Structure of the reaction center from *Rhodobacter sphaeroides* R-26. The cofactors. *Proc. Natl. Acad. Sci. U.S.A.* 84: 5730–5734, 1987.
2. Allen, J. P., G. Feher, T. O. Yeates, H. Komiya, and D. C. Rees. Structure of the reaction center from *Rhodobacter sphaeroides* R-26. Protein subunits. *Proc. Natl. Acad. Sci. U.S.A.* 84: 6162–6166, 1987.
3. Branden, C., and J. Tooze. *Introduction to Protein Structure.* New York: Garland, 1991.
4. Brändén, C. I. Founding fathers and families. *Nature* 346: 607–608, 1990.
5. Cantor, C. R. and P. R. Schimmel. *Biophysical Chemistry.* San Francisco: W. H. Freeman, 1980.
6. Catterall, W. A. Structure and function of voltage-sensitive ion channels. *Science* 242: 50–61, 1988.
7. Chothia, C. One thousand families for the molecular biologist. *Nature* 357: 543–544, 1992.
8. Cowan, S. W., T. Schirmer, G. Rummel, M. Steiert, R. Ghosh, R. A. Pauptit, J. N. Jansonius, and J. P. Rosenbusch. Crystal structures explain functional properties of two *E. coli* porins. *Nature* 358: 727–733, 1992.
9. Diesenhofer, J., O. Epp, K. Miki, R. Huber, and H. Michel. X-ray structure analysis of a membrane protein complex—electron density map of the chromophores of the photosynthetic reaction center from *rhodopseudomonas viridis*. *J. Mol. Biol.* 180: 385–398, 1984.
10. Diesenhofer, J., O. Epp, K. Miki, R. Huber, and H. Michel. Structure of the protein subunits in the photosynthetic reaction center of *Rhodopseudomonas viridis*. *Nature* 318: 618–623, 1985.
11. Garavito, M. and J. P. Rosenbusch. Three-dimensional crystals of an integral membrane protein: an initial x-ray analysis. *J. Cell Biol.* 86: 327–329, 1980.
12. Henderson, R., J. M. Baldwin, T. A. Ceska, E. Zemlin, E. Beckmann, and K. H. Downing. Model for the structure of bacteriorhodopsin based on high resolution electron cryo-microscopy. *J. Mol. Biol.* 213: 899–929, 1990.
13. Henderson, R. and P.N.T. Unwin. Three dimensional model of the purple membrane obtained by electron microscopy. *Nature* 257: 28–32, 1975.
14. Hendrickson, W. A., and K. Wütrich. *Macromolecular Structures. Curr. Biol.,* New York: published annually.
15. Kabsch, W., H. G. Mannherz, D. Suck, E. F. Pai, and K. C. Holmes. Atomic structure of the actin:DNase I complex. *Nature* 347: 37–44, 1990.
16. Kühlbrandt, W. and D. N. Wang. Three-dimensional structure of plant light-harvesting complex determined by electron crystallography. *Nature* 350: 130–4, 1991.
17. Langone J. J. (ed.). *Methods in Enzymology.* vol. 202, New York: Academic Press, 1991.
18. Lesk, A. M. *Protein Architecture: A Practical Approach.* New York: Oxford University Press, 1991.
19. Matthews, I., R. P. Sharma, A. G. Lee, and J. M. East. Trans-membranous organization of (Ca-Mg) ATPase from sarcoplasmic reticulum; evidence for lumenal location of residues 877–888. *J. Biol. Chem.* 265: 18737–18740, 1990.

20. Nielsen, T. B. and J. A. Reynolds. Measurements of molecular weights by gel electrophoresis. *Meth. Enzymol.* 48: 1–20. New York: Academic Press, 1976.

21. Reynolds, J. A., and D. R. McCaslin. The role of detergents in membrane reconstitution. *Subcell. Biochem.* 14: 1–23, 1989.

22. Reynolds, J. A., and C. Tanford. Determination of molecular weight of the protein moiety in protein–detergent complexes without direct knowledge of detergent binding. *Proc. Natl. Acad. Sci. U.S.A.* 73: 4467–4470, 1976.

23. Stokes, D. L., and N. M. Green. Structure of Ca-ATPase: electron microscopy of frozen-hydrated crystals at 6 A resolution in projection. *J. Mol. Biol.* 213: 529–38, 1990.

24. Tanford, C., J. A. Reynolds, and E. A. Johnson. Sarcoplasmic reticulum calcium pump: a model for Ca^{2+} binding and Ca^{2+} coupled phosphorylation. *Proc. Natl. Acad. Sci. U.S.A.* 84: 7094–8, 1987.

25. Watt, R. M., and J. A. Reynolds. Interaction of apolipoprotein B from human serum low-density lipoprotein with egg yolk phosphatidylcholine. *Biochemistry* 20: 3897–3901, 1981.

26. Wütrich, K. Protein structure determination in solution by NMR spectroscopy. *J. Biol. Chem.* 265: 22059–22062, 1990.

4. Energy transduction mechanisms (animals and plants)

VLADIMIR SKULACHEV | *A. N. Belozersky Institute of Physico-Chemical Biology, Moscow State University, Moscow, Russia*

CHAPTER CONTENTS

Abbreviations. $\Delta\bar{\mu}_{H^+}$ and $\Delta\bar{\mu}_{Na^+}$, transmembrane differences in electrochemical potentials of H$^+$ and NA$^+$ ions, respectively; ΔpH, ΔpK and ΔpNA, transmembrane concentration differences of H$^+$, K$^+$, and Na$^+$ ions, respectively; $\Delta\Psi$, transmembrane electric potential difference; CCCP, m-chlorocarbonlycyanide phenylhydrazone; DCCD, N,N′dicyclohexylcarbodiimide.

TRADITIONALLY, THE CHEMICAL WORK (biosyntheses) AND THE ATP-DRIVEN MECHANICAL WORK occurring in non-membranous parts of the cell are assumed to be beyond the scope of classical bioenergetics. In this chapter we consider the bioenergetic mechanisms responsible for *(1)* the production of convertible energy currencies that is, adenosine triphosphate [ATP] protonic and sodium potentials and face *(2)* the utilization of these currencies to perform membrane-linked work in animal and plant cells. Bacterial systems are mentioned only where necessary to explain eukaryotic mechanisms. Mechanical work is reviewed elsewhere in this volume, and biosynthetic mechanisms are the domain of molecular biologists and biochemists rather than physiologists.

THE ANIMAL CELL

General Pattern of the Energy Transductions

In animal cells, utilization of energy sources, that is, substrates of glycolysis or respiration, results in the formation of ATP or electrochemical differences in H$^+$ potentials ($\Delta\bar{\mu}_{H^+}$) that are used to perform different types of work.

$\Delta\bar{\mu}_{H^+}$ consists of electric and chemical components; the electric potential difference ($\Delta\Psi$) and the pH difference (ΔpH). Figure 4.1 illustrates what we mean by $\Delta\Psi$ and ΔpH. If there is an electric potential difference across the membrane, generated, for example, by a battery, then H$^+$ ions tend to move from the positive to the negative compartment (from left to right in Fig. 4.1*A*). The same tendency appears when the concentration of H$^+$ ions in the left compartment is higher than in the right (Fig. 4.1*B*). This potential energy may be utilized if the membrane has a device capable of coupling the downhill H$^+$ movement to the performance of useful work.

The energy stored in $\Delta\bar{\mu}_{H^+}$ can be calculated from eq. 4.1:

$$\Delta\bar{\mu}_{H^+} = F\,\Delta\Psi + R\,T\,\ln[H^+]_p/[H^+]_n \qquad (4.1)$$

where R is the gas constant, T is the absolute temperature, F is Faraday's constant, $[H^+]_p$ is the molar concentration of H$^+$ ions in a more positive (or more acidic) compartment, and $[H^+]_n$ is the molar concentration in a more negative (or more alkaline) compartment.

$\Delta\bar{\mu}_{H^+}$ has units of J·mole^{-1}. To express it in volts *(V)*, one should divide it by F. For this quantity, Mitchell (119) introduced the term "proton motive force" (Δp) which, at 25°C, may be calculated according to Eq. 4.2:

$$\Delta p = \Delta\bar{\mu}_{H^+}/F = \Delta\Psi - 0.06\Delta pH \qquad (4.2)$$

The difference between $\Delta\Psi$ and ΔpH results because pH is a negative logarithm of H^+ concentration. Indeed, $\Delta\bar{\mu}_{H^+}$ increases when the left compartment in Figure 4.1 becomes more positive or its pH lowers.

According to Eq. 4.2, $\Delta pH = 1$ is equivalent to $\Delta\Psi = 0.06$ V. The same value expressed in $kJ \cdot mole^{-1}$ will be 5.7, and that in $kcal \cdot mole^{-1}$ will be 1.37.

ATP and all the substances capable of forming it via group-transfer enzymatic reactions are defined as "high-energy compounds." The free energy of hydrolysis of the high-energy compounds is not lower than that of ATP; for example, under physiological conditions this value for ATP is 10 $kcal \cdot mole^{-1}$, (whereas under standard conditions it is about 7 $kcal \cdot mole^{-1}$).

It should be stressed that the term "high-energy compound" is of biological rather than chemical significance. For example, creatine phosphate should be regarded as a high-energy compound in vertebrates (possessing creatine phosphate kinase to phosphorylate ADP), but not in those invertebrates in which arginine phosphate kinase substitutes for the above-mentioned enzyme.

There are two ways to hydrolyze ATP:

$$ATP + H_2O \rightarrow ADP + P_i \qquad (4.3)$$

$$ATP + H_2O \rightarrow AMP + PP_i \qquad (4.4)$$

Hydrolysis of terminal phosphate in ATP and formation of ADP and inorganic phosphate (P_i) is shown to take place when the energy-consuming process driven by the ATP hydrolysis requires energy less than or equal to 10 $kcal \cdot mole^{-1}$. This is also the case when the energy requirement is much larger than 10 $kcal \cdot mole^{-1}$. Here a special mechanism (for example, the actomyosin filament) makes it possible to use the energy of many ATP molecules to perform a given function.

If the energy requirement is only slightly higher than 10 $kcal \cdot mole^{-1}$ (that is, by several $kcal \cdot mole^{-1}$), then ATP is hydrolyzed to AMP and inorganic pyrophosphate (PP_i). Under standard conditions, the energy of hydrolysis is almost equal for two anhydride bonds in the ATP molecule. However, in the living cell the energy release is usually several kcal higher when AMP and PP_i are formed instead of ADP and P_i. This is because of a much lower cytosolic concentration of PP_i than that of P_i. Such an effect is the result of PP_i hydrolysis by soluble pyrophosphatase. This may explain why adenosine triphosphate, rather than adenosine diphosphate (also a high-energy compound), is employed as convertible energy currency.

The following major types of useful work are inherent in the animal cell:

1. Chemical work, that is syntheses of biological compounds, transfer of reducing equivalents from positive to negative redox potentials, and electroneutral processes of uphill transport of solutes

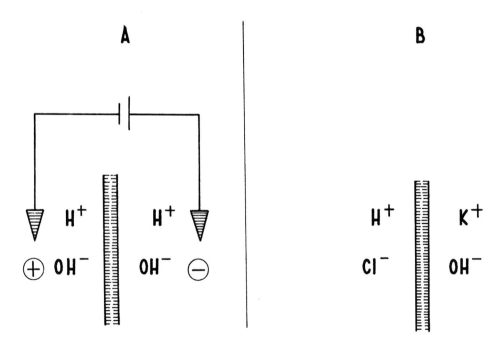

FIG. 4.1. $\Delta\Psi$ (A) and ΔpH (B) between two compartments separated by a membrane (vertical septum).

2. Electrical work, that is electrophoretic movement of proteins and low-molecular compounds across the membranes or alongside them

3. Mechanical work, that is processes of motility supported by ATP hydrolysis

4. Thermoregulatory heat production

Uphill transports of solutes mentioned in *(2)* and *(3)* are conventionally defined as "osmotic work." In animal cells, osmotic work is associated with *(i)* the outer cell membrane (plasma membrane), *(ii)* the inner mitochondrial membrane, and *(iii)* membranes of intracellular secretory granules—endosomes, lysosomes, etc. To support uphill transport, transmembrane electrochemical Na^+ potential differences ($\Delta\bar{\mu}_{Na^+}$ on plasma membrane—or $\Delta\bar{\mu}_{H^+}$ on other membranes of the animal cell) are employed. It is Na^+/K^+-ATPase that produces $\Delta\bar{\mu}_{Na^+}$, whereas $\Delta\bar{\mu}_{H^+}$ is generated by respiration (mitochondria) or ATP hydrolysis (secretory granules, etc.).

The main energy flow in the animal cell can be described by Eq. 4.5 as it was originally proposed by Mitchell (119) within the framework of his famous chemiosmotic hypothesis:

$$\text{Respiration} \rightarrow \Delta\bar{\mu}_{H^+} \rightarrow \text{ATP} \rightarrow \text{work} \quad (4.5)$$

$\Delta\bar{\mu}_{H^+}$ is formed by the respiratory chain, the enzyme sequence catalyzing oxidation of NADH by oxygen. The respiratory chain is localized in the inner mitochondrial membrane. The same membrane contains H^+-ATP-synthase, the enzyme carrying out ATP formation from ADP and inorganic phosphate at the expense of the respiratory chain-produced $\Delta\bar{\mu}_{H^+}$. This process belongs to the category of membrane-linked phosphorylation.

In the animal cell, there are also ATP-forming processes of another type, that is, substrate-level phosphorylations. They proceed in the water phase of the cell and do not require association with membranes. It is in this manner that ATP and GTP (guanosine triphosphate) synthases are coupled to glycolysis and α-ketoglytarate decarboxylation, respectively. Their contribution to the total energy production is small under aerobic conditions, whereas glycolysis appears to be the only energy-yielding mechanism under anaerobiosis.

The energy-transduction highways of the animal cell are summarized in Figures 4.2 and 4.3.

Substrate-Level Phosphorylations

Glycolytic Phosphorylations. The glycolytic chain includes two ATP production mechanisms of the substrate-level type (Fig. 4.4). In the first, called *glycolytic oxidoreduction* (Fig. 4.4, reactions 1–4), the energy transduction

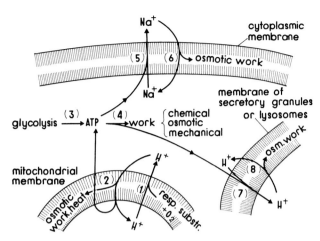

FIG. 4.2. Localization of the energy transduction processes in the animal cell, *(1)* Respiratory chain pumps H^+ from mitochondria. *(2)* H^+ comes back performing chemical work (ATP synthesis) or "osmotic work" (uphill transport of metabolites). *(3)* ATP is formed by glycolysis. *(4)* Chemical, "osmotic," and mechanical work is driven by ATP hydrolysis. *(5)* Na^+ is extruded from the cell by Na^+/K^+-ATPase. *(6)* Na^+ comes into the cell via Na^+, solute-symporters of the outer cell membrane. *(7)* H^+ is pumped to the secretory granules, lysosomes, etc. by H^+-ATPase of vacuolar type. *(8)* The H^+ efflux from these vesicles supports "osmotic work."

is localized at the step of an aldehyde-to-carboxylic acid oxidation, coupled to formation of a high-energy thioester bond (reaction 2). In the preceding step, the semialdehyde compound is formed as a result of combination of the substrate, glyceraldehyde 3-phosphate, and the SH-group of cysteine-149 of glyceraldehyde 3-phosphate dehydrogenase (reaction 1). Semialdehyde oxidation is coupled to reduction of the enzyme-bound NAD^+. The next step represents phosphorolysis of the thioester so that 1,3-diphosphoglycerate is formed (reaction 3). Reaction 3 completes that span of glycolysis which is catalyzed by glyceraldehyde 3-phosphate dehydrogenase. This enzyme is a 146 kD protein composed of four equal subunits. Its concentration in the aqueous phase of the cell is very high (in muscle and yeast, it amounts to about 10% and 20%, respectively, of the total water-soluble proteins). In fact, the enzyme concentration is somewhat higher than that of coenzyme (NAD^+). As to the concentration of the substrate, glyceraldehyde 3-phosphate, it is one or two orders of magnitude lower than the enzyme concentration. Such an unusual relationship might be explained by high chemical reactivity of the aldehyde, which can easily modify amino and SH-groups, among others. It is possible that a very high concentration of the enzyme metabolizing the aldehyde results in immediate conversion of the substrate to product wherever the substrate appears.

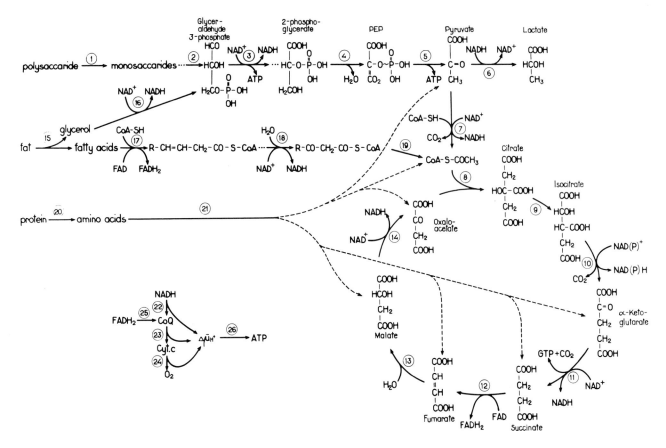

FIG. 4.3. Energy-yielding metabolic pathways of the animal cell. 1–6, glycolysis (3-PGA, 2-PGA and PEP, 3-phosphoglyceraldehyde, 2-phosphoglycerate and phospho*enol*pyruvate, respectively); 7, oxidative decarboxylation of pyruvate; 8–14, the citric acid cycle; 15, lipolysis of the neutral fat; 16, glycerol phosphorylation and oxidation; 17–19, β-oxidation of fatty acid (R, remainder of the fatty acid hydrocarbon chain); 20, proteolysis; 21, conversion of various amino acids to pyruvate, acetyl CoA and the citric acid cycle intermediates; 22–24, the respiratory chain; 25, reduction of CoQ by FADH₂ produced in reactions 12 and 17; 26, ATP formation by H⁺-ATP-synthase. Energy transductions, namely formation of $\Delta\bar{\mu}_{H^+}$ or high-energy ATP precursors takes place at stages 3, 4, 11, 22, 23, and 24.

Glyceraldehyde 3-phosphate dehydrogenase forms complexes with the preceding and the succeeding enzymes of the glycolytic chain—aldolase and 1,3-diphosphoglycerate kinase. The latter catalyzes reaction 4 (see Fig. 4.4), that is, ADP phosphorylation by 1,3-diphosphoglycerate. Glyceraldehyde 3-phosphate dehydrogenase can also bind some ATP-consuming enzymes and, in particular, those involved in protein synthesis. In the nucleus, glyceraldehyde 3-phosphate dehydrogenase was shown to be bound to chromosomes (24).

The enolase reaction (Figs. 4.4, reaction 6) is the second energy transduction step of glycolysis. It consists in dehydration of 2-phosphoglycerate (formed from 3-phosphoglycerate by phosphoglyceromutase, reaction 5) so that phospho*enol*pyruvate is produced. Enolase represents the 88 kD protein composed of two identical subunits. Mg^{2+} or Mn^{2+} serve as cofactors.

To form ATP, phospho*enol*pyruvate phosphorylates ADP, which is catalyzed by phospho*enol*pyruvate kinase (reaction 7). The obtained *enol*pyruvate is spontaneously equilibrated with the keto-form of pyruvate (reaction 8).

Oxidative Decarboxylation-Driven Phosphorylation. Oxidative decarboxylation of α-ketoglutarate, a reaction of the citric acid cycle, is found to be coupled to phosphorylation. The process is shown in Fig. 4.5.

The process starts with the combination of α-ketoglutarate and thiamine pyrophosphate (TPP) (reaction 1). As a result of decarboxylation of the reaction product, a succinyl semialdehyde derivative is formed (reaction 2). Reaction 3 is the energy-yielding step.

Here semialdehyde is oxidized by the enzyme-bound lipoate so that dihydrolipoate and a very short-lived high-energy transient, succinyl $\sim$ TPP, are produced. It is assumed that the TPP amino group is essential to stabilize, to some degree, succinyl $\sim$ TPP. Two reaction products, when formed, immediately react with each other, resulting in appearance of TPP and succinyl lipoate (reaction 4). The latter reacts with CoA, giving succinyl CoA and dihydrolipoate (reaction 5). Reactions 1–4 and reaction 5 are catalyzed by α-ketoglutarate decarboxylase and succinyl CoA transacylase, respectively.

It is succinate thiokinase that is responsible for the ensuing energy transfer from succinyl CoA to GTP (reactions 6–8). At first, phosphorolysis of the enzyme-bound succinyl CoA occurs (reaction 6). Formed succinyl phosphate phosphorylates an imidazole residue of one of the enzyme histidines (reaction 7). After this, phosphoryl is transferred to GDP so that GTP is produced. (For reviews, see refs. 144, 215).

GTP may be utilized by GTP-consuming processes in the mitochondrial matrix where all these events are localized.

It seems remarkable that one of the mitochondrial energy-yielding reactions results primarily in formation of GTP, not ATP. In fact, ATP produced inside the mitochondrion is expelled to the cytosol by the $\Delta\Psi$-driven ATP^{4-}/ADP^{3-} antiporter (see later under $\Delta\bar{\mu}_{H^+}$- and $\Delta\bar{\mu}_{Na^+}$-Driven Transports of Solutes). The antiporter shows very high specificity for adenine nucleotides. This means that energy conserved in the form of GTP cannot be exported from the matrix of intact mitochondria. Thus, in spite of the presence of the antiporter, the mitochondrion always has at its disposal small but available portions of the conserved energy. It should be mentioned, however, that under some special conditions, the ATP/ADP antiporter can be modified in such a way that it forms nonspecific pores that are permeable to low molecular mass solutes and impermeable to macromolecules. Apparently, the pore allows GTP to be released from the matrix and to reach the mitochondrial intermembrane space, where nucleotide diphosphate kinase is localized. The latter can catalyze the phosphoryl transfer from GTP to ADP.

Besides GTP and succinate, α-ketoglutarate oxidative decarboxylation results in reduction of lipoate to dihydrolipoate (reaction 5), which is reoxidized to

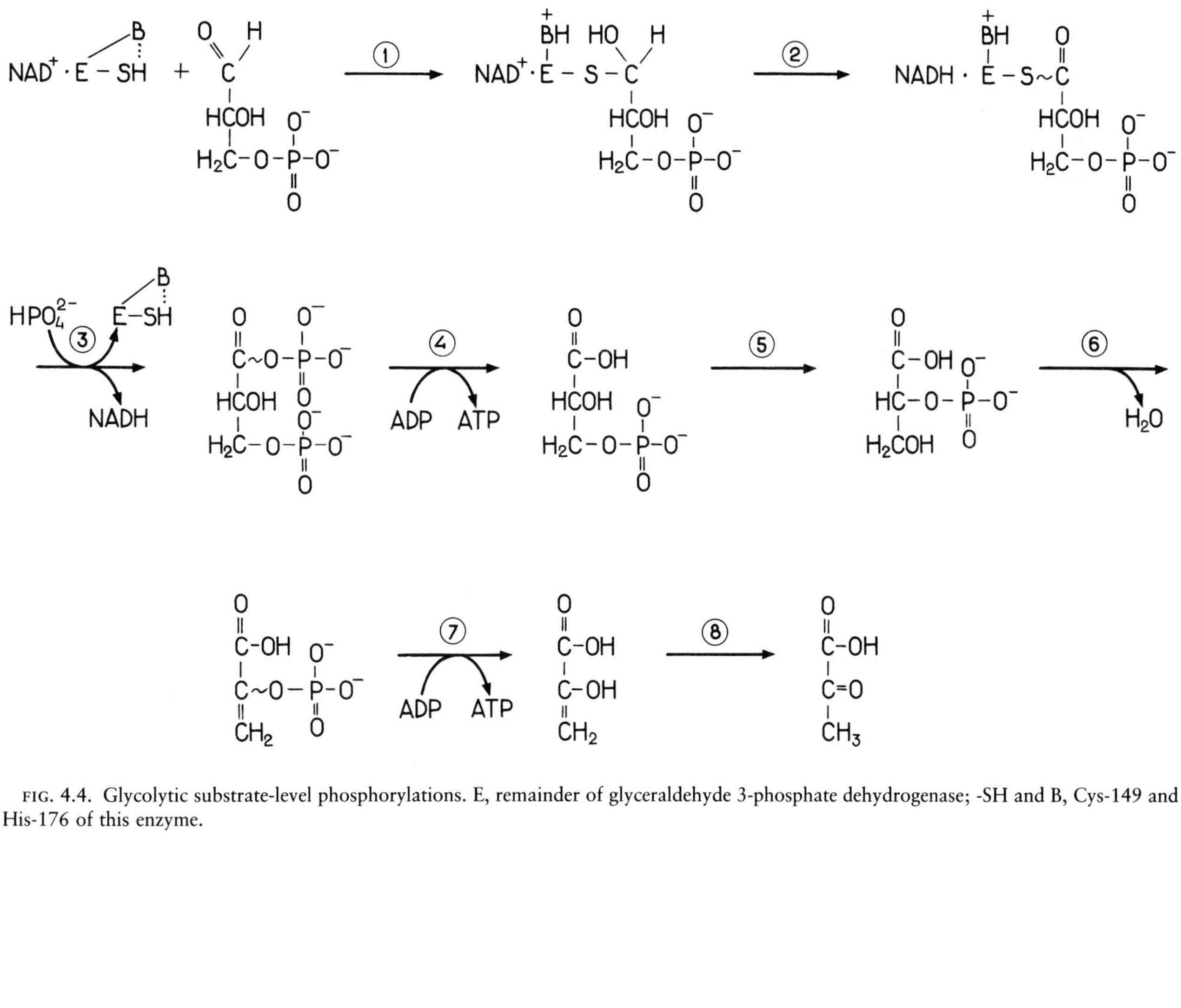

FIG. 4.4. Glycolytic substrate-level phosphorylations. E, remainder of glyceraldehyde 3-phosphate dehydrogenase; -SH and B, Cys-149 and His-176 of this enzyme.

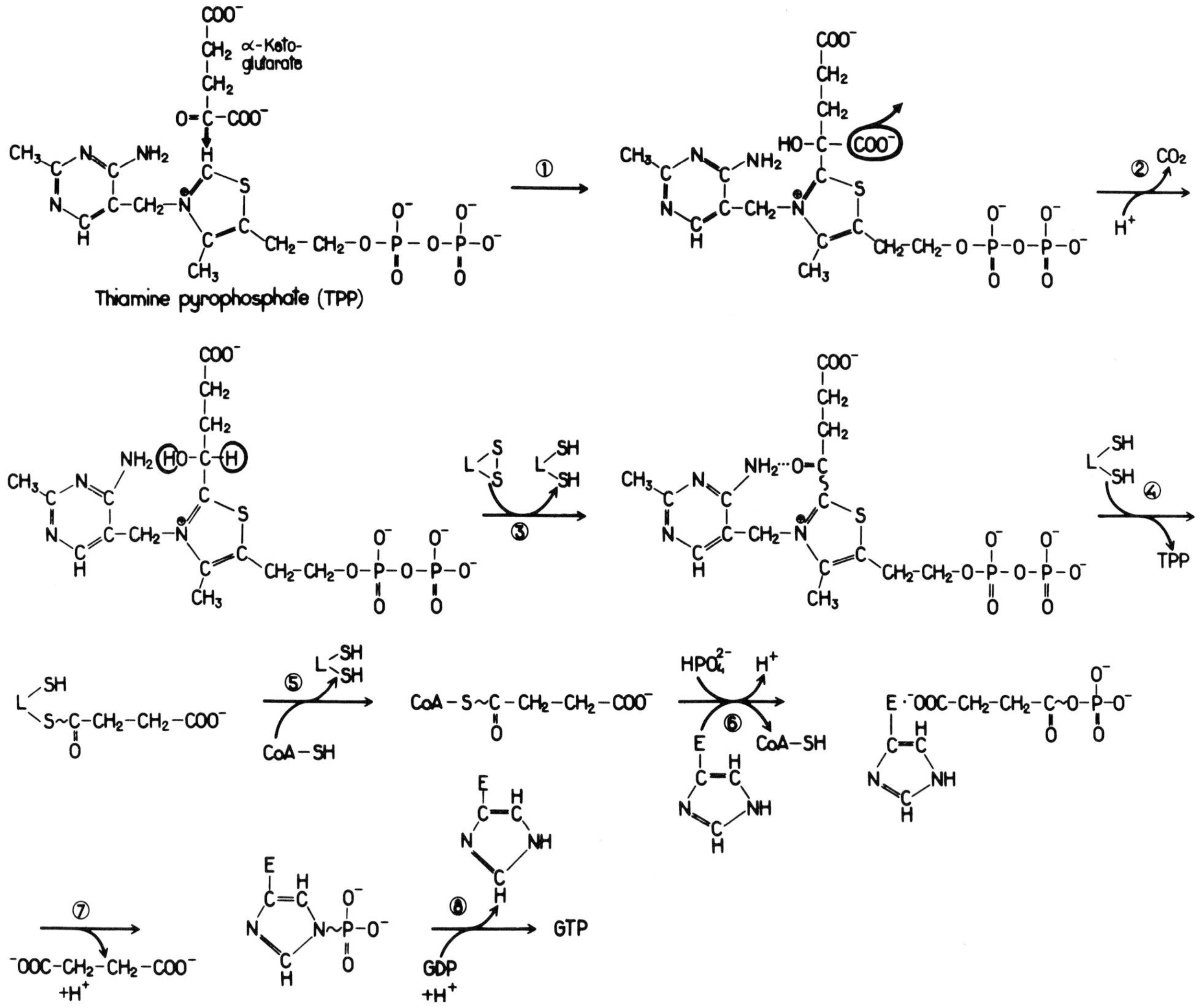

FIG. 4.5. Phosphorylation coupled to the oxidative decarboxylation of α-ketoglutarate. TPT, thiamine pyrophosphate; L, remainder of lipoate; E, succinate thiokinase.

lipoate by dihydrolipoate dehydrogenase, a FAD-containing enzyme, thus reducing NAD^+.

The initial steps of oxidative decarboxylation of pyruvate are quite similar to those for α-ketoglutarate. The main difference between the two processes is that phosphorylation does not occur in the case of pyruvate. Here, acetyl CoA, which is formed instead of succinyl CoA, combines with oxaloacetate to produce citrate (see Fig. 4.3, reaction 8). Moreover, acetyl CoA is used in fatty acid biosynthesis and in some other synthetic processes. It should be noted, that succinyl CoA is sometimes utilized for purposes other than succinate + GTP formation—for example, for heme synthesis.

It has been shown that several enzymes involved in the ketoacid oxidation are organized in a spherical particle, the α-ketoglutarate or pyruvate dehydroge-nase complex, which can be as large as 300 Å in diameter. A single complex contains many copies of each complex-containing enzyme; for example the pyruvate dehydrogenase complex (MW = 4800 kD) is composed of 16 molecules of pyruvate decarboxylase (produces acetyl CoA), 64 copies of transacetylase, and 8 copies of dihydrolipoate dehydrogenase. It is transacetylase that forms the core of the complex.

Respiratory Chain

The simplest scheme for the mitochondrial respiratory chain has already been shown in Figure 4.3, reactions 22–24. In Figure 4.6, a detailed version of the same scheme is presented.

The respiratory chain is composed of three multisubunit complexes, NADH-CoQ reductase (complex I), CoQH$_2$-cytochrome c reductase (bc_1 complex or complex III), and cytochrome oxidase (complex IV). An additional complex, succinate-CoQ reductase (succinate dehydrogenase or complex II) transfers reducing equivalents from succinate to CoQ with no NAD$^+$ involved. This isoenergetic reaction *per se* does not produce energy. Therefore it will not be considered further in this context.

As shown in Figure 4.6, the respiratory chain starts with NADH oxidation, CoQ being the low molecular weight oxidant. In this process, several redox centers of the NADH-CoQ reductase complex are involved, namely, flavine mononucleotide (FMN) and four iron–sulfur components designated FeS$_I$1, FeS$_I$2, FeS$_I$3, and FeS$_I$4. The NADH-CoQ reductase produces CoQH$_2$, which is then oxidized by the Q-cycle system of complex III. The latter includes the FeS cluster of complex III (FeS$_{III}$) and two cytochrome b hemes, one of high potential (b_h) and one of low potential (b_l). As a result of Q-cycle operation, iron–sulfur protein III (FeS$_{III}$) is reduced. From FeS$_{III}$ the electrons are transferred to cytochrome c_1 and then to cytochrome c, the reductant of cytochrome oxidase that includes hemes a and a_3, and three atoms of Cu. Finally, cytochrome oxidase reduces molecular oxygen.

The process of NADH oxidation by oxygen is coupled to the translocation of ten H$^+$ ions from the mitochondrial matrix to the intermembrane space of the mitochondrion. Each of the three main respiratory chain complexes, I, III, and IV, has been shown to be competent in H$^+$ translocation and hence may be regarded as a $\Delta\bar{\mu}_{H^+}$ generator.

Sources of Reducing Equivalents. Reducing equivalents can enter the respiratory chain at different levels, depending upon the redox potential of the oxidized substrate.

If the potential is lower than, equal to, or slightly higher than that of the NADH/NAD$^+$ pair (-0.3 V), then the entire respiratory chain sequence can be involved in the oxidation process. This is the case for the majority of respiratory substrates. NAD$^+$ frequently serves as the immediate oxidant for the substrates, for example: in two of the four dehydrogenations in the citric acid cycle (NAD-linked oxidation of isocitrate and malate), in the second dehydrogenation in the β-oxidation system of fatty acids (oxidation of 3-hydroxy fatty acyl CoA), in glycolytic oxidoreduction (oxidation of 3-phosphoglyceraldehyde), as well as in oxidation of lactate, β-hydroxybutyrate, and some other substrates.

In rare cases, NADP$^+$ appears to be the substrate oxidant instead of NAD$^+$. In principle, NADPH formed may then reduce NAD$^+$ in the transhydrogenase reaction (see later under *H$^+$-Nicotinamide Nucleotide Transhydrogenase*). In energized membranes, however, it does not occur, because transhydrogenase

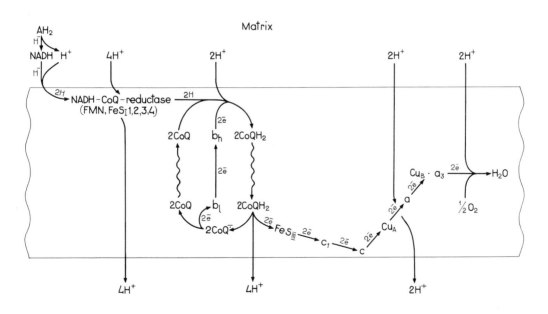

FIG. 4.6. Respiratory chain of mitochondria. AH$_2$, A respiratory substrate; FeS$_I$1,2,3,4, corresponding iron-sulfur centers of NADH-CoQ reductase; b_h and b_l, high- and low-potential hemes of cytochrome b; FeS$_{III}$, iron-sulfur protein of the complex III; c_1, c, a, and a_3, corresponding cytochromes.

is organized as a $\Delta\bar{\mu}_{H^+}$ consumer operating in the NADH $\rightarrow$ NADP$^+$ direction. Therefore, NADPH is usually not oxidized by the respiratory chain, being used instead as a source of reducing equivalents for biosyntheses.

If the redox potential of the oxidation substrate is much more negative than that of the NADH/NAD$^+$ pair, an energy conservation mechanism may be included even before the respiratory chain. This is the case for oxidative decarboxylation of ketoacids that is, α-ketoglutarate and pyruvate (see earlier under Oxidative Decarboxylation-Driven Phosphorylation). Redox potential of α-ketoglutarate/succinate pair is equal to -0.67 V, whereas that of pyruvate/acetyl CoA is equal to -0.7 V.

If the redox potential of the substrate is significantly lower than that of NAD, the reducing equivalents are transferred to the middle or terminal segments of the respiratory chain with no NAD involved. This is true for one of the citric acid cycle substrates, succinate (redox potential, 0.03 V), and for fatty acyl CoA, the substrate of the first oxidoreduction in the β-oxidation system. Succinate, as well as fatty acyl CoA dehydrogenase, feeds the respiratory chain with electrons transferred to the bc_1 complex.

In very rare cases, when the redox potential of the oxidized substrate (for example, ascorbate) is more positive than that of CoQ, reducing equivalents enter the respiratory chain at the cytochrome c level, so only the cytochrome oxidase mechanism proves to be involved in energy transduction.

NADH-CoQ Reductase. Mitochondrial NADH-CoQ reductase (reviewed in refs. 148, 196, 209) comprises approximately 30 different subunits. The molecular mass of the complex is estimated to be 700–900 kD. It contains several redox groups, that is, one FMN

and about 14 iron–sulfur clusters (non-heme irons) of complex I, abbreviated as FeS$_I$.

Two subcomplexes can be distinguished in the NADH-CoQ reductase. One of them is more peripheral and bears FMN and three FeS clusters of redox potential around -0.25 V; namely, one binuclear center (that is, containing 2 Fe atoms), FeS$_I$1, and two tetranuclear centers, FeS$_I$3 and FeS$_I$4. Another (hydrophobic) subcomplex is inlaid into the membrane. It contains a tetranuclear center, FeS$_I$2 (redox potential around -0.1 V). The peripheral subcomplex is responsible for the NADH binding. It is assumed that FMN plays the role of the NADH oxidant. The reaction between NADH and FMN is specifically inhibited by rhein (4,5-dihydroxyanthraquinone-2-carboxylate).

FeS$_I$2 presumably serves as the CoQ reductant. CoQ reduction is specifically inhibited by rotenone, piericidin, barbiturates, and some CoQ analogues. Complex I has been shown to contain two different CoQ binding sites.

Mitochondrial NADH-CoQ-reductase contains six types of mitochondrial DNA-encoded polypeptides, which are synthesized inside mitochondria. Devoid of redox centers, they play a role in the self-assembly of the complex by being incorporated into the membrane before other polypeptides of the complex, which are nuclear DNA-encoded and synthesized in the cytoplasm. One of the mitochondrion-synthesized subunits binds DCCD and might be, therefore, involved in the H$^+$ pumping. Electron microscopy of the membrane crystals of mitochondrial NADH-CoQ reductase showed that the NADH-binding peripheral subcomplex protrudes from the membrane into the matrix space (208) (Fig. 4.7).

The assumption that the NADH–CoQ reductase complex is competent in $\Delta\bar{\mu}_{H^+}$ generation initially advanced by Mitchell (120), is based on the fact

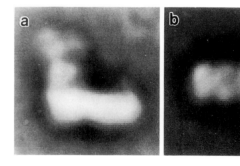
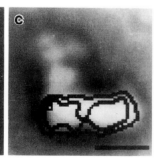

FIG. 4.7 Electron microscopic analysis of single NADH-CoQ reductase molecule from *Neurospora crassa* and its hydrophobic membranous fraction. *(a)* The average particle of the whole enzyme; *(b)* an average of the hydrophobic fraction; *(c)* the averaged particle of the whole enzyme superimposed by the averaged particle of the hydrophobic fraction. Bar, 20 nm. It is assumed that the hydrophobic subcomplex is plugged through the membrane whereas the hydrophilic subcomplex is protruded to the mitochrondrial matrix space [From Weiss et al. (208) with permission].

that electron transfer via this segment of the respiratory chain can be coupled to ADP phosphorylation. In particular, Schatz and Racker (166) showed phosphorylation coupled to NADH oxidation by CoQ_1 in submitochondrial vesicles. In our group, it was found that oxidation of NADH by fumarate generates $\Delta\Psi$ across the membrane of submitochondrial vesicles (177).

In the NADH–CoQ reductase proteoliposomes, H^+ pumping coupled to NADH oxidation has been shown (150). If H^+-ATP synthase were present in the same proteoliposomes, coupled ATP formation could be demonstrated.

In early studies, it was found that the transfer of one electron from NADH to CoQ is coupled to the transmembrane movement of one proton. In more recent studies, an $H^+/\bar{e}$ value equal to 2 was reported (25).

The mechanism of $\Delta\bar{\mu}_{H^+}$ generation by NADH–CoQ reductase remains obscure (for discussion see refs. 95, 149, 179, 208, 209.)

CoQH$_2$–Cytochrome c Reductase. NADH–CoQ reductase converts CoQ to $CoQH_2$ which donates electrons to the next respiratory chain segment called $CoQH_2$–cytochrome c reductase (the bc_1 complex, complex III, reviewed in refs. 179, 194). The purified enzyme complex has been reconstituted with phospholipids to form proteoliposomes, which proved to be competent in the formation of $\Delta\bar{\mu}_{H^+}$ coupled to electron transfer from synthetic quinol derivatives to cytochrome c. If proteoliposomes also contained H^+-ATP synthase, coupled ATP formation could be shown (34).

Complex III contains four redox centers: hemes b_l and b_h associated with cytochrome b, a binuclear iron-sulfur cluster (FeS_{III}) bound to the corresponding apoprotein, and heme c combined with cytochrome c_1 apoprotein. The complex forms dimers which protrude from both sides of the membrane (Fig. 4.8).

The polypeptide composition of the complex from beef heart mitochondria is shown in Table 4.1. A similar polypeptide pattern was revealed when yeast and *Neurospora* mitochondria were analyzed, although some variations in molecular masses of subunits were detected. A comparison with the next respiratory chain complex, cytochrome oxidase, showed that they probably contain no subunits in common.

Besides the redox carriers, the mitochondrial bc_1 complex contains eight polypeptides without prosthetic groups (Table 4.1). One of them (subunit VI) binds CoQ. It consists of 110 amino acid residues and a high percentage (about 50) of α-helical structures.

There are some indications that subunits VIII and X form a subcomplex with cytochrome c_1. One of these

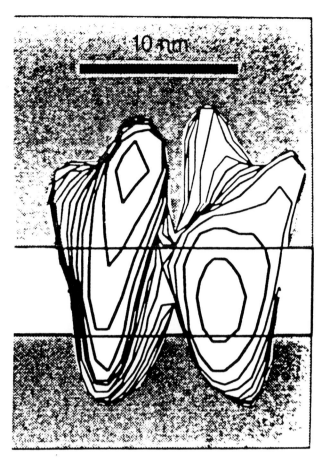

FIG. 4.8. Projections of $CoQH_2$-cytochrome c reductase reconstructed by means of an electron microscopic study of membrane crystals. The white horizontal section represents the membrane bilayer, the upper dark area the matrix space of mitochondria and the lower dark area the intermembrane space [From Leonard et al. (105) with permission].

polypeptides was shown to facilitate the binding of cytochrome c_1 to cytochrome c. In this subunit, 27% of the total amino acid residues are glutamic acid and glutamine. An acidic cluster of eight consecutive glutamic acid residues is located near the N-terminus. One may speculate that this negatively charged sequence interacts with cytochrome c, which contains many positively charged amino acid residues. It was suggested that the smallest (XI) subunit takes part in the binding of the FeS_{III} protein.

As for the other polypeptides possessing no redox centers, their functions remain obscure. Some of them might participate in the organization of proton, rather than electron, transfer pathways, or in the arrangement of the complex in the membrane. The latter role was tentatively attributed to the largest subunits I and II called "core proteins." Recently however, it was found that at least core protein I is incorporated into the bc_1

TABLE 4.1. *Subunit Composition of the* b-c$_1$ *Complex from Beef Heart Mitochondria*

Subunit	Molecular mass (kD)	Designation	Function	Reference
I	49.5	Core protein I	Unknown	(113)
II	47	Core protein II	Unknown	(113)
III	44	Cytochrome *b*	Redox	(163)
IV	28	Cytochrome c_1	Redox	(200)
V	21.5	FeS$_{III}$ protein	Redox	(69)
VI	13.5	Q$_{III}$ protein	CoQ binding	(165)
VII	9.5	—	Unknown	(165)
VIII	9	—	Cyt. c_1 binding?	(165)
IX	8	—	Subunit VI binding?	(165)
X	7	—	Cyt. c_1 binding?	(165)
XI	6.5	—	FeS protein binding?	(165)

complex during its biogenesis much later than cytochromes *b* and c_1, thus providing an indication against the organizing function of the core protein in the assembly of $CoQH_2$–cytochrome *c* reductase. In any case, core proteins are in some way essential for the functioning of bc_1 complex since their removal results in the loss of redox activity, which can be reactivated by adding core proteins. It was found that the bc_1 complex forms the dimers shown in Fig. 4.8. Each 15 nm–long monomer is oriented perpendicular to the plane of the membrane, the largest (7 nm) part being exposed to the matrix aqueous phase, while the smallest (3 nm) part, is exposed to the intermembrane water. The middle 5 nm part is immersed into the membrane, in contact with the other monomer in the dimer.

We now consider subunits that are equipped with redox groups, beginning with cytochrome *b*.

Cytochrome *b* is the only complex III polypeptide encoded by mitochondrial DNA. It contains nine hydrophobic regions long enough to form transmembrane α-helical columns. However, only eight of them are actually oriented across the membrane, whereas one α helix is arranged alongside.

The apocytochrome *b* binds hemes b_h and b_l in such a way that the planes of *b* hemes are oriented perpendicular to the membrane surface. Each heme is bound to apoprotein by at least three bonds. Two of them are formed by histidine imidazoles and the heme iron, and the third by a positively charged arginine or lysine residues of the protein and a negatively charged propionate residue of the heme. The distance between the hemes is about 20 Å (212).

FeS$_{III}$ protein (Slater's factor, BAL-sensitive factor, or Rieske's protein) contains one 2Fe-2S cluster as the redox center. FeS$_{III}$ protein (molecular mass 21.5 kD) contains 199 amino acids. It is characterized by a small N-terminal hydrophilic part (29 residues), a hydrophobic α-helical column (25 residues) and a C-terminal domain that is relatively hydrophilic (145 residues).

The FeS$_{III}$ cluster is formed by two irons, two S^{2-} residues, and four cysteine residues. Fortunately, in the FeS$_{III}$ polypeptide there are only four cysteins (positions 142, 147, 161, and 163), so it is clear that just this part of the amino acid sequence is involved in the construction of the redox center (69).

Cytochrome c_1 is a single polypeptide composed of 241 amino acids. It contains one heme *c* covalently attached to cysteine residues 37 and 40. Together with some other subunit(s) of complex III, cytochrome c_1 organizes the binding site for cytochrome *c*, which has been shown to interact with complex III by the same region with which it interacts with cytochrome oxidase. This fact implies movement of cytochrome *c* during its functioning.

There is no doubt that the cytochrome *c*-binding domain of cytochrome c_1 is localized on the outer surface of the inner mitochondrial membrane. The protein may be anchored in the membrane by its single hydrophobic segment localized close to the C-terminus. This segment can be detached from the heme-containing hydrophilic part of cytochrome c_1 by means of limited proteolysis.

The distance from the heme to the surface of the protein globule in isolated cytochrome c_1 is about 1 nm, whereas such a distance for the FeS cluster in the FeS$_{III}$ protein is 1.9 nm (for review, see ref. 211).

There is an indication that the above-described complexity of structure of $CoQH_2$–cytochrome *c* reductase is not the obligatory property of this $\Delta\bar{\mu}_{H^+}$ generator. In *Paracoccus denitrificans*, the same enzyme competent in energy transduction was found to contain only three polypeptides, namely 62 kD (cytochrome c_1),

39 kD (two-heme cytochrome b), and 20 kD (FeS protein) (214).

A functional scheme for the $CoQH_2$–cytochrome c reductase complex is shown in Figure 4.6. There are several independent pieces of evidence confirming the cyclic operation of CoQ, postulated by Mitchell (121), including the action of inhibitors. Duel and Thorn (53) reported in 1962 on the need to combine the two different inhibitors, BAL (British Anti-Lewisite, the name for 2,3-dimercaptopropanol) and antymicin A, to block the reduction of mitochondrial cytochrome b. As found later, this is due to the existence of two distinct pathways for electrons to cytochrome b. One of them, sensitive to antymicin, as well as to other inhibitors such as funiculosin and 2-n-heptyl-4-hydroxyquinoline N-oxide, was shown to be the result of the interaction between CoQ and heme b_h, which, according to Konstantinov's group (101), is localized closer to the matrix side of the mitochondrial membrane. The other pathway is apparently localized closer to the outer membrane side. It leads to heme b_l, requires FeS_{III}, and is inhibited by BAL, which destroys the FeS_{III} cluster in the presence of oxygen. The same effect can be achieved by removing the FeS_{III} protein from the membrane. Besides BAL, a group of inhibitors affect electron transfer between $CoQH_2$ and FeS_{III}, namely, myxothiazol, 5-n-undecyl-6-hydroxy-2, 7-dioxobenzothiazol (UHDBT), and mucidin (for review, see ref. 211). An inhibiting effect of low concentrations of Zn^{2+} seems to be directed toward FeS_{III} (183).

The redox potentials of the electron carriers involved are compatible with the Q-cycle scheme, that is, -0.04 V for heme b_l, $+0.04$ V for heme b_h, about $+0.28$ V for FeS_{III}, and $+0.22$ V and $+0.25$ V for cytochromes c_1 and c, respectively. Thus, electrogenic stages of the electron transfer directed across the membrane (from b_l to b_h), being partially responsible for $\Delta\Psi$ formation by complex III, proved to be energy-yielding. Moreover, energy is accumulated at those stages of the Q-cycle when the electrogenic transfer of protons occurs.

At the same time, electron transfer from FeS_{III} to cytochrome c_1 and further to cytochrome c, which is postulated to be directed along the membrane, seems to occur without any significant changes in the redox potential and, hence, in the free energy.

One of requirements of the Q-cycle scheme shown in Figure 4.6 is the existence of the stable semiquinone anion ($CoQ^{\cdot-}$) which is formed on FeS_{III} and then is oxidized by heme b_l. This form of CoQ has been described (94). There is some evidence that one of the small subunits of the $CoQH_2$– cytochrome c-reductase has a role to play in stabilizing $CoQ^{\cdot-}$.

Summing up the data on $\Delta\bar{\mu}_{H^+}$ generation by the bc_1 complex, we should emphasize that in spite of many uncertainties in the details of this mechanism, one essential point is clear: The transfer of one reducing equivalent from $CoQH_2$ to cytochrome c results in the translocation of one net negative charge across the membrane, the process being accompanied by the release of two H^+ ions on the outer membrane surface (Fig. 4.6).

Cytochrome Oxidase. Cytochrome oxidase catalyzes the oxidation of reduced cytochrome c by molecular oxygen. The reaction is coupled to the formation of $\Delta\bar{\mu}_{H^+}$, as shown both in intact and in reconstituted membrane systems (for reviews, see 9, 10, 211).

The cytochrome oxidase reductant cytochrome c represents a very stable, small, heme-containing protein composed of 104 amino acids and heme c. The heme is covalently bound to the apoprotein (via SH groups of Cys-14 and Cys-17), a feature distinguishing c-type cytochromes from those of a- and b-types, where noncovalent bonds are involved in the heme binding.

The fifth and sixth ligands of the iron are formed by the His-18 imidazole and the Met-80 sulfur. Both of these amino acid residues are invariant in all the known cytochromes c, while some others may vary in different kinds of organisms. For instance, there are no differences in the amino acid sequence of cytochrome c from man and chimpanzee; between man and horse there are 12 differences, and between man and *Neurospora*, 44 differences. The three-dimensional structure of cytochrome c is known from X-ray studies (0.2 nm resolution; Fig. 4.9).

Cytochrome c is found to combine with the cytochrome c_1 region of $CoQH_2$-cytochrome c reductase or, alternatively, with subunit II of cytochrome oxidase. In these interactions, a lysine-rich part (Lys-13, 72, 86, possibly 79, and 27) near the solvent-accessible heme edge is involved. In subunit II, Asp-158, 112 and Glu-198, 114 carboxyl anions seem to form salt bridges with cationic amino groups of the above-mentioned lysines of cytochrome c.

It is generally agreed that reductants and oxidants attack that edge of the cytochrome c heme which is exposed to water. It is also assumed that cytochrome c shuttles between cytochrome c_1 and cytochrome c oxidase by means of lateral diffusion along the membrane surface. In $CoQH_2$–cytochrome c reductase proteoliposomes, cardiolipin was shown to be specifically involved in cytochrome c binding to membrane surfaces.

Mitochondrial cytochrome oxidase contains four redox centers, that is, two hemes of the a-type differing in their properties (a and a_3), and two copper atoms (Cu_A and Cu_B).

Mammalian cytochrome oxidase ($\sim$200 kD) con-

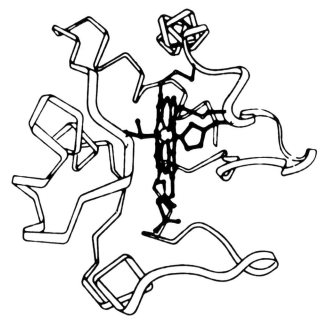

FIG. 4.9. Structure of cyochrome c from tuna muscle mitochondria. Heme and its ligands are shown by black lines [From Salamma (160) with permission].

tains three major subunits I, II, and III of 57, 26, and 30 kD, respectively, coded by the mitochondrial DNA, and at least nine polypeptides of lower molecular mass coded by nuclear DNA: IV (17 kD), V (12.5 kD), VIa (10.5 kD), VIb (9.5 kD), VIc (8.5 kD), VII (10 kD), VIIIa (5.5 kD), VIIIb (5 kD), and VIIIc (6 kD) (82, 212). The enzyme forms dimers resembling in this respect NADH–CoQ reductase and $CoQH_2$–cytochrome reductase.

X-ray data at 2.8Å resolution obtained by Yoshikawa and coworkers (195, 195a) are shown in Figure 4.10. Data on the primary structure of all the cytochrome c oxidase subunits are available (reviewed in ref. 213). The alternations of hydrophilic and hydrophobic segments are consistent with the assumption that subunits I, II, and III form 12, 2, and 7 α-helical transmembrane columns, respectively.

It is generally accepted that all the redox centers of cytochrome oxidase are localized in subunits I and II. Among them, only two Cu_A are in subunit II. This subunit, near the C-terminus, has a domain homologous to the Cu-binding site of other copper proteins (213).

It was revealed that hemes are oriented perpendicular to the plane of the membrane. There are indications that heme a is localized closer to the outer surface of the membrane (212), whereas heme a_3 was found to be immersed to the membrane core. The distance between the two hemes proved to be about 1.4 nm. It is about 1 nm when projected normal to the plane of the membrane.

The arrangement of heme a_3 is very similar to that of hemoglobins and myoglobins. The fifth ligand is formed by the histidine imidazole, while O_2 may serve as the sixth ligand. Cu_B is shown to reside within 0.3–0.5 nm from the heme a_3 iron (212). It was suggested that Cu and Fe of heme a_3 can be connected by a common ligand that can be replaced by O_2.

Subunit III, like subunit I, is very hydrophobic. Subunits of a smaller molecular mass, encoded by a nuclear genome, are relatively hydrophilic.

According to the data of Kuhn and Kadenbach (100), the mitochondrion-encoded subunits I–III are identical in all rat tissues studied. At the same time, all the nuclear-encoded subunits showed immunological differences between two or more tissues; moreover, differences were observed between these subunits isolated from fetal and adult tissues (54). These observations, as well as the fact that bacterial cytochrome oxidases contain only three polypeptides corresponding to mitochondrial subunits I, II, and III, may indicate that nuclear-encoded subunits are necessary for the regulation of cytochrome oxidase in the eukaryotic cell rather than for its catalytic and energy-transducing functions (81). A direct influence of ATP, ADP, and phosphate on activity of cytochrome oxidase reconstituted into proteoliposomes has been reported (75, 76, 111). Such an effect was absent from the P. denitrificans enzyme, being associated with small subunits of the mammalian cytochrome oxidase (75). Stoichiometric binding of ATP to cytochrome oxidase was reported recently (153).

It is interesting that mammalian cytochrome oxidase is less active than those from bacteria. Dissociation of some nuclear-coded subunits increases the activity to a value similar to that of the bacterial enzymes (83). This illustrates a principle that the role of enzymes is not only to reach the maximal rate of the catalyzed reaction but also to allow the activity to be controlled by a set of regulatory mechanisms (175).

It is usually assumed that an electron removed from cytochrome c is immediately transferred to Cu_A. This cytochrome oxidase redox center occupies the closest position to the enzyme surface exposed to the intermembrane space. Moving in the direction from the surface to the membrane interior, the electron is proposed to be translocated from Cu_A to heme a and further to the heme a_3-Cu_B complex, the terminal component of respiratory chain that reduces O_2 (see Fig. 4.6).

It is suggested that the O_2 molecule, when it is bound to cytochrome oxidase and reduced, forms a peroxy-intermediate where a —O—O— group plays the role

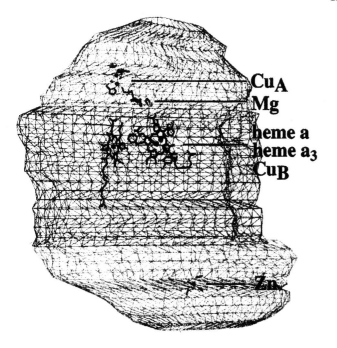

FIG. 4.10. The X-ray structure of bovine cytochrome c oxidase at 2.8 Å resolution. Schematic representation of metal and heme location [From Tsukihara et al. (195) with permission.]

of the ligand between the heme a_3 iron and Cu_B (10, 213).

In the original presentation of the chemiosmotic hypothesis, Mitchell (119) assumed that membrane charging by redox $\Delta\bar{\mu}_{H^+}$ generators is a result of the transmembrane movement of electrons (the loop scheme). Later, however, it appeared that the loop scheme proves insufficient to account for all the energy-transducing events in this enzyme. In particular, it does not explain an observation of Wikström (210) that the oxidation of an electron carrier, for example, ferrocyanide or reduced cytochrome c, by mitochondria (or cytochrome oxidase proteoliposomes), results in a transient acidification of the incubation medium. It has also been shown in proteoliposome experiments that the released H^+ ions originate from the intravesicular water space since the effect could be demonstrated only if there was a pH buffer inside proteoliposomes. The addition of a protonophore completely arrested the acidification (212, 213).

This qualitative experiment was also supported by a quantitative estimation of the number of protons released from vesicles by each electron donor oxidized by cytochrome c oxidase. In mitochondria, an $H^+/\bar{e}$ ratio equal to 1 was found. In inside-out submitochondrial vesicles, alkalization was shown to be induced by a cytochrome oxidase reaction, the $H^+/\bar{e}$ ratio being

close to 2. Both in right-side-out and inside-out vesicles, cytochrome oxidase proved to be operating in an electrogenic fashion, as if two charges were transported across the membrane for each electron transferred from cytochrome c to oxygen (212).

All these relationships can be explained by the scheme shown in Figure 4.6. As one can appreciate from this scheme, the oxidation of one cytochrome c molecule by oxygen causes a transmembrane movement of two electric charges due to cooperation of the following processes: *(1)* transfer of one electron over a half-membrane distance (from cytochrome c to a_3); *(2)* transfer of one proton over a half-membrane distance (from the matrix to heme a_3); *(3)* transfer of one more proton all through the entire membrane thickness (from the matrix to the intermembrane space). This system is usually called a "cytochrome oxidase proton pump."

Any scheme for the cytochrome oxidase mechanism requires H^+-conducting paths crossing a significant part of the hydrophobic barrier. The first indication of this kind was obtained by the Konstantinov group in our laboratory (8). It was revealed that a pH dependence of the redox potential is inherent not only in cytochrome a_3 (which transfers electrons to O_2 with subsequent addition of H^+ ions to form H_2O), but also in cytochrome a. It was shown that the cytochrome a redox potential can respond to pH changes in the matrix space. In other experiments, the pH dependence of the cyanide binding to heme a_3 was studied. It was shown that a $\Delta\Psi$ imposed across the membrane changes this binding in such a way as if heme a_3 were localized in the middle part of the membrane, being connected with the matrix by a "proton well" (4).

The exact mechanism of the cytochrome oxidase proton pump remains obscure. Here the situation resembles that in NADH–CoQ reductase and is in contrast to the $CoQH_2$–cytochrome c reductase, where the Q-cycle scheme seems to be sufficient to account for all available information.

There have been several attempts to explain both NADH–CoQ reductase and cytochrome oxidase mechanisms in terms of direct coupling when H^+ pumping appears to be the inevitable consequence of the chemical reaction (for example, addition of two electrons to CoQ inevitably results in binding of two protons, etc.— see Fig. 4.6). To explain why two H^+ are translocated across the membrane for one electron, cyclic schemes of the electron transfer were postulated for initial and terminal spans of the respiratory chain.

Alternative mechanisms assume that electron transfer and H^+ pumping are coupled indirectly (conformationally) like that in H^+-nicotinamide nucleotide transhydrogenase, where H^+ ions pumped through the membrane are not identical to those involved in the oxidoreduction (see later under H^+-Nicotinamide Nucleotide Transhydrogenase). In the direct coupling mechanism, the $H^+/\bar{e}$ ratio is fixed, having stoichiometric characteristics of the chemical process catalyzed by a given enzyme. As to the indirect mechanism, here the $H^+/\bar{e}$ ratio may be related to the number of H^+ ions transported by the enzyme, and is hence under some form of control (for reviews, see refs. 10, 104, 122, 179). In this context, it should be mentioned that the electron flow in cytochrome oxidase can be decoupled from the pump mechanism. Such an effect results in decrease in $H^+/\bar{e}$ ratio from 2 to 1. This seems to happen at certain critical levels of $\Delta\bar{\mu}_{H^+}$ and is related to the electron flow rate (31, 140).

Similar decreases in the $H^+/\bar{e}$ stoichiometry may also occur in the NADH–CoQ reductase span. Therefore, a $H^+/\bar{e}$ ratio equal to 5 for the entire respiratory chain (NADH oxidation by oxygen) should be regarded as the maximal value, which may lower under certain conditions.

Estimations of exact $H^+/\bar{e}$ value is difficult due to recirculation of the pumped H^+ ions. It appears to be easier to measure another parameter related to $H^+/\bar{e}$, namely the ADP/O ratio of the oxidative phosphorylation process. In this case, $\Delta\bar{\mu}_{H^+}$ generated by respiration is utilized by H^+-ATP synthase to form ATP from ADP and phosphate. Measuring oxygen consumption induced by addition of a known amount of ADP to respiring mitochondria, it is possible to estimate the ADP/O ratio. In some cases (for example, in mitochondria isolated from 14- to 21-day-old chicks), respiration without ADP may be as low as 1% of that in the presence of ADP (193). If the ADP stimulation of respiratory rate is not strong, the phosphorylating efficiency of respiration may be estimated by measuring the P/O ratio. To this end, the incubation mixture with mitochondria is supplemented with glucose and hexokinase, which utilize the formed ATP to produce glucose 6-phosphate. The latter cannot be metabolized by isolated mitochondria.

ADP/O and P/O ratios depend first upon the number of H^+ ions which are (1) exported from mitochondria by the respiratory chain for each atom of consumed oxygen and (2) imported to mitochondria for each ATP synthesized by H^+-ATP synthase. At least one more parameter should also be taken into account, namely import of ADP and phosphate and export of ATP, since it is known that H^+-ATP synthase uses intramitochondrial ADP and P_i to form intramitochondrial ATP. ADP import and ATP export are shown to occur electrophoretically, being catalyzed by the

ATP^{4-}/ADP^{3-} antiporter, whereas phosphate is taken up by mitochondria in an electroneutral, ΔpH-driven fashion by the $H_2PO_4^-$, H^+-symporter (see later under $\Delta\bar{\mu}_{H^+}$- and $\Delta\bar{\mu}_{Na^+}$-Driven Transports of Solutes). This means that exchange of extramitochondrial ADP and phosphate for intramitochondrial ATP is coupled with electrophoretic uptake of one H^+ ion.

The maximal ADP/O and P/O ratios measured in succinate-oxidizing mitochondria proved to be equal to 2, which can be explained assuming that (1) six H^+ ions are expelled from mitochondria when two electrons are transferred from succinate to oxygen (see Fig. 4.6), (2) two H^+ ions per one ATP enter the mitochondrial matrix via H^+-ATP synthase and (3) one H^+ ion is taken up due to cooperation of the $ATP^{4-}_{in}/ADP^{3-}_{out}$ antiporter and the $(H_2PO_4^-)_{out}$, H^+_{out} symporter.

Following this logic, one may predict that the maximal ADP/O ratios for oxidation of NADH and ascorbate should be equal to 3.3 and 1.3, respectively, since ten H^+ ions are exported by the entire respiratory chain and four H^+ ions by its cytochrome oxidase span. These values are close to the highest ones observed experimentally (193). Lower values, which are often reported in the literature, may result from (1) increase in the passive H^+ conductance of the membrane (H^+ leakage and uncoupling), (2) decrease in the $H^+/\bar{e}$ ratio in NADH–CoQ reductase and/or cytochrome oxidase (slips of the H^+ pumps or decoupling), or (c) activation of electron transfer(s) bypassing the energy-coupling site(s) of the respiratory chain.

H^+-ATP-Synthase and H^+-ATPases

H^+-ATP synthase is the enzyme catalyzing $\Delta\bar{\mu}_{H^+}$-driven ADP phosphorylation by inorganic phosphate in mitochondria, chloroplasts, and respiring or photosynthetic bacteria.

The H^+-ATP synthase complex can be dissociated into two subcomplexes, one integrated into the membrane and responsible for H^+ transport (factor F_0), and the other peripheral and loosely bound to the membrane and catalyzing ATP synthesis and hydrolysis (factor F_1).

Mitochondrial H^+-ATP synthase is composed of thirteen types of subunits. Among them there are the F_1 subcomplex (five types of subunits), the F_0 complex (four types of subunits), and four additional subunits (169, 202).

The F_1 complex has six major subunits (3α and 3β) and three minor ones (γ, δ, and ϵ). The major subunits resemble their bacterial and chloroplast analogues in this respect: It was found that the β-subunit from yeast mitochondria exhibits a greater than 70% conservation of protein sequence when compared with E. coli, beef heart mitochondria, and chloroplasts. A similar situation was revealed in the α-subunit.

A comparison of minor F_1 subunits from E. coli and animal mitochondria carried out by Walker et al. (203, 204) showed that δ- and ϵ-subunits are different, while the γ-subunits are similar. Taking the sequence analysis data into account, the authors concluded that the mitochondrial δ-subunit is homologous to the E. coli ϵ-subunit.

The so-called oligomycin sensitivity-conferring protein (OSCP), a 21 kD subunit found in mitochondria, shows sequence homologies with the δ- and b-subunits of the E. coli enzyme.

The hydrophobic multicopy subunit 9 of mitochondrial factor F_0 (Beechey proteolipid, or DCCD-binding subunit) is homologous to subunit c of E. coli. Each F_0 is probably composed of twelve copies of subunit 9. Sequences resembling those in the E. coli subunit a were found in mitochondrial 22 kD subunit 6 (factor F_2 and factor B).

Five subunits in mitochondrial H^+-ATP synthase are absent from E. coli: (1) 5.5 kD ϵ-subunit of F_1; (2) 8 kD subunit A6L of F_0; (3) 9 kD factor F_6; (4) 9.5 kD protein inhibitor of F_1 ATPase activity; (5) 18.5 kD subunit d which, together with F_6 and OSCP, is required for correct binding of F_1 and F_0 (179, 202).

A subunit composition of mitochondrial F_0F_1 is shown in Table 4.2.

In animal cells and yeast, all the F_1 subunits are encoded by the nuclear genome and synthesized in the cytoplasm as somewhat larger precursors. In the DNA of animal mitochondria, there are two overlapping genes encoding subunits 6 and A6L (201).

Subunits 6, 9, and A6L are all proteolipids. They are soluble in $CHCl_3$:methanol (2:1) and can be extracted with this solvent from mitochondrial H^+-ATP synthase.

The H^+-ATP synthase complex is so large that it protudes from the membrane surface for long distances. The protruding part (factor F_1) faces the matrix space of mitochondria (65, 146).

Negative staining of inside-out submitochondrial vesicles reveals knobs 9 nm in diameter covering the entire outer membrane surface (Fig. 4.11). Racker (46) showed that the knobs vanished after mechanical agitation, accompanied by the disappearance of ATPase and ATP synthase activity of the vesicles and the appearance of 9 nm spherical particles and ATPase in the supernatant. Reconstitution of these spheres with vesicles resulted in the reappearance of knobs and ATPase synthase activity in vesicles. From these observations, Racker concluded that knobs represent the catalytic part of the H^+-ATP synthase, that is factor F_1.

Experiments showed that some parts of H^+-ATP synthase, other than F_1, are still in the membrane

TABLE 4.2. *Comparison of the Subunit Composition of the* E. coli *and Animal Mitochondrial* H^+-ATP Synthases *

Complex	E. coli (kD)	Animal mitochondria (kD)
F_1	α (55) $\longleftrightarrow$	α (55)
	β (50) $\longleftrightarrow$	β (51)
	γ (31.5) $\longleftrightarrow$	γ (30
	δ (19.5)	
	ϵ (15) $\longleftrightarrow$	δ (21)
	No analogue	ϵ (5.5)
F_0	a (30)	Subunit 6 = F_2 = F_B = a (22)
	b (17)	b = F_0I = PVP (24.5)
	c (8) $\longleftrightarrow$	Subunit 9 = DCCD-binding proteolipid = c (7.5)
	No analogue	A6L (8)
		OSCP (21)
Other subunits	No analogue	F_6 (9)
	No analogue	d (18.5)
	No analogue	Inhibitor protein (9.5)

* Arrows show the homologies in amino acid sequences.
Source: Ovchinnikov et al. (139); Walker et al. (201, 202, 204).

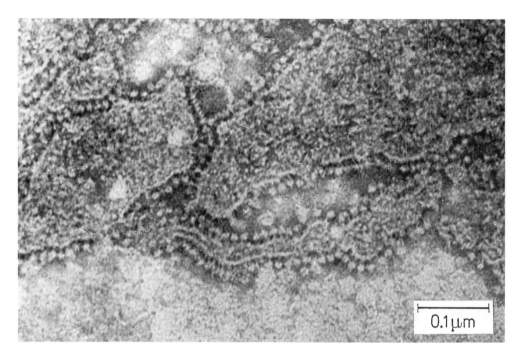

FIG. 4.11. The negative staining of the inside-out submitochrondrial vesicles. As can be seen, there are numerous knobs on the outer side of the membrane. These knobs are factors F_1 of H^+-ATP synthase, protruding from the membrane (negative staining, electron microphotograph by L. E. Bakeeva).

despite the detachment of knobs. It was found that it is the H^+-conducting F_0 complex that remains membrane-bound. Both F_1 and F_0 proved to be necessary for reconstituting ATP synthase activity at the proteoliposomal level (147).

The results of a computer analysis of electron micrographs of the negatively stained mitochondrial F_1 are shown in Figure 4.12. It was found that the six protein masses, most probably major subunits (3α and 3β), are arranged in two layers approximately at the vertices of a triangular antiprism. The frontal projection was shown to be about 10 nm in diameter. Occasionally one may detect yet another (seventh) protein mass localized in the central part of the mitochondrial factor F_1, which may represent a complex of minor subunits (19).

In an x-ray study at 3.6 Å resolution, it was found that the mitochondrial F_1 molecule may be approximated by an ellipsoid with axes of $12 \times 12 \times 7.5$ nm (17).

Only indirect information is available concerning the three-dimensional structure of factor F_0. It is assumed that there are six and two α-helical transmembrane rods respectively in the hydrophobic subunits 6 and 9.

Factor F_1, when detached from the membrane sector of H^+-ATP synthase, retains the ability to hydrolyze

ATP to ADP and phosphate. The reaction proceeds vigorously but loses sensitivity to olygomycin, diethylstilbestrol, and low concentrations of DCCD, differing in this respect from ATP hydrolysis by the membrane-bound F_0F_1 complex.

It has been established that adenine nucleotides can be bound to six binding sites of the F_1 molecule, that is, one site per major subunit. Both α- and β-subunits of F_1 contain conserved sequences similar to those in other ATP-binding proteins. Three of the binding sites have a much higher affinity for nucleotides than the other three sites. Thus, tightly and loosely bound nucleotides are contained in F_1 (23, 99).

When only one nucleotide binding site is saturated with ATP, factor F_1 hydrolyzes ATP very slowly (so-called unisite catalysis) (67). ATP binding at a second site caused a manifold increase in the reaction rate as a result of a significant acceleration of the release of the products (ADP and P_i).

δ and ϵ subunits proved not to be needed for maximal ATPase activity of isolated F_1. At the same time, the γ subunit was shown to stimulate fourfold activity of the $\alpha_3\beta_3$ subcomplex (86).

Vinogradov and coworkers (118) reported that ATP hydrolysis by isolated mitochondrial F_1 is strongly inhibited by ADP. According to the data obtained in our group by Milgrom and Murataliev (117), this inhibition is due to ADP binding in the noncatalytic nucleotide binding site of factor F_1. In the mitochondrial membrane, ADP elimination from the inhibited factor F_1 was found to be $\Delta\bar{\mu}H$-dependent (97). This means that ADP inhibition can occur only when $\Delta\bar{\mu}_{H+}$ is low, that is, when there is a danger of the cellular ATP pool being exhausted due to ATPase activity of F_0F_1 complex.

In mitochondria, a function similar to that of the tightly bound ADP is performed by a special protein inhibitor (58). The chloroplast and bacterial protein inhibitors seem to be identical to the ϵ-subunit of factor F_1. Like ADP, the protein inhibitor suppresses ATPase activity of isolated F_1. It also inhibits the membrane-bound F_0F_1 complex when $\Delta\bar{\mu}_{H+}$ lowers (for review see ref. 97).

Several independent lines of evidence indicate that in the F_1 catalytic site, hydrolysis of ATP proceeds with no covalent intermediates involved, as shown by Eq. (4.6):

$$\underset{\begin{subarray}{c}|\\|\end{subarray}}{\overset{\begin{subarray}{c}|\\|\end{subarray}}{\text{ADPO}}}\overset{\displaystyle\text{HOH}}{\underset{\displaystyle\overset{\displaystyle\text{OH}}{\underset{\displaystyle\text{OH}}{|}}}{\underset{}{-}\text{P}=\text{O}}} \longrightarrow \text{ADPOH} + \overset{\displaystyle\text{OH}}{\underset{\displaystyle\text{OH}}{\text{HO}-\text{P}=\text{O}}} \qquad (4.6)$$

This conclusion is based upon data of isotope analysis of the hydrolytic process as well as its resistance to hydroxylamine, vanadate and other reagents inhibiting the ATPases that catalyze ATP cleavage via a high-energy phosphoenzyme intermediate (22, 98, 206).

Feldman and Sigman (55) showed that factor CF_1, isolated from chloroplasts, can synthesize ATP from tightly bound ADP in response to the addition of inorganic phosphate. The synthesized ATP, tightly bound to CF_1, could be released into the aqueous phase only after denaturation of CF_1. Later, the same authors demonstrated a synthesis of bound ATP from bound ADP and exogenous P_i in completely uncoupled chloroplasts (56), confirming a suggestion made 10 years earlier by Cross and Boyer, who studied uncoupled mitochondria (45). In 1983 Sakamoto and Tonomura (159) reported on the synthesis of bound ATP by isolated mitochondrial F_1 from a medium containing ADP and P_i. The process was stimulated significantly by 30% dimethylsulfoxide. These observations indicate that $\Delta\bar{\mu}_{H+}$ is required for ATP removal from the F_1 active site (and/or for tight ADP and P_i binding to this site) rather than for ATP synthesis per se.

Calculations made in our group by Kozlov (99) indicate that the equilibrium constant of ATP hydrolysis in the F_1 catalytic site is no more than 100 and, in all likelihood, is much smaller. This means that the synthesis of bound ATP required a much smaller portion of energy than that of free ATP in solution. It is possible that practically all of the energy is expended at the stage of the ATP release and/or ADP + P_i binding, as first suggested by Boyer (22).

Removal of F_1 from mitochondrial membranes always results in H^+ leakage through F_0. F_0-containing proteoliposomes are also leaky to H^+. The effect is abolished by the specific F_0 inhibitors, oligomycin, and DCCD. Reconstitution of the F_0F_1 complex suppresses the leakage as well (23). Modification of a free carboxylic group of a dicarboxylic amino acid residue in

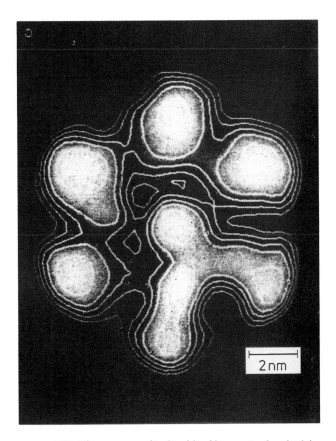

FIG. 4.12. The structure of isolated beef heart mitochondrial factor F_1. Final result of the image analysis of 379 projections of negatively stained F_1 molecules [From Boekema et al. (19) with permission].

subunit 6 by DCCD abolishes both H^+ transport and $\Delta\bar{\mu}_{H^+}$-powered ATP synthesis. This residue is localized in the middle of the hydrophobic core of the membrane, that is, halfway from one membrane surface to the other. To inhibit H^+-ATPase activity or H^+ conductance, it is sufficient to modify only one or two of twelve subunits 9 with DCCD (63).

As was stressed by Kagawa, "F_0 . . . is unlikely to be a simple hole through which H^+ flows, because the rate of H^+ translocation through F_0 obeys saturation kinetics with K_m values of about 10^{-7} M, indicating the presence of H^+ binding site(s)" (84).

The H^+-conductance of the chloroplast factor F_0 was found to be about $2H^+ \cdot \mu S^{-1} \cdot CF_0^{-1}$ at 100 mV $\Delta\Psi$ (108). The selectivity of the conductance toward H^+ is as high as 10^7 in comparison to Na^+ or K^+.

H^+-ATP synthase incorporated into proteoliposomes can (1) generate $\Delta\bar{\mu}_{H^+}$ when ATP is hydrolyzed and (2) synthesize ATP when $\Delta\bar{\mu}_{H^+}$ is artificially imposed across the proteoliposomal membrane. Kagawa and Racker pioneered in the studies on F_0F_1 proteoliposomes (87).

An important question concerning the F_0F_1 mechanism is how many H^+ must be transported downhill to form one ATP molecule from ADP and P_i. ADP/O measurements indicate that the H^+/ATP ratio for synthesis of intramitochondrial ATP is probably equal to 2 but a higher value is not excluded.

In plants, the ADP photophosphorylation occurs on the outer membrane surface of thylakoids. The produced ATP is consumed in large amounts in the chloroplast stroma to support the synthesis of glucose. In this case, the ATP^{4-}/ADP^{3-} antiporter and $H_2PO_4^+$, H^+ symporters cannot be used to add energy to ATP formation. Here, the H^+/ATP ratio is usually assumed to be 3 (192).

The principles of the functioning of H^+-ATP synthase may be formulated as follows: (1) ADP and P_i bound to factor F_1 can produce bound ATP in an energy-independent fashion; (2) energy is required to transport the bound ATP from the F_1 catalytic site to water and/or to transport ADP and P_i from water to the catalytic site; (3) this energy should be in the form of $\Delta\bar{\mu}_{H^+}$ (i) produced by a natural $\Delta\bar{\mu}_{H^+}$ generator or (ii) artificially imposed across the F_0F_1-containing membrane (ΔpH or $\Delta\Psi$ are equally active in supporting the process); (4) it is F_0 that is involved in translocation of H^+ through H^+-ATP synthase.

As already mentioned, ADP can inhibit ATPase activity of factor F_1. This effect was studied in detail by Vinogradov et al. (118). In particular, it was found that the ADP inhibition was greatly potentiated by azide, and the effect was reversed by sulphite. Azide as

well as the F_1 protein inhibitor did not affect respiratory ATP synthesis. Vinogradov and co-workers (118) concluded that (1) there are two conformers of factor F_1, one for ATP synthesis and the other for ATP hydrolysis, and (2) tightly bound ADP stabilizes the synthetic conformer.

In his original version of the chemiosmotic hypothesis, Mitchell (120) suggested that the role of F_0 consists in H^+ translocation between the outer aqueous phase and the F_1 catalytic site, assuming that the translocated protons are consumed in the ATP synthase reaction:

$$ADP^{3-} + PO_4^{3-} + 2H^+ \rightarrow ATP^{4-} + H_2O \quad (4.7)$$

Alternative versions assume that the H^+ ions translocated across the membrane via F_0 are not involved per se in H_2O formation by F_1 (Fig. 4.13). As for $\Delta\bar{\mu}_{H^+}$ utilization, it may occur in such a way that H^+ transport across the F_0 part of the enzyme induces a confor-

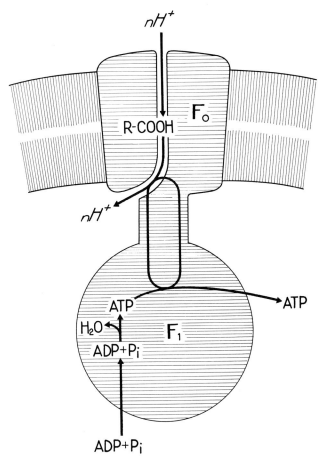

FIG. 4.13. H^+-ATP synthase mechanism: translocation of two or three hydrogen ion (nH^+) via F_0 results in a conformation change in F_1 which facilitates the release of the tightly-bound ATP from the F_1 catalytic site. Hydrogen ions passing F_0 are not involved in H_2O formation accompanying the ATP synthesis from ADP and P_1.

mational change in F_0 and, in turn, in F_1, causing an ATP release from the catalytic site. This conformational change may be triggered by protonation or deprotonation of some group(s) in the F_0 molecule, for example, the DCCD-sensitive carboxylate (21).

Evidence for the long-range conformational interaction between the DCCD-sensitive site of F_0 and the F_1 catalytic site is summarized by Senior (169).

The strongest indication that F_0-translocated protons are not identical with those incorporated into H_2O in the ATP synthase reaction is furnished by studies on Na^+-driven phosphorylation. It was found that in some bacteria, one and the same F_0F_1-type ATP synthase can pump either H^+ or Na^+. It is obvious that such an effect could not occur if the pumped ion were involved in the H_2O formation (48).

In the animal cell, the ATP–$\Delta\bar{\mu}_{H^+}$ interconversion can be catalyzed not only by mitochondrial H^+-ATP-synthase but also by H^+-ATPases of the vacuolar type. This enzyme was found to be localized in various intracellular vesicles other than mitochondria that is, lysosomes, endosomes, secretory granules, etc., as well as in the plasma membrane of some types of tissues. Its function consists in H^+-motive ATP hydrolysis rather than in $\Delta\bar{\mu}_{H^+}$-driven ATP synthesis. H^+-ATPase is similar, but not identical, to the F_0F_1 complex. In animals, H^+-ATPase of chromaffin granules seems to be the first sufficiently well-characterized specimen of this type of enzyme.

Chromaffin granules are vesicles suspended in the cytoplasm of the cells of the adrenal medulla. They accumulate catecholamines (adrenaline and noradrenaline) to a final concentration of 0.55 M, or about 25,000-fold higher than in the cytosol (134). Catecholamines represent the major cationic species inside these secretory granules. The corresponding anion proved to be ATP, the concentration of which is slightly above 0.1 M. There are some indications that both catecholamine and ATP are sympathetic cotransmitters interacting with corresponding receptors.

Catecholamine accumulation inside chromaffin granules is mediated by catecholamine cation/nH^+ antiporter and powered by the H^+-ATPase–generated $\Delta\bar{\mu}_{H^+}$. H^+-ATPase catalyzes H^+-influx into vesicles, supported by the energy ATP (134).

The subunit composition of this ATPase is the following (kD): 115; $(72)_3$, $(57)_3$, 41, 34, 33 (F_1); 20 and $(16)_n$ (F_0) when n is most probably 6. This is in fact similar to that of the tonoplast ATPase of plants-and-fungi. The difference is that the chromaffin granule ATPase contains a 115 kD subunit which is absent from tonoplasts. Some sequence homologies between major subunits of this ATPase and those of F_1 were revealed. Minor subunits show no such homology. Studies at the proteoliposomal level showed that the enzyme catalyzes H^+ transport in an electrogenic way (37, 38, 39).

Inhibitor analysis of this H^+-ATPase revealed that, in contrast to mitochondrial H^+-ATP synthase, it is resistant to oligomycin. Located in the membrane, it was inhibited by DCCD at a 50-fold higher concentration of the inhibitor than the mitochondrial H^+-ATP synthase. It is the 16 kD subunit that combines with DCCD. Detachment from the membrane desensitizes the enzyme to DCCD (37). Vanadate, unlike plant and fungal plasmalemmal H^+-ATPases, fails to block ATP-hydrolysis and the ATPase-linked H^+ transport in chromaffin granules. On the other hand, it is very sensitive to SH reagents such as N-ethylmaleimide (39).

Various types of animal cells employ H^+-ATPase when a hormone or neurotransmitter is stored before being excreted from the cell. It is firmly established that this mechanism, besides catecholamine, applies to serotonin in platelets, acetylcholine in synaptic vesicles, a group of peptide hormones in the hypophysis, and insulin in pancreatic islet cells. Moreover, H^+-ATPase was described in lysosomes, endosomes, clathrin-coated vesicles, phagosomes, and the outer cell membrane of some animal tissues, such as in the bladder epithelium and certain types of kidney cells. In the literature, one finds some indications of the functioning of H^+-ATPase in the plasma membranes of brain cells and Ehrlich tumor cells, as well as in the Golgi and endoplasmic reticulum membranes (109, 112, 179).

The subunit composition of H^+-ATPase from clathrin-coated vesicles, which was purified 200-fold, proved to be similar to that from chromaffin granules that is, 100, $(73)_3$, $(58)_3$, 40, 38, 34, 33, 19, and $(17)_6$ kD (7). The 17 kD subunit proteoliposomes were shown to possess a DCCD-sensitive H^+ permeability (192). Practically the same subunits were found in renal tubules (62). On the other hand, lysosomal H^+-ATPase was shown to lack the 100 kD polypeptide (64). All these ATPases proved to be (1) resistant to oligomycin, vanadate, and diethylstilbestrol; (2) sensitive to DCCD but usually at concentrations higher than mitochondrial H^+-ATP synthase; and (3) sensitive to N-ethylmaleimide. These properties are similar to those of the plant tonoplast H^+-ATPase. On the other hand, the tonoplast ATPase is completely inhibited by NO_3^- which is without any effect on ATPase of clathrin-coated vesicles.

In chromaffin and other secretory granules, the H^+-ATPase–generated $\Delta\bar{\mu}_{H^+}$ supports accumulation of the secreted compounds inside the granule. On the other hand, the function of some other H^+-ATPases consists of acidification of the vesicle interior rather than in the membrane energizations. This is inherent in lysosomal, endosomal, and possibly Golgi apparatus H^+-ATPases. In these systems, $\Delta\Psi$ formed by H^+-ATPase is converted to ΔpH due to electrophoretic influx of an anion, for example, Cl^-. In lysosomes, acidification of the inner solution is required to create optimal conditions for operation of the lytic enzymes having acidic pH optima. In fact, lysosomal enzymes are inactive at neutral pH typical of the cytosol. As a result, the cytosolic components cannot be digested when these enzymes are occasionally released from lysosomes. Acidic pH also seems necessary for the functioning of endosomes and some other cytoplasmic inclusions possessing H^+-ATPases.

An important role is played by H^+-ATPases in the plasma membrane of some types of renal cells and bladder epithelium: they are involved in the transcellular transport of acidic equivalents. Here again, $\Delta\Psi \rightarrow \Delta pH$ transition seems to be necessary.

In the gastric mucosa, H^+/K^+-ATPase is present in the outer cell membrane. The enzyme is specialized in acidification of the lumen of the stomach. The problem of $\Delta\bar{\mu}_{H^+} \rightarrow \Delta pH$ transduction does not arise since H^+/K^+-ATPase exchanges H^+ ions for an equal number of K^+ ions. Apparently this ATPase transfers only one H^+ per ATP. As a result, huge pH gradients can, in principle, be formed.

It was found that H^+/K^+-ATPase falls into the so-called P-category, resembling H^+-ATPase of the outer cell membrane of plants and fungi, as well as animal Ca^{2+} and Na^+/K^+ ATPases in the following respects: (1) a phosphorylated intermediate can be identified, (2) it is a carboxylate belonging to the 100 kD polypeptide that is phosphorylated, (3) vanadate and hydroxylamine are inhibitory. The enzyme is probably composed of four 100 kD subunits (158, 187). Similar H^+/K^+-ATPases seem to be present in the colonic mucosa (88).

Na+/K+-ATPase and Ca2+-ATPase

Na^+/K^+-ATPase is localized in the outer membrane of animal cells, and is responsible for $\Delta\Psi$, ΔpNa, and ΔpK generations across this membrane. Na^+/K^+-ATPase catalyzes the exchange of $2K^+_{out}$ for $3Na^+_{in}$ per molecule of hydrolyzed ATP. The enzyme with a molecular mass of about 550 kD is composed of four subunits: two α and two β. Molecular masses of the subunits are about 95 and 45 kD, respectively (115, 167). Electron microscopy study of two-dimensional crystals of the enzyme (137) showed that the vertical dimension of the $\alpha\beta$-protomer is about 10 nm, that is, larger than

the lipid bilayer thickness. The α-subunit was found to protrude on the inner, as well as on the outer membrane surfaces, while the β-subunit is exposed on the outer surface only. The C-terminal region of the β-subunit is glycosylated. The mass of carbohydrate moiety is 7 kD. The chemical modification and antibody technique data showed that almost two-thirds of the α-subunit is localized outside the membrane, its cytoplasmic domain being three-fold larger than that located on the outer surface. The α-subunits are somewhat narrowed in the central part and contact each other in the intramembrane region. The height of the contact region is about 3 nm.

Na^+/K^+-ATPase has been reconstituted into proteoliposomes which are competent in energy transduction. The freeze-etching technique showed that proteoliposome membrane contains 9–12 nm particles, the size corresponding to the $\alpha_2\beta_2$ complex.

Analysis of the hydropathy profiles of the amino acid sequence indicates the presence of eight and one α-helical columns in α- and β-subunits, respectively (138).

The N-terminal domain of the α-subunit was found to face the cytoplasmic membrane side. The carboxyl of an Asp-96 localized in this domain is phosphorylated by ATP when a phosphoenzyme is formed (167).

The Na^+/K^+-ATPase mechanism is described in terms of the following scheme.

$$E_1 + ATP \xrightarrow{Na^+_{in}} E_1P + ADP \qquad (4.8)$$

$$E_1P \longrightarrow E_2P \qquad (4.9)$$

$$E_2P \xrightarrow{K^+_{out}} E_2 + P_i \qquad (4.10)$$

$$E_2 \longrightarrow E_1 \qquad (4.11)$$

According to the scheme, Na^+ (presumably Na^+_{in}) is required for enzyme phosphorylation. The phosphoenzyme undergoes a conformational change ($E_1 \rightarrow E_2$ transition) and then decomposes to form E_2 and P_i in a K^+_{out}-dependent fashion. E_2 relaxes to E_1, ready to enter the next cycle. It is the E_2P state that is fixed by ouabain, the specific Na^+/K^+-ATPase inhibitor. Oligomycin was found to affect the cycle at the E_1 and E_1P steps, stimulating Na^+ binding to the enzyme (K_d for Na^+ decreases from 340 to 60 μM). The K_d for K^+ was found to be the same (6 μM) with and without oligomycin (115).

Rb^+, Cs^+, NH_4^+, Tl^+, Li^+, and Na^+ were reported to substitute to some degree for K^+ at the K^+ sites. On the other hand, the Na^+ binding sites are known to have relatively high specificity. Among the above-mentioned cations, only Li^+ was found to partially substitute for Na^+. The Li^+/K^+-ATPase activity is, however, very small, and the affinity for Li^+ is much lower than for Na^+. It is interesting that H^+ can substitute for Na^+ and that the affinity for H^+ is, in fact, at least two orders of magnitude higher than for Na^+. Nevertheless at physiological pH, $[Na^+]$ is much higher than $[H^+]$ so that Na^+, rather than H^+, is translocated by Na^+/K^+-ATPase. However, under slightly acidic conditions (pH = 5.7), ATPase activity and H^+/K^+ antiport activity were reported in Na^+/K^+-ATPase proteoliposomes (68; for reviews, see refs. 125, 167).

Ca^{2+}-ATPase is present on all plasma membrane and, in some tissues, in the endoplasmic reticulum of the animal cells. This P-type enzyme pumps Ca^{2+} from cytosol to the extracellular medium or to the endoplasmic reticulum interior. There are no indications that the formed electrochemical Ca^{2+} potential difference is utilized to support any useful work. Most probably the only function of Ca^{2+}-ATPase is to regulate cytosolic Ca^{2+} levels (32, 33).

$\Delta\bar{\mu}_{H^+}$- and $\Delta\bar{\mu}_{Na^+}$-Driven Transports of Solutes

H^+-ATPase, H^+/K^+-ATPase, Na^+/K^+-ATPase, and Ca^{2+}-ATPase (considered earlier under H^+-ATP-Synthase and H^+-ATPases and Na^+/K^+-ATPase and Ca^{2+}-ATPase) exemplify systems utilizing ATP to transport solute(s) namely H^+, K^+, Na^+, and Ca^{2+}. However, the more typical situation is when uphill transport of a solute is supported by $\Delta\bar{\mu}_{H^+}$ or $\Delta\bar{\mu}_{Na^+}$. The former and the latter are used in the inner mitochondrial membrane and in the plasma membrane of animal cells, respectively.

For solute S, accumulated by a $\Delta\bar{\mu}_{H^+}$-driven system in a negatively charged compartment (for example, in the mitochondrial matrix), the accumulation ratio ($[S]_{in}/[S]_{out}$) will obey the following equation:

$$RT\ln[S]_{in}/[S]_{out} = -(n + Z)F\Delta\Psi - nRT\ln[H^+]_{in}/[H^+]_{out} \qquad (4.12)$$

where n is the number of H^+ ions symported with one S molecule, and Z is the number of positive charges on S (96).

If a transport system accumulates a cation of a strong base inside a negatively charged compartment, or extrudes an anion from such a compartment, $\Delta\Psi$ may be the only driving force with no ΔpH involved.

For a monovalent cation C^+, the accumulation ratio will be:

$$RT\ln[C^+]_{in}/[C^+]_{out} = -F\Delta\Psi \qquad (4.13)$$

It is easy to calculate that the tenfold gradient of C^+ requires $\Delta\Psi$ of about 60 mV to be formed ($t° = 25°$). Such simple electrophoretic behavior is inherent in the valinomycin-mediated transport of K^+ ions. A similar

mechanism is responsible for Ca^{2+} accumulation by mitochondria. The only difference is that in this case coefficient Z (see Eq. 4.12) is equal to 2, so that 30 mV $\Delta\Psi$ appears to be sufficient to maintain the tenfold gradient.

Both valinomycin and the mitochondrial Ca^{2+} uniporter have a rather simple problem to solve. They should recognize the transported ion and facilitate its diffusion across the hydrophobic membrane core. No special device to transduce $\Delta\Psi$ energy into an ion concentration gradient is necessary, since the overall process consists in the movement of the ion in the electric field (electrophoresis). An ATP^{4-} efflux from mitochondria is also electrophoretic. However, here the system is complicated by the fact that the $ATP^{4-}_{in}/ADP^{3-}_{out}$ antiport, rather than ATP^{4-} uniport occurs: the purpose is not to extrude ATP from mitochondria but to substitute inner ATP for outer ADP. The antiport is catalyzed by a 32 kD protein forming dimers. In each monomer, six α-helical columns were found. The turnover number is rather low (25 s^{-1}), whereas the amount of the protein in the membrane is large (up to 10% of the inner membrane protein).

The ATP/ADP-antiporter is specifically inhibited by the glycosides atractylate and carboxyatractylate, affecting the protein from the extramitochondrial side, and by bongkrekic acid from the matrix side (reviewed in refs. 90, 198).

The electrophoretic mechanism of solute transport results in cations and anions accumulating in negatively and positively charged compartments, respectively. It is obvious, however, that separation of solutes into cations and anions is hardly the aim of a living system. One way to overcome, at least partially, the limitation of a simple electrophoretic principle is to convert the $\Delta\Psi$ component of the protonic potential to ΔpH. In fact, $\Delta\Psi$ is the primary form of $\Delta\bar{\mu}_{H^+}$ because the electric capacity of the membrane is much lower than the pH buffer capacity of membrane-washing solutions. To convert $\Delta\Psi$ to ΔpH, it is necessary to discharge the electric potential difference by an electrophoretic flux of ions other than H^+ and OH^-.

If $\Delta\Psi = 0$, then Eq. 4.12 may be simplified as follows:

$$[S]_{in}/[S]_{out} = ([H^+]_{out}/[H^+]_{in})^n \qquad (4.14)$$

The simplest case is when a weak acid or a weak base is accumulated in the more alkaline or more acidic compartment, respectively. To do this, only two conditions need to be fulfilled: the membrane should be permeable *(1)* to a noncharged form of the acid (base) and *(2)* it must be impermeable to a charged form of this acid (base). Accordingly, the following simple events take place, resulting in the accumulation, for example, of anion A^- of weak acid AH in the alkaline mitochondrial matrix:

$$AH_{out} \rightarrow AH_{in}; \qquad (4.15)$$

$$AH_{in} + OH^-_{in} \rightarrow A^-_{in} + H_2O_{in}. \qquad (4.16)$$

Consumed OH^-_{in} is regenerated by the respiratory chain removing H^+ from the mitochondrial matrix.

Using this mechanism, acetate is taken up by energized mitochondria. Since acetic acid in its protonated form, CH_3COOH, is a small noncharged molecule, it enters mitochondria without any carriers, moving down ΔpH, neutralizes intramitochondrial OH^- ions, and accumulates inside in the form of acetate anion. The latter is charged and thus cannot cross the mitochondrial membrane and escape from the matrix.

A similar mechanism, but operating in the opposite direction, may give rise to NH_3 efflux:

$$(NH_3)_{in} \rightarrow (NH_3)_{out} \qquad (4.17)$$

$$(NH_3)_{out} + H^+_{out} \rightarrow (NH^+_4)_{out} \qquad (4.18)$$

Thus, the existence of $\Delta\bar{\mu}_{H^+}$ in its secondary form (ΔpH) allows weak acid and bases to be involved in the uphill transport in the direction determined by the pH gradient. Yet just as in the case of the $\Delta\Psi$-driven transport of cations and anions, it is hardly possible to obtain a reasonable composition of, for example the mitochondrial matrix by accumulation of anions of weak acids and cations of strong bases, and by extrusion of cations of weak bases and anions of strong acids. It seems obvious that in some cases the direction of the solute fluxes must be opposite.

For cations of strong bases, the problem may be solved by using cation/H^+ antiporters. For example, mitochondria, besides the electrophoretic system for Ca^{2+} accumulation, have also an electroneutral $Ca^{2+}/2H^+$ antiporter extruding Ca^{2+} from the mitochondrial matrix down ΔpH. Thus, mitochondria can not only accumulate Ca^{2+} but also actively release it. It is clear that these two systems must be alternatively actuated since their cooperation results in Ca^{2+} circulation dissipating $\Delta\bar{\mu}_{H^+}$ energy.

Similarly, anions of strong acids can be accumulated inside the matrix in a ΔpH-dependent fashion if they are symported with H^+. This may be exemplified by phosphate transport systems of mitochondria and bacteria. Electroneutral H_3PO_4 is practically absent from water solution at physiological pH. Phosphate anions must be extruded from the mitochondrial matrix and bacterial cytoplasm down $\Delta\Psi$. Nevertheless phosphate is accumulated, not extruded. This results from the operation of the phosphate carrier which symports $H_2PO_4^-$ with H^+ (or antiports $H_2PO_4^-$ against OH^-). The carrier was shown to be a 30 kD protein

structurally resembling ATP/ADP-antiporter (for review, see ref. 179).

Sometimes both $\Delta\Psi$ and ΔpH are the driving forces for uphill solute transports. This is the case, for instance, when the transported compound does not contain either ionized atoms or mobile protons. Such a compound may be symported with H^+ (or antiported against H^+) down total $\Delta\bar{\mu}_{H^+}$.

An interesting example of the $\Delta\bar{\mu}_{H^+}$-driven uphill transport represents accumulation of the positively charged catecholamines (RN^+H_3) in chromaffin granules of the adrenal medulla. Here, H^+-ATPase (see earlier under H^+-ATP-Synthase and H^+-ATPases) pumps H^+ into the granules, so that their interior is positively charged and acidified. When accumulated in granules, RN^+H_3 is antiported against $2H^+$. Thus, two protons and only one net positive charge are released from the granule per imported RN^+H_3 molecule. As a result, at equal $\Delta\Psi$ and ΔpH, the contribution of ΔpH appears to be twofold larger than that of $\Delta\Psi$ according to Eq. 4.12.

There are several systems where $\Delta\bar{\mu}_{H^+}$ powers the formation of a gradient of a solute (S_1) which then is used for uphill transport of another solute (S_2).

In mitochondria, a long $\Delta\bar{\mu}_{H^+}$-dependent cascade was described when the transport of phosphate and citric cycle intermediates was studied. The initial event is $\Delta\Psi$ formation by the respiratory chain. $\Delta\Psi$ is then discharged by a flux of ionized penetrants (electrophoretic $ATP^{4-}_{in}/ADP^{3-}_{out}$ exchange, Ca^{2+} or K^+ uniport, etc.) and ΔpH is formed. ΔpH is utilized to form ΔpP_i via $H_2PO_4^-$, H^+ symport directed from outside to inside. ΔpP_i powers the accumulation of malate by means of $HPO_4^{2-}/malate^{2-}$ antiporter. The last step of the cascade is citrate import via $malate^{2-}/citrate^{3-}$, H^+ antiporter (Fig. 4.14). Such a complicated system allows a very sophisticated regulation of the metabolic pattern to be organized (130).

Uphill transport of solutes across the outer membrane of the animal cell is supported by the Na^+/K^+-ATPase–produced $\Delta\bar{\mu}_{Na^+}$. For this purpose, Na^+, solute-symporters are used. This is the way by which amino acids, sugars, fatty acids, and some other compounds are imported by cells.

H^+–Nicotinamide Nucleotide–Transhydrogenase

The reversible transfer of hydride ion (H^-) between NAD and NADP (Eq. 4.19) is catalyzed by enzymes called "transhydrogenases."

$$NADH + NADP^+ \rightleftharpoons NAD^+ + NADPH \qquad (4.19)$$

Such an activity was first described by Colowick et al. (42). Later Daniellson and Ernster (46) discovered

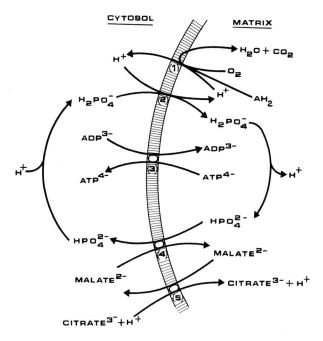

FIG. 4.14. Mitochondrial—cytosolic relationships: the transport cascade of the inner mitochondrial membrane. Respiratory chain pumps H^+ from the matrix (1). $H_2PO_4^-$, H^+ symporter (2) and ADP^{3-}/ATP^{4-} antiporter (3) mediate the influx of $H_2PO_4^-$ and ADP and the efflux of ATP. $HPO_4^{2-}/malate^{2-}$ antiport (4) results in malate influx. $Malate^{2-}/(citrate^{3-}+H^+)$ antiport (5) gives rise to citrate accumulation in the matrix [After Pedersen and Wehrle, (142) with permission].

that in energized submitochondrial vesicles, the equilibrium of this reaction is strongly shifted to NADPH formation. The [NADPH] × [NAD$^+$]/[NADP$^+$] × [NADH] ratio attained was 500, whereas in nonenergized vesicles it was close to 1 (the latter is in agreement with the almost equal redox potentials of NADH/ NAD$^+$ and NADH/NADP$^+$ couples, differing in no more than 5 mV). Mitchell (120) proposed that transhydrogenase represented an additional energy-coupling site in the respiratory chain. It was suggested that oxidation of NADPH by NAD$^+$ generated $\Delta\bar{\mu}_{H+}$ in the same direction as respiration, whereas the reverse process consumed respiration-produced $\Delta\bar{\mu}_{H+}$. Within the framework of this concept, a shift in the equilibrium of the transhydrogenase reaction under energized conditions is a simple consequence of the energetics of transhydrogenase and other $\Delta\bar{\mu}_{H+}$ generators.

Free energy change (ΔG) in the process of conversion of substrate (S) to product (P) can be calculated as:

$$\Delta G = \Delta G_o + RT\ln\,[S]/[P], \qquad (4.20)$$

where ΔG_o is the energy change when $[S] = [P]$. In all the $\Delta\bar{\mu}_H+$ generators other than transhydrogenase, ΔG_o is rather large, as a rule, about 7–10 kcal·mol^{-1}. In the transhydrogenase reaction, however, ΔG_o is

close to zero. Therefore this enzyme can effectively compete with respiration or ATPase in $\Delta\bar{\mu}H$ formation only if [NADPH] × [NAD$^+$] were several orders of magnitude higher than [NADP$^+$] × [NADH], a most unlikely situation in the living cell.

Mitchell's hypothesis concerning the transhydrogenase mechanism presupposes that (1) NADPH $\rightarrow$ NAD$^+$ transhydrogenation must result in the generation of $\Delta\bar{\mu}_{H+}$ in the same direction as respiration or ATP hydrolysis, and (2) NADH $\rightarrow$ NADP$^+$ transhydrogenation must generate a $\Delta\bar{\mu}_{H+}$ of opposite direction, both these effects were decreased by the reaction products. Experiments performed in Liberman's laboratory and in our group confirmed the above assumptions (49, 50, 66, 107). It was found that inside-out submitochondrial vesicles and proteoliposomes with transhydrogenase from beef heart mitochondria generated $\Delta\Psi$, inside positive, when oxidizing NADPH by NAD$^+$. Oxidation of NADH by NADP$^+$ also generated $\Delta\Psi$, but now it was inside negative. In both cases, $\Delta\Psi$ disappeared when the reaction products were added to equalize concentrations of four nicotinamide nucleotides in the reaction mixture. The transhydrogenase-produced $\Delta\Psi$ was the linear function of ln[NADPH] × [NAD$^+$]/[NADP$^+$] × [NADH]. Thus it was concluded that transhydrogenase in inside-out submitochondrial vesicles catalyzes the following process:

$$\begin{aligned}NADPH + NAD^+ + nH^+_{out} &\leftrightarrow \\ NADP^+ + NADH + nH^+_{in}\end{aligned} \qquad (4.21)$$

If a $\Delta\bar{\mu}_{H+}$ promoting H^+_{in} efflux is formed by other $\Delta\bar{\mu}_{H+}$ generators, the reaction is shifted to the left. The n value in Eq. 4.21 seems to be equal to 0.5.

To understand the H^+-transhydrogenase mechanism, it is essential to take into account the fact that a direct stereospecific transfer of H^- ion takes place between the A-position of the nicotinamide ring in NAD and the B-position of that in NADP, with no proton exchange with water (77). Therefore, protons translocated by the enzyme through the membrane are not identical to those transferred from one nicotinamide nucleotide to the other in a transhydrogenase reaction.

Both NAD(H) and NADP(H) are known to interact with the matrix (in bacteria, cytoplasmic) side of the membrane, bound to two different sites. It was shown that labeling of the enzyme in proteoliposomes by an intravesicular membrane-impermeable photoprobe was enhanced by NADP$^+$, decreased by NADH, and not significantly affected by NADPH. It was concluded that substrate binding alters the conformation of the transhydrogenase domain protruding from the opposite surface of the membrane (143).

It is known that $\Delta\bar{\mu}_{H^+}$ not only shifts the equilibrium in the direction of $NADP^+$ reduction, but also greatly accelerates this process. Moreover, $\Delta\bar{\mu}_{H^+}$ decreases the K_m of transhydrogenase for $NADP^+$ from 40 μM down to 6.5 μM and increases the K_m for NAD^+ from 28 μM to 43.5 μM. Thus, $\Delta\bar{\mu}_{H^+}$ favors $NADP^+$ reduction and hinders NAD^+ reduction (157).

The principle of the H^+-transhydrogenase mechanism remains obscure (for discussion, see refs. 77, 156, 157). Structurally, the mitochondrial enzyme proved to be a dimer composed of two identical 120 kD subunits (3, 77), whereas three domains were identified in the transhydrogenase monomer. One of them is embedded in the membrane (20% of the entire protein molecule); 50% and 30% were exposed to the matrix and intermembrane spaces, respectively (207).

The master function of H^+-transhydrogenase undoubtedly consists of the maintenance of the high [NADPH]/[NADP$^+$] ratio, essential for supporting reductive syntheses, which usually include NADPH-oxidizing step(s). In fact, $\Delta\bar{\mu}_{H^+}$ produced by a respiratory chain and consumed by NADH → NADPH transhydrogenation, is ultimately utilized as an additional driving force of reductive syntheses.

The high level of NADPH is also necessary to provide reducing equivalents (1) for the cytochrome P-450–mediated detoxication of xenobiotics and (2) for the SH-glutathione–mediated reduction of SS-bonds in proteins. Moreover, there is an indication obtained in studying isolated hepatocytes (73) that transhydrogenase operating in the NADPH → NADH direction sustains for some time $\Delta\Psi$ at about 130 mV in the absence of respiration and at very low [ATP]. This $\Delta\Psi$ may be essential for maintaining the proper ion distribution between mitochondria and cytosol, as well as for correct arrangement of proteins in the inner mitochondrial membrane.

$\Delta\bar{\mu}_{H^+}$ as an Energy Source for Heat Production

In warm-blooded animals, thermoregulatory heat production may be due to operation of three main types of mechanisms: (1) ATP hydrolysis, (2) $\Delta\bar{\mu}_{H^+}$ dissipation, and (3) substrate oxidation with no ATP and $\Delta\bar{\mu}_{H^+}$ involved. Of these, the $\Delta\bar{\mu}_{H^+}$ dissipation (uncoupling) mechanism seems to be the most important. In 1960, the first experiments demonstrating the uncoupling effect of short-term cold exposure of animals were carried out. It was found that pigeons which were shorn before the experiments (to exclude physical thermoregulation) survived at 20°C in the air for 15–20 min when exposed to cold for the first time. The pigeons cooled for the second time the next day acquired the ability to survive for hours.

The study of oxidative phosphorylation in mitochondria isolated from the breast muscle of pigeons cooled for the first time revealed a decrease in the P/O ratio by a factor of 1.6. The second cold exposure resulted in about a sixfold lowering of the P/O ratio, that is, an almost complete uncoupling of respiration.

Further studies of this phenomenon revealed that the uncoupling attains its maximum within 15 min of the second cold exposure. This can be shown not only in birds but also in mammals (174, 184).

The same experiments showed that concentration of free fatty acids increases significantly, while that of triglycerides decreases in the muscle of an animal repeatedly subjected to short cold exposure. Addition of the delipidized serum albumin which binds free fatty acids increased the P/O ratio in mitochondria from the cold-treated animals. On the other hand, free fatty acids isolated from muscle of cold-treated animals and added to mitochondria from nontreated animals was found to induce uncoupling (106). It was shown that palmitate releases respiratory control if added at rather low concentration (1×10^{-5} M). This effect was found to be inhibited specifically by 5×10^{-7} M carboxyatractyloside, pointing to the involvement of the ATP/ADP antiporter (see earlier under $\Delta\bar{\mu}_{H^+}$- and $\Delta\bar{\mu}_{Na^+}$-Driven Transports of Solutes). ADP added with or without oligomycin also decreased the palmitate-induced stimulation of respiration (5, 6, 168). As was shown in our group by Brustovetsky et al. (26), low palmitate concentrations increase mitochondrial H^+ conductivity. Carboxyatractyloside strongly inhibited the palmitate action.

According to Brustovetsky et al. (27), uncoupling in skeletal muscle and liver mitochondria from ground squirrels arousing from hibernation is abolished by carboxyatractylate and serum albumin. A carboxyatractylate-sensitive increase in the H^+ permeability was also revealed in liver mitochondria of the newborn rat immediately after birth (196).

To explain the ATP/ADP antiporter-mediated uncoupling by fatty acids, it was postulated (181) that the antiporter consists of (1) adenine nucleotide-specific, carboxyatractyloside and bongkrekic acid–sensitive gates facing the intermembrane and matrix water spaces, respectively, and (2) rather nonspecific anion-translocating machinery embedded in the membrane core. Anions of fatty acids (A^-) are assumed to reach the anion-translocating part with no specific gate involved. Then they are translocated across the membrane-like adenine nucleotides, that is, in a $\Delta\Psi$-driven fashion. Facilitation of the transmembrane movement of A^- may be of principal importance for the fatty acid–mediated uncoupling, since the anionic form of fatty acids tends to occupy a position in the interface with a —COO$^-$ group facing the water and the hydrocarbon "tail" penetrating into the membrane core. Therefore, lipid membranes are of low permeability for fatty acid anions and of high permeability for protonated fatty acids (AH), as well as acetic acid. If the ATP/ADP antiporter can carry A$^-$, it must strongly increase the uncoupling efficiency of fatty acids.

In agreement with the above reasoning, palmitate did not increase the H$^+$ conductance of planar phospholipid membranes and failed to abolish $\Delta\Psi$ generation in cytochrome oxidase proteoliposomes. Under the same conditions, gramicidin and protonophorous uncoupler CCCP abolished $\Delta\Psi$ in both mitochondria and proteoliposomes. In mitochondria, gramicidin and CCCP effects proved to be carboxyatractyloside-resistant. As for dinitrophenol, carboxyatractyloside partially inhibited the action of its low concentrations on mitochondria. This may be explained by involvement of ATP/ADP antiporter in the transfer of the dinitrophenol anions (pK = 4) (5, 6).

Horvath et al. (74) reported that stearic acid and other acidic lipids compete for approximately 50 sites at the intramembrane surface of the ATP/ADP antiporter dimer. This number of first-shell lipid sites is unusually large for a protein of this size. In the same group, it was found that the pK_a of stearic acid at the lipid–water interface increases up to 8.0. At the lipid–peripheral protein interface, this value varied from 8.6 to 9.6, depending on the type of added protein (161).

The fatty acid–induced uncoupling mediated by the ATP/ADP antiporter can be produced by either saturated or nonsaturated fatty acids. Another potentially thermogenic effect inherent only in saturated fatty acids was described by Azzi and co-workers (102). It was found that very low (submicromolar) palmitate concentrations decrease respiratory control in cytochrome oxidase proteoliposomes from 8 to 4.5, without any increase in the H$^+$ conductance or decrease in the H$^+$/O ratio. The mechanism of this allosteric effect remains unclear.

A very important characteristic of the fatty acid–mediated uncoupling is the fact that there is no need for a special recoupling mechanism. The uncoupling effect must disappear spontaneously when the delivery rate of fatty acids to mitochondria becomes slower than that of their oxidation. These relationships seem to take place if the ambient temperature increases, the cold receptors of the skin stop working, and the system resulting in the release of fatty acids from triglycerides is switched off. The pool of free fatty acids in mitochondria is quickly oxidized by the potent β-oxidation mechanism. A decrease in the concentration of free fatty acids results inevitably in a recoupling of oxida-tion and phosphorylation. Thus, the fact that fatty acids are both substrates of oxidation and regulators of this process automatically prevents the system from being uncoupled for a time longer than necessary. This is a remarkable feature since the uncoupled respiration is thermodynamically more favorable than the coupled one. Therefore, a mitochondrion may, in principle, encounter some problems when returning from the uncoupled state to the coupled state.

All the features described thus far are characteristic of the response of muscle to repeated cold exposure. There is, however, a fast thermogenic response in the muscle that does not require cold pretreatment. This is shivering, which, as for any muscle contraction, must be supported by ATP hydrolysis. In this case, the respiratory energy is first converted into $\Delta\bar{\mu}_{H^+}$, then into intramitochondrial ATP which exchanges for cytoplasmic ADP and is hydrolyzed by actomyosin ATPase. Apparently, this pathway is too long and too slow to be effective for urgent heat production. Moreover, it requires muscle contraction, that is, the specific function of this tissue in order to be strongly activated (oxygen consumption by the organism upon cold exposure is as high as during hard muscle work). Therefore, the main principle of thermoregulation, that is, to make the function independent of the surrounding temperature, would not be achieved when muscle contraction is involved in additional heat production in the cold. Therefore, shivering decreases with the development of cold adaptation. The decrease in shivering coincides with the appearance of the metabolic response to noradrenaline, the mediator of cold-induced lypolysis (172, 186). That the muscle is still involved in thermogenesis when shivering disappears has been proved by the observation that at this stage of adaptation the efficiency of respiration in supporting muscle contraction of cold-exposed animals appears to be lower than that of the control animals (16). This can be easily explained by uncoupling.

Many features of the fatty acid–mediated thermoregulatory uncoupling first discovered in skeletal muscle were then observed in brown fat, a mammalian tissue specializing in additional heat production in the cold. Although brown fat accounts for at most 1%–2% of the total body mass, sympathetic nerve stimulation of this tissue in cold-acclimated animals can increase its heat production until it becomes an important source of thermogenesis (131, 186) and is responsible for up to one-third of the overall metabolic rate (57). Under these conditions, brown fat was found to be capable of producing as much as 400 W heat·kg^{-1}, that is, several orders of magnitude more than the thermogenic capacity of mammalian tissues in general (about 1 W·kg^{-1} body wt in a resting adult male) (126).

Brown fat is localized in the upper part of the back, where it envelopes blood vessels that transport blood to the brain. Thus, the heat produced by brown fat may be of great significance for the organism to survive at strong cooling.

The brown color of this tissue comes from a very high content of mitochondria that are well equipped for heat production. In particular, these mitochondria have a significant excess of respiratory chain enzymes over H^+–ATP synthase. Yet most importantly they contain a special protein, thermogenin, which mediates a transmembrane downhill H^+ flow as was suggested by Nicholls (129).

Thermogenin, a 32 kD protein, can amount to 10%–15% of the total brown fat mitochondrial protein (89, 132). Its sequence is similar to that of the ATP/ADP antiporter. Like the antiporter, thermogenin is synthesized without formation of a higher molecular weight precursor (154), differing in this respect from the $H_2PO_4^-$, H^+ symporter, which also resembles the antiporter in the domain composition.

Just as in the case of the ATP/ADP antiporter in muscle mitochondria, H^+ conductance by thermogenin in brown fat mitochondria is activated by 10^{-6}–10^{-5} M free fatty acids. Purine nucleotides inhibit this conductance (132). Both of these phenomena have been reproduced in thermogenin proteoliposomes (20, 173, 190).

It may be that thermogenin represents a monofunctional derivative of the ATP/ADP antiporter that has lost the main function of the latter, that is, the capability of exchanging adenine nucleotides. Instead, it specializes in an additional function of the antiporter, namely, fatty acid–mediated uncoupling. Within the framework of our hypothesis, thermogenin facilitates transmembrane flow of the fatty acid anions (181).

An interesting problem is the relation of thermogenin to Cl^- permeability which is enormously high in brown fat mitochondria (132) and in thermogenin proteoliposomes (190). Cl^- transport does not require fatty acids; it is blocked by purine nucleotides. All of these effects may be explained if we assume that Cl^- substitutes, certainly at much higher concentrations, for fatty acid anions in combining with cationic ligands in the thermogenin molecule to be transported across the membrane.

The conclusion about the non-specificity of the translocating machinery of thermogenin to the transported anions was recently confirmed by Garlid and coworkers (59, 78, 79). It was shown that brown fat mitochondria and thermogenin proteoliposomes demonstrate the purine nucleotide–sensitive transport of hexane sulfonate, as well as of a large group of monovalent, monopolar anions (alkylsulfonates, benzenesul-

fonate, oxohalogenides, hypophosphate, hexafluorophosphate, and pyruvate).

Several studies have been undertaken in order to learn the role of thermogenin in cold adaptation. It was found that the adaptive process is accompanied by an increase in the thermogenin mRNA level in the cytoplasm (154) and in the concentration of this protein in the mitochondria of brown fat (132).

Studies on whole animals, brown adipose tissue in situ, brown adipocytes, and isolated mitochondria revealed the following chain of events involved in a thermogenic response of this tissue to cold.

1. Cold receptors in the skin are activated and send signals to the thermoregulatory center in the hypothalamus.

2. From the hypothalamus, the signals are transmitted to brown fat via sympathetic neurons. Noradrenaline is released from nerve terminals into the intercellular spaces of the brown adipose tissue.

3. Noradrenaline binds to β-adrenergic receptors, which are localized on the outer surface of the brown fat cell plasma membrane.

4. β-adrenergic receptors, combining with noradrenaline, activate adenylate cyclase.

5. Adenylate cyclase produces cAMP from ATP.

6. cAMP switches on the protein kinase cascade, resulting in the activation of lipase.

7. Lipase releases fatty acids and glycerol when triglycerides in intracellular neutral fat droplets are hydrolyzed.

8. Fatty acids perform two functions: they are the respiratory substrates and the uncouplers dissipating the respiration-produced $\Delta\bar{\mu}_{H^+}$ in a thermogenin-mediated fashion.

Intracellular Power Transmission

It is generally believed that ATP diffusion is the mechanism that allows energy to be transported between various intracellular regions. In very large muscle cells, this mechanism is assumed to be supplemented with diffusion of creatine phosphate. However, some studies carried out in our group indicate that the intracellular power transmission can also proceed in the form of $\Delta\bar{\mu}_{H^+}$, spreading along membranes of extended mitochondria.

It is obvious that both the $\Delta\Psi$ and ΔpH constituents of $\Delta\bar{\mu}_{H^+}$ once they are formed across the membrane, immediately spread along it. The electrical conductance of the media on both sides of the coupling membrane is very high since these media are aqueous solutions of electrolytes. At the same time, conductance of the membrane may be extremely low. This means that $\Delta\Psi$, if produced by the $\Delta\bar{\mu}_{H^+}$ generator, cannot avoid fast irradiation over the membrane surface. ΔpH also must irradiate rather quickly because of the high rate of H^+ diffusion in water and the high concentration of mobile pH buffers. This means that the $\Delta\bar{\mu}_{H^+}$ produced by a $\Delta\bar{\mu}_{H^+}$ generator in a certain area of the membrane can, in principle, be transmitted, as such, along the membrane and transduced into work when used in another region of the same membrane. In 1969–1971 we extended this line of reasoning to the hypothesis that coupling membranes act as power-transmitting cables at the cellular level (175, 176).

Translated from Greek, the word "mitochondrion" means "thread-grain." This term was introduced by cytologists who used the light microscope. The first students of mitochondria always indicated that mitochondria may exist in two basic forms: *(1)* filamentous and *(2)* spherical or ellipsoid.

The development of the electron microscope and thin section techniques changed this opinion in such a way that filamentous mitochondria came to be regarded as a very rare exception, while the spherical shape was assumed to be canonical. This change of views stemmed from the fact that single section electron microscopy deals with a two-dimensional, rather than a three-dimensional picture of the cell. A structure viewed this way may be erroneously interpreted as one of a number of small grains unless it is reconstituted with the aid of many serial sections.

It is clear that if $\Delta\bar{\mu}_{H^+}$ transmission is confined to a single spherical mitochondrion, then this mechanism cannot be used to transport power for a distance comparable to the size of an eukaryotic cell. However, if mitochondria are, at least under certain conditions, filamentous, then the role of mitochondria as intracellular protonic cables may be discussed as a realistic hypothesis.

Three approaches have shaken the dogma of spherical mitochondria, namely: *(1)* the reconstitution of three-dimensional electron micrographs of the whole cell with the aid of serial thin sections; *(2)* high-voltage electron microscopy allowing one to increase the thickness of the studied preparation; and *(3)* the staining of mitochondria with fluorescent penetrating cations, making it possible to return to studies of mitochondria in the living cell by means of light microscopy.

The serial section method was applied first in studies on unicellular eukaryotes. Here, very large and complicated mitochondrial structures were detected. For example, in a flagellate, *Polytomella agilis,* Burton and Moore (29) described a single (!) mitochondrion which looked like a hollow, perforated sphere arranged immediately below the outer cell membrane so the cytoplasm and the nucleus proved to be packed into a mitochondrial "stringbag."

Later a single giant mitochondrion was described in some yeast cells. Threadlike, 10 μm–long mitochondria were found in exocrine cells of the pancreas. A chain of the long end-to-end arranged mitochondria was revealed in spermatozoa (179).

Muscle tissue seems to be one of the most interesting objects for studying intracellular power transmission. Multicellular muscle cells (fibers) are very large, and their energy requirements are extremely high. In a hardworking muscle, gradients of oxygen and substrates between the periphery and the core of the cell will arise, the effect limiting the scope of the work performed. $\Delta\bar{\mu}_{H^+}$ transmission from the muscle cell edge to its core along mitochondrial membranes might solve this problem. If so, the dimensions of muscle mitochondria will be particularly large.

The first indications of the existence of a mitochondrial system penetrating the muscle fibers of higher animals were obtained by Bubenzer (28), Gauthier and Padykula (60, 61), and Ogata and Murata (136). The authors studied random sections of rat diaphragmatic muscle, *Musculi semitendinosus,* and human intercostal muscle, respectively. Bakeeva et al. (11, 15) in our laboratory undertook a systematic investigation of serial sections of rat diaphragm (151). It was shown that in this tissue, the mitochondrial material is organized into networks piercing the I-band regions of the muscle cell near the Z-discs. The networks are connected with columns oriented perpendicular to their plane, that is, parallel to myofibrils. Moreover, there are branches, arranged parallel to Z-discs, connecting the networks with mitochondrial clusters at the fiber periphery. Such a system, defined as the mitochondrial reticulum *(Reticulum mitochondriale),* is found to be characteristic of the diaphragm of adult animals. It is absent from the diaphragm of rat embryos and newborn rats (11). Further study (12) revealed the time course of postnatal development of the mitochondrial framework in the diaphragm. Its formation was shown to be completed during the first 2 months after birth.

Generally, the mitochondrial reticulum may be organized in two cardinally different modes: as a giant organelle surrounded by continuous outer and inner membranes or, alternatively, as an assembly of many end-to-end associated mitochondria. An intermediate version provides for the case when the common outer membrane covers many mitoplasts, each of them being surrounded by its own inner membrane.

It should be mentioned in this context that sometimes the reticulum-forming mitochondrial filaments form dark partitions built of four membranes. The intermembrane spaces are filled with osmiophilic material. Apparently, these partitions represent junctions of two branches of the mitochondrial reticulum. This structure, discovered in our laboratory by Bakeeva et al. (11) in diaphragm, was later studied in detail by the same group in heart muscle since here mitochondrial junctions proved to be especially numerous (13, 14).

It was found that in this tissue, mitochondria also form a three-dimensional system, but instead of a thin, filamentous mitochondrial network found in diaphrgm and skeletal muscle, the heart has a multitude of thick, poorly branched organelles. However, all of them are coupled to each other by numerous junctions. The junction zone was found to represent a disk with a diameter of 0.1–1.0 μm. In this zone, membranes and intermembrane spaces are of higher density. Two outer membranes of contacting mitochondria appear to maximally approach each other in a manner similar to that observed in tight junctions between cell membranes. Each mitochondrion was shown to be connected with its neighbors by several such junctions. They were found in myocardiocytes with both contracted and relaxed myofibrils. On the other hand, we could not detect mitochondrial contacts in the hearts of 3-day-old rats. Later, mitochondrial contacts were described in the heart muscle of an invertebrate (135).

The electron microscopic method employed in the above studies deals with fixed material so it is not possible to follow directly the functioning of mitochondria. To overcome this limitation, we decided to return to light microscopy. Here, functional analysis is quite possible; but, the great disadvantage is low contrast, insufficient to see with certainty thin mitochondrial filaments and networks. The low resolution of the transmission light microscope, due to the wavelength of the visible light, can be overcome if we deal with light emission instead of light absorption. If the light emission is sufficiently strong, the light source will be seen in the dark, even if it cannot be observed as a light-absorbing body.

Thus the problem was how to obtain light emission from mitochondria. To do this, we applied the approach we introduced earlier to detect $\Delta\Psi$ formation across the mitochondrial membrane, that is, synthetic penetrating cations (107). The only modification of the method was that fluorescent cations were used. The idea was that fluorescent cations, electrophoretically moving into mitochondria, may be accumulated in the matrix space. A concentration of the cation inside may be 10^4 higher than outside mitochondria provided that $\Delta\bar{\mu}_{H^+}$ is in the form of $\Delta\Psi$. If one succeeds in finding such a cation, "invisible" mitochondrial filaments may be seen under the fluorescent microscope. In search of penetrating fluorescent cations, we turned our attention to rhodamines. These compounds are (1) positively charged, (2) somewhat hydrophobic, and (3) have a

high quantum yield of fluorescence. Moreover, rhodamine derivatives are known to specifically stain mitochondria in the living cell. This was shown more than 50 years ago in light transmission microscopy studies (80, 189).

As was found in our group by Severina (171), ethyl-substituted rhodamine, added to solutions washing the planar phospholipid membrane, generates a diffusion potential predicted by the Nernst equation. This means that ethylrhodamine may be regarded as a penetrating cation. In other experiments, carried out in our group by Zorov, ethylrhodamine was employed for detecting a mitochondrial reticulum in the diaphragmatic muscle.

Independently, Chen and coworkers found that the treatment of cultured fibroblasts with commercially available rhodamine-B–conjugated IgG antibody resulted in the staining of "snakelike structures" inside the cell (36), which later were identified as mitochondria (205). Systematic investigation of rhodamines 3B, 6G, and 123 and other fluorescent hydrophobic cations, such as cyanine dyes and safranin O, showed that they too are selectively accumulated in energized mitochondria inside the cell so that mitochondria become fluorescent. Discharging $\Delta\bar{\mu}_{H^+}$ by uncouplers or $\Delta\Psi$ by valinomycin + K^+, makes it possible to abolish mitochondrial fluorescence, while nigericin, converting ΔpH into $\Delta\Psi$, increases it. Not only fibroblasts but also primary cultures of bladder epithelium and of myocardiocytes were found to respond in such a way.

The main result of the fluorescent studies of mitochondria in the living cell is that very frequently they have the form of filaments, tens of micrometers in length. The filaments may be so long as to connect the cell core and the cell periphery, or even cross the cell from edge to edge. In some cases, mitochondrial networks were detected. A typical case in point is that two mitochondrial populations: (1) filamentous or network-forming and (2) small spherical, oval or rodlike organelles coexist in the same cell. Such a pattern was revealed in our laboratory by Zorov when the primary culture of human fibroblasts was studied (Fig. 4.15).

It should be mentioned that the preferable orientation of the filamentous mitochondria along the long axis of the cell, as seen in Figure 4.15, was shown to require microtubules. The addition of agents inducing decomposition of microtubules prevents this orientation, although the length of the mitochondrial filaments does not decrease.

At the same time, some drugs and other in vivo treatments were shown to provoke fragmentation of long mitochondria in the cell (for example, see ref. 199). In fact, fluorescent studies of mitochondria have confirmed the data obtained by old methods, indicating

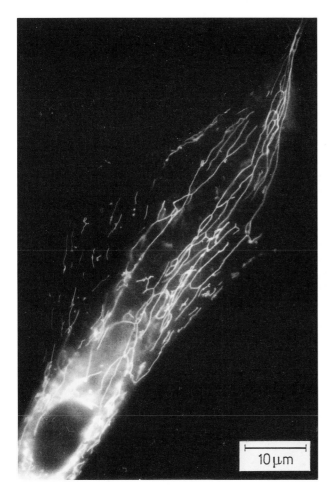

FIG. 4.15. Mitochondria revealed in a living human fibroblast by means of the penetrating fluorescent cation ethylrhodamine (Photograph by D. B. Zorov)

that thread-grain transition is a typical feature of these organelles (179, 180).

An obvious prediction of the hypothesis considering a filamentous mitochondrion as a power-transmitting cable is that the whole filament must be de-energized when any part becomes leaky. This prediction was verified in our group. Combining a laser and a fluorescent microscope, Zorov (1, 2, 51) succeeded in illuminating a single mitochondrial filament in a human fibroblast cell stained with ethylrhodamine. A very narrow laser beam (the diameter of the light spot was commensurate with the thickness of the mitochondrial filaments) was used to cause localized damage of the filament. As shown in Figure 4.16, the laser treatment resulted in the disappearance of the rhodamine fluorescence in the entire 50 μm–long filament. It is essential that (1) other filaments remain fluorescent so that the laser effect is not the result of the nonspecific damage of the cell and (2) the illuminated filament

retains its continuity when scrutinized under a phase-contrast or electron microscope (Fig. 4.16). In fact, no trace of laser-induced damage was found.

A similar study was performed on the myocardiocyte cell culture, showing that the illumination of a single mitochondrion gives rise to a quenching of a cluster composed of many mitochondria. Electron microscopic analysis revealed that mitochondria, which become quenched after such laser treatment, are connected with the illuminated mitochondria by intermitochondrial contacts, whereas those retaining fluorescence are not.

The simplest explanation of these findings is this: In a myocardiocyte, there are several "octopuslike" mitochondrial clusters formed by many mitochondria joined by mitochondrial junctions (we called such a cluster "Streptio mitochondriale" [2]). These junctions are of high electric conductance, and so the cluster is de-energized when at least one of the cluster-composing mitochondria becomes leaky (Fig. 4.17).

The operation of a filamentous mitochondrion or mitochondrial cluster as an H^+-conducting cable allows the long-distance diffusion of metabolites, ATP, ADP, and phosphate to be replaced with proton movement along mitochondrial membranes (178). Such a substitution may (1) accelerate power transmission and (2) direct the delivery of energy to a certain intracellular area.

On a cellular scale, the H^+ movement along the mitochondrial filament may be approximated as one-dimensional diffusion. This means that replacement of, for example, ATP diffusion in the cytosol with the diffusion of H^+ and mobile pH buffers along filamentous mitochondria is equivalent to the replacement of a process occurring in the three-dimensional space with that in the one-dimensional space. Such an effect undoubtedly saves time when energy should be delivered to a certain place inside the cell.

THE PLANT CELL

General Pattern of the Energy Transductions

Figure 4.18 shows the plant cell energy transduction pattern. In the light, energy is accumulated by the $\Delta\bar{\mu}_{H^+}$-generating photoredox chain (reaction 1), which is also reduced $NADP^+$ by electrons removed from H_2O (not shown in Fig. 4.18; see Fig. 4.19). These processes are localized in the thylakoid membrane of chloroplasts. In the same membrane, there is only one $\Delta\bar{\mu}_{H^+}$-consuming enzyme, namely, H^+-ATP synthase. Another H^+-ATP synthase is operating in the inner membrane of mitochondria. It utilizes $\Delta\bar{\mu}_{H^+}$ generated

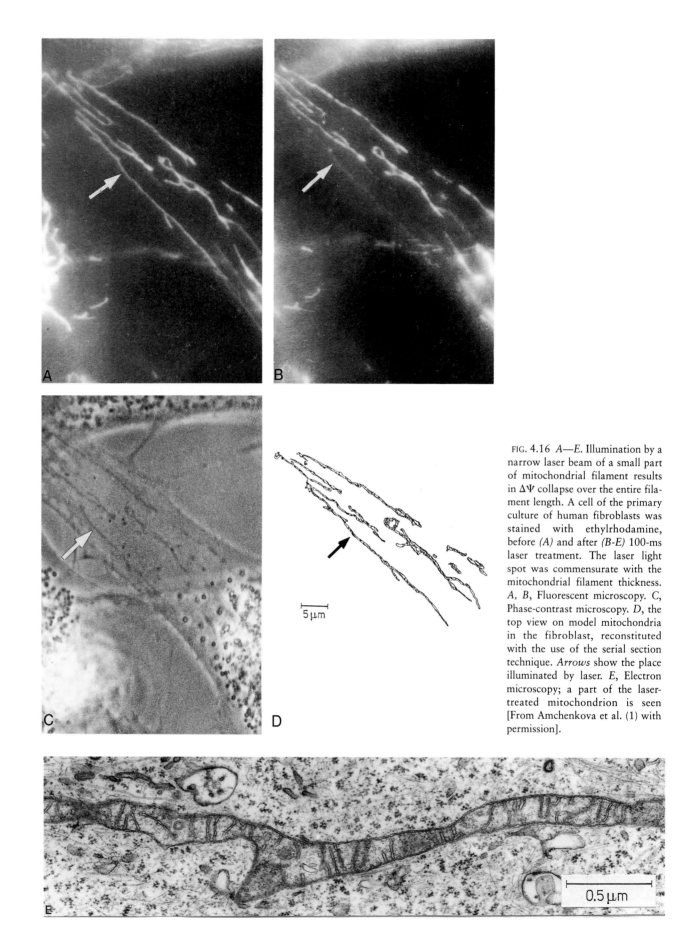

FIG. 4.16 *A—E*. Illumination by a narrow laser beam of a small part of mitochondrial filament results in $\Delta\Psi$ collapse over the entire filament length. A cell of the primary culture of human fibroblasts was stained with ethylrhodamine, before *(A)* and after *(B-E)* 100-ms laser treatment. The laser light spot was commensurate with the mitochondrial filament thickness. *A, B,* Fluorescent microscopy. *C,* Phase-contrast microscopy. *D,* the top view on model mitochondria in the fibroblast, reconstituted with the use of the serial section technique. *Arrows* show the place illuminated by laser. *E,* Electron microscopy; a part of the laser-treated mitochondrion is seen [From Amchenkova et al. (1) with permission].

5 μm

0.5 μm

FILAMENTOUS MITOCHONDRION

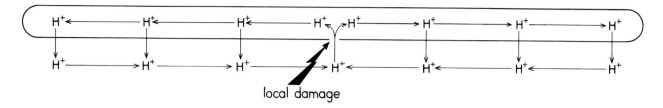

MITOCHONDRIAL CLUSTER (Streptio mitochondriale)

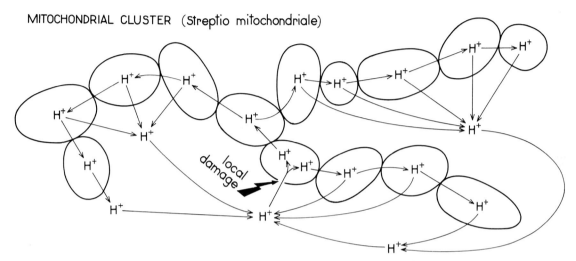

FIG. 4.17. De-energization of filamentous mitochondrion (fibroblast) or mitochondrial cluster (heart muscle cell) induced by local damage of the mitochondrial membrane. In the latter case, it is assumed that the mitochondrial junctions are H^+ permeable [From Skulachev, (180) with permission].

by the respiratory chain. ATP can also be formed by glycolysis.

In the outer cell membrane, an H^+-ATPase is found that hydrolyzes ATP and forms the $\Delta\bar{\mu}_{H^+}$ required to import substrates by H^+, substrate-symporters, and export sucrose by the H^+/sucrose-antiporter. The Na^+/H^+ antiporter forms $\Delta\bar{\mu}_{Na^+}$ which probably serves as a $\Delta\bar{\mu}_{H^+}$ buffer.

One more H^+-ATPase is localized in the membrane-surrounding vacuole (tonoplast). The $\Delta\bar{\mu}_{H^+}$ formed is used by some porters to create gradients of substances between cytosol and vacuole. In the same membrane, there is H^+-pyrophosphatase (H^+-PPase) which produces $\Delta\bar{\mu}_{H^+}$ at the expense of hydrolysis of the inorganic pyrophosphate.

Photoredox Chain

The flow diagram of the photosynthetic redox chain is shown in Fig. 4.19. It is based on the so-called Z-scheme postulated by Hill and Rendall (72). The mechanism includes two types of photosynthetic reaction center complexes (photosystems I and II) and several enzymes, catalyzing the dark reactions of water oxida-

tion, Q-cycle, and $NADP^+$ reduction. Water and $NADH^+$ serve as the initial reductant and final oxidant of this thylakoid membrane–linked system, respectively (the redox potential difference, 1.2 V).

The process is initiated by absorption of a photon by antenna chlorophyll (not shown in Fig. 4.19). Excitation migrates from the antenna chlorophyll to a photosystem chlorophyll. If this is the chlorophyll dimer of photosystem II ($\lambda_{max} = 680$ nm), it is oxidized by pheophytin via a Chl_{680} monomer. From pheophytin, an electron is transferred to the bound plastoquinone (PQ_A), connected in some way with a non-heme iron. The electron then reaches another bound plastoquinone molecule (PQ_B). The chlorophyll cation radical $(Chl_{680})^{\ddagger}_2$ accepts an electron from a tyrosine residue of a photosystem II subunit, which, in turn, is reduced by electrons removed from the water by means of a complicated water-splitting system containing four atoms of manganese. Electrons, protons, and molecular oxygen are released when the water molecule is split. The released protons appear in the intrathylakoid space.

As to the reduced PQ, it is oxidized by the PQH_2-plastocyanin reductase (cytochrome b_6f complex),

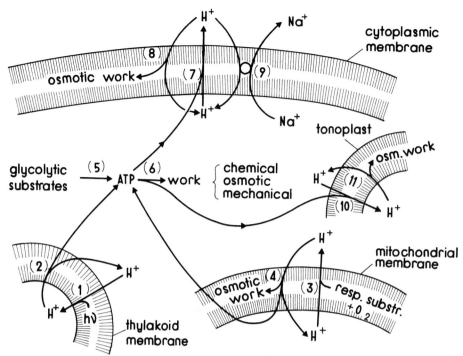

FIG. 4.18. General pattern of the energy transductions in the plant cell. *(1)* Photoredox chain pumps H^+ to the thylakoid interior. *(2)* H^+-ATP-synthase forms ATP, which is coupled to the downhill H^+ efflux from thylakoid. *(3)* H^+ is pumped from the mitochondrial matrix space by the respiratory chain. *(4)* Downhill H^+ influx to matrix is coupled to ATP synthesis or to performance and "osmotic" work (for example, uptake of solutes by mitochondria via H^+, solute-symporters). *(5)* ATP is formed by glycolysis. *(6)* ATP is utilized to perform chemical, "osmotic," and mechanical work. *(7)* ATP is hydrolyzed by the plasma membrane H^+-ATPase which pumps H^+ from the cell. *(8)* Downhill H^+ movement supports "osmotic" work of the outer cell membrane. *(9)* Na^+ is pumped from the cell by Na^+/H^+-antiporter. *(10)* H^+ is pumped to vacuole by the tonoplast H^+-ATPase. *(11)* Downhill H^+ efflux from vacuole supports "osmotic" work [From Skulachev, (182) with permission].

which is an analogue of the bc_1 complex. According to the scheme, the complex is comprised of two hemes of cytochrome b_6 (low and high potential), FeS_{III}, and cytochrome f that functions in much the same way as cytochrome c_1. The b_6f complex is assumed to organize the Q-cycle in a manner similar to that already described in mitochondria. As a result, the transport of one electron from PQ to cytochrome f is coupled to *(1)* the translocation of one charge across the thylakoid membrane, *(2)* the release of two H^+ ions into the thylakoid interior, and *(3)* the absorption of two H^+ ions from the chloroplast stroma.

From cytochrome f, the electron moves to the Cu-containing protein plastocyanin and reduces the cation radical of photosystem I chlorophyll dimer ($\lambda_{max} = 700$ nm). $(Chl_{700})^+_2$ is formed as a result of the photo-oxidation of the excited $(Chl_{700})_2$. From $(Chl_{700})_2$ the electrons are transferred via Chl_{693} and vitamin K_1 to a primary stable acceptor, the four-nuclear (4Fe–4S) iron–sulfur component (FeS_x). At the next step, the electrons reach four other nuclear clusters (FeS_B, FeS_A),

to the binuclear cluster of the 11 kD protein ferredoxin, and further on to a flavoprotein containing FAD as the prosthetic group. From FAD, the reducing equivalents reach $NADP^+$. The formed NADPH is oxidized by a water-soluble enzyme system reducing CO_2 to carbohydrates, $(—CHOH—)_n$.

Taking into account the operation of the water-splitting system and the Q-cycle, one can estimate the quantum efficiency of the transmembrane H^+ pumping by this redox chain as $6H^+$ per $4h\nu$. This corresponds to the $H^+/\bar{e}$ ratio equal to 3 (the number of pumped H^+ ions per each electron transferred from H_2O to CO_2).

Within the framework of the scheme in Figure 4.19, there are three $\Delta\bar{\mu}_{H^+}$-generators in the photosynthetic redox chain of chloroplast: two light-dependent generators (the transmembrane electron transports from Chl_{680} to the bound PQ and from Chl_{700} to Fd, called photosystem I and photosystem II) and one light-independent generator (the b_6f-complex) (for reviews, see refs. 18, 91, 92, 124, 133, 155, 214).

Photosystem I. Photosystem I was shown to contain at least five different polypeptides with apparent molecular masses of 83, 82, 22, 19, and 9 kD. There are some indications that Chl_{700}, vitamin K_1, and FeS_x are bound to two large subunits, whereas FeS_A and FeS_B are associated with 9 kD subunit (Fig. 4.20). The primary structures of the large photosystem I polypeptides have already been established. They proved to show 45% homology (for reviews, see refs. 91, 124).

Besides the Z-scheme, there is an alternative way the photosynthetic redox chain functions in chloroplasts, namely, the cyclic electron transport around photosystem I with no photosystem II involved. In such a case, reducing equivalents are transported from ferredoxin back to $(Chl_{700})^+_2$ instead of being accepted by $NADP^+$. It is suggested that ferredoxin reduces PQ so that photosystem II appears to be unnecessary to regenerate $(Chl_{700})_2$ from $(Chl_{700})^+_2$ (for reviews, see refs. 18, 44, 91).

The $\Delta\bar{\mu}_{H^+}$ formation mediated by photosystem I includes very fast ($\tau < 100$ ns) $\Delta\Psi$ generation. This process is much faster than any reaction other than electron transfer from $(Chl_{700})_2$ to FeS_X. The generated $\Delta\Psi$ constitutes a major part of the total electrogenesis linked with the operation of photosystem I. There is nothing surprising about these relationships: it is known that

plastocyanin faces the thylakoid interior, whereas ferredoxin and ferredoxin-$NADP^+$ reductase face the thylakoid exterior. Therefore, the photosystem I–mediated transport of reducing equivalents must cross the membrane. Measurements showed that the transmembrane movement of electrons is electrically not compensated and is therefore electrogenic. Thus, photosystem I may roughly be described in terms of the Mitchell electron-transferring half-loop (120).

Photosystem II. There are nine types of polypeptides in the photosystem II complex (Fig. 4.21). Their molecular masses are 56, 52, 39.5, 39, 33, 23, 17, 9, and 4.5 kD. Amino acid sequences of 39.5 and 39 kD polypeptides are similar to the L- and M-subunits of the bacterial reaction center complex (70, 105). The 39.5 and 39 kD polypeptides have sites for binding the primary and secondary quinones, PQ_A and PQ_B. The smallest subunits (9 and 4.5 kD) were shown to form cytochrome b_{559}. Each of them contains one histidine involved in heme coordination (124).

Chl_{680} most probably forms a dimer. The redox potential of the Chl_{680} dimer in the ground state is extremely high (about $+1.1$ V), which explains why $(Chl_{68})^+_2$ can serve as an oxidant for water (the redox potential of the H_2O/O_2 pair at neutral pH is $+0.82$

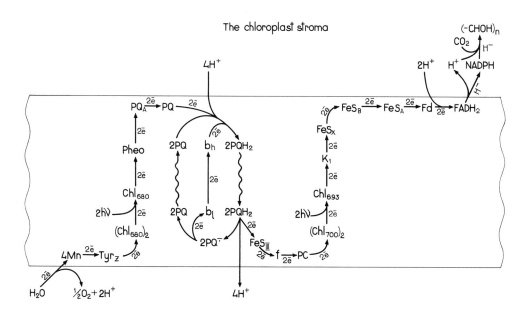

FIG. 4.19. The non-cyclic photosynthetic redox chain of chloroplasts. *Tyr_z*, tyrosine-161, electron donor of the photosystem II; *Chl_680*, chlorophyll of the photosystem II; *Pheo*, pheophytin; *PQ_A* and *PQ_B*, plastoquinones bound to photosystem II; PQ, plastoquinone pool; *b_h* and *b_l*, high- and low-potential hemes of cytochrome b_6; *FeS_III*, iron-sulfur cluster of b_6f complex; *f*, cytochome *f*; *PC*, plastocyanin; *Chl_700* and *Chl_693*, chlorophylls of photosystem I; *FeS_X*, the primary stable electron acceptor of photosystem I; *FeS_A* and *FeS_B*, two iron-sulfur clusters tightly bound to the photosystem I; *Fd*, ferredoxin.

The chloroplast stroma

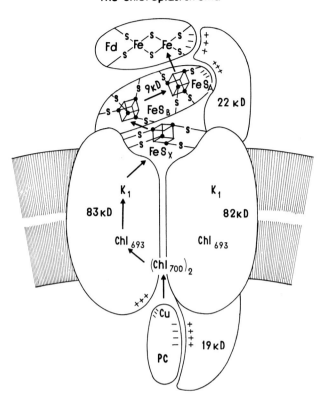

The thylakoid lumen

FIG. 4.20. A tentative scheme of the photosystem I reaction center complex. *83, 82, 22, 19,* and *9 kD,* subunits of corresponding molecular masses; PC, water-soluble protein plastocyanin; *(Chl$_{700}$)$_2$* and *Chl$_{693}$*, the photosystem I chlorophyll dimer and monomer, respectively; *K$_1$*, tightly bound vitamin K$_1$ (phylloquinone); *FeS$_x$*, the 4Fe–4S cluster operating as the primary stable electron acceptor; *FeS$_B$* and *FeS$_A$*, the 4Fe–4S clusters taking part in the electron transfer from FeS$_x$ to Fd; *Fd*, water-soluble protein ferredoxin containing 2Fe–2S clusters; *arrows*, the electron transfer pathway [Adapted from Knaff (91) with permission].

V). In contrast, the redox potential of the excited (Chl$_{680}$)$_2$ is more negative than −0.6 *V*. Intermediate acceptors of the electron removed from Chl$_{680}$ are probably the Chl$_{680}$ monomer and pheophytin (the redox potential of the latter is −0.61 *V*).

The primary stable electron acceptor was shown to be tightly bound plastoquinone PQ$_A$. An additional bound plastoquinone, PQ$_B$ participates in the further electron transfer along the photosynthetic redox chain. Between PQ$_A$ and PQ$_B$ there is a non-heme iron. Histidine residues serve the Fe^{2+} as ligands (155).

The role of the primary electron donor for (Chl$_{680}$)$_2$$^+$ is played by the 39 kD subunit tyrosine-161 (116). The half-time of the Tyr-(Chl$_{680}$)$_2$ oxidoreduction varies, depending upon conditions, from 50 ns to 500 ns.

The water-splitting system is used as the reductant for Tyr-161. It is composed of 33, 23, and 17 kD polypeptides and an additional protein containing four Mn^{2+}. It has been suggested that the protein in question is identical to the 32 kD subunit carrying tyrosine-161 (for review, see ref. 155).

The role of cytochrome b_{559} found in the photosystem II reaction center complex is not clear. It has been speculated that it participates in a hypothetical system for the cyclic transport of reducing equivalents around photosystem II. This may protect photosystem II against photodamage at high light intensities (30).

At present, the mechanism of $\Delta\bar{\mu}_{H^+}$ generation by photosystem II can be pictured by analogy with the more elaborate photosynthetic reaction centers of the purple bacteria. Apparently it consists of the light-dependent transmembrane movement of an electron from the excited pigment (Chl$_{680}$)$_2$ to the primary stable acceptor (PQ$_A$) localized on the opposite sides of the membrane hydrophobic barrier. Just this mechanism was demonstrated for *Rhodopseudomonas viridis* reaction centers by means of structural (x-ray study of atomic resolution) and functional (direct measurements of the $\Delta\Psi$ generation by a voltmeter) investigations (47, 52, 179). As was shown in these studies, about 90% of accumulated energy is stored due to transmem-

The chloroplast stroma

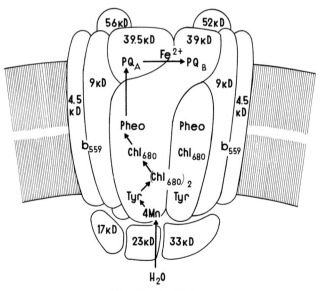

The thylakoid lumen

FIG. 4.21. A tentative scheme of the photosystem II reaction center complex. *56, 52, 39.5, 39, 33, 23, 17, 9,* and *4.5 kD,* subunits of corresponding molecular mass; *(Chl$_{680}$)$_2$* and *Chl$_{680}$*, the photosystem II chlorophyll dimer and monomer, respectively; *Pheo*, pheophytin; *PQ$_A$* and *PQ$_B$*, bound plastoquinones; *arrows*, the electron transfer pathway [Adapted from Rutherford (155) with permission].

brane movement of electron jumping from one redox group to another (direction of the electron flow was found to be from the outer membrane surface to the cytoplasmic membrane surface). The rest (10%) are accumulated because of movement of H^+ ions in the opposite direction. The secondary quinone is the point at which electron and proton meet to form quinol.

A mysterious feature common to the bacterial photoredox center complex, as well as to photosystems I and II, is the presence of a nonfunctional second branch (in Figs. 4.20 and 4.21, the right-side chlorophyll monomers and pheophytin or vitamin K_1) (for discussion see ref. 179).

Plastoquinol-Plastocyanin Reductase. The Q-cycle system is organized in chloroplasts by PQH_2-plastocyanin reductase (the b_6f complex), which is a functional analogue of mitochondrial complex III. The b_6f complex is composed of at least four polypeptides, three of which contain redox centers, namely, cytochrome f (32 kD), two-heme cytochrome b_6 (23 kD), and FeS protein (20 kD). The fourth subunit (17 kDa) does not directly participate in the electron transfer (40, 145). Both hemes in cytochrome b_6 have α-maxima at 563 nm. They differ in their redox potentials, which are -0.15 V and -0.03 V for b_1 and b_h, respectively (41). Cytochrome b_6 also contains five histidines, four of which participate in the coordination of two iron ions in two hemes. Both hemes appear to be arranged in the hydrophobic membrane core perpendicular to the plane of the thylakoid membrane. It is essential that the distance between the heme edges, according to the model, is 12 Å and that between the Fe atoms is 20 Å. This means that an electron moving from b_1 to b_h crosses about half of the hydrophobic membrane core. This is in agreement with the measurements performed on the bc_1 complex in mitochondria, which showed that $b_1 \rightarrow b_h$ electron transfer is responsible for about 40% of $\Delta\bar{\mu}_{H^+}$ generation by the Q-cycle systems (93). The rest (60%) of $\Delta\bar{\mu}_{H^+}$ is apparently formed as a result of the proton movement, as shown in Figure 4.19. It is important that the redox potential difference between hemes b_1 and b_h corresponds to half of the $\Delta\bar{\mu}_{H^+}$ value produced by $\Delta\bar{\mu}_{H^+}$ generators.

A putative partner of cytochrome b_6 in PQH_2 oxidation is the iron–sulfur protein, similar to that found in the mitochondrial complex III (FeS$_{III}$). There is little available information on this electron carrier. However, the next component of the redox chain, cytochrome f, has been studied in detail.

The amino acid sequence of the 285 residue cytochrome f protein shows that the Cys-X-Y-Cys-His sequence, which is characteristic of covalent heme binding by a c-type cytochrome, occurs at residues 21–25 in the N-terminal segment. His-25 and Lys-145 or Lys-222 serve as the heme ligands. The heme is localized in the large (residues 1–250) N-terminal domain exposed to the water in the thylakoid lumen. The residues 251–270 are hydrophobic and presumably form an α-helical column that functions as a membrane-spanning anchor.

The functionally analogous mitochondrial cytochrome c_1 shares a relatively small sequence homology with cytochrome f, but it also has a heme-binding site and a hydrophobic anchor not far from the N- and the C-ends, respectively (43). The cytochrome f spectrum shows an α-maximum at 555 nm. The redox potential is $+365$ mV.

The average diameter of the b_6f complex is about 8.5 nm. The complexes tend to form dimers.

H^+-ATP-Synthase, H^+-ATPases and H^+-Pyrophosphatase

The photoredox chain-produced $\Delta\bar{\mu}_{H^+}$ is utilized by H^+-ATP-synthase of F_0F_1 type, which resembles its mitochondrial analogue (see earlier under H^+-ATP-Synthase and H^+-ATPases). Factor F_1 (called here CF_1) has the $\alpha_3\beta_3\gamma\delta\epsilon$ structure. The ϵ-subunit is homologous to the bacterial protein and operates like the protein inhibitor of mitochondrial F_1. Factor CF_0 contains four types of subunits, called I, II, III, and IV. Subunit III is homologous to the DCCD-binding subunit 9 of mitochondrial factor F_0 (71). The I, III, IV, α-, β-, and ϵ-subunits of the chloroplast F_0F_1 complex were shown to be synthesized in chloroplasts, with β- and ϵ-genes being interconnected (85, 202). It was found that F_1 subunits of mitochondria, coexisting with chloroplasts within the same plant cell, differ in their antigenic sites from homologous CF_1 subunits (188).

The outer cell membrane (plasmalemma) of plants and fungi contains H^+-ATPase, the enzyme extruding H^+ ions from the cytoplasm at the expense of ATP energy. The nature of H^+ transport is electrogenic. The enzyme belongs to the P-type of ATPases forming a vanadate- and hydrohylamine-sensitive aspartyl phosphate intermediate when ATP is hydrolyzed. Diethylstilbestrol and DCCD are also inhibitory (114, 170).

This H^+-ATPase is a single polypeptide of the 100 kD molecular mass (for the primary structure, see ref. 170). Its content in the plasmalemma reaches 15% of the total protein. The mechanisms of ATP hydrolysis was suggested to be the same as in other P-ATPase (see earlier under Na^+/K^+-ATPase and Ca^{2+}-ATPase).

There must be some physiological significance to the fact that the pH optimum of the plasmalemmal H^+-ATPase is between 6.0 and 7.0. At pH greater than 7.0, the activity of the enzyme decreases considerably. It could be hypothesized that the enzyme is used to prevent acidification of the cytoplasm by removing excess H^+ ions when the intracellular pH drops below the neutral level. Apparently, pH regulation is a function of the outer cell membrane H^+-ATPase. At the same time, it is obvious that this ATPase also performs another function, that of formation of $\Delta\bar{\mu}_{H^+}$, which is then utilized by H^+ solute symporters and H^+/solute antiporters. In this way, uphill importation of metabolites into the cell is organized. Sucrose produced in green tissues of the plant as a result of photosynthesis is exported from the cells by the sucrose/H^+ antiporter.

In several laboratories, indications have been obtained that a vanadate-sensitive ATPase is also located in the outer membrane of the chloroplast (reviewed in ref. 179).

Vacuolar H^+-ATPase is localized in tonoplasts, the membrane of the plant (or fungal) vacuole. It is well known that there are some solute gradients between the cytosol and the vacuole. A special H^+-ATPase was identified as a device that creates a driving force for uphill solute transport across tonoplast. The ATP-binding sites of this ATPase face the cytosol. ATP hydrolysis is coupled to an electrogenic H^+ influx into the vacuole.

Vacuolar ATPase consists of detachable (F_1-like) and membranous (F_0-like) subcomplexes. The former is composed of 67, 55, 52, 44, and 32 kD subunits whereas the latter consists of subunits of 100 and 16 kD. The smallest subunit binds DCCD- and N-ethylmaleimide (123, 127, 141). The enzyme is resistant to vanadate, diethylstibestrol, oligomycin, and azide. Sequence homology with V-type H^+-ATPase of the animal secretory granules and lysosomes has been demonstrated (103, 127).

An additional $\Delta\bar{\mu}_{H^+}$ generator recently described in plant tonoplasts was H^+-pyrophosphatase (35, 152, 162). This enzyme is composed of two 73 kD subunits (110, 164).

As for $\Delta\bar{\mu}_{H^+}$ consumers other than H^+-ATP-synthase, they are presented in the plant cell by previously mentioned H^+/solute-symporters or H^+/solute-antiporters in the plasma membrane and tonoplast. In rare cases, $\Delta\bar{\mu}_{H^+}$ dissipation is used for thermoregulatory heat production. For instance, in some plants fluorescence accompanied strong increases in the temperature in flowers, resulting in volatization of attractants. There are some indications that fatty acids are also involved in this process (for review, see ref. 179).

CONCLUSION

Consideration of energy transduction mechanisms in animal and plant cells allows us to draw some general conclusions concerning bioenergetics:

1. The cell avoids direct utilization of external energy sources in the performance of useful work. It transforms the energy of these sources to a convertible energy currency, that is, ATP, $\Delta\bar{\mu}_{H^+}$ or $\Delta\bar{\mu}_{Na^+}$ which is then expended to support various types of energy-consuming processes.

2. Cells always possess at least two energy currencies, one that is water soluble (ATP); the other, membrane-linked ($\Delta\bar{\mu}_{H^+}$ and/or $\Delta\bar{\mu}_{Na^+}$). In animals, intracellular membranes (inner membrane of mitochondria, membranes of secretory granules, endosomes, lysosomes, etc.) serve $\Delta\bar{\mu}_{H^+}$ while the outer cell membrane serves $\Delta\bar{\mu}_{Na^+}$ as the convertible energy currency. In plant cells, it appears that all coupling membranes deal through $\Delta\bar{\mu}_{H^+}$.

3. It is $\Delta\bar{\mu}_{H^+}$ that is primarily produced by respiration and photosynthesis, the main energy-yielding processes of eukaryotic cells. Glycolysis and oxidative decarboxylation of α-ketoglutarate that primarily produce ATP and GTP, respectively, are quantitatively supplementary. However, they appear to be of vital importance under conditions when the main processes cease (anaerobiosis and, in plants, darkness).

4. To perform mechanical work, ATP is utilized in eukaryotic cells. To support uphill transport of solutes, $\Delta\bar{\mu}_{H^+}$ is usually used in the plant cell and in intracellular organelles of the animal cell, whereas $\Delta\bar{\mu}_{Na^+}$ serves this purpose in the animal cell plasma membrane. Biosyntheses, as a rule, utilize nucleotide triphosphate energy. However, in some reductive biosyntheses, a high NADPH/$NADP^+$ ratio is used as an additional driving force. The H^+-nicotinamide nucleotide transhydrogenase keeps the NADPH/$NADP^+$ ratio at a high level.

REFERENCES

1. Amchenkova, A. A., L. E. Bakeeva, Yu. S. Chentsov, and V. P. Skulachev. Coupling membranes as energy-transmitting cables. I. Filamentous mitochondria in fibroblasts and mitochondrial clusters in cardiomyocytes. *J. Cell. Biol.* 107: 481–495, 1988.

2. Amchenkova, A. A., L. E. Bakeeva, V. A. Drachev, D. B. Zorov, V. P. Skulachev, and Yu. S. Chentsov. Mitochondrial electric cable. *Vestnik MGU. ser. biol.* 3: 3–15, 1986 (Russ.).

3. Anderson, W. M. and R. R. Fisher. The subunit structure of bovine heart mitochondrial transhydrogenase. *Biochim. Biophys. Acta* 635: 194–199, 1981.

4. Andreev, I. M., V. Yu. Artzatbanov, A. A. Konstantinov, and V. P. Skulachev. Cyanide binding with ferricytochrome a_3 in rat liver mitochondria. *Dokl. Acad. Nauk SSSR* 244: 1013–1018, 1979 (Russ.).

5. Andreyev, A. Yu., T. O. Bondareva, V. I. Dedukhova, E. N. Mokhova, V. P. Skulachev, and N. I. Volkov. Carboxyatractylate inhibits the uncoupling effect of free fatty acids. *FEBS Lett.* 226: 265–269, 1988.

6. Andreyev, A. Yu., T. O. Bondareva, V. I. Dedukhova, E. N.

Mokhova, V. P. Skulachev, L. M. Tsofina, N. I. Volkov, and T. V. Vygodina. The ATP/ADP-antiporter is involved in the uncoupling effect of fatty acids on mitochondria. *Eur. J. Biochem.* 182: 585–592, 1989.

7. Arai, H., G. Terres, S. Pink, and M. Forgas. Topography and subunit stoichiometry of the coated vesicle proton pump. *J. Biol. Chem.* 263: 8796–8802, 1988.

8. Artzatbanov, V. Yu., A. A. Konstantinov, and V. P. Skulachev. Involvement of intramitochondrial protone in redox reactions of cytochrome *a*. *FEBS Lett.* 87: 180–185, 1978.

9. Azzi, A. and M. Müller. Cytochrome *c* oxidases: polypeptide composition, role of subunits, and location of active metal centers. *Arch. Biochem. Biophys.* 280: 242–251, 1990.

10. Babcock, G. T. and M. Wikström. Oxygen activation and the conservation of energy in cell respiration. *Nature* 356: 301–309, 1992.

11. Bakeeva, L. E., Yu. S. Chentsov, and V. P. Skulachev. Mitochondrial framework *(Reticulum mitochondriale)* in rat diaphragm muscle. *Biochim. Biophys. Acta* 501: 349–369, 1978.

12. Bakeeva, L. E., Yu. S. Chentsov, and V. P. Skulachev. Ontogenesis of mitochondrial reticulum in rat diaphragm muscle. *Eur. J. Cell Biol.* 25: 175–181, 1981.

13. Bakeeva, L. E., Yu. S. Chentsov, and V. P. Skulachev. Intermitochondrial contacts in myocardiocytes. *J. Mol. Cell. Cardiol.* 15: 413–420, 1983.

14. Bakeeva, L. E., A. A. Shevelev, Yu. S. Chentsov, and V. P. Skulachev. A freeze-fracture study of mitochondrial junctions in rat cardiomyocytes. *Biol. membrany* 2: 133–143, 1985 (Russ.).

15. Bakeeva, L. E., V. P. Skulachev, and Yu. S. Chentsov. Mitochondrial reticulum: organization and possible functions of a novel intracellular structure in the muscle tissue. *Vestnik MGU, ser. biol.* 3: 23–28, 1977 (Russ.).

16. Bazhenov, Yu. I. *Thermogenesis and Muscle Activity in the Cold Adaptation.* Leningrad: Nauka, 1981 (Russ.).

17. Bianchet, M., X. Ysern, J. Hullihen, P. L. Pedersen, and L. M. Amzel. Mitochondrial ATP synthase. *J. Biol. Chem.* 266: 21197–21201, 1991.

18. Blankenship, R. E. and R. C. Prince. Excited-state redox potentials and the Z scheme of photosynthesis. *Trends Biochem. Sci.* 10: 382–383, 1985.

19. Boekema, E. J., J. A. Berden, and M. G. Van Heel. Structure of mitochondrial F$_1$-ATPase studied by electron microscopy and image processing. *Biochim. Biophys. Acta* 851: 353–360, 1986.

20. Bouilland, F., D. Ricquier, T. Gulik-Krzywicki, and C. M. Gary-Bobo. The possible proton translocating activity of the mitochondrial uncoupling protein of brown adipose tissue. *FEBS Lett.* 164: 272–276, 1983.

21. Boyer, P. D. A perspective of the binding change mechanism for ATP synthesis. *FASEB J.* 3: 2164–2178, 1989.

22. Boyer, P. D., R. L. Cross, and W. Momsen. A new concept for energy coupling in oxidative phosphorylation based on a molecular explanation of the oxygen exchange reactions. *Proc. Natl. Acad. Sci. USA* 70: 2837–2839, 1973.

23. Bragg, P. D. The ATPase complex of *Escherichia coli. Can. J. Biochem. Cell Biol.* 62: 1190–1197, 1984.

24. Bränden, C. I. and H. Eklund. Structure and mechanism of liver alcohol dehydrogenase, lactate dehydrogenase and glyceraldehyde-3-phosphate dehydrogenase. In: *Dehydrogenases Requiring Nicotinamide Coenzymes,* edited by J. Jeffery. Basel: Birkhäuser Verlag, 1990, pp. 41–84.

25. Brown, G. C. and M. D. Brand. Proton/electron stoichiometry of mitochondrial complex I estimated from the equilibrium thermodynamic force ratio. *Biochem. J.* 252: 473–479, 1988.

26. Brustovetsky, N. N., V. I. Dedukhova, M. V. Yegorova, E. N. Mokhova, and V. P. Skulachev. Inhibitors of the ATP/ADP antiporter suppress stimulation of mitochondrial respiration and H$^+$ permeability by palmitate and anionic detergents. *FEBS Lett.* 272: 187–189, 1990a.

27. Brustovetsky, N. N., Z. G. Amerkanov, M. V. Yegorova, E. N. Mokhova, and V. P. Skulachev. Carboxylate-sensitive uncoupling in liver mitochondria from ground squirrels during hibernation and arousal. *FEBS Lett.* 272: 190–192, 1990b.

28. Bubenzer, H.-J. Die dünnen und die dicken Muskelfasern des Zwerchfells der Ratte. *Z. Zellforsch.* 69: 520–550, 1966.

29. Burton, M. and J. Moore. The mitochondrion of the flagellate, *Polymella agilis. Ultrastruct. Res.* 48: 414–419, 1974.

30. Canaani, O. and M. Havaux. Evidence for a biological role in photosynthesis for cytochrome *b-559*—component of photosystem II reaction center. *Proc. Natl. Acad. Sci. USA* 87: 9295–9299, 1990.

31. Capitanio, N., G. Capitanio, E. De Nitto, G. Villani, and S. Papa. H$^+$/$\bar{e}$ stoichiometry of mitochondrial cytochrome complexes reconstituted in liposomes. *FEBS Lett.* 288: 179–182, 1991.

32. Carafoli, E. Intracellular calcium regulation, with special attention to the role of the plasma membrane calcium pump. *J. Cardiovasc. Pharmacol* 12: S77–S84, 1988.

33. Carafoli, E., G. Inesi, and B. P. Rosen. Calcium transport across biological membranes. In: *Metal Ions in Biological Systems,* edited by H. Sigel. New York: Marcel Dekker Inc., 1984, v. 17, pp. 130–186.

34. Casey, R. P. Membrane reconstitution of the energy-conserving enzymes of oxidative phosphorylation. *Biochim. Biophys. Acta* 768: 319–347, 1984.

35. Chanson, A., J. Fichmann, D. Spear, and L. Taiz. Pyrophosphate-driven proton transport by microsomal membranes of corn coleoptiles. *Plant Physiol.* 79: 159–164, 1985.

36. Chen, L. B., I. C. Summerhayes, L. V. Johnson, M. L. Walsh, S. D. Bernal, and T. J. Lampidis. Probing mitochondria in living cells with rhodamine 123. *Cold Spring Harbor Symp. Quant Biol.* 46: 141–155, 1982.

37. Cidon, S., H. Ben-David, and N. Nelson. ATP-driven proton fluxes across membranes of secretory organelles. *J. Biol. Chem.* 258: 11684–11688, 1983.

38. Cidon, S., and N. Nelson. Novel ATPase in the chromaffin granule membrane. *J. Biol. Chem.* 258: 2892–2898, 1983.

39. Cidon, S., and N. Nelson. Purification of N-ethylmaleimide-sensitive ATPase from chromaffin granule membranes. *J. Biol. Chem.* 261: 9222–9227, 1986.

40. Clark, R. D., and G. Hind. Isolation of a five-polypeptide cytochrome *b–f* complex from spinach chloroplasts. *J. Biol. Chem.* 258: 10348–10354, 1983.

41. Clark, R. D., and G. Hind. Spectrally distinct cytochrome *b-563* components in a chloroplast cytochrome *b–f* complex: interaction with a hydroxyquinoline N-oxide. *Proc. Natl. Acad. Sci. USA* 80: 6249–6253, 1983.

42. Colowick, S. P., N. O. Kaplan, E. F. Neufield, and M. M. Ciotti. Pyridine nucleotide transhydrogenase. I. Indirect evidence for the reaction and purification of the enzyme. *J. Biol. Chem.* 195: 95–106, 1952.

43. Cramer, W. A., W. R. Widger, R. G. Herrmann, and A. Trebst. Topography and function of thylakoid membrane proteins. *Trends Biochem. Sci.* 10: 125–129, 1985.

44. Crofts, A. R. and Wraight, C. A. The electrochemical domain of photosynthesis. *Biochim. Biophys. Acta* 426: 149–185, 1983.

45. Cross, R. L. and P. D. Boyer. Evidence for detection of AT^{32}P bound at the coupling sites of mitochondrial oxidative phosphorylation. *Biochem. Biophys. Res. Commun.* 51: 56–59, 1973.

46. Danielson, L. and L. Ernster. Energy-dependent reduction of triphosphopyridine nucleotide by reduced diphosphopyridine nucleotide, coupled to the energy-transfer system of the respiratory chain. *Biochem. Z.* 338: 188–205, 1963.

47. Deisenhofer, J., O. Epp, K. Miki, R. Huber, and H. Michel. X-ray structure analysis of a membrane protein complex. Electron density map at 3 Å resolution and a model of the chromophores of the photosynthetic reaction center from *Rhodopseudomonas viridis. J. Mol. Biol.* 180: 385–398, 1984.

48. Dimroth, P. Sodium ion transport decarboxylases and other aspects of sodium ion cycling in bacteria. *Microbiol. Rev.* 51: 320–340, 1987.

49. Dontsov, A. E., L. L. Grinius, A. A. Jasaitis, I. I. Severina, and V. P. Skulachev. A study on the mechanism of energy coupling in the redox chain. I. Transhydrogenase: the fourth site of the redox chain energy coupling. *J. Bioenerg. Biomembr.* 3: 277–303, 1972.

50. Drachev, L. A., A. A. Kondrashin, A. Yu. Semenov, and V. P. Skulachev. Reconstitution of biological molecular generators of electric current. Transhydrogenase. *Eur. J. Biochem.* 113: 213–218, 1980.

51. Drachev, V. A., and D. B. Zorov. Mitochondrion as an electric cable. Experimental verification of the hypothesis. *Dokl. Akad. Nauk. SSSR* 287: 1237–1238, 1986 (Russ.).

52. Dracheva, S. M., L. A. Drachev, A. A. Konstantinov, A. Yu. Semenov, V. P. Skulachev, A. M. Arutjunjan, V. A. Shuvalov, and S. Zaberezhnaya. Electrogenic steps in the redox reactions catalyzed by photosynthetic center complexes from *Rhodopseudomonas viridis. Eur. J. Biochem.* 171: 253–264, 1988.

53. Duel, D. H., and M. B. Thorn. Effects of 2,3-dimercaptopropanol and antimycin on absorption spectra of heart-muscle preparations. *Biochim. Biophys. Acta* 59: 426–436, 1962.

54. Ewart, G. D., Y.-Z. Zhang, and R. A. Capaldi. Switching of bovine cytochrome *c* oxidase subunit VIa isoforms in skeletal muscle during development. *FEBS Lett.* 292: 79–84, 1991.

55. Feldman, R. I. and D. S. Sigman. The synthesis of enzyme-bound ATP by a soluble chloroplast coupling factor 1. *J. Biol. Chem.* 257: 1676–1683, 1982.

56. Feldman, R. I. and D. S. Sigman. The synthesis of ATP by the membrane-bound ATP synthase complex from medium ^{32}P$_i$ under completely uncoupled conditions. *J. Biol. Chem.* 258: 12178–12183, 1983.

57. Foster, D. O. Quantitative contribution of brown adipose tissue thermogenesis to overall metabolism. *Can. J. Biochem.* 62: 618–622, 1984.

58. Frangione, B., E. Rosenwasser, H. S. Penefsky, and M. E. Pullman. Amino acid sequences of the protein inhibitor of mitochondrial adenosine triphosphatase. *Proc. Natl. Acad. Sci. USA* 78: 7403–7407, 1981.

59. Garlid, K. D. New insights into mechanisms of anion uniport through the uncoupling protein of brown adipose tissue mitochondria. *Biochim. Biophys. Acta* 1018: 151–154, 1990.

60. Gauthier, G. F. On the relationship of ultrastructural and cytochemical features to color in mammalian skeletal muscle. *J. Zellforsch. Mikrosk. Anat.* 95: 462–482, 1969.

61. Gauthier, G. F. and H. A. Padykula. Cytological studies of fiber types in skeletal muscle. A comparative study of the mammalian diaphragm. *J. Cell Biol.* 28: 333–354, 1966.

62. Gillespie, J., S. Ozanne, J. Percy, M. Warren, J. Haywood, and D. Apps. The vacuolar H$^+$-translocating ATPase of renal tubules contains a 115-kDa glycosylated subunit. *FEBS Lett.* 282: 69–72, 1991.

63. Glaser, E., B. Norling, J. Kopecky, and L. Ernster. Comparison of the effects of oligomycin and dicyclohexylcarbodiimide on mitochondrial ATPase and related reactions. *Eur. J. Biochem.* 121: 525–531, 1982.

64. Gluck, S., and J. Caldwell. Proton-translocating ATPase from bovine kidney medulla: partial purification and reconstitution. *Am. J. Physiol.* 254 (*Renal Fluid Electrolyte Physiol.* 23): F71–F79, 1988.

65. Gogol, E. P., U. Lucken, and R. A. Capaldi. The stalk connecting the F$_1$ and F$_0$ domains of ATP synthase visualized by electron microscopy of unstained specimens. *FEBS Lett.* 219: 274–278, 1987.

66. Grinius, L. L., A. A. Jasaitis, J. P. Kadziauskas, E. A. Liberman, V. P. Skulachev, V. P. Topali, and M. A. Vladimirova. Conversion of biomembrane-produced energy into electric form. I. Submitochondrial particles. *Biochim. Biophys. Acta* 216: 1–12, 1970.

67. Grubmeyer, C., H. S. Penefsky. Cooperativity between catalytic sites in the mechanism of action of beef heart mitochondrial adenosine triphosphatase. *J. Biol. Chem.* 256: 3728–3734, 1981.

68. Hara, Y., J. Yamada, and M. Nakao. Proton transport catalyzed by the sodium pump. Ouabain-sensitive ATPase activity and the phosphorylation of Na,K-ATPase in the absence of sodium ions. *J. Biochem.* 99: 531–539, 1986.

69. Harnisch, U., H. Weiss, and W. Sebald. The primary structure of the iron-sulfur subunit of ubiquinol-cytochrome *c* reductase from *Neurospora* determined by cDNA and gene sequencing. *Eur. J. Biochem.* 149: 95–99, 1985.

70. Hearst, L. E. and K. Sauer. Protein sequence homologies between portions of the L and M subunit of reaction centers of *Rhodopseudomonas capsulata* and the Q$_B$-protein of chloroplast thylakoid membranes; a proposed relation to quinone-binding sites. *Z. Naturforsch.* 39c: 421–424, 1984.

71. Hennig, J. and R. C. Herrmann. Chloroplast ATP synthase of spinach contains nine nonidentical subunit species, six of which are encoded by plastid chromosomes in two operon in a phylogenetically conserved arrangement. *Mol. Gen. Genet.* 203: 117–128, 1986.

72. Hill, R. and F. Rendall. Function of the two cytochrome components in chloroplasts: a working hypothesis. *Nature* 186: 136–137, 1960.

73. Hoek, J. B. and J. Rydström. Physiological roles of nicotinamide nucleotide transhydrogenase. *Biochem. J.* 254: 1–10, 1988.

74. Horvath, L. I., M. Drecs, M. Klingenberg, and D. Marsh. Lipid-protein interactions in ADP-ATP carrier/egg phosphatidylcholine recombinants: studies by spin-label ESP spectroscopy. *Biochemistry* 29: 10664–10669, 1990.

75. Hüther, F.-J. and B. Kadenbach. Specific effects of ATP on the kinetics of reconstituted bovine heart cytochrome-c oxidase. *FEBS Lett.* 207: 89–94, 1986.

76. Hüther, F.-J., B. Kadenbach. Intraliposomal nucleotides change the kinetics of reconstituted cytochrome *c* oxidase from bovine heart but not from *Paracoccus denitrificans. Biochem. Biophys. Res. Commun.* 153: 525–534, 1988.

77. Jackson, J. B. The proton-translocating nicotinamide adenine dinucleotide transhydrogenase. *J. Bioenerg. Biomembr.* 23: 715–741, 1991.

78. Jezek, P. and Garlid, K. D. New substrates and competitive inhibitors of the Cl$^-$ translocating pathway of the uncoupling

protein of brown adipose tissue mitochondria. *J. Biol. Chem.* 265: 19303–19311, 1990.

79. Jezek, P., D. E. Orosz, and K. D. Garlid. Reconstitution of the protein of brown adipose tissue mitochondria. *J. Biol. Chem.* 265: 19296–19302, 1990.

80. Johannes, H. Beitrage zur Vitalfarbung von Pilzmycelien. II. Die Inturbanz der Farbung mit Rhodaminen. *Protoplama* 36: 181–194, 1941.

81. Kadenbach, B. Regulation of respiration and ATP synthesis in higher organisms: hypothesis. *J. Bioenerg. Biomembr.* 18: 39–54, 1986.

82. Kadenbach, B., M. Ungibauer, J. Jarausch, U. Bugé, and L. Kuhn-Nantwig. The complexity of respiratory complexes. *Trends Biochem. Sci.* 7: 398–400, 1983.

83. Kadenbach, B., A. Stroh, F.-J. Hüther, A. Reimann, and D. Steverding. Evolutionary aspects of cytochrome c oxidase. *J. Bioenerg. Biomembr.* 23: 321–334, 1991.

84. Kagawa, Y. H^+-ATP synthetase from a thermophilic bacterium In: *Chemiosmotic Proton Circuits in Biological Membranes,* edited by V. P. Skulachev and P. C. Hinkle, London: Addison-Wesley, 1981, pp. 421–434.

85. Kagawa, Y. Proton motive ATP synthesis. In: *Bioenergetics,* edited by Ernster, Amsterdam: Elsevier Science Publishers, 1984, pp. 149–157.

86. Kagawa, Y., S. Ohta, and Y. Otawara-Hamamoto. $\alpha_3\beta_3$ complex of thermophilic ATP synthase. Catalysis without the γ-subunit. *FEBS Lett.* 249: 67–69, 1989.

87. Kagawa, Y. and E. Racker. Partial resolution of the enzymes catalyzing oxidative phosphorylation. XXV. Reconstitution of vesicles catalyzing $^{32}P_1$-adenosine triphosphate exchange. *J. Biol. Chem.* 246: 5477–5487, 1971.

88. Kaunitz, J. D., G. Sachs. Identification of a vanadate-sensitive potassium dependent proton pump from rabbit colon. *J. Biol. Chem.* 26: 14005–14010, 1986.

89. Klaus, S., L. Casteilla, F. Bouillaud, and D. Ricquier. The uncoupling protein UCP: a membraneous mitochondrial ion carrier exclusively expressed in brown adipose tissue. *Int. J. Biochem.* 23: 791–801, 1991.

90. Klingenberg, M. The ADP-ATP translocation in mitochondrial membrane potential controlled transport. *J. Membr. Biol.* 56: 97–105, 1980.

91. Knaff, D. B. The photosystem I reaction centre. *Trends Biochem. Sci.* 13: 460–461, 1988.

92. Knaff, D. B. and M. Hirasawa. Ferredoxin-dependent chloroplast enzymes. *Biochim. Biophys. Acta.* 1056: 93–125, 1991.

93. Konstantinov, A. A. Vectorial electron and proton transfer steps in the cytochrome bc_1 complex. *Biochim. Biophys. Acta* 1018: 138–141, 1990.

94. Konstantinov, A. A. and E. K. Ruuge. Semoquinone Q in the respiratory chain of electron transfer particles. *FEBS Lett.* 81: 137–141, 1977.

95. Kotlyar, A. B., V. D. Sled, D. Sh. Burbaev, I. A. Moroz, and A. D. Vinogradov. Coupling site I and the rotenone-sensitive ubisemiquinone in tightly coupled submitochondrial particles. *FEBS Lett.* 264: 17–20, 1990.

96. Kotyk, A. Coupling of secondary active transport with $\Delta\mu_H^+$. *J. Bioenerg. Biomembr.* 15: 307–319, 1983.

97. Kozlov, I. A. and B. V. Chernyak. Regulation of H^+-ATPases in oxidative and photophosphorylation. *Trends Biochem. Sci.* 11: 32–35, 1986.

98. Kozlov, I. A., V. P. Skulachev. H^+-ATPase and membrane energy coupling. *Biochim. Biophys. Acta* 463: 29–89, 1977.

99. Kozlov, I. A. and V. P. Skulachev. An H^+-ATP synthase: a substrate translocation concept. *Curr. Top. Membr. Trans.* 16: 285–301, 1972.

100. Kuhn, N., B. Kadenbach. Isolation and properties of cytochrome c oxidase from rat liver and quantification of immunological differences between isozymes from various rat tissues with subunit-specific antisera. *Eur. J. Biochem.* 149: 147–158, 1985.

101. Kunz, W. S., A. A. Konstantinov, L. M. Tsofina, E. A. Liberman. Localization of a ferricyanide-reactive site of cytochrome b–c_1 complex, possibly of cytochrome b or ubisemiquinone, at the outer face of submitochondrial particles. *FEBS Lett.* 172: 261–266, 1984.

102. Labonia, N., M. Müller, and A. Azzi. The effect of non-esterified fatty acids on the proton-pumping cytochrome c oxidase reconstituted into liposomes. *Biochem. J.* 254: 139–145, 1988.

103. Lay, S., J. C. Watson, J. N. Hansen, and H. Sze. Molecular cloning and sequencing of cDNAs encoding the proteolipid subunit of the vacuolar H^+-ATPase from a higher plant. *J. Biol. Chem.* 266: 16078–16084, 1991.

104. Larsen, R. W., L.-P. Pan, S. M. Musser, Z. Li, and S. I. Chan. Could Cu_B be the site of redox linkage in cytochrome c oxidase? *Proc. Natl. Acad. Sci. USA* 89: 723–727, 1992.

105. Leonard, K., H. Haiker, and H. Weiss. Three-dimensional structure of NADH: ubiquinone reductase (complex I) from *Neurospora* mitochondria determined by electron microscopy of membrane crystals. *J. Mol. Biol.* 194: 277–286, 1987.

106. Levachev, M. M., E. A. Mishukova, V. G. Sivkova, and Skulachev. V. P. Energetics of pigeon at self-warming after hypothermia. *Biokhimiya* 30: 864–874, 1965 (Russ.).

107. Liberman, E. A. and V. P. Skulachev. Conversion of biomembrane-produced energy into electric form. IV. General discussion. *Biochim. Biophys. Acta* 216: 30–42, 1970.

108. Lill, H., G. Althoff, and W. Junge. Analysis of ionic channels by a flash spectrometric technique applicable to thylakoid membranes: CF_0, the proton channel of the chloroplast ATP synthase, and, for comparison, gramicidin. *J. Membr. Biol.* 98: 69–78, 1987.

109. Lukacs, G. L., O. D. Rotstein, and S. Grinstein. Phagosomal acidification is mediated by a vacuolar-type H^+-ATPase in murine macrophages. *J. Biol. Chem.* 265: 21099–21107, 1990.

110. Maeshima, M. Oligomeric structure of H^+-translocating inorganic pyrophsphatase of plant vacuoles. *Biochim. Biophys. Res. Commun.* 168: 1157–1162, 1990.

111. Malatesta, F., G. Antonini, P. Sarti and M. Brunoru. Modulation of cytochrome oxidase activity by inorganic and organic phosphate. *Biochem. J.* 248: 161–165, 1987.

112. Maloney, P. C., and T. H. Wilson. The evolution of ion pumps. *Bioscience* 35: 43–48, 1985.

113. Mandel-Hartvig, I. and B. D. Nelson. Comparative study of the peptide composition of complex III (quinol-cytochrome c reductase). *J. Bioenerg. Biomembr.* 15: 289–299, 1983.

114. Marra, E. and A. Ballarin-Denti. The proton pumps of the plasmalemma and the tonoplast of higher plants. *J. Bioenerg. Biomembr.* 17: 1–21, 1985.

115. Matsui, H. Na^+, K^+-ATPase: conformational change and interaction with sodium and potassium ions. In: *Transport and Bioenergetics in Biomembranes,* edited by R. Sato and Y. Kagawa, Tokyo: Japan. Sci. Soc. Press, 1982, pp. 165–187.

116. Metz, J. G., P. J. Nixon, M. Rogner, G. W. Brudvig, B. A. Dinner. Directed alteration of the D1 polypeptide of photosystem II: evidence that tyrosine-161 is the redox component, Z, connecting the oxygen-evolving complex to the primary electron donor, P680. *Biochemistry* 28: 6960–6969, 1989.

117. Milgrom, Ya. M. and M. B. Murataliev. Mitochondrial F_1-ATPase: the alterations of P_i-binding properties and catalytic activity are caused by ADP redistribution between nucleotide binding sites, the process including the nucleotide release and rebinding steps. *Biol. membrany* 3: 781–791, 1986 (Russ.).

118. Minkov, I. B., E. A. Vasilyeva, A. F. Fitin, and A. D. Vinogradov. Differential effects of ADP on ATPase and oxidative phosphophorylation in submitochondrial particles. *Biochem. Intern.* 1: 478–485, 1980.

119. Mitchell, P. Coupling of phosphorylation to electron and hydrogen transfer by a chemiomostic type of mechanism. *Nature* 191: 144–148, 1961.

120. Mitchell, P. Chemiosmotic coupling in oxidative and photosynthetic phosphorylation. *Biol. Rev.* 41: 445–502, 1966.

121. Mitchell, P. Possible molecular mechanisms of the proton motive function of cytochrome systems. *J. Theor. Biol.* 62: 327–367, 1976.

122. Mitchell, P., R. Mitchell, A. J. Moody, I. C. West, H. Baum, and J. M. Wrigglesworth. Chemiosmotic coupling in cytochrome oxidase. Possible protonmotive O loop and O cycle mechanisms. *FEBS Lett.* 188: 1–7, 1985.

123. Moriyama, Y. and N. Nelson. Lysosomal H^+-translocating ATPase has a similar subunit structure to chromaffin granule H^+-ATPase complex. *Biochim. Biophys. Acta* 980: 241–247, 1989.

124. Murphy, D. J. The molecular organization of the photosynthetic membranes of higher plants. *Biochim. Biophys. Acta* 864: 33–94, 1986.

125. Nakao, M. Molecular biological approaches in Na^+,K^+-ATPase and H^+,K^+-ATPase pump studies. In: *New Era in Bioenergetics,* edited by Y. Mukohata, Tokyo: Acad. Press, 1991, p. 1.

126. Nedergaard, J., B. Cannon. Thermogenic mitochondria. In: *Bioenergetics,* edited by L. Ernster, Amsterdam: Elsevier Science Publishers, 1984, pp 291–314.

127. Nelson, N. Structure, function, and evolution of proton-ATPases. *Plant Physiol.* 86: 1–3, 1988.

128. Nelson, H., S. Mandiyan, T. Noumi, Y. Moriyama, M. C. Miedel, and N. Nelson. Molecular cloning of cDNA encoding the c subunit of H^+-ATPase from bovine chromaffin granules. *J. Biol. Chem.* 265: 20390–20393, 1990.

129. Nicholls, D. G. Brown adipose tissue mitochondria. *Biochim. Biophys. Acta* 549: 1–22, 1979.

130. Nicholls, D. G. *Bioenergetics. Introduction to the Chemiosmotic Theory.* London: Academic Press, 1982.

131. Nicholls, D. G. and R. M. Locke. Heat generation by mitochondria. In: *Chemiosmotic Proton Circuits in Biological Membranes,* edited by V. P. Skulachev and P. C. Hinkle, London: Addison-Wesley, 1981, pp. 567–576.

132. Nicholls, D. G. and R. M. Locke. Thermogenic mechanisms in brown fat. *Physiol. Rev.* 64: 1–64, 1984.

133. Nitschke, W. and A. W. Rutherford. Photosynthetic reaction centers: variations on a common structural theme? *Trends Biochem. Sci.* 16: 241–245, 1991.

134. Njus, D. The chromaffin vesicle and the energetics of storage organelles. *J. Auton. Nerv. Syst.* 7: 35–40, 1983.

135. Nylund, A., N. J. Komarova, P. R. Rumyanstsev, A. Tjonneland, and S. Okland. Heart ultrastructure in *Petrobius brevistylis* (Archaeognatha: Microcoryphia) *Entomol. Gener.* 11: 263–272, 1986.

136. Ogata, T. and F. Murata. Cytological features of three fiber types in human striated muscle. *Tohoku J. Exp. Med.* 99: 225–245, 1969.

137. Ovchinnikov, Yu. A., V. V. Demin, A. N. Barnakov, A. P. Kuzin, A. V. Lunev, N. N. Modyanov, and K. N. Dzhandzhugazyan. Three-dimensional structure of $(Na^+ + K^+)$-ATPase revealed by electron microscopy of two-dimensional crystals. *FEBS Lett.* 190: 73–76, 1985.

138. Ovchinnikov, Yu. A., N. N. Modyanov, N. Broude, K. E. Petrukhin, A. V. Grishin, N. M. Arzamazova, N. A. Aldanova, G. F. Monastyrskaya, and E. D. Sverdlov. Pig kidney Na^+,K^+-ATPase. Primary structure and spatial organization. *FEBS Lett.* 201: 237–245, 1986.

139. Ovchinnikov, Yu. A., N. N. Modyanov, V. A. Grinkevich, N. A. Aldanova, O. E. Trubetskaya, I. V. Nazimov, T. Hundal, and L. Ernster. Amino acid sequence of the oligomycin sensitivity-conferring protein (OSCP) of beef-heart mitochondria and its homology with the δ-subunit of the F_1-ATPase of *Escherichia coli. FEBS Lett.* 166: 19–22, 1984.

140. Papa, S., N. Capitanio, G. Capitanio, E. De Nitto, and M. Minuto. The cytochrome chain of mitochondria exhibits variable $H^+/\bar{e}$ stoichiometry. *FEBS Lett.* 288: 183–186, 1991.

141. Parry, R. V., J. C. Turner, and P. A. Rea. High purity preparations of higher plant vacuolar H^+-ATPase reveal additional subunits. *J. Biol. Chem.* 264: 20025–20032, 1989.

142. Pedersen, P. L. and J. P. Wehrle. Phosphate transport processes of animal cells. In: *Membranes and Transport,* edited by A. N. Martonosi, New York: Plenum Press, 1982, Vol. 1, pp. 645–669.

143. Pennington, R. M. and R. R. Fisher. Reconstituted mitochondrial transhydrogenase is a transmembrane protein. *FEBS Lett.* 164: 345–349, 1983.

144. Perham, R. N. Domains, motifs, and linkers in 2-oxo acid dehydrogenase multienzyme complexes: a paradigm in the design of a multifunctional protein. *Biochemistry* 30: 8501–8512, 1991.

145. Phillips, A. L. and J. C. Gray. Isolation and characterization of a cytochrome *b–f* complex from pea chloroplasts. *Eur. J. Biochem.* 137: 553–560, 1983.

146. Racker, E. *Mechanisms in Bioenergetics.* London: Academic Press. 1965.

147. Racker, E. *A New Look at Mechanisms in Bioenergetics.* New York: Academic Press, 1976.

148. Ragan, C. I. Structure of NADH-ubiquinone reductase (complex I). In: *Curr. Top. Bioenerg.,* edited by C. P. Lee, San Diego: Academic Press, 1987, pp. 1–36.

149. Ragan, C. I. Structure and function of an archetypal respiratory chain complex: NADH-ubiquinone reductase. *Biochem. Soc. Trans.* 18: 515–516, 1990.

150. Ragan, I. and P. C. Hinkle. Ion transport and respiratory control in vesicles formed from reduced nicotinamide adenine dinucleotide coenzyme Q reductase and phospholipids. *J. Biol. Chem.* 250: 8472–8476, 1975.

151. Rambourg, A. and D. Segretain. Three-dimensional electron microscopy of mitochondria and endoplasmic reticulum in the red muscle fiber of the rat diaphragm. *Anat. Rec.* 197: 33–48, 1980.

152. Rea P. A. and R. J. Poole. Proton-translocating inorganic pyrophosphatase in red beet (*Beta vulgaris* L.) tonoplast vesicles. *Plant Physiol.* 77: 46–52, 1985.

153. Reimann, A. and B. Kadenbach. Stoichiometric binding of 2′ (or 3′)-O-(2,4,6-trinitrophenyl)-adenosine 5′-triphosphate to bovine heart cytochrome *c* oxidase. *FEBS Lett.* 307: 294–296, 1992.

154. Ricquier, D., J. Thibault, F. Bouilland, and Y. Kuster. Molecular approach to thermogenesis in brown adipose tissue. *J. Biol. Chem.* 258: 6675–6677, 1983.

155. Rutherford, A. W. Photosystem II, the water-splitting enzyme. *Trends Biochem. Sci.* 14: 227–232, 1989.

156. Rydström, J., C.-P. Lee, and L. Ernster. Energy-linked nicotinamide nucleotide transhydrogenase. In: *Chemiosmotic Proton Circuits in Biological Membranes,* edited by V. P. Skulachev and P. C. Hinkle, London: Addison-Wesley, 1981, pp. 483–508.

157. Rydström, J., B. Persson, and H.-I. Tang. Mitochondrial nicotinamide nucleotide transhydrogenase. In: *Bioenergetics,* edited by L. Ernster, Amsterdam: Elsevier Science Publishers, 1984, pp. 297–319.

158. Sachs, G., H. R. Koelz, T. Berglindh, E. Rabon and G. Saccomani. Aspects of gastric proton-transport ATPase. In: *Membranes and Transport,* edited by A. N. Martonosi, N.Y.-London: Plenum Press, 1982, Vol. 1, pp. 633–643.

159. Sakamoto, J. and Y. Tonomura. Synthesis of enzyme-bound ATP by mitochondrial soluble F_1-ATPase in the presence of dimethylsulfoxide. *J. Biochem.* 93: 1601–1614, 1983.

160. Salamma, F. R. Structure and function of cytochromes. *Annu. Rev. Biochem.* 46: 299–330, 1977.

161. Sankaram, M. B., P. J. Brophy, W. Jordi, and D. Marsh. Fatty acid pH titration and the selectivity of interaction with extrinsic proteins in dimyristoylphosphatidylglycerol dispersions. *Biochim. Biophys. Acta* 1021: 63–69, 1990.

162. Sarafian, V. and R. J. Poole. Putification of an H^+-translocating inorganic pyrophosphatase from vacuole membranes of red beet. *Plant Physiol.* 91: 34–38, 1989.

163. Saraste, M. Location of haem-binding sites in the mitochondrial cytochrome *b. FEBS Lett.* 166: 367–372, 1984.

164. Sato, M. H., M. Maeshima, Y. Ohsumi, and M. Yoshida. Dimeric structure of H^+-translocating pyrophosphatase from pumpkin vacuolar membranes. *FEBS Lett.* 290: 177–180, 1991.

165. Schagger H., H. Borcheert, H. Aquila, T. A. Link, and G. Von Jagow. Isolation and amino acid sequence of the smallest subunit of beef heart bc_1 complex. *FEBS Lett.* 190: 89–94, 1985.

166. Schatz, G. and E. Racker. Partial resolution of the enzymes catalyzing oxidative phosphorylation. VII. Oxidative phosphorylation in the diphosphopyridine nucleotide-cytochrome *b* segment of the respiratory chain: assay and properties of submitochondrial particles. *J. Biol. Chem.* 241: 1429–1437, 1966.

167. Schwartz, A. and J. H. Collins. Na^+/K^+-ATPase. Structure of the enzyme and mechanism of action of digitalis. In: *Membranes and Transport,* edited by A. N. Martonosi, New York: Plenum Press, 1982, Vol. 1, pp. 521–527.

168. Schönfeld, P. Does the function of adenine nucleotide translocase in fatty acid uncoupling depend on the type of mitochondria? *FEBS Lett.* 264: 246–248, 1990.

169. Senior, A. E. ATP synthesis by oxidative phosphorylation. *Physiol. Rev.* 68: 177–231, 1988.

170. Serrano, R. Structure and function of proton translocating ATPase in plasma membranes of plants and fungi. *Biochim. Biophys. Acta* 947: 1–28, 1988.

171. Severina, I. I. and V. P. Skulachev. Ethylrhodamine as a fluorescent penetrating cation and a membrane potential-sensitive probe in cyanobacterial cells. *FEBS Lett.* 165: 67–71, 1984.

172. Shiota, M. and S. Masumi. Effect of norepinephrine on consumption of oxygen in perfused skeletal muscle from cold-exposed rats. *Am. J. Physiol.* 245 (Endocrinol. Metab. 17): E482–E489, 1988.

173. Shrago, E., J. McTigue, S. Kathiyar, and G. Woldegiorgis. Preparation of a highly purified reconstituted uncoupling protein to study biochemical mechanism(s) of proton conductance. In: *Hormones, Thermogenesis, and Obesity,* edited by H. Lardy and F. Stratman, Amsterdam: Elsevier, 1989, pp. 129–136.

174. Skulachev, V. P. Regulation of the coupling of oxidation and phosphorylation. *Proc. 5th Intern. Biochem. Congr.* 5: 365–374, 1963.

175. Skulachev, V. P. *Energy Accumulation Process in the Cell.* Moscow: Nauka, 1969 (Russ.).

176. Skulachev, V. P. Energy transformation in the respiratory chain. *Curr. Top. Bioenerg.* 4: 127–190, 1971.

177. Skulachev, V. P. *Energy Transduction in Biomembranes.* Moscow: Nauka, 1972 (Russ.).

178. Skulachev, V. P. Integrating functions of biomembranes. Problems of lateral transport of energy, metabolites and electrons. *Biochim. Biophys. Acta* 604: 297–320, 1980.

179. Skulachev, V. P. *Membrane Bioenergetics,* Berlin: Springer-Verlag, 1988.

180. Skulachev, V. P. Power transmission along biological membranes. *J. Membr. Biol.* 114: 97–112, 1990.

181. Skulachev, V. P. Fatty acid circuit as a physiological mechanism of uncoupling of oxidative phosphorylation. *FEBS Lett.* 294: 158–162, 1991.

182. Skulachev, V. P. The laws of cell energetics. *Eur. J. Biochem.* 208: 203–209, 1992.

183. Skulachev, V. P., V. V. Chistyakov, A. A. Jasaitis, and E. G. Smirnova. Inhibition of the respiratory chain by zinc ions. *Biochim. Biophys. Res. Commun.* 26: 1–6, 1967.

184. Skulachev, V. P. and S. P. Maslov. Role of non-phosphorylating oxidation in thermoregulation. *Biokhimiya* 25: 1058–1064, 1960 (Russ.).

185. Skulachev, V. P., S. P. Maslov, V. G. Sivkova, L. P. Kalinichenko, and G. M. Maslova. Cold-induced incoupling of oxidation and phosphorylation in the muscles of white mice. *Biokhimiya* 28: 70–79, 1963 (Russ.).

186. Smith, R. E. and B. A. Horwitz. Brown fat and thermogenesis. *Physiol. Rev.* 49: 330–425, 1969.

187. Soumarmon, A. and M.J.M. Lewin. Gastric (H^+,K^+)-ATPase. *Biochemie* 68: 1287–1291, 1986.

188. Spitsberg, V. L., N. E. Pfeiffer, B. Partridge, D. E. Wylie, S. M. Schuster. Isolation and antigenic characterization of corn mitochondrial F_1-ATPase. *Plant Physiol.* 77: 339–345, 1986.

189. Srügger, S. Die vitalfarbung des Protoplasmas mit Rhodamine B und 6G. *Protoplasma* 30: 85–100, 1938.

190. Strieleman, P. J., K. L. Schalinske, and E. Shrago. Fatty acid activation of the reconstituted brown adipose tissue mitochondria uncoupling protein. *J. Biol. Chem.* 260: 13402–13405, 1985.

191. Strotmann, H., and D. Lohse. Determination of the H^+/ATP ratio of the H^+ transport-coupled reversible chloroplast ATPase reaction by equilibrium studies. *FEBS Lett.* 229: 308–312, 1988.

192. Sun, S.-Z., X.-S. Xie, and D. K. Stone. Isolation and reconstitution of the dicyclohexylcarbodiimide-sensitive proton pore of the clathrin-coated vesicle proton translocating complex. *J. Biol. Chem.* 262: 14790–14794, 1987.

193. Toth, P. P., K. J. Sumerix, S. Ferguson-Miller, and C. H. Suelter. Respiratory control and ADP: O coupling ratios of isolated chick heart mitochondria. *Arch. Biochim. Biophys.* 276: 199–211, 1990.

194. Trumpower, B. L. Cytochrome bc_1 complexes of microorganisms. *Microbiol. Rev.* 54: 101–129, 1990.

195. Tsukihara, T., H. Aoyama, E. Yamashita, T. Tomizaki, H. Yamaguchi, K. Shinzawa-Itoh, R. Nakashima, R. Yaono, and S. Yoshikawa. Structure of metal sites of oxidized bovine heart cytochrome c oxidase at 2.8 Å. *Science* 269: 1069–1074, 1995.

196. Valcarce, C. and M. Cuezva. Interaction of adenine nucleotides with the adenine nucleotide translocase regulates the developmental changes in proton conductance of the inner mitochondrial membrane. *FEBS Lett.* 294: 225–228, 1991.

197. Van Belzen, R. and S.P.J. Albracht. The pathway of electron transfer in NADH:Q oxidoreductase. *Biochim. Biophys. Acta* 974: 311–320, 1989.

198. Vignais, P. V., M. R. Block, F. Boulay, G. Brandolin, and G.J.M. Lauquin. Functional and topological aspects of the mitochondrial adenine-nucleotide carrier. In: *Membranes and Transport,* edited by A. N. Martonosi, New York: Plenum Press, 1982, Vol. 1, pp. 405–414.

199. Vorobjev, I. A. and D. B. Zorov. Diazepam inhibits cell respiration and induces fragmentation of mitochondrial reticulum. *FEBS Lett.* 163: 311–314, 1983.

200. Wakabayashi, S., H. Matsubara, C. H. Kim, K. Kawai, T. E. King. The complete amino acid sequence of bovine heart cytochrome c_1. *Biochem. Biophys. Res. Commun.* 97: 1548–1554, 1980.

201. Walker, J. E., A. L. Cozens, M. Dyer, I. M. Fearnley, S. Powell, M. J. Runswick, and L. J. Tybulewicz. Genes for ATP synthases from photosynthetic bacteria, chloroplasts and mitochondria. *EBEC* 4: 1, 1986.

202. Walker, J. E., M. J. Runswick, and L. Poulter. ATP synthase from bovine mitochondria. The characterization and sequence analysis of two membrane-associated subunits and of the corresponding cDNAs. *J. Mol. Biol.* 197: 89–100, 1987.

203. Walker, J. E., M. J. Runswick, M. Saraste. Subunit equivalence in *Escherichia coli* and bovine heart mitochondrial F_1F_0 ATPases. *FEBS Lett.* 146: 393–396, 1982.

204. Walker, J. E., I. M. Fearnley, N. J. Gay, B. W. Gibson, F. D. Northrop, S. J. Powell, M. J. Runswick, M. Saraste, and L. J. Tybulewicz. Primary structure and subunit stoichiometry of F_1-ATPase from bovine mitochondria. *J. Mol. Biol.* 184: 677–701, 1985.

205. Walsh, M. L., J. Jen, and L. B. Chen. Transport of serum components into structures similar to mitochondria. *Cold Spring Harbor Conf. Proliferation* 6: 513–520, 1979.

206. Webb, M. R., C. Grubmeyer, H. S. Penefsky, and D. R. Trenthams. The stereochemical course of phosphoric residue transfer catalyzed by beef heart mitochondrial ATPase. *J. Biol. Chem.* 255: 11637–11639, 1980.

207. Weis, J. K., L.N.Y. Wu, and R. R. Fisher. The orientation of transhydrogenase in the inner mitochondrial membrane in rat liver. *Arch. Biochem. Biophys.* 257: 424–429, 1987.

208. Weiss, H. and T. Friedrich. Redox-linked proton translocation by NADH-ubiquinone reductase (complex I). *J. Bioenerg. Biomembr.* 23: 743–754, 1991.

209. Weiss, H., T. Friedrich, G. Hofhaus, and D. Preis. The respiratory-chain NADH dehydrogenase (complex I) of mitochondria. *Europ. J. Biochem.* 197: 563–576, 1991.

210. Wikström, M. Proton pump coupled to cytochrome c oxidase in mitochondria. *Nature* 266: 271–273, 1977.

211. Wikström, M., K. Krab, and M. Saraste. *Cytochrome Oxidase—A Synthesis.* London: Academic Press, 1981.

212. Wikström, M., and M. Saraste. The mitochondrial respiratory chain. In: *Bioenergetics.* edited by L. Ernster, Amsterdam: Elsevier Science Publishers, 1984, pp. 49–94.

213. Wikström, M., M. Saraste, and T. Penttila. Relationships between structure and function in cytochrome oxidase. In: *The Enzymes of Biological Membranes,* edited by A. N. Martonosi, London: Plenum Press, 1985, Vol 4, pp. 111–148.

214. Witt, H. T. Functional mechanism of water splitting photosynthesis. *Photosynt. Res.* 29: 55–77, 1991.

215. Yang, Z., and B. L. Trumpower. Protonmotive Q cycle pathway of electron transfer and energy transduction in the three-subunit ubiquinol-cytochrome c oxidoreduction complex of *Paracoccus denitrificans. J. Biol. Chem.* 263: 11962–11970, 1988

216. Yeaman, S. J. The 2-oxo acid dehydrogenase complexes: recent advances. *Biochem. J.* 257: 625–632, 1989.

5. Principles of regulation and control in biochemistry: a pragmatic, flux-oriented approach

B. CRABTREE | Rowelt Research Institute, Aberdeen, Scotland

E. A. NEWSHOLME | Merton College, University of Oxford, Oxford, England

N. B. REPPAS | Balliol College, University of Oxford, Oxford, England

CHAPTER CONTENTS

THE ADVENT OF RECOMBINANT DNA TECHNOLOGY, AND THE RESULTING SURGE IN MOLECULAR BIOLOGY have made it all too easy to believe that the study of metabolic pathways is outdated and/or complete. "Anatomically" there may be some truth in this, but the way in which the rates of these pathways are regulated is of crucial importance for the whole of biology; indeed Koshland (89) has pointed out that the processes of molecular biology are themselves rather complex pathways! Moreover, transferring enzymes between tissues or species can be regarded as very naive (perhaps dangerously so) if no account is taken of the regulation of these enzymes in the originating cell.

Regulation of enzymes and pathways in situ should therefore be seen as an essential part of biochemical knowledge. Unfortunately, the subject has been hampered by both practical and theoretical difficulties, which have perhaps tended to put off experimental workers. As will be discussed later, however, many of the theoretical problems (if not the terminology) have now been clarified and, although technical difficulties remain, recent advances in non-invasive techniques such as NMR (17, 28, 29, 75, 80, 96, 97, 124, 149, 150) should help to provide methods for testing the various models proposed for regulation of metabolism. The ultimate aim is to develop quantitative kinetic models of the pathways (or, more generally, fluxes or systems). In many cases this ideal is a long way off, and a basically qualitative or algebraic approach is all that is possible. Such an approach should not be scorned: it has given valuable insight into regulatory systems, and we believe that it will continue to do so. Most importantly, it provides a firm foundation for more complex quantitative models, as will be seen throughout the chapter.

This chapter takes a somewhat pragmatic approach that we have developed over several years, referred to as *flux-oriented theory* (FOT) (44–46, 109, 111). This provides a systematic organization of the kinetic data needed to understand metabolic regulation and indicates the types of experimental investigations that would be useful. Two other important approaches, *biochemical systems theory* (BST) (74, 138–141, 175) and *metabolic control theory* (MCT), alternatively known as *control analysis,* (71, 73, 76–79, 182) are discussed later.

Most sections of the chapter are based on hypothetical, abstract systems since our aim is to highlight the general principles involved in regulation. Real metabolic pathways or systems, of course, are more complex. In our experience, however, the extra complexity is superimposed on relatively simple basic structures, and these structures provide important insights into the control and regulation of metabolism.

FLUX-ORIENTED THEORY OF REGULATION

This approach is based on the concept of a metabolic flux, which extends the familiar concept of a pathway (44, 45, 109). A hypothetical example of a simple flux is given in Figure 5.1. The initial intermediate (or pathway substrate) is converted into an end product, D, via a series of enzyme-catalyzed reactions (E_1–E_3): D then leaves the system via the final reaction, E_4.

The entire system is assumed to be in a steady state; this means that all the reactions are proceeding at constant rates, equal to the rate of conversion of A into D, and all the substrate concentrations (except for A) do not vary with time. This equality of the rates of the component reactions is achieved by a balancing provided by the concentrations of intermediates B–D, as a result of the rates of E_2–E_4 being a function of their substrate concentrations. For example, if the rate of E_1 were to exceed that of E_2, there would be a net increase in B; this would increase the rate of E_2 until it became equal to that of E_1. Similarly, C balances E_3–E_2 and D balances E_4–E_3. The resulting steady state gives a defined and hence controllable system which is insulated from the fluctuations in the environ-

ment. It must be stressed, however, that the steady state is an abstraction, or model, to which few if any systems conform exactly. Nevertheless, it is sufficiently close to the physiological state of most pathways to allow a satisfactory modelling of their control systems. A great simplification is achieved by the absence of time as a variable.

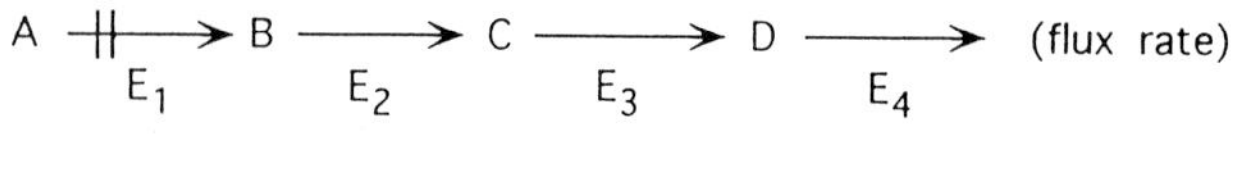

Flux-Generating Steps

The situation with the initial substrate in Figure 5.1, A, is of some interest since, during the operation of the system, the concentration of A will decline continuously. Therefore, if E_1 responded to A (as the other reactions do to their substrates), the rate of E_1 and hence all the other reactions would also decrease continuously making a steady state impossible. Consequently, it has been suggested (41, 44, 107, 110) that the initial reaction of a flux (which need not be in the same cell or tissues as the rest of it: see Fig. 5.2) is made unresponsive to its pathway substrate, A, (Fig. 5.1) by being saturated with A. The decreasing concentration of the initial substrate would then not affect the steady state flux until its concentration became so low that the system had effectively consumed most of it (an example is the near-depletion of glycogen in some muscles after a period of activity). The flux in Figure 5.1 can therefore be considered as being generated at E_1; this has been termed the flux-generating reaction (or step) for the entire flux (44, 109, 110).

The flux-generating step, which with the final reaction of the flux provides the insulation from environmental fluctuations, is somewhat analogous to a battery (accumulator) in an electrical circuit since it effectively provides a constant head of flux (44). However, as with the steady state itself, this concept is an abstraction that may not apply in all cases: for example, if only a small proportion of A (Fig. 5.1) is used relatively slowly (as is the case for protein degradation to provide fuel for short periods of starvation) the resulting disruption of the steady state would not be serious if E_1 responded to A. Moreover, with particulate fuels, unresponsiveness (or zero order kinetics) may sometimes be produced without saturating E_1 with A. These points are discussed in more detail in APPENDIX 1. For the remainder of this chapter, the flux-generating step is simply assumed to be saturated with the initial substrate.

Branched Fluxes: Reversible Reactions

In the system outlined in Figure 5.1, the flux, J, is generated at reaction E_1 and transmitted along the rest of the system via the substrate-dependence of the

$$A \xrightarrow{\;\;\;\;} B \xrightarrow{\;\;\;\;\;\;} C \xrightarrow{\;\;\;\;\;\;} D \xrightarrow{\;\;\;\;\;\;} \text{(flux rate)}$$
$$\;\;\; E_1 \quad\quad\quad E_2 \quad\quad\quad E_3 \quad\quad\quad E_4$$

FIG. 5.1. Simple linear flux. *A–D* represent metabolic intermediates, E_1–E_4 represent enzyme-catalyzed reactions and symbol ⇥ denotes saturation (here, with A).

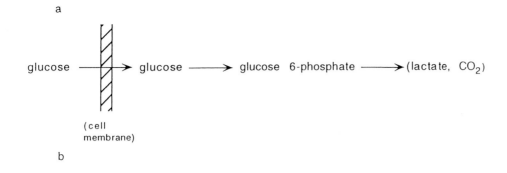

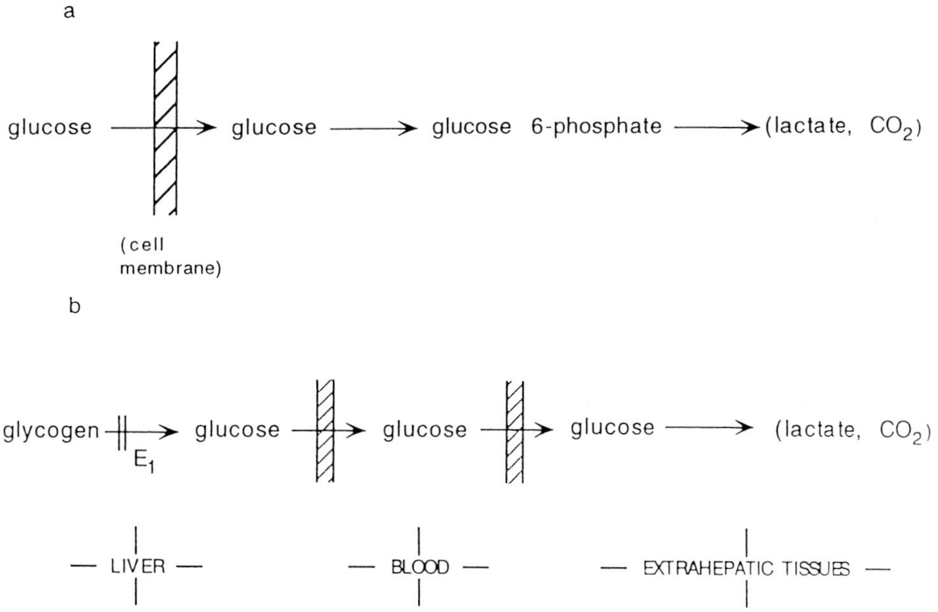

FIG. 5.2. Metabolic carbohydrate flux. (a) Conventional pathway of glycolysis, originating with glucose entry into a cell, (b) whole-body flux, of which (a) is a component. E_1 = glycogen phosphorylase. In this, and other figures the cofactors are omitted, unless relevant to the discussions.

subsequent reactions, E_2–E_4.[1] However, this is a very simple system indeed. Most metabolic systems have extensive branching and not all the component reactions are unidirectional (that is, kinetically irreversible).

A simple example of a branched system is shown in Figure 5.3. Here a flux, J, generated at E_1, branches into two fluxes, K and L, at the branch-point, S. This system therefore consists of three fluxes, each generated at E_1. This concept of branched fluxes is useful in explaining the control of glycolysis in vivo by treating the overall flux as the sum of component fluxes such as glycolysis from glucose and glycolysis from glycogen (109).

The system in Figure 5.4a contains a kinetically reversible reaction, E_2. This means that this reaction now responds to its product in addition to its substrate. Such reversible (near-equilibrium) reactions are very common in metabolism: for example, many reactions of glycolysis are reversible in situ (Fig. 5.4b and Table 5.1). Their regulatory significance is considerable and will be discussed later under Effect of the Kinetic Reversibility of the Reaction, R_v. In fact, unidirectional reactions are needed only to generate the flux, and to ensure the removal of end products. However, these

too have important regulatory properties and are therefore frequently found in the middle of a flux. A notable example is the reaction catalyzed by phosphofructokinase, which neither generates nor removes the end products of a flux (Fig. 5.4b), but is a key regulatory site for glycolysis (see later).

Other Types of Flux: Energy Fluxes and Equivalence

So far, only simple carbon-based (C) fluxes have been considered. However, the generalized concept of a flux can be extended to other types of intermediates, notably cofactors such as ATP and the pyridine nucleotides (NADH, NADPH), which link conventional pathways into fluxes relating directly to their physiological functions (2, 44). Perhaps the best example is the transmission of a biological energy flux via the adenine nucleotides ATP and ADP; which enables chemical energy to be transferred from one pathway to another to provide for biosyntheses, heat production and the performance of mechanical work (most notably in muscle).

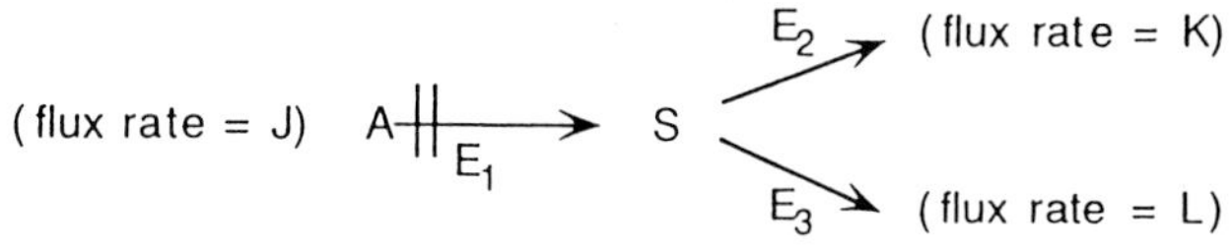

[1]In Flux-Oriented Theory the term *flux* is used both as a general descriptive term for an extended pathway or system (as above), and as a measure of the rate of flow of material through that pathway. This should not cause any serious confusion, since the exact usage should be apparent from its context.

(flux rate = J)   A —||—→   S E_2 ↗ (flux rate = K)
              E_1 E_3 ↘ (flux rate = L)

FIG. 5.3. Branched flux.

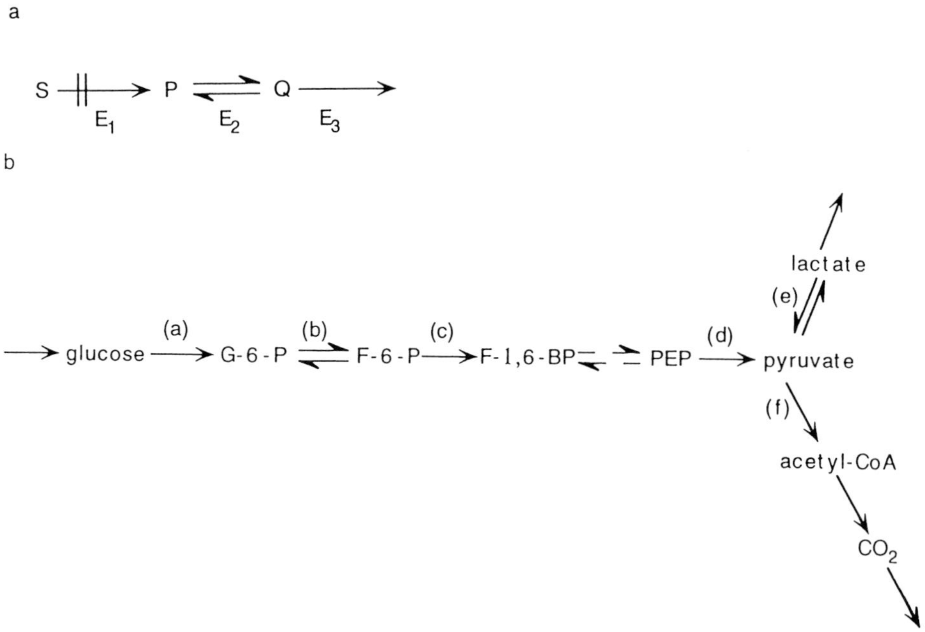

a

$$S \xrightarrow[E_1]{\;\;||\;\;} P \underset{E_2}{\overset{}{\rightleftharpoons}} Q \xrightarrow[E_3]{}$$

b

$$\xrightarrow{} \text{glucose} \xrightarrow{(a)} \text{G-6-P} \underset{}{\overset{(b)}{\rightleftharpoons}} \text{F-6-P} \xrightarrow{(c)} \text{F-1,6-BP} \rightleftharpoons \rightleftharpoons \text{PEP} \xrightarrow{(d)} \text{pyruvate}$$

lactate

(e)

(f)

acetyl-CoA

CO_2

FIG. 5.4. Fluxes with reversible steps. *(a)* Hypothetical linear flux, *(b)* glycolysis: *a* = hexokinase (HK); *b* = phosphoglucose isomerase (PGI); *c* = phosphofructokinase (PFK); *d* = pyruvate kinase (PK); *e* = lactate dehydrogenase (LDH); *f* = pyruvate dehydrogenase (PDH).

An example of such an energy flux is illustrated in Figure 5.5*a*. Here the C-flux of lactate production (J_l) is part of a wider E-flux (J_E) which involves the continuous synthesis and hydrolysis of ATP. J_E could represent the flux of energy needed for muscular contraction. Since two moles of ATP are synthesized per mole of glucose converted to lactate, J_E is equal to $2 \times J_l$. In this case, when one flux is a multiple of another, they can be regarded as equivalent in a regulatory sense: because, for example, a two-fold increase of J_l will be accompanied by a two-fold increase in J_E.

However, the system in Figure 5.5*b* is somewhat

TABLE 5.1. *Status of Glycolytic Reactions in Heart Muscle In Vivo*

Reaction Catalyzed By	K	Γ	K/Γ	Nature
Hexokinase	4700	0.08	59000	Irreversible
Glucosephosphate isomerase	0.4	0.24	1.7	Reversible
6-Phosphofructokinase	1050	0.03	35000	Irreversible
Aldolase	10^{-4}	9×10^{-6}	11	Reversible
Glyceraldehyde phosphate dehydrogenase plus phosphoglycerate kinase	850	9	94	Reversible (see text)
Phosphoglyceromutase	0.14	0.12	1.2	Reversible
Enolase	3.6	1.4	2.6	Reversible
Pyruvate kinase	2000	40	500	Irreversible
Glycogen phosphorylase	0.3	0.003	100	Irreversible

Source: data from (114). *K* denotes the equilibrium constant and Γ the mass action ratio, both measured in the direction of the glycolytic flux: for example, with hexokinase,

$$\Gamma = \frac{[glucose\text{-}6\text{-}phosphate] \cdot [ADP]}{[glucose] \cdot [ATP]}$$

where the concentrations refer to those in vivo.

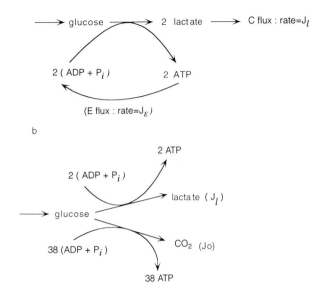

FIG. 5.5. Energy fluxes. (a) Equivalent fluxes; $J_{E} = 2 \times J_l$, (b) non-equivalent fluxes; total energy flux $= 2 \times J_l + 38 \times J_0$, which is not proportional to, and thus not equivalent to, the total carbon flux, $J_0 + J_l$.

different. Although the oxidation of glucose to CO_2 (J_o) is equivalent to an energy flux of $38 \times J_o$, the total energy flux of the system is now $(2 \times J_l + 38 \times J_o)$. This cannot be expressed as a multiple of the total carbon (glycolytic) flux, J_g, that is, $(J_l + J_o)$. Thus, in this system the glycolytic flux, J_g, is not equivalent to the energy flux, J_E.

This means that, for example, a reduced demand for ATP, and hence a reduced J_E, need not be met by a corresponding reduction in the rate of glycolysis, J_g. It could instead be met by switching flux from oxidation (J_o) to lactate formation (J_l); this could, under some conditions, result in less ATP production without changing the total rate of glycolysis. Physiologically, this would mean using extra energy to convert the lactate back to glucose (for example, in the liver); and this may be the reason mammalian tissues use the powerful regulator, citrate (which interacts with phosphofructokinase: Fig. 5.4b), to inhibit glycolysis during starvation (see refs. 11, 12, 114, 117, 169), thereby ensuring that a reduction of the ATP demand from glycolysis is met by a corresponding decrease in the total carbon flux. In this context it is interesting that, in insect flight muscle, there is little if any lactate formation, so that the total flux of glycolysis, J_g, will now be equal to J_o (Fig. 5.5b) and hence equivalent to the total energy flux, $38 \times J_o$. The total carbon flux of glycolysis will now follow any changes in the energy flux, so that a specific regulator of PFK will not be

needed. Indeed, the phosphofructokinase in insect flight muscle does not respond to citrate.

Internal and External Effectors

In the system outlined in Figure 5.6 the concentration of B is such that E_2 is balanced to E_1; likewise the concentration of C is such that E_3 is balanced to E_2. The concentrations of both B and C are thus determined entirely by the reactions of this system and cannot therefore change except as a result of some prior interaction with one or more of these reactions. B and C are therefore referred to as internal effectors of this flux. In contrast, the concentrations of metabolites X and Y are not determined by flux J and are therefore referred to as external effectors of this flux (44, 45, 109).

It is important to note that, in FOT, terms such as internal and external effectors (as well as others; see later under Control and Regulation of Metabolic Fluxes and Concentrations: Communication and Regulatory Sequences) are defined only relative to a given flux: they should never be used in an absolute sense. Thus, although X and Y are external effectors of flux J, they will be determined by, and hence internal effectors of, some other flux system.

A system such as that in Figure 5.6, in which the metabolites split neatly into internal and external effectors, has been referred to as an open system or structure (45, 46). However, such systems are not very common in metabolism, because of the linking of fluxes via cofactors and other shared intermediates. For example, in Figure 5.7, metabolites B and C are internal effectors of flux J and X is a totally external effector. However, the concentrations of M and N, which at first sight may look like totally external effectors, are partially determined by flux J: they are internal effectors for the wider system consisting of $J + K$. Consequently, metabolites such as M and N (which are often cofactors such as the adenine and pyridine nucleotides) are referred to as partially external effectors of J, and their interaction with flux J represents a 'closed'

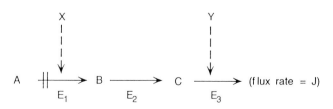

FIG. 5.6. Linear flux with external effectors. The broken lines denote regulatory (for example, allosteric) interactions.

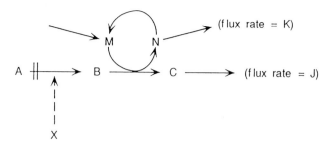

FIG. 5.7. Partially external regulators. The concentrations of metabolites M and N are only partially determined by flux J. On the other hand, they are totally determined by the wider system, $J + K$. Regulator X (totally external for J) communicates with M, but M (partially external for J) does not communicate with X.

structure (45, 46). Qualitatively, partially external effectors can be treated as if they were totally external, but the distinction becomes extremely important when analyzing systems quantitatively (see later under Z Is a Mass-Action Effector of E_2: Analysis of the Effects of a Partially External Regulator).

This concept of partially external effectors enables the analysis of truncated fluxes. For example, the system in Figure 5.8 is a truncated form of that in Figure 5.6: the system now starting with B, the product of the generating reaction E_1. C and Y are still internal and external effectors, respectively, for this truncated flux (J_t). However, B is now partially external, relative to this flux, since its concentration is only partially determined by J_t. In this way, fluxes can be split up and analyzed systematically, without having to include all the reactions back to the flux generating step. This is especially important when considering fluxes which span the whole body (for example, postabsorptive glycogenolysis and lipolysis).

Control and Regulation of Metabolic Fluxes and Concentrations: Communication and Regulatory Sequences

In a simple system such as that in Figure 5.6, a change in concentration of metabolite X will change the activity of E_1 and hence the rate of conversion of A into B.

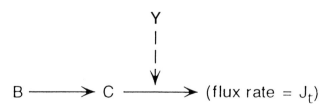

FIG. 5.8. Truncated flux. The reactions producing B are omitted, so that B is a partially external regulator of the truncated flux, J_t.

This will increase B and hence increase the rate of E_2; which, in turn, will increase C and thus increase the rate of E_3. For a defined change of X, these ensuing changes in concentrations and rates will proceed until the system is balanced and a new steady state has therefore been attained: if X activates E_1 the flux, J, will have increased: if X inhibits E_1 it will have decreased. Thus X can be regarded as being able to communicate with the concentrations of B and C, and with reactions E_1–E_3. The above effects can be summarized as the following communication sequence (44, 45, 109):

$$X \sim E_1 \sim B \sim E_2 \sim C \sim E_3,$$

where the symbol $\sim$ denotes a communication (from left to right).

Now this type of sequence will apply to any external effector of E_1, so that we can write a general communication sequence for this reaction as,

$$E_1 \sim B \sim E_2 \sim C \sim E_3$$

or

$$E_1 \sim [B,C,J].$$

Reactions such as E_1, which communicate with all the reactions of J, provide potential regulatory (or control) sites for J; in addition, E_1 communicates with, and hence is potentially regulatory for, B and C. External (including partially external) effectors of such reactions are then said to be potential regulators for J, B, and C (44, 109).

In contrast, the other external effector in this system, Y, can only communicate with C. For example, if Y is an activator of E_3, an increased Y will increase E_3 and thus tend to decrease C. However, there is no further communication from C to any of the other reactions or intermediates of this system, so that the net effect of the change of Y is just a decreased C, which proceeds until the rate of E_3 is once more equal to that of E_2. Y, and hence reaction E_3, is potentially regulatory only for C in this system.

Similarly, E_2 is potentially regulatory only for B, since there is no communication from B to E_1, and C is fixed by the need to balance E_3 effectively to E_1 (which is not altered by effectors of E_2). Thus E_2 is potentially regulatory only for B. (Although there will be an initial increase of C, after activation of E_2, this will not be maintained and the concentration of C will return to its original value in the new steady state.)

The above analysis shows that a potentially regulatory site for any defined variable (that is, a flux or concentration) is one that can communicate with that variable; and this is the definition of the term regulatory adopted in FOT (44). It should be stressed that, as with the terms internal and external effector, this term must relate to a defined variable and should never be used in an absolute sense.

With branched systems the regulation of the component fluxes need not involve a communication with the flux-generating step. For example, in Figure 5.3, an external effector of E_2 (or E_3) could change fluxes K and L reciprocally, by a switching at the branch point (S), without affecting the total flux, J, and hence the flux-generating step, E_1. Thus, in this system reactions E_2 and E_3 are regulatory sites for fluxes K and L, but not for J; once again illustrating that this term must refer to a defined flux.

In Figure 5.6, the only way of communicating with, and hence regulating, the flux is via an external interaction at E_1 which is therefore the sole regulatory site for the system. However, this would be a most ineffective control system for most pathways, because an increased demand for C (for example, via an increased activity of E_3) could not be met by an increased supply from A. However, the system in Figure 5.9a does enable the supply of C to be regulated by demand, as a result of feedback inhibition in the system.

In Figure 5.9a this feedback is of two types. First, a reversible reaction at E_2 enables changes of C to be communicated to B. Second, an inhibition of E_1 by its product, B, enables B to communicate with E_1. An increased activity of E_3 (due to a change in the external effector Y) will now decrease C which, via the reversible reaction at E_2, will then decrease B, and this will increase E_1. Thus, E_3 communicates with all the reactions of the system, via the sequence

$$E_3 \sim C \sim E_2 \sim B \sim E_1$$

so that E_3 is now regulatory for J. As a result, any increased demand for C can be met by an increased supply.

This system also illustrates that the term regulatory is crucially dependent on the internal kinetic structure

of the system: if this structure were to be changed (either by design or in a diseased state), the regulatory structure would change. For example, if reaction E_2 (Fig. 5.9a) were to become irreversible, the feedback from C to B would disappear and E_3 would be unable to regulate the flux of A to C (that is, J).

The effect of C on E_2 in Figure 5.9a results from a mass-action effect, which need not affect the activity of the catalyst (enzyme). However, that of B on E_1 results from a direct inhibition of the catalytic activity of this enzyme. In theory, any product of an enzyme reaction will inhibit the catalytic activity because the catalytic mechanism will automatically enable the product to bind at the active site, thereby resulting in competitive inhibition relative to the corresponding substrate (49). Therefore, the absence of product inhibition at the irreversible reactions in several of the systems discussed hitherto is, strictly speaking, invalid. However, with many irreversible reactions in metabolism, the concentration of the product in vivo is too small for this competitive effect to be significant. This conclusion is supported by several instances where product inhibition at irreversible reactions is needed for control purposes, and yet does not proceed via the competitive mechanism at the active site. For example, the competitive product inhibition of hexokinase by glucose-6-phosphate would require a glucose-6-phosphate concentration of approximately 65mM for half-saturation (177). At physiological concentrations of glucose-6-phosphate (approximately 0.5 mM) there would be no significant effect. However, product inhibition of hexokinase provides an important control link in the regulation of glycolysis (52) and occurs via a noncompetitive mechanism (with respect to glucose) characterized by half-saturation at a glucose-6-phosphate concentration of approximately 0.4 mM (177). Similarly, the competitive inhibition of phosphofructokinase by fructose-1,6-bisphosphate (FBP) is not physiologically significant; on the contrary, an activating effect of FBP is probably important under physiological conditions (169).

On the other hand, in some tissues the concentration of citrate may be high enough to enable its competitive effect on citrate synthase to be physiologically significant (8).

Thus, at irreversible reactions, product inhibition should not be automatically assumed to be present. As with every other proposed communication, its significance needs to be assessed by comparing the physiological concentration with the affinity of the enzyme for it.

The above specific effects of glucose-6-phosphate and FBP are examples of allosteric effects since they are exerted by binding to sites other than the active one (102, 114, 118, 163). Such control of enzyme

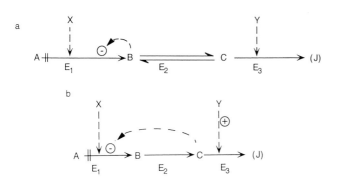

FIG. 5.9. Flux with feedback inhibition. (a) Simple linear system, (b) system with extended feedback (end product) inhibition. ⊖ denotes an inhibition, ⊕ denotes an activation.

activity enables a wide range of communications to occur because the system is not limited by the need to use metabolites chemically resembling the substrates, cofactors and products. Indeed, metabolic control systems could not be understood, even qualitatively, until the advent of this concept. Allosteric effects also allow the operation of feedforward mechanisms, such as the activation of pyruvate kinase in liver by FBP (53). The net result is a complex, tightly organized, yet flexible control system which has enormous benefits to the cell; but which can soon give a headache to the experimental worker trying to unravel it!

As a result of allosteric inhibition, the communication between reaction E_3 and E_1 in Figure 5.9a could be made directly by C, as in Figure 5.9b. Here, E_3 is now regulatory for J via the communication sequence

$$E_3 \sim C \sim E_1 \sim B \sim E_2.$$

This is an example of end-product inhibition of a pathway, a study of which in bacteria paved the way for the development of the allosteric concept (171).

Finally, the presence of several potential control sites in a system (for example, in Figure 5.9a, all three reactions are potentially regulatory for J) means that, in general, a flux does not have a unique control site, or pacemaker (77, 78). Indeed, a single regulator may interact simultaneously at more than one such site. An example is the interaction of the glycolytic regulator, AMP, with both phosphofructokinase and phosphorylase, both of which are regulatory for glycogenolysis (Fig. 5.10).

Control Versus Regulation

In Figure 5.11 both E_1 and E_2 are potentially regulatory for flux J (and metabolite S). However, since there are no external regulators of E_2, only E_1 is actually used to change J and S. The potential regulatory sequence from E_2,

$$E_2 \sim S \sim E_1,$$

is therefore said to be silent (46). However, for any given change in the external regulator, Z, the actual values of J and S will be influenced by the kinetics of

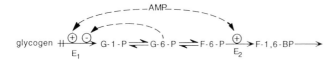

FIG. 5.10. Outline of the control of glycolysis by ATP. $E_1 =$ phosphorylase b; $E_2 =$ phosphofructokinase (PFK). $\oplus$ denotes an activation.

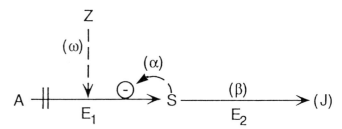

FIG. 5.11. System with two potential control sites, only one of which is used. Here, as in subsequent figures, the greek letters denote relative-change intrinsic or component sensitivities.

reaction E_2. For example, if this reaction has a low affinity for S, a greater concentration of S will be needed to balance the reactions in the steady state so that S will be greater the lower the affinity of E_2 for S. Thus, although E_2 does not play any role in changing S in this system, it does play a role in determining the steady state concentration of S. Indeed, in general, any potential communication with a variable such as J or S will help to determine the steady state value of that variable.

Recently, it has been suggested that the factors determining the value of a variable (that is, all the factors that can potentially communicate with it) be defined as controlling that variable; that is, helping to constrain its value to that observed in the steady state. In contrast, factors that actually change the value of a variable from one steady state to another are defined as regulating that variable (73). Thus, in Figure 5.11, Z would be defined as regulating J and S, whereas E_2, E_1, and Z would be defined as controlling J and S. Therefore regulation becomes a subset of control, at least qualitatively!

This qualitative distinction between control and regulation is quite helpful, and will be adopted in this chapter. As will be seen later, one of its effects is to highlight and resolve much of the confusion that has surrounded the subject in recent years. However, attempts to use this definition as a means of quantifying control (as opposed to regulation) are far less satisfactory: as will be shown later under Metabolic Control Theory.

Identification of Communication Sequences In Vivo

From the above discussions, it is evident that, to understand the control or regulation of any pathway, it is necessary to identify both the external and internal communications. This means establishing the reversibility of the component reactions, identifying allosteric interactions, and deciding whether such interactions (as well as those of substrates, cofactors, and products) are physiologically significant.

There are at least two methods for establishing the kinetic reversibility of a reaction in situ. First, since a reversible reaction is one that is not far from equilibrium, they can be identified by comparing the mass-action ratio, Γ, with the corresponding equilibrium constant, K (114, 118). The mass-action ratio is the same ratio as that for the equilibrium constant, but with concentrations at equilibrium replaced by those in situ. Thus, for a reaction at equilibrium, K and Γ are equal, and the ratio, K/Γ, is equal to 1. Consequently, if K/Γ is close to 1 the reaction will be near equilibrium, and hence reversible, in situ: if K/Γ is very much greater than 1, the reaction will be far from equilibrium and hence irreversible. Values of K/Γ for the reactions of glycolysis are shown in Table 5.1, from which it can be seen that the reactions catalyzed by phosphorylase, hexokinase, phosphofructokinase, and pyruvate kinase are almost certainly irreversible. The others are reversible, with glyceraldehyde-3-phosphate dehydrogenase plus phosphoglycerate kinase, a borderline case (discussed later).

The kinetic reversibility of a reaction may also be established by measuring the fluxes through it, a technique usually requiring isotopes. For example, in perfused rat heart, $^{14}CO_2$ was readily incorporated into citrate, thereby establishing the reversibility of isocitrate dehydrogenase. However, no label appeared in acetyl-CoA or malate, indicating that, in this particular tissue, the reaction catalyzed by citrate synthase is probably irreversible (129).

For quantitative studies, the extent of the reversibility can be measured by the function R_v (formerly, R [44]: see later under Effect of the Kinetic Reversibility of the Reaction, R_v), so that this problem does not arise (at least in theory!). However, experimental determinations of reversibility in situ are fraught with problems, and these may be such that the information is at best qualitative only (for example, the data in Table 5.1). First, the value of the equilibrium constant, K, is not usually an absolute thermodynamic constant, but is affected by such factors as ionic strength, pH, and metal ion concentrations. Consequently, values of K determined in vitro may not be applicable at the site of the reaction. Secondly, the metabolite concentrations required for Γ (and for calculating specific activities needed for isotope incorporation studies) are usually based on tissue contents (as in Table 5.1). Quite often, such contents do not reflect concentrations as a result of (1) the existence of the metabolite in more than one cell compartment (for example, acetyl-CoA, malate, and oxaloacetate occur both in the cytoplasm and in

the mitochondrial matrix [1]); (2) binding of significant amounts of metabolite to cell proteins and other structures (for example, in muscle significant amounts of both ADP and AMP are bound [136]); (3) direct passage of some intermediates (channelling) from one enzyme to the next in the pathway, resulting in concentrations at the active sites that are different from those in the bulk phase of the cell (6, 85, 122, 160, 168).

For example, binding of glyceraldehyde-3-phosphate to aldolase and other proteins is most probably the reason for the apparent irreversibility of glyceraldehyde-3-phosphate dehydrogenase plus phosphoglycerate kinase in Table 5.1). This step is almost certainly reversible in situ because no separate enzyme system is required to carry the reverse flux of gluconeogenesis in tissues such as liver and kidney cortex.

Some of these problems (especially those due to compartmentation of the metabolites) may be resolved, at least partially, although the techniques can be quite difficult and cumbersome (1). There are also specific techniques which can measure the concentrations of certain metal ions, notably Ca^{2+} (66, 100, 189). However, the most promising set of general techniques are those which involve direct, non invasive, determinations of intracellular concentrations: most notable amongst these is nuclear magnetic resonance, NMR, mentioned earlier. Although the sensitivity of the latter method is often too low to measure metabolic intermediates below concentrations of approximately 0.2 mM (75), one would expect this problem to be resolved eventually and enable more accurate intracellular concentrations to be obtained.

The above problems also bedevil the establishment of other communications in situ. For example, deciding whether a reaction is flux-generating requires a comparison of the concentration of its substrate with the kinetic response of the enzyme. The revelant kinetic function here is the $S_{0.5}$ value, which is the concentration of substrate required to give half-maximal activation (for simple hyperbolic responses, $S_{0.5}$ is the Michaelis constant, K_m). Thus, for a reaction to be flux-generating, it must be unresponsive to the concentration of its substrate, S, in situ, so that S must be much greater than $S_{0.5}$. However, this comparison is made even more difficult by the particulate nature of many initial substrates (for example, glycogen and triacylglycerol), which makes even the very definition of the term concentration somewhat dubious (see APPENDIX 1). In addition, the $S_{0.5}$ value has to be deduced from enzymological studies in vitro, and there is no guarantee that it applies in vivo.

Similarly, the establishment of whether an irreversible reaction responds to its substrate(s) in situ involves comparisons of their concentrations with the relevant $S_{0.5}$ values, again leading to the possibility of significant errors. The same conclusions apply to the interactions of possible allosteric effectors (both activators and inhibitors) and to possible inhibition by products at irreversible reactions. (In contrast with substrates and activators, an inhibition is significant only if the concentration of inhibitor is much greater than the $I_{0.5}$ value, which, in the classical hyperbolic case, is the familiar K_i value [49].)

With all these potentially depressing pitfalls it may seem surprising that so much has actually been achieved in studying metabolic regulation! However, the major problems occur when dealing with any specific detailed quantitative analysis of systems. Qualitative modelling is less of a problem. All that is needed for this is an identification of whether the component reactions are reversible or not; and whether any metabolite interactions with the enzymes are likely to be of physiological significance (for example, whether a substrate concentration is similar to or much greater than the relevant $S_{0.5}$ value). Accurate values of metabolite concentrations, kinetic responses, etc. are not required. This enables a very crude control model to be set up (as in the hypothetical schemes discussed hitherto) and from this any potential control sites can be identified. With simple systems the identification of control sites can be made by using regulatory sequences, but with more complex systems this procedure becomes potentially misleading and is replaced by more general methods based on either graph theory (APPENDIX 3) or the qualitative manipulation of matrices (38). The conclusions from such models can then be tested by using specific activators and inhibitors of the enzymes to see whether communication sequences predicted by the model actually operate in vivo (for example, ref. 134).

Qualitative modelling can give a great deal of information about the operation of control systems. Indeed, much of our current knowledge of metabolic control systems has been deduced from a predominantly qualitative type of analysis (see refs. 41, 107, 114, 118, 132). Unfortunately, with the current emphasis on detailed quantitation, it is all too easy to scorn such analyses nowadays. Although a detailed kinetic modelling is the ultimate aim for understanding fully the details of metabolic control systems, especially the relative importance of different control sites, such modelling is very time consuming; and there are consequently relatively few such models available (examples include glycolysis in erythrocytes [71, 130], and various pathways in bacteria and other unicellular organisms [190]). For most pathways, a predominantly qualitative modelling is all that is possible with current analytical data.

Need for Sensitive Regulation In Vivo

It is often observed experimentally that quite large changes of flux can be produced with quite small changes in the concentrations of regulators and internal metabolites. For example, the rate of glycogenolysis in muscle (Fig. 5.10) may increase several hundredfold, but this is accompanied by changes in concentrations of the glycolytic intermediates of less than 4 fold (41, 107, 114, 118). Why should this be so?

First, the concentrations of most metabolic intermediates are in the range, 0.01–1 mM; a hundred fold increase in these would give final concentrations in the range 1–100 mM. Changes of this magnitude would not only strain the solvent capacity of the cell water, but would also precipitate out salts of Ca^{2+} and other metal ions, as well as deplete the limited amounts of inorganic phosphate, CoA, and other cofactors (3).

Secondly, many metabolic intermediates are involved in, or communicate with, equilibria or reactions that are essential for providing the kinetic structure of metabolism and hence its control; large changes in these concentrations could seriously damage this structure. For example, the reaction catalyzed by glycogen phosphorylase,

$$glycogen_n + P_i \rightarrow glucose\text{-}1\text{-}phosphate + glycogen_{n-1}$$

is kinetically irreversible in vivo, despite an equilibrium constant of approximately 0.3 (that is, in the direction of glycogen synthesis, Table 5.1). This irreversibility results from the concentration of glucose-1-phosphate (approx. 0.03 mM) being maintained much lower than that of inorganic phosphate (P$_i$, approx. 10mM), so that the mass action ratio, Γ, is approximately 0.003. The value of K/Γ is therefore 100 and the reaction is irreversible (Table 5.1). If the concentration of glucose-1-phosphate were to increase, and that of P$_i$ were to decrease one hundredfold, the value of Γ would become approximately 30 and the reaction would proceed in the direction of glycogen synthesis. Since phosphorylase catalyses a reaction that serves to generate a steady state flux of glycogenolysis in situ (APPENDIX 1; Fig. 5.2b), it must remain irreversible in the direction of glycogenolysis at all times. Consequently, only small changes of glucose-1-phosphate and/or P$_i$ can be tolerated (133).

However, these conclusions only apply to metabolites which transmit important metabolic fluxes of energy and biosynthesis. They need not apply to those whose role is solely as controlling agents, for example second messengers such as cyclic AMP, Ca^{2+}, and inositol phosphates (4, 15, 27, 69, 100, 127, 184). For such regulatory metabolites the resting concentrations may be very low indeed, perhaps less than 1 μM. Thus even a thousandfold increase will not produce final concentrations that are deleterious to the cell. Furthermore, since they participate in no metabolic reactions other than those synthesizing and degrading them, large changes in their concentrations do not disrupt the overall kinetic and regulatory structure of metabolism.

Metabolic Sensitivities

The mechanisms which provide a large or sensitive response of a flux to its regulators can be analyzed qualitatively (for example, refs. 41, 107), but this soon becomes difficult when, as is usually the case, several mechanisms act together. Therefore we need methods for (1) quantifying the effects of each mechanism and (2) combining these effects to produce an overall response.

Let us consider a general communication, $X \sim Y$, where X is a stimulus (for example, a regulator) and Y the ensuing response (for example, a flux, concentration or reaction rate). If a given change of stimulus, ΔX, produces a response of ΔY, the strength of the communication can be expressed as the ratio of the absolute changes, $\Delta Y / \Delta X$, or of the relative changes, $(\Delta Y/Y)/(\Delta X/X)$. (A relative change is equivalent to a percentage change and its usefulness in quantifying metabolic regulatory systems will be seen later.)

However, to enable the strengths of complex systems to be calculated from those of their components, it is necessary to work with the mathematical limits of these ratios, dY/dX and $(dY/Y)/(dX/X)$: in FOT these functions are termed absolute-change and relative-change sensitivities, respectively (44, 45).

Thus, for the communication $X \sim Y$, the absolute-change sensitivity, a_X^Y, is defined by the equation,

$$a_X^Y = dY/dX \qquad (5-1)$$

and the corresponding relative-change sensitivity, s_X^Y, is defined by the equation,

$$s_X^Y = (dY/Y)/(dX/X) \qquad (5-2)$$

$$= d\ln Y/d\ln X \qquad (5-3)$$

The equation for s_X^Y can be developed further, as follows:

$$d\ln Y = s_X^Y \cdot d\ln X \qquad (5-4)$$

so that, upon integration,

$$\ln Y = s_X^Y \cdot \ln X + \ln k \qquad (5-5)$$

where k is a constant.
Therefore,

$$\ln Y = \ln(k \cdot X^s) \qquad (5-6)$$

or,

$$Y = k \cdot X^s \qquad (5-7)$$

Equation 5–7 is a power equation representing the communication, $X \sim Y$, and has been used as the basis for a quantitative algebraic analysis of control systems (44). In this respect the approach is analogous to biochemical systems theory (see later under Biochemical Systems Theory). However, there is one important difference: equation 5–7 refers to infinitesimal changes whereas the power equations of BST refer to large finite changes ($\Delta X, \Delta Y$). Consequently, the value of the integration constant is not needed in FOT, whereas it is in BST. Moreover, although the use of finite power equations has been justified as a reasonable approximation in several cases (41, 107, 139, 158, 175), it always suffers from a potential drawback of having to assume that the index (that is, s_X^Y in equation 5–7) is effectively constant over the response. This assumption is never guaranteed and, in our opinion, it is better to avoid having to make it by basing the analysis on infinitesimals.

However, a major problem with using infinitesimal power equations such as equation 5–7 is that they cannot be applied to systems where Y changes sign, for example if Y represented a glycolytic flux that reversed to become a gluconeogenic one (157). Therefore the analysis in FOT is now based on equations analogous to equation 5–4, which apply under all conditions, including a reversal of the sign of Y (37, 46). To avoid excessive notation, a term such as $d\ln X$ ($= dX/X$) is written as X_ρ (previously $\overset{r}{X}$); so that equation 5–4 would be written as,

$$s_X^Y = Y_\rho/X_\rho \qquad (5-8)$$

or,

$$Y_\rho = s_X^Y \cdot X_\rho \qquad (5-9)$$

Combining Sensitivities: Product and Addition Rules. Since both absolute-change and relative-change sensitivities are differential coefficients, they can be combined by the customary rules of calculus. Two such rules are of fundamental importance (44).

First, if two or more communications are linked together consecutively, that is, head to tail, as in the following system,

$$X \overset{a_X^Y}{\sim} Y \overset{a_Y^Z}{\sim} Z,$$

the absolute-change sensitivity of the overall change, dZ/dX, is the product $(dY/dX) \cdot (dZ/dY)$, that is,

$a_X^Y \cdot a_Y^Z$. The same conclusions apply to the corresponding relative-change sensitivities, so that

$$s_X^Z = s_X^Y \cdot s_Y^Z \qquad (5-10)$$

This product rule is supplemented by a corresponding addition rule which applies to communications which operate independently. For example, the following overall communication between X and Y,

$$X \overset{a_1}{\sim} Y \overset{a_2}{\sim} X$$

consists of two independent or distinct routes. For a given absolute change of X, dX, the responses of Y via each route are,

$$(dY)_1 = a_1 \cdot dX \qquad (5-11)$$

$$(dY)_2 = a_2 \cdot dX \qquad (5-12)$$

Since the changes are infinitesimal,

$$dY = (dY)_1 + (dY)_2 \qquad (5-13)$$

$$= (a_1 + a_2) \cdot dX \qquad (5-14)$$

whence,

$$a_X^Y = dY/dX$$
$$= a_1 + a_2 \qquad (5-15)$$

The same rule applies to the corresponding relative-change sensitivities, so that,

$$s_X^Y = s_1 + s_2 \qquad (5-16)$$

Like the product rule, equation 5–13, and hence the addition rule, applies only to infinitesimal changes; this is the reason infinitesimally based functions are used to evaluate the strengths of complex systems.

These two rules enable the overall sensitivity of any complex system to be evaluated. For example, with the system,

$$X \overset{a_1}{\sim} Z \overset{a_2}{\sim} Y \overset{a_3}{\sim} X$$

applying the product rule to the left-hand sequence gives,

$$(a_X^Y)_{left} = a_1 . a_2,$$

so that, by the addition rule,

$$a_X^Y = a_3 + a_1 . a_2 \qquad (5-17)$$

Correspondingly, for the relative-change functions,

$$s_X^Y = s_3 + s_1 \cdot s_2 \qquad (5-18)$$

These two rules are the fundamental basis of the quantitation of metabolic regulatory or control sys-

tems, and all subsequent derivations in this chapter are based on them, albeit implicitly in many complex situations.

Rate Equations and Conservations Expressed as Sensitivities. One important consequence of the addition rule is that it allows a rate equation to be expressed in terms of the sensitivities of the rate to the effectors (21, 44). For example, consider a general reaction whose rate, v, depends on the concentrations of several effectors, $S_1, S_2, \ldots S_n$ (which may be substrates, cofactors, products, or allosteric effectors). The effect of each effector on v can be diagrammed as follows,

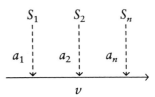

This represents a set of distinct communications, so that from equation 5–13,

$$dv = a_1 \cdot dS_1 + a_2 \cdot dS_2 + \ldots a_n \cdot dS_n \qquad (5-19)$$

Likewise, for the corresponding relative-change sensitivities,

$$d\ln v = s_1 \cdot d\ln S_1 + s_2 \cdot d\ln S_2 + \ldots s_n \cdot d\ln S_n \qquad (5-20)$$

or,

$$v_\rho = s_1 \cdot S_{1\rho} + s_2 \cdot S_{2\rho} + \ldots s_n \cdot S_{n\rho} \qquad (5-21)$$

Equation 5–21 may be expressed as an infinitesimal power equation (analogous to equation 5–7),

$$v = k \cdot S_1^s \cdot S_2^s \cdot \ldots S_n^s \qquad (5-21a)$$

which may be used as a basis for algebraic and other quantitative analyses (44); unless, as mentioned above, v changes sign.

In addition to rate equations, many metabolic systems contain branched fluxes (for example, Fig. 5.3) and may also have conserved pools of metabolites: for example, under many circumstances the sum of NAD and NADH is constant. For regulatory analysis, branched systems and conservations are equivalent, because a branch-point equation such as,

$$J = K + L \qquad (5-22)$$

(Fig. 5.3) is equivalent to a conservation,

$$J - K - L = 0 \text{ (that is, a constant)} \qquad (5-23)$$

Therefore, let us consider a general conservation,

$$A + B = k \qquad (5-24)$$

where k is a constant (including zero). For infinitesimal absolute changes,

$$dA + dB = dk$$
$$= 0 \qquad (5\text{--}25)$$

since k is constant.

Thus, for an analysis based on absolute changes, any conservations can be written by analogy with equation 5–25.

For relative changes, equation 5–25 can be expressed as,

$$dA \cdot (A/A) + dB \cdot (B/B) = 0$$

whence, using the ρ subscript to denote dA/A and dB/B,

$$A \cdot A_\rho + B \cdot B_\rho = 0 \qquad (5\text{--}26)$$

Consequently, for a system based on absolute changes, equation 5–23 for the branched system would be written as,

$$dJ - dK - dL = 0 \qquad (5\text{--}27)$$

and, for one based on relative changes, as (44),

$$J \cdot J_\rho - K \cdot K_\rho - L \cdot L_\rho = 0 \qquad (5\text{--}28)$$

Equations analogous to equations 5–21 and 5–26 form the basis of most of the algebraic and quantitative analyses of FOT, allowing complex sensitivities to be calculated or derived systematically from the component rate and conservation equations of any metabolic system (37, 45, 46). However, it must be stressed that these sensitivities, whether based on absolute or relative changes, refer only to infinitesimal changes. Because their numerical values vary continuously during a large, or finite, change, they cannot be used to calculate the response to such changes directly (although they may be used indirectly, as shown later under Calculation of Responses to Large (Finite) Changes of Regulators). However, a most useful property of these net sensitivity functions is that they allow an algebraic analysis of the effects of proposed control mechanisms on the overall response of a system; several examples of this approach will be given in subsequent sections of this chapter.

Which Type of Sensitivity Should Be Used? So far, absolute-change and relative-change sensitivities have been treated with equal importance, since they both obey the fundamental combination rules. However, each type has its own particular uses and, as will be seen later under Analysis of Non-Steady States, the absolute-change function is of great importance for analyzing and calculating time-dependent responses. Also, as shown later under Calculating the Relative Importance of Regulators and Regulatory Sequences During a Finite Response: Mean-Value Sensitivities, it is the most appropriate function to use when computing the relative importance of different control or regulatory sites in situ.

However, when considering the overall changes in concentrations and fluxes from one steady state to another, the relative-change function is the most appropriate one to use for at least two reasons. First, it is dimensionless (unlike the absolute-change function) and is therefore algebraically simpler. Secondly, it relates more directly to the general effect of changes on metabolic reactions or equilibria. For example, a given absolute change of 0.1 mM would have little effect on an enzyme or equilibrium if the initial concentration were 10 mM. In this case the relative change would be 0.01, which represents a 1.01-fold increase: a drop in the ocean. However, if the initial concentration were 0.01 mM, the relative change would be 10, which represents an 11-fold increase which would have a much more noticeable effect. Thus, relative-changes and sensitivities based upon them are the most useful functions for describing overall steady-state responses. All the major approaches use them for this purpose, either explicitly (as in FOT and MCT) or implicitly (as the index in the power equations for BST).

Intrinsic and Component Sensitivities. Each term on the right-hand side (RHS) of equation 5–21 refers to the effect on v of an infinitesimal change of S when all the other effector concentrations remain constant. Thus, if the concentrations of $S_2 \ldots S_n$ remain constant, so that $S_{2\rho} \ldots S_{n\rho}$ are all zero,

$$v_\rho = s_1 \cdot S_{1\rho}$$

so that,

$$s_1 = v_\rho / S_{1\rho} \qquad (5\text{--}29)$$

This type of sensitivity is referred to, in FOT, as an intrinsic sensitivity, and given the symbol s_i (34, 35, 42, 44, 45, 108). Consequently, equation 5–21 may be written in terms of intrinsic sensitivities, as follows:

$$v_\rho = s^v_{i(S_1)} \cdot S_{1\rho} + s^v_{i(S_2)} \cdot S_{2\rho} + \ldots s^v_{i(S_n)} \cdot S_{n\rho} \quad (5\text{--}30)$$

However, with some rate equations in complex systems it is often simpler to combine the effects of several of the effectors into a single sensitivity function (44). For example, the adenine nucleotides, ATP, ADP, and AMP, often act together at a reaction: at phosphofructokinase, their action can be summarized as follows:

Since ADP and AMP activate this enzyme and since their concentrations change in a direction opposite to that of ATP, the net effect of the intercommunication between the nucleotides is to reinforce the inhibitory effect of ATP. Moreover, at cellular concentrations of these nucleotides, the adenylate kinase equilibrium $(ATP + AMP \rightleftharpoons 2ADP)$ and the $ATP + ADP + AMP$ $(+IMP)$ conservation act together to convert a small change of ATP concentration into a much larger change in that of AMP (107, 114, 118). Thus ADP and AMP act as auxiliary regulators (as do other important regulators of this enzyme, such as F-2,6-BP, P_i and citrate [167]), enabling large changes in the activity of phosphofructokinase to result from relatively small changes in ATP concentration. As discussed earlier under Need for Sensitive Regulation In Vivo, small changes of ATP are desirable since ATP is a key intermediate participating directly in many reactions and equilibria. The use of auxiliary or secondary regulators is a common method for reducing changes in concentrations of key intermediates and regulators such as ATP (35, 41, 109).

The full rate equation for phosphofructokinase,

$$v_\rho = ATP_\rho \cdot s_{i(ATP)}^v + ADP_\rho \cdot s_{i(ADP)}^v + AMP_\rho \cdot s_{i(AMP)}^v + \text{other terms} \tag{5-31}$$

may be written in terms of ATP alone. Thus, by using the product rule (equation 5−10) for the communications between ATP and ADP, and between ATP and AMP, the terms in ADP_ρ and AMP_ρ can be eliminated to produce the equation,

$$v_\rho = ATP_\rho \cdot s_{i(ATP)}^v + ATP_\rho \cdot s_{i(ADP)}^v \cdot s_{ATP}^{ADP} + ATP_\rho \cdot s_{i(AMP)}^v \cdot s_{ATP}^{AMP} + \text{other terms} \tag{5-32}$$

$$= ATP_\rho \cdot [s_{i(ATP)}^v + s_{i(ADP)}^v \cdot s_{ATP}^{ADP} + s_{i(AMP)}^v \cdot s_{ATP}^{AMP}] + \text{other terms} \tag{5-33}$$

$$= ATP_\rho \cdot s_{ATP}^v + \text{other effector terms} \tag{5-34}$$

where s_{ATP}^v is a complex function consisting of the three intrinsic sensitivities and two sensitivities relating to the internucleotide communications.

Since we are often dealing with energy fluxes, for which ATP is a major pathway substrate (see earlier under Other Fluxes: Energy Fluxes and Equivalence), it is often more convenient to write complex rate equations such as equation 5−31 in terms of ATP alone (as in equation 5−34), thereby reducing the algebraic complexity of the system. However, when his is done, the resulting sensitivity, s_{ATP}^v, is not an intrinsic one because the intrinsic sensitivity to ATP refers to the effects of ATP acting alone. It is therefore referred to as a component sensitivity.

By analogy with the definition for intrinsic sensitivity, a component sensitivity of v to an effector, S_j, (equation 5−21) is the response of v_ρ to $S_{j\rho}$ when all other effectors of the reaction, as it is written in the system under analysis, remain constant. If a system is written in terms of equation 5−31, the coefficient of ATP_ρ will be the intrinsic sensitivity; if the system is written in terms of ATP alone (equation 5−34), the coefficient will be the component sensitivity, whose value is given by equation 5−33.

For the rest of this chapter, the sensitivity terms in the component rate equations will be assumed to represent component sensitivities, unless otherwise stated. Moreover, to reduce the amount of cumbersome notation, greek letters will be used to denote component (or intrinsic) sensitivities. Any such letter will refer to a defined system, but will inevitably be used several times in different contexts.

Expressing Intrinsic and Component Sensitivities in Terms of Reaction Kinetics.

Up to now, sensitivities have referred to general communications with no specified mechanisms. However, in this form, they would be of little practical use in analyzing control systems. Therefore we now have to incorporate the effects of several important kinetic mechanisms that are believed to occur in cells. For this, the classical equations of enzyme kinetics and reaction rates will be applied. Although this approach has been criticized (9, 179, 181), there is currently no other satisfactory way; and the results have, in any case, provided too many valuable insights into the kinetics and responses of organized systems to be dismissed as irrelevant. As with so many other approaches to metabolic regulation, it should be seen as useful approximation.

Effect of enzyme kinetic response. Let us first consider a simple enzyme-catalyzed reaction whose rate, v, is a saturable function of an effector, S. S may be a substrate, cofactor, product, allosteric activator, or inhibitor. In many situations the rate may be approximated by the following Hill equation (5−2):

$$v = \frac{V_m \cdot S^n}{K^n + S^n} \tag{5-35}$$

where V_m is the maximum rate when S is saturating and K is a parameter whose units are those of concentration. In fact, K is equal to $S_{0.5}$, the concentration of S that produces a half-maximal rate $(V_m/2)$. In general, both V_m and K depend on the concentrations of other effectors so that they are more properly regarded as parameters than constants.

n is an index and, when its value is unity, equation 5−35 becomes the familiar Michaelis-Menten equation describing a hyperbolic response,

$$v = \frac{V_m \cdot S}{K_m + S} \qquad (5\text{-}36)$$

where K_m is the Michaelis constant. Curves showing typical responses for $n = 1$ and $n = 4$ are shown in Figure 5.12.

Equations 5–35 and 5–36 refer to an irreversible (that is, unidirectional) reaction, but this is no problem because the effects of kinetic reversibility can be incorporated at a later stage (see later under Effect of the Kinetic Reversibility of the Reaction, R_v).

To calculate the effect of the kinetic response on the intrinsic sensitivity of v to S, equation 5–2 is used in the form,

$$r_S^v = \frac{dv}{dS} \cdot \frac{S}{v} \qquad (5\text{-}37)$$

where r denotes a kinetic-response component of the intrinsic sensitivity, s_i (44). If no other mechanisms operate, $r = s_i$.

Differentiating equation 5–35 with respect to S,

$$\frac{dv}{dS} = \frac{n \cdot K^n \cdot V_m \cdot S^{n-1}}{(K^n + S^n)^2} \qquad (5\text{-}39)$$

representing the absolute-change sensitivity of this response. Therefore, from equation 5–37,

$$r_S^v = \frac{n \cdot K^n}{(K^n + S^n)} \qquad (5\text{-}41)$$

so that, for a Michaelis-Menten response, when $n = 1$,

$$r_S^v = \frac{K_m}{K_m + S} \qquad (5\text{-}42)$$

Equations 5–41 and 5–42 show that the value of r, and hence s_i, varies continuously with the concentration

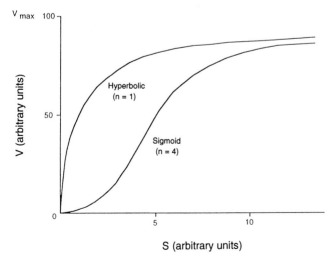

V max   100

V (arbitrary units)

Hyperbolic
(n = 1)

50

Sigmoid
(n = 4)

0

5   10

S (arbitrary units)

FIG. 5.12. Hill responses. The curves are calculated from equation 5–35, with $K = 1$ and $V_m = 100$.

of S. As mentioned previously, this variability is a characteristic of sensitivity functions in general. They are not constants and refer only to specific infinitesimal changes. Indirect methods must be used to calculate larger changes (see later under Calculation of Responses to Large [Finite] Changes of Regulators).

Equations 5–41 and 5–42 also show that, when $S << K$, the value of r approaches n; when $S >> K$, the value of r approaches zero. Consequently, to make use of the sensitivity inherent in a sigmoid response (that is, when $n > 1$: Fig. 5.12), the value of S must not be too much greater than K (or $S_{0.5}$).

For an inhibitor, the value of n is negative. Thus, when $n = -1$ (that is, classical linear inhibition), replacing S by I and K by K_i, equation 5–41 becomes,

$$r_I^v = \frac{-K_i^{-1}}{K_i^{-1} + I^{-1}} \qquad (5\text{-}43)$$

or, more familiarly,

$$r_I^v = \frac{-I}{I + K_i} \qquad (5\text{-}44)$$

This equation shows that, as with an activator (S), the value of r varies continuously with I. However, there is an important difference. When $I << K_i$ the value of r tends to zero, and when $I >> K_i$, the value of r tends to -1 (in the general case, $-n$). These results are the inverse of those for an activator and show that, whereas S must not be too much greater than K ($S_{0.5}$) for an activation to be physiologically significant, I must not be too much lower than K ($I_{0.5}$) for the inhibition to be significant.

Mass-action and catalytic effects. If S is an allosteric effector, r describes an effect purely on the catalyst (that is, enzyme). This catalytic effect has been denoted by the symbol, r_c (44); so that in this case, $r = r_c$. However, if S takes part in the reaction, that is, is a substrate, cofactor, or product, there is a mass-action component of r, r_m, in addition to the catalytic one, r_c. Since these sensitivities represent distinct communications between S and the rate, v, application of the addition rule (equation 5–16) gives,

$$r = r_m + r_c \qquad (5\text{-}45)$$

The importance of splitting r into its mass-action and catalytic components will become clear when considering the effects of reversibility on the intrinsic sensitivity.

For example, if S is a substrate with a Michaelis-Menten response, so that, by equation 5–42,

$$r = K_m / (K_m + S)$$

and if one molecule of S participates in the reaction, so that the mass-action response is linear in S and therefore $r_m = 1$,

$$r_c = r - r_m$$

$$= K_m/(K_m + S) - 1 \qquad (5-46)$$

$$= -S/(K_m + S) \qquad (5-47)$$

Equation 5–47 shows that, when S is much less than K_m, the value of r_c approaches zero, as expected since few binding sites will be occupied and most of the catalyst will be active. However, as S increases, the value of r_c becomes more and more negative, and approaches -1 as S becomes very large. This negativity reflects the continuously decreasing number of binding sites available as S increases. Eventually the negative catalystic effect cancels out the mass-action component, r_m, to give the zero order kinetics associated with saturation (when $r = 0$).

Finally, although the general effects of the kinetic response have been illustrated by using a Hill equation, the calculations of r can be applied to any enzyme-kinetic equation. For example, if the equation is a two-substrate response of the type,

$$v = \frac{V_m \cdot A \cdot B}{A \cdot B + B \cdot K_a + A \cdot K_b + K_{ab}} \qquad (5-48)$$

the value of r for A (r_A^v) is obtained by differentiating the equation with respect to A, and treating the other substrate, B, as a constant. Similarly the value of r_B^v is obtained by differentiating equation 5–48 with respect to B, treating A as a constant. These r values are therefore functions of the other substrate(s), adding a further contribution to the variability of these sensitivities during any noninfinitesimal response.

Effect of the kinetic reversibility of the reaction, R_v. When a reaction is kinetically reversible (near equilibrium), for example,

$$S \underset{v_r}{\overset{v_f}{\rightleftharpoons}} P$$

the net rate, v, is equal to $v_f - v_r$; where v_f and v_r are the rates of the forward and reverse components of the reaction, respectively (if the reaction is at equilibrium, $v_f = v_r$ and v is zero).

Let us initially (for simplicity) assume that a substrate, S, interacts linearly with the enzyme catalyzing the reaction, so that the value of r is unity. In this situation,

$$v_f = k \cdot S$$

where k is a rate constant, so that

$$v_{f\rho} = S_\rho \qquad (5-49)$$

Since

$$v = v_f - v_r \qquad (5-50)$$

and we are evaluating a component of the intrinsic sensitivity to S, that is, when P and hence v_r is constant, differentiating equation 5–50 gives,

$$dv = dv_f \qquad (5-51)$$

which can be written as,

$$dv/v = (dv_f/v) \cdot (v_f/v_f) \qquad (5-52)$$

or,

$$v_\rho = (v_f)_\rho \cdot (v_f/v) \qquad (5-53)$$

Thus, from equation 5–49,

$$v_\rho = S_\rho \cdot (v_f/v) \qquad (5-54)$$

The component of the intrinsic sensitivity of v to S due to the reversibility of the reaction is given by the ratio, v_ρ/S_ρ, and is therefore equal to the function v_f/v, which is termed the reversibility, R_v (formerly, R (44)).

R_v is related to the ratio K/Γ (see earlier under Identification of Communication Sequences In Vivo) by the equation (44, 108),

$$R_v = \frac{K/\Gamma}{K/\Gamma - 1} \qquad (5-55)$$

and may therefore be calculated from values of K/Γ, such as those in Table 5.1. R_v may also be measured directly from the net rate, v, and a determination of either v_f or v_r by using isotopic methods. For example, the reversibility of isocitrate dehydrogenase in the perfused rat heart has been measured from the incorporation of $^{14}CO_2$ into citrate (thereby measuring v_r); from the data in Scheme 1 of Randle et al. (129), the value of R_v for the isocitrate dehydrogenase step is approximately 4. However, as discussed earlier, the accuracy of any determination of R_v will be limited by the difficulties of measuring K, Γ, and isotopic fluxes in situ.

By similar arguments, but keeping S and hence v_f constant, the component of the intrinsic sensitivity to the product, P, is,

$$s_{i(P)}^v = -(R_v - 1) \qquad (5-56)$$

Equations 5–55 and 5–56 show that, when a reaction is completely irreversible, so that $v_r = 0$ and $v = v_f$, the value of $R_v = 1$. In this situation the value of $s_{i(S)}^v = 1$ (that is, there is just the linear response between S and the enzyme activity) and $s_{i(P)}^v = 0$ (since, in this simple example, there is no effect of P on v when $v_r = 0$).

At the other extreme, when the reaction approaches equilibrium, v_f becomes much greater than v, so that R_v becomes very large. Consequently, both $s_{i(S)}^v$ and $s_{i(P)}^v$ become very large, showing that, as a reaction approaches equilibrium, the intrinsic sensitivity to its substrates, products, and cofactors becomes very large (in agreement with earlier qualitative predictions (41 107)). However, as will be seen later, this may not always be the case for the net sensitivity in situ.

A major problem with the derivation of equations 5–55 and 5–56 is that both S and P were assumed to interact only with v_f and v_r, respectively. In other words, the effects are purely mass-action and the catalystic components, r_c, are zero. This is most unlikely (and impossible for allosteric effectors) and we must therefore consider the more general situation where S also interacts with the catalyst.

In the general situation, the communication between S and v can be written as,

$$
S \underset{r_c \quad v_r \quad -(R_v-1)}{\overset{r_m+r_c \quad v_f \quad R_v}{\rightleftharpoons}} v
$$

where the bottom sequence includes the catalystic effect of S on the reverse process. This must be present, and be of sensitivity r_c, because any effect on the catalyst changes v_f and v_r by the same factor (103).

Applying the product rule (equation 5–10) to each individual sequence and combining the results by the addition rule (equation 5–16),

$$s_{i(S)}^v = R_v \cdot (r_m + r_c) - r_c \cdot (R_v - 1) \qquad (5\text{–}57)$$

$$= R_v \cdot r_m + r_c \qquad (5\text{–}58)$$

A similar analysis for P shows that, in general,

$$s_{i(P)}^v = -(R_v - 1) \cdot r_m + r_c \qquad (5\text{–}59)$$

where the r terms now refer to the kinetic response of the enzyme to P.

Equations 5–58 and 5–59 show that R_v only increases the mass-action component, r_m, and thus only increases the intrinsic sensitivity to substrates, products, and cofactors: the response to allosteric effectors (for which $r_m = 0$) is not affected (34, 44). This means that any allosteric effects on reactions known to be near equilibrium (reversible) in vivo may be physiologically unimportant since these effects will be greatly opposed by the large effect of R_v on the response to the substrates and products (34). This point is considered in

more detail later (see under Effect of Reaction Reversibility (R_v) In Situ on the Net Sensitivity to External Regulators).

However, these conclusions about allosteric effects on near equilibria need not apply to a reaction which spans a membrane; consider, for example, the transport system for metabolite X in Figure 5.13. This is a kinetically reversible system with M an allosteric regulator. However, since M is present in only one of the compartments, it can communicate only with one process, here v_f. Consequently, the response to M is the same as that of a purely mass-action substrate, via the sequence,

$$
M \overset{r}{\rightleftharpoons} v_f \overset{R_v}{\rightleftharpoons} v
$$

so that,

$$s_{i(M)}^v = R_v \cdot r \qquad (5\text{–}60)$$

Therefore, unlike an enzyme reaction in free solution, the intrinsic sensitivity to the allosteric effector, M, now increases as the reaction approaches equilibrium. Allosteric effects on reversible transport systems (for example, due to metal ions such as Ca^{2+} and Na^+) may thus have important regulatory roles.

So far in this section, an algebraic approach has been used to examine the effects of kinetic response and reversibility on intrinsic sensitivities; this type of approach will be used extensively in later sections. It can be applied to infinitesimally based sensitivities because the general equations apply at each point of a finite (large) response. For example, although a numerical value for $s_{i(S)}^v$ (calculated from equation 5–58) would apply only to a specific infinitesimal change (since s varies continuously during a larger response), the algebraic conclusion that an increased R_v will increase the value of this sensitivity applies to all values of s, and hence to an entire finite response. Conse-

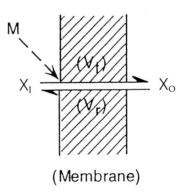

(Membrane)

FIG. 5.13. Reversible system spanning a membrane. X_I and X_O denote concentrations inside and outside a cell compartment, respectively. M is a metabolite present only on the inside.

quently, this type of algebraic approach is not limited by the infinitesimally based nature of sensitivities and can be used to provide valuable insights into the workings of metabolic regulatory systems.

Effect of substrate cycling. A substrate cycle (sometimes referred to as a futile cycle) occurs when two separate reactions oppose each other (81, 106, 108). For example, in Figure 5.14*a* the conversion of *S* to *P* (by E_2) is opposed by the conversion of *P* back to *S* (by E_4), despite there being a net flux (*J*) of *S* to *P* across the cycle. As shown, such cycling is accompanied by the consumption of energy, here, as in many systems, represented by the net hydrolysis of ATP to ADP plus P_i. Such cycles may therefore play an important role in the production of heat and/or dissipation of excess fuel (106, 108). However, only their role in metabolic regulation will be considered here.

From Figure 5.14*a* it can be seen that the cycle between *S* and *P* is analogous to a kinetically reversible reaction; so that one regulatory consequence of substrate cycling is the provision of feedback from *P* to *S*. Indeed, this has been shown for the cycle between fatty acids and triacylglycerol in dogs (95). However, the crucial difference between a substrate cycle and a reversible reaction is that the forward and reverse reactions (*F* and *C*, respectively) are distinct chemical reactions, catalyzed by different enzymes. This has important consequences, one of which is to enable a wider regulation of the net flux, because effectors of E_4 can be used in addition to those of E_2. Moreover, as shown later, it can also increase the sensitivity of the response to a regulator, and decrease the effectiveness of internal oppositions.

The analogy with reversible reactions can be used to derive the effect of the cycle on the sensitivity of the net flux across the cycle, v_c, to *S* and *P* (108). Thus, if enzyme E_2 responds linearly to *S* (so that $r = 1$) and reaction *F* is completely irreversible (so that $R_v = 1$), the intrinsic sensitivity of the net rate to *S* is the function equivalent to "v_f/v" in equation 5–55, that is, F/v_c ($= F/J$ in the steady state). Therefore, since

$$F = J + C,$$
$$s_{i(S)}^{v_C} = F/J$$
$$= (J + C)/J$$
$$= 1 + C/J \qquad (5\text{–}61)$$

Similarly, if enzyme E_4 responds linearly to *P*, and *C* is completely irreversible, the intrinsic sensitivity of v_c to *P* is the function equivalent to $-(v_f/v - 1)$ (equation 5–56) such that,

$$s_{i(P)}^{v_C} = -(1 + C/J - 1)$$
$$= -C/J \qquad (5\text{–}62)$$

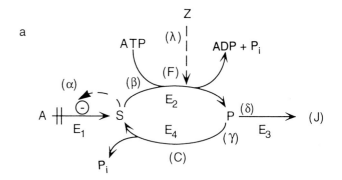

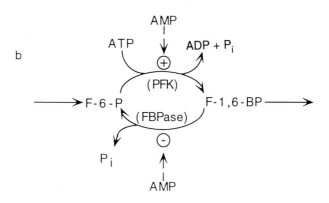

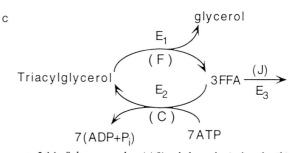

FIG. 5.14. Substrate cycles. *(a)* Simple hypothetical cycle, *(b)* cycle between fructose-6-phosphate (F-6-P) and fructose-1,6-bisphosphate (FBP), *(c)* cycle between triacylglycerol and FFA; the 7 ATP molecules consumed per turn comprise two each for the conversion of the three molecules of FFA to their CoA esters, and one for the formation of glycerol phosphate.

The functions $1 + C/J$ and $-C/J$, in equations 5–61 and 5–62, are actually the sensitivities of the elementary communications, $v_{E_2} \curvearrowright v_C$ and $v_{E_4} \curvearrowright v_C$, respectively; just as R_v and $-(R_v - 1)$ are the sensitivities of the elementary communications, $v_f \curvearrowright v$ and $v_r \curvearrowright v$, respectively (44). Thus, if the responses of *S* and *P* with E_2 and E_4 are not linear (as will usually be the case in vivo), these effects can be incorporated by using the product and addition rules (equations 5–10 and 5–16). For example, if

$$r_S^{E2} = K/(K+S) \text{ (see equation 5–42)}$$

$$s_{i(S)}^{v_C} = \frac{K \cdot (1+C/J)}{(K+S)} \qquad (5\text{–}63)$$

This is because the intrinsic sensitivity of v_C to S refers to the consecutive communication sequence,

$$S \overset{(r_S)}{\sim} \text{enzyme } E_2 \overset{(R_v=1)}{\sim} \text{rate of } E_2 \overset{(1+C/J)}{\sim} \text{net flux}$$

and the overall intrinsic sensitivity is therefore the product of the components.

The above procedures effectively convert the step represented by the cycle into a single reaction of net flux v_C. Although this is not absolutely necessary, it can often result in a great simplification of the algebra (see later under Effect of a Substrate Cycle In Situ on the Net Sensitivity to External Regulators).

Since reactions E_2 and E_4 are catalyzed by separate chemical reactions, an allosteric effector (for example, Z in Fig. 5.14a) need only interact with one of the reactions of the cycle (in contrast with its interaction with a reversible reaction in free solution). Consequently, the functions $1+C/J$ and $-C/J$ apply to the interaction of all effectors of the forward and reverse reactions, respectively. Thus, applying the product rule (equation 5–10) as above,

$$s_{i(Z)}^{v_C} = \lambda \cdot (1+C/J) \qquad (5\text{–}64)$$

where λ is the intrinsic sensitivity of E_2 to Z (Fig. 5.14a).

Indeed, an effector may have opposite effects on the component reactions of a substrate cycle; for example AMP activates phosphofructokinase but inhibits fructose-1,6-bisphosphatase (118) (Fig. 5.14b). In this example there are two communications between AMP and the net flux across the cycle, v_C,

$$\text{AMP} \overset{\alpha}{\sim} \text{PFK} \overset{(1+C/J)}{\sim} v_C \text{ and}$$

$$\text{AMP} \overset{\beta}{\sim} \text{FBPase} \overset{-C/J}{\sim} v_C$$

Therefore, by using the product rule on each sequence and combining them by the addition rule (equation 5–16),

$$s_{i(AMP)}^{v_C} = \alpha \cdot (1+C/J) - \beta \cdot C/J \qquad (5\text{–}65)$$

However, since AMP inhibits FBPase, β is negative and can therefore be replaced by $-\beta$, where β is a positive number. Equation 5–65 then becomes,

$$s_{i(AMP)}^{v_C} = \alpha \cdot (1+C/J) + \beta \cdot C/J \qquad (5\text{–}66)$$

Since all the factors on the right-hand side of equation 5–66 are now positive numbers, this equation shows that the opposite effects of AMP on the component reactions gives a further increased sensitivity due to the term $\beta \cdot C/J$. Physiologically, this enables the net flux of glycolysis (equivalent to J) to be regulated by even smaller changes of AMP.

Further examples of calculating complex intrinsic sensitivities at substrate cycles, using the product and addition rules, are given by Crabtree and Newsholme (44).

Furthermore, and again in contrast with a reversible reaction, it is possible to saturate one or more of the component reactions of a substrate cycle (for example, the intrinsic sensitivities α and/or γ in Figure 5.14a may be zero). This is impossible with a reversible reaction since the mass-action conditions automatically imply that v_f and v_r respond to the substrates and products, respectively. As will be seen later, the possibility of saturating the component reactions of a substrate cycle may be of considerable regulatory importance in increasing the sensitivity provided by such cycles in situ.

Equations 5–61 to 5–66 show that the value of the intrinsic sensitivity increases without limit as the rate of cycling, C, increases (although there is usually a limit in situ: see later under Effect of a Substrate Cycle In Situ on the Net Sensitivity to External Regulators). Consequently, a high sensitivity may require a high rate of cycling; this may cause problems physiologically because of the accompanying heat production, which may be considerable (60, 108). Indeed, the condition of malignant hyperthermia (hyperpyrexia), often seen in pigs and humans during anaesthesia, appears, at least in part, to be the result of excessive substrate cycling between fructose-6-phosphate and fructose-1,6-bisphosphate (25). Therefore, it has been suggested that, under resting conditions when there is no demand for a high sensitivity, the rates of cycling are maintained as low as possible. However, when pathway activity is anticipated (for example, to provide ATP for muscular activity), the rates of cycling, and hence the sensitivities, are increased. This increase only lasts for a short time, however, until the activity occurs or the resting state is returned to. The increased heat production does not last for long enough to damage the animal (108, 114). Catecholamines may play a role in varying the rates of cycling according to anticipated demand, since these hormones have been observed to increase the rates of several substrate cycles, both in vivo and in vitro (5, 22, 23, 106, 116); such a role would also be consistent with their physiological importance for flight or fight.

Finally, as with reversible reactions, the rates of substrate cycling are extremely difficult to measure in vivo. Radio- or stable isotopes have to be used and there are many problems in interpreting the results (40, 81, 176). Therefore, current values for substrate cycling in vivo have to be treated with caution; a caveat that unfortunately applies to many quantitative analyses of metabolic regulatory systems.

Effect of interconvertible enzyme forms (interconversion cycles). Many enzymes exist in two forms, one catalytically active (a form) and the other (b form) either inactive or much less active under cellular conditions (Fig. 5.15). Often the a form results from the removal of an inhibitory effect on the b form; for example, phosphorylase a is much less sensitive to inhibition by glucose-6-phosphate and ATP (56, 163).

The important point is that the changes from a to b and vice versa involve covalent changes to the protein, often involving the addition of P_i from ATP and its subsequent removal by phosphatases. (Other covalent changes include the addition of an adenylate (AMP) residue to the protein, as in glutamine synthetase from *E. coli* [161].)

Consequently, as with a substrate cycle, there is the potential for a continuous cycling between the a and b forms enabling regulators of both E_1 and E_2 (Fig. 5.15) to change the proportion of enzyme in the active, a form and thus control the enzyme activity. However, in contrast with a substrate cycle, there is no net flux across the cycle (for example, from a to b) in the steady state. Such continuous cycling was first suggested by Sols and Gancedo (156) and by Newsholme and Start (117), who also suggested that such a cycle could provide large changes in enzyme activity (that is, the a form) for small changes in concentrations in regulators

of the interconverting enzymes, E_1 or E_2 (for example, regulator Z in Fig. 5.15). Indeed Newsholme and Crabtree (41, 107, 36) showed that such a cycle, termed an interconversion cycle, could result in a reversible near-total inhibition or activation of the enzyme activity. This resulted from the interconverting enzymes being unresponsive to changes in the concentrations of the a and b forms, although this was not stated explicitly.

However, there was no detailed quantitation of such interconversion cycles until the detailed analyses of Stadtman and Chock (161), followed by those of Koshland and coworkers (59, 92, 93). The latter group showed explicitly that, if the converting enzymes were almost saturated with, and hence unresponsive to, their substrates, the a and b forms, (that is, if the values of sensitivities α and β, Fig. 5.15, were very small) there could be a very large change in the a form (and hence active enzyme) in response to very small changes in regulators such as Z. They have termed this effect zero order ultrasensitivity and have obtained evidence that it operates to control isocitrate dehydrogenase in *E. coli* as well as in the phosphorylase interconversion cycles.

Let us therefore analyze this effect algebraically, using rate equations for E_1 and E_2 (Fig. 5.15) based on equation 5–21. (A similar analysis, based on infinitesimal power equations, has been given previously [44].)

For reaction E_1,

$$F_\rho = \beta \cdot b_\rho + \omega \cdot Z_\rho \qquad (5\text{–}67)$$

For reaction E_2,

$$C_\rho = \alpha \cdot a_\rho \qquad (5\text{–}68)$$

Since, in the steady state, $F = C$ (so that $F_\rho = C_\rho$), the right-hand sides of these equations can be equated to give,

$$\beta \cdot b_\rho + \omega \cdot Z_\rho = \alpha \cdot a_\rho \qquad (5\text{–}69)$$

The conservation relationship,

$$a + b = \text{constant},$$

is written, by analogy with equation 5–26, as,

$$a \cdot a_\rho + b \cdot b_\rho = 0 \qquad (5\text{–}70)$$

Combining equations 5–69 and 5–70 to eliminate the inactive, b form,

$$a_\rho = Z_\rho \cdot \left[\frac{\omega}{\alpha + \beta \cdot \dfrac{a}{b}} \right] \qquad (5\text{–}71)$$

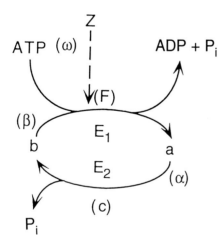

FIG. 5.15. Interconversion cycle. Enzyme E exists in two forms, a and b, which are interconverted by E_1 and E_2. The energy consumption here is represented by a net hydrolysis of ATP to ADP and P_i. In the steady state the rates of E_1 and E_2 are equal, and there is no net flux across the cycle.

so that the intrinsic sensitivity of a to Z, that is a_ρ/Z_ρ, is given by the equation,

$$(s_{i(Z)}^a)_{cycle} = \frac{\omega}{\alpha + \beta \cdot \dfrac{a}{b}} \qquad (5-72)$$

Equation 5–72 shows that, in agreement with the earlier work outlined above, the value of this intrinsic sensitivity can become very large indeed, if both α and β approach zero (that is, E_1 and E_2 become saturated with the b and a forms, respectively). Even the large increase in the a/b ratio, which will occur as the b form is converted into the a form, will have little effect under these conditions.

This mechanism could therefore have a powerful effect on the response of enzymes such as E_1 to regulators such as Z, and there is experimental evidence for its operation during the regulation of isocitrate dehydrogenase in *E. coli* (92, 93) and glycogen breakdown in muscle (101).

From equation 5–72 it should be noted that, for zero order ultrasensitivity to be effective, both component sensitivities, α and β, must be small. If one is large, it will dominate the denominator and there will be no ultrasensitivity. This can be understood by considering the response to Z (Fig. 5.15) when β is large, so that E_1 responds readily to the b form. Then, since any increased conversion of b into a must, by the conservation, be accompanied by a fall in b, this fall will quickly lower the rate of E_1 and hence oppose any further conversion of b into a. Thus, the overall conversion of b into a will be much less than if β were small, whatever the value of the other component sensitivity, α.

The conditions for zero order ultrasensitivity also require that there is little if any deleterious interaction among the component sensitivities, α, β, and ω. As shown by Cardenas and Cornish-Bowden (18), this can occur if Z interacts with E_1 competitively, by changing the affinity of E_1 for the b form. In this case, as Z increases, the value of β also increases; and, if the enzyme follows classical inhibition patterns, the net effect is an intrinsic sensitivity (equation 5–72) of less than ω! In other words the interconversion cycle has now attenuated the effect of Z instead of amplifying it. (This attenuation may be physiologically important in some cases, but no examples are known.) Consequently it has been suggested that noncompetitive or uncompetitive effectors of interconverting enzymes are the most likely candidates for producing an ultrasensitive response. However, because of the overall ultrasensitivity, these component interactions may be weak, (for example, in Fig. 5.15 the value of ω may be

small) and therefore easily overlooked in enzymological observations (18).

Since an interconversion cycle effectively changes the activity of the catalyst, it belongs to the catalystic component, r_c, of the kinetic response (see earlier under Mass-Action and Catalystic Effects). Moreover, since it is a direct communication with the enzyme activity, by the addition rule, it will be added to any other terms of r_c. For example, consider the regulation of pyruvate dehydrogenase by its substrate, pyruvate. As shown in Figure 5.16, pyruvate interacts via an interconversion cycle as well as directly (128). Thus, applying the product rule to the interaction via the cycle, and then combining both routes by the addition rule,

$$s_{pyruvate}^{PDH} = r_m + r_c + \omega \cdot s_{phosphatase}^a \qquad (5-73)$$

Since ω and $s_{phosphatase}^a$ are both negative, their product will be positive and the net activation via the cycle will therefore reinforce that via the direct route. Indeed, if there is zero order ultrasensitivity, the value of $\omega \cdot s_{phosphatase}^a$ may be very large indeed, so that small changes in pyruvate concentration could produce large changes in the enzyme activity. Moreover, since the values of $ws_{phosphatase}^a$ can be varied, by changing the component sensitivities of the cycle, the entire intrinsic sensitivity of the enzyme to pyruvate may be under metabolic control, and varied according to anticipated demand.

Overall Sensitivities of Regulatory Mechanisms In Situ: Internal Oppositions. In previous sections it has been shown how intrinsic and component sensitivities can be built up from elementary communications, using the product and addition combination rules. However, it is now necessary to analyze how this intrinsic sensitivity is

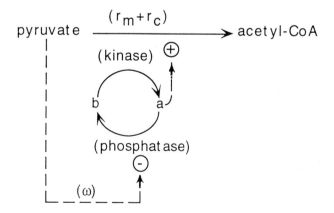

FIG. 5.16. Regulation of pyruvate dehydrogenase (PDH) by pyruvate. Pyruvate has a direct effect $(r_m + r_c)$ and an indirect one, via inhibition of the phosphatase component of the interconversion cycle.

modified when the mechanisms are part of a wider system. As a simple example, consider the system in Figure 5.17a (which was analyzed by Crabtree and Newsholme [42] using infinitesimal power equations and, more recently, by Snell and Fell [55, 155], in relation to serine metabolism: Figure 5.17b). This is a truncated flux, J_t, with S a partially external regulator of E_1. Assuming that S interacts by purely mass-action effects, the intrinsic sensitivity of E_1 to S is equal to the reversibility function, R_v, for E_1 (see earlier under Effect of the Kinetic Reversibility of the Reaction, R_v). However, this describes the response when the concentrations of all the other effectors of E_1 are constant; and, in situ, that of the product P will vary. To obtain the net sensitivity to S it is therefore necessary to combine several equations of the system, as with those of the interconversion cycle (equations 5–67 to 5–72) in the previous section.

By analogy with equation 5–21, the rate equation for E_1 can be written as,

$$J_\rho = R_v \cdot S_\rho - (R_v - 1) \cdot P_\rho \qquad (5-74)$$

assuming that S interacts linearly with E_1 and using the functions in equations 5–55 and 5–56. Similarly, for E_2,

$$J_\rho = \delta \cdot P_\rho \qquad (5-75)$$

since this reaction is irreversible.
Replacing P_ρ in equation 5–74 by using equation 5–75,

$$J_\rho = S_\rho \cdot \left[\frac{\delta \cdot R_v}{\delta + R_v - 1} \right] \qquad (5-75a)$$

so that the overall sensitivity of J to S, J_ρ/S_ρ, is given by the equation,

$$s_S^J = \frac{\delta \cdot R_v}{\delta + R_v - 1} \qquad (5-76)$$

This equation shows that, although the intrinsic sensitivity of E_1 to S is equal to R_v for E_1 (equation 5–55), the net sensitivity in situ is a much more complicated function which includes the effects of P. Moreover, as R_v increases, so that E_1 moves closer to equilibrium,

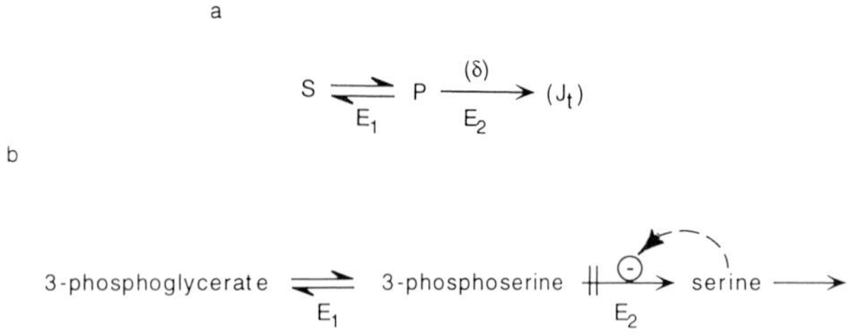

FIG. 5.17. Reversible reactions in situ. (a) Simple hypothetical system, (b) flux of serine biosynthesis in vivo. E_1 = phosphoglycerate dehydrogenase plus phosphoserine aminotransferase; E_2 = phosphoserine phosphatase. The symbol, ╫ denotes saturation.

the value of s_S^J does not become infinitely large (as in equation 5–58), but approaches an upper limit of δ. Therefore, when E_1 is reversible, the net sensitivity of the flux to S depends critically on the response of the subsequent reaction, E_2, to P: if this is large, s_S^J will be large, but if it is small, then s_S^J will be small.

This can be appreciated qualitatively by noting that, when E_1 is reversible, P is effectively an inhibitor of E_1. Thus, as S increases, so that the rate of E_1 and hence P increases, the increased P will oppose the action of S, thus reducing the net response to S. If δ is large, only a small increase in P will be needed to produce large changes in the rate of the subsequent reaction, E_2; so that the opposition due to P will be small. On the other hand, if δ is small, then larger changes of P are needed to increase the rate of E_2 and there will be a greater opposition to the effect of S. As will be seen in later sections, such oppositions are very common in metabolic systems and are often the price that must be paid to provide a flexible regulatory system.

Consequently, to make use of the high intrinsic sensitivity conferred by reversibility, the irreversible reactions of the system (for example, E_2 in Fig. 5.17a) must have intrinsic or component sensitivities that are as large as possible. As shown in earlier sections, this can be provided by non-Michaelis-Menten (sigmoid) kinetics (that is, $n > 1$ in equation· 5–35), substrate cycling, interconvertible forms, and auxiliary effectors, each mechanism acting either singly or in concert.

The role and significance of metabolic oppositions can be illustrated further by calculating the net sensitivities of flux J to regulator Z in the system outlined in Figure 5.11. The overall sensitivities are derived, as previously, by writing the component rate equations in the form of equation 5–21, so that, for E_1,

$$J_\rho = \alpha \cdot S_\rho + \omega \cdot Z_\rho \qquad (5-77)$$

and for E_2,

$$J_\rho = \beta \cdot S_\rho \qquad (5-78)$$

Eliminating S_ρ between these two equations shows that,

$$J_\rho = \frac{\beta \cdot \omega}{\beta - \alpha} \cdot Z_\rho$$

so that J_ρ/Z_ρ, that is, the net sensitivity of J to Z, is given by the equation,

$$s_Z^J = \frac{\beta \cdot \omega}{\beta + \alpha} \qquad (5-79)$$

where, as previously, the negative α (due to inhibition of E_1 by S) is replaced by $-\alpha$, where α is a positive number.

Equation 5–79 shows that, for any value of ω (that is, strength of interaction of Z with E_1), the value of s_Z^J increases with β, to a limiting value of ω. However the value of s_Z^J decreases to zero as α increases. This effect of α results from an opposition by S on the effect of regulator Z. Assuming that Z is an activator, so that ω is positive, an increased Z increases S. However, since S inhibits E_1, this increased S will oppose the activating effect of Z and result in a smaller increase in the net flux, J. As the value of α increases, the strength of this opposition also increases until it is so strong that there is little if any effect of Z on J. Why, therefore, are such oppositions tolerated?

The answer to this question is that, in the system in Figure 5.11, there is only one regulatory site (at E_1) and in this case it would make sense to keep the value of α, and hence the opposition to Z, as low as possible. However, as stressed earlier, most systems have multiple control sites; so that, in general, reaction E_2 of Figure 5.11 would be used in addition to E_1, thereby allowing the system to be regulated by demand for S as well as by supply. For this feedback mechanism to operate effectively, the effect of S on E_1 must be large: indeed, if a hypothetical regulator, Q, interacted with E_2 in Figure 5.11, an analysis similar to that above would show that,

$$s_Q^J = \frac{\lambda \cdot \alpha}{\beta + \alpha} \qquad (5\text{–}80)$$

where λ is the intrinsic sensitivity of E_2 to Q.

This equation shows that, for this feedback (demand) system, a high value of β would be deleterious because the activation of E_2 by S, which was essential for the interaction of Z, now opposes that of Q. Indeed, this is an example of a general result: regulatory sequences acting in opposite directions will oppose each other (44). The cell therefore, as a price for flexibility of control, has to adapt to potentially significant oppositions to its regulatory mechanisms in situ. It may do this be providing high values of external-interaction sensitivities, such as ω and λ, or it may vary the values of α and β according to the sequence being used.

For example, in muscle, glycolysis is controlled principally if not exclusively by the demand for ATP as illustrated schematically in Figure 5.18. For this reason, the feedback sequence, and hence α, is very powerful; so much so that it is difficult to increase the rate of glycolysis simply by increasing the supply of glucose. This is because the ATP concentrations in situ (approx. $5mM$) effectively saturate the myofibrillar ATPase (188) so that β, and hence the opposition due to changes of ATP, is very small. However, in predominantly biosynthetic tissues, such as liver and white adipose tissue, the need to regulate the supply of inter-

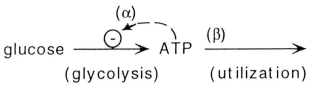

FIG. 5.18. Basic control of glycolysis by ATP. This diagram illustrates the general principles: the presence of auxiliary regulators and other mechanisms makes the actual system much more complicated.

mediates from glycolysis by the availability of glucose means that the forward (supply) regulatory sequence is at least as important as the feedback (demand) one (114). Therefore, the component sensitivity β is much larger than in muscle, and α much smaller, enabling the rate of glycolysis to be regulated by the availability of glucose in addition to the demand for ATP.

Finally, let us consider the response of the internal effector, S, to regulator, Z, in Figure 5.11. Solving the rate equations (equations 5–77 and 5–78) for S_ρ, instead of J_ρ,

$$s_Z^S = \frac{\omega}{\beta + \alpha} \qquad (5\text{–}81)$$

Similarly, for the hypothetical regulator of E_2, Q, (not shown in Fig. 5.11),

$$s_Q^S = \frac{\lambda}{\beta + \alpha} \qquad (5\text{–}82)$$

These equations show that, for any given values of the interaction sensitivities, ω or λ, the response of S is smaller the greater the values of the component sensitivities, β and/or α. Thus large values of these component sensitivities enable the flux, J, to be changed with relatively small changes in the concentrations of its intermediates, thereby minimizing any potentially deleterious effects on the kinetic structure of the regulatory system (see earlier under Need for Sensitive Regulation In Vivo). Consequently, those mechanisms that provide high intrinsic sensitivities (reversibility, sigmoid responses, substrate cycles, interconversion cycles) may also play a very important role in stabilizing the regulatory structures of metabolism, in addition to providing a sensitive response of fluxes to their regulators.

Deriving Net Sensitivities by Using Matrix Algebra. The derivation of net sensitivities by solving simultaneous equations directly, as adopted in the previous section, is quite satisfactory for small systems. However, with larger ones, especially those with extensive branching and containing conserved pools, it soon becomes very difficult and cumbersome. In this case, the set of simultaneous equations is more readily and systematically solved by using the techniques of matrix algebra (37, 46). This type of analysis has also been used by other workers, using approaches other than FOT (21, 24, 54, 57, 72, 87, 131, 154, 183); but the method outlined below is, in our opinion, the most direct and straightforward way of solving metabolic systems that exist in a steady state.

Let us illustrate the method by considering the system in Figure 5.19, which comprises a linear (that is, unbranched) flux consisting of three reactions, with an external effector, Z, interacting with E_2. As before, the rate equations are written in the form of equation 5–21,

For E_1,

$$J_\rho = \alpha \cdot S_\rho \qquad (5\text{–}83)$$

for E_2,

$$J_\rho = \beta \cdot S_\rho + \gamma \cdot P_\rho + \lambda \cdot Z_\rho \qquad (5\text{–}84)$$

for E_3,

$$J_\rho = \delta \cdot P_\rho \qquad (5\text{–}85)$$

Now the term $\lambda \cdot Z_\rho$ in equation 5–84 represents the effect of the specific regulator, Z, on the activity of E_2. Its operation can be written as the successive communications,

$$Z \overset{\lambda}{\frown} E_2, \quad E_2 \overset{s_2^{Var}}{\frown} variable,$$

so that by the product rule (equation 5–10),

$$s_Z^{Var} = \lambda \cdot s_2^{Var} \qquad (5\text{–}86)$$

The sensitivity of the second sequence, s_2^{Var}, is the effect on the variable (for example, flux, J) of an infinitesimal relative change of the activity of E_2, caused by the interaction of some (unspecified) totally external regulator. (It does not refer to the interaction of partially external regulators: see later under Z Is a Mass-Action Effector of E_2: Analysis of the Effects of a Partially External Regulator.) Therefore, a function such as s_2^{Var} represents a core sensitivity for the system,

from which the effects of any specific external regulator can be derived by using the product rule, as in equation 5–86. (In several previous publications we have referred to these core sensitivities as control coefficients. However, as shown later under Metabolic Control Theory, this is a misnomer that we no longer wish to use.)

Consequently, equations 5–83 to 5–85 can be generalized to allow for the interactions of any totally external regulators, by including a general term, E_ρ, to denote the effects of the enzyme activities. (This type of direct analysis was first used by Kacser [76] to analyze a simple branched system, but was not further developed because of a preference for control theorems. See later under Metabolic Control Theory.)

For E_1,

$$J_\rho = \alpha \cdot S_\rho + E_{1\rho} \qquad (5\text{–}87)$$

for E_2,

$$J_\rho = \beta \cdot S_\rho + \gamma \cdot P_\rho + E_{2\rho} \qquad (5\text{–}88)$$

for E_3,

$$J_\rho = \delta \cdot P_\rho + E_{3\rho} \qquad (5\text{–}89)$$

Here, E_ρ terms have been included for each enzyme of the system, and not just for E_2, to derive a completely general set of core sensitivities which apply to any system having the same structure as that in Figure 5.19: for example, the effects of a regulator of E_1, let us say X, could easily be derived, without resolving the system, by forming the product,

$$s_X^{Var} = s_{i(X)}^{E_1} \cdot s_1^{Var} \qquad (5\text{–}90)$$

(However, if only the response to $E_{2\rho}$ (that is, Z) is required, the system can be written and solved with $E_{1\rho}$ and $E_{3\rho}$ both equal to zero.)

To derive the net or core sensitivities of the variables to the enzyme activities, equations 5–87 to 5–89 are written as follows:

$$J_\rho - \alpha \cdot S_\rho - 0 \cdot P_\rho = E_{1\rho} \qquad (5\text{–}91)$$

$$J_\rho - \beta \cdot S_\rho - \gamma \cdot P_\rho = E_{2\rho} \qquad (5\text{–}92)$$

$$J_\rho - 0 \cdot S_\rho - \delta \cdot P_\rho = E_{3\rho} \qquad (5\text{–}93)$$

where all the variables have been transferred to the left-hand side and the externally modifiable enzyme activities are on the right-hand side. This is then expressed as the matrix equation:

$$\underset{\mathbf{N}}{\begin{pmatrix} 1 & -\alpha & 0 \\ 1 & -\beta & -\gamma \\ 1 & 0 & -\delta \end{pmatrix}} \underset{\mathbf{v}}{\begin{pmatrix} J_\rho \\ S_\rho \\ P_\rho \end{pmatrix}} = \underset{\mathbf{p}}{\begin{pmatrix} E_{1\rho} \\ E_{2\rho} \\ E_{3\rho} \end{pmatrix}} \qquad (5\text{–}94)$$

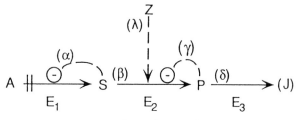

FIG. 5.19. Linear flux with external regulator of E_2.

which can be generalized as,

$$N \cdot v = p \qquad (5-95)$$

Here N is a square (in this example 3×3) matrix containing component sensitivities (and, in branched and conserved systems, fluxes and concentrations: see later); v is a vector (that is a single-column matrix) containing infinitesimal relative changes in the variables; and p is a vector containing infinitesimal relative changes in the enzyme activities (due to the action of unspecified totally external regulators).

Each row of this matrix equation consists of one of the component reactions of the system and it should be noted that equation 5–94 is therefore not the only way of writing this system in matrix form. For example, if E_1 and E_3 were interchanged in p, the rows 1 and 3 of N would be correspondingly interchanged; the rows of v would not, however, be affected. On the other hand, if the variables of v are interchanged, there would be a corresponding interchange of the columns of N, but the rows of p would be unaffected.

The general solution of equation 5–95 can be expressed algebraically as,

$$v = N^{-1} \cdot p \qquad (5-96)$$

where N^{-1} is the inverse matrix of N, defined by the equations,

$$N \cdot N^{-1} = N^{-1} \cdot N \qquad (5-97)$$
$$= I$$

where I is the identity matrix for the above 3×3 system,

$$I_3 = \begin{pmatrix} 1 & 0 & 0 \\ 0 & 1 & 0 \\ 0 & 0 & 1 \end{pmatrix}$$

If N^{-1} can be calculated, equation 5–96 shows that it can be postmultiplied by p (which in this system contains just enzyme activities) to give a matrix that relates changes in the variables to externally induced changes of enzyme activity. Thus, writing N^{-1} symbolically as,

$$N^{-1} = \begin{pmatrix} n_{11} & n_{12} & n_{13} \\ n_{21} & n_{22} & n_{23} \\ n_{31} & n_{32} & n_{33} \end{pmatrix} \qquad (5-98)$$

where the first and second numbers in the subscript denote the row and column position, respectively, equation 5–96 can be written as,

$$\begin{pmatrix} J_\rho \\ S_\rho \\ P_\rho \end{pmatrix} = \begin{pmatrix} n_{11} & n_{12} & n_{13} \\ n_{21} & n_{22} & n_{23} \\ n_{31} & n_{32} & n_{33} \end{pmatrix} \begin{pmatrix} E_{1\rho} \\ E_{2\rho} \\ E_{3\rho} \end{pmatrix}$$

$$= \begin{pmatrix} n_{11} \cdot E_{1\rho} & + & n_{12} \cdot E_{2\rho} & + & n_{13} \cdot E_{3\rho} \\ n_{21} \cdot E_{1\rho} & + & n_{22} \cdot E_{2\rho} & + & n_{23} \cdot E_{3\rho} \\ n_{31} \cdot E_{1\rho} & + & n_{32} \cdot E_{2\rho} & + & n_{33} \cdot E_{3\rho} \end{pmatrix} \quad (5-99)$$

Thus, considering the top row of equation 5–99,

$$J_\rho = n_{11} \cdot E_{1\rho} + n_{12} \cdot E_{2\rho} + n_{13} \cdot E_{3\rho}$$

it can be seen that n_{11} is the overall sensitivity of J to the activity of E_1, that is, s_1^J; n_{12} is the overall sensitivity of J to E_2 and n_{13} is the overall sensitivity of J to E_3. Analogous conclusions for the other variables, S and P, follow from considering the other rows of equation 5–99. Therefore equation 5–98, and hence N^{-1}, can be written in terms of core sensitivities as follows,

$$N^{-1} = \begin{pmatrix} s_1^J & s_2^J & s_3^J \\ s_1^S & s_2^S & s_3^S \\ s_1^P & s_2^P & s_3^P \end{pmatrix} \qquad (5-100)$$

In general, the elements of N^{-1} give the core sensitivities of the system. A general core sensitivity, s_j^{Var}, will be found in the row of N^{-1} corresponding to variable, Var (for example, the core sensitivities of J are in the 1st row of N^{-1} in equation 5–100), and in the column of N^{-1} corresponding to E_j (for example, the core sensitivities for E_2 are in the second column of N^{-1}).

The identification of core sensitivities can be made more easily if N^{-1} is augmented to include the elements of vectors v and p. Thus, equation 5–100 can be augmented as follows:

$$\begin{array}{cc} \begin{matrix} E_1 & E_2 & E_3 \end{matrix} & \\ \begin{pmatrix} s_1^J & s_2^J & s_3^J \\ s_1^S & s_2^S & s_3^S \\ s_1^P & s_2^P & s_3^P \end{pmatrix} \!\! \begin{matrix} J \\ S \\ P \end{matrix} & \begin{matrix} (\mathbf{p}^T) \\ \\ \end{matrix} \\ (N^{-1}) \qquad (\mathbf{v}) & \end{array} \qquad (5-101)$$

(from which the ρ subscripts have been omitted, for simplicity). Here the elements of v in equation 5–95 have been placed in their corresponding rows, and those of p in their corresponding columns. (In matrix algebra terminology, this top row is the transpose of p, denoted as $\mathbf{p}^T$.) Identification of the core sensitivities is now straightforward.

Calculation of the inverse matrix, N^{-1}. If numerical values for the component sensitivities in matrix N are known or can be calculated, the elements of N^{-1}, and hence numerical values for the core sensitivities, can be calculated by using a commercial computer program. However, numerical data for metabolic systems are often too inaccurate to enable such accurate quantitation and usually an algebraic solution of N^{-1} is required. Although some advanced commercial programs (for example, Mathematica: Wolfram Research, Inc.,

USA) may enable this to be done by computer, in most cases a manual method is adopted.

For this, the general matrix equation,

$$N \cdot v = p,$$

is written as,

$$N \cdot v = I \cdot p \qquad (5-102)$$

where I is the identity matrix corresponding to the size of N.

The rows of N are then manipulated (by adding and subtracting multiples of the rows to or from each other) until N is transformed into I. Corresponding row manipulations on the right-hand side identity matrix will then have converted it into N^{-1}; so that equation 5–102 has become,

$$I \cdot v = N^{-1} \cdot p \qquad (5-103)$$

or,

$$v = N^{-1} \cdot p$$

With the system in Figure 5.19, solution of its matrix equation (equation 5–94) by this method gives,

$$N^{-1} = (1/\det N) \begin{pmatrix} \beta \cdot \delta & \alpha \cdot \delta & \alpha \cdot \gamma \\ (\delta + \gamma) & -\delta & -\gamma \\ \beta & \alpha & -(\beta + \alpha) \end{pmatrix} \qquad (5-104)$$

where, as before, $\alpha = -\alpha$, $\gamma = -\gamma$, so that all the components of N^{-1} are positive numbers.

Det N is known as the determinant of matrix N. Methods for evaluating determinants are given in most standard mathematical texts (for example, ref. 162). In this particular example, its value is given by the equation,

$$\det N = \beta \cdot \delta + \alpha \cdot \delta + \alpha \cdot \gamma \qquad (5-105)$$

From equation 5–104, it can be seen that det N is the common denominator of all the core sensitivities of the system. This is a general result, whose significance is discussed in APPENDIX 2.

Thus, augmenting equation 5–104 as in 5–101,

$$N^{-1}_{(aug)} = (1/\det N) \left(\begin{array}{ccc|c} E_1 & E_2 & E_3 & \\ \beta \cdot \delta & \alpha \cdot \delta & \alpha \cdot \gamma & J \\ (\delta + \gamma) & -\delta & -\gamma & S \\ \beta & \alpha & -(\beta + \alpha) & P \end{array} \right)$$

$$(5-106)$$

The core sensitivity of, for example, S to E_2 is seen, from inspection of the matrix, to be equal to $-\delta/(\det N)$. Consequently, the net sensitivity of S to regulator Z (Fig. 5.19) is, from the product rule (equation 5–10),

$$s^S_Z = -\lambda \cdot \delta/(\det N) \qquad (5-107)$$

The net sensitivities of J and P to Z can be calculated in a similar way and are,

$$s^J_Z = \lambda \cdot \alpha \cdot \delta/(\det N) \qquad (5-108)$$

$$s^P_Z = \lambda \cdot \alpha/(\det N) \qquad (5-109)$$

If Z also interacts with another regulatory site (for example at E_1, with intrinsic sensitivity ω, see Fig. 5.26), then the net sensitivity of the interaction with J via this route is,

$$\omega \cdot s^J_1 = \omega \cdot \beta \cdot \delta/(\det N) \qquad (5-110)$$

using the results in equation 5–106. Since the overall communication of Z with J (or any other variable) is now the result of two distinct communication sequences, $Z \sim E_1 \sim J$ and $Z \sim E_2 \sim J$, the overall sensitivity of J to Z is given by the addition rule (equation 5–16); so that,

$$\begin{aligned} s^J_Z &= (s^J_Z)_1 + (s^J_Z)_2 \\ &= (\omega \cdot \beta \cdot \delta)/\det N + (\lambda \cdot \alpha \cdot \delta)/\det N \quad (5-111) \\ &= \delta \cdot (\omega \cdot \beta + \lambda \cdot \alpha)/\det N \end{aligned}$$

In this way the core sensitivities for a system can be used to evaluate the net sensitivity of any variable to any specified external regulator, acting singly or in combination with others. (The situation with a partially external regulator is discussed later under Z Is a Mass-Action Effector of E_2: Analysis of the Effects of a Partially External Regulator.)

The general matrix equation (equation 5–95) may also be solved by using graph theory. Indeed, with this technique it is often possible to obtain the solution directly from the systems, without writing out the matrix. A short account of graph theory, applied to FOT, is given in APPENDIX 3.

The above analysis and resultant core sensitivities will now be used to analyze, algebraically, the resultant effect of certain control mechanisms in situ.

Effect of Reaction Reversibility (R_v) In Situ on the Net Sensitivity to External Regulators. As shown earlier, the kinetic reversibility of a reaction, R_v, increases the intrinsic sensitivity to a substrate, product, or cofactor (mass-action effectors); but, in free solution, it has no effect on the intrinsic sensitivity to an allosteric effector. However, it was also shown earlier that this high sensitivity to a substrate may be effectively opposed in situ, if the sensitivity of the next reaction is low. In this section we shall examine the effect of reaction reversibility on the sensitivity of a flux, J, to an external regulator, Z, of a reversible reaction, E_2. This system

(Fig. 5.20) has the same structure as that analyzed in the previous section (Fig. 5.19) so that the core sensitivities in equation 5–106 can be used directly.

Therefore, from equation 5–108,

$$s_Z^J = \lambda \cdot \alpha \cdot \delta / (\det N)$$

which, using equation 5–105, becomes,

$$s_Z^J = \lambda \cdot \alpha \cdot \delta / (\beta \cdot \delta + \alpha \cdot \delta + \alpha \cdot \gamma) \qquad (5-112)$$

Assuming that S and P both interact linearly with E_2 (so that the r component of the intrinsic sensitivity is 1: see earlier under Effect of Enzyme Kinetic Response), then from equations 5–55 and 5–56,

$$s_{i(S)}^{E_2} = \beta$$
$$= R_v \qquad (5-113)$$

and,

$$s_{i(P)}^{E_2} = \gamma \qquad (5-114)$$
$$= -(R_v - 1)$$

whence,

$$\gamma = (R_v - 1) \qquad (5-115)$$

where R_v is the reversibility of reaction E_2.

Replacing β and γ in equation 5–112 by these functions of R_v,

$$s_Z^J = \frac{\lambda \cdot \alpha \cdot \delta}{R_v \cdot (\delta + \alpha) + \alpha \cdot (\delta - 1)} \qquad (5-116)$$

Let us now examine the effects of R_v on this overall sensitivity when Z is an allosteric or a mass-action effector of E_2.

Z is an allosteric regulator of E_2. If Z is an allosteric regulator of E_2 (that is, a totally external one), the intrinsic sensitivity will not be affected by R_v (see earlier under Effect of the Kinetic Reversibility of the Reaction, R_v) and will thus remain equal to λ (as in equation 5–116). Therefore, as R_v increases, that is, reaction E_2 becomes more reversible or closer to equilibrium, equation 5–116 shows that the sensitivity of J to Z tends to zero. In other words, an allosteric effector will be a very ineffective regulator of J if it

interacts at a reversible reaction, as predicted qualitatively (107) and shown algebraically by using infinitesimal power equations (44).

Moreover, from the functions given in equation 5–106, it can be seen that there are no terms in either β or γ (and hence R_v) in the numerators of either s_Z^S or s_Z^P; so that the net sensitivities of S and P to Z also tend to zero as R_v increases.

These effects of R_v can be understood by noting that an increased activity of E_2 (due to Z) will be opposed *in situ* by the ensuing fall in S and rise in P (Fig. 5.20). Since the strengths of these oppositions depend upon $s_{i(S)}^{E_2}$ and $s_{i(P)}^{E_2}$, both of which increase with R_v (equations 5–58, 5–59), the opposition to the unchanged interaction of Z with E_2 (λ) will become stronger as R_v increases, making Z less and less effective.

However, if E_2 is vectorially organized (that is, spans two distinct compartments—see Fig. 5.13) and if Z can only interact from one side (for example, that which contains S), then, as shown earlier under Effect of the Kinetic Reversibility of the Reaction, R_v, the interaction of Z with E_2 will be a function of R_v (see equation 5–60). Assuming a linear response of E_2 to Z, the intrinsic sensitivity, λ, is equal to R_v; so that equation 5–116 becomes,

$$s_Z^J = \frac{\alpha \cdot \delta \cdot R_v}{R_v \cdot (\delta + \alpha) + \alpha \cdot (\delta - 1)} \qquad (5-117)$$

As R_v increases, the value of s_Z^J now tends to a limit, not of zero, but,

$$(s_Z^J)_\infty = \frac{\alpha \cdot \delta}{\delta + \alpha} \qquad (5-118)$$

If $R_v = 1$, so that E_2 is completely irreversible, from equation 5–117,

$$(s_Z^J)_{irrev} = \frac{\alpha}{\alpha + 1} \qquad (5-119)$$

Denoting the ratio $(s_Z^J)_\infty / (s_Z^J)_{irrev}$ by the symbol ρ, equations 5–118 and 5–119 show that,

$$\rho = \frac{\delta \cdot (\alpha + 1)}{\delta + \alpha} \qquad (5-120)$$

Thus, if $\delta = 1$, $\rho = 1$, and there is no effect of R_v on the net response of J to Z. If δ is greater than 1, ρ is also greater than 1 so that an increased reversibility of E_2 now increases the response of J to Z. If δ is less than 1, ρ is also less than 1 and an increased reversibility of E_2 then progressively decreases (attenuates) the response of J to Z.

Therefore, in this situation, the effect of the reversibility of the regulatory reaction on the overall response to Z in situ cannot be predicted, even qualitatively,

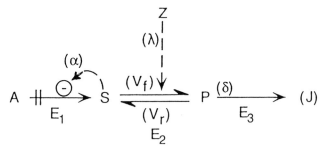

FIG. 5.20. Linear flux with reversible reaction at E_2.

from its effects on the intrinsic sensitivity (42, 44). As with the net sensitivity of flux J to substrate S in Figure 5.17a, it depends critically on the sensitivity of the subsequent step(s).

Z is a mass-action effector of E_2: analysis of the effects of a partially external regulator. If Z is a mass-action effector of E_2 (that is, a substrate or cofactor), the structure of the system is no longer the same as that in Figure 5.20. This is because the concentration of Z is now partially determined by E_2, and hence by flux J. Therefore, Z is now a partially external regulator of J and the correct structure is shown in Figure 5.21.

Qualitatively, Z can still be regarded as if it were totally external, but quantitatively there is an important difference: the core sensitivities now have to include terms relating to the interaction of Z, which are not included in equation 5–106. For example, the interaction of the totally external regulator of E_2, W, (Fig. 5.21) will result in a change of Z. Therefore, the appropriate value for s_2^j will have to include this change of Z and consequently will not be equal to the corresponding function in equation 5–106; using the latter to describe the response to W will therefore give erroneous results.

Nevertheless, the interaction of Z itself can be evaluated by using the core sensitivities in equation 5–106, since this interaction does not produce changes in the concentrations of other partially external regulators in this particular system. It can, under these rather special conditions, therefore be treated as if it were totally external, that is, as if the system were as outlined in Figure 5.20. However, if this is done, it is important to stress that any core sensitivities, such as s_2^j, are now specific for the interaction of Z, and do not apply to any other regulator (for example, W). They are therefore termed pseudo core sensitivities (formerly pseudo control coefficients) and given symbols, for example, $s_{2(Z)}^j$, to denote that they refer to a specific (partially external) regulator, Z (37, 46).

Therefore, assuming that Z interacts linearly with E_2 (so that the r component is 1), $s_{i(Z)}^{E_2}$ is equal to R_v. Thus the equation for the response of J to Z is the same as equation 5–117 and the same conclusions

apply. Consequently, when Z is a mass-action effector of E_2, an increased reversibility of E_2 may increase the net sensitivity of J to Z, decrease it, or have no effect at all, depending on the value of δ. Thus, here also, the effects of reversibility on the net response in situ cannot be predicted from the effects on the intrinsic sensitivity, and each case must be treated on its own merits (42, 44, 55, 155).

Finally, when a system contains partially external regulators, such as Z in Figure 5.21, (and most metabolic systems do) it becomes very difficult to solve the system algebraically; that is, to obtain all the core sensitivities required to evaluate the responses to totally external regulators such as W. This is because Z is now a variable of the system and belongs to the vector, $\mathbf{v}$, of equation 5–95. As a result, there are now more variables than equations in the system as written (that is, Fig. 5.21) and an algebraic solution is impossible (37, 46).

Sometimes this problem can be overcome by making some simplifying assumptions. For example, if the flux being analyzed (that is, J in Fig. 5.21) is only a small fraction of the total flux through Z, it is possible to assume that Z_ρ is zero during the response to regulators such as W, allowing the core sensitivities in equation 5–106 to be used.

However, this approximation will only occasionally be valid and, in general, it is necessary to remove the closed component (represented by the interaction of Z), by widening the system so that Z becomes totally internal. This can be done by adding arbitrary fluxes entering and leaving Z with appropriate component sensitivities from Z. Thus, in Figure 5.22, the system in Figure 5.21 has been opened by adding fluxes I and O with component sensitivities to Z of ψ and ζ, respectively. This adds two new variables (I and O) and three new equations (the rate equations for I and O and the branch conservation, $I = J + O$). There is now the same number of variables and equations and the system can therefore be solved algebraically to give the core sensitivities. Although these contain terms in I, O, ψ and ζ, they can still provide valuable algebraic expressions for examining the effects of control mechanisms on the response of the system. For numerical calculations however, the values of these additional terms will need to be determined, and this is not always easy, or sometimes even possible!

Effect of a Substrate Cycle In Situ on the Net Sensitivity to External Regulators.
Figure 5.14a shows a system containing a substrate cycle in place of reaction E_2 of Figures 5.19 and 5.20. By analogy with equations 5–87 to 5–89, the component equations for the system are as follows:

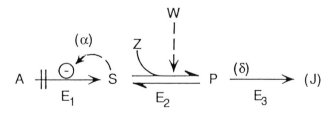

FIG. 5.21. Linear flux with partially external regulator at E_2.

For E_1,

$$J_\rho = \alpha \cdot S_\rho + E_{1\rho} \qquad (5\text{--}121)$$

for E_2,

$$F_\rho = \beta \cdot S_\rho + E_{2\rho} \qquad (5\text{--}122)$$

for E_3,

$$J_\rho = \delta \cdot P_\rho + E_{3\rho} \qquad (5\text{--}123)$$

for E_4,

$$C_\rho = \gamma \cdot P_\rho + E_{4\rho} \qquad (5\text{--}124)$$

There is now an additional equation resulting from the flux conservation,

$$J = F - C$$

which becomes (equation 5–28),

$$J \cdot J_\rho - F \cdot F_\rho + C \cdot C_\rho = 0 \qquad (5\text{--}125)$$

Equations 5–121 to 5–125 are then written, as previously, in matrix form to give equation 5–126,

$$\begin{pmatrix} 1 & 0 & 0 & -\alpha & 0 \\ 0 & 1 & 0 & -\beta & 0 \\ 0 & 0 & 1 & 0 & -\gamma \\ 1 & 0 & 0 & 0 & -\delta \\ J & -F & C & 0 & 0 \end{pmatrix} \begin{pmatrix} J_\rho \\ F_\rho \\ C_\rho \\ S_\rho \\ P_\rho \end{pmatrix} = \begin{pmatrix} E_{1\rho} \\ E_{2\rho} \\ E_{4\rho} \\ E_{3\rho} \\ 0 \end{pmatrix} \qquad (5\text{--}126)$$

$$(N) (v) (p)$$

Here, unlike the system described by equation 5–94, the E_ρ terms in vector **p** are not in numerical order. Also, **p** contains a zero element (from the conservation relationship in the bottom row). This means that the inverse matrix, N^{-1}, contains more elements than the total number of core sensitivities. Thus, writing N^{-1} in terms of core sensitivities (as in equation 5–101), we obtain the augmented equation 5–127,

$$N^{-1} = \begin{pmatrix} E_1 & E_2 & E_4 & E_3 & 0 \\ s_1^J & s_2^J & s_4^J & s_3^J & a \\ s_1^F & s_2^F & s_4^F & s_3^F & b \\ s_1^C & s_2^C & s_4^C & s_3^C & c \\ s_1^S & s_2^S & s_4^S & s_3^S & d \\ s_1^P & s_2^P & s_4^P & s_3^P & e \end{pmatrix} \begin{pmatrix} (p^T) \\ J \\ F \\ C \\ S \\ P \end{pmatrix} \qquad (5\text{--}127)$$

$$(N^{-1}) (v)$$

in which the order of the E elements in p^T is the same as in **p** and is also reflected in the subscripts of the core sensitivities in N^{-1}.

It can also be seen that N^{-1} contains an extra column which is annihilated by the zero element in **p** when the product, $N^{-1} \cdot p$, is formed. Thus the elements in this column, a–e, do not appear in the product, $N^{-1} \cdot p$, and hence do not represent core sensitivities. For the present discussions they can therefore be ignored.

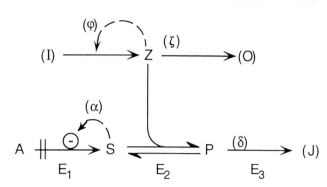

FIG. 5.22. Opening the system in Figure 5.21.

In general, if there are m E_ρ terms in vector **p**, out of a total of n elements, the last $(n-m)$ columns of N^{-1} will be annihilated when forming the product, $N^{-1} \cdot p$. Therefore, when calculating core sensitivities, it is only necessary to compute the first m columns of N^{-1}; and this can be done automatically by writing equation 5–95 as,

$$N \cdot v = I_{(n,m)} \cdot p \qquad (5\text{--}128)$$

where $I_{(n,m)}$ is a matrix consisting of only the first m columns of the identity matrix I_n: for the system in Figure 5.14a, this is expressed as shown in equation 5–129.

$$I_{(n,m)} = I_{(5,4)}$$

$$= \begin{pmatrix} 1 & 0 & 0 & 0 \\ 0 & 1 & 0 & 0 \\ 0 & 0 & 1 & 0 \\ 0 & 0 & 0 & 1 \\ 0 & 0 & 0 & 0 \end{pmatrix} \qquad (5\text{--}129)$$

Carrying out the row manipulations on equation 5–128 to convert N into I then converts $I_{(n,m)}$ into the matrix of core sensitivities, without having to evaluate the unnecessary columns of N^{-1}.

However, further consideration of the solution of systems containing conservations will be deferred until the next section since there is an important short cut. For now, let us just consider the effects of regulator, Z, on the net flux, J, of Figure 5.14a. To do this, we do not need to solve equation 5–126, because the structure of Figure 5.14a is the same as in Figure 5.19, so that the core sensitivities given in equation 5–106 can be applied directly.

Using the equations describing the effects of a substrate cycle on the intrinsic sensitivity of the net flux across the cycle (equations 5–68, 5–69), the sensitivity functions in equation 5–106 are transformed as follows:

$$\beta \text{ (Fig. 5.19)} \rightarrow \beta \cdot (1 + C/J) \qquad (5-130)$$
$$\gamma \text{ (Fig. 5.19)} \rightarrow -\gamma \cdot C/J \qquad (5-131)$$

so that,

$$\gamma \text{ (Fig. 5.19)} \rightarrow \gamma \cdot C/J \qquad (5-131a)$$

(NB In Fig. 5.14a γ is positive, whereas in Fig. 5.19 it is negative.)

$$\lambda \text{ (Fig. 5.19)} \rightarrow \lambda \cdot (1 + C/J) \qquad (5-132)$$

(Equation 5-132 applies because the cycle increases the intrinsic sensitivity to all effectors of E_2 and E_4, not just their substrates, products, and cofactors.)

Equation 5-112 therefore becomes,

$$s_Z^J = \frac{\alpha \cdot \delta \cdot \lambda \cdot \left(1 + \dfrac{C}{J}\right)}{\alpha \cdot \delta + \beta \cdot \delta \cdot \left(1 + \dfrac{C}{J}\right) + \alpha \cdot \gamma \cdot \dfrac{C}{J}} \qquad (5-133)$$

This equation shows that, as C increases, the value of s_Z^J approaches a limiting value (35, 54), given by the equation,

$$(s_Z^J)_\infty = \frac{\alpha \cdot \delta \cdot \lambda}{\beta \cdot \delta + \alpha \cdot \gamma} \qquad (5-134)$$

This contrasts with the effects of C on the intrinsic sensitivity of the net flux to Z, which increases without limit (see earlier under Effect of Substrate Cycling).

In the absence of cycling (that is, when $C = 0$), equation 5-133 shows that,

$$(s_Z^J)_0 = \frac{\alpha \cdot \lambda}{\alpha + \beta} \qquad (5-135)$$

so that, defining ρ as the ratio $(s_Z^J)_\infty/(s_Z^J)_0$ (see earlier under Z Is an Allosteric Regulator of E_2) and combining equations 5-134 and 5-135,

$$\rho = \frac{\delta \cdot (\alpha + \beta)}{\beta \cdot \delta + \alpha \cdot \gamma} \qquad (5-136)$$

Therefore, as with a reversible reaction (see earlier under Effect of Reaction Reversibility (R_v) In Situ on the Net Sensitivity to External Regulators), if $\delta = 1$, $\rho = 1$, and the cycle has no effect on the net sensitivity of J to Z. If δ is less than 1, the cycle reduces the sensitivity to Z. Only if δ is greater than 1 does the cycle actually increase the net sensitivity of J to Z, in line with its effects on the intrinsic sensitivity although this reaches a limiting value, given by equation 5-134. As before, these effects are due to internal oppositions (via changes in S and P) to the effect of Z, and these oppositions are represented by the terms in β and γ in the denominator of equation 5-133.

Consequently, as with a reversible reaction, the effect of substrate cycling on the sensitivity of a regulator in

situ cannot generally be predicted from the intrinsic sensitivity alone. However, unlike a reversible reaction, the effects of oppositions due to S and P can be eliminated by saturating E_2 with S and E_4 with P (Fig. 5.14a), so that β and γ become zero. Equation 5-133 then becomes,

$$s_Z^J = \lambda \cdot (1 + C/J) \qquad (5-137)$$

from which it can be seen that the net sensitivity is now equal to the intrinsic sensitivity, the oppositions having been eliminated. (Such saturation is not possible for the substrates, products, and cofactors of a kinetically reversible reaction, because any attempt to saturate a mass-action effect results in the reaction becoming kinetically irreversible [26].)

However, improving the sensitivity by saturating out the oppositions may not always be possible in situ (35). For example, in the substrate cycle between F-6-P and FBP (Fig. 5.14b), FBPase must be able to respond to FBP in liver (that is, γ cannot be zero) because this communication is needed to transmit the flux of glucose formation from lactate, alanine, and other gluconeogenic precursors. Similarly, PFK cannot be saturated with F-6-P in this tissue because this communication is needed to transmit the fluxes of glucose and glycogen to lipids and other biosynthetic products. Consequently, in tissues such as liver, the oppositions caused by F-6-P and FBP to the interaction of regulators of the cycle (for example, AMP, Fig. 5.14b) cannot be removed, and can be seen as another example of the price to be paid for a flexible and varied control of the pathway.

In muscle, the situation with the F-6-P/FBP cycle is somewhat different, since there is no significant gluconeogenesis, and glycolysis is effectively controlled by the demand for ATP (see earlier under Overall Sensitivities of Regulatory Mechanisms In Situ: Internal Oppositions). Therefore, FBPase may be saturated with FBP in vivo so that $\gamma = 0$ and the opposition due to FBP removed (35). However, the other opposition to regulators of the cycle, caused by the effect of F-6-P on PFK, cannot be removed by saturation, because all the allosteric regulatory effects of intermediates such as ATP, AMP, F-2,6-BP, etc. operate ultimately by changing the affinity of PFK for F-6-P, that is, the $S_{0.5}$ value (12, 61, 169). Therefore, saturating PFK with F-6-P would reduce or eliminate these effects and thereby destroy the control systems for glycolysis. Consequently, even in muscle, there must always be some opposition to the effects of AMP and other regulators of the cycle; but the precise extent of this will have to await a detailed measurement of the rates of cycling and enzyme kinetic responses in situ.

Regulatory Analysis of a Branched System: Matrix Partitioning. Figure 5.23 shows a simple branched system containing three fluxes, J, K, and L, and one internal metabolite, S; there are also two external regulators, X and Y. By analogy with equations 5–87 to 5–89, the rate equations for this system can be written as, For E_1,

$$J_\rho = \alpha \cdot S_\rho + E_{1\rho} \qquad (5\text{–}138)$$

for E_2,

$$K_\rho = \beta \cdot S_\rho + E_{2\rho} \qquad (5\text{–}139)$$

for E_3,

$$L_\rho = \gamma \cdot S_\rho + E_{3\rho} \qquad (5\text{–}140)$$

and there is one branch-point conservation, J-K-L = 0, which may be written as (equation 5–28),

$$J \cdot J_\rho - K \cdot K_\rho - L \cdot L_\rho = 0$$

As in previous examples, these four equations can be combined to give matrix equation 5–141.

$$\underbrace{\begin{pmatrix} 1 & 0 & 0 & | & -\alpha \\ 0 & 1 & 0 & | & -\beta \\ 0 & 0 & 1 & | & -\gamma \\ \hline J & -K & -L & | & 0 \end{pmatrix}}_{(\mathbf{N})} \underbrace{\begin{pmatrix} J_\rho \\ K_\rho \\ L_\rho \\ \hline S_\rho \end{pmatrix}}_{(\mathbf{v})} = \underbrace{\begin{pmatrix} E_{1\rho} \\ E_{2\rho} \\ E_{3\rho} \\ \hline 0 \end{pmatrix}}_{(\mathbf{p})} \qquad (5\text{–}141)$$

Here, as in the previous example, vector $\mathbf{p}$ has a zero element, so that only the first three columns of the inverse matrix, $\mathbf{N}^{-1}$, are required for evaluating the core sensitivities of this system. However, it is possible to solve such branched systems more easily by a preliminary partitioning of the matrix equation, indicated by the broken lines in equation 5–141 (39). This splits the matrices into the system of submatrices shown in equation 5–142,

$$\begin{pmatrix} \mathbf{I} & | & \mathbf{A} \\ \hline \mathbf{B} & | & \mathbf{C} \end{pmatrix} \begin{pmatrix} \mathbf{j} \\ \hline \mathbf{m} \end{pmatrix} = \begin{pmatrix} \mathbf{x} \\ \hline \mathbf{y} \end{pmatrix} \qquad (5\text{–}142)$$

where $\mathbf{I}$ is the identity matrix ($\mathbf{I}_3$ in this particular example).

Multiplying out the submatrices and simplifying results in the equations,

$$[\mathbf{C} - \mathbf{B} \cdot \mathbf{A}] \cdot \mathbf{m} = [\mathbf{y} - \mathbf{B} \cdot \mathbf{x}] \qquad (5\text{–}143)$$

$$\mathbf{j} = [\mathbf{x} - \mathbf{A} \cdot \mathbf{m}] \qquad (5\text{–}144)$$

where the multiplications represent matrix multiplication (162).

Equations 5–143 and 5–144 have effectively separated the fluxes (in $\mathbf{j}$) from the internal effectors (in $\mathbf{m}$). Moreover, the only matrix equation requiring inversion (and hence row manipulations) is equation 5–143, and this is of size $m \times m$, where m is the number of internal effectors. This is often much smaller than the original matrix, N, and is therefore much easier to invert: for

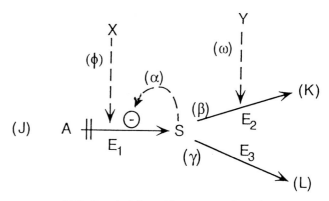

FIG. 5.23. Branched flux with two external regulators.

example, in Figure 5.23, there is just one internal effector, the value of m is therefore 1, and the matrix $[\mathbf{C} - \mathbf{B} \cdot \mathbf{A}]$ is of size 1×1 (that is, is a single element), compared with the original 4×4 matrix in equation 5–141. Partitioning can therefore save a great deal of algebra. Equation 5–143 is analogous to the S-matrix systems, characteristic of BST (see later under Biochemical Systems Theory), except that infinitesimal rather than finite quantities are used.

After solving equation 5–143 to obtain the net or core sensitivities of the internal metabolites, those of the fluxes are then calculated by using equation 5–144. This requires no matrix inversions, and hence no further row manipulations.

The detailed calculations proceed as follows. From equation 5–143,

$$\mathbf{m} = [\mathbf{C} - \mathbf{B} \cdot \mathbf{A}]^{-1} \cdot [\mathbf{y} - \mathbf{B} \cdot \mathbf{x}] \qquad (5\text{–}145)$$

which may be solved by using row manipulations, after writing it in the form (equation 5–102),

$$[\mathbf{C} - \mathbf{B} \cdot \mathbf{A}] \cdot \mathbf{m} = \mathbf{I}_m \cdot [\mathbf{y} - \mathbf{B} \cdot \mathbf{x}] \qquad (5\text{–}146)$$

where $\mathbf{I}_m$ is the identity matrix of size m. Manipulating $[\mathbf{C} - \mathbf{B} \cdot \mathbf{A}]$ on the left-hand side to $\mathbf{I}_m$ then converts $\mathbf{I}_m$ on the right-hand side to $[\mathbf{C} - \mathbf{B} \cdot \mathbf{A}]^{-1}$.

The product, $[\mathbf{C} - \mathbf{B} \cdot \mathbf{A}]^{-1} \cdot [\mathbf{y} - \mathbf{B} \cdot \mathbf{x}]$, is then expressed as $\mathbf{S}_m \cdot \mathbf{e}$, where $\mathbf{e}$ is a vector containing all the enzyme activity terms (unlike $\mathbf{p}$, it does not contain any zero elements), so that,

$$\mathbf{m} = \mathbf{S}_m \cdot \mathbf{e} \qquad (5\text{–}147)$$

This is then used in equation 5–144 to give the equation,

$$\mathbf{j} = [\mathbf{x} - \mathbf{A} \cdot \mathbf{S}_m \cdot \mathbf{e}] \qquad (5\text{–}148)$$

which, after carrying out the matrix operations on the right-hand side, can be written as,

$$\mathbf{j} = \mathbf{S}_j \cdot \mathbf{e} \qquad (5\text{–}149)$$

The rows of matrices S_m and S_j contain the core sensitivities for the internal effectors and the fluxes, respectively; so that the entire solution of the equation, $N \cdot v = p$, can be written as,

$$v = \begin{pmatrix} S_j \\ S_m \end{pmatrix} e \qquad (5\text{-}150)$$

which can then be augmented to identify the core sensitivities for each variable.

Thus, from equation 5–141, the submatrices are as follows:

$$A = -\begin{pmatrix} \alpha \\ \beta \\ \gamma \end{pmatrix}; \quad B = (J \ -K \ -L); \quad C = (0)$$

$$j = \begin{pmatrix} J_\rho \\ K_\rho \\ L_\rho \end{pmatrix}; \quad m = (S_\rho)$$

$$x = \begin{pmatrix} E_{1\rho} \\ E_{2\rho} \\ E_{3\rho} \end{pmatrix}; \quad y = (0)$$

Therefore, equation 5–143 becomes,

$$\left[(0) + (J - K - L) \cdot \begin{pmatrix} \alpha \\ \beta \\ \gamma \end{pmatrix} \right] [(S_\rho)]$$

$$= \left[(0) - (J - K - L) \cdot \begin{pmatrix} E_{1\rho} \\ E_{2\rho} \\ E_{3\rho} \end{pmatrix} \right]$$

whence,

$$(\alpha J - \beta K - \gamma L) \cdot (S_\rho) = -(J \cdot E_{1\rho} - K \cdot E_{2\rho} - L \cdot E_{3\rho})$$

This is a simple equation in S_ρ (corresponding to $m = 1$ in this example), so that the inverse of $[C - B \cdot A]$ is just the reciprocal of $(\alpha J - \beta K - \gamma L)$. Therefore, by analogy with equation 5–145,

$$m = S_\rho$$
$$= \frac{-(J \cdot E_{1\rho} - K \cdot E_{2\rho} - L \cdot E_{3\rho})}{(\alpha \cdot J - \beta \cdot K - \gamma \cdot L)} \qquad (5\text{-}151)$$

By analogy with equation 5–147, this can be written as,

$$m = (s_1^S \ s_2^S \ s_3^S) \begin{pmatrix} E_{1\rho} \\ E_{2\rho} \\ E_{3\rho} \end{pmatrix} \qquad (5\text{-}152)$$

where, for example,

$$s_1^S = \frac{-J}{\alpha \cdot J - \beta \cdot K - \gamma \cdot L}$$

From equation 5–148,

$$j = \begin{pmatrix} J_\rho \\ K_\rho \\ L_\rho \end{pmatrix}$$
$$= \begin{pmatrix} E_{1\rho} \\ E_{2\rho} \\ E_{2\rho} \end{pmatrix} + \begin{pmatrix} \alpha \\ \beta \\ \gamma \end{pmatrix} (s_1^S \ s_2^S \ s_3^S) \begin{pmatrix} E_{1\rho} \\ E_{2\rho} \\ E_{3\rho} \end{pmatrix} \qquad (5\text{-}153)$$

After combining the matrices (equation 5–149), equation 5–153 takes the form shown in equation 5–153a.

$$\begin{pmatrix} J_\rho \\ K_\rho \\ L_\rho \end{pmatrix} = \begin{pmatrix} 1 + \alpha \cdot s_1^S & \alpha \cdot s_2^S & \alpha \cdot s_3^S \\ \beta \cdot s_1^S & 1 + \beta \cdot s_2^S & \beta \cdot s_3^S \\ \gamma \cdot s_1^S & \gamma \cdot s_2^S & 1 + \gamma \cdot s_3^S \end{pmatrix} \begin{pmatrix} E_{1\rho} \\ E_{2\rho} \\ E_{3\rho} \end{pmatrix} \qquad (5\text{-}153a)$$

Therefore, from equation 5–150, the full solution of equation 5–141 is as shown in equation 5–154. The values of each s_j^S are obtained from equation 5–151.

$$v = \begin{pmatrix} J_\rho \\ K_\rho \\ L_\rho \\ S_\rho \end{pmatrix}$$
$$= \begin{pmatrix} 1 + \alpha \cdot s_1^S & \alpha \cdot s_2^S & \alpha \cdot s_3^S \\ \beta \cdot s_1^S & 1 + \beta \cdot s_2^S & \beta \cdot s_3^S \\ \gamma \cdot s_1^S & \gamma \cdot s_2^S & 1 + \gamma \cdot s_3^S \\ s_1^S & s_2^S & s_3^S \end{pmatrix} \begin{pmatrix} E_{1\rho} \\ E_{2\rho} \\ E_{3\rho} \end{pmatrix} \qquad (5\text{-}154)$$

This procedure has automatically produced the solution matrix containing just the core sensitivities, that is, just the first 3 columns of N^{-1}. Moreover, since the denominator of equation 5–151, $(\alpha J - \beta K - \gamma L)$, is the common denominator of all the core sensitivities, it is not only the determinant of the reduced matrix, $[C - B \cdot A]$, but also that of the full matrix, N (47). (If α is replaced by $-\alpha$ this determinant becomes $-(\alpha J + \beta K + \gamma L)$ and is therefore negative. However, this is just the result of the particular ordering of the component equations in equation 5–141 and has no physiological significance; a different order of equations would produce positive values for det N.)

Using the functions in equation 5–151, the augmented solution of equation 5–141, and hence the system in Figure 5.23, is,

$$N_{(aug)}^{-1} =$$

$$(1/\text{det } N) \begin{pmatrix} & E_1 & E_2 & E_3 & \\ \hline -(\beta \cdot K + \gamma \cdot L) & -\alpha \cdot K & -\alpha \cdot L & J \\ -\beta \cdot J & -(\gamma \cdot L + \alpha \cdot J) & \beta \cdot L & K \\ \gamma \cdot J & \gamma \cdot K & -(\beta \cdot K + \alpha \cdot J) & L \\ -J & -K & -L & S \end{pmatrix}$$
$$\qquad (5\text{-}155)$$

where, as before, $\alpha = -\alpha$. (There is no need to include the zero in $\mathbf{p}$ on the top row, since only the first three rows of N^{-1} have been generated.)

Branch-point sensitivities. Let us use the results in equation 5–155 to examine the response of flux K to the external regulator of E_2, Y (Fig. 5.23). From the product rule (equation 5–10),

$$s_Y^K = \omega \cdot s_2^K$$

and from equation 5–155,

$$s_2^K = \frac{\gamma \cdot L + \alpha \cdot J}{\alpha \cdot J + \beta \cdot K + \gamma \cdot L} \qquad (5\text{–}156)$$

Therefore,

$$s_Y^K = \frac{\omega \cdot (\gamma \cdot L + \alpha \cdot J)}{\alpha \cdot J + \beta \cdot K + \gamma \cdot L} \qquad (5\text{–}157)$$

This equation can be simplified, by replacing L with $(J-K)$ and dividing through by J, to give,

$$s_Y^K = \frac{\omega \cdot \left(\alpha + \gamma - \gamma \cdot \dfrac{K}{J}\right)}{(\alpha + \gamma) + (\beta - \gamma) \cdot \dfrac{K}{J}} \qquad (5\text{–}158)$$

where K/J must always be less than 1.

Equation 5–158 shows that, as K/J becomes very small, so that flux K is only a small fraction of the total flux through S (that is, flux J), the value of s_Y^K tends to a limiting value of ω, the intrinsic sensitivity of the interaction of Y with E_2. Thus, under these conditions, there is effectively no opposition to the effect of Y. This can be understood qualitatively by noting that, if K is only a small fraction of J, changes in K will hardly disturb S, which will therefore remain effectively constant. Thus an increase in E_2, due to the interaction of Y, will not be accompanied by any significant fall in S and there will consequently be no oppositions to Y in this system.

This property of a branch point in reducing or eliminating internal oppositions, thereby increasing the net sensitivity of the response to external regulators was reported independently by Koshland and co-workers (94) and by Newsholme and co-workers (44, 112). It may be of considerable physiological importance in rapidly-proliferating tissues, such as tumors, lymphocytes, and intestinal cells, which require a non-limited supply of biosynthetic precursors, especially those for nucleic acid synthesis.

For example, glutamine may be of especial importance for supplying nitrogen for biosynthesis in such tissues (Fig. 5.24), and it is known to be metabolized at a rate far in excess of that needed for supporting

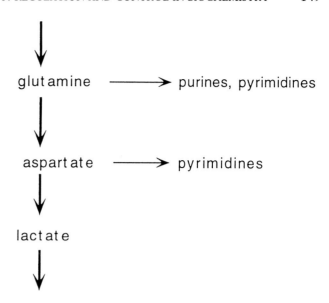

FIG. 5.24. The pathway of glutaminolysis.

biosynthesis (68, 98, 173, 186, 187). Newsholme and co-workers (105, 112, 113, 115, 119, 125) have therefore proposed that this high rate of glutaminolysis serves, by the branch-point effect outlined above, to enable a sensitive regulation of the flux of nitrogen from glutamine by regulators of purine and/or pyrimidine biosynthesis. Similarly, high rates of glycolysis in many tumors (10, 90) and the high rate of lactate production in brain (30) may also enable a sensitive regulation of biosynthetic fluxes for which glycolytic intermediates are the precursors (for example, glucose-6-phosphate is a precursor of the ribose and hence deoxyribose moieties of nucleic acids [105]).

However, the role of the branch-point effect may not always be to provide a high sensitivity of one of the fluxes to regulation. For example, Snell and Fell (55, 155) have shown that the pathway of serine biosynthesis from 3-phosphoglycerate in rabbit liver represents only a small fraction of the total glycolytic flux, and should therefore be highly sensitive to its external regulators. Nevertheless, the reaction forming serine (phosphoserine phosphatase) is effectively saturated by phosphoserine under most physiological conditions (Fig. 5.25). This saturation alone would be sufficient to eliminate any oppositions due to changes in 3-phosphoglycerate and phosphoserine. The branch point effect is therefore superfluous as far as a sensitive regulation of this flux is concerned, and other roles must be sought.

Perhaps the major role of the branch-point effect in this example is to stabilize the concentration of the branch intermediate (which would be the initial precursor for a biosynthetic pathway). Using an analysis

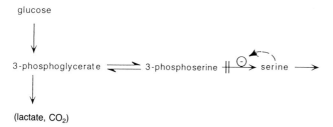

FIG. 5.25. Relationship of serine biosynthesis to glycolysis.

similar to that earlier (equations 5–156–5–158), the sensitivity of the branch-point intermediate, S, (Fig. 5.23) to Y is given by the equation,

$$s_Y^S = \frac{-\omega \cdot \dfrac{K}{J}}{(\alpha + \gamma) + (\beta - \gamma) \cdot \dfrac{K}{J}} \qquad (5{-}159)$$

which shows that, as K/J becomes very small, the value of the net sensitivity of S to Y tends to zero. Thus, as predicted qualitatively, S becomes stable to regulators of flux K when K is a small fraction of the total flux, J. With reference to serine biosynthesis, the resulting stabilization of the glycolytic intermediate, 3-phosphoglycerate, could be important because this intermediate is in communication with many metabolic equilibria, and large fluctuations in its concentration could therefore damage the kinetic structure of the control systems (see earlier under Need for Sensitive Regulation In Vivo). Such stabilization would be produced by saturating phosphoserine phosphatase with phosphoserine and may therefore be the main role of the branch point effect in this system (55, 155).

Whether the branch-point effect operates to stabilize intermediates, or to provide a high sensitivity of a flux, or both, will depend on the system under discussion. However, in either case, there is an apparently large wastage of flux through the entire system (that is, flux L in Fig. 5.23). As with substrate cycling, this may represent a price worth paying for providing a sensitive and stable response.

Calculation of Responses to Large (Finite) Changes of Regulators.

In previous sections, attention has been concentrated on the use of sensitivity functions for analyzing the algebraic properties of the regulatory systems, thereby obtaining a great deal of useful information about the physiological importance of regulatory mechanisms, such as kinetic reversibility and substrate cycling. As stated earlier, for such analyses, it does not matter that the sensitivities refer to infinitesimal changes since the algebraic properties apply generally

to the entire response of any finite (physiological) change.

However, it is possible to use infinitesimally based sensitivities for numerical calculations of finite changes by an indirect method known as finite decomposition (44, 47). In this method a given large fold change in the regulator concentration (or, more generally, the stimulus) is divided into n successive, equal, fold changes, each sufficiently small to apply the infinitesimal sensitivity equations. A satisfactory value of n for many metabolic responses is 100. (If X changes from X_1 to X_2, the fold change is X_2/X_1. The corresponding relative change is $(X_2 - X_1)/X_1$, so that a fold change = 1 + relative change.)

To illustrate the principles of this method, let regulator X (Fig. 5.23) change from X_1 to X_2. The finite fold change, X_f, is therefore X_2/X_1 and the infinitesimal fold change, X_ϕ, is obtained from the equation,

$$X_\phi = X_f^{(1/n)} \qquad (5{-}160)$$

(which divides X_f into n successive and equal fold changes, X_ϕ, each of which is small enough to apply the sensitivity equations). X_ϕ is then converted into the equivalent (infinitesimally based) relative change, X_ρ, using the equation,

$$X_\rho = X_\phi - 1 \qquad (5{-}161)$$

The matrix equations are now written specifically for the interaction of X, so that the term $E_{1\rho}$ in vector $\mathbf{p}$ of equation 5–149, or its partitioned equivalent, is replaced by $\phi \cdot X_\rho$ and the other E_ρ terms are set to zero. Thus, equation 5–141 takes the form shown in equation 5–162.

$$\begin{pmatrix} 1 & 0 & 0 & -\alpha \\ 0 & 1 & 0 & -\beta \\ 0 & 0 & 1 & -\gamma \\ J & -K & -L & 0 \end{pmatrix} \begin{pmatrix} J_\rho \\ K_\rho \\ L_\rho \\ S_\rho \end{pmatrix} = \begin{pmatrix} \phi \cdot X_\rho \\ 0 \\ 0 \\ 0 \end{pmatrix} \qquad (5{-}162)$$

The initial values for the component sensitivities are calculated from the initial values of the fluxes and concentrations in the system, as outlined earlier (see earlier under Expressing Intrinsic and Component Sensitivities in Terms of Reaction Kinetics). The equation is then solved for J_ρ, K_ρ, L_ρ, and S_ρ. (Since this is a numerical solution, computer programs can be used. Therefore, a preliminary partitioning of the matrix equation is of less importance than for algebraic analyses, although it may sometimes help to reduce computing time with large branched systems.)

The new values for each variable are then calculated from the relation, new value = previous value·[1 + infinitesimal relative change] and the values of the component sensitivities in equation 5–162 are read-justed to these new values.

The entire process is repeated n times and the final values of the variables are those resulting from the application of the finite stimulus, X_f.

If more than one regulator changes, the method can be applied in two ways. Again with reference to Figure 5.23, if regulators X and Y both change, a calculated response to X (with no change of Y) can be followed by a calculated response to Y (with no further change of X); the final values of the variables then represent the combined effect of the two regulators. Alternatively, equation 5–141 can be written in terms of both infinitesimal relative changes (X_ρ and Y_ρ) and the combined response calculated directly. In this case vector $\mathbf{p}$ in equation 5–141 becomes,

$$\mathbf{p} = \begin{pmatrix} \phi \cdot X_\rho \\ \omega \cdot Y_\rho \\ 0 \\ 0 \end{pmatrix}$$

where, as before $X_\rho = [X_f^{(1/n)} - 1]$ and $Y_\rho = [Y_f^{(1/n)} - 1]$.

This method can be used to calculate the steady state response to any finite change of stimulus even for systems which show a net flux reversal (for example, the change of a net glycolytic flux into a net gluconeogenic one). Moreover, since it does not involve time derivatives, it is a very fast method for calculating the theoretical steady state response of any proposed kinetic control model to changes in its regulator concentrations. These predicted responses may then be compared with those measured experimentally in order to test the model.

Unfortunately, the numerical data needed to calculate the values of the component sensitivities, and hence these finite responses, are usually not very accurate. Consequently, a full application of this, or any other, method for numerical computation must await the arrival of much better experimental data.

Also, the method assumes that the system remains stable throughout the response, that is, no points of instability or system breakdown are encountered. Such continued stability is not guaranteed with some systems, especially those containing positive feedback (for example, the product activation of PFK by FBP) or negative feedforward mechanisms (for example, substrate inhibition of an enzyme at physiological concentrations). However, system breakdown or instability is readily tested for by calculating the sign of the determinant of matrix N, det N, (or, after partitioning, by the sign of det $[C - B \cdot A]$) (47). If the system is stable to begin with, it will remain so if the sign of the determi-

nant does not change after each step of the method. If it does change, the system has encountered an unstable region, in which case the method cannot be used beyond that point and alternative procedures, using time-dependent functions, must be used instead. (A full discussion of the role of det N in determining the stability of a steady state system is given in APPENDIX 2.)

However, the major problem with the method of finite decomposition, as outlined above, is that a relative change (dX/X) is not defined when the variable, X, is zero. Thus the method requires the initial values of all the variables.

This problem can be circumvented by basing the method on absolute changes (dX) instead of relative ones, since these are defined at $X = 0$. In this modification, the vectors $\mathbf{p}$ and $\mathbf{v}$ of equation 5–95 contain infinitesimal absolute changes in place of the relative ones, and the component sensitivities in matrix N are replaced by the corresponding absolute-change functions, which may be derived from the relative-change functions by the relation,

$$a_X^Y = (s_X^Y) \cdot X/Y \quad \text{(see equations 5–1 and 5–2)}$$

(This conversion must be done algebraically, not numerically, since X and Y may be initially zero.)

The situation with conservations is slightly different, however: for example, the flux conservation (Fig. 5.23),

$$J - K - L = 0$$

becomes, in terms of absolute changes,

$$dJ - dK - dL = 0 \quad (5\text{–}163)$$

Thus the flux terms in matrix N are replaced by $+1$ or -1 and the matrix equation, analogous to equation 5–162, takes the form shown in equation 5–164, where a, b, g, and p are the absolute-change sensitivities corresponding to α, β, γ, and ϕ, respectively.

$$\begin{pmatrix} 1 & 0 & 0 & -a \\ 0 & 1 & 0 & -b \\ 0 & 0 & 1 & -g \\ 1 & -1 & -1 & 0 \end{pmatrix} \begin{pmatrix} dJ \\ dK \\ dL \\ dS \end{pmatrix} = \begin{pmatrix} p \cdot dX \\ 0 \\ 0 \\ 0 \end{pmatrix} \quad (5\text{–}164)$$

As with relative changes, the finite absolute change, ΔX, is divided into n successive small changes, dX, and equation 5–164 is solved n successive times, adjusting the values of the component sensitivities, and checking the sign of det N, after each step.

Calculating the Relative Importance of Regulators and Regulatory Sequences During a Finite Response: Mean-Value Sensitivities. Previous sections have considered the question,

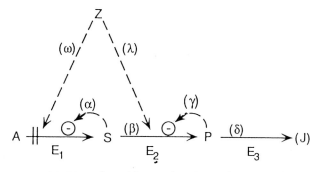

FIG. 5.26. Linear flux with a regulator interacting at two sites.

"How strong is the response to a given regulator and how can this strength be modified in vivo?". This question can be answered algebraically, using infinitesimally based sensitivities, and numerically, using finite decomposition. In this section another important question is considered: "How important is any given regulator or regulatory sequence during a metabolic change in vivo?".

To answer this latter question, we need to consider the contribution of a regulatory sequence to an overall response. For example, in Figure 5.26 a single regulator Z interacts with two regulatory reactions, E_1 and E_2. (This system is analogous to the effect of AMP on glycolysis: Fig. 5.10.) Since the system is structurally the same as that in Figure 5.19, the core sensitivities in equation 5–106 can be used here also. Therefore, by the product rule,

$$(s_Z^J)_{via\,E_1} = \omega \cdot s_1^J$$
$$= \frac{\omega \cdot \beta \cdot \delta}{\beta \cdot \delta + \alpha \cdot \delta + \alpha \cdot \gamma} \qquad (5\text{--}165)$$

Likewise,

$$(s_Z^J)_{via\,E_2} = \lambda \cdot s_2^J$$
$$= \frac{\lambda \cdot \alpha \cdot \delta}{\beta \cdot \delta + \alpha \cdot \delta + \alpha \cdot \gamma} \qquad (5\text{--}166)$$

The net effect of Z on J is therefore given by the addition rule (equation 5–16),

$$s_Z^J = (s_Z^J)_{via\,E1} + (s_Z^J)_{via\,E2}$$
$$= \frac{\omega \cdot \beta \cdot \delta}{\beta \cdot \delta + \alpha \cdot \delta + \alpha \cdot \gamma} + \frac{\lambda \cdot \alpha \cdot \delta}{\beta \cdot \delta + \alpha \cdot \delta + \alpha \cdot \gamma} \qquad (5\text{--}167)$$

Using the identity, $s_Z^J = J_\rho/Z_\rho$, equation 5–167 can be written as,

$$J_\rho = \frac{\omega \cdot \beta \cdot \delta}{\beta \cdot \delta + \alpha \cdot \delta + \alpha \cdot \gamma} \cdot Z_\rho$$
$$+ \frac{\lambda \cdot \alpha \cdot \delta}{\beta \cdot \delta + \alpha \cdot \delta + \alpha \cdot \gamma} \cdot Z_\rho \qquad (5\text{--}168)$$

The two terms in Z_ρ on the right-hand side of equation 5–168 represent the contributions of the two regulatory sequences, $Z \sim E_1 \sim J$ and $Z \sim E_2 \sim J$, to J_ρ. The relative importance of these sequences can therefore be measured as the ratio of the two terms. Thus,

$$\frac{contribution\;via\;E_1}{contribution\;via\;E_2} = \frac{\omega \cdot \beta}{\lambda \cdot \alpha} \qquad (5\text{--}169)$$

Since equations 5–168 and 5–169 refer to infinitesimal changes, they cannot be used to calculate the contributions of each regulatory sequence to a finite change of J (although, as in previous sections, they can be used to examine general algebraic properties). However, it is possible to extend these questions to cover finite changes by integrating the component equations of the system. Thus, for the system in Figure 5.26, the rate of E_1 can be written, by analogy with equation 5–21, as,

$$v_{1\rho} = \alpha \cdot S_\rho + \omega \cdot Z_\rho \qquad (5\text{--}170)$$

Integrating both sides, to simulate a finite (non-infinitesimal) response,

$$\int v_{1\rho} = \int dlnv_1$$
$$= \int \alpha \cdot S_\rho + \int \omega \cdot Z_\rho \qquad (5\text{--}171)$$

Now,

$$\int dlnv_1 = ln(v_1)_{final} - ln(v_1)_{initial} \qquad (5\text{--}172)$$
$$= ln\,J_{final} - ln\,J_{initial}$$
$$= ln(J_{final}/J_{initial})$$
$$= ln\,J_f \qquad (5\text{--}173)$$

where the subscript, f, denotes a fold change.

Thus, equation 5–171 becomes,

$$ln\,J_f = \int \alpha \cdot S_\rho + \int \omega \cdot Z_\rho \qquad (5\text{--}174)$$
$$= \int \alpha \cdot dlnS + \int \omega \cdot dlnZ \qquad (5\text{--}175)$$

The integrals on the right-hand side of this equation cannot be evaluated algebraically because both α and ω vary continuously during a finite response to Z. However, they can be expressed as mean-value functions, defined by the general equation,

$$(s_X^Y)_m = \frac{\int s_X^Y \cdot dlnX}{lnX_{final} - lnX_{initial}} \qquad (5\text{--}176)$$

$$= \frac{\int s_X^Y \cdot d\ln X}{\ln X_f} \qquad (5-177)$$

where the subscript, m, denotes a mean value.

Therefore, equation 5–175 can be written as,

$$\ln J_f = \alpha_m \cdot \ln S_f + \omega_m \cdot \ln Z_f \qquad (5-178)$$

$$= \ln S_f^{\alpha_m} + \ln Z_f^{\omega_m} \qquad (5-179)$$

$$= \ln(S_f^{\alpha_m} \cdot Z_f^{\omega_m}) \qquad (5-180)$$

so that,

$$J_f = S_f^{\alpha_m} \cdot Z_f^{\omega_m} \qquad (5-181)$$

which represents a finite power equation,

$$J = k \cdot S^{\alpha_m} \cdot Z^{\omega_m} \qquad (5-182)$$

This type of equation forms the basis of the approach known as biochemical systems theory (discussed in detail later); the above analysis shows that the index of such equations is a mean-value relative-change sensitivity, defined by equation 5–177.

Consequently, all the equations for the system in Figure 5.26 could be written in terms of $\ln J_f$ (as in equation 5–178) and the system solved to give, by analogy with equation 5–168, the equation,

$$\ln J_f = \frac{\omega_m \cdot \beta_m \cdot \delta_m}{\beta_m \cdot \delta_m + \alpha_m \cdot \delta_m + \alpha_m \cdot \gamma_m} \cdot \ln Z_f$$
$$+ \frac{\lambda_m \cdot \alpha_m \cdot \delta_m}{\beta_m \cdot \delta_m + \alpha_m \cdot \delta_m + \alpha_m \cdot \gamma_m} \cdot \ln Z_f \quad (5-183)$$

Each term of the right-hand side can be regarded as a contribution of each regulatory sequence to $\ln J_f$. The mean-value sensitivities can be evaluated from a numerical simulation of the time course of the system between the two steady states (see later under Analysis of Non-Steady States), using a numerical integration based on equation 5–177. The contributions of each sequence to the finite response, $\ln J_f$, then calculated and compared by using the respective terms in $\ln Z_f$ in equation 5–183.

Unfortunately, this method of comparing regulatory sequences has two major problems. First, a contribution to a function such as $\ln J_f$ is rather obscure. Secondly, and potentially more serious, is that functions such as $\ln J_f$ are not defined if the flux, J, reverses in direction (157).

As with calculations of finite responses (see earlier under Calculation of Responses to Large (Finite) Changes of Regulators), these difficulties may be overcome by basing the calculations on absolute-change

sensitivities instead of the relative change functions. For this, equation 5–170 is written as,

$$dv_1 = a \cdot dS + o \cdot dZ \qquad (5-184)$$

where a and o are the absolute-change sensitivities corresponding to α and ω, respectively [so that $a = \alpha \cdot (v_1/S)$; $o = \omega \cdot (v_1/Z)$]. Integrating equation 5–184,

$$\int dv_1 = \Delta J$$

$$= \int a \cdot dS + \int o \cdot dZ \qquad (5-185)$$

As before, the integrals on the right-hand side can be replaced by mean-value functions, defined (by analogy with equation 5–177) as follows:

$$(a_X^Y)_m = \frac{\int a_X^Y \cdot dX}{\Delta X} \qquad (5-186)$$

so that, equation 5–185 becomes,

$$\Delta J = a_m \cdot \Delta S + o_m \cdot \Delta Z \qquad (5-187)$$

Consequently, equation 5–183 can be written in terms of mean-value absolute-change sensitivities as follows:

$$\Delta J = \frac{o_m \cdot b_m \cdot d_m}{b_m \cdot d_m + a_m \cdot d_m + a_m \cdot g_m} \cdot \Delta Z$$
$$+ \frac{l_m \cdot a_m \cdot d_m}{b_m \cdot d_m + a_m \cdot d_m + a_m \cdot g_m} \cdot \Delta Z \quad (5-188)$$

where the roman letters correspond to the greek ones in equation 5–168.

As with equation 5–168, each term on the right-hand side of equation 5–188 represents the contribution of each regulatory sequence to the absolute change of J, ΔJ, a much more appropriate function than $\ln J_f$. Moreover, ΔJ is defined even if J reverses, so these contributions are quite general.

Therefore, the contribution of any regulatory sequence to a change in a variable can be quantified by calculating the contribution to the absolute change in that variable; using functions corresponding to those in equations 5–165 to 5–168, but with infinitesimal relative-change sensitivities replaced by mean absolute-change sensitivities. (As with the use of absolute-change sensitivities for calculating finite responses, any conservations in the system, including branch points, are accommodated by replacing the appropriate variables in matrix N by +1 or −1, depending on their sign in N.)

Mean-value sensitivities are calculated numerically from a kinetic model describing the change from the initial to the final steady state. Usually the regulator (for example, Z, Fig. 5.26) is assumed to change instan-

TABLE 5.2. *Contributions of Regulatory Interactions to ΔJ*

	$Z_f = 2$	$Z_f = 10$
Initial value of J	1	1
Final value of J	1.457	1.92
ΔJ	0.457	0.92
ΔZ	1.0	9.0
Initial value of α	-0.5	-0.5
Mean value of a (a_m)	-0.73	-0.96
Initial value of β	0.5	0.5
Mean value of b (b_m)	0.606	0.876
Initial value of ω	1	1
Mean value of o (o_m)	0.6	0.112
Initial value of λ	0.5	0.5
Mean value of l (l_m)	0.335	0.093
Contribution to ΔJ of the interaction of Z with E_1 (($Cn^{\Delta J}$)$_1$)	0.274	0.482
Contribution to ΔJ of the interaction of Z with E_2 (($Cn^{\Delta J}$)$_2$)	0.183	0.437
$(Cn^{\Delta J})_1 + (Cn^{\Delta J})_2$	0.457	0.919
$\dfrac{(Cn^{\Delta J})_1}{(Cn^{\Delta J})_2}$	1.5	1.1

The values refer to the system in Figure 5.26, with roman letters denoting absolute-change sensitivities corresponding to the relative-change functions (greek letters). For simplicity, the value of γ, and hence g, is zero (which also means that the value of d is not required: see equation 5–188). Each contribution was calculated, using the functions on the right-hand side of equation 5–188, for a 2- or a 10-fold instantaneous increase in the concentration of the regulator Z, via a numerically generated (Euler-based) time course (see later under Analysis of Non-Steady States) and evaluation of the mean-value sensitivities as in equation 5–189.

The rates of both E_1 and E_2 are assumed to be described by a general two-substrate, Hill-type, equation,

$$v = \frac{V_m \cdot S^\phi \cdot X^\mu}{(S^\phi \cdot X^\mu + K_1 \cdot S^\phi + K_2 \cdot X^\mu + K_3)}$$

where μ and ϕ are powers, and the K values are functions of the individual rate constants of the catalytic mechanism. The absolute-change (intrinsic) sensitivities of $E_{(1or2)}$ to S and Z, were calculated as $(dv/dS)_{Zconstant}$ or $(dv/dZ)_{Sconstant}$, respectively. For both E_1 and E_2 the value of V_m is 4 units and all the K values are 1 unit. The initial values of S and Z are each 1 unit.

For E_1, $\phi = -1$ (product inhibition by S) and $\mu = 2$; for E_2, both ϕ and $\mu = 1$.

taneously, that is, much faster than the variables, S and P. A time course of S and P is then generated by using a numerical procedure, as outlined later under Analysis of Non-Steady States. At each point of the time course the values of the infinitesimally based sensitivities, for example, b_S^{E2}, are calculated and the integrals for equation 5–186 obtained as the sum of each of these values multiplied by dS for that point of the time course. Thus, for b_m, using equation 5–186,

$$b_m = \frac{\int b_S^{E2} \cdot dS}{\Delta S}$$
$$= \frac{\sum (b_S^{E2} \cdot dS)}{\Delta S} \qquad (5–189)$$

where the sum is taken over the time taken to change from the initial to the final (or, in practice, 99% of the final) steady state.

An example of such a calculation, for the interactions of Z with flux J via E_1 and E_2 (Fig. 5.26), is given in Table 5.2. The results show that each of the two contributions to ΔJ each depends on the size of the stimulus, that is, the change in Z. This is a general property of functions based on mean-value sensitivities, which, unlike most of the other functions used in this chapter, do not just depend on the initial and final steady conditions. Instead they are path-dependent. For example, since the value of b_S^{E2} (the absolute-change sensitivity corresponding to β in Fig. 5.26) depends on the other effectors of E_2, Z and P, in addition to S, the value of the sum in equation 5–189 depends on the values of Z and P at each point of the time course. Hence the corresponding mean-value sensitivity will depend on both the overall changes, ΔZ and ΔP, and the rates at which Z and P change during the approach to the new steady state. Although, for simplicity, the interaction of P with E_2 was assumed to be zero in Table 5.2, the effect of Z on the mean-value sensitivities, b_m and a_m, was still present; this is the reason each contribution in Table 5.2 depends on the size of the stimulus, ΔZ.

The path-dependence of the mean-value sensitivities (which applies to both relative-change and absolute-change types) means that, whilst a general algebraic analysis of the relative importance of regulator sites can be made by examining the infinitesimally based functions (for example, equation 5–169), a precise evaluation of their relative importance under physiological conditions cannot be made in the absence of a suitable kinetic model of the system. Moreover, in contrast with the method of finite decomposition for calculating specific responses, a time course must be generated. At present, a major problem with such

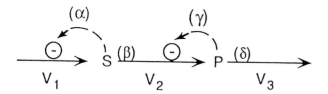

FIG. 5.27. Linear non-steady state system. Here v_1, v_2 and v_3 are not mutually equal, so that S and P vary with time and there is no defined system flux; α–γ denote absolute-change sensitivities.

calculations, as with those for finite responses, is the scarcity of accurate numerical data. Nevertheless, the above analysis does at least indicate the type of functions that will be needed when techniques improve, as they almost certainly will in the future.

Analysis of Non-Steady States

Although the steady state assumption provides a valuable and relatively simple structure on which to base discussions of regulatory mechanisms, it is an abstraction which cannot always be applied. The following are cases when time becomes an important variable: oscillations, pulses, and the transient changes that occur when one steady changes into another as a result of regulation. On a wider plane, cell division and growth are also important examples of time-dependent, and hence non-steady state, systems.

The entry of time into the equations creates major problems for any detailed analysis since we now have to solve sets of differential equations and this is very difficult to do analytically, that is, to provide an algebraic solution, for all but the simplest systems. (For an excellent review of the analysis of non-steady state systems, see ref 71.) To outline the principles and problems, let us consider the system in Figure 5.27, which is structurally the same as that in Figure 5.19. However, the system is no longer in a steady state, so that S and P vary continuously with time. Most importantly, this means that the rates of the component reactions (v_1–v_3) are not equal to each other such that there is no defined system flux, that is, J. Consequently, a set of simple simultaneous equations (for example, equations 5–91 to 5–93) cannot now be written—each reaction must be considered individually.

For the system in Figure 5.27, the differential equations are as follows:
For the net flow into S,

$$dS/dt = v_1 - v_2 \qquad (5\text{–}190)$$

For the net flow into P,

$$dP/dt = v_2 - v_3 \qquad (5\text{–}191)$$

These equations can be differentiated once more to give,

$$d^2S/dt^2 = dv_1/dt - dv_2/dt \qquad (5\text{–}192)$$

$$d^2P/dt^2 = dv_2/dt - dv_3dt \qquad (5\text{–}193)$$

Each differential coefficient on the right-hand side of equations 5–192 and 5–193 can be written as a product, $(dv/dX) \cdot (dX/dt)$, where X denotes a general effector and the first term in the product is an absolute-change sensitivity. With this substitution, equations 5–192 and 5–193 become,

$$d^2S/dt^2 = (dv_1/dS) \cdot (dS/dt) - (dv_2/dS) \cdot (dS/dt) \\ - (dv_2/dP) \cdot (dP/dt) \qquad (5\text{–}194)$$

$$d^2P/dt^2 = (dv_2/dS) \cdot (dS/dt) + (dv_2/dP) \cdot (dP/dt) \\ - (dv_3/dP) \cdot (dP/dt) \qquad (5\text{–}195)$$

or, in terms of absolute-change sensitivities (see Fig. 5.27),

$$d^2S/dt^2 = (a-b) \cdot (dS/dt) - g \cdot (dP/dt) \qquad (5\text{–}196)$$

$$d^2P/dt^2 = b \cdot (dS/dt) + (g-d) \cdot (dP/dt) \qquad (5\text{–}197)$$

Writing S'' and P'' for the second differentials, and S' and P' for the first differentials, we can express equations 5–196 and 5–197 in matrix form as,

$$\begin{pmatrix} S'' \\ P'' \end{pmatrix} = - \begin{pmatrix} (b-a) & g \\ -b & (d-g) \end{pmatrix} \begin{pmatrix} S' \\ P' \end{pmatrix} \qquad (5\text{–}198)$$

which can be written generally as,

$$\mathbf{m}'' = -Q \cdot \mathbf{m}' \qquad (5\text{–}199)$$

where $\mathbf{m}$ is a vector containing the internal effectors of the system.

Equation 5–199 can be integrated to give,

$$\mathbf{m}' = -Q \cdot \mathbf{m} + \mathbf{n} \qquad (5\text{–}200)$$

where $\mathbf{n}$ is a vector corresponding to a constant of integration. By analyzing a system about its steady-state value (provided that this exists!), vector $\mathbf{n}$ can be incorporated into $\mathbf{m}$ to give the equation,

$$\mathbf{m}'_s = -Q \cdot \mathbf{m}_s \qquad (5\text{–}201)$$

where the subscript, s, denotes changes around the steady state values.

Equation 5–201 can, in theory, be solved analytically for a system by using functions termed the eigenvalues and eigenvectors of matrix Q (104). Thus, since a particular solution of equation 5–201 will have the form,

$$\mathbf{m_p} = \mathbf{a} \cdot e^{-\lambda t} \qquad (5\text{–}201a)$$

where **a** is a vector and λ a constant,

$$\mathbf{m_p'} = -\lambda \cdot \mathbf{a} \cdot e^{-\lambda t} \qquad (5\text{-}201b)$$

Putting equations 5–201a and 5–201b, into equation 5–201 gives,

$$-\lambda \cdot \mathbf{a} \cdot e^{-\lambda t} = -Q \cdot \mathbf{a} \cdot e^{-\lambda t}$$

so that,

$$Q \cdot \mathbf{a} = \lambda \cdot \mathbf{a} \qquad (5\text{-}201c)$$

A vector, **a**, with this property, is termed an eigenvector of matrix Q and λ is a constant known as the corresponding eigenvalue, and both can be found by standard algebraic or numerical procedures (104, 162). The complete solution of equation 5–201, and hence the dynamical properties of the system, is the sum of all the particular solutions (that is, $\Sigma(\mathbf{m_p})$):

$$\mathbf{m} = \sum_{}^{n} (\mathbf{a}_j \cdot e^{-\lambda_j t})$$

where n is the total number of eigenvalues and corresponding eigenvectors of matrix Q.

The eigenvalues may be either real or complex numbers (that is, have the form $w \pm i \cdot z$, where w is the real component, z the imaginary component, and $i^2 = -1$). If λ is a simple real number (that is, $z = 0$), it must be positive for a steady state to be attained. This makes the coefficient of t in equation 5–201a negative, so that the exponential approaches zero as time increases, and **m** becomes a constant (steady-state) vector. Consequently, for a straightforward approach to a steady state, all the eigenvalues of Q must be real and positive. If any are negative, the exponential will increase with time and the system will break down completely.

If λ is a complex number, the system will oscillate with a frequency determined by the value of the imaginary component, z. If the real component, w, is zero, the oscillations will be sustained, whereas if w is positive, they will fade with time and a steady state will eventually be attained. However, if w is negative, the amplitude or size of the oscillations will increase with time until the system breaks down.

Consequently, the stability of a system can be determined if the eigenvalues and eigenvalues of matrix Q can be calculated. Unfortunately, even determining the nature and signs of the eigenvalues is not easy, especially as the systems become large; and there is, as yet, no simple graphical method, analogous to the use of graph theory for steady states (APPENDIX 3) for making these assessments. Moreover, the calculations of eigenvalues and eigenvectors requires the elements of matrix Q to be constants. Although this may be assumed for simplicity with hypothetical abstract systems, it cannot apply generally to real metabolic systems because the

elements of Q include sensitivities, whose values change continuously during any physiological response. Therefore, with real metabolic systems, any calculated eigenvalues of matrix Q will apply only to a specific infinitesimal response and will therefore be of little, if any, use for predicting the stability of the system generally. (It should be noted that, with non-steady state systems, the values of the elements of matrix Q may affect the qualitative nature of the dynamics, for example whether a system oscillates or not. This contrasts with the elements of the steady state matrix, N (equation 5–95), that only affect the size, not the nature, of the response, which is always a smooth approach to a new steady state.)

Therefore, for real metabolic non-steady states, specific numerical solutions must be used in which a succession of small time intervals (dt) is taken, beginning at some appropriate time (usually zero) and the differential equations (for example, equations 5–190 and 5–191 for Fig. 5–27) solved for dS and dP as follows:

$$dS = dt \cdot (v_1 - v_2) \qquad (5\text{-}202)$$

$$dP = dt \cdot (v_2 - v_3) \qquad (5\text{-}203)$$

The new values of S and P are then calculated, as $S + dS$ and $P + dP$, respectively, and the reaction rates, $v_1 - v_3$, recalculated from the kinetic rate equations; equations 5–202 and 5–203 are then solved for the next time interval. In this way, S and P (as well as $v_1 - v_3$) are obtained as functions of time. (In practice, this simple Euler method works only with relatively simple systems and, for more complex systems, more sophisticated Runge-Kutta methods, for example, would be used instead [104]. However, the general principles are the same.)

As with many attempts to compute the behavior of metabolic systems, this type of numerical simulation is bedeviled by the lack of accurate numerical data and hence accurate kinetic models. For the time being therefore, we need an adequate descriptive method to understand the factors involved in the production and maintenance of non-steady state systems. A suitable elementary function is the speed of response of a metabolite or reaction to a change in its state.

Speed of Response. Consider a simple isolated reaction,

$$S \xrightarrow[v_t]{}$$

The behavior of S with time, S_t, is given by the differential equation,

$$dS_t/dt = -v_t \qquad (5\text{-}204)$$

If v is proportional to S (that is, first order), equation 5–204 can be written as,

$$dS_t/dt = -a \cdot S_t$$

or,

$$dS_t/dt + a \cdot S_t = 0 \qquad (5\text{–}205)$$

where a is the first order rate constant. (Since $a = dv/dS$, it will, in general, be the absolute-change sensitivity of v to S.)

The solution of equation 5–205 is,

$$S_t = S_0 \cdot e^{-a \cdot t} \qquad (5\text{–}206)$$

so that,

$$t = \frac{1}{a} \cdot \ln \frac{S_0}{S_t} \qquad (5\text{–}207)$$

Equation 5–207 shows that, for any given amount of S consumed, the time taken is inversely proportional to a: so that we can consider the speed of response of S to be directly proportional to a.

Although this quantitative relationship applies to first order responses, it can be applied usefully and descriptively to any response by considering a to be equal to the ratio of finite absolute changes, $\Delta v / \Delta S$, that is, treating the finite response as if it were linear in S.

Therefore, in general (44),

$$\textit{speed of response of } S \propto \frac{\Delta v}{\Delta S} \qquad (5\text{–}208)$$

$$\propto \frac{\Delta J}{\Delta S} \qquad (5\text{–}209)$$

if we are considering transient changes between steady states.

From this function it can be predicted that the speed of response of S will be increased by increasing Δv and/or by decreasing ΔS. As will be shown below, both these mechanisms are used to vary the speeds of response of metabolic intermediates and hence reactions.

Transient Changes Between Steady States: Functional Readiness and Pulsed Stimuli. When the concentration of a regulator (such as Z in Fig. 5.26) changes, the concentrations of both S and P change, transiently, until the new steady state becomes established. The duration of this transient state will, in most cases, be required to be as short as possible, especially with an important ATP-producing pathway such as glycolysis in muscle. Therefore, the speeds of response of S and P should be as large as possible.

From equation 5–209, these speeds are proportional to $\Delta J / \Delta S$ and $\Delta J / \Delta P$. Now ΔJ, the absolute change of the flux, is determined by the metabolic role of the pathway and cannot be varied independently. For example, the change of glycolysis in muscle will be determined by the change in demand for ATP. However, the other component of the ratio, ΔS or ΔP, can be made as small as possible by ensuring high intrinsic or component sensitivities. From the definition of a relative change for S,

$$\Delta S = S_{initial} \cdot S_{rel} \qquad (5\text{–}210)$$

where $S_{initial}$ is the initial value of S and S_{rel} denotes a relative change of S. Thus, for any given initial concentration, the value of ΔS is smaller, the smaller the value of S_{rel}. As shown earlier under Overall Sensitivities of Regulatory Mechanisms In Situ: Internal Oppositions, S_{rel} will be smaller, the greater the values of the component sensitivities, α and β (Fig. 5.27). Likewise, ΔP will be smaller, the greater the values of δ and γ.

Consequently, those mechanisms that provide a high intrinsic or component sensitivity to the internal effectors (for example, kinetic reversibility, interconvertible enzyme forms, substrate cycling, and sigmoid kinetics) not only create a stable kinetic control structure, but also speed up the response of the system to its regulators. Indeed the role of kinetic reversibility in speeding up the response of glycolysis was proposed by Bucher and Russman (14), and termed functional readiness.

However, there is an additional way of increasing the speed of response of S (or P). From equation 5–210, for any give relative change, ΔS becomes smaller as the initial value of S becomes smaller. In a system where the enzymes and metabolites are in free solution, a low initial S could be achieved by a high catalytic activity (or kinetic power [83, 84, 86]) of the enzymes removing S: for example, $S_{initial}$ will be smaller, the greater the ratio $V_m / S_{0.5}$ for E_2. Whilst increasing the kinetic power may be possible in some cases, in general it would increase the amount of protein in the cell, which could seriously affect the solvent capacity (see earlier under Need for Sensitive Regulation In Vivo). However, by confining S to a small compartment near the active site of E_2, the value of $S_{initial}$ at the active site could be maintained very low: indeed, in the limit, S could be passed directly from E_1 to E_2 without ever becoming free in solution (83, 85, 123).

There is now strong, although not conclusive, evidence that such channelling does occur in vivo (122, 160) and, as discussed above, one of its roles could be to provide a fast response of a pathway to its regulators. However, such a tight channelling may not always be advantageous because metabolites such as S and P would not be able to communicate with other reactions and fluxes. This may not be a problem for the intermediates of fatty acid synthesis, which are passed directly from one enzyme site to another (163), but it would cause problems for regulators such as the adenine and pyridine nucleotides, as well as acetyl-CoA and other intermediates. These need to be free in order to communicate between pathways and thereby exercise their regulatory functions. Therefore, it is of interest that, in the span of the Krebs cycle from malate to citrate, there may be a tight channelling of intermediates during a state of high flux; but this is relaxed under other conditions, allowing oxaloacetate and other intermediates to communicate with other systems (168).

In any case, a simple restricted diffusion of intermediates, such as S and P in Figure 5.26, from the active sites of enzymes E_1–E_3 could provide a transient channelling during a short-lasting change of flux; since, for such a short period, S could be directed towards the active site of E_2 and P towards the active site of E_3, thereby providing a fast response. However, this type of leaky channelling could only operate for a short time, because eventually S and P would diffuse away from the channel, until their steady state concentrations were the same as in its absence (32). Consequently, this type of channelling is likely to be of value only for improving the speed of short-lasting, that is, pulsed, changes of flux (for example, those caused by pulses of Ca^{2+} or other secondary cellular messengers).

A similar situation may underlie the different anomeric specificities of the glycolytic enzymes, phosphofructokinase (PFK) and fructose-1,6-bisphosphatase (FBPase) to fructose-1,6-bisphosphatase (FBP). PFK generates the β anomeric form of FBP, whereas FBPase requires the α form (88). Since the rate of conversion of the β form into the α one is relatively slow, a short-lasting increase of PFK will produce FBP that will preferentially be used by aldolase (also β specific) and will not enter the reverse reaction of the substrate cycle between F6P and FBP (Fig. 5.14b). Thus, for a short period, the component sensitivity, γ, (Fig. 5.14a) will effectively be zero and there will be no oppositions to the increased PFK activity. The net sensitivity of glycolysis to regulators of PFK is thereby increased. However, as with a leaky channel, this can only last for a relatively short time, because the gradual conversion of the β into the α form will eventually increase the rate of FBPase and bring in the opposition due to γ.

Modification of the Speed of Response of an Interconversion Cycle.
In the previous section, the speed of response was analyzed with respect to changes of ΔS, since the ΔJ term in equation 5–209 was considered to be

unavailable for independent control. However, with an interconversion cycle (Fig. 5.15) the term, ΔJ, is equal to ΔC (or ΔF since C and F are equal in the steady state). Moreover, since this is an independent cycle, its flux is not constrained to any other metabolic system (unlike the flux through glycolysis), so that F (and C) can take any values.

As with metabolite S (equation 5–208),

$$\Delta C = C_{initial} \cdot C_{rel},$$

so that, for any given chance of regulator, Z (Fig. 5.15) (and hence C_{rel}), ΔC and hence the speed of response of the a and b forms to Z will be increased by an increased initial rate of cycling. Therefore, the speed of response of the system to its external regulators can be varied by varying the initial rate of cycling (43, 44).

This may be important in regulating the response of protein systems, for example those involving G-proteins (58, 151), to pulses of a regulator. As shown in Figure 5.28, a given pulse of a regulator (lasting for time, τ) will produce a pulsed response of the a and b forms. If τ is smaller than the time taken for the cycle to attain a new steady state, the pulse height of the response depends on the speed of response of the system, and can therefore be varied by varying the initial rate of cycling, C (44). Such a variable response may play a role in modifying the transmission of nervous and chemical impulses at synapses, as well as the response of cellular systems to pulses of Ca^{2+} that result from the action of several hormones (189).

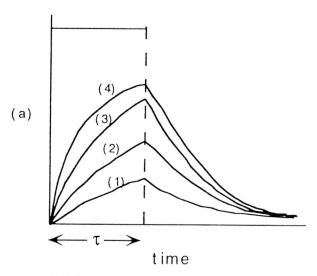

(a)

(4)
(3)
(2)
(1)

τ

time

FIG. 5.28. Effect of the speed of response on a pulsed system. The pulsed stimulus lasts for time τ. Curves 1–4 show the effect of a coordinated increase of the initial rate of cycling between the b and a forms of an interconversion cycle. In curve 4 a steady state has been reached within time τ and any further increase in the rate of initial cycling has no further effect.

However, this alerting effect of an interconversion cycle will usually require a coordinated increase of both E_1 and E_2; otherwise the initial distribution of the a and b forms would be changed—greatly so, if the cycle operates with zero order ultrasensitivity. Such changes may have undesirable consequences for metabolism. It is therefore of interest that several enzymes, such as E_1 and E_2 in Figure 5.15, which produce metabolic cycles (either between different enzyme forms, or between metabolites such as F-6-P and F-2, 6-BP) are found on the same protein molecule (16, 51, 170, 172).

In most examples of these bifunctional enzymes, regulators have been observed to produce reciprocal regulation; for example, with the PFK-2/FBPase-2 enzyme, phosphorylation of a serine residue activates the FBPase but inhibits the PFK component (16, 62, 172). Nevertheless, it is possible that such bifunctional proteins may have regulators which activate both E_1 and E_2 to the same extent, thereby giving a coordinated change of fluxes F and C without significantly affecting the distributions of the a and b forms.

A similar type of coordinated regulation may occur with some substrate cycles; for example, that involving the hydrolysis and reesterification of triacylglycerol (Fig. 5.14c). Several hormones increase both the rate of this cycle (C) and the net flux of FFA mobilization (J), in such a way that the ratio C/J is approximately constant (see discussion after ref. 113). Indeed, the elevation of both C and J in this cycle may continue for several hours after a period of exercise, with the increased cycling accounting for as much as 50% of the extra uptake of O_2 under these conditions (5).

However, a proportionate increase in C and J is equivalent to the situation when δ (Fig. 5.14a) = 1, (see earlier under Effect of a Substrate Cycle In Situ on the Net Sensitivity to External Regulators) and therefore completely opposes the effects of the cycle on the response of J to hormonal regulators of either E_1 or E_2 (Fig. 5.14c). On the other hand, by maintaining the C/J ratio under all conditions of hormonal mobilization, the feedback sensitivity of FFA production to changes in FFA, as a result of changes in demand for FFA (reaction E_3, Fig. 5.14c), is also maintained. This latter sensitivity may be the more important because of the need to avoid large changes of FFA. Changes in hormone concentrations can be much larger (see earlier under Need for Sensitive Regulation In Vivo) so that the mobilization sensitivity conferred by the cycle may be of lesser importance than the feedback sensitivity to FFA. For example, a five-fold increase in mobilization, J, at constant cycling, C, would reduce the C/J ratio by 80%, necessitating a larger change of FFA to balance the supply of FFA to demand. Mobilization sensi-

tivity may therefore be sacrificed to provide a high feedback sensitivity at all times, at least with the FFA mobilizing system (Fig. 5.14c).

Metabolic Buffers. Sometimes it may be necessary to decrease the speed of response to provide a buffer against sudden changes: for example, in muscle, the ATP concentration must not suffer large changes or else the kinetic control structure could be damaged (see earlier under Need for Sensitive Regulation In Vivo).

Buffering by an equilibrium. A simple way of buffering an intermediate is via an equilibrium, as shown in Figure 5.29a. Here, metabolite S is in equilibrium with Q, so that,

$$Q = K_{eq} \cdot S \qquad (5-211)$$

S and Q now represent a single pool of size $(S + Q)$ or, using equation 5-211, $S(1 + K_{eq})$. Therefore, changes in this pool, due to ΔS, are now equal to $\Delta S(1 + K_{eq})$; so that, from equation 5-209, the speed of response of S will be decreased by the factor $(1 + K_{eq})$, as shown mathematically (107).

Consequently, a large value of K_{eq} slows down changes in S and therefore smooths out any sudden changes resulting from changes in the fluxes I and/or O (Fig. 5.29a). Indeed, for very short-lasting changes, the value of S may hardly change at all. For the buffering of ATP in muscle, the equilibrium between ATP/ADP and creatine phosphate/creatine provides a satisfactory buffer for short times (185), and the acetyl-carnitine/carnitine couple may provide a similar buffering for the acetyl-CoA/CoA system (135).

However, this type of buffering is only transient. If the changes in fluxes I and/or O persist, the final steady state value of S will be the same as in the absence of the buffer (107). Thus, as with a leaky channel, equilibrium buffering only slows down the change in S: it does not affect the final steady state value.

Problems with a tight binding of ligands to proteins. Slowing down a change of S will also occur if the equilibrium represents the binding of S to a protein or enzyme (as a regulator) and can be very pronounced when the concentration of S is similar to that of the protein. Consequently, if S is a second messenger (for example, cyclic AMP) produced by hormone action and binds tightly to its receptor protein, it could take a long time to remove S from the protein after cessation of the hormonal stimulus; the response to the hormone would therefore take a long time to switch off (7, 107, 117).

For this reason it was proposed (177) that the binding of such regulators would be relatively weak, and the correspondingly weak response amplified by a system of interconversion cycles (that is, a cascade): an example being the response of glycogen breakdown to cyclic

a

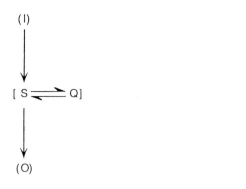

b

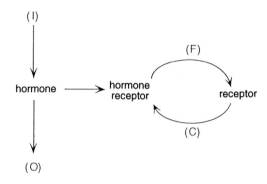

c

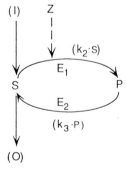

FIG. 5.29. Metabolic buffers. *(a)* Equilibrium. $Q = K_{eq}.S$ and the resulting larger pool of S is shown in parentheses. Fluxes I and O are equal in the steady state. *(b)* Receptor recycling. The hormone binds tightly to its receptor, but the complex is destroyed when it is internalized (via F); the free receptor is then recycled (via C). Here, fluxes I and O represent the delivery and removal of the hormone, respectively. *(c)* Substrate cycle. Here, in the steady state there is no net flux across the cycle, that fluxes I and O are equal. S and P may denote *(1)* glucose and glycogen, *(2)* amino acids and protein, *(3)* fatty acids and triacylglycerol, in which case the cycles may be regarded as storage cycles. E_1 and E_2 are assumed to have first-order responses to S and P, respectively.

AMP and Ca^{2+}. The net result is a strong, but rapidly reversible, response of the system to the regulator.

Alternatively, if S represents a hormone and Q its complex with a receptor, the problems caused by a tight binding of S could be overcome by recycling the receptor complex (7152), thereby generating free receptor continuously (Fig. 5.29*b*). The off response (to a decreased hormone production, flux I) is then determined, not by the slow dissociation of hormone from the receptor, but by the rate of receptor recycling. As with other similar cycles, this rate (and hence the speed of response to changes in hormone concentrations) may be under cellular control and therefore be varied according to physiological requirements.

Buffering by a substrate cycle. Buffering may also be provided by a substrate cycle, since this has the same kinetic structure as an equilibrium (Fig. 5.29*c*). Examples of such buffering may be the storage cycles between glucose and glycogen, between amino acids and protein in the liver, and between fatty acids and triacylglycerols in adipose tissue and liver (Fig. 5.29*c*). These may serve to buffer blood glucose, amino acids, and fatty acids against large changes after absorption from the gut.

Moreover, with a substrate cycle, the function equivalent to K_{eq} is the ratio k_2/k_3 (Fig. 5.29*c*) and, unlike K_{eq}, this function can be varied by changing the activities of enzymes E_1 and E_2 (for example, via regulator, Z). Thus, unlike equilibrium buffering, the speed of response of a substrate cycle and hence its buffering capacity, can be under cellular control, the price being an increased use of energy by the cycle. However, as with the sensitivity conferred by substrate cycles (see earlier under Effect of Substrate Cycling), the rate of cycling and hence the energy consumption may be kept low when there is little demand for buffering, but increased when demand is anticipated. For example, with the storage cycles in liver and other tissues, the ratio k_2/k_3 might be increased after the ingestion of a meal in order to prepare these cycles to buffer the entry of fuels into the blood.

Metabolic Oscillations. If there is a delay between the production of an internal effector, X, and its removal, X may sometimes overshoot the balance point (that is, the steady state value) after the interaction of a regulator, resulting in oscillations of X around the balance point (Fig. 5.30). For example, in Figure 5.31, metabolite S is removed directly by E_2, but also indirectly via the communication sequence, $S \sim M \sim E_2$. This latter sequence introduces a delay between changes of S (due to the interaction of Z) and its rate of removal via E_2. If this delay is substantial, then, despite the presence of the direct communication of S with E_2, S will overshoot the balance point (at which

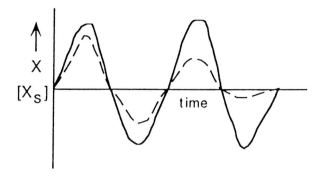

FIG. 5.30. Oscillations. X_s is the steady state value, or balance point around which X oscillates. The broken line shows a fading, or damped, oscillation.

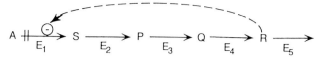

FIG. 5.32. System with a long feedback loop.

the rates of E_1 and E_2 are equal) and oscillations of S will occur.

Usually such oscillations fade with time (that is, are damped: Fig. 5.30), but under some conditions they may be sustained, and be physiologically important. The rhythmic contraction of the heart is an obvious example, but pulses or oscillations of Ca^{2+} are now recognized to play an important role in cellular signalling. However, when they occur during the transition of one steady state into another, oscillations can be a nuisance and their occurrence under such conditions needs to be minimized.

One way of minimizing the occurrence of oscillations is to ensure that any indirect communication sequences such as that in Figure 5.31 respond rapidly to S. From previous discussions (see earlier under Transient Changes Between Steady States: Functional Readiness and Pulsed Stimuli) this means that the component sensitivities of such sequences should be large. Another way would be to have a buffering system at S, thus reducing the rate of change of S and ensuring that the indirect systems have sufficient time to operate before S overshoots the balance point.

Oscillations may also occur if a feedback loop spans several reactions (63); for example the loop from me-

tabolite R to reaction E_1 in Figure 5.32. This is because, as in Figure 5.31, S is now removed via an indirect communication sequence, $S \sim P \sim Q \sim R \sim E_1$. Since the delay between changes of S and the response of E_1 increases with the number of reactions in this sequence, as the feedback loop becomes longer, the delay will eventually be sufficient to result in (damped) oscillations of S.

Consequently, long feedback loops may present stability problems and, although the occurrence of such oscillations could be minimized by having high component sensitivities in the sequence, an alternative method is to reduce them by using branched systems, such as those in Figures 5.23–5.25. Referring to Figure 5.23, changes in flux K can be achieved by switching fluxes between K and L, with no need to communicate further upstream to E_1. If the span represented by E_1 consists of several consecutive enzymes (as in Figs. 5.24 and 5.25), this switch avoids a long feedback loop from S, and thus improves the stability of the entire system. This may be another reason for the use of branched systems such as those in Figures 5.24 and 5.25. As shown previously under Branch-Point Sensitivities, a large flux in the other branch, L, would be needed to ensure a sensitive response of K and/or a stabilized concentration of S; but this may be an acceptable price for the ensuing stability of the system.

TWO OTHER IMPORTANT APPROACHES TO METABOLIC CONTROL AND REGULATION

Throughout this chapter we have used our flux-oriented theory (FOT), which we believe to be the simplest and most flexible way of discussing and analyzing metabolic regulatory systems. However, there are two other important alternative approaches: biochemical systems theory, developed by Savageau and coworkers (139, 141, 159, 175), and metabolic control theory (or control analysis), developed by Kacser and other European scientists (31, 73, 77, 79, 182). Each of these will now be considered in turn.

Biochemical Systems Theory (BST)

This is the oldest of the three approaches considered in this chapter and, as in FOT, rate equations are set up

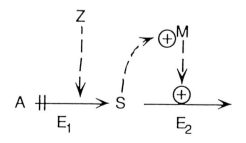

FIG. 5.31. System with indirect removal of an intermediate. The indirect communication, $S \sim M \sim E_2$, introduces a delay between the production of S (by E_1) and its removal (by E_2).

for each component reaction and a set of simultaneous equations solved. However, in BST, all the rate equations are described by finite power equations. Thus, for the following reaction, using the terminology of BST,

$$X_1 \xrightarrow[v_2]{(g_{21})\ (g_{22})} X_2$$

the rate of X_2 formation, v_2, is given by the equation,

$$v_2 = dX_2/dt = \alpha_2 \cdot X_1^{g_{21}} \cdot X_2^{g_{22}} \qquad (5\text{-}212)$$

α_2 is the X_2 formation rate constant. (Degradation rate constants are indicated by the symbol β.) g_{21} and g_{22} are formation indices (as opposed to degradation indices which are denoted by h_{ij}), with the first and second subscripts denoting the number of the reaction and effector, respectively. These indices are equivalent to the component sensitivities of FOT. However, the major difference between BST and FOT is that equation 5-212 describes a noninfinitesimal response, whereas a power equation in FOT represents an infinitesimal response. Consequently, the value of the rate constant, α, is needed in BST, but not in FOT.

As a simple example of the use of S-System BST, consider the system in Figure 5.33, which is the BST equivalent of Figure 5.23. One first writes out a node equation for the dependent variable X_2 in terms of the fluxes that contribute to its steady state value,

$$dX_2/dt = v_1 - (v_2 + v_3) \qquad (5\text{-}213)$$

These elementary fluxes are aggregated into net fluxes for synthesis (V_{+2}) and for degradation (V_{-2}):

$$dX_2/dt = V_{+2} - V_{+2} \qquad (5\text{-}214)$$

where,

$$V_{+2} = v_1 \qquad (5\text{-}215)$$

and,

$$V_{-2} = v_2 + v_3 \qquad (5\text{-}216)$$

Each aggregate rate law is then replaced by the corresponding power law representation as follows,

$$V_{+2} = \alpha_2 \cdot X_1^{g_{21}} \cdot X_2^{g_{22}} \qquad (5\text{-}217)$$

$$V_{-2} = \beta_2 \cdot X_2^{h_{22}} \qquad (5\text{-}218)$$

Thus,

$$dX_2/dt = \alpha_2 \cdot X_1^{g_{21}} \cdot X_2^{g_{22}} - \beta_2 \cdot X_2^{h_{22}} \qquad (5\text{-}219)$$

At a steady state, the time derivative can be set to zero, so that equation 5-219 becomes,

$$\alpha_2 \cdot X_1^{g_{21}} \cdot X_2^{g_{22}} = \beta_2 \cdot X_2^{h_{22}} \qquad (5\text{-}220)$$

By taking logarithms of both sides of equation 5-220, $\ln X_2$ can be expressed in terms of the external regulator X_1 as,

$$\ln X_2 = \frac{(\ln\alpha_2 - \ln\beta_2) + g_{21} \cdot \ln X_1}{(h_{22} - g_{22})} \qquad (5\text{-}221)$$

This equation describes a finite change of X_2. However, BST does consider infinitesimals; and in its terms "logarithmic gain" (L), defined for the effect of X_1 on X_2, as,

$$L(X_2, X_1) = d\ln X_2/d\ln X_1 \qquad (5\text{-}222)$$

is the same as the net sensitivity of X_2 to X_1 ($s_{X_1}^{X_2}$) in FOT.

Although the analysis of BST is concentrated on determining the response of the internal effectors to external regulation, the response of the fluxes (that is, $v_1 - v_3$ in Fig. 5.33) can be calculated by a procedure analogous to that in FOT (equation 5-148) (158). Thus this approach, as with FOT, calculates responses directly and systematically from the rate equations of the system.

However, there are two problems with the use of BST compared with FOT. First, there is the need to aggregate fluxes at a branch point, as in the above example. With simple systems, this is not too difficult (175), but it becomes more so as the complexity increases. In FOT, such aggregations are either not required (as, for example, in the solution of equation 5-141) or are included automatically as a result of matrix partitioning (equations 5-142 to 5-144). Indeed, matrix partitioning in FOT results in a matrix equation (equation 5-143) which is an infinitesimally based equivalent of the S-matrix of BST.

The second problem with BST is its use of finite, rather than infinitesimal, power equations. Although Savageau and co-workers have provided much evidence that this approximation is satisfactory in many cases (139, 158, 175), it can never be guaranteed to apply in all cases. This is because the index g is a mean-value relative-change sensitivity (equation 5-182) and its

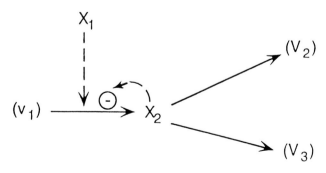

FIG. 5.33. A branched system in BST formulation.

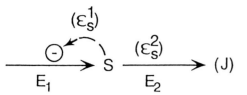

FIG. 5.34. Linear flux in MCT formulation. A comparison with Figure 5.11 shows that $\epsilon_S^1 = \alpha$; $\epsilon_S^2 = \beta$.

value therefore depends on the way in which other effectors of the reaction change during the approach to a steady state. If these changes are ever large and/or the reaction is very sensitive to them, the actual value of the index may be significantly different from that assumed in the finite approximation. Since (as in FOT) finite responses can always be calculated from infinitesimally based equations using a technique such as finite decomposition this will in general provide a more accurate method than the assumption of finite power equations. Moreover, FOT is also a simpler approach in that it does not require the values of the rate constants.

Metabolic Control Theory (MCT)

Both BST and FOT are approaches which are directly concerned with the regulation of metabolic systems, as defined earlier under Control Versus Regulation: that is, the response of the system variables (fluxes and intermediate concentrations) to changes in the concentrations of external regulators. However, MCT is predominantly concerned with attempting to measure the control of the variables: that is, the relative importance of those factors that determine the steady state values of the variables. It therefore considers regulation only indirectly.

As shown earlier under Control Versus Regulation, the distinction between control and regulation can be quite useful, and we have adopted it in this chapter. To recapitulate, a variable is controlled by all the reactions and regulators which *can* communicate with it, whereas it is regulated only by those reactions and regulators which *do* communicate with it. Thus, in Figure 5.11, S is controlled by Z, E_1, and E_2, but is regulated only by Z, since E_2 has no regulators. In other words, of the two potential regulatory sequences, $E_1 \sim S \sim E_2$ and $E_2 \sim S \sim E_1$, only the former is used. The latter is silent in this system.

MCT takes this argument a stage further, by assuming that these two regulatory sequences actually operate in the steady state to constrain, and hence control, S. This control is then quantified as the effect of infinitesimal changes of the enzyme activities on the vari-

ables (73, 79). Thus, for metabolite S in Figure 5.11 we have, using the core sensitivities of FOT,

$$d\ln S/d\ln E_1 = s_1^S \qquad (5-223)$$

$$d\ln S/d\ln E_2 = s_2^S \qquad (5-224)$$

Simultaneous infinitesimal changes of E_1 and E_2 therefore produce a change in S given by the addition rule (equation 5-16); so that,

$$d\ln S = (d\ln S)_1 + (d\ln S)_2 \qquad (5-225)$$

$$= s_1^S \cdot d\ln E_1 + s_2^S \cdot d\ln E_2 \qquad (5-226)$$

Control is then defined in MCT, and hence quantified, as the separate contributions of each reaction to $d\ln S$ in equation 5-226. Thus the control of S by E_1 is defined and measured by the function $s_1^S \cdot d\ln E_1$; and the control of S by E_2 is defined and measured by the function $s_2^S \cdot d\ln E_2$. Therefore, for the same infinitesimal relative changes in the enzyme activities (that is, when $d\ln E_1 = d\ln E_2$) the control of S by E_1 and E_2 is defined and measured by the values of s_1^S and s_2^S, respectively. Consequently, in MCT, these functions are referred to as control coefficients, denoted by the symbol C. Thus the above core sensitivities are denoted as C_1^S and C_2^S.

Similar conclusions apply to any variable. Thus for flux, J, the core sensitivities, s_1^J and s_2^J are denoted as C_1^J and C_2^J, respectively, and defined as measuring the control of J by E_1 and E_2, respectively.

Since core sensitivities, and thus control coefficients, can often be measured experimentally (for example, by using inhibitor titrations [48, 65, 182], or a top-down approach [13]), MCT appears to provide an attractive experimental approach for measuring control in metabolic systems. For example, by measuring C_1^J and C_2^J under several different physiological conditions, it is possible to examine whether control shifts from one site to another. This determination and interpretation of control coefficients has therefore become widespread in recent years (55, 64, 65, 91, 126, 137, 153, 155).

Unfortunately, although this definition of control gives a mathematically consistent theory, it is almost certainly meaningless physiologically. This is because control is defined in terms of the operation of regulatory sequences (Equation 5-226) and there seems to be no justification for this. Qualitatively, it is satisfactory, because a zero control coefficient indicates no communication from the enzyme activity to the variable; the enzyme does not control that variable. However, it is difficult to understand why, for example, the effect of an infinitesimal change of E_2 on S (Fig. 5.11) should measure the contribution of the regulatory sequence,

$E_2 \sim S \sim E_1$, in the steady state, when no such changes are occurring!

Indeed, the control of S by E_2 (which, in Fig. 5.11, has no external regulators) results from the kinetic response of E_2 to S (β), after Z changes, and not the response of S to E_2, as implied by the regulatory sequence defining the control of S by E_2. Defining the control of S by E_2 in terms of an external change of E_2 on S is therefore unphysiological and potentially misleading.

These problems can be highlighted more directly by evaluating the steady state concentration of S (Fig. 5.11). Since this is a finite quantity, we shall use finite power equations (as in BST) in the form of equation 5–182; so that,

$$v_1 = k_1 \cdot S^{\alpha_m} \qquad (5\text{--}227a)$$

$$v_2 = k_2 \cdot S^{\beta_m} \qquad (5\text{--}227b)$$

where the index denotes a mean relative-change sensitivity. k_1 and k_2 are rate constants (with k_1 incorporating the effects of regulator Z) and, since this is a finite equation, their values cannot be ignored.

In the steady state, when $v_1 = v_2$,

$$k_1 \cdot S^{\alpha_m} = k_2 \cdot S^{\beta_m} \qquad (5\text{--}227c)$$

so that, taking logarithms of both sides and solving for S,

$$\ln S = \frac{1}{\beta_m - \alpha_m} \cdot \ln k_1 - \frac{1}{\beta_m - \alpha_m} \cdot \ln k_2 \qquad (5\text{--}227d)$$

The functions on the right-hand side of equation 5–227d are the contributions of each reaction to $\ln S$ and therefore represent the physiological control of S by E_1 and E_2. Each of these contributions has a factor, $\pm 1/(\beta_m - \alpha_m)$, which is the same function as the corresponding control coefficient, $\pm(1/(\beta - \alpha)$. However, equation 5–227d shows that, for the control of S, mean-value sensitivities and not the infinitesimal ones need to be used. As shown earlier, these are path dependent and cannot be derived purely from steady state analyses.

Furthermore, the $\ln k$ terms in equation 5–227d show that, to evaluate the physiological control of S by E_1 and E_2, the values of the rate constants (that is, the residual activity when the terms in S have been divided out of v) are also needed; so that even mean-value based control coefficients are not sufficient.

Most importantly, as would be expected for the control of a steady state variable (see earlier), the functions in equation 5–227d do not refer to changes,

and therefore apply without any problems to silent control sites. Thus, the contribution of E_2 to $\ln S$, and hence the physiological control of S in Figure 5–11, is described by the function, $-[1/(\beta_m - \alpha_m)] \cdot \ln k_2$, which does not involve or imply any changes of E_2.

Consequently, the definition of control adopted in MCT (equation 5–226), involving infinitesimal changes, is not satisfactory for measuring physiological control. This, in turn, means that the numerical values of control coefficients are only qualitative reflections of the control potential of each site; they do not measure physiological control. For these reasons, we now prefer to use the term core sensitivity in place of control coefficient.

Although the terms on the right-hand side of equation 5–227d suggest a physiological measurement of control, this would not be practical because, as a result of their path dependence (see earlier under Calculating the Relative Importance of Regulators and Regulatory Sequences During a Finite Response: Mean-Value Sensitivities), calculating the mean-value sensitivities would require a knowledge of exactly how the steady-state value of S was produced, starting from zero! However, the factors controlling a variable are always included in the sensitivity of that variable to external regulators. For example, the net sensitivity of S to Z (Fig. 5.11) is given by the equation,

$$s_Z^S = \frac{\omega}{\alpha + \beta}$$

and the control of S by E_2 is represented by the β term in the denominator. Since such regulatory sensitivities automatically include the control factors, there is no need for a separate measure of control anyway; and we would suggest that this term should only be used descriptively.

Another problem with MCT is its quantitation, again stemming from its preoccupation with measuring control. Unlike FOT or BST, it is not founded directly on rate equations, but indirectly via the use of control theorems, of which there are two main types: summation and connectivity theorems (134, 918, 197). Thus, referring to Figure 5.34, the summation theorems are, using MCT nomenclature,

$$C_1^J + C_2^J = 1 \qquad (5\text{--}228)$$

$$C_1^S + C_2^S = 0 \qquad (5\text{--}229)$$

and the connectivity theorems are,

$$\epsilon_S^1 \cdot C_1^J + \epsilon_S^2 \cdot C_2^J = 0 \qquad (5\text{--}230)$$

$$\epsilon_S^1 \cdot C_1^S + \epsilon_S^2 \cdot C_2^S = -1 \qquad (5\text{--}231)$$

where ϵ is termed an elasticity coefficient and is equivalent to a component or intrinsic sensitivity in FOT. (No distinction has so far been made between an intrinsic and a component sensitivity in MCT.)

Equations 5–228 to 5–231 can be solved by using a matrix method which generates the control coefficients in terms of the elasticity coefficients. For the system in Figure 5.34, the matrix is given in equation 5–232,

$$\begin{pmatrix} 1 & 1 \\ -\epsilon_S^1 & -\epsilon_S^2 \end{pmatrix} \begin{pmatrix} C_1^J & C_1^S \\ C_2^J & C_2^S \end{pmatrix} = \begin{pmatrix} 1 & 0 \\ 0 & 1 \end{pmatrix} \quad (5\text{–}232)$$

from which it can be seen that the square matrix on the left-hand side is the transpose of the corresponding matrix in FOT (that is, the rows and columns of matrix N are interchanged).

This type of matrix, based on theorems rather than explicit rate equations, is quite acceptable for very simple fluxes since the theorems can usually be written by inspection. However, for branched systems, such as Figure 5.23, it becomes much more difficult because, in addition to the summation and connectivity theorems, additional theorems are needed for the branch points. Although such theorems can, in theory, be generated for systems of any complexity (215), they soon become complex. Moreover, special theorems are also required to incorporate conserved pools (169) and interconversion cycles (185), and these also soon become quite complicated.

Consequently, a quantitative analysis based on theorems soon becomes cumbersome, far more so than the aggregation processes needed for BST. For these and other reasons we prefer and recommend analyses based on the direct and straightforward use of rate and conservation equations, for example equation 5–94.

Finally, a great deal of confusion has arisen because MCT is often presented and used as if it applied directly to physiological regulation, but it does not. Thus, the numerical values of control coefficients do not indicate the relative importance of potential regulatory sites in vivo; since, amongst other things, the site may not be used for regulation (for example, E_2 in Fig. 5.11), and no account is taken of the interaction of regulators, such as Z (Fig. 5.11). Also, infinitesimally based sensitivities do not measure the strengths of the corresponding physiological responses. Indeed, we wonder how many experimental investigators would have adopted MCT so readily had they realized that it does not apply directly to regulation or even measure control in a physiological sense.

Of course, since the infinitesimal sensitivities of MCT are often equivalent to those of FOT (for example,

core sensitivities and control coefficients, component sensitivities and elasticity coefficients), the MCT-based functions can be applied to regulation in the same way as those of FOT (134); and this has been advocated. However, because of the unsatisfactory basis of MCT, both qualitatively and quantitatively, its nomenclature is unnecessarily complicated (and, in the case of control coefficients, seriously misleading). Neither is it as flexible as the relatively simple 's'-based system adopted in this chapter, a summary of which is given in Table 5.3. For these reasons we do not recommend that MCT functions should be used to describe metabolic regulatory systems.

A Note on Control Theorems. The reason why control theorems are used in MCT is that they are regarded as fundamental relationships (31). However, this is not the case, and several workers (20, 21, 31, 142, 157) have shown that they are derived relationships based on standard matrix identities which, in FOT nomenclature are as follows,

$$N \cdot N^{-1} = I \quad (5\text{–}233)$$

$$N^{-1} \cdot N = I \quad (5\text{–}234)$$

where N is the matrix in equation 5–95 and N^{-1} is its inverse. I is the corresponding identity matrix.

For the system in Figure 5.34, using the functions of MCT,

$$N = \begin{pmatrix} 1 & -\epsilon_S^1 \\ 1 & -\epsilon_S^2 \end{pmatrix} \quad (5\text{–}235)$$

and, by analogy with equation 5–100 (replacing 's' by 'C'), we can write N^{-1} as,

$$N^{-1} = \begin{pmatrix} C_1^J & C_2^J \\ C_1^S & C_2^S \end{pmatrix} \quad (5\text{–}236)$$

Equation 5–233 then becomes the matrix equation shown in equation 5–237 and,

$$\begin{pmatrix} 1 & -\epsilon_S^1 \\ 1 & -\epsilon_S^2 \end{pmatrix} \begin{pmatrix} C_1^J & C_2^J \\ C_1^S & C_2^S \end{pmatrix} = \begin{pmatrix} 1 & 0 \\ 0 & 1 \end{pmatrix} \quad (5\text{–}237)$$

after multiplying out the matrices on the left-hand side, equation 5–238.

$$\begin{pmatrix} C_1^J - \epsilon_S^1 \cdot C_1^S & C_2^J - \epsilon_S^1 \cdot C_2^S \\ C_1^J - \epsilon_S^2 \cdot C_1^S & C_2^J - \epsilon_S^2 \cdot C_2^S \end{pmatrix} = \begin{pmatrix} 1 & 0 \\ 0 & 1 \end{pmatrix} \quad (5\text{–}238)$$

TABLE 5.3. 's'-Based Nomenclatue of FOT *

Symbol	Name	Definition
$s_j^{variable}$	Core sensitivity	Net (relative-change) sensitivity of variable to reaction E_j
$s_{j(P)}^{variable}$	Pseudo core sensitivity	Net (relative-change) sensitivity of variable to E_j, mediated by the specific partially external regulator, P
$s_X^{variable}$	Net sensitivity	Net (relative-change) sensitivity of variable to regulator X
$s_S^{E_j}$ or $s_S^{v_j}$ or s_S^j	Component sensitivity	(Relative-change) sensitivity of reaction E_j to S at constant concentrations of all the other effectors of E_j, *in the system as written*
$s_{j(S)}^{E_j}$	Intrinsic sensitivity	(Relative-change) sensitivity of reaction E_j to S at constant concentrations of all other possible effectors of E_j, whether written out in the system or not

* Corresponding absolute-change sensitivities are denoted by the symbol *a* in place of *s*. As in the other approaches, it would be possible to simplify the nomenclature of FOT further by referring to all fluxes as F_1 to F_n, all internal effectors as S_1 to S_n, all external effectors as X_1 to X_n and all partially external effectors as P_1 to P_n. However, we have retained the familiar notations here.

By equating corresponding elements on both sides of equation 5–238, it can be seen to have generated four theorems relating to the same enzyme activity, a characteristic of theorems generated from the identity, $N \cdot N^{-1} = 1$. For example, one such theorem, relating to E_1, is,

$$C_1^J - \epsilon_S^1 \cdot C_1^S = 1 \qquad (5\text{–}239)$$

or, in FOT nomenclature,

$$s_1^J - s_S^1 s_1^S = 1 \qquad (5\text{–}240)$$

Such relationships have a practical use (64, 78). Thus, if the activity of E_1 is varied by a specific activator or inhibitor and the (infinitesimal) responses of J and S extrapolated from the measurements (48, 65, 166), the resulting values of C_1^J (s_1^J) and C_1^S (s_1^S) will enable the value of ϵ_S^1 (s_S^1) to be calculated from equation 5–240. This value can then be compared with that predicted from a kinetic model of the system (see earlier under Expressing Intrinsic and Component Sensitivities in Terms of Reaction Kinetics), thereby testing the assumptions of the model. Indeed, we believe that the only practical value of numerically evaluated sensitivities (especially control coefficients) is their use in setting up and testing proposed kinetic models.

When the right-hand side of the theorem is 1 (as in equation 5–239), it is necessary to determine the values of the core sensitivities or control coefficients; and, if this involves the use of a specific (or hopefully specific) inhibitor, the kinetic response of the enzyme to the inhibitor must be known (48, 65). However, if the right-hand side is zero, as, for example, with the following theorem from equation 5–238,

$$C_1^J - \epsilon_S^2 \cdot C_1^S = 0$$

the kinetic response to the inhibitor need not be known. This is because,

$$\epsilon_S^2 = C_1^J / C_1^S$$
$$= (\phi \cdot C_1^J)/(\phi \cdot C_1^S)$$

where ϕ is the intrinsic sensitivity (or elasticity coefficient) of E_1 to the inhibitor, I. Thus, by the product rule (equation 5–10),

$$\epsilon_S^2 = s_I^J / s_I^S$$

Therefore ϵ_S^2 (or, in FOT nomenclature, s_S^2) can be determined directly from the infinitesimal responses of J and S to I without having to know the value of ϕ.

In contrast, theorems based on the other identity, $N^{-1} \cdot N = 1$ (equation 5–234), relate to the same variable (that is, S or J). Thus, for the system in Figure 5.34, equation 5–234 becomes equation 5–241,

$$\begin{pmatrix} C_1^J & C_2^J \\ C_1^S & C_2^S \end{pmatrix} \begin{pmatrix} 1 & -\epsilon_S^1 \\ 1 & -\epsilon_S^2 \end{pmatrix} = \begin{pmatrix} 1 & 0 \\ 0 & 1 \end{pmatrix} \quad (5{-}241)$$

which, after multiplying out the matrices on the left-hand side, becomes,

$$\begin{pmatrix} (C_1^J + C_2^J) & -(\epsilon_S^1 \cdot C_1^J + \epsilon_S^2 \cdot C_2^J) \\ (C_1^S + C_2^S) & -(\epsilon_S^1 \cdot C_1^S + \epsilon_S^2 \cdot C_2^S) \end{pmatrix} = \begin{pmatrix} 1 & 0 \\ 0 & 1 \end{pmatrix} \quad (5{-}242)$$

Equating corresponding elements on both sides of equation 5–242 gives the summation and connectivity theorems (equations 5–228–5–231), each of which relates to the same variable.

This analysis shows that control theorems are certainly not fundamental: in fact they are a very indirect and inefficient way of generating net sensitivities, compared with the direct analyses of FOT and BST. Moreover, these are occasions when the summation theorem cannot be applied easily, for example when a system contains partially external effectors; if one enzyme catalyses more than one reaction of a pathway, some of the connectivity theorems do not apply either (19). Such theorems should therefore be regarded as providing potentially useful algebraic short-cuts, rather than as fundamental relationships of metabolic control.

We believe that the real fundamental relationships of metabolic control and regulation are those represented by the product and addition rules (equations 5–10 and 5–16). As stated earlier, all subsequent quantitation is derived from them; and therefore so are the theorems. Indeed, it is quite easy to see this with the connectivity theorem. Thus, in Figure 5.34, the net communication between the internal effector, S, and the flux, J, results from the sequences, $S \frown E_1 \frown J$ and $S \frown E_2 \frown J$. By applying the product rule, the overall sensitivities via these routes are, $\epsilon_S^1 \cdot C_1^J$ and $\epsilon_S^2 \cdot C_2^J$, respectively. The overall sensitivity of J to S is therefore, by the addition rule, $\epsilon_S^1 \cdot C_1^J + \epsilon_S^2 \cdot C_2^J$. However, this must be zero, since S is an internal effector and therefore cannot regulate J. The result is the connectivity theorem in equation 5–230. Similarly, S can be regarded as being able to communicate with itself via three sequences, $S \frown E_1 \frown S$, $S \frown E_2 \frown S$ and $S \frown S$, whose overall sensitivities are, by the product rule, $\epsilon_S^1 \cdot C_1^S$, $\epsilon_S^2 \cdot C_2^S$ and 1, respectively. Therefore, by the addition rule, the net sensitivity of S to S is equal to $\epsilon_S^1 \cdot C_1^S + \epsilon_S^2 \cdot C_2^S + 1$, which must also be zero, or S would change spontaneously! The result is the connectivity theorem in equation 5–231.

SUMMARY AND FUTURE PROSPECTS

The usefulness of the approach described in this chapter (FOT) is that it provides a rigorous description of metabolic regulation at all levels. These range from a qualitative identification of potential control sites (by using graph theory or a qualitative diagonalization of the matrix equation) to a detailed numerical calculation of finite responses (via finite decomposition) and the relative importance of control sites in vivo.

The approach can also be extended to deal with non-steady states—an area which has only been outlined briefly here and where much more work is still needed.

Given the current inaccuracy of much kinetic data in situ, perhaps the most useful aspect of this (and other) approaches is the ability to derive general algebraic equations for complex responses in situ. As shown throughout this chapter, these equations can give useful insights into the role of control mechanisms embedded in a pathway when their regulatory properties may be very different from those in isolation. This approach, based on hypothetical systems, may also be extended to include numerical simulations (for example, refs. 32, 70, 73), again providing useful information about control systems in general.

An important area of research where this type of approach could be valuable in organizing the data currently available is that involving the interactions and interrelations between inositol phosphates and other messenger systems (for example, Ca^{2+} and cyclic AMP). The inositol phosphates, in particular, participate in many cyclic fluxes; and, from the discussions earlier under Effect of a Substrate Cycle In Situ on the Net Sensitivity to External Regulators and Regulatory Analysis of a Branched System: Matrix Partitioning, these could provide important regulatory mechanisms, in both steady and non-steady states.

When dealing with specific pathways (for example, glycolysis) the situation becomes much more complicated because a satisfactory kinetic control model is required. This, in turn, requires more accurate kinetic data in situ which, as stated throughout this chapter, are difficult to obtain with many pathways. Nevertheless, a surprising amount of information can be obtained from relatively simple models. Indeed, much of our knowledge of control systems has been derived in this way.

Even with simple models of pathways, it must be stressed that any numerical predictions or conclusions will depend on the validity of the model, and hence the reliability of the kinetic data. Independent experimental testing of the models is therefore essential. One way of doing this is to compare predicted responses (generated by either a time course or, if only steady states are involved, by finite decomposition) with those measured experimentally. Another way is to compare experimentally determined core sensitivities (that is, control coefficients) with those predicted from the kinetics of the model. (It must be reemphasized that numerical values of sensitivities [whether they are core or component sensitivities] have little if any physiological meaning and should be used only as a means of testing a proposed control model. In particular, they should never be used as measures of control.)

Another approach, which will almost certainly become more important in the near future, is the application of recombinant DNA technology to metabolic control systems (120). This may help not only to identify metabolic communications, and hence potential control sites in vivo, but also to provide an experimental test of some or all of the quantitative predictions of the control model.

In our opinion, the experimental approach to metabolic regulation should initially consist of identifying the basic kinetics of the component reactions of the flux (pathway) to identify the internal communications that make up potential regulatory sequences. This involves identifying any flux-generating steps, reversible and irreversible reactions, significant product inhibitions, and external (allosteric) interactions. From this, a predominantly qualitative model may be built, allowing the identification of potential regulatory sites. The existence of these sites may then be tested experimentally, for example by seeing whether a specific activator or inhibitor of the proposed regulatory site does indeed change the flux or whatever variable is under consideration. This type of experimental approach has recently been summarized in a six-point strategy (50).

The next steps would be to refine the kinetic data to allow for more detailed quantitative investigations of the model. Even then, it must be remembered that the detailed predictive powers of the model may be very limited. For example, no kinetic model of glycolysis before 1980 indicated the presence of the important regulator fructose-2,6-bisphosphate! However, for many practical purposes, even a relatively simple model may often contain enough information to enable an approximate, but satisfactory, understanding of how and when a pathway is regulated in situ; and how strong the regulatory interactions are likely to be. Such information is obtained quite readily from the model, by using the approach described in this chapter.

We conclude this chapter by considering an important theoretical and practical problem faced by all the main quantitative approaches to metabolic regulation, that is, FOT, BST, or MCT. All these are based on the concepts of reaction rates and concentrations; in other words, they assume that the law of mass-action, or its equivalent, applies in vivo. This creates two main problems. First, as mentioned above, it is very difficult to measure reaction rates, enzyme kinetic responses and, especially, concentrations in cells so that kinetic models of metabolic regulatory systems have to be deduced from quite inaccurate data, making their detailed predictive powers very limited. However, as techniques improve, as they surely will, more accurate kinetic information should become available.

The second problem, which is more abstract and potentially more serious, questions whether the law of mass-action really applies in living cells, and, in particular, whether the term concentration has any meaning (67, 181). This is because the size of many intracellular compartments is so small that they may contain relatively few molecules. Since concentration is a statistical effect of a large number of molecules, it may not be applicable in such small compartments, and other functions may be required. Some theoretical work has been done on this problem, notably several attempts to apply the principles of irreversible or non-equilibrium thermodynamics (86, 164, 165, 180), but they have so far given little if any theoretical insight into the subject, nor are they yet supported by conclusive experimental evidence. There are also strong arguments against any physiological significance of thermodynamic, as opposed to kinetic, coupling (82). Therefore, despite criticisms that biochemists are using 19th century methods of analysis (see ref. 179), it seems reasonable to persist with the law of mass-action, or its equivalents, until there is definite experimental evidence that it cannot be used.

However, one area where there will almost certainly be potential problems with the classical analysis is in the interconversion of small sets of cellular proteins, for example G-proteins and those interacting with DNA and RNA. In such assemblies there may be only a few molecules of protein and this will make it difficult to apply the classical concepts of concentration, except in certain circumstances (67).

To see what may happen in such a situation, let us consider the simple interconversion cycle in Figure 5.35a. Here a single protein molecule alternates between the a and b forms; for example, a phosphate group may be added and then removed continuously to the molecule. If this cyclical addition and removal of phosphate is low, this system would have to be analyzed, not by classical rate equations, but by stochastic techniques, involving random walks (33). Needless to say, such an analysis would be extremely complicated.

However, if the rate of addition and removal of the

a

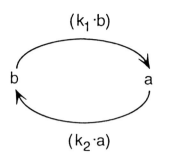

$(k_1 \cdot b)$

b   a

$(k_2 \cdot a)$

b

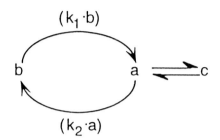

$(k_1 \cdot b)$

b   a ⇌ c

$(k_2 \cdot a)$

FIG. 5.35. Interconversion cycle with a single molecule of protein. a, b, and c refer to the same molecule and hence cannot coexist. k_1 and k_2 are classical rate constants that would give the observed rates as a function of the fractional times spent as a or b. The conversion of a into c takes a time τ.

phosphate group is rapid, there is then a definite average time spent in each form, and this is equivalent to the classical concentration: the rate equations, equations 5–67 and 5–68, can then be applied. To see this, let us assume that Figure 5.35a represents a classical system consisting of a large number of molecules. Each rate equation would be characterized by a rate constant, k_1 and k_2, and the average times spent in each form (121) would be $1/k_1$ (for the b form) and $1/k_2$ (for the a form). Thus the fraction of the total time $(1/k_1 + 1/k_2)$ spent in the a form, τ_a, would be given by the equation,

$$\tau_a = \frac{1/k_2}{1/k_1 + 1/k_2} \qquad (5\text{–}243)$$

$$= \frac{k_1}{k_1 + k_2} \qquad (5\text{–}244)$$

If the rates of addition and removal of phosphate are rapid, the fraction of the total time spent as a, τ_a, will be a defined (that is, non-fluctuating) function for a system containing just one molecule of protein; equation 5–244 shows that it can be expressed as a ratio of (in this case imaginary) rate constants. More-

over, from classical kinetics, if $(a + b)$ is denoted by S, so that $b = S - a$, in the steady state,

$$k_1 \cdot (S - a) = k_2 \cdot a \qquad (5\text{–}245)$$

whence,

$$a/S = k_1/(k_1 + k_2) \qquad (5\text{–}246)$$

Consequently, from equation 5–244,

$$\tau_a = a/S \qquad (5\text{–}247)$$

Therefore, if the rate of cycling between a and b is sufficient to give a defined average time spent in each form, the fraction of the total time spent as a is proportional to the classical concentration of a, and can be used in its place for such small assemblies.

Indeed, kinetic analyses, based on time rather than rate constants, have been advocated by some workers (99, 121). Although they can be quite complicated compared with classical kinetics, since they involve probability functions, they may be more applicable to systems in which time is an important parameter (for example, the movement of RNA along ribosomes, and the conformational changes that occur in proteins).

Let us now consider what may happen in the single-molecule protein cycle (Fig. 5.35) when the rates of interconversion are increased in a coordinate fashion so that k_1/k_2 remains constant. In the classical situation, such a coordinated increase in the rates does not change the ratio a/b and, earlier under Modification of the Speed of Response of an Interconversion Cycle, was proposed as a means of alerting the cycle to obtain a fast response during a subsequent stimulus. Moreover, if a is in equilibrium with a further form c, (Fig. 5.36b) the amount of this form is also unaffected by a coordinate increase in the rate constants.

However, when the system consists of a single protein molecule, alternating now between forms a, b, and c (Fig. 5.35b), the situation following a coordinated increase in the rate constants, k_1 and k_2, may be very different. As the rate of interconversion of the b and a forms is increased, the protein molecule spends less and less time in each form (although the fraction of the total time spent in each form, and hence the concentration of a and b will not be affected: equation 5–246). If it takes a fixed time, τ, to produce the conformational change converting a to c, then, as the rate of cycling between a and b increases, there will be a point at which the a form lasts for a time less than τ. If any conformational changes of a (that is, those which would have eventually resulted in c) are reversed completely when it is converted to the b form, the c form will no longer be formed. Therefore, its effective concentration will fall to zero at a critical coordinated rate of cycling. If c is the enzymically active species, it means that the activity will be lost at some critical rate of cycling, even though the b/a ratio is unaffected. Such an effect would not be predicted from a classical kinetic analysis, based on an assembly containing a large number of molecules, and, if it can occur physiologically, it may play an important role in controlling enzyme activity. In any case, it is a salutary reminder that, as far as the analysis of metabolic regulatory systems is concerned, there is still much to be learned.

Finally, one of the most important principles to have emerged from studying metabolic regulatory systems is that cells are prepared to use energy to provide regulatory mechanisms in the form of substrate and interconversion cycles. There is also much excess capacity in that most enzymes seldom become saturated with their substrates and hence operate well below their maximum capacities. Thus, metabolism would probably be regarded as being quite inefficient by an economist, especially one of a monetarist disposition! However, as we have seen, such an opinion would be most short-sighted because the result of all this "inefficiency" is a stable yet flexible system that responds to changes in the environment without being at the mercy of them. Consequently, energy that is not used for seemingly useful purposes is not necessarily wasted. Moreover, treating such energy usage as an unacceptable wastage, and taking steps to reduce or eliminate it (from an animal or an economic system), may produce a system so inflexible that it is damaged or destroyed by even slightly adverse changes in its environment when a more expensive, but flexible, system would have survived.

APPENDIX 1. FLUX-GENERATING STEPS REAPPRAISED

Let the system in Figure 5.1 represent a whole-body pathway which originates from a defined initial substrate (A), which will usually be a storage form of a fuel like glycogen, triacylglycerol, or protein. Reaction E_1 may also represent the uptake of nutrients from the gut.

As stated in the text (see the section titled Flux-Generating Steps), if the rate of E_1 (v_1) responds to the continuously decreasing concentration of A, this rate, and hence that of the system flux, J, will decline continuously, making a steady state impossible to attain. However, for a certain time (for example, before the rate has fallen by more than approx. 10%), the system will be in an approximately steady state.

Assuming a first-order response of E_1 to A, so that $v_1 = a \cdot A$, the time taken to reduce the fluxes of utilization of the major fuels by 10% in man can be calculated from their initial contents and rates of utilization (Ta-

ble 5.4). It can be seen that these times vary widely: from a few seconds for muscle glycogen, to several days for the body stores of protein and fat.

Thus, when considering the short-term regulation of the systems utilizing fat or protein, they can be assumed to be in a steady state even if the response of E_1 to A is first order.

At the other extreme, under first-order conditions, the steady state for muscle glycogenolysis would last for only a few seconds. Such a rapid breakdown of the steady state would cause physiological problems, because, in muscle, a near-maximum rate of phosphorylase activity (equivalent to E_1) is needed to supply sufficient ATP to support intense activity, such as sprinting (114), and a continuous reduction in the steady state flux would seriously interfere with this. Therefore, muscle glycogenolysis needs to be effectively insulated from the decrease in glycogen, for example, by saturating phosphorylase with glycogen. There is then no effect on the flux as the concentration of glycogen falls until it has fallen to such an extent that it is practically all used up. At this point, a near-constant rate of glycogenolysis will have been maintained for approx 100/60, i.e. 1.7 min (Table 5.4), by which time either the intense activity has ceased or the ATP requirements have been met by other fuels (for example, blood glucose or fatty acids).

Saturation of phosphorylase with glycogen (deduced by comparing the glycogen "concentration" with the K_m value of phosphorylase [114]) also seems to occur in the liver and, as in muscle, enables the whole-body flux of liver glycogen to glucose to be unaffected by the falling concentration of glycogen in the liver (as during short-term fasting). The time scale in liver is longer than in muscle (Table 5.4) and saturation enables the liver glycogen stores to maintain a near-constant production of glucose from glycogen for approx 97/5.5, i.e. 18 h (Table 5.4). This is sufficient to provide glucose during a short-term (overnight) fast. In longer-term starvation, the glucose requirements are reduced and then are supplied mainly by protein for the first few days and, for long-term fasting, by glycerol produced from the hydrolysis of triacylglycerol.

Similarly, for fat utilization, "saturation" of the lipase in adipose tissue with triacylglycerol enables fat to supply the energy requirements of starvation at a near constant rate for approx. 337,000/10050, i.e. 34 days (Table 5.4).

Another example of a saturated reaction serving as an important flux-generating step is the release of glutamine from skeletal muscle. This amino acid, which may be derived from protein breakdown as well as de novo synthesis from glutamate and glutamate precursors, is transported out of the muscle via an irreversible

TABLE 5.4. *Rates of Fuel Utilization by Human Tissues In Vivo* *

Fuel	Muscle Glycogen	Liver Glycogen	Body Triacylglycerol	Body Protein
Content (A)	100^a	97^b	$337{,}000^c$	$150{,}000^c$
Rate of use (v_1)	60^d	5.5^e	$10{,}050^f$	$10{,}050^f$
$a\ (=v_1/A)$	0.5 min^{-1}	0.06 hr^{-1}	0.03 day^{-1}	0.07 day^{-1}
$t_{10\%}$	12 sec	1.9 hr	3.4 days	1.5 days

Units of A: $^a\mu$mol/g; ^{b}g/liver; ^{c}kJ/man.
Units of v_1: $^d\mu$mol/min/g (during sprinting); ^{e}g/hr/liver (during fasting); ^{f}kJ/day/man (during starvation).

* Values refer to humans and are from (114), pp 337, 362, 370, 538, 541: a is the effective first order rate constant if the fuel were used via a first order process. $t_{10\%}$ is the time taken to reduce the (first order) flux by 10%; and, from equation 5–207, is equal to $(1/a).\ln(0.9)$, or approx $0.11/a$.

carrier-mediated system which seems to be saturated with intracellular glutamine. For example, in rats, the concentration of intracellular glutamine, approx 4 mM, is approximately tenfold greater than the appropriate K_m (125). This transport system can therefore serve as the flux-generating step for the important flux of muscle glutamine to cells of the intestine and kidney, and for the immune system, for which glutamine is an important fuel (10, 98, 113, 115, 119, 174, 186, 187).

There are, however, at least two exceptions to the saturation principle for flux-generating steps: general protein breakdown to supply energy (as during starvation) and nutrient uptake from the gut.

From the data in Table 5.4, if general protein degradation (equivalent to E_1 in Fig. 5.1) were saturated with protein, the protein stores could supply energy at a near-constant rate for ~150,000/10,050, i.e. 15 days. However, saturation at this step is not possible, for reasons similar to those preventing saturation of FBPase with FBP in liver. Protein degradation must respond rapidly to the concentration of protein to provide a cycle (referred to as protein turnover) for the rapid regulation of the concentrations of enzymes and other proteins (178); and this takes precedence over insulating the steady state flux. In this case, the lack of "insulation" is not a serious problem because *(1)* the time scale is quite large (approx 1.5 days to reduce the steady-state flux by 10%, Table 5.4) and *(2)* only approx. 20% of the body's protein can be used as a fuel anyway (any greater use would be severely disabling, or even fatal; therefore physiological mechanisms operate to spare protein after a few days of fasting [114]). Thus, by the time the reduction in steady state flux becomes significant, protein is no longer needed as a fuel. Consequently, because of the long time scale and the use of only a relative small fraction of the total amount of the fuel, the flux-generating step for the

process of protein degradation to provide energy need not be saturated. (It is not known whether general amino-acid transport out of muscle is saturated, as for glutamine. If this is the case, these transporters could provide flux-generating steps for each amino acid (or group of amino acids), reducing or eliminating the problems associated with nonsaturating protein degradation. However, as with protein degradation, saturation of these transporters with their intracellular amino acids would not be of critical importance during extreme conditions such as starvation; indeed, even glutamine transport may become unsaturated during severe sepsis [115].)

A similar situation arises with the uptake of nutrients from the gut. This process, which supplies a significant amount of fuel to the body after feeding, cannot be saturated; if it could be, an increased amount of ingested food would result in significant quantities of fuel passing through the gut without being absorbed. Consequently, insulation here would be wasteful. Moreover, much of the fuel is taken to the liver (via the hepatic portal vein), where storage cycles (for example, those between glucose and glycogen, proteins and amino acids) act as buffers to prevent large fluctuations in the rate of fuel absorption from affecting the corresponding steady state fluxes in the systemic circulation (see the section titled Metabolic Buffers; Fig. 5.29c). Therefore, during intestinal absorption, it is not possible to define a unique "flux-generating step"; it is represented instead by a complex system comprising gut uptake (flux I, Fig. 5.29c) and the net flux across the storage cycle (between S and P), both of which will change continuously to provide a smooth and steady output, O. A system, such as that in Fig. 5.2b, therefore refers to postabsorptive metabolism.

Finally, the concept of "saturation" as regards storage forms such as glycogen and triacylglycerol may not be quite the same as for an enzyme whose substrate is in true solution. Thus, in Figure 5.36, a particle of glycogen is acted on by a molecule of phosphorylase (and related enzymes), E; and E does not dissociate as the glucose units are removed in the form of glucose-1-phosphate. In this situation, the glycogen particle will shrink as glucose units are removed, but the rate of production of the units will not decrease. Thus the system will behave as if glycogen were saturating and a constant flux will be generated despite the presence of free (unbound) enzyme. Moreover, the presence of free enzyme means that an increased glycogen content (that is, number of particles) could increase the rate of glycogenolysis. Thus this particulate system could allow a flux-generating step to respond to an increase in its substrate concentration! Evidence for such glycogen particles was obtained many years ago (56), and it has since been found that elevating the glycogen content of skeletal muscle increases the rate of glycogenolysis during subsequent muscular activity (133).

A similar situation may also apply to triacylglycerol by lipase, since this fuel exists (in adipose tissue) in the form of large fat droplets to which lipase and other enzymes are bound (118).

APPENDIX 2. STABILITY OF THE STEADY STATE

Most of the systems discussed in this chapter have been assumed to exist in a steady state. As stated earlier, this is a satisfactory model of many pathways. However, whether a system exists in a steady state or not depends on the values of the component sensitivities in the matrix N (equation 5–95). As stressed in earlier sections, the numerical values of these sensitivities will usually vary continuously during any finite response. Under some conditions these variations may cause the system to become so unstable that the steady state

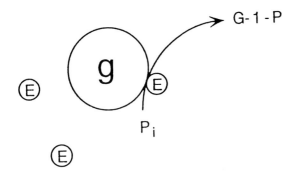

FIG. 5.36. Flux-generating step with a particulate system. G is a glycogen particle, to which phosphorylase and other enzymes *(E)* are tightly bound. E does not dissociate as glucose-1-phosphate is released.

breaks down. It is therefore essential that any steady state analysis (for example, finite decomposition) is able to determine if and when this point is reached. Fortunately, the criterion for such a "global" stability is simply the determinant of matrix N (equation 5–95), referred to as det N (47).

As shown in the text, all the net sensitivities of a system have a common denominator equal to det N. Thus, if det N ever becomes zero, no steady state can exist. Moreover, if det N changes sign, the system moves from stability to instability or vice versa. This can be appreciated by considering the response of flux J to regulator Z in Figure 5.11. If Z is an activator of E_1, an increased Z will increase J. Even though the associated increase of S inhibits E_1, and hence opposes the action of Z, there is always a net increase of E_1 and hence J. Thus the net sensitivity, s_Z^J, is positive. Since det N is the denominator of s_Z^J, a change in its sign will make s_Z^J negative. However, a negative response of J to Z is impossible with this system and therefore indicates that stability has been lost. The system then either breaks down completely, or, as a result of further changes in the values of the component sensitivities in matrix N, moves into a new region of stability where det N becomes positive again.

Such instability followed by a restoration of stability can produce "multiple" steady states. This means that, for any given concentrations of external regulators (parameters), there are several possible steady state values for each flux and internal metabolite, and not just one (71). These states clearly cannot co-exist, and whichever state does exist is determined by the way in (or path by) which the stimuli are applied. One potentially important role of multiple steady states is to provide a switch mechanism, whereby a system in one steady state is changed, by a stimulus, into a new steady state that persists when the stimulus is removed (36, 62, 132, 143). This type of mechanism (which, if produced by a system of interconvertible forms of an enzyme, as in Figure 5.37a–c, is energy dependent) may be important in producing the persistent changes characteristic of cell differentiation or even memory.

Unfortunately, stability cannot simply be identified with a predetermined sign of det N because this sign depends on the way in which the equations are ordered in the matrix equations (for example, equation 5–94). If two equations are interchanged, the sign of det N changes! However, for any given structure of N, the sign representing stability can be derived by noting that a steady state will always be stable *if all the feedback effects are negative and all the feedforward ones are positive*. Referring to Figure 5.11, this means that positive values of the feedforward component sensitivi-

a

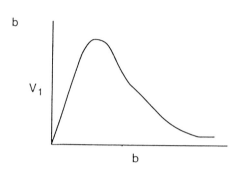

b

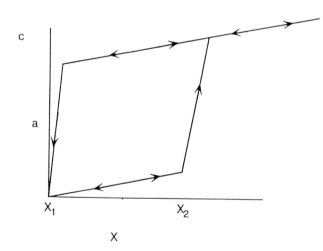

FIG. 5.37. Switch mechanism at an interconversion cycle. *(a)* Cycle, *(b)* response of E_1 to b: substrate inhibition, *(c)* response of active, a form to regulator X. At X_2 there is a large increase of a, which is not reversed until X is lowered to X_1, a value much lower than X_2. This system before shows hysteresis and enables a response to be maintained in the absence of the original stimulus (36).

ties (β) and negative values of the feedback sensitivity (α) will give the sign for det N that guarantees stability. (As can be seen from Figure 5.11, the terms *feedforward* and *feedback* refer to communications in the same and opposite directions to the flux, respectively.)

In this example,

$$N = \begin{pmatrix} 1 & -\alpha \\ 1 & -\beta \end{pmatrix}$$

whence,

$$\det N = -\beta + \alpha \qquad (5-248)$$

To make the feedback sensitivity negative, α is replaced by $-\alpha$ where α is a positive number, so that,

$$\det N = -\beta - \alpha$$

Since all the component terms are now positive numbers, det N is negative. Consequently, for this representation of matrix N, a negative value for det N guarantees stability. In contrast, a positive value guarantees instability. Thus, if the sign of det N changes from negative to positive during a finite decomposition of this system, a region of instability has been reached and the method must then be terminated at that point. (It should be noted, from equation 5–248, that a positive value for α, i.e. an activating effect of S on E_1 [Fig. 5.11], can give a negative value for det N, and hence a stable steady state, but only if α is less than β.)

However, the above test for stability applies only if none of the fluxes in the vector of variables, **v** (equation 5–95), changes sign. This would not happen with any of the hypothetical systems illustrated in this chapter, but it would be the case with the system in Figure 5.38, in which the net flux, J, can reverse according to the relative magnitudes of the other component fluxes, $A-E$. If J were included in the matrix equation for the system, it can be shown that J is a factor of det N; thus det N would change sign with J. However, this would not change the signs of the net sensitivities of S and P, since J is also a factor of the numerator of these sensitivities. Consequently, when analyzing any potentially reversible flux by finite decomposition, only the irreversible fluxes ($A-E$ in Fig. 5.38) and not the reversible ones (i.e. J) should be included in vector v. The values of the reversible fluxes can then be calcu-

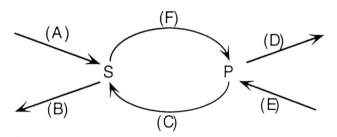

FIG. 5.38. A fully reversible system. The net flux J ($=A-B$; $=F-C$; $=D-E$) can reverse and hence change sign.

lated from the irreversible ones at the end of the procedure. (This procedure is an example of the "reversible" strategy described and advocated by Sorribas and Savageau in relation to BST [159].)

APPENDIX 3. SOLVING METABOLIC SYSTEMS BY USING GRAPH THEORY

Graph theory, which has recently been applied to metabolic systems by Sen (144–148), is a very interesting and potentially useful method for solving matrix equations like equation 5–96. Indeed, using this method, it is possible to solve the system without having to write the matrix at all! Full details of this approach can be found in the papers by Sen; here we will illustrate the principles of the approach by considering the simple system in Figure 5.11, but ignoring the regulator Z.

The basic matrix equation for this system,

$$\begin{pmatrix} 1 & -\alpha \\ 1 & -\beta \end{pmatrix} \begin{pmatrix} J_\rho \\ S_\rho \end{pmatrix} = \begin{pmatrix} E_{1\rho} \\ E_{2\rho} \end{pmatrix}$$

is transformed by placing all the functions on the right-hand side, so that,

$$\begin{pmatrix} \dfrac{J_\rho}{-1} & \dfrac{S_\rho}{\alpha} \\ -1 & \beta \end{pmatrix} \begin{pmatrix} J_\rho \\ S_\rho \end{pmatrix} + \begin{pmatrix} E_{1\rho} \\ E_{2\rho} \end{pmatrix} = 0 \qquad (5\text{–}249)$$

As with the equation for N^{-1} (equation 5–101), the matrix N has been augmented, but this time by the transpose of $\mathbf{v}$ ($\mathbf{v}^T$).

Equation 5–249 allows a "digraph" of the system to be drawn (Fig. 5.39a), in which the variables, J and S are nodes. Reactions E_1 and E_2 are connected to their corresponding variables (in equation 5–249) by broken lines, each of which is given a "weight" of $+1$.

The rest of the digraph is constructed by drawing arrows from the variables in $\mathbf{v}^T$ to those in $\mathbf{v}$ with weights equal to the corresponding elements in the matrix N. For example, there is an arrow from S to J with weight α, and one from J to J of weight -1 (known as a *self-loop*). A zero element in matrix N indicates no corresponding arrow.

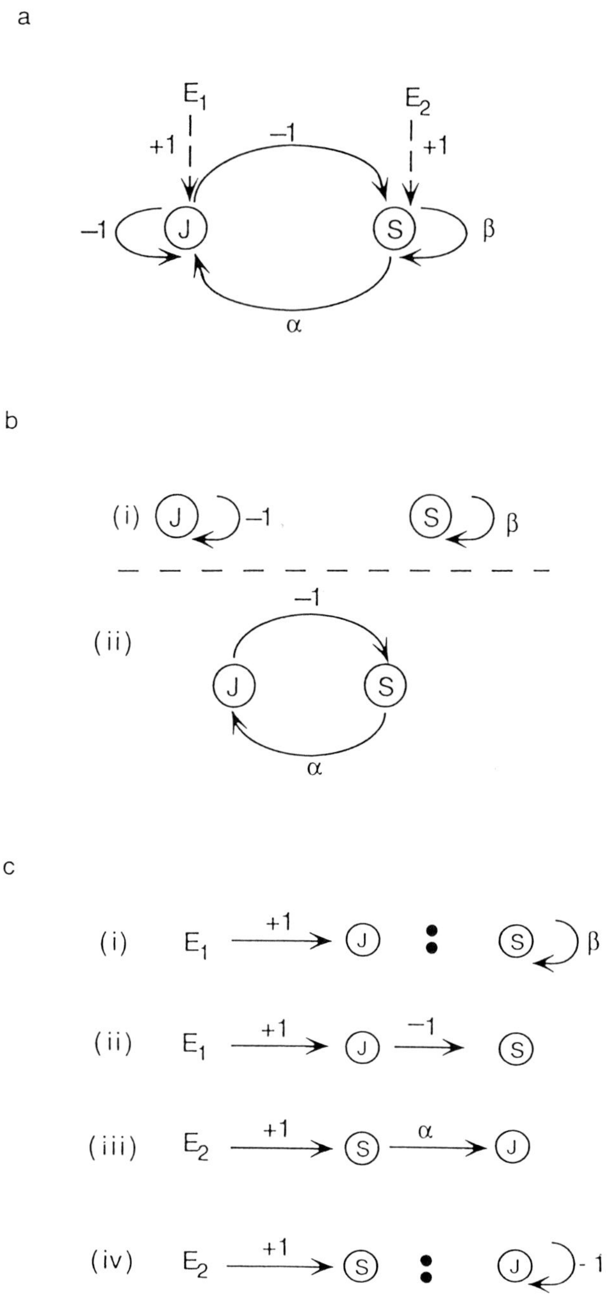

Calculation of Det N

To calculate the determinant of N (see the text section titled Deriving Net Sensitivities by Using Matrix Algebra and APPENDIX 2), the terms in E_1 and E_2 are ignored, and attention is focused on the variables, J and S. The digraph is split into "directed circuits", that is paths beginning and ending on the same node

FIG. 5.39. Evaluating core sensitivities by graph theory. *(a)* Digraph for the system in Figure 5.11 (omitting Z). Variables J and S are nodes, *(b)* directed circuits and connections for the digraph, *(c)* directed paths and one-connections for each variable. For further details see text.

(variable) and passing no more than once through any component node. These circuits are shown in Figure 5.39b, which shows that a self-loop also represents a directed circuit. A complete set of directed circuits constitutes a "connection"; so that, from Figure 5.39b, there are two separate connections for this system.

Each connection is then given a "gain", which is the product of all the weights on the arrows (or edges) of the connection. If the number of circuits in the connection is zero or even, the gain is multiplied by $+1$; if the number of circuits is odd, the gain is multiplied by -1.

Thus, for the top connection (i) in Figure 5.39b,

$$\text{gain} = (-1) \cdot (\beta) \cdot (+1) \text{ (since there are two circuits)}$$
$$= -\beta$$

For the bottom connection (ii),

$$\text{gain} = (-1) \cdot (\alpha) \cdot (-1) \text{ (since there is just one circuit)}$$
$$= \alpha$$

Det N is then equal to the sum of the gains, so that

$$\det N = \alpha - \beta$$

which can be verified by inspection of matrix N.

Det N is the common denominator of all the core sensitivities. To obtain the numerators, we consider the "directed paths" from E_1 and E_2 to the required variable. A directed path is one that passes from E to the variable, either directly or via several nodes, but does not pass through any given node more than once. Thus, to calculate the numerator of s_1^J we form the directed path $E_1 \longrightarrow J$ (weight $+1$) and, for that of s_1^S, the directed path $E_1 \longrightarrow J \longrightarrow S$ (combined weight $+1 \cdot -1$) [Fig. 5.39c (i) and (ii)].

However, these paths need to form part of a "one-connection" in which any nodes not used in the path must be part of separate directed circuits, either as the result of a self-loop or in combination with other nodes. Thus, for s_1^J, the path $E_1 \longrightarrow J$ does not include node S. However, this node has a self-loop, thereby ensuring a one-connection [Fig. 5.39c(i)]. The gain of this connection is, as before, the product of all the weights and is multiplied by -1 if the number of *directed circuits* is odd. For the numerator of s_1^J, the total gain is therefore $(+1) \cdot (\beta) \cdot (-1)$, since there is only one circuit, so that, using the value for det N,

$$s_1^J = -\beta/(\alpha - \beta)$$

Similarly, the other one-connections in Figure 39c (ii–iv) allow calculations of the numerators of s_1^S, s_2^J and s_2^S, respectively.

Although this is only a simple example that could easily be solved by other means, graph theory provides a systematic method of solution that can be applied to any system, no matter how complex. As mentioned above, the method can also be used without having to write down the matrix (148).

A major advantage of using graph theory is that it provides both a qualitative and a quantitative analysis of a system. For a qualitative identification of potential control sites, all that is needed is to identify a directed path from the site to the variable, and then check to ensure that the nodes not used in this path can form directed circuits. The quantitative analysis (sensitivity functions) is then obtained, if required, from the gains of each connection, as shown above.

However, this use of what are termed "signal flow" graphs can become cumbersome with complex systems as a result of having to check that each node not used in path can form a directed circuit. This problem may be significantly improved by applying another type of graph, "spanning trees", to this analysis. This has been done with MCT, based on theorems (147), but has so far not been attempted with FOT. Future work in this area should be most interesting, especially if this type of graphical analysis could be applied to matrix equations for non-steady states (equations 5–200 and 5–201) to enable predictions of stability (see text section titled Analysis of Non-Steady States).

REFERENCES

1. Akerboom, T. P. M., R. Van Der Meer, and J. M. Tager. Techniques for the investigation of intracellular compartmentation. *Tech. Metabol. Res.* B205:1–33, 1979.

2. Atkinson, D. E. Regulation of enzyme activity. *Annu. Rev. Biochem.* 35: 85–124, 1966.

3. Atkinson, D. E. Limitation of metabolite concentrations and the conservation of solvent capacity in the living cell. *Curr. Top. Cell. Regul.* 1: 29–43, 1969.

4. Axelrod, J. Receptor-mediated activation of phospholipase A_2 and arachidonic acid release in signal transduction. *Biochem. Soc. Trans.* 18: 503–507, 1990.

5. Bahr, R., P. Hansson, and O. M. Sejersted. Triglyceride/fatty acid cycling is increased after exercise. *Metabolism* 39: 993–999, 1990.

6. Batke, J. Channeling of glycolytic intermediates by temporary, stationary bi-enzyme complexes is probable *in vivo*. *Trends Biochem. Sci.* 14: 481–482, 1989.

7. Beck, J. S. On internalization of hormone-receptor complex and receptor cycling. *J. Theor. Biol.* 132: 263–276, 1988.

8. Beeckmans, S. Some structural and regulatory aspects of citrate synthase. *Int. J. Biochem.* 16: 341–351, 1984.

9. Berry, M. N., R. B. Gregory, A. R. Grivell, D. C. Henley, J. W. Phillips, P. G. Wallace, and G. R. Welch. Linear relationships between mitochondrial forces and cytoplasmic flows argue for the organized energy-coupled nature of cellular metabolism. *FEBS Lett.* 224: 201–207, 1987.

10. Board, M., S. Humm, and E. A. Newsholme. Maximum activities of key enzymes of glycolysis, glutaminolysis, pentose phosphate pathway and tricarboxylic acid cycle in normal, neoplastic and suppressed cells. *Biochem J.* 265: 503–509, 1990.

11. Bowman, R. H. Inhibition of citrate metabolism by sodium fluoroacetate in the perfused rat heart and the effect on phosphofructokinase activity and glucose utilization. *Biochem. J.* 93: 13C–15C, 1964.

12. Bristow, J., D. M. Bier, and L. G. Lange. Regulation of adult and fetal myocardial phosphofructokinase. Relief of cooperativity and competition between fructose-2,6-bisphosphate, ATP and citrate. *J. Biol. Chem.* 262: 2171–2175, 1987.

13. Brown, G. C., R. P. Hafner, and M. D. Brand. A 'top-down' approach to the determination of control coefficients in metabolic control theory. *Eur. J. Biochem.* 188: 321–325, 1990.

14. Bucher, T. and W. Russmann. Equilibrium and nonequilibrium in the glycolysis system. *Angew. Chem. Internat. Edit.* 3: 426–439, 1964.

15. Burgoyne, R. D. and A. Morgan. The control of free arachidonic acid levels. *Trends Biochem. Sci.* 15: 364–366, 1990.

16. Burnell, J. N. and M. D. Hatch. Activation and inactivation of an enzyme catalyzed by a single, bifunctional protein: a new example and why. *Arch. Biochem. Biophys.* 245: 297–304, 1986.

17. Burt, C. T., T. Glonek, and M. Barany. Analysis of phosphate metabolites, the intracellular pH, and the state of adenosine triphosphate in intact muscle by phosphorus nuclear magnetic resonance. *J. Biol. Chem.* 251: 2584–2591, 1976.

18. Cardenas, M. L. and A. Cornish-Bowden. Characteristics necessary for an interconvertible enzyme cascade to generate a highly sensitive response to an effector. *Biochem. J.* 257: 339–345, 1989.

19. Cascante, M., E. I. Canela, and R. Franco. Control analysis of systems having two steps catalysed by the same protein molecule in unbranched chains. *Eur. J. Biochem.* 192: 369–371, 1990.

20. Cascante, M., R. Franco, and E. I. Canela. Use of implicit methods from general sensitivity theory to develop a systematic approach to metabolic control. II. complex systems. *Math. Biosci.* 94: 289–309, 1989.

21. Cascante, M., R. Franco, and E. I. Canela. Use of implicit methods from general sensitivity theory to develop a systematic approach to metabolic control. I. Unbranched pathways. *Math. Biosci.* 94: 271–288, 1989.

22. Challis, R. A. J., B. Crabtree, and E. A. Newsholme. Hormonal regulation of the rate of the glycogen/glucose-1-phosphate cycle in skeletal muscle. *Eur. J. Biochem.* 163: 205–210 1987.

23. Challiss, R.J.A., J.R.S. Arch, and E. A. Newsholme. The rate of substrate cycling between fructose-6-phosphate and fructose-1,6-bisphosphate in skeletal muscle. *Biochem. J.* 221: 153–161, 1984.

24. Chen, Y-D. and H. V. Westerhoff. How do inhibitors and modifiers of individual enzymes affect steady state fluxes and concentrations in metabolic systems? *Math. Modelling* 7: 1173–1180, 1986.

25. Clark, M. G., C. H. Williams, W. F. Pfeifer, D. P. Bloxham, P. C. Holland, C. A. Taylor, and H. A. Lardy. Accelerated substrate cycling of fructose-6-phosphate in the muscles of malignant hyperthermic pigs. *Nature* 245: 99–101, 1973.

26. Cleland, W. W. The kinetics of enzyme-catalysed reactions with two or more substrates and products. III. Prediction of initial velocity and inhibition patterns by inspection. *Biochim. Biophys. Acta* 67: 188–196, 1963.

27. Cohen, P. The role of protein phosphorylation in the hormonal control of enzyme activity. *Eur. J. Biochem.* 151: 439–448, 1985.

28. Cohen, S. M. Simultaneous ^{13}C and ^{31}P NMR studies of perfused rat liver: effects of insulin and glucagon and a ^{13}C NMR assay of free Mg^{2+}. *J. Biol. Chem.* 258: 14294–14308, 1983.

29. Cohen, S. M., R. G. Shulman, and A. C. McLaughlin. Effects of ethanol on alanine metabolism in perfused mouse liver studied by ^{13}C NMR. *Proc. Natl. Acad. Sci. USA* 76: 4808–4812, 1979.

30. Cohen, S. R. Why does brain make lactate? *J. Theor. Biol.* 112: 429–432, 1985.

31. Cornish-Bowden, A. Metabolic control theory and biochemical systems theory: different objectives, different assumptions, different results. *J. Theor. Biol.* 136: 365–377, 1989.

32. Cornish-Bowden, A. Failure of channelling to maintain low concentrations of metabolic intermediates. *Eur. J. Biochem.* 195: 103–108, 1991.

33. Cox, D. R. and H. D. Miller. *The Theory of Stochastic Processes.* London: Methuen, 1965.

34. Crabtree, B. Reversible (near equilibrium) reactions and substrate cycles. *Biochem. Soc. Trans.* 4: 1046–1048, 1976.

35. Crabtree, B. Theoretical considerations of the sensitivity conferred by substrate cycles *in vivo. Biochem. Soc. Trans.* 4: 999–1002, 1976.

36. Crabtree, B. A metabolic switch produced by enzymically interconvertible forms of an enzyme. *FEBS Lett.* 187: 193–195, 1985.

37. Crabtree, B. Metabolic regulation. In: *Quantitative Aspects of Ruminant Digestion and Metabolism,* edited by J. M. Forbes and J. France. Butterworths, 1992.

38. Crabtree, B. A qualitative method for identifying external control sites in metabolic systems. *Am. J. Physiol.* 262 (*Regulatory Integrative Comp Physiol.* 31): R806–R812.

39. Crabtree, B., G. Collins, and M. F. Franklin. A simplified method for calculating complex metabolic sensitivites by using matrix partitioning. *Biochem. J.* 263: 289–292, 1989.

40. Crabtree, B. and G. E. Lobley. Measuring metabolic fluxes in organs and tissues with single and multiple tracers. *Proc. Nutr. Soc.* 47: 353–364, 1988.

41. Crabtree, B. and E. A. Newsholme. Comparative aspects of fuel utilization and metabolism by muscle. In: *Insect Muscle,* edited by P. N. R. Usherwood. London & New York: Academic Press, 1975, p. 405–500.

42. Crabtree, B. and E. A. Newsholme. Sensitivity of a near-equilibrium reaction in a metabolic pathway to changes in substrate concentration. *Eur. J. Biochem.* 89: 19–22, 1978.

43. Crabtree, B. and E. A. Newsholme. Reply to letter from Susan Moore on substrate cycling. *Trends Biochem. Sci.* 10: 387, 1985.

44. Crabtree, B. and E. A. Newsholme. A quantitative approach to metabolic control. *Curr. Top. Cell. Regul.* 25: 21–76, 1985.

45. Crabtree, B. and E. A. Newsholme. A systematic approach to describing and analysing metabolic control systems. *Trends Biochem. Sci.* 12: 4–12, 1987.

46. Crabtree, B. and E. A. Newsholme. The derivation and interpretation of control coefficients. *Biochem. J.* 247: 113–120, 1987.

47. Crabtree, B. and E. A. Newsholme. A method for testing the stability of a steady state system during the calculation of a response to large changes in regulator concentration. *FEBS Lett.* 280: 329–331, 1991.

48. Derr, R. F. Modern metabolic control theory II. Determination of flux-control coefficients. *Biochem. Archiv.* 2: 31–44, 1986.

49. Dixon, M. and E. C. Webb. *Enzymes.* New York: Longmans, 1979.

50. Easterby, J. S. *Biochem. J.* 199: 155–161, 1981.

51. El-Maghrabi, M., T. Claus, T. Pilkis, E. Fox, and S. Pilkis. *J. Biol. Chem.* 257: 7603–7607, 1982.

52. England, P. J. and P. J. Randle. Effectors of rat-heart hexokinases and the control of rates of glucose phosphorylation in the perfused rat heart. *Biochem. J.* 105: 907–920, 1967.

53. Engstrom, L. The regulation of liver pyruvate kinase by phosphorylation-dephosphorylation. *Curr. Top. Cell. Regul.* 13: 29–52, 1978.

54. Fell, D. A. and H. M. Sauro. Metabolic control and its analysis. *Eur. J. Biochem.* 148: 555–561, 1985.

55. Fell, D. A. and K. Snell. Control analysis of mammalian serine biosynthesis. *Biochem. J.* 256: 97–101, 1988.

56. Fischer, E. H., A. Pocker, and J. C. Saari. The structure, function

and control of glycogen phosphorylase. *Essays Biochem.* 6: 23–68, 1970.

57. Giersch, C. Control analysis of biochemical pathways: a novel procedure for calculating control coefficients, and an additional theorem for branched pathways. *J. Theor. Biol.* 134: 451–462, 1988.

58. Gilman, A. G. G-proteins: transducers of receptor-generated signals. *Annu. Rev. Biochem.* 56: 615–649, 1987.

59. Goldbeter, A. and D. E. Koshland. An amplified sensitivity arising from covalent modification in biological systems. *Proc. Nat. Acad. Sci. U.S.A.* 78: 6840–6844, 1981.

60. Goldbeter, A. and D. E. Koshland, Jr. Energy expenditure in the control of biochemical systems by covalent modification. *J. Biol. Chem.* 262: 4460–4471, 1987.

61. Goldhammer, A. R. and H. H. Paradies. Phosphofructokinase: structure and function. *Curr. Top. Cell. Regul.* 15: 109–141, 1979.

62. Goldstein, B. M. and A. N. Ivanova. Hormonal regulation of 6-phosphofructo-2-kinase/fructose-2,6-bisphosphatase: kinetic models. *FEBS Lett.* 217:212–215, 1987.

63. Goldstein, B. N. and E. L. Shevelev. Stability of multienzyme systems with feedback regulation: a graph-theoretical approach. *J. Theor. Biol.* 112: 493–503, 1985.

64. Groen, A. K., C.W.T. Van Roermudn, R. C. Vervoorn, and J. M. Tager. Control of gluconeogenesis in rat liver cells. *Biochem. J.* 237: 379–389, 1986.

65. Groen, A. K., R.J.A. Wanders, H.V. Westerhoff, R. Van Der Meer, and J. M. Tager. Quantification of the contribution of various steps to the control of mitochondrial respiration. *J. Biol. Chem.* 257: 2754–2757, 1982.

66. Grynkiewicz, G., M. Poenie, and R. Y. Tsien. A new generation of Ca^{2+} indicators with greatly improved fluorescence properties. *J. Biol. Chem.* 260: 3440–3450, 1985.

67. Halling, P. J. Do the laws of chemistry apply to living cells? *Trends Biochem. Sci.* 14: 317–318, 1989.

68. Hanson, P. J. and D. S. Parsons. Metabolism and transport of glutamine and glucose in vascularly perfused small intestine rat. *Biochem. J.* 166: 509–519, 1977.

69. Hawthorne, J. N. Phosphoinositides and metabolic control: how many messengers? *Biochem. Soc. Trans.* 16: 657–660, 1988.

70. Heinrich, R. and S. M. Rapoport. The utility of mathematical models for the understanding of metabolic systems. *Biochem. Soc. Trans.* 11: 31–35, 1983.

71. Heinrich, R., S. M. Rapoport, and T. A. Rapoport. Metabolic regulation and mathematical models. *Prog. Biophys. Molec. Biol.* 32: 1–82, 1977.

72. Heinrich, R. and T. A. Rapoport. A linear steady state treatment of enzymatic chains: general properties, control and effector strength. *Eur. J. Biochem.* 42: 89–95, 1974.

73. Hofmeyr, J-H. S. and A. Cornish-Bowden. Quantitative assessment of regulation in metabolic systems. *Eur. J. Biochem.* 200: 223–236, 1991.

74. Irvine, D. H. Objectives, assumptions and results of metabolic control theory and biochemical systems theory. *J. Theor. Biol.* 143: 139–143, 1990.

75. Jeffrey, F.M.H., A. Rajagopal, C. R. Malloy, and A. D. Sherry.[13] C NMR: a simple yet comprehensive method for analysis of intermediary metabolism. *Trends Biochem. Sci.* 16: 5–10, 1991.

76. Kacser, H. The control of enzyme systems *in vivo:* elasticity analysis of the steady state. *Biochem. Soc. Trans.* 11: 35–40, 1983.

77. Kacser, H. and J. A. Burns. The control of flux. *Symp. Soc. Exp. Biol.* 27: 65–104, 1973.

78. Kacser, H. and J. A. Burns. Molecular democracy: who shares the controls? *Biochem. Soc. Trans.* 7: 1149–1160, 1979.

79. Kacser, H. and J. W. Porteous. Control of metabolism: what do we have to measure? *Trends Biochem. Sci.* 12: 5–14, 1987.

80. Katz, J. An n.m.r. study of the tricarboxylic acid cycle. *Biochem. J.* 263: 997, 1989.

81. Katz, J. and R. Rognstad. Futile cycles in the metabolism of glucose. *Curr. Top. Cell. Regul.* 10: 238–287, 1976.

82. Keizer, J. L. Thermodynamic coupling in chemical reactions. *J. Theor. Biol.* 49: 323–335, 1975.

83. Keleti, T. Kinetic power: basis of enzyme efficiency, specificity and evolution. *J. Molec. Catal.* 47: 271–279, 1988.

84. Keleti, T. and Vértessy, B. Kinetic power, and theory of control. In *Dynamics of Biochemical Systems,* edited by S. Damjanovich, T. Keleti and L. Trón) Amsterdam: Elsevier, 1996, p. 3–10.

85. Keleti, T. and J. Ovadi. Control of metabolism by dynamic macromolecular interactions. *Curr. Top. Cell. Regul.* 29: 1–33, 1988.

86. Keleti, T. and G. R. Welch. The evolution of enzyme kinetic power. *Biochem. J.* 223: 299–303, 1984.

87. Kell, D. B. and H. V. Westerhoff. Metabolic control theory: its role in microbiology and biotechnology. *FEMS Microbiol. Rev.* 39: 305–320, 1986.

88. Koerner, T.A.W., R. J. Voll, and E. S. Younathan. A proposed model for the regulation of phosphofructokinase and frustose-1,6-bisphosphatase based on their reciprocal anomeric specificities. *FEBS Lett.* 84: 207–213, 1977.

89. Koshland, D. R., Jr. Control of enzyme activity and metabolic pathways. In: *Metabolic Regulation,* edited by R. S. Ochs, R. W. Hanson, and J. Hall. Amsterdam: Elsevier, 1985, p. 1–8.

90. Krebs, H. A. The Pasteur effect and the relations between respiration and fermentation. *Essays Biochem.* 8: 1–34, 1972.

91. Kunz, W., F. N. Gellerich, L. Schild, and P. Schonfeld. Kinetic limitations in the overall reactions of mitochondrial oxidative phosphorylation accounting for flux-dependent changes in the apparent ratio. *FEBS Lett.* 233: 17–21, 1988.

92. La Porte, D. C. and D. E. Koshland. A protein with kinase and phosphatase activities involved in regulating the tricarboxylic acid cycle. *Nature,* 300: 458–460, 1982.

93. La Porte, D. C. and D. E. Koshland. Phosphorylation of isocitrate dehydrogenase as a demonstration of enhanced sensitivity in covalent regulation. *Nature,* 305: 286–290, 1983.

94. LaPorte, D. C., K. Walsh, and D. E. Koshland, Jr. The branch point effect: ultrasensitivity and subsensitivity to metabolic control. *J. Biol. Chem.* 259: 14068–14075, 1984.

95. Madsen, J., J. Bulow, and N. E. Nielsen. Inhibition of fatty acid mobilization by arterial free fatty acid concentration. *Acta Physiol. Scand.* 127: 161–166, 1986.

96. Malloy, C. R., A. D. Sherry, and F.M.H. Jeffrey. Carbon flux through citric acid cycle pathway in perfused heart by ^{13}C NMR spectroscopy. *FEBS Lett.* 212: 58–62, 1987.

97. Malloy, C. R., A. D. Sherry, and F.M.H. Jeffrey. Evaluation of carbon flux and substrate selection through alternate pathways involving the citric acid cycle of the heart by ^{13}C NMR spectroscopy. *J. Biol. Chem.* 263: 6964–6971, 1988.

98. Matsuno, T. Bioenergetics of tumor cells: glutamine metabolism in tumor cell mitochondria. *Int. J. Biochem.* 19: 303–307, 1987.

99. Mazur, A. K. A probabilistic view on steady-state enzyme reactions. *J. Theor. Biol.* 148: 229–242, 1991.

100. McCormack, J.G., A. Halestrap, and R. M. Denton. Role of calcium ions in the regulation of mammalian intramitochondrial metabolism. *Physiol. Rev.* 70: 391–425, 1990.

101. Meinke, M. H. and R. D. Edstrom. Muscle glycogenolysis:

regulation of the cyclic interconversion of phosphorylase *a* and phosphorylase *b*. *J. Biol. Chem.* 266: 2259–2266, 1991.

102. Monod, J., J-P. Changeux, and F. Jacob. Allosteric proteins and cellular control systems. *J. Mol. Biol.* 6: 306–329, 1963.

103. Morris, J. G. *A Biologist's Physical Chemistry.* London: Edward Arnold, 1974.

104. Morris, J. L. *Computational Methods in Elementary Numerical Analysis.* Chichester: Wiley & Sons, 1983.

105. Newsholme, E. A. and M. Board. Application of metabolic-control logic to fuel utilization and its significance in tumor cells. *Adv. Enzyme Regul.* 31: 225–246, 1991.

106. Newsholme, E. A., R. A. J. Challiss, and B. Crabtree. Substrate cycles; their role in improving sensitivity in metabolic control. *Trends Biochem. Sci.* 9: 277–280, 1984.

107. Newsholme, E. A. and B. Crabtree. Metabolic aspects of enzyme activity regulation. *Symp. Soc. Exp. Biol.* 27: 429–460, 1973.

108. Newsholme, E. A. and B. Crabtree. Substrate cycles in metabolic regulation and in heat generation. *Biochem. Soc. Symp.* 41: 61–109, 1976.

109. Newsholme, E. A. and B. Crabtree. Theoretical principles in the approaches to the control of metabolic pathways, and their application to the control of glycolysis in muscle. *J. Mol. Cell. Cardiol.* 11: 839–856, 1979.

110. Newsholme, E. A. and B. Crabtree. Flux-generating and regulatory steps in metabolic control. *Trends Biochem. Sci.* 6: 53–55, 1981.

111. Newsholme, E. A. and B. Crabtree. Qualitative and quantitative approaches for control: Their current value. In: *Regulation of Hepatic Function*, edited by N. Grunnet and B. Quistorff. Copenhagen: Munksgaard, 1991, p. 214–226.

112. Newsholme, E. A., B. Crabtree, and M.S.M. Ardawi. The role of high rates of glycolysis and glutamine utilization in rapidly dividing cells. *Biosci. Rep.* 5: 393–400, 1985.

113. Newsholme, E. A., B. Crabtree, and M. Parry-Billings. The energetic cost of regulation: an analysis based on the principles of metabolic-control-logic. In: *Energy Metabolism: tissue determinants and cellular corollaries*, edited by J. M. Kinney and H. M. Tucker. Raven Press, New York, 1992, p. 467–493.

114. Newsholme, E. A. and A. R. Leech. *Biochemistry for the Medical Sciences.* Chichester: John Wiley & Sons, 1983.

115. Newsholme, E. A., P. Newsholme, R. Curi, E. Challoner, and M.S.M. Ardawai. A role for muscle in the immune system and its importance in surgery, trauma, sepsis and burns. *Nutrition.* 4: 261–268, 1988.

116. Newsholme, E. A. and J. C. Stanley. Substrate cycles: their role in control of metabolism with specific references to the liver. *Diabetes/Metabolism Reviews.* 3: 295–305, 1987.

117. Newsholme, E. A. and C. Start. General aspects of the regulation of enzyme activity and the effects of hormones. In: *Handbook of Physiology. Section 7: Endocrinology*, edited by D. F. Steiner and N. Freinkel. Washington, D.C.: American Physiological Society, 1972, p. 369–383.

118. Newsholme, E. A. and C. Start. *Regulation in Metabolism.* Chichester: Wiley & Sons, 1973.

119. Newsholme, P., S. Gordon, and E. A. Newsholme. Rates of utilization and fates of glucose, glutamine, pyruvate, fatty acids and ketone bodies by mouse macrophages. *Biochem. J.* 242: 631–636, 1987.

120. Nimmo, H. G. and P. Cohen. Applications of recombinant DNA technology to studies of metabolic regulation. *Biochem. J.* 247: 1–13, 1987.

121. Ninio, J. Alternative to the steady state method: derivation of reaction rates from first-passage times and pathway probabilities. *Proc. Natl. Acad. Sci. USA* 84: 663–667, 1987.

122. Ovadi, J. Physiological significance of metabolic channelling. *J. Theor. Biol.* 152: 1–22, 1991.

123. Ovadi, J., P. Tompa, B. Vertessy, F. Orosz, T. Keleti, and G. R. Welch. The perfection of substrate-channelling in interacting enzyme systems: transient-time analysis. *Biochem. J.* 1991.

124. Pahl-Wostl, C. and J. Seelig. Metabolic pathways for ketone body production. ^{13}C NMR spectroscopy of rat liver in vivo using ^{13}C-multilabeled fatty acids. *Biochemistry* 25: 6799–6807, 1986.

125. Parry-Billings, M. *Studies in glutamine metabolism in muscle.* Oxford: Thesis, Oxford University, 1989.

126. Pryor, H. J., J. E. Smyth, P. T. Quinlan, and A. N. Halestrap. Evidence that the flux control coefficient of the respiratory chain is high during gluconeogenesis from lactate in hepatocytes from starved rats. *Biochem. J.* 247: 449–457, 1987.

127. Putney, J. W., Jr. Formation and actions of calcium-mobilizing messenger inositol-1,4,5-trisphosphate. *Am. J. Physiol.* 252 (*Gastrointest. Liver Physiol. 15*): G149–G157, 1987.

128. Randle, P. J. Fuel selection in animals. *Biochem. Soc. Trans.* 14: 799–806, 1986.

129. Randle, P. J., P. J. England, and R. M. Denton. Control of the tricarboxylate cycle and its interactions with glycolysis during acetate utilization in rat heart. *Biochem. J.* 117: 677–695, 1970.

130. Rapoport, T. A., R. Heinrich, and S. M. Rapoport. The regulatory principles of glycolysis in erythrocytes *in vivo* and *in vitro*: a minimal comprehensive model describing steady states, quasi-steady states and time-dependent processes. *Biochem. J.* 154: 449–469, 1976.

131. Reder, C. Metabolic control theory: a structural approach. *J. Theor. Biol.* 135: 175–201, 1988.

132. Ricard, J. Dynamics of multi-enzyme reactions, cell growth and perception of ionic signals from the external milieu. *J. Theor. Biol.* 128: 253–278, 1987.

133. Richter, E. A. and H. Galbo. High Glycogen levels enhance glycogen breakdown in isolated contracting skeletal muscle. *J. Appl. Physiol.* 61: 827–831, 1986.

134. Rognstad, R. Rate-limiting steps in metabolic pathways. *J. Biol. Chem.* 254: 1875–1878, 1979.

135. Sacktor, B. Regulation of intermediary metabolism, with special reference to the control mechanisms in insect flight muscle. *Adv. Insect Physiol.* 7: 267–347, 1970.

136. Sahlin, K. Control of energetic processes in contracting human skeletal muscle. *Biochem. Soc. Trans.* 19: 353–358, 1991.

137. Salter, M., R. G. Knowles, and C. I. Pogson. Quantification of the importance of individual steps in the control of aromatic amino acid metabolism. *Biochem. J.* 234: 635–647, 1986.

138. Savageau, M. A. Biochemical systems analysis III. Dynamic solutions using a power-law approximation. *J. Theor. Biol.* 26: 215–226, 1970.

139. Savageau, M. A. The behavior of intact biochemical control systems. *Curr. Top. Cell. Regul.* 6: 63–130, 1972.

140. Savageau, M. A. A theory of alternative designs for biochemical control systems. *Biomed. Biochim. Acta* 6: 875–880, 1985.

141. Savageau, M. A. Mathematics of organizationally complex systems. *Biomed. Biochim. Acta* 6: 839–844, 1985.

142. Savageau, M. A., E. O. Voit, and D. H. Irvine. Biochemical systems theory and metabolic control theory: 2. he role of summation and connectivity relationships. *Math. Biosci.* 86: 147–169, 1987.

143. Schiffmann, Y. Self-organization in biological membranes. *Biochem. Soc. Trans.* 14: 1195–1196, 1986.

144. Sen, A. K. Topological analysis of metabolic control. *Math. Biosci.* 102: 191–223, 1990.

145. Sen, A. K. Metabolic control analysis. An application of signal flow graphs. *Biochem. J.* 269: 141–147, 1990.

146. Sen, A. K. A graph-theoretic analysis of metabolic regulation in linear pathways with multiple feedback loops and branched pathways. *Biochim. Biophys. Acta* 1059: 293–311, 1991.

147. Sen, A. K. Quantitative analysis of metabolic regulation: a graph-theoretic analysis using spanning trees. *Biochem. J.* 275: 253–258, 1991.

148. Sen, A. K. Calculation of control coefficients of metabolic pathways: a flux-oriented graph-theoretic approach. *Biochem. J.* 279: 55–65, 1991.

149. Shulman, G. I. and L. Rossetti. Influence of the route of glucose administration on hepatic glycogen repletion. *Am. J. Physiol.* 257 *(Endocrinol. Metab. 20)* E681–E685, 1989.

150. Shulman, G. I., D. L. Rothman, T. Jue, P. Stein, R. A. Defronzo, and R. G. Shulman. Quantitation of glycogen synthesis in normal subjects and subjects with non insulin dependent diabetes by ^{13}C nuclear magnetic resonance spectroscopy. *New Engl. J. Med.* 322: 223–228, 1990.

151. Sibley, D. R. and R. J. Lefkowitz. Molecular mechanisms of receptor desensitization using the β-adrenergic receptor-coupled adenylate cyclase system as a model. *Nature* 317: 124–129, 1985.

152. Sibley, D. R., R. H. Strasser, J. L. Benovic, K. Daniel, and R. J. Lefkowitz. Phosphorylation/ dephosphorylation of the β-adrenergic receptor regulates its functional coupling to adenylate cyclase and subcellular distribution. *Proc. Natl. Acad. Sci. USA* 83: 9408–9412, 1986.

153. Small, J. R. and D. A. Fell. Responses of metabolic systems: application of control analysis to yeast glycolysis. *Biochem. Soc. Trans.* 15: 238, 1987.

154. Small, J. R. and D. A. Fell. The matrix method of metabolic control analysis: its validity for complex pathway structures. *J. Theor. Biol.* 136: 181–197, 1989.

155. Snell, K. and D. A. Fell. Metabolic control analysis of mammalian serine metabolism. *Adv. Enzyme Regul.* 30: 13–32, 1990.

156. Sols, A. and C. Gancedo. Primary regulatory enzymes and related proteins. In: *Biochemical Regulatory mechanisms in eukaryotic cells*, edited by E. Kun and S. Grisolia. New York: Wiley-Interscience, 1972, p. 85–114.

157. Sorribas, A. and M. A. Savageau. A comparison of variant theories of intact biochemical systems. II. Flux-oriented and metabolic control theories. *Math. Biosci.* 94: 195–238, 1989.

158. Sorribas, A. and M. A. Savageau. A comparison of variant theories of intact biochemical systems. I. Enzyme-enzyme interactions and biochemical systems theory. *Math. Biosci.* 94: 161–193, 1989.

159. Sorribas, A. and M. A. Savageau. Strategies for representing metabolic pathways within biochemical systems theory: reversible pathways. *Math. Biosci.* 94: 239–269, 1989.

160. Srivastava, D. K. and S. A. Bernhard. Metabolite transfer via enzyme-enzyme complexes. *Science.* 234: 1081–1086, 1986.

161. Stadtman, E. R. and P. B. Chock. Interconvertible enzyme cascades in metabolic regulation. *Curr. Top. Cell. Regul.* 13: 53–95, 1978.

162. Stephenson, G. *Mathematical Methods for Science Students.* London: Longmans, 1961.

163. Stryer, L. *Biochemistry.* New York: Freeman, 1988.

164. Stucki, J. W. The optimal efficiency and the economic degrees of coupling of oxidative phosphorylation. *Eur. J. Biochem.* 109: 269–283, 1980.

165. Stucki, J. W. The thermodynamic-buffer enzymes. *Eur. J. Biochem.* 109: 257–267, 1980.

166. Tager, J. M., A. K. Groen, R. J. A. Wanders, J. Duszynski, H. V. Westerhoff, and R. C. Vervoorn. Control of mitochondrial respiration. *Biochem. Soc. Trans.* 11:40–43, 1983.

167. Tejwani, G. A., A. Ramaiah, and M. Ananthanaryana. Regulation of glycolysis in muscle. The role of ammonium and synergism among the positive effectors of phosphofructokinase. *Arch. Biochem. Biophys.* 158: 195–199, 1973.

168. Tompa, P., J. Batke, J. Ovadi, G. R. Welch, and P. A. Srere. Quantitation of the interaction between citrate synthase and malate dehydrogenase. *J. BIol. Chem.* 262: 6089–6092, 1987.

169. Tornheim, K. Activation of muscle phosphofructokinase by fructose-2,6-bisphosphate and fructose-1,6-bisphosphate is differently affected by other regulatory metabolites. *J. Biol. Chem.* 260: 7985–7989 1985.

170. Traut, T. W. Uridine-5′-phosphate synthesis: evidence for substrate cycling involving this bifunctional protein. *Arch. Biochem. Biophys.* 268: 108–115, 1989.

171. Umbarger, H. E. Amino acid biosynthesis and its regulation. *Annu. Rev. Biochem.* 47: 533–606, 1978.

172. Van Schaftingen, E. Fructose-2,6-bisphosphate. *Adv. Enzymol.* 59: 315–395, 1987.

173. Vinay, P., G. Lemieux, and a Gougoux. Characteristics of glutamine metabolism by rat kidney tubules: a carbon and nitrogen balance. *Can. J. Biochem.* 57: 346–356, 1979.

174. Vinay, P., J. P. Mapes, and H. A. Krebs. Fate of glutamine carbon in renal metabolism. *Am. J. Physiol.* 234 *(Renal Fluid Electrolyte Physiol. 3)*: F123–F129, 1978.

175. Voit, E. O. and M. A. Savageau. Accuracy of alternative representations for integrated biochemical systems. *Biochemistry* 26: 6869–6880, 1987.

176. Wajngot, A., V. Chandramouli, W. C. Schumann, K. Kumaran, S. Efendic, and B. R. Landau. Testing of the assumptions made in estimating the extent of futile cycling. *Am. J. Physiol.* 256 *(Endocrinol. Metab. 19)*: E668–E6765, 1989.

177. Walker, D. G. The nature and function of hexokinases in animal tissues. *Essays Biochem.* 2: 33–67, 1966.

178. Waterlow, J. C., P. J. Garlick, and D. J. Millward. *Protein Turnover in Mammalian Tissues and in the Whole Body.* Amsterdam: Elsevier, 1978.

179. Welch, G. R. Some problems in the usage of Gibbs free energy in biochemistry. *J. Theor. Biol.* 114: 433–446, 1985.

180. Welch, G. R. and T. Keleti. On the "cytosociology" of enzyme action *in vivo*: a novel thermodynamic correlate of biological evolution. *J. Theor. Biol.* 93: 701–735, 1981.

181. Welch, G. R., T. Keleti, and B. Vertessy. The control of cell metabolism for homogeneous vs. heterogeneous enzyme systems. *J. Theor. Biol.* 130: 407–422, 1988.

182. Westerhoff, H. V., A. K. Groen, and R.J.A. Wanders. Modern theories of metabolic control and their applications. *Biosci. Rep.* 4: 1–22, 1984.

183. Westerhoff, H. V. and D. B. Kell. Matrix method for determining steps most rate-limiting to metabolic fluxes in biotechnological processes. *Biotechnol. Bioeng.* 30: 101–107, 1987.

184. Williamson, J. R., R. H. Cooper, S. K. Joseph, and A. P. Thomas. Inositol trisphosphate and diacylglycerol as intracellular second messengers in liver. *Am. J. Physiol.* 248 *(Cell Physiol. 17)*: C203–C216, 1985.

185. Wilson, D. F, K. Nishiki, and M. Erecinska. Energy metabolism in muscle and its regulation during individual contraction-

relaxation cycles. In: *Metabolic Regulation,* edited by R. S. Ochs, R. W. Hanson, and J. Hall. Amsterdam: Elsevier, 1985, p. 77–86.

186. Windmueller, H. G. and A. E. Spaeth. Uptake and metabolism of plasma glutamine by the small intestine. *J. Biol. Chem.* 249: 5070–5079, 1974.

187. Windmueller, H. G. and A. E. Spaeth. Identification of ketone bodies and glutamine as the major respiratory fuels *in vivo* for postabsorptive rat small intestine. *J. Biol. Chem.* 253: 69–76, 1978.

188. Woledge, R. C., N. A. Curtin, and E. Homsher. *Energetic Aspects of Muscle Contraction.* London: Academic Press, 1985.

189. Woods, N. M., S. R. Cuthbertson, and P. H. Cobbold. Repetitive transient rises in cytoplasmic free calcium in hormone-stimulated hepatocytes. *Nature* 319: 600–602, 1986.

190. Wright, B. E. and P. J. Kelly. Kinetic models of metabolism in intact cells, tissues and organisms. *Curr. Top. Cell. Regul.* 19: 103–158, 1981.

6. Basic principles of transport

ROBERT I. MACEY | *Department of Molecular and Cell Biology, University of California, Berkeley, California*

TERESA F. MOURA | *Faculdade de Ciências e Tecnologia UNL, and Instituto de Tecnologia Química e Biológica, Universidade Nova de Lisboa, Lisbon, Portugal*

CHAPTER CONTENTS

DEVELOPMENT OF KINETIC MODELS has gone hand in hand with empirical measurements of membrane transport since the earliest recognition of the physiological importance of membrane permeability. The wide interest in these models reflects their utility in the interpretation and planning of experimental investigations. In addition, kinetic models provide clues about mechanism. They can be expected to supply physiological significance to molecular structural detail as it becomes available, and conversely they can be expected to furnish clues about the higher-order membrane structure of a sequenced membrane transport protein. Finally, kinetic models are indispensable for the integration of large amounts of experimental data and for embedding the details of a particular transport system

into a larger physiological context. Examples of these include reconstruction of axonal excitation and conduction from voltage clamp data (53), volume regulation in epithelia (93, 129), fate of red cells in the renal medulla (99) and the lung (141), generation of irreversibly sickled blood cells (94), cardiac excitation (33, 110), and renal physiology (128, 139, 140).

We review some basic principles of equilibrium thermodynamics and kinetics that have found extensive application, and that have formed the starting point for more advanced, detailed work. As such our treatment is elementary; for the most part we prefer simple intuitive examples to more general, rigorous proofs. We have made liberal use of a number of excellent books and reviews (35, 36, 38, 51, 77, 79, 80, 90, 125, 126, 127).

THERMODYNAMICS

Thermodynamics deals with the possibility of a system changing from one state to another without any work being done on the system by an external agency. If work is required, thermodynamic calculations can predict the minimal amount of energy required for the transition. Further, thermodynamics provides the foundation for comparing the magnitudes of important physiological driving forces arising from concentration, osmotic pressure, hydrostatic pressure, and voltage gradients as well as their relation to metabolic energy.

A thermodynamic state is fully defined by specifying the temperature, pressure, volume and chemical composition of each phase in the system. Although thermodynamics deals directly with differences in energy between different states, it is not concerned with how the system changes from one state to another. Its strength lies in the fact that its results are independent of the pathway or mechanism of the change so that thermodynamic predictions can be relied on with confidence. On the other hand, thermodynamics predictions reveal little if anything about mechanistic details.

Maximal Work Is Attained on Reversible Paths

Spontaneity, maximal work, equilibrium, and *reversible paths* are essential concepts for thermodynamics. We illustrate them with the simple mechanical example shown in Figure 6.1.

A linear spring is attached to a weightless pan containing sand with mass m_o. The pan is arrested at an equilibrium height at $x = 0$ where the force exerted by the stretched spring is just balanced by the weight of the sand. When a portion of sand with mass Δm is removed, the spring contracts, raising the new lighter

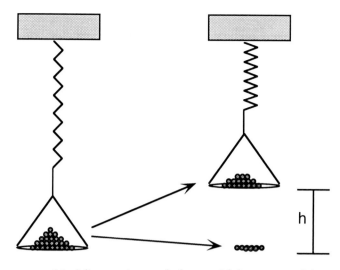

FIG. 6.1. A linear spring attached to a weightless pan containing sand rises a distance h when sand is removed.

mass, $m_o - \Delta m$, to a new equilibrium position at $x = h$. The work done by the spring in this one-step process, $w(1)$, is given by the weight times the distance moved, i.e.,

$$w(1) = (m_o - \Delta m)gh = m_o gh - \Delta mgh \quad (6-1)$$

where g represents the gravitational constant. Now suppose we repeat the process of moving the pan from $x = 0$ to $x = h$, but this time we do it in two steps. First we remove half the mass, $\Delta m/2$. The pan will move, say, 1/2 the distance to a new position at $h/2$. Its work during this first stage is $w_1 = (m_o - \Delta m/2)g(h/2)$. In the second stage we again remove $\Delta m/2$. The new mass is $m_o - \Delta m$ and the pan moves an additional distance $h/2$ to $x = h$ with work done $w_2 = (m_o - \Delta m)g(h/2)$. The total work in this process is

$$w(2) = w_1 + w_2 = m_o gh - (3/4)\Delta mgh$$
$$= w(1) + (1/4)\Delta mgh \quad (6-2)$$

Clearly $w(2)$, the work done along the two-step path, is greater than $w(1)$; a larger weight was lifted along the first stage of this path. If we repeat the process by removing $\Delta m/4$ in four steps, the work done $w(4)$ will be larger than $w(2)$. As we introduce more and more steps the work done becomes larger and larger as we lift more weight to higher intermediate levels between $x = 0$ and $x = h$. The greatest amount of work is obtained when we remove one grain of sand at a time. If we started with smaller grains, we could extract even more work. In the limiting case we let the mass of each grain approach zero as the number of removal steps becomes infinite to find W_{max}, the ideal maximal amount of work possible.

The example illustrates that there is no limit on the possible paths. However, there is an ideal pathway called the reversible path. On a reversible path, any *infinitesimal* change in the direction of force changes the direction of the path. (This means that at each stage along the path, the system is in "equilibrium.") All other paths are called irreversible. In our example, the reversible pathway would be closely approached if we remove one grain of sand at a time. If, at any stage, we add a grain of sand instead of removing it (i.e., we create an "infinitesimal" change in direction of the force), we change the direction of the path—the spring extends rather than contracts. The smaller the grains of sand, the closer we approach an ideal reversible path. Although we can approximate the reversible path in experiments, it does not exist in practice; real paths are irreversible.

Our conclusion that work obtained along a reversible path is greater than work done on any other path is very general. Using the second law of thermodynamics it can be shown in general that at *constant temperature and pressure,* the work done by the system will be the same for any reversible path and that $w_{max} = w_{rev} > w_{irrev} = w_{actual}$, or

$$w_{rev} - w_{actual} > 0 \quad (6-3)$$

Further, for any spontaneous change, w_{actual} must be positive or zero. If it was negative then we would need to supply work to the system; the process would not be spontaneous. Thus

$$w_{rev} \geq w_{actual} \geq 0 \quad (6-4)$$

Any transition for which this inequality holds is thermodynamically feasible; its spontaneous occurrence is not ruled out. Conversely when the inequality does not hold the transition cannot occur spontaneously. An experimental result that seems to contradict this inequality simply means that the entire "system" has not been identified—the transition is not completely described. The system must be coupled to another energy source that remains to be specified.

"Useful" work = work − PdV. It is convenient to refine this definition of spontaneity by excluding work done by any volume expansion. If the transition (again at constant temperature and pressure = P) involves a volume change ΔV, then mechanical work = $P\Delta V$ will be done against the prevailing, usually atmospheric, pressure. Since both pressure and net ΔV depend only on the state of the system, they will be the same no matter what path is traversed—i.e., $P\Delta V_{rev} = P\Delta V_{irrev}$ so that $(w_{rev} - P\Delta V_{rev}) - (w_{actual} - P\Delta V_{irrev}) = w_{rev} - w_{actual} \geq 0$. If we define "useful work" as $w' \equiv w - P\Delta V$, we retain the same formal criterion for a spontaneous process:

$$w'_{max} \geq w'_{actual} \qquad (6\text{-}5)$$

This criterion states that the process will take place only if the maximal work that can be extracted from the system is greater than work the system is asked to perform. The introduction of "useful work" simply makes incidental work done by or against the atmosphere irrelevant to our criterion for spontaneity. When a spontaneous process takes place, it is as though the capacity or potential of the system to do useful work has diminished. At constant T and P we call this capacity or potential the Gibbs free energy (sometimes just "free energy") and we denote it by G.

Change in Gibbs Free Energy Reflects Reversible w'. Thermodynamic applications are generally carried out by calculating ΔG, the change in G of the system as it moves from one state to another. These changes in energy depend only on the initial and final state of the system and are independent of the transition path. Changes in G are in fact simply equal to $-w'_{max}$; we write

$$\Delta G = G_{final} - G_{initial} = -w'_{rev} \qquad (6\text{-}6)$$

ΔG is another way of writing $-w'_{max}$ (which is equal to $-w'_{rev}$). The negative sign expresses the fact that the system's free energy decreases each time a spontaneous change takes place. For a spontaneous change, $w'_{rev} \geq w'_{actual}$. This implies that $-w'_{rev} \leq -w'_{actual}$, so

$$\Delta G = -w'_{rev} \leq -w'_{actual} \qquad (6\text{-}7)$$

Equilibrium at Constant T and P. In the equilibrium state if the system is left to itself no further spontaneous change is possible. Therefore there can be no conceivable change for which $w'_{rev} > w'_{actual}$. The only alternative is $w'_{rev} = w'_{actual}$. This reflects our earlier statement that a reversible path consists of a sequence of equilibrium states. The free energy change becomes

$$\Delta G_{EQUIL} = -w'_{rev} = -w'_{actual} \qquad (6\text{-}8)$$

If the system does no work on the environment, $w'_{actual} = 0$ so that

$$\Delta G_{EQUIL} = 0 \qquad (6\text{-}9)$$

Changes in Free Energy Are Obtained from Chemical Potentials: Free Energy per Mole

Applications of thermodynamics to transport problems necessarily deal with changes in composition of multiphase systems (e.g., two solutions separated by a membrane). At constant T and P, when the composition of a solution is changed by addition of dn moles of a substance—the change in G will be given by

$$dG = (dG/dn)\, dn = \mu\, dn \qquad (6\text{-}10)$$

where $\mu \equiv (dG/dn)$, the free energy per mole (more precisely, the partial molar free energy) is called the *chemical potential.* Each substance in each phase will have its own chemical potential. If more than one component changes, then we add the contribution of each to find the total change in G.

For example, assume two substances, **A** and **B**, are constrained to move together across the membrane from a solution on the "in" side to the "o" side. Each time one mole of **A** moves, b moles of **B** move. We represent this as though it were a chemical reaction:

$$\mathbf{A_{in}} + b\mathbf{B_{in}} \longrightarrow \mathbf{A_o} + b\mathbf{B_o} \qquad (6\text{-}11)$$

Each time a small number, dn, of molecules of **A** move, side "o" gains dn molecules and side "in" gains $-dn$ (i.e., it loses dn). The change in G_o due to **A** alone is $\mu_{A\_o}dn$, while the change in G_{in} due to **A** alone is $-\mu_{A\_in}dn$. Also, every time dn molecules of **A** move, $b\, dn$ molecules of **B** move, so that **B** contributes $\mu_{B\_o}bdn$ to G_o and $-\mu_{B\_in}bdn$ to G_{in}. As a result we have

$$dG = dG_o + dG_{in}$$
$$= (\mu_{A\_o} + b\mu_{B\_o})dn - (\mu_{A\_in} + b\mu_{B\_in})dn \leq -w' \qquad (6\text{-}12)$$

We have calculated the free energy change per dn moles. The dn is presumed to be sufficiently small so that no significant changes in concentrations, etc., will take place during the indicated "reaction." It is conventional to divide both sides by dn and express this result as a change in free energy per mole ΔG, with a corresponding work per mole, W'.

$$\Delta G = (\mu_{A\_o} + b\mu_{B\_o}) - (\mu_{A\_in} + b\mu_{B\_in}) \leq -W' \qquad (6\text{-}13)$$

Pursuing the analogy with chemical reactions, note that ΔG is simply the difference of the chemical potentials of the final state minus those of the initial state, just as it is the corresponding difference of the products and reactants in a chemical reaction. More precisely, we can generalize this to cover any transport and/or chemical reaction by

$$\Delta G = \left(\sum_i \nu_i \mu_i\right)_{FINAL} - \left(\sum_i \nu_i \mu_i\right)_{INITIAL} \leq -W' \qquad (6\text{-}14)$$

where the summation is carried out over all participants in the "reaction" and the ν_i are the corresponding stoichiometric coefficients. In our example shown in Equation 6–11, we identify side "in" with the initial state and "o" with the final; in a homogenous chemical reaction the initial state corresponds to reactants, the final state to products.

Electrochemical Potentials Incorporate Electrical Work. Finally, we make one further definition to accommodate the electric field acting on ions as they pass through a membrane. When an ion moves from a region where the electrical potential is Ψ_{in} to one where the potential is Ψ_o, then the electrical work done equals the charge transported times the difference in electrical potential. If dn moles of ion moves, and there are z charges per ion, then the total charge transported (in coulombs) $= zFdn$, where $F =$ the Faraday constant (96,494 coul./equiv.). If z includes the algebraic sign of the charge, then $w_{electrical} = zF(\Psi_o - \Psi_{in})dn$. We denote any remaining work (e.g., gravitational) by the symbol w''—i.e., $w' = w'' + w_{electrical}$, and apply this to Equation 6–12 to obtain

$$dG = (\mu_{A\_o} + b\mu_{B\_o})dn - (\mu_{A\_in} + b\mu_{B\_in})dn$$
$$\leq -z_A F(\Psi_o - \Psi_{in})dn - bz_B F(\Psi_o - \Psi_{in})dn - w'' \quad (6\text{–}15)$$

where w'' denotes the non-PdV, nonelectrical work. Adding the electrical work term to both sides of the inequality leaves

$$d\tilde{G} = (\mu_{A\_o} + z_A F\Psi_o)dn + b(\mu_{B\_o} + z_B F\Psi_o)dn$$
$$- (\mu_{A\_in} + z_A F\Psi_{in})dn - b(\mu_{B\_in} + z_B F\Psi_{in})dn \leq w'' \quad (6\text{–}16)$$

Since $w'' = 0$ in most transport applications, this suggests a new definition: the electrochemical potential, $\tilde{\mu}$, is defined as

$$\tilde{\mu} \equiv \mu + zF\Psi \quad (6\text{–}17)$$

With $w'' = 0$, the criterion for spontaneity becomes

$$d\tilde{G} = \tilde{\mu}_{A\_o}dn + b\tilde{\mu}_{B\_o}dn - \tilde{\mu}_{A\_in}dn - b\tilde{\mu}_{B\_in}dn \leq 0 \quad (6\text{–}18)$$

where the electrochemical free energy, $\tilde{G}$, is defined by

$$\tilde{G} \equiv G + \Sigma z_i F\Psi \quad (6\text{–}19)$$

Note that for a nonelectrolyte $z = 0$, so that $\tilde{G} = G$. Further, only differences in electrical potential have significance. As a result in a simple one-phase system, we may set $\psi = 0$ (i.e., consider the solution grounded) so that $\tilde{G} = G$.

Our principal result follows by repeating the arguments used for chemical potentials. Namely, with $w'' = 0$,

$$\Delta\tilde{G} = \sum_i \nu_i\left(\tilde{\mu}_i\right)_{\text{FINAL}} - \sum_i \nu_i\left(\tilde{\mu}_i\right)_{\text{INITIAL}} \leq 0 \quad (6\text{–}20)$$
$$\text{at constant } T \text{ and } P$$

where the inequality holds for any real change, and the equality implies equilibrium. We shall use this relation repeatedly.

Uncoupled Transport Equilibrium Is Obtained When Electrochemical Potentials Are Equal in Each Phase. Consider the equilibrium distribution of substances on two sides ("in" and "o") of a membrane. Using Equation 6–20 with subscripts "in" and "o" replacing "initial" and "final," we have

$$\Delta\tilde{G} = \sum_i \nu_i\tilde{\mu}_{i\_o} - \sum_i \nu_i\tilde{\mu}_{i\_in} = \sum_i \nu_i\left(\tilde{\mu}_{i\_o} - \tilde{\mu}_{i\_in}\right) \leq 0 \quad (6\text{–}21)$$

Clearly, if $\tilde{\mu}_{i\_o} - \tilde{\mu}_{i\_in} = 0$ for all i, then $\Delta\tilde{G} = 0$. If the solutes move independently, then the converse can also be shown; if $\Delta\tilde{G} = 0$, then $\tilde{\mu}_{i\_o} - \tilde{\mu}_{i\_in} = 0$ for each i (147). Equality of electrochemical potentials is a necessary and sufficient condition for equilibrium of independently moving solutes.

If a single substance moves independently of other solutes from "in" to "o," then its associated free energy change is also independent and we can predict the direction of its movement in general by calculating what happens in the special case where it is the only solute that moves. In this case, $\Delta\tilde{G} = \tilde{\mu}_{i\_o} - \tilde{\mu}_{i\_in}$ will be negative if and only if $\tilde{\mu}_{i\_in} > \tilde{\mu}_{i\_o}$. If $\tilde{\mu}_{i\_o} > \tilde{\mu}_{i\_in}$, then $\Delta\tilde{G}$ would be positive and transport could only occur in the reverse direction, from "o" to "in." We conclude that, given a path, an independently moving substance will abandon regions of high electrochemical potential in favor of regions with lower potential. In other words, independently moving substances move down their electrochemical gradient.

Explicit Form of the Electrochemical Potential. Physiological transport is generally driven by gradients of concentration, pressure, and electrical potential. Accordingly it is expedient to obtain the explicit dependence of the electrochemical potential on these variables. For dilute solutions it can be shown (34, 123) that the electrochemical potential of a solute S can be split into the following terms:

$$\tilde{\mu}_s = \overline{\mu}_s^{\,oo}(T) + PV_s + RT\ln\overline{n}_s + z_s F\Psi \quad (6\text{–}22)$$

where V_s is the molar volume of S, and $\overline{n}_s$ is its mole fraction; i.e., $\overline{n}_s \equiv n_s/(n_w + \Sigma n_i)$, where n_s is the number of moles of S, n_w is the number of moles of water and the summation is carried over all solutes, including S. The first term of Equation 6–22 contains energy specific to the chemical species (e.g., its bond energies); it also depends on the solvent and on temperature. The second term shows the pressure dependence; the third term, sometimes called the "osmotic energy," depends solely on temperature and concentration; and the last term is the electrical energy. A similar expression (with $z = 0$) is also valid for the chemical potential of the solvent of the dilute solution.

The mole fraction $\bar{n}_s$ can be converted to the more convenient molar concentration C_s by noting that in dilute solutions the number of moles of any solute, n_i, is much smaller than n_w, the number of moles of water. Letting V_w denote the molar volume of water

$$\bar{n}_s = \frac{n_s}{n_w + \Sigma n_i} \approx \frac{n_s}{n_w} = V_w \left(\frac{n_s}{n_w V_w} \right) \approx V_w C_s \quad (6-23)$$

Using this expression in Equation 6–22, $\bar{\mu}_s$ can be expressed in any of the following equivalent forms:

$$\begin{aligned} \bar{\mu}_s &= \bar{\mu}_s{}^{oo}(T) + P V_s + RT \ln V_w C_s + z_s F \Psi \\ &= \bar{\mu}_s{}^{oo}(T) + P V_s + RT \ln V_w + RT \ln C_s + z_s F \Psi \\ &= \mu_s{}^{oo}(T) + P V_s + RT \ln C_s + z_s F \Psi \\ &= \mu_s^o(T,P) + RT \ln C_s + z_s F \Psi \quad (6-24) \end{aligned}$$

where $\mu_s^{oo}(T) \equiv \bar{\mu}_s{}^{oo}(T) + RT \ln V_w$ and $\mu_s^o(T,P) \equiv \mu_s^{oo}(T) + P V_s$ depends on temperature, pressure, the nature of the solvent, but not on the concentration.

When solutions are not sufficiently dilute, solutes interact, and the concentration term is replaced by an "effective concentration" called the activity, usually denoted by "a_s." More conveniently, we define an activity coefficient γ_s by $a_s = \gamma_s C_s$. In this case

$$\bar{\mu}_s \equiv \mu_s^o(T,P) + RT \ln \gamma_s C_s + z_s F \Psi \quad (6-25)$$

and the departure of γ_s from unity measures the departure of the solution from our dilute solution assumption.

Molecular Interactions Can Cause Departure from Ideal Behavior.

When can we expect significant departures of γ from unity? The derivation of Equation 6–22 requires that the partial vapor pressure of any component in a solution, say **S**, is proportional to the mole fraction of **S**. These solutions are called ideal. We can imagine two cases where this assumption should hold:

1. The intermolecular forces between all solute and solvent species are essentially the same. Then the tendency of **S** to rise into the vapor phase will be independent of its neighbors and will depend only on the proportion of **S** molecules present (mole fraction). This case does not apply to most aqueous solutions of physiological importance.

2. Solutions are sufficiently dilute so that each molecule of **S** is effectively and completely surrounded by solvent. **S** is then in a uniform environment, which will not change if the solution is diluted even further. This means we can ignore interactions between solutes and, again in this instance, the tendency of **S** to escape into the vapor will depend only on the relative proportion (mole fraction). Just how dilute the solution has to be

for this to hold depends on the solute and the solvent. If the interactive forces between solute molecules fall off rapidly with the distance between the molecules, then the solution can become more concentrated before these interactions become important. We can expect this to hold for nonpolar nonelectrolytes. On the other hand, forces between ions or dipolar molecules decrease slowly with distance so that the solution has to become quite dilute to be ideal. As a general rule, for the solution to be ideal, the mean interaction energy between solute molecules must be small compared to their thermal energy, kT, the Boltzmann constant times the absolute temperature.

In more practical terms, we turn to empirical measurements of activity. These can be measured by several methods, including vapor pressure, freezing points, and electrochemical measurements. For most gases activity coefficients are very close to unity up until about 1 to 2 atm. For most nonelectrolytes deviations of activity from concentration is quite small up to about 0.5 M. For electrolytes, the situation is not nearly so convenient. Activity coefficients for uniunivalent salts like NaCl or KCl show 10% deviations from ideal in solutions as dilute as 0.01 M. In the physiological range between 0.1 and 0.2 M their activity is reduced to about three-fourths the ideal concentration. Deviations of multivalent ions are stronger. Useful tables of activity coefficients can be found in the text by Robinson and Stokes (117).

In most electrolyte solutions that are not too concentrated, the most significant solute interaction is due to electrostatic charge. Each ion is subject to the electric field produced by all other ions, and to a first approximation it depends on the net charge of the ion but not on the chemical species that carries the charge. (A K^+ ion is influenced almost as much by another K^+ ion as by an Na^+ ion.) Hence the activity of a given ion depends largely on the total concentration of charge—more specifically, on the ionic strength. Results of empirical investigations led to the formulation of Lewis's principle of ionic strength: "In dilute solutions, the activity coefficient of a given strong electrolyte is the same in all solutions of the same ionic strength" (44). This notion gives some rationalization for approximating activities with concentration, because very often applications only involve ratios of concentrations, say C_{in}/C_o. Strictly speaking, it should be $\gamma_{in} C_{in}/\gamma_o C_o$. However, if the ionic strength is similar in both the "in" and "o" solutions, then the activity coefficients may also be similar and cancel. Although we shall follow the common practice of using concentrations rather than activities in theoretical expositions, more exact thermodynamic calculations should always replace ionic concentrations with activities whenever they are known.

Solutes in Phase Equilibrium

The Nernst Potential and the Donnan Ratio. Ions will diffuse down concentration gradients and will also flow in response to an electrical potential gradient (electric field). These two sources of movement can oppose each other and in some instances balance so that no net movement takes place. How can we compare concentration gradients with electric gradients? We apply Equation 6–24 to the equilibrium of ions across a membrane separating solutions "in" and "o." At equilibrium $\tilde{\mu}_{i\_o} = \tilde{\mu}_{i\_in}$ for all permeable ions so that

$$\tilde{\mu}_{i\_o}^o(T,P) + RT \ln C_{i\_o} + z_i F \Psi_o$$
$$= \tilde{\mu}_{i\_in}^o(T,P) + RT \ln C_{i\_in} + z_i F \Psi_{in} \quad (6\text{--}26)$$

Since the solvent (water) as well as T and P are the same on both sides of the membrane, it follows that $\tilde{\mu}_{i\_o}^o(T,P) = \tilde{\mu}_{i\_in}^o(T,P)$ will cancel. Rearranging terms, we arrive at the following expression for the membrane potential required to prevent movement of an ion down its given concentration gradient. This potential that will maintain equilibrium is called the Nernst potential, or sometimes the equilibrium potential of the ion in question.

$$(\Psi_i)_{EQUIL} = \Psi_{in} - \Psi_o = -\frac{RT}{z_i F} \ln \frac{C_{i\_in}}{C_{i\_o}} = -\frac{RT}{F} \ln \left(\frac{C_{i\_in}}{C_{i\_o}} \right)^{1/z_i}$$
$$(6\text{--}27)$$

Alternatively, we may rewrite Equation 6–27 as

$$\frac{C_{i\_in}}{C_{i\_o}} = e^{-\frac{z_i F}{RT}(\Psi_{in} - \Psi_o)} \quad (6\text{--}28)$$

If there are several ion species, all in equilibrium, the concentration distribution of each ion will satisfy Equation 6–27. Using the last equality of this expression we rewrite Equation 6–28 for each permeable ion species $i, j, k \ldots$, as

$$\left(\frac{C_{i\_in}}{C_{i\_o}} \right)^{1/z_i} = \left(\frac{C_{j\_in}}{C_{j\_o}} \right)^{1/z_j} = \left(\frac{C_{k\_in}}{C_{k\_o}} \right)^{1/z_k} = \ldots$$
$$= e^{-\frac{F}{RT}(\Psi_{in} - \Psi_o)} \equiv r_{DONNAN} \quad (6\text{--}29)$$

where the Donnan ratio defined above shows the relation between cations (with positive z) and anions (negative z) of any valence at equilibrium.

Partition Coefficients. Results of the last paragraph are easily generalized to include cases where ions or non-electrolytes are equilibrated between dissimilar phases (e.g., oil and water), which we designate by subscripts 1 and 2. Again we set $\tilde{\mu}_{i\_1} = \tilde{\mu}_{i\_2}$, but in this case $\tilde{\mu}_{i\_1}^o(T,P) \neq \tilde{\mu}_{i\_2}^o(T,P)$, so that the equilibrium condition Equation 6–26 leads to

$$\frac{C_1}{C_2} = e^{-\frac{\mu_1^o - \mu_2^o}{RT}} e^{-\frac{zF}{RT}(\Psi_1 - \Psi_2)} = e^{-\frac{\Delta G^o}{RT}} e^{-\frac{zF}{RT}(\Psi_1 - \Psi_2)}$$
$$= K e^{-\frac{zF}{RT}(\Psi_1 - \Psi_2)} \quad (6\text{--}30)$$

where $\Delta G^o \equiv \mu_1^o - \mu_2^o$ and the partition coefficient K (more specifically, this is called the *intrinsic partition coefficient*) is defined by $K \equiv e^{-\Delta G^o/RT}$. For nonelectrolytes $z = 0$ so that the concentration ratio is simply equal to the partition coefficient.

Osmotic Equilibrium

Van't Hoff's Law. The chemical potential for water requires separate treatment; when solutes are dilute, water is concentrated. It is expedient to write the chemical potential in terms of the *solute* concentrations rather than water concentration. We begin with an ideal solution where we relate μ to the mole fraction of water, $\bar{n}_w$, as

$$\mu_w \equiv \bar{\mu}_w^{oo}(T) + PV_w + RT \ln \bar{n}_w \quad (6\text{--}31)$$

$$\bar{n}_w \equiv \frac{n_w}{n_w + \Sigma n_i} = 1 - \frac{\Sigma n_i}{n_w + \Sigma n_i} \approx 1 - \frac{\Sigma n_i}{n_w} \quad (6\text{--}32)$$

where the last expression takes advantage of the fact that $n_w \gg \Sigma n_i$ in dilute solutions (cf. Equation 6–22). Inserting Equation 6–32 into 6–31 and applying the mathematical approximation that $\ln(1-x) \approx -x$ when $x \ll 1$,

$$\mu_w = \bar{\mu}_w^{oo}(T) + PV_w + RT \ln \left(1 - \frac{\Sigma n_i}{n_w} \right)$$
$$\approx \bar{\mu}_w^{oo}(T) + PV_w - RT \frac{\Sigma n_i}{n_w} \quad (6\text{--}33)$$

Finally, we note that $n_w V_w =$ the total volume of water, so that the molal concentration of the ith solute is simply $C_i = n_i/(n_w V_w)$. Substituting this into Equation 6–33 we obtain the following useful approximation to the chemical potential of water in terms of the concentrations of solutes

$$\mu_w \equiv \bar{\mu}_w^{oo}(T) + PV_w - V_w RT \Sigma C_i = \bar{\mu}_w^{oo}(T) + V_w (P - \Pi)$$
$$(6\text{--}34)$$

where, by definition $\Pi \equiv RT \Sigma C_i$.

For nonideal solutions, we again use an "effective concentration" by multiplying each C_i with an osmotic coefficient, ϕ. Ions and proteins, in particular, have osmotic coefficients that deviate significantly from unity. Although both activity and osmotic coefficients reflect nonideal behavior, they are not identical. Activity coefficients are defined to account for the chemical potential of the solute. Osmotic coefficients are applied to the solute to account for the chemical potential of the solvent.

Osmotic equilibrium between two solutions can be obtained by equating μ_w in the two solutions. This often involves a difference in hydrostatic pressure. When one of the solutions is pure water the difference in P is called the osmotic pressure. Consider two solutions "in" and "o" where we allow each solution to be at a different P. This does not violate our general restriction to changes at constant T and P. This constraint allows for differences in pressure in different phases; it requires only that pressure in each phase remain constant (137). Setting $\mu_{w\_in} = \mu_{w\_o}$, and canceling the $\bar{\mu}_w^{oo}$ that are the same in both solutions, we have

$$P_{in} - P_o = RT\Sigma(C_{i\_in} - C_{i\_o}) = (\Pi_{in} - \Pi_o) = \Delta\Pi$$
(6–35)

which is a statement of van't Hoff's law.

Osmotic Balance in Animal Cells. Animal cells are unable to withstand significant hydrostatic pressure differences ($P_{in} \approx P_o$) and the ionic and protein components are often far from ideal, so that osmotic balance is generally achieved by setting $\Sigma(\phi_{i\_in}C_{i\_in} - \phi_{i\_o}C_{i\_o}) = 0$. A good discussion and experimental demonstration of osmotic equilibrium in cells is given by Freedman and Hoffman (42).

Chemiosmotic Coupling

Thermodynamics of the Na-K Pump. To illustrate the incorporation of chemical coupling into a transport system, we utilize the Na^+-K^+ pump, whose stoichiometry is $mNa^+/nK^+/1ATP$ as follows:

$$m\,Na_{in} + n\,K_o + ATP_{in} \longrightarrow m\,Na_o + n\,K_{in}$$
$$+ ADP_{in} + P_{i\_in}$$
(6–36)

where we have assumed that hydrolysis of ATP and liberation of ADP and P_i take place in the "inside solution." Applying Equation 6–21 we have

$$\Delta\tilde{G} = m\tilde{\mu}_{Na\_o} + n\tilde{\mu}_{K\_in} + \tilde{\mu}_{ADP\_in} + \tilde{\mu}_{P_i\_in}$$
$$- (m\tilde{\mu}_{Na\_in} + n\tilde{\mu}_{K\_o} + \tilde{\mu}_{ATP\_in})$$
(6–37)

Substitute Equation 6–24 and note that $\mu_{Na\_o}^o = \mu_{Na\_in}^o$ and $\mu_{K\_o}^o = \mu_{K\_in}^o$ because the solvent, T, and P are the same in both solutions. It follows that:

$$\Delta\tilde{G} = mRT\ln[Na_o] + mz_{Na}F\Psi_o + nRT\ln[K_{in}] + nz_KF\Psi_{in}$$
$$+ \mu_{ADP}^o + RT\ln[ADP_{in}] + \mu_{P_i}^o + RT\ln[P_{i\_in}]$$
$$- mRT\ln[Na_{in}] - mz_{Na}F\Psi_{in} - nRT\ln[K_o]$$
$$- nz_KF\Psi_o - \mu_{ATP}^o - RT\ln[ATP_{in}]$$
(6–38)

or collecting terms

$$\Delta\tilde{G} = \Delta G_{ATP}^o + RT\ln\frac{[ADP_{in}][P_{i\_in}]}{[ATP_{in}]} + RT\ln\frac{[Na_o]^m[K_{in}]^n}{[Na_{in}]^m[K_o]^n}$$
$$- (mz_{Na} - nz_K)F(\Psi_{in} - \Psi_o) \leq 0$$
(6–39)

where $\Delta G_{ATP}^o \equiv \mu_{ADP}^o + \mu_{P_i}^o - \mu_{ATP}^o$ is the standard free energy of ATP hydrolysis ≈ -7.30 kcal mole^{-1} at pH = 7.0, and $z_{Na} = z_K = 1$ are the valences of Na^+ and K^+ (including the algebraic sign of their charge).

Equation 6–39 shows the thermodynamic constraints between chemical, osmotic, and electrical energies. The inequality is based on the assumption that the pump is running in the forward direction, e.g., energy of ATP hydrolysis is used to move Na^+ and K^+ against their electrochemical gradients. In any case where $\Delta\tilde{G} > 0$, the reaction would proceed in the reverse direction, energy from electrochemical gradients being utilized for synthesis of ATP.

If the pump could operate optimally (100% efficiency), maximal (reversible) work could be extracted and $\Delta\tilde{G} = 0$. This condition allows us to extract an expression for the maximal concentration ratios that the pump could provide in terms of the chemical energy available as well as the electrical work it is called on to perform.

$$\frac{[Na_o]^m[K_{in}]^n}{[Na_{in}]^m[K_o]^n} = \frac{[ATP_{in}]}{[ADP_{in}][P_{i\_in}]}e^{-\frac{\Delta G_{ATP}^o - (mz_{Na} - nz_K)F(\Psi_{in} - \Psi_o)}{RT}}$$
(6–40)

This maximal efficiency is an idealization; it is accomplished at zero rate because the system is at equilibrium. At constant membrane potential, tipping the scale by increasing the ratio on the left-hand side would begin to synthesize ATP (43). Physiologically, this rarely occurs; instead of utilizing Na^+ or K^+ gradients, the energy source for ATP synthesis usually comes from electrochemical gradients of hydrogen ions.

Generalized Transport Scheme. The method utilized for the Na^+-K^+ pump is easily generalized to a more complex system that will be useful in later discussions. Consider the general reaction shown below, where transport of solutes **A, B, C,** . . . are coupled to a chemical reaction where the substrate **S** is converted to product **P**.

$$n_A\mathbf{A_{in}} + n_B\mathbf{B_{in}} + n_C\mathbf{C_o} + \ldots + \mathbf{S} \longrightarrow n_A\mathbf{A_o} + n_B\mathbf{B_o}$$

$$+\, n_C\mathbf{C_{in}} + \ldots + \mathbf{P} \qquad (6\text{--}41)$$

Just as in the Na-K pump example, we apply Equation 6–21 to arrive at

$$\Delta\tilde{G} = \Delta G^o_{S\to P} + RT\ln\frac{[\mathbf{P}]}{[\mathbf{S}]} + RT\ln\frac{[\mathbf{A_o}]^{n_A}[\mathbf{B_o}]^{n_B}[\mathbf{C_{in}}]^{n_C}\ldots}{[\mathbf{A_{in}}]^{n_A}[\mathbf{B_{in}}]^{n_B}[\mathbf{C_o}]^{n_C}\ldots}$$

$$-\,(n_A z_A + n_B z_B - n_C z_C + \ldots)F(\Psi_{in} - \Psi_o) \le 0 \qquad (6\text{--}42)$$

Cycles, Rate Constants, and Detailed Balance. If, in analogy with Equation 6–19, we define $\Delta\tilde{G}^o \equiv \Delta G^o_{S\to P} + \Sigma z_i F\Delta\Psi$, then it follows from Equation 6–42 that

$$K^{eq} = \mathrm{e}^{\frac{-\Delta\tilde{G}^o}{RT}} \qquad (6\text{--}43)$$

where

$$K^{eq} \equiv \left(\frac{[\mathbf{A_o}]^{n_A}[\mathbf{B_o}]^{n_B}[\mathbf{C_{in}}]^{n_C}\ldots[\mathbf{P}]}{[\mathbf{A_{in}}]^{n_A}[\mathbf{B_{in}}]^{n_B}[\mathbf{C_o}]^{n_C}\ldots[\mathbf{S}]}\right)_{EQUILIBRIUM}$$

$$(6\text{--}44)$$

is the formal equilibrium constant of the reaction as written in Equation 6–41.

Like all thermodynamic calculations, Equation 6–43 is independent of the path; details of the pathway always cancel. We can exploit this fact to gain some general insight into properties that must be shared by all paths. Figure 6.2 shows a cyclic reaction where a charged solute **C** reacts with membrane carrier **Y** on the inner surface of the membrane (where $\Psi = \Psi_{in}$) and is pumped out of the cell (where $\Psi = \Psi_o$). Energy is supplied by the substrate **S**, which is degraded to **P**.

The overall reaction for one cycle is

$$\mathbf{C_{in}} + \mathbf{S} + \mathbf{Y_4} \to \mathbf{C_o} + \mathbf{P} + \mathbf{Y_4} \qquad (6\text{--}45)$$

We have included the redundant term $\mathbf{Y_4}$ to emphasize that the reaction refers to a cycle. We consider the equilibrium state of the system when $\Delta\tilde{G} = 0$. The $\Delta\tilde{G}^o$ for this net reaction is given by

$$\Delta\tilde{G}^o = \Delta\tilde{G}^o_{S\to P} + z_c F(\Psi_o - \Psi_{in}) \qquad (6\text{--}46)$$

This $\Delta\tilde{G}^o$ can also be broken into the sum of free energies for the individual steps. Letting $\Delta\tilde{G}^o_i$ represent the standard free energy change for the ith step of the cycle (leading to $\mathbf{Y_i}$), we have

$$\Delta\tilde{G}^o = \Delta\tilde{G}^o_1 + \Delta\tilde{G}^o_2 + \Delta\tilde{G}^o_3 + \Delta\tilde{G}^o_4 \qquad (6\text{--}47)$$

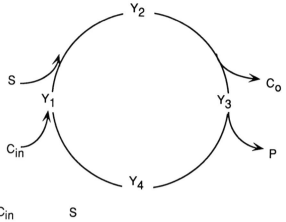

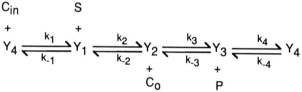

FIG. 6.2. Two ways of depicting a cyclic reaction. After reacting with membrane carrier **Y** on the inner surface of the membrane, solute **C** is pumped out of the cell. Energy is supplied by the substrate **S**, which is degraded to **P**.

In addition, we note from the principle of detailed balance (34, 65) that for any reaction sequence to be in equilibrium, each individual step must be in equilibrium. Applying Equation 6–43 to each individual step, we have $K^{eq}_i = \mathrm{e}^{-\Delta\tilde{G}^o_i/RT}$. Applying it to the entire cycle yields

$$K^{eq} = \mathrm{e}^{\frac{-\Delta\tilde{G}^o}{RT}} = \mathrm{e}^{\frac{-(\Delta\tilde{G}^o_1 + \Delta\tilde{G}^o_2 + \Delta\tilde{G}^o_3 + \Delta\tilde{G}^o_4)}{RT}} = K^{eq}_1 K^{eq}_2 K^{eq}_3 K^{eq}_4$$

$$(6\text{--}48)$$

Finally, we recognize that each equilibrium constant is the ratio of forward-to-backward rate constants, i.e., $K^{eq}_i = k_i/k_{-i}$ so that

$$\frac{k_1 k_2 k_3 k_4}{k_{-1} k_{-2} k_{-3} k_{-4}} = \mathrm{e}^{\frac{-\Delta\tilde{G}^o}{RT}} \qquad (6\text{--}49)$$

Utilization of Equation 6–49 in the development of kinetic models of channels and carriers requires the evaluation of $\Delta\tilde{G}^o$. For transport between aqueous phases (say, from inside to outside the cell) at constant temperature and pressure, substances that are chemically transformed will contribute to the ΔG^o portion of $\Delta\tilde{G}^o$ while transported charged substances will contribute to the Ψ terms. In our example, **C** contributes to the electrical term $\Delta\Psi$, while **S** and **P** contribute to the chemical term ΔG^o. If **S** and **P** were charged, and if their entrance and exit from the cycle were spatially separate, then they would also contribute to the electrical term. If the system was uncoupled from an energy source (**S** and **P** are omitted from Equation 6–45), leaving only the electrical term, Equation 6–43 would reduce to the Nernst equation Equation 6–28. If **C** was uncharged it would degenerate to the equilibrium state where $C_{in} = C_o$.

DIFFUSION

Flux Is Proportional to the Concentration Gradient: Fick's First Law

We begin with the simplest transport process, diffusion of a single solute down its concentration gradient, where net movement occurs as a result of the random motion of individual solute particles. Diffusion is described by Fick's law, which assumes that the flux J, defined as the number of moles of substance *crossing a unit area* in a unit time (moles cm^{-2} s^{-1}), is proportional to the concentration gradient. Letting C denote the concentration (moles cm^{-3}) of the solute **C**, and x represent the distance (cm) from a fixed reference point along the diffusion path, we have [*]

$$J = -D\frac{dC}{dx} \qquad (6-50)$$

where D is the diffusion coefficient, which will be assumed to be a constant. Equation 6–50 can be derived in several ways. We use the following kinetic approach because it is often extended to include diffusion through membranes. Assume the solute particles are embedded in a lattice of solvent molecules. For simplicity, we analyze one-dimensional diffusion along the x axis where the flux is given by the net rate at which solutes pass over a series of equal-height energy barriers that separate each position (Fig. 6.3).

Let λ be the distance between positions in the lattice (this will correspond to submolecular dimensions), C_i the concentration of solute at a particular position $x = x_i$, C_{i+1}, the concentration at x_{i+1}, and let k be half the average number of jumps each molecule makes per second. On average each particle makes an equal number of forward and backward jumps so that k represents the average number of forward jumps. We will calculate the flux J through a 1 cm^2 surface perpendicular to the plane of the page and located midway between x_i and x_{i+1} (Fig. 6.3). The volume bounded by this surface and projected backward a distance λ is

[*] For simpler notation, we omit brackets in the molar concentration terms. The symbol for the substance is written in bold type to distinguish it from its concentration.

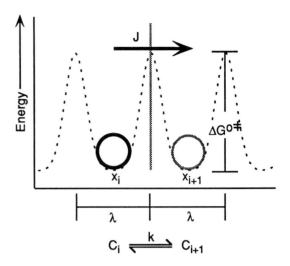

FIG. 6.3. Lattice model of diffusion. The solute at position x_i is "caged" by surrounding water molecules. It may advance to position x_{i+1} when it acquires sufficient thermal energy to break out of its cage. Energy barriers separating the solute from adjacent positions are illustrated, with average distance between activation energy peaks of height $\Delta G^{o\ddagger}$ given by λ. J represents net flux from x_i to x_{i+1}.

given by λ cm $\times$ 1 $cm^2 = \lambda$ cm^3, while the number of solute molecules within this volume is $C_i \lambda$. During each second, the number of moles jumping through the surface and going from x_i to x_{i+1} is given by $(C_i \lambda)k$. Similarly, the number jumping in the reverse direction (x_{i+1} to x_i) is given by $(C_{i+1} \lambda)k$, so that

$$J = \lambda k C_i - \lambda k C_{i+1} = \lambda k (C_i - C_{i+1}) \qquad (6-51)$$

Let $\Delta C = C_{i+1} - C_i$, and $\Delta x = x_{i+1} - x_i = \lambda$. It follows that $C_i - C_{i+1} = -\lambda \Delta C / \Delta x$, and substituting this result into Equation 6–51 leaves

$$J = -\lambda^2 k \frac{\Delta C}{\Delta x} \qquad (6-52)$$

Remembering that $\Delta x = \lambda$ is very small (of the order of 10^{-8} cm), we can approximate $\Delta C / \Delta x$ by dC/dx to arrive at

$$J = -\lambda^2 k \frac{dC}{dx} \qquad (6-53)$$

Equation 6–53 is consistent with Equation 6–50 if

$$D = \lambda^2 k \qquad (6-54)$$

The jumping frequency k has the dimensions s^{-1}, while λ has dimensions in cm. It follows from Equation 6–54 that the dimensions of D will be cm^2 s^{-1}. The jumping frequency k, and consequently D, both depend

TABLE 6.1. *Diffusion Coefficients of Simple Solutes in Water*

Substance	Molecular Weight (g·mole^{-1})	$D \times 10^5$ (cm$^2 \cdot$s^{-1})
O_2	32	1.98
CO_2	44	1.77
NaCl	58.5	1.39
KCl	74.5	1.68
Urea	60	1.18
Glycerol	92	0.83
Glucose	180	0.60
Lactose	342	0.43
Myoglobin	17,500	0.11

on temperature and on the mass (molecular weight) of the solute. Typical values of D for biological molecules (at 20°C) are listed in Table 6.1.

Most molecules involved in diffusional transport across biomembranes have molecular weights less than 200 g mole^{-1}, and it can be seen from Table 6.1 that their aqueous diffusion coefficients will be on the order of 10^{-5} cm^2/s.

A popular interpretation of the jumping frequency k is provided by the extension of chemical reaction rate theory to problems in viscosity and diffusion. The theory assumes each solute molecule remains in its position restrained by forces of attraction to neighboring solvent molecules. The solute oscillates back and forth about its average position at the bottom of an energy well and occasionally acquires enough energy through thermal interactions to tear itself loose and advance to the next position in the lattice.

Formal development and application of the theory assumes that molecules advance from one position to the next, a distance roughly equivalent to the molecular size, after passing through a high-energy transition state corresponding to the energy at the peak of the barrier (45). At any given time molecules residing in the energy wells are thought to be in quasi-equilibrium with a small number of activated molecules in the transition state, and further, it is assumed that a molecule jumps directly from well to peak in essentially one step. With these assumptions it can be shown that

$$k = k^o e^{-\frac{\Delta G^{o\ddagger}}{RT}} \qquad (6\text{--}55)$$

where $\Delta G^{o\ddagger}$ represents the standard free energy difference between peak and well of the energy barrier and k^o is a constant. (k^o is generally assumed to be $k_B T/h = 6.11 \times 10^{12}$ s^{-1} where k_B is the Boltzmann constant and h is Planck's constant).

These assumptions, leading to Equation 6–55, are reasonable for diffusion in solids, but they do not appear to be valid for diffusion in liquids (24, 134). In particular, empirical estimates show that values of $\Delta G^{o\ddagger}$ in liquids are too low to exclude a sizable proportion of molecules from the activated state, and experimental evidence indicates that individual molecular motions are much smaller than intermolecular distances. Attractive as the rate theory molecular picture is, it cannot be taken literally for diffusion in liquids. However, the analysis leading to Equation 6–53 is independent of these assumptions and will hold no matter what distance we ascribe to λ, provided it is comparable or smaller than distances where concentration changes are negligible. We can interpret our rate constant k as the reciprocal of ½ the average time a molecule spends in a specific position. Since molecules diffuse independently of one another, this average time will be constant, independent of concentration.

Conservation of Matter: Fick's Second Law

For most diffusion problems Equation 6–50 is insufficient, because it is only one equation in the two unknowns J and C. An additional condition is required, and this is provided by the conservation of matter. In the following examples we use this condition to eliminate J so that C may be obtained as an explicit function of x and t. Once this has been accomplished, the flux can always be determined by Equation 6–50.

First consider the spherical symmetry shown in Figure 6.4A. Diffusion takes place in the radial direction only, so that C is a function of the radial coordinate r and the time t. We apply conservation of matter to the spherical shell that lies between two imaginary concentric spheres of radius r and $r + dr$. The inner sphere has volume V and surface area A, while the larger exterior sphere has volume $V + dV$ and area $A + dA$. If C denotes the concentration within the shell and m denotes the total number of moles of solute within the shell, then, since the volume of the shell is dV, it follows that $m = C\, dV$. Now follow how m changes as solute flows into the shell through the inner surface at a rate JA and out of the outer surface at a rate $JA + d(JA)$. We have

$$\frac{dm}{dt} = dV \frac{dC}{dt} = JA - (JA + d(JA))$$

$$= -d(JA) = -J\,dA - A\,dJ \qquad (6\text{--}56)$$

Since $V = (4/3)\pi r^3$, $dV = 4\pi r^2 dr$; similarly $A = 4\pi r^2$, so $dA = 8\pi r\, dr$. Substituting these values into Equation 6–56 and dividing by dr leaves

A. Sphere

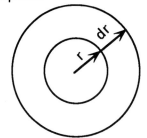

B. Cylinder

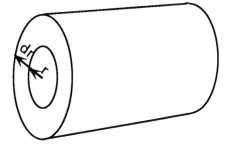

C. Plane

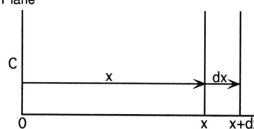

FIG. 6.4. Common types of symmetry encountered in diffusion problems: *A.* spherical symmetry; *B.* cylindrical symmetry; *C.* planar symmetry.

$$\frac{\partial C}{\partial t} = -\left(\frac{2}{r}J + \frac{\partial J}{\partial r}\right) \qquad (6\text{--}57)$$

Equation 6–57, called the *continuity equation*, holds whenever matter is continuously distributed regardless of the cause of J. In our case, J is due to diffusion. We substitute Equation 6–50 (with r replacing x) into Equation 6–57 to eliminate J and arrive at *Fick's second law* for spherical symmetry:

$$\frac{\partial C}{\partial t} = D\left(\frac{\partial^2 C}{\partial r^2} + \frac{2\,\partial C}{r\,\partial r}\right) = \frac{D}{r^2}\frac{\partial}{\partial r}\left(r^2\frac{\partial C}{\partial r}\right) \qquad (6\text{--}58)$$

For cylindrical symmetry (Fig. 6.4*B*), we replace the spherical shells by cylindrical shells of arbitrary length *l*. Repeating the same arguments, we arrive at an expression identical to Equation 6–56. However, for cylinders $V = \pi r^2 l$, $dV = 2\pi r l\,dr$, $A = 2\pi r l$, $dA = 2\pi l\,dr$.

Using these values to arrive at an analogous continuity equation for cylinders and then using Equation 6–50 yields

$$\frac{\partial C}{\partial t} = D\left(\frac{\partial^2 C}{\partial r^2} + \frac{1}{r}\frac{\partial C}{\partial r}\right) = \frac{D}{r}\frac{\partial}{\partial r}\left(r\frac{\partial C}{\partial r}\right) \qquad (6\text{--}59)$$

For the final case, planar symmetry (Fig. 6.4*C*), replace the shells with a planar region bounded by two planes perpendicular to the direction of flow (the *x* axis). Again we arrive at Equation 6–56, but now $dV = A\,dx$ and A is constant. Instead of Equation 6–57, we have

$$\frac{\partial C}{\partial t} = -\frac{\partial J}{\partial x} \qquad (6\text{--}60)$$

and using Equation 6–50,

$$\frac{\partial C}{\partial t} = D\frac{\partial^2 C}{\partial x^2} \qquad (6\text{--}61)$$

Equations 6–58, 6–59, and 6–61 are special cases of Fick's second law, each one appropriate for diffusion with specific geometrical symmetry. It is convenient to write these three equations in the composite form:

$$\frac{\partial C}{\partial t} = D\left(\frac{\partial^2 C}{\partial r^2} + \frac{n-1}{r}\frac{\partial C}{\partial r}\right) \quad n = 1, 2, 3 \quad (6\text{--}62)$$

where $n = 1$, 2, or 3 for planar (one dimension, with x replacing r), cylindrical (two dimensions), and spherical (three dimensions) symmetry, respectively. These equations are the beginning points for solution of the many time-dependent diffusion problems in biology. Each equation has an infinite number of solutions. Finding the solution that fits a particular case requires the specification of initial and boundary conditions, that is, the concentration at all points for a particular time (usually $t = 0$) and the concentration or flux at a particular place (usually a boundary) for all times. For the more general three-dimensional case with *no symmetry*, it can be shown that

$$\frac{\partial C}{\partial t} = D\left(\frac{\partial^2 C}{\partial x^2} + \frac{\partial^2 C}{\partial y^2} + \frac{\partial^2 C}{\partial z^2}\right) \qquad (6\text{--}63)$$

where x, y, and z are the usual three Cartesian coordinates. A large number of solutions for various conditions can be found in the textbooks by Crank (27) and by Carslaw and Jaeger (19).

Progress of Diffusion: Mean Square Displacement = 2 Dt

In simple diffusion, each molecule moves about at random, independent of all the others. How fast is diffusion? Suppose we consider the N molecules occupying a specific position, and follow the x coordinate of these molecules as they jump about at random. At some later time, some of the molecules will have moved a large distance, but others will have hardly moved at all. How far do they travel on average? The answer is zero because positive displacements (to the right) and negative displacements (to the left) occur with the same frequency, and they cancel. A more informative answer is obtained if we first square each displacement and then take the average of the squared displacements. To this end we continue with our lattice model illustrated in Figure 6.3 and make the simplifying assumption that jumps are all the same size and equal to λ. Let $x_i(n)$ denote the position of the ith molecule after n jumps. This molecule arrives at its position after making a jump of $\pm\lambda$ from its last position at $x_i(n-1)$. We have

$$x_i(n) = x_i(n-1) \pm \lambda \qquad (6\text{--}64)$$

Squaring both sides,

$$x_i(n)^2 = x_i(n-1)^2 \pm 2\lambda\ x_i(n-1) + \lambda^2 \qquad (6\text{--}65)$$

Now consider all N molecules. After n steps, the average squared displacement $\overline{x(n)^2}$, will be

$$\overline{x(n)^2} = \frac{1}{N}\sum_{i=1}^{N} x_i(n)^2 = \frac{1}{N}\sum_{i=1}^{N}(x_i(n-1)^2 \pm 2\lambda x_i(n-1) + \lambda^2)$$

$$= \frac{1}{N}\sum_{i=1}^{N} x_i(n-1)^2 \pm \frac{2\lambda}{N}\sum_{i=1}^{N} x_i(n-1) + \lambda^2 \qquad (6\text{--}66)$$

Since $+\lambda$ and $-\lambda$ will occur with equal frequency, the sum of $\pm 2\lambda x_i(n-1)$ will cancel and we are left with

$$\overline{x(n)^2} = \frac{1}{N}\sum_{i=1}^{N} x_i(n-1)^2 + \lambda^2 = \overline{x(n-1)^2} + \lambda^2 \qquad (6\text{--}67)$$

The fact that all N molecules start from the same position at $x=0$ means that $\overline{x(0)^2}=0$, i.e., before any jumps occur they are still at $x=0$. We use this trivial result in Equation 6–67 to calculate that $\overline{x(1)^2}=\lambda^2$, use this result to calculate $\overline{x(2)^2}=2\lambda^2$, and keep repeating the argument to finally arrive at

$$\overline{x(n)^2} = n\lambda^2 \qquad (6\text{--}68)$$

But n is simply equal to the number of jumps per second times t, the total time that has elapsed. Since by definition $2k$ = the number of jumps per second (both forward and backward) we can write,

$$\left.\begin{array}{l} \overline{x^2} = 2\ k\ \lambda^2 t = 2Dt \\[2mm] \text{and} \\[2mm] x_{rms} \equiv \sqrt{\overline{x^2}} = \sqrt{2Dt} \end{array}\right\} \quad \text{for 1 dimension} \quad (6\text{--}69)$$

where $\overline{x^2}$ denotes the mean squared displacement at any time t, and x_{rms} denotes (square) root of the mean squared displacement.

This result is modified when diffusion takes place in more than one dimension. For two dimensions where the solute diffuses in the x-y plane, random displacements in the x coordinate are completely independent of y coordinate displacements. Arguments given above may be applied to each coordinate so that the mean squared displacement along each axis is given by $\overline{x^2}=2Dt$ and $\overline{y^2}=2Dt$. Since the actual displacement is given by $r^2=x^2+y^2$, we have $\overline{r^2}=4Dt$ for two dimensions. Similarly, in three dimensions, $r^2=x^2+y^2+z^2$ so that $\overline{r^2}=6Dt$. In general,

$$\overline{r^2} = 2nDt, \qquad n = 1, 2, 3 \qquad (6\text{--}70)$$

A lucid discussion of diffusion transit times is given by Hardt (49) and by Berg (15).

Randomly Diffusing Molecules Spread Out in a Normal Distribution

Equation 6–68 provides a simple estimate of how far diffusion has progressed at any time; in general, the root mean squared distance, x_{rms}, increases with the square root of t. But this is a statistical average. As time increases, molecules that originated at a fixed position (say, the y-z plane located at $x=0$) will be distributed throughout space. An explicit form for the distribution can be obtained from the solution of Equation 6–61 that matches the relevant initial/boundary conditions. This solution is given by

$$C(x,t) = \frac{B_1}{(4\pi Dt)^{1/2}}\exp\left(\frac{-x^2}{4Dt}\right) \qquad (6\text{--}71)$$

where B_1 is a constant that equals the total number of molecules originally present at $t=0$ in each cm^2 of the y-z plane at $x=0$. This function is illustrated by the "bell-shaped" curves in Figure 6.5, where the concentration at each position is plotted for three different values of t.

To verify Equation 6–71 we need to prove that it satisfies Equation 6–61 as well as the two initial/boundary conditions listed below as Equation 6–72. Satisfaction of Equation 6–61 is established by computing $\partial C/\partial t$ and $\partial^2 C/\partial x^2$ directly from Equation 6–71 and substituting them into Equation 6–61. The two initial/boundary conditions require that $C(x,0)=0$ for any $x \neq 0$ and that total number of molecules, summed over all positions, remains constant at all times. If A is the total constant cross-sectional area under consideration, then since we began with AB_1 molecules in our original plane, conservation requires that the space available for diffusion continues to have exactly AB_1

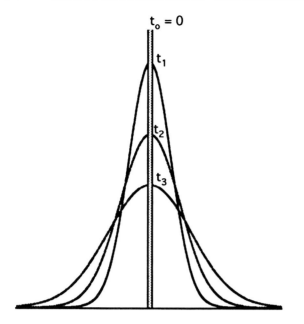

FIG. 6.5. Concentration profiles, described by Equation 6–71, show the fate of solute molecules that originate (time $t=t_o=0$) within the thin shaded rectangle as they diffuse through the solution. The y axis represents the concentration at each position. As time progresses (e.g., at $t=t_1$) the solutes diffuse in both directions, depleting the rectangle and creating a bell-shaped (Gaussian) curve. The curve flattens more and more with time (e.g., $t=t_2$) until at $t=\infty$ the concentration profile becomes uniformly flat.

molecules at all times. Mathematically these conditions are

$$\lim_{t\to 0} C(x,t)=0 \qquad \text{for} \qquad x\neq 0$$

$$\int_{-\infty}^{\infty} C(x,t)dV=\int_{-\infty}^{\infty} C(x,t)Adx=AB_1 \quad (6\text{–}72)$$

The first condition can be verified by application of L'Hopital's rule, while the integral can be evaluated after substituting $x^2/4Dt=v^2$ and noting from standard tables of integrals that (see ref. 1, for example)

$$\int_{-\infty}^{\infty} \exp(-v^2)dv=\sqrt{\pi}$$

If we normalize Equation 6–71 by dividing by B_1, we are left with the following function designated by $Pr(x,t)$:

$$Pr(x,t)=\frac{C(x,t)}{B_1}=\frac{1}{(4\pi Dt)^{1/2}}\exp\left(\frac{-x^2}{4Dt}\right) \quad (6\text{–}73)$$

$Pr(x,t)$ is the normal or Gaussian distribution with a variance (i.e., mean squared displacement from the origin) equal to $2Dt$. A variance $=2Dt$ implies that the standard deviation $=\sqrt{2Dt}$, and it follows from the properties of the normal distribution that for any time

t, 68% of the molecules will lie within a region defined by $x=\pm\sqrt{2Dt}$, and 32% will have diffused beyond those limits.

It is no odd coincidence that our statistical calculation, Equation 6–68, of the mean square displacement of diffusing molecules is also equal to $2Dt$. $Pr(x,t)$ is a probability distribution; the quantity $Pr(x,t)dx$ equals the probability of finding a given molecule at some point between x and $x+dx$ at any time t (15). This interpretation suggests an easy way to generalize our results from one-dimensional (x) to two- (x, y) and three- (x, y, z) dimensional diffusion corresponding to cylindrical and spherical symmetry. In the two-dimensional case we consider the fate of all molecules that originate on a line (the z axis) perpendicular to the x-y plane, lying at $x=y=0$. Since random displacements in the x and y direction are completely independent, Equation 6–73, with y replacing x, also yields the correct probability function for finding a given molecule at y and t. With independent movements the probability of finding the molecule at the more specific position x,y,t will be given by $Pr(x,y,t)=Pr(x,t)$ $Pr(y,t)$, the product of the two independent probabilities. Letting B_2 denote total number of molecules originally present at $t=0$ in each cm of the line on the z axis (at $x=0$, $y=0$), we have

$$C(x,y,t)=B_2Pr(x,t)Pr(y,t)$$

$$=B_2\left(\frac{1}{(4\pi Dt)^{1/2}}\exp\left(\frac{-x^2}{4Dt}\right)\right)\left(\frac{1}{(4\pi Dt)^{1/2}}\exp\left(\frac{-y^2}{4Dt}\right)\right)$$

$$=\frac{B_2}{4\pi Dt}\exp\left(\frac{-(x^2+y^2)}{4Dt}\right)=\frac{B_2}{4\pi Dt}\exp\left(\frac{-r^2}{4Dt}\right)$$

$$(6\text{–}74)$$

Similarly in three dimensions (spherical symmetry), letting B_3 denote total number of molecules originally present at $t=0$ at the point at the origin ($x=0$, $y=0$, $z=0$), we have

$$C(r,t)=\frac{B_3}{(4\pi Dt)^{3/2}}\exp\left(\frac{-r^2}{4Dt}\right) \quad (6\text{–}75)$$

The expressions Equation 6–74 and Equation 6–75 can also be obtained as solutions to Equations 6–58 and 6–59 with appropriate initial/boundary conditions analogous to Equation 6–72. Equations 6–71, 6–74, and 6–75 can be written in the composite form as

$$C(r,t)=\frac{B_n}{(4\pi Dt)^{n/2}}\exp\left(\frac{-r^2}{4Dt}\right) \qquad n=1, 2, 3$$

$$(6\text{–}76)$$

Note that the dependence of Equations 6–76 on the spatial variable (x or r) is independent of n. It follows that, aside from a time-dependent scaling factor, the plots of C versus r will be the same for $n = 1$, 2, or 3.

The Diffusion Front, Where Solute Depletion Ends and Accumulation Begins, Moves as $\sqrt{2nDt}$

Solutions represented by Equation 6–76 yield important insights to problems where the molecules diffuse independently of one another, within spaces that do not contain reflecting barriers, and where the molecules are neither created nor consumed. This follows because we can begin at $t = 0$ with any arbitrary molecular distribution, select source molecules in any plane, line, or point (corresponding to $n = 1$, 2, 3), define a coordinate system placing the source at the origin, and the subsequent fate of the source molecules will be completely determined by Equation 6–76. Since all molecules are assumed to move independently, this description will hold independently of the initial position and subsequent behavior of any nonsource molecule.

Equation 6–76 shows that at $r = 0$, $C(0,t) = B_n (4\pi Dt)^{-n/2}$ so that the original source concentration at $r = 0$ is depleted and decreases as $t^{-n/2}$. This depletion is fastest for the point source (spherical case with $n = 3$), and slowest for the plane source with $n = 1$. Tracking of how adjacent areas accumulate solute and then become depleted as more remote areas begin to acquire solute can be accomplished taking the first and second partial derivatives of Equation 6–76 with respect to r. These are:

$$\frac{\partial C}{\partial r} = -\frac{B_n}{(4\pi Dt)^{n/2}}\left(\frac{r}{2Dt}\right)\exp\left(\frac{-r^2}{4Dt}\right)$$

$$\frac{\partial^2 C}{\partial r^2} = \frac{B_n}{(4\pi Dt)^{n/2}}\left(\frac{r^2 - 2Dt}{(2Dt)^2}\right)\exp\left(\frac{-r^2}{4Dt}\right) \quad (6\text{–}77)$$

and substituting them into Equation 6–62,

$$\frac{\partial C}{\partial t} = D\frac{B_n}{(4\pi Dt)^{n/2}(2Dt)^2}(r^2 - 2nDt)\exp\left(\frac{-r^2}{4Dt}\right)$$
$$n = 1, 2, 3 \quad (6\text{–}78)$$

The negative of Equation 6–77, $-\partial C/\partial r$ is proportional to the flux at each position while $\partial C/\partial t$, given by Equation 6–78, represents the rate of accumulation of solute at each position. These are plotted in Figure 6.6 for $n = 1$, where $\partial^2 C/\partial r^2 = (1/D)/(\partial C/\partial t)$. From Equation 6–78 it is apparent that $(\partial^2 C)/(\partial r^2)$ vanishes whenever $r = \pm\sqrt{2Dt}$, indicating that the flux J will be at a maximum when $r = \pm\sqrt{2Dt}$. However, $\partial C/\partial t$ vanishes at $r = \pm\sqrt{2nDt}$. $\partial C/\partial t$ will be positive for $|r| > |\pm\sqrt{2nDt}|$ and it will be negative for $|r| < |\pm\sqrt{2nDt}|$. Hence $r = \pm\sqrt{2Dt}$ indicates that the position separat-

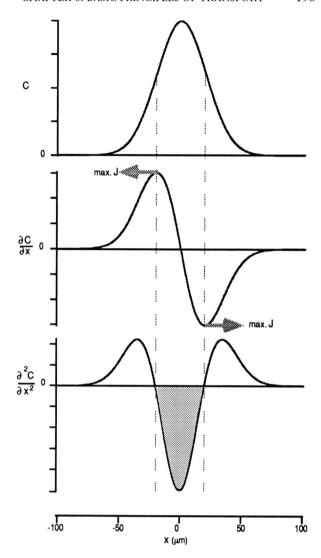

FIG. 6.6. Plots illustrating time course of diffusion from a source originally located at $x = 0$. The top plot is similar to Figure 6.5. The second and third plots illustrate the corresponding flux and solute accumulation respectively ($n = 1$; see text). The shaded area distinguishes regions of solute accumulation from regions of depletion.

ing the region that is losing solute (gray areas in Fig. 6.6) from those regions that are gaining solute. In Figure 6.6, the position $r = \pm\sqrt{2Dt}$ is indicated by the vertical lines on each side of $r = 0$.

In all cases the root mean square displacement of molecules from their original position agrees precisely with the transition position separating regions of accumulation from regions of depletion, giving further credence to the use of $r = \pm\sqrt{2nDt}$ as a measure of the progress of diffusion. In the one-dimensional case, $n = 1$, this position also agrees with the position of maximum flux, but this correspondence disappears for the higher dimensions $n = 2$ and 3.

TABLE 6.2. *Examples of Diffusion Transit Time with* $D = 10^{-5}$ *cm^2 s^{-1}*

Distance x_{rms}	Time for Diffusion $t = \frac{(x_{rms})^2}{2D}$	Example
50 Å	12.5 ns	Thickness of lipid bilayer
200 Å	0.2 μs	Synaptic cleft
1 μm	0.5 ms	Size of mitochondrion
10 μm	50 ms	Radius of mammalian cell
100 μm	5 s	Diameter of large muscle fiber
250 μm	31 s	Radius of giant squid axon
1 mm	8.3 min	1/2 thickness of frog sartorius muscle
2 mm	33.3 min	1/2 thickness of lens (eye)
5 mm	3.5 h	Radius of mature ovarian follicle
2 cm	2.3 days	Thickness of ventricular myocardium

Although use of $r = \pm\sqrt{2nDt}$ to provide estimates of the diffusion distance in time t is a rough approximation, it allows us to distinguish cases where the time dependence of diffusion is significant from those cases where diffusion occurs so rapidly that its time dependence can be ignored. Some simple examples are shown in Table 6.2 for $n = 1$.

Diffusion Transients in Thin Membranes Are Very Rapid

Most treatments of membrane permeability rely on the assumption that the flux of solute is constant at each position within the membrane. We clarify this assumption with the simple case shown in Figure 6.7, where a homogeneous membrane d cm thick separates two well-stirred large baths of water.

At the beginning ($t = 0$), permeable solute is suddenly introduced on the left side (side *in*), raising the concentration from $C = 0$ to $C = C_{in}$. If the baths are sufficiently large these initial concentrations will remain constant at C_{in} on the left where $x < 0$, and virtually 0 on the right where $x > d$. At $t = 0$, solute immediately diffuses into the membrane through the surface facing side *in* ($x = 0$), but little if any leaves the membrane at $x = d$. This follows because it takes time for an appreciable amount of solute to diffuse through the membrane. As time progresses, the concentration of solute within the membrane builds up and solute begins to leave the membrane at $x = d$ at an increasing rate. Eventually solute leaves membrane at the same rate that it enters, and the concentration within the membrane will show no further change. The system has

entered a steady state where the flux through the membrane remains constant.

How long does it take the membrane to reach a steady state? Using Equation 6–71, we estimate that 32% of the molecules that begin at $x = 0$ will have diffused at least as far as $\pm d$ cm by $t = d^2/2D$ s (i.e., 16% will have moved past $x = -d$ to the left, and 16% will have moved past the membrane surface at $x = +d$ on the right). If d is of the order of 10^{-6} cm and D is the same in the membrane as in water, then $t =$

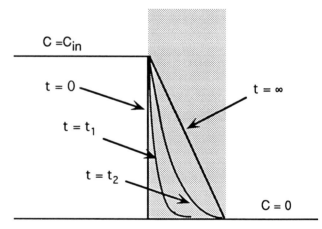

FIG. 6.7. Concentration profiles within a homogeneous membrane following sudden addition of solute to side "in" on the left. The concentration on side "o" at $x > d$ is maintained at zero. The y axis represents concentration, the x axis distance. At time $t = 0$ solute is added at concentration $C = C_{in}$ to the left-hand side. At that instant there is no solute within the membrane: the concentration profile is represented by the black line. As time progresses, the membrane fills with solute and the concentration profile quickly approaches the steady state shown by the straight line at $t = \infty$. Intermediate profiles are represented by the curves labeled t_1 and t_2. For most practical purposes, the steady state is reached "instantly" (see text).

$(10^{-6})^2/2 \times 10^{-5} = 5 \times 10^{-8}$ s, a very short time! Diffusion coefficients within membranes are likely to be smaller than in aqueous solution. Levitt (90), for example, indicates a loss of one order of magnitude within the narrow confines of a gramicidin channel. A more extreme example is provided by the very large hydrophobic anion tetraphenylboron. Using its mobility estimate given by Andersen and Fuchs (6), we calculate a diffusion coefficient of 4×10^{-8} cm^2 s^{-1} for tetraphenylboron in lipid bilayers. However, even if D in the membrane is much smaller, the same qualitative conclusion will hold: e.g., let $D = 10^{-9}$ cm^2 s^{-1} and we still have t smaller than 10^{-3} s. This calculation is only approximate because it ignores the fact that the baths are well stirred, but this will only hasten the transport process so that t will be even smaller than our estimate. The estimate just given strongly suggests that diffusion transients in a thin membrane are negligible: when the concentrations of solute in the bathing solutions change, the membrane shifts from one steady state to the next almost instantly. This conjecture can be verified by a more precise analysis, which solves Equation 6–61 with the appropriate initial and boundary conditions. It can be shown that the flux J approximates its constant steady-state value (error $\leq$ 1%) in a time given by $t = 0.54\, d^2/D$. Again using values of D and d given in the preceding paragraphs, we see that for practical purposes it appears as though the membrane is in a steady state at all times; more succinctly, the membrane is in a "quasi steady state." The quasi steady state implies that the concentration within the membrane does not vary with time. Setting $\partial C/\partial t = 0$ in Equation 6–60, we find $\partial J/\partial x = 0$ so that J can be considered independent of position within the membrane. This assumption is common practice, it has been justified by other arguments for electrodiffusion of ions (21), and we shall make it repeatedly in sections that follow.

Permeation through Membranes Takes Place in At Least Three Steps

To apply these results to cell membranes, we account for at least three processes:

1. Passage of the solute from an aqueous medium into the membrane
2. Diffusion through the interior of the membrane
3. Passage of the solute out of the membrane into the aqueous medium on the opposite side

These three processes are shown in Figure 6.8 in terms of energy barriers.

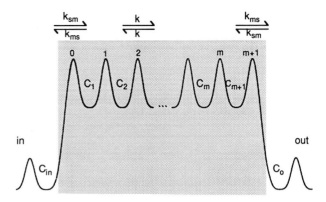

FIG. 6.8. A symmetric homogeneous membrane is represented by a series of energy barriers. Interfacial barriers (barriers 0 and $m+1$) are encountered by the solute upon entering or leaving the membrane. Internal barriers (1 to m) are encountered while passing through the membrane.

There are two interfacial barriers, 0 and $m+1$, which correspond to processes 1 and 3. Rate constants for passage across these barriers are k_{sm} (solution to membrane) and k_{ms} (membrane to solution). The membrane interior is assumed to be homogeneous so that all rate constants for jumping across internal barriers (either forward or backward) are the same and equal to k. The distance between barriers is given by λ, and there are a total of m barriers in the membrane interior. Let C_i equal the concentration in front of the ith barrier and let C_{in} and C_o equal the concentrations in the media adjacent to the two sides of the membrane, as illustrated in Figure 6.8. Further, assuming a quasi steady state within the membrane implies that the flux J through each barrier is equal and, following the development of Equation 6–51, we proceed to account for each barrier as:

	Barrier	
interfacial	0	$J = k_{sm}\lambda C_{in} - k_{ms}\lambda C_1$
internal	1	$J = k\lambda C_1 - k\lambda C_2$
	2	$J = k\lambda C_2 - k\lambda C_3$
	.	. . .
	.	. . .
	.	. . .
	m	$J = k\lambda C_m - k\lambda C_{m+1}$
interfacial	$m+1$	$J = k_{ms}\lambda C_{m+1} - k_{sm}\lambda C_o$ (6–79)

Multiplying both sides of the expressions that involve internal barriers (1 to m) by k_{ms}/k and adding the results for all barriers yields

$$2J + m\frac{k_{ms}}{k}J = k_{sm}\lambda(C_{in} - C_o) \qquad (6\text{–}80)$$

Solving for J,

$$J = \frac{k_{sm}k\lambda}{2k+mk_{ms}}(C_{in}-C_o) = P(C_{in}-C_o) \quad (6-81)$$

where

$$P \equiv \frac{k_{sm}k\lambda}{2k+mk_{ms}} \quad (6-82)$$

is defined as the permeability coefficient. Note from Equation 6-82 that the dimensions of P are the same as $k\lambda$—i.e., cm s^{-1} (like a velocity).

The Exchange Time for Filling or Emptying of Cells Equals V/PA

How long does it take a cell to equilibrate solutes across its membrane? Suppose the cell is suddenly placed in a large volume of a well-stirred solution, where the solute concentration is C_o. If the solute is not metabolized and simply diffuses through the membrane, then we can apply Equation 6-81. Let V equal the cell volume and A the surface area of the cell membrane. Let $n = C_{in}V$ equal the number of moles of solute within the cell at any time, and let $\Delta C = C_{in} - C_o$. Then, by Equation 6-81, the total flux *into* the cell is given by

$$JA = \frac{dn}{dt} = \frac{d(C_{in}V)}{dt} = -PA\Delta C \quad (6-83)$$

Assuming both C_o and V remain constant, Equation 6-83 can be written as

$$\frac{d(C_{in}V)}{dt} = V\frac{dC_{in}}{dt} = V\frac{d\Delta C}{dt} = -PA\Delta C$$

or

$$\frac{d\Delta C}{\Delta C} = -\frac{PA}{V}dt \quad (6-84)$$

Integrating Equation 6-84 and using the condition that at $t=0$, $\Delta C = \Delta C(0)$, we have

$$\Delta C = \Delta C(0)e^{-\frac{PA}{V}t} = \Delta C(0)e^{-t/\tau} \quad (6-85)$$

where

$$\tau = \frac{V}{PA} \quad (6-86)$$

is called the exchange time. Since $e^{-3} = 0.05$, after three exchange times have elapsed the concentration gradient will have collapsed to less than 5% of its initial value. When τ is small, equilibration is fast. τ depends on the (1) membrane permeability; (2) the shape of the cell

(the more spherical, the larger V/A); and (3) the size of the cell. For a given shape, the smaller the cell the smaller the V/A. For example, an isolated cell organelle can be expected to equilibrate faster than a whole cell, even if both membranes were identical. The half-time $t_{1/2}$ for equilibration can be obtained by substituting $t = t_{1/2}$ when $\Delta C(t) = \Delta C(0)/2$ into Equation 6-85, yielding $t_{1/2} = 0.69\ \tau$.

In Simple Exponential Processes, the Time Constant = the Mean Residence Time

For simple processes where a reservoir fills or empties exponentially the time constant is equal to the residence time, i.e., it equals the average time a molecule spends within the reservoir. This can be shown via our last example of a cell emptying solute by simple diffusion through its membrane. Suppose at $t=0$ the solute inside the cell consisting of N_o molecules is suddenly labeled. We follow each labeled molecule, counting time in equal intervals of Δt, and noting how many molecules leave within each interval. If Δn_i leave in the interval between t_i and $t_i + \Delta t$, then the time these Δn_i molecules spend within the cell is approximately t_i, and the average time spent by *all* molecules within the cell is given by

$$t_{ave} \approx \frac{\Delta n_0 t_{0+}\Delta n_1 t_1 + \ldots}{\Delta n_0 + \Delta n_1 + \ldots} = \frac{\sum\limits_{i=0}^{\infty}\Delta n_i t_i}{N_o} \quad (6-87)$$

As Δt becomes smaller, our approximation becomes better. To apply Equation 6-85 note that there are no labeled molecules in the external solution, so that $C_o = 0$ and Equation 6-85 reduces to $C_{in} = C_{in}(0)e^{-t/\tau}$. Multiplying both sides by the constant V and letting $N_t = C_{in}V$ denote the total number of molecules inside the cell at any time t, we have

$$N_t = N_o e^{-t/\tau} \quad (6-88)$$

With a very small Δt, we can use this to approximate Δn_i because the number that leave in any time interval between t and $t+dt$ (i.e., those whose "age" lies between t and $t+dt$) is given by $-dN_{t_i}$ (the minus sign occurs because we count the number of molecules that *leave* the cell for the environment—each time the environment gains one, the cell loses one). After differentiating Equation 6-88, we have

$$\Delta n_i = -dN_{t_i} \approx -\frac{dN_{t_i}}{dt}\Delta t = \frac{1}{\tau}N_o e^{-\frac{t_i}{\tau}}\Delta t \quad (6-89)$$

Using this result together with our expression for t_{ave}, we have

$$t_{ave} \approx \sum_{i=0}^{\infty} \frac{t_i}{\tau} e^{-\frac{t_i}{\tau}} \Delta t \qquad (6-90)$$

As Δt gets smaller and smaller, our approximation becomes more exact and the summation approaches an integral. We obtain our result $t_{ave} = \tau$ by writing

$$t_{ave} = \int_0^{\infty} \frac{t}{\tau} e^{-t/\tau} dt = \tau \int_0^{\infty} x e^{-x} dx = \tau \qquad (6-91)$$

where we have evaluated the integral by making the substitution $x = t/\tau$, and consulting standard tables of integrals.

Cytoplasmic Diffusion Transients Are Rapid

The analysis in the preceding section assumed that solute entering the cell behaved as though it mixed instantaneously with the contents of the cytoplasm; the time required for diffusion within the cytoplasm was ignored so that the rate of solute release (or uptake) was determined solely by the properties of the membrane. To test this assumption, we consider the opposite extreme by assuming transport through the membrane is rapid compared to cytoplasmic diffusion. In this case, the release is determined almost entirely by diffusion within the cell. Consider a spherical cell of radius R releasing solute into a well-stirred infinite medium. The concentration profile within the cell at all times can be obtained by solving Equation 6–58 with the conditions that $C = 0$ at all times in the bath (where $r > R$) and that the initial concentration within the cell is $C = $ constant at all points where $r < R$. Integration of this profile throughout the cell interior provides an expression for the average intracellular concentration $\bar{C}$. It can be shown (50) that the time $(t_{1/2})_D$ taken for $\bar{C}$ to reach half its initial concentration is given by

$$(t_{1/2})_D = 0.03 \frac{R^2}{D} \qquad (6-92)$$

Compare this with the analogous result Equation 6–86 calculated for the case where membrane permeability was rate limiting. Using a spherical cell, $V/A = R/3$ so that the corresponding half-time for the permeability limiting case is

$$(t_{1/2})_P = 0.69\tau = 0.23 \frac{R}{P} \qquad (6-93)$$

If $(t_{1/2})_P >> (t_{1/2})_D$, then membrane permeation is much slower than cytoplasmic diffusion. In this case, transport is rate limited by membrane permeation and the solute concentration will be virtually uniform through-

out the cytoplasm. Using Equations 6–92 and 6–93, this criterion for permeation limited transport can be written as

$$\frac{D}{RP} >> 0.13 \qquad (6-94)$$

This inequality is satisfied for water-soluble solutes permeating most cells. Other things being equal, the smaller the cell, the larger D/RP—i.e., the easier it is to satisfy Equation 6–94. If diffusion in the cytoplasm is similar to water (with $D \approx 10^{-5}$ cm·s^{-1}), then even if we take an unusually large cell, with $R = 100$ μm $= 10^{-2}$ cm, Equation 6–94 will be satisfied for any solute with $P << 7.7 \times 10^{-3}$ cm s^{-1}, and we may assume that C is uniform throughout the cytoplasm. If cytoplasmic D is significantly less than 10^{-5} cm s^{-1}, then our conclusion will have to be modified. However, for most cells R is an order of magnitude smaller than 100 μm and for most solutes P is also very much smaller than 7.7×10^{-3} cm s^{-1}, so there is room for D to be reduced without violating the inequality.

Unstirred Layers Can Be a Significant Barrier

Mechanical stirring is never perfect, because there is always a thin film of fluid adhering to the membrane where turbulent movements produced by stirring are very sparse or absent. Effects of this stagnant layer on permeability is estimated by assuming that stirring is perfect in the bulk of the solution but totally absent at a distance δ from the membrane. Thus, the membrane is surrounded by an unstirred layer δ cm thick where transport can occur only by diffusion. Actually, turbulent stirring does not suddenly stop at a specific distance; rather, it diminishes erratically as the membrane is approached. The concept of an unstirred layer is a simple idealization, that has proved to be very useful and fairly accurate in practice (50).

To show how the unstirred layer affects permeability measurements, we develop the model illustrated in Figure 6.9, which shows three membranes in series.

The two outer membranes with permeabilities P_1 and P_2 represent water films (unstirred layers) with thicknesses δ_1 and δ_2 cm. The middle shaded membrane (with permeability P_m) represents the "real" membrane. Solute on the left-hand side is present in a stirred bath at concentration C_{in}. It diffuses through the first unstirred layer dropping to C_{m_in}, then through the membrane where it drops to C_{m_o}, and finally through the second unstirred layer to reach the stirred bath concentration C_o on the right.

Estimates of P_m are commonly obtained by dividing the flux J through the system by the appropriate con-

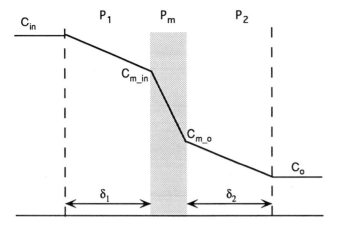

FIG. 6.9. Effects of unstirred layers on permeability. The illustration, consisting of three "membranes" in series, shows the concentration profiles through each of these. Concentration is plotted on the y axis, distance on the x axis. P_1 and P_2 represent the permeabilities of the two unstirred layers; P_m represents the permeability of the membrane.

centration difference ΔC. If we were unaware of the unstirred layers, we would set $\Delta C = C_{in} - C_o$ and erroneously assign the value of the composite permeability, P_{comp}, of the three membranes in series to P_m. A correct value for P_m is obtained only when we use $\Delta C = C_{m\_in} - C_{m\_o}$, but these concentrations are rarely available to the experimenter. We use a less direct approach by analyzing the composite system as the sum of its parts as follows. In a steady state the flux through each membrane will be equal. For the composite system we have

$$\frac{J}{P_{comp}} = C_{in} - C_o \qquad (6-95)$$

while for each component

$$\frac{J}{P_1} = C_{in} - C_{m\_in}$$

$$\frac{J}{P_m} = C_{m\_in} - C_{m\_o}$$

$$\frac{J}{P_2} = C_{m\_o} - C_o \qquad (6-96)$$

Adding these three, we have

$$J\left(\frac{1}{P_1} + \frac{1}{P_m} + \frac{1}{P_2}\right) = C_{in} - C_o \qquad (6-97)$$

and comparing this sum with the expression for the composite yields

$$\frac{1}{P_{comp}} = \frac{1}{P_1} + \frac{1}{P_m} + \frac{1}{P_2} \qquad (6-98)$$

Equation 6–98 reflects the intuitive result that the observed permeability P_{comp} will equal P_m whenever the unstirred-layer permeabilities P_1 and P_2 are much higher than P_m. Estimates for P_1 and P_2, can be obtained from Equation 6–82. If a membrane is simply a film of water, then $k_{sm} = k_{ms} = k$ and Equation 6–82 becomes

$$P = \frac{k\lambda^2}{(2+m)\lambda} \qquad (6-99)$$

Noting that the total number of barriers $(2+m)$ times the barrier width λ is simply the thickness δ and, from Equation 6–54, using $k\lambda^2 = D$, we find that Equation 6–99 becomes $P = D/\delta$ so that Equation 6–98 yields

$$\frac{1}{P_{comp}} = \frac{1}{P_m} + \frac{\delta_1}{D} + \frac{\delta_2}{D} \qquad (6-100)$$

or

$$P_m = \frac{1}{1 - (\delta_1 + \delta_2)\dfrac{P_{comp}}{D}} P_{comp} \qquad (6-101)$$

It follows that the criterion for ignoring unstirred layers is

$$(\delta_1 + \delta_2)P_{comp} << D \approx 10^{-5} \text{ cm}^2 \text{ s}^{-1} \qquad (6-102)$$

If inequality 6–102 is not satisfied, then Equation 6–101 can be used to calculate P_m.

Both Equations 6–101 and 6–102 require the value of $\delta_1 + \delta_2$. This can be estimated in several ways. The simplest is to measure the composite permeability to a small lipid-soluble molecule that is known to have a very high P_m. If P_m is sufficiently large, the term $1/P_m$ can be ignored in Equation 6–100 so that P_{comp} is determined primarily by the unstirred layer and Equation 6–100 can be used to estimate $\delta_1 + \delta_2$. Once the value of $\delta_1 + \delta_2$ is known for any system, it can be used repeatedly in Equation 6–101 or 6–102 for any other solute. Even if the lipid-soluble permeability is not fast enough to ignore $1/P_m$, the method can be used for an upper-limit estimate of the size of the unstirred layer, and in many instances that is all that is required.

Another method used in ion-transport studies for estimating unstirred layer size is to suddenly change the concentration of a permeable ion in the stirred bath while continuously monitoring the membrane potential. The time required to reach a new steady value of the membrane potential is then used to calculate the size of the unstirred layer (28, 144). Measurements of unstirred layers range all the way from 1 to 500 μm, depending on the system. They are especially large in chamber systems that involve large areas of membrane, e.g., in black lipid bilayers and epithelial sheets. (For a more detailed discussion of unstirred layers—e.g., in the vicinity of channels—see 12, 20, 51, 76.)

Membrane Diffusion Is Assumed to Be Rate-Limiting for Plasma Membranes

Returning to our more detailed analysis of plasma membrane permeability, we examine the relative magnitude of individual barriers by taking the reciprocal of Equation 6–82.

$$\frac{1}{P}=\frac{2}{k_{sm}\lambda}+\frac{mk_{ms}}{k_{sm}k\lambda}=\frac{2\lambda}{k_{sm}\lambda^2}+\frac{mk_{ms}\lambda}{k_{sm}k\lambda^2} \quad (6-103)$$

To interpret this expression, first note that k_{sm}/k_{ms} is the rate constant for migration of solute from solution to membrane, divided by the rate constant for migration from membrane to solution. This ratio is the solubility or partition coefficient (it is the kinetic equivalent of Equation 6–30) of the solute in the membrane, which we denote by

$$K=\frac{k_{sm}}{k_{ms}} \quad (6-104)$$

Further, letting d denote the thickness of the membrane, we have $d=(2+m)\lambda$. If d is of the order of 50–100 Å and λ is simply a few angstroms at most, it follows that $m >> 2$ and $d=(2+m)\lambda \approx m\lambda$. Finally, note that the diffusion coefficient in the membrane is given by $D_m=k\lambda^2$. By analogy we also set $D_{sm}=k_{sm}\lambda^2$ so that Equation 6–103 becomes

$$\frac{1}{P}=\frac{2\lambda}{D_{sm}}+\frac{d}{KD_m}=\frac{\lambda}{D_{sm}}+\frac{d}{KD_m}+\frac{\lambda}{D_{sm}} \quad (6-105)$$

Comparing with Equation 6–98, we can interpret the three terms in Equation 6–105 as arising from transport through three membranes in series. The first represents passage through the solution–membrane interface with "permeability" D_{sm}/λ, the second arises from diffusion through the membrane with permeability KD_m/d, and the third arises from passage through the membrane-solution interface, again with "permeability" D_{sm}/λ. Most applications assume that diffusion through the membrane is rate limiting, so that $KD_m/d << D_{sm}/2\lambda$, allowing Equation 6–105 to be replaced by

$$P=\frac{KD_m}{d} \quad (6-106)$$

The same result can be obtained directly from Equation 6–82 by assuming that $mk_{ms} >> 2k$. It can also be derived by integrating Equation 6–50 with the assumption that the interfacial processes are fast enough to establish a quasi-equilibrium at the solu-

tion–membrane interface at all times. Although independent estimates of the components of Equation 6–106 give an accurate description of water diffusion through planar lipid bilayers (37), in most cases the validity of the underlying assumptions are difficult to test. Nevertheless, Equation 6–106 provides the point of departure for many studies and has led to useful results, as shown below.

Selective Permeability of Lipid Bilayers Is Determined Primarily by Solubility

It is apparent from Equation 6–106 that P will increase with lipid solubility (because of K) and decrease with molecular size (because of D_m; see Table 6.1) of the solute. Both factors vary from solute to solute. However, for lipid bilayer membranes the variation of lipid solubility is orders of magnitude times greater than corresponding variations in molecular size (and presumably D_m). Thus, variations in permeability from solute to solute is determined primarily by lipid solubility.

This is illustrated in Figure 6.10, which shows a plot of nonelectrolyte permeability in egg phosphatidylcho-

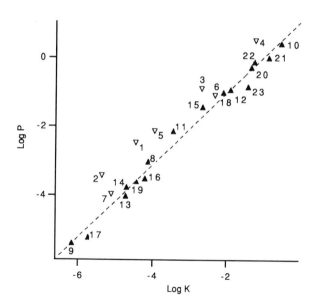

FIG. 6.10. A plot of nonelectrolyte permeability in egg phosphatidylcholine-decane bilayers as a function of their partition coefficient in hexadecane redrawn from the paper of Walter and Gutknecht (138). Numbers in the figure correspond to the solutes as follows: 1. water; 2. hydrofluoric acid; 3. ammonia; 4. hydrochloric acid; 5. formic acid; 6. methylamine; 7. formamide; 8. nitric acid; 9. urea; 10. thiocyanic acid; 11. acetic acid; 12. ethylamine; 13. ethanediol; 14. acetamide, 15. propionic acid; 16. 1,2-propanediol; 17. glycerol; 18. butyric acid; 19. 1,4-butanediol; 20. benzoic acid; 21. hexanoic acid; 22. salicylic acid; 23. codeine.

line-decane bilayers as a function of their partition coefficient in hexadecane taken from the paper of Walter and Gutknecht (138). The correlation is excellent even with permeabilities covering a range of 10^6. Similar results were obtained using olive oil instead of hexadecane, but poorer correlations occurred with the more polar solvents octanol and ether. Those solutes that fall furthest from the regression line (open triangles) all have molecular weights less than 50. But even for these, the permeabilities cover a range of 10,000 while the deviations from the regression line are maximally 2- to 15-fold. This result underscores the fact that lipid bilayer permeability is determined primarily by lipid solubility, with other factors playing a secondary role.

In attempting to evaluate these secondary effects, Walter and Gutknecht showed that the deviations of the small solutes from the regression line, in contrast to the others (with molecular weights ranging from 50 to 300), show a steep dependence on molecular volume. They interpret this in terms of the dependence of membrane diffusion coefficient on molecular size. After comparing this size dependence of small and large solutes with comparable studies of diffusion in artificial media, they conclude that the membrane barrier behaves more like a soft polymer (e.g., rubber, polyurethane) than a hydrocarbon liquid (see also 96, 142).

A qualitative description of the dependence of lipid solubility and corresponding biological membrane permeability on specific chemical structure is given by Overton's empirical rules. Examples of these rules cited by Wright and Diamond (143) are:

1. Permeability and lipid solubility will increase with:

    a. increased alkylation
    b. increased chain length
    c. replacement of oxygen by sulfur
    d. halogenation

2. Permeability and lipid solubility will decrease with the introduction of:

    a. oxygen groups (hydroxyl, ketone, aldehyde, ether, and carboxylic acid)
    b. amino groups

Details of how these rules follow from considerations of intermolecular forces between solute and membrane hydrocarbon (or lipid solvents) are discussed by Diamond and Wright (32). A more recent review on predicting partition coefficients from molecular structure has been written by Leo (82).

WATER TRANSPORT

Water Transport Can Be Driven by Three Different Gradients

In principle water permeation through a membrane can be measured in three different ways, corresponding to the three driving forces hydraulic, osmotic, and diffusional gradients (see Fig. 6.11).

Figure 6.11A depicts a "filtration" experiment where a hydraulic piston pushing on side "in" forces water through the membrane to side "o." A very permeable mechanical support is provided to prevent mechanical deformation (in plants and bacteria this corresponds to a cell wall). In part B an osmotic solute on side "o" draws the water toward it, and in part C labeled water is placed on side "in," and we simply measure the appearance of labeled water on side "o" as it diffuses down its concentration gradient.

Although hydraulic or osmotic experiments usually express water flux in terms of volume flow, J_{vol} ($cm^3 \ s^{-1}$), while diffusional flux is generally measured in molar terms, J_{molar} (moles s^{-1}), these are easily interconverted. If V_w is the molar volume of water, we use $J_{molar} = J_{vol}/V_w$ to convert all fluxes to molar terms and define three permeability coefficients by the following relations:

$$J_f \equiv P_f \frac{\Delta P}{RT}, \qquad J_{osm} \equiv P_{osm} \frac{-\Delta \Pi}{RT}, \qquad J_d \equiv P_d \Delta C^*$$

$$(6-107)$$

where J_f, J_{osm}, and J_d are molar fluxes measured under hydrostatic pressure, P, osmotic, Π, and labeled water concentration, C^*, gradients, respectively. The term RT is included in the first two definitions so that the driving force units will all be consistent in moles cm^{-3}.

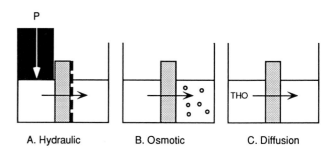

A. Hydraulic   B. Osmotic   C. Diffusion

FIG. 6.11. Three ways of measuring water flow through a membrane. A. A hydraulic piston pushing on side "in" forces water through the membrane to side "o." A permeable mechanical support is provided to prevent mechanical deformation of the membrane. B. An impermeable solute on side "o" draws the water toward it. C. Flux of labeled water placed on side "in."

The minus sign appears in the osmotic flux equation because water flows *up* an osmotic gradient, from low to high osmotic pressures.

Since all of these experiments deal with water flow through the same membrane, it is not unreasonable to expect the permeability coefficients to be related. Empirical results show that all of these coefficients are identical in nonporous membranes where water molecules permeate individually, independently of one another. However, in porous membranes where water may permeate in bulk, results show that while $P_f = P_{osm}$, P_f and P_{osm} are distinctly greater than P_d. The following paragraphs examine the basis for these results.

In Lipid Membranes Water Transport Occurs by Solubility-Diffusion Mechanism, with $P_f = P_{osm} = P_d$

The demonstration that $P_f = P_{osm} = P_d$ in nonporous membranes is based on the fact that water is sparingly soluble and can be treated as a dilute solute in the membrane phase. In each of the three cases, examination of the water partitioning at each membrane interface will show that a concentration gradient for water is established within the membrane, implicating a common mechanism; water simply diffuses down its concentration gradient with the same diffusion coefficient.

To analyze water partitioning, we assume equilibrium at the interfaces. Letting μ_{aq} denote the chemical potential for water in the aqueous phase just adjacent to the membrane and μ_m its chemical potential just within the membrane, we have $\mu_{aq} = \mu_m$. An explicit expression for μ_{aq} can be obtained directly from Equation 6–34. However, since water is a dilute *solute* within the membrane phase the corresponding expression for μ_m is obtained from Equation 6–22. (In this case it will be expedient to use the mole fraction form of the chemical potential to avoid problems with dimensions.) If C_i denotes the concentration of an impermeable solute dissolved in the aqueous phases only, we write from Equation 6–34 and Equation 6–22:

$$\mu_{aq} = \overline{\mu}_{aq}^{oo}(T) + PV_{w\_aq} - V_{w\_aq}RT\Sigma C_i$$
$$= \overline{\mu}_{aq}^{oo}(T) + PV_{w\_aq} - \Pi V_{w\_aq}$$
$$\mu_m = \overline{\mu}_m^{oo}(T) + PV_{w\_m} + RT \ln (\overline{n}_m)$$

$$(6\text{--}108)$$

where $\Pi = RT\Sigma C_i$ and V_{w_aq} and V_{w_m} are the molar volumes of water in the aqueous and membrane phase, respectively.

Setting $\mu_{aq} = \mu_m$ on the interface facing side "in" and solving for $\overline{n}_{m\_in}$, the mole fraction of water leaves:

$$\overline{n}_{m\_in} =$$
$$\exp\left(\frac{(\overline{\mu}_{w\_aq}^{oo} - \overline{\mu}_{w\_m}^{oo}) + (P_{aq\_in}V_{w\_aq} - P_{m\_in}V_{w\_m}) - \Pi_{in}V_{w\_aq}}{RT}\right)$$
$$= K^o \exp\left(\frac{P_{aq\_in}V_{w\_aq} - P_{m\_in}V_{w\_m} - \Pi_{in}V_{w\_aq}}{RT}\right)$$

$$(6\text{--}109)$$

where $K^o = \exp\left\{\dfrac{-(\overline{\mu}_{w\_m}^{oo} - \overline{\mu}_{w\_aq}^{oo})}{RT}\right\}$

Let V_m denote the partial molar volume of the major membrane component (e.g., oil), and let C denote water concentration. Then the concentration of water in the membrane at the inner interface C_{m_in} is given by

$$C_{m\_in} = \frac{\overline{n}_{m\_in}}{V_m}$$

so that

$$C_{m\_in} = \frac{K^o}{V_m} \exp\left\{\frac{P_{aq\_in}V_{w\_aq} - P_{m\_in}V_{w\_m} - \Pi_{in}V_{w\_aq}}{RT}\right\}$$

$$(6\text{--}110)$$

Repeating the argument for side "o" and taking the difference yields the water concentration difference, ΔC_m, across (but within) the membrane.

$$\Delta C_m \equiv C_{m\_in} - C_{m\_o}$$
$$= \frac{K^o}{V_m}\left(\exp\left\{\frac{P_{aq\_in}V_{w\_aq} - P_{m\_in}V_{w\_m} - \Pi_{in}V_{w\_aq}}{RT}\right\}\right.$$
$$\left. - \exp\left\{\frac{P_{aq\_o}V_{w\_aq} - P_{m\_o}V_{w\_m} - \Pi_{o}V_{w\_aq}}{RT}\right\}\right)$$

$$(6\text{--}111)$$

If we restrict our attention to reasonable pressure gradients—say, less than one or two atmospheres—then since $RT \approx 22.4$ L atm mole^{-1} and V_w is of the order of 0.018 L mole^{-1} the PV terms in the exponential will be much less than unity. Further, in dilute solutions osmotic gradients are never greater than, say, 1 Osm so that, with $V_{w\_aq} = 0.018$ L mole^{-1}, $(\Pi V_{w\_aq})/(RT) = V_{w\_aq}\Sigma C_i$ is also much less than unity. It follows that to a very good approximation the exponential terms can be expanded in linear form (i.e., $e^x \approx 1 + x$), leaving

$$\Delta C_m$$
$$= \frac{K^o}{V_m}\left(\frac{(P_{aq\_in} - P_{aq\_o})V_{w\_aq} - (P_{m\_in} - P_{m\_o})V_{w\_m} - (\Pi_{in} - \Pi_o)V_{w\_aq}}{RT}\right)$$

$$(6\text{--}112)$$

Finally because of the mechanical support there will be no hydrostatic pressure drop across the membrane;

it occurs entirely across the support. Accordingly, $P_{m\_in} = P_{m\_o}$ and we have

$$\Delta C_m = \frac{V_{w\_aq}}{V_m} K^o \left(\frac{(P_{aq\_in} - P_{aq\_o}) - (\Pi_{in} - \Pi_o)}{RT} \right)$$

$$= \frac{V_{w\_aq}}{V_m} K^o \left(\frac{\Delta P}{RT} - \frac{\Delta \Pi}{RT} \right) \qquad (6\text{--}113)$$

Let D_m denote the diffusion coefficient for water through the membrane. Using Equations 6–81 and 6–106 (with $K = 1$ because we have already accounted for interfacial partitioning) we arrive at

$$J = \frac{V_{w\_aq} K^o D_m}{V_m d} \left(\frac{\Delta P}{RT} - \frac{\Delta \Pi}{RT} \right) \qquad (6\text{--}114)$$

which holds for either the hydraulic experiment, the osmotic experiment, or both. Setting the osmotic gradient = 0 and comparing Equations 6–107 and 6–114 shows that

$$P_f = \frac{V_{w\_aq} K^o D_m}{V_m d}$$

A similar comparison with $\Delta P = 0$ shows that

$$P_{osm} = \frac{V_{w\_aq} K^o D_m}{V_m d}$$

Thus $P_f = P_{osm}$.

To show that $P_d = P_f$, we modify the same argument by noting that THO is a dilute *solute* in *both* aqueous and membrane phases. Hence we now use Equation 6–22 for both μ_{aq} and μ_m. (Also note that there is no pressure difference in the two solutions so that $P_{aq\_in} = P_{aq\_o} = P_{atm}$.) Instead of Equation 6–109, the mole fraction, $\bar{n}^*$, of THO is given by

$$\bar{n}^*_{m\_in} = \bar{n}^*_{aq\_in} K^o \exp \left\{ \frac{P_{atm} V_{w\_aq} - P_{m\_in} V_{w\_m}}{RT} \right\}$$

$$\approx \bar{n}^*_{aq\_in} K^o \left(1 + \frac{P_{atm} V_{w\_aq} - P_{m\_in} V_{w\_m}}{RT} \right)$$

$$\bar{n}^*_{m\_o} = \bar{n}^*_{aq\_o} K^o \exp \left\{ \frac{P_{atm} V_{w\_aq} - P_{m\_o} V_{w\_m}}{RT} \right\}$$

$$\approx \bar{n}^*_{aq\_o} K^o \left(1 + \frac{P_{atm} V_{w\_aq} - P_{m\_o} V_{w\_m}}{RT} \right)$$

$$(6\text{--}115)$$

where expansion of the exponential is again justified by the small values of $P_x V_x$ compared to RT. Repeating the argument yields a corresponding expression for

side "o." Letting C^* denote the concentration of THO, we substitute

$$C^*_m = \frac{\bar{n}^*_m}{V_m}, \quad C^*_{aq} = \frac{\bar{n}^*_{aq}}{V_{w\_aq}}$$

and recalling that $P_{m\_in} = P_{m\_o}$ we find ΔC^* given by

$$\Delta C^*_m = \frac{V_{w\_aq} K^o (C^*_{aq\_in} - C^*_{aq\_o})}{V_m} \left(1 + \frac{P_{atm} V_{w\_aq} - P_{m\_in} V_{w\_m}}{RT} \right)$$

$$\approx \frac{V_{w\_aq} K^o (C^*_{aq\_in} - C^*_{aq\_o})}{V_m} \qquad (6\text{--}116)$$

where we have used the fact that $(P_{atm} V_{w\_aq} - P_{m\_in} V_{w\_m})/RT \ll 1$. Applying Equation 6–106 shows that

$$J = \frac{V_{w\_aq} K^o D_m}{V_m d} (C^*_{aq\_in} - C^*_{aq\_o})$$

with

$$P_d = \frac{V_{w\_aq} K^o D_m}{V_m d} = P_f = P_{osm} \qquad (6\text{--}117)$$

and thus the driving force for water movement in all three cases is simple diffusion.

Osmotic Gradients Generate Hydraulic Pressure Gradients in Aqueous Channels

Aqueous channels formed by transmembrane proteins consist of a continuum of water stretching through the membrane, forming a water bridge between the two bathing solutions. When impermeable solute is added to one side of the membrane it induces an "osmotic" flow of water. What generates this flow? Again, our analysis is based on an examination of the water partitioning at each membrane interface. In this case, however, it is apparent that water cannot be considered a dilute solute within the membrane; it remains a solvent within the channel as well as in the two bathing solutions. As a consequence we shall show that an impermeable solute induces a hydrostatic pressure gradient that generates hydraulic flow in the channel (101,102).

Consider the porous membrane illustrated in Figure 6.12. Side "o" contains an impermeable solute that cannot enter the channel, while side "in" is pure water. We assume rapid equilibrium at the membrane interfaces. Letting subscript m refer to the channel and using Equation 6–34 in all cases, we set $\mu_{aq} = \mu_m$ at each interface. Since there is no solute within the channel or in the bathing medium on side "in," we have:

$$\mu_{aq\_in} = \bar{\mu}^{oo}_{aq}(T) + P_{aq\_in} V_{w\_aq}$$

$$\mu_{m\_in} = \bar{\mu}^{oo}_m(T) + P_{m\_in} V_{w\_m} \qquad (6\text{--}118)$$

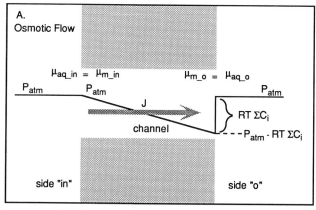

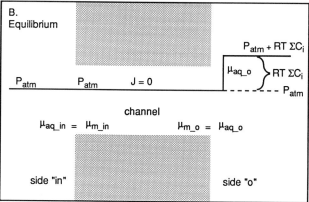

FIG. 6.12. A membrane water channel forms a continuum bridging the two bathing solutions. Both side "in" and the channel contain pure water, while side "o" contains an impermeable solute. Rapid equilibrium at the interfaces requires continuity of the chemical potential for water, but the impermeable solute on side "o" creates a discontinuity in its concentration at the "o" interface. These boundary conditions are satisfied when there is an equal and opposite discontinuity in hydrostatic pressure at the "o" interface. *Panel A:* Plot of hydrostatic pressure profile through the membrane. The pressure drop across the membrane creates a *hydraulic* flow of water through the membrane even though both bathing media are at atmospheric pressure. *Panel B:* Equilibrium ($J=0$) ensues when the cryptic pressure drop within the membrane is wiped out by the application of a hydrostatic pressure $(P-P_{atm})=\Delta\Pi=RT\Sigma C_i$ to side "o."

If we assume that channel water is no different from water in the bathing media, then $\overline{\mu}_{aq}^{oo}(T)=\overline{\mu}_m^{oo}(T)$ and $V_{w\_aq}=V_{w\_m}$, so that after setting $\mu_{aq\_in}=\mu_{m\_in}$ we are left with a simple equality of pressure across the solute-free interface, i.e.,

$$P_{m\_in}=P_{aq\_in} \qquad (6-119)$$

For interface "o" μ_m does not change, but μ_{aq} must account for the impermeable solute. We have

$$\mu_{aq\_o}=\overline{\mu}_{aq}^{oo}(T)+P_{aq\_o}V_{w\_aq}-V_{w\_aq}RT\sum C_i$$

$$\mu_{m\_o}=\overline{\mu}_m^{oo}(T)+P_{m\_o}V_{w\_m} \qquad (6-120)$$

Again with $\overline{\mu}_{aq}^{oo}(T)=\overline{\mu}_m^{oo}(T)$ and $V_{w\_aq}=V_{w\_m}$ and setting $\mu_{aq\_o}=\mu_{m\_o}$, we now find the following discontinuity of pressure across the interface on side "o":

$$P_{m\_o}=P_{aq\_o}-RT\sum C_i=P_{aq\_o}-\Pi_o \qquad (6-121)$$

Subtracting Equations 6–119 and 6–121 shows that $\Delta\Pi P_m$, the hydrostatic pressure gradient within the membrane, induced by the osmotic solute is given by

$$\Delta P_m \equiv P_{m\_in}-P_{m\_o}=P_{aq\_in}-(P_{aq\_o}-\Pi_o)=\Delta P_{aq}+\Pi_o$$
$$(6-122)$$

where $\Delta P_{aq} \equiv P_{aq\_in}-P_{aq\_o}$ is the imposed hydrostatic pressure difference between the two aqueous bathing solutions. Most often the bathing solutions are both exposed to the same atmosphere—in this case, $\Delta P_{aq}=0$. Even in this case where there is no external evidence of a hydrostatic pressure gradient, Equation 6–122 shows that a hydraulic gradient equal to Π_o will be induced *within* the membrane, with the pressure just within the membrane at the solute side (side "o" in this case) reduced Π_o units below atmospheric (Equation 6–122). It is this hydraulic pressure gradient that drives the osmotic flow. With porous membranes osmosis no longer has the characteristics of water diffusion; it moves in bulk, more like water driven through a pipe.

The hydraulic pressure drop across a porous membrane induced by an osmotic gradient is absent during simple diffusion and thus P_{osm} and P_d are not equal. For example, compare the solute flux through a single channel when the channel cross section is sufficiently large to treat its water contents as a continuum. Osmotically induced hydraulic flow will resemble flow through a pipe and will follow Poiseuille's law*; the flow (and consequently the hydraulic permeability) will be proportional to the fourth power of the channel radius. On the other hand, diffusional flux through the same channel will be simply proportional to the cross-sectional channel area—i.e., it will be proportional to the radius squared. Thus there is no a priori basis to predict the relation between P_f and P_d; it will depend on channel geometry.

At osmotic equilibrium there can be no driving forces and ΔP_m must vanish. With a nonzero Π_o Equation 6–122 shows that this can be attained only with a hydrostatic pressure gradient between aqueous phases that is equal and opposite to Π_o—i.e., $\Delta P_{aq} \equiv P_{aq\_in}$

---

* Poiseuille's law relates the fluid flow F (cm³/sec), $=J_fV_wA$, in a cylindrical pipe, of cross-sectional area A, radius r and length L, to ΔP, the pressure gradient driving the flow, as $F=(\pi r^4/8\eta L)\Delta P$, where η is the fluid viscosity (145). In a large pipe $K=1$ and using Equations 6–106 and 6–107 the total diffusional flux through the same pipe is given by $J_dA=P_dA\Delta C=(\pi r^2D/L)\Delta C$.

$-P_{aq\_o} = -\Pi_o$. This is simply the definition of osmotic pressure. These two cases are illustrated in Figure 6.12.

In Narrow Channels the Ratio
$P_{osmotic}/P_{diffusion}$ = the Number of Water Molecules Contained within the Channel

The other extreme, narrow channels that constrain water molecules to move in a single file and prevent them from passing one another, is probably a closer approximation to natural channels found in cell membranes. Analysis of these channels leads to a remarkable relation that allows the average number of water molecules occupying a channel to be calculated directly from macroscopic measurements; the number of molecules is simply equal to P_{osm}/P_d, the ratio of osmotic to diffusional permeability. This observation was originally derived for arbitrary channel occupancies by Levitt (85) and further modified by Finkelstein and Rosenberg (39).

To derive the relation we first consider a general mechanical model where the flux is induced by some unspecified force F; e.g., F could arise from osmotic or hydraulic pressure gradients or from a gravitational field. The molecular velocity v generated by F builds up almost instantaneously to the point where frictional retarding forces just balance the applied force F. If the frictional coefficient of each molecule is given by ξ, and if each channel contains $n_{w/c}$ water molecules, then the total frictional force within the channel is $n_{w/c}\xi v$ and setting this equal to F yields

$$F = n_{w/c}\xi v \qquad (6-123)$$

Because of the single-file (no pass) constraint, each time a molecule moves from one end of channel to the solution on the opposite side, all of the $n_{w/c} - 1$ channel molecules lying ahead of it will also move, delivering a total of $n_{w/c}$ molecules to the opposite side. If L denotes the length of a channel, then L/v = the time for the molecule to travel from one end to the other so that the molecular flux of water (molecules s^{-1}) through the channel will be given by $j = n_{w/c}(v/L)$. If N_A denotes Avogadro's number, and if there are ρ channels per cm^2 of membrane, then the molar flux (moles cm^{-2} sec^{-1}) will be $J_F = \rho j/N_A = \rho n_{w/c}v/N_A L$ and using Equation 6–123 to eliminate v:

$$J_F = \frac{\rho F}{N_A \xi L} \qquad (6-124)$$

The flux of tracer water, J_F^*, is obtained by multiplying J_F by the very small fraction of channel water that carries the label. It is convenient to express this fraction in terms of concentrations, i.e.,

$$J_F^* = \frac{\rho F}{N_A \xi L}\left(\frac{\overline{C}^*}{C}\right) \qquad (6-125)$$

where C denotes the concentration of water and $\overline{C}^*$ denotes the mean concentration of labeled tracer in the membrane channels.

Equations 6–124 and 6–125 hold regardless of the nature of F; it may be osmotic, hydraulic, gravitational, etc. Choosing osmotic, we estimate F_{osm} by noting that the work done by F_{osm} in moving $n_{w/c}$ molecules across the channel (a distance L) is PV work. Since the volume of water contained within a single channel equals $n_{w/c} V_w/N_A$, the PV work w is given by

$$w = \Delta P\, n_{w/c}\frac{V_w}{N_A} = \Delta\Pi\, n_{w/c}\frac{V_w}{N_A} = F_{osm}L \qquad (6-126)$$

Solving Equation 6–126 for F_{osm}, substituting its value for F in Equation 6–124, and comparing with Equation 6–107, we have

$$P_{osm} = \frac{\rho\, RT\, V_w}{N_A^2\, \xi\, L^2}\, n_{w/c} \qquad (6-127)$$

Finding the corresponding expression for P_d is more complex; we resort to a useful artifice and consider a hypothetical equilibrium where the diffusive flux of a tracer of labeled water molecules is just balanced by an opposing external gravitational force F_g. We choose gravity to take advantage of the slight difference in mass between ordinary and labeled water. We imagine water diffusing through a membrane in a vertical direction—against gravity. Since our final result will involve a relation solely between *physical constants*, this will be independent of the conceptual device used to derive it, just as it would not matter what particular "experiment" was utilized to measure the constants. Using Equations 6–107 and 6–125 to balance the two fluxes and remembering that at equilibrium the gravitational flux is directed downward while the diffusive flux is directed up, we have

$$J_{F_g} + J_d = -\frac{\rho\, F_g}{N_A \xi L}\,\frac{\overline{C}^*}{C} + P_d\,\Delta C^* = 0 \quad (6-128)$$

This last condition is useful only if we can relate $\overline{C}^*$ and ΔC^*. The relation can be obtained from the thermodynamic equilibrium condition (Equation 6–8) $\Delta G_{equil} = -w'_{rev}$. Here ΔG refers to the free energy change in transporting a mole of tracer a distance L cm from the lower side "in" to the upper side "o" and the molar work $w' = N_A F_g L$. Since labeled water is simply a dilute solute dissolved in ordinary water, we use Equation 6–24, with $z = 0$ and $\mu^o_{in} = \mu^o_o$, to obtain

$$\Delta G_{equil} = \mu_o - \mu_{in} = RT \ln C^*_o - RT \ln C^*_{in}$$

$$= -w'_{rev} = -N_A F_g L \qquad (6\text{--}129)$$

Solving for the concentrations

$$\frac{C^*_o}{C^*_{in}} = \exp\left(\frac{-N_A F_g L}{RT}\right) \qquad (6\text{--}130)$$

or

$$\Delta C^* \equiv (C^*_{in} - C^*_o) = C^*_{in}\left(1 - \exp\left(\frac{-N_A F_g L}{RT}\right)\right)$$

$$\approx \frac{N_A F_g L}{RT} C^*_{in} \qquad (6\text{--}131)$$

where the very small value of L (of the order of 10^{-6} cm) easily justifies the linear approximation to the exponential. For example, if $N_A F_g$ represents the gravitational force acting on one mole of water, then $N_A F_g = 18$ grams $\times$ 980 cm s^{-2}, the mass times the gravitational constant. With $L = 10^{-6}$ cm and with $RT = 2.48 \times 10^{10}$ erg mol^{-1} (at $T = 298°$ K), the value of $N_A F_g L / RT$ is of the order of 7×10^{-13}. This very small value also implies that $C^*_{in} \approx C^*_o$. On the other hand, since $\overline{C}^*$ must lie somewhere between C^*_{in} and C^*_o, we can write

$$\overline{C}^* = C^*_{in}(1 - \varepsilon) \quad \text{where } 0 < \varepsilon < \frac{\Delta C^*}{C^*_{in}} = \frac{N_A F_g L}{RT}$$

$$(6\text{--}132)$$

From our numerical example, it is clear that $\varepsilon << 1$.

Solving Equation 6–132 for C^*_{in} and using Equation 6–131 leaves

$$\Delta C^* = \frac{N_A F_g L}{(1 - \varepsilon)RT} \overline{C}^* \qquad (6\text{--}133)$$

Substituting this into Equation 6–128, solving for P_d, and neglecting ε in comparison to 1, we have

$$P_d = \frac{\rho\, RT}{N_A^2\, \xi L^2}\, \frac{1}{C} \qquad (6\text{--}134)$$

Note that the solution within the channel is virtually pure water at concentration $C = N_A n_{w/c}/N_A n_{w/c} V_w = 1/V_w$. Thus

$$P_d = \frac{\rho\, RTV_w}{N_A^2\, \xi L^2} \qquad (6\text{--}135)$$

and taking the ratio of Equation 6–127 and Equation 6–135 establishes the result

$$\frac{P_{osm}}{P_d} = n_{w/c} \qquad (6\text{--}136)$$

Using experimental estimates of this ratio, typical values for $n_{w/c}$ are reported to be 9 for the gramicidin channel (29) and 10 for red blood cells (105).

Coupling of Solute and Solvent Transport Is Described by the Kedem-Katchalsky Equations

The preceding analyses of water transport assumed that all solutes (other than labeled water) were impermeable. In contrast, many of the solutes in physiological systems are permeable, and when there are solute gradients we can expect both solute and water to move. In most cases water permeation is far more rapid than solute permeation. As a result the solutes appear to behave as though they were impermeable, allowing water to be treated as though it is in equilibrium (see earlier, under Osmotic Balance in Animal Cells). When a rapidly permeating solute is present, this quasi-equilibrium condition cannot be applied, and interactions between rates of water and solute transport become significant. The Kedem-Katchalsky (KK) equations (67) were developed to provide a kinetic description of simultaneous solute and water transport that takes these interactions into account. These are written in terms of a volume flux J_V, which represents the total volume of *solution* (water plus solutes) that flows across a unit surface per unit time, as well as a new parameter (to be discussed) called the reflection coefficient, denoted by σ. Restricting our discussion to a single permeable solute S, the equations are:

$$J_v = \frac{P_{osm} V_w}{RT} (\Delta P - \sigma_s \Delta\Pi_s - \Delta\Pi_{imp})$$

$$= P_{osm} V_w\left(\frac{\Delta P}{RT} - \Sigma\Delta C_i - \sigma_s \Delta C_s\right)$$

$$J_s = (1 - \sigma)\overline{C}_s J_V + P_s \Delta C_s \qquad (6\text{--}137)$$

where J_s denotes the molar flux of S, $\Delta\Pi_s = RT\Delta C_s$ denotes the ideal osmotic pressure contribution of S, P_s represents its permeability, $\Delta C_s = C_{s\_in} - C_{s\_o}$, $\overline{C}_s$ represents a "mean" concentration within the membrane, and $\Delta\Pi_{imp} = RT\,\Sigma\Delta C_i$ denotes the osmotic pressure contribution of all impermeable solutes.

Detailed derivation of these equations from irreversible thermodynamics is given in the text of Katchalsky and Curran (65) as well as the papers of Kedem and Katchalsky (67, 68); however, their general applicability has drawn considerable criticism from several authors (e.g., [66, 113]). Therefore, we will not attempt to repeat the derivation. Instead we present the following heuristic interpretations of a few simple cases.

First consider the wide channel under circumstances illustrated in Figure 6.12. As before, side "in" contains pure water, but now the impermeable solutes are replaced by a single permeable solute S. When the solute diameter is greater than the channel diameter it will be reflected off the membrane walls and not gain entrance to the channel. If the solute diameter is slightly less than channel diameter a large fraction will also be reflected, but occasionally some will be admitted, and as solute diameter becomes smaller its frequency of successful entries into the channel will become larger. Thus we might anticipate that the concentration within the channel of larger solute molecules is less than the corresponding concentration in the bathing medium. (How this inequality can be realized even under equilibrium conditions is discussed by Finkelstein [36]). Let the fraction of solute molecules that approach the channel entrance only to be reflected backwards be denoted by σ_s. Then the fraction that does enter will be given by $1 - \sigma_s$ and $1 - \sigma_s$ acts as a partition coefficient, where the concentration at the membrane interface, just inside the channel, $C_{s\text{-}channel}$, will equal $(1 - \sigma_s) C_s$. Now let us repeat the arguments embodied in Equations 6–118 to 6–122, only this time we allow solute into the channel. For side "in" Equation 6–118 are still applicable, but for side "o" Equation 6–120 are replaced by:

$$\mu_{aq\_o} = \overline{\mu}_{aq}^{oo}(T) + P_{aq\_o} V_{w\_aq} - V_{w\_aq} RTC_s$$
$$\mu_{m\_o} = \overline{\mu}_{m}^{oo}(T) + P_{m\_o} V_{w\_aq} - V_{w\_aq} RT(1 - \sigma_s) C_s$$
$$(6\text{--}138)$$

and using the same arguments with no further modification we find a pressure discontinuity at the "o" interface given by $P_{m\_o} = P_{aq\_o} - \sigma_s RTC_s$. Thus, the intramembrane pressure gradient is now

$$\Delta P_m \equiv P_{m\_in} - P_{m\_o} = \Delta P_{aq} + \sigma_s \Pi_{s\_o} \quad (6\text{--}139)$$

Equation 6–139 shows that in a porous membrane, the "effective" osmotic pressure exerted by a permeable solute is only a fraction of the pressure that would have been exerted by an equivalent concentration of impermeable solute. That fraction, beginning with a value of unity for an impermeable solute, gets less and less the more permeable the solute, until finally when solute enters the channel as easily as water, its "partition coefficient" $1 - \sigma_s = 1$, so that $\sigma_s = 0$. The development of Equation 6–139 also illustrates how the KK equation for J_v (see Equation 6–137 with $\sigma_s \Pi_{s\_o} = \sigma_s \Delta \Pi_s$) arises in a porous system. A more complicated, three-dimensional treatment of the above (7) is intuitively described by Finkelstein (36).

The equation for J_s contains two terms; when $J_v = 0$, it degenerates into the second term, which is the familiar diffusional form $J_s = P_s \Delta C_s$. To illustrate the first term, consider the same porous membrane when $C_{s\_in} = C_{s\_o} \equiv C_s$ so that $\Delta C_s = 0$, but $J_v \neq 0$. (This can

be accomplished by simply adding impermeable solute to one side of the membrane.) In this case, even though $\Delta C_s = 0$, there will be a nonzero solute flux because the solvent flow through the channel carries solute by convection. This flux will equal J_v times the solute concentration within the membrane. But the solute concentration within the membrane is given by $(1 - \sigma_s) C_s$. It follows that $J_s = (1 - \sigma_s) C_s J_v$ and we have recovered the first term in the expression for J_s (see Equation 6–137), where $\overline{C}_s = C_s$. In this simple case where $\Delta C_s = 0$ it is easy to assign a value for $\overline{C}_s$. In the more general case a "mean value" for C_s within the channel is assigned. Although $\overline{C}_s$ must lie somewhere between the concentrations on the two sides of the membrane, its value depends on further assumptions about the membrane. The KK analysis requires that $\overline{C}_s = (\Delta C_s)/[\ln(C_{s\_in}/C_{s\_o})]$, but this assignment has been criticized repeatedly (reviewed by Katz and Bresler [66]). In view of the ambiguity, some authors prefer the simpler arithmetic mean and set $\overline{C}_s \approx \frac{1}{2}(C_{s\_in} + C_{s\_o})$.

Other examples are lucidly discussed by Finkelstein (36). For example, after accounting for the difference in water flux, J_w, and volume flux, J_v (i.e., $J_v \equiv V_w J_w + V_s J_s$), it is shown how reflection coefficients for permeable solutes in oil-like membranes are given by

$$\sigma_s = 1 - \frac{P_s V_s}{P_d V_w} \quad (6\text{--}140)$$

where V_s is the molar volume of the solute and P_d is the diffusional permeability of water. For most polar solutes $P_s << P_d$, so that $\sigma_s \approx 1$.

Finkelstein also shows how a corresponding expression for a single-file channel can be written in terms of the channel volumes. In cases where the channels contain at most one solute molecule, let V_{ch_solute} equal the volume of a solute-containing channel (e.g., the solution volume occupied by the solute plus water) and let V_{ch_water} equal the volume of the water-filled, solute-free channel. Then

$$\sigma_s = \left(1 - \frac{P_s V_{ch\_solute}}{P_d V_{ch\_water}}\right) \quad (6\text{--}141)$$

Other properties of coupling of water and ion movements in narrow ion channels are discussed by Levitt (89).

IONIC DIFFUSION

Diffusion with Superimposed Drift Due to External Forces

Electrolyte transport is complicated by the fact that ion movements caused by electrical forces are superimposed upon ordinary diffusion. To treat this, first consider only those solute motions that are generated by an external force f. When the force is first applied in a liquid medium, solutes will accelerate quickly building up to a terminal velocity v where acceleration due to the applied force is just balanced by frictional forces encountered by the solute as it displaces and slides past molecules of the solvent. Due to the small mass (low inertia) of the solute the terminal velocity v is reached almost instantaneously, so that in general we can equate the applied and frictional forces. Since v is proportional to the frictional force it will also be proportional to the applied force f. We write

$$v = uf \qquad (6\text{--}142)$$

where u is a proportionality constant, the particle mobility, which is the inverse of the frictional coefficient. To relate the particle flux j generated by f to v, first note that if dx is the distance traveled by a solute particle in time dt, then $v = dx/dt$. Now construct the volume element shown in Figure 6.13, with dimension dx, dy, and dz.

Let n denote the concentration in terms of the number of particles per unit volume (i.e., $n = N_A C$, where $N_A =$ Avogadro's number). If motion is constrained to the x direction, then after an elapsed time dt all solute particles that were contained within the volume (given by $n\ dx\ dy\ dz$) will have crossed the shaded plane

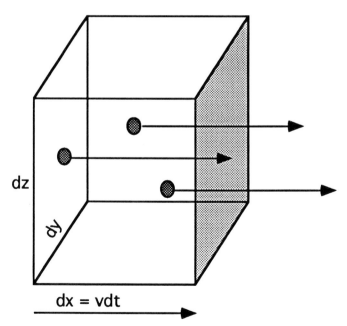

FIG. 6.13. Efflux of solute out of an elementary volume where motion is constrained to the x direction. If the solute velocity $= dx/dt$, then all solutes contained within the box defined by $dxdydz$ at time t will have passed through the shaded plane in time $t+dt$.

whose area $dA = dy\ dz$. The number of particles passing through the plane of area dA in time dt equals $j\ dA\ dt$. Since this flux equals the number of particles within the volume, which equals $n\ dx\ dy\ dz$, we have

$$j\ dA\ dt = j\ dy\ dz\ dt = n\ dx\ dy\ dz \qquad (6\text{--}143)$$

Dividing both sides by $dydzdt$, we obtain

$$j = n\frac{dx}{dt} = nv = unf \qquad (6\text{--}144)$$

By definition the molar flux $J = j/N_A$, the force per mole $\bar{f} = N_A f$, the molar mobility $U = u/N_A$, and the molar concentration $C = n/N_A$, so that substituting these quantities for j, $\bar{f}$, u, and n we obtain

$$J = UC\bar{f} \qquad (6\text{--}145)$$

Equation 6–145 is unambiguous as long as the forces (e.g., electrical, gravitational, etc.) are well defined. But how do we deal with diffusion? It is only the combination of a nonuniform concentration distribution together with the random chaotic movements generated by thermal energy that leads to net diffusive movement. There is no conventional "diffusion force" driving individual molecules in the direction of decreasing concentration gradients. A satisfactory description of diffusion in the presence of external forces requires development at the microscopic level where statistics of individual molecules are taken into account (127).

On a macroscopic level we have already applied thermodynamics to the equilibria of ions where the tendency to diffuse in one direction is counterbalanced by an oppositely directed electrical force. Work done or energy difference is the unifying concept of thermodynamics that allows the magnitudes of diffusion and electrical migration to be compared. We extend this development in a nonrigorous manner to kinetics by introducing a generalized "thermodynamic force." We begin by noting that the classical work done is given by force times distance—i.e., $w = \bar{f}\ dx$. From thermodynamics the negative of work done per mole of transported solute is given by the difference in electrochemical potential—i.e., $w = -d\tilde{\mu}$. If we equate these two work terms and divide by dx, we obtain

$$\bar{f} = -\frac{d\tilde{\mu}}{dx} \qquad (6\text{--}146)$$

Substituting this into Equation 6–145 and expanding with the aid of Equation 6–24 we arrive at the Nernst-Planck equation

$$J = -UC\frac{d\tilde{\mu}}{dx} = -UC\left(\frac{d\mu^o}{dx} + RT\frac{d\ln C}{dx} + zF\frac{d\Psi}{dx}\right) \qquad (6\text{--}147)$$

This general form of the Nernst-Planck equation can also be derived from statistical theory without recourse to thermodynamic "forces" (127).

In many applications diffusion occurs in a homogeneous medium and we can set $d\mu^o/dx = 0$. Using this and setting $z = 0$, we should recover Fick's First Law

$$J = -URTC \frac{d \ln C}{dx} = -URT \frac{dC}{dx} \quad (6-148)$$

Comparing Equation 6–148 and Equation 6–50 implies that $D = URT$, an equation known as Einstein's relation. In addition to recovering Fick's First Law from Equation 6–147 we can set $J = 0$, $d\mu^o/dx = 0$, and integrate the resulting expression to recover the Nernst equation Equation 6–27 for electrochemical equilibrium. Finally, if we assume both μ^o and C are constant everywhere so that $d\mu^o/dx = dC/dx = 0$ in Equation 6–147, we have

$$J = -UCzF \frac{d\Psi}{dx} = UqE \quad (6-149)$$

where $q = CzF$ is the charge per unit volume and

$$E = -\frac{d\Psi}{dx}$$

is the electrical field. Multiplying J by zF changes the molar flux into an equivalent current, I. Applying this to Equation 6–149 yields the equivalent of Ohm's law—i.e., $I = -(z^2F^2UC/dx)d\Psi$. Note that z^2F^2UC/dx, the "conductance" (inverse of "resistance") is proportional to concentration.

Using $D = URT$, we can rewrite the Nernst-Planck equation in any of the following equivalent forms

$$J = -UC \left(\frac{d\mu^o}{dx} + RT \frac{d \ln C}{dx} + zF \frac{d\Psi}{dx} \right)$$

$$= -U \left(RT \frac{dC}{dx} + C \left(\frac{d\mu^o}{dx} + zF \frac{d\Psi}{dx} \right) \right)$$

$$= -D \left(\frac{dC}{dx} + \frac{C}{RT} \left(\frac{d\mu^o}{dx} + zF \frac{d\Psi}{dx} \right) \right)$$

$$= -D \left(\frac{dC}{dx} + C \frac{d\varphi}{dx} \right)$$

$$\varphi \equiv ((\mu^o(x) - \mu_o^o) + zF(\Psi(x) - \Psi_o))/RT \quad (6-150)$$

where $\mu^o(x)$ and $\Psi(x)$ denote μ^o and Ψ at any position x while μ_o^o and Ψ_o represent constant reference values for μ^o and Ψ in the aqueous solution on side "o."

The Nernst-Planck equation relates ionic flux to the derivatives (gradients) of concentration and electrical

potential. More-useful expressions require a relation between flux and the actual concentrations and electrical potentials existing on each side of the membrane that are measurable. These are obtained by integrating the Nernst-Planck equations. In our models of nonelectrolyte permeability we made the tacit assumption that the membrane was homogeneous, i.e., that μ^o and D, as well as the area available for diffusion, did not depend on position (x) within the membrane. The latter two items can be accommodated formally, replacing D with $D(x)a(x)$, where $a(x)$ denotes the area within 1 cm^2 of membrane surface that is available for diffusion. Note that $a(x)$ is a simple fraction having dimensions of cm^2/cm^2 of membrane—i.e., it is dimensionless.

It is possible to obtain a partial integral of Equation 6–150 for a nonhomogeneous membrane. For the problem shown in Figure 6.14, first note that the equality

$$\frac{d}{dx} (Ce^\varphi) = \left(\frac{dC}{dx} + C \frac{d\varphi}{dx} \right) e^\varphi \quad (6-151)$$

is easily verified by formal differentiation of the left-hand side. We use this result after multiplying both sides of Equation 6–150 by $(e^\varphi)/[D(x)a(x)]$ to find

$$\frac{Je^\varphi}{D(x)a(x)} = -\left(\frac{dC}{dx} + C \frac{d\varphi}{dx} \right) e^\varphi = -\frac{d}{dx} (Ce^\varphi) \quad (6-152)$$

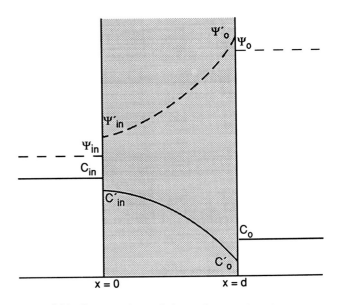

FIG. 6.14. Concentration and electrical potential profiles through a membrane.

In a steady state where J is constant, Equation 6–152 can be integrated between the point $x=0$ where $C=C'_{in}$, $\varphi=\varphi'_{in}$, and the point x where $C=C'(x)$, $\varphi=\varphi'(x)$

$$J\int_0^x \frac{e^{\varphi(x)}dx}{D(x)a(x)} = \frac{J}{\Lambda(x)} = C'_{in}e^{\varphi'in} - C'(x)e^{\varphi'(x)}, \quad (6–153)$$

$$\Lambda(x) \equiv \frac{1}{\displaystyle\int_0^x \frac{e^{\varphi(x)}dx}{D(x)a(x)}} \quad (6–154)$$

and where C'_{in}, and φ'_{in} refer to quantities at $x=0$ just *within* the membrane. (Discontinuities may occur in going from bulk aqueous solution to the membrane because of differences in solubility and the presence of surface charge; e.g., see 38 and 104.)

Substituting $x=d$ into Equation 6–153 and solving for J leaves

$$J = \Lambda(d)(C'_{in}e^{\varphi'in} - C'_o e^{\varphi'o}) \quad (6–155)$$

where C'_o and φ'_o refer to quantities at $x=d$ just *within* the membrane.

It remains to relate C'_{in}, C'_o, φ'_{in}, and φ'_o to corresponding quantities in the aqueous media. For simplicity, assume that ions at the membrane interfaces ($x=0$, $x=d$) equilibrate very rapidly compared to diffusion through the membrane so that boundary ions are equilibrated across the interfaces. The criterion for ionic equilibria between two phases (primed and unprimed) is found by setting $\tilde{\mu}' = \tilde{\mu}$ for each ion at the boundary. In our case (see Fig. 6.14)

$$\frac{C'_{in}}{C_{in}} = e^{-(\varphi'in - \varphi_{in})}, \qquad \frac{C'_o}{C_o} = e^{-(\varphi'o - \varphi_o)} \quad (6–156)$$

and using these in Equation 6–155

$$J = \Lambda(d)\left(C_{in}e^{\varphi_{in}} - C_o e^{\varphi_o}\right) \quad (6–157)$$

To express Equation 6–157 in more conventional terms we adopt the standard convention that the external solution is grounded ($\Psi_o = 0$), making the *trans* membrane potential $\Psi_m \equiv \Psi_{in} - \Psi_o = \Psi_{in}$. Using these conventions together with the definition of φ (Equation 6–150) and the fact that $\mu^o_{in} = \mu^o_o$, we have

$$\varphi_{in} = ((\mu^o_{in} - \mu^o_o) + zF\Psi_m)/RT = zF\Psi_m/RT,$$

$$\varphi_o = ((\mu^o_o - \mu^o_o) + zF(\Psi_o - \Psi_o))/RT = 0 \quad (6–158)$$

Combining Equations 6–157 and 6–158 yields

$$J = \Lambda(d)\left(C_{in}e^{\frac{zF\Psi_m}{RT}} - C_o\right) \quad (6–159)$$

with

$$\Lambda(d) = \frac{1}{\displaystyle\int_o^d \frac{e^{\varphi(x)}}{D(x)a(x)}dx} = \frac{1}{\displaystyle\int_o^d \frac{e^{\frac{zF\Psi(x)}{RT}}}{D(x)a(x)\,K(x)}dx}$$

$$K(x) \equiv e^{\frac{-(\mu^o(x) - \mu^o_o)}{RT}} \quad (6–160)$$

where $K(x)$ is a local partition coefficient, defined above. This equation can also be written in equivalent but more symmetric form by factoring $e^{zF\Psi_m/2RT}$ to obtain

$$J = \Lambda(d)e^{\frac{zF\Psi_m}{2RT}}\left(C_{in}e^{\frac{zF\Psi_m}{2RT}} - C_o e^{-\frac{zF\Psi_m}{2RT}}\right)$$

$$= \Lambda_s(d)\left(C_{in}e^{\frac{zF\Psi_m}{2RT}} - C_o e^{-\frac{zF\Psi_m}{2RT}}\right) \quad (6–161)$$

with

$$\Lambda_s(d) = \Lambda(d)e^{\frac{zF\Psi_m}{2RT}} \quad (6–162)$$

Equation 6–159 is only a partial solution to problems of ion diffusion, because evaluation of $\Lambda(d)$ depends on how Ψ is distributed through the membrane. Nevertheless it can yield important results (see below). Its significance is that it separates the concentration terms from other factors in a fairly general nonhomogeneous membrane. Equations 6–159 and 6–160 have also been used for discussions of nonelectrolyte ($z=0$) permeability in nonhomogeneous membranes (3, 31).

Ions Transported by Simple Diffusion Follow the Ussing Flux Ratio Relation

Expressions for influx, J_{influx}, and efflux, J_{efflux}, of an ion can be obtained from Equation 6–159 by setting the appropriate concentration equal to zero:

$$J_{influx} = \Lambda(d)C_o$$

$$J_{efflux} = \Lambda(d)C_{in}e^{\frac{zF\Psi_m}{RT}} \quad (6–163)$$

Taking the ratio of these two leaves Ussing's flux ratio (136)—i.e.,

$$\frac{J_{influx}}{J_{efflux}} = \frac{C_o}{C_{in}}e^{-\frac{zF\Psi_m}{RT}} = e^{-\frac{zF(\Psi_m - \Psi_{EQUIL})}{RT}} \quad (6–164)$$

where the last equality follows from Equation 6–27. Equation 6–164 is a remarkable result because it contains only quantities that are accessible from the bathing solutions and it applies to a wide variety of membranes where the nature of the nonhomogeneity ($D(x)$, $a(x)$, $\mu^o(x)$, and $\Psi(x)$) has been unspecified. The derivation shows that any ion that obeys the Nernst-Planck equation will also obey Equation 6–164. However, we have not shown the converse; in fact, it does not necessarily follow. Further generalizations of the Ussing flux ratio are discussed by Patlak and Rapoport (114).

Bulk Solutions Carry No Net Charge

Whereas electric forces and potentials are generated by the separation of charge, the amount of charge required to create substantial voltages is very small. In any small macroscopic volume of a neutral solution of, say, 0.1 mM NaCl, the number of Na^+ ions virtually equals the number of Cl^- ions. It can be shown, for example, that the number of excess Na^+ ions that would have to be added to that volume to create even a large electric field (say, 100 V/cm) would be negligible when compared to the enormous number of Na^+ ions already present. For this reason, when accounting for the concentrations of ions, any macroscopic volume of an electrolyte solution is treated as though it is electrically neutral. Even when electric fields are present, the total number of positive charges are virtually equal to the total number of negative charges. In the case of membrane potentials, this principle of macroscopic electrical neutrality is easily illustrated by considering net charge distribution across the membrane. Most biological membranes have an electrical capacitance of the order of 10^{-6} farad cm^{-2}. By definition, capacitance (farads) $\equiv$ net charge (Coulombs)/voltage (volts); for a membrane this translates to net charge on the membrane divided by the membrane potential. For 1 cm^2 of membrane with a substantial membrane potential of the order of 100 mV $= 10^{-1}$ V, the net charge separated by the membrane would equal $10^{-1} \times 10^{-6} = 10^{-7}$ Coulombs. Using the Faraday constant (96,485 Coulombs equiv^{-1}), we find that 10^{-7} Coulombs is the charge carried by approximately 10^{-12} moles of univalent ions. Thus, charging a whole square centimeter of membrane to 100 mV is accomplished by the separation of only 10^{-12} mole of univalent ions. Further, more-detailed analyses show that the charge separation is confined to regions that are less than 100 Å from the membrane. It follows that we can treat both intra- and extracellular fluids as though they are electrically neutral even in the presence of membrane potentials. At rest the intracellular side of the membrane is charged negatively, but this charge, contained entirely within an extremely thin film next to the membrane, is too small to significantly alter bulk ionic concentrations. It follows from macroscopic electrical neutrality that there is no transport of net charge through a membrane in a steady state unless there is a source (and sink) of electrical charge. In voltage-clamp experiments, the source and sink is provided by an external power supply. In an action potential, the depolarized section on the membrane serves as a partial source (or sink) for other sections. A steady state with no external source (or sink) requires that the sum of charges carried by ions through the membrane must be zero

$$\sum_i z_i J_i = 0 \qquad (6\text{--}165)$$

where z_i is the valence (including algebraic sign) and J_i the flux of the ith ion species.

Combining Equations 6–159 and 6–160 with Equation 6–165 yields a very simple result for a homogeneous membrane when all permeating ions have the same sign and valence. In a homogeneous membrane D, a, and K are constants and Equations 6–160 reduces to

$$\Lambda(d) = \frac{D\,a\,K}{d} \frac{d}{\int_o^d e^{\frac{zF\Psi(x)}{RT}}\,dx} = P\,Q(z,d) \qquad (6\text{--}166)$$

with

$$P \equiv \frac{D\,a\,K}{d}, \qquad Q(z,d) \equiv \frac{d}{\int_o^d e^{\frac{zF\Psi(x)}{RT}}\,dx} \qquad (6\text{--}167)$$

When all permeating ions have the same sign and valence, the function $Q(z,d)$ will have the same value for all ions and it will cancel. For example, suppose the membrane is permeable only to Na^+ and K^+. In addition, assume there are no active transport systems that transport net charge. In this case Equation 6–165 reduces to

$$J_{Na} + J_K = PNaQ(+1,d)\left([Na]_{in}e^{\frac{zF\Psi_m}{RT}} - [Na]_o\right)$$
$$+ P_K Q(+1,d)\left([K]_{in}e^{\frac{zF\Psi_m}{RT}} - [K]_o\right) = 0 \qquad (6\text{--}168)$$

Solving for Ψ_m yields

$$\Psi_m = \frac{RT}{F}\ln\left(\frac{P_K[K]_o + P_{Na}[Na]_o}{P_K[K]_{in} + P_{Na}[Na]_{in}}\right)$$

$$= \frac{RT}{F}\ln\left(P_K[K]_o + P_{Na}[Na]_o\right)$$

$$- \frac{RT}{F}\ln\left(P_K[K]_{in} + P_{Na}[Na]_{in}\right) \qquad (6\text{--}169)$$

a result that is easily generalized to any number of ions of the same valence.

The last equality allows a simple interpretation of Equation 6–169: Ψ_m arises from the difference of two terms: the tendency for positive charge to flow in minus the tendency of positive charge to flow out. In our example, the steady state is possible only because the flow of Na^+ in is balanced by the flow of K^+ out (or vice versa). If only one ion was permeable, say $P_K = 0$, then the new steady state Equation 6–169 degenerates into the Nernst equation for Na^+ equilibrium.

The Constant Field Is a Convenient Idealization

To develop more general cases where both cation and anion diffusion are significant, it is necessary to evaluate $Q(z,d)$. This requires specification of how the electric field varies at each position in the membrane. For example, we could assume that electrical neutrality holds everywhere, not only in bulk solutions, but even within the membrane. This is a good approximation, provided the electrical fields ($-d\Psi/dx$) are of reasonable size. It holds for many thick artificial membranes that can be treated as bulk phases. However, it is not a good approximation for lipid bilayers or cell membranes. A typical cell membrane 50 Å $= 5 \times 10^{-7}$ cm thick with a membrane potential of 100 mV sustains an electric field on the order of 0.10 (volts)/ 5×10^{-7} (cm) $= 2 \times 10^5 = 200,000$ V cm^{-1}! More-elaborate analyses show that electrical neutrality will not be maintained within these intense fields.

It is a given that Ψ moves from Ψ'_{in} to Ψ'_{o} as we traverse the membrane from side "in" to "o," but just how Ψ makes the transition between these endpoints depends on the unknown details of the path taken by the ions as they cross the membrane. Lacking this information, we adopt the simplest case, the constant field approximation that assumes that $-d\Psi/dx =$ constant so that Ψ changes linearly from Ψ'_{in} to Ψ'_{o} as x goes from 0 to d. The assumption is easily justified in lipid bilayers by appeal to the Poisson equation (115), which requires $d^2\Psi/dx^2$ at any point x to be proportional to the concentration of electrical charge at x. Since low ion solubility ensures that the concentration of electrical charge within the interior of the bilayer is very small, $d^2\Psi/dx^2 \approx 0$ and it follows that $-d\Psi/dx \approx$ constant. However, the same argument is not applicable to channels. The validity of the constant field assumption for bilayers and for more general cases has been discussed extensively (2, 21, 48, 83, 84, 100, 111).

Constant-Field Flux. A linear Ψ implies that

$$\Psi = \Psi'_{in} + \frac{\Psi'_{o} - \Psi'_{in}}{d} x \qquad (6–170)$$

Assuming D, a, and K are constants and using Equation 6–170 in Equation 6–167 yields

$$Q(z,d) = \frac{zF \dfrac{\Psi'_{in} - \Psi'_{o}}{RT}}{e^{\frac{zF}{RT}\Psi'_{in}} - e^{\frac{zF}{RT}\Psi'_{o}}} \qquad (6–171)$$

where the Ψ'_{in} and Ψ'_{o} refer to Ψ within the membrane. To relate these potentials to Ψ_{in} and Ψ_{o}, the measurable potentials in the two baths surrounding the membrane, we need values for interfacial potentials ($\Psi'_{o} - \Psi_{o}$) and ($\Psi'_{in} - \Psi_{in}$). Unfortunately, these values are rarely at hand, and for simplicity it is common practice to ignore them by setting $\Psi'_{o} - \Psi_{o} = \Psi'_{in} - \Psi_{in} = 0$. In this case $\Psi'_{in} = \Psi_m$, $\Psi'_{o} = 0$, and Equation 6–171 becomes

$$Q(z,d) = \frac{zF \dfrac{\Psi_m}{RT}}{e^{\frac{zF}{RT}\Psi_m} - 1} \qquad (6–172)$$

Finally, from Equations 6–159 and 6–166 we arrive at the constant-field flux equation

$$J = P \frac{\dfrac{zF}{RT}\Psi_m}{e^{\frac{zF}{RT}\Psi_m} - 1} (C_{in} e^{\frac{zF}{RT}\Psi_m} - C_o) \qquad (6–173)$$

The corresponding symmetric forms of Equations 6–173 and 6–172 are

$$J = P\, Q_s(z,d) \left(C_{in} e^{\frac{zF\Psi_m}{2RT}} - C_o e^{-\frac{zF\Psi_m}{2RT}} \right) \quad (6–174)$$

with

$$Q_s(z,d) = \frac{\dfrac{zF}{RT}\Psi_m}{e^{\frac{zF}{RT}\Psi_m} - 1} e^{\frac{zF\Psi_m}{2RT}} = \frac{\dfrac{zF}{2RT}\Psi_m}{\sinh\left(\dfrac{zF}{2RT}\Psi_m\right)} \qquad (6–175)$$

Constant-Field Membrane Potential. A corresponding expression for Equation 6–169 can be obtained for cases in which both cation and anion diffusion are significant. When $|z|$ is the same for all ions, the results are simplified because substitution of $+z$ and $-z$ into Equation 6–172 results in the simplification

$$Q(-z,d) = Q(z,d) e^{\frac{zF}{RT}\Psi_m} \qquad (6–176)$$

A good example is provided by the important case where only univalent ions K^+, Na^+, and Cl^- are significant. Using Equations 6–165, 6–173, and 6–176 and solving for Ψ_m, we have

$$\Psi_m = \frac{RT}{zF} \ln \left(\frac{P_K[K]_o + P_{Na}[Na]_o + P_{Cl}[Cl]_{in}}{P_K[K]_{in} + P_{Na}[Na]_{in} + P_{Cl}[Cl]_o} \right) \quad (6-177)$$

Equation 6–177 has the same loose interpretation as Equation 6–169, i.e., Ψ_m arises from the difference of two terms: the tendency of positive charge to flow in minus the tendency of positive charge to flow out.

Equations 6–173 or 6–174 and 6–177 are the commonly used equations of the constant field theory. Note that when $P_{Cl} = 0$, Equation 6–177 becomes identical to Equation 6–169, as it should. Further, if both P_{Na} and P_{Cl} are zero, Equation 6–177 reduces to the Nernst equation for K^+. Other conditions for the validity of Equation 6–177 have been discussed by several authors (11, 112, 123, 131). In particular, note that Equation 6–177 follows from Equations 6–159, 6–166, and 6–165 whenever $Q(z) = Q(-z)$; a constant field is only one special case.

Analogous expressions can be derived for permeating mixtures of univalent and divalent ions by making use of the following relation obtained from Equations 6–172:

$$Q(2,d) = \frac{\frac{2F\Psi_m}{RT}}{e^{\frac{2F\Psi_m}{RT}} - 1} = \frac{\frac{2F\Psi_m}{RT}}{\left(e^{\frac{F\Psi_m}{RT}} + 1\right)\left(e^{\frac{F\Psi_m}{RT}} - 1\right)}$$

$$= \frac{2}{\left(e^{\frac{F\Psi_m}{RT}} + 1\right)} Q(1,d) \quad (6-178)$$

For example, consider the case where K^+ and Ca^{2+} are the primary permeating ions. Substituting Equation 6–178 into Equation 6–173 and noting from Equation 6–165 that $J_K + 2J_{Ca} = 0$, we find that $Q(1,d)$ cancels and we are left with the following quadratic equation in $e^{F\Psi_m/RT}$.

$$(P_K[K]_{in} + 4P_{Ca}[Ca]_{in})e^{\frac{2F\Psi_m}{RT}} + (P_K[K]_{in} - P_K[K]_o)e^{\frac{F\Psi_m}{RT}}$$

$$- (P_K[K]_o + 4P_{Ca}[Ca]_o) = 0 \quad (6-179)$$

After solving for $e^{\frac{zF\Psi_m}{RT}}$ and taking ln of both sides, we have

$$\Psi_m = \frac{RT}{F} \ln \frac{-(P_K[K]_{in} - P_K[K]_o) + \sqrt{a}}{2(P_K[K]_{in} + 4P_{Ca}[Ca]_{in})}$$

$$a = (P_K[K]_{in} - P_K[K]_o)^2 + 4(P_K[K]_{in} + 4P_{Ca}[Ca]_{in})$$

$$(P_K[K]_o + 4P_{Ca}[Ca]_o) \quad (6-180)$$

When $P_{Ca} = 0$, Ψ_m calculated from Equation 6–180 reduces to the Nernst potential for K^+. Similarly, when $P_K = 0$, Ψ_m reduces to the Nernst potential for Ca^{2+}.

Using the same Equations 6–178, 6–173, and 6–165, similar expressions can be obtained for more complicated mixtures (73, 124).

Conductance Depends on Ionic Concentrations

In addition to the molar flux J, ion movements can also be interpreted and measured in terms of charge flux—i.e., in terms of electrical current I. The relation between I and J for any ion, i, is given by

$$I_i = z_i F J_i \quad (6-181)$$

Applying this to Equation 6–173, we have

$$I_i = \frac{z_i^2 F^2 \frac{\Psi_m}{RT}}{e^{\frac{zF\Psi_m}{RT}} - 1} P_i (C_{in} e^{\frac{zF\Psi_m}{RT}} - C_o) \quad (6-182)$$

In the simplest case where $C_{in} = C_o = C$, Equation 6–182 leaves

$$(I_i)_{C_o = C_{in}} = \frac{z_i^2 F^2 P_i C_i}{RT} \Psi_m \quad (6-183)$$

Here, current is directly proportional to voltage, a plot of I vs. Ψ_m is a straight line passing through the origin, and the membrane behaves as an ideal ohmic resistor. In this case the constant of proportionality, g_i, relating current to voltage is called conductance; it is given by

$$(g_i)_{C_o = C_{in}} = \frac{z_i^2 F^2 P_i C_i}{RT} \quad (6-184)$$

Although both g_i and P_i reflect the ease of membrane permeation by a particular ion, they are not equivalent. The conductance is proportional to P_i, the permeability, *times* C_i, the concentration of ions (number of charge available to carry the current). The value of F^2/RT at 20°C is approximately 3.8×10^6 coul equiv^{-1} volt^{-1}, so that if C_i is measured in millimoles liter^{-1} and P in cm s^{-1}, then for this special symmetric case $(C_o = C_{in} = C_i)$ $g_i = 3.8 z_i C_i P_i$ with the dimensions of Siemen cm^{-2}. Note that each ion species is characterized by its own conductance. The total membrane conductance will equal the sum of the individual or *partial* conductances.

In the more general case, where C_o is not necessarily equal to C_{in}, the partial conductance is defined as the slope of the current-vs.-voltage curve.

$$g_i \equiv \frac{\partial I_i}{\partial \Psi_m} \quad (6-185)$$

Inspection of Equation 6–182 shows that the current–voltage relation is generally nonlinear, so that g_i will depend on both voltage and concentration. Behavior of the current–voltage relation is most easily seen by examining its asymptotic behavior as the membrane potential becomes large:

$$(I_i)_{\psi_m \to +\infty} = \frac{z_i^2 F^2 P_i C_{in}}{RT} \Psi_m, \quad (g_i)_{\psi_m \to +\infty} = \frac{z_i^2 F^2 P_i C_{in}}{RT}$$

$$(I_i)_{\psi_m \to -\infty} - \frac{z_i^2 F^2 P_i C_o}{RT} \Psi_m, \quad (g_i)_{\psi_m \to -\infty} = \frac{z_i^2 F^2 P_i C_o}{RT}$$

$$(6-186)$$

At high positive membrane potentials, positive current is directed outward and g_i approaches a constant value proportional to C_{in}. This reflects the fact that when the driving force is large, conductance is limited by the charge available to move from inside to out. Similarly, at high negative membrane potentials, positive current is directed inward and g_i approaches a constant value proportional to C_o.

The general constant-field conductance calculated from Equations 6–182 and 6–185 can be seen in Equation 6–187 below. It moves between the two limits defined in Equation 6–186, as shown in Figure 6–15. A nonconstant conductance that depends on the direction of current flow is often referred to as *rectification.*

An additional useful limiting case is obtained by taking the limit of Equation 6–187 as Ψ_m approaches zero and

$$g_i^o \equiv \left(\frac{\partial I}{\partial \Psi_m}\right)_{\psi_m \to 0} = \frac{z_i^2 F^2 P_i}{RT} \left(\frac{C_{in} + C_o}{2}\right)$$

$$(6-188)$$

where g_i^o as defined above is called the *small signal conductance.* It is the arithmetic average of the two extreme limits shown in Equation 6–186. Because of its simplicity, the small signal conductance is a popular experimental measurement.

Permeability Ratios Can Be Measured by Changes in Membrane Potential

Absolute permeability coefficients can be calculated from conductance measurements (cf. Equation 6–184) as well as flux measurements. In addition, relative permeabilities can be conveniently obtained from reversal potentials ("resting," zero-current potentials). For example, assume the membrane is only permeable to univalent cations A^+, B^+, C^+, etc. First we bathe the cell in a solution containing only A^+, measure Ψ_m, and then quickly replace A^+ with B^+ and repeat the

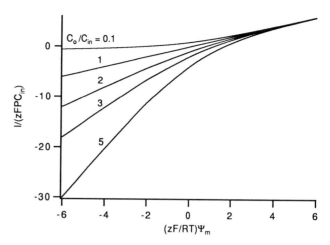

FIG. 6.15. Normalized I vs. Ψ_m plot obtained from Equation 6–182 for various values of C_o while C_{in} is held fixed. Values for C_o/C_{in} are indicated on the graph. Note that the slopes (conductances) of all curves approach the same limiting value (determined by Equation 6–186) at infinite Ψ_m.

measurement. We utilize Equation 6–169 to describe the measurements as follows:

$$\Psi_{m1} = \frac{RT}{F} \ln \frac{P_A[A_o]}{P_A[A_{in}] + P_B[B_{in}] + P_C[C_{in}] + \ldots}$$

$$\Psi_{m2} = \frac{RT}{F} \ln \frac{P_B[B_o]}{P_A[A_{in}] + P_B[B_{in}] + P_C[C_{in}] + \ldots}$$

$$\Delta\Psi_m = \Psi_{m1} - \Psi_{m2} = \frac{RT}{F} \ln \frac{P_A[A_o]}{P_B[B_o]}$$

$$(6-189)$$

Since both $[A_o]$ and $[B_o]$ are controlled by the experimenter, the ratio P_A/P_B is easily obtained from measurements of $\Delta\Psi_m$, provided the internal concentrations of **A**, **B**, **C**, etc. do not change during the experiment.

An Electrogenic Pump Contributes to Ψ_m

In the preceding discussion we assumed that ion transport occurs only by diffusion. The fact that cell membranes also contain ion pumps that actively transport ions need not affect our results for ionic fluxes provided we interpret J as only part of the flux, that part that is due to diffusion. However, ion pumps do affect expressions for Ψ_m because they may contribute to the net charge carried across the membrane. When ion pumps are present, Equation 6–165 is modified by adding another term, J_{pump}, i.e.,

$$g_i = \frac{\partial I_i}{\partial \Psi_m} = \frac{z_i^2 F^2 P_i \left(\left[e^{\frac{zF\Psi_m}{RT}} - \left(1 + \frac{zF\Psi_m}{RT}\right) \right] C_{in} e^{\frac{zF\Psi_m}{RT}} - \left[\left(1 - \frac{zF\Psi_m}{RT}\right) e^{\frac{zF\Psi_m}{RT}} - 1 \right] C_o \right)}{RT \left(e^{\frac{zF\Psi_m}{RT}} - 1 \right)^2}$$

$$(6-187)$$

$$J_{pump} + \sum_i z_i J_i = 0 \qquad (6-190)$$

where J_{pump} represents the *net* positive charge carried by *all* pumps,

$$J_{pump} \equiv \sum_i z_i J_{pump\_i}.$$

If the pump is electrically neutral—e.g., if it pumps 1 K^+ in exchange for 1 Na^+—then J_{pump} will be zero and our expressions for Ψ_m remain unmodified. With an electrogenic pump (e.g., exchange of 2 K^+ for 3 Na^+) added to a membrane that is only permeable to K^+, Na^+, and Cl^-, we use Equations 6–173 and 6–190 to solve for $e^{(zF/RT)\Psi_m}$ and, after taking logarithms, we have the following extended version of Equation 6–177:

$$\Psi_m =$$
$$\frac{RT}{F} \ln \left(\frac{P_K[K]_o + P_{Na}[Na]_o + P_{Cl}[Cl]_{in} - \dfrac{RT}{F}\dfrac{J_{pump}}{\Psi_m}}{P_K[K]_{in} + P_{Na}[Na]_{in} + P_{Cl}[Cl]_o - \dfrac{RT}{F}\dfrac{J_{pump}}{\Psi_m}} \right)$$
$$(6-191)$$

Although Equation 6–191 is a transcendental equation with Ψ_m appearing on both sides, explicit solutions for Ψ_m can be obtained by numerical methods. If J_{pump} is known, its contribution to Ψ_m can be estimated using Equation 6–191.

When the entire system, consisting of cytoplasm, external medium, as well as membrane, is in a steady state, then pumps will balance leaks and Equation 6–191 takes on a very simple, useful, form (56, 107). For example, suppose there is an exchange pump for Na^+ and K^+, but no pump for Cl^-. If J_{Nap} and J_{Kp} represent the Na^+ and K^+ fluxes through the pump, then in a steady state where there is no net flux of any ion

$$J_{Na} = -J_{Nap}, \; J_K = -J_{Kp}, \; J_{Cl} = 0,$$
$$J^+_{pump} = J_{Nap} + J_{Kp} \qquad (6-192)$$

Denoting the stoichiometry of the $Na^+ - K^+$ pump by n_p, we also have

$$J_{Kp} = -n_p J_{Nap} \qquad (6-193)$$

Using these steady-state conditions together with Equation 6–173 in Equation 6–190 and solving for Ψ_m,

$$\Psi_m = \frac{RT}{zF} \ln \left(\frac{P_K[K]_o + n_p P_{Na}[Na]_o}{P_K[K]_{in} + n_p P_{Na}[Na]_{in}} \right) \qquad (6-194)$$

The only difference between Equation 6–194 and the corresponding Equations 6–169 and 6–177 (derived by assuming an electrically neutral pump) is that P_{Na} is replaced by $n_p P_{Na}$. In many cells, the pump exchanges 2 K^+ for 3 Na^+, making $n_p = 2/3$. Further, in many cases $P_K \gg P_{Na}$. If this holds, Equation 6–194 implies that the electrogenic pump's contribution to the normal steady-state membrane potential will be small.

Channels Can Be Incorporated into the Nernst-Planck Formulation

Without any further assumptions, the continuum approach exemplified by the constant-field equations is not sufficient to account for a number of passive ion-transport properties found in natural membranes and in bilayers that are doped with channel-forming ionophores. These discrepancies have been summarized by Andersen (4) and by Hille (51). They include:

1. Ion fluxes may level off and "saturate" as the driving forces increase. Sometimes they go through a maximum and then actually decrease as the driving force increases.

2. Many ion channels show rectification in symmetric solutions where $C_o = C_{in}$ in contrast with Equation 6–183.

3. Unidirectional ion fluxes do not obey Ussing's flux ratio Equation 6–164. This, despite the fact that the relation stems almost directly from the Nernst-Planck equation and does not depend on the constant field or other assumptions.

4. Many ion channels show complex relations between conductance and permeability. Under some conditions a channel may have high conductance with low permeability and vice versa under other conditions.

5. In some channels, the permeability ratios found by reversal or bi-ionic potentials (Equation 6–189) do not agree with permeabilities measured with ionic fluxes.

The Nernst-Planck equation (Equation 6–147) or its integral (Equation 6–159) implies that the membrane has an unlimited capacity to hold ions and to transport them. From this perspective, ions move independently and the flux will keep increasing without limit as long as the driving force gets larger and larger.

We can begin to address some of these problems with the realization that most physiological ion transport takes place via a limited number of ion channels (with no contribution from the lipid bilayer) and that once an ion enters a channel it will electrostatically repel further entry by another ion of similar sign. In the simplest case this repulsion is sufficiently large so that occupation of a channel by more than one ion is highly unlikely and the Nernst-Planck formulation can be extended in a simple way to include saturation kinetics. We illustrate with an adaptation of Levitt's original, more general, development (90). For simplicity, we adopt the constant-field assumption together with a constant D, channel cross-sectional area a, and μ^o (see Equation 6–160), as well as the neglect of boundary potentials so that $\Psi'_{in} = \Psi_{in}$, $\Psi'_o = \Psi_o$. Analytically, the single-occupancy assumption can be introduced through the boundary conditions relating concentrations at the interface between the aqueous solution and the channel. These can be approached by considering the simple chemical equilibrium

$$C_{in} + Y^o \rightarrow C'_{in}, \qquad C_o + Y^o \rightarrow C'_o$$

where C' represents the concentration of ion C just within the membrane and Y^o represents the density of unoccupied channels (moles of channel per cm^2 of membrane). We have

$$C'_{in} = C_{in} Y^o K, \qquad C'_o = C_o Y^o K \qquad (6\text{--}195)$$

where K is an equilibrium constant. Note that the product KY^o is dimensionless; like a partition coefficient, it equals the ratio of membrane to aqueous concentrations.

To derive the flux, we begin with Equation 6–153 for intramembrane diffusion and apply the usual constant-field assumptions, $\Psi(x) = \Psi_m(1 - x/d)$, $\Psi'_{in} = \Psi_m$, $\Psi'_o = 0$, and with D, a, and μ^o held constant. Then, with $\varphi \equiv (\mu^o - \mu_o^o + zF\Psi)/RT$, and the aid of Equations 6–153 and 6–154, we have

$$J = \Lambda'(x)(C'_{in} e^{\varphi'_{in}} - C'(x) e^{\varphi'(x)})$$

$$= \frac{D\,a}{d} \frac{\frac{zF}{RT}\Psi_m}{e^{\frac{zF}{RT}\Psi_m} - e^{\frac{zF\Psi_m(1-x/d)}{RT}}}$$

$$\times \left(C'_{in} e^{\frac{zF}{RT}\Psi_m} - C'(x) e^{\frac{zF\Psi_m(1-x/d)}{RT}} \right)$$

$$(6\text{--}196)$$

Letting $x = d$ in the above equation yields

$$J = \frac{D\,a}{d} \frac{\frac{zF}{RT}\Psi_m}{e^{\frac{zF}{RT}\Psi_m} - 1} \left(C'_{in} e^{\frac{zF}{RT}\Psi_m} - C'_o \right) \qquad (6\text{--}197)$$

Inserting the boundary conditions (Equation 6–195), we have

$$J = Y^o \frac{D\,a\,K}{d} \frac{\frac{zF}{RT}\Psi_m}{e^{\frac{zF}{RT}\Psi_m} - 1} \left(C_{in} e^{\frac{zF}{RT}\Psi_m} - C_o \right)$$

$$= Y^o \frac{D\,a\,K}{d} Q(z,d) \left(C_{in} e^{\frac{zF}{RT}\Psi_m} - C_o \right) \qquad (6\text{--}198)$$

Equation 6–198 differs from the corresponding constant-field flux (Equation 6–173) by the appearance of Y^o, the density of unoccupied channels. An expression for Y^o can be found from conservation of channels. If Y_t represents the total channel density (occupied or not) and Y^C represents the density of occupied channels, then

$$Y^o = Y_t - Y^C \qquad (6\text{--}199)$$

Note that Y^C is not simply equal to C'. Rather, it equals the total number of moles of C ions per unit area of membrane, regardless of the position of the ion within a given channel (one and only one ion per channel; each ion blocks its channel). To find this number, we consider the ensemble of all channels within a given membrane area, α, and let x denote the distance coordinate along the axis of the channel, starting with $x = 0$ on side "in" and ending at $x = d$ on side "o." Then the number of moles lying within the membrane between x and $x + dx$ will be given by $C'(x)$ (αdx) and the total number of moles contained in the ensemble of channels, defined by our unit area, $\alpha = 1$, of membrane, is given by

$$Y^C = \int_0^d C'(x)\,dx \qquad (6\text{--}200)$$

Evaluation of this integral requires an explicit expression for C' as a function of x. This can be obtained by using Equation 6–197 to eliminate J from Equation 6–196, and solving for $C'(x)$.

$$C'(x) = \frac{\left(C'_{in} e^{\frac{zF\Psi_m}{RT}} - C'_o \right) - (C'_{in} - C'_o) e^{\frac{zF\Psi_m}{RT}\frac{x}{d}}}{e^{\frac{zF\Psi_m}{RT}} - 1}$$

$$(6\text{--}201)$$

Inserting the boundary conditions (Equation 6–195) into the right-hand side of the above, we have our final expression for intramembrane concentration profile in terms of the aqueous concentrations:

$$C'(x) = KY^o \frac{\left(C_{in} e^{\frac{zF\Psi_m}{RT}} - C_o \right) - (C_{in} - C_o) e^{\frac{zF\Psi_m}{RT}\frac{x}{d}}}{e^{\frac{zF\Psi_m}{RT}} - 1}$$

$$(6\text{--}202)$$

Using Equation 6–202 to evaluate the integral in Equation 6–200, the result can be inserted into Equation 6–199 to arrive at

$$Y^o = \frac{Y_t}{1 + \dfrac{C_{in}}{K_{in}} + \dfrac{C_o}{K_o}} \qquad (6\text{--}203)$$

$$K_{in} = \frac{\dfrac{zF\Psi_m}{RT}\left(e^{\frac{zF\Psi_m}{RT}} - 1\right)}{K\,d\left[\left(\dfrac{zF\Psi_m}{RT} - 1\right)\left(e^{\frac{zF\Psi_m}{RT}} - 1\right) + \dfrac{zF\Psi_m}{RT}\right]}$$

$$K_o = \frac{\dfrac{zF\Psi_m}{RT}\left(e^{\frac{zF\Psi_m}{RT}} - 1\right)}{K\,d\left(\left(e^{\frac{zF\Psi_m}{RT}} - 1\right) - \dfrac{zF\Psi_m}{RT}\right)}$$

$$(6\text{--}204)$$

Finally, substituting Equation 6–203 into Equation 6–198, we have

$$J = P \ Q(z, d) \frac{\left(C_{in}e^{\frac{zF\Psi_m}{RT}} - C_o\right)}{1 + \frac{C_{in}}{K_{in}} + \frac{C_o}{K_o}}$$

$$P \equiv \frac{D \ a \ K \ Y_t}{d} \qquad (6-205)$$

Single-Occupancy Channels Show Saturation Kinetics. Notice from Equation 6–205 that at constant voltage and at low concentrations with $C_{in} << K_{in}$, $C_o << K_o$, then $J \approx P \ Q(z, d)(C_{in}e^{zF\Psi_m/RT} - C_o)$ and, at fixed membrane potential, the flux varies linearly with concentration. However, as the concentration increases, competition for channels develops and the flux begins to level off. If we remove the solute from one side—say, side "o" so that $C_o = 0$—then the expression for J resembles the Michaelis-Menten formulation for a simple enzyme:

$$(J)_{C_o=0} = \frac{J^{max\_o}C_{in}}{K_{in} + C_{in}} \qquad (6-206)$$

where $J^{max\_o} = P \ Q(z, d) \ K_{in}e^{zF\Psi_m/RT}$ is the limiting maximal flux as C_{in} becomes large, and K_{in} defined above in Equation 6–204 is the concentration of C_{in} where the flux equals 1/2 its maximal rate. The corresponding expression when $C_{in} = 0$ is

$$(J)_{C_{in}=0} = \frac{J^{max\_in}C_o}{K_o + C_o} \qquad (6-207)$$

where $J^{max\_in} = P \ Q(z, d) \ K_o$, and K_o is defined in Equation 6–204.

Even though the channels show saturation kinetics, they retain a number of properties of the older free diffusion theory embodied by Equation 6–173. For example, the flux and/or current in symmetric solutions with $C_{in} = C_o$ is given by

$$J = \frac{I}{zF} = P \ \frac{zF}{RT}\left(\frac{C}{1 + Kd \ C}\right)\Psi_m \qquad (6-208)$$

which, like Equation 6–183, still shows a linear "Ohm's Law" behavior but has a more complex concentration dependence.

Competition. The same analysis can be employed for two different ion species, **B** and **C**, competing for the same channels. Flux of each species is treated independently as shown above, but the calculation of Y^O now takes into account that occupancy by either ion, **B** or **C**, completely blocks the channel. Instead of Equation 6–199 we now have $Y_t = Y^O + Y^C + Y^B$. Carrying out the analysis yields

$$J_C = P_C \ Q(z_C, d) \frac{\left(C_{in}e^{\frac{z_CF\Psi_m}{RT}} - C_o\right)}{1 + \frac{C_{in}}{K_{C\_in}} + \frac{C_o}{K_{C\_o}} + \frac{B_{in}}{K_{B\_in}} + \frac{B_o}{K_{B\_o}}}$$

$$J_B = P_B \ Q(z_B, d) \frac{\left(B_{in}e^{\frac{z_BF\Psi_m}{RT}} - B_o\right)}{1 + \frac{C_{in}}{K_{C\_in}} + \frac{C_o}{K_{C\_o}} + \frac{B_{in}}{K_{B\_in}} + \frac{B_o}{K_{B\_o}}} \qquad (6-209)$$

where the constants P_C, P_B, K_{C_in}, K_{C_o}, K_{B_in}, and K_{B_o} are obtained directly from their analogs, Equations 6–204 and 6–205, simply by assigning the appropriate subscripts to z, P, and K. The inhibitory effect of **B** on J_C can be seen by the fact that B only occurs in the denominator of the expression for J_C. Similarly, C only occurs in the denominator of J_B.

Ussing Flux Ratio. Measurements of unidirectional fluxes generally place a tracer, or label, on one side of the membrane—say, on side "in"—and measure its appearance on the other, side "o." Suppose solute **B** in Equation 6–209 was a tracer for **C**. We emphasize this by replacing B with C^* (labeled C). Since an ideal tracer does not disturb the system, all of the **C** and **C*** (and **B**) parameters must be equal, i.e.,

$$K_{in}^B \equiv K_{in}^* = K_{in}^C = K_{in}, \quad K_o^B \equiv K_o^* = K_o^C = K_o,$$

$$P^B \equiv P^* = P^C = P \qquad (6-210)$$

and the superscript is no longer necessary. Further, C^* is utilized in tracer quantities, which means that

$$C_{in} + C_{in}^* \approx C_{in}, \qquad C_o + C_o^* \approx C_o \qquad (6-211)$$

Substituting these quantities in Equation 6–209 and letting J^*_{efflux} represent the efflux of the tracer

$$J^*_{efflux} = \frac{P\left[C_{in}^*e^{\frac{zF}{RT}\Psi_m}\right]}{1 + \frac{C_{in}}{K_{in}} + \frac{C_o}{K_o}} \qquad (6-212)$$

The denominator shows the competitive influence of the unlabeled **C** on the efflux of the labeled tracer **C***.

If J_{efflux} represents the unidirectional efflux of unlabeled **C**, then since $J^*_{efflux} = J_{efflux} \ (C^*_{in}/C_{in})$

$$J_{efflux} = J^*_{efflux} \frac{C_{in}}{C^*_{in}} = \frac{P\left[C_{in}e^{\frac{zF}{RT}\Psi_m}\right]}{1 + \frac{C_{in}}{K_{in}} + \frac{C_o}{K_o}} \qquad (6-213)$$

Using a similar argument for the unlabeled influx, J_{influx}, obtained by placing a tracer on side "o" and measuring its appearance on side "in," yields

$$J_{influx} = \frac{P[C_o]}{1 + \dfrac{C_{in}}{K_{in}} + \dfrac{C_o}{K_o}} \qquad (6\text{–}214)$$

Taking the ratio leads to the same Ussing relation, Equation 6–164, characteristic of formulations based on free diffusion.

$$\frac{J_{efflux}}{J_{influx}} = \frac{C_{in}}{C_o} \, e^{\frac{zF}{RT}\Psi_m} \qquad (6\text{–}215)$$

This follows even though the channels can be saturated.

Membrane Potentials: $z_C = z_B$. Finally, if **B** and **C** are the only permeating ions, and assuming $z_C = z_B$, Equation 6–209 can be used to derive a solution for the membrane potential. Using Equation 6–165, we set $J_C + J_B = 0$ and solve for Ψ_m as

$$\Psi_m = \frac{RT}{F} \ln \left(\frac{P_C C_o + P_B B_o}{P_C C_{in} + P_B B_{in}} \right) \qquad (6\text{–}216)$$

which is the analog of Equation 6–169.

Generalizations. Single ion channels are tractable because all ion interactions are confined to the membrane–solution interface; ions within the channels move independently. This is no longer the case when more than one ion is allowed into the channel, and the analysis is correspondingly more complex (90, 92).

More general cases can also include nonhomogeneous, asymmetric, charged channels. These can be treated by dropping the assumption that $d\mu^o/dx = 0$, or by introducing a new potential energy function (90). Figure 6.16 shows an example of a model channel discussed by Dani and Levitt (30), where $a(x)$ varies from the wide-mouthed vestibule on the left to a narrow, negatively charged aperture that approximates the size of an ion on the right. The fact that an ion within the channel loses much of its interaction with water dipoles is accounted for by the low dielectric constant ($\varepsilon = 2$) in the nearby channel walls. The dotted lines show equipotential lines generated by the channel charge, the full lines show lines of force.

In the next section, we discuss some of the elementary factors that contribute to energy barriers.

ENERGY BARRIERS

The Born Energy Estimates the Work Required to Transfer an Ion from One Medium to Another

Following Born's classic calculation of hydration energies (17), we begin with an estimation of the work done ($\Delta\tilde{G}^o$) in transferring an ion from one medium (water)

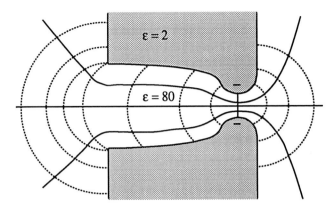

FIG. 6.16. A model channel illustrating the continuum approach with a macroscopic dielectric constant. Channel cross-sectional area, $a(x)$, varies from the wide-mouthed vestibule on the left to a narrow, negatively charged, aperture that approximates the size of an ion on the right. The *dotted lines* show equipotential lines generated by the channel charge; the *solid lines* show lines of force. (Redrawn from Dani and Levitt [30].)

to another (e.g., a hydrocarbon medium similar to the interior of a lipid bilayer):

$$C^+_{water} \longrightarrow C^+_{hydrocarbon}$$

We assume that electrostatic energy is the only significant work done in this process. This allows us to carry out the calculation in the following two stages: First, we assume that an *uncharged* ion is introduced into the hydrocarbon phase; this requires no electrical work. Second, we introduce charge bit by bit until the final ionic charge, q_o, is obtained. The work done in this process is the electrostatic energy of the ion in the hydrocarbon. Now repeat the process in the water. The difference in these two energies represents the work required to transport the ion from hydrocarbon to water.

To calculate the charging energy, we represent the ion by a rigid sphere of radius r_i bearing a charge q. Using Gauss's law it can be shown (145) that at any distance $r > r_i$ from the center, the electric field (and potential) generated by a spherically symmetric charge is identical to the field generated by the same total charge concentrated at a point at the center of the sphere. In a vacuum the potential Ψ_r at a point r distant from a single point charge (or from the center of the corresponding spherical "ion") is given by $\Psi_r = q/4\pi\varepsilon_o r$, where Ψ_r is measured in volts, q in Coulombs, r in meters, and $\varepsilon_o = 8.85 \times 10^{-12}$ Coulomb^2 joule^{-1} meter^{-1} is the permittivity of free space. If the sphere is immersed in any material (say, water or hydrocarbon), the electrical force and potential is reduced by a factor ε_r, the relative permittivity of the medium, which is also known as its dielectric constant. Taking this into account, we now have

$$\Psi_r = q/4\pi\varepsilon_o\varepsilon_r \quad \text{for} \quad r > r_i \qquad (6\text{--}217)$$

The dielectric constant, a dimensionless number, is a measure of the ability of the medium to shield the electrical charge.

By definition, the work done in adding a small charge dq to any point at potential Ψ is given by Ψdq. Starting with an uncharged sphere in hydrocarbon, the work done per ion in charging it up to $q_o = 1.6 \times 10^{-19}$ Coulomb (the charge of an electron or proton) would be

$$W^o_{hydrocarbon} = \int_0^{q_o} \Psi_{r_i} dq$$

$$= \frac{1}{4\pi\varepsilon_o\varepsilon_{r\_hydrocarbon}r_i} \int_0^{q_o} q\,dq = \frac{q_o^2}{8\pi\varepsilon_o\varepsilon_{r\_hydrocarbon}r_i}$$

$$(6\text{--}218)$$

A similar calculation can be made for the water phase. After taking the difference of the two results (hydrocarbon − water) and multiplying by Avogadro's number, N_A, to convert to work per mole of univalent ions, we are left with the $\Delta\tilde{G}^o$ required to move the ion from water to hydrocarbon which is often called the Born energy; i.e.,

$$\Delta\tilde{G}^o_{Born} = \Delta\tilde{G}^o_{water \rightarrow hydrocarbon} = \frac{N_A q_o^2}{8\pi\varepsilon_o r_i}\left(\frac{1}{\varepsilon_{r\_hydrocarbon}} - \frac{1}{\varepsilon_{r\_water}}\right)$$

$$= \frac{166}{r_i^o}\left(\frac{1}{\varepsilon_{r\_hydrocarbon}} - \frac{1}{\varepsilon_{r\_water}}\right) \text{ kcal mol}^{-1}$$

$$(6\text{--}219)$$

where r_i^o is the ionic radius in Angstroms ($r_i^o = r_i \times 10^{-10}$). The partition coefficient for a univalent ion is then given by (see Equation 6–30)

$$\frac{C_{hydrocarbon}}{C_{water}} = e^{\frac{-\Delta\tilde{G}^o}{RT}} = e^{\frac{\frac{166}{r_i^o}\left(\frac{1}{\varepsilon_{r\_hydrocarbon}} - \frac{1}{\varepsilon_{r\_water}}\right)}{RT}}$$

$$(6\text{--}220)$$

The dielectric constant for lipid bilayers is generally around 2, while that for water is very large, about 78.5, reflecting the large dipole moment of water molecules. Inserting these values with $r_K^o = 1.5$ Å for K^+ gives a $\Delta\tilde{G}^o = 53.95$ kcal mol^{-1}. With $RT = 0.592$ kcal mol^{-1} at 25°C, we arrive at a concentration distribution ratio (solubility) of 2.6×10^{-40}. Note that for ions with smaller radii (e.g., Na$^+$) the solubility will be even less. It is no wonder that lipid bilayers are virtually impermeable to small ions. Since the ionic radius appears in the exponential of Equation 6–220, the solu-

bility is highly sensitive to ionic size. If we increase r_i^o by a factor of 3—say, to 4.5 Å—the distribution ratio increases some 26 orders of magnitude, from 2.6×10^{-40} to 6.5×10^{-14}! Large ions ($r_i > 4$Å) begin to show some solubility and are often able to permeate bilayers. This is particularly true for uncouplers of (1) oxidative phosphorylation, e.g., FCCP$^-$ (carbonylcyanideptrifluoromethoxyphenylhydrazone), DNP$^-$ (2,4-dinitrophenol), (2) potential sensitive dyes ANS$^-$ (1-analino-8-napthalenesulfonate), and (3) insertion of protein-charged groups into membranes.

The same theory can be used to estimate the free energy of ion hydration (C$^+_{vacuum} \rightarrow$ C$^+_{water}$); we repeat the same arguments, but replace $\varepsilon_{r\_hydrocarbon}$ by $\varepsilon_{r\_vacuum} = 1$. This allows the general approach to be tested by comparing it to empirical data. Results show that the unmodified Equation 6–219 substantially overestimates hydration energies. The theory can be improved considerably and brought into agreement with the data by making small increments to the ionic radii (14, 17). This is justified because at atomic dimensions water can no longer be treated as a continuum; the finite size of oxygen atoms contained in water molecules restricts their electrical centers from direct contact with the ion. As a result, the electrically equivalent "hole" within the water created by space occupied by the ion is larger than its crystal radius would suggest. Although Equations 6–219 and 6–220 are based on a primitive model, they capture the main feature of ion solubility: partitioning between water and other nonpolar media is determined largely by the strong interaction of ions with water dipoles.

Born Energy, Image Forces, Dipole Potentials, and Hydrophobic Interactions Contribute to the Energy Barriers of Lipid Bilayers

Application of Equation 6–219 to membranes tacitly assumes that ions are sufficiently far from the membrane water interface so that they no longer interact with water dipoles and that the membrane is homogeneous with little or no structure. The first assumption if faulty because bilayers are very thin; even in their center, an ion will not be isolated from induced electrical interactions with water. Forces that arise from these interactions are called image forces, because they are commonly computed by a mathematical artifice called the "method of images." The Born energy corrected for image forces has been derived for any position x within a homogeneous lipid bilayer of thickness d by Neumcke and Lauger (109) in the form of an infinite series. If the membrane thickness equals d, then for $r < x < d/2$, the series is given by

$$\Delta G^o_{Born\_Image} = \Delta G^o_{Born} - \frac{N_A q_o^2 \alpha}{16\pi\varepsilon_o \varepsilon_{r\_hydrocarbon}}$$

$$\times \left\{ \frac{1}{x} + \frac{1}{d} \sum_{n=1}^{\infty} \left(\frac{\alpha^{2n}}{n+x/d} + \frac{\alpha^{2n-2}}{n-x/d} + \frac{\alpha^{2n}}{n+r_i/d} + \frac{\alpha^{2n-2}}{n-r_i/d} \right) \right\}$$

(6–221)

This series can be approximated to within 4% by the following (41):

$$\Delta G^o_{Born\_Image} \approx$$

$$\frac{N_A q_o^2}{8\pi\varepsilon_o \varepsilon_{r\_hydrocarbon} r_i} \left(1 - \frac{r_i}{2x} - 1.2 \left(\frac{r_i}{d}\right)\left(\frac{x}{d}\right)^2 \right) \quad r_i < x < d/2$$

(6–222)

where it is assumed that $\varepsilon_{r\_water} >> \varepsilon_{r\_hydrocarbon}$.

Plots comparing Equation 6–219 and Equation 6–221 are shown in Figure 6.17, where the striking effect of ion size is clearly shown.

The permeability of bilayers to large anions is dramatically larger than the permeability to cations of corresponding size and hydrophobicity. The hydrophobic anion tetraphenylboron ($z = -1$), for example, permeates bilayers as much as 10^9 times faster than its cationic analog tetraphenylarsonium ($z = +1$) (5, 6), despite the fact that both ions are univalent, have similar radii ($r_i = 4.2$ Å), and have identical hydrophobic groups. This dramatic selectivity is not predicted by the Born-image theory. Rather, it appears to be due to electrostatic forces generated by dipoles oriented perpendicular to the plane of the membrane and giving rise to a "dipole potential." The source of these dipoles is apparently the carbonyl and ester groups that link the two fatty acid chains to the glycerol backbone of the phospholipid molecule. Oriented water molecules at the membrane interface may also contribute, but zwitterionic phospholipid head groups probably do not, because they lie flat within the plane of the membrane so that their dipole moment is directed perpendicular to the transport path. The implicated ester oxygens direct their negativity toward the aqueous phase, making the membrane interior positive. But the hydrophobic membrane interior is the bottleneck for permeation by a polar molecule. If it is electropositive it will attract anions, repel cations, and thus account for a selective permeability of the bilayer.

A quantitative, self-consistent construction of energy barriers relevant to bilayer binding as well as permeability of hydrophobic ions developed by Flewelling and Hubbell (40, 41) is shown in Figure 6.18. In addition to the electrostatic forces described above,

they also include neutral hydrophobic forces. Plot A shows the individual components drawn for univalent hydrophobic cations and anions with a radius of 4.2 Å corresponding to tetraphenylboron and tetraphenylarsonium. The upper trace shows a composite consisting of the Born image neutral hydrophobic energies. It is similar to Figure 5.17, but the addition of hydrophobic energies estimated at around -4 kcal mol^{-1} within the membrane lowers the composite and is responsible for the dip in the curve that occurs within the membrane close to the interface regions where the Born-image energy has not risen substantially. The two lower curves of plot A, "+ Dipole" and "− Dipole," are plots of dipole energies for cations and anions, respectively. Plot B shows the sum of these energies. Note that the differences between anion and cation is due to the dipole term. The energy barrier pattern shows two energy wells within the membrane but near the interfaces. It suggests a three-step transport model (aqueous → energy well(1)→ energy well(2) → aqueous) that is similar to one developed and successfully applied by Ketterer et al. (69).

Solvation Energies Are Important Determinants of Channel Accessibility

Most physiological ions are virtually excluded from the bilayer. Instead they are transported through water-filled protein channels. Although these channels provide lower energy barriers for ion transport, they are narrow, sometimes allowing only single-file passage, and the environment presented to an ion contained within them is complex and not equivalent to bulk

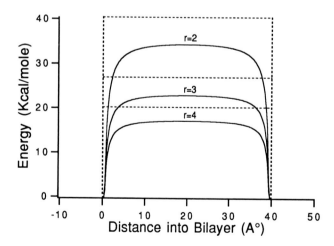

FIG. 6.17. Born energy *(dashed lines)* for three different ionic radii are obtained from Equation 6–219. Corresponding Born energy corrected for "image forces" and shown as *solid lines* are obtained from Equation 6–221 using the first 20 terms of the series.

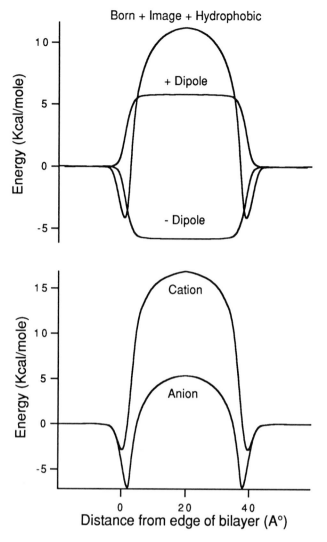

FIG. 6.18. Lipid bilayer energy barriers for a hydrophobic ion with radius = 4.2 Å. *Top panel:* The upper trace shows a composite consisting of the Born image plus neutral hydrophobic energies. The two lower curves of plot A, "+ Dipole" and "– Dipole," are plots of dipole energies for cations and anions respectively. *Bottom panel* shows the composite sum of these energies, one for anions, the other for cations. The negative dipole potential lowers the energy barrier for anions, making them very permeable. (Redrawn from Flewelling and Hubbel [41].)

water. Just as in bilayers, problems of solute distribution between bathing solution and channel interior is an important determinant of permeability, and similar methods of analysis have been employed.

Within the channel, in the radial direction an ion is close to the channel wall, an interface of relatively low dielectric constant due to hydrophobic parts of the channel protein and the lipid bilayer. In the longitudinal direction it is adjacent to water molecules that have also been displaced from their normal environment, making it difficult to assign a local dielectric constant

within the channel. In addition, an ion, even in the center of the channel, may be sufficiently close to the edge for some interaction with the bulk water of the bathing solutions. These interfaces give rise to a complex set of forces. They have been analyzed by several authors (61, 86).

Dipole potentials arising within the lipid bilayer appear to be shielded by channel proteins, substantially reducing any influence on ion transport through the channel (58). However, there are polar groups (often containing oxygen) that line the pathway through the channel. Although these groups are not sufficiently oriented to summate and produce a large field (like the bilayer dipole potential), they interact with channel ions, providing local energy wells that influence the advancement of ions into and through the channel. Ions, for example, may leave the polar environment of the water because polar oxygens within the channel can substitute for the water oxygens. Energy changes, which occur as water (solvent) molecules associated with the solute are replaced by the polar groups of the membrane, have been called "solvation energy."

Relevant forces governing ion permeation have been studied most intensively in the gramicidin channel. This simple channel is formed in bilayers by a dimer of gramicidin A, a linear pentadecapeptide. Transport occurs via a narrow cylindrical pore approximately 26 Å long and 4 Å in diameter produced by two single-stranded right-handed β helices. Hydrogen-bonded carbonyl groups line the interior of the pore, while hydrophobic amino acid side chains extend into the bilayer. Structural constraints appear to restrict permeation to single-file transport of both partially dehydrated ions and water.

The most detailed investigations of this channel utilize the method of molecular dynamics simulations (121). In this computer-generated approach, the history of an initial arrangement of several hundred to a thousand particles is followed by tracking the trajectory of each particle as it moves under the influence of assumed intermolecular forces (or potentials). Using Newton's laws, individual positions are estimated in a very short time interval (shorter than the time between collisions), and this process is repeated tens of thousands of times, generating the desired trajectories over a short but significant time span.

The initial system used by Roux and Karplus (120) in their molecular dynamics model consisted of 314 peptide atoms for the gramicidin dimer, 190 water molecules, 85 spherical models of CH_3 groups (representing the hydrocarbon membrane containing the channel), and one Na^+. Their reconstruction of the energy barriers for the entire system is shown in Figure 6.19.

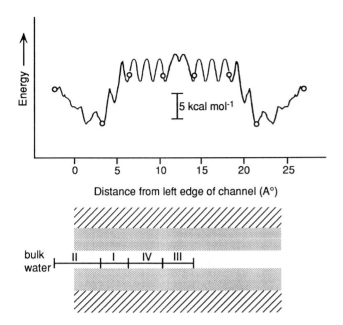

FIG. 6.19. Energy-barrier profile of gramicidin channel constructed from molecular dynamics simulations by Roux and Karplus. Simulations were accomplished by dividing the channel into four sections labeled *I* to *IV* *(lower figure)*, corresponding to sections separated by open circles in energy profile *(upper figure)*. Significance of the different sections: II—transition from bulk water to single file solvation; I—the beginning of single-file transport; III—single-file transport; IV—intermolecular region. (Redrawn from Roux and Karplus [120].)

As illustrated in the figure, the free energy profile consists of 14 barriers and 15 wells with barriers separated by approximately 1.5 Å, corresponding to the helical length of a periodic β helix. The region where the two monomers join is labeled III in the lower diagram.

Beginning on the left in Figure 6.19, a Na^+ located in the bulk water has five to six water molecules in its first hydration shell. As it begins to enter the channel in region II (which stretches for about 5 Å), it *progressively* dehydrates as water is replaced by channel carbonyls. As the ion finally reaches the beginning of the single-file region, its nearest neighbors consist of two water molecules and four channel carbonyl groups. Note that during this passage there is no substantial energy barrier that can be identified with a dehydration step. The dehydration appears to be more characteristic of a continuous transformation, with carbonyl groups compensating for loss of water, rather than an abrupt step. This conclusion is consistent with the analysis of Jakobsson and Chiu (57), who found their experimental data best fit by energy profiles that show no significant energy barrier near the channel entrance. It is at variance with earlier molecular dynamics simulations reviewed in Roux and Karplus (121).

The lowest energy wells are located at the end of region II and the beginning of I. These are thought to correspond to the major Na^+ binding sites that have been detected experimentally with NMR. Region I represents the beginning of single-file diffusion. It contains the largest activation barriers, and it is tempting to ascribe these large barriers to image forces, smaller but analogous to the large internal barrier found in bilayers. However, closer analysis by Roux and Karplus shows that effects of water molecules within the bulk solution (the origin of the image forces) is swamped by the forces arising from electrostatic interactions with eight water molecules pointing toward the ion in the channel. Long-range image forces do not play a significant role. Instead, the barrier appears to be due to local interactions between the ion, the two neighboring waters, and the channel.

Region IV, representing a region of fully developed single-file diffusion within the interior of the channel, was modeled with an infinite periodic poly(L,D)-alanine β helix (119). The "interior" lies deep within the channel where boundary effects due to entrance barriers (accounted for in regions II and I) are negligible. This region contains energy barriers approximately 4.5 kcal mol^{-1} for Na^+, but only about 1 kcal mol^{-1} for the larger K^+, and this corresponds to the faster transport rates of the larger cations. The larger barrier (and lower transport rates) of the smaller cations is due to their stronger association with carbonyl oxygens. Within the energy wells in the channel interior Na^+ lies off center, while the large K^+ lies closer to the channel axis. The motion of Na^+, composed of several jumps over small energy barriers between discrete states, is controlled by local interactions arising from collisions with neighboring carbonyls and the two nearest waters. The local flexibility of the helix plays an important role by helping Na^+ maintain its close contact with the carbonyls; during its transition from site to site the ion is always in contact with at least two carbonyls and two water molecules. In contrast, K^+ transport seems to be controlled by diffusion of water; it has much smaller barriers and motion proceeds by continuous diffusion as described by the Nernst-Planck diffusion approximation. Predicted transport rates for Na^+ are in good agreement with experimental data, but predicted K^+ and water transport are too slow (118). (The authors suggest that channel nonhomogeneities not incorporated into the model may be responsible.)

Surface Potentials Modify the Transmembrane Potential as Well as Local Ion Concentrations

Negative electrical charges arising from ionized groups on the phospholipids and proteins that make up biological membranes lower the electrical potential of the membrane surface relative to the potential in the bulk solution far from the surface. They attract compensatory (counter) cations from the solution, concentrating them near the surface. A complex cation distribution results from the balance of electrical attraction toward the surface and the tendency for diffusion in the opposite direction. Anions, repelled by the surface charge and tending to diffuse toward it, find themselves in a similar but oppositely oriented distribution where they are partially excluded from regions near the surface.

The importance of surface charge for ion transport is reflected by the fact that flux equations require concentrations and the electrical potential at the membrane surfaces. With significant surface charge, neither the ionic concentration nor the electrical potential in the bulk phase represents the corresponding items at the surface. Within this context, for a given surface charge we would like to know: *(1)* What is the potential difference between the surface and the bulk phase? *(2)* What is the relation between ionic concentrations at the surface and in the bulk? *(3)* How far out into the solution do surface effects extend? *(4)* How does the actual potential drop across the membrane relate to the measured potential difference between the bathing solutions? *(5)* How are the above affected by the ionic composition of the bathing solution?

Approximate answers to these questions can be obtained for a simple Guoy-Chapman model of the membrane surface. The model assumes the surface charge is uniformly spread out in a continuum over the membrane surface, and that the aqueous solution, also treated as a continuum, has a uniform, well-defined dielectric constant extending all the way up to the mathematical interface representing the membrane–aqueous solution boundary. Further, it assumes that the ions can be treated as point charges. Analysis of the Guoy-Chapman model (10, 17, 103) shows that when the surface on side "in" is bathed by a mixture of ions, all of the same valence, the surface potential Ψ_s (in millivolts) on side "in" at 25°C is given by

$$\Psi_s \equiv \Psi_{surface} - \Psi_{in} = \frac{2RT}{zF} \sinh^{-1}\left(\frac{\theta\sigma_s}{C^{1/2}}\right)$$

$$= \frac{51}{z} \sinh^{-1}\left(\frac{136\,\sigma_s}{C^{1/2}}\right) \qquad (6\text{--}223)$$

where σ_s is the surface charge density in electronic charges per Å^2, C is the concentration of salts (moles per liter) in the bulk solution, and $\theta = 1/(8\varepsilon\varepsilon_o RT)^{1/2}$. If, for example, 20% of the phospholipids have a single net negative charge (i.e., 20% are either phosphatidyl serine, glycerol, or inositol), then, assuming each phospholipid molecule occupies 60 Å^2, the surface charge

density will be 1 charge per 300 Å^2 (= 0.0033 charges per Å^2). It follows that in solution of, say, 100 mM KCl the electrical potential at the surface will lie 59 millivolts below the potential in the bulk phase.

Given the surface potential, the relation between ionic concentrations at the surface and in the bulk is given by the equilibrium condition Equation 6–28, i.e.,

$$C_{surface} = C_{bulk}\, e^{\frac{-zF\Psi_s}{RT}} \qquad (6\text{--}224)$$

In our example with $\Psi_s = -59$ mV, we expect the univalent cations at the surface to be approximately 1000 mM, ten times more concentrated than in the bulk solution. Anions will be 10 mM, ten times less concentrated.

A further result of the Guoy-Chapman theory shows how the potential changes as we move from the surface toward the bulk solution. Letting x denote the distance of any point in the aqueous solution from the membrane surface (where $x = 0$), it can be shown that

$$\Psi(x) - \Psi_{in} = \frac{2RT}{zF} \ln\left(\frac{1 + \alpha e^{-\kappa x}}{1 - \alpha e^{-\kappa x}}\right)$$

$$\alpha \equiv \frac{e^{\frac{zF\Psi_s}{2RT}} - 1}{e^{\frac{zF\Psi_s}{2RT}} + 1}, \qquad \kappa \equiv \left(\frac{2z^2 F^2 C}{\varepsilon_r \varepsilon_o}\right)^{\frac{1}{2}} \qquad (6\text{--}225)$$

where $1/\kappa$ is the *Debye length*, an important distance that is often used to characterize the spatial dependence of electrical potentials in electrolyte solutions.

Equation 6–225 can be simplified by expanding α in a Taylor's series about the point $\Psi_s = 0$. Using a fifth-order expansion yields

$$\alpha = 0.2500\,\frac{zF\Psi_s}{RT}$$

$$- 0.0052\left(\frac{zF\Psi_s}{RT}\right)^3 + 0.0001\left(\frac{zF\Psi_s}{RT}\right)^5 - \ldots$$

$$(6\text{--}226)$$

For $|zF\Psi_s/2RT| < 1$, we can ignore the higher-order terms and insert $\alpha \approx zF\Psi_s/4RT$ into Equation 6–225 to arrive at

$$\Psi(x) - \Psi_{in} = (\Psi_{surface} - \Psi_{in})e^{-\kappa x} = \Psi_s e^{-\kappa x} \qquad (6\text{--}227)$$

showing that the difference in electrical potential between the surface and the bulk phase will die away exponentially as we move away from the surface with a characteristic length given by the Debye length, $1/\kappa$. This provides a simple estimate of how far surface effects extend out into the solution. For example, again using 100 mM KCl, the Debye length obtained from its definition in Equation 6–225 will be about 10 Å and, since $e^{-3} = 0.05$, after moving 30 Å away from

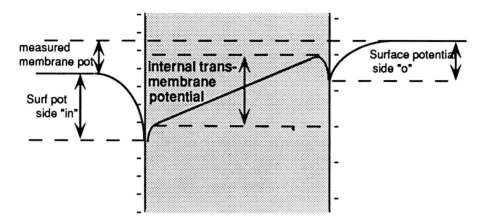

FIG. 6.20. Electrical potential profiles in the presence of surface charge. Surface-charge density on inner surface at left is assumed larger than the corresponding charge density on outer surface at the right. The fast rise in the potential on entering the membrane from either side "in" or "o" is due to a small dipole potential. Notice that the internal transmembrane potential difference is larger than the measured membrane potential.

the membrane surface the potential will be within 5% of its value in bulk solution. Although Equation 6–227 is most accurate for low surface potentials it also provides a good estimate for reasonable values. For example, letting $\Psi_s = -100$ mv and using the exact Equation 6–225, the distance where the potential reaches 5% of its value in bulk solution is 29.4 Å (compared to 30 Å calculated with Equation 6–227).

Figure 6.20 illustrates how the internal membrane potential may differ from the measured potential difference between the two bathing solutions. The solution on the right (side "o") is grounded; its potential is zero. However, at the negatively charged surface on side "o" the potential drops to Ψ_{s_o}. This change in potential begins to be apparent at about 30 Å from the surface. Similarly on the right, side "in," the potential changes from Ψ_m, the measured membrane potential in the bulk solution, to Ψ_{s_in} at the surface. The figure illustrates the common case where the inner surface carries a larger negative charge than the outer surface; accordingly, $\Psi_{s\_in} < \Psi_{s\_o}$. This establishes the potential at each interface and therefore the potential drop across the membrane. Details of the drop (e.g., constant field) have been discussed earlier. The figure assumes a linear profile superimposed on a symmetric dipole potential depicted by the sudden increase in potential that occurs just after entering the membrane from either side.

Ions in the bulk solution redistribute near the surface and partially compensate, or "screen," the net charge fixed to the membrane. The more concentrated the ions, the more effective the screening. Further, the more charge carried by each ion (the higher the z), the more effective the screening. These conclusions are borne out by analysis. Increasing either C or z will

decrease Ψ_s (Equation 6–223), as well as the Debye length, $1/\kappa$ (Equation 6–226). Thus, the surface potential and the distance that surface effects extend into the medium (Equation 6–225), are both depressed.

Quantitative results summarized by Equations 6–223, 6–225, 6–226, and 6–227 are limited to cases where ions in the aqueous media all have the same valence, i.e., $|z_{cation}| = |z_{anion}|$. For more general mixtures of ions with no restrictions on the valence, Equation 6–223 is replaced by the Grahame equation (10):

$$\sigma_s = \left[2\varepsilon_r\varepsilon_o RT\Sigma C_i \left(e^{-\frac{z_i F\Psi_s}{RT}} - 1 \right) \right] \quad (6\text{–}228)$$

where the summation is carried out over all ions in the bulk solution. The Grahame equation must be solved numerically for Ψ_s.

One of the chief criticisms of literal applications of the Guoy-Chapman model arises from the assumption that charge is uniformly smeared out over large areas of the membrane. Even in uniform lipid bilayers the charge is discrete and the charge provided by the mixture of phospholipids and proteins typical of cell membranes is certainly not uniform. Nevertheless, the theory provides useful insights as a simple first approximation; it has been successfully applied in many cases (103, 104). Reviews of more recent developments, as well as their application to bilayers and channels, are given in several articles (18, 63, 64, 74, 104, 108).

Transport across Energy Barriers

Energy barriers provide a mechanism for including more detailed membrane properties into Nernst-Planck transport equations. Energy profiles (as a function of x) for image forces, dipole potentials, solvation, etc., such as those illustrated above, can be incorporated into the term for μ^o of Equation 6–150 (e.g., see 31 for nonelectrolytes, and 90 for ions). The development and evaluation of more realistic, detailed energy profiles is a focus of intensive research (18, 30, 59, 60, 62, 63, 64, 90, 91, 120, 121).

Eyring Rate Theory: Rate Constants Depend on Ψ

Rate theory provides an alternative to the Nernst-Planck continuum. It assumes that the transported particle occupies only a finite number of sites during its passage and characterizes the transport in terms of jumping over energy barriers that exist between sites. This approach is attractive because it has a self-contained method to deal with heterogeneous transport pathways. It was used to derive the permeability equations for nonelectrolytes, Equations 6–82 and 6–106. Adopting the same model for ions, a similar analysis applies, only now each rate constant must be modified by the electrical energy of the ion at each site in the electrical field. The following outlines the Eyring method, the simplest and most popular approach for incorporating Ψ_m into rate constants.

At the ith site, let $\Psi = \Psi_i$, and at site $i+1$, let $\Psi = \Psi_{i+1}$. Assume the energy barrier is symmetric with its peak located midway between the two sites, and that electrical potential at this peak is given by the mean of the two site potentials, i.e., $\Psi = 1/2\,(\Psi_{i+1} + \Psi_i)$. The electrical work done in moving a mole of ions from site i to the peak, $w_{i,p}$, will be given by

$$w_{i,p} = zF\left(\frac{1}{2}(\Psi_{i+1} + \Psi_i) - \Psi_i\right) = \frac{1}{2}zF(\Psi_{i+1} - \Psi_i)$$

$$(6\text{--}229)$$

while the work done in moving from $i+1$ to the peak will be given by

$$w_{i+1,p} = zF\left(\frac{1}{2}(\Psi_{i+1} + \Psi_i) - \Psi_{i+1}\right) = -\frac{1}{2}zF(\Psi_{i+1} - \Psi_i)$$

$$(6\text{--}230)$$

The corresponding activation energies in the presence of the electric field for the forward and reverse direction are

$$\Delta\tilde{G}^{o\ddagger}_{+i} = \Delta G^{o\ddagger}_{+i} + \frac{1}{2}zF(\Psi_{i+1} - \Psi_i)$$

$$\Delta\tilde{G}^{o\ddagger}_{-i} = \Delta G^{o\ddagger}_{-i} - \frac{1}{2}zF(\Psi_{i+1} - \Psi_i) \qquad (6\text{--}231)$$

where $\Delta G^{o\ddagger}$ is the activation free energy, in the absence of an electric field. Recalling Equation 6–55, the rate

constants for electrolytes in the forward and backward direction are

$$k_{+i} = k^o_{+i}e^{-\frac{\Delta\tilde{G}^{o\ddagger}_{+i}}{RT}} = k^o_{+i}e^{-\frac{\Delta G^{o\ddagger}_{+i}}{RT} - \frac{zF}{2RT}(\Psi_{i+1} - \Psi_i)}$$

$$= \gamma_{+i}e^{-\frac{zF}{2RT}(\Psi_{i+1} - \Psi_i)}$$

$$k_{-i} = k^o_{-i}e^{-\frac{\Delta\tilde{G}^{o\ddagger}_{-i}}{RT}} = k^o_{-i}e^{-\frac{\Delta G^{o\ddagger}_{-i}}{RT} + \frac{zF}{2RT}(\Psi_{i+1} - \Psi_i)}$$

$$= \gamma_{-i}e^{+\frac{zF}{2RT}(\Psi_{i+1} - \Psi_i)}$$

$$(6\text{--}232)$$

where $\gamma_{\pm i} \equiv k^o_{\pm i}\,e^{-(\Delta G^{o\ddagger}_{\pm i}/RT)}$ and the subscripts " $+$ " and " $-$ " refer to rate constants in the forward $(i \to i+1)$ and back $(i+1 \to i)$ directions, respectively.

Taking a constant field example, we have $\Psi = \Psi_m(1 - x/d)$. If the ith site is located at x_i and letting the fractional distance between sites be $\delta_i = (x_{i+1} - x_i)/d$, we have:

$$\Psi_{i+1} - \Psi_i = -\Psi_m(x_{i+1} - x_i)/d = -\Psi_m\delta_i$$

$$k_{+i} = \gamma_{+i}e^{+\frac{zF}{2RT}\delta_i\Psi_m} \qquad k_{-i} = \gamma_{-i}e^{-\frac{zF}{2RT}\delta_i\Psi_m} \qquad (6\text{--}233)$$

More-general treatments omit any reference to constant fields; we drop the middle expression in Equation 6–233 and simply accept the definition, $\delta_i \equiv (\Psi_{i+1} - \Psi_i/ - \Psi_m)$, as the fractional change in potential between sites i and $i+1$ (normalized by the total trans-membrane potential). In these cases, we refer to δ_i as an "electrical distance," but it is often simply an empirical constant. Strictly speaking, k is an electrochemical rate constant that for consistency should be written with a tilde: $\tilde{k}$. To simplify notation we will omit the tilde, with the understanding that all rate constants for ions may be dependent on electrical potential.

The conventional treatment, given above, focuses entirely on the effect of the electric field on the transported ion and completely neglects its effect on charges or dipoles associated with the channel. These can be substantial even in the case of an electrically neutral channel. As an ion progresses from energy well to peak, carbonyl dipoles reorient, and this accounts for part of the work in moving the ion. But this work will depend on the electric field generated by the membrane potential. Assuming a constant field in a gramicidin channel, it can be shown (119) that the electrical work contains an additional term, equal to minus the electric field times the average change in the x component of the dipole of the channel-water system when the ion goes from the bottom of the energy well to the peak. Letting $\Delta\overline{m}_{dipole}$ represent this average change in dipole in molar terms (i.e., the molecular dipole times Avogadro's number), the result given by Roux and Karplus, expressed in our notation, replaces Equations 6–229 and 6–233 with

$$w_{i,p} = \frac{1}{2} zF(\Psi_{i+1} - \Psi_i) - \Delta\overline{m}_{dipole}\left(\frac{\Psi_m}{d}\right)$$

$$= -\left(\frac{x_{i+1} - x_i}{2d} + \frac{\Delta\overline{m}_{dipole}}{zFd}\right) zF\Psi_m.$$

$$k_{+i} = \gamma_{+i} \exp\left[\frac{zF}{2RT}\left(\frac{x_{i+1} - x_i}{d} + \frac{2\Delta\overline{m}_{dipole}}{zFd}\right)\Psi_m\right]$$

$$(6\text{--}233a)$$

Comparing Equation 6–233a with Equation 6–233 shows that the apparent "electrical distance" is now given by

$$\frac{x_{i+1} - x_i}{d} + \frac{2\Delta\overline{m}_{dipole}}{zFD}.$$

The distance between sites in the helical model was $x_{i+1} - x_i = 1.5$ Å, and, from their simulations with the Na$^+$ ion, Roux and Karplus estimate that $2\Delta\overline{m}_{dipole}/zF = 0.8$ Å. Hence, in the interior of the gramicidin channel, neglect of the channel dipole contribution to the electrical work produced by the advance of an Na$^+$ leads to a 53% error in the apparent "electrical distance."

Kinetic Approaches

At present there is no single kinetic method that fulfills all criteria for detail, precision, feasibility, and convenience. Different methods lend themselves to different problems. The relative merits and disadvantages of rate theory vs. a continuum approach have been compared in several reviews (4, 23, 25, 30, 88, 90, 118). Mathematically, the continuum approach is more subtle; conceptual as well as formal difficulties appear when mutual interactions of ions (e.g., more than one ion in the same channel) are taken into account (23, 90). Further, molecular dynamics simulations (120) show that differences in orientational freedom of bulk water and channel water make it difficult for a continuum electrostatic treatment, based on an isotropic dielectric constant, to approximate the channel environment. However, a continuum approach has the important advantage that it provides a model with a few adjustable parameters that relates permeation data to physical features of the channel. This approach will probably find more application as details of channels are uncovered.

The chief advantage of rate theory is that it provides a simple, straightforward means of dealing with a multitude of complexities, including ion interactions and the fluctuations of energy barriers, that are bound to arise with channels and carriers. Further, the correspondence of rate theory transport with enzyme kinetics provides a wealth of analytical, graphical, and computer techniques developed over the years that can be applied directly to transport problems. However, the simple formalism of rate theory is somewhat misleading; when more than a few barriers are considered, the number of unknown parameters as well as the algebraic tedium in dealing with them quickly become excessive. A more important objection resides in attempts to interpret the rate constants. The theory, as developed by Eyring (45), provides a means for resolving a rate constant into two factors: an activation energy term, $e^{-\Delta G^{\circ\ddagger}/RT}$, and a jumping frequency factor, k^o. The frequency factor is generally assumed to be given by $(k_BT)/(h) = 6.11 \times 10^{12}$ s^{-1}, where k_B is the Boltzmann constant and h is Planck's constant. Using this value for k^o allows $\Delta\tilde{G}^{o\ddagger}$ to be calculated directly from the rate constant, and presumably this energy can then be related to features of the transport protein. However, the use of Eyring rate theory with molar energy barriers below 10 RT is questionable at best and, unlike many chemical reactions, estimates of energy barriers for channels lie significantly below that level (23). Application of the theory to the gramicidin channel, for example, overestimates the transport rate of Na$^+$ by one order of magnitude (118). Without an unambiguous interpretation of rate constants, we are left with descriptive parameters that are not easily related to structural detail or to the membrane potential.

Models based on Eyring rate theory assume that barrier crossings occur in a single "ballistic" jump, where rate constants depend primarily on the height of the barrier. In contrast, alternative, more realistic, rate-constant formulations (24, 25) based on stochastic methods focus on frictional processes arising from frequent collisions and on the shape of the energy barrier. These details are not currently available for most channels.

It is generally agreed that molecular dynamics simulations will provide the most detailed and precise analysis of transport energetics and kinetics that is available. Its limited use to date stems not only from the fact that it requires a detailed description of channel structure; it has also been severely constrained by the limitations of currently available computers. Transit times for ions through channels are measured in terms of microseconds, while realistic molecular dynamic simulations are usually less than a few nanoseconds, making it necessary to break simulations into segments. The simulations leading to regions I, II, and III of Figure 6.19, for example, are reported to have taken on the order of 300 Cray YMP hours. Other problems related to the accuracy of computations of molecular dynamics simulations are reviewed in Roux and Karplus (121). Attempts to find some compromise between the rigorous demands of molecular dynamics and macroscopic approaches introduce a rapidly fluctuating random force into the equations of motion (Langevin equation [116]) to substitute for many of the environmental collisions (23, 118).

Despite its limitations, rate theory has been used more extensively than any other approach with considerable success for bilayers, channels, and carriers (13, 51, 52, 69, 77, 79, 80). It has been particularly useful in dealing with saturation, competition, ion interactions, selectivity, multiple binding sites, and fluctuating barriers. It is an effective tool for dealing with the concentration dependence of transport, and we shall use it extensively in that context. While recognizing that each rate constant depends on the membrane potential, we circumvent some of the problems discussed above by refraining from any explicit interpretation of the rate constants. In the next section, we apply rate theory to simple channels suggested by qualitative features of Figure 6.19.

CHANNELS

Channels are transport pathways mediated by proteins that usually contain "binding" sites for the transported solutes. The defining feature of a channel is that the binding sites are simultaneously accessible to solutes on either side of the membrane. Where details are available, their most outstanding property seems to be the very high transport capability (turnover number) of each channel, which may be as high as 10^8 s^{-1} (77, 79, 80).

A channel containing two binding sites that can bind a maximum of one solute C at a time is illustrated in Figure 6.21. Each channel can exist in three states: an empty channel Y^{oo}, a channel with C on the inside Y^{co}, and a channel with C on the outside Y^{oc}. Inclusion of doubly occupied channel Y^{cc} is excluded from this example by assuming that two C ions would repel each other, or that the concentrations are so low as to make its occurrence improbable. Our analysis begins with a calculation of how these states are distributed throughout the membrane in the steady state. Once this is determined, the steady-state solute flux across the membrane can be found from the net transition rate between any two states—say, from Y^{CO} to Y^{OC}. Letting Y_t represent the channel density—i.e., the number of channels per cm^2 membrane—and letting Y^{oo}, Y^{co}, and Y^{oc} represent the number of channels per cm^2 that are in each of the states Y^{oo}, Y^{co}, and Y^{oc}, respectively, we can write

$$J = k_1 Y^{co} - k_{-1}Y^{oc} = k_o Y^{oc} - k_{-o}C_o Y^{oo}$$

$$= k_{in}C_{in}Y^{oo} - k_{-in}Y^{co} \qquad (6-234)$$

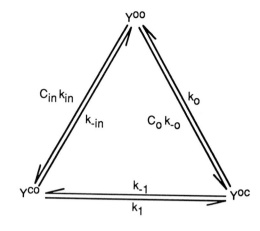

$$C_{in} + Y^{oo} \underset{k_{-in}}{\overset{k_{in}}{\rightleftharpoons}} Y^{co} \underset{k_{-1}}{\overset{k_1}{\rightleftharpoons}} Y^{oc} \underset{k_{-o}}{\overset{k_o}{\rightleftharpoons}} Y^{oo} + C_o$$

FIG. 6.21. Illustration of a simple channel with two binding sites. Each channel can exist in three states: an empty channel Y^{oo}, a channel with C on the inside Y^{co}, and a channel with C on the outside Y^{oc}. The upper triangular sketch shows the transitions among the states. The lower "reaction" scheme shows the reversible transport of C_{in} to C_o.

Relevant solutions for Y^{oo}, Y^{co}, and Y^{oc} can be obtained from the condition that they do not change in the steady state (see discussion of Fig. 6.7). From Figure 6.21 we have:

$$\frac{dY^{oo}}{dt} = -(k_{in}C_{in} + k_{-o}C_o)Y^{oo} + k_{-in}Y^{co} + k_o Y^{oc} = 0$$
$$(6-235)$$

$$\frac{dY^{co}}{dt} = (k_{in}C_{in})Y^{oo} - (k_1 + k_{-in})Y^{co} + k_{-1}Y^{oc} = 0$$
$$(6-236)$$

$$\frac{dY^{oc}}{dt} = (k_{-o}C_o)Y^{oo} + k_1 Y^{co} - (k_o + k_{-1})Y^{oc} = 0$$
$$(6-237)$$

These are not independent equations; the last can be obtained as the sum of the first two. A third independent condition, the site conservation equation, results from the fact that any channel must be in one and only one of the three states, so that

$$Y^{oo} + Y^{co} + Y^{oc} = Y_t \qquad (6-238)$$

Equations 6–236, 6–237, and 6–238 comprise a set of three independent equations in three unknowns that can be represented in more compact matrix notation as

$$\begin{bmatrix} 1 & 1 & 1 \\ k_{in}C_{in} & -(k_1 + k_{-in}) & k_{-1} \\ k_{-o}C_o & k_1 & -(k_o + k_{-1}) \end{bmatrix} \begin{bmatrix} Y^{oo} \\ Y^{co} \\ Y^{oc} \end{bmatrix} = \begin{bmatrix} Y_t \\ 0 \\ 0 \end{bmatrix}$$
$$(6-239)$$

Solutions for the three unknowns, Y^{oo}, Y^{co}, and Y^{oc}, can be solved by standard determinant methods (97). The solution for each unknown is given by the ratio of two 3×3 determinants. In our case, let d_{coef} denote the determinant defined by the 3×3 matrix of Equation 5–239, and let d_{oo}, d_{co}, and d_{oc} represent the determinants formed by respectively replacing the first, second, and third column in d_{coef} with the column formed from the elements Y_t, 0, 0 (i.e., by the column on the right-hand side of Equation 6–239). The solutions for the channel states can then be written as

$$Y^{oo} = \frac{d_{oo}}{d_{coef}}, \quad Y^{co} = \frac{d_{co}}{d_{coef}}, \quad Y^{oc} = \frac{d_{oc}}{d_{coef}} \quad (6\text{–}240)$$

Each of these determinants can be "expanded by columns" (97) and written as a linear combination of three additional 2×2 determinants. For example, d_{oc}, the numerator in the ratio for Y^{oc}, is formed by the determinant of Equation 6–239, with the last column having been replaced by Y_t, 0, 0. This determinant, shown below, is then expanded by selecting the first column and writing

$$d_{oc} = \begin{vmatrix} 1 & 1 & Y_t \\ k_{in}C_{in} & -(k_1 + k_{-in}) & 0 \\ k_{-o}C_o & k_1 & 0 \end{vmatrix}$$

$$= 1 \begin{vmatrix} -(k_1 + k_{-in}) & 0 \\ k_1 & 0 \end{vmatrix} - k_{in}C_{in} \begin{vmatrix} 1 & Y_t \\ k_1 & 0 \end{vmatrix}$$

$$+ k_{-o}C_o \begin{vmatrix} 1 & Y_t \\ -(k_1 + k_{-in}) & 0 \end{vmatrix}$$

$$= y_t(k_{in}k_1 C_{in} + k_{-o}(k_1 + k_{-in})C_o) \quad (6\text{–}241)$$

which is linear in both C_{in} and C_o. The coefficient determinant, d_{coef}, of Equation 6–239) is the denominator of all the channel densities (Y^{oo}, Y^{co}, or Y^{oc}) and is equal to:

$$d_{coef} = \begin{vmatrix} 1 & 1 & 1 \\ k_{in}C_{in} & -(k_1 + k_{-in}) & k_{-1} \\ k_{-o}C_o & k_1 & -(k_o + k_{-1}) \end{vmatrix}$$

$$= (k_1 k_o + k_o k_{-in} + k_{-1} k_{-in}) + k_{in}(k_1 + k_o + k_{-1})C_{in}$$
$$+ k_{-o}(k_1 + k_{-1} + k_{-in})C_o \quad (6\text{–}242)$$

which is also a linear function of C_{in} and C_o. So, Y^{oc} is:

$$Y^{oc} = \frac{Y_t(k_{in}k_1 C_{in} + k_{-o}(k_1 + k_{-in})C_o)}{d_{coef}} \quad (6\text{–}243)$$

Expanding the other determinants for the other channel densities leads to analogous expressions, and using Equation 6–234 with $J = k_1 Y^{co} - k_{-1} Y^{oc}$, we finally arrive at

$$J = \frac{Y_t(k_{in}k_o k_1 C_{in} - k_{-in}k_{-o}k_{-1}C_o)}{d_{coef}} \quad (6\text{–}244)$$

Although there are six rate constants, they are not all independent. At equilibrium $J = 0$ and according to Equation 6–28, $C_o = C_{in}e^{zF/RT\Psi_m}$. Using this condition together with $J = 0$ in Equation 6–244 yields

$$k_{in}k_o k_1 = k_{-in}k_{-o}k_{-1}e^{\frac{zF}{RT}\Psi_m} \quad (6\text{–}245)$$

Substituting Equation 6–242 and 6–245 into Equation 6–244 and rearranging terms, we have

$$J = \frac{P[C_{in}e^{\frac{zF}{RT}\Psi_m} - C_o]}{1 + \frac{C_{in}}{K_{in}} + \frac{C_o}{K_o}} \quad (6\text{–}246)$$

where

$$P = \frac{Y_t k_{-in}k_{-o}k_{-1}}{k_1 k_o + k_o k_{-in} + k_{-1}k_{-in}} \quad (6\text{–}247)$$

$$K_{in} = \frac{k_1 k_o + k_o k_{-in} + k_{-1}k_{-in}}{k_{in}(k_1 + k_o + k_{-1})} \quad K_o = \frac{k_1 k_o + k_o k_{-in} + k_{-1}k_{-in}}{k_{-o}(k_1 + k_{-1} + k_{-in})}$$

$$(6\text{–}248)$$

Many features of this derivation are applicable in more complex cases. Note that each channel density (Y^{oo}, Y^{co}, Y^{oc}) is given by the ratio of two determinants and that the net flux is given by the difference of these ratios (after multiplying by the appropriate rate constant). These ratios all have a common denominator, which equals the determinant of the coefficient matrix shown in Equation 6–239; this will also be the denominator in the expression for J. Further, this coefficient determinant is more complex than any of the others, because they are formed from it by replacing a column by a simpler one consisting of one constant, Y_t, and the rest zeros. It follows in many cases that the formal structure of the flux equation can be deduced simply by inspecting the coefficient matrix. In our above example, the concentrations C_o and C_{in} occur only in the first column of the matrix. Selecting the first column to expand the determinant by columns, it is apparent that expansion of the determinant can be at most a linear combination of the concentrations. This also holds for any of the other determinants utilized in the derivation. It follows that the expression for J is the ratio of two linear combinations (bilinear).

Single-Occupancy Channels with Binding Sites Show Saturation Kinetics

The concentration dependence of these simple channels is similar to the concentration dependence of channels derived from the Nernst-Planck continuum (Equations

6–206 and 6–207). At constant voltage and at low concentrations ($C_{in} \ll K_{in}$, $C_o \ll K_o$), it follows from Equation 6–246 that $J = P[C_{in} e^{zF/RT \Psi_m} - C_o]$ and the flux varies linearly with concentration. As the concentration increases, competition for channels develops and the flux begins to level off. With $C_o = 0$, J follows Michaelis-Menten kinetics,

$$(J)_{C_o = 0} = \frac{J^{max-0} C_{in}}{K_{in} + C_{in}} \qquad (6-249)$$

where $J^{max-0} = P \, K_{in} e^{\frac{zF}{RT} \Psi_m}$

The corresponding expression when $C_{in} = 0$ is

$$(J)_{C_{in} = 0} = \frac{J^{max-in} C_o}{K_o + C_o} \qquad (6-250)$$

where $J^{max-in} = P \, K_o$. Notice the asymmetry indicated by the fact that both the J^{max} and the concentrations for 1/2 maximal flux, K_o and K_{in}, differ in the two cases.

Competition, Unidirectional Flux, and the Ussing Flux Ratio

Figure 6.22 shows that the kinetics of two solute species, **C** and **B**, competing for a single-occupancy channel, adds two more channel states, $\mathbf{Y^{BO}}$ and $\mathbf{Y^{OB}}$, to the three already considered.

Using superscripts to identify parameters with corresponding solutes, the steady-state channel states are determined by:

$$\frac{dY^{CO}}{dt} = (k^C_{in} C_{in}) Y^{OO} - (k^C_1 + k^C_{-in}) Y^{CO} + k^C_{-1} Y^{OC} = 0$$

$$\frac{dY^{OC}}{dt} = (k^C_{-o} C_o) Y^{OO} + k^C_1 Y^{CO} - (k^C_o + k^C_{-1}) Y^{OC} = 0$$

$$\frac{dY^{BO}}{dt} = (k^B_{in} B_{in}) Y^{OO} - (k^B_1 + k^B_{-in}) Y^{BO} + k^B_{-1} Y^{OB} = 0$$

$$\frac{dY^{OB}}{dt} = (k^B_{-o} B_o) Y^{OO} + k^B_1 Y^{BO} - (k^B_o + k^B_{-1}) Y^{OB} = 0$$

$$\frac{dY^{OO}}{dt} = -(k^C_{in} C_{in} + k^C_{-o} C_o + k^B_{in} B_{in} + k^B_{-o} B_o) Y^{OO}$$
$$+ k^C_{-in} Y^{CO} + k^C_o Y^{OC} + k^B_{-in} Y^{BO} + k^B_o Y^{OB} = 0 \qquad (6-251)$$

Again the differential equations are not all independent, and we use the conservation equation

$$Y^{OO} + Y^{CO} + Y^{OC} + Y^{BO} + Y^{OB} = Y_t \qquad (6-252)$$

to replace the expression for dY^{OO}/dt to arrive at a system of five independent equations represented by

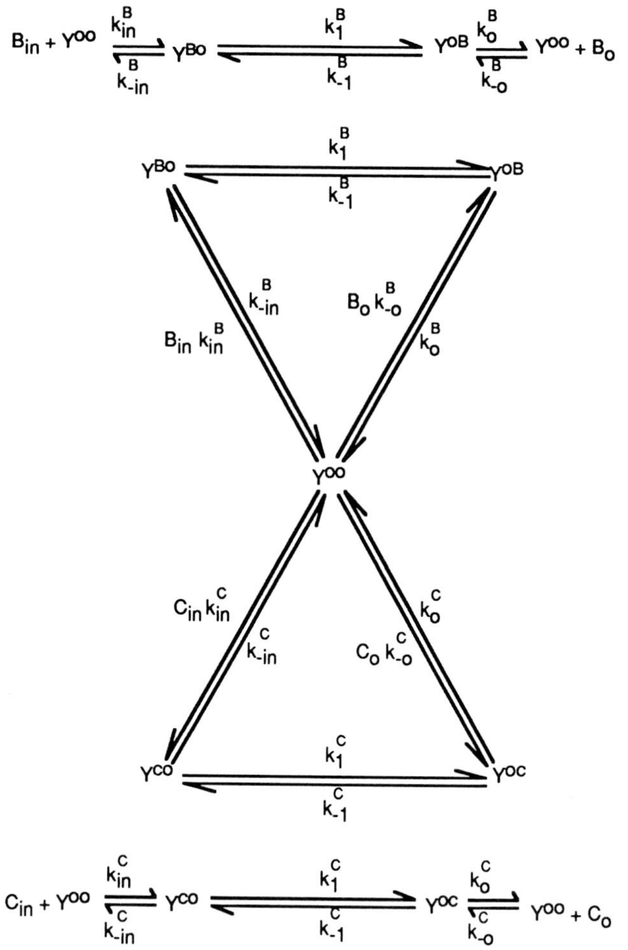

FIG. 6.22. Reaction scheme illustrating a single-occupancy channel and two binding sites, where two solutes, **C** and **B**, compete. See Fig. 6.21 for details.

$$\begin{bmatrix} 1 & 1 & 1 & 1 & 1 \\ k^C_{in} C_{in} & -(k^C_1 + k^C_{-in}) & k^C_1 & 0 & 0 \\ k^C_{-o} C_o & k^C_1 & -(k^C_o + k^C_{-1}) & 0 & 0 \\ k^B_{in} B_{in} & 0 & 0 & -(k^B_1 + k^B_{-in}) & k^B_1 \\ k^B_{-o} B_o & 0 & 0 & k^B_1 & -(k^B_o + k^B_{-1}) \end{bmatrix}$$

$$\times \begin{bmatrix} Y^{OO} \\ Y^{CO} \\ Y^{OC} \\ Y^{BO} \\ Y^{OB} \end{bmatrix} = \begin{bmatrix} Y_t \\ 0 \\ 0 \\ 0 \\ 0 \end{bmatrix} \qquad (6-253)$$

These can be resolved using the same determinant methods as in Equations 6–239. Notice that in setting up each determinant the solute concentrations C_{in}, C_o, B_{in}, and B_o will appear only in the first column in a product with Y^{oo}. This reflects the assumption that each channel can contain at most one solute molecule; the only possible reaction of free solute with membrane occurs when the channel is coming from the vacant state. Since the C_{in}, C_o, B_{in}, and B_o occur only in the first column, we can expand each determinant by columns and remove these terms as factors of the corresponding minor determinants. It follows that both the numerator and the denominator in the explicit expression for each Y will consist at most of a linear combination of C_{in}, C_o, B_{in}, and B_o. As a result, the net flux for **C** as given by any of the relations in Equations 6–234 will also be determined by a ratio of two linear combinations of the solutes. Similar remarks hold for **B**. Finally, notice that the determinant of the coefficient matrix in Equation 6–253 will be a common denominator in all expressions. It follows that we can write

$$J_C = \frac{a_0 + a_1 C_{in} + a_2 C_o + b_1 B_{in} + b_2 B_0}{b_0 + a_3 C_{in} + a_4 C_o + b_3 B_{in} + b_4 B_0} \qquad (6\text{–}254)$$

$$J_B = \frac{\alpha_0 + \alpha_1 C_{in} + \alpha_2 C_o + \beta_1 B_{in} + \beta_2 B_0}{b_0 + a_3 C_{in} + a_4 C_o + b_3 B_{in} + b_4 B_0} \qquad (6\text{–}255)$$

where all of the a_i, b_i, α_i, β_i are constants that depend on voltage but have no dependence on C_{in}, C_o, B_{in}, and B_o. Since J_C vanishes whenever $C_{in} = C_o = 0$ for all possible values of B_{in}, and B_o, and all of the infinite values that Ψ_m can take independent of the concentrations (e.g., with an external source of current as in a voltage clamp), it follows that

$$a_0 = b_1 = b_2 = 0 \qquad (6\text{–}256)$$

Further, whenever $B_{in} = B_o = 0$, Equation 6–254 must reduce to Equation 6–246 for all values of C_{in}, C_o and Ψ_m. This requires that

$$a_1 = P^C e^{\frac{zF}{RT}\Psi_m}, \quad a_2 = P^C, \quad a_3 = \frac{1}{K_{in}^C}, \quad a_4 = \frac{1}{K_o^C} \qquad (6\text{–}257)$$

where P^C, K_{in}^C, K_o^C are defined by applying the superscript C to Equation 6–247, and 6–248.

The kinetic problem is symmetric; provided we take proper account of the superscripts, solute **C** and **B** are mathematically equivalent and we can apply the same arguments to solute **B** with the result that

$$J_C = \frac{P^C\left[C_{in} e^{\frac{z_C F}{RT}\Psi_m} - C_o\right]}{1 + \dfrac{C_{in}}{K_{in}^C} + \dfrac{C_o}{K_o^C} + \dfrac{B_{in}}{K_{in}^B} + \dfrac{B_o}{K_o^B}}, \quad J_B = \frac{P^B\left[B_{in} e^{\frac{z_B F}{RT}\Psi_m} - B_o\right]}{1 + \dfrac{C_{in}}{K_{in}^C} + \dfrac{C_o}{K_o^C} + \dfrac{B_{in}}{K_{in}^B} + \dfrac{B_o}{K_o^B}}$$

$$(6\text{–}258)$$

where the parameters with superscript B are also defined by Equation 6–247 and 6–248. The membrane current is the simple sum

$$I = z_C F J_C + z_B F J_B \qquad (6\text{–}259)$$

Notice that Equations 6–258 have the same formal dependency on *concentrations* as Equation 6–209. The

same arguments used in Equation 6–209 can be used to translate the results of Equation 6–258 into unidirectional fluxes that would be measured by tracers. Thus, if **B** is an isotope of **C**, then $P_C = B_B$, $z_C = z_B$, and setting $C_o = 0$ and $B_{in} = 0$, we measure efflux with J_C and influx with J_B. Expressions similar to Equation 6–213 and 6–214 readily emerge, and taking the ratio, we have $J_{influx}/J_{efflux} = J_B/J_C = (B_o/C_{in})e^{-zF\Psi_m/RT}$, which, because **B** is a tracer of **C**, is the same as Equation 6–164. This follows even though the channels' binding sites can be saturated.

Single-Occupancy Channels: Voltage Dependence in Symmetric Channels

Discussion of voltage dependence is complex, because each of the rate constants includes a voltage-dependent term whose magnitude depends on the position of the barrier within the membrane. To simplify matters, we restrict our discussion to symmetric channels. If identical binding sites are located very close to the inner and outer membrane surfaces, it is reasonable to assume that the entire voltage drop across the membrane occurs across the middle barrier between the two sites. If interfacial potentials are ignored, the voltages at the first and third barrier are approximately equal to those in their neighboring bathing solutions, and it follows that k_{in}, k_{-in}, k_o, and k_{-o}, are voltage-independent. Further, if the sites are identical, then the binding and dissociation rate constants will also be identical:

$$k_{in} = k_{-o} \qquad k_o = k_{-in} \qquad (6\text{–}260)$$

Finally, symmetry requires that the middle barrier is located halfway through the membrane so that, recalling Equation 6–232,

$$k_1 = \gamma_1 e^{\frac{1}{2}\frac{zF}{RT}\Psi_m} \quad k_{-1} = \gamma_{-1} e^{-\frac{1}{2}\frac{zF}{RT}\Psi_m} \quad (6\text{–}261)$$

where $\gamma_1 = \gamma_{-1}$ is the rate constant (either forward or backward) when $\Psi_m = 0$.

Using Equations 6–260 and 6–248, we have

$$K_o = \frac{k_{-in}}{k_{in}} = \frac{k_o}{k_{-o}} = K \quad K_{in} = \frac{k_o}{k_{-o}} = \frac{k_{-in}}{k_{in}} = K \qquad (6\text{–}262)$$

i.e., in this symmetric model the half-maximal concentrations are independent of voltage; they are equal to each other and both equal K, the dissociation constant for solute distribution between membrane binding site and aqueous solution.

Substituting Equations 6–261 and 6–262 into 6–247 and 6–246, and writing the flux in terms of current by using the relation $I = zFJ$, we separate the current into two factors; the first depends only on voltage, the second on concentration.

$$I = zFJ$$

$$= \frac{zFY_t\gamma_1 k_{in} e^{-\frac{1}{2}\frac{zF}{RT}\Psi_m}}{\left(\gamma_1\left[e^{-\frac{1}{2}\frac{zF}{RT}\Psi_m} + e^{\frac{1}{2}\frac{zF}{RT}\Psi_m}\right] + k_{-in}\right)} \times \frac{\left(C_{in}e^{\frac{zF}{RT}\Psi_m} - C_o\right)}{\left(1 + \frac{C_{in}+C_o}{K}\right)}$$

$$(6\text{--}263)$$

Although the concentration dependence is familiar, the voltage terms are not; setting $C_{in} = C_o$, for example, no longer reduces to a simple Ohmic (linear) dependence on Ψ_m. Also, channel saturation is revealed as the current reaches a maximal value when the voltage becomes large.

$$\lim_{\Psi_m \to +\infty} I = \frac{I_{max}C_{in}}{K + (C_{in} + C_o)} \quad \lim_{\Psi_m \to -\infty} I = \frac{-I_{max}C_o}{K + (C_{in} + C_o)}$$

$$(6\text{--}264)$$

where I_{max}, the absolute maximum current obtained when concentration gradient as well as voltage becomes infinite, is given by $I_{max} = zFk_{in}KY_t = zFk_{-in}Y_t$.

Just as in the case of the Nernst-Plank formulation (Equation 6–208), the symmetric two-site channel shows rectification based on an asymmetry in concentrations on the two sides of the membrane. However, in this case the limiting conductances implied by Equation 6–264 are zero. There is an absolute maximum current that can flow (see Fig. 6.23). In either direction, it is dependent only on the number of channels and on the dissociation rate constant, i.e., $I_{max} = zFk_{-in}Y_t = zFk_oY_t$. This simple dependence arises because a large concentration gradient—say, $C_{in} = \infty$, $C_o = 0$—will ensure that all channels contain an ion while $\Psi_m = \infty$ drives all the membrane states into the $\mathbf{Y}^{oc}$ configuration (i.e., $Y^{oc} = Y_t$). As a result, the rate-limiting step is simply dissociation of the ion from the membrane binding site, which we have *assumed* to be voltage independent.

Single-Occupancy-Channel Results
Can Be Generalized to N Sites

Both the numerator and the denominator in the expression for J for a two-site single-occupancy model consist of a linear combination of C_{in} and C_o. This form arises in all single-occupancy models regardless of the number of membrane sites. Consider the more general case where we introduce n intermediate sites denoted by

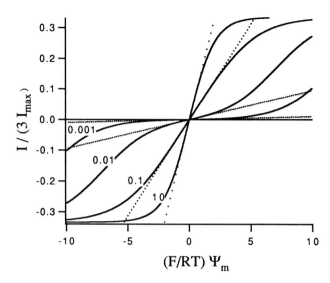

FIG. 6.23. Normalized plot of I vs. Ψ_m (Equation 6–263) with $C_{in} = C_o = K$. Values of λ/k_{-in} are indicated in the graph. Channel saturation becomes more evident at lower voltages the larger the λ/k_{-in} (translocation rate compared to dissociation rate). *Dotted straight lines* show the corresponding linear conductance predicted by corresponding constant-field equations.

$\mathbf{Y}_i (i = 1, 2, \ldots n)$ within the membrane as in Figure 6.24.

Let k_i denote the rate constant between $\mathbf{Y}_i$ and $\mathbf{Y}_{i+1}$. Expressions for the channel densities Y_i, Y^{OO}, and Y^{OC} and the flux take the form:

$$\frac{dY^{OO}}{dt} = -(k_{in}C_{in} + k_{-o}C_o)Y^{OO} + k_{-in}Y^{CO} + k_oY^{OC} = 0$$

$$\frac{dY^{CO}}{dt} = (k_{in}C_{in})Y^{OO} - (k_1 + k_{-in})Y^{CO} + k_{-1}Y_1 = 0$$

$$\vdots$$

$$\frac{dY_i}{dt} = k_iY_{(i-1)} - (k_{i+1} + k_{-i})Y_i + k_{-(i+1)}Y_{(i+1)} = 0$$

$$\vdots$$

$$\frac{dY^{OC}}{dt} = (k_{-o}C_o)Y^{OO} + k_{n+1}Y_n - (k_o + k_{-(n+1)})Y^{OC} = 0$$

$$(6\text{--}265)$$

$$Y^{OO} + Y^{CO} + Y^{OC} + \sum_{i=1}^{n} Y_i = Y_t \qquad (6\text{--}266)$$

$$J = k_iY_{i-1} - k_{-i}Y_i \qquad (6\text{--}267)$$

$$\begin{bmatrix} 1 & 1 & 1 & 1 & \ldots & 1 & 1 & 1 \\ k_{in}C_{in} & -(k_1 + k_{-in}) & k_{-1} & & \ldots & 0 & 0 & 0 \\ 0 & k_1 & -(k_{-1} + k_2) & k_{-2} & \ldots & 0 & 0 & 0 \\ \vdots & \vdots & \vdots & \vdots & \ddots & \vdots & \vdots & \vdots \\ 0 & 0 & 0 & 0 & \ldots & k_n & -(k_{n+1} + k_{-n}) & k_{-(n+1)} \\ k_{-o}C_o & 0 & 0 & 0 & \ldots & 0 & k_{n+1} & -(k_{-(n+1)} + k_o) \end{bmatrix} \begin{bmatrix} Y^{OO} \\ Y^{CO} \\ Y_1 \\ \vdots \\ Y_n \\ Y^{OC} \end{bmatrix} = \begin{bmatrix} Y_t \\ 0 \\ 0 \\ \vdots \\ 0 \\ 0 \end{bmatrix}$$

$$(6\text{--}268)$$

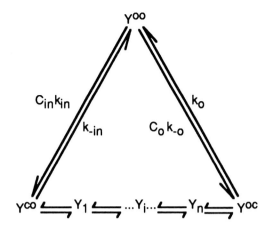

$$C_{in} + Y^{oo} \underset{k_{-in}}{\overset{k_{in}}{\rightleftharpoons}} Y^{co} \underset{k_{-1}}{\overset{k_1}{\rightleftharpoons}} Y_1 \rightleftharpoons ...Y_i...\rightleftharpoons Y_n \underset{k_{-(n+1)}}{\overset{k_{n+1}}{\rightleftharpoons}} Y^{oc} \underset{k_{-o}}{\overset{k_o}{\rightleftharpoons}} Y^{oo} + C_o$$

FIG. 6.24. Reaction scheme illustrating a single-occupancy channel with $n+2$ binding sites. See Figure 6.21 for details.

We employ the same arguments used in the discussion of the determinants of competitive ions in a two-site channel. Again, in each determinant in the n site channel the solute concentrations C_{in} and C_o will appear only in the first column in a product with Y^{oo}. Again, this reflects the fact that the only possible reaction of free solute with membrane occurs when the channel is coming from the vacant state. Since the C_{in} and C_o occur only in the first column, we can expand each determinant by columns and remove these terms as factors of the corresponding minor determinants. It follows that both the numerator and the denominator in the explicit expression for each Y will consist at most of a linear combination of C_{in} and C_o. As a result in the steady state, the net flux as given by any of the relations stated in Equation 6–267 will also be determined by the ratio of two linear combinations of C_{in} and C_o. Finally, using the equilibrium condition Equation 6–28, which is independent of path, we find that J is again given by

$$J = \frac{P\left[C_{in}e^{\frac{zF}{RT}\Psi_m} - C_o\right]}{1 + \frac{C_{in}}{K_{in}} + \frac{C_o}{K_o}} \qquad (6\text{–}269)$$

where P, K_{in}, and K_o have the same phenomenological significance, but they are made of different composites of the rate constants k_i. Equation 6–269 is formally

equivalent to Equation 6–246; in other words, observing the dependence of flux on solute concentration cannot distinguish between one or many sites. However, their dependence on voltage will be different.

Multiple Occupancy

Multiple-occupancy models allow more than one solute to occupy the channel. Channels may contain a fixed ionized charge, which would attract the first ion and make the channel appear relatively neutral to the second ion. Even in uncharged channels such as those formed by gramicidin, the electrostatic repulsion is insufficient to prevent channel occupation by more than one ion, especially at high concentrations (86, 87).

Evidence for natural occurrence of multi-ion channels has been described by Hille and others (13, 51, 52). Multiple-occupancy criteria include: *(1)* failure of the flux (or current or conductance) to show the concentration dependence predicted by Equation 6–269; *(2)* failure to find that permeability ratios determined by zero current potentials in asymmetric solutions (e.g., Equation 6–189) are independent of concentration (75); *(3)* failure of flux, inhibited by a blocking ion, to show the concentration dependencies predicted by Equation 6–258; and *(4)* a flux ratio that is more sensitive to the voltage than predicted by the Ussing relation.

Flux ratio deviations can often be conveniently described by raising the Ussing prediction to the *n*th power, i.e.,

$$J_{influx}/J_{efflux} = \left(\frac{C_o}{C_{in}} e^{-\frac{zF\Psi_m}{RT}} \right)^n \qquad (6\text{--}270)$$

where *n* is a constant undetermined parameter to be obtained by fitting experimental data. The value of *n* can often serve as an indicator of the number of ions that the channel contains. When $n = 1$ the Equation 6–270 reduces to the original Ussing equation and, as we have seen, this is characteristic of single-occupancy channels as well as free diffusion. Finding $n > 1$ is generally taken as evidence that the channel contains more than one ion. If the channel contains *m* sites and if the concentration of the single permeant species is sufficiently high so that all *m* sites are always occupied, then for narrow single-file channels (that do not allow one ion to pass another) it can be shown that $n = m + 1$ (54, 89, 98). In the more general case of high concentrations where a single vacancy is allowed (i.e., channels contain either *m* or $m - 1$ ions) it can be shown that $m - 1 < n < m$ (72, 89).

More-detailed kinetics of multi-ion channels can become formidable even in simple cases. Hille, for example, points out that the complete general steady-state analysis of a two-site channel with two permeant ion species results in an expression for *J* containing 5760 different voltage-dependent terms in the denominator. Multiple-occupancy models have been developed by several authors (e.g., 51, 52, 90, 92, 135).

SIMPLE CARRIERS

Net Flux

Most proteins can undergo conformational changes at room temperature. When these occur in transport proteins they may give rise to kinetics that are often interpreted in terms of carrier models. The simple carrier model, illustrated in Figure 6.25, assumes that a mobile component **Y** within the membrane facilitates permeation by combining with a permeant **C** on one side, discharging it on the opposite side, and then completing a cycle by returning in the discharged from to the original side.

The distinctive property of a carrier is that at any instant its reactive site is exposed only to one side of the membrane; exposure to the opposite side requires passage to a new conformational state. This contrasts with channels where reactive sites are exposed simulta-

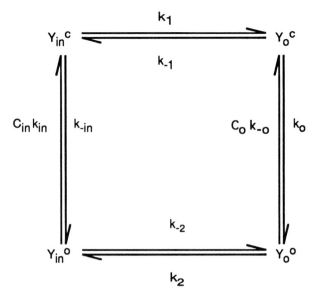

FIG. 6.25. Illustration of a simple carrier where a mobile component **Y** within the membrane facilitates permeation. Beginning on side "in" the solute C_{in} combines with an empty reactive carrier site that is exposed on side "in" and is designated as Y_{in}^o. The reaction forms a carrier-solute complex Y_{in}^c. Y_{in}^c then undergoes a translocation step (conformation change) that exposes the reactive site to side "o," where it is designated as Y_o^c. The complex Y_o^c then dissociates, releasing C_o to side "o." The free carrier Y_o^o with the reactive site still facing side "o" will then complete the cycle by translocating its reactive site to side "in," where it again becomes Y_{in}^o.

neously to both sides of the membrane. Compared to channels, carriers appear to be very slow. Turnover rates for ion channels can be as high as 10^8 s^{-1}, while maximal rates for pumps and carriers appear to fall within the range of 10 to 1000 s^{-1}. This slow turnover for pumps and carriers is believed to be rate limited by membrane protein conformational changes involved in transport (80, 126). Conformational changes in channels and their relation to carriers is discussed later, under FLUCTUATING BARRIERS: CHANNELS AND CARRIERS.

We begin the analysis by noting that the solute flux is equal to the flux of carrier complex, and, in a steady state, the flux of free carrier must be equal and opposite to the flux of solute-carrier complex, i.e.,

$$J = k_1 Y_{in}^C - k_{-1} Y_o^C = k_2 Y_o^O - k_{-2} Y_{in}^O$$
$$= k_{in} C_{in} Y_{in}^O - k_{-in} Y_{in}^C = k_o Y_o^C - k_{-o} C_o Y_o^O \qquad (6\text{--}271)$$

Expressions for the four forms of **Y** are obtained from solutions to the kinetic equations derived from Figure 6.25. These equations are:

$$\frac{dY_{in}^O}{dt} = -(k_{-2} + C_{in}k_{in})Y_{in}^O + k_{-in}Y_{in}^C + 0 \ Y_o^C + k_2 Y_o^O = 0$$

$$\frac{dY_{in}^C}{dt} = C_{in}k_{in} \ Y_{in}^O - (k_1 + k_{-in})Y_{in}^C + k_{-1}Y_o^C + 0 \ Y_o^O = 0$$

$$\frac{dY_o^C}{dt} = 0 \ Y_{in}^O + k_1 \ Y_{in}^C - (k_{-1} + k_o)Y_o^C + C_o k_{-o} \ Y_o^O = 0$$

$$\frac{dY_o^O}{dt} = k_{-2} \ Y_{in}^O + 0 \ Y_{in}^C + k_o \ Y_o^C - (k_2 + C_o k_{-o})Y_o^O = 0$$

$$(6\text{--}272)$$

Only three of these four expressions are independent; we obtain the fourth condition from the conservation of carriers, i.e.,

$$Y_{in}^O + Y_{in}^C + Y_o^C + Y_o^O = Y_t \qquad (6\text{--}273)$$

Using the first three expressions in Equation 6–272 together with Equation 6–273 yields four independent equations for the four unknown states of **Y**.

$$\begin{bmatrix} -k_{-2}-C_{in}k_{in} & k_{-in} & 0 & k_2 \\ C_{in}k_{in} & -k_1-k_{-in} & k_{-1} & 0 \\ 0 & k_1 & -k_{-1}-k_o & C_o k_{-o} \\ 1 & 1 & 1 & 1 \end{bmatrix} \begin{bmatrix} Y_{in}^O \\ Y_{in}^C \\ Y_o^C \\ Y_o^O \end{bmatrix} = \begin{bmatrix} 0 \\ 0 \\ 0 \\ Y_t \end{bmatrix}$$

$$(6\text{--}274)$$

Unlike the matrices in simple channels, concentration-dependent terms in the above coefficient matrix are not confined to a single column or a single row. Consequently, we cannot expect its determinant to be a simple linear function of the concentrations; inspection of the matrix shows that in addition to the linear terms we can anticipate products of C_{in} and C_o that signify a more complex kinetics for carriers.

Solutions of Equation 6–274 for the distribution of carrier states can be obtained using determinant methods similar to those employed for channels. These are easily evaluated on a desktop computer using any one of a number of commercial algebraic processors (e.g., Math Cad, Mathematica, Maple, Theorist). The solutions for Y_{in}^O and Y_o^O are:

$$Y_{in}^O = \frac{P_o(k_2[k_{-1}k_{-in} + k_1k_o + k_{-in}k_o] + k_{-in}k_{-o}k_{-1}C_o)}{k_{-in}k_{-o}k_{-1}k_{-2}\left(1 + \dfrac{C_{in}}{K_{in}} + \dfrac{C_o}{K_o} + \dfrac{C_{in}C_o}{K_{ino}}\right)}$$

$$(6\text{--}275)$$

$$Y_o^O = \frac{P_o(k_{-2}[k_{-1}k_{-in} + k_1k_o + k_{-in}k_o] + k_{in}k_o k_1 C_{in})}{k_{-in}k_{-o}k_{-1}k_{-2}\left(1 + \dfrac{C_{in}}{K_{in}} + \dfrac{C_o}{K_o} + \dfrac{C_{in}C_o}{K_{ino}}\right)}$$

$$(6\text{--}276)$$

Substituting Equation 6–275 and Equation 6–276 into Equation 6–271, we have

$$J = \frac{P_o(k_{in}k_o k_1 k_2 C_{in} - k_{-in}k_{-o}k_{-1}k_{-2}C_o)}{k_{-in}k_{-o}k_{-1}k_{-2}\left(1 + \dfrac{C_{in}}{K_{in}} + \dfrac{C_o}{K_o} + \dfrac{C_{in}C_o}{K_{ino}}\right)} \qquad (6\text{--}277)$$

where

$$K_{in} = \frac{(k_2 + k_{-2})(k_{-1}k_{-in} + k_1k_o + k_{-in}k_o)}{k_{in}(k_2 k_1 + k_2 k_{-1} + k_2 k_o + k_1 k_o)}$$

$$K_o = \frac{(k_2 + k_{-2})(k_{-1}k_{-in} + k_1k_o + k_{-in}k_o)}{k_{-o}(k_1 k_{-2} + k_{-2}k_{-1} + k_{-2}k_{-in} + k_{-1}k_{-in})}$$

$$(6\text{--}278)$$

$$K_{ino} = \frac{(k_2 + k_{-2})(k_{-1}k_{-in} + k_1k_o + k_{-in}k_o)}{(k_1 + k_{-1})k_{-o}k_{in}}$$

$$P_o = \frac{(k_{-in}k_{-o}k_{-1}k_{-2}Y_t)}{(k_2 + k_{-2})(k_{-1}k_{-in} + k_1k_o + k_{-in}k_o)}$$

$$(6\text{--}279)$$

At equilibrium, $J = 0$, and according to Equation 6–28, $C_o = C_{in}e^{zF/RT\Psi_m}$. It follows from Equation 6–277 that

$$k_{in}k_o k_1 k_2 = k_{-in}k_{-o}k_{-1}k_{-2}\,e^{\frac{zF}{RT}\Psi_m} \qquad (6\text{--}280)$$

a relation that can also be obtained directly from application of the principle of detailed balance Equation 6–49. Inserting Equation 6–280 into Equation 6–277, we arrive at the following expression for the net flux:

$$J = \frac{P_o(C_{in}\,e^{\frac{zF}{RT}\Psi_m} - C_o)}{1 + \dfrac{C_{in}}{K_{in}} + \dfrac{C_o}{K_o} + \dfrac{C_{in}C_o}{K_{ino}}} \qquad (6\text{--}281)$$

This expression is very similar to the corresponding flux for a simple channel. The parameters P_o, K_{in}, and K_o have the same interpretation; P_o corresponds to permeability coefficients at low concentrations. K_{in} is the concentration of C_{in} that will result in half the maximal transport rate when $C_o = 0$. Conversely, K_o is the "half-maximal concentration" for C_o when $C_{in} = 0$. Note the asymmetry implied by the fact that K_o and K_{in} are not equal. The only difference between Equation 6–281 and the corresponding expression for simple channels is the appearance of the nonlinear term $C_{in}C_o/K_{ino}$ in the denominator. In fact, the appearance of this term can form the basis for an experimental test to distinguish between simple channel or simple carrier (146).

Unidirectional Flux

Unidirectional efflux is restricted to that solute transport where the solute: *(1)* arises in side "in," and *(2)* makes the dissociation transition from Y_o^c to C_O. This is not simply equal to $k_o Y_o^c$, because this term includes solutes that may have arisen on either side of the membrane. To obtain the unidirectional flux, we follow the efflux J_{efflux}^* of labeled tracer C* when $C_{in}^* \neq C_o^* = 0$, so that any tracer arriving on side "o" can only have originated on side "in." In a steady state the flux of tracer combining with unoccupied carrier equals the flux of tracer-carrier complex, Y*, across the membrane, which equals the discharge of tracer on side "o," i.e.,

$$J_{efflux}^* = k_{in} C_{in}^* Y_{in}^O - k_{-in} Y_{in}^* \quad (6\text{-}282)$$

$$= k_1 Y_{in}^* - k_{-1} Y_o^* \quad (6\text{-}283)$$

$$= k_o Y_o^* \quad (6\text{-}284)$$

These three equations can be used to eliminate the labeled carrier species Y_{in}^* and Y_o^* to arrive at

$$J_{efflux}^* = \frac{k_1 k_{in} k_o C_{in}^* Y_{in}^O}{k_{-1} k_{-in} + k_1 k_o + k_{-in} k_o} \quad (6\text{-}285)$$

Using Equation 6–275 together with Equation 6–280 yields

$$J_{efflux}^* = \frac{P_o \left(\dfrac{C_o}{K_4} + 1 \right) C_{in}^* e^{\frac{zF}{RT} \Psi_m}}{1 + \dfrac{C_{in}}{K_{in}} + \dfrac{C_o}{K_o} + \dfrac{C_{in} C_o}{K_{ino}}} \quad (6\text{-}286)$$

where

$$K_4 \equiv \frac{k_2 (k_{-1} k_{-in} + k_1 k_o + k_{-in} k_o)}{k_{-1} k_{-in} k_{-o}} \quad (6\text{-}287)$$

Since $J_{efflux}^* = J_{efflux}(C_{in}/C_{in}^*)$

$$J_{efflux} = \frac{P_o \left(\dfrac{C_o}{K_4} + 1 \right) C_{in} e^{\frac{zF}{RT} \Psi_m}}{1 + \dfrac{C_{in}}{K_{in}} + \dfrac{C_o}{K_o} + \dfrac{C_{in} C_o}{K_{ino}}}$$

$$= \frac{J_{efflux}^{max} C_{in}}{K_{efflux}^M + C_{in}} \quad (6\text{-}288)$$

$$J_{efflux}^{max} = \frac{K_{ino} K_{in} P_o e^{\frac{zF}{RT} \Psi_m}}{K_4} \left(\frac{K_4 + C_o}{K_{ino} + K_{in} C_o} \right)$$

$$K_{efflux}^M = \frac{K_{ino} K_{in}}{K_o} \left(\frac{K_o + C_o}{K_{ino} + K_{in} C_o} \right) \quad (6\text{-}289)$$

where J_{efflux}^{max} and K_{efflux}^M are Michaelis-Menten parameters discussed below.

For J_{influx}, similar arguments show that

$$J_{influx} = \frac{P_o \left(\dfrac{C_{in} e^{\frac{zF}{RT} \Psi_m}}{K_4} + 1 \right) C_o}{1 + \dfrac{C_{in}}{K_{in}} + \dfrac{C_o}{K_o} + \dfrac{C_{in} C_o}{K_{ino}}} = \frac{J_{influx}^{max} C_o}{K_{influx}^M + C_o} \quad (6\text{-}290)$$

$$J_{influx}^{max} = \frac{K_{ino} K_o P_o}{K_4} \left(\frac{K_4 + C_{in} e^{\frac{zF}{RT} \Psi_m}}{K_{ino} + K_o C_{in}} \right)$$

$$K_{influx}^M = \frac{K_{ino} K_o}{K_{in}} \left(\frac{K_{in} + C_{in}}{K_{ino} + K_o C_{in}} \right) \quad (6\text{-}291)$$

At first sight, the appearance of a new parameter K_4 in the unidirectional flux seems to suggest the availability of another measurable parameter that would provide information beyond that obtainable from net flux measurements. However, K_4 is not an independent parameter. Tedious substitutions using Equations 6–278, 6–279, and 6–280 show that K_4 is related to K_{in}, K_o, and K_{ino}, through the expression:

$$\frac{e^{\frac{zF}{RT} \Psi_m}}{K_o} + \frac{1}{K_{in}} = \frac{K_4}{K_{ino}} + \frac{e^{\frac{zF}{RT} \Psi_m}}{K_4} \quad (6\text{-}292)$$

In principle, no new information is obtained from unidirectional flux measurements; they can be predicted from a comprehensive knowledge of the kinetics of net flux and vice versa. Both net and unidirectional fluxes are fully described by the same four parameters, P_o, K_{in}, K_o, and K_{ino}. However, in practice, accurate measurements of these parameters may be formidable, and recourse to redundant measurements is often highly desirable to ensure reliability (95, 125, 126).

Dependence on *cis* Concentration: Michaelis-Menten Kinetics. We analyze the details of unidirectional fluxes by using influx as an example. The nomenclature *cis* and *trans* are used to refer to the compartment from *(cis)* and toward *(trans)* which the flux originated or is directed. First consider the dependence of J_{influx} on C_o as illustrated by the last equality in Equation 6–290. If we hold C_{in} and Ψ_m constant, then J_{influx}^{max} and K_{influx}^M are constant, making J_{influx} a rectangular hyperbola with the usual Michaelis-Menten interpretations: *(1)* When C_o is small J_{influx} increases linearly with a slope = $J_{influx}^{max}/K_{influx}^M$. *(2)* When C_o becomes very large J_{influx} saturates approaching an asymptote = J_{influx}^{max}. *(3)* When $C_o = K_{influx}^M$, $J_{influx} = 1/2\ J_{influx}^{max}$. The fact that J_{influx}^{max} and K_{influx}^M are themselves functions of C_{in} raises the more interesting question: How does J_{influx} depend on C_{in}?

Dependence on *trans* Concentration: *Trans*-stimulation/inhibition. The dependence of J_{influx} on C_{in} is shown by fixing the value of C_o and taking the derivative of J_{influx} in Equation 6–290 with respect to C_{in}

$$\frac{\partial J_{influx}}{\partial C_{in}} = \frac{P_o\left(\dfrac{1}{K_4 e^{\frac{-zF}{RT}\Psi_m}} - \dfrac{1}{K_{in}} + \dfrac{C_o}{K_o K_4 e^{\frac{-zF}{RT}\Psi_m}} - \dfrac{C_o}{K_{ino}}\right)}{\left(1 + \dfrac{C_{in}}{K_{in}} + \dfrac{C_o}{K_o} + \dfrac{C_{in}C_o}{K_{ino}}\right)^2} C_o$$

$$(6\text{–}293)$$

Since the denominator is always positive, the sign of the derivative depends only on the sign of the numerator. More specifically, the derivative will be positive whenever

$$\frac{1}{K_4 e^{\frac{-zF}{RT}\Psi_m}} - \frac{1}{K_{in}} + \frac{C_o}{K_o K_4 e^{\frac{-zF}{RT}\Psi_m}} - \frac{C_o}{K_{ino}} > 0 \qquad (6\text{–}294)$$

Equation 6–294 can be simplified by using Equation 6–292 to eliminate K_{ino}, leaving

$$\frac{(K_{in} e^{\frac{zF}{RT}\Psi_m} - K_4)(C_o + K_4)}{K_4 K_{in}} > 0 \qquad (6\text{–}295)$$

Whenever $K_{in} > K_4 e^{-zF/RT\Psi_m}$ the derivative is positive; increasing C_{in}, the concentration on the side that is *receiving* the influx will stimulate J_{influx}. This phenomenon is called *trans*-stimulation. When $K_{in} < K_4 e^{-zF/RT\Psi_m}$, J_{influx} will decrease with C_{in}, a phenomenon called *trans*-inhibition. A simple carrier can show either *trans*-inhibition or *trans*-stimulation. This contrasts with a simple single-occupancy channel, which shows only *trans*-inhibition.

To interpret *trans*-stimulation, substitute the definitions of K_{in} and K_4, Equations 6–278 and 6–287, into the criterion $K_{in} > K_4 e^{-zF/RT\Psi_m}$. The inequality reduces to

$$k_1 > \left(1 + \frac{(k_1 + k_{-1})}{k_o}\right)k_{-2} \qquad (6\text{–}296)$$

and we conclude that for *trans*-stimulation it is necessary (but not sufficient) that $k_1 > k_{-2}$; i.e., the translocation of carrier complex from "in" to "o" is faster than the corresponding translocation of free carrier (see Fig. 6.25). In this case, adding solute to side "in" (increasing C_{in}) creates more carrier complex, which can return faster to side "o" where it can pick up more tracer. In order for this to be effective, it is important that once the unlabeled carrier complex reaches side "o" it dwells there long enough to dissociate before "turning around" and moving back to side "in." A fast dissociation step is promoted by large values of

k_o, while a long dwell time is promoted by low values of k_{-1}. Both of these factors are reflected in our formal criterion Equation 6–296, which is independent of concentration.

Using similar arguments, corresponding criteria for *trans* stimulation of *efflux* are:

$$K_0 > K_4 \quad \text{or} \quad k_{-1} > \left(1 + \frac{(k_1 + k_{-1})}{k_{-in}}\right)k_2 \qquad (6\text{–}297)$$

Michaelis-Menten Parameters Are Bilinear Functions of *trans* Concentration. Interpreting the concentration dependence of J^{max} and K^M is facilitated by the recognition that Equations 6–289 and 6–291 are both *bilinear* in C_{in}. A bilinear function $y(x)$ together with its derivative has the form

$$y(x) = \frac{a_1 + a_2 x}{a_3 + a_4 x} \quad \frac{dy}{dx} = \frac{a_2 a_3 - a_1 a_4}{(a_3 + a_4 x)^2} \qquad (6\text{–}298)$$

where a_1, a_2, a_3, and a_4 are constants. From the derivative, it is seen that whenever

$$a_2 a_3 - a_1 a_4 > 0 \qquad (6\text{–}299)$$

y will increase *monotonically* with x, and when $a_2 a_3 - a_1 a_4 < 0$, y will decrease *monotonically* with x. This criterion can be applied to any of the Michaelis-Menten parameters to ferret out their qualitative concentration dependencies by inspection. For example, using Equations 6–289 and 6–291, we have

J^{max}_{influx} increases with C_{in} when $K_{ino} e^{\frac{zF}{RT}\Psi_m} > K_o K_4$

J^{max}_{efflux} increases with C_o when $K_{ino} > K_{in} K_4$

K^M_{influx} increases with C_{in} when $K_{ino} > K_o K_{in}$

K^M_{efflux} increases with C_o when $K_{ino} > K_o K_{in}$

Simple Carriers Do Not Follow the Ussing Flux Ratio. Taking the ratio of Equation 6–288 to Equation 6–290, we have

$$\frac{J_{efflux}}{J_{influx}} = \left(\frac{K_4 + C_o}{K_4 + C_{in} e^{\frac{zF}{RT}\Psi_m}}\right)\left(\frac{C_{in} e^{\frac{zF}{RT}\Psi_m}}{C_o}\right) \qquad (6\text{–}300)$$

showing that the Ussing flux ratio is not valid for simple carriers. To express the deviation as in Equation 6–270, we first define an Ussing ratio

$$U_r \equiv \frac{C_{in} e^{\frac{zF}{RT}\Psi_m}}{C_o} \qquad (6\text{–}301)$$

and write

$$\frac{J_{efflux}}{J_{influx}} = \left(\frac{K_4 + C_o}{K_4 + C_o U_r}\right)U_r = U_r^n \qquad (6\text{–}302)$$

Solving for n

$$n = 1 + \frac{\log\left(\dfrac{K_4 + C_o}{K_4 + C_o U_r}\right)}{\log(U_r)} \qquad (6\text{-}303)$$

If $U_r > 1$, then the numerator in Equation 6–303 is negative while the denominator is positive, making $n < 1$. If $U_r < 1$, then the signs of the numerator and denominator are reversed; we still have $n < 1$. Finally if $U_r = 1$, then the system is in equilibrium and direct substitution into Equation 6–302 shows that influx = efflux, as it should and n is indeterminate.

Comparing these conclusions with corresponding results from channels provides a criterion that may help discriminate between kinetic behavior of channels from that of simple carriers; i.e., $n > 1$ for channels and $n < 1$ for simple carriers. However, this result for simple carriers cannot be generalized. For example, Horowicz et al. (55) show how a simple bivalent carrier model that requires binding two ions before translocation can yield an $n = 2$.

Equilibrium at the Boundaries

The influence of individual rate constants on transport kinetics is illustrated by considering the two limiting cases: *(1)* where transport is rate limited by the translocation of the carrier-solute complex (equilibrium at the boundaries), and *(2)* where transport is rate limited by reactions at the interfaces of free carrier with solute (translocation equilibrium). Equilibrium at the boundaries is a very common assumption. In this case, reactions of the carrier with solute at the interfaces is very rapid compared to the translocation step; i.e., the interfacial rate constants k_{in}, k_{-in}, k_o, and k_{-o} are all much larger than translocation rate constants k_1, k_{-1}, k_2, and k_{-2}. We designate this condition and define the chemical equilibrium constants at the interfaces by:

$$\left.\begin{array}{l} k_{in} \\ k_{-in} \\ k_o \\ k_{-o} \end{array}\right\} >> \left\{\begin{array}{l} k_1 \\ k_{-1} \\ k_2 \\ k_{-2} \end{array}\right. \qquad \begin{array}{l} K_{in}^{eq} = \dfrac{k_{-in}}{k_{in}} \\[2ex] K_o^{eq} = \dfrac{k_o}{k_{-o}} \end{array} \qquad (6\text{-}304)$$

Using these inequalities in our condition for *trans*-stimulation, Equation 6–296, we are left with a simpler criterion,

$$k_1 > k_{-2} \qquad (6\text{-}305)$$

which is now both a necessary *and sufficient* condition for *trans*-stimulation. A large value of k_o now guarantees that there will be ample time for dissociation of unlabeled solute-carrier complex at interface "o" so

that the sole requisite for *trans*-stimulation is that the solute-carrier complex translocates faster than the free carrier.

Using Equations 6–304, 6–278 to 6–280, 6–287, and 6–291, J_{influx}^{max} is now given by

$$J_{influx}^{max} = Y_t e^{\frac{zF}{RT}\Psi_m} k_{-1} \times \frac{K_{in}^{eq} k_{-2} + k_1 C_{in}}{K_{in}^{eq}(k_{-2} + k_{-1}) + (k_1 + k_{-1})C_{in}} \qquad (6\text{-}306)$$

where K_{in}^{eq} denotes the equilibrium constant between carrier and solute on side "in." Again, J_{influx}^{max} is bilinear in C_{in}. Applying 6–299 shows that J_{influx}^{max} will increase with C_{in} when $k_1 > k_{-2}$; i.e., with equilibrium at the boundaries *trans*-stimulation is always accompanied by an increase in J_{influx}^{max}.

When $C_{in} = 0$, $J_{influx}^{max} = Y_t e^{zF/RT\Psi_m} \times k_{-1} k_{-2}/(k_{-1} + k_{-2})$, a function that increases with both k_{-1} and k_{-2}. As shown in Figure 6.25, these are the rate constants promoting delivery of carrier-complex to side "in" and return of empty carrier to side "o." The rate constant k_1, for return of the loaded carrier from side "in" to "o," does not appear because the density of loaded carriers on side "in" is virtually zero when $C_{in} = 0$.

When C_{in} becomes infinite, $J_{influx}^{max} = Y_t e^{zF/RT\Psi_m} \times k_{-1} k_1/(k_{-1} + k_1)$, a function that is independent of k_2 and k_{-2}. This reflects the fact that both C_{in} and C_o are very large so that the carriers are saturated, leaving no empty carrier on either side of the membrane. There is no traffic of empty carriers; all transport takes place by exchange.

Turning to K_{influx}^M, we find a more complex concentration dependence. Using Equations 6–291, 6–278, 6–279, and 6–304 yields

$$K_{influx}^M = \frac{(k_2 + k_{-2})K_{in}^{eq} + (k_2 + k_1)C_{in}}{(k_{-2} + k_{-1})K_{in}^{eq} + (k_1 + k_{-1})C_{in}} K_o^{eq} \qquad (6\text{-}307)$$

Applying Equation 6–299 shows that K_{influx}^M will increase with C_{in} when $(k_1 - k_{-2})(k_{-1} - k_2) > 0$. Thus, *trans*-stimulation ($k_1 > k_{-2}$) will be accompanied by an increase in K_{influx}^M provided $k_{-1} > k_2$; otherwise K_{influx}^M will decrease. Conversely, *trans*-inhibition ($k_1 < k_{-2}$) will be accompanied by an increase in K_{influx}^M provided $k_{-1} < k_2$; otherwise it will decrease.

Rate-Limiting Steps at the Boundaries

Transport is rate limited by reactions at the interfaces when reactions of the carrier with solute at the interfaces is very slow compared to the translocation step. In other words, interfacial rate constants k_{in}, k_{-in}, k_o, and k_{-o} are all much smaller than translocation rate constants k_1, k_{-1}, k_2, and k_{-2}. We designate this condition, and define the translocation equilibrium constants K_1 and K_2, by

$$\left.\begin{matrix} k_{in} \\ k_{-in} \\ k_o \\ k_{-o} \end{matrix}\right\} << \left\{\begin{matrix} k_1 \\ k_{-1} \\ k_2 \\ k_{-2} \end{matrix}\right. \qquad \begin{matrix} K_1 \equiv \dfrac{k_{-1}}{k_1} \\[2mm] K_2 \equiv \dfrac{k_2}{k_{-2}} \end{matrix} \qquad (6-308)$$

J_{influx}^{max} now becomes

$$J_{influx}^{max} = \frac{Y_t k_{in} k_o K_2}{k_o(K_1+1)(K_1 k_{-in} + k_o)}$$

$$\times \frac{k_2(K_1 k_{-in} + k_o) + k_{in} k_o K_2 C_{in}}{k_2 + K_2 k_{in} C_{in}} \qquad (6-309)$$

As might be expected J_{influx}^{max} is bilinear in C_{in}. The condition that J_{influx}^{max} increase with C_{in} is given by

$$k_{in} k_o K_2 k_2 - (K_1 k_{-in} + k_o)k_2(K_2 k_{in})$$
$$= -K_2 K_1 k_2 k_{-in} k_{in} > 0 \qquad (6-310)$$

a condition that clearly never exists. It follows that *trans*-inhibition will always occur when rate-limiting steps are at the boundaries. Under these conditions, the translocation rates are irrelevant so that stimulatory effects of solute on translocation steps are insignificant. In any particular simple carrier system, the experimental demonstration of *trans*-stimulation can be taken as proof that the boundary reactions are not rate limiting in the simple carrier model.

Using Equations 6–308, 6–278, 6–279, and 6–291, K_{influx}^M becomes the bilinear

$$K_{influx}^M = \frac{k_2}{k_o(K_1+1)}$$

$$\times \frac{(K_2+1)(K_1 k_{-in} + k_o) + (K_1+1)K_2 k_{in} C_{in}}{k_2 + K_2 k_{in} C_{in}}$$

$$(6-311)$$

Energy-driven Simple Carrier Systems

Participation of metabolic reactions in one or more transport steps may generate an active transport of solute against its apparent electrochemical gradient. The kinetics of this coupling is formally very similar to the above discussions, with the major caveat that the "rate constants" k_i are no longer independent of concentration. In addition to their dependence on membrane potential, they now depend on concentrations of metabolic substrates and products that are coupled to the transport process.

To illustrate, consider the transport of **C** coupled to an energy source (**S** → **P**). We represent the net reaction by

$$C_{in} + S \rightleftharpoons C_o + P \qquad (6-312)$$

and assume that the entire reaction (**S** → **P**) takes place on side "in." Although the explicit concentration dependence of each rate constant will depend on the entry/exit position of each metabolic constituent to/from the transport cycle, thermodynamic constraints at equilibrium provide one important general relation that is independent of pathway details. At equilibrium the simple Nernst equation no longer applies, making Equation 6–280 invalid. However, in place of the Nernst equation, we obtain from Equations 6–42, 6–43, and 6–44,

$$(C_o)_{EQUIL} = \left(C_{in}\frac{S}{P}e^{-\frac{\Delta G_{SP}^o - zF\Psi_m}{RT}}\right)_{EQUIL} \qquad (6-313)$$

Setting $J = 0$ (equilibrium) in Equation 6–277 and utilizing Equation 6–313, we have

$$k_{in}k_o k_1 k_2 = k_{-in}k_{-o}k_{-1}k_{-2}\frac{S}{P}e^{-\frac{\Delta G_{SP}^o - zF\Psi_m}{RT}}$$

$$(6-314)$$

Using Equation 6–314 in place of 6–280 and repeating the arguments that led to Equations 6–281, 6–288, and 6–290, we rewrite these flux equations as

$$J = \frac{P_o\left(C_{in}\dfrac{S}{P}e^{-\frac{\Delta G_{SP}^o - zF\Psi_m}{RT}} - C_o\right)}{1 + \dfrac{C_{in}}{K_{in}} + \dfrac{C_o}{K_o} + \dfrac{C_{in}C_o}{K_{ino}}} \qquad (6-315)$$

$$J_{influx} = \frac{P_o\left(\dfrac{C_{in}\dfrac{S}{P}e^{-\frac{\Delta G_{SP}^o - zF\Psi_m}{RT}}}{K_4} + 1\right)C_o}{\dfrac{C_{in}}{K_{in}} + \dfrac{C_o}{K_o} + \dfrac{C_{in}C_o}{K_{ino}} + 1} \qquad (6-316)$$

and

$$J_{efflux} = \frac{P_o\left(\dfrac{C_o}{K_4} + 1\right)C_{in}\dfrac{S}{P}e^{-\frac{\Delta G_{SP}^o - zF\Psi_m}{RT}}}{1 + \dfrac{C_{in}}{K_{in}} + \dfrac{C_o}{K_o} + \dfrac{C_{in}C_o}{K_{ino}}} \qquad (6-317)$$

where the parameters are also functions of the S and/or P concentrations as well as Ψ_m.

The explicit concentration dependence of the parameters can be obtained in any particular case by simple substitution. For example, when conversion of substrate **S** to **P** occurs in a single step that is coupled to the binding of solute C_{in} to the carrier component Y_{in}^o, then the concentration S will always be associated with the forward rate constant of the reaction, k_{in}, while P will be associated with the reverse rate constant, k_{-in}.

The implication is that in all flux parameters, k_{in} is replaced by the product $k_{in}S$ and k_{-in} is replaced by the product $k_{-in}S$. The resulting expressions for these parameters are:

$$K_{in} = \frac{(k_2 + k_{-2})(k_{-1}k_{-in}P + k_1 k_o + k_{-in}Pk_o)}{k_{in}S(k_2 k_1 + k_2 k_{-1} + k_2 k_o + k_1 k_o)} \quad (6\text{--}318)$$

$$K_o = \frac{(k_2 + k_{-2})(k_{-1}k_{-in}P + k_1 k_o + k_{-in}Pk_o)}{k_{-o}(k_1 k_{-2} + k_{-2}k_{-1} + k_{-2}k_{-in}P + k_{-1}k_{-in}P)} \quad (6\text{--}319)$$

$$K_{ino} = \frac{(k_2 + k_{-2})(k_{-1}k_{-in}P + k_1 k_o + k_{-in}Pk_o)}{(k_1 + k_{-1})k_{-o}k_{in}S} \quad (6\text{--}320)$$

$$P_o = \frac{k_{-in}Pk_{-o}k_{-1}k_{-2}y_t}{(k_2 + k_{-2})(k_{-1}k_{-in}P + k_1 k_o + k_{-in}Pk_o)} \quad (6\text{--}321)$$

and

$$K_4 = \frac{k_2(k_{-1}k_{-in}P + k_1 k_o + k_{-in}Pk_o)}{k_{-1}k_{-in}Sk_{-o}} \quad (6\text{--}322)$$

On the other hand, if the hydrolysis of S occurs in two steps, with the first step (the binding of S) coupled to the binding of C_{in}, and the second step (release of P) coupled with the translocation step k_1, then, for all the flux parameters, k_{in} will be replaced by the product $k_{in}S$ and k_1 will be replaced by $k_1 P$.

These two examples illustrate our primary point: For any solute transport that is coupled with an energy source, the numerator of the net flux equation will reflect the equilibrium condition when the net flux is zero, and the parameters of the flux equations will reflect their dependence on S and P.

COTRANSPORT

A system that rigidly couples the movement of two solutes and constrains them to move in the same direction is called cotransport (also referred to as symport). Its physiological significance is that it allows the gradient of one solute to drive another solute uphill against its electrochemical gradient. Cotransport systems are ubiquitous, often using the gradient of Na^+ to drive the accumulation of metabolically important substrates like glucose and amino acids into the cell. The kidney and intestinal tract (70) provide examples where cotransport has been extensively studied. A table with many examples may be found in Stein (125).

Thermodynamics: Cotransport Can Move Solutes "Uphill"

We consider a cotransport system of two solutes, A and B, whose stoichiometry is $mA : nB$. For generality, we include the possibility that the system is coupled to a chemical energy source, and represent the net reaction by

$$mA_{in} + nB_{in} + S_{in} \longrightarrow mA_o + nB_o + P_{in} \quad (6\text{--}323)$$

where it is assumed that the reaction $(S \rightarrow P)$ takes place in the "inside solution." Applying Equation 6–20, we have

$$\Delta \tilde{G} = \tilde{\mu}_o^A + \tilde{\mu}_o^B + \tilde{\mu}_{in}^P - (\tilde{\mu}_{in}^A + \tilde{\mu}_{in}^B + \tilde{\mu}_{in}^S) \quad (6\text{--}324)$$

Using Equation 6–25, noting that $\mu_{A_o}^o = \mu_{A_{in}}^o$ and $\mu_{B_o}^o = \mu_{B_{in}}^o$, and collecting terms yields

$$\Delta \tilde{G} = \Delta G_{SP} + RT \ln \frac{A_o^m B_o^n}{A_{in}^m B_{in}^n}$$
$$- (mz_A + nz_B)F(\Psi_{in} - \Psi_o) \leq 0 \quad (6\text{--}325)$$

where ΔG_{SP} is the free energy of $S \rightarrow P$ and z_A and z_B are the valences of A and B (including the algebraic sign of their charge).

Equation 6–324 shows the thermodynamic constraints between chemical, osmotic, and electrical energies. The inequality is based on the assumption that the system is running in the indicated (forward) direction from "in" to "o." The maximal concentration ratios that the cotransport can provide is found at equilibrium where $\Delta \tilde{G} = 0$ and

$$\frac{A_o^m B_o^n}{A_{in}^m B_{in}^n} = e^{-\frac{\Delta G_{SP} - (mz_A + nz_B)F(\Psi_{in} - \Psi_o)}{RT}} \quad (6\text{--}326)$$

When the reaction $S \rightarrow P$ is not coupled to any process,

$$\frac{A_o^m B_o^n}{A_{in}^m B_{in}^n} = e^{\frac{(mz_A + nz_B)F(\Psi_{in} - \Psi_o)}{RT}} = e^{\frac{(mz_A + nz_B)F\Psi_m}{RT}} \quad (6\text{--}327)$$

An example of uphill cotransport, as well as the significance of its stoichiometry, is readily provided by the common case where one solute, say A, is uncharged so that $z_A = 0$ and where $m = 1$ (e.g., Na^+-glucose cotransport). Rearranging the ratio leaves:

$$\frac{A_{in}}{A_o} = \frac{B_o^n}{B_{in}^n} e^{-\frac{nz_B F\Psi_m}{RT}} \quad (6\text{--}328)$$

Letting B represent Na^+, then $z_B = 1$ and with the nominal values $B_o/B_{in} = 10$, $\Psi_m = -60$ mV, so that $e^{-nz_B F\Psi_m/RT} \approx 10$, we find for a 1:1 stoichiometry ($m = n = 1$) that the intracellular accumulation of solute A, A_{in}/A_o can be as high as 100-fold. However, for a 1:2 stoichiometry ($m = 1$, $n = 2$) this accumulation increases to 10,000. A 1:1 stoichiometry has the advantage of being less costly, dissipating only one Na^+ for each **A** that is transported. It serves very well as long as it is not required to "pump" against steep gradients. This implicit strategy of using different transport stoichiometries for different tasks is apparently exploited by the mammalian kidney as it reabsorbs virtually all of the glucose that is filtered. In the early parts of the proximal tubule where the glucose concentration in the lumen is still relatively high, the stoichiometry is 1:1, while in the more distal parts where glucose concentrations begin to approach zero, the stoichiometry changes to 1:2 (132, 133).

Kinetic Description

The cotransport carrier illustrated in Figure 6.26 has the following properties:

1. Each carrier molecule has two reactive sites; one site reacts with **A**, the other with **B**. The stoichiometry is $m = n = 1$; there are no metabolic energy coupled reactions.

2. The sites always face the same side of the membrane.

3. By changing conformation the carrier can reorient the exposure of both sites from one side of the membrane to the other.

4. Conformation changes leading to site reorientation can take place only when the carrier is completely free of **A** and **B** or when it becomes a ternary complex having bound *both* **A** and **B**.

In Figure 6.26 we utilize superscripts to identify binding sites: "o" implies an empty binding site for **A**, while "•" represents an empty site for **B**. Subscripts are used to express carrier conformation. For example, $Y_{in}^{o•}$ shows a carrier with the two vacant sites facing side in, while $Y_{o}^{o•}$ shows the allowed conformation change where both sites are now facing the outside. Superscripts A and B denote sites that are occupied by **A** and/or **B**. For example, $Y_{in}^{A•}$ denotes a carrier with only **A** bound on the inside site, Y_{in}^{oB} has **B** bound on the inside with a vacant **A**, and Y_{in}^{AB} has both **A** and **B** bound to the inside sites.

To avoid an algebraic nightmare, we assume equilibrium at each boundary (i.e., at cycles $Y_{in}^{o•} \rightarrow Y_{in}^{A•} \rightarrow Y_{in}^{AB} \rightarrow Y_{in}^{oB} \rightarrow Y_{in}^{o•}$ and $Y_{o}^{o•} \rightarrow Y_{o}^{A•} \rightarrow Y_{o}^{AB} \rightarrow Y_{o}^{oB} \rightarrow Y_{o}^{o•}$). This will follow when conformation steps are rate limiting, as appears to be the case for most pumps and carriers (see Lauger [80] as well as discussion following Equation 6–310). Within these reservations the scheme is fairly general; in particular, it allows for any of the following three different reaction sequences at the boundaries:

1. Random Sequence. The random carrier has specific reactive sites for each solute. These sites are exposed, simultaneously, at only one side of the membrane at a time. In this case all boundary reactions take place independently. At boundary "in," $Y_{in}^{o•}$ can react with either A_{in} or B_{in}, and similarly Y_{in}^{AB} can dissociate into

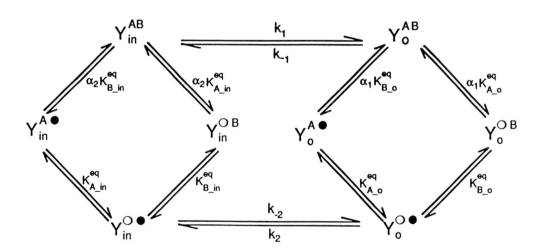

FIG. 6.26. Cotransport random carrier ($m = n = 1$). The cycle begins on side "in" where A_{in} or B_{in} combine with its reactive site on component $Y_{in}^{o•}$, forming the complex $Y_{in}^{A•}$ or Y_{in}^{oB}. This complex reacts with the other solute, forming the ternary complex Y_{in}^{AB}. Y_{in}^{AB} undergoes a translocation step, exposing the reactive sites to side "o," and becomes Y_{o}^{AB}. Y_{o}^{AB} dissociates, releasing in a random order A_{o} or B_{o} to side "o," going through the intermediate complexes $Y_{o}^{A•}$ or Y_{o}^{oB}, before becoming totally in the undissociated form $Y_{o}^{o•}$. $Y_{o}^{o•}$ will then translocate them to side "in," completing the cycle. The association-dissociation reactions at each boundary are assumed to be in equilibrium. At these boundaries only three out of the four cyclic dissociation constants, K^{eq}, are independent. This is reflected in the terms containing α_1 and α_2 (see discussion of Equations 6–331, 6–332, and 6–333).

either $Y_{in}^{\circ B}$ or $Y_{in}^{A\bullet}$. Analogous remarks hold for boundary "o."

2. Ordered Sequence. This mechanism refers to the case where the order of binding of the solutes is always the same, with one solute species always binding before the other.

3. Simultaneous Sequence. In this mechanism both solutes bind simultaneously to the carrier. In this case, only the totally free and totally bound forms of the carrier exist.

Our objectives in this section on kinetics of cotransport are: *(1)* Derive a formal expression for net flux that may be used for describing the physiological operation of the transport system. *(2)* Estimate the relevant parameters from unidirectional flux measurements. *(3)* Provide graphical criteria for distinguishing between the three different reaction sequences.

General Net Flux

According to Figure 6.26, the steady state, net solute flux of **A** or **B** equals the flux of the ternary carrier complex, because this is the only form of the carrier complexed with solutes that can translocate the membrane.

$$J_A = J_B = k_1 Y_{in}^{AB} - k_{-1} Y_o^{AB} \quad (6\text{--}329)$$

Progression to an explicit solution for this net flux requires explicit solutions for Y_o^{AB} and Y_{in}^{AB}. Rather than continuing to formulate steady-state differential equations for individual carrier species, it is much simpler to proceed directly to the equilibrium conditions at the boundaries.

Equilibrium at the Boundaries and Application of Detailed Balance. Assume that the reactions at the boundaries of the solutes **A** and **B** with the reactive sites of component **Y** are in quasi-equilibrium at all times, with the following *dissociation* constants, K_x^{eq} and L_x^{eq}, for the reactions at side "in" and at side "o."

Examine the four-state cycle $Y_{in}^{\circ\bullet} \rightarrow Y_{in}^{A\bullet} \rightarrow Y_{in}^{AB} \rightarrow Y_{in}^{\circ B} \rightarrow Y_{in}^{\circ\bullet}$ shown in Figure 6.26 by noting that the overall reaction $Y_o^{AB} \rightarrow Y_o^{\circ\bullet}$ can proceed along two

$$K_{A\_in}^{eq} = \frac{Y_{in}^{\circ\bullet} A_{in}}{Y_{in}^{A\bullet}} \qquad K_{A\_o}^{eq} = \frac{Y_o^{\circ\bullet} A_o}{Y_o^{A\bullet}}$$

$$K_{B\_in}^{eq} = \frac{Y_{in}^{\circ\bullet} B_{in}}{Y_{in}^{\circ B}} \qquad K_{B\_o}^{eq} = \frac{Y_o^{\circ\bullet} B_o}{Y_o^{\circ B}}$$

$$L_{A\_in}^{eq} = \frac{Y_{in}^{\circ B} A_{in}}{Y_{in}^{AB}} \qquad L_{A\_o}^{eq} = \frac{Y_o^{\circ B} A_o}{Y_o^{AB}} \quad (6\text{--}330)$$

$$L_{B\_in}^{eq} = \frac{Y_{in}^{A\bullet} B_{in}}{Y_{in}^{AB}} \qquad L_{B\_o}^{eq} = \frac{Y_o^{A\bullet} B_o}{Y_o^{AB}}$$

different pathways, $Y_{in}^{AB} \rightarrow Y_{in}^{\circ B} \rightarrow Y_{in}^{\circ\bullet}$ and $Y_{in}^{AB} \rightarrow Y_{in}^{A\bullet} \rightarrow Y_{in}^{\circ\bullet}$. Following the arguments used to derive Equations 6–48 and 6–49, the dissociation constant for the first path is $L_{A\_in}^{eq} K_{B\_in}^{eq}$ while the constant for the second path is $L_{B\_in}^{eq} K_{A\_in}^{eq}$. But there can be only one dissociation constant for the overall reaction (there is only one $\Delta\tilde{G}^o$), and it follows that $L_{A\_in}^{eq} K_{B\_in}^{eq} = L_{B\_in}^{eq} K_{A\_in}^{eq}$. If we define

$$\alpha_2 \equiv \frac{L_{B\_in}^{eq}}{K_{B\_in}^{eq}} = \frac{L_{A\_in}^{eq}}{K_{A\_in}^{eq}} \quad (6\text{--}331)$$

then

$$L_{B\_in}^{eq} = \alpha_2 K_{B\_in}^{eq} \quad \text{and} \quad L_{A\_in}^{eq} = \alpha_2 K_{A\_in}^{eq} \quad (6\text{--}332)$$

Similar arguments for the cycle $Y_o^{\circ\bullet} \rightarrow Y_o^{A\bullet} \rightarrow Y_o^{AB} \rightarrow Y_o^{\circ B} \rightarrow Y_o^{\circ\bullet}$ yield

$$L_{B\_o}^{eq} = \alpha_1 K_{B\_o}^{eq} \quad \text{and} \quad L_{A\_o}^{eq} = \alpha_1 K_{A\_o}^{eq} \quad (6\text{--}333)$$

where

$$\alpha_1 \equiv \frac{L_{B\_o}^{eq}}{K_{B\_o}^{eq}} = \frac{L_{A\_o}^{eq}}{K_{A\_o}^{eq}} \quad (6\text{--}334)$$

Now consider the net transport cycle illustrated in Figure 6.27.

Reading from left to right, note that the forward direction of the first two reactions on side "in" are association. Accordingly, we represent their (association) equilibrium constants by the reciprocal of the corresponding dissociation constants. Reactions on side "o" are dissociations. Applying Equation 6–48, we equate the product of the individual equilibrium constants to the exponential of the change in standard electrochemical free energy,

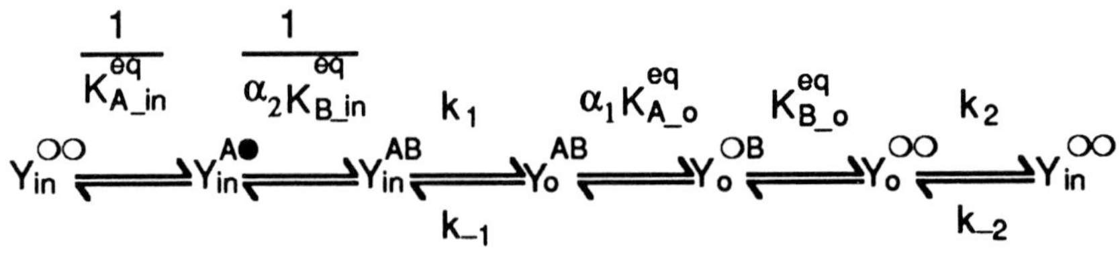

FIG. 6.27. Reaction cycle used to derive Equation 6–335. See text.

$$\left(\frac{1}{K_{A\_in}^{eq}}\right)\left(\frac{1}{\alpha_2 K_{B\_in}^{eq}}\right)\left(\frac{k_1}{k_{-1}}\right)\alpha_1 K_{A\_o}^{eq} K_{B\_o}^{eq}\left(\frac{k_2}{k_{-2}}\right)=e^{-\frac{\Delta\bar{G}_o}{RT}}$$

$$=e^{\frac{(z_A+z_B)F\Psi_m}{RT}} \tag{6-335}$$

Steady-state Translocation and Conservation of Carrier. In contrast to boundary reactions, the "translocation" steps are assumed to be rate limiting. Rather than at equilibrium, they are in a steady state. This implies that the net flux of loaded carriers from "in" to "o" equals the net flux of unloaded carriers from "o" to "in."

$$k_1 Y_{in}^{AB} - k_{-1} Y_o^{AB} = k_2 Y_o^{\circ\bullet} - k_{-2} Y_{in}^{\circ\bullet} \tag{6-336}$$

Finally, in addition to the above conditions we also require conservation of total carrier (Y_T) in the membrane

$$Y_T = Y_o^{\circ\bullet} + Y_o^{A\bullet} + Y_o^{\circ B} + Y_o^{AB} + Y_{in}^{\circ\bullet} + Y_{in}^{A\bullet} + Y_{in}^{\circ B} + Y_{in}^{AB} \tag{6-337}$$

Solving this system of Equations 6–330 to 6–337 for Y_o^{AB} and Y_{in}^{AB}, and substituting their values in the net flux equation, Equation 6–329, we obtain:

$$J = P_{AB}\frac{(A_{in}B_{in}e^{\frac{(z_A+z_B)F\Psi_m}{RT}} - A_o B_o)}{denom} \tag{6-338}$$

with

$$denom = 1 + \frac{A_{in}}{K_{A\_in}} + \frac{B_{in}}{K_{B\_in}} + \frac{A_o}{K_{A\_o}} + \frac{B_o}{K_{B\_o}}$$
$$+ A_o B_o\left(\frac{1}{M_{in}} + \frac{A_{in}}{M_{A\_in}} + \frac{B_{in}}{M_{B\_in}}\right)$$
$$+ A_{in}B_{in}\left(\frac{1}{M_o} + \frac{A_o}{M_{A\_o}} + \frac{B_o}{M_{B\_o}}\right)$$
$$+ \frac{A_{in}B_{in}A_o B_o}{K_{ino}} \tag{6-339}$$

where the parameters P_{AB}, K_x, and M_x are defined in Table 6.3.

Unidirectional Fluxes

To obtain the unidirectional flux, we follow the efflux J_{efflux}^* of labeled tracer, say A^*, when $A_{in}^* \neq A_o^* = 0$. From Equation 6–329 we have $J_{A\,efflux}^* = k_1 Y_{in}^{A^*B} - k_{-1} Y_o^{A^*B}$, but since $A_o^* = 0$ it follows from the equilibrium conditions at the boundaries, Equation 6–330, that $Y_o^{A^*B} = 0$. Further, rapid equilibrium at the boundaries implies that $Y_{in}^{A^*B} = Y_{in}^{AB}(A_{in}^*/A_{in})$, and finally, as usual, $J_{A\,efflux}^* = J_{A\,efflux}(A_{in}^*/A_{in})$. It follows that

$$J_{A\_efflux} = k_1 Y_{in}^{AB} \quad \text{and} \quad J_{A\_efflux} = J_{B\_efflux} \tag{6-340}$$

Using corresponding arguments for influx,

$$J_{A\_influx} = k_{-1} Y_o^{AB} \quad \text{and} \quad J_{A\_influx} = J_{B\_influx} \tag{6-341}$$

Using the solutions of Equations 6–330 to 6–337 for Y_o^{AB} and Y_{in}^{AB}, in the influx equation for **A** or **B** we obtain

$$J_{A\_influx} = J_{B\_influx}$$
$$= \frac{P_{AB}\left(\frac{K_{B\_o}}{M_{B\_o}}A_{in}B_{in}e^{\frac{(z_A+z_B)F\Psi_m}{RT}}+1\right)A_o B_o}{denom} \tag{6-342}$$

The efflux of the solutes has a similar expression as long as we interchange in the numerator $A_o B_o \to A_{in}B_{in}e^{(z_A+z_B)F\Psi_m/RT}$.

$$J_{A\_efflux} = J_{B\_efflux}$$
$$= \frac{P_{AB}\left(\frac{K_{B\_o}}{M_{B\_o}}A_o B_o+1\right)A_{in}B_{in}e^{\frac{(z_A+z_B)F\Psi_m}{RT}}}{denom} \tag{6-343}$$

TABLE 6.3. *Cotransport Parameters*

$$P_{AB} = \frac{k_{-1}k_{-2}}{(k_{-2}+k_2)(\alpha_1 K_{A\_o}^{eq} K_{B\_o}^{eq})}Y_T$$

$$K_{A\_in} = K_{A\_in}^{eq}\frac{k_2+k_{-2}}{k_2}$$

$$K_{B\_in} = K_{B\_in}^{eq}\frac{k_2+k_{-2}}{k_2}$$

$$M_{in} = \alpha_1 K_{A\_o}^{eq} K_{B\_o}^{eq}\frac{k_{-2}+k_2}{k_{-2}+k_{-1}}$$

$$M_{A\_in} = \alpha_1 K_{A\_o}^{eq} K_{B\_o}^{eq} K_{A\_in}^{eq}\frac{k_{-2}+k_2}{k_{-1}}$$

$$M_{B\_in} = \alpha_1 K_{A\_o}^{eq} K_{B\_o}^{eq} K_{B\_in}^{eq}\frac{k_{-2}+k_2}{k_{-1}}$$

$$K_{ino} = \alpha_1 K_{A\_o}^{eq} K_{B\_o}^{eq}\alpha_2 K_{A\_in}^{eq} K_{B\_in}^{eq}\frac{k_{-2}+k_2}{k_{-1}+k_1}$$

$$K_{A\_o} = K_{A\_o}^{eq}\frac{k_{-2}+k_2}{k_{-2}}$$

$$K_{B\_o} = K_{B\_o}^{eq}\frac{k_{-2}+k_2}{k_{-2}}$$

$$M_o = \alpha_2 K_{A\_in}^{eq} K_{B\_in}^{eq}\frac{k_{-2}+k_2}{k_2+k_1}$$

$$M_{A\_o} = \alpha_2 K_{A\_in}^{eq} K_{B\_in}^{eq} K_{A\_o}^{eq}\frac{k_{-2}+k_2}{k_1}.$$

$$M_{B\_o} = \alpha_2 K_{A\_in}^{eq} K_{B\_in}^{eq} K_{B\_o}^{eq}\frac{k_{-2}+k_2}{k_1}$$

Unidirectional Fluxes Can Be Expressed by Michaelis-Menten Kinetics with Concentration-dependent Parameters. Michaelis-Menten formulations for the analysis of the unidirectional fluxes are a useful tool for resolving parameters, exploring mechanisms, and generally guiding experimental design. For simplicity, we restrict our discussion to a common experimental condition where $\Psi_m = 0$ (e.g., valinomycin doped membranes with $[K^+]_{in} = [K^+]_o$). Relations between net flux parameters and the Michaelis-Menten parameters will be made explicit in a later section.

The Michaelis-Menten expressions for the influx of **A** are

$$J_{A\_influx} = \frac{J_{A\_influx}^{max} A_o}{K_{A\_influx}^M + A_o} \qquad (6\text{-}344)$$

$J_{A\_influx}^{max}$ and $K_{A\_influx}^M$ are a function of co-solute concentration, B_o, as well as the "in" concentrations A_{in} and B_{in}. For expediency their dependence on A_{in}, B_{in} is collected into a common function $X = f(A_{in}, B_{in})$, and is defined as

$$X \equiv \frac{1 + \dfrac{A_{in}}{K_{A\_in}^{eq}} + \dfrac{B_{in}}{K_{B\_in}^{eq}} + \dfrac{A_{in}B_{in}}{\alpha_2 K_{A\_in}^{eq} K_{B\_in}^{eq}}}{k_{-2} + k_1 \dfrac{A_{in}B_{in}}{\alpha_2 K_{A\_in}^{eq} K_{B\_in}^{eq}}} \qquad (6\text{-}345)$$

Note that X is constant for fixed values of A_{in} and B_{in}. Using X, we have

$$J_{A\_influx}^{max} = \frac{Y_T k_{-1} B_o}{\alpha_1 K_{B\_o}^{eq} + (1 + k_{-1}X)B_o}$$

$$K_{A\_influx}^M = \alpha_1 K_{A\_o}^{eq} \frac{(1 + k_2 X)K_{B\_o}^{eq} + B_o}{\alpha_1 K_{B\_o}^{eq} + (1 + k_{-1}X)B_o} \qquad (6\text{-}346)$$

Similar expressions for **B** are:

$$J_{B\_influx} = \frac{J_{B\_influx}^{max} B_o}{K_{B\_influx}^M + B_o}$$

$$J_{B\_influx}^{max} = \frac{Y_T K_{-1} A_o}{\alpha_1 K_{A\_o}^{eq} + (1 + k_{-1}X)A_o} \qquad (6\text{-}347)$$

$$K_{B\_influx}^M = \alpha_1 K_{B\_o}^{eq} \frac{(1 + k_2 X)K_{A\_o}^{eq} + A_o}{\alpha_1 K_{A\_o}^{eq} + (1 + k_{-1}X)A_o} \qquad (6\text{-}348)$$

Linearization of the Michaelis-Menten Equation. In order to estimate $J_{A\_influx}^{max}$ and $K_{A\_influx}^M$ (or corresponding expressions for **B**) and to determine the mechanism under study (i.e., random, sequential, or simultaneous binding), it is common practice to use linear plots of the Michaelis-Menten equations. Some of the most popular include the Lineweaver Burk, $1/J_{A\_influx} = f(1/A_o)$, the Hanes, $A_o/J_{A\_influx} = f(A_o)$, the Eadie-Hofstee, $J_{A\_influx} = f(J_{A\_influx}/A_o)$, and the Cornish-Bowden, $J_{A\_influx}^{max}$

$= f(K_{A\_influx}^M)$. Any of these provide estimates of parameters and of the mechanism involved (26).

Lineweaver-Burk Plots. We use the Lineweaver-Burk plot as an example. It is obtained by inverting the flux equation Equation 6–344

$$\frac{1}{J_{A\_influx}} = \left(\frac{1}{J_{A\_influx}^{max}}\right) + \left(\frac{K_{A\_influx}^M}{J_{A\_influx}^{max}}\right)\frac{1}{A_o}$$

$$\frac{1}{J_{B\_influx}} = \left(\frac{1}{J_{B\_influx}^{max}}\right) + \left(\frac{K_{B\_influx}^M}{J_{B\_influx}^{max}}\right)\frac{1}{B_o} \qquad (6\text{-}349)$$

For each value of B_o, the plot of $y = 1/J_{A\_influx}$ vs. $x = 1/A_o$ will give a straight line with a y-axis intercept, b_A, equal to $1/J_{A\_influx}^{max}$, an x-axis intercept, x_A, equal to $-1/K_{A\_influx}^M$, and a slope m_A equal to $K_{A\_influx}^M/J_{A\_influx}^{max}$. Similar remarks hold for plots of $1/J_{B\_influx}$ vs. $1/B_o$. More explicitly, we have for the intercepts, b_x, and slopes, m_x:

$$b_A \equiv \frac{1}{J_{A\_influx}^{max}} = \left(\frac{\alpha_1 K_{B\_o}^{eq}}{k_{-1}Y_T}\right)\frac{1}{B_o} + \left(\frac{1 + k_{-1}X}{k_{-1}Y_T}\right)$$

$$b_B \equiv \frac{1}{J_{B\_influx}^{max}} = \left(\frac{\alpha_1 K_{A\_o}^{eq}}{k_{-1}Y_T}\right)\frac{1}{A_o} + \left(\frac{1 + k_{-1}X}{k_{-1}Y_T}\right) \qquad (6\text{-}350)$$

and

$$m_A \equiv \frac{K_{A\_influx}^M}{J_{A\_influx}^{max}} = \left(\frac{\alpha_1 K_{A\_o}^{eq}(1 + k_2 X)}{k_{-1}Y_T}\right)\frac{1}{B_o} + \left(\frac{\alpha_1 K_{A\_o}^{eq}}{k_{-1}Y_T}\right)$$

$$m_B \equiv \frac{K_{B\_influx}^M}{J_{B\_influx}^{max}} = \left(\frac{\alpha_1 K_{B\_o}^{eq} K_{A\_o}^{eq}(1 + k_2 X)}{k_{-1}Y_T}\right)\frac{1}{A_o} + \left(\frac{\alpha_1 K_{B\_o}^{eq}}{k_{-1}Y_T}\right)$$

$$(6\text{-}351)$$

Note that for $1/J_{A\_influx}$ vs. $1/A_o$ plots, Equation 6–349, as B_o increases both the slope m_A (Equation 6–351) and the intercept b_A (Equation 6–350) decrease. Since the intercepts equal the reciprocal of $J_{A\_influx}^{max}$, it follows that $J_{A\_influx}^{max}$ increases with B_o. Similar remarks hold for $1/J_{B\_influx}$ vs. $1/B_o$ plots.

The Family of Straight Lines Generated by Lineweaver-Burk Plots Intersect at a Common Point. Under certain conditions the position of intersection of these straight line plots provides a convenient diagnostic test for specific mechanisms. The two Lineweaver Burk plots ($1/J_{A\_influx}$ vs. $1/A_o$) obtained at two different values of B_o, say B_{o1} and B_{o2}, will have different slopes, m_{A1} and m_{A2}, as well as two different intercepts, b_{A1} and b_{A2}. These two straight lines will intersect at a point whose x and y coordinates are denoted here as x_A and y_A, respectively. These coordinates are found by the condition that the intersection point is shared by each line so that $y_A = m_{A1}x_A + b_{A1} = m_{A2}x_A + b_{A2}$. Solving for x_A

with the aid of Equations 6–350 and 6–351, and then using this value in the last expression for y_A, yields

$$x_A = -\frac{b_{A2}-b_{A1}}{m_{A2}-m_{A1}} = \frac{-1}{K^{eq}_{A\_o}(1+k_2X)}$$

$$y_A = m_{A1}x_A + b_{A1} = \frac{(1+k_{-1}X)(1+k_2X)-\alpha_1}{k_{-1}Y_T(1+k_2X)}$$

$$(6\text{–}352)$$

Note that although m_A and b_A depend on B_o, x_A and y_A do not. In other words, the intersection point is independent of B_o; all straight lines in the family, constructed by altering B_o, intersect at a common point. Further, since the value of x_A is negative this intersection point will lie to the left of the y axis in either the second quadrant (if y_A is positive), the third quadrant if y_A is negative), or on the x axis (if y_A vanishes). The sign of y_A is determined by the sign of its numerator, $(1+k_{-1}X)(1+k_2X)-\alpha_1$.

Similar remarks hold for families of straight lines generated by plotting $1/J_{B\_influx}$ vs. $1/B_o$ for different values of A_o. In this case, we obtain the corresponding expressions

$$x_B = -\frac{b_{B2}-b_{B1}}{m_{B2}-m_{B1}} = \frac{-1}{K^{eq}_{B\_o}(1+k_2X)}$$

$$y_B = m_{B1}x_B + b_{B1} = \frac{(1+k_{-1}X)(1+k_2X)-\alpha_1}{k_{-1}Y_T(1+k_2X)}$$

$$(6\text{–}353)$$

Note that $y_A = y_B$; the vertical position of the intersection point of the family of linear plots will be identical for both types of influx plots ($1/J_{A\_influx}$ vs. $1/A_o$ and $1/J_{B\_influx}$ vs. $1/B_o$). This is not true for the horizontal coordinate, $x_A \neq x_B$. Properties of these plots are illustrated in Figure 6.28 (see legends for more detail).

On Influx, the Random Sequence Becomes Ordered as $\alpha_1 \rightarrow$ 0. The ordered sequential mechanism appears when

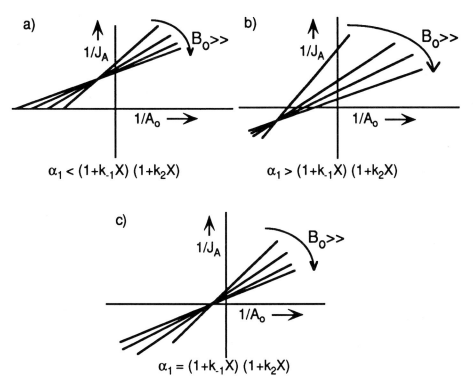

FIG. 6.28. Cotransport random: Families of Lineweaver-Burk plots of $1/J_{A\_influx}$ versus $1/A_o$ for *increasing* values of B_o indicated by the *curved arrow*. All plots intersect in the second or third quadrant, depending on the relative values of α_1 and the product $(1+k_{-1}X)(1+k_2X)$. To interpret plots, recall that for each value of B_o, the plot shows a straight line with a y-axis intercept equal to $1/J^{max}_{A\_influx}$ and an x-axis intercept equal to $-1/K^M_{A\_influx}$ *a*: $\alpha_1 < (1+k_{-1}X)(1+k_2X)$. From Equations 6–352 and 6–353, $y_A = y_B > 0$ and the intersection lies in the second quadrant.. From the changes in the x and y intercepts, it follows that $J^{max}_{A\_influx}$ increases with B_o while $K^M_{A\_influx}$ decreases. *b*: If $\alpha_1 > (1+k_{-1}X)$ $(1+k_2X)$ then $y_A = y_B < 0$. The intersection lies below the x axis, in the third quadrant. Both $J^{max}_{A\_influx}$ and $K^M_{A\_influx}$ will increase with B_o. *c*: If $\alpha_1 = (1+k_{-1}X)(1+k_2X)$ then $y_A = y_B = 0$. The intersection lies on the x axis. In addition, $J^{max}_{A\_influx}$ increases with increasing B_o, while $K^M_{A\_influx}$ remains constant, independent of B_o.

the solute binding to the carrier occurs in lockstep fashion, with one solute always binding before the other. When solute **A**, for example, always binds to the carrier before solute **B**, then the carrier form $Y_o^{\circ B}$ does not exist. To accommodate this within the scheme of Figure 6.26, the dissociation constant for this complex has to be very high, i.e., $K_{B\_o}^{eq} \to \infty$. But an infinite $K_{B\_o}^{eq}$ will drive Y_o^{AB} to zero unless $\alpha_1 \to 0$ so that the product $\alpha_1 K_{B\_o}^{eq}$ is finite while $\alpha_1 K_{A\_o}^{eq} \to 0$. ($K_{A\_o}^{eq}$ must remain finite in order not to drive $Y^{A\bullet}$ to zero). If $\alpha_1 \to 0$, then $\alpha_1 < (1 + k_{-1}X)(1 + k_2X)$, and the intersection point of the families of lines of $1/J_{A\_influx}$ vs. $1/A_o$ lies in the second quadrant. The only distinguishing characteristic of this plot is that as $B_o \to \infty$, the slope m_A (Equation 6–351) vanishes, leaving a horizontal flat line that may be difficult to demonstrate experimentally (Fig. 6.29). However, plots of $1/J_{B\_influx}$ vs. $1/B_o$ yield a more useful diagnostic. As shown in Equation 6–353, when $K_{B\_o}^{eq} \to \infty$, $x_B = 0$, and the family of straight lines intersect on the y axis (see Fig. 6.29). The combination of these two plots, but particularly the latter, will identify the sequential ordered mechanism with solute **A** always binding first.

Kinetics of Simultaneous Binding of mA and nB Resembles Simple Carrier Kinetics When Carrier Concentrations Are Replaced with the Product A^mB^n

The generalized reaction scheme for complex stoichiometries (beyond 1:1) becomes very diffuse and falls outside of the scope of this chapter. However, in the special case where the binding of the solutes is simultaneous, the kinetics are greatly simplified. For in this case m**A** and n**B** act as a unit; all binding and release

reactions are proportional to A^mB^n. If we sketch the reaction scheme for this particular case, it will be identical to Figure 6.25, the scheme for the simple carrier with C_{in} replaced by $A_{in}^m B_{in}^n$ and C_o replaced by $A_o^m B_o^n$. It follows that all the simple carrier kinetic equations, steady state as well as equilibrium at the boundaries, will apply to this particular case of cotransport as long as we make the replacements $C_{in} \to A_{in}^m B_{in}^n$, $C_o \to A_o^m B_o^n$, and $e^{zF\Psi_m/RT} \to e^{[(mz_A + nz_B)F\Psi_m]/(RT)}$. In addition, since each time the carrier cycles (corresponding to the transport of 1 **C**) it delivers m**A** and n**B**, we replace J_C, the flux of **C**, by $J_C \to J_A/m = J_B/n$.

We illustrate with the Michaelis-Menten equations. From Equation 6–290 we have

$$J_{C\_influx} = \frac{J_{C\_influx}^{max} C_o}{K_{C\_influx}^M + C_o} \qquad (6\text{–}354)$$

where $J_{C\_influx}^{max}$ and $K_{C\_influx}^M$ are defined in Equation 6–291 and with the C subscript added for clarity in the following. Making the replacements listed above, we have

$$J_{A\_influx} = m\,J_{C\_influx} = \frac{m\,J_{C\_influx}^{max} A_o^m B_o^n}{K_{C\_influx}^M + A_o^m B_o^n}$$

$$= \frac{(m\,J_{C\_influx}^{max})A_o^m}{(K_{C\_influx}^M/B_o^n) + A_o^m} = \frac{J_{A\_influx}^{max} A^m}{(K_{A\_influx}^M)^m + A^m}$$

$$(6\text{–}355)$$

where the maximal influx of **A**, $J_{A\_influx}^{max} = m J_{C\_influx}^{max}$, can be obtained directly from Equation 6–291 and the concentration of A_o required for 1/2 maximal influx, $K_{A\_influx}^M = (K_{C\_influx}^M/B^n)^{1/m}$, can also be obtained directly from Equation 6–291.

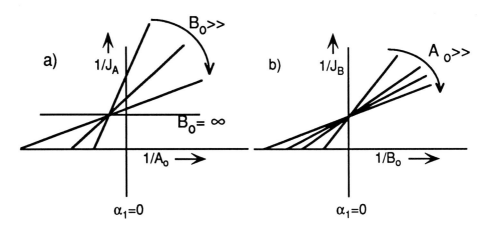

FIG. 6.29. Cotransport sequential. Families of Lineweaver-Burk plots illustrating graphical criteria for an ordered sequence mechanism for influx with A_o always binding first. *a*: Plot of $1/J_{A\_influx}$ versus $1/A_o$ for increasing values of B_o indicated by the *curved arrow*. As $B_o \to \infty$, the slope $\to 0$. *b*: Plot of $1/J_{B\_influx}$ versus $1/B_o$ for increasing values of A_o indicated by the *curved arrow*. All lines intersect on the y axis.

The equations for double reciprocal plots for **A** and **B** are:

$$\frac{1}{J_{A\_influx}} = \frac{1}{J_{A\_influx}^{max}} + \frac{(K_{A\_influx}^{M})^{m}}{J_{A\_influx}^{max}} \frac{1}{A_{o}^{m}}$$

$$\frac{1}{J_{B\_influx}} = \frac{1}{J_{B\_influx}^{max}} + \frac{(K_{B\_influx}^{M})^{n}}{J_{B\_influx}^{max}} \frac{1}{B_{o}^{n}} \qquad (6\text{--}356)$$

Estimates of the values of m, n, and the respective parameters can be obtained by rearranging the equations into log-log plots, i.e., Hill plots (130).

As an example of graphical analysis, suppose that A_{in} and B_{in} are held constant while influx is measured as a function of A_{o}. It follows from Equation 6–291 that $J_{A\_influx}^{max} = m\,J_{C\_influx}^{max}$ *will be independent of* B_{o}. But $1/J_{A\_influx}^{max}$ is the *y*-intercept of the plot of $1/J_{A\_influx}$ vs. $1/A_{o}^{m}$, so that the intercept is also independent of B_{o}. In other words, a necessary condition for simultaneous binding is that all straight lines generated by plots of $1/J_{A\_influx}$ vs. $1/A_{o}^{m}$ for various values of B_{o} intercept at the same point on the *y* axis.

Relations between the Net Flux Equation Parameters and the Michaelis-Menten–type Parameters for 1:1 Stoichiometry

Table 6.4 shows the relations that can be established between the net flux equation parameters (which are constants, independent of experimental conditions) and the concentration-dependent Michaelis-Menten param-

eters. These relations, shown in the first column, refer to Michaelis-Menten parameters obtained by corresponding experimental conditions listed in the second column.

COUNTERTRANSPORT

While cotransport systems couple the movement of two solutes constraining them to move in the *same* direction, countertransport systems (also called antiport, or exchange) catalyze their exchange by constraining them to move in the *opposite* direction. The physiological significance of countertransport is similar to cotransport; it provides a mechanism for the gradient of one solute to drive another solute uphill against its electrochemical gradient. Countertransport systems are very common; some examples are the ATP/ADP antiport in the mitochondrion (71), the Na^{+}/Ca^{+2} exchanger in muscle cells and other plasma membranes (16, 106), the H^{+}/Na^{+} exchanger (8, 9), and the Cl^{-}/HCO_{3}^{-} exchanger (122). Countertransport systems may be directly coupled to an energy source like ATP; the most obvious example is the $Na^{+}\text{-}K^{+}$ pump (43, 46, 47).

Thermodynamics

We represent a countertransport system of two solutes **A** and **B** (with valences z_{A} and z_{B}) whose stoichiometry

TABLE 6.4. *Relation of Cotransport Parameters to the Concentration-Dependent MM (Michaelis-Menten) Parameters for* $m = n = 1$

Relation	Experimental Conditions ($\Psi_{m} = 0$ assumed in all cases)
$K_{A\_o} = K_{A\_influx}^{M}$	$A_{in} = B_{in} = B_{o} = 0$
$K_{B\_o} = K_{B\_influx}^{M}$	$A_{in} = B_{in} = A_{o} = 0$
$K_{A\_in} = K_{A\_efflux}^{M}$	$A_{o} = B_{o} = B_{in} = 0$
$K_{B\_in} = K_{B\_efflux}^{M}$	$A_{o} = B_{o} = A_{in} = 0$
$M_{in} = K_{B\_o}K_{A\_influx}^{M}$	$A_{in} = B_{in} = 0$ and $B_{o} \to \infty$
$M_{A\_in} = M_{in}K_{A\_efflux}^{M}$	$A_{o} = B_{o} \to \infty$ and $B_{in} = 0$
$M_{B\_in} = M_{in}K_{B\_efflux}^{M}$	$A_{o} = B_{o} \to \infty$ and $A_{in} = 0$
$M_{o} = K_{B\_in}K_{A\_efflux}^{M}$	$A_{o} = B_{o} = 0$ and $B_{in} \to \infty$
$M_{o}^{A} = M_{o}K_{A\_influx}^{M}$	$A_{in} = B_{in} \to \infty$ and $B_{o} = 0$
$M_{B\_o} = M_{o}K_{B\_influx}^{M}$	$A_{in} = B_{in} \to \infty$ and $A_{o} = 0$
$K_{ino} = M_{A\_o}K_{B\_influx}^{M}$	$A_{in} = B_{in} \to \infty$ and $A_{o} \to \infty$
$P = \dfrac{J_{A\_influx}^{max}}{K_{A\_o}K_{B\_influx}^{M}}$	$J_{A\_influx}^{max} \to A_{in} = B_{in} = 0$ and $B_{o} \to \infty$ $K_{B\_influx}^{M} \to A_{in} = B_{in} = 0$ and $A_{o} \to \infty$

is mA/nB, and which is coupled to an energy source ($S_{in} \longrightarrow P_{in}$) as follows:

$$mA_{in} + nB_o + S_{in} \longrightarrow mA_o + nB_{in} + P_{in} \quad (6\text{--}357)$$

Applying Equation 6–20 and using arguments similar to those employed in the discussion of cotransport, we obtain:

$$\Delta \tilde{G} = \Delta G_{SP} + RT \ln \frac{A_o^m B_{in}^n}{A_{in}^m B_o^n} - (mz_A - nz_B)F(\Psi_{in} - \Psi_o) \le 0$$
$$(6\text{--}358)$$

The maximal concentration ratios that the countertransport can provide are given by

$$\frac{A_o^m B_{in}^n}{A_{in}^m B_o^n} = e^{-\frac{\Delta G_{SP} - (mz_A - nz_B)F(\Psi_{in} - \Psi_o)}{RT}} \quad (6\text{--}359)$$

Note that Equation 6–359 can be obtained from Equation 6–326 as long as we interchange $B_{in} \leftrightarrow B_o$ and $z_B \leftrightarrow -z_B$. Both modifications reflect the directional change in the flux of **B** as we go from co- to countertransport.

A well-known example of countertransport, the H^+/Na^+ exchanger, couples an inward flow of 1 Na^+ to an outward flow of 1H^+ with no other energy source. Here, $m = n = 1$, $z_A = z_B = +1$, so that the exponential term $= 1$ and the maximal concentration ratio will be obtained when

$$\frac{[Na_o^+]}{[Na_{in}^+]} = \frac{[H_o^+]}{[H_{in}^+]}$$

For this case, assuming a very rapid exchange where no other transport process of H^+ or Na^+ is significant, the extracellular/intracellular concentration ratio for H^+ equals the corresponding ratio for Na^+; the H^+ ratio is clamped to the Na^+ ratio.

Another important example is provided by the Na^+/Ca^{2+} exchanger in muscle cells; in parallel with the Ca^{2+} pump, it keeps the intracellular level of Ca^{2+} very low. In this case, the influx of 3 Na^+ is coupled with the efflux of 1 Ca^{2+}. If we let Na^+ be solute **B**, and Ca^{2+} be solute **A**, then $m = 3$ and $n = 1$, while $z_B = +1$ and $z_A = +2$. If

$$\Psi_m = -60 \text{ mV}, [Na_o^+]/[Na_{in}^+] = 10$$

then the maximum ratio becomes

$$\frac{[Ca_{in}^{2+}]}{[Ca_o^{2+}]} = \frac{[Na_{in}^+]^3}{[Na_o^+]^3} e^{\frac{F\Psi_m}{RT}} = 10^{-4} \quad (6\text{--}360)$$

showing how cells bathed in normal Ca^{2+} concentrations of the order of 2 mM can maintain submicromolar intracellular Ca^{2+} levels.

Kinetic Models

We discuss countertransport kinetics by considering two different reaction schemes:

1. **Ping-Pong Model.** The carrier system **Y** binds only one solute at a time and the ternary complex does not exist (Fig. 6.30). This system can transport a maximum of one solute species for each translocation step.

2. **Sequential Model.** Both solutes bind to the carrier system **Y**, and only the free and the carrier complexed with both solutes can translocate (see Fig. 6.32). In contrast with the ping-pong model, this system transports both solutes during each translocation. This model is very similar to the cotransport model already described, and the same three special cases—random, ordered, and simultaneous sequences—can be considered.

As in cotransport, we assume equilibrium at the boundaries and no energy-coupled reactions. Our objectives are to derive a formal expression for net flux

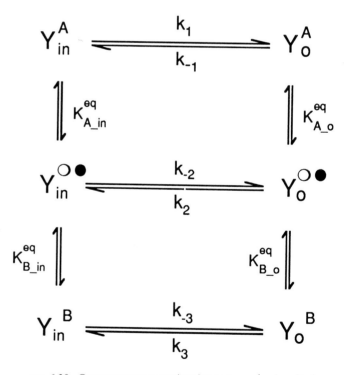

FIG. 6.30. Countertransport carrier ping-pong mechanism. Beginning on side "in" the solute A_{in} or B_{in} combine with the reactive site of component $Y_{in}^{\circ \bullet}$. The complex Y_{in}^A or Y_{in}^B will then undergo a translocation step, moving the reactive site to side "o." The complex then dissociates, releasing A_o or B_o to side "o." The undissociated carrier with the reactive site still facing side "o" will then either translocate its reactive site in the free ($Y_o^{\circ \bullet}$) or bound form (Y_o^A or Y_o^B) to side "in." Reactions of the solute **A** and **B** with the reactive site of component **Y** are assumed to be in equilibrium.

used to describe the operation of the carrier under physiological conditions, to estimate the relevant parameters from unidirectional flux measurements, and to provide graphical criteria for distinguishing between the different reaction sequences.

The countertransport models presented in Figures 6.30 and 6.32 have the following common properties: *(1)* Each carrier molecule has two reactive sites; one site reacts with **A**, the other with **B**. The stoichiometry is $m=n=1$. More complex stoichiometries are taken up later. *(2)* By changing conformation, the carrier changes the orientation of the two sites. We again utilize superscripts to identify binding sites: "o" implies an empty binding site for **A**, while "•" represents an empty site for **B**. Superscript A or B denotes nonvacant sites that are bound with **A** and/or **B**. Subscripts are used to express carrier conformation.

Ping-Pong Model

Figure 6.30 shows the schematics of a countertransport ping-pong system, where only three forms of the carrier system can occur on each side of the membrane. $Y^{\circ\bullet}$ denotes the free carrier with two vacant binding sites, one for **A** and one for **B**; Y^A denotes that once **A** is bound to its binding site, **B** can no longer bind to its vacant binding site. Conversely, Y^B denotes that when **B** is bound, **A** is excluded. All the forms of the carrier can reorient their sites in the membrane. By allowing the free carrier $Y^{\circ\bullet}$ to reorient, we have relaxed the restriction of a rigid exchange coupling.

Net Flux. From Figure 6.30 the net solute flux of **A** and **B** is given by

$$J_A = k_1 Y_{in}^A - k_{-1} Y_o^A \qquad J_B = k_{-3} Y_{in}^B - k_3 Y_o^B$$
$$(6\text{--}361)$$

and using arguments similar to those that led to Equation 6-340,

$$J_{A\_influx} = k_{-1} Y_o^A \qquad J_{B\_influx} = k_3 Y_o^B$$
$$J_{A\_efflux} = k_1 Y_{in}^A \qquad J_{B\_efflux} = k_{-3} Y_{in}^B \qquad (6\text{--}362)$$

Explicit solutions for these net and unidirectional flux equations in terms of the solute concentrations requires resolution of Y_{in}^A, Y_o^A, and Y_{in}^B, Y_o^B as functions of solute concentration.

Equilibrium at the boundaries and detailed balance. The assumption of local equilibrium at the solute-carrier interface implies that

$$K_{A\_in}^{eq} = \frac{Y_{in}^{\circ\bullet} A_{in}}{Y_{in}^A} \qquad K_{A\_o}^{eq} = \frac{Y_o^{\circ\bullet} A_o}{Y_o^A}$$

$$K_{B\_in}^{eq} = \frac{Y_{in}^{\circ\bullet} B_{in}}{Y_{in}^B} \qquad K_{B\_o}^{eq} = \frac{Y_o^{\circ\bullet} B_o}{Y_o^B} \qquad (6\text{--}363)$$

where the K's represent dissociation constants. The relation for detailed balance (see Equations 6–48 and 6–49 applied to the cycles $Y_{in}^{\circ\bullet} \to Y_{in}^A \to Y_o^A \to Y_o^{\circ\bullet} \to Y_{in}^{\circ\bullet}$ and $Y_{in}^{\circ\bullet} \to Y_{in}^B \to Y_o^B \to Y_o^{\circ\bullet} \to Y_{in}^{\circ\bullet}$ of Fig. 6.30) is:

$$\left(\frac{1}{K_{A\_in}^{eq}}\right)\left(\frac{k_1}{k_{-1}}\right) K_{A\_o}^{eq}\left(\frac{k_2}{k_{-2}}\right) = e^{\frac{z_A F}{RT}\Psi_m}$$

$$\left(\frac{1}{K_{B\_in}^{eq}}\right)\left(\frac{k_2}{k_{-2}}\right) K_{B\_o}^{eq}\left(\frac{k_{-3}}{k_3}\right) = e^{\frac{z_B F}{RT}\Psi_m} \qquad (6\text{--}364)$$

Steady-state translocation and conservation of carrier. The steady-state translocation step requires that the net flux of loaded carriers from "in" to "o" equals the net flux of unloaded carriers returning from "o" to "in,"

$$(k_1 Y_{in}^A - k_{-1} Y_o^A) + (k_{-3} Y_{in}^B - k_3 Y_o^B) = (k_2 Y_o^{\circ\bullet} - k_{-2} Y_{in}^{\circ\bullet})$$
$$(6\text{--}365)$$

while the conservation of carriers requires

$$Y_T = Y_o^{\circ\bullet} + Y_o^A + Y_o^B + Y_{in}^{\circ\bullet} + Y_{in}^A + Y_{in}^B \qquad (6\text{--}366)$$

Using Equation 6–363 to eliminate Y_{in}^A, Y_o^A, Y_{in}^B, and Y_o^B in Equation 6–365 and Equation 6–366 leaves two equations in the two unknowns $Y_o^{\circ\bullet}$ and $Y_{in}^{\circ\bullet}$. Solving for these and resubstituting their solutions back into Equation 6–363 yields the desired explicit expressions for Y_{in}^A, Y_o^A, Y_{in}^B, and Y_o^B, which can be utilized directly in the net flux equations (Equation 6–361) and also in expressions for unidirectional fluxes (Equation 6–362). For solute **A** we obtain:

$$J_A = \frac{P_o(A_{in} e^{\frac{z_A F}{RT}\Psi_m} - A_o) + P_{AB}(A_{in} e^{\frac{z_A F}{RT}\Psi_m} B_o - A_o B_{in} e^{\frac{z_B F}{RT}\Psi_m})}{denom}$$
$$(6\text{--}367)$$

$$J_{A\_influx} = \frac{P_o\left(1 + \frac{A_{in} e^{\frac{z_A F}{RT}\Psi_m}}{K_4}\right) A_o + P_{AB}(B_{in} e^{\frac{z_B F}{RT}\Psi_m}) A_o}{denom}$$
$$(6\text{--}368)$$

$$J_{A\_efflux} = \frac{P_o\left(1 + \frac{A_o}{K_4}\right) A_{in} e^{\frac{z_A F}{RT}\Psi_m} + P_{AB} B_o A_{in} e^{\frac{z_A F}{RT}\Psi_m}}{denom}$$
$$(6\text{--}369)$$

$$denom = 1 + \frac{A_{in}}{K_{A\_in}} + \frac{B_{in}}{K_{B\_in}} + \frac{A_o}{K_{A\_o}} + \frac{B_o}{K_{B\_o}}$$
$$+ \frac{A_{in} A_o}{K_{AinAo}} + \frac{A_{in} B_o}{K_{AinBo}} + \frac{B_{in} A_o}{K_{BinAo}} + \frac{B_{in} B_o}{K_{BinBo}} \qquad (6\text{--}370)$$

where the parameters are as shown in Table 6.5.

TABLE 6.5. *Countertransport Ping-pong Parameters*

$$P_{AB} = \frac{k_3 k_{-2} k_{-1}}{k_2(k_2 + k_{-2})K_{A\_o}^{eq} K_{B\_o}^{eq}} Y_T \qquad P_o = \frac{k_{-2} k_{-1} Y_T}{(k_2 + k_{-2})K_{A\_o}^{eq}}$$

$$K_{A\_in} = K_{A\_in}^{eq} \frac{k_2 + k_{-2}}{k_2 + k_1} \qquad K_{A\_o} = K_{A\_o}^{eq} = \frac{k_{-2} + k_2}{k_{-2} + k_{-1}}$$

$$K_{B\_in} = K_{B\_in}^{eq} \frac{k_2 + k_{-2}}{k_2 + k_{-3}} \qquad K_{B\_o} = K_{B\_o}^{eq} \frac{k_{-2} + k_2}{k_{-2} + k_3}$$

$$K_{AinAo} = K_{A\_in}^{eq} K_{A\_o}^{eq} \frac{k_{-2} + k_2}{k_{-1} + k_1} \qquad K_{AinBo} = K_{A\_in}^{eq} K_{B\_o}^{eq} \frac{k_{-2} + k_2}{k_3 + k_1}$$

$$K_{BinAo} = K_{B\_in}^{eq} K_{A\_o}^{eq} \frac{k_{-2} + k_2}{k_{-1} + k_{-3}} \qquad K_{BinBo} = K_{B\_in}^{eq} K_{B\_o}^{eq} \frac{k_{-2} + k_2}{k_3 + k_{-3}}$$

$$K_4 = \frac{k_2 K_{A\_o}^{eq}}{k_{-1}}$$

The first term in the numerator on the left in the net flux equation (6–367) represents transport through the slippage pathway, cycle $Y_{in}^{\circ\bullet} \rightarrow Y_{in}^A \rightarrow Y_o^A \rightarrow Y_o^{\circ\bullet} \rightarrow Y_{in}^{\circ\bullet}$. It is formally identical to a simple carrier; it remains intact while the second term vanishes as Equation 6–367 reduces to the expression for the net flux of a simple carrier when $B_{in} = B_o = 0$. The second term represents the exchange flux. When $z_A = z_B$ the exponential term factors out of the exchange flux, leaving the difference in the concentration products as the simple driving force for exchange.

Similar remarks apply to unidirectional fluxes. The first term in the numerator of Equations 6–368 and 6–369 represents the "slippage," while the second term accounts for the exchange. These expressions reduce to corresponding expressions for a simple carrier when the concentration of **B** equals zero on both compartments.

The expressions for the net and unidirectional fluxes of **B** are similar to Equations 6–367, 6–368, and 6–369, with the same denominator. The numerator, however, is altered by the following replacements: $k_1 \rightarrow k_{-3}$, $k_{-1} \rightarrow k_3$, $K_{A\_o}^{eq} \rightarrow K_{B\_o}^{eq}$, $K_{A\_in}^{eq} \rightarrow K_{B\_in}^{eq}$, $A_o \rightarrow B_o$, $A_{in} \rightarrow B_{in}$, and $z_A \rightarrow z_B$.

Flux equations with no slippage. Slippage disappears whenever $P_o = 0$. This is equivalent to setting $k_{\pm 2} = 0$. However, straightforward substitution into Equations 6–367, 6–368, or 6–370, and equations in Table 6.5 leads to indeterminate forms. This is easily resolved by multiplying the numerator and denominator of Equations 6–367, 6–368, and 6–369 by $(k_2 + k_{-2})$. The parameters P_{AB} and K_{x_y} are modified as shown below, and the resulting fluxes become:

$$J_A = \frac{P'_{AB}(A_{in}e^{\frac{z_A F}{RT}\Psi_m}B_o - A_o B_{in}e^{\frac{z_B F}{RT}\Psi_m})}{denom'} \qquad (6\text{–}371)$$

$$J_{A\_influx} = \frac{P'_{AB}B_{in}e^{\frac{z_B F}{RT}\Psi_m}A_o}{denom} \qquad J_{A\_efflux} = \frac{P'_{AB}B_o A_{in}e^{\frac{z_A F}{RT}\Psi_m}}{denom'}$$
$$(6\text{–}372)$$

$$denom' = \frac{A_{in}}{K'_{A\_in}} + \frac{B_{in}}{K'_{B\_in}} + \frac{A_o}{K'_{A\_o}} + \frac{B_o}{K'_{B\_o}}$$
$$+ \frac{A_{in}A_o}{K'_{AinAo}} + \frac{A_{in}B_o}{K'_{AinBo}} + \frac{B_{in}A_o}{K'_{BinAo}} + \frac{B_{in}B_o}{K'_{BinBo}}$$
$$(6\text{–}373)$$

$$K'_{X\_y} = \frac{K_{X\_y}}{(k_2 + k_{-2})}$$

$$P'_{AB} = (k_2 + k_{-2}) P_{AB} = \frac{k_3}{K_{B\_o}^{eq}} \frac{k_1}{K_{A\_in}^{eq}} e^{\frac{-z_A F}{RT}\Psi_m} Y_T$$
$$= \frac{k_{-3}}{K_{A\_o}^{eq}} \frac{k_{-1}}{K_{B\_in}^{eq}} e^{\frac{-z_B F}{RT}\Psi_m} Y_T$$
$$(6\text{–}374)$$

where we have also used Equation 6–364 to eliminate the indeterminate ratio k_{-2}/k_2 from $(k_2 + k_{-2}) P_{AB}$.

In this pure countertransport system with a 1/1 stoichiometry, P'_{AB} is a measure of the joint, low-concentration, permeability of **A** and **B**, as the carrier traverses the membrane complexed with one solute, releases it, and returns loaded with the other. Conditions for zero net flux as well as the nonequilibrium flux direction are in complete agreement with the thermodynamic description given above.

Unidirectional Flux. For simplicity, we assume $\Psi_m = 0$. In general, Equation 6–368 can be rewritten as

$$J_{A\_influx} = \frac{J_{A\_influx}^{max}A_o}{K_{A\_influx}^M + A_o} \qquad (6\text{–}375)$$

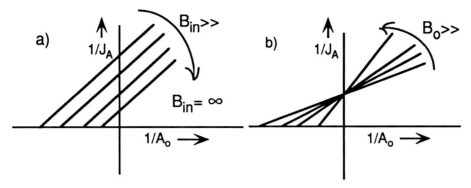

FIG. 6.31. Countertransport ping-pong. Lineweaver-Burk plot for the influx of a solute **A** for different concentrations of **B**. *a*: Plot of stimulation of the influx of **A** by the counter solute (**B**_{in}) on the *trans* side of the membrane where it binds to the free carrier when there is no slippage. *b*: Plot of a competitive inhibition. Influx of **A** is decreased by **B**_o, which competes for the same carrier site on the same side "o" (*cis* side) of the membrane.

$$J_{A\_influx}^{max} = \frac{k_{-1} Y_T}{(1 + k_{-1}X)}$$

$$K_{A\_influx}^{M} = K_{A\_o}^{eq} \frac{(1 + k_2 X)\, K_{B\_o}^{eq} + (1 + k_3 X)\, B_o}{K_{B\_o}^{eq}(1 + k_{-1}X)}$$

$$(6\text{--}376)$$

where $X = f(A_{in}, B_{in})$ shows the dependence of these parameters on the internal concentrations and is given by

$$X = \frac{1 + \dfrac{A_{in}}{K_{A\_in}^{eq}} + \dfrac{B_{in}}{K_{B\_in}^{eq}}}{k_{-2} + k_1 \dfrac{A_{in}}{K_{A\_in}^{eq}} + k_{-3} \dfrac{B_{in}}{K_{B\_in}^{eq}}} \qquad (6\text{--}377)$$

Lineweaver-Burk Plots Can Test for Ping-pong Kinetics

B_{in} *stimulates the influx of* A_o. Activation of the influx of **A**_o by the countersolute **B**_{in} is easily illustrated by holding A_{in} constant, setting $B_o = 0$ (or if $K_{B\_o}^{eq} = \infty$), and plotting $1/J_{A\_influx}$ vs. $1/A_o$. By varying the concentration of **B**_{in} a family of straight lines is obtained where the slope reduces to

$$\frac{K_{A\_influx}^{M}}{J_{A\ influx}} = \frac{K_{A\_o}^{eq}}{k_{-1}\, Y_T}(1 + k_2 X) \qquad (6\text{--}378)$$

Figure 6.31*a* shows this activation of the influx of **A**_o by increasing values of B_{in}, when the slippage component is negligible, $k_2 = 0$, and the slope becomes constant, independent of X and thus independent of B_{in}. These Lineweaver Burk plots consist of a family of parallel lines, an identifying criterion for a pure ping-pong mechanism ($k_2 = k_{-2} = 0$) whenever $B_o = 0$ (or $K_{B\_o}^{eq} = \infty$).

B_o *competitively inhibits the influx of* A_o. Although **B** on the *trans* side has a stimulating effect, **B** on the *cis* side of the membrane competitively inhibits the flux of **A**.

In this case, plotting $1/J_{A\_influx}$ vs. $1/A_o$ (Fig. 6.31*b*) while holding A_{in} and B_{in} constant (i.e., $X = $ constant) for various values of B_o yields a family of straight lines, all crossing at $(0, 1/J_{A\_influx}^{max})$ on the y axis. From Equations 6–376 and 6–377, we see that $J_{A\_influx}^{max}$ is independent of B_o. Accordingly, all lines intercept the y axis at a common point. On the other hand, the slope of each line is given by $K_{A\_influx}^{M}/J_{A\_influx}^{max}$ and since, from Equation 6–376, $K_{A\_influx}^{M}$ increases linearly with B_o, the slopes of the $1/J_{A\_influx}$ plots also increase with B_o. Although J_{A_influx} decreases with B_o, the fact that $J_{A\_influx}^{max}$ is not inhibited implies that B_o acts as competitive inhibitor.

Sequential Model

Unlike the ping-pong model, the counter sequential carrier illustrated in Figure 6.32 permits simultaneous biding of both **A** and **B** and allows translocation only when the carrier is completely free of **A** and **B** or when it becomes a ternary complex having bound *both* **A** and **B**. Further, the binding sites always face opposite sides of the membrane, e.g., when **A** site (o) faces inside, **B** site (•) faces outside. By changing conformation (i.e., translocation) the carrier can interchange the orientation of the two sites.

In Figure 6.32, subscripts express carrier conformation and superscripts indicate binding sites. For example, $_o Y_{in}^{\bullet}$ shows a carrier with vacant **A** site facing out, while vacant **B** site faces in. $_{in} Y_o^{\bullet}$ shows the allowed conformation change where the vacant **A** site now faces in and the vacant **B** site faces out. $_{in}^{A} Y_o^{B}$ denotes a carrier with **A** bound on the inside site and **B** bound on the outside, $_o Y_{in}^{B}$ has **B** bound on the inside with a vacant **A** site facing out, and $_{in}^{A} Y_o^{\bullet}$ has **A** bound on the inside with a vacant **B** site facing out.

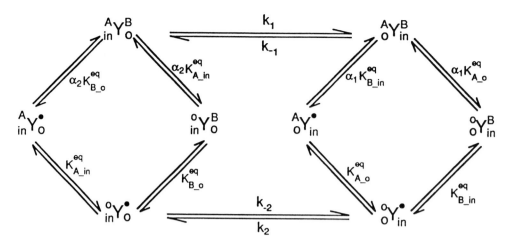

FIG. 6.32. Countertransport sequential model with $m = n = 1$. When **A** binds to $_{in}^{\circ}Y_o^{\bullet}$ or $_o^{\circ}Y_{in}^{\bullet}$, it forms the complex $_{in}^{A}Y_o^{\bullet}$ or $_o^{A}Y_{in}^{\bullet}$. This complex will then bind **B** on the empty reactive site located on the opposite side of the membrane, forming $_{in}^{A}Y_o^{B}$ or $_o^{A}Y_{in}^{B}$. On the other hand, if **B** binds first, then the complex $_{in}^{\circ}Y_o^{B}$ or $_o^{\circ}Y_{in}^{B}$ would be formed before $_o^{A}Y_o^{B}$ or $_o^{A}Y_{in}^{B}$. Only three out of the four cyclic dissociation constants, K^{eq}, are independent (see Equations 6–48 and 6–49). For each step, the dissociation constant is located closest to the dissociation arrow.

Note that the reaction scheme in Figure 6.32 is similar to the scheme of cotransport of Figure 6.26. The main difference is that here, the **B** moves in an opposite direction to solute **A**, while in cotransport they move in the same direction. In Figure 6.32, when **A** moves from in to out, **B** moves from out to in. In cotransport (Fig. 6.26), **B** always follows **A**.

Net Flux. In the steady state the net flux of **A** and **B** equals the flux of the ternary carrier complex and is equal to:

$$J_A = J_B = k_1 \, _{in}^{A}Y_o^{B} - k_{-1} \, _o^{A}Y_{in}^{B} \qquad (6\text{–}379)$$

The influx (J_{influx}) and efflux (J_{efflux}) of **A** and **B** can be expressed as:

$$J_{A\_influx} = k_{-1} \, _o^{A}Y_{in}^{B} \quad \text{where} \quad J_{A\_influx} = J_{B\_efflux}$$

$$J_{A\_efflux} = k_1 \, _{in}^{A}Y_o^{B} \quad \text{where} \quad J_{A\_efflux} = J_{B\_influx} \qquad (6\text{–}380)$$

Equilibrium at the boundaries and detailed balance. We assume that the reactions of free solutes **A** and **B** with the reactive sites are in fast equilibrium with the dissociation constants $K_{A\_o}^{eq}$, $K_{B\_o}^{eq}$, $\alpha_1 K_{A\_o}^{eq}$, $\alpha_2 K_{B\_o}^{eq}$ for the reactions at side "o" of the membrane and $K_{A\_in}^{eq}$, $K_{B\_in}^{eq}$, $\alpha_2 K_{A\_in}^{eq}$, and $\alpha_1 K_{B\_in}^{eq}$ for the reactions at side "in." These dissociation constants are:

$$K_{A\_in}^{eq} = \frac{_{in}^{\circ}Y_o^{\bullet} A_{in}}{_{in}^{A}Y_o^{\bullet}} \qquad K_{A\_o}^{eq} = \frac{_o^{\circ}Y_{in}^{\bullet} A_o}{_o^{A}Y_{in}^{\bullet}}$$

$$K_{B\_in}^{eq} = \frac{_o^{\circ}Y_{in}^{\bullet} B_{in}}{_o^{\circ}Y_{in}^{B}} \qquad K_{B\_o}^{eq} = \frac{_{in}^{\circ}Y_o^{\bullet} B_o}{_{in}^{\circ}Y_o^{B}} \qquad (6\text{–}381)$$

$$\alpha_2 K_{A\_in}^{eq} = \frac{_{in}^{\circ}Y_o^{B} A_{in}}{_{in}^{A}Y_o^{B}} \qquad \alpha_1 K_{A\_o}^{eq} = \frac{_o^{\circ}Y_{in}^{B} A_o}{_o^{A}Y_{in}^{B}}$$

$$\alpha_1 K_{B\_in}^{eq} = \frac{_o^{A}Y_{in}^{\bullet} B_{in}}{_o^{A}Y_{in}^{B}} \qquad \alpha_2 K_{B\_o}^{eq} = \frac{_{in}^{A}Y_o^{\bullet} B_o}{_{in}^{A}Y_o^{B}}$$

The relation for detailed balance (see Equations 6–48 and 6–49), applied to the cycle $_{in}^{\circ}Y_o^{\bullet} \to \, _{in}^{A}Y_o^{\bullet} \to \, _{in}^{A}Y_o^{B} \to \, _o^{A}Y_{in}^{B} \to \, _o^{\circ}Y_{in}^{B} \to \, _o^{\circ}Y_{in}^{\bullet} \to \, _{in}^{\circ}Y_o^{\bullet}$, is:

$$\left(\frac{1}{K_{A\_in}^{eq}}\right)\left(\frac{1}{\alpha_2 K_{B\_o}^{eq}}\right)\left(\frac{k_1}{k_{-1}}\right)\alpha_1 K_{A\_o}^{eq} \, K_{B\_in}^{eq} \left(\frac{k_2}{k_{-2}}\right) = e^{-\frac{\Delta \tilde{G}^{\circ}}{RT}}$$

$$= e^{\frac{(z_A - z_B)F\Psi_m}{RT}} \qquad (6\text{–}382)$$

Steady-state translocation and carrier conservation. The translocation steps are in a steady state. This implies

$$k_1 \, _{in}^{A}Y_o^{B} - k_{-1} \, _o^{A}Y_{in}^{B} = k_2 \, _o^{\circ}Y_{in}^{\bullet} - k_{-2} \, _{in}^{\circ}Y_o^{\bullet} \qquad (6\text{–}383)$$

The conservation of total carrier (Y_T) implies

$$Y_T = \, _{in}^{\circ}Y_o^{\bullet} + \, _{in}^{A}Y_o^{\bullet} + \, _{in}^{\circ}Y_o^{B} + \, _{in}^{A}Y_o^{B} + \, _o^{A}Y_{in}^{B}$$

$$+ \, _o^{A}Y_{in}^{\bullet} + \, _o^{\circ}Y_{in}^{B} + \, _o^{\circ}Y_{in}^{\bullet} \qquad (6\text{–}384)$$

Kinetics of sequential co- and countertransport can be interchanged. Equations 6–381 and 6–384 are sufficient to determine all of the carrier states (the *Y*'s) in terms of rate constants, concentrations, and Y_t. Substituting these into Equation 6–379 and 6–380 completes the kinetic characterization of our model. Notice that all of these equations are identical to corresponding equations for the cotransport provided we make the following substitutions:

$$B_o \to B_{in}, \quad K^{eq}_{B\_o} \to K^{eq}_{B\_in}, \quad \text{and} \quad z_B \to -z_B \qquad (6\text{--}385)$$

Further, making the same substitutions the solutions for the flux equations will be identical to Equations 6–338, 339, 342, and 343, with parameters similar to the ones in Table 6.3. Finally, using the substitutions from Equation 6–385, the same Lineweaver-Burk plots and the same discussion is applicable.

Generalization to m-n Stoichiometry for Simultaneous Binding

All of the countertransport results can be easily recast for the more general case where m molecules of **A** and/or n molecules of **B** are involved in each cycle. Derivations and equations and plots can be literally repeated after making the following substitutions:

$$A \to A^m, \quad B \to B^n, \quad z_A \to m z_A, \quad z_B \to n z_B$$

Further, letting $J(m)_A$, $J(m)_{A\_efflux}$, $J(m)_{A\_influx}$ denote the net flux, influx, and efflux, of **A**, we have

$$J(m)_A = m J(1)_A \equiv m J_A, \quad J(m)_{A\_influx} = m J(1)_{A\_influx}$$
$$\equiv m J_{A\_influx}, \quad J(m)_{A\_efflux} = m J(1)_{A\_efflux}$$
$$\equiv m J_{A\_efflux}$$

Similarly

$$J(n)_B = n J(1)_B \equiv n J_B, \quad J(n)_{B\_influx} = n J(1)_{B\_influx}$$
$$\equiv n J_{B\_influx}, \quad J(n)_{B\_efflux} = n J(1)_{B\_efflux} \equiv n J_{B\_efflux}$$

FLUCTUATING BARRIERS: CHANNELS AND CARRIERS

Up to this point we have considered channels and carriers as two distinct objects. Channels were static structures containing solute-binding sites that are simultaneously accessible to solutes on either side of the membrane. On the other hand, carriers allowed conformational changes to take place within the membrane so that binding sites were exposed alternately to one side of the membrane or to the other, but never simultaneously to both. Both channels and carriers are composed of integral membrane proteins, and examination of proteins as dynamic structures as well as single-channel electrical recordings of electrolyte transport suggests that we relax our restrictive definition of channels as rigid static structures.

Physical measurements show that proteins commonly fluctuate between conformational states at room temperature. Transition times for some of these states have been estimated to lie within the picosecond range, whereas others may last as long as seconds. Thus, it is plausible that many of these transitions take place at rates comparable and even slower than solute transit through the membrane, so that we can expect that energy barriers which fluctuate in confluence with conformation transitions will influence transport kinetics in channels as well as carriers. In fact, as we shall see, rigid channels and ideal carriers can be considered as opposite poles of a continuum provided by channels with fluctuating energy barriers. The following discussion treats transport in an ensemble of channels with fluctuating barriers (79). An excellent exposition of the stochastic principles for interpreting single channels is given by Colquhoun and Hawkes (22).

Consider a simple channel containing a single solute-binding site. The channel fluctuates between two states, a "ground" state **Y** and a "perturbed" state **Y**$_*$, each with a single vacant binding site denoted by **Y**$^\circ$ and **Y**$_*^\circ$. Bound states are denoted by **Y**^{C} and **Y**$_*^C$. The transport scheme for a channel in the ground state can be represented by the following reaction scheme:

$$\mathbf{C_{in}} + \mathbf{Y^\circ} \underset{k_{-in}}{\overset{k_{in}}{\rightleftharpoons}} \mathbf{Y^C} \underset{k_{-o}}{\overset{k_o}{\rightleftharpoons}} \mathbf{Y^\circ} + \mathbf{C_o}$$
$$(6\text{--}386)$$

with a similar diagram for the perturbed state. Thus there are four channel states, two in the ground state, **Y**$^\circ$ (unoccupied) and **Y**^{C} (occupied), and two in a perturbed state, **Y**$_*^\circ$ (unoccupied) and **Y**$_*^C$ (occupied). Total channel behavior, including fluctuations between states, is illustrated in Figure 6.33

Rate constants are different in the two states because the energy barriers have been perturbed, as indicated in Figure 6.33. In general, the fluctuation frequencies $k_{\pm 1} \neq k_{\pm 2}$; they depend on whether the channel is occupied or not. The distribution of channel states is derived in the usual manner by assuming a steady state. To reduce algebraic clutter, we make use of the following substitutions:

$$\alpha \equiv k_{in} C_{in} + k_{-o} C_o \qquad \alpha^* \equiv k^*_{in} C_{in} + k^*_{-o} C_o$$
$$\beta \equiv k_{-in} + k_o \qquad \beta^* \equiv k^*_{-in} + k^*_o$$
$$(6\text{--}387)$$

Steady-state equations are:

$$\frac{dY^\circ}{dt} = -(\alpha + k_{-2})Y^\circ + \beta Y^C + k_2 Y_*^\circ + 0\ Y_*^C = 0$$

$$\frac{dY^C}{dt} = \alpha\ Y^\circ - (k_1 + \beta) Y^C + 0\ Y_*^\circ + k_{-1} Y_*^C = 0$$

$$\frac{dY_*^\circ}{dt} = k_{-2} Y^\circ + 0\ Y^C - (\alpha^* + k_2)Y_*^\circ + \beta^* Y_*^C = 0$$

$$Y^\circ + Y^C + Y_*^\circ + Y_*^C = Y_t \qquad (6\text{--}388)$$

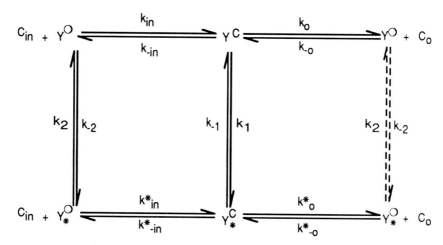

FIG. 6.33. Reaction scheme for single-site channel with energy barriers that can fluctuate between two conformations, **Y** and **Y**₊. The *dashed arrows* are redundant; they have been drawn to facilitate an overview of all transport pathways.

where Y_t denotes the density of channels. Additional conditions can be obtained by application of detailed balance restrictions to the following set of three complete membrane cycles, each of which is accompanied by the net transport of one mole from "in" to "o":

$$\mathbf{Y}^\circ \underset{k_{-in}}{\overset{k_{in}}{\rightleftharpoons}} \mathbf{Y}^C \underset{k_{-o}}{\overset{k_o}{\rightleftharpoons}} \mathbf{Y}^\circ, \quad \mathbf{Y}^\circ_* \underset{k^*_{in}}{\overset{k^*_{-in}}{\rightleftharpoons}} \mathbf{Y}^C_* \underset{k^*_{-o}}{\overset{k^*_o}{\rightleftharpoons}} \mathbf{Y}^\circ_*$$

(6–389)

and

$$\mathbf{Y}^\circ \underset{k_{-in}}{\overset{k_{in}}{\rightleftharpoons}} \mathbf{Y}^C \underset{k_{-1}}{\overset{k_1}{\rightleftharpoons}} \mathbf{Y}^C_* \underset{k^*_{-o}}{\overset{k^*_o}{\rightleftharpoons}} \mathbf{Y}^\circ_* \underset{k_{-2}}{\overset{k_2}{\rightleftharpoons}} \mathbf{Y}^\circ$$

(6–390)

Applying detailed balance Equation 6–49 to Equations 6–389 and 6–390, we have

$$\frac{k_{in}k_o}{k_{-in}k_{-o}} = \frac{k^*_{in}k^*_o}{k^*_{-in}k^*_{-o}} = \frac{k_{in}k_1k^*_ok_2}{k_{-in}k_{-1}k^*_{-o}k_{-2}} = e^{\frac{zF}{RT}\Psi_m}$$

(6–391)

Finally, the net flux does not distinguish between channels; it is given by the sum of the fluxes into the membrane's ground and perturbed channels. In steady state this is equal to the corresponding sum out of the membrane, i.e.,

$$J = (k_{in}C_{in}Y^\circ - k_{-in}Y^C) + (k^*_{in}C_{in}Y^\circ_* - k^*_{-in}Y^C_*)$$
$$= (k_oY^C - k_{-o}C_oY^\circ) + (k^*_oY^C_* - k^*_{-o}C_oY^\circ_*)$$

(6–392)

An explicit expression for J in terms of C_{in} and C_o is obtained by using Equations 6–387, 6–388, and 6–

391 to obtain explicit expressions for the Y's and substituting these into Equation 6–392. The result is a ratio of two quadratic functions of C_{in} and C_o. Using Equation 6–391 to eliminate any three rate constants in the numerator allows the driving force to be factored as shown (cf. 78, 79, 81).

$$J = \frac{(P_o + Q_{in}C_{in}e^{\frac{zF}{RT}\Psi_m} + Q_oC_o)(C_{in}e^{\frac{zF}{RT}\Psi_m} - C_o)}{1 + \frac{C_{in}}{K_{in}} + \frac{C_o}{K_o} + \frac{C_{in}C_o}{K_{ino}} + \frac{C_{in}^2}{K_{inin}} + \frac{C_o^2}{K_{oo}}}$$

(6–393)

where the parameters are as listed in Table 6.6.

Channel Transport Properties Depend on the Rate of Transition between the Conformations

Our model consists of two channels, one in the ground state, the other perturbed, operating in parallel. Let us assume that the perturbed channel has a higher permeability than the ground-state channel. Then the observed permeability of the composite will depend on the relative fraction that is in each state. These two channels are distinct from two similar parallel static channels because they are interconvertible, so it is reasonable to assume that their unique behavior will be related to their rates of interconversion. Suppose the interconversion rate is very rapid, much more rapid than the average dwell time of a solute in the channel. Then each solute "sees" an average barrier structure and the channel behaves as though it were a single static channel with rate constants that are a composite of the two states.

TABLE 6.6. *Parameters for Channels with Fluctuation Barriers*

$$P_o = \frac{Y_t([k_{-1}k_o + k_1k_o^*]k_{-in}k_{-in}^* + [k_{-1}k_{-in} + k_1k_{-in}^*][k_{-1}k_o + k_1k_o^* + k_ok_o^*])k_{-2}k_{-o}^*}{k_1k_o^*m_o}$$

$$Q_{in} = \frac{Y_t(k_{-1}k_o + k_1k_o^*)k_{in}k_{-in}^*k_{-o}^*}{k_o^*m_o} \qquad Q_o = \frac{Y_t(k_{-1}k_{-in} + k_1k_{-in}^*)k_{-o}k_{-o}^*}{m_o}$$

$$K_{in} = \frac{m_o}{(k_2 + k_1)(k_{-in}^* + k_o^*)k_{in} + (k_{-2} + k_{-1})(k_{-in} + k_o)k_{in}^* + (k_1 + k_{-1})(k_2k_{in} + k_{-2}k_{in}^*)}$$

$$K_o = \frac{m_o}{(k_2 + k_1)(k_{-in}^* + k_o^*)k_{-o} + (k_{-2} + k_{-1})(k_{-in} + k_o)k_{-o}^* + (k_1 + k_{-1})(k_2k_{-o} + k_{-2}k_{-o}^*)}$$

$$K_{ino} = \frac{m_o}{(k_1 + k_{-1})(k_{in}k_{-o}^* + k_{-o}k_{in}^*)} \qquad K_{inin} = \frac{m_o}{(k_1 + k_{-1})k_{in}k_{in}^*} \qquad K_{oo} = \frac{m_o}{(k_1 + k_{-1})k_{-o}k_{-o}^*}$$

$$m_o = (k_2 + k_{-2})([k_{-in}^* + k_o^*]k_1 + [k_{-1} + k_{-in}^* + k_o^*][k_{-in} + k_o])$$

Now consider the other extreme; the interconversion rate between states is slow, compared to the average dwell time of a solute in the channel. Also assume that the presence of the solute within the channel promotes conversion to the faster perturbed state, and this state persists for some time after the solute leaves before it relaxes to the slower ground state. The next solute will then find a distribution of available (unoccupied) channels, some fast, some slow. Further, this distribution will depend on concentration (e.g., when the concentration is near zero more of the channels will be in the slow state). As a result the channel flux will have a much more complex concentration dependence, as illustrated by Equation 6–393.

Channels with Fluctuating Barriers Do Not Show Michaelis-Menten Kinetics

Consider the simple case with $C_{in} = 0$. In our singly occupied, static structures, the influx follows Michaelis-Menten kinetics increasingly monotonically with C_o to an asymptote. This does not necessarily hold in dynamic channels. With $C_{in} = 0$, Equation 6–393 reduces to

$$-J = \frac{(P_o + Q_oC_o)C_o}{\dfrac{C_o^2}{K_{oo}} + \dfrac{C_o}{K_o} + 1} \qquad (6\text{–}394)$$

which is plotted in Figure 6.34 for two different sets of parameters. Although the lower plot rises monotonically to an asymptote, neither curve is a rectangular hyperbola. The upper curve obtained by simply changing P_o from 1 to 3 rises to a maximum and then falls; it shows qualitative differences from typical Michaelis-Menten behavior.

The same conclusions hold for small signal ohmic

conductance (81). Finding a maximum in the concentration dependence of conductance is often regarded as a sign of multiple occupancy as the first occupant electrostatically repels the second. However, the maximum is just as easily predicted from dynamic channels with a single occupant.

Channels with Fluctuating Barriers Can Show Carrier Kinetics

If we allow the barrier heights on both sides of the binding site to fluctuate between nominal values, where transport occurs, to very large values that effectively close the channel, then we can construct a channel that is indistinguishable from a carrier (79). It is entirely possible that rigid channels and conventional carriers

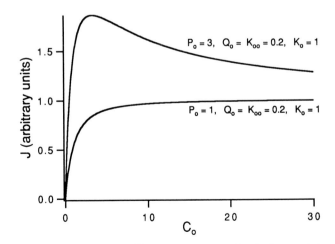

FIG. 6.34. Two plots of flux vs. concentration for singly occupied dynamic channels, with different P_o, obtained from Equation 6–394. Values of the parameters are shown on each curve. Neither curve is a rectangular hyperbola. Note the maximum in the upper plot.

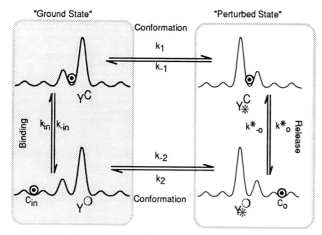

FIG. 6.35. Operation of cycle (Equation 6–390) shows how channel with very large fluctuating barriers behaves like conventional carrier.

are merely opposite ends of a continuous spectrum of channel behavior. As illustrated in Figure 6.35, we require only that the channel alternates from the ground state where it is open on the left (closed on the right, which implies $k_o = k_{-o} = 0$) to the perturbed state where it is open on the right (closed on the left with $k^*_{in} = k^*_{-in} = 0$). The "translocation" step is simply the transition from ground to perturbed state.

Setting $k_o = k_{-o} = k^*_{in} = k^*_{-in} = 0$ in Equation 6–392 and Table 6.6, we find that Q_{in}, Q_o, C_{in}^2/K_{inin}, and C_o^2/K_{oo} all vanish so that Equation 6–393 reduces to

$$J = \frac{P_o(C_{in} e^{\frac{zF}{RT}\Psi_m} - C_o)}{1 + \frac{C_{in}}{K_{in}} + \frac{C_o}{K_o} + \frac{C_{in}C_o}{K_{ino}}} \qquad (6\text{–}395)$$

which is formally identical to the simple carrier flux equation Equation 6–281. Further, the parameters also reduce to the simple carrier. For example,

$$P_o = \frac{k_{-2}k_{-1}k_{-in}k^*_{-o}Y_t}{(k_2 + k_{-2})(k_{-1}k_{-in} + k_1 k^*_o + k_{-in}k^*_o)} \qquad (6\text{–}396)$$

is virtually identical to the corresponding Equation 6–279. The only difference is notation for k_o and k_{-o}; Equation 6–396 indicates that transport between binding site and solution "o" takes place in the "perturbed" state, while Equation 6–279 does not. Similar remarks hold for the remaining parameters K_{in}, K_o, and K_{ino}.

We thank the Junta Nacional de Investigação Científica e Tecnológica and the Cigarette and Tobacco Surtax Fund of California (31T-0302) for financial support and the Instituto de Tecnologia Química e Biológica, Universidade Nova de Lisboa for a Visiting Professorship to R. I. Macey.

REFERENCES

1. Abramowitz, M., and Stegun, I. A. *Handbook of Mathematical Functions with Formulas Graphs and Mathematical Tables.* National Bureau of Standards Applied Mathematics Series 55, 1964.
2. Agin, D. Electroneutrality and electrodiffusion in the squid axon. *Proc. Natl. Acad. Sci. USA* 57: 1232–1238, 1967.
3. Andersen, O. S. Permeability properties of unmodified lipid bilayer membranes. In: *Membrane Transport in Biology,* edited by G. Giebisch, D. C. Tosteson, and H. H. Ussing. Berlin: Springer-Verlag, 1978, p. 369–446.
4. Andersen, O. S. Kinetics of ion movements mediated by carriers and channels. *Methods Enzymol.* 171: 62–112, 1989.
5. Andersen, O. S., A. Finkelstein, I. Katz, and A. Cass. Effect of phloretin on the permeability of thin lipid membranes. *J. Gen. Physiol.* 67: 749–771, 1976.
6. Andersen, O. S., and M. Fuchs. Potential energy barriers to ion transport within lipid bilayers: studies with tetraphenylborate. *Biophys. J.* 15: 795–830, 1975.
7. Anderson, J. L., and D. M. Malone. Mechanism of osmotic flow in porous membranes. *Biophys. J.* 14: 957–982, 1974.
8. Aronson, P. S. Kinetic properties of the plasma membrane Na$^+$-H$^+$ exchanger. *Annu. Rev. Physiol.* 47: 546–560, 1985.
9. Aronson, P. S., and P. Igarashi. Molecular properties and physiological roles of the renal Na$^+$-H$^+$ exchanger. *Curr. Top. Membr. Transp.* 26: 57–75, 1986.
10. Aveyard, R., and D. A. Haydon. *An Introduction to the Principles of Surface Chemistry.* Cambridge: Cambridge University Press, 1973.
11. Barr, L. Membrane potential profiles and the Goldman equation. *J. Theor. Biol.* 9: 351–359, 1965.
12. Barry, P. H., and J. M. Diamond. Effects of unstirred layers on membrane phenomena. *Physiol. Rev.* 64: 763–872, 1984.
13. Begenisich, T., and C. Smith. Multi-ion nature of potassium channels in squid axons. *Curr. Top. Membr. Transp.* 22: 353–369, 1984.
14. Benedek, G. B., and F. M. H. Villars. *Physics with Illustrative examples from Biology and Medicine.* Reading, MA: Addison-Wesley, 1979.
15. Berg, H. C. *Random Walks in Biology.* Princeton: Princeton University Press, 1983.
16. Blaustein, M. P. Sodium-calcium exchange in cardiac, smooth, and skeletal muscle: Key to contractility. *Curr. Top. Membr. Transp.* 34: 289–330, 1989.
17. Bockris, J. O., and A. K. N. Reddy. *Modern Electrochemistry.* New York: Plenum Press, 1970.
18. Cai, M., and P. C. Jordan. How does vestibule surface charge affect ion conduction and toxin binding in a sodium channel? *Biophys. J.* 57: 883–891, 1990.
19. Carslaw, H. S., and J. C. Jaeger. *Conduction of Heat in Solids.* London: Oxford University Press, 1959.
20. Chiu, S. W., and E. Jakobsson. Stochastic theory of singly occupied ion channels. II. Effects of access resistance and potential gradients extending into the bath. *Biophys. J.* 55: 147–157, 1989.
21. Cole, K. S. Electrodiffusion models for the membrane of squid giant axon. *Physiol. Rev.* 45: 340–379, 1965.
22. Colquhoun, D., and A. G. Hawkes. The principles of the sto-

chastic interpretation of ion-channel mechanisms. In: *Single Channel Recording,* edited by B. Sakmann and E. Neher. New York: Plenum, 1983.

23. Cooper, K., E. Jakobsson, and P. Wolynes. The theory of ion transport through membrane channels. *Prog. Biophys. Mol. Biol.* 46: 51–96, 1985.

24. Cooper, K. E., P. Y. Gates, and R. S. Eisenberg. Diffusion theory and discrete rate constants in ion permeation. *J. Membr. Biol.* 106: 95–105, 1988a.

25. Cooper, K. E., P. Y. Gates, and R. S. Eisenberg. Surmounting barriers in ionic channels. *Q. Rev. Biophys.* 21: 331–364, 1988b.

26. Cornish-Bowden, A., and C. W. Wharton. *Enzyme Kinetics.* Oxford: IRL Press, 1988.

27. Crank, J. *The Mathematics of Diffusion.* Oxford: Clarendon Press, 1975.

28. Dainty, J., and C. R. House. "Unstirred layers" in frog skin. *J. Physiol. (Lond.)* 182: 66–78, 1966.

29. Dani, J. A., and D. G. Levitt. Water transport and ion-water interaction in the gramicidin channel. *Biophys. J.* 35: 501–508, 1981.

30. Dani, J. A., and D. G. Levitt. Diffusion and kinetic approaches to describe permeation in ionic channels. *J. Theor. Biol.* 146: 289–301, 1990.

31. Diamond, J. M., G. Szabo, and Y. Katz. Theory of nonelectrolyte permeation in a generalized membrane. *J. Membr. Biol.* 17: 148–152, 1974.

32. Diamond, J. M., and E. M. Wright. Molecular forces governing non-electrolyte permeation through cell membranes. *Proc. R. Soc. Lond. [Biol.]* 172: 273–316, 1969.

33. DiFrancesco, D., and D. Noble. A model of cardiac electrical activity incorporating ionic pumps and concentration changes. *Philos. Trans. R. Soc. Lond. Biol.* 307: 353–98, 1985.

34. Eisenberg, D., and D. Carothers. *Physical Chemistry.* Menlo Park, CA.: Benjamin Cummings, 1979.

35. Ferreira, H. G., and M. W. Marshall. *The Biophysical Basis of Excitability.* New York: Cambridge University Press, 1985.

36. Finkelstein, A. *Water Movement Through Lipid Bilayer, Pores, and Plasma Membranes.* New York: John Wiley & Sons, 1987.

37. Finkelstein, A., and A. Cass. Permeability and electrical properties of thin lipid membranes. *J. Gen. Physiol.* 52: 145s, 1968.

38. Finkelstein, A. and A. Mauro. Physical principles and formalisms of electrical excitability. In: *Handbook of Physiology. The Nervous System,* edited by Eric R. Kandel, Baltimore, Maryland: Am. Physiol. Soc., sect 1, vol. I, 1977, p. 161–213.

39. Finkelstein, A., and P. A. Rosenberg. Single file transport: Implications for ion and water movement through gramidicin A channels. In: *Membrane Transport Processes,* edited by C. F. Stevens and R. W. Tsien. New York: Raven, 1979, p. 73–88.

40. Flewelling, R. F., and W. L. Hubbell. Hydrophobic ion interactions with membranes. *Biophys. J.* 49: 531–540, 1986a.

41. Flewelling, R. F., and W. L. Hubbell. The membrane dipole potential in a total membrane potential model. *Biophys. J.* 49: 541–552, 1986b.

42. Freedman, J. C., and J. F. Hoffman. Ionic and osmotic equilibria of human red blood cells treated with nystatin. *J. Gen. Physiol.* 74: 157–185, 1979.

43. Garrahan, P. J., and I. M. Glynn. Driving the sodium pump backwards to form adenosine triphosphate. *Nature* 211: 1414–1415, 1966.

44. Glasstone, S. *Thermodynamics for Chemists.* Huntington, NY: R. E. Krieger, 1972.

45. Glasstone, S., K. J. Laidler, and H. Eyring. *The Theory of Rate Processes.* New York: McGraw-Hill, 1941.

46. Glynn, I. M. Sodium and potassium movements in human red cells. *J. Physiol.* 34: 278–310, 1956.

47. Glynn, I. M. The action of cardiac glycosides on sodium and potassium movements in the human red cells. *J. Physiol.* 136(1): 148–173, 1957.

48. Goldman, D. Potential, impedance, and rectification in membranes. *J. Gen. Physiol.* 27: 37–60, 1944.

49. Hardt, S. L. The diffusion transit time: a simple derivation. *Bull. Math. Biol.* 41: 89–99, 1981.

50. Helfferich, F. *Ion Exchange.* New York: McGraw-Hill, 1962.

51. Hille, B. *Ionic Channels of Excitable Membranes.* Sunderland, MA: Sinauer Associates, 1992.

52. Hille, B., and W. Schwarz. Potassium channels as multi-ion single file pores. *J. Gen. Physiol.* 72: 409–442, 1978.

53. Hodgkin, A. L., and A. F. Huxley. A quantitative description of membrane current and its application to conduction and excitation in nerve. *J. Physiol.* 117: 500–544, 1952.

54. Hodgkin, A. L., and R. D. Keynes. The potassium permeability of a giant nerve fibre. *J. Physiol.* 128: 61–68, 1955.

55. Horowicz, P., P. W. Gage, and R. S. Eisenberg. The role of the electrochemical gradient in determining potassium fluxes in frog striated muscle. *J. Gen. Physiol.* 51: 193s–203s, 1968.

56. Jacquez, J. A. A general relation between membrane potential, ion activities, and pump fluxes for nonsymmetric cells in a steady state. *Math. Biosci.* 53: 53–57, 1981.

57. Jakobsson, E., and S. W. Chiu. Stochastic theory of ion movement in channels with single-ion occupancy. *Biophys. J.* 52: 33–45, 1987.

58. Jordan, P. C. Electrostatic modeling of ion pores. II. Effects attributable to the membrane dipole potential. *Biophys. J.* 41: 189–195, 1983.

59. Jordan, P. C. Effect of pore structure on energy barriers and applied voltage profiles. I. Symmetrical channels. *Biophys. J.* 45: 1091–1100, 1984a.

60. Jordan, P. C. Effect of pore structure on energy barriers and applied voltage profiles. II. Unsymmetrical channels. *Biophys. J.* 45: 1101–1107, 1984b.

61. Jordan, P. C. The total eletrostatic potential in a gramicidin channel. *J. Membr. Biol.* 78: 91–102, 1984c.

62. Jordan, P. C. Ion channel electrostatics and the shapes of channel proteins. In: *Ion Channel Reconstitution,* edited by C. Miller. New York: Plenum Press, 1986, p. 37–55.

63. Jordan, P. C. How pore mouth charge distributions alter the permeability of transmembrane ionic channels. *Biophys. J.* 51: 297–311, 1987.

64. Jordan, P. C., R. J. Bacquet, J. A. McCammon, and P. Tran. How electrolyte shielding influences the electrical potential in transmembrane ion channels. *Biophys. J.* 55: 1041–1052, 1989.

65. Katchalsky, A. and P. F. Curran. *Nonequilibrium Thermodynamics in Biophysics.* Cambridge, MA: Harvard University Press, 1965.

66. Katz, M. A., and E. H. Bresler. Osmosis. In: *Edema,* edited by N. C. Staub and A. E. Taylor, New York: Raven Press, 1984, p. 39–60.

67. Kedem, O., and A. Katchalsky. Thermodynamic analysis of the permeability of biological membranes to non-electrolytes. *Biochim. Biophys. Acta* 27: 229–246, 1958.

68. Kedem, O., and A. Katchalsky. A physical interpretation of the phenomenological coefficients of membrane permeability. *J. Gen. Physiol.* 45(1): 143–179, 1961.

69. Ketterer, B., B. Neumcke, and P. Lauger. Transport mechanism of hydrophobic ions through lipid bilayer membranes. *J. Membr. Biol.* 5: 225–245, 1971.

70. Kinne, R. K. H. Selectivity and direction: plasma membranes in

renal transport. *Am. J. Physiol.* 260 (*Renal Fluid Electrolyte Physiol.* 29): F153–F162, 1991.

71. Klingenberg, M. The ADP-ATP translocation in mitocondria, a membrane potential controlled transport. *J. Membr. Biol.* 56: 97–105, 1980.

72. Kohler, H. H., and K. Heckmann. Unidirectional fluxes in saturated single-file pores of biological and artificial membranes. I. Pores containing no more than one vacancy. *J. Theor. Biol.* 79: 381–401, 1979.

73. Kotyk, A., K. Janacek, and J. Koryta. *Biophysical Chemistry of Membrane Functions.* New York: John Wiley & Sons, 1988.

74. Latorre, R., P. Labarca, and D. Naranjo. Surface charge effects on ion conduction in ion channels. *Methods Enzymol.* 207: 471–501, 1992.

75. Lauger, P. Ion transport through pores: a rate theory analysis. *Biochim. Biophys. Acta* 311: 423–441, 1973.

76. Lauger, P. Diffusion-limited ion flow through pores. *Biochim. Biophys. Acta* 455: 493–509, 1976.

77. Lauger, P. Kinetic properties of ion carriers and channels. *J. Membr. Biol.* 57: 163–178, 1980.

78. Lauger, P. Conformational transitions of ionic channels. In: *Single Channel Recording,* edited by B. Sakmann and E. Neher. New York: Plenum Press, 1983.

79. Lauger, P. Dynamics of ion transport systems in membranes. *Physiol. Rev.* 67: 1296–1331, 1987.

80. Lauger, P. *Electrogenic Ion Pumps.* Sunderland, MA: Sinauer Associates, 1991.

81. Lauger, P., W. Stephan, and E. Frehland. Fluctuation of barrier structure in ionic channels. *Biochim. Biophys. Acta* 602: 167–180, 1980.

82. Leo, A. Hydrophobic parameter: measurement and calculation. *Methods Enzymol.* 202: 544–591, 1991.

83. Levie, R. Mathematical modeling of transport of lipid soluble ions and ion carrier complexes through lipid bilayer membranes. *Adv. Chem. Phys.* 37: 99–137, 1978.

84. Levie, R., and H. Moreira. Transport of ions of one kind through thin membranes. *J. Membr. Biol.* 9: 241–260, 1972.

85. Levitt, D. G. A new theory of transport for cell membrane pores. I. General theory and application to red cell. *Biochim. Biophys. Acta* 373: 115–131, 1974.

86. Levitt, D. G. Electrostatic calculations for an ion channel. I. Energy and potential profiles and interactions between ions. *Biophys. J.* 22: 209–219, 1978a.

87. Levitt, D. G. Electrostatic calculations for an ion channel. II. Kinetic behavior of the gramicidin A channel. *Biophys. J.* 22: 221–248, 1978b.

88. Levitt, D. G. Comparison of Nernst-Plank and reaction rate models for multiply occupied channels. *Biophys. J.* 37(3): 575–587, 1982.

89. Levitt, D. G. Kinetics of movement in narrow channels. *Curr. Topic. Membr. Transp.* 21: 182–197, 1984.

90. Levitt, D. G. Interpretation of biological ion channel flux data—Reaction-rate versus continuum theory. *Annu. Rev. Biophys. Biophys. Chem.* 15: 29–57, 1986.

91. Levitt, D. G. General continuum theory for multiion channel. I. Theory. *Biophys. J.* 59: 271–277, 1991a.

92. Levitt, D. G. General continuum theory for multiion channel. II. Application to acetylcholine channel. *Biophys. J.* 59: 278–288, 1991b.

93. Lew, V. L., H. G. Ferreira, and T. Moura. The behaviour of transporting epithelial cells. I. Computer analysis of a basic model. *Proc. R. Soc. Lond [Biol.]* 206: 53–83, 1979.

94. Lew, V. L., C. J. Freeman, O. E. Ortiz, and R. M. Bookchin. A mathematical model of the volume, pH, and ion content regulation in reticulocytes. Application to the pathophysiology of sickle cell dehydration. *J. Clin. Invest.* 87: 100–112, 1991.

95. Lieb, W. R. A kinetic approach to transport studies. In: *Red Cell Membranes—A Methodological Approach,* edited by J. C. Ellory and J. D. Young. New York: Academic Press, 1982, p. 135–154.

96. Lieb, W. R., and W. D. Stein. Non-Stokesian nature of transverse diffusion within human red cell membranes. *J. Membr. Biol.* 92: 111–119, 1986.

97. Lipschutz, S. *Schaum's Outline of Theory and Problems of Linear Algebra.* New York: McGraw-Hill, 1991.

98. Macey, R. I. Mathematical models of membrane transport processes. In: *Physiology of Membrane Disorders,* edited by T. E. Andreoli, J. F. Hoffman, D. D. Fanestil, and S. G. Schultz. New York: Plenum, 1986, p. 111–131.

99. Macey, R. I., and L. W. Yousef. Osmotic stability of red cells in renal circulation requires rapid urea transport. *Am. J. Physiol.* 254 (*Cell Physiol* 23): C699–C674, 1988.

100. Macgillivary, A. D., and D. Hare. Applicability of Goldman's constant field assumption to biological systems. *J. Theor. Biol.* 25: 113–126, 1969.

101. Mauro, A. Some properties of ionic and nonionic semipermeable membranes. *Circulation* 21: 845–854, 1960.

102. Mauro, A. The role of negative pressure in osmotic equilibrium and osmotic flow. In: *Water Transport across Epithelia, Alfred Benzon Symposium 15,* edited by H. H. Ussing, N. A. Lassen, and O. Sten Knudsen. Copenhagen: Munksgaard, 1981, p. 107–119.

103. McLaughlin, S. Electrostatic potentials at membrane-solution interfaces. *Curr. Top. Membr. Transp.* 9: 71–95, 1977.

104. McLaughlin, S. The electrostatic properties of membranes. *Annu. Rev. Biophys. Biophys. Chem.* 18: 113–136, 1989.

105. Moura, T. F., R. I. Macey, D. Y. Chien, D. M. Karan, and H. Santos. Thermodynamics of all-or-none water channel closure in red cells. *J. Membr. Biol.* 81: 105–111, 1984.

106. Mullins, L. J. An electrogenic saga: Consequences of sodium-calcium exchange in cardiac muscle. In: *Electrogenic Transport: Fundamental Principles of Physiological Implications,* edited by M. P. Blaustein and M. Lieberman. New York: Raven Press, 1984, p. 161–179.

107. Mullins, L. J., and K. Noda. The influence of sodium-free solutions on the membrane potential of frog muscle fibers. *J. Gen. Physiol.* 47: 117–132, 1963.

108. Naranjo, D., R. Latorre, D. Cherbavaz, P. McGill, and M. F. Schumaker. A simple model for surface charge on ion channel proteins. *Biophys. J.* 66: 59–70, 1994.

109. Neumcke, B., and P. Lauger. Nonlinear electrical effects in lipid bilayer membranes. II. Integration of the generalized Nernst-Planck equations. *Biophys. J.* 9: 1160–1170, 1969.

110. Noble, D., and G. Bett. Reconstructing the heart: a challenge for integrative physiology, *Cardiovasc. Res.* 27: 1701–1712, 1993.

111. Patlak, C. S. Contributions to the theory of active transport. *Bull. Math. Biophys.* 18: 271–315, 1956.

112. Patlak, C. S. Derivation of an equation for the diffusion potential. *Nature* 188(4754): 944–945, 1960.

113. Patlak, C. S., D. A. Goldstein, and J. F. Hoffman. The flow of solute and solvent across a two-membrane system. *J. Theor. Biol.* 5: 426–442, 1963.

114. Patlak, C. S., and S. I. Rapoport. Use of transient and steady-state measurements of the unidirectional flux ratio for the determination of the free energy change of chemical reactions and active transport systems. *Bull. Math. Biol.* 42: 529–537, 1980.

115. Purcell, E. M. *Electricity and Magnetism.* New York: McGraw-Hill, 1985.

116. Reif, F. *Fundamentals of Statistical and Thermal Physics.* New York: McGraw-Hill, 1965.

117. Robinson, R. A., and R. H. Stokes. *Electrolyte Solutions.* London: Butterworth, 1965.

118. Roux, B., and M. Karplus. Ion transport in a gramicidin-like channel: dynamics and mobility. *J. Phys. Chem.* 95: 4856–4868, 1991a.

119. Roux, B., and M. Karplus. Ion transport in a model gramicidin channel. Structure and thermodynamics. *Biophys. J.* 59: 961–981, 1991b.

120. Roux, B., and M. Karplus. Ion transport in the gramicidin channel: Free energy of the solvated right-handed dimer in a model membrane. *J. Am. Chem. Soc.* 115: 3250–3262, 1993.

121. Roux, B., and M. Karplus. Molecular dynamics simulations of the gramicidin channel. *Annu. Rev. Biophys. Biomol. Struct.* 23: 731–761, 1994.

122. Salhany, J. M. 1990. *Erythrocyte Band 3 Protein.* Boca Raton, FL: CRC Press, 1990.

123. Schultz, S. G. *Basic Principles of Membrane Transport.* London: Cambridge University Press, 1980.

124. Spangler, S. G. Expansion of the constant field equation to include both divalent and monovalent ions. *Alabama J. Med. Sci.* 9: 218–223, 1972.

125. Stein, W. D. *Transport and Diffusion across Cell Membranes.* San Diego: Academic Press, 1986.

126. Stein, W. D. *Channels, Carriers, and Pumps: An Introduction to Membrane Transport.* San Diego: Academic Press, 1990.

127. Sten-Knudsen, O. Passive transport processes. In: *Membrane Transport in Biology,* edited by G. Giebisch, D. C. Tosteson, and H. H. Ussing. Berlin: Springer-Verlag, 1978, p. 5–114.

128. Strieter, J., J. L. Stephenson, G. Giebisch, and A. M. Weinstein, A mathematical model of the rabbit cortical collecting tubule. *Am. J. Physiol.* 263 (*Renal Fluid Electrolyte Physiol.* 32): F1063–F1075, 1992.

129. Strieter, J., J. L. Stephenson, L. G. Palmer, and A. M. Weinstein. Volume-activated chloride permeability can mediate cell volume regulation in a mathematical model of a tight epithelium. *J. Gen. Physiol.* 96(2): 319–344, 1990.

130. Stryer, L. *Biochemistry.* New York: W. H. Freeman, 1988.

131. Teorell, T. Transport processes and electrical phenomena in ionic membranes. *Prog. Biophys. Biophys. Chem.* 3: 305–369, 1953.

132. Turner, R. J., and A. Moran. Futher studies of proximal tubular brush border membrane. D-glucose transport heterogeneity. *J. Membr. Biol.* 70: 37–45, 1982a.

133. Turner, R. J. and A. Moran. Heterogeneity of sodium dependent D-glucose transport sites along the proximal tubule: evidence from vesicle studies. *Am. J. Physiol* 242 (*Renal Fluid Electrolyte Physiol.* 11): F406–F414, 1982b.

134. Tyrrell, H. J. V., and K. R. Harris, *Diffusion in Liquids: A Theoretical and Experimental Study.* London, Butterworths, 1984.

135. Urban, B. W., and S. B. Hladky. Ion transport in the simplest single file pore. *Biochim. Biophys. Acta* 554: 410–429, 1979.

136. Ussing, H. H. Distinction by means of tracers between active transport and diffusion. *Acta Physiol. Scand.* 19: 43–56, 1949.

137. Wall, F. T. *Chemical Thermodynamics: A Course of Study.* San Francisco: W. H. Freeman, 1974.

138. Walter, A., and J. Gutknecht. Permeability of small nonelectrolytes through lipid bilayer membranes. *J. Membr. Biol.* 90: 207–217, 1986.

139. Weinstein, A. M. Chloride transport in a mathematical model of the rat proximal tubule. *Am. J. Physiol.* 263 (*Renal Fluid Electrolyte Physiol.* 32): F784–F798, 1992.

140. Wexler, A. S., R. E. Kalaba, and D. J. Marsh. Three-dimensional anatomy and renal concentrating mechanism. I. Modeling results. *Am. J. Physiol.* 260 (*Renal Fluid Electrolyte Physiol.* 29): F368–F383, 1991.

141. Wieth, J. O., O. S. Andersen, J. Brahm, P. J. Bjerrum, and J. R. Borders. Chloride-bicarbonate exchange in red blood cells: physiology of transport and chemical modification of binding sites. *Philos. Trans. R. Soc. Lond. Biol.* 299: 383–399, 1982.

142. Wright, E. M., and N. Bindslev. Thermodynamic analysis of nonelectrolyte permeation across the toad urinary bladder. *J. Membr. Biol.* 29: 289–312, 1976.

143. Wright, E. M., and J. M. Diamond. Patterns of nonelectrolyte permeability. *Proc. R. Soc. Lond. [Biol.]* 172: 227–271, 1969.

144. Wright, E. M., A. P. Smulders, and J. M. Tormey. The role of the lateral intracellular spaces and solute polarization effect in the passive flow of water across the rabbit gallbladder. *J. Membr. Biol.* 7: 198–219, 1972.

145. Young, H. D. *University Physics.* Reading, MA: Addison-Wesley, 1992.

146. Yousef, L. W. and R. I. Macey. A method to distinguish between pore and carrier kinetics applied to urea transport across the erythrocyte membrane. *Biochim. Biophys. Acta* 984: 281–288, 1989.

147. Zemansky, M. W., and R. H. Dittman, *Heat and Thermodynamics: An Intermediate Textbook.* New York: McGraw-Hill, 1981.

7. Membrane transport in single cells

MICHAEL L. JENNINGS | Department of Physiology and Biophysics, University of Arkansas for Medical Sciences, Little Rock, Arkansas

MARK A. MILANICK | Department of Physiology, University of Missouri School of Medicine, and Dalton Cardiovascular Research Center, Columbia, Missouri

THE STUDY OF SOLUTE TRANSPORT IN SINGLE CELLS is inherently simpler than in intact tissues (for example, muscle, epithelia, brain). Intact tissues in general consist of multiple cell types with differing transport characterics. In addition, unstirred layers or other trapped extracellular compartments can make analysis of transport data in intact tissues far more complex than in single cells. For these reasons, single cells have been used extensively in studies of transport across the plasma membrane. Our purpose is to summarize conceptual and experimental approaches to studying transport in single cells, illustrated by two specific transport systems (the Na^+, K^+-ATPase and the Cl^--HCO_3^- exchanger) that have been thoroughly characterized in single cells.

Our emphasis is on aspects of transport in single cells that are not covered in detail elsewhere in this volume. For example, the authors have attempted to minimize overlap between this chapter and the chapter by Reuss on epithelial transport, which covers some aspects of the sodium pump and Cl^--HCO_3^- exchanger. Single cells are widely used for voltage-clamp studies using microelectrodes and patch electrodes. This chapter contains reference to measurements of membrane potential but does not include the sizable literature on voltage-clamp studies of channels in individual cells.

Our main emphasis is on how single cells have been used for studies of the catalytic mechanism of individual transport systems, illustrated by the Na^+-K^+-ATPase and the Cl^--HCO_3^- exchanger. In addition to studies of mechanisms of transport, single cells have also been used extensively for the study of the regulation of transport. As an illustration of approaches to the study of transport regulation, we include a discussion of cell-volume regulation in single cells.

EXAMPLES OF PREPARATIONS OF SINGLE CELLS

Large Cells that Can Be Studied Individually

Figure 7.1 shows some examples of preparations of isolated cells that have been used for transport studies. Very few cells are sufficiently large that transport measurements can be made in a single individual cell by tracer methods. In order for a single cell to be large enough for a transport measurement, the cell must contain a quantity of the solute of interest that is in the analytically available range. For example, a single cell of volume $1000 \ \mu^3$ (that is, a cuboid cell that is $10 \ \mu$ on each side) contains about 0.1 pmol K^+ and much smaller amounts of all other solutes. Even with radionuclides, these amounts are very hard to detect. For most cells, however, solute transport is measured in a large number of functionally identical cells in suspension or attached to a culture dish.

A few cells are large enough to be used individually for transport measurements. The *Xenopus laevis* oocyte has a diameter of about $1000 \ \mu m$ and a volume of $5 \times 10^8 \ \mu^3$ (0.5 μl), that is, a factor of 10^5–10^6 larger than most mammalian cells. The *Xenopus* oocyte can be voltage clamped quite readily and has been used for numerous electrical measurements of expressed ion channels (231). Measurement of electric current in *Xenopus* oocytes can also be used to quantify coupled transport processes that result in net charge movement (180, 200). In addition, single oocytes have been used to measure tracer efflux of $^{36}Cl^-$; the *Xenopus* oocyte is sufficiently large that the radioactivity in a single cell can be monitored continuously (171). Water transport in single oocytes has been measured by optical determination of oocyte swelling (194).

The barnacle giant muscle fiber and squid or *Myxicola* giant axons are large enough to be internally dialyzed, which makes it possible not only to measure tracer fluxes but also to clamp the cytoplasm to a desired solute (salt, pH, metabolite) composition. Methods for internal dialysis of these preparations have been described by Brinley and Mullins (20), Russell (211), Russell and Brodwick (212), and Abercrombie and Sjodin (2). Either influx or efflux can be measured. There are other examples of large single cells that can be used for transport studies, including epithelial cells that have been fused to form a large internal compartment (176). The principles governing the use of these cells are the same as those for other large cells.

Blood Cells

Blood is a rich source of free-living single cells that are well-suited for transport experiments. In terms of mass, the predominant blood cell is of course the erythrocyte, which, because of its abundance and simplicity, has been used more extensively for transport studies than any other cell. The mature mammalian red cell has only one membrane and therefore only one internal compartment, thus simplifying kinetic analysis of transport data. Red blood cells can be prepared from whole blood by centrifugation, with aspiration of white cells from the top of the red cell pellet. For most transport applications, it is not necessary to use more elaborate methods to completely remove white cells (for example, alpha cellulose or density centrifugation), because the internal volume of contaminating white cells is an extremely small fraction of the volume of red cells, and most white cells are removed by aspiration.

Lymphocytes and neutrophils can be prepared from whole blood by differential centrifugation on Ficoll-Hypaque (17). Once isolated, neutrophils and lymphocytes can be used for transport studies by methods similar to those used for red cells (for example, 223, 232). Platelets can also be separated from whole blood by differential centrifugation and used in transport studies.

Cells Prepared by Dissociating Intact Tissue

Numerous methods have been published for isolating cells from intact tissues. The methods for preparing dissociated cells in general include some combination of enzymatic treatment and mechanical disruption. The major considerations in these kinds of isolations are to preserve the characteristics of the original cell as closely as possible. The criteria that have been used to evaluate the quality of a preparation of dissociated cells include

FIG. 7.1. Some examples of preparations of individual cells that have been used extensively for transport studies. *A.* Cardiac myocytes from guinea pig. *B.* Endothelial cells from cow. *C.* Enterocytes from chick. *D.* Oocytes from Xenopus. *E.* Erythrocytes from human. *F.* Smooth muscle cells from cow.

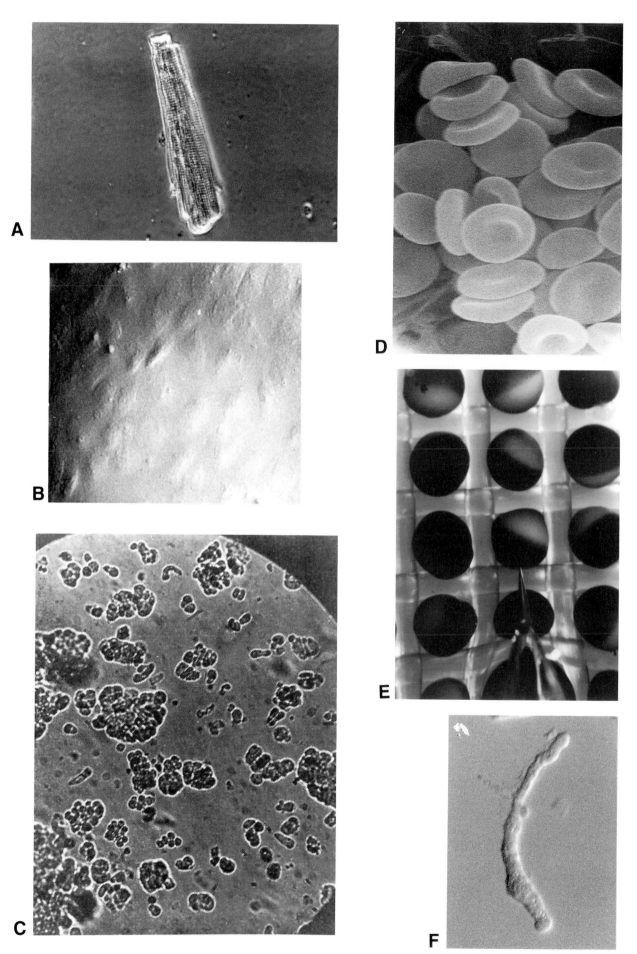

A

B

C

D

E

F

263

morphology, membrane resistance, immunocytochemical markers, contractility, and response to hormonal stimulation. It is beyond the scope of this discussion to review methods for preparing cells from various organs. Examples of established methods for preparing dissociated cells include those for preparing enterocytes (126), muscle cells (152, 210, 238), tubule cells from defined nephron segments (127), endothelial cell (239), hepatocytes (229), and adipocytes (207).

Permanent Cell Lines and Expression Systems

There are many examples of primary cell cultures and permanent cell lines that have been used extensively in transport studies (for example, 71, 113, 151, 160). Most of the cell lines were derived either from a tumor or from tissue of a fetal or very young animal. Some cell lines are grown in suspension, some attached to solid supports (culture dish, coverslip, microcarrier beads, porous filter), and some are propagated in an animal host. Mammalian cells are often used following transfection of the cDNA encoding a transporter of interest (for example, 114, 148). Microorganisms such as bacteria (154, 208) and *Neurospora* (235) are single cells that are well-suited for transport measurements. Yeast is becoming more commonly used for functional expression of mammalian transport proteins (101). Our emphasis is on studies of intact cells rather than organelles or isolated plasma membrane vesicles. However, the principles that apply to the study of transport in single cells can also be used to study transport in organelles such as mitochondria or isolated plasma-membrane vesicles.

METHODS FOR MEASURING TRANSPORT

Tracer Influx and Efflux: Definition of Flux

The application of radionuclides to the study of membrane transport has a long history. Theoretical considerations are described by Dawson in the *Handbook of Physiology, Gastrointestinal Physiology,* and by Macey and Mara in this edition of the *Handbook* (Chapter 6). The interested reader should consult these sources for mathematical derivations of the equations governing the use of radioactive tracers in membrane-transport experiments. The present treatment is intended to provide an intuitive understanding that should be sufficient to allow the design and interpretation of simple quantitative experiments, particularly as applied to single cells.

The same principles apply to individual single cells (for example, *Xenopus* oocyte or barnacle muscle fiber)

or an ensemble of functionally identical cells (for example, fibroblasts, erythrocytes, adipocytes, yeast, or bacteria). The following introduction is intended to provide definitions of unidirectional flux and net flux, with only minimal reference to mechanisms of particular transporters. Later in this chapter, the use of tracer flux measurements to analyze transport mechanisms is discussed in connection with specific transport systems: the Na^+-K^+-ATPase and the Cl^--HCO_3^- exchanger.

Populations of single cells are often considered to consist of two well-stirred compartments (extracellular and intracellular) separated by a single barrier. The intracellular compartment is assumed to be well-stirred simply because diffusional equilibration within the cell is usually rapid compared with transport across the cell membrane. This assumption can be an oversimplification if there are slowly exchanging compartments within the cell or if the solute in question is transported extremely rapidly across the plasma membrane (see later under Unstirred Layers). For the present purpose, it is assumed that the intracellular compartment is homogeneous. The effects of unstirred layers on transport measurements are discussed briefly later in this chapter.

Consider first a solute S that is neither consumed nor produced by the cell and is distributed in a steady state, that is, the amounts of intracellular and extracellular solute are constant over the time of the experiment. By definition, there is no overall net flux. There are, however, unidirectional fluxes. The unidirectional influx is defined as the quantity of S (moles) that crosses 1 cm^2 of the membrane in the inward direction per unit time. The unidirectional efflux is defined in the same way, and the net flux (same units) is the difference between unidirectional influx and efflux. The most straightforward dimensional units for flux are moles/area-time. In practice, many different units are used, all of which could in principle be converted to moles/area-time. Some examples of flux units are nmol/10^6 cells-min; nmol/mg protein-min; μmol/ml cells-min; and μmol/cm^2-min.

If the solute of interest is initially only on one side of the membrane, then the initial unidirectional flux is equal to the initial net flux, because the initial unidirectional flux in the opposite direction is zero. If the solute of interest is on both sides of the membrane, the only way to measure a unidirectional flux is with a tracer. The tracer is often a radioisotope of the solute of interest, but one element can sometimes be used as an appropriate tracer for another (for example, rubidium as a tracer for potassium for studies of the Na^+-K^+-ATPase). In a tracer flux experiment it is assumed that the tracer is chemically indistinguishable from the bulk solute, that is, the movement of the tracer is representative of that of the whole population of solute molecules or ions.

For example, in an influx experiment, if extracellular tracer initially represents one part in 10^6 of the solute, and if 100 molecules of tracer are found to cross 1 cm^2 of membrane per minute, then the unidirectional influx of the solute is 100×10^6 molecules per cm^2 per minute. The number of molecules of tracer is measured as the number of disintegrations per minute (DPM). Therefore, in order to measure the initial unidirectional influx, it is necessary to know the extracellular specific activity, that is, the disintegrations per minute (DPM) per mole of solute. The flux is then calculated from the initial increase in intracellular radioactivity (for example, DPM/mg protein-min), the tracer influx, divided by the initial extracellular specific activity (DPM/nmol).

The influx measured in these units can be compared among different laboratories and among different cell preparations. Unfortunately, authors sometimes express transport data in units that are difficult or impossible to compare with those from other laboratories. For example, suppose the initial influx of an amino acid into a cultured fibroblast is expressed as CPM per culture dish per minute. This information cannot be compared with that from other laboratories without knowing other information (CPM/nmol, cells/dish, area/cell) that would make it possible to convert the data into flux units. Data of this kind are useful mainly for determining the effect of an intervention (for example, hormone, inhibitor) in a single experiment. Transport data are of much more lasting value when presented in units that can easily be converted into a genuine flux (amount/area-time).

Unidirectional Influx. A unidirectional influx of S can be measured as follows. At $t = 0$, a radioactive form of solute S is added to the extracellular medium, and, at various subsequent times, an aliquot of suspension is removed, and the intracellular radioactivity is determined after separation of cells from extracellular medium. The methods for separating cells from medium are diverse and include centrifugation and washes in cold medium; centrifugation through a nonaqueous layer (oil or butylphthalate); or washes of cells on filters. The criteria for adequate separation are the following. There should be no movement of tracer in either direction during the separation. No cells should be lysed or otherwise lost. If cells are lost, there should be some way (protein determination or cell count) to determine how many cells are lost. Finally, the extracellular radioactivity should be removed, that is, the residual extracellular radioactivity should not contribute significantly to the measured intracellular

activity. If complete removal of extracellular radioactivity is not possible, an estimate of residual extracellular space can be made by including a separate tracer that does not enter the cells (for example, inulin or sucrose).

After separating cells from extracellular medium, the cell-associated radioactivity is determined. The initial rate of increase in intracellular radioactivity is then used to calculate the unidirectional influx J_{oi} (nmol/mg protein-min):

$$J_{oi} = \{d[\text{DPM}_i/\text{mg protein}]/ dt\}/[\text{Extracellular DPM/nmol}] \quad (7-1)$$

The important point is that, in order for an influx measurement to be meaningful, the extracellular specific activity must be known (in order to convert DPM to moles), the initial rate should be measured, and the influx must refer to some specified number of cells (measured as cell number, mg protein, ml cells, or cm^2 surface area).

If the sampling rate is not sufficiently rapid to measure true initial influx, then it is necessary to correct for tracer "backflux," which becomes increasingly important as the intracellular specific activity increases. Influx data can be corrected for backflux by fitting the influx data to an exponential time course. Under conditions of no net flux, the time course of intracellular radioactivity has the following form (see, for example, Lepke and Passow [150]):

$$y(t) - y(0) = [y(\infty) - y(0)][1 - e^{-kt}] \quad (7-2)$$

where $y(t)$, $y(0)$, and $y(\infty)$ are respectively the intracellular radioactivity at time t, at $t = 0$, and at tracer equilibrium, that is, when the intracellular specific activity equals the extracellular specific activity. The rate constant k (min^{-1}) is the rate constant for tracer exchange. Under most conditions the size of the intracellular compartment is much smaller than the size of the extracellular compartment, implying that the final specific activity is essentially the same as the initial extracellular specific activity. Under these conditions the unidirectional flux is equal to the rate constant k times the intracellular content of the solute of interest. Note that in the limit of short time points, $e^{-kt} = 1 - kt$, and the time course of tracer influx is approximately linear—that is, Equation 7–2—for short time periods, is approximated by Equation 7–1.

Unidirectional Efflux. Unidirectional efflux can be measured by preloading cells with a fixed amount of radioactive tracer, washing off the extracellular tracer, and then measuring the rate of loss of tracer from cells into a medium that is initially free of radioactivity. The simplest way to do this is to centrifuge the cells periodically and measure the radioactivity in a volume of supernatant. It is also possible to measure tracer efflux by washing off the extracellular medium and measuring cell-associated radioactivity, but, as a practical matter it is more convenient to measure efflux as an increase in extracellular radioactivity rather than a decrease in intracellular radioactivity. If time points must be taken rapidly, inhibitor stop or filtration methods are used (for example, 84, 136). Very rapid sampling with a continuous-flow filtration tube allows effluxes with half-times less than 0.1 sec to be measured (18).

As is true of influx, the time course of tracer efflux under conditions of no net solute flux is a single exponential (31) described by Equation 7–2. Under conditions in which the size of the intracellular compartment is much smaller than the extracellular compartment, the unidirectional efflux is calculated as the rate constant (min^{-1}) times the intracellular solute content (nmol/10^6 cells, nmol/mg protein, umol/ml cells, or any other units that specify an amount of solute per amount of cells). To determine the rate constant k for efflux, the extracellular radioactivity is compared with the radioactivity at infinite time (determined either by incubation for several half-times of tracer exchange or by lysing the cells with trichloroacetic acid, detergent, or strong base).

Under conditions of no net flux, the time course of tracer influx and efflux is always a single exponential, as long as there is a single well-stirred extracellular compartment and the intracellular compartment consists of a single cell or a homogeneous population of cells. The single exponential time course of tracer flux has no implication regarding the number of transport pathways that mediate the flux of a given solute. There could be a single pathway or several different transport pathways contributing to the observed tracer flux; when there is no net flux, the time course of tracer exchange is a single exponential, no matter how many different transport pathways are operative.

Example of a Tracer Exchange Experiment. The following example illustrates the relationships among unidirectional influx, unidirectional efflux, concentration, and specific activity (Fig. 7.2). Consider a suspension of cells (0.1% cytocrit) in a steady state with a particular extracellular medium. The low cytocrit is chosen here so that the total amount of intracellular solute is much smaller than the total amount of extracellular solute. Under these conditions, the extracellular specific activity (DPM/nmol) of solute does not change significantly during the influx. The only parameter that changes is the intracellular specific activity. The intracellular K^+ concentration is 120 mM, which in these cells corresponds to 500 nmol K^+/mg protein. The extracellular

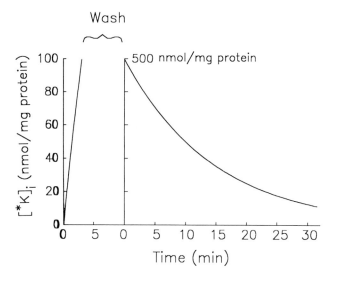

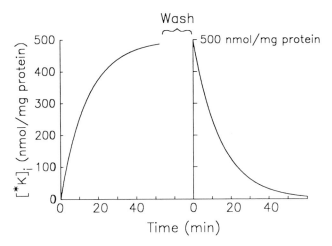

Total cellular K^+ = 500 nmol/mg protein

Initial *K influx = 35 nmol/mg protein—min

Influx rate constant = 0.07/min

Initial *K efflux = 35 nmol/mg protein—min

Efflux rate constant = 0.07/min

FIG. 7.2. Unidirectional influx and efflux under conditions of no net flux. *Lower graph:* Tracer is allowed to equilibrate before efflux is measured. *Upper graph:* Tracer influx reaches only 20% of the extracellular specific activity before the cells are washed to begin efflux measurement. In both cases, unidirectional influx and efflux are equal. The tracer influx appears to be more rapid than efflux in the lower part of the figure because the initial intracellular specific activity (DPM/nmol) is much less than the original extracellular specific activity in the influx experiment. Therefore, each DPM that crosses the membrane at the beginning of the efflux experiment represents more K^+ ions than in the beginning of the influx experiment. See text for further discussion.

K^+ concentration is 5 mM. At $t = 0$, $^{42}K^+$ is added to the medium, and aliquots are removed periodically for determination of intracellular radioactivity. Over the first ten minutes, the increase of intracellular radioactivity with time is nearly linear, and a unidirectional influx of 35 nmol/mg protein-min is calculated.

Alternatively, the time course could be measured for times sufficient to reach tracer equilibrium (equal specific activities) between the cells and medium, in which case a rate constant of 0.07/min is measured. From this rate constant, and the intracellular contents (500 nmol/mg protein), an efflux of 35 nmol/mg protein is calculated, in agreement with that determined from the initial influx measurement.

It is important to understand that tracer equilibrium, that is, the situation at $t >> 60$ min in Figure 7.2, means simply that the intracellular and extracellular specific activities (DPM/nmol) are equal. The intracellular tracer concentration (DPM/μl cell water) is much higher than the extracellular concentration (DPM/μl extracellular water). At tracer equilibrium there will be the same gradient of tracer as there is of nonradioactive K^+. The K^+ gradient is maintained by a balance of active inward transport by the Na^+,K^+-ATPase and downhill outward transport by any of a number of possible channels or transporters (later under Pump-Leak Paradigm).

It is instructive to consider an experiment in which the tracer K^+ influx is measured for a period of time, after which the tracer is removed from the extracellular medium and the efflux of tracer from the cells is measured. Consider first an experiment in which the tracer K^+ influx is allowed to continue until the intracellular and extracellular specific activities (DPM/nmol) are equal. The time course of influx will be a single exponential with a characteristic rate constant as defined in Equation 7–2. If the cells are then washed and the efflux measured, the efflux will also be a single exponential, with the same rate constant as that for influx (Fig. 7.2). In other words, as long as cells and medium are in a steady state with constant intracellular and extracellular concentrations, then the rate constants for influx and efflux are equal, as are the unidirectional fluxes. The unidirectional influx and efflux must be equal, because there is no net flux.

Consider now the same experiment, but now the influx is measured for a shorter time, so that the intracellular specific activity reaches only 20% that of the extracellular solute (Fig. 7.2). Again, the cells are washed without loss of radioactivity, and the time course of efflux is measured. In this case, it takes much longer for the radioactivity to leave the cells than it took to load them initially. This kind of result could be incorrectly interpreted as evidence that the unidirec-

tional influx is much larger than the unidirectional efflux. In fact, the unidirectional influx and efflux are still equal to each other.

The reason that the tracer efflux appears to be slower than influx is that the *intracellular* specific activity at the beginning of the efflux is much smaller than the extracellular specific activity had been at the beginning of influx, because the cells and medium had not reached tracer equilibrium before the efflux. In other words, each tracer ion during the efflux experiment represents 5 times more K^+ ions than had been true during influx. The point of this example is that the number of radioactive molecules or ions that cross a membrane per unit time can be interpreted quantitatively only if it is known how many total ions or molecules each DPM represents. As a practical matter, it is perfectly valid to measure a unidirectional efflux without insisting that the cells are initially loaded to the same specific activity as the external medium. As long as the rate constant for efflux and the total cellular contents of the solute are known, a unidirectional efflux can be calculated.

Unidirectional Flux under Conditions of Net Flux. The above example is relatively simple because the concentrations of solute inside and outside the cell are constant, and the unidirectional influx must equal the unidirectional efflux. The same definitions of unidirectional influx and efflux apply when there is a net flux, but the situation is more complicated because the chemical concentration of solute changes with time. Therefore, it is crucial to measure true initial fluxes.

The simplest example of a tracer flux measurement under conditions of net flux is when the substrate (tracer and bulk) is initially only present on one side of the membrane. Under these conditions the unidirectional flux must equal the net flux, because there is initially no flux in the opposite direction. In this kind of experiment the specific activity is the same on both sides of the membrane and does not change throughout the experiment. Therefore, the radioactivity is a measure of the chemical amount of the substrate, as long as the specific activity is known. Under conditions of net flux, the time course of tracer movement may not be an exponential, because the flux may be a saturable function of substrate concentration. For example, the net efflux of glucose from a human red blood cell into a glucose-free medium is constant for more than half the total time course (230). The same is true of the efflux of Cl^- from red cells suspended in an $SO_4^=$ medium (107). In both cases the efflux is not strongly dependent on the intracellular substrate concentration, because some other step in the catalytic cycle limits the overall transport rate.

Definition of Permeability Coefficient

The permeability coefficient (cm/s) is a useful parameter for comparing the transport of various solutes in different membranes. The permeability coefficient was originally defined for solutes that cross membranes by simple diffusion. For an uncharged substance that is distributed in equal concentrations on both sides of the membrane,

$$P \ (\text{cm/s}) = J \ (\mu\text{mol/cm}^2 - \text{s})/[S](\mu\text{mol/cm}^3) \quad (7\text{–}3)$$

where J is the unidirectional influx or efflux (they are equal under these conditions), and $[S]$ is the substrate concentration. In a net flux experiment, the permeability coefficient is defined similarly:

$$P \ (\text{cm/sec}) = J \ (\mu\text{mol/cm}^2 - \text{sec})/ \\ ([S]_i - [S]_o)(\mu\text{mol/cm}^3) \quad (7\text{–}4)$$

where J is now the net efflux.

Unstirred Layers

As mentioned earlier, one of the attractions of studying transport in single cells is that unstirred layers are much less of a problem than in intact tissue. Nonetheless, the issue of unstirred layers can arise in tracer flux measurements with single cells. The problem is most significant when solute permeability is high and diffusion through an unstirred layer immediately outside the cell is a significant barrier compared with transport through the membrane. The approximate effect of an unstirred layer on the apparent permeability coefficient is as follows (see, for example, 251):

$$[P_{\text{app}}]^{-1} = [P_{\text{real}}]^{-1} + \delta/D \quad (7\text{–}5)$$

where P_{app} (cm/s) is the apparent (measured) permeability coefficient, P_{real} (cm/s) is the actual permeability coefficient, δ is the thickness (cm) of the unstirred layer, and D is the diffusion coefficient (cm^2/s) of the solute in the extracellular medium. For example, if there is an unstirred layer of 10 μ (10^{-3} cm), and if the diffusion coefficient of the solute is 10^{-6} cm^2/s, a real permeability coefficient of 10^{-4} cm/s will be overestimated by 10%. The permeability coefficients for most solutes through most membranes are smaller than this, and unstirred layers are not a major problem for studies of suspensions of cells.

Multiple Intracellular Compartments

In very large cells such as the *Xenopus* oocyte, intracellular unstirred layers can in principle cause errors in measurements of permeability, because the thickness of the cytoplasmic unstirred layer can be much larger

than 10 μ. An intracellular unstirred layer has the same effect as an extracellular unstirred layer: It causes an underestimate of the permeability of the membrane itself because the unstirred layer constitutes a significant barrier. Again, these errors are most significant for rapidly transported solutes.

For most solute transport studies in "small" cells, that is, volume of 1000 μ^3 or less and shortest linear dimensions of 10 μ or less, the intracellular space is considered a single well-stirred compartment. Cells of course have organelles, and some solutes are not distributed uniformly between cytoplasm and various organelles. The best-characterized example is calcium, which is sequestered in endoplasmic reticulum of many cells and sarcoplasmic reticulum of muscle cells. It is important to consider the possible effects of sequestration in organelles when interpreting transport data in intact cells.

Optical Methods for Measuring Transport

In addition to the commonly used tracer methods, it is also possible to use optical methods to measure changes in intracellular concentrations of various substances. For example, either absorbance or fluorescence indicators are commonly used to measure the time course of changes in intracellular free Cl$^-$ (252), Ca^{2+} (255), and pH (241). The absorbance or fluorescence of the substrate itself has been used to study the function of the red blood cell anion transporter (32, 174) and the multiple drug resistance P-glycoprotein (3). Another optical method for studying transport in suspensions of cells is light scattering. Light scattering has been used for many years as a continuous measure of cell volume during net transport experiments in single cells (67, 129, 230). The principle of the method is that light scattering depends on the size of the cells as well as the refractive index difference between cells and medium. During net solute transport, if the solute concentration is high (>50 mM), the cell volume and refractive index difference will change with time as the solute is transported, and light scattering will allow an estimate of the time course of transport. Another way to measure volume changes during the net movement of solute and water is by image analysis.

Ion-Sensitive Microelectrodes

In relatively large cells such as those of *Necturus* epithelia (4, 256), frog muscle (1), sheep Purkinje fibers (250), or snail neurons (241, 243), it is possible to measure net ion transport with cytoplasmic ion-sensitive electrodes. Ion-sensitive electrodes can also be used to provide a rapid measure of cell volume. Cells are loaded with a quaternary ammonium ion (tetramethylammonium; TMA), and an ion-selective electrode monitors changes in concentration (activity) of the ion (206). A rapid influx or efflux of water (caused by an osmotic gradient) causes dilution or concentration of the TMA.

Membrane Potential

Quantitative interpretation of ion-transport measurements requires knowledge of the membrane potential (although some ion-transport processes are not strongly dependent on potential, as discussed later under ANION EXCHANGER: Electrical Properties of the Exchange Pathway and under SODIUM PUMP: Electrical Properties of the Na Pump). There are several methods for measuring and/or controlling the membrane potential. The most direct method is to make direct electrical contact with cytoplasm with microelectrodes or patch electrodes. Ion channels are of course commonly studied in this way, but it is also possible to study pumps, exchangers, and cotransporters with the membrane potential clamped with microelectrodes. This approach is especially useful if the transporter or pump mediates net current flow, so that electric current can be used as a measure of the ion fluxes through the transporter. Examples of transporters that have been characterized by direct electrical measurements include the Na^+, K^+-ATPase (38), the Na^+-Ca^{2+} exchanger (159), and the Na^+-glucose cotransporter (180). Electric current is most often used as the measure of ion flux in cells that are voltage-clamped, but it is also possible to measure tracer fluxes in voltage-clamped squid axons (66, 198) or *Xenopus oocytes* (82). In many cells it is impractical to use microelectrodes to manipulate the membrane potential and simultaneously measure solute transport, because most individual cells are too small for tracer measurements on a single cell, and some transporters are electroneutral and therefore cannot be assayed as a net current. In suspensions of cells the membrane potential can be estimated by optical methods, using potential-sensitive dyes (96, 140). It is sometimes possible to manipulate the membrane potential with ion gradients and/or ionophores like valinomycin and, in the same cells, measure the flux of a solute mediated by a particular transporter. This is possible only if the method used to vary the membrane potential does not otherwise interfere with the flux measurement. For example, it would not be useful to vary the membrane potential with a K^+ ionophore and measure the effect on the K^+ flux through a specific transporter, because the K^+ flux mediated by valinomycin would contribute too much to the total K^+ flux. Often, however, it is possible to use ion gradients and ionophores to vary

the membrane potential without interfering with the flux measurement (for example, 111, 165).

Nuclear Magnetic Resonance

Nuclear magnetic resonance has been used to measure solute transport in suspensions of cells. The basis for most NMR methods for measuring transport is that a chemical shift reagent can be confined to one side of the membrane, and the intracellular pool can be distinguished from the extracellular pool (45). NMR can also be used to measure the net influx of a solute (for example, ^{31}P PO_4) in samples that have been separated from the extracellular medium (137).

TRANSPORT MECHANISMS

Rationale for Investigating Transport Mechanisms

This chapter describes methods for investigating transport mechanisms in single cells. In this context the term *mechanism* refers to a specific sequence of elementary rate processes, for example, binding, release, and unimolecular conformational changes. Each of these events represents the transition from one physical structure to another, for example, a free intracellular substrate combining with an empty intracellular site to become an inward-facing protein-substrate complex. There are several motivations for determining the mechanism of a transport process: *(1)* The kinetic mechanism can place physical constraints on the transporter. For example, if the anion transport system exchanges Cl^- for Cl^- with a 1:1 stoichiometry with ping-pong kinetics (15, 57), there is no conformation of the protein in which two Cl^- ions are bound at the same time to transport sites. *(2)* A quantitative understanding of the kinetic mechanism allows one to model the response of a cell to a given set of conditions. For example, the function of cardiac muscle depends on how ion fluxes through the Na^+/Ca^{2+} exchanger respond to changes in membrane potential and ion concentrations. *(3)* An understanding of kinetic mechanism can lead to insights regarding the ways in which transport is regulated. The substrate itself is a regulator of some transporters, for example, intracellular Cl^- inhibits the $Na^+/K^+/Cl^-$ cotransporter (19); recognition of this kind of regulation arose from the study of the kinetic mechanism of the transporter.

In trying to predict the contribution of a particular transporter to the integrative behavior of a living cell, tissue, or organism, a genuine mechanistic understanding of the transporter is far more useful than a limited set of empirical relationships. Such an understanding

TABLE 7.1. *Properties of the Na Pump and the Anion Exchanger*

Property	Anion Exchanger, AE1	Na Pump
Current kinetic model	Ping-pong	Ping-pong
Stoichiometry	1:1	3Na:2K
Modes	Self-exchange (Cl:Cl)	3 Na efflux:2 K influx
	Heteroexchange (Cl:HCO3)	3 Na efflux:3 Na influx
	Cl: (1 SO4 = +1 H+)	3 Na efflux:2 Na influx
		3 na efflux: no influx
		2 K efflux:3 Na influx
		2 K efflux:2 K influx
		2 K efflux: no influx
Substrates	Most small inorganic anions	Efflux: H, Li, Na
	Some organic anions	Influx: H, Li, Na, K, Rb, Cs, NH4+
Electrically silent?	Yes, normally	No; 3 Na efflux/2 K influx
		Yes; 3 Na efflux/3 Na influx
Electrogenic	When chemically modified	3 Na efflux/2 K influx and some other modes
Em effects on	Slight on Cl/Cl On sulfate binding	Under many, but not all conditions
	On modified band 3	
ATP as energy source	No	Yes
Covalent modification during catalytic cycle	No	Yes

has not been achieved for any transporter. However, significant progress has been made in the understanding of transport mechanisms, largely as a result of studying individual cells or populations of identical cells. Dozens of different transport processes are known to exist in cells, each catalyzed by a different membrane protein. In addition, the diffusional movement of some solutes across the lipid bilayer is physiologically significant. It is not the purpose of this chapter to present a catalogue of different transport systems. Instead, two examples of well-studied transporters are presented in detail. The goal of these examples is to illustrate the kind of mechanistic insights that can be attained from a quantitative analysis of transport data in single cells. The two transporters to be considered are an electrogenic active transporter (the Na$^+$,K$^+$-ATPase), and an electroneutral obligatory ion exchanger (red blood cell Cl$^-$-HCO$_3^-$ exchanger). Some properties are summarized in Table 7.1. For these two well-characterized transporters, the same general themes are discussed: the catalytic cycle, electrical properties, and structure-function relations.

Catalytic Cycles

Transport proteins, even those that are relatively abundant, are always present in catalytic amounts. That is,

the number of moles of transporter in a cell is in general much smaller than the number of moles of substrate to be transported. In order for the transporter to move a significant amount of substrate, it must operate in a repetitive catalytic cycle. A central goal of the kinetic analysis of transport proteins is to define what the catalytic cycle is for a given transporter, estimate the rates of the elementary processes, and, ultimately, correlate elementary events with physical conformations of the transport protein (134, 135).

In transporter proteins having a single substrate (simple carriers), there are only a small number of possible catalytic cycles. For example, net influx consists of external binding, inward translocation, internal release, and return of the transporter to a state in which it can once again bind external substrate. Inward translocation may not be a single step. There may or may not be a large conformational change associated with substrate translocation, but, if there is, the conformational change must be reversed in the absence of substrate in order for net substrate movement to occur.

Many transporters have multiple substrates that move in either the same or opposite directions. For these, there are many more possible types of catalytic cycle than for single-substrate transporters (222). For multiple-substrate transporters, the kinetics are often described in terminology developed for enzymes having multiple substrates (25).

An exchange of ions could take place by a variety of mechanisms. In a ping-pong mechanism the two exchanging ions take turns crossing the membrane. A complete catalytic cycle consists of external binding of a substrate anion, inward translocation, and internal release, followed by internal binding of the other substrate, outward translocation, and external release (Fig. 7.3A). In a ping-pong mechanism, the tight coupling between influx and efflux is a consequence of the requirement that the transport protein be occupied with a transportable substrate at the substrate binding site in order for the inward or outward translocation event to take place. That is, there is no "slippage," which is a term used to describe a translocation event in the absence of bound substrate; such an event, if it takes place at all, is orders of magnitude slower than in the presence of Cl^- or HCO_3^-. If there is no *trans* substrate (on the opposite side of the membrane from the labeled substrate) that can be transported, then the catalytic cycle will not be completed and therefore does not repeat rapidly. This is a major difference between a coupled transport system and a simple carrier in which a single substrate can be transported repeatedly in the same direction without anything else being transported.

Another possible mechanism for ion exchange involves a complex among the transport protein and both internal and external substrates. The terminology describing such mechanisms is confusing. The term *sequential* is sometimes used, because the reaction

⟶

FIG. 7.3. *A.* Schematic model of a ping-pong system that exchanges squares for circles. The outward-facing conformations are striped. Starting at the top and proceeding clockwise: empty transporter with site facing outside. Outside square binding to transporter. Translocation of the square so that the square now has access to the inside. Release of the square to the inside. The steps for the circle are similar. Note that there is no conformation that has both a square and a circle bound. The dashed line in the middle illustrates slippage; this allows for movement of the square in the absence of the circle by allowing the empty transporter to move the empty site. More details on the specifics of each conformation are presented in Figure 7.4. *B.* Schematic model of a non-ping-pong system that exchanges squares for circles. The outward-facing conformations are striped. Starting at the right top and proceeding counterclockwise: empty transporter with both sites facing outside. Outside square binding to transporter. Translocation of the sites and the square to the inside. Binding of inside circle to the transporter, which promotes the release of the square to the inside. Translocation of the sites and the circle to the outside, with subsequent release of circle to the outside. Note that there is a conformation that has both a square and a circle bound *(lower left)*, in contrast to the ping-pong model. Other non-ping-pong models are possible by varying the order of addition of the substrates, of release of the substrates, and of translocation of one or the other sites. In all cases, the essential feature is the existence of at least one conformation with both substrates bound.

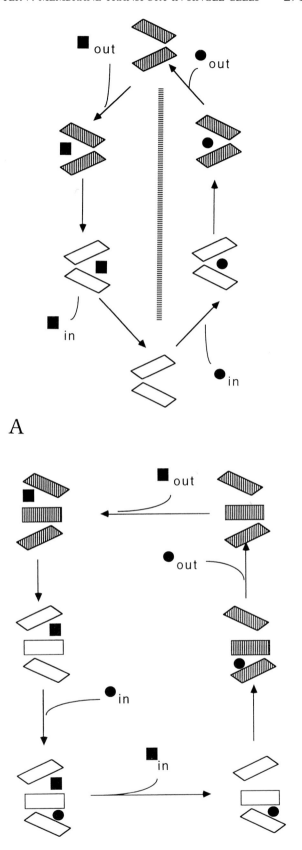

mechanism requires the binding of internal and external substrate (in either ordered or random sequence) before any of them is translocated. The term *simultaneous* has also been applied to such mechanisms, because the translocation event simultaneously moves both substrates. To try to minimize confusion related to terminology, we will use the term "non-ping-pong" for this kind of mechanism. The important distinction between ping-pong and non-ping-pong is that, in the latter, states of the protein exist in which both internal and external substrates are bound.

There are many possible variations on each of the above mechanisms, including hybrids in which there is a conformational change between inward-facing and outward-facing states (as in ping-pong) but in which external binding precedes external release, resulting in a ternary complex (as in non-ping-pong). For both the Na^+ pump and the anion exchanger, it is instructive to discuss some approaches that have been taken to determining the catalytic cycle. In discussing these approaches, it is important to reiterate that kinetic tests cannot prove a mechanism to be correct. The value of a kinetic test is to rule out specific hypotheses.

The determination of the kinetic mechanism can be an attempt to determine which conformational *states* of the protein are allowed and which are forbidden. Some physical implications can be drawn from such results. Additional kinetic information is provided by determining which conformational *changes* are allowed and which are forbidden, that is, what are the possible paths between states of the protein.

There are many tests that help one decide if a system is ping-pong or non-ping-pong (see Table 7.2). One is to measure the ratio of V_{max}/K_m. This is done by varying the concentration of one substrate (for example, extracellular K^+ for the Na^+,K^+-ATPase or extracellular Cl^- for the anion exchanger), while keeping the concentrations of all other substrates constant. The maximal flux (V_{max}) and substrate concentration that gives half-maximal flux (K_m) under this set of conditions is then measured. The same kind of measurement is then done at a different concentration of *trans* substrate (that is, the substrate on the opposite side of the membrane from the substrate whose flux is being measured). For ping-pong mechanisms, the ratio of V_{max}/K_m is independent of the concentration of *trans* substrate.

Other useful approaches to distinguishing ping-pong from non-ping-pong kinetics are to use inhibitors and alternate substrates. Both these approaches are illustrated by the Na^+,K^+-ATPase and $Cl^--HCO_3^-$ exchanger. For both transporters that will be discussed in detail here, several different modes of operation of the transporter have been studied. For example, the Na^+ pump can mediate the normal $3 Na^+/2 K^+$ exchange as well as Na^+/Na^+ and K^+/K^+ exchange, or, in some cells, Na^+ efflux that is not coupled to the influx of a cation. The anion exchanger of red blood cells can mediate Cl^--Cl^- exchange, net anion exchange (exchange of one anion for a different anion), or conductive anion flux. Studies of the different modes of operation of a transport system are important for at least four reasons:

1. The results can provide critical tests of kinetic models and certainly place constraints on acceptable models.

2. The information obtained about the different rate-limiting steps under a variety of conditions provides an important basis for sensitive comparisons between the normal protein in its native environment and proteins reconstituted in different lipid environments, or proteins (perhaps mutated) expressed in a different cell type. If the only quantitative comparison is the

TABLE 7.2. *Methods for Distinguishing Ping-pong from Non-ping-pong Mechanisms*

Method	Ping-pong Prediction	Non-ping-pong Prediction	Where Discussed
V_{max}/K_m	Constant	Depends upon trans ion until saturation	AE and pump
Inhibitor that binds to TA	Denominator of velocity equation has A_iB_oI term and A_iA_oI term	Depends upon particular model	Pump
Inhibitor that competes with B_o	Denominator of velocity equation has Cli I term	Denominator of velocity equation has Cli I and I term	AE
A/A, B/B exchange rates predict A/B exchange rate	Yes	No	AE * and pump
Half-cycle	Yes	No	AE and pump

*The special case where A/A is much faster than B/B is discussed in detail in the anion exchange section. Some of the predictions for a ping-pong model with slippage are different and are discussed in the pump section; see also Sachs (214) and Restrepo et al. (204).

turnover number of the normal cycle, then important changes in individual steps in the cycle could be overlooked. A more sensitive comparison would be the ratio of the maximal velocities of different modes of operation, because different steps will be rate-limiting for different modes of operation.

3. Knowledge of other modes of operation is useful for determining whether structural alterations affect specific steps or sites or generally disable the enzyme; alterations that render the enzyme completely inert are often not as revealing as alterations that only prevent one or two previously allowed conformational changes.

4. In addition, the results of the experiments on the normal enzyme should provide guidance on the types of experiments to attempt with the modified enzyme. As the understanding of kinetics and structure-function relationships improves, it will be possible to utilize the information from initial functional studies of modified proteins to make new predictions. Experiments based on the new predictions will provide a much better test of the models proposed.

ANION EXCHANGER

Red Blood Cell Anion Transport

One of the best-characterized transport proteins is the red blood cell anion exchanger. The physiological function of this protein is to facilitate CO_2 transport and excretion. The mechanism of CO_2 uptake by blood in a systemic capillary is as follows (209): CO_2 enters the blood by simple diffusion and is very slowly hydrated in the plasma, where there is very little carbonic anhydrase activity. The bulk of the CO_2 diffuses into the red cell, in which the high concentration of carbonic anhydrase very rapidly converts CO_2 into HCO_3^-, which then leaves the red cell in exchange for extracellular Cl^-. The same events take place in reverse in pulmonary capillaries, where the situation is somewhat more complex because of the presence of extracellular carbonic anhydrase activity (13). The physiological advantage conferred by red cell Cl^--HCO_3^- exchange is that it allows the rapid conversion of CO_2 to plasma HCO_3^-, with the bulk of the actual hydration taking place in the heavily buffered red cell, thereby minimizing the change in extracellular pH. In order for red cell Cl^--HCO_3^- exchange to be effective, it must be complete in the time the cell spends in a capillary (one second or less). Accordingly, anion exchange in red cells is very rapid. The high rate of anion transport in red cells is a consequence of a large number of copies (10^6 per cell) of the anion exchange protein known as band 3, capnophorin, or AE1, which catalyzes anion

transport with a high turnover rate (50,000 ions per transporter per second; see 18, 186).

Electrically Silent Obligatory Exchange

Until about 25 years ago it was widely assumed that the high rate of Cl^- transport in red cells was a consequence of a high Cl^- conductance (see 185). However, tracer flux studies with the potassium ionophore valinomycin (102, 129) as well as estimates of the effect of valinomycin on the membrane potential (96) showed that the Cl^- conductance of the red cell membrane is much smaller than would be expected on the basis of tracer exchange measurements. It is instructive to reiterate the arguments that led to this conclusion.

The distribution of Cl^-, HCO_3^-, and OH^- across the red blood cell membrane is very close to that expected from a Donnan equilibrium. The intracellular concentration of each anion is about 0.7 times the extracellular concentration (for example, 83), and membrane potential in red cells is about -8 mV, which is close to the value expected if Cl^-, H^+, and OH^- are all distributed at electrochemical equilibrium:

$$[Cl^-]_{in}/[Cl^-]_{out} = [HCO_3^-]_{in}/[HCO_3^-]_{out} = exp\ [FV_m/RT]$$

$$(7-6)$$

In human red blood cells, the rate constant for exchange of intracellular $^{36}Cl^-$ for extracellular Cl^- is about 800/min at 37°C (18). The rate constant for efflux of K^+ from the same cells is about 0.0002/min (21), that is, Cl^- tracer flux is over 10^6 times as rapid as that of K^+. The tracer fluxes, steady-state distributions, and membrane potential are completely consistent with the idea that the Cl^- tracer flux is through a current-carrying channel and that the Cl^- conductance is several orders of magnitude higher than that of Na^+ and K^+. However, several experimental observations are not consistent with this idea. The discovery of cation ionophores (gramidicin, valinomycin) made it possible to increase the permeability of the red cell membrane to alkali cations (see 88, 102, 129). For example, valinomycin is a cyclic depsipeptide (alternating peptide and ester bonds) that forms a lipid-soluble complex with K^+, Rb^+, and Cs^+ but not with Na^+ (5). Valinomycin induces a rapid conductive K^+ permeability in lipid bilayers and red blood cell membranes. Harris and Pressman (88) found that the rate of net loss of KCl from red cells treated with valinomycin is relatively slow unless a proton ionophore is also present to allow exchange of K^+ for H^+. If the tracer Cl^- flux were through a conductive pathway, there should be no effect of a proton ionophore on valinomycin-mediated K^+ efflux.

Subsequent quantitative tracer and net efflux studies by Hunter (102) and Knauf et al. (129) confirmed that the conductive Cl^- permeability (P_{Cl}) is about 10,000-fold smaller than expected from $^{36}Cl^-$ efflux measurements. In both studies the cation fluxes were analyzed with the assumption that they follow the constant field equation (78, 94). The following analysis is that used by Knauf et al. (129) for the net efflux of K^+ and Cl^- in cells treated with valinomycin. The net flux of Cl^-, according to the constant field equation is:

$$J_{Cl} = P_{Cl} \, [\ln B][Cl_i B - Cl_o]/[B - 1] \quad (7\text{--}7)$$

where $B = \exp[-FV_m/RT]$. A similar equation can be written for the net efflux of K^+.

$$J_K = P_K \, [\ln B][K_i - K_o B]/[B - 1] \quad (7\text{--}8)$$

The only net fluxes are ionic currents carried by K^+ and Cl^-; therefore the expressions for each flux can be equated and the equation can be solved for B:

$$B = [P_i K_i + P_{Cl} Cl_o]/[P_K K_o + P_{Cl} Cl_i] \quad (7\text{--}9)$$

The net flux of K^+ and Cl^- was measured as a function of the valinomycin concentration and was found to reach a maximum at about 1 μM, even though the K^+ permeability continued to increase at higher valinomycin concentrations. That is, the net efflux at high valinomycin concentrations is limited by Cl^-, not K^+. Quantitative analysis of the data according to the constant field equation showed that the ionic permeability coefficient for Cl^- is about 2.5×10^{-8} cm/s at 37°C, or 10,000-fold less than that expected if the tracer Cl^--Cl^- exchange flux were through a conductive pathway.

The above analysis depends on the assumption that valinomycin-mediated K^+ transport can be described by the constant field equation. Valinomycin is a carrier, and the kinetics of valinomycin-mediated K^+ transport are not simple. Tracer K^+ fluxes in the presence of valinomycin overestimate the actual P_K (11). However, the conclusion that P_{Cl} is much lower than that expected from Cl^--Cl^- exchange measurements is verified by studies using gramicidin, a channel-forming antibiotic in which estimates of P_K are more straightforward (56).

Direct measurement of the electrical resistance of the large red cells of *Amphiuma* (141) indicated that the membrane resistance is higher than would be expected if the $^{36}Cl^-$ flux took place through a conductive channel. Measurement of membrane potential of human red cells with fluorescent dyes led to the same conclusion (96), that is, that over 99% of the Cl^--Cl^- exchange flux does not contribute to the conductance of the membrane.

Kinetic Mechanism of Anion Exchange

Rate-Limiting Step. A good starting point for kinetic analysis of a transport process is to determine whether the rate-limiting step in the overall catalytic cycle is binding/release or translocation. Nuclear magnetic resonance experiments indicate that, at 0°C, the rate of exchange of Cl^- with the aqueous medium is much more rapid than the overall rate of anion exchange (46). At higher temperature, it is possible that binding may not be much faster than translocation, but, under the conditions used for most kinetic analysis of red cell anion exchange (low temperature), it is likely that translocation rather than binding/release is the rate-limiting event.

Tests of the Ping-pong Mechanism. The ping-pong mechanism implies that there must be two structurally distinct forms of the transporter: inward-facing and outward-facing (Fig. 7.4). Since the transporter is a protein, these forms must represent different conformational states. The proportion of transporters in each form

PING–PONG MECHANISM

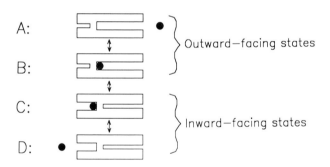

FIG. 7.4. Schematic representation of the ping-pong mechanism for coupled exchange. The four states of the system are shown. All transitions between these states are reversible. The following is the sequence of states in the influx of a substrate, that is, half of a complete catalytic cycle. *A*. The empty outward-facing state. This state has no substrate bound at the transport site, but an extracellular substrate has diffusional access to the site. Intracellular substrate cannot bind to this state. *B*. The loaded outward-facing state. This state is formed when a substrate moves from the bulk extracellular medium and binds to the outward-facing transport site. At present this site has a kinetic rather than a physical definition, because the detailed structure of the protein is unknown. The outward-facing site is wherever the substrate is immediately before the rate-limiting inward translocation step. *C*. The loaded inward-facing state. After formation of the loaded outward-facing state, the protein-substrate complex undergoes a conformational change that allows the substrate diffusional access to the intracellular medium. The structural differences between the inward-facing and outward-facing states are unknown. *D*. The unloaded inward-facing state. This state is formed when the substrate moves from the inward facing transport site to the bulk intracellular solution. Extracellular substrate can no longer reach the transport site of this state of the protein.

depends on the intrinsic free energies of each form (which need not be equal) as well as the intracellular and extracellular concentrations of substrate anions and inhibitors.

Most tests of the transport mechanisms are experiments carried out in the steady state (30, 84, 130–135, 166, 221, 222). The steady state here is defined in the same way as it is defined in enzymology. The overall system may not be at equilibrium; there could be a net exchange of one anion for another. However, over the time course of the measurement, the rate of exchange is constant, that is, the number of turnovers of the catalytic cycle per second is constant, and the amounts of all intermediates are constant. In the steady state, the mass flow through each stage of the catalytic cycle is the same; otherwise there would be buildup or depletion of one of the intermediates.

Consider a steady state in which intracellular anion **A** is exchanging for extracellular anion **B**. For simplicity, assume that there are initially no competing anions, that is, **A** is the only transportable intracellular anion, and **B** is the only transportable extracellular anion.

The expression for the initial exchange flux can be derived as follows. There are four unknown variables: $[E]_i$, the concentration of free inward-facing transporters (no bound anions); $[E]_o$, the concentration of free outward-facing state; $[EA]_i$ the concentration of the inward-facing state with anion **A** bound at the substrate binding site; and $[EB]_o$, the concentration of outward-facing state with anion **B** bound at the substrate binding site. The binding/release reactions to inward-facing and outward-facing states are described by the following equations:

$$[EA]_i = [E]_i[A]_i/K_{Ai} \qquad (7-10)$$

$$[EB]_o = [E]_o[B]_o/K_{Bo} \qquad (7-11)$$

where K_{Ai} and K_{Bo} are respectively the dissociation constants for the binding of **A** and **B** to the inward- and outward-facing states. Equations 7–10 and 7–11 describe the equilibrium relationships for the intracellular and extracellular binding events. With the assumption that binding is much more rapid than translocation, these equilibrium relationships are approximately correct even in the presence of net exchange of **A** for **B**.

In the steady state, the efflux of **A** is exactly balanced by the influx of **B**. In terms of the reaction mechanism, the number of outward translocation events per second must be equal to the number of inward translocation events per second. The inward and outward translocation events are each characterized by a unimolecular rate constant (units s^{-1}). The net efflux of **A** is given by the following expression:

$$J_A = k_{Aio}[EA]_i - k_{Aoi}[EA]_o \qquad (7-12)$$

where k_{Aio} and k_{Aoi} are the unimolecular rate constants for the outward and inward translocation events for anion **A**. An analogous expression can be written for the net influx of **B**:

$$J_B = k_{Boi}[EB]_o - k_{Bio}[EB]_i \qquad (7-13)$$

To simplify the algebra, consider only the initial flux, during which there is no significant accumulation of either intracellular **B** or extracellular **A**. Accordingly, the second terms in each of the equations for the fluxes of **A** and **B** are negligible.

In the steady state, the influx of **B** is equal to the efflux of **A**:

$$k_{Aio}[EA]_i = k_{Boi}[EB]_o \qquad (7-14)$$

In order to solve for the four unknowns ($[E]_i$, $[E]_o$, $[EA]_i$, and $[EB]_o$), four equations are required. The first three are the equilibrium binding equations, 7–10 and 7–11, and the steady-state requirement that influx equals efflux (Equation 7–14). The fourth equation is conservation of mass: the sum of all the forms of the transporter is constant:

$$[E]_i + [E]_o + [EA]_i + [EB]_o = [E]_t \qquad (7-15)$$

These four equations can be solved to give an expression for the efflux of **A** per transporter:

$$J_A = J_B = \frac{k_{Aio}k_{Boi}E_t}{k_{Aio}(1 + K_{Bo}/[B]_o) + k_{Boi}(1 + K_{Ai}/[A]_i)} \qquad (7-16)$$

It is instructive to consider a limiting case for Equation 7–16. The exchange flux in general depends on the translocation rate constants for both exchange partners. However, if the translocation of one of the anions is much faster than that of the other, Equation 7–16 becomes considerably simpler. For example, if the rate constant k_{Aio} for efflux of **A** is much faster than that for the influx of **B**, and if the concentration of **A** is not vanishingly small, then the second term in the denominator is much smaller than the first term and Equation 7–16 becomes:

$$J_A = J_B = k_{Boi}E_t/(1 + K_{Bo}/[B]_o) \qquad (7-17)$$

Equation 7–17 is the equivalent of the Michaelis-Menten equation for the rate of an enzyme reaction:

$$J_B = J_{Bmax}[B]_o/([B]_o + K_{1/2}) \qquad (7-18)$$

with $J_{Bmax} = k_{Boi}$ and $K_{1/2} = K_{Bo}$. That is, the concentration of extracellular **B** at which the influx is half maximal is equal to the dissociation constant for the binding of **B** to the outward-facing transport site. This is true *only* under the specialized condition in which the translocation of **B** is much slower than that of **A**, that is, when the catalytic cycle is limited by **B**.

Next consider an experiment in which the concentration of extracellular **B** is varied at a fixed concentration of intracellular **A**, but now the translocation rate constant for **A** is not much faster than that of **B**. Equation 7–16 can be rewritten in Michaelis-Menten form:

$$J_B = J_{Bmax}[B_o]/([B_o + K_{1/2})] \qquad (7\text{--}19)$$

where

$$J_{Bmax} = \frac{k_{Aio}k_{Boi}E_t}{(k_{Aio} + k_{Boi} + k_{Boi}K_{Ai}/[A_i])} \qquad (7\text{--}20)$$

and

$$K_{1/2} = \frac{k_{Aio}K_{Bo}}{(k_{Aio} + k_{Boi} + k_{Boi}K_{Ai}/[A_i])} \qquad (7\text{--}21)$$

So in the more general case in which **A** is not transported much faster than **B**, the exchange flux is still a simple, saturable function of the concentration of extracellular **B**, but the half-maximal concentration of **B** depends not only on the dissociation constant K_{Bo} but also on the concentration of intracellular **A**, the dissociation constant for intracellular **A** binding, and the translocation rate constants for **A** and **B**.

The ping-pong model implies that, in an experiment in which the intracellular concentration of **A** is fixed and the extracellular concentration of **B** is varied, the apparent J_{max} and $K_{1/2}$ for **B** each depend on the concentration of intracellular **A** as well as on the translocation rate constants, but the ratio $J_{max}/K_{1/2}$ is independent of intracellular **A**. This can be shown by dividing Equation 7–11 by Equation 7–12, resulting in the following:

$$J_{max}/K_{1/2} = k_{Boi}E_t/K_{Bo} \qquad (7\text{--}22)$$

Predictions of Non-ping-pong Models. The catalytic cycle for a simple non-ping-pong mechanism involves the binding of intracellular substrate and extracellular substrate before the rate-limiting exchange event can take place (Fig. 7.5). There are various versions of the model, in which substrates bind in one or the other order or in random order. In the simplest non-ping-pong model the substrates bind in random order, and the binding of one substrate does not influence the binding of the other. The next-simplest model is binding in random order, but with the provision that binding to the inward-facing site may affect the affinity for binding to the outward-facing site.

For non-ping-pong models of the kind we are considering here, the catalytic cycle is complete after the rate-limiting translocation event and release of the substrates (Fig. 7.5). That is, the protein is immediately ready for another cycle of binding from both surfaces,

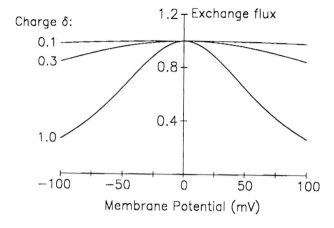

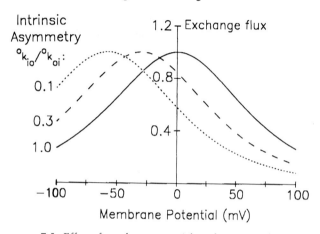

FIG. 7.5. Effect of membrane potential on electroneutral 1:1 anion exchange. The anion exchange flux is plotted as a function of membrane potential. *Upper:* In this example, the transporter is intrinsically symmetric and the membrane potential is varied between −100 mV and +100 mV. The three curves represent three different amounts of charge translocated through the membrane potential difference in the rate-limiting step. Note that, because the transporter is intrinsically symmetric, it is not possible to distinguish between positive and negative charge movement in the translocation step. *Lower:* In this example, it is assumed that a full net negative charge moves through the transmembrane potential difference in the translocation step. The three curves refer to different intrinsic asymmetries, represented as the ratio of outward to inward translocation rate constants at a membrane potential of zero. The curves are normalized to the same maximum exchange flux. Note that the sign of the net charge translocated can be determined only if there is independent knowledge of the intrinsic asymmetry of the exchanger. For example, if a full net positive charge were translocated, and if $^{o}k_{io}/^{o}k_{oi} = 10$, the flux would have exactly the same dependence on potential as it would if there were a full net negative charge translocated and $^{o}k_{io}/^{o}k_{oi} = 0.1$ *(dotted curve).*

exchange, and release; the protein does not exist in distinct inward-facing and outward-facing states. (There would almost certainly be some sort of conformational change during the translocation event, but the important feature of a non-ping-pong model is that both substrates participate in the translocation event.) The algebraic expression describing the flux for such a mechanism is derived as follows:

As was done for the ping-pong model, assume that intracellular **A** exchanges for extracellular **B**. Let E be the empty transporter; AE is the transporter with **A** bound to an inward-facing site; EA is the transporter with **A** bound to an outward-facing site; EB is the transporter with **B** bound to an outward-facing site; BE is the transporter with **B** bound to an inward-facing site. From these definitions, the equilibrium binding equations can be written:

$$[AE] = [A]_i[E]/K_{Ai} \qquad (7-23)$$

$$[BE] = [B]_i[E]/K_{Bi} \qquad (7-24)$$

$$[EA] = [A]_o[E]/K_{Ao} \qquad (7-25)$$

$$[EB] = [B]_o[E]/K_{Bo} \qquad (7-26)$$

The ternary complex AEB has **A** bound to the inward-facing site and **B** bound to the outward-facing site. The other ternary complexes are AEA, BEB, and BEA, with the corresponding definitions. The equilibrium between AEB and EB is

$$[AEB] = [A]_i[EB]/\alpha K_{Ai} = [A][A]_i[B]_o/\alpha K_{Ai}K_{Bo} \qquad (7-27)$$

where α is the factor by which binding of extracellular **B** to the outward-facing site alters the dissociation constant for the binding of intracellular **A** to the inward-facing site. This factor could be either greater or less than one.

An equivalent equation is

$$[AEB] = [B]_o[AE]/\alpha K_{Bo} = [E][A]_i[B]_o/\alpha K_{Ai}K_{Bo} \qquad (7-28)$$

The factor α must be the same in both expressions, because the free energy of the doubly loaded transporter must be the same irrespective of whether **A** or **B** bound first.

The net exchange flux of intracellular **A** for extracellular **B** is given by the following expression:

$$J_{AB} = k_{AB}[AEB] - k_{BA}[BEA] \qquad (7-29)$$

where k_{AB} is the unimolecular rate constant for the transition from $[AEB]$ to $[BEA]$, and k_{BA} is the rate constant for the reverse transition. As with the ping-pong model, consider only the simple case in which the initial exchange of intracellular **A** for extracellular **B** is measured before there is significant buildup of either extracellular **A** or intracellular **B**. Under these conditions, the only forms of the transporter are $[E]$,

$[AE]$, $[EB]$, and $[AEB]$. The sum of the concentrations of these must be the total E_t:

$$E_t = [E] + [EB] + [AEB] + [AE] \qquad (7-30)$$

With the equilibrium equations substituted for each form of the transporter, this equation becomes:

$$[E] = E_t/\{1 + [A]_i/K_{Ai} + [B]_o/K_{Bo} + [A]_i[B]_o/\alpha K_{Ai}K_{Bo}\} \qquad (7-31)$$

The initial exchange flux (per transporter) of intracellular **A** for extracellular **B** is then given by the following:

$$J_A = J_B = \frac{k_{AB}[A]_i[B]_oE_t}{(\alpha K_{Ai}K_{Bo} + \alpha[A]_iK_{Bo} + \alpha[B]_oK_{Ai} + [A]_i[B]_o)} \qquad (7-32)$$

Now consider the experiment in which the concentration of extracellular B is varied at a fixed concentration of intracellular A. As was shown above for a ping-pong mechanism, the flux will be a hyperbolic function of extracellular **B**:

$$J_B = J_{Bmax}[B_o]/([B_o + K_{1/2}) \qquad (7-33)$$

where

$$J_{Bmax} = \frac{k_{AB}[A]_iE_t}{\alpha K_{Ai} + [A]_i} \qquad (7-34)$$

and

$$K_{1/2} = \alpha K_{Bo}([A]_i + K_{Ai})/(\alpha K_{Ai} + [A]_i) \qquad (7-35)$$

The ratio $J_{Bmax}/K_{1/2}$ is then given by:

$$J_{Bmax}/K_{1/2} = k_{AB}[A]_i/[\alpha K_{Bo}([A]_i + K_{Ai})] \qquad (7-36)$$

Therefore, the ratio of maximum flux divided by the half-saturation constant ($J_{max}/K_{1/2}$) should depend on the concentration of *trans* anion in this kind of non-ping-pong model. This dependence is very different from that of a ping-pong model.

The ratio $J_{max}/K_{1/2}$ is the initial slope of a plot of the flux as a function of concentration at very low concentration ($B_o << K_{1/2}$). In a ping-pong mechanism, this slope should not depend on the *trans* anion, because the concentration of *cis* anion is so low that it is rate-limiting. In a non-ping-pong model, the limiting slope at low concentrations depends on the *trans* anion because of the need to form a ternary complex.

Gunn and Fröhlich (84) measured the exchange of Cl^- for Cl^- or Br^- across the red blood cell membrane, at various concentrations of anions on each side of the membrane. The data indicate that the ratio of J_{max} to $K_{1/2}$ is independent of the concentration of *trans* anion, as predicted by the ping-pong mechanism. Restrepo et al. (205) did the same kind of experiment for $Cl^- - Cl^-$ exchange in HL60 cells (a cell line derived from a human myeloid leukemia) and found the opposite result: The ratio J_{max} to $K_{1/2}$ is a hyperbolic function of the *trans* anion, as predicted by a non-ping-pong model. Anion exchange in these cells is catalyzed by the AE2 protein, which has a membrane domain with very similar sequence to that of red cell band 3. Despite this similarity, the kinetics of anion exchange in the two kinds of cell appear to be fundamentally different.

Other Ways to Distinguish Ping-pong from Non-ping-pong.

Another prediction of the ping-pong mechanism is that the rate of net exchange of a rapidly transported anion for a slowly transported anion should be determined by the slow anion (Equation 7–16 above). In a non-ping-pong mechanism (in which a ternary complex is involved in the rate-limiting step), the flux could depend on both anions. Experimentally, the self-exchange fluxes of Cl^-, Br^-, and I^- in media containing 120 mM NaCl, NaBr, or NaI are 260, 32, and 1, respectively (31). All three halide self-exchange fluxes are still much larger than that of $SO_4^=$ at neutral pH. Despite the large differences among the halide exchange rates, the net exchange fluxes of intracellular Cl^-, Br^-, or I^- for extracellular $SO_4^=$ are all very similar (106). A related finding that is consistent with a ping-pong mechanism is that, when intracellular Cl^- exchanges for either $SO_4^=$ or phosphate, the half-maximal intracellular Cl^- concentration is very low, as expected if the $SO_4^=$ half-cycle rather than the Cl^- half-cycle is rate-limiting. A non-ping-pong mechanism, without additional assumptions, predicts similar $K_{1/2}$ for intracellular Cl^-, whether it is exchanging for Cl^- or $SO_4^=$ (107).

Use of inhibitors. As is true of the Na^+ pump, inhibitors have been used extensively to investigate the kinetics of anion exchange. The ping-pong mechanism predicts that, in the absence of other rapidly penetrating anions, the proportion of inward-facing and outward-facing states of the transporter should be strongly influenced by the Cl^- gradient (128). If an inhibitor binds more strongly to the outward-facing than to the inward-facing conformation, then the inhibitory potency should be influenced in a predictable way by the Cl^- gradient. For competitive inhibitors this kind of study must be designed to eliminate effects of competition. It is possible, however, to find conditions in which the effects of the Cl^- gradient on inhibitory potency can be measured without major confounding effects of Cl^- competition.

Knauf and co-workers (59, 130, 131) have shown that the Cl^- gradient strongly affects the potency of stilbenedisulfonate derivatives and several other inhibitors of anion exchange. In agreement with this idea,

the inhibitory potency of the stilbenedisulfonate DNDS is affected by the Cl^- gradient (107).

An interesting example of the use of inhibitors was the study of Restrepo et al. (204) of the effect of DIDS on $Cl^- - Cl^-$ exchange in HL-60 cells (a human cell line derived from a myeloid leukemia). These cells express the protein AE2 (33), which has a membrane domain that is highly homologous to that of red cell band 3. Restrepo et al. (204) showed that DIDS is a competitive inhibitor of Cl^- transport in HL60 cells, but the slope of the Dixon plot (1/Uninhibited flux × half-maximal inhibitory DIDS concentration) is not consistent with a ping-pong model, even if slippage (translocation of unloaded site) is allowed. The result is, however, consistent with a non-ping-pong mechanism of transport. This result indicates that two similar proteins (red cell and HL60 anion exchangers) can have very different kinetics.

Single turnover experiments. The number (1.2×10^6) of copies of band 3 polypeptide per human red cell is the equivalent of roughly 20 μ moles band 3 per liter of packed cells. It is possible to load resealed red blood cell ghosts with low concentrations of Cl^- (for example, 60 μM), with citrate as the predominant intracellular anion. When these ghosts are suspended in a medium containing no rapidly penetrating anion, there is a rapid efflux of about 800,000 Cl^- ions per cell, followed by a much slower further efflux (107). This efflux represents a half-turnover of the catalytic cycle: internal binding, translocation, and external release of Cl^-. This result is in complete agreement with a ping-pong mechanism but would not be expected from the simplest non-ping-pong mechanism, because it demonstrates that efflux is possible in the absence of influx. It should be pointed out that the experiment was performed in the presence of extracellular $SO_4^=$, which can bind to band 3 but is not transported appreciably at 0°C. Therefore, it is possible that the release of extracellular Cl^- takes place only after extracellular $SO_4^=$ binding (220). Nonetheless, the single turnover experiment is good evidence that anion translocation involves a conformational change between inward-facing and outward-facing states, as in the ping-pong mechanism.

Intrinsic Asymmetries

Some transport systems have obvious kinetic asymmetries. For a simple ping-pong mechanism there are two possible kinds of kinetic asymmetry. First, intracellular and extracellular binding affinities may not be equal. Second, inward and outward translocation events may not take place at the same rates. For a 1:1 exchanger, there is a priori no reason to expect that the exchange is asymmetric. Physiologically, net Cl^-–HCO_3^- exchange across the red cell membrane needs to work rapidly in both directions. There is considerable evidence, however, that the transporter itself is asymmetric, and it is instructive to discuss various approaches to determining the nature of this asymmetry.

Consider an experiment in which Cl^-–Cl^- exchange is measured at a fixed, high intracellular concentration of Cl^- and a varying extracellular Cl^- concentration. The flux is a hyperbolic function of the extracellular $[Cl^-]$; the concentration at which the flux is half-maximal is a function not only of the true extracellular binding affinity but also of the inward and outward translocation rates for Cl^-. This may be seen by rewriting Equation 7–21:

$$K_{1/2out}=\frac{k_{Aio}K_{Bo}}{(k_{Aio}+k_{Boi}+k_{Boi}K_{Ai}/[A_i])} \quad (7\text{--}37)$$

For the present example, the subscripts A and B can be dropped with the understanding that A is intracellular Cl^- and B is extracellular Cl^-. Equation 7–37 then becomes:

$$K_{1/2out}=\frac{k_{io}K_o}{(k_{io}+k_{oi}+k_{oi}K_i/[Cl_i])} \quad (7\text{--}38)$$

The intracellular Cl^- concentration is assumed to be present at a high enough concentration that the third term in the denominator may be neglected. Equation 7–38 then becomes:

$$K_{1/2out}=\frac{k_{io}K_o}{(k_{io}+k_{oi})} \quad (7\text{--}39)$$

The $K_{1/2}$ for external Cl^- at high intracellular Cl^- concentration, then, depends not only on the true extracellular dissociation constant but also on the translocation rate constants for influx and efflux of Cl^-. The only case in which $K_{1/2}$ is equal to the true dissociation constant K_o is if the outward translocation rate constant (k_{io}) is much larger than that (k_{oi}) for inward translocation. The corresponding expression for intracellular Cl^- is:

$$K_{1/2in}=\frac{k_{oi}K_i}{(k_{io}+k_{oi})} \quad (7\text{--}40)$$

Experimentally, the $K_{1/2}$ for extracellular Cl^- is considerably smaller than the $K_{1/2}$ for intracellular Cl^-, indicating an asymmetry in either the binding affinities, translocation rate constants, or both (84, 89). It is not possible to distinguish between binding and translocation asymmetry without additional information. Knauf et al. (133) took advantage of their finding that flufenamic acid, a noncompetitive inhibitor, binds more strongly to the outward-facing than to the inward-

facing state of red cell band 3. A careful series of experiments in which the inhibitory potency of flufenamic acid was measured at different Cl^- concentrations allowed the authors to estimate the proportions of Cl^--loaded and free inward- and outward-facing transporter in the presence of symmetric 150 mM Cl^-. Most of the asymmetry in the $K_{1/2}$ for Cl^- is a consequence of asymmetric translocation rate constants rather than true asymmetric binding affinities.

The above example is an illustration of how difficult it is to determine fundamental kinetic parameters (dissociation constants, translocation rate constants) from transport measurements. The emphasis here has been on theoretical considerations, that is, the relation between the observed $K_{1/2}$ and the true dissociation constant. In addition, there are several practical difficulties that should be mentioned. In varying the Cl^- concentration over a wide range, it is necessary to vary either the ionic strength or to use a substitute anion. No true "spectator" anion (that does not interact at all with the transporter) is known to exist for band 3 (see 108, 132). In addition, at high substrate concentrations, the substrate anion itself can bind to a separate site, denoted the modifier site (30 128, 132), which causes inhibition. Finally, the apparent substrate affinity for ion binding to a membrane protein can be influenced by surface charge. For these reasons, in addition to the theoretical kinetic considerations, it is exceedingly difficult to measure a true Cl^- or HCO_3^- affinity for band 3.

Electrical Properties of the Exchange Pathway

One of the fundamental questions about transport systems involves the nature of the conformational changes of a transport cycle and, in particular, the nature of the conformation change that leads to ion translocation. For many, translocation suggests that the ion moves, but in a general sense, translocation refers to the event that changes the access of the ion. At some point, the ion has diffusional access from the internal solution and at another point the ion has access to the external solution. (This is probably a fundamental difference between transporters and ion channels.) Clearly, the movement of an ion can change its access, but also a change of access can be achieved if the part of the protein that forms a gate moves. In either case (ion movement or segment of protein movement) there may be charge movement through the membrane electric field. In this case, membrane potential will alter the rate of charge movement. Thus, there has been a concerted effort for many years to measure the influence of membrane potential on the rate of ion movement. Initially the studies are simply concerned with which, if any, steps in the cycle are charge moving, but ultimately one of the questions is whether the charge moving is ion movement or gate movement. A thorough theoretical description of possible I–V curves from single kinetic models has been described (35, 87, 143, 145, 146). The following discussion on the case of electroneutral anion exchange illustrates many of the basic principles, which also apply to the discussion of membrane potential effects on the Na^+ pump.

Studies summarized above established that over 99% of the $^{36}Cl^-$ efflux into a Cl^--containing medium represents an obligatory, tightly coupled exchange. The simplest such exchange is a 1:1 exchange of monovalent anions. It is possible in principle that the exchanger could require two Cl^- to exchange for two Cl^- (or two other anions), but there is no evidence for such exchange, and the kinetics of transport (see 57, 128) are consistent with a 1:1 exchange. A complete catalytic cycle for this exchange results in zero net movement of charge through the transmembrane field. However, one or more of the elementary steps in the catalytic cycle may involve a net movement of an unpaired ion through part of the transmembrane potential.

In order to address the question of whether a net charge moves in one of the elementary steps, it is necessary to define the most likely catalytic cycle. As discussed above under Kinetic Mechanism of Anion Exchange, the kinetics of both monovalent and divalent anion exchange in red blood cells are consistent with a ping-pong mechanism, in which the pair of exchanging anions do not cross the rate-limiting barrier simultaneously. Rather, the exchanging anions take turns crossing the membrane. Although the kinetics are complex and cannot be explained entirely by the simplest form of a ping-pong mechanism, it is likely that the basic features of the mechanism are correct; each half of the catalytic cycle consists of binding, translocation, and release of a single anion. The working assumption is that the rate-limiting step is translocation rather than binding.

Consider first the question of whether the rate-limiting translocation step is associated with net charge movement. At saturating substrate concentrations, the expression for the exchange of intracellular tracer anion with extracellular (nontracer) anion is the following (see above):

$$J = k_{io}k_{oi}E_t/(k_{io} + k_{oi}) \qquad (7-41)$$

where J is the flux (ions/transport protein-sec), k_{io} is the outward translocation rate constant, and k_{oi} is the inward translocation rate constant. Let δ = the net charge moved through the transmembrane potential

during the translocation step. For example, $\delta = 0.3$ is the equivalent of 0.3 charges moving through the entire potential change, or 0.6 charges moving through half the transmembrane potential, or any combination of charge and fractional potential whose product is 0.3.

In order to write an expression for the effect of membrane potential on either k_{io} or k_{oi}, it is necessary to make an assumption about the shape of the energy barrier between the two states (146). Because nothing is known about the nature of this barrier, it is assumed that the barrier is symmetric, that is, half the potential change is experienced on each side of the transition state. Accordingly, the effect of potential on the rate constants can be written as follows:

$$k_{io} = {}^ok_{io} \exp(-\delta FV/2RT) \qquad (7-42)$$

$$k_{oi} = {}^ok_{oi} \exp(\delta FV/2RT) \qquad (7-43)$$

where ${}^ok_{io}$ and ${}^ok_{oi}$ are the respective rate constants at zero membrane potential. With these expressions for the rate constants, the tracer flux at saturating substrate concentrations is:

$$J = {}^ok_{io}{}^ok_{oi}/({}^ok_{io}\exp[-\delta FV/2RT] + {}^ok_{oi}\exp[\delta FV/2RT])$$

$$(7-44)$$

A plot of J as a function of the membrane potential is expected to show a maximum, with a lower flux at either very positive or very negative potentials. The potential at which the flux is maximum depends on the intrinsic symmetry of the exchange process. If ${}^ok_{io} = {}^ok_{oi}$, that is, if the translocation event is symmetric, then the flux will be maximal at zero membrane potential, with inhibition at either positive or negative potentials. If, on the other hand, the translocation event is asymmetric—for example, if ${}^ok_{io}$ is 10 times ${}^ok_{oi}$—then the exchange will have an optimal membrane potential other than zero. Figure 7.6 shows the theoretical exchange flux as a function of membrane potential for varying degrees of intrinsic asymmetry and varying net charges moved in the translocation event. An important point to be inferred from these curves is that the exchange flux is at a maximum when the rate constants for influx and efflux are equal, that is, when the transport is symmetric. If there is net charge transport during the translocation event, then any electric field that tends to make the rate constants more asymmetric will lower the overall exchange flux.

The relationships shown in Figure 7.6 illustrate that it is difficult to interpret the effect (or lack of effect) of membrane potential on a tracer exchange flux without independent information about the relative rates of inward and outward translocation. For example, if the tracer flux were measured at 0 and −75 mV, and if no effect of potential were detected, it would be tempt-

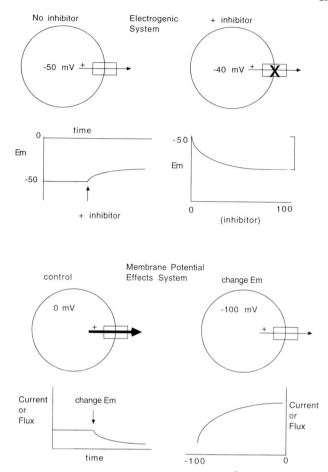

FIG. 7.6. The difference between an electrogenic system and a system that responds to membrane potential. *Upper panel:* Addition of an inhibitor to an electrogenic system will alter the membrane potential. *Lower panel:* In contrast, changes of membrane potential can alter the rate of a transport system (as indicated by the change in arrow size). Electrogenicity and voltage dependence are two different characteristics. For some range of membrane potentials, all electrogenic systems will be altered by membrane potential. However, as discussed in the text, there are conditions where electrogenic systems may be membrane potential independent and electroneutral systems can be membrane potential dependent.

ing to conclude that essentially no net charge is moved relative to the membrane potential during the rate-limiting translocation event. However, it is possible that the transporter has an intrinsic asymmetry such that the flux is the same at 0 and -75 mV even if the translocation step is associated with substantial net charge movement. Accordingly, in order to study the electrical properties of an electrically silent exchange process, it is important to vary the membrane potential over a very wide range, and it is also helpful to have independent information about the rates of influx and efflux.

Grygorczyk et al. (82) varied the membrane potential

in *Xenopus* oocytes expressing mouse band 3 protein. A significant effect of potential on the tracer Cl^- flux was observed, consistent with the idea that the value of δ in this system is about 0.2. The effect of potential on the mouse protein expressed in oocytes is in contrast with observations in intact human red cells. For example, Fröhlich et al. (58) found that gramicidin has no significant effect on either the maximal exchange or the half-maximal external Cl^- concentration for Cl^--Cl^- exchange. Milanick and Gunn (165) found no effect of membrane potential on $SO_4^=$ influx at low pH, under conditions in which the $SO_4^=$ influx consists of $SO_4^=$-H^+ cotransport (105, 164). Jennings et al. (110) measured the effect of potential on anion exchange under a variety of conditions, including those in which the influx half-cycle is rate-limiting. The only detectable effect of potential is on the inhibitory potency of divalent anions, for example, $SO_4^=$ as competitors for Cl^-. This effect was interpreted as evidence that the site at which extracellular $SO_4^=$ competes with Cl^- is within the transmembrane field; the ions move through about 15% of the potential on the way from the bulk extracellular medium to this site. In summary, in intact red cells there is no detectable effect of membrane potential on anion translocation through the band 3 protein. Anion exchange through the same protein (mouse rather than human), expressed in *Xenopus* oocytes, does exhibit a slight dependence on potential (82). The reason for the difference between the expressed and red cell protein is not known. In any case, all the data indicate that, to a first approximation, the rate-limiting translocation step is electrically neutral, that is, it does not involve unpaired charge moving through the transmembrane field. A detailed interpretation of this finding cannot be made in the absence of further understanding of the structure of the protein.

Although the translocation step is electrically neutral for both Cl^- and $SO_4^=(+H^+)$ in native band 3, the situation is changed by chemical modification. Treatment of intact cells with Woodward's reagent K and BH_4^- converts a particular glutamate side chain (Glu 681) into an alcohol, thus removing the negative charge that is normally on this residue at neutral pH (112). In modified cells, net Cl^--$SO_4^=$ exchange is electrogenic and is no longer associated with the cotransport of H^+ (109). The $SO_4^=$ translocation event appears to be nearly electroneutral in the modified cells, but the Cl^- translocation event is associated with the movement of net positive charge. This suggests that, in normal band 3, the translocation event involves the movement of two protein-bound positive charges and two negative charges through the transmembrane field, resulting in an electroneutral translocation event. The two negative charges are both on the substrate in the case of $SO_4^=$. For monovalent substrates such as Cl^-, the second translocated negative charge is believed to be the side chain of glutamate 681. Not enough is known about the structure of the protein to make a more detailed statement about what really happens during the translocation step.

SODIUM PUMP

The physiological function of the Na pump is to create and maintain low intracellular Na and high intracellular K levels. Low intracellular Na levels are important for generation of action potentials in excitable cells, for the uptake of nutrients in epithelial cells, and for the acute maintenance of cell volume in many mammalian cells. High K levels are permissive for protein synthesis. Virtually all mammalian cells have an Na pump; red cells of animals of the order Carnivora are the primary exception (162, 163, 167, 181, 183). Red cells are not required to generate action potentials, have modest nutrient needs that are easily met by equilibration with plasma nutrients, and do not synthesize proteins. Mammalian steady-state cell volume (see later under Pump-Leak Paradigm) generally requires maintaining intracellular Na at less than electrochemical equilibrium values. In red cells of animals of the order Carnivora, the intracellular Na concentration is similar to the extracellular Na concentration; since the membrane potential is -10 mV, the Na is not at equilibrium. Instead of using an Na pump to maintain this Na level, these cells use Na/Ca exchange to extrude Na and the Ca pump to maintain the Ca gradient (162, 163, 181, 183).

The Na^+-K^+-ATPase (Na^+ pump) uses the free energy of ATP hydrolysis to perform the uphill transport of Na^+ outward and K^+ inward across the plasma membranes of most animal cells (35, 38, 61, 72, 233). Under nearly all conditions the stoichiometry of the pump is 3 Na^+ and 2 K^+ transported per ATP (27, 34–38, 40, 51, 53–54, 61, 72–75, 115, 116, 119, 120, 124, 143, 153, 192, 193, 196, 233, 234, 247). Our discussion is confined mainly to the nature of the catalytic cycle, that is, the steps involved in Na^+ and K^+ transport.

Electrically Silent and Electrogenic Na Pump Modes

Electrogenic transport modes result in the movement of net charge. Modes that result in no net charge movement (after completion of a full cycle) are electroneutral. The Na pump has both electrogenic and electroneutral modes, for example, 3Na/2K exchange is electrogenic and 3Na/3Na or 2K/2K exchanges are

electroneutral (7, 14, 27, 28, 34, 44, 72, 119, 120, 124). As discussed above, the anion exchanger is electroneutral (1CL/1HCO₃), but chemical modification can convert it into an electrogenic transporter (1Cl/1 $SO_4^=$).

The determination of whether a transport system is electrogenic or electroneutral could be considered as a subset of determining the substrates and stoichiometry of a system. The determination that the Na pump is electrogenic added support to the 3Na:2K stoichiometry model. Without proof of electrogenicity, models with additional counter-ions are often difficult to exclude—for example, the Na pump could have mediated the exchange of 3Na + 1Cl for 2 K or 3 Na for 2 K and 1 H. The finding that the anion exchanger is electroneutral supports the model of 1:1 anion/anion exchange.

The most obvious method to prove electrogenicity is to show that when the transporter is activated, the membrane potential changes (which is precisely the definition of electrogenic). While measurement of changes in E_m is most direct, it is not always practical. In order to measure a change of E_m in response to a particular transport system, the system must be a major pathway for net charge movement. In many cells, this can be difficult to achieve.

If operation of the transport system does not lead to a change of membrane potential, there are two possible explanations. One is that the system is electroneutral under those conditions. The other is that the membrane conductance is so high that it short-circuits (or masks) any influence of the transport system on the membrane potential. Operation of the Na pump in untreated red cells results in essentially no change of membrane potential, because the anion conductance is large compared to the current generated by the Na pump (96). A measurable signal is obtained in sulfate-loaded, DIDS-treated red cells (98). An even larger signal can be generated in ghosts loaded with tartrate and treated with DIDS (43).

In many cases, the effect of membrane potential on a transport system is determined; in many cell types this is a straightforward experiment and it can lead to informative results (examples are presented in the next section). However, in general, it does *not* distinguish an electroneutral from an electrogenic system (Fig. 7.6). The Na pump provides the best example of this case. The forward mode of the Na pump is electrogenic over a large potential range but is not potential dependent (in the absence of extracellular Na) (for example, 35, 37). So here is a case of a system that is clearly electrogenic (ouabain-sensitive current is measured) but that is membrane-potential independent over most of the experimentally accessible voltage range. In contrast, an alternative mode of operation of the Na pump is Na/Na exchange. This process is electroneutral, as there is no measurable ouabain-sensitive current (66). Nevertheless, the process is membrane-potential dependent (66). So here is a case of a system that is clearly electroneutral but is potential dependent. Because so many have inferred electrogenicity from membrane potential dependence, perhaps a more familiar, but analogous, case should be considered. Consider the question of how to distinguish a transport system that transports protons from one that is pH dependent. Clearly there are systems that do not transport protons (for example, Na, K, Cl cotransport) and that are pH dependent, and there are systems that transport protons (for example, H + amino acid cotransport) that are pH independent (over some pH range).

There is one circumstance when membrane potential dependence can prove electrogenicity. If the membrane potential alters the direction of the *net* movement of the ions, then membrane potential has altered the direction of the reaction and must therefore have altered the driving force or free energy. This implies that the reaction involves net charge movement. An analogous situation occurs for Na/H exchange. The fact that the pH gradient alters the net movement of Na implies that H is transported during the reaction (8). Reeves and Hale (202) showed that the membrane potential alters the direction of net movement of Ca mediated by the Na/Ca exchanger, thus showing that the Na/Ca exchanger in cardiac membrane was electrogenic.

There are three general methods for the measurement of changes of E_m:

1. Direct electrical measurement (voltage clamp)
2. Fluorescent probes (membrane potential sensitive, pH probes and probes for other ions [cf. 3])
3. Movement of ions in response to E_m

Na/K Exchange: Electrode Measurements. The electric measurement of Na-pump activity is usually determined as pump current in voltage-clamp mode. In this situation, the electrodes are used to detect a small change in membrane potential, and a feedback system supplies current to maintain the membrane potential at the original setting.

Extensive studies on nerve and muscle preparations were performed in the 1960s; in the early part of the decade, it was presumed that the Na pump was electroneutral, but by the time of Thomas's review (242), it had been shown in many preparations of excitable cells that the Na pump was electrogenic (see also 37, 72, 75).

As a modern example of this approach, we review the evidence that the Na pump is electrogenic in *Xeno-*

pus oocytes (138, 245). In order to study Na pump current it is necessary to find conditions in which the pump current is a substantial amount of the current that is measured. The conditions usually involve ion substitution and addition of channel inhibitors. Pump current is identified as Na_i-dependent, K_o-dependent, ouabain-inhibitable current. Lafaire and Schwarz (138) showed that addition of dihydroouabain depolarized oocytes substantially, as would be expected if dihydroouabain inhibited a process that resulted in positive charge efflux. This depolarization was rapidly reversible upon washout of dihydroouabain. The changes in membrane potential upon the addition of cardiac glycosides required the presence of extracellular K and intracellular Na. In many preparations it is also possible to show that the cardiac glycoside-sensitive current requires ATP.

It is important to point out that activation or inhibition of the pump should result in fairly rapid changes of E_m, as only a few pump cycles are required to move enough charge to change the potential. Slow changes in E_m (for example, on the order of minutes) are more likely to reflect changes of ion concentrations due to pump inhibition and probably do not reflect the direct effect of the Na pump on membrane potential (Fig. 7.7).

Ouabain-induced changes in current need not be directly due to the Na pump but could be due to changes of ion concentrations following pump inhibition. In large cells such as oocytes, it is unlikely that the bulk concentration will change in a few minutes; however, local ion concentrations may change. In particular, because of the large K gradient and the extracellular geometry, there may be appreciable K accumulation just outside the membrane that would alter the membrane potential (60, 61, 199). Also, local changes in cytosolic Na may alter the Ca influx through the Na/Ca exchanger, which is electrogenic and which may subsequently alter cytosolic Ca. In many cells cytosolic Ca opens K and Cl channels. Thus ouabain treatment may indirectly lead to an alteration in E_m due to the change in K or Cl conductances.

Na/Na Exchange: Electrode Measurements. In squid axons, it is possible to measure efflux of radioactive Na and to measure current from the same area of the voltage-clamped axon. The pump will primarily perform Na/Na exchange in the absence of potassium and the presence of Na, ADP, ATP, and Mg. Gadsby et al. (66) showed that there is a cardiac glycoside-sensitive Na flux under these conditions; in the absence of ADP or Mg or extracellular Na, no cardiac glycoside–sensitive flux was observed. The cardiac glycoside-sensitive Na efflux averaged 10 pmol/cm²/s at 0 mV. If this flux

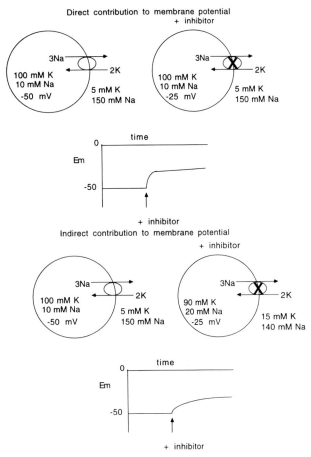

FIG. 7.7. Inhibition of the Na pump could alter membrane potential for two different reasons. *Upper panel:* The Na pump directly contributes to the membrane potential and thus inhibition of the pump rapidly leads to dissipation of the membrane potential. *Lower panel:* The Na pump does not directly contribute to the membrane potential. However, inhibition of the pump will lead to alterations in the Na and K gradients, and these will alter the membrane potential. Since the changes in ion gradients are relatively slow, the time course for membrane potential change will be relatively slow. Thus, when attempting to determine if a new system is electroneutral or electrogenic, one needs to determine if changes in membrane potential reflect direct operation of the transporter or are secondary to ion changes caused by the transporter.

involved the movement of 1 net charge per 3 Na ions effluxed (as expected for a 3 Na/2 Na exchange), then there should be a cardiac glycoside–sensitive current of 333 nA/cm², as shown by Equation 7–45.

$$1/3 \times 10 \text{ pmol/cm}^2/\text{s} \times 10^5 \text{pC/pmol} \times 1 \text{ pA/pC/s}$$
$$= 333,000 \text{ pA/cm}^2 \qquad (7\text{–}45)$$

In contrast, under the same conditions, and from the same region of each axon, the cardiac glycoside–sensitive current was 34 nA/cm². Thus less than 10% of the Na flux would represent electrogenic 3 Na/2 Na exchange. This is a similar argument to that developed above (see Tests of the Ping-Pong Mechanism) to con-

clude that anion exchange was electrically silent. Clearly, the majority of the flux under these conditions is electrically silent.

Na/K Exchange: Fluorescence Measurements. Mammalian red blood cells are small; it is difficult to obtain electrode measurements of membrane potential (or voltage-clamped current) (96). As an alternative, fluorescent dyes have been used to measure the membrane potential (140). Why are fluorescent dyes apparently more sensitive than electrode techniques? Note that it is possible to measure pump current in cardiac cells and oocytes, where the density of pump is high or where the surface area is large. But in red cells, with current technology, measurement of pump current with electrodes is not possible. This conclusion can be reached by two different calculations: The conductance of a whole red cell (sulfate-loaded, DIDS-treated) is about 100 times smaller than the conductance of a 10 Gohm whole-cell patch (the resistance has been estimated to be 10^6 ohm cm² and the surface area is about 100 μ^2, and thus the resistance is 1000 gigaohms) (96). Alternatively, one can estimate the pump current from the Na flux measurement (and the conversion of red cell volume to surface area); the pump flux is approximately 1 mmol of charge/liter packed cells per hour (97). One liter of red cells contains 10^{13} cells, one mole of charge is 10^5 coulombs, and 1 amp is a coulomb per second. Thus the current is approximately 0.0028 pA, which is not measurable by current techniques.

Hoffman et al. (97) studied the pump current using voltage-sensitive fluorescent dyes in red blood cells. In the initial studies, the resistance of the red cell was increased by using sulfate-loaded red cells and treating with DIDS. This increased the resistance by approximately an order of magnitude. In addition, the Na concentration was increased so that the Na pump current was increased. Under these conditions, addition of K to the extracellular medium caused a hyperpolarization (as expected for net positive charge efflux). The potential returned to the basal value upon the addition of ouabain. The ability to both turn on and turn off the transport system, and thus see both an increase and decrease in dye fluorescence, allows one to eliminate some artifacts as explanations for the changes in dye fluorescence. It is important to point out that the activation or inhibition of the pump should result in fairly rapid changes of E_m, as only a few pump cycles are required to move enough charge to change the potential (Fig. 7.7). Because the movement of many dyes is slower than an electrode response, pump-induced changes in E_m may require a minute or two for detection of the steady-state fluorescence value; any further changes in fluorescence more likely reflect changes in ion concentrations (and subsequent E_m changes) or in cell volume (and subsequent changes in dye-quenching and fluorescent properties).

In order to calculate a stoichiometry from the changes in membrane potential, one needs to estimate the membrane conductance (= 1/membrane resistance) and then one can apply Ohm's law, I = V/R, to calculate the pump current. The pump current can then be compared to pump-mediated Na efflux. In order to optimize the membrane potential changes, red cells were loaded with sulfate and treated with DIDS (both processes increasing R) and loaded with Na (to increase pump current, I) (97). If one assumes that sulfate is the most permeant ion and that the DIDS-insensitive sulfate flux is conductive, then measurement of tracer sulfate efflux can be used to estimate membrane resistance. These calculations are consistent with a 3:2 stoichiometry (97). (Note that the fact that the lack of an ouabain-sensitive sulfate flux under these conditions [42, 43] suggests that there is another conductance that is larger than the sulfate conductance.)

Note that the operation of the Na pump of untreated cells (containing chloride and approximately 10 mM Na and not treated with DIDS) results in almost no change of membrane potential (97). This does not imply that the Na pump is only electrogenic in the presence of sulfate and DIDS, but rather that the cell conductance short-circuits the pump current. Thus, in general, if one desires to conclude that a given transport system is electroneutral because its operation does not alter membrane potential, it is necessary to have an independent measurement of cell conductance and to quantitatively estimate the expected change of potential given the measured values of membrane conductance and estimate current as if all the flux were conductive.

Uncoupled Na Efflux: Fluorescence Measurements. In the absence of extracellular Na and K, there is an ouabain-inhibitable Na efflux that has the properties expected for Na movement through the pump; this is termed the *uncoupled mode* (70, 76). Under the conditions of uncoupled Na flux, in red blood cells, it is possible to measure an ouabain-sensitive efflux of sulfate and of phosphate under appropriate conditions (42, 43, 155–157). The existence of an ouabain-sensitive anion flux under these conditions is not sufficient information to conclude that the anion movements are mediated directly by the pump. An important alternative is that the ouabain-sensitive anion flux occurs secondary to an ouabain-sensitive change in membrane potential, owing to an electrogenic uncoupled Na flux. This putative change in membrane potential could increase the driving force of movement of compensating charge;

under these conditions anions could be the most conductive ion. In order to rule out this alternative explanation, Dissing and Hoffman (43) showed that under the conditions of an uncoupled Na efflux and ouabain-sensitive sulfate efflux, there was no detectable fluorescence change when using a membrane potential sensitive fluorescent dye. Therefore, the ouabain-sensitive anion flux was not an indirect effect of membrane potential but rather reflected sulfate flux mediated by the Na pump. Since the uncoupled Na flux is about one-sixth of the Na flux in the Na/K mode in red cells, one would expect a substantial change in fluorescence if there were no compensating charge movement through the Na pump. Thus this mode of the pump is electroneutral in intact red blood cells (43).

Electrogenic Na Pump Modes: Transported Protons and Charge Compensating Proton Movements. A particularly intriguing set of experiments were performed in Blostein's lab (190, 191). One of the questions that sparked this line of inquiry was whether the Na-pump stoichiometry was fixed at 3 Na per 2 K. Blostein reasoned that at low Na there would be a substantial fraction of Na pumps with less than 3 Na ions bound; if the translocation (and release) conformational changes occurred at a reasonable rate, then one should be able to measure a stoichiometry of 1 or 2 Na per 2 K. This was in fact observed, particularly if the cytoplasmic proton concentration was increased. The stoichiometry of 1 Na per 2 K could result from translocation of 1 cation (Na) out of the cell and 2 K ions in, in which case the current flow (and membrane potential change) would be the opposite direction from that observed when the pump mediates 3 Na/2 K. An alternative explanation is that the pump still translocates 3 cations out of the cell; for example, 2 H+ could accompany the 1 Na. These two possibilities result in different models of how the Na pump works. In the second case, apparently the translocation or release conformational change requires a precise amount of charge (3+). In the first case, the site apparently excludes protons and does not require a precise amount of charge; the optimal charge (3) does lead to the optimal rate of transport.

To distinguish between these models, an inside-out vesicle preparation was used (14). Before making the vesicles, the intact red cells were treated with DIDS to inhibit the anion exchanger and decrease the anion conductance (190). An operating anion exchanger would be capable of moving protons that would interfere with the measurement of net proton movements. After treating the red cells with DIDS, the cells were washed extensively to remove any noncovalently attached DIDS (190, 191). This is important because intracellular DIDS inhibits the Na pump (187). Fluorescein-dextran was incorporated inside; the fluorescence of fluorescein is H-dependent.

Operation of the Na pump, in the absence of cytosolic Na, led to changes in the free [H+] inside the vesicles (that is, at the extracellular surface), as illustrated in Figure 7.8 (190, 191). This could have been the result of H+ movements in place of Na movements. (Since the vesicles are inside out, Na or H would move into the vesicles and K out of the vesicles.) Alternatively, the change of free [H+] could have occurred secondary to a change of membrane potential (see Fig. 7.8). If the pump had been mediating the exchange of 1 Na per 2 K, then the inside of the vesicle would have become more negative with respect to the medium. The change in membrane potential would have altered the driving forces for all ions; if H+ were the most permeant or if H+ moved though a conducting pathway, then H+ would have entered the vesicles, leading to a change of E_m.

To determine if the H movements were directly through the pump, or indirect due to changes of E_m, Polvani and Blostein added lipophilic ions (190, 191). These ions can easily cross the membrane and would redistribute rapidly if there were a change of membrane potential. At very low pH (approximately 6.2) the addition of lipophilic ions had no effect on the proton movement. The results of these experiments clearly indicated that at low pH (approximately 6.2) the Na pump could mediate H/K and Na/H exchange with extracellular H acting as a surrogate K and intracellular H acting as a surrogate Na. At more modest acidic pH values (6.8), the addition of lipophilic ions altered the proton movement. This shows that the pump no longer moved protons directly, but instead was electrogenic with net charge movement in the opposite direction of the normal mode. This electrogenic component at low Na is consistent with a 1 Na/2 K stoichiometry (190, 191).

Pump Stoichiometry

The determination that the Na pump was electrogenic added support to the 3Na/2K stoichiometry model. In contrast, stoichiometry implies electrogenicity only if (1) it is possible to exclude all other ion movements (particularly difficult are protons and bicarbonate), (2) if the measurements are sufficiently precise to eliminate 2:2 stoichiometries, and (3) the measurements are of net ion movement and are not complicated by exchanges (see, for example, 189). The measurement of the ratio of pump-mediated net Na and K movements in red cells requires determining the ratio of the difference in total flux minus ouabain-insensitive flux.

Clearly, even precise measurements of the fluxes can lead to imprecise estimates of the ratio due to the compounding of the errors. Thus, while the 3:2 ratio in red blood cells was well accepted since the early 1970s, the study of Hoffman et al. (97) was the definitive proof that the red cell pump was electrogenic. Some of the most highly accurate measurements of Na pump stoichiometry are those of Rakowski et al. (198) in voltage-clamped squid axons; that paper considers all possible stoichiometries up to 11:10, and the only ratio statistically in agreement with the data was 3:2.

Kinetic Mechanism of the Na Pump

Transport vs. Biochemistry. The Na-pump cycle involves the transport of Na and K and the transient, covalent, association of the gamma phosphate of ATP with the pump protein. This latter aspect allows for detailed biochemical analysis that is not possible with simple carriers or transporters in which covalent bonds are not formed or broken during the transport cycle. The measurement of the properties of the phosphointermediate of the pump led very early to the development of a ping-pong model for the movement of Na and K (see, for example, 73, 75). Briefly, it was found that, in the presence of Na+ and ATP, a phosphorylated intermediate is formed that is sufficiently stable to quantify. The gamma phosphate from ATP is transferred to an aspartate residue to form an acyl phosphate that retains much of the free energy that was originally in the terminal phosphate of ATP. This phosphorylated intermediate breaks down rapidly in the presence of K+. A simplified scheme of the essential features of the kinetics of the Na pump is shown in Figure 7.9.

The establishment of an Na+-dependent phosphorylation and K+-dependent dephosphorylation led to the development of a model (the Post-Albers model) in which the Na translocation steps (and enzyme phosphointermediate formation) occurred separately from the K translocation steps (and enzyme dephosphorylation). In the nomenclature of Cleland (25) for enzymes with multiple substrates, this is a ping-pong mechanism because Na+ is bound and transported, followed by K+ binding and translocation. A detailed review of the properties of the phosphointermediate, including whether it really occurs during the normal pump cycle, is beyond the scope of this chapter (51, 77, 175, 233, 234).

In this section we focus solely on the transport evidence that the Na pump mediates a ping-pong exchange of Na and K. The mechanism of the Na pump has been determined by several complimentary techniques, including V_{max}/K_m measurements, inhibitor

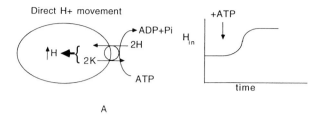

Direct H+ movement

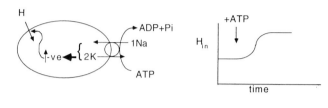

Indirect H movement

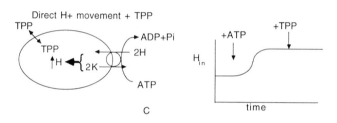

Direct H+ movement + TPP

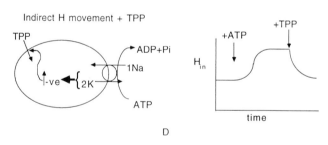

Indirect H movement + TPP

FIG. 7.8. Possible changes of proton concentration in response to a coupled process that directly moves protons in exchange for K *(A and B)* and directly moves Na in exchange for K *(C and D)*. The experimental preparation is inside-out vesicles from red blood cells treated so that the primary conductive pathway is protons. The Na pump mediates K efflux under these conditions in response to ATP in the bath. In *A*, Na pump mediates an electroneutral K/H exchange. H influx will directly elevate the intravesicular H. The increase in intracellular H will lead to an increased H efflux from the vesicles. Eventually, the proton influx through the Na pump will be balanced by the increased passive H efflux and there will be no further change of H. In *B*, the Na pump mediates an electrogenic 2K/1Na exchange and does not directly move H. Operation of the pump leads to an inside negative potential. Because the primary conductive pathway in these vesicles is H, the inside negative potential will lead to a conductive influx of H. The H influx will elevate the intravesicular H and the measured result will be similar to that observed in *A*. The increase in intracellular H will lead to an increased H efflux from the vesicles. Eventually, the proton influx in response to the electro-

studies, transient kinetics, and rates of alternative transport modes.

V_{max}/K_m Studies. This discussion of the Na pump will consist of two parts: the studies examining the ratio of V_{max} to K_m in red cells and the evidence for slippage of the Na pump in red cells. The pump flux is measured as the difference in the unidirectional Na^+ or K^+ (Rb^+) flux in the presence and absence of ouabain or other cardiac glycosides, which are potent inhibitors of the Na^+,K^+-ATPase (53, 54, 73). Even for systems with complex kinetics such as the Na^+ pump, it is possible to measure the unidirectional pump flux as a function of the *cis* substrate concentration, with all other variables held reasonably constant. For each combination of all the other variables (for example, *trans* ion concentrations), there will be some maximum flux (V_{max}) and some concentration of *cis* substrate that gives a half-maximal flux (K_m). The ratio of V_{max} to K_m as a function of the *trans* ion can be used to distinguish ping-pong from non-ping-pong models. It is important to point out that an exchange mechanism that is not fully coupled complicates the analysis. A fully coupled process is one in which efflux of ion X requires the presence and influx of ion Y. The anion exchanger (see earlier under Electrically Silent Anion Transport) is, to a very good approximation, completely coupled, as the rate of net Cl flux is at least 10,000 times slower than the exchange rate. The movement of ion X without the movement of ion Y has been called "slippage" or "uncoupled flux" (50, 56). The Na pump does allow some slippage, and thus this kinetic analysis will include the complications that are associated with slippage. A kinetic scheme for the Na pump that incorporates the uncoupled Na flux and additional ATP effects is shown in Figure 7.10. For the following kinetic analysis, the ATP concentration will be fixed and thus we will not be concerned with the details of how ATP interacts with pump. Below is presented a more complete discussion of whether the uncoupled pathway shown in Figure 7.10 is an appropriate model for the uncoupled flux.

One can generalize the above discussion about $V_{max}/$

genic Na pump will be balanced by the increased passive H efflux and there will be no further change of H. In *C*, addition of TPP will not alter the H concentration. Since operation of the pump has not changed the membrane potential, addition of TPP should not alter the H flux mediated by the pump nor the passive pathway. Thus, there will be no change in H. In *D*, addition of TPP will alter the H concentration. TPP will short-circuit the membrane potential change. Thus, the membrane potential will return to its value in the absence of ATP. Thus, there will be no driving force for passive H efflux and the H concentration will return toward basal values.

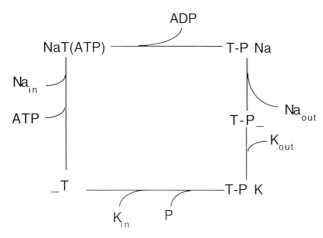

FIG. 7.9. Simplied kinetic scheme for Na/K exchange by the Na pump. T is the pump. The conformation with the empty transport site having access to the intracellular solution is _T and having access to the extracellular solution is T—P_. After addition of ATP and Na$_{in}$, the terminal phosphate is covalently attached to the pump, ADP is released, and Na is translocated. Following Na release, K$_{out}$ can bind. The transport site returns to the inside to release K and the covalent P$_i$ bond is broken, releasing P$_i$ (to the inside). Not explicitly shown are the occluded states (51, 77). During translocation, there is a conformation where K ions are trapped inside the pump and have access to neither the inside nor the outside.

K$_m$ ratio to state that, in a fully coupled ping-pong system, the ratio of V$_{max}$ to K$_m$ is independent of the concentration of the *trans* ion. In contrast, for non-ping-pong systems and for ping-pong systems with slippage, the ratio of V$_{max}$/K$_m$ for one substrate is a function of the other substrate concentration (see equations in Table 7.3). This fact was not widely appreciated in the early 1970s, and thus several studies on Na pump kinetics were used to rule out a ping-pong model. In light of our current knowledge, one could say that the observation that the ratio V$_{max}$/K$_m$ depends on the second substrate rules out fully coupled ping-pong models but is consistent with two types of models: non-ping-pong and ping-pong with slippage.

In the 1970s several groups studied the effect of *trans* ions on V$_{max}$ and K$_{1/2}$ (68, 69, 99, 213). For these studies, it was necessary to vary the intracellular Na concentration; several techniques for varying Na in red cells were available. The early studies (68, 69, 99) were done in the presence of products (that is, Na$_0 \neq 0$); in the presence of products it is difficult to distinguish between ping-pong and non-ping-pong models. Sachs (213) found that intracellular Na changed the V$_{max}$ for extracellular K but did not change Km. As discussed above, a ping-pong mechanism predicts that the ratio V$_{max}$/K$_m$ for K is independent of Na. Since the data did not fit the prediction, these results ruled out fully coupled ping-pong models. Al-

though the uncoupled Na pump flux was appreciated at the time, the models constructed were all non-ping-pong. These models, based upon Na and K transport data, were surprising because it was easiest to describe the biochemical data by a ping-pong (Post-Albers) model. Garay and Garrahan (69) do hint at the possibility that uncoupled fluxes may be involved, but they clearly favor a non-ping-pong model. A resolution of the pump kinetic mechanism awaited a better appreciation of the kinetic consequences of an uncoupled mode, and alternative methods for distinguishing ping-pong from non-ping-pong models, which came in 1979 (214).

Uncoupled Na Efflux. In red blood cells, Garrahan and Glynn (70) observed an ouabain-sensitive Na efflux in the absence of extracellular potassium. Control experiments were performed to be sure that there was no contaminating extracellular K. Even at that time, ouabain was thought to be a highly specific inhibitor of the Na pump; however, to their credit Garrahan and Glynn realized that there were two interpretations to their result: *(1)* The pump was not obligated to exchange Na for K, for example, there was some slippage, and *(2)* ouabain inhibited another process in addition to the Na pump. Other characteristics of the uncoupled Na efflux were also consistent with it being mediated by the Na pump. (7, 27, 76)

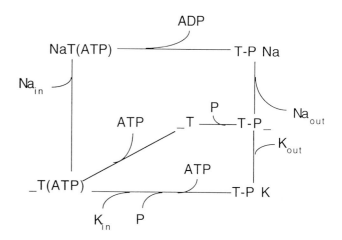

FIG. 7.10. Slightly less simple kinetic scheme for the Na pump. This scheme incorporates two additional points. *(1)* The middle pathway from T—P_ to _T(ATP) can occur in the absence of K$_{out}$ and is the pathway for uncoupled Na efflux. See text for a more complete discussion of the uncoupled Na pathway. *(2)* ATP with low affinity (and without hydrolysis) accelerates the K translocation steps. Thus, ATP can bind to the lower pathway. Whether this ATP effect occurs at the separate physical site from where the terminal phosphate of ATP is transferred to the pump remains to be determined. Most kinetic results are consistent with only one physical site for ATP, but do not rule out models of two physically distinct ATP sites.

TABLE 7.3. *Exchange Equations with $A_o = B_i = O$*

Model	Numerator	Denominator
Ping-pong	$V_{max} A_i B_o$	$X A_i + Y B_o + A_i B_o$
Non-ping-pong (#1 and #2)	$V_{max} A_i B_o$	$X A_i + Y B_o + A_i B_o + Z$
Ping-pong with slippage	$V_{max} A_i B_o + W_{max} A_i$	$X A_i + Y B_o + A_i B_o + Z$

A_i and B_o are the substrates for A_{efflux}/B_{influx} exchange process. V_{max} is the maximal flux for exchange; W_{max} is the maximal efflux of A_i in the absence of B_o. X, Y, and Z are sums of products of rate constants and are independent of substrate concentration. In some cases they are the maximal $K_{1/2}$ values, for example, for the ping-pong, X is the maximal $K_{1/2}$ for B_o (obtained at saturating A_i). The random-ordered, not rapid equilibrium nonping-pong model equation is more complicated.

Inhibitor Studies. It is difficult, if not impossible, to distinguish between a ping-pong model with slippage and a non-ping-pong model by measuring the substrate dependence of the flux. This is because the kinetic equations for these two classes of models are nearly identical (see Table 7.3). Thus, another approach is required. It turns out that certain types of inhibitors can be used to distinguish between these models. This approach was first applied to the Na pump by Sachs (214); his presentation of the argument is elegant, and we encourage the reader to read Sachs's account.

Consider a ping-pong and a non-ping-pong scheme and consider an inhibitor that binds to only one conformation. Which conformation would produce different results (kinetic equations) for the two models? An inhibitor that binds to the conformation TNa (Fig. 7.11) will generate different kinetic equations for the ping-pong model and for non-ping-pong model #1. The equations can be derived by standard algebraic techniques or by the King-Altman graphical approach. For the models presented here, it is possible to determine the difference between ping-pong and non-ping-pong model terms by inspection using the King-Altman approach. The interested reader is encouraged to read the approach in Segel (228) or the original King-Altman papers. This graphical approach may be particularly appealing to those who are not mathematically inclined. In the following analysis, we assume that $K_i = 0$.

In a ping-pong model, an inhibitor that binds to TNa will generate two denominator terms that contain I as a factor: $Na_i K_o I$ and $Na_i Na_o I$ (see Table 7.4). The non-ping-pong model #1 also generates two denominator terms that contain I as a factor, but the terms are different: $Na_i I$ and $Na_o I$. When $Na_o = 0$, the two models are distinguished by the fact that the ping-pong model has a term with $K_o I$ but the non-ping-pong

model lacks such a term. (In the non-ping-pong model, K_o and I are mutually exclusive and so there can be no term that contains their product.) A ping-pong model with slippage generates the additional term $Na_i I$; clearly the presence of the term $Na_i K_o I$ still distinguishes between a ping-pong model with slippage and non-ping-pong model #1. The requirement for a term that contains $Na_i K_o I$ can be determined by experiments in which K_o is varied. A non-least-squares fitting procedure can be used to determine if the term $Na_i K_o I$ is well defined, for example, that the presence of the term reduces the sum of the squares of the differences and that the standard error for the term is <25% of its value (26).

Another important distinction between ping-pong (with slippage) and non-ping-pong model #1 is that Na_o will change the inhibition pattern for the former but not the latter. As stated above, the ping-pong equation includes the term $Na_i Na_o I$, while the non-ping-pong model does not include this term. (In nonping-pong model #1, Na_i and Na_o are mutually exclusive so there can be no term that contains their product.) While the non-ping-pong model has the term $Na_o I$, this term can be neglected if Na_i is saturating: The velocity equation will be of the form $V_{max} Na_i K_o / (A + Na_i B)$, where A is all terms without Na_i (including $Na_o I$) and B is the sum of the coefficients of all the terms that include Na_i as a factor. The definition of saturating Na_i is that $Na_i * B >> A$, so that the equation reduces to $V_{max} K_o / (B)$. The inhibitor terms for ping-pong and non-ping-pong model #1 are given in Table 7.4.

Sachs (214) showed that oligomycin had the properties expected of an inhibitor that binds to TNa. For example, oligomycin had no effect on the K/K exchange mode of the pump, observed in the presence of ATP and Pi and the absence of Na. This suggests that

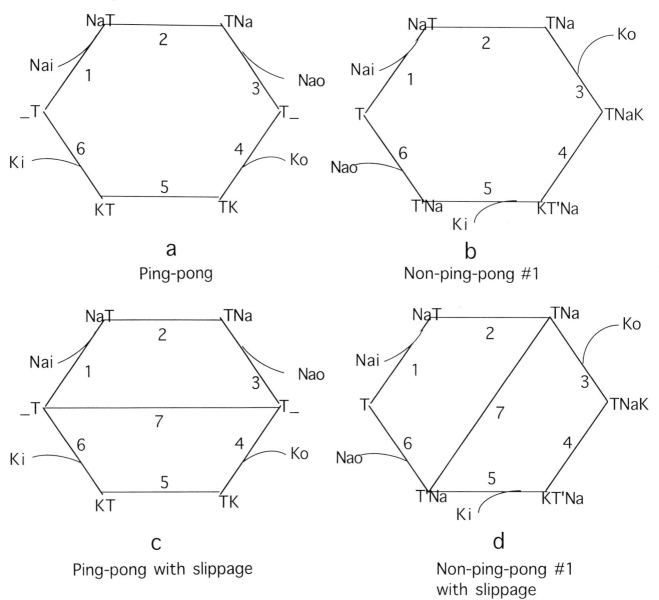

FIG. 7.11. Cyclic diagrams for different transport mechanisms for the exchange of Na_i for K_o. For all diagrams the rate constants from k_1 to k_6 are in the clockwise direction and the rate constants for k_{-1} to k_{-6} are in the counterclockwise direction. T is the transporter protein. *a.* A ping-pong mechanism. The conformation with the empty transport site having access to the intracellular solution is __T and having access to the extracellular solution is T__. Steps 3 and 6 are the release of product. *b.* Non-ping-pong mechanism no. 1. A key feature of non-ping-pong mechanisms is the formation of a ternary complex, TNak, which has both substrates bound. There are a variety of possible non-ping-pong models based on different (or random) ordered binding and release. In this model, steps 1 and 3 are the addition of substrate. Step 4 represents the translocation of k, and step 2 the translocation of Na. In this model, steps 5 and 6 are release of product. A major difference in the cyclic diagrams between this model and that shown in *a* is that K_o adds before Na_i is released in this model. *c.* A ping-pong mechanism with slippage. The addition of step 7 allows the interconversion of T__ to __T in the absence of K_o; thus the transporter can cycle (and transport) Na in the absence of K_o or K in the absence of Na. This step is referred to as uncoupled Na flux in the Na pump field and is called slippage in the anion exchanger field. *d.* Non-ping-pong mechanism no. 1 with slippage. The addition of step 7 allows transporter to cycle (and transport) Na in the absence of K_o.

conformations with K bound do not bind oligomycin. Also, oligomycin was uncompetitive with respect to extracellular Na, intracellular Na, and ATP for Na/Na exchange, as expected if oligomycin binds to the Na-loaded form, TNa. (Uncompetitive inhibition means that the term S I is required to fit the data, but not a term that contains I but is independent of S, where S is a substrate.)

With this background, Sachs was then able to test the predictions of the ping-pong and non-ping-pong models. He found that oligomycin inhibition (in the absence of Na_o) was consistent with the presence of a term K_oI (214). (Since Na_i was fixed, it was not used in the analysis; both models predict that Na_i will effect the inhibition by I.) Furthermore, the addition of Na_o altered the type of inhibition. These results are consistent with the ping-pong model (with slippage) and are not consistent with the non-ping-pong model.

It is important at this stage in the investigation to ask if other types of non-ping-pong models are consistent with the data. Indeed, Sachs has constructed such a model. The requirement is that the non-ping-pong model generate a term of K_oI. Non-ping-pong model #1 did not generate such a term because K_o bound to the same form as I. If we alter the order of binding, so that K_o binds before Na_i and allow K_i to be released before Na_o, then I and K_o are not mutually exclusive (Fig. 7.12). Non-ping-pong model #2 is also consistent with the oligomycin results, but as Sachs points out, is not easily reconciled with the cation dependence of the Na/Na and K/K exchange modes (214). In contrast, the ping-pong model predicts the observed properties of Na/Na and K/K exchange modes (214). This work of Sachs represents outstanding kinetic analysis and shows the importance of kinetics: it can rule out models. The limitations are also shown: More than one model is consistent with any set of data.

These results on the Na pump illustrate an important point about kinetic (and scientific) approaches: The analysis does not prove a model, but can be used to rule out models. It is likely that more than one model will provide an equivalently good fit to the data; the final determination of which model is preferred involves the often subjective determination of which one is simpler or more appealing.

Exchange Modes. The finding that the Na pump can mediate Na/Na exchange, K/K exchange, and ATP/ADP exchange are all consistent with a ping-pong model, and the ping-pong model provides for definite constraints on the relation between the rates of Na/Na, K/K, and Na/K exchange, as was illustrated above for the anion exchanger. Sachs, in a series of papers (215–217) has found that the exchange modes, in general, have the properties expected from a ping-pong exchange of Na for K for the forward pump mode.

Transients. A very clear difference in the behavior of a ping-pong model and non-ping-pong models is that in a ping-pong model the protein is capable of transporting one substrate in the absence of the other; this can only occur once, because the pump then requires the other substrate to complete the cycle. So observation of a half-cycle provides strong support for a ping-pong model. Forbush (49) used kidney vesicles, which have a high concentration of Na pumps, and a turntable-based collection apparatus to provide fast time resolution (see also 50), to determine if the Na pump could perform a half-cycle. Indeed, he found that there was a half-cycle of Na efflux in the absence of extracellular K. The rate of this burst was faster than the overall Na/K exchange rate (at pH values < 7.5), consistent with the half-cycle being a part of the normal pump cycle.

Summary of Na Pump Mechanism. The success of the ping-pong model with slippage (Post-Albers model for the Na pump), in describing the results of the V_{max}/K_m

TABLE 7.4. *Inhibitor Terms for Different Kinetic Mechanisms when the Inhibitor Binds to TNa*

Mechanism	$Na_o = O$	With Na_o
Ping-pong	Na_iK_oI	$Na_iK_oI + Na_iNa_oI$
Ping-pong with slippage	$Na_iK_oI + Na_iI$	$Na_iK_oI + Na_iNa_oI + Na_iI$
Non-ping-pong #1	Na_iI	Na_iI
Non-ping-pong #1 with slippage	Na_iI	$Na_iI + Na_oI$
Non-ping-pong #2	Na_iK_oI	$Na_iK_oI + Na_iNa_oI$

The coefficients (which are a function of the rate constants and independent of substrate, product, or inhibitor concentrations) are not shown. These terms were derived using the King-Altman approach, as described in the text, and can also be derived by standard algebraic techniques (see text for references).

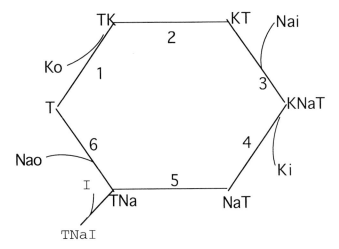

Non-ping-pong model #2

FIG. 7.12. Cyclic diagram for non-ping-pong model #2. A key feature of non-ping-pong mechanisms is the formation of a ternary complex, TNaK, which has both substrates bound. In this model, K_o adds before Na_i. (See Fig. 7.11 for more details.) This model fits with the oligomycin data because K_o and I are not mutually exclusive and Na_o and Na_i are not mutually exclusive.

studies, the oligomycin studies, the exchange modes, and the half-cycle measurements, accounts for a great part of its popularity (55). It is important to point out that there are results that are not easily accommodated by this simple model; most of these can be accommodated by ad hoc additions, for example, modifier Na binding sites. Whether or not these modified ping-pong models remain a better description of pump behavior than do non-ping-pong models awaits further experimental results.

We would like to examine two other aspects: (1) the uncoupled Na flux and (2) rate of Na pump reversal.

Uncoupled Na flux. There are currently three different explanations for what occurs during the uncoupled pump flux, that is, Na efflux in the absence of extracellular Na and K:

1. Na/H exchange
2. Na-anion cotransport
3. purely electrogenic Na efflux

Only the latter case was postulated when Sachs did the oligomycin experiments. First we review the information for each case, and then we discuss whether Na/H exchange or Na-anion transport would allow alternative explanations for the oligomycin experiments.

At high extracellular proton concentrations (that is, low pH), three different groups have shown that there

is Na/H exchange. This was elegantly shown by Polvani and Blostein (190, 191) in inside-out vesicles from red blood cells (see earlier under Electrogenic Na Pump Modes: Transported Protons and Charge Compensating Proton Movements). Goldshleger et al. (79; see also 123) reached a similar conclusion based on slightly less direct measurements: that the uncoupled flux was not electrogenic and that it was not anion dependent; therefore it must involve the exchange of a cation and the only ones present were protons. Dissing and Hoffman (43) clearly measured an ouabain-sensitive H influx in red cells at pH values less than 7. Thus, three different preparations show Na/H exchange at pH < 7; however, these results are also irrelevant for this discussion of mechanism, as the experiments of Sachs and others were all done above pH 7.

Above pH 7, the red cell inside-out vesicles became too leaky for Polvani and Blostein's measurements. But at pH 6.8, Blostein and Polvani clearly demonstrated (190, 191) an electrogenic component to the uncoupled Na flux. Goldshleger et al. (79) clearly observed an electrogenic current in reconstituted kidney pump liposomes at alkaline pH values. So in these cases, the uncoupled Na efflux behaves as originally suggested by Garrahan and Glynn (70) and as modeled by Sachs (214).

Dissing and Hoffman (43) found that intact red blood cells mediated an ouabain-sensitive sulfate efflux. (Marin and Hoffman [155–157] have also observed an ouabain-sensitive phosphate efflux that is consistent with the notion that the terminal phosphate from ATP accompanies Na efflux along with the sulfate in the uncoupled mode.) These data appear convincing but have not been widely accepted in the Na pump field, probably for a combination of three reasons. (1) The pump mediated anion fluxes have only been observed in red cells; it remains possible that the red cell is peculiar in this regard. (2) The pump-mediated anion fluxes have been reported by only one laboratory. Although it is important to note that no one has reported an inability to measure the fluxes nor the finding that the uncoupled flux is electrogenic in intact red cells. (3) The pump-mediated anion fluxes would require two important modifications to the current theory. (1) How does one construct a protein that sometimes carries sulfate out and sometimes carries K in and in both cases resets to carry sodium out? (2) How does the pump know when to take the sulfate? In the current ping-pong model with slippage, it is not until the pump releases the transported Na that it is aware of whether or not extracellular K is present. These reasons are not compelling, but given the general inertia of science and the otherwise solid (but not complete) success of the Post-Albers model, the Post-

Albers model remains the most popular explanation of how the Na pump functions.

The most convenient way to accommodate the Na-anion cotransport and to maintain a ping-pong model is to suppose that the T-to-T conformational change is associated with anion efflux (at low H concentrations). This merely modifies Figure 7.11C, by including intracellular anion concentration with rate constant k_7.

Rate of Pump Reversal. The rates and substrate dependencies of Na/Na and K/K exchange on the Na pump are

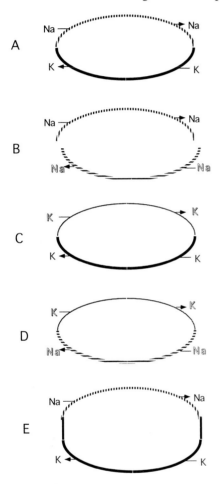

FIG. 7.13. The rate of some exchanges predict the rate of other exchanges for a ping-pong model of the Na pump. *A.* The Na efflux/K influx exchange mode. *B.* Na efflux / Na influx exchange mode (which involves some of the same steps as in Na/K exchange, as indicated by the similar type of line). *C.* K efflux / K influx exchange mode (which involves some of the same steps as in Na/K exchange, as indicated by the similar type of line). *D.* K efflux /Na influx exchange (pump reversal). For the simple scheme diagrammed, the slowest steps from the other three exchange modes must govern the rate of the reverse mode. *E.* Possible alternative explanation for why K/Na exchange is slow is that there is a conformational change between the Na binding forms and the K binding forms. Since Na and K compete on both sides, this explanation is less likely than the possibility that it is difficult to achieve optimal concentrations of ADP and P_i for pump reversal (see text).

exactly as expected from a ping-pong model. However, there is one alternative mode whose rate does not seem to fit a simple ping-pong model, pump reversal, that is, Na influx/K efflux. The rates of Na efflux/K influx and Na efflux/Na influx and K efflux/K influx can be used to predict the rate of K efflux/Na influx; in theory, K efflux/Na influx should not be slower than the slowest of the other three exchange modes. In fact, in the few cases where K efflux/Na influx has been measured, it is much slower than the other three exchanges. The other three exchanges occur at approximately the same rate, under optimal substrate conditions.

Why is the K efflux/Na influx data not easily compatible with a ping-pong model? The rate of K efflux/Na influx will be governed by the slower of the two half-cycles: K efflux or Na influx. Suppose K efflux is slower. Then K efflux/K influx should be as slow as the reverse mode (or even slower if K influx were slower than K efflux). There are two ways out of this dilemma: either K efflux/Na influx has not been measured under optimal substrate conditions or the two exchange reactions do not quite overlap, that is, there is a finite conformational change that must occur following the release of Na before the pump can bind K (but no such change is required to bind Na). Thus, the exchange reactions do not encompass the full cycle (see Fig. 7.13). This explanation is unlikely for the Na pump, because extracellular Na and K compete and such a model does not predict competition. On the other hand, achieving optimal substrate concentrations may be very difficult: High ADP is required to reverse the Na steps, but ADP is quite likely to bind to the ATP site on the K form of the enzyme as well (29, 121).

Electrical Properties of the Na Pump

The studies summarized above established that the Na pump mediates an exchange of 3 Na ions for 2 K ions with the net movement of one charge per cycle. Thus a complete catalytic cycle for the pump results in net movement of charge through the transmembrane field. Thus, at least one of the elementary steps in the catalytic cycle must involve a net movement of an unpaired ion through part of the transmembrane potential. The rate of any conformational change that moves net charge through the transmembrane potential will depend on the size of the transmembrane potential; qualitatively, inside negative potentials will slow steps that involve net positive charge efflux and inside positive potentials will accelerate net positive charge efflux. A discussion of the quantitative nature of the dependence of rate on membrane potential was presented earlier under Electrical Properties of the Exchange Pathway.

To address the question of whether a net charge

moves in one of the elementary steps, it is necessary to define the most likely catalytic cycle. As discussed earlier under Kinetic Mechanism of the Na Pump, the kinetics of the Na pump are consistent with a ping-pong mechanism, in which Na and K do not cross the rate-limiting barrier simultaneously. Rather, the Na transport steps and the potassium transport steps are separate. Although the kinetics are complex and cannot be explained entirely by the simplest form of a ping-pong mechanism, it is likely that the basic features of the mechanism are correct: half of the catalytic cycle consists of binding, translocation, and release of three sodium ions; the other half of similar steps for potassium.

It should be pointed out that, unlike the anion exchanger case, ion dissociation may be rate limiting for some modes of the Na pump. Ion binding can only be rate limiting at nonsaturating concentrations; a saturating concentration implies that an increase in concentration (and therefore rate since k = concentration × rate constant) has no effect on the overall cycle.

In some preparations, the pump current is essentially voltage independent from -100 mV to $+50$ mV, in the absence of extracellular Na and with saturating extracellular K concentrations (for the results in oocytes, see ref. 199, Fig. 5; ref. 226, Fig. 2; and ref. 249). (A thorough review of the initial work on voltage dependence of the pump has been presented [40].) The pump is clearly electrogenic under these conditions, because pump current was being measured. The lack of effect of membrane potential implies that under these conditions the rate-limiting step is not moving net charge through the electric field.

The lack of effect of membrane potential on the electrogenic Na pump may seem surprising at first. Thus, we briefly review the thermodynamic constraints on potential dependence. The Na pump moves ions against an electrochemical gradient using the energy from ATP hydrolysis. Suppose the concentrations of ATP, ADP, Pi, Na_i, Na_o, K_i, and K_o are fixed. Then there exists a potential where the free energy from ATP hydrolysis is equal and opposite to the electrochemical free energy for Na and K transport. This point is the reversal potential for the pump. For physiologically relevant ion concentrations, E_{rev} is approximately -250 mV (35, 40). At the reversal potential, there is no net movement of Na or K (or net ATP hydrolysis) so there is no current, but the unidirectional fluxes may be sizable. As the membrane potential becomes more positive than the reversal potential, there will be net Na efflux and net K influx (and net ATP hydrolysis) and the pump will generate a current of net positive charge efflux. As the membrane potential becomes more negative than the reversal potential, the pump will mediate net Na influx, K efflux, and ATP synthesis

and the pump current will be inward. Clearly the membrane current is membrane potential dependent near the reversal potential, and this fact is independent of the kinetic model of the pump (so long as the stoichiometry of Na:K:ATP is fixed at all potentials). However, the potential dependence of the Na pump at other potentials depends on the pump mechanism and cannot be predicted a priori as it is dependent on the kinetics of the pump (see for example, 87, 143, and also 147). Figure 7.14 shows several possible I–V curves for a pump with a reversal potential of -250 mV (35). When $Na_o = 0$, the pump cannot go in reverse and E_{rev} is not defined.

A constant current–vs.–membrane potential curve does place important constraints on any model. It clearly implies that, under the conditions studied, the rate-limiting step does not carry net charge. Clearly, however, more information can be obtained from non-constant current (or flux) vs. membrane potential curves in the range of -100 mV to $+50$ mV. There are three different steady-state approaches:

1. One could vary the free energy of the system so that the reversal potential is in the range of -100 mV to $+50$ mV; this can be difficult, because the current (or flux) may be quite small under these conditions.

2. Once can vary the substrate, product, or temperature in order to make the potential dependent step rate determining.

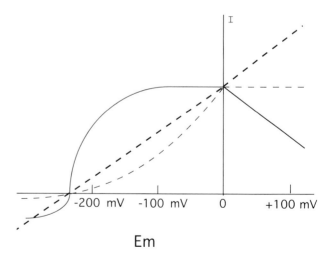

FIG. 7.14. Some examples of shapes of I–V (current vs. membrane potential) curves with $E_{rev} = -250$ mV. (Adapted from DeWeer [35]). These curves were drawn manually to illustrate the following possibilities: (1) There can be a large range of voltage where current is independent of membrane potential (see text for more details). (2) Near E_{rev} the current may be very small. (3) Away from E_{rev} there are no thermodynamic constraints on the shape of the curve (though it can't cross the voltage axis again). For a more detailed discussion of I–V curve shapes, see, for example, Hansen et al. (87), DeWeer (35), and Läuger (143).

3. One can measure different modes of the pump, in which the potential dependent step is rate determining. (These approaches are not mutually exclusive.)

These are all steady-state approaches with the common theme that one manipulates the concentrations of substrates, products, and temperature and measures a steady-state flux (or current) that reflects a particular cycle of the pump. If the rate-determining step of this mode is the conversion from conformation A to conformation B, and if this step moves net charge, then the steady-state flux or current measurement will depend on the applied membrane potential. This is a powerful technique but suffers from the usual disadvantage of steady-state measurements: Because a number of different conformational changes often contribute to the overall rate, there is often ambiguity in assigning the potential dependence to a particular step. Thus, many investigators try to measure different modes of the pump, and to manipulate substrate and product concentrations so that several different experimental approaches can be used to measure the putative same step (6, 34–41, 61–66, 80, 125, 138, 158, 172, 173, 177, 195–200, 203, 225, 226, 227, 246, 248, 249). An alternative technique for determining which step moves net charge involves transient electrical measurements (16, 47, 48, 63, 90–93, 100, 144, 219, 240).

Both steady-state and transient electrical techniques have been used to identify the charge-moving steps of the sodium pump. Because steady-state measurements are more accessible to most investigators, we will concentrate on those results, but it must be stressed that the transient measurements of Na-pump charge movement have represented an important technical advance and have provided important new information (6, 9, 16, 47, 48, 92, 93, 100, 144, 145, 172, 219, 240).

As mentioned above, there are conditions where a particular steady-state result can be fit by more than one model; occasionally results are obtained that are clearly best fit by one simple model. In order to make the discussion easier to follow, we list the four different experimental conditions that have been well studied, with a brief comment on the interpretation. This will be followed by a more detailed discussion of the Na/Na exchange data and the Na/K data at low K_o.

1. Na/K current ($Na_0 = 0$, $K_0 > 5$ mM)

The voltage independence of the steady-state Na/K current implies that the rate-limiting step does not carry charge (see 12, 104, 201 for discussion of rate-limiting). Thus, if the rate-limiting step is K deocclusion, then it does not carry charge. Alternatively, if Na dissociation carries charge, it is not rate limiting. (Note that Forbush and Klodos [52] have shown that pH alters the rate-limiting step. It may be that the rate-limiting step is different for Na pumps from different species [90, 226, 247, 249, 254]).

2. Na/Na exchange, an example of approach 3

The fact that steady-state Na/Na exchange is voltage dependent implies that the rate-determining step for this mode does carry charge (66). We should point out that the term "rate determining" is ambiguous (201). Membrane potential predominantly affects the $K_{1/2}$ of Na/Na exchange, not the V_{max} (at saturating Na_i). Thus, the results are qualitatively consistent with membrane-potential-altering steps involved with Na binding. A more detailed analysis is presented below.

3. Na/K current in the presence of Na_0, an example of approach 2

In the presence of extracellular Na, the Na/K current is potential dependent (62, 80, 138, 158, 172, 173, 198, 219, 226, 227). Extracellular Na is an inhibitor of Na/K current. Negative potentials make extracellular Na a more potent inhibitor. A simple interpretation of this result is that negative potentials attract extracellular Na to its binding site. Clearly, this picture is consistent with the Na/Na exchange result. There are alternative explanations of the result that negative potentials make extracellular Na a more potent inhibitor; most of these are not consistent with Na/Na exchange data. For example, membrane potential could alter other steps so that negative potentials increase the amount of time that the pump spends in the conformation that binds extracellular Na. In this case, extracellular Na would be a better inhibitor even though there is no direct effect of membrane potential on the Na ions.

4. Na/K current, $K_o < 5$ mM, an example of approach 2

At low extracellular K concentrations, the Na/K current is decreased by positive potentials (138, 198, 219, 226, 227). This may seem surprising: Positive potentials might have been expected to increase the rate of an enzyme that results in positive charge efflux. This result once again shows the danger of using changes in the magnitude of the driving force to predict kinetic changes. An informative method to analyze this result is to state the results as follows: the I–V curve is potential independent at saturating K and is potential dependent at low K. At low K, K binding becomes partially rate limiting; if K binding involves net charge transfer through part of the electric field, it would be potential dependent, and inside positive potentials would be expected to decrease the rate of K binding. This result is also discussed in more detail below.

These results are consistent with a model in which extracellular Na and K binding steps are potential dependent. We discuss two aspects of these results in more detail: Na/Na exchange and Na/K exchange at low extracellular K concentrations.

Detailed Analysis of Na/Na Exchange. Na/Na exchange and ATP/ADP exchange are alternative modes of the Na pump that have been well studied (see for example 15, 34, 118, 120, 122). Two types of Na/Na exchange have been described: *(1)* ADP-dependent Na/Na exchange represents the reverse of the Na efflux steps, requires ADP as well as ATP, and results in no net ATP hydrolysis (15, 34, 119, 120, 122). It was predicted to be electrically silent because of the 1Na:1Na stoichiometry (70). *(2)* Na/Na ATPase represents extracellular Na acting like K so the Na movements are accompanied by ATP hydrolysis and the process is electrogenic (15, 149). Only ADP-dependent Na/Na exchange will be considered here, since the Na movements were electrically silent. Figure 7.15 shows a diagram of Na/Na exchange where intracellular Na and extracellular Na are considered substrates. Chemically, protein-Na complexes behave the same whether ^{22}Na, ^{23}Na, or ^{24}Na is bound; however, if the state $E^{22}Na$ releases Na to the outside, that will contribute to the measured ^{22}Na efflux rate, but if the state $E^{23}Na$ releases Na to the outside solution, it will not be detected. This diagram is consistent with the result that ATP/ADP exchange can occur in the absence of Na_0

As described above for Cl/Cl exchange, one expects an electroneutral exchange process, in which each half-cycle moves charge through the electric field, to result in a biphasic flux vs. E_m curve. The shape of the curve and the point of maximal flux provide important information about the asymmetry of the system and the amounts of charge moved in the rate-limiting step.

Gadsby et al. (66) have studied the voltage dependence of Na/Na exchange in squid axons. This has required impressive technical developments in order to prove that the measured Na flux comes from the voltage-clamped portion of the axon (198).

Na/Na exchange, though electroneutral, was potential dependent (66). Surprisingly, the curve was not biphasic, but rather appeared to saturate, that is, between -30 mV and -90 mV, the flux is essentially constant (66). The data are of very high quality. However, there remains the possibility that the curve may not actually saturate; rather, it may have a broad plateau. In terms of model building, saturation is an interesting observation. Unfortunately, it was not experimentally possible to extend the voltage range. Thus, models that predict biphasic responses with a broad plateau are not completely eliminated.

Let us assume that the flux-vs.-voltage curve for Na/Na exchange is monotonic. How can the flux-vs.-E_m

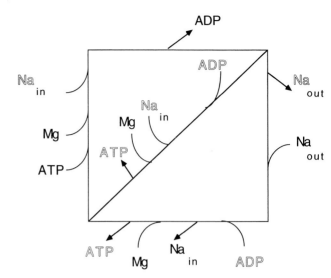

FIG. 7.15. Cyclic diagram of electroneutral Na/Na exchange mode of the Na pump. This model provides explicit consideration of radioactive tracers, ADP (for example, 3H or 14C, ADP) and Na (for example, 22Na or 24Na). The ATP/ADP exchange rate is determined by the rate of appearance of ATP from ADP. Na/Na exchange is measured by the appearance of Na_o. While the protein is assumed to treat Na22, Na23, and Na24 exactly the same, this explicit treatment of the tracers allows for easier consideration of some properties of the cycle. The outer loop *(square)* is the cycle required for Na/Na exchange. The outer loop is also one possible cycle that gives ATP/ADP exchange. The inner loop *(upper triangle)* is the cycle that allows for ATP/ADP exchange in the absence of Na_o.

curve for an electroneutral exchange be monotonic when the theory developed above predicted a biphasic curve? There are two different theoretical interpretations. The theory developed in the anion section had two important assumptions: that the effect of the field was symmetric and that the voltage-dependent step was rate limiting at $E_m = 0$.

Gadsby et al. (66) realized that the mathematics would generate monotonic curve if the voltage dependence (reflected in delta) were completely asymmetric—that is, $\delta = 1$. Shown in Figure 7.16 are flux-vs.-potential curves for different asymmetry factors. How can one construct a physical model in which delta is so asymmetric? It is difficult to construct a conformational change with $\delta = 1$. Examination of the data from a different view will provide additional insight. The data are well fit by a model in which the membrane potential alters the $K_{1/2}$ for Na but does not affect the V_{max}. Here V_{max} refers to the flux at saturating Na_o. This also suggests that the voltage dependence occurs in the steps of binding and release. Mathematically, this might be expressed by stating that the rate of Na_o binding = $Na_0 \times$ the association rate constant (at Em

= 0) × exp δFV/2RT. The natural tendency is to associate the factor δFV/2RT with the association rate constant (and therefore claim that dissociation is voltage independent because of the complete asymmetry). Alternatively, and this is an important insight of Gadsby et al., one could group the factors as [Na$_0$ × exp δFV/2RT] × the association rate constant. In this case, the association and dissociation rates are both membrane potential independent. The effect of membrane potential is essentially to alter the probability of Na$_0$ binding to the transport site. This statement is better understood if we now consider a physical model: the access well model (66, 143, 219). The idea is that the Na ion must pass through the electric field (and therefore part of the protein) on its way to the binding site. The effect of a negative inside potential is to increase the probability that an Na ion has transversed this access channel and therefore increase the probability that the Na ion will bind to the transport site. Clearly, when Na$_0$ is not saturating, this will increase the flux. This access well model provides a pleasing picture for the observations. One way to view the asymmetry is that negative potentials effectively increase the probability of Na$_0$ binding (and thus increase the overall flux) but do not alter the probability of Na$_0^+$ binding, since the effective concentration of Na$_0^*$ is zero. The stimulatory effect of extracellular Na and of negative potential on Na/Na exchange appear equivalent.

Gadsby et al. (66) clearly point out that this asymmetry may be only apparent. This raises the question: Is it possible to generate a broad plateau in the flux-vs.-potential curve with a symmetric electric field effect? The answer is yes, as shown by the simulations in Figure 7.17. In this figure, the rate constants for extracellular Na binding and dissociation are varied. This is a simplified version of the description of Hilgemann (92, 93); Hilgemann developed his model to fit his transient data. When the Na concentration is not saturating, then Na binding is partially rate limiting. Thus inside negative potentials, which increase the rate of Na binding, would speed up the overall cycle. At more negative potentials the rate of Na release will become rate limiting, and thus making the potential more negative will eventually slow the overall cycle. When the rate constant for Na release is of the same order as other steps in the cycle, there will be a relatively sharp peak in the flux-vs.-potential curve (Fig. 7.17). However, if the rate constant for Na release is much greater than the rate of other steps in the cycle (for example; the rate of Na release is much faster than the conformational change that precedes release, at E$_m$ = 0), there will be broad plateau. This occurs because there is a voltage range where neither binding nor release is rate limiting and thus potential will have no effect on the overall cycle. But because binding and release are both voltage dependent, there will be voltages where they eventually become rate limiting.

Detailed Analysis of Na/K Exchange at Low K$_o$.
In 1986, Lafaire and Schwarz (138) reported that the I–V curve for the Na pump in oocytes was biphasic. A negative slope was observed at low extracellular potassium in the presence of extracellular Na from 0 to +50 mV. Biphasic curves were not observed by other investigators (who worked at high K concentrations), and they were theoretically surprising. Also, experimentally they were observed at low K and positive potentials, conditions in which possible artifacts due to local K accumulation or time-dependent contaminating currents are difficult to exclude. Thus, substantial work was required to convince some other investigators that the effects were due to Na-pump mediated current (199, 200). In fact, there have now been several independent studies supporting the finding that the Na/K pump

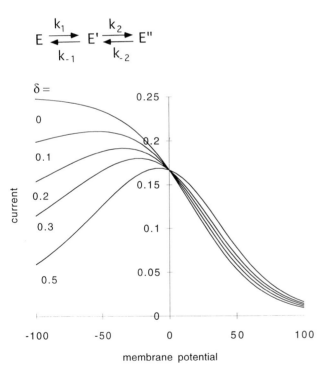

FIG. 7.16. Flux vs. membrane potential simulations for different values of δ, the electric field asymmetry factor. These cures were drawn for a simple two-step reaction scheme shown at the tip of the figure. The rate equation is Flux = 1/[1/k$_1$ + 1/k$_{-1}$ + 1/k$_2$ + 1/k$_{-2}$ + k$_{-1}$/(k$_1$k$_2$) + k$_2$/(k$_{-1}$k$_{-2}$)], see Gadsby et al., 1993 (66). For this simulation, all the rate constants were set to 1 at E$_m$ = O. The voltage dependence (of step 2) was modeled as k$_2$ = k$_o$2 exp((δ$_s$/25) and k$_{-2}$ = k$_{o-2}$ exp (−(1−δ$_s$)/25). As δ tend toward zero, a plateau is achieved at negative potentials. The δ values are indicated.

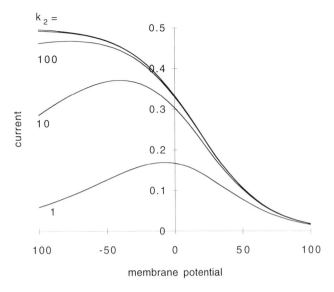

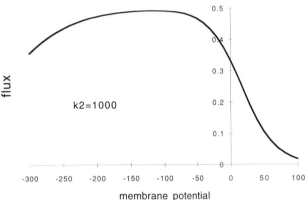

FIG. 7.17. Flux vs. membrane potential simulations for different values of rate constants $k_2 = k_{-2}$. *Upper:* These curves were drawn for a simple two-step reaction scheme shown at the top of Figure 7.16, with the rate equation Flux $= 1/[1/k_1 + 1/k_{-1} + 1/k_2 + 1/k_{-2} + k_{-1}/(k_1 k_2) + k_2/(k_{-1} k_{-2})]$. For this simulation, $\delta = 0.5$, $k_1 = k_{-1} = 1$ and the k_2 and k_{-2} values at $E_m = 0$ are as indicated. *Lower:* When $k_2 = k_{-2} = 1000$ plateaus in the range from -50 to -100 mV, over a large range the flux eventually declines. This is because no matter how fast k_2 and k_{-2} are at $E_m = 0$ (that is, k_{o2} and k_{o-2}), at extreme potentials k_2 and k_{-2} must become rate limiting; compare Hilgemann (93).

current is biphasic at low K (in the presence of extracellular Na).

Lafaire and Schwarz (138) found that at low extracellular potassium, the current was inversely related to the membrane potential in the absence of extracellular Na. Thus, making the membrane potential more negative actually speeded up the rate of the net positive charge efflux generated by the Na pump. This effect was observed only at low K and not at saturating K. At low K, one of the rate-limiting steps is K binding, and this step is the primary candidate for the potential dependent step under these conditions.

The biphasic I–V curve for the Na pump observed in the presence of Na and at low K (138, 200, 219, 225–227) is now thought to result from two potential dependent steps: from approximately 0 mV to 50 mV, the increasingly positive membrane potential decreases the probability of extracellular K binding and thus decreases the overall pump rate. From 0 mV to −100 mV, the increasingly negative potential increases the probability of extracellular Na binding, which inhibits the overall pump rate. Figure 7.18 illustrates a simple example in which there are two voltage-dependent steps. The voltage dependence of these steps as well as the net flux are shown. Branched models can also give rise to biphasic curves, though they do not adequately describe the Na-pump data (161, 222).

Consistent with this picture are the results obtained at saturating extracellular K concentrations. Here the predominant effect is that negative potentials inhibit the pump. Negative potentials and extracellular Na have similar effects in inhibiting the pump (172, 198, 226); in squid axons a twofold change in Na_o is equivalent to a 26 mV change in membrane potential (66).

Transients. An alternative technique for determining which step moves net charge involves transient electrical measurements. In this technique, conditions are manipulated so that all of the pump molecules are in conformation A. Then a parameter is abruptly altered

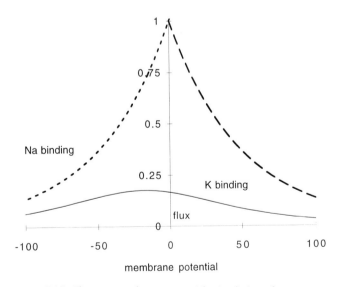

FIG. 7.18. Flux vs. membrane potential stimulations for a two-step model in which both steps are voltage dependent. *Solid line:* the overall flux as a function of membrane potential. *Dashed lines:* the rates of the voltage-dependent steps. Note that at extreme potentials, the voltage-dependent steps become rate limiting and therefore dominate the rate of the overall cycle.

(temperature, membrane potential, ATP concentration, etc.) and the conformational change from A to B is monitored. If this conformational change is associated with net charge movement through the transmembrane field, an electrical signal may be generated. This signal may be detected as a transient current (47, 48, 172, 197) or its integral, a displacement of charge (92, 93), or a change in the fluorescence of an electrochromic dye (6, 16, 145, 240). While in theory transient and steady-state methods provide similar information about the charge-moving step (143), there are advantages and disadvantages to both approaches. The transient charge movement measurement is more direct but is technically more challenging, and measurement requires a preparation with a sufficiently high density of pumps so that the pump charge movement is detectable. A knowledge of the kinetics of the pump allows the investigator to manipulate ionic and substrate/ product conditions so that the pump can be fixed in one conformation for a substantial amount of time. This does create a small amount of ambiguity in the interpretation. In cells (and in steady-state measurements), the pumps are constantly cycling and spend very little time in any particular conformational state (that is, for a turnover of 100 per second, no state is occupied for more than approximately 10 msec). However, in the transient measurements, the pump cycle is often interrupted so that the pump can be maintained in conformation A for many seconds during experimental setup before the perturbation occurs. It is conceivable that during this time the pump has changed conformation from A to A′, and that A′ is not part of the normal pump cycle. Thus, most investigators try to determine whether the properties of the conformational change are consistent with a change of A to B—that is, are consistent with the steady-state rate for the conversion.

Using whole-cell patch-clamp techniques, Nakao and Gadsby (172) showed that a transient current was observed under conditions of Na/Na exchange but not K/K exchange (at saturating K concentrations) (9), thus providing some of the early evidence that Na steps involve net charge transfer. Other evidence was obtained using a bilayer technique (16, 47, 48) and using electrochromic dyes (6, 90, 240).

Recently, Hilgemann (92, 93), Heyse et al. (90), and Holmgren and Rakowski (100) have measured transients under a variety of conditions using the giant patch technique, electrochromic dyes, and high-time resolution two-electrode voltage clamp, respectively. In all three cases, release of the first Na to the extracellular solution is the major electrogenic step. However, the models of Heyse et al. (90) and Hilgemann (92) differ in some respects from the model of Holmgren and

Rakowski (100), which is very similar to the steady-state model proposed by Gadsby et al. (66) for Na/Na exchange. In Hilgemann's model, the release step is rapid at $E_m = 0$ and involves a symmetric electric field ($\delta = 0.5$). The studies by these four groups involve different species, different tissues, and different techniques. The fact that different models are used to fit the data in these studies may reflect differences in pump behavior. It will be interesting to see what model best fits transient charge movements and steady-state Na/ Na exchange data obtained on the same tissue from the same species.

Summary of Transport Mechanisms

The Na pump and the anion exchanger have been extensively studied. We hope we have provided examples that indicate some of the value of quantitative analysis of transporter function. This quantitative analysis has provided a wealth of important information about the electrical and kinetic properties of the Na pump and the anion exchanger. We have attempted to indicate the strengths and weaknesses of the approaches as well as some of the present ambiguities in interpretation, because we hope that the questions asked about the anion exchanger and the Na pump will be asked about other membrane-transport proteins.

The determination of fundamental kinetic parameters from a transport experiment is quite difficult, but the information is essential to the long-term goal of understanding the molecular mechanism of transport. As more is learned about the structure of transport proteins (for example, 253), it will become important to try to make a connection between various structural states and the kinetic events associated with the function of the transport protein. This can only be done quantifying the rates and energetics of the elementary events of binding and translocation.

INTEGRATIVE TRANSPORT STUDIES IN SINGLE CELLS: CELL-VOLUME REGULATION

We have focused on the study of the individual transport mechanisms in single cells. In addition to being useful for the study of transport kinetics and catalytic mechanism, single cells have also been used extensively in the study of the regulation of transport. Some aspects of transport regulation (for example, stimulation of transport by growth factors, or the regulation of transporters during cellular differentiation) are beyond the scope of our discussion. However, there is one area of transport regulation that has been studied so extensively in single cells that it should be mentioned here, and that is cell-volume regulation. The discussion of cell-volume regulation here is intended to be an introduction to the subject and not a comprehensive treatment. Interested readers are referred to the recent book on the subject edited by Strange (237).

With the exception of a few specialized tissues, cell membranes in general have a high permeability to water, mediated in part by the lipid bilayer and to varying degrees by the family of water channels, or aquaporins, that have recently been described (23). Because the water permeabilities of cell membranes are relatively high, the total amount of cell water depends on the total amount of intracellular osmotically active solute. The intracellular solute in turn depends on various transport processes, and, in some cells, the synthesis and degradation of osmolytes.

Pump-Leak Paradigm

The pump-leak concept of homeostasis is a recurring theme in cell physiology (167). In most cells the gradients of Na^+, K^+, and Ca^{2+} are maintained in a steady state by way of a balance between uphill fluxes driven by the ATP-driven pumps and downhill fluxes that are mediated by a variety of channels and transporters. The downhill pathways (those that do not use metabolic energy directly) were once collectively called "leaks," but it is now recognized that the leak pathways are complex, diverse, and in some cases highly regulated. The pump-leak idea also applies to epithelial cells that exhibit transcellular transport, except that the pumps and leaks are on different sides of the cell (see Reuss, this volume).

An important advance in the development of concepts related to ion homeostasis was the work of Tosteson and Hoffman (244) on sheep red blood cells. The steady-state cation composition of the red cells of some sheep (HK) is similar to that of other (noncarnivore) red cells: 110 mM K^+ and 30 mM Na^+. However, other sheep (LK) have red cells that have lower K^+ (20 mM) and higher Na^+ (130 mM). In both HK and LK sheep red cells, the intracellular cation contents are in a steady state in which the net uphill (pump-mediated) Na^+ efflux and K^+ influx are balanced by net downhill ("leak") Na^+ influx and K^+ efflux. However, the number of functioning Na^+ pumps is much higher in HK than in LK cells. Although the cation concentrations are very different in HK and LK cells, the same principle of a steady-state balance of pump and leak is operative in both kinds of cell. Moreover, the pump-leak model, along with the measured pump and leak fluxes, can account quantitatively for the different compositions of HK and LK cells.

Volume-Sensitive Transport Processes

The earlier formulations of the pump-leak hypothesis for cell-volume regulation did not include provision for acute regulation of either the active or downhill transport processes in response to changes in conditions. In the past 20 years, however, it has become clear that many cells have transport systems (ion channels, exchangers, or cotransporters) whose activity is sensitive to cell volume.

In general, an acute increase in cell volume (caused experimentally by exposure to a hypotonic medium or by preloading a cell with solute) will activate transport systems that cause cell shrinkage back toward the normal volume. This regulatory volume decrease (RVD) is caused by different transport processes in different cells. For example, in lymphocytes the RVD is mediated by separate K^+ and Cl^- channels (223). In many species of red cells, the RVD is mediated by the coupled cotransport of K^+ and Cl^- (22, 142).

Conversely, exposure of cells to hypertonic medium causes shrinkage followed (in some but not all cells) by a regulatory volume increase (RVI). RVI is mediated by Na^+-H^+ exchange in some cells (81) and Na^+-K^+-Cl^- cotransport (85, 95) in others. For both RVI and RVD, the regulatory volume change takes place by way of thermodynamically downhill transport processes. The Na^+-K^+ ATPase provides the gradients of Na^+ and K^+, and the volume-responsive transporters provide downhill pathways for either the gain or loss of solute as needed to maintain cell volume.

Detection of Cell Volume

Considerable recent work has been devoted to the question of how a change in cell volume causes an increase or decrease in the activity of particular transport processes (86, 224, 237). The problem is sometimes divided into two broad questions: How does a cell sense its volume, and what are the signaling mechanisms by which a change in cell volume results in the activation or inactivation of a particular transporter?

Cells could sense volume by way of mechanical deformations of the membrane and/or cytoskeleton. There is ample evidence for activation of volume-sensitive channels by membrane stretch (10, 24, 218). There is also very good evidence for a physical connection between various transporters in many tissues to the cytoskeleton. Readers are referred to the recent chapter by Mills et al. (168) on the role of the cytoskeleton in cell-volume regulation.

Another possible mechanism that cells may use to detect the concentration of cytoplasm involves the concept of macromolecular crowding (169, 170). In cytoplasm, the activity coefficient for a given protein is highly dependent on the total concentration of macromolecules as a consequence of the limited availability of free solvent volume. Because of cytoplasmic crowding, enzyme activities can be highly dependent on the total protein concentration. This dependence of enzyme activity on total protein concentration provides a potential mechanism for regulation of transport systems that control cell volume.

Transduction of Volume Signal to Transport Proteins

Numerous cell-signaling pathways have been proposed to mediate cell-volume regulation in various cells. Intracellular Ca^{2+} has a role in the activation of some, but certainly not all, swelling-activated channels. Arachidonic acid metabolites are believed to have a role in the regulatory volume decrease in Ehrlich ascites tumor cells (139).

There is considerable evidence that phosphorylation-dephosphorylation cycles are of importance in controlling at least some volume-sensitive transport processes. For example, the serine-threonine protein phosphatase inhibitors okadaic acid and calyculin are potent inhibitors of swelling-stimulated K^+/Cl^- cotransport in both mammalian and avian red cells (111, 117, 178, 236), suggesting that a protein dephosphorylation event is one of the steps associated with activation of transport by cell swelling. In avian erythrocytes, which exhibit shrinkage-stimulated $Na^+/K^+/Cl^-$ cotransport, okadaic acid and calyculin stimulate the cotransporter in the absence of cell shrinkage (178, 179, 188). These findings suggest that a phosphorylation event may activate shrinkage-stimulated $Na^+/K^+/Cl^-$ cotransport.

The proposal that a dephosphorylation activates swelling-sensitive transport, whereas phosphorylation activates shrinkage-stimulated transport, is attractive because it suggests that there is at least some overlap in the mechanisms for regulating swelling-activated and shrinkage-activated transporters. The best experimental support for coordinated regulation is the work by Parker and coworkers using dog red cells, which have a swelling-activated K^+/Cl^- cotransporter and a shrinkage-activated Na^+/H^+ exchange. Various manipulations (for example, changes in cell volume, intracellular Mg^{2+}, phosphatase inhibitors) that activate one system will tend to inactive the other, so that either one system or the other, but not both, is activated under a given set of conditions (182, 184).

In summary, in the past 10 to 20 years there have been major advances in the understanding of cell-volume regulation. The identities of many (but certainly not all) of the transport systems that mediate RVD and RVI in various cells are known. A major challenge for the future is to establish, at both the cellular and molecular level, the mechanisms by which cell volume activates and inactivates these transporters.

We thank Drs. Paul De Weer, David Gadsby, Phil Knauf, and Robert Rakowski for very helpful comments about the manuscript; Drs. George Kimmich, Owen Hamill, Leona Rubin, and Mike Sturek for the micrographs shown in Figure 7.1; and Ms. Elizabeth Kendrick for typing the manuscript. M.L.J. has been supported by NIH grant GM 26861; M.A.M., by NIH grant DK37512.

REFERENCES

1. Abercrombie, R. F., R. W. Putnam, and A. Roos. The intracellular pH of frog skeletal muscle: Its regulation in isotonic solutions. *J. Physiol. (Lond.)* 345: 175–187, 1983.
2. Abercrombie, R. F., and R. A. Sjodin. Sodium efflux in Myxicola giant axons. *J. Gen. Physiol.* 69: 765–778, 1977.
3. Altenberg, G. A., C. G. Vanoye, J. K. Horton, and L. Reuss. Unidirectional fluxes of rhodamine 123 in multidrug-resistant cells: Evidence against direct drug extrusion from the plasma membrane. *Proc. Natl. Acad. Sci. U.S.A.* 91: 4654–4657, 1994.
4. Altenberg, G. A., J. Copello, C. Cotton, K. Dawson, Y. Segal, F. Wehner, and L. Reuss. Electrophysiological methods for studying ion and water transport in *Necturus* gall bladder epithelium. *Methods Enzymol.* 192: 650–683, 1990.
5. Andreoli, T. E., M. Tieffenberg, and D. C. Tosteson. The effect of valinomycin on the ionic permeability of thin lipid membranes. *J. Gen. Physiol.* 50: 2527–2545, 1967.
6. Apell, H.-J. Electrogenic properties of the Na,K pump. *J. Membr. Biol.* 110: 103–114, 1989.
7. Apell, H.-J., V. Haring, and M. Roudna. Na,K-ATPase in artificial lipid vesicles. Comparison of Na,K and Na-only pumping mode. *Biochim. Biophys. Acta* 1023: 81–90, 1990.
8. Aronson, P. S. Electrochemical driving forces for secondary active transport: Energetics and kinetics of Na^+-H^+ exchange and Na^+-glucose cotransport. *Soc. Gen Physiol. Ser.* 38: 49–70, 1984.
9. Bahinski, A., M. Nakao, and D. C. Gadsby. Potassium translocation by the Na/K pump is voltage insensitive. *Proc. Natl. Acad. Sci. U.S.A.* 85: 3412, 1988.
10. Bear, C. E. A nonselective cation channel in rat liver cells is activated by membrane stretch. *Am. J. Physiol.* 258(*Cell Physiol.* 27): C421–C428, 1990.
11. Bennekou, P., and P. Christophersen. Flux ratio of vaolinomycin-mediated K^+ fluxes across the human red cell membrane in the presence of the protonophore CCCP. *J. Membr. Biol.* 93: 221–227, 1986.
12. Benson, S. W. *The Foundations of Chemical Kinetics,* edited by S. W. Benson. New York: McGraw-Hill, 1960.
13. Bidani, A., and E. D. Crandall. Velocity of CO_2 exchanges in the lungs. *Annu. Rev. Physiol.* 50: 639–652, 1988.
14. Blostein, R. Measurement of Na and K transport and Na,K-ATPase activity in inside-out vesicles from mammalian erythrocytes. *Methods. Enzymol.* 156: 171, 1988.
15. Blostein, R. Sodium pump-catalyzed sodium-sodium exchange associated with ATP hydrolysis. *J. Biol. Chem.* 258(13): 7948–7953, 1983.
16. Borlinghaus, R., H.-J. Apell, and P. Läuger. Fast charge translocations associated with partial reactions of the Na,K-pump: I.

Current and voltage transients after photochemical release of ATP. *J. Membr. Biol.* 97: 161–178, 1987.

17. Boyum, A. Isolation of mononuclear cells and granulocytes from human blood. *Scand. J. Clin. Lab. Invest.* 21 (Suppl. 97): 77–89, 1968.

18. Brahm, J. Temperature-dependent changes of chloride transport kinetics in human red cells. *J. Gen. Physiol.* 70: 283–306, 1977.

19. Breitwieser, G. E., A. A. Altamirano, and J. M. Russell. Osmotic stimulation of Na(+)-K(+)-Cl-cotransport in squid giant axon is [Cl−]i dependent. *Am. J. Physiol.* 258 (*Cell Physiol.* 27): C749–753, 1990.

20. Brinley, F. J., Jr., and L. J. Mullins. Sodium extrusion by internally dialyzed squid axons. *J. Gen. Physiol.* 50: 2303–2331, 1967.

21. Brugnara, C., H. F. Bunn, and D. C. Tosteson. Regulation of erythrocyte cation and water content in sickle cell anemia. *Science* 232: 388–390, 1986.

22. Brugnara, C., A. S. Kopin, H. F. Bunn, D. C. Tosteson. Regulation of cation content and cell volume in hemoglobin erythrocytes from patients with homozygous hemoglobin disease. *J. Clin. Invest.* 75: 1608–1617, 1985.

23. Chrispeels, M. J., and P. Agre. Aquaporins: Water channel proteins of plant and animal cells. *Trends Biol. Sci.* 19: 421–425, 1994.

24. Christensen, O. Mediation of cell volume regulation by Ca influx through stretch-activated channels. *Nature* 330: 66–68, 1987.

25. Cleland, W. W. The kinetics of enzyme-catalyzed reactions with two or more substrates or products. I. Nomenclature and rate equations. *Biochim. Biophys. Acta* 67: 104–137, 1963.

26. Cleland, W. W. Statistical analysis of enzyme kinetic data. in: *Methods Enzymol.*, S. P. 63: 103–137, 1988.

27. Cornelius, F. Functional reconstitution of the sodium pump. Kinetics of exchange reactions performed by reconstituted Na/K-ATPase. *Biochim. Biophys. Acta* 1071: 19–66, 1991.

28. Cornelius, F., and J. C. Skou. Na+-Na+ exchange mediated by (Na+ + K+)-ATPase reconstituted into lipsomes. Evalaution of pump stoichiometry and response to ATP and ADP. *Biochim Biophys. Acta* 818(2): 211–221, 1985.

29. Covarrubias, Y.G.-M., and P. De Weer. Kinetics of magnesium interaction with (Na+ + K+)-ATPase. *J. Gen. Physiol.* 96: 48a–49a, 1990.

30. Dalmark, M. Effects of halides and bicarbonate on chloride transport in human red blood cells. *J. Gen. Physiol.* 67: 223–234, 1976.

31. Dalmark, M., and J. O. Weith. Temperature dependence of chloride, bromide, iodide, thiocyanate and salicylate transport in human red cells. *J. Physiol. (Lond.)* 224: 583–610, 1972.

32. Darmon, A., O. Eidelman, and Z. I. Cabantchik. A method for measuring anion transfer across membranes of hemoglobin-free cells and vesicles by continuous monitoring of fluorescence. *Anal. Biochem.* 119: 313–321, 1982.

33. Demuth, D. R., L. C. Showe, M. Ballantine, A. Palumbo, P. J. Fraser, L. Cioe, G. Rovera, and P. J. Curtis. Cloning and structural characterization of a human non-erythroid band 3-like protein. *EMBO J.* 5: 1205–1214, 1986.

34. De Weer, P. Na,K ATPase: Reaction mechanisms and ion translocating steps. *Curr. Top. Membr. Transp.* 19: 599–623, 1983.

35. De Weer, P. Electrogenic pumps: Theoretical and practical considerations. *Soc. Gen. Physiol. Ser.* 38: 1–15, 1984.

36. De Weer, P. The Na/K pump: A current generating enzyme. In: *Regulation of Potassium Transport Across Biological Membranes*, edited by L. Reuss, J. M. Russell, and G. Szabo. Austin: University of Texas Press 1990, p. 5–28.

37. De Weer, P., D. C. Gadsby, and R. F. Rakowski. Voltage

dependence of the Na-K pump. *Annu. Rev. Physiol.* 50: 225, 1988.

38. De Weer, P., D. C. Gadsby, and R. F. Rakowski. Overview: Stoichiometry and voltage dependence of the Na/K pump. *Prog. Clin. Biol. Res.* 268A: 421–434, 1988.

39. De Weer, P., R. F. Rakowski. Current generated by backward-running electrogenic Na pump in squid giant axons. *Nature* 309(5967): 450–452, 1984.

40. De Weer P., D. C. Gadsby, and R. F. Rakowski. Voltage dependence of the Na-K pump. *Annu. Rev. Physiol.* 50: 225–41, 1988.

41. De Weer, P., R. F. Rakowski, and D. C. Gadsby. Voltage sensitivity of the Na+/K+ pump: Structural implications. In: *The Sodium Pump*, edited by E. Bamberg and W. Schoner. New York: Springer, 1994, p. 472–481.

42. Dissing, S., and J. F. Hoffman. Anion-coupled Na efflux mediated by the Na/K pump in human red blood cells. *Curr. Top. Membr. Transp.* 19: 693, 1983.

43. Dissing, S., and J. F. Hoffman. Anion-coupled Na efflux mediated by the human red blood cell Na/K pump. *J. Gen. Physiol.* 96: 167–193, 1990.

44. Esmann, M., and J. C. Skou. Temperature-dependencies of various catalytic activities of membrane-bound Na+/K+-ATPase from ox brain, ox kidney and shark rectal gland and of C12E8-solubilized shark Na+/K+-ATPase. *Biochim. Biophys. Acta* 944(3): 344–350, 1988.

45. Fabry, M. E., and M. Eistenstadt. Water exchange across red cell membranes: II. Measurement by nuclear magnetic resonance T1, T2 and T12 hydrid relaxation. The effects of osmolarity, cell volume, and medium. *J. Membr. Biol.* 42: 375–398, 1978.

46. Falke, J. J., K. J. Kanes, and S. I. Chan. The kinetic equation for the chloride transport cycle of band 3. *J. Biol. Chem.* 260: 9545–9551, 1985.

47. Fendler, K., E. Grell, E. Bamberg. Kinetics of pump currents generated by the Na+, K+-ATPase. *FEBS Lett.* 224: 83–88, 1987.

48. Fendler, K., S. Jaruschewski, A. S. Hobbs, R. W. Albers, J. P. Froehlich. Presteady state charge translocation in NaK-ATPase from eel electric organ. *J. Gen. Physiol.* 102: 631–666, 1993.

49. Forbush III, B. Na+ movement in a single turnover of the Na pump. *Proc. Natl. Acad. Sci. U.S.A.* 81: 5310–5314, 1984.

50. Forbush III, B. Rapid release of K and Rb from an occluded state of the Na,K-pump in the presence of ATP or ADP. *J. Biol. Chem.* 262: III04, 1987.

51. Forbush III, B. Overview: Occluded ions and Na,K-ATPase. In: *The Na+, K+-pump, Part A: Molecular Aspects*, edited by J. C. Skou et al. Alan R. Liss, Inc., p. 1988, 229–248.

52. Forbush III, B., and I. Klodos. Rate-limiting steps in Na translocation by the Na/K pump. *Soc. Gen. Physiol. Ser.* 46: 210–225, 1991.

53. Fozzard, H. A. Cellular basis for inotropic changes in the heart. *Am. Heart J.* 116(1): 230–235, 1988.

54. Fozzard, H. A., and M. F. Sheets. Cellular mechanism of action of cardiac glycosides. *J. Am. Coll. Cardiol.* 5(5): 10A–15A, 1985.

55. Froehlich, J. P. and Fendler K. The partial reactions of the Na+- and Na+ + K+-activated adenosine triphosphatases. *Soc. Gen. Physiol. Ser.* 46: 227–247, 1991.

56. Fröhlich, O. Relative contributions of the slippage and tunneling mechanisms to anion net efflux from human erythrocytes. *J. Gen. Physiol.* 84: 887–893, 1984.

57. Fröhlich, O., and R. B. Gunn. Erythrocyte anion transport: The kinetics of a single-site obligatory exchange system. *Biochim. Biophys. Acta* 864: 169–194, 1986.

58. Fröhlich, O., C. Liebson, and R. B. Gunn. Chloride net efflux from intact erythrocytes under slippage conditions. Evidence for a positive charge on the anion binding/transport site. *J. Gen. Physiol.* 81: 127–152, 1983.

59. Furuya, W., T. Tarshis, F.-Y. Law, and P. A. Knauf. Transmembrane effects of intracellular H2DIDS. Evidence for two conformations of the transport site of human erythrocyte anion exchange protein. *J. Gen. Physiol.* 83: 657–681, 1984.

60. Gadsby, D. C. Hyperpolarization of frog skeletal muscle fibers and of canine cardiac purkinje fibers during enhanced Na$^+$-K$^+$ exchange: Extracellular K$^+$ depletion or increased pump current? *Cur. Top. Membr. Transp.* 16: 17–33, 1982.

61. Gadsby, D. C. The Na/K pump of cardiac cells. *Annu. Rev. Biophys. Bioeng.* 13: 373–398, 1984.

62. Gadsby, D. C., and M. Nakao. Steady-state current-voltage relationship of the Na/K pump in guinea pig ventricular mycoytes. *J. Gen. Physiol.* 94(3): 511–537, 1989.

63. Gadsby, D. C., M. Nakao, and A. Bahinski. Voltage dependence of transient and steady-state Na/K pump currents in myocytes. *Mol. Cell. Biochem.* 89(2): 141–146, 1989.

64. Gadsby, D. C., M. Nakao, A. Bahinski, G. Nagel, and M. Suenson. Charge movements via the cardiac Na,K-ATPase. *Acta Physiol. Scand.* 146: 111–123, 1992.

65. Gadsby, D. C., J. Kumura, and A. Noma. Voltage dependence of Na/K pump current in isolated heart cells. *Nature* 315(6104): 63–65, 1985.

66. Gadsby, D. C., R. F. Rakowski, and P. De Weer. Extracellular access to the Na,K pump: Pathway similar to ion channel. *Science* 260: 100–103, 1993.

67. Galey, W. R., J. D. Owen, and A. K. Solomon. Temperature dependence of nonelectrolyte permeation across red cell membranes. *J. Gen. Physiol.* 61: 727–746, 1973.

68. Garay, R. P., and P. J. Garrahan. The interaction of sodium and potassium with the sodium pump in red cells. *J. Physiol.* 231: 297–325, 1973.

69. Garrahan, P. J., and R. P. Garay. The distinction between sequential and simultaneous models for sodium and potassium transport. *Curr. Top. Membr. Transp.* 8: 19–97, 1976.

70. Garrahan, P. J., and I. M. Glynn. The behavior of the sodium pump in red cells in the absence of external potassium. *J. Physiol. (Lond.)* 192: 159–174, 1967.

71. Geck, P., C. Pietrzyk, B.-C. Burckhardt, B. Pfeiffer, and E. Heinz. Electrically silent cotransport of Na, K, and Cl in Ehrlich cells. *Biochim. Biophys. Acta* 600: 432–447, 1980.

72. Glynn, I. M. The electrogenic sodium pump. *Soc. Gen. Physiol. Ser.* 38: 33–48, 1984.

73. Glynn, I. M. The Na$^+$,K$^+$-transporting adenosine triphosphatase. In: *The Enzymes of Biological Membranes*, edited by A. N. Martonosi. New York: Plenum Press, 1985, p. 35–114.

74. Glynn, I. M. How does the sodium pump pump? *Soc. Gen. Physiol. Ser.* 43: 1–17, 1988.

75. Glynn, I. M. Overview: The coupling of enzymatic steps to the translocation of sodium and potassium. *Prog. Clin. Biol. Res.* 268A: 435–460, 1988.

76. Glynn, I. M., and S.J.D. Karlish. ATP hydrolysis associated with an uncoupled sodium flux through the sodium pump: Evidence for allosteric effects of intracellular ATP and extracellular sodium. *J. Physiol.* 256: 465–496, 1976.

77. Glynn, I. M., and S.J.D. Karlish. Occluded cations in active transport. *Annu. Rev. Biochem.* 59: 171–205, 1990.

78. Goldman, D. E. Potential, impedence, and rectification in membranes. *J. Gen. Physiol.* 27: 37–60, 1943.

79. Goldshleger, R., Y. Shahak, and S.J.D. Karlish. Electrogenic and electroneutral transport modes of renal Na/K ATPase reconstituted into proteoliposomes. *J. Membr. Biol.* 113: 139–154, 1990.

80. Goldshleger, R., S.J.D. Karlish, A. Rephaeli, and W. D. Stein. The effect of membrane potential on the mammalian sodium-potassium pump reconstituted into phospholipid vesicles. *J. Physiol.* 387: 331–355, 1987.

81. Grinstein, S., S. Cohen, J. D. Goetz, and A. Rothstein. Osmotic and phorbol ester–induced activation of Na/H exchange: Possible role of protein phosphorylation in lymphocyte volume regulation. *J. Cell Biol.* 101: 269–276, 1985.

82. Grygorczyk, R., W. Schwarz, and H. Passow. Potential dependence of the "electrically silent" anion exchange across the plasma membrane of *Xenopus* oocytes mediated by the band 3 protein of mouse red blood cells. *J. Membr. Biol.* 99: 127–136, 1987.

83. Gunn, R. B., M. Dalmark, D. C. Tosteson, and J. O. Wieth. Characteristics of chloride transport in human red blood cells. *J. Gen. Physiol.* 61: 185–206, 1973.

84. Gunn, R. B., and O. Fröhlich. Asymmetry in the mechanism for anion exchange in human red blood cell membranes (Evidence for reciprocating sites that react with one transported anion at a time). *J. Gen. Physiol.* 74: 351–374, 1979.

85. Haas, M. Properties and diversity of (Na-K-Cl) cotransporters. *Annu. Rev. Physiol.* 51: 443–457, 1989.

86. Hallows, K. R., and P. A. Knauf. Principles of cell volume regulation. In: *Cellular and Molecular Physiology of Cell Volume Regulation*, edited by K. Strange. Boca Raton, FL: CRC Press, 1994, p. 3–29.

87. Hansen, U.-P., D. Gradmann, D. Sanders, and C. L. Slayman. Interpretation of current-voltage relationships for "active" ion transport systems: I. Steady-state reaction-kinetic analysis of class-I mechanisms. *J. Membr. Biol.* 63: 165–190, 1981.

88. Harris, E. J., and B. C. Pressman. Obligate cation exchanges in red cells. *Nature* 216: 918–920, 1967.

89. Hautmann, M., and K. F. Schnell. Concentration dependence of the chloride self-exchange and homoexchange fluxes in human red cell ghosts. *Pflugers Arch.* 405: 193–201, 1985.

90. Heyse, S., I. Wuddel, H.-J. Apell, and W. Stürmer. Partial reactions of the Na,K-ATPase: Determination of rate constants. *J. Gen. Physiol.* 104: 197–240, 1994.

91. Hilgemann, D. W. Giant excised cardiac sarcolemmal membrane patches: Sodium and sodium-calcium exchange currents. *Pflugers Arch.* 415: 247–249, 1989.

92. Hilgemann, D. W. Flexibility and constraint in the interpretation of Na$^+$/K$^+$ pump electrogenicity: What is an access channel? In: *The Sodium Pump: Structure Mechanism, Hormonal Control and Its Role in Disease*, edited by E. Bamberg and W. Schoner. New York: Springer, 1994, p. 507.

93. Hilgemann, D. W. Channel-like function of the Na,K pump probed at microsecond resolution in giant membrane patches. *Science* 263: 14, 1994.

94. Hodgkin, A. L., and B. Katz. The effect of sodium ions on the electrical activity of the giant axon of the squid. *J. Physiol. (Lond.)* 108: 37–77, 1949.

95. Hoffman, E. K. Anion transport systems in the plasma membrane of vertebrate cells. *Biochim. Biophys. Acta* 684: 1–31, 1986.

96. Hoffman, J. F., and P. C. Laris. Determination of membrane potential in human and Amphiuma red blood cells by means of a fluorescent probe. *J. Physiol. (Lond.)* 239: 519–552, 1974.

97. Hoffman, J. F., J. H. Kaplan, T. J. Callahan. The Na/K pump in red cells is electrogenic. *Federation Proc.* 38: 2440, 1979.

98. Hoffman, J. F., J. H. Kaplan, T. J. Callahan, and J. C. Freedman. Electrical resistance of the red cell membrane and the relation

between net anion transport and the anion exchange mechanism. *Ann. NY Acad. Sci.* 314: 357–360, 1980.

99. Hoffman, P. G., and D. C. Tosteson. Active sodium and potassium transport in high potassium and low potassium sheep red cells. *J. Gen. Physiol.* 58: 438–466, 1971.

100. Holmgren, M., and R. F. Rakowski. Pre–steady state transient currents mediated by the Na/K pump in internally perfused *Xenopus* oocytes. *Biophys. J.* 66: 912–922, 1994.

101. Horowitz, B., K. A. Eakle, G. Scheiner-Bobis, G. R. Randolph, C. Y. Chen, R. A. Hitzeman, and R. A. Farley. Synthesis and assembly of functional mammalian Na,K-ATPase in yeast. *J. Biol. Chem.* 265: 4189–4192, 1990.

102. Hunter, M. J. A quantitative estimate of the non-exchange restricted chloride permeability of the human red cell. *J. Physiol. (Lond.)* 218: 49P–50P, 1971.

103. Jencks W P. Utilization of binding energy and coupling rules for active transport and other coupled vectorial processes. *Methods Enzymol.* 171: 145–64, 1989.

104. Jenks, W. P. *Catalysis in Chemistry and Enzymology.* New York: McGraw-Hill, 1969.

105. Jennings, M. L. Proton fluxes associated with erythrocyte membrane anion exchange. *J. Membr. Biol.* 28: 187–205, 1976.

106. Jennings, M. L. Apparent "recruitment" of sulfate transport sites by the Cl gradient across the human erythrocyte membrane. In: *Membrane Transport in Erythrocytes,* edited by U. V. Lassen, H. H. Ussing, and J. O. Wieth. Copenhagen: Munksgaard, 1980, p. 450–463.

107. Jennings, M. L. Stoichiometry of a half-turnover of band 3, the chloride transport protein of human erythrocytes. *J. Gen. Physiol.* 79: 169–185, 1982.

108. Jennings, M. L. Structure and function of the red blood cell anion transport protein. *Annu. Rev. Biophys. Chem.* 18: 397–430, 1989.

109. Jennings, M. L. Rapid electrogenic sulfate-chloride exchange mediated band 3 in human erythrocytes. *J. Gen. Physiol.* 105: 21–47, 1995.

110. Jennings, M. L., M. Allen, and R. K. Schulz. Effects of membrane potential on electrically silent transport. *J. Gen. Physiol.* 96: 991–1012, 1990.

111. Jennings, M. L., and R. K. Schulz. Okadaic acid inhibition of KCl cotransport. Evidence that protein dephosphorylation is necessary for activation of transport by either cell swelling or N-ethylmaleimide. *J. Gen. Physiol.* 97: 799–817, 1991.

112. Jennings, M. L., and J. S. Smith. Anion-proton co-transport through the human red blood cell band 3 protein. Role of glutamate 681. *J. Biol. Chem.* 267: 13964–13971, 1992.

113. Jentsch, T. J., T. R. Stahlknecht, H. Hollwede, D. G. Fischer, S. K. Keller, and M. Widerholt. A bicarbonate-dependent process inhibitable by disulfonic stilbenes and a sodium/hydrogen exchange mediate $^{22}Na^+$ uptake into cultured bovine corneal endothelium. *J. Biol. Chem.* 260: 795–801, 1985.

114. Jewell, E. A., and J. B. Lingrel. Comparison of the substrate dependence properties of the rat, Na,K-ATPase $\alpha1$, $\alpha2$, and $\alpha3$ isoforms expressed in HeLa cells. *J. Biol. Chem.* 266: 16925–16930, 1991.

115. Jørgensen, P. L. Conformational transitions in the alpha-subunit and ion occlusion. *Soc. Gen. Physiol. Ser.* 46: 189–200, 1991.

116. Jørgensen, P. L., and J. P. Andersen. Structural basis for E_1-E_2 conformational transitions in Na,K-pump and Ca-pump proteins. *J. Membr. Biol.* 103: 95–120, 1988.

117. Kaji, D., and Y. Tsukitani. Role of protein phosphatase in activation of KCl cotransport in human erythrocytes. *Am. J. Physiol.* 260 (*Cell Physiol.* 29): C178–C182, 1991.

118. Kaplan, J. H. Sodium pump-mediated ATP:ADP exchange: The sided effect of sodium and potassium ions. *J. Gen. Physiol.* 80: 915, 1982.

119. Kaplan, J. H. Sodium ions and the sodium pump: Transport and enzymatic activity. *Am. J. Physiol.* 245 (*Gastrointest. Liver Physiol.* 5): G327–G333, 1983.

120. Kaplan, J. H. Ion movements through the sodium pump. *Annu. Rev. Physiol.* 47: 535, 1985.

121. Kaplan, J. H., and L. J. Kenney. ADP supports ouabain-sensitive K-K exchange in human red blood cells. *Ann. N.Y. Acad. Sci.* 402: 292, 1982.

122. Kaplan, J. H., and L. J. Kenney. Temperature effects of sodium pump phosphoenzyme distribution in human red blood cells. *J. Gen. Physiol.* 85: 123–136, 1985.

123. Karlish S.J.D. Measurement of active and passive Na^+ and K^+ fluxes in reconstituted vesicles. *Methods Enzymol.* 156: 179–88, 1988.

124. Karlish S J. The mechanism of active cation transport by the Na/K-pump. *Prog. Clin. Biol. Res.* 273: 207–216, 1988.

125. Karlish, S.J.D., R. Goldschleger, Y. Shanak, and A. Rephaeli. Charge transfer by the Na/K pump. In: *The Na,K pump, Part A: Molecular Aspects,* edited by J. C. Skou, J. G. Nørby, A. B. Maunsbach, and M. Esmann. New York: Alan R. Liss, Inc., 1988, p. 519–524.

126. Kimmich, G. A. Preparation and properties of mucosal epithelial cells isolated from small intestine. *Biochemistry* 9: 3659–3668, 1970.

127. Kinne, R.K.H. Transport in isolated cells from defined nephron segments. *Methods Enzymol.* 191: 380–409, 1990.

128. Knauf, P. A. Erythrocyte anion exchange and the band 3 protein: Transport kinetics and molecular structure. *Curr. Top. Membr. Transp.* 12: 249–363, 1979.

129. Knauf, P. A., G. F. Fuhrmann, S. Rothstein, and A. Rothstein. The relationship between exchange and net anion flow across the human red blood cell membrane. *J. Gen. Physiol.* 69: 363–386, 1977.

130. Knauf, P. A., F.-Y. Law, T. Tarshis, and W. Furuya. Effects of the transport site conformation on the binding of external NAP-tauring to the human erythrocyte anion exchange system. Evidence for intrinsic asymmetry. *J. Gen. Physiol.* 83: 683–701, 1984.

131. Knauf, P. A., and N. A. Mann. Use of niflumic acid to determine the nature of the asymmetry of the human erythrocyte anion exchange system. *J. Gen. Physiol.* 83: 703–725, 1984.

132. Knauf, P. A., and N. A. Mann. Location of the chloride self-inhibitory site of the human erythrocyte anion exchange system. *Am. J. Physiol.* 251 (*Cell Physiol.* 20): C1–C9, 1986.

133. Knauf, P. A., L. J. Spinelli, and N. A. Mann. Flufenamic acid senses conformation and asymmetry of human erythrocyte band 3 anion transport protein. *Am. J. Physiol.* 257 (*Cell Physiol.* 26): C277–C289, 1989.

134. Krupka, R. M. Role of substrate binding forces in exchange-only transport systems: I. Transition-state theory. *J. Membr. Biol.* 109: 151–158, 1989.

135. Krupka, R. M. Role of substrate binding forces in exchange-only transport systems: II. Implications for the mechanism of the anion exchanger of red cells. *J. Membr. Biol.* 109: 159–171, 1989.

136. Ku, C.-P., M. L. Jennings, and H. Passow. A comparison of the inhibitory potency of reversibly acting inhibitors of anion transport on chloride and sulfate movements across the human red cell membrane. *Biochim. Biophys. Acta.* 553: 132–141, 1979.

137. Labotka, R. J., and A. Omachi. Erythrocyte anion transport

of phosphate analogus. *J. Biol. Chem.* 262: 305–311, 1987.

138. Lafaire, A. V., and W. Schwarz. Voltage dependence of the rheogenic Na$^+$/K$^+$-A ATPase in the membrane of oocytes of *Xenopus laevis. J. Membl. Biol.* 91(1): 43–51, 1986.

139. Lambert, I. H., E. K. Hoffmann, and P. Christensen. Role of prostaglandins and leukotrienes in volume regulation in Ehrlich ascites tumor cells. *J. Membr. Biol.* 98: 247–256, 1987.

140. Laris, P. C., and J. F. Hoffman. Optical determination of electrical properties of red blood cell and Ehrlich ascites tumor cell membranes with fluorescent dyes. In: *Optical Methods in Cell Physiology, Vol. 40,* edited by P. De Weer and B. M. Salzberg. New York: John Wiley and Sons, 1986, p. 199.

141. Lassen, U. V. Membrane potential and membrane resistance of red blood cells. In: *Oxygen Affinity of Hemoglobin and Red Cell Acid Base Status,* edited by M. Rorth and P. Astrup. Copenhagen: Munksgaard, 1972, p. 291–306.

142. Lauf, P. K. K$^+$:Cl$^-$ cotransport: sulfhydryls, divaltne cations, and the mechanism of volume activation in a red cell. *J. Membr. Biol.* 88: 1–13, 1985.

143. Läuger, P. *Electrogenic Ion Pumps.* Sunderland, MA: Sinauer Associates, Inc., 1991, p. 313.

144. Läuger, P., and H.-J. Apell. Transient behavior of the Na$^+$/K$^+$-pump: Microscopic analysis of nonstationary ion-translocation. *Biochim. Biophys. Acta* 944: 451–464, 1988.

145. Läuger, P., and H.-J. Apell. Voltage dependence of partial reactions of the Na$^+$/K$^+$ pump: Predictions from microscopic models. *Biochim. Biophys. Acta* 945: 1–10, 1988.

146. Läuger P., and P. Jauch. Microscopic description of voltage effects on ion-driven cotransport systems. *J. Membr. Biol.* 91: 275–284, 1986.

147. Läuger, P. Voltage dependence of sodium-calcium exchange: Prediction from kinetic models. *J. Membr. Biol.* 99: 1–11, 1987.

148. Lee, B. S., R. B. Gunn, and R. R. Kopito. Functional differences among nonerythroid anion exchangers expressed in a transfected human cell line. *J. Biol. Chem.* 266: 11448–11454, 1991.

149. Lee, K. H., and R. Blostein. Red cell sodium fluxes catalysed by the sodium pump in the absence of K$^+$ and ADP. *Nature* 285: 338–339, 1980.

150. Lepke, S., and H. Passow. The permeability of the human red blood cell to sulfate ions. *J. Membr. Biol.* 6: 158–182, 1971.

151. Levinson, C. Sodium-dependent ion cotransport in a steady-state Ehrlich ascites tumor cells. *J. Membr. Biol.* 87: 121–130, 1985.

152. Lieberman, M., S. D. Hauschka, Z. W. Hall, B. R. Eisenberg, R. Horn, J. V. Walsh, R. W. Tsien, A. W. Jones, J. L. Wlaker, M. Poenie, F. Fay, F. Fabiato, and C. C. Ashley. Isolated muscle cells as a physiological model. *Am. J. Physiol.* 253 (*Cell Physiol.* 22): C349–C363, 1987.

153. Lingrel, J. B., and T. Kuntzweiler. Na$^+$, K($+$)-ATPase. *J. Biol. Chem.* 269(31): 19659–19662, 1994.

154. Maloney, P. C., S. V. Ambudkar, V. Anantharam, L. A. Sonna, and A. Varadhachary. Anion-exchange mechanisms in bacteria. *Microbiol. Rev.* 54: 1–17, 1990.

155. Marin, R., and J. F. Hoffman. Cytoplasmic anions and substrate-derived PO4 are simultaneously transported with Na in "uncoupled" Na efflux mediated by the red cell Na/K pump. *J. Gen. Physiol.* 88: 37a, 1986.

156. Marin, R., and J. F. Hoffman. Phosphate from the phosphointermediate (EP) of the human red blood cell Na/K pump is coeffluxed with Na, in the absence of external K. *J. Gen. Physiol.* 104: 1–32, 1994.

157. Marin, R., and J. F. Hoffman. ADP + orthophosphate (P$_1$) stimulates an Na/K pump-mediated coefflux of P$_1$ and Na in human red blood cell ghosts. *J. Gen. Physiol.* 104: 33–55, 1994.

158. Marx, A., J. P. Rubbersberg, and R. Rudel. Dependence of the electrogenic pump current of *Xenopus* oocytes on external potassium. *Pflusers Arch.* 408(5): 537–539, 1987.

159. Matsuoka, S., and D. W. Hilgemann. Steady state and dynamic properties of cardiac sodium-calcium exchange. Ion and voltage dependencies of the transport cycle. *J. Gen. Physiol.* 100: 963–1001, 1992.

160. McRoberts, J. A., C. T. Tran, and M. H. Saier, Jr. Characterization of low potassium-resistant mutants of the Madin-Darby canine kidney cell line with defects in NaCl/KCl symport. *J. Biol. Chem.* 258: 12320–12326, 1983.

161. Milanick, M. A. A branched reaction mechanism for the Na/K pump as an alternative explanation for a non-monotonic current vs. membrane potential response. *J. Membr. Biol.* 119: 33–39, 1991.

162. Milanick, M. A. Na-Ca exchange: Evidence against a ping-pong mechanism and against a Ca pool in ferret red blood cells. *Am. J. Physiol.* 261 (*Cell Physiol. 30*): 261: C185–193, 1991.

163. Milanick, M. A., and M.D.S. Frame: Intracellular Na, extracellular Ca, Cd and Mn: Implications for kinetic models of Na/Ca exchange in ferret red blood cells. *Ann. NY Acad. Sci.* 639: 604–615, 1991.

164. Milanick, M. A., and R. B. Gunn. Proton-sulfate cotransport: Mechanism of H$^+$ and sulfate addition to the chloride transporter of human red blood cells. *J. Gen. Physiol.* 79: 87–113, 1982.

165. Milanick, M. A., and R. B. Gunn. Proton-sulfate cotransport: External proton activation of sulfate influx into human red blood cells. *Am. J. Physiol.* 247 (*Cell Physiol. 16*): 247: C247–C259, 1984.

166. Milanick, M. A., and R. B. Gunn. Proton inhibition of chloride exchange: A synchrony of band 3 proton and anion transport sites. *Am. J. Physiol.* 250 (*Cell Physiol. 19*) C955, 1986.

167. Milanick, M. A. and J. F. Hoffman. Ion transport and volume regulation in red blood cells. *Ann. NY Acad. Sci.* 25: 174–186, 1987.

168. Mills, J. W., E. M. Schwiebert, and B. A. Stanton. The cytoskeleton and cell volume regulation. In: *Cellular and Molecular Physiology of Cell Volume Regulation,* edited by R. Strange. Ann Arbor, MI: CRC Press, Chap. 14, p. 241–253.

169. Minton, A. P. The effect of volume occupancy upon the thermodynamic activity of proteins: Some biochemical consequences. *Mol. Cell. Biochem.* 55: 119–140, 1992.

170. Minton, A. P., G. C. Colclasure, and J. C. Parker. Model for the role of macromolecular crowding in the regulation of cellular volume. *Proc. Natl. Acad. Sci. U.S.A.* 89: 10504–10506, 1992.

171. Morgan, M., P. Hanke, R. Grygorczyk, A. Tintschl, H. Fasold, and H. Passow. Mediation of anion transport in oocytes of *Xenopus laevis* by biosynthetically inserted band 3 protein from mouse spleen erythroid cells. *EMBO J.* 4: 1927–1931, 1985.

172. Nakao, M., and D. C. Gadsby. Voltage dependence of Na translocation by the Na/K pump. *Nature* 323: 628–630, 1986.

173. Nakao, M., and D. C. Gadsby. [Na] and [K] dependence of the Na/K pump current-voltage relationship in guinea pig ventricular myocytes. *J. Gen. Physiol.* 94(3): 539–565, 1989.

174. Newton, A. C., and W. H. Huestis. Efflux of dipicolinic acid from human erythrocytes, sealed membrane fragments, and

band 3-liposome complexes: A fluorescence probe for the erythrocyte anion transporter. *Anal. Biochem.* 156: 56–60, 1986.

175. Norby, J. G., and I. Kodos. Overview: The phosphointermediates of Na,K-ATPase. *Prog. Clin. Biol. Res.* 263A: 249–70, 1988.

176. Oberleithner, H., U. Kersting, and M. Hunter. Cytoplasmic pH determines K^+ conductance in fused renal epithelial cells. *Proc. Natl. Acad. Sci. U.S.A.* 85: 8345–8349, 1988.

177. Omay, H. S., and W. Schwarz. Voltage-dependent stimulation of Na + /K(+)-pump current by external cations: Selectivity of different K + cogeners. *Biochim. Biophys. Acta* 1104(1): 167–173, 1992.

178. Palfrey, H. C. Protein phosphorylation control in the activity of volume-sensitive transport systems. In: *Cellular and Molecular Physiology of Cell Volume Regulation,* edited by K. Stunga. Boca Raton, IL: CRC Press. Chap. 12 1994.

179. Palfrey, H. C., and E. B. Pewitt. The ATP and Mg^{2+} dependence of Na^+-K^+-^2Cl-cotransport reflects a requirement for protein phosphorylation: Studies using calyculin A. *Pflugers Arch.* 425: 321–328, 1993.

180. Parent, L., S. Supplisson, D.D.F. Loo, and E. M. Wright. Electrogenic properties of the cloned Na/glucose cotransporter: I. Voltage-clamp studies. *J. Membr. Biol.* 125: 49–62, 1992.

181. Parker, J. C. Sodium and calcium movements in dog red blood cells. *J. Gen. Physiol.* 71: 1–17, 1978.

182. Parker, J. C., C. Colclasure, and T. J. McManus. Coordinated regulation of shrinkage-induced Na/H exchange and swelling-induced [K-Cl] cotransport in dog red cells. Further evidence from activation kinetics and phosphatase inhibition. *J. Gen. Physiol.* 98: 869–892, 1991.

183. Parker, J. C., H. J. Gitelman, P. S. Glosson, and D. L. Leonard. Role of calcium in volume regulation by dog red blood cells. *J. Gen. Physiol.* 65: 84–96, 1975.

184. Parker, J. C., T. J. McManus, L. C. Starke, and J. H. Gitelman. Coordinated regulation of Na/H exchange and [K-Cl] cotransport in dog red cells. *J. Gen. Physiol.* 96: 1141–1152, 1990.

185. Passow, H. Passive ion permeability of the erythrocyte membrane. *Prog. Biophys. Mol. Biol.* 19: 425–467, 1969.

186. Passow, H. Molecular aspects of band 3 protein-mediated anion transport across the red blood cell membrane. *Rev. Physiol. Biochem. Pharmacol.* 103: 62–203, 1986.

187. Pedemonte, C. H., and Kaplan, J. H. Chemical modification as an approach to elucidation of sodium pump structure-function relations. *Am. J. Physiol.* 258 (*Cell Physiol.* 27): C1–C23, 1990.

188. Pewitt, E. G., R. S. Hegde, M. Haas, and H. C. Palfrey. The regulation of Na/K/^{2}Cl cotransport and bumetanide binding in avian erythrocytes by protein phosphorylation and dephosphorylation. *J. Biol. Chem.* 265: 20747–20756, 1990.

189. Plesner I. W. Use of experimental isotope-exchange fluxes in reversible enzyme and membrane transport models, assessed by simultaneous computer simulation of unidirectional and net chemical rates. *Biochem. J.* 286: 295–303, 1992.

190. Polvani, D., and R., Blostein. Protons as substitutes for sodium and potassium pump reaction. *J. Biol. Chem.* 263: 16757, 1988.

191. Polvani, D., and R. Blostein. Effects of cytoplasmic sodium concentration on the electrogenicity of the sodium pump. *J. Biol. Chem.* 264(26): 15182–15185, 1989.

192. Pratap, P. R., and J. D. Robinson. Rapid kinetic analyses of the Na + /K(+)-ATPase distinguish among different criteria for conformational change. *Biochim. Biophys. Acta* 1151(1): 89–98, 1993.

193. Pratap, P. R., J. D. Robinson, and M. I. Steinberg. The reaction sequence of the Na^+/K^+-ATPase: Rapid kinetic measurements distinguish between alternative schemes. *Biochim. Biophys. Acta* 1069: 288, 1991.

194. Preston, G. M., T. Piazza Carroll, W. B. Guggino, and P. Agre. Appearance of water channels in *Xenopus* oocytes expressing red cell CHIP28 protein. *Science* 256: 385–387, 1992.

195. Rakowski, R. F. Simultaneous measurement of changes in current and tracer flux in voltage-clamped squid giant axon. *Biophys. J.* 55(4): 663–71, 1989.

196. Rakowski, R. F. Stoichiometry and voltage dependence of the Na + /K + pump in squid giant axons and *Xenopus* oocytes. *Soc. Gen. Physiol. Ser.* 46: 339–353, 1991.

197. Rakowski, R. F. Charge movement by the Na/K pump in *Xenopus* oocytes. *J. Gen. Physiol.* 101: 117–144, 1993.

198. Rakowski, R. F., D. C. Gadsby, and P. De Weer. Stoichiometry and voltage dependence of the sodium pump in voltage-clamped, internally dialyzed squid giant axons. *J. Gen. Physiol.* 93(5): 903–941, 1989.

199. Rakowski, R. F., and C. L. Paxson. Voltage dependence of Na/K pump current in *Xenopus* oocytes. *J. Membr. Biol.* 106: 173–182, 1988.

200. Rakowski, R. F., L. A. Vasilets, J. LaTona, and W. Schwarz. A negative slope in the current-voltage relationship of the Na/K pump in *Xenopus* oocytes produced by reduction of external K. *J. Mem. Biol.* 121: 177–187, 1991.

201. Ray, W. J., Jr. Rate-limiting step: A quantitative definition. Application to steady-state enzymic reactions. *Biochemistry* 22(20): 4625–4637, 1983.

202. Reeves, J. P., and C. C. Hale. The stoichiometry of the cardiac sodium-calcium exchange system. *J. Biol. Chem.* 259(12): 7733–7739, 1984.

203. Rephaeli, A., D. E. Richards, and S.J.D. Karlish. Electrical potential accelerates the $E_1P(Na) \rightarrow E_2P$ conformational transition of (Na,K)-ATPase in reconstituted vesicles. *J. Biol. Chem.* 261: 12437–12440, 1986.

204. Restrepo, D., B. L. Cronise, R. B. Snyder, L. J. Spinelli, and P. A. Knauf. Kinetics of DIDS inhibition of HL-60 cell anion exchange rules out ping-pong model with slippage. *Am. J. Physiol.* 260 (*Cell Physiol.* 29): C535–C544, 1991.

205. Restrepo, D., D. J. Kozody, L. J. Spinelli, P. A. Knauf. Cl-Cl exchange in promyelocytic HL-60 cells follows simultaneous rather than ping-pong kinetics. *Am. J. Physiol.* 257 (*Cell Physiol.* 26): C520–527, 1989.

206. Reuss, L. Changes in cell volume measured with an electrophysiologic technique. *Proc. Natl. Acad. Sci. U.S.A.* 82: 6014–6018, 1985.

207. Rodbell, M. Metabolism of isolated fat cells. I. Effects of hormones on glucose metabolism and lipolysis. *J. Biol. Chem.* 239: 375–380, 1964.

208. Rosen, B. P. Recent advances in bacterial ion transport. *Annu. Rev. Microbiol.* 40: 263–286, 1986.

209. Roughton, F.J.W. Recent work on carbon dioxide transport by the blood. *Physiol. Rev.* 15: 241–296, 1935.

210. Rubin, L. J., R. S. Keller, J. L. Parker, and H. R. Adams. Contractile dysfunction of ventricular myocytes isolated from endotoxemic guinea pigs. *Shock* 2: 113–120, 1994.

211. Russell, J. M. ATP-dependent chloride influx into internally dialyzed squid giant axons. *J. Membr. Biol.* 28: 335–349, 1976.

212. Russell, J. M., and M. S. Brodwick. Properties of chloride transport in barnacle muscle fibers. *J. Gen. Physiol.* 73: 343–368, 1979.

213. Sachs, J. R. Kinetics of the inhibition of the Na-K pump

by external sodium. *J. Physiol. (Lond.)* 264: 449–470, 1977.

214. Sachs, J. R. The order of release of sodium and addition of potassium in the sodium-potassium pump reaction mechanism. *J. Physiol. (Lond.)* 302: 219–240, 1980.

215. Sachs, J. R. Potassium exchange as part of the overall reaction mechanism of the sodium pump of the human red blood cell. *J. Physiol. (Lond.)* 374: 221–244, 1981.

216. Sachs, J. R. The order of addition of sodium and release of potassium at the inside of the sodium pump of the human red cell. *J. Physiol. (Lond.)* 381: 149–168, 1986.

217. Sachs, J. R. Successes and failures of the Albers-Post model in predicting ion flux kinetics. In: *The Sodium Pump; Structure, Mechanism, and Regulation,* edited by J. H. Kaplan and P. De Weer. New York: Rockefeller University Press, 1991, p. 249–266.

218. Sackin, H. A stretch-activated K^+ channel sensitive to cell volume. *Proc. Natl. Acad. Sci. USA* 86: 1731–1735, 1992.

219. Sagar, A., and R. F. Rakowski. Access channel model for the voltage dependence of the forward-running Na^+/K^+ pump. *J. Gen. Physiol.* 103: 869–894, 1994.

220. Salhany, J. M., and P. B. Rauenbuehler. Kinetics and mechanism of erythrocyte anion exchange. *J. Biol. Chem.* 258: 245–249, 1983.

221. Salhany, J. M., and J. C. Swanson. Kinetics of passive anion transport across the human erythrocyte membrane. *Biochemistry* 17: 3354–3362, 1978.

222. Sanders, D. Generalized kinetic analysis of ion-driven cotransport systems: II. Random ligand binding as a simple explanation for non-michaelian kinetics. *J. Membr. Biol.* 90: 67–87, 1986.

223. Sarkadi, B., E. Mack, and A. Rothstein. Ionic events during the volume response of human peripheral blood lymphocytes to hypotonic media. I. Distinctions between volume-activated Cl^- and K^+ conductance pathways. *J. Gen. Physiol.* 83: 497–512, 1984.

224. Sarkadi, B., and J. C. Parker. Activation of ion transport pathways by changes in cell volume. *Biochim. Biophys. Acta* 1071: 407–427, 1991.

225. Schwarz, W., and L. A. Vasilets. Variations in voltage-dependent stimulation of the $Na+/K+$ pump in *Xenopus* oocytes by external potassium. *Soc. Gen. Physiol. Ser.* 46: 327–338, 1991.

226. Schwarz, W., L. A. Vasilets, H. Omay, A. Efthymiadis, J. Rettinger, and S. Elsner. Electrogenic properties of the endogenous and of modified *torpedo* Na^+/K^+-pumps in *Xenopus* oocytes: The access channel for external cations. In: *The Sodium Pump,* edited by E. Bomberg and W. Schoner. Darmstadt, Germany: Steinkopff, 1994, p. 482–494.

227. Schweigert, B., A. V. Lafaire, and W. Schwarz. Voltage dependence of the Na-K ATPase: Measurements of ouabain-dependent membrane current and ouabain binding in oocytes of *Xenopus laevis. Pflugers. Arch.* 412(6): 579–588, 1988.

228. Segel, I. H. *Enzyme Kinetics: Behavior and Analysis of Rapid Equilibrium and Steady-State Enzyme Systems,* edited by I. H. Segel. New York: John Wiley and Sons, 1975.

229. Seglen, P. O. Preparation of isolated rat liver cells. *Methods Cell. Biol.* 13: 29–84, 1976.

230. Sen, A. K., and W. F. Widdas. Determination of the temperature and pH dependence of glucose transfer across the human erythrocyte membrane measured by glucose exit. *J. Physiol. (Lond.)* 160: 392–403, 1962.

231. Sigel, E. Use of *Xenopus* oocytes for the functional expression of plasma membrane proteins. *J. Membr. Biol.* 117: 201–221, 1990.

232. Simchowitz, L., and P. DeWeer. Chloride movements in human neutrophils. Diffusion, exchange, and active transport. *J. Gen. Physiol.* 88: 167–194, 1986.

233. Skou, J. C. The fourth Datta lecture. The energy coupled exchange of Na^+ for K^+ across the cell membrane. The Na^+, $K(+)$-pump. *FEBS Lett.* 268(2): 314–324, 1990.

234. Skou, J. C., and M. Esmann. The Na, K-ATPase. *J. Bioenerg. Biomembr.* 24(3): 249–261, 1992.

235. Slayman, C. L., H. Kuroda, and A. Ballarin-Denti. Cation effluxes associated with the utpake of TPP^+, TPA^+, and $TPMP^+$ by *Neurospora:* Evidence for a predominantly electroneutral influx process. *Biochim. Biophys. Acta* 1190: 57–71, 1994.

236. Starke, L. C., and M. J. Jennings. [K-Cl] cotransport in rabbit red cells: Further evidence for regulation by protein phosphatase type 1. *Am. J. Physiol.* 244 (*Cell Physiol.* 33): C118–C124, 1993.

237. Strange, K., ed. *Cellular and Molecular Physiology of Cell Volume Regulation.* Ann Arbor, MI. CRC Press., 1994.

238. Sturek, M., K. Kunda, and Q. Hu. Sarcoplasmic reticulum buffering of myoplasmic calcium in bovine coronary artery smooth muscle. *J. Physiol.* 451: 25–48, 1992.

239. Sturek, M., P. Smith, and L. Stehno-Bittel. In vitro models of vascular endothelial cell calcium regulation. In: *Ion Channels of Vascular Smooth Muscle Cells and Endothelial Cells,* edited by N. Sperelakis and H. Kuriyama. New York: Elsevier p. 1991, 349–364.

240. Stürmer, W., R. Bühler, H.-J. Apell, and P. Läger. Charge translocation by the Na,K-pump: II. Ion binding and release at the extracellular face. *J. Membr. Biol.* 121: 163–176, 1991.

241. Thomas, J. A., R. N. Buchsbaum, A. Zimniak, and E. Racker. Intracellular pH measurement in Ehrlich asictes tumor cells utilizing spectroscopic probes generated in situ. *Biochemistry* 18: 2210–2218, 1979.

242. Thomas, R. C. Electrogenic sodium pump in nerve and muscle cells. *Physiol. Rev.* 52(3): 563–594, 1972.

243. Thomas, R. C. The effect of carbon dioxide on the intracellular pH and buffering power of snail neurones. *J. Physiol. (Lond.)* 255: 715–735, 1976.

244. Tosteson, D. C., and J. F. Hoffman. Regulation of cell volume by active cation transport in high and low potassium sheep red cells. *J. Gen. Physiol.* 44: 169–194, 1960.

245. Turin, L. Electrogenic sodium pumping in *Xenopus* blastomeres: Apparent pump conductance and reversal potential. *Soc. Gen. Physiol. Ser.* 38: 345–351, 1984.

246. Vasilets, L. A., and W. Schwarz. Regulation of endogenous and expressed Na^+/K^+ pumps in *Xenopus* oocytes by membrane potential and stimulation of protein kinases. *J. Membr. Biol.* 125(2): 119–132, 1992.

247. Vasilets, L. A., and W. Schwarz. Structure-function relationships of cation binding in the Na^+/K^+-ATPase. *Biochim. Biophys. Acta* 1154: 201–222, 1993.

248. Vasilets, L. A., H. S. Omay, T. Ohata, S. Noguchi, M. Kawamura, and W. Schwarz. Stimulation of the Na^+/K^+ pump by external $[K^+]$ is regulated by voltage-dependent gating. *J. Biol. Chem.* 266: 16285–16288, 1991.

249. Vasilets, L. A., T. Ohta, S. Noguchi, M. Kawamura, and W. Schwarz. Voltage-dependent inhibition of the sodium pump by external sodium: Species differences and possible role of the N-terminus of the α-subunit. *Eur. Biophys. J.* 21: 433–443, 1993.

250. Vaughan-Jones, R. D. Regulation of chloride in quiescent sheep heart Purkinje fibers studied using intracellular chloride and

pH-sensitive microelectrodes. *J. Physiol. (Lond.)* 137: 19203, 1979.

251. Verkman, A. S., and J. A. Dix. Effect of unstirred layers on binding and reaction kinetics at a membrane surface. *Anal. Biochem.* 142: 109–116, 1984.

252. Verkman, A. S., M. C. Sellers, A. C. Chao, T. Leung, and R. Ketcham. Synthesis and characterization of improved chloride-sensitive fluorescent indicators for biological applications. *Anal. Biochem.* 178: 355–361, 1989.

253. Wang, D. N., V. E. Sarabia, R.A.F. Reithmeier, and W. Kuhlbrandt. Three-dimensional map of the dimeric membrane domain of the human erythrocyte anion exchanger, Band 3. *EMBO J.* 13: 3230–3235, 1994.

254. White, B., and R. Blostein. Comparison of red cell and kidney (Na$^+$, K$^+$)-ATPase at 0°C. *Biochim. Biophys. Acta* 688: 685–690, 1982.

255. Williams, D. A., and F. S. Fay. Intracellular calibration of the fluorescent calcium indicator Fura-2. *Cell Calcium* 11: 75–83, 1990.

256. Zeuthen, T. The effects of chloride ions on electrodiffusion on the membrane of a leaky epithelium. Studies of intact tissue by microelectrodes. *Pflugers Arch.* 408: 267–274, 1987.

8. Epithelial transport

L U I S R E U S S | Department of Physiology and Biophysics, The University of Texas Medical Branch, Galveston, Texas

A TRANSPORTING EPITHELIUM is a sheet of closely apposed cells that can perform vectorial transport of solutes and/or water from one bathing solution to the other. Epithelia are barriers between the body and the external world, or between fluid compartments within the body, but they perform also transport, which eventually contributes to determining the volume and composition of the fluid compartments on both sides of the epithelial sheet. A very important function of transporting epithelia is homeostatic, i.e., designed to regulate the volume and composition of the body fluids. Other functions are absorption of nutrients (small intestine), excretion of catabolites (kidney, liver), and separation from the environment by a low-permeability barrier (epidermis). The aim of this chapter is to provide a general view of epithelial transport function from a cell-physiology viewpoint, focusing on the epithelial cell as the functional unit in the epithelium. The emphasis is on structure–function relationships, basic transport mechanisms, and transport regulation, using examples from specific epithelia, but with little discussion of specific epithelial organs. The discussion is addressed to advanced students, postdoctoral trainees, and newcomers to the field, not to established investigators. My intention is to present a balanced view of epithelial physiology, with an emphasis on ion and water transport, including discussion of classical studies, recent developments, and areas of incomplete knowledge.

TRANSPORTING EPITHELIA ARE SHEETS OF POLAR CELLS

Epithelia can have one or more cell types. They are remarkably heterogeneous in structural and functional properties, as best demonstrated by the features of different segments of the renal tubule, or of the intestine epithelium. Epithelia exemplify well the importance of structural organization in cell physiology. *Polarity* of epithelial cells includes structural, biochemical, and functional properties. To understand epithelial transport we must understand first the basics of epithelial morphology and cell biology (reviewed in 89 and 409).

Epithelial Structure Involves Specialized Cell–Cell and Cell–Matrix Junctions

Epithelial cells contact their neighbors via *junctional complexes* (see below), *desmosomes,* which include both spot desmosomes and belt desmosomes (or zonula adherens of the junctional complexes), and *gap junctions.* Desmosomes provide mechanical attachment of the cells; the attachment between the adjoining cells is due to cell-adhesion molecules (CAMs). CAMs can be Ca^{2+}-independent or Ca^{2+}-dependent *(cadherins).* Only the latter have been identified in epithelial cells. Cadherins are a family of integral membrane glycoproteins of 700–750 amino acid residues, with a single

transmembrane domain. They exhibit homophilic binding—that is, cadherins of the same type, from neighboring cells, bind to each other. The isoform expressed in epithelial cells is called *E-cadherin* or *uvomorulin.* It is located in the adhesion belts of the junctional complexes (see below) and effectively connects the cortical actin cytoskeleton of neighboring cells. The intracellular domain of E-cadherin interacts with cytoskeletal actin via several proteins (α, β, γ catenins, vinculin, α actinin, and plakoglobin). The CAMs of desmosomes are desmogleins and desmocollins, which attach to intermediate filaments in the cytoskeleton via specific proteins (desmoplakins, plakoglobin) (see ref. 80).

Gap junctions are large channels that communicate the cytosolic compartments of the neighboring cells. A complete gap junction consists of two connexons (one for each cell), each having six identical transmembrane subunits arranged parallel to each other, perpendicularly to the plasma membrane. Gap junctions are permeable to large molecules (up to about 1 kDa). Their conductance is regulated by phosphorylation, intracellular pH, intracellular pCa, and membrane voltage (46). The main functional consequence of this high-permeability cell-to-cell pathway is to make the epithelial layer into a functional syncytium.

A functional epithelium requires not only cell–cell contacts, but also contacts with the *basal lamina,* which is a specialized region at the interface between epithelia and underlying connective tissue. Epithelial cells are attached to the basal lamina by *hemidesmosomes* and *focal contacts.* Both forms of attachment involve *integrins,* which are transmembrane proteins that bind to specific motifs in extracellular matrix components. The hemidesmosomes attach to intermediate filaments in the cytoskeleton via a desmoplakin-like protein. Focal contacts attach to actin filaments via a complex of proteins that includes talin, α-actinin, and vinculin. It is becoming clear that integrins play a variety of roles in regulation of cell function. In epithelial cells, integrins appear to be involved in the initiation of structural and functional polarization following plating epithelial cells dissociated from monolayer cultures (528, 533, 534). They may be also involved in transducing extracellular mechanical energy into internal energy to reshape the cytoskeleton (657). Epithelial cell junctions are shown schematically in Figure 8.1.

Epithelial Polarity Is Essential for Vectorial Transport

The plasma membrane of the epithelial cell has two domains, *apical* and *basolateral,* separated by the *junctional complexes* (also named tight junctions, although in many epithelia they are leaky, not tight). In order

APICAL

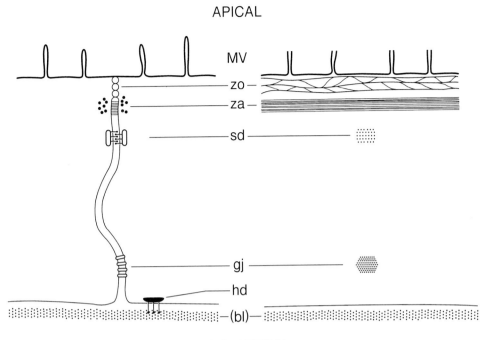

BASOLATERAL

FIG. 8.1. Epithelial cell junctions. *Left:* two adjacent epithelial cells viewed in a section normal to the apical surface. *Right:* lateral view of one cell. Abbreviations denote features shown in both diagrams: *MV* = microvilli; *zo* = zonula occludens; *za* = zonula adherens (or belt desmosome); *sd* = spot desmosome; *gj* = gap junction; *hd* = hemidesmosome; *bl* = basal lamina. Tight junction (junctional complex) includes *zo* and *za*. Modified from Cereijido (97), with permission.

for epithelia to perform vectorial transport between solutions of the same or similar composition, the transport properties of apical and basolateral membranes must be different. This results from differences in lipid composition and, most importantly, in expression of transport proteins among the two membrane domains (89, 409). Polarity also involves differences in cytoskeletal structure and organelle distribution. The polarity of the cytoskeleton is important in establishing and maintaining polar expression of transport proteins (for reviews, see refs. 414, 434). We will briefly discuss some transport-relevant structural aspects of epithelial cell polarity, concentrating on four issues: the junctional complexes, the features of the apical membrane domain, the features of the basolateral membrane domain, and the distribution of organelles. This follows closely the recent article by Matlin and Caplan (409). For further details, see Caplan and Rodríguez-Boulan (89).

The *junctional complexes* include the *zonula occludens* ("tight junction") and the *zonula adherens*. The latter appears to have as a major function its association with the cytoskeleton and hence influences the shape of regions of the epithelial cell (for instance, microvilli) and of the entire epithelial sheet (planar

and tubular epithelia). The *zonula occludens* is the transport-relevant junction. It functions mostly as a barrier between the two fluid compartments separated by the epithelium, but in many epithelia it is also an important permeation pathway, certainly for ions and arguably for water. Junctional transport is passive, secondary to other (transcellular) transport events.

The salient features in junctional structure are the contacts (fusion) of the outer leaflets of the plasma membranes of the adjacent cells observed in thin sections by electron microscopy, and the strands (ridges in P faces, grooves in E faces) visible in freeze-fracture replicas: the strands are arranged in roughly parallel arrays that branch and anastomose. The number of parallel strands encountered in a trajectory perpendicular to the epithelial plane appears to correlate with the degree of "tightness" of the junctional pathway, assessed by the electrical resistance (107; see also 14 and 512). Based on these structural observations, current models for the junctions imply that the outer leaflets of the plasma membranes are fused. This would explain the barrier function defined above—that is, the junctions limit the intercellular (i.e., junctional) permeation of polar substances in the transepithelial direction. The fusion of the outer leaflets of the membranes of adjacent cells also explains the *fence* function of the junctions (146)—that is, molecules that span both leaflets of the plasma membrane do not translocate by lateral diffusion between apical and basolateral membrane domains. In contrast, molecules that are contained within the inner leaflet of one of the plasma membrane domains move freely between the two domains (152). The molecular details of junctional structure are not clear; there is still disagreement on whether they are made solely of lipids or whether proteins are directly involved in the cell-to-cell contact. Studies in avian epithelia strongly suggest that the protein *occludin* may be the junctional protein accounting for cell–cell contact (203, 252). Occludin is a 65 kDa integral membrane protein which by immunofluorescence and immunoelectron microscopy has been shown to be expressed exclusively at junctional domains of epithelia and endothelial cells, precisely at the membrane-contact points. Following the molecular identification of chicken occludin, the sequences of human, mouse, dog, and rat-kangaroo isoforms were obtained (17b). Human, mouse, and dog occludins are ca 90% identical, but only ca 50% identical with the chicken and rat-kangaroo isoforms. Other proteins are important in assembly and function of the junctions (e.g., ZO-1; see 14 and 31), but, in contrast to occludin, they are peripheral and hence do not form part of the junctions themselves. Recent studies indicate that occludin targeting to the junctional region involves binding of its C-terminal domain to ZO-1 (204). Occludin, ZO-1, actin, and other proteins participate in the regulation of paracellular permeability (see 17a for review).

The *apical membrane domain* in many epithelia is characterized by the presence of microvilli in association with a specialized region of the cytoskeleton, the terminal web. For review of structural and biochemical aspects of the apical cytoskeleton of epithelial cells, see Mooseker (426). The major function of microvilli is to increase the effective apical surface area. In the mammalian renal proximal tubule, the amplification factor is ca. 20-fold (412). In addition, the microvilli constitute an anatomic unstirred layer in series with the apical surface, a fact that may have functional significance. Solute transport within an unstirred layer is exclusively by diffusion, and hence slow compared to convective (bulk flow) transport. This facilitates the generation of local concentration gradients of solutes transported in either direction. For instance, the action of digestive enzymes at the surface of the microvilli causes high concentration of products (e.g., hexoses) in the immediate vicinity of the absorptive membrane; the unstirred layer tends to preserve this high solute concentration because of lack of convective mixing. Microvilli are microscopic folds of the apical membrane domain around a cytoskeletal core. The core is composed of a bundle of 20 to 30 parallel thin actin filaments that are linked among themselves, and also to the plasma membrane and to the terminal web, by several proteins, of which villin and fimbrin are the principal ones. Villin is particularly interesting, because it regulates, in a [Ca^{2+}]-dependent fashion, the association and length of the actin filaments. Further, expression of villin cDNA in nonepithelial cells results in the formation of microvilli. The *terminal web* anchors the microvilli and also has connections to desmosomes present in the lateral surface of the epithelial cells and to the cytoskeleton associated with the zonula adherens. It contains *spectrin* overlying intermediate filaments. Spectrin provides rigidity to the cortical cytoskeleton. This arrangement stiffens the microvilli and keeps them perpendicular to the cell surface (7). Inasmuch as the terminal web contains myosin and tropomyosin in addition to actin, it has been speculated that it may be contractile and generate or influence motion of microvilli. In addition, the terminal web could control the tension on the junctions by circumferential contraction of the zonula adherens. The contraction of this intercellular cytoskeletal network could mediate epithelial morphogenesis—for example, folding of the epithelial sheet into tubes (7). It is possible that the putative role of intracellular [Ca^{2+}] in junctional permeability involves terminal-web proteins, but this is speculative.

The *basolateral membrane domain* does not contain microvilli, but in many epithelia is endowed with deep infoldings that substantially increase its surface area. In virtually all ion-transporting epithelia (the choroid plexus epithelium is a conspicuous exception) the Na^+ pump is expressed only in the basolateral membrane. Regardless of the targeting mechanism, which remains obscure, it appears that retention of the pumps in this membrane domain requires binding to ankyrin; ankyrin in turn binds to spectrin, and this protein anchors the complex to the cytoskeleton by binding to actin. Furthermore, at least in some epithelial cells the sodium pump is expressed in the lateral, but not in the basal region of the basolateral membrane (13). Recent studies suggest that E-cadherin, present only in lateral regions of the basolateral membrane, promotes the insertion of Na^+ pumps (419). Furthermore, Na^+, K^+-ATPase, E-cadherin, ankyrin, and fodrin colocalize in renal epithelial cells, indicating that anchoring of Na^+, K^+-ATPase to the cytoskeleton plays a major role in maintaining its polar expression (484a). Some proteins are apparently expressed only in basolateral membrane domains, some of them exclusively in the basal region (e.g., integrins) or exclusively in the lateral region (e.g., E-cadherin).

The *polarized distribution of organelles* is well exemplified by the familiar elongated mitochondria parallel to basolateral membrane infoldings in renal proximal tubule cells. The proximity of membrane proteins responsible for active transport (i.e., Na$^+$ pumps) and metabolic energy sources (i.e., mitochondria) is obviously convenient for metabolic coupling of transport. The mechanism of this distribution is not clear, but microtubules, which are polarized in epithelial cells (435), transport mitochondria in other cells (561, 639). The positions of other organelles are determined by the cytoskeleton, in particular the arrays of microtubules (e.g., Golgi apparatus around the nucleus). Of major importance for regulation of transepithelial transport is the polarized distribution of intracellular vesicular compartments containing transporters that can be inserted or retrieved from the plasma membrane in regulated fashion. Examples are the water pores inserted in the apical membranes of some tight epithelia following stimulation with vasopressin (reviewed in 266 and 654), and the regulated fusion of vesicles or tubule-vesicles containing H$^+$ pumps to the apical membranes of H$^+$-secreting epithelia (229, 538). Figure 8.2 illustrates the main features of epithelial cell polarity.

Until very recently, transport studies in isolated epithelial cells were limited because of loss of polarity following isolation. The development of enzymatic treatments that result in isolated epithelia cells with preserved polarity (578a, 613a, 622a, 622b) allows for optical and electrophysiological studies with access to both membrane domains, in cells with virtually complete preservation of the polarity observed in situ (613a, 622a), and exhibiting normal regulatory responses (622b).

TRANSPORTING EPITHELIA GENERATE AND MAINTAIN DIFFERENCES IN CHEMICAL COMPOSITION BETWEEN FLUID COMPARTMENTS

Epithelia separate the body from the outside world and also separate fluid compartments within the body. Epithelia are barriers and also transport ions, inorganic solutes, and water between these compartments. Transepithelial transport is translocation of matter *across* the epithelial sheet. This translocation can be in two directions: *absorption* (toward the internal medium—i.e., the extracellular fluid of the animal), and *secretion* (toward the external medium, which in most cases is

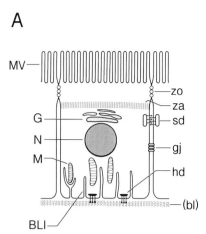

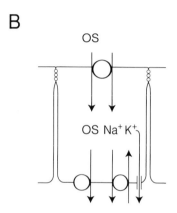

FIG. 8.2. Main features of epithelial cell polarity. *A.* Diagram depicting structural polarity: *MV* = microvilli; *G* = Golgi apparatus; *N* = nucleus; *M* = mitochondria; *BLI* = basolateral-membrane infoldings; *zo* = zonula occludens; *za* = zonula adherens; *sd* = spot desmosome; *gj* = gap junction; *hd* = hemidesmosome; *(bl)* = basal lamina. Positions of organelles are typical of most epithelial cells. *B.* Specific transport proteins are confined to different domains (apical or basolateral) of the plasma membrane. In example depicted, organic solute (*OS*, e.g., glucose) is transported into cell (across apical membrane) by secondary-active transport via Na$^+$-OS cotransporters and out of cell (across basolateral membrane) by uncoupled passive transport via OS carriers. Na$^+$ transport is downhill at apical membrane (via the cotransporters) and uphill at basolateral membrane (via the Na$^+$,K$^+$-ATPase). The K$^+$ channels in basolateral membrane mediate efflux of the K$^+$ that enter the cell via the Na$^+$,K$^+$-ATPase.

the lumen of the epithelial organ—e.g., renal tubule or gastrointestinal tract). Transepithelial *active transport* is a transcellular process, primarily dependent upon the operation of so-called ion pumps *(primary active transport)*. This creates passive driving forces (dependent on chemical or electrochemical potential differences) that may result in diffusive or electrodiffusive *passive transport* of other substances, or in coupled active transport of other substances *(secondary active transport)*.

Transepithelial Transport Involves Active Ion Transport

The thermodynamic definition of active transport is simple: it is the form of transport that occurs in the absence of or against the prevailing electrochemical gradient, i.e., it is energetically uphill transport. There are two kinds of active transport:

Primary active transport is characterized by direct use of metabolic energy, i.e., that released by the hydrolysis of ATP; the transporter responsible for primary active transport is a *pump*. Cells of transporting epithelia can express one or more of four transporting ATPases, namely the Na^+, K^+-ATPase, the H^+-ATPase, the H^+, K^+-ATPase, and the Ca^{2+}-ATPase.

Secondary active transport is characterized by the utilization of energy prestored in the electrochemical gradient for one substrate to translocate another species. In animal cells in general, and in epithelial cells in particular, secondary active transport is generally coupled to Na^+ transport, i.e., the energy stored in the Na^+ electrochemical gradient (produced by the Na^+ pump) is utilized to transport other substrates. The transport process can be cotransport (also called symport, in which Na^+ moves down its electrochemical gradient and the actively transported substrate moves in the same direction) or countertransport (also called antiport and exchange, in which Na^+ and the actively transported substrate move in opposite directions).

Passive Transport Processes also Contribute to Transepithelial Transport

Passive transport is energetically *downhill*, that is, driven by the existing chemical or electrochemical gradient. In isothermal and isobaric systems, when the driving forces can be only those derived from differences in concentration and/or electrical potential, the "driving force" for passive transport is given by

$$\Delta \bar{\mu}_j = z_j V_m F + RT \ln (C_j^i / C_j^o) \qquad (8-1)$$

where $\Delta \bar{\mu}_j$ is the electrochemical potential difference (the subscript j denotes the transported ion), z is the valence, V_m is the membrane voltage, F is the Faraday, R is the gas constant, T is the absolute temperature, C is concentration, and the superscripts i and o denote the two sides of the membrane (inside and outside, respectively). Based on this equation, Ussing derived a general expression that provides an excellent test for passive transport (633). This is the *flux ratio equation*, which establishes that

$$J_{in}/J_{out} = (C_i/C_o) \exp (z V_m F/RT) \qquad (8-2)$$

where J denotes flux (the subscripts "in" and "out" denote influx and efflux, respectively), and the other parameters have been defined. If the experimentally determined ratio of unidirectional fluxes (J_{in}/J_{out}) deviates from the prediction given by the flux ratio equation, which describes the passive driving forces, active transport must be suspected. Other explanations for deviation are exchange diffusion and single-file diffusion. See Schultz (563) for a detailed treatment.

Passive transport across biological membranes can be of two general types, namely *diffusion* across the lipid bilayer or *mediated transport*, which is via membrane transporters, that is, carriers or channels. Carriers are transporters that facilitate overall downhill transmembrane translocation of one or more species by a process that involves substrate binding on one side, change in conformation of the transporter, and translocation of the substrate and the binding site to the opposite side of the membrane. Carriers can transport either one substrate (facilitated diffusion, uniport) or two or more: in the same direction (cotransport, symport) or in opposite direction (countertransport, exchange or antiport). Transport via uniporters is always passive, whereas transport via symporters and antiporters involves passive translocation of at least one substrate (always) and may include secondary-active translocation of one or more additional substrates, provided that the net driving force is favorable.

Coupling is a term that has been used ambiguously in the epithelial transport literature. Thermodynamic coupling is different from molecular coupling. Molecular coupling implies mediated transport of more than one species by a transporting molecule, i.e, a carrier or a pump: the Na^+ and K^+ fluxes via the Na^+ pump are coupled, as are the Na^+, K^+, and Cl^- fluxes via the Na^+-K^+-$2Cl^-$ cotransporter. In addition to this form of coupling, there is thermodynamic coupling. Transepithelial Na^+ transport via an apical membrane channel in series with a basolateral membrane pump generates a transepithelial voltage that in turn can drive passive Cl^- transport via a different pathway. In this case, the "coupling" does not involve the transport molecule itself, but instead involves a more complex transfer of energy between two different transport processes.

Transepithelial Transport Involves Transcellular and Paracellular Pathways

Transepithelial transport of a specific substrate can be in principle *transcellular* or *paracellular*. In a normal, undamaged epithelium the paracellular pathway is intercellular (junctions in series with lateral intercellular spaces). Transcellular transport is via the cells, i.e., must involve transport across one membrane, distribution of the transported substance in the intracellular compartment, and transport across the opposite membrane. In epithelia with more than one cell type, net transport can be the complex result of transport via multiple parallel pathways, transcellular and paracellular, with different transport properties. Damage of an epithelial preparation, for instance by manipulation in vitro, can create artifactual paracellular pathways. Edge damage in preparations mounted between half-chambers is a recognized example of such artifacts (236, 277).

It has been occasionally suggested that transcellular vectorial transport might involve a path connecting the two cell-membrane domains, bypassing the bulk cytoplasm (e.g., 424). This hypothesis is based on structural observations suggesting the existence of intracellular tubular formations that could subserve this function. Experiments with invasive techniques that allow for measurements of ion activities and cell water content in transporting epithelial cells appear to rule out this theory and demonstrate instead that the "transport pool" for ions undergoing transepithelial transport is the entire cell interior. Illustrating this point, Figure 8.3 shows changes in electrochemical signals proportional to intracellular ion concentrations following experimental maneuvers that change transepithelial transport rates. These results are inconsistent with the notion of a small transport compartment bypassing the cytosol. However, several experimental studies suggest that inhomogeneous distribution of transporters in the cell-membrane domains, in combination with specific geometric factors, can result in compartmentalized transport. For instance, in fused renal epithelial cells it has been proposed that Na^+ entry results in uptake by the nucleus before distribution in the cytoplasm (443).

During steady-state transepithelial ion transport the fluid compartments delimited by the epithelial membranes (two extracellular and one intracellular compartment) must remain electroneutral. This fact imposes constraints on the transport processes. In the steady state and in the absence of external current application, spontaneous transport must obey the principle of macroscopic electroneutrality. In other words, net transepithelial transport of a cation must be balanced by net transport of anion(s) in the same direction and/or net transport of other cation(s) in the opposite direction. The balancing flux is generally driven by the transepithelial voltage generated by the primary transport process. Separation of charge in biological systems does occur, but only microscopically. It can be shown that the measured voltages across cell mem-

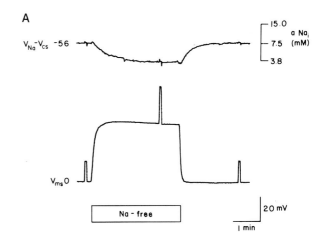

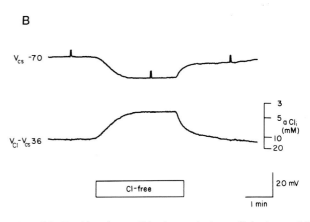

FIG. 8.3. Rapid and reversible changes in intracellular ion activities following removal of Na^+ (replaced with tetramethylammonium, TMA^+) or Cl^- (replaced with cyclamate) from apical bathing solution. Numbers on the left denote voltages (in mV) at the beginning of each record. A. Top, voltage measured by intracellular Na^+-selective electrode, referenced to the membrane voltage ($V_{Na} - V_{cs}$), intracellular Na^+ activity (aNa_i) scale on the right; bottom, transepithelial voltage (V_{ms}). Mucosa-positive change in V_{ms} was caused by $Na^+ - TMA^+$ paracellular bi-ionic potential. Na^+ removal from apical bathing solution causes a rapid, reversible fall in aNa_i. B. Top, basolateral-membrane voltage (V_{cs}); bottom, voltage measured by intracellular Cl^--selective electrode, referenced to the membrane voltage ($V_{Cl} - V_{cs}$), intracellular Cl^- activity (aCl_i) scale on the right. During exposure to Cl^--free mucosal medium, cell hyperpolarizes and aCl_i falls rapidly and reversibly. These results indicate that the Na^+ and Cl^- transport pools are accessible to the ion-sensitive microeletrode and hence involved the entire cytoplasm. From Reuss (511), with permission.

branes involve only minute excesses of charges of opposite polarities at the two surfaces of the phospholipid bilayer (686). Net transport of charge across an epithelium can occur in the steady state if a current is experimentally applied across the epithelium; under these conditions, charge balance occurs by ion transport at the electrodes in contact with the bathing solutions (see Fig. 8.7 and accompanying text). The parallel fluxes providing charge compensation in transporting epithelia are not necessarily via the same transport molecule or even via the same transepithelial pathway. For example, K^+ and Cl^- can be transported across the same membrane via separate channels; in Cl^--secreting epithelia it is widely accepted that Cl^- is actively transported via the epithelia cells, while Na^+ follows passively via the paracellular pathway.

Transepithelial transport is predominantly a transcellular process. Transporting epithelial cells are unique in that there is a continuous traffic of solute and/or water across them. These fluxes can be very large relative to the cell contents; e.g., in a normal mammalian renal proximal tubule, cell water is completely replaced every ~20 s (596). In the steady state, while cell volume and composition remain constant, the net amount of transported materials entering the cell across one membrane per unit time must be equal to the net amount leaving across the other membrane in the same time. In addition, when the transport rate at one cell membrane is altered, for instance by a hormone, this change must be rapidly and effectively followed by a matching change in transport rate at the other membrane, so that the steady-state transport rates match each other. This important adaptive mechanism, called *cross talk*, was first recognized by Schultz (564, 565). Much experimental work during the past decade has demonstrated that there are several forms of cross talk that account for the adjustment of transport rate across one cell membrane following primary changes in transport rate across the other one.

Chemical and Electrical Gradients Couple Ion Fluxes in Epithelia

Primary ion transport across epithelia cell membranes can result in changes in thermodynamic driving forces, which in turn may drive transport of other substances. A frequent way in which this happens in ion-transporting epithelia is by electrical coupling between the two transport events—that is, the first one alters the membrane voltage, and the change in membrane voltage in turn provides the driving force for transport of the second species. This process is better understood by an elementary treatment of epithelial electrophysiology (equivalent circuit analysis). Detailed discussions of basic epithelial electrophysiology and circuit analysis can be found in several excellent reviews (11, 64, 108, 130, 278, 374, 464, 483, and 562).

An epithelium consisting of one cell type and a finite paracellular pathway can be represented by the equivalent circuit shown in Figure 8.4. The main concept conveyed in this figure is that the voltage across each of the three components (cell membranes and

paracellular pathway) depends on all circuit elements. The reason is that there is intraepithelial current flow. The loop current (i_e) is given by:

$$i_e = \frac{E_b + E_s - E_a}{R_a + R_b + R_s} \quad (8-3)$$

and the membrane voltages (e.g., for the apical membrane) are given by:

$$V_a = E_a - i_e R_a \quad (8-4)$$

Inserting the expression for i_e into each membrane voltage equation yields the following three equations:

$$V_a = \frac{E_a(R_a + R_s) + (E_b - E_s)R_a}{R_a + R_b + R_s} \quad (8-5)$$

$$V_b = \frac{E_b(R_a + R_s) + (E_a + E_s)R_b}{R_a + R_b + R_s} \quad (8-6)$$

$$V_s = \frac{(E_b - E_a)R_s + E_s(R_a + R_b)}{R_a + R_b + R_s} \quad (8-7)$$

These equations demonstrate that, in principle, the voltage of a given membrane domain in an epithelial cell depends on all parameters in the circuit and not on the properties of that membrane alone, which are denoted by the so-called zero-current voltage (or equivalent electromotive force, EMF) and resistance. To show the differences, in Figure 8.4 we show the solution of Equations 8–5 through 8–7 for realistic assumed values of EMFs and resistances. The membrane voltages are in every case different from the respective EMFs. In conclusion: *(1)* the voltages of the two epithelial cell membrane domains depend on each other because they are connected by the paracellular pathway, which acts as an electrical shunt, and *(2)* changes in the properties of one of the membranes (EMF or electrical resistance) alter not only its own voltage, but also the voltage across the opposite membrane. These two points underscore the importance of membrane voltage as a mechanism of thermodynamic coupling in steady-state ion transport, and in intermembrane cross talk (464). The "shunting" process described requires a low R_s. If the value of R_s approaches infinity, the cell-membrane voltages approach the values of the respective EMFs, as can be seen by introducing $R_s \to \infty$ in Equations 8–5 and 8–6.

THE BUILDING BLOCKS OF EPITHELIAL FUNCTION ARE MEMBRANE TRANSPORTERS

As explained above, transepithelial transport can take place across the transcellular pathway, the paracellular pathway, or both. In most epithelia, biophysical studies of paracellular transport suggest a watery, electrically charged pathway that can be well-described as con-

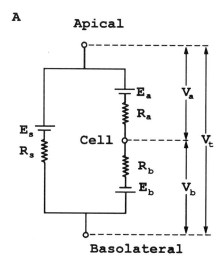

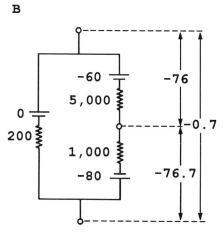

FIG. 8.4. *A*. Steady-state equivalent circuit of an epithelium with one cell type and a paracellular pathway of finite conductance. Each element in the circuit (*a*: apical membrane, *b*: basolateral membrane, *s*: paracellular pathway) is represented by a Thévenin electrical equivalent, i.e., an electromotive force (EMF), *E*, in series with a resistance, *R*. The Thévenin equivalent is an acceptable representation of any combination of linear electrical elements at the steady state. *B*. Membrane voltages for an epithelium with the indicated values of EMFs (in mV) and resistances (in $\Omega \cdot cm^2$). Polarities referred to basolateral solution (transepithelial values) or to adjacent solution (cell membrane values). Note that all voltages differ from the respective EMF values. Modified from Reuss et al. (515), with permission.

sisting of pores. Paracellular transport in high-permeability epithelia involves small solutes and possibly water and is always a passive process. Transcellular transport involves substrate translocation across both membrane domains and distribution of the substrate in an intracellular transport pool. In the case of highly liposoluble substrates, passive transport occurs predominantly via the lipid bilayer, by a *solubility-diffusion* process. An example of this kind of transport is the permeation of CO_2. Hydrophilic solutes, which

are less soluble in lipids, undergo passive transport via membrane proteins, i.e., mediated transport. Mediated transport can be primary active, secondary active, or passive. Examples of these modes of transport involve ions and most nutrients.

Transepithelial transport of ions, and to a certain extent of water, requires membrane transporters because the permeability of the lipid bilayer for water and hydrophilic substances is very low. These transporters are integral membrane proteins. Both their individual properties and their concerted operation must be unraveled in order to understand the transport function of epithelial cells. The transporters expressed in epithelial-cell membranes have the potential to perform two kinds of functions, namely transepithelial—*vectorial*—transport (the substrate enters the cell at one membrane and leaves the cell across the other membrane) and simple—*housekeeping*—transport (the substrate either enters or leaves the cell across a single membrane domain, but does not cross the cell). The function of vectorial transport is homeostatic (regulation of volume and/or composition of body-fluid compartments by absorption or secretion), either absorptive (transport of nutrients into the extracellular fluid) or excretory (elimination of certain solutes from the body). In contrast, the function of housekeeping transport is to regulate the volume and/or composition of the epithelial cell itself.

The transporters expressed in cell membranes can be classified, functionally, in three groups: *carriers*, *channels*, and *pumps*.

Ion pumps perform primary active transport. In the case of transporting epithelia, ion pumps are ATPases that use the energy liberated by ATP hydrolysis to translocate the substrate in an energetically uphill fashion. The energy is thought to be directly used in changing the conformation of the protein in a way that accounts for the substrate translocation, i.e., by making the bound substrate accessible to the opposite side of the membrane. For an excellent review of ion pumps, see Läuger (352). Only a handful of ion pumps have been identified in epithelial cells, namely the Na^+, K^+-ATPase, the Ca^{2+}-ATPase, The H^+-ATPase, and the H^+, K^+-ATPase (Table 8.1). Other plasma-membrane epithelial ion pumps have been proposed, but either they are expressed in unique epithelia or the evidence for their existence is controversial. Some structural and functional features of epithelial cell-membrane ion pumps are illustrated in Figure 8.5.

Channels and carriers are integral membrane proteins that can perform only *overall* passive transport; however, in the cases of carriers that transport more than one solute the electrochemical potential of one of these can be used to perform secondary active transport of another one. Both channels and carriers undergo

TABLE 8.1. *Plasma-Membrane Ion Pumps in Epithelial Cells*

Pump	Epithelium	Membrane Domain	Function	Subunits	References
P-type					
Na$^+$,K$^+$-ATPase	Ubiquitous	BL	Basolateral membrane Na$^+$ and K$^+$ transport Lower [Na$^+$]$_i$, raise [K$^+$]$_i$, increase V$_m$.	$(\alpha\beta)_2$	382, 588, 589
Ca^{2+}-ATPase	SIE Renal cortex*	BL	Basolateral membrane Ca^{2+} transport	$(\alpha\beta)_2$	90, 91
H$^+$,K$^+$-ATPase	GE, CE, CCD	A	H$^+$ secretion, K$^+$ absorption	$(\alpha\beta)_2$	319, 539
V-type					
H$^+$-ATPase	CCD, MCD	A or BL	H$^+$ or HCO$_3^-$ secretion	At least 9	225

For an authoritative reference on ion pumps, see Läuger (352). Abbreviations: GE = gastric epithelium, SIE = small intestine epithelium, CE = colon epithelium, CCD = cortical collecting duct, MCD = medullary collecting duct, A = apical, BL = basolateral. *Tubule segment not identified.

changes in conformation during or in relation to transport. Carriers and channels exhibit typical substrate selectivities, pharmacological inhibition, and saturation (although this is more evident in carriers). It is thought that carriers have substrate-binding sites that are accessible from one of the solutions adjacent to the membrane at a time; the conformational change required for translocation makes the binding site accessible to the other side. Channels differ in that when they are open their interior is accessible to both adjacent solu-

tions at the same time. An ion channel can exist in open (permeable, conductive) or closed (impermeable, nonconductive) states. The transition from the closed to the open state is called *gating*, where the gate is the part of the channel protein that "moves" during the transition. There is an essential difference between the transport-related conformational changes of channels and carriers. One gating episode allows the channel to be in the transporting mode for as long as it remains open, while a carrier must undergo the conformational

A

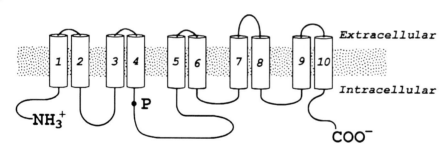

B

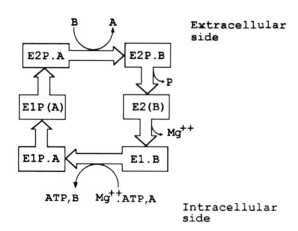

FIG. 8.5. Ion pumps in epithelial cells. *A.* Predicted general structure of α-subunit of P-type pump (representative of Na$^+$,K$^+$-, Ca^{2+}- and H$^+$,K$^+$-ATPases). The α-subunit contains binding site(s) for transported substrate(s), ATPase activity, regulatory domains, and inhibitor binding sites. It is a single peptide chain, with amino and carboxyl termini in the cytoplasmic side, probably ten transmembrane α-helices and ATP binding site in the second cytoplasmic loop. Based on Lingrel and Kunzweiler (382). *B.* Generalized catalytic cycle for P-type ATPases modeled as countertransport pumps exchanging Na$^+$ *(A)* for K$^+$ *(B)*, Ca^{2+} *(A)* for H$^+$ *(B)*, and H$^+$ *(B)* for K$^+$ *(A)* (Na$^+$,K$^+$-, Ca^{2+}- and H$^+$, K$^+$-ATPase, respectively. *E1* = conformation with ion-binding sites accessible from cytoplasm, *E2* = conformation with ion-binding sites accessible from the extracellular face, *E(A)* or *E(B)* = ion "occluded" in the protein. Modified from Sachs and Munson (542), with permission.

change in every transporting cycle. This difference is the basis for the principal criterion to distinguish channels and carriers, namely the *turnover number:* open channels will typically allow for ion permeation at rates of 10^6-10^8 s^{-1}, whereas carriers can transport at rates of the order of only 10^2-10^4 s^{-1} (601). Hence, carriers are slower transporters than channels. However, carriers can have multiple binding sites of different selectivities and thus perform secondary active transport by molecular flux coupling, a process that is not possible in channels. While channels are always *uniporters,* carriers can be *uniporters, symporters,* or *antiporters* (definitions and synonyms for these terms are given earlier in this chapter). For excellent, comprehensive reviews on channels and carriers, see Hille (282) and Stein (601), respectively.

Table 8.2 lists some of the main ion channels expressed in plasma membranes of epithelial cells. These channels can contribute to homocellular regulation (e.g., the basolateral K$^+$ channel recycles the K$^+$ taken up by the Na$^+$, K$^+$-ATPase) or vectorial transport (e.g., Na$^+$ channels in the apical membrane can be the pathway for Na$^+$ absorption; K$^+$ channels in the apical membrane can be the pathway for K$^+$ secretion). To contribute to active vectorial transport, the channels must be in series with a pump (apical Na$^+$ channel and basolateral Na$^+$, K$^+$-ATPase in some Na$^+$-absorptive epithelia).

Table 8.3 lists some of the carriers expressed in plasma membranes of cells from transporting epithelia. They are classified according to the number of substrates and the direction of transport. As in the case of

TABLE 8.2. *Examples of Plasma-Membrane Ion Channels in Epithelial Cells*

Channel, γ (pS)	Epithelia	Domain	Function	Regulation	Blocker	Special Features
Sodium						
Na$^+$, 5	FSE, AUB CCD, MCD	A	Electrodiffusive Na$^+$ entry	(+) aldo, PKA (−) Ca$^{2+}{}_i$, H$^+{}_i$	Amiloride	3 subunits cloned
Potassium						
MKC, 100–200	CCD, GBE SIE, CE, MUB	A BL	K$^+$ loss, cell shrinkage	(+) Ca$^{2+}{}_i$, ↓ V$_m$ (−) H$^+{}_i$	TEA$^+$	Not cloned from epithelia
IRKC, 2–60	PT, CCD, TALH PT, AE, FSE, CCD	A BL	K$^+$ secretion K$^+$ recycling	(+) ATP, PKA	Ba$^{2+}{}_o$	ROMK1 cloned, 391 aa (Figure 8.14)
Chloride						
CFTR, 7–10	AE, SIE, CE, PDE, GBE	A	Cl$^-$ secretion	(+) ATP, PKA	DPC	Cloned, 1480 aa (Fig. 16), P$_{Cl}$>P$_I$, linear, multi-ion pore
CSCC	AE, CE	A	Cl$^-$ secretion	(+) Ca$^{2+}{}_i$	DIDS	Not cloned, P$_I$>P$_{Cl}$, linear
SSCC, 30–50	AE, SIE, CE	A/BL	Cl$^-$ secretion	(−) ↓ V$_m$	DIDS	Not cloned, P$_I$>P$_{Cl}$, linear
ORCC, 40–50	AE, CE	A	Cl$^-$ secretion	(+) ↓ V$_m$, ATP$_o$	DIDS	Not cloned, outwardly- rectifying
ClC-K1, 40 (?)	TALH	BL	Cl$^-$ absorption	(+) PKA, Cl$_i$, ↓ V$_m$		ClC-K1 cloned, 687 aa, TALH channel?
DBCC, 40–50	CCD	BL	Volume regu- lation?	(+) ↑ V$_m$		Not cloned

For an authoritative reference on ion channels, see Hille (282). For reviews on epithelial ion channels, see text. Abbreviations: γ = single-channel conductance, *aa* = amino-acid residues, (+) = activation, (−) = inhibition, ↓ V$_m$ = membrane depolarization. *Channels:* MKC = maxi K$^+$ channel, IRKC = inwardly-rectifying K$^+$ channel, CFTR = cystic fibrosis transmembrane conductance regulator, CSCC = Ca^{2+}-sensitive Cl$^-$ channel, SSCC = swelling-sensitive Cl$^-$ channel, ORCC = outwardly-rectifying Cl$^-$ channel, ClC-K1 = Cl$^-$ channel from kidney thin ascending loop of Henle, DBCC = double-barrel Cl$^-$ channel; *gastrointestinal-tract epithelia:* GE = gastric epithelium, SIE = small intestine epithelium, CE = colon epithelium, SGE = salivary gland epithelium, GBE = gallbladder epithelium, PDE = pancreatic-duct epithelium; *renal-tubule epithelia:* PT = proximal tubule, TALH = thick ascending segment of Henle's loop, DT = distal tubule, CCD = cortical collecting duct, MCD = medullary collecting duct; *other epithelia:* AE = airway epithelium, LGE = lacrimal-gland epithelium, FSE = frog skin epithelium, AUB = amphibian urinary bladder epithelium, MUB = mammalian urinary bladder epithelium.   ↓ V$_m$: depolarization; ↑ V$_m$: hyperpolarization.

TABLE 8.3. *Examples of Plasma-Membrane Ion Carriers in Epithelial Cells*

Carrier	Isoform	Epithelium	Domain	Function	Features
SYMPORTERS					
Na^+-glucose	SGLT1,2	SIE, PT	A	Na^+ and glucose influx*	Cloned (Fig. 8.13); SGLT1: 662 aa, $2Na^+$: 1 gluc; SGLT2, 672 aa, 1 Na^+: 1 gluc
Na^+-amino acid	SAAT1	SIE, PT	A	Na^+ and amino acid influx	Cloned [333], 660 aa, transports A, S, C, P, G
Na^+-phosphate	NPT1	PT	A	Na^+ and phosphate influx	Cloned [677], 465 aa
Na^+-Cl^-	CCC-3	DT, WFUB	A	Na^+ and Cl^- influx	Cloned, 1002, 1023 aa, respectively
Na^+-K^+-$2Cl^-$	CCC-1 (NKCC1)	SRG	BL	Na^+, K^+ and Cl^- influx	Cloned, 1191 aa (Fig. 8.17)
	CCC-2 (NKCC2)	TALH	A	Na^+, K^+ and Cl^- influx	Cloned, 1099 aa
K^+-Cl^-		GBE, PT, TALH	BL	K^+ and Cl^- efflux*	Cloned, 1085 aa
Na^+-$(HCO_3^-)_3$		PT, TALH, GE	BL	Na^+ and HCO_3^- efflux	Cloned [528a]
ANTIPORTERS					
Anion Exchangers					
Cl^-/HCO_3^-	AE1b	CCD, MCD	BL	HCO_3^- efflux	Cloned, 850 residues, truncated AE1, allosteric internal H^+ site
		SIE, GBE	A	Cl^- influx	Not cloned, allosteric internal H^+ site
Na^+-dependent Cl^-/HCO_3^-		PT	BL	HCO_3^- efflux	Not cloned, minor role, $Na^+(HCO_3^-)_2$/Cl^-
Cl^-/formate		PT	A	Cl^- influx	Not cloned
Cation Exchangers					
Na^+/H^+	NHE2	SIE, GE, CE	?	Na^+ influx, H^+ efflux	Cloned, 813 residues, allosteric internal H^+ site, high sensitivity to amiloride
	NHE3	PT, SIE, GE	A	Na^+ influx, H^+ efflux	Cloned, 831 residues, allosteric internal H^+ site, low sensitivity to amiloride
Na^+/Ca^{2+}	NACA2,3	kidney	BL	Ca^{2+} efflux $[Ca^{2+}]_i$ regulation	Cloned, 1966 aa, 1945 aa, respectively [332]

For reviews on carriers, see Stein (601) and Reithmeier (507). References not given in text are in brackets.   Abbreviations: *aa*: = amino-acid residues. *gastrointestinal-tract epithelia*: GE = gastric epithelium, SIE = small intestine epithelium, CE = colon epithelium, SGE = salivary gland epithelium, GBE = gallbladder epithelium, PDE = pancreatic-duct epithelium, SRG = shark rectal gland; *renal-tubule epithelia*: PT = proximal tubule, TALH = thick ascending segment of Henle's loop, DT = distal tubule, CCD = cortical collecting duct, MCD = medullary collecting duct; *other epithelia*: AE = airway epithelium, LGE = lacrimal-gland epithelium, FSE = frog skin epithelium, AUB = amphibian urinary bladder epithelium, MUB = mammalian urinary bladder epithelium, WFUB = winter flounder urinary bladder epithelium.   *Influx and efflux denote flux into the cell and out of the cell, respectively. For the apical membrane, influx is in the absorptive direction and efflux in the secretory direction; for the basolateral membrane, influx is in the secretory direction and efflux is in the absorptive direction.

channels, for carriers to contribute to active transepithelial transport their function must be coordinated with that of a pump. However, carriers may be in series with a pump (e.g., apical Na^+-glucose cotransporter in series with the basolateral Na^+, K^+-ATPase in some Na^+-absorptive epithelia) or in parallel with a pump (e.g., basolateral Na^+, K^+, $2Cl^-$ cotransporter and Na^+, K^+-ATPase in Cl^--secreting epithelia). In the latter case, a third transporter is needed at the opposite membrane. Hence, both carriers and channels in epithelial-cell membranes can be thought of as "leak" pathways coordinated with primary active transporters. This "pump-leak" scheme is of the greatest importance in the functional organization of ion-transporting epithelia.

MECHANISMS OF ION TRANSPORT

The Two-Membrane Hypothesis: A General Epithelial-Transport Model

Many, but not all, epithelia share striking similarities in their functional organization. The basic scheme, formulated by Ussing and coworkers in the fifties (330), is the *two-membrane hypothesis*, originally developed to explain the mechanism of Na^+ transport in frog skin epithelium but successfully applied to a number of other transporting epithelia. As illustrated in Figure 8.6, the epithelium is represented by two barriers (or cell-membrane domains) with distinct transport features: the *outer* (apical) membrane is Na^+ permeable; the *inner* (basolateral) membrane is K^+ permeable and contains also the transporter responsible for the active transport step, namely the Na^+, K^+-ATPase. This scheme elegantly explains the basic features of Na^+ transport by frog skin and similar so-called tight epithelia and constitutes the starting point for other models. Let us dissect the elements of the system: *(1)* The transported ion must cross two membranes (and implicitly is distributed in the intracellular compartment). *(2)* For transepithelial ion transport to occur in the absence of a favorable external driving force, appropriate transport pathways must exist in both mem-

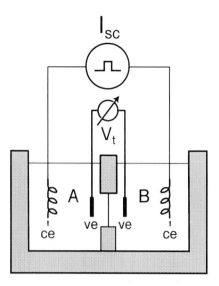

FIG. 8.7. Short-circuit current technique. Epithelium (vertical line between ve's) is tightly mounted in the middle partition of an Ussing chamber, separating solutions of identical composition, at the same pressure. Current is applied across the epithelium with an external circuit (*ce* = current electrode) and the transepithelial voltage (V_t) is measured (*ve* = voltage electrode). The short-circuit current (I_{sc}) is the current needed to make $V_t = 0$. This current, in the conditions defined, is equivalent to the sum of active ion fluxes across the epithelium. Pitfalls in the use of this technique include: *(1)* improper short-circuiting: series resistances in preparation produce voltage drops and the true transepithelial voltage is not zero; *(2)* the compositions of the bathing solution are different, i.e., there are passive (electrodiffusive or osmotic) driving forces that can account for net ion fluxes.

branes, and at least one of the transport steps must be active. In the case of frog epidermis and similar epithelia, the "pump" is in the basolateral membrane and the "leak" is in the apical membrane. *(3)* Active transport across one membrane (in this case, the basolateral one) changes the chemical or electrochemical gradient across the apical membrane, and hence "drives" passive transport across this membrane. *(4)* In the steady state, transepithelial transport must be electroneutral. Hence, in the absence of external current, if only one ion (e.g., Na^+) is actively transported, a passive permeation pathway ("shunt") must exist for at least one other ion (e.g., Cl^-).

The basic studies conducive to the formulation of the two-membrane hypothesis were: first, the invention of the short-circuit current technique (638; see Fig. 8.7), second, the assessment of the ionic permeabilities of the apical and basolateral surfaces of the epidermis (330), and third, the demonstration of the existence of a Na^+, K^+-ATPase at the basolateral barrier (329). Using the short-circuit current technique, it was demonstrated that the only ion actively transported across the epithelium was Na^+. The permeability studies con-

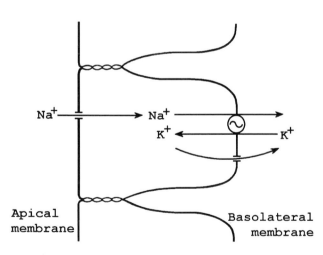

FIG. 8.6. Transport mechanism in Na^+-absorptive epithelia. The two-membrane hypothesis for Na^+-absorbing "tight" epithelia (330). At the apical-cell membrane, Na^+ entry is via an Na^+-channel blockable by amiloride. At the basolateral cell membrane, extrusion of Na^+ is mediated by an Na^+, K^+-ATPase inhibitable by ouabain. K^+ "recycles" across the basolateral cell membrane via a K^+ channel. Modified from Reuss and Cotton (518), with permission.

sisted of reducing anion permeation (by replacing Cl^- with the impermeant SO_4^{2-} or adding the Cl^- transport blocker Cu^{2+}) and then replacing mol-by-mol cations on one bathing solution at a time, observing the effect on the transepithelial voltage. The results revealed a striking difference: the apical membrane exhibited Na^+ selective permeability, whereas the basolateral membrane was K^+ selective. The basolateral location of the Na^+, K^+-ATPase was initially argued from the inhibition of transepithelial Na^+ transport brought about by K^+ removal from the adjacent bathing solution. The demonstration of inhibition of transport by pump inhibitors such as ouabain added support for this idea (329). Metabolic poisons also inhibited active Na^+ transport and hence reduced the short-circuit current (291).

Confirmation of the basic tenets of the two-membrane hypothesis has been obtained in a variety of studies in frog skin epithelium, amphibian and mammalian urinary bladder, colon, renal collecting tubule, and cultured epithelia of renal and intestinal origin (reviewed in 105, 166, 181, 263, 276, 354, 368, 380, 393, 457, 634, and 636).

The Original Two-Membrane Hypothesis Must Be Modified to Explain Transport in Other Epithelia. The success of the two-membrane hypothesis in explaining active Na^+ transport in simple epithelia is undeniable, and the elegance and simplicity of the model are widely appreciated. Although the basic tenets of this hypothesis are applicable to virtually every epithelium, it does not provide a complete explanation for the mechanisms of transepithelial ion transport in epithelia that exhibit very different properties from those of the frog skin epithelium.

The two-membrane hypothesis could not be extended to observations in gallbladder epithelium (143, 144, 145). The difficulties were that the transported species in the gallbladder appeared in fact to be NaCl instead of Na^+ and that the overall ion permeability of the epithelium was orders of magnitude higher than that of the frog skin. Together, these features result in electroneutral transport, i.e., the gallbladder transports at a rate comparable to that of the frog skin epithelium, but the short-circuit current is very low or nil. Further investigations of ion-transport mechanisms in renal proximal tubule (288, 682) and gallbladder (199, 201) culminated in the development of two important new ideas: first, the notion of ion transport pathways in parallel with the cells (paracellular or "shunt" pathway), which had already been suggested by Ussing and Windhager (637) from experiments in frog skin epithelia; second, the idea of electroneutral ion entry in absorptive epithelia. The latter notion, developed

from studies in small intestine (431) and gallbladder (197), proposes that the entry step in absorptive epithelia is not necessarily via ion channels, but can be via carriers (later identified as either Na^+-Cl^- or Na^+-K^+-$2Cl^-$ cotransporters, or Na^+/H^+ and Cl^-/HCO_3^- exchangers operating in parallel).

A second limitation of the original two-membrane hypothesis is that it did not incorporate internal feedback control between the transporters—that is, no functional regulation within a membrane (518) or functional regulation from one membrane to the other (564, 565, 568). Hence, adaptive changes in transport rate across a membrane domain in response to alterations in transport across the other domain would occur solely in response to changes in intracellular substrate concentration. In other words, cell volume and/or composition would be necessarily perturbed following changes in transport rate at one of the membranes. The experimental work that resulted in the cross-talk concept had not yet been carried out at the time that the two-membrane hypothesis was formulated. Nevertheless, Ussing and his associates made the first experimental observation showing a functional link between the two barriers of the frog skin epithelium (397).

Regardless of these shortcomings, virtually all cell models of epithelial transport processes start with the two-membrane model. With minimal reformulation, the paradigm is as follows: The initial step is active transport of one or more ions by an ATPase *(pump)*. This causes changes in the electrochemical potential differences between the cell and the extracellular solutions. Hence a driving force is generated, which results in net flux(es) via an overall passive, downhill transport mechanism *(leak)*. The nature of this secondary process (flux via channels or carriers) is what determines different functional transport modalities, as illustrated in the next section for Na^+-pump based systems. When the active step is mediated by the Na^+, K^+-ATPase, K^+ imported by the pump can either be transported across the epithelium in the opposite direction to Na^+ transport, if a K^+ transporter exists at the contralateral membrane, or recycles across the ipsilateral membrane, if this membrane expresses both Na^+ pumps and one or more K^+ transport pathways.

Tight and Leaky Epithelia: Junctional Ion Permeability Varies among NaCl-Absorptive Epithelia. The studies of Diamond in the early sixties (143, 144, 145) questioned the generality of the bioelectric concepts implied by the two-membrane hypothesis. The basic observation of Diamond was that the fish gallbladder epithelium, exposed to identical physiological salt solutions, could perform transepithelial fluid transport at a reasonable rate without generating a measurable transepithelial voltage. Soon thereafter, electrophysiological studies in renal tubules and in amphibian gallbladder proved that the high transepithelial conductance of these preparations is due to a high electrodiffusive permeability of the junctional complexes. Further, these results were confirmed by ultrastructural studies revealing high junctional permeability to markers such as lanthanum in leaky epithelia (392), a feature that is absent in native tight epithelia (637). The essential studies are illustrated in Figure 8.8. The notion of "tight" and "leaky" epithelia derived from this work.

The strategy of the electrophysiological studies was to estimate the electrical resistances of the cell membranes (by intracellular cable analysis or other techniques) and compare these values with the measured transepithelial resistance. Inasmuch as the transcellular and the paracellular pathway are in parallel (see Fig.

8.4), the transepithelial resistance (R_t) is related to transcellular $(R_a + R_b)$ and paracellular resistance (R_s) according to

$$\frac{1}{R_t} = \frac{1}{R_a + R_b} + \frac{1}{R_s} \qquad (8-8)$$

which clearly shows that the low-resistance pathway determines the value of R_t. The geometry of epithelial cells and epithelial tissues complicates this kind of analysis. First, the electrical data are usually related to the dimensions of the whole epithelium, which does not necessarily represent true membrane surface area because of apical-membrane microvilli and basolateral membrane infoldings that can amplify the surface by an order of magnitude or more. Further, cell width can explain large differences in paracellular conductance, because it determines the total horizontal length of

A

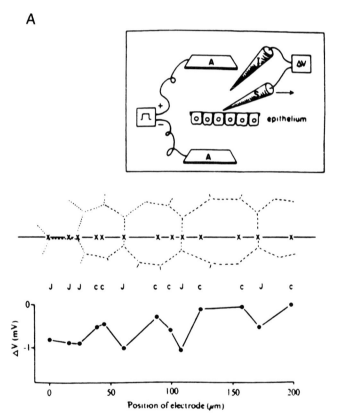

B

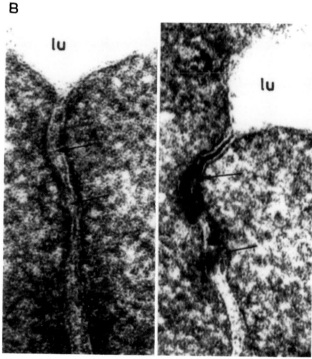

FIG. 8.8. Demonstration that some epithelia have "leaky" (high-permeability) junctions. *A.* Voltage scanning in *Necturus* gallbladder epithelium. High-conductance pathways for transepithelial current flow were detected with microelectrode *(S)* exploring electric field with respect to distant electrode in apical bathing solution *(I)*. S was moved along apical surface while large current was passed across the tissue (electrodes *A* in the inset). Microelectrode path is shown with respect to cell borders; x = tip position during voltage measurement, over junction *(J)* or over cell *(C)*. In voltage scan, downward

deflections mean current sinks (high-conductance pathway). This experiment demonstrates junctional location of a high-conductance transepithelial pathway. From Frömter and Diamond (201), with permission. *B.* Junctional region of rabbit gallbladder epithelial cells after exposure to La[3+] on the apical side. Lanthanum is clearly seen in intercellular pathway, indicating junctional permeability. Specimen fixed in glutaraldehyde, postfixed in OsO_4, and stained on grid with uranyl acetate and lead citrate. X 250,000. From Machen et al. (392), with permission.

the junctional network per cm^2 of epithelium. Hence, absolute resistance or conductance values for individual epithelia must be interpreted together with structural data. Reasonable estimates of membrane surface areas can be obtained from measurements of capacitance (109, 373), a method based on the observation that cell membranes have capacitances of the order of 1 μFcm^{-2}, or from morphometric analysis (412). The best criterion to assess the degree of epithelial leakiness is not the value of R_t, or even the value of R_s, but the ratio R_c/R_s (where R_c is the transcellular resistance = $R_a + R_b$). The larger this ratio, the leakier the epithelium. In most cases, differences in R_t values between epithelia can be largely explained by differences in the value of R_s, but an exception has been claimed, i.e., a "leaky" epithelium with a high-permeability transcellular pathway and a low-permeability paracellular pathway (28).

Studies in many epithelia have clearly shown that the degree of epithelial leakiness correlates with the mechanism of Na$^+$ transport. Ion transport in leaky epithelia is generally electroneutral, whereas in tight epithelia it is electrogenic, and the electrodiffusive permeability of the paracellular pathway is not the only explanation for this observation. A large fraction of ion entry across the apical membrane in epithelia such as those of the proximal tubule, the small intestine, and the gallbladder is electroneutral (in proximal tubule and small intestine a significant fraction of Na$^+$ entry is electrogenic, via cotransport with hexoses, amino acids and other organic substrates). In contrast, in epithelia such as those of the frog skin, toad urinary bladder, and renal collecting duct, ion entry across the apical membrane is entirely electrogenic—that is, via channels.

The paracellular pathway may have several functional roles in leaky epithelia. The functional role of the paracellular pathway in leaky epithelia is not entirely clear. It has been suggested that this pathway may be of importance for transepithelial transport of both ions and water.

A function of the paracellular shunt is to provide electrical coupling between the two cell membranes, as explained earlier (see 562). In many leaky epithelia, the relative K$^+$ permeability of the basolateral membrane is greater than that of the apical membrane, and hence the basolateral-membrane zero-current voltage is greater. Current flow between the two membranes results in hyperpolarization of the apical one, which in turn increases the driving force for inflow of positively charged permeable molecules or ions (e.g., Na$^+$ via channels or electrogenic cotransporters). This is the case in the proximal tubule and small intestine, where there are electrogenic Na$^+$-nutrient cotransporters (see Table 8.3), but not in epithelia that lack these carriers.

In the latter, it is possible that a function of intraepithelial current flow is to hyperpolarize the apical membrane to prevent K$^+$ loss via K$^+$ channels.

A second role for the paracellular pathway is passive ion transport, driven by differences in electrical potential, concentration, or both. For instance, if transcellular transport is electrogenic, a transepithelial voltage will develop, which can drive paracellular transport of other ion(s). Examples of this phenomenon are clear in Cl$^-$-transporting epithelia, both absorptive (e.g., thick ascending segment of the loop of Henle) and secretory (e.g., intestinal crypts). The transport of Cl$^-$ in these epithelia is transcellular and electrogenic, creating a transepithelial voltage with negative polarity on the side to which Cl$^-$ is transported. In parallel with Cl$^-$ there is an Na$^+$ flux in the same direction, thought to occur predominantly or entirely via the paracellular pathway (245, 586, reviewed in 241, 503, and 612). Another mechanism by which transcellular transport can result in paracellular passive ion transport is by changes in substrate concentration, instead of changes in transepithelial voltage. This process is favored when one of the fluid compartments is small, and therefore the best examples are found in proximal tubule. The arrangement *in series* of segments with different permeability properties also facilitates this process. A good example is that of anion transport in the renal proximal tubule. In the early segments, Na$^+$ reabsorption occurs in parallel with HCO$_3^-$ rather than Cl$^-$ reabsorption (237, 501). Inside the cell, H$_2$CO$_3$, whose production is catalyzed by carbonic anhydrase, partially dissociates in H$^+$ and HCO$_3^-$. The H$^+$ is secreted across the apical membrane in exchange for Na$^+$ (Na$^+$/H$^+$ exchanger) titrates lumen HCO$_3^-$, and the cell HCO$_3^-$ is cotransported across the basolateral membrane with Na$^+$. The predominant reabsorption of NaHCO$_3$ results in water reabsorption by osmosis (see 680), and less-permeable solutes are concentrated in the lumen. Among these solutes is Cl$^-$. Therefore, in the late proximal tubule the Cl$^-$ concentration in the lumen rises to a value higher than that in the interstitial fluid, which generates a driving force for passive Cl$^-$ reabsorption via the paracellular pathway.

A third possible function of the paracellular pathway is water transport. The net transepithelial solute transport tends to reduce the osmolality of the fluid in contact with the apical membrane and to increase the osmolality of the fluid adjacent to the basolateral membrane. This difference in osmolality causes water flow via the cells and the paracellular pathway according to their respective osmotic water permeabilities. The magnitude of the osmotic water flow across the paracellular pathway in leaky epithelia is controversial. An argument against a significant role is that the total surface area of the junctions is extremely small compared to that of the cell membranes (596, and see later, under MECHANISMS OF TRANSEPITHELIAL WATER TRANSPORT).

Finally, related to the previous one, a fourth possible function of the paracellular pathway is to permit solvent drag (180, 289). This phenomenon consists of net solute flux in the same direction as the net osmotic (or hydrostatic) water flow, and is interpreted as due to frictional interaction between solvent and solute during viscous flow. Solvent drag requires net water flow and a finite permeability of the water pathway for the solute. The effect of solvent drag would be to enhance transepithelial solute transport by coupling active salt transport to osmotic water flow and to passive solute transport. The magnitude of solvent drag in epithelial preparations is uncertain, because the presence of un-stirred fluid layers in series with the epithelial membranes makes such calculations difficult (289, and see later, under MECHANISMS OF TRANSEPITHELIAL WATER TRANSPORT).

In the Steady State, Transepithelial NaCl Absorption Satisfies the Electroneutrality Principle and Involves Coordination between Transporters. In epithelial cells performing vectorial transport, in the steady state both cell volume and ionic composition remain constant, indicating that the rates of entry and exit of each transported species must be the same. Deviations from this balance are only transiently tolerable; persistent lack of adjustment of the two rates results in a net gain or a net loss of solute by the cell, as shown in Figure 8.9. If at least one of the cell membranes is water permeable, then the change in cell solute content is accompanied by a change in cell water volume in the same direction. This under-scores the need for a mechanism of adjustment of the transport rates at the two membranes ("cross talk"), a concept that will be discussed in detail later, under MECHANISMS OF REGULATION OF TRANSEPITHELIAL TRANSPORT.

In addition, steady-state salt transport under spontaneous conditions (i.e., without experimental application of current) is an electroneutral process. In other words, the net charge balance of the cell must be zero, and the net transfer of charge across the epithelium (from one bathing solution into the other) must also be zero. These constraints are satisfied by different epithelia in different ways, as illustrated in Figure 8.10.

In some epithelia, the dominant or sole transfer is transcellular, and electroneutrality is satisfied at each membrane by either direct or indirect flux coupling. Direct or *molecular* coupling denotes transport of the two species via the same protein, i.e., cotransport or exchange. Indirect coupling involves transport via dif-

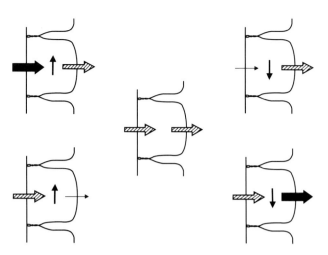

FIG. 8.9. Expected changes in epithelial-cell solute and/or water content by lack of adjustment of solute transport rates at the two membranes. *Center:* salt-absorbing epithelial cell at steady state, i.e., apical solute entry equals basolateral solute exit and cell solute content remains constant. *Left:* increases in solute content, by primary increase in entry *(top)* or primary decrease in exit *(bottom)*. *Right:* decreases in solute content, by primary decrease in entry *(top)* or primary increase in exit *(bottom)*. In all four instances, cell water volume would change in same direction as cell solute content, because of osmotic water flow (at least one of the cell membranes has a high water permeability). From Reuss (510), with permission.

ferent pathways. The coupling can be *thermodynamic* (for instance, ion flux through one transporter changes the membrane voltage and therefore the driving force for transport of the other ion through another transporter) or *allosteric* (for example, transport via a pathway changes the intracellular or extracellular level of a substrate, which in turn activates the other transporter by binding to an allosteric site). In recent years, several examples have been proposed that illustrate the complexity of the interrelationships between transporters expressed in the same membrane domain. Understanding these processes has become a major task in epithelial-cell physiology, as discussed in more detail later (see MECHANISMS OF REGULATION OF TRANSEPITHELIAL TRANSPORT).

In other epithelia, net charge transfer across the cells may occur in the steady state, if it is balanced by a paracellular net charge transfer of opposite sign. For instance, let us consider the isolated frog skin epithelium incubated with identical physiologic salt solutions on both sides. In the absence of external current application, the transcellular Na^+ flux can be accompanied by passive, paracellular Cl^- transport (driven by the transepithelial voltage). The condition of zero transepithelial current is referred to as "open circuit." In contrast, if a current equivalent to the net transepithelial Na^+ flux is applied, then the transepithelial voltage is

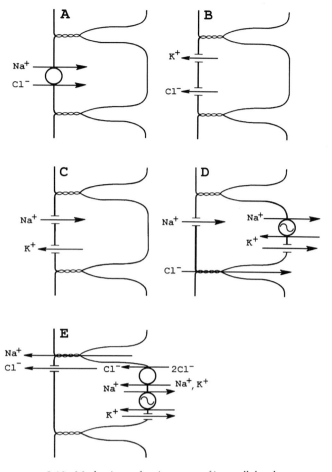

FIG. 8.10. Mechanisms of maintenance of intracellular electroneutrality and solution electroneutrality during steady-state transepithelial ion transport. In all diagrams, Na$^+$ or K$^+$ transport is the primary event, indirectly or directly coupled to anion flux in same direction or to cation flux in the opposite direction. Examples of epithelia in which the scheme depicted has been proven or is suspected are denoted in parentheses. *A.* Na$^+$ and Cl$^-$ fluxes are directly coupled by cotransport (mammalian renal distal tubule). *B.* K$^+$ and Cl$^-$ fluxes occur via independent channels; the fluxes could be "coupled" by the membrane voltage, both as driving force for electrodiffusion and as gating mechanism (amphibian gallbladder stimulated by cAMP). *C.* Na$^+$ and K$^+$ fluxes occur in opposite directions, via separate channels; they may be "coupled" by the membrane voltage (mammalian renal cortical collecting data). *D.* Absorptive Na$^+$ and Cl$^-$ transepithelial fluxes occur via different pathways, transcellular and paracellular, respectively (frog skin epithelium). *E.* Secretory Cl$^-$ and Na$^+$ fluxes occur via different pathways, transcellular and paracellular, respectively (airway epithelium).

canceled, the driving force for passive Cl$^-$ transport becomes zero, and only Na$^+$ is transported (the charge balance is provided by electron flow in the external circuit and ion flows at the electrode-solution interfaces, as explained in Fig. 8.7).

Paracellular anion transport is indispensable in most electrogenic Na$^+$-transporting epithelia to maintain

absorption. Otherwise, the transepithelial voltage generated by active Na$^+$ transport would reach a level equivalent to the maximum capacity of the Na$^+$ pump and transport would cease ("static head"). Under these conditions, basolateral net Na$^+$ transport is abolished, intracellular [Na$^+$] falls, and across the apical membrane the Na$^+$ chemical driving force is equal and opposite to the electrical driving force and Na$^+$ entry also ceases. The transepithelial voltage at which net Na$^+$ transport ceases (if [Na$^+$] is the same in both solutions) is sometimes referred to as the "electromotive force of the Na$^+$ pump," E_{Na} (633). If the tissue is experimentally subjected to a transepithelial voltage opposite to and greater than E_{Na}, then the Na$^+$ pump "reverses": Na$^+$ enters the cell across the basolateral membrane and ATP is synthesized (136).

Mechanisms of Transepithelial NaCl Transport in Absorptive Epithelia

In NaCl-absorptive epithelia, Na$^+$ translocation always involves an active transport step at the membrane adjacent to the *trans* solution (i.e., the solution toward which Na$^+$ is transported). In most epithelia, this is the basolateral membrane; in the choroid plexus and the retinal pigment epithelium, the pump is located on the apical membrane and the direction of transport is reversed; hereafter we will ignore this exception. For clarity I will classify NaCl-transporting epithelia in two groups: *(1) Na$^+$-transporting epithelia,* which are electrogenic, characterized by the presence of Na$^+$ channels in the apical membrane and generally tight, i.e., with high paracellular electrical resistance; examples are the renal collecting duct, frog skin, and both amphibian and mammalian urinary bladder. *(2) NaCl-transporting epithelia,* which are electroneutral (or nearly electroneutral), do not express a significant number of apical membrane Na$^+$ channels (Na$^+$ entry is carrier-mediated), and are generally leaky, i.e., with low paracellular electrical resistance; the simplest example is the gallbladder epithelium; epithelia from the small intestine and renal proximal tubule are also included in this group for simplicity, although Na$^+$ entry mechanisms are partly electroneutral and partly electrogenic, but nevertheless not channel-mediated.

The functional organization of the basolateral membrane is similar for both Na$^+$-transporting and NaCl-transporting epithelia. As stated in the two-membrane hypothesis, at this membrane Na$^+$ is pumped out of the cell, whereas K$^+$ is pumped into the cell by the Na$^+$, K$^+$-ATPase and generally recycles via a leak pathway in the same membrane, although in some cases it is also transported across the opposite membrane. The nature of the K^+ leak has been clarified during the last decade. It consists of K^+ channels, probably of several types, and, in some epithelia, an electroneutral K^+-Cl^- cotransporter. The principal difference between Na^+-transporting and NaCl-transporting epithelia is in the mechanism of Na^+ entry across the apical membrane. Na^+ entry can be channel-mediated or carrier-mediated, and the latter can be electrogenic or electroneutral. Examples of the transporters involved in these processes are listed in Table 8.2 and Table 8.3.

The Distinctive Feature of Electrogenic, Na^+-Transporting Epithelia Is the Presence of Amiloride-Sensitive Na^+ Channels in the Apical Membrane. The apical-membrane channels accounting for the electrodiffusive Na^+ permeability of frog skin epithelium and other tight epithelia, and predicted from the Ussing model, are the so-called amiloride-sensitive epithelial Na^+ channels (reviewed in 46a, 47, 216, and 458).

Epithelial Na^+ channels have variable selectivity for Na^+ over K^+, exhibit little if any voltage sensitivity, and are sensitive to the pyrazine diuretic amiloride (and some of its analogs). Palmer (457) has classified these channels, according to their Na^+ selectivity, in three groups: *(1)* High selectivity: $P_{Na}/P_K \geq 10$, low conductance (~ 5 pS), long open and closed times (0.5–5.0s), found in tissues and in epithelial cells in culture. *(2)* Moderate selectivity: $P_{Na}/P_K = 3$–4, higher conductance (7–15 pS), much shorter open and closed times (≤ 50 ms), found only in cultured cells. *(3)* Nonselective: $P_{Na}/P_K \leq 1.5$, either high (23–28 pS) or low (≥ 3 pS) conductance, found only in cultured cells. Channels in the three groups have similar affinity for amiloride ($K_i \leq 0.5$ μM). The existence of moderate and low-selectivity channels in native tissues is questionable. A possibility is that they represent degraded high-selectivity channels (370, 371, 384).

Amiloride is an effective blocker of the high-selectivity channel. The apparent affinity for amiloride and the pattern of effects of amiloride analogs are typical and distinct from the patterns of block of other transporters by these drugs, e.g., Na^+/H^+ and Na^+/Ca^{2+} exchangers (232, 323). The amiloride binding site has not been identified, but the mechanism of block appears to be of the plug type, sensing about 12% of the membrane electric field (259, 455; see also 216).

The identification of the molecular components of the amiloride-sensitive Na^+ channel has been a difficult process, including purification and reconstitution (33, 48, 49, 446, 552; see 216 for review), as well as molecular cDNA cloning methods. The answer is not yet complete, but major progress has been made. Recently, two groups reported the molecular cloning of αrENaC (α-subunit of rat Epithelial Na^+ Channel) (83, 383). Following expression in *Xenopus laevis* oocytes, the currents had the properties expected if they were underlain by high-selectivity amiloride-sensitive channels, but small currents were observed, suggesting lack of other factors in the oocyte expression system. αrENaC mRNA levels are high in the expected tissues (colon, kidney) and absent in other organs. αrENaC has a predicted molecular mass of 78 kDa, in contrast with the typical 150 kDa amiloride-binding proteins found by protein-purification techniques. αrENaC has significant homology with invertebrate touch-sensitive genes, but not with genes coding for other channels or amiloride-sensitive transporters (83). Two additional cDNAs with significant degree of homology with αrENaC (β and γrENaC) have been recently cloned (85). Coexpression experiments revealed that large amiloride-sensitive currents (100-fold that elicited by αrENaC alone) were only observed with the three transcripts. This result strongly supports the notion that these three homologous cDNAs encode subunits of the high-selectivity, amiloride-sensitive Na^+ channel. A recent study of the membrane topology suggests that all three subunits have two transmembrane domains, intracellular N and C termini, and large extracellular loops with multiple N-linked glycosylation sites (84). Conceivably, other subunits may contribute to the channel itself or its regulation. One candidate is the ~ 160 kDa protein Apx (600). In conclusion, the high-selectivity, amiloride-sensitive epithelial channel is likely to be a heteromultimeric membrane protein some of whose subunits (including pore-forming ones) have been cloned. The results obtained from protein purification and reconstitution yield a ~ 700 kDa protein whose Na^+-conducting, amiloride-binding subunit is 130–180 kDa. The molecular-biological studies have so far identified α, β, and γrENaC, with molecular masses ranging from 72 to 78 kDa. Epithelial Na^+ channel isoforms of human (416b) and bovine origin (201a) have also been cloned.

These studies have defined a new class of Na^+ channel that is unrelated to the voltage-sensitive Na^+ channel from excitable cells. The discovered peptides have considerable degree of homology with mechanosensitive channels and with proteins involved in neurodegenerative disorders, suggesting that they all belong to a new family of ion channels (reviewed in 47, and 458). Other channel subunits with only two transmembrane domains are inwardly rectifying K^+ channels, a mechanosensitive channel of *E. coli*, and an ATP-gated cation channel (see ref. 440a for review).

Amiloride-Sensitive Epithelial Na$^+$ Channels Are Tightly Regulated by Hormonal and Nonhormonal Mechanisms

Aldosterone. Aldosterone stimulates Na$^+$ absorption in electrogenic Na$^+$-transporting epithelia in a biphasic fashion. The early effects are evident in minutes and its action persists for hours. The mechanisms of action of aldosterone remain controversial after decades of experimental work. Renal epithelial cells express three classes of corticosteroid receptors whose functional roles are not entirely clear (407). There seems to be agreement in that the main effect is to increase the Na$^+$ permeability of the apical membrane, but this is not the sole effect (47, 215, 457). It is not entirely clear whether aldosterone elevates the number of "operational" high-selectivity channels and/or their open probability (311, 453, 461). These effects require de novo protein synthesis (214), but it is not known whether the proteins in question are channels, components thereof, or regulatory factors. The time course of the effect of aldosterone is complex, and it has been suggested that early events involve activation of channels already expressed in the membrane, while slow effects would require synthesis and insertion of new channels or some of their subunits (25, 26). Brefeldin A (an inhibitor of intracellular vesicle traffic) inhibits Na$^+$ transport in A6 monolayers by reducing the density of apical-membrane channels and abolishes the stimulating effect of aldosterone (183b). Both effects suggest a regulatory role for Na$^+$ channel insertion and removal.

Antidiuretic hormones. Vasopressin and other antidiuretic peptides stimulate apical-membrane Na$^+$ permeability in electrogenic, Na$^+$-transporting epithelia. The effect is quite rapid (minutes) and involves vasopressin binding to V2 receptors expressed in the basolateral membrane, activation of adenylate cyclase, and increase in cAMP levels (reviewed in 47 and 216). The increase in permeability appears to be due at least in part to fusion of channel-containing subapical membrane vesicles with the apical membrane (220), although in toad urinary bladder epithelium the full effect of vasopressin on Na$^+$ transport can occur without an increase in capacitance (604). Other mechanisms are possible (see 47, 558; also, see later, under MECHANISMS OF REGULATION OF TRANSEPITHELIAL TRANSPORT). G-proteins and the degree of actin polymerization may be involved in the effect of vasopressin (490, 491).

Feedback inhibition. Negative-feedback inhibition couples the apical-membrane Na$^+$ permeability, that is, elevations in [Na$^+$]$_i$ inhibit P$_{Na}$ and decreases in [Na$^+$]$_i$ stimulate P$_{Na}$ (373, 397). There is indirect evidence suggesting that the mechanism of this inhibition involves an elevation in [Ca^{2+}]$_i$, probably mediated by changes in the rate of basolateral Na$^+$/Ca^{2+} exchange. This hypothesis is supported by patch-clamp studies correlating Ca^{2+} levels and Na$^+$-channel activity (195, 587). The effect of [Ca^{2+}]$_i$ appears to be mediated by activation of PKC (381, 447).

Atrial natriuretic peptide. ANP, a natriuretic hormone released by the atria of the heart, promotes natriuresis primarily by increasing glomerular filtration rate, but also by inhibiting Na$^+$ reabsorption (325). The latter mechanism is by inhibition of the low-selectivity Na$^+$ channels of the inner-medullary collecting duct via elevation of cell levels of cGMP (377, 378, 379).

Other factors. The amiloride-sensitive channel is inhibited by elevations in [Ca^{2+}]$_i$ and decreases in pH$_i$ (218, 460), activated by transmethylation (553), activated by phosphorylation, both by PKA and Ca^{2+}-dependent kinases (381, 447), inhibited by PKC (29a), and regulated by G-proteins (87, 221, 377) and actin (491). Short actin filaments activate expressed α,β,γ-rENaC, and PKA appears to potentiate this effect by reducing elongation of actin filaments (50a). None of these mechanisms has been unequivocally demonstrated to mediate the effects of aldosterone or vasopressin. Interestingly, activation of apical-membrane adenosine receptors has a biphasic effect on amiloride-sensitive channels, i.e., stimulation at lower agonist concentrations and inhibition at high agonist concentrations (390a). With normal urine concentrations of adenosine, Na$^+$ reabsorption is favored; high filtered NaCl load and renal ischemia would elicit natriuresis (390a).

Liddle's disease: a genetic disorder of the amiloride-sensitive Na$^+$ channel. Liddle's disease is a hereditary disease characterized by hypertension associated with positive Na$^+$ balance, expansion of extracellular fluid volume, and decreased levels of plasma renin. This disorder has been traced to a mutation of the gene encoding the β subunit of an amiloride-sensitive Na$^+$ channel, which truncates the C-terminal domain. Lymphocytes from affected individuals exhibit maximal activation of whole-cell Na$^+$ currents, which are abolished by addition of a peptide mimicking the C-terminus (74a).

In Electroneutral NaCl-Transporting Epithelia, Na$^+$ Entry Is Not via Apical-Membrane Channels, but Carrier-Mediated.

In epithelia such as those of small intestine, renal proximal tubule, and gallbladder, all of which absorb Na$^+$ salts, there has been no conclusive demonstration of apical-membrane Na$^+$ channels, although the possibility has been raised from observations such as the effect of low amiloride concentrations on electrical parameters (see 123) It is clear that Na$^+$ entry across the apical membrane is, at least predominantly, carrier-mediated. Four kinds of transporters have been demonstrated to contribute to Na$^+$ entry in NaCl-absorptive epithelia, namely the Na$^+$-organic substrate cotransporters, the Na$^+$/H$^+$ exchanger, the Na$^+$-K$^+$-2Cl$^-$ cotransporter, and the thiazide-sensitive Na$^+$-Cl$^-$ cotransporter (Figs. 8.11 and 8.12). In this section I will discuss the first two; the last two will be discussed in the next section, MECHANISMS OF ION TRANSPORT IN PRIMARY CL$^-$-TRANSPORTING EPITHELIA.

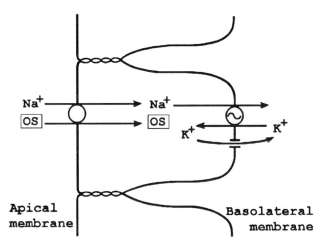

FIG. 8.12. Transport mechanisms in a "leaky" rheogenic Na$^+$-absorbing epithelium. Na$^+$ influx is by cotransport with organic solutes (*OS*, e.g., sugar or amino acid). The transporters at the basolateral cell membrane are as shown in Figure 8.11. Modified from Reuss and Cotton (518), with permission.

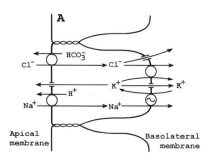

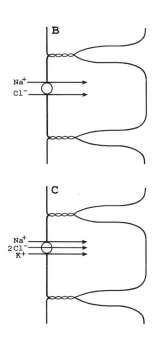

FIG. 8.11. Transport mechanisms in "leaky" electroneutral NaCl-absorbing epithelia. *A.* Na$^+$ and Cl$^-$ influxes across apical cell membrane via exchanges with H$^+$ and HCO$_3^-$, respectively. Cl$^-$ exit across basolateral cell membrane is via Cl$^-$ channels and/or via K$^+$-Cl$^-$ cotransport. Na$^+$ extrusion is mediated by Na$^+$,K$^+$-ATPase. K$^+$ recycles via channels (basolateral or both cell membranes). This model accounts for NaCl transport in most "leaky" NaCl-absorptive epithelia. *B.* The mechanism of salt entry is Na$^+$-Cl$^-$ cotransport (demonstrated for flounder urinary bladder and mammalian renal distal tubule and proposed for rabbit and *Necturus* gallbladder under certain conditions). Basolateral transport mechanisms as in top diagram. *C.* Mechanism of salt entry is Na$^+$-K$^+$-2Cl$^-$ cotransport (demonstrated for flounder intestine). Basolateral transport mechanisms as in top diagram. Modified from Reuss and Cotton (518), with permission.

Sodium organic solute cotransporters. These transporters account for large fractions of Na$^+$ entry in small intestine and renal proximal tubule. Quantitatively, the main ones are Na$^+$ glucose and Na$^+$-amino acid cotransporters. Both are electrogenic, and hence addition of the cotransported solute to the apical solution results in depolarization of the apical membrane, in small intestine (530, 678) as well as in renal proximal tubule (200, 406). Several cotransporters have been cloned in recent years. As an example, I will discuss the Na$^+$-glucose cotransporters (reviewed in 273, 274, 507, 621, 687, and 688). The first eukaryote Na$^+$ cotransporter cloned was one isoform of the Na$^+$/glucose cotransporter (272). Two gene families have been so far identified, homologous to the Na$^+$/glucose cotransporter (SGLT1) and the Na$^+$/Cl$^-$/GABA cotransporter (GAT-1,) respectively.

The Na$^+$ cotransporters perform secondary active transport, i.e., promote intracellular accumulation of nutrients, neurotransmitters, vitamins, bile salts, and anions. The energy for uphill transport of these solutes is provided by the Na$^+$ electrochemical gradient across the cell membrane, generated by the operation of the Na$^+$, K$^+$ pump. This essential discovery, which emanated from Crane's studies of glucose transport by small intestine (121), is referred to as the *sodium-gradient hypothesis*. Some Na$^+$ cotransporters that are functionally important in epithelial cells, and their basic properties, are listed in Table 8.3.

Transepithelial glucose absorption in small intestine and renal proximal tubule involves influx across the apical (luminal) membrane via Na$^+$-glucose cotransporters (SGLT gene family) (687) and efflux across the basolateral membrane via glucose uniporters (GLUT gene family) (621). Two SGLT isoforms are expressed in the renal proximal tubule (SGLT1 and SGLT2), with 59% amino acid identity among the human gene products (673). SGLT1 has a high affinity for glucose (K_m = 0.35 mM), low capacity, and a 2:1 Na^+:glucose stoichiometry. It is expressed in the late proximal tubule (S2 and S3 segments). SGLT2 has a lower affinity for glucose (K_m = 1.6 mM), high capacity, and 1:1 Na^+:glucose stoichiometry (305). It is expressed in the early proximal tubule (S1 segment). SGLT2 reabsorbs most of the filtered glucose, while SGLT1 transports a smaller amount but reduces the glucose concentration in the lumen to near zero. SGLT1 is the only isoform found in enterocytes. In addition to their role in the absorption of sugar, the apical-membrane Na^+/glucose cotransporters contribute to salt and water absorption of sugar, the apical-membrane Na^+/glucose cotransporters contribute to salt and water absorption, both in intestine and renal proximal tubule.

The basolateral-membrane glucose transporters are part of a family of facilitated glucose transporters (GLUT), of which there are several isoforms. GLUT1, GLUT2, and GLUT5 are the main glucose uniporters expressed in transporting epithelial cells. GLUT2 transports fructose, glucose, and galactose across the basolateral membrane (75, 622). The expression of GLUT2 is restricted to surface enterocytes and hence it is also a developmental marker (621). In contrast, GLUT5 is expressed solely on the apical membrane of intestinal cells (134, 401). Its function is the Na^+-independent transport of fructose across the brush-border membrane.

SGLT1 is a 664-residue protein with significant N-linked glycosylation; hydropathy plots predict 12 transmembrane domains (Fig. 8.13). Aspects of this membrane topology have been confirmed by protease digestion of membrane vesicles followed by immunologic analysis and sequencing of the resulting peptides (283). The sequence of SGLT1 is highly conserved in mammals. Regulation by phosphorylation has not been demonstrated, although the protein has several consensus phosphorylation sites (687). In contrast with other carriers, SGLT1 does not possess a large cytosolic domain, which suggests that another subunit has a regulatory role. Although SGLT1 alone is able to translocate glucose together with sodium, it has been found that the membrane-anchored protein RS1, also expressed in apical membranes, increases significantly the V_{max} of the uptake expressed by SGLT1 (331). Hence, RS1 may be a regulatory subunit of SGLT1.

Functional studies of SGLT1 expressed in several cell types (53, 272, 296, 469, 470) yielded substrate specificity and kinetic properties consistent with those measured in native cells. Electrophysiological studies of SGLT1 expressed in *Xenopus laevis* oocytes are consistent with a narrowly defined kinetic model whose rate-limiting step at 0 mV is the return of the empty carrier to the external surface, but in a polarized membrane it shifts to the Na^+ dissociation at the cytosolic surface (469, 470). In the absence of sugar, membrane depolarization was shown to cause large transient currents (charge movement) attributable to voltage-induced changes in the orientation of one or more cotransporter domains in the membrane. Electrophysiological studies of SGLT1 expressed in LLC-PK cells (45a) yielded results compatible with two possible kinetic models, one of them consistent with the one described above.

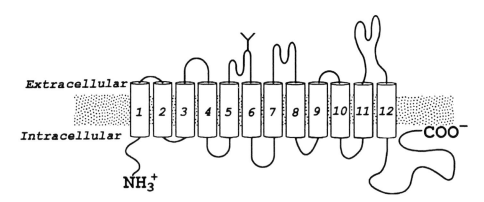

FIG. 8.13. Predicted secondary structure of human SGLT1. The protein consists of single chain of 664 amino acids, 12 putative transmembrane α-helices, N-glycosylation at Asn248, and cytoplasmic location of both amino and carboxyl termini. The carboxyl terminus is closely associated to the plasma membrane. Model for low-affinity Na^+-glucose cotransporter SGLT2 is almost identical. Modified from Hediger and Rhoads (274) with permission.

Na⁺/H⁺ exchanger. The Na^+/H^+ exchanger (NHE) is a carrier that exchanges Na^+ and H^+ electroneutrally. The chemical gradients for Na^+ and H^+ are such that in most cells and under most conditions the exchanger catalyzes Na^+ influx and H^+ efflux. Five isoforms have been identified at the molecular level: NHE1 (551), NHE2 (626, 663), NHE3 (451, 624), NHE4 (451), and NHE5 (322). From the point of view of their role in epithelial transport, we are interested in two potential functions of NHE isoforms: *(1)* involvement in *transepithelial transport*, i.e., Na^+ absorption and H^+ secretion mediated by apical membrane Na^+/H^+ exchange in epithelial cells such as those of the proximal tubule or the small intestine, and *(2)* regulation of intracellular pH (and in some instances cell volume), i.e., *housekeeping* function of NHE expressed in the basolateral membranes of most epithelial cells. The Na^+/H^+ exchanger contributes to Na^+ entry across the apical membrane of proximal tubule (20, 60), small intestine (612) and gallbladder (509, 511, 668). In the thick ascending segment of the loop of Henle, Na^+/H^+ exchangers are expressed in both apical and basolateral membranes, and they appear to be functionally coupled (235b).

The NHE molecules contain 717 to 831 amino acids, with identity ranging from 35% to 54%. Both N and C termini are intracellular and the protein is predicted to have 12 membrane-spanning domains and some are glycosylated (for further details, see 489, 507, and 625). The primary sequences of the isoforms reveal greater degree of homology of the membrane domains, which suggests the same transport mechanism; less-conserved domains could account for the differences in targeting, pharmacological properties, and regulation. The isoforms exhibit large differences in sensitivity to blockers (amiloride and its analogs) but have similar substrate affinities. NHE1 is the housekeeping exchanger, expressed in most cells. NHE5 is expressed in brain, testes, spleen, and skeletal muscle. NHE2, -3, and -4 are expressed in epithelial cells (see Table 8.3). It is clear that NHE3 is targeted only to the apical membrane domain (12a, 52) and NHE2 and NHE4 (in kidney) are targeted to the basolateral membrane domain (487, 594a). From the mRNA levels, the organs with greater expression are as follows (reviewed in 440, 507, and 625): *(1)* NHE2 (amiloride sensitive): rat small intestine and gastric mucosa, rabbit renal medulla, and descending colon; *(2)* NHE3 (amiloride resistant): renal cortex (proximal convoluted tubule, and which ascending limb of Henle's loop [12a, 52]), large and small intestine, and gastric mucosa; *(3)* NHE4: rat gastric mucosa and large and small intestine. It appears that NHE3 alone or together with NHE2 underlies exchanger-mediated Na^+ entry across

the apical membrane of epithelia such as those of the proximal tubule, intestine, and gallbladder.

The Na^+/H^+ exchanger is a dimer (170, 187) and it has been suggested that there is cooperative interaction between two Na^+-binding sites (452), probably one belonging to each subunit (507). Coexpression experiments indicate that there is no formation of heterodimers.

The regulation of NHE is complex and largely isoform dependent. A common feature of all NHE isoforms appears to be the existence of an internal H^+-binding site that activates exchange when $[H^+]_i$ rises (22) (for review of regulation of NHE1, see 440 and 489). In transfected fibroblasts, PKC activation by phorbol esters stimulates both NHE1 and NHE2 but inhibits NHE3 (364). Dopamine inhibits proximal tubule Na^+/H^+ exchange via parallel pathways, cAMP-dependent and cAMP-independent, respectively (173). In any event, the effects PKA, PKC, and tyrosine kinases can be stimulatory or inhibitory depending on isoform and cell type (235a, 304b, 625, 690a). Activation of tyrosine kinase inhibits the apical-membrane NHE3 in thick ascending loop of Henle in isosmotic solutions and underlies the inhibition elicited by hyposmotic solutions (235a). In transformed mouse proximal tubule cells, medium acidification activates transcription of immediate early genes, an effect that is blocked by tyrosine kinase inhibitors (691). Inhibition of apical-membrane Na^+/H^+ exchange by cAMP has been demonstrated in *Necturus* gallbladder epithelium (519) and proximal tubule cells (666). The cAMP effect is mediated by PKA activation and appears to involve a cofactor (NHE-RF) expressed in kidney (including proximal tubule), small intestine and liver (667a). Glucocorticoids increase NHE3 mRNA levels in small-intestine cells (695) and in renal cortex (37). Both acute and chronic treatment with glucocorticoids stimulate Na^+/H^+ exchange, but only chronic administration elevates NHE3 mRNA levels, suggesting two mechanisms of stimulation of the exchanger (37). The chronic regulation appears to be by a transcriptional mechanism (304a). Activation by hypertonicity has been demonstrated for NHE1 (51b) and NHE2 (587a). In contrast, NHE3 from thick ascending loop of Henle is inhibited by hypertonicity (235a). In NHE1, the volume and growth-factor sensitivities appear to reside in different regions of the molecule (51b). Chimeras of NHE1 and NHE3 indicate that the Ca^{2+} sensitivity of NHE1 resides in its carboxy-terminal cytoplasmic domain (654a).

Cl⁻ Transport Mechanisms Are Different in Na⁺- and NaCl-Transporting Epithelia. The mechanisms of Cl^- transport in NaCl-absorptive epithelia are not as well understood

as the mechanisms of Na^+ transport. In the case of electrogenic Na^+-transporting epithelia, such as frog skin, the presence of Cl^- is not required for Na^+ transport, but Cl^- undergoes passive transport, secondary to transepithelial Na^+ transport. The Cl^- transport pathway may be transcellular or paracellular. Transcellular transport can occur via the Na^+-transporting cells or via a different cell type, and paracellular transport involves permeation of the junctions and the lateral intercellular spaces (Fig. 8.12). In high-resistance, Na^+-transporting epithelia Cl^- transport is by electrodiffusion and hence electrogenic. An early experimental observation suggesting this process is that the transepithelial electrical resistance increases dramatically following replacement of external-solution Cl^- with other anions (330), and also with low concentrations of Cu^{2+}, thought to reduce electrodiffusive Cl^- permeability (174, 330).

The simple basolateral membrane transport scheme described above, namely Na^+,K^+-ATPase in parallel with K^+ channels, is complicated in some epithelia by the existence of additional ion-transport pathways. Some pathways involving Na^+, Cl^-, and K^+, and representative examples, are listed in Table 8.2 and 8.3. Exchange and cotransport systems contribute to the regulation of intracellular pH and cell volume (reviewed in 59, 182, 425, 510, 597, 635, and 651). Cl^- channels have been found in basolateral membranes of several electrogenic Na^+-transporting epithelia, such as rabbit urinary bladder (261), renal cortical collecting tubule (550), and colon (149). Their function is not entirely clear. They are probably not involved in transepithelial Cl^- transport (inasmuch as the apical membrane is virtually Cl^--impermeable), but they could contribute to maintenance of membrane voltage and also to cell-volume regulation following cell swelling.

In *electroneutral or near-electroneutral, NaCl-transporting epithelia,* Cl^- entry across the apical membrane is a carrier-mediated process, either exchange (Cl^-/anion, e.g., Cl^-/HCO_3^-) or cotransport (Na^+-Cl^- or Na^+-K^+-$2Cl^-$). Examples of epithelia endowed with apical-membrane anion exchangers include renal proximal tubule (21), small intestine (375, 376, 612), and gallbladder (509, 511, 516). In the case of renal proximal tubule, anions other than HCO_3^- exchange for Cl^-, most prominently formate (306) (see Table 8.3). In contrast with Na^+-transporting epithelia, in which Na^+ transport is Cl^--independent, in NaCl-transporting epithelia steady-state transport requires the presence of both ions, which are translocated in the same direction. Both Na^+-Cl^- and Na^+-K^+-$2Cl^-$ cotransport involve direct (molecular) coupling between the ion fluxes, whereas the case of Cl^-/

HCO_3^- exchange in parallel with Na^+/H^+ exchange involves indirect coupling. This point can be clarified by considering why removal of Na^+ from the apical bathing solution causes abolishment of Cl^- transport: Na^+ removal reverses the fluxes via the Na^+/H^+ exchanger; this acidifies the cytoplasm, i.e., intracellular $[HCO_3^-]$ falls and hence: *(1)* the driving force for anion exchange decreases, and *(2)* the anion exchanger is inhibited by an effect of pH_i on an allosteric site (see later, under MECHANISMS OF REGULATION OF TRANSEPITHELIAL TRANSPORT).

In most mammalian NaCl-absorptive epithelia, NaCl entry occurs by the parallel operation of Na^+/H^+ and Cl^-/HCO_3^- exchangers, but Na^+-K^+-$2Cl^-$ cotransport has been demonstrated in flounder intestine (445, 612) and Na^+-Cl^- cotransport has been claimed to occur in *Necturus* and rabbit gallbladder epithelia (124, 125, 133, 165), in mammalian distal tubule (488), and in flounder urinary bladder (606). Some of the transporters involved in these processes will be discussed in later sections: for Na^+-K^+-$2Cl^-$ cotransporter see later, under MECHANISMS OF ION TRANSPORT IN PRIMARY CL$^-$-TRANSPORTING EPITHELIA and for Cl^-/HCO^-_3 exchanger see under MECHANISMS OF ION TRANSPORT IN H^+ AND HCO_3^--TRANSPORTING EPITHELIA.

In NaCl-Absorptive Epithelia, a Cl^- Efflux Mechanism Is Necessary at the Basolateral Membrane. As evident from comparing Figures 8.6, 8.11, and 8.12, basolateral-membrane Cl^--transport pathways are not necessary in electrogenic Na^+-absorptive epithelia, but are necessary in NaCl-absorptive epithelia, in which entry of Na^+ and Cl^- occur in parallel, either by cotransport or by the coordinated operation of Na^+/H^+ and Cl^-/HCO_3^- (or Cl^-/anion) exchanges. The two pathways identified by transport studies in intact cells and in membrane vesicles are Cl^- channels (230, 502) and a K^+-Cl^- cotransporter (115, 169, 351, 508, 555). Studies of these transporters have been difficult because of the poor accessibility of this membrane to electrophysiologic techniques and the difficulties involved in purifying basolateral membrane. Basolateral Cl^- channels have not been identified at the molecular level, but the experimental evidence for their existence is convincing. The K^+-Cl^- cotransporter has been recently cloned (222a). Details will be discussed later under the Cation–Chloride Cotransporters. In renal proximal tubule, the Cl^- conductance of the basolateral membrane appears to be negligible under control conditions (reviewed in 502 and 503), but is increased by cell swelling (396, 672), suggesting that K^+-Cl^- cotransport is the main steady-state Cl^- transport mechanism in normal conditions. In small intestine, basolateral-membrane Cl^- channels have been found in patch-clamp studies (223, 582, 583), and it is unclear whether K^+-Cl^- cotransport contributes to Cl^- transport across this membrane. In the *Necturus* gallbladder, there is evidence for both a basolateral-membrane Cl^- conductance and K^+-Cl^- cotransport (511). The Cl^- conductive pathway (one or more channels) is activated by HCO_3^-/CO_2 by a mechanism independent of sustained changes in intracellular pH (509, 605). Bicarbonate stimulation of NaCl absorption has been also reported in gallbladders of other species (122, 275, 403), but it has been less studied in other epithelia (reviewed in 523).

A potential problem with the identification of basolateral-membrane ion channels in epithelial cells is the fact that detachment of the cells from the basal lamina may result in abnormal channel expression (114). Hence, comparison of the single-channel data with the properties of the membrane assessed in situ (e.g., with intracellular-microelectrode techniques) is necessary to verify whether the channels found are expressed in the cell in situ.

In NaCl-Absorptive Epithelia, K^+ Undergoes Transport, Either Transepithelially or across the Membrane Expressing the Na^+, K^+-ATPase. In absorptive epithelia, K^+ may undergo net absorption, net secretion, or no net transepithelial transport, but it is partly recycled across the basolateral membrane. An example of a Na^+-absorptive epithelium that does not perform substantial transepithelial K^+ transport is the frog skin, in which basolateral K^+ entry via the Na^+, K^+-ATPase is balanced by electrodiffusive K^+ exit through basolateral membrane K^+ channels (166, 638). A simple case of K^+ secretion in parallel with Na^+ absorption is that of the principal cell of the renal cortical collecting tubule in mammals, in which basolateral uptake of K^+, again via the Na^+, K^+-pump, elevates intracellular $[K^+]$ to levels above electrochemical equilibrium; because of the presence of K^+-conductive channels in the apical membrane, part of the K^+ is secreted to the lumen (463, 662). Finally, in a few instances K^+ can be absorbed in parallel with Na^+. This requires an active transport step inward across the apical membrane or outward across the basolateral membrane. An example is the renal proximal tubule, where basolateral-membrane K^+ channels have been identified (548), but the mechanism of active K^+ absorption is unclear (689).

K^+ Channels in NaCl-Absorptive Epithelia. Among epithelial cells, K^+ channels have been best studied in renal-tubule cells, in situ or in culture (see 346, 662). Several channels, with distinct properties, have been found in apical and/or basolateral membranes of most tubule segments, in some instances including demonstrations of functional significance, regulation, and molecular identity. In this section I will focus the discussion on two kinds of channels, namely the renal ATP-sensitive, inwardly rectifying K^+ channel and the "maxi" K^+ channel expressed in several epithelia. The latter channel is activated by membrane depolarization and elevations of $[Ca^{2+}]_i$, while the former one is voltage- and Ca^{2+}-insensitive. Maxi K^+ channels have been observed in other absorptive epithelia, such as those from *Necturus* gallbladder (579), rat colon (77, 321), human airway epithelium (344), and rabbit small intestine (Copello, Peerce, and Reuss, unpublished). In renal tubule, the available information suggests that maxi K^+ channels are not involved in transepithelial transport, and hence their function is unknown (662). In contrast, in secretory epithelia maxi K^+ channels are activated by secretagogues, via elevation in $[Ca^{2+}]_i$ (308, 483) and play a role in fluid secretion. Recent studies in some absorptive epithelia, however, suggest that these channels may also be activated by extracellular messengers and contribute to regulation of ion and fluid absorption (see later, under MECHANISMS OF REGULATION OF TRANSEPITHELIAL TRANSPORT).

Apical membrane K^+ channels. The inwardly rectifying, ATP-sensitive K^+ channels, expressed in apical membranes of several renal-tubule segments, differ from ATP-sensitive channels expressed in other cells such as pancreatic β-cell and cardiomyocytes (see 24) in their insensitivity to TEA^+, lower or nil sensitivity to sulfonylureas, and lower sensitivity to ATP (662). In cortical collecting ducts, ATP-sensitive, apical membrane K^+ channels contribute to transepithelial transport (662). In the thick ascending limb of Henle, they allow for efflux of the K^+ taken up via the Na^+-K^+-$2Cl^-$ cotransporter also expressed in the apical membrane. This recycling is essential for net NaCl absorption. The high K^+ selectivity of the apical membrane, underlain by this channel, contributes to the lumen-positive voltage that drives paracellular Na^+ reabsorption. Inwardly rectifying ATP-sensitive K^+ channels are expressed in the apical membrane of principal cells of the cortical collecting duct, where they are the apical-membrane pathway for K^+ secretion (662, 689). In the basolateral membrane of the proximal tubule, a similar channel is thought to accomplish recycling of K^+ taken up by the Na^+, K^+-ATPase (627), analogously to the apical membrane K^+ channel in the loop of Henle. The apical-membrane K^+ channels in collecting duct and ascending limb of Henle's loop appear to correspond to the ROMK channels (or homologous proteins) described in the next paragraph.

A rat-kidney inwardly rectifying, ATP-sensitive K^+ channel and homologues have been recently cloned (286). The ROMK1 protein (ROM = renal outer medulla) is ca. 45 kDa and has only two potential-membrane-spanning segments (M1 and M2) flanking an amino acid sequence that exhibits 59% similarity to the 17-amino-acid "P" or pore-forming segment of voltage-gated K^+ channels (in the latter channels, the segment in question is flanked by transmembrane domains 5 and 6). In ROMK1 there is no distinct homology to the S4 region (putative voltage sensor) of voltage-gated channels. An ATP-binding motif (Walker type A) is present in a 27-amino-acid stretch following M1 (Fig. 8.14). The same region contains potential sites for PKA- and PKC-mediated phosphorylation. Both Mg-ATP and PKA-mediated phosphorylation restore the channel activity after run-down (420). The channel has high K^+/Na^+ selectivity, high open probability (P_0) at V_m above -60 mV, moderate inward rectification, and single-channel conductance ca. 39 pS (inward current). These properties match those of the native renal ATP-sensitive K^+ channels. Northern blot analysis revealed ROMK1 expression in renal cortex and outer medulla, but not in inner medulla (286). Recent studies by Hebert and associates have demonstrated that mRNAs corresponding to three isoforms of ROMK channels are expressed in all tubule segments from the thick ascending limb of the loop of Henle to the end of the cortical collecting tubule, each isoform with a specific distribution (56, 357). In conclusion, the functional characteristics of ROMK1, together with its intrarenal distribution, suggest that ROMK channels underlie the low-conductance, ATP-sensitive K^+

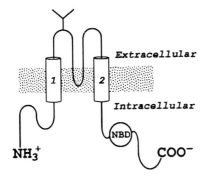

FIG. 8.14. Predicted secondary structure of ROMK1. In contrast with other cloned ion channels, the protein has only two transmembrane α-helices. The extracellular loop has an N-glycosylation site and appears to be responsible for formation of the conductive pathway. Amino and carboxyl termini are on the cytoplasmic side. The carboxyl-terminus region is long and contains the nucleotide-binding domain (NBD). Modified from Hebert and Ho (271), with permission.

channels found in the distal nephron and responsible for K^+ secretion. An important feature of these gene products is their small size and apparently simple membrane topology, which should make them extremely valuable for studies of structure-function relationships.

Mechanisms of regulation of ROMK channels are under active investigation. Other inwardly rectifying K^+ channels have been shown to exist as membrane complexes of the subunits described above (two membrane-spanning domains), and the sulfonyl urea receptor (SUR, a membrane protein belonging to the ATP-binding cassette—ABC—superfamily). Coexpression of SUR is necessary for full channel activity and pharmacological sensitivity (296a). ROMK2, an alternatively spliced isoform of ROMK1 (truncated amino terminus) encodes an ATP-sensitive K^+ channel. Although SUR is not expressed in the kidney, coexpression of ROMK2 with CFTR, another ABC protein, changes the pharmacological properties of the channel, making it sensitive to glibenclamide, a sulfonyl urea (419a).

Basolateral membrane K^+ channels. The functions of K^+ channels expressed in basolateral membranes of Na^+ or NaCl-transporting epithelia are: *(1)* to recycle the K^+ taken up by the Na^+, K^+-ATPase, and *(2)* to hyperpolarize the basolateral membrane (and if the paracellular pathway has a finite conductance, to hyperpolarize the apical membrane as well). Intracellular-microelectrode experiments demonstrate high basolateral-membrane K^+ conductance and electrodiffusive permeability, and patch-clamp studies reveal K^+ channels. It is not always clear that the channels observed in patch-clamp studies underlie the membrane conductance, either because of lack of parallel studies or lack of demonstration of preservation of cell polarity and normal targeting of membrane proteins. The literature on basolateral-membrane K^+ channels is vast; I will review only a few examples of Na^+- and NaCl-transporting epithelia.

In electrogenic Na^+-transporting epithelia, the principal basolateral-membrane K^+ channel appears to be inwardly rectifying. In rabbit cortical collecting tubule, the single-channel conductances are 2 and 20 pS, at negative and positive voltages, respectively, and the P_o increases with hyperpolarization (550a). In rat cortical collecting duct, a low-conductance (28 pS) and an intermediate-conductance K^+ channel (50–85 pS) have been identified (284, 661). The intermediate-conductance K^+ channel is voltage-dependent, i.e., hyperpolarization increases the P_o (661). Both channels appear to be regulated by cGMP-dependent protein kinase (285); the low-conductance channel is activated by nitric oxide (384a). In frog skin, the channel has an inward conductance of 15 pS, is blocked by Ba^{2+} and low pH_i, the P_o increases with depolarization, and the rectification is due to Mg^{2+} block. This channel accounts for the K^+ conductance of the basolateral membrane (632).

In NaCl-transporting epithelia, basolateral-membrane K^+ channels also accomplish the functions stated above. Of special importance is their role in the restoration of the apical-membrane voltage following the depolarization elicited by Na^+-organic solute cotransport (253). Fusion of basolateral plasma membrane vesicles from *Necturus* intestine with planar lipid bilayers demonstrated a large-conductance K^+ channel (116). The P_o is Ca^{2+}-independent, but increases with hyperpolarization. It was suggested that following increased apical-membrane Na^+ entry by cotransport the basolateral membrane resistance increases because of closure of K^+ channels; the parallel increase in pump current would then hyperpolarize the (high-resistance) basolateral membrane helping restore the membrane voltage (116). In whole-cell studies on enterocytes isolated from guinea-pig villi, the K^+ current, supposed but not proven to be basolateral in origin, was inwardly rectifying, Ca^{2+}-independent, sensitive to Ba^2 and scorpion venom, but insensitive to tetraethylammonium (TEA^+) (582). However, in another study in the same preparation, the K^+ current was activated by depolarization and showed complex inactivation. Single-channel studies revealed an 8 pS channel with similar inactivation kinetics to the current (619). In the latter study, inwardly rectifying currents or channels were not observed. The voltage dependence of these channels would result in their activation after the depolarization produced by apical-membrane Na^+-organic solute cotransport, and hence contribute to the ensuing repolarization. The reasons for the differences between these results and those obtained both in *Necturus* intestine and in the earlier studies in guinea-pig enterocytes are not clear. In conclusion, the basolateral-membrane K^+ channels of enterocytes have not been definitively identified.

In rabbit proximal tubule, basolateral-membrane K^+ channels are inwardly rectifying, with inward conductances of ca. 50 pS, in the straight segment (231) and the convoluted (S_2) segment (468). The P_o is increased by depolarization, decreased by internal acidification, and decreased by millimolar concentrations of ATP in the cytoplasmic surface (627). The K^+ conductance of the basolateral membrane of rabbit proximal tubule is elevated by cell swelling (43, 396, 672). The inwardly rectifying channel may contribute to the coupling between basolateral pump-mediated Na^+ transport and basolateral K^+ conductance, likely to be mediated by

cell swelling, intracellular ATP levels, and intracellular pH (42, 627).

Basolateral-Membrane K^+-Cl^- Cotransporter. The K^+-Cl^- cotransporter mediates coupled, electroneutral fluxes of K^+ and Cl^- across plasma membranes of epithelial (115, 169, 508, 555) and nonepithelial cells (351). Under physiological conditions, the fluxes are outward; the coupling ratio $(K^+:Cl^-)$ is 1. The cotransporter is inhibited by loop diuretics (but with lower affinities than those for the Na^+-K^+-$2Cl^-$ cotransporter) and by N-ethylmaleimide. In some cells, cotransport is activated by swelling. The K^+-Cl^- cotransporter (KCC) has been recently identified at the molecular level in rabbit, rat, and human (222a). It belongs to the family of cation–chloride cotransporters, which includes the Na^+-K^+-$2Cl^-$ cotransporters (NKCC) and the Na^+-Cl^- cotransporter (NCC). The molecular characteristics of this family of transporters are discussed later under The Cation–Chloride Cotransporters.

Mechanisms of Ion Transport in Primary Cl^--Transporting Epithelia

The first demonstration of primarily active transepithelial Cl^- transport was provided by Zadunaisky in studies of the amphibian corneal epithelium (696). Since this pioneering work, active transepithelial Cl^- transport has been identified in a number of epithelia. These epithelia are electrogenic, displaying a short-circuit current mainly or entirely attributable to net Cl^- transport (241, 696). The net Cl^- flux depends on the Na^+ gradient, developed and maintained by primary active transport (586). In fact, Cl^- transport can be abolished by Na^+ removal or by dissipating the Na^+ gradient between the extracellular fluid and the cell interior. In most instances Cl^--transporting epithelia are secretory (e.g., exocrine glands such as salivary, pancreatic, eccrine, lacrimal and shark rectal gland, airway epithelium, intestinal crypts, corneal epithelium), but in a few instances they are absorptive (e.g., thick ascending segment of the loop of Henle).

Cellular Mechanisms of Transepithelial Cl^- Transport. The main ion-transport mechanisms in secretory and absorptive Cl^--transporting epithelia are depicted in Figure 8.15A and 8.15B, respectively. In both instances, a two-membrane scheme with pump and leak transporters explains well the experimental observations, but there are some subtle differences between the absorptive and secretory schemes.

In secretory epithelia (Fig 8.15A), the essential com-

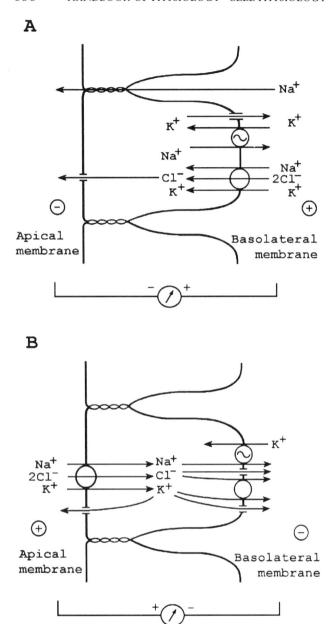

FIG. 8.15. Mechanisms of ion transport in Cl$^-$-transporting epithelia. *A.* Transport model for exocrine-gland Cl$^-$-secreting epithelium. Intracellular [Cl$^-$] is maintained above electrochemical equilibrium by electroneutral Na$^+$-K$^+$-2Cl$^-$ cotransport across the basolateral cell membrane. Na$^+$ influx is balanced by efflux via Na$^+$,K$^+$-ATPase; K$^+$ influxes are balanced by efflux via K$^+$ channels. Transepithelial secretion is induced by activation of apical-membrane Cl$^-$ channel(s). Modified from Reuss and Cotton (518), with permission. *B.* Model of Cl$^-$ transport in thick ascending limb of Henle's loop, a Cl$^-$-absorptive epithelium. Cl$^-$ entry across the apical membrane is mediated by Na$^+$-K$^+$-2Cl$^-$ cotransporter. Most basolateral Cl$^-$ efflux proceeds through Cl$^-$ channels. Modified from Reeves and Andreoli (503), with permission.

ponents are a Cl$^-$ channel in the apical membrane and a three-component mechanism of active influx of Cl$^-$ across the basolateral membrane. The three components operate together to cause net Cl$^-$ uptake across the basolateral membrane. They are the Na$^+$-K$^+$-2Cl$^-$ cotransporter, the Na$^+$,K$^+$-ATPase, and a K$^+$ channel. The operation of the Na$^+$ pump causes a fall in intracellular Na$^+$ and a rise in intracellular K$^+$, as well as a cell-negative EMF, because of the large K$^+$ gradient and the K$^+$-selectivity of the basolateral membrane and the electrogenicity of the pump itself. In the steady state, the prevailing ion-concentration gradients favor inward fluxes through the cotransporter, elevating both [Cl$^-$] and [K$^+$] above equilibrium levels in the cell. K$^+$ exits via the channels in the basolateral membrane, whereas Cl$^-$ exits via the channels in the apical membrane. The ion selectivities of the two membrane domains, in combination with the activities of the principal ions in the cell and the extracellular fluids, determine cell-negative EMFs at both membranes. However, in Cl$^-$-secretory epithelia (e.g., corneal epithelium) the EMF of the basolateral membrane ($\approx E_K$) is greater than that of the apical membrane ($\approx E_{Cl}$) (520). This has two effects: first, in the transporting epithelium exposed to the same solution on both sides there will be an apical-negative transepithelial voltage; second, there will be a loop current tending to depolarize the basolateral membrane and to hyperpolarize the apical membrane, with both voltages being displaced away from E_K and E_{Cl}, respectively. Transcellular transport, as depicted in Figure 8.15A, involves only Cl$^-$; both Na$^+$ and K$^+$ recycle across the basolateral membrane, via the Na$^+$ pump and the K$^+$ channel, respectively. The transepithelial voltage, oriented apical-negative, constitutes a driving force for paracellular Na$^+$ transport, so that the net result is secretion of NaCl. This in turn results in net fluid secretion, probably by transcellular osmotic water flow. This model was first proposed by Silva et al. (586) to explain the mechanism of Cl$^-$ secretion by the shark rectal gland and has since been applied to a number of Cl$^-$-secretory epithelia (415, 612). A major component of the response of Cl$^-$-secretory epithelia to agonists that stimulate secretion is the coordinated activation of apical-membrane Cl$^-$ channels and basolateral-membrane K$^+$ channel. The latter effect tends to hyperpolarize both cell membranes and hence maintain the driving force for Cl$^-$ secretion. The channels responsible for K$^+$ recycling across the basolateral membrane have been well studied in the human-colon cell line T84. Carbachol activates inwardly rectifying (ca. 55 pS) K$^+$ channels (617). These channels are both cAMP- and Ca^{2+}-sensitive, and their control involves also PKC and ATP binding (615). In other words, they appear to be regulated in coordination with the Ca^{2+}- and/or cAMP-mediated stimulation of the apical membrane Cl^- channels, discussed in the next section.

The transport mechanisms in Cl^--absorptive epithelia are depicted schematically in Figure 8.15B. There are major similarities with the Cl^--secretion model depicted in Figure 8.15A. In both instances, the Cl^- channel is expressed at the *trans* membrane (i.e., the membrane across which Cl^- *leaves* the cell). Also, in both cases the entry pathway at the *cis* membrane involves both Na^+-K^+-$2Cl^-$ cotransport and K^+ recycling via K^+ channels, and the membrane voltages are similar in the two systems. However, in the Cl^--absorptive epithelium the Na^+,K^+-ATPase is located at the *trans* membrane, so that Na^+ entering the cell via cotransport at the *cis* membrane is not recycled, but instead undergoes transcellular transport. In the thick ascending limb of the loop of Henle, Cl^- absorption is entirely transcellular, whereas Na^+ absorption is partly transcellular and partly paracellular. With identical solutions on both sides of the epithelium, the paracellular Na^+ flux is driven by the transepithelial voltage, which is lumen-positive by a similar mechanism to that explaining the lumen-negative polarity in Cl^--secreting epithelia (241, 245, 503).

In the thick ascending limb of the loop of Henle K^+ enters the cells across the apical membrane by secondary active transport (Na^+-K^+-$2Cl^-$ cotransport). Most of the K^+ taken up recycles across the apical membrane via K^+ channels, which share at least some characteristics with ROMK1, although other K^+ channels have been also found (see 244). In the rat thick ascending limb there are intermediate-conductance (ca. 70 pS) and low-conductance (ca. 30 pS) K^+ channels. The intermediate-conductance channel is blocked by arachidonic acid (660a); the low-conductance channel is activated by PKA-mediated phosphorylation (658). Another moiety of K^+ entering the cells via cotransport is passively transported across the basolateral membrane, via K^+ channels (293, 694) and/or K^+-Cl^- cotransport (241, 245, 248). However, this is controversial, i.e., there seem to be differences in basolateral-membrane K^+ channel activity that depend on species, segment, and experimental conditions. In addition, the basolateral membrane of thick ascending limb cells contains nonselective cation channels (476) and probably a K^+-HCO_3^- cotransporter (362). The latter transporter has been demonstrated in suspensions of rat medullary thick ascending limbs, and hence it is not certain that it is expressed in the basolateral membrane.

The Essential Transporters in Cl^--Transporting Epithelia Are the Cl^- Channels and the Na^+-K^+-$2Cl^-$ Cotransporter. As pointed out by several authors (187b, 196, 639a), there

are no clear patterns for definitive classification of epithelial Cl^- channels; i.e., the primary sequences and the functional characteristics, in particular the mechanisms of regulation, differ widely (see Table 8.2). Typically, Cl^--secreting epithelial cells express in their apical membrane one or more of the following Cl^- channels: *(1)* the so-called *cystic fibrosis transmembrane conductance regulator* (CFTR), *(2)* outwardly rectifying channels, *(3)* Ca^{2+}-sensitive channels, and *(4)* swelling-activated channels. Our current knowledge of epithelial Cl^- channels, with emphasis on CFTR, as well as the identification and initial characterization of the Na^+-K^+-$2Cl^-$ cotransporter, are presented in the remainder of this section.

Epithelial anion channels. Epithelial-cell Cl^- channels are important for transepithelial transport, but in addition participate in the control of cell volume and composition and may influence the function of other Cl^- channels. The only Cl^- channels that have been definitively demonstrated at the molecular level are those of the ClC family and CFTR. Other channels probably exist, but await identification; several proteins that have been suggested to underlie Cl^--channel activity could be channel regulators instead (301, 495). Some channels well-characterized from a biophysical point of view have not been cloned. Finally, several studies have been carried out on isolated cells, and it is not always certain that channel expression in the surface membrane of these cells was equivalent to the situation in situ.

ClC-2 and ClC-K. ClC-2, a member of the ClC family, is expressed in many cell types, including several epithelial organs (620). This channel is an excellent candidate for one or more functional roles in epithelial cells, but knowledge of its expression in defined cells and membrane domains is incomplete. The channel is activated by cell swelling and appears to contribute to cell-volume regulation in heterologous expression systems, but this is not entirely clear in vivo (495). ClC-K1 and ClC-K2 (first cloned in rat) are closely related genes homologous to other ClC family members, and are expressed exclusively in the kidney. ClC-K1 was shown by RT-PCR of isolated nephron segments to be expressed predominantly in the thin ascending segment of the loop of Henle (630). Immunohistochemistry and immunoelectron microscopy demonstrated its presence in both apical and basolateral membranes (631), and its expression in *Xenopus* oocytes resulted in a Cl^- current with functional and pharmacological properties consistent with those of the tubule segment studied in vitro (631). These studies provide a molecular explanation for the high Cl^- permeability of the thin ascending segment of the loop of Henle. A very similar gene product is expressed in the thick segment of the ascending limb of Henle's loop and the distal convoluted tubule, and less in the late proximal tubule (S3 segment) and the cortical collecting tubule (314). ClC-K2 is expressed in all tubule segments as well as in the glomerulus (314). Dehydration induces transcription of rClC-K1 (630). Two related isoforms are expressed in human kidney (314, 618a) and two additional members of the ClC family (ClC-K2L and ClC-K2S) have been identified in rat kidney (1). Both are predominantly expressed in the thick-ascending loop of Henle and in the cortical collecting duct. Expression in *Xenopus laevis* oocytes yielded outwardly rectifying anion currents ($Br^- > Cl^- > I^-$ selectivity), blocked by DIDS and DPC (1). Recent patch-clamp studies of basolateral membranes from mouse cortical thick ascending limbs revealed two Cl^- channels with conductances of 45 and 7–9 pS, both stimulated by forskolin. The low-conductance channel is more active under control conditions and has properties consistent with the basolateral membrane Cl^- conductance (251). ClC-3 (310), found in kidney and other organs, is less than 30% identical to other ClC proteins. Another ClC molecule, ClC-5, recently cloned from rat kidney, is very nearly homologous with ClC-3. It is expressed in kidney and colon, underlies an outwardly rectifying current, and is blocked by disulfonic stilbene derivatives (548a).

The molecular weight of ClC gene products ranges from ≈75 to ≈110 kDa. Hydropathy analyses suggest 12 transmembrane domains, but the topology is not known precisely (247, 314). Most ClC channels have the halide selectivity sequence $Cl^- > Br^- > I^-$, except ClC-K1 (630), ClC-K2L and ClC-K2S (1), and ClC-3 (310). Another Cl^- channel belonging to the ClC family was cloned and sequenced from a rabbit gastric cDNA library and expressed in oocytes (402). The predicted protein is highly homologous to rat brain ClC-2. The channel is linear, has a conductance of 30 pS with 800 mM Cl^- on both sides, permeability sequence $I^- > Cl^- > NO_{3-}$, and is activated by PKA and ATP (there are two consensus sequences for PKA-mediated phosphorylation). A unique property is that the channel is active at an extracellular pH of 3.0. All these features are consistent with those of the native gastric epithelial Cl^- channel, studied by fusion of purified membrane vesicles coexpressing H^+,K^+-ATPase (128). These studies support the conclusion that this is the Cl^- channel of the apical membrane of HCl-secreting (i.e., parietal) gastric epithelial cells.

In most other studies in epithelial cells, the functional role of putative ClC-family channels is less clear. The conductances of ClC-2 and ClC-K channels have not been published. Although it is likely that these gene products are anion channels involved in transepithelial Cl^- transport, the channels identified with molecular-

biological techniques have not yet been matched with those studied with biophysical approaches. However, further progress in this area is fast and answers can be expected in the near future.

CFTR. CFTR has received considerable attention during the last few years, and it is beyond the scope of this chapter to cover in detail any of the several areas of rapid progress. Several excellent recent reviews are available (95, 202, 208, 525, and 674). CFTR belongs to the superfamily of ATP-binding cassette (ABC) proteins, also referred to as traffic ATPases. Another member of this family is the P-glycoprotein of multidrug-resistant cancer cells. CFTR was cloned and sequenced (313, 526, 529). The similarity with previously sequenced ABC transporters suggested the possibility that CFTR might not be a Cl^- channel per se, but a regulator. However, at present there is solid information supporting the conclusion that CFTR is an ATP-dependent, cAMP-regulated Cl^- channel. First, the protein has been purified to virtual homogeneity and reconstituted, yielding channel activity consistent with that observed in native CFTR-expressing cells and in transfected cells (39). Second, experimental mutations involving charged residues in putative transmembrane domains resulted in changes in ion selectivity, indicating that these residues are part of the ion-conductive pore (16). CFTR is a linear (ohmic) channel in symmetric Cl^--containing solutions, has a conductance of 7–10 pS, and in most instances appears to be more permeable to Cl^- than to I^- (for further details, see 17).

The predicted topology of CFTR is shown in Figure 8.16. The protein has two halves, each including six transmembrane domains, some of which appear to form the pore (16), and two nucleotide-binding consensus sequences (NBDs). The two halves are bridged by a large hydrophilic moiety called the regulatory domain ("R-domain"), which must be phosphorylated for normal gating. Studies of interferences between Cl^- and SCN^- permeation indicate that CFTR has a multi-ion pore (618). Replacement of a single arginine with glutamine abolished this effect (618). Mutagenesis studies and reconstitution of partial constructs in planar lipid bilayers suggest that transmembrane segments 2 and 6 are involved in forming the pore (444). Surprisingly, the R-domain alone has been reported to cause anion-channel activity when inserted in planar lipid bilayers (18). The significance of the latter observations is unclear.

Cystic fibrosis is the result of CFTR mutations. The most common mutation is a phenylalanine deletion in the first NBD (ΔF508). This mutation leads to defective processing and poor transfer of the protein to the cell membrane (103, 142; reviewed in 127). ΔF508 CFTR

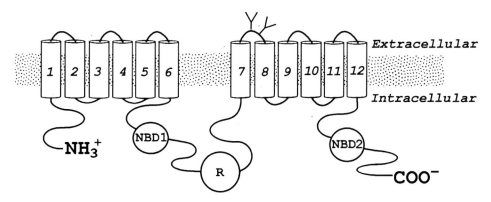

FIG. 8.16. Predicted secondary structure of CFTR. The protein consists of single peptide chain with two halves, each compromising six putative transmembrane α-helices and a nucleotide-binding domain (NBD). The two halves are linked by the R domain, a highly charged regulatory region. R domain possesses consensus phosphorylation sites for protein kinases A and C and Ca^{2+}-calmodulin kinase. $\Delta F508$ deletion found in 70% of all CF cases is located in first NBD. Two consensus sites for N-linked glycosylation are present between transmembrane domains 7 and 8. Modified from Welsh et al. (674), with permission.

is recognized by hsp70 (a heat shock protein) and degraded in the ER (692). A few copies are inserted in the cell membrane, but they have low stability (385). Both this mutant and the wild-type CFTR function as cAMP-regulated Cl^- channels in the endoplasmic reticulum, as demonstrated by patch-clamp studies of outer membranes of isolated nuclei (475). Interestingly, targeting of $\Delta F508$ expressed in nonepithelial cells to the plasma membrane is corrected by low temperature (142, 156). Decreased incubation temperature of airway epithelial cells expressing $\Delta F508$ CFTR not only corrects CFTR channel function but also reduces the number of copies of plasma-membrane outwardly rectifying Cl^- channels (163).

Many other CFTR mutations are found, which account for varying degrees of severity of the disease (676). Mutations of positively charged residues (arginines) in TM 2 and TM 6 result in correctly processed proteins but small Cl^- currents, in part because of lower single-channel conductance. Charged residues determining multi- or single-ion occupancy may account for these differences in channel conductance (618).

Regulation of CFTR is a complicated process requiring PKA-mediated phosphorylation and binding of ATP or hydrolyzable ATP analogs (15, 96, 208). PKA activates CFTR even after elimination by mutation of all consensus sites in the R-domain (98, 524), likely due to cryptic PKA-phosphorylation sites, one of which has been recently identified (580). The effect of PKA appears to be modulated by the actin cytoskeleton (183a). The notion that CFTR gating requires ATP hydrolysis is based on the demonstration that ortho-

vanadate and BeF_3 stabilize the open conformation of phosphorylated, ATP-gated CFTR channels (36). The first nucleotide-binding domain catalyzes ATP hydrolysis (327). There is a complicated relationship between PKA-mediated phosphorylation and the function of the two NBDs (36, 295). However, deletion of the entire C-terminal half of CFTR does not abolish Cl^--channel activity or alter pore properties, although channel regulation is perturbed (584). It is possible that in these deletion experiments there is low-level, residual expression of full-length CFTR. Gadsby and his associates (208, 295), working with cardiac CFTR, have proposed a gating mechanism that involves *(1)* PKA-mediated phosphorylation of two distinct sets of sites (first P_1 and then P_2), *(2)* change in conformation of the R-domain, activating the nucleotide-binding domains, and *(3)* ATP binding and hydrolysis. P_1 and P_2 are dephosphorylated by different phosphatases. P_1 and $P_1 P_2$ phosphorylation allow for ATP hydrolysis at one NBD (NBDA), causing brief channel openings. Only following $P_1 P_2$ phosphorylation and channel opening by ATP hydrolysis at NBDA, NBDB can interact with ATP, which stabilizes the channel in a long open state until ATP hydrolysis and $ADP + P_i$ dissociation. This and other gating schemes are under intense investigation.

Recent studies indicate that reductions in PKA activity result in decreased expression of CFTR; conversely, cAMP induces CFTR in a cell-specific fashion (416a). The quantitative significance of these results is uncertain.

It has also been proposed that defective acidification of intracellular organelles plays an important role in the pathogenesis of CF (32a). However, endosomal and trans-Golgi pH in fibroblasts transfected with wild-type or ΔF508-CFTR did not differ significantly, with or without CFTR activation (581a).

CFTR is also a regulator. An exciting recent hypothesis is that CFTR not only is a Cl^- channel by itself, but also a regulator of other Cl^- channels and of Na^+ channels. This hypothesis arose from studies in CFTR knockout mice and from experiments in which CFTR and outwardly rectifying Cl^- channels were coexpressed. In transgenic CFTR animals, outwardly rectifying Cl^- channels are expressed but inactive (207). Further, CFTR transfection in CF cell lines restores the activity of the outwardly rectifying channel (162). In addition, extracellular ATP activates outwardly rectifying Cl^- channels in cultured airway epithelial cells and stimulates Cl^- secretion (611). From other experiments, it was suggested that CFTR expression results in ATP extrusion, either because CFTR is ATP-conducting (505) or because it controls ATP transport via another pathway. Putting these results together, Guggino and associates speculated that CFTR expression controls the activity of outwardly rectifying Cl^- channels by an autocrine mechanism involving ATP efflux and ATP activation of the coexpressed channels (162, 249, 576). Recently, Schwiebert et al. (575) provided strong evidence supporting the above hypothesis: *(1)* the activation of outwardly rectifying Cl^- channels requires CFTR expression and intracellular ATP, *(2)* it is abolished (in whole cells and excised patches) by ATP scavenging with either glucose and hexokinase or apyrase, *(3)* extracellular ATP is effective at low concentrations, probably acting via P_{2U} receptors, *(4)* both ATP unidirectional-efflux and single-channel studies support cAMP-stimulated, glibenclamide-blocked ATP release via CFTR itself or a closely related pathway, and *(5)* with expression of CFTR mutants the activation of outwardly rectifying Cl^- channels is abolished or reduced. Hence, the regulatory mechanism would involve CFTR-mediated ATP efflux, and autocrine activation of outwardly rectifying Cl^- channels via P_{2U} purinergic receptors and unidentified G-protein(s) (575). However, studies from other groups (374a, 501a) dispute the contention that CFTR is an ATP-conductive channel. An alternative possibility is that CFTR directly activates outwardly rectifying Cl^- channels, as proposed from studies in which tracheal-membrane proteins were immunopurified and co-reconstituted in planar lipid bilayers. Interestingly, the outwardly rectifying Cl^- channel was insensitive to ATP in these studies (303).

Na^+ absorption in airway epithelia involves apical-membrane Na^+ channels. In cells expressing normal CFTR, these channels are insensitive to aldosterone,

vasopressin, and cAMP (458). In airway epithelia from CF individuals, however, Na^+ absorption is stimulated by cAMP (63). Recent studies have shown that the Na^+ channel in normal and CF airway epithelia is the "regular" amiloride-sensitive epithelial Na^+ channel (ENaC), and the mRNA levels in CF cells are not increased, although Na^+ transport is stimulated (76). Studies of expression of ENaC and coexpression of both CFTR and ENaC in a heterologous system demonstrated that when ENaC alone is expressed, cAMP stimulates amiloride-sensitive Na^+ transport, while in cells coexpressing both channels cAMP inhibits Na^+ transport (610). The mechanism of the effect of CFTR on ENaC has not been established.

In addition to secretory epithelia, CFTR is expressed in myocardium (207a) and kidney (426a). Renal expression involves both the full-length gene product and an isoform lacking the second set of transmembrane domains and the second NBD. Nevertheless, this isoform was functional when expressed in oocytes.

Other epithelial Cl^- channels. Several other Cl^- channels expressed in secretory epithelial cells have been identified by cell-physiological, biophysical or pharmacological studies, but lack molecular identification (reviewed in 17, 81, 187b, 242, 608a). The main channels are the following: *(1)* Ca^{2+}-activated Cl^- channel, which is expressed both in apical membranes of airway and colonic epithelial cells, and is also linear, but more permeable to I^- than to Cl^- and, in contrast with CFTR, is blocked by the disulfonic stilbene derivative DIDS. *(2)* Swelling-activated Cl^- channel, also expressed in airway and intestinal epithelia, with single-channel conductance of 35–50 pS, $P_I > P_{Cl}$, and DIDS sensitivity. *(3)* Outwardly rectifying Cl^- channel, which in contrast with all of the above is activated by depolarization, has a conductance of 40–50 pS, $P_I > P_{Cl}$, and DIDS sensitivity. This is the channel claimed to be activated by external ATP and abnormally regulated, together with CFTR, in cystic fibrosis (see earlier in this section, under Cystic Fibrosis).

The main basolateral-membrane Cl^- transport pathway in the thick ascending limb of the loop of Henle is a Cl^- conductance (i.e., one or more channels) whose properties have been assessed from patch-clamp (243, 251, 477) and lipid-bilayer reconstitution studies (502). Initial patch-clamp studies revealed a 42 pS linear (477) or outwardly rectifying channel (243), insensitive to Ca^{2+}, but activated by depolarization. In the bilayer-reconstitution studies the channel was linear in symmetric Cl^- solutions and also activated by depolarization. The channel is activated by $[Cl^-]_i$ in the physiological range (502); in addition, the Cl^- conductance (560) and the Cl^- channel (477, 502) are stimulated by cAMP, by a PKA-mediated mechanism (502).

Both cAMP and changes in $[Cl^-]_i$ could contribute to modulation of basolateral membrane Cl^- conductance—for example, following stimulation of NaCl reabsorption by vasopressin. This picture has been recently complicated by the finding, in basolateral membranes of mouse cortical thick-ascending limbs, of two Cl^- channels of ca. 45 and 7–9 pS. The smaller channel has features consistent with the function of the segment studied in vitro, from the viewpoints of biophysics, pharmacology, and regulation (251). As stated above, under CFTR, renal expression of both full-length and a truncated CFTR isoform has recently been demonstrated (426a).

Proteins that appear to regulate Cl^- channels. Several other proteins have been suspected to be Cl^- channels. The group includes phospholemman (425a), pI_{Cln}, (340) IsK, Cl (27), and P-glycoprotein (the protein upregulated in multi-drug-resistant cancer cells 639a). However, the experimental data available do not conclusively prove that these molecules are Cl^- channels, and the possibility exists that they are instead regulators of the Cl^- channels themselves (see 301 and 495 for discussions of this issue).

The cation–chloride cotransporters. The Na^+-K^+-$2Cl^-$ cotransporter (NKCC) is a carrier protein that, although not exclusive to epithelial cells, has a major functional role in primary Cl^--transporting epithelia. NKCC performs electroneutral translocation of Cl^-, Na^+, and K^+ with the stoichiometry 2:1:1 (reviewed in 255), but other stoichiometries have been proposed (535). Transport is inhibited by high-ceiling diuretics (furosemide, bumetanide, and others).

NKCC is expressed in Cl^--transporting epithelia in the membrane domain opposite to that in which the Cl^- channel is present. In the case of Cl^- absorption (e.g., in thick ascending limb of the loop of Henle in some species), NKCC is present in the apical membrane (305a) and contributes to net Na^+ and Cl^- reabsorption. In the case of Cl^- secretion (e.g., airway epithelia), NKCC is present in the basolateral membrane. In both instances, the NKCC performs *secondary-active* Cl^- transport, which raises intracellular $[Cl^-]$ to levels above those predicted for electrochemical equilibrium. Thus, Cl^- can then exit the cell by electrodiffusion via channels across the opposite membrane (i.e., basolateral membrane in the loop of Henle cell, apical membrane in the airway epithelial cell). The operation of the NKCC is thus coordinated with those of the Na^+,K^+-ATPase, ultimately responsible for the main component of the thermodynamic driving force for influx of Na^+,K^+, and Cl^-, namely the low $[Na^+]_i$, and the K^+ channels that allow for K^+ recycling. The three transporters are expressed in the same membrane domain and operate in concert.

NKCC is a 195 kDa glycoprotein in shark rectal gland (188, 190), where immunolocalization studies demonstrate its presence in the basolateral membrane of Cl^--secreting cells (390). A homologous 150 kD protein has been identified in other species (255). At least three isoforms, demonstrated immunologically, exist in shark. The Na^+-K^+-$2Cl^-$ cotransporter from shark rectal gland (NKCC1) has been cloned, sequenced, and functionally expressed in human cells (690); using probes from this gene product, mRNA was found in several organs, including liver, kidney, stomach, and colon. The human colon cotransporter is 74% identical to shark NKCC1 and has been functionally expressed in human embryonic kidney cells (478b). A second isoform (NKCC2, 61% identity with NKCC1) was found in rabbit kidney (478); independently, another Na^+-K^+-$2Cl^-$ cotransporter (rBSC) was cloned from rat kidney medulla and functionally expressed in *Xenopus* oocytes (210). NKCC2 can be alternatively spliced to yield three isoforms with differential expression within the kidney (478).

The NKCC genes appear to be related to the cDNA encoding the thiazide-sensitive (Na^+-Cl^-) cotransporter (TSC) in flounder urinary bladder and mammalian distal tubule (211), suggesting a large gene family. Haas (256) has proposed the name cation-chloride cotransporters (CCC), with the following grouping: CCC1 (shark rectal gland NKCC, NKCC1), CCC2 (mammalian kidney NKCCs: NKCC2, rBSC), CCC3 (winter-flounder urinary bladder and mammalian distal tubule thiazide-sensitive Na^+-Cl^- cotransporter). Recently, mammalian K^+-Cl^- cotransporters (KCC1) have been cloned (222a, 478a). Potential additional members of the family include other Na^+-K^+-$2Cl^-$ and Na^+-Cl^- cotransporters.

The NKCC-1 gene product is 1191 amino acids long, with long N- and C-terminal domains located in the cytoplasmic side (each with a threonine phosphorylation site) and twelve predicted transmembrane helices; TM 7 and 8 are separated by a large extracellular loop with several glycosylation consensus sites (Fig. 8.17). There is considerable homology of transmembrane domains among the Na^+-K^+-$2Cl^-$ and Na^+-Cl^- cotransporters identified so far (454), in particular in the transmembrane domains, the second cytoplasmic-side loop, and a C-terminus region (256). Elegant studies of the kinetics of transport by the Na^+-K^+-$2Cl^-$ cotransporter suggest a minimum model with ordered binding (Na^+, Cl^-, K^+, Cl^-) with glide symmetry (389). Bumetanide is thought to compete for the second Cl^- site (see 256). No structure-function relationships have been established.

Regulation of NKCC (see 256 for a detailed review) appears to involve changes in the number of *active*

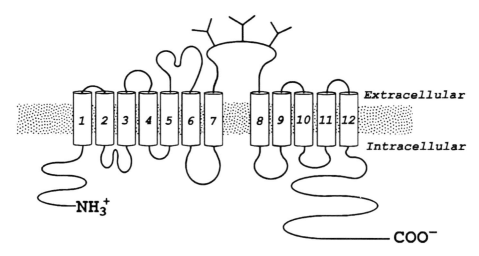

FIG. 8.17. Predicted secondary structure for shark rectal gland Na^+-K^+-$2Cl^-$ cotransporter (NKCC1). The protein comprises 1,191 amino acid residues. Amino and carboxyl termini are on the cytoplasmic side. There are two domains of seven and five putative transmembrane α-helices, respectively, joined by an extracellular loop containing several N-glycosylation sites (branched lines). This predicted secondary structure is shared by other members of putative caution-Cl^- cotransporter family (see text). Modified from Xu et al. (690), with permission.

copies expressed in the membrane, whose assessment is possible because bumetanide and analogs bind only to the activated NKCC (257). Secretagogues that elevate cAMP, as well as cell shrinkage, increase the NKCC activity (189, 386). The activation involves serine- and threonine-residue phosphorylation (387). Direct phosphorylation by PKA appears to have been ruled out because the sites are not within PKA consensus domains and cAMP is not elevated by reductions in cell volume that also activate NKCC (386, 387). An appealing possibility is that the intracellular $[Cl^-]$ plays a pivotal role in the activation of NKCC, thus linking the activity of the apical-membrane Cl^- channels to the activity of the cotransporter expressed in the opposite membrane domain (257a, 388, 527). The sequence of events following agonist-mediated stimulation of transport would be as follows: activation of Cl^- channels, fall of $[Cl^-]_i$, stimulation of the cotransporter expressed in the opposite membrane domain, stimulation of Cl^- entry. Several lines of evidence support this hypothesis, but direct proof is lacking.

Recent studies starting with rabbit kidney and rat brain cDNA libraries resulted in the cloning of the K^+-Cl^- cotransporters, rbKCC1 and rtKCC1 (222a). A human cotransporter (hKCC1), 97% identical with rbKCC1, was also obtained (222a). KCC1 consists of 1085 amino-acid residues, with 24%–25% identity with the Na^+-K^+-$2Cl^-$ cotransporter (NKCC) and the Na^+-Cl^- cotransporter (NCC). The membrane topology, predicted from hydropathy analysis, is similar to that of NKCC (Fig. 8.17), i.e., cytoplasmic N-

and C-termini, 12 membrane-spanning domains, and a large extracellular loop with N-linked glycosylation sites. Functional expression was accomplished in human embryonic kidney (HEK293) cells.

Mechanisms of Ion Transport in H^+- and HCO_3^--Transporting Epithelia

Transepithelial acid/base transport is present in a variety of epithelia, including, for example, virtually every segment of the renal tubule. As is the case with transport of other ions, transepithelial H^+ transport may have a homeostatic function, i.e., contribute to the regulation of the pH of body fluids, or an organ-specific function, e.g., activation of pepsinogen in the gastric fluid. In addition, H^+ transport regulates the intracellular pH (pH_i) of the epithelial cell itself, but this does not necessarily result in transepithelial transport.

We will restrict our discussion to epithelial cells contributing to net transepithelial H^+ or HCO_3^- transport. Examples include renal-tubule segments, gastric mucosa, and excretory ducts of exocrine gland (e.g., pancreas).

Epithelial Cells Transport H^+ via H^+ Pumps or Na^+/H^+ Exchanger. Net transepithelial H^+ transport is most frequently secretory and involves an active step across the apical membrane. Two basic models exist in epithelial cells, depending on the molecule responsible for active H^+ transport: (1) a Na^+/H^+ exchanger, as in the renal proximal tubule and small intestine (20, 541); in this case, H$^+$ transport is secondary active, depending on the Na$^+$ chemical gradient across the apical membrane. *(2)* a H$^+$- or H$^+$, K$^+$-ATPase, as in distal segments of the renal tubule, urinary bladder of reptiles, and gastric epithelium; in these instances, H$^+$ transport is primary active. Typically, only some specialized cells in the epithelium express H$^+$ pumps and can therefore accomplish H$^+$ secretion; in the distal segments of the nephron the α-intercalated (mitochondria-rich) cells (3, 573, 602), and in the gastric glands, the parietal cells (192, 496, 538).

Epithelial Cells Transport HCO$_3^-$ by Cotransport with Na$^+$ or Exchange for Cl$^-$. From a homeostatic point of view, H$^+$ secretion or HCO$_3^-$ absorption are equivalent processes. Ultimately, the proton secreted by an epithelial cell comes from the ionization of an intracellular water molecule, a reaction catalyzed by carbonic anhydrase. Thus, for every H$^+$ secreted there is an OH$^-$ "left behind" in the cell. This OH$^-$ reacts with CO$_2$, becoming HCO$_3^-$, which can be transported independently of H$^+$ transport. The two principal mechanisms of HCO$_3^-$ transport across plasma membranes of epithelial cells are cotransport with Na$^+$ and exchange with Cl$^-$. In the case of H$^+$ secretory cells mentioned above, HCO$_3^-$ is transported across the basolateral membrane, from cell to interstitial fluid. Other cells, such as pancreatic duct cells (343), enterocytes (186, 591, 612), and distal nephron cells of alkali-loaded animals (569, 598), secrete HCO$_3^-$ and absorb (or reabsorb) H$^+$.

The Cell Models of Transepithelial H$^+$ and HCO$_3^-$ Transport Are Mirror Images of Each Other. Transepithelial acid or base transport can occur when the H$^+$ and HCO$_3^-$ transporters are expressed in opposite membrane domains, as shown in Figure 8.18. If H$^+$ is translocated to the lumen, the overall process includes H$^+$ secretion and HCO$_3^-$ absorption (or reabsorption). If H$^+$ is transported to the basolateral bathing solution, the process is reversed, i.e., there are H$^+$ absorption and HCO$_3^-$ secretion. Under most circumstances CO$_2$ is present in all three compartments (cell and solutions in contact with apical and basolateral surfaces of the epithelial cells). In addition, buffering of the transported species (H$^+$ or HCO$_3^-$) results in CO$_2$ production. Finally, CO$_2$ is highly permeable across lipid membranes, so that it tends to equilibrate by diffusion. This provides continuous replenishment of the CO$_2$ lost from the intracellular compartment because of the reaction with OH$^-$ (see Fig. 8.18).

H$^+$ secretion. The cell models accounting for H$^+$ secretion are shown in Figure 8.19, panels *A, B,* and *C.*

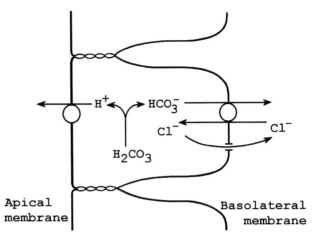

FIG. 8.18. Basic mechanisms of ion transport in H$^+$ and HCO$_3^-$ secretory epithelia. Carbonic anhydrase catalyzes formation of H$^+$ and OH$^-$ from water; HCO$_3^-$ is then formed by reaction of OH$^-$ and CO$_2$. Net result, shown in figure, is production of H$^+$ and HCO$_3^-$. In H$^+$-secreting epithelium there is apical-membrane H$^+$ extrusion and basolateral-membrane HCO$_3^-$ extrusion. In HCO$_3^-$ secreting epithelium, polar expression of these transporters is reversed. Cl$^-$ channel in parallel with Cl$^-$/HCO$_3^-$ exchanger maintains intracellular [Cl$^-$]. Precise basolateral-membrane transporters vary among H$^+$-secreting epithelia (see text).

The essential difference is the *predominant* mechanism of H$^+$ translocation across the apical membrane. In A, model representing proximal renal tubule and small intestine, H$^+$ exits the cell mainly by Na$^+$/H$^+$ exchange. In B, model representing the H$^+$-secreting, α-intercalated cell of the renal collecting tubule, H$^+$ is secreted predominantly by an electrogenic H$^+$-ATPase. In C, model representing the oxyntic cell of the gastric epithelium, H$^+$ is extruded by an electroneutral H$^+$,K$^+$-ATPase. HCO$_3^-$ transport across the basolateral membrane occurs in all three instances, either in exchange for Cl$^-$ or via cotransport with Na$^+$.

HCO$_3^-$ secretion. The mechanism of HCO$_3^-$ secretion is depicted in Figure 8.20. This scheme represents the situation in excretory ducts and base-secreting intercalated cells of the renal cortical collecting tubule (β-intercalated cells), which express similar transporters as the α-intercalated cells, but with opposite polarity. An interesting feature in the regulation of acid/base transport in the collecting tubule of the kidney is that following changes in systemic acid base status of the animal the number of H$^+$-secreting cells appears to increase and the number of HCO$_3^-$ secreting cells appears to decrease. It has been suggested that the mechanism could be a change in polarized expression of the relevant transporters (4). This is discussed further under MECHANISMS OF REGULATION OF TRANSEPITHELIAL TRANSPORT.

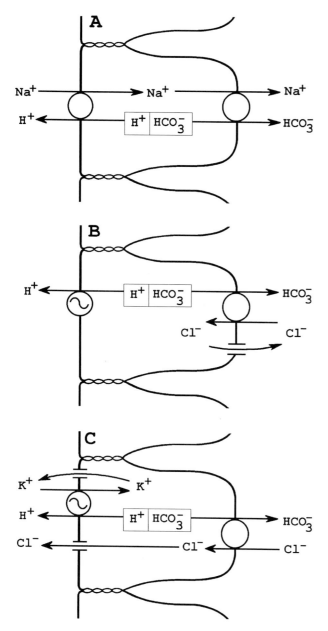

FIG. 8.19. Specific transport mechanisms in H$^+$-secreting epithelia: *A.* Renal proximal tubule. *B.* α-intercalated cell of the renal collecting tubule. *C.* Oxyntic cell of the gastric mucosa. See text for details.

Membrane Transporters in H$^+$- and HCO$_3^-$-Transporting Epithelia. In addition to the Na$^+$/H$^+$ exchanger, already discussed (see under Na$^+$/H$^+$ Exchanger) the transporters involved specifically in transepithelial H$^+$ and HCO$_3^-$ transport are the H$^+$- and H$^+$,K$^+$-ATPases, and the HCO$_3^-$ transporters, namely anion exchangers and the Na$^+$-HCO$_3^-$-cotransporters. Information on their structure and function is succinctly presented in the sections that follow.

Epithelial H$^+$ pumps. Two kinds of H$^+$-transporting ATPases are expressed in plasma membranes of epithe-

lial cells, namely vacuolar H$^+$-ATPases in the kidney (reviewed in 225) and H$^+$,K$^+$-ATPases, in gastric mucosa, colon epithelium, and renal tubule epithelial cells (reviewed in 280, 319, 539, and 543).

The vacuolar H$^+$-ATPase contributes to H$^+$ secretion in proximal and distal nephron. H$^+$ transport mediated by this pump is regulated by and contributes to the maintenance of systemic acid-base homeostasis. Vacuolar H$^+$-ATPases are a family of transporters differing in the composition of specific subunits and are regulated by cytosolic enzymes in a compartment-selective way, i.e., plasma membrane and lysosomal enzymes are regulated differently. The vacuolar H$^+$-ATPase in intercalated cells of the renal collecting duct is found in an intracellular membrane pool that can be inserted by exocytosis in the plasma membrane following appropriate stimuli, in particular intracellular acidification (reviewed in 5 and 225).

In the collecting duct, H$^+$ secretion is carried out by specialized cells rich in carbonic anhydrase, the α-intercalated cells (70, 400). The vacuolar H$^+$-ATPase from bovine kidney has been isolated and characterized (227, 228, 664). It is a multimeric protein of molecular weight ~580 kDa. SDS-PAGE of immunopurified enzyme yielded over ten polypeptides, of apparent molecular weights ranging from ~12 to ~70 kDa (227, 664). Immunolocalization studies revealed abundant expression in apical membrane of the proximal tubule, moderate expression in the thick ascending limb and distal convoluted tubule and abundant expression in the intercalated cells of the connecting tubule and collecting duct (72, 73). There is evidence suggesting that this kind of ATPase is involved in both H$^+$ secretion and HCO$_3^-$ secretion in α- and β-intercalated cells of the cortical collecting tubule, respectively. This re-

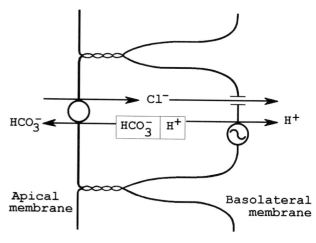

FIG. 8.20. Transport mechanisms in HCO$_3^-$-secreting epithelia. Net operation is HCO$_3^-$ secretion and Cl$^-$ absorption. Note that polarized expression of H$^+$ and HCO$_3^-$ transporters is opposite to that in H$^+$-secreting epithelia.

sults from pump expression in either the apical membrane (α cells) or the basolateral membrane (β cells). For further details, see later, under MECHANISMS OF REGULATION OF TRANSEPITHELIAL TRANSPORT.

The gastric H$^+$,K$^+$-ATPase is an α,β heterodimer (585a) that belongs to the P-type ion-transporting ATPases. The α-subunit, responsible for the transport function and ATP hydrolysis, consists of 1033 amino acids (MW 114 kDa), whereas the β-subunit is composed of 291 amino acids (MW 60–90 kDa) and is heavily glycosylated. Both subunits are quite homologous to the corresponding peptides of the Na$^+$,K$^+$-ATPase. The membrane topology of the α-subunit has not been definitively established. (For reviews, see 280, 319, 539).

The enzyme is expressed in the tubulovesicular system and in the apical membrane of the parietal cells of the gastric epithelium, where its function is secretion of HCl. It does this by an uphill, electroneutral, obligatory exchange of H$^+$ for K$^+$, coupled to Mg^{2+}-dependent ATP hydrolysis. It can generate transmembrane H$^+$ concentration ratios greater than 10^6 (540), which makes it the most "powerful" ion pump. Stimulation of the parietal cell with agonists results in translocation of H$^+$,K$^+$-ATPase molecules from tubulovesicular (cytoplasmic) structures to the apical membrane domain, as well as in increased apical membrane conductance for both K$^+$ and Cl$^-$, which could result from activation and/or insertion of channels (191). The pump-mediated H$^+$ and K$^+$ fluxes, together with the channel-mediated K$^+$ and Cl$^-$ fluxes, result in net HCl secretion.

The deduced amino-acid sequence of the α-subunit yields a molecular weight of 114 kDa with three potential N-glycosylation sites. α-Subunits of several mammalian species are over 95% identical. The homology of α-subunits of porcine H$^+$,K$^+$-ATPase and Na$^+$,K$^+$-ATPase is ca. 60% (ca. 40% in the N-terminal domain). (For further details, see 280, 319, 497, and 539.) Mouse and frog α-subunits have been coexpressed with β-subunits of rabbit H$^+$,K$^+$-ATPase in *Xenopus* oocytes, and both assembly and functional expression were demonstrated (408).

The apparent molecular weight of the glycosylated β-subunit is 60–90 kDa. As is the case with the Na$^+$, K$^+$-ATPase, this subunit has no catalytic function, but is necessary for expression and activity of the enzyme (see 319).

The colon and kidney H$^+$,K$^+$-ATPases are expressed in apical membranes. Colonocytes express an Na$^+$-independent, K$^+$-stimulated ATPase activity sensitive to vanadate and SCH 28080 (138, 254). This ATPase appears to perform active H$^+$ secretion and is both functionally and immunologically similar to the gastric H$^+$,K$^+$-ATPase (309). The colonic H$^+$,K$^+$-ATPase

has been cloned (126). Its α-subunit is ca. 60% identical to the catalytic subunit of gastric H$^+$,K$^+$-ATPase and to the Na$^+$,K$^+$-ATPase isoforms (319).

The mRNA for the colonic H$^+$,K$^+$-ATPase is also expressed in kidney. The intercalated cells in the cortical collecting duct and other distal nephron segments express this pump, which can be inhibited by omeprazole and SCH28080 and is insensitive to ouabain (see 684 for a review). This pump is located in the apical membrane and contributes to K$^+$ reabsorption and H$^+$ secretion. It was suggested that its main physiologic function is K$^+$ reabsorption, which is stimulated in animals subjected to low-K$^+$ diets (151, 684), but recent studies suggest a role in H$^+$ secretion in animals in normal K$^+$ balance (19). Recent studies in rat kidney suggest that both gastric and colonic H$^+$,K$^+$-ATPases are expressed at the mRNA level (341). Interestingly, the colonic isoform appears to be upregulated by hypokalemia, but not by acidosis, whereas the gastric isoform is not regulated by either condition (156a).

Epithelial HCO$_3^-$ transporters. Anion exchangers (AE family) carry out electroneutral exchange of Cl$^-$ and HCO$_3^-$ (506), and hence contribute to the maintenance and regulation of intracellular pH and [Cl$^-$], and cell volume. The prototype is the red cell band 3 (AE1). The AE family includes the related genes AE2 and AE3. The cDNA's encoding for AE1 (335) and other members of the family (reviewed in 506) have been cloned. AE1b (67, 342) is a truncated version of AE1, consisting of 850 instead of 911–929 amino acid residues. Its predicted membrane topology includes intracellular N and C termini, probably 12 transmembrane domains, and N-linked glycosylation on the fourth extracellular loop. AE1b is expressed in the renal collecting duct (basolateral membrane of intercalated cells). Na$^+$-independent Cl$^-$/HCO$_3^-$ exchange has also been demonstrated in basolateral membranes of rabbit late proximal tubule (430) and rabbit medullary collecting duct (267), but the molecular identity of the transporters involved is unknown.

The proteins of the AE family carry out anion exchange that is electroneutral and passive. The fluxes are dictated by the chemical gradients for Cl$^-$ and HCO$_3^-$; since in most cells the Cl$^-$ gradient is much larger, the net fluxes are inward for Cl$^-$ and outward for HCO$_3^-$, i.e., the anion exchangers are base extruders. Disulfonic stilbene derivatives inhibit anion exchange and are thus valuable tools in studying this process (82). However, these agents also inhibit Na$^+$-HCO$_3^-$ cotransport (61) and block certain anion channels (432).

Recombinant AE2 expressed in *Xenopus* oocytes is activated by intracellular alkalinization (291a), a feature that can coordinate fluxes via Na$^+$/H$^+$ and Cl$^-$/HCO$_3^-$ exchangers. It has also been proposed that changes in the polar expression of anion exchangers, as well as in expression of the vacuolar H^+ pump (see above), contribute to the regulation of H^+ and HCO_3^- transport in the cortical collecting duct. This hypothesis is discussed later, under MECHANISMS OF REGULATION OF TRANSEPITHELIAL TRANSPORT.

Other base extruders expressed in epithelial cells are the Na^+-HCO_3^- cotransporter and possibly the Na^+-dependent Cl^-/HCO_3^- exchanger ($Na^+(HCO_3^-)_2/Cl^-$). The existence of the Na^+-HCO_3^- cotransporter is firmly established on the bases of studies in cells and basolateral-membrane vesicles from renal tubules and other epithelia, whereas the expression of Na^+-dependent Cl^-/HCO_3^- exchanger in epithelial cells is likely but not proven (see below). The Na^+-HCO_3^- cotransporter is electrogenic, whereas the Na^+-dependent Cl^-/HCO_3^- exchanger would be electroneutral.

Biophysical studies of the Na^+-HCO_3^- cotransporter suggest Na^+:HCO_3^- stoichiometries of at least 1:2 (9, 62, 492, 581, 693). The best assessments of the stoichiometry (Na^+:HCO_3^- = 1:3) were obtained in membrane vesicles of renal origin (593) and in rat proximal tubules in vivo (581). Given the inwardly directed Na^+ gradient in most cells, base extrusion will take place when the membrane voltage (cell interior negative) is rather large. The $Na^+HCO_3^-$ cotransporter was described in *Ambystoma* proximal tubule (61) and subsequently demonstrated in mammalian proximal tubule as well (78). It is independent of the presence of Cl^- (61, 239) and sensitive to disulfonic stilbene derivatives (61, 79). Its properties, well studied in amphibian and mammalian proximal tubules (reviewed in 492), include voltage sensitivity of transport, Na^+ dependence of the HCO_3^- equivalent permeability, direct (molecular) coupling between Na^+ and HCO_3^- fluxes, and "paradoxical" effect of membrane voltage on tracer Na^+ uptake (in the presence of HCO_3^-, Na^+ uptake is stimulated by an inside-positive voltage of basolateral membrane vesicles). Transport studies in isolated membrane vesicles (592) and in vivo studies in proximal tubules (581) suggest that the transported species is $NaHCO_3$ + CO_3^{2-}. In the proximal tubule, the Na^+-HCO_3^- cotransporter accounts for >90% of the HCO_3^- flux across the basolateral membrane (492). Other epithelia in which this cotransporter is expressed are thick ascending limb of Henle's loop (339), monkey kidney cells in culture (302), and gastric oxyntic cells (129). The cotransporter is subject to acute and chronic regulation. Acutely, it appears to respond to changes in pH_i by an effect on a putative modifier site (594); chronically, its activity is increased by systemic acidosis, either metabolic or respiratory (492).

The Na^+-HCO_3^- cotransporter (NBC) from *Ambystoma triginum* was recently cloned by expression in *Xenopus laevis* oocytes (528a). The aNBC is a 4.3-kb cDNA insert encoding a protein with several (probably 12) transmembrane domains and no substantial homology with sequenced proteins. Expression in oocytes revealed the main features demonstrated in epithelial cells, i.e., electrogenicity, Na^+ dependency, HCO_3^- dependency, Cl^- independence, and disulfonic-stilbene block (528a).

The Na^+-dependent Cl^-/HCO_3^- exchanger has been conclusively demonstrated in invertebrate cells (see 58 for review) and has been proposed to be expressed in basolateral membranes of proximal tubule cells, on the bases of pH_i changes in response to changes in external $[Cl^-]$ and $[Cl^-]_i$ changes elicited by alterations in external HCO_3^-/CO_2 (10, 250, 297, 556). These effects are Na^+-dependent and it has been claimed that they cannot be attributed to thermodynamic coupling of a Na^+-HCO_3^- cotransporter and a Cl^- conductance, because the latter is deemed negligible (see discussion in 492). The data are consistent with electroneutral exchange involving $Na^+(HCO_3^-)_2/Cl^-$ or an equivalent process. However, the studies on which these interpretations are based are technically difficult, and the possibility of conductive contributions to transport has not been definitively excluded. Further, Na^+-dependent Cl^-/HCO_3^- exchange is not present in membrane vesicles (21). These arguments certainly do not rule out the hypothesis that this transporter is expressed in epithelial cells, but suggest a need for additional studies. In proximal tubule, the contribution of this transporter to HCO_3^- transport across the basolateral membrane would be small in comparison to that of the Na^+-HCO_3^- cotransporter. Expression of $Na^+(HCO_3^-)_2/Cl^-$ in epithelial cells appears to have been proposed only for renal proximal tubule.

MECHANISMS OF TRANSEPITHELIAL WATER TRANSPORT

The study of the mechanisms of transepithelial water transport has progressed considerably in the last decade because of two major experimental developments—namely, accurate measurements of water-permeability coefficients, which allowed for identification of transepithelial water-transport pathways and estimation of osmotic driving forces, and the identification of the molecules responsible for water-pore behavior in certain epithelial-cell membranes.

Transepithelial Water Transport Is Linked to Transepithelial Salt Transport

In epithelia, vectorial water transport is a process of great homeostatic importance. Water balance is regulated to a great extent by changes in the rate of water loss in the urine, by mechanisms to be discussed in this section from a cell physiology point of view. Further, the renal proximal tubules and the small intestine absorb large volumes of water, in a human adult about 130 and about 8 liters per day, respectively. Two distinct types of transepithelial water transport can be observed in epithelia in situ: *(1)* net transport between nominally isosmotic fluids (e.g., in the renal proximal tubule), and *(2)* net transport down a preexisting osmotic gradient (e.g. in the collecting tubule, from a low-osmolality lumen fluid to a high-osmolality peritubular fluid). There is considerable experimental evidence indicating that the mechanism of water transport is largely passive and driven by osmotic forces. In other words, the water moves across the epithelial barriers by a downhill mechanism, because of a difference in water-chemical potential. However, there is evidence supporting the possibility of secondary-active water transport (see the section that follows). Other theories, such as pinocytosis and electroosmosis, have been proposed, but their experimental support is poor (reviewed in 680).

Transepithelial water transport can occur in either the absorptive or the secretory direction. Although most epithelia are either absorptive or secretory, in some epithelial organs absorption and secretion can be performed by different regions, and a change in net direction from absorption to secretion can occur as well, e.g., in the small intestine.

Secondary-Active Water Transport? Recent experiments in *Necturus* choroid plexus epithelium (699a) and frog retinal pigment epithelium (699b) indicate that water transport may be coupled to solute transport across individual cell-membrane domains, and that under favorable conditions water can be moved uphill, driven by the net ion gradients, a process *like* cotransport. In choroid plexus and pigment epithelium, the transporters involved would be the Na^+-K^+-$2Cl^-$ cotransporter and the H^+-lactate-cotransporter, respectively. These observations suggest that secondary-active water transport may, after all, occur in biological membranes (513a).

Epithelia Are Widely Diverse in Their Water-Transport Characteristics

Different epithelia have very different osmotic water permeabilities. The osmotic water permeability of cell membranes (P_f, also P_{os}) varies greatly, from virtually nil at the apical membrane of cells of the thick ascending loop of Henle (and of distal and collecting tubule cells in the absence of antidiuretic hormone), to 400–600 $\mu m \cdot s^{-1}$ in mammalian red blood cells and renal proximal tubules. In pure lipid bilayers, P_f values range from <1 to ca. 100 $\mu m \cdot s^{-1}$. Most epithelial cell membranes have P_f values near the high range of lipid-bilayer P_f. In some cases, however, they are either very tight to water or highly water permeable (Table 8.4; see also 623). Recent experimental results indicate that membranes with *very high* P_f values express pores in their plasma membranes, and these pores appear to be specific for water. There is no explanation for the *very low* P_f values of other cell membranes.

A simple classification of epithelia distinguishes three groups, depending on the value of the osmotic permeability coefficient and on whether it is regulated or not:

TABLE 8.4. *Osmotic Water Permeability of Epithelial Cell Membranes*

Epithelium	Apical	Basolateral	Transepithelial	References
ADH-Insensitive				
Rabbit PST	4500	5000	4280	94
Necturus gallbladder	640	460	350	119
Rabbit/Rat TALH	—	—	0–20*	503
ADH-Sensitive				
Rabbit CCD (−) ADH	70	450	—	609
(+) ADH	310	490	—	
Rabbit IMCD (−) ADH	70	480	—	184
(+) ADH	260	390	—	

Abbreviations: ADH = antidiuretic hormone (vasopressin); PST = proximal straight tubule; TALH = thick ascending limb of the loop of Henle; CCD = cortical collecting duct; IMCD = inner-medullary collecting duct. P_f values are rounded and expressed in $\mu m \cdot s^{-1}$, without correction for membrane folding factors; i.e., they are referred to "idealized" epithelial surface. For P_f values for membrane folding, see ref. 623. * Range of several studies.

(1) Epithelia with high osmotic water permeability: In this group the P_f is always high and there is no regulation. Examples include virtually all so-called leaky epithelia, namely renal proximal tubule, descending limb of the loop of Henle, small intestine, gallbladder, choroid plexus, and others. In many of these epithelia the high water permeability of the cell membranes is attributable to the presence of proteinaceous pores in the cell membranes, both apical and basolateral. The contribution of the junctional pathway to water flow is controversial and will be discussed below.

(2) Epithelia with low osmotic water permeability: The only example in mammals is the ascending limb of the loop of Henle, in which the P_f is extremely low and insensitive to antidiuretic hormone. The barriers to water permeation are the apical membranes and the junctional complexes. The basolateral membrane is quite permeable to water, so that changes in the osmolality of the basolateral solution cause rapid changes in cell volume (268, 613), whereas changes in the osmolality of the apical bathing solution do not alter cell volume. It is interesting to note that the transepithelial P_f is low although the junctions are quite permeable to ions. This is one of many instances in which water- and ion-permeability coefficients correlate poorly. Even when ion channels are expressed in cell membranes and are permeable to water, the number of channel molecules is relatively low, so that the P_f is still small. The apical membrane P_f of the cells of the ascending segments of the loop of Henle is among the lowest in nature. The explanation for this unusually low P_f value is unknown. It is also unclear whether the junctional complexes of this tubule segment have different properties from those of tubule segments with different water permeabilities.

(3) Epithelia with regulated water permeability: Epithelia of the collecting segments of the renal tubule and the anuran epidermis and urinary bladder exhibit a low baseline P_f which is increased in response to changes in plasma osmolality. The regulation of water permeability in these epithelia is essential for water homeostasis. Increases in plasma osmolality cause secretion of antidiuretic hormone (hereafter ADH, in mammals vasopressin) from the neurohypophysis. Vasopressin binds to V_2 receptors in the basolateral membrane of the target cells, and by a G protein–mediated mechanism activates adenylyl cyclase. The resulting increase in intracellular cAMP levels promotes exocytotic fusion with the apical membrane, of subapical tubulovesicles containing preformed water pores, thus increasing its P_f (70, 260, 262, 646). Cessation of stimulation by ADH results in removal of the pores by endocytosis (reviewed in 654).

Transepithelial Water Transport in Leaky Epithelia Is Nearly Isosmotic

In so-called leaky epithelia (e.g., renal proximal tubule, small intestine, gallbladder) under most physiologic conditions the fluid compartments on both sides of the epithelium are isosmotic or nearly isosmotic.

A pivotal observation in early studies of transepithelial water absorption was the demonstration that it can take place *against* the osmotic gradient, assessed from the osmolalities of the bulk bathing solutions (132). This is the case in both intestine and renal proximal tubule, and clearly indicates that water transport is uphill, although not necessarily a primary active phenomenon. Other studies in small intestine demonstrated that uphill water absorption occurs only in the direction of net solute absorption, and hence suggested coupling between solute and water fluxes. The task, then, was to explain how such coupling can occur and result in net water flow from a more-concentrated to a less-concentrated solution.

Three-Compartment Model. An epithelial-cell layer and the surrounding fluids constitute a multibarrier, multicompartment system. The properties of the barriers and compartments can explain how the coupling between transepithelial solute and water transport occurs and results in *apparently active* water transport. The two main proposals developed around this idea were the three-compartment model of Curran and MacIntosh (131) and the standing-gradient hypothesis of Diamond and Bossert (148).

The three-compartment model (Fig. 8.21*A*) accounts for water transport from *cis*-solution to *trans*-solution *against* the osmotic pressure difference between the solutions bathing the epithelium. Transport of salt into an unstirred compartment within the epithelium renders this compartment hyperosmotic to the *cis* solution (*hyperosmotic middle-compartment*). Water flows from the *cis* solution (A) into the middle compartment (M) by osmosis, and *solution* flows from the middle compartment (M) to the *trans* solution (B) by bulk flow, because of the increase in hydrostatic pressure in the middle compartment (M). The barrier separating *cis* solution and middle compartment is semipermeable (high reflection coefficient, low solute permeability), whereas the barrier separating middle compartment and *trans* solution is porous (low reflection coefficient, high solute permeability). In the epithelium, the *cis*-side membrane would be formed by the two cell membranes in series and the *trans*-side membrane by the basal end of the lateral intercellular spaces in series with the basement membrane. Water would be transported against the osmotic pressure difference for as long as the middle compartment remains hyperosmotic to the *cis* compartment. A clear prediction from the model is that the fluid emerging from the epithelium must be hyperosmotic to the solution in the *cis* side. This prediction was not borne out by experimental observations, which showed that the osmolality of the emerging fluid was not significantly different from that of the *cis* solution (145). The standing-gradient hypothesis is a refinement of the Curran and MacIntosh model that takes care of this difficulty.

Standing-Gradient Hypothesis. This mechanism, proposed by Diamond and Bossert (148), is depicted in Figure 8.21*B*. Its main tenets are: *(1)* The hyperosmotic middle compartment is the lateral intercellular space. *(2)* The junctions are assumed to be effectively impermeable to water and solutes. *(3)* Solute is transported from the cell to the trans solution across the apical-most region of the lateral membranes. *(4)* In contrast, osmotic water flow occurs throughout the lateral membrane. *(5)* The solution in the lateral intercellular space is unstirred. *(6)* Longitudinal solute diffusion in the lateral intercellular space is restricted by the geometry (the spaces are long and narrow). These conditions result in an osmotic gradient from blind end to open end of the lateral intercellular space, with progressive lowering of the osmolality. The mathematical model shows that near-isosmolality of the emerging fluid is more likely if the channel is long and narrow, the solute diffusion coefficient is small, and the cell membrane P_f is high relative to the rate of solute transport.

Experimental work after the formulation of the hypothesis showed that some of the assumptions listed above are not valid. First, the junctions (in the case of leaky epithelia) have high ionic permeability (199, 288, 679, 682). Second, active ion transport across the basolateral membrane is not restricted to the apical regions of the spaces, but homogeneously distributed, as demonstrated by the distribution of Na^+, K^+-ATPase molecules by ouabain binding (421). Third, morphometric analysis and electrophysiologic calculations suggest that the lateral-space diffusion time constants are short, and hence that longitudinal standing osmotic gradients are unlikely (281, 315, 547, 670, 671). In addition, it would seem that only if the spaces are very long and narrow and the cell membrane P_f very high would there be osmotic equilibration of the transported fluid. Realistic geometry and reasonable values for cell-membrane P_f predict a hyperosmotic emerging fluid (547). The standing-gradient hypothesis also involves predictions of the magnitude of the hyperosmolality of the fluid in the lateral intercellular spaces.

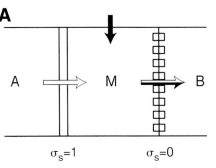

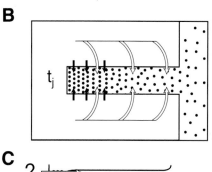

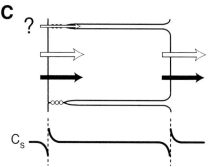

FIG. 8.21. Mechanisms of water transport in leaky epithelia. *Panel A.* Three-compartment model of Curran and MacIntosh (131). Because of solute entry *(solid arrow)* $C_s(M) > C_s(A)$, where C_s is total solute concentration or osmolality. This causes osmotic water flow from compartment *A* to compartment *M (open arrow)*. The elevation of the hydrostatic pressure in compartment *M* causes solution flow from compartment *M* into compartment *B (open/solid arrow),* regardless of the solute concentration (and hence osmotic pressure) in the latter. Modified from Whittembury and Reuss (680), with permission. *Panel B.* Standing-gradient hypothesis of Diamond and Bossert (148). Solute transport *(solid arrows)* into channel *(lateral intercellular space)* causes a local increase in osmolality; water flows osmotically across bounding membranes *(open arrows),* "diluting" the solution in the channel. Transport toward open end is by bulk flow and diffusion. Modified from Reuss and Cotton (517), with permission. *Panel C.* Near-isosmotic transport model. Because of high osmotic water permeability of cell membranes, differences in solution osmolality (C_s) needed to account for fluid transport are small, probably localized at epithelium-solution interfaces. Salt transport causes dilution of solution on *cis* side and concentration of solution on *trans* side; these differences cause osmotic water flow from *cis* to *trans*. Magnitude of paracellular water flow is uncertain. Because of high surface area of lateral membranes and small volume of lateral intercellular spaces, space osmolality is "clamped" by cell osmolality, making longitudinal gradients small at most.

The lower the P_f of the cell membranes, the greater the driving force necessary to account for a certain rate of water transport. Early measurements of transepithelial P_f values were gross underestimations because of lack of consideration of unstirred-layer artifacts (289).

Near-Isosmotic Fluid Transport. Current P_f values for high-permeability membranes range from 100 to 400 $\mu m \cdot s^{-1}$, one to two orders of magnitude higher than early estimates (reviewed in 680). Therefore, instead of a mean difference in osmolality of the order of 20 $mosmol \cdot kg^{-1}$, a difference of only 1–2 $mosmol \cdot kg^{-1}$ may suffice to account for the rates of transepithelial transport (119, 480, 680). Transepithelial salt transport in leaky epithelia should result in finite, albeit small, changes in the osmolalities of *cis* and *trans* solutions in the immediate vicinity of the membranes, namely *cis*-solution hyposmolality and *trans*-solution hyperosmolality (669). Such changes have been measured in fluid samples of isolated-perfused renal tubules (34, 240). Hence, the accumulating evidence suggests that so-called isosmotic transepithelial water transport is not isosmotic, but near-isosmotic, that the osmotic gradients required to account for the measured water transport rates are very small, and that substantially hyperosmotic compartments or large standing gradients do not exist (Fig. 21C; [596, 680]).

Paracellular Recirculation of Solute. Ussing and his co-workers have proposed an interesting mechanism that can explain truly isosmotic transepithelial water transport (341a, 636a). After arguing that transcellular net water transport must be negligible under truly isosmotic conditions, they bring the paracellular pathway into consideration. For instance, in frog-skin glands, Cl^- secretion in transcellular and Na^+ secretion is paracellular but involves *cation recirculation*: paracellular Na^+ flux into the gland lumen is followed by its partial reuptake by the acinar cells (e.g., apical-membrane Na^+ channel, basolateral membrane Na^+, K^+-ATPase) and resecretion via the paracellular pathway. The recirculation of Na^+ would ultimately account for the net water flux, which would be paracellular (for additional details, see 341a).

Role of Other Driving Forces. In vivo, and to a certain extent in vitro, other "forces" than the difference in osmolality between the solutions on each side of the barrier can contribute to transepithelial water transport between isosmotic or near-isosmotic solutions. In this context, one must consider three possibilities (for details, see 680): *(1)* Differences in protein concentration that result in a colloid-osmotic pressure difference: In proximal tubule a small fraction of fluid reabsorption

can be attributed to this mechanism. *(2)* Differences in effective osmolality because of asymmetric distribution of solutes with high reflection coefficient: If the osmolalities of the two solutions are the same but the reflection coefficients of one or more solutes differ there will be a difference in effective osmotic pressure, which will tend to drive a net water flux toward the less-permeant solute(s). This is the case in the late renal proximal tubule by the complete reabsorption of organic solutes and the preferential reabsorption of HCO_3; these solutes have higher reflection coefficients than Cl^-. *(3)* Differences in hydrostatic pressure: In contrast with the regime in the cardiovascular system, the hydrostatic pressures across epithelial layers are generally small and do not contribute importantly to net fluid transport.

Transepithelial Water Transport in Leaky Epithelia Can Be Transcellular and/or Paracellular

In principle, transepithelial osmotic water flow can take place across the transcellular pathways (cell membranes in series with the cytoplasm) and/or across the paracellular pathway (junctional complexes in series with the lateral intercellular spaces). While there is solid experimental evidence in favor of transcellular water flow, paracellular water flow remains controversial.

Arguments for Transcellular Osmotic Water Transport. The principal argument for transcellular water transport is the direct demonstration of high P_f of the cell membranes. In leaky epithelia with high water permeability such as mammalian renal proximal tubule and *Necturus* gallbladder, the P_f values of apical and basolateral membrane are typically several hundred $\mu m \cdot s^{-1}$ (renal tubule: see 92, 93, 94, 234, 235, 680; gallbladder: see 119, 480, 699). Luminal-membrane microvilli and basolateral membrane infoldings amplify membrane surface area severalfold, in particular in mammalian proximal tubule, so that the P_f values corrected per unit membrane area are far less than the values relative to idealized cylindrical geometry of the tubule. Nevertheless, the presence of pores in certain epithelial-cell membranes has been demonstrated by both biophysical and molecular-biological methods (see below).

Arguments for Paracellular Osmotic Water Transport. The contribution of the paracellular pathway to transepithelial osmotic water flow cannot be measured directly at the present time. Instead, it can be estimated from calculations based on a comparison between the P_f values of the entire epithelium and those of the cell membranes. Inasmuch as the pathways are in parallel, the transepithelial P_f (P_f^t) is the sum of the cellular and paracellular components ($P^t_f = P^c_f + P^p_f$). In proximal tubule, the data support the notion that P^t_f is significantly greater than P^c_f and hence suggest that P^p_f is sizable (see 680). However, the errors in the measurements of both transepithelial and cell-membrane P_f are considerable, limiting the value of these calculations.

A second argument supporting paracellular water transport is based on the blocking effect of organomercurial compounds (see later, under Molecular Identity of Water Pores in Epithelial-Cell Membranes). The inhibition of cell-membrane P_f by pCMBS is large (>90%), yet the transepithelial P_f falls only to about half, suggesting a parallel pathway for osmotic water flow (680). However, these calculations assume a single site of action of pCMBS (the ipsilateral membrane domain) and other effects are possible.

A third kind of argument in support of paracellular osmotic water flow is based on experimental observations indicative of coupling of water and solute fluxes, namely *solvent drag* and *electrokinetic phenomena*. The notion is that if osmotic water flow occurs through relatively large pores (with finite solute permeability), then viscous drag will carry solute in the same direction as the water, generating a flux that is not accounted for by the difference in solute chemical or electrochemical potential between the bulk solutions. Apparent solvent drag can be measured using hydrophilic nonelectrolytes (typically low-molecular-weight sugars); a net flux occurs in the same direction as the water flow, and there is a linear correlation between the two. Positive results have been obtained in several flat and tubule-shaped epithelia (see 680). Among electrokinetic phenomena the easiest to determine is the *streaming potential* elicited by ion flow in the same direction as the water flow, due to the ionic selectivity of the water-permeation pathway. Transepithelial streaming potentials have been measured for a couple of decades and purported to denote coupling of water and ion transport in the same pathway, supposedly paracellular (reviewed in 623). An interesting recent hypothesis is that the paracellular permeability of small-intestine epithelium can be regulated in response to changes in apical-solution osmolality. Following digestion by membrane-bound hydrolytic enzymes the large increase in osmolality at the cell surface and/or the increase in solute entry would result in cytoskeletal changes conducive to an increase in junctional permeability. Thus, a large fraction of the solute and fluid absorption would be paracellular, by solvent drag (466). This issue is discussed further under MECHANISMS OF REGULATION OF TRANS-EPITHELIAL TRANSPORT.

A problem with solvent-drag and streaming-potential experiments is that in many instances the results can be explained by the existence of unstirred fluid layers adjacent to the epithelial surfaces. Solute transport in these layers is exclusively by diffusion (there is no convection), and therefore water transport (from *cis* to *trans* side) changes the solute concentrations in the vicinity of the cell-membrane surfaces, increasing the concentration in the cis compartment and decreasing it in the trans compartment. Hence, the flux of nonelectrolyte can be the result of these changes in concentration (or *unstirred-layer polarization*) instead of flux coupling in the permeation pathway (pseudo solvent drag instead of "true" solvent drag). Similarly, the change in transepithelial voltage could be due to the changes in ion concentrations at the membrane surfaces generated by the water flux, i.e., a simple diffusion potential (pseudo streaming potential). In many instances, the experimental results are amenable to both interpretations, but in some cases it has been argued that only unrealistic unstirred-layer thicknesses would invalidate the conclusion of true solvent drag (see 680). In recent studies of apparent streaming potentials in *Necturus* gallbladder epithelium it was shown that comparison of the time courses of the changes in transepithelial voltage produced by *(1)* adding osmolyte, and *(2)* ionic substitution permits differentiation of true and apparent streaming potentials (513, 521, 522).

In conclusion, there is solid evidence that the cellular pathway contributes to transepithelial osmotic water transport and there are experimental observations supporting the possibility of significant paracellular water flow. However, the evidence for the latter is indirect and sometimes questionable because of potential artifacts resulting from the presence of unstirred layers. Definitive resolution of this problem must await direct assessments of transcellular and paracellular water transport. This is a difficult task, considering the limited technical options available to measure water flow.

Water Permeation across Cell Membranes of Some Leaky Epithelia Is via Constitutive Pores

Osmotic water flow across a cell membrane could be by solubility diffusion via the lipid moiety of the membrane and/or by single-file or laminar flow via pores piercing the membrane. It is clear that most lipid membranes have sizable P_f values, and hence the existence of pores is not necessary to account for some osmotic water permeation. However, cell membranes with high P_f (such as those of leaky epithelia and ADH-stimulated tight epithelia) do contain pores. The existence of these pores was first proven with biophysical techniques. Recently, their molecular identity and functional properties have started to be unraveled (reviewed in 2, 140, 164, and 647a). An excellent review of the biophysics of water transport across artificial, cellular, and epithelial membranes was written by Finkelstein (180).

Biophysical Criteria for Water Pores

High P_f/P_d ratio. If water permeates through pores, osmotic water flow will involve some kind of interaction among permeating water molecules, either frictional interaction (viscous, laminar, or Poiseuille flow, as in thin capillaries) or single-file transport. In either case, the mechanism of water permeation will not be simple diffusion (as in the case of permeation across the lipid bilayer) because the water molecules do not move independently. The result is that the ratio of osmotic (P_f) to diffusive water-permeability coefficient (P_d) is significantly greater than unity (the value of P_f/P_d is proportional to the square of the pore radius for a pore allowing for viscous flow and to the number of water molecules in the pore for a single-file pore). However, the presence of unstirred layers can lead to spuriously high values of this ratio because the value of P_d is underestimated more than the value of P_f. Hence, unstirred-layer corrections are necessary for the interpretation of P_f/P_d (reviewed in 180). In sum, a valid demonstration that $P_f/P_d > 1$ is evidence for pore-mediated osmotic water flow. Whether the value of P_f/P_d denotes the radius of a rather wide pore or the number of water molecules in a single-filing pore must be determined by independent criteria.

Low activation energy. The Arrhenius activation energy (E_a) for water permeation via pores is close to the E_a for water self-diffusion (<5 kcal·mol^{-1}), which is much lower than the E_a for water diffusion across a phospholipid bilayer (ca. 12 kcal·mol^{-1}).

Hg sensitivity. When cell membranes contain water pores, the fraction of water flow via the pores is sensitive to HgCl$_2$ and organomercurial compounds. This feature suggested the presence of a critical SH group in the pore protein (391), which has been confirmed in molecular-biological studies.

Interactions between water and solute fluxes. If the pores are large enough that solutes present in the solution partition into the pore, then there will be frictional interactions between the solutes and the water fluxes in the pore, as discussed above in the context of paracellular water flow. The two main forms of interaction are *solvent drag* and *electrokinetic phenomena*. Although both have been demonstrated for other membrane pores, it appears that water pores in epithelial cells are impermeable to other molecules, i.e., probably too narrow for either solvent drag or electrokinetic phenomena to occur.

As stated above, one should suspect water pores in cell membranes with high P_f values. Of the epithelial ones, the best studied is the apical membrane of the renal proximal tubule, for which all the above-listed criteria have been demonstrated (see 680). Epithelial water pores have been identified in a number of organs and cell types (Table 8.5). In the apical membrane of renal proximal tubules, the pores are present in the membrane and active (permeable) under control conditions, whereas in the renal collecting tubule they are contained in a cytoplasmic vesicular pool in the resting condition and are inserted in the apical membrane following stimulation with ADH (71).

Mechanisms of Transepithelial Water Transport in ADH-Sensitive Epithelia

In contrast with the constitutive high P_f of the proximal tubule and most other epithelia, the P_f of mammalian collecting tubule and amphibian epithelia such as the frog epidermis and urinary bladder is under hormonal regulation. In the absence of ADH, the apical membrane is virtually impermeable to water, whereas the basolateral membrane is highly permeable (557). Further, increases in apical-solution osmolality, which in-

TABLE 8.5. *Expression of Water Pores in Epithelial Cell Membranes*

Aquaporin	Epithelium	Membrane Domain	AA Residues/TMS	Hg Sensitivity
AQP1 (AQP-CHIP, CHIP 28)	PT, tDLH, CPE, BDE, GBE, ESGE, CEnd	A/BL	269/6	Yes
AQP2 (AQP-CD, WCH-CD)	CCD, MCD	A	271/6	Yes
AQP3	MCD	BL	285/6	Yes
AQP5	SG, LG, lung	?	265/6	Yes

From references 2, 324, and 498. Abbreviations: *AA* = amino acid; *TMS* = transmembrane segments. AQP 3 is also permeable to urea and glycerol. AQP4 does not appear to be expressed in epithelial cells. *Epithelia:* PT = proximal tubule, tDLH = thin descending limb of Henle's loop, CPE = choroid plexus epithelium, BDE = biliary-duct epithelium, GBE = gallbladder epithelium, ESGE = eccrine sweat gland epithelium, CEnd = corneal endothelium, CCD = cortical collecting duct, MCD = medullary collecting duct, SG = salivary gland, LG = lacrimal gland.

crease junctional permeability, result in a small increase in P_f compared to that elicited by ADH, although the junctional permeability to urea increases (559). These results support the notion that the effect of ADH on P_f is exclusively or mainly at the apical membrane.

The available experimental evidence indicates that apical-membrane P_f of ADH-sensitive epithelia is regulated by insertion and retrieval of intracellular vesicles containing water pores (70, 260, 262, 646). The existence of such pores is supported by measurements of the ratio of osmotic and diffusive water-permeability coefficients ($P_f/P_d > 1$), but unstirred layers complicate the interpretation of such studies and there were errors in the early literature (see 180). The problem has been reexamined in both toad urinary bladder (365, 366, 471) and cortical collecting tubule (269, 557), with appropriate correction for unstirred-layer effects. In both cases a high value of P_f/P_d was found. Further, there is evidence that hydrophilic nonelectrolytes do not permeate the ADH-induced pathway (see below). Taken together, these results indicate that osmotic water flow across apical membranes of ADH-sensitive epithelia is mediated by narrow, single-file pores. A strong additional argument against osmotic water flow by solubility diffusion is the lack of parallelism between the effects of ADH on water and nonelectrolyte permeability (179). While in artificial lipid bilayers membrane fluidity changes alter proportionally the permeabilities for water and hydrophobic nonelectrolytes (178), in amphibian urinary bladder ADH increases P_f 50-fold with practically no change in nonelectrolyte permeability (179, 486). Observations in membrane vesicles support this conclusion; in fact, increasing the fluidity of urinary bladder endocytic vesicles decreased, instead of increasing, the P_f (206).

The water pores are thought to correspond to structures visible as small membrane particles in freeze-fracture electron microscopy. The particles are grouped in aggregates that inside the cell are contained in tubule vesicles named *aggrephores* (70, 260, 262). ADH stimulates the fusion of aggrephores with the apical membrane by a process involving changes in the actin and spectrin cytoskeleton in the subapical region of the cell. The breakdown and reorganization of the actin network is thought to permit vesicle fusion (reviewed in 266). The fusion has been demonstrated by both electron-microscopic techniques and measurements of membrane electrical capacitance, which increases by 10%–30% with ADH stimulation (462, 465, 604).

Whether the intra-membrane particle aggregates suffice to account for the P_f has been extensively discussed (365, 366, 652, 653; see also 180). Levine et al. (366) have argued that water pores in parallel occupying only the aggregates in the apical membrane surface cannot account for the ADH-induced P_f and suggested the possibility of a series-parallel "showerhead" model (366), i.e., a long and wide channel (stem) in series with numerous short, single-file pores (in the cap). Alternatively, the aggregates in the fused tubules may contribute to the total P_f. Wade (653) has calculated that such a model would account for the measured P_f value following ADH stimulation.

An ingenious method to assess apical membrane osmotic water permeability in renal tubule segments was developed by Verkman and coworkers (647). A fluorescent marker was injected in the circulation of ADH-deficient rats. The marker filtered in the glomerulus and was taken up by endocytosis across apical membranes of renal-tubule cells. In the renal cortex, most of the vesicles were from proximal tubules, whereas in the papilla they were from collecting tubules (358, 359). Hence, it was possible to obtain fractions containing endocytic vesicles from either proximal tubule or collecting tubule. The P_f of the membranes of these vesicles was measured by stop-flow quenching upon raising the external osmolality (102). The P_f of vesicles from renal papilla of ADH-deficient rats was very low; ADH treatment resulted in an increase in P_f with low activation energy (359, 647; see also Molecular Identity of Water Pores in Epithelial-Cell Membranes). Biophysical studies of endocytic vesicles from amphibian urinary bladder indicate that the water pores are fully functional and not regulated (206).

The selectivity of the ADH-sensitive water pores has been controversial. In some cells, ADH increases both P_f and the permeability of small hydrophilic nonelectrolytes, such as urea. In toad urinary bladder, phloretin inhibits solute permeability but not P_f (363), whereas general anesthetics inhibit P_f but not solute permeability (367). In addition, studies in endocytic vesicles revealed that the increase in urea permeability brought about by ADH appears to be independent from the regulation of P_f (326, 585), and in cortical collecting tubule ADH appears to increase only water permeability (238, 270). These results indicate that the water and solute permeation pathways stimulated by ADH are separate.

The H^+ permeability of apical membrane of toad urinary bladder epithelium was found to increase upon treatment with ADH (226), but this result has been disputed (472). The H^+ permeability of endocytic vesicles from renal medulla is about the same, independent of whether they have high or low P_f (585).

There is no definitive evidence for recycling of the water channels in mammalian collecting tubule or amphibian urinary bladder, but it is clear that the vesicles containing the water channels, in contrast to vesicles from proximal tubule cells, are not acidified in the presence of ATP (360). This suggests that there is no degradation of the water pores, supporting the possibility of recycling.

Molecular Identity of Water Pores in Epithelial-Cell Membranes

From radiation inactivation studies the apparent molecular mass of the water pore in cells of the renal proximal tubule was estimated to be ca. 30 kDa (643), but its molecular identification was serendipitous, i.e., the result of a search for a function of a membrane protein found in the red cell membrane (2). A family of water channel proteins has been identified, and the generic name *aquaporin* has been proposed (2).

CHIP28 is a 28 kDa channel-forming integral protein (acronym CHIP) found in the red blood cell membrane with a density of ~150,000 copies per cell (141, 590). CHIP was cloned in 1991 (493) and later shown to correspond to the water pore previously identified by biophysical methods. The protein exists in two forms, namely 28 kDa nonglycosylated (CHIP28) and N-glycosylated (glyCHIP) of molecular weight of 40–60 kDa (590). The assembled membrane protein is a tetramer with one glyCHIP per tetramer (304, 590). Negative-stain electron microscopy suggests that the structure of CHIP is that of a square-shaped oligomer with 7 nm side and a central pit (655, 656) (Fig. 8.22). For further details, see the reviews by Agre et al (1a, 2).

The experimental evidence supporting the notion that CHIP is a water pore includes: (1) injection of CHIP RNA in *Xenopus* oocytes elicits an increase in P_f that has a low E_a, is sensitive to $HgCl_2$ and inhibited by antisense oligonucleotides (159, 494, 700), (2) CHIP reconstitution in liposomes confers an osmotic water permeability with all the characteristics of pores, and (3) CHIP is expressed in the membranes of cells that demonstrably have a high osmotic water permeability mediated by pores.

In the kidney, CHIP protein is expressed in apical and basolateral membranes of renal proximal tubule cells and descending thin limbs of the loop of Henle (141, 437, 438, 536). Other sites of expression are epithelial cells of choroid plexus, ciliary body, lens, biliary tract, eccrine sweat gland and male reproductive tract, and capillary and venular endothelial cells (2, 74, 439).

The primary sequence and membrane-topology studies suggest the structure depicted in Figure 8.22 (see 2 and 164). N- and C-termini are cytoplasmic and there are six transmembrane domains, with an internal repeat present in each half of the molecule. Sequence analysis and site-directed mutagenesis experiments strongly suggest that the first cytoplasmic loop (loop

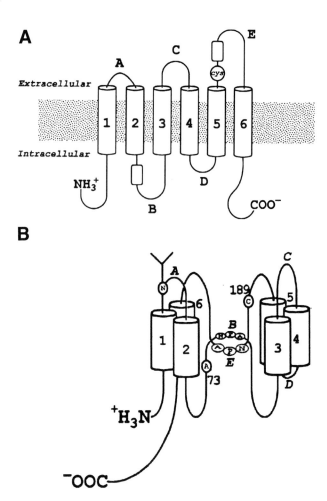

FIG. 8.22. Membrane topology and putative structure of aquaporin 1 (CHIP28). *A.* Predicted secondary structure: CHIP comprises two repeats of three putative transmembrane domains each. Amino and carboxyl termini are on the cytoplasmic side. There is a single N-glycosylation site in loop *A* (not shown). Loops *B* and *E* contain highly conserved Asn-Pro-Ala motifs thought to contribute to the formation of the water pore. Loop *E* contains Cys189, the residue conferring Hg-sensitivity to water permeation. *B.* Proposed structure of the water pore of a CHIP monomer. α-helical transmembrane domains form the peripheral wall, whereas loops *B* and *E* assemble in the membrane to form the thin barrier accounting for the selectivity for water. The diagram shows a monomer that is functional per se, but CHIP monomers assemble into tetrameric complexes. Modified from Engel et al. (164), with permission.

B) and the third extracellular loop (loop E) contribute to the pore structure, spanning the lipid bilayer from the opposite surfaces. The motif Asn-Pro-Ala, present in loops B and E in all members of the gene family, is essential for function, as is Cys189, which confers $HgCl_2$ sensitivity to the pore. Ala73 and Cys189, which occupy mirror positions in loops B and E, respectively, are critical for pore function as revealed by site-directed mutagenesis.

CHIP is a tetramer shaped as an hourglass, with a

deep cavity facing the outside (extracellular medium) and a shallow cavity facing the inside (cytoplasm). The two halves of the molecule contribute loops B and E, and the motifs Asn-Pro-Ala form a shelflike structure adjacent to the transmembrane domains. This structure would be a thin barrier separating the two cavities of the tetramer and would be perforated by four narrow pores, one for each CHIP subunit (Fig. 8.22) (164).

The selectivity of these pores for water is very high, i.e., there is negligible permeation of ions or small nonelectrolytes. This is consistent with a narrow aqueous pore that would hinder sterically (size selection) the permeation of uncharged solutes, but the explanation for the apparent lack of permeation of H^+ (or OH^-) equivalents is less clear, perhaps implying charge barriers in the permeation pathway. Recent studies suggest that cAMP stimulation makes aquaporin 1 cation permeable (see the section that follows).

A Water-Pore Gene Family. cDNAs encoding several proteins homologous to CHIP have been identified in animal and plant cells. Some of these proteins confer high osmotic water permeability (aquaporins, AQP). These include (in addition to CHIP, called AQP1). *(1)* the ADH-regulated water channel of the renal collecting tubule (205, 437a), also called WCH-CD, AQP-CD and AQP_2; *(2)* the basolateral-membrane water channel expressed in cells from renal collecting tubule and intestine (160, 298), also called AQP3; *(3)* the mercury-insensitive water channel (264) expressed in renal vasa recta, cells lining subarachnoid space, and ventricles in the central nervous system and conjunctiva, called AQP4; *(4)* a recently cloned water channel expressed in salivary and lacrimal glands and in lung (356, 498), called AQP5; *(5)* MIP26, a lens protein that has recently shown to have water-pore properties (428); and *(6)* plant homologues (413, 450), called TIPs (tonoplast intrinsic proteins). Nomenclature and properties of aquaporins found in epithelial cells are summarized in Table 8.5. Consistent with the notion that some aquaporins confer constitutive osmotic water permeability and some are regulated, two recent studies demonstrated that AQP-CD, the water channel underlying the vasopressin-sensitive osmotic water permeability of apical membrane of renal collecting-duct cells, is upregulated by the hormone via its effect on V_2 receptors (150, 265). The acute effect of vasopressin on collecting-duct water permeability involves translocation of AQP-CD pores from a vesicular pool to the apical membrane (436).

It has been recently proposed that AQP1 expressed in *Xenopus* oocytes is stimulated by cAMP (692a). The effects, mediated by PKA, would involve both an increase in water permeability and the acquisition of cation permeability. In the absence of cAMP stimulation, AQP1 has been claimed to be impermeable to ions (494, 700).

MECHANISMS OF REGULATION OF TRANSEPITHELIAL TRANSPORT

Regulation of transepithelial ion transport involves changes in the transport rate of one or more transporters in response to signals that can be physical (e.g., membrane voltage) or chemical (e.g., binding of a hormone to a receptor). At the level of single transport proteins, changes in transport rate can result from *(1)* changes in activity (i.e., increase or decrease in the turnover rate) with a constant number of operative transporters in the plasma membrane, or *(2)* increase or decrease in the number of operative transporters in the membrane by changes in their degree of activation or in the rate of incorporation or retrieval. For didactic reasons one can classify mechanisms of regulation of mass transport across epithelial membranes as *extrinsic* and *intrinsic,* depending on the origin of the stimulus.

Extrinsic regulation results from an external influence on the epithelial cells—for example, the effect of a hormone. In the simplest case, the hormone binds to a plasma-membrane receptor, which is activated, elevating the level of an intracellular second messenger, which in turn acts on one or more effector molecules, changing their activity and hence the rate of transport. Intrinsic regulation encompasses mechanisms that occur within the epithelium itself and usually involves a functional relationship between two or more transporters. These transporters may be present in the same membrane domain (*intramembrane regulatory mechanisms*) or in the opposite membrane domains (*crosstalk mechanisms*). Intrinsic regulation per se does not necessarily result in changes in the rate of transepithelial transport. Instead, it provides adjustments of transport rates within a cell-membrane domain (518) or among the opposite membrane domains in order to preserve cell volume and composition during changes in the rate of transepithelial transport (564, 565, 568). Intrinsic regulation can result from a primary change in transport rate (e.g., increased Na^+-glucose entry in enterocytes causes changes in the activity of basolateral-membrane K^+ channels) or occur as a secondary effect of an extrinsic influence on transport rate (e.g., hormonal activation of Na^+ channels in the cortical collecting duct indirectly activates the basolateral-membrane Na^+,K^+-ATPase).

The time course of regulatory events is generally a good indication of the mechanism involved (4). Ion channels can respond to gating factors (e.g., membrane

TABLE 8.6. *Classification of Mechanisms of Extrinsic Regulation of Epithelial Transport*

RAPID REGULATION

Rapid Change in the Activity of Individual Transporters

    Membrane voltage

    Membrane stretch

    Reversible binding of regulatory ligands: Ca^{2+}, H^+, ATP, cyclic nucleotides

    Reversible covalent modification (phosphorylation/dephosphorylation): protein kinases (PKA, cGMP-dependent PK, PKC, Ca^{2+} − calmodulin activated PK), tyrosine kinases, phosphoprotein phosphatases

    Protein-protein interaction: calmodulin, G-proteins, others

Rapid Change in the Number of Transporters

    Exocytosis/endocytosis

Rapid Regulation of the Paracellular Conductance

LONG-TERM REGULATION

Change in Synthesis Rate

    Aldosterone, glucocorticoids, thyroid hormone, $1,25(OH)_2$-cholecalciferol

Change in Degradation Rate

    Proteases

Molecular Replacement

Change in Polar Expression

voltage) in milliseconds, while activation, inactivation, insertion or removal of carriers or pumps takes from milliseconds to seconds, and protein synthesis is followed by a significant lag time, of the order of several hours. A classification of mechanisms of regulation of transepithelial transport based on this tenet is presented in Table 8.6.

Rapid Regulation

Rapid regulation occurs in time frames ranging from milliseconds to a few minutes. Two general mechanisms can be distinguished, namely *a change in the activity of a fixed number of transporters* or *a change in the number of transporters expressed in the membrane.*

Rapid Change in the Activity of Individual Transporters. Channels, carriers, and pumps can be in principle regulated by physical or chemical factors. Among the physical factors, there are well-demonstrated effects of changes in *membrane voltage* and circumstantial evidence has been provided for *membrane stretch*. The variety of chemical mechanisms of regulation is greater,

including *ligand binding, reversible covalent modification,* and *protein-protein interactions.*

Membrane voltage. A regulatory role of membrane voltage has been clearly demonstrated for epithelial ion channels, in particular voltage-gated K^+ channels. The best-studied examples are so-called maxi-K^+ channels, which are also Ca^{2+}-sensitive. Membrane depolarization increases the open probability (P_o) of these channels. Most epithelial cells have a K^+ equilibrium potential (E_K) greater than the membrane voltage (V_m), so that the net driving force favors K^+ efflux from the cell. Therefore, membrane depolarization tends to result in K^+ efflux by the change in driving force brought about by membrane depolarization and the increase in P_K brought about by the increase in P_o. Two examples are the response of *Necturus* gallbladder epithelium to cAMP (117, 481) and the changes in basolateral-membrane K^+ conductance in villus enterocytes upon the depolarization elicited by coupled entry of Na^+ and organic solute across the apical membrane (568).

Membrane stretch. There is abundant evidence supporting the notion that some ion channels are sensitive to membrane stretch—i.e., mechanosensitive. This has been demonstrated in nonepithelial cells (484) as well as in epithelial cells (175, 176, 544, 545, 546). P_o increases in response to membrane stretch imposed by changes in pipette pressure or by cell swelling (176, 545, 629; reviewed in 372, 537 and 546). The effect of membrane stretch is reversible (537). A question is the physiological significance of some of these observations, i.e., to what extent is the membrane stretch produced by changing hydrostatic pressure in the patch pipette relevant to changes in cell volume in situ? In this context, demonstrations of effects of changes in pipette pressure as well as effects of osmotic cell swelling on channel activity are more convincing of a physiological significance. There is abundant experimental data supporting a role of membrane stretch in cell-volume regulation upon swelling but the precise sensors and transduction mechanisms of cell-volume regulatory responses have not been identified, and the question remains as to whether the parameter sensed is of physical or chemical nature.

Stretch-activated channels can be of three types, namely cation selective, K^+ selective, and Cl^- selective (546). A mode of activation would be as follows: A primary increase in cell volume (e.g., by increased salt entry causing osmotic water inflow) would result directly in activation of K^+ and/or Cl^- channels, or in activation of nonselective cation channels; the latter would cause Ca^{2+} influx, in turn activating K^+ and/or Cl^- channels (104). There is considerable evidence favoring a role of $[Ca^{2+}]_i$ in regulatory volume decrease in epithelia (193, 416, 485).

Recent work on mechanosensitive channels demonstrates that their activation by a step change in tension is followed by rapid adaptation (258). Hence, instead of eliciting a sustained change in electrical properties of the membrane or in transmembrane ion fluxes, these channels may rather have the role of *switches* in a signaling cascade. For example, the main consequence of activation of nonselective cation channels by membrane stretch could be a brief Ca^{2+}-influx, which in turn could activate other (Ca^{2+}-sensitive) ion channels, e.g., K^+ and/or Cl^- channels, with a time course dependent on their own properties. However, some K^+ channels appear to be mechanosensitive by themselves and do not require the second-messenger role of Ca^{2+} for their activation (546). Mechanosensitive cation channels exhibit amiloride sensitivity (532) and primary-structure homology with the epithelial Na^+ channel (83, 85, 383). It has been proposed that they are part of a superfamily that also includes epithelial Na^+ channels and *C. elegans* touch-sensitive channels (155, 287, 290).

The cytoskeleton appears to play a major role in conferring, modulating or transducing membrane tension (484, 657), and molecular components of the cytoskeleton (e.g., actin) can modulate ion-channel function by chemical mechanisms (see Protein–protein interactions, below).

There is no conclusive evidence for mechanosensitivity of other membrane transporters (carriers or pumps). The activity of the Na^+,K^+-pump in skeletal muscle is modulated by cell-volume changes (645), but whether membrane stretch is involved in this effect is not clear.

Regulatory ligands. Numerous second messengers regulate transporter activity by direct binding, which is thought to result in a change in conformation altering the functional properties of the transporter.

Calcium ions can regulate the activity of transport proteins by both direct and indirect mechanisms (reviewed in 482). Direct effects have been best characterized in the case of certain channels. Maxi-K^+ channels are activated by Ca^{2+} from the cytosolic side; the activation is represented as a shift to the left of the function relating P_o to membrane voltage, i.e., as $[Ca^{2+}]_i$ is elevated, the channel gates at progressively more negative membrane voltages. The effect is reversible and involves competition with H^+ binding, which is inhibitory. These channels have been found in numerous epithelial cells (57, 113, 292, 320, 321, 579), and their function is not always clear. In cortical collecting tubule, apical-membrane maxi-K^+ channels are closed under physiological conditions, and hence do not contribute to K^+ secretion. In *Necturus* gallbladder, in contrast, the channel is activated by the depolar-

ization elicited by an increase in cAMP level, which activates an apical membrane CFTR-like Cl^- channel; the coordinated activation of Cl^- and K^+ channels results in KCl loss and cell shrinkage (117, 514). This cAMP-regulated Cl^- channel is not dependent on intracellular $[Ca^{2+}]$ (279, 336). Similar effects of cAMP, involving loss of cell K^+ and Cl^- are observed in the colon cell line T84 (417), and activation of Cl^- secretion by cAMP has been reported in small-intestine absorptive cells (328).

The amiloride-sensitive Na^+ channel in some epithelia is inhibited by elevations in $[Ca^{2+}]_i$. Such results have been obtained in both amphibian epithelia such as that of the toad urinary bladder (101, 217, 218) and mammalian epithelia such as the cortical collecting duct (195, 587). However, there appears to be no effect of $[Ca^{2+}]_i$ on Na^+ channels in excised apical membrane patches (460), suggesting that the effect is indirect. It is possible that $[Ca^{2+}]_i$ has a role in feedback control of apical membrane permeability in tight epithelia (683; see later, under Cross-Talk Mechanisms).

Ca^{2+} regulation of epithelial transport also involves indirect effects that are mediated by cytoskeletal alterations, exocytosis, and changes in the activity of cytosolic enzymes. In the latter category, calpain (a Ca^{2+}-dependent protease) has been shown to activate protein kinases by partial proteolysis (317, 318).

Lowering *pH* on the cytosolic side of the membrane reduces the P_o of amiloride-sensitive Na^+ channels in the cortical collecting tubule (218, 219, 456), as well as the P_o of maxi-K^+ (see above) and other epithelial K^+ channels (40, 99, 441, 660, 672). It has been suggested that the increase in apical membrane conductance elicited by aldosterone in the collecting tubule is due to intracellular alkalinization (442), but this effect could not be proven during chronic stimulation (453). Lowering the pH_i also inhibits the activity of the Na^+,K^+-ATPase (65, 158).

The anion ATP^{4-} binds to purinergic receptors present in the plasma membranes of many cell types. ATP may cause an increase in plasma membrane permeability to large molecules (209), but extracellular ATP can also activate specific conductances, including nonselective cation channels (45), K^+, and Cl^- channels (118, 194, 595). The latter point is quite interesting because it has been proposed that the outwardly rectifying Cl^- channel frequently coexpressed with CFTR is activated by extracellular ATP, and ATP may be secreted via CFTR (162, 249, 505, 575, 576, but see also 374a, 501a).

ATP-sensitive K^+ channels are expressed in many cells. For instance, islet β cells express K^+ channels inhibited by intracellular ATP and involved in insulin secretion (see 23). Ion-transporting epithelia can express K$^+$ channels that are stimulated by ATP, inhibited by ATP, or either stimulated or inhibited depending on the ATP concentration. The K$^+$ channel expressed in basolateral membrane of rabbit proximal tubule is inhibited by millimolar ATP concentrations on the cytosolic surface, whereas in the cortical collecting duct of the same animal the apical-membrane low-conductance K$^+$ channel is activated by PKA plus low (ca. 0.1 mM) ATP concentrations, but inhibited by millimolar ATP or ADP concentrations (662).

Cyclic nucleotides modulate channel function in many cells via activation of specific protein kinases, which in turn cause protein phosphorylation. In one documented instance, however, cGMP was shown to reduce the P$_o$ of amiloride-sensitive Na$^+$ channels from renal medullary collecting tubule by a direct effect (378), in addition to activation of cGMP-dependent protein kinase.

Eicosanoids (arachidonic-acid metabolites) have regulatory roles both as intercellular and intracellular messengers (reviewed in 35a). Arachidonic acid is released from the plasma membrane by the action of phospholipase A$_2$ (PLA$_2$), a cytosolic enzyme that associates with the plasma membrane following elevation of cytosolic [Ca^{2+}]. Arachidonic acid is metabolized by (1) the cyclooxygenase pathway, resulting in production of prostaglandins, thromboxanes, and prostacyclin, (2) the lipoxygenase pathway, producing leuko-

trienes and other agents; and (3) the cytochrome P-450 epoxygenase pathway, producing epoxyeicosatrienoic acids (EETs) and hydroxyeicosatetraenoic acids (HEPEs). From the point of view of regulation of epithelial transport, prostaglandins, in particular PGE$_2$, are the best studied eicosanoids. PGE$_2$ stimulates intestinal Cl$^-$ secretion and duodenal HCO$^-_3$ secretion, and inhibits renal collecting duct Na$^+$ reabsorption (reviewed in 35a). The mechanisms of the effects of prostaglandins and other eicosanoids are under active investigation.

Although Cl$^-$ has not been conclusively demonstrated to act as a regulator by binding to transporters, some instances of modulation of epithelial-cell ion transport by changes in intracellular [Cl$^-$] have been reported. As explained above, intracellular [Cl$^-$] appears to modulate Na$^+$-K$^+$-2Cl$^-$ (386, 527). In addition, in HT29 (colon epithelium) cells, intracellular [Cl$^-$] appears to affect electrodiffusive Cl$^-$ permeability (177).

Protein covalent modification. Covalent modifications of proteins such as *phosphorylation, methylation,* and *ADP ribosylation* have been suspected or demonstrated to be involved in the regulation of transepithelial transport. These processes will be discussed in this section, with an emphasis on phosphorylation/dephosphorylation.

Protein kinases catalyze the transference of the γ-phosphate from ATP to specific amino-acid residues

TABLE 8.7. *Examples of Regulation of Epithelial Transport Processes by Protein Phosphorylation and Dephosphorylation*

Enzyme	Tissue	Agonist	Target	Effect	References
cAMP-dependent PK (PKA)	CCD	Vasopressin	Na$^+$ channel	Insertion; activation?	220, 405, 554
	AE	Isoproterenol	CFTR	Activation (requires ATP)	15
cGMP-activated PK	MCD	ANF	Na$^+$ channel	Inactivation	697
	WFIE		Na$^+$-K$^+$-2Cl$^-$ cotransporter	Inhibition	499
	CE		Cl$^-$ channel	Activation	137
Protein kinase C (PKC)	AE	Phorbol esters	CFTR	Activation	675
	HF		CFTR	Activation	51, 616
	GBE	Phorbol esters	CFTR-like channel	Activation	279
	PT		Na$^+$-H$^+$ exchanger	Activation	665, 667
Ca^{2+}-CAL PK	PT		Na$^+$-H$^+$ exchanger	Small activation	665
Protein phosphatases	A6	Insulin	Apical Cl$^-$ channel	Activation	404
	AE		CFTR	Inactivation	44
	SRG		CFTR	Inactivation	345

Abbreviations: ANF = atrial natriuretic factor; CAL = calmodulin. *Epithelia*: CCD = cortical collecting duct, AE = airway epithelium, MCD = medullary collecting duct, WFIE = Winter flounder intestine epithelium, CE = colon epithelium, HF = human fibroblasts (transfected), GBE = gallbladder epithelium, PT = proximal tubule, A6 = frog-kidney cell line, SRG = shark rectal gland.

of target proteins (312). Specific phosphatases cause dephosphorylation, making the reaction reversible. Several protein kinases contribute to the regulation of transepithelial transport. Examples of their effects are listed in Table 8.7. As an example, I discuss the role of cAMP-dependent protein kinase A (PKA).

PKA consists of regulatory and catalytic subunits. cAMP binding to the regulatory subunit dissociates the two peptides, releasing the constitutively active catalytic subunit, which then catalyzes protein phosphorylation of threonine or serine residues in consensus sequences of the form R-X-S(T).

There are numerous instances of PKA-mediated regulation of epithelial transport (51a, 66, 298a), two of which are briefly discussed here. Vasopressin increases Na^+ absorption in toad urinary bladder, frog skin, and renal collecting ducts. Activation of V_2 receptors by vasopressin stimulates adenylate cyclase and elevates cAMP levels in the target cells, increasing the Na^+ permeability of the apical membrane. The effect is probably mediated by phosphorylation of a subunit of the amiloride-sensitive Na^+ channel (554). Whether the putative phosphorylation of the subunit is a direct effect of PKA is not certain, because the hormone also elevates phospholipase C activity, which in turn generates IP_3 and raises $[Ca^{2+}]_i$. Hence, the phosphorylation may also occur through the action of a calcium-activated kinase. The most studied and best-understood effect of PKA on channel function is on CFTR, where PKA-mediated phosphorylation and ATP binding appear to be essential for channel activity. Purified PKA activates apical Cl^- channels of absorptive epithelia. In situ, this leads to a decrease in net Cl^- absorption or to net Cl^- secretion (112, 117, 198, 279). As explained earlier (see Epithelial Anion Channels) PKA-mediated phosphorylation of CFTR is necessary for channel gating by ATP hydrolysis cycles (208). There are two sets of PKA phosphorylation sites that are sensitive to different phosphatases.

Another form of covalent modification of membrane transporters is methylation, claimed to contribute to the activation of amiloride-sensitive Na^+ channels (553, 681), as explained under MECHANISMS OF TRANSEPITHELIAL NaCl TRANSPORT IN ABSORPTIVE EPITHELIA.

Cholera toxin and pertussis toxin catalyze ADP ribosylation and hence modulate G-protein activity. Hence, there has been a search for cytosolic ADP-ribosyl transferases that might act as regulators of cellular processes such as ion transport. However, no conclusive demonstration of their presence or effect is presently available.

Protein–protein interactions. Ion transporters are often protein complexes including peptides that do not participate in ion translocation but facilitate assembly or membrane insertion of the complex, or regulate its

function by other mechanisms. Such peptides may themselves be targets for kinases or other regulators.

A target of *calmodulin* is the plasma-membrane Ca^{2+}-ATPase (479), involved in transepithelial Ca^{2+} transport. Calmodulin activates the Ca^{2+}-ATPase in enterocytes and in other epithelial cells (644).

G proteins directly activate certain ion channels (68, 69, 106). In the realm of epithelial-cell channels, G-protein involvement has been demonstrated in amiloride-sensitive Na^+ channels. Purified $G\alpha_{i-3}$ activated the channel when added to the cytoplasmic side of excised patches from apical membrane of A6 cells (29, 87) or inner-medullary collecting duct cells (377). It has also been shown that a G protein activates Cl^- channels in the apical membrane of intercalated cells of renal collecting tubule (577). For other effects of G proteins on epithelial as well as on nonpolar-cell ion channels, see the review by Clapham (106).

Actin binding has been shown to modulate epithelial channels. In the case of Na^+ channels in A6 cells (88, 491) actin polymerization is necessary for PKA-mediated activation. Effects on Cl^- channels (578) and on K^+ channels (659) have been demonstrated in renal cortical collecting tubule. In addition, actin polymerization appears to be necessary for cAMP-mediated activation of the basolateral-membrane Na^+-K^+-2Cl^- cotransporter in a colon epithelial cell line (410), and unpolymerized actin stimulates rat kidney Na^+,K^+-ATPase (86a). The mechanism of these effects is unknown.

Several *transmembrane proteins,* some of them initially considered to be ion channels, may in fact be regulators. Most of the relevant results are very recent and their interpretations are not always clear (see 495). The prevailing view is that these molecules, either soluble, mostly cytoplasmic proteins (e.g., I_{Cln}), or transmembrane proteins (e.g., phospholemman and P-glycoprotein) may be modulators of preexisting Cl^- channels. The effectors would be voltage-sensitive or swelling-activated Cl^- channels constitutively expressed in the recipient cells. Transfection of another small protein (called IsK, or more recently, IsKCl) appears to cause activation of both K^+ and Cl^- channels. In several studies, point mutations of some of these proteins changed the properties of channels expressed in transfected cells. This is not conclusive demonstration that the mutated protein is a channel per se, i.e., mutation of a regulator could change kinetic or pharmacological properties of the effector. It is too early to draw definitive conclusions from these studies, but the possibility exists of multiple mechanisms of regulation of transporter function via protein-protein interactions, a topic currently undergoing intense scrutiny.

Rapid Change in the Number of Transporters. The mechanism discussed in this section is the translocation of assembled transporters from intracellular vesicular compartments to the plasma membrane or vice versa. Exocytosis and endocytosis are constitutive phenomena that in the steady state tend to maintain constant the number of copies of each transport protein in the plasma membrane. Newly synthesized units are added to the membrane by exocytosis and "aged" membrane proteins—e.g., proteins altered by partial proteolysis or other processes—are removed by endocytosis. In certain epithelia, exocytosis and endocytosis are *regulated* and changes in their rates can result in changes in the number of transporters present in the plasma membrane. The transporter molecules are preassembled and incorporated in membrane vesicles or tubules near the relevant membrane domain. Hence, the change in number of copies in the plasma membrane is independent of changes in the rate of synthesis of transport proteins. A physical or chemical stimulus causes fusion of the vesicles or tubules with the membrane, increasing both the membrane area and the number of transporters. The change in membrane area can be estimated by morphometric analysis of tissue sections (e.g., 412) or by measuring the electrical capacitance of the membrane (e.g., 109). Upon cessation of the stimulus, the transporters are retrieved by specific endocytosis. They can be either transferred to lysosomes and degraded or recycled to the plasma membrane during another stimulation cycle. Exocytosis occurs in the time frame of milliseconds to seconds, involves major cytoskeletal rearrangements, and is regulated by intracellular mediators.

Studies of ion-transport regulation by exocytotic insertion of transporters in epithelial-cell membranes have been carried out for both apical- and basolateral-membrane transporters. Apical-membrane examples are the regulation of osmotic water permeability and Na^+ permeability in vasopressin-sensitive epithelia, the change in apical-membrane surface area with distention in the mammalian urinary bladder (369), and the fusion of membrane vesicles containing H^+-ATPases or H^+,K^+-ATPases in H^+-secreting epithelia. A basolateral-membrane example is the regulation of Na^+ transport by insertion of Na^+,K^+-ATPase molecules (35, 55).

In gastric mucosa the transporter involved in exocytotic regulation is the H^+,K^+-ATPase and one of the stimuli is histamine (191, 538, 540), whereas in renal H^+-secreting epithelia (collecting tubule, reptilian urinary bladder) the transporter involved is a vacuolar-type H^+-ATPase and the stimulus is cell acidification, e.g., by increasing P_{CO_2} (572).

The cellular mechanisms of the effect of vasopressin on water permeability are discussed earlier (see MECHANISMS OF TRANSEPITHELIAL WATER TRANSPORT). In contrast with the effect of aldosterone on apical-membrane Na^+ permeability, i.e., activation of channels already present in the apical membrane of tight epithelia, the effect of vasopressin appears to be mediated by insertion of Na^+ channels from a cytoplasmic vesicular pool (214, 504). Both aldosterone and vasopressin stimulate apical-membrane Na^+ permeability in epithelia such as that of the toad urinary bladder (e.g., 214), but the mechanisms appear to be radically different. Treatment of the apical surface of the epithelium with trypsin irreversibly inhibits Na^+ transport. After trypsin removal, aldosterone does not stimulate Na^+ transport, but vasopressin does (220). These data strongly support the view that aldosterone stimulates channels already present in the apical membrane, whereas vasopressin induces the insertion of new channels (see 558). However, in studies in toad urinary bladder it was shown that stimulation by vasopressin was not always accompanied by a change in membrane capacitance (604).

As an additional example of regulation of transepithelial transport by exocytosis/endocytosis, I summarize studies of the regulation of H^+-ATPase function in renal epithelia. For reviews of the analogous process in the parietal cell of the gastric epithelium, see Wolosin (685) and Sachs (538).

In renal collecting tubule and in turtle urinary bladder, both epithelia that acidify the lumen solution, H^+ secretion is rapidly and reversibly stimulated by cell acidification, for instance by raising solution P_{CO_2} (229, 418, 572). The increase in transport rate occurs without change in the maximum electrochemical gradient generated by the H^+ pump. Instead, it involves an increase in the number of H^+-ATPase molecules in the apical membrane by exocytotic fusion of cytoplasmic vesicles containing functional pumps (86, 229). The exocytosis was elegantly demonstrated by loading the H^+-secretory cells with a fluorescent probe (by constitutive endocytosis) and then observing extrusion of fluorophore following cell acidification, followed by sustained secretion of H^+ equivalents (229). The role of intracellular acidification is supported by the observation that similar results are elicited by salts of permeant short-chain fatty acids such as butyrate and acetate (4, 86, 640). The link between the fall in pH_i and the exocytosis appears to be an increase in $[Ca^{2+}]_i$ by entry from the external solution, probably because of membrane depolarization and Ca^{2+} channel activation, without release from stores. If the $[Ca^{2+}]_i$ elevation is experimentally prevented, both the exocytosis and the increase in H^+ pumping are prevented as well (86, 640).

In addition to constitutive endocytosis, the epithelia of renal collecting tubules and reptilian urinary bladders are endowed with regulated (pH$_i$-dependent) endocytosis. The rate of internalization of fluorescent probes is stimulated by cell alkalinization, either by lowering PCO_2 or adding inhibitors of carbonic anhydrase. Endocytosis is much slower than exocytosis.

Studies in renal collecting duct have revealed that aldosterone promotes synthesis of a latent Na$^+$,K$^+$-ATPase pool whose insertion in the basolateral membrane would be stimulated by an elevation in intracellular [Na$^+$] (35, 55) or by vasopressin (120). The mechanism by which the elevation of intracellular [Na$^+$] causes selective Na$^+$-pump exocytotic insertion is unknown.

Rapid Regulation of the Paracellular Conductance. Changes in the electrical resistance of the junctional complexes and/or the lateral intercellular spaces may have major effects on epithelial function, in particular in so-called leaky epithelia, in which the paracellular resistance is low (see Chemical and Electrical Gradients Couple Ion Fluxes in Epithelia). The main effect of an increase in paracellular resistance is to reduce the cell-membrane voltage drops produced by intraepithelial current flow. This may alter the driving forces for ion transport across one or both cell membranes.

cAMP. It has been suggested that cAMP increases the junctional resistance and alters the structure of the junctions in gallbladder epithelium (50, 157), and hence that the effects of cyclic nucleotides on transepithelial ion transport may involve junction-related mechanisms. Changes in junctional properties by cAMP were also described in intestinal epithelia from flounder (307) and goldfish (30). However, detailed electrophysiological studies in *Necturus* gallbladder strongly suggest that the increase in transepithelial resistance elicited by cAMP is caused by a decrease in the width of the lateral intercellular spaces secondary to the reduction in transepithelial fluid absorption, and not by an increase in junctional resistance (337, 338). The membrane-voltage changes elicited by cAMP in *Necturus* gallbladder epithelium are dominated by a large increase in apical-membrane Cl$^-$ conductance that determines the change in transepithelial voltage (112, 117, 481). This of course does not exclude the possibility that in other epithelia cAMP may in fact alter junctional selectivity, but rigorous proof has not been provided.

Organic solutes. Experiments in mammalian small intestine suggest that large increases in apical-solution concentration of organic solutes that are cotransported with Na$^+$ produce an increase in paracellular permeability concomitant with structural changes in the junc-

tions (399, 466, 467). This suggests an adaptive increase in paracellular fluid absorption when transcellular absorption is increased, i.e., it would be possible to absorb organic solute (e.g., glucose) at rates well above the maximum expected from transcellular carrier-mediated transport. The proposed mechanism is an alteration in the cytoskeleton brought about by solute uptake or the resulting change in cell volume; this would cause structural changes in the junctional complexes and increase their permeability to water and polar solutes. Solvent drag from apical to basolateral solution would be driven by the hyperosmolality of the fluid transported across the cells, if the reflection coefficient for the luminal solute (e.g., glucose) is low (466). There is structural evidence for junctional alterations (398) and it has been suggested that the paracellular conductance and the basolateral-membrane capacitance increase (466, 467), but the latter determinations are questionable (see below and discussion in ref. 467). These are provoking observations, but, as in other studies (reviewed in 32), there is no conclusive proof that the change in paracellular conductance is junctional and generalized—i.e., involving the entire junctional domain—and not localized to specific areas. Such a proof is very difficult, requiring the complicated electrophysiological analysis performed by Kottra et al. in *Necturus* gallbladder epithelium (337, 338).

Other factors. There is no doubt that Ca^{2+} and PKC play important roles in tight-junction assembly, but regulation of the function and properties of assembled junctions by these agents has not been conclusively demonstrated. Other interesting possibilities of regulation of junctional permeability have been proposed from studies of Caco cells and of rat inner-medullary collecting duct. In Caco cells, nitric oxide has been reported to reversibly decrease the transepithelial electrical resistance, and possibly the paracellular permeability, together with alterations in junctional and cytoskeletal structure (549). These changes do not involve cGMP. In rat inner medullary collecting duct, dehydration of the animal appears to cause an increase in paracellular permeability to water and nonelectrolytes that would contribute to the generation of a maximally concentrated urine (185).

Long-term Regulation

Following Al-Awqati's classification (4), we distinguish four groups of possible mechanisms of long-term regulation: change in the rate of synthesis, change in the rate of degradation, molecular replacement, and change in polar expression.

Change in the Rate of Synthesis. Broadly considered, regulation of gene expression involves the following potential control levels: *(1)* transcription, *(2)* RNA processing and/or modification, *(3)* mRNA transport, *(4)* mRNA degradation and storage, *(5)* translation, and *(6)* posttranslational modulation of protein activity (429). Well-documented examples of regulation of gene expression in epithelia are scarce, because of the difficulties imposed by the complexity of the tissues and their functions, but major progress in the use of molecular-genetics approaches to understand epithelial transport function is already taking place. An example is the demonstration of an increase in mRNA levels of the H^+-ATPase 31 kDa subunit in cortical collecting duct cells isolated from rabbits subjected to an acid load (172a). Another example pertains to the effects of medium hyperosmolality on transcription of membrane transporters, stress proteins, and metabolic enzymes (251a). What follows is a discussion of the effect of aldosterone on transepithelial Na^+ transport, as an example of long-term regulation by change in the rate of synthesis.

Steroid hormones and other hormones bind to receptors belonging to the same gene family. These receptors are transcription factors, i.e., proteins having three domains, whose function are hormone-binding, DNA-binding, and regulatory, respectively (167, 168). Deletion of the regulatory domain activates the receptor. The gene-regulatory function is contained in a "zinc-finger" structure in the DNA-binding domain of the receptor.

Examples of this mechanism are the effects of mineralocorticoids, glucocorticoids, thyroid hormone, and active vitamin D_3 on transepithelial ion transport. The best-studied one is the effect of the mineralocorticoid aldosterone on Na^+ transport in tight epithelia such as the renal collecting duct (reviewed in 169a, 531, and 558).

Aldosterone binds to cytoplasmic receptors and the complex is transported to the nucleus where it stimulates transcription. Several proteins are synthesized, none of which have been identified (54, 614). The effects of aldosterone on Na^+ transport are entirely dependent on mRNA and protein synthesis (161, 214).

The main effect of aldosterone is to stimulate transepithelial Na^+ transport, by one or more of the following mechanisms: *(1)* stimulation of apical-membrane Na^+ entry (by an increase in Na^+ permeability), *(2)* stimulation of the Na^+,K^+-ATPase expressed at the basolateral membrane (by an increase in the number of pumps), and *(3)* increase in ATP supply to the pump. The available data appear to support all three possibilities and suggest a pleiotropic effect of the hormone, but the picture is complicated because inasmuch as aldosterone increases Na^+ entry and hence $[Na^+]_i$, it is possible that some of its effects are indirect.

The effect of the hormone is biphasic. The early effect is explained by activation of preexisting, high-selectivity apical-membrane Na^+ channels (26, 220, 453), but nevertheless requires protein synthesis. The simplest interpretation is that the de novo synthesized protein is a regulatory subunit of the Na^+ channel, but there is no direct support for this notion. Methylation of one or more subunits has also been proposed (553, 681) but remains controversial (see 558).

The late effect of aldosterone involves principally Na^+,K^+-ATPase activation (213, 222, 224, 459, 648, 649). In toad urinary bladder, aldosterone stimulates the synthesis of pump α and β subunits, apparently by an Na^+-independent mechanism, but activation of Na^+,K^+-ATPase molecules may require an elevation in $[Na^+]_i$ entry (35, 55). In mammalian cortical collecting tubule, the late effect of aldosterone involves activation of the Na^+,K^+-ATPase, but this effect is secondary to Na^+ entry (459), i.e., a high intracellular $[Na^+]_i$ is necessary for surface expression of pumps synthesized following stimulation with aldosterone (449). Increases in $[Ca^{2+}]_i$ stimulate transcription of both α and β subunits of the Na^+,K^+-ATPase (500), but this is probably a slow process.

The metabolic theory of the effect of aldosterone is supported by demonstrations of increased citrate synthesis in the kidney medullary region following exposure to the hormone (353). Whether this effect is primary or secondary is not clear.

Other effects of aldosterone are to increase apical membrane P_K in the collecting tubule (660) and to stimulate urine acidification. With respect to the effect on P_K, an interesting recent development is the observation that aldosterone stimulates Na^+/H^+ exchange and thus elevates the pH_i of its target cells (442). The alkalinization might explain, at least in part, the effects of the hormone on apical-membrane P_{Na} and P_K, inasmuch as the P_o's of apical membrane Na^+ and K^+ channels increase with cytosolic alkalinization (442, 660). However, during long-term aldosterone treatment this mechanism could not be demonstrated (453). Aldosterone has also been claimed to stimulate H^+ secretion by the V-type H^+-ATPase (6), by mechanisms not fully understood. DOCA stimulates HCO_3^- secretion in rat cortical collecting duct (212).

Na^+,K^+-ATPase gene regulation is emerging as highly complex in epithelial cells. Some recent observations illustrate the variety of regulatory mechanisms that can operate in different animal species and cell types. For instance, exposure of immortalized rat hepatocytes to serum causes Na^+ entry and results in stimulation of the synthesis of $\alpha 1$ and $\beta 1$ subunits. Only the transcription of the $\alpha 1$ gene increases, although both $\alpha 1$ and $\beta 1$ mRNA levels rise; the increase in $\beta 1$ mRNA requires Na^+ in the medium, whereas the transcriptional effect is Na^+-independent (316). This indicates that the mechanisms of regulation of synthesis of the two subunits are entirely different. The importance of gene regulation on developmental expression is well illustrated by studies in rat small intestine showing that the expression of $\alpha 1$ and $\beta 1$ Na^+,K^+-ATPase subunits, assessed by mRNA levels, increases threefold from neonate to adult animals (698). It has been proposed that the difference in levels of expression of the pump can explain the high susceptibility of immature animals to cholera toxin.

Change in the Rate of Degradation. Transport proteins have a finite lifetime and are degraded mostly by endocytosis and hydrolysis by proteases in lysosomes. Epithelial cells from small intestine and proximal tubule express proteases in the apical membrane or secrete them. The amiloride-sensitive Na^+ channel in rabbit urinary bladder has been shown to undergo partial digestion by an enzyme secreted across the same membrane (370, 371, 384). The channel loses Na^+/K^+ selectivity but remains virtually impermeable to anions. The generality and quantitative importance of this kind of mechanism are uncertain.

Na^+,K^+-ATPase methionine-labeling studies in a pig-kidney cell line indicate that shortly after synthesis β subunits are in excess with respect to α subunits, and that this is rapidly compensated by rapid degradation, while later on degradation of α and β subunits (presumably heterodimers) proceeds at a much slower rate (361).

Molecular Replacement. Molecular replacement as a form of regulation of transepithelial transport is suggested by the existence of gene families and superfamilies of transporters, multisubunit transport proteins and isoforms with tissue-specific expression, developmental expression, and/or distinct functional characteristics. Hence, changes in the nature of the transporter—e.g., by a change in isoform expressed instead of in the number of copies—can modulate the transport rate. There are well-documented examples of molecular replacement as part of normal development of several cell types. A possible example of this kind of mechanism in mature epithelia is the higher K_i for ouabain of Na^+-glucose cotransport by ileal mucosa from streptozocin diabetic rats compared to controls (171). A similar difference was elicited in vitro, in mucosae from normal animals, by glucagon pretreatment. The effect of glucagon had a significant latency and was abolished by inhibition of protein synthesis. Finally, mRNA encoding the $\alpha 1$-isoform of the Na^+,K^+-ATPase (which has a low sensitivity to ouabain) was twofold higher in intestines from diabetic rats compared to control animals. These studies suggest that the difference in ouabain sensitivity is due to a change in pump isoform expression brought about by an increase in glucagon levels (171).

Several mechanisms could potentially result in changes in transporter function by "replacement" of either the main subunit or accessory subunits. For instance, a modulating factor could exert its effect on RNA splicing and hence elicit synthesis of an alternate-splicing product with different functional characteristics. Also, changing the isoform of a single subunit can confer different properties to a multisubunit transporter, and the level of production of each isoform can be subject to control. In the case of transport proteins including numerous subunits, the scope of potential functional regulation is enormous, as are the possibilities for genetic abnormalities. Finally, we must consider modulation by regulatory proteins, which can also be subject to alternate splicing and isoform-specific regulation. It is likely that in the coming years we will witness major developments in this field. At present, the catalog of these regulatory mechanisms is virtually empty.

Change in Polar Expression. The hypothesis that epithelial cells can reverse their polarity was proposed by Al-Awqati and coworkers as an explanation for the change of collecting tubule intercalated cells from H^+-secreting to HCO_3^--secreting, or vice versa, following changes in the diet that alter the acid-base status of the experimental animals (574). As explained above, there are two types of intercalated cells: H^+-secreting, or α, and HCO_3^--secreting, or β. The definition of a cell as either α or β is based on a combination of morphological, immunocytochemical, and functional criteria that are not always the same for all investigators. As shown in Figures 8.18 and 8.19, both the H^+-ATPase and the Cl^-/HCO_3^- exchanger are expressed at opposite membranes in these two cell types.

The basic experimental observation is that a mild metabolic acidosis elicits a change in collecting-duct function from HCO_3^--secreting to H^+-secreting (see 570). After rabbits are acid loaded, the relative numbers of α- and β-intercalated cells in the cortical collecting tubule were reported to change in favor of the α type (574). From these and other studies it was postulated that the disappearance of β cells and the appearance of α cells are due to a *transformation* of the former cell type into the latter, i.e., to a redistribution of transporters so that the essential molecules, namely the H^+-ATPase and the Cl^-/HCO_3^- exchanger, disappear from one membrane domain and appear in the opposite one. This is an attractive hypothesis, but several recent studies suggest a far more complicated picture. This scenario would apply only to the cortical collecting tubule; in the outer medullary collecting duct, where most of the regulation of H^+ excretion seems to occur, the intercalated cells secrete H^+ in a regulated fashion but do not secrete HCO_3^- (see 570 for review).

In the rat, H^+-ATPase molecules are expressed exclusively in the apical membrane of α cells and exclusively in the basolateral membrane of β cells (73, 433). It is possible that the H^+-ATPases expressed in α and β cells are not identical, but there is no evidence supporting this. In the rabbit, β cells have been operationally defined by peanut-lectin agglutinin binding to the apical membrane. These cells appear to have diffuse cytoplasmic staining with anti-H^+ pump antibodies without clear membrane staining (571). The facts are more complicated when it comes to the Cl^-/HCO_3^- exchanger. In rat and rabbit α cells, the AE1 isoform has been shown by immunological methods to be expressed only in the basolateral membrane (8, 153, 154, 334), and the same antibody fails to label β cells in either rat or rabbit (153, 571). However, it was recently claimed that rabbit cortical collecting tubule β-intercalated cells do express AE1 gene-product immunoreactivity in purified apical membranes, but that the (intracellular) epitope is inaccessible in situ (641). Further, studying immortalized cultures of β-intercalated cells, the same group has concluded that the plating density controls the targeting of AE1 gene product, because of secretion of a specific extracellular-matrix protein only in cells plated at high density; under these conditions, the AE1 is targeted to the basolateral membrane (642, see also 4a). The difference in experimental preparations, and the membrane-purification procedure needed to demonstrate apical-membrane expression of AE1, make these results difficult to interpret. Further, it has also been reported that AE1 mRNA levels in cells from rabbit cortical collecting ducts are about tenfold higher in β cells, and that metabolic acidosis increases mRNA levels three- to fivefold in the aggregate of both cell types (172). Both results argue against apical membrane targeting of the AE1 gene product in β-intercalated cells. Hence, at the present time it is unclear what isoform of the Cl^-/HCO_3^- exchanger is expressed in apical membranes of β cells. In conclusion, the proposal that intercalated cells can reverse their polarity and thus regulate transepithelial ion transport remains both provoking and controversial and is in need of further investigation.

In other systems, reversal of epithelial polarity has

been well-documented, although none of these instances can be unambiguously claimed to constitute physiological regulatory phenomena. One example is that of thyroid cells, which when cultured at low concentration of serum form follicles with inward-facing apical membranes and junctional complexes. If the serum concentration is elevated, polarity is reversed (410a, see also 603). MDCK cells grown in suspension tend to form cysts with polarized membrane domains. When the cysts are incubated in collagen gel, polarity reverses, and this process appears to involve β_1 integrins (448). Alterations in polar expression of K^+ channels have been observed in *Necturus* gallbladder epithelium following physical separation of the cells from the underlying basal lamina without loss of the junctional complexes. Under these conditions, maxi-K^+ channels normally expressed only in the apical membrane can be demonstrated in the basolateral membrane domain (114). Finally, a redistribution of integrins in damaged renal tubule cells has been implied as an important pathogenic mechanism in acute renal failure. The suggestion is that loss of basolateral-membrane integrins causes cell exfoliation and integrin redistribution may facilitate adhesion of exfoliated cells and hence tubule obstruction (233).

Intramembrane Regulation and Cross-Talk Mechanisms

Transepithelial transport of NaCl or another salt must satisfy the electroneutrality principle and hence involves coordination among transporters. Further, in the steady state, for both cell volume and ionic composition to remain constant, the rates of entry and exit of the transported species must be the same. When only one ion is transported, intrinsic regulation implies adjustments between the rates at the two membrane domains—i.e., influx and efflux must be equal. When two or more ions are transported, intrinsic regulation may involve: *(1) intramembrane regulation:* adjustment of the rates of net transport at the same membrane, e.g., apical-membrane NaCl entry has the stoichiometry $Na^+:Cl^- = 1:1$, and *(2) cross talk:* adjustment of the transport rates at the two membrane domains, e.g., apical NaCl entry = basolateral NaCl exit.

Intramembrane Regulation. Intramembrane regulation involves adjustments of fluxes via carriers, via channels, or between pumps and other transporters. The case of adjustment between transport mediated by the Na^+,K^+-ATPase and the basolateral membrane K^+ channel is referred to as pump-leak parallelism (566). In this section I shall discuss one example of intramembrane regulation involving two carriers, another one involving two channels, and a case of pump-leak parallelism.

Adjustment between the transport rates of carriers expressed in the same membrane. In electroneutral NaCl-absorptive epithelia, changes in the rate of Na$^+$ entry are matched, in the steady state, by changes in the rate of Cl$^-$ entry, and vice versa. These effects were initially interpreted as indicative of apical membrane Na$^+$-Cl$^-$ cotransport (197, 431), but later studies demonstrated that in most of these tissues NaCl entry is in fact the result of apical membrane Na$^+$/H$^+$ and Cl$^-$/HCO$_3^-$ exchanges (375, 376, 509, 516, 612, 668). Experiments using intracellular ion-selective microelectrodes demonstrated unambiguously that Na$^+$ and Cl$^-$ transport across the apical membrane could be *transiently* dissociated (509). Nevertheless, in the steady state the transport rates match each other because of changes in intracellular pH. For instance, in *Necturus* gallbladder epithelium, if Cl$^-$ is removed from the apical bathing solution, intracellular [Cl$^-$] falls rapidly, and intracellular Na$^+$ also falls, but more slowly. The mechanism of the latter effect is as follows: *(1)* removal of lumen Cl$^-$ elevates pH$_i$ because the fluxes via the Cl$^-$/HCO$_3^-$ exchanger are reversed (516); *(2)* the lowering of [H$^+$]$_i$ reduces Na$^+$/H$^+$ exchange because H$^+$ is a substrate and its intracellular concentration is reduced and also because the Na$^+$/H$^+$ exchanger undergoes allosteric activation by internal H$^+$, and hence its activity decreases at high pH$_i$ (12, 22). A similar mechanism is likely to operate when the primary change is in Na$^+$ entry; the apical-membrane ileal Cl$^-$/HCO$_3^-$ exchanger is inhibited by low pH$_i$ and activated by high pH$_i$ (291a, 427).

Adjustment between the activities of channels expressed in the same membrane. Changes in membrane voltage can result in adjustments of the fluxes via "electrogenic" membrane transporters by two mechanisms, namely a change in electrochemical driving force and a change in kinetics of voltage-sensitive transporters. The second mechanism appears to be significant only in the case of ion channels.

In *Necturus* gallbladder epithelium, cAMP activates an apical membrane Cl$^-$ conductance (481). This depolarizes the membrane by ca. 20 mV, which increases the driving force for K$^+$ exit and increases ca. fivefold the P$_o$ of maxi-K$^+$ channels expressed in the same membrane. The end result is conductive loss of K$^+$ and Cl$^-$ and cell shrinkage (117). Coordination of the function of two ion channels by changes in membrane voltage can also be a form of cross talk in leaky epithelia, since the two cell membranes are electrically connected via the low-resistance paracellular pathway (see later, under Cross-Talk Mechanisms).

Pump-leak parallelism. Pancreatic beta cells, cardiomy-

ocytes, neurons, and other cells express [ATP]$_i$-sensitive K$^+$ channels whose P$_o$ is modulated by the nucleotide by a purinergic-receptor mechanism, with no ATP hydrolysis (38). This mechanism could play a role in epithelia in coupling the rate of the pump to the basolateral membrane P$_K$. Compelling evidence favoring this hypothesis has been obtained in rabbit proximal tubule cells (627). These cells have [ATP]$_i$-sensitive K$^+$ channels in the basolateral membrane. These channels are the main determinants of the K$^+$ conductance of the basolateral membrane and are activated by depletion of cell ATP. Further, [ATP]$_i$ decreases when transepithelial Na$^+$ transport is stimulated. Finally, when [ATP]$_i$ is experimentally raised, the effect of Na$^+$ transport on basolateral-membrane K$^+$ conductance is abolished. Consistent results have been obtained by other investigators (42). Since ATP-sensitive K$^+$ channels are expressed in several epithelia, this mechanism may explain pump-leak parallelism (627). Coupling between the activity of the Na$^+$,K$^+$-ATPase and the TEA$^+$-sensitive K$^+$ channels has also been shown in *Xenopus* oocytes (290a). These results question estimates of pump current from use of cardioactive steroids in the absence of K$^+$-channel blockers. In epithelial cells expressing ATP-sensitive K$^+$ channels in the apical membrane, [ATP]$_i$ may couple the rate of basolateral pumping to apical K$^+$ secretion. This would be a form of cross talk.

Cross-Talk Mechanisms. The mechanisms accounting for adjustment of the rates of ion transport between the two cell-membrane domains (e.g., the increase in Na$^+$ pumping across the basolateral membrane following an increase in Na$^+$ entry via apical membrane channels) have been collected under the generic name of *cross-talk mechanisms* (147, 518, 564, 565, 568), and are thought to respond to the following general scheme: *(1)* there is a primary change in the rate of transport at one of the membranes; *(2)* this results in a change in the level of a signal (in the membrane or in the cell); *(3)* this signal, which can be physical or chemical, is sensed and transduced; and *(4)* the effect is a change in the activity of one or more transporters at the opposite membrane.

Cross talk in tight epithelia. The first experimental observation supporting the idea of cross talk in epithelial cells was made by MacRobbie and Ussing in frog skin epithelium (397). They observed that inhibition of the basolateral Na$^+$,K$^+$-ATPase with G-strophanthidin, or by lowering the pH of the inner bathing solution, caused a decrease in the passive permeability of the apical cell membrane to Na$^+$ and of the basolateral cell membrane to K$^+$ (397). In the toad urinary bladder, changes in the osmolality of the serosal bathing solution caused parallel changes in transport properties of both apical and basolateral cell membranes (183). Exposure of tissues to metabolic inhibitors or a K^+-free serosal bathing solution, maneuvers expected to lead to pump inhibition, also resulted in a decrease in apical membrane P_{Na} (100, 373).

Cell swelling following exposure of isolated, perfused rabbit renal cortical collecting ducts to ouabain was followed by partial volume recovery (607). Also, when Na^+ was restored after a period of removal from the solution bathing the apical surface of rabbit colonic epithelium, a large transepithelial Na^+ current was observed, which relaxed slowly (567, 628). The time course of the current was accounted for by changes in apical membrane amiloride-sensitive P_{Na} permeability. Similar conclusions were reached in studies on frog skin (355). These results suggest that the rate of transepithelial Na^+ absorption by "tight" epithelia is dependent upon apical cell membrane P_{Na}, which in turn can be affected by changes in $[Na^+]_i$, either directly or via alterations in $[Ca^{2+}]_i$ and/or pH_i. The demonstration of Na^+/Ca^{2+} exchange in basolateral membranes of epithelial cells (683) suggests that Ca^{2+} could mediate cross talk between the two cell membranes. Following a reduction of the basolateral Na^+ efflux, for instance by pump inhibition, $[Na^+]_i$ would rise, which should reduce or even reverse Ca^{2+} transport via the Na^+/Ca^{2+} exchanger (683). The elevation in $[Ca^{2+}]_i$ in turn would reduce apical membrane P_{Na}, and hence Na^+ entry, completing the cross-talk mechanism. Patch-clamp studies of renal collecting duct cells indicate that increases in both intracellular $[H^+]$ and $[Ca^{2+}]$ inhibit amiloride-sensitive Na^+ channels by reducing P_o (195, 587), but the effect of Ca^{2+} appears to be indirect, i.e., it was not present in excised membrane patches (460).

The simplest signal linking the rate of Na^+ entry (as cause) and the pump activity (as effect) is the change in $[Na^+]_i$. However, studies of the relationship between $[Na^+]_i$ and transepithelial net Na^+ flux (J_{Na}) in rabbit colon, rabbit urinary bladder and toad urinary bladder do not support this simple notion (reviewed in 565 and 568).

Amiloride addition to the solution bathing the apical surface of toad or frog urinary bladder epithelium induced a rapid (seconds) fall in basolateral membrane conductance with no change in cell volume (135) and the rate of volume regulatory decrease (RVD) after transient exposure to hyposmotic solutions was slowed by amiloride pretreatment. It was suggested that inhibition of apical membrane P_{Na} resulted in inhibition of basolateral membrane K^+ permeability (P_K) and thereby slowed K^+ and Cl^- effluxes during RVD (135). Conversely, exposure of *Necturus* urinary blad-

der to Ba^{2+} on the basolateral surface to reduce K^+ conductance caused a near-simultaneous fall in apical membrane P_{Na} (139).

Cross talk in leaky epithelia. In small intestine and renal proximal tubule, entry of organic solute coupled to Na^+ depolarized the apical membrane voltage and reduced the ratio of cell-membrane resistances (R_a/R_b). These changes were followed by a slower repolarization, concomitant with an increase in R_a/R_b. Phloridzin (blocker of Na^+-sugar cotransport) raised R_a/R_b to a value greater than that observed prior to exposure to the sugar, and Ba^{2+} addition to the basolateral bathing solution prevented the secondary rise in R_a/R_b. These data suggest that there is an increase in basolateral K^+ conductance in response to the increased solute influx across the apical-cell membrane (347, 564, 565, 568), probably due to cell swelling. Consistent with this interpretation, exposure to a hyposmotic bathing solution increased the Ba^{2+}-sensitive basolateral membrane K^+ conductance in small intestine (350). Further, isolated guinea-pig enterocytes swelled and subsequently regulated their volume following exposure to alanine and glucose (395). The regulation involved activation of K^+ and Cl^- conductances.

In leaky epithelia, maneuvers that increased Na^+ absorption by two- to fourfold did not elevate $[Na^+]_i$ (568), denoting Na^+ pump activation. The mechanism of this effect is unknown. Possibilities include changes in the number of operational pump molecules or in the turnover number.

In renal proximal tubule, basolateral-membrane K^+ conductance also changed in proportion to the rate of Na^+ absorption, and it was shown that this is not coupled to $[Ca^{2+}]_i$ (349). Hyposmotic swelling was followed by partial regulatory volume decrease, due to activation of Cl^- and K^+ conductances and absolutely dependent upon extracellular Ca^{2+} (416). In perfused rabbit proximal tubules, cell swelling elicited by either hyposmotic solutions or entry of Na^+ and organic solute increased basolateral-membrane K^+ conductance (43); $[Ca^{2+}]_i$ increased with hyposmotic swelling, but not with Na^+-organic solute influx and RVD occurred only with the former perturbation (41). Hence, it seems clear that the mechanisms linking these two mechanisms of cell swelling to the changes in basolateral ionic conductances are different. Strange (608) has pointed out that the functional consequences of cell swelling vary depending on the swelling mechanism.

Inhibition of the Na^+,K^+-ATPase elevates $[Na^+]_i$, decreases $[K^+]_i$, and causes membrane depolarization and increase in $[Cl^-]_i$. In proximal tubules, the swelling is very slow and depends on the presence of cotransported organic solutes in the lumen. In their absence, V_m and cell volume remained stable for hours (348,

394, 650). Cells from some epithelia express inwardly rectifying K$^+$ channels in the basolateral membrane (116, 568). It has been suggested that these channels are involved in membrane repolarization (after the depolarization produced by entry of Na$^+$ and cotransported solutes), but there is no agreement in some experimental observations (619; see also Basolateral-Membrane K$^+$ Channels in NaCl-Absorptive Epithelia).

Signaling mechanisms in cross talk. There are many forms of membrane cross talk in epithelial cells, and no single, unifying mechanism accounting for all of these has emerged. The information available suggests roles of cell volume and cytoplasmic ion concentrations as primary signals.

The mechanisms by which changes in cell volume alter membrane-transport processes are beginning to be understood. A reasonable hypothesis is a dual control system involving a physical sensing mechanism—i.e., membrane/cytoskeleton tension—and a chemical sensing mechanism, either specific or nonspecific. Cytoplasmic [Ca^{2+}] and protein phosphorylation may be involved in the transduction of the signals. In addition, intracellular pH, intracellular [ATP], membrane voltage, and other factors can alter the rate of transport via plasma membrane channels, carriers, and pumps. As suggested before (568) the signals and the transduction mechanisms could be multiple and interactive.

Here, I shall discuss the hypothesis of cross talk mediated by changes in cell volume. Other possible mechanisms have been discussed in detail in recent reviews (193, 416, 518, 546, 565, 568).

A change in cell volume would be the initial signal sensed by the cell and transduced to physical or chemical events that ultimately alter membrane transport. A mechanical-strain sensor in the membrane and/or the cytoskeleton would detect changes in tension. Mechanosensitive channels have been identified in epithelial and other cell types, and their properties have been discussed above (see Rapid Change in the Activity of Individual Transporters). An argument against a membrane-associated mechanical sensor is the demonstration that KCl cotransport in rabbit red blood cells is activated by increases in cell volume, but not by changes in cell shape (299). The alternative hypothesis postulates a chemical sensor of cell-volume changes. This was considered unlikely because cell-volume regulation occurs following very small changes in cell volume (568). However, the signal may be amplified—by an enzymatic cascade, for instance, perhaps including phosphorylation and/or dephosphorylation. The end result would be chemical modification of one or more transporters (246, 300, 474). This would produce a change in solute transport proportionally greater than the changes in cell volume. The signal could be the concentration of one or more cytoplasmic molecules. The process can be *specific*, e.g., the sensed parameter is a second messenger or an enzyme, or *nonspecific*, e.g., the sensed quantity is the total concentration of macromolecules. A candidate for specific chemical sensing mechanism is the intracellular [Mg^{2+}] (300, 599), but definitive proof is lacking. An interesting proposal for nonspecific sensing is the total protein concentration, which within certain ranges can exert large effects on the chemical activities of specific solutes (macromolecular-crowding hypothesis; see 110, 111). Further details can be found in recent reviews (422, 423, 473). The role of changes in ionic strength on the activity of transporters involved in cell-volume regulation has been discussed by Parker et al. (474a).

I am grateful to Drs. G. A. Altenberg, P. S. Aronson, C. U. Cotton, S. A. Lewis, S. G. Schultz and K. R. Spring for constructive comments on sections of this chapter; to Drs. V. L. Schuster, M. J. Stutts and W.-H. Wang for discussions; to Drs. D. Stubbs and R. Torres for invaluable help with the figures and the bibliography, respectively; and to L. Durant and A. Lucas for secretarial help. This work was supported in part by grant DK38734 of the National Institutes of Health.

REFERENCES

1. Adachi, S., S. Uchida, H. Ito, M. Hata, M. Hiroe, F. Marumo, and S. Sasaki. Two isoforms of a chloride channel predominantly expressed in thick ascending limb of Henle's loop and collecting ducts of rat kidney. *J. Biol. Chem.* 269: 17677–17683, 1994.

1a. Agre, P., D. Brown, and S. Nielsen. Aquaporin water channels: unanswered questions and unresolved controversies. *Curr. Opin. Cell Biol.* 7: 472–483, 1995.

2. Agre P., G. M. Preston, B. L. Smith, J. S. Jung, S. Raina, C. Moon, W. B. Guggino, and S. Nielsen. Aquaporin CHIP: the archetypal molecular water channel. *Am. J. Physiol.* 265: (*Renal Fluid Electrolyte Physiol.* 34): F463–F476, 1993.

3. Al-Awqati, Q. H$^+$ transport in urinary epithelia. *Am. J. Physiol.* 235: F77–F88, 1978.

4. Al-Awqati, Q. Cellular and molecular mechanisms of regulation of ion transport in epithelia. In: *The Kidney: Physiology and Pathophysiology*, edited by D. W. Seldin and G. Giebisch. New York: Raven, 1992, p. 625–644.

4a. Al-Awqati, Q. Plasticity in epithelial polarity of renal intercalated cells: targeting of the H$^+$-ATPase and band 3. *Am. J. Physiol.* 270 (*Cell Physiol* 39): C1571–C1580.

5. Al-Awqati, Q., and R. Beauwens. Cellular mechanisms of H$^+$ and HCO$_3^-$ transport in tight urinary epithelia. In: *Handbook of Physiology. Renal Physiology*, edited by E. E. Windhager. New York: Oxford University Press for the American Physiological Society, 1992, sect. 8, vol. 1, chapt. 8, p. 323–350.

6. Al-Awqati, Q., L. Norby, A. Mueller, and P. R. Steinmetz. Characteristics of stimulation of H$^+$ transport by aldosterone in turtle urinary bladder. *J. Clin. Invest.* 58: 351–358, 1976.

7. Alberts, B., D. Bray, J. Lewis, M. Raff, K. Roberts, and J. D. Watson. *Molecular Biology of the Cell*, 3rd Edition. New York: Garland Publishing, Inc., 1994.

8. Alper, S. L., J. Natale, S. Gluck, H. F. Lodish, and D. Brown. Subtypes of intercalated cells in rat kidney collecting duct defined by antibodies against the erythroid band 3 and renal vacuolar H^+-ATPase. *Proc. Natl. Acad. Sci. USA.* 86: 5429–5433, 1989.

9. Alpern, R. J. Mechanism of basolateral membrane $H^+/OH^-/HCO_3^-$ transport in the rat proximal convoluted tubule. A sodium-coupled electrogenic process. *J. Gen. Physiol.* 86: 613–636, 1985.

10. Alpern, R. J., and M. Chambers. Basolateral membrane Cl/HCO_3 exchange in rat proximal convoluted tubule. Na-dependent and -independent modes. *J. Gen. Physiol.* 89: 581–598, 1987.

11. Altenberg, G. A., J. Copello, C. Cotton, K. Dawson, Y. Segal, F. Wehner, and L. Reuss. Electrophysiological methods for studying ion and water transport in *Necturus* gallbladder epithelium. *Methods Enzymol.* 192: 650–683, 1990.

12. Altenberg, G. A., and L. Reuss. Apical membrane Na^+/H^+ exchange in *Necturus* gallbladder epithelium. Its dependence on extracellular and intracellular pH and on external Na^+ concentration. *J. Gen. Physiol.* 95: 369–392, 1990.

12a. Amemiya, M., J. Loffing, M. Lotscher, B. Kaissling, R. J. Alpern, and O. W. Moe. Expression of NHE-3 in the apical membrane of rat renal proximal tubule and thick ascending limb. *Kidney Int.* 48: 1206–1215, 1995.

13. Amerongen, H. H., J. A. Mack, J. M. Wilson, and M. R. Neutra. Membrane domains of intestinal epithelial cells: distribution of Na,K-ATPase and the membrane skeleton in adult rat intestine during fetal development and after epithelial isolation. *J. Cell Biol.* 109: 2129–2138, 1989.

14. Anderson, J. M., M. S. Balda, and A. S. Fanning. The structure and regulation of tight junctions. *Curr. Opin. Cell Biol.* 5: 772–778, 1993.

15. Anderson, M. P., H. A. Berger, D. P. Rich, R. J. Gregory, A. E. Smith, and M. J. Welsh. Nucleoside triphosphates are required to open the CFTR chloride channel. *Cell* 67: 775–784, 1991.

16. Anderson, M. P., R. J. Gregory, S. Thompson, D. W. Souza, P. Paul, R. C. Mulligan, A. E. Smith, and M. J. Welsh. Demonstration that CFTR is a chloride channel by alteration of its ion selectivity. *Science* 253: 202–205, 1991.

17. Anderson, M. P., D. N. Sheppard, H. A. Berger, and M. J. Welsh. Chloride channels in the apical membrane of normal and cystic fibrosis airway and intestinal epithelial. *Am. J. Physiol.* 263 (*Lung Cellular and Molecular Physiology* 32): L1–L14, 1992.

17a. Anderson, J. M., and C. M. Van Itallie. Tight junctions and the molecular basis for regulation of paracellular permeability. *Am. J. Physiol.* 269 (*Gastrointest. Liver Physiol.* 32): G467–G475, 1995.

17b. Ando-Akatsuka, Y., M. Saitou, T. Hirase, M. Kishi, A. Sakaki-bara, M. Itoh, S. Yonemura, M. Furuse, and S. Tsukita. Interspecies diversity of the occludin sequence: cDNA cloning of human, mouse, dog, and rat-kangaroo homologues. *J. Cell Biol.* 133: 43–47, 1996.

18. Arispe, N., E. Rojas, J. Hartman, E. J. Sorscher, and H. B. Pollard. Intrinsic anion channel activity of the recombinant first nucleotide binding fold domain of the cystic fibrosis transmembrane regulator protein. *Proc. Natl. Acad. Sci. U.S.A.* 89: 1539–1543, 1992.

19. Armitage, F. E., and C. S. Wingo. Luminal acidification in K-replete OMCD$_i$: contributions of H-K-ATPase and bafilomycin-A$_1$-sensitive H-ATPase. *Am. J. Physiol.* 267 (*Renal Fluid Electrolyte Physiol.* 36): F450–F458, 1994.

20. Aronson, P. S. Mechanisms of active H^+ secretion in the proxi-mal tubule. *Am. J. Physiol.* 245 (*Renal Fluid Electrolyte Phsyiol.* 14): F647–F659, 1983.

21. Aronson, P. S. The renal proximal tubule. A model for diversity of anion exchangers and stilbene-sensitive anion transporters. *Annu. Rev. Physiol.* 51: 419–441, 1989.

22. Aronson, P. S., J. Nee, and M. A. Suhm. Modifier role of internal H^+ in activating the Na^+-H^+ exchanger in renal microvillus membrane vesicles. *Nature* 299: 161–163, 1982.

23. Ashcroft, F. M. Adenosine 5'-triphosphate-sensitive potassium channels. *Annu. Rev. Neurosci.* 11: 97–118, 1988.

24. Ashcroft, S. J. H., and F. M. Ashcroft. Properties and functions of ATP-sensitive K channels. *Cell Signal.* 2: 197–214, 1990.

25. Asher, C., R. Eren, L. Kahn, O. Yeger, and H. Garty. Expression of the amiloride-blockable Na^+ channel by RNA from control vs. aldosterone stimulated tissue. *J. Biol. Chem.* 267: 16061–16065, 1992.

26. Asher, C., and H. Garty. Aldosterone increases the apical Na^+ permeability of toad bladder by two different mechanisms. *Proc. Natl. Acad. Sci. U.S.A.* 85: 7413–7417, 1988.

27. Attall, B., E. Gullemare, F. Lesage, E. Honore, G. Romey, M. Lazdunski, and J. Barhanin. The protein IsK is a dual activator of K^+ and Cl^- channels. *Nature* 365: 850–852, 1993.

28. Augustus, J., J. Bijman, C. H. van Os, and J. F. G. Slegers. High conductance in an epithelial membrane not due to extracellular shunting. *Nature* 268: 657–658, 1977.

29. Ausiello, D. A., J. L. Stow, H. F. Cantiello, J. B. de Almeida, and D. J. Benos. Purified epithelial Na^+ channel complex contains the pertussis toxin-sensitive $G\alpha_{i-3}$ protein. *J. Biol. Chem.* 267: 4759–4765, 1992.

29a. Awayda, M. S., I. I. Ismailov, B. K. Berdiev, C. M. Fuller, and D. J. Benos. Protein kinase regulation of a cloned epithelial Na^+ channel. *J. Gen. Physiol.* 108: 49–65, 1996.

30. Bakker, R., and J. A. Groot. Further evidence for the regulation of the tight junction ion selectivity by cAMP in goldfish intestinal mucosa. *J. Membr. Biol.* 111: 25–35, 1989.

31. Balda, M. S., and J. M. Anderson. Two classes of tight junctions are revealed by ZO-1 isoforms. *Am. J. Physiol.* 264 (*Cell Physiol.* 33): C918–C924, 1993.

32. Balda, M. S., M. B. Fallon, C. M. Van Itallie, and J. M. Anderson. Structure, regulation, and pathophysiology of tight junctions in the gastrointestinal tract. *Yale J. Biol. Med.* 65: 725–735, 1992.

32a. Barasch, J., B. Kiss, A. Prince, L. Saiman, D. Gruenert, and Q. Al-Awqati. Defective acidification of intracellular organelles in cystic fibrosis. *Nature* 352: 70–73, 1991.

33. Barbry, P., O. Chassande, R. Marsault, M. Lazdunski, and C. Frelin. [^{3}H]phenamil binding protein of the renal epithelium Na^+ channel. Purification, affinity labeling, and functional reconstitution. *Biochemistry* 29: 1039–1045, 1990.

34. Barfuss, D. W., and J. A. Schafer. Hyperosmolarity of absorbate from isolated rabbit proximal tubules. *Am. J. Physiol.* 247 (*Renal Fluid Electrolyte Physiol.* 16): F130–F139, 1984.

35. Barlet-Bas, C., C. Khadouri, S. Marsy, and A. Doucet. Enhanced intracellular sodium concentration in kidney cells recruits a latent pool of Na-K-ATPase whose size is modulated by corticosteroids. *J. Biol. Chem.* 265: 7799–7809, 1990.

35a. Barrett, K. M., and T. D. Bigby. New roles for eicosanoids as regulators of epithelial function and growth. *News Physiol. Sci.* 10: 153–159, 1995.

36. Baukrowitz, T., T.-C. Hwang, A. C. Nairn, and D. C. Gadsby. Coupling of CFTR Cl^- channel gating to an ATP hydrolysis cycle. *Neuron* 12: 473–482, 1994.

37. Baum, M., O. W. Moe, D. L. Gentry, and R. J. Alpern. Effect of glucocorticoids on renal cortical NHE-3 and NHE-1 mRNA.

Am. J. Physiol. 267 (*Renal Fluid Electrolyte Physiol.* 36): F437–F442, 1994.

38. Bean, B. P. Pharmacology and electrophysiology of ATP-activated ion channels. *Trends Pharmacol. Sci.* 13: 87–90, 1992.

39. Bear, C. E., C. Li, N. Kartner, R. J. Bridges, T. J. Jensen, M. Ramjeesingh, and J. R. Riordan. Purification and functional reconstitution of the cystic fibrosis transmembrane conductance regulator (CFTR). *Cell* 68: 809–818, 1992.

40. Beck, J. S., S. Breton, G. Giebisch, and R. Laprade. Potassium conductance regulation by pH during volume regulation in rabbit proximal convoluted tubules. *Am. J. Physiol.* 263 (*Renal Fluid Electrolyte Physiol.* 32): F453–F458, 1992.

41. Beck, J. S., S. Breton, R. Laprade, and G. Giebisch. Volume regulation and intracellular calcium in the rabbit proximal convoluted tubule. *Am. J. Physiol.* 260 (*Renal Fluid Electrolyte Physiol.* 29): F861–F867, 1991.

42. Beck, J. S., R. Laprade, and J.-Y. Lapointe. Coupling between transepithelial Na transport and basolateral K conductance in renal proximal tubule. *Am. J. Physiol.* 266 (*Renal Fluid Electrolyte Physiol.* 35): F517–F527, 1994.

43. Beck, J. S., and D. J. Potts. Cell swelling, co-transport activation and potassium conductance in isolated perfused rabbit kidney proximal tubules. *J. Physiol. (Lond.)* 425: 369–378, 1990.

44. Becq, F., T. J. Jensen, X.-B. Chang, A. Savoia, J. M. Rommens, L.-C. Tsui, M. Buchwald, J. R. Riordan, and J. W. Hanrahan. Phosphatase inhibitors activate normal and defective CFTR chloride channels. *Proc. Natl. Acad. Sci. U.S.A.* 91: 9160–9164, 1994.

45. Benham, C. D., and R. W. Tsien. A novel receptor-operated Ca^{2+}-permeable channel activated by ATP in smooth muscle. *Nature* 328: 275–278, 1987.

45a. Bennett, E., and G. A. Kimmich. The molecular mechanism and potential dependence of the Na^+/glucose cotransporter. *Biophys. J.* 70: 1676–1688.

46. Bennett, M.V.L., L. C. Barrio, T. A. Bargiello, D. C. Spray, E. Hertzberg, and J. C. Saez. Gap junctions: new tools, new answers, new questions. *Neuron* 6: 305–320, 1991.

46a. Benos, D. J., M. S. Awayda, B. K. Berdiev, A. L. Bradford, C. M. Fuller, O. Senyk, and I. I. Ismailov. Diversity and regulation of amiloride-sensitive Na^+ channels. *Kidney Int.* 49: 1632–1637, 1996.

47. Benos, D. J., M. S. Awayda, I. I. Ismailov, and J. P. Johnson. Structure and function of amiloride-sensitive Na^+ channels. *J. Membr. Biol.* 143: 1–18, 1995.

48. Benos, D. J., G. Saccomani, B. M. Brenner, and S. Sariban-Sohraby. Purification and characterization of the amiloride-sensitive sodium channel from A6 cultured cells and bovine renal papilla. *Proc. Natl. Acad. Sci. U.S.A.* 83: 8525–8529, 1986.

49. Benos, D. J., G. Saccomani, and S. Sariban-Sohraby. The epithelial sodium channel. Subunit number of location of the amiloride binding site. *J. Biol. Chem.* 262: 10613–10618, 1987.

50. Bentzel, C. J., and K. Loeschke. The tight junction-intercellular space: an integrated transport path. In: *Isotonic Transport in Leaky Epithelia, Alfred Benzon Symposium 34*, edited by H. H. Ussing, J. Fischbarg, O. Sten-Knudsen, E. H. Larsen, and N. J. Willumsen. Copenhagen: Munksgaard, 1993, p. 233–243.

50a. Berdiev, B. K., A. G. Prat, H. F. Cantiello, D. A. Ausiello, C. M. Fuller, B. Jovov, D. J. Benos, and I, I. Ismailov. Regulation of epithelial sodium channels by short actin filaments. *J. Biol. Chem.* 271: 17704–17710, 1996.

51. Berger, H. A., S. M. Travis, and M. J. Welsh. Regulation of the cystic fibrosis transmembrane conductance regulator Cl^- channel by specific protein kinases and protein phosphatases. *J. Biol. Chem.* 268: 2037–2047, 1993.

51a. Bertorello, A. M., and A. I. Katz. Regulation of Na^+-K^+ pump activity: pathways between receptors and effectors. *News Physiol. Sci.* 10: 253–259, 1995.

51b. Bianchini, L., A. Kapus, G. Lukacs, S. Wasan, S. Wakabayashi, J. Pouysségur, F. H. Yu, J. Orlowski, and S. Grinstein. Responsiveness of mutants of NHE1 isoform of Na^+/H^+ antiport to osmotic stress. *Am. J. Physiol.* 269 (*Cell Physiol.* 38): C998–C1007, 1995.

52. Biemesderfer, D., J. Pizzonia, A. Abu-Alfa, M. Exner, R. Reilly, P. Igarashi, and P. S. Aronson. NHE3: a Na^+/H^+ exchanger isoform of renal brush border. *Am. J. Physiol.* 265 (*Renal Fluid Electrolyte Physiol.* 34): F736–F742, 1993.

53. Birner, B., D.D.F. Loo, and E. M. Wright. Voltage-clamp studies of the intestinal Na^+/glucose cotransporter cloned from rabbit small intestine. *Pflugers Arch.* 418: 79–85, 1991.

54. Blazer-Yost, B., and M. Cox. Aldosterone-induced proteins: characterization using lectin-affinity chromatography. *Am. J. Physiol.* 249 (*Cell Physiol.* 18): C215–C225, 1985.

55. Blot-Chabaud, M., F. Wanstok, J. P. Bonvalet, and N. Farman. Cell sodium-induced recruitment of Na^+-K^+ ATPase pumps in rabbit cortical collecting tubules is aldosterone dependent. *J. Biol. Chem.* 265: 11676–11681, 1990.

56. Boim, M. A., K. Ho, M. E. Shuck, M. J. Bienkowski, J. H. Block, J. L. Slightom, Y. Yang, B. M. Brenner, and S. C. Hebert. ROMK inwardly rectifying ATP-sensitive K^+ channel. II. Cloning and distribution of alternative forms. *Am. J. Physiol.* 268 (*Renal Fluid Electrolyte Physiol.* 37): F1132–F1140, 1995.

57. Bolívar, J. J., and M. Cereijido. Voltage and Ca^{2+}-activated K^+ channel in cultured epithelial cells. *J. Membr. Biol.* 97: 43–51, 1987.

58. Boron, W. F. Transport of H^+ and of ionic weak acids and bases. *J. Membr. Biol.* 72: 1–16, 1983.

59. Boron, W. F. Intracellular pH regulation in epithelial cells. *Annu. Rev. Physiol.* 48: 377–388, 1986.

60. Boron, W. F., and E. L. Boulpaep. Intracellular pH regulation in the renal proximal tubule of the salamander. Na-H exchange. *J. Gen. Physiol.* 81: 29–52, 1983.

61. Boron, W. F., and E. L. Boulpaep. Intracellular pH regulation in the renal proximal tubule of the salamander. Basolateral HCO_3^- transport. *J. Gen. Physiol.* 81: 53–94, 1983.

62. Boron, W. F., and E. L. Boulpaep. The electrogenic Na/HCO_3 cotransporter. *Kidney Int.* 36: 392–403, 1989.

63. Boucher, R. C., M. J. Stutts, M. R. Knowles, L. Cantley, and J. T. Gatzy. Na^+ transport in cystic fibrosis respiratory epithelia. Abnormal basal rate and response to adenylate cyclase activation. *J. Clin. Invest.* 78: 1245–1252, 1986.

64. Boulpaep, E. L., and H. Sackin. Electrical analysis of intraepithelial barriers. *Curr. Top. Membr. Transp.* 13: 169–197, 1980.

65. Breitwieser, G. E., A. A. Altamirano, and J. M. Russell. Effects of pH changes on sodium pump fluxes in squid giant axon. *Am. J. Physiol.* 253 (*Cell Physiol.* 22): C547–C554, 1987.

66. Breton, S., J. S. Beck, and R. Laprade. cAMP stimulates proximal convoluted tubule Na^+-K^+-ATPase activity. *Am. J. Physiol.* 266 (*Renal Fluid Electrolyte Physiol.* 35): F400–F410, 1994.

67. Brosius, F. C., S. L. Alper, A. M. García, and H. F. Lodish. The major kidney band 3 transcript predicts an amino-terminal truncated band 3 polypeptide. *J. Biol. Chem.* 264: 7784–7787, 1989.

68. Brown, A. M. Ionic channels and their regulation by G protein subunit. *Annu. Rev. Physiol.* 52: 197–213, 1990.

69. Brown, A. M., and L. Birnbaumer. Direct G-protein gating of ion channels. *Am. J. Physiol.* 254 (*Heart Circ. Physiol.* 23): H401–H410, 1988.

70. Brown, D. Membrane recycling and epithelial cell function.

Am. J. Physiol. 256 (*Renal Fluid Electrolyte Physiol.* 25): F1–F12, 1989.

71. Brown, D. Structural-functional features of vasopressin-induced water flow in the kidney collecting duct. *Semin. Nephrol.* 11: 478–501, 1991.

72. Brown, D., S. Hirsch, and S. Gluck. An H^+-ATPase in opposite plasma membrane domains in kidney epithelial cell subpopulations. *Nature* 331: 622–624, 1988.

73. Brown, D., S. Hirsch, and S. Gluck. Localization of a proton-pumping ATPase in rat kidney. *J. Clin. Invest.* 82: 2114–2126, 1988.

74. Brown, D., J.-M. Verbavatz, G. Valenti, B. Lui, and I. Sabolic. Localization of the CHIP28 water channel in reabsorptive segments of the rat male reproductive tract. *Eur. J. Cell. Biol.* 61: 264–273, 1993.

74a. Bubien, J. K., I. I. Ismailov, B. K. Berdiev, T. Cornwell, R. P. Lifton, C. M. Fuller, J.-M. Achard, D. J. Benos, and D. G. Warnock. Liddle's disease: abnormal regulation of amiloride-sensitive Na^+ channels by β-subunit mutation. *Am. J. Physiol.* 270 (*Cell Physiol* 39): C208–C213, 1996.

75. Burant, C. F., J. Takeda, E. Brot-Laroche, G. I. Bell, and N. O. Davidson. Fructose transporter in human spermatozoa and small intestine in GLUT5. *J. Biol. Chem.* 267: 14523–14526, 1992.

76. Burch, L. H., C. R. Talbot, M. R. Knowles, C. M. Canessa, B. C. Rossier, and R. C. Boucher. Relative expression of the human epithelial Na^+ channel subunits in normal and cystic fibrosis airways. *Am. J. Physiol.* 269 (*Cell Physiol.* 38): C511–C518, 1995.

77. Burckhardt, B.-C., and H. Gögelein. Small and maxi K^+ channels in the basolateral membrane of isolated crypts from rat distal colon: single-channel and slow whole-cell recordings. *Pflugers Arch.* 420: 54–60, 1992.

78. Burckhardt, B.-Ch., K. Sato, and E. Frömter. Electrophysiological analysis of bicarbonate permeation across the peritubular cell membrane of rat kidney proximal tubule. I. Basic observations. *Pflugers Arch.* 401: 34–42, 1984.

79. Burckhardt, B.-Ch., K. Sato, and E. Frömter. Electrophysiological analysis of bicarbonate permeation across the peritubular cell membrane of rat kidney proximal tubule. II. Exclusion of HCO_3^- effects on other ion permeabilities and on coupled electroneutral HCO_3^- transport. *Pflugers Arch.* 401: 43–51, 1984.

80. Buxton, R. S. Nomenclature of the desmosomal cadherins. *Cell Biol.* 121: 481–483, 1993.

81. Cabantchik, Z. I., and R. Greger. Chemical probes for anion transporters of mammalian cell membranes. *Am. J. Physiol.* 262 (*Cell Physiol.* 31): C803–C827, 1992.

82. Cabantchik, Z. I., and A. Rothstein. Membrane proteins related to anion permeability of human red blood cells. I. Localization of disulfonic stilbene binding sites in proteins involved in permeation. *J. Membr. Biol.* 15: 207–226, 1974.

83. Canessa, C. M., J.-D. Horisberger, and B. C. Rossier. Epithelial sodium channel related to proteins involved in neurodegeneration. *Nature* 361: 467–470, 1993.

84. Canessa, C. M., A.-M. Mérillat, and B. C. Rossier. Membrane topology of the epithelial sodium channel in intact cells. *Am. J. Physiol.* 267 (*Cell Physiol.* 36): C1682–C1690, 1994a.

85. Canessa, C. M., L. Schild, G. Buell, B. Thorens, I. Gautschil, J.-D. Horisberger, and B. C. Rossier. Amiloride-sensitive epithelial sodium channel is made of three homologous subunits. *Nature* 367: 463–467, 1994.

86. Cannon, C., J. van Adelsberg, S. Kelly, and Q. Al-Awqati. Carbon dioxide induced exocytotic insertion of H^+ pumps in turtle bladder luminal membrane: role of cell pH and calcium. *Nature* 314: 443–446, 1985.

86a. Cantiello, H. F. Actin filaments stimulate the Na^+-K^+-ATPase. *Am. J. Physiol.* 269 (*Renal Fluid Electrolyte Physiol.* 38): F637–F643, 1995.

87. Cantiello, H. F., C. R. Patenaude, and D. A. Ausiello. G-protein subunit, αi-3, activates a pertussis toxin-sensitive Na^+ channel from the epithelial cell line, A6. *J. Biol. Chem.* 264: 20867–20870, 1989.

88. Cantiello, H. F., J. L. Stow, A. G. Prat, and D. A. Ausiello. Actin filaments regulate epithelial Na^+ channel activity. *Am. J. Physiol.* 261 (*Cell Physiol.* 30): C882–C888, 1991.

89. Caplan, M. J., and Rodríguez-Boulan, E., chapt. 17, this volume.

90. Carafoli, E. Calcium pump of the plasma membrane. *Physiol. Rev.* 71: 129–153, 1991.

91. Carafoli, E. Biogenesis: Plasma membrane calcium ATPase: 15 years of work on the purified enzyme. *FASEB J.* 8: 993–1002, 1994.

92. Carpi-Medina, P., E. González, B. Lindemann, and G. Whittembury. The continuous measurement of tubular volume changes in response to step changes in contraluminal osmolality. *Pflugers Arch.* 400: 343–348, 1983.

93. Carpi-Medina, P., E. González, and G. Whittembury. Cell osmotic water permeability of isolated rabbit proximal convoluted tubules. *Am. J. Physiol.* 244 (*Renal Fluid Electrolyte Physiol.* 13): F554–F563, 1984.

94. Carpi-Medina, P, and G. Whittembury. Comparison of transcellular and transepithelial water osmotic permeabilities in the isolated proximal straight tubule of the rabbit kidney. *Pflugers Arch.* 412: 66–74, 1988.

95. Carroll, T. P., E. M. Schwiebert, and W. B. Guggino. CFTR: Structure and function. *Cell Physiol. Biochem.* 3: 388–399, 1993.

96. Carson, M. R., S. M. Travis, and M. J. Welsh. The two nucleotide-binding domains of cystic fibrosis transmembrane conductance regulator (CFTR) have distinct functions in controlling channel activity. *J. Biol. Chem.* 270: 1711–1717, 1995.

97. Cereijido, M. Evolution of ideas on the tight junction. In: *Tight Junctions,* edited by M. Cereijido. Boca Raton, FL: CRC, 1992, p. 1–13.

98. Chang, X. B., J. A. Tabcharani, Y. X. Hou, A.T.J. Jensen, N. Kartner, N. Alon, J. W. Hanrahan, and J. R. Riordan. Protein kinase A (PKA) still activates CFTR chloride channel after mutagenesis of all 10 PKA consensus phosphorylation sites. *J. Biol. Chem.* 268: 11304–11311, 1993.

99. Chang, D., N. L. Kushman, and D. C. Dawson. Intracellular pH regulates basolateral K^+ and Cl^- conductances in colonic epithelial cells by modulating Ca^{2+} activation. *J. Gen. Physiol.* 98: 183–196, 1991.

100. Chase, H. S., Jr., and Q. Al-Awqati. Regulation of the sodium permeability of the luminal border of toad bladder by intracellular sodium and calcium. *J. Gen. Physiol.* 77: 693–712, 1981.

101. Chase, H. S., Jr., and Q. Al-Awqati. Calcium reduces the sodium permeability of luminal membrane vesicles from toad bladder: Studies using a fast reaction apparatus. *J. Gen. Physiol.* 81: 643–665, 1983.

102. Chen, P.-Y., D. Pearce, and A. S. Verkman. Membrane water and solute permeability determined quantitatively by self-quenching of an entrapped fluorophore. *Biochemistry* 27: 5713–5719, 1988.

103. Cheng, S. H., R. Gregory, J. Marshall, S. Paul, D. W. Souza, G. A. White, C. R. O'Riordan, and A. E. Smith. Defective intracellular transport and processing of CFTR is the molecular basis of most cystic fibrosis. *Cell* 63: 827–834, 1990.

104. Christensen, O. Mediation of cell volume regulation by Ca^{2+}

influx through stretch-activated channels. *Nature* 330: 66–68, 1987.

105. Civan, M. M, and H. Garty. Toad urinary bladder as a model for studying transepithelial sodium transport. *Methods Enzymol.* 192: 683–697, 1990.

106. Clapham, D. E. Direct G protein activation of ion channels? *Annu. Rev. Neurosci.* 17: 441–464, 1994.

107. Claude, P. Morphological factors influencing transepithelial permeability: a model for the resistance of the zonula occludens. *J. Membr. Biol.* 39: 219–232, 1978.

108. Clausen, C. Impedance analysis in tight epithelia. *Methods Enzymol.* 171: 628–641, 1989.

109. Clausen, C., T. E. Machen, and J. M. Diamond. Use of AC impedance analysis to study membrane changes related to acid secretion in amphibian gastric mucosa. *Biophys. J.* 41: 167–178, 1993.

110. Colclasure, G. C., and J. C. Parker. Cytosolic protein concentration is the primary volume signal in dog red cells. *J. Gen. Physiol.* 98: 881–892, 1991.

111. Colclasure, G. C., and J. C. Parker. Cytosolic protein concentration is the primary volume signal for swelling-induced [K-Cl] cotransport in dog red cells. *J. Gen. Physiol.* 100: 1–10, 1992.

112. Copello, J., T. A. Heming, Y. Segal, and L. Reuss. cAMP-activated apical membrane chloride channels in *Necturus* gallbladder epithelium. Conductance, selectivity, and block. *J. Gen. Physiol.* 102: 177–199, 1993.

113. Copello, J., Y. Segal, and L. Reuss. Cytosolic pH regulates maxi K^+ channels in *Necturus* gall-bladder epithelial cells. *J. Physiol. (Lond.)* 434: 577–590, 1991.

114. Copello, J., F. Wehner, and L. Reuss. Artifactual expression of maxi-K^+ channels in basolateral membrane of gallbladder epithelial cells. *Am. J. Physiol.* 264 (*Cell Physiol.* 33): C1128–C1136, 1993.

115. Corcia, A., and W. McD. Armstrong. KCl cotransport. A mechanism for basolateral chloride exit in *Necturus* gallbladder. *J. Membr. Biol.* 76: 173–182, 1983.

116. Costantin, J., S. Alcalen, A. S. Otero, W. P. Dubinsky, and S. G. Schultz. Reconstitution of an inwardly rectifying potassium channel from the basolateral membranes of *Necturus* enterocytes into planar lipid bilayers. *Proc. Natl. Acad. Sci. U.S.A.* 86: 5212–5216, 1989.

117. Cotton, C. U., and L. Reuss. Effects of changes in mucosal solution Cl^- or K^+ concentration on cell water volume of *Necturus* gallbladder epithelium. *J. Gen. Physiol.* 97: 667–686, 1991.

118. Cotton, C. U., and L. Reuss. Electrophysiological effects of extracellular ATP on *Necturus* gallbladder epithelium. *J. Gen. Physiol.* 97: 949–972, 1991.

119. Cotton, C. U., A. M. Weinstein, and L. Reuss. Osmotic water permeability of *Necturus* gallbladder. *J. Gen. Physiol.* 93: 649–679, 1989.

120. Coutry, N., N. Farman, J. P. Bonvalet, and M. Blot-Chabaud. Synergistic action of vasopressin and aldosterone on basolateral Na^+-K^+-ATPase in the cortical collecting duct. *J. Membr. Biol.* 145: 99–106, 1995.

121. Crane, R. K. Intestinal absorption of sugars. *Physiol. Rev.* 40: 789–825, 1960.

122. Cremaschi, D., S. Hénin, and G. Meyer. Stimulation by HCO_3^- of Na^+ transport in rabbit gallbladder. *J. Membr. Biol.* 47: 145–170, 1979.

123. Cremaschi, D., and G. Meyer. Amiloride-sensitive sodium channels in rabbit and guinea-pig gall-bladder. *J. Physiol. (Lond.)* 326: 21–34, 1982.

124. Cremaschi, D., G. Meyer, C. Rossetti, G. Bottà, and P. Palestini. The nature of the neutral Na^+-Cl^- coupled entry at the apical membrane of rabbit gallbladder epithelium. I. Na^+/H^+, Cl^-/HCO_3^- double exchange and Na^+/Cl^- symport. *J. Membr. Biol.* 95: 209–218, 1987.

125. Cremaschi, D., C. Porta, G. Bottà, and G. Meyer. Nature of the neutral Na^+-Cl^- coupled entry at the apical membrane of rabbit gallbladder epithelium: IV. Na^+/H^+, Cl^-/HCO_3^- double exchange, hydrochlorothiazide-sensitive Na^+-Cl^- symport and Na^+-K^+-$2Cl^-$ cotransport are all involved. *J. Membr. Biol.* 129: 221–235, 1992.

126. Crowson, M. S., and G. E. Shull. Isolation and characterization of a cDNA encoding the putative distal colon H^+,K^+-ATPase. *J. Biol. Chem.* 267: 13740–13748, 1992.

127. Cunningham, S. A., R. A. Frizzell, and A. P. Morris. Vesicular targeting and the control of ion secretion in epithelial cells: implications for cystic fibrosis. *J. Physiol. (Lond.)* 482: 27S–30S, 1995.

128. Cuppoletti, J., A. M. Baker, and D. H. Malinowska. Cl^- channels of the gastric parietal cell that are active at low pH. *Am. J. Physiol.* 264 (*Cell Physiol.* 33): C1609–C1618, 1993.

129. Curci, S., L. Debellis, and E. Frömter. Evidence for rheogenic sodium bicarbonate cotransport in the basolateral membrane of oxyntic cells of frog gastric fundus. *Pflugers Arch.* 408: 497–504, 1987.

130. Curci, S., and E. Frömter. Electrophysiological techniques in the analysis of ion transport across gastric mucosa. *Methods Enzymol.* 192: 82–93, 1990.

131. Curran, P. F., and J. R. MacIntosh. A model system for biological water transport. *Nature* 193: 347–348, 1962.

132. Curran, P. F., and A. K. Solomon. Ion and water fluxes in the ileum of rats. *J. Gen. Physiol.* 41: 143–168, 1957.

133. Dausch, R., and K. R. Spring. Regulation of NaCl entry into *Necturus* gallbladder epithelium by protein kinase C. *Am. J. Physiol.* 266 (*Cell Physiol.* 35): C531–C535, 1994.

134. Davidson, N. O., A.M.L. Hausman, C. A. Ifkovits, J. B. Buse, G. W. Gould, C. F. Burant, and G. I. Bell. Human intestinal glucose transporter expression and localization of GLUT5. *Am. J. Physiol.* 262 (*Cell Physiol.* 31): C795–C800, 1992.

135. Davis, C. W., and A. L. Finn. Sodium transport inhibition by amiloride reduces basolateral membrane potassium conductance in tight epithelia. *Science* 216: 525–527, 1982.

136. Dawson, D. C., and Q. Al-Awqati. Induction of reverse flow of Na^+ through the active transport pathway in toad urinary bladder. *Biochim. Biophys. Acta* 508: 413–417, 1978.

137. de Jonge, H. R., N. Van Den Berghe, B. Tilly, M. Kansen, and J. Bijman. (Dys)regulation of epithelial chloride channels. *Biochem. Soc. Trans.* 17: 816–818, 1989.

138. del Castillo, J. R., V. M. Rajendran, and H. J. Binder. Apical membrane localization of ouabain-sensitive K^+-activated ATPase activities in rat distal colon. *Am. J. Physiol.* 261 (*Gastrointest. Liver Physiol.* 24): G1005–G1011, 1991.

139. Demarest, J. R., and A. L. Finn. Interaction between the basolateral K^+ and apical Na^+ conductances in *Necturus* urinary bladder. *J. Gen. Physiol.* 89: 563–580, 1987.

140. Dempster, J. A., A. N. Van Hoek, and C. H. van Os. The quest for water channels. *News Physiol. Sci.* 7: 172–176, 1992.

141. Denker, B. M., B. L. Smith, F. P. Kuhajda, and P. Agre. Identification, purification, and partial characterization of a novel M_r 28,000 integral membrane protein from erythrocytes and renal tubule. *J. Biol. Chem.* 263: 14634–15642, 1988.

142. Denning, G. M., M. P. Anderson, J. F. Amara, J. Marshal, A. E. Smith, and M. J. Welsh. Processing of mutant cystic

fibrosis transmembrane conductance regulator is temperature-sensitive. *Nature* 358: 761–764, 1992.

143. Diamond, J. M. The reabsorptive function of the gall-bladder. *J. Physiol. (Lond.)* 161: 442–473, 1962.

144. Diamond, J. M. The mechanism of solute transport by the gall-bladder. *J. Physiol. (Lond.)* 161: 474–502, 1962.

145. Diamond, J. M. The mechanism of water transport by the gall-bladder. *J. Physiol. (Lond.)* 161: 503–527, 1962.

146. Diamond, J. M. The epithelial junction: bridge, gate, and fence. *Physiologist* 20: 10–17, 1977.

147. Diamond, J. M. Transcellular cross-talk between epithelial cell membranes. *Nature* 300: 683, 1982.

148. Diamond, J. M., and W. H. Bossert. Standing-gradient osmotic flow: A mechanism for coupling of water and solute transport in epithelia. *J. Gen. Physiol.* 50: 2061–2083, 1967.

149. Diener, M., M. Nobles, and W. Rummel. Activation of basolateral Cl$^-$ channels in the rat colonic epithelium during regulatory volume decrease. *Pflugers Arch.* 421: 530–538, 1992.

150. DiGiovanni, S. R., S. Nielsen, E. I. Christensen, and M. A. Knepper. Regulation of collecting duct water channel expression by vasopressin in Brattleboro rat. *Proc. Natl. Acad. Sci. U.S.A.* 91: 8984–9888, 1994.

151. Doucet, A., and S. Marsy. Characterization of K-ATPase activity in distal nephron: stimulation by potassium depletion. *Am. J. Physiol.* 253 (*Renal Fluid Electrolyte Physiol.* 22): F418–F423, 1987.

152. Dragsten, P. R., R. Blumenthal, and J. S. Handler. Membrane asymmetry in epithelia: is the tight junction a barrier to diffusion in the plasma membrane? *Nature* 294: 718–722, 1981.

153. Drenckhahn, D., K. Schluter, D. P. Allen, and V. Bennett. Co-localization of band 3 with ankyrin and spectrin at the basal membrane of intercalated cells in the rat kidney. *Science* 230: 1287–1289, 1985.

154. Drenckhahn, D., M. Oelmann, P. Schaef, M. Wagner, and S. Wagner. Band 3 is the basolateral anion exchanger of dark epithelial cells of turtle urinary bladder. *Am. J. Physiol.* 252 (*Cell. Physiol.* 21): C570–C574, 1987.

155. Driscoll, M., and M. Chalfie. The mec-4 gene is a member of a family of *Caenorhabditis elegans* genes that can mutate to induce neuronal degeneration. *Nature* 849: 588–593, 1991.

156. Drumm, M. L., A. H. Pope, W. H. Cliff, J. M. Rommens, S. A. Marvins, L. Tsui, F. Collins, R. A. Frizzell, and J. M. Wilson. Correction of cystic fibrosis defect in vitro by retrovirus-mediated gene transfer. *Cell* 62: 227–1233, 1990.

156a. DuBose, T. D., Jr., J. Codina, A. Burges, and T. A. Pressley. Regulation of H$^+$-K$^+$-ATPase expression in kidney. *Am. J. Physiol.* 269 (*Renal Fluid Electrolyte Physiol.* 38): F500–F507, 1995.

157. Duffey, M. E., B. Hainau, S. Ho, and C. J. Bentzel. Regulation of epithelial tight junction permeability by cyclic AMP. *Nature* 294: 451–453, 1981.

158. Eaton, D. C., K. L. Hamilton, and K. E. Johnson. Intracellular acidosis blocks the basolateral Na-K pump in rabbit urinary bladder. *Am. J. Physiol.* 247 (*Renal Fluid Electrolyte Physiol.* 16): F946–F954, 1984.

159. Echevarría, M., G. Frindt, G. M. Preston, M. Snezana, P. Agre, J. Fischbarg, and E. Windhager. Expression of multiple water channel activities in *Xenopus* oocytes injected with mRNA from rat kidney. *J. Gen. Physiol.* 101: 827–841, 1993.

160. Echevarría, M., E. E. Windhager, S. S. Tate, and G. Frindt. Cloning and expression of AQP3, a water channel from the medullary collecting duct of rat kidney. *Proc. Natl. Acad. Sci. U.S.A.* 91: 10997–11001, 1994.

161. Edelman, I. S. Receptors and effectors in hormone action on

the kidney. *Am. J. Physiol.* 245 (*Renal Fluid Electrolyte Physiol.* 14): F333–F339, 1981.

162. Egan, M., T. Flotte, S. Afione, R. Solow, P. L. Zeitlin, B. J. Carter, and W. B. Guggino. Defective regulation of outwardly rectifying Cl$^-$ channels in cystic fibrosis by protein kinase A corrected by insertion of CFTR. *Nature* 358: 581–584, 1992.

163. Egan, M. E., E. M. Schwiebert, and W. B. Guggino. Differential expression of ORCC and CFTR induced by low temperature in CF airway epithelial cells. *Am. J. Physiol.* 268 (*Cell Physiol.* 37): C243–C251, 1995.

164. Engel, A., T. Walz, and P. Agre. The aquaporin family of membrane water channels. *Curr. Opin. Struct. Biol.* 4: 545–553, 1994.

165. Ericson, A.-C., and K. R. Spring. Coupled NaCl entry into *Necturus* gallbladder epithelial cells. *Am. J. Physiol.* 243 (*Cell Physiol.* 12): C140–C145, 1982.

166. Erlij, D., and H. H. Ussing. Transport across amphibian skin. In: *Membrane Transport in Biology, Vol. III,* edited by G. Giebisch, D. C. Tosteson, and H. H. Ussing. Berlin: Springer-Verlag, 1978, p. 175–208.

167. Evans, R. M. The steroid and thyroid hormone receptor superfamily. *Science* 240: 889–895, 1988.

168. Evans, R. M. Molecular characterization of the glucocorticoid receptor. *Recent Prog. Horm. Res.* 45:1–22, 1989.

169. Eveloff, J., and D. G. Warnock. K-Cl transport systems in rabbit renal basolateral membrane vesicles. *Am. J. Physiol.* 252 (*Renal Fluid Electrolyte Physiol.* 21): F883–F889, 1987.

169a. Ewart, H. S., and A. Klip. Hormonal regulation Na$^+$-K$^+$-ATPase: mechanisms underlying rapid and sustained changes in pump activity. *Am. J. Physiol.* 269 (*Cell Physiol.* 38): C295–C311, 1995.

170. Fafournoux, P., J. Noël, and J. Pouysségur. Evidence that Na$^+$/H$^+$ exchanger isoforms NHE1 and NHE3 exist as stable dimers in membranes with a high degree of specificity for homodimers. *J. Biol. Chem.* 269: 2589–2596, 1994.

171. Fedorak, R. N., N. Cortas, and M. Field. Diabetes mellitus and glucagon alter ouabain-sensitive Na$^+$-K$^+$-ATPase in rat small intestine. *Diabetes* 40: 1603–1610, 1991.

172. Fejes-Tóth, G., W.-R. Chen, E. Rusvai, T. Moser, and A. Náray-Fejes-Tóth. Differential expression of AE1 in renal HCO$_3$-secreting and -reabsorbing intercalated cells. *J. Biol. Chem.* 269: 26717–26721, 1994.

172a. Fejes-Tóth, G., and A. Náray-Fejes-Tóth. Effect of acid/base balance on H-ATPase 31kD subunit mRNA levels in collecting duct cells. *Kidney Int.* 48: 1420–1426, 1995.

173. Felder, C. C., F. E. Albrecht, T. Campbell, G. M. Eisner, and P. A. José. cAMP-independent, G protein-linked inhibition of Na$^+$/H$^+$ exchange in renal brush border by D1 dopamine agonists. *Am. J. Physiol.* 264 (*Renal Fluid Electrolyte Physiol.* 33): F1032–F1037, 1993.

174. Ferreira, K.T.G. The effect of copper on frog skin. The role of sulphydryl groups. *Biochim. Biophys. Acta* 510: 298–304, 1978.

175. Filipovic, D., and H. Sackin. A calcium-permeable stretch-activated cation channel in renal proximal tubule. *Am. J. Physiol.* 260 (*Renal Fluid Electrolyte Physiol.* 29): F119–F129, 1991.

176. Filipovic, D., and H. Sackin. Stretch- and volume-activated channels in isolated proximal tubule cells. *Am. J. Physiol.* 262 (*Renal Fluid Electrolyte Physiol.* 31): F857–F870, 1992.

177. Fine, D. M., C. F. Lo, L. Aguillar, D. L. Blackmon, and M. H. Montrose. Cellular chloride depletion inhibits cAMP-activated electrogenic chloride fluxes in HT29-18-C$_1$ cells. *J. Membr. Biol.* 145: 129–141, 1995.

178. Finkelstein, A. Water and nonelectrolyte permeability of lipid bilayer membranes. *J. Gen. Physiol.* 68: 127–135, 1976.

179. Finkelstein, A. Nature of the water permeability increase induced by antidiuretic hormone (ADH) in toad urinary bladder and related tissues. *J. Gen. Physiol.* 68: 137–143, 1976.

180. Finkelstein, A. *Water Movements through Lipid Bilayers, Pores and Plasma Membranes: Theory and Reality.* New York: Wiley, 1987.

181. Finn, A. L. Transport across amphibian urinary bladder. In: *Membrane Transport in Biology, Vol. III,* edited by G. Giebisch, D. C. Tosteson, and H. H. Ussing. Berlin: Springer-Verlag, 1978, p. 209–238.

182. Finn, A. L. Symports and antiports in epithelial cell volume regulation. In: Na^+/H^+ *Exchange,* edited by S. Grinstein. Boca Raton, FL: CRC, 1988. p. 191–200.

183. Finn, A. L., and L. Reuss. Effects of changes in the composition of the serosal solution on the electrical properties of the toad urinary bladder epithelium. *J. Physiol. (Lond.)* 250: 541–558, 1975.

183a. Fischer, H., B. Illek, and T. E. Machen. The actin filament disrupter cytochalasin D activates the recombinant cystic fibrosis transmembrane conductance regulator Cl^- channel in mouse 3T3 fibroblasts. *J. Physiol. (Lond.)* 489: 745–754, 1995.

183b. Fisher, R. S., F. G. Grillo, and S. Sariban-Sohraby. Brefeldin A inhibition of apical Na^+ channels in epithelia. *Am. J. Physiol.* 270 (*Cell Physiol.* 39): C138–C147, 1996.

184. Flamion, B., and K. R. Spring. Water permeability of apical and basolateral cell membranes of rat inner medullary collecting duct. *Am. J. Physiol.* 259 (*Renal Fluid Electrolyte Physiol.* 28): F986–F999, 1990.

185. Flamion, B., K. R. Spring, and M. Abramow. Adaptation of inner medullary collecting duct to dehydration involves a paracellular pathway. *Am. J. Physiol.* 268 (*Renal Fluid Electrolyte Physiol.* 37): F53–F63, 1995.

186. Flemström, G., and A. Garner. Gastroduodenal HCO_3^- transport: characteristics and proposed role in acidity regulation and mucosal protection. *Am. J. Physiol.* 242 (*Gastrointest. Liver Physiol.* 5): G183–G193, 1982.

187. Fliegel, L., R. S. Haworth, and J.R.B. Dyck. Characterization of the placental brush border membrane Na^+/H^+ exchanger: identification of thiol-dependent transitions in apparent molecular size. *J. Biochem.* 289: 101–107, 1993.

187a. Folkesson, H. G., M. A. Matthay, A. Frigeri, and A. S. Verkman. Transepithelial water permeability in microperfused distal airways. Evidence for channel-mediated water transport. *J. Clin. Invest.* 97: 664–671, 1996.

187b. Fong, P., and T. J. Jentsch. Molecular basis of epithelial Cl channels. *J. Membr. Biol.* 144: 189–197, 1995.

188. Forbush, B., III, and M. Haas. Identification of a 200 K Dalton component of the Na,K,Cl-cotransport system in membranes from dogfish shark rectal gland. *Biophys. J.* 53: 222a, 1988.

189. Forbush, B. III, M. Haas, and C. Lytle. The Na-K-Cl cotransporter in shark rectal gland. I. Regulation in the intact perfused gland. *Am. J. Physiol.* 262 (*Cell Physiol.* 31): C1000–C1008, 1992.

190. Forbush, B., III, C. Lytle, C. Xu, J. A. Payne, and D. Biemesderfer. The Na,K,Cl cotransporter of shark rectal gland. *Renal Physiol. Biochem.* 17: 201–204, 1994.

191. Forte, J. G., D. K. Hanzel, T. Urushidani, and J. M. Wolosin. Pumps and pathways for gastric HCl secretion. *Ann. NY Acad. Sci.* 574: 145–158, 1989.

192. Forte, J. G., T. E. Machen, and K. J. Obrink. Mechanisms of gastric H^+ and Cl^- transport. *Annu Rev. Physiol.* 42: 111–126, 1980.

193. Foskett, J. K. The role of calcium in the control of volume regulatory transport pathways. In: *Cellular and Molecular Physiology of Cell Volume Regulation,* edited by K. Strange, Boca Raton: CRC, 1994, p. 259–278.

194. Friedrich, F., H. Weiss, M. Paulmichl, and F. Lang. Activation of potassium channels in renal epithelioid cells (MDCK). *Am. J. Physiol.* 256 (*Cell Physiol.* 25): C1016–1021, 1989.

195. Frindt, G., R. B. Silver, E. E. Windhager, and L. G. Palmer. Feedback regulation of Na channels in rat CCT. II. Effects of inhibition of Na entry. *Am. J. Physiol.* 264 (*Renal Fluid Electrolyte Physiol.* 33): F565–F574, 1993.

196. Frizzell, R. A., and W. H. Cliff. Chloride channels. No common motif. *Curr. Biol.* 2: 285–287, 1992.

197. Frizzell, R. A., M. C. Dugas, and S. G. Schultz. Sodium chloride transport by rabbit gallbladder direct evidence for a coupled NaCl influx process. *J. Gen. Physiol.* 65: 769–795, 1975.

198. Frizzell, R. A., G. Reckhammer, and R. L. Shoemaker. Altered regulation of airway epithelial chloride channels in cystic fibrosis. *Science* 233: 558–560, 1986.

199. Frömter, E. The route of passive ion movement through the epithelium of *Necturus* gallbladder. *J. Membr. Biol.* 8: 259–301, 1972.

200. Frömter, E. Electrophysiological analysis of rat renal sugar and amino acid transport. I. Basic Phenomena. *Pflugers Arch.* 393: 179–189, 1982.

201. Frömter, E., and J. Diamond. Route of passive ion permeation in epithelia. *Nature [New Biol.]* 235: 9–13, 1972.

201a. Fuller, C. M., M. S. Awayda, M. P. Arrate, A. L. Bradford, R. G. Morris, C. M. Canessa, B. C. Rossier, and D. J. Benos. Cloning of a bovine renal epithelial Na^+ channel subunit. *Am. J. Physiol.* 269 (*Cell Physiol.* 38): C641–654, 1995.

202. Fuller, C. M., and B. J. Benos. CFTR! *Am. J. Physiol.* 263 (*Cell Physiol.* 32): C267–C286, 1992.

203. Furuse, M., T. Hirase, M. Itoh, A. Nagafuchi, S. Yonemura, S. Tsukita, and S. Tsukita. Occludin: a novel integral membrane protein localizing at tight junctions. *J. Cell Biol.* 123: 1777–1788, 1993.

204. Furuse, M., M. Itoh, T. Hirase, A. Nagafuchi, S. Yonemura, S. Tsukita, and S. Tsukita. Direct association of occludin with ZO-1 and its possible involvement in the localization of occludin at tight junctions. *J. Cell Biol.* 127: 1617–1626, 1994.

205. Fushimi, K., S. Uchida, Y. Hara, Y. Hirata, F. Marumo, and S. Sasaki. Cloning and expression of apical membrane water channel of rat kidney collecting tubule. *Nature* 361: 549–552, 1993.

206. Fushimi, K., and A. S. Verkman. Relationship between vasopressin-sensitive water transport and plasma membrane fluidity in kidney collecting tubule. *Am. J. Physiol.* 260 (*Cell Physiol.* 29): C1–C8, 1991.

207. Gabriel, S. E., L. L. Clarke, R. C. Boucher, and M. J. Stutts. CFTR and outward rectifying chloride channels are distinct proteins with a regulatory relationship. *Nature* 363: 263–266, 1993.

207a. Gadsby, D. C., G. Nagel, and T.-C. Hwang. The CFTR chloride channel of mammalian heart. *Annu. Rev. Physiol.* 57: 387–416, 1995.

208. Gadsby, D. C., and A. C. Nairn. Regulation of CFTR channel gating. *Trends Biochem. Sci.* 19: 513–518, 1994.

209. Gallacher, D. V. Are there purinergic receptors on parotid acinar cells? *Nature* 296: 83–86, 1982.

210. Gamba, G., A. Miyanoshita, M. Lombardi, J. Lytton, W.-L. Lee, M. A. Hediger, and S. C. Hebert. Molecular cloning,

primary structure, and characterization of two members of the mammalian electroneutral sodium-(potassium)-chloride co-transporter family expressed in kidney. *J. Biol. Chem.* 269: 17713–17722, 1994.

211. Gamba, G., S. N. Saltzberg, M. Lambardi, A. Miyanoshita, J. Lytton, M. A. Hediger, B. M. Brenner, and S. C. Hebert. Primary structure and functional expression of a cDNA encoding the thiazide-sensitive, electroneutral sodium-chloride co-transporter. *Proc. Natl. Acad. Sci. U.S.A.* 90: 2749–2753, 1994.

212. Garcia-Austt, J., D. W. Good, M. B. Burg, and M. A. Knepper. Deoxycorticosterone-stimulated bicarbonate secretion in rat cortical collecting ducts: effects of luminal chloride removal and in vivo acid loading. *Am. J. Physiol.* 249 (*Renal Fluid Electrolyte Physiol.* 18): F205–F212, 1985.

213. Garg, L. C., M. A. Knepper, and M. B. Burg. Mineralocorticoid effects on Na-K-ATPase in individual nephron segments. *Am. J. Physiol.* 240 (*Renal Fluid Electrolyte Physiol.* 9): F536–F544, 1981.

214. Garty, H. Mechanisms of aldosterone action in tight epithelia. *J. Membr. Biol.* 90: 193–205, 1986.

215. Garty, H. Regulation of Na^+ permeability by aldosterone. *Semin. Nephrol.* 12: 24–29, 1992.

216. Garty, H. Molecular properties of epithelial, amiloride-blockable Na^+ channels. *FASEB J.* 8: 0522–0528, 1994.

217. Garty, H., and C. Asher. Ca^{2+}-induced down-regulation of Na^+ channels in toad bladder epithelium. *J. Biol. Chem.* 261: 7400–7406, 1986.

218. Garty, H., C. Asher, and O. Yeger. Direct inhibition of epithelial Na^+ channels by Ca^{2+} and other divalent cations. *J. Membr. Biol.* 95: 151–162, 1987.

219. Garty, H., E. D. Civan, and M. M. Civan. Effects of internal and external pH on amiloride-blockable Na^+ transport across toad urinary bladder vesicles. *J. Membr. Biol.* 87: 67–75, 1985.

220. Garty, H., and I. S. Edelman. Amiloride-sensitive trypsinization of apical sodium channels. Analysis of hormonal regulation of sodium transport in toad bladder. *J. Gen. Physiol.* 81: 785–803, 1983.

221. Garty, H., O. Yeger, A. Yanovsky, and C. Asher. Guanosine nucleotide-dependent activation of the amiloride-blockable Na^+ channel. *Am. J. Physiol.* 256 (*Renal Fluid Electrolyte Physiol.* 25): F965–F969, 1989.

222. Geering, K., M. Girardet, C. Bron, J.-P. Kraehenbuhl, and B. C. Rossier. Hormonal regulation of (Na^+,K^+)-ATPase biosynthesis in the toad bladder. Effect of aldosterone and 3,5,3′-triiodo-1-thyronine. *J. Biol. Chem.* 257: 10338–10343, 1982.

222a. Gillen, C. M., S. Brill, J. A. Payne, and B. Forbush, III. Molecular cloning and functional expression of the K-Cl co-transporter from rabbit, rat, and human. *J. Biol. Chem.* 271: 16237–16244.

223. Giraldez, F., K. J. Murray, F. V. Sepúlveda, and D. N. Sheppard. Characterization of a phosphorylation-activated Cl^--selective channel in isolated *Necturus* enterocytes. *J. Physiol. (Lond.)* 416: 517–538, 1989.

224. Glick, G. G., F. Ismail-Beigi, and I. S. Edelman. Hormonal regulation of Na,K-ATPase. *Prog. Clin. Biol. Res.* 2688: 277–295, 1988.

225. Gluck, S. L. The structure and biochemistry of the vacuolar H^+ ATPase in proximal and distal urinary acidification. *J. Bioenerg. Biomembr.* 24: 351–358, 1992.

226. Gluck, S., and Q. Al-Awqati. Vasopressin increases water permeability by inducing pores. *Nature* 284: 631–632, 1980.

227. Gluck, S., and J. Caldwell. Immunoaffinity purification and characterization of vacuolar H^+ ATPase from bovine kidney. *J. Biol. Chem.* 262: 15780–15789, 1987.

228. Gluck, S., and J. Caldwell. Proton-translocating ATPase from bovine kidney medulla: partial purification and reconstitution. *Am. J. Physiol.* 254 (*Renal Fluid Electrolyte Physiol.* 23): F71–F79, 1988.

229. Gluck, S., C. Cannon, and Q. Al-Awqati. Exocytosis regulates urinary acidification in turtle bladder by rapid insertion of H^+ pumps into the luminal membrane. *Proc. Natl. Acad. Sci. U.S.A.* 79: 4327–4331, 1982.

230. Gögelein, H. Chloride channels in epithelia. *Biochim. Biophys. Acta* 947: 521–547, 1988.

231. Gögelein, H., and R. Greger. Properties of single K^+ chanels in the basolateral membrane of rabbit proximal straight tubules. *Pflugers Arch.* 410: 288–295, 1987.

232. Goldstein, O., C. Asher, P. Barbryk, E. J. Cragoe, Jr., W. Clauss, and H. Garty. An epithelial high-affinity amiloride binding site, different from the Na^+ channel. *J. Biol. Chem.* 268: 7856–7862, 1993.

233. Goligorsky, M. S., W. Lieberthal, L. Racusen, and E. E. Simon. Integrin receptors in renal tubular epithelium: new insights into pathophysiology of acute renal failure. *Am. J. Physiol.* 264 (*Renal Fluid Electrolyte Physiol.* 33): F1–F8, 1993.

234. González, E., P. Carpi-Medina, H. Linares, and G. Whittembury. Water osmotic permeability of the apical membrane of proximal straight tubular (PST) cells. *Pflugers Arch.* 402: 337–339, 1984.

235. González, E., P. Carpi-Medina, and G. Whittembury. Cell osmotic water permeability of isolated rabbit proximal straight tubules. *Am. J. Physiol.* 242 (*Renal Fluid Electrolyte Physiol.* 11): F321–F330, 1982.

235a. Good, D. G. Hyperosmolality inhibits bicarbonate absorption in rat medullary thick ascending limb via a protein-tyrosine kinase-dependent pathway. *J. Biol. Chem.* 270: 9883–9889, 1995.

235b. Good, D. W., T. George, and B. A. Watts, III. Basolateral membrane Na^+/H^+ exchange enhances HCO_3^- absorption in rat medullary thick ascending limb: Evidence for functional coupling between basolateral and apical membrane Na^+/H^+ exchangers. *Proc. Natl. Acad. Sci. U.S.A.*

236. Gordon, L.G.M., G. Kottra, and E. J. Frömter. Methods to detect, quantify, and minimize edge leaks in Ussing chambers. *Methods Enzymol.* 192: 697–710, 1990.

237. Gottschalk, C. W., W. E. Lassiter, and M. Mylle. Localization of urine acidification in the mammalian kidney. *Am. J. Physiol.* 198: 581–585, 1960.

238. Grantham J. J., and M. B. Burg. Effect of vasopressin and cyclic AMP on permeability of isolated collecting tubules. *Am. J. Physiol.* 211: 255–259, 1966.

239. Grassl, S. M., and P. S. Aronson. Na^+/HCO_3^- co-transport in basolateral membrane vesicles isolated from rabbit renal cortex. *J. Biol. Chem.* 261: 8778–8873, 1986.

240. Green, R., and G. Giebisch. Luminal hypotonicity: A driving force for fluid absorption from the proximal tubule. *Am. J. Physiol.* 246 (*Renal Fluid Electrolyte Physiol.* 15): F167–F174, 1984.

241. Greger, R. Ion transport mechanisms in thick ascending limb of Henle's loop of mammalian nephron. *Physiol. Rev.* 65: 760–797, 1985.

242. Greger, R. Chloride channel blockers. *Methods Enzymol.* 191: 793–810, 1990.

243. Greger, R., M. Bleich, and E. Schlatter. Ion channels in the

thick ascending limb of Henle's loop. *Renal Physiol. Biochem.* 13: 37–50, 1990.

244. Greger, R., M. Bleich, and E. Schlatter. Ion channel regulation in the thick ascending limb of the loop of Henle. *Kidney Int.* 40: S119–S124, 1991.

245. Greger, R., and E. Schlatter. Properties of the basolateral membrane of the cortical thick ascending limb of Henle's loop of rabbit kidney. A model for secondary active chloride transport. *Pflugers Arch.* 396: 325–334, 1983.

246. Grinstein, S., W. Furuya, and L. Bianchini. Protein kinases, phosphatases, and the control of cell volume. *News Physiol. Sci.* 7: 232–237, 1992.

247. Grunder, S., A. Thiemann, M. Pusch, and J. T. Jentsch. Regions involved in the opening of CLC-2 chloride channel by voltage and cell volume. *Nature* 360: 759–762, 1992.

248. Guggino, W. B. Functional heterogeneity in the early distal tubule of the *Amphiuma* kidney: evidence for two modes of Cl^- and K^+ transport across the basolateral cell membrane. *Am. J. Physiol.* 250 (*Renal Fluid Electrolyte Physiol.* 19) F430–F440, 1986.

249. Guggino, W. B. Outwardly rectifying chloride channels and CF: a divorce and remarriage. *J. Bioenerg. Biomembr.* 25: 27–35, 1993.

250. Guggino, W. B., R. London, E. L. Boulpaep, and G. Giebisch. Chloride transport across the basolateral cell membrane of the *Necturus* proximal tubule: Dependence on bicarbonate and sodium. *J. Membr. Biol.* 71: 227–240, 1983.

251. Guinamard, R., A. Chraïbi, and J. Teulon. A small-conductance Cl^- channel in the mouse thick ascending limb that is activated by ATP and protein kinase A. *J. Physiol. (Lond.)* 485: 97–112, 1995.

251a. Gullans, S. R., D. M. Cohen, R. Kojima, J. Randall, B. M. Brenner, B. Santos, and A. Chevaile. Transcriptional responses to tubule challenges. *Kidney Int.* 49: 1578–1681, 1996.

252. Gumbiner, B. M. Breaking through the tight junction barrier. *J. Cell Biol.* 123: 1631–1633, 1993.

253. Gunter-Smith, P. J., E. Grasset, and S. G. Schultz. Sodium-coupled amino acid and sugar transport by *Necturus* small intestine. *J. Membr. Biol.* 66: 25–39, 1982.

254. Gustin, M. C., and D. B. P. Goodman. Characterization of the phosphorylated intermediate of the K^+-ouabain-insensitive ATPase of the rabbit colon brush-border membrane. *J. Biol. Chem.* 257: 9629–9633, 1982.

255. Haas, M. Properties and diversity of (Na-K-Cl) cotransporters. *Annu. Rev. Physiol.* 51: 443–457, 1989.

256. Haas, M. The Na-K-Cl cotransporters. *Am. J. Physiol.* 267 (*Cell Physiol.* 36): C869–C885, 1994.

257. Haas, M., B. Forbush, III. [^{3}H]bumetamide binding to duck red cells. Correlation with inhibition of ($Na^+K^+2Cl^-$) cotransport. *J. Biol. Chem.* 261: 8434–8441, 1986.

257a. Haas, M., D. McBrayer, and C. Lytle. [Cl^-]$_i$-dependent phosphorylation of the Na-K-Cl cotransport protein of dog tracheal epithelial cells. *J. Biol. Chem.* 270: 28955–28961, 1995.

258. Hamill, O. P., and D. W. McBride, Jr. Molecular mechanisms of mechanoreceptor adaptation. *News Physiol. Sci.* 9: 53–59, 1994.

259. Hamilton, K. L., and D. C. Eaton. Single-channel recordings from amiloride-sensitive epihelial sodium channel. *Am. J. Physiol.* 249 (*Cell Physiol.* 18): C200–C207, 1985.

260. Handler, J. S. Antidiuretic hormone moves membranes. *Am. J. Physiol.* 255 (*Renal Fluid Electrolyte Physiol.* 24): F375–F382, 1988.

261. Hanrahan, J. W., W. P. Alles, and S. A. Lewis. Single anion-selective channels in basolateral membrane of a mammalian tight epithelium. *Proc. Natl. Acad. Sci. U.S.A.* 82: 7791–7795, 1985.

262. Harris, H. W., and J. S. Handler. The role of membrane turnover in the water permeability response to antidiuretic hormone. *J. Membr. Biol.* 103: 207–216, 1988.

263. Harvey, B. J. Cellular mechanisms of regulation of ion and water channels and pumps in high-resistance epithelia. In: *Isotonic Transport in Leaky Epithelia,* Alfred Benzon Symposium 34, edited by H. H. Ussing, J. Fischbarg, O. Sten-Knudsen, E. H. Larsen, and N. J. Willumsen. Copenhagen: Munskgaard, 1993, p. 312–332.

264. Hasegawa, H., T. Ma, W. Skach, M. A. Matthay, and A. S. Verkman. Molecular cloning of a mercurial-insensitive water channel expressed in selected water-transporting tissues. *J. Biol. Chem.* 269: 5497–5500, 1994.

265. Hayashi, M., S. Sasaki, H. Tsuganezawa, T. Monkawa, W. Kitajima, K., Konishi, K. Fushimi, F. Marumo, and T. Saruta. Expression and distribution of aquaporin of collecting duct are regulated by vasopressin V_2 receptor in rat kidney. *J. Clin. Invest.* 94: 1778–1783, 1994.

266. Hays, R. M., N. Franki, H. Simon, and Y. Gao. Antidiuretic hormone and exocytosis: lessons from neurosecretion. *Am. J. Physiol.* 267 (*Cell Physiol.* 36): C1507–C1524, 1994.

267. Hays, S. R., and R. J. Alpern. Basolateral membrane Na^+-independent Cl^-/HCO_3^- exchange in the inner stripe of the outer medullary collecting tubule. *J. Gen. Physiol.* 95: 347–368, 1990.

268. Hebert, S. C. Hypertonic cell volume regulation in mouse thick limbs. I. ADH dependency and nephron heterogeneity. *Am. J. Physiol.* 250 (*Cell Physiol.* 19): C907–C919, 1986.

269. Hebert, S. C., and T. E. Andreoli. Interactions of temperature and ADH on transport in cortical collecting tubules. *Am. J. Physiol.* 238 (*Renal Fluid Electrolyte Physiol.* 9): F470–F480, 1980.

270. Hebert, S. C., and T. E. Andreoli. Water transport and osmoregulation by terminal nephron segments. In: *The Kidney: Physiology and Pathophysiology,* edited by D. W. Seldin and G. Giebisch. New York: Raven, 1985, p. 933–949.

271. Hebert, S. C., and K. Ho. Structure and functional properties of an inwardly rectifying ATP-regulated K^+ channel from rat kidney. *Renal Physiol. Biochem.* 17: 143–147, 1994.

272. Hediger, M. A., M. J. Coady, T. S. Ikeda, and E. M. Wright. Expression cloning and cDNA sequencing of the Na^+/glucose cotransporter. *Nature* 330: 379–381, 1987.

273. Hediger, M. A., Y. Kanai, G. You, and S. Nussberger. Mammalian ion-coupled solute transporters. *J. Physiol. (Lond.)* 482: 7S–17S, 1995.

274. Hediger, M. A., and D. B. Rhoads. Molecular physiology of sodium-glucose cotransporters. *Physiol. Rev.* 74: 993–1026, 1994.

275. Heintze, K., K.-U. Petersen, P. Olles, S. H. Saverymuttu, and J. R. Wood. Effects of bicarbonate on fluid and electrolyte transport by the guinea pig gallbladder: a bicarbonate-chloride exchange. *J. Membr. Biol.* 45: 43–59, 1979.

276. Helman, S. I., and N. L. Kizer. Apical sodium ion channels of tight epithelia as viewed from the perspective of noise analysis. *Curr. Top. Membr. Transp.* 37: 117–155, 1990.

277. Helman, S. I., and D. A. Miller. Edge damage effect on measurements of urea and sodium flux in frog skin. *Am. J. Physiol.* 226: 1198–1203, 1974.

278. Helman, S. I., and S. M. Thompson. Interpretation and use of electrical equivalent circuit in studies of epithelial tissues. *Am.*

J. Physiol. 243 (*Renal Fluid Electrolyte Physiol.* 12): F519–F531, 1982.

279. Heming, T. A., J. Copello, and L. Reuss. Regulation of cAMP-activated apical membrane chloride conductance in gallbladder epithelium. *J. Gen. Physiol.* 103: 1–18, 1994.

280. Hersey, S. J., and G. Sachs. Gastric acid secretion. *Physiol. Rev.* 75: 155–189, 1995.

281. Hill, A. E. Solute-solvent coupling in epithelia: a critical examination of the standing-gradient osmotic flow theory. *Proc. R. Soc. Lond. (Biol.)* 190: 99–114, 1975.

282. Hille, B. Ionic channels of excitable membranes. Sunderland: Sinauer, 1992.

283. Hirayama, B. A., C. D. Smith and E. M. Wright. Secondary structure of the Na$^+$/glucose cotransporter. *J. Gen. Physiol.* 100: 19–20 (Abst.), 1992.

284. Hirsch, J., and E. Schlatter. K$^+$ channels in the basolateral membrane of rat cortical collecting duct. *Pflugers Arch.* 424: 470–477, 1993.

285. Hirsch, J., and E. Schlatter. K$^+$ channels in the basolateral membrane of rat cortical collecting duct are regulated by a cGMP-dependenet protein kinase. *Pflugers Arch.* 429: 338–344, 1995.

286. Ho, K., C. G. Nichols, W. J. Lederer, J. Lytton, P. M. Vassiley, M. V. Kanazirska, and S. C. Hebert. Cloning and expression of an inwardly rectifying ATP-regulated potassium channel. *Nature* 362: 31–38, 1993.

287. Hong, K., and M. Driscoll. A transmembrane domain of the putative channel subunit MEC-4 influences mechanotransduction and neurodegeneration in *C. elegans. Nature* 367: 470–473, 1994.

288. Hoshi, T., and F. Sakai. A comparison of the electrical resistances of the surface cell membrane and cellular wall in the proximal tubule of the newt kidney. *Jpn. J. Physiol.* 17: 627–637, 1967.

289. House, C. R. *Water Transport in Cells and Tissues.* London: Arnold, 1974.

290. Huang, M., and M. Chalfie. Gene interactions affecting mechanosensory transduction in *Caenorhabditis elegans. Nature* 367: 467–470, 1994.

290a. Huang, H., H. St.-Jean, M. J. Coady, J.-Y. Lapointe. Evidence for coupling between Na$^+$ pump activity and TEA-sensitive K$^+$ currents in *Xenopus laevis* occytes. *J. Membr. Biol.* 143: 29–35, 1995.

291. Huf, E. G., N. S. Doss, and J. P. Wills. Effects of metabolic inhibitors and drugs on ion transport and oxygen consumption in isolated frog skin. *J. Gen. Physiol.* 41: 397–417, 1957.

291a. Humphreys, B.D.L. Jiang, M. N. Chernova, and S. L. Alper. Hypertonic activation of AE3 anion exchanger in *Xenopus* oocytes via NHE-mediated intracellular alkalinization. *Am. J. Physiol.* 268 (*Cell Physiol.* 37): C201–C209, 1995.

292. Hunter, M., K. Kawahara, and G. Giebisch. Calcium-activated epithelial potassium channels. *Miner. Electrolyte Metab.* 14: 48–57, 1988.

293. Hurst, A. M., M. Duplain, and J.-Y. Lapointe. Basolateral membrane potassium channels in rabbit cortical thick ascending limb. *Am. J. Physiol.* 263 (*Renal Fluid Electrolyte Physiol.* 32): F262–F267, 1992.

294. Hwang, T. C., L. Lu, P. L. Zeitlin, D. C. Gruenert, R. Huganir, and W. B. Guggino. Cl$^-$ channels in CF: lack of activation by protein kinase C and cAMP-dependent protein kinase. *Science* 244: 1351–1353, 1989.

295. Hwang, T. C., G. Nagel, A. C. Nairn, and D. C. Gadsby. Regulation of the gating of cystic fibrosis transmembrane conductance regulator Cl channels by phosphorylation and ATP

296. Ikeda, T. S., E.-S. Hwang, M. J. Coady, B. A. Hirayama, M. A. Hediger, and E. M. Wright. Characterization of a Na$^+$/glucose cotransporter cloned from rabbit intestine. *J. Membr. Biol.* 110: 87–95, 1989.

296a. Inagaki, N., T. Gonoi, J. P. Clement, IV, N. Namba, J. Inazawa, G. González, L. Aguilar-Bryan, S. Seinio, and J. Bryan. Reconstitution of I$_{KATP}$: an inward rectifier subunit plus the sulfonylurea receptor. *Science* 270: 1166–1170, 1995.

297. Ishibashi, K., F. C. Rector, and C. A. Berry. Role of Na-dependent Cl/HCO$_3$ exchange in basolateral Cl transport of rabbit proximal tubules. *Am. J. Physiol.* 264 (*Renal Fluid Electrolyte Physiol.* 33): F251–F258, 1993.

298. Ishibashi, K., S. Sasaki, K. Fushimi, S. Uchida, K. Nakajima, Y. Yamaguchi, T. Gojobori T, and F. Marumo. Molecular cloning and expression of a new member of the aquaporin family (AQP3) with permeability to glycerol and urea in addition to water expressed at the basolateral membrane of kidney collecting duct cells. *Proc. Natl. Acad. Sci. U.S.A.* 91: 6269–6273, 1994.

298a. Ismailov, I. I. and D. J. Benos. Effects of phosphorylation on ion channel function. *Kidney Int.* 48: 1167–1179, 1995.

299. Jennings, M. L., and R. K. Schulz. Swelling-activated KCl cotransport in rabbit red cells: flux is determined mainly by cell volume rather than shape. *Am. J. Physiol.* 259 (*Cell Physiol.* 28): C960–C967, 1990.

300. Jennings, M. L., and R. K. Schulz. Okadaic acid inhibition of KCl cotransport. Evidence that protein dephosphorylation is necessary for activation of transport by either cell swelling or N-ethylmaleimide. *J. Gen. Physiol.* 97: 799–818, 1991.

301. Jentsch, T. J. Molecular physiology of anion channels. *Curr. Opin. Cell Biol.* 6: 600–606, 1994.

302. Jentsch, T. J., P. Schwartz, B. S. Schill, B. Langer, A. P. Lepple, S. K. Keller, and M. Wiederholt. Kinetic properties of the sodium bicarbonate (carbonate) symport in monkey kidney epithelial cells (BSC-1). *J. Biol. Chem.* 261: 10673–10679, 1986.

303. Jovov, B., I. I. Ismailov, and D. J. Benos. Cystic fibrosis transmembrane conductance regulator is required for protein kinase A activation of an outwardly rectified anion channel purified from bovine tracheal epithelia. *J. Biol. Chem.* 270: 1521–1528, 1995.

304. Jung, J. S., G. M. Preston, B. L. Smith, W. B. Guggino, and P. Agre. Molecular structure of the water channel through aquaporin CHIP: the tetrameric-hourglass model. *J. Biol. Chem.* 269: 14648–14654, 1994.

304a. Kandasamy, R. A., and J. Orlowski. Genomic organization and glucocorticoid transcriptional activation of the rat Na$^+$/H$^+$ exchanger Nhe3 gene. *J. Biol. Chem.* 271: 10551–10559, 1996.

304b. Kandasamy, R. A., F. H. Yu, R. Harris, A. Boucher, J. W. Hanrahan, and J. Orlowski. Plasma membrane Na$^+$/H$^+$ exchanger isoforms (NHE-1, -2, and -3) are differentially responsive to second messenger agonists of the protein kinase A and C pathways. *J. Biol. Chem.* 270: 9209–29216, 1995.

305. Kanai, Y., W. S. Lee, G. You, D. Brown, and M. A. Hediger. The human kidney low affinity Na$^+$/glucose cotransporter SGLT2. Delineation of the major renal reabsorptive mechanism for D-glucose. *J. Clin. Invest.* 93: 397–404, 1994.

305a. Kaplan, M. R., M. D. Plotkin, W.-S. Lee, Z.-C. Xu, J. Lytton, and S. C. Hebert. Apical localization of the Na-K-Cl cotransporter, rBSCl, on rat thick ascending limbs. *Kidney Int.* 49: 40–47, 1996.

306. Karniski, L. P., and P. S. Aronson. Formate: a critical interme-

diate for chloride transport in the proximal tubule. *News Physiol. Sci.* 2: 160–164, 1987.

307. Karsny, E. J., J. L. Madara, D. R. DiBona, and R. A. Frizzell. Cyclic AMP regulates tight junction permselectivity in flounder intestine. *Federation Proc.* 42: 1100–1108, 1983.

308. Kasai, H., and G. J. Augustine. Cytosolic Ca^{2+} gradients triggering unidirectional fluid secretion from exocrine pancreas. *Nature* 348: 735–738, 1990.

309. Kaunitz, J. D., and G. Sachs. Identification of a vanadate-sensitive potassium-dependent proton pump from rabbit colon. *J. Biol. Chem.* 261: 14005–14010, 1986.

310. Kawasaki, M., S. Uchida, T. Monkawa, A. Miyawaki, K. Mikoshiba, F. Marumo, and S. Sasaki. Cloning and expression of a protein kinase C-regulated chloride channel abundantly expressed in rat brain neuronal cells. *Neuron* 12: 597–604, 1994.

311. Kemendy, A. E., T. R. Kleyman, and D. C. Eaton. Aldosterone alters the open probability of amiloride-blockable sodium channels in A6 epithelia. *Am. J. Physiol.* 263 (*Cell Physiol.* 32): C825–C837, 1992.

312. Kennelly, P. J., and E. G. Krebs. Consensus sequences as substrate specificity determinants for protein kinases and protein phosphatases. *J. Biol. Chem.* 266: 15555–15558, 1991.

313. Kerem, B., J. M. Rommens, J. A. Buchanan, D. Markiewicz, T. K. Cox, A. Chakravarti, M. Buchwald, and L.-C. Tsui. Identification of the cystic fibrosis gene: genetic analysis. *Science* 245: 1073–1080, 1989.

314. Kieferle, S., P. Fong, M. Bens, A. Vandewalle, and T. J. Jentsch. Two highly homologous members of the ClC chloride channel family in both rat and human kidney. *Proc. Natl. Acad. Sci. U.S.A.* 91: 6943–6947, 1994.

315. King-Hele, J. A. Approximate analytical solutions for water and solute flow in intercellular spaces with a leaky tight junction. *J. Theor. Biol.* 80: 451–465, 1979.

316. Kirtane, A., N. Ismail-Beigi, and F. Ismail-Beigi. Role of enhanced Na^+ entry in the control of Na,K-ATPase gene expression by serum. *J. Membr. Biol.* 137: 9–15, 1994.

317. Kishimoto, A., N. Kajikawa, M. Shiota, and Y. Nishizuka. Proteolytic activation of calcium-activated, phospholipid-dependent protein kinase by calcium-dependent neutral protease. *J. Biol. Chem.* 258: 1156–1164, 1983.

318. Kishimoto, A., K. Mikawa, K. Hashimoto, I. Yasuda, S. Tanaka, M. Tominaga, T. Kuroda, and Y. Nishizuka. Limited proteolysis of protein kinase C subspecies by calcium-dependent neutral protease (calpain). *J. Biol. Chem.* 264: 4088–4092, 1989.

319. Klaassen, C.H.W., and J.J.H.H.M. De Pont. Gastric H^+/K^+-ATPase. *Cell Physiol. Biochem.* 4: 115–134, 1994.

320. Klaerke, D. A., and P. L. Jorgensen. Role of Ca^{2+}-activated K^+ channel in regulation of NaCl reabsorption in thick ascending limb of Henle's loop. *Comp. Biochem. Physiol.* [A] 904: 757–765, 1988.

321. Klaerke, D. A., H. Wiener, T. Zeuthen, and P. L. Jorgensen. Ca^{2+} activation and pH dependence of a maxi K^+ channel from rabbit distal colon epithelium. *J. Membr. Biol.* 136: 9–21, 1993.

322. Klanke, C. A., Y. R. Su, D. F. Callen, Z. Wang, P. Meneton, N. Baird, R. A. Kandasamy, J. Orlowski, B. E. Otterud, M. Leppert, G. E. Shull, and A. G. Menon. Molecular cloning and physical and genetic mapping of a novel human Na/H exchanger (NHE5/SLC9A5) to chromosome 16q22.1. *Genomics* 25: 615–622, 1995.

323. Kleyman, T. R., and E. J. Cragoe, Jr. Cation transport probes: the amiloride series. *Methods Enzymol.* 191: 739–755, 1990.

324. Knepper, M. A. The aquaporin family of molecular water channels. *Proc. Natl. Acad. Sci. U.S.A.* 91: 6255–6258, 1994.

325. Knepper, M. A., S. P. Lankford, and Y. Terada. Renal tubular actions of ANF. *Can. J. Physiol. Pharmacol.* 69: 1537–1545, 1991.

326. Knepper, M. A., and R. A. Star. The vasopressin-regulated urea transporter in renal inner medullary collecting duct. *Am. J. Physiol.* 259 (*Renal Fluid Electrolyte Physiol.* 28): F393–F401, 1990.

327. Koo, Y. H., and P. L. Pedersen. The first nucleotide binding fold of the cystic fibrosis transmembrane conductance regulator can function as an active ATPase. *J. Biol. Chem.* 270: 22093–22096, 1995.

328. Köckerling, A., and M. Fromm. Origin of cAMP-dependent Cl^- secretion from both crypts and surface epithelial of rat intestine. *Am. J. Physiol.* 264 (*Cell Physiol.* 33): C1294–C1301, 1993.

329. Koefoed-Johnson, V. The effect of g-strophanthin (ouabain) on the active transport of sodium through the isolated frog skin. *Acta Physiol. Scand.* 42 (Suppl. 145): 87–88, 1957.

330. Koefoed-Johnson, V., and H. H. Ussing. The nature of the frog skin potential. *Acta Physiol. Scand.* 42: 298–308, 1958.

331. Koepsell, H., and M. Veyh. Structure of Na$^+$-D-glucose cotransport system. *Cell Physiol. Biochem.* 4: 206–216, 1994.

332. Kofuji, P., W. J. Lederer, and D. H. Schulze. Na/Ca exchanger isoforms expressed in kidney. *Am. J. Physiol.* 265 (*Renal Fluid Electrolyte Physiol.* 34): F598–F603, 1993.

333. Kong, C.-T., S.-F. Yet, and J. E. Lever. Cloning and expression of a mammalian Na^+/amino acid cotransporter with sequence similarity to Na^+/glucose cotransporters. *J. Biol. Chem.* 268: 1509–1512, 1993.

334. Kopito R. R., M. M. Andersson, D. A. Herzlinger, Q. Al-Awqati, and H. F. Lodish. Structure and tissue-specific expression of the mouse anion-exchanger gene in erythroid and renal cells. In: *Cell Physiology of Blood*, edited by R. B. Gunn and J. C. Parker. New York: Rockefeller, 1988, p. 151–162.

335. Kopito, R. R., and H. F. Lodish. Primary structure and transmembrane orientation of the murine anion exchange protein. *Nature* 316: 234–238, 1985.

336. Kottra, G. Calcium is not involved in the cAMP-mediated stimulation of Cl^- conductance in the apical membrane of *Necturus* gallbladder epithelium. *Pflugers Arch.* 429: 647–658, 1995.

337. Kottra, G., W. Haase, and E. Frömter. Tight-junction tightness of *Necturus* gall bladder epithelium is not regulated by cAMP or intracellular Ca^{2+}. I. Microscopic and general electrophysiological observations. *Pflugers Arch.* 425: 528–534, 1993.

338. Kottra, G., and E. Frömter. Tight-junction tightness of *Necturus* gall bladder epithelium is not regulated by cAMP or intracellular Ca^{2+}. II. Impedance measurements. *Pflugers Arch.* 425: 535–545, 1993.

339. Krapf, R. Basolateral membrane $H/OH/HCO_3$ transport in the rat cortical thick ascending limb. *J. Clin. Invest.* 82: 234–241, 1988.

340. Krapivinsky, G. B., M. J. Ackerman, E. A. Gordon, L. D. Krapivinsky, and D. E. Clapham. Molecular characterization of a swelling-induced chloride conductance regulatory protein, pl_{Cln}. *Cell* 76: 439–448, 1994.

341. Kraut, J. A., F. Starr, G. Sachs, and M. Reuben. Expression of gastric and colonic H^+-K^+-ATPase in the rat kidney. *Am. J. Physiol.* 268: (*Renal Fluid Electrolyte Physiol.* 37): F581–F587, 1995.

341a. Kristensen, P., and H. H. Ussing. Epithelial organization. In: *The Kidney. Physiology and Pathophysiology*, Chapter 10, 2nd edition, edited by D. W. Seldin, and G. Giebisch. New York: Raven Press, p. 265–285, 1992.

342. Kudrycki, K. D., and G. E. Shull. Primary structure of rat kidney band 3 anion exchange protein deduced from a cDNA. *J. Biol. Chem.* 264: 8185–8192, 1989.

343. Kuijpers, G.A.J., and J.J.H.H.M. De Pont. Role of proton and bicarbonate transport in pancreatic cell function. *Annu. Rev. Physiol.* 49: 87–104, 1987.

344. Kunzelmann, K., H. Pavenstadt, C. Beck, O. Unal, P. Emmrich, H. J. Arndt, and R. Greger. Characterization of potassium channels in respiratory cells. I. General Properties. *Pflugers Arch.* 414: 291–296, 1989.

345. La, B.-Q., S. L. Carosi, J. Valentich, S. Shenolikar, and S. C. Sansom. Regulation of epithelial chloride channels by protein phosphatase. *Am. J. Physiol.* 260 (*Cell Physiol.* 29): C1217–C1223, 1991.

346. Lang, F., F. Friedrich, M. Paulmichl, W. Schobersberger, A. Jungwirth, M. Ritter, M. Steidl, H. Weiss, E. Wöll, E. Tschernko, R. Paulmichl, and C. Hallbrucker. Ion channels in Madin-Darby canine kidney cells. *Renal Physiol. Biochem.* 13: 82–93, 1990.

347. Lang, F., G. Messner, and W. Rehwald. Electrophysiology of sodium-coupled transport in proximal renal tubules. *Am. J. Physiol.* 250 (*Renal Fluid Electrolyte Physiol.* 19): F953–962, 1986.

348. Lang, F., G. Messner, W. Wang, M. Paulmichl, H. Oberleithner, and P. Deetjen. The influence of intracellular sodium activity on the transport of glucose in proximal tubule of frog kidney. *Pflugers Arch.* 401: 14–21, 1984.

349. Lapointe, J., L. Garneau, P. D. Bell, and J. Cardinal. Membrane crosstalk in the mammalian proximal tubule during alterations in transepithelial sodium transport. *Am. J. Physiol.* 258 (*Renal Fluid Electrolyte Physiol.* 27): F339–F345, 1990.

350. Lau, K. R., R. L. Hudson, and S. G. Schultz. Cell swelling induces a barium-inhibitable potassium conductance in the basolateral membrane of *Necturus* small intestine. *Proc. Natl. Acad. Sci. U.S.A.* 81: 3591–3594, 1984.

351. Lauf, P. K., J. Bauer, N. C. Adragna, H. Fujise, A.M.M. Zade-Oppen, K. H. Ryu, and E. Delpire. Erythrocyte K-Cl cotransport: proteins and regulation. *Am. J. Physiol.* 263 (*Cell Physiol.* 32): C917–C932, 1992.

352. Läuger, P. *Electrogenic Ion Pumps*. Sunderland, MA: Sinauer, 1991.

353. Law, P. Y., and I. S. Edelman. Induction of citrate synthase by aldosterone in the rat kidney. *J. Membr. Biol.* 41: 41–64, 1978.

354. Leaf, A. From toad bladder to kidney. *Am. J. Physiol.* 242 (*Renal Fluid Electrolyte Physiol.* 11): F103–F111, 1982.

355. Leblanc, G., and F. Morel. Na and K movements across the membranes of frog skin epithelia associated with transient current changes. *Pflugers Arch.* 358: 159–177, 1975.

356. Lee, M. D., K. Y. Bhakta, S. Raina, R. Yonescu, C. A. Griffin, N. G. Copeland, D. J. Gilbert, N. A. Jenkins, G. M. Preston, and P. Agre. The human aquaporin-5 gene. *J. Biol. Chem.* 271: 8599–8604, 1996.

357. Lee, W.-S., and S. C. Hebert. ROMK inwardly rectifying ATP-sensitive K+ channel. I. Expression in rat distal nephron segments. *Am. J. Physiol.* 268 (*Renal Fluid Electrolyte Physiol.* 37): F1124–F1131, 1995.

358. Lencer, W. I., D. Brown, D. A. Ausiello, and A. S. Verkman. Endocytosis of water channels in rat kidney: cell specificity and correlation with in vivo antidiuretic states. *Am. J. Physiol.* 259 (*Cell Physiol.* 28): C920–C932, 1990.

359. Lencer, W. I., A. S. Verkman, D. A. Ausiello, A. Arnaout, and D. Brown. Endocytic vesicles from renal papilla which retrieve the vasopressin-sensitive water channel do not contain an H+ ATPase. *J. Cell Biol.* 111: 379–389, 1990.

360. Lencer, W. I., P. Weyer, A. S. Verkman, D. A. Ausiello, and D. Brown. FITC-dextran as a probe for endosome function and localization in kidney. *Am. J. Physiol.* 258 (*Cell Physiol.* 27): C309–C317, 1990.

361. Lescale-Matys, L., D. S. Putnam, and A. A. McDonough. Na+-K+-ATPase α_1- and β_1-subunit degradation: evidence for multiple subunit specific rates. *Am. J. Physiol.* 264 (*Cell Physiol.* 33): C583–C590, 1993.

362. Leviel, F., P. Borensztein, P. Houillier, M. Paillard, and M. Bichara. Electroneutral K+/HCO$_3^-$ cotransport in cells of medullary thick ascending limb of rat kidney. *J. Clin. Invest.* 90: 869–878, 1992.

363. Levine, S., N. Franki, and R. M. Hays. Effect of phloretin on water and solute movement in the toad bladder. *J. Clin. Invest.* 52: 1435–1442, 1973.

364. Levine, S. A., M. H. Montrose, C. M. Tse, and M. Donowitz. Kinetics and regulation of three cloned mammalian Na+/H+ exchangers stably expressed in a fibroblast cell line. *J. Biol. Chem.* 268: 25527–25535, 1993.

365. Levine, S. D., M. Jacoby, and A. Finkelstein. The water permeability of toad urinary bladder. I. Permeability of barriers in series with the luminal membrane. *J. Gen. Physiol.* 83: 529–542, 1984.

366. Levine, S. D., M. Jacoby, and A. Finkelstein. The water permeability of toad urinary bladder. II. The value of Pf/Pd(w) for the antidiuretic hormone-induced water permeation pathway. *J. Gen. Physiol.* 83: 543–562, 1984.

367. Levine, S. D., R. D. Levine, R. E. Worthington, and R. M. Hays. Selective inhibition of osmotic water flow by general anesthetics in toad urinary bladder. *J. Clin. Invest.* 58: 980–988, 1976.

368. Lewis, S. A. A reinvestigation of the function of the mammalian urinary bladder. *Am. J. Physiol.* 232 (*Renal Fluid Electrolyte Physiol.* 3): F187–F195, 1977.

369. Lewis, S. A. The mammalian urinary bladder: it's more than accommodating. *News Physiol. Sci.* 1: 61–65, 1986.

370. Lewis, S. A., and W. P. Alles. Urinary kallikrein: a physiological regulator of epithelial Na+ absorption. *Proc. Natl. Acad. Sci. U.S.A.* 83: 5345–5348, 1986.

371. Lewis, S. A., and C. Clausen. Urinary proteases degrade epithelial sodium channels. *J. Membr. Biol.* 122: 77–88, 1991.

372. Lewis, S. A., and P. Donaldson. Ion channels and cell volume regulation: chaos in an organized system. *News Physiol. Sci.* 5: 112–119, 1990.

373. Lewis, S. A., D. C. Eaton, and J. M. Diamond. The mechanism of Na+ transport by rabbit urinary bladder. *J. Membr. Biol.* 28: 41–70, 1976.

374. Lewis, S. A., and J. W. Hanrahan. Physiological approaches for studying mammalian urinary bladder epithelium. *Methods Enzymol.* 192: 632–650, 1990.

374a. Li, C., M. Ramjeesingh, and C. E. Bear. Purified cystic fibrosis transmembrane conductance regulator (CFTR) does not function as an ATP channel. *J. Biol. Chem.* 271: 11623–11626, 1996.

375. Liedtke, C. M., and U. Hopfer. Mechanism of Cl− translocation across small intestinal brush-border membrane. I. Absence of Na+-Cl− cotransport. *Am. J. Physiol.* 242 (*Gastrointest. Liver Physiol.* 5): G263–G271, 1982.

376. Liedtke, C. M., and U. Hopfer. Mechanism of Cl− translocation across small intestinal brush-border membrane. II. Demonstration of Cl−-OH− exchange and Cl− conductance. *Am. J.*

Physiol. 242 (*Gastrointest. Liver Physiol.* 5): G272–G280, 1982.

377. Light, D. B., D. A. Ausiello, and B. A. Stanton. Guanine nucleotide-binding protein, αi-3, directly activates a cation channel in rat renal inner medullary collecting duct cells. *J. Clin. Invest.* 84: 352–356, 1989.

378. Light, D. B., J. D. Corbin, and B. A. Stanton. Dual ion-channel regulation by cyclic GMP and cyclic GMP-dependent protein kinase. *Nature* 344: 336–339, 1990.

379. Light, D. B., E. M. Schwiebert, K. H. Karlson, and B. A. Stanton. Atrial natriuretic peptide inhibits a cation channel in renal inner medullary collecting duct cells. *Science* 243: 383–385, 1989.

380. Lindemann, B. Fluctuation analysis of sodium channels in epithelia. *Annu. Rev. Physiol.* 46: 497–515, 1984.

381. Ling, B. N., and D. C. Eaton. Effects of luminal Na^+ on single Na^+ channels in A6 cells, a regulatory role of protein kinase C. *Am. J. Physiol.* 256 (*Renal Fluid Electrolyte Physiol.* 25): F1094–F1103, 1989.

382. Lingrel, J. B., and T. Kuntzweiler. Na^+,K^+-ATPase. *J. Biol. Chem.* 269: 19659–19662, 1994.

383. Lingueglia, E., N. Voilley, R. Waldmann, M. Lazdunski, and P. Barbry. Expression cloning of an epithelial amiloride-sensitive Na^+ channel—a new channel type with homologies to *Caenorhabditis elegans* degenerins. *FEBS Lett.* 318: 95–99, 1993.

384. Loo, D. D., S. A. Lewis, M. S. Ifshin, and J. M. Diamond. Turnover, membrane insertion, and degradation of sodium channels in rabbit urinary bladder. *Science* 221: 1288–1290, 1983.

384a. Lu, M., and W.-H. Wang. Nitric oxide regulates the low-conductance K^+ channel in basolateral membrane of cortical collecting duct. *Am. J. Physiol.* 270 (*Cell Physiol.* 39): C1336–C134, 1996.

385. Lukacs, G. L., X. B. Chang, C. Bear, N. Kartner, A. Mohamed, J. R. Riordan, and S. Grinstein. The ΙF508 mutation decreases the stability of cystic fibrosis transmembrane conductance regulator in the plasma membrane. *J. Biol. Chem.* 268: 21592–21598, 1994.

386. Lytle, C., and B. Forbush, III. Na-K-Cl co-transport in the shark rectal gland. II. Regulation in isolated tubules. *Am. J. Physiol.* 262 (*Cell Physiol.* 31): C1009–C1017, 1992.

387. Lytle, C., and B. Forbush, III. The Na-K-Cl cotransport-protein of shark rectal gland. II. Regulation by direct phosphorylation. *J. Biol. Chem.* 267: 25438–25443, 1992.

388. Lytle, C., and B. Forbush, III. Regulatory phosphorylation of the secretory Na-K-Cl cotransporter: modulation by cytoplasmic Cl. *Am. J. Physiol.* 270 (*Cell Physiol.* 39): C437–C448, 1996

389. Lytle, C., and T. J. McManus. A minimal kinetic model of (Na+K+2Cl) cotransport with ordered binding and glide symmetry. *J. Gen. Physiol.* 88: 36a (Abstr.), 1986.

390. Lytle, C., J.-C. Xu, D. Biemesderfer, M. Haas, and B. Forbush, III. The Na-K-Cl cotransport protein of shark rectal gland. I. Development of monoclonal antibodies, immunoaffinity purification, and partial biochemical characterization. *J. Biol. Chem.* 267: 25428–25437, 1992.

390a. Ma, H. and B. N. Ling. Luminal adenosine receptors regulate amiloride-sensitive Na^+ channels in A6 distal nephron cells. *Am. J. Physiol.* 270 (*Renal Fluid Electrolyte Physiol.* 39): F798–F805, 1996.

391. Macey, R. I., and R.E.I. Farmer. Inhibition of water and solute permeability in human red cells. *Biochim. Biophys. Acta* 211: 104–106, 1970.

392. Machen, T. E., D. Erlij, and F.B.P. Wooding. Permeable junctional complexes. The movement of lanthanum across rabbit gallbladder and intestine. *J. Cell Biol.* 54: 302–312, 1972.

393. Macknight, A.D.C. Ion and water transport in toad urinary epithelia. In: *Handbook of Physiology: Renal Physiology,* edited by E. E. Windhager. New York: Oxford University Press for the American Physiological Society, 1992, sect. 8, vol. 1, chapt. 8, p. 271–320.

394. Macknight, A.D.C., and A. Leaf. Regulation of cellular volume. *Physiol. Rev.* 57: 510–573, 1977.

395. MacLeod, R. J., and J. R. Hamilton. Separate K^+ and Cl^- transport pathways are activated for regulatory volume decrease in jejunal villus cells. *Am. J. Physiol.* 260 (*Gastrointest. Liver Physiol.* 23): G405–G415, 1991.

396. Macri, P., S. Breton, J. S. Beck, J. Cardinal, and R. Laprade. Basolateral K^+, Cl^- and HCO_3^- conductances and cell volume regulation in rabbit PCT. *Am. J. Physiol.* 264 (*Renal Fluid Electrolyte Physiol.* 33): F365–F374, 1993.

397. MacRobbie, E.A.C., and H. H. Ussing. Osmotic behavior of the epithelial cells of frog skin. *Acta Physiol. Scand.* 53: 348–365, 1961.

398. Madara, J. M. Functional morphology of epithelium of the small intestine. In: *Handbook of Physiology. The Gastrointestinal System. Intestinal Absorption and Secretion,* edited by S. G. Schultz, M. Field and R. A. Frizzell. Bethesda, MD Am. Physiol. Soc., 1991, sect. 6, vol. IV, chapt. 3, p. 83–120.

399. Madara, J. L., and J. R. Pappenheimer. Structural basis for physiological regulation of paracellular pathways in intestinal epithelia. *J. Membr. Biol.* 100: 149–164, 1987.

400. Madsen, K. M., and C. C. Tisher. Structure-function relationships in H^+-secreting epithelia. *Federation Proc.* 44: 2704–2709, 1985.

401. Mahraoui, L., M. Rousset, E. Dussaulx, A. Darmoul, A. Zweibaum, and E. Brot-Laroche. Expression and localization of GLUT5 in Caco-2 cells in human small intestine and colon. *Am. J. Physiol.* 263 (*Gastrointest. Liver Physiol.* 26): G312–G318, 1993.

402. Malinowska, D. H., E. Y. Kupert, A. Bahinshi, A. M. Sherry, and J. Cuppoletti. Cloning, functional expression, and characterization of a PKA-activated gastric Cl^- channel. *Am. J. Physiol.* 268 (*Cell Physiol.* 37): C191–C200, 1995.

403. Martin, D. W. The effect of the bicarbonate ion on the gallbladder salt pump. *J. Membr. Biol.* 18: 219–230, 1974.

404. Marunaka, Y., and D. C. Eaton. Effects of insulin and phosphatase on a Ca^{2+}-dependent Cl-channel in a distal nephron cell line (A6). *J. Gen. Physiol.* 95: 773–789, 1990.

405. Marunaka, Y., and D. C. Eaton. Effects of vasopressin and cAMP on single amiloride-blockable Na channels. *Am. J. Physiol.* 260 (*Cell Physiol.* 29): C1071–C1084, 1991.

406. Maruyama, T., and T. Hoshi. The effect of D-glucose on the electrical potential profile across the proximal tubule of newt kidney. *Biochim. Biophys. Acta* 282: 214–225, 1972.

407. Marver, D. Evidence of corticosteroid action along the nephron. *Am. J. Physiol.* 246 (*Renal Fluid Electrolyte Physiol.* 15): F111–F123, 1984.

408. Mathews, P. M., D. Claeys, F. Jaisser, K. Geering, J.-D. Horisberger, J.-P. Kraehenbuhl, and B. C. Rossier. Primary structure and functional expression of the mouse and frog α-subunit of the gastric $H^+ - K^+$-APase. *Am. J. Physiol.* 268 (*Cell Physiol.* 37): C1207–C1214, 1995.

409. Matlin, K. S., and M. J. Caplan. Epithelial cell structure and polarity. In: *The Kidney: Physiology and Pathophysiology,* edited by D. W. Seldin and G. Giebisch. New York: Raven, 1992, p. 447–473.

410. Matthews, J. B., C. S. Awtrey, and J. L. Madara. Microfilament-dependent activation of Na$^+$/K$^+$/2Cl$^-$ cotransport by cAMP in intestinal epithelial monolayers. *J. Clin. Invest.* 90: 1608–1613, 1992.

410a. Mauchamp, J., A. Margotat, M. Chambard, B. Charrier, L. Remy, and M. Michel-Bechet. 1979. Polarity of three dimensional structures derived from isolated hog thyroid cells in primary culture. *Cell Tissue Res.* 204: 417–430, 1979.

411. Maunsbach, A. E. Cellular mechanism of tubular protein transport. *Int. Rev. Physiol.* 2: 145–167, 1976.

412. Maunsbach, A. B., and E. L. Boulpaep. Quantitative ultrastructure and functional correlates in proximal tubule of *Ambystoma* and *Necturus. Am. J. Physiol.* 246 (*Renal Fluid Electrolyte Physiol.* 15): F710–F724, 1984.

413. Maurel, C., J. Reizer, J. I. Schroeder, M. J. Chrispeels. The vacuolar membrane protein γ-TIP creates water specific channels in *Xenopus* oocytes. *EMBO J.* 12: 2241–2247, 1993.

414. Mays, R. W., K. A. Beck, and W. J. Nelson. Organization and function of the cytoskeleton in polarized epithelial cells: a component of the protein sorting machinery. *Curr. Opin. Cell Biol.* 6: 16–24, 1994.

415. McCann, J. D., and M. J. Welsh. Regulation of Cl$^-$ and K$^+$ channels in airway epithelium. *Annu. Rev. Physiol.* 52: 115–136, 1990.

416. McCarty, N. A., and R. G. O'Neil. Calcium signalling in cell volume regulation. *Physiol. Rev.* 72: 1037–1061, 1992.

416a. McDonald, R. A., R. P. Matthews, R. L. Idzerda, and G. S. McKnight. Basal expression of the cystic fibrosis transmembrane conductance regulator gene is dependent on protein kinase A activity. *Proc. Natl. Acad. Sci. U.S.A.* 92: 7560–7564, 1995.

416b. McDonald, F. J., P. M. Snyder, P. B. McCray, and M. J. Welsh. Cloning, expression, and tissue distribution of a human amiloride-sensitive Na$^+$ channel. *Am. J. Physiol.* 266 (*Lung Cell. Mol. Physiol.* 10): L78–L734, 1994.

417. McEwan, G. T., C. D. Brown, B. H. Hirst, and N. L. Simmons. Characterization of volume-activated ion transport across epithelial monolayers of human intestinal T84 cells. *Pflugers Arch.* 423: 213–220, 1993.

418. McKinney, T. D., and K. K. Davidson. Effects of respiratory acidosis on HCO$_3^-$ transport by rabbit collecting tubules. *Am. J. Physiol.* 255: (*Renal Fluid Electrolyte Physiol.* 24): F656–F665, 1988.

419. McNeill, H., M. Ozawa, R. Kemler, and W. J. Nelson. Novel function of the cell adhesion molecule uvomorulin as an inducer of cell surface polarity. *Cell* 62: 309–316, 1990.

419a. McNicholas, C.M.W.B. Guggino, E. M. Schwiebert, S. C. Hebert, G. Giebisch, and M. E. Egan. Sensitivity of a renal K$^+$ channel (ROMK2) to the inhibitory sulfonylurea compound glibenclamide is enhanced by coexpression with the ATP-binding cassette transporter cystic fibrosis transmembrane regulator. *Proc. Natl. Acad. Sci. U.S.A.* 93: 8083–8088, 1996.

420. McNicholas, C. M., W. Wang, K. Ho, S. G. Hebert, and G. Giebisch. Regulation of ROMK1 K$^+$ channel activity involves phosphorylation processes. *Proc. Natl. Acad. Sci. U.S.A.* 91: 8077–8081, 1994.

421. Mills, J. W., and D. R. DiBona. Distribution of Na$^+$ pump sites in the frog gallbladder. *Nature* 271: 273–275, 1978.

422. Minton, A. P. Influence of macromolecular crowding on intracellular association reactions: possible role in volume regulation. In: *Cellular and Molecular Physiology of Cell Volume Regulation*, edited by K. Strange. Boca Raton, FL: CRC, 1994, p. 181–190.

423. Minton, A. P., G. C. Colclasure, and J. C. Parker. Model for the role of macromolecular crowding in regulation of cellular volume. *Proc. Natl. Acad. Sci. U.S.A.* 89: 10504–10506, 1992.

424. Møllgård, K., and J. Rostgaard. Morphological aspects of some sodium transporting epithelia suggesting a transcellular pathway *via* elements of endoplasmic reticulum. *J. Membr. Biol.* 40: 71–89, 1978.

425. Montrose-Rafizadeh, C., and W. B. Guggino. Cell volume regulation in the nephron. *Annu. Rev. Physiol.* 52: 761–772, 1990.

425a. Moorman, J. R., S. J. Ackerman, G. C. Kowdley, M. P. Griffin, J. P. Mounsey, Z. Chen, S. E. Cala, J. J. O'Brian, G. Szabo, and L. R. Jones. Unitary anion currents through phospholemman channel molecules *Nature* 377: 737–740, 1995.

426. Mooseker, M. S. Organization, chemistry and assembly of the cytoskeletal apparatus of the intestinal brush border. *Annu. Rev. Cell Biol.* 1: 209–241, 1985.

426a. Morales, M. M., T. P. Carroll, T. Mortia, E. M. Schwiebert, O. Devuyst, P. D. Wilson, A. G. Lopes, B. A. Stanton, H. C. Dietz, G. R. Cutting, and W. B. Guggino. Both the wild type and a functional isoform of CFTR are expressed in kidney. *Am. J. Physiol.* 270 (*Renal Fluid Electrolyte Physiol.* 39): F1038–F1048, 1996.

427. Mugharbil, A., R. G. Knickelbein, P. S. Aronson, and J. W. Dobbins. Rabbit ileal brush-border membrane Cl-HCO$_3$ exchanger is activated by an internal pH sensitive modifier site. *Am. J. Physiol.* 259 (*Gastrointest. Liver Physiol.* 22): G666–G670, 1990.

428. Mulders, S. M., G. M. Preston, P. M. T. Deen, W. B. Guggino, C. H. van Os, and P. Agre. Water channel properties of major protein of lens. *J. Biol. Chem.* 270: 9010–9016, 1995.

429. Müller, H.-P., and W. Schaffner. Regulation of gene activity in eukaryotes. In: *Essentials of Molecular Biology*, edited by D. Freifelder and G. M. Malacinski. Boston: Jones and Bartlett, 1993, p. 359–389.

430. Nakhoul, N. L., L. K. Chen, and W. F. Boron. Intracellular pH regulation in rabbit S3 proximal tubule: basolateral Cl-HCO$_3$ exchange and Na-HCO$_3$ cotransport. *Am. J. Physiol.* 258 (*Renal Fluid Electrolyte Physiol.* 27): F371–F381, 1990.

431. Nellans, H. N., R. A. Frizzell, and S. G. Schultz. Coupled sodium-chloride influx across the brush border of rabbit ileum. *Am. J. Physiol.* 225: 467–475, 1973.

432. Nelson, D. J., J. M. Tang, and L. G. Palmer. Single-channel recordings of apical membrane chloride conductance in A6 epithelial cells. *J. Membr. Biol.* 80: 81–89, 1984.

433. Nelson, R. D., X.-L. Guo, K. Masood, D. Brown, M. Kalkbrenner, and S. Gluck. Selectively amplified expression of an isoform of the vacuolar H$^+$ ATPase 56 kilodalton subunit in renal intercalated cells. *Proc. Natl. Acad. Sci. U.S.A.* 89: 3541–3545, 1992.

434. Nelson, W. J. Cytoskeleton functions in membrane traffic in polarized epithelial cells. *Semin. Cell Biol.* 2: 375–385, 1991.

435. Nelson, W. J. Regulation of cell surface polarity from bacteria to mammals. *Science* 258: 948–955, 1992.

436. Nielsen, S., C.-L. Chou, D. Marples, E. I. Christensen, B. K. Kishore, and M. A. Knepper. Vasopressin increases water permeability of kidney collecting duct by inducing translocation of aquaporin-CD water channels to plasma membrane. *Proc. Natl. Acad. Sci. U.S.A.* 92: 1013–1017, 1995.

437. Nielsen, S., S. R. DiGiovanni, E. I. Christensen, M. A. Knepper, and H. W. Harris. Cellular and subcellular immunolocalization of vasopressin-regulated water channel in rat kidney. *Proc. Natl. Acad. Sci. U.S.A.* 90: 11663–11667, 1993.

437a. Nielsen, S., D. Marples, J. Frøkiær, M. Knepper, and P. Agre. The aquaporin family of water channels in kidney. An update on physiology and pathophysiology of aquaporin-2. *Kidney Int.* 49: 1718–1723, 1996

438. Nielsen S., B. L. Smith, E. I. Christensen, M. A. Knepper, and P. Agre. CHIP28 water channels are localized in constitutively water-permeable segments of the nephron. *J. Cell Biol.* 120: 371–383, 1993.

439. Nielsen, S., B. L. Smith, E. I. Christensen, and P. Agre. Distribution of the aquaporin CHIP in secretory and resorptive epithelia and capillary endothelia. *Proc. Natl. Acad. Sci. U.S.A.* 90: 7272–7279, 1993.

440. Noël, J., and J. Pouysségur. Hormonal regulation, pharmacology, and membrane sorting of vertebrate Na^+/H^+ exchanger isoforms. *Am. J. Physiol.* 268 (*Cell Physiol.* 37): C283–C296, 1995.

440a. North, R. A. Families of ion channels with two hydrophobic segments. *Curr. Opin. Cell Biol.* 9: 474–483, 1996.

441. Novak, I., and R. Greger. Effect of bicarbonate on potassium conductance of isolated perfused rat pancreatic ducts. *Pflugers Arch.* 419: 76–83, 1991.

442. Oberleithner, H., U. Kersting, S. Silbernagl, W. Steigner, and U. Vogel. Fusion of cultured dog kidney (MDCK) cells. II. Relationship between cell pH and K^+ conductance in response to aldosterone. *J. Membr. Biol.* 111: 49–56, 1989.

443. Oberleithner, H., S. Wunsch, and S. Schneider. Patchy accumulation of apical Na^+ transporters allows cross talk between extracellular space and cell nucleus. *Proc. Natl. Acad. Sci. U.S.A.* 89: 241–245, 1992.

444. Oblatt-Montal, M., G. L. Reddy, T. Iwamoto, J. M. Tomich, and M. Montal. Identification of an ion channel-forming motif in the primary structure of CFTR, the cystic fibrosis chloride channel. *Proc. Natl. Acad. Sci. U.S.A.* 91: 1495–1499, 1994.

445. O'Grady, S. M., M. W. Musch, and M. Field. Stoichiometry and ion affinities of the Na-K-Cl cotransport system in the intestine of the winter flounder (*Pseudopleuronectes americanus*). *J. Membr. Biol.* 91: 33–41, 1986.

446. Oh, Y., and D. J. Benos. Single-channel characteristics of a purified bovine renal amiloride-sensitive Na^+ channel in planar lipid bilayers. *Am. J. Physiol.* 264 (*Cell Physiol.* 33): C1489–C1499, 1993.

447. Oh, Y., P. R. Smith, A. L. Bradford, D. Keeton, and B. J. Benos. Regulation by phosphorylation of purified epithelial Na^+ channels in planar lipid bilayers. *Am. J. Physiol.* 265 (*Cell Physiol.* 34): C85–C91, 1993.

448. Ojakian, G. K., and R. Schwimmer. Regulation of epithelial cell surface polarity reversal by β_1 integrins. *J. Cell Sci.* 107: 561–576, 1994.

449. O'Neil, R. G., and R. A. Hayhurst. Sodium-dependent modulation of the renal Na-K-ATPase: influence of mineralocorticoids on the cortical collecting duct. *J. Membr. Biol.* 85: 169–179, 1985.

450. Opperman, C. H., C. G. Taylor, and M. Conkling. Root-knot nematode-directed expression of a plant root-specific gene. *Science* 263: 221–223, 1994.

451. Orlowski, J., R. A. Kandasamy, and G. E. Shull. Molecular cloning of putative members of the Na/H exchanger gene family. *J. Biol. Chem.* 267: 9331–9339, 1992.

452. Otsu, K., J. L. Kinsella, P. Heller, and J. P. Froehlich. Sodium dependence of the Na^+-H^+ exchanger in the pre-steady state. *J. Biol. Chem.* 268: 3184–3193, 1993.

453. Pácha, J., G. Frindt, L. Antonian, R. B. Silver, and L. G. Palmer. Regulation of Na channels of the rat cortical collecting tubule of aldosterone. *J. Gen. Physiol.* 102: 25–52, 1993.

454. Palfrey, C., and A. Cossins. Cell physiology: fishy tales of kidney function. *Nature* 371: 377–378, 1994.

455. Palmer, L. G. Voltage-dependent block by amiloride and other monovalent cations of apical Na channels in the toad urinary bladder. *J. Membr. Biol.* 80: 153–165, 1984.

456. Palmer, L. G. Ion selectivity of epithelial Na channels. *J. Membr. Biol.* 96: 97–106, 1987.

457. Palmer, L. G. Epithelial Na channels: function and diversity. *Annu. Rev. Physiol.* 54: 51–66, 1992.

458. Palmer, L. G. Epithelial Na channels and their kin. *News Physiol. Sci.* 10: 61–67, 1995.

459. Palmer, L. G., L. Antonian, and G. Frindt. Regulation of the Na-K pump of the rat cortical collecting tubule by aldosterone. *J. Gen. Physiol.* 102: 43–57, 1993.

460. Palmer, L. G., and G. Frindt. Effects of cell Ca and pH on Na channels from rat cortical collecting tubule. *Am. J. Physiol.* 253 (*Renal Fluid Electrolyte Physiol.* 22): F333–F339, 1987.

461. Palmer, L. G., J.H.-Y. Li, B. Lindemann, and I. S. Edelman. Aldosterone control of the density of sodium channels in the toad urinary bladder. *J. Membr. Biol.* 64: 91–102, 1982.

462. Palmer, L. G., and M. Lorenzen. Antidiuretic hormone-dependent membrane capacitance and water permeability in the toad urinary bladder. *Am. J. Physiol.* 244 (*Renal Fluid Electrolyte Physiol.* 13): F195–F204, 1983.

463. Palmer, L. G., and H. Sackin. Regulation of renal ion channels. *FASEB J.* 2: 3061–3065, 1988.

464. Palmer, L. G., and H. Sackin. Electrophysiological analysis of transepithelial transport. In: *The Kidney: Physiology and Pathophysiology,* edited by D. W. Seldin and G. Giebisch. New York: Raven, 1992, p. 361–405.

465. Palmer, L. G., and N. Speez. Modulation of antidiuretic hormone-dependent capacitance and water flow in toad urinary bladder. *Am. J. Physiol.* 246 (*Renal Fluid Electrolyte Physiol.* 15): F501–F508, 1984.

466. Pappenheimer, J. R. On the coupling of membrane digestion with intestinal absorption of sugars and amino acids. *Am. J. Physiol.* 265 (*Gastrointest. Liver Physiol.* 28): G409–G417, 1993.

467. Pappenheimer, J. R., and J. L. Madara. Role of active transport in regulation of junctional permeability and paracellular absorption of nutrients by intestinal epithelia. In: *Isotonic Transport in Leaky Epithelia,* Alfred Benzon Symposium 34, edited by H. H. Ussing, J. Fischbarg, O. Sten-Knudsen, E. H. Larsen and N. J. Willumsen. Copenhagen: Munksgaard, 1993, p. 221–230.

468. Parent, L., J. Cardinal, and R. Sauvé. Single-channel analysis of a K channel at basolateral membrane of rabbit proximal convoluted tubule. *Am. J. Physiol.* 254 (*Renal Fluid Electrolyte Physiol.* 23): F105–F113, 1988.

469. Parent, L., S. Supplisson, D.D.F. Loo, and E. M. Wright. Electrogenic properties of the cloned Na^+/glucose cotransporter. I. Voltage-clamp studies. *J. Membr. Biol.* 125: 49–62, 1992.

470. Parent, L., S. Supplisson, D.D.F. Loo, and E. M. Wright. Electrogenic properties of the cloned Na^+/glucose cotransporter. II. A transport model under nonrapid equilibrium conditions. *J. Membr. Biol.* 125: 63–79, 1992.

471. Parisi, M., and J. Bourguet. The single file hypothesis and the water channels induced by antidiuretic hormone. *J. Membr. Biol.* 71: 189–193, 1983.

472. Parisi, M., J. Wietzerbin, and J. Bourguet. Intracellular pH, transepithelial pH gradients and ADH-induced water channels. *Am. J. Physiol.* 244 (*Renal Fluid Electrolyte Physiol.* 13): F712–F718, 1983.

473. Parker, J. C. In defense of cell volume? *Am. J. Physiol.* 265 (*Cell Physiol.* 34): C1191–C1200, 1993.

474. Parker, J. C., G. C. Colclasure, and T. J. McManus. Coordinated regulation of shrinkage-induced Na/H exchange and swelling-induced [K-Cl] cotransport in dog red cells. Further evidence from activation kinetics and phosphatase inhibition. *J. Gen. Physiol.* 98: 869–880, 1991.

474a. Parker, J. C., P. B. Dunham, and A. P. Minton. Effects of ionic strength on the regulation of Na/H exchange and K-Cl cotransport in dog red blood cells. *J. Gen. Physiol.* 105: 677–699, 1995.

475. Pasyk, E. A., and K. Foskett. Mutant (F508) cystic fibrosis transmembrane conductance regulator Cl^- channel is functional when retained in endoplasmic reticulum of mammalian cells. *J. Biol. Chem.* 270: 12347–12350, 1995.

476. Paulais, M., and J. Teulon. A cation channel in the thick ascending limb of Henle's loop of the mouse kidney: inhibition by adenine nucleotides. *J. Physiol. (Lond.)* 413: 315–327, 1989.

477. Paulais, M., and J. Teulon. cAMP-activated chloride channel in the basolateral membrane of the thick ascending limb of the mouse kidney. *J. Membr. Biol.* 113: 253–260, 1990.

478. Payne, J. A., and B. Forbush, III. Alternatively spliced isoforms of the putative renal Na-K-Cl cotransporter are differentially distributed within the rabbit kidney. *Proc. Natl. Acad. Sci. U.S.A.* 91: 4544–4548, 1994.

478a. Payne, J. A., T. J. Stevenson, and L. F. Donaldson. Molecular characterization of a putative K-Cl cotransporter in rat brain. A neuronal-specific isoform. *J. Biol. Chem.* 271: 16245–16252.

478b. Payne, J. A., J.-C. Xu, M. Haas, C. Y. Lytle, D. Ward, and B. Forbush, III. Primary structure, functional expression, and chromosomal localization of the bumetanide-sensitive Na-K-Cl cotransporter in human colon. *J. Biol. Chem.* 270: 17977–17985, 1995.

479. Penniston, J. T., and A. Enyedi. Plasma membrane Ca^{2+} pump: recent developments. *Cell Physiol. Biochem.* 4: 148–159, 1994.

480. Persson, B. E., and K. R. Spring. Gallbladder epithelial cell hydraulic water permeability and volume regulation. *J. Gen. Physiol.* 79: 481–505, 1982.

481. Petersen, K.-U., and L. Reuss. Cyclic AMP-induced chloride permeability in the apical membrane of *Necturus* gallbladder epithelium. *J. Gen. Physiol.* 81: 705–729, 1983.

482. Petersen, O. H. New aspects of cytosolic calcium signaling. *News Physiol. Sci.* 11: 13–17, 1996.

483. Petersen, O. H., M. Wakui, Y. Osipchuk, D. Yule, and D. V. Gallacher. Electrophysiology of pancreatic acinar cells. *Methods Enzymol.* 192: 300–308, 1990.

484. Petrov, A. G., and P.N.R. Usherwood. Mechanosensitivity of cell membranes. Ion channels, lipid matrix and cytoskeleton. *Eur. Biophys. J.* 23: 1–19, 1994.

484a. Piepenhagen, P. A., L. L. Peters, S. E. Lux, and J. Nelson. Differential expression of Na^+-K^+-ATPase, ankyrin, fodrin, and E-cadherin along the kidney nephron. *Am. J. Physiol.* 269 (*Cell Physiol.* 38): C1417-C1432, 1995.

485. Pierce, S. K., and A. D. Politis. Ca^{2+}-activated cell volume recovery mechanisms. *Annu. Rev. Physiol.* 52: 27–42, 1990.

486. Pietras, R. J., and E. M. Wright. The membrane action of antidiuretic hormone (ADH) on toad urinary bladder. *J. Membr. Biol.* 22: 107–123, 1975.

487. Pizzonia, J. H., M. Exner, D. Biemesderfer, A. K. Abu-Alfa, M. S. Wu, P. Igarashi, and P. S. Aronson. Immunochemical characterization of the Na/H exchanger isoform NHE4 in rat kidney. *J. Am. Soc. Nephrol.* 5: 297, (Abstr.), 1994.

488. Planelles, G., and T. Anagnostopoulos. Thiazide-sensitive Na-Cl cotransport mediates NaCl absorption in amphibian distal tubule. *Pflugers Arch.* 421: 307–313, 1992.

489. Pouysségur, J. Molecular biology and hormonal regulation of vertebrate Na^+/H^+ exchanger isoforms. *Renal Physiol. Biochem.* 17:190–193, 1994.

490. Prat, A. G., D. A. Ausiello, and H. F. Cantiello. Vasopressin and protein kinase-A activate G-protein-sensitive epithelial Na^+ channels. *Am. J. Physiol.* 265 (*Cell Physiol.* 34): C218–C223, 1993.

491. Prat, A. G., A. M. Bertorello, D. A. Ausiello, and H. F. Cantiello. Activation of epithelial Na^+ channels by protein kinase-A requires actin filaments. *Am. J. Physiol.* 265 (*Cell Physiol.* 34): C224–C233, 1993.

492. Preisig, P. A., and R. J. Alpern. Basolateral membrane H/HCO_3 transport in renal tubules. *Kidney Int.* 39: 1077–1086, 1991.

493. Preston, G. M., and P. Agre. Molecular cloning of the red cell integral membrane protein of M_r 28,000: a member of an ancient channel family. *Proc. Natl. Acad. Sci. U.S.A.* 88: 11110–111114, 1991.

494. Preston, G. M., T. P. Carroll, W. B. Guggino, and P. Agre. Appearance of water channels in *Xenopus* oocytes expressing red cell CHIP28 protein. *Science* 256: 385–387, 1992.

495. Pusch, M., and T. J. Jentsch. Molecular physiology of voltage-gated chloride channels. *Physiol. Rev.* 74: 813–828, 1994.

496. Rabon, E., J. Cuppoletti, D. Malinowska, A. Smolke, H. F. Helander, J. Mendlein, and G. Sachs. Proton secretion by the gastric parietal cell. *J. Exp. Biol.* 106: 119–133, 1983.

497. Rabon, E. C., and M. A. Reuben. The mechanism and structure of the gastric H,K-ATPase. *Annu. Rev. Physiol.* 52: 321–344, 1990.

498. Raina, S., G. M. Preston, W. B. Guggino, and P. Agre. Molecular cloning and characterization of an aquaporin cDNA from salivary, lacrimal, and respiratory tissues. *J. Biol. Chem.* 270: 1908–1912, 1995.

499. Rao, M. C., N. T. Nash, and M. Field. Differing effects of cGMP and cAMP on ion transport across flounder intestine. *Am. J. Physiol.* 246 (*Cell Physiol.* 15): C167–C171, 1984.

500. Rayson, B. M. $[Ca^{2+}]_i$ regulates transcription rate of the Na^+/K^+-ATPase α_1 subunit. *J. Biol. Chem.* 266: 21335–21338, 1991.

501. Rector, F. C., N. W. Carter, and D. W. Seldin. The mechanism of bicarbonate reabsorption in the proximal and distal tubules of the kidney. *J. Clin. Invest.* 44: 278–290, 1965.

501a. Reddy, M. M., P. M. Quinton, C. Haws, J. J. Wine, R. Grygorczyk, J. A. Tabcharani, J. W. Hanrahan, K. L. Gunderson, and R. R. Kopito. Failure of the cystic fibrosis transmembrane conductance regulator to conduct ATP. *Science* 271: 1876–1879, 1996.

502. Reeves, W. B., and T. E. Andreoli. Renal epithelial chloride channels. *Annu. Rev. Physiol.* 54: 29–50, 1992.

503. Reeves, W. B., and T. E. Andreoli. Sodium chloride transport in the loop of Henle. In: *The Kidney, Physiology and Pathophysiology*, edited by D. W. Seldin and G. Giebisch. New York: Raven, 1992, p. 1975–2001.

504. Reif, M. C., S. L. Troutman, and J. A. Schafer. Sodium transport by rat cortical collecting tubule. Effects of vasopressin and deoxycorticosterone in rat and rabbit CCD. *Am. J. Physiol.* 259 (*Renal Fluid Electrolyte Physiol.* 28): F147–F156, 1990.

505. Reisin, I. L., A. G. Prat, E. H. Abraham, J. F. Amara, R. J. Gregory, D. A. Ausiello, and H. F. Cantiello. The cystic fibrosis

transmembrane conductance regulator is a dual ATP and chloride channel. *J. Biol. Chem.* 269: 20584–20591, 1994.

506. Reithmeier, R. A. F. The erythrocyte anion transporter (Band 3). *Curr. Opin. Struct. Biol.* 3: 515–523, 1993.

507. Reithmeier, R.A.F. Mammalian exchangers and co-transporters. *Curr. Opin. Cell Biol.* 6: 583–594, 1994.

508. Reuss, L. Basolateral KCl co-transport in a NaCl-absorbing epithelium. *Nature* 305: 723–726, 1983.

509. Reuss, L. Independence of apical membrane Na^+ and Cl^- entry in *Necturus* gallbladder epithelium. *J. Gen. Physiol.* 84: 423–445, 1984.

510. Reuss, L. Cell volume regulation in nonrenal epithelia. *Renal Physiol. Biochem.* 11: 187–201, 1988.

511. Reuss, L. Ion transport across gallbladder epithelium. *Physiol. Rev.* 69: 503–545, 1989.

512. Reuss, L. Tight junction permeability to ions and water. In: *The Tight Junction,* edited by M. Cereijido, Boca Raton, FL: CRC, 1991, p. 49–66.

513. Reuss, L. Pathways for osmotic water transport in gallbladder epithelium. In: *Isotonic Transport in Leaky Epithelia,* Alfred Benzon Symposium 34, edited by H. H. Ussing, J. Fischbarg, O. Sten-Knudsen, E. H. Larsen and N. J. Willumsen. Copenhagen: Munksgaard, 1993, p. 181–200.

513a. Reuss, L. Active water transport? *J. Physiol.,* 497: 1, 1996.

514. Reuss, L. and G. A. Altenberg. cAMP-activated Cl^- channels: regulatory role in gallbladder and other absorptive epithelia. *News Physiol. Sci.* 10: 86–91, 1995.

515. Reuss, L., E. Bello-Reuss, and T. P. Grady. Effects of ouabain on fluid transport and electrical properties of *Necturus* gallbladder. Evidence in favor of a neutral basolateral sodium transport mechanism. *J. Gen. Physiol.* 73: 385–402, 1979.

516. Reuss, L., and J. L. Costantin. Cl^-/HCO_3^- exchange at the apical membrane of *Necturus* gallbladder. *J. Gen. Physiol.* 83: 801–818, 1984.

517. Reuss, L., and C. U. Cotton. Isosmotic fluid transport across epithelia. *Contemp. Nephrol.* 4: 1–37, 1988.

518. Reuss, L., and C. U. Cotton. Volume regulation in epithelia: transcellular transport and cross-talk. In: *Cellular and Molecular Physiology of Cell Volume Regulation,* edited by K. Strange. Boca Raton, FL: CRC, 1994, p. 31–48.

519. Reuss, L., and K.-U. Petersen. Cyclic AMP inhibits Na^+/H^+ exchange at the apical membrane of *Necturus* gallbladder epithelium. *J. Gen. Physiol.* 85: 409–429, 1985.

520. Reuss, L., P. Reinach, S. A. Weinman, and T. P. Grady. Intracellular ion activities and Cl^- transport mechanisms in bullfrog corneal epithelium. *Am. J. Physiol.* 244 (*Cell Physiol.* 13): C336–C347, 1983.

521. Reuss, L., B. Simon, and C. U. Cotton. Pseudo-streaming potentials in *Necturus* gallbladder epithelium. II. The mechanism is a junctional diffusion potential. *J. Gen. Physiol.* 99: 317–338, 1992.

522. Reuss, L., B. Simon, and Z. Xi. Pseudo-streaming potentials in *Necturus* gallbladder epithelium. I. Paracellular origin of the transepithelial voltage changes. *J. Gen. Physiol.* 99: 297–316, 1992.

523. Reuss, L., and J. S. Stoddard. Role of H^+ and HCO_3^- in salt transport in gallbladder epithelium. *Annu. Rev. Physiol.* 49:35–49, 1987.

524. Rich, D. P., H. A. Berer, S. H. Cheng, S. M. Travis, M. Saxena, A. E. Smith, and M. J. Welsh. Regulation of the cystic fibrosis transmembrane conductance regulator Cl^- channel by negative charge in the R domain. *J. Biol. Chem.* 268: 20259–20267, 1993.

525. Riordan, J. R. The cystic fibrosis transmembrane conductance regulator. *Annu. Rev. Physiol.* 55:609–630, 1993.

526. Riordan, J. R., J. M. Rommens, B. Kerem, N. Alon, R. Rozmahel, Z. Grzelczak, J. Zielenski, S. Lok, N. Plvsic, J.-L. Chou, M. L. Drumm, M. C. Iannuzzi, F. S. Collins, and T. C. Tsui. Identification of the cystic fibrosis gene: cloning and characterization of complementary DNA. *Science* 245: 1066–1074, 1989. Erratum: *Science* 245: 1437, 1989.

527. Robertson, M. A., and J. K. Foskett. Na^+ transport pathways in secretory acinar cells: membrane cross talk mediated by $[Cl^-]_i$. *Am. J. Physiol.* 267 (*Cell Physiol.* 36): C146–C156, 1994.

528. Rodríguez-Boulan, E., and E. Nelson. Morphogenesis of the polarized epithelial cell phenotype. *Science* 245: 718–725, 1989.

528a. Romero, M. F., M. A. Hediger, E. L. Boulpaep, and W. F. Boron. Physiology of the cloned renal electrogenic Na^+/HCO_3^- cotransporter, NBC. *J. Gen. Physiol.* 108: 16–17a, 1996.

529. Rommens, J. M., M. C. Iannuzzi, B. S. Kerem, M. L. Drumm, G. Melmer, M. Dean, R. Rozmahel, J. L. Cole, D. Kennedy, N. Hidaka, M. Zsiga, M. Buchwald, J. R. Riordan, L.-C., Tsui, and F. S. Collins. Identification of the cystic fibrosis gene: chromosome walking and jumping. *Science* 245: 1059–1065, 1989.

530. Rose, R. C., and S. G. Schultz. Studies on the electrical potential profile across rabbit ileum. Effects of sugars and amino acids on transmural and transmucosal electrical potential differences. *J. Gen. Physiol.* 57: 639–663, 1971.

531. Rossier, B. C., and L. G. Palmer. Mechanisms of aldosterone action on sodium and potassium transport. In: *The Kidney: Physiology and Pathophysiology,* edited by D. W. Seldin and G. Giebisch. New York: Raven, 1992, p. 1373–1409.

532. Ruesch, A., C. J. Kros, and G. P. Richardson. Block by amiloride and derivatives of mechano-electrical transduction in outer hair cells of mouse cochlear cultures. *J. Physiol. (Lond.)* 474: 75–86, 1994.

533. Ruoslahti, E. Integrins. *J. Clin. Invest.* 87: 1–5, 1991.

534. Ruoslahti, E., N. A. Noble, S. Kagami, and W. A. Border. Integrins. *Kidney Int.* 44: S17–S22, 1994.

535. Russell, J. M. Cation-coupled chloride influx in squid axon. Role of potassium and stoichiometry of the transport process. *J. Gen. Physiol.* 81: 909–926, 1983.

536. Sabolic, I., G. Valenti, J. M. Verbavatz, A. N. van Hoek, A. S. Verkman, D. A. Ausiello, and D. Brown. Localization of the CHIP28 water channel in rat kidney. *Am. J. Physiol.* 263 (*Cell Physiol.* 32): C1225–C1233, 1992.

537. Sachs, F. Stretch-sensitive ion channels. An update. In: *Sensory Transduction,* edited by D. P. Corey and S. D. Roper. New York: Rockefeller, 1992, p. 241–260.

538. Sachs, G. The gastric pump: the H^+, K^+-ATPase. In: *Physiology of the Gastrointestinal Tract,* edited by L. R. Johnson. New York: Raven, 1987, p. 865–881.

539. Sachs, G., M. Besancon, J. M. Shin, F. Mercier, K. Munson, and S. Hersey. Structural aspects of the gastric H,K-ATPase. *J. Bioenerg. Biomembr.* 24: 301–308, 1992.

540. Sachs, G., H. H. Chang, E. Rabon, R. Schackmann, M. Lewis, and G. Saccomani. A nonelectrogenic H^+ pump in plasma membranes of hog stomach. *J. Biol. Chem.* 251: 7690–7698, 1976.

541. Sachs, G., L. D. Faller, and E. Rabon. Proton/hydroxyl transport in gastric and intestinal epithelia. *J. Membr. Biol.* 64: 123–135, 1982.

542. Sachs, G., and K. Munson. Mammalian phosphorylating ionmotive ATPASes. *Curr. Opin. Cell Biol.* 3: 685–694, 1991.

543. Sachs, G., J. M. Shin, M. Besancon, K. Munson, and S. Hersey. Topology and sites in the H,K-ATPase. *Ann. N.Y. Acad. Sci.* 671: 204–216, 1992.

544. Sackin, H. Stretch-activated potassium channels in renal proximal tubules. *Am. J. Physiol.* 253 (Renal Fluid Electrolyte Physiol. 22): F1253–F1262, 1987.

545. Sackin, A. A stretch-activated H^+ channel sensitive to cell volume. *Proc. Natl. Acad. Sci. U.S.A.* 86: 1731–1735, 1989.

546. Sackin, H. Stretch-activated ion channels. In: *Cellular and Molecular Physiology of Cell Volume Regulation*, edited by K. Strange. Boca Raton, FL: CRC, 1994, p. 259–278.

547. Sackin, H., and E. L. Boulpaep. Models for coupling of salt and water transport. Proximal tubular reabsorption in *Necturus* kidney. *J. Gen. Physiol.* 66: 671–733, 1975.

548. Sackin, H., and L. G. Palmer. Basolateral potassium channels in renal proximal tubule. *Am. J. Physiol.* 253: (Renal Fluid Electrolyte Physiol. 23): F476–F487, 1987.

548a. Sakamoto, H., M. Kawasaki, S. Uchida, S. Sasaki, and F. Marumo. Identification of a new outwardly rectifying Cl^- channel that belongs to a subfamily of the ClC Cl^- channels. *J. Biol. Chem.* 271: 10210–10216, 1996.

549. Salzman, A. L., M. J. Menconi, N. Unno, R. M. Ezzell, D. M. Casey, P. K. González, and M. P. Fink. Nitric oxide dilates tight junctions and depletes ATP in cultured Caco-2BBe intestinal epithelial monolayers. *Am. J. Physiol.* 268 (Gastrointest. Liver Physiol. 31): G361–G373, 1995.

550. Sansom, S. C., B.-Q. La, and S. L. Carosi. Double-barreled chloride channels of collecting duct basolateral membrane. *Am. J. Physiol.* 259 (Renal Fluid Electrolyte Physiol. 28): F46–F52, 1990.

550a. Sansom, S. C., B. Q. La, and S. L. Carosi. Potassium and chloride channels of the basolateral membrane (BLM) of the rabbit cortical collecting duct (CCD) (Abstract). *Kidney Int.* 37 (Supp. 1): 570, 1990.

551. Sardet, C., A. Franchi, and J. Pouysségur. Molecular cloning, primary structure, and expression of the human growth factor-activatable Na^+/H^+ antiporter. *Cell* 56: 271–280, 1989.

552. Sariban-Sohraby, S., M. Abramov, and R. S. Fisher. Single channel behavior of a purified epithelial Na^+ channel subunit that binds amiloride. *Am. J. Physiol.* 263 (Cell Physiol. 32): C1111–C1117, 1992.

553. Sariban-Sohraby, S., M. Burg, W. P. Wiesmann, P. K. Chiang, and J. P. Johnson. Methylation increases sodium transport into A6 apical membrane vesicles: possible mode of aldosterone action. *Science* 225: 745–746, 1984.

554. Sariban-Sohraby, S., E. J. Sorscher, B. M. Brenner, and D. J. Benos. Phosphorylation of a single subunit of the epithelial Na^+ channel protein following vasopressin treatment of A6 cells. *J. Biol. Chem.* 263: 13875–13879, 1988.

555. Sasaki, S., K. Ishibashi, N. Yoshiyama, and T. Shiigai. KCl cotransport across the basolateral membrane of rabbit renal proximal straight tubules. *J. Clin. Invest.* 81: 194–199, 1988.

556. Sasaki, S., and N. Yoshiyama. Interaction of chloride and bicarbonate transport across the basolateral membrane of rabbit proximal straight tubule. Evidence for sodium coupled chloride/bicarbonate exchange. *J. Clin. Invest.* 81: 1004–1011, 1988.

557. Schafer, J. A., and T. E. Andreoli. Cellular constraints to diffusion. The effect of antidiuretic hormone on water flows in isolated mammalian collecting ducts. *J. Clin. Invest.* 51: 1264–1278, 1972.

558. Schafer, J. A., and C. T. Hawk. Regulation of Na^+ channels in the cortical collecting duct by AVP and mineralocorticoids. *Kidney Int.* 41: 255–268, 1992.

559. Schafer, J. A., S. L. Troutman, and T. E. Andreoli. Osmosis in cortical collecting tubules. ADH-independent osmotic flow rectification. *J. Gen. Physiol.* 64: 228–240, 1974.

560. Schlatter, E., and R. Greger. cAMP increases the basolateral chloride conductance in the isolated perfused medullary thick ascending limb. *Pflugers Arch.* 405: 367–376, 1985.

561. Schnapp, B. J., R. D. Vale, M. P. Sheetz, and T. S. Reese. Single microtubules from squid axoplasm support bidirectional movement of organelles. *Cell* 40: 455–462, 1985.

562. Schultz, S. G. Application of equivalent electrical circuit models to study of sodium transport across epithelial tissues. *Federation Proc.* 38: 2024–2029, 1979.

563. Schultz, S. G. *Basic Principles of Membrane Transport.* Cambridge: Cambridge University Press, 1980.

564. Schultz, S. G. Homocellular regulatory mechanisms in sodium-transporting epithelia: avoidance of extinction by "flush-through." *Am. J. Physiol.* 241 (Renal Fluid Electrolyte Physiol. 10): F579–590, 1981.

565. Schultz, S. G. Membrane cross-talk in sodium-absorbing epithelial cells. In: *The Kidney: Physiology and Pathophysiology,* edited by D. W. Seldin and G. Giebisch. New York: Raven, 1992, p. 287–299.

566. Schultz, S. G. The "pump-leak" parallelism in *Necturus* enterocytes: some cellular and molecular insights. *Renal Physiol. Biochem.* 17: 134–137, 1994.

567. Schultz, S. G., R. A. Frizzell, and H. N. Nellans. Active sodium transport and the electrophysiology of rabbit colon. *J. Membr. Biol.* 33: 351–384, 1977.

568. Schultz, S. G., and R. L. Hudson. Biology of sodium-absorbing epithelial cells: dawning of a new era. In: *Handbook of Physiology. The Gastrointestinal System. Intestinal Absorption and Secretion,* edited by S. G. Schultz, M. Field, and R. A. Frizzell. Betuesda, MD: Am. Physiol. Soc., 1991, sect. 6, vol. IV, chapt. 2, p. 45–81.

569. Schuster, V. L. Cortical collecting duct bicarbonate secretion. *Kidney Int.* 40: S47–S50, 1991.

570. Schuster, V. L. Function and regulation of collecting duct intercalated cells. *Annu. Rev. Physiol.* 55: 267–288, 1993.

571. Schuster, V. L., G. Fejes-Tóth, A. Náray-Fejes-Tóth, and S. Gluck. Co-localization of H^+-ATPase and band 3 anion exchanger in rabbit collecting duct intercalated cells. *Am. J. Physiol.* 260 (Renal Fluid Electrolyte Physiol. 29): F506–F517, 1991.

572. Schwartz, G. J., and Q. Al-Awqati. Carbon dioxide causes exocytosis of vesicles containing H^+ pumps in isolated perfused proximal and collecting tubules. *J. Clin. Invest.* 75: 1638–1644, 1985.

573. Schwartz, G. J., and Q. Al-Awqati. Regulation of transepithelial H^+ transport by exocytosis and endocytosis. *Annu. Rev. Physiol.* 48: 153–161, 1986.

574. Schwartz, G. J., J. Barasch, and Q. Al-Awqati. Plasticity of functional epithelial polarity. *Nature* 318: 368–371, 1985.

575. Schwiebert, E. M., M. E. Egan, T.-H. Hwang, S. B. Fulmer, S. A. Allen, G. R. Cutting, and W. B. Guggino. CFTR regulates outwardly rectifying chloride channels through an autocrine mechanism involving ATP. *Cell* 81: 1063–1073, 1995.

576. Schwiebert, E. M., T. Flotte, G. R. Cutting, and W. B. Guggino. Both CFTR and outwardly rectifying chloride channels contribute to cAMP-stimulated whole cell chloride currents. *Am. J. Physiol.* 266 (Cell Physiol. 35): C1464–C1477, 1994.

577. Schwiebert, E. M., D. B. Light, G. Fejes-Tóth, A. Naray-Fejes-Tóth, and B. A. Stanton. A GTP-binding protein activates

chloride channels in a renal epithelium. *J. Biol. Chem.* 265: 7725–7728, 1990.

578. Schwiebert, E. M., J. W. Mills, and B. A. Stanton. Actin-based cytoskeleton regulates a chloride channel and cell volume in a renal cortical collecting duct cell lines. *J. Biol. Chem.* 269: 7081–7089, 1994.

578a. Segal, A. S., E. L. Boulpaep, and A. B. Maunsbach. A novel preparation of dissociated renal proximal tubule cells that maintain epithelial polarity in suspension. *Am. J. Physiol.* 270 (*Cell Physiol.* 39): C1843–C1863, 1996.

579. Segal, Y., and L. Reuss. Maxi K$^+$ channels and their relationship to the apical membrane conductance in *Necturus* gallbladder epithelium. *J. Gen. Physiol.* 95: 791–818, 1990.

580. Seibert, F. S., J. A. Tabcharani, X.-B. Chang, A. M. Dulhanty, C. Matthews, J. W. Hanrahan, and J. R. Riordan. cAMP-dependent protein kinase-mediated phosphorylation of cystic fibrosis transmembrane conductance regulator residue Ser-753 and its role in channel activation. *J. Biol. Chem.* 270: 2158–2162, 1995.

581. Seki, G., S. Coppola, K. Yoshitomi, B.-C. Burckhardt, I. Samarzijia, S. Müller-Berger, and E. Frömter. On the mechanism of bicarbonate exit from renal proximal tubular cells. *Kidney Int.* 49: 1671–1677, 1996.

581a. Seksek, O., J. Biwersi, and A. S. Verkman. Evidence against defective trans-Golgi acidification in cystic fibrosis. *J. Biol. Chem.* 271: 15542–15548, 1996.

582. Sepúlveda, F. V., F. Fargon, and P. A. McNaughton. K$^+$ and Cl$^-$ currents in enterocytes isolated from guinea-pig small intestinal villi. *J. Physiol.* 434: 351–367, 1991.

583. Sepúlveda, F. V., and W. T. Mason. Single channel recordings obtained from basolateral membranes of isolated rabbit enterocytes. *FEBS Lett.* 191: 87–91, 1985.

584. Sheppard, D. N., L. S. Ostedgaard, D. P. Rich, and M. J. Welsh. The amino-terminal portion of CFTR forms a regulated Cl$^-$ channel. *Cell* 76: 1091–1098, 1994.

585. Shi, L.-B., D. Brown, and A. S. Verkman. Water, urea and proton transport properties of endosomes containing the vasopressin-sensitive water channel from toad bladder. *J. Gen. Physiol.* 95: 941–960, 1990.

585a. Shin, J. M., and G. Sachs, Dimerization of the gastric H$^+$,K$^+$-ATPase. *J. Biol. Chem.* 271: 1904–1908, 1996.

586. Silva, P., J. Stoff, M. Field, L. Fine, J. N. Forrest, and F. H. Epstein. Mechanism of active chloride secretion by shark rectal gland: role of Na-K-ATPase in chloride transport. *Am. J. Physiol.* 233 (*Renal Fluid Electrolyte Physiol.* 4): F298–F306, 1977.

587. Silver, R. B., G. Frindt, E. E. Windhager, and L. G. Palmer. Feedback regulation of Na channels in rat CCT. I. Effects of inhibition of Na pump. *Am. J. Physiol.* 264 (*Renal Fluid Electrolyte Physiol.* 33): F557–F564, 1993.

587a. Singh, G., J. Orlowski, and M. Soleimani. Transient expression of Na$^+$/H$^+$ exchanger isoform NHE-2 in LLC-PK$_1$ cells: Inhibition of endogenous NHE-3 and regulation by hypertonicity. *J. Membr. Biol.* 151: 261–268, 1996.

588. Skou, J. C. The Na-K Pump. *NIPS* 7: 95–100, 1992.

589. Skou, J. C., and M. Esmann. The Na,K-ATPase. *J. Bioenerg. Biomembr.* 24: 249–261, 1992.

590. Smith, B. L., and P. Agre. Erythrocyte M$_r$ 28,000 transmembrane protein exists as a multisubunit oligomer similar to channel proteins. *J. Biol. Chem.* 266: 6407–6415, 1991.

591. Smith, P. L., M. A. Cascairo, and S. K. Sullivan. Sodium dependence of luminal alkalinization by rabbit ileal mucosa. *Am. J. Physiol.* 249 (*Gastrointest. Liver Physiol.* 12): G358–G368, 1985.

592. Soleimani, M., and P. S. Aronson. Ionic mechanism of Na$^+$/HCO$_3^-$ contransport in rabbit renal basolateral membrane vesicles. *J. Biol. Chem.* 264: 18302–18308, 1989.

593. Soleimani, M., S. M. Grassl, and P. S. Aronson. Stoichiometry of Na$^+$-HCO$_3^-$ cotransport in basolateral membrane vesicles isolated from rabbit renal cortex. *J. Clin. Invest.* 79: 1276–1280, 1987.

594. Soleimani, M., G. A. Lesoine, J. A. Bergman, and T. D. McKinney. A pH modifier site regulates activity of the Na$^+$/HCO$_3^-$ cotransporter in basolateral membranes of kidney proximal tubule. *J. Clin. Invest.* 88: 1135–1140, 1991.

594a. Soleimani, M., G. Singh, G. L. Bizal, S. R. Gullans, and J. A. McAteer. Na$^+$/H$^+$ exchanger isoforms NHE-2 and NHE-1 in inner medullary collecting duct cells: expression, functional localization, and differential regulation. *J. Biol. Chem.* 269: 27973–27978, 1994.

595. Soltoff, S. P., M. K. McMillan, E. J. Cragoe Jr., L. C. Cantley, and B. R. Talamo. Effects of extracellular ATP on ion transport systems and [Ca^{2+}]$_i$ in rat parotid acinar cells. Comparison with the muscarinic agonist carbachol. *J. Gen. Physiol.* 95: 319–346, 1990.

596. Spring, K. R. Mechanism of fluid transport by epithelia. In: *Handbook of Physiology. The Gastrointestinal System. Intestinal Absorption and Secretion*, edited by S. G. Schultz, M. Field, and R. A. Frizzell. Bethesda, MD: Am. Physiol. Soc., 1991, sect. 6, vol. IV, chapt. 0, p. 195–207.

597. Spring, K. R., and A. W. Siebens. Solute transport and epithelial cell volume regulation. *Comp. Biochem. Physiol.* 90A: 557–560, 1988.

598. Star, R. A., M. B. Burg, and M. A. Knepper. Bicarbonate secretion and chloride absorption by rabbit cortical collecting ducts. *J. Clin. Invest.* 76: 1123–1130, 1985.

599. Starke, L. C. Regulation of Na/K/2Cl and KCl cotransport in duck red cells. Ph.D. Thesis, Durham, NC: Duke University, 1989.

600. Staub, O., F. Verrey, T. R. Kleyman, D. J. Benos, B. C. Rosier, and J. P. Kraehenbuhl. Primary structure of an apical protein from *Xenopus laevis* that participates in amiloride-sensitive sodium channel activity. *J. Cell Biol.* 119: 1497–1506, 1992.

601. Stein, W. D. *Channels, Carriers, and Pumps. An Introduction to Membrane Transport*. San Diego: Academic, 1990.

602. Steinmetz, P. R. Cellular organization of urinary acidification. *Am. J. Physiol.* 251 (*Renal Fluid Electrolyte Physiol.* 20): F173–F186, 1986.

603. Stern, C. D., and D. O. MacKenzie. Sodium transport and the control of epiblast polarity in the early chick embryo. *J. Embryol. Exp. Morphol.* 77: 73–98, 1983.

604. Stetson, D. L., S. A. Lewis, W. Alles, and J. B. Wade. Evaluation by capacitance measurements of antidiuretic hormone induced membrane area changes in toad bladder. *Biochim. Biophys. Acta* 689: 267–274, 1982.

605. Stoddard, J., and L. Reuss. Dependence of cell membrane conductances on bathing solution HCO$_3^-$/CO$_2$ in *Necturus* gallbladder. *J. Membr. Biol.* 102: 163–174, 1988.

606. Stokes, J. B. Passive NaCl transport in the flounder urinary bladder: predominance of a cellular pathway. *Am. J. Physiol.* 254 (*Cell Physiol.* 23): C229–C237, 1988.

607. Strange, K. Ouabain-induced cell swelling in rabbit cortical collecting tubule: NaCl transport by principal cells. *J. Membr. Biol.* 107: 249–261, 1989.

608. Strange, K. Are all cell volume changes the same? *News Physiol. Sci.* 9: 223–227, 1994.

608a. Strange, K., F. Emma, and P. S. Jackson. Cellular and molecular physiology of volume-sensitive anion channels. *Am. J. Physiol.* 270 (*Cell Physiol.* 39): C711–C730, 1996.

609. Strange, K., and K. R. Spring. Cell membrane permeability of rabbit cortical collecting duct. *J. Membr. Biol.* 96: 27–43, 1987.

610. Stutts, M. J., C. Canessa, J. C. Olson, M. Hamrick, J. A. Cohn, B. C. Rossier, and R. C. Boucher. CFTR as a cAMP-dependent regulator of sodium channels. *Science* 269: 847–850, 1995.

611. Stutts, M. J., T. C. Chinet, S. J. Mason, J. M. Fullton, L. L. Clarke, and R. C. Boucher. Regulation of Cl⁻ channels in normal and cystic fibrosis airway epithelial cells by extracellular ATP. *Proc. Natl. Acad. Sci. U.S.A.* 89: 1621–1625, 1992.

612. Sullivan, S. K., and M. Field. Ion transport across mammalian small intestine. In: *Handbook of Physiology. The Gastrointestinal System. Intestinal Absorption and Secretion*, edited by S. G. Schultz, M. Field and R. A. Frizzell. Bethesda MD: Am. Physiol. Soc., 1991, sect. 6, vol. IV, chapt. 10, p. 287–301.

613. Sun, A. M., S. N. Saltzbert, D. Kikeri, and S. C. Hebert. Mechanisms of cell volume regulation by the mouse medullary thick ascending limb of Henle. *Kidney Int.* 38: 1019–1029, 1990.

613a. Supplisson, S., D. D. Loo, and G. Sachs. Diversity of K⁺ channels in the basolateral membrane of resting *Necturus* oxyntic cells. *J. Membr. Biol.* 123: 209–221, 1991.

614. Szerlip, H. M., L. Waisberg, M. Clayman, E. Neilson, J. B. Wade, and M. Cox. Aldosterone-induced proteins: purification and localization of GP65, 70. *Am. J. Physiol.* 256 (*Cell Physiol.* 25): C865–C872, 1989.

615. Tabcharani, J. A., A. Boucher, J.W.L. Eng, and J. W. Hanrahan. Regulation of an inwardly rectifying K channel in the T₈₄ epithelial cell line by calcium, nucleotides and kinases. *J. Membr. Biol.* 142: 255–266, 1994.

616. Tabcharani, J. A., X.-B. Chang, J. R. Riordan, and J. W. Hanrahan. Phosphorylation-regulated Cl⁻ channel in CHO cells stably expressing the cystic fibrosis gene. *Nature* 352: 628–631, 1991.

617. Tabcharani, J. A., R. A. Harris, A. Boucher, J.W.L. Eng, and J. W. Hanrahan. Basolateral K channel activated by carbachol in the epithelial cell line T₈₄. *J. Membr. Biol.* 142: 241–254, 1994.

618. Tabcharani, J. A., J. M. Rommens, Y. X. Hou, X. B. Chang, L. Tsui, J. R. Riordan, and J. W. Hanrahan. Multi-ion pore behavior in the CFTR chloride channel. *Nature* 366: 79–82, 1993.

618a. Takeuchi, Y., S. Uchida, F. Marumo, and S. Sasaki. Cloning, tissue distribution, and intrarenal localization of ClC chloride channels in human kidney. *Kidney Int.* 48: 1497–1503.

619. Tatsuta, H., S. Ueda, S. Morishima, and Y. Okada. Voltage- and time-dependent K⁺ channel currents in the basolateral membrane of villus enterocytes isolated from guinea pig small intestine. *J. Gen. Physiol.* 103: 429–446, 1994.

620. Thiemann, A., S. Grunder, M. Pusch, and T. J. Jentsch. A chloride channel widely expressed in epithelial and non-epithelial cells. *Nature* 356: 57–60, 1992.

621. Thorens, B. Facilitated glucose transporters in epithelial cells. *Annu. Rev. Physiol.* 55: 591–608, 1993.

622. Thorens, B., Z.-Q. Cheng, D. Brown, and H. F. Lodish. Liver glucose transporter: a basolateral protein in hepatocytes and intestine and kidney epithelial cells. *Am. J. Physiol.* 259 (*Cell Physiol.* 28): C279–C285, 1990.

622a. Torres, R. J., G. A. Altenberg, J. A. Copello, G. Zampighi, and L. Reuss. Preservation and structural and functional polarity in isolated epithelial cells. *Am. J. Physiol.* 270 (*Cell Physiol.* 39): C1864–C1874, 1996.

622b. Torres, R. J., G. A. Altenberg, J. A. Cohn, and L. Reuss. Polarized expression of cyclic AMP-activated chloride channels in isolated epithelial cells. *Am. J. Physiol.* 271 (*Cell Physiol.* 40): C1574–C1582, 1996.

623. Tripathi, S., and E. L. Boulpaep. Mechanisms of water transport by epithelial cells. *Q. J. Exp. Physiol.* 74: 385–417, 1989.

624. Tse, C.-M., S. R. Brant, M. S. Walker, J. Pouysségur, and M. Donowitz. Cloning and sequencing of a rabbit cDNA encoding an intestinal and kidney-specific Na⁺/H⁺ exchanger isoform [NHE-3]. *J. Biol. Chem.* 267:9340–9346, 1992.

625. Tse, M., S. Levine, C. Yun, S. Brant, L. Counillon, J. Pouysségur, and M. Donowitz. Structure/function studies of the epithelial isoforms of the mammalian Na⁺/H⁺ exchanger gene family. *J. Membr. Biol.* 135: 93–108, 1993.

626. Tse, C. M., S. A. Levine, C.H.C. Yun, M. H. Montrose, P. J. Little, J. Pouysségur, and M. Donowitz. Cloning and expression of a rabbit cDNA encoding a serum-activated ethylisopropyl amiloride resistant epithelial Na⁺/H⁺ exchanger isoform (NHE-2). *J. Biol. Chem.* 268: 11917–11924, 1993.

627. Tsuchiya, K., W. Wang, G. Giebisch, and P. A. Welling. ATP is a coupling modulator of parallel Na,K-ATPase-K-channel activity in the renal proximal tubule. *Proc. Natl. Acad. Sci. U.S.A.* 89: 6418–6244, 1992.

628. Turnheim, K., R. A. Frizzell, and S. G. Schultz. Interaction between cell sodium and the amiloride-sensitive sodium entry step in rabbit colon. *J. Membr. Biol.* 39: 233–256, 1978.

629. Ubl, J., H. Murer, and H.-A. Kolb. Ion channels activated by osmotic and mechanical stress in membranes of opossum kidney cells. *J. Membr. Biol.* 104: 223–232, 1988.

630. Uchida, S., S. Sasaki, T. Furukawa, M. Hiraoka, T. Imai, Y. Hirata, and F. Marumo. Molecular cloning of a chloride channel that is regulated by dehydration and expressed predominantly in kidney medulla. *J. Biol. Chem.* 268: 3821–3824, 1993.

631. Uchida, S., S. Sasaki, K. Nitta, K. Uchida, S. Horita, H. Nihei, and F. Marumo. Localization and functional characterization of rat kidney-specific chloride channel, ClC-K1. *J. Clin. Invest.* 95: 104–113, 1995.

632. Urbach, V., E. van Kerkhove, and B. J. Harvey. Inward-rectifier potassium channels in basolateral membranes of frog skin epithelium. *J. Gen. Physiol.* 103: 583–604, 1994.

633. Ussing, H. H. The distinction by means of tracers between active transport and diffusion. *Acta. Physiol. Scand.* 19: 43–56, 1949.

634. Ussing, H. H. *The Alkali Metal Ions in Biology. I. The Alkali Metal Ions in Isolated Systems and Tissues*. Berlin: Springer-Verlag, 1960.

635. Ussing, H. H. Epithelial cell volume regulation illustrated by experiments in frog skin. *Renal Physiol.* 9: 38–46, 1986.

636. Ussing, H. H. Epithelial transport: frog skin as a model system. In: *Membrane Transport. People and Ideas*, edited by D. C. Tosteson. New York: Oxford, 1989, p. 337–362.

636a. Ussing, H. H., and K. Eskesen. Mechanism of isotonic water transport in glands. *Acta Physiol. Scand.* 136: 443–454, 1989.

637. Ussing, H. H, and E. E. Windhager. Nature of shunt path and active sodium transport path through frog skin epithelium. *Acta Physiol. Scand.* 61: 484–504, 1964.

638. Ussing, H. H., and K. Zerahn. Active transport of sodium as the source of electric current in the short-circuited isolated frog skin. *Acta Physiol. Scand.* 23: 110–127, 1951.

639. Vale, R. D., B. J. Schnapp, T. S. Reese, and M. P. Sheetz. Movement of organelles along filaments dissociated from the axoplasm of the squid giant axon. *Cell* 40: 449–454, 1985.

639a. Valverde, M.A.S.P. Hardy, and F. V. Sepúlveda. Chloride channels: a state of flux. *FASEB J.* 9: 509–515, 1995.

640. van Adelsberg, J., and Q. Al-Awqati. Regulation of cell pH by calcium-mediated exocytosis of H⁺ ATPases. *J. Cell Biol.* 102: 1638–1645, 1986.

641. van Adelsberg J., J. C. Edwards, and Q. Al-Awqati. The apical Cl/HCO₃ exchanger of β-intercalated cells. *J. Biol. Chem.* 268: 11283–11289, 1993.

642. van Adelsberg, J., C. Edwards, J. Takito, B. Kiss, and Q. Al-Awqati. An induced extracellular matrix protein reverses the polarity of band 3 in intercalated epithelial cells. *Cell* 76: 1053–1061, 1994.

643. van Hoek, A. N., M. L. Hom, L. H. Luthjens, M. D. deJone, J. A. Dempster, and C. H. van Os CH. Functional unit of 30 kDa for proximal tubule water channels as revealed by radiation inactivation. *J. Biol. Chem.* 266: 16633–16635, 1991.

644. van Os, C. H. Transcellular calcium transport in intestinal and renal epithelial cells. *Biochim. Biophys. Acta* 906: 195–222, 1987.

645. Venosa, R. A. Hypo-osmotic stimulation of active Na transport in frog muscle: apparent upregulation of Na⁺ pumps. *J. Membr. Biol.* 120:97–104, 1991

646. Verkman, A. S. Mechanisms and regulation of water permeability in renal epithelia. *Am. J. Physiol.* 257 (*Cell Physiol.* 26): C837–C850, 1989.

647. Verkman, A. S., W. Lencer, D. Brown, and D. A. Ausiello. Endosomes from kidney collecting tubule contain the vasopressin-sensitive water channel. *Nature* 333: 268–269, 1988.

647a. Verkman, A. S., A. N. van Hoek, T. Ma, A. Frigeri, W. R. Skach, A. Mitra, B. K. Tamarappoo, and J. Farinas. Water transport across mammalian cell membranes. *Am. J. Physiol.* 270 (*Cell Physiol.* 39): C12–C30, 1996.

648. Verrey, F., J.-P. Kraehenbuhl, and B. C. Rossier. Aldosterone induces a rapid increase in the rate of Na,K-ATPase gene transcription in cultured kidney cells. *Mol. Endocrinol.* 3: 1367–1376, 1989.

649. Verrey, F., E. Schaerer, P. Zoerkler, M. P. Paccolat, K. Geering, J. P. Kraehenbuhl, and B. C. Rossier. Regulation by aldosterone of Na⁺,K⁺-ATPase mRNAs, protein synthesis, and sodium transport in cultured kidney cells. *J. Cell Biol.* 105: 1231–1237, 1987.

650. Völkl, H., J. Geibel, R. Greger, and L. Lang. Effects of ouabain and temperature on cell membrane potentials in isolated perfused straight proximal tubules of the mouse kidney. *Pflugers Arch.* 407: 252–257, 1986.

651. Völkl, H., M. Paulmichl, and F. Lang. Cell volume regulation in renal cortical cells. *Renal Physiol. Biochem.* 3–5: 158–173, 1988.

652. Wade, J. B. Hormonal modulation of epithelial structures. *Curr. Top. Membr. Transp.* 13: 124–147, 1980.

653. Wade, J. B. Membrane structural studies of the action of vasopressin. *Federation Proc.* 44: 2687–2692, 1985.

654. Wade, J. B. Role of membrane traffic in the water and Na⁺ responses to vasopressin. *Semin. Nephrol.* 14: 322–332, 1994.

654a. Wakabayashi, S., T. Ikeda, J. Noël, B. Schmitt, J. Orlowski, J. Pouysségur, M. Shigekawa. Cytoplasmic domain of the ubiquitous Na⁺/H⁺ exchanger NHE1 can confer Ca²⁺ responsiveness to the apical isoform NHE3. *J. Biol. Chem.* 270: 26460–26465, 1995.

655. Walz, T., B. L. Smith, P. Agre, and A. Engel. The 3-D structure of human erythrocyte aquaporin CHIP. *EMBO J.* 13: 2985–2993, 1994.

656. Walz, T., B. L. Smith, M. L. Zeidel, A. Engel, and P. Agre. Biologically active two-dimensional crystals of aquaporin CHIP. *J. Biol. Chem.* 269: 1583–1586, 1994.

657. Wang, N., J. P. Butler, and D. E. Ingber. Mechanotransduction across the cell surface and through the cytoskeleton. *Science* 260: 1124–1127, 1993.

658. Wang, W.-H. Two types of K⁺ channels in thick ascending limb of rat kidney. *Am. J. Physiol.* 267 (*Renal Fluid Electrolyte Physiol.* 36): F599–F605, 1994.

659. Wang, W.-H., A. Cassola, and G. Giebisch. Involvement of actin cytoskeleton in modulation of apical K channel activity in rat collecting duct. *Am. J. Physiol.* 267 (*Renal Fluid Electrolyte Physiol.* 36): F592–F598, 1994.

660. Wang, W.-H., R. M. Henderson, J. Geibel, S. White, and G. Giebisch. Mechanism of aldosterone-induced increase of K⁺ conductance in early distal renal tubule cells of the frog. *J. Membr. Biol.* 111: 277–289, 1989.

660a. Wang, W.-H., and M. Lu. Effect of arachidonic acid on activity of the apical K⁺ channel in the thick ascending limb of the rat kidney. *J. Gen. Physiol.* 106: 727–743, 1995.

661. Wang, W.-H., C. M. McNicholas, A. S. Segal, and G. Giebisch. A novel approach allows identification of K channels in the lateral membrane of rat CCD. *Am. J. Physiol.* 266 (*Renal Fluid Electrolyte Physiol.* 33): F813–F822, 1994.

662. Wang, W.-H., H. Sackin, and G. Giebisch. Renal potassium channels and their regulation. *Annu. Rev. Physiol.* 54: 81–96, 1992.

663. Wang, Z., J. Orlowski, and G. E. Shull. Primary structure and functional expression of a novel gastrointestinal isoform of the rat Na⁺/H⁺ exchanger. *J. Biol. Chem.* 268: 11925–11928, 1993.

664. Wang, Z.-Q., and S. Gluck. Isolation and properties of bovine kidney brush border vacuolar H⁺-ATPase. A proton pump with enzymatic and structural differences from kidney microsomal H⁺-ATPase. *J. Biol. Chem.* 265: 21957–21965, 1990.

665. Weinman, E. J., W. P. Dubinsky, K. Fisher, D. Steplock, Q. Dinh, L. Chang, and S. Shenolikar. Regulation of reconstituted renal Na⁺/H⁺ exchanger by calcium-dependent protein kinases. *J. Membr. Biol.* 103: 237–244, 1988.

666. Weinman, E. J., and S. Shenolikar. Regulation of the renal brush border membrane Na⁺/H⁺ exchanger. *Annu. Rev. Physiol.* 55: 289–304, 1993.

667. Weinman, E. J., and S. Shenolikar. Protein kinase C activates the renal apical membrane Na⁺/H⁺ exchanger. *J. Membr. Biol.* 93: 133–139, 1986.

667a. Weinman, E. J., D. Steplock, Y. Wang, and S. Shenolikar. Characterization of a protein cofactor that mediates protein kinase A regulation of the renal brush border membrane Na⁺-H⁺ exchanger. *J. Clin. Invest.* 95: 2143–2149.

668. Weinman, S. A., and L. Reuss. Na⁺-H⁺ exchange and Na⁺ entry across the apical membrane of *Necturus* gallbladder. *J. Gen. Physiol.* 83: 57–74, 1984.

669. Weinstein, A. M., and J. L. Stephenson. Electrolyte transport across a simple epithelium. Steady-state and transient analysis. *Biophys. J.* 27: 165–186, 1979.

670. Weinstein, A. M., and J. L. Stephenson. Coupled water transport in standing gradient models of the lateral intercellular space. *Biophys. J.* 35: 167–191, 1981.

671. Weinstein, A. M., and J. L. Stephenson. Models of coupled salt and water transport across leaky epithelia. *J. Membr. Biol.* 60: 1–20, 1981.

672. Welling, P. A., and R. G. O'Neil. Ionic conductive properties of rabbit proximal straight tubule basolateral membrane. *Am. J. Physiol.* 258 (*Renal Fluid Electrolyte Physiol.* 27): F940–F950, 1990.

673. Wells, R. G., A. M. Pajor, Y. Kanai, E. A. Turk, E. M. Wright, and M. A. Hediger. Cloning of a human kidney cDNA with similarity to the sodium-glucose cotransporter. *Am. J. Physiol.* 263 (*Renal Fluid Electrolyte Physiol.* 32): F459–F465, 1992.

674. Welsh, M. J., M. P. Anderson, D. P. Rich, H. A. Berger, G. M. Denning, L. S. Ostedgaard, D. N. Sheppard, S. H. Cheng, R. J. Gregory, and A. E. Smith. Cystic fibrosis transmembrane conductance regulator: a chloride channel with novel regulation. *Neuron* 8: 821–829, 1992.

675. Welsh, M. J. Effect of phorbol ester and calcium ionophores of chloride secretion in canine tracheal epithelium. *Am. J. Physiol.* 253 (*Cell Physiol.* 22): C828–C834, 1987.

676. Welsh, M. J., and A. E. Smith. Molecular mechanisms of CFTR chloride channel dysfunction in cystic fibrosis. *Cell* 73: 1251–1254, 1993.

677. Werner, A., M. L. Moore, N. Mantei, J. Biber, G. Semenza, and H. Murer. Cloning and expression of cDNA for a Na/Pi cotransport system of kidney cortex. *Proc. Natl. Acad. Sci. U.S.A.* 88: 9608–9612, 1991.

678. White, J. F., and W. McD. Armstrong. Effect of transported solutes on membrane potentials in bullfrog small intestine. *Am. J. Physiol.* 221: 194–201, 1971.

679. Whittembury, G., and F. A. Rawlins. Evidence of a paracellular pathway for ion flow in the kidney proximal tubule: Electron microscopic demonstration of lanthanum precipitate in the tight junction. *Pflugers Arch.* 330: 302–309, 1971.

680. Whittembury, G., and L. Reuss. Mechanisms of coupling of solute and solvent transport in epithelia. In: *The Kidney: Physiology and Pathophysiology*, edited by D. W. Seldin and G. Giebisch. New York: Raven, 1992, p. 317–360.

681. Wiesmann, W. P., J. P. Johnson, G. A. Miura, and P. K. Chiang. Aldosterone-stimulated transmethylations are linked to sodium transport. *Am. J. Physiol.* 248 (*Renal Fluid Electrolyte Physiol.* 17): F43–F47, 1985.

682. Windhager, E. E., E. L. Boulpaep, G. Giebisch. Electrophysiological studies in single nephrons. In: *Proceedings of the Third International Congress on Nephrology*, edited by G. E. Schreiner: Washington, D.C., 1966, Vol. 1. New York: Karger, 1967, p. 35–47.

683. Windhager, E. E., and A. Taylor. Regulatory role of intracellular calcium ions in epithelial Na transport. *Annu. Rev. Physiol.* 45: 519–532, 1983.

684. Wingo, C. S., and B. D. Cain. The renal H-K-ATPase: Physiological significance and role in potassium homeostasis. *Annu. Rev. Physiol.* 55: 323–347, 1993.

685. Wolosin, J. M. Ion transport studies with H^+-K^+-ATPase-rich vesicles: implications for HCl secretion and parietal cell physiology. *Am. J. Physiol.* 248 (*Gastrointest. Liver Physiol.* 11): G595–G607, 1985.

686. Woodbury, W. F. The cell membrane: ionic and potential gradients and active transport. In: *Physiology and Biophysics*, edited by T. C. Ruch and H. D. Patton. Philadelphia: Saunders, 1965, p. 1–25.

687. Wright, E. M. The intestinal Na^+/glucose cotransporter. *Annu. Rev. Physiol.* 55: 575–589, 1993.

688. Wright, E. M., K. M. Hager, and E. Turk. Sodium cotransport proteins. *Curr. Opin. Cell Biol.* 4: 696–702, 1992.

689. Wright, F. S., and G. Giebisch. Regulation of potassium excretion. In: *The Kidney: Physiology and Pathophysiology*, edited by D. W. Seldin and G. Giebisch. New York: Raven, 1992, p. 2209–2247.

690. Xu, J.-C., L. Lytle, T. T. Zhu, J. A. Payne, E. Benz, Jr., and B. Forbush III. Molecular cloning and functional expression of the bumetanide-sensitive Na-K-Cl cotransporter. *Proc. Natl. Acad. Sci. U.S.A.* 91: 2201–2205, 1994.

690a. Yamaji, Y., M. Amemiya, A. Cano, P. A. Preisig, R. T. Miller, O. W. Moe, and R. J. Alpern. Overexpression of csk inhibits acid-induced activation of NHE-3. *Proc. Natl. Acad. Sci. U.S.A.* 92: 6274–6278, 1995.

691. Yamaji, Y., O. W. Moe, R. T. Miller, and R. J. Alpern. Acid activation of immediate early genes in renal epithelial cells. *J. Clin. Invest.* 94: 1297–1303, 1994.

692. Yang, Y., S. Janich, J. A. Cohn, and J. M. Wilson. The common variant of cystic fibrosis transmembrane conductance regulator is recognized by hsp70 and degraded in a pre-Golgi nonlysosomal compartment. *Proc. Natl. Acad. Sci. U.S.A.* 90: 9480–9484, 1993.

692a. Yool, A. J., W. D. Stamer, J. W. Regan. Forskolin stimulation of water and cation permeability in aquaporin 1 water channels. *Science* 273: 1216–1218, 1996.

693. Yoshitomi, K., B.-Ch. Burckhardt, and E. Frömter. Rheogenic sodium-bicarbonate cotransport in the peritubular cell membrane of rat renal proximal tubule. *Pflugers Arch.* 406: 360–366, 1985.

694. Yoshitomi, K., C. Koseki, J. Taniguchi, and M. Imai. Functional heterogeneity in the hamster medullary thick ascending limb of Henle's loop. *Pflugers Arch.* 408: 600–608, 1987.

695. Yun, C. H. C., S. Gurubhagavatula, S. Levine, J. L. M. Montgomery, S. R. Brant, M. E. Cohen, E. J. Cragoe, J. Pouysségur, C. M. Tse, and M. Donowitz. Glucocorticoid stimulation of ileal Na^+ absorptive cell brush border Na^+/H^+ exchange and association with an increase in message for NHE-3, an epithelial Na^+/H^+ exchanger isoform. *J. Biol. Chem.* 268: 206–211, 1993.

696. Zadunaisky, J. A. Active transport of chloride in frog cornea. *Am. J. Physiol.* 211: 506–512, 1966.

697. Zeidel, M. L., D. Kikeri, P. Silva, M. Burrowes, and B. M. Brenner. Atrial natriuretic peptides inhibit conductive sodium uptake by rabbit inner medullary collecting duct cells. *J. Clin. Invest.* 82: 1067–1074, 1988.

698. Zemelman, B. V., W. A. Walker, and S.-H. W. Chu. Expression and developmental regulation of Na^+,K^+ adenosine triphosphatase in the rat small intestine. *J. Clin. Invest.* 90: 1016–1022, 1992.

699. Zeuthen, T. Relations between intracellular ion activities and extracellular osmolarity in *Necturus* gallbladder epithelium. *J. Membr. Biol.* 66: 109–121, 1982.

699a. Zeuthen, T. Cotransport of K^+, Cl^- and H_2O by membrane proteins from choroid plexus epithelium of *Necturus maculosus*. *J. Physiol. (Lond.)* 478: 203–219, 1994.

699b. Zeuthen, T., S. Hamann, and M. la Cour. Cotransport of H^+, lactate and H_2O by membrane proteins in retinal pigment epithelium of bullfrog. *J. Physiol. (Lond.),* 497: 3–17, 1996.

700. Zhang, R., W. Skach, H. Hasegawa, A. N. van Hoek, and A. S. Verkman. Cloning, functional analysis and cell localization of a kidney proximal tubule water transporter homologous to CHIP28. *J. Cell Biol.* 120: 359–369, 1993.

9. Intracellular pH

ROBERT W. PUTNAM | *Department of Physiology and Biophysics, Wright State University School of Medicine, Dayton, Ohio*

ALBERT ROOS | *Department of Cell Biology and Physiology, Washington University School of Medicine, St. Louis, Missouri*

THE PH OF THE CYTOSOL (intracellular pH, pH_i) is a major component of a cell's "climate" and a pervasive factor in cellular events. It is specified or measured whenever they are discussed or examined. The variety of events in which pH_i is now known to play a role and the complexity of the membrane-bound proteins that control pH_i are so great that reviews like this one may soon be replaced by surveys of more limited scope, focusing on those areas that are unresolved and remain under active investigation. Among them are the role of H^+ in the physiology of channels and transporters of all kinds and the nature of their conformational changes induced by H^+, the anatomy and physiology of the subunits that make up the pH_i-regulating proteins, and regional pH differences in the cytosol which thus far have received little attention. Some of these areas of study are taken up by us in this review. Finally, the acid–base status, extracellular and intracellular, of the human organism, especially in disease, continues to be of vital interest to clinicians.

Studies extending over nearly a century, but especially those of the last three decades, have led to a rather detailed understanding of the mechanisms that account for the steady-state value of pH_i and its course in response to perturbations. We shall briefly list some of the steps that have led to our present knowledge. Early work was aimed at defining the nature of the plasma membrane; it used changes in pH_i principally as a tool to this end. Thus, the experiments on plant and animal cells by Overton (283, 284) convinced him of the membrane's largely lipid composition, a discovery of prime importance to cell physiology. Only the uncharged (lipid-soluble) forms of weak acids and bases, he found, gained entry and thereby acidified or alkalinized the cells, whereas strong acids and bases had no effect. Subsequent studies confirmed these findings (190, 191).

An accurate way to measure pH_i had to be devised before further progress could be made (see TECHNIQUES). Kite (209) made an early attempt, and concluded from the color changes of drops of dye injected into *Amoeba proteus* that the organism's pH_i was "neutral to slightly alkaline." Overton's observations were later applied to derive pH_i from the distribution of weak acids and bases between the extracellular and intracellular compartment, on the assumptions that their uncharged forms equilibrate across the plasma membrane, and that the charged forms are impermeant. A large literature sprung up on pH_i measurements with weak electrolytes, and valuable results were obtained in many cell types. Most of this work was done on tissues or entire organs; some of it on suspensions of cells or organelles. It has been reviewed in detail by Roos and Boron (317). The assumption of impermeability to ions is not always justified: NH_4^+ and HCO_3^-, for instance, can pass the membrane of many cells (for examples, see later under SOME OBSERVATIONS ON INTRACELLULAR PH TRANSITS). However, there is adequate theoretical and experimental evidence that, at least in the case of weak acids such as the commonly used DMO (5,5-dimethyloxazolidine-2,4 dione), ionic permeability can be neglected without introducing significant errors in the computed pH_i (62, 203, 315). Even so, these methods have serious limitations. For instance, the relative slowness of the indicators' distribution limits their use to steady-state conditions, and the cells must be destroyed for analysis so that only one measurement can be made.

Rapid and continuous measurements of pH_i in single cells became possible when electrodes were introduced whose voltage output was pH-sensitive, with tips small enough ("microelectrodes") to enter the cell without much damage. The technique requires cell impalement by a second, reference microelectrode. It took some 30 years before Caldwell and others replaced the original hydrogen (Pt/H_2) (372) and antimony (72) electrodes, which had serious drawbacks, with electrodes made of pH-sensitive, closed-tip glass capillaries (78, 79, 216, 361). This was a considerable achievement, and proper pH-sensitive glass, not susceptible to interference by other cell constituents, became the material of choice. The Caldwell electrode does not lend itself to easy reproduction and is rather large; the design was subsequently improved by Hinke (172) and by Thomas (377). Much of our present knowledge on the subject has been obtained with these two types of electrodes. More recently, open-tipped microelectrodes of ordinary glass, their tips filled with an H^+-sensitive liquid ion exchanger, have become popular because of their ease of construction. Fluorescent pH-sensitive dyes that enter cells by diffusion and then undergo modification so that they are trapped inside are important additions to pH_i instrumentation. Finally, there is the nuclear magnetic resonance (NMR) technique, which permits simultaneous measurement of many cell constituents besides H^+. These methods will be examined in this chapter.

It had been recognized 60 years ago by Fenn and his coworkers that the intracellular H ion concentration

(they derived it from the CO_2 distribution), at least in resting frog muscle, was too low to conform to a passive (Donnan) transmembrane H^+ distribution. We now know that this is the case in nearly all cells, except non-nucleated red cells; it is compatible with later observations that, in rat muscle cells, even very high concentrations of weak acids—DMO or lactic acid—only slightly reduce intracellular pH (315a, 315b). Fenn and Cobb (121) concluded that "some continuous supply of energy would be necessary" to maintain H ion concentration away from equilibrium; Fenn and Maurer (122) suggested phosphocreatine and lactic acid metabolism as the energy source. Twenty years later, Hill (169) proposed an H^+ extruding (and, of course, energy-requiring) mechanism. It became important to identify the modes by which pH_i is stabilized: acid accumulates not just by passive entry of H^+, a slow process, but by other, often faster events such as metabolic generation of H^+, especially during hypoxia and cellular activity, and the entry of the protonated forms of weak acids. Buffers can only temporarily protect the cell from changes in acidity, and it is now clear that, following Hill, H ions (more precisely, H^+ equivalents) are ejected from the cell through special membrane-spanning proteins. These molecules employ ATP as the "continuous supply of energy," either directly (primary active transport; see later, under $H^+-ATPases$ (Proton Pumps)) or indirectly (secondary active transport). The latter way is the more frequent one; here, H^+ extrusion (or inward movement of HCO_3^-) is tightly coupled to the movement of another ion, often Na^+. Energy is expended through a membrane-bound Na^+/K^+ ATPase to keep the intracellular concentration of this ion far below equilibrium, and it is the resulting inwardly directed Na^+ electrochemical gradient that powers H^+ extrusion.

Another twenty years passed before Thomas (377) demonstrated this by direct measurement with intracellular microelectrodes. When (nominally) CO_2-free snail neurons were acid-loaded by exposure to CO_2, pH_i first fell as CO_2 entered the cells, but then briskly recovered. Upon CO_2 removal, pH_i rose to a value that exceeded the initial one ("overshoot"). Both recovery and overshoot could be explained by acid extrusion during CO_2 exposure; the overshoot was greater the more prolonged the exposure (58). These transients are examined later, under SOME OBSERVATIONS ON INTRACELLULAR PH TRANSITS. HCO_3^- entry was apparently responsible for the recovery (57); when acidification was achieved through HCl injection in the absence of CO_2/HCO_3^-, recovery was very slow (380). The process involves the exit of internal Cl^- (324) as well as the entry of Na^+ (380). It is electroneutral (379, 380), and is inhibitable by the disulfonic stilbene derivatives DIDS and SITS (379). It comprises either exchange of external Na^+ + HCO_3^- for internal Cl^- + H^+, or external Na^+ + 2 HCO_3^- for internal Cl^- (a choice between the two is difficult to make), the energy being provided by the Na^+ gradient. In the squid giant axon, however, pH_i recovery is achieved through entry of the ion pair $NaCO_3^-$ in exchange for Cl^- (53). (See MECHANISMS OF PH REGULATION).

A second H ion extruder was demonstrated by Murer et al. (261) in rat intestine and kidney brush border membrane vesicles and by Johnson et al. (195) in sea urchin eggs. This transporter, called the Na^+/H^+ exchanger, involves an exchange of internal H^+ for external Na^+; it is also energized by the Na^+ gradient, but does not require HCO_3^-. Na^+/H^+ exchangers, in a range of isoforms, are now known to be present in nearly every cell. These transporters, which are inhibited to varying degrees by amiloride and its derivatives, will be described in more detail later, under MECHANISMS OF PH REGULATION. A useful way of studying them is by observing pH_i recovery from an acid load in the (nominal) absence of CO_2. In the technique of CO_2-free acid loading, first described by Boron and De Weer (57, 58), the cells are transiently exposed to NH_4^+/NH_3 in the form of, for instance NH_4Cl. Upon its removal, pH_i falls to values below the initial one. The reason for this "undershoot" (which increases with the duration of exposure) is discussed later, under SOME OBSERVATIONS ON INTRACELLULAR PH TRANSIENTS.

The significance of pH_i for many cellular events has already been alluded to. To a large extent, this can be ascribed to the effect of pH_i changes on proteins within the cytosol and on proteins that span membranes. Because of the imidazole groups of the histidine residues whose pKs are in the physiological range, the local charge of these proteins, and thus probably their conformation, will vary with pH_i. This is a principal reason why enzymatic activity and channel properties are often modified by changes in pH_i. (The pKs of the imidazole groups, it should be remembered, can vary rather widely, owing to the presence of other charged groups or dipoles in their vicinity.)

The earliest observations on the role of pH_i in cell physiology were almost certainly made by Isaac Newton, and are recorded in the *Journal Books of the Royal Society* for November 13, 1712, and March 31, 1720. In the latter entry, Newton, "the President in the Chair," reminisced about "a very remarkable Experiment he made formerly [this must have been around 1664] in Trinity Colledge Kitchin at Cambridge upon the heart of an Eel which he Cutt into three pieces and observed every One of them Beat at the same Instant & Interval putting Spittle upon any of the Sections had no Effect but a Drop of Viniger utterly Extinguished its Motion." In this breathless description we have the negative inotropic effect of intracellular acidity produced by external application of a weak acid, abundantly confirmed since then and controlled by the application of Sir Isaac's spittle which, like that of ordinary mortals, must have been slightly alkaline. The uniform rhythmicity of the eel's heart was duly noticed, as well as the myogenic nature of the beat. All in Trinity Colledge Kitchin!

Later, under SOME CELLULAR PROCESSES AFFECTING OR AFFECTED BY pH_i, we describe some of the events that can be influenced by pH_i. The reader may be impressed by their diversity. Truly, intracellular pH is a prominent component of the cell's climate.

TECHNIQUES

Several methods are in current use for the measurement of intracellular pH. They employ a diversity of tools, the most prominent of which are microelectrodes, intracellular dyes, and nuclear magnetic resonance (NMR). We will summarize the principles and some of the limitations of each method, and give examples of their application.

pH-Sensitive Microelectrodes

It has been more than 40 years since the first measurement of intracellular pH with a glass microelectrode (78). The pH-sensitive glass of these electrodes was protected along most of its length; its tip was unshielded. The pH could only be measured in large cells such as crab muscle fibers; the tip's diameter was about 600 μm (79). Since this early work, there has been much improvement in electrode design (for reviews, see 148, 173, 217, 317, and 381). At present, three different types of pH-sensitive microelectrodes are in use: glass microelectrodes, either with protruding tip (172) or with recessed tip (377); liquid ion exchange (LIX) microelectrodes (20, 84); and patch or suction microelectrodes (314).

Before examining the properties of the electrode types, some general comments are in order. All three have in common the presence of an H^+-selective barrier at the tip that separates the buffered salt solution in the electrode from the unknown external solution. To the extent that the barrier is permeable solely to H ions, the electrode's potential difference, ΔV, between its internal solution and the external solution is linearly related to the pH difference between these solutions, ΔpH. That is, the Nernst equation is obeyed: $\Delta V =$

$(2.303RT/F)\Delta pH$, where the constant between parentheses amounts to 59 mV at 25°C; ΔV (ideally) changes by 59 mV per unit of pH change in the unknown solution. When the electrode tip is placed in a cell, its signal, with reference to ground, will be due both to the pH-sensitive voltage and the membrane potential (V_m). To derive a signal that is linearly related solely to cellular pH, the membrane potential, which is often measured independently with a conventional V_m microelectrode, is subtracted from the pH electrode's output. Although impaling a cell with two electrodes is rather invasive, obtaining simultaneous measurements of both intracellular pH and V_m is a major advantage. Damage can be limited by combining the two into a double-barreled electrode (see pH-Sensitive Liquid Ion Exchange (LIX) Microelectrodes). The general theory of pH and its measurement with electrodes has been discussed by Bates (32).

pH-Sensitive Glass Microelectrodes. The H^+-selective barrier in these electrodes is the special glass itself. Permeable only to H^+, the capillaries are pulled to a fine tip that is then sealed. The shaft not inside the cell must be shielded. To this end, Caldwell (78) employed pH-insensitive glass, wax, and shellac, which made his electrode rather awkward to use. The Hinke electrode (172) was a considerable improvement. Here, the tip of a sealed pH-sensitive glass capillary protrudes from a slightly larger, pH-insensitive glass capillary. The two are sealed with either cement or by glass–glass seal. The newer model has a smaller exposed tip than the original one: tip diameter ranges from 1 to 20 μm, exposed length from 10 to 100 μm. These electrodes are well suited for measurements on barnacle muscle fibers and squid axons where they are introduced parallel to the cells' axis (59, 61, 327).

The protruding-tip glass microelectrode has several advantages: fast response time (about 1–2 s) (58), high selectivity for H^+ (slope about 57 mV/pH unit), and low resistance (1–10 GΩ), which minimizes noise. It can be used for many months. A disadvantage is that its application is limited to long and large cells.

A different design was introduced by Thomas in 1974: the recessed-tip microelectrode (377, 381). Here, the sealed pH-sensitive capillary is contained within an open tip pH-insensitive shielding capillary that slightly extends beyond the pH-sensitive tip. The inner glass is heated under high pressure, which establishes a seal between the two capillaries. Upon cooling, the inner shaft separates and is removed, leaving the pH-sensitive tip sealed within the tip of the shielding electrode. This results in a recess volume between the outer glass and the tip. When a cell is impaled, the cytosol exchanges with the recess volume. Since only the open tip (often

about 1 μm in diameter) needs to enter the cell, even small cells such as mammalian skeletal muscle fibers and amphibian proximal tubule cells (diameter 25–35 μm) can be impaled (9, 55). The electrodes are also in present use in modest-sized cells (see, for instance, 300, 380, and 401).

Recessed-tip microelectrodes, like other glass electrodes, are selective for H^+ and can be used repeatedly for many experiments, as long as their tips remain sharp enough to impale a cell. A disadvantage is that they have a long response time since the recess volume must fill with cytosol by diffusion. The time for 90% response can be as long as 1–2 min (381), but it can be reduced to about 15 s by carefully matching the tapers of the inner and outer electrodes, thus reducing recess volume (55).

pH-Sensitive Liquid Ion Exchange (LIX) Microelectrodes. The introduction of LIX microelectrodes (20, 239) represented another advance, not least because of their relative ease of construction. Fundamentally, these electrodes share the same principle with electrodes made from pH-sensitive glass. The selective H^+ barrier in the open, non-pH-sensitive glass tip is, however, formed by a liquid ion exchanger dissolved in an organic solvent, the LIX mixture. The end of a conventional open-tip microelectrode is filled with a small amount of this mixture, and the electrode is then back-filled with a buffered electrolyte solution. As with the glass electrode, a potential difference is established across the ion exchanger that, ideally, is linearly related to the difference in pH. A major advantage of these electrodes is that their tip is no larger than that of conventional V_m microelectrodes, thus allowing their introduction into small cells. Of course, an electrode for measuring V_m must be introduced as well. Both electrodes can be combined into one double-barreled microelectrode system; this has the advantage that pH_i and V_m are measured at almost exactly the same locus, although these electrodes obviously have larger tips than do single-barreled electrodes. Both single- and double-barreled LIX microelectrodes are much simpler to construct than electrodes in which the sensor is pH-sensitive glass.

The originally proposed liquid ion exchanger, tridodecylamine (TDDA), and its solvent (20), have certain technical disadvantages, such as loss of sensitivity in electrodes with small tips and the requirement for equilibration with CO_2. The recently introduced LIX, 4-nonadecylpyridine (ETH 1907) seems to obviate these problems (84). This LIX is dissolved in potassium tetrakis(4-chlorophenyl)borate (KTpClPB) in o-nitrophenyl octylether (o-NPOE). It is effective over the pH range 2 to 9, does not require equilibration with CO_2, shows little interference from Cl^-, and maintains its pH sensitivity even with very small electrode tips; all this is in contrast to the original carrier (84). LIX electrodes have a fast response time (1–15 s) (19) and, in general, lower resistance than the glass microelectrodes. Thus, they are less noisy, except when the tips are very fine. Their major disadvantages are limited useful life, often of only a few days, occasional responsiveness to other ions, and a tendency for the calibration slope to change. Upon initial construction, they will often exhibit a super-Nernstian slope (in excess of 59 mV/pH unit), which sometimes changes during the course of an experiment. It follows that they exhibit more drift than glass electrodes (6, 173). Also, drugs may interact with the exchanger and thus affect electrode response. Despite these drawbacks, LIX electrodes are widely used, especially in small cells (27, 206, 228). Aickin (6) successfully applied them to mammalian smooth muscle cells (length 10 μm, width 3 μm).

pH-Sensitive Patch or Suction Microelectrodes. Recently, a novel pH-sensitive electrode has been described (314) that is composed of a suction or patch electrode in combination with a liquid ion exchanger. The electrode is constructed with a rather large (1–2 μm diameter) and blunt tip. When the tip is filled with one of the liquid ion exchangers, it becomes a pH-sensitive microelectrode. It is pressed against the cell membrane and negative pressure is applied within the electrode, which causes the cell membrane to break and form a tight seal. Thus, the tip is exposed to the cell's interior. To measure V_m simultaneously either a conventional V_m microelectrode or a second patch electrode is used.

The microelectrode becomes stable shortly after contact with the cytosol has been established, and remains useful for up to several hours of continuous recording. These electrodes have a near-Nernstian slope; their minimum useful pH range is between 6 to 8.

pH-sensitive patch microelectrodes have advantages but also drawbacks. Their large tip size results in low resistance (400 MΩ) for a pH electrode; thus they are quiet and stable and respond quickly to changes in pH (time constant about 50 ms). They can be used in any cell that can be patch clamped. On the other hand, they can only be used once, even though they maintain their selectivity for several days; after initial use, they will no longer form a tight seal with the cell membrane (314). Furthermore, since they have a higher initial resistance than the conventional patch microelectrode, due to the LIX column in the tip, the increase in resistance when the tip touches the cell membrane is not as striking. It is, therefore, difficult to determine

exactly when the cell membrane is touched and when to apply suction. In addition, the electrodes are prone to the same problems as are the regular LIX electrodes. It also must be kept in mind that if V_m is measured with an additional, conventional, patch microelectrode, the latter can dialyze the cytoplasm: its filling solution will replace the cytosol by diffusion through its large open tip. The effect of this on pH_i may not be negligible.

pH-Sensitive Fluorescent Indicators

Optical methods that use pH-sensitive colored molecules for the evaluation of pH of plant and animal cells have been used for nearly 80 years. A historical survey is given in the review by Roos and Boron (317). Once the problem of their intracellular incorporation has been overcome, the advantages of their use are considerable. The molecules respond rapidly to pH changes and, most important, can be used in small cells, either singly or in groups, and in organelles where more invasive techniques are impractical. In addition, cells loaded with fluorescent pH-sensitive probes can be visualized with fluorescence microscopy (for review, see ref. 114a) in order to study pH_i regulation within single cells in a population or within various compartments in a cell (for example, ref. 144a). However, in contrast to microelectrode methods, membrane voltage need not be measured when pH_i is determined optically, and usually is not. This is a drawback, since membrane voltage is a useful index of the state of the cell and, in combination with pH_o, allows derivation of the electrochemical H^+ gradient. The present popularity of optical methods is largely due to the introduction of new compounds that fluoresce, have excellent pH sensitivity, are available as neutral derivatives that can diffuse across membranes, and, most important, once inside the cell can be rendered impermeant by esterases that convert them into their charged fluorescent parents.

This is not to overlook some significant work with colored, nonfluorescent dyes. For example, Baylor et al. (33) used the dye phenol red, a weak acid, iontophoretically injected into single frog twitch fibers. After correction for absorbance by dye-free fibers, the absorbance ratio of the dye-loaded fiber at 560 nm over that at 480 nm (the isosbestic point, that is, the wavelength at which absorbance is independent of pH) was used as a measure of pH_i. Calibration was done with dye solutions in cuvettes. The resting pH_i, 6.9, in reasonable agreement with data obtained with microelectrodes (1), rose by 0.004 following an action potential, and then was maintained at 0.002 above resting value.

The following summary will be limited to fluorescent dyes. Fluorescence may be described operationally as the "immediate" (about 10^{-8} s delay) re-emission of absorbed light energy; this is in contrast with phosphorescence, where the delay may be as much as 1 s. The wavelength of the emitted light is always longer than that of the exciting (absorbed) light. The absorption and fluorescence emission spectra, when plotted on a frequency (rather than wavelength) scale, are mirror images of each other.

The intensity of fluorescence emission depends on the wavelength and intensity of the incident light, the number of fluorescent molecules, the extinction coefficient, and the quantum yield (number of quanta emitted/number of quanta absorbed), the latter two varying with wavelength.

Optical Principles of Fluorescence Measurements for pH$_i$.

In the simplest arrangement, the single cells, groups of cells, or organelles containing the pH$_i$-sensitive fluorescent probe are excited at a single wavelength. The intensity of the emitted light, again monochromatic but of longer wavelength, is monitored by a photometric device and recorded. It is used as an index of pH$_i$. This approach has the disadvantage that unsuspected changes in the intensity of the light source or in the preparation (such as variations in optical path length, probe concentration, or photobleaching) produce changes in emitted light intensity that would erroneously be ascribed to changes in pH$_i$. For this reason, measurements at two wavelengths are preferable, in which the ratio of the two intensities is used as a measure of pH$_i$. The ratio should not be affected by these interfering factors. One of the two wavelengths may be that of the isoexcitation point (at which the emitted light is insensitive to pH changes), but this is not mandatory. Dual measurements can be done in one of two ways: (1) by exciting at two different wavelengths (achieved, for instance, by a rotating wheel containing two filters and interposed between light source and preparation) and measuring the emitted light at a single wavelength; or (2) by exciting at one wavelength and measuring the emitted light at two wavelengths (for instance, by two photomultiplier tubes in parallel). The choice of method depends on the particular fluorescent dye used; for some dyes, either method is applicable. It should be added that measurements of *absorbance* to evaluate pH$_i$ can also be made with these fluorescent probes, exactly as with nonfluorescent colored compounds. The preparation is illuminated at one or two wavelengths, and the absorbance measured at these same wavelengths (82, 375). A correction must be made for the intrinsic absorbance of probe-free cells.

Types of Indicators.

There are two general types of pH-sensitive fluorescent indicators: (1) strongly negatively charged molecules that in esterified, uncharged, colorless form readily diffuse into cells, where they are cleaved by esterases, the resulting charged molecules remaining "trapped" inside; (2) molecules that are weak bases whose distribution across the membranes of cells or organelles is determined by the transmembrane pH gradient. These indicators must, therefore, remain present in the external fluid. Two papers, some twenty years ago, initiated the use of these two types (108, 376).

Fluorescent trapped indicators. Of the several charged groups of this type of pH-sensitive dye, one is a phenolic group with pK 6–7 that is partially dissociated at the cell's pH. The difference in fluorescence of the deprotonated and protonated forms is used to evaluate pH$_i$. The other ionized groups are carboxyl groups (pK around 4) that remain largely in the deprotonated, negatively charged form. The indicator is added to the external fluid in esterified form in which most of the charges are neutralized. The colorless, lipid-soluble compound diffuses into the cell where esterase(s) release the fluorophore. Its now exposed charges retard or prevent it from leaving the cell ("trapped indicator"). The parent of one series of such compounds is the dye fluorescein. Fluorescein is one of the brightest dyes known, with a large extinction coefficient and a high quantum yield (0.9). However, it has only one carboxyl group in addition to its phenolic group and, therefore, rapidly leaks out of the cell. It also can enter the relatively alkaline mitochondrial compartment where, like other weak acids, it is highly concentrated. Thus, the signal emitted by the cell is contaminated by a mitochondrial contribution (375).

A derivative, 6-carboxylfluorescein, has one extra carboxyl group, resulting in a lower but still significant rate of leakage. Also, its pK (about 6.5) is rather low for intracellular measurements. Moreover, it requires lowering pH$_o$ to 6.2 to facilitate entry, since even in esterified form it still has one free carboxyl group, which must be in the protonated form for the molecule to diffuse into the cell (375). Another derivative, 5,6-dicarboxy fluorescein, has a similarly low pK and poor retention rate (313). The compound 4',5'-dimethyl-5(and -6)carboxyfluorescein (pK about 7.3) has also been used (82). The most attractive derivative is 2',7'-bis(2-carboxyethyl)-5(and -6) carboxyfluorescein (BCECF), which has four carboxyl groups (313). The pK of its phenolic group is about 7, and its retention is excellent though not perfect. Its quantum yield is as high as that of fluorescein. For these reasons it is the compound of choice of this fluorescent family (65, 302, 313, 415). In order to obtain a signal ratio,

BCECF is excited alternately at two excitation wavelengths (around 500 and 440 nm); the emission is monitored at around 530 nm. At 440 nm, the emitted light intensity is insensitive to pH.

With one exception, esterification of these derivatives is achieved by acetylation. At the low dye concentrations, the fall in pH_i due to the intracellular release of acetic acid that accompanies de-esterification is not significant (82). In the case of BCECF, where the acetoxymethyl ester is used, not only acetate and H^+ but also formaldehyde are released.

Fluorescein derivatives can also be retained in the cell by covalent binding to macromolecules such as dextran or proteins. In that case, they must be placed in the cell either by microinjection or by allowing them to be endocytosed. Endocytosed compounds have been used to measure the pH of lysosomes (277), of the Golgi complex (144a, 336a), and of individual isolated endocytic vesicles (282a, 338).

A problem in the use of fluorescein-derived indicators is that fluorescent drugs such as amiloride, SITS, or cinnamate and their derivatives, which are often used in the study of pH_i may interfere with the measurements. For this reason, indicators formed by combining a naphthalene moiety with rhodamine (so-called SNARFs; see 413) may sometimes be preferable because of their longer excitation wavelengths. One of this group, carboxy-SNARF-1 (carboxy-seminaphtorhodafluor-1) is excited at 540 nm and its emission measured at 590 nm and 640 nm. Most of the drugs mentioned above are not markedly excited at these wavelengths (67). Another group of such indicators is formed by combining a naphthalene moiety with fluorescein (SNARLs; see 413).

Details of other, less frequently used probes can be found in the reviews by Tsien (394), Kotyk and Slavik (217), and Haugland and Minta (162).

Fluorescent weak bases. The second type of fluorescent molecules that have been used to estimate the relative pH of the cytosol with respect to the external fluid, but especially the pH of acidic organelles, is a group of amines: 9-aminoacridine, atebrin, acridine orange, quinine, and acridine (227). They are weak bases whose distribution between cellular compartments depends on the pH difference of the compartments. Their charged (protonated) form is impermeant, while their uncharged (deprotonated) form, being lipid-soluble, rapidly equilibrates across membranes. This leads to the more acidic compartment containing a high total concentration of the probe. The probes must, of course, remain present in the external fluid during measurement; in this respect they differ from the first type of fluorescent indicators. The general subject of the distribution of weak bases (and acids) between com-

partments of different acidity has been discussed by Roos and Boron (317).

These fluorescent amines can respond to pH differences in several ways: *(1)* by a decrease in quantum yield in the presence of an acidic compartment; *(2)* by a pH-dependent shift in their fluorescence spectra; or *(3)* by a concentration-dependent shift. A few examples of these applications will be given. *(1)* When gastric microsomal vesicles suspended in an alkaline medium (pH about 8) containing 9-aminoacridine are artificially acidified (pH 4), they promptly reduce ("quench") dye fluorescence by as much as 70%. This seems to be due to self-interaction of the concentrated probe in the vesicles (108). Since the probe can bind to components of the cell, quantification in cellular systems is difficult (227). *(2)* Quinine and acridine show a red shift in their emission spectra and an increase in quantum yield in going from an alkaline to an acidic environment. By measuring the shift, the pH difference between vesicles and the medium in which they are suspended can be calculated by comparison with a standard curve (fluorescence vs. pH) (226). Much lower probe concentrations can be used than under *(1)*. *(3)* At very high concentrations of acridine orange, multimeric dye aggregates are formed and a new fluorescence peak appears (at about 640 nm), but the quantum yield is much lower than that of the monomer fluorescence. The most acidic compartments of a cell can accumulate enough dye to produce the polymeric red fluorescence, while the rest appears green. Thus, sea urchin eggs upon fertilization display red granules. Quantification is hardly possible (227).

Calibration of Trapped Indicators. Calibration of the probes trapped in the cell should ideally be done with the probes in situ. Their optical properties in the cell and in solutions are somewhat different due to possible probe binding to cell constituents or to variations in ionic environment. Intracellular calibration is achieved by collapsing the transmembrane pH gradient with high $[K^+]_o$ in the presence of the K^+/H^+ exchanger nigericin or a K^+-selective ionophore such as valinomycin plus a protonophore such as carbonyl cyanide m-chlorophenyl hydrazone (CCCP), while exposing the cell to solutions of different pH_o. The emission ratio–pH_i ($\simeq pH_o$) relationship is then measured over the pertinent pH_i range. Since calibration requires several preparations (of the same cell type), the results are normalized by comparing the ratios with that at a "standard" pH_i, for example, 7.0. The data can be fitted to a sigmoid curve that, in essence, is the titration curve of the weakly acid phenolic groups of the probe. At the inflection point, pH_i equals the probe's apparent intracellular pK. It should be emphasized that, since quantum yield varies with wavelength, the apparent pK depends on the particular wavelength pair chosen (29).

In a less elaborate way of calibration, the emission ratios of high $[K^+]_o$ nigericin-treated cells at very low and very high pH are first obtained; at these two pHs the probe should largely be in the protonated or deprotonated form, respectively. The unknown pH_i can then be obtained by a comparison with each of these two ratios (67, 152). In principle, all calibration approaches are combinations of the Henderson-Hasselbalch equation applied to the probe's pH-sensitive phenolic group, and Beer's law applied to the fluorescence emitted by each of the partners of the conjugate pair of this group.

Calibration, applicable to cell suspensions in cuvettes, can also be achieved by lysing the cells with digitonin. The pH of the medium is determined at which lysis produces no spectral change ("null method"; see 29). Corrections may have to be made to compensate for the small (about 5 nm) blue shift that occurs when the dye enters the medium from the intracellular environment (82, 313, 374).

Nuclear Magnetic Resonance Spectroscopy

Nuclear magnetic resonance (NMR) spectroscopy occupies a somewhat unusual position among the methods of measuring intracellular pH. It requires considerable theoretical and practical expertise and elaborate equipment, and is, in general, not employed if determining the H ion concentration is the only objective. Rather, its advantage in biological applications lies in that it permits the simultaneous acquisition of qualitative and quantitative information on a large number of cell constituents besides H^+.

The technique observes the magnetic character of atomic nuclei through the resonant response to a high-frequency magnetic field, when the nucleus is placed in a strong static magnetic field. Most nuclear species possess angular momentum, called nuclear spin. Its magnitude follows from quantum mechanics and is either a half-integral or integral multiple of $h/2\pi$ (h is Planck's constant). The abbreviated expression is "spin $+\frac{1}{2}$, $-\frac{1}{2}$, 1, 3/2," etc. Accompanying the spin is a magnetic dipole moment, as if electric charge were circulating around the spin's axis. In the static magnetic field, a torque acts on the nucleus through its magnetic moment and produces a precession about the direction of the field, like the precession of a spinning top. The frequency of this precession is determined not only by the strength of the static field, but, most important, also by the local fields generated by neighboring electrons circulating through the bonding structure. Thus, the precession frequency is characteristic not only of the

identity of the nucleus (^1H vs. ^2H, for example), in which case the frequency differences are large, but also of the chemical environment of the nucleus ($-CH_3$ vs. $-CH_2$, for example), in which case the frequency differences are small. When now the nucleus is irradiated with pulses containing a range of radio frequencies, each nucleus absorbs or emits that frequency which corresponds to its precession frequency: it is in resonance. Between pulses, the ground-state and excited-state spin populations return to their thermal equilibrium distribution: the system of spins relaxes with exchange of energy with the surrounding "lattice." At thermal equilibrium, the ground-state spin population is in slight excess over that of the excited state, and there is a net absorption of energy. It is this absorption that is detected at the resonant frequency. Pulse duration and time between pulses are critical parameters in the arrangement. For the majority of experiments, nuclei with spin $\frac{1}{2}$ are utilized; this includes the nuclei of ^{31}P, ^{19}F, ^{13}C, ^{15}N, and ^1H. The NMR spectrum obtained by observing, for instance, ^{31}P contains peaks for each chemically distinct nucleus of a ^{31}P atom. Their intensity (integrated area) is a measure of the amount of the particular chemical group or compound in the sample. The separation of the peaks is called the chemical shift. The peaks must be identified by appropriate calibration; their location in the NMR spectrum is given with reference to that of a well-known compound. Because the natural abundances of ^1H and ^{31}P are nearly 100%, ^1H and ^{31}P are the most commonly observed nuclei in biological work; the low concentrations of ^{13}C and ^{15}N (abundances are 1% or less) make them less suitable. The enormously intense ^1H resonance of ubiquitous water requires selective suppression techniques for ^1H detection of low concentrations of biomolecules.

The use of ^{31}P NMR spectroscopy for measuring pH_i is based on the chemical shift of certain intracellular phosphorus-containing, pH-sensitive compounds, the most common of which is inorganic phosphate. If there were no exchange of P between $H_2PO_4^{-1}$ and HPO_4^{-2} (the principal inorganic phosphates of the cell), each would show up as a separate peak. In fact, the exchange is so fast that the conjugate pair appears as one single peak whose chemical shift is their mole fraction-weighted average, and is thus a function of pH_i. The chemical shift of the reference compound should ideally be pH-independent (3), at least over the pH_i range that is being studied; phosphocreatine is often used. An in vitro titration of a phosphate-containing mock-cytosolic solution, in which pH and chemical shift are simultaneously measured, allows the assignment of absolute pH_i values to the experimentally observed shift. In cases in which phosphate is present in too small amounts, pH-sensitive compounds such as phos- phonium derivatives or 2-deoxyglucose (which phos- phorylates intracellularly in the 6 position) can be added. However, the lack of effect on pH_i of such extraneous substances must first be established.

One problem with the NMR method is the low signal-to-noise ratio. This can be compensated for by increasing magnetic field strength, sample volume, or cell concentration. It follows that the method cannot be applied to very small amounts of tissue or to single cells, although as little as 50 mg of muscle tissue has been successfully studied by ^{31}P NMR methods (156). Signal averaging is another way to obtain a better signal-to-noise ratio, at the expense, of course, of time resolution. Under favorable conditions it is now possi- ble to measure changes in pH_i (and of other cell constit- uents) with a time resolution of 15 s or even less (135). The studies should be carried out under proper physiological conditions, such as adequate oxygen- ation. This was not always observed in some of the older work in which the physical complexities of the arrangement received most attention.

The NMR method can be used (with caution) on compartments of intact cells such as mitochondria (135, 341), on concentrated suspensions of microor- ganisms or cells (143, 193, 369), or on tissues (130, 295). Of course, problems of interpretation multiply with the complexity of the preparation; identification of the various phosphate peaks and, especially, their assignment to specific compartments or subcompart- ments often requires separate procedures, and are sometimes speculative. Even the distinction between signals generated by intracellular and extracellular phosphate may give rise to some uncertainty. Relatively recent studies in which pH_i measurements and pH_i changes obtained with NMR were compared with re- sults from more conventional approaches (for instance, microelectrodes) have established the NMR technique as a valid and accurate method to measure pH_i (143, 156, 193, 305, 369).

For more details, the reader is referred to the physical literature (a good start is Ackerman and d'Avignon, [2]), and to summaries of the biological application of NMR (for instance, 130 and 135).

SOME OBSERVATIONS ON INTRACELLULAR pH TRANSIENTS

The mechanisms that contribute to pH_i regulation, discussed later, under MECHANISMS OF PH REGULATION are, in general, studied by first perturbing the steady conditions through a sudden change in the composition of the extracellular or intracellular fluid, and then

observing the time course of pH_i as it moves toward a new steady state. We shall give here five examples of the transients induced when an acid or alkaline load is imposed on a cell by exposing it to CO_2 (H_2CO_3) or to NH_3, the substances most commonly used for this purpose. Other weak acids, such as butyric or propi- onic acid, or bases such as methylamine are sometimes used, and acidification has also been achieved by direct intracellular injection of H^+ or NH_4^+ (377, 380).

Figures 9.1–9.4 illustrate responses to CO_2, obtained on different preparations. In the first three, pH_i was measured with a glass microelectrode; in the fourth, with a liquid ion exchanger (see earlier, under pH- Sensitive Microelectrodes). In Figure 9.1 (52), the course of pH_i is shown when a single giant barnacle muscle fiber, initially in CO_2-free artificial seawater (pH_o 7.6), was exposed at room temperature to seawa- ter of the same pH_o but containing 50 mM HCO_3^- and equilibrated with 5% CO_2. In the absence of CO_2, pH_i was stable at ~7.3; V_m was ~ -60 mV. This pH_i is significantly higher than the equilibrium value of ~6.6, that is, there is an inwardly directed electrochemical H^+ gradient. In the first transient, the CO_2 exposure lasted for only 12 min. The pH_i rapidly fell, leveled off at about 7.03, and, upon return to CO_2-free solution,

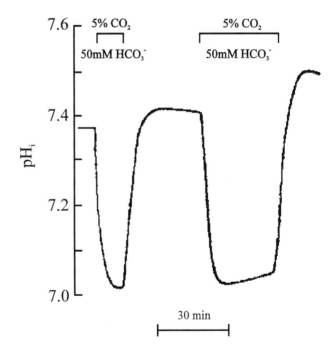

FIG. 9.1. pH_i transients of a barnacle muscle fiber in CO_2-free artificial seawater (pH 7.6, room temperature) upon exposure to a solution containing 5% CO_2/50 mM HCO_3^- of the same pH, mea- sured with a glass pH-sensitive microelectrode of the Hinke type (52). Reprinted with the permission of the American Physiological Society. For interpretation of this and the following four figures, see text under SOME OBSERVATIONS ON INTRACELLULAR pH TRANSIENTS.

slightly "overshot" the control pH value. The second, longer (33 min) CO_2 exposure produced nearly the same initial fall, but a greater overshoot. In both cases, the fall results from entry of CO_2 that, after hydration to H_2CO_3, yields H ions upon dissociation. Assuming that at the end of the fall in pH_i the CO_2 had (nearly) equilibrated across the membrane, the equilibrium potentials of H^+, OH^-, and HCO_3^- are all about $59(7.03-7.6) = -34$ mV. Since, at this point, V_m was -49 mV, passive fluxes (H^+ inward, OH^- and HCO_3^- outward) would produce further acidification, assuming the ions to be permeant. Instead, the fiber alkalinized ("recovered"). This indicates an active, energy-requiring acid-extruding mechanism, involving either one or a combination of these ions. From separate experiments it was concluded that influx of HCO_3^- was at least in part responsible. Other ionic requirements (Na^+ and Cl^-) of the acid extrusion will be examined later, under Na^+-Dependent Cl^-/HCO_3^- Exchanger. The magnitude of the initial fall in pH_i is dependent on the efficacy of H^+ buffering (see later, under Physicochemical Buffers), and also on the starting pH_i; the lower this value, the less the fall (mass action law).

Upon CO_2/HCO_3^- removal, CO_2 passively diffuses out of the fiber, and most of the intracellular HCO_3^- combines with H^+ and also leaves as CO_2. This not only cancels nearly the entire intracellular acid load brought about by CO_2 entry, but also "uncovers" the effect of extrusion of acid equivalents (entry of HCO_3^-) that took place during the exposure. This is the reason for the overshoot, and for its being greater, the longer the exposure to CO_2. It is blunted by any passive exit of HCO_3^-, but this should play only a minor roll.

Figure 9.2 (380) illustrates a similar experiment in a snail neuron that was, in succession, acidified by three decreasing CO_2 pulses at constant pH_o (7.5). Recovery (to ~7.2) from acidification was nearly complete at the end of about 20 min of CO_2 exposure. As in barnacle muscle, HCO_3^- influx against its gradient was found to be at least partly responsible. The details of the mechanism are probably similar to those in barnacle muscle. The record also shows that the cell promptly and nearly completely recovered from the alkaline load (overshoot) that was imposed by CO_2 withdrawal. The mechanism of this process was not further studied here; some general aspects of recovery from alkaline loading will be mentioned under Na^+-Independent Cl^-/HCO_3^- Exchanger. The shorter time frame of the events in snail neuron, as compared with those in barnacle muscle, can in part be ascribed to the greater surface/volume ratio of the neurons.

The result of a similar acid-loading experiment in frog muscle at pH_o 7.35 is shown in Figure 9.3 (1).

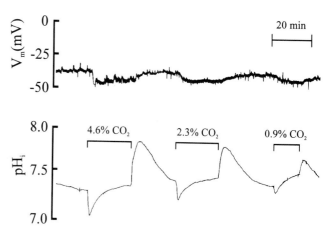

FIG. 9.2. V_m and pH_i transients in snail neurons in CO_2-free buffer (pH 7.5, room temperature) upon successive exposures to three different levels of CO_2/HCO_3^- at the same pH of 7.5. V_m was measured with a standard 3 M KCl-filled glass microelectrode and pH_i with a glass pH-sensitive microelectrode of the Thomas type. Modified from Thomas (380). Reprinted with the permission of the *Journal of Physiology* (London). The original pH_i trace has been refined by inverting the ordinate to make the figure intelligible to New World scientists.

No recovery at all was observed during the more than 20 min of exposure to CO_2, and thus there was no overshoot upon removal (Fig. 9.3A). But when the experiment was repeated in the absence of Na^+ (Fig. 9.3B), the initial rapid pH_i fall was followed by further, slower acidification, and CO_2 removal now produced an undershoot. Evidently, acid extrusion requiring Na^+ did take place in Figure 9.3A but was masked by acidification, shown (in other experiments) to be due to HCO_3^- efflux down its electrochemical gradient.

The fourth example of the course of pH_i in response to acid loading with CO_2 is given in Figure 9.4 (6). A guinea pig smooth muscle cell from vas deferens was initially in a solution containing 3% CO_2, 14 mM HCO_3^-; pH_o was 7.35 throughout. When CO_2/HCO_3^- was removed from the medium, pH_i after a transient overshoot, did not just return to its level in 3% CO_2, but actually fell to a value more than 0.2 unit lower. A comment on this rather surprising observation is given later, under Determinants of Steady-State pH_i.

A frequently observed response to an alkaline load is illustrated in Figure 9.5 (Putnam, unpublished observations). A monolayer culture of primary rat astrocytes was pulsed with 30 mM NH_4Cl (pH_o 7.4) in the nominal absence of CO_2. The fluorescent indicator BCECF (see earlier, under pH-Sensitive Fluorescent Indicators) was used to monitor pH_i. The resulting pH_i transient was the reverse of that resulting from CO_2 exposure: initial rapid alkalinization followed by slower acidification. The initial rise in pH_i is due to

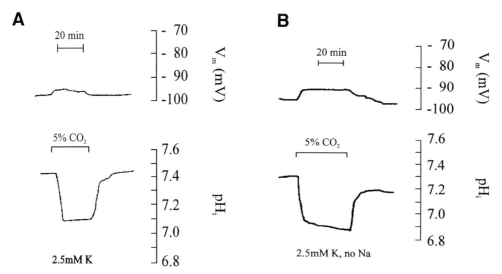

FIG. 9.3. V_m and pH_i transients in semitendinosus muscle fibers from the frog, *Rana pipiens*, in CO_2-free HEPES buffer (pH 7.35, room temperature), upon exposure to a solution containing 5% CO_2/24 mM HCO_3^- of the same pH. V_m was measured with a standard 3M KCl-filled glass microelectrode and pH_i with a glass pH-sensitive microelectrode of the Thomas type. Fibers were in solutions containing 2.5 mM K^+ either with *(A)* or without *(B)* external Na^+ (Na^+ completely substituted by NMDG) (1). Reprinted with the permission of the American Physiological Society.

diffusional influx of NH_3. At the end of the rise, NH_3 should (nearly) have equilibrated across the membrane. The subsequent downward-slanting pH_i course ("plateau acidification") is interpreted as due to the slower inward electrodiffusion of NH_4^+; both chemical and electrical gradients favor this movement (even though V_m was not measured as is customary in fluorometric studies, it almost certainly was negative). NH_4^+ entry drives the intracellular equilibrium $NH_4^+ \rightleftarrows NH_3 + H^+$ to the right, resulting in NH_3 leaving the cell. Thus, the NH_3/NH_4^+ pair functions as an inward shuttling system of H ions. The gain of H^+ equivalents during plateau acidification results in a pH_i undershoot when NH_4Cl is removed. Figure 9.5 also shows that without Na^+, no recovery from the undershoot took place, and that drugs such as amiloride similarly prevent recovery. This will be examined later, under Cation/H^+ Exchangers. It should be added that, in the presence of CO_2/HCO_3^-, the Na^+-independent Cl^-/HCO_3^- exchanger (mentioned under that heading) might also play a role in plateau acidification during NH_4Cl exposure.

The large permeability of most plasma membranes to NH_3 and CO_2 and the lower permeabilities to HCO_3^- and NH_4^+ are the bases of transients such as those shown in Figures 9.1 to 9.5. The pH_i of a few cells, however, behaves differently to exposure to NH_3/NH_4^+ or CO_2/HCO_3^-. For example, mouse and rat renal tubule cells from the medullary thick ascending limb (MTAL) of Henle's loop *acidify* rather than alka-

linize when their apical membrane is exposed to NH_3/NH_4^+ (141, 208). An unusually low (or even absent) apical-membrane permeability to NH_3, combined with a high permeability to NH_4^+, could be responsible. Indeed, when NH_4^+ entry is blocked, NH_3/NH_4^+ causes no pH_i change (208). Yet, under alkaline luminal conditions (which enhance the inward-driving force on NH_3), cell alkalinization does take place when NH_4^+ entry is blocked; an apical NH_3 permeability of 7×10^{-3} cm s^{-1} was measured (141). A low membrane permeability to NH_3 has also been observed in apical membrane of mammalian bladder (83a) and colonic crypt cells (346a). Parietal and chief cells of rabbit gastric glands, which secrete acid and pepsin, are another example of unorthodox pH_i transients. Exposure of their apical membranes to either NH_3/NH_4^+ or CO_2/HCO_3^- results in no change at all of pH_i, strongly suggesting that these membranes are nearly impermeable to both NH_3 and NH_4^+, and to both CO_2 and HCO_3^- (405). These gland cells are also resistant to extreme luminal acidification (406).

Our discussion of the expected pH_i transients upon exposure of cells to CO_2 is based on the assumption that CO_2 and HCO_3^- are in equilibrium, in both the extracellular and intracellular fluids. This is a reasonable assumption since in most cells, carbonic anhydrase (for review, see ref. 348a) is present, which greatly speeds up the conversion of CO_2 to HCO_3^- (actually, the hydration of CO_2); even the uncatalyzed equilibrium is achieved in only a few seconds. Thus, an equilibrated

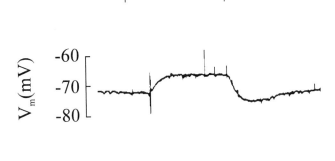

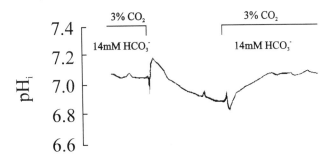

FIG. 9.4. V_m and pH_i transients in a guinea-pig smooth muscle cell from vas deferens. The cells were first exposed to 3% CO_2/14 mM HCO_3^-, then to CO_2-free solution, and finally again to 3% CO_2 solution. Temperature was 35°C. V_m and pH_i were measured with a double-barreled microelectrode using a reference liquid ion exchanger for the V_m barrel and a pH-sensitive liquid ion exchanger for the pH barrel. Modified from Aickin (6). Reprinted with the permission of the *Journal of Physiology* (London).

solution cannot contain only CO_2 and no HCO_3^-, nor only HCO_3^- and no CO_2. A recent technique (421c) describes the making of solutions that are out-of-equilibrium with respect to CO_2/HCO_3^-. It involves the rapid mixing of two solutions: one, an acid solution equilibrated with CO_2, the other a CO_2/HCO_3^--free solution buffered to a high pH. The two are designed so that their mixture has the desired pH (~7.4 at 37°C for mammalian cells) and will contain CO_2 but very little HCO_3^-. Conversely, a mixture containing HCO_3^- but little CO_2 can be made by the confluence of an alkaline solution containing HCO_3^- with a CO_2/HCO_3^--free solution buffered to a low pH. The mixtures contain a carbonic anhydrase inhibitor (348a), and should promptly (within 1 s) be applied to the preparation. It will now be possible to differentiate the actions of CO_2 and HCO_3^- separately in processes where, thus far, their relative roles could not be disentangled.

MECHANISMS OF pH REGULATION

The cytosol and the plasma membrane of every cell contain systems that serve to minimize changes of

the cytosolic pH. Such changes are brought about by metabolic events that release or consume H or OH ions, by movements of these ions across plasma membranes or organellar membranes, or by the protonated or deprotonated forms of weak acids and bases that may donate or accept protons upon entering the cytosol. The stabilization of pH_i is of vital importance to the life of the cell. Even small changes can have striking effects, especially on charge and conformation of proteins that contain ionizable groups, thereby modifying their functions. Examples of such functional effects will be given later, under SOME CELLULAR PROCESSES AFFECTING OR AFFECTED BY pH_i.

Stability requires that the disturbing H (or OH) ions are extruded from the cell, ideally as soon and as rapidly as they are produced. However, the membrane-bound transporters through which this removal takes place are relatively slow; they will be discussed later in this section. We shall first examine the intracellular mechanisms that can promptly take up (or release) H^+ and thereby minimize pH_i disturbances. By far the most important of these mechanisms is physicochemical buffering, the subject of the discussion that follows. It must be emphasized that the role of buffers can only be a limited and temporary one because of their finite capacity for H ions. Eventually, their original condition must be restored through removal from the cell of the acid or alkaline loads. Metabolic processes that can also play a buffering role, especially in mammalian brain and liver, have been summarized in two previous reviews (90, 317).

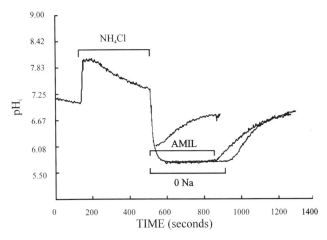

FIG. 9.5. pH_i transients in cultured monolayers of primary rat astrocytes upon exposure to a CO_2-free solution containing 30 mM NH_4Cl, measured with the pH-sensitive fluorescent dye BCECF. The pH_o was 7.4 throughout; temperature was 37°C. The superimposed pH_i courses are shown when, upon NH_4Cl removal, the solution was temporarily changed either to an Na^+-free solution (substitution by N-methyl-D-glucamine), or to one containing 1 mM amiloride (Shrode and Putnam, unpublished observations).

Physicochemical Buffers

A pH buffer, more precisely buffer pair, is a proton donor ("acid"), $(HA)^z$, in equilibrium with its conjugate proton acceptor ("base"), A^{z-1}.

$$(HA)^z \rightleftarrows A^{z-1} + H^+ \qquad (9-1)$$

where the exponents indicate the net charges of the molecules. Acids can be electroneutral ($z = 0$) or they can carry net positive ($z > 0$) or negative ($z < 0$) charges of any magnitude. For an acid to be an effective buffering partner in an aqueous medium such as the cytosol, it should be able to oppose the physiologically encountered alkaline loads (reductions in $[H^+]$) by yielding graded amounts of H^+, that is, by increasing its degree of dissociation (shift to the right of Equation 9-1). Its conjugate partner should be able to oppose acid loads (elevations in $[H^+]$) by combining with (nearly) all of the added H^+ (shift to the left of Equation 9-1). The partial and variable dissociation of the acids qualifies them as "weak." The buffer acid's dissociation constant, K (or dissociation exponent, pK $= -\log K$) is a crucial determinant of the conjugate pair's efficacy. In a cytosol whose pH usually ranges between 6 and 8, acids with pKs of, say, 2 will remain nearly completely deprotonated, whereas acids with pKs of 10 will remain protonated; such acids are, therefore, ineffective in buffering the cytosol.

The index of buffering power or "buffer value," β, is defined as the ratio of the small quantity of base added to a liter of water, $d[OH^-]$ (mmoles l^{-1}), over the resulting rise in pH: $\beta = d[OH^-]/dpH$. It is identical with the ratio of the small quantity of added acid, $d[H_3O^+]$, commonly written as $d[H^+]$, over the resulting fall in pH. Addition of acid represents "negative addition" of base; thus, β is always a positive number. It should be obvious that β represents the slope of the titration curve (OH^- vs. pH) when the conjugate acid is titrated with, say, NaOH, and varies with pH, even in simple solutions with only one buffer.

The buffer value of a single buffer pair can be evaluated from the mass action law applied to the equilibrium in Equation 9-1. If it is assumed, for simplicity, that the acid is electroneutral, then:

$$K' = ([H^+][A^-])/[HA] \qquad (9-2)$$

where K' is the acid's apparent dissociation constant at a specified temperature and ionic strength. To a close approximation, $[A^-]$ can be equated with the amount of OH^-—for example, in the form of NaOH—added to a liter of the weak acid. Equation 9-2, with this approximation, was first stated by Henderson (166), except that, in modern use, the bracketed H^+ connotes activity rather than concentration.

Closed Buffer Systems. Buffers whose total concentration, $[C]$, remains unchanged with pH changes are called "closed." Because $[C] = [A^-] + [HA]$, Equation 9-2 can be written:

$$K' = ([H^+][A^-]/([C] - [A^-])) \qquad (9-3)$$

Since, by definition, $dpH = -d\log[H^+] = -d[H^+]/2.3[H^+]$, β can now be obtained from equation 9-3:

$$\beta = d[OH^-]/dpH = d[A^-]/dpH \qquad (9-4)$$
$$= 2.3 [H^+] K'[C]/([H^+] + K')^2$$

From Equation 9.4, it can be seen that β approaches zero both when $[H^+]$ approaches zero and infinity. The conditions of maximum buffering, derived from Equation 9-4 by setting $d\beta/dpH$ to zero, are $[H^+] = K'$ or pH = pK', and (using Equation 9-3) $[A^-] = 0.5 [C]$, that is, half of the total buffer is in the form of the deprotonated partner. The maximum buffer value of any buffer, obtained from Equation 9-4 by equating $[H^+]$ and K', is 0.58 [C] (215, 316, 400). At pHs half a unit higher or lower than pK', β is 0.42 [C] (Fig. 9.6); at pHs 1 unit higher or lower, β is 0.19 [C] (Fig. 9.6); 2 units higher or lower, 0.022 [C].

It is to be pointed out that the conditions of maximal buffering power are those at which, upon addition of some base, the smallest change in *pH* occurs, and *not* the smallest change in $[H^+]$. The smallest change in $[H^+]$, that is, the maximal value of $-d[A^-]/d[H^+]$, occurs when (practically) all of the acid has been converted into its salt, that is, under very alkaline conditions. This is evident by observing that the value of this derivative, $K'[C]/([H^+] + K')^2$ (from Equation 9-3), approaches its maximum (namely $[C]/K'$) as $[H^+]$ approaches zero. On the other hand, the conventional expression for β, $d[A^-]/dpH$, is maximal when the *normalized* change of $[H^+]$, that is $d[H^+]/[H^+]$, is minimal, because, as pointed out above, $d[H^+]/[H^+]$ represents dpH.

Even low concentrations of buffer offer significant resistance to an imposed acid or alkaline load. For example, addition of 0.5 ml of a 1 millimolar HCl or NaOH solution to 1 liter of pure water of pH 7 changes pH by about 0.7 unit, but the presence of as little as 1 m*M* buffer of pK' 7 reduces this pH change to less than 0.001. Even under unfavorable buffering conditions—for example, pK' differing from pH by a full unit—the pH change is less than 0.003 and the free $[H^+]$ change is only about 0.1% of what it would have been without the buffer.

From the definition of β it is clear that the total buffer value of a mixture of buffer species is the sum of the individual values at the particular pH, each expressed as in Equation 9-4.

Finally, even without weak electrolytes water has a finite buffering power. This is not surprising, since H_2O itself is both a weak acid, OH^- being its conjugate base, and a weak base, the hydrated proton, H_3O^+, being its conjugate acid. The buffering power of water, β_{H_2O}, takes these two properties into account:

$$\beta_{H_2O} = d[OH^-]/dpH - d[H^+]/dpH \quad (9-5)$$
$$= 2.3 \,([OH^-] + [H^+])$$

This holds also for water containing strong acids or bases. The buffering power of water is least at neutral pH ($[H^+] = [OH^-] = 10^{-7}\,M$), namely 0.00046 m$M$. It assumes high values at great acidity and alkalinity; at pH 2 and 12, $\beta_{H_2O} = 23$ mM, at pH 1 and 13, 230 mM, etc. The buffering power of water can also be obtained in a slightly different way (400). For example, when the concentration of a $10^{-2}\,M$ solution of HCl (pH 2) is raised to $10.1 \times 10^{-3}\,M$ (pH 1.9956 . . .), β_{H_2O} is $0.1/(2 - 1.9956 . . .) = 23.1$ mM, in agreement with β_{H_2O} from Equation 9–5.

Application to Intracellular Conditions. Unfortunately, a reasonably complete picture of the various cytosolic buffer species, their pKs, and their concentrations is available for only a few cells. Of the mobile buffers present in nominally CO_2-free frog skeletal muscle, carnosine (β-alanylhistidine, pK ~6.78) is by far the most important one (9.1–19.5 mmoles per liter cell water), followed by inorganic phosphate (1.4–6 mM, pK_2 of phosphoric acid ~6.8) (140). Immobile, fixed buffers in the form of myofibrillar and other proteins are present in much higher concentrations. Their histidine residues, which are mainly responsible for their buffering power, may come to about 62 mM (99). Both carnosine and histidine owe their buffering ability to the imidazole group. All these pKs are at 25°C but usually unspecified ionic strength. Assuming pH_i to be uniform throughout the cell water (this is almost certainly not the case; see later, under Cytosolic H Ion Movements and the Role of Buffers), the combined buffering power due to these three categories amounts to 34–40 mM at pH_i 6.6, 22–28 mM at pH_i 7.0, and 12–16 mM at pH_i 7.4, of which 26, 14, and 7 mM, respectively, is due to proteins. This is in reasonable agreement with direct measurements (21) over the pH_i range 6.0 to 7.4 of the non-CO_2, so-called intrinsic, buffering power. Not included in the calculation are the terminal α-amino groups whose pKs are below 8 and sometimes below 7, due to the acidifying effect of neighboring peptide and disulfide linkages (115); their buffer contribution is difficult to evaluate. The increase of intrinsic β with decreasing pH_i over the physiological range has also been found in other cell types whose precise buffer compositions are unknown (for references, see 21).

Almost equally detailed cytosolic analyses have been made on giant axons of *Myxicola infundibulum* (134), a marine polychaete, and of the squid (*Loligo pealii*; see 111, 213). Both contain large amounts of free amino acids, ~400 mM total, nearly half of the axons' osmotic requirements. Glycine, cysteic and aspartic acids, and, in squid axon, isethionic acid are predominant. Small amounts of histidine residues are present in the neurofilaments. Nonbicarbonate β of the squid axon (pH_i ~7.3), measured by pulsing with NH_3/NH_4^+ (53) (see later, under Measurement of Intracellular Buffering Power) is 9–11 mM owing to the free amino acids (pKs of the amino and sulfhydryl groups is ~10, of the carboxyl groups, ~2), β (obtained by titration of the extruded axoplasm) increases to more than 200 mM both at unphysiologically high pH (9.5–10.5) and low pH (2–4) (361). In *Myxicola* extruded axoplasm (pH_i 7.5 at 10°C), β, measured by titration, similarly rises steeply from ~18 mM at pH 7 to more than 80 mM at pH 8 (11).

Open Buffer Systems. Whereas in closed buffers their *total concentration* remains unchanged with pH_i changes, in a second group, the so-called open buffers, it is the intracellular concentration of the *uncharged partner*, protonated or deprotonated, that does not change. This is the case when two conditions are fulfilled: *(1)* the uncharged partner is also present in the external fluid, and rapidly equilibrates across the plasma membrane, while the charged partner permeates the membrane, at best, only very slowly. Carbonic acid and a number of weak acids such as propionic acid are examples of highly permeant uncharged protonated forms; bases such as NH_3 and procaine are examples of permeant uncharged deprotonated forms (membrane permeability to, for instance, HCO_3^- and NH_4^+ may be significant under certain conditions—see earlier, under SOME OBSERVATIONS ON INTRACELLULAR PH TRANSIENTS); *(2)* volume and/or turnover rate of the external fluid is sufficiently great and mixing is sufficiently efficient for its composition (pH and buffer concentration) to remain unaltered over a period of time. Under most in vivo and in vitro circumstances, this condition is satisfied. (An exception are the studies of Dubuisson [114], in which the external CO_2/HCO_3^--containing fluid of in vitro frog muscle was reduced to a thin film.)

When, under open buffer conditions, the uncharged partner is the *protonated* form, for example, $H_2CO_3(CO_2)$, the transmembrane ratio of its charged partner is *inversely* proportional to the H^+ ratio, provided K' is the same inside and outside of the cell. This follows from the mass action law applied to both sides of the plasma membrane: $[CO_2]K' = [H^+]_i[HCO_3^-]_i$ = $[H^+]_o[HCO_3^-]_o$. On the other hand, when the uncharged partner is the *deprotonated* form, for example, NH_3, the transmembrane ratio of its charged partner is *directly* proportional to that of H^+: $K'/[NH_3]$ = $[H^+]_i/[NH_4^+]_i = [H^+]_o/[NH_4^+]_o$. The transmembrane distributions of the protonated and deprotonated partners of weak acids and bases have been discussed in detail (317).

The buffering power of an open buffer system can easily be derived. When the protonated form, HA, of an acid remains constant with pH changes, K' = $([H^+][A^-])/[HA]$ (Equation 9–2) simplifies to $[A^-]$ = $M/[H^+]$, where the constant $M = K'[HA]$. This gives $\beta = d[A^-]/dpH = 2.3 M/[H^+]$ or $2.3 [A^-]$. Thus, for the open HCO_3^-/CO_2 buffer system, $\beta = 2.3 [HCO_3^-]$. When the deprotonated form, B, of a base ($HB^+ \rightleftarrows B + H^+$) remains constant, $[HB^+] = [H^+]/N$, where the constant $N = K'/[B]$ and $\beta = -d[HB^+]/dpH$ = $2.3 [H^+]/N$ or $2.3 [HB^+]$. Thus, for the open NH_3/NH_4^+ buffer system, $\beta = 2.3[NH_4^+]$.

The buffer value of an open system can greatly exceed the corresponding closed-system value. For instance, in a resting in vivo frog skeletal muscle fiber (pH_i ~7.1), $[CO_2]_i$ is about 0.9 mM (P_{CO_2} 30 mm Hg, CO_2 solubility 0.03 mmole l^{-1} mm Hg^{-1}). The pK'_1 ("overall" first ionization exponent) of CO_2 is 6.1 (304), and thus $[HCO_3^-]_i$ comes to 9 mM. If the fiber were closed to CO_2, β_{CO_2} at this pH_i would be small, only 1.9 mM (Equation 9–4), because pH_i and pK'_1 differ by a full unit. But with the system open to CO_2, which is the usual situation, $\beta_{CO_2} = 2.3 \times 9 = 20.7$ mM, about the same as the non-CO_2 buffering power (21). The stabilizing effect on pH_i of this open system is further illustrated by the following example. If no intracellular buffers other than HCO_3^-/CO_2 were to be present, one millimole of H^+ added to a liter of fiber water would, under closed conditions, reduce $[HCO_3^-]_i$ and raise $[CO_2]_i$ each by 1 mM, raise P_{CO_2} from 40 to 73 mm Hg and lower pH_i by 0.30. Under open conditions, $[HCO_3^-]$ is again reduced by 1 mM, but the generated CO_2 rapidly leaves the cell so that P_{CO_2} remains unchanged, and pH_i falls by less than 0.04. The advantage of the open system over the closed one for a weak acid such as CO_2 (H_2CO_3) becomes more striking the higher the pH. On the other hand, for a weak base like NH_3 the advantage is more striking the lower the pH. This is shown in Figure 9.6, which compares the buffering powers of open and closed systems as a function of pH.

It was pointed out many years ago by Van Slyke (400) that the intact organisms of higher animals can be considered open to CO_2: the CO_2 concentration (proportionate to the partial pressure of CO_2) in the

intracellular and extracellular fluids, and thus in the blood, is usually kept within narrow limits by lung ventilation and its regulatory mechanisms. This results in the HCO_3^-/CO_2 buffering system being of great importance in pH stabilization of the blood and all other compartments, even though the average pH (~7.4) in the body fluids is more than a unit higher than the "overall" pK'_1 of carbonic acid.

Effect of Temperature. In general, the pK of a weak acid or base increases as the temperature is lowered according to the integrated Van't Hoff equation (see 317). The change in pH with temperature of a buffer solution is practically equal to the change in pK; the electrolyte's degree of dissociation remains unchanged. That is, the concentration of each of the partners of the buffer pair is unaffected by temperature. These relations hold only for closed systems. For instance, in a closed CO_2/HCO_3^- buffer, the CO_2 concentration (product of P_{CO_2} and CO_2 solubility) remains unchanged when temperature is lowered: P_{CO_2} is reduced while solubility increases. But when the system is open, it is P_{CO_2} that remains unchanged, and thus CO_2 concentration increases. The result is that pH of the open buffer increases less than pK. Even though, in

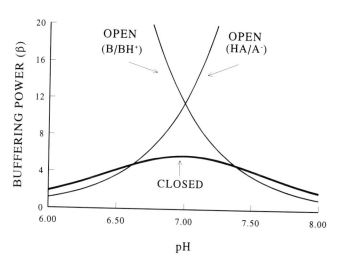

FIG. 9.6. Comparison of the buffering power, β, under closed and open conditions of each of two monovalent buffers; a weak acid, HA/A^-, and a weak base, B/BH^+. "Closed" means fixed total buffer concentration, [C]. "Open" means fixed concentration of the uncharged partner, [HA] or [B]. For simplicity, the buffers are assumed to have the same pK' of 7.0 and, at pH = pK', the same [C], 10 mM, both when open and closed. The β-pH relationship of the two closed buffers is the same; β is maximal at pH = pK', namely 5.76 mM/pH unit (that is, 0.576[C]). For the open buffers, β at pH = pK' is twice this value, or 11.52 mM/pH unit (that is, 2.303[A$^-$] or 2.303[HB$^+$]), but their β-pH relationships have no maximum. Instead, β increases exponentially with pH in the case of the acid, and decreases exponentially with pH in the case of the base. See text under Physicochemical Buffers for further discussion.

general, an inverse relationship has been found between pH_i and temperature (317), it is not a straightforward matter to apply these theoretical considerations to physiological conditions in which pH_o and HCO_3^- are often manipulated, and rates of acid extrusion and metabolism may vary with temperature.

Cytosolic H Ion Movements and the Role of Buffers. Thus far, the pH of the cell has been assumed to be uniform throughout the cell water. However, regional differences constantly arise, brought on by a variety of events. The production or removal of H^+ by metabolic processes in the cytosol leads to local acidification or alkalinization close to the pertinent substrates and enzymes. The activity of the membrane-bound exchangers that extrude H^+ equivalents will raise pH_i in their immediate vicinity, whereas mechanisms that involve extrusion of base equivalents acidify adjacent cytosolic regions. In the case of the mitochondria (see later, under Mitochondrial H^+ Movements), proton extrusion from their matrix may acidify the cytosol next to these organelles. Also, protons are taken up by the sarcoplasmic reticular cisternae of skeletal muscle fibers when they release calcium in response to depolarization (33, 185); this produces regional alkalinization of the cytosol. Such H ion differences generate diffusional movements in which pH buffers play an important role. Mobile buffers greatly enhance the *steady-state* flux of H^+ (151, 154). Since free H ions are present in only micromolar amounts, their flux is driven by very small concentration gradients and thus provides a slow and ineffective mechanism for dissipating differences in $[H^+]$. On the other hand, the accompanying gradients of H ions bound to buffers are in the millimolar range. For this reason, mobile buffers greatly facilitate H ion transport, often by several orders of magnitude. Their large gradients more than make up for their slower diffusion (117). If the buffers are immobilized, their facilitating role should be abolished. Indeed, when phosphate buffers are attached to large cellulose particles in in vitro experiments, they have no effect on H^+ diffusion (151). As an example of in vivo facilitation, buffers may rapidly replenish the store of free H ions at the inner mouth of H^+ channels (see later, under Passive Transmembrane Proton Movements). The number of ions that leave a cell such as a salamander oocyte through these channels in a fraction of a second can equal or surpass the total number of intracellular free H ions (31). It might be added that facilitation of H^+ diffusion by mobile buffers is entirely comparable to facilitation of oxygen diffusion by oxyhemoglobin (251, 419).

A second aspect of H^+ kinetics is also affected by buffers: *the rate of propagation* of an H ion disturbance

originating as a local change in pH_i. Buffers, either mobile or fixed, slow this rate. In other words, they reduce the apparent diffusion constant of H^+ (D_H^{app}). This will intuitively be apparent: the diffusion constant of any buffer is always less than that of H^+. The original papers (11, 186, 198) should be consulted for the rigorous derivation of D_H^{app}. What is most useful is that the resulting expression (which is based on Fick's second law of diffusion) can to a close approximation be simplified (186) to:

$$D_H^{app} = D_M \beta_M/(\beta_M + \beta_F) \qquad (9-6)$$

where D_M and β_M are the diffusion constant and buffering power of the mobile buffer, and β_F is the buffering power of the fixed buffer (in general, proteins). It is assumed that (1) the pH differences are small (186) (a less realistic assumption that $[H^+]$ is much less than the buffers' dissociation constants leads to the same end result [198]); (2) the diffusion constants of both conjugate partners of the mobile buffer are the same; (3) H^+ binds rapidly to the buffers; and (4) the buffers do not exchange with the extracellular environment. For simplicity, only one species of each of the two buffer types, each with a single dissociable group, is considered in equation 9–6, which can, however, easily be generalized (186). The equation shows that in the presence of only mobile buffer, the apparent H ion diffusion constant approaches the diffusion constant of the buffer. According to the more complete expression for D_H^{app} (not shown), a fixed buffer by itself slows H^+ diffusion by 4 or 5 orders of magnitude. Thus, adding a mobile buffer to a system containing only immobile buffers will speed up diffusion of H^+; the rate will, however, remain below that of free diffusion. Equation 9–6 might explain why in *Myxicola* axoplasm the apparent H ion diffusion constant at pH_i 8–9 is several times that at pH_i 5–6 (11). At the high pH_i's, the amino groups of the large amount of free amino acids (about 400 mM) importantly contribute to β_M, while the contribution of the fixed buffers presumably remains relatively unchanged and might even fall.

Equation 9–6 has been applied (186) to intact frog twitch muscle fibers (pH_i ~7) whose buffer composition is reasonably well known (140). With β_M (mostly due to carnosine) = 12 mM, D_M = 5 × 10^{-6} cm^{2}s^{-1}, and β_F (due to proteins) = 21 mM, D_H^{app} comes to 1.8 × 10^{-6} cm^2 s^{-1}. Compared with this, D_H in aqueous solution (18°C) is 8 × 10^{-5} cm^2 s^{-1}, nearly 50 times as great. A cytosolic pH gradient established by cisternal Ca^{2+}/H^+ exchange and extending over half a sarcomere's length (~1 μm) would decay in about 3 ms (186).

Recently, the apparent H^+ diffusion constant in the

presence of buffers has been computed without the assumption that the pH differences are small. Although the work (404) actually addresses the diffusion of Ca^{2+}, it is applicable to H^+ as well. In this more complete expression, a second term, which is a nonlinear function of the H^+ gradient, is added to the equation of Irving et al. (186); this term can be significant when the H^+ gradient is large.

Finally, there is the matter of H^+ diffusion along membrane surfaces, which becomes especially important in the case of the mitochondrion (see later, under Mitochondrial H^+ Movements). How do H ions, pumped across the inner membrane of this organelle, reach other proteins in the same membrane, such as F_0-F_1 H^+ ATPase? The pumped protons might equilibrate with the aqueous bulk phase before being consumed, or they could laterally diffuse in the membrane interface compartment, at a rate faster than the surface to bulk transfer rate (164). Lateral diffusion would probably be mediated by surface-bound buffer groups (acidic and basic surface residues such as lipid head groups). The dwell time of protons on these groups is known to decrease with lower pKs. Indeed, head groups with low pKs enhance proton movement, an observation that strengthens the lateral-movement option (334).

Measurement of Intracellular Buffering Power. Measurements of β are, by definition, performed by observing the displacement of pH_i when small known quantities of H^+ or OH^- equivalents are added to known volumes of cell water. The most direct techniques of addition are electrophoresis and pressure injection. The former employs two micropipettes through which H or HCO_3^- ions are electrically injected. The total injected charge is measured, and from their previously determined transference numbers the current fractions carried by these ions are obtained. An estimate of cell volume and of percentage cell water is required. With pressure injection of HCl, Cl^- activity can be monitored with a Cl^--sensitive microelectrode. The presumed identity of intracellular H^+ and Cl^- spaces obviates knowing cell volume. However, the Cl^- activity coefficient must be known. These approaches, applied to snail neurons in the nominal absence of CO_2, have yielded values for the intrinsic (that is, non-CO_2) buffering power of the order of 10 mM (242, 378).

A second method for measuring β is indirect but less difficult; it is the most commonly used approach (52, 60). It takes advantage of the rapid equalization, discussed above, of the uncharged conjugate partner of a weak acid or base across the plasma membrane, the charged partner being very much less permeant if at all. Before the test, the cell should not contain the

acid or base. The cell is exposed to a known total concentration of, say, propionic acid + propionate in a bathing solution of known pH. The Henderson-Hasselbalch equation, pH = pK' + log ($[A^-]/[HA]$), gives the external, and thus intracellular, concentration of propionic acid if its pK' is known. Some of the entering molecules dissociate; one propionate ion is formed for each proton released. Once propionic acid equilibrium is reached, the reduced pH_i, monitored by microelectrode or optically, is used to calculate intracellular propionate concentration, again using the Henderson-Hasselbalch equation, on the assumption that the intracellular and extracellular pKs are the same. Buffering power is thus obtained: $\beta = -\Delta[\text{propionate}]_i/\Delta pH_i$. CO_2 can equally be used as long as the cell initially is (nominally) CO_2-free; the external (and thus intracellular) CO_2 concentration is the product of P_{CO_2} of the bathing fluid and CO_2 solubility, assumed to be the same on both sides of the membrane. Weak bases such as NH_3 can similarly be used. The cell is exposed to a known concentration of, say, NH_3/NH_4^+ at known pH, and the change in pH_i is measured. Some of the entering NH_3 accept a proton, one NH_4 ion being formed for each proton removed from the cell water. The calculated $[NH_4^+]_i$ provides the value for β: $\beta = \Delta[NH_4^+]_i/\Delta pH_i$. Since buffering power varies with pH_i, these βs refer only to a cytosolic pH intermediate between the initial and final values. It is, therefore, desirable that the pH_i changes be kept as small as is compatible with the resolving power of the measurements.

The weak-acid method is vitiated by acid extrusion that is stimulated by the fall in pH_i; this would lead to an overestimate of β. Extrusion can be blocked by agents such as stilbene derivatives or amiloride or its derivatives. The NH_3/NH_4^+ method suffers from the permeability—often not inconsiderable—of the plasma membrane to the NH_4 ion, which also leads to an overestimation of β. One way to minimize this effect is to calculate β from the fall in pH_i when NH_3/NH_4^+ is removed from the bath: most of the intracellular NH_4^+ will leave as NH_3 because the electric field opposes NH_4^+ exit. Other ways to minimize complications in the use of these methods have been discussed by Roos and Boron (317).

Membrane Transport Systems

Over 100 years ago, Nasse (266) found that exposing red blood cells to CO_2 resulted in a loss of Cl^- (he called it "kitchen salt") from the serum. This represents the first evidence of the existence of a transporter, the Cl^-/HCO_3^- exchanger. The stoichiometry of a 1-for-1 exchange required an additional 80 years to be discovered (384). It was not until the early 1970s that the protein responsible for this anion exchange was identified. SDS gel electrophoresis of the membrane proteins of red blood cell ghosts yielded six major bands (120, 212). It is the protein of the third band (molecular weight 89 kDa) that mediates Cl^-/HCO_3^- exchange. Shortly thereafter, two additional membrane-bound pH-regulating transport systems were described: an Na^+/H^+ exchanger in rat intestinal and renal proximal tubule cells (261) and sea urchin eggs (195), and a mechanism that exchanges external Na^+ and HCO_3^- (or CO_3^{2-}) for internal Cl^- in invertebrate nerve cells (squid giant axons and snail and crayfish neurons; see 253, 325, 380) and barnacle muscle cells (60, 61). Further experiments have shown that many cells have multiple pH-regulating transport systems; also for each type of transport system, there appear to be several distinct isoforms (proteins with similar function but somewhat different molecular structure and possibly different ligand affinities and kinetics), sometimes even in the same cell. These transporters may contribute to other cellular processes in addition to regulation of pH_i, such as regulation of intracellular Na^+ and Cl^- concentrations, maintenance of cell volume, and transcellular movements across epithelia of ions such as Na^+, Cl^-, and H^+. We shall now briefly review the various exchangers.

Cation/H$^+$ Exchangers. Two different types of cation/ H^+ exchangers have been described: the alkalinizing Na^+/H^+ exchanger (NHE) (Na^+ entering, H^+ leaving), and the acidifying K^+/H^+ exchanger (KHE) (H^+ entering, K^+ leaving). The NHE is electroneutral (but see 5, 371). It is inhibited to varying degrees by the diuretic amiloride and its analogs (see reviews of Benos [37] and of Kleyman and Cragoe [211]) and is nearly ubiquitous in vertebrate cells (124, 273a). An analogous Na^+-for-H^+ transport is seen in yeast and bacteria, but the proteins that mediate this exchange are not homologous to vertebrate NHEs (273a). We will therefore restrict our comments to NHEs from vertebrate cells. The KHE is also electroneutral and is inhibitable by amiloride. It has been reported in nucleated red blood cells (76) and corneal epithelial cells (48). Both NHE and KHE are involved in cell-volume regulation (see later, under CELL-VOLUME REGULATION).

Thermodynamically, the NHE under most conditions is poised far from equilibrium: the inward electrochemical Na^+ gradient greatly exceeds the inward H^+ gradient. (When the Na^+ gradient is reversed, the exchanger can actually acidify the cell.) Kinetically, the most important intracellular stimulus to its activity is cell acidification, especially since the H ion, independent of its role as substrate, also binds allosterically.

This results in a greater-than-linear dependence of NHE on intracellular H^+ (24, 26). Most NHEs normally operate at near saturation with respect to external Na^+, so that changes in Na^+ concentration hardly change their activity. The effect of external H^+ depends on whether or not it is a purely competitive inhibitor of Na^+ (24).

Two additional observations about the exchanger's function should be mentioned: (1) Kinetic analysis of amiloride-sensitive ^{22}Na uptake (which is indicative of NHE activity) in renal brush border membrane vesicles at 0°C reveals a biphasic uptake. The initial uptake is very rapid with a Hill coefficient greater than 1, and is followed by a linear uptake phase (Hill coefficient = 1). The initial phase suggests that the transporter forms oligomers (282). The formation of homodimers of NHE has also been observed in fibroblasts (119a). (2) Exposure of cells to chronic acidosis can result in increased transcription and in increased number and overall activity of NHE (184), which, thus, may in part be determined by the previous acid/base history of a cell.

A number of substances (for example, hormones, growth factors, neurotransmitters) can activate NHE by a variety of means, including phosphorylation of an internal domain of the exchanger, by the binding of Ca^{2+}/calmodulin to a putative autoinhibitory domain on NHE (thus releasing inhibition) and possibly by the interaction of a regulatory protein with NHE (for review see ref. 273a and 421a). This process is often mediated by phosphoinositide breakdown leading to increased intracellular Ca^{2+} and increased activity of protein kinase C (38). Activation involves an increase in the affinity of NHE for H^+ either at the transport site or the allosteric modifying site (96, 142, 407).

The Na^+/H^+ exchanger is a glycoprotein of 80–110 kDa (about 820 amino acids), with 10–12 putative membrane-spanning regions (N terminal domain of about 500 amino acids) that include both the transport sites for Na^+ and H^+ and the H^+_i allosteric modulatory site (407). There are putative glycosylation sites on the first external loop of this domain (96, 273a). NHE also contains a cytoplasmic domain (C terminal end of over 300 amino acids) that contains the major regulatory and phosphorylation sites of the exchanger, including serine (330). Phosphorylation of the cytoplasmic domain can modify the H^+ affinity of sites within the membrane spanning domain. The roles of the membrane spanning domain in transport and the cytoplasmic domain in regulation are clarified by experiments where NHE mutants, lacking the cytoplasmic domain, are expressed in fibroblasts that have been made to lack a functional NHE. In such cells, the expressed NHE is capable of Na^+/H^+ exchange and contains the H^+ allosteric modulatory site, but it can no longer be activated by the usual NHE-activating agents such as growth factors (407).

Not all Na$^+$/H$^+$ exchangers are the same (86). A few examples of functional differences will be listed. *(1)* It was thought that all NHEs are amiloride-sensitive, but an amiloride-insensitive NHE (K_I for amiloride 10–25 times higher than amiloride-sensitive NHE) has been demonstrated in several cell types, including hippocampal neurons (307), porcine kidney epithelial cells (155), and brush border membranes from placenta (221) and kidney (319). *(2)* Some NHEs apparently lack the internal H$^+$ modifier site (306). *(3)* Some NHEs have an ion selectivity that is less restricted than that of the "common" NHE (NHE-1) (293).

Progress has been made in correlating some of these

functional differences with different NHE isoforms (273a, 421a). At least five isoforms of NHE have been described (Table 9.1). The first, NHE-1, is found in virtually all cells and is believed to be the "house-keeper," responsible for pH$_i$ regulation. NHE-2 is predominantly localized in the apical membranes of villus epithelial cells of gastrointestinal tract (jejunum, ileum, and large intestine) and kidney medulla, where it may play a role in the reabsorption and transcellular movement of Na$^+$ (49). NHE-3 resides in NaCl-absorbing epithelia of intestine (especially ascending colon, jejunum, and ileum) and kidney cortex (proximal tubule brush border membranes; see 43 and 351). There is good evidence that this isoform mediates Na$^+$ influx across the apical membranes of these reabsorbing epi-

TABLE 9.1. *Properties of the Isoforms of Vertebrate Na$^+$/H$^+$ Exchange*

Isoform	M.W. (kDa)	% Homology with NHE-1	Amiloride Sensitivity[b]	Localization[c]	Effectors[d]	References
NHE-1	91	>90[a]	+ (AMIL)	Ubiquitous	Growth factors (K_m effect)[e]	(329)
			+ (EIPA)		Cell shrinkage	(330)
			+ (MPA)		PKC[f]	(388)
					Phorbol esters	(158)
NHE-2	91	42–50	+ (AMIL)	*GI tract* (apical membranes)	Growth factors (V_{max} effect)[g]	(183)
			− (EIPA)	Liver	PKC	(91)
			− (MPA)	Uterus	Phorbol esters	(96)
						(179)
				Kidney		
				Adrenal gland		(390)
				Skeletal & cardiac muscle		(409)
				Brain		
NHE-3	93	38–41	− (AMIL)	*GI tract* (mostly intestine)	Serum (V_{max} effect)	(389)
			− (EIPA)	*Kidney*	Glucocorticoids	(97)
			− (MPA)	Heart	PKC–Inhibition	(391)
				Brain	Phorbol ester–Inhibition	(420)
						(421)
NHE-4	81	43	?	*GI tract* (mostly stomach)	Cell shrinkage	(281)
				Kidney		(50)
				Brain		(392)
				Uterus		
				Skeletal muscle		
β-NHE	85	64	+ (AMIL)	*Trout RBC*	cAMP	(51)
			+ (MPA)		Arrestin-desensitization	(258)

[a] NHE-1 isoforms from different tissues and species have more than 90% amino acid homology.
[b] AMIL—amiloride; EIPA—5-N-ethyl-N-isopropylamiloride; MPA—5-N-(methylpropyl) amiloride. A + indicates that amiloride or its derivative inhibits the exchanger; a − indicates that amiloride or its derivative does not inhibit the exchanger.
[c] Italicized tissues represent major sites of localization of isoform.
[d] An effector is any factor that can either stimulate or inhibit. Factor is stimulatory unless indicated otherwise.
[e] Indicates that the growth factor increases the affinity of the exchanger for internal H$^+$.
[f] Protein kinase C.
[g] Indicates that the growth factor increases the maximal velocity of the exchanger.

thelia (393, 420). In addition to the characteristics listed in Table 9.1, these three isoforms can be pharmacologically differentiated based on their inhibition by a new NHE inhibitor, HOE694 (96a, 97). HOE694 [(3-methylsulphonyl-4-piperidinobenzoyl) guanidine methanesulphonate] has a guanidinium group that is similar to that of amiloride but has a very different K_I value for NHE1, 2, and 3 (0.16, 5, and 650 μM, respectively). The function of NHE-4, which is localized largely in stomach (especially antrum) and intestine (proximal small intestine, cecum, proximal colon), is not known. It may be activated by cell shrinkage (50), which would suggest that it plays a role in volume regulation, especially in the renal medullar environment, which experiences high osmotic pressures. Finally, isoform β-NHE is present in the membrane of trout red blood cells (51). It is activated by cAMP but unaffected by cell acidification except when already activated by cAMP or phorbol esters (258). Catecholamines released by hypoxia activate this isoform; the resulting red blood cell alkalinization helps compensate for the hypoxia by increasing the O_2 content of the hemoglobin (Bohr effect).

Very little is known about the structure and distribution of the K^+/H^+ exchanger. Under normal conditions, it exchanges intracellular K^+ for an external H^+ and thus achieves net solute efflux and cellular acidification. It has been found in nucleated red blood cells from *Amphiuma* (76), where in conjunction with Cl^-/HCO_3^- exchange it mediates the efflux of KCl (and therefore of cell water) during regulatory volume decrease in response to cell swelling (see later, under Cell-Volume Regulation), and in corneal epithelial cells (48), where it may mediate pH$_i$ recovery from an alkaline load.

HCO_3^--Dependent Transporters. These comprise several systems that are involved in HCO_3^- (or CO_3^{2-}) transport. They are almost always (for an exception, see later under Na$^+$-HCO_3^- Cotransporter) inhibited by disulphonic stilbene derivatives such as SITS, DIDS, and DNDS (see review of Cabantchik and Greger [75]), and include: *(1)* electroneutral Na$^+$-independent Cl^-/HCO_3^- exchange (band 3 in red blood cells); *(2)* electroneutral Na$^+$-dependent Cl^-/HCO_3^- exchange (that is, Na$^+$ + HCO_3^-/Cl^- exchange); *(3)* electrogenic or electroneutral Na$^+$-HCO_3^- cotransport; and *(4)* electroneutral K$^+$-HCO_3^- cotransport.

1. The Na$^+$-independent Cl^-/HCO_3^- exchanger is found in a wide variety of cells, including red blood cells and neutrophils, cardiac Purkinje fibers, renal mesangial cells, and gallbladder epithelial cells. It normally mediates the electroneutral exchange of an external Cl^- for an internal HCO_3^-, and thus results in net uptake of solute and acidification of the cytoplasm

(but see 310). In blood it enhances the CO_2-carrying capacity by exchanging red cell HCO_3^- (formed in abundance because of the high buffer, that is, hemoglobin, concentration) for plasma Cl^-. It plays an important role in pH$_i$ recovery from an alkaline load when CO_2/HCO_3^- is present (132, 344, 345, 402). In human neutrophils, cinnamate derivatives inhibit the exchanger; SITS has little or no effect (345). The exchanger also contributes to transepithelial HCO_3^- movements and to regulation of cell volume and internal Cl^- (13). In rabbit ileal brush-border membrane, it is activated by an internal pH-sensitive modifier site: when both internal pH and HCO_3^- are increased, Cl^- uptake is stimulated in a sigmoidal fashion. OH^- by itself is a poor substrate for the exchanger, but it "activates" the transport of HCO_3^- (259).

This exchanger (AE, for anion exchanger) has three isoforms (13, 14, 323), AE 1, 2, and 3. They are all glycoproteins and exhibit about 80%–90% homology. AE 1, the smallest (MW $\approx$ 115 kDa, 800–900 amino acids), is the same as the band 3 protein from red cells (the name is derived from its relative position on gels after electrophoresis). It allows hemoglobin to contribute to the buffering of blood plasma: CO_2, released in the tissues, is converted within the red cells to H^+ and HCO_3^-; the H^+ is buffered, while HCO_3^- is exchanged with plasma Cl^- through the band 3 protein (so-called chloride shift). In its complete absence, plasma [HCO_3^-] and pH in vivo are reduced, the latter by 0.15 unit (184b, 371a), a surprisingly small amount. This protein is also found on the basolateral membranes of type A renal intercalated cells (323). It has 14 membrane spanning regions. The large cytoplasmic domain (about 400 amino acids), which is on the N-terminal end (214), contains regulatory sites, including a binding site for the cytoskeletal protein ankyrin. The middle 400 amino acids (containing some of the membrane-spanning regions) form the transport domain (234). The AE 2 and 3 isoforms are larger (MW $\approx$ 145–165 kDa). AE 2 is found in a wide variety of cells, including gastric parietal cells, the basolateral membrane of choroid plexus cells, and the contralacunar membrane of osteoclasts (13). It is believed to be the isoform that returns pH$_i$ to its normal value in response to an alkaline load, to mediate transpithelial HCO_3^- movements, and probably to aid in the regulation of cell volume (323). AE 2 has an internal pH-sensitive allosteric regulatory binding site, which activates it at alkaline pH$_i$ (259). AE 3 has a far more restricted distribution than AE 2. It has predominantly been found in heart and central nervous system (13), as well as the fore-stomach and large intestine (219). Its specific function is unknown.

2. The Na$^+$-dependent Cl^-/HCO_3^- exchanger was originally described as the pH-regulating transport

system in invertebrate nerve and muscle cells (52, 60, 61, 253, 325, 380). It has since been found in a wide variety of cells (177), and its kinetics have been studied extensively. It mediates the net influx of HCO_3^- and Na^+ (or sometimes $NaCO_3^-$) and efflux of Cl^- in an electroneutral fashion (63, 317). The exchange is driven by the inwardly directed Na^+ gradient, and results in cellular alkalinization; 2 acid equivalents per cycle are removed, while 1 Na^+ enters and 1 Cl^- leaves. Numerous kinetic models for interaction of the participating ions are possible (61), in which acid extrusion is achieved in different ways: *(1)* 2 HCO_3^-'s enter, *(2)* 1 HCO_3^- enters while 1 H^+ leaves, or *(3)* one ion pair, $NaCO_3^-$ enters. Kinetic analysis indicates that in squid giant axon the third variant is most likely (53, 59). On the other hand, in barnacle muscle $NaCO_3^-$ is definitely not involved in the exchange (61). (There are other differences between the barnacle and squid giant axon transporter. For instance, the former is easily reversible [326], in contrast with the latter [53]). Thus, while the transporter has not been cloned, it is likely that at least 2 isoforms exist.

3. The Na^+-HCO_3^- cotransporter was originally described in renal proximal tubule cells (56). In these cells, it is localized in the basolateral membrane, is electrogenic, and mediates the net efflux of Na^+ and HCO_3^-. A similar transport has since been reported in several other cell types, including hepatocytes (123, 137), glial cells (111a, 112, 268), renal medullary thick limb cells (207), retinal pigment epithelium (181), cardiac Purkinje fibers (100), and smooth muscle cells (7, 8). These cotransporters have stoichiometries (base equivalents: Na^+) that vary from 3:1 to 1:1.

In the case of the basolateral membrane of renal proximal tubule cells, 3 base equivalents, namely 1 HCO_3^- and 1 CO_3^{2-}, are transported with 1 Na^+ (352). This is illustrated in Figure 9.7A. A stoichiometry of 3:1 was also found in salamander glial cells (268); the nature of the base equivalents was not determined. In contrast, in frog retinal pigment epithelium (181) and leech glial cells (112, 260), the stoichiometry is 2:1. It is not known whether 2 HCO_3^- ions or 1 CO_3 ion per Na ion are involved (see Fig. 9.7B).

Finally, an electroneutral Na^+-HCO_3^- cotransporter has been identified in sheep cardiac Purkinje cells. It is not affected by, and does not affect, membrane potential (100; see Fig. 9.7C). A similar mechanism is present in guinea pig ureter smooth muscle cells (7, 8). Interestingly, the latter is not inhibited by DIDS (8), a feature

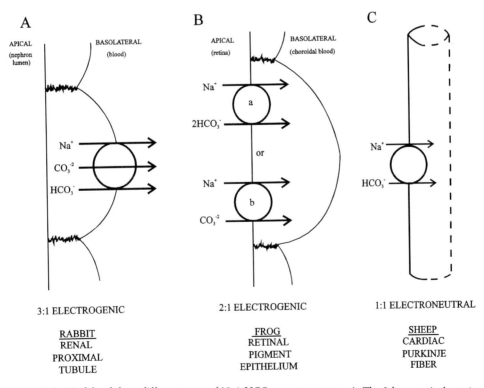

FIG. 9.7. Models of three different types of Na^+-HCO_3^- cotransporters. *A.* The 3 base equivalents:1 Na^+ electrogenic cotransporter present in rabbit renal proximal tubule cells. The transport involves the efflux of 1 HCO_3^- and 1 CO_3^{-2} along with a single Na^+. *B.* The 2 base equivalents:1 Na^+ electrogenic cotransporter. It is not certain whether this mechanism in frog retinal pigment epithelium involves the influx of each Na^+ with 2 HCO_3^- (Model a) or 1 CO_3^{-2} (Model b). *C.* The 1 HCO_3^-:1 Na^+ electroneutral cotransporter in sheep cardiac Purkinje fibers. It involves the influx of a single HCO_3^- per Na$^+$.

that distinguishes it from all other Na^+-HCO_3^- cotransporters. Based on these differences in stoichiometry and inhibitor sensitivity, the various Na^+-HCO_3^- cotransporters may represent different isoforms or even distinct transport proteins.

The Na^+-HCO_3^- cotransporter is believed to serve at least two different functions, depending on the cell in which it is located. In certain epithelial cells, it is part of the mechanism that mediates transepithelial HCO_3^- transport (56, 181). In other cells, it mediates cellular HCO_3^- influx and contributes to pH regulation upon intracellular acidification (8, 100, 137). Finally, in at least one cell type, leech glial cells (260), the membrane potential and the reversal potential for the Na^+-HCO_3^- cotransporter are similar, suggesting that pH_i in glial cells could be a function of the membrane potential.

The Na^+-HCO_3^- cotransporter from salamander (*Ambystoma tigrinum*) renal proximal tubule has now been cloned (314a, 314b). This large protein (~1025 amino acids) shows no substantial homology with any known protein and thus appears to be a unique transport protein. Its sequence suggests ten putative membrane spanning regions and large cytoplasmic domains at both the C- and N-terminal ends. Several possible phosphorylation sites are present within these cytoplasmic domains. The protein also contains one very large extracellular loop and several potential glycosylation sites, and is found in the kidney, bladder, intestines, brain, and eye. It is not known what the structural relationship is between this cotransporter from salamander (3:1 stoichiometry) and those cotransporters with stoichiometries of 2:1 and 1:1.

4. The K^+-HCO_3^- cotransporter has been described in rat renal medullary thick ascending limb cells (44a, 227a) and in squid axon (178a, 178b). Little is known about the structure and regulation of this transporter, but it appears to function in the recovery of cell pH from alkaline loads and in the renal reabsorption of HCO_3^-. It may also contribute to the regulation of cell $[K^+]$ or volume (178b).

H^+-ATPases (Proton Pumps).

Proton pumps transport H^+ using the energy directly from ATP hydrolysis. Three different types of H^+ ATPases exist: (1) F_0-F_1; (2) E_1-E_2 or P-type; and (3) vacuolar or V-type (10, 138, 267, 322).

1. The F_0-F_1 *ATPase* or ATP synthase is a most important enzyme, instrumental in providing nearly every cell except red blood cells with the main supply of ATP. It is localized in the inner membrane of mitochondria and chloroplasts, and in the plasma membranes of aerobic bacteria. It mediates the formation of ATP from ADP, phosphate, and H^+, using as energy source the H^+ electrochemical gradient that, in the case of mitochondria, has been established across the inner membrane during the electron-transfer steps of oxidative phosphorylation (see later, under Mitochondrial H^+ Movements). The protons return from the external fluid to the mitochondrion through this enzyme, down their electrochemical gradient. The F_0 domain of the enzyme is large (380 kDa), with five subunits that span the membrane and form an H^+ pore. The F_1 cytoplasmic domain is smaller (100 kDa), has four subunits, and contains the ATPase activity. These two domains are joined by a stalk or neck region so that the entire ATPase has a lollipop shape (350). The stalk contains binding sites for the inhibitor oligomycin. F_0-F_1 ATPases can also be inhibited by azide and N,N'-dicyclohexylcarbodi-imide (DCCD).

2. The *P-type ATPases* form phosphorylated intermediates in the process of ATP hydrolysis (10). P-type H^+ ATPases have been demonstrated in the apical membrane of gastric gland cells (126), and in the surface membranes of some yeast cells, protozoans, and higher plants (264). They are responsible for acid secretion into the stomach, where they exchange internal H^+ for external K^+ in an electroneutral fashion. In plants, yeast, and protozoans, they extrude H ions electrogenically. Again, the cytoplasmic domain of these enzymes contains the ATP hydrolysis site. They are inhibited by vanadate and have molecular weights of 100–110 kDa.

3. The *vacuolar or V-type ATPases* have been described in the vacuolar membranes of yeast and plant cells, and in the vacuolar and also surface membranes of various eukaryotic cells (42, 66, 139, 161, 307a, 322, 355, 367). Their major function is the acidification of intracellular organelles (for example, lysosomes, secretory granules, endosomes; see later, under Endocytosis/Exocytosis). Their presence on the surface membrane suggests a role in pH_i regulation (macrophages) or in acidification of the external milieu (kidney collecting duct cells, osteoclasts). V-type ATPases are similar in structure to F_0-F_1 ATPases in that they are large (200–400 kDa), have a lollipop shape, and are electrogenic. They do not seem to form phosphorylated intermediates (322), are insensitive to oligomycin and vanadate, but can be inhibited by DCCD, N-ethylmaleimide (NEM), and bafilomycin (367). It is likely that V-type ATPases evolved more recently than the other types of ATPases since they are found in eukaryotic cells only, whereas F_0-F_1 and P-type ATPases are present in prokaryotes as well.

Other Transporters that Can Affect pH_i.

Some transport systems can affect pH_i even though pH_i regulation may not be their primary function. For example, the Ca^{2+}

concentration in the cell body of the snail neuron is largely maintained by a plasma membrane–bound ATP-dependent Ca^{2+} pump that exchanges extracellular H^+ (not Na^+!) for intracellular Ca^{2+}, and is inhibited by vanadate (336). (Ca^{2+}-H^+ exchange in these cells may also take place across the mitochondrial membrane; see later, under Mitochondrial H^+ Movements.) Another example is the *Na+-organic anion cotransporter* of renal proximal tubule cells that mediates the combined influx of Na^+ with an organic anionic weak base, such as lactate (342), acetate (265), or succinate and citrate (285a), and thereby alkalinizes the cells (25). A third example is the *Cl−/formate exchanger* of proximal tubule cells (201). It is comparable to the Cl^-/HCO_3^- exchanger in that it mediates Cl^- influx in exchange for formate efflux, thereby acidifying the cell. The purpose of these transporters seems to be principally the transepithelial transport of organic anions and the reabsorption of, respectively, Na^+ and Cl^-. This subject has been reviewed by Aronson (25).

Passive Transmembrane Proton Movements. In some cells H^+ can cross the membrane through what has been postulated to be H ion channels. Such movements, that is, H^+ currents, have been demonstrated in snail neurons (73, 74, 382), immature oocytes of the axolotl (31), macrophages (200), stimulated neutrophils (110), alveolar cells (84a, 109), osteoclasts (274a), and enterocytes (79a). The channels are quite specific for H ions and allow outward current under conditions of membrane depolarization (voltage gating) (110a). The current is promoted by intracellular acidification and extracellular alkalinization (31). Externally applied divalent cations such as Cu^{2+}, Zn^{2+}, Cd^{2+}, and Ni^{2+} block the current. Unequivocal proof for the existence of H^+ channels is difficult to establish: the very small size of the currents estimated from noise analysis (unitary conductance probably < 0.4 pS, turnover number ~ 10^4 s^{-1}) makes differentiation between channel and carrier impossible (74). Nevertheless, the mechanism, whether channel or carrier can contribute to significant acid removal, especially during prolonged depolarization such as occurs in active cardiac myocytes and in sea urchin and *Xenopus* oocytes upon fertilization (362). Intracellular buffers would make up most of the H^+ loss (see Physicochemical Buffers), but there may be local alkalinization next to the inner surface of the membrane (382). The subject has been reviewed by Lukacs et al. (233); the effects of external and internal buffer concentrations on H^+ current have more recently been examined (110b).

Artificial phospholipid bilayers are demonstrably permeant to H ions, very much more so than to other ions (107, 153). The mechanism is under dispute (153, 263). Two quite different modes have been proposed (290a). At membrane thickness of less than about 25 Å, pores may be the dominant mechanism. In thicker membranes, the proton permeability coefficient decreases by 2 orders of magnitude, from 10^{-2} to about 10^{-4} cm·s^{-1}, and permeation seems to proceed mainly by partitioning into the bilayer's hydrophobic phase, followed by diffusion. This holds for K^+ as well as for H^+, but the K^+ permeability coefficient remains 7 or 8 orders less than that of H^+. At least in the case of pores, this striking difference most likely is due to the protons in the form of hydronium ions, H_3O^+, or even $H_5O_2^+$, being shuttled through the hydrogen-bonded network of water molecules, "hopping" from molecule to molecule (295a, 327a). Because of the low [H^+], the H^+ movements are so small that the H^+ contribution to membrane conductance is negligible. Nevertheless, the H^+ permeation could lead to a slow inward "leak" down the H^+ electrochemical gradient, and thus to some degree of cell acidification.

Basis for Multiple Transporters in the Same Cell. We return now to a consideration of why most cells have two or even more pH-regulating transporters. One explanation is that each transporter is specialized to regulate pH$_i$ under specific conditions. For example, many cells have both Na^+/H^+ and Cl^-/HCO_3^- exchangers. Na^+/H^+ exchange will return pH$_i$ toward normal in response to cytoplasmic *acidification*, while Cl^-/HCO_3^- exchange will return it in response to *alkalinization* (13). A more subtle example of this specialization is seen in vascular smooth muscle cells, which contain the Na^+/H^+ and the Na^+-dependent Cl^-/HCO_3^- exchanger (199), both of which are activated by cell acidification and drive the pH$_i$ in an alkaline direction. Kinetic analysis shows that the Na^+-dependent Cl^-/HCO_3^- exchanger is activated by small acid disturbances; further acidification does not greatly increase its activity. In contrast, the Na^+/H^+ exchanger is most active when the cell is very acid (Fig. 9.8). Thus, in smooth muscle cells the Na^+/H^+ exchanger seems to provide an "emergency" system, while the Na^+-dependent Cl^-/HCO_3^- exchanger mainly corrects the more common small acid disturbances.

Another reason for having a variety of pH-regulating transporters in one cell is that they may respond differently to externally applied substances such as hormones, mitogens, and second messengers. Some substances will activate all pH-regulating transporters in a particular cell (132), but it is far more common that each transporter is affected differently (88, 287, 301, 320). This selectivity of activation makes for a more differentiated pH$_i$ response to the agents.

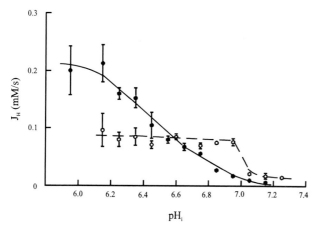

FIG. 9.8. Dependence of rate of acid efflux, J_H (product of rate of pH_i recovery and buffering power), upon pH_i during pH_i recovery from cell acidification in cultured BC3H-1 myocytes. ●, Acid efflux mediated by the Na^+/H^+ exchanger; ○, acid efflux mediated by the Na^+-dependent Cl^-/HCO_3^- exchanger. The values for the Na^+/H^+ exchanger were collected in the absence of CO_2/HCO_3^- in cells acidified to varying degrees (pH_i 5.95 to 7.15) after exposure to 15 mM NH_4Cl (for example, see Fig. 9.5). The rate of recovery was determined from the slope of the pH_i vs. time trace during the first minute of pH_i recovery from acidification and has been shown to be due to Na^+/H^+ exchange only (299). The values for the Na^+-dependent Cl^-/HCO_3^- exchanger were collected in the presence of 5% $CO_2/24$ mM HCO_3^- and 1 mM amiloride (to inhibit Na^+/H^+ exchange). These cells were acidified to varying degrees (pH_i 6.15 to 7.25) after exposure to either 15 or 30 mM NH_4Cl. Again, the rate of recovery was determined from the slope of the pH vs. time trace during the first minute of pH_i recovery from acidification. The recovery under these conditions has been shown to be entirely due to Na^+-dependent Cl^-/HCO_3^- exchange (299). Each experiment yielded only one recovery value. Each point represents the mean ± 1 standard error of from 3–8 separate experiments. The curves were fitted by eye. Note that the activity of the Na^+-dependent Cl^-/HCO_3^- exchanger predominates in the pH_i range of 7.2 to about 6.7. Below 6.6, the Na^+/H^+ exchanger increasingly predominates (Putnam, unpublished observations).

Regulation of cellular properties other than pH_i may be a primary purpose for the presence of some of the "pH-regulating" transporters. For example, in vascular smooth muscle cells, where the activity of the Na^+-dependent Cl^-/HCO_3^- exchanger is nearly pH_i-independent over a broad range, the exchanger may serve to regulate Cl_i (199). In the case of a similar transporter in squid axon, its K_m values for external HCO_3^- (2.3 mM) and Na^+ (77 mM) are considerably below normal seawater concentrations of these ions (about 12 and 425 mM, respectively), whereas the K_m for internal Cl^- (84 mM) is rather close to the normal Cl_i concentration (100–150 mM) (63). Under physiological conditions, the external Na^+ and HCO_3^- transport sites should thus be close to saturation, but the rate of the exchanger would be responsive to changes in Cl_i^-. It would seem, therefore, that the control of

Cl_i^- may well be an important function of this exchanger. As another example, the Na^+/H^+ exchanger in cardiac muscle cells, which contributes to the regulation of pH_i, has been shown also to play a major role in controlling intracellular Na^+ activity in these cells (127). Therefore, the presence of multiple pH-regulating transporters within a cell combines pH stabilization with regulation of cell-volume, V_m, $[Cl^-]_i$ and $[Na^+]_i$.

A particularly good example of the role of H^+-transporting carriers in cell function other than pH_i regulation is provided by some observations on barnacle muscle fibers. Here pH_i regulation is (nearly) exclusively mediated by the Na^+-dependent Cl^-/HCO_3^- exchanger (317). The Na^+/H^+ exchanger, though present, is only weakly activated by cell acidification (105), but its activity is markedly increased in hypertonic solutions, apparently in response to cell shrinkage (104). The major role for the Na^+/H^+ exchanger in this preparation therefore seems to be cell-volume regulation (see later, under Cell-Volume Regulation). It is not known whether hypertonicity activates the transporters that are already present in the membrane, whether it induces recruitment of more of them, or both.

Basis for Multiple Isoforms of a Particular Transporter. The basis for multiple isoforms of a given transporter may well be the same as that for multiple transporters in a cell, namely that different isoforms may regulate pH_i under different conditions, respond differently to externally applied agents, or serve different functions within a cell. Their locations in the membrane may also be different. Information is only available about isoforms of the Na^+/H^+ exchanger. Thus, NHE-1 is restricted to the basolateral membranes while NHE-2 and NHE-3 are restricted to the apical membranes of kidney and gastrointestinal epithelial cells (80, 183, 393). NHE-1 in these cells is undoubtedly involved in the regulation of pH_i, while NHE-3 is probably involved in the reabsorption and transcellular movement of Na^+ (393). In nucleated red blood cells of certain fish, β-NHE is activated by low partial pressures of O_2, but not by cell acidification or shrinkage. The activated β-NHE will alkalinize red cells, thereby increasing the O_2 affinity of hemoglobin through Root and Bohr shifts. This increases the oxygen-transport ability of blood and opposes the effect of hypoxia (258). Thus, these three different NHE isoforms serve different functions.

Differential responses of various NHE isoforms to externally applied agents have also been demonstrated. Vasopressin and calcitonin, when applied to cultured renal epithelial cells, inhibit apical NHE-2 but activate basolateral NHE-1 (80). Increased cAMP, which in general either inhibits or has no effect on Na^+/H^+ exchange, activates β-NHE in fish red cells (258).

It is likely that differences in function and regulation of the isoforms of other pH-regulating transporters will be found in the future.

Determinants of Steady-State pH_i

The pH of a cell will be in a steady state when the rate of acidification equals that at which acid is removed. The acidification rate represents the total effect of metabolic acid production, influx of H^+ and of protonated molecules such as propionic acid or NH_4^+, and efflux of deprotonated molecules such as NH_3 or HCO_3^-. Sometimes these fluxes are mediated by transporters such as the Cl^-/HCO_3^- exchanger. Processes that remove acid include metabolic reactions that consume H ions, and transporters such as Na^+/H^+ and $Na^+ + HCO_3^-/Cl^-$ exchangers that remove H^+ or H^+ equivalents from the cell.

The effect on pH_i of changes in pH_o should be briefly examined, especially since this effect varies among different cell types. Thus, in the nominal absence of CO_2, the acidity of frog skeletal muscle is only slightly affected by pH_o changes; pH_i changes by less than 0.1 when pH_o is changed by a full unit ($\Delta pH_i/\Delta pH_o \sim 0.1$; see 47 and 303). At the other extreme, in carotid body glomus cells (which are assumed to play an important role in chemoreception of the carotid body), this ratio is as high as 0.85: their pH_i is very sensitive to external pH alterations (415). Cultured mammalian myocytes with a ratio of 0.7 display an intermediate sensitivity (302). In the presence of CO_2, there are also marked differences in pH_i sensitivity to pH_o. In mammalian smooth muscle (6), skeletal muscle (9), Purkinje fibers of the heart (116), and neutrophils (344) the $\Delta pH_i/\Delta pH_o$ ratio amounts to 0.3–0.4, while in glomus cells pH_i sensitivity to pH_o is again high, with a ratio of 0.6–0.8, irrespective of whether pH_o is changed at constant pCO_2 or constant $[HCO_3^-]_o$ (68, 415). This high sensitivity is nearly matched by that of human red cells (128; for a comparison see 415). In red cells, the H^+ distribution is determined by the net charge concentrations of the impermeant ions on either side of the plasma membrane (Donnan equilibrium); this is not the case in the other cells. The basis for the different effects of pH_o on steady-state pH_i in these cells is not fully understood.

Attention should be drawn to the varying effects on steady-state pH_i of exposing originally CO_2-free cells to media containing CO_2/HCO_3^-, a common experimental procedure. The immediate acidification due to CO_2 influx is sometimes followed by at most a partial recovery; a steady-state pH_i value is reached that is lower than the original one. This occurs, for instance, in frog skeletal muscle fibers (1, 47). In other cells, such as the snail neuron (380), pH_i returns nearly to its original value (see Fig. 9.2). The steady-state pH_i in CO_2-free medium can actually be lower than in CO_2-containing medium, as illustrated in Figure 9.4 for guinea pig vas deferens smooth muscle (6). This relationship is also seen in guinea pig ureter smooth muscle (7), and in other cell types such as renal mesangial cells (65), cultured fibroblasts (44), glial cells (112), and cultured myocytes (302). The precise pH_i at which the rates of the two opposing processes, acid production and acid removal, are equal is determined by the sensitivities to pH_i of each of the many events that make up these two processes, and cannot be predicted a priori. There is no reason why these sensitivities, including sensitivities to CO_2, should all be the same in every cell and, thus, why balance should always be reached at the same pH_i.

A practical lesson is that the steady-state pH_i values in vivo—that is, with CO_2 present—may be significantly lower or higher than those of the same cells in vitro under the artificial CO_2-free conditions in which many past experiments were carried out.

SOME CELLULAR PROCESSES AFFECTING OR AFFECTED BY pH_i

Even though the amount of free H ions in the cytosol is extremely small compared to that of nearly all other cell constituents, alterations in pH_i can have striking effects, especially on charge and conformation of proteins that contain ionizable groups, and thus on their functions. We shall give several examples of how changes in pH_i can affect cellular processes and functions.

Metabolism

We have selected two metabolic processes to illustrate the interrelationships between metabolism and cytosolic pH: the production of lactate through glycolysis, especially in skeletal muscle, and the activation of neutrophils.

1. A few words should first be said about the conditions under which lactate is generated. It has been known for many years that hypoxia stimulates lactate production in working muscle. The classic work of Hill and his coworkers in the early 1920s (159, 170) confirmed this relationship, and seemed to firmly establish the concept that an increase in lactate production (lactate concentration) indicates inadequate O_2 supply to the muscles. The present view is less rigid. Experi-

ments in vivo and in vitro strongly suggest that lactate can be formed by active muscle even when the oxygen supply to all parts of the muscle must be judged to be adequate. Factors other than hypoxia apparently play a role, but the precise nature of these factors is by no means clear (136). We now examine the relationship between lactate and H ions.

The rate of production and appearance of lactate ions through glycolysis is almost always greater than the concurrent rate of production of H ions. The precise relationship between these two ions is very much dependent on pH. The production of 2 moles of lactate ions from 1 mole of glucose is always associated with the production of 2 moles of ATP (from ADP and inorganic phosphate). But at, say, pH of 6.8 (a reasonable value for pH_i), only 1.15 equivalent of H^+ is simultaneously formed, at pH 7.4, only 0.4 equivalent, and at pH > 8, none at all. Only at unphysiologically low pH (<6) will the amounts of lactate and H ions formed be the same. The precise stoichiometry follows from the pKs of the participating species, namely ADP, ATP, phosphate and their respective Mg^{2+} complexes. For completeness' sake it should be added that if increased free energy demand were to require hydrolysis of all the ATP generated by glycolysis, the overall rate of H^+ production would then become independent of pH and equal to the rate of lactate production. The reason is that H^+ production from ATP hydrolysis is also pH-sensitive, but in a direction opposite to that of lactate formation (176).

The interaction between hypoxia, pH_i, and the amounts of various metabolites has been studied in the isolated ferret heart with ^{31}P nuclear magnetic resonance (349). Cyanide, which inhibits mitochondrial oxidation, was used to mimic hypoxia. The results indicate that the intracellular acidosis produced by cyanide is largely due to lactic acid production, as is to be expected. On the other hand, when glycolysis is partly inhibited by removing glucose from the perfusate, cyanide-induced acidosis is partly due to protons released by ATP hydrolysis. When glycolysis is completely inhibited by iodoacetate, cyanide-induced acidosis is entirely due to ATP hydrolysis.

In working skeletal muscle the demand for ATP as an energy source may be so great as to threaten its supply. This tendency is opposed by hydrolysis of another high-energy compound, creatine phosphate, which, through its hydrolysis catalyzed by creatine kinase, supplies its phosphate group to restore the threatened ATP level. (Later, the creatine is rephosphorylated by ATP.) From the point of cellular H ion concentration it is important that creatine phosphate hydrolysis is associated with removal of some H ions, because of the partial conversion of hydrolysis-derived

HPO_4^{-2} to $H_2PO_4^{-1}$ (apparent pK of $H_2PO_4^-$ is ~6.8). The fraction of HPO_4^{-2} that is converted depends, of course, on pH_i. At any rate, this alkalinization opposes acidification due to glycolysis and comes into play especially during severe and prolonged muscle activity. This is important for muscle function, since acidity is known to impair both Ca^{2+} release (321) and mechanical efficacy (119), probably through substitution of Ca^{2+} by H^+ on troponin C (279). The reader is reminded that Ca^{2+} release from the sarcoplasmic reticulum in response to depolarization both initiates contraction and stimulates glycolysis (89). Two examples illustrate the role played by creatine phosphate hydrolysis in affecting the course of pH_i: (1) In isolated, artificially acidified (pH_i 5.90–6.69) frog muscle fibers, depolarization by 25 mM K^+ strikingly raises pH_i and reduces creatine phosphate concentration; ATP remains unchanged (21). When creatine phosphate hydrolysis is blocked by 2,4-dinitrofluorobenzene, no alkalinization is seen. On the other hand, normal fibers (pH_i ~7.37) are not alkalinized by depolarization, and creatine phosphate remains unchanged. What we may see here is depression of glycolysis by acidification and, thus, of the rate of ATP generation by glycolysis; creatine phosphate hydrolysis is, therefore, called upon. Indeed, the measured changes in concentration of various glycolytic intermediates in the acid fibers point to reduced activity of phosphofructokinase (21). This is an exquisitely pH-sensitive controlling enzyme in the glycolytic pathway (387). (2) Normal liver contains only a small amount of creatine kinase. Transgenic introduction of creatine kinase into mouse liver allows the perfused liver to maintain a normal ATP level and a normal pH_i (~7.5) for more than 15 min of hypoxia (5% O_2/95% N_2 in perfusate). In normal liver, on the other hand, hypoxia depletes ATP, while pH_i falls to below 7 (246). These results again show the supportive role of creatine phosphate hydrolysis in pH_i regulation.

2. A second metabolic process in which intracellular H ions play a significant role is the series of events associated with activation of (human) neutrophils. These cells react to a wide range of soluble agents and particulate matter such as bacteria by chemotaxis, phagocytosis, degranulation, and complicated biochemical transformations that take place only to a minor extent in resting neutrophils. The transformations that take place only to a minor extent in resting neutrophils. The transformations following exposure to the phorbol ester, 12-O-tetradecanoylphorbol-13 acetate (TPA), have been studied by Grinstein and Furuya (147). The neutrophils respond with a "metabolic burst" in which molecular O_2 is reduced to the superoxide radical O_2^-; in the process, NADPH is oxidized to $NADP^+$ with the production of one H ion

per NAPD$^+$ formed. (The radical is "dismutated" to H_2O_2 upon leaving the cell, a slow process that consumes some H$^+$.) Through the activated hexose monophosphate shunt, in which glucose-6-phosphate is oxidized to CO_2 and water, NADP$^+$ is promptly converted back to NADH with production of another H ion. Lactate production is not increased. The pH$_i$ falls by about 0.1 unit, then returns approximately to its resting value (≈ 7.2). In the presence of amiloride or in Na-free medium, this course is replaced by a fall in pH$_i$ of about 0.25. Thus, the pH$_i$ recovery is most likely due to activation of Na$^+$/H$^+$ exchange (366). The pH$_i$ response seems to be mediated by protein kinase C, which is believed to be the cellular receptor of the phorbol ester.

Simultaneously with cell acidification, the acidity of the medium increases, somewhat faster in the presence of Na$^+$ than in its absence, as would be expected from the presence of an active Na$^+$/H$^+$ exchange. It is of interest that phagocytosis and chemotaxis can be inhibited (by cytochalasin B) without interference with internal or external acidification.

The total rate of appearance of H$^+$ (intracellular plus extracellular) in a suspension of activated cells is in approximate agreement with theoretical predictions, taking into account the measured rate of O_2^- production (about 3 mmoles per 10^6 cells in 5 min), its stoichiometry (presumably 1 or 2 moles of H$^+$ per mole of O_2^-), intracellular and extracellular buffering powers, and fractional cell volume (147).

The factors that lead to external acidification are not clear. Even with Na$^+$/H$^+$ exchange blocked, H ions appear in the external fluid during the metabolic burst. The outward diffusion of generated CO_2 could make a contribution. Another pathway might be H$^+$-specific "channels" that open upon membrane depolarization, conduct only in the outward direction, and are blocked by external Cd^{2+} and Zn^{2+}. They have been described in resting neutrophils (110) where they resemble those in other cell types (see earlier, under Passive Transmembrane Proton Movements). No firm data are available on V$_m$ of activated neutrophils; their resting V$_m$ ranges between -50 and -60 mV (343).

Cell–Cell Coupling: Gap Junctions

In nearly all embryonic and many adult tissues, the plasma membranes of neighboring cells come in close approximation with each other in several places to form intercellular bridges insulated from the extracellular space. These channels provide for direct cell-to-cell communication. They are permeable to molecules up to a molecular weight of 800–1200 Da (diameter 1.2–2.0 nm), including ions, and thus permit diffusion

of metabolites and low-resistance electrical coupling between cells. Such channels and the bimembranous structures containing them are called *gap junctions*. Furshpan and Potter (129) were the first to propose electrical coupling from the very short (~ 0.1 msec) delay between the presynaptic spike and the postsynaptic potential, when the crayfish giant motor synapse (now called *electrical synapse*) was stimulated presynaptically. Low-resistance connections were soon found in nonexcitable cells (220, 232) and since then have been demonstrated in many other cell types. They are now ascribed to gap junctions, and electrical measurements are often used to evaluate them. Junctional structure, which varies in different cells, has been studied by physical and chemical methods (see, for example, 41, 237, 238, 285, 290, 312, 398, 399; for further references see Liu et al. [229]).

Gap junctions deserve a place in this chapter because their conductance can often be reversibly reduced by lowering pH$_i$. This was first demonstrated by Turin and Warner (395, 396), who applied solutions equilibrated with 100% CO_2 to amphibian embryonic cells; external acidification by itself was without effect. This effect has since been observed in cells as widely divergent as fish blastomeres (359) and mammalian liver (358) and heart (357). It is referred to as "gating" of the junctions by H ions. It is now realized that junctional conductance responds in a complicated way to acidification, and that the steady-state relationship between conductance and pH$_i$ varies among cell types. The immediate result of acidification is often a rise in conductance (36, 229), which is converted to a fall as acidification proceeds. The normalized steady-state relationship between conductance and pH$_i$ usually is S-shaped, which suggests that the response to H$^+$ is a cooperative process. A curve is fitted to the data based on the Hill equation, whose coefficient (the greater the coefficient, the steeper the plot—that is, the greater the sensitivity to pH$_i$) varies with cell type, as does the "pK" (that is, the pH$_i$ at which half-maximal conductance is achieved) (356, 357).

Cloning techniques have been used to express major gap junction proteins (subunits) in *Xenopus* oocytes (229). These subunits are called *connexins*. A gap junctional channel consists of six of these connexins (237). Each connexin has a cytoplasmic N- and C-terminus, four membrane-spanning domains, and two extracellular loop domains, which are all highly conserved. The cytoplasmic domains of each protein are connexin-specific, and may provide regulatory and channel properties, which will now be examined. The major connexins in liver (connexin 32) and in heart and between glial cells in brain (connexin 43), when expressed in the oocytes, display different relationships between their conductance and pH_i—that is, their pKs and Hill coefficients are different. These differences resemble those of the respective natural junctions, which suggests that the difference in pH_i sensitivity of heart and liver junctions is due to inherent variations in their chemical and/or physical properties and not to variations in cytoplasmic composition. There is evidence (229) that the difference is largely due to the long (149 residues) carboxyl cytoplasmic tail of cardiac connexin. Cleaving 125 amino acids from this tail converts the cardiac Hill coefficient and its pK to those of liver connexin; that is, the two Hill curves become identical. In these studies, pH_i was changed by Na acetate addition to the superfusate. It is important to note that the S-shaped relationship between conductance and pH_i held only at rather low pH_i values (5.8–6.8). At pH_i = pK (6.6 for connexin 43, 6.1 for connexin 32), conductance fell by a factor of 2 to 3 per unit fall in pH_i. At higher, more physiologic pH_i, conductance actually rose with acidification; this biphasic response has already been mentioned above.

A role of histidine 95 in the pH gating of connexin 43 has recently been suggested (115a). Moreover, it has been proposed that, at acid pH_i, two regions of the carboxyl cytoplasmic tail of connexin 43 form a "gating particle" that interacts with a separate region of the connexin protein to bring about closure of the channel (115b, 257a). Questions arise as to whether or not the effect of acidity is uniform on all junctions of a particular cell. A few limited data have been published from single-channel recordings on neonatal rat myocytes (69; replotted in 357). After CO_2 uncoupling, the approximately Gaussian unitary conductance distribution is not any different from that in the control, with a peak at ~50 pS (other gap junctions seem to have higher normal conductances of 120–150 pS; for example, see 269). This suggests that the fall in single-channel current is due to a fall in mean open time, rather than in conductance, in response to acidification.

The concept that intracellular H^+ changes directly affect gap junction conductance has not gone unchallenged, mainly because of the finding that raising intracellular Ca^{2+} concentration can also reduce conductance. This was first shown (in fact, even before the pH_i effect was observed) by Loewenstein (231) and Rose and Loewenstein (318) in salivary gland cells of insect larvae. Could it be that $[H^+]_i$ exerts its effect via an effect on Ca^{2+}? This possibility, of course, opens the reciprocal one of Ca^{2+} exerting its effect via changes in H^+. The option that Ca^{2+} and H^+ each affect the gap junction directly can also not be ruled out. Spray et al. (359) propose that the changes in conductance upon acidification are due only to direct

action of H ions on the junction. Their view is based on experiments with amphibian embryonic cells, and has been expanded to other preparations that do display junctional insensitivity to changes in $[Ca^{2+}]_i$ (see review by Spray and Bennett [356]). On the other hand, Peracchia (291, 292) points to conflicting data in different cell systems, and supports the hypothesis of a Ca^{2+}-mediated effect of H^+ on the junctions. Arellano et al. (23) made conductance measurements on coupled lateral axons from crayfish internally perfused with solutions containing various amounts of H^+, Ca^{2+}, cAMP, and a cAMP-dependent protein kinase. In the presence of a phosphorylating internal perfusate (pH 7), conductance was reduced by about 30% when pCa was lowered from 6.4 to 6.0. Since low pH and phosphorylation did not alter gap junction conductance by themselves, but only modulated the Ca^{2+} effect, the authors propose that Ca^{2+} regulates cell-to-cell communication "probably through direct interaction with the channel protein." This is quite different from the view of Spray and Bennett (356). At any rate, the variability of results in different preparations seems well established. A uniform mechanism may not exist (see review by Bennett and Verselis [35]). To add another layer of complexity, the ionic sensitivity of gap junctional conductance of neonatal rat heart cells is significantly different from that of adult cells; the reason is unknown (122a).

For the sake of completeness, some other variables should be mentioned that can influence intercellular communication. Among them are transjunctional voltage (157), phosphorylation (353, 368), certain anesthetics (196), cyclic AMP (360), low ATP (363), and arachidonic acid (125). Synergistic actions of various such regulators (including Ca^{2+} and H^+) have also been suggested (414). Attempts have been made to interpret the action of some of these agents on cardiac gap junctions in terms of the physiology of the heart as a whole (229, 357); such attempts are in the early stage.

DNA Synthesis and Cell Growth

That cell activation is associated with a rise in pH_i was first observed by Johnson et al. (195): fertilization of sea urchin eggs is associated with cellular alkalinization. Since then, many other cell-activating agents that stimulate DNA synthesis and cell growth have been found to have a similar effect on pH_i in a variety of cells. For example, serum, epidermal growth factor (EGF), platelet-derived growth factor (PDGF), insulin, thrombin, vasopressin, and phorbol esters all alkalinize (by about 0.1–0.2 pH unit) cells ranging from frog skeletal muscle fibers to cultured fibroblasts (for review, see 256). In all cases, alkalinization is the result of activation of Na^+/H^+ exchange. In mutant fibroblasts that lack the Na^+/H^+ exchanger, serum does not increase either pH_i or DNA synthesis, but is effective if pH_i has been artificially raised (297). (An elevated pH_i by itself does not stimulate DNA synthesis [296]). Data of this kind have led to the suggestion that an increase in pH_i must be part of an early signaling pathway for growth.

Subsequent considerations have led to a somewhat different view. Most of the above studies were done in the nominal absence of CO_2. In CO_2-containing medium (the natural environment), quite a few cells have a higher pH_i than in the absence of CO_2 (see before, under Determinants of Steady-State pH_i), and the activating agents often do not elicit an elevation of pH_i but still stimulate DNA synthesis and growth (40, 44, 369). In fact, A431 carcinoma cells (81) and mesangial cells (131, 132) increase DNA synthesis in response to epidermal growth factor and arginine vasopressin, respectively, even though these agents produce a *decrease* in pH_i. Also, in lymphocytes, interleukin-2, a growth factor, stimulates DNA synthesis even when Na^+/H^+ exchange is blocked by amiloride (247). Thus, an increase in pH_i is not essential for stimulation of DNA synthesis or growth. There may be a "permissive" pH_i (223) or a "threshold" pH_i (44, 296), defined as the minimum value required for cell responsiveness to growth factors. As long as the pH_i remains above this threshold, the cell would respond even when the growth factor acidifies it.

If increased pH_i is not essential for growth, what is the "purpose" of the increased activity of transporters? It has been suggested that the transporters' response enhances the "housekeeping ability" of the cell (44, 132). It might simply be one more manifestation of heightened cellular activity.

In some instances large pH_i changes do appear to play a determining role in cell activation. Two specific case will be highlighted. In many oocytes, fertilization results in a significant rise in pH_i, from 6.8 to about 7.2 (71, 195, 337), with resulting increased rates of DNA and protein synthesis and changes in shape, ultimately resulting in cell division (71). Alkalinization by itself has been shown to activate unfertilized eggs (but see 307b), as evidenced by: increased protein synthesis (118), DNA synthesis (241), and chromosome condensation (240). An increase in pH_i can even activate protein synthesis in a cell-free system derived from sea urchin eggs (418).

A second example of the role of a pH_i rise in cell activation is the arousal from dormancy in the embryos of the brine shrimp, *Artemia*. The encysted embryo remains in a hypometabolic state until it is subjected to a cycle of dehydration and rehydration. If O_2 is available, pH_i increases by as much as 1.6 units (from 6.3 to 7.9) and the embryo continues its development (70, 71), while undergoing a series of metabolic changes. All of these changes are prevented by acidification, even in the presence of oxygen. Even without O_2, alkalinization can initiate these changes (71, 178). On the other hand, in the nonfeeding larval stage of the nematode *Caenorhabditis elegans*, initiation of development appears to be accompanied by a fall in pH_i by as much as 1 unit (403). It is not known whether acidification by itself can trigger development in *C. elegans*.

Multiple Drug Resistance. Recently, another possible way in which pH_i can affect cell growth has been suggested. Cells cultured in the presence of growth-inhibiting chemotherapeutic agents such as adriamycin, daunomycin, or vincristine often develop resistance to these drugs and, interestingly, even to drugs to which the cell never was exposed. This phenomenon is referred to as *multiple drug resistance* (mdr). It is usually correlated with the appearance of a membrane-bound phosphorylated glycoprotein of 170–180 kilodaltons, often referred to as P-glycoprotein (197), or to the appearance of a similar transporter (141a, 259a, 412a). P-glycoprotein is believed to act as a drug-efflux pump for positively charged drug molecules; it uses ATP as an energy source.

In some cells exhibiting multiple drug resistance, pH_i is elevated (64, 204, 346, 373), although not in all mdr cells (16). Several mechanisms have been proposed to account for the elevation of pH_i. The efflux of the protonated (that is, cationic) form of chemotherapeutic drugs, many of which are weak bases (pKs between 7.4 and 8.2), could raise pH_i (373). Alternatively, the pump itself may act as a proton pump, resulting in the direct efflux of H^+ (373). P-glycoprotein has also been proposed to activate other pH-regulating transporters, such as the Na^+/H^+ exchanger (64) and the vacuolar H^+ ATPase (236), which could lead to cellular alkalinization.

An elevated pH_i could contribute to multiple drug resistance in a number of ways. Chemotherapeutic drugs are believed to enter through passive drug influx. Therefore, the more alkaline a cell, the less should be the accumulation of these weak bases. Indeed, daunomycin accumulation in non-mdr cells can be markedly reduced by raising pH_i by about 0.3 unit (346). Drug effectiveness should be correspondingly reduced. A change in pH_i could also modify the interaction between a drug and its intracellular binding site(s), and in this way alter effectiveness (346). Finally, a change in pH_i could affect the accumulation of chemotherapeutic drugs in acidic intracellular organelles (257).

The role of acidic intracellular compartments in multiple drug resistance has been implicated by recent studies on human breast cancer cells (MCF-7) (334a). Drug resistant MCF-7 cells have a pH_i that is elevated only about 0.2 pH unit compared to drug-sensitive cells (16, 334a). This difference is believed to be too small to account for the difference in the levels of drug accumulation in sensitive vs. resistant cells. However, measurements of pH within acidic compartments involved in exocytosis (the *trans* Golgi network and the pericentriolar recycling compartment; see later under Endocytosis/Exocytosis) show that these compartments are more acidic (pH ~5.9) in resistant than in sensitive (pH ~6.5) cells (334a). These data suggest that drug resistance may involve maintenance of an acid pH in intracellular compartments capable of accumulating and exporting chemotherapeutic drugs and that it may be the pH gradient between these compartments and the cytoplasm that determines the degree of drug resistance.

Recently, cells overexpressing the cystic fibrosis transmembrane conductance regulator (CFTR), a transport protein homologous to P-glycoprotein, have been found to exhibit multiple drug resistance (412a). In cystic fibrosis, where CFTR is abnormal, the pH of acidic intracellular compartments (such as the *trans* Golgi network) is elevated (10a). Thus, it appears that both P-glycoprotein and CFTR play a crucial role in maintaining a low pH in certain intracellular compartments and that this function may be important for the expression of the multiple drug–resistance phenotype.

Multiple drug resistance is a complex phenomenon. It must be admitted that, just as P-glycoprotein expression does not always correlate with drug resistance (28, 180, 271), in some cells there is a poor correlation between pH_i and drug resistance (16, 64). Important issues with respect to the role of pH_i remain uncertain. They include its relative importance compared to that of P-glycoprotein in drug resistance, the mechanism responsible for the cellular alkalinization in some mdr cells, and the importance of pH changes within acidic intracellular organelles.

Membrane Channels

Membrane channels are formed by membrane-spanning proteins. They allow the movement of solutes between the intracellular and extracellular compartments through conducting pathways. The function of a variety of ion channels can be modified by changes in pH_i or pH_o (see 254) through changes in the degree of protonation of ionizable groups in these proteins (for example, 202), and also through other mechanisms. The effects of pH have been most thoroughly studied in potassium channels. At least four different types are affected by pH: *(1)* inward rectifier K^+ channels, *(2)* delayed rectifier K^+ channels, *(3)* Ca^2-activated maxi-K^+ channels (maxi-K(Ca)); and *(4)* ATP-regulated K^+ channels (K(ATP)). Intracellular acidification inhibits inward rectifier, delayed rectifier, and maxi-K(Ca) channels, but may activate K(ATP) channels (Table 9.2).

Inward rectifier K^+ channels exhibit increased conductance upon hyperpolarization and decreased conductance upon depolarization (171). They are inhibited by extracellular Ba^{2+}, by tetraethylammonium (TEA), and by intracellular but not extracellular acidification (45, 182, 188, 255, 278). In many cells, the pH_i effect on inward rectifier K^+ channels is sigmoidal (pK 6.1–6.3 for most cells, 7.3 for kidney cells) with a Hill coefficient of 2–3, suggesting multiple H^+ binding sites. Between pH_i 7 and 6, single-channel conductance is reduced only slightly, but open probability is reduced markedly (188). These authors consider three possible mechanisms to explain the effect of decreased pH_i: *(1)* reduction of the negative surface potential through increased binding of H ions to negative surface charges, thereby reducing the concentration of cations near the membrane; *(2)* increased binding of H ions to a regulatory site on the channel that, through a conformational change, alters the channel's conductance; or *(3)* direct interference of H ions with the movement of K^+ through the channel by binding to a site in the channel. None of these models fits the data completely, but the suggestion of multiple H^+ binding sites leaves open the possibility that a combination of these mechanisms may be involved in H ion inhibition.

Delayed rectifier K^+ channels are activated by depolarization and can be inhibited by TEA, Ba^{2+}, and quinidine (171). As is the case with inward rectifier K^+ channels, a decrease in pH_i inhibits the channel. The relationship is sigmoidal (pK 6.3) with a Hill coefficient >1 (87, 252, 410). The pH effect is complicated: it is dependent on the initial holding potential (87). Only at a potential between −20 and −70 mV is channel inactivation sensitive to pH_i (although channel activation is pH_i insensitive under these conditions). When this voltage effect is taken into consideration, the sigmoidal relationship between inactivation and pH_i becomes linear between pH_i 6 and 10 (87). Such a nonsaturating effect of pH_i on channel function is usually indicative of H ion screening of or binding to membrane surface charges. However, screening or binding should affect channel activation as well as inactivation. Clay (87) offers the alternative possibility that H ions inhibit these channels by binding to multiple sites with different pK values.

The maxi-K(Ca) channel is characterized by a very large conductance (about 200 pS). It is voltage sensitive

TABLE 9.2. *Summary of pH$_i$ Effects on K$^+$ Channels*

Channel	pH$_i$ Effect	Comments	References
Inward rectifier	Inhibited by acidification	Hill coefficient = 2–3	(255)
		Acidification reduces open probability	(45)
			(182)
			(278)
			(188)
Delayed rectifier	Inhibited by acidification	Hill coefficient >1	(410)
		Inhibition dependent on holding potential	(252)
			(87)
		Acidification causes inactivation at more negative potentials	
Maxi-K$^+$ (Ca)	Inhibited by acidification	Hill coefficient = 1	(92)
		H ions compete with Ca for binding	(85)
			(222)
			(94)
			(225)
ATP-regulated	Activated by acidification*	H ions interfere with ATP binding	(98)
			(103)
		H ions increase single channel open probability	

*In pancreatic β cells, acidification inhibits ATP-regulated K$^+$ channels (248).

and is blocked by TEA and charybdotoxin (171). A fall in pH$_i$ inhibits this channel, as has been shown in a number of cells (85, 92, 94, 222, 225). This may be due to competition between H$^+$ and Ca^{2+} for the same binding site. It would account for the increased channel closed time and the decreased open time as pH$_i$ is reduced. H ions do not affect single-channel conductance. Thus, it appears that H ions bind to the Ca^{2+} sites on maxi-K(Ca) channels, thereby reducing the effect of Ca^{2+}.

The ATP-regulated K$^+$ channel also exhibits a pH$_i$ effect. These channels are voltage insensitive but are inhibited by ATP. In fact, at physiological ATP concentrations (>1 mM) they are quiescent (171). The effect of pH$_i$ is complex (98). For example, in cardiac myocytes, reducing pH$_i$ from 7.4 to 6.5 inhibited single-channel conductance by 30% (both in the presence and absence of 0.2 mM Mg-ATP) but increased mean open probability (P$_o$) eight-fold (in the presence of 0.2 mM Mg-ATP). This suggests that H ions reduce ATP binding to the channels and increase P$_o$ despite the direct H$^+$ effect on single-channel conductance. A qualitatively similar ability of H ions to interfere with ATP binding to the channel has been shown in skeletal muscle (103). In contrast, in pancreatic β-cells K(ATP) channels are *inhibited* by decreased pH$_i$ (248).

The ability of pH$_i$ to affect K$^+$ channels has several physiological implications. In *excitable* cells, the effects on stimulus-contraction and stimulus-secretion coupling may be profound. Three examples will be given: *(1)* In slow crayfish muscle, intracellular acidification inhibits delayed-rectifier K$^+$ channels. This results in membrane depolarization and, hence, leads to Ca^{2+} entry and the appearance of Ca^{2+} spikes (252). *(2)* Cardiac ischemia produces cell acidification and reduction of ATP, both of which lead to opening of ATP-regulated K channels (98). This would shorten the plateau phase of the action potential, thus reducing the magnitude of Ca^{2+} influx, and therefore the development of twitch tension. On the other hand, ischemia-induced arrhythmias may be triggered by ventricular depolarization resulting from inhibition of inward-rectifier K$^+$ channels by intracellular acidosis (188). *(3)* Acidification of islet β cells in the presence of low glucose concentration inhibits ATP-regulated K$^+$ channels (as was stated above). This leads successively to membrane depolarization, Ca^{2+} entry, and insulin secretion (248).

In *nonexcitable* cells such as kidney cells, low pH$_i$-induced depolarization results from inhibition of the inward rectifier (278). Such depolarization could reduce the normal acidifying efflux of HCO$_3^-$ on the electrogenic Na$^+$-HCO$_3^-$ cotransporter (see earlier, under Membrane Transport Systems) and result in cellular alkalinization (408). Cellular acidification, in addition to depolarizing these cells, would reduce intracellular HCO$_3^-$, and the combination of reduced V_m and intracellular HCO$_3^-$ could actually reverse the cotransporter, resulting in HCO$_3^-$ influx. This influx would result in cellular alkalinization and thus a recovery of pH$_i$ back toward its initial value.

The activity of Na$^+$ channels can also be affected by changes in pH$_i$. In squid axons, H ions can directly block certain Na$^+$ channels by binding to sites within the channels themselves (411), whereas in skeletal muscle H ions prevent Na$^+$ channel inactivation (274). It must be admitted that both of these effects occur at physiologically unrealistic values of pH$_i$ (5.1–5.3). In frog skin epithelium, decreasing pH$_i$ from 8 to 7 results in a large decrease in the apical-membrane Na$^+$ conductance. This pH$_i$ effect has a Hill coefficient of 2 and a pK of 7.25 (160). Further, lowering pH from 7.4 to 6.4 in the solution bathing the cytoplasmic face of apical Na$^+$ channels from rat cortical collecting tubules resulted in an eight-fold decrease in the probability that the channels would be open (286). In other epithelial cells, intracellular Ca^{2+} has been shown to inhibit Na$^+$ conductance that is relieved by intracellular acidification; this suggests competition between H$^+$ and Ca^{2+} for binding sites on the Na$^+$ channel (133). Ca^{2+} channels also can be affected by changes in pH$_i$. The conductances of both the dihydropyridine-sensitive Ca^{2+} channel and the sarcoplasmic reticulum Ca^{2+} release channel of skeletal muscle are reduced by intracellular acidification (235, 298, 321). Finally, a Ca^{2+}-activated nonselective cation channel from mammalian kidney tubules is markedly affected by pH$_i$, having an optimal open probability between pH 6.8 to 7.0 that decreases sharply at both acid and alkaline values of pH$_i$ (84b). These findings suggest that pH$_i$ inhibits the nonselective cation channel through two distinct mechanisms.

Cell-Volume Regulation

The regulation of cell volume and pH$_i$ was recently reviewed (177). There is a good reason for linking these two aspects of cellular physiology. Several of the transport systems concerned with pH$_i$ regulation (see earlier under Membrane Transport Systems) turn out to be involved, under certain circumstances and in certain cell types, with volume regulation. The following discussion will briefly examine this involvement.

It must be pointed out that, even without alterations in the external milieu such as changes in osmolarity, cells require mechanisms for volume regulation. The

high intracellular concentration of impermeant anionic macromolecules would, according to the Gibbs-Donnan equilibrium, lead to passive entry of permeant ions (especially Na$^+$) and water; this eventually would lyse the cells. The system of ion-extruding (and, of course, energy-requiring) pumps and leaks proposed more than 30 years ago (385) obviates this intolerable outcome. The model, it will be noted, dates from about the time of discovery of the membrane-bound Na$^+$-K$^+$-ATPase and its ion-transposing properties (see later, under Mitochondrial H$^+$ Movements). In some cells, however, blocking Na$^+$-K$^+$-ATPase actually results in shrinkage, possibly due to K$^+$ efflux (17, 18).

Cells usually respond to imposed volume changes by activating mechanisms that return their volume toward the initial value. Chamberlin and Strange (83) and Hoffmann and Simonsen (177) have reviewed this subject. These compensatory responses most often involve net movement of ions, accompanied by water, across the plasma membrane. The rapid regulatory volume decrease (RVD) in response to swelling (usually due to exposure to hypotonic media) is often achieved by net efflux of K$^+$ and/or Cl$^-$ (in high Na$^+$ dog red cells by Na$^+$/Ca^{2+} exchange); the rapid regulatory volume increase (RVI) in response to shrinkage in hypertonic media, by net influx of Na$^+$ and Cl$^-$. The pathways which mediate these ionic movements vary. In some cells, ion-conductive channels are utilized, in others electroneutral cotransporters, in still others electroneutral pH$_i$-regulating exchangers (288). Here we shall only examine the contributions of the latter. The work of Cala (76, 77) on *Amphiuma* nucleated red cells may serve as an example. When these cells are exposed to a hypertonic solution, RVI is achieved by enhanced activity of two parallel mechanisms. One exchanges external Na$^+$ for internal H$^+$, the other external Cl$^-$ for internal HCO$_3^-$. These pathways are independent: DIDS, an inhibitor of anion exchange (see earlier, under HCO$_3^-$-dependent Transporters) prevents Cl$^-$ entry but not Na$^+$ entry. Since both H$^+$ and HCO$_3^-$ leave the cell, the medium becomes enriched with H$_2$CO$_3$ and acidifies.

In hypotonic solutions, RVD in the *Amphiuma* red cells is achieved by activation of K$^+$/H$^+$ exchange (K$^+$ leaving) and Cl$^-$/HCO$_3^-$ exchange (Cl$^-$ leaving). The medium becomes alkaline. DIDS blocks the anion exchange; there is then little volume change.

What sets off these events is not clear, but most likely the Na$^+$ and Cl$^-$ fluxes or the K$^+$ and Cl$^-$ fluxes are coupled as a result of the changes in pH$_i$ (and thus of [HCO$_3^-$]$_i$) that result from the Na$^+$/H$^+$ or K$^+$/H$^+$ exchanges. Cala (77) provided circumstantial evidence that these two exchanges are mediated by the same transporting pathway. The electroneutrality of the exchanges has been concluded from the absence of significant V_m changes, as measured directly with microelectrodes (76).

The magnitude of the osmotic effects of the Cl^-/HCO_3^- and Na^+/H^+ exchangers requires some discussion. Even though equal numbers of particles are transported across the plasma membrane, neither exchange is "osmotically silent." With the help of the HCO_3^--pH diagram (Fig. 9.9), made popular by Davenport (102), we will illustrate the osmotic consequences of 5 mmoles of HCO_3^- leaving the cytosol in exchange for Cl^-. Let the initial conditions (point A) be: $pH_i = 7.3$, $[HCO_3^-]_i = 16$ mM, and $Pco_2 = 33$ mm Hg and constant (open system; see earlier, under Physicochemical Buffers). This point is located at the intersection of the iso-Pco_2 curve, which represents the $[HCO_3^-]_i$-pH_i relationship at constant $Pco_2 = 33$ mm Hg (calculated from the Henderson-Hasselbalch equation), and the straight line ("CO_2 absorption curve"), which gives the $[HCO_3^-]_i$-pH_i relationship when the cytosol is titrated with CO_2 (follow the line from right to left). Its slope is a measure of the *non*-HCO_3^- buffering power, as was first recognized by Peters and Van Slyke (294): H ions and HCO_3 ions are generated in equal amounts. (This is the basis of the non-HCO_3^- buffering power

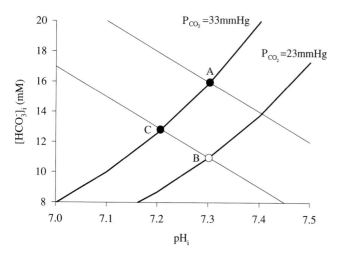

FIG. 9.9. Effect on cell acid-base balance of removal of 5 mmoles of HCO_3^- per liter cytosolic water (in exchange for 5 mmoles of Cl^-). Point *A* represents initial conditions: $pH_i = 7.3$, $[HCO_3^-]_i = 16$ mM, $Pco_2 = 33$ mm Hg. Point *B* represents the hypothetical case in which the reduction of pH_i resulting from HCO_3^- removal is canceled by simultaneous reduction of Pco_2: $pH_i = 7.3$, $[HCO_3^-]_i = 11$ mM, $Pco_2 = 23$ mm Hg. Point *C* represents the actual conditions when HCO_3^- is removed at unchanged Pco_2: $pH_i = 7.2$, $[HCO_3^-]_i = 13$ mM, $Pco_2 = 33$ mm Hg. The iso-Pco_2 curves are constructed taking the "overall" pK_1' of CO_2 as 6.1 and CO_2 solubility as 0.03 mmole l^{-1} mm Hg^{-1}. The slope of the parallel lines is a measure of the non-HCO_3^- buffering power, which remains the same in the three conditions at an assumed value of 20 mM per pH unit. For details, see text.

measurement with a CO_2 pulse; see earlier, under Measurement of Intracellular Buffering Power.) The non-HCO_3^- buffering power is assumed to remain constant at 20 mM over the narrow pH range under consideration. The removal of 5 mM HCO_3^- will shift to the right the intracellular equilibrium $CO_2 + H_2O \rightleftarrows H^+ + HCO_3^-$, with the generation of some H^+ and HCO_3^- from CO_2. The pH_i will be somewhat reduced, but most of the generated H ions are consumed by the non-HCO_3^- buffers, thereby shifting the equilibrium further to the right. Thus, the generated HCO_3^- will partially make up for the HCO_3^- removal. To what extent? Imagine pH_i being returned to its original value (7.3) by reducing Pco_2—that is, by driving the above equilibrium to the left. The non-HCO_3^- buffers will then be titrated back, and the HCO_3^- generation canceled. At this point (B), $[HCO_3^-]_i$ is 11 mM, exactly 5 mM less than the original amount. Non-HCO_3^- buffering power has not been affected by this maneuver; therefore, B is located at the intersection of a buffer line of the original slope but displaced downward by 5 mM HCO_3^-, and an iso-CO_2 curve at the reduced Pco_2 (23 mm Hg). The actual acid–base conditions when 5 mM HCO_3^- are removed at unaltered Pco_2 (33 mm Hg) are thus given by point C, which is located at the intersection of the lower buffer line and the original iso-CO_2 curve. These conditions are $pH_i = 7.2$ and $[HCO_3^-]_i = 13$ mM. Therefore, removal of 5 mM HCO_3^- reduces $[HCO_3^-]_i$ by only 3 mM. The reduction will be less the greater the buffering power, as inspection of the diagram readily shows. The total osmotic effect of the 5-for-5 Cl^-/HCO_3^- exchange is thus an enrichment of osmotically active particles by $5 - 3 = 2$ mmoles (Cl ions) per liter of cell water.

The osmotic consequences of 5 mmoles of Na^+ entering the cytosol in exchange for 5 mmoles of H^+ can be calculated in a comparable way, again using the Davenport diagram. Assuming the same initial conditions as before and Pco_2 again constant at 33 mm Hg, the loss of 5 mM H^+ will increase pH_i from 7.30 to 7.39, and $[HCO_3^-]_i$ from 16 to 19.3 mM. Thus, the Na^+/H^+ exchanger raises the number of osmotically active particles by $5 + 3.3 = 8.3$ mmoles per liter of cell water. If the exchange takes place in the nominal absence of CO_2, the calculation is simpler. The pH_i increase will be $5/20 = 0.25$, from 7.30 to 7.55. This corresponds to a reduction in free H ions of about 2×10^{-5} mM, a trivial amount. Thus, the cytosol will be enriched with 5 mmole particles (Na ions) per liter cell water.

Finally, a word should be said about the osmotic consequences when both the Cl^-/HCO_3^- and Na^+/H^+ exchangers are active in the same cell. When Pco_2 is maintained constant and the two exchangers are functioning at the same rate, then equal amounts of H^+ and HCO_3^- will be extruded. No change in pH_i will occur; the net osmotic gain of the cell will solely be due to the sum of the entries of Na^+ and of Cl^-.

It follows from the above that pH_i perturbations that activate exchangers should produce cell-volume changes, but few data on this subject are available. C6 glioma cells respond to exposure (and uptake) of a weak acid by swelling as expected; the subsequent, slower increase in volume may be ascribed to the Na^+/H^+ exchanger being activated by the cell's acidification (192).

Shrinkage-induced activation of the common exchanger isoform, NHE-1, requires the presence of a site on the cytoplasmic domain between amino acid 566 and the N terminus, that is distinct from the regulatory site activated by growth factors (41a). The mechanism by which cell shrinkage activates Na^+/H^+ exchange is not fully understood. It is known that shrinkage increases the affinity of the exchanger for H^+ (146, 194, 340) and that this activation is dependent on cellular ATP (149, 339), suggesting a phosphorylation step in the activation mechanism, even though the exchanger itself does not seem to be phosphorylated directly (150, 273a). None of the well-known pathways for activation of Na^+/H^+ exchange (for example, those mediated by changes in Ca^{2+}, cAMP, or protein kinases; see earlier, under Cation/H^+ Exchangers) seems to be involved in shrinkage-induced activation of the exchanger (105, 145, 194, 339, 417). Shrinkage-induced activation of Na^+/H^+ exchange can be inhibited by calmodulin inhibitors (101, 146, 339), cytochalasin B (101), and inhibitors of myosin light chain kinase (339). Finally, activation may require a GTP-binding protein (105). A model for activation of Na^+/H^+ exchange by shrinkage has been proposed that involves phosphorylation of myosin light chain and possible altered stress fiber activity (339).

Shrinkage has recently been shown to activate Na^+-dependent Cl^-/HCO_3^- exchange in Chinese hamster ovary cells (310a). This finding is of interest since this exchanger should not mediate any net solute transport and these cells exhibit no volume regulation in response to shrinkage (330a). As with Na^+/H^+ exchange, the activation of Na^+-dependent Cl^-/HCO_3^- exchange by cell shrinkage occurred in cells depleted of protein kinase C and in cells in which Ca^{2+} had been chelated, indicating that neither of these signaling systems were involved in the activation pathway (310a). These data suggest that pH-regulating transporters may be activated upon cell shrinkage even if they are not involved in cell volume regulation, indicating that alkalinization per se may serve some function in shrunken cells.

Mitochondrial H^+ Movements

In aerobic cell respiration, oxygen is used as the terminal oxidant in a reaction that forms half of the water-oxygen cycle in nature, the other half being photosynthesis:

$$\frac{1}{2}O_2 + 2H^+ + 2\ e^- \xrightleftharpoons[\text{photosynthesis}]{\text{respiration}} 1H_2O \quad (9-7)$$

Here we are concerned with the respiration half of the cycle, especially with respect to the fate of H ions. Respiration takes place in the mitochondrion. In the chain of respiratory events, high-energy electrons derived from substrate H atoms are transferred, two at a time, through each of three large (300–800 kDa) respiratory enzyme complexes in succession. The enzymes are located in the inner membrane of the mitochondrion. The electrons lose their energy step by step as they pass down this respiratory chain, while the affinity for electrons of the three complexes (their redox potential) increases pari passu, and reaches a maximum in molecular oxygen, the final oxidant. Each of the complexes uses a large part of the energy released by the electrons to extrude protons from the aqueous matrix of the mitochondrion across the inner membrane into the space between inner and outer membrane. In this way, a transmembrane electrochemical proton gradient is created. Some protons may diffuse into the cytosol across the outer membrane, which has wide pores (354). There is some uncertainty about the H^+-electron stoichiometry at the first two complexes; probably a total of less than 4 protons are extruded per electron pair that passes through them. The last complex, cytochrome c oxidase, extrudes two protons (30), while it consumes another two, also derived from the matrix, plus the two electrons in the reduction of half an O_2 molecule to one molecule of water (see equation 9–7). Its structure has recently been described, and the path taken by protons in the conversion of oxygen to water is now known with reasonable certainty (189). However, the mechanism by which protons are pumped from the aqueous matrix through the enzyme across the membrane is still open to discussion.

A word should be said about the origins of the high-energy electrons. They are extracted in the form of NADH and $FADH_2$ from acetyl coenzyme A in the course of its stepwise degradation in the citric acid cycle (which takes place in the mitochondrial matrix). Carbohydrates and fats, the primary energy suppliers, are first broken down in the cytosol to pyruvate and fatty acids, which then pass into the mitochondrial matrix. There they are converted to acetyl CoA. Most of the energy of the foodstuffs is conserved as high-energy electrons.

The proton gradient across the inner membrane of the mitochondrion has an electrical component, ΔV, the voltage difference across the membrane (inside negative), and a chemical component, $(2.303RT/F)$ $(pH_{cyt} - pH_{mit})$, which also has a negative value (that is, $pH_{cyt} < pH_{mit}$) [mit = mitochondrial matrix, cyt = cytosol]. The relative contributions of the two components to the total gradient depend on the properties of the inner membrane. At negative ΔV this membrane is nearly impermeable to ions (354). If there were no other ion-translocating mechanisms, electrostatic repulsion forces would prevent a chemical gradient from being established, and the energy of proton extrusion would be stored solely in the form of a voltage difference. Carrier proteins are, however, embedded in the inner membrane that can translocate Cl^-, Na^+, and various anions (see below; also see 224). Of the total electrochemical proton gradient of about -170 mV across an average cell's inner mitochondrial membrane, the electrical component may amount to -150 mV (230), the chemical component to -20 mV. That is, in this example the pH of matrix water is about 0.3 unit higher than that of cytosolic water, which has, indeed, been observed (341).

A principal function of the proton gradient is to provide the energy for ATP synthesis from ADP and phosphate, which are present in the matrix. This process is mediated by a special H^+ ATPase, also called ATP synthase or F_0-F_1 ATPase (see earlier, under Membrane Transport Systems), a major component of the inner mitochondrial membrane. The external protons, two or three per molecule of ATP formed, pass back through this enzyme into the mitochondrial matrix. Under suitable energetic conditions that are determined by the relative concentrations of the reactants (ADP, phosphate and ATP), ATP synthase can mediate the reverse reaction: hydrolysis of ATP to ADP and phosphate. The energy liberated is then used to establish an electrochemical gradient across the inner membrane of the same polarity as that created by the respiratory enzyme chain.

The process whereby respiration (oxidation) is coupled to ATP formation is called *oxidative phosphorylation*. The existence of this linkage had been known for some years, but its precise mode—namely, through proton extrusion across a membrane with the creation of an electrochemical gradient as the penultimate energy repository (the ultimate one being ATP)—was proposed and demonstrated later (249, 250). It was Mitchell who named the sum total of these events "chemiosmotic coupling." The concept is now generally accepted, but it was initially received with skepti-

cism. This is not surprising: "coupling" of biochemical and biophysical (membrane-translocating) processes introduced a novel feature into what up to that time had been a biochemical approach to mitochondrion-based ATP formation. It must be admitted that the dynamics of proton movements in the vicinity of biological membranes is by no means agreed upon; a variety of models have been proposed, which are discussed by Kasianowicz and Bezrukov (202). The work by Heberle et al. (164) and by Scherrer (334) mentioned earlier under Cytosolic H Ion Movements and the Role of Buffers is pertinent in this respect.

Besides its role in ATP synthesis, the proton electrochemical gradient subserves transport functions. Some of these functions have already been alluded to. At least ten different carrier proteins have been identified in the inner membrane, and their molecular aspects characterized (218). Several of these systems are electrogenic and are driven by the electrical component of the proton gradient, that is, the voltage difference across the inner membrane. Among them are the ADP/ATP exchanger, which allows matrix ADP to be replenished and ATP to be moved into the cytosol, and the aspartate/glutamate exchanger. On the other hand, ions such as phosphate, citrate, pyruvate, and glutamate are transported inward accompanied by protons (224).

The relationship between Ca^{2+} and H^+ movements at the mitochondrion deserves mention here. The free Ca^{2+} concentration in the matrix water is uncertain, but surely the driving force on the ion is overwhelmingly inward from cytosol to matrix because of the greatly negative value of ΔV and the divalent nature of the ion. On the other hand, the inner membrane has a very low ionic permeability. It has been proposed (242) that in snail neurons some of the Ca^{2+} injected into the cytosol finds its way into the mitochondrion in exchange for H^+, possibly on a carrier: the injection was accompanied by a transient fall in cytosolic pH. In agreement, the mitochondrial pH of isolated respiring mitochondria increased when Ca^{2+} in the medium was raised by 0.5 mM (admittedly a highly unphysiological concentration); simultaneously, the pH in the weakly buffered medium fell (4).

The extrusion of protons can be shunted by substances that make the inner membrane permeable to protons. These "uncouplers" belong to the class of special weak acids, such as 2,4-dinitrophenol, of which both the protonated and deprotonated forms are lipid-soluble. The partners provide a shuttling system for the inward movement of protons down their electrochemical gradient. This mode of action is comparable to that of the NH_3–NH_4^+ conjugate pair; its proton-shuttling ability was discussed earlier, under SOME

OBSERVATIONS ON INTRACELLULAR pH TRANSIENTS. The uncouplers are different in that they do not leave the inner membrane, but go back and forth. These compounds deplete the energy supply (proton gradient) available for ATP synthesis, while the oxidative process accelerates, uncoupled from the "braking" effect of the gradient. The oxidative energy is dissipated as heat.

Physiologically increased heat production resulting from uncoupled oxidation is put to good use in young mammals under conditions of cold stress, where it contributes to the maintenance of body temperature (nonshivering thermogenesis). The extra heat is generated by a special organ, brown fat, that is dispersed over the animals' body and usually is found in immediate contact with major vessels such as the aorta (386). This fat has a rich supply of blood vessels and sympathetic fibers. Brown fat's enormous heat-producing capacity (it can generate up to half of the total heat of the animal even though it constitutes only 1%–2% of body weight) is due to the uncoupling ability of its huge number of mitochondria that contain an unusually large concentration of respiratory enzyme complexes in their inner membranes. The ATP synthase activity, on the other hand, is low. Especially in cold-adapted animals, these membranes contain a special 32 kDa protein, called uncoupling protein, which represents about 15% of the inner membrane protein. Its synthesis is essentially controlled by norepinephrine released from sympathetic endings. Through this protein, H^+ can enter the mitochondrial matrix, thus shunting the proton-pumping mechanism.

The uncoupling protein of brown fat is one of the best characterized of the family of mitochondrial carrier proteins. Its sequence of 306 amino acids has been determined directly, and indirectly by sequencing its cDNA in a variety of mammals. In the membrane it functions as a dimer. Thus far, the data reveal considerable homologies with the ADP/ATP carrier, the phosphate carrier, and the glutamate/malate exchanger (210). Yet, in spite of these advances, the mode by which it translocates H ions remains a mystery. This protein can also translocate chloride, bromide, nitrate, and thiocyanate ions (270), but whether the same domain is involved in the transport of these ions as in that of H^+ is uncertain. Purine nucleotides such as ATP and GDP are powerful blockers of H^+ translocation, and it is generally agreed that fatty acids are the physiologic agents that abolish this block, and so "open" the uncoupling pathway.

The series of events in in vivo brown adipocytes following cold stress may be as follows: The stress releases norepinephrine. Its binding to β receptors on the cell membrane sets in motion a lipolytic cascade.

The resulting free fatty acids serve two functions: to provide oxidizable substrate to the mitochondria, and to act as second messenger in converting the uncoupling protein from the nonconducting to the conducting state. Several observations confirm individual steps of this sequence. Norepinephrine greatly increases respiration and heat production of isolated brown fat cells, as do fatty acids, especially when the cells are taken from cold-adapted animals. On the other hand, pyruvate (normally derived from glycolysis), which is an excellent respiratory substrate for the brown adipocytes, fails to uncouple the mitochondria. It is hypothesized that "fatty acids do not simply introduce an ohmic conductance, but clamp the electrochemical proton gradient at a level that is not only largely independent of the respiratory rate, but also is sufficiently high to allow continued transport of metabolites into the matrix [see above], and even a measure of ATP synthesis" (270).

Cytoskeleton

The cytoskeleton consists of protein polymers, which form a network of filaments and tubules in the cell. The cytoskeleton serves a variety of functions. It plays an important role in cell division, providing a pathway for the movement of chromosomes and organelles, it determines the shape of the cell, and it operates in secretion. The two main cytoskeletal structures are microtubules and microfilaments, which are predominantly composed of polymers of tubulin and actin, respectively. Since these proteins contain certain amino acids whose local charges are pH dependent, it can be expected that changes in pH_i will affect cytoskeletal structure and function. We will discuss two examples of the influence of pH_i on the cytoskeleton: (1) the effect on microtubule assembly/disassembly and vesicle transport on microtubules, and (2) the effect on actin assembly into microfilaments.

1. Tubulin can assemble into two types of microtubules. Purified tubulin forms free microtubules. Most microtubules, however, form in vivo in a directed fashion from a nucleation site (for example, a centrosome or an axoneme, referred to as a microtubule-organizing center, MTOC). In an early study, assembly of purified microtubule proteins (from bovine brain or clam oocytes) was shown to be dependent on pH_i: below pH 7 free microtubule assembly was favored, and above pH 7 disassembly was favored (309, 365). A similar effect of pH_i was observed in cultured mammalian cells (106). In contrast, the pH_i effect on MTOC-nucleated microtubule assembly is just the opposite: acidic pH_i reduces assembly while alkaline pH_i increases it (106, 333, 364, 365). Furthermore, these MTOC-nucleated

microtubules are polarized, with a plus end and a minus end. Purified tubulin grows in both directions and this growth is faster at acidic pH, whereas nucleated growth occurs only at the plus end and is enhanced at alkaline pH. This implies the presence of minus-end inhibiting factors in the cytosol as well as pH-dependent plus-end microtubule growth promotors in the cytosol (365).

An example may illustrate the physiological significance of pH_i in microtubule assembly. Unfertilized sea urchin eggs lack an apparent cytoskeleton. Within 5 min of fertilization, microtubules are formed. The microtubules are nucleated by the sperm centrioles and form a radial structure, the *sperm aster*. The sperm aster is responsible for the movement of both the incorporated sperm nucleus and the egg nucleus (nuclear migration) toward the center of the cell (332). The appearance of the sperm aster is preceded by a fertilization-induced rise in pH_i from 6.8 to 7.2 (337), which is both necessary and sufficient for nuclear migration (333). When, in fertilized eggs, the pH_i changes are prevented, no sperm aster is formed and no nuclear migration occurs. On the other hand, when, in unfertilized eggs, pH_i is raised, both sperm aster formation and nuclear migration are seen. Thus, pH-dependent microtubule formation is essential for proper egg development upon fertilization.

In polarized cells (for example, nerves and epithelial cells), microtubules growing from nucleating sites are polarized. In nerve cells, the plus end is oriented away from the cell body and into the axon; in epithelial cells it is oriented toward the basolateral membrane (205). These microtubules are associated with "microtubule motors" (mechanochemical ATPases such as kinesin and dynein) that move vesicles in either the minus or plus direction. Movement of certain vesicles along these microtubules is pH-dependent. For instance, late endosomes (see later, under Endocytosis/Exocytosis) move toward the plus end upon acidification and toward the minus end upon alkalinization (168, 289). Thus, at pH_i of about 6.5, late endosomes move into the axons of nerves or toward the basolateral region of epithelial cells, while at 7.5 the late endosomes are largely localized in the neuronal cell body or in the apical-membrane region of epithelial cells. Not all vesicles, however, are similarly affected. The movement of lysosomes on microtubules is not affected as dramatically by changes in pH_i as that of late endosomes, and early endosomal distribution is unaffected by pH_i (289). Thus, large changes of pH_i could affect the microtubule motors and alter the direction of transport and cellular localization of some, but not all, transported vesicles.

In summary, changes of pH_i can affect both the assembly of and direction of transport on microtubules

and thereby potentially alter cytoskeletal-dependent cellular processes.

2. In nonmuscle cells, actin fibers (F-actin) form by polymerization of globular actin monomers (G-actin) into microfilaments. These microfilaments are dynamic structures, existing only for short periods and continually being re-formed and broken down, often appearing in different regions of a cell and at different developmental stages. Actin polymerization/depolymerization reactions are involved in such cell functions as motility, shape changes, and formation of microvilli.

The polymerization of purified G-actin into F-actin is pH-dependent, decreasing with increasing pH_i over the range 6–8. An alkaline pH results in a higher critical actin concentration for polymerization (328, 422) and a higher dissociation rate constant (328), both of which contribute to actin depolymerization. These data on purified actin suggest that cell alkalinization should be accompanied by reduction in the number of microfilaments, but often just the opposite is seen. For example, sea urchin eggs alkalinize from 6.6 to 7.4 upon fertilization, and this alkalinization is necessary for the formation of actin filament bundles in microvilli (34). Similarly, an alkalinization is necessary for actin polymerization as part of the acrosomal reaction in sperm (383). It should be noted, however, that not all actin polymerization events are under the influence of pH_i. Activation of neutrophils by chemotactic factors, which initiate directed cell motility, is accompanied by an increase in actin polymerization, which is pH independent (113, 262). The contrast between the effect of pH on polymerization of purified G-actin and its in vivo effects indicates that stimulation of actin polymerization by cell alkalinization cannot be attributed to a direct pH effect on actin monomers. It has been hypothesized that increased cell pH reduces the binding of actin monomers to an associated protein. This binding would block polymer formation (34, 383). Thus, according to this theory, pH_i alters actin polymerization by affecting its interaction with regulatory proteins.

Indeed, the interaction of actin with several proteins in the cell can be pH sensitive. For instance, talin is a protein that can bind to actin, facilitate actin crosslinking, and help to attach actin filaments to the cell membrane at points of attachment to the extracellular matrix (focal adhesions). The binding between talin and actin is pH dependent, increasing as pH_i is lowered from 7.4 to 6.6 (335, 421b). Another protein, actin depolymerizing factor (ADF), isolated from chick brains, catalyzes the rapid depolymerization of actin filaments. The ability of ADF to depolymerize actin is maximal at pH 8 and falls fivefold between 7.7 and 7.1 (163). This pH_i effect on ADF may underlie the shape or motility changes that are seen when certain cells are alkalinized (163). Finally, F-actin bundling in the slime mold, *Dictyostelium discoideum,* which is mediated by elongation factor 1α, is reduced as pH_i is increased from 6.2 to 7.8 (114b).

In summary, the ability of changes of pH_i to alter microfilament formation is a complex process involving both effects on actin–actin interactions and effects on interactions between actin and regulatory binding proteins. While actin filament formation is often increased by cell alkalinization, the crosslinking, bundling, and forming of these filaments into networks is often reduced at alkaline values of pH_i due to reduced interactions between actin and the regulatory binding proteins. These latter effects may be related to increased cytoplasmic solation, protein synthesis, and cell motility with increased pH_i (114b).

Endocytosis/Exocytosis

Macromolecules can enter and leave cells through endocytosis and exocytosis. Endocytosis can be nonselective (fluid-phase) or selective (receptor-mediated). In the latter case, specific macromolecules (ligands) bind to specific surface receptors and are internalized, after which the ligand dissociates from the receptor and the receptor is recycled to the surface. Internally synthesized proteins, on the other hand, are packaged into vesicles in the *trans* Golgi network and are then directed to various sites, including vesicles from which proteins can be secreted by exocytosis. Thus, both endocytic and exocytic pathways involve the intracellular movement of macromolecules and membranes through the vesicular system of the cell. The pHs of some components of this vesicular system are substantially lower than cytoplasmic pH. At least four acidic vesicle pools have been identified—namely, endosomes, lysosomes, vesicles formed from the *trans* Golgi network, and secretory vesicles (for reviews, see 22, 244, and 370).

Endocytic vesicles, formed by pinching off of clathrin-coated pits in the surface membrane, initially contain ligands bound to receptors and are filled with extracellular solution. Thus the pH of these initial vesicles is very similar to pH_o about 7.4. However, within seconds, these vesicles fuse with early endosomes, whose pH is more acid, around 5.5–6.3 (133a, 133b, 243, 282a). Over the next several minutes, the endocytosed material moves to a compartment (the late endosomes) whose pH is as low as 5.5 (this was measured by incorporating macroglobulin labeled with the pH-sensitive fluorescent dye fluorescein [see pH-Sensitive Fluorescent Indicators] into the endosomes; [397]). Later, the contents of these late endosomes are found within the lysosomes (pH 4.5–5; see 165). Thus,

early and late endosomes and lysosomes are acidic compartments.

The exocytic pathway begins with the synthesis of new proteins in the endoplasmic reticulum. The proteins are then transported through the Golgi apparatus. The pH within the Golgi apparatus is about 6.2 to 6.5 (144a, 336a). The proteins that are to be secreted are concentrated by binding to specialized regions of the *trans* Golgi network membrane (signal patch) contained within clathrin-coated membrane buds (144). These buds form secretory vesicles. The protein dissociates from the patch region and is condensed and stored in these vesicles. Upon appropriate stimulation, the vesicles fuse with the surface membrane (exocytosis) and the protein is secreted. After secretion, the remaining membrane components are recycled to the *trans* Golgi network. In endocrine but not exocrine cells (370), the condensing and storage vesicles are acidic. For instance, insulin-containing vesicles in pancreatic islet cells have an internal pH between 5 and 5.7 (280).

Acidification of these various vesicular compartments is achieved by a vacuolar or V-type ATPase (see earlier, under Vacuolar or V-type ATPases), which resides in the vesicular membrane. These H^+ pumps are thought to be electrogenic and will thus create a *trans*-vesicular electrical gradient (vesicle interior positive). Inhibition of V-type H^+ATPases results in an increase in vesicle pH and a loss of vesicular transport (endocytosis/exocytosis) (202a, 286a).

Acidification serves several key roles in vesicular function. In some vesicles, especially lysosomes and late endosomes, a low pH is essential for proper function of the degradative enzymes that are contained in them, such as lysosomal hydrolases and endoproteases (243). During receptor-mediated endocytosis, the low pH in the early endosomes facilitates the dissociation of ligand from receptor, which is necessary for receptor recycling. Further, endocytosed substances like viruses (348) or bacterial toxins (46) undergo conformational changes that are facilitated by low pH. These changes enable them to fuse and thus penetrate the vesicle membranes. Finally, during exocytosis, acidic storage vesicles often form in endocrine cells. For example, catecholamines in adrenal medullary cells are stored as chromaffin granules (273). The low pH (~5.7) of these granules protects the amines from degradation, is conducive to their passive accumulation of catecholamines (which are weak bases) and, most important, appears to be required for enormously concentrating the catecholamines through an active, electrogenic transporter that exchanges protons for catecholamine ion (273): the catecholamine concentration in the granules is some 30,000 times higher than in the cytosol.

It is now clear that cytoplasmic pH can affect the very mechanism of endocytosis and exocytosis. Reduction of cytoplasmic pH to below 6.8 in a mutant fibroblast cell line reversibly blocks both fluid-phase and receptor-mediated endocytosis as well as the exocytic movement of a viral glycoprotein from the *trans* Golgi network to the surface membrane (95). A number of weak bases such as chloroquine, methylamine, and ammonia, which accumulate in acid compartments, also inhibit endocytosis and exocytosis (79b, 370). These observations suggest that the pH *difference* across the membranes of vesicles may be a significant factor in their movement and their behavior at the plasma membrane.

Other Cellular Processes

Finally, a few of the many other cellular processes in which pH_i plays a role will be briefly mentioned.

Changes in pH_i affect cell motility. The function of pH_i in actin polymerization has already been mentioned (see earlier, under Cytoskeleton). An elevated pH_i is also required for flagellar beating in sperm (331), suggesting that pH_i may, in general, affect flagellar-mediated cell movements.

H ions and intracellular messengers such as Ca^{2+}, cAMP, cGMP, and G proteins can interact. Thus, Ca^{2+} and H^+ can interact by sharing intracellular buffer sites (39), by transmembrane Ca^{2+}/H^+ exchange (272), by Ca^{2+} activation of pH-regulating transporters (for example, [167]), by competition for binding to calmodulin (71), and by H^+ effects on Ca^{2+} channels (see earlier, under Membrane Channels). H^+ and cAMP and cGMP also interact in various ways: cAMP-metabolizing enzymes (adenylate cyclase and cyclic nucleotide phosphodiesterase) have distinct pH dependencies (71); elevation of pH_i decreases cGMP by inhibiting guanylyl cyclase (93, 347), and increases the cAMP and cGMP response to adrenergic stimulation (174, 175).

The regulation of pH_i in epithelial cells has been reviewed (54, 184a, 218a, 232a). The pH_i-regulating transporters in these cells can also participate in transepithelial transport of various solutes such as K^+ (416), Na^+ (234a, 311) or organic ions (25). The pH-regulating transporters often have specialized localizations in the epithelial membrane and precise control mechanisms to enable them to achieve these various functions.

When certain cells are grown under alkaline conditions, they may be transformed both in appearance and properties. This suggests that gene expression is altered by the change in pH_i. One possible way in which this may be brought about is through an alteration of the interaction between DNA and histones. Histones are positively charged molecules that bind to DNA, reducing the repulsion of negatively charged phosphate groups on DNA and allowing the DNA to adopt a tightly packed structure. The negatively charged phosphate groups in DNA will remain charged as pH_i increases, but the positive charge on the histones will decrease, reducing the binding of histones to DNA. This will tend to loosen the DNA packing, allowing it to be more easily transcribed (276).

Changes in pH_i have also been shown to alter the expression of specific genes in various cells: acidification of hamster embryo cells triggers immediate early-gene expression (187); growing cells in acid medium increases the mRNA for NHE-1 and the synthesis of this Na^+/H^+ exchange isoform (15); alkaline pH reduces the expression of a gene-regulating protein in *Staphylococcus* (308); and low external pH can induce the expression of the multidrug-resistance protein (see earlier, under DNA Synthesis and Cell Growth) in humans (412).

We end this review where we began, with the inotropic effect of pH_i changes on muscle. As was stated at the beginning of the chapter, it has been known for a long time that a decrease in cell pH reduces the contractile capabilities of heart muscle and of skeletal muscle. A variety of mechanisms have been proposed to explain this phenomenon. Prominent among them are intracellular Ca^{2+}-related factors such as acidosis-induced reduction in the response of the contractile proteins to Ca^{2+}, due to reduced binding to troponin C. This reduction occurs even though the size of the Ca^{2+} transient that activates contraction is increased (12). On the basis of work on skinned, fast-twitch skeletal muscle fibers it has been proposed that $H_2PO_4^-$, which increases as pH_i falls, may also be a factor in the negative inotropic effect of acidosis (275). This seems not to be the case in skinned heart muscle or in intact heart (see review by Orchard and Kentish [279]). The developed tension of heart muscle fibers is far more sensitive to pH_i than that of skeletal muscle fibers; studies in transgenic mice have shown that this greater sensitivity resides in troponin C (245). Thus, it is likely that multiple pathways exist by which changes in pH_i can affect muscle contractility. It is a testament to the significance of pH_i, as well as to the unending search for scientific understanding, that this question is still under active study, some 330 years after Isaac Newton made the original observations.

We thank Joseph Ackerman for comments on the section *Nuclear Magnetic Resonance Spectroscopy*, Stanley Misler for comments on the section *Membrane Channels*, Ron Abercrombie for enlightening discussions on diffusion, and Luis Reuss for critically reading this chapter. We also thank Phyllis Douglas for producing the figures and for secretarial help. This work was supported in part by the NIH (grant number AR38881 to R.W.P.) and the Juvenile Diabetes Foundation, International (to R.W.P.).

REFERENCES

1. Abercrombie, R. F., R. W. Putnam, and A. Roos. The intracellular pH of frog muscle: its regulation in isotonic solutions. *J. Physiol. (Lond.)* 345: 175–187, 1983.

2. Ackerman, J.J.H., and D. A. d'Avignon. Basic concepts of high resolution nuclear magnetic resonance spectroscopy. In: *Medical Magnetic Resonance: a Primer*, edited by T. F. Budinger and A. R. Margolis. Berkeley, CA: Society of Magnetic Resonance in Medicine, 1988, p. 249–262.

3. Ackerman, J.J.H., G. E. Soto, W. M. Spees, Z. Zhu, and J. L. Evelhoch. The NMR chemical shift pH measurement revisited: analysis of error and modeling of pH-dependent reference. *Magn. Reson. Med.* 36: 674–683, 1996.

4. Addanki, S., F. D. Cahill, and J. F. Sotos. Determination of intramitochondrial pH and intramitochondrial-extramitochondrial pH gradient of isolated heart mitochondria by the use of 5,5-dimethyl-2,4-oxazolidine-dione. *J. Biol. Chem.* 243: 2337–2348, 1968.

5. Ahearn, G. A., P. Franco, and L. P. Clay. Electrogenic 2 Na^+/1 H^+ exchange in crustaceans. *J. Membr. Biol.* 116: 215–226, 1990.

6. Aickin, C. C. Direct measurement of intracellular pH and buffering power in smooth muscle cells of guinea-pig vas deferens. *J. Physiol. (Lond.)* 349: 571–585, 1984.

7. Aickin, C. C. Regulation of intracellular pH in the smooth muscle of guinea-pig ureter: Na^+ dependence. *J. Physiol. (Lond.)* 479: 301–316, 1994.

8. Aickin, C. C. Regulation of intracellular pH in the smooth muscle of guinea-pig ureter: HCO_3 dependence. *J. Physiol. (Lond.)* 479: 317–329, 1994.

9. Aickin, C. C., and R. C. Thomas. Micro-electrode measurement of the intracellular pH and buffering power of mouse soleus muscle fibres. *J. Physiol. (Lond.)* 267: 791–810, 1977.

10. Al-Awqati, Q. Proton-translocating ATPases. *Annu. Rev. Cell Biol.* 2: 179–199, 1986.

10a. Al-Awqati, Q., J. Barasch, and D. Landry. Chloride channels of intracellular organelles and their potential role in cystic fibrosis. *J. Exp. Biol.* 172: 245–266, 1992.

11. Al-Baldawi, N. F., and R. F. Abercrombie. Cytoplasmic hydrogen ion difusion coefficient. *Biophys. J.* 61: 1470–1479, 1992.

12. Allen, D. G., and C. H. Orchard. The effect of changes of pH on intracellular calcium transients in mammalian cardiac muscle. *J. Physiol. (Lond.)* 335: 555–567, 1983.

13. Alper, S. L. The band 3-related anion exchanger (AE) gene family. *Annu. Rev. Physiol.* 53: 549–564, 1991.

14. Alper, S. L. The band 3-related AE anion exchanger gene family. *Cell Physiol. Biochem.* 4: 265–281, 1994.

15. Alpern, R. J., Y. Yamaji, A. Cano, S. Horie, R. T. Miller, O. W. Moe, and P. A. Preisig. Chronic regulation of the Na/H antiporter. *J. Lab. Clin. Med.* 122: 137–140, 1993.

16. Altenberg, G. A., G. Young, J. K. Horton, D. Glass, J. A. Belli, and L. Reuss. Changes in intra- or extracellular pH do not mediate P-glycoprotein-dependent multidrug resistance. *Proc. Natl. Acad. Sci. USA* 90: 9735–9738, 1993.

17. Alvarez-Leefmans, F. J., S. M. Gamino, and L. Reuss. Cell volume changes upon sodium pump inhibition in *Helix aspera* neurons. *J. Physiol. (Lond.)* 458: 603–619, 1992.

18. Alvarez-Leefmans, F. J., H. Cruzblanca, S. M. Gamino, J. Altamirano, A. Nani, and L. Reuss. Transmembrane ion movements elicited by sodium pump inhibition in *Helix aspera* neurons. *J. Neurophysiol.* 71: 1787–1796, 1994.

19. Ammann, D. *Ion-Selective Microelectrodes. Principles, Design and Application.* New York: Springer Verlag, 1986.

20. Ammann, D., F. Lanter, R. Steiner, P. Schulthess, Y. Shijo, and W. Simon. Neutral carrier based hydrogen ion selective microelectrode for extra- and intracellular studies. *Anal. Chem.* 53: 2267–2269, 1981.

21. Amorena, C. E., T. J. Wilding, J. K. Manchester, and A. Roos. Changes in intracellular pH caused by high K in normal and acidified frog muscle. *J. Gen. Physiol.* 96: 959–972, 1990.

22. Anderson, R.G.W., and L. Orci. A view of acidic intracellular compartments. *J. Cell Biol.* 106: 539–543; 1988.

23. Arellano, R. O., A. Rivera, and F. Ramon. Protein phosphorylation and hydrogen ions modulate calcium-induced closure of gap junction channels. *Biophys. J.* 57: 363–367, 1990.

24. Aronson, P. S. Kinetic properties of the plasma membrane Na^+-H^+ exchanger. *Annu. Rev. Physiol.* 47: 545–560, 1985.

25. Aronson, P. S. The renal proximal tubule: a model for diversity of anion exchangers and stilbene-sensitive anion transporters. *Annu. Rev. Physiol.* 51: 419–441, 1989.

26. Aronson, P. S., J. Nee, and M. Suhm. Modifier role of internal H^+ in activating the Na^+-H^+ exchanger in renal microvillus membrane vesicles. *Nature* 299: 161–163, 1982.

27. Astion, M. L., A. Chvatal, and R. K. Orkand. Further studies of electrogenic Na^+/HCO_3^- cotransport in glial cells of *Necturus* optic nerve: regulation of pH_i. *Glia* 4: 461–468, 1991.

28. Baas, F., A.P.M. Jongsma, H. J. Broxterman, R. J. Arceci, D. Housman, G. L. Scheffer, A. Riethorst, M. van Groenigen, A.W.M. Nieuwint, and H. Joenje. Non-P-glycoprotein mediated mechanism for multidrug resistance precedes P-glycoprotein expression during *in vitro* selection for doxorubicin resistance in a human lung cancer cell line. *Cancer Res.* 50: 5392–5398, 1990.

29. Babcock, D. F. Examination of the intracellular ionic environment and of ionophore action by null point measurements employing the fluorescein chromophore. *J. Biol. Chem.* 258: 6380–6389, 1983.

30. Babcock, G. T., and M. Wikström. Oxygen activation and the conservation of energy in cell respiration. *Nature* 356: 301–309, 1992.

31. Barish, M. E., and C. Baud. A voltage-gated hydrogen ion current in the oocyte membrane of the axolotl, *Ambystoma*. *J. Physiol. (Lond.)* 352: 243–263, 1984.

32. Bates, R. G. *Determination of pH. Theory and Practice*, 2nd edition. New York: John Wiley and Sons, Inc., 1973.

33. Baylor, S. M., W. K. Chandler, and M. W. Marshall. Optical measurements of intracellular pH and magnesium in frog skeletal muscle fibres. *J. Physiol. (Lond.)* 331: 105–137, 1982.

34. Begg, D. A., and L. I. Rebhun. pH regulates the polymerization of actin in the sea urchin egg cortex. *J. Cell Biol.* 83: 241–248, 1979.

35. Bennett, M.V.L., and V. K. Verselis. Biophysics of gap junctions. *Cell Biol.* 3: 29–47, 1992.

36. Bennett, M.V.L., V. K. Verselis, R. L. White, and D. C. Spray. Gap junctional conductance: gating. In: *Gap Junctions*. E. L. Hertzberg and R. G. Johnson. New York: Alan R. Liss, Inc., 1988, p. 287–304.

37. Benos, D. J. Amiloride: a molecular probe of sodium transport in tissues and cells. *Am. J. Physiol.* 242 (*Cell Physiol.* 11): C131–C145, 1982.

38. Berridge, M. J. Inositol trisphosphate and diacylglycerol as second messengers. *Biochem. J.* 220: 345–360, 1984.

39. Bers, D. M., and D. Ellis. Intracellular calcium and sodium activity in sheep heart Purkinje fibres. Effect of changes of external sodium and intracellular pH. *Pflugers Arch.* 393: 171–178, 1982.

40. Besterman, J. M., S. J. Tyrey, E. J. Cragoe, Jr., and P. Cuatrecasas. Inhibition of epidermal growth factor-induced mitogenesis by amiloride and an analog: Evidence against a requirement for Na^+/H^+ exchange. *Proc. Natl. Acad. Sci. USA* 81: 6762–6766, 1984.

41. Beyer, E. C., D. L. Paul, and D. A. Goodenough. Connexin 43: a protein from rat heart homologous to a gap junction protein from liver. *J. Cell Biol.* 105: 2621–2629, 1987.

41a. Bianchini, L., A. Kapus, G. Lukacs, S. Wasan, S. Wakabayashi, J. Pouysségur, F. H. Yu, J. Orlowski, and S. Grinstein. Responsiveness of mutants of NHE1 isoform of Na^+/H^+ antiport to osmotic stress. *Am. J. Physiol.* 269 (*Cell Physiol.* 38): C998–C1007, 1995.

42. Bidani, A., and S.E.S. Brown. ATP-dependent pH_i recovery in lung macrophages: evidence for a plasma membrane H^+-ATPase. *Am. J. Physiol.* 259 (*Cell Physiol.* 28): C586–C598, 1990.

43. Biemesderfer, D., J. Pizzonia, A. Abu-Alfa, M. Exner, R. Reilly, P. Igarashi, and P. S. Aronson. NHE3: a Na^+/H^+ exchanger isoform of renal brush border. *Am. J. Physiol.* 265 (*Renal Fluid Electrolyte Physiol.* 34): F736–F742, 1993.

44. Bierman, A. J., E. J. Cragoe, Jr., S. W. de Laat, and W. H. Moolenaar. Bicarbonate determines cytoplasmic pH and suppresses mitogen-induced alkalinization in fibroblastic cells. *J. Biol. Chem.* 263: 15253–15256, 1988.

44a. Blanchard, A., F. Leviel, M. Bichara, R.-A. Podevin, and M. Paillard. Interactions of external and internal K^+ with K^+-HCO_3^- cotransporter of rat medullary thick ascending limb. *Am. J. Physiol.* 271 (*Cell Physiol.* 40): C218–C225, 1996.

45. Blatz, A. L. Asymmetric proton block of inward rectifier channels in skeletal muscle. *Pflugers Arch.* 401: 402–407, 1984.

46. Blewitt, M. G., L. A. Chung, and E. London. Effect of pH on the conformation of diphtheria toxin and its implications for membrane penetration. *Biochemistry* 24: 5458–5464, 1985.

47. Bolton, T., and R. D. Vaughan-Jones. Continuous direct measurement of intracellular chloride and pH in frog skeletal muscle. *J. Physiol. (Lond.)* 270: 801–833, 1977.

48. Bonanno, J. A. K^+-H^+ exchange, a fundamental cell acidifier in corneal epithelium. *Am. J. Physiol.* 260 (*Cell Physiol.* 29): C618–C625, 1991.

49. Bookstein, C., A. M. DePaoli, Y. Xie, P. Niu, M. W. Musch, M. C. Rao, and E. B. Chang. Na^+/H^+ exchangers, NHE-1 and NHE-3, of rat intestine. Expression and localization. *J. Clin. Invest.* 93: 106–113, 1994.

50. Bookstein, C., M. W. Musch, A. DePaoli, M. Villereal, B. Bacallao, and E. B. Chang. Characterization of a unique sodium-hydrogen exchange isoform (NHE-4) in inner medulla of the rat kidney and in transfected NHE-deficient murine fibroblasts. *Mol. Biol. Cell* 4: 311a, 1993.

51. Borgese, F., C. Sardet, M. Cappadoro, J. Pouysségur, and R. Motais. Cloning and expression of a cAMP-activated Na^+/H^+ exchanger: evidence that the cytoplasmic domain mediates hormonal regulation. *Proc. Natl. Acad. Sci. USA* 89: 6765–6769, 1992.

52. Boron, W. F. Intracellular pH transients in giant barnacle muscle fibers. *Am. J. Physiol.* 233 (*Cell Physiol.* 2): C61–C73, 1977.

53. Boron, W. F. Intracellular pH-regulating mechanism of the squid axon. Relation between the external Na^+ and HCO_3^- dependences. *J. Gen. Physiol.* 85: 325–346, 1985.

54. Boron, W. F. Intracellular pH regulation in epithelial cells. *Annu. Rev. Physiol.* 48: 377–388, 1986.

55. Boron, W. F., and E. L. Boulpaep. Intracellular pH regulation in the renal proximal tubule of the salamander. Na^+-H^+ exchange. *J. Gen. Physiol.* 81: 29–52, 1983.

56. Boron, W. F., and E. L. Boulpaep. Intracellular pH regulation in the renal proximal tubule of the salamander. Basolateral HCO_3^- transport. *J. Gen. Physiol.* 81: 53–94, 1983.

57. Boron, W. F., and P. De Weer. Active proton transport stimulated by CO_2/HCO_3^-, blocked by cyanide. *Nature* 259: 240–241, 1976.

58. Boron, W. F., and P. De Weer. Intracellular pH transients in squid giant axons caused by CO_2, NH_3, and metabolic inhibitors. *J. Gen. Physiol.* 67: 91–112, 1976.

59. Boron, W. F., and R. C. Knakal. Intracellular pH-regulating mechanism of the squid axon. Interaction between DNDS and extracellular Na^+ and HCO_3^-. *J. Gen. Physiol.* 93: 123–150, 1989.

60. Boron, W. F., W. C. McCormick, and A. Roos. pH regulation in barnacle muscle fibers: dependence on intracellular and extracellular pH. *Am. J. Physiol.* 237 (*Cell Physiol.* 6): C185–C193, 1979.

61. Boron, W. F., W. C. McCormick, and A. Roos. pH regulation in barnacle muscle fibers: dependence on extracellular sodium and bicarbonate. *Am. J. Physiol.* 240 (*Cell Physiol.* 9): C80–C89, 1981.

62. Boron, W. F., and A. Roos. Comparison of microelectrode, DMO and methylamine methods for measuring intracellular pH. *Am. J. Physiol.* 231: 799–809, 1976.

63. Boron, W. F., and J. M. Russell. Stoichiometry and ion dependencies of the intracellular-pH-regulating mechanism in squid giant axons. *J. Gen. Physiol.* 81: 373–399, 1983.

64. Boscoboinik, D., R. S. Gupta, and R. M. Epand. Investigation of the relationship between altered intracellular pH and multidrug resistance in mammalian cells. *Br. J. Cancer* 61: 568–572, 1990.

65. Boyarsky, G., M. B. Ganz, R. B. Sterzel, and W. F. Boron. pH regulation in single glomerular mesangial cells. I. Acid extrusiion in absence and presence of HCO_3^-. *Am. J. Physiol.* 255 (*Cell Physiol.* 24): C844–C856, 1988.

66. Brown, D., S. Hirsch, and S. Gluck. An H^+-ATPase in opposite plasma membrane domains in kidney epithelial cell subpopulations. *Nature* 331: 622–624, 1988.

67. Buckler, K. J., and R. D. Vaughan-Jones. Application of a new pH-sensitive fluoroprobe (carboxy-SNARF-1) for intracellular pH measurement in small isolated cells. *Pflugers Arch.* 417: 234–239, 1990.

68. Buckler, K. J., R. D. Vaughan-Jones, C. Peers, D. Lagadic-Gossmann, and P.C.G. Nye. Effects of extracellular pH, P_{CO_2} and HCO_3^- on intracellular pH in isolated type-I cells of the neonatal rat carotid body. *J. Physiol. (Lond.)* 444: 703–721, 1991.

69. Burt, J. M., and D. C. Spray. Single channel events and gating behavior of the cardiac gap junction channel. *Proc. Natl. Acad. Sci. USA* 85: 3431–3434, 1988.

70. Busa, W. B., and J. H. Crowe. Intracellular pH regulates transitions between dormancy and development of brine shrimp (*Artemia salina*) embryos. *Science* 221: 366–368, 1983.

71. Busa, W. B., and R. Nuccitelli. Metabolic regulation via intracellular pH. *Am. J. Physiol.* 246 (*Regulatory Integrative Comp. Physiol.* 15): R409–R438, 1984.

72. Buytendijk, F.J.J., and M. W. Woerdeman. Die physico-chemischen Erscheinungen während der Entwicklung. I. Die Messung der Wasserstoffionenkonzentration. *Wilhelm Roux Arch. Entwicklungsmech. Org.* 112: 387–410, 1927.

73. Byerly, L., R. Meech, and W. Moody, Jr. Rapidly activating hydrogen ion currents in perfused neurones of the snail, *Lymnaea stagnalis. J. Physiol. (Lond.)* 351: 199–216, 1984.

74. Byerly, L., and Y. Suen. Characterization of proton currents in neurones of the snail. *J. Physiol. (Lond.)* 413: 75–89, 1989.

75. Cabantchik, Z. I., and R. Greger. Chemical probes for anion transporters of mammalian cell membranes. *Am. J. Physiol.* 262 (*Cell Physiol.* 31): C803–C827, 1992.

76. Cala, P. M. Volume regulation by *Amphiuma* red blood cells: The membrane potential and its implications regarding the nature of the ion-flux pathways. *J. Gen. Physiol.* 76: 683–708, 1980.

77. Cala, P. M. Volume regulation by *Amphiuma* red blood cells: the role of Ca^{2+} as modulator of alkali metal/H^+ exchange. *J. Gen. Physiol.* 82: 761–784, 1983.

78. Caldwell, P. C. An investigation of the intracellular pH of crab muscle fibres by means of micro-glass and micro-tungsten electrodes. *J. Physiol. (Lond.)* 126: 169–180, 1954.

79. Caldwell, P. C. Studies on the internal pH of large muscle and nerve fibres. *J. Physiol. (Lond.)* 142: 22–62, 1958.

79a. Calonge, M. L., and A. A. Hundáin. PKC activators stimulate H^+ conductance in chicken enterocytes. *Pflugers Arch.* 431: 594–598, 1996.

79b. Carnell, L., and H.-P.H. Moore. Transport via the regulated secretory pathway in semi-intact PC12 cells: role of intracisternal calcium and pH in the transport and sorting of secretogranin II. *J. Cell Biol.* 127: 693–705, 1994.

80. Casavola, V., S. J. Reshkin, H. Murer, and C. Helmle-Kolb. Polarized expression of Na$^+$/H$^+$ exchange activity in LLC-PK$_1$/PKE$_{20}$ cells: II. hormonal regulation. *Pflugers Arch.* 420: 282–289, 1992.

81. Cassel, D., B. Whiteley, J. X. Zhuang, and L. Glaser. Mitogen-independent activation of Na$^+$/H$^+$ exchange in human epidermoid carcinoma A431 cells: Regulation by medium osmolarity. *J. Cell. Physiol.* 122: 178–186, 1985.

82. Chaillet, J. R., and W. F. Boron. Intracellular calibration of a pH-sensitive dye in isolated, perfused salamander proximal tubules. *J. Gen. Physiol.* 86: 765–794, 1985.

83. Chamberlin, M. E., and K. Strange. Anisosmotic cell volume regulation: a comparative view. *Am. J. Physiol.* 257 (*Cell Physiol.* 26): C159–C173, 1989.

83a. Chang, A., T. G. Hammond, T. T. Sun, and M. L. Zeidel: Permeability properties of the mammalian bladder apical membrane. *Am. J. Physiol.* 267 (*Cell Physiol.* 36): C1483–C1492, 1994.

84. Chao, P., D. Ammann, A. Oesch, W. Simon, and F. Lang. Extra- and intracellular hydrogen ion-selective microelectrodes based on neutral carriers with extended pH response range in acid media. *Pflügers Arch.* 411: 216–219, 1988.

84a. Cherney, V. V., V. S. Markin, and T. E. DeCoursey. The voltage-activated hydrogen ion conductance in rat alveolar epithelial cells is determined by the pH gradient. *J. Gen. Physiol.* 105: 861–896, 1995.

84b. Chraïbi, A., R. Guinamard, and J. Teulon. Effects of internal pH on the nonselective cation channel from the mouse collecting tubule. *J. Membr. Biol.* 148: 83–90, 1995.

85. Christensen, O., and T. Zeuthen. Maxi K$^+$ channels in leaky epithelia are regulated by intracellular Ca^{2+}, pH and membrane potential. *Pflugers Arch.* 408: 249–259, 1987.

86. Clark, J. D., and L. E. Limbird. Na$^+$-H$^+$ exchanger subtypes: a predictive review. *Am. J. Physiol.* 261 (*Cell Physiol.* 30): C945–C953, 1991.

87. Clay, J. R. I$_K$ inactivation in squid axons is shifted along the voltage axis by changes in the intracellular pH. *Biophys. J.* 58: 797–801, 1990.

88. Clemens, N., W. Siffert, and P. Scheid. Thrombin-induced cytosolic alkalinization in human platelets occurs without an apparent involvement of HCO$_3^-$/Cl$^-$ exchange. *Pflügers Arch.* 416: 68–73, 1990.

89. Cohen, P., C. B., Klee, C. Picton, and S. Shenolikar. Calcium control of muscle phosphorylaase kinase through the combined action of calmodulin and troponin. *Ann. NY Acad. Sci.* 356: 151–161, 1980.

90. Cohen, R. D., and R. A. Iles. Intracellular pH: measurement, control and metabolic interrelationships. *Crit. Rev. Clin. Lab. Sci.* 6: 101–143, 1975.

91. Collins, J. F., T. Honda, S. Knobel, N. M. Bulus, J. Conary, R. DuBois, and F. K. Ghishan. Molecular cloning, sequencing, tissue distribution, and functional expression of a Na$^+$/H$^+$ exchanger (NHE-2). *Proc. Natl. Acad. Sci. USA* 90: 3938–3942, 1993.

92. Cook, D. L., M. Ikeuchi, and W. Y. Fujimoto. Lowering of pH$_i$ inhibits Ca^{2+}-activated K$^+$ channels in pancreatic B-cells. *Nature* 311: 269–271, 1984.

93. Cook, S. P., and D. F. Babcock. Selective modulation by cGMP of the K$^+$ channel activated by speract. *J. Biol. Chem.* 268: 22402–22407, 1993.

94. Copello, J., Y. Segal, and L. Reuss. Cytosolic pH regulates maxi K$^+$ channels in *Necturus* gall-bladder epithelial cells. *J. Physiol. (Lond.)* 434: 577–590, 1991.

95. Cosson, P., I. de Curtis, J. Pouysségur, G. Griffiths, and J. Davoust. Low cytoplasmic pH inhibits endocytosis and transport from the *trans*-Golgi network to the cell surface. *J. Cell Biol.* 108: 377–387, 1989.

96. Counillon, L., and J. Pouysségur. Molecular biology and hormonal regulation of vertebrate Na$^+$/H$^+$ exchanger isoforms. In: *Molecular Biology and Function of Carrier Proteins*, edited by L. Reuss, J. M. Russell, Jr., and M. L. Jennings. New York: Rockefeller University Press, 1993, p. 169–185.

96a. Counillon, L., and J. Pouysségur. Structure-function studies and molecular regulation of the growth factor activatable sodium-hydrogen exchanger (NHE-1). *Cardiovas. Res.* 29: 147–154, 1995.

97. Counillon, L., W. Scholz, H. J. Lang, and J. Pouysségur. Pharmacological characterization of stably transfected Na$^+$/H$^+$ antiporter isoforms using amiloride analogs and a new inhibitor exhibiting anti-ischemic properties. *Mol. Pharmacol.* 44: 1041–1045, 1993.

98. Cuevas, J., A. L. Bassett, J. S. Cameron, T. Furukawa, R. J. Myerburg, and S. Kimura. Effect of H$^+$ on ATP-regulated K$^+$ channels in feline ventricular myocytes. *Am. J. Physiol.* 261 (*Heart Circ. Physiol.* 30): H755–H761, 1991.

99. Curtin, N. A. Buffer power and intracellular pH of frog sartorius mucle. *Biophys. J.* 50: 837–841, 1986.

100. Dart, C., and R. D. Vaughan-Jones. Na$^+$-HCO$_3^-$ symport in the sheep cardiac Purkinje fibre. *J. Physiol. (Lond.)* 451: 365–385, 1992.

101. Dascalu, A., A. Nevo, and R. Korenstein. Hyperosmotic activation of the Na$^+$/H$^+$ exchanger in a rat bone cell line: temperature dependence and activation pathways. *J. Physiol. (Lond.)* 456: 503–518, 1992.

102. Davenport, H. W. *The ABC of Acid–Base Chemistry.* Chicago University of Chicago Press, 1947.

103. Davies, N. W., N. B. Standen, and P. R. Stanfield. The effect of intracellular pH on ATP-dependent potassium channels of frog skeletal muscle. *J. Physiol. (Lond.)* 445: 549–568, 1992.

104. Davis, B. A., E. M. Hogan, and W. F. Boron. Intracellular Cl$^-$

dependence of hypertonically activated Na-H exchange activity in barnacle muscle. *Biophys. J.* 57: 186a, 1990.

105. Davis, B. A., E. M. Hogan, and W. F. Boron. Activation of Na-H exchange by intracellular lithium in barnacle muscle fibers. *Am. J. Physiol.* 263 (*Cell Physiol.* 32): C246–C256, 1992.

106. De Brabander, M., G. Geuens, R. Nuydens, R. Willebrords, and J. De Mey. Microtubule stability and assembly in living cells: the influence of metabolic inhibitors, taxol and pH. *Cold Spring Harb. Symp. Quant. Biol.* 45: 227–240, 1982.

107. Deamer, D. W., and J. W. Nicholls. Proton flux mechanisms in model and biological membranes. *J. Membr. Biol.* 107: 91–103, 1989.

108. Deamer, D. W., R. C. Prince, and A. R. Crofts. The response of fluorescent amines to pH gradient across liposomes membranes. *Biochim. Biophys. Acta* 274: 323–335, 1972.

109. DeCoursey, T. E. Hydrogen ion currents in rat alveolar epithelial cells. *Biophys. J.* 60: 1243–1253, 1991.

110. DeCoursey, T. E., and V. V. Cherny. Potential, pH, and arachidonate gate hydrogen ion currents in human neutrophils. *Biophys. J.* 65: 1590–1598, 1993.

110a. DeCoursey, T. E., and V. V. Cherney. Voltage-activated hydrogen ion currents. *J. Membr. Biol.* 141: 203–223, 1994.

110b. DeCoursey, T. E., and V. V. Cherney. Effects of buffer concentration on voltage-gated H^+ currents: does diffusion limit the conduction? *Biophys. J.* 71: 182–193, 1996.

111. Deffner, G.G.J., and R. E. Haftner. Chemical investigations of the giant nerve fibres of squid. I. Fractionation of dialyzable constituents of axoplasm and quantitative determination of the free amino acids. *Biochim. Biophys. Acta* 32: 362–374, 1959.

111a. Deitmer, J. W., and C. R. Rose. pH regulation and proton signalling by glial cells. *Prog. Neurobiol.* 48: 73–103, 1996.

112. Deitmer, J. W., and W.-R. Schlue. An inwardly directed electrogenic sodium-bicarbonate cotransport in leech glial cells. *J. Physiol. (Lond.)* 411: 179–194, 1989.

113. Downey, G. P., and S. Grinstein. Receptor-mediated actin assembly in electropermeabilized neutrophils: role of intracellular pH. *Biochem. Biophys. Res. Commun.* 160: 18–24, 1989.

114. Dubuisson, M. A discussion on muscular contraction and relaxation: their physical and chemical basis, introduced by A. V. Hill. *Proc. R. Soc. Lond. [Biol.]* 137: 40–87, 1950.

114a. Dunn, K. W., S. Mayor, J. N. Myers, and F. R. Maxfield. Applications of ratio fluorescence microscopy in the study of cell physiology. *FASEB J.* 8: 573–582, 1994.

114b. Edmonds, B. T., J. Murray, and J. Condeelis. pH regulation of the F-actin binding properties of *Dictyostelium* elongation factor 1α. *J. Biol. Chem.* 270: 15222–15230, 1995.

115. Edsall, J. T., and J. J. Wyman. *Biophysical Chemistry, Vol. I.* New York: Academic Press, 1958.

115a. Ek, J. F., M. Delmar, R. Perzova, and S. M Taffet. Role of histidine 95 on the pH gating of the cardiac gap junction protein connexin-43. *Circ. Res.* 74: 1058–1064, 1994.

115b. Ek-Vitorín, J. F., G. Calero, G. E. Morley, W. Coombs, S. M. Taffet, and M. Delmar. pH regulation of connexin-43: molecular analysis of the gating particle. *Biophys. J.* 71: 1273–1284, 1996.

116. Ellis, D., and R. C. Thomas. Direct measurement of the intracellular pH of mammalian cardiac muscle. *J. Physiol. (Lond.)* 262: 755–771, 1976.

117. Engasser, J.-M., and C. Horvath. Buffer-facilitated proton transport pH profile of bound enzymes. *Biochim. Biophys. Acta* 358: 178–192, 1974.

118. Epel, D., R. A. Steinhardt, T. Humphreys, and D. Mazia. An analysis of the partial metabolic derepression of sea urchin eggs by ammonia: the existence of independent pathways. *Dev. Biol.* 40: 245–255, 1974.

119. Fabiato, A., and F. Fabiato. Effects of pH on the myofilaments and the sarcoplasmic reticulum of skinned cells from cardiac and skeletal muscle. *J. Physiol. (Lond.)* 276: 233–255, 1978.

119a. Fafournoux, P., J. Noël, and J. Pouysségur. Evidence that Na^+/H^+ exchanger isoforms NHE-1 and NHE-3 exist as stable dimers in membranes with a high degree of specificity for homodimers. *J. Biol. Chem.* 269: 2589–2596, 1994.

120. Fairbanks, G., T. L. Steck, and D.F.H. Wallach. Electrophoretic analysis of the major polypeptides of the human erythrocyte membrane. *Biochemistry* 10: 2606–2617, 1971.

121. Fenn, W. O., and D. M. Cobb. The potassium equilibrium in muscle. *J. Gen. Physiol.* 17: 629–656, 1934.

122. Fenn, W. O., and F. W. Maurer. The pH of muscle. *Protoplasma* 24: 337–345, 1935.

122a. Firek, L., and R. Weingart. Modification of gap junction conductance by divalent cations and protons in neonatal rat heart cells. *J. Mol. Cell Cardiol.* 27: 1633–1643, 1995.

123. Fitz, J. G., S. D. Lidofsky, M.-H. Xie, and B. F. Scharschmidt. Transmembrane electrical potential difference regulates Na^+/HCO_3^- cotransport and intracellular pH in hepatocytes. *Proc. Natl. Acad. Sci. USA* 89: 4197–4201, 1992.

124. Fliegel, L., and O. Fröhlich. The Na^+/H^+ exchanger: an update of structure, regulation and cardiac physiology. *Biochem. J.* 296: 273–285, 1993.

125. Fluri, G. S., A. Rudisuli, M. Willi, S. Rohr, and R. Weingart. Effects of arachidonic acid on the gap junctions of neonatal rat heart cells. *Pflügers Arch.* 417: 149–156, 1990.

126. Forte, J. G., D. K. Hanzel, T. Urushidani, and J. M. Wolosin. Pumps and pathways for gastric HCl secretion. *Ann. NY Acad. Sci.* 574: 145–1158, 1989.

127. Frelin, C., P. Vigne, and M. Lazdunski. The role of the Na^+/H^+ exchange system in cardiac cells in relation to the control of the internal Na^+ concentration. A molecular basis for the antagonistic effect of ouabain and amiloride on the heart. *J. Biol. Chem.* 259: 8880–8885, 1984.

128. Funder, J., and J. O. Wieth. Chloride and hydrogen ion distribution between human red cells and plasma. *Acta Physiol. Scand.* 68: 234–245, 1966.

129. Furshpan, E. J., and D. D. Potter. Transmission at the giant motor synapses of the crayfish. *J. Physiol. (Lond.)* 145: 289–325, 1959.

130. Gadian, D. G., G. K. Radda, M. J. Dawson, and D. R. Wilkie. pH measurements of cardiac and skeletal muscle using ^{31}P-NMR. In: *Intracellular pH: Its Measurement, Regulation, and Utilization in Cellular Functions,* edited by R. Nuccitelli and D. W. Deamer. New York: Alan R. Liss, Inc., 1982, p. 60–70.

131. Ganz, M. B., G. Boyarsky, W. F. Boron, and R. B. Sterzel. Effects of angiotensin II and vasopressin on intracellular pH of glomerular mesangial cells. *Am. J. Physiol.* 254 (*Renal Fluid Electrolyte Physiol.* 23): F787–F794, 1988.

132. Ganz, M. B., G. Boyarsky, R. B. Sterzel, and W. F. Boron. Arginine vasopressin enhances pH_i regulation in the presence of HCO_3 by stimulating three acid–base transport systems. *Nature* 337: 648–651, 1989.

133. Garty, H., C. Asher, and O. Yeger. Direct inhibition of epithelial Na^+ channels by a pH-dependent interaction with calcium, and by other divalent ions. *J. Membr. Biol.* 95: 151–162, 1987.

133a. Gekle, M., S. Mildenberger, R. Freudinger, and S. Silbernagl. Endosomal alkalinization reduces J_{max} and K_m of albumin receptor-mediated endocytosis in OK cells. *Am. J. Physiol.* 268 (*Renal Fluid Electrolyte Physiol.* 37): F899–F906, 1995.

133b. Gekle, M., and S. Silbernagl. Comparison of the buffer capacity of endocytotic vesicles, lysosomes and cytoplasm in cells derived from the proximal tubule of the kidney (opossum kidney cells). *Pflugers Arch.* 429: 452–454, 1995.

134. Gilbert, D. S. Axoplasm chemical composition in *Myxicola* and solubility properties of its structural proteins. *J. Physiol. (Lond.)* 253: 303–319, 1975.

135. Gillies, R. J., J. R. Alger, J. A. den Hollander, and R. G. Shulman. Intracellular pH measured by NMR: methods and results. In: *Intracellular pH: Its Measurement, Regulation, and Utilization in Cellular Functions.* edited by R. Nuccitelli and D. W. Deamer. New York: Alan R. Liss, Inc., 1982, p. 79–104.

136. Gladden, L. B. Current "anaerobic threshold" controversies. *Physiologist* 27: 312–318, 1984.

137. Gleeson, D., N. D. Smith, and J. L. Boyer. Bicarbonate-dependent and -independent intracellular pH regulatory mechanisms in rat hepatocytes. Evidence for Na^+-HCO_3^- cotransport. *J. Clin. Invest.* 84: 312–321, 1989.

138. Gluck, S. L. The structure and biochemistry of the vacuolar H^+ ATPase in proximal and distal urinary acidification. *J. Bionerg. Biomembr.* 24: 351–359, 1992.

139. Gluck, S. L. The vacuolar H^+-ATPases: versatile proton pumps participating in constitutive and specialized functions of eukaryotic cells. *Int. Rev. Cytol.* 137C: 105–137, 1993.

140. Godt, R. E., and D. W. Maughan. On the composition of the cytosol of relaxed skeletal muscle of the frog. *Am. J. Physiol.* 254 (*Cell Physiol.* 23): C591–C604, 1988.

141. Good, D. W. Ammonium transport by the thick ascending limb of Henle's loop. *Annu. Rev. Physiol.* 56: 623–647, 1994.

141a. Grant, C. E., G. Valdimarsson, D. R. Hipfner, K. C. Almquist, S.P.C. Cole, and R. G. Deeley. Overexpression of multidrug resistance-associated protein (MRP) increases resistance to natural product drug. *Cancer Res.* 54: 357–361, 1994.

142. Green, J., and S. Muallem. A common mechanism for activation of the Na^+/H^+ exchanger by different types of stimuli. *FASEB J.* 3: 2408–2414, 1989.

143. Greenfield, N. J., M. Hussain, and J. Lenard. Effect of growth state and amines on cytoplasmic and vacuolar pH, phosphate and polyphosphate levels in *Saccharomyces cerevisiae*: a ^{31}P-nuclear magnetic resonance study. *Biochim. Biophys. Acta* 926: 205–214, 1987.

144. Griffiths, G., and K. Simons. The *trans* Golgi network: sorting at the exit site of the Golgi complex. *Science* 234: 438–443, 1986.

144a. Grinstein, S. Non-invasive measurement of the luminal pH of compartments of the secretory pathway. *Physiologist* 39: 144, 1996.

145. Grinstein, S., and S. Cohen. Cytoplasmic [Ca^{2+}] and intracellular pH in lymphocytes. Role of membrane potential and volume-activated Na^+/H^+ exchange. *J. Gen. Physiol.* 89: 185–213, 1987.

146. Grinstein, S., S. Cohen, J. D. Goetz, and A. Rothstein. Osmotic and phorbol ester-induced activation of Na^+/H^+ exchange: possible role of protein phosphorylation in lymphocyte volume regulation. *J. Cell Biol.* 101: 269–276, 1985.

147. Grinstein, S., and W. Furuya. Cytoplasmic pH regulation in phorbol ester-activated human neutrophils. *Am. J. Physiol.* 251 (*Cell Physiol.* 20): C55–C65, 1986.

148. Grinstein, S., and R. W. Putnam. Measurement of intracellular pH. In: *Methods in Membrane and Transporter Research.* edited by J. Schafer, G. Giebisch, P. Kristensen, and H. Ussings. Austin, TX: R. G. Landes Company, 1994, p. 113–141.

149. Grinstein, S., and A. Rothstein. Mechanisms of regulation of the Na^+/H^+ exchanger. *J. Membr. Biol.* 90: 1–12, 1986.

150. Grinstein, S., M. Woodside, C. Sardet, J. Pouysségur, and D. Rotin. Activation of the Na^+/H^+ antiporter during cell volume regulation. Evidence for a phosphorylation-independent mechanism. *J. Biol. Chem.* 267: 23823–23828, 1992.

151. Gros, G., W. Moll, H. Hoppe, and H. Gros. Proton transport by phosphate diffusion—a mechanism of facilitated CO_2 transfer. *J. Gen. Physiol.* 67: 773–790, 1976.

152. Grynkiewicz, G., M. Poenie, and R. Y. Tsien. A new generation of Ca^{2+} indicators with greatly improved fluorescence properties. *J. Biol. Chem.* 260: 3440–3450, 1985.

153. Gutknecht, J. Proton conductance through phospholipid bilayers: water wires or weak acids? *J. Bioenerg. Biomembr.* 19: 427–442, 1987.

154. Gutknecht, J., and D. C. Tosteson. Diffusion of weak acids across lipid bilayer membranes: effect of chemical reactions in the unstirred layers. *Science* 182: 1258–1261, 1973.

155. Haggerty, J. G., N. Agarwal, R. F. Reilly, E. A. Adelberg, and C. W. Slayman. Pharmacologically different Na/H antiporters on the apical and basolateral surfaces of cultured porcine kidney cells (LLC-PK₁). *Proc. Natl. Acad. Sci. USA* 85: 6797–6801, 1988.

156. Hamm, J. R., and G. M. Yue. ^{31}P nuclear magnetic resonance measurements on intracellular pH in giant barnacle muscle. *Am. J. Physiol.* 252 (*Cell Physiol.* 21): C30–C37, 1987.

157. Harris, A. L., D. C. Spray, and M.V.L. Bennett. Kinetic properties of a voltage-dependent junctional conductance. *J. Gen. Physiol.* 77: 95–117, 1981.

158. Harris, S. P., N. W. Pichards, C. D. Logsdon, J. Pouysségur, and D. C. Dawson. Cloning of partial length cDNAs homologous to the human NHE-1 antiporter from reptilian [turtle] colon. *J. Gen. Physiol.* 100: 34a, 1992.

159. Hartree, W., and A. V. Hill. The recovery heat production in muscle. *J. Physiol. (Lond.)* 56: 367–381, 1922.

160. Harvey, B. J., S. R. Thomas, and J. Ehrenfeld. Intracellular pH controls cell membrane Na^+ and K^+ conductances and transport in frog skin epithelium. *J. Gen. Physiol.* 92: 767–791, 1988.

161. Harvey, W. R., and N. Nelson, eds. V-ATPases. *Journal of Experimental Biology,* vol. 172. Cambridge, U.K.: The Company of Biologists Ltd., 1992.

162. Haugland, R. P., and A. Minta. Design and application of indicator probes. In: *Non-invasive Techniques in Cell Biology,* edited by J. K. Foskett and S. Grinstein. New York: Wiley-Liss, 1990, p. 1–20.

163. Hayden, S. M., P. S. Miller, A. Brauweiler, and J. R. Bamburg. Analysis of the interactions of actin depolymerizing factor with G- and F-actin. *Biochemistry* 32: 9994–10004, 1993.

164. Heberle, J., J. Riesle, G. Thiedemann, D. Oesterhelt, and N. A. Dencher. Proton migration along the membrane surface and retarded surface to bulk transfer. *Nature* 370: 379–382, 1994.

165. Heilmann, P., W. Beisker, U. Miaskowski, P. Camner, and W. G. Kreyling. Intraphagolysosomal pH in canine and rat alveolar macrophages: flow cytometric measurements. *Environ. Health Perspect.* 97: 115–120, 1992.

166. Henderson, L. J. Concerning the relationship between the strength of acids and their capacity to preserve neutrality. *Am. J. Physiol.* 21: 173–179, 1908.

167. Hendey, B., M. D. Mamrack, and R. W. Putnam. Thrombin induces a calcium transient that mediates an activation of the Na^+/H^+ exchanger in human fibroblasts. *J. Biol. Chem.* 264: 19540–19547, 1989.

168. Heuser, J. Changes in lysosome shape and distribution correlated with changes in cytoplasmic pH. *J. Cell Biol.* 108: 855–864, 1989.

169. Hill, A. V. The influence of the external medium on the internal pH of muscle. *Proc. R. Soc. Lond. [Biol]* 144: 1–22, 1955.

170. Hill, A. V., and H. Lupton. Muscular exercise, lactic acid, and the supply and utilization of oxygen. *Q. J. Med.* 16: 135–171, 1923.

171. Hille, B. *Ion Channels of Excitable Membranes,* 2nd edition. Sunderland, MA: Sinauer Assoc., Inc., 1992.

172. Hinke, J.A.M. Cation-selective microelectrodes for intracellular use. In: *Glass Electrodes for Hydrogen and Other Cations,* edited by G. Eisenman. New York: Dekker, 1967, p. 467–477.

173. Hinke, J.A.M. Thirty years of ion-selective microelectrodes: disappointments and successes. *Can. J. Physiol. Pharmacol.* 65: 873–878, 1987.

174. Ho, A. K., M. Girard, I. Young, and C. L. Chik. Intracellular pH on adrenergic-stimulated cAMP and cGMP production in rat pinealocytes. *Am. J. Physiol.* 261 (*Cell Physiol.* 30): C642–C649, 1991.

175. Ho, A. K., L. O'Brien, M. Girard, and C. L. Chik. Intracellular pH on protein kinase C and ionomycin potentiation of isoproterenol-stimulated cyclic AMP and cyclic GMP production in rat pinealocytes. *J. Neurochem.* 59: 2304–2310, 1992.

176. Hochachka, P. W., and T. P. Mommsen. Protons and anaerobiosis. *Science* 219: 1391–1397, 1983.

177. Hoffmann, E. K., and L. O. Simonsen. Membrane mechanisms in volume and pH regulation in vertebrate cells. *Physiol. Rev.* 69: 315–382, 1989.

178. Hofmann, G. E., and S. C. Hand. Subcellular differentiation arrested in *Artemia* embryos under anoxia: Evidence supporting a regulatory role for intracellular pH. *J. Exp. Zool.* 253: 287–302, 1990.

178a. Hogan, E. M., M. A. Cohen, and W. F. Boron. K^+- and HCO_3^--dependent acid–base transport in squid giant axons. I. Base efflux. *J. Gen. Physiol.* 106: 821–844, 1995.

178b. Hogan, E. M., M. A. Cohen, and W. F. Boron. K^+- and HCO_3^--dependent acid–base transport in squid giant axon. II. Base influx. *J. Gen. Physiol.* 106: 845–862, 1995.

179. Honda, T., S. M. Knobel, N. M. Bulus, and F. K. Ghishan. Kinetic characterization of a stably expressed novel Na^+/H^+ exchanger (NHE-2). *Biochim. Biophys. Acta* 1150: 199–202, 1993.

180. Huet, S., B. Schott, and J. Roberts. P-glycoprotein overexpression cannot explain the complete doxorubicin-resistance phenotype in rat glioblastoma cell lines. *Br. J. Cancer* 65: 538–544, 1992.

181. Hughes, B. A., J. S. Adorante, S. S. Miller, and H. Lin. Apical electrogenic $NaHCO_3$ cotransport. A mechanism for CO_3 absorption across the retinal pigment epithelium. *J. Gen. Physiol.* 94: 125–150, 1989.

182. Hunter, M., H. Oberleithner, R. M. Henderson, and G. Giebisch. Whole-cell potassium currents in single early distal tubule cells. *Am. J. Physiol.* 255 (*Renal Fluid Electrolyte Physiol.* 24): F699–F703, 1988.

183. Igarashi, P., R. F. Reilly, D. Hildebrandt, D. Biemesderfer, N. A. Reboucas, C. W. Slayman, and P. S. Aronson. Molecular biology of renal Na^+-H^+ exchangers. *Kidney Int.* 40 (Suppl. 33): S84–S89, 1991.

184. Igarashi, P., M. I. Freed, M. B. Ganz, and R. F. Reilly. Effects of chronic metabolic acidosis on Na^+-H^+ exchangers in LLC-PK$_1$ renal epithelial cells. *Am. J. Physiol.* 263 (*Renal Fluid Electrolyte Physiol.* 32): F83–F88, 1992.

184a. Ilundáin, A. Intracellular pH regulation in intestinal and renal epithelial cells. *Comp. Biochem. Physiol.* 101A: 413–424, 1992.

184b. Inaba, M., A. Yawata, I. Koshino, K. Sato, M. Takeuchi, Y. Takakuwa, S. Manno, Y. Yawata, A. Kanzaki, J. Sakai, A. Ban, K. Ono, and Y. Maede. Defective anion transport and marked spherocytosis with membrane instability caused by hereditary total deficiency of red cell band 3 in cattle due to a nonsense mutation. *J. Clin. Invest.* 97: 1804–1817, 1996.

185. Irving, M., J. Maylie, N. L. Sizto, and W. K. Chandler. Simultaneous monitoring of changes in magnesium and calcium concentrations in frog cut twitch fibres containing antipyrlazo III. *J. Gen. Physiol.* 93: 585–608, 1989.

186. Irving, M., J. Maylie, N. L. Sizto, and W. K. Chandler. Intracellular difusion in the presence of mobile buffers. Application to proton movement in muscle. *Biophys. J.* 57: 717–721, 1990.

187. Isfort, R. J., D. B. Cody, T. N. Asquith, G. M. Ridder, S. B. Stuard, and R. A. Leboeuf. Induction of protein phosphorylation, protein synthesis, immediate-early-gene expression and cellular proliferation by intracellular pH modulation. Implications for the role of hydrogen ions in signal transduction. *Eur. J. Biochem.* 213: 349–357, 1993.

188. Ito, H., J. Vereecke, and E. Carmeliet. Intracellular protons inhibit inward rectifier K^+ channel of guinea-pig ventricular cell membrane. *Pflugers Arch.* 422: 280–286, 1992.

189. Iwata, S., C. Ostermeier, B. Ludwig, and H. Michel. Structure of 2.8 Å resolution of cytochrome c oxidase from *Paracoccus denitrificans*. *Nature* 376: 660–669, 1995.

190. Jacobs, M. H. The production of intracellular acidity by neutral and alkaline solutions containing carbon dioxide. *Am. J. Physiol.* 53: 457–463, 1920.

191. Jacobs, M. H. The influence of ammonium salts on cell reaction. *J. Gen. Physiol.* 5: 181–188, 1922.

192. Jakubovicz, D. E., S. Grinstein, and A. Klip. Cell swelling following recovery from acidification in C6 glioma cells: an in vitro model of postischemic brain edema. *Brain Res.* 435: 138–146, 1987.

193. Jans, A.W.H., E. S. Krijnen, J. Luig, and R.K.H. Kinne. A ^{31}P-NMR study on the recovery of intracellular pH in LLC-PK1/Cl$_4$ cells from intracellular alkalinization. *Biochim. Biophys. Acta* 931: 326–334, 1987.

194. Jean, T., C. Frelin, P. Vigne, and M. Lazdunski. The Na^+/H^+ exchange system in glial cell lines. Properties and activation by an hyperosmotic shock. *Eur. J. Biochem.* 160: 211–219, 1986.

195. Johnson, J. D., D. Epel, and M. Paul. Intracellular pH and activation of sea urchin eggs after fertilisation. *Nature* 262: 661–664, 1976.

196. Johnston, M. F., S. A. Simon, and F. Ramon. Interaction of anaesthetics with electrical synapses. *Nature* 286: 498–500, 1980.

197. Juliano, R. L., and V. Ling. A surface glycoprotein modulating drug permeability in Chinese hamster ovary cell mutants. *Biochim. Biophys. Acta* 455: 152–162, 1976.

198. Junge, W., and S. McLaughlin. The role of fixed and mobile buffers in the kinetics of proton movement. *Biochim. Biophys. Acta* 890: 1–5, 1987.

199. Kahn, A. M., E. J. Cragoe, Jr., J. C. Allen, C. L. Seidel, and H. Shelat. Effects of pH$_i$ on Na^+-H^+, Na^+-dependent, and Na^+-independent Cl-HCO_3^- exchangers in vascular smooth muscle. *Am. J. Physiol.* 261 (*Cell Physiol.* 30): C837–C844, 1991.

200. Kapus, A., R. Romanek, A. Q. Yi, O. D. Rotstein, and S. Grinstein. A pH-sensitive and voltage-dependent proton conductance in the plasma membrane of macrophages. *J. Gen. Physiol.* 102: 729–760, 1993.

201. Karniski, L. P., and P. S. Aronson. Chloride/formate exchange with formic acid recycling: a mechanism of active chloride

transport across epithelial membranes. *Proc. Natl. Acad. Sci. USA* 82: 6362–6365, 1985.

202. Kasianowicz, J. J., and S. M. Bezrukov. Protonation dynamics of the α-toxin ion channel from spectral analysis of pH-dependent current fluctuations. *Biophys. J.* 69: 94–105, 1995.

202a. Kataoka, T., M. Muroi, S. Ohkuma, T. Waritani, J. Magae, A. Takatsuki, S. Kondo, M. Yamasaki, and K. Nagai. Prodigiosin 25-C uncouples vacuolar type H⁺-ATPase, inhibits vacuolar acidification and affects glycoprotein processing. *FEBS Lett.* 359: 53–59, 1995.

203. Keifer, D. W., and A. Roos. Membrane permeability to the molecular and ionic forms of DMO in barnacle muscle. *Am. J. Physiol.* 240 (*Cell Physiol.* 9): C73–C79, 1981.

204. Keizer, J. G., and H. Joenje. Increased cytosolic pH in multidrug-resistant human lung tumor cells: Effect of verapamil. *J. Natl. Cancer Inst.* 81: 706–709, 1989.

205. Kelly, R. B. Microtubules, membrane traffic and cell organization. *Cell* 61: 5–7, 1990.

206. Kettenmann, H., B. R. Ransom, and W.-R. Schlue. Intracellular pH shifts capable of uncoupling cultured oligodendrocytes are seen only in low HCO₃⁻ solution. *Glia* 3: 110–117, 1990.

207. Kikeri, D., S. Azar, A. Sun, M. L. Zeidel, and S. C. Hebert. Na⁺-H⁺ antiporter and Na⁺-(HCO₃⁻)ₙ symporter regulate intracellular pH in mouse medullary thick limbs of Henle. *Am. J. Physiol.* 258 (*Renal Fluid Electrolyte Physiol.* 27): F45–F456, 1990.

208. Kikeri, D., A. Sun, M. L. Zeidel, and S. C. Hebert. Cell membranes impermeable to NH₃. *Nature* 339: 478–480, 1989.

209. Kite, G. L. Studies on the physical properties of the protoplasm. I. The physical properties of the protoplasm of certain animal and plant cells. *Am. J. Physiol.* 32: 146–164, 1913.

210. Klaus, S., L. Casteilla, F. Bouillaud, and D. Riequier. The uncoupling protein UCP: a membraneous mitochondrial ion carrier exclusively expressed in brown adipose tissue. *Int. J. Biochem.* 23: 791–801, 1991.

211. Kleyman, T. R., and E. J. Cragoe, Jr. Amiloride and its analogs as tools in the study of ion transport. *J. Membr. Biol.* 105: 1–21, 1988.

212. Knauf, P. A. Erythrocyte anion exchange and the band 3 protein: transport kinetics and molecular structure. *Curr. Top. Membr. Transp.* 12: 249–363, 1979.

213. Koechlin, B. A. On the chemical composition of the axoplasm of squid giant nerve fibres with particular reference to its ion pattern. *J. Biophys. Biochem. Cytol.* 1: 511–529, 1955.

214. Kopito, R. R., and H. F. Lodish. Primary structure and transmembrane orientation of the murine anion exchange protein. *Nature* 316: 234–238, 1985.

215. Koppel, M., and K. Spiro. Über die Wirkung von Moderatoren (Puffern) bei der Verschiebung des Säure-Basengleichwichtes in biologischen Flüssigkeiten. *Biochem. Ztschr.* 65: 409–439, 1914.

216. Kostyuk, P. G., and G. A. Sorokina. On the mechanism of hydrogen ion distribution between cell protoplasm and the medium. In: *Membrane Trasnport and Metabolism,* edited by A. Kleinzeller and A. Kotyk. London: Academic Press, 1961, p. 193–203.

217. Kotyk, A., and J. Slavik. *Intracellular pH and Its Measurement.* Boca Raton: CRC Press, 1989.

218. Krämer, K., and F. Palmieri. Molecular aspects of isolated and reconstituted carrier proteins from animal mitochondria. *Biochim. Biophys. Acta* 974: 1–23, 1989.

218a. Krapf, R., and R. J. Alpern. Cell pH and transepithelial H/HCO₃ transport in the renal proximal tubule. *J. Membr. Biol.* 131: 1–10, 1993.

219. Kudrycki, K. E., P. R. Newman, and G. E. Shull. cDNA cloning and tissue distribution of mRNAs for two proteins that are related to the band 3 Cl⁻/HCO₃⁻ exchanger. *J. Biol. Chem.* 265: 462–471, 1990.

220. Kuffler, S. W., and D. D. Potter. Glia in the leech contral nervous system: physiological properties and neuron-glia relationships. *J. Neurophysiol.* 27: 290–320, 1964.

221. Kulanthaivel, P., F. H. Leibach, V. B. Mahesh, E. J. Cragoe, Jr., and V. Ganapathy. The Na⁺-H⁺ exchanger of the placental brush-border membrane is pharmacologically distinct from that of the renal brush-border membrane. *J. Biol. Chem.* 265: 1249–1252, 1990.

222. Kume, H., K. Takagi, T. Satake, H. Tokuno, and T. Tomita. Effects of intracellular pH on calcium-activated potassium channels in rabbit tracheal smooth muscle. *J. Physiol. (Lond.)* 424: 445–457, 1990.

223. Lagarde, A. E., and J. Pouysségur. The Na⁺/H⁺ antiport in cancer. *Cancer Biochem. Biophys.* 9: 1–14, 1986.

224. LaNoue, K. F., and A. C. Schoolwerth. Metabolic transport in mitochondria. *Annu. Rev. Biochem.* 48: 871–922, 1979.

225. Laurido, C., S. Candia, D. Wolff, and R. Latorre. Proton modulation of a Ca²⁺-activated K⁺ channel from rat skeletal muscle incorporated into planar bilayers. *J. Gen. Physiol.* 98: 1025–1043, 1991.

226. Lee, H. C., and J. G. Forte. A novel method for measurement of intravesicular pH using fluorescent probes. *Biochim. Biophys. Acta* 601: 152–166, 1980.

227. Lee, H. C., J. G. Forte, and D. Epel. The use of fluorescent amines for the measurement of pHᵢ: Applications in liposomes, gastric microsomes, and sea urchin gametes. In: *Intracellular pH: Its Measurement, Regulation, and Utilization in Cellular Functions,* edited by R. Nuccitelli and D. Deamer. New York: Alan R. Liss, Inc., 1982, p. 135–160.

227a. Leviel, F., P. Borensztein, P. Houillier, M. Paillard, and M. Bichara. Electroneutral K⁺/HCO₃⁻ cotransport in cells of medullary thick ascending limb of rat kidney. *J. Clin. Invest.* 90: 869–878, 1992.

228. Liu, S., D. Piwnica-Worms, and M. Lieberman. Intracellular pH regulation in cultured embryonic chick heart cells. Na⁺-dependent Cl⁻/HCO₃⁻ exchange. *J. Gen. Physiol.* 96: 1247–1269, 1990.

229. Liu, S., S. Taffet, L. Stoner, M. Delmar, M. L. Vallano, and J. Jalife. A structural basis for the unequal sensitivity of the major cardiac and liver gap junctions to intracellular acidification: The carboxyl tail length. *Biophys. J.* 64: 1422–1433, 1993.

230. Loew, L. M., R. A. Tuft, W. Carrington, and F. S. Fay. Imaging in five dimensions: time-dependent membrane potentials in individual mitochondria. *Biophys. J.* 65: 2396–2407, 1993.

231. Loewenstein, W. R. Permeability of membrane junctions. *Ann. NY Acad. Sci.* 137: 441–472, 1966.

232. Loewenstein, W. R., and Y. Kanno. Studies on an epithelial (gland) cell junction. I. Modification of surface membrane permeability. *J. Cell Biol.* 22: 565–586, 1964.

232a. Lubman, R. L., and E. D. Crandall. Regulation of intracellular pH in alveolar epithelial cells. *Am. J. Physiol.* 262 (*Lung Cell. Mol. Physiol.* 6): L1–L14, 1992.

233. Lukacs, G. L., A. Kapus, A. Nanda, R. Romanek, and S. Grinstein. Proton conductance of the plasma membrane: properties, regulation, and functional role. *Am. J. Physiol.* 265 (*Cell Physiol.* 34): C3–C14, 1993.

234. Lux, S. E., K. M. John, R. R. Kopito, and H. F. Lodish. Cloning and characterization of band 3, the human erythrocyte anion-exchange protein (AE1). *Proc. Natl. Acad. Sci. USA* 86: 9089–9093, 1989.

234a. Lyall, V., G. M. Feldman, and T.U.L. Biber. Regulation of apical Na$^+$ conductive transport in epthelia by pH. *Biochim. Biophys. Acta* 1241: 31–44, 1995.

235. Ma, J., and J-Y. Zhao. Highly cooperative and hysteresis response of the skeletal muscle ryanodine receptor to changes in proton concentrations. *Biophys. J.* 66: A19, 1994.

236. Ma, L., and M. S. Center. The gene encoding vacuolar H$^+$-ATPase subunit C is overexpressed in multidrug-resistant HL60 cells. *Biochem. Biophys. Res. Commun.* 182: 675–681, 1992.

237. Makowski, L., D.L.D. Caspar, W. C. Phillips, and D. A. Goodenough. Gap junction structures II. Analysis of the X-ray diffraction data. *J. Cell Biol.* 74: 629–645, 1977.

238. Makowski, L., D.L.D. Caspar, W. C. Phillips, and T. S. Baker. Gap junction structure. VI. Variation and conservation in connexon conformation and packing. *Biophys. J.* 45: 208–218, 1984.

239. Matsumura, Y., S. Aoki, K. Kajino, and M. Fujimoto. The double-barreled microelectrode for the measurement of intracellular pH, using liquid ion-exchanger, and its biological application. *Proc. Int. Cong. Physiol. Sci.* 14: 572, 1980.

240. Mazia, D. Chromosome cycles turned on in unfertilized sea urchin eggs exposed to NH$_4$OH. *Proc. Natl. Acad. Sci. USA* 71: 690–693, 1974.

241. Mazia, D., and A. Ruby. DNA synthesis turned on in unfertilized sea urchin eggs by treatment with NH$_4$OH. *Exp. Cell Res.* 85: 167–172, 1974.

242. Meech, R. W., and R. C. Thomas. Effect of measured calcium chloride injections on the membrane potential and internal pH of snail neurones. *J. Physiol. (Lond.)* 298: 111–129, 1980.

243. Mellman, I. The importance of being acid: the role of acidification in intracellular membrane traffic. *J. Exp. Biol.* 172: 39–45, 1992.

244. Mellman, I., R. Fuchs, and A. Helenius. Acidification of the endocytic and exocytic pathways. *Annu. Rev. Biochem.* 55: 663–700, 1986.

245. Metzger, J. M., M. S. Parmacek, E. Barr, K. Pasyk, W.-I. Lin, K. L. Cochrane, L. J. Field, and J. M. Leiden. Skeletal troponin C reduces contractile sensitivity to acidosis in cardiac myocytes from transgenic mice. *Proc. Natl. Acad. Sci. USA* 90: 9036–9040, 1993.

246. Miller, K., J. Halow, and A. P. Koretsky. Phosphocreatine protects transgenic mouse liver expressing creatine kinase from hypoxia and ischemia. *Am. J. Physiol.* 265 (*Cell Physiol.* 34): C1544–C1551, 1993.

247. Mills, G. B., E. J. Cragoe, Jr., E. W. Gelfand, and S. Grinstein. Interleukin 2 induces a rapid increase in intracellular pH through activation of a Na$^+$/H$^+$ antiport. Cytoplasmic alkalinization is not required for lymphocyte proliferation. *J. Biol. Chem.* 260: 12500–12507, 1985.

248. Misler, S., K. Gillis, and J. Tabcharani. Modulation of gating of a metabolically regulated, ATP-dependent K$^+$ channel by intracellular pH in B cells of the pancreatic islet. *J. Membr. Biol.* 109: 135–143, 1989.

249. Mitchell, P. Coupling of phosphorylation to electron and hydrogen transfer by a chemiosmotic type of mechanism. *Nature* 191: 144–148, 1961.

250. Mitchell, P. Compartmentation and communication in living systems. Ligand conduction: a general catalytic principle in chemical, osmotic and chemiosmotic reaction systems. *Eur. J. Biochem.* 95: 1–20, 1979.

251. Moll, W. Measurements of facilitated diffusion of oxygen in red blood cells at 37°C. *Pflugers Arch.* 305: 269–278, 1969.

252. Moody, W. J., Jr. Appearance of calcium action potentials in crayfish slow muscle fibres under conditions of low intracellular pH. *J. Physiol. (Lond.)* 302: 335–346, 1980.

253. Moody, W. J., Jr. The ionic mechanism of intracellular pH regulation in crayfish neurones. *J. Physiol. (Lond.)* 316: 293–308, 1981.

254. Moody, W. J., Jr. Effects of intracellular H$^+$ on the electrical properties of excitable cells. *Annu. Rev. Neurosci.* 7: 257–278, 1984.

255. Moody, W. J., Jr., and S. Hagiwara. Block of inward rectification by intracellular H$^+$ in immature oocytes of the starfish *Mediaster aequalis*. *J. Gen. Physiol.* 79: 115–130, 1982.

256. Moolenaar, W. H. Effect of growth factors on intracellular pH regulation. *Annu. Rev. Physiol.* 48: 363–376, 1986.

257. Moriyama, Y., T. Manabe, Y. Yoshimori, Y. Tashiro, and M. Futai. ATP-dependent uptake of anti-neoplastic agents by acidic organelles. *J. Biochem.* 115: 213–218, 1994.

257a. Morley, G. E., S. M. Taffet, and M. Delmar. Intramolecular interactions mediate pH regulation of connexin-43 channels. *Biophys. J.* 70: 1294–1302, 1996.

258. Motais, R., F. Borgese, B. Fievet, and F. Garcia-Romeu. Regulation of Na$^+$/H$^+$ exchange and pH in erythrocytes of fish. *Comp. Biochem. Physiol.* 102A: 597–602, 1992.

259. Mugharbil, A., R. G. Knickelbein, P. S. Aronson, and J. W. Dobbins. Rabbit ileal brush-border membrane Cl-HCO$_3$ exchanger is activated by an internal pH-sensitive modifier site. *Am. J. Physiol.* 259 (*Gastrointest. Liver Physiol.* 22): G666–G670, 1990.

259a. Mülder, H. S., J. Lankelma, H. Dekker, H. J. Broxterman, and H. M. Pinedo. Daunorubicin efflux against a concentration gradient in non-P-glycoprotein multidrug-resistant lung-cancer cells. *Int. J. Cancer* 59: 275–281, 1994.

260. Munsch, T., and J. W. Deitmer. Sodium-bicarbonate cotransport current in identified leech glial cells. *J. Physiol. (Lond.)* 474: 43–53, 1994.

261. Murer, H., U. Hopfer, and R. Kinne. Sodium/proton antiport in brush-border membranes isolated from rat small intestine and kidney. *Biochem. J.* 154: 597–604, 1976.

262. Naccache, P. H., S. Therrien, A. C. Caon, N. Liao, C. Gilbert, and S. R. McColl. Chemoattractant-induced cytoplasmic pH changes and cytoskeletal reorganization in human neutrophils. Relationship to the stimulated calcium transients and oxidative burst. *J. Immunol.* 142: 2438–2444, 1989.

263. Nagle, J. F. Theory of passive proton conductance in lipid bilayers. *J. Bioenerg. Biomembr.* 19: 413–426, 1987.

264. Nakamoto, R. K., R. Rao, and C. W. Slayman. Transmembrane segments of the P-type cation-transporting ATPases. *Ann. NY Acad. Sci.* 574: 165–179, 1989.

265. Nakhoul, N. L., and W. F. Boron. Acetate transport in the S3 segment of the rabbit proximal tubule and its effect on intracellular pH. *J. Gen. Physiol.* 92: 395–412, 1988.

266. Nasse, H. Untersuchungen über den Austritt und Eintritt von Stoffen (Transsudation and Diffusion) durch die Wand der Haargefässe. *Arch Gesamte Physiol. Menschen Tiere* 16: 604–634, 1878.

267. Nelson, N. Evolution of organellar proton-ATPases. *Biochim. Biophys. Acta* 1100: 109–124, 1992.

268. Newman, E. A., and M. L. Astion. Localization and stoichiometry of electrogenic sodium bicarbonate cotransport in retinal glial cells. *Glia* 4: 424–428, 1991.

269. Neyton, J., and A. Trautmann. Single channel currents of an intercellular junction. *Nature* 317: 331–335, 1985.

270. Nicholls, D. G., S. A. Cunningham, and E. Rial. The bioenergetic mechanisms of brown adipose tissue mitochondria. In:

Brown Adipose Tissue, edited by P. Trayhurn and D. G. Nicholls. London: E. Arnold, 1986, p. 52–85.

271. Nieuwint, A.W.M., F. Baas, J. Wiegant, and H. Joenje. Cytogenetic alterations associated with P-glycoprotein- and non-P-glycoprotein-mediated multidrug resistance in SW-1573 human lung tumor cell lines. *Cancer Res.* 52: 4361–4371, 1992.

272. Niggli, V., E. Sigel, and E. Carafoli. Inhibition of the purified and reconstituted Ca^{2+} pump of erythrocytes by μM levels of DIDS and NAP-taurine. *FEBS Lett.* 138: 164–166, 1982.

273. Njus, D., J. Knoth and M. Zallakian. Proton-linked transport in chromaffin granules. *Curr. Top. Bioeng.* 11: 107–147, 1981.

273a. Noël, J., and J. Pouysségur. Hormonal regulation, pharmacology, and membrane sorting of vertebrate Na^+/H^+ exchanger isoforms. *Am. J. Physiol.* 268 (*Cell Physiol.* 37): C283–C296, 1995.

274. Nonner, W., B. C. Spalding, and B. Hille. Low intracellular pH and chemical agents slow inactivation gating in sodium channels of muscle. *Nature* 284: 360–363, 1980.

274a. Nordström, T., O. D. Rotstein, R. Romanek, S. Asotra, J.N.M. Heersche, M. F. Manolson, G. F. Brisseau, and S. Grinstein. Regulation of cytoplasmic pH in osteoclasts. Contribution of proton pumps and a proton-selective conductance. *J. Biol. Chem.* 270: 2203–2212, 1995.

275. Nosek, T. M., K. Y. Fender, and R. E. Godt. It is diprotonated inorganic phosphate that depresses force in skinned skeletal muscle fibers. *Science* 236: 191–193, 1987.

276. Oberleithner, H., and A. Schwab. Alkaline stress-induced cell transformation. *News Physiol. Sci.* 9: 165–168, 1994.

277. Ohkuma, S., and B. Poole. Fluorescence probe measurement of the intralysosomal pH in living cells and the perturbation of pH by various agents. *Proc. Natl. Acad. Sci. USA* 73: 3327–3331, 1978.

278. Ohno-Shosaku, T., T. Kubota, J. Yamaguchi, and M. Fujimoto. Regulation of inwardly rectifying K^+ channels by intracellular pH in opossum kidney cells. *Pflugers Arch.* 416: 138–143, 1990.

279. Orchard, C. H., and J. C. Kentish. Effects of changes of pH of contractile function of cardiac muscle. *Am. J. Physiol.* 258 (*Cell Physiol.* 27): C967–C981, 1990.

280. Orci, L., M. Ravazzola, M. Amherdt, O. Madsen, A. Perrelet, J.-D. Vassalli, and R.G.W. Anderson. Conversion of proinsulin to insulin occurs coordinately with acidification of maturing secretory vesicles. *J. Cell Biol.* 103: 2273–2281, 1986.

281. Orlowski, J., R. A. Kandasamy, and G. E. Shull. Molecular cloning of putative members of the Na/H exchanger gene family. cDNA cloning, deduced amino acid sequence, and mRNA tissue expression of the rat Na/H exchanger NHE-1 and two structurally related proteins. *J. Biol. Chem.* 267: 9331–9339, 1992.

282. Otsu, K., J. Kinsella, B. Sacktor, and J. P. Froehlich. Transient state kinetic evidence for an oligomer in the mechanism of Na^+-H^+ exchange. *Proc. Natl. Acad. Sci. USA* 86: 4818–4822, 1989.

282a. Overly, C. C., K.-D. Lee, E. Berthiaume, and P. J. Hollenbeck. Quantitative measurement of intraorganelle pH in the endosomal-lysosomal pathway in neurons by using ratiometric imaging with pyranine. *Proc. Natl. Acad. Sci. U.S.A.* 92: 3156–3160, 1995.

283. Overton, E. Über die osmotischen Eigenschaften der Zelle in ihrer Bedeutung fur die Toxicologie und Pharmacologie. *Z. Phys. Chem.* 22: 189–209, 1897.

284. Overton, E. Beiträge zur allgemeinen Muskel und Nervenphysiologie. *Pflugers Arch.* 92: 115–280, 1902.

285. Page, E., and Y. Shibata. Permeable junctions between cardiac cells. *Annu. Rev. Physiol.* 43: 431–441, 1981.

285a. Pajor, A. M., and H. G. Valmonte. Expression of the renal Na^+/dicarboxylate cotransporter, NaDC-1, in COS-7 cells. *Pflugers Arch.* 431: 645–651, 1996.

286. Palmer, L. G., and G. Frindt. Effects of cell Ca and pH on Na channels from rat cortical collecting tubule. *Am. J. Physiol.* 253 (*Renal Fluid Electrolyte Physiol.* 22): F333–F339, 1987.

286a. Palokangas, H., K. Metsikkö, and K. Väänänen. Active vacuolar H^+ ATPase is required for both endocytic and exocytic processes during viral infection of BHK-21 cells. *J. Biol. Chem.* 269: 17577–17585, 1994.

287. Paradiso, A. M., M. C. Townsley, E. Wenzl, and T. E. Machen. Regulation of intracellular pH in resting and in stimulated parietal cells. *Am. J. Physiol.* 257 (*Cell Physiol.* 26): C554–C561, 1989.

288. Parker, J. C. In defense of cell volume? *Am. J. Physiol.* 265 (*Cell Physiol.* 34): C1191–C1200, 1993.

289. Parton, R. G., C. G. Dotti, R. Bacallao, I. Kurtz, K. Simons, and K. Prydz. pH-induced microtubule-dependent redistribution of late endosomes in neuronal and epithelial cells. *J. Cell Biol.* 113: 261–274, 1991.

290. Paul, D. Molecular cloning of cDNA for rat liver gap junction protein. *J. Cell Biol.* 103: 123–134, 1986.

290a. Paula, S., A. G. Volkov, A. N. Van Hoek, T. H. Haines, and D. W. Deamer. Permeation of protons, potassium ions, and small polar molecules through phospholipid bilayers as a function of membrane thickness. *Biophys. J.* 70: 339–348, 1996.

291. Peracchia, C. Increase in gap junction resistance with acidification in crayfish septate axons is closely related to changes in intracellular calcium but not hydrogen ion concentration. *J. Membr. Biol.* 113: 75–92, 1990.

292. Peracchia, C. Effects of caffeine and ryanodine on low pH$_i$-induced changes in gap junction conductance and calcium concentration in crayfish septate axons. *J. Membr. Biol.* 117: 79–89, 1990.

293. Periyasamy, S. M., S. S. Kakar, K. D. Garlid, and A. Askar. Ion specificity of cardiac sarcolemmal Na^+/H^+ antiporter. *J. Biol. Chem.* 265: 6035–6041, 1990.

294. Peters, J. P., and D. D. Van Slyke. *Quantitative Clinical Chemistry, Vol. I: Interpretations.* Baltimore: Williams and Wilkins, 1931, p. 909–912.

295. Petroff, O.A.C., J. W. Pritchard, K. L. Behar, J. R. Alger, J. A. den Hollander, and R. G. Shulman. Cerebral intracellular pH by ^{31}P nuclear magnetic resonance spectroscopy. *Neurology* 35: 781–788, 1985.

295a. Pomès, R., and B. Roux. Structure and dynamics of a proton wire: a theoretical study of H^+ translocation along the single-file water chain in the Gramicidin A channel. *Biophys. J.* 71: 19–39, 1996.

296. Pouysségur, J., A. Franchi, G. L'Allemain, and S. Paris. Cytoplasmic pH, a key determinant of growth factor-induced DNA synthesis in quiescent fibroblasts. *FEBS Lett.* 190: 115–119, 1985.

297. Pouysségur, J., C. Sardet, A. Franchi, G. L'Allemain, and S. Paris. A specific mutation abolishing Na^+/H^+ antiport activity in hamster fibroblasts precludes growth at neutral and acidic pH. *Proc. Natl. Acad. Sci. USA* 81: 4833–4837, 1984.

298. Prod'hom, B., D. Pietrobon, and P. Hess. Direct measurement of proton transfer rates to a group controlling the dihydropyridine-sensitive Ca^{2+} channel. *Nature* 329: 243–246, 1987.

299. Putnam, R. W. pH regulatory transport systems in a smooth muscle-like cell line. *Am. J. Physiol.* 258 (*Cell Physiol.* 27): C470–C479, 1990.

300. Putnam, R. W. Localization of pH-regulating transport systems to the surface membrane of semitendinosus muscle from the frog, *Rana pipiens. J. Physiol. (Lond.)* 427: 56P, 1990.

301. Putnam, R. W., P. B. Douglas, and D. Dewey. Control of HCO_3-dependent exchangers by cyclic nucleotides in vascular smooth muscle cells. In: *Regulation of Smooth Muscle Contraction,* edited by R. S. Moreland. New York: Plenum Press, 1991, p. 461–472.

302. Putnam, R. W., and R. D. Grubbs. Steady-state pH_i, buffering power, and effect of CO_2 in a smooth-muscle like cell line. *Am. J. Physiol.* 258 (*Cell Physiol.* 27): C461–C469, 1990.

303. Putnam, R. W., and A. Roos. Aspects of pH_i regulation in frog skeletal muscle. *Curr. Top. Membr. Transp.* 26: 35–56, 1986.

304. Putnam, R. W., and A. Roos. Which value for the first dissociation constant of carbonic acid should be used in biological work? *Am. J. Physiol.* 260 (*Cell Physiol.* 29): C1113–C1116, 1991.

305. Raffin, J. P., M. T. Thebault, and J. Y. Le Gall. Changes in phosphmetabolites and intracellular pH in the tail muscle of the prawn *Palaemon serratus* as shown by in vivo ^{31}P-NMR. *J. Comp. Physiol. B* 158: 223–228, 1988.

306. Rajendran, V. M., and H. J. Binder. Characterization of Na-H exchange in apical membrane vesicles of rat colon. *J. Biol. Chem.* 265: 8408–8414, 1990.

307. Raley-Susman, K. M., E. J. Cragoe, Jr., R. M. Sapolsky, and R. R. Kopito. Regulation of intracellular pH in cultured hippocampal neurons by an amiloride-insensitive Na^+/H^+ exchanger. *J. Biol. Chem.* 266: 2739–2745, 1991.

307a. Ravesloot, J. H., T. Eisen, R. Baron, and W. F. Boron. Role of Na-H exchangers and vacuolar H^+ pumps in intracellular pH regulation in neonatal rat osteoclasts. *J. Gen. Physiol.* 105: 177–208, 1995.

307b. Rees, B. B., C. Patton, J. L. Grainger, and D. Epel. Protein synthesis increases after fertilization of sea urchin eggs in the absence of an increase in intracellular pH. *Dev. Biol.* 169: 683–698, 1995.

308. Regassa, L. B., and M. J. Betley. Alkaline pH decreases expression of the accessory gene regulator *(agr)* in *Staphylococcus aureus. J. Bacteriol.* 174: 5095–5100, 1992.

309. Regula, C. S., J. R. Pfeiffer, and R. D. Berlin. Microtubule assembly and disassembly at alkaline pH. *J. Cell Biol.* 89: 45–53, 1981.

310. Restrepo, D., D. J. Kozody, L. J. Spinelli, and P. A. Knauf. pH homeostasis in promyelocytic leukemic HL60 cells. *J. Gen. Physiol.* 92: 489–507, 1988.

310a. Reusch, H. P., J. Lowe, and H. E. Ives. Osmotic activation of a Na^+-dependent Cl^-/HCO_3^- exchanger. *Am. J. Physiol.* 268 (*Cell Physiol.* 37): C147–C153, 1995.

311. Reuss, L., Y. Segal, and G. Altenberg. Regulation of ion transport across gallbladder epithelium. *Annu. Rev. Physiol.* 53: 361–373, 1991.

312. Revel, J.-P., B. J. Nicholson, and S. B. Yancey. Chemistry of gap junctions. *Annu. Rev. Physiol.* 47: 263–279, 1985.

313. Rink, T. J., R. Y. Tsien, and T. Pozzan. Cytoplasmic pH and free Mg^{2+} in lymphocytes. *J. Cell Biol.* 95: 189–196, 1982.

314. Rodrigo, G. C., and R. A. Chapman. A novel resin-filled ion-sensitive micro-electrode suitable for intracellular measurements in isolated cardiac myocytes. *Pflugers Arch.* 416: 196–200, 1990.

314a. Romero, M. F., M. A. Hediger, E. L. Boulpaep, and W. F. Boron. Expression cloning of the renal electrogenic Na/HCO_3 cotransporter (NBC) from *Ambystoma tigrinum. The Physiologist* 39: 144, 1996.

314b. Romero, M. F., M. A. Hediger, E. L. Boulpaep, and W. F. Boron. Physiology of the cloned renal electrogenic Na/HCO_3 cotransporter. *J. Gen. Physiol.* 108: 16–17a, 1996.

315. Roos, A. Intracellular pH and intracellular buffering power of the cat brain. *Am. J. Physiol.* 209: 1233–1246, 1965.

315a. Roos, A. Intracellular pH and buffering power of rat muscle. *Am. J. Physiol.* 221: 182–188, 1971.

315b. Roos, A. Intracellular pH and distribution of weak acids across cell membranes. A study of D- and L-lactate and of DMO in rat diaphragm. *J. Physiol. (Lond.)* 249: 1–25, 1975.

316. Roos, A., and W. F. Boron. The buffer value of weak acids and bases: origin of the concept, and first mathematical derivation and application to physico-chemical systems. The work of M. Koppel and K. Spiro (1914). *Respir. Physiol.* 40: 1–32, 1980.

317. Roos, A., and W. F. Boron. Intracellular pH. *Physiol. Rev.* 61: 296–434, 1981.

318. Rose, B., and W. R. Loewenstein. Permeability of a cell junction and the local cytoplasmic free ionized calcium concentration: A study with aequorin. *J. Membr. Biol.* 28: 87–119, 1976.

319. Ross, W., W. Bertrand, and A. Morrison. A photoactivatable probe for the Na^+/H^+ exchanger cross-links a 66-kDa renal brush border membrane protein. *J. Biol. Chem.* 265: 5341–5344, 1990.

320. Rothenberg, P., L. Glaser, P. Schlesinger, and D. Cassel. Activation of Na^+/H^+ exchange by epidermal growth factor elevates intracellular pH in A431 cells. *J. Biol. Chem.* 258: 12644–12653, 1983.

321. Rousseau, E., and J. Pinkos. pH modulates conducting and gating behaviour of single calcium release channels. *Pflugers Arch.* 415: 645–647, 1990.

322. Rudnick, G. ATP-driven H^+ pumping into intracellular organelles. *Ann. Revu. Physiol.* 48: 403–413, 1986.

323. Ruetz, S., A. E. Lindsey, and R. R. Kopito. Function and biosynthesis of erythroid and nonerythroid anion exchangers. In: *Molecular Biology and Function of Carrier Proteins,* edited by L. Reuss, J. M. Russell, Jr. and M. L. Jennings. New York: Rockefeller University Press, 1993, p. 193–200.

324. Russell, J. M., and W. F. Boron. Role of chloride in regulation of intracellular pH. *Nature* 264: 73–74, 1976.

325. Russell, J. M., and W. F. Boron. Intracellular pH regulation in squid giant axons. In: *Intracellular pH: Its Measurement, Regulation, and Utilization in Cellular Functions,* edited by R. Nuccitelli and D. Deamer. New York: Alan R. Liss, 1982, p. 221–237.

326. Russell, J. M., W. F. Boron, and M. S. Brodwick. Intracellular pH and Na fluxes in barnacle muscle with evidence for reversal of the ionic mechanism of intracellular pH regulation. *J. Gen. Physiol.* 82: 47–78, 1983.

327. Russell, J. M., and M. S. Brodwick. The interaction of intracellular Mg^{2+} and pH on Cl^- fluxes associated with intracellular pH regulation in barnacle muscle fibers. *J. Gen. Physiol.* 91: 495–513, 1988.

327a. Sagnella, D. E., and G. A. Voth. Structure and dynamics of hydronium in the ion channel Gramicidin A. *Biophys. J.* 70: 2043–2051, 1996.

328. Sampath, P., and T. D. Pollard. Effects of cytochalasin, phalloidin, and pH on the elongation of actin filaments. *Biochemistry* 30: 1973–1980, 1991.

329. Sardet, C., A. Franchi, and J. Pouysségur. Molecular cloning, primary structure, and expression of the human growth factor-activatable Na^+/H^+ antiporter. *Cell* 56: 271–280, 1989.

330. Sardet, C., L. Counillon, A. Franchi, and J. Pouysségur. Growth factors induce phosphorylation of the Na^+/H^+ antiporter, a glycoprotein of 110 kD. *Science* 247: 723–7726, 1990.

330a. Sarkadi, B., L. Attisano, S. Grinstein, M. Buchwald, and A. Rothstein. Volume regulation of Chinese hamster ovary cells in anisoosmotic media. *Biochim. Biophys. Acta* 774: 159–168, 1984.

331. Sato, F., J. Mogami, and S. A. Baba. Flagellar quiescence and transience of inactivation induced by rapid pH drop. *Cell Motil. Cytoskeleton* 10: 374–379, 1988.

332. Schatten, G. The supramolecular organization of the cytoskeleton during fertilization. In: *Subcellular Biochemistry, Vol. 1,* edited by D. B. Roodyn. New York: Plenum Press, 1984, p. 359–453.

333. Schatten, G., T. Bestor, R. Balczon, J. Henson, and H. Schatten. Intracellular pH shift leads to microtubule assembly and microtubule-mediated motility during sea urchin fertilization: correlations between elevated intracellular pH and microtubule activity and depressed intracellular pH and microtubule disassembly. *Eur. J. Cell Biol.* 36: 116–127, 1985.

334. Scherrer, P. Proton movement on membranes. *Nature* 374: 222, 1995.

334a. Schindler, M., S. Grabski, E. Hoff, and S. M. Simon. Defective pH regulation of acidic compartments in human breast cancer cells (MCF-7) is normalized in Adriamycin-resistant cells (MCF-7adr). *Biochemistry* 35: 2811–2817, 1996.

335. Schmidt, J. M., R. M. Robson, J. Zhang, and M. H. Stromer. The marked pH dependence of the talin-actin interaction. *Biochem. Biophys. Res. Commun.* 197: 660–666, 1993.

336. Schwiening, C. J., H. J. Kennedy, and R. C. Thomas. Calcium-hydrogen exchange by the plasma membrane Ca-ATPase of voltage-clamped snail neurones. *Proc. R. Soc. Lond. [Biol.]* 253: 285–289, 1993.

336a. Seksek, O., J. Biwersi, and A. S. Verkman. Direct measurement of *trans*-Golgi pH in living cells and regulation by second messengers. *J. Biol. Chem.* 270: 4967–4970, 1995.

337. Shen, S. S., and R. A. Steinhardt. Direct measurement of intracellular pH during metabolic derepression of the sea urchin egg. *Nature* 272: 253–254, 1978.

338. Shi, L. B., K. Fushimi, H. R. Bae, and A. S. Verkman. Heterogeneity in ATP-dependent acidification in endocytic vesicles from kidney proximal tubule. Measurement of pH in individual endocytic vesicles in a cell-free system. *Biophys. J.* 59: 1208–1217, 1991.

339. Shrode, L. D., J. D. Klein, W. C. O'Neill, and R. W. Putnam. Shrinkage-induced activation of Na^+/H^+ exchange in primary rat astrocytes: role of myosin light chain kinase. *Am. J. Physiol.* 268 (Cell Physiol. 37): C257–C266, 1995.

340. Shrode, L. D., and R. W. Putnam. Intracellular pH regulation in primary rat astrocytes and C6 glioma cells. *Glia* 12: 196–210, 1994.

341. Shulman, R. G., T. R. Brown, K. Ugurbil, S. Ogawa, S. M. Cohen, and J. A. den Hollander. Cellular applications of ^{31}P and ^{13}C nuclear magnetic resonance. *Science* 205: 160–166, 1979.

342. Siebens, A. W., and W. F. Boron. Effect of electroneutral luminal and basolateral lactate transport on intracellular pH in salamander proximal tubules. *J. Gen. Physiol.* 90: 799–831, 1987.

343. Simchowitz, L., I. Spillberg, and P. De Weer. Sodium and potassium fluxes and membrane potential of human neutrophils. *J. Gen. Physiol.* 79: 453–479, 1982.

344. Simchowitz, L., and A. Roos. Regulation of intracellular pH in human neutrophils. *J. Gen. Physiol.* 85: 443–470, 1985.

345. Simchowitz, L., and A. O. Davis. Intracellular pH recovery from alkalinization. *J. Gen. Physiol.* 96: 1037–1059, 1990.

346. Simon, S., D. Roy, and M. Schindler. Intracellular pH and the control of multidrug resistance. *Proc. Natl. Acad. Sci. USA* 91: 1128–1132, 1994.

346a. Singh, S. K., H. J. Binder, J. P. Geibel, and W. F. Boron. An apical permeability barrier to NH_3/NH_4^+ in isolated, perfused colonic crypts. *Proc. Natl. Acad. Sci. USA* 92: 11573–11577, 1995.

347. Siskind, M. S., E. A. Alexander, and J. H. Schwartz. Regulation of cGMP production by intracellular alkalinization in cultured rat inner medullary collecting duct cells. *Biochem. Biophys. Res. Commun.* 170: 860–866, 1990.

348. Skehel, J. J., P. M. Bayley, E. B. Brown, S. R. Martin, M. D. Waterfield, J. M. White, I. A. Wilson, and D. C. Wiley. Changes in the conformation of influenza virus hemagglutinin at the pH optimum of virus-mediated membrane fusion. *Proc. Natl. Acad. Sci. USA* 79: 968–972, 1982.

348a. Sly, W. S., and P. Y. Hu. Human carbonic anhydrases and carbonic anhydrase deficiencies. *Annu. Rev. Biochem.* 64: 375–401, 1995.

349. Smith, G. L., P. Donoso, C. J. Bauer, and D. A. Eisner. Relationship between intracellular pH and metabolite concentrations during metabolic inhibition in isolated ferret heart. *J. Physiol. (Lond.)* 472: 11–22, 1993.

350. Soboll, S., T. A. Link, and G. Von Jagow. Proton transport across the mitochondrial membrane and subcellular proton concentrations. In: *pH Homeostasis. Mechanisms and Control,* edited by D. Häussinger. New York: Academic Press, 1988, p. 97–122.

351. Soleimani, M., C. Bookstein, J. A. McAteer, Y. J. Hattabaugh, G. L. Bizal, M. W. Musch, M. Villereal, M. C. Rao, R. L. Howard, and E. B. Chang. Effect of high osmolality on Na^+/H^+ exchange in renal proximal tubule cells. *J. Biol. Chem.* 269: 15613–15618, 1994.

352. Soleimani, M., and P. S. Aronson. Ionic mechanism of Na^+-HCO_3^- cotransporter in rabbit renal basolateral membrane vesicles. *J. Biol. Chem.* 264: 18302–18308, 1989.

353. Somogyi, R., A. Batzer, and H. A. Kolb. Inhibition of electrical coupling in pairs of murine pancreatic acinar cells by OAG and isolated protein kinase C. *J. Membr. Biol.* 108: 273–283, 1989.

354. Sorgato, M. C., and O. Moran. Channels in mitochondrial membranes: knowns, unknowns, and prospects. *Crit. Rev. Biochem. Mol. Biol.* 28: 127–171, 1993.

355. Spanswick, R. M. Vacuolar and cell membrane H^+-ATPases of plant cells. *Ann. NY Acad. Sci.* 574: 180–188, 1989.

356. Spray, D. C., and M.V.L. Bennett. Physiology and pharmacology of gap junctions. *Annu. Rev. Physiol.* 47: 281–303, 1985.

357. Spray, D. C., and J. M. Burt. Structure-activity relations of the cardiac gap junction channel. *Am. J. Physiol.* 258 (Cell Physiol. 27): C195–C205, 1990.

358. Spray, D. C., R. D. Ginzberg, E. A. Morales, Z. Gatmaitan, and I. M. Arias. Electrophysiological properties of gap junctions between dissociated pairs of rat hepatocytes. *J. Cell Biol.* 101: 135–144, 1986.

359. Spray, D. C., A. L. Harris, and M.V.L. Bennett. Gap junctional conductance is a simple and sensitive function of intracellular pH. *Science* 211: 712–715, 1981.

360. Spray, D. C., and J. C. Saez. Agents that affect gap junctinal conductance: sites of action and specificities. In: *Biochemical Regulation of Intercellular Communication,* edited by M. A. Mehlman, New York: Alan R. Liss, 1988, p. 1–26.

361. Spyropoulos, C. S. Cytoplasmic pH of nerve fibres. *J. Neurochem.* 5: 185–194, 1960.

362. Steinhardt, R. A., S. Shen, and D. Mazia. Membrane potential, membrane resistance and an energy requirement for the devel-

opment of potassium conductance in the fertilization reaction of echinoderm eggs. *Exp. Cell Res.* 72: 195–203, 1972.

363. Sugiura, H., J. Toyama, N. Tsuboi, K. Kamiya, and I. Kodama. ATP directly affects junctional conductance between paired ventricular myocytes isolated from guinea pig heart. *Circ. Res.* 66: 1095–1102, 1990.

364. Suprenant, K. A. Alkaline pH favors microtubule self-assembly in surf clam, *Spisula solidissima*, oocyte extracts. *Exp. Cell Res.* 184: 167–180, 1989.

365. Suprenant, K. A. Unidirectional microtubule assembly in cell-free extracts of *Spisula solidissima* oocytes is regulated by subtle changes in pH. *Cell Motil. Cytoskeleton* 19: 207–220, 1991.

366. Swallow, C. J., S. Grinstein, and O. D. Rotstein. Regulation and functional significance of cytoplasmic pH in phagocytic leukocytes. *Curr. Top. Membr. Transp.* 35: 227–247, 1990.

367. Swallow, C. J., S. Grinstein, and O. D. Rotstein. A vacuolar type H^+-ATPase regulates cytoplasmic pH in murine macrophages. *J. Biol. Chem.* 265: 7645–7654, 1990.

368. Swenson, K. I., H. Piwnica-Worms, H. McNamee, and D. L. Paul. Tyrosine phosphorylation of the gap junction protein connexin43 is required for the pp60[v-src]-induced inhibition of communication. *Cell Regul.* 1: 989–1002, 1990.

369. Szwergold, B. S., T. R. Brown, and J. J. Freed. Bicarbonate abolishes intracellular alkalinization in mitogen-stimulated 3T3 cells. *J. Cell. Physiol.* 138: 227–235, 1989.

370. Tager, J. M., J.M.F.G. Aerts, R.J.A. Oude Elferink, A. K. Groen, M. Hollemans, and A. W. Schram. pH regulation of intracellular membrane flow. In: *pH Homeostasis. Mechanisms and Control,* edited by D. Häussinger. New York: Academic Press, 1988, p. 123–162.

371. Taglicht, D., E. Padan, and S. Schuldiner. Proton-sodium stoichiometry of NhaA, an electrogenic antiporter from *Escherichia coli. J. Biol. Chem.* 268: 5382–5387, 1993.

371a. Tanner, M.J.A. The acid test for band 3. *Nature* 382: 209–210, 1996.

372. Taylor, C. V., and D. M. Whitaker. Potentiometric determinations in the protoplasm and cell-sap of nitella. *Protoplasma* 3: 1–6, 1927.

373. Thiebaut, F., S. J. Currier, J. Whitaker, R. P. Haugland, M. M. Gottesman, I. Pastan, and M. C. Willingham. Activity of the multidrug transporter results in alkalinization of the cytosol: Measurement of cytosolic pH by microinjection of a pH-sensitive dye. *J. Histochem. Cytochem.* 38: 685–690, 1990.

374. Thomas, J. A. Intracellularly trapped pH indicators. In: *Optical Methods in Cell Physiology,* P. De Weer and B. M. Salzberg. New York: John Wiley and Sons, 1986, p. 311–325.

375. Thomas, J. A., R. N. Buchsbaum, A. Zimniak, and E. Racker. Intracellular pH measurements in Ehrlich ascites tumor cells utilizing spectroscopic probes generated *in situ. Biochemistry* 18: 2210–2218, 1979.

376. Thomas, J. A., R. E. Cole, and T. A. Langworthy. Intracellular pH measurements with a spectroscopic probe generated *in situ. Federation Proc.* 35: 1455, 1976.

377. Thomas, R. C. Intracellular pH of snail neurones measured with a new pH-sensitive glass micro-electrode. *J. Physiol. (Lond.)* 238: 159–180, 1974.

378. Thomas, R. C. The effect of carbon dioxide on the intracellular pH and buffering power of snail neurones. *J. Physiol. (Lond.)* 255: 715–735, 1976a.

379. Thomas, R. C. Ionic mechanism of the H^+ pump in a snail neurone. *Nature* 262: 54–55, 1976b.

380. Thomas, R. C. The role of bicarbonate, chloride and sodium ions in the regulation of intracellular pH in snail neurones. *J. Physiol. (Lond.)* 273: 317–338, 1977.

381. Thomas, R. C. *Ion-sensitive Intracellular Microelectrodes: How to Make and Use Them.* San Francisco: Academic Press, 1978.

382. Thomas, R. C., and R. W. Meech. Hydrogen ion currents and intracellular pH in depolarized voltage-clamped snail neurones. *Nature* 299: 826–828, 1982.

383. Tilney, L. G., D. P. Kiehart, C. Sardet, and M. Tilney. Polymerization of actin. IV. Role of Ca^{++} and H^+ in the assembly of actin and in membrane fusion in the acrosomal reaction of echinoderm sperm. *J. Cell Biol.* 77: 536–550, 1978.

384. Tosteson, D. C. Halide transport in red blood cells. *Acta Physiol. Scand.* 46: 19–41, 1959.

385. Tosteson, D. C., and J. F. Hoffman. Regulation of cell volume by active cation transport in high and low potassium sheep red cells. *J. Gen. Physiol.* 44: 169–194, 1960.

386. Trayhurn, P., and D. G. Nicholls, eds. *Brown Adipose Tissue.* London: Edward Arnold, 1986.

387. Trivedi, B., and W. H. Danforth. Effect of pH on the kinetics of frog muscle phosphofructokinase. *J. Biol. Chem.* 241: 4110–4112, 1966.

388. Tse, C. M., A. I. Ma, V. W. Yang, A.J.M. Watson, S. Levine, M. H. Montrose, J. Potter, C. Sardet, J. Pouysségur, and M. Donowitz. 1991. Molecular cloning and expression of a cDNA encoding the rabbit ileal villus cell basolateral membrane Na^+/H^+ exchanger. *EMBO J.* 10: 1957–1967, 1991.

389. Tse, C. M., S. R. Brant, M. S. Walker, J. Pouysségur, and M. Donowitz. Cloning and sequencing of a rabbit cDNA encoding an intestinal and kidney-specific Na^+/H^+ exchanger isoform (NHE-3). *J. Biol. Chem.* 267: 9340–9346, 1992.

390. Tse, C. M., S. A. Levine, C.H.C. Yun, M. H. Montrose, P. J. Little, J. Pouysségur, and M. Donowitz. Cloning and expression of a rabbit cDNA encoding a serum-activated ethylisopropylamiloride-resistant epithelial Na^+/H^+ exchanger isoform (NHE-2). *J. Biol. Chem.* 268: 11917–11924, 1993.

391. Tse, C. M., S. A. Levine, C.H.C. Yun, S. R. Brant, J. Pouysségur, M. H. Montrose, and M. Donowitz. Functional characteristics of a cloned epithelial Na^+/H^+ exchanger (NHE3): Resistance to amiloride and inhibition by protein kinase C. *Proc. Natl. Acad. Sci. USA* 90: 9110–9114, 1993.

392. Tse, M., S. Levine, C. Yun, S. Brant, L. T. Counillon, J. Pouysségur, and M. Donowitz. Structure/function studies of the epithelial isoforms of the mammalian Na^+/H^+ exchanger gene family. *J. Membr. Biol.* 135: 93–108, 1993.

393. Tse, C.-M., S. A. Levine, C.H.C. Yun, S. R. Brant, S. Nath, J. Pouysségur, and M. Donowitz. Molecular properties, kinetics and regulation of mammalian Na^+/H^+ exchangers. *Cell Physiol. Biochem.* 4: 282–300, 1994.

394. Tsien, R. Y. 1989. Fluorescent indicators of ion concentrations. In: *Fluorescence Microscopy of Living Cells in Culture,* edited by Y. L. Wang and D. L. Taylor. New York: Academic Press, 1989, p. 127–156.

395. Turin, L., and A. E. Warner. Carbon dioxide reversibly abolishes ionic communication between cells of early amphibian embryo. *Nature* 270: 56–57, 1977.

396. Turin, L., and A. E. Warner. Intracellular pH in early *Xenopus* embryos: Its effect on current flow between blastomeres. *J. Physiol. (Lond.)* 300: 489–504, 1980.

397. Tycko, B., and F. R. Maxfield. Rapid acidification of endocytic vesicles containing α_2-macroglobulin. *Cell* 26: 643–651, 1982.

398. Unwin, P.N.T. Gap junction structure and the control of cell-to-cell communication. In: *Junctional Complexes of Epithelial*

Cells, edited by G. Bock and S. Clark (Ciba Foundation Symp. 125). New York: Wiley, 1987, p. 78–91.

399. Unwin, P.N.T., and G. Zampighi. Structure of the gap junction between communicating cells. *Nature* 283: 545–549, 1980.

400. Van Slyke, D. D. On the measurement of buffer values and on the relationship of buffer value to the dissociation constant of the buffer and the concentration and reaction of the buffer solution. *J. Biol. Chem.* 52: 525–570, 1922.

401. Vanheel, B., A. De Hemptinne, and I. Leusen. Acidification and intracellular sodium ion activity during simulated myocardial ischemia. *Am. J. Physiol.* 259 (*Cell Physiol.* 28): C169–C179, 1990.

402. Vaughan-Jones, R. D. Chloride activity and its control in skeletal and cardiac muscle. *Philos. Trans. R. Soc. Lond. Biol.* 299: 537–548, 1982.

403. Wadsworth, W. G., and D. L. Riddle. Acidic intracellular pH shift during *Caenorhabditis elegans* larval development. *Proc. Natl. Acad. Sci. USA* 85: 8435–8438, 1988.

404. Wagner, J., and J. Keizer. Effects of rapid buffers on Ca^{2+} diffusion and Ca^{2+} oscillations. *Biophys. J.* 67: 447–456, 1994.

405. Waisbren, S. J., J. P. Geibel, I. M. Modlin, and W. F. Boron. Unusual permeability properties of gastric gland cells. *Nature* 368: 332–335, 1994.

406. Waisbren, S. J., J. P. Geibel, W. F. Boron, and I. M. Modlin. Luminal perfusion of isolated gastric glands. *Am. J. Physiol.* 266 (*Cell Physiol.* 35): C1013–C1027, 1994.

407. Wakabayashi, S., P. Fafournoux, C. Sardet, and J. Pouysségur. The Na^+/H^+ antiporter cytoplasmic domain mediates growth factor signals and controls "H^+-sensing." *Proc. Natl. Acad. Sci. USA* 89: 2424–2428, 1992.

408. Wang, W., Y. Wang, S. Siblernagl, and H. Oberleithner. Fused cells of frog proximal tubule: II. Voltage-dependent intracellular pH. *J. Membr. Biol.* 101: 259–265, 1988.

409. Wang, Z., J. Orlowski, and G. E. Shull. Primary structure and functional expression of a novel gastrointestinal isoform of the rat Na/H exchanger. *J. Biol. Chem.* 268: 11925–11928, 1993.

410. Wanke, E., E. Carbone, and P. L. Testa. K^+ conductance modified by a titratable group accessible to protons from the intracellular side of the squid axon membrane. *Biophys. J.* 26: 319–324, 1979.

411. Wanke, E., E. Carbone, and P. L. Testa. The sodium channel and intracellular H^+ blockage in squid axons. *Nature* 287: 62–63, 1980.

412. Wei, L.-Y., and P. D. Roepe. Low external pH and osmotic shock increase the expression of human MDR protein. *Biochemistry* 33: 7229–7238, 1994.

412a. Wei, L. Y., M. J. Stutts, M. M. Hoffman, and P. D. Roepe. Overexpression of the cystic fibrosis transmembrane conductance regulator in NIH 3T3 cells lowers membrane potential and intracellular pH and confers a multidrug resistance phenotype. *Biophys. J.* 69: 883–895, 1995.

413. Whitaker, J. E., R. P. Haugland, and F. G. Pendergast. Spectral and photophysical studies of Benzo{c}xanthene dyes: dual emission pH sensors. *Anal. Biochem.* 194: 330–344, 1991.

414. White, R. L., J. E. Doeller, V. K. Verselis, and B. A. Wittenberg. Gap junctional conductance between pairs of ventricular myocytes is modulated synergistically by H^+ and Ca^{++}. *J. Gen. Physiol.* 95: 1061–1075, 1990.

415. Wilding, T. J., B. Cheng, and A. Roos. pH regulation in adult rat carotid body glomus cells. Importance of extracellular pH, sodium and potassium. *J. Gen. Physiol.* 100: 593–608, 1992.

416. Wingo, C. S., and B. D. Cain. The renal H-K-ATPase: Physiological significance and role in potassium homeostasis. *Ann. Revu. Physiol.* 55: 323–347, 1993.

417. Winkel, G. K., C. Sardet, J. Pouysségur, and H. E. Ives. Role of cytoplasmic domain of the Na^+/H^+ exchanger in hormonal activation. *J. Biol. Chem.* 268: 3396–3400, 1993.

418. Winkler, M. M., and R. A. Steinhardt. Activation of protein synthesis in a sea urchin cell-free system. *Dev. Biol.* 84: 432–439, 1981.

419. Wittenberg, J. B. The molecular mechanism of hemoglobin-facilitated oxygen diffusion. *J. Biol. Chem.* 241: 104–114, 1966.

420. Yun, C.H.C., S. Gurubhagavatula, S. A. Levine, J.L.M. Montgomery, S. R. Brant, M. E. Cohen, E. J. Cragoe, Jr., J. Pouysségur, C. M. Tse, and M. Donowitz. Glucocorticoid stimulation of ileal Na^+ absorptive cell brush border Na^+/H^+ exchange and association with an increase in message for NHE-3, an epithelial Na^+/H^+ exchanger isoform. *J. Biol. Chem.* 268: 206–211, 1993.

421. Yun, C.H.C., P. J. Little, S. K. Nath, S. A. Levine, J. Pouysségur, C. M. Tse, and M. Donowitz. Leu143 in the putative fourth membrane spanning domain is critical for amiloride inhibition of an epithelial Na^+/H^+ exchanger isoform (NHE-2). *Biochem. Biophys. Res. Commun.* 193: 532–539, 1993.

421a. Yun, C.H.C., C.-M. Tse, S. K. Nath, S. A. Levine, S. R. Brant, and M. Donowitz. Mammalian Na^+/H^+ exchanger gene family: structure and function studies. *Am. J. Physiol.* 269 (*Gastrointest. Liver Physiol.* 32): G1–G11, 1995.

421b. Zhang, J., R. M. Robson, J. M. Schmidt, and M. H. Stromer. Talin can crosslink actin filaments into both networks and bundles. *Biochem. Biophys. Res. Commun.* 218: 530–537, 1996.

421c. Zhao, J., E. M. Hogan, M. O. Bevensee, and W. F. Boron. Out-of-equilibrium CO_2/HCO_3^- solutions and their use in characterizing a new K/HCO_3 cotransporter. *Nature* 374: 636–639, 1995.

422. Zimmerle, C. T., and C. Frieden. Effect of pH on the mechanism of actin polymerization. *Biochemistry* 27: 7766–7772, 1988.

10. Osmolytes and cell-volume regulation: physiological and evolutionary principles

GEORGE N. SOMERO | Hopkins Marine Station, Stanford University, Pacific Grove, California

PAUL H. YANCEY | Department of Biology, Whitman College, Walla Walla, Washington

CHAPTER CONTENTS

ONE OF THE GREAT QUESTS OF BIOLOGY, INDEED OF ALL THE SCIENCES, is the discovery of unifying principles that can interrelate and explain what previously had appeared to be complex, unrelated, or even arbitrary phenomena. The study of cellular osmotic agents, osmolytes, the inorganic ions and low-molecular-weight organic molecules that constitute the major share of the cell's osmotic content, exemplifies this type of scientific achievement. What once appeared to be a widely diverse array of solutions to the problems posed by osmotic stress in bacteria, fungi, algae, higher plants, and animals is coming increasingly to be seen instead as a reflection of a narrow and highly constrained pattern of evolution, one based on a very small number of physical–chemical principles that appear to have been applicable to all cell types since the dawn of cellular evolution.

Philosophically and historically, our appreciation for the evolutionary principles underlying the evolution of osmolyte systems stems from the seminal 1913 treatise of L. J. Henderson (108), which demonstrated how different components of the environment—notably, water and the ions accumulated in biological fluids—provide a "fit" environment for life as we know it. Some 70 to 80 years later, cell physiologists working with taxonomically diverse organisms that encounter water stress, and biophysicists studying water structure and the effects of small solutes and macromolecules on the organization of water, have converged in their thinking concerning what is required for establishing a hospitable solvent environment in the cell. The internal milieu must facilitate the attainment and retention of physiologically appropriate properties of macromolecular systems and membranes, properties that in almost all cases are strongly affected by the types of solutes present in the cellular water and by their actual concentrations. These properties include protein folding, assembly, and function; regulation of gene expression; and conservation of appropriate membrane structure and function. As more has been learned about the physics and chemistry of water, and how the properties of water are influenced by small solutes and macromolecules, we have come to understand how these ubiquitous interactions among solvent and solutes establish the basic ground rules of osmolyte system design. Establishing a "fit" intracellular solution through selection of appropriate osmotic solutes was a primordial evolutionary task. Thus, one finds in comparative analysis of diverse organisms a common set of solutions to osmotic stress. The evolution of osmolyte systems occurred hand-in-hand with the evolution of macromolecular and cellular systems. The most basic characteristics of macromolecular design, and many of the key properties of the cellular level of biological organization, reflect the pivotal roles of osmotic systems in biological evolution.

The convergence that has occurred in the evolutionary development of intracellular osmolyte systems is, not surprisingly, paralleled by a high level of convergence in the research programs of microbiologists, botanists, and zoologists interested in osmotic phenomena (197, 237). During the past two decades workers in all these disciplines have reached a consensus concerning the basic and evolutionarily common principles entailed in the development of osmolyte systems.

To emphasize the near universality of these principles—principles that seem to be strongly violated only by the extremely halophilic archaebacteria that may contain K^+ at concentrations of up to 5 M in their cells (134)—we use examples from diverse procaryotic and eucaryotic organisms to illustrate the major theme developed in this chapter. One purpose for doing this is to emphasize the similarities in osmolyte systems across almost all species. Another important reason for adopting this broadly comparative approach is to emphasize the similar responses of macromolecules to a given type and concentration of solute regardless of taxon or degree of osmotic stress experienced by the organism. These common properties of diverse cell types show that in order to cope with osmotic stress,

most of the evolutionary "work," so to speak, has been borne by the osmolytes; proteins, and probably nucleic acids as well, generally are not in any way uniquely adapted to the particular degree of osmotic stress faced by the organism. The primary adaptations found in macromolecular systems are largely those that effect the regulation of osmolyte composition and concentration. Seen from a broad evolutionary perspective, then, the adaptations that characterize osmolyte systems of most procaryotic and eucaryotic cells represent a highly efficient mode of evolution, one in which the required adaptive changes in proteins are localized in a few systems, rather than distributed throughout most, if not all, proteins—as in the case of the aberrant halophilic archaebacteria that use high concentrations of inorganic ions rather than organic osmolytes to adjust the osmolarity of the cellular water.

WHAT MAKES A SOLUTE AN OSMOLYTE?

Because of the enormous number of types of inorganic ions and small organic molecules in the cell, it is important to define as rigorously as possible what makes a solute an osmolyte. As a general definition, we propose the following: *an osmolyte is any inorganic or organic species whose concentration is regulated during the process of cell volume regulation in parallel with the osmotic stress imposed on the cell.* By "in parallel" we do not imply that a similar slope exists for the changes in total environmental (extracellular) osmolarity and the concentration of a putative osmolyte. We imply only that an increase in extracellular osmolarity will be followed by an increase in the intracellular concentration of the solute in question, and that decreases in extracellular osmolarity will be followed by a decrease in osmolyte concentration. We emphasize that volume regulation must be occurring for this definition to be valid, because withdrawal or addition of water to the cell without volume regulation taking place will obviously lead to the concentration or dilution, respectively, of most if not all intracellular solutes.

We leave open the question of a minimal or threshold level response in the sense of how large a change in the concentration of the solute in question is necessary before the solute is truly acting as an osmolyte. Often there is no ambiguity about whether a solute plays a significant osmotic role: glycerol concentrations may rise by one to three molal during osmotic stress in certain algae (24). Free amino acid concentrations in the cells of euryhaline marine invertebrates may change by a few tenths M during osmotic adjustment (237). Urea concentrations, and the concentrations of other organic osmolytes that co-occur with urea in urea-rich cells, may rise to levels of several hundred mmoles/kg water under water stress (232). However, physiologically significant changes may be much smaller in some cases. For instance, in mammalian brain, the osmotic stress imposed by hypernatremia led to statistically significant increases in the concentrations of a number of organic osmolytes (myoinositol [65%], glycine betaine [54%], glycerophosphorylcholine [GPC] [132%], glutamine [143%], glutamate [84%], and taurine [78%]), most of which are known to be critically important in other organs (for example, the kidney). However, in each case the actual concentration changes in brain were in the range of only one to a few millimoles per kilogram of cellular water (141). These changes are much smaller than those found in the kidney (34, 233), yet qualitatively both organs show a similar pattern of osmolyte accumulation under water stress. Mammalian brain may be one of the organs most sensitive to osmotic stress, and even minor changes in osmolyte concentrations may yield important physiological benefits to brain function and, thereby, the organism's survival (201). It is noteworthy—in the context of determining what the minimal physiologically important changes in osmolyte concentrations might be—that the sensors and transducers involved in osmotic regulation may have sensitivities to transmembrane osmotic gradients on the order of a few milliosmoles per kilogram of water, and to changes in cell volume of <3% (reviewed by Chamberlin and Strange [45].)

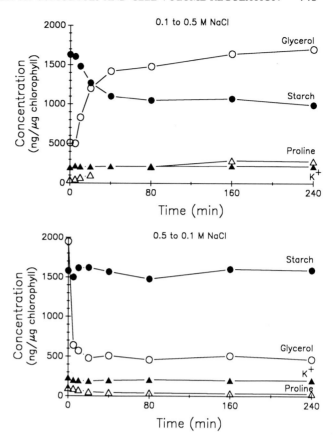

FIG. 10.1. Time courses of change in intracellular contents of starch, glycerol, proline, and potassium ion in the unicellular green alga *Chlorococcum submarinium* exposed to rapid changes in salinity. *Upper panel.* Shift from 0.1 M to 0.5 M NaCl. *Lower panel.* Shift from 0.5 M to 0.1 M NaCl. [Redrawn from Blackwell and Gilmour (22).]

THE BASIC OSMOREGULATORY RESPONSE: CONSERVATION PAIRED WITH CHANGE

Figure 10.1 illustrates one of the most basic characteristics of the osmoregulatory responses of cells: during volume regulation, the concentrations of major inorganic ions like K$^+$ generally are strongly conserved whereas the concentrations of organic osmolytes may change enormously (202, 237). This generalization applies to most types of procaryotic and eucaryotic cells, and it is clearly manifested by the unicellular green alga *Chlorococcum submarinum*, subjected to large and rapid changes in external salinity (22). Upon transfer from 0.1 M to 0.5 M NaCl (Fig. 10.1; *upper panel*), the cells displayed a rapid increase in the concentrations of glycerol and proline, two dominant organic osmolytes; a rapid and parallel fall in the content of starch, the primary source of the glycerol; and only a minimal increase in intracellular K$^+$. The reverse osmotic stress, transfer from 0.5 M to 0.1 M NaCl (Fig. 10.1, *lower panel*), led to a very rapid loss of glycerol

and proline, and a very minor change in K$^+$. The time courses for organic osmolyte accumulation and removal are seen to differ significantly: during hyperosmotic stress, glycerol and proline concentrations reach maximal values only after 160–240 min, whereas during hypoosmotic stress, both osmolytes reach new, lower steady-state concentrations within 40 min.

Paired closely with the kinetics of changes in organic osmolyte concentration were the responses of two key physiological functions, respiration rate and photosynthetic activity (22). Perturbation of these functions by large and rapid shifts in external osmolarity, which initially led to large shifts in cellular water content, was eliminated in concert with the attainment of a new osmotic steady-state effected by shifting organic osmolyte concentrations.

The patterns of osmotic adaptation observed in this unicellular alga are paradigmatic in several important ways, and they provide a very appropriate conceptual backdrop for the central questions addressed in this chapter. Of primary importance is the observation that inorganic ions like K^+ typically undergo only relatively small and transient changes in concentration, whereas certain small organic molecules, often polyhydric alcohols like glycerol and certain free amino acids like proline, account for virtually all of the adaptive change in intracellular osmolarity. The kinetics of organic osmolyte accumulation and removal may differ. These adjustments in osmolyte concentrations often are closely paralleled by changes in major cellular activities that are perturbed by changes in solute concentration and composition.

These observations raise the following questions. First, what classes of solutes tend to vary in concentration in parallel with changes in external osmolarity? That is, what are the major classes of osmolytes in different organisms? Second, what solutes tend to be highly conserved in concentration regardless of osmotic stress or taxon? Third, how are the concentrations of osmolytes regulated to effect the correct adjustments of cellular osmolarity in as rapid a manner as possible? What roles are played by intracellular storage molecules that can be rapidly converted to low-molecular-weight osmolytes, as seen in the conversion of starch to glycerol? How important is de novo biosynthesis of organic molecules, such as the formation of free amino acids via transamination of alpha-keto acids and the accumulation of "dead-end" metabolites like the nitrogenous waste product urea? What role does the uptake of organic osmolytes from the environment, molecules termed *osmoprotective solutes* (59), play in different cell types and different organisms? Fourth, what are the kinetics of osmolyte regulation, and how do the kinetics of hyperosmotic regulation differ from hypoosmotic regulation? Do all osmolytes exhibit similar kinetics, or do they differ in how rapidly, or in what sequence, they accumulate or decrease in concentration? Fifth, what are the fundamental physical and chemical principles that underlie the observed preferences of virtually all cell types for only a few classes of osmolytes? How have these basic principles shaped the design of the cellular solution in widely diverse procaryotic and eucaryotic species?

OSMOLYTE TAXONOMY: EVOLUTIONARY CONVERGENCE AND CONSERVATION

The Discovery of Organic Osmolytes

Early in the twentieth century, L. Frédericq (77) discovered that marine cartilaginous fishes (elasmobranchs) and invertebrates had insufficient concentrations of inorganic ions in their cells to account for their approx-imate osmotic balance with seawater [reviewed by Smith (194), and Clark (48)]. The osmotic difference was soon accounted for in the elasmobranchs by the high quantities of a nitrogenous waste product, urea, already known in these animals, and by trimethylamine-N-oxide (TMAO), discovered in 1909 (194). However, not until the 1950s and later were other types of organic osmolytes identified in cells exposed to high osmolalities. Among the first additional organic osmolytes to be discovered were free amino acids in marine invertebrates (48), glycerophosphorylcholine (GPC) in the mammalian renal medulla (83, 211, 212), and polyhydric sugar alcohols (polyols) like glycerol in osmophilic yeasts and marine algae (29, 30).

Although these early studies suggested that very different adaptations to osmotic stress had evolved in different phylogenetic lineages, it is now recognized that the types of organic osmolytes used are very widely distributed taxonomically, and the organic solutes found among diverse procaryotic and eucaryotic species are universally restricted to about four major classes of compounds: free amino acids and derivatives, methylated ammonium and sulfonium compounds, sugars and polyols, and urea. Figure 10.2 shows the chemical structures, and Table 10.1 lists the major taxonomic distributions, of representative organic osmolytes of these four major classes. These data demonstrate clearly that some organic osmolytes are found in an extreme diversity of organisms. For example, the methylammonium solute glycine betaine is a major osmolyte in many bacteria, higher plants, marine elasmobranch fishes, the "living fossil" coelacanth (*Latimeria chalumnae*), marine invertebrates, and the mammalian kidney (Table 10.1). Similar taxonomically wide occurrences are noted for some polyols (sugar alcohols with an hydroxyl group present on each carbon) and free amino acids. Whether a particular instance of taxonomically wide exploitation of an organic osmolyte represents evolutionary convergence or, instead, conservation of an ancient mechanism is not always clear. It should also be noted that whereas some cells may accumulate preferentially only a single type of organic osmolyte, some groups of organisms are found to have all categories of osmolytes in one tissue. The mammalian kidney accumulates urea, the polyols *myo*-inositol and sorbitol, the methylamines glycerophosphorylcholine (GPC) and betaine (83), and the amino acid taurine (160) (Fig. 10.2).

Polyols and Sugars

First studied in algae where glycerol and related compounds often accumulate to concentrations in excess

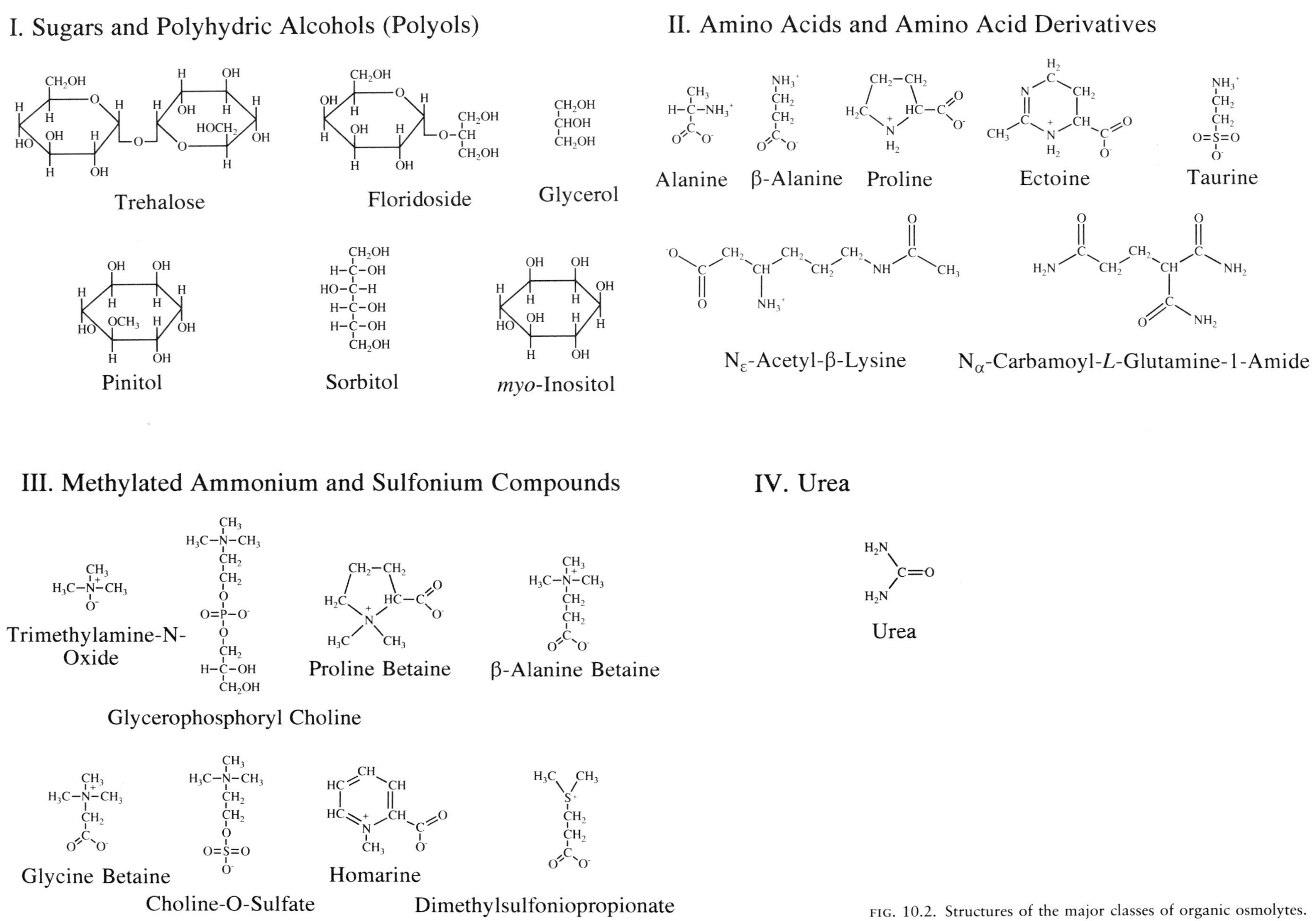

I. Sugars and Polyhydric Alcohols (Polyols)

Trehalose

Floridoside

Glycerol

Pinitol

Sorbitol

myo-Inositol

II. Amino Acids and Amino Acid Derivatives

Alanine

β-Alanine

Proline

Ectoine

Taurine

N_ϵ-Acetyl-β-Lysine

N_α-Carbamoyl-*L*-Glutamine-1-Amide

III. Methylated Ammonium and Sulfonium Compounds

Trimethylamine-N-Oxide

Glycerophosphoryl Choline

Proline Betaine

β-Alanine Betaine

Glycine Betaine

Choline-O-Sulfate

Homarine

Dimethylsulfoniopropionate

IV. Urea

Urea

FIG. 10.2. Structures of the major classes of organic osmolytes.

TABLE 10.1. *Major Classes of Organic Solutes: Taxonomic Distributions of Representative Organic Osmolytes*

Organism	Sugars and Polyhydric Alcohols	References
Osmolytes		
Eubacteria	Trehalose, sucrose, mannitol, glucosylglycerol (heteroside), mannosucrose	59, 144, 196, 207
Algae	Glycerol, mannitol, sorbitol, volemitol, altritol, sucrose, mannose, glucose, heterosides (for example, floridoside), cyclohexane	24, 29, 31, 66, 67, 108, 217
Fungi, yeast	Glycerol, erythritol, arabitol, mannitol, etc.	20, 24
Vascular plants	*Myo*-inositol, sorbitol, glucose, sucrose, mannitol, methyl-inositols (pinitol, L-quebrachitol, O-methyl-muco-inositol, etc.)	73, 76, 94, 181, 200, 228
Animals		
Artemia embryo	Glycerol, trehalose	103
Insects—freezing	Glycerol, sorbitol, erythritol, ribitol, sucrose	69, 84
—salinity	Trehalose	
Marine elasmobranchs	*Myo*-inositol, *scyllo*-inositol	127, 187
Bony fish—freezing	Glycerol	172
Amphibia—freezing	Glucose, glycerol	182, 203
Mammals—kidney	Sorbitol, *myo*-inositol	83, 106, 141
—brain	*Myo*-inositol	

Organism	Amino Acids and Derivatives	References
Archaebacteria (methanogens)	Glutamine, glutamate, proline, β-glutamate, β-glutamine, N_ϵ-acetyl-β-lysine	178
Eubacteria	Glutamine, glutamate, proline, ectoine, hydroxyectoine, N_α-carbamoyl-L-glutamate 1-amide, N-acetylglutaminyl-glutamine amide, pipecolate, γ-amino butyrate	82, 95, 186, 195, 207
Algae	Proline, alanine, glycine, glutamate	31, 71, 107
Protozoa	Glycine, alanine, proline	121
Vascular plants	Proline	228
Animals		
Marine invertebrates (various phyla)	Glycine, alanine, proline, serine, taurine, strombine, etc.	48, 142, 237
Insects (brackish)	Proline, serine	84

(continued)

of 1–2 M (24, 29), a wide variety of sugars and polyols (Fig. 10.2) have subsequently been characterized in a phylogenetically diverse array of species (Table 10.1). The sugar and polyol osmolytes include monosaccharides such as glucose, disaccharides such as sucrose and mannosucrose (196), heterosides [combinations of sugars and glycerol (24)], linear polyols with three to six carbons, and cyclic polyols or cyclitols (Fig. 10.2). There are also methylated cyclitol osmolytes including pinitol (O-methyl-chiro-inositol) (Fig. 10.2), ononitol (O-methyl-myo-inositol), O-methyl-scyllo-inositol (187), and O-methyl-muco-inositol (76, 200).

The sugar and polyol osmolytes occur in the greatest variety in photosynthetic organisms and fungi (24), but distribution of some types is scattered among distantly related groups and seems clearly the result of evolutionary convergence. For example, *myo*-inositol and sorbitol are used as osmolytes in marine algae, some angiosperms, and mammals (Table 10.1).

Sugars and polyols play important roles in adaptation to water stress arising from freezing and cellular desiccation. They are the dominant osmotic solutes used in organisms adapting to freezing temperatures, for example, freeze-tolerant species (69, 182, 203). In addition to lowering freezing points by colligative means and by enhancing supercooling, the sugars and polyols serve as important osmolytes in freeze-tolerant organisms by reducing cell shrinkage as extracellular

TABLE 10.1. *Continued*

Organism	Sugars and Polyhydric Alcohols	References
Marine cyclostomes	Glycine, alanine, proline, etc.	124
Marine elasmobranchs	Taurine β-alanine, glycine, alanine, etc.	124
Amphibia	Various α-amino acids	14
Mammals—kidney, heart, brain	Taurine, glutamine	136, 141, 159, 207

Organism	Methylated Ammonium and Sulfonium Solutes	References
Archaebacteria (methanogens)	Glycine betaine	178
Eubacteria	Glycine betaine, choline-O-sulfate, proline betaine, taurine betaine, β-alanine betaine, glutamate betaine, pipecolate betaine, DMSP	59, 135, 144, 208
Algae	Glycine betaine, proline betaine, homarine, choline-O-sulfate, dimethyl-taurine, taurine betaine, DMSP	67, 71, 118, 123, 125, 148, 173
Vascular plants	Glycine betaine, proline betaine, β-alanine betaine, choline-O-sulfate, DMSP	94, 228
Animals		
Marine invertebrates	Glycine betaine, TMAO, proline betaine	48, 166
Marine cyclostomes	TMAO	124
Marine elasmobranchs	Glycine betaine, TMAO, sarcosine	124, 231
Coelacanth	Glycine betaine, TMAO	99
Amphibia *(Xenopus)*	GPC	220
Mammals—kidney, brain	Glycine betaine, GPC	83, 106, 141
Avians (erythrocytes)	Taurine	189

Organism	Urea	References
Gastropods—estivating		115
Marine elasmobranchs	(With methylamines)	124, 194, 231
Coelacanth	(With methylamines)	99
Lungfish—estivation		194
Amphibia—salinity	(With methylamines in *Xenopus*)	14, 149
—estivation		
Mammals—kidney	(With methylamines)	83

Abbreviations: DMSP, dimethylsulfoniopropionate; TMAO, trimethylamine-N-oxide, GPC, glycerophosphorylcholine.

fluids freeze and create high osmotic pressure. In a limited number of marine teleosts from high latitudes, glycerol has been discovered at high enough concentrations to make the fish approximately iso-osmotic with seawater, a highly unusual characteristic for a teleost fish (172). Glycerol concentrations are regulated seasonally; winter fish had high glycerol levels compared with summer fish.

Disaccharides, especially trehalose, are commonly accumulated in anhydrobiotic organisms (57). Here they may serve a number of functions, including the stabilization of cellular membranes and proteins against damage due to withdrawal of virtually all of the cellular water.

Free Amino Acids and Their Derivatives

Free amino acids were first discovered to be osmolytes in several marine invertebrate phyla, with the predominant forms being small, polar, zwitterionic molecules, especially glycine, alanine, proline, glutamine, β-alanine and the β-aminosulfonic acid taurine (48, 51). These solutes are now recognized as important osmolytes in a great diversity of organisms (Table 10.1; Fig. 10.2). Negatively charged amino acids are occasionally found as osmolytes, but at least in the case of glutamate in bacteria, electroneutrality is achieved with elevated intracellular K^+ (68). When combinations of glutamate and K^+ are accumulated in salt-stressed bacteria, this may be only a temporary adaptation. Positively charged amino acids, lysine and arginine, tend not to be used as osmolytes.

Recently, a number of other amino acids and amino acid derivatives have been found as osmolytes in eubacteria and methanogenic archaebacteria. These solutes include β-glutamate and β-glutamine (178), γ-amino butyrate, cyclic forms such as pipecolate (208) and ectoine (120) (Fig. 10.2), and modified types including N_ϵ-acetyl-β-lysine (an acetylated form of the amino acid that converts a positively charged molecule to a zwitterion (132) and a dipeptide, N-acetyl-glutaminylglutamine amide (195) (Fig. 10.2). The osmolyte that is exploited depends on the species of bacterium, the access to the osmolyte (or a precursor) in the medium, and the external salinity (178).

Methylated Ammonium and Sulfonium Compounds

Another widespread type of organic osmolyte is zwitterions distinguished by having charged methylated nitrogen or sulfur atoms. The former, generally referred to as methylamines, are far more common (Fig. 10.2). Most methylated amines and sulfonium compounds are fully substituted and thus are also called quaternary ammonium compounds (QACs) or tertiary sulfonium compounds (TSCs). However, there are exceptions to this general rule such as sarcosine (N-methylglycine) in some cartilaginous fishes (124) and N,N-dimethyl-taurine in marine red algae (118). By far the most widespread methylamine is glycine betaine (Fig. 10.2) (N,N,N,-trimethylglycine), one of several betaines such as proline betaine and β-alanine betaine (Fig. 10.2), β-glutamate betaine (144), γ-amino butyrate betaine, pipecolate betaine (208), and taurine betaine (148).

Urea and Urea with Methylamines

Less common than the types of organic osmolyte patterns discussed above is the occurrence of high levels of the nitrogenous waste product urea (Fig. 10.2). Initially found as an osmolyte in marine elasmobranchs (at an average intracellular and blood concentration of ~400 mM) for balancing body-fluid osmolality with seawater (194), it is used similarly in some amphibians adapting to salinity (14), and it accumulates to high levels in some estivating vertebrates (for example, lungfish) (64, 149) and gastropods (115), where it may help conserve body water. In the mammalian kidney, where urea is concentrated as a waste product, its concentration may rise to several molar in some xeric species such as desert rodents (183, 232).

In nonestivating organisms, high urea concentrations are associated with high concentrations of other classes of osmolytes, most notably methylamines such as trimethylamine-N-oxide (TMAO) (Fig. 10.2; Table 10.1). In cartilaginous fishes, the concentrations of these two solute classes are typically found at a 2:1 or 3:2 ratio of urea to total methylamines (231, 241). The significance of this pattern, and the mechanisms by which it is attained and regulated, will be discussed below (see later under MECHANISMS OF SOLUTE EFFECTS—AND NONEFFECTS)

In summary, broad surveys of numerous procaryotic and eucaryotic species have shown that only four major classes of organic osmolytes are accumulated within cells. In most cases these organic solutes make up the bulk of the osmotically active solutes. Inorganic ions generally are regulated to relatively stable levels, especially in eucaryotic cells, whereas the concentrations of organic osmolytes may rise to levels of 2–3 M, especially in the case of sugars and polyols. We now examine the effects of inorganic ions and the primary classes of organic osmolytes on selected in vitro and in vivo systems to ascertain the selective factors responsible for the patterns of osmolyte evolution and regulation found in the different kingdoms.

OSMOLYTE EFFECTS: PERTURBATION, STABILIZATION, AND COMPATIBILITY

The occurrence of only a small number of classes of organic osmolytes among all procaryotic and eucaryotic taxa (Fig. 10.2; Table 10.1), and the conservation of inorganic ion concentrations within narrow ranges in the intracellular fluids (37, 202, 237) leads to the question of how these different solutes influence essential cellular processes and structures. Can one explain the evolutionary trends and the physiological regulatory patterns observed in osmolyte systems in terms of how different osmolytes affect—or, indeed, fail to affect—such critical cellular functions as enzymatic activity, rates of protein synthesis, and control of gene expression?

Changes in Concentrations of Inorganic Ions Are Generally Perturbing of Biochemical Systems

We consider first the effects of changes in inorganic ion concentration and composition on selected in vitro systems. This is a logical starting point for an analysis of osmolyte system design because it allows us to address a central question concerning the "economics" of osmolyte utilization. If, as has been estimated for photosynthetic cells, the metabolic cost of using organic osmolytes is up to ten times the cost of using inorganic ions like K^+ (171), why do nearly all types of cells undergoing volume regulation employ organic solutes, many of which play important additional roles in the cell as substrates for adenosine triphosphate (ATP) generation or biosynthetic precursors? Is it not possible to design macromolecular systems that can maintain appropriate functional and structural characteristics in the face of widely varying concentrations of inorganic ions?

The data presented in Figure 10.3 provide a partial answer to these questions by illustrating several common features of inorganic ion effects on biochemical reactions. For most biochemical systems so examined in vitro, inorganic ions are found to affect strongly the system's functional and/or structural characteristics (237). The activities of most biochemical processes show a relatively sharp response to variations in the concentration of any type of inorganic ion, and the strength of the effect differs among ions and salts. Homologous (or analogous) systems in widely different organisms typically display similar responses to different types of inorganic ions and to variations in the concentrations of a given type of ion. Organic osmolytes generally are much less perturbing, even at extremely high concentrations; yet, like inorganic ions, they have similar effects on a given type of structure or process from taxonomically diverse organisms.

The effects of two potassium salts, potassium chloride (KCl) and potassium acetate, on translation by a cell-free system from mouse L cells illustrate commonly observed in vitro responses to ionic composition and concentration (Fig. 10.3, *upper left panel*). The rate of protein synthesis is strongly affected by changes in salt concentration for both potassium salts. For KCl and K-acetate, increases in salt concentration up to 75 m*M* or 125 m*M*, respectively, activate translation; further increases in salt concentration are inhibitory. Thus, whereas a certain level of potassium ion appears to be

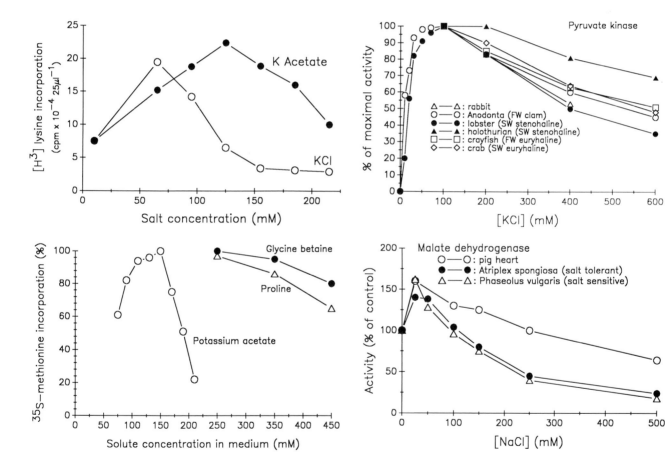

FIG. 10.3. Effects of salts on in vitro activities of protein translation systems and enzymes. *Upper left panel.* Effects of KCl and K-acetate on protein synthesis by a cell-free translation system from mouse L-cells. [Redrawn from Weber et al. (216)]. *Lower left panel.* Effects of K-acetate, glycine betaine, and proline on protein synthesis by a cell-free translation system from wheat germ, using wheat leaf mRNA. [Redrawn from Wyn-Jones (228).] *Upper right panel.* Effects of variations in KCl concentration on the maximal velocity of the pyruvate kinase reactions of diverse animals. [Redrawn from Bowlus and Somero (25).] *Lower right panel.* Effects of variations in NaCl concentration on activities of malate dehydrogenases from pig heart, a salt-sensitive plant (*Phaseolus vulgaris*), and a salt-tolerant plant (*Atriplex spongiosa*). [Redrawn from Greenway and Osmond (97).]

required for translational activity, a level that coincides with intracellular K^+ concentrations, concentrations in excess of physiological K^+ inhibit this key process. These data show why adjusting the intracellular osmolarity with inorganic ions is apt to be disadvantageous for an osmoconforming species or cell: during cell volume regulation, the rates of essential metabolic processes such as protein synthesis will be governed not only by the cell's demands for the products of the metabolic pathway, but also, and perhaps chiefly, by the effects of fluctuations in intracellular osmolarity. It is unlikely that the perturbations caused by shifts in inorganic ion concentrations will be in concert with the requirements of the cell for the pathway's products. For this reason alone, inorganic ions appear to be a poor type of osmolyte in osmoconforming organisms.

These data also provide an initial perspective on the varying effects of different classes of solutes on cellular processes. The chloride ion and acetate ion exhibit very different influences on translation: the inorganic anion, Cl^-, is significantly more inhibitory at high concentrations than the organic anion, acetate. These differences are fully in agreement with the Hofmeister series effects discussed below (see later under MECHANISMS OF SOLUTE EFFECTS—AND NONEFFECTS). Furthermore, as a caveat for experimental protocol design, the differences seen between KCl and K-acetate show that the use of an inorganic salt like KCl may be very inappropriate if one is attempting to establish in vitro conditions intended to simulate the intracellular milieu, which contains low concentrations of chloride ion relative to the summed concentration of small organic anions.

The solute effects observed for a particular macromolecular system from one species generally are found as well in studies of the homologous (or analogous) system in other species, even very distantly related species. This observation illustrates one of the central points of this chapter: the universality of physical-chemical principles governing osmolyte system design. The lower left panel of Figure 10.3 shows the effects of K-acetate and two organic osmolytes, glycine betaine and proline, on translation by a cell-free system using wheat germ translational apparatus and wheat leaf messenger RNA (mRNA). The wheat translation system, like that from mouse L-cells, exhibits a sharp response to variations in K-acetate concentration, and an optimum near 125–150 mM is found for both the plant and animal systems. In contrast to the effects of K-acetate, the addition of much higher concentrations of the two commonly occurring organic osmolytes glycine betaine and proline have relatively small effects on the rate of translation.

The responses to inorganic ions, and the interspecific similarities in these responses, noted in comparison of

the two in vitro translation systems are mirrored in a large number of comparative studies of solute effects on enzyme function (198, 237). For example, skeletal muscle pyruvate kinases (PK) from animals differing in osmotic characteristics (total intracellular osmolarity and degree of euryhalinity) are generally similar in their responses to variations in KCl concentration (Fig. 10.3, *upper right panel*). Pyruvate kinases from an osmotically dilute freshwater clam (*Anodonta* sp.), stenohaline and euryhaline marine invertebrates, and a mammal are all activated by increasing KCl concentrations up to approximately 75 mM–100 mM. Further increases in KCl concentration are inhibitory. The PK showing the highest tolerance of elevated KCl concentrations is from a stenohaline echinoderm, the holothurian *Parastichopus parvimensis*. Although there is an apparently lower K^+ requirement for maximal activity by PK of the freshwater clam, and an enhanced salt tolerance by the echinoderm PK, there is nonetheless a very high degree of similarity in the responses to changes in KCl concentration among all homologues of the enzyme from invertebrates and vertebrates (25). However, a salt-insensitive form of PK has been reported in a salt-adapted plant (218), which indicates that there exist exceptions to the general rule stating that all variants of a given type of protein respond similarly to a particular type of osmolyte.

Similarity in response to salt concentration is shown in comparisons of animal and plant malate dehydrogenases (MDH) (Fig. 10.3, *lower right panel*). All three MDHs are activated by low concentrations of NaCl and inhibited by concentrations above approximately 25 mM. The porcine MDH is somewhat less sensitive to rising NaCl concentration than the two plant MDHs. The MDH of the salt-tolerant plant *Atriplex spongiosa* responds to changes in NaCl concentration in parallel with the enzyme from the salt-sensitive plant *Phaseolus vulgaris*. These data thus provide further support for the hypotheses that inorganic ions are poorly suited for osmolyte function, especially in osmoconforming euryhaline species, and that homologous (or analogous) systems from species with different intracellular osmolarities and tolerance of osmotic stress are generally very similar in their responses to inorganic ions.

Organic Osmolyte Compatibility with Biochemical Functions In Vitro

The advantages to osmoconforming species of using organic osmolytes are illustrated by the data of Figure 10.4, which show the "compatibility" (28–30) of certain organic osmolytes with a variety of biochemical functions. Compatible solutes are those organic solutes

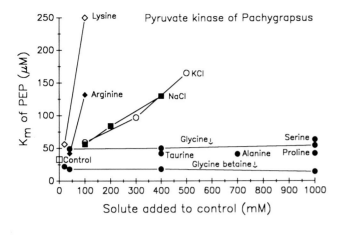

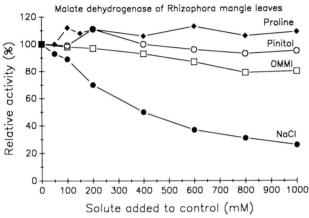

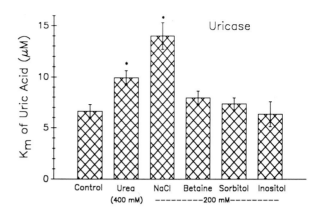

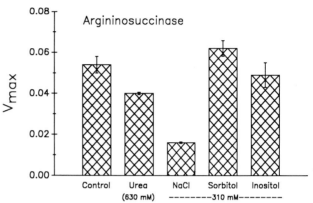

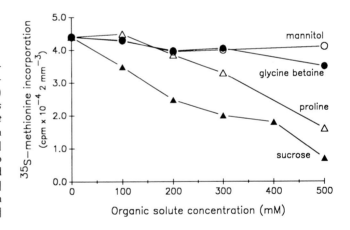

FIG. 10.4. Compatibility of organic osmolytes with in vitro biochemical functions. *Upper left panel:* Effects of solutes on the apparent Michaelis–Menten constant (K_m) of phosphoenolpyruvate (PEP) for the pyruvate kinase reaction of the marine crab *Pachygrapsus crassipes*. [Redrawn from Bowlus and Somero (25).] *Lower left panel.* Effects of solutes on activity of malate dehydrogenase from leaves of *Rhizophora mangle*. [Redrawn from Sommer et al. (200).] *Upper right and middle right panels.* Effects of solutes on two mammalian renal enzymes, uricase (K_m of uric acid) *(upper)* and argininosuccinase (V_{max}) *(middle)*. [Redrawn from Yancey (233).] *Lower right panel.* Effects of solutes on translation of wheat germ RNA by wheat germ ribosomes. [Redrawn from Gibson et al. (86).]

that can be accumulated to high concentrations and employed at widely different concentrations during volume regulation without perturbing cellular processes or structures.

Enzyme–ligand interactions typically are very sensitive to chemical and physical perturbation (114, 198, 199), and solute effects on these interactions serve to illustrate very clearly the differences between compatible and noncompatible solutes. The upper left panel of Figure 10.4 demonstrates for pyruvate kinase from the marine crab *Pachygrapsus crassipes* how the Michaelis-

Menten constant (K_m) of the substrate phosphoenolpyruvate (PEP) is affected by a number of inorganic and organic solutes. In common with the patterns shown in Figure 10.3, both NaCl and KCl are perturbing of PEP binding to PK, as measured by the increases in K_m of PEP as salt concentration increases. These large increases in K_m of PEP would be expected to inhibit catalytic activity and disrupt the regulatory function of the enzyme. For most of the organic solutes tested, either no significant effect or only small changes in K_m of PEP were observed, even at solute concentrations of up to

1 *M*. The commonly occurring amino acid osmolytes, glycine, serine, proline, and alanine, and the sulfonic amino acid taurine, a breakdown product of amino acid catabolism, had only small effects on the K_m of PEP. Glycine betaine consistently reduced the K_m of PEP below the control value. Two of the amino acids tested, lysine and arginine, were extremely perturbing of PEP–PK interactions. Neither of these two basic amino acids serves as a major osmolyte in cells. The perturbation of K_m of PEP by these net positively charged amino acids could be due to the formation of PEP–amino acid complexes (25, 60); see later under MECHANISMS OF SOLUTE EFFECTS—AND NONEFFECTS).

Sugar and polyhydric alcohol (polyol) osmolytes exhibit compatibility with biochemical function that is similar to that observed with the compatible amino acid osmolytes. Sugar and polyol concentrations can be raised to very high levels (up to 2–3 molal in halophilic algae; 24, 29) without major perturbation of physiological function. Malate dehydrogenase of the plant *Rhizophora mangle* is inhibited by rising concentrations of NaCl, but is affected only slightly by the organic osmolytes proline, pinitol, and 1D-1-O-methyl-mucoinositol (OMMI), even at concentrations up to 1 *M* (Fig. 10.4, *lower left panel*). It bears noting that OMMI not only may function as an osmolyte, but additionally may serve other key roles: a scavenger of oxygen free radicals (193) and a cryoprotectant during freezing stress (203). Thus, selection for a given type of organic osmolyte may reflect advantages of the solute in addition to those of compatibility as an osmolyte (see later under EVOLUTIONARY PERSPECTIVES).

Enzymes of the mammalian kidney, like the enzymes of euryhaline osmoconforming species living in habitats with high and varying external osmolarities, must function in the presence of high and fluctuating concentrations of organic solutes, including urea, glycine betaine, glycerophosphorylcholine, sorbitol, inositol and taurine (34, 157, 158, 160, 233). The effects of these four organic solutes and NaCl on two renal enzymes, uricase and argininosuccinase, are illustrated in the upper and middle panels, respectively, of Figure 10.4. Urea and NaCl significantly increase the K_m of uric acid for uricase and reduce the maximal velocity (V_{max}) of the argininosuccinase reaction. The effects of urea on the two renal enzymes indicate that at least this particular organic osmolyte is highly incompatible with enzyme function at physiological concentrations. The other major organic osmolytes of the mammalian kidney that were tested, glycine betaine, sorbitol, and inositol had no effect on either K_m or V_{max} of either of these enzymes.

The perturbation by salts of translation in vitro (Fig. 10.3, *left panels*) contrasts with the effects of organic osmolytes on this process (Fig. 10.4, *bottom right panel*). Mannitol and glycine betaine were without substantial effects on translation at concentrations as high as 500 m*M*. Proline appears generally compatible with translation up to concentrations of almost 200 m*M*, a concentration that seldom is reached by any *individual* organic osmolyte. Sucrose, however, is inhibitory of translation, although it is not as inhibitory as inorganic salts. Some of the inhibitory effects of high concentrations of sucrose could be the result of viscosity increases.

These four sets of data illustrate the following important aspects of organic osmolyte compatibility with macromolecular function. First, the lack of effect of organic osmolytes at physiologically realistic concentrations enables the cells of osmoconforming organisms (or the cells of organs like the mammalian kidney that encounter large changes in osmolarity due to state of hydration, nitrogen excretion requirements, and salt loading) to increase and decrease osmolyte concentrations without perturbing metabolic function. This is the defining feature of osmolyte compatibility as first emphasized by Brown and Simpson (30). Second, the compatibility of organic osmolytes belonging to all of the major classes shown in Figure 10.2, with the exception of urea, can be generally similar, although exceptions can be identified (for example, glycine betaine may significantly reduce K_m values, and high sucrose concentrations may be inhibitory to some processes). Third, urea is highly incompatible with biochemical function, even at the concentrations found, for example, in mammalian kidneys and in the body fluids of cartilaginous fishes. Fourth, the effects—or noneffects—of organic osmolytes are neither species- nor system-specific to any large extent, which is a reflection of the universality of the mechanisms governing osmolyte–macromolecular interactions.

Organic Osmolyte Compatibility with Protein Structure In Vitro

The commonly occurring organic osmolytes generally have been found to favor the stabilization of protein structure and the integrity of protein–nucleic acid complexes (237, 43). Some, but not all, of the solutes shown to be perturbing of in vitro biochemical function have been found to perturb protein structure. The monovalent ions K^+, Na^+, and Cl^-, which typically perturb function, are generally nonperturbing of protein structure up to relatively high concentrations.

These relationships are illustrated in Figure 10.5. Figure 10.5*A* shows the effects of KCl, NaCl, and several compatible or perturbing organic solutes on the thermal stability of bovine pancreatic ribonuclease (RNase), an enzyme frequently employed as a model protein in studies of protein structural stability. All of the common organic osmolytes, with the exception of urea, increase the denaturation temperature of RNase. Urea is a strong protein denaturant, and its effects on RNase stability are noted at concentrations of $0.4–0.8\ M$; that is, at concentrations in the physiological range for the inner medulla of the mammalian kidney, and the body fluids of urea-rich species like elasmobranch fishes, the coelacanth, and certain estivating amphibians. Arginine and guanidinium-Cl are also strongly perturbing. KCl and NaCl at concentrations up to $1\ M$ are without a significant effect on ribonuclease thermal stability. In terms of protein structure, then, these salts are compatible solutes, at least at common physiological concentrations. Their perturbing effects on enzymatic activity at low salt concentrations such as those used in the studies shown in Figure 10.3 are more likely to result from disruption of enzyme–ligand interactions than from perturbation of enzyme conformation or assembly.

Although organic osmolytes are generally stabilizing of protein structure, not all these osmolytes have equivalent stabilizing effects. Note in the solute effects on RNase, for example, that the free amino acid osmolytes are differentially effective in stabilizing this protein's structure. Glutamate, glycine, and alanine are significant stabilizers, whereas proline has no effect on RNase stability. The methylamines trimethylamine-N-oxide (TMAO) and glycine betaine are very effective stabilizers of RNase structure.

Differential effects of organic osmolytes have been observed in many other in vitro studies. Figure 10.5B shows the effects of osmolytes belonging to three of the major classes on the resistance of malate dehydrogenase (MDH) from leaves of the plant *Salsola soda* to thermal inactivation (incubation at 45°C for 40 mins). The amino acid (proline), the polyol (sorbitol), and the methylamine (glycine betaine) all increased the stability of the protein. The solutes differed, however, in the degree of protection provided. Sorbitol and glycine betaine were significantly more effective than proline in protecting the enzyme.

As in the case of amino acid osmolytes, substantial differences have been discovered among polyol osmolytes in their capacities for protein stabilization. Figure 10.5C shows the effects of six polyols on the stability of chymotrypsin at 44°C. All the polyols except ethylene glycol increase thermal stability in a concentration-dependent manner, as gauged by an increase in the standard free energy change of the denaturation reaction. However, the stabilizing effects of the polyols differ substantially, with the order of effectiveness: inositol > mannitol = sorbitol > erythritol > glycerol.

Very similar trends were found in studies of the thermal stability of lysozyme (Fig. 10.5D) and the rate of reassociation of urea-denatured asparaginase (Fig. 10.5E). The ranking of the polyols in terms of effectiveness in stabilizing lysozyme structure was: inositol > sorbitol > xylitol > erythritol > glycerol > ethylene glycol. For lysozyme, even the latter polyol exerted a minor stabilizing influence. Ethylene glycol also favored a significant increase in the rate of reassociation of subunits of urea-denatured asparaginase, but was much less effective than the other polyols, which displayed structure-stabilizing effects in the order: inositol > erythritol > glucose > sucrose > glycerol. These generally consistent trends in the relative stabilizing effects of different polyols among three different proteins is another illustration of the commonality of solute effects observed from protein to protein and from organism to organism.

In this treatment of solute effects in vitro we have focused on proteins, because of the large amount of information on solute-protein effects. Solute effects on membrane systems may also be important, although there is less detailed information available to support this conclusion. Organic osmolytes may influence the structure of the lipid bilayer. The osmolytes glycine, proline, and glycine betaine increased the phase transition temperatures of vesicles of dipalmitoylphosphatidylcholine (DPPC) (180). The magnitude of this effect was proportional to the degree of nitrogen methylation, with the strongest effect being found for glycine betaine. The finding that solutes that perturb protein structure also perturb phospholipid bilayers, as evidenced by increases in leakiness of vesicles (3), indicates that many of the principles that underlie osmolyte selection for compatibility with protein function also may apply in the case of membrane systems.

The osmolarity encountered by a cell can also affect the fatty acid composition of its cellular membranes. When the salt-tolerant yeast *Zygosaccharomyces rouxii* was grown in the presence of 15% NaCl, the fatty acid composition differed significantly from control cells grown in low-salt medium (116). The effect of these shifts in membrane phospholipid fatty acid composition was to reduce the fluidity of the membranes of the high-salt yeast cells. It was hypothesized, but not shown, that one advantage of this reduced membrane fluidity could be a reduction in permeability to glycerol, a compatible osmolyte accumulated in this species (116).

Compatibility of Organic Osmolytes: In Vivo and Cell Culture Studies

Osmoprotection in Bacteria and Yeasts. Although tests in vitro with isolated proteins and other biochemical preparations provide solid evidence for compatibility

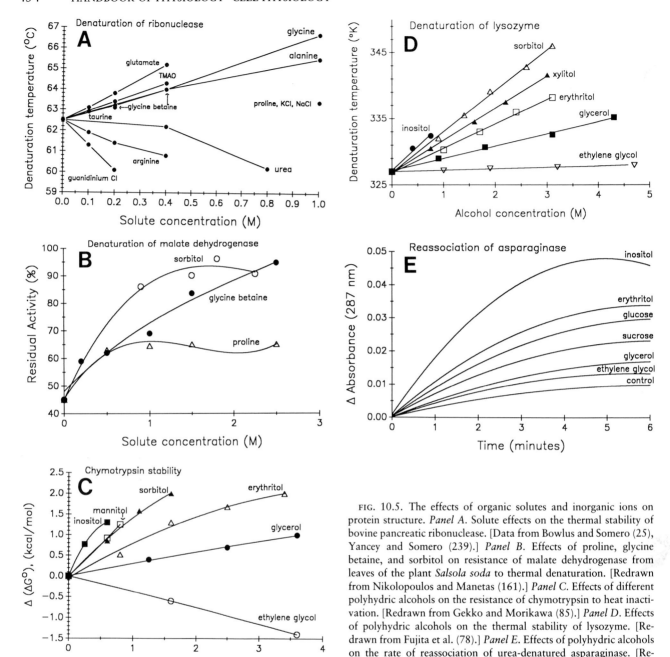

FIG. 10.5. The effects of organic solutes and inorganic ions on protein structure. *Panel A.* Solute effects on the thermal stability of bovine pancreatic ribonuclease. [Data from Bowlus and Somero (25), Yancey and Somero (239).] *Panel B.* Effects of proline, glycine betaine, and sorbitol on resistance of malate dehydrogenase from leaves of the plant *Salsola soda* to thermal denaturation. [Redrawn from Nikolopoulos and Manetas (161).] *Panel C.* Effects of different polyhydric alcohols on the resistance of chymotrypsin to heat inactivation. [Redrawn from Gekko and Morikawa (85).] *Panel D.* Effects of polyhydric alcohols on the thermal stability of lysozyme. [Redrawn from Fujita et al. (78).] *Panel E.* Effects of polyhydric alcohols on the rate of reassociation of urea-denatured asparaginase. [Redrawn from Shifrin and Parrot (188).]

as a selective factor in osmolyte evolution, experiments on compatibility in living systems are more direct. Experimentation on living systems requires defined manipulation of intracellular osmolyte contents. Such controlled variation in intracellular osmolyte concentrations can be achieved in some cases by variation in medium composition and in others by selection of mutant cells that differ in osmoregulatory properties.

One testable prediction is that cells maintained under

hyperosmotic stress will do better (for example, will grow more rapidly) if they are provided in the medium with either compatible osmolytes or required precursors that can be taken up by the cell and used in osmolyte synthesis. In such tests, the added solute is often termed an *osmoprotectant* (59). Osmoprotectants thus are solutes that can be transported into the cell, where they either can serve immediately as compatible osmolytes or can be further metabolized to one or more compatible osmolytes. In bacteria, the phenomenon of osmoprotection is readily demonstrated by altering the composition of laboratory media. Most experiments have been done with *Escherichia coli*, which can survive in elevated salinities of several hundred m*M* NaCl. However, growth in unsupplemented medium is quite slow at higher osmolalities (Fig. 10.6). Without the presence of osmoprotectants in the medium, the cells use K$^+$ and glutamate as their major osmolytes, at least during the initial phases of osmotic stress (see later under REGULATION OF OSMOLYTE CONCENTRATIONS). Addition of extracellular betaine or one of its precursors (for example, choline), or other compatible osmolytes such as choline-o-sulfate or β-alanine betaine, greatly increases cell division rate (Fig. 10.6) and extends the range of survival. As the osmoprotective solutes are accumulated in the cells, K$^+$ and glutamate concentrations are reduced (59, 133).

Osmoprotectants also affect bacterial respiration. Halotolerant bacteria exposed to 1.8 *M* KCl have suppressed respiration, which can be restored with exogenous solutes with effectiveness in the order: dimethylsulfonioacetate > glycine betaine > dimethylglycine > β-alanine betaine = sarcosine > glycine (191).

Another approach to altering cell organic osmolytes is through mutation. *E. coli* mutants incapable of replacing K$^+$ and glutamate with trehalose are impaired in osmoadaptation (68); other mutants that overproduce proline or increase glycine betaine uptake have enhanced osmotolerance (96, 145). A halotolerant yeast, *Debaryomyces hansenii*, accumulates glycerol and arabinitol when exposed to high NaCl medium. A mutant strain that accumulates much less of these polyols grows more slowly in that medium. Addition of exogenous glycerol, arabinitol, or mannitol (but not

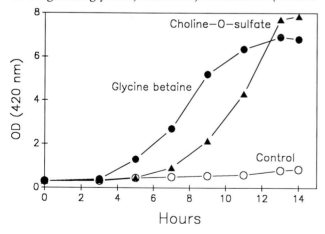

FIG. 10.6. The effects of the osomoprotectants choline-O-sulfate and glycine betaine on growth of *E. coli* cells in medium containing 0.65 M NaCl. [Redrawn from Hanson (105).] β-alanine betaine showed osmoprotection similar to that afforded by glycine betaine (not shown).

erythritol) restores growth (1). Finally, transfer of the gene for choline oxidase, the complex that produces glycine betaine from choline, has been completed from a soil bacterium (*Arthrobacter pascens*) to an *E. coli* mutant defective in glycine betaine synthesis. The result is enhanced salinity tolerance in the latter bacterium (179).

Osmoprotection and Osmolyte Compatibility in Vertebrate Cells. Osmoprotection by exogenous solutes is not restricted to cases where the osmoprotectants are found in the external environment. The accumulation of osmoprotectants from the extracellular fluids is an analogous process, and is important in osmoregulatory adjustments in multicellular organisms (for example, in the mammalian kidney). For example, glucose taken up from the extracellular fluids by cells of the renal medulla can be converted to the compatible solute sorbitol, a major kidney osmolyte (34).

Studies of osmoprotectant and compatible solute effects in the mammalian kidney medulla have, in effect, tested the opposite prediction to that described above for bacteria: cells will do worse in hyperosmotic medium if deprived of compatible osmolytes. In standard growth medium made hyperosmotic with NaCl, a line of rabbit inner medullary epithelial cells (PAP-HT25) relied mainly on sorbitol for cell volume maintenance, at least in normal growth medium. These cells took up glucose, which they converted to sorbitol through the aldose reductase reaction. The sorbitol-accumulating cells grew well (as measured by ability to form clonal colonies in culture) in hyperosmotic media (for example, with 250 mM NaCl compared to normal nonrenal blood at about 150 mM). Addition of aldose reductase inhibitors (Fig. 10.7A, tolrestat groups), a class of drugs that prevent the conversion of glucose to sorbitol, resulted in drastically reduced cell growth in parallel with declining cell sorbitol content (Fig. 10.7A). In contrast tolrestat had no effect at 150 mM NaCl (control), a condition under which the cells had little or no sorbitol (236). In the presence of the aldose reductase inhibitors, the lack of sorbitol synthesis (paired with the failure of other organic solutes to increase sufficiently to compensate) would result in higher concentrations of noncompatible solutes, either by cell shrinkage or by ions entering from the medium. The elevated concentrations of noncompatible solutes probably caused the observed low growth rates through inhibition of various biochemical processes. Thus the hypothesis of sorbitol compatibility is clearly supported by these in vitro studies.

These initial cell culture experiments were followed by two additional lines of study. First, in vivo testing was done by feeding an aldose reductase inhibitor to rats for 10, 21, and 72 days (70, 238) and analyzing their kidneys for osmolytes. Like the PAP cells (Fig. 10.7A), sorbitol content of the inner medulla declined greatly with drug treatment in 10 and 72 days (Fig. 10.7B, sorbinil group). In contrast to the PAP cell study, however, glycine betaine contents in the 10- and 72-day treatments rose in almost perfect compensation for the reduced sorbitol concentrations, such that there was no noticeable osmotic damage, and the relationship to Na^+ remained constant (238). In the 21-day group, neither sorbitol nor betaine contents changed, a pattern as yet unexplained (238).

Second, later studies on PAP cells confirmed the results from the in vivo studies performed with rats: if PAP cells were provided glycine betaine (normally absent) in growth medium, they accumulated this osmoprotectant when sorbitol production was suppressed

by an aldose reductase inhibitor. The large decrease in cell viability previously seen (Fig. 10.7A) was reversed by betaine accumulation (Fig. 10.7A, betaine group). Thus, betaine and sorbitol may serve as interchangeable compatible osmolytes in the kidney.

Other vertebrate cells have been examined for growth in hyperosmotic conditions with organic solutes. Growth of early mouse embryos in vitro is inhibited by increased NaCl concentration in the medium, but addition of glutamine reversed this inhibition. Lawitts and Biggers (137) propose that glutamine is acting as a compatible osmolyte. SV-3T3 and chick fibroblasts cells have inhibited protein synthesis and growth rates, plus reduced cell volumes, in 0.5 osmolar medium. Addition of glycine betaine to the medium restored these properties to normal; addition of N,N-dimethylglycine, sarcosine (N-methylglycine), glycine,

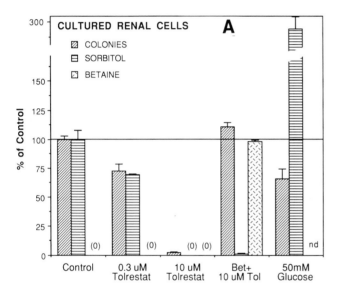

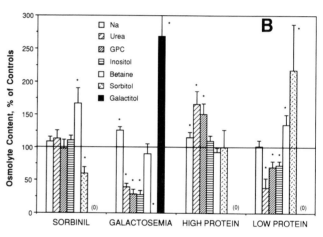

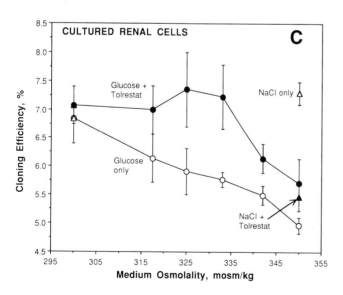

FIG. 10.7. A. Effects of exogenous tolrestat (Tol, an aldose reductase inhibitor), glycine betaine (Bet, 5 mM), or glucose on the number of successful colonies formed (from seeded individual cells) and cellular osmolyte contents (mmol/kg protein, normalized to control amount of sorbitol) of mammalian renal cells (PAP-HT25) after 7–14 days in culture, in 500–550 mosmolar media (* = significant change). [Redrawn from Yancey et al. (236) and Moriyama et al. (154).] B. Effects of various dietary manipulations on osmolytes of rat renal inner medulla (contents in mmol/kg wet wet or mmol/kg protein, expressed as percent of appropriate control contents). Sorbinil: animals treated for 10 days with the aldose reductase inhibitor sorbinil in the diet at 40 mg/kg/d [data from Yancey et al. (238)]; Galactosemia: animals with 50% dietary galactose for 10 days [data from Bondy et al. (23)]; High Protein and Low Protein: animals on 50% or 4.5% protein, respectively (compared to 20% for controls) [data from Peterson et al. (163)]. (* = significant change.) C. Effects of high glucose and the aldose reductase inhibitor tolrestat on cloning efficiency of PAP-HT25 cells. At near normal osmolality high glucose inhibits cloning efficiency, but tolrestat offsets this effect through limiting the accumulation of sorbitol. [Redrawn after Yancey et al. (236).]

proline, taurine, or glycerol had weaker but nonetheless some restoring effects as well (164).

Another approach to the examining of solute compatibility in vivo involves exposing cultured cells to high concentrations of extracellular solutes added to initially isosmotic medium. For solutes that penetrate cell membranes and thus directly alter the intracellular milieu, compatible solutes should have fewer disruptive effects than solutes predicted to be perturbing of cellular function. This approach has been used with clonal colony growth of Madin-Darby canine kidney (MDCK) cells. The studies used two osmolytes that readily cross cell membranes: glycerol, the archetypical compatible solute in protists, and urea, a well-known (noncompatible) protein destabilizer. When MDCK cells were grown with these osmolytes in the medium, glycerol was somewhat inhibitory, but clearly was far more compatible than urea at higher osmolalities (see Fig. 10.9A).

Excess Intracellular Accumulation; Diabetes. Finally, in contrast to compatibility of osmolytes when used for appropriate osmotic balance, excess accumulation of an osmolyte intracellularly should be detrimental by causing swelling. This is the process hypothesized to occur in one of the side effects of hyperglycemia in diabetes mellitus: in organs having aldose reductase (such as lens, retina, nerves, kidney), excessive blood glucose drives the reaction to sorbitol such that excessive amounts accumulate in cells (36), which may explain the occurrence of the ruptured cells seen in some organs (36). In vitro PAP cells have been tested to investigate this process. Cells grown at 550 mOsm with 60 mM glucose have excessively high contents of sorbitol and reduced growth (Fig. 10.7A, glucose group). Similarly, in cells grown in medium initially isosmotic but supplemented with increasing amounts of glucose, growth was inhibited in correlation with the amount of glucose (Fig. 10.7C) and cell sorbitol content (not shown). Addition of an aldose reductase inhibitor restored both cell sorbitol content and growth in the glucose-exposed cells. In contrast, other cells exposed to added NaCl (with control glucose levels) grew at control rates, in correlation with increased cell sorbitol (not shown), while the drug tolrestat reduced sorbitol content below normal and inhibited growth (Fig. 10.7C).

ORGANIC OSMOLYTE EFFECTS: COUNTERACTING SOLUTE SYSTEMS

In contrast to the usually benign effects of most organic osmolytes, the waste product urea is a well-known perturbant of macromolecules. At concentrations found in marine elasmobranchs and mammalian kidneys, a wide variety of biological functions are inhibited or disrupted, including alteration (usually inhibition) of enzyme kinetics (V_{max} or K_m), destabilization of protein folding and assembly, inhibition of muscle contraction and tissue respiration, and death of exposed cells and organisms (231). For example, bony fish exposed to seawater containing 400 mM urea begin dying as their blood urea levels pass 200 mM (98). Occasionally, major effects of low urea concentrations (for example, 25 mM) (104) are seen on proteins, but most effects are noted only when urea concentrations rise above approximately 100 mM.

A clue about the resolution of the "urea paradox"—that is, the puzzle as to why a perturbing solute has been selected as a major organic osmolyte in diverse species—is offered by the observation that in urea-rich organisms and tissues, urea is not the only organic osmolyte that is accumulated. Methylamines like trimethylamine-N-oxide (TMAO), glycine betaine, and GPC co-occur with urea (220, 231, 237). This observation prompted studies of urea and methylamine effects on proteins of marine elasmobranch fishes (240, 241). These studies showed that the methylamines can offset the effects of urea in many cases. These studies showed that, in addition to organic osmolyte systems that rely strictly on solutes that are individually nonperturbing (that is, on compatible osmolytes), other systems employ a family of solutes that may be individually perturbing, but which have no *net* effect. These latter systems are termed "counteracting solute systems" to emphasize the algebraic additivity of solute effects.

Counteracting Solute Effects In Vitro

Protein Function. Figure 10.8 shows some of the typical effects of urea and urea-counteracting methylamines on kinetic parameters for several enzymes from urea-rich and non–urea-accumulating species. Urea at physiological concentrations usually raises K_m values (Fig. 10.8A: stingray PK) and inhibits maximal velocities (Fig. 10.8B: pig kidney argininosuccinase; Fig. 10.8C: kidney c-AMP phosphodiesterase activity). In another case, enzyme phosphorylation level was found to be reduced by urea (Fig. 10.8D). These urea effects are typically not species-specific: enzymes from organisms containing high urea levels generally are no less sensitive to urea than the homologues present in species without high intracellular urea (239, 240, 241). It should be noted that urea-insensitivity in protein function (for example, with hemoglobins) may be a primitive trait, rather than a derived adaptation to elevated urea concentrations (184).

Methylamines generally have effects on protein func-

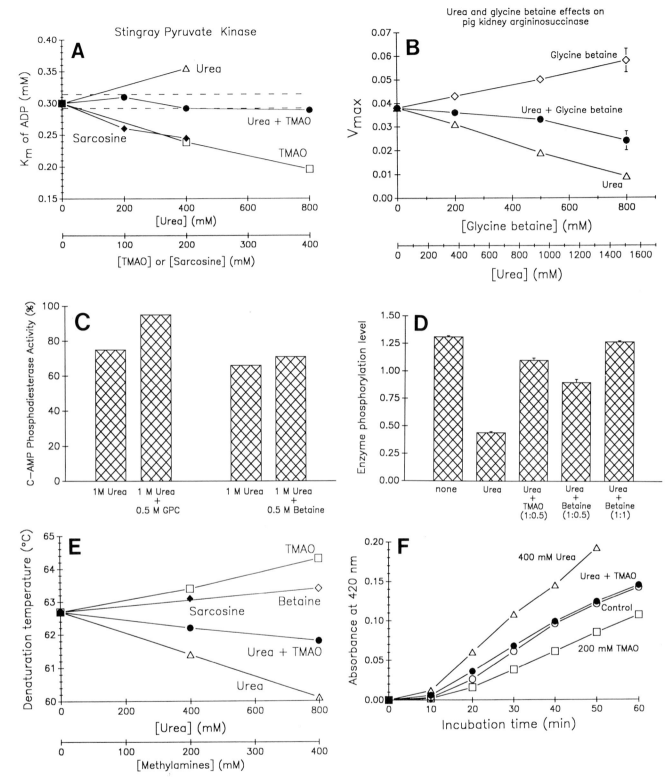

FIG. 10.8. Counteracting effects of urea and methylamines on in vitro biochemical systems. *Panel A*. Effects on K_m of ADP for pyruvate kinase of the stingray *(Urolophis halleri)*. Horizontal dashed line shows the 95% confidence interval around the control (no urea or methylamines) K_m value. [Redrawn after Yancey and Somero (241).] *Panel B*. Effects on V_{max} of the porcine kidney argininosuccinase reaction (unpublished data of T. Arnell, M. Blykowski, and P. H. Yancey). *Panel C*. Effects on activity of MDCK cell c-AMP-phosphodiesterase. [Data from Subramaniam and Jackson (204).] Rates are expressed relative to control (no added urea or methylamine). For all but the treatment "1 M urea + 0.5 M GPC" rates were significantly lower than control rates. *Panel D*. Effects on degree of phosphorylation of sarcoplasmic reticulum ATPase. [Data from deMeis and Inesi (61).] *Panel E*. Effects on thermal denaturation of bovine pancreatic ribonuclease. [Redrawn after Yancey and Somero (240)]. *Panel F*. Labeling of sulfhydryl (−SH) groups of glutamate dehydrogenase (mammalian) by 4-chloro-7-nitrobenzofurazan. [Redrawn after Yancey and Somero (240).]

tion and structure that are opposite to those of urea, such that when the two types of solutes are both present in solution they offset each other. This counteraction is typically maximal at about a 2:1 urea:methylamine ratio, which is the approximate ratio found in elasmobranch cells. This ratio works well for offsetting urea effects on the K_m and V_{max} values of many enzymes besides those shown in Fig. 10.8A,B: elasmobranch and mammalian glutamate dehydrogenase (241), elasmobranch and teleost actomyosin ATPase (231), elasmobranch lactate dehydrogenase (K_m of NADH, but not K_m of pyruvate), argininosuccinase, creatine kinase (231), carbamoyl phosphate synthetase (4), yeast alcohol dehydrogenase (147), teleost glycine cleavage complex (155), and mammalian (renal) Ca^{2+}-ATPase (214).

Counteraction has also been seen with more complex systems, including contraction of skinned muscle fibers (elasmobranch) (2), and various aspects of mitochondrial respiration. Some of the latter show urea activation and TMAO inhibition (5, 15), a less common pattern that has also been noted for one enzyme (241). In these cases, TMAO may make some enzymes too rigid or too aggregated for proper function, whereas urea may restore flexibility and the proper aggregation state (147, 237, 241).

Protein Stability. Counteracting effects of urea and methylamine osmolytes also are seen on various aspects of protein structure, including thermal denaturation of bovine ribonuclease (Fig. 10.8E), catalase (147), and lysozyme (7), renaturation rate of elasmobranch and mammalian lactate dehydrogenase (240), and unfolding of bovine glutamate dehydrogenase in the presence of a thiol reagent, 4-chloro-7-nitrobenzofurazan (Fig. 10.8F). In each case the effect of urea was to destabilize protein structure, and the effects of the methylamines alone was to enhance structure. Together, at concentration ratios found physiologically, the solutes were counteracting, as in their effects on protein function. Recently, betaine at 2.5 M was found to reverse the large increase in the dielectric increment and relaxation time of bovine serum albumin in 5 M urea (17). A non-natural urea counteractant may be dimethylsulfoxide (DMSO), a commonly used cryoprotectant, which was found to protect Na^+-K^+ ATPase from inactivation by urea (152).

Urea Counteraction in Living Systems

Tests of the counteracting solutes hypothesis in vivo are more difficult than in vitro, but are necessary for direct demonstration that urea–methylamine counteraction is physiologically important. One testable prediction is that, like proteins, cells will function better with an intracellular mixture of urea and methylamines than with either urea or methylamine alone (237). Urea and betaine in the growth medium will penetrate MDCK cells, such that intracellular contents of both solutes can be readily manipulated (235). As shown in Figure 10.9A, addition of urea or betaine alone in the medium greatly reduced cell colony growth, except for some stimulation by both solutes at lower concentrations. However, addition of betaine with high urea partly or fully restored normal growth (Fig. 10.9A), in accordance with the counteracting-osmolytes hypothesis. Addition of the compatible solutes sorbitol and glycine had no urea-counteracting effects, although *myo*-inositol did counteract urea somewhat (Fig. 10.9A).

The counteracting-osmolytes hypothesis also predicts that urea-counteracting osmolytes should be regulated specifically in response to urea concentration, and not to the concentration of NaCl or osmotic pressure per se. That is, this hypothesis predicts that the concentrations of urea counteractants should maintain a consistent ratio to urea concentrations inside cells. Experiments with both MDCK and PAP-HT25 mammalian renal cells in culture support this prediction. Raising medium osmolality with NaCl triggers an increase in all intracellular organic osmolytes (including GPC); however, urea addition alone results in increased GPC (a methylamine) only. In PAP-HT25 cells, the urea effect is stronger than for NaCl at equal osmolalities.

In vivo, the mammalian kidney also fits the prediction well for the methylamine GPC, as revealed in tests with animals in which renal contents of urea were altered largely independently of sodium through variable dietary protein. In high-protein diets, renal urea (and, to a lesser extent, sodium) contents increase, with only GPC increasing among the non-urea organic osmolytes (Fig. 10.7B). In low-protein diets, urea (but not sodium) contents decreased greatly, concomitant with decrease in GPC (and, to a lesser extent, inositol) content (Fig. 10.7B). Across all diets tested, only renal contents of urea and GPC maintained a strong linear correlation (Fig. 10.9B).

Results with the other renal methylamine, glycine betaine, appear paradoxical. Glycine betaine, shown to counteract urea effects on renal cells in vitro (above), is also found in a linear correlation with urea in antidiuretic and diuretic animals, but the ratio differs in the two states: betaine is not down-regulated in diuresis in proportion to urea decrease (232, 234). Further, whereas urea stimulates GPC accumulation in cultured renal cells and probably in vivo (above; Fig. 10.7B), betaine (and sorbitol) contents are increased in apparent compensation for reduced GPC (and inositol) when medullary urea content is lowered in vivo by low dietary protein (Fig. 10.7B).

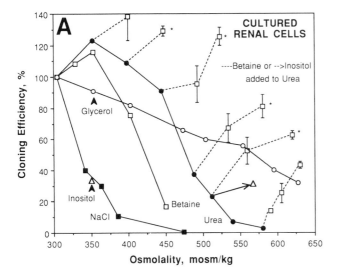

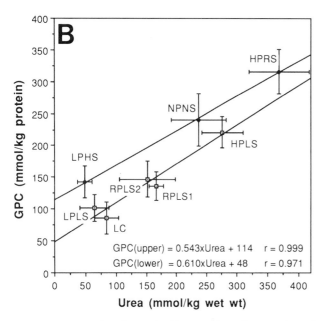

FIG. 10.9. *A.* The effects of additions of NaCl, urea, glycerol, inositol, and glycine betaine to the growth medium on the colony-forming efficiency of MDCK cells. Colony-forming efficiency values are the number of colonies formed per number of single cells seeded on the growth plates, as normalized to control values (no test solutes added to the growth medium, which had an osmolality of 305 mosm/kg). Error bars represent SDs. *Open squares* connected by *dashed lines* represent glycine betaine added to cultures containing urea. (* = 2:1 ratio of urea to glycine betaine.) [Redrawn from Yancey and Burg (235).] *B.* GPC and urea contents in renal inner medulla of rats subjected to various combinations of dietary protein and salt load. *Upper regression line:* Diets are low protein/high salt (LPHS), normal protein/normal salt (NPNS), and high protein/reduced salt (HPRS). *Lower regression line:* Diets are all low salt (salt load lower than HS, NS, and RS diets on upper line). LP = low protein, RP = reduced protein (two separate experiments), HP = high protein, LC = low calorie diet (animals restricted to maintenance level of food). [Redrawn from Peterson et al. (163).]

Overall, then, the link between renal urea and GPC may fit the counteracting-osmolytes hypothesis. However, a constant urea:methylamine (GPC plus betaine) ratio is clearly not maintained in all conditions, because of the lack of—or inverse correlation between—urea and betaine. If the counteracting-osmolytes hypothesis has validity for the mammalian kidney, there are several possible explanations for the failure of *total* methylamines to decrease in the low-protein diets. First, it may be that GPC is a stronger (thus more critical) counteractant toward urea than is glycine betaine. This possibility is suggested by the results shown in Fig. 10.8*C:* GPC was more effective than glycine betaine in offsetting the perturbation of c-AMP phosphodiesterase activity. Furthermore, as noted in the discussion on enzymes in the previous section, betaine is generally not as effective in counteraction as is TMAO (the major methylamine of cartilaginous fishes) (240, 241). This and the fact that betaine is a widespread osmolyte in organisms with little or no urea (Table 10.1) and may functionally substitute for sorbitol as a compatible osmolyte (above, Fig. 10.4C), suggests it may be more compatible than urea-counteracting, despite its methylamine structure. Second, it may be that a precise urea:methylamine ratio is not crucial for adequate protein functions at low urea levels, especially if betaine acts more as a compatible osmolyte. Finally, inositol may be a weak urea-counteractant, as discussed above. However, the role of glycine betaine (other than as an osmotic effector) remains uncertain.

Counteraction in living systems may have medical applications. The possible involvement of methylamines in human uremia and its treatment have been suggested by Lee et al. (139). In an interesting extrapolation of the counteracting solutes hypothesis, Kloiber et al. (128) found that methylamine injections in rats provided strong protection against hyperammonemia. In another study, *E. coli* were found to use trace amounts of human urinary osmolytes (including glycine betaine) to grow effectively in the presence of salts and urea. Glycine betaine alone did not provide protection, but it did so in combination with salts, sugars, or mannitol. This phenomenon could play a role in urinary tract infections (46).

Salt Counteraction (Haloprotection)

The best example of another naturally occurring form of counteracting solute effects again lies in the methylamines, which in some cases are able to offset perturbation by inorganic salts of protein function and structure. This mechanism, the "haloprotective effect" (59), could be relevant when the concentrations of intracellular inorganic ions rise; for example, during the early stages of hyperosmotic regulation (see later under REGULATION OF OSMOLYTE CONCENTRATIONS).

Pollard and Wyn Jones (168) first showed that NaCl inhibition of barley MDH activity could be reversed by several organic osmolytes, with the order of effectiveness glycine betaine > dimethylglycine > sarcosine (N-methylglycine) > glycine. The efficacy of counteraction of KCl inhibition of a cyanobacterial ribulosebisphosphate carboxylase-oxygenase (RuBisCO) by organic osmolytes (Fig. 10.10) reveals a similar dependence on degree of osmolyte methylation. TMAO and glycine betaine are more effective at reversing salt inhibition than sarcosine or glycine.

Although methylamine counteraction of salt effects is not as general as reversal of urea effects, methylamine–salt counteractions have been reported in a number of different systems, including bacterial trehalase activity (110), the stability of barnacle myofilament architecture (50), inhibition of force development in rabbit and lobster muscle fibers (90), and the stability of the spinach photosystem complex (156). In the latter study, salt-perturbation of the association of extrinsic membrane proteins with the phospholipid bilayer was examined. One molar NaCl caused the dissociation of extrinsic proteins from membranes of the photosystem 2 complex of spinach, but 1 M glycine betaine prevented this disruption of membrane structure.

Other organic osmolytes besides the methylamines have also been reported to offset salt perturbation of biochemical systems. For example, the aggregation of chick erythrocyte chromatin induced by NaCl was counteracted by several organic osmolytes, with the order of effectiveness: taurine > glycine > proline > mannitol > sorbitol (32). Interestingly, TMAO was found to have no effect. Glycerol has been shown to offset NaCl inhibition of some enzymes of the halotolerant green alga *Dunaliella*, but other enzymes were not protected by glycerol (88). Sorbitol reverses the inhibition by NaCl and KCl of myosin denaturation and ATPase activity (130).

An unusual type of counteracting osmoprotection has been reported for cultures of mammalian renal medulla cells: primary explants and permanent cell lines survive better in a medium with high urea and NaCl (two perturbants) than with either alone (157, 225). It is not known whether this is a true counteraction, however; perhaps urea and salt exposure triggers a more optimal combination of intracellular osmolytes than either perturbant does alone.

Exceptions to Counteraction

Despite the widespread occurrence of methylamine-urea counteraction in diverse biochemical and physiological systems, this algebraic additivity of solute effects is not universal. Some enzymes that are perturbed by urea are not restored by the addition of TMAO to the in vitro medium (104, 147, 233, 241). For urea-rich species there may be one or more additional ways of offsetting these urea effects. Selection could favor the evolution of urea-insensitive proteins (231), albeit this strategy would seem to be of limited importance because of the common bases for urea effects on proteins (see later under MECHANISMS OF SOLUTE EFFECTS—AND NON-EFFECTS). One mechanism for enhancing tolerance of urea for a protein kinetic property that is urea-sensitive but not restored by methylamines has been shown for the skeletal muscle isozyme of lactate dehydrogenase (239). The K_m of pyruvate is increased by physiological concentrations of urea but is not sensitive to methylamines, unlike the K_m of NADH, which exhibits urea–methylamine counteraction. For the LDHs of elasmobranch fishes the K_m of pyruvate measured in the absence of urea is low relative to the values found for LDHs of other vertebrates (239). However, when physiological concentrations of urea are added to the in vitro assay system, the K_m of pyruvate is increased to a value similar to that found at a common temperature for the LDHs of other vertebrates. Thus, the set points for protein functions can be adjusted evolutionarily to values that lead to an actual *requirement* for the presence of urea for physiologically appropriate function by the protein. The extent to which proteins that are urea-sensitive but not restored by methylamines are so modified is not known.

It may also be the case that some degree of urea perturbation is tolerable for systems that are non-limiting, for example, enzymes that play at most a

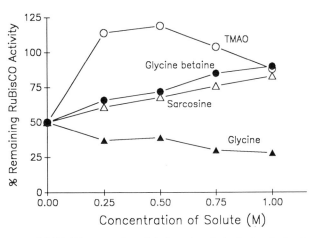

FIG. 10.10. The protection of *Aphanothece halophytica* ribulosebisphosphate carboxylase/oxygenase activity from inhibition by KCl by organic solutes with different degrees of methylation. [Redrawn from Incharoensakdi et al. (119).]

minimal role in governing flux through a pathway. Furthermore, the requirements for counteraction of urea effects may depend on the particular physiological and environmental relationships of the organism. Hibernators and estivators, for example, may accumulate high concentrations of urea during periods of inactivity, and in these instances the inhibiting effects of urea on metabolic activity could be adaptive (104, 237).

REGULATION OF OSMOLYTE CONCENTRATIONS

Interspecific Similarities in Basic Regulatory Strategies

Much as similar types of osmolytes are used in phylogenetically diverse species (Table 10.1) so are there many similarities among species in the basic types of regulatory mechanisms used to adjust osmolyte concentrations in the cell. In cell types as diverse as bacteria and mammalian kidney cells, two principal mechanisms are used to elevate the concentrations of osmolytes under conditions of hyperosmotic stress: uptake from the medium of both inorganic ions and organic molecules, and de novo synthesis of the latter. Down-regulation of osmolyte concentration also is achieved via similar mechanisms in many different cell types. Under conditions of hypoosmotic stress, inorganic ions and some organic osmolytes move out of the cell, either passively or through active transport mechanisms, and other organic osmolytes are either catabolized or added to osmotically inactive polymers. The kinetics of osmolyte regulation differ considerably among species and between the processes of osmolyte accumulation and decrease in cellular osmolarity. The rate at which the cell is able to decrease its osmolarity during hypoosmotic regulation may greatly exceed the rate at which osmolytes are built up during hyperosmotic stress; this difference has been noted in taxonomically diverse species.

The extracellular solutes that are accumulated for osmotic purposes (osmoprotectants), whether found in the external environment (for example, pore water in soil or in the ocean) or in the extracellular fluids of multicellular species, fulfill two functions: either they can be used directly as osmolytes (for example, glycine betaine and proline taken up from soils by bacteria and plants) or they can be metabolized, usually via relatively few enzymatic steps, to compatible solutes. The uptake of choline and subsequent conversion to glycine betaine in bacteria and plants is a good illustration of this second type of osmoprotectant (59). Only two enzymatic reactions, those catalyzed by choline mono-oxygenase (plants) or choline dehydrogenase (bacteria) and betaine aldehyde dehydrogenase, are

needed to convert choline to glycine betaine. As a general rule, the uptake of osmoprotectants can be energetically less costly than the de novo synthesis of organic osmolytes by the cell (170). The osmoprotectants that are accumulated under certain laboratory conditions may be solutes not found in the natural environment of the organism. For example, MOPS buffer was found to accumulate in osmotically stressed E. coli cells when no other organic osmoprotectants were available (44). Cayley and colleagues (44) referred to MOPS as a "gratuitous" osmolyte.

Osmolyte Regulation in Bacteria and Plants

Sequential Changes in Osmolyte Utilization during Salinity Acclimation. The time-dependent changes in intracellular osmolyte composition observed in plants and bacteria during salinity acclimation afford an especially good perspective concerning several of the key attributes of osmoregulation: the relative advantages of different types of osmolytes, the complementary roles of osmoprotectants and de novo synthesized osmolytes, and the relationship between environmental availability of osmoprotectants and the type of osmolyte(s) serving as the dominant contributors to intracellular osmoregulation.

In a variety of types of cells, hyperosmotic stress leads to an initial rise in the concentration of inorganic ions, notably, K^+. This is shown for the green alga *Chlorococcum submarinium* in Figure 10.1. Increases in ambient salinity led to slight increases in intracellular K^+ concentration, and decreases in salinity led to modest decreases in K^+ (22). Much larger increases in intracellular inorganic ion concentrations accompany hyperosmotic stress in many bacteria. For example, in cultures of growing *E. coli* K-12 cells in which 0.5 M NaCl was added to the growth medium, large amounts of K^+ were rapidly taken up from the medium (68). Concomitantly, the cells synthesized large amounts of glutamate, with final glutamate concentrations approximating one-half the $[K^+]_i$. The pairing of K^+ uptake and de novo synthesis of glutamate effected stabilization of net charge in the cell. The maintenance of charge neutrality by simultaneous accumulation of an organic solute with a net charge and an inorganic ion with the opposite charge may be common in bacteria and archaebacteria. A methanogenic archaebacterium, *Methanothermus sociabilis,* accumulated K^+ and cyclic-2,3-diphosphoglycerate (which bears three negative charges) at a 1:3 ratio during growth at high temperature (109).

Although the initial response to hyperosmotic shock in growing *E. coli* K-12 cells involves elevations in K^+ and glutamate concentrations, within 30 minutes the cells began to synthesize the compatible sugar trehalose (68). Within approximately 2 h, trehalose had replaced K^+ and glutamate as the dominant osmolyte in the cell. However, if proline was available in the medium, it was preferentially accumulated and served as the major intracellular osmolyte. Thus, the choice of organic osmolyte is seen to reflect in part the environmental availability of compatible solutes (or their precursors).

Further insights into the relative advantages of different types of organic osmolytes have come from studies in which more than one class of organic osmolyte is present in the medium as a potential osmoprotectant. Such studies have shown that a clear preference exists in the type of solute accumulated, preferences that reflect the effects observed in in vitro studies of osmolyte effects. In several different species of bacteria glycine betaine was accumulated in preference to any of the other potential osmoprotectants included in the medium (proline, ectoine, mannosucrose, and N_e-acetyl-β-lysine (59). Glycine betaine is typically the most effective osmoprotectant in reducing osmotic stress in bacteria (138). In studies of osmolyte effects in vitro glycine betaine is seen to be effective in stabilizing protein structure and in counteracting the effects of salts and urea on protein structure and function. The osmoprotective capacities of glycine betaine are further reflected in the occurrence of effective, salt-regulated uptake systems for this osmolyte (see later under Regulation of Uptake Systems for Osmoprotectants).

The progressive shift during osmotic regulation from less to more compatible osmolytes has been shown very clearly for the euryhaline cyanobacterium *Synechocystis* PCC6714 (176). When cells were shifted from freshwater to 0.5 *M* NaCl, a rapid influx of NaCl occurred in approximately 2 min. The increased intracellular Na^+ allowed maintenance of cell volume. Over the next 20 min, Na^+ was exchanged for K^+, a more compatible inorganic ion. The two organic osmolytes sucrose and glucosylglycerol then began to build up in the cell, and K^+ simultaneously declined. Over time, the latter, more compatible osmolyte became the dominant osmotic agent in the cells. Results like these show that during the initial period of cell volume regulation it may be important to use less than optimally compatible solutes to ensure that the cell is not damaged by the hyperosmotic stress. Later, when the uptake and synthetic pathways required for accumulation of compatible solutes are activated, the intracellular milieu can be restored to a state more conducive for optimal macromolecular structure and function.

Nutritional Determinants of Osmolyte Selection. Although the selective accumulation of certain types of organic osmoprotectants from the medium reflects their favorable effects on cellular function, other factors can militate against the use of favorable osmoprotectants under certain nutritional circumstances. In media where nitrogen is limiting, sugar or polyol synthesis may play a dominant role in hyperosmoregulation in bacteria, such that the nitrogen found in choline, glycine betaine, or proline betaine can be exploited for other purposes (81). Nutritional effects on choice of osmolyte also apply in the case of organisms that can synthesize both nitrogen-containing and non-nitrogenous osmolytes. The halophilic phototrophic bacterium *Ectothiorhodospira halochloris* is capable of synthesizing ectoine, glycine betaine, and trehalose (81). When cells were grown under nitrogen-limiting conditions, trehalose partially replaced glycine betaine in the intracellular osmolyte pool, and ectoine was catabolized. The fall in glycine betaine concentration appeared to stem from reduced synthesis; betaine seemed not to serve as a nitrogen source for other processes. Under hyperosmotic conditions in some bacteria the catabolism of the nitrogen-containing osmolytes choline, glycine betaine, proline, and proline betaine is inhibited (89, 167). Hypoosmotic shock can lead to the rapid excretion of glycine betaine (59, 174, 175).

There may be complex interactions between the transport, de novo synthesis, and catabolic processing of different organic osmolytes. For example, in *Ectothiorhodospira halochloris* the enzyme trehalase, which hydrolyzes trehalose to glucose, is activated by glycine betaine and inhibited by salts (110). When glycine betaine is present the K_m of the enzyme for trehalose is reduced from approximately 0.5 *M* to 0.16 *M*, and the inhibitory effects of salts are partially reversed. In this organism, then, the presence of glycine betaine may facilitate the degradation of trehalose to glucose. This type of regulation could be important if the cell were first to accumulate trehalose and, subsequently, be capable of taking up glycine betaine as an osmoprotectant from the medium. The two glucose moieties of trehalose could be added to a storage polysaccharide, for use in subsequent osmotic responses or for nutrient purposes.

The interplay between selection of osmolytes and nutritional conditions observed in many bacteria is also seen in comparisons of different plants. Plants, as photoautotrophs, typically rely heavily on sugar and polyhydric alcohol osmolytes under conditions of hyperosmotic stress. These osmolytes can be readily generated, either directly or indirectly, from photosynthetic products. Starch reserves can quickly be converted into osmotically active solutes like glycerol, as shown for the unicellular green alga *C. submarinum* (Fig. 10.1). In another unicellular green alga, *Dunalie-*

lla tertiolecta, glycerol concentrations also reach maximal levels within about 30 min due to a rapid breakdown of starch reserves and subsequent metabolism of the released sugar to the polyol glycerol (21). During hypoosmotic conditions, glycerol may be catabolized or released to the medium. Note that in *C. submarinum* the reciprocal change in starch and glycerol concentrations seen during hyperosmotic stress was not found under hypoosmotic stress (Fig. 10.1).

Other storage compounds in plants may also yield osmotically active solutes. In the alga *Poterioochromonas malhamensis* the storage polysaccharide chrysolaminarin is rapidly converted to isofloridoside (galactosylglycerol) (122). A key regulatory enzyme in this process is isofloridoside phosphate synthetase. This enzyme normally exists in an inactive state in the cell, and requires limited proteolysis to be catalytically active. Hyperosmotic stress causes proteolytic activation of the enzyme, and, thus, the production of isofloridoside.

One of the principal nutrient-related differences in osmoregulatory strategies found among plant species involves the role of nitrogenous osmolytes like glycine betaine and proline. Aquatic plants, especially nitrogen-limited marine species, tend not to use nitrogen-containing osmolytes, but instead rely largely on sugars and polyols. Unicellular algae and macroalgae may also accumulate high concentrations of β-dimethylsulfoniopropionate (DMSP) (71, 123, 173). This sulfonium compound bears a strong structural similarity to methylated ammonium solutes like glycine betaine and β-alanine betaine (Fig. 10.2). The availability of sulfur in seawater (sulfate is near 28 mM in seawater) may favor synthesis of this solute in place of a methylated ammonium osmolyte. DMSP has been shown to be a compatible solute (162). However, the compatibility of DMSP is temperature-dependent: although compatible at low temperatures, at high temperatures DMSP can denature proteins, possible because of its ability to interact with hydrophobic groups (6, 162).

Vascular plants may accumulate substantial amounts of nitrogenous osmolytes (for example, glycine betaine) if the soil contains ample nitrogen, particularly if this nitrogen is in the form of osmoprotectants like betaine and choline (105, 151). The energetic advantages to plants of exploiting soil osmoprotectants rather than photosynthetic products are thought to be large, perhaps up to tenfold (171). It should be noted that organic osmolytes are thought to be compartmentalized almost entirely in the cytoplasmic compartments of plant cells; osmolyte concentrations in vacuoles are likely to be low. This uneven distribution of osmolytes in different cellular compartments creates experimental difficulties in determining precise physiological concentrations of osmolytes in plant cells (that is, in the cytoplasm) (151). Thus, estimates of organic osmolyte concentrations based on whole cell water or volume are apt to seriously underestimate cytoplasmic osmolyte concentrations.

Regulation of Uptake Systems for Osmoprotectants. Although in both bacterial and plant cells it is energetically less costly to transport an osmoprotectant into the cell than to synthesize compounds like trehalose or glycerol, there are nonetheless some costs entailed in osmoprotectant uptake. One of the initial costs of osmoprotectant accumulation may be the synthesis of the transporters needed to effect rapid uptake of the solute(s) in question. In bacterial systems several transport systems for osmoprotectants have been identified and partially characterized in terms of their localization, kinetic properties, and responsiveness to changes in extracellular osmolarity. The *proU* operon of *E. coli* and *Salmonella typhimurium* encodes a low-affinity transport system for proline and a high-affinity system for glycine betaine (59). This system is strongly (200--700-fold) induced by increases in salinity, and the induction is observed within a few minutes of the osmotic shock (58, 59). Although the mechanisms by which the *proU* system is induced by high salinity are not fully understood, there is evidence that the high concentrations of K-glutamate that *E. coli* builds up to maintain turgor pressure during the early phases of osmotic adaptation can stimulate enhanced transcription of the *proU* operon and, thereby, uptake of glycine betaine to replace the K-glutamate. In an in vitro study of regulation of the *proU* operon that employed a DNA template and RNA polymerase, elevated concentrations of K-glutamate increased transcription of genes regulated by the *proU* promoter (170); transcription of other genes was not enhanced.

Transport of osmoprotectants into the cell need not entail transcription and synthesis of new protein. The ProP transport system of *E. coli*, which is a permease for both glycine betaine and proline, is synthesized constitutively, but its activity is modulated by external salinity (59). As in the case of salinity-dependent gene expression, the mechanism of salt modulation of enzymatic activity is not fully understood. Based on available data, a model has been developed that incorporates the effects of membrane stretch and ligand modulation of the protein (59). Stretch-activated membrane transport systems may be of widespread importance in regulating osmolyte transport in many types of procaryotic and eucaryotic cells (45).

Osmolyte Regulation in Invertebrates

The two primary sources of osmolytes described for bacterial and plant cells, uptake from the medium and synthesis within the cell, exist as well for osmoconforming marine invertebrates, which use free amino acids as the principal type of organic osmolyte (Table 10.1). Exogenous sources of free amino acids are of two major types. Soft-bodied invertebrates (but not hard-bodied species like crustaceans) are highly effective at taking up dissolved free amino acids from seawater (146, 226, 227). To some degree these free amino acids must contribute to the osmotic regulatory abilities of the organism (56), much as they may provide a substantial amount of the organisms' nutrition (for example, in larval stages) (146). Dietary sources of amino acids and other organic osmolytes are also common, notably for carnivorous species that feed on prey rich in protein, glycine betaine, or urea.

The production of free amino acid osmolytes within cells occurs via several processes. Protein turnover is one. Although it is debatable whether invertebrates rely significantly on special osmolyte-generating polymers (45), molecules that would serve a function analogous to that served by starch in unicellular green algae, there is evidence to suggest that protein degradation contributes importantly to the production of free amino acids. Normal protein turnover would, of course, lead to the availability of substantial quantities of free amino acids which, if not reused for protein synthesis or otherwise metabolized, could assume an osmolyte role. In the bivalve mollusc *Geukensia demissa*, when proteolysis was inhibited free amino acid accumulation also decreased (62). Aminopeptidase activity was found to correlate with accumulation of free amino acids. In *Mytilus edulis,* allozyme variants of aminopeptidase I that differed in catalytic activity were correlated with differences in ability to accumulate free amino acids (63, 112). Populations of *M. edulis* found in high salinity habitats had a higher frequency of the gene encoding the catalytically more active aminopeptide I allozyme (129).

De novo synthesis of amino acids is also known to play an important role in hyperosmotic regulation (38–40, 45). Glycogen could serve as a type of osmolyte storage compound in the sense that activation of glycogen catabolism through the glycolytic sequence could yield large amounts of the alpha-keto acid pyruvic acid, which can be converted to alanine through transamination, or to larger amino acids such as proline, glutamate, and aspartate through carboxylation by the reaction catalyzed by malic enzyme (38). Alanine, along with other nonessential amino acids such as proline and glycine, is commonly one of the amino acids exhibiting the most rapid and largest response to salinity change (62, 63).

The full complement of enzymatic machinery involved in synthesizing free amino acids during hyperosmotic stress remains uncertain. Earlier studies focused on the role of glutamate dehydrogenase (GDH) (see ref. 87 for review). It was postulated that activation of GDH by rising intracellular concentrations of sodium ion could supply increased amounts of glutamate which would facilitate transamination reactions, which use glutamate as a source of amino groups, and serve as a precursor for the synthesis of proline. More recent studies have cast doubt on a pivotal role for GDH, although glutamate does serve as a precursor for proline during hyperosmotic stress (for example, in the copepod *Tigriopus californicus*) (39). Further studies of proline synthesis in this species showed that proline accumulation could be blocked with inhibitors of protein synthesis (40). These results suggest that one or more of the enzymes required for synthesis of proline must be induced by salinity stress for enhanced synthesis of proline to be possible.

The suite of free amino acids accumulated in marine invertebrates varies considerably from species to species and from organ to organ within a species (25, 87). The choice of free amino acid accumulated may also be influenced by dietary status of the organism. Goolish and Burton (93) reported that specimens of *T. californicus* with high lipid contents accumulated significantly more proline than alanine, whereas animals with low lipid contents or animals exposed to anoxia accumulated alanine in preference to proline. These data were interpreted in terms of the higher energy costs of proline synthesis relative to synthesis of alanine (93). These authors calculated that hyperosmotic stress of this copepod (a shift from 50% to 100% seawater) necessitated approximately 12% of daily energy use for osmoregulation.

Under hypoosmotic stress, some or all of the organic osmolytes decrease in concentration as the result of several mechanisms, including transport out of the cell and catabolism (45, 87). As in the case of the process of osmolyte accumulation, different species show different patterns of osmolyte loss from the intracellular pool during hypoosmotic regulation. In general, loss of organic osmolytes during hypoosmotic stress occurs more rapidly than their accumulation during hyperosmotic stress.

Osmolyte Regulation in Lower Vertebrates

The regulation of cell volume with organic osmolytes is prevalent in euryhaline elasmobranchs and teleosts, and in some amphibian species, notably, frogs and toads that undergo long periods of hibernation or estivation (92). In most of these species nitrogenous solutes that are dead-end metabolites (for example, taurine, urea, and TMAO) play dominant roles in cell volume regulatory responses to osmotic stress. Characteristically, the organic osmolytes used in volume regulation in a particular cell type are not metabolized by the cell (92). Thus, many of the most important volume-regulatory mechanisms entail the uptake into, or release from, the cell of osmolytes in response to changes in external osmolarity.

The regulatory mechanisms responsible for adjusting organic osmolyte concentrations in lower vertebrates are best understood in the case of volume regulatory processes in fish blood cells. In both euryhaline elasmobranchs and teleosts, amino acids, including β-alanine and amino acid derivatives like taurine, are regulated in concert with changes in extracellular osmolarity (45, 91). In erythrocytes of the skate *Raja erinacea* hypoosmotic stress led to an efflux of β-alanine and taurine (91). Hypoosmotic stress elicited a rapid increase in the production of diacylglycerol, an activator of protein kinase C (150). Further evidence that protein kinase C plays an important role in governing osmolyte efflux came from the demonstration that addition of activators of the enzyme such as phorbol ester or elevation of intracellular Ca^{2+} mimicked the effect of hypoosmotic stress on taurine efflux.

In teleost erythrocytes too, taurine transport is modulated by the osmolarity of the medium. In erythrocytes of an eel and a flounder, hypoosmotic stress led to an increase in amino acid permeability and loss of solutes such as taurine from the cell (75). Under these conditions a NaCl-dependent amino acid uptake pathway is inhibited (75).

The fates of the organic osmolytes that are released into the blood during hypoosmotic volume regulation include metabolism in the liver (β-alanine) and elimination via the kidneys (140). In both cases the processes appear to be self-regulating because of the relationships between the blood concentrations of the solutes and the K_m values of the transport systems. Thus, normal plasma concentrations of β-alanine in the skate *R. erinacea* are near 0.2–0.4 m*M*, and the K_m value of the hepatocyte uptake system is approximately 0.2 m*M* (140). Excretion of taurine by the teleost kidney involves both filtration and secretion (185).

In urea-rich organisms such as marine elasmobranchs, estivating lungfishes, and hibernating or estivating amphibians, urea is generally distributed similarly in different body compartments because of its ability to permeate most membranes (74). The gills of elasmobranchs represent a notable exception (26).

In urea-rich species, then, urea does not play a significant role in establishing osmotic balance between the intra- and the extracellular fluids.

The Mammalian Kidney

Among vertebrates, our knowledge of organic osmolyte regulation is greatest in the case of the mammalian kidney. The complex and medically important dynamics of osmotic regulation and excretion have led to an enormous literature and, of late, a strong focus on the regulatory systems that govern the intracellular concentration of urea and the organic solutes potentially capable of protecting the cell from the toxic effects of urea.

The regulation of osmolyte concentrations in the mammalian kidney must satisfy two requirements: the total osmolarity of the cellular fluids must be regulated to effect volume regulation, and the types of osmolytes accumulated must achieve the required counteraction of urea effects. Despite the large amount of study devoted to mammalian kidney physiology, it has only been during the past ten years that the roles of organic osmolytes, and the mechanisms by which their concentrations are regulated, have been clarified (34, 35, 233).

Osmolyte Distributions in the Kidney. For the mammalian renal medulla, early studies asked whether its major organic solutes were, in fact, of critical importance in maintaining renal osmotic balance in vivo (13). Several observations suggest that organic osmolytes are important in achieving both osmotic balance (volume regulation) and urea counteraction. First, NaCl and urea are generally found in an increasing gradient from renal cortex to inner medulla in antidiuretic animals (211). Because urea is thought to equilibrate across membranes, the main cause of osmotic imbalance for a medullary cell should be variations in NaCl, which is thought to remain largely extracellular (12, 18, 19). To maintain cell volume, the total intracellular concentration of all osmolytes other than urea should exhibit a gradient similar to that for extracellular NaCl. In general, this prediction is met by the renal contents of polyols and methylamines, as illustrated for the kidney of antidiuretic rabbit (Fig. 10.11; *left panels*): total sorbitol + inositol + glycine betaine + GPC follows a gradient quite similar to that for NaCl, but not for urea, in kidneys sectioned into seven regions from cortex to papillary tip (in these studies, the data are for total contents of extracted kidneys, rather than intra- or extracellular concentrations). Several other studies, in which Na^+ was not measured, also show similar gradients for organic osmolytes (9, 102, 222).

As another test of the prediction that total non-urea osmolytes should increase in parallel with sodium chloride concentration, osmolyte contents have been determined in kidneys of the laboratory rat and in three wild rodent species ranging in water adaptation from the xeric pocket mouse (which never drinks water) to the mesic montane vole. In Figure 10.11 *(upper right panel),* total contents of methylamines plus polyols are plotted against sodium contents for kidneys of antidiuretic animals (held 2 or more days without water). The correlation is remarkably similar across all kidney regions (cortex to papilla) and species.

To function as osmolytes above the 150 mM Na$^+$ found in "normal" mammalian plasma (202), concentrations of methylamines plus polyols should be held at about 1.7 times that of Na$^+$ (0.85 for Na$^+$ plus Cl$^-$) not 1.0, because inorganic ions have lower activity coefficients. To estimate this ratio, the slope of 0.66 found for contents (Fig. 10.11, *upper right*) must be corrected for differing extracellular and intracellular spaces, which are unknown for most species. However, in the rat, cells of cortex and outer medulla occupy about 60% of the space, but only about 33% in inner medulla (165). If these values are used to estimate concentrations in Figure 10.11 (for example, for inner medulla by dividing total methylamines plus polyols by 0.33 and Na by .67), the slope becomes 1.5 ($r^2 = 0.70$) not far from the 1.7 predicted (see also ref. 232), with a Na$^+$ intercept of 177, not far from "normal" blood at about 150 mM Na$^+$.

Changes in Osmolyte Concentrations. A second prediction about the ratios of Na$^+$ and non-urea organic osmolytes can be made: as renal sodium levels vary within an animal due to changing water or salt intake, polyols and methylamines should be up- or down-regulated to maintain the same ratio as seen in the antidiuretic animals. Several studies have shown this prediction to be met: animals given access to water (sometimes sweetened) or in other ways made diuretic generally have reduced contents of renal polyols and methylamines compared to water-deprived animals (8, 9, 83, 102, 222, 224, 232, 234, 238). Results for various mammals are plotted in Figure 10.11 *(lower right panel);* again, the correlation to [Na$^+$] is very similar. Recently, taurine has been confirmed to act as a variable osmolyte in rat kidneys in similar experiments (160).

In vitro preparations confirm these results: in freshly isolated tubules from rat inner medulla, cell contents of GPC, sorbitol and inositol (betaine was not measured) varied directly with extracellular NaCl, but not with urea (223). Similar results were found for sorbitol (no other osmolytes were measured) in freshly isolated cells from rat inner medullary ducts (101).

Other disturbances to renal osmolytes also reveal

that the medullary cells appear to sense cell volume changes and to regulate organic osmolyte concentrations appropriately. As discussed in the section on osmoprotection and osmolyte compatibility in vertebrate cells, suppression of one osmolyte, sorbitol, by aldose reductase inhibition triggers the renal medulla to up-regulate glycine betaine in apparent compensation for reduced sorbitol levels (Fig. 10.7C, sorbinil group). In another in vivo study, the polyol accumulation of diabetes was simulated by galactosemia, in which high dietary galactose results in elevated galactitol accumulation in tissues with aldose reductase activity. In galactosemic rats, the high galactitol in the renal inner medulla resulted in reduced contents of sorbitol, inositol, and GPC in apparent osmotic compensation, and in greatly reduced urea contents (23, Fig. 10.7C). In galactosemic rats given an aldose reductase inhibitor, urea content of medulla remained low, galactitol content was reduced such that the sum of galactitol + sorbitol was the same as sorbitol in control animals, but inositol and GPC contents remained depressed while betaine contents doubled. Whether osmotic balance was maintained is not clear, and interpretation is confounded by the apparent need for the kidney to maintain a constant urea:GPC ratio for counteraction. It is likely, therefore, that renal cell volume is held roughly constant by regulation of total polyols and methylamines. However, individual osmolytes may vary in different, poorly understood ways. In particular, sorbitol is almost absent from outer medulla, while *myo*-inositol is highest in outer medulla in rabbit (Fig. 10.11, *left panel*) and all other species examined (19, 83, 223, 232). The reason for these patterns is not known; however, the sum of the two polyols follows a gradient very similar to that for Na$^+$.

Despite these differences in individual osmolytes, total cellular osmolytes must be regulated by the kidney to track extracellular osmotic changes. How this is accomplished is unknown. The regulatory mechanisms for individual osmolytes in vivo are also poorly understood, but a general pattern is emerging: the inner medulla can rapidly release organic osmolytes when exposed to a sudden decrease in osmolality, but can only slowly accumulate them over hours to days when exposed to a rapid increase (19, 224). This pattern and regulatory processes involved have been more thoroughly elucidated in vitro. Figure 10.12 summarizes the known regulatory processes in renal cells.

Sorbitol has been most thoroughly studied. It is produced from glucose by aldose reductase and metabolized to fructose by sorbitol dehydrogenase. In vivo, aldose reductase (54) and its mRNA (55) are present in a renal gradient matching that of sorbitol content; sorbitol dehydrogenase is found in an inverse pattern

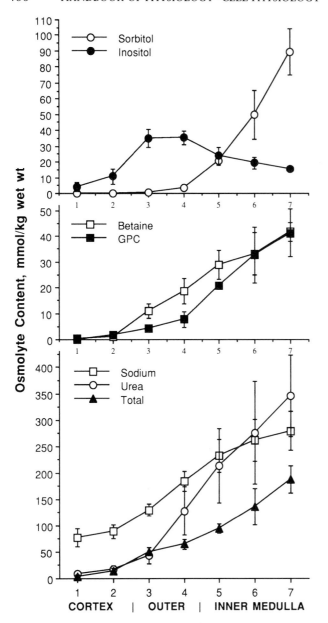

FIG. 10.11. *Left panels.* Distributions of osmolytes in kidneys of antidiuretic rabbits (held without water for 2 days). The kidneys were cut into seven sections, from the outer cortex *(section 1)* to the tip of the inner medulla *(section 7),* as shown on the abscissa. "Total" in the bottom frame indicates the sum of methylamines + polyols. [Redrawn after Yancey and Burg (234).] *Right panels.* Correlations between renal contents of sodium and total methylamines and polyols for rat, rabbit, and wild rodents (mesic montane

vole, *Microtus montanus;* deer mouse, *Peromyscus m. gambeli;* and xeric desert pocket mouse, *Perognathus parvus).* Each point represents four to seven animals. The upper points in each frame represent values for the inner medulla, the middle points are for the outer medulla, and the lowest points are for cortex. The upper panel is for antidiuretic animals and the lower panel is for diuretic animals (given water for 3 or more days); "x" values in lower panel are data replotted from upper panel. [Redrawn after Yancey (233).]

(highest in cortex). Antidiuretic treatment (of diabetes insipidus rats) results in increased aldose reductase mRNA, enzyme and product (sorbitol) in inner medulla, changes that take several days to complete. Reduction of sorbitol in normal rats made diuretic may result from cellular release, a more rapid process than synthesis and accumulation (55).

In vitro, sorbitol accumulation in PAP-HT25 cells is induced by hypertonic media (but not if a membrane-penetrating solute such as urea or glycerol is used). The accumulation is again slow, taking 4 days to reach a maximum (10). This is preceded by an increase in aldose reductase mRNA content, which takes 1 day to reach a maximum, and in aldose reductase enzyme content, which takes 3 days to reach a maximum. Control appears to be largely at the level of aldose reductase gene transcription (measured by nuclear run-on), which is activated 17-fold 12 h after switching cells from isotonic to hypertonic medium (192).

The trigger for transcriptional control is not known. However, in time course and ouabain-utilizing (that is, Na^+-K^+-ATPase inhibition) studies on adaptation to hyperosmotic media, the amount of aldose reductase activity in PAP cells correlated perfectly with cell ionic strength measured as the sum of $K^+ + Na^+$ (a weaker correlation was found with K^+ alone, and no correlation was found with Na^+ alone) (209, 210). Thus, as suggested by studies of salt effects on gene transcription in bacteria, a direct effect of changes in ionic strength during cell volume regulation on gene expression might occur (111).

In contrast to the slow up-regulation during hyperosmotic adaptation, PAP cells suddenly exposed to lower osmolality increase their sorbitol permeability within 30s (190), and 30% of cell sorbitol is released within 10 min (153). Permeability appears to be through a specific apical polyol "permease" with specificity: sorbitol > mannitol > inositol >> L-glucose (190). The permeability is calcium-dependent in primary renal cultures and may involve fusion of channel-laden vesicles with the plasma membrane (125). In PAP cells, permeability is inhibited by inhibitors of arachidonic acid metabolism and of cytochrome P450 (80), which implicates an arachidonic acid metabolite of the cytochrome P450 pathway as a second messenger (80, 83).

Inositol in the mammalian kidney is taken in by a basolateral, Na^+-dependent membrane transporter in renal slices and MDCK cells (83) and may also be synthesized. It is obtained from the diet and by synthesis from glucose in the liver (83). Hypertonically stressed MDCK cells show an increase in the V_{max} of the transporter, which may be the result of transcrip-

tional control. Recently, the cDNA for the MDCK transporter was cloned, shown to induce inositol transport in *Xenopus* oocytes, and used as a probe to show increase of transcript following hypertonic shock (131). Again, accumulation of inositol following the shock takes several days, while upon return to lower osmolality, inositol is released rapidly (20% within 1 h) (159). This release is through a channel whose activation is blocked by Ca^{2+}-channel blockers and inhibitors of cytochrome P450 (11).

Glycine betaine is also taken up from the blood by a basolateral Na^+-dependent transporter (100), obtained from the diet or liver, or synthesized in the kidney itself by the choline oxidase system (83). Patterns of up-regulation of the transporter in hyperosmotically shocked MDCK cells are virtually identical to those for inositol (83, 230). The cloning of the cDNA for the transporter and its enhanced expression under osmotic stress has been done as for the inositol transporter (230). There is no evidence for increased synthesis of glycine betaine under hyperosmotic conditions, so uptake may be the most important means for increasing glycine betaine concentrations. The time courses of uptake in hyperosmolality and release after switching to hypoosmolality are similar to those for polyols (83, 157); efflux is also dependent on calcium and cytochrome P450 (11).

Glycerophosphorylcholine is probably regulated by control of the degradation pathway, at least during up-regulation. It is an intermediate in the membrane phospholipid pathways, and probably is derived in vivo from phosphatidylcholine and hydrolyzed by GPC:choline phosphodiesterase (242). Regulation in the renal medulla is unclear, but MDCK cells require choline in the medium (159). Exposure to elevated NaCl or urea concentrations suppresses the MDCK cells' GPC: choline phosphodiesterase without altering GPC synthesis (242). GPC rapidly decreases after hypoosmotic shock, probably by efflux via an unknown membrane pathway (159).

Taurine also uses a basolateral Na^+-dependent transporter, with a V_{max} that is increased in hyperosmotically shocked MDCK cells (210). Transcription and/or RNA regulation are again involved: the transporter gene has been cloned and used to show that the taurine transporter mRNA increases in MDCK cells exposed to hypertonicity (210).

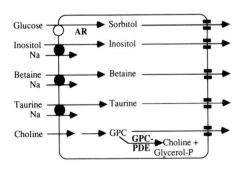

FIG. 10.12. Summary model of organic osmolyte accumulation and loss in mammalian renal medullary cells. The kinetics of the different processes differ. When extracellular osmolality is changed, activities of transporters and enzymes change slowly *(left side of model)*, but permeability to solutes changes rapidly *(right side of model)*. [Model after Burg (34).]

Stress Protein Induction in Hyperosmotic Stress

In Madin-Darby kidney cells hyperosmotic NaCl (but not glycerol or urea) led to a large (four- to five-fold) increase in the mRNA for heat shock protein 70 (Hsp70) (52). This observation suggests that during the early stages of adaptation to hyperosmotic stress, one or more of the heat shock (stress) proteins, a group of proteins with varied functions, including the control of protein folding and aggregation, may play an important role in protecting the cell from the effects of high solute concentrations. These stress proteins may be critical during the early phases of osmotic regulation—that is, during the period when compatible solutes are still being accumulated to restore cell volume.

Because stress proteins are ubiquitous, it is possible that they are important not only in mammalian kidney cells, but also in a wide variety of other species that experience osmotic stress. Stress proteins may be most important in species or cells that show substantial, but transient, increases in inorganic ion concentrations during osmotic stress, because of the perturbing effects of these ions on macromolecular systems. Broader study of the role of stress proteins in osmotic adaptation is clearly warranted.

MECHANISMS OF SOLUTE EFFECTS—AND NON-EFFECTS

The Hofmeister Series and Organic Osmolyte Structures

The fact that a given type of organic osmolyte typically has effects on proteins that are independent of both the type of protein being examined and the species from which the protein was isolated suggests the following important conclusion: osmolyte effects—or, as more commonly found, non-effects—on protein structure and function derive from common and fundamental aspects of protein structure and from basic interactions between proteins and the solvent and solute. The basic features of protein–water–solute interactions are now fairly clear, and they provide a broad and firm conceptual foundation for interpreting osmolyte effects. Studies during the past two decades have shown, first, what the important chemical structures are that make an organic solute "fit" for use as an osmolyte and, second, what the occurrence of these chemical features means in terms of establishing favorable or unfavorable interactions among proteins, solutes, and water.

M. E. Clark (48, 51) and Wyn Jones et al. (229) were the first to propose a basis in structural chemistry for the natural selection of organic osmolytes. They drew attention to the similarities in structure between the compatible free amino acid and methylamine osmolytes, on the one hand, and the protein structure–stabilizing ions of the Hofmeister series (112), on the other (Fig. 10.2 and Fig. 10.13). The Hofmeister series is an empirical ranking of solutes based on their dif-

THE HOFMEISTER SERIES

$(\text{---Stabilizing}$ $\text{Destabilizing---})$

ANIONS F^- PO_4^{3-} SO_4^{2-} CH_3COO^- Cl^- Br^- I^- SCN^-

CATIONS $(CH_3)_4N^+$ $(CH_3)_2NH_2^+$ NH_4^+ K^+ Na^+ Cs^+ Li^+ Mg^{2+} Ca^{2+} Ba^{2+}

FIG. 10.13. Hofmeister series of ions.

fering tendencies to stabilize (precipitate) or destabilize (solubilize) proteins (53, 114, 215).

Hofmeister series effects bear strong resemblance to several of the key properties of organic osmolytes. The Hofmeister series ranking of solute effects is generally independent of the protein being studied, a phenomenon observed in the study of in vitro effects of organic osmolytes. The effects of Hofmeister solutes show algebraic additivity: addition of a destabilizing ion plus a stabilizing ion, at a precise ratio of concentrations, will lead to no net effect on protein structure. Thus, solute counteraction, like solute compatibility, may find explanation in terms of the Hofmeister series.

As is apparent from comparing the structures shown in Figures 10.2 and 10.13, the commonly occurring free amino acid and methylamine osmolytes bear groups that are structurally similar to the stabilizing ions of the Hofmeister series: methylammonium ions, NH_3, COO^-, and $PO_4^=$. To a first approximation, the free amino acid and methylammonium (and methylsulfonium) osmolytes will favor protein structural stability. However, among the organic osmolytes there are clear differences in structure-stabilizing ability. For example, the extent of nitrogen methylation influences the stabilizing effect of a solute (Figs. 10.7 and 10.10), with stabilization having the ranking: glycine betaine > sarcosine > glycine (Fig. 10.9) and $CH_3)_4N^+$ > $(CH_3)_2 H_2N^+$ > NH_4^+ (Fig. 10.13). The selection of a particular type or the set of organic osmolytes present in a cell thus may entail the question of how much stabilizing effect on proteins is advantageous relative to the chemical and physical conditions faced by the cell (see later under Favorable Effects of Compatible Solutes Not Related to Osmoregulation).

Differential stabilizing effects have also been observed in comparisons of nonionic solutes. Sugars and polyols may be stabilizing of protein structure (Fig. 10.5), and these solutes differ among themselves in a consistent fashion in their effects on protein stability. Thus the selective basis for their accumulation may be the same as that for amino acid, methylammonium, and methylsulfonium osmolytes: osmolytes are accumulated that do not disrupt protein structure, where "disruption" must be appreciated as including overstabilization as well as destabilization of structure.

Strong structure-destabilizing groups or molecules (for example, urea and the guanidinium group) are not common contributors to the organic osmolyte pool, except for cases in which the destabilizing compound (for example, urea) is paired with a counteractant like TMAO or GPC.

Before considering the physicochemical basis for the favorable properties of compatible organic osmolytes, it is critical to emphasize that the accumulation of solutes that are stabilizing of protein structure should not be taken to mean that a primary and common objective of organic osmolyte accumulation is the enhancement of intracellular protein stability. This erroneous interpretation of osmolyte adaptation (221) may be based on the failure to realize that at the relatively low concentrations at which organic osmolytes typically are found under most physiological conditions (for example, at concentrations no higher than about 300 mmoles/kg cellular water^{-1} in most cases) (237), all but the most stabilizing of compatible organic osmolytes are not apt to have a significant effect on protein structural stability or function (*cf.* Fig. 10.3). Thus, despite the fact that compatible solutes can effect substantial stabilization of protein structure at high concentrations, under most physiological conditions there is little reason to expect that an increase in intracellular osmolarity due to organic osmolyte accumulation will lead to more rigid proteins or more stable protein–protein interactions. In fact, the organic osmolytes that accumulate to the highest concentrations in the most osmotically concentrated cells are solutes that have relatively low stabilizing effects on proteins. For example, in the unicellular green alga *Dunaliella,* in which glycerol accumulates to concentrations in excess of 2 molal under osmotic stress, the effect of this buildup in organic osmolyte concentration on protein structure is apt to be small (see Fig. 10.5). As emphasized earlier, a primary adaptive advantage of using compatible organic osmolytes is the ability of the cell to vary their concentrations widely without significantly affecting the structures of functions of macromolecules.

Cases where protein stabilization is favored by organic osmolyte accumulation include counteracting (or haloprotective) osmolyte systems, in which perturbation by one type of solute, urea or high salt, is offset by the presence of another type of solute that has the opposite effect on protein structure or function. The algebraic additivity principle found with Hofmeister series ions is illustrated by the counteracting solute mechanism. Organic molecules may also contribute to protein and membrane stabilization in contexts different from those of cell volume regulation—for example, during extreme desiccation and in the face of freezing or heat stress (see later under Favorable Effects of Compatible Solutes Not Related to Osmoregulation).

Preferential Exclusion of Compatible Osmolytes from the Protein Surface

Timasheff and colleagues (7, 205, 206) have discovered the physicochemical basis of the varied effects of different solutes, including organic osmolytes and ions of the Hofmeister series, on protein structure. Their hypothesis, the preferential exclusion (or preferential hydration) hypothesis, accounts for the differential effects of a wide variety of stabilizing and destabilizing solutes in terms of the tendency of the solute either to be excluded from, or to interact with, the surface of a protein. They have demonstrated that the similar effects on protein function and structure seen with the stabilizing ions of the Hofmeister series and the compatible organic osmolytes are due to this common physicochemical mechanism. Thus, the hypothesis of Clark (48) and Wyn Jones (229) that the "fitness" of organic osmolytes is a consequence of their physicochemical similarities to stabilizing Hofmeister series ions has been substantiated.

The fundamental discovery of Timasheff and colleagues is that both the inorganic ions and the organic solutes that stabilize protein structure are preferentially excluded from the immediate hydration shell around a protein. Thus, the protein is preferentially hydrated, and it accumulates an adjacent layer of water molecules in which stabilizing organic solutes are less concentrated than in the water that is distant from the protein surface (Fig. 10.14*B*). Destabilizing solutes such as SCN^-, urea, and guanidinium bind effectively to proteins and facilitate their unfolding. They do not show preferential exclusion from the water adjacent to the protein's surface, and may be more concentrated near the protein than in the bulk cell water because of their protein-binding abilities (Fig. 10.14*A*).

Why should the degree to which a solute is excluded from the immediate hydration layer around a protein be correlated closely with the solute's effect on protein stability? The thermodynamic hypothesis developed by Timasheff and coworkers focuses on the entropy of the entire system: the protein, the solute, plus the water. If a solute is excluded from the water immediately adjacent to the protein surface, a local decrease in system entropy takes place because the solute will be distributed less randomly throughout the entire aqueous phase. Thermodynamically, this is an unfavorable situation. However, if the protein were to unfold and expose even more surface area, the situation would be even worse from a thermodynamic perspective. With the exposure to solvent of a larger area of protein

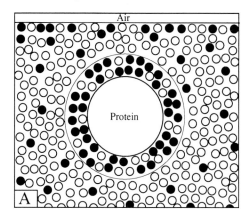
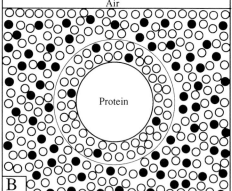

FIG. 10.14. Preferential hydration model of Timasheff, illustrating distribution of water *(open circles)* and solute *(closed circles)* following equilibrium dialysis of a protein. Solutes that show preferential binding to a protein (that is, destabilizing solutes) occur at higher concentration inside dialysis bag than in solution outside bag *(frame A)*. Structure-stabilizing solutes are excluded from the water near protein and thus occur at higher concentrations outside dialysis bag *(frame B)*. Because structure-stabilizing solutes increase the surface tension of water, they also tend to be excluded from the water–air interface. [Redrawn after Timasheff (205).]

surface, even more water would be adjacent to protein surface and a larger fraction of total solvent would be unable to accommodate the solute effectively. A bad situation thermodynamically would be made even worse (Fig. 10.14). Therefore, solutes that are excluded preferentially from water near the protein surface favor the compact, folded state of the protein, and the aggregation of protein subunits. At sufficiently high concentrations, protein stabilizers favor protein precipitation, the maximal reduction in protein surface in contact with water, as shown by Hofmeister more than a century ago.

Structure-destabilizing solutes such as SCN^-, urea, and guanidinium bind effectively to proteins, for example, to peptide backbone linkages (Fig. 10.15). Structure-destabilizing solutes exhibit what is termed "preferential interaction" with proteins, as opposed to the "preferential exclusion" shown by stabilizing solutes (206). Because they are not excluded from the hydration layer around the protein, the preferential exclusion stabilizing mechanism outlined above does not apply. In fact, because denaturants bind well to proteins, the unfolding of the protein is favored because more surface area is made available for interaction with the denaturant. Hydrophobic solutes denature proteins because of their abilities to bind strongly to nonpolar side chains that normally may be buried within the interior of the protein.

The underlying physicochemical basis of the preferential exclusion of stabilizing solutes appears to involve several factors (205). Some solutes are extremely bulky and are excluded from the protein surface largely by geometrical (steric) factors. In some cases the charge

borne by a solute may lead to its repulsion from the protein surface. However, probably the most general mechanism responsible for preferential exclusion of stabilizing solutes involves changes in water structure induced both by the proteins and the stabilizing solutes. Most structure-stabilizing solutes enhance the structure of the water in their immediate vicinity. These structure-enhancing effects are manifested in several physical properties of water, for example, by increases in the surface tension of water and alterations in the mobility of water molecules, as reflected by decreased self-diffusion coefficients (49, 143). To a first approximation, then, if both protein and stabilizing solute are increasing the structure of the water immediately adjacent to them, it will be difficult for a stabilizing solute to interact with the same group of water molecules whose structure is being influenced by the surface of a protein. The stabilizing solute will be accumulated preferentially in the "bulk" solution, where surfaces are not already having a strong influence on water structure.

It is instructive to examine the structures of certain organic osmolytes to see how natural selection has led to particular chemical features that appear consistent with the preferential exclusion model. Most organic osmolytes are either uncharged polar molecules (for example, sugars and polyols) or zwitterionic solutes that bear no net charge. These molecules will not bind strongly to charged groups on proteins, nor will they bind to ligands in solution. A particularly interesting example of solute evolution is seen with N_ϵ-acetyl-β-lysine (Fig. 10.2). Lysine bears a net positive charge and is highly disruptive of protein function (Fig. 10.3) and structure. By removing one positive charge from the amino acid through acetylation, lysine can be converted to a compatible solute, N_ϵ-acetyl-β-lysine, which is found as an osmolyte in a methanogenic archaebacterium (132, Table 10.1).

Solutes with a large hydrophobic moment (for example, solutes with long alkyl chains like those of valine and isoleucine) are disruptive of protein structure because of their abilities to bind to nonpolar regions on proteins and thereby favor protein unfolding. Strongly hydrophobic amino acids are not found as organic osmolytes, partly because of their potential for binding to proteins, and also because of their limited solubilities relative to simple amino acids like glycine and alanine. The structures of several organic osmolytes appear to reflect selection for decreased hydrophobic moment. Despite having a five-membered carbon chain, proline has a small hydrophobic moment because of its ring

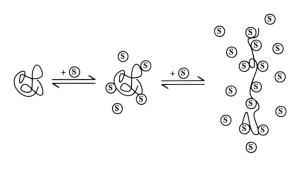

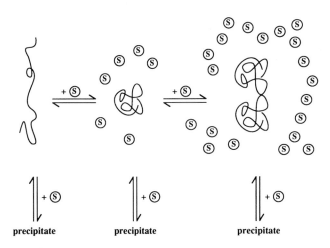

FIG. 10.15. The opposite effects of binding solutes and preferentially excluded solutes on protein conformation, assembly, and solubility. *Upper panel.* Binding solute *(S)* favors unfolding of the protein (native coil) to the denatured state (random, extended coil). *Lower panel.* Preferentially excluded solute *(S)* favors conversion of unfolded protein *(left)* into compact, folded conformation of the protein *(center)*. Higher concentrations of excluded solute favor enhanced protein–protein interactions *(right)*, and, at sufficiently high solute concentrations, protein precipitation. [Redrawn from Low (143).]

structure. Ectoine (Fig. 10.2) likewise is a cyclical amino acid derivative with a low hydrophobic moment. Hydrophobic moment can also be reduced by shifting the amino group from the alpha to the beta carbon, as found in β-alanine.

Solute Interactions with Ligands in Solution

The full range of solute effects that are involved in the selection of organic osmolytes includes more than protein–solute interactions. Some solutes appear to be incompatible with cellular function not because (or not only because) they interact with proteins, but because they interact strongly with ligands in solution. As shown in Figure 10.3, both lysine and arginine are strongly inhibitory of the interactions between pyruvate kinase and phosphoenolpyruvate. The large increases in apparent K_m of PEP induced by arginine and lysine are likely the consequence of binding of PEP by these two basic (that is, net positively charged) amino acids in solution (60). Complexes between PEP (or other phosphorylated ligands) and arginine or lysine would effectively reduce the concentration of free ligand available for binding to the enzyme. Thus, apparent K_m values will be high, and catalysis may be inhibited. In this context it is pertinent to note that the active sites of enzymes that bind phosphorylated ligands like PEP typically contain arginyl residues to stabilize the ligand's binding (177). The zwitterionic and uncharged compatible osmolytes thus are seen to be unlikely to perturb enzymatic function for a second reason: they will not interact with free ligands in solution, much as they fail to interact with the protein surface.

Nonreactivity of Modified Amino Acid Osmolytes

Another aspect of solute nonreactivity pertains in the case of the amino acid osmolytes that have structures that appear to reduce or preclude their interacting with amino acid-utilizing systems. The β-amino acids are unlikely to interfere with the metabolism of the α-isomers of the same amino acid. The β-amino acids also will not affect protein synthesis reactions. In some invertebrates D-amino acids may be important in osmotic regulation (169). The D-isomers will again be relatively nonreactive with the enzymes that are involved in amino acid metabolism and protein synthesis.

Monosaccharide Reactivity with Proteins

Another criterion for solute compatibility pertains in the case of sugars. Monosaccharides that exist largely or entirely in the open (carbonyl) configuration can interact nonenzymatically with the terminal amino groups of proteins to form Schiff bases. These covalent modifications may perturb protein function. Glucose has the least tendency to occur in the open configuration of any common hexose, yet even high glucose concentrations can lead to substantial protein glycosylation (33).

The common sugar and polyol osmolytes (Fig. 10.2) generally have ring structures. For instance, glucose may occur in the disaccharide trehalose. When free glucose does occur at high concentrations, it is in the case of freezing stress, for example, in hibernating frogs (69, 182, 203). At these low temperatures, the rate of nonenzymatic glycosylation is apt to be very low.

Favorable Effects of Compatible Solutes Not Related to Osmoregulation

Although the primary focus of this chapter is the role of organic osmolytes in volume regulation, many of the same classes of solutes that are accumulated to high levels during osmotic stress are also of importance in adapting cells to other types of stress, including freezing temperatures, extremes of desiccation as found in anhydrobiotic species, and extremes of high temperature.

The organic solutes accumulated in freeze-tolerant organisms appear to protect proteins during freeze–thaw cycles by their preferential exclusion from the protein surface (42). Solutes such as glycerol that accumulate to high concentrations in all body fluid compartments in freeze-tolerant insects (69) may thus prevent denaturation of proteins in the extracellular compartments; for example, the hemolymph, where ice formation can be tolerated in freeze-tolerant species.

In the case of extreme desiccation the preferential exclusion of solutes is not the key mechanism for stabilizing proteins and membranes. Organisms capable of undergoing extreme desiccation (for example, *Artemia* embryos, nematodes, and pollen grains) accumulate either trehalose or sucrose (41, 57). These polyhydroxy solutes are hypothesized to hydrogen bond to dried proteins and membranes as water substitutes, and thereby preserve the native structures of the molecules.

A role for compatible organic osmolytes in the stabilization of protein structure in thermophilic organisms or in heat-shocked cells of mesophiles has been suggested by several recent studies. Although as discussed above the use of compatible solutes under most physiological conditions is not based on a requirement for enhancement of protein stability, under extremes of high temperature the use of structure-stabilizing organic molecules may be an important mechanism for preventing thermal inactivation of proteins. In spores of the protist *Dictyostelium discoideum* elevation in intracellular trehalose content enhanced the heat stability of the cells (72). Heat shock of the yeast *Saccharomyces cerevisiae* led to a massive accumulation of trehalose and increased thermal tolerance (117). The potential roles that trehalose might serve in adapting cells to stress is reviewed by Van Laere (213).

An unusual example of how a dominant intracellular organic solute may contribute to thermal resistance has been reported for methanogenic archaebacteria of the genus *Methanothermus*. Culturing of cells at different temperatures led to large effects on intracellular K^+ concentrations and on the levels of the major organic solute, cyclic 2,3-diphosphoglycerate (cDPG) (109). In the most thermotolerant species examined, *Mt. sociabilis*, concentrations of K^+ and cDPG rose to 1060 and 320 mM, respectively, in cells cultured at 88°C. The potassium salt of cDPG was shown to be highly stabilizing of two thermally labile proteins, purified malate dehydrogenase and glyceraldehyde-3-phosphate dehydrogenase, of a *Methanothermus* species. Thus, the abilities of cells of these methanogens to grow at high temperatures appears to be due, at least in some important measure, to the stabilization by organic solutes of inherently thermally labile proteins. The protein-stabilizing effects of cDPG were not very different from the effects of potassium phosphate, a stabilizing Hofmeister salt, so cDPG appears to lack unusual stabilizing effectiveness. Surprisingly, it was found that cDPG did not stabilize the glyceraldehyde-3-phosphate dehydrogenase from a mammal. This finding suggests that, unlike almost all of the solute effects found for compatible solutes, the characteristics of the protein may help influence the effects of the solute.

In another extremely thermophilic archaebacterium, *Methanococcus jannaschii*, more than 90% of the soluble organic material was the β-amino acid β-glutamate (178). In these cells, too, a high concentration of a compatible solute may be serving dual roles: osmotic agent and thermal stability enhancer.

Inorganic Ions: Perturbation and Compatibility

Some of the basic principles that apply in establishing the "fitness" of organic osmolytes pertain as well for inorganic ions. Potassium ion is the dominant monovalent cation in almost all cells. Its selective accumulation may require considerable metabolic energy. These facts suggest that K^+ may be a more compatible inorganic ion than, say, Na^+. Wiggins (219) has hypothesized that the potassium ion's tendencies to be accommodated differentially into the different forms of cellular water underlie the ion's "fitness." K^+ has a lower tendency than Na^+ to interact with the water adjacent to surfaces, for example, protein filaments and enzymes. Thus K^+ is a better preferentially excluded ion than Na^+, as shown by its position in the Hofmeister series. Because of its low tendency to interact with water adjacent to proteins, K^+ can readily neutralize fixed charges, for example, carboxylate groups.

The effects of ions on protein function also differ, as shown by the Hofmeister series effects on enzymatic activity (215). For some enzymes (for example, pyruvate kinase) K^+ is required for activity, and Na^+ is inhibitory (25). Thus, K^+ can be regarded as a more compatible monovalent cation than Na^+.

In the context of the effects of inorganic ions on cellular function, even compatible ions such as K^+ can be perturbing at sufficiently high concentrations—for example, as shown by increases in K_m values (Fig. 10.3). It has been argued, however, that in bacteria grown under high osmolarities the accumulation of high concentrations of K^+ with the concomitant loss of cellular water may not always be perturbing because of the concentrating effect that dehydration of the cell will have on cellular macromolecules (43, 44). Cayley et al. (43, 44) grew *E. coli* in media with osmolarities ranging from 0.1 to 1.02 osm. The thermodynamic activity of K^+ increased from 0.14 M to 0.76 M over this range of media osmolarities, and cellular water decreased proportionately. The high intracellular activity of potassium ion would be expected to decrease the binding of DNA regulatory proteins to gene regulatory elements. However, this did not occur. These authors argued that the increased activity of the DNA binding proteins with loss of cellular water compensated for the increased activity of K^+ (43). The enhanced "molecular crowding" induced by dehydration would be expected to enhance association of cellular macromolecules (79).

Despite the apparent ability of some molecular systems (that is, DNA–protein interactions) to be sustained in the presence of high intracellular activities of K^+, it seems unlikely that the high K^+ levels induced by dehydration would be nonperturbing to most biochemical systems. Indeed, Cayley et al. (43) found that growth was slowed in the strongly dehydrated cells, perhaps as a result of salt perturbation of numerous salt-sensitive enzymes. The compensatory effects of molecular crowding thus may apply for some types of macromolecular association processes, but they appear unable to free the cell fully from perturbation by high intracellular ion activities.

EVOLUTIONARY PERSPECTIVES

Macromolecular vs. "Micromolecular" Evolution

Faced with osmotic stress and the need for volume regulation, organisms have several options available

for maintaining the structural and functional integrity of the cell. They may closely defend the osmotic composition and concentration of the cells, thereby maintaining cell volume and avoiding perturbation of macromolecular constituents that would result from shifts in intracellular solute and water content. Most vertebrates follow this general pattern, although the mammalian kidney displays a marked divergence from this strategy. The costs of defending the internal milieu via a close defense of the solute status quo include substantial energy expenditures for ion transport.

A second option is to conform osmotically by allowing large changes in the intracellular concentrations of inorganic ions, chiefly K^+. This pattern of adaptation to osmotic stress is rare. Although hyperoosmotic stress does lead to transient increases in inorganic ion concentrations in many species, for the most part these changes are relatively minor and short-lived. As compatible organic solutes are added to the cell, inorganic ion concentrations tend to return to prestress values. It is only in the extremely halophilic archaebacteria that inorganic ions play a dominant role in building up the intracellular osmolarity (134). These extremely halophilic bacteria may accumulate K^+ to concentrations of approximately 5 M in the cellular water (134). From what has been shown about salt effects on protein function and structure (for example, Figs. 10.3 and 10.4), it is apparent that enzymes of extremely halophilic archaebacteria must be very different from the analogous or homologous proteins of other organisms. In fact, the majority of proteins from extremely halophilic archaebacteria that have been studied not only function at vastly higher salt concentrations than those tolerated by enzymes from other species, but actually require these high salt concentrations for the acquisition of native structure and normal physiological function (134). The halophilic nature of these proteins is due to pervasive differences in amino acid composition relative to proteins of all other taxa. Proteins of extremely halophilic archaebacteria contain high percentages of acidic (glutamyl and aspartyl) residues, reduced percentages of strongly hydrophobic residues, and increased percentages of weakly hydrophobic residues. This highly unusual amino acid composition renders these proteins absolutely dependent on high intracellular cation (chiefly K^+) concentrations for titrating charged residues and for effecting the burial of weakly hydrophobic groups within the protein interior. Although these adaptations enable these cells to withstand extremely high external and intracellular salinities, they also lock the cells into a requirement for these conditions.

Species that are euryhaline osmoconformers, and organs such as the mammalian kidney that tolerate large changes in osmolarity, exploit the organic osmolytes that have been the primary focus of this chapter. Unlike the pervasive changes in protein structure that are necessitated by the use of high intracellular salt concentrations, modifications of the cellular biochemistry are quite minimal in the case of species that exploit compatible organic osmolytes. By using the appropriate "micromolecules," then, the species avoids the necessity of pervasive alteration of its macromolecular systems. The required adaptations for effectively controlling the concentrations of organic osmolytes are restricted largely to transport and synthesis systems that are involved in regulating the intracellular concentrations of these solutes. Membrane transporters for solute uptake, regulatory systems for controlling the activities of these transporters, and enzymatic reactions for generating or catabolizing organic osmolytes are the primary types of adaptations required for effective regulation of organic osmolyte concentrations.

Evolution of Osmolyte Molecules: An Overview of Principles of Selection

The vast majority of studies of molecular evolution have focused on macromolecules—DNA, RNA, and proteins. Relatively little attention has been paid to the selective advantages or disadvantages of different types of low-molecular-weight species ("micromolecules") such as osmolytes, and to the ways in which the basic physicochemical properties of macromolecules may have helped dictate what types of "micromolecules" are "fit" for accumulation in cells (16, 205, 237). Likewise, with a few exceptions (108, 219), the significance of the properties of water in governing the evolution of macromolecular systems has received relatively little attention. We have attempted to show that a comprehensive picture of biological evolution, as well as a firm understanding of critical aspects of cellular volume and osmotic regulation, can be obtained only by examining the interactions of all three of these critical cellular components, macromolecules, "micromolecules," and water.

Selection Based on Preferential Hydration Characteristics.
These complex and biologically critical interactions have been most cogently investigated in studies of the effects of different solutes on protein stability and water structure. Analysis by Timasheff and colleagues (205, 206) has provided a thermodynamic model, the preferential hydration theory, which explains the basis of the Hofmeister series effects on proteins and, thus, solute compatibility and solute counteraction. This model provides a general framework for analyzing how natural selection has shaped osmolyte systems, and for

explaining many of the specific osmolyte patterns found in cells—for example, the mixture of urea, methylamines, and polyols found in the inner medulla of the mammalian kidney. Differences among organic osmolytes likewise can be rationalized in terms of Timasheff's theory. For example, different degrees of structure-stabilizing activity may be advantageous in different contexts. Glycerophosphorylcholine in the mammalian kidney is a case in point. Structurally, GPC would appear to be the optimal type of stabilizing osmolyte from the standpoint of Hofmeister series properties. Glycerophosphorylcholine contains three structure-stabilizing groups: a fully methylated nitrogen (choline), a phosphate group, and a glycerol moiety. In the inner medulla of the mammalian kidney, GPC appears to be the principal solute used to counteract the effects of high urea concentrations. The concentration of GPC tracks changes in urea concentration more fully than the concentration of any other organic osmolyte. GPC is also found to be more effective than another methylammonium compound, glycine betaine, in counteracting the effect of urea on protein function (Fig. 10.8).

Similar considerations apply in the choice of polyols used in osmoregulation. In the mammalian kidney sorbitol and inositol are the chief polyols accumulated. Unlike glycerol, sorbitol and inositol can offset urea perturbation of some proteins. Sorbitol and inositol have especially strong protein-stabilizing effects (Figs. 10.8).

Although the physicochemical principles developed by Timasheff and others to account for osmolyte compatibility are generally applicable, physiological and environmental factors can also influence the types of organic osmolytes accumulated by a given organism, or by a particular type of cell.

Selection Based on Carbon and Nitrogen Sources.
As already discussed, one of the primary determinants of the types of organic osmolytes found in a cell is the organism's source of reduced (fixed) carbon. Photoautotrophic species typically rely on osmolytes that derive from sugars (Table 10.1). Storage polysaccharides are converted to osmotically active units (glycerol) when hyperosmotic stress occurs. However, not all photoautotrophic species rely exclusively on sugars and polyols. Although many aquatic species use sugars and polyols as the primary osmolytes under salinity stress, terrestrial plants with better access to nitrogen also use nitrogen-containing osmolytes such as glycine betaine.

Nitrogen availability in the diet is also reflected in the osmolyte composition of animal cells. In animals the use of sugars and polyols is much less common (Table 10.1), and is found chiefly in species facing freezing stress or extreme desiccation (anhydrobiosis). Because many animals have a diet rich in protein, the exploitation of nitrogenous waste products (for example, urea, glycine betaine, GPC and TMAO, and free amino acids) is easily rationalized. The use of nitrogenous waste products appears to be a very efficient mechanism for osmotic regulation because these terminal metabolites have no other metabolic function, and represent no further demand on the cell's energy metabolism. Nitrogenous end-products may be toxic, however, as the counteracting solute mechanism found in urea–TMAO, urea–GPC, and urea–sorbitol/inositol systems shows.

Selection Based on Cell Temperature. Temperature plays at least two roles in helping to govern the composition of organic osmolyte systems. First, considerations of solubility come into play when osmolytes having generally similar effects on macromolecules nonetheless differ in their solubilities. For example, the solubility of inositol (ring structure) is less than that of sorbitol (chain structure), so the latter solute is more fit for osmolyte function under conditions of low temperature.

A second temperature effect involves the temperature dependence of hydrophobic interactions (7). Hydrophobic interactions are enhanced as temperature is raised, at least up to temperatures of $50°–60°C$. The enhancement of hydrophobic interactions at elevated temperatures means that a solute with some hydrophobic character (hydrophobic moment) may be compatible at low temperatures, but disruptive of protein function at higher temperatures. For example, dimethylsulfoxide (DMSO) and the osmolyte DMSP both appear to disrupt protein structure at high temperatures, whereas at low temperatures they are stabilizing (7, 162). In view of the potentially negative effects of organic solutes with large hydrophobic moments, it is not surprising to find that the commonly occurring organic osmolytes are generally lacking in strongly hydrophobic regions (Fig. 10.2). Thus, organic osmolytes are not only beneficial for euryhaline osmoconformers because they can be varied widely in concentration without affecting cellular structures and functions, but they also are advantageous for eurythermal organisms whose cell temperatures span a wide range, the upper limits of which may establish dangerous potentials for disruptive hydrophobic interactions.

An organism's body temperature may influence the choice of osmolyte when solute perturbation of proteins is a threat. We hypothesize that the use of GPC as a urea counteractant in the mammalian kidney may in part reflect the high cellular temperatures of these species relative to those of urea-rich fishes, which use

TMAO as the primary counteractant of urea effects. To offset the effects of high urea concentrations, which may be more than twice those found in urea-rich fishes, plus high temperatures at which urea perturbation will be enhanced, a particularly powerful urea-counteractant may be required in the mammalian kidney's inner medulla.

Similar considerations may pertain in the case of selection of sugar and polyol osmolytes. In mammalian kidneys, sorbitol and inositol, which are especially strong structure stabilizers relative to other sugar and polyol osmolytes (Fig. 10.5), are accumulated. Algae found in high temperature brine pools accumulate polyols and sugars, which have negligible hydrophobic moments.

Because most studies of preferential hydration have been done near $20°C$ (7), there needs to be additional investigation of the temperature dependence of solute binding and exclusion to clarify the roles of adaptation temperature in the development of organic osmolyte systems.

Summary: The Adaptive Significance of Osmolyte System Evolution

The evolutionary patterns observed in macromolecular and "micromolecular" systems can only be understood when placed in a broad and synthetic frame of reference that includes the properties of water. Water not only provides the solvent medium for most metabolic transformations and transport processes; the physicochemical properties of water are a major factor governing the folding and assembly of proteins, nucleic acids, and membranes. Macromolecular surfaces in turn strongly affect the properties of the nearby water molecules (205, 206, 219). To a considerable extent these influences of macromolecules on water structure may determine the evolutionary fitness of different types of "micromolecules" for use as osmotic agents.

Physiological regulation of osmolyte composition and concentration must be viewed in an evolutionary context that incorporates several inherent properties of the proteins of a species. In a very general sense, the properties of proteins, including structural stability, subunit assembly characteristics, ligand binding ability, and catalytic activity, are set within ranges appropriate for physiological function through evolutionary changes in amino acid composition. The values for these parameters are, however, highly susceptible to perturbation by solutes, especially inorganic ions (113, 198). The threats to protein systems posed by water stress, whether this arises from changes in ambient salinity, from the synthesis of high concentrations of metabolic products like urea, or from other sources, are generally thwarted by the regulation of the osmotic composition and concentration of the intracellular milieu, such that the essential characteristics of proteins established through evolutionary processes are defended. The use of appropriate osmolytes to defend the evolutionarily selected characteristics of macromolecules and membranes can be viewed as a highly efficient evolutionary strategy in the sense that only a relatively small fraction of the cell's macromolecules, chiefly those responsible for regulating the synthesis and transport of osmolytes, need to be specifically adapted for the osmotic challenges faced by the cell. For the vast majority of proteins, no adaptation to osmotic conditions is necessitated.

The significance of this freedom of most proteins from a requirement for adaptation to osmotic conditions is perhaps no better illustrated than by the mammalian kidney. Here, because of the use of counteracting and compatible solutes, the same types of proteins found in other organs of the animal are capable of maintaining their structures and functions in the face of widely varying osmolarity and high urea concentrations—conditions that clearly pose a threat to protein function and structure, yet are effectively neutralized through the use of appropriate organic osmolytes.

Portions of this work were supported by National Science Foundation grants DCB88-12180 and IBN92-06660.

REFERENCES

1. Adler, L., and L. Gustafsson. Polyhydric alcohol production and amino acid pool in relation to halotolerance of the yeast *Debaryomyces hansenii*. *Arch. Microbiol.* 124: 123–130, 1980.
2. Altringham, J. D., P. H. Yancey, and I. A. Johnston. The effects of osmoregulatory solutes on tension generation by dogfish skinned muscle fibres. *J. Exp. Zool.* 96: 443–445, 1982.
3. Anchordoguy, T. J., J. F. Carpenter, C. A. Cecchine, J. H. Crowe, and L. M. Crowe. Effects of protein perturbants on phospholipid bilayers. *Arch. Biochem. Biophys.* 283: 356–361, 1990.
4. Anderson, P. M. Purification and properties of the glutamine- and N-acetyl-L-glutamate-dependent carbamoyl phosphate synthetase from liver of *Squalus acanthias*. *J. Biol. Chem.* 256: 12228–12238, 1981.
5. Anderson, P. M. Ketone body and phosphoenolpyruvate formation by isolated hepatic mitochondria from *Squalus acanthias* spiny dogfish. *J. Exp. Zool.* 254: 144–154, 1990.
6. Arakawa, T., J. F. Carpenter, Y. A. Kita, and J. H. Crowe. The basis for toxicity of certain cryoprotectants: a hypothesis. *Cryobiology* 27: 401–415, 1990.
7. Arakawa, T., and S. N. Timasheff. The stabilization of proteins by osmolytes. *Biophys. J.* 47: 411–414, 1985.
8. Avison, M. J., D. L. Rothman, T. W. Nixon, W. S. Long, and N. J. Siegel. ^{1}H NMR study of renal trimethylamine responses to dehydration and acute volume loading in man. *Proc. Natl. Acad. Sci. U.S.A.* 88: 6053–6057, 1991.
9. Bagnasco, S., R. Balaban, H. Fales, Y.-M. Yang, and M. Burg. Predominant osmotically active organic solutes in rat and rabbit renal medullas. *J. Biol. Chem.* 261: 5872–5877, 1986.
10. Bagnasco, S., S. Uchida, R. Balaban, P. Kador, and M. Burg. Induction of aldose reductase and sorbitol in renal inner medullary cells by elevated extracellular NaCl. *Proc Natl Acad Sci U.S.A.* 84: 1718–1720, 1987.
11. Bagnasco, S. M., M. H. Montrose, and J. S. Handler. Role of calcium in organic osmolyte efflux when MDCK cells are shifted from hypertonic to isotonic medium. *Am. J. Physiol.* 264 (*Cell Physiol.* 33): C1165–1170, 1993.
12. Balaban, R., and M. B. Burg. Osmotically active organic solutes in the renal inner medulla. *Kidney Int.* 31: 562–564, 1987.
13. Balaban, R., and M. A. Knepper. Nitrogen-14 nuclear magnetic resonance spectroscopy of mammalian tissues. *Am. J. Physiol.* 245 (*Cell Physiol.* 14): C439–C444, 1983.
14. Balinsky, J. B. Adaptation of nitrogen metabolism to hyperosmotic environment in amphibia, *J. Exp. Zool.* 215: 350–356, 1981.
15. Ballantyne, J. S., and T. W. Moon. The effects of urea, trimethylamine oxide and ionic strength on the oxidation of acyl carnitines by mitochondria isolated from the liver of the little skate *Raja erinacea*. *J. Comp. Physiol.* 156B: 845–851, 1986.
16. Ballantyne, J. S., C. D. Moyes, and T. W. Moon. Compatible and counteracting solutes and the evolution of ion and osmoregulation in fishes. *Can. J. Zool.* 65: 1883–1888, 1987.
17. Bateman, J. B., G. F. Evans, P. R. Brown, C. Gabriel, and E. H. Grant. Dielectric properties of the system bovine albumin:urea:betaine in aqueous solution. *Phys. Med. Biol.* 37: 175–182, 1992.
18. Beck, F., A. Dorge, R. Rick, and K. Thurau. Intra- and extracellular elemental concentrations of rat inner papilla in antidiuresis. *Kidney Int.* 25: 397–403, 1984.
19. Beck, F. X., M. Schmolke, W. G. Guder, A. Dorge, and K. Thurau. Osmolytes in renal medulla during rapid changes in papillary tonicity. *Am. J. Physiol.* 262 (*Renal Fluid Electrolyte Physiol.* 31): F849–F856, 1992.
20. Beever, R. E., and E. P. Laracy. Osmotic adjustment in the filamentous fungus *Aspergillus nidulans*. *J. Bacteriol.* 168: 1358–1365, 1986.
21. Beltmans, D., and A. Van Laere. Glycerol cycle enzymes and intermediates during adaptation of *Dunaliella teriolecta* cells to hyperosmotic stress. *Plant Cell Environ.* 10: 185–190, 1987.
22. Blackwell, J. R., and Gilmour, D. J. Physiological response of the unicellular green alga *Chlorococcum submarinum* to rapid changes in salinity. *Arch. Microbiol.* 157: 86–91, 1991.
23. Bondy, C., B. D. Cowley, Jr., S. L. Lightman, and P. F. Kador. Feedback inhibition of aldose reductase gene expression in rat renal medulla:galactitol accumulation reduces enzyme messenger RNA levels and depletes cellular inositol content. *J. Clin. Invest.* 86: 1103–1108, 1990.
24. Borowitzka, L. Glycerol and other carbohydrate osmotic effectors. In: *Transport Processes, Iono- and Osmoregulation*, edited by R. Gilles and M. Gilles-Ballien, Berlin: Springer-Verlag, 1985, p. 437–453.
25. Bowlus, R. D., and G. N. Somero. Solute compatibility with enzyme function and structure: rationales for the selection of osmotic agents and end-products of anaerobic metabolism in marine invertebrates. *J. Exp. Zool.* 208: 137–152, 1979.
26. Boylan, J. W. Gill permeability in *Squalus acanthia*. In: *Sharks, Skates and Rays*, edited by P. W. Gilbert, R. F. Mathewson, and D. P. Rall. Baltimore, MD: Johns Hopkins Press, 1967, p. 197–206.

27. Bradley, T. J. Physiology of osmoregulation in mosquitoes. *Annu. Rev. Entomol.* 32: 439–462, 1987.

28. Brown, A. D. Compatible solutes and extreme water stress in eukaryotic micro-organisms. *Adv. Microbial Physiol.* 17: 181–242, 1978.

29. Brown, A. D. *Microbial Water Stress Physiology: Principles and Perspectives.* New York: Wiley, 1990.

30. Brown, A. D., and J. Simpson. Water relations of sugar-tolerant yeasts: the role of intracellular polyols. *J. Gen. Microbiol.* 72: 589–591, 1972.

31. Brown, L. M., and J. A. Hellebust. The contribution of organic solutes to osmotic balance in some green and eustigmatophyte algae. *J. Phycol.* 16: 265–270, 1980.

32. Buche, A., P. Colson, and C. Houssier. Organic osmotic effectors and chromatin structure. *J. Biomol. Struct. Dynam.* 8: 601–618, 1990.

33. Bunn, H. F., and P. J. Higgins. Reaction of monosaccharides with proteins: possible evolutionary significance. *Science* 213: 222–224, 1982.

34. Burg, M. B. Molecular basis for accumulation of compatible osmolytes in mammalian renal cells. In: *Water and Life: A Comparative Analysis of Water Relationships at the Organismic, Cellular, and Molecular Levels,* edited by G. N. Somero, C. B. Osmond, and C. L. Bolis, Berlin: Springer-Verlag, 1992, p. 33–51.

35. Burg, M. B., and A. Garcia-Perez. How tonicity regulates gene expression. *J. Am. Soc. Nephrol.* 3: 121–127, 1992.

36. Burg, M. B., and P. F. Kador. Sorbitol, osmoregulation, and the complications of diabetes. *J. Clin. Invest.* 81: 635–640, 1988.

37. Burton, R. F. The composition of animal cells: solutes contributing to osmotic pressure and charge balance. *Comp. Biochem. Physiol.* 76B: 663–671, 1983.

38. Burton, R. S. Incorporation of ^{14}C-bicarbonate into the free amino acid pool during hyperosmotic stress in an intertidal copepod. *J. Exp. Zool.* 238: 55–61, 1986.

39. Burton, R. S. Regulation of proline synthesis during osmotic stress in the copepod *Tigriopus californicus. J. Exp. Zool.* 259:166–173, 1991.

40. Burton, R. S. Regulation of proline synthesis in osmotic reponse: effects of protein synthesis inhibitors. *J. Exp. Zool.* 259: 272–277, 1991.

41. Carpenter, J. F., and J. H. Crowe. Modes of stabilization of a protein by organic solutes during desiccation. *Cryobiology* 25: 459–470, 1988.

42. Carpenter, J. S., S. J. Prestrelski, and T. Arakawa. Separation of freezing- and drying-induced denaturation of lyophilized proteins using stress-specific stabilization: I. Enzyme activity and calorimetric studies. *Arch. Biochem. Biophys.* 303: 456–464, 1993.

43. Cayley, S., B. A. Lewis, H. J. Guttman, and M. T. Record. Characterization of the cytoplasm of *Escherichia coli* K-12 as a function of external osmolarity. Implications for protein-DNA interactions in vivo. *J. Mol. Biol.* 222: 281–300, 1991.

44. Cayley, S., B. A. Lewis, and M. T. Record. Origins of the osmoprotective properties of betaine and proline in *Escherichia coli* K-12. *J. Bacteriol.* 174: 1586–1595, 1992.

45. Chamberlin, M. E., and K. Strange. Anisosmotic cell volume regulation: a comparative view. *Am. J. Physiol.* 257 (*Cell Physiol.* 26): C159–173, 1989.

46. Chambers, S. T., and C. M. Kunin. The osmoprotective properties of urine for bacteria: the protective effect of betaine and human urine against low pH and high concentrations of electrolytes, sugars and urea. *J. Infect. Dis.* 152: 1308–1315, 1985.

47. Chauncey, B., M. V. Leite, and L. Goldstein. Renal sorbitol accumulation and associated enzyme activities in diabetes. *Enzyme* 39: 231–234, 1988.

48. Clark, M. E. The osmotic role of amino acids: discovery and function. In: *Transport Processes, Iono- and Osmoregulation,* edited by R. Gilles and M. Gilles-Ballien, Berlin: Springer-Verlag, 1985, p. 412–423.

49. Clark, M. E. Non-Donnan effects of organic osmolytes in cell volume changes. *Curr. Top. Membr. Transp.* 30: 251–271, 1987.

50. Clark, M. E., J.A.M. Hinke, and M. E. Todd. Studies on water in barnacle muscle fibres. II. Role of ions and organic solutes in swelling of chemically-skinned fibres. *J. Exp. Biol.* 90: 43–63, 1981.

51. Clark, M. E., and M. Zounes. The effects of selected cell osmolytes on the activity of lactate dehydrogenase from the euryhaline polychaete, *Nereis succinea. Biol. Bull. Woods Hole* 153: 468–484, 1977.

52. Cohen, D. M., J. C. Wasserman, and S. R. Gullans. Immediate early gene and HSP70 expression in hyperosmotic stress in MDCK cells. *Am. J. Physiol.* 261 (*Cell Physiol.* 30): C594–C601, 1991.

53. Collins, K. D., and M. W. Washabaugh. The Hofmeister effect and the behavior of water at interfaces. *Q. Rev. Biophys.* 18: 323–422, 1985.

54. Corder, C. N., J. G. Collins, T. S. Brannan, and J. Sharma. Aldose reductase and sorbitol dehydrogenase distribution in rat kidney. *J. Histochem. Cytochem.* 25: 1–8, 1977.

55. Cowley, B. D. Jr, J. D. Ferraris, D. Carper, and M. B. Burg. In vivo osmotic regulation of aldose reductase mRNA, protein, and sorbitol content in rat renal medulla. *Am. J. Physiol.* 258 (*Renal Fluid Electrolyte Physiol.* 27): F154–F161, 1990.

56. Crowe, J. Transport of exogenous substrate and cell volume regulation in bivalve molluscs. *J. Exp. Zool.* 233: 21–28, 1977.

57. Crowe, J. H., F. A. Hoekstra, and L. M. Crowe. Anhydrobiosis, *Annu. Rev. Physiol.* 54: 579–599, 1992.

58. Csonka, L. N. Physiological and genetic responses of bacteria to osmotic stress. *Microbiol. Rev.* 53: 121–147, 1989.

59. Csonka, L. N., and A. D. Hanson. Prokaryotic osmoregulation: genetics and physiology. *Annu. Rev. Microbiol.* 45: 569–606, 1991.

60. Cumme, G. A., A. Horn and W. Achilles. Metal complex formation and its importance for enzyme regulation. *Acta Biol. Med. Ger.* 31: 349–363, 1973.

61. De Meis, L., and G. Inesi. Effects of organic solvents, methylamines, and urea on affinity for P_i of the Ca^{2+}-ATPase of sarcoplasmic reticulum. *J. Biol. Chem.* 263: 157–161, 1988.

62. Deaton, L. E. Hyperosmotic cellular volume regulation in the ribbed mussel *Geukensia demissa:* inhibition by lysosomal and proteinase inhibitors. *J. Exp. Zool.* 244: 375–382, 1987.

63. Deaton, L. E., T. J. Hilbish, and R. K. Koehn. Protein as a source of amino nitrogen during hyperosmotic volume regulation in the mussel *Mytilus edulis. Physiol. Zool.* 57: 609–619, 1984.

64. DeLaney, R. G., Lahiri, S., Hamilton, R., and Fishman, A. P., Acid-base balance and plasma composition in the aestivating lungfish (*Protopterus*), *Am. J. Physiol.* 232 (*Regulatory Integrative Comp. Physiol.* 3): R10–R17, 1977.

65. Dickinson, D.M.J., and G. O. Kirst. Osmotic adjustment in marine eukaryotic algae: the role of inorganic ions, quaternary ammonium, tertiary sulphonium and carbohydrate solutes. I. Diatoms and a Rhodophyte. *New Phytol.* 106: 645–655, 1987.

66. Dickinson, D.M.J., and G. O. Kirst. Osmotic adjustment in marine eukaryotic algae: the role of inorganic ions, quaternary ammonium, tertiary sulphonium and carbohydrate solutes. II.

Prasinophytes and Haptophytes. *New Phytol.* 106: 657–666, 1987.

67. Dickinson, D.M.J., and G. O. Kirst. The role of β-dimethylsulphoniopropionate, glycine betaine and homarine in the osmoacclimation of *Platymonas subcordiformis. Planta* 167: 536–543, 1986.

68. Dinnibier, U., E. Limpinsel, R. Schmid, and E. P. Bakker. Transient accumulation of potassium glutamate and its replacement by trehalose during adaptation of growing cells of *Escherichia coli* K-12 to elevated sodium chloride concentrations, *Arch. Microbiol.* 150: 348–357, 1988.

69. Duman, J. G., D. W. Wu, L. Xu, D. Tursman, and T. M. Olsen. Adaptations of insects to subzero temperatures, *Q. Rev. Biol.* 66: 387–410, 1991.

70. Edmands, S., and P. H. Yancey. Effects on rat renal osmolytes of extended treatment with an aldose reductase inhibitor. *Comp. Biochem. Physiol.* 103C: 499–502, 1992.

71. Edwards, D. M., R. H. Reed, and W.D.P. Stewart. Osmoaccumulation in *Enteromorpha intestinalis:* long-term effects of osmotic stress on organic solute accumulation. *Mar. Biol.* 98: 467–476, 1988.

72. Emyanitoff, R. G. and B. E. Wright. Effect of intracellular carbohydrates on heat resistance of *Dictyostelium discoideum* spores. *J. Bacteriol.* 140: 1008–1012, 1979.

73. Evans, R. D., R. A. Black, W. H. Loescher, and R. J. Fellows. Osmotic relations of the drought-tolerant shrub *Artemisia tridentata* in response to water stress. *Plant Cell Environ.* 15: 49–59, 1992.

74. Fenstermacher, J., F. Sheldon, J. Ratner, and A. Roomet. The blood to tissue distribution of various polar materials in the dogfish, *Squalus acanthias. Comp. Biochem. Physiol.* 42A: 195–204, 1972.

75. Fincham, D. A., M. W. Wolowyk, and J. D. Young. Volume-sensitive taurine transport in fish erythrocytes. *J. Membr. Biol.* 96: 45–56, 1987.

76. Ford, C. W. Accumulation of O-methyl-inositols in water-stressed *Vigna* species. *Phytochemistry* 21: 1149–1151, 1982.

77. Frédericq, L. Sur la concentration moleculaire du sang et des tissus chez les animaux aquaticques. *Bull. Acad. Belg. Cl. Sci.* 1901: 428–454, 1901.

78. Fujita, Y., Y. Iwasa, and Y. Noda. The effect of polyhydric alcohols on the thermal denaturation of lysozyme as measured by differential scanning calorimetry. *Bull. Chem. Soc. Jpn.* 55: 1896–1900, 1982.

79. Fulton, A. B. How crowded is the cytoplasm? *Cell* 30: 345–347, 1982.

80. Furlong, T. J., T. Moriyama, J. Capdevila, D. T. Rice, and K. R. Spring. Activation of osmolyte efflux from cultured renal papillary epithelial cells. *J. Membr. Biol.* 123: 269–277, 1991.

81. Galinski, E. A., and R. M. Herzog. The role of trehalose as a substitute for nitrogen-containing compatible solutes in *Ectothiorhodospira halochloris. Arch. Microbiol.* 153: 607–613, 1990.

82. Galinski, E. A., and A. Oren. Isolation and structure determination of a novel compatible solute from the moderately halophilic purple sulfur bacterium *Ectothiorhodospira marismortui. Eur. J. Biochem.* 198: 593–598, 1991.

83. Garcia-Perez, A., and M. B. Burg. Renal medullary organic osmolytes. *Physiol. Rev.* 71: 1081–1114, 1991.

84. Garret, M. A., and T. J. Bradley. Extracellular accumulation of proline, serine and trehalose in the haemolymph of osmoconforming brackish-water mosquitoes. *J. Exp. Biol.* 129: 231–238, 1987.

85. Gekko, K., and T. Morikawa. Thermodynamics of polyol-induced thermal stabilization of chymotrypsinogen. *J. Biochem.* 90: 51–60, 1981.

86. Gibson, T. S., J. Speirs, and C. J. Brady. Salt-tolerance in plants. II. *In vitro* translation of m-RNAs from salt-tolerant and salt-sensitive plants on wheat germ ribosomes. Responses to ions and compatible organic solutes. *Plant Cell Environ.* 7: 579–587, 1984.

87. Gilles, R. Volume regulation in cells of euryhaline invertebrates. *Curr. Top. Membr. Transp.* 30: 205–247, 1987.

88. Ginzburg, M. *Dunaliella:* a green alga adapted to salt. *Adv. Bot. Res.* 14: 93–183, 1987.

89. Glous, K., and D. Le Rudulier. Transport and catabolism of proline betaine in salt-stressed *Rhizobium meliloti. Arch. Microbiol.* 151: 143–148, 1989.

90. Godt, R. E., R.T.H. Fogaca, M.A.W. Andrews, and T. M. Nosek. Influence of ionic strength on contractile force and energy consumption of skinned fibers from mammalian and crustacean striated muscle. In: Advances in Experimental Medicine and Biology, vol. 332. *Mechanism of Myofilament Sliding in Muscle Contraction,* edited by H. Sugi, G. H. Pollack. New York: Plenum, 1993, p. 763–774.

91. Goldstein, L. Volume regulation in the erythrocyte of the little skate, *Raja erinacea. J. Exp. Zool. Suppl.* 2: 136–142, 1989.

92. Goldstein, L., and A. Kleinzeller. Cell volume regulation in lower vertebrates. *Curr. Top. Membr. Transp.* 30: 181–204, 1987.

93. Goolish, E. M., and R. S. Burton. Energetics of osmoregulation in an intertidal copepod: effects of anoxia and lipid reserves on the pattern of free amino acid accumulation. *Func. Ecol.* 3: 81–89, 1989.

94. Gorman, J., R. G. Wyn Jones, and E. McDonnell. Some mechanisms of salt tolerance in crop plants. *Plant Soil* 89: 15–40, 1985.

95. Gouesbet, G., Blanco, C., Hamelin, J., and Bernard, T. Osmotic adjustment in *Brevibacterium ammoniagenes:* pipecolic acid accumulation at elevated osmolalities. *J. Gen. Microbiol.* 138: 959–965, 1992.

96. Gowrishankar, J., P. Jayashree, and K. Rajkumari. Molecular cloning of an osmoregulatory locus in *Escherichia coli:* increased *proU* gene dosage results in enhanced osmotolerance. *J. Bacteriol.* 168: 1197–1204, 1986.

97. Greenway, H., and C. B. Osmond. Salt responses of enzymes from species differing in salt tolerance. *Plant Physiol.* 49: 256–259, 1972.

98. Griffith, R. W., P.K.T. Pang, and L. A. Benedetto. Urea tolerance in the killifish, *Fundulus heteroclitus. Comp. Biochem. Physiol.* 62A: 327–330, 1979.

99. Griffith, R. W., B. L. Umminger, B. F. Grant, P.K.T. Pang, and G. E. Pickford. Serum composition of the coelacanth *Latimeria chalumnae* Smith. *J. Exp. Zool.* 187: 87–102, 1974.

100. Grossman, E. B., and S. C. Hebert. Renal inner medullary choline dehydrogenase activity: characterization and modulation. *Am. J. Physiol.* 256 (*Renal Fluid Electrolyte Physiol.* 25): F107–F112, 1989.

101. Grunewald R. W., and R.K.H. Kinne. Intercellular sorbitol content in isolated rat renal inner medullary collecting duct cells: regulation by extracellular osmolality. *Pflugers Arch.* 414: 178–184, 1989.

102. Gullans, S. R., J. D. Blumenfeld, J. A. Balschi, M. Kaleta, R. M. Brenner, C. W. Heilig, and S. C. Hebert. Accumulation of major organic osmolytes in rat renal medulla in dehydration. *Am. J. Physiol.* 255 (*Renal Fluid Electrolyte Physiol.* 24): F626–F634, 1988.

103. Hand, S. C. Water content and metabolic organization in

anhydrobiotic animals, In: *Water and Life: A Comparative Analysis of Water Relationships at the Organismic, Cellular, and Molecular Levels,* edited by G. N. Somero, C. B. Osmond, and C. L. Bolis, Berlin: Springer-Verlag, 1992, p. 104–127.

104. Hand, S. C., and G. N. Somero. Urea and methylamine effects on rabbit muscle phosphofructokinase, *J. Biol. Chem.* 257: 734–741, 1982.

105. Hanson, A. Compatible solute synthesis and compartmentation in plants. In: *Water and Life: A Comparative Analysis of Water Relationships at the Organismic, Cellular, and Molecular Levels,* edited by G. N. Somero, C. B. Osmond, and C. L. Bolis, Berlin: Springer-Verlag, 1992, p. 52–60.

106. Heilig, C. W., M. E. Stromski, J. D. Blumenfeld, J. P. Lee, and S. R. Gullans. Characterization of the major brain osmolytes that accumulate in salt-loaded rats. *Am. J. Physiol.* 257 (*Renal Fluid Electrolyte Physiol.* 26) 257: F1108–1116, 1989.

107. Hellebust, J. A. Mechanisms of response to salinity in halotolerant microalgae. *Plant Soil* 89: 69–81, 1985.

108. Henderson, L. J. *The Fitness of the Environment.* New York: Macmillan, 1913.

109. Henzel, R., and H. Konig. Thermoadaptation of methanogenic bacteria by intracellular ion concentration. *FEMS Microbiol. Lett.* 49: 75–79, 1988.

110. Herzog, R. M., E. A. Galinski, and H. G. Truper. Degradation of the compatible solute trehalose in *Ectothiorhodospira halochloris:* isolation and characterization of trehalase. *Arch. Microbiol.* 153: 600–606, 1990.

111. Higgins, C. F., J. Cairney, D. A. Stirling, L. Sutherland, and I. R. Booth. Osmotic regulation of gene expression: ionic strength as an intracellular signal? *Trends Biochem. Sci.* 12: 339–344, 1987.

112. Hilbish, T. J., L. E. Deaton and R. K. Koehn. Effect of allozyme polymorphism on regulation of cell volume. *Nature* 298: 688–689, 1982.

113. Hochachka, P. W. and G. N. Somero. *Biochemical Adaptation.* Princeton, NJ: Princeton Univ. Press, 1984.

114. Hofmeister, F. On the understanding of the effect of salts. Second report. On regularities in the precipitating effects of salts and their relationship to their physiological behavior. *Naunyn-Schmiedebergs Arch. Exp. Pathol. Pharmakol.* 24: 247–260, 1888.

115. Horne, F. R. Accumulation of urea by a pulmonate snail during estivation. *Comp. Biochem. Physiol.* 38A: 565–570, 1971.

116. Hosono, K. Effect of salt stress on lipid composition and membrane fluidity of the salt-tolerant yeast *Zygosaccharomyces rouxii. J. Gen. Microbiol.* 138: 91–96, 1992.

117. Hottiger, T., T. Boller, and A. Wiemken. Rapid changes of heat and desiccation tolerance correlated with changes in trehalose content in *Saccharomyces cerevisiae* cells subjected to temperature shifts. *FEBS Lett.* 220: 113–115, 1987.

118. Impellizzeri, G., S. Mangiafico, G. Oriente, M. Piattelli, and S. Sciuto. Amino acids and low-molecular-weight carbohydrates of some marine red algae. *Phytochemistry* 14: 1549–1557, 1975.

119. Incharoensakdi, A., T. Takabe, and T. Akazawa. Effect of betaine on enzyme activity and subunit interaction of ribulose-1, 5-bisphosphate carboxylase/oxygenase from *Aphanothece halophytica. Plant Physiol.* 81: 1044–1049, 1986.

120. Jebbar, M., T. Talibart, K. Glous, T. Bernard, and C. Blanco. Osmoprotection of *Escherichia coli* by ectoine: uptake and accumulation characteristics. *J. Bacteriol.* 174: 5027–5035, 1992.

121. Kaneshiro, E. S., Holz, G. G. Jr., and Dunham, P. B. Osmoregulation in a marine ciliate, *Miamiensis avidus.* II. Regulation of intracellular free amino acids. *Biol. Bull. Woods Hole* 137: 161–169, 1969.

122. Kauss, H., K. S. Thomson, M. Thomson and W. Jeblick. Osmotic regulation. Physiological significance of proteolytic and nonproteolytic activation of isofloridoside-phosphate synthetase. *Plant Physiol.* 63: 455–459, 1979.

123. Keller, M. D. Dimethyl sulfide production and marine phytoplankton: the importance of species composition and cell size. *Biol. Oceanogr.* 6: 375–382, 1988.

124. King, P. A., and L. Goldstein. Organic osmolytes and cell volume regulation in fish. *Mol. Physiol.* 4: 53–66, 1983.

125. Kinne, R.K.H. The role of organic osmolytes in osmoregulation: from bacteria to mammals. *J. Exp. Zool.* 265: 346–355, 1993.

126. Kirst, G. O. Salinity tolerance of eukaryotic marine algae. *Annu. Rev. Plant Physiol. Plant Mol. Biol.* 40: 21–53, 1989.

127. Kleinzeller, A. Trimethylamine oxide and the maintenance of volume of dogfish shark rectal gland cells. *J. Exp. Zool.* 236: 11–17, 1985.

128. Kloiber, O., B. Banjac, and L. R. Drewes. Protection against acute hyperammonemia: the role of quaternary amines. *Toxicology* 49: 83–90, 1988.

129. Koehn, R. K., B. L. Bayne, M. N. Moore, and J. F. Siebenaller. Salinity related physiological and genetic differences between populations of *Mytilus edulis. Biol. J. Linn. Soc.* 14: 319–334, 1980.

130. Kumazawa, Y., and Arai, K. Suppressive effect of sorbitol on denaturation of carp myosin B induced by neutral salts. *Nippon Sui. Gak.* 56: 679–686, 1990.

131. Kwon, H. M., A. Yamauchi, S. Uchida, A. S. Preston, A. Garcia-Perez, M. B. Burg, and J. S. Handler. CLoning of the cDNA for a Na^+/myo-inositol cotransporter, a hypertonicity stress protein. *J. Biol. Chem.* 267: 6297–6310, 1992.

132. Lai, M.-C., K. R. Sowers, D. E. Robertson, M. F. Roberts, and R. P. Gunsalus. Distribution of compatible solutes in the halophilic methanogenic archaebacteria. *J. Bacteriol.* 173: 5352–5358, 1991.

133. Landfald B., and A. R. Strom. Choline-glycine betaine pathway confers a high level of osmotic tolerance in *Escherichia coli. J. Bacteriol.* 165: 849–855, 1986.

134. Lanyi, J. Salt-dependent properties of proteins from extremely halophilic bacteria. *Bacteriol. Rev.* 38: 272–290, 1974.

135. Larsen, P. I., L. K. Syndes, B. Landfald, and A. R. Strom. Osmoregulation in *Escherichia coli* by accumulation of organic osmolytes: betaines, glutamic acid, and trehalose. *Arch. Microbiol.* 147: 1–7, 1987.

136. Law, R. O. Amino acids as volume-regulatory osmolytes in mammalian cells. *Comp. Biochem. Physiol.* 99A: 263–277, 1991.

137. Lawitts, J. A., and J. D. Biggers. Joint effects of sodium chloride, glutamine, and glucose in mouse preimplantation embryo culture media. *Mol. Reprod. Dev.* 31: 189–194, 1992.

138. Le Rudulier, D., and L. Bouillard. Glycine betaine, an osmotic effector in *Klebsiella pneumoniae* and other members of the Enterobacteriaceae. *Appl. Environ. Microbiol.* 46: 152–159, 1983.

139. Lee, J. A., H. A. Lee, and P. J. Sadler. Uraemia: is urea more important than we thought? *Lancet* 338: 1438–1440, 1991.

140. Leech, A. R., and L. Goldstein. β-alanine oxidation in the liver of the little skate, *Raja erinacea. J. Exp. Zool.* 225: 9–14, 1983.

141. Lien, Y-H, H., J. I. Shapiro, and L. Chan. Effects of hypernatremia on organic brain osmoles. *J. Clin. Invest.* 85: 1427–1435, 1990.

142. Loomis, S. H., J. F. Carpenter, and J. H. Crowe. Identification

of strombine and taurine as cryoprotectants in the intertidal bivalve *Mytilus edulis. Biochim. Biophys. Acta.* 943: 113–118, 1988.

143. Low, P. S. Molecular basis of the biological compatibility of nature's osmolytes. In: *Transport Processes, Iono- and Osmoregulation,* edited by R. Gilles and M. Gilles-Ballien, Berlin: Springer-Verlag, Berlin, 1985, p. 469–477.

144. MacKay, M. A., R. S. Norton, and L. J. Borowitzka. Organic osmoregulatory solutes in cyanobacteria. *J. Gen. Microbiol.* 130: 2177–2191, 1984.

145. Mahan, M. J., and L. N. Csonka. Genetic analysis of the *proBA* genes of *Salmonella typhimurium:* physical and genetic analyses of the cloned *proB+A+* genes of *Escherichia coli* and of a mutant allele that confers proline overproduction and enhanced osmotolerance. *J. Bacteriol.* 156:1249–???, 1983.

146. Manahan, D. T. Adaptations by invertebrate larvae for nutrient acquisition from seawater. *Am. Zool.* 30: 147–160, 1990.

147. Mashino, T., and I. Fridovich. Effects of urea and trimethylamine-N-oxide on enzyme activity and stability. *Arch. Biochem. Biophys.* 258: 356–360, 1987.

148. Mason, T. G., and G. Blunden. Quaternary ammonium and tertiary sulfonium compounds of algal origin as alleviators of osmotic stress. *Bot. Mar.* 32: 313–316, 1989.

149. McClanahan, L. Adaptations of the spadefoot toad, *Scaphiopus couchi,* to desert environments. *Comp. Biochem. Physiol.* 20: 73–99, 1967.

150. McConnell, F. M., and L. Goldstein. Intracellular signals and volume regulatory response in skate erythrocytes. *Am. J. Physiol.* 255 (*Regulatory Integrative Comp. Physiol.* 24): R982–R987, 1988.

151. McCue, K. F., and A. D. Hanson. Drought and salt tolerance: towards understanding and application. *Trends Biotechnol.* 8: 358–362, 1990.

152. Mirsalikhova, N. M. Stabilization of Na$^+$, K$^+$-adenosine triphosphatase by dimethyl-sulfoxide in the case of inactivation by urea. *Biokhimiya* 43: 43–39, 1978.

153. Moriyama, T., A. Garcia-Perez, and M. B. Burg. Factors affecting the ratio of different organic osmolytes in renal medullary cells. *Am. J. Physiol.* 259 (*Renal Fluid Electrolyte Physiol.* 28): F847–F858, 1990.

154. Moriyama, T., A. Garcia-Perez, A. D. Olson, and M. B. Burg. Intracellular betaine substitutes for sorbitol in protecting renal medullary cells from hypertonicity. *Am. J. Physiol.* 260 (*Renal Fluid Electrolyte Physiol.* 29): F494–F497, 1991.

155. Moyes, C. D., and T. W. Moon. Solute effects on the glycine cleavage system of two osmoconformers *Raja erinacea* and *Mya arenaria* and an osmoregulator *Pseudopleuronectes americanus. J. Exp. Zool.* 242: 1–8, 1987.

156. Murata, N., P. S. Mohanty, H. Hayashi, G. C. Papageorgiou. Glycinebetaine stabilizes the association of extrinsic proteins with the photosynthetic oxygen-evolving complex. *FEBS Lett.* 296: 187–189, 1992.

157. Nakanishi, T., R. S. Balaban, and M. B. Burg. Survey of osmolytes in renal cell lines. *Am. J. Physiol.* 255 (*Cell Physiol.* 24): C181–C191, 1988.

158. Nakanishi, T., and M. B. Burg. Osmoregulatory fluxes of *myo*-inositol and betaine in renal cells. *Am. J. Physiol.* 257 (*Cell Physiol.* 26): C964–C970, 1989.

159. Nakanishi, T., and M. B. Burg. Osmoregulation of glycerophosphorylcholine content of mammalian renal cells. *Am. J. Physiol.* 257 (*Cell Physiol.* 26): C795–C801, 1989.

160. Nakanishi, T., O. Uyama, and M. Sugita. Osmotically regulated taurine content in the rat renal inner medulla. *Am. J.*

161. *Physiol.* 261 (*Renal Fluid Electrolyte Physiol.* 30): F957–962, 1991.

161. Nikolopoulos, D., and Y. Manetas. Compatible solutes and *in vitro* stability of *Salsola soda* enzymes: proline incompatibility. *Phytochemistry* 30: 411–413, 1991.

162. Nishiguchi, M. K. and G. N. Somero. Temperature- and concentration-dependence of compatibility of the organic osmolyte β-dimethylsulfoniopropionate. *Cryobiology* 29: 118–124, 1992.

163. Peterson, D. P., K. M. Murphy, R. Ursino, K. Streeter, and P. H. Yancey. Effects of dietary protein and salt on rat renal osmolytes: co-variation in urea and glycerophosphorylcholine contents. *Am. J. Physiol.* 263 (*Renal Fluid Electrolyte Physiol.* 32): F594–F600, 1992.

164. Petronini, R. G., E. M. De Angelis, P. Borghetti, A. F. Borghetti, K. P. Wheeler. Modulation by betaine of cellular responses to osmotic stress. *Biochem. J.* 282: 69–73, 1992.

165. Pfaller, W. Structure function correlation on rat kidney. In: *Advances in Anatomy, Embryology and Cell Biology,* edited by W. Hild, J. van Limborgh, R. Ortmann, J. E. Pauly, and T. H. Schiebler, Berlin: Springer-Verlag, 1982, vol. 70, p. 22.

166. Pierce, S. K., S. C. Edwards, P. H. Mazzocchi, L. J. Klinger, M. K. Warren. Proline betaine: a unique osmolyte in an extremely euryhaline osmoconformer. *Biol. Bull. Woods Hole* 167: 495–500, 1984.

167. Pocard, J.-A., T. Bernard, L. Smith and D. Le Rudulier. Characterization of three choline transport activities in *Rhizobium meliloti:* modulation by choline and osmotic stress. *J. Bacteriol.* 171: 531–537, 1989.

168. Pollard, A., and R. G. Wyn Jones. Enzyme activities in concentrated solutions of glycinebetaine and other solutes. *Planta* 144: 291–298, 1979.

169. Preston, R. L. Occurrence of D-amino acids in higher organisms: a survey of the distribution of D-amino acids in marine invertebrates. *Comp. Biochem. Physiol.* 87B: 55–62, 1987.

170. Prince, W. S., and M. R. Villarejo. Osmotic control of proU transcription is mediated through direct action of potassium glutamate on the transcription complex. *J. Biol. Chem.* 265: 17673–17679, 1990.

171. Raven, J. A. Regulation of pH and generation of osmolarity in vascular plants: a cost-benefit analysis in relation to efficiency of use of energy, nitrogen and water. *New Phytol.* 101: 25–77, 1985.

172. Raymond, J. A. Glycerol is a colligative antifreeze in some northern fishes. *J. Exp. Zool.* 262: 347–352, 1992.

173. Reed, R. H. Measurement and osmotic significance of β-dimethylsulphonioproionate in marine macroalgae. *Mar. Biol. Lett.* 4: 173–181, 1983.

174. Reed, R. H., and W.D.P. Stewart. Physiological responses of *Rivularia atra* to salinity: osmotic adjustment in hypersaline media. *New Phytol.* 95: 595–603, 1983.

175. Reed, R. H., S.R.C. Warr, N. W. Kirby and W.D.P. Stewart. Osmotic shock-induced release of low molecular weight metabolites from free-living and immobilized cyanobacteria. *Enzyme Microb. Technol.* 8: 101–104, 1986.

176. Reed, R. H., S.R.C. Warr, D. L. Richardson, D. J. Moore, W. D. P. Stewart. Multiphasic osmotic adjustment in a euryhaline cyanobacterium. *FEMS Microbiol. Lett.* 28: 225–229, 1985.

177. Riordan, J. F., K. D. McElvany and C. L. Borders, Jr. Arginyl residues: anion recognition sites in enzymes. *Science,* 195: 884–885, 1977.

178. Robertson, D. E., and M. F. Roberts. Organic osmolytes in methanogenic archaebacteria. *Biofactors* 3: 1–9, 1991.

179. Rozwadowski, K. L., G. G. Khachatourians, and G. Selvaraj. Choline oxidase, a catabolic enzyme in *Arthrobacter pascens*, facilitates adaptation to osmotic stress in *Escherichia coli*. *J. Bacteriol.* 173: 472–478, 1991.

180. Rudolph, A. S., and B. Goins. The effect of hydration stress solutes on phase behavior of hydrated dipalmitoylphosphatidylcholine. *Biochim. Biophys. Acta* 1066: 90–94, 1991.

181. Sacher, R. F., and R. C. Staples. Inositol and sugars in adaptation of tomato to salt. *Plant Physiol.* 77: 206–210, 1985.

182. Schmid, W. D. Survival of frogs in low temperature. *Science* 215: 697–698, 1982.

183. Schmidt-Nielsen, B., K. Schmidt-Nielsen, A. Brokaw, and H. Schneiderman. Water conservation in desert rodents. *J. Cell. Comp. Physiol.* 32: 331–360, 1948.

184. Scholnick, D. A., and C. P. Mangum. Sensitivity of hemoglobins to intracellular effectors: primitive and derived features. *J. Exp. Zool.* 259: 32–42, 1991.

185. Schrock, H., R., Forster, and L. Goldstein. Renal handling of taurine in marine fish. *Am. J. Physiol.* 242 (*Regulatory Integrative Comp. Physiol.* 13): R64–R69, 1982.

186. Schuh, W., and H. Puff. The crystal structure of ectoine, a novel amino acid of potential osmoregulatory function. *Z. Naturforsch.* 40c: 780–784, 1985.

187. Sherman, W. R., P. C. Simpson, and S. L. Goodwin. *Scyllo*-inositol and *myo*-inositol levels in tissues of the skate *Raja erinacea*. *Comp. Biochem. Physiol.* 59B: 201–202, 1978.

188. Shifrin S., and C. L. Parrott. Influence of glycerol and other polyhydric alcohols on the quaternary structure of an oligomeric protein. *Arch. Biochem. Biophys.* 166: 426–432, 1975.

189. Shihabi, Z. K., H. O. Goodman, R. P. Holmes, and M. I. O'Connor. The taurinecontent of avian erythrocytes and its role in osmoregulation. *Comp. Biochem. Physiol.* 92A: 545–549, 1989.

190. Siebens, A. and K. Spring. A novel sorbitol transport mechanism in cultured renal papillary epithelial cells. *Am. J. Physiol.* 257 (*Renal Fluid Electrolyte Physiol.* 26): F937–F946, 1989.

191. Skedy-Winkler, C., and Y. Avi-Dor. Betaine: induced stimulation of respiration of high osmolarities in halotolerant bacterium. *Biochem. J.* 150: 219–226, 1975.

192. Smardo, F. L. Jr., M. B. Burg, and A. Garcia-Perez. Kidney aldose reductase gene transcription is osmotically regulated. *Am. J. Physiol.* 262 (*Cell Physiol.* 31): C776–C782, 1992.

193. Smirnoff, N., and Q. I. Cumbes. Hydroxyl radical scavenging activity of compatible solutes. *Phytochemistry* 28: 1057–1060, 1989.

194. Smith, H. W. The retention and physiological role of urea in the elasmobranchii. *Biol. Rev.* 11: 49–82, 1936.

195. Smith, L. T., and G. M. Smith. An osmoregulated dipeptide in stressed *Rhizobium meliloti*. *J. Bacteriol.* 171: 4714–4717, 1989.

196. Smith, L. T., G. M. Smith, and M. A. Madkour. Osmoregulation in *Agrobacterium tumefaciens*: accumulation of a novel disaccharide is controlled by osmotic strength and glycine betaine. *J. Bacteriol.* 172: 6849–6855, 1990.

197. Somero, G. N. Adapting to water stress: Convergence on common solutions. In: *Water and Life: A Comparative Analysis of Water Relationships at the Organismic, Cellular, and Molecular Levels*. edited by G. N. Somero, C. B. Osmond, and C. L. Bolis, Berlin: Springer-Verlag, 1992, p. 3–18, 1992.

198. Somero, G. N. Protons, osmolytes, and fitness of internal milieu for protein function. *Am. J. Physiol.* (*Regulatory Integrative Comp. Physiol.* 20). 251: R197–R213, 1986.

199. Somero, G. N., C. B. Osmond, and C. L. Bolis (Eds). *Water and Life: A Comparative Analysis of Water Relationships at the Organismic, Cellular, and Molecular Levels*, Berlin: Springer-Verlag, 1992.

200. Sommer, C., B. Thonke, and M. Popp. The compatibility of D-pinitol and 1D-1-0-methyl-muco-inositol with malate dehydrogenase activity. *Bot. Acta* 103: 270–273, 1990.

201. Star, R. A. Hyperosmolar states. *Am. J. Med. Sci.* 300: 402–412, 1990.

202. Steinbach, H. B. The prevalence of K. *Perspect. Biol. Med.* 5: 338–355, 1962.

203. Storey, K. B., and J. M. Storey. Freeze tolerance in animals. *Physiol. Rev.* 68: 27–84, 1988.

204. Subramaniam, S., and B. A. Jackson. Differential effects of trimethylamines (TMA) on enzyme activity in MDCK cells. *FASEB J.* 6: A958, 1992.

205. Timasheff, S. A physicochemical basis for the selection of osmolytes by nature. In: *Water and Life: A Comparative Analysis of Water Relationships at the Organismic, Cellular, and Molecular Levels*, edited by G. N. Somero, C. B. Osmond, and C. L. Bolis, Berlin: Springer-Verlag, 1992, p. 70–84.

206. Timasheff, S. The control of protein stability and association by weak interactions with water: How do solvents affect these processes? *Annu. Rev. Biophys. Biomol. Struct.* 22: 67–97, 1993.

207. Trachtman, H., R. Barbour, J. A. Sturman, and L. Finberg. Taurine and osmoregulation: taurine is a cerebral osmoprotective molecule in chronic hypernatremic dehydration. *Pediatr. Res.* 23: 35–39, 1988.

208. Truper, H. G., and E. A. Galinski. Concentrated brines as habitats for microorganisms. *Experientia* 42: 1182–1187, 1986.

209. Uchida, S., A. Garcia-Perez, H. Murphy, and M. B. Burg. Signal for induction of aldose reductase in renal medullary cells by high external NaCl. *Am. J. Physiol.* 256 (*Cell Physiol.* 25): C614–C620, 1989.

210. Uchida, S., H. M. Kwon, A. Yamauchi, A. S. Preston, F. Marumo, and J. S. Handler. Molecular cloning of the cDNA for an MDCK cell Na^+ and Cl^--dependent taurine transporter that is regulated by hypertonicity. *Proc. Natl. Acad. Sci. U.S.A.* 89: 8230–8234, 1992.

211. Ullrich, K. J. Uber das Vorkommen von Phosphoverbindungen in verschieden Nierenabschnitten und Anderungen ihrer Konzentration in Abhangigkeit vom Divresezustand. *Pflugers Arch.* 262: 551–561, 1956.

212. Ullrich, K. J. Glycerophosphorylcholinumsatz und glycerophosphorylcholindiesterase in der Saugetier-Niere. *Biochem. Z.* 331: 98–102, 1959.

213. Van Laere, A. Trehalose, reserve and/or stress metabolite? *FEMS Microbiol. Rev.* 63: 201–210, 1989.

214. Vieyra, A., C. Caruso-Neves, J. R. Meyer-Fernandes. ATP—^{32}P exchange catalyzed by plasma membrane Ca^{2+}-ATPase from kidney proximal tubules. *J. Biol. Chem.* 266: 10324–10330, 1991.

215. Von Hippel, P. H., and T. Schleich. The effects of neutral salts on the structure and conformational stability of macromolecules in solution. In: *Structure and Stability of Biological Macromolecules*, edited by S. N. Timasheff and G. D. Fasman, New York: Marcel Dekker, 1969, p. 417–574.

216. Weber, L. A., E. D. Hickey, P. A. Maroney, and C. Baglioni. Inhibition of protein synthesis by Cl^-. *J. Biol. Chem.* 252: 4007–4010, 1977.

217. Wegmann, K. Osmoregulation in eukaryotic algae. *FEMS Microbiol. Rev.* 39: 37–43, 1986.

218. Wiencke, C. The response of pyruvate kinase from the inter-

tidal red alga *Porphyra umbilicalis* to sodium and potassium ions. *J. Plant Physiol.* 116: 447–453, 1984.

219. Wiggins, P. M. Role of water in some biological processes. *Microbiol. Rev.* 54: 432–449, 1991.

220. Wilkie, D. R., and S. Wray. The response of the African frog *Xenopus laevis* to increased salinity demonstrated by ^{31}P-NMR spectroscopy. *J. Physiol.* 381: 30P, 1986.

221. Winzor, C. L., D. J. Winzor, L. G. Paleg, G. P. Jones, and B. P. Naidu. Rationalization of the effects of compatible solutes on protein stability in terms of thermodynamic nonideality. *Arch. Biochem. Biophys.* 296: 102–107, 1992.

222. Wirthensohn G, F. Beck, and W. G. Guder. Role and regulation of glycerophosphorylcholine in rat renal papilla. *Pflugers Arch.* 409: 411–415, 1987.

223. Wirthensohn G, S. Lefrank, M. Schmolke, and W. Guder. Regulation of organic osmolyte concentrations in tubules from rat renal inner medulla. *Am. J. Physiol.* 256 (*Renal Fluid Electrolyte Physiol.* 25): F128–F135, 1989.

224. Wolff, S. D., T. S. Stanton, S. L. James, and R. S. Balaban. Acute regulation of the predominant organic osmolytes of the rabbit renal inner medulla. *Am. J. Physiol.* 257 (*Renal Fluid Electrolyte Physiol.* 26): F676–F681, 1989.

225. Woolverton, W., S. Githens, R. O'Dell-Smith, and C. Bartell. Rat renal papillary tissue explants survive and produce epithelial monolayers in culture media made hyperosmotic with sodium chloride and urea. *J. Exp. Zool.* 256: 189–199, 1990.

226. Wright, S. H., and D. T. Manahan. Integumental nutrient uptake by aquatic organisms. *Annu. Rev. Physiol.* 51: 585–600, 1989.

227. Wright, S. H., T. M. Wunz, and A. L. Silva. Betaine transport in the gill of a marine mussel, *Mytilus californianus. Am. J. Physiol.* 263 (*Regulatory Integrative Comp. Physiol.* 32): R226–R232, 1992.

228. Wyn Jones, R. G. Phytochemical aspects of osmotic adaptation. *Rec. Adv. Phytochem.* 18: 55–78, 1984.

229. Wyn Jones, R. G., R. Storey, R. A. Leigh, N. Ahmad, and A. Pollard. A hypothesis on cytoplasmic osmoregulation. In: *Regulation of Cell Membrane Activities in Plants,* edited by E. Marre and O. Ciferri, Amsterdam: Elsevier, 1977, p. 121–136.

230. Yamauchi, A., S. Uchida, H. M. Kwon, A. S. Preston, R. B. Robey, A. Garcia-Perez, M. B. Burg, and J. S. Handler. Cloning of a Na^+- and Cl^--dependent betaine transporter that is regulated by hypertonicity. *J. Biol. Chem.* 267: 649–652, 1992.

231. Yancey, P. H. Organic osmotic effectors in cartilaginous fishes.

In: *Transport Processes, Iono- and Osmoregulation,* edited by R. Gilles and M. Gilles-Ballien, Berlin: Springer-Verlag, 1985, p. 424–436.

232. Yancey P. H. Osmotic effectors in kidneys of xeric and mesic rodents: corticomedullary distributions and changes with water availability. *J. Comp. Physiol.* 158B: 369–380, 1988.

233. Yancey, P. H. Compatible and counteracting aspects of organic osmolytes in mammalian kidney cells *in vivo* and *in vitro.* In: *Water and Life: A Comparative Analysis of Water Relationships at the Organismic, Cellular, and Molecular Levels,* edited by G. N. Somero, C. B. Osmond, and C. L. Bolis, Berlin: Springer Verlag, 1992, p. 19–32.

234. Yancey P. H., and M. B. Burg. Distributions of major organic osmolytes in rabbit kidneys in diuresis and antidiuresis. *Am. J. Physiol.* 257 (*Renal Fluid Electrolyte Physiol.* 26): F602–607, 1989.

235. Yancey, P. H., and M. B. Burg. Counteracting effects of urea and betaine on colony-forming efficiency of mammalian cells in culture. *Am. J. Physiol.* 258 (*Regulatory Integrative Comp. Physiol.* 27): R198–204, 1990.

236. Yancey P. H., M. B. Burg, and S. M. Bagnasco. Effects of NaCl, glucose and aldose reductase inhibitors on cloning efficiency of renal cells. *Am. J. Physiol.* 258: (*Cell Physiol.* 27): C156–C163, 1990.

237. Yancey, P. H., M. E. Clark, S. C. Hand, R. D. Bowlus, and G. N. Somero. Living with water stress: the evolution of osmolyte systems. *Science* 217: 1212–1222, 1982.

238. Yancey, P. H., R. G. Haner, and T. Freudenberger. Effects of an aldose reductase inhibitor on osmotic effectors in rat renal medulla. *Am. J. Physiol.* 259 (*Renal Fluid Electrolyte Physiol.* 28): F733–F738, 1990.

239. Yancey, P. H., and G. N. Somero. Urea-requiring lactate dehydrogenases of marine elasmobranch fishes. *J. Comp. Physiol.* 125: 135–141, 1978.

240. Yancey, P. H., and G. N. Somero. Counteraction of urea destabilization of protein structure by methylamine osmoregulatory compounds of elasmobranch fishes. *Biochem. J.* 182: 317–323, 1979.

241. Yancey, P. H., and G. N. Somero. Methylamine osmoregulatory compounds in elasmobranch fishes reverse urea inhibition of enzymes. *J. Exp. Zool.* 212: 205–213, 1980.

242. Zablocki, K., S.P.F. Miller, A. Garcia-Perez and M. B. Burg. Accumulation of glycerophosphorylcholine (GPC) by renal cells: osmotic regulation of GPC: choline phosphodiesterase. *Proc. Natl. Acad. Sci. USA.* 88: 7820–7824, 1991.

11. Of membrane stability and mosaics: the spectrin cytoskeleton

JON S. MORROW

DAVID L. RIMM

SCOTT P. KENNEDY

CAROL D. CIANCI

JOHN H. SINARD

SCOTT A. WEED

Department of Pathology, Yale University School of Medicine, New Haven, Connecticut

CHAPTER CONTENTS

THE PRESENCE OF A SUPPORTING INFRASTRUCTURE BENEATH ERYTHROCYTE MEMBRANES was surmised in 1968 by Vincent Marchesi and Edward Steers, working at the National Institutes of Health, who observed fragmentation and vesiculation of erythrocyte membranes following the removal of protein (311). The high-molecular-weight protein responsible for stabilization of the erythrocyte membrane was identified and named *spectrin* (312, 508) because it was derived from hemoglobin-depleted erythrocyte "ghosts." Now, more than a quarter of a century later and after intense investigation by laboratories throughout the world, the nature of the erythrocyte *membrane skeleton* (also variously called the *cortical cytoskeleton* or the *spectrin skeleton*) is understood in reasonable detail. Spectrin, and most major proteins with which it associates, has been cloned and sequenced; interactions between components of the membrane skeleton have been studied by quantitative in vitro assay and visualized in situ by sophisticated ultrastructural techniques; and many forms of inherited hemolytic disease are now understood as arising from specific mutations in spectrin or in one of its associated proteins (301, 542, for other topical reviews). In addition, beginning in the early 1980s, proteins related to the spectrin skeleton of erythrocytes were identified in a variety of nonerythroid tissues. These studies demonstrated a remarkable degree of sequence conservation between spectrin from different species and tissues but increasingly highlighted an unexpected complexity in non-erythroid spectrin skeletal composition, developmental regulation, and intracellular sorting. These observations challenged paradigms based solely on the erythrocyte spectrin skeleton and suggested that the spectrin membrane skeleton contributes to cellular function in ways that not even the most enthusiastic of the early investigators imagined.

This chapter focuses primarily on spectrin, its established role in the erythrocyte, and the emerging evidence for isoform and compositional complexity in the spectrin skeleton of non-erythroid cells. The mechanisms by which this complexity is achieved and maintained are reviewed. The structural model that has emerged from studies of the erythrocyte is reviewed in the light of current knowledge, and this model is extended to include the way that spectrin may interact with the membrane in non-erythroid cells. A unifying hypothesis of spectrin skeletal function is proposed, called the *linked mosaic model*. This concept proposes a universal role of spectrin as a multifunctional membrane-associated organizing center about which macromolecular complexes containing membrane proteins and receptors, cytoplasmic signaling molecules, and structural elements are gathered and coordinately regulated. Evidence that this membrane-organizing role of the spectrin skeleton also forms the basis for several aspects of its contributions to the stability and shape of the erythrocyte membrane is reviewed.

THE RED CELL MEMBRANE SKELETON

Our understanding of the cortical membrane cytoskeleton comes largely from studies of mammalian erythrocytes, where it bestows the structural and functional adaptability needed for the red cell to withstand the dynamic shear stresses of blood circulation. Its primary role in the red cell is to provide global membrane support, a function contributed to by its intrinsic tensile properties, regulation of the lateral mobility and organization of integral membrane proteins (see later under THE TRILAYER COUPLE . . .), and possibly even interactions with phospholipids (31, 91, 301, 542). Besides spectrin, an ensemble of peripheral and integral membrane proteins contribute to the erythrocyte skeleton, either by forming lateral associations within the plane of the cortical cytoskeleton or by firmly anchoring it to at least three distinct integral membrane proteins. These features of the red cell cytoskeleton have been reviewed extensively (31, 91, 168, 296, 297, 301, 542). Their basic layout is depicted in Figure 11.1, along

with views of the in situ membrane skeleton as discerned from complementary techniques (Fig. 11.2.)

Approximately 16 major proteins contribute to either the attachment or stability of the erythrocyte membrane skeleton (Table 11.1). Many additional proteins are also present at lower levels, including transport proteins and blood group antigens, but the nature of most minor components or their function remains obscure. The cortical cytoskeletal proteins are all peripheral, meaning that they can be extracted from the membrane by chaotropic agents without detergents. Conversely, proteins that anchor the cytoskeleton to the bilayer are integral, the two most abundant being the anion exchanger (AE1 or band 3) and glycophorins A and C.

With the discovery of multiple points of skeletal attachment to the membrane (Fig. 11.1 and Table 11.1), a central question has been, which of these many interactions guide the assembly of the skeleton during erythropoiesis and which are primarily responsible for stabilizing the mature skeleton? This has turned out to be a surprisingly difficult question to answer. Almost all studies clearly indicate that the β-subunit of spectrin is required for membrane assembly and anchoring of the spectrin skeleton; hence any required interaction(s) must be functions bestowed by this subunit. Evidence indicating that α-spectrin is usually (but not always, 271) synthesized in excess of β-spectrin in most cells, but that its assembly into the mature skeleton is limited by the amount of β-spectrin, also supports the primacy of β-spectrin in the process of assembly and stabilization (181, 267, 345). Early on, it was demonstrated that a specific site within β-spectrin interacts with ankyrin (357), the first linking protein identified (34). This fact, together with the observation that the conditions (0.1 mM EDTA) needed to release spectrin from ghost membranes (Fig. 11.1B) closely paralleled those that inhibit the binding of β-spectrin to ankyrin in vitro, and that ankyrin could mediate the reassociation of spectrin to inside-out-vesicles (IOVs) depleted of ankyrin and protein 4.1 (29, 34), suggested that the ankyrin linkage to AE1 guided attachment of the skeleton to the membrane in erythrocytes (reviewed in 28). Supporting this notion was the observation that in some cell lines stable spectrin assembly is not observed until the onset of AE1 synthesis, suggesting that AE1 (via ankyrin) initiated skeletal assembly (179, 180). It is now apparent that the answer to what initiates and guides spectrin assembly at the membrane is much more complicated. Several other membrane-linking proteins have emerged, such as protein 4.1 (503), α- and β-adducin (335, 465), and actin (85, 297, for review). In addition, direct interactions between β-spectrin and unknown integral membrane proteins that

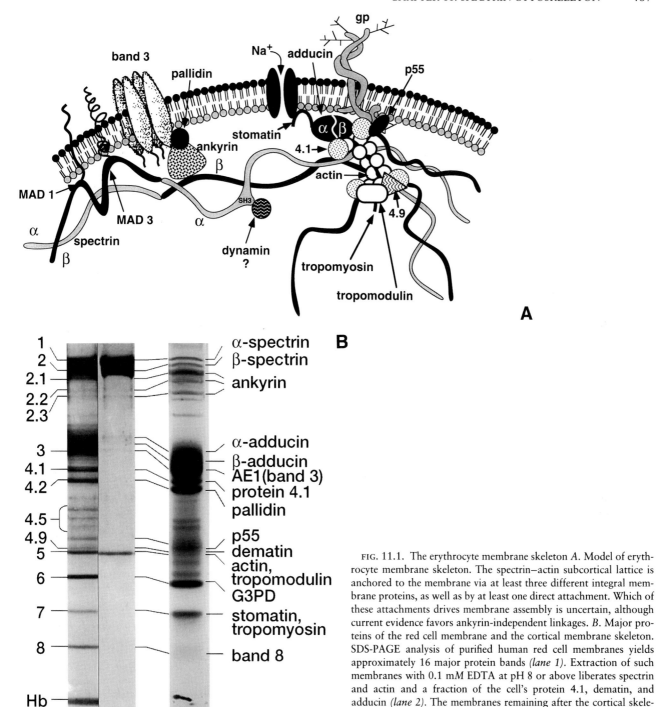

FIG. 11.1. The erythrocyte membrane skeleton A. Model of erythrocyte membrane skeleton. The spectrin–actin subcortical lattice is anchored to the membrane via at least three different integral membrane proteins, as well as by at least one direct attachment. Which of these attachments drives membrane assembly is uncertain, although current evidence favors ankyrin-independent linkages. B. Major proteins of the red cell membrane and the cortical membrane skeleton. SDS-PAGE analysis of purified human red cell membranes yields approximately 16 major protein bands *(lane 1)*. Extraction of such membranes with 0.1 mM EDTA at pH 8 or above liberates spectrin and actin and a fraction of the cell's protein 4.1, dematin, and adducin *(lane 2)*. The membranes remaining after the cortical skeleton is removed retain all of the integral proteins and many of the linker proteins *(lane 3)*. The nomenclature according to their order on SDS-PAGE gels as well as their assigned names is given.

do not require ankyrin or any other linking protein also exist. These *ankyrin-independent* interactions are capable of mediating the attachment of spectrin to membranes both in vitro (109, 207, 287, 481) and in vivo (479). When the ankyrin-to-AE1 linkage is experimentally dissociated without disrupting other skeletal associations, the skeleton remains assembled, albeit more fragile (292), and at least limited assembly of the nascent spectrin skeleton proceeds before the synthesis of AE1 (255, 271). Very recent data also indicating that linkage via ankyrin to AE1 is not required for membrane skeletal assembly come from two

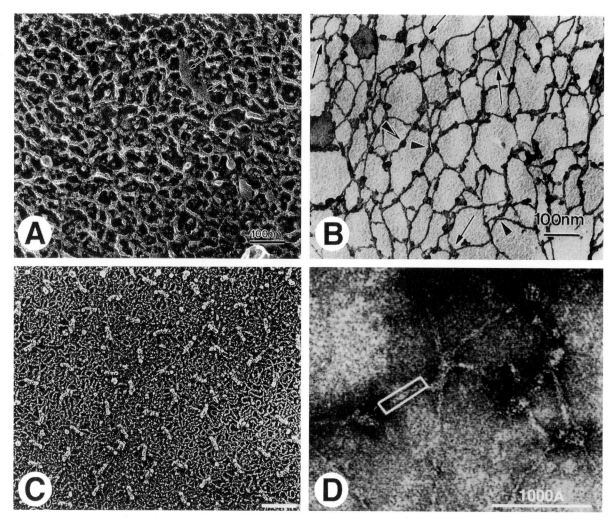

FIG. 11.2. Four views of the spectrin membrane skeleton. A. The erythrocyte cytoskeleton in situ. This sample was prepared from paraformaldehyde-fixed ghosts attached to coated coverslips and sheered to expose the skeleton, followed by quick-freeze, deep-etch, rotary-replication (QFDERR). The arrows point to unquenched water left in the sample. The average filament length is 29–37 nm, which is about a third of the extended length of the spectrin heterodimer. This information together with immunogold labeling has suggested that in situ the skeleton is highly compact, with many side-to-side associations of spectrin as well as a large fraction of spectrin hexamers and octamers as well as tetramers (518). B. A view of the spread isolated skeleton prepared by extraction of ghosts with 2.5% Triton X-100, followed by spreading onto a formvar film followed by QFDERR. After this treatment, the compact skeleton seen in situ is highly expanded. Small arrows point to apparent actin sites; large arrows point to spectrin hexamer junctions. Arrowheads point to laterally associated filaments. The double arrowhead desig-

nates an ankyrin–AE1 complex (519). (Panels A and B were adapted with permission from the references cited.) C. Quick-freeze deep-etch replica of the inner face of a red cell membrane in situ that has been extracted with NP-40 and NaCl to remove most of the ankyrin and protein 4.1, exposing the network of intertwined spectrin and 67 nm actin nuggets. Compared with (A), this view depicts a much looser skeleton. It is likely that the salt and detergent treatment used in this preparation has loosened the side-to-side associations of spectrin as well as removed many associated proteins. [Micrograph provided compliments of Dr. J. Heuser and reprinted with permission from Coleman et al. (91).] D. Image of an expanded skeleton prepared by negative staining. The stretched spectrin filaments (tetramers) predominate in these preparations, although hexamers are seen (279). The preparation shown here was used to determine by Fourier analysis the supercoiling of the spectrin heterotetramer (see Fig. 11.4). [Adapted with permission from McGough and Josephs (323).]

additional observations. Mice in which the AE1 gene has been deleted by targeted disruption display marked anion transport anomalies and a hemolytic anemia, yet have a spectrin skeleton that is assembled and apparently intact (402, 558). Related observations have also been made in a cow with a spontaneous mutation

that results in an absence of AE1 (550). The erythrocytes of these animals display marked instability with endo- and exo-vesiculation and transport anomalies, but nevertheless an assembled albeit less stable spectrin membrane skeleton.

These studies, although still preliminary, are consis-

TABLE 11.1. *Major Proteins Involved with the Human Erythrocyte Membrane Skeleton*

SDS Gel Band[a]	Protein	SDS-PAGE $M_r \div 10^3$	Calculated[b] MW $\div 10^3$	Copies[c] Cell $\div 10^3$	Copies/ Spectrin	Copies/ Adducin	Type[d] Protein	GenBank[e] Accession No.
1	αI-spectrin	240	281	240	1.0	8	C	M61877
2	βI-spectrin	220	246	240	1.0	8	C	J05500
2.1-2.3	Ankyrin[f]	210–190	210	120	0.5	4	L	X16609
3′ (prime)[g]	α-Adducin	105	81	30	0.1	1	C, L	X58141
	β-Adducin	100	80	30	0.1	1	C, L	X58199
3[h]	Anion exchanger 1	90–100	102	1,200	5.0	40	T	M27819
4.1[i]	Protein 4.1a	80	66	200	0.8	7	C, L	M14933
	Protein 4.1b	78						
4.2	Pallidin	72	77	250	1.0	8	C, L	M30646
4.5	p55	55	53	80	0.3	3	L	M64925
4.9	Dematin[j]	48 & 52	43 & 46	140	0.6	5	C	L19713 & U28389
5	β-Actin	43	42	500	2.1	17	C	X00351
	Tropomodulin	43	41	30	0.1	1	C	M77016
6	G3PD[k]	35	36	500	2.1	17	L	M17851
7.1	Tropomyosin[l]	27 & 29	28[l]	70	0.3	2	C	X04588
7.2	Stomatin	31	32	200	0.8	7	T	M81635
PAS[m]-1	Glycophorin A	36	14	1,000	4.2	33	R	X51798
PAS-2	Glycophorin C[n]	32	14	200 (C + D)	0.8	7	R	M36284
PAS-3	Glycophorin B	20	8	200	0.8	7	R	J02982
	Glycophorin D	23	11				R	M36284
	Glycophorin E[p]		6				R	M29610

Adapted and modified from (301).

[a] Based on original numbering system of Fairbanks and Steck (134).

[b] All calculated MW's do not include any contributions from carbohydrate (i.e., for AE1 and all of the glycophorins).

[c] Based on densitometric scanning of Coomassie blue stained SDS-PAGE gels or, in the case of spectrin and ankyrin, by radioimmunoassay (301).

[d] Classification of protein as: (C), cytoskeletal component; (L), a linking protein that joins the skeleton to an integral membrane protein; (T), a transporter related protein; or (R) a receptor type protein. T and R are both integral membrane proteins, while C and L are peripheral membrane proteins. Typically, transporter proteins (T) are type II integral membrane proteins with considerable mass within the bilayer; receptor proteins (R) are usually type I integral membrane proteins with only a single bilayer.

[e] The GenBank accession number allows easy retrieval of the completed cDNA sequence and derived protein sequence. In cases where more than one sequence has been accessioned, the most complete sequence is listed.

[f] Band 2.1 represents a full-length transcript of ankyrin. Band 2.2 (195 Mr) arises by alternative transcription of the ANK1 gene (261). Smaller ankyrins (band 2.3, 175 Mr; band 2.6, 145 Mr) are also detectable in erythrocytes. Whether these arise from proteolysis or represent additional alternative transcripts, as in non-erythroid cells (117, 249, 400), is unknown.

[g] Band 3 obscures the α & β adducin bands on SDS gels. Adducin was first detected in membranes in which band 3 had been digested from the outside, and was revealed as band "3 prime" (298).

[h] Band 3 migrates on SDS-PAGE as a broad band due to heterogeneous glycosylation.

[i] Two protein 4.1 bands may be discerned, 4.1a and 4.1b. The larger band arises from the smaller band by a process of slow deamidation of glutamine and aparagine, a process reflecting RBC age.

[j] Two protein bands are apparent, one at 48 Mr and one at 52 Mr. Both arise from the same gene; the larger subunit contains an internal insertion that arises by alternative mRNA splicing.

[k] Only a small fraction of G3PD exists on the membrane in vivo. Under the low-ionic strength conditions of red cell membrane preparation, it is partitioned selectively to its binding site on AE1. This binding has a down-regulatory effect on its enzymatic activity (291).

[l] Tropomyosin has been identified by immunologic criteria only; the red cell protein has not been cloned. The data shown are for fibroblast tropomyosin.

[m] PAS refers to periodic acic–Schiff stained gels. All of the glycophorins are poorly stained and inapparent on Coomassie blue stained gels.

[n] Glycophorins C&D.

[p] Glycophorin E is related to C&D. Its message has been identified, but the protein product has not (65, 505).

tent with an emerging body of other data suggesting that integral membrane proteins per se play a major role in determining erythrocyte shape and stability (69, 257, 341, 448), and that ankyrin is just one of several membrane linkages that can mediate membrane assembly (109, 194, 195, 284, 285, 288, 313, 349, 420, 421, 465, 484). In this sense, a more accurate view of ankyrin's role is as a mechanism for attaching certain integral membrane proteins to the underlying skeletal lattice, rather than as an attachment of the skeleton to the membrane. While these different viewpoints might seem merely semantic, as described below an alternative view of spectrin skeletal function is as a membrane-organizing center, and that its contributions to membrane stability follow directly from this organizing role. By viewing its ankyrin-mediated linkages as points of membrane organization rather than as centers of cytoskeletal attachment, a unifying view of both erythroid and non-erythroid spectrin function emerges.

HOW DOES THE SPECTRIN MEMBRANE SKELETON STABILIZE THE RED CELL?

Three features of the red cell determine its shape and stability: (1) its surface area to volume ratio (SA/V); (2) the viscosity of its cell contents (mainly hemoglobin), which limits the rate at which it may deform; and (3) the intrinsic viscoelastic properties of the bilayer membrane with its associated spectrin skeleton. The spectrin skeleton contributes to the first and third of these features. Typically, a mature red cell has a generous surface area to volume ratio, allowing large deformations at a constant volume. This ratio is perturbed either by loss of membrane surface area or by a failure of the cell to regulate its internal volume. Examples of both conditions are common. A variety of lipophilic compounds can intercalate into either face of the lipid bilayer, expanding one or both and disrupting cell shape in accord with classical bilayer couple theory (340, 351, 454, 496). Hemolytic conditions associated with instability of the spectrin skeleton or an inadequate amount of skeletal proteins lead to membrane vesiculation and the loss of membrane surface area, reducing the SA/V and resulting in fragile, small, spherocytic red cells (4, 67). Similarly, conditions in which the control of cell volume is abnormal, most typically manifested by a failure to control Na^+/K^+ ratios properly across the membrane, lead either to cell dehydration or overhydration and other shape changes, such as those seen in hereditary stomatocytosis (262, 484). Interestingly, the spectrin skeleton, acting through linkages to adducin and stomatin, may also play a role in the regulation of such fluxes (see later under stomatin).

The membrane skeleton also contributes to membrane shape and stability by modifying the viscoelastic properties of the membrane. The removal of spectrin by low ionic strength media induces membrane instability and vesiculation. Manipulations that induce an aggregation of the spectrin skeleton, such as low pH or Ca^{2+} loading, lead to marked shaped changes such as echinocytes and often the loss of spectrin-free membrane blebs (132). In mice with congenital absence of α-spectrin (sph/sph), the severe fragility and spherocytic shape of their red cells is ameliorated by returning spectrin to the membrane (101). Similarly, mutations in spectrin or any several of its associated proteins are now recognized as the basis of a wide array of inherited hemolytic disease (Fig. 11.3) (see reviews by 22, 301, 309, 401).

Less clear is how the spectrin skeleton actually contributes to the stability and shape of the membrane. Several features of the skeleton play a role (reviewed in 338, 339, 495). The membrane bestows an element of elasticity to the cell, facilitating the recovery of normal red cell shape after transient deformations such as those that occur during passage through capillaries or splenic sinusoids. In its native state, the spectrin skeleton also appears to be under a slight tension, yet with the spectrin molecules still highly condensed relative to their fully extended contour length (Figs. 2 and 4) (323, 519). This feature, together with the highly anastamosing yet isotropic arrangement of the cortical skeleton (Fig. 2), contributes to membrane elasticity, especially under conditions of large deformation. Consonant with this notion, several investigators have hypothesized that spectrin functions like an ionic gel or thermodynamic spring, with the state of spectrin condensation (that is, length) mediated by electrostatic interactions within the protein and sensitive to the ionic milieu (486–488, 498). In vitro, both the isolated spectrin subunits as well as the heterotetramer have a scaled flexural rigidity similar to that of an actin filament (90). Fourier reconstructions of the spectrin heterotetramer in situ demonstrate how its length and dimensions change as the molecule is stretched (Fig. 11.4).

However, recent observations suggest that the elastic properties of the skeleton per se may be less important to membrane shape and rigidity than its ability to control the organization of integral membrane proteins (256, 257, 337–339, 495). Recent studies reveal the area expansivity modulus of the spectrin skeleton to be orders of magnitude weaker (described as 100-fold softer than "soft latex") than the expansivity modulus of the bilayer to which it is joined (339). It is therefore hard to attribute more than a negligible direct contribution of the skeleton to the shape, stretch, or bending properties of the membrane, except at extremes of

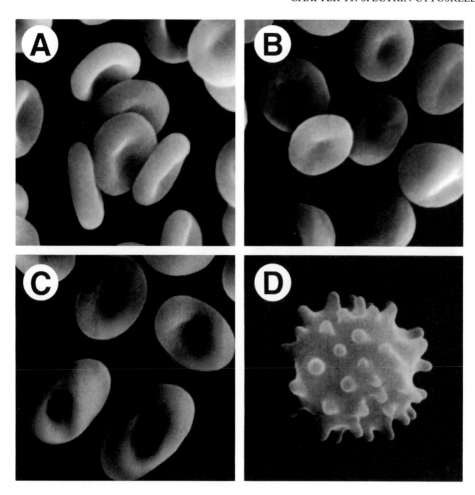

FIG. 11.3. Alterations in the spectrin skeleton alter the shape and stability of erythrocytes. Scanning electron microscopic images: *A*. Normal red cell. *B*. Defects in ankyrin or AE1 often lead to spherocy-tosis. *C*. Inherited defects in the self-association of spectrin often lead to hereditary elliptocytosis. *D*. Increased intracellular Ca^{2+}, lowered pH, or ATP depletion induce echinocytic transformations.

deformation or under conditions of rapid deformation when elastic limits are approached and the sheer modulus of the skeleton becomes important. Other studies accent the importance of the integral membrane proteins in controlling the cell's shape and deformation properties. This is most dramatically demonstrated in a condition called hereditary ovalocytosis, in which red cells are axially deformed and 10 to 20 times more rigid than normal (341, 448). The spectrin–actin skeleton of these cells appears to be normal. Their only defect is a mutation in AE1, in which nine amino acids (residues 400–408) are deleted adjacent to the beginning of AE1's first transmembrane domain. It has been hypothesized that this change renders the cytoplasmic domain of AE1 susceptible to entanglement in the underlying spectrin skeleton, thereby rigidifying the cell (341). Consistent with this is the observation that AE1 in these cells displays a markedly restricted rotational mobility that is relieved by the removal of ankyrin and

pallidin (509). However, a more general explanation may be that this mutation alters the conformation of AE1 in a way that substantively changes the energetics of its interaction with other integral proteins both within and at the surface of the bilayer. How this might happen and the potential importance of controlling integral membrane protein organization even in a red cell, derives from a consideration of bilayer couple theory and the distributional free energy of embedded membrane molecules (131, 220, 257, 493, 494, 497).

THE TRILAYER COUPLE—SPECTRIN AS A MEMBRANE ORGANIZER

If integral proteins modulate the membrane's shape and deformability, and if the direct contributions of the spectrin skeleton to the stretch and bending energies of the membrane are negligible, then it is fair to ask

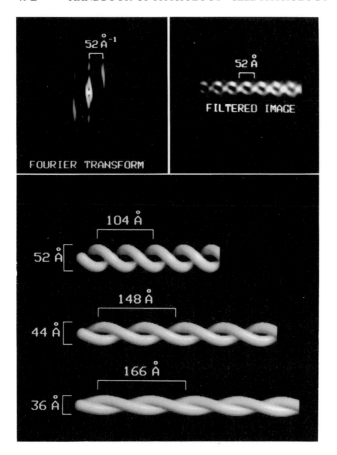

FIG. 11.4. Fourier filtered images of the spectrin heterotetramer in situ. These images were derived from negative stained electron micrographs of the stretched spectrin cortical skeleton such as the micrograph shown in Figure 11.2D. This analysis revealed a coiled-coil quaternary conformation of spectrin in situ. Note that the spectrin heterotetramer can be linearly deformed, accounting for its proposed "spring-like" properties. Shown are three filtered images demonstrating how the pitch and length of the spectrin helix change with length. Each turn of the helix is estimated to contain four spectrin repeat units. Upon extraction from the membrane, spectrin heterodimers and heterotetramers assume a much more open and loose conformation (cf. Fig. 11.12). [Adapted with permission from McGough and Josephs (323).]

why membrane instability and shape changes appear whenever the spectrin skeleton is perturbed or defective. Part of the answer may lie in the impact of integral protein distributions on the bending energies of a bilayer, and the influence of the spectrin skeleton on these distributions. In a simplified way, the impact that the distribution of asymmetric integral proteins have on the shape and stability of a membrane bilayer is illustrated in Figure 11.5 (316, 494). These considerations have been theoretically explored in detail, leading to some surprising insights (131, 219, 220, 257, 316, 493, 494, 497).

In these treatments, following bilayer couple theory,

the membrane is considered as a trilayer, consisting of the phospholipid bilayer and a third layer consisting of the skeleton. Hydrophobic forces join the two lipid layers, and the skeleton is attached via its interactions with integral proteins. The three layers are considered to be in close contact, but otherwise laterally unconstrained. If the energy of each is considered independently, the energy of the trilayer system (W) is obtained as the sum of the elastic energies of the bilayer (W_b), the elastic energy of the skeleton (W_s), the interaction (binding) energy between the skeleton and the bilayer (W_i), and the energy of interaction of the embedded molecules with the bilayer (W_{mol}) (220, 493):

$$W = W_b + W_s + W_i + W_{mol} \qquad (11–1)$$

Neglecting for the moment the sheer elasticity of the skeleton, the interaction energy term, and the embedded protein energy term, the elastic energy of each

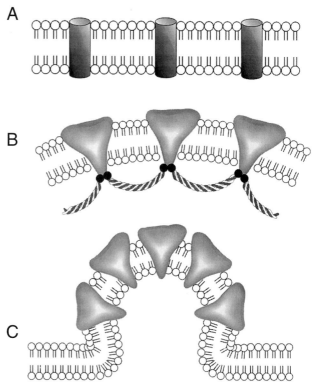

FIG. 11.5. Effects of embedded proteins as predicted by the bilayer couple hypothesis. Integral membrane proteins can have a major effect on the shape and stability of the bilayer. Control by a cortical skeleton ameliorates this potential effect. A. Cylindrical embedded proteins can impart curvature to the membrane by expanding one face preferentially. They thus will tend to cluster in regions of curvature that reduce their net interaction energy with the bilayer. A cortical skeleton aids in maintaining a homogeneous distribution of these molecules, and thereby contributes to membrane stability. C. Released of the constraints of a cortical skeleton, asymmetric embedded proteins can induce acute membrane curvatures and instability.

layer can be described either individually or collectively as the sum of two deformational modes involving isotropic membrane area expansion and membrane bending. The membrane bending term includes both nonlocal bending and local bending terms. This approximation yields three terms that describe the overall elastic energy of the trilayer considered as a unit. [Note: the following derivation follows closely the analysis set forth by Svetina and colleagues, to whom full credit is given (493).] Specifically, the energy of the trilayer is given by:

$$W = \frac{1}{2}\frac{K}{A_o}(A - A_o)^2 + \frac{1}{2}\frac{k_r}{A_o}(C - C_o)^2 \quad (11\text{-}2)$$
$$+ \frac{1}{2}k_c \int (c_1 + c_2 - c_o)^2 \, dA$$

In this equation, A_o is the equilibrium value of the area expansivity term A, and K is the area expansivity modulus of the overall trilayer. C_o is the equilibrium term for the integral of all curvatures over the membrane, and k_r is the nonlocal bending elastic modulus. In the last term, k_c is the local membrane bending modulus and c_o is the spontaneous local curvature. Collectively, the first term in Equation 11-2 accounts for the energy of stretching of the membrane considered as a trilayer of lipid and cortical skeleton, while the second two terms account for the energy of bending of the trilayer.

Following classic generalized bilayer couple theory, the shape of the closed membrane (and cell) at a given volume and surface area will be determined by the minimum of the sum of the local and nonlocal bending energies in Equation 11-2. Such treatments predict a variety of shapes as a function of the relative volumes and surface area differences between bilayers (256), including ellipsoidal shapes and spherical shapes with budding vesicles (493). Of particular interest is what happens to the cortical skeletal layer under cell deformation, and what is the impact of embedded protein on the energetics, shape, and stability of the bilayer with or without a tethering skeleton.

It is plausible that the skeleton seeks to reduce its energy under deformation, and that this energy (W_{skel}) is the sum of its elastic and interaction energies:

$$W_{skel} = W_s + W_i \quad (11\text{-}3)$$
$$W_{skel} = \frac{1}{2}K_s(A_s - A_{so})^2/A_{so} - \gamma(A - A_f)$$

where K_s is the area expansivity modulus of the skeleton, A_s is the area of the skeleton, A_f is the area of the bilayer that is not underlain with the cortical skeleton, and A_{so} is the areas of the relaxed skeleton. The attachment energy of the skeleton per unit area of the membrane (γ) arises from the total of all binding interactions of the skeleton with the bilayer and its embedded proteins. Because the area of the relaxed skeleton A_{so} is less than that of the bilayer (498), under normal circumstances the skeleton is slightly expanded and under some tension. Thus, the determination of whether the skeleton will stretch with the bilayer under deformation, or detach and collapse, leaving the bilayer without an underlying skeleton, will derive from the balance between the favorable energetics of decreasing skeletal expansion versus the loss of binding energy that accompanies detachment, as defined in Equation 11-3. Calculations based on this model suggest that detachment may be relatively rare except under extremes of deformation (493), although this tendency would be expected to increase whenever there was a weakening in the strength of skeletal-to-membrane attachment. Such weakening occurs in some forms of hereditary spherocytosis, and the considerations above may offer insights into why these cells bud off skeletal-free microvesicles (for review, see 301).

The effect of the integral membrane proteins themselves on membrane shape and stability may also be significant (Fig. 11.5) (316, 494). Assuming that all embedded molecules are equivalent, axisymmetric, and perpendicular to the membrane, and taking the energy of interaction for each molecule to be approximated as a second order polynomial of the two local membrane curvatures at the site of the molecule, the energy of the embedded molecules W_{mol} is (257, 493):

$$\frac{W_{mol}}{8\pi k_c} = \kappa p \int [(c_1 + c_2 - c_s)^2 - \frac{4}{3}c_1 c_2] \quad (11\text{-}4)$$
$$(n/\bar{n}) \, da + p \int (n/\bar{n}) \ln(n/\bar{n}) \, da$$

where the constant c_s represents the ideal curvature (that is, minimal energy) for the embedded molecules; $\kappa = K_e/kTR_o^2$, K_e is a constant proportional to the strength of the interaction of the embedded molecules with the bilayer; k is the Boltzmann constant, T the temperature, $p = NkT/8\pi k_c$; N is the total number of embedded molecules; n is the surface density of the molecules; $\bar{n}$ is the uniform density of the same number of molecules in the membrane. Minimizing Equation 11-1 for a fixed number of embedded molecules, with W_{mol} included (but neglecting W_s and W_i), reveals that the distribution of the embedded molecules and cell shape and curvature are strongly interdependent (257, 316, 493).

The presence of an underlying cortical skeleton attached to the integral proteins will resist displacement of the integral proteins proportional to the amount the laterally homogeneous skeleton is deformed (Fig. 11.5). The effect of a homogeneous skeleton is thus generally

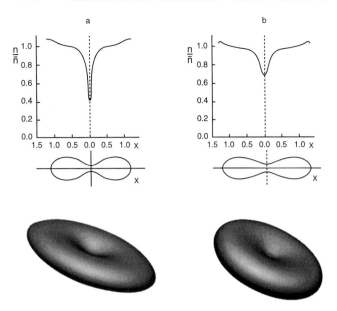

FIG. 11.6. By controlling the distribution of embedded proteins, the cortical skeleton controls cell shape. The equilibrium erythrocyte shape and the lateral density of the embedded proteins and cortical skeleton has been calculated *(A)* without the stabilizing effect of the skeleton or *(B)* with a cortical skeleton. In these calculations, the relative volume was held constant at 0.6, and the relative surface areas of the two leaves of the bilayer was set at 1.0465; $c_s = 4$; $k = 0.005$; $p = 500$. Note that in case *A*, the equilibrium shape is elliptical, whereas in *B* the equilibrium shape is axisymmetric and biconcave. Also note the markedly inhomogeneous distribution of protein in the case without skeletal stabilization. Views are presented along the minor axis; the ellipse semiaxis ratio in *A* is 1.18. [Adapted with permission from Svetina et al. (493).]

to resist inhomogeneities in integral membrane protein distributions, although it can also be expected that an inhomogeneous skeleton may favor localized inhomogeneities in membrane proteins. Such locally inhomogeneous skeletal arrays have been observed at points of cell–cell contact, receptor clusters, or on internal membrane compartments (see Fig. 11.20). The elastic energy of skeletal deformation, now expressed in terms of the number of embedded molecules, all of which are assumed to be bound directly or indirectly to the cortical skeleton, is:

$$\frac{W_s}{8\pi k_c} = \alpha_s \int (\bar{n}/n - 1)^2 \, (n/\bar{n}) \, da \qquad (11-5)$$

where $\alpha_s = K_s R_o^2 / 4k_c$. The predicted consequences for cell shape as a function of protein distribution is presented in Figure 11.6 (493). In this figure, the predicted cells are viewed along their minor axis, and the calculated integral and skeleton protein lateral density is plotted. Since the assumption has been made that all integral proteins are tethered to the skeleton, the skele-

tal profile and the integral membrane protein profile are synonymous. In Figure 11.6A, the effects of having integral membrane proteins alone in the bilayer is plotted. Note that the cell becomes elliptical, and that the protein distribution within the bilayer is markedly inhomogeneous, with the center part of the dimple having a protein density only approximately 40% of that at the periphery. Conversely, in Figure 11.6B, the stabilizing effect of the skeleton is considered (Equation 11–5). Now the shape becomes axisymmetric, and the protein distribution much more uniform, with the dimple protein density being approximately 70% of that at the periphery.

It can also be demonstrated by the above treatment that as the integral protein density rises, the lack of a cortical skeleton–bilayer interaction may lead to membrane instability and vesiculation (493). This is demonstrated in Figure 11.7. Two cell shapes are compared with respect to their membrane free energy (Equations 11–2 and 11–4) as a function of the number of embedded molecules *(p)* in the absence of a cortical skeleton. In the absence of any embedded molecules, the discoid shape is favored. However, as the number of embedded molecules (not stabilized by skeletal attachment) exceeds approximately 125/cell, membrane vesiculation

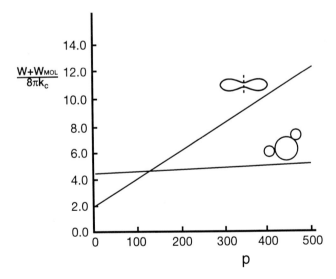

FIG. 11.7. Increasing the number of embedded molecules may induce instability in the cell membrane. The relative membrane free energy is calculated as a function of the number of embedded proteins for two different shapes, the discocyte and a shape consisting of a spherical mother cell and two spherical daughter vesicles. For both cells, $v = 0.6$; $k_r/k_c = 3$; $c_s = 2$, $k = 0.005$. For the discocyte, $\Delta a_o = 1.038$; for the mother cell and vesicles, net $\Delta a = 1.7105$, $r_{mother} = 0.6985$, and $r_{daughter} = 0.5060$. Note that as the number of embedded proteins *(p)* rises, the cell tends toward vesiculation, a consequence resisted by the presence of a cortical skeleton (cf. Fig. 11.6). [Adapted with permission from Svetina et al. (493).]

and fragmentation of the cell becomes energetically preferred. These analyses, while obviously making many assumptions, nevertheless illustrate the fundamental factors that determine membrane stability and morphology, and the putative role of the skeleton in these processes.

Collectively, these considerations challenge many common perceptions of the cortical spectrin cytoskeleton, and suggest that even in the erythrocyte it functions not so much as a structural scaffold of springs and girders, but more like a delicate and pliant fabric that by tethering membrane proteins sustains lateral order in the membrane. As discussed below, this notion fits well with the putative role of spectrin in non-erythroid cells, and offers a unifying and parsimonious concept of the spectrin skeleton that transcends cell type.

COMPONENTS OF THE ERYTHROCYTE MEMBRANE SKELETON

As outlined in Table 11.1, many proteins contribute to the spectrin-based skeleton of the mammalian red cell, and homologues of these exist in other cells as well. These proteins are discussed individually in greater detail in this section. Other proteins, many not generally recognized to exist in the red cell but that contribute to the spectrin skeleton in non-erythroid cells, are summarized in Table 11.2. Emphasis in the sections that follow is on spectrin and those proteins actually considered most central to the cortical skeleton; the integral proteins to which the skeleton may anchor but which do not actually form part of the skeleton are discussed only briefly and selectively.

Spectrin

Structure. Spectrin is the major component of the erythrocyte membrane skeleton, constituting over 20% of the structure by weight. The functional unit is a heterodimer, composed of two subunits (Fig. 11.8). In the red cell, each functional unit contains one alpha-subunit and one beta-subunit arranged antiparallel, although in other cells homodimers and multimers of just beta-type spectrins may exist (see later under Spatial and Temporal Polarization—Muscle). With the recognition of spectrin in non-erythroid cells and the cloning of several spectrin genes has come the realization that spectrin is actually part of a gene superfamily of actin-binding proteins. Other members of this family include α-actinin and dystrophin (189, 529, 542). Spectrin is most closely related to α-actinin; dystrophin is a distant member of the family (Fig. 11.9). All are characterized by the presence of multiple homologous repetitive units of about 106 amino acids, interspersed with regions of nonrepetitive sequence that mediate a variety of protein interactions (Fig. 11.8). All spectrins also contain nonhomologous sequences at their ends, leading to a tripartite structure in which nonhomologous amino and carboxy terminal domains (I and III respectively) flank a largely repetitive central domain II. This is most evident in spectrin and dystrophin. Other homologies exist independent of the repeat structure. The amino-terminal domain I of β-spectrin, α-actinin, and dystrophin is unrelated to the spectrin repeat structure, but does show strong homology to a diverse family of actin-binding proteins, including APB 120 (gelation factor) and ABP 280 (filamin), adducin, and fimbrin (19, 61, 112, 124, 171, 202, 247, 319, 531). As noted below, these proteins share a conserved 27-residue sequence required for actin binding (re-

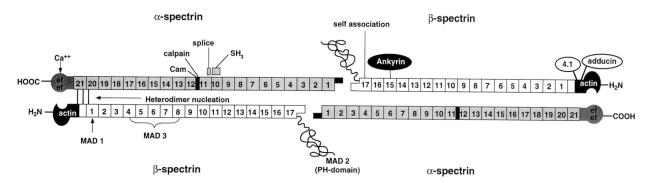

FIG. 11.8. Spectrin typically is a multifunctional heterodimer that self-associates. The functional unit is most often an $\alpha\beta$-heterodimer. Each subunit contains non-homologous NH$_2$- and COOH-terminal domains (domains I and III, respectively), joined by a central domain (domain II) composed of multiple homologous repeats of $\approx$106 residues each. Within the molecules are many sites that interact with other proteins or regulator molecules.

TABLE 11.2. *Interactions Involving Spectrin*

Interaction	Linker/Cofactor	Tissue	Selected References
Membrane Interactions with Spectrin			
Band 3 (anion exchangers)	Ankyrin	RBC, kidney, brain	29, 33, 34, 348
Na/K-ATPase	Ankyrin	Kidney	248, 352, 370, 375
Amiloride-sensitive Na$^+$ channel	Ankyrin	Kidney	469
Voltage-gated Na$^+$ channel	Ankyrin	Brain	476, 477
ABGP 205	Ankyrin	Brain	297
Inositol triphosphate receptor	Ankyrin	Brain	226
CD44 (gp85)	Ankyrin	Lymphoma	280, 283
CD44-like 116kD protein	Ankyrin	Endothelium	49
CAMS related to L1/neurofascin	Ankyrin	Brain	105, 107
H/K-ATPase	Ankyrin	Gastric cells	468
Glycophorin C	Protein 4.1	Erythrocytes	195
Thy-1	Protein 4.1	Lymphocytes	50
Cadherin	α(E)-catenin	Kidney	286
Stomatin	Adducin	RBC and others	465
$\beta\gamma$ Subunits of trimeric G-proteins	Direct		406, 511
CD45 (gp180) (tyrosine phosphatase)	Direct	Lymphocytes	51, 282
cGMP-gated cation channel	Direct	Retina (rods)	342
Dynamin (GTPase)	Direct	MEL cells, brain	72
Epithelial Na$^+$ Channel	Direct	MDCK cells	433
N-CAM 180	Direct	Brain	409
Phospholipid	Direct	RBC et al.	118, 343, 344, 414
GP lb-IX—to actin	ABP-280	Platelets	9, 289
Cadherin—to actin	α(E)-catenin	MDCK cells	428
Cytoskeletal Interactions with Spectrin			
Actin	Adducin	RBC	155, 335
	Protein 4.1	RBC	84, 85, 87, 546
Intermediate filaments	Direct	Brain, RBC (avian)	148, 149, 198, 264, 308
Tubulin	Direct	Brain	133, 426
Plectin	Direct	Glioma cells	198
Synapsin	Direct	Brain	463
A60	Direct	Brain (axons)	192
Other Interactions			
Synapsin to actin	Direct	Brain	16, 403
Tropomyosin to actin	Direct	RBC, muscle	144, 146
Dematin (protein 4.9) to actin	Actin	RBC	216, 461
Protein 4.1 and glycophorin to p55	Protein 4.1	RBC	195, 313, 314
Ankyrin to vimentin	Ankyrin	RBC (avian)	159, 160
Spectrin to IP3	Direct	Brain, etc.	217, 526

viewed in 189, 190). Similarly, the COOH-terminal region (domain III) of α-spectrin contains EF-hand type Ca^{2+} binding sequences that share strong homology to the COOH-region of α-actinin, as well as to regions within hundreds of other proteins that bind Ca^{2+}, including calmodulin, troponin C, and myosin light chain kinase (25, 258, 367, 512). Even within domain II, regions of divergent sequence and specialized bind-

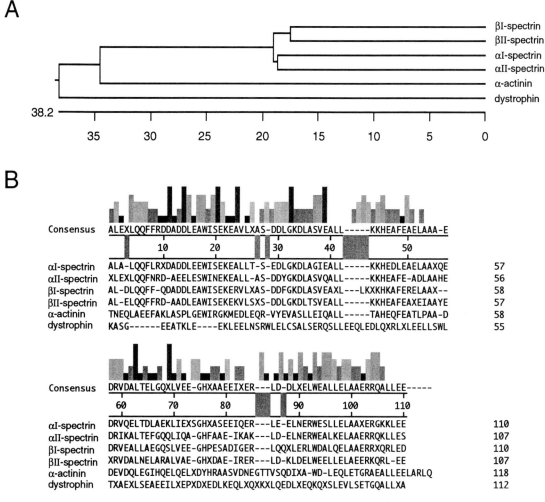

FIG. 11.9. The spectrins are the prototypical members of a gene superfamily that includes α-actinin and dystrophin. The derived protein sequences of the repetitive unit found in the central domain of the four documented spectrin genes, α-actinin, and dystrophin were first compared within each protein to arrive at a consensus repeat structure for each protein. These consensus repeats were then compared with each other to estimate their overall degree of relatedness. *A*. A relatedness tree in which the aggregate distance separating the different proteins is proportional to their similarity. This analysis indicates that the superfamily has three major branches, with the spectrins being most related to each other, then to α-actinin, and only distantly to dystropin. *B*. Sequence comparison of the consensus repeat units for each protein in the superfamily. Note the strong conservation of the trp at position 18 in the repetitive unit. A similar residue occurs at position 21 in dystrophin, although the alignment algorithm does not score this in this comparison. [Morrow and Rimm, unpublished obervations.]

ing domains exist (Fig. 11.10). The *raison d'être* of only a few of these sequence divergences is known, yet they tend to be conserved across species. Examples would include the *src* homology domain (SH3) found at the center of both αI and αII spectrin (530), or the specialized sequence responsible for ankyrin binding found within the 15th repeat of β-spectrin. The role of these and other specialized regions within spectrin are discussed in greater detail below.

A point of some confusion with respect to the repeti-

tive structure of spectrin has been defining the precise boundaries of the true repetitive unit. Based on the primary structure of αI-spectrin, sequence repeat units of approximately 106 amino acids were identified beginning with the first identified repeated residue in α-spectrin at residue 24 (472). When this phasing is extended over the entire molecule (439), αI spectrin was found to have an excess of approximately one-third of a repeat unit. Beta spectrin also displayed a similar structure, with larger nonhomologous domains

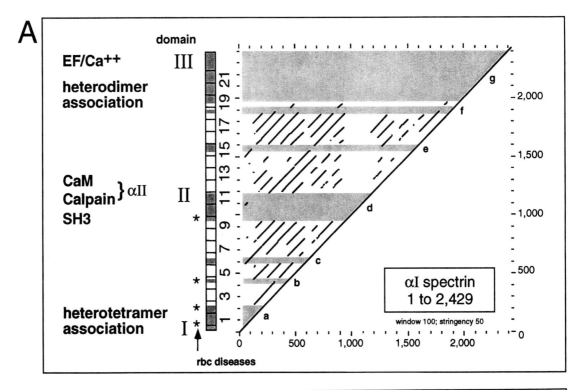

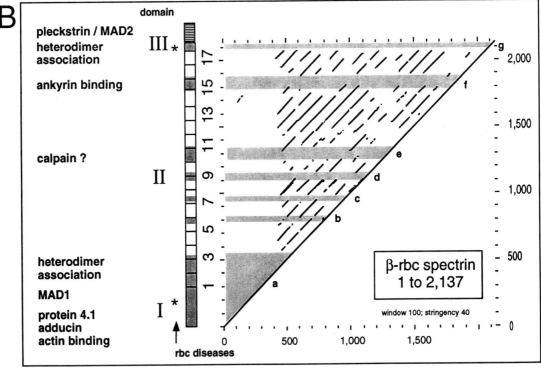

FIG. 11.10. Spectrin contains many regions of sequence divergence associated with specialized function. The sequences of either *(A)* αI spectrin or *(B)* βI spectrin were compared with themselves by "dot blot" analysis. The parallel lines spaced at 106 residues are indicative of the repetitive structure of spectrin. Note the nonrepetitive structure in domains I and III of both spectrins, as well as the many regions of divergence even with domain II. Sites marked by * indicate the approximate location of several mutations that cause hemolytic disease (reviewed in 301). Note that all known functional domains (labeled), as well as disease causing mutations, are located in regions of sequence divergence. The role of many such regions remains undetermined, although it is likely that they may harbor activities necessary for some of spectrin's myriad functions (see Table 11.2).

I and III, and an excess of approximately two-thirds of a repeat unit. This ambiguity was resolved by recognition that the true phasing of the repeat unit is displaced by approximately 27–29 residues toward the COOH-terminus (Fig. 11.9) (54), and ultimately by the resolution of the crystal structure of a spectrin repeat (54) (Fig. 11.11). Based on this phasing, α-spectrin typically has one-third of a repeat unit at its amino-terminus; β-spectrin typically has an incomplete 17th repeat unit that contains only two-thirds of a full repeat sequence. As has been demonstrated directly or indirectly in many studies, the ability of spectrin

A

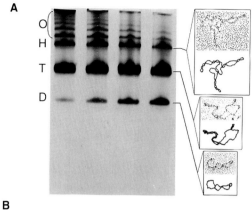

B

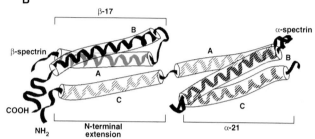

FIG. 11.12. The self-association of $\alpha\beta$-spectrin heterodimers to form tetramers is mediated by domain I of α-spectrin and repeat 17 of β-spectrin. *A.* Spectrin $\alpha\beta$ heterodimers undergo a concentration-dependent self-association to form tetramers and larger oligomers. This process can be most clearly demonstrated by non-denaturing gel electrophoresis, in which each oligomeric species migrates as a separate band. The different forms of spectrin can also be demonstrated by electron microscopy after rotary shadowing. As the concentration of spectrin incubated in vitro is diminished from 24 mg/ml *(left lane)* to 12, 6, and 3 mg/ml respectively, spectrin's state of self-association changes from larger oligomeric species towards the dimer. Oligomerization to species beyond the tetramer is a property more pronounced in αI/βI spectrin than in αII/βII spectrin. [Adapted with permission from Morrow and Marchesi (356).] *B.* The joining of two $\alpha\beta$-heterodimers to form a spectrin tetramer involves the paired association of the amino-terminal extension of α-spectrin that is homologous to helix C with the incomplete repeat 17 of β-spectrin. This association process thus essentially recreates a complete spectrin triple helical repeat motif (cf. Fig. 11.11) (235).

A

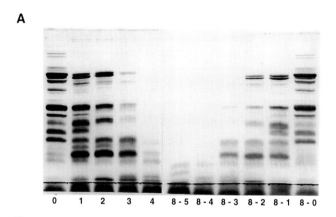

B

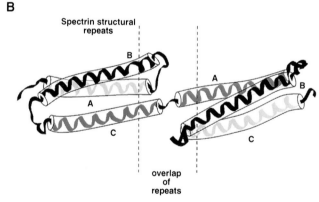

FIG. 11.11. The repetitive unit of spectrin is a stable triple helix, capable of refolding spontaneously. *A.* Spectrin's characteristic pattern of resistance to proteolytic digestion by trypsin is recovered spontaneously after denaturation with urea. This resistance derives from spectrin's secondary and tertiary structure, which limits access of the protease to over 300 potential sites in each subunit (473). In these experiments, carried out by Dr. William Knowles and Vincent Marchesi at Yale, αIβI spectrin was denatured with increasing amounts of urea; then, for half the samples, the urea was diluted or dialyzed away, and the protein subjected to 60 min of trypsin proteolysis at 0°C. Note the spontaneous recovery of its resistance to trypsin, indicative that the protein can faithfully recover at least most aspects of its secondary and tertiary structure after denaturation (Drs. Knowles and Marchesi, personal communication). *B.* The putative structure of two spectrin repeat units as derived from the crystal coordinates of *Drosophila* αI spectrin's 14th repeat unit. It is probable that helix C joins helix A with minimal discontinuity.

heterodimers to self-associate to form tetramers and larger oligomers is mediated by the association of the one-third repeat unit at the amino-terminus of α-spectrin with the two-thirds of a repeat unit at the COOH-terminal end of domain II in β-spectrin (Fig. 11.12) (235, 357, 471, 515).

Spectrin's repetitive unit is able to fold spontaneously to a putative correct conformation both as an isolated subunit (234, 235, 54) as well as when it is part of the intact native protein (62) (Fig. 11.11), a property that has facilitated the determination of the loci of multiple functional domains within spectrin (Fig. 11.8) (23, 90, 92, 103, 109, 114, 125, 184–186, 188, 232, 234, 236,

253, 287, 355–357, 471, 473–475, 523). A surprising finding with respect to spectrin structure is that the 106 residue sequence repeat units do not correspond to the exon/intron structure of the spectrin gene (6, 272). For example, the two largest exons in β (erythroid) spectrin, exon 14 (871 bp) and 16 (757 bp), extend over four and three repeat units respectively, while βI spectrin's tenth repeat unit is encoded by four exons. These observations indicate that evolutionarily the 106–amino acid repetitive unit characteristic of the spectrin gene superfamily did not derive from a primordial mini-gene that underwent repeated duplication to form the present-day members of this family (6). Alternatively, a comparison of homologies within its various repeat units suggests that spectrin has evolved from a smaller protein that has been duplicated at least once, because the strongest internal homology exists between the two halves of the protein (60, 439). It is also noteworthy that the repeat-to-repeat homology across species or between different spectrin isoforms (for example, αI vs. αII) is highly conserved, and is much better than the homology between neighboring repeats. This strong conservation of each repeat throughout evolution may indicate that the repeats serve a more complex role than simply linking specialized functional domains together.

Limited Proteolysis of Spectrin–Trypsin Resistant Domains. Because of the very large size of spectrin, the first successful attempts to characterize the structural features and functional domains of spectrin utilized its reproducible resistance to digestion by trypsin *in vitro* (Fig. 11.11) (243, 357, 473). By combining limited digestion under carefully controlled conditions with two-dimensional IEF-SDS-PAGE analysis, characteristic peptide maps can be obtained that allow the identification and characterization of not only specialized functional domains (21, 23, 357), but also the sites of post-translational modification (336) and the identification of a wide array of mutations in either α- or β-spectrin (see reviews 301, 309). A typical two-dimensional tryptic map and the nomenclature of the assigned peptide fragments is shown in Figure 11.13. Because of the emergence of a new spectrin nomenclature that has evolved from an understanding of the sequence and genes of spectrin, these peptide fragments should now be referenced with the prefix "T," to designate them as tryptic fragments. This convention will avoid future confusion with the now standard nomenclature used to describe the different spectrin isoforms. Thus, the previously designated αI fragment is now termed Tα–I to distinguish it from the αI isoforms of spectrin. Similar maps have also been generated for αII spectrin (188).

Isoform Diversity and Nomenclature. The increasingly recognized diversity of the spectrins, as well as species differences, has led to a confusing array of different names for similar molecules, and has also obscured real differences in molecules (for example the differences in distribution and presumably function of the general beta spectrin first identified in brain and the beta spectrin found in mammalian erythrocytes). A more rational nomenclature has been proposed to rectify this confusion (307, 54). This nomenclature is similar to earlier suggestions (55), and is based on the mammalian (human) spectrins (Fig. 11.14). Non-human spectrins are referenced to their human analogues by homology or, if not homologous, by a descriptive notation. Briefly, the basic heterodimer of spectrin (usually) contains one α- and one β-subunit. There are currently two recognized (human) genes for each subunit. These are numbered (by Roman numerals) in their order of discovery: αI and βI encoding for erythrocyte spectrin, αII and βII encoding for the general forms of non-erythrocyte spectrin. Species arising by alternative transcription (whether confirmed or predicted) are designated by the symbol Σ (for "subtype"), followed by a numeric designation. Uncertain transcripts are designated with a "*". Thus, the 270 kDa transcript of erythrocyte β-spectrin is properly designated βIΣ2, for it is the second transcript (subtype) of the first β-spectrin (gene) to be discovered. The alpha subunit of what was previously called fodrin (transcript unspecified) is now properly designated αIIΣ*, or just αII spectrin.

Recent cloning of human fetal brain αII spectrin, together with earlier work and analysis of the transcriptional diversity in different cells and tissues, has begun to reveal the full isoform diversity that appears to characterize the spectrins (76, 77, also Cianci and Morrow, unpublished data, 346, 478). At least three regions of alternative transcription exist in αII spectrin. In accord with the principles for the nomenclature of spectrin set forth above, the lung fibroblast sequence (346) is properly called spectrin αIIΣ1. A form identical to lung fibroblast spectrin, but lacking the 60 (bp) (20 aa) insert (insert "1") near the calpain cleavage site also exists (Fig. 11.14), and is properly designated αIIΣ2. A recently identified fetal form that incorporates a novel 15 bp (5 residue) insert (insert "2") as well as the 60 bp insert is designated αIIΣ3 (77). The spectrin found in brain with insert "2" but not insert "1" is αIIΣ4. Additional isoforms also exist, because some spectrins lack insert 3. Although these isoform variations are relatively modest from a protein sequence standpoint, they appear to be important from the standpoint of neuronal ecology. For example, reverse transcriptase–polymerase chain reaction (RT-PCR) analysis of a variety of rat tissues using "intron jump-

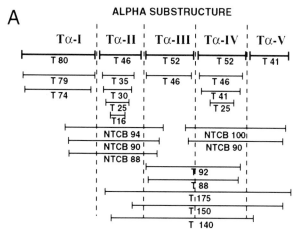

ALPHA SUBSTRUCTURE

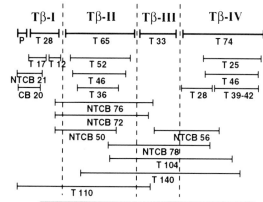

BETA SUBSTRUCTURE

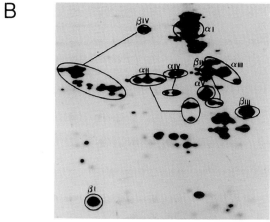

FIG. 11.13. Domain maps of spectrin generated by trypsin digestion. αI/βI spectrin extracted from erythrocytes was subjected to mild trypsin digestion at 0°C, cleaving the protein into a reproducible number of well-characterized large fragments. *A.* The alignment of the various tryptic fragments, as determined by high-resolution peptide mapping (473). Also shown is the alignment of some of the fragments generated by 2-nitro-5-thiocyanobenzoic acid (NTCB) digestion of αI/βI spectrin. *B.* The pattern following limited trypsin digestion. Peptide fragments are resolved by two-dimensional IEF-SDS-PAGE analysis, and visualized by Coomassie blue. The pattern that results is of limited complexity, facilitating the identification of the sites of functional domains, post-translational modifications, and inherited mutations in this very large protein. [Adapted with permission from Speicher et al. (473).]

ing" primers demonstrates that αIIΣ3 and αIIΣ4 are confined almost exclusively to brain, whereas αIIΣ1 and αIIΣ2 are not expressed in brain (76, also Cianci and Morrow, unpublished). Both forms of insert 1 are expressed in brain. This is the insert that flanks the SH3 domain and the calpain cleavage site in αII spectrin.

Most spectrins are composed of two nonidentical high-molecular-weight subunits. Avian species express only a single α-subunit in both erythroid and nonerythroid cells. In mammalian cells, both subunits may vary. Sequence analysis of α-spectrins from different species has shown that the nonerythroid α-chains are strictly conserved, with greater than 90% amino acid identity across species. In contrast, erythroid α-spectrins are 50% to 60% identical. Less is known about conservation of β-spectrin. Mammalian erythrocytes express only βIΣ1 spectrin, which may be a spliciform confined to erythroid lineage cells. Most other tissues express transcripts of the βII spectrin gene, and many also express alternative βI spectrin transcripts such as βIΣ2.

A summary of the calculated size and other properties of the human spectrins is presented in Table 11.3.

αI-Spectrin. The αI spectrin gene resides on chromosome 1 in both humans and mice (212, 439) and its genomic organization has been determined (254). While its mobility on SDS-PAGE has been estimated at approximately 240 kDa, the true calculated molecular weight (MW) of this protein is in excess of 280 kDa (Table 11.3). αI-Spectrin has been extensively studied in the context of a variety of hemolytic disorders, initially by the limited tryptic digestion method (Fig. 11.13) and more recently by molecular genetic approaches. These studies have identified over nine distinct tryptic cleavage abnormalities that relate to instabilities in the spectrin skeleton (113, 301). The most severe of these involve defects in αI-spectrin's self-association domain near the amino-terminus (Fig. 11.12), and affect the pattern of digestion of the Tα-I peptide (also called the 80k peptide based on its mobility in SDS-PAGE, 357). This tryptic fragment harbors the self-association site of αI spectrin (356, 357). Changes in the tryptic resistance of this peptide led to the first discovery of a mutation in spectrin that caused a hemolytic disease (242, 243).

Of the human spectrins that have been well characterized, αI appears to have the most restricted distribution. It has been convincingly demonstrated only in erythrocytes or erythrocyte precursors. However, isolated reports of its appearance in other tissues have appeared. By Northern blotting, transcripts larger than erythrocyte αI spectrin mRNA that hybridize with αI spectrin probes have been reported in some nonerythroid cells (415). Using antibodies reactive to αI spectrin, large immunoreactive bands of approximately

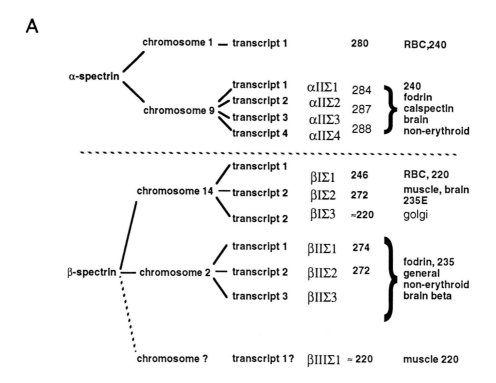

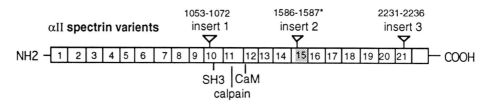

FIG. 11.14. The nomenclature of the spectrins. *A.* There are four known genes for spectrin; these encode spectrins αI, αII, βI, and βII. Multiple isoforms arise from each of these genes except αI (at least none have so far been discovered). The various isoforms for each spectrin are annotated by the addition of a Σ followed by arabic numbers. Non-mammalian spectrins are referenced to this nomenclature by homology or, if not homologous, by special notation (307, 543). *B.* Alternative transcripts in αII spectrin. Three sites of alternative mRNA splicing have been identified in αII spectrin (76, 77, 346). Note: numbering is based on the lung fibroblast sequence, which does not contain insert 2.

280 and 270 kDa have been detected in skeletal muscle (522). Using the Mab IID2 (552), which is highly specific for the Tα-I peptide of αI spectrin, convincing immunoreactivity has been demonstrated along the stress fibers and near the nucleus in primary cultures of sertoli cells (557). Most convincingly, a recent report has identified αIΣ1 reactivity in the cerebellum in a pattern coincident with βIΣ2, and has demonstrated

that these two forms can be coprecipitated from this tissue (80). Thus, while all of these observations await verification, collectively they raise the specter that even αI spectrin has a role to play in non-erythroid cells. Its association with cytoplasmic structures in the study of Sertoli cells is particularly intriguing, given the now documented association of βI spectrin and ankyrin homologues with Golgi membranes (20, 117), and the

TABLE 11.3. *Summary of Calculated Size and Other Properties of Human Spectrins*

Spectrin Isoform	αIΣ1		αIIΣ1		βIΣ1		βIIΣ1	
M W	281085		284310		246346		274658	
# A.A.	2429		2472		2137		2364	
1 μg	3.558 pMoles		3.517 pMoles		4.059 pMoles		3.641 pMoles	
Molar Ext.	304100 ± 5%		280480 ± 5%		290700 ± 5%		326140 ± 5%	
1 A(280)	0.92 mg/ml		1.01 mg/ml		0.85 mg/ml		0.84 mg/ml	
Isoelectric	4.91		5.18		5.08		5.43	
Charge @ pH 7	−137.8		−101.12		−98.22		−77.16	
Amino Acid(s)	Number	% by Frequency	Number	% by Frequency	Number	% by Frequency	Number	% by Frequency
Charged (RKHYCDE)	938	38.62	928	37.54	776	36.31	895	37.86
Acidic (DE)	475	19.56	458	18.53	383	17.92	425	17.98
Basic (KR)	326	13.42	345	13.96	274	12.82	336	14.21
Polar (NCQSTY)	563	23.18	583	23.58	488	22.84	577	24.41
Hydrophobic (AILFWV)	828	34.09	841	34.02	738	34.53	766	32.4
Ala	214	8.81	251	10.15	184	8.61	188	7.95
Cys	20	0.82	14	0.57	15	0.7	14	0.59
Asp	174	7.16	177	7.16	149	6.97	156	6.6
Glu	301	12.39	281	11.37	234	10.95	269	11.38
Phe	73	3.01	84	3.4	66	3.09	61	2.58
Gly	88	3.62	96	3.88	98	4.59	78	3.3
His	68	2.8	71	2.87	65	3.04	71	3
Ile	96	3.95	90	3.64	88	4.12	116	4.91
Lys	186	7.66	196	7.93	131	6.13	179	7.57
Leu	299	12.31	268	10.84	255	11.93	246	10.41
Met	39	1.61	50	2.02	43	2.01	65	2.75
Asn	97	3.99	99	4	66	3.09	93	3.93
Pro	42	1.73	28	1.13	48	2.25	46	1.95
Gln	172	7.08	192	7.77	162	7.58	153	6.47
Arg	140	5.76	149	6.03	143	6.69	157	6.64
Ser	128	5.27	148	5.99	118	5.52	153	6.47
Thr	97	3.99	90	3.64	88	4.12	115	4.86
Val	104	4.28	108	4.37	103	4.82	109	4.61
Trp	42	1.73	40	1.62	42	1.97	46	1.95
Tyr	49	2.02	40	1.62	39	1.82	49	2.07

Values calculated using the program Protean™, from DNAstar, Inc.

association of 4.1 homologues with stress fibers and the nucleus (86, 503). It will be interesting to see if there is also an αI spectrin counterpart associated with these components.

αII-spectrin. The gene for αII spectrin resides on chromosomes 9 (272). αII spectrin is widely distributed throughout most cells, with the exception of the mature erythrocyte. It is abundant in brain, and this is the most common source from which it is purified. Together with βII spectrin it forms a relatively stable αII/βII heterodimer and heterotetramer. It is this form, without reference to the isoform subtypes, that is most often been referred to as fodrin or brain spectrin. αII spectrin has been fully characterized in *Drosophila* (126), chicken brain (530), human lung fibroblasts (346), and most recently human fetal brain (77). Collectively, these studies have identified at least three sites of alternative transcription (Fig. 11.14). In many respects αII spectrin is similar to αI spectrin, and overall it shares nearly 60% similarity with αI spectrin.

However, despite these similarities, αII displays many functional specializations not found in αI spectrin. Unlike αI, it contains a region of nonhomologous sequence near the predicted center of helix C in its 11th structural repeat. This is the site where calmodulin binds and also is a proteolytically hypersensitive region where micro-calpain most readily cleaves spectrin (184, 185, 187, 188). Interestingly, *Drosophila* αI spectrin does not have a calmodulin binding site within the 11th repeat unit, but a sequence with similar activity does appear after the 14th repeat unit (125). These findings suggest that calmodulin binding is an activity required of αII spectrin only late in its evolutionary history, and that this activity has been independently acquired by both flies and humans in a similar but slightly different central region of the protein. Other differences also exist: αII spectrin is more rigid than αI spectrin (90), presumably due to tighter packing between the slightly overlapping subunits (Fig. 11.11) (548). The ligand affinity of their SH3 domains differs, as does their self-association affinities. These functions are discussed in greater detail below.

βI spectrin. βI spectrin is the only β-spectrin found in the mammalian red cell. It arises from a single gene on chromosome 14 in the human (541, 544) and chromosome 12 in the mouse (79, 265). The first non-erythroid βI spectrin identified was in skeletal muscle and brain (170, 307, 427, 54, 55, 555). This βIΣ2 spectrin (formerly called muscle spectrin or 235E spectrin) results from differential processing of βI spectrin pre-mRNA, resulting in the insertion of four additional exons (71). The resulting protein differs from erythroid βI spectrin (βIΣ1) in its COOH-terminal domain III, where alternate exon usage generates an open reading frame by eliminating the last 22 amino acids in βIΣ1 spectrin and extends the C-terminus an additional 213 residues to generate βIΣ2 spectrin (54). This analysis assumes that the N-terminal half of the muscle βI spectrin is identical to that in the red cell, for the complete muscle transcript has yet to be cloned to verify this assumption. The new domain III inserted into βIΣ2 spectrin displays high homology with domain III of βII spectrins and contains unique functional sites for binding to G proteins and membranes. The switch in gene processing, from βII (the more general form) to βIΣ1 spectrin, occurs early in the erythroid differentiation program, and appears to be a pathway unique to the erythroid lineage (71). βIΣ2 spectrin is expressed abundantly in muscle, where it appears late in development (412, 534); in granular cells of the cerebellum, where it is concentrated at the postsynaptic density (307); and in selected regions of the neocortex (260). A similar spectrin also appears to be expressed in chicken muscle and in chicken cerebellum based on immunologic criteria (266, 371). Other possible isoforms of βI spectrin have been identified immunologically in association with the ACH receptor in skeletal muscle (41) and with the Golgi apparatus in epithelial cells (20, 117). Whether these forms are identical to βIΣ2, or represent additional splice forms or gene products remains to be determined.

Evidence from both brain (307) and muscle (41, 412) suggests that at least some of these non-erythroid βI spectrin forms may form homodimers or homotetramers, or at least not associate in vivo with any immunologically recognizable α-spectrin. Similar findings were also found in avian tissues. However, it is clear that all of these β-spectrins retain the capacity to associate with α-spectrins (90), and α-spectrin has been identified in association with βI spectrin by co-localization in both cardiac and skeletal muscle (522), in cultured C2C12 cells transfected with βI spectrin constructs (Weed and Morrow, unpublished), and associated with an αI immunoreactive spectrin in brain (80). It remains to be determined whether other isoforms of α-spectrin associate with βIΣ* transcripts, and what factors determine whether this spectrin will assemble as a heterodimer/tetramer or as a homopolymer.

βII spectrin. The gene for βII spectrin resides on chromosome 2 in humans (208). The mouse form has also been characterized (302) and is found on mouse chromosome 11 (42). βII is the most prevalent β-spectrin, and appears to be present in most but not all tissues (for example, it is absent in mature skeletal muscle cells [534]). βII spectrin is the only spectrin so far with documented alternative transcripts arising in the amino-terminal domain I (287, 543; also B. Forget, unpublished). As with βIΣ2 spectrin, βII spectrin has 17 typical repeat units, an actin-binding motif in domain I, and an extended domain III that is similar to domain III of βIΣ2 spectrin. Most of the binding interactions of spectrin reside in β-spectrin (Fig. 11.8 and below). The gamma spectrin of avians probably is equivalent to βII spectrin based on its distribution and immunologic cross-reactivity (266). A βII spectrin homologue exists in *Drosophila* (60). βII spectrin is approximately 60% similar to βI spectrin overall, although on a repeat-to-repeat comparison the similarities are higher.

Other β-spectrins. A unique isoform of β-spectrin has been described in the terminal web of avian intestinal epithelial cells (TW260). This spectrin, which has not been cloned or fully characterized, is unusual because it is associated with the intestinal terminal web microfilaments in avians but not with membranes (201). It also does not appear to bind any of the conventional ankyrins (207) and is expressed late in development on embryonic day 15 (166). In mammals, a conventional βII spectrin assumes the role of TW260 in the avian brush border (63). The nature of the differences between TW260 and mammalian βII spectrin is unknown.

In *Drosophila*, besides a conventional β-spectrin (60), there also exists an unusual high molecular weight β-spectrin (430 kDa) termed β_H spectrin (127). This spectrin has no known human homologue. Because it is complexed with a conventionally sized α-spectrin it cannot be an exact counterpart to human dystrophin, which it otherwise resembles. β_H is also not an alternative transcript of *Drosophila* βII spectrin, for it arises from a unique gene (127). Its functional significance remains unknown. Other β-spectrins also undoubtedly exist but remain poorly characterized. For example, spectrin has been identified in amoebae (408), and a 220 kDa protein immunoreactive with β-spectrin antibodies appears to be a component of a plant membrane skeleton (136, 138, 290, 333, 462).

Functional Domain Structure of Spectrin. Spectrin is a large multifunctional molecule. A summary of the binding functions associated with spectrin is depicted in Fig. 11.8, and they are described briefly here.

Self-association. Individual spectrin subunits assemble in the cytosol to form heterodimers and heterotetramers. This process requires two separate assembly events: *(1)* the antiparallel assembly of an α-spectrin with a β-spectrin and *(2)* the joining of two heterodimers to form the heterotetramer or even larger oligomeric complexes. These two processes proceed by distinct mechanisms.

The joining of the individual α- and β-subunits to form the heterodimer appears to be a *nucleation-propagation* phenomenon that is largely independent of isoform type. Weak interactions over several regions of both subunits appear to stabilize the mature heterodimer. These weak associations alone are not strong enough to individually bring the two subunits together, although when a strongly interacting site does bind the two subunits (nucleation), then the weak interactions collectively will stabilize the complex (propagation). The sites of inter-subunit association were initially mapped to regions near both ends as well as the middle of the heterodimer by noting which peptides remained complexed under non-denaturing conditions after tryptic digestion (357). Subsequently, it was found that only peptides encompassing the amino-terminal third of βI-spectrin would reassemble in a side-side fashion with isolated αI-spectrin, suggesting that the strong interchain binding site (the *nucleation* site) resides near the actin binding end of the two subunits (450, 475). Subsequent studies have narrowed the regions involved to repeats 1–2 of β-spectrin and repeats 20 and 21 of α-spectrin, with contributions from domain I of β-spectrin also involved (413, 520, 521). (Note: in the studies of *Drosophila* spectrin, the authors have chosen to call domain I of β-spectrin "repeat 1"; therefore, the "beta 2" segment referred to in these studies (520, 521) and other studies with *Drosophila* spectrin is equivalent to the mammalian and avian β-spectrin repeat 1 as referenced by most other authors and as used in this chapter.) Whether the nucleation event between spectrin subunits is regulated in any way remains an open question. In vivo, as already discussed, α-spectrin is usually synthesized in excess of β-spectrin, with the consequence that αβ heterodimers rapidly form even in the absence of membrane assembly.

The second type of subunit association is the joining of two heterodimers to form tetramers and larger oligo-mers (89, 196, 354, 356, 418). As noted above, this process occurs by an association involving the amino-terminal domain of α-spectrin with sequences within the 17th repeat of β-spectrin (Fig. 11.12) (114, 235, 357, 471). The mechanism of this association reaction is now well understood. In essence, self-association occurs because the 17th repeat unit of β-spectrin is incomplete and lacks helix C (235). Correspondingly, the amino-terminal domain of α-spectrin begins with a sequence analogous to helix C of a repeat unit. Thus, the joining of the two heterodimers recreates a complete triple helical spectrin repeat unit. Proof of this mechanism comes from the observation that deletion of the 17th repeat along with helix C of βI spectrin's 16th repeat unit recreates a bonafide self-association domain with affinity for αI-spectrin equal to that of the native site (235). Interestingly, the presence of an extended domain III, as occurs in βIΣ2 and βIIΣI spectrin, does not seem to affect the accessibility or affinity of the self-association site in β-spectrin (235). The self-association reactions of spectrin also seem to respond to seemingly remote changes elsewhere in the spectrin molecule. For example, the propensity to form heterotetramers is enhanced when phosphorylated ankyrin binds to β-spectrin, even though the two binding sites are separated by at least one full structural repeat (74, 72, 294, 532). Mutations in αI spectrin can also diminish self-association, even when they are more than four repeat units away (301, for review). And perhaps most dramatically, the only known physiologic effect that will dissociate αIIβII spectrin heterotetramers (but not the heterodimer) in vitro without strong chaotropic agents, denaturants, or detergents is the cleavage of αII spectrin by calpain in the presence of calmodulin, both acting almost at the precise center of the αII spectrin subunit over 100 kDa from the self-association site (Fig. 11.15, also see section below) (187). How these effects are transmitted through the extended spectrin molecule is not understood.

Ankyrin binding. β-spectrin binds ankyrin with 10–100 n*M* affinities through a specialization of its 15th repeat unit (234, 357). The minimal site encompasses all of the 15th sequence repeat, and extends approximately 25 residues into the 16th sequence repeat (βI spectrin residues 1,768–1,898). When one examines this site with respect to the phasing of the structural repeat units, one finds that the downstream limit of the ankyrin binding site in βI-spectrin correlates almost precisely with the anticipated extent of the phased structural unit. It follows that a small portion of the upstream repeat unit (structural repeat 14) must also be present for ankyrin binding activity, presumably for reasons of stability. The 15th repeat of βI spectrin lacks the highly conserved tryptophan at position 18

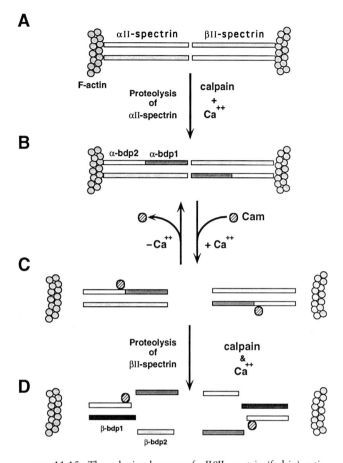

FIG. 11.15. The calpain cleavage of αIIβII spectrin (fodrin) activates its susceptibility to regulation by Ca^{2+} and calmodulin. Diagram depicting the synergistic action of calpain and calmodulin on the self-association and actin-binding properties of spectrin. A. Intact αIIβII spectrin binds F-actin and cross-links actin filaments by forming a stable heterotetramer. Calcium and calmodulin are without effect on these interactions. B. Proteolysis of the α subunit, creating the breakdown products α-bdp1 and α-bdp2, has no immediate effect on the actin binding of spectrin or its self-associative properties, and the spectrin molecule remains assembled and tetrameric under native conditions. The exact (calculated) size of the major proteolytic products of αII spectrin, α-bdp1 and α-bdp2, are 135,832 and 148,495 Da respectively. C. The binding of calmodulin to α-cleaved spectrin dissociates the tetramer to dimers and reduces its ability to bind actin. This process is rapidly reversible and requires calcium. (Note: this is the only physiologic condition yet identified that will lead to dissociation of αIIβII spectrin heterotetramers.) D. Continued action of calpain in the presence of calmodulin leads to cleavage of the β-subunit, generating β-bdp1 and β-bdp2. This cleavage causes dissociation of the heterodimer and irreversible loss of actin binding and crosslinking properties. The placement of these fragments relative to the intact molecule is as indicated. Note that the two fragments comprising the actin-binding end of the molecule remain non-covalently associated, while the other fragments dissociate. [Adapted from Harris and Morrow (187) with permission.]

within the phased structural repeat unit (Fig. 11.9). This residue is required for optimal subunit stability (303). The specialized function of this repeat (vis-à-vis ankyrin binding) probably arises from a unique 43 residue sequence in the latter third of this repeat, which must be presented in the proper conformational context for ankyrin binding to occur (234). However, additional direct contribution of the needed upstream sequences (which are highly conserved) to binding still cannot be excluded.

Protein 4.1 binding. Protein 4.1 stabilizes the interaction of spectrin with F-actin in a tertiary complex (84, 87, 121, 380, 516). It also binds directly to both βI and βII spectrin with submicromolar affinity (23, 24, 92). The binding site for protein 4.1 is within the amino terminal domain I of β-spectrin, near the site of actin binding (Fig. 11.8) (23). Although the precise boundaries of the site have not been defined, it appears to be very near codon 202, based on the loss of 4.1 binding activity in spectrin kissimmee (24). The replacement of tryptophan by arginine at this site, well within the 272-residue domain I of βI spectrin, renders the domain unstable and susceptible to oxidative and proteolytic damage (21). However, it should be noted that the kissimmee mutation per se does not diminish the ability of recombinant peptides to bind protein 4.1 and actin complexes (24), suggesting that residue 202 is not critical for binding.

Adducin binding. Adducin binds in binary interactions with the β subunit of spectrin and F-actin, and more strongly in a ternary interaction involving all three proteins (90, 335). Beyond this, the binding site has not been more closely defined, although very recent studies implicate the N-terminal actin-binding domain (domain I) and portions of β-spectrin repeat domains I and 2 (276).

Direct membrane association domains (MADs). As discussed above, spectrin has the capacity to associate directly with membranes in an ankyrin-independent fashion (207, 481). Such interactions may be the major factor in the topographic assembly of the spectrin skeleton in many cells (353, 479). Ankyrin-independent binding is mediated by β-spectrin (109, 207, 287). Three domains have been recognized from in vitro studies. MAD1 is localized to repeat 1 in both βI and βII spectrin, and recombinant peptides encompassing this repeat both bind to stripped brain membranes and can inhibit the binding of full-length native αIIβII spectrin to such membranes (287). This extended repeat unit contains two regions of nonhomologous sequence that are believed to account for its special membrane-binding activities (287). The identification of membrane-binding activity in this repeat unit was a surprising finding, for several other functions have also been attributed to this repeat [for example, adducin–actin binding (276, 475) and heterodimer association (520, 521)]. How a single repeat unit of spectrin can participate in so many activities remains to be resolved.

A second ankyrin-independent membrane association domain (MAD2) resides in the COOH-terminal domain III of $\beta I\Sigma 2$ and βII spectrin (109, 287). MAD2 is not found in domain III of $\beta I\Sigma I$ (erythrocyte) spectrin. This site binds strongly to brain membranes in vitro, and also strongly inhibits the binding of native $\alpha II \beta II$ spectrin to these membranes (287). Contained within MAD2 is a so-called pleckstrin homology (PH) domain. The three-dimensional structure of this domain in βII spectrin has been solved (304). Pleckstrin is the major protein kinase C substrate in platelets, and the PH domain of this protein is a sequence motif found in a wide array of signaling proteins and some cytoskeletal proteins (191, 322). It is likely that MAD2 gains its membrane-binding activity from its PH domain (109, 287, 526).

Two classes of ligands have been identified for proteins containing PH domains. One ligand, as exemplified by β-adrenergic receptor kinase (β ark), is targeted to its membrane-bound receptor substrate by binding to the dissociated, prenylated, membrane-anchored $\beta\gamma$-subunits of heterotrimeric G proteins (511). Prior to receptor activation, the $\beta\gamma$-subunits are apparently inaccessible for binding by the PH domain because they are complexed with the α-subunit of the trimeric receptor complex. The PH domain in MAD2 of β-spectrin also binds to the β-subunit of heterotrimeric G proteins (511), suggesting either a mechanism by which spectrin binding to the membrane is regulated, or else a mechanism by which dissociated G protein subunits might be organized by binding to the cortical skeleton. The second ligand identified that binds to the PH domain at least in some proteins is phosphatidylinositol bisphosphate (IP3) (217, 526). The structure of the PH domain with bound IP3 has been determined (217). It also appears that IP3 mediates the binding of MAD2 to brain membranes (526). Unresolved is whether this is the only site of direct membrane interaction in MAD2, and if so, why this binding is sensitive to protease treatment of the membranes (109, 287, 481).

Spectrin's third direct membrane association domain, which we propose to call MAD3, is found in domain II of β-spectrin, between repeat units 2 and 7 (109). Although sequences in this region will bind directly as recombinant peptides to synaptosomal membranes, unlike MAD1 and MAD2 they will not inhibit the binding of intact native $\alpha II \beta II$ spectrin to these membranes in vitro (109, 287). MAD3 appears to interact with a 64 kDa calmodulin (CaM) sensitive membrane glycoprotein that may also interact with

laminin in a Ca^{2+} dependent fashion (108). If these finding are confirmed, they would represent a direct molecular linkage between the extracellular matrix and the spectrin skeleton.

The physiologic role of these direct membrane association domains, as well as the other linkages of spectrin to the membrane (for example, ankyrin, adducin, protein 4.1, actin), remains uncertain. As already discussed, neither ankyrin nor MAD2 (the PH domain) appears to mediate the assembly of spectrin to the membrane in vivo, at least in epithelial cells (353, 479) or skeletal muscle (535). Conversely, sequences encompassing MAD1 and portions of MAD3 are sufficient for assembly in epithelial cells (479).

SH3 domain interactions. Both αI and αII spectrin contain a region between repeats 10 and 11 with homology to the amino terminal half of the *src* tyrosine kinase protein (SH3). This so-called SH3 domain, approximately 60 amino acids in length, is found in a variety of proteins involved with signal transduction as well as in a variety of cytoskeletal proteins (reviewed in 360, 393). Its structure has been determined in αII spectrin to 1.8 Å resolution (361). The domain is composed of five antiparallel beta-strands arranged as a compact beta-barrel. Ligands for the SH3 domain in general share a common proline-rich binding motif (422) and include such molecules as 3BP-1 (with homology to Bcr), small GTPase activating proteins (for example, GAP-rho) (78, 425), the GTPase dynamin (172), and several components of the Grb-2, SOS, and ras signaling cascades (18, 130, 293, 434, 467). At least five distinct ligands that bind the SH3 domain of spectrin have been detected (73), only two of which have so far been characterized. One ligand is the alpha subunit of the epithelial sodium channel (α-rENaC) (433); the other is dynamin (73). The relationship of dynamin to spectrin binding is of some interest, for it highlights a divergent specificity in SH3 domain binding between different spectrin isoforms, and also calls into question the possible role of dynamin, a GTPase involved with endocytosis, in the mature erythrocyte. Dynamin does not bind the SH3 domain of αII spectrin very well (73, 172). Conversely, it does bind the SH3 domain of αI spectrin, and both proteins are up-regulated during erythropoiesis. Dynamin can be found in both the cytosol and on the membrane of mature erythrocytes (73). While the role of dynamin in the mature red cell (if any) remains enigmatic, it is intriguing that αI spectrin, the most divergent of the spectrins, has *gained* the capacity to bind via its SH3 domain to this GTPase.

Ca^{2+}, calmodulin, and calpain interactions. Domain I of both αI and αII spectrin contains sequences highly homologous to the Ca^{2+} binding motifs in calmodulin and other Ca^{2+} binding proteins, as noted above.

These "EF-hand" motifs are responsible for spectrin's micromolar affinity for Ca^{2+} (125, 299, 523). The physiological significance of Ca^{2+} binding in mammalian αII and αI spectrin remains uncertain. In sea urchins, Ca^{2+} binding directly modulates the affinity of spectrin for actin (141), but apparently not in mammalian spectrin (187).

The calcium-regulated protein calmodulin (CaM) binds with submicromolar affinity to a site within the 11th repeat unit of αII spectrin (184) or the 14th repeat unit of *Drosophila* spectrin (125). CaM also binds with lesser affinity to a sequence within βI spectrin sequence repeat units 2 and 3 (8, 450, 472). Both interactions appear to have the potential to regulate spectrin's interaction with F-actin (8, 187), although only the interaction with αII spectrin may be physiologically relevant given the weak affinity of the βI spectrin site for CaM (8). Alternatively, it has been proposed that the interaction of CaM with βI spectrin at a site adjacent to the heterodimer nucleation site might regulate the assembly of the heterodimer (450); however in vivo evidence for such regulation or even the need for such a regulatory mechanism is lacking. Of particular interest is the coordinated interaction of the calcium-activated protease micro-calpain (also known as calcium-dependent proteinase I or CDP-I) and calmodulin on αII spectrin (183–185, 451) (Fig. 11.15). Although αII spectrin binds CaM in a Ca^{2+} dependent fashion, this binding per se has no apparent effect on spectrin's properties or state of association. However, CaM accelerates the rate of calpain proteolysis of αII spectrin, yielding a single cleavage of αII spectrin after tyr 1177 (184, 185). The presence of CaM on αII spectrin also enhances the rate of calpain cleavage of βII spectrin, even though CaM does not bind with high affinity to βII spectrin (185). A second calpain cleavage of βII spectrin can also occur under some circumstances (482). Collectively, these specific and targeted protease cleavages have a remarkable effect on spectrin, at least in vitro (Fig. 11.15). Whereas CaM binding is without effect on spectrin's actin binding and self-associative properties before cleavage, after the cleavage of αII spectrin by calpain, these properties become subject to reversible regulation by CaM and Ca^{2+} (187). However, if the βII subunit is cleaved by calpain (an enzyme with many regulatory inputs, reviewed in 359), the entire spectrin heterotetrameric complex dissociates irreversibly. Because βII spectrin cleavage at approximately 10 micromolar levels of Ca^{2+} occurs only in the presence of active CaM and is kinetically slower (185), spectrin together with CaM and micro-calpain conceptually form a system that: *(1)* in the absence of calcium, constituitively cross-links actin filaments; *(2)* when activated by calpain cleavage, exhibits reversible Ca^{2+}-regulated actin cross-linking; and *(3)* with sustained Ca^{2+} and calpain activity, irreversibly disassembles. A now considerable body of evidence indicates that the calpain cleavage of αII spectrin at tyr_{1177} is a common sequel of many physiologic and pathologic processes, from NMDA receptor stimulation to secretion (reviewed in 165). The significance of these observations *vis-à-vis* receptor organization or function remains to be elucidated.

Other spectrin interactions. Spectrin has also been reported to interact with adenosine triphosphate (ATP), intermediate filaments, tubulin, plectin, synapsin I, and a variety of other proteins, either directly or indirectly (see Table 11.2). The sites of binding and significance of these putative interactions remains in most instances to be determined.

Actin

Actin is one of the most abundant proteins on earth and is the major component of the microfilamentous cytoskeleton. The actin-based cytoskeleton courses throughout the cytoplasm of most cells (except the mammalian red cell), and is concentrated in the cortical region of the cell, in stress fibers, and in other specialized regions. In non-nucleated red cells, all of the filamentous actin exists only in the submembraneous compartment as a component of the spectrin–actin cortical cytoskeleton.

Filamentous actin (F-actin) is composed of helically polymerized actin monomers (globular or G-actin, MW = 41,800 Da). The filament is typically double helical with seven monomers per turn and a distinct polarity, as determined by the way that the head (S1) fragment of myosin uniformly decorates an actin filament when viewed by electron microscopy. The so-called barbed or fast-growing end undergoes rapid exchange with G-actin monomers; at the pointed or slow-growing monomer, exchange is more leisurely. Actin filaments form the connecting link between a range of diverse actin cross-linking, bundling, capping, and severing proteins that collectively organize and structure the actin cytoskeleton. Similarly, other actin-binding proteins use actin as an organizing center to localize or optimize specific functions. It is also the basis for motility and is the substrate for myosin-based contractile activity.

Three major isoforms of actin have been described (α, β, and γ)(197). The cortical cytoskeleton is composed predominantly of β actin; stress fibers and sarcomeres are predominantly α actin. These are coded by separate genes and have unique patterns of expression, differing albeit incompletely understood functions, and preferences for different sets of binding partners (460).

Actin has been extensively visualized by both fluorescence and electron microscopy, and it has generally been attributed a structural or supportive role. However, these analyses lack the sensitivity to detect the full spectrum of isoform diversity or other functions of actin. For example, there are unique types of actin filaments in the cortex of the cell (see below). Actin also forms specialized structures in the core of microvilli (166, 201). Recently, a series of *actin-related* proteins have been described, markedly expanding the diversity of actin types and presumably actin's functions (449). These actin-related proteins (ARPs) usually form part of larger structures and have only recently been appreciated (81, 82). It is now clear that a whole family of actin-related proteins exist (for review of nomenclature, see 449). Most convincingly, ARP-1 (actin-related protein–1), a component of the centromere, has been purified and shown competent to form filaments like conventional F-actin (446).

In the red cell, actin exists primarily in the form of short F-actin oligomeric double filaments, 12–14 monomers long (55, 88, 455, 517), although longer filaments have been observed under some conditions (11, 536). The way the length of these filaments is determined and maintained remains incompletely understood. Several proteins in the red cell (for example, spectrin, protein 4.1, adducin, dematin, tropomyosin, and tropomodulin) all interact directly with actin and collectively must contribute to its dynamics and organization. The importance of these proteins is reinforced by early studies demonstrating that even mild proteolysis of erythrocyte ghosts (presumably damaging proteins responsible for controlling actin filament structure) leads to uncontrolled polymerization of actin and the appearance of large microfilamentous arrays (310). Recent data suggest that a major factor determining actin filament length in the red cell may be its interaction with tropomyosin and tropomodulin (173, 176), both of which are present at the junctional complexes of the in situ cortical cytoskeleton (517). Tropomodulin is a protein that binds to tropomyosin and caps the "pointed" or slowly growing end of actin filaments (146, 533). In skeletal muscle, antibodies that block tropomodulin binding lead to an uncontrolled growth in actin filament lengths (176). It is believed that tropomodulin plays a similar role in erythrocytes.

The reason that actin filament size needs to be so closely regulated in the erythrocyte is unclear, but it is clear that the process is important. Agents that disrupt actin filaments such as DNAase treatment dissociate the cortical cytoskeleton (453), whereas agents such as phalloidin that stabilize filaments enhance the rigidity of the membrane (365).

Ankyrin

Ankyrins are a family of large proteins that have emerged as important adapter molecules mediating linkages between several integral membrane proteins and the underlying spectrin-based cytoskeleton (reviewed in 28, 32). First described in erythrocytes (298), ankyrin-like proteins have now been discovered in virtually all vertebrate cells (300) and even in invertebrate tissues (13, 128, 382). Multiple ankyrins have arisen by both gene duplication and by alternative mRNA splicing, resulting in a protein family with enormous complexity. The nomenclature of the ankyrins remains in flux. A simple scheme has been proposed that follows conventions similar to those now used to name the spectrins (Table 11.4) (115). This nomenclature is based on the three known human ankyrin genes; nonhuman ankyrins are denoted by their homologies to the human ankyrins (or by separate notation if nonhomologous). The human ankyrins are denoted by the capital letters ANK (and the mouse ankyrins by the letters Ank (see 231, 400), followed by Roman numerals in their order of discovery. Transcripts arising by alternative mRNA splicing are denoted by the symbol Σ followed by Arabic numerals; unconfirmed transcripts are designated with a "*" until their identity is ascertained. Thus, as shown in Table 11.4, human erythrocyte ankyrins (Ank$_R$, arising from chromosome 8) are properly termed ANKIΣ1 (for protein 2.1), ANKIΣ2 (for protein 2.2), and so forth; human brain ankyrins (Ank$_B$, mapped to chromosome 4) are referred to as ANKIIΣ1 (440 kDa) and ANKIIΣ2 (220 kDa); and the general isoforms of ankyrin (Ank$_G$, located on chromosome 10) are named ANKIIIΣ1 (480 kDa), ANKIIIΣ2 (270 kDa), ANKIIIΣ3 (190 kDa), and ANKIIIΣ4 (119 kDa, Ank$_{G119}$), and so on.

Most ankyrins described to date (28, 38, 153, 178, 250, 259, 261, 300, 383) contain three independently folded domains (Fig. 11.16): *(1)* a highly conserved N-terminal 89 kDa domain of repeats that associates with several membrane proteins; *(2)* a well-conserved central 62–67 kDa domain that binds spectrin; and *(3)* a variably sized but ≈55 kDa C-terminal "regulatory" domain that can modulate the activities of the first two domains. Several novel ankyrin isoforms, some of which lack all or part of the repetitive or regulatory domain, have been recently elucidated (117, 250, 400). A striking feature of the 89 kDa domain is the presence

TABLE 11.4. *Proposed Nomenclature for Human Ankyrins*

Old Name	Chromosome	Transcript	Tissues	New Name	Reference
Ank$_R$, ANK1	8	1 (7.2 Kb)	Red Cells (206 kDa); Cerebellum; Purkinje	ANKIΣ1	261, 300
		2 (6.8 Kb)	Red Cells	ANKIΣ2	178, 300
Ank$_B$, ANK2	4	1 ($\approx$ 13 Kb)	Neurons; Glia (440 kDa)	ANKIIΣ1	383
		2 ($\approx$ 9 Kb)	Unmyelinated axon (220 kDa)	ANKIIΣ2	259, 383
Ank$_G$, ANK3	10	1 (15 Kb)	Nodes of Ranvier; axonal initial segments (480 kDa)	ANKIIIΣ1	231, 259
		2 (10 Kb)	Nodes of Ranvier; axonal initial segments (220 kDa)	ANKIIIΣ2	231, 259
		3 ($\approx$ 7 Kb)	Kidney; other epithelial tissues (190 kDa)	ANKIIIΣ3	117, 400
		4 ($\approx$ 6 Kb)	Kidney, placenta, muscle, MDCK Golgi (119 kDa)	ANKIIIΣ4	117

of 24 tandemly arrayed repeats, each consisting of 33 amino acids. Ankyrin-like repeats are present in an intriguing number of apparently unrelated proteins (28, 300), including membrane proteins involved in cell differentiation such as *Drosophila* Notch and *Caenorhabditis elegans* Glp-1; cytoplasmic proteins that regulate the cell cycle such as *S. cerevisiae* SW16 and SW14; and nuclear proteins such as the transcription factors Nf-Kappa B and GABP-beta. These findings suggest a role for ankyrin well beyond simply linking cytoskeletal proteins to the membrane.

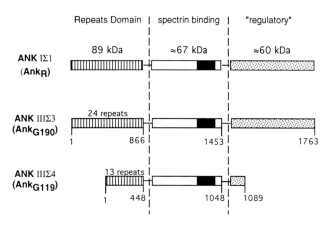

FIG. 11.16. Tripartite domain structure of ankyrin. Ankyrin isolated from erythrocytes (ANKI or Ank$_R$) contains three independently folded domains. The ankyrin repeats domain (ARD) consists of 24 nonidentical repeats of 33 residues each. The ARD motif is a common feature of many proteins. This domain in ankyrin is responsible for binding to AE1, Na,K-ATPase, and typically other membrane proteins. The central domain is involved with ankyrin's interactions with spectrin. The region of the most critical residues are shaded black. The least well understood of ankyrin's domains is the putative regulator domain. Alternative transcripts involving this domain alter ankyrin's other interactions. Also shown is the derived structure of a recently cloned ankyrin with wide tissue distribution (Ank G, ANK3, ANKIII) (117, 250, 400).

The 89 kDa ARD region appears to be globular and composed of four independently folded subdomains of six repeats each (329). All four sets of six repeats appear to be required for proper folding into a membrane-binding globular configuration, with the removal of even a single repeat reducing α-helicity by 40%. All four subdomains may participate in binding to AE1 (band 3), although the most amino-terminal 12 repeats alone retain most of the activity of the whole protein (104, 110). However, the isoform of the band 3 (AE3) in kidney does not bind ankyrin (120), presumably because of differences in its cytoplasmic domain (524). The voltage-dependent sodium channel in the brain has been shown to associate primarily with the terminal 11 repeats (476), and α-Na$^+$-K$^+$ATPase requires both the repeats domain and the spectrin-binding domain to achieve the highest affinity interaction with ankyrin in erythrocytes and epithelial cells (104). Other ligands include gastric H$^+$,K$^+$-ATPase (468), the renal amiloride-sensitive sodium channel (469), and the cardiac Na$^+$ Ca^{2+} exchanger (277) (Table 11.2). Besides integral membrane transport proteins and spectrin, ankyrin also binds vimentin (159), tubulin (111), the inositol triphosphate receptor (46, 47), and the cell adhesion molecule CD44 (283). The marked diversity between these proteins suggests that a complex evolutionary process has generated individualized binding sites within ankyrin that mediate its attachment to multiple ligands.

The central 62–67 kDa spectrin-binding domain is conserved in all human isoforms of ankyrin characterized to date. Within this region, the "minimal" spectrin-binding region has recently been identified as encompassing residues 1136–1160 of red cell ANKIΣ1 (407) and residues 669–860 of renal ANKIIIΣ4 (117). However, the N-terminal region of the spectrin-binding domain, which is less well conserved, also appears to play a role (110, 139). ANKιΣ1 (red cell) and ANKιιΣ2 (brain) interact with βIΣ1 and βIIΣ* spectrin with much different affinities (186), and it has been postulated that the divergent N-terminal sequence within the larger spectrin-binding domain of ankyrin may impart on different members of the ankyrin family their characteristic spectrin-binding affinities (407). Other factors that influence spectrin–ankyrin interactions are the state of spectrin self-association and the phosphorylation state of ankyrin (74, 72, 294, 532). Unphosphorylated ankyrin preferentially binds to spectrin tetramers rather than to dimers with a tenfold greater affinity; this preferential binding is abolished by phosphorylation, although the sites involved and the physiologic significance of this interaction remain unclear. In addition, the affinity of ankyrin for spectrin oligomers is enhanced by concomitant binding of the anion exchanger with the repeats domain (74). Finally, spectrin–ankyrin interactions are influenced by alternative splicing of ANKιΣ1, which deletes 163 residues within the regulatory domain to produce a smaller "activated" ANKιΣ2 (protein 2.2) with an increase in affinity for spectrin, anion exchanger, and tubulin (111, 178). Collectively, these in vitro studies suggest that the interaction between ankyrin, spectrin, and its membrane-binding sites is finely tuned and subject to many levels of regulation, the purpose of which remains enigmatic.

Protein 4.1

Protein 4.1 is an 80 kDa sulfhydryl-rich phosphoprotein originally described on the basis of its electrophoretic mobility in erythrocyte membranes. It is now recognized as the prototypical member of a larger family of proteins involved with mediating the interaction of actin filaments with membranes (147, 151, 443). The other members of this family include ezrin, moesin, and radixin. All members of this protein 4.1 superfamily share a conserved 30 kDa N-terminal domain responsible for membrane binding (284) but otherwise poorly understood (Fig. 11.17). Immunoreactive forms of protein 4.1 have since been found in most tissues, where they exhibit a complex heterogeneity in molecular weights, subcellular localization, and primary amino acid sequence (95, 504). A rich isoform and functional diversity of 4.1 forms exists due to posttranslational modifications such as phosphorylation and glycosylation, as well as tissue-specific and developmentally regulated alternative splicing of the pre-mRNA derived from the gene on chromosome 1p33–p34.2 (209). Immunoreactive 4.1 polypeptides range in size from 30 to 210 kDa, and in distribution from membrane-associated in red cells and polarized epithelial cells; nuclear staining in cultured leukemic and Madin-Darby Canine Kidney (MDCK) cells (97, 503); perinuclear staining in endothelial cells (273); and on stress fibers in yet other cells (86). A unified nomenclature is not available.

In its most abundant form, protein 4.1 is an 80 kDa monomer in solution, folded into four distinct domains (Fig. 11.17) (273): (1) a hydrophobic 30 kDa N-terminal domain conserved between all members of the protein 4.1 superfamily; this region contains the membrane association site (via the anion exchanger and glycophorin c and d in red cells) (194, 195, 390);

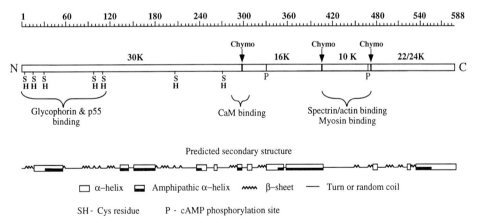

FIG. 11.17. Domain structure of protein 4.1. Many isoforms of protein 4.1 exist, all arising by a complex pattern of alternative mRNA spicing over a region encompassing 23 exons (94). The functional domain structure of the erythroid protein is shown here, based on several studies (75, 93, 94, 98, 203–205, 272, 284, 391). The 10 kDa spectrin-actin binding region appears to be an erythroid-specific spiciform, suggesting that this activity may not be fundamental to protein 4.1's function. The other members of the 4.1 family (ezrin, moesin, radixin) share their strongest homologies in the 30 kD NH$_2$-terminal region. The predicted secondary structure is derived from the algorithms of Garnier et al. (157).

(2) a hydrophilic 16 kDa region containing a protein kinase C phosphorylation site (204); *(3)* a highly charged 10 kDa domain that mediates the interaction with spectrin and actin (98); and *(4)* an acidic 12.6 kDa C-terminal domain. Protein 4.1 also binds calmodulin at the junction between the 30 kDa and 16 kDa domains (75, 233, 502), with a corresponding reduction in its membrane affinity (284).

Because spectrin–actin complexes precede expression of protein 4.1 in maturing erythrocytes (68), and because protein 4.1 directly binds both spectrin and actin (23, 350), protein 4.1 no doubt plays a role in the stabilization of the spectrin–actin network after it has formed. However, a large fraction of protein 4.1 associates with red cell membranes independent of spectrin, suggesting that the two most obvious functions of protein 4.1 (i.e., spectrin–actin binding vs. membrane binding) are distinct. The major integral membrane protein ligands for protein 4.1 in red cells are the anion exchanger and glycophorins C and D (194, 195, 390). Its binding to glycophorin also involves p55. Isoforms of AE1 are present in most human tissues, and the glycophorins represent a typical type I membrane protein. Thus, membrane-binding sites for 4.1 probably exist in all tissues. However, it has been recently shown that although the kidney anion exchanger isoform (AE1) is restricted to the basolateral domain and colocalizes with ankyrin, it does not bind either renal protein 4.1 or ankyrin (120, 524). Protein 4.1 in cultured epithelial cells, as with other components of this gene family, often localizes along zones

of cell–cell contact (56, 147, 353), raising the possibility that it associates with components of the cell–cell adhesion apparatus.

Adducin

Adducin was first identified as a substrate for protein kinase C (387) and later purified and characterized using its calmodulin-binding activity (156). There are approximately 30,000 copies per red blood cell. Red cell adducin is composed of two polypeptide subunits (α, β) (Fig. 11.18). These subunits display aberrant mobility on polyacrylamide gels, with apparent molecular weights of 103–120 kD and 97–110 kD, respectively (156, 335, 528). The genes for both chains have been cloned and sequenced, yielding predicted molecular weights for the two subunits of 81 kD and 80 kD respectively (228). The predicted protein sequences show 49% overall identity and 66% similarity between the two chains, with the greatest degree of similarity residing in the N-terminal half of the proteins. Electron microscopy, hydrodynamic studies, and protease digestion experiments all suggest that the protein is highly asymmetric (156, 213, 227, 228, 445, 528), with a protease resistant N-terminal 38–40 kD globular domain, a 8–9 kD "neck" domain, and a highly protease sensitive C-terminal tail domain. Near the very C-terminus of both chains is an identical 22 amino acid highly basic domain (11 lysines) that is similar to the functional domain of the MARCKS (myristoylated alanine-rich C kinase substrate) protein.

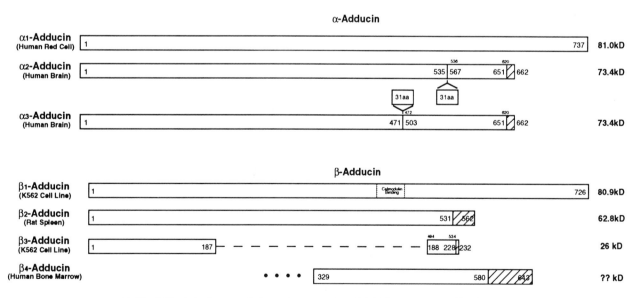

FIG. 11.18. Adducin isoforms and alternative transcripts. Multiple transcripts of both α-adducin and β-adducin have been identified. All of these so far involve alternative transcripts involving the COOH-terminal third of the molecule (aka the "tail region"). This region is very proteolytically sensitive.

Protein kinase C can phosphorylate each adducin chain at a serine within the highly basic domain C-terminal domain, and there appears to be a second phosphorylatable serine at the beginning of this domain in each chain (320). The beta chain of adducin can bind calmodulin in a calcium-dependent manner, and this binding site has been localized to amino acid residues 425–461 within the tail of the beta subunit (445).

Isoform Diversity and Nomenclature. Multiple isoforms of adducin exist (Fig. 11.18). The human α-adducin gene is located on chromosome 4p16.3 (278, 507). It spans 80 kb and contains 17 exons. The red cell sequence, designated α-1, encodes a 737 amino acid protein (228). The α-2 sequence, cloned from brain, utilizes two alternate exons: the first resulting in a 31 amino acid insertion between residues 535 and 536 of the red cell sequence and the second adding a different 11 amino acid C-terminus after residue 621 of the alpha-1 sequence (507). A third alpha isoform, α-3, also isolated from brain, contains a different 31 amino acid insertion between residues 471 and 472 of the red cell sequence and the same C-terminal alteration seen in the α-2 sequence (167). The 4.0 kb message for alpha adducin is seen in essentially all tissues examined (228).

The human beta adducin gene is located on chromosome 2p14 (164). The red cell cDNA (β-1) was cloned and sequenced from the K562 human erythroleukemia cell line and encodes a 726 amino acid protein (228). The β-2 isoform was the first beta adducin sequenced, and was cloned from rat spleen (440, 441). It is very similar to the red cell sequence until codon 530, after which the sequence diverges for 32 amino acids prior to terminating. A similarly spliced isoform has not yet been found in any human tissues. A β-3 adducin isoform has also been cloned from the K562 cell line. The smallest adducin is β-3; it is composed of the first 187 residues of the red cell sequence, followed immediately by residues 494–534 of the red cell sequence, plus four more amino acids before the stop codon (162, 163). A fourth isoform (β-4) has also been identified that contains an 86 base pair exon inserted after residue 580 of the β-1 sequence (464, 466). The remainder of the red cell coding sequence is then joined, but in a different frame, resulting in a 63 amino acid long novel and highly basic COOH-terminal sequence. A full-length clone has not yet been isolated for the β-4 isoform, so it is not yet known if this isoform shows other alterations in the NH$_2$-terminal half. β-Adducin message also varies significantly in size from 3.0 to 8.1 kb in different tissues, consonant with the many isoforms that have been identified, and appears to be absent from the liver and kidney (228).

Recently, a third isoform of adducin, designated γ-adducin, has been isolated from rat brain and kidney (123). The predicted sequence is 50%–60% homologous to the human α- and β-adducin sequences and has a nearly identical MARCKS-like highly basic C-terminal domain. It appears to be the product of a different gene than those encoding either the alpha or beta isoforms. Probes specific for the γ-adducin message revealed a 4.5 kb message in multiple human tissues.

Function and Functional Domains. Adducin subunits appear to exist as either α β heterodimers, or $(\alpha \beta)_2$ tetramers, with the latter probably representing the predominant species (213). Subunit association is mediated predominantly by the "head" regions of the molecules, but there are also inter-chain interactions within the tail sequences. In addition, antibodies specific for the rat γ-adducin protein co-precipitate rat α-adducin (123), suggesting that $\alpha\gamma$ oligomers can form as well as $\alpha\beta$ oligomers.

Adducin can bundle actin filaments (335) and binds tightly to spectrin/actin complexes (155, 335), suggesting that like protein 4.1 it may facilitate the association of these proteins in the cytoskeleton. The binding sites in adducin for spectrin and actin have been localized to the tails of both the α- and β-subunits. Phosphorylation by protein kinase C reduces adducin's affinity for spectrin (320), an interaction mediated by spectrin's β-subunit (90). Adducin has been localized to the sites of cell–cell contact in cultured epithelial cells. This process requires productive cell–cell contact as mediated by extracellular Ca^{2+}. Phorbol esters induce the phosphorylation of adducin and lead to a redistribution of the adducin away from contact sites (230). The highly basic COOH-termini of both α- and β-adducin also each contain a binding site for the integral membrane protein stomatin (band 7.2b), creating a direct membrane anchor for adducin. The possible significance of this interaction for the regulation of membrane ion permeability is discussed below.

Adducin has also been localized to the chromosomes of mouse oocytes undergoing meiosis, perhaps playing a role in asymmetric chromosomal segregation or imprinting (405); the message for adducin is also polarized in *Drosophila* oocytes, suggesting a role for this protein in the establishing the anteroposterior axis in *Drosophila* embryos (119).

An adducin abnormality has been identified in the Milan hypertensive strain (MHS) of rats, a strain that spontaneously develops hypertension and thus has been used to investigate the pathogenesis of hypertension (441). Hypertension in these rats has been genetically linked to abnormal cation transport properties both in the proximal tubule cells of the kidneys and in the red blood cells (35, 140). The transport abnormalities are linked to a cytoskeletal defect, because stripping of the cytoskeleton to form inside-out-vesicles eliminates the differences seen between MHS rats and their normotensive controls (140). Cross-immunization experiments between the hypertensive MHS rats and their closely related control normotensive strain have resulted in the identification of a single amino acid difference in the β-adducin (β-2 isoform) sequence between these strains (440, 441, 513). Another point mutation has also been identified in the alpha adducin gene of these rats (36, 66). These associations are particularly intriguing, given adducin's association with stomatin, a putative regulator of transmembrane ion transport (211, 465).

Dematin (Protein 4.9)

Dematin, originally termed protein 4.9, is an actin-bundling and cross-linking protein (461). The functional unit is a trimer composed of subunits of either 48 or 52 kDa. There are approximately 43,000 copies of dematin trimer per cell, roughly equivalent to one trimeric unit for every short actin oligomeric filament (216). The two sized subunits arise from the same gene by differential mRNA splicing, and may be developmentally regulated and tissue type specific (137, 432). Human dematin is located on chromosome 8, near the locus of the ankyrin gene (14). Mouse dematin is located on chromosome 14 (398). Dematin displays homology to the amino-terminal 30 kDa domain of the actin bundling and severing protein villin (419). Its interaction with actin is regulated by protein kinase A phosphorylation, but not by protein kinase C (216). In the erythrocyte, dematin together with tropomyosin and tropomodulin may participate in the regulation of actin filament structure and dynamics.

Pallidin (Protein 4.2)

Pallidin (band 4.2) is a 77 kDa myristolylated protein that is highly homologous to transglutaminase, although it lacks the critical cys residue at the active site of transglutaminase, rendering it devoid of transglutaminase activity (reviewed in 83, 252). It derives its name from recognition that mutations in this protein, found on chromosome 2 in the mouse (251), account for one of the 12 genetically independent mouse mutations that map to unique chromosomal loci but have similar phenotypes (539). The pallid phenotype in the mouse includes hypopigmentation, abnormal melanosome structure, prolonged bleeding times due to abnormal platelet dense granules, and increased lysosomal

enzyme content in the kidneys. Collectively, these findings suggest that mutations in pallidin lead to a complex defect in granule and/or organelle formation (539). Human disorders of similar phenotype include Hermansky-Pudlak syndrome and Chédiak-Higashi syndrome (301). It is also noteworthy that another mouse mutation called *mocha*, which has a similar phenotype, results from a defect in the Ank3 gene (399), suggesting that at least one of the alternative transcripts of Ank3 interacts with pallidin to mediate storage granule structure or function. The precise role of pallidin in the red cell remains a mystery, although it binds to AE1 and ankyrin, and may form either an additional link between the skeleton and the membrane or else augment the attachment of ankyrin with AE1. Mutations in pallidin have been associated with some forms of hereditary spherocytosis (113, 193, 218, 499).

p55 (an Erythrocyte Membrane-Associated Guanylate Kinase)

p55 is a 55 kDa protein originally found to copurify with the erythrocyte actin-bundling protein dematin (protein 4.9) (215, 216, 461). It is a peripheral membrane protein that remains associated with the membrane under conditions that remove most other peripheral proteins (437). Its sequence suggests that p55 is largely a hydrophilic globular protein with no transmembrane component; it is constitutively expressed during erythropoiesis in human reticulocytes (437). There are about 80,000 copies of p55 monomer per cell, and its copy number remains constant from reticulocytes to mature erythrocytes. It is the major palmitoylated protein in the human erythrocyte membrane (437). Its precise function in the red cell is unknown.

Its membrane-anchoring sites, in addition to its palmitoyl linkage, are protein 4.1 and glycophorin C. The importance of these linkages has been established by the absence of p55 in patients with hereditary deficiencies of either protein 4.1 or glycophorin C (5), and also by in vitro binding studies localizing p55 and glycophorin C binding to the NH$_2$ terminal 30 kDa domain of protein 4.1 (314). Another study has identified a 12 amino acid region in the cytoplasmic domain of glycophorin C and a positively charged motif flanked by the SH3 and guanylate kinase domains of p55 as the sites responsible for binding to the 30 kDa (COOH-terminal) domain of protein 4.1 (313).

Functional Domain Structure. The predicted primary structure of p55 contains a conserved *src* homology 3 (SH3) (437) found in several signaling proteins (360). Sequence analysis of protein p55 also identifies it as member of a larger superfamily of proteins recently termed MAGuKs, for *membrane-associated guanylate kinases* (7, 57, 547, 548). This family includes a *Drosophila* discs-large tumor suppressor gene (Dlg) (547) and its human homologue (Hdlg) (295); the tight junction proteins Z0–1 (540) and Z0–2 (for review 7, 224); the rat synaptic proteins SAP-90/PSD-95 (70, 240) and SAP97 (358); and the *lin-2* gene required for vulval signaling in *C. elegans* (reviewed in 239).

Sequence comparisons among the MAGuKs reveal three major domains: *(1)* an N-terminal DHR (*disc-large* homology region) domain (548), containing from one to three 90–amino acid repeat domains (a repeat domain of unknown function that is also found in non-MAGuK proteins; for review, see 410); *(2)* a *src* homology 3 (SH3) domain that mediates specific protein–protein interactions (for review, see 393); and *(3)* a sequence similar to that found in the enzymatic site of guanylate kinase. This enzyme catalyzes synthesis of GDP from GMP. p55 is the smallest member of this family, composed of only a single DHR repeat, one SH3 domain, and the guanylate kinase region.

The best characterized member of this family is the septate junction protein *discs-large* (dlg-A) in *Drosophila* (547). *dlg* mutations interfere in the imaginal discs with proliferation control, apical to basal cell polarity, cell adhesion, and the ability of cells to differentiate. Recent studies have identified the putative roles of the various domains. The human homologue of *hdlg*, as with p55, binds in vitro to the 30 kDa domain of protein 4.1 (295) via two distinct sites on *hdlg*, the DHR domain and a region between the SH3 domain and the guanylate kinase region. In addition, other studies have demonstrated that DHR regions one and two of PSD-95 associate with the C-terminus of the Shaker-type K$^+$ channel (238). These results indicate that the DHR region may function in mediating protein–protein interactions and in directing the MAGuKs to specialized membrane sites involved in cell signaling.

Stomatin

Stomatin, also known as erythrocyte membrane protein band 7.2b (EPB72), is a 31 kD integral membrane protein present in the red cell membrane at approximately 200,000 copies per red cell (199). It has been purified, although in a denatured state (525), and is insoluble in most detergents. Although strongly associated with the membrane, the protein is present in Triton extracted cytoskeletal protein preparations, indicating interaction with the cytoskeleton (483).

The stomatin coding sequence is 288 amino acids

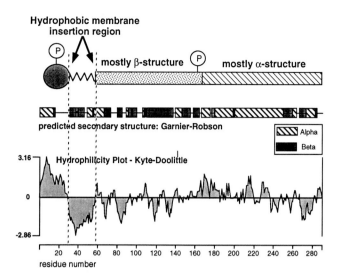

FIG. 11.19. Stomatin contains a hydrophobic 28-residue sequence thought to intercalate into the bilayer. A schematic view of stomatin suggests a typical transmembrane protein. However, both the NH2- and COOH- termini are accessible from the cytoplasmic side of the membrane (199, 442). Therefore, stomatin may assume an unusual structure with a putative hairpin loop inserting into the bilayer. Stomatin appears to regulate Na$^+$/K$^+$ leaks, and binds to adducin (465).

long and contains a 29 amino acid–long highly hydrophobic region near the N-terminus (residues 26–54, Fig. 11.19 (200, 485)). The large C-terminal domain has been established to be cytoplasmic by protease digestion experiments in sealed and unsealed ghosts (199). cAMP stimulation of sealed red cells has been shown to result in phosphorylation near the N-terminus at serine 9, indicating that this region is also cytoplasmic in orientation (442). This predicts a highly unusual monotopic structure with a hairpin loop insertion into the bilayer. Additional experiments with the mec-2 gene product in *C. elegans,* a protein very similar in sequence to stomatin, supports this conclusion about the overall topology of the stomatin protein (211). In addition to phosphorylation in response to cAMP stimulation (182, 442, 525), stomatin also appears to be palmitylated (525). Cross-linking studies suggest that stomatin may exist as a homodimer or trimer (527).

Isoform Diversity and Nomenclature. Only a single isoform of stomatin has been identified to date. The gene is located on chromosome 9q34.1 (154, 538) and contains seven exons distributed over 40 kb of DNA (152). Two polyadenylation sequences are present in the 3′ untranslated region, generating transcripts of 2.2 and 3.1 kb (152). Stomatin has been detected by Western blotting not only in the red cell but also in the liver, kidney, and a number of hematopoietic cell lines (199, 200, 525). Despite some inconsistencies, the tissue distribution appears to be wide, with message detected in fibroblasts, liver, kidney, brain, gut, and heart in one study (485), and in all tissues examined except brain, colon, and ovary in another study (152).

Function. Clues for the function of stomatin (as well as the name for this protein) have come from investigation of the autosomal dominant hemolytic anemia *hereditary stomatocytosis*. On peripheral smear, the red blood cells of affected patients show swelling and a single deep central "mouth-shaped" depression (hence the term stomatocytosis). In addition to this architectural change, these red cells have a transmembrane leak of the monovalent cations Na^+ and K^+ (483, 553). This leak can be "plugged" by treatment of the cells with the cross-linker dimethyl adipimate (327), strongly suggesting that a transmembrane protein channel is responsible for this leak. Hereditary stomatocytosis patients have no detectable levels of the stomatin protein in their red cells (129, 263). Thus, the absence of stomatin appears to correlate with a transmembrane cation link, suggesting the stomatin is a regulator of some ion channel. This channel has yet to be identified.

Stomatin also binds to the cytoskeleton (483), and the binding site for stomatin has been recently identified as the highly basic C-terminus of adducin (465). As discussed below, this interaction of stomatin and adducin, together with evidence emerging from investigations in other organisms, suggests a novel role for the cortical cytoskeleton and these proteins in mechanosensory signal transduction.

Tropomyosin and Tropomodulin

As already discussed, tropomyosin and tropomodulin play a major role in determining the state of actin in the red cell skeleton, and together with dematin are probably responsible for the control of actin filament length in these cells (146, 173, 176, 517). Tropomyosins are a family of actin filament binding proteins that exhibit extensive cell-type specific isoform diversity. Erythrocyte tropomyosin is a heterodimer with subunits of 27 and 29 kDa (144, 145). Together with stomatin, these two subunits migrate on SDS-PAGE as band 7 (Fig. 11.1B). Compared to other tropomyosins, the erythrocyte protein self-polymerizes more poorly (305), but binds more avidly to tropomodulin (15, 492). Tropomodulin is a 41 kDa-protein that binds with a 2:1 stoichiometry to one end of tropomyosin and associates with the pointed end of actin filaments

(533). It migrates with the actin band on SDS-PAGE, and is present in diverse tissues (491).

Dynamin

Dynamin was originally discovered as a 100 kDa nucleotide-dependent microtubule-binding protein purified from calf brain (444, 458, 459). It binds microtubules in vitro (444, 459), although its association with microtubules in intact cells has not been reproducibly demonstrated. Dynamin displays a tripartite GTP binding consensus sequence (378) isoforms exist. A form confined primarily in neuronal tissue and displaying a developmental time course of expression is commonly referred to as dynamin (135, 366). Subsequently, other isoforms were identified including a gene with wider tissue distribution (96, 470). This form is now referred to as dynamin 2. Rat brain dynamin and rat liver dynamin 2 share an overall sequence identity of 79% (470); are highly homologous (> 80%) in their N-terminus; and diverge significantly in their C-terminus (which contains 28–30% proline residues). Both forms of dynamin are subject to differential splicing, with each dynamin exhibiting at least four protein forms (430, 470). The functional significance of the alternative splicing remains unclear. The nature of the dynamin found in red cells has not been determined (73). There is substantial evidence that dynamin is involved in vesicle trafficking in other cells, but the relationship of this function to its role in the red cell or its association with αI spectrin, if any, remains undetermined.

Interactions with Phospholipids

Spectrin can interact directly with phospholipids in vitro and penetrate unilamellar and multilamellar liposomes (118, 229, 306, 315, 330–332). Spectrin's interaction with phosphatidylcholine is especially intriguing given that this lipid is a cofactor in activation of protein kinase C, and protein kinase C co-localizes with spectrin in lymphocytes (175). However, this interaction is weak (39). Perhaps the best understood interaction of a phospholipid with spectrin at the structural level occurs via spectrin's PH domain, which as described above binds IP3 (217, 526). However, despite these interactions, spectrin does not appear to be responsible for the transbilayer asymmetry in phospholipid composition (414). It also remains uncertain whether any of these putative lipid binding sites actually anchor spectrin to the membrane. Given the sensitivity of spectrin membrane binding to proteases (207, 481), it is probably more likely that spectrin's lipid binding activity acts to maintain the lateral order of certain lipid domains at the membrane surface.

THE SPECTRIN SKELETON OF NON-ERYTHROID CELLS

The non-erythroid spectrin skeleton, first appreciated in the early 1980s, displays a remarkable diversity in composition and distribution. Typically, its localization at the plasma membrane is dynamic and developmentally dependent; it interacts with an even wider array of partners than found in the erythrocyte (Table 11.2), and it functions not only on the plasma membrane but also on internal membranes and other cytoskeletal filament systems as well. These observations, accumulated over the past decade, have forced a reevaluation of the lessons learned from study of the red cell skeleton. Some of these features are summarized here.

Spatial and Temporal Polarization

A key feature distinguishing the spectrin skeleton in non-erythroid cells is its localization both temporally and spatially to specialized surface domains. Some of these features are illustrated in Figure 11.20.

Neurons. All of the various spectrin isoforms so far discovered can be found in nervous tissue (reviewed in 170). The most common spectrins in brain are derived from αII and βII, with diversity arising from various combinations of the alternative transcripts of these genes. These forms are most abundant in axons and in presynaptic terminals, and to a lesser degree in the soma and PSDs throughout the cortex (274, 275). Conversely, βIΣ2 spectrin is concentrated in cerebellar granule cells and in regions of the frontal cortex (169, 260, 307). It is localized to cell bodies and dendrites,

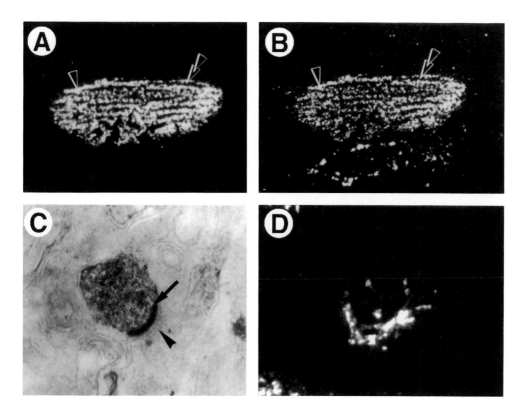

FIG. 11.20. Spectrin in non-erythroid cells is often highly organized. A. Rat myotubes after extraction with saponin and stained with rhodamine-bungarotoxin, which stains the acetylcholine receptor (AChR) clusters. B. Same preparation stained with Mab VIIF7, which reacts with βI spectrin. Note the coincident distribution, indicating that this spectrin is codistributed with the AChR,. [Adapted with permission from Bloch and Morrow (41).] C. βIΣ2 spectrin is concentrated at the postsynaptic density in the molecular layer of the rat cerebellum. Immunoperoxidase labeling EM. Bar = 0.5. μm [Adapted with permission from (307).] D. Immunofluorescent image of MDCK cell stained with an antibody to Ank G119(ANKIIIΣ4). The punctate eccentric cytoplasmic staining adjacent to the nucleus marks a form of βI spectrin associated with the Golgi apparatus (20, 117).

but is concentrated at the plasma membrane almost exclusively at the PSD (Fig. 11.20) (307). Both forms are found associated with organelles, synaptic vesicles, and plasma membranes.

Neutrophils. Spectrin associates in neutrophils with chemotactic receptors subsequent to their activation with formyl peptides (222, 223, 381, 417). This association is thought to play a role in the segregation of receptor plasma membrane domains from their signal transduction partners, thereby establishing a molecular basis for the homologous desensitization of cells after ligand binds receptor (reviewed in 241). Such a mechanism occurs with the NADPH-dependent superoxide complex that associates with the cytoskeleton upon activation, laterally organizing into a functional plasma membrane domain (417).

Lymphocytes. Perhaps the most dramatic example of a dynamic rearrangement of spectrin during a physiologic process occurs in lymphocytes. Here, ligand or IgG-induced cell surface receptor capping and interleukin 2 secretion induce a redistribution of spectrin (αII-βII?) from a cytoplasmic meshwork of spectrin filaments and membranous vesicles to the plasma membrane (40). When spectrin migrates to the plasma membrane in these cells, it is localized to regions occupied by activated surface receptors (40, 268, 270, 392). The immunofluorescence spectrin staining pattern is heterogenous in lymph node T cells, displaying either a cortical or diffuse pattern. In addition, multiple small and large aggregates of spectrin are located in the trans-Golgi region or near the nucleus. Lymphocytes displaying aggregated spectrin have more tightly organized membrane lipids than those with diffuse staining. Protein kinase C (the PKC βII isoform) is translocated to the cytoplasmic spectrin aggregates upon lymphocyte activation by phorbol esters or by antigen-specific receptors, indicating that the positioning of the spectrin-based cytoskeleton is sensitive to activation signals and may play a role in the function or positioning of PKC βII (174). Analogues of ankyrin have also been described in these cells, including a small 72 kDa ankyrin (a possible splice variant of ANK3; see discussion in 117). These ankyrins, as with spectrin, were found to dynamically redistribute between cytosolic and organized membrane associated pools (52, 175).

Both spectrin, ankyrin, and protein 4.1 homologues associate with a variety of receptors and other proteins in lymphocytes, many of which have now well recognized roles (Table 11.2). Examples of these include Thy-1 (gp25), CD44 (a transmembrane adhesion molecule), CD45 (a tyrosine phosphatase), the IP3 receptor,

and the ryanodine receptor (44–49, 221, 281–283, 537).

Secretory Cells. Evidence for a role for αIIβII spectrin in regulated secretion has come from studies in adrenal chromaffin, gastric parietal and parotid acinar cells, and in lung alveolar epithelial cells (12, 328, 396, 556). Following stimulation of chromaffin cells, a dramatic redistribution of spectrin occurs, with the protein migrating from a continuous ring in the subplasmalemmal region to patches (396). Introduction of spectrin antibodies into permeabilized cells reduces catecholamine secretion by 50%, and a portion of the cytoplasmic spectrin may be redistributed to intracellular vesicles during secretion. It has been proposed that spectrin inhibits the docking and fusion of secretory vesicles in cells with regulated secretion (12).

Muscle. Skeletal muscle is unusual that in addition to expressing a single form of αII spectrin, it expresses three known isoforms of β spectrin at different stages of development. Adult mammalian skeletal muscle contains αIIΣ* spectrin (415, 522) and two erythroid-like forms of the β subunit: a 235k Da form termed βIΣ2 (412, 522, 541) and a 220 kDa form termed β_{AChR} spectrin (41). While all of these βI isoforms were initially identified by immuno-cross-reactivity with antibodies raised against βIΣ1 (erythrocyte) spectrin (10, 41), they appear to be distinct molecules based on immunolocalization studies. βIΣ2 spectrin is found in a continuous pattern along the inner surface of the sarcolemma and is enriched in costameres, the regions overlying the Z-line and I band of the sarcomere (371, 412). Although the exact molecular organization of costameres is not yet known, costameres have been shown to contain also several other cytoskeletal proteins. Those so far identified include ankyrin (372), vinculin (100, 388, 389, 412), talin (27), γ actin (100), and dystrophin (318, 334, 412, 489).

β_{AChR} spectrin (probably a βIΣ3 spectrin, Fig. 11.14) is localized exclusively in mature muscle at the neuromuscular junction, where it is found in regions enriched with acetylcholine receptors (AChRs) (Fig. 11.20) (41, 122). High-resolution electron microscopy (EM) analysis of deep-etched membranes from the neuromuscular junction has demonstrated that β_{AChR} spectrin exists as a homopolymer and forms a lattice-like array nearly identical to that found in the erythrocyte (416). These βI isoforms also display differential associations with the αII subunit: βIΣ2 spectrin, as in the cerebellum (307), is associated at least partially with αII spectrin whereas the β_{AChR} spectrin is not (41, 534).

The developmental changes in β spectrin composition during myogenesis are also of some interest. Evidence that spectrin plays an important role in early myogenesis has come largely from studies on avian skeletal muscle. Avian muscle differs from mammalian muscle in that fully differentiated avian muscle contains homologues of both $\beta I\Sigma 1$ and $\beta I\Sigma 2$ spectrin (originally termed "β" and "β'," 371). In primary avian myoblasts, the predominant β spectrin isoform is a homologue of βII spectrin (termed "γ" spectrin in that study) that upon myotube fusion in vitro gradually switches from this spectrin isoform to βI spectrin isoform (374). Switching of β spectrin subunit composition occurs shortly after myoblast fusion (within 1–2 days postfusion) as determined by both protein (373) and mRNA (347) analysis. The switching of these isoforms is triggered by post-translational events, for they require the presence of extracellular Ca^{2+} for the switch to occur (373). While all three of these β spectrins ("γ", "β," and "β'") were detected in equal amounts beginning in early (4-day) myotubes, "β" spectrin synthesis increased tenfold in 15-day myotubes (374). During the same period the amounts of "γ" and "β'" spectrin remained unchanged, indicating that the production of "β'" spectrin had yet to achieve the levels found in mature skeletal muscle (373). In mammalian systems, the switching of β spectrin isoforms occurs later in development. The overall profile is similar, with βII spectrin diminishing until it is absent at birth, and $\beta I\Sigma 2$ spectrin increasing from day E12 onward (246, 534). Thus the βII spectrin subunit can be thought of as the "fetal" β spectrin isoform during myogenesis and the $\beta I\Sigma 2$ subunit is the "adult" β spectrin isoform. Earlier studies that documented βII spectrin in adult skeletal muscle (166, 411) were probably detecting βII spectrin in the endothelial cells and stroma that contaminate all muscle tissue preparations. Conversely, αII appears to be the only α spectrin expressed in skeletal muscle (374, 415, 522, 534).

Epithelial Cells. The participation of spectrin in the establishment and/or maintenance of specialized membrane domains has been studied predominantly in kidney epithelium where $\alpha II\beta II$ spectrin, ankyrin, and the basolateral protein Na^+-K^+ ATPase are co-localized in the basolateral domain (352, 375, 376). In cultures of Madin-Darby canine kidney (MDCK) epithelial cells these proteins become colocalized upon the induction of cell–cell contact. This process appears to be initiated by engagement at the cell surface of E-cadherin (see below) (325, 326, 353). Polarization of spectrin to the basolateral domain results in its incorporation into a Triton X-100 insoluble matrix and increased metabolic stability. ANKI binds directly to the cytoplasmic domain of the Na^+-K^+ ATPase α-subunit, at a site in-

volving the second and third putative cytoplasmic domains (116, 352, 375). Evidence for the important role of E-cadherin in this assembly process comes from the observation that expression of full length but not a truncated cytoplasmic domain of E-cadherin in fibroblasts induces the redistribution of Na^+-K^+ ATPase, spectrin, and ankyrin to sites of cadherin-mediated cell–cell contacts (325). The effect of cadherin is even more clearly illustrated in studies on retinal pigment epithelium. In contrast to most transport epithelial where Na^+-K^+ ATPase is expressed basolaterally, retinal pigment epithelium presents this enzyme on the apical membrane. This expression is accompanied by a reversal of the ankyrin and spectrin polarity (compared to other epithelia) (177). Interestingly, the expression of transfected E-cadherin in these cells overrides the endogenous signals, induces the appearance of a new ankyrin, and imposes a conventional polarity of Na^+-K^+ ATPase, spectrin, and ankyrin on the cell [see discussion on cadherins (317)].

Development. Spectrin and ankyrin levels and distribution are regulated in the complex series of events that accompany *Drosophila* development (128, 397). In the early stages of development spectrin is concentrated in the cortex and in cytoplasmic pools surrounding the rapidly dividing nuclei. As the embryo develops, spectrin migrates with the nuclei to the surface and is concentrated in the supranuclear caps. During mitotic divisions the cytoskeletal caps elongate (interphase and prophase) and divide (metaphase and anaphase). During cellularization spectrin concentrates in the cortex and in the lateral margins of columnar epithelial cells, extending to a position apical of the furrow canals. In the final stages of development spectrin leaves the lateral margins and localizes almost exclusively to apical boundaries. Mutations in α-spectrin in *Drosophila* cause first- to second-instar lethality (269). Spectrin-deficient larvae lose contact between epithelial cells of the gut, and display loss of cell–substratum interactions. These changes suggest that a spectrin skeleton is required to stabilize the cell–cell interactions required for maintenance of cell shape and subcellular organization within tissues. Similar patterns of embryonic lethality involving developmental disorganization but not cellular lethality have also been noted with anti-sense suppression of protein 4.1 in *Xenopus* embryos (161). Dynamic patterns of spectrin expression and distribution during development have also been shown in mouse and sea urchin embryos where spectrin rapidly accumulates on newly synthesized organelles and vesicle membranes that undergo polarized movements during oocyte maturation and early embryonic development (43, 142). In unfertilized sea urchin eggs, spectrin specifically associates with cortical granules, acidic vesicles, and yolk platelets at relatively the same level as it associates with the plasma membrane (143).

PROTEINS INTERACTING WITH SPECTRIN IN NON-ERYTHROID CELLS

In addition to the proteins known to participate in the spectrin skeleton in erythrocytes, several additional proteins not found in red cells have been implicated in the spectrin skeleton of non-erythroid cells (see Table 11.2). Although these are too numerous to describe in detail, some stand out as illustrative of the way the spectrin skeleton may interact with other filament systems or with components of the cell's adhesion apparatus. These are described in greater detail below.

Cytoskeletal Elements

The spectrin cytoskeleton has been shown to interact in vitro and colocalize in vivo with several major structural protein systems including actin, intermediate filaments, and microtubules.

Intermediate Filaments. Spectrin associates directly or indirectly with at least three classes of intermediate filaments: desmin, vimentin and neurofilaments. The localization of the $\beta I\Sigma 2$ isoform of spectrin in striated muscle to the junction between the Z-line and the sarcolemma is believed to provide support to the sarcolemma by its interaction with the intermediate filaments in this region (373). In other cells, the microinjection of spectrin-specific antibodies causes spectrin and vimentin filaments to aggregate near the nucleus, presumably because of a direct interaction (308). Interestingly, this aggregation does not affect cell shape or the distribution of actin filaments at the membrane. Spectrin also interacts directly with plectin and IFAP-300, two related intermediate filament proteins found in glial and kidney cells (198). Both β I and βII spectrin bind to the rod-like 20 kDa domain of the light subunit of neurofilaments (NF-L) (149, 150). This binding involves the amino terminal region (domain I) of β spectrin at a site coincident with the actin binding domain (148), and interacts with the same (aminoterminal) region of NF-L that binds MAP-2 (148, 150). Spectrin from mammalian erythrocytes ($\alpha I\beta I$) also binds desmin in co-sedimentation assays (264), and a spectrin ($\alpha II\beta II$) is closely associated with intermediate filaments in a polarized distribution in lymphocytes (270). Besides these direct interactions, the aminoterminal "head" domains of desmin and vimentin filaments also bind end-on to ankyrin (159, 160). Taken together, these interactions indicate that spectrin tethers selected proteins to the intermediate filament as well as the actin cytoskeleton.

Microtubules. $\alpha II\beta II$ spectrin purified from brain interacts directly with and co-purifies with microtubules in vitro (133, 426). Spectrin may also interact indirectly with microtubules via microtubule-associated proteins such as MAP-2 or Tau, or with ankyrin, protein 4.1, or synapsin, each of which independently binds microtubules (17, 30, 64, 106, 225, 404). The physiologic significance of spectrin—microtubule interactions is unknown.

Adhesion Proteins

Work in a variety of cells as described above has identified the interaction of the spectrin—actin skeleton with cell—cell adhesion components as central to the topographic assembly of the skeleton, especially at points of cell—cell contact. Some of the key proteins involved in this process that interact either directly or indirectly with spectrin or other components of the cortical skeleton are described here.

Cadherins. The cadherin family represents a group of type I transmembrane proteins involved with mediating cell—cell adhesion in a Ca^{2+} dependent fashion. As noted above, at least in epithelial cells, the nascent spectrin skeleton and several of its associated proteins (4.1, adducin, ankyrin, etc.) assemble at points of cell—cell contact coincident with the clustering of E-cadherin. The cadherins and their associated cytoplasmic complexes are thus believed to play a major role in guiding the assembly of the cortical spectrin skeleton to topographically defined regions of the plasma membrane, at least in epithelial cells (for review, see 353, 369, 431). Structurally the cadherins are defined by an extracellular domain composed of five repeated units that bestow a capacity for calcium-sensitive, homotypic adhesion (for review, see 501). Each cadherin also has a hydrophobic membrane-spanning region and a cytoplasmic tail. Several cadherins have been described, which collectively make up a gene superfamily (reviewed in 429). These have been named on the basis of their tissue of origin. E-cadherin (E for epithelial) and N-cadherin (neural) are the best studied of the group. Other cadherins, including M-cadherin (muscle), and cadherin-5 or VE-cadherin (for vascular endothelial cells) are also classic cadherins. The desmosomal cadherins demonstrate homotypic adhesion but display less homology in both the cytoplasmic as well as their extracellular domains (59). Other more distantly related cadherin molecules have also been described. Examples include the *ret* protein with homology in the extracellular domain (447), and protein tyrosine phosphatase *mu,* with homology only in the cytoplasmic domain (53).

The extra-cellular domain of the cadherins is their most characteristic feature. Synthesized as a precursor, 156 residues are cleaved from the NH_2-terminus during maturation, and the protein is also glycosylated (363). Cadherins typically have five extracellular domains, of which the first three are well conserved. In the first domain is a highly conserved sequence (HAV) that is required for homotypic adhesion (377). Each extracellular domain also contains a less well conserved sequence (LDRE·(X)n·DXND) implicated in calcium binding (for review, see 500). The extracellular domain also contains consensus sites for glycosylation and four conserved cysteines; their pattern of cross-linkage is not known. Based on crystal studies of the extracellular domain of N-cadherin (452), the molecule appears to be a dimer with stabilizing calcium molecules nested into the repeated structure. It has been proposed that two cadherins interact *via* their first extracellular domains, interacting with each other on opposing cadherin dimers in a zipperlike fashion (452). Although this is an attractive model, especially with respect to dimerization, the interaction of the first extra-cellular domain with itself is less well accepted. Electron microscopic studies suggest that the distance between the two membrane faces of a cadherin-based junction are too short for an end-to-end apposition and may be more consistent with the first repeat interacting with the third repeat on the opposing face (551).

The cytoplasmic face of the cadherin molecule is a critical region for both adhesion and regulation. Even though adhesion takes place outside the cell, the cytoplasmic domain of cadherin, together with an ensemble of two or three associated proteins called catenins, is required for functional adhesion (364, 384). A conserved region within the last 37 amino acids (but not including the last seven) of the cytoplasmic domain of cadherin interacts with β-catenin (386, 480). A second highly conserved region is also present just inside the membrane. The function of this region is not known, although some hints of its function have emerged. The transfection of E-cadherin into a retinal cell line that normally contains only B-cadherin (a neural subtype) redistributes its Na^+-K^+ ATPase from the apical to the basolateral surface (317). The sequence of B- and E-cadherin differs most in the region of their cytoplasmic domain that is closest to the membrane. This region may also be the site at which p120cas binds (423, 456) and possibly the protein tyrosine phosphatase *mu* (PTPase-μ) (54). Thus, cadherin's second highly conserved region may act to regulate the composition and assembly of other components of the complex. The connection of cadherin to the cytoskeleton is medi-

ated by α- and β-catenin, as discussed below. This linkage forges a connection to the spectrin–actin cytoskeleton but leaves many unanswered questions, such as how interactions at the cytoplasmic face mediate extracellular adhesion or what are the processes that stabilize the adhesion complex once nascent contact has been made (326). Figure 11.21 presents a cartoon depicting some of these interactions and our current state of knowledge.

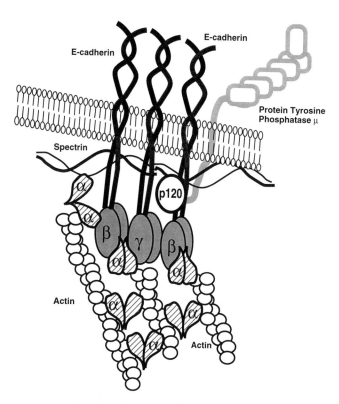

FIG. 11.21. A hypothetical model of the cadherin-based transmembrane adhesion complex and its relationship to the associated cortical cytoskeleton. The Greek letters α and β indicate the corresponding catenins. Although there is evidence that α-catenin can bind both spectrin (286) and actin (428) and that is exists as a dimer, the linkages shown are hypothetical. It is not known if catenin/spectrin and catenin/actin interactions can occur simultaneously. It is possible that α-catenin not only links the cadherin complex to the cytoskeleton but also participates in actin bundling in other parts of the cell or in other cell types. There is no good evidence for a direct interaction between E-cadherin and spectrin, but they co-localize and are suspected to be joined by α-catenin. The interactions between β- and α-catenin are presented as an example of how β-catenin (a member of the *arm* family) binds directly to both E-cadherin and α-catenin. It is not yet known which, if any, other members of the *arm* family can participate in this interaction. Plakoglobin, previously known as γ-catenin, may participate, but cannot bind the same cadherin molecule (although theoretically both could bind a cadherin dimer). α-Actinin, another member of the spectrin superfamily (see Fig. 11.9), has also been implicated in this complex, but its exact role remains uncertain; and hence it is not included in this figure. Both p120 (CAS) and Ptpaseμ also bind to cadherin, but neither appears to bind in same region as β-catenin or to interact directly with spectrin.

β-catenin. β-catenin is perhaps the most complex member of the cadherin-based membrane adhesion complex. It is a member of the *arm* family, a group of signal transduction proteins first recognized in *Drosophila* and defined by an internal 42 amino acid–repeated structure typically flanked by unique N- and C-terminal domains (395). The 42 amino acid repeat occurs 13 times in β-catenin (394). The repeat region of this molecule contains the binding site for E-cadherin (214). Some members of the *arm* family, such as plakoglobin, are very similar to β-catenin (395); others are more divergent and may bind the cadherins at different sites (for example, p120cas)(423, 456) or not at all (smgGDS)(237). β-catenin also binds directly to α-catenin through its N-terminal domain (1). Its binding to both E-cadherin and α-catenin appear to be independent events, and a pool of β-catenin complexed with α-catenin but not cadherin has been demonstrated (368).

β-catenin interacts with a number of other proteins, including APC (the product of a gene that is mutated in many cases of colon cancer) (435, 490), the EGF receptor (206), the erbB2 (HER2) non-receptor tyrosine kinases (379) and the actin-bundling protein, fascin (506). These interactions suggest a more complex and diverse role for this molecule than simply one of linking cell–cell adhesion to the cytoskeleton. The regulation of adhesion may be controlled by β-catenin phosphorylation. The protein is tyrosine phosphorylated in *v-src* transformed cells (26, 321). The kinase responsible for this modification is not known, although candidates exist, including the oncogenes *trk*, *met* (the HGF/scatter factor receptor) and *src* (for review, see 37). Serine and threonine phosphorylation may also mediate the binding specificity of β-catenin. When phosphorylated on serine by glycogen synthase kinase 3 (GSK3), β-catenin binds APC (436, 510). Conversely, when dephosphorylated it binds cadherin and forms functional adhesive junctions. This mechanism has been proposed as a means of modulating the cytoplasmic pool of β-catenin and a way to regulate adhesion.

Plakoglobin. Plakoglobin is another member of the *arm* family indirectly associated with the adhesion complex. This protein has been independently described by two groups, and has also been named γ-catenin (99, 384). Like β-catenin, it tightly associates with the C-terminal cytoplasmic domain of cadherins (245) and interacts with α-catenin (438), which in turn links it to the actin skeleton. Plakoglobin, through other mechanisms, links the cadherin complex to intermediate filaments (514) It is concentrated at both conventional and desmosomal type cadherin junctions (368). Some studies suggest that cadherins can interact with either β-catenin or plakoglobin, but not both at the same time (1).

p120cas. The most recent catenin-type molecule to be identified is p120cas(CAS). Initially described as a 120 kDa substrate of the *src* tyrosine kinase (424), it is now known to be phosphorylated in response to growth factors EGF, PDGF, and others stimulating receptor tyrosine kinase activity (324). This protein appears to be an integral part of the cadherin cell adhesion complex (3, 423). However, despite its similarities to β-catenin, CAS binds only to cadherins (102), possibly at a different site than that utilized by β-catenin and plakoglobin (456). It does not bind α-catenin (102).

α-Catenin. The linkage of cadherin to the cytoskeleton is primarily mediated by α-catenin. This conclusion is now well supported. α-catenin is less tightly associated with cadherin and β-catenin in extraction assays (385); α-catenin binds directly to and bundles F-actin (428); the COOH-terminal domain of α-catenin is needed for full actin-binding activity (428); and the COOH-terminal (but not the NH2-terminal) domain alone is sufficient to mediate productive adhesion when covalently joined as a chimeric molecule to E-cadherin (362). α-Catenin is related to vinculin, another protein involved with the attachment of the filamentous cytoskeleton and α-actinin to the membrane (58).

α-Catenin appears to interact with members of the cortical cytoskeleton beyond just actin. Two other interactions have been proposed. α-actinin, a member of the spectrin superfamily (see above), colocalizes with α-catenin both at cellular junctions and along stress fibers in fibroblasts (244). α-actinin also coprecipitates with the cadherin/catenin complex under some conditions (244). Spectrin also interacts with α-catenin, at least in vitro (286), and co-localizes by immunofluorescence at nascent cell–cell junctions (353).

EVOLVING CONCEPTS

The major role of the spectrin cytoskeleton in erythrocytes is to bestow global structural support and deformability; however, the temporal and spatial polarization of the non-erythroid spectrins suggest that their primary role is not simply one of plasma membrane support. Spectrin is often observed in a polarized phenotype associated with intracellular vesicles, transmembrane receptors, and specific membrane domains in a dynamic state influenced by the level of cell activation, development, or polarity. Even erythrocyte spectrin, which in the mature erythrocyte exists as a two-dimensional homogeneous network on the cytoplasmic face of the membrane, displays a polarized phenotype during hematopoiesis when the spectrin, AE1, and ankyrin of the orthochromatic erythroblast are polarized in the newly forming reticulocyte and are segregated away from the ejecting nucleus (158). Collectively, these observations and those described in previous sections imply a significant role for the spectrin skeleton in membrane assembly, vesicle trafficking, and the organization of membrane signaling systems, at least at zones of cell–cell contact or following lymphocyte activation.

Recent data also suggest yet another unusual way that the spectrin skeleton may participate in the ecology of the cell, and that is as a *mechanosensory signal transduction system*. There are multiple roles for mechanosensory signal transduction at the cellular level. Cell volume regulation, blood pressure regulation, and cell–cell contact sensation are a few examples. An effective system must have two components: an ability to sense mechanical deformation, and the ability to transduce that mechanical signal into a chemical or electrical signal within the cell. The recent identification (465) of a specific interaction between the cytoskeletal protein adducin and a regulator of transmembrane ion transport, stomatin, suggests that these proteins may form the core of a mechanosensory system. The spectrin–actin cytoskeletal network has the necessary elasticity to "sense" mechanical deformation of the cell. Stomatin appears to regulate cation transport and therefore is in a position to create a chemical or electrical signal. Adducin links stomatin to the cytoskeleton. The ability of both stomatin and adducin to be phosphorylated suggests that this interaction may be regulated.

Although circumstantial, the evidence for the role of adducin and stomatin in forming a mechanosensory system is becoming compelling. The binding site for stomatin on adducin is the region homologous to the MARCKS protein (465), and the MARCKS protein is believed to serve a signaling role in cells (2). Work with the Milan hypertensive strain of rats has linked mutations in the adducin genes to abnormalities of blood pressure regulation (36). A stomatin analogue in *C. elegans* (mec-2) is directly involved in mechanosensation in these worms (211). Mec-2 has been shown to interact with and regulate mec-4, mec-6, and mec-10, all homologous to subunits of the epithelial sodium channel (210), the same sodium channel that when mutated produces Liddle's syndrome, a heritable form of human hypertension (457). Adducin has been shown to be present in many cell types and the message for stomatin has been identified in a variety of human

tissues; therefore, further characterization of this interaction is certain to have relevance far beyond its role in a rare hemolytic disease, including a better understanding of mechanisms underlying hypertension and the unusual ways that the spectrin skeleton has been utilized to mediate surface membrane events.

CONCLUSIONS: THE LINKED MOSAIC MODEL

Observations on the spectrin skeleton in erythroid and non-erythroid cells collectively lead to the following conclusions:

1. Spectrin is a large multifunctional flexible protein highly conserved through evolution.
2. Spectrin binds directly or indirectly to an array of integral membrane proteins.
3. Spectrin binds to an array of cytoplasmic linker and regulatory proteins.
4. Spectrin binds to all major filaments: actin, intermediate filaments, and tubulin.
5. A spectrin skeleton (of some isotype) may be found wherever there are membranes.
6. Spectrin lends shape and dynamic stability to membranes.
7. Spectrin's interactions are subject to many levels of post-translational control.

In addition, the energetics of the bilayer with embedded proteins suggest that the control of integral membrane protein distributions is a fundamental and crucially important task, especially in areas of the membrane that are laterally inhomogeneous (e.g., polarized cells, PSD), but even in areas of the membrane expected to be homogeneous (e.g., the red cell membrane). Taken together, these considerations suggest a unifying model of the spectrin skeleton. This model is termed the *linked mosaic*.

In the *linked mosaic* model, spectrin's fundamental role is to locally tether integral membrane and cytosolic proteins into a semi-organized array. The intrinsic flexibility of spectrin allows the array considerable local freedom of motion, but limits wider excursions beyond the limits of the mosaic. A series of mosaics are linked by spectrin's interactions with other cytoskeletal elements; most commonly the linker is actin, but intermediate filaments and tubulin may also serve. Such filament interactions may also serve to constrain isotypes of spectrin and certain associated proteins (for example, ankyrin) to specific cellular compartments (such as a "vesicular spectrin skeleton" vs. the "cortical spectrin skeleton") (115). It is envisioned that integral proteins tethered to a given mosaic may be exchanged with other mosaics, such as during membrane assembly or lymphocyte receptor activation ("vesicular mosaics" exchanging with "cortical mosaics") or during lateral migration of an integral membrane protein (for example, diffusion of AE1 in the plane of the erythrocyte membrane or the process of receptor patching and capping in lymphocytes). These concepts and this model is depicted in Figure 11.22.

Regardless of the merits of this model or others, it is clear that the spectrin-based skeleton is a central and crucially important component of the machinery by which a cell organizes its affairs. The full range of its

interactions is only now being appreciated, despite decades of study. It would be wise for active cell biologists and physiologists to "stay tuned."

The kind assistance and insights of Drs. Prasad Devarajan, Zhushan Zhang, Susan Glantz, Narla Mohandas, Christian Lombardo, Alan Harris, David Speicher, and Vincent T. Marchesi are gratefully acknowledged. Dr. Sása Svetina is especially thanked for his discussions on the energetics of bilayer membranes, and for providing several manuscripts in advance of their publication. Work reviewed in this chapter that was the authors' was supported by grants from the National Institutes of Health, United States of America.

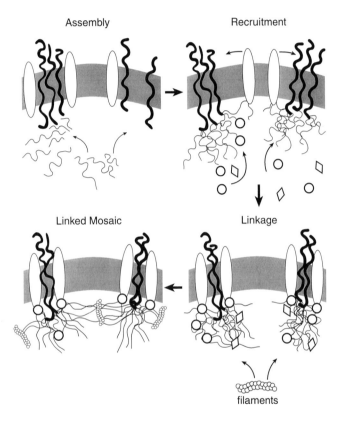

Assembly   Recruitment

Linked Mosaic   Linkage

filaments

FIG. 11.22. The linked mosaic model of spectrin action. The fundamental role of the spectrin skeleton is to control lateral order in the plane of the membrane. Through its capacity to bind multiple ligands selectively, and to self-associate through hetero-and homotypic interactions, end-on and side to side, spectrin alone can form ordered arrays of mosaics of limited size. Associated with these nascent arrays are various embedded and soluble proteins. It is hypothesized that the spectrin mosaics with their associated proteins are joined by linking interactions, most commonly involving actin or other filament systems, to form larger arrays. Collectively, these *linked mosaics* serve to control the lateral organization of integral membrane proteins, and to recruit selected cytoplasmic components (e.g., signal transduction molecules, kinases) in a way that enhances receptor efficiencies, creates signal transduction systems, and stabilizes the membrane against spontaneous and deleterious changes in the distribution of its integral membrane proteins. This proposed role applies with equal validity to internal as well as surface membranes. Perhaps for this reason spectrin and ankyrin are also associated with internal membranes.

REFERENCES

1. Aberle, H., S. Butz, J. Stappert, H. Weissig, R. Kemler, and H. Hoschuetzky. Assembly of the cadherin-catenin complex in vitro with recombinant proteins. *J. Cell Sci.* 107: 3655–3663, 1994.
2. Aderem, A. Signal transduction and the actin cytoskeleton: the roles of MARCKS and profilin. *Trends Biochem. Sci.* 17: 438–43, 1992.
3. Aghib, D. F., and P. D. McCrea. The E-cadherin complex contains the src substrate p120. *Exp. Cell Res.* 218: 359–69, 1995.
4. Agre, P., E. P. Orringer, and V. Bennett. Deficient red-cell spectrin in severe, recessively inherited spherocytosis. *N. Engl. J. Med.* 306: 1155–1161, 1982.
5. Alloisio, N., N. Dalla Venezia, A. Rana, K. Andrabi, P. Texier, F. Gilsanz, J. P. Cartron, J. Delaunay, and A. H. Chishti. Evidence that red blood cell protein p55 may participate in the skeleton-membrane linkage that involves protein 4.1 and glycophorin C. *Blood* 82: 1323–1327, 1993.
6. Amin, K. M., A. Scarpa, J. C. Winkelmann, P. J. Curtis, and B. G. Forget. The exon-intron organization of the human erythroid beta-spectrin gene. *Genomics* 18: 118–125, 1993.
7. Anderson, J. M., A. S. Fanning, L. Lapierre, and C. M. Van Itallie. Zonula occludens ZO-1 and ZO-2: membrane-associated guanylate kinase homologues (MAGuks) of the tight junction. *Biochem. Soc. Trans.* 23: 470–475, 1995.
8. Anderson, J. P., and J. S. Morrow. The interaction of calmodulin with human erythrocyte spectrin: inhibition of protein 4.1 stimulated actin binding. *J. Biol. Chem.* 262: 6365–6372, 1987.
9. Andrews, R. K., and J. E. Fox. Identification of a region in the cytoplasmic domain of the platelet membrane glycoprotein Ib-IX complex that binds to purified actin-binding protein. *J. Biol. Chem.* 267: 18605–18611, 1992.
10. Appleyard, S. T., M. J. Dunn, V. Dubowitz, M. L. Scott, S. L. Pittman, and D. M. Shotton. Monoclonal antibodies detect a spectrin-like protein in normal and dystrophic human skeletal muscle. *Proc. Natl. Acad. Sci. U.S.A.* 81: 776–780, 1984.
11. Atkinson, M.A.L., J. S. Morrow, and V. T. Marchesi. The polymeric state of actin in the human erythrocyte cytoskeleton. *J. Cell Biochem.* 18: 493–505, 1982.
12. Aunis, D., and M.-F. Bader. The cytoskeleton as a barrier to exocytosis in secretory cell. *J. Exp. Biol.* 139: 253–266, 1988.
13. Axton, J. M., F. L. Shamanski, L. M. Young, D. S. Henderson, J. B. Boyd, and W. T. Orr. The inhibitor of DNA replication encoded by the Drosophila gene plutonium is a small, ankyrin repeat protein. *EMBO J.* 13: 462–70, 1994.
14. Azim, A. C., J. H. Knoll, A. H. Beggs, and A. H. Chishti.

Isoform cloning, actin binding, and chromosomal localization of human erythroid dematin, a member of the villin superfamily. *J. Biol. Chem.* 270: 17407–17413, 1995.

15. Babcock, G. G., and V. M. Fowler. Isoform-specific interaction of tropomodulin with skeletal muscle and erythrocyte tropomyosins. *J. Biol. Chem.* 269: 27510–27518, 1994.

16. Bähler, M. F., and P. Greengard. Synapsin I bundles F-actin in a phosphorylation-dependent manner. *Nature* 326: 704–707, 1987.

17. Baines, A. J., and V. Bennett. Synapsin I is a microtubule-bundling protein. *Nature* 319: 145–147, 1986.

18. Baltensperger, K., L. M. Kozma, A. D. Cherniack, J. K. Klarlund, A. Chawla, U. Banerjee, and M. P. Czech. Binding of the ras activator son of sevenless to insulin receptor substrate-1 signaling complexes. *Science* 260: 1950–1952, 1993.

19. Baron, M. D., M. D. Davison, P. Jones, and D. R. Critchley. The sequence of chick a-actinin reveals homologies to spectrin and calmodulin. *J. Biol. Chem.* 262: 17623–17630, 1987.

20. Beck, K. A., J. A. Buchanan, V. Malhotra, and W. J. Nelson. Golgi spectrin: identification of an erythroid beta-spectrin homolog associated with the Golgi complex. *J. Cell Biol.* 127: 707–723, 1994.

21. Becker, P. J., J. S. Morrow, and S. E. Lux. Abnormal oxidant sensitivity and beta chain structure of spectrin in hereditary spherocytosis associated with defective spectrin-protein 4.1 binding. *J. Clin. Invest.* 80: 557–565, 1987.

22. Becker, P. S., and S. E. Lux. Hereditary spherocytosis and related disorders. *Clin. Haematol.* 14: 15–43, 1985.

23. Becker, P. S., M. A. Schwartz, J. S. Morrow, and S. E. Lux. Radiolabel-transfer cross-linking demonstrates that protein 4.1 binds to the N-terminal region of beta spectrin and to actin in binary interactions. *Eur. J. Biochem.* 193: 827–836, 1990.

24. Becker, P. S., W. T. Tse, S. E. Lux, and B. G. Forget. Beta spectrin kissimmee: a spectrin variant associated with autosomal dominant hereditary spherocytosis and defective binding to protein 4.1. *J. Clin. Invest.* 92: 612–616, 1993.

25. Beggs, A. H., T. J. Byers, J. H. Knoll, F. M. Boyce, et al. Cloning and characterization of two human skeletal muscle alpha-actinin genes located on chromosomes 1 and 11. *J. Biol. Chem.* 267: 9281–9288, 1992.

26. Behrens, J., L. Vakaet, R. Friis, E. Winterhager, R. F. Van, M. M. Mareel, and W. Birchmeier. Loss of epithelial differentiation and gain of invasiveness correlates with tyrosine phosphorylation of the E-cadherin/beta-catenin complex in cells transformed with a temperature-sensitive v-SRC gene. *J. Cell Biol.* 120: 757–766, 1993.

27. Belkin, A. M., N. I. Zhidkova, and V. E. Koteliansky. Localization of talin in skeletal and cardiac muscles. *FEBS Lett.* 200: 32–36, 1986.

28. Bennett, V. Ankyrins. Adaptors between diverse plasma membrane proteins and the cytoplasm. *J. Biol. Chem.* 267: 8703–8706, 1982.

29. Bennett, V., and D. Branton. Selective association of spectrin with the cytoplasmic surface of human erythrocyte plasma membranes. Quantitative determination with purified (32P) spectrin. *J. Biol. Chem.* 252: 2753–2763, 1977.

30. Bennett, V., and J. Davis. Erythrocyte ankyrin: immunoreactive analogues are associated with mitotic structures in cultured cells and with microtubules in brain. *Proc. Natl. Acad. Sci. U.S.A.* 78: 7550–7554, 1981.

31. Bennett, V., and D. M. Gilligan. The spectrin-based membrane skeleton and micron-scale organization of the plasma membrane. *Annu. Rev. Cell Biol.* 9: 27–66, 1993.

32. Bennett, V., and S. Lambert. The spectrin skeleton: from red cells to brain. *J. Clin. Invest.* 87: 1483–1489, 1991.

33. Bennett, V., and P. J. Stenbuck. Identification and partial purification of ankyrin, the high affinity membrane attachment site for human erythrocyte spectrin. *J. Biol. Chem.* 254: 2533–2541, 1979.

34. Bennett, V., and P. J. Stenbuck. The membrane attachment protein for spectrin is associated with band 3 in human erythrocyte membranes. *Nature* 280: 468–73, 1979.

35. Bianchi, G., P. Ferrarri, D. Trizio, M. Ferrandi, L. Torielli, B. R. Barber, and E. Polli. Red blood cell abnormalities and spontaneous hypertension in the rat: a genetically determined link. *Hypertension* 7: 319–325, 1985.

36. Bianchi, G., G. Tripodi, G. Casari, S. Salardi, B. R. Barber, R. Garcia, P. Leoni, L. Torielli, D. Cusi, M. Ferrandi, et al. Two point mutations within the adducin genes are involved in blood pressure variation. *Proc. Natl. Acad. Sci. U.S.A.* 91: 3999–4003, 1994.

37. Birchmeier, W., K. M. Weidner, and J. Behrens. Molecular mechanisms leading to loss of differentiation and gain of invasiveness in epithelial cells. *J. Cell Sci. Suppl.* 17: 159–164, 1993.

38. Birkenmeier, C. S., R. A. White, L. L. Peters, E. J. Hall, S. E. Lux, and J. E. Barker. Complex patterns of sequence variation and multiple 5′ and 3′ ends are found among transcripts of the erythroid ankyrin gene. *J. Biol. Chem.* 268: 9533–9540, 1993.

39. Bitbol, M., C. Dempsey, A. Watts, and P. F. Devaux. Weak interaction of spectrin with phosphatidylcholine-phosphatidylserine. *FEBS Lett.* 244: 217–222, 1989.

40. Black, J. D., S. T. Koury, R. B. Bankert, and E. A. Repasky. Heterogeneity in lymphocyte spectrin distribution: ultrastructural identification of a new spectrin-rich cytoplasmic structure. *J. Cell Biol.* 106: 97–109, 1988.

41. Bloch, R. J., and J. S. Morrow. An unusual β-spectrin associated with clustered acetylcholine receptors. *J. Cell Biol.* 108: 481–493, 1989.

42. Bloom, M. L., B. K. Lee, C. S. Birkenmeier, Y. Ma, W. E. Zimmer, S. R. Goodman, E. M. Eicher, and J. E. Barker. Brain beta spectrin isoform 235 (Spnb-2) maps to mouse chromosome 11. *Mamm. Genome* 3: 293–295, 1992.

43. Bonder, E. M., D. J. Fishkind, N. M. Cotran, and D. A. Begg. The cortical actin-membrane cytoskeleton of unfertilized sea urchin eggs: analysis of the spatial organization and relationship of filamentous actin, nonfilamentous actin, and egg spectrin. *Dev. Biol.* 134: 327–341, 1989.

44. Bourguignon, L. Y., A. Chu, H. Jin, and N. R. Brandt. Ryanodine receptor-ankyrin interaction regulates internal Ca^{2+} release in mouse T-lymphoma cells. *J. Biol. Chem.* 270: 17917–17922, 1995.

45. Bourguignon, L. Y., N. Iida, and H. Jin. The involvement of the cytoskeleton in regulating IP3 receptor-mediated internal Ca^{2+} release in human blood platelets. *Cell Biol. Int.* 17: 751–758, 1993.

46. Bourguignon, L. Y., and H. Jin. Identification of the ankyrin-binding domain of the mouse T-lymphoma cell inositol 1,4,5-trisphosphate (IP3) receptor and its role in the regulation of IP3-mediated internal Ca^{2+} release. *J. Biol. Chem.* 270: 7257–60, 1995.

47. Bourguignon, L. Y., H. Jin, N. Iida, N. R. Brandt, and S. H. Zhang. The involvement of ankyrin in the regulation of inositol 1,4,5-trisphosphate receptor-mediated internal Ca^{2+} release from Ca^{2+} storage vesicles in mouse T-lymphoma cells. *J. Biol. Chem.* 268: 7290–7297, 1993.

48. Bourguignon, L. Y., E. L. Kalomiris, and V. B. Lokeshwar.

Acylation of the lymphoma transmembrane glycoprotein, GP85, may be required for GP85-ankyrin interaction. *J. Biol. Chem.* 266: 11761–11765, 1991.

49. Bourguignon, L. Y., V. B. Lokeshwar, J. He, X. Chen, and G. J. Bourguignon. A CD44-like endothelial cell transmembrane glycoprotein (GP116) interacts with extracellular matrix and ankyrin. *Mol. Cell Biol.* 12: 4464–4471, 1992.

50. Bourguignon, L. Y., S. J. Suchard, and E. L. Kalomiris. Lymphoma Thy-1 glycoprotein is linked to the cytoskeleton via a 4.1-like protein. *J. Cell Biol.* 103: 2529–2540, 1986.

51. Bourguignon, L. Y., S. J. Suchard, M. L. Nagpal, and J. J. Glenney. A T-lymphoma transmembrane glycoprotein (gp180) is linked to the cytoskeletal protein, fodrin. *J. Cell Biol.* 101: 477–487, 1985.

52. Bourguignon, L. Y., G. Walker, S. J. Suchard, and K. Balazovich. A lymphoma plasma membrane-associated protein with ankyrin-like properties. *J. Cell Biol.* 102: 2115–2124, 1986.

53. Brady-Kalnay, S. M., A. J. Flint, and N. K. Tonks. Homophilic binding of PTP mu, a receptor-type protein tyrosine phosphatase, can mediate cell-cell aggregation. *J. Cell Biol.* 122: 961–972, 1993.

54. Brady-Kalnay, S. M., D. L. Rimm, and N. K. Tonks. Receptor protein tyrosine phosphatase PTPmu associates with cadherins and catenins in vivo. *J. Cell Biol.* 130: 977–986, 1995.

55. Branton, D., C. M. Cohen, and J. Tyler. Interaction of cytoskeletal proteins on the human erythrocyte membrane. *Cell* 24: 24–32, 1981.

56. Bretscher, A. Rapid phosphorylation and reorganization of ezrin and spectrin accompany morphological changes induced in A-431 cells by epidermal growth factor. *J. Cell Biol.* 108: 921–930, 1989.

57. Bryant, P. J., K. L. Watson, R. W. Justice, and D. F. Woods. Tumor suppressor genes encoding proteins required for cell interactions and signal transduction in Drosophila. *Dev. Suppl.* 239–249, 1993.

58. Button, E., C. Shapland, and D. Lawson. Actin, its associated proteins and metastasis. [Review]. *Cell Motil. Cytoskeleton* 30: 247–251, 1995.

59. Buxton, R. S., and A. I. Magee. Structure and interactions of desmosomal and other cadherins. *Semin. Cell Biol.* 3: 157–167, 1992.

60. Byers, T., E. Brandin, R. Lue, E. Winograd, and D. Branton. The complete sequence of Drosophila beta-spectrin reveals supra-motifs. *Proc. Nat. Acad. Sci. U.S.A.* 89: 6187–6191, 1992.

61. Byers, T. J., A. Husain-Chishti, R. R. Dubreuil, D. Branton, and L. S. Goldstein. Sequence similarity of the amino-terminal domain of Drosophila beta spectrin to alpha actinin and dystrophin. *J. Cell Biol.* 109: 1633–1641, 1989.

62. Calvert, R., P. Bennett, and W. Gratzer. Properties and structural role of the subunits of human spectrin. *Eur. J. Biochem.* 107: 355–361, 1980.

63. Carboni, J., C. L. Howe, A. B. West, K. W. Barwick, M. S. Mooseker, and J. S. Morrow. Characterization of intestinal brush border cytoskeletal proteins of normal and neoplastic human epithelial cells: a comparison with the avian brush border. *Am. J. Pathol.* 129: 589–600, 1987.

64. Carlier, M. F., C. Simon, R. Cassoly, and L. A. Pradel. Interaction between microtubule-associated protein tau and spectrin. *Biochimie* 66: 305–311, 1984.

65. Cartron, J. P., and C. Rahuel. Human erythrocyte glycophorins: protein and gene structure analyses. *Transfus. Med. Rev.* 6: 63–92, 1992.

66. Casari, G., C. Barlassina, D. Cusi, L. Zagato, R. Muirhead, M. Righetti, P. Nembri, K. Amar, M. Gtti, F. Macciardi, G. Binelli, and G. Bianchi. Association of the alpha adducin locus with essential hypertension. *Hypertension* 25: 320–326, 1995.

67. Chasis, J. A., P. Agre, and N. Mohandas. Decreased membrane mechanical stability and in vivo loss of surface area reflect spectrin deficiencies in hereditary spherocytosis. *J. Clin. Invest.* 82: 617–623, 1988.

68. Chasis, J. A., L. Coulombel, J. Conboy, S. McGee, K. Andrews, Y. W. Kan, and N. Mohandas. Differentiation-associated switches in protein 4.1 expression. Synthesis of multiple structural isoforms during normal human erythropoiesis. *J. Clin. Invest.* 91: 329–438, 1993.

69. Chasis, J. A., M. E. Reid, R. H. Jensen, and N. Mohandas. Signal transduction by glycophorin A: role of extracellular and cytoplasmic domains in a modulatable process. *J. Cell Biol.* 107: 1351–1357, 1988.

70. Cho, K. O., C. A. Hunt, and M. B. Kennedy. The rat brain postsynaptic density fraction contains a homolog of the Drosophila discs-large tumor suppressor protein. *Neuron* 9: 929–942, 1992.

71. Chu, Z. L., A. Wickrema, S. B. Krantz, and J. C. Winkelmann. Erythroid-specific processing of human beta spectrin I premRNA. *Blood* 84: 1992–1999, 1994.

72. Cianci, C. D., E.-T. Chen, and J. S. Morrow. Cooperative interactions in the spectrin-based cortical cytoskeleton: a kinetic study. *Federation Proc.* 240, 1987.

73. Cianci, C. D., P. G. Gallagher, B. G. Forget, and J. S. Morrow. Unique subsets of proteins bind the SH3 domain of αI spectrin and Grb2 in differentiating mouse erythroleukemia (MEL) cells. *Mol. and Cell. Biol.* 5: 271a (abstract), 1995.

74. Cianci, C. D., M. Giorgi, and J. S. Morrow. Phosphorylation of ankyrin down-regulates its cooperative interaction with spectrin and protein 3. *J. Cell Biochem.* 37: 301–315, 1988.

75. Cianci, C. D., A. S. Harris, S. Mische, and J. S. Morrow. The calmodulin binding site of protein 4.1. submitted, 1996.

76. Cianci, C. D., J. C. Kim, J. McLaughlin, P. R. Stabach, C. R. Lombardo, and J. S. Morrow. Spectrin isoform diversity and assembly in non-erythroid cells. *Cell. Mol. Biol. Lett.* 1: 79–99, 1996.

77. Cianci, C. D., and J. S. Morrow. Cloning and characterization of human fetal brain alpha II spectrin. GenBank Accession number U26396, 1995.

78. Cicchetti, P., B. J. Mayer, G. Thiel, and D. Baltimore. Identification of a protein that binds to the SH3 region of *abl* and is similar to *bcr* and *GAP-rho*. *Science* 257: 803–806, 1992.

79. Cioe, L., P. Laurila, P. Meo, K. Krebs, S. Goodman, and P. J. Curtis. Cloning and nucleotide sequence of a mouse erythrocyte beta-spectrin cDNA. *Blood* 70: 915–920, 1987.

80. Clark, M. B., Y. Ma, M. L. Bloom, J. E. Barker, I. S. Zagon, W. E. Zimmer, and S. R. Goodman. Brain alpha erythroid spectrin: identification, compartmentalization, and beta spectrin associations. *Brain Res.* 663: 223–236, 1994.

81. Clark, S. W., and D. I. Meyer. Centractin is an actin homologue associated with the centrosome [see comments]. *Nature* 359: 246–250, 1992.

82. Clark, S. W., O. Staub, I. B. Clark, E. L. Holzbaur, B. M. Paschal, R. B. Vallee, and D. I. Meyer. Beta-centractin: characterization and distribution of a new member of the centractin family of actin-related proteins. *Mol. Biol. Cell* 5: 1301–1310, 1994.

83. Cohen, C. M., E. Dotimas, and C. Korsgren. Human erythrocyte membrane protein band 4.2 (pallidin). *Semin. Hematol.* 30: 119–137, 1993.

84. Cohen, C. M., and S. F. Foley. The role of band 4.1 in the association of actin with erythrocyte membranes. *Biochim. Biophys. Acta* 688: 691–701, 1982.

85. Cohen, C. M., and S. F. Foley. Spectrin-dependent and -independent association of F-actin with the erythrocyte membrane. *J. Cell Biol.* 86: 694–698, 1980.

86. Cohen, C. M., S. F. Foley, and C. Korsgren. A protein immunologically related to erythrocyte band 4.1 is found on stress fibres on non-erythroid cells. *Nature* 299: 648–650, 1982.

87. Cohen, C. M., and C. Korsgren. Band 4.1 causes spectrin-actin gels to become thixiotropic. *Biochem. Biophys. Res. Commun.* 97: 1429–1435, 1980.

88. Cohen, C. M., J. M. Tyler, and D. Branton. Spectrin-actin associations studied by electron microscopy of shadowed preparations. *Cell* 21: 875–883, 1980.

89. Cole, N., and G. B. Ralston. Enhancement of self-association of human spectrin by polyethylene glycol. *Int. J. Biochem.* 26: 799–804, 1994.

90. Coleman, T. R., D. J. Fishkind, M. S. Mooseker, and J. S. Morrow. Contributions of the beta-subunit to spectrin structure and function. *Cell Motil. Cytoskeleton* 12: 248–263, 1989.

91. Coleman, T. R., D. J. Fishkind, M. S. Mooseker, and J. S. Morrow. Functional diversity among spectrin isoforms. *Cell Motil. Cytoskeleton* 12: 225–247, 1989.

92. Coleman, T. R., A. S. Harris, S. M. Mische, M. S. Mooseker, and J. S. Morrow. Beta spectrin bestows protein 4.1 sensitivity on spectrin-actin interactions. *J. Cell Biol.* 104: 519–526, 1987.

93. Conboy, J., Y. W. Kan, S. B. Shohet, and N. Mohandas. Molecular cloning of protein 4.1, a major structural element of the human erythrocyte membrane skeleton. *Proc. Natl. Acad. Sci. U.S.A.* 83: 9512–9516, 1986.

94. Conboy, J. G. Structure, function, and molecular genetics of erythroid membrane skeletal protein 4.1 in normal and abnormal red blood cells. *Semin. Hematol.* 30: 58–73, 1993.

95. Conboy, J. G., J. Y. Chan, J. A. Chasis, Y. W. Kan, and N. Mohandas. Tissue- and development-specific alternative RNA splicing regulates expression of multiple isoforms of erythroid membrane protein 4.1. *J. Biol. Chem.* 266: 8273–8280, 1991.

96. Cook, T. A., R. Urrutia, and M. A. McNiven. Identification of dynamin 2, an isoform ubiquitously expressed in rat tissue. *Proc. Natl. Acad. Sci. U.S.A.* 91: 644–648, 1994.

97. Correas, I. Characterization of isoforms of protein 4.1 present in the nucleus. *Biochem. J.* 279: 581–585, 1991.

98. Correas, I., T. L. Leto, D. W. Speicher, and V. T. Marchesi. Identification of the functional site of erythrocyte protein 4.1 involved in spectrin-actin associations. *J. Biol. Chem.* 261: 3310–3315, 1986.

99. Cowin, P., H. P. Kapprell, W. W. Franke, J. Tamkun, and R. O. Hynes. Plakoglobin: a protein common to different kinds of intercellular adhering junctions. *Cell* 46: 1063–1073, 1986.

100. Craig, S. W., and J. V. Pardo. Gamma actin, spectrin, and intermediate filament proteins colocalize with vinculin at costameres, myofibril-to-sarcolemma attachment sites. *Cell Motil.* 3: 449–462, 1983.

101. Dahl, S. C., R. W. Geib, M. T. Fox, M. Edidin, and D. Branton. Rapid capping in alpha-spectrin-deficient MEL cells from mice afflicted with hereditary hemolytic anemia. *J. Cell Biol.* 125: 1057–1065, 1994.

102. Daniel, J. M., and A. B. Reynolds. The tyrosine kinase substrate p120cas binds directly to E-cadherin but not to the adenomatous polyposis coli protein or alpha-catenin. *Mol. Cell Biol.* 15: 4819–4824, 1995.

103. Davis, J., and V. Bennett. Brain spectrin. Isolation of subunits and formation of hybrids with erythrocyte spectrin subunits. *J. Biol. Chem.* 258: 7757–7766, 1983.

104. Davis, J. Q., and V. Bennett. The anion exchanger and Na+K(+)-ATPase interact with distinct sites on ankyrin in in vitro assays. *J. Biol. Chem.* 265: 17252–17256, 1990.

105. Davis, J. Q., and V. Bennett. Ankyrin-binding activity of nervous system cell adhesion molecules expressed in adult brain. *J. Cell Sci. Suppl.* 17: 109–117, 1993.

106. Davis, J. Q., and V. Bennett. Brain ankyrin. A membrane-associated protein with binding sites for spectrin, tubulin, and the cytoplasmic domain of the erythrocyte anion channel. *J. Biol. Chem.* 259: 13550–13559, 1984.

107. Davis, J. Q., T. McLaughlin, and V. Bennett. Ankyrin-binding proteins related to nervous system cell adhesion molecules: candidates to provide transmembrane and intercellular connections in adult brain. *J. Cell Biol.* 121: 121–33, 1993.

108. Davis, L., and V. Bennett. Purification of a calmodulin-sensitive spectrin-binding membrane glycoprotein from brain. *Mol. Biol. Cell* 6: 1566A, 1995.

109. Davis, L. H., and V. Bennett. Identification of two regions of beta G spectrin that bind to distinct sites in brain membranes. *J. Biol. Chem.* 269: 4409–4416, 1994.

110. Davis, L. H., and V. Bennett. Mapping the binding sites of human erythrocyte ankyrin for the anion exchanger and spectrin. *J. Biol. Chem.* 265: 10589–10596, 1990.

111. Davis, L. H., J. Q. Davis, and V. Bennett. Ankyrin regulation: an alternatively spliced segment of the regulatory domain functions as an intramolecular modulator. *J. Biol. Chem.* 267: 18966–18972, 1992.

112. de Arruda, M. V., S. Watson, C. S. Lin, J. Leavitt, and P. Matsudaira. Fimbrin is a homologue of the cytoplasmic phosphoprotein plastin and has domains homologous with calmodulin and actin gelation proteins. *J. Cell Biol.* 111: 1069–1070, 1990.

113. Delaunay, J., N. Alloisio, L. Morle, and F. Baklouti. The genetic disorders of the red cell skeleton. *Nouv. Rev. Er. Hematol.* 33: 63–70, 1991.

114. DeSilva, T. M., K. C. Peng, K. D. Speicher, and D. W. Speicher. Analysis of human red cell spectrin tetramer (head-to-head) assembly using complementary univalent peptides. *Biochemistry* 31: 10872–10878, 1992.

115. Devarajan, P., and J. S. Morrow. The spectrin cytoskeleton and organization of polarized epithelial cell membranes. In: *Membrane Protein–Cytoskeleton Interactions,* edited by W. J. Nelson. New York: Academic Press, 1996, Vol. 43, p. 97–128.

116. Devarajan, P., D. A. Scaramuzzino, and J. S. Morrow. Ankyrin binds to two distinct cytoplasmic domains of Na,K-ATPase α subunit. *Proc. Natl. Acad. Sci. U.S.A.* 91: 2965–2969, 1994.

117. Devarajan, P., P. R. Stabach, A. S. Mann, T. Ardito, M. Kashgarian, and J. S. Morrow. Identification of a small cytoplasmic ankyrin (Ank$_{G119}$) in kidney and muscle that binds $\beta I\varsigma^*$ spectrin and associates with the Golgi apparatus. *J. Cell Biol.* 133: 819–830, 1996.

118. Diakowski, W., and A. F. Sikorski. Brain spectrin interacts with membrane phospholipids. *Acta Biochim. Pol.* 41: 153–154, 1994.

119. Ding, D., S. Parkhurst, and H. Lipshitz. Different genetic requirements for anterior RNA localization revealed by the distribution of adducin-like transcripts during Drosophila oogenesis. *Proc. Natl. Acad. Sci. U.S.A.* 90: 2512–2516, 1993.

120. Ding, Y., J. R. Casey, and R. R. Kopito. The major kidney AE1 isoform does not bind ankyrin (Ank1) in vitro. An essential role for the 79 NH2-terminal amino acid residues of band 3. *J. Biol. Chem.* 269: 32201–32208, 1994.

121. Discher, D. E., R. Winardi, P. O. Schischmanoff, M. Parra, J. G. Conboy, and N. Mohandas. Mechanochemistry of protein 4.1's spectrin-actin-binding domain: ternary complex interactions, membrane binding, network integration, structural strengthening. *J. Cell Biol.* 130: 897–907, 1995.

122. Dmytrenko, G. M., D. W. Pumplin, and R. J. Bloch. Dystrophin in a membrane skeletal network: localization and comparison to other proteins. *J. Neurosci.* 13: 547–558, 1993.

123. Dong, L., C. Chapline, B. Mousseau, L. Fowler, K. Ramsay, J. Stevens, and S. Jaken. 35H, a sequence isolated as a protein kinase C binding protein, is a novel member of the adducin family. *J. Biol. Chem.* 270: 25534–25540, 1995.

124. Dubreuil, R. R. Structure and evolution of the actin crosslinking proteins. *Bioessays* 13: 219–26, 1991.

125. Dubreuil, R. R., E. Branton, J. H. Reisberg, L. S. Goldstein, and D. Branton. Structure, calmodulin-binding, and calcium-binding properties of recombinant alpha spectrin polypeptides. *J. Biol. Chem.* 266: 7189–7193, 1991.

126. Dubreuil, R. R., T. J. Byers, A. L. Sillman, D. Bar-Zvi, L. S. Goldstein, and D. Branton. The complete sequence of Drosophila alpha-spectrin: conservation of structural domains between alpha-spectrins and alpha-actinin. *J. Cell Biol.* 109: 2197–2205, 1989.

127. Dubreuil, R. R., T. J. Byers, C. T. Stewart, and D. P. Kiehart. A beta-spectrin isoform from Drosophila (beta H) is similar in size to vertebrate dystrophin. *J. Cell Biol.* 111: 1849–1858, 1990.

128. Dubreuil, R. R., and J. Yu. Ankyrin and beta-spectrin accumulate independently of alpha-spectrin in Drosophila. *Proc. Natl. Acad. Sci. U.S.A.* 91: 10285–10289, 1994.

129. Eber, S. W., W. M. Lande, T. A. Iarocci, W. C. Mentzer, P. Hohn, J. S. Wiley, and W. Schroter. Hereditary stomatocytosis: consistent association with an integral membrane protein deficiency. *Br. J. Haematol.* 72: 452–455, 1989.

130. Egan, S. E., B. W. Giddings, M. W. Brooks, L. Buday, A. M. Sizeland, and R. A. Weinberg. Association of Sos Ras exchange protein with Grb2 is implicated in tyrosine kinase signal transduction and transformation [see comments]. *Nature* 363: 45–51, 1993.

131. Elgsaeter, A., and A. Mikkelsen. Shapes and shape changes in vitro in normal red blood cells. *Biochim. Biophys.* Acta 1071: 273–290, 1991.

132. Elgsaeter, A., D. M. Shotton, and D. Branton. Intramembrane particle aggregation in erythrocyte ghosts. II. The influence of spectrin aggregation. *Biochim. Biophys.* Acta 426: 101–122, 1976.

133. Fach, B. L., S. F. Graham, and R. A. Keates. Association of fodrin with brain microtubules. *Can. J. Biochem. Cell Biol.* 63: 372–381, 1985.

134. Fairbanks, G., T. L. Steck, and D. F. Wallach. Electrophoretic analysis of the major polypeptides of the human erythrocyte membrane. *Biochemistry* 10: 2606–2617, 1971.

135. Faire, K., F. Trent, J. M. Tepper, and S. M. Bonder. Analysis of dynamin isoforms in mammalian brain: dynamin-1 expression is spatially and temporally regulated during postnatal development. *Proc. Natl. Acad. Sci. U.S.A.* 89: 8376–8380, 1992.

136. Falzone, C. J., Y. H. Kao, J. Zhao, D. A. Bryant, and J. T. Lecomte. Three-dimensional solution structure of PsaE from the cyanobacterium Synechococcus sp. strain PCC 7002, a photosystem I protein that shows structural homology with SH3 domains. *Biochemistry* 33: 6052–6062, 1994.

137. Faquin, W. C., C. A. Husain, and D. Branton. Expression of dematin (protein 4.9) during avian erythropoiesis. *Eur. J. Cell Biol.* 53: 48–58, 1990.

138. Faraday, C. D., and R. M. Spanswick. Evidence for a membrane skeleton in higher plants. A spectrin-like polypeptide co-isolates with rice root plasma membranes. *FEBS Lett.* 318: 313–6, 1993.

139. Fearon, E. R., T. Finkel, M. L. Gillison, S. P. Kennedy, J. F. Casella, G. F. Tomaselli, J. S. Morrow, and C. V. Dang. Karyoplasmic interaction selection strategy (KISS): A general strategy to detect protein-protein interactions in mammalian cells. *Proc. Natl. Acad. Sci. U.S.A.* 89: 7958–7962, 1992.

140. Ferrarri, P., L. Torielli, M. Ferrandi, and G. Bianchi. Volumes and Na transports in intact red blood cells, resealed ghosts, and inside-out vesicles of Milan-hypertensive rats. In: *Membrane Pathology,* edited by G. Bianchi, E. Carafoli and A. Scarpe, New York: Ann NY Acad Sci, 1986, p. 561–566.

141. Fishkind, D. J., E. M. Bonder, and D. A. Begg. Isolation and characterization of sea urchin egg spectrin: calcium modulation of the spectrin-actin interaction. *Cell Motil. Cytoskeleton* 7: 304–314, 1987.

142. Fishkind, D. J., E. M. Bonder, and D. A. Begg. Sea urchin spectrin in oogenesis and embryogenesis: a multifunctional integrator of membrane-cytoskeletal interactions. *Dev. Biol.* 142: 453–464, 1990.

143. Fishkind, D. J., E. M. Bonder, and D. A. Begg. Subcellular localization of sea urchin egg spectrin: evidence for assembly of the membrane-skeleton on unique classes of vesicles in eggs and embryos. *Dev. Biol.* 142: 439–452, 1990.

144. Fowler, V. M., and V. Bennett. Erythrocyte membrane tropomyosin. Purification and properties. *J. Biol. Chem.* 259: 5978–5989, 1984.

145. Fowler, V. M., and V. Bennett. Tropomyosin: a new component of the erythrocyte membrane skeleton. *Prog. Clin. Biol. Res.* 159: 57–71, 1984.

146. Fowler, V. M., M. A. Sussmann, P. G. Miller, B. E. Flucher, and M. P. Daniels. Tropomodulin is associated with the free (pointed) ends of the thin filaments in rat skeletal muscle. *J. Cell Biol.* 120: 411–420, 1993.

147. Franck, Z., R. Gary, and A. Bretscher. Moesin, like ezrin, colocalizes with actin in the cortical cytoskeleton in cultured cells, but its expression is more variable. *J. Cell Sci.* 105: 219–231, 1993.

148. Frappier, T., J. Derancourt, and L. A. Pradel. Actin and neurofilament binding domain of brain spectrin beta subunit. *Eur. J. Biochem.* 205: 85–91, 1992.

149. Frappier, T., F. Regnouf, and L. A. Pradel. Binding of brain spectrin to the 70-kDa neurofilament subunit protein. *Eur. J. Biochem.* 169: 651–657, 1987.

150. Frappier, T., F. Stetzkowski-Marden, and L. A. Pradel. Interaction domains of neurofilament light chain and brain spectrin. *Biochem. J.* 275 (Pt 2): 521–527, 1991.

151. Funayama, N., A. Nagafuchi, N. Sato, S. Tsukita, and S. Tsukita. Radixin is a novel member of the band 4.1 family. *J. Cell Biol.* 115: 1039–1048, 1991.

152. Gallagher, P. G., and B. G. Forget. Structure, organization and expression of the human band 7.2b gene, a candidate gene for hereditary hydrocytosis. *J. Biol. Chem.* 270: 26358–26363, 1995.

153. Gallagher, P. G., W. T. Tse, A. L. Scarpa, S. E. Lux, and B. G. Forget. Large numbers of alternatively spliced isoforms

of the regulatory region of human erythrocyte ankyrin. *Trans. Assoc. Am. Physicians* 105: 268–277, 1992.

154. Gallagher, P. G., M. Upender, D. C. Ward, and B. G. Forget. The gene for human erythrocyte membrane protein band 7.2 (EPB72) maps to 9q33-q34 centromeric to the Philadelphia chromosome translocation breakpoint region. *Genomics* 18: 167–169, 1993.

155. Gardner, K., and V. Bennett. Modulation of spectrin-actin assembly by erythrocyte adducin. *Nature* 328: 359–362, 1987.

156. Gardner, K., and V. Bennett. A new erythrocyte membrane-associated protein with calmodulin binding activity. Identification and purification. *J. Biol. Chem.* 261: 1339–1348, 1986.

157. Garnier, J., D. J. Osguthorpe, and B. Robson. Analysis of the accuracy and implications of simple methods for predicting the secondary structure of globular proteins. *J. Mol. Biol.* 120: 97–120, 1978.

158. Geiduschek, J. B., and S. J. Singer. Molecular changes in the membrane of mouse erythroid cells accompanying differentiation. *Cell* 16: 149–163, 1979.

159. Georgatos, S. D., and V. T. Marchesi. The binding of vimentin to human erythrocyte membranes: a model system for the study of intermediate filament-membrane interactions. *J. Cell Biol.* 100: 1955–1961, 1985.

160. Georgatos, S. D., D. C. Weaver, and V. T. Marchesi. Site specificity in vimentin-membrane interactions: intermediate filament subunits associate with the plasma membrane via their head domains. *J. Cell Biol.* 100: 1962–1967, 1985.

161. Giebelhaus, D. H., D. W. Eib, and R. T. Moon. Antisense RNA inhibits expression of membrane skeleton protein 4.1 during embryonic development of Xenopus. *Cell* 53: 601–615, 1988.

162. Gilligan, D., and V. Bennett. Alternative splicing of adducin beta subunit. *J. Cell Biol.* 115: 42a, 1991.

163. Gilligan, D. M., and V. Bennett. Localization of adducin isoforms in brain and kidney. *Mol. Biol. Cell* 5: 421 (abstract), 1994.

164. Gilligan, D. M., J. Lieman, and V. Bennett. Assignment of the human beta-adducin gene (ADD2) to 2p13-p14 by in situ hybridization. *Genomics* 28: 610–612, 1995.

165. Glantz, S. B., and J. S. Morrow. The spectrin-actin cytoskeleton and membrane integrity in hypoxia. In: *Tissue Oxygen Deprivation: Developmental, Molecular and Integrated Function*, edited by G. G. Haddad and G. Lister, New York: Marcel Dekker, 1995, p. 153–192.

166. Glenney, J. R., Jr., and P. Glenney. Fodrin is the general spectrin-like protein found in most cells whereas spectrin and the TW protein have a restricted distribution. *Cell* 34: 503–512, 1983.

167. Goldberg, Y. P., B. Y. Lin, S. E. Andrew, J. Nasir, R. Graham, M. L. Glaves, G. Hutchinson, J. Theilmann, D. G. Ginzinger, K. Schappert, L. Clarke, J. M. Rommens, and M. R. Hayden. Cloning and mapping of the alpha-adducin gene close to D4S95 and assessment of its relationship to Huntington disease. *Hum. Mol. Genet.* 1: 669–675, 1992.

168. Goodman, S. R., K. E. Krebs, C. F. Whitfield, B. M. Riederer, and I. S. Zagon. Spectrin and related molecules. *CRC Crit. Rev. Biochem.* 23: 171–234, 1988.

169. Goodman, S. R., I. S. Zagon, C. F. Whitfield, L. A. Casoria, P. J. McLaughlin, and T. L. Laskiewicz. A spectrin-like protein from mouse brain membranes: immunological and structural correlations with erythrocyte spectrin. *Cell Motil.* 3: 635–647, 1983.

170. Goodman, S. R., W. E. Zimmer, M. B. Clark, I. S. Zagon, J. E. Barker, and M. L. Bloom. Brain spectrin: of mice and men. *Brain Res. Bull.* 36: 593–606, 1995.

171. Gorlin, J. B., R. Yamin, S. Egan, M. Stewart, T. P. Stossel, D. J. Kwiatkowski, and J. H. Hartwig. Human endothelial actin-binding protein (ABP-280, nonmuscle filamin): a molecular leaf spring. *J. Cell. Biol.* 111: 1089–1105, 1990.

172. Gout, I., R. Dhand, I. D. Hiles, M. J. Fry, G. Panayotou, P. Das, O. Truong, N. F. Totty, J. Hsuan, G. W. Booker, et al. The GTPase dynamin binds to and is activated by a subset of SH3 domains. *Cell* 75: 25–36, 1993.

173. Gregorio, C. C., and V. M. Fowler. Mechanisms of thin filament assembly in embryonic chick cardiac myocytes: tropomodulin requires tropomyosin for assembly. *J. Cell Biol.* 129: 683–695, 1995.

174. Gregorio, C. C., R. T. Kubo, R. B. Bankert, and E. A. Repasky. Translocation of spectrin and protein kinase C to a cytoplasmic aggregate upon lymphocyte activation. *Proc. Natl. Acad. Sci. U.S.A.* 89: 4947–4951, 1992.

175. Gregorio, C. C., E. A. Repasky, V. M. Fowler, and J. D. Black. Dynamic properties of ankyrin in T lymphocytes: colocalization with spectrin and protein kinase C beta. *J. Cell Biol.* 125: 345–358, 1994.

176. Gregorio, C. C., A. Weber, M. Bondad, C. R. Pennise, and V. M. Fowler. Requirement of pointed-end capping by tropomodulin to maintain actin filament length in embryonic chick cardiac myocytes. *Nature* 377: 83–86, 1995.

177. Gundersen, D., J. Orlowski, and E. Rodriquez-Boulan. Apical polarity of Na, K-ATPase in retinal pigment epithelium is linked to a reversal of the ankyrin-fodrin submembrane cytoskeleton. *J. Biol. Chem.* 112: 863–872, 1991.

178. Hall, T. G., and V. Bennett. Regulatory domains of erythrocyte ankyrin. *J. Biol. Chem.* 262: 10537–10545, 1987.

179. Hanspal, M., J. S. Hanspal, R. Kalraiya, S. C. Liu, K. E. Sahr, D. Howard, and J. Palek. Asynchronous synthesis of membrane skeletal proteins during terminal maturation of murine erythroblasts. *Blood* 80: 530–539, 1992.

180. Hanspal, M., J. S. Hanspal, R. Kalraiya, and J. Palek. The expression and synthesis of the band 3 protein initiates the formation of a stable membrane skeleton in murine Rauscher-transformed erythroid cells. *Eur. J. Cell Biol.* 58: 313–318, 1992.

181. Hanspal, M., R. Kalraiya, J. Hanspal, K. E. Sahr, and J. Palek. Erythropoietin enhances the assembly of alpha, beta spectrin heterodimers on the murine erythroblast membranes by increasing beta spectrin synthesis. *J. Biol. Chem.* 266: 15626–15630, 1991.

182. Harell, D., and M. Morrison. Two-dimensional separation of erythrocyte membrane proteins. *Arch. Biochem. Biophys.* 193: 158–168, 1979.

183. Harris, A. S., C. D. Cianci, S. M. Mische, and J. S. Morrow. Characterization of human erythrocyte protein 4.1-calmodulin (CAM) interactions. In: *Proceedings of the Symposium on Cytoskeleton and Cell Regulation.* Steamboat Springs, CO: 237, 1990.

184. Harris, A. S., D. Croall, and J. S. Morrow. The calmodulin-binding site in fodrin is near the site of calcium-dependent protease-I cleavage. *J. Biol. Chem.* 263: 15754–15761, 1988.

185. Harris, A. S., D. E. Croall, and J. S. Morrow. Calmodulin regulates fodrin susceptibility to cleavage by calcium-dependent protease I. *J. Biol. Chem.* 264: 17401–17408, 1989.

186. Harris, A. S., L. A. D. Green, K. J. Ainger, and J. S. Morrow. Mechanisms of cytoskeletal regulation (1): Functional differences correlate with antigenic dissimilarity in human brain

and erythrocyte spectrin. *Biochim. Biophys. Acta* 830: 147–158, 1985.

187. Harris, A. S., and J. S. Morrow. Calmodulin and calcium-dependent protease I coordinately regulate the interaction of fodrin with actin. *Proc. Natl. Acad. Sci. U.S.A.* 87: 3009–3013, 1990.

188. Harris, A. S., and J. S. Morrow. Proteolytic processing of human brain alpha spectrin (Fodrin): Identification of a hypersensitive site. *J. Neurosci.* 8: 2640–2651, 1988.

189. Hartwig, J. H. Spectrin superfamily, subfamily 1: The spectrin family. *Protein Profile* 1: 715–749, 1994.

190. Hartwig, J. H., and D. J. Kwiatkowski. Actin-binding proteins. *Curr. Opin. Cell Biol.* 3: 87–97, 1991.

191. Haslam, R. J., H. B. Koide, and B. A. Hemmings. Pleckstrin domain homology [letter]. *Nature* 363: 309–310, 1993.

192. Hayes, N. V., F. E. Holmes, J. Grantham, and A. J. Baines. A60, an axonal membrane-skeletal spectrin-binding protein. *Biochem. Soc. Trans.* 23: 54–58, 1995.

193. Hayette, S., D. Dhermy, M. E. dos Santos, M. Bozon, D. Drenckhahn, N. Alloisio, P. Texier, J. Delaunay, and L. Morle. A deletional frameshift mutation in protein 4.2 gene (allele 4.2 Lisboa) associated with hereditary hemolytic anemia. *Blood* 85: 250–256, 1995.

194. Hemming, N. J., D. J. Anstee, W. J. Mawby, M. E. Reid, and M. J. Tanner. Localization of the protein 4.1-binding site on human erythrocyte glycophorins C and D [published erratum appears in *Biochem. J.* 1994 Jun 15;300(Pt 3):920]. *Biochem. J.* 299: 191–196, 1994.

195. Hemming, N. J., D. J. Anstee, M. A. Staricoff, M.J.A. Tanner, and N. Mohandas. Identification of the membrane attachment sites for protein 4.1 in the human erythrocyte. *J. Biol. Chem.* 270: 5360–5366, 1995.

196. Henniker, A., and G. B. Ralston. Reinvestigation of the thermodynamics of spectrin self-association. *Biophys. Chem.* 52: 251–258, 1994.

197. Herman, I. M. Actin isoforms. [Review]. *Curr. Opin. Cell Biol.* 5: 48–55, 1993.

198. Herrmann, H., and G. Wiche. Plectin and IFAP-300K are homologous proteins binding to microtubule-associated proteins 1 and 2 and to the 240-kilodalton subunit of spectrin. *J. Biol. Chem.* 262: 1320–1325, 1987.

199. Hiebl-Dirschmied, C. M., G. R. Adolf, and R. Prohaska. Isolation and partial characterisation of the human band 7 integral membrane protein. *Biochim. Biophys. Acta* 1065: 195–202, 1991.

200. Hiebl-Dirschmied, C. M., B. Entler, C. Glotzmann, I. Maurer-Fogy, C. Stratowa, and R. Prohaska. Cloning and nucleotide sequence of cDNA encoding human erythrocyte band 7 integral membrane protein. *Biochim. Biophys. Acta* 1090: 123–124, 1991.

201. Hirokawa, N., R. E. Cheney, and M. Willard. Location of a protein of the fodrin-spectrin-TW260/240 family in the mouse intestinal brush border. *Cell* 32: 953–65, 1983.

202. Hock, R. S., G. Davis, and D. W. Speicher. Purification of human smooth muscle filamin and characterization of structural domains and functional sites. *Biochemistry* 29: 9441–9451, 1990.

203. Horne, W. C., S. C. Huang, P. S. Becker, T. K. Tang, and E. J. Benz, Jr. Tissue-specific alternative splicing of protein 4.1 inserts an exon necessary for formation of the ternary complex with erythrocyte spectrin and F-actin. *Blood* 82: 2558–2563, 1993.

204. Horne, W. C., T. L. Leto, and V. T. Marchesi. Differential phosphorylation of multiple sites in protein 4.1 and protein 4.9 by phorbol ester-activated and cyclic AMP-dependent protein kinases. *J. Biol. Chem.* 260: 9073–9076, 1985.

205. Horne, W. C., W. C. Prinz, and E. K. Tang. Identification of two cAMP-dependent phosphorylation sites on erythrocyte protein 4.1. *Biochim. Biophys. Acta* 1055: 87–92, 1990.

206. Hoschuetzky, H., H. Aberle, and R. Kemler. Beta-catenin mediates the interaction of the cadherin-catenin complex with epidermal growth factor receptor. *J. Cell Biol.* 127: 1375–1380, 1994.

207. Howe, C. L., L. M. Sacramone, M. S. Mooseker, and J. S. Morrow. Mechanisms of cytoskeletal regulation: modulation of membrane affinity in avian brush border and erythrocyte spectrins. *J. Cell Biol.* 101: 1379–1385, 1985.

208. Hu, R. J., S. Moorthy, and V. Bennett. Expression of functional domains of beta G-spectrin disrupts epithelial morphology in cultured cells. *J. Cell Biol.* 128: 1069–80, 1995.

209. Huang, J. P., C. J. Tang, G. H. Kou, V. T. Marchesi, E. J. Benz, Jr., and T. K. Tang. Genomic structure of the locus encoding protein 4.1. Structural basis for complex combinational patterns of tissue-specific alternative RNA splicing. *J. Biol. Chem.* 268: 3758–66, 1993.

210. Huang, M., and M. Chalfie. Gene interactions affecting mechanosensory transduction in Caenorhabditis elegans [see comments]. *Nature* 367: 467–470, 1994.

211. Huang, M., G. Gu, E. L. Ferguson, and M. Chalfie. A stomatin-like protein necessary for mechanosensation in C. elegans. *Nature* 378: 292–295, 1995.

212. Huebner, K., A. P. Palumbo, M. Isobe, C. A. Kozak, S. Monaco, G. Rovera, C. M. Croce, and P. J. Curtis. The alpha-spectrin gene is on chromosome 1 in mouse and man. *Proc. Natl. Acad. Sci. U.S.A.* 82: 3790–3793, 1985.

213. Hughes, C. A., and V. Bennett. Adducin: a physical model with implications for function in assembly of spectrin-actin complexes. *J. Biol. Chem.* 270: 18990–18996, 1995.

214. Hulsken, J., W. Birchmeier, and J. Behrens. E-cadherin and APC compete for the interaction with beta-catenin and the cytoskeleton. *J. Cell Biol.* 127: 2061–2069, 1994.

215. Husain-Chishti, A., W. Faquin, C. C. Wu, and D. Branton. Purification of erythrocyte dematin (protein 4.9) reveals an endogenous protein kinase that modulates actin-bundling activity. *J. Biol. Chem.* 264: 8985–8991, 1989.

216. Husain-Chishti, A., A. Levin, and D. Branton. Abolition of actin-bundling by phosphorylation of human erythrocyte protein 4.9. *Nature* 334: 718–721, 1988.

217. Hyvonen, M., M. J. Macias, M. Nilges, H. Oschkinat, M. Saraste, and M. Wilmanns. Structure of the binding site for inositol phosphates in a PH domain. *EMBO J.* 14: 4676–4685, 1995.

218. Ideguchi, H., J. Nishimura, H. Nawata, and N. Hamasaki. A genetic defect of erythrocyte band 4.2 protein associated with hereditary spherocytosis. *Br. J. Haematol.* 74: 347–353, 1990.

219. Iglic, A., S. Svetina, and B. Zeks. Depletion of membrane skeleton in red blood cell vesicles. *Biophys J.* 69: 274–279, 1995.

220. Iglic, A., S. Svetina, and B. Zeks. A role of membrane skeleton in discontinuous red blood cell shape transformations. *Cell Mol. Biol. Lett.* 1: 137–144, 1996.

221. Iida, N., V. B. Lokeshwar, and L. Y. Bourguignon. Mapping the fodrin binding domain in CD45, a leukocyte membrane-associated tyrosine phosphatase. *J. Biol. Chem.* 269: 28576–28583, 1994.

222. Jesaitis, A. J., G. M. Bokoch, J. O. Tolley, and R. A. Allen. Lateral segregation of neutrophil chemotactic receptors into

actin- and fodrin-rich plasma membrane microdomains depleted in guanyl nucleotide regulatory proteins. *J. Cell Biol.* 107: 921–928, 1988.

223. Jesaitis, A. J., J. O. Tolley, and R. A. Allen. Receptor-cytoskeleton interactions and membrane traffic may regulate chemoattractant-induced superoxide production in human granulocytes. *J. Biol. Chem.* 261: 13662–13669, 1986.

224. Jesaitis, L. A., and D. A. Goodenough. Molecular characterization and tissue distribution of ZO-2, a tight junction protein homologous to ZO-1 and the Drosophila discs-large tumor suppressor protein. *J. Cell Biol.* 124: 949–961, 1994.

225. Johnson, G. V., J. M. Litersky, and R. S. Jope. Degradation of microtubule-associated protein 2 and brain spectrin by calpain: a comparative study. *J. Neurochem.* 56: 1630–1638, 1991.

226. Joseph, S. K., and S. Samanta. Detergent solubility of the inositol trisphosphate receptor in rat brain membranes. Evidence for association of the receptor with ankyrin. *J. Biol. Chem.* 268: 6477–6486, 1993.

227. Joshi, R., and V. Bennett. Mapping the domain structure of human erythrocyte adducin. *J. Biol. Chem.* 265: 13130–13136, 1990.

228. Joshi, R., D. M. Gilligan, E. Otto, T. McLaughlin, and V. Bennett. Primary structure and domain organization of human alpha and beta adducin. *J. Cell Biol.* 115: 665–675, 1991.

229. Kahana, E., J. C. Pinder, K. S. Smith, and W. B. Gratzer. Fluorescence quenching of spectrin and other red cell membrane cytoskeletal proteins. Relation to hydrophobic binding sites. *Biochem. J.* 282: 75–80, 1992.

230. Kaiser, H. W., E. O'Keefe, and V. Bennett. Adducin: Ca^{++}-dependent association with sites of cell-cell contact. *J. Cell Biol.* 109: 557–569, 1989.

231. Kapfhamer, D., D. E. Miller, S. Lambert, V. Bennett, T. W. Glover, and M. Burmeister. Chromosomal localization of the ankyrinG gene (ANK3/Ank3) to human 10q21 and mouse 10. *Genomics* 27: 189–91, 1995.

232. Karinch, A. M., W. E. Zimmer, and S. R. Goodman. The identification and sequence of the actin-binding domain of human red blood cell beta-spectrin. *J. Biol. Chem.* 265: 11833–11840, 1990.

233. Kelly, G. M., B. D. Zelus, and R. T. Moon. Identification of a calcium-dependent calmodulin-binding domain in Xenopus membrane skeleton protein 4.1. *J. Biol. Chem.* 266: 12469–12473, 1991.

234. Kennedy, S. P., S. L. Warren, B. G. Forget, and J. S. Morrow. Ankyrin binds to the 15th repetitive unit of erythroid and non-erythroid β-spectrin. *J. Cell Biol.* 115: 267–277, 1991.

235. Kennedy, S. P., S. A. Weed, B. G. Forget, and J. S. Morrow. A partial structural repeat forms the heterodimer self-association site of all β-spectrins. *J. Biol. Chem.* 269: 11400–11408, 1994.

236. Kennedy, S. P., S. A. Weed, J. Winkleman, and J. S. Morrow. The self-association site of recombinant human erythroid and muscle β-spectrin. *J. Cell Biol.* 115: 42a, 1991.

237. Kikuchi, A., K. Kaibuchi, Y. Hori, H. Nonaka, T. Sakoda, M. Kawamura, T. Mizuno, and Y. Takai. Molecular cloning of the human cDNA for a stimulatory GDP/GTP exchange protein for c-Ki-ras p21 and smg p21. *Oncogene* 7: 289–93, 1992.

238. Kim, E., M. Niethammer, A. Rothschild, Y. N. Jan, and M. Sheng. Clustering of Shaker-type K^+ channels by interaction with a family of membrane-associated guanylate kinases. *Nature* 378: 85–88, 1995.

239. Kim, S. K. Tight junction, membrane-associated guanylate kinases and cell signaling. *Curr. Opin. Cell Biol.* 7: 641–649, 1995.

240. Kistner, U., B. M. Wenzel, R. W. Veh, C. Cases-Langhoff, A. M. Garner, U. Appeltauer, B. Voss, E. D. Gundelfinger, and C. C. Garner. SAP90, a rat presynaptic protein related to the product of the *Drosophila* tumor suppressor gene *dlg*-A. *J. Biol. Chem.* 268: 4580–4583, 1993.

241. Klotz, K. N., and A. J. Jesaitis. Neutrophil chemoattractant receptors and the membrane skeleton. *Bioessays* 16: 193–198, 1994.

242. Knowles, W. J., W. J. Knowles, J. S. Morrow, D. W. Speicher, H. S. Zarkowsky, N. Mohandas, W. C. Mentzer, S. B. Shohet, and V. T. Marchesi. Spectrin from patients with hereditary pyropoikilocytosis have self-association defects and an abnormal tryptic digestion map. In: *ICN/UCLA Symposium on Differentiation and Function of Hematopoietic Cell Surfaces,* Keystone, CO, February. Keystone: ICN/UCLA, 1981.

243. Knowles, W. J., J. S. Morrow, D. W. Speicher, H. S. Zarkowsky, N. Mohandas, W. C. Mentzer, S. B. Shohet, and V. T. Marchesi. Molecular and functional changes in spectrin from patients with hereditary pyropoikilocytosis. *J. Clin. Invest.* 71: 1867–1877, 1983.

244. Knudsen, K. A., A. P. Soler, K. R. Johnson, and M. J. Wheelock. Interaction of alpha-actinin with the cadherin/catenin cell-cell adhesion complex via alpha-catenin. *J. Cell. Biol.* 130: 67–77, 1995.

245. Knudsen, K. A., and M. J. Wheelock. Plakoglobin, or an 83-kD homologue distinct from beta-catenin, interacts with E-cadherin and N-cadherin. *J. Cell Biol.* 118: 671–679, 1992.

246. Kobayashi, T., S. Ohno, Y. C. Park-Matsumoto, N. Kameda, and T. Baba. Developmental studies of dystrophin and other cytoskeletal proteins in cultured muscle cells. *Microsc. Res. Tech.* 30: 437–457, 1995.

247. Koenig, M., A. P. Monaco, and L. M. Kunkel. The complete sequence of dystrophin predicts a rod-shaped cytoskeletal protein. *Cell* 53: 219–226, 1988.

248. Koob, R., M. Zimmermann, W. Schoner, and D. Drenckhahn. Colocalization and coprecipitation of ankyrin and Na+,K+-ATPase in kidney epithelial cells. *Eur. J. Cell Biol.* 45: 230–237, 1988.

249. Kordeli, E., J. Davis, B. Trapp, and V. Bennett. An isoform of ankyrin is localized at nodes of Ranvier in myelinated axons of central and peripheral nerves. *J. Cell Biol.* 110: 1341–1352, 1990.

250. Kordeli, E., S. Lambert, and V. Bennett. AnkyrinG. A new ankyrin gene with neural-specific isoforms localized at the axonal initial segment and node of Ranvier. *J. Biol. Chem.* 270: 2352–2359, 1995.

251. Korsgren, C., and C. M. Cohen. cDNA sequence, gene sequence, and properties of murine pallidin (band 4.2), the protein implicated in the murine pallid mutation. *Genomics* 21: 478–485, 1994.

252. Korsgren, C., J. Lawler, S. Lambert, D. Speicher, and C. M. Cohen. Complete amino acid sequence and homologies of human erythrocyte membrane protein band 4.2. *Proc. Natl. Acad. Sci. U.S.A.* 87: 613–617, 1990.

253. Kotula, L., T. M. DeSilva, D. W. Speicher, and P. J. Curtis. Functional characterization of recombinant human red cell α-spectrin polypeptides containing the tetramer binding site. *J. Biol. Chem.* 268: 14788–14793, 1993.

254. Kotula, L., L. D. Laury-Kleintop, L. Showe, K. Sahr, A. J. Linnenbach, B. Forget, and P. J. Curtis. The exon-intron orga-

nization of the human erythrocyte alpha-spectrin gene. *Genomics* 9: 131–140, 1991.

255. Koury, M. J., M. C. Bondurant, and S. S. Rana. Changes in erythroid membrane proteins during erythropoietin-mediated terminal differentiation. *J. Cell Physiol.* 133: 438–448, 1987.

256. Kralj-Iglic, V., S. Svetina, and B. Zeks. The existence of non-axisymmetric bilayer vesicle shapes predicted by the bilayer couple model. *Eur. Biophys. J.* 22: 97–103, 1993.

257. Kralj-Iglic, V., S. Svetina, and B. Zeks. Lateral distribution of membrane constituents and cellular shapes. *Eur. Biophys. J.* submitted, 1996.

258. Kretsinger, R. H., and S. Nakayama. Evolution of EF-hand calcium-modulated proteins. IV. Exon shuffling did not determine the domain compositions of EF-hand proteins. *J. Mol. Evol.* 36: 477–488, 1993.

259. Kunimoto, M., E. Otto, and V. Bennett. A new 440-kD isoform is the major ankyrin in neonatal rat brain. *J. Cell Biol.* 115: 1319–1331, 1991.

260. Lambert, S., and V. Bennett. Postmitotic expression of ankyrinR and beta R-spectrin in discrete neuronal populations of the rat brain. *J. Neurosci.* 13: 3725–3735, 1993.

261. Lambert, S., H. Yu, J. T. Prchal, J. Lawler, P. Ruff, D. Speicher, M. C. Cheung, Y. W. Kan, and J. Palek. cDNA sequence for human erythrocyte ankyrin. *Proc. Natl. Acad. Sci. U.S.A.* 87: 1730–1734, 1990.

262. Lande, W. M., and W. C. Mentzer. Haemolytic anaemia associated with increased cation permeability. *Clin. Hematol.* 14: 89–103, 1985.

263. Lande, W. M., P. W. Thiemann, and W. M. Mentzer. Missing band 7 membrane protein in two patients with high Na, low K erythrocytes. *J. Clin. Invest.* 70: 1273–1280, 1982.

264. Langley, R. C., Jr., and C. M. Cohen. Association of spectrin with desmin intermediate filaments. *J. Cell Biochem.* 30: 101–109, 1986.

265. Laurila, P., L. Cioe, C. A. Kozak, and P. J. Curtis. Assignment of mouse beta-spectrin gene to chromosome 12. *Somat. Cell Mol. Genet.* 13: 93–97, 1987.

266. Lazarides, E., and W. J. Nelson. Erythrocyte and brain forms of spectrin in cerebellum: distinct membrane-cytoskeletal domains in neurons. *Science* 220: 1295–1296, 1983.

267. Lazarides, E., and C. Woods. Biogenesis of the red blood cell membrane-skeleton and the control of erythroid morphogenesis. *Annu. Rev. Cell Biol.* 5: 427–452, 1989.

268. Lee, J. K., J. D. Black, E. A. Repasky, R. T. Kubo, and R. B. Bankert. Activation induces a rapid reorganization of spectrin in lymphocytes. *Cell* 55: 807–816, 1988.

269. Lee, J. K., R. S. Coyne, R. R. Dubreuil, L. S. Goldstein, and D. Branton. Cell shape and interaction defects in alpha-spectrin mutants of Drosophila melanogaster. *J. Cell Biol.* 123: 1797–809, 1993.

270. Lee, J. K., and E. A. Repasky. Cytoskeletal polarity in mammalian lymphocytes in situ. *Cell Tissue Res.* 247: 195–202, 1987.

271. Lehnert, M. E., and H. F. Lodish. Unequal synthesis and differential degradation of alpha and beta spectrin during murine erythroid differentiation. *J. Cell Biol.* 107: 413–426, 1988.

272. Leto, T. L., D. Fortugno-Erikson, D. Barton, T. L. Yang-Fong, U. Francke, A. S. Harris, J. S. Morrow, V. T. Marchesi, and E. J. Benz, Jr. Comparison of non-erythroid α-spectrin genes reveals strict homology among diverse species. *Mol. Cell. Biol.* 8: 1–9, 1988.

273. Leto, T. L., and V. T. Marchesi. A structural model of human erythrocyte protein 4.1. *J. Biol. Chem.* 259: 4603–4608, 1984.

274. LeVine, H. I., and N. E. Sayoun. Involvement of fodrin-binding proteins in the structure of the neuronal postsynaptic density and regulation by phosphorylation. *Biochem. Biophys. Res. Commun.* 138: 59–65, 1986.

275. Levine, J., and M. Willard. Fodrin: axonally transported polypeptides associated with the internal periphery of many cells. *J. Cell Biol.* 90: 631–643, 1981.

276. Li, X.-L., and V. Bennett. Identification of the spectrin subunit and domains required for formation of adducin/spectrin/actin complexes. *Mol. Biol. Cell* 6: 269a (abstract), 1995.

277. Li, Z. P., E. P. Burke, J. S. Frank, V. Bennett, and K. D. Philipson. The cardiac Na^+-Ca^{2+} exchanger binds to the cytoskeletal protein ankyrin. *J. Biol. Chem.* 268: 11489–11491, 1993.

278. Lin, B., J. Nasir, H. McDonald, R. Graham, J. M. Rommens, Y. P. Goldberg, and M. R. Hayden. Genomic organization of the human alpha-adducin gene and its alternately spliced isoforms. *Genomics* 25: 93–99, 1995.

279. Liu, S. C., L. H. Derick, and J. Palek. Visualization of the hexagonal lattice in the erythrocyte membrane skeleton. *J. Cell Biol.* 104: 527–536, 1987.

280. Lokeshwar, V. B., and L. Y. Bourguignon. The lymphoma transmembrane glycoprotein GP85 (CD44) is a novel guanine nucleotide-binding protein which regulates GP85 (CD44)-ankyrin interaction. *J. Biol. Chem.* 267: 22073–22078, 1992.

281. Lokeshwar, V. B., and L. Y. Bourguignon. Post-translational protein modification and expression of ankyrin-binding site(s) in GP85 (Pgp-1/CD44) and its biosynthetic precursors during T-lymphoma membrane biosynthesis. *J. Biol. Chem.* 266: 17983–17989, 1991.

282. Lokeshwar, V. B., and L. Y. Bourguignon. Tyrosine phosphatase activity of lymphoma CD45 (GP180) is regulated by a direct interaction with the cytoskeleton. *J. Biol. Chem.* 267: 21551–21557, 1992.

283. Lokeshwar, V. B., N. Fregien, and L. Y. Bourguignon. Ankyrin-binding domain of CD44(GP85) is required for the expression of hyaluronic acid-mediated adhesion function. *J. Cell Biol.* 126: 1099–1109, 1994.

284. Lombardo, C. R., and P. S. Low. Calmodulin modulates protein 4.1 binding to human erythrocyte membranes. *Biochim. Biophys. Acta* 1196: 139–144, 1994.

285. Lombardo, C. R., D. L. Rimm, S. P. Kennedy, B. G. Forget, and J. S. Morrow. Ankyrin independent membrane sites for non-erythroid spectrin. *Mol. Biol. Cell* 4(suppl) 57a, 1993.

286. Lombardo, C. R., D. L. Rimm, E. Koslov, and J. S. Morrow. Human recombinant alpha-catenin binds to spectrin. *Mol. Biol. Cell* 5 (suppl): 47a, 1994.

287. Lombardo, C. R., S. A. Weed, S. P. Kennedy, B. G. Forget, and J. S. Morrow. βII-Spectrin (fodrin) and βIς2-spectrin (muscle) contain NH2- & COOH-terminal membrane association domains (MAD1 & MAD2). *J. Biol. Chem.* 269: 29212–29219, 1994.

288. Lombardo, C. R., B. M. Willardson, and P. S. Low. Localization of the protein 4.1-binding site on the cytoplasmic domain of erythrocyte membrane band 3. *J. Biol. Chem.* 267: 9540–9546, 1992.

289. López, J. A., B. Leung, C. C. Reynolds, C. Q. Li, and J.E.B. Fox. Efficient plasma membrane expression of a functional platelet glycoprotein Ib-IX complex requires the presence of its three subunits. *J. Biol. Chem.* 267: 12851–12859, 1992.

290. Lorenz, M., B. Bisikirska, B. Hanus-Lorenz, K. Strzalka, and A. F. Sikorski. Proteins reacting with anti-spectrin antibodies are present in Chlamydomonas cells. *Cell Biol Int.* 19: 83–90, 1995.

291. Low, P. S., D. P. Allen, T. F. Zioncheck, P. Chari, B. M.

Willardson, R. L. Geahlen, and M. L. Harrison. Tyrosine phosphorylation of band 3 inhibits peripheral protein binding. *J. Biol. Chem.* 262: 4592–4596, 1987.

292. Low, P. S., B. M. Willardson, N. Mohandas, M. Rossi, and S. Shohet. Contribution of the band 3-ankyrin interaction to erythrocyte membrane mechanical stability. *Blood* 77: 1581–1586, 1991.

293. Lowenstein, E. J., R. J. Daly, A. G. Batzer, W. Li, B. Margolis, R. Lammers, A. Ullrich, E. Y. Skolnik, D. Bar-Sagi, and J. Schlessinger. The SH2 and SH3 domain-containing protein GRB2 links receptor tyrosine kinases to ras signaling. *Cell* 70: 431–442, 1992.

294. Lu, P. W., C. J. Soong, and M. Tao. Phosphorylation of ankyrin decreases its affinity for spectrin tetramer. *J. Biol. Chem.* 260: 14958–14964, 1985.

295. Lue, R. A., S. M. Marfatia, D. Branton, and A. H. Chishti. Cloning and characterization of hdlg: the human homologue of the Drosophila discs large tumor suppressor binds to protein 4.1. *Proc. Natl. Acad. Sci. U.S.A.* 91: 9818–9822, 1994.

296. Luna, E. J. Molecular links between the cytoskeleton and membranes. *Curr. Opin. Cell Biol.* 3: 120–126, 1991.

297. Luna, E. J., and A. L. Hitt. Cytoskeleton–plasma membrane interactions. *Science* 258: 955–964, 1992.

298. Luna, E. J., G. H. Kidd, and D. Branton. Identification by peptide analysis of the spectrin-binding protein in human erythrocytes. *J. Biol. Chem.* 254: 2526–2532, 1979.

299. Lundberg, S., V. P. Lehto, and L. Backman. Characterization of calcium binding to spectrins. *Biochemistry* 31: 5665–5671, 1992.

300. Lux, S. E., K. M. John, and V. Bennett. Analysis of cDNA for human erythrocyte ankyrin indicates a repeated structure with homology to tissue-differentiation and cell-cycle control proteins. *Nature* 344: 36–42, 1990.

301. Lux, S. E., and J. Palek. Disorders of the red cell membrane. In: *Blood: Principles and Practice of Hematology,* edited by R. I. Handin, S. E. Lux and T. P. Stossel, Philadelphia: JB Lippincott, 1995, p. 1701–1816.

302. Ma, Y., W. E. Zimmer, B. M. Riederer, M. L. Bloom, J. E. Barker, and S. R. Goodman. The complete amino acid sequence for brain beta spectrin (beta fodrin): relationship to globin sequences [published errata appear in *Brain Res. Mol. Brain Res.* 1993 Oct;20(1–2):179 and 1994 Jan;21(1–2):181]. *Brain Res. Mol. Bain Res.* 18: 87–99, 1993.

303. MacDonald, R. I., A. Musacchio, R. A. Holmgren, and M. Saraste. Invariant tryptophan at a shielded site promotes folding of the conformational unit of spectrin. *Proc. Natl. Acad. Sci. U.S.A.* 91: 1299–1303, 1994.

304. Macias, M. J., A. Musacchio, H. Ponstingl, M. Nilges, M. Saraste, and H. Oschkinat. Structure of the pleckstrin homology domain from beta-spectrin. *Nature* 369: 675–677, 1994.

305. Mak, A. S., G. Roseborough, and H. Baker. Tropomyosin from human erythrocyte membrane polymerizes poorly but binds F-actin effectively in the presence and absence of spectrin. *Biochim. Biophys. Acta* 912: 157–166, 1987.

306. Maksymiw, R., S. F. Sui, H. Gaub, and E. Sackmann. Electrostatic coupling of spectrin dimers to phosphatidylserine containing lipid lamellae. *Biochemistry* 26: 2983–2990, 1987.

307. Malchiodi-Albedi, F., M. Ceccarini, J. C. Winkelmann, J. S. Morrow, and T. C. Petrucci. The 270 kDa splice variant of erythrocyte β-spectrin (βIς2) segregates *in vivo* and *in vitro* to specific domains of cerebellar neurons. *J. Cell Sci.* 106: 67–78, 1993.

308. Mangeat, P. H., and K. Burridge. Immunoprecipitation of noneryhrocyte spectrin within live cells following microinjection of specific antibodies: relation to cytoskeletal structures. *J. Cell Biol.* 98: 1363–1377, 1984.

309. Marchesi, S. L. Mutant cytoskeletal proteins in hemolytic disease. In: *Ordering the membrane-cytoskeleton trilayer,* edited by M. S. Mooseker and J. S. Morrow, New York: Academic Press, 1991, p. 155–174.

310. Marchesi, V. T., and G. E. Palade. Inactivation of adenosine triphosphatase and disruption of red cell membranes by trypsin: protective effect of adenosine triphosphate. *Proc. Natl. Acad. Sci. U.S.A.* 58: 991–995, 1967.

311. Marchesi, V. T., and E. Steers. Selective solubilization of a protein component of the red cell membrane. *Science* 159: 203–204, 1968.

312. Marchesi, V. T., T. W. Tillack, R. L. Jackson, J. P. Segrest, and R. E. Scott. Chemical characterization and surface orientation of the major glycoprotein of the human erythrocyte membrane. *Proc. Natl. Acad. Sci. U.S.A.* 69: 1445–1449, 1972.

313. Marfatia, S. M., R. A. Leu, D. Branton, and A. H. Chishti. Identification of the protein 4.1 binding interface on glycophorin C and p55, a homologue of the Drosophila discs-large tumor suppressor protein. *J. Biol. Chem.* 270: 715–719, 1995.

314. Marfatia, S. M., R. A. Lue, D. Branton, and A. H. Chishti. In vitro binding studies suggest a membrane-associated complex between erythroid p55, protein 4.1, and glycophorin C. *J. Biol. Chem.* 269: 8631–8634, 1994.

315. Mariani, M., D. Maretzki, and H. U. Lutz. A tightly membrane-associated subpopulation of spectrin is 3H-palmitoylated. *J. Biol. Chem.* 268: 12996–3001, 1993.

316. Markin, V. S. Lateral organization of membranes and cell shapes. *Biophys. J.* 36: 1–19, 1981.

317. Marrs, J. A., C. Andersson-Fisone, M. C. Jeong, L. Cohen-Gould, C. Zurzolo, I. R. Nabi, E. Rodriguez-Boulan, and W. J. Nelson. Plasticity in epithelial cell phenotype: modulation by expression of different cadherin cell adhesion molecules. *J. Cell Biol.* 129: 507–519, 1995.

318. Masuda, T., N. Fujimaki, E. Ozawa, and H. Ishikawa. Confocal laser microscopy of dystrophin localization in guinea pig skeletal muscle fibers. *J. Cell Biol.* 119: 543–548, 1992.

319. Matsudaira, P. Modular organization of actin crosslinking proteins. *Trends Biochem. Sci.* 16: 87–92, 1991.

320. Matsuoka, Y., C. A. Hughes, and V. Bennett. Spectin/actin assembly activity of adducin is regulated by phosphorylation of the MARCKS-related domain by protein kinase C. *Mol. Cell. Biol.* 6: 269a (abstract), 1995.

321. Matsuyoshi, N., M. Hamaguchi, S. Taniguchi, A. Nagafuchi, S. Tsukita, and M. Takeichi. Cadherin-mediated cell-cell adhesion is perturbed by v-src tyrosine phosphorylation in metastatic fibroblasts. *J. Cell Biol.* 118: 703–714, 1992.

322. Mayer, B. J., R. Ren, K. L. Clark, and D. Baltimore. A putative modular domain present in diverse signaling proteins [letter]. *Cell* 73: 629–630, 1993.

323. McGough, A. M., and R. Josephs. On the structure of erythrocyte spectrin in partially expanded membrane skeletons. *Proc. Natl. Acad. Sci. U.S.A.* 87: 5208–5212, 1990.

324. McManus, M. J., D. C. Connolly, and N. J. Maihle. Tissue- and transformation-specific phosphotyrosyl proteins in v-erbB-tranformed cells. *J. Virol.* 69: 3631–3638, 1995.

325. McNeill, H., M. Ozawa, R. Kemler, and W. J. Nelson. Novel function of the cell adhesion molecule uvomorulin as an inducer of cell surface polarity. *Cell* 62: 309–316, 1990.

326. McNeill, H., T. A. Ryan, S. J. Smith, and W. J. Nelson.

Spatial and temporal dissection of immediate and early events following cadherin-mediated epithelial cell adhesion. *J. Cell Biol.* 120: 1217–1226, 1993.

327. Mentzer, W. C., B. H. Lubin, and S. Emmons. Correction of the permeability defect in hereditary stomatocytosis by dimethyl adipimate. *N. Engl. J. Med.* 294: 1200–1205, 1976.

328. Mercier, F., H. Reggio, G. Devilliers, D. Bataille, and P. Mangeat. Membrane-cytoskeleton dynamics in rat parietal cells: mobilization of actin and spectrin upon stimulation of gastric acid secretion. *J. Biol. Chem.* 108: 441–453, 1989.

329. Michaely, P., and V. Bennett. The membrane-binding domain of ankyrin contains four independently folded subdomains, each comprised of six ankyrin repeats. *J. Biol. Chem.* 268: 22703–22709, 1993.

330. Michalak, K., M. Bobrowska, K. Bialkowska, J. Szopa, and A. F. Sikorski. Interaction of erythrocyte spectrin with some nonbilayer phospholipids. *Gen. Physiol. Biophys.* 13: 57–62, 1994.

331. Michalak, K., M. Bobrowska, and A. F. Sikorski. Interaction of bovine erythrocyte spectrin with aminophospholipid liposomes. *Gen. Physiol. Biophys.* 12: 163–170, 1993.

332. Michalak, K., M. Bobrowska, and A. F. Sikorski. Investigation of spectrin binding to phospholipid vesicles using a isoindole fluorescent probe. Thermal properties of the bound and unbound protein. *Gen. Physiol. Biophys.* 9: 615–624, 1990.

333. Michaud, D., G. Guillet, P. A. Rogers, and P. M. Charest. Identification of a 220 kDa membrane-associated plant cell protein immunologically related to human beta-spectrin. *FEBS Lett.* 294: 77–80, 1991.

334. Minetti, C., F. Bentrame, G. Marcenaro, and E. Bonilla. Dystrophin at the plasma membrane of human muscle fibers shows a costameric localization. *Neuromusc. Dis.* 2: 99–109, 1992.

335. Mische, S. M., M. S. Mooseker, and J. S. Morrow. Erythrocyte adducin: a calmodulin regulated actin bundling protein that stimulates spectrin-actin binding. *J. Cell Biol.* 105: 2837–2845, 1987.

336. Mische, S. M., and J. S. Morrow. Post-translational regulation of the erythrocyte cortical cytoskeleton. *Protoplasma* 145: 167–175, 1988.

337. Mohandas, N. Molecular basis for red cell membrane viscoelastic properties. *Biochem. Soc. Trans.* 20: 776–782, 1992.

338. Mohandas, N., and J. A. Chasis. Red blood cell deformability, membrane material properties and shape: regulation by transmembrane, skeletal and cytosolic proteins and lipids. *Semin Hematol.* 30: 171–192, 1993.

339. Mohandas, N., and E. Evans. Mechanical properties of the red cell membrane in relation to molecular structure and genetic defects. *Annu. Rev. Biophys. Biomol. Struct.* 23: 787–818, 1994.

340. Mohandas, N., A. C. Greenquist, and S. B. Shohet. Bilayer balance and regulation of red cell shape changes. *J. Supramolecular Structure Cellular Biochem.* 9: 453–458, 1978.

341. Mohandas, N., R. Winardi, D. Knowles, A. Leung, M. Parra, E. George, J. Conboy, and J. Chasis. Molecular basis for membrane rigidity of hereditary ovalocytosis: A novel mechanism involving the cytoplasmic domain of band 3. *J. Clin. Invest.* 89: 686–692, 1992.

342. Molday, L. L., N. J. Cook, U. B. Kaupp, and R. S. Molday. The cGMP-gated cation channel of bovine rod photoreceptor cells is associated with a 240-kDa protein exhibiting immunochemical cross-reactivity with spectrin. *J. Biol. Chem.* 265: 18690–18695, 1990.

343. Mombers, C., P. W. van Dijck, L. L. van Deenen, J. de Gier, and A. J. Verkleij. The interaction of spectrin–actin and synthetic phospholipids. *Biochim. Biophys. Acta* 470: 152–160, 1977.

344. Mombers, C., A. J. Verkleij, J. de Gier, and L. L. van Deenen. The interaction of spectrin–actin and synthetic phospholipids. II. The interaction with phosphatidylserine. *Biochim. Biophys. Acta* 551: 271–281, 1979.

345. Moon, R. T., and E. Lazarides. beta-Spectrin limits alpha-spectrin assembly on membranes following synthesis in a chicken erythroid cell lysate. *Nature* 305: 62–65, 1983.

346. Moon, R. T., and A. P. McMahon. Generation of diversity in nonerythroid spectrins. Multiple polypeptides are predicted by sequence analysis of cDNAs encompassing the coding region of human nonerythroid alpha-spectrin. *J. Biol. Chem.* 265: 4427–4433, 1990.

347. Moon, R. T., J. Ngai, B. J. Wold, and E. Lazarides. Tissue-specific expression of distinct spectrin and ankyrin transcripts in erythroid and nonerythroid cells. *J. Cell Biol.* 100: 152–160, 1985.

348. Morgans, C. W., and R. R. Kopito. Association of the brain anion exchanger, AE3, with the repeat domain of ankyrin. *J. Cell Sci.* 105: 1137–1142, 1993.

349. Moriyama, R., C. R. Lombardo, R. F. Workman, and P. S. Low. Regulation of linkages between the erythrocyte membrane and its skeleton by 2,3-diphosphoglycerate. *J. Biol. Chem.* 268: 10990–10996, 1993.

350. Morris, M. B., and S. E. Lux. Characterization of the binary interaction between human erythrocyte protein 4.1 and actin. *Eur. J. Biochem.* 231: 644–650, 1995.

351. Morrow, J. S., and R. A. Anderson. Shaping the too fluid bilayer. *Lab. Invest.* 54: 237–240, 1986.

352. Morrow, J. S., C. Cianci, T. Ardito, A. Mann, and M. T. Kashgarian. Ankyrin links fodrin to alpha Na/K ATPase in Madin-Darby canine kidney cells and in renal tubule cells. *J. Cell Biol.* 108: 455–465, 1989.

353. Morrow, J. S., C. D. Cianci, S. P. Kennedy, and S. L. Warren. Polarized assembly of spectrin and ankyrin in epithelial cells. In: *Ordering the Membrane Cytoskeleton Trilayer*, edited by M. S. Mooseker and J. S. Morrow, New York: Academic Press, 1991, p. 227–244.

354. Morrow, J. S., W. Haigh, and V. T. Marchesi. Spectrin oligomers: a structural feature of the erythrocyte cytoskeleton. *J. Supra. Struc. Cell Biochem.* 17: 275–287, 1981.

355. Morrow, J. S., A. S. Harris, P. E. Shile, L. A. D. Green, and K. J. Ainger. Structural and functional domains of human brain and muscle spectrin: relationship to human erythrocyte spectrin. *Proceedings of the International Symposium on Contractile Proteins in Muscle and Non-Muscle Cell Systems and Their Morphophysiopathology (Sassari, Italy).* 1: 125–133, 1985.

356. Morrow, J. S., and V. T. Marchesi. Self-assembly of spectrin oligomers *in vitro*: a basis for a dynamic cytoskeleton. *J. Cell Biol.* 88: 463–468, 1981.

357. Morrow, J. S., D. W. Speicher, W. J. Knowles, C. J. Hsu, and V. T. Marchesi. Identification of functional domains of human erythrocyte spectrin. *Proc. Natl. Acad. Sci. U.S.A.* 77: 6592–6596, 1980.

358. Muller, B. M., U. Kistner, R. W. Veh, C. Cases-Langhoff, B. Becker, E. D. Gundelfinger, and C. C. Garner. Molecular characterization and spatial distribution of SAP97, a novel presynaptic protein homologous to SAP90 and the *Drosophila* discs-large tumor suppressor protein. *J. Neurosci.* 15: 2354–2366, 1995.

359. Murachi, T. Intracellular regulatory system involving calpain and calpastatin. *Biochem. Int.* 18: 263–294, 1989.

360. Musacchio, A., T. Gibson, V. P. Lehto, and M. Saraste. SH3—

an abundant protein domain in search of a function. *FEBS Lett.* 307: 55–61, 1992.

361. Musacchio, A., M. Noble, R. Pauptit, R. Wierenga, and M. Saraste. Crystal structure of a Src-homology 3 (SH3) domain. *Nature* 359: 851–855, 1992.

362. Nagafuchi, A., S. Ishihara, and S. Tsukita. The roles of catenins in the cadherin-mediated cell adhesion: functional analysis of E-cadherin-alpha catenin fusion molecules. *J. Cell Biol.* 127: 235–245, 1994.

363. Nagafuchi, A., Y. Shirayoshi, K. Okazaki, K. Yasuda, and M. Takeichi. Transformation of cell adhesion properties by exogenously introduced E-cadherin cDNA. *Nature* 329: 341–343, 1987.

364. Nagafuchi, A., and M. Takeichi. Cell binding function of E-cadherin is regulated by the cytoplasmic domain. *Eur. Mol. Biol. Organization J.* 7: 3679–3684, 1988.

365. Nakashima, K., and E. Beutler. Comparison of structure and function of human erythrocyte and human muscle actin. *Proc. Natl. Acad. Sci. U.S.A.* 76: 935–938, 1979.

366. Nakata, T., A. Iwamoto, Y. Noda, R. Takemura, H. Yoshikura, and N. Hirokawa. Predominant and developmentally regulated expression of dynamin in neurons. *Neuron* 7: 461–469, 1991.

367. Nakayama, S., N. D. Moncrief, and R. H. Kretsinger. Evolution of EF-hand calcium-modulated proteins. II. Domains of several subfamilies have diverse evolutionary histories. *J. Mol. Evol.* 34: 416–48, 1992.

368. Nathke, I. S., L. Hinck, J. R. Swedlow, J. Papkoff, and W. J. Nelson. Defining interactions and distributions of cadherin and catenin complexes in polarized epithelial cells. *J. Cell Biol.* 125: 1341–1352, 1994.

369. Nelson, W. J. Regulation of cell surface polarity in renal epithelia. *Pediatr. Nephrol.* 7: 599–604, 1993.

370. Nelson, W. J., and R. W. Hammerton. A membrane-cytoskeletal complex containing Na^+,K^+-ATPase, ankyrin, and fodrin in Madin-Darby canine kidney (MDCK) cells: implications for the biogenesis of epithelial cell polarity. *J. Cell Biol.* 108: 893–902, 1989.

371. Nelson, W. J., and E. Lazarides. Expression of the beta subunit of spectrin in nonerythroid cells. *Proc. Natl. Acad. Sci. U.S.A.* 80: 363–367, 1983.

372. Nelson, W. J., and E. Lazarides. Goblin (ankyrin) in striated muscle: identification of the potential membrane receptor for erythroid spectrin in muscle cells. *Proc. Natl. Acad. Sci. U.S.A.* 81: 3292–3296, 1984.

373. Nelson, W. J., and E. Lazarides. Posttranslational control of membrane-skeleton (ankyrin and alpha beta-spectrin) assembly in early myogenesis. *J. Cell Biol.* 100: 1726–1735, 1985.

374. Nelson, W. J., and E. Lazarides. Switching of subunit composition of muscle spectrin during myogenesis in vitro. *Nature* 304: 364–368, 1983.

375. Nelson, W. J., and P. J. Veshnock. Ankyrin binding to (Na^+ + K^+) ATPase and implications for the organization of membrane domains in polarized cells. *Nature* 328: 533–536, 1987.

376. Nelson, W. J., and P. J. Veshnock. Dynamics of membrane-skeleton (fodrin) organization during development of polarity in Madin-Darby canine kidney epithelial cells. *J. Cell Biol.* 103: 1751–1765, 1986.

377. Nose, A., K. Tsuji, and M. Takeichi. Localization of specificity determining sites in cadherin cell adhesion molecules. *Cell* 61: 147–155, 1990.

378. Obar, R. A., C. A. Collins, J. A. Hammarback, H. S. Shpetner, and R. B. Vallee. Molecular cloning of the microtubule-associated mechanochemical enzyme dynamin reveals homol-ogy with a new family of GTP-binding proteins. *Nature* 347: 256–261, 1990.

379. Ochiai, A., S. Akimoto, Y. Kanai, T. Shibata, T. Oyama, and S. Hirohashi. c-erbB-2 gene product associates with catenins in human cancer cells. *Biochem. Biophys. Res. Commun.* 205: 73–78, 1994.

380. Ohanian, V., L. C. Wolfe, K. M. John, J. C. Pinder, S. E. Lux, and W. B. Gratzer. Analysis of the ternary interaction of the red cell membrane skeletal proteins spectrin, actin, and 4.1. *Biochemistry* 23: 4416–4420, 1984.

381. Omann, G. M., W. N. Swann, Z. G. Oades, C. A. Parkos, A. J. Jesaitis, and L. A. Sklar. N-formylpeptide-receptor dynamics, cytoskeletal activation, and intracellular calcium response in human neutrophil cytoplasts. *J. Immunol.* 139: 3447–3455, 1987.

382. Otsuka, A. J., R. Franco, B. Yang, K. H. Shim, L. Z. Tang, Y. Y. Zhang, P. Boontrakulpoontawee, A. Jeyaprakash, E. Hedgecock, V. I. Wheaton, et al. An ankyrin-related gene (unc-44) is necessary for proper axonal guidance in Caenorhabditis elegans. *J. Cell Biol.* 129: 1081–1092, 1995.

383. Otto, E., M. Kunimoto, T. McLaughlin, and V. Bennett. Isolation and characterization of cDNAs encoding human brain ankyrins reveal a family of alternatively spliced genes. *J. Cell Biol.* 114: 241–253, 1991.

384. Ozawa, M., H. Baribault, and R. Kemler. The cytoplasmic domain of the cell adhesion molecule uvomorulin associates with three independent proteins structurally related in different species. *Eur. Mol. Biol. Org. J.* 8: 1711–1717, 1989.

385. Ozawa, M., and R. Kemler. Molecular organization of the uvomorulin-catenin complex. *J. Cell Biol.* 116: 989–996, 1992.

386. Ozawa, M., M. Ringwald, and R. Kemler. Uvomorulin-catenin complex formation is regulated by a specific domain in the cytoplasmic region of the cell adhesion molecule. *Proc. Natl. Acad. Sci. U.S.A.* 87: 4246–4250, 1990.

387. Palfrey, H. C., and A. Waseem. Protein kinase C in the human erythrocyte. Translocation to the plasma membrane and phosphorylation of bands 4.1 and 4.9 and other membrane proteins. *J. Biol. Chem.* 260: 16021–16029, 1985.

388. Pardo, J. V., J. D'Angelo Siliciano, and S. W. Craig. A vinculin-containing cortical lattice in skeletal muscle: transverse lattice elements ("costameres") mark sites of attachment between myofibrils and sarcolemma. *Proc. Natl. Acad. Sci. U.S.A.* 80: 1008–1012, 1983.

389. Pardo, J. V., J. D. Siliciano, and S. W. Craig. Vinculin is a component of an extensive network of myofibril-sarcolemma attachment regions in cardiac muscle fibers. *J. Cell Biol.* 97: 1081–1088, 1983.

390. Pasternack, G. R., R. A. Anderson, T. L. Leto, and V. T. Marchesi. Interactions between protein 4.1 and band 3. An alternative binding site for an element of the membrane skeleton. *J. Biol. Chem.* 260: 3676–3683, 1985.

391. Pasternack, G. R., and R. H. Racusen. Erythrocyte protein 4.1 binds and regulates myosin. *Proc. Natl. Acad. Sci. U.S.A.* 86: 9712–9716, 1989.

392. Pauly, J. L., R. B. Bankert, and E. A. Repasky. Immunofluorescent patterns of spectrin in lymphocyte cell lines. *J. Immunol.* 136: 246–253, 1986.

393. Pawson, T., and J. Schlessinger. SH2 and SH3 domains. *Current Bio.* 3: 434–442, 1993.

394. Peifer, M., S. Berg, and A. B. Reynolds. A repeating amino acid motif shared by proteins with diverse cellular roles [letter]. *Cell* 76: 789–791, 1994.

395. Peifer, M., P. D. McCrea, K. J. Green, E. Wieschaus, and B. M. Gumbiner. The vertebrate adhesive junction proteins

beta-catenin and plakoglobin and the Drosophila segment polarity gene armadillo form a multigene family with similar properties. *J. Cell Biol.* 118: 681–691, 1992.

396. Perrin, D., and H. D. Soling. No evidence for calpain I involvement in fodrin rearrangements linked to regulated secretion. *FEBS Lett.* 311: 302–304, 1992.

397. Pesacreta, T. C., T. J. Byers, R. Dubreuil, D. P. Kiehart, and D. Branton. Drosophila spectrin: the membrane skeleton during embryogenesis. *J. Cell Biol.* 108: 1697–709, 1989.

398. Peters, L. L., E. M. Eicher, A. C. Azim, and A. H. Chishti. The gene encoding the erythrocyte membrane skeleton protein dematin (Epb4.9) maps to mouse chromosome 14. *Genomics* 26: 634–635, 1995.

399. Peters, L. L., E. M. Eicher, T. C. Hoock, K. M. John, M. Yialamas, and S. E. Lux. Evidence that the mouse platelet storage pool deficiency mutation *mocha* is a defect in a new member of the ankyrin gene family. *Blood* 82: 340a (abstract), 1993.

400. Peters, L. L., K. M. John, F. M. Lu, E. M. Eicher, A. Higgins, M. Yialamas, L. C. Turtzo, A. J. Otsuka, and S. E. Lux. Ank3 (epithelial ankyrin), a widely distributed new member of the ankyrin gene family and the major ankyrin in kidney, is expressed in alternatively spliced forms, including forms that lack the repeat domain. *J. Cell Biol.* 130: 313–330, 1995.

401. Peters, L. L., and S. E. Lux. Ankyrins: structure and function in normal cells and hereditary spherocytes. *Semin. Hematol.* 30: 85–118, 1993.

402. Peters, L. L., R. A. Shivdasani, S.-C. Liu, M. Hanspal, K. M. John, J. M. Gonzalez, C. Brugnara, B. Gwynn, N. Mohandas, S. L. Alper, S. H. Orkin, and S. E. Lux. Anion exchanger 1 (Band 3) is required to prevent erythrocyte membrane surface loss but not to form the membrane skeleton. *Cell* 86: 917–927, 1996.

403. Petrucci, T. C., and J. S. Morrow. Synapsin-I: An actin bundling protein under phosphorylation control. *J. Cell Biol.* 105: 1355–1363, 1987.

404. Petrucci, T. P., and J. S. Morrow. The actin and tubulin binding domains of Synapsin Ia and Ib. *Biochemistry* 30: 413–422, 1990.

405. Pinto-Correia, C., E. G. Goldstein, V. Bennett, and J. S. Sobel. Immunofluorescence localization of an adducin-like protein in the chromosomes of mouse oocytes. *Dev. Biol.* 146: 301–311, 1991.

406. Pitcher, J. A., J. Inglese, J. B. Higgins, J. L. Arriza, P. J. Casey, C. Kim, J. L. Benovic, M. M. Kwatra, M. G. Caron, and R. J. Lefkowitz. Role of beta gamma subunits of G proteins in targeting the beta-adrenergic receptor kinase to membrane-bound receptors. *Science* 257: 1264–1267, 1992.

407. Platt, O. S., S. E. Lux, and J. F. Falcone. A highly conserved region of human erythrocyte ankyrin contains the capacity to bind spectrin. *J. Biol. Chem.* 268: 24421–24426, 1993.

408. Pollard, T. D. Purification of a high molecular weight actin filament gelation protein from Acanthamoeba that shares antigenic determinants with vertebrate spectrins. *J. Cell Biol.* 99: 1970–1980, 1984.

409. Pollerberg, G. E., K. Burridge, K. E. Krebs, S. R. Goodman, and M. Schachner. The 180-kD component of the neural cell adhesion molecule N-CAM is involved in a cell-cell contacts and cytoskeleton-membrane interactions. *Cell Tissue Res.* 250: 227–236, 1987.

410. Ponting, C. P., and C. Philips. DHR domains in syntrophins, neuronal NO synthases and other intracellular proteins. *Trends Biochem. Sci.* 20: 102–103, 1995.

411. Porter, G. A. The membrane skeleton of costameres. In: *Physiology*. Baltimore: University of Maryland School of Medicine, 1993.

412. Porter, G. A., G. M. Dmytrenko, J. C. Winkelmann, and R. J. Bloch. Dystrophin colocalizes with β-spectrin in distinct subsarcolemmal domains in mammalian skeletal muscle. *J. Cell Biol.* 117: 997–1005, 1992.

413. Pradhan, D., C. Lombardo, P. R. Stabach, and J. S. Morrow. Surface plasmon resonance detects a strong heterodimer nucleation site in domain I of beta-spectrin. in preparation, 1996.

414. Pradhan, D., P. Williamson, and R. A. Schlegel. Bilayer/cytoskeleton interactions in lipid-symmetric erythrocytes assessed by a photoactivatable phospholipid analogue. *Biochemistry* 30: 7754–7758, 1991.

415. Prchal, J. T., T. Papayannopoulou, and S.-H. Yoon. Patterns of spectrin transcripts in erythroid and non-erythroid cells. *J. Cell Physiol.* 144: 287–294, 1990.

416. Pumplin, D. W. The membrane skeleton of acetylcholine receptor domains in rat myotubes contains antiparallel homodimers of β-spectrin in filaments quantitatively resembling those of erythrocytes. *J. Cell Sci.* 108: 3145–3154, 1995.

417. Quinn, M. T., C. A. Parkos, and A. J. Jesaitis. The lateral organization of components of the membrane skeleton and superoxide generation in the plasma membrane of stimulated human neutrophils. *Biochim. Biophys. Acta* 987: 83–94, 1989.

418. Ralston, G. B. The concentration dependence of the activity coefficient of the human spectrin heterodimer. A quantitative test of the Adams-Fujita approximation. *Biophys. Chem.* 52: 51–61, 1994.

419. Rana, A. P., P. Ruff, G. J. Maalouf, D. W. Speicher, and A. H. Chishti. Cloning of human erythroid dematin reveals another member of the villin family. *Proc. Natl. Acad. Sci. U.S.A.* 90: 6651–6655, 1993.

420. Reid, M. E., Y. Takakuwa, J. Conboy, G. Tchernia, and N. Mohandas. Glycophorin C content of human erythrocyte membrane is regulated by protein 4.1. *Blood* 75: 2229–2234, 1990.

421. Reid, M. E., Y. Takakuwa, G. Tchernia, R. H. Jensen, J. A. Chasis, and N. Mohandas. Functional role for glycophorin C and its interaction with the human red cell membrane skeletal component, protein 4.1. *Prog. Clin. Biol. Res.* 319: 553–571; discussion 572–578, 1989.

422. Ren, R., B. J. Mayer, P. Cicchetti, and D. Baltimore. Identification of a ten-amino acid proline-rich SH3 binding site. *Science* 259: 1157–1161, 1993.

423. Reynolds, A. B., J. Daniel, P. D. McCrea, M. J. Wheelock, J. Wu, and Z. Zhang. Identification of a new catenin: the tyrosine kinase substrate p120cas associates with E-cadherin complexes. *Mol. Cell Biol.* 14: 8333–8342, 1994.

424. Reynolds, A. B., L. Herbert, J. L. Cleveland, S. T. Berg, and J. R. Gaut. p120, a novel substrate of protein tyrosine kinase receptors and of p60v-src, is related to cadherin-binding factors beta-catenin, plakoglobin and armadillo. *Oncogene* 7: 2439–2445, 1992.

425. Ridley, A. J., and A. Hall. The small GTP-binding protein rho regulates the assembly of focal adhesions and actin stress fibers in response to growth factors. *Cell* 70: 389–399, 1992.

426. Riederer, B. M., and S. R. Goodman. Association of brain spectrin isoforms with microtubules. *FEBS Lett.* 277: 49–52, 1990.

427. Riederer, B. M., I. S. Zagon, and S. R. Goodman. Brain spectrin (240/235) and brain spectrin (240/235E): two distinct spectrin subtypes with different locations within mammalian neural cells. *J. Cell Biol.* 102: 2088–2096, 1986.

428. Rimm, D. L., E. R. Koslov, P. Kebriaei, C. D. Cianci, and J. S. Morrow. α_1(E)-catenin is a novel actin binding and bundling protein mediating the attachment of F-actin to the membrane adhesion complex. *Proc. Natl. Acad. Sci. U.S.A.* 92: 8813–8817, 1995.

429. Rimm, D. L., and J. S. Morrow. Molecular cloning of human E-cadherin suggests a novel subdivision of the cadherin superfamily. *Biochem. Biophys. Res. Commun.* 200: 1754–1761, 1994.

430. Robinson, M. S. The role of clathrin, adaptors and dynamin in endocytosis. *Curr. Opin. Cell Biol.* 6: 538–544, 1994.

431. Rodriguez-Boulan, E., and W. J. Nelson. Morphogenesis of the polarized epithelial cell phenotype. *Science* 245: 718–725, 1989.

432. Roof, D., A. Hayes, G. Hardenbergh, and M. Adamian. A 52 kD cytoskeletal protein from retinal rod photoreceptors is related to erythrocyte dematin. *Invest. Ophthalmol. Vis. Sci.* 32: 582–593, 1991.

433. Rotin, D., D. Bar-Sagi, H. O'Brodovich, J. Merilainen, V. P. Lehto, C. M. Canessa, B. C. Rossier, and G. P. Downey. An SH3 binding region in the epithelial Na$^+$ channel (alpha rE-NaC) mediates its localization at the apical membrane. *EMBO J* 13: 4440–4450, 1994.

434. Rozakis, A. M., J. McGlade, G. Mbamalu, G. Pelicci, R. Daly, W. Li, A. Batzer, S. Thomas, J. Brugge, P. G. Pelicci, et al. Association of the Shc and Grb2/Sem5 SH2-containing proteins is implicated in activation of the Ras pathway by tyrosine kinases. *Nature* 360: 689–692, 1992.

435. Rubenfeld, B., B. Souza, I. Albert, O. Muller, S. H. Chamberlain, F. R. Masiarz, S. Munemitsu, and P. Polakis. Association of the APC gene product with beta-catenin. *Science* 262: 1731–1734, 1993.

436. Rubinfeld, B., I. Albert, B. Souza, S. Munemitsu, and P. Polakis. Interaction of the APC tumor suppressor protein with catenins. *Mol. Biol. Cell* 6: 117a, 1995.

437. Ruff, P., D. W. Speicher, and A. Husain-Chishti. Molecular identification of a major palmitoylated erythrocyte membrane protein containing the src homology 3 motif. *Proc. Natl. Acad. Sci. U.S.A.* 88: 6595–6599, 1991.

438. Sacco, P. A., T. M. McGranahan, M. J. Wheelock, and K. R. Johnson. Identification of plakoglobin domains required for association with N-cadherin and alpha-catenin. *J. Biol. Chem.* 270: 20201–20206, 1995.

439. Sahr, K. E., P. Laurila, L. Kotula, A. L. Scarpa, E. Coupal, T. L. Leto, A. J. Linnenbach, J. C. Winkelmann, D. W. Speicher, V. T. Marchesi, P. J. Curtis, and B. G. Forget. The complete cDNA and polypeptide sequences of human erythroid α-spectrin. *J. Biol. Chem.* 265: 4434–4443, 1990.

440. Salardi, S., R. Modica, M. Ferrandi, P. Ferrari, L. Torielli, and G. Bianchi. Characterization of erythrocyte adducin from the Milan hypertensive strain of rats. *J. Hypertens.* 64: S196–S198, 1988.

441. Salardi, S., B. Saccardo, G. Borsani, R. Modica, M. Ferrandi, M. G. Tripodi, M. Soria, P. Ferrari, F. E. Baralle, A. Sidoli, and G. Bianchi. Erythrocyte adducin differential properties in the normotensive and hypertensive rats of the milan strain. Characterization of spleen adducin m-RNA. *Am. J. Hypertens.* 2: 229–237, 1989.

442. Salzer, U., H. Ahorn, and R. Prohaska. Identification of the phosphorylation site on human erythrocyte band 7 integral membrane protein: implications for a monotopic protein structure. *Biochim. Biophys. Acta* 1151: 149–152, 1993.

443. Sato, N., N. Funayama, A. Nagafuchi, S. Yonemura, S. Tsukita, and S. Tsukita. A gene family consisting of ezrin, radixin and moesin. Its specific localization at actin filament/plasma membrane association sites. *J. Cell Sci.* 103: 131–143, 1992.

444. Scaife, R., and R. L. Margolis. Biochemical and immunochemical analysis of rat brain dynamin interaction with microtubules and organelles in vivo and in vitro. *J. Cell Biol.* 111: 3023–3033, 1990.

445. Scaramuzzino, D. A., and J. S. Morrow. Calmodulin-binding domain of recombinant erythrocyte beta-adducin. *Proc. Natl. Acad. Sci. U.S.A.* 90: 3398–3402, 1993.

446. Schafer, D. A., S. R. Gill, J. A. Cooper, J. E. Heuser, and T. A. Schroer. Ultrastructural analysis of the dynactin complex: an actin-related protein is a component of a filament that resembles F-actin. *J. Cell Biol.* 126: 403–412, 1994.

447. Schneider, R. The human protooncogene ret: a communicative cadherin. *Trends Biochem.* 17: 468–469, 1992.

448. Schofield, A. E., M. J. Tanner, J. C. Pinder, B. Clough, P. M. Bayley, G. B. Nash, A. R. Dluzewski, D. M. Reardon, T. M. Cox, R. J. Wilson, et al. Basis of unique red cell membrane properties in hereditary ovalocytosis. *J. Mol. Biol.* 223: 949–958, 1992.

449. Schroer, T. A., E. Fyrberg, J. A. Cooper, R. H. Waterston, D. Helfman, T. D. Pollard, and D. I. Meyer. Actin-related protein nomenclature and classification. *J. Cell Biol.* 127: 1777–1778, 1994.

450. Sears, D. E., V. T. Marchesi, and J. S. Morrow. A calmodulin and α-subunit binding domain in human erythrocyte beta-spectrin. *Biochim. Biophys. Acta* 870: 432–442, 1986.

451. Seubert, P., M. Baudry, S. Dudek, and G. Lynch. Calmodulin stimulates the degradation of brain spectrin by calpain. *Synapse* 1: 20–24, 1987.

452. Shapiro, L., A. M. Fannon, P. D. Kwong, A. Thompson, M. S. Lehmann, G. Grubel, J. F. Legrand, N. J. Als, D. R. Colman, and W. A. Hendrickson. Structural basis of cell-cell adhesion by cadherins. *Nature* 374: 327–337, 1995.

453. Sheetz, M. P. DNase-I-dependent dissociation of erythrocyte cytoskeletons. *J. Cell Biol.* 81: 266–270, 1979.

454. Sheetz, M. P., and S. J. Singer. Biological membranes as bilayer couples. A molecular mechanism of drug-erythrocyte interactions. *Proc. Natl. Acad. Sci. U.S.A.* 71: 4457–4461, 1974.

455. Shen, B. W., R. Josephs, and T. L. Steck. Ultrastructure of the intact skeleton of the human erythrocyte membrane. *J. Cell Biol.* 102: 997–1006, 1986.

456. Shibamoto, S., M. Hayakawa, K. Takeuchi, T. Hori, K. Miyazawa, N. Kitamura, K. R. Johnson, M. J. Wheelock, N. Matsuyoshi, M. Takeichi, et al. Association of p120, a tyrosine kinase substrate, with E-cadherin/catenin complexes. *J. Cell Biol.* 128: 949–957, 1995.

457. Shimkets, R. A., D. G. Warnock, C. M. Bositis, C. Nelson-Williams, J. H. Hansson, M. Schambelan, J. R. Gill, Jr., S. Ulick, R. V. Milora, J. W. Findling, R. Lifton, et al. Liddle's syndrome: heritable human hypertension caused by mutations in the beta subunit of the epithelial sodium channel. *Cell* 79: 407–414, 1994.

458. Shpetner, H. S., and R. B. Vallee. Dynamin is a GTPase stimulated to high levels of activity by microtubules. *Nature* 355: 733–735, 1992.

459. Shpetner, H. S., and R. B. Vallee. Identification of dynamin, a novel mechanochemical enzyme that mediates interactions between microtubules. *Cell* 59: 421–432, 1989.

460. Shuster, C. B., and I. M. Herman. Indirect association of ezrin with F-actin: isoform specificity and calcium sensitivity. *J. Cell Biol.* 128: 837–848, 1995.

461. Siegel, D. L., and D. Branton. Partial purification and characterization of an actin-bundling protein, band 4.9, from human erythrocytes. *J. Cell Biol.* 100: 775–785, 1985.

462. Sikorski, A. F., W. Swat, M. Brzezinska, Z. Wroblewski, and B. Bisikirska. A protein cross-reacting with anti-spectrin antibodies is present in higher plant cells. *Z Naturforsch [C]* 48: 580–583, 1993.

463. Sikorski, A. F., G. Terlecki, I. S. Zagon, and S. R. Goodman. Synapsin I-mediated interaction of brain spectrin with synaptic vesicles. *J. Cell Biol.* 114: 313–318, 1991.

464. Sinard, J. H., G. W. Stewart, A. C. Argent, D. M. Gilligan, and J. S. Morrow. A novel isoform of beta adducin utilizes an alternatively spliced exon near the C-terminus. *Mol. Biol. Cell* 6: 269a, 1995.

465. Sinard, J. H., G. W. Stewart, A. C. Argent, and J. S. Morrow. Stomatin binding to adducin: a novel link between transmembrane ion transport and the cytoskeleton. *Mol. Biol. Cell* 5 (suppl): 421a, 1994.

466. Sinard, J. H., G. W. Stewart, A. C. Argent, P. R. Stabach, and J. S. Morrow. Characterization of a novel alternative transcript of beta adducin. *Genbank* #U43959, 1995.

467. Skolnik, E. Y., A. Batzer, N. Li, C.-H. Lee, E. Lowenstein, M. Mohammadi, B. Margolis, and J. Schlessinger. The function of GRB2 in linking the insulin receptor to ras signaling pathways. *Science* 260: 1953–1955, 1993.

468. Smith, P. R., A. L. Bradford, E. H. Joe, K. J. Angelides, D. J. Benos, and G. Saccomani. Gastric parietal cell H($^+$)-K($^+$)-ATPase microsomes are associated with isoforms of ankyrin and spectrin. *Am. J. Physiol.* 264: C63–70, 1993.

469. Smith, P. R., G. Saccomani, E. H. Joe, K. J. Angelides, and D. J. Benos. Amiloride-sensitive sodium channel is linked to the cytoskeleton in renal epithelial cells. *Proc. Natl. Acad. Sci. U.S.A.* 88: 6971–6975, 1991.

470. Sontag, J.-M., E. M. Fykse, Y. Ushkaryov, J.-P. Liu, P. J. Robinson, and T. C. Sudhof. Differential expression and regulation of multiple dynamins. *J. Biol. Chem.* 269: 4547–4554, 1994.

471. Speicher, D. W., T. M. DeSilva, K. D. Speicher, J. A. Ursitti, P. Hembach, and L. Weglarz. Location of the human red cell spectrin tetramer binding site and detection of a related "closed" hairpin loop dimer using proteolytic footprinting. *J. Biol. Chem.* 268: 4227–4235, 1993.

472. Speicher, D. W., and V. T. Marchesi. Erythrocyte spectrin is comprised of many homologous triple helical segments. *Nature* 311: 177–180, 1984.

473. Speicher, D. W., J. S. Morrow, W. J. Knowles, and V. T. Marchesi. Identification of proteolytically resistant domains of human erythrocyte spectrin. *Proc. Natl. Acad. Sci. U.S.A.* 77: 5673–5677, 1980.

474. Speicher, D. W., J. S. Morrow, W. J. Knowles, and V. T. Marchesi. A structural model of human erythrocyte spectrin: alignment of chemical and functional domains. *J. Biol. Chem.* 257: 9093–9101, 1982.

475. Speicher, D. W., L. Weglarz, and T. M. DeSilva. Properties of human red cell spectrin heterodimer (side-to-side) assembly and identification of an essential nucleation site. *J. Biol. Chem.* 267: 14775–14782, 1992.

476. Srinivasan, Y., M. Lewallen, and K. J. Angelides. Mapping the binding site on ankyrin for the voltage-dependent sodium channel from brain. *J. Biol. Chem.* 267: 7483–7489, 1992.

477. Srinivasan, Y. L., E. J. Davis, V. Bennett, and K. Angelides. Ankyrin and spectrin associate with voltage-dependent Na channels in brain. *Nature* 333: 177–180, 1988.

478. Stabach, P. R., C. D. Cianci, S. B. Glantz, Z. Zhang, and J. S.

479. Stabach, P. R., S. P. Kennedy, and J. S. Morrow. Polarized assembly of the spectrin skeleton is ankyrin-independent in MDCK cells. *Mol. Biol. Cell* 4: 58a, 1993.

480. Stappert, J., and R. Kemler. A short core region of E-cadherin is essential for catenin binding and is highly phosphorylated. *Cell Adhes. Commun.* 2: 319–327, 1994.

481. Steiner, J. P., and V. Bennett. Ankyrin-independent membrane protein-binding sites for brain and erythrocyte spectrin. *J. Biol. Chem.* 263: 14417–14425, 1988.

482. Steiner, J. P., H. T. J. Walke, and V. Bennett. Calcium/calmodulin inhibits direct binding of spectrin to synaptosomal membranes. *J. Biol. Chem.* 264: 2783–2791, 1989.

483. Stewart, G. W., and A. C. Argent. The integral band 7 membrane protein of the human erythrocyte membrane. *Biochem. Soc. Trans.* 20: 785–790, 1992.

484. Stewart, G. W., A. C. Argent, and B. C. Dash. Stomatin: a putative cation transport regulator in the red cell membrane. *Biochim. Biophys. Acta* 1225: 15–25, 1993.

485. Stewart, G. W., B. E. Hepworth-Jones, J. N. Keen, B.C.J. Dash, A. C. Argent, and C. M. Casimir. Isolation of cDNA coding for a ubiquitous membrane protein deficient in high Na, low K stomatocytic erythrocytes. *Blood* 79: 1593–1601, 1992.

486. Stokke, B. T., A. Mikkelsen, and A. Elgsaeter. The human erythrocyte membrane skeleton may be an ionic gel. I. Membrane mechanochemical properties. *Eur. Biophys. J.* 13: 203–218, 1986.

487. Stokke, B. T., A. Mikkelsen, and A. Elgsaeter. The human erythrocyte membrane skeleton may be an ionic gel. II. Numerical analyses of cell shapes and shape transformations. *Eur. Biophys. J.* 13: 219–233, 1986.

488. Stokke, B. T., A. Mikkelsen, and A. Elgsaeter. The human erythrocyte membrane skeleton may be an ionic gel. III. Micropipette aspiration of unswollen erythrocytes. *J. Theor. Biol.* 123: 205–211, 1986.

489. Straub, V., R. E. Bittner, J. J. Leger, and T. Voit. Direct visualization of the dystrophin network on skeletal muscle fiber membrane. *J. Cell Biol.* 119: 1183–1191, 1992.

490. Su, L. K., B. Vogelstein, and K. W. Kinzler. Association of the APC tumor suppressor protein with catenins. *Science* 262: 1734–1737, 1993.

491. Sung, L. A., V. M. Fowler, K. Lambert, M. A. Sussman, D. Karr, and S. Chien. Molecular cloning and characterization of human fetal liver tropomodulin. A tropomyosin-binding protein. *J. Biol. Chem.* 267: 2616–2621, 1992.

492. Sussman, M. A., and V. M. Fowler. Tropomodulin binding to tropomyosins. Isoform-specific differences in affinity and stoichiometry. *Eur. J. Biochem.* 205: 355–362, 1992.

493. Svetina, S., A. Iglic, V. Kralj-Iglic, and B. Zeks. Cytoskeleton and red cell shape. *Cell Mol. Biol. Lett.* 1: 67–75, 1996.

494. Svetina, S., V. Kralj-Iglic, and B. Zeks. Cell shape and lateral distribution of mobile membrane constituents. In: *Proceedings X. School on Biophysics of Membrane Transport, Part II,* edited by J. Kuczera and S. Przestalski, Wroclaw: Agricultural University of Wroclaw, 1990, p. 139–155.

495. Svetina, S., S. Vrhoec, M. Gros, and B. Zeks. Red blood cell membrane as a possible model system for studying mechanisms of chlorpromazine action. *Acta Pharmacologica* 42: 25–35, 1992.

496. Svetina, S., and B. Zeks. Bilayer couple as a possible mechanism of biological shape formation. *Biomed. Biochim. Acta* 44: 979–986, 1985.

Morrow. Site-directed mutagenesis of αII spectrin (fodrin) reveals determinants of its μ-calpain susceptibility. *Biochemistry.* In press, 1996.

497. Svetina, S., and B. Zeks. Elastic properties of layered membranes and their role in transformations of cellular shapes. In: *Biomechanics of Active Movement and Division of Cells,* edited by N. Akkas, Berlin: Springer-Verlag, 1994, p. 479–486.

498. Svoboda, K., C. F. Schmidt, D. Branton, and S. M. Block. Conformation and elasticity of the isolated red blood cell membrane skeleton. *Biophys. J.* 63: 784–793, 1992.

499. Takaoka, Y., H. Ideguchi, M. Matsuda, N. Sakamoto, T. Takeuchi, and Y. Fukumaki. A novel mutation in the erythrocyte protein 4.2 gene of Japanese patients with hereditary spherocytosis (protein 4.2 Fukuoka). *Br. J. Haematol.* 88: 527–533, 1994.

500. Takeichi, M. Cadherin cell adhesion receptors as a morphogenetic regulator. *Science* 251: 1451–1455, 1991.

501. Takeichi, M. Morphogenetic roles of classic cadherins. *Curr. Opin. Cell Biol.* 7: 619–627, 1995.

502. Tanaka, T., K. Kadowaki, E. Lazarides, and K. Sobue. Ca^{2+}-dependent regulation of the spectrin/actin interaction by calmodulin and protein 4.1. *J. Biol. Chem.* 266: 1134–1140, 1991.

503. Tang, T. K., T. L. Leto, V. T. Marchesi, and E. J. Benz. Expression of specific isoforms of protein 4.1 in erythroid and non-erythroid tissues. *Adv. Exp. Med. Biol.* 241: 81–95, 1988.

504. Tang, T. K., Z. Qin, T. Leto, V. T. Marchesi, and E. J. Benz, Jr. Heterogeneity of mRNA and protein products arising from the protein 4.1 gene in erythroid and nonerythroid tissues. *J. Cell Biol.* 110: 617–624, 1990.

505. Tanner, M. J. The major integral proteins of the human red cell. *Baillieres Clin. Haematol.* 6: 333–356, 1993.

506. Tao, Y.-S., R. Edwards, J. Bryan, and P. D. McCrea. The actin bundling protein fascin associates with beta-catenin. *Mol. Biol. Cell* 6: 300a, 1995.

507. Taylor, S. A., R. G. Snell, A. Buckler, C. Ambrose, M. Duyao, D. Church, C. S. Lin, M. Altherr, G. P. Bates, N. Groot, et al. Cloning of the alpha-adducin gene from the Huntington's disease candidate region of chromosome 4 by exon amplification. *Nat. Genet.* 2: 223–227, 1992.

508. Tillack, T. W., S. L. Marchesi, V. T. Marchesi, and E. Steers, Jr. A comparative study of spectrin: a protein isolated from red blood cell membranes. *Biochim. Biophys. Acta* 200: 125–131, 1970.

509. Tilley, L., R. A. McPherson, G. L. Jones, and W. H. Sawyer. Structural organisation of band 3 in Melanesian ovalocytes. *Biochim. Biophys. Acta* 1181: 83–89, 1993.

510. Torres, M., C. Yost, D. Kimelman, and R. T. Moon. Phosphorylation of beta-catenin by glycogen synthase kinase-3 increases its interaction with cadherin and decreases its axis inducing activity in xenopus. *Mol. Biol. Cell* 6: 117a, 1995.

511. Touhara, K., J. Inglese, J. A. Pitcher, G. Shaw, and R. J. Lefkowitz. Binding of G protein beta gamma-subunits to pleckstrin homology domains. *J. Biol. Chem.* 269: 10217–10220, 1994.

512. Trave, G., A. Pastore, M. Hyvonen, and M. Saraste. The C-terminal domain of alpha-spectrin is structurally related to calmodulin. *Eur. J. Biochem.* 227: 35–42, 1995.

513. Tripodi, G., A. Piscone, G. Borsani, S. Tisminetzky, S. Salardi, A. Sidoli, P. James, S. Pongor, G. Bianchi, and F. E. Baralle. Molecular cloning of an adducin-like protein: evidence of a polymorphism in the normotensive and hypertensive rats of the milan strain. *Biochem. Biophys. Res. Commun.* 177: 939–947, 1991.

514. Troyanovsky, S. M., R. B. Troyanovsky, L. G. Eshkind, V. A. Krutovskikh, R. E. Leube, and W. W. Franke. Identification of the plakoglobin-binding domain in desmoglein and its role in plaque assembly and intermediate filament anchorage. *J. Cell. Biol.* 127: 151–160, 1994.

515. Tse, W. T., M. C. Lecomte, F. F. Costa, M. Garbarz, C. Feo, P. Boivin, D. Dhermy, and B. G. Forget. Point mutation in the beta-spectrin gene associated with alpha 1/74 hereditary elliptocytosis. Implications for the mechanism of spectrin dimer self-association. *J. Clin. Invest.* 86: 909–916, 1990.

516. Ungewickell, E., P. M. Bennett, R. Calvert, V. Ohanian, and W. B. Gratzer. In vitro formation of a complex between cytoskeletal proteins of the human erythrocyte. *Nature* 280: 811–814, 1979.

517. Ursitti, J. A., and V. M. Fowler. Immunolocalization of tropomodulin, tropomyosin and actin in spread human erythrocyte skeletons. *J. Cell Sci.* 107: 1633–1639, 1994.

518. Ursitti, J. A., D. W. Pumplin, J. B. Wade, and R. J. Bloch. Ultrastructure of the human erythrocyte cytoskeleton and its attachment to the membrane. *Cell Motil. Cytoskeleton* 19: 227–243, 1991.

519. Ursitti, J. A., and J. B. Wade. Ultrastructure and immunocytochemistry of the isolated human erythrocyte membrane skeleton. *Cell Motil. Cytoskeleton* 25: 30–42, 1993.

520. Viel, A., and D. Branton. Interchain binding at the tail end of the Drosophila spectrin molecule. *Proc. Natl. Acad. Sci. U.S.A.* 91: 10839–10843, 1994.

521. Viel, A., M. Gee, and D. Branton. Molecular analysis of interchain binding at the tail end of Drosophila spectrin. *Mol. Biol. Cell* 6: 270a (abstract), 1995.

522. Vybiral, T., J. C. Winkelmann, R. Roberts, E. H. Joe, D. L. Casey, J. K. Williams, and H. F. Epstein. Human cardiac and skeletal muscle spectrins: differential expression and localization. *Cell Mot. Cytoskeleton* 21: 293–304, 1992.

523. Wallis, C., E. Wenegieme, and J. Babitch. Characterization of calcium binding to brain spectrin. *J. Biol. Chem.* 267: 4333–4337, 1992.

524. Wang, C. C., R. Moriyama, C. R. Lombardo, and P. S. Low. Partial characterization of the cytoplasmic domain of human kidney band 3. *J. Biol. Chem.* 270: 17892–17897, 1995.

525. Wang, D., W. C. Mentzer, T. Cameron, and R. M. Johnson. Purification of band 7.2b, a 31-kDa integral phosphoprotein absent in hereditary stomatocytosis. *J. Biol. Chem.* 266: 17826–17831, 1991.

526. Wang, D. S., and G. Shaw. The association of the C-terminal region of beta I sigma II spectrin to brain membranes is mediated by a PH domain, does not require membrane proteins, and coincides with a inositol-1,4,5 triphosphate binding site. *Biochem. Biophys. Res. Commun.* 217: 608–615, 1995.

527. Wang, K., and F. M. Richards. An approach to nearest neighbor analysis of membrane proteins. *J. Biol. Chem.* 249: 8005–8018, 1974.

528. Waseem, A., and H. C. Palfrey. Erythrocyte adducin. Comparison of the alpha- and beta-subunits and multiple-site phosphorylation by protein kinase C and cAMP-dependent protein kinase. *Eur. J. Biochem.* 178: 563–573, 1988.

529. Wasenius, V. M., O. Narvanen, V. P. Lehto, and M. Saraste. Alpha-actinin and spectrin have common structural domains. *FEBS Lett.* 221: 73–76, 1987.

530. Wasenius, V. M., M. Saraste, P. Salven, M. Eramaa, L. Holm, and V. P. Lehto. Primary structure of the brain alpha-spectrin. *J. Cell Biol.* 108: 79–93, 1989.

531. Way, M., B. Pope, and A. Weeds. Molecular biology of actin binding proteins: evidence for a common structural domain in the F-actin binding sites of gelsolin and alpha-actinin. *J. Cell Sci. Suppl.* 14: 91–94, 1991.

532. Weaver, D. C., G. R. Pasternack, and V. T. Marchesi. The

structural basis of ankyrin function II. Identification of two functional domains. *J. Biol. Chem.* 259: 6170–6175, 1984.

533. Weber, A., C. R. Pennise, G. G. Babcock, and V. M. Fowler. Tropomodulin caps the pointed ends of actin filaments [see comments]. *J. Cell Biol.* 127: 1627–1635, 1994.

534. Weed, S. A., P. R. Stabach, and J. S. Morrow. Developmental expression of βIς2 spectrin in developing rat brain. in preparation, 1996.

535. Weed, S. A., P. R. Stabach, and J. S. Morrow. The MAD2 (pleckstrin homology) domain of βIΣ2 spectrin is not required for cortical membrane assembly or myotube formation in skeletal muscle. in preparation, 1996.

536. Weinstein, R. S., H. D. Tazelaar, and J. M. Loew. Red cell comets: ultrastructure of axial elongation of the membrane skeleton. *Blood Cells* 11: 343–366, 1986.

537. Welsh, C. F., D. Zhu, and L. Y. Bourguignon. Interaction of CD44 variant isoforms with hyaluronic acid and the cytoskeleton in human prostate cancer cells. *J. Cell Physiol.* 164: 605–612, 1995.

538. Westberg, J. A., B. Entler, R. Prohaska, and J. P. Schroder. The gene coding for erythrocyte protein band 7.2b (EPB72) is located in band q34.1 of human chromosome 9. *Cytogenet. Cell Genet.* 63: 241–243, 1993.

539. White, R. A., L. L. Peters, L. R. Adkison, C. Korsgren, C. M. Cohen, and S. E. Lux. The murine pallid mutation is a platelet storage pool disease associated with the protein 4.2 (pallidin) gene. *Nat Genet.* 2: 80–3, 1992.

540. Willott, E., M. S. Balda, A. S. Fanning, B. Jameson, C. Van Itallie, and J. M. Anderson. The tight junction protein ZO-1 is homologous to the Drosophila discs-large tumor suppressor protein of septate junctions. *Proc. Natl. Acad. Sci. U.S.A.* 90: 7845–7838, 1993.

541. Winkelmann, J. C., J. G. Chang, W. T. Tse, A. L. Scarpa, V. T. Marchesi, and B. G. Forget. Full-length sequence of the cDNA for human erythroid beta-spectrin. *J. Biol. Chem.* 265: 11827–11832, 1990.

542. Winkelmann, J. C., F. F. Costa, B. L. Linzie, and B. G. Forget. Beta spectrin in human skeletal muscle. Tissue-specific differential processing of 3′ beta spectrin pre-mRNA generates a beta spectrin isoform with a unique carboxyl terminus. *J. Biol. Chem.* 264: 20449–54, 1990.

543. Winkelmann, J. C., and B. G. Forget. Erythroid and nonerythroid spectrins. *Blood* 81: 3173–3185, 1993.

544. Winkelmann, J. C., T. L. Leto, P. C. Watkins, R. Eddy, T. B. Shows, A. J. Linnenbach, K. E. Sahr, N. Kathuria, V. T. Marchesi, and B. G. Forget. Molecular cloning of the cDNA for human erythrocyte beta-spectrin. *Blood* 72: 328–334, 1988.

545. Winograd, E., D. Hume, and D. Branton. Phasing the conformational unit of spectrin. *Proc. Natl. Acad. Sci. U.S.A.* 88: 10788–10791, 1991.

546. Wolfe, L. C., K. M. John, J. C. Falcone, A. M. Byrne, and S. E. Lux. A genetic defect in the binding of protein 4.1 to spectrin in a kindred with hereditary spherocytosis. *N Engl. J. Med.* 307: 1367–1374, 1982.

547. Woods, D. F., and P. J. Bryant. The discs-large tumor suppressor gene of Drosophila encodes a guanylate kinase homolog localized at septate junctions. *Cell* 66: 451–464, 1991.

548. Woods, D. F., and P. J. Bryant. ZO-1, DigA and PSD-95/SAP-90; homologous proteins in tight, septate and synaptic cell junctions. *Mech. Devel.* 44: 85–89, 1993.

549. Yan, Y., E. Winograd, A. Viel, T. Cronin, S. C. Harrison, and D. Branton. Crystal structure of the repetitive segments of spectrin. *Science* 262: 2027–2030, 1993.

550. Yawata, Y., M. Inaba, A. Yawata, A. Kanzaki, K. Ono, M. Takeuchi, K. Sato, Y. Maede, and Y. Takakuwa. Complete band 3 deficiency in cattle: a model for hereditary sherocytosis with striking instability of cytoskeletal network with marked exo- and endocytosis. *Blood* 86 (suppl): 468a (abstract), 1995.

551. Yonemura, S., M. Itoh, A. Nagafuchi, and S. Tsukita. Cell-to-cell adherens junction formation and actin filament organization: similarities and differences between non-polarized fibroblasts and polarized epithelial cells. *J. Cell Sci.* 108: 127–142, 1995.

552. Yurchenco, P. D., D. W. Speicher, J. S. Morrow, W. J. Knowles, and V. T. Marchesi. Monoclonal antibodies as probes of domain structure of the spectrin alpha subunit. *J. Biol. Chem.* 257: 9102–9107, 1982.

553. Zarkowsky, H. S., F. A. Oski, R. Shaafi, S. B. Shohet, and D. G. Nathan. Congenital hemolytic anemia with high sodium, low potassium red cells. I Studies of membrane permeability. *N. Engl. J. Med.* 278: 573–581, 1968.

554. Zimmer, W. E., Y. Ma, I. S. Zagon, and S. R. Goodman. Developmental expression of brain β-specrin isoform messenger RNAs. *Brain Res.* 594: 75–83, 1992.

555. Zimmer, W. E., Y. P. Ma, and S. R. Goodman. Identification of a mouse brain beta-spectrin cDNA and distribution of its mRNA in adult tissues. *Brain Res. Bull.* 27: 187–193, 1991.

556. Zimmerman, U. J., D. W. Speicher, and A. B. Fisher. Secretagogue-induced proteolysis of lung spectrin in alveolar epithelial type II cells. *Biochim. Biophys. Acta* 1137: 127–134, 1992.

557. Ziparo, E., B. M. Zani, A. Filippini, M. Stefanini, and V. T. Marchesi. Proteins of the membrane skeleton in rat Sertoli cells. *J. Cell Sci.* 86: 145–154, 1986.

558. Southgate, C. D., A. H. Chishti, B. Mitchell, S. J. Yi, and J. Palek. Targeted disruption of the murine erythroid band 3 gene results in spherocytosis and severe haemolytic anaemia despite a normal membrane skeleton. *Nature Genet.* 14: 227–230, 1996.

12. Extracellular matrix: a solid-state regulator of cell form, function, and tissue development

DONALD E. INGBER | *Departments of Pathology and Surgery, Children's Hospital and Harvard Medical School, Boston, Massachusetts*

CHAPTER CONTENTS

THE ROLE OF EXTRACELLULAR MATRIX (ECM) IN THE REGULATION OF CELL FORM AND FUNCTION is coming into sharper focus. In the not so distant past, ECM was viewed as a poorly defined structural lattice that surrounded cells and linked neighboring tissues within organs. Extracellular matrix was thought to act primarily to maintain normal tissue boundaries. In recent years, it has become clear that the molecules that make up ECM control a variety of normal cell behaviors, including growth, differentiation, morphogenesis, and tissue repair. Extracellular matrix also has been found to play a critical role in pathological processes such as cancer. The purpose of this chapter is not to present an exhaustive review of the structure–function relations of different ECM molecules or an in-depth analysis of their cell surface receptors. This information can be found in a variety of excellent reviews (3, 38, 41, 86, 113, 114, 121, 127), including Chapter 22 in this book. Rather, I hope to emphasize new concepts in matrix biology and to review recent advances that have been made in terms of understanding how ECM molecules exert their effects on cell form and function. Special emphasis will be placed on the role of mechanical forces in cell regulation and on the mechanism of transmembrane signaling across ECM receptors.

EXTRACELLULAR MATRIX: COMPOSITION AND STRUCTURE

Extracellular matrix may be defined as a macromolecular complex that contains collagenous molecules (types I–XVIII), glycoproteins (fibronectin, laminin, entactin, vitronectin, thrombospondin, chondronectin, osteonectin, fibrin), elastin, proteoglycans (containing heparan sulfate, chondroitin sulfate, dermatan sulfate, or keratin sulfate), and/or glycosaminoglycans (hyaluronic acid) as its major constituents. Extracellular matrix has been referred to as the "cellular glue" that holds cells and tissues together. Others view ECM as "cellular slime" upon which cells attach and move. Still others think of ECM as a tissue boundary that limits tissue intermixing. In reality, ECM is all of these and more, as will be described below.

Extracellular matrix occurs in higher animals in two forms: it is referred to as *interstitial matrix* in connective tissues and as *basement membrane* (BM) within epithelium. In interstitial matrix, ECM molecules fill the tissue space and surround individual cells (Fig. 12.1A,B). In BM, ECM molecules are deposited only along the basal surface of the epithelium and do not extend between cells of the same type (Fig. 12.1C,D). Interstitial matrix molecules tend to form a three-dimensional lattice, whereas BM components become organized within a flat "chicken wire" like mesh. This difference in organization is most likely due to differences in the self-assembly properties of interstitial versus BM collagens (41, 108).

Extracellular matrix molecules are always produced by cells that secrete them into the surrounding tissue space. Organized ECMs form in the extracellular milieu through a combination of self-assembly (for example, collagen fibril formation, BM collagen network assembly) and specific binding interactions between different ECM components. Fibronectin, one of the most ubiquitous ECM molecules, contains multiple functional domains that have distinct binding properties for different elements of the ECM (for example; collagen, fibrin, heparan sulfate) (113). Binding interactions between fibronectin and specific cell surface receptors also appear to play a critical role in ECM assembly (91). In embryological development, the epithelium is primarily responsible for deposition of its BM, although mesenchyme can also contribute to its formation (8, 29). Thus, the same ECM often can contain components derived from different cell types.

The three-dimensional organization of the ECM lattice, combined with differences in its collagen composition (41), is largely responsible for the biomechanical properties characteristic of each tissue. For example, the elasticity of skin results from the presence of a loose hydrated gel composed of interstitial collagens, proteoglycans, glycoproteins, and an especially large complement of elastin. In contrast, collagen fibrils are tightly woven together into cables within tendons, where they are designed to carry large tensile loads. Combinations of collagens, proteoglycans, and glycoproteins are organized in a different manner within cartilage so that they maximize the compressive load-bearing capabilities of the highly hydroscopic proteoglycan constituents. In bone, cell-derived ECM serves as a nidus for precipitation of inorganic ions, such as calcium phosphate, and results in formation of a rigid trabecular lattice with even greater load-bearing capabilities.

It is important to emphasize that ECM is not a static structure, nor is it deposited in a random manner. In the case of bone, it has been known for over 75 years that ECM is deposited in specific patterns that correspond precisely to engineering lines of tension and compression that are characteristic for any structure of that size and shape with similar load-bearing characteristics (70). It is also clear that changes in the way one carries weight, due to either injury or design (for example; binding of infants' feet in China), result in bone matrix remodeling until that new mechanical stress is minimized. Thus, bone matrix is a dynamic structure that can spontaneously remodel in response to force. This is less easy to visualize in other tissues. However, it is likely that all ECMs exhibit similar properties. Perhaps one of the most common examples in pathology is the change in structure of the vasculature that is observed in patients with systemic hypertension. The force sensitivity of ECM also has been recently taken advantage of by plastic surgeons who now

mechanically expand regions of skin to increase the surface area of skin flaps for use in reconstructive surgery. For these reasons and others that will be discussed below (see later under Scaffolding for Orderly Tissue Renewal and Repair), ECM is viewed as the site at which mechanical forces are transmitted to and from cells (50).

EXTRACELLULAR MATRIX AND TISSUE ARCHITECTURE

Changes in ECM are observed throughout all stages of embryonic development. Individual ECM molecules (laminin) are detected in a disorganized array as early as the eight-cell stage of the embryo (74, 75). Later, the first BM forms—separating the ectoderm from the newly formed endoderm—when type IV collagen, heparan sulfate–proteoglycan, and fibronectin are also deposited. When the third germ layer forms (mesoderm), a new BM is created. In contrast, interstitial types I and III collagens first appear in mesodermal tissues on day 8 in the mouse (75). A similar sequence of random deposition of laminin molecules, followed by appearance of other BM components and their collective reorganization into a planar BM array, occurs when primary mesenchyme transforms into secondary epithelium during kidney tubulogenesis (30). Thus, BM is always deposited beneath an epithelium at the time of its emergence.

Subsequent tissue growth and morphogenesis are made possible through ECM extension and remodeling. Specific patterns of histogenetic organization (for example, acinar versus tubular) result from establishment of differentials in ECM turnover and cell growth at selective sites (Fig. 12.2). These growth differentials are often regulated by complex epithelial–mesenchymal interactions. For example, in developing salivary gland, the epithelium imparts morphologic stability by producing BM whereas the mesenchyme drives histogenetic remodeling by altering BM degradation at different sites (8, 12, 13, 119). Mesenchyme stimulates BM dissolution along the tips of the growing lobule by secreting proteolytic enzymes (for example, hyaluronidase) whereas it slows BM breakdown within the quiescent clefts at the base of the expanding lobules by depositing interstitial fibrillar collagen. Adjacent epithelial cells respond to BM dissolution by depositing new ECM components at even higher levels, such that net lateral extension of BM results in regions of tissue expansion. In contrast, the fibrillar collagen that is deposited in the clefts inhibits BM breakdown by binding BM heparan sulfate proteoglycan and interfering with its degradation (26). Regions that exhibit the highest rate of ECM turnover also display the most

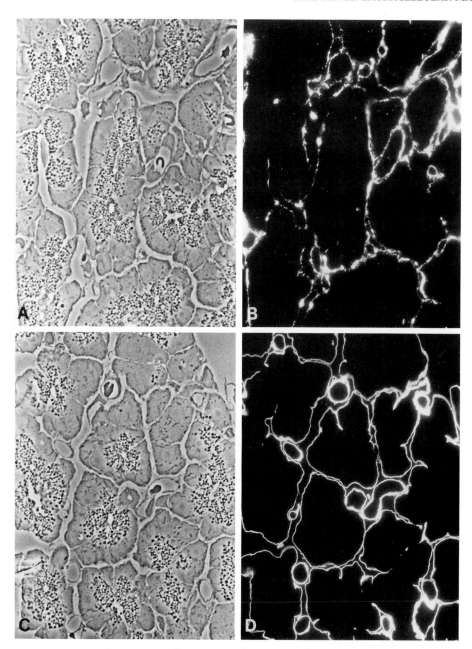

FIG. 12.1. Interstitial matrix versus basement membrane (BM). Phase contrast *(A,C)* and immunofluorescence *(B,D)* micrographs of frozen sections of normal rat exocrine pancreas stained with antibodies against interstitial type III collagen *(A,B)* or BM type IV collagen *(C,D)*. Interstitial fibrillar collagen is distributed in an amorphous pattern filling the connective tissue space. Basement membrane collagen appears in a linear pattern (planar in three dimensions) and therefore forms a physical tissue boundary separating the epithelium and vascular endothelium from surrounding connective tissue.

rapid rate of cell proliferation. In this manner, the growing tips of expanding lobules bud outward from the clefts, which are relatively fixed in place, and thus glandular morphogenesis ensues (Fig. 12.2). Similar coupling between ECM turnover, cell growth, and tissue expansion is observed during vascular morphogenesis (35); however, the endothelium itself is apparently responsible for both elaboration of degradative enzymes and simultaneous deposition of new ECM components (Fig. 12.2).

Interestingly, when embryonic epithelia and mesenchyme are isolated from different tissues and then combined heterotypically, it is the tissue source of the mesenchyme (for example, kidney versus mammary)

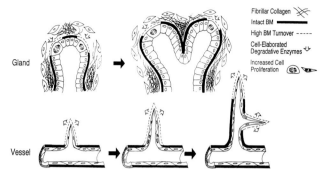

FIG. 12.2. Extracellular matrix (ECM) and tissue development. Pattern formation during morphogenesis results from establishment of local differentials of cell growth, ECM turnover, the tissue expansion. Regions of tissue expansion correspond to areas in which basement membrane (BM) turnover and cell proliferation are the highest. Importantly, the rate of BM deposition must exceed its degradation in regions of rapid growth and turnover because net growth (lateral extension) of BM is observed during tissue development. Differences in tissue pattern (for example, glandular acini versus vascular tubes) result from differences in the number and distribution of the growth centers. Glandular morphogenesis involves complex epithelial–mesenchymal interactions whereas angiogenesis (capillary morphogenesis) appears to be driven largely by the endothelial cell itself (see text for appropriate references).

that determines the specific three-dimensional pattern (tubular versus acinar) that results (103). Although the basis of this pattern-forming behavior is unknown, mechanical forces generated by the mesenchyme are likely to play an important regulatory role (44, 57, 92).

In the adult, rates of ECM turnover and cell proliferation are normally low. However, they may return to high levels in response to injury, physiological growth stimulation (for example, mammary gland development, altered hemodynamic forces), or as a result of pathology (for example, arthritis, inflammation, neoplasia). Maintenance of stable tissue form requires a fine balance between ECM synthesis and degradation. For example, involution of developing tissues, such as Müllerian duct and mammary gland, appears to be triggered by induction of BM breakdown (59, 126, 144). A similar link between ECM turnover and tissue regression is also observed during angiogenesis [Fig. 12.3; (53, 59)]. These studies support the possibility that changes in ECM turnover may actually control cell growth and viability rather than vice versa (57). The mechanism by which changes in cell–ECM interactions regulate cell growth will be discussed in more detail later under Solid State Growth Regulator.

Finally, it is important to point out that tissue growth and morphogenesis can be inhibited by interfering with either the synthetic (53, 83) or degradative (90) limbs of the ECM turnover loop. The reason for this is that BM extension as well as cell growth and migration

likely requires continued coupling between ECM synthesis and breakdown. In this manner, BM extension may be similar to glycogen chain extension in the liver: preexisting polymers must by "clipped" by enzymes before new structural subunits can be added. However, the actual mechanism of BM extension in living tissues remains essentially unexplored. This may represent an exciting area for future research that could have wider implications than previously recognized, given that homologues of ECM molecules and their receptors appear to play a role in morphogenesis of plants (135, 138) as well as animals.

FUNCTIONS OF THE EXTRACELLULAR MATRIX

Physiological Cell Attachment Foundation

The first and foremost function of ECM, as it relates to cells, is its role as a physiological substratum for

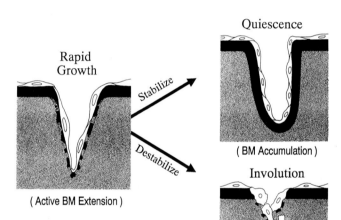

FIG. 12.3. Extracellular matrix (ECM) and tissue remodeling. Changes in basement membrane (BM) turnover that accompany tissue growth and involution during capillary development are shown here. Regions of the growing vessel that undergo the most active remodeling exhibit high rates of ECM turnover. However, BM synthesis must exceed BM degradation to produce net BM extension in these regions. Formation of new stable tissue form correlates with a decrease in BM degradation and accumulation of an intact BM. Once fully formed, BM synthesis and degradation are maintained at very low and approximately equal levels throughout adult life. In contrast, growing tissues may be induced to regress by rapidly increasing ECM degradation or by inhibiting ECM deposition, such that total dissolution of BM results. Normal cells rapidly lose viability when they are freed from anchorage and thus progressive tissue involution ensues. During involution, cells at the leading edge that sit on the BM with the highest turnover rate are the first to lose their adhesions and retract. This proceeds backwards along the growing sprout in a progressive manner until the entire vessel involutes (59). In this sense, controlled tissue involution is essentially normal tissue development in reverse.

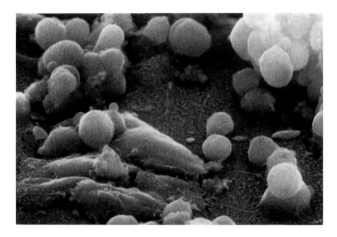

FIG. 12.4. Basement membranes (BM) supports cell attachment and spreading. Scanning electron micrograph showing that exogenous intact BM (isolated from human amnion) can promote cell attachment and spreading, even in the presence of the protein synthesis inhibitor cycloheximide (experimental details are described in ref. 62).

cell anchorage. This concept was initially based on the experimental observation that treatment of whole tissues with ECM-degrading enzymes (collagenase, proteases) resulted in tissue breakdown and release of individual cells. Conversely, attachment of dispersed cells could be rapidly promoted, even in the absence of new protein synthesis, by providing cells with an intact ECM (Fig. 12.4). Purified ECM molecules can also support cell adhesion (69) and serum promotes cell attachment because it contains high concentrations of fibronectin and vitronectin, two highly adhesive matrix components. In fact, tissue culture dishes are essentially bacteriological (nonadhesive) plastic dishes that have been chemically treated in order to enhance adsorption of serum proteins and cell-elaborated matrix. (Interestingly, net surface charge is not the critical determinant for cell adhesion, as often assumed in the past; attachment proteins and hence cells will adhere to both positively and negatively charged dishes.)

Early studies with dishes coated with purified ECM molecules suggested that epithelial cells prefer to use BM proteins (for example, laminin, type IV collagen) for their attachment whereas mesenchymal cells adhere primarily to interstitial ECM components (for example, fibronectin, types I, III collagens). For example, the attachment and spreading of primary mammary epithelia require de novo collagen synthesis; however, this requirement can be circumvented only by providing BM (type IV) collagen and not fibrillar (type I) collagen substrata (143). Yet, more recent studies show that many epithelial and mesenchymal cells can use multiple

ECM components, from both BM and interstitial matrix, for their attachment (60, 69, 88).

Spatial Organizer of Polarized Epithelium

One of the most common characteristics of epithelial cells is that they exhibit polarized form as well as function. An excellent example of this is the consistent cell orientation that is found within a secretory epithelium normally surrounded by an intact BM (Fig. 12.5). Dissociated epithelial cells commonly lose this normal orientation when cultured on standard tissue culture substrata. However, they restore their normal polarized form if allowed to contact exogenous BM (31, 62) or if they are able to synthesize and accumulate their own matrix (125). Binding interactions between ECM molecules and distinct cell surface receptors also mediate epithelial polarization during development (120). Thus, BM normally may serve to integrate and maintain individual cells within a polarized epithelium. Clearly, there are many intracellular and intercellular determinants of polarized cell function (for example, cytoskeletal organization, organelle movement, junctional complex formation). However, anchorage to BM appears to provide an initial point of orientation and stability upon which additional steps in the epithelial organization cascade can build (62, 100, 101).

Solid-State Growth Regulator

Clearly, a major advance in the understanding of growth control relates to the discovery of soluble growth factors that induce cell division. However, the growth-promoting action of a mitogen by itself is not sufficient to explain how normal tissues develop. For correct histogenesis to occur, local growth differentials must be established in a tissue microenvironment that is saturated with soluble mitogens (Fig. 12.2). Recent in vitro studies demonstrate that although ECM and diffusible growth factors act in concert to control morphogenesis, ECM molecules are actually the dominant regulators (54). This is because ECM molecules dictate whether individual cells will proliferate, involute, or differentiate in response to soluble cues (51, 53, 55, 88).

Cell attachment to ECM is also the basis of anchorage-dependent growth control. It has been known for many years that normal cells will grow only if attached and spread on a solid substrate and that many rapidly lose viability when maintained in a round form or in suspension (10, 37). In contrast, loss of anchorage-dependence is a critical hallmark of transformation (116, 147). As described above, cells commonly attach and spread either by depositing new

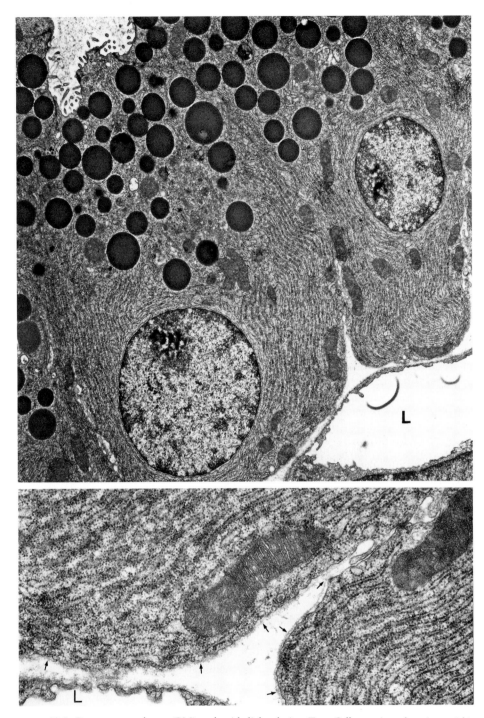

FIG. 12.5. Basement membrane (BM) and epithelial polarity. *Top:* Cells consistently orient within polarized epithelium, nucleus and rough endoplasmic reticulum (ER) at the base, Golgi complex above, and secretory granules in the apex, as shown in this electron micrograph of rat exocrine pancreas. *Bottom:* A higher magnification image from the same view showing that the highly polarized epithelium sits atop a continuous BM that separates the cells from the surrounding connective tissue space. Tips of the *small arrows* abut on the epithelial BM; *L,* lumen of a small capillary.

matrix components or by binding to exogenous ECM. Furthermore, pharmacological inhibitors of ECM deposition (for example, proline analogues) inhibit cell spreading and growth in vitro (143) and induce growing tissues to regress in vivo (53, 144).

Additional studies with cultured cells have shown that cell growth varies depending on the type of ECM component used for cell attachment (for example, collagen versus fibronectin) (60, 69) and that ECM molecules regulate cell growth at least in part by modulating cell shape (51, 60, 88). Thus, these findings suggest that local changes of ECM composition and integrity, such as those that might be observed in regions of increased ECM turnover during morphogenesis (Fig. 12.2), may serve to regulate cell sensitivity to soluble mitogens. The molecular and biomechanical basis for this type of growth control will be discussed later under TRANSMEMBRANE SIGNALING ACROSS MATRIX RECEPTORS.

Inducer of Differentiation

Differentiated cells usually lose their specialized functions when cultured on tissue culture plastic. In contrast, cells maintain high levels of differentiated functions (for example, albumin secretion by hepatocytes, milk secretion by mammary cells, capillary tube formation by endothelial cells) when cultured on complex ECM substrata (intact BM, native collagen gels, BM gels) (11, 31, 123; see also Chapter 22 in this book). or on dishes coated with purified matrix molecules (55, 88). ECM apparently can regulate differentiation by altering gene transcription, mRNA stability, protein synthesis, and/or secretory rates (11, 15, 88, 99, 124).

In certain tissues, the specific spatial organization of different ECM components appears to be critical for maintenance of the differentiated phenotype. For example, the composition of skeletal muscle fiber BM differs between synaptic and extrasynaptic regions (104). Synaptic BM retained after cell injury directs placement of new acetylcholine receptors on the surfaces of regenerating muscle cells (18). This effect appears to be mediated by specific ECM molecules, such as heparan sulfate–proteoglycan (5) and laminin (132), that are concentrated within the neuromuscular BM. Accumulation of BM, which is regulated by the functional activity of these cells (105), also results in limitation of lateral receptor mobility within the plasma membrane (6).

Tissue Boundary

BM establishes a physical tissue boundary between epithelium and underlying connective tissue and thereby normally restricts mixing between these two cell populations (Fig. 12.1C,D). Basement membrane also may regulate macromolecular transport between tissues, given that the BM functions as a semipermeable filtration barrier in the kidney glomerulus (32). The filtration function in other tissues remains to be explored.

Storage Site for Soluble Regulators

Extracellular matrices may function in the storage and release of soluble mitogens. The mesenchymal mitogen and angiogenic factor, basic fibroblast growth factor, has been localized within BM in normal tissues (36). Cell proliferation may be suppressed as a result of local sequestration of this mitogen whereas release of stored factors (stormones) may turn on growth in response to injury. On the other hand, substrate-bound platelet derived growth factor is mitogenic in vitro (118) and associates with collagen fibers within healing wounds (81). Other types of regulatory molecules, such as cell surface receptor–bound urokinase-type plasminogen activator and its type I inhibitor, also appear to be contained within ECM (96).

Scaffolding for Orderly Tissue Renewal and Repair

All tissues are dynamic structures. Thus, maintenance of specialized tissue form requires that cells lost due to injury or aging be replaced in an organized fashion. Orderly tissue renewal requires the continued presence of BM as an extracellular scaffolding or template that maintains the original architectural form and assures for accurate regeneration of preexisting structures. Basement membrane assures for (1) correct repositioning of cells, (2) specificity for cell type, (3) reestablishment of polarized cell orientation, and (4) stimulation of cell migration and growth that are required for tissue repair (134). For example, after ischemic injury to muscle in the rat, muscle cells grow only along the inside of the muscle BM, endothelial cells along capillary BM, and connective tissue remains limited to its normal circumferential distribution (134). In contrast, loss of BM integrity during wound healing results in loss of normal tissue pattern or what is known as "scar" formation. Cell migration and growth also require de novo deposition of ECM proteins (82, 122, 143), which in turn mediate cell attachment and spreading. The general role of ECM in cell motility has been previously reviewed (151).

Another point that often is not discussed is that mechanical force transmission across ECM is critical for control of tissue repair. During wound healing, fibroblasts produce wound contraction and connective tissue remodeling by depositing ECM proteins and exerting tension on their ECM adhesions. Recent studies suggest that transmembrane force transmission and wound reorganization depend on whether or not specific ECM receptors (for example, $\alpha5\beta1$ integrins) are expressed on the fibroblast cell surface (140). Cell-generated forces are transmitted over large distances over fibronectin-containing microfibrils (ligands for $\alpha5\beta1$ integrins), which act as cables linking fibroblasts to each other and to collagenous matrix. Tension exerted by cells leads to realignment of collagen bundles, which in turn determines the spatial orientation and morphology of associated connective tissue cells (44, 68). Transmission of cell tension to ECM is also required for fibroblast migration (33, 44) and extension of nerve growth cones (45).

In this manner, the morphogenetic response observed at the tissue level is orchestrated in time and space by mechanical forces that are generated within individual cells and transmitted across their interconnecting ECM adhesions. Mechanical forces may play a similar role during embryonic development (44, 57, 92).

Target for Deregulation during Tumor Formation and Invasion

In the past, BM was viewed as a host barrier through which a malignant tumor must gain the ability to invade. In fact, BM is normally a specialized product of overlying epithelial cells that also plays a central role as a stabilizer of epithelial form and orientation. Thus, changes in ECM may play an important role during neoplastic transformation, both before and after the onset of malignant invasion.

As described above, an epithelium is a dynamic structure and its orderly renewal requires the continued presence of BM as an extracellular template that retains the original tissue form. Tissue regeneration requires increased growth and migration; however, it is a normal process because it is highly organized and reversible. If normal cells lose contact with ECM under normal conditions, they will rapidly lose viability and die because they depend on anchorage for their survival (Fig. 12.3). On the other hand, an epithelium may come to escape its normally tight growth constraints as well as its requirement for anchorage (for example, due to infection with an oncogenic virus). Cell multiplication without concomitant extension of new BM results in a piling up of cells and hence formation of a local swelling or "tumor" (61).

Continued contact between cells and intact BM is required for maintenance of epithelial polarity (62). Thus, the loss of cell orientation and resultant disorganization of tissue form observed during dysplasia and early stages of tumor formation may be partly due to tumor cells gaining the ability to survive and proliferate free of normal contact with BM (61, 130). The importance of ECM in the earliest phases of tumor formation is also supported by the finding that tumor suppressor genes have been identified that code for an ECM protein (thrombospondin; [98]) and an enzyme involved in ECM synthesis (lysyl oxidase; [66]).

When the balance between ECM synthesis and degradation can no longer be maintained, local breakdown of BM is observed. The tumor is now malignant because it is free to invade neighboring tissues and to metastasize throughout the body. The importance of ECM in tumor invasion and metastasis has been long recognized. While past excitement centered on the identification of specific collagenases that appear to be required for malignant invasion (79), more recent work has focused on the importance of a "proteolytic balance" involving both enzymes and enzyme inhibitors, such as members of the TIMP family (67, 78). Cell–ECM binding interactions as well as specific integrins also have been found to be important for cancer metastasis (150).

TRANSMEMBRANE SIGNALING ACROSS MATRIX RECEPTORS

Cell Surface Matrix Receptors

All of the cell-modulating effects of ECM molecules are believed to be produced by binding to specific cell surface ECM receptors. The best-characterized ECM receptors are members of a family of transmembrane proteins known as *integrins* (3, 121). Integrins were first isolated based on their ability to recognize a common amino acid sequence, arginine–glycine–aspartate (RGD), found within many ECM molecules, although integrins with different binding specificities have also been identified. These cell surface receptors share the following properties (Fig. 12.6): *(1)* they are heterodimers composed of non-covalently coupled α and β subunits; *(2)* their ECM specificity depends on the combination of α and β chains; *(3)* the cytoplasmic portion of some β chains contains the consensus tyrosine phosphorylation site that is also found in receptors for EGF and insulin; and *(4)* they appear to be concentrated within cell adhesion sites in close association with intracellular microfilaments, actin-associated molecules (for example; talin, α-actinin, paxillin), and a variety of non-receptor protein kinases (19, 129).

Cell-surface heparan sulfate–proteoglycans also function as ECM receptors (14). The core protein of at least one of these proteoglycans (Syndecan) spans

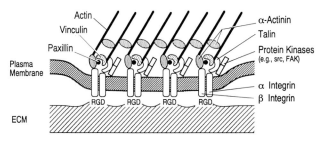

FIG. 12.6. Diagram of the focal adhesion complex. The extracellular portion of integrin receptors recognize specific amino acid binding sites (for example, RGD) within extracellular matrix (ECM) molecules. The intracellular portion of these receptors physically interlinks with the actin cytoskeleton via binding interactions with actin-associated proteins, such as talin, α-actinin, vinculin, and paxillin. This localized membrane/cytoskeletal microdomain also has been found to contain molecules that are thought to mediate chemical signaling by ECM (for example, the protein kinases, *c-src* and FAK kinase). The precise location of these proteins within the complex is currently unknown.

the cell surface and reportedly interacts with actin-containing microfilaments (106). Cell-surface heparan sulfate proteoglycans may play an additional role in growth control by binding soluble growth factors and presenting them to their own high-affinity growth factor receptors (148). Other ECM receptor proteins have been reported (for example, 67 Kda laminin receptor; [17]); however, their molecular identity remains to be determined.

Chemical Signaling

A major new advance in matrix biology is the demonstration that ECM binding to cell surface integrin receptors can act directly to activate intracellular chemical signaling pathways and induce gene expression (48, 110). Integrin receptors do not themselves have any known endogenous protein kinase activity. However, it has been shown that cell binding to fibronectin recruits tyrosine kinases (19), stimulates phosphorylation of specific proteins (9, 34, 149), increases inositol lipid synthesis (85) and breakdown (16, 24, 85) releases intracellular calcium (64), activates the Na^+H^+ antiporter (63) and induces early immediate growth response genes, such as c-fos (28) in the absence of growth factors. In other words, cell binding to ECM has been found to activate signal transduction mechanisms similar to those used by soluble mitogens.

Integrins clearly mediate chemical signaling. Integrin $\alpha IIb\beta3$ (also known as glycoprotein IIb/IIIa) has been shown to regulate Na^+H^+ exchange as well as tyrosine-specific protein phosphorylation in platelets (9, 34). Clustering of integrin $\beta1$ or $\alpha5$ chains, but not receptor occupancy alone, is sufficient to activate the

Na^+H^+ antiporter in fibroblasts and endothelial cells (111). Furthermore, a variety of different ECM molecules and integrin receptors share the ability to stimulate this signaling pathway (111, 112), a pathway whose activation is required for growth in these cells (63). In other cell types, either occupancy alone (1) or clustering (141) of specific integrins is sufficient to induce gene expression. However, these genes are associated with specialized cell functions (for example, differentiation, motility) rather than cell proliferation. Recent work also suggests that specific ECM-responsive gene regulatory elements may exist (109). It is important to emphasize that these signaling events occur as a result of a direct ECM–receptor interaction and are independent of subsequent changes of cell shape (1, 112, 141).

Mechanical Signaling

Ligation and clustering of specific transmembrane ECM receptors by itself cannot explain how ECM molecules switch cells between differentiation and growth. For example, while activation of chemical signaling pathways by ECM binding is necessary for growth (63, 111), it is not sufficient. Large-scale tension-dependent changes of cell shape are also required for subsequent cell cycle progression and entry into S phase (42, 51, 88, 117a). It is also known that the same ECM substratum conveys different regulatory signals depending on whether or not it can resist cell-generated forces and thereby alter cell form (Fig. 12.7). For example, fibroblasts commonly spread, secrete ECM proteins, and grow when cultured on attached collagen gels. However, these cells round, decrease matrix synthesis, and become quiescent within one hour after the gels are released and stress is allowed to dissipate (87). The flexibility achieved when collagen gels are floated also provides a suitable context for cell retraction, expression of differentiation-specific genes, and deposition of BM by mammary epithelial cells (31, 77, 124). In fact, the mechanical properties of ECM gels, such as Matrigel, appear to be a critical determinant of morphogenesis in many cell types (72, 94, 123, 131). Interestingly, malleable BM gels (for example, Matrigel) lose their ability to induce cell rounding and differentiation when they are made rigid (94).

Another way to vary the amount of tension that cells experience, as well as cell spreading, is to change the number of potential attachment points that can resist cell tractional forces (Fig. 12.7). This can be accomplished by varying the density of a single type of purified ECM molecule (for example, fibronectin) that is adsorbed onto planar, rigid surfaces (51, 55, 88, 128). Cell spreading and growth are inhibited in capil-

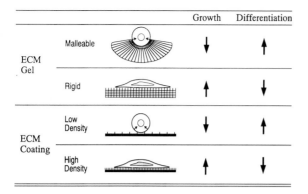

FIG. 12.7. Extracellular matrix (ECM) mechanics and control of cell function. Cells exert tension on their ECM adhesions. Highly adhesive ECM substrata that are malleable (for example, native collagen gels, Matrigel, thin silicone rubber) can not resist cytoskeletal tension and thus they promote cell rounding. However, if the same substrata are made rigid (for example, due to chemical cross-linking or immobilizing the gel on a dish), they support cell extension. Similar changes in cell shape can be induced by varying the ability of a rigid substratum to resist cell tension. This can be done by changing the density of a single type of purified ECM molecule (for example, fibronectin, laminin, or different collagen types) that is pre-adsorbed on otherwise nonadhesive dishes. For many cell types, round cells tend to maintain high levels of tissue-specific functions, whereas spread cells turn off differentiation and switch on growth, regardless of the method of cell shape control (see text for appropriate references).

lary endothelial cells (51, 55) and hepatocytes (88) when ECM coating densities are lowered, cell tension is decreased, and rounding is promoted, even when cells are cultured in the presence of saturating amounts of soluble mitogens. Interestingly, differences in the ability of different ECM components to support cell growth also correlate directly with their effects on cell shape (60). Furthermore, when growth is turned off, both morphological differentiation (capillary tube formation) and expression of tissue-specific genes (albumin in liver) are switched on (55, 88). Hepatocytes also maintain high levels of differentiated function and remain quiescent when cultured on dishes that are coated with a high density of synthetic RGD-peptide that ligates cell surface integrin receptors and supports cell attachment, but does not promote cell spreading (89). Artificially altering the total cell–ECM contact area over which forces can be transmitted (that is, focal contact area) (93) or the topography of the adhesive surface (23, 117a) provides similar control over the shape, orientation, and growth of fibroblasts.

In summary, these studies suggest that cells can be switched between genetic programs associated with growth and differentiation simply by altering ECM mechanics and thereby changing cell shape (Fig. 12.7). Cell attachment to ECM in the absence of cell spread-

ing (for example, on a malleable ECM) results in switching on of what is essentially a "default" differentiation pathway. For cells to turn off differentiation and proliferate, the ECM foundation must exhibit mechanical properties that can physically resist cell tractional forces and thus support cell extension. In other words, the same ECM molecule may induce either growth or differentiation depending on its ability to sustain high levels of cytoskeletal tension (54). However, it is important to point out that a cell's sensitivity to tension and to cell shape control may vary greatly from one tissue to another (for example, chondrocytes versus endothelial cells).

Mechanical force transmission across cell surface receptors may be critical for other cell functions as well. As described above, fibroblasts must exert tension on their ECM adhesions to contract wounds. Leukocytes move out of blood vessels by physically pulling on cell surface receptors that line the surface of the endothelium (39). Cytotoxic T cells also must apply tension to receptors on the surface of their target cells in order to accomplish cell killing (47). Interestingly, recent estimates of the force exerted on a single cell by "large-scale" forces such as gravity (1 μdyne/cell) (4) reveal that it is many times less than that generated within the cytoskeleton and resisted by the ECM (33, 80, 128). Thus, adherent cells within tissues normally exist in a state of isometric tension. Activation of a cell response by exogenous forces (for example, releasing an attached collagen gel, application of fluid shear stress) may therefore result from a change in the cellular mechanical force balance (50, 117).

Transmembrane integrin receptors are excellent candidates for molecules that transmit forces to and from the cytoskeleton (50) because they physically link actin-associated proteins (for example, talin, vinculin, α-actinin, paxillin) to ECM molecules (129) and to adhesion receptors on the surface of other cells (121). Other transmembrane receptors (for example, syndecans, cadherins) that interconnect with the cytoskeleton may mediate force transmission as well.

Clear demonstration that specific integrins mediate force transmission comes from analysis of *Drosophila* embryos that express the *myospheroid* mutation (133). Cells in the developing muscles of these embryos fail to express integrin β_{PS}, the *Drosophila* homologue of the β-integrin of vertebrate fibronectin receptors. These embryos develop normally until the first muscle contractions begin. Once tension is exerted, the muscle cells retract from their ECM insertion sites and become spheroidal. Disruption of force transmission across integrins results in both loss of muscle function and disorganization of intracellular cytoarchitecture in this system. Integrins also have been shown to mediate outward force transmission in fibroblasts (107) and inward force transfer in endothelial cells (136) as well as gravity sensation in plants (138).

Different integrin subunits may differ in their ability to mediate force transfer across the cell surface and to the cytoskeleton. Clustering of β_1 integrins by immobilized ligands or soluble antibodies is sufficient to induce cytoskeletal associations in fibroblasts (46) and endothelial cells (136). In contrast, phosphorylation of the α_6 subunit is required to promote its association with the cytoskeleton and to induce macrophage attachment and spreading on laminin (115). In epidermal cells, integrin $\alpha_3\beta_1$ interlinks ECM molecules with actin filaments in focal contacts, whereas integrin $\alpha_6\beta_4$ interconnects with intermediate filaments in hemidesmosome-like structures (21). Human melanoma and lung carcinoma cells can use either $\alpha_V\beta_3$ or $\alpha_V\beta_5$ integrin to attach to vitronectin, but only $\alpha_V\beta_3$ colocalizes with vinculin, talin, and the ends of actin filaments (139). In contrast, two different integrins ($\alpha_5\beta_1$ and $\alpha_V\beta_3$) have to act in concert to induce cell spreading and associated cytoskeletal alterations in melanoma cells (22). Force transmission by integrins also may be modulated by altering the exposure or function of ligand binding sites on the outer surface of the cell (2, 145).

Signal Integration

How do mechanical forces produce a biochemical response once they are transmitted across integrins? One mechanism for transducing mechanical signals into biochemical information is through force-dependent release of chemical second messengers. The fact that cell binding to immobilized ECM molecules activates chemical signaling pathways, suggests that integrins may convey chemical signals much like growth factor receptors. However, in contrast to growth factor receptors, ECM ligands have to both cluster integrins and physically prevent their internalization in order to produce sustained activation of the Na^+H^+ antiporter (112). Furthermore, to induce growth in anchorage-dependent cells, immobilized ECM molecules must be presented on a surface that can also support large-scale changes of cell shape (51, 88, 89). Thus, binding of integrins may elicit different chemical signals depending on whether or not their ligands can physically resist cytoskeletal tension. Interestingly, while both soluble and immobilized ECM molecules stimulate cell migration, they use different signaling pathways (7).

Mechanochemical transduction by integrins also may be mediated by force-induced changes in cytoskeletal organization that accompany cell shape changes. Mechanical forces are balanced between tensile actin

filaments, microtubular struts, and ECM anchoring supports (45, 52, 56, 57); integrins are responsible for maintaining the stability of this equilibrium (117, 136). The importance of mechanical forces in cell shape control can be easily visualized using three-dimensional tension-dependent ("tensegrity") cell models that are constructed with sticks and elastic string (52, 56, 57). These models, like living cells, spread and flatten when attached to rigid surfaces; however, they pull flexible substrata up into folds and remain round. In other words, studies with these purely mechanical cell models suggest that ECM may convey regulatory information to the cell by modulating the cellular force balance and thereby changing cytoskeletal form.

Changes in the cytoskeleton may result directly from mechanical deformation at the site of force application or through global rearrangements within the cytoskeletal lattice (52, 136). When the cytoskeletal force balance is perturbed, thermodynamic parameters are also altered (20, 57) and coordinated changes in actin bundle assembly and microtubule polymerization ensue (25, 45, 71). Resultant changes in cytoskeletal organization may modulate cell function by altering binding interactions between cytoskeletal filaments and different elements of the cell's metabolic machinery (for example, membrane receptors, enzymes, polyribosomes, mRNA, organelles) or by changing nuclear structure (15, 43, 57, 58). Cytoskeletal changes also could be transduced directly into a chemical response by activating "stretch-activated" membrane ion channels on the cell surface (102), altering associations between tubulin and G proteins (97), or perturbing interactions between actin-binding proteins (gelsolin and profilin) and phosphoinositides (40, 65). In addition, protein phosphorylation cascades can be initiated directly within focal contacts (27, 149).

A possible linkage between integrins, physical forces, cytoskeletal changes, and chemical signal generation is supported by the finding that adenylate cyclase and release of intracellular calcium can be stimulated by physical extension of the cytoskeleton caused by cell spreading on ECM (64), mechanical stretching of the substratum (76, 146), or hyposmotic swelling (137). Interestingly, force-induced release of chemical messengers may, in turn, feed back to regulate force transmission across the cell surface by modulating exposure of extracellular binding sites on integrins (145) or by altering their ability to anchor to the cytoskeleton (73). ECM-induced changes in inositol lipid metabolism (85) could also influence cytoskeletal organization directly by altering the availability of free actin monomer (40, 65). In other words, the chemical and mechanical signaling mechanisms used by ECM are most likely interdependent. Interestingly, recent studies using a newly developed method to isolate intact focal adhesion complexes suggest that much of this integration may occur locally at the site of integrin binding (84, 95).

CONCLUSIONS

It has long been recognized that ECM plays a major role in the regulation of cell form and function as well as tissue development. Over the past decade, great advances have been made in terms of identifying and characterizing different types of ECM molecules and their receptors. Nevertheless, the molecular mechanism by which cells sense and respond to changes in ECM remains unclear. Recent work confirms that ECM acts in a structural role to stabilize normal tissue architecture and maintain tissue function, as assumed in the past. In addition, however, it has revealed that ECM acts as a critical morphogenetic regulator by modulating cell sensitivity to soluble mitogens within the local tissue microenvironment. Specific transmembrane receptors, such as integrins, have been identified and shown to mediate the effects of ECM on cell behavior. Still, the specific series of molecular events that begins with ECM binding and ends with changes in gene expression and growth remains to be deciphered. Elucidation of this mechanism will require understanding of how these receptors transmit chemical and mechanical signals across the cell surface as well as how different types of signals are integrated inside the cell.

REFERENCES

1. Adams, J. C., and F. M. Watt. Fibronectin inhibits terminal differentiation of human keratinocytes. *Nature* 340: 307–309, 1989.
2. Adams, J. C., and F. M. Watt. Changes in keratinocyte adhesion during terminal differentiation: reduction in fibronectin binding precedes $\alpha5\beta1$ integrin loss from the cell surface. *Cell* 63: 425–435, 1990.
3. Albelda, A. M., and C. A. Buck. Integrins and other cell adhesion molecules. *FASEB J.* 4: 2868–2880, 1990.
4. Albrecht-Buehler, G. In defense of "nonmolecular" cell biology. *Int. Rev. Cytol.* 120: 191–241, 1990.
5. Anderson, M. J., and D. M. Fambrough. Aggregates of acetylcholine receptors are associated with plaques of basal lamina heparan sulfate proteoglycan on the surface of skeletal muscle fibers. *J. Cell Biol.* 97: 396–1411, 1983.
6. Axelrod, D., P. Raudin, D. E. Koppel, J. Schlessinger, W. W. Webb, E. L. Elson, and T. R. Podleski. Lateral motion of fluorescently labeled acetylcholine receptors in membranes of developing muscle fibers. *Proc. Natl. Acad. Sci. U.S.A.* 73: 4594–4598, 1976.
7. Aznavoorian, S., M. L. Stracke, H. Krutzsch, E. Schiffman, and L. A. Liotta. Signal transduction for chemotaxis and haptotaxis by matrix molecules in tumor cells. *J. Cell Biol.* 110: 1427–1438, 1990.

8. Banerjee, S. D., R. H. Cohn, and M. R. Bernfield. Basal lamina of embryonic salivary epithelia. Production by the epithelium and role in maintaining lobular morphology. *J. Cell Biol.* 73: 445–463, 1977.
9. Banga, H. S., E. R. Simons, L. F. Brass, and S. E. Rittenhouse. Activation of phospholipases A and C in human platelets exposed to epinephrine: role of glycoproteins IIb/IIIa and dual role of epinephrine. *Proc. Natl. Acad. Sci. U.S.A.* 83: 9197–9201, 1986.
10. Ben Ze'ev, A., S. R. Farmer, and S. Penman. Protein synthesis requires cell-surface contact while nuclear events respond to cell shape in anchorage-dependent fibroblasts. *Cell* 21: 365–372, 1980.
11. Ben Ze'ev, A. G., S. Robinson, N. L. Bucher, and S. R. Farmer. Cell-cell and cell-matrix interactions differentially regulate the expression of hepatic and cytoskeletal genes in primary cultures of rat hepatocytes. *Proc. Natl. Acad. Sci. U.S.A.* 85: 1–6, 1988.
12. Bernfield, M. R., and S. D. Benerjee. The basal lamina in epithelial-mesenchymal interactions. In: *Biology and Chemistry of Basement Membranes*, edited by N. Kefalides, New York: Academic Press, 1978, p. 137–148.
13. Bernfield, M. R., S. D. Banerjee, and R. H. Cohn. Dependence of salivary epithelial morphology and branching morphogenesis upon acid mucopolysaccharide-protein proteoglycan at the epithelial cell surface. *J. Cell Biol.* 52: 674–689, 1972.
14. Bernfield, M. R., R. Kokenyesi, M. Kato, M. T. Hinkes, J. Spring, R. L. Gallo, and E. J. Lose. Biology of the syndecans: a family of transmembrane heparan sulfate proteoglycans. *Annu. Rev. Cell Biol.* 8: 365–393, 1992.
15. Bissell, M. G., H. G. Hall, and G. Parry. How does extracellular matrix direct gene expression? *J. Theor. Biol.* 99: 31–68, 1982.
16. Breuer, D., and C. Wagener. Activation of the phosphatidylinositol cycle in spreading cells. *Exp. Cell Res.* 182: 659–663, 1989.
17. Brown, S. S., H. L. Malinoff, and M. S. Wicha. Connectin: cell surface protein that binds both laminin and actin. *Proc. Natl. Acad. Sci. U.S.A.* 80: 5927–5930, 1983.
18. Burden, S. J., P. B. Sargent, and U. J. McMahan. Acetylcholine receptors accumulate at original synaptic sites in the absence of nerve. *J. Cell Biol.* 82: 412–425, 1979.
19. Burridge, K. Substrate adhesion in normal and transformed fibroblasts: organization and regulation of cytoskeletal, membrane, and extracellular matrix components at focal contacts. *Cancer Rev.* 4: 18–78, 1986.
20. Buxbaum, R. E., and S. R. Heidemann. A thermodynamic model for force integration and microtubule assembly during axonal elongation. *J. Theor. Biol.* 134: 379–390, 1988.
21. Carter, W. G., P. Kauer, S. G. Gil, P. J. Gahr, and E. A. Wayner. Distinct functions for integrins $\alpha3\beta1$ in focal adhesions and $\alpha_6\beta_4$/bullous pemphigoid antigen in a new stable anchoring contact (SAC) of keratinocytes: relation to hemidesmosomes. *J. Cell Biol.* 111: 3141–3154, 1990.
22. Charo, I. F., L. Nannizzi, J. W. Smith, and D. A. Cheresh. The vitronectin receptor $\alpha_v\beta_3$ binds fibronectin and acts in concert with $\alpha_5\beta_1$ in promoting cellular attachment and spreading on fibronectin. *J. Cell Biol.* 111: 2795–2800, 1990.
23. Clark, P., P. Connolly, A.S.G. Curtis, J.A.T. Dow, and C.D.W. Wilkinson. Topographical control of cell behaviour: II. Multiple grooved substrata. *Development* 108: 635–644, 1990.
24. Cybulsky, A. V., J. V. Bonventre, R. F. Quigg, L. S. Wolfe, and D. J. Salant. Extracellular matrix regulates proliferation and phospholipid turnover in glomerular epithelial cells. *Am. J. Physiol.* 259 (*Renal Fluid Electrolyte Physiol.* 28): F326–F337, 1990.
25. Danowski, B. Fibroblast contractility and actin organization are

stimulated by microtubule inhibitors. *J. Cell Sci.* 93: 255–266, 1989.

26. David, G., and M. R. Bernfield. Collagen reduces glycosaminoglycan degradation by cultured mammary epithelial cells: possible mechanism for basal lamina formation. *Proc. Natl. Acad. Sci. U.S.A.* 76: 786–790, 1979.

27. Davis, S., M. L. Lu, S. H. Lo, S. Lin, J. A Butler, B. J. Drucker, T. M. Roberts, Q. An, and L. B. Chen. Presence of an SH2 domain in the actin-binding protein tensin. *Science* 252: 712–715, 1991.

28. Dike, L., and S. Farmer. Cell adhesion induces expression of growth associated genes in suspension-arrested fibroblasts. *Proc. Natl. Acad. Sci. U.S.A.* 85: 5792–6796, 1988.

29. Dodson, J. W., and E. D. Hay. Control of corneal differentiation by extracellular materials. Collagen as promotor and stabilizer of epithelial stroma production. *Dev. Biol.* 38: 249–270, 1971.

30. Ekblom, P., K. Alitalo, A. Vaheri, R. Timpl, and L. Saxen. Induction of a basement membrane glycoprotein in embryonic kidney: possible role of laminin in morphogenesis. *Proc. Natl. Acad. Sci. U.S.A.* 77: 485–489, 1980.

31. Emerman, J. T., and D. R. Pitelka. Maintenance and induction of morphological differentiation in dissociated mammary epithelium on floating collagen membranes. *In Vitro* 13: 316–328, 1977.

32. Farquhar, M. G. Structure and function in glomerular capillaries: role of the basement membrane in glomerular filtration. In: *Biology and Chemistry of Basement Membranes*, edited by N. Kefalides, New York: Academic Press, 1978, p. 137–148.

33. Felder, S., and E. L. Elson. Mechanics of fibroblast locomotion: quantitative analysis of forces and motions at the leading lamella of fibroblasts. *J. Cell Biol.* 6: 2415–2526, 1990.

34. Ferrell, Jr, J. R., and G. S. Martin. Tyrosine-specific protein phosphorylation is regulated by glycoprotein IIb/IIIa in platelets. *Proc. Natl. Acad. Sci. U.S.A.* 86: 2234–2238, 1989.

35. Folkman, J. Angiogenesis: initiation and control. *Ann. N.Y. Acad. Sci.* 401: 212–227, 1982.

36. Folkman, J., M. K. Klagsbrun, J. Sasse, M. Wadzinski, D. E. Ingber, and I. Vlodavsky. A heparin-binding angiogenic protein—basic fibroblast growth factor—is stored within basement membrane. *Am. J. Pathol.* 130: 393–400, 1988.

37. Folkman, J., and A. Moscona. Role of cell shape in growth control. *Nature* 273: 345–349, 1978.

38. Frazier, W. A. Thrombospondins. *Curr. Opin. Cell Biol.* 3: 792–799, 1991.

39. Gallick, S., S. Usami, K. M. Jan, and S. Chien. Shear stress-induced detachment of human polymorphonuclear leukocytes from endothelial cell monolayers. *Biorheology* 26: 823–834, 1989.

40. Goldschmidt-Clermont, P. J., L. M. Machesky, J. J. Baldassare, and T. D. Pollard. The actin-binding protein profilin binds to PIP$_2$ and inhibits its hydrolysis by phospholipase C. *Science* 247: 1575–1578, 1990.

41. Gordon, M. K., and B. R. Olsen. The contribution of collagenous proteins to tissue-specific matrix assemblies. *Curr. Opin. Cell Biol.* 2: 833–838, 1990.

42. Guadagno, T. M., and R. K. Assoian. G1/S control of anchorage-independent growth in the fibroblast cell cycle. *J. Cell Biol.* 115: 1419–1425, 1991.

43. Hansen, L. K., and D. E. Ingber. Regulation of nucleocytoplasmic transport by mechanical forces transmitted through the cytoskeleton. In: *Nuclear Trafficking*, edited by C. Feldherr, San Diego: Academic Press, 1992, p. 71–86.

44. Harris, G. K., D. Stopak, and P. Wild. Fibroblast traction as a mechanism for collagen morphogenesis. *Nature* 290: 249–251, 1981.

45. Heidemann, S. R., and R. E. Buxbaum. Tension as a regulator and integrator of axonal growth. *Cell Motil. Cytoskeleton* 17: 6–10, 1990.

46. Horvath, A., and S. Kellie. Regulation of integrin motility and cytoskeletal association in normal and RSV-transformed chick embryo fibroblasts. *J. Cell Sci.* 97: 307–315, 1990.

47. Hubbard, B. B., M. W. Glacken, J. R. Rodgers, and R. R. Rich. The role of physical forces on cytotoxic T cell-target cell conjugate stability. *J Immunol* 144: 4129–4138, 1990.

48. Hynes, R. O. Integrin signaling. *Cell* 69: 11–25, 1992.

49. Ikawa, H., R. L. Trelstad, J. M. Hutson, T. F. Manganara, and P. K. Donahoe. Changing patterns of fibronectin, laminin, type IV collagen, and a basement membrane proteoglycan during rat Mullerian duct regression. *Dev. Biol.* 102: 260–263, 1984.

50. Ingber, D. E. Integrins as mechanochemical transducers. *Curr. Opin. Cell Biol.* 3: 841–848, 1991.

51. Ingber, D. E. Fibronectin controls capillary endothelial cell growth by modulating cell shape. *Proc. Natl. Acad. Sci. U.S.A.* 87: 3579–3583, 1990.

52. Ingber, D. E. Cellular tensegrity: defining new rules of biological design that govern the cytoskeleton. *J. Cell Sci.* 104: 613–627, 1993.

53. Ingber, D. E., and J. Folkman. Inhibition of angiogenesis through inhibition of collagen metabolism. *Lab. Invest.* 59: 44–51, 1988.

54. Ingber, D. E., and J. Folkman. How does extracellular matrix control capillary morphogenesis? *Cell* 58: 803–805, 1989.

55. Ingber, D. E., and J. Folkman. Mechanochemical switching between growth and differentiation during fibroblast growth factor-stimulated angiogenesis in vitro: role of extracellular matrix. *J. Cell Biol.* 109: 317–330, 1989.

56. Ingber, D. E., and J. Folkman. Tension and compression as basic determinants of cell form and function: utilization of a cellular tensegrity mechanism. In: *Cell Shape: Determinants, Regulation and Regulatory Role*, edited by W. Stein and F. Bronner, Orlando: Academic Press, 1989, p. 1–32.

57. Ingber, D. E., and J. D. Jamieson. Cells as tensegrity structures: architectural regulation of histodifferentiation by physical forces tranduced over basement membrane. In: *Gene Expression During Normal and Malignant Differentiation*, edited by L. C. Andersson, C. G. Gahmberg, and P. Ekblom, Orlando: Academic Press, 1985, p. 13–32.

58. Ingber, D. E., S. Karp, G. Plopper, L. Hansen, and D. Mooney. Mechanochemical transduction across extracellular matrix and through the cytoskeleton. In: *Physical Forces and the Mammalian Cell*, edited by J. A. Frangos and C. L. Ives, San Diego: Academic Press, 1993, p. 61–79.

59. Ingber, D. E., J. A. Madri, and J. Folkman. A possible mechanism for inhibition of angiogenesis by angiostatic steroids: induction of capillary basement membrane dissolution. *Endocrinology* 119: 1768–1775, 1986.

60. Ingber, D. E., J. A. Madri, and J. Folkman. Extracellular matrix regulates endothelial growth factor action through modulation of cell and nuclear expansion. *In Vitro Cell Dev. Biol.* 23: 387–394, 1987.

61. Ingber, D. E., J. A. Madri, and J. D. Jamieson. Neoplastic disorganization of pancreatic epithelial cell-cell relations: role of basement membrane. *Am. J. Pathol.* 121: 248–260, 1985.

62. Ingber, D. E., J. A. Madri, and J. D. Jamieson. Basement membrane as a spatial organizer of polarized epithelia: exogenous basement membrane reorients pancreatic epithelial tumor cells in vitro. *Am. J. Pathol.* 122: 129, 1986.

63. Ingber, D. E., D. Prusty, J. Frangione, E. J. Cragoe Jr, C. Lechene, and M. A. Schwartz. Control of intracellular pH and growth by fibronectin in capillary endothelial cells. *J. Cell Biol.* 110: 1803–1812, 1990.

64. Jaconi, M. E., J. M. Theler, W. Schlegel, R. D. Appel, S. D. Wright, and P. D. Lew. Multiple elevations of cytosolic-free Ca^{2+} in human neutrophils: initiation by adherence receptors of the integrin family. *J. Cell Biol.* 112: 1249–1257, 1991.

65. Janmey, P. A., and T. P. Stossel. Modulation of gelsolin functions by phosphatidylinositol 4,5-biphosphate. *Nature* 325: 362–364, 1987.

66. Kenyon, K., S. Contente, P. C. Trackman, J. Tang, H. M. Kagan, and R. M. Friedman. Lysyl oxidase and *rrg* messenger RNA. *Science* 253: 802, 1991.

67. Khokha, R., and D. T. Denhardt. Matrix metalloproteinases and tissue inhibitor of metalloproteinases: a review of their role in tumorigenesis and tissue invasion. *Invasion and Metastasis* 9: 391–405, 1989.

68. Klebe, R. J., H. Caldwell, and S. Milam. Cells transmit spatial information by orienting collagen fibers. *Matrix* 9: 451–458, 1989.

69. Kleinman, H. K., R. J. Klebe, and G. R. Martin. Role of collagenous matrices in the adhesion and growth of cells. *J. Cell Biol.* 88: 473–485, 1981.

70. Koch, J. The laws of bone architecture. *Am. J. Anat.* 21: 177–298, 1917.

71. Kolodney, M. S., and R. B. Wyslomerski. Isometric contraction by fibroblasts and endothelial cells: a quantitative study. *J. Cell Biol.* 117: 73–82, 1992.

72. Kubota, Y., H. K. Kleinman, G. R. Martin, and T. J. Lawley. Role of laminin and basement membrane in the morphological differentiation of human endothelial cells into capillary-like structures. *J. Cell Biol.* 107: 1589–1598, 1988.

73. Lampugnani, M. G., M. Giorgi, E. Dejana, and P. C. Marchisio. Endothelial cell motility, integrin receptor clustering, and microfilament organization are inhibited by agents that increase intracellular cAMP. *Lab. Invest.* 63(4): 521–531, 1990.

74. Leivo, I. Structure and composition of early basement membranes: studies with early embryos and teratocarcinoma cells. *Med. Biol.* 61: 1–30, 1983.

75. Leivo, I., A. Vaheri, R. Timpl, and J. Wartiovaara. Appearance and distribution of collagens and laminin in the early mouse embryo. *Dev. Biol.* 76: 1000–1114, 1980.

76. Letsou, G. V., O. Rosales, S. Maitz, A. Vogt, and B. E. Sumpio. Stimulation of adenylate cyclase activity in cultured endothelial cells subjected to cyclic stretch. *J. Cardiovasc. Surg.* 31: 634–639, 1990.

77. Li, M. L., J. Aggeler, D. A. Farson, C. Hatier, J. Hassell, and M. J. Bissell. Influence of a reconstituted basement membrane and its components on casein gene expression and secretion in mouse mammary epithelial cells. *Proc. Natl. Acad. Sci. U.S.A.* 84: 136–140, 1987.

78. Liotta, L. A., P. S. Steeg, and W. G. Stetler-Stevenson. Cancer metastasis and angiogenesis: an imbalance of positive and negative regulation. *Cell* 64: 327–336, 1991.

79. Liotta, L. A., K. Tryggvason, S. Garbisa, I. Hart, C. M. Foltz, and S. Shafie. Metastatic potential correlates with enzymatic degradation of basement membrane collagen. *Nature* 284: 67–68, 1980.

80. Lotz, M. M., C. A. Burdal, H. P. Erickson, and D. R. McClay. Cell adhesion to fibronectin and tenascin: quantitative measurements of initial binding and subsequent strengthening strenghtening response. *J. Cell Biol.* 109: 1795–1805, 1989.

81. Lynch, S. E., J. C. Nixon, R. B. Colvin, and H. N. Antoniades. The role of platelet-derived growth factor in wound healing: synergistic effects with other growth factors. *Proc. Natl. Acad. Sci. U.S.A.* 84: 7696–7700, 1987.

82. Madri, J. A., and K. S. Stenn. Aortic endothelial cell migration: I. Matrix requirements and composition. *Am. J. Pathol.* 106: 180–186, 1982.

83. Maragoudakis, M. E., M. Sarmonika, and M. Panoutsacopoulou. Inhibition of basement membrane biosynthesis prevents angiogenesis. *J. Pharm. Exp. Ther.* 244: 729–733, 1988.

84. McNamee, H., H. G. Liley, and D. E. Ingber. Integrin-dependent control of inositol lipid synthesis in vascular endothelial cells and smooth muscle cells. *Exp. Cell Res.* 224: 116–122, 1996.

85. McNamee, H., D. E. Ingber, and M. A. Schwartz. Adhesion to fibronectin stimulates inositol lipid synthesis and enhances PDGF-induced inositol lipid breakdown. *J. Cell Biol.* 121: 673–678, 1993.

86. Mercurio, A. M. Laminins: multiple forms, multiple receptors. *Curr. Opin. Cell Biol.* 2: 845–849, 1990.

87. Mochitate, K., P. Pawelek, and F. Grinnell. Stress relaxation of contracted collagen gels: Disruption of actin filament bundles, release of cell surface fibronectin, and down-regulation of DNA and protein synthesis. *Exp. Cell Res.* 193: 198–207, 1991.

88. Mooney, D., L. Hansen, S. Farmer, J. Vacanti, R. Langer, and D. Ingber. Switching from differentiation to growth in hepatocytes: control by extracellular matrix. *J. Cell Physiol.* 151: 497–505, 1992.

89. Mooney, D., R. Langer, L. K. Hansen, J. P. Vacanti, and D. E. Ingber. Induction of hepatocyte differentiation by the extracellular matrix and an RGD-containing synthetic peptide. *Proc. Mat. Res. Soc. Symp. Proc.* 252: 199–204, 1992.

90. Moses, M. A., and R. Langer. A metalloproteinase inhibitor as an inhibitor of neovascularization. *J. Cell. Biochem.* 47: 230–235, 1991.

91. Mosher, D. F., J. Suttile, C. Wu, and J. A. McDonald. Assembly of extracellular matrix. *Curr. Opin. Cell Biol.* 4: 810–818, 1992.

92. Nogawa, H., and Y. Nakanishi. Mechanical aspects of the mesenchymal influence on epithelial branching morphogenesis of mouse salivary gland. *Development* 101: 491–500, 1987.

93. O'Neill, C., P. Jordan, and P. Riddle. Narrow linear strips of adhesive substratum are powerful inducers of both growth and total focal contact area. *J. Cell Sci.* 95: 577–586, 1990.

94. Opas, M. Expression of the differentiated phenotype by epithelial cells in vitro is regulated by both biochemistry and mechanics of the substratum. *Dev. Biol.* 131: 281–293, 1989.

95. Plopper, G., H. P. McNamee, L. E. Dike, K. Bojanowski, and D. E. Ingber. Convergence of integrin and growth factor receptor signaling pathways within the focal adhesion complex. *Mol. Biol. Cell* 6: 1349–1365, 1995.

96. Pollanen, J., O. Saksela, E. M. Salonen, P. Andreasen, L. Nielsen, K. Dano, and A. Vaheri. Distinct localizations of urokinase-type plasminogen activator and its type 1 inhibitor under cultured human fibroblasts and sarcoma cells. *J. Cell Biol.* 104: 1085–1096, 1987.

97. Rasenick, M. M., N. Wang, and K. Yan. Specific associations between tubulin and G proteins: participation of cytoskeletal elements in cellular signal transduction. *Biol. Med. Signal Trans.* 381–386, 1990.

98. Rastinejad, F., P. F. Polverini, and N. P. Bouck. Regulation of the activity of a new inhibitor of angiogenesis by a cancer suppressor gene. *Cell* 56: 345–355, 1989.

99. Reid, L. M. Stem cell biology, hormone/matrix synergies and liver differentiation. *Curr. Opin. Cell Biol.* 2: 121–130, 1990.

100. Rodriguez-Boulan, E., and W. J. Nelson. Morphogenesis of the polarized epithelial cell phenotype. *Science* 245: 718–725, 1989.

101. Rodriguez-Boulan, E., K. T. Paskiet, and D. D. Sabatini. Assembly of enveloped viruses in Madin-Darby canine kidney cells: polarized budding from single attached cell and from clusters of cells in suspension. *J. Cell Biol.* 96: 866–874, 1983.

102. Sachs, F. Mechanical transduction in biological systems. *Crit. Rev. Biomed. Eng.* 16: 141–169, 1988.

103. Sakakura, T., Y. Nishizura, and C. Dawe. Mesenchyme-dependent morphogenesis and epithelium-specific cytodifferentiation in mouse mammary gland. *Science* 194: 1439–1441, 1976.

104. Sanes, J. R., E. Engvall, R. Butowski, and D. Hunter. Molecular heterogeneity of basal laminae: isoforms of laminin and collagen IV at the neuromuscular junction and elsewhere. *J. Cell Biol.* 111: 1685–1699, 1990.

105. Sanes, J. R., and J. C. Lawrence Jr. Activity-dependent accumulation of basal lamina by cultured rat myotubules. *Dev. Biol.* 97: 123–136, 1983.

106. Saunders, S., M. Jalkanen, S. O'Farrell, and M. R. Bernfield. Molecular cloning of syndecan, an integral membrane proteoglycan. *J. Cell Biol.* 108: 1547–1056, 1989.

107. Schiro, J. A., B. M. Chan, W. T. Roswit, P. D. Kassner, A. P. Pentland, M. E. Hemler, A. Z. Eisen, and T. S. Kupper. Integrin $\alpha2\beta1$ (VLA 2) mediates reorganization and contraction of collagen matrices by human cells. *Cell* 67: 403–401, 1991.

108. Schittny, J. C., and P. Yurchenco. Basement membranes: molecular organization and function in development and disease. *Curr. Opin. Cell Biol.* 1: 983–988, 1989.

109. Schmidhauser, C., M. J. Bissell, C. A. Myers, and G. F. Casperson. Extracellular matrix and hormones transcriptionally regulate bovine beta-casein 5′ sequences in stably transfected mouse mammary cells. *Proc. Natl. Acad. Sci. U.S.A.* 87: 9118–9122, 1990.

110. Schwartz, M. A. Transmembrane signaling by integrins. *Trends Cell Biol.* 2: 304–308, 1992.

111. Schwartz, M. A., D. E. Ingber, M. Laurence, T. A. Springer, and C. Lechene. Multiple integrins share the ability to induce elevation of intracellular pH. *Exp. Cell Res.* 195: 533–535, 1991.

112. Schwartz, M. A., C. Lechene, and D. E. Ingber. Insoluble fibronectin activates the Na^+H^+ antiporter by inducing clustering and immobilization of its receptor, independent of cell shape. *Proc. Natl. Acad. Sci. U.S.A.* 88: 7849–7853, 1991.

113. Schwartzbauer, J. E. Fibronectin: from gene to protein. *Curr. Opin. Cell Biol.* 3: 786–791, 1991.

114. Seyedin, S. M., and D. M. Rosen. Matrix proteins of the skeleton. *Curr. Opin. Cell Biol.* 2: 914–919, 1990.

115. Shaw, L. M., J. M. Messier, and A. M. Mercurio. The activation dependent adhesion of macrophages to laminin involves cytoskeletal anchoring and phosphorylation of the $\alpha_6\beta_1$ integrin. *J. Cell Biol.* 110: 2167–2174, 1990.

116. Shin, S., V. H. Freedman, R. Risser, and R. Pollack. Tumorignicity of virus-transformed cells in nude mice is correlated specifically with anchorage-independent growth in vitro. *Proc. Natl. Acad. Sci. U.S.A.* 72: 4435–4439, 1975.

117. Sims, J., S. Karp, and D. E. Ingber. Altering the cellular mechanical force balance results in integrated changes in cell, cytoskeletal, and nuclear shape. *J. Cell Sci.* 103: 1215–1222, 1992.

117a. Singhvi, R., A. Kumar, G. P. Lopez, G. N. Stephanopoulos, D. I. C. Wang, G. M. Whitesides, and D. E. Ingber. Engineering cell shape and function. *Science* 264: 696–698, 1994.

118. Smith, J. C., J. P. Singh, J. S. Lillquist, D. S. Goon, and C. D. Stiles. Growth factors adherent to cell substrate are mitogenically active in situ. *Nature* 296: 154–156, 1982.

119. Smith, R. L., and M. R. Bernfield. Mesenchyme cells degrade epithelial basal lamina glycosaminoglycan. *Dev. Biol.* 94: 378–390, 1982.

120. Sorokin, L., A. Sonnenberg, M. Aumailley, R. Timpl, and P. Ekblom. Recognition of the laminin E8 cell-binding site by an integrin possessing the $\alpha6$ subunit is essential for epithelial polarization in developing kidney tubules. *J. Cell Biol.* 111: 1265–1273, 1990.

121. Springer, T. A. The sensation and regulation of interactions with the extracellular environment: the cell biology of lymphocyte adhesion receptors. *Annu. Rev. Cell Biol.* 6: 359–402, 1990.

122. Stenn, K. S., J. A. Madri, and F. J. Roll. Migrating epidermis produces AB2 collagen and requires continual collagen synthesis for movement. *Nature* 277: 229–232, 1979.

123. Stoker, A. W., C. H. Streuli, M. Martins-Green, and M. J. Bissell. Designer microenvironments for the analysis of cell and tissue function. *Curr. Opin. Cell Biol.* 2: 864–874, 1990.

124. Streuli, C. H., and M. J. Bissell. Expression of extracellular matrix components is regulated by substratum. *J. Cell Biol.* 110: 1405–1415, 1990.

125. Sugrue, S. P., and E. D. Hay. Response of basal epithelial cell surface and cytoskeleton to solubilized extracellular matrix molecules. *J. Cell Biol.* 91: 45–54, 1981.

126. Talhouk, R. S., M. J. Bissell, and Z. Werb. Coordinate expression of extracellular matrix-degrading proteinases and their inhibitors regulate mammary epithelial function during involution. *J. Cell Biol.* 118: 1271–1282, 1992.

127. Toole, B. P. Hyaluronan and its binding proteins, the hyaladherins. *Curr. Opin. Cell Biol.* 2: 839–844, 1990.

128. Truskey, G. A., and J. S. Pirone. The effect of fluid shear stress upon cell adhesion to fibronectin-treated surfaces. *J. Biomed. Mater. Res.* 24: 1333–1353, 1990.

129. Turner, C. E., and K. Burridge. Transmembrane molecular assemblies in cell-extracellular matrix interactions. *Curr. Opin. Cell Biol.* 849–853, 1991.

130. Vembu, D., L. A. Liotta, M. Paranjpe, and C. W. Boone. Correlation of tumorigenicity with resistance to growth inhibition by cis-hydroxyproline. *Exp. Cell Res.* 124: 247–252, 1979.

131. Vernon, R. B., J. C. Angello, L. Iruela-Arispe, T. F. Lane, and E. H. Sage. Reorganization of basement membrane matrices by cellular traction promotes the formation of cellular networks in vitro. *Lab. Invest.* 66: 536–547, 1992.

132. Vogel, Z., C. N. Christian, M. Vigny, H. C. Bauer, P. Sonderegger, and M. P. Daniels. Laminin induces acetylcholine receptor aggregation on cultured myotubes and enhances the receptor aggregation activity of a neuronal factor. *J. Neurosci.* 3: 1058–1068, 1983.

133. Volk, T., L. I. Fessler, and J. H. Fessler. A role for integrin in the formation of sarcomeric cytoarchitecture. *Cell* 63: 525–536, 1990.

134. Vracko, R. Basal lamina scaffold-anatomy and significance for maintenance of orderly tissue structures. *Am. J. Pathol.* 77: 314–346, 1974.

135. Wagner, V., C. Brian, and R. S. Quatrano. Role of a vitronectin-like molecule in embryo adhesion of the brown alga Fucus. *Proc. Natl. Acad. Sci. U.S.A.* 89: 3644–3648, 1992.

136. Wang, N., J. P. Butler, and D. E. Ingber. Mechanotransduction across the cell surface and through the cytoskeleton. *Science* 260: 1124–1127, 1993.

137. Watson, P. Direct stimulation of adenylate cyclase by mechanical forces in S49 mouse lymphoma cells during hyposomotic swelling. *J. Biol. Chem.* 265: 6569–6675, 1990.

138. Wayne, R., M. P. Staves, and A. C. Leopold. The contribution of the extracellular matrix to gravisensing in characean cells. *J. Cell Sci.* 101: 611–623, 1992.

139. Wayner, E. A., R. A. Orlando, and D. A. Cheresh. Integrins $\alpha v\beta 3$ and $\alpha v\beta 5$ contribute to cell attachment to vitronectrin but differentially distribute on the cell surface. *J. Cell Biol.* 113: 919–929, 1991.

140. Welch, M. P., G. F. Odland, and R. F. Clark. Temporal relationships of F-actin bundle formation, collagen and fibronectin matrix assembly, and fibronectin receptor expression to wound contraction. *J. Cell Biol.* 110: 133–145, 1990.

141. Werb, Z., P. M. Tremble, O. Behrendtsen, E. Crowley, and C. H. Damsky. Signal transduction through the fibronectin receptor induces collagenase and stromelysin gene expression. *J. Cell Biol.* 109: 877–889, 1989.

142. Wessells, N. K. *Tissue Interactions and Development.* New York: Benjamin Press, 1977.

143. Wicha, M. S., L. A. Liotta, G. Garbisa, and W. R. Kidwell. Basement membrane collagen requirements for attachment and growth of mammary epithelium. *Exp. Cell Res.* 124: 181–190, 1979.

144. Wicha, M. S., L. A. Liotta, B. K. Vonderhaar, and W. R. Kidwell. Effects of inhibition of basement membrane collagen deposition on rat mammary gland development. *Dev. Biol.* 80: 253–261, 1980.

145. Willigen, G. V., and Akkerman, J. N. Protein kinase C and cyclic AMP regulate reversible exposure of binding sites for fibrinogen on the glycoprotein IIB–IIIA complex of human platelets. *Biochem. J.* 273: 115–120, 1991.

146. Wirtz, H.R.W., and L. G. Dobbs. Calcium mobilization and exocytosis after one mechanical stretch of lung epithelial cells. *Science* 250: 1266–1269, 1990.

147. Wittelsberger, S. C., K. Kleene, and S. Penman. Progressive loss of shape-responsive metabolic controls in cells with increasingly transformed phenotype. *Cell* 24: 859–866, 1981.

148. Yayon, A., M. Klagsbrun, J. D. Esko, P. Leder, and D. M. Ornitz. Cell surface heparin-like molecules are required for binding of basic fibroblast growth factor to its high affinity receptors. *Cell* 64: 841–848, 1991.

149. Zachary, I., and E. Rozengurt. Focal adhesion kinase (p125[fak]): a point of convergence in the action of neuropeptides, integrins, and on cogenes. *Cell* 71: 891–894, 1992.

150. Zetter, B. R. The cellular basis of site-specific tumor metastasis. *N. Engl. J. Med.* 322: 605–612, 1990.

151. Zetter, B. R., and S. E. Brightman. Cell motility and the extracellular matrix. *Curr. Opin. Cell Biol.* 2: 850– 56, 1990.

13. Microtubule motors in cell and tissue function

MICHAEL P. SHEETZ | *Department of Physiology, Duke University Medical Center, Durham, North Carolina*

THE FIRST MICROTUBULE MOTOR TO BE DESCRIBED was dynein of sperm axonemes. That description included an in vitro motility assay of the sliding of axonemal microtubules past one another in partially digested axonemes [reviewed in (7)]. For nearly 25 years dynein remained the only microtubule motor and its analysis was hampered by the large size of the major dynein heavy chain (>300 kDa) and the large number of subunits in the protein (>10 in a complex of 1300 kDa).

Because microtubules were a prominent structural component in virtually all eukaryotic cells, it was natural to postulate that dynein-like motors might be present in other cells as the motors moving organelles in such processes as mitosis and axonal transport. Using antibodies and ATPase activity as the major criteria, many searched for cytoplasmic analogues of the axonemal dynein. Many of those studies were in eggs and although dynein was found, it was difficult to rule out the possibility that it was an axonemal dynein precursor or a storage form. Work then focused on the squid giant axon, which had no axonemes; significant amounts of the cytoplasm could be extruded free of a plasma membrane.

Two technical developments aided in the search for microtubule motors in the squid axoplasm. First, video-enhanced microscopy was developed by the Allens (Nina and Robert) (1) and by Shinya Inoue (12). The concept was similar to that used by satellites that had to expand the contrast of a small portion of the gray scale to detect low-contrast structures. The Allens introduced the additional feature of digital image subtraction to remove the pattern noise from the lens imperfections and this made it possible to observe structures that were otherwise below the detectable limit of the eye. Collaborations between Allen and a group of neurobiologists at the Marine Biological Labs produced the first complete views of the vesicular axoplasmic transport (2). Video-enhanced differential interference contrast microscopy (VE-DIC) became the technology of choice for most subsequent motility assays. Second, the conceptual basis of in vitro motility assays was provided by the development of a quantitative in vitro assay for myosin motility using latex beads coated with myosin (32). Microtubule motors were purified and defined on the basis of their activity in in vitro motility assays as observed by video-enhanced differential interference contrast microscopy.

From the early description of the movement of axoplasmic vesicles on transport fibers that could be teased from axoplasm, it was clear that a mechanism similar to that of myosin-coated bead movement was responsible for the transport of axoplasmic vesicles. Namely, a motor on the vesicles was moving on the transport filaments. Subsequent studies defined the transport filaments that supported bidirectional movement as microtubules (24). It was then found that microtubules were transported across the glass (18) and that the microtubule transport could be catalyzed by a soluble fraction of axoplasm lacking vesicles (38). Using microtubule transport as an assay, it was possible to purify the factor from axoplasm that catalyzed the movement. To purify kinesin, a microtubule affinity step was employed in which adenylylimidodiphosphate (AMP-PNP) (16) instead of ATP depletion (subsequently it was found that ADP decreased kinesin's affinity for microtubules) was used to increase microtubule-motor affinity; the complex was separated from other supernatant proteins and ATP was employed to release the motor from the microtubules. When the ATP release fraction was separated on a sizing gel column, one fraction catalyzed microtubule gliding, but that fraction had a very small microtubule-activated ATPase activity (37). Another fraction had a high ATPase activity but did not catalyze microtubule gliding (it was cytoplasmic dynein but the concentration was insufficient for catalyzing motility). When purified, the factor that catalyzed motility was a protein that contained 110, 65, and 67 kDa subunits in squid axoplasm, squid neural tissue, and bovine brain (37). Because of its motile activity, it was named kinesin. Shortly thereafter, a microtubule-activated ATPase activity was reported for a 110 kDa protein from porcine brain (3a) and a sea urchin analogue of kinesin was identified (25). In a centrosome motility assay, kinesin was shown to move only toward the plus ends of the microtubules, which is opposite to the direction of dynein movement (38a).

In crude axoplasmic supernatants, there was an activity that produced minus-end directed motility analogous to dynein (38a). A microtubule affinity protocol was used to isolate a dynein-like protein from *Caenorhabditis elegans* (the worm has no flagella and no regular dynein) but by an unusual circumstance the direction of movement was reported to be the same as that of kinesin (polymerization of microtubules from axonemes in the presence of cytoplasmic dynein was preferentially from the minus rather than the plus end as usual) (19a). Other investigators using a similar protocol isolated a minus-end directed motor from bovine brain that had been previously identified as a microtubule-associated protein (Map 1C) (21). Because of the many similarities between dynein from flagella and the brain protein, it was called cytoplasmic dynein.

The criterion of ATP-dependent microtubule binding was used to isolate several other proteins and at least one of those had an ATPase activity (GTPase activity was greater). That protein, called dynamin, was subsequently shown to be the bovine analogue of the shibiri protein from flies (involved in endocytosis) and the microtubule binding has not been correlated with a motor activity. Most analogues of the motor proteins have been identified using sequence similarities or antibody cross-reactivity.

WHY MOTORS—AND WHERE—IN TISSUES?

Within the past ten years there has been a great proliferation of the identified microtubule-based motors [see reviews (5, 8, 9, 11, 27, 33, 39, 41)]. Although the estimates may be slightly inflated, there are reports of 35 kinesin-like and more than 5 cytoplasmic dynein-like motors in the *Drosophila* genome. This raises the obvious question of why so many different microtubule motors are present. The answer in part lies in the need for cells to transport many different intracellular components rapidly from one site to another in a vectorial fashion, often in opposite directions under different control signals. From chromosomes to mRNAs and whole nuclei to the smallest membrane

vesicles, cellular structures must be transported at the right time after the proper signal in a specific direction. Diffusion-based transport is sufficiently rapid to drive some movements, particularly the transport of globular actin subunits to the leading edges of cells. For structures larger than 25 nm, including membrane vesicles and most mRNAs, the rates of diffusion are considerably slower and there are even major barriers for their movement to certain regions of cytoplasm (19). Further, there is often a need for force generation and for concentration of components that can be driven easily by motors. Force is clearly needed during mitosis for the separation of chromosomes (6, 20, 40) or for the stretching of membrane tubes within cytoplasm to form the endoplasmic reticulum (ER) (4). Concentration of the Golgi components is clearly needed as well as a number of larger structures that are organized in specific cell regions (28). Bulk transport of vesicles within axons or proteinaceous material across epithelia all require motors for speed and efficiency. With the vast proliferation of motors it is clear that directed transport of many different materials could be independently controlled and directed. A growing body of literature indicates that separate motors may be moving distinct vesicle subpopulations (13, 14, 30, 42). Because many studies involve the knockout of specific motors, there is a rapidly expanding literature on the effects of motor depletion. Surprisingly, many of the phenotypes are quite mild and that raises the questions of whether the motors are weak catalysts (increasing rates of transport only two to threefold over diffusive mechanisms) or whether there are overlapping functions of the many motors such that other motors compensate for the loss of one.

In worms the loss of kinesin-like proteins is not a tight lethal mutation and often animals survive to a late stage (22, 31). Neurological phenotypes are common. In yeast, the loss of a kinesin analogue is not lethal until a myosin analogue is also deleted (17). As a result, interest has been revived in the possible role for actin-based motors in vesicle transport. In some systems such as squid axoplasm, there is direct evidence of an actin-based vesicle movement. The exact role that actin-based motility plays in organelle transport is obscure, although an obvious possibility is that it catalyzes the movement from the ends of microtubules through the actin cortex to the plasma membrane.

BASIC CELLULAR MOTOR MECHANISMS

How do motors move in general? In all systems defined so far, the filamentous substrate defines the direction of movement and overall length of movement. Regula-

MICROTUBULE ORGANIZATION

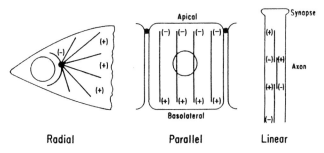

Radial   **Parallel**   **Linear**

FIG. 13.1. This illustrates three types of microtubule organization. The single microtubule organizing center *(centrosome)* in fibroblasts produces a radial array of microtubules with their plus ends toward the periphery. In epithelial cells, multiple microtubule organizing centers in the apical region give rise to a parallel array of microtubules. Within axons there is a third pattern of overlapping microtubules with all of their plus ends toward the periphery, thus forming a continuous path from the cell body to the synapse.

tion of motor activity determines whether or not the motor will move and vesicle or other cargo attachment sites will determine which motor binds to the cargo. This is all very similar to the transport systems we experience in daily life. Roadways are the analogues of the filamentous substrates. Rotary versus piston engines are analogues of two different types of motor proteins. The different vehicles that are powered by the same motor types clearly have different linkages to the motors and move under different conditions. From the distribution and polarity of the substrate filaments we can know which direction motor proteins will move. For example, in Figure 13.1 the orientation of the axonal microtubules is shown to be linear with the plus ends toward the synapse. Under these circumstances, kinesin will move materials from the cell body to the synapse. Conversely, cytoplasmic dynein will move materials from the synapse to the cell body. A number of obvious problems arise in such a system. First, because there is no protein synthesis in the synapse, cytoplasmic dynein must be carried in an inactive form to the synapse. Second, a switch must exist to convert the motor anchors from a kinesin-binding complex to a cytoplasmic dynein binding complex. In other systems the problems relate to how vesicle formation is coordinated with movement. Most of these questions will await a better understanding of the organelle–motor complexes.

ORGANIZATION OF MICROTUBULES IN TISSUES

Within many different types of cells the microtubules form structurally rigid elements that shape the cyto-

plasm and cell boundaries and provide roadways to form and maintain cellular asymmetries. In tensegrity models of cytoplasm, the microtubules are the rigid structural rods upon which are hung the more flexible actin filaments and membranes. Because there are few microtubule-organizing centers within cells, they can form the major organizing sites for cellular polarity generation. Such asymmetries as apical–basolateral, cell body–synapse or front–back are all correlated with different microtubule organizations that logically correlate with different transport functions. In many highly specialized tissues such as the sertoli cells the microtubules are organized such that they can support many of the motile events involved in sperm assembly. One of the most confusing microtubule arrangements is the antiparallel array found in dendritic processes (3). Because microtubules of opposite polarity are next to one another in dendrites, a kinesin-driven vesicle should move in two directions equally well and have no overall transport. We do know that MAP-2 is present in dendrites selectively and it inhibits motor movement on microtubules (18). Thus, MAP2 coating of one polarity of microtubules would block movement on those microtubules and allow unidirectional transport on the other microtubules. Alternative mechanisms include compartmentalization of the dendritic cytoplasm or the action of other motors that somehow exploit the bidirectional microtubule arrays.

ORGANELLE–MOTOR COMPLEXES

Within a cell, a complex of a motor and several other proteins links the organelle to the microtubule and is responsible for catalyzing movement. Several proteins have been identified that are involved in one way or another in the process of organelle movement through the use of in vitro assays of organelle movement on microtubules [reviewed in (28)]. In most instances, the mixing of organelles, supernatant, and motors is required for motility, although cases have been reported where the motors are not removed from vesicles by alkali extraction (23). Where it has been studied in detail, two soluble factors cause a dramatic increase in the level of motility in vitro. One of those, the dynactin complex, is specific for cytoplasmic dynein (28) and the *Drosophila* analogue is responsible for the glued phenotype (10), which has major neurological abnormalities. Dynactin is particularly interesting because it contains an actin-like protein, centractin, and could constitute a link between the actin and microtubule motor systems (26). The other factor has not been identified at the molecular level but both kinesin and cytoplasmic dynein-dependent activity are promoted by the same fractions (29). On the organelle surface is a protein receptor and one member of this family of proteins, the kinectins, has been defined (36). The kinectins have interesting characteristics, including a very large mass of alpha helical coiled-coil on the cytoplasmic vesicle surface, a single N-terminal hydrophobic sequence, proximity to both cytoplasmic dynein and kinesin binding sites, and evidence for alternatively spliced forms (42).

The types of vesicles moved by kinesin and cytoplasmic dynein have not been totally defined. With the definition of motor receptors in the future, it will be possible to determine which populations of organelles are capable of moving by kinesin and cytoplasmic dynein or by other motors. Those organelles include the ER, Golgi, and endosomes. Mitochondria in brain rely upon a kinesin-like protein that is distinct from the most abundant kinesin form (13, 14). In the cases of *Drosophila* and *C. elegans,* the mutation of kinesin does not block mitosis nor does it block axonal transport of synaptic vesicles. In *C. elegans,* the loss of a kinesin analogue does block synaptic vesicle accumulation, which indicates that some different vesicle populations are transported by different motors (22). Thus, although some different populations of vesicles may be transported by the same motors through modified receptors, other vesicle populations appear to rely on specialized motors presumably binding to specialized receptors.

The linkages between the motors and the vesicles are critical in control of movement. The vesicle contains the information necessary to direct its movement and the cell can generally control the level of movement, that is, when to activate or inactivate the motility. For example, an endocytic vesicle is moved from a peripheral location to the center of the cell (typically by cytoplasmic dynein) and the cell will choose when to activate the transport, such as in response to a hormone. Once the vesicle reaches the Golgi region it fuses with the trans-Golgi network (TGN) and then the motor is released or otherwise inactivated. Further processing occurs in the TGN and an exocytic vesicle is formed by budding. That exocytic vesicle must contain the signal for the outward movement (typically kinesin driven) and it will wait for the proper cellular signal to move. The basic elements of the system needed for motility are the cargo vesicle, the linkage protein, the cellular regulatory signal, and finally the motor. In such a model (see Fig. 13.2), the vesicles need to be formed with a motor receptor on their surface for future movement. The receptor on the surface must contain information about the direction of movement. If the cell is in the proper state for movement, then the proper motor will bind and the cell will activate

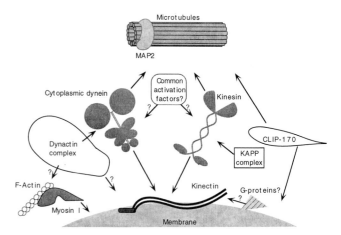

FIG. 13.2. This diagram illustrates the basic structure of the motors and the known components of the organelle motor complex. Kinectin is a membrane-attached protein and appears to bind or interact with both motors. Dynactin activates organelle motility for only cytoplasmic dynein and has centractin bound to it which could interact with other myosin motility systems. There are probably many additional components that contribute to the regulation of organelle motility, these include the MAPs on the microtubule, which could block motor interactions with the microtubule.

movement. Once at the periphery, the vesicle will fuse with the proper target membrane and the motor will be inactivated. With further processing, a new endocytic vesicle will form and start its trek back to the cell center.

MICROTUBULE MOTOR STRUCTURE

The two major classes of microtubule motors, kinesin and cytoplasmic dynein, share a number of structural similarities. They both are dimeric, with two ATPase-containing heads that are connected to a common tail or stalk structure. In the conventional paradigm, the stalk is involved in linking the motors to the cargo and the two heads are involved in walking along the microtubule in a hand-over-hand or similar fashion. Although the exact number of genes in each class has not been defined, many more members of the kinesin family appear to be in the *Drosophila* genome. There is considerable diversity in the kinesin family, with several monomeric forms and unusual tail structures. The motor families have been defined on the basis of sequence similarities [see reviews (5, 8, 9, 41)]. The conserved motor domain of kinesin is about 350 amino acids long and typical homologies are greater than 50% between members. Classifications are normally made on the basis of similarities in the nonmotor domains.

In the kinesin family are the kar3, ncd, or nod proteins that move toward the minus end of microtubules like cytoplasmic dynein. Naturally, these reverse kinesins raise important questions about mechanism—but also the question, why use such a large protein as cytoplasmic dynein when a smaller one will do? Perhaps the important functions that require cytoplasmic dynein relate to aspects of its motility such as the mechanism of cargo loading or regulation that are not well understood. Because of the large number of motors, it might be expected that several operate at once in a given organism. There are several examples where multiple motors are expressed simultaneously in a single cell. A simple rationale for the multiple motors is that they will be under different control mechanisms that are obviously required for multiple transport phenomena to occur within the same cell.

Neuronal kinesin was defined as a tetrameric (A2B2) protein with both heavy (115 kDa) and light chains (72 kDa) on the basis of its ability to move microtubules along glass surfaces. Such in vitro assays have been essential in defining the motors because they have such low microtubule-activated ATPase activities. The motors appear to be tightly down-regulated normally and only with proper activation in vivo or with anionic beads in vitro will a microtubule-activated ATPase be observed with an intact motor. This is critical, for at the normal cellular concentrations of the motors the total ATP content of the cell would be utilized within a matter of seconds.

MECHANISM OF CONVERSION OF ATP ENERGY INTO MOTILITY

In recent years several labs have looked at the elements of motor movement on microtubules at a molecular level that was only possible because the microtubule motors work as single motors. The force of a single kinesin has been measured to be two picoNewtons (pN) in an isometric condition (on a > 30 sec. time scale) (15) while values of about 5 pN have been measured on a shorter time scale (34, 35). The ability to detect single molecule movements and to determine the mechanics of single molecules makes it now possible to test models of the molecular events involved in energy conversion. The major models include the swinging crossbridge and a ratcheting or rocking crossbridge. From the work that has been done, it is clear that there are only minor differences between the basic mechanisms of motor function, and the models are moving closer rather than diverging because of the increased experimental data. The major differences for the microtubule motors compared to myosin relate to

the lifetime of the tightly bound state. For a single motor to move along a microtubule, the motor must keep at least one head on the microtubule at all times, otherwise it would diffuse away into solution. One way to accomplish this is for the binding of a second head to catalyze the release of the first head, perhaps through a mechanical coupling. This is an important aspect of the process to consider in terms of the basic process of converting chemical into mechanical energy.

SUMMARY

There are a large number of microtubule motors within the cytoplasm of cells that appear to catalyze the directed movement of many different intracellular organelles ranging from proteins to nuclei. The two-headed structure of many of the motors is sufficient for a single molecule to carry a cargo many microns along a microtubule. Within the cell and presumably within many tissues as well these motors increase the rate of cellular processes by severalfold. Thus some cells and organisms can survive without specific motors but the rate of functions are often significantly compromised. We expect to find that the motors play critical roles in many physiological processes at the level of accelerating rates and creating asymmetries within cells and tissues.

REFERENCES

1. Allen, R. D., N. S. Allen, and J. L. Travis. Video-enhanced contrast, differential interference contrast (AVEC-DIC) microscopy: a new method capable of analyzing microtubule related motility in the reticulopodial network of *Allogromia laticollans*. *Cell Motil.* 1: 291–302, 1981.
1a. Allen, R. D., D. G. Weiss, J. H. Hayden, D. T. Brown, H. Fujiwake, M. Simpson. Gliding movement of and bidirectional transport along single native microtubules from squid axoplasm: evidence for an active role of microtubules in cytoplasmic transport. *J. Cell Biol.* 100: 1736–1752, 1985.
2. Allen, R. D., J. Metuzals, I. Tasaki, S. T. Brady, and S. P. Gilbert. Fast axonal transport in squid giant axon. *Science* 218: 1127–1129, 1982.
3. Baas, P. W., J. S. Deitch, M. M. Black, and G. A. Banker. Polarity orientation of microtubules in hippocampal neurons: uniformity in the axon and nonuniformity in the dendrite. *Proc. Natl. Acad. Sci. U.S.A.* 85: 8335–8339, 1988.
3a. Brady, S. T. A novel brain ATPase with properties expected for the fast axonal transport motor. *Nature* 317: 73–75, 1985.
4. Dabora, S. L., and M. P. Sheetz. The microtubule-dependent formation of a tubulovesicular network with characteristics of the ER from cultured cell extracts. *Cell* 54: 27–35, 1988.
5. Endow, S. A., and M. A. Titus. Genetic approaches to molecular motors [Review]. *Annu. Rev. Cell Biol.* 8: 29–66, 1992.
6. Fuller, M. T., and P. G. Wilson. Force and counterforce in the mitotic spindle [Review]. *Cell* 71: 547–50, 1992.
7. Gibbons, I. R. Dynein ATPases as microtubule motors [Review]. *J. Biol. Chem.* 263: 15837–15840, 1988.

8. Goldstein, L. S. With apologies to Scheherazade: tails of 1001 kinesin motors [Review]. *Annu. Rev. Genet.* 27: 319–351, 1993.

9. Hirokawa, N. Mechanism of axonal transport. Identification of new molecular motors and regulations of transports [Review]. *Neurosci. Res.* 18: 1–9, 1993.

10. Holzbaur, E. L., J. A. Hammarback, B. M. Paschal, N. G. Kravit, K. K. Pfister, and R. B. Vallee. Homology of a 150K cytoplasmic dynein-associated polypeptide with the Drosophila gene Glued [published erratum appears in *Nature* 1992 Dec 17;360(6405):695]. *Nature* 351: 579–583, 1991.

11. Hoyt, M. A. Cellular roles of kinesin and related proteins [Review]. *Curr. Opin. Cell Biol.* 6: 63–68, 1994.

12. Inoue, S. Video image processing greatly enhances contrast, quality, and speed in polarization-based microscopy. *J. Cell Biol.* 89: 346–356, 1981.

13. Jellali, A., B. M. Metz, I. Surgucheva, V. Jancsik, C. Schwartz, D. Filliol, V. I. Gelfand, and A. Rendon. Structural and biochemical properties of kinesin heavy chain associated with rat brain mitochondria. *Cell Motil. Cytoskeleton* 28: 79–93, 1994.

14. Kondo, S., Y. R. Sato, Y. Noda, H. Aizawa, T. Nakata, Y. Matsuura, and N. Hirokawa. KIF3A is a new microtubule-based anterograde motor in the nerve axon. *J. Cell Biol.* 125: 1095–1107, 1994.

15. Kuo, S. C., and M. P. Sheetz. Force of single kinesin molecules measured with optical tweezers. *Science* 260: 232–234, 1993.

16. Lasek, R. J., and S. T. Brady. Adenylyl imidodiphosphate (AMP-PNP), a non-hydrolyzable analogue of ATP produces a stable intermediate in the motility cycle of fast axonal transport. *Nature* 316: 645–647, 1985.

17. Lillie, S. H., and S. S. Brown. Immunofluorescence localization of the unconventional myosin, Myo2p, and the putative kinesin-related protein, Smy1p, to the same regions of polarized growth in Saccharomyces cerevisiae. *J. Cell Biol.* 125: 825–842, 1994.

18. Lopez, L. A., and M. P. Sheetz. Steric inhibition of cytoplasmic dynein and kinesin motility by MAP2. *Cell Motil. Cytoskeleton* 24: 1–16, 1993.

19. Luby-Phelps, K., and D. L. Taylor. Subcellular compartmentalization by local differentiation of cytoplasmic structure. *Cell Motil. Cytoskeleton* 10: 28–37, 1988.

19a. Lye, R. J., M. E. Porter, J. M. Scholey, and J. R. McIntosh. Identification of a microtubule-based cytoplasmic motor in the nematode C. elegans. *Cell* 51: 309–318, 1987.

20. McIntosh, J. R., and C. M. Pfarr. Mitotic motors [Review]. *J. Cell Biol.* 115: 577–585, 1991.

21. Paschal, B. M., H. S. Shpetner, and R. B. Vallee. MAP 1C is a microtubule-activated ATPase which translocates microtubules in vitro and has dynein-like properties. *J. Cell Biol.* 105: 1273–1282, 1987.

22. Patel, N., M. D. Thierry, and J. R. Mancillas. Cloning by insertional mutagenesis of a cDNA encoding Caenorhabditis elegans kinesin heavy chain. *Proc. Natl. Acad. Sci. U.S.A.* 90: 9181–9185, 1993.

23. Schnapp, B. J., T. S. Reese, and R. Bechtold. Kinesin is bound with high affinity to squid axon organelles that move to the plus-end of microtubules. *J. Cell Biol.* 119: 389–399, 1992.

24. Schnapp, B. J., R. D. Vale, M. P. Sheetz, and T. S. Reese. Single microtubules from squid axoplasm support bidirectional movement of organelles. *Cell* 40: 455–462, 1985.

25. Scholey, J. M., M. E. Porter, P. M. Grissom, and J. R. McIntosh. Identification of kinesin in sea urchin eggs, and evidence for its localization in the mitotic spindle. *Nature* 318: 483–486, 1985.

26. Schroer, T. A. New insights into the interaction of cytoplasmic dynein with the actin-related protein, Arp 1 [Review]. *J. Cell Biol.* 127: 1–4, 1994.

27. Schroer, T. A. Structure, function and regulation of cytoplasmic dynein. [Review]. *Curr. Opin. Cell Biol.* 6: 69–73, 1994.

28. Schroer, T. A., and M. P. Sheetz. Functions of microtubule-based motors. *Annu. Rev. Physiol.* 53: 629–652, 1991.

29. Schroer, T. A., and M. P. Sheetz. Two activators of microtubule-based vesicle transport. *J. Cell Biol.* 115: 1309–1318.

30. Sekine, Y., Y. Okada, Y. Noda, S. Kondo, H. Aizawa, R. Takemura, and N. Hirokawa. A novel microtubule-based motor protein (KIF4) for organelle transports, whose expression is regulated developmentally. *J. Cell Biol.* 127: 187–201, 1994.

31. Shakir, M. A., T. Fukushige, H. Yasuda, J. Miwa, and S. S. Siddiqui. C. elegans osm-3 gene mediating osmotic avoidance behaviour encodes a kinesin-like protein. *Neuroreport* 4: 891–894, 1993.

32. Sheetz, M. P., and J. A. Spudich. Movement of myosin-coated fluorescent beads on actin cables in vitro. *Nature,* 303:31–35, 1983.

33. Skoufias, D. A., and J. M. Scholey. Cytoplasmic microtubule-based motor proteins [Review]. *Curr. Opin. Cell Biol.* 5: 95–104, 1993.

34. Svoboda, K., and S. M. Block. Force and velocity measured for single kinesin molecules. *Cell* 77: 773–84, 1994.

35. Svoboda, K., C. F. Schmidt, B. J. Schnapp, and S. M. Block. Direct observation of kinesin stepping by optical trapping interferometry [see comments]. *Nature* 365: 721–727, 1993.

36. Toyoshima, I., H. Yu, E. R. Steuer, and M. P. Sheetz. Kinectin, a major kinesin-binding protein on ER. *J. Cell Biol.* 118: 1121–1131, 1992.

37. Vale, R. D., T. S. Reese, and M. P. Sheetz. Identification of a novel force-generating protein, kinesin, involved in microtubule-based motility. *Cell* 42: 39–50, 1985.

38. Vale, R. D., B. J. Schnapp, T. S. Reese, and M. P. Sheetz. Organelle, bead, and microtubule translocations promoted by soluble factors from the squid giant axon. *Cell* 40: 559–569, 1985.

38a. Vale, R. D., B. J. Schnapp, T. Mitchison, E. Steuier, T. S. Reese, M. P. Sheetz. Different axoplasmic proteins generate movement in opposite directions along microtubules in vitro. *Cell* 43: 623–632, 1985.

39. Vallee, R. Molecular analysis of the microtubule motor dynein [Review]. *Proc. Natl. Acad. Sci. U.S.A.* 90: 8769–8772, 1993.

40. Wadsworth, P. Mitosis: spindle assembly and chromosome motion [Review]. *Curr. Opin. Cell Biol.* 5: 123–128, 1993.

41. Walker, R. A., and M. P. Sheetz. Cytoplasmic microtubule-associated motors [Review]. *Annu. Rev. Biochem.* 62: 429–451, 1993.

42. Yu, H., C. V. Nicchitta, J. Kumar, M. Becker, I. Toyoshima, and M. P. Sheetz. Characterization of kinectin, a kinesin-binding protein: primary sequence and N-terminal topogenic signal analysis. *J. Cell Biol.* 6: 171–183, 1995.

14. Membrane fusion

R. BLUMENTHAL
D. S. DIMITROV

Section on Membrane Structure and Function, National Cancer Institute,
National Institutes of Health, Bethesda, Maryland

THE FIRST MEMBRANE FUSION REACTION probably occurred a billion years ago. How could it have happened and why was it so important? One possible scenario is the following. The formation of membranes accelerated the evolution of life because of compartmentalization of the biochemical reactions. However, development of the prebiotic "cells" and reproduction of the nucleic acids were limited by increasing the cell volume, which minimizes the surface-to-volume ratio, allowing low exposure of nonpolar moieties to water. This decreases the efficiency of the membrane-bound biochemical reactions. Growth of prebiotic cells might also be limited by "barriers to fusion" established by electrostatic, hydration-repulsion and/or mechanical forces (24). The lack of mechanical stability of large membranes was presumably a crucial factor in providing the appropriate conditions for cell division, which is critically dependent on the ability of the membranes to fuse. One can imagine that strong forces (for example; ocean waves) led to contact of parts of membrane from the same cell or from other cells, and mechanical destabilization. This may have resulted in local fusion, cell division, or cell fusion. The two phenomena, cell division and cell fusion, provided the critical events needed for the rapid evolution of life. Processes involving fusion of parts of the same membrane leading to separation of compartments (for example, cell division) are called membrane fission to distinguish them from the process of membrane fusion, which leads to joining of compartments enclosed by different membranes. Because the physical mechanisms of fission and fusion appear to be the same, and both involve membrane merging, we will use the term membrane fusion for both of them.

There are two fundamental characteristics of any membrane fusion reaction that are important for understanding the mechanisms and biological importance of membrane fusion. First, the membrane fusion reaction requires molecular contact and destabilization. Membranes were designed by nature to be stable and resist external forces. In order to fuse they must be forced to change their structure to conformations appropriate for their merging. The contact of membranes and their destabilization were presumably induced originally by mechanical forces. In living organisms specialized proteins are predominantly used for that purpose.

Second, the membrane fusion reaction leads to either separation (division) of one compartment (cell) into two or several compartments or to joining of two or more compartments into one. The separation into compartments by the fusion (fission) reaction allows replication and transfer of substances that originally belonged to the parent compartment. Examples are cell division, endocytosis, and budding of transport vesicles in the secretory and endocytic pathway. The joining of two compartments into one leads to mixing of the membrane and aqueous contents of the two compartments. Examples are cell fusion in fertilization and hybridization, exocytosis, entry of enveloped viruses into cells, myotube formation, and intracellular transfer of material by vesicles.

The progress in studying and understanding the mechanisms of membrane fusion was largely dependent on the knowledge of membrane structure (see chapter 2 by Thompson et al. and chapter 3 by Tanford and Reynolds, this volume). Membrane fusion involves rapid localized rearrangements of the lipid matrix that are difficult to monitor. For almost two centuries fusion was monitored by observing the consequences of fusion rather than the very act of fusion. Formation of multinucleated cells (syncytia) is one of the best known and easy to observe phenomena that have been used as indication of plasma membrane fusion. Syncytia were observed for the first time by Muller in the early decades of the nineteenth century [see (267)]. Membrane fusion was suggested (173) and later clearly established (177) as a mode of their formation. It was not until the 1960s, however, that the studies of fusion "exploded." During a relatively short period of time a number of interesting fusion phenomena were discovered and characterized by using light microscopy [for history see (267)]: (1) viruses can induce formation of giant multinucleated cells (syncytia) (135, 227, 228, 269), (2) during fertilization the acrosomal membrane interdigitates and then coalesces with the egg membrane (64), (3) mononucleated myoblasts fuse to form myotubes, at least in vitro (43, 136, 167), and (4) cell hybrids can be formed in vitro by spontaneous cell fusion (19, 301). In the early 1970s liposomes were discovered (16) and used as a model system to study fusion (240). Polyethylene glycol (PEG) was introduced as a fusion agent for plant protoplasts (33, 157) and

animal cells (254). About a decade ago it was reported that cell fusion can be also induced by external electric fields (electrofusion) (217, 294, 334, 358) and even by laser beams (289). [For review of older literature on fusion, see (257) and for more recent reviews see the books (21, 49, 97, 104, 219, 203, 224, 304, 350).]

Although light microscopy provides a great deal of information about the phenomenology of fusion, more advanced physical techniques are required to elucidate mechanisms of membrane fusion. During the past three decades mechanisms of membrane fusion were extensively studied by using a wide variety of such techniques, including electron microscopy, electron spin resonance, fluorescence video microscopy, spectrofluorometry and electrophysiological methods. In the next section these techniques are discussed in more detail.

OBSERVATION OF FUSION REQUIRES PHYSICAL TECHNIQUES FOR MONITORING MIXING OF MEMBRANES AND THE COMPARTMENTS THEY ENCLOSE

The criterion for fusion of membranes is the merging of their lipid bilayer and cytoplasmic continuity. This leads to mutual diffusion of the lipid molecules and eventually to intermixing of water-soluble substances bounded by the membranes. The majority of the assays for monitoring fusion are, therefore, based on measuring intermixing of lipid- or water-soluble molecules by using a wide variety of physical techniques. Most of them use light microscopy, fluorescence marker redistribution [for review see (96, 186)] electrophysiological techniques [for reviews, see (8, 181)] and rapid freezing (cryofixation) electron microscopy techniques [for review see (163)].

Morphological Changes Following Fusion Are Observed by Light Microscopy but Membrane Fusion May Occur without such Changes

Light microscopy has been widely used to study fusion. Fusion is detected as a morphological change leading to formation of larger fused membrane products. The most dramatic visual manifestation of fusion activity is the formation of syncytia (Fig. 14.1). Syncytia are commonly defined as giant cells having a diameter equal to or larger than four diameters of the original unfused cells (85, 180). A major problem with syncytia formation as an assay for fusion is that fusion may occur but remain undetected because of formation of small fusion junctions of size smaller than the resolution of the light microscopy (85). This is, for example, the case with electrofusion of red blood cells at low electric field strength (90). Formation of lumen and

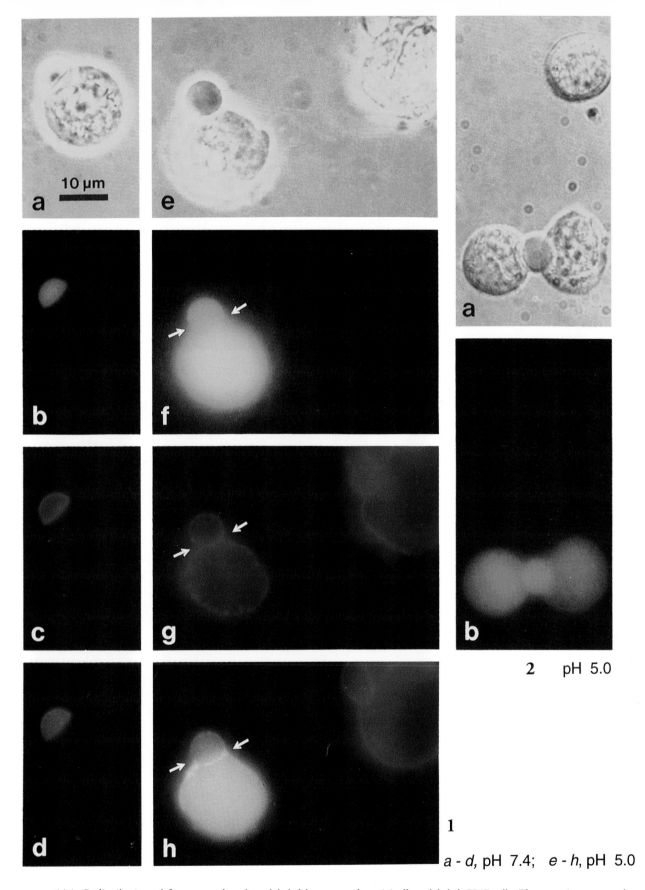

PLATE 14.1. Redistribution of fluorescent dyes from labeled human red blood cells (RBC) to unlabeled GP4F cells following fusion as detected by fluorescence microscopy. *Panel 1.* (*a*) shows a RBC double-labeled with a water-soluble fluorescent dye, NBD, and a membrane-soluble dye, R18, and attached to a GP4F cell in phase contrast incubated at pH 7.4 for 90 sec. (b–d) show pictures taken under fluorescence of NBD, R18, and both, respectively. Note that the fluorophores are confined to the RBC cytoplasm and membrane, respectively; there is no transfer to either label to the fibroblast. (*e–h*) are photos of GP4F cells decorated with double-labeled RBC after 90 s at pH 5.0 and 37°C. Note redistribution of the two labels

to the originally unlabeled GP4F cells. The opposing arrows show that the membrane between the fusing cells is still visible with the membrane label (*g* and *h*), but not in (*f*) where the cytoplasmic dye has redistributed. The red rhodamine fluorescence appears yellow in *h*, because of the overlying green of the NBD. *Panel 2* shows that hemoglobin does not move from RBC to GP4F after fusion detected by NBD redistribution. (*a*) is taken under phase contrast and shows that the hemoglobin, which appears pink, is still contained inside the RBC; (*b*) is taken under fluorescence and shows the redistribution of the NBD. [From (280).]

Top View

Side View Cross Section

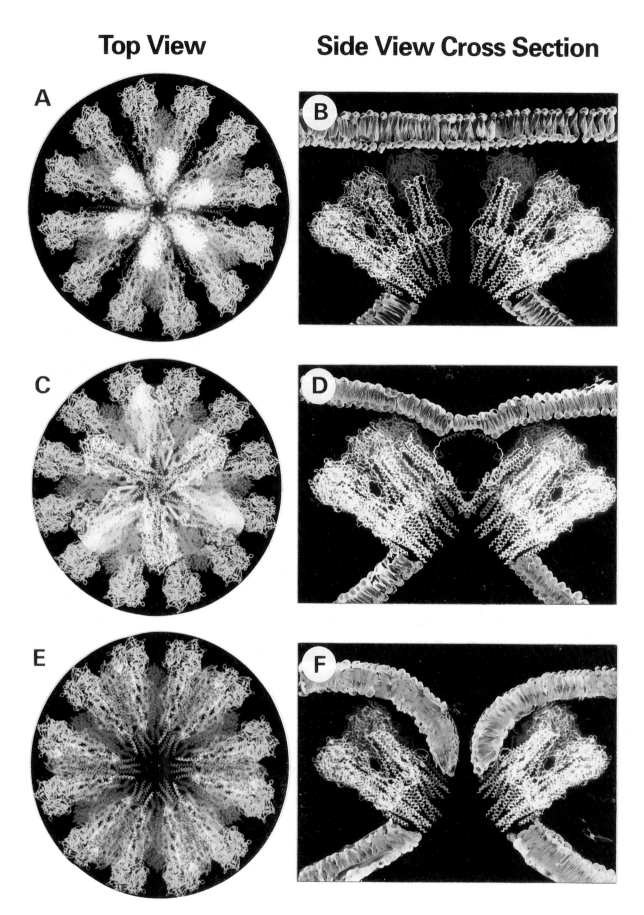

PLATE 14.2. A structural model of HA-mediated membrane fusion. The left panels show a view from the top; the right panels—a side view cross section. *A* and *B* represent the initial assembly of HA trimers around a fusion site. *C* and *D* show the initial interaction of fusion peptides with target membranes. *E* and *F* show formation of a fusion pore. Note the bending of the membrane that is needed to get the fusion peptides (in red in the right side panels) at close proximity to the target membrane (127). [Kindly provided by Dr. H. R. Guy.]

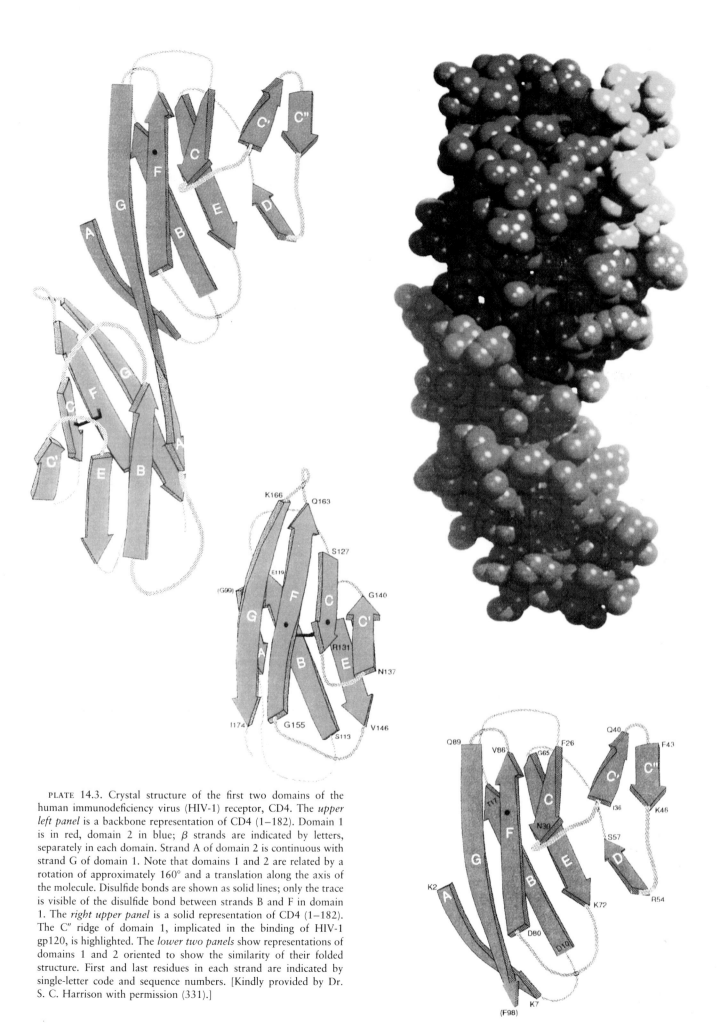

PLATE 14.3. Crystal structure of the first two domains of the human immunodeficiency virus (HIV-1) receptor, CD4. The *upper left panel* is a backbone representation of CD4 (1–182). Domain 1 is in red, domain 2 in blue; β strands are indicated by letters, separately in each domain. Strand A of domain 2 is continuous with strand G of domain 1. Note that domains 1 and 2 are related by a rotation of approximately 160° and a translation along the axis of the molecule. Disulfide bonds are shown as solid lines; only the trace is visible of the disulfide bond between strands B and F in domain 1. The *right upper panel* is a solid representation of CD4 (1–182). The C″ ridge of domain 1, implicated in the binding of HIV-1 gp120, is highlighted. The *lower two panels* show representations of domains 1 and 2 oriented to show the similarity of their folded structure. First and last residues in each strand are indicated by single-letter code and sequence numbers. [Kindly provided by Dr. S. C. Harrison with permission (331).]

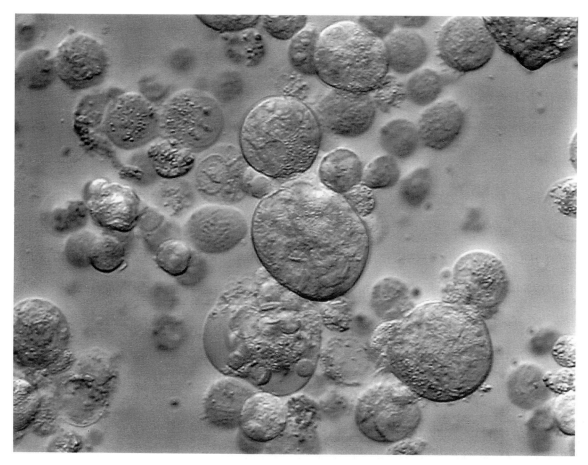

FIG. 14.1. Formation of syncytia mediated by the human immunodeficiency virus (HIV-1) envelope glycoprotein. CD4-negative T lymphocytes (12E1), expressing the HIV-1 envelope glycoprotein encoded by a recombinant vaccinia virus (vPE16), were mixed with CD4-positive T cells (Molt3). The pictures were taken under DIC 16 h after mixing.

rounding of cells may occur at high strengths of the fusogenic electric field or when the spectrin network is disrupted. It is also commonly observed that combination of viral and host factors contribute to syncytia formation by viruses. Some influenza (119), herpes (268), and paramyxoviruses (211) do not lead to syncytia formation even though they are fusion competent. With the advance of the light microscopy techniques, such as differential interference contrast (DIC), video-enhanced microscopy and confocal laser microscopy, it is now possible to observe also fusion of small vesicles with cells, as in exocytosis (293).

Fluorescence Microscopy and Spectrofluorometry Allow Quantitation of Membrane Fusion Events in Living Cells

The use of fluorescence microscopy is based on redistribution of fluorescent molecules following fusion. The fluorescent markers can be incorporated either in the membranes or in the compartments bounded by the membranes. In the first case the membranes are labeled with membrane-soluble dyes, which commonly consist of hydrophobic tail(s) and a fluorescence reporter group. The labeling procedure is one of the critical steps for successful use of fluorescent techniques to monitor fusion. Commonly the membranes are labeled by rapidly mixing them with a solution containing the dye. To avoid artifacts, dyes that are loosely associated with the surface need to be separated from membrane-embedded fluorescent molecules and a variety of controls need to be performed for nonspecific dye transfer in the absence of fusion. Cells or vesicles can also be labeled with water-soluble dyes. The fusion is then indicated by the appearance of the water-soluble dye in the unlabeled compartment.

A very reliable way of incorporating fluorescent phospholipid molecules into one of the fusion partners is by reconstitution. This can easily be done when liposomes are used as targets for fusion. Alternatively, reconstituted vesicles containing dye and fusion protein (for example, reconstituted viral envelopes) are excellent vehicles for monitoring fusion events (26).

The next step is the mixing of labeled with unlabeled membranes and monitoring the dye redistribution to the unlabeled membranes. The membranes can either be mixed at the conditions of fusion (commonly 37°C) or prebound and then transfered to fuse. Because the membranes need some time to adhere before fusion, the measured fusion kinetics may be different when using the two protocols. Whenever possible, it is preferable to prebind the membranes and then to transfer them at fusion conditions. The dye redistribution from the labeled to the unlabeled membrane, which is an indication for fusion, can be observed visually (see Plate 14.1, facing page 564), and the number of fused cells counted or recorded by using video recorders. In the latter case the kinetics of dye transfer for single cells can be analyzed quantitatively.

The possibility of recording the kinetics of fusion of single cells is a major advantage of fluorescence microscopy for monitoring fusion. However, this technique is very time consuming and it is difficult to perform statistical analyses. To measure the kinetics of fusion for a population of cells methods based on spectrofluorometry are more appropriate. One of the most popular assays is based on the dequenching of the octadecylrhodamine (R18) (132, 160) (see Fig. 14.2). This fluorescent dye has decreased fluorescence (quenches) at high concentrations. After fusion it dilutes into the target membrane and its fluorescence increases. The increase in fluorescence is an indication for fusion and proportional to the number of fused membranes. This assay is very convenient and has been extensively used during the past decade. Another interesting assay for measuring fusion in populations of cells is based on the phenomenon of resonance energy transfer. In this case the membrane partners are labeled with two different fluorescent molecules—donor and acceptor. Before fusion the energy transfer is minimal because of the physical separation of the two dyes. After fusion the dyes intermix and the fluorescence due to the energy transfer increases. The fusion yield can then be quantitated by measuring the increase in fluorescence [for a review of the resonance energy transfer and other fluorescence-based assays see (96, 207)].

Quantitation of fusion involves measuring yields, rates and delays. Fusion yields are commonly defined as the number of fused labeled cells normalized to the total number of labeled cells (302). They can be measured as function of time. The rate of fusion is then the derivative of the fusion yield with respect to time. Delays (lag times) are the time periods between the application of the fusion trigger and the onset of

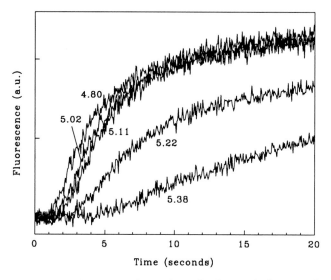

FIG. 14.2. Fluorescence dequenching after fusion of influenza with erythrocyte ghosts as measured by a spectrofluorometer. The rapid kinetics of fluorescence changes (in arbitrary units [a.u.]) upon fusion was triggered by mixing (using a stop-flow technique) of equal volumes of an R18-influenza virus–ghost suspension and a PBS-citrate solution to reach the indicated pH. Nine data sets were averaged for each pH. The temperature was 37°C. [From (56).]

membrane merging. In biophysical assays the onset of fusion occurs when the average signal exceeds the level of noise. Because of the stochastic nature of the reactions leading to the onset of fusion, the values of delay times for single fusion events (158, 185) will be different from one (pre-fusion) complex to another. Therefore, the rate of fusion of a population of prefusion complexes is determined by the distribution of delay times (50).

The spectrofluorimetric assays measure rates of fusion by monitoring the fluorescence increase over time of a population of complexes (25, 56, 57, 186, 209, 261, 280, 315). In this case, in addition to the contribution of variations in delays, there will be also an effect of the rate of dye transfer to the unlabeled cells (50, 272). In measurements of fusion of single influenza virions with erythrocytes, it was found that the measured rate of R18 redistribution was much slower than expected from simple lipid diffusion calculations (185). On the other hand it has been shown that water-soluble dye can be transferred through pores by electroosmosis at a much faster rate than expected from diffusion (89). Therefore, dye redistribution rates need to be considered in estimating rates of fusion (50).

Electron Microscopy Provides Direct Observation of Structural Rearrangements Due to Fusion

Electron microscopy (EM) is the only technique that allows direct visualization of the structure of fusion

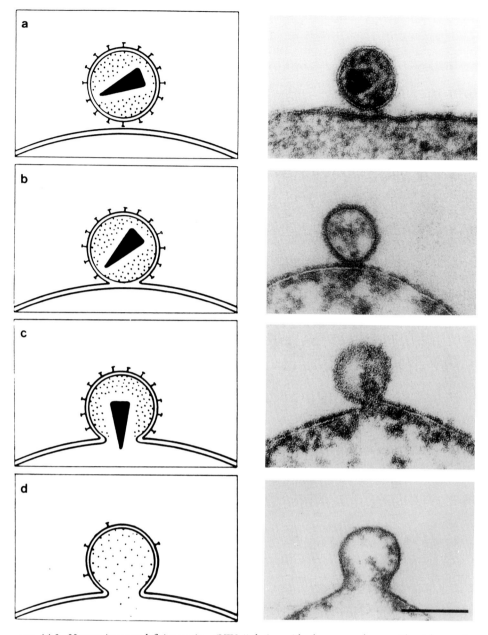

FIG. 14.3. Human immunodeficiency virus (HIV-1) fusion with plasma membranes. The right panels show ultrathin sections of H9 cells incubated with purified HIV-1 and embedded in Epon. The left panels represent schematically stages in the fusion process. *(a)* Adsorption of HIV-1 to the cell membrane after 2 min at 4°C. *(b,c)* Fusion after 1–3 min at 37°C *(d)* Empty viral envelope after fusion with the cell membrane and release of the viral ribonucleoprotein complex. Bar = 150 nm. Kindly provided by Dr. C. Grewe. [With permission from (125).]

intermediates. However, fusion is a very rapid and localized phenomenon, that involves molecular rearrangements of small molecules (lipids). This makes the use of electron microscopy for studying mechanisms of fusion very difficult. Major problems are the possibility of artifacts due to the preparation of specimen, misinterpretations of the EM pictures, and difficulties in quantitation. One approach to avoiding

artifacts is to freeze the specimen rapidly [for review see (163)].

Both thin-sectioning and freeze-fracture were used to observe fusion (47, 164–166, 229, 231). The series of electron micrographs presented in Figure 14.3 show in detail a sequence of events occurring during the initial stages of entry of human immunodeficiency virus 1 (HIV-1) into cells (125). The lipid bilayers of the viral envelope and of the target membrane appear to merge into a pentalaminar structures before the membrane diaphragm ruptures leading to entry of the nucleocapsid into the cell (see below for a discussion of virus–cell fusion). Such merging of the two leaflets in contact before cytoplasmic continuity (so-called pentalaminar structures) has originally been described in exocytosis of secretory granules of pancreatic acinar cells (233). In an early study (234) parts of the pentalaminar sandwich seemed to merge into two dark lines, which presumably implied one single bilayer membrane. Such a "bilayer diaphragm" has also been seen in a freeze-fracture study of exocytosis in zoospores (248). According to Chandler and Heuser (47), however, such "intermediates" seem to be an artifact, resulting from exposure to high concentrations of glycerol. Pentalaminar structures were not observed during fusion of vacuoles processed by rapid freezing but were found in chemically fixed specimens (47, 113). Membrane fusion appeared to occur in a specific membrane site, where a fusion pore was formed immediately after the two membranes came into contact with each other (113).

Another interesting phenomenon, revealed by thin-sectioning EM, is the existence of regular periodic patterns of local contacts when erythrocytes were adhered by polyionic macromolecules (60, 61). Such regular contacts may lead to the formation of localized fusion junctions involving small areas (probably nanometers or tens of nanometers) (314, 316), which then expand resulting in small intracellular vesicles as observed by EM in electrofusion of erythrocyte ghosts (54).

Thin sections do not allow detection of structural changes in the membrane involved in fusion. Use of electron-dense probes for carbohydrate residues or ferritin-labeled antibodies to specific surface components [for example; viral membrane proteins, see (188)] reveal redistribution of membrane components concomitant with fusion. For instance, it has been shown that lectin-binding sites are absent from the plasma membrane involved in a pentalaminar association with the membrane of a secretory granule of mast cells (174). Thin sections have also revealed electron-dense bridges, which are interpreted as microfilamentous elements of the cell web, between two interacting membranes (231).

Freeze-fracture EM reveals more elaborate structural organization of membranes before, during, and after fusion. In aldehyde-fixed, glycerol-impregnated mammalian cells, exocytosis is accompanied by the formation of particle-free patches at the fusion sites. Similar areas have been seen in viral-plasma membrane fusion (164). Thus the process of fusion has been interpreted

as involving "bare lipid patches" (189). In protozoan exocytosis, however, the membranes that are about to fuse appear instead to be decorated by a specific particle arrangement, the "rosette" (281). Moreover, the existence of particle-free patches in aldehyde-fixed, glycerol-impregnated mammalian cells has been ascribed to an artifact, because they were not found in unfixed, unimpregnated cells fractured after rapid freezing (47). Possible artifacts in observing fusion by EM was analyzed in detail by Knoll and Plattner (163), who argue that today cryofixation is the most reliable method known.

Figure 14.4 shows quick-freezing and freeze-fracture applied to the study of early stages of exocytosis in rat peritoneal mast cells (45, 46, 48, 71). Mast cells briefly stimulated with 48/80 (a synthetic polycation and well-known histamine-releasing agent) at 22°C displayed single, narrow-necked pores (some as small as 0.05 μm in diameter) joining single granules with the plasma membrane. Pores that had become as large as 0.1 μm in diameter were clearly etchable and thus represented aqueous channels connecting the granule interior with the extracellular space. Granules exhibiting pores usually did not have wide areas of contact with the plasma membrane, and clearings of intramembrane particles, seen in chemically fixed mast cells undergoing exocytosis, were not present on either plasma or granule membranes. Fusion of interior granules later in the secretory process also appeared to involve pores. Also found were groups of extremely small, etchable pores on granule membranes that may represent the earliest aqueous communication between fusing granules.

Freeze-fracture also allows us to observe possible intermediates in fusion. It has been suggested that some of the small globules (7–12 nm in diameter) observed in freeze-fractures of membranes and commonly attributed to membrane proteins might represent "lipid particles" or inverted micelles (326). A study of Hui and coworkers (147) and a theoretical analysis of Siegel (295) indicate that the existence of such inverted micelles within a single bilayer is quite implausible. However, such structures could very well exist as interbilayer fusion intermediates (295), which seem to be observed by cryomicroscopy (297).

Patch-Clamp Techniques Allow the Monitoring of Very Fast Openings of Fusion Pores

Electrophysiologic assays are based on measuring the postsynaptic potentials due to the secretion of transmitter by a nerve terminal. In some synapses, miniature postsynaptic potentials can be recorded with a submillisecond time resolution probably at the level of single

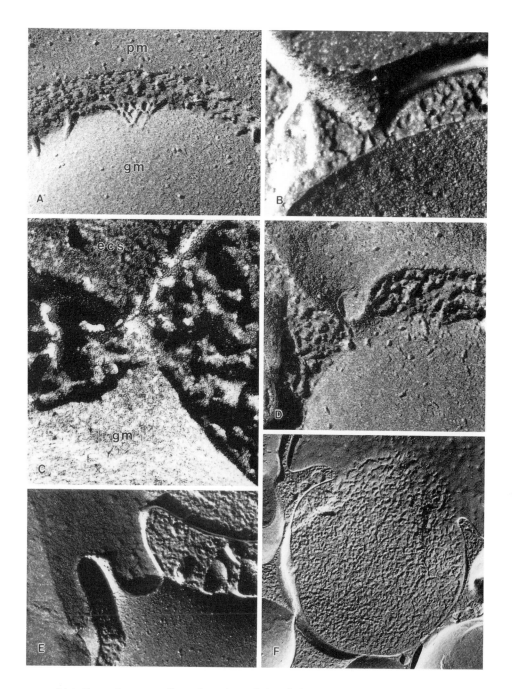

FIG. 14.4. Freeze-fracture replicas of membrane fusion during exocytosis in rat peritoneal mast cells. Mast cells were stimulated with 8 μg/ml of the histamine releaser 48/80, then rapidly frozen 15 s later. A. In the unstimulated cell the plasma membrane *(pm)* is separated from underlying granule membrane *(gm)* by a layer of cytoplasm. Magnification ×270,000. B. Prior to membrane fusion and pore formation, the plasma membrane dimples inward toward the granule. Initial pore formation is thought to occur within a highly localized area of contact between the two membranes probably no greater than 50 nm in diameter. ×220,000. C. A long narrow pore (10 nm inner diameter by 40 nm in length) connects the plasma membrane with the granule membrane *(gm)* bringing the extracellular space *(ecs)* into continuity with the granule interior. At this early stage of growth, pores are quite variably in morphology. ×200,000. D,E. Even as the pore widens to 30 nm *(D)* and 200 nm *(E)* the remainder of the granule membrane is well separated from the plasma membrane. ×190,000 *(D)* and ×100,000 *(E)*. F. Finally, the pore grows to produce the typical omega figure characteristic of exocytosis in its final stages. Note the etched matrix of the granule emerging into the extracellular space. ×56,000. [Micrographs were reproduced by permission from (48) *(A and D)*: (46) *(B and E)*: (71) *(C)*; and (45) *(F)*, and kindly provided by Dr. D. E. Chandler.]

569

synaptic vesicles (75, 159, 212). While of extremely high time resolution, this technique is applicable only to systems where postsynaptic membrane potentials can be measured. This limitation is overcome by a new technique for measuring delays and rates of fusion, which is based on electrochemical monitoring of single secretory events (55). Another technique, which has been applied to a wider variety of systems, is based on measurements of the capacitance increase due to the increase in membrane area following fusion (153, 182, 214). By using this method one can monitor fusion of single secretory vesicle reported as capacitance increase in small steps (Fig. 14.5). Such steps have been seen in adrenal chromaffin cells (214), pancreatic acinar cells (195), neutrophils (220) and mast cells (106). Mast cells from a defective strain of beige mice are particularly convenient because of their giant secretory vesicles (1–5 μm in diameter) which give very large electric signals upon secretion, allowing us to detect details that are hardly seen during exocytosis of smaller vesicles (37, 357). Analysis of these currents allows estimation of the fusion pore size during its formation and expansion (306) and the rate of fusion (10).

While it is believed that fusion is an irreversible phenomenon, Fernandez and coworkers (106) found that occasionally the step increase in capacitance, which reports formation of a fusion pore, fluctuates for a while (a second or less) and then returns to baseline, which indicates closing of the fusion pore. The electrical conductance of a flickering pore (0.7–4 nS in giant vesicles) is consistent with a 4nm–10 nm diameter pore of 150 nm length. Fusion pores formed between erythrocytes and fibroblasts, expressing the influenza hemagglutinin, also nearly always show prolonged and extensive conductance fluctuations reminiscent of the flicker phenomenon seen in exocytosis (308) (see below).

It was found that the activation energy for opening of exocytotic fusion pores is higher (23 kcal/mol) than that for the pore closing (6.95 kcal/mol), which was interpreted as an indication for involvement of proteins in the fusion pore opening and lipid in its closing (221). Recently, stochastically occurring signals were recorded from adrenal chromaffin cells using a carbon-fiber electrode as an electrochemical detector (55). These signals obey statistics characteristic for quantal release; however, in contrast to neuronal transmitter release, secretion occurred with a significant delay after short step depolarizations. Furthermore, a pedestal or "foot" at the onset of unitary events, which may represent the slow leak of catecholamine molecules out of a narrow "fusion pore" before the pore, dilates for complete exocytosis.

WHAT DO WE LEARN FROM "NONBIOLOGICAL" FUSION PROCESSES?

Until fairly recently the study of mechanisms of membrane fusion has been dominated by the use of model systems. Particularly, artificial lipid membranes in the form of liposomes or of planar bilayers have been used extensively as tools for those studies because they provide a clear way to look at the influence of lipid state on fusion (24). The chemical composition of liposomes can be varied considerably from pure one-component systems to mixtures of various phospholipids differing in head-group or hydrocarbon chain configuration. Thus questions relating to head-group specificity, bilayer packing, phase transitions and separations can be studied in detail. Other membrane components such as cholesterol, glycoproteins, and various proteins or glycoproteins can be incorporated into the phospholipid bilayer, and their effects on fusion studied under controlled conditions. Membrane-active molecules and drugs that are known to alter fusion of natural membranes can be added externally to a well-defined system, and their influence on the mechanism of fusion assessed in detail. The results from these simple systems can then be correlated with the evidence on fusion with natural membranes. There are a number of objections leveled against artificial vesicles: first, in the early studies on vesicle–vesicle interaction it was not clear that fusion, rather than other forms of interaction such as lipid exchange, was taking place. Second, some liposomes represent thermodynamically metastable forms following sonication or other procedures to form vesicles. Third, the mechanism of lipid membrane fusion might be totally different from that of biological membrane fusion.

A case in point is the fusion of liposomes with planar phospholipid bilayers as a model system for exocytosis (62). Whereas osmotic swelling of vesicles, attached to the planar bilayer in a "pre-fusion" state, is a necessary and sufficient condition for fusion in this system (62), it is not required for exocytosis in secretion (357) (see Fig. 14.5). Moreover, the lipid membrane fusion is characterized by a rapid diaphragmatic breakage, where initial pores cannot be detected within microseconds, whereas the exocytotic membrane fusion is characterized by the formation of a semi-stable intermediate, the fusion pore (see earlier under Electron Microscopy. . . .) The discovery of fusion pores, and of a variety of specialized proteins that mediate membrane fusion has led to a "paradigm shift" (170) in the field that relegates the lipid to a secondary role. This statement is not intended to slight the lipidologists: it is still important to learn about shapes, forces, transi-

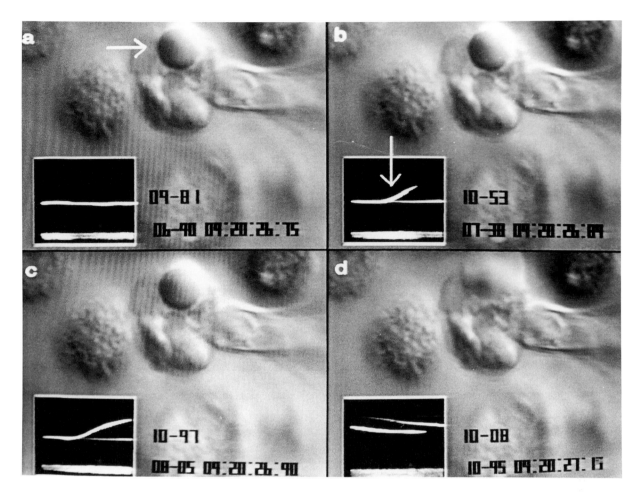

FIG. 14.5. Capacitance increase following fusion of secretory vesicles during exocytosis of beige mice mast cells. The capacitance numbers 6.90, 7.30, 8.05, and 10.95 (in pF) are next to the oscilloscope graphs and left to the eight-digit timer with hr, min, sec and 0.01 sec. The upper oscilloscope trace shows capacitance; the lower, conductance. The oscilloscope parameters are 50 ms/trace, persistence of 50 ms. The microscopy images were taken with a video camera and stored on tape for later analysis. Four sequential frames are shown. In (a) no activity occurs and it provides a reference for the granule size. The arrow indicates the granule. In (b) the capacitance increased from 6.90 to 7.30. The arrow indicates the initial opening of a fusion pore. The capacitance further increased in (c) and (d) indicating widening of the fusion pore(s). Swelling of the granule (d) was observed after fusion (357). Conductance, which is indicated with numbers above the capacitance, transiently changes during the experiment, but no average change corresponds to the fusion event. [The photo was kindly provided by Dr. J. Zimmerberg.]

tions of lipids to find out how easily they can be coaxed by the fusion proteins to undergo the catastrophic rearrangements necessary for fusion. In the next sections, lessons learned from several model systems are discussed in detail.

Ca^{2+} Induces Aggregation Destabilization, and Fusion of Liposomes Containing Phospholipids with Negatively Charged Head-groups

Calcium ions induce aggregation and fusion of a variety of lipid bilayers containing acidic components. Papa-hadjopoulos and coworkers found that addition of Ca^{2+} to suspension of phosphatidylserine (PS) vesicles causes fusion and formation of large structures, called "cochleate" cylinders (238; 240–243). Fusion was accompanied by extensive leakage of the vesicle aqueous contents. The Ca^{2+}-induced fusion can be modulated by annexins and depends on the nature of the lipid molecules (201). Parsegian and Rand (245) suggested that interaction of calcium ions with the negatively charged phospholipids results in a very strong inter-membrane attraction, which leads to an increase of the membrane tension. The increase in tension results in localized destabilization and rupture of the membrane. Those regions of destabilized membranes which are in contact can fuse, while the rupture of the rest of the vesicle membrane leads to leakage. An analysis of the role of calcium ions in fusion can be found in recent reviews (239, 348).

Although Ca^{2+}-induced fusion of PS-containing liposomes might be unrelated to biological membrane fusion, important lessons from the system can be learned because it shows how membranes destined to fuse must overcome barriers due to intermembrane repulsion. In biological systems the aggregation and fusion functions are separated and fulfilled by different proteins (see later under VIRAL ENVELOPE PROTEINS . . .). However, in some cases (for example, influenza hemagglutinin), they are part of the same molecule. In the Ca^{2+}–PS system the bivalent ion acts both as a cross-linking agent and a fusogen. In any theory of fusion the aggregation step should be explicitly accounted for. Fortunately, what makes lipid membranes come close together or what holds them apart can at this point be fitted into the framework of a theory. The likelihood of surfaces coming in close contact has been estimated in the framework of the Derjaguin-Landau-Verwey-Overbeek (DLVO) theory (327). This theory has been developed for aggregation of colloidal particles and considers the interplay of attractive van der Waals and repulsive electric double layer forces between charged surfaces in liquids. It allows calculation of the free energy of interaction between the particles, G, as a function of the separation distance between the surfaces, h. The ratio of the number of particles at a distance h to that at an infinite separation, where G is zero, is given by the Boltzmann factor, $\exp[-G(h)/kT]$, where k is the Boltzmann constant and T is absolute temperature. The aggregates of particles are stable—the particles are adhered when $-G/kT \gg 1$. For fusion to occur the membranes should be in a stable apposition, that is, the intermembrane energy of interaction G should be sufficient to overcome the Brownian motion.

The DLVO theory fails to describe interactions between lipid bilayers at short separations (of the order of nanometers). Parsegian, Rand and their collaborators discovered a surprisingly strong, exponentially growing (characteristic decay length of 0.1–0.2 nm) repulsive force between phospolipid bilayers [for reviews see (245, 266)] (Fig. 14.6). This force has been measured for a variety of lipid systems and used for analysis of membrane adhesion. Because of the insensitivity of this force to bilayer electric charges and salts in the solution, it was concluded that it is due to hydration of the polar lipid groups by water. In spite of the numer-

ous studies, the nature of the hydration force is still poorly understood.

The existence of a "hydration" barrier allows a straightforward explanation of the mechanism of action of Ca^{2+} in the PS system as a dehydrating agent. However, dehydration alone is not sufficient to cause fusion, unless it also leads to membrane destabilization. For example, charge neutralization by some monovalent and divalent (Mg^{2+}) ions can give rise to aggregation of bilayers but not to fusion (349). Dehydration by dextran, sucrose, or glycerol commonly does not cause fusion (29). The crucial event in fusion of lipid bilayers seems to be their destabilization or/and formation of defects. It is reasonable to assume that following membrane destabilization or defect formation the nonpolar fatty acid chains will be transiently exposed to water. This can induce hydrophobic attraction between membranes that have such defects. According to Ohki (222) the hydrophobicity of the membrane surface is related to the surface free energy of monolayers. He found that the concentration of divalent cations required to change significantly the surface pressure of lipid monolayers correlates with the threshold concentration needed for fusion of vesicles. Bilayer defects can be formed by Ca^{2+}-induced lateral phase separation, which results in the formation of rigid crystalline domains of acid lipids within a mixed lipid membrane (238, 242). PEG also induces defects. Hui and coworkers (148) have observed "point defects" in bilayers from a mixture of egg PC and soybean PE, which were fused by freeze-thawing. A variety of fusogenic chemicals can fuse membranes by creating defects (189). Commonly the concentrations of those fusogens needed for fusion correlates with their lytic activity. The notion that destabilization is a critical factor in fusion is also consistent with the observations that small unilamellar vesicles (SUV) that have higher curvature and are unstable have an intrinsic higher capacity for fusion (178, 208, 222, 348). The correlation between formation of pores by high voltage electric pulses and electrofusion provides another striking example (354) (see later under Destabilization by High . . .).

The notion that adhesion of membranes is necessary but not sufficient for fusion has a notable exception. This is the case when the adhesion is very strong. As shown by Parsegian and Rand (244) the membrane tension T is proportional to the interaction energy G. With an increase in the absolute value of G, the tension T increases until reaching a critical value (about 3 dynes/cm) where the membranes rupture. Parsegian and Rand suggested that the ruptured membranes can fuse at the contact area [see, for example; (245)]. An example is Ca^{2+}-induced fusion of PS vesicles, where

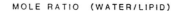

MOLE RATIO (WATER/LIPID)

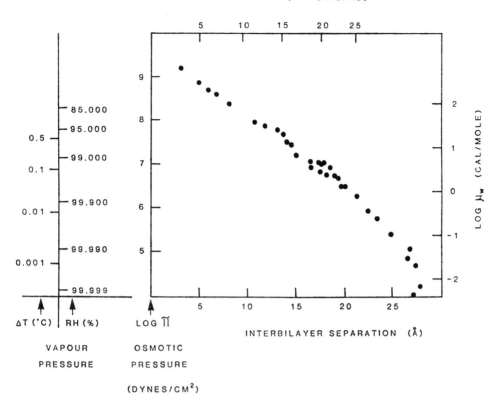

FIG. 14.6. Intermembrane forces between egg phosphatidylcholine lipid bilayers as function of the interbilayer separation. The bilayer separation is also expressed as molar ratio of water to lipid. The intermembrane forces are expressed in four different ways: *(i)* osmotic stress π; *(ii)* chemical potential μ_w; *(iii)* equivalent relative humidity, and *(iv)* temperature increase that would correspond to the same changes in equivalent relative humidity and osmotic stress. Note the tremendous change in the intermembrane repulsion with a decrease in the intermembrane separation: more than 10^5 times for an about ten-fold decrease in the separation. Note also the almost linear dependence of the logarithm of the repulsive force on the separation. [Kindly provided by Dr. V. A. Parsegian, with permission.]

the interaction energy becomes so high (>27 dynes/cm^2) on addition of Ca^{2+} that membranes rupture and fuse extensively. The increase in membrane tension as a mechanical destabilizing agent was found to promote or enhance fusion in a variety of systems, as PEG-induced fusion of cell membranes (166) (see below) and liposome-planar bilayer membrane fusion (62), and in electrofusion (359).

Experiments showing that exocytosis can be inhibited by increasing the osmotic pressure of the medium led to the suggestion that osmotic swelling might be an important step in the granule–plasma membrane fusion (100, 253). Moreover, in mast cells secretory vesicle swelling concomitant with exocytosis has been visualized directly (72). However, a careful examination of the sequence of events in exocytosis of secretory granules in mast cells demonstrate that swelling follows fusion and cannot be causal to fusion (357) (see Fig.

14.5). Similar observations have been reported in the field of viral fusion in that neither intact virus–cell fusion, nor viral envelope protein-induced cell fusion appears to depend on osmotic forces (24, 280).

Fusion of Lipid Membranes by Amphipathic and Nonpolar Molecules Correlates with Their Lytic and Aggregational Activity

A variety of chemicals can fuse membranes (189). Most of them are able to induce either aggregation or lysis or both. Lipids or lipid components fuse cells. Among surface-active lipoidal substances, one of the best studied with regard to its fusing capabilities is lysolecithin (1, 255). It is very lytic and at low concentrations (0.1 mg/mL–1 mg/mL) can induce fusion of cells. Recently, it was found, however, that lysophosphatidylcholine also inhibits fusion of cell membranes (52). Lucy and his colleagues (2, 3, 189) studied a number of other fusogenic agents, as oleoylglycerol and oleic acid, which destabilize membranes and induce fusion. They suggested that when a fusogen such as oleoylglycerol penetrates a cell membrane, the membrane structure is altered leading to an increase in permeability. The increase in permeability is an indication of membrane destabilization, but also may lead to an increase in the intracellular concentration of calcium ions and activation of cellular proteinases. It was hypothesized that the cleavage of intracellular proteins and production of fusogenic peptides then may promote or enhance fusion (190).

Hydrophobic peptides are able to induce fusion of small lipid vesicles [see (24) for a review]. The most interesting peptides in this context are those corresponding to the viral "fusion peptide" sequences (see later under VIRAL ENVELOPE PROTEINS . . .) (226, 265, 339). Synthetic peptides corresponding to the sequence of the amino terminus of the HA2 subunit of influenza virus hemagglutinin (see below) induced fusion of liposomes in a pH-dependent fashion. Protonation of glutamic acid residues in the fusion peptide sequence presumably lowered the barrier to insertion of the peptide into the bilayer (225). The fusion process caused leakage of aqueous liposomal contents. A correlation between the alpha-helical content of peptides and their fusogenicity was noted, but this was not absolute (225, 265, 339). One characteristic of fusion-active peptides appears to be their uneven distribution of side chain bulkiness over the peptides (35, 213). The "hydrophobic and rough" peptides may cause vesicle-aggregation and perturb the closely apposing membranes more effectively to cause fusion. Although lessons about mechanisms of membrane fusion may be learned from studies of interactions of synthetic peptides with lipid bilayers, a note of caution is in order. In the viral fusion sections we discuss studies of cell fusion mediated by influenza hemagglutinin which indicate that peptide–peptide interactions rather than peptide bilayer interactions need to be considered as crucial elements of viral fusion.

Polypeptides—for example, polylysine (115, 329) and apocytochrome C (328)—can also induce fusion of liposomes containing phospholipids with negatively charged headgroups. The extent of polylysine-induced fusion was dependent on the charge ratio between bound polylysine and phosphatidylserine (PS) in the outer monolayer and excess polylysine inhibited fusion (329). The data indicate that there must be sufficient sites on the vesicles and sufficient polypeptide to achieve effective aggregation. For fusion to occur after aggregation, charges on the vesicles must be neutralized either by polypeptide–PS interaction or by protonation

of the PS carboxyl groups that lead to destabilization. We learn from these studies that a delicate balance must be struck between requirements for the polypeptide to cross-link apposing membranes to cause fusion, and the opposing tendency to limit fusion by steric hindrance when the polypeptide is in excess.

Dehydration, Aggregation, and Destabilization of Membranes by Polyethelene Glycol Are Essential for Fusion of Lipid Membranes

Polyethylene glycol (PEG) of molecular weight 1 kDa or higher strongly binds and structures water. Nuclear magnetic resonance (NMR) measurements indicate that all water molecules change mobility in a solution of > 13% PEG (17) and no mobile water is detectable above 48% 6 kDa PEG (323). The dielectric constant of PEG–water solutions decreases from 80 to 50 with an increase in the PEG concentration from 10%–50%, but the polarity of the solution is still higher than the solvent polarity for formation of lipid bilayers (dielectric constant < 35) (11).

Polyethyle glycol induces aggregation and destabilization of lipid bilayers, predominantly by dehydration and decreasing the polarity of the solution. Even very small concentrations of PEG (2.5%) lead to aggregation of egg phosphatidylcholine (PC) small unilamellar vesicles (SUV) (30). The aggregation is reversible for PEG concentrations below 20% (30). Transfer of lipid molecules between SUV (145) and various ultrastructural defects (31, 32) was observed at higher PEG concentrations. The ultrastructural defects may indicate exposure of hydrophobic portions of the lipid bilayer.

PEG has been widely used to fuse a variety of cells (4, 34, 157, 254, 257). While fusion of cells requires incubation of cells with relatively high PEG concentrations (40%–50%) and subsequent dilution, fusion of lipid vesicles can be achieved at lower concentrations (smaller than 40%) and without dilution [see the review (145)]. It was proposed that high PEG concentrations are needed to provide regions of cell membranes free of proteins, probably by protein aggregation. This is consistent with EM observations showing lack of intramembrane particles (IMP) and enhanced PEG-induced fusion after protease treatment of cells. The dilution of cells after treatment with PEG leads to osmotic swelling, which can further destabilize the membranes (24). This is consistent with observations that incubation of erythrocyte ghosts in hypotonic solutions after PEG treatment enhanced fusion (146).

The observation that PEG-induced fusion of cell membranes requires removal of the proteins from the sites of fusion, as well as an additional destabilizing stress, shows that membrane proteins provide barriers for fusion in this particular case, probably because of steric hindrance. Therefore the lipid composition of the membranes is the critical factor in the efficiency of PEG-mediated fusion. It was found that mouse fibroblasts resistant to PEG fusion have elevated levels of neutral lipids, particularly triglycerides and an ether-link diacylglycerol, and a higher percentage of saturated fatty acyl chains (269, 270). It has been shown that rates of influenza fusion with membranes of liposomes (325) and erythrocyte ghosts (129) also decrease with an increase in the degree of saturation of acyl chains in phospholipids. Enhanced rates of fusion with lipids containing small headgroups and *cis*-unsaturated acyl chains are attributed to reduction of the dehydration energy of the membrane surface (266), or the increased propensity to form curved structures (53). However, there was no difference in fusion of unmodified cells with intact influenza virions compared to fusion with PEG-resistant cell lines (262). We learn from these observations that the virus chooses its appropriate domain of lipids from the plethora of lipid species in the mammalian plasma membrane.

Destabilization by High-Voltage Electric Pulses Leads to Fusion of Adjoining Membranes

During the past decade numerous studies have shown that external electric fields can induce fusion of a wide variety of cell and artificial membranes [for review see (49, 81, 219)]. This experimental observation is a demonstration of an inherent ability of membranes to fuse if appropriate conditions are provided and indicates existence of properties of membrane systems, related to fusion, that are largely independent of the type of membranes. These properties include membrane stability and adhesion.

External direct current (DC) fields can destabilize membranes and induce formation of pores (electroporation) [for recent review see (49)]. In the late 1960s and the early 1970s it was found that application of high-voltage direct current (DC) pulses to cell suspensions leads to killing of bacteria and yeasts (277), lysis of erythrocytes and protoplasts (279), release of catecholamines and ATP from chromaffin granules (216), and transcellular ion flow in bacteria (361) [for early electroporation data see (279)]. This was originally attributed to electric breakdown of the cell membrane, which implies irreversible rupture of the membrane (278). Later it was shown that the membrane permeability changes can be transient in nature (216) and that they can be analyzed in terms of reversible dielectric breakdown (360). The first electroporative gene transfer into living cells with the subsequent actual expression of the foreign gene (218) led to explosive development of the studies on interactions of membranes with high voltage pulses. The term electroporation was introduced (218) and presently generally accepted to refer not only to the phenomenon of formation of pores but also to all pore-related events caused by exposure of membranes to high field strengths (333).

In 1979 Senda and coworkers published a paper describing an interesting observation (294). When an electric field was applied by microelectrodes to two plant protoplasts, brought into close contact by using a micromanipulator, the cells underwent morphological changes until they formed a single fusion product as observed by light microscopy. At about the same time three other research groups fused cells by high-voltage electric pulses but using different approaches to bring the cells into close contact. Berg's group used polyethylene glycol (PEG) to aggregate cells and then stimulated yeast protoplast fusion with an electric field (334). The evidence for fusion was genetic because the pulses led to formation of viable hybrids that grew on minimal media to form prototrophic colonies. Neumann and colleagues (217) achieved cell agglutination by rolling the cell suspension in plastic tubes, while Zimmermann and Scheurich (358) used alternating current (AC) fields to bring plant protoplasts at close approach by dielectrophoresis.

The common and striking feature of these first observations of electrofusion was that the magnitude of the transmembrane voltage needed for fusion was about the same (of the order of several hundred millivolts to 1 V) as that for electroporation, and electrofusion was insensitive to the way the cells were brought into contact. Later it was demonstrated that not only the absolute value of the transmembrane voltage correlates with that needed for electroporation, but also the entire functional dependence of the pulse voltage on its duration is the same (354) (Fig. 14.7). It was also found that the membranes still fuse if they are electroporated first and then brought into contact (303, 320), which indicates existence of long-lived fusogenic states. These observations led to the important conclusions that destabilization of membranes is critical for fusion, that pores may be involved as fusion intermediates, and that the fusion is largely independent on the way the membranes are brought at close contact.

Molecular Rearrangements in the Lipid Bilayers during the Very Act of Fusion May Involve Intermediate Structures

Membrane contact is required for fusion and membrane destabilization is the critical event in triggering it; formation of intermediate structures, however, may

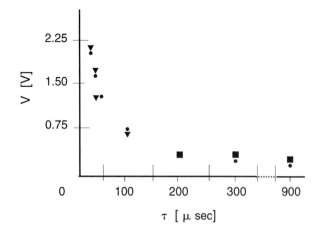

FIG. 14.7. Correlation between the threshold transmembrane voltage V of electrofusion *(triangles)*, of electroporation *(circles)*, and of cell destruction *(squares)* as function of the pulse duration τ. The square of the threshold voltage is inversely proportional to the pulse duration in agreement with the fluctuation wave mechanism of electroporation. [From (354).]

be needed to decrease the energy barriers of fusion. Because of the very small amount of membrane lipid involved in fusion pores and their very short lifetimes, direct measurement of the intermediate structures in fusion of lipid bilayers is very difficult, if not impossible. Some progress was made by using NMR for "bulk" fusion of membranes (171, 172, 319) and cryomicroscopy (297). The lack of direct experimental measurements of lipid bilayer fusion hinders the understanding of the molecular mechanisms of fusion. In contrast, there are several theoretical models that try to calculate intermediate structures requiring minimal activation energy for their formation (169, 296, 318). These models are helpful in the design of new experiments for better understanding of the molecular rearrangements during fusion. Figure 14.8 shows hypothetical intermediates in bilayer fusion as suggested by a theoretical analysis (53, 296).

SPECIALIZED PROTEINS MEDIATE FUSION IN LIFE PROCESSES

Occasionally, cell membranes can fuse spontaneously (267), most of the physiological processes involving fusion require specific recognition and control, which implies participation of proteins. Numerous studies have shown that proteins mediate viral fusion and are involved in a variety of other life processes involving fusion [see the books (21, 97, 224, 304, 350) and the review articles (24, 27, 134, 194, 249, 340, 342, 345, 353). While viral fusion proteins are the best character-

ized, recently a number of proteins were implicated in fusion during fertilization, myotube formation, exocytosis, and intracellular fusion. In other processes such as cell division and formation of osteoclasts, the role of specific proteins remains to be elucidated.

VIRAL ENVELOPE PROTEINS CONTAIN HYDROPHOBIC "FUSION PEPTIDE" SEQUENCES

Many enveloped viruses assemble their outermost coat in the plasma membrane of the host cell by a budding process of progeny particles. The viruses vary in appearance from rod-shaped to spherical, with diameters of 100 nm–200 nm. The lipid components of these enveloped viruses are nearly identical to those found in the plasma membrane of the host cell; the proteins, however, are virus-specific. Significant advances have been made in recent years in elucidating the role of viral spike proteins in inducing membrane fusion. They include: *(i)* The first high-resolution image of a membrane fusion protein by x-ray crystallography (352); *(ii)* the elucidation of the primary sequence of a large number of viral membrane proteins using DNA sequencing techniques (120); *(iii)* the development of genetic and chemical methods for site-specific alterations of viral protein structure (34, 110, 119, 141, 142, 309); *(iv)* the recognition of the essential role of specific receptors for fusion in the target membrane (65); *(v)* functional reconstitution of viral spike glycoproteins into lipid vesicles (26); and *(vi)* development of biophysical techniques such as fluorescence and electron spin resonance and electrophysiological techniques to monitor fusion [for a review see (21)].

Some of the enveloped viruses enter cells via the endocytic pathway (342), and subsequently fuse with the membrane of the endocytic vesicle. In the case of influenza and many other viruses the fusion is triggered by the low pH in the endosome. With other virions (e.g., human immunodeficiency virus [HIV], see below) the fusion is triggered at neutral pH by a viral envelope protein–cell surface receptor interaction. In the latter case it has not been established whether the infectious route of entry is via the endosome or directly at the plasma membrane.

The fusion process is mediated by a wide variety of envelope glycoproteins. There is no sequence homology between fusion proteins of different families, which argues for the lack of a single ancestral gene (194). Perhaps they evolved from cellular recognition and receptor molecules that later evolved to fusion proteins. All viral fusion proteins are integral membrane proteins with their N-termini external and their C-termini inter-

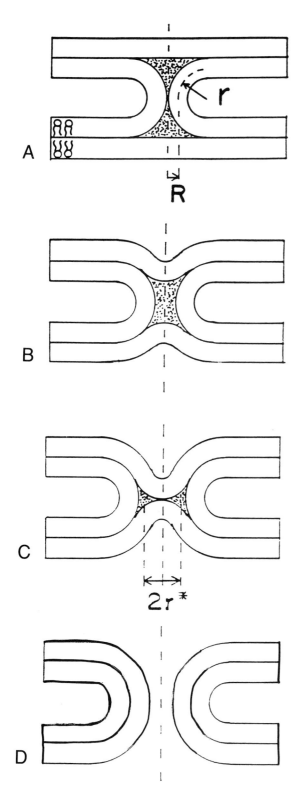

nal to the membrane, and 90%–95% of the protein mass is external to the membrane.

The viral envelope proteins are usually activated by post-translational proteolytic cleavage that releases a hydrophobic peptide, the so-called fusion peptide, which is conserved within, but not between, virus families. Some virus families (for example; rhabdoviruses) do not have a cleaved N-terminal fusion peptide. In the case of the vesicular stomatitis virus (VSV) envelope glycoprotein an internal sequence of 19 uncharged amino acids (residues 118–136) has been proposed as a possible "fusion peptide" (225, 343).

To Enter a Cell a Virus Must Find the Receptor That Invites It In

Because virus infection is in many cases selective with respect to the host, the virus must interact with the cell surface in a specific way. The specificity of viral entry is commonly achieved by specific ligand–receptor interactions. Viruses exploit cellular surface molecules as receptors for binding and entry into host cells. The binding of viruses has been studied extensively in relation to fusion. Viruses utilize a variety of proteins, lipids, or oligosaccharides as receptors [for review see (65, 70, 195, 216, 313, 338). Two approaches have been particularly productive in identifying such viral receptors: antibodies to cell surface antigens have been generated to block viral attachment or infectivity, and DNA transfer has been used to recover genes that confer infectivity by the virus to otherwise nonpermissive cells. Thus CD4 was first identified as the receptor for HIV by screening antibodies to leukocyte antigens, followed by cloning the cDNA for CD4 and transfection into non-permissive cells (see below). Likewise, further receptors have been identified, for example CR2 for the Epstein-Barr virus (203), and the laminin receptor for Sindbis virus (332). Receptors for human viruses have been identified that belong to the immunoglobulin superfamily [for example; CD4 for HIV (see later under The Receptor CD4 Plays . . .), ICAM-1 for rhinoviruses (183), and an immunoglobulin-like receptor for poliovirus (11)]. Recently an integrin called VLA-2 has been identified as the receptor for echovirus-1 (22). (It should be noted that rhinovirus, poliovirus, and echo-1 are non-enveloped viruses.) A

FIG. 14.8. Hypothetical intermediates in fusion of lipid bilayers. The monolayers of the membranes are drawn as slabs, and all the structures are cylindrically symmetric about the vertical axis. A. The stalk intermediate (53). The structure is a catenoidal monolayer sandwiched between two flat monolayers. r is the minor radius of the catenoid. B. Evolution of the stalk toward an activated stalk intermediate. The flat monolayers start to dimple, and the radius of the catenoid in the plane of the membranes expands. C. Activated stalk. The dimples of the trans monolayers meet, and the two axially symmetric voids transform into a ring of trigonally symmetric void of radius r^*. The high curvature stress, and the stress due to stabilization of the voids, is concentrated in a small monolayer area of the two dimples, driving rupture of these two monolayers to form a fusion pore (D) [From (296).]

striking finding during the past years was the identification of membrane transporters as receptors for the ecotropic murine leukemia virus (330) and gibbon ape leukemia viruses (155), which turned out to be an amino acid and phosphate permease, respectively. This finding opens up the possibility of an alliance between cellular physiologists and virologists, which might pay handsome dividends in unveiling the dual roles of particular cell surface molecules. To allow entry into the cell the receptor plays a passive role, ligand binding, and in some cases an active role, triggering a conformational change in the envelope protein that sets the fusion cascade in motion. Below we shall discuss in more detail the dual role played by CD4 in mediating entry of human immunodeficiency virus type 1 (HIV-1) into cells.

Some Viruses Require More Than One Type of Envelope Protein for Entry

For a number of virus families it was demonstrated that a single protein is sufficient to induce membrane fusion as indicated by syncytia formation following cloning and expression of the gene products of viral envelope proteins on the plasma membrane of cells. This is the case for orthomyxoviruses, rhabdoviruses, and retroviruses.

In the case of paramyxoviruses that contain two glycoproteins (HN, hemagglutinin-neuraminidase, and F, fusion factor) in the outer leaflet of the lipid bilayer (228), the situation is not so clear. The F protein is absolutely required for the entry of viruses into the host cells and for cell–cell fusion (287). The HN binds to terminal sialic acid containing receptors and is a neuraminidase (228). A variety of fusion studies indicate that both HN and F are required for paramyxovirus-induced cell fusion (99, 144, 187, 211, 228, 276, 344), although the F protein alone has been reported as sufficient to mediate membrane fusion (6, 142, 143, 192, 246, 292). In the latter case the binding function was provided by adding an external ligand, such as wheat germ agglutinin (143) or antibody (292), or by using HepG2 cells as targets, which contain a cell surface receptor (the asialoglycoprotein receptor (14), which binds to the F protein (193)). A recent study with viral envelopes reconstituted with F + HN or F alone indicates synergy between F and HN in mediating membrane fusion (15). This result might form the basis for an explanation of why different laboratories obtain different results on membrane fusion mediated by the F protein alone. It will be of interest to learn what parts of the proteins are responsible for the specific interactions (144) that lead to synergy between the proteins.

The Semliki Forest virus (SFV) envelope consists of three glycoproteins, E1, E2, and E3 [for review and references see (194, 342)]. E1 (50 kDa) and E2 (50 kDa) are transmembrane proteins; E3 (10 kDa) is noncovalently associated with these on the outer side of the viral membrane. E2 and E3 are derived from a precursor, p62, which is cleaved before the protein reaches the plasma membrane, but this cleavage is not required for fusion activity. There is no obvious hydrophobic fusion sequence, but E1 includes a conserved uncharged domain. Both E1 and E2 undergo a conformational change at low pH, which is required for fusion. E1 and E2 both may be involved in the fusion reaction; E3 may not be essential. The three envelope glycoproteins form a noncovalently associated heterononamer. Fusion of SFV requires cholesterol in the target membrane, which interacts with E1. Fusion is pH-dependent, very rapid, and efficient.

The envelope of herpes simplex virus (HSV), which fuses at neutral pH at the plasma membrane (305), consists of six membrane glycoproteins. Fuller and Lee (114) have suggested that specific interactions of one of those components of the virion surface and a cell surface receptor may trigger a conformational change that exposes fusogenic domains of other viral proteins. Thus, the fusion process might be the result of a cooperative sequential chain of events requiring several different viral proteins.

In the following we will describe in detail two examples of fusion mediated by single viral envelope proteins, whose mechanism have been most extensively studied: influenza hemagglutinin (HA) and HIV-1 gp120-gp41. They represent two different categories of triggering mechanism; HA by low pH and gp120-gp41 by a receptor–ligand interaction, which sets the fusion cascade in motion. Many of the experimental approaches, hypotheses, and putative mechanisms could be applied to the larger family of fusogenic envelope glycoproteins.

Influenza Hemagglutinin Was the Only Fusion Protein with Known Three-Dimensional Structure

The influenza hemagglutinin protein (HA) is the best-known among the viral envelope proteins, its structure having been elucidated over ten years ago (352). Moreover, a lot of knowledge has been gained about its cell biology, receptor interactions, and pH-regulated conformational changes. Practical advantages such as its stable expression in cell lines and the ease of regulating its fusion activity (with protons) combine to make it an appropriate tool for a detailed study of the molecular mechanism of viral envelope protein–mediated fusion.

Cloning of HA and expression of the gene product on the plasma membrane of cells demonstrated that this protein was sufficient to induce polykaryon formation with a pH dependence-similar to that observed for fusion of intact virus with cells (120). Hemagglutinin is a homotrimer (3 × 84 kDa) in which every subunit consists of two disulfide-linked glycoproteins, HA1 and HA2, derived by proteolytic cleavage of a common precursor HA0 (Fig. 14.9) (352). A change in tertiary structure of the HA molecule (or a soluble form, BHA1) in response to a decrease in pH has since been demonstrated by a number of techniques. These include the morphology of spike glycoproteins observed by

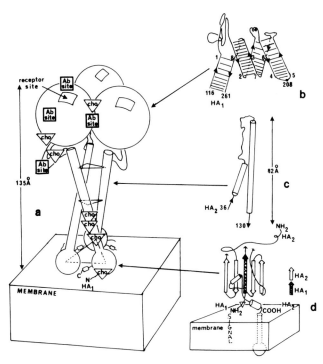

FIG. 14.9. Crystal structure of soluble hemagglutinin (HA). *a.* Schematic diagram of the 1968 hemagglutinin trimer showing carbohydrate attachment sites *(CHO)*, antigenic sties *(Ab site)*, the host-receptor binding site and the arrangement relative to the membrane. *b.* The eight-stranded β-sheet structure and looped-out region in the globular domain. *c.* HA$_2$ residues 36–130, including two α-helices (cylinders), form part of the fibrous region. Three of the long helices, one from each monomer, pack together as a triple-stranded coiled-coil that stabilizes the timer. *d.* The membrane end of the molecule contains a five-stranded β-sheet. The central strand is the N-terminal of HA$_1$ and the adjacent strands come from the C-terminal chain of HA$_2$. Broken lines suggest the path of the hydrophobic anchoring peptide cleaved by bromelain. The possible attachment of the N-terminus to the membrane by the signal sequence is also indicated. The site of the first oligosaccharide chain (residue 8) is shown as a triangle. Note the large separation (more than 8.2 nm) of the end of the fusion peptide from the globular domain that contains the receptor binding sites. [Kindly provided by Dr. D. C. Wiley with permission (352).]

cryo-electron microscopy, susceptibility to proteases, tryptophan fluorescence spectroscopy, and immunoprecipitation (93, 273, 299, 341). The interrelationships between the conformation detected by those techniques and the fusogenic conformation is unclear. There is, however, broad agreement that exposure of the HA2 amino terminal domain is somehow crucial to fusion (119, 122, 161). The first 20 or so residues of HA2 are highly conserved, relatively hydrophobic, and considered analogous to putative fusion peptide sequences found in other viral spike glycoproteins (36). This "fusion" peptide is buried at the interface between monomers in the neutral pH trimer structure, 35 Å from the virus bilayer surface and 100 Å from the tip of the trimer. White and Wilson (341) used a panel of monoclonal antibodies to probe the low pH induced conformational change of BHA and showed that this peptide becomes rapidly exposed, as well as demonstrating dissociation of the globular top domains. These studies inspired many cartoons showing the HA trimer unfurling like a flower to expose a fusion peptide stamen. Later studies have shown that neither fusion nor exposure of the fusion peptide require the extensive conformational change associated with top dissociation (260, 310). However, recent studies with cysteine mutants that link the tops of the trimer indicate that some movement of the tops is required to unleash the fusion peptide (122, 161). The extensive conformational change has been associated with low pH-induced inactivation of fusion activity rather than with fusion per se (260). For several strains of influenza A viruses it has been shown that incubation of virus alone under low pH conditions eliminates their fusion capacity over time (minutes) (291). This phenomenon is not observed with the A/Japan/305/57 strain of the virus, nor is the more extensive conformational change, as defined by an unaltered spike morphology observed by cryo-electron microscopy (Fig. 14.10) (260). Until very recently the HA was the only fusion protein with a known 3-D structure. Recently, however, Rey et al. (266a) determined the structure of the external glycoprotein of tick-borne encephalitis virus. They proposed that, unlike the spikes seen on influenza and many other viruses, the flavivirus envelope protein is lying down. This shows the variety of structures viruses can use to get into their prey. The number of conformations available to HA under low pH conditions when virus is bound to biological membranes appears to be quite high, which leads to complicated kinetic schemes to account for experimental observations. The receptor for influenza virus can be any cell surface molecule with oligosaccharides terminating in N-acetylneuraminic acid (sialic acid). Sialic acid residues fit into a shallow pocket on HA1 at the tip of the molecule (336). Although the binding constants are in the millimolar range (286) statistical co-operativity leads to tight association of virus with cells. In the following we shall describe how triggering of viral fusion may set a complex set of events in motion that finally terminates in the coalescence of membranes and aqueous spaces.

The Process of HA-mediated Membrane Fusion Can Be Dissected into a Number of Elementary Steps

Figure 14.11 shows the various steps in the pathway that have been resolved kinetically using a variety of biophysical techniques (28, 56, 158, 209, 260, 280, 290). It should be noted that the steps depicted in Figure 14.11 have been defined operationally, in terms of experimental maneuvers to identify them, and they do not represent a detailed molecular description of intermediates in membrane fusion. Although most of the data on fusion pores are derived from fusion of

erythrocytes loaded with fluorescent dyes with influenza hemagglutinin–expressing cells, the general conclusions were corroborated by experiments of fusion of single virions with cells (185). Low pH–induced conformational changes produce activated HA that has been examined by a variety of biochemical, biophysical, and morphological techniques (260, 299, 340, 341). Although many states of activated HA have been proposed (260, 341), it is generally agreed that liberation of the fusion peptide is a minimum requirement for fusion (122, 161). The appearance of the fusion peptide is depicted in activated HA in Figure 14.11. The states following HA activation are depicted as aggregates of trimers, based on evidence that more than one trimer is needed to induce HA-mediated fusion (102, 209). However, because the delay in HA-mediated fusion was not dependent on a higher power of HA surface density (56), it is unlikely that aggregation of HA trimers forms part of the rate-limiting step in stages subsequent to activated HA.

pH 7.4   pH 4.9

JAPAN

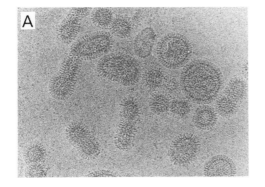
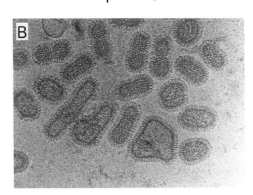

x31

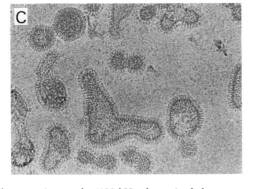
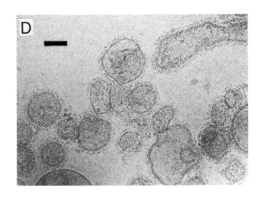

FIG. 14.10. Electron micrographs (100 kV) of unstained, frozen, hydrated influenza virus. *A* and *B*. Virus from the Japan strain. *C* and *D*. Virus from the X:31 strain. Panels A and C contain virus at pH 7.4 and 37°C; panels B and D contain virus incubated for 15 min at pH 4.9 and 37°C and subsequently neutralized. Note the morphological changes in the envelope spikes of the X:31 strain at the low pH and the lack of visible changes in the spikes of the Japan strain. [From (260).]

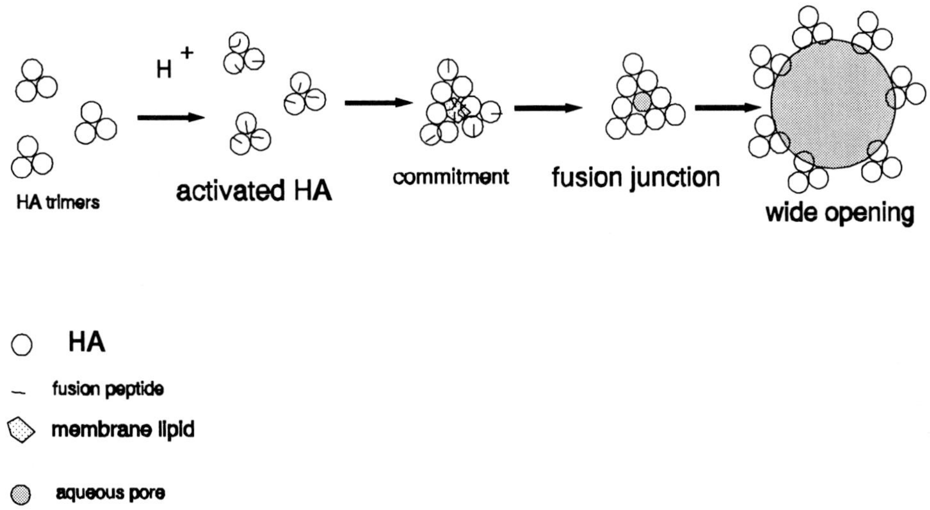

FIG. 14.11. Cartoon of steps in HA-mediated fusion. The HA trimers are activated by the hydrogen ions. The activated HA trimers undergo a complex cascade of phenomena to form a committed state after which the fusion proceeds even if the pH is reversed back to neutral. Then fusion junctions and pores form that further expand to a wide opening.

Commitment is defined as the step in HA fusion initiated by low pH triggering, but which cannot be stopped by returning to neutral pH (158, 209). At 37°C and optimal pH the $t_{1/2}$ for commitment is about one-third of the delay time. However, at low temperatures the time differential is larger, in that commitment can occur within 1 to 2 minutes (290), whereas the subsequent delay will take about 10 minutes (290, 311). The committed state is insensitive to biochemical maneuvers (e.g., treatment with neuraminidase, trypsin, or DTT) that affect HA1–target membrane interactions, but is inactivated by treatment with enzymes (for example, thermolysin, proteinase K) that act on HA2 (290). Moreover, following low pH activation at low temperatures but prior to lipid mixing, HA2 is labeled by photoreactive lipid probes inserted into the target membrane (310). Therefore, the committed state represents interaction of HA2 (most likely the fusion peptide) with the target membrane. Dissociation between conformational changes and commitment on the one hand, and the lag time leading to fusion pore formation on the other hand was seen with the X31 strain of influenza virus. With other stains [for example, PR/8 of influenza A (232), and influenza C (108)] the conformational change appears to be rate-limiting.

A striking observation concerning intermediates in the process of HA fusion is the formation of small fusion junctions (see earlier under Electron Microscopy . . .) that allow movement of lipid and small molecules but that restrict the movement of large molecules (280, 308). The flow of lipid and of protein is also restricted in the intermediate stage (184, 185). The fusion junc-

tion has been probed in more detail by capacitance patch clamp studies of fusion between HA-expressing cells and erythrocytes (307, 308). Changes in capacitance (a membrane area) have been monitored to detect fusion and in conductance properties to estimate the size of an aqueous pore linking the two cytoplasms. An initial conductance jump of about 150 pS is observed, which grows over subsequent tens of milliseconds where it is temporarily arrested at a conductance of about 600 pS. The conductance jump appears to precede lipid movement (323a, 357a). This conductance is consistent with a pore of molecular dimensions (no more than twice the diameter of a gap junction channel). The pore is by no means a stable structure: frequently "flickering" is observed between conductances of 0 and 600 pS as if the pores are opening and closing repeatedly. No direct evidence exists for involvement of HA proteins as structural elements of the initial fusion junction, although membrane-associated proteins are presumably playing some role in limiting the expansion of the pore. Both the capacitance measurements and the fact that lag times obtained with small water-soluble dyes are within the range obtained with lipid fluorophores (158, 280) indicate that the fusion junction is a small pore and not merely a lipid mixing intermediate.

The wide opening associated with nucleocapsid delivery can be measured in the erythrocyte-HA-expressing cell system by transfer of large molecules. Fusion of dextran-loaded erythrocyte ghosts with HA-expressing cells leads to delivery of 30% to 40% fluorescently labeled dextran under conditions where all bound erythrocytes transfer a lipid dye (290, 308). Moreover, transition from fusion junction, which allows all of the lipid dye to move, to the formation of the large opening was aborted with HA carrying a single site mutation in the fusion peptide (290). This is consistent with the notion that interaction of the fusion peptide with the target membrane can give rise to lipid perturbations and transient pores, sufficient to produce full lipid redistribution. This interaction does not appear to be very dependent on the structure of the fusion peptide. Surprisingly, however, the fusion peptide seems to play an important role in later processes related to pore widening, which eventually results in delivery of the nucleocapsid into the cell.

These finding are consistent with a model for HA-mediated membrane fusion which assigns an important role for the fusion peptide in stabilizing HA trimer assemblies by interactions of some of the fusion peptides with the transmembrane portion of the viral envelope protein while simultaneously other fusion peptides interact with the target membrane to induce a curved membrane structure (127) (see Plate 14.2, facing page 565. Whereas the initial fusion peptide–lipid interactions, which give rise to fusion pores and lipid mixing, are relatively nonspecific, stabilization of the pore and its widening might require more specific peptide–peptide interactions that are determined by the fusion peptide sequence. Recently, an interesting "spring-loaded" model was suggested as a mechanism of influenza fusion (43a). This model involves formation of a coiled-coil structure which was also demonstrated by solving the 3-D structure of the HA2 at low "fusion" pH (39a). Only further experiments will show whether this model can explain the characteristic features of HA-mediated fusion. It is now possible to test such models by the combination of recombinant DNA and biophysical techniques, and thereby gain further insight into the mechanism of HA-mediated membrane fusion.

Human Immunodeficiency Virus Type 1 (HIV-1), the Primary Etiological Agent of the Acquired Immunodeficiency Syndrome (AIDS), Enters Cells by Membrane Fusion at Neutral pH

Entry of HIV-1 into cells is initiated by the interaction of the envelope glycoprotein gp120–gp41 with its receptor, CD4, on the target cell membrane (18, 116, 125, 176, 191, 197, 247, 298, 312). Infected cells frequently form giant multinucleated cells (syncytia) (162, 256) by fusion of their membranes (179, 180). Cell fusion is a potentially important mechanism of the cytopathic effects of HIV (179, 180, 300), which are considered to be the main cause of the defects of the immune system leading to AIDS (105), and is involved

in HIV-1 transmission (92, 282). The *env* gene of HIV-1 encodes a polypeptide (gp160) of about 840–860 amino acids, the precise length depending on the isolate. This is folded in the endoplasmic reticulum into a complex, disulphide-linked structure that is maintained in the endoplasmic reticulum membrane by virtue of the hydrophobic, membrane-spanning domain in the gp41 moiety. During post-translational processing, the precursor glycoproteins oligomerise and are extensively glycosylated before a cellular proteinase cleaves the precursor at a characteristic sequence to create the mature glycoproteins gp120 and gp41. Several post-translational modifications to the *env* precursor—disulphide-bonding, glycosylation, cleavage, and oligomerisation—are necessary for attainment of full biological activity of the mature virus, although uncleaved and partially glycosylated gp160 can bind CD4 (204). In endoplasmic reticulum and plasma membranes, the carboxyl-terminus of gp41 is intracellular and the amino-terminus extracellular (168). Gp120 is attached onto the extracellular domain of gp41 by noncovalent bonds whose strength is altered by single site mutations in the binding domains of gp120 and gp41 (168). The *env* glycoproteins are packaged onto virions by a process that is imperfectly understood. The virions bud out from the cell membrane and vesicle membranes, which form the surface membrane of the mature virion (118). The surface of the mature virion is studded with up to 72 glycoprotein knobs or spikes (118) that are, at least for the commonly studied laboratory-adapted isolates of HIV-1, prone to dissociation from the virus (204).

A high affinity interaction between the viral glycoprotein (gp120–gp41 complex) (131, 346) and the cellular receptor CD4 (74, 162) is necessary for both virus–cell (175, 191, 198) and cell–cell fusion (13, 179, 300). In vitro, HIV-1 can also infect cells of certain lineages that do not express detectable levels of CD4 on their surfaces, and in which HIV-1 infection is not blocked by anti-CD4 antibodies (58). In general, these routes of infection are inefficient and their significance for HIV-1 infection in vivo is uncertain. Because even less is known about the mechanism of CD4-independent infection than about the CD4-dependent route, we will limit our discussion to the CD4-dependent entry. Although HIV-1–cell fusion can be detected by electron microscopy and fluorescence dye redistribution within a few minutes after formation of a stable CD4–gp120–gp41 complex (91, 125, 298), virus entry is generally considered to take place relatively slowly over a period of several hours, the rate depending on the virus and cell type (107) with virus binding being the rate-limiting step (230). The HIV-1–cell fusion reaction is activated at neutral pH by CD4 binding, which may have an analogous function to the low-pH activation of influenza virus fusion. Although electron microscopic observations indicate that HIV-1 virus–cell fusion occurs at the cell surface as well as in internal vesicles (125), it is not yet clear what the infectious route of entry is.

The Receptor CD4 Plays Both a Passive and an Active Role in Allowing Entry of the Virus into the Cell

CD4 is a cell surface glycoprotein of 55kDa molecular weight. It is a member of the immunoglobulin superfamily and consists of four extracellular immunoglobulin-like domains anchored to the cell surface by a transmembrane region followed by a short cytoplasmic domain (204). CD4 is associated predominantly with the lineage of T-lymphocytes known as helper T cells, but has also been detected on other cells of the immune system, and in certain nonhematopoietic tissues (285). Genetically engineered CD4 molecules have been expressed either as membrane-anchored proteins or as truncated, soluble derivatives (sCD4). SCD4 has proved to be a useful reagent to probe the details of viral envelope protein–membrane receptor interaction that fits the ligand–receptor paradigm (204). A plethora of studies have been carried of binding of sCD4 to intact virus as well as to cells expressing the env glycoprotein. Equilibrium dissociation constants fall in the range of 1—10 nM, with a forward rate constant of 10^5 M^{-1} s^{-1} and dissociation rate constant of 3.3×10^{-4} s^{-1} at 37°C (87). Surprisingly, there is a break in the forward rate constant at about 18°C (87).

Initial studies with monoclonal antibodies (mAbs) localized the region of CD4 interactive with gp120 to that recognised by mAbs clustering with Leu3a and OKT4a (12). Mutational analysis together with antibody epitope mapping have been integrated into the known structure of the first two domains of CD4, defined by X-ray crystallography (274, 331) (see Plate 14.3, facing page 564). The high-affinity binding site for gp120 lies within the first, immunoglobulin-like domain (D1) of CD4 (204). It consists primarily of a stretch of about 20 amino acids (residues 40–60) in which a β-strand (the C″ strand) is linked at one end by the CDR-2 homologous loop (CDR-2 loop) to the C′ strand, and at the other end to the D strand. This structure composed of three β-strands probably forms the backbone of a ridge that may bind into a cleft in gp120.

Regions outside the high-affinity gp120 binding site have been reported to play a role in HIV-1–env–mediated fusion. Specifically a second site on CD4/D1, corresponding to the CDR-3 homologous region (CDR-3 loop), has been identified based on inhibition studies using peptides synthesized from the CD4 sequences between 81 and 101, site-directed mutagenesis, and studies with mAbs that had been epitope mapped to that region (204). However, recent data from Broder and Berger (38) indicate that CD4 molecules with a diversity of mutations encompassing the CDR3 region efficiently support HIV-1–env–mediated cell fusion.

Expression at the surface of CD4 cells of chimeric molecules consisting of domains 1 and 2 of CD4 fused either to the hinge, transmembrane region, and cytoplasmic tail of CD8 or to a glycolipid anchor conferred susceptibility to HIV-1 infection (154, 258). However, the efficiency of infection and rate of syncytia formation with the CD4/CD8 chimera was greatly reduced (123). Therefore, it appears that regions beyond the first domain are important for fusion although they do not participate in gp120 binding. Further evidence for this contention comes from studies with mAbs that inhibit HIV-1 infection and fusion of CD4$^+$ cells without interfering with the binding of virus or recombinant gp120 (41, 44, 128, 283). Thus the entire extracellular complement of CD4 is probably required for efficient HIV-1 infection and syncytium formation, the membrane proximal domains being involved subsequently to virus binding to D1. It seems that in addition to its induction of essential conformational changes in the HIV-1 envelope glycoproteins, CD4 may also need to alter its cell surface configuration to efficiently mediate HIV-1-membrane fusion. Since sCD4 has been shown to be an extended, rod-like molecule of about 125Å in length (33), and fusion of two membranes requires greater proximity than this, CD4 may need to "collapse" onto the cell surface after HIV-1 binding. In addition, clustering of CD4 may be required at the contact point between the virion and the cell in order to form a fusion complex capable of initiating membrane coalescence (see later under Stable Envelope . . .).

Because human CD4 inserted into nonhuman cells will not support HIV-1 entry into those cells (13, 39, 94, 191), additional components in the target membrane appear to be required for fusion. Experiments with human–nonhuman hybrid cells indicate that components of the human cell membrane are necessary to overcome the block in HIV-1–env glycoprotein-mediated fusion of nonhuman cells (39, 94). Candidates for molecules in the target membrane accessory to CD4 are membrane-embedded proteins, glycoconjugates and/or phospholipids. Several cell surface proteins have been proposed to interact with specific regions of gp120 and gp41 (51, 263). Recently, fusin (105a) and CC-CKR-5 (6a, 75a, 93a), were identified as coreceptors for T-cell line and macrophage tropic HIV-1 isolates, respectively. They belong to the family of G-protein–coupled seven-transmembrane-domain proteins and have short extracellular portions. One might speculate that the coreceptors interact with the CD4-gp120-gp41 complex, helping to expose the gp41 fusion peptide which, in concert with fusion peptides from other gp41 molecules, induces fusion [Fig. 14.12; (81a)].

The notion that CD4 binding to virions leads to activation of the fusion pathway via the induction of conformational changes in the envelope glycoproteins was based on the observation of sCD4 enhancement of SIV_{agm} infection (7), and of sCD4-induced shedding of gp120 from gp41 (204). Although such shedding represents a gross conformational change in the HIV-1 envelope, it does not appear to be necessary for fusion, for a sCD4-resistant HIV-1 mutant retaining gp120 after sCD4 binding is highly fusogenic (199, 321). Furthermore, primary isolates that have reduced affinity for sCD4 only shed gp120 slowly compared to the rate of shedding from laboratory-adapted strains (204). Because the primary viruses obviously can fuse with peripheral blood lymphocytes at 37°C, complete dissociation of gp120 from gp41 may not be required for fusion. One caveat is that we do not know whether sCD4 and cellular CD4 behave equivalently; a second is that only a small percentage of gp120 molecules in a fusion complex may need to dissociate from gp41.

Other conformational changes in the HIV-1 (IIIB) envelope have also been identified that are independent of gp120 shedding. Exposure of the V3 loop, a domain of gp120 that is the principal neutralizing determinant, is increased after sCD4 binding, measured both by antibody binding and by enhanced cleavage by an exogenous proteinase (284). CD4 binding to gp120 also causes the exposure of previously occult epitopes on gp41 (284). Although the gp41 fusion peptide has not yet been detected immunologically, either before or after sCD4 binding, mAbs have revealed CD4-induced exposure of two regions of gp41. One of these regions is close to the 5-amino disulfide bonded loop on gp41 (284) and could be a contact region for gp120. Alternatively, this epitope corresponds closely to a region of gp41 reported to be a binding site for a 45kDa cell membrane protein (51, 263) but the significance of this is not yet known. The other epitope is within a proposed α helix close to the transmembrane region of gp41 (284). CD4 appears to induce multiple conformational changes, with differing temperature dependences. Establishing how these changes are coordinated, and which are necessary for fusion, will be essential for understanding the virus–cell fusion reaction. Depending on the strain of virus used, sCD4 may either enhance or neutralize HIV-1 infectivity. This is very similar to the effect of pH on different strains of influenza virus: pH-induced conformational changes may lead, to either an activated fusogenic state or a desensitized state, depending on circumstances and strain of virus (260).

Stable Envelope Glycoprotein–Receptor Complex Formation Is Rate-limiting in the Overall Fusion Process

Entry of HIV-1 into cells requires binding of gp120 to cellular CD4, putative conformational changes in the *env* protein and CD4, and formation of fusion pores between viral and host membranes, which develop into large openings allowing release of the nucleocapsid. As

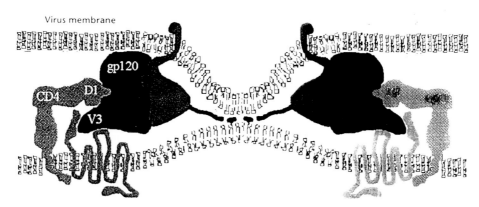

FIG. 14.12. A model of possible interactions between HIV-1 coreceptors, viral gp120-gp41, and CD4 resulting in fusion of the viral and host cell membranes. The coreceptors are depicted to interact with the V3 loop of gp120, which determines the viral tropism. They may also interact, albeit weakly, with CD4. Those interactions lead to exposure of the fusion peptide, which induces fusion of the viral and cell membranes. While two HIV-1 coreceptor-CD4-gp120-gp41 complexes are shown, their actual number in the fusion complex is unknown. [Modified from 81a.]

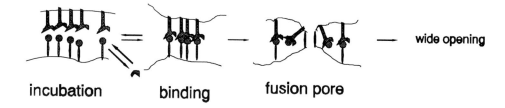

FIG. 14.13. Human immunodeficiency virus (HIV-1) envelope glycoprotein-mediated cell fusion cartoon. The initial binding of gp120 to CD4 leads to close membrane contact and conformational changes that result in formation of fusion pores and their wide opening.

a model system to dissect those events, interactions between HIV-1 *env*–expressing cells, infected with recombinant vaccinia viruses encoding the genes for gp120–gp41, and CD4$^+$ target cells are studied. The scheme is shown in Figure 14.13. Formation of fusion pores can be assessed by lipid mixing and cytoplasmic continuity using fluorescent dyes (85). Formation of multinucleated giant cells is indicative of formation of the large openings that allow nucleocapsid entry. It was found that 50% of the cells exhibited fusion pore formation within 15 to 60 minutes whereas 50% syncytia formation is slower and takes hours to accomplish. Those differences could be resolved by taking into account that at least four cell fusion events need to occur before syncytia can be observed, which leads to the conclusion of a relatively short lag between pore formation and syncytia formation. That conclusion was corroborated using video microscopy with an image-enhanced Nomarski differential interference contrast optics to follow fusion of individual cells in real time (82a). The analysis of the videotape recordings showed several characteristic features of the HIV-1 envelope–mediated cell fusion, representative prints of which are shown in Figure 14.14: *(i)* cells made contact relatively rapidly (within minutes), in many cases by using microspikes to "touch" and adhere to the neighboring cells, *(ii)* the adhered cells fused after a relatively long "waiting" period, which varied from 15 minutes to hours, *(iii)* the morphological changes after membrane fusion, which led to disappearance of the interface separating the two cells, were rapid (within a minute), and *(iv)* the process of syncytia formation involved subsequent fusion with other cells and not simultaneous fusion of many cells. In some cases it appeared that fusion was accompanied by formation of intracellular vesicles. Fifty percent escape

from sCD4 neutralization of HIV-1–env–mediated cell fusion occurs in 30 minutes, which is about the t$_{1/2}$ for cell fusion based on dye redistribution (86, 282). This indicates that interactions leading to a stable CD4–gp120–gp41 complex are rate limiting in the case of cell fusion. Similar observations were made with fusion of intact HIV-1 with cells (107). However, there are exceptions to this pattern: *(i)* HIV-1 envelope–mediated fusion of cells lacking the surface adhesion molecule LFA-1 proceeds to fusion pore formation, but establishment of syncytia was significantly delayed and their number decreased (124), and *(ii)* when CD4 was expressed in the "wrong" (that is, nonhuman) target membrane (39, 204), or in the absence of Ca^{2+} in the medium, fusion pore formation was blocked although gp120–CD4 interactions were not affected (84).

Multiple Copies of the HIV-1 Envelope Glycoprotein May Be Required for Fusion Pore Formation

From the above discussion it appears that HIV-1-env–mediated membrane fusion occurs rapidly once the slower binding stage is complete. Studies with influenza hemagglutinin (HA) had suggested that the fusion requires the association of multiple viral envelope proteins at the plasma membrane (28a) (see earlier under Influenza Hemagglutinin). Based on such findings a model for HA-mediated membrane fusion has been proposed that assigns an important role for the fusion peptide in stabilizing HA trimer assemblies by interactions of some of the fusion peptides with the transmembrane portion of the viral envelope protein while simultaneously, others fusion peptides interact with the target membrane to induce a curved membrane structure (see Plate 14.3, facing page 564). Although the

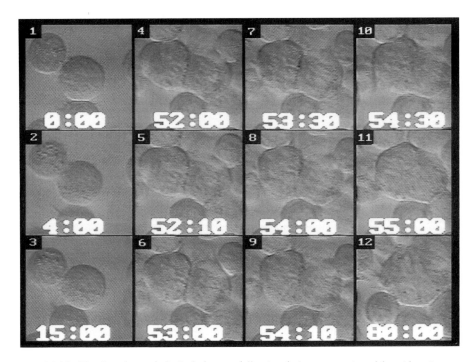

FIG. 14.14. Kinetics of morphological changes following fusion as monitored by videomicroscopy under DIC. Cells expressing human immunodeficiency virus (HIV-1) envelop glycoprotein encoded by a vaccinia recombinant were mixed with CD4-expressing cells at time 0:00 (min:sec). The cells stayed in contact for 52 min. The contact area expanded (compare 15:00 with 52:00 to 53:00) and within about a minute (53:30 to 54:10) the visible boundary between the cells disappeared. The cells then underwent morphological changes leading to rounding of the fusion product (54:30, 55:00) and formation of small syncytium after fusion with other cells (80:00).

model is based on HA, which contains more structural information, based on high-resolution crystal structure we surmise that similar considerations hold for gp120–gp41. On the assumption that the major fusion peptide is on gp41, this may be inserted into the host cell membrane or may provide a hydrophobic interface across which membrane lipids might flow. Other potential cell surface binding sites for gp41 (263) could be involved at this point in the fusion pathway. Studies with fusion peptide analogues of envelope proteins of HIV-1 and SIV suggest that the fusion peptide may insert obliquely into the cell membrane, the precise angle being critical (35, 36). Presumably multiple interactions are necessary for the membranes to come together, peptides from all components of an oligomer and many such oligomers forming the final complex as has been proposed for influenza virus. A prediction of the model is that fusion requires cooperation between a number of fusion peptides and thus fusion activity will depend on env glycoprotein surface density.

Electron micrographs (118) and theoretical studies indicate (175) that multiple contacts form between the virion and the cell surface (204). Moreover, it appears that the number of syncytia formed after cell fusion mediated by the gp120–gp41 molecules and their rate of formation decrease with decreasing the level of surface expression of gp120–gp41 (Dimitrov and Blumenthal, unpublished observations). Clustering of CD4 molecules at the site of virion binding might be required at this stage for a fusion complex to be formed. Uncertainties about the structure of the glycoprotein spikes include the number of gp120–gp41 molecules that comprise an oligomer; whether the components are arranged symmetrically within a spike; and whether the spikes are rigidly anchored onto the *gag* shell protein p17 or are free to diffuse on the virus surface. The mature HIV-1 and HIV-2 spikes have been suggested to be trimeric (117, 337). However, the balance of evidence favors a tetrameric structure for the spike that is assembled as a pair of dimers held together by noncovalent bonds between gp41 molecules (204). Recently it has been shown that HIV-1 envelope glycoprotein dimers can bind two CD4 molecules simultaneously (98).

In another approach to study cooperative assembly of a fusion complex, Freed and coworkers (109) constructed gp41 mutant with a polar substitution at N-terminal amino acid 2 (the 41.2 mutation) that dominantly interferes with both syncytium formation and infection mediated by the wild-type HIV-1 envelope glycoprotein. The interference by the 41.2 mutant is not a result of aberrant envelope glycoprotein synthesis, processing, or transport. The 41.2 mutant elicits a dominant interfering effect even in the presence of excess wild-type glycoprotein, suggesting that a higher-order envelope glycoprotein complex is involved in membrane fusion.

SPERM MEMBRANE PROTEINS INVOLVED IN SPERM–EGG FUSION MAY RESEMBLE VIRAL FUSION PROTEINS

The sperm membrane is specialized for fusion with other cell membranes (see chapter 23 by Epel, this volume). Fusion is restricted to specific regions of the spermatozoon. In sea urchins and other marine animals the site of fusion is the inner surface of the acrosomal granule membrane, which is exposed following exocytosis of the granule to become the external surface of the sperm acrosomal process. The acrosomal process membrane is coated with lysin proteins (103), which are needed for the penetration of the sperm through the egg extracellular envelopes, and adhesive proteins (bindins) (324). Fusion of phospholipid vesicles can be induced by both lysin from abalone sperm (140) and bindin from sea urchin sperm (121). The analysis of the abalone lysin sequence (112) suggests that the amino-terminal half of the molecule may be arranged in four amphipathic helical segments that may be critical for its binding and possibly fusion activity. However, the sperm fusion rate in vivo is much faster than the observed rate of bindin-induced vesicle fusion and faster than lysin-mediated fusion. In addition, the association of bindin with vesicles is reversible and therefore it associates peripherally rather than integrating within the bilayer. As discussed in previous sections, many different kinds of peptides and proteins can induce liposome fusion in a nonspecific way. Therefore, while the hypothesis for the possible role of lysin and bindin in the acrosomal membrane fusion is interesting, it remains to be proved.

The mammalian sperm membrane also has regions implicated in fusion (the equatorial segment and the post-acrosomal segment of acrosome-reacted sperm), which show a great diversity in antigenic and morphological specialization. Some of the antibodies raised against those antigens displayed significant fertilization inhibitory activity (259, 279). For example, preincubation of acrosome-reacted guinea pig sperm with saturating levels of an antibody raised against the PH-30 antigen inhibits the fertilization of zona-free eggs and decreases the average number of fusing sperm cells by 75% (259).

The primary sequence of the PH-30 was recently determined (23). It has two subunits. The α subunit consists of 289 amino acids containing one predicted transmembrane domain. It is heavily glycosylated. The predicted molecular weight of unmodified mature PH-30 α is 29,700; the glycosylated protein on reducing SDS-polyacrylamide gels has a relative molecular mass of 60 kDa. The PH-30 β subunit consists of 353 amino acids with two potential sites for N-linked glycosylation. Its calculated molecular mass (39 kDa) is slightly smaller than that observed on reducing SDS gels (44 kDa). Both subunits have predicted transmembrane domains, large N-terminal extracellular domains and short cytoplasmic tails. Computer-aided sequence and structure analysis indicated that the α subunit contains a putative fusion peptide typical of viral fusion proteins (see earlier under VIRAL ENVELOPE . . .), and that the β subunit has a domain related to a family of soluble integrin ligands found in snake venoms. Based on these findings and the fusion inhibitory activity of the antibody raised against the PH-30, it was suggested that the PH-30 α/β complex resembles many viral fusion proteins in their membrane topology, binding, and fusion (23). Despite the lack of direct experimental proof of its fusion activity, PH-30 may be the first recognized cellular analogue of a viral fusion protein, which opens new exciting possibilities for understanding how cell membranes fuse.

TOWARD A RESOLUTION OF FUSION PROTEINS IN EXOCYTOSIS

Exocytosis involves fusion of a secretory organelle with the cell plasma membrane (see chapter 16 by Sengupta et al., this volume). It is probably the most actively studied fusion event. Electron microscopic analysis (see Fig. 14.4) and patch clamp studies (see Fig. 14.5) of the various stages of the exocytotic fusion process have inspired our thinking about fusion mechanisms. Those studies revealed that biological membrane fusion is quite different from lipid membrane fusion, which is characterized by membrane diaphragms as intermediated stages that then rupture to provide continuity of aqueous spaces. Exocytotic membrane fusion, on the other hand, is characterized by the formation of semi-stable intermediates, fusion pores (see earlier under Electron Microscopy. . .). However, the determination of the specific molecular constituents of those fusion pores is still elusive. The fusion pores have been postulated to be assemblies of lipid and proteins (9).

Several classes of proteins have been proposed as components of the fusion assembly. They include members of the annexin family (67), phosphoproteins, and synaptophysin. One member of the annexin family, synexin (47 kDa) or annexin VII, was isolated from secretory cells and it induces aggregation of chromaffin granules in the presence of low concentrations of calcium ions (68, 69). The aggregated granules can be fused by addition of small amounts (5 μM) of arachidonic acid (66). Synexin also enhances the aggregation and fusion of negatively charged liposomes in the presence of Ca^{2+}, and reduces the threshold calcium ions concentration required for fusion of liposomes of certain types to 10 μM (137–139). Addition of arachidonic acid enhances the rate of fusion but not aggregation (200). Its effect, however, is not specific for synexin. Recently, it was found that arachidonic acid is not essential for Ca^{2+}-dependent exocytosis in adrenal chromaffin cells (205). Human synexin contains a unique 167–amino acid N-terminal segment, which is highly hydrophobic (42, 251). The existence of such hydrophobic segment and the extensive hydrophobic character of a 36-kDa tetrad repeat support the hypothesis for membrane fusion occurring through a hydrophobic bridge (251, 252). Although the hydrophobic bridge hypothesis is interesting, it needs further experimental proof.

Another proposal of a potentially fusogenic protein comes from the observation that during synchronous (1 s) exocytosis one phosphoprotein band (PP65, 63–65 kDa) was selectively dephosyphorylated for less than 1 s and then phosphorylated for about 10 s (355). The dephosphorylation can transiently increase the protein hydrophobicity and cause membrane fusion. Indeed, fusion can be induced by exogenous phosphatases and inhibited by mAb against PP65 [see in (249)].

Yet another potential fusogenic protein is the neuronal protein synaptophysin (p38, 38 kDa), which is a well-characterized integral membrane protein of synaptic vesicles. This protein can be phosphorylated by an intravesicular kinase (237). Synaptophysin mediates fusion in vitro after reconstitution in liposomes, and its hexamers form channels (322) that could be intermediates in fusion. It is also necessary for calcium-dependent glutamate secretion in a reconstituted oocyte expression system (5). Many other proteins are implicated in the regulation of exocytosis, including the recently discovered Exo1 and Exo2 proteins, which stimulate calcium-dependent exocytosis in permeabilized adrenal chromaffin cells (206). Whether these proteins can regulate exocytosis at the level of fusion is unknown.

Although the participation of membrane-associated and partly integrated proteins in membrane fusion during exocytosis appears quite certain, key questions as to whether there exists at all any fusogenic protein specific for exocytosis and what kind of modification

it would have to undergo to become fusogenic, remain to be answered [see the recent reviews (9, 40, 101, 201, 249, 252, 356)].

MULTIPLE PROTEINS MAY BE REQUIRED FOR INTRACELLULAR FUSION

The eukaryotic cells are composed of compartments enclosed by membranes that are responsible for the biosynthesis, processing, and transfer of proteins and lipids. The specific transfer of material is mediated by transport vesicles (233) that bud from the donor compartment and then fuse with the acceptor compartment (see chapter 15 by Hauri and Schweizer and chapter 16 by Sengupta et al., this volume). The specific and tightly controlled fusion event results in delivery of soluble and membrane components of the vesicles to the next organelle in the pathway. The recent studies in several laboratories have led to identification and characterization of components of the transport vesicle fusion apparatus, which is highly conserved between widely diverged species and at different locations within the cell [for review and references see (271)]. They include components of the proteinaceous coat on some transport vesicles, regulatory elements belonging to the GTP-binding protein families, and N-ethylmaleimide sensitive "fusion" protein (NSF) (192, 351). NSF turns out to be a homotetramer of 76 kDa subunits, found both in a soluble form and as a Golgi peripheral membrane protein. Thus, fusion protein might be a misnomer. The NSF release from membranes seems to require ATP hydrolysis. The amino acid sequence of the NSF contains consensus motifs for ATP binding and soluble NSF becomes rapidly and irreversibly inactivated in the absence of ATP or ADP, which indicate that NSF itself hydrolyzes ATP.

The binding of NSF to membranes is specific and saturable, and requires a heat-sensitive integral membrane receptor and one or more soluble NSF attachment proteins (SNAPs) (59, 335) SNAPs appear to be peripheral membrane components. Their activity is attributed to three proteins (35, 36, and 39 kDa), each able to some extent to mediate NSF binding to Golgi membranes. By using purified proteins it was shown that SNAPs first bind to Golgi membranes and then NSF interacts with SNAP. It was suggested that one or more SNAPs bind to the integral membrane receptor(s) and undergo a chemical or conformational change, exposing a binding site for NSF (271). The NSF–SNAP-receptor complex may then interact with fatty acyl CoA and additional soluble or membrane-bound proteins that trigger the fusion of adjacent membrane bilayers. ATP hydrolysis might provide energy to drive the fusion reaction and to release soluble components back into the cytosol.

The amino acid sequence of NSF is 48% identical to the predicted sequence of the SEC18 gene product (Sec18p). NSF and Sec18p appear to be true functional homologues. The gene encoding NSF was cloned from a CHO cDNA library; SEC18 is from *Saccharomyces cerevisiae*. The remarkable similarity between NSF and Sec18p extends also to similarity between SNAPs and another yeast gene product—Sec17p. These findings indicate that proteins required for fusion of transport vesicles are conserved between species. It also seems likely that the fusion proteins are conserved not only between diverged species but also at different locations within the cell. A number of in vitro studies have shown a requirement for NSF in transport from the endoplasmic reticulum to the Golgi (20), in transport between the later compartments of the Golgi apparatus (192), and for endocytosis and fusion of endocytotic vesicles (73). Whether other components specific for different locations inside the cell are required remains to be elucidated.

Evidence that GTP-binding proteins take part in vesicular transport comes from in vitro assays that reconstitute fusion between intracellular compartments and from secretion-deficient mutants of yeast (126). In yeast, mutations in the sec4 or ypt1 genes encoding small GTP-binding proteins inhibit constitutive membrane flow at the plasma membrane or Golgi complex, respectively. It has been suggested that membrane fusion–fission events are regulated by cycling of small GTP-binding proteins between a membrane-bound and free state.

An indication that heterotrimeric GTP-binding proteins ($G\alpha\beta\gamma$) are involved in intracellular membrane fusion comes from inhibition studies with mastoparan and other compounds that increase nucleotide exchange by G proteins (63). Moreover, addition of beta gamma subunits of G proteins to the fusion assay antagonized the stimulatory effect of GTP-γ-S (63). It has also been shown that [ALF$_4$] blocks several steps in in vitro transport along secretory (202) and endocytic pathways (196). This reagent does not affect small GTP-binding proteins, but it activates $G\alpha\beta\gamma$ (156). Evidence was also obtained by showing that overexpression of an inhibitory $G\alpha$ retards the secretion of constitutively secreted proteoglycan (317). Given the function of G proteins in signal transduction (see chapter 5 by Crabtree et al., this volume), these findings may provide insight into mechanisms by which fusion pores are assembled and disassembled.

TOWARD A PHYSICOCHEMICAL ANALYSIS OF FUSION KINETICS

Fusion is a dynamic process that develops in time. To fully understand its mechanism one needs to know the time sequence of events leading to fusion. In what follows we present a physicochemical analysis of the kinetic pathways of fusion. We begin with a characterization of fusion of monolayers. This then leads to a plausible qualitative description of fusion of lipid bilayers and cell membranes. Although we have repeatedly pointed out the difference between lipid bilayer and cell membrane fusion, the kinetics of approach of membranes bear common features. It is after the initial contacts are made that the differences show up.

To understand kinetic mechanisms of fusion one needs to know the interplay between the attractive forces, which drive the fusion process, and the repulsive forces, which pose barriers to fusion. It is reasonable to assume that the system utilizes the *fastest* pathway to reach equilibrium. Thirty years ago, Scheludko [see in (288)] suggested that the most rapid kinetic pathway leading to fast coalescence in colloid systems driven by attraction between surfaces is the growth of thermal fluctuations of the shape of the interacting surfaces, represented as a superposition of surface (fluctuation) waves. The fluctuation wave mechanism of bubble and drop coalescence was further developed by Ivanov, Scheludko, and their collaborators (150, 152) to include the viscous resistance of the liquid film as a whole and the physical properties of the surfactants. It was found (150) that the viscosity of the monolayer and the surface diffusion of its molecules (264) are critical for the rate of growth of the fluctuation waves. This finding explained why the critical separation at which the bubbles fuse is affected by the monolayer density, a phenomenon observed in a variety of foam and emulsion systems. For very high monolayer densities, the energy barriers determine the lifetime of the film separating the bubbles. [For the current concepts of thin film dynamics and instability, see (95, 149, 151).]

Further studies of the dynamic properties of lipid monolayers have shown that in addition to the surface diffusion, there is another process by which the lipid molecules can move very rapidly under the action of differences in surface pressures arising from differences in the monolayer densities; this is the so-called Marangoni effect [see, e.g., (88)]. The Marangoni effect leads to much faster lateral motion of the lipid molecules than the surface diffusion and is strongly affected by the type of lipid molecules, their escape into the bulk of the liquid, the liquid viscosity and the presence of protein molecules (88, 235, 236). For example, for dipalmitoyllecithin monolayers the Marangoni "diffusion" coefficient was measured to be in the range of 350 to 3400 cm^2/s (88), which is about nine orders of magnitude higher than the respective surface diffusion coefficient. However, because the Marangoni effect depends on the thickness of the liquid support, for approaching bubbles at separations of about 10 nm it can result in only 10^3-fold faster rate of lipid molecules motion than the diffusion.

The lessons to be learned from the studies of these "simple" systems are (1) there are two major kinetic mechanisms for the fusion process to develop—growth of fluctuational waves and overcoming energy barriers; (2) in both cases the properties of the thin liquid layer between the surfaces, and those of the monolayers, are critical for the kinetics of fusion; and (3) the fusion per se is diffusion of the lipid molecules and Marangoni effect and therefore is very fast (μs) and determined by the type of the lipid molecules and the interactions with their environment, especially by their phase state.

While fusion of lipid monolayers is "simply" intermixing of their molecules driven by concentration and surface pressure gradients (that is, by lateral diffusion and Marangoni effect), fusion of lipid bilayers is more complicated because of their three-dimensional structure. Cell membranes are even more complicated. They are highly heterogeneous; contain significant amounts of proteins, glycoproteins, and glycolipids; are surrounded by surface layers of adsorbed molecules; and are attached to cytoskeleton [for a current view on the structure and dynamics of lipid bilayers and cell membranes, see the recent review of Sackmann (275) and chapter 2 by Thompson et al. and chapter 3 by Tanford and Reynolds, this volume]. The strong mutual attraction of the two leaflets in the lipid bilayer and the steric hindrance by surface nonlipid components lead to restrictions in the ability of the outer leaflets to intermix. Therefore, formation of intermediate structures may be needed to overcome the energy barrier due to the internal bilayer interactions (296) and the steric barriers imposed by surface proteins. While this fundamental difference between monolayers and bilayer membranes may lead to different molecular mechanisms, the major kinetic pathways of fusion of monolayers and bilayer membranes may be similar. The liquid layer between the approaching surfaces and the surface properties of monolayers and bilayers are essentially the same. Even the van der Waals forces and the bending elasticity, which depend on the thickness of the bilayer, may not differ significantly, provided that the physicochemical environment is the same. These similarities indicate that the kinetics of the intermembrane interactions leading to fusion is qualitatively the same as for monolayers covering bubble (or drop) surfaces. The fusion per se ultimately requires intermixing of the lipid molecules and therefore should involve "monolayer type" of lateral diffusion and Marangoni effect. Based on these arguments and the similarity of the phenomenological equations that describe the mechanics of monolayers and bilayers, it was proposed that fusion of bilayer membranes may follow similar kinetic pathways as fusion of monolayers in colloid systems, and that the knowledge gained in colloid and surface chemistry should be used to describe some features of fusion of bilayer and cell membranes (76–79, 82, 83). What are possible kinetic pathways of bilayer membrane fusion is shown schematically in Figure 14.15.

Fusion kinetics mediated by viral glycoproteins might then be described in the following hypothetical way. The exposure of the viral hydrophobic fusion peptide to the water environment results in an increase in the free energy of the peptide and its immediate environment, possibly including the surfaces of the interacting membranes. This leads to a high free energy state probably similar to that of a fusogenic electropore in electrofusion. Therefore, the kinetic pathways of viral fusion after activation of the viral proteins could be very similar to those of electrofusion and other types of fusion. The existence of a high energy state of the viral fusion protein leads to local approach of the lipid bilayer and destabilization of the membranes and the liquid layer between them. The lipid matrix of the membrane then undergoes localized structural rearrangements resulting in lipid intermixing. An indication that this stage of viral fusion is also similar to that in electrofusion is the magnitude of the activation energy, which in both cases is of the order of the activation energy for the lateral mobility of the lipid molecules (15–25 kcal/mol) (90, 280). After the formation of fusion junctions and pores, they can either expand and lead to formation of giant cells or stabilize and preserve the morphology of aggregated cells (54, 90).

The comparison between electrofusion and virus fusion as well as with other types of fusion may lead to discovery of new important parameters affecting the kinetics of biological fusion. For example, by analogy with electrofusion one might predict that delays in viral fusion should increase with increasing the viscosity of the medium. This was confirmed experimentally for fusion of Sendai virus (82). A detailed comparison between electrofusion and viral fusion is made elsewhere (80). These considerations could be applied to a wide variety of biological processes involving fusion, including exo- and endocytosis, fertilization, cell division, intracellular transport and myotube formation. In the following sections we describe data on fusion

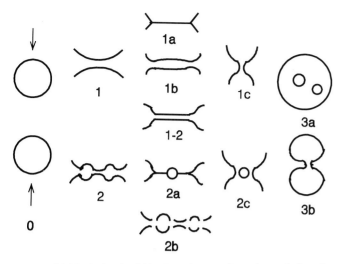

FIG. 14.15. A sketch of kinetic pathways of membrane fusion. *0.* Two membranes approach each other under the action of a driving force due to intermembrane attraction or external fields (e.g., electric). *1.* The membranes approach each other by gradually changing their shapes to almost flat membranes that surround a liquid layer of almost uniform thickness. *2.* Fast localized membrane approach due to unstable shape undulations. *1a.* A single bilayer membrane formed by "semi-fusion" of two membranes that can expand to reach an equilibrium state. *1b.* Nonuniform "nonwavy" liquid layer between approach membranes. *1–2.* A liquid layer between membranes that has an uniform thickness. *2a.* Localized "semifusion" of membranes and formation of "lenses." *2b.* Destabilization of the membranes and formation of pores. *1c,2c.* Fusion of membranes and formation of intracellular vesicles of other structures. *3a,b.* Post–membrane fusion phenomena leading to final fusion products. [Modified from (80).]

kinetics, which can be interpreted by physicochemical principles discussed above.

Delays in Fusion Are Proportional to the Fusion Barriers and Decrease with an Increase in the Strength of the Fusogen

Kinetics of fusion, measured by fluorescent video microscopy of individual membranes (28a, 90, 209, 280) and spectrofluorimetry of populations of membranes (57, 133), consist of three major parts: *(i)* lag times (delays), *(ii)* fluorescence changes, and *(iii)* plateaus that give the maximum fusion yields. Figure 14.2 shows an example of a spectrofluorimetric record of virus–cell fusion kinetics. Similar dependencies were observed by fluorescence video microscopy in electrofusion of erythrocyte ghosts (90) (Fig. 14.16) and for individual cell fusion events induced by viral fusion proteins [see, for example; Figure 5 in (209) for fusion induced by the influenza hemagglutinin]. In a previous section we have shown that the process of HA-mediated membrane fusion can be dissected into a number of elemen-

tary steps, which include triggering, commitment, fusion pore formation, and wide opening. In the following we will present a physicochemical analysis of the delay times, which can provide important information about lifetimes of fusion intermediates.

Delays in fusion vary widely for different systems. They can be in the millisecond range for the neurotransmitter release (9, 130) or minutes and hours for cell fusion induced by the HIV envelope proteins (85). For planar–planar bilayer fusions (53) the lifetimes of the events leading to fusion ("waiting time" for fusion) are in the range of seconds to minutes. Even for the same system, but for different experimental conditions, the delays can vary in orders of magnitude. For instance, delays in electrofusion of erythrocyte ghosts are in the range of milliseconds to minutes (90). Delays in viral fusion are in the range of tens of milliseconds to minutes (57, 133, 209, 308).

A study of delays in electrofusion of erythrocyte ghosts (90) showed that they decreased over the range 4 to 0.3 s with an increase in *(i)* the pulse strength from 1 to 0.25 kV/mm, *(ii)* the pulse duration in the range 0.073 to 1.8 ms, *(iii)* the dielectrophoretic force that brings the membranes at close apposition before triggering fusion. They increased proportionally to the increase in the medium viscosity. The delays decreased 2 to 3 times with an increase in temperature from 21° to 37°C. The Arrhenius plot yielded straight lines. The calculated activation energy, 17 kcal/mol, does not depend on the pulse strength (90).

The delays in fusion of fibroblasts expressing influenza hemagglutinin with erythrocytes decreased with decreasing the pH. They were of the order of seconds for pH in the range 5.4 to 4.9 (280). The activation energy is 18 kcal/mol for temperatures between 27° and 50°C (209). The delays in fusion of intact viruses with erythrocyte ghosts showed similar trends but were shorter than for cell fusion (57). The delay in fusion of Sendai virus with erythrocyte ghosts increased with an increase in medium viscosity (82).

The data for delays in electrofusion of erythrocyte ghosts (90) can be described by an empirical formula (80):

$$\text{Delay} = C\mu\exp(E_a/RT)$$

where C is a constant that does not depend on the activation energy and the viscosity of the liquid between the membranes, μ, E_a—activation energy, R—gas constant, and T—absolute temperature. The constant C is inversely proportional to the driving force of the fusion reaction and proportional to the system resistance to fusion. Therefore, it depends on all the factors that determine the driving force and the fusion resistance. The activation energy does not depend on

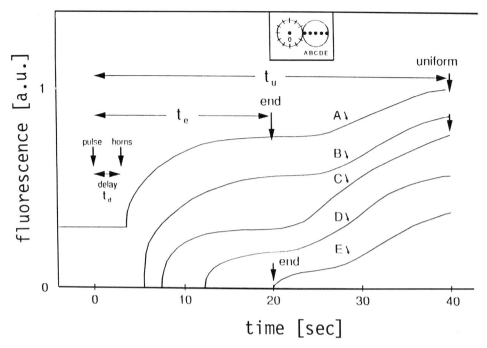

FIG. 14.16. Delays and kinetics of fluorescent dye redistribution from a labeled to an originally unlabeled erythrocyte ghost after fusion induced by a high-voltage electric pulse. The space locations of the points of measurement are shown in the inset. The distances, from the contact with the labeled membrane are (in μm) $A,0$ (the contact); B, 1.7; C, 3.3 (the center of the erythrocyte ghost); D, 5; and E, 6.6 (the far end of the membrane). The fluorescence intensity is normalized to that at the center of the labeled membrane (point 0 in the inset). The fluorescence intensity at the membrane contact (point A in the inset) is higher than zero even before the pulse because of the close proximity of the labeled membrane. The increase in fluorescence begins after a delay at a point in time that coincides with a characteristic appearance of the fluorescence as "horns" (see Fig. 14.13). The dye diffuses until reaching the far ends of the membranes (after time t_e) and an almost uniform distribution (after time t_u). [From (90) with permission.]

the strength of the fusogen. Typical values of C are in the range 10 to 1000 Pa^{-1} and of the activation energy E_a − 10 to 30 kcal/mol. A theoretical analysis of the delays in fusion is presented elsewhere (82).

Rates of Fusion Can Provide Information for the Time Course of Membrane Merging and Fusion Pore Expansion

While the delays reflect the lifetime of the intermediates before the actual intermixing of the membrane components, the rates of fusion provide information for the subsequent stages of the fusion process. When the fusion rate is measured by fluorescent dye redistribution assays in individual cells, the rate of fluorescence changes reflects the rate of diffusion of the dye through the intermembrane junction after fusion. There are two limiting cases: (1) the dye transfer through the intermembrane junction is fast compared to the lateral

diffusion in the originally unlabeled membrane and (2) the transfer through the intermembrane junction is slower than the diffusion (50).

In the first case the rate of dye transfer is entirely determined by the diffusion coefficient of the dye in the originally unlabeled membrane and the membrane geometry. This is the case, for example, shown in Figure 14.16, borrowed from the study of electrofusion of erythrocyte ghosts (90). The plot of the square of the distance between the diffusion front and the membrane contact zone versus time is linear. The slope gives 4D, D being the diffusion coefficient, which for this particular case is equal to 0.55×10^{-8} cm^2/s. This is one method of measuring the lateral diffusion of fluorescent dyes. Therefore, in this case the increase in fluorescence in the initial stage of the membrane transfer is not a measure of the rate of fusion. It only shows that the rate of dye transfer through the intermembrane junctions is faster than the rate of diffusion. After the diffusion front reaches the far end of the cell, the fluorescence intensity continues to increase until reaching the final equilibrium state. This stage also depends on the diffusion. The case of fast dye transfer through intermembrane junctions occurs when the total length of membrane contact by junctions is larger than the contact perimeter. For example, if the intermembrane junctions are channels, fast dye transfer occurs when the product of the number of channels and their average perimeter is larger than the perimeter of the membrane contact.

When the rate of dye transfer through the fusion junction is lower than the diffusion rate, the increase of fluorescence is uniform. If depends only on the time and the properties of the fusion junctions. The rate of increase reflects the number and size of the fusion junctions and their change with time. When the number of size of the fusion junctions are constant, the fluorescence increase should be linear with time in the beginning of the dye transfer. Any differences from the linear relationship could be attributed to changes of the size and number of the fusion junctions, including creation of new or closing of old junctions. Therefore, fluorescence changes after delays may contain important information about intermediates in fusion and their evolution with time. Unfortunately, this important process has not yet been rigorously analyzed.

Fusion Yields and Delays Are Related but May Reflect Different Properties of the Fusing Membranes

For a population of cells the change of the parameters of the system, which leads to decrease of the delay, causes commonly an increase in fusion yields. For example, in electrofusion of erythrocyte ghosts, the increase in pulse strength and duration decreases the delay and increases the fusion yield (90). This means that under stronger pulses more cells acquire the property to fuse. Those that would have fused anyway do it faster. Several experimental results, however, indicate that while fusion yields and delays are related they may reflect different physical parameters (90): *(1)* Under strong pulses practically all cells fuse (that is, the fusion yield is near 100%) and further increase in the pulse strength does not lead to an increase in the fusion yield. The delay can be decreased, however, severalfold with the further increase of the pulse strength. *(2)* For weak pulses the delay does not change significantly. The fusion yield, however, increases severalfold with the increase of the pulse strength. *(3)* The fusion yield changes significantly, but the delay does not, in buffers of different ionic strength. Rigorous interpretation of these findings based on physicochemical principles is still lacking.

DOES UNDERSTANDING MEMBRANE FUSION NEED NEW BREAKTHROUGHS IN METHODOLOGY?

The very act of fusion involves fast (microsecond to millisecond) molecular rearrangements of small numbers (several to several hundreds) of molecules. The majority of the membrane molecules do not participate in the fusion reaction. From an experimental point of view, the fast small signal of few molecules on the huge background of nonparticipating molecules is a major problem—one that does not allow us to monitor the molecular conformational changes during the membrane merging. It seems, therefore, that at the current level of development of the experimental techniques, the molecular mechanisms of fusion cannot be elucidated. Is there any fundamental limit to our knowledge of fusion mechanisms imposed by the current experimental techniques, and do we need new breakthroughs in methodology to discover the fusion mechanisms? It is difficult to give a definite answer to this question. Recently, new techniques, including photosensitized labeling and site-specific mutagenesis, are being successfully used in fusion research. While those and other new methods are promising and have yielded important information, techniques of the future may lead to major developments in our understanding of fusion mechanisms.

NOTE ADDED IN PROOF

Very recently we found that the gp120 of the HIV-1 envelope glycoprotein forms a trimeric complex with the primary receptor, CD4, and the coreceptor molecule, CXCR-4 (formerly called fusin). This is the first experimental evidence for physical association between a part of a virus envelope glycoprotein and two receptor molecules (C. K. Lapham, J. Ouyang, B. Chandrasekar, N. Y. Nguyen, D. S. Dimitrov, and H. Golding. *Science* 274: 602–605, 1996).

We thank Drs. C. Grewe, D. E. Chandler, J. Zimmerberg, V. A. Parsegian, D. C. Wiley, H. R. Guy and S. C. Harrison for kindly providing prints of Figures 14.3, 14.4, 14.5, 14.6, and Plates 14.2 and 14.3, respectively, and Drs. H. Golding, H. Elson, A. Puri, C. Schoch, C. Pack, M. Krumbiegel, C. Broder, and E. Berger for helpful comments and providing unpublished data.

REFERENCES

1. Ahkong, Q. F., F. C. Cramp, D. Fisher, J. I. Howell, and J. A. Lucy. Studies on chemically induced fusion. *J. Cell Sci.* 10: 769–787, 1972.
2. Ahkong, Q. F., D. Fisher, W. Tampion, and J. A. Lucy. The

fusion of erythrocytes by fatty acids, retinol, and alpha-tocopherol. *Biochem. J.* 136: 147–155, 1973.

3. Ahkong, Q. F., D. Fisher, W. Tampion, and J. A. Lucy. Mechanisms of cell fusion. *Nature* 253: 194–195, 1975.

4. Ahkong, Q. F., J. I. Howell, J. A. Lucy, F. Safwat, M. R. Davey, and E. C. Cocking. Fusion of hen erythrocytes with yeast protoplasts induced by polyethylene glycol. *Nature* 255: 266–267, 1975.

5. Alder, J., B. Lu, F. Valtorta, P. Greengard, and M-M. Poo. Calcium-dependent transmitter secretion reconstituted in Xenopus oocytes: requirement for synaptophysin. *Science* 257: 657–661, 1992.

6. Alkhatib, G., C. Richardson, and S. H. Shen. Intracellular processing, glycosylation, and cell-surface expression of the measles virus fusion protein (F) encoded by a recombinant adenovirus. *Virology.* 175: 262–270, 1990.

6a. Alkhatib, G., C. Combadiere, C. C., Broder, Y. Feng, P. E., Kennedy, P. M. Murphy, and E. A. Berger. CC CKR5: a RANTES, MIP-1alpha, MIP-1beta receptor as a fusion cofactor for macrophage-tropic HIV-1. *Science* 272: 1955–1958, 1996.

7. Allan, J. S., J. Strauss, and D. W. Buck. Enhancement of SIV infection with soluble receptor molecules. *Science* 247: 1084–1088, 1990.

8. Almers, W. Exocytosis. *Annu. Rev. Physiol.* 52: 607–624, 1990.

9. Almers, W., L. J. Breckenridge, A. Iwata, A. K. Lee, A. E. Spruce, and F. W. Tse. Millisecond studies of single membrane fusion events. *Ann. NY Acad. Sci.* 318: 327, 1991.

10. Almers, W., and F. W. Tse. Transmitter release from synapses: does a preassembled fusion pore initiate exocytosis? *Neuron* 4: 813–818, 1900.

11. Arnold, K., A. Hermann, L. Pratsch, and K. Gawrisch. The dielectric properties of aqueous solutions of polyethylene glycol and their influence on membrane structure. *Biochim. Biophys. Acta* 815: 515–518, 1985.

12. Arthos, J., K. C. Deen, M. A. Chaikin, J. A. Fornwald, G. Sathe, Q. J. Sattentau, P. R. Clapham, R. Weiss, J. S. McDougal, C. Pietropaolo, R. Axel, A. Truneh, P. J. Maddon, and R. W. Sweet. Identification of the residues in human CD4 critical for the binding of HIV. *Cell* 57: 469–481, 1989.

13. Ashorn, P. A., E. A. Berger, and B. Moss. Human immunodeficiency virus envelope glycoprotein/CD4-mediated fusion of non-primate cells with human cells. *J. Virol.* 64: 2149–2156, 1990.

14. Ashwell, G., and J. Harford. Carbohydrate-specific receptors of the liver. *Annu. Rev. Biochem.* 51: 531–554, 1982.

15. Bagai, S., A. Puri, R. Blumenthal, and D. P. Sarkar. Hemagglutinin-neuraminidase enhances F protein-mediated membrane fusion of reconstituted Sendai virus envelopes with cells. *J. Virol.* 67: 3312–3318, 1994.

16. Bangham, A. D., M. M. Standish, and J. C. Watkins. Diffusion of univalent ions across lamellae of swollen phospholipids. *J. Mol. Biol.* 13: 238–252, 1968.

17. Baran, A. A., I. M. Solomentseva, V. V. Mank, and O. D. Kurilenko. Role of the solvation factor in stabilizing disperse systems containing water-soluble polymers. *Dokl. Akad. Nauk SSSR*, 1972, p. 363–366.

18. Barre-Sinoussi, F., J. C. Chermann, F. Rey, M. T. Nugeyre, S. Chamaret, J. Gruest, C. Dauguet, C. Axler-Blin, F. Vezinet-Brun, C. Rouzioux, W. Rozenbaum, and L. Montagnier. Isolation of a T-lymphotropic retrovirus from a patient at risk for Acquired Immune Deficiency Syndrome (AIDS). *Science* 220: 868–871, 1983.

19. Barski, G., S. Sorieul, and F. Cornefert. "Hybrid" type cells in combined cultures of two different mammalian cell strains. *J. Natl. Cancer Inst.* 26: 1269–1291, 1961.

20. Beckers, C. J., M. R. Block, B. S. Glick, J. E. Rothman, and W. E. Balch. Vesicular transport between the endoplasmic reticulum and the Golgi stack requires the NEM-sensitive fusion protein. *Nature* 339: 397–398, 1989.

21. Bentz, J. *Viral Fusion Mechanisms.* Boca Raton, Florida: CRC Press, 1993.

22. Bergelson, J. M., M. P. Shepley, B. M. Chan, M. E. Hemler, and R. W. Finberg. Identification of the integrin VLA-2 as a receptor for echovirus 1 [see comments]. *Science* 255: 1718–1720, 1992.

23. Blobel, C. P., T. G. Wolfsberg, C. W. Turck, D. G. Myles, P. Primakoff, and J. M. White. A potential fusion peptide and an integrin ligand domain in a protein active in sperm-egg fusion. *Nature* 356: 248–252, 1992.

24. Blumenthal, R. Membrane fusion. *Curr. Top. Membr. Transp.* 29: 203–254, 1987.

25. Blumenthal, R., A. Bali-Puri, A. Walter, D. Covell, and O. Eidelman. pH-Dependent fusion of vesicular stomatitis virus with Vero cells: measurement by dequenching of octadecylrhodamine fluorescence. *J. Biol. Chem.* 262: 13614–13619, 1987.

26. Blumenthal, R., and A. Loyter. Reconstituted viral envelopes—"Trojan Horses" for drug delivery and gene therapy. *TibTech* 9: 41–45, 1991.

27. Blumenthal, R., A. Puri, and D. S. Dimitrov. Initial steps of enveloped virus entry into animal cells. In: *Biotechnology of Cell Regulation*, edited by R. Verna and Y. Nishizuka, New York: Raven Press, 1991, p. 115–133.

28. Blumenthal, R., C. Schoch, A. Puri, and M. J. Claque. A dissection of steps leading to viral envelope protein-mediated membrane fusion. *Ann. NY Acad. Sci.* 635: 285–296, 1991.

28a. Blumenthal, R., D. P. Sarkar, S. Durell, D. E. Howard, and S. J. Morris. Dilation of the influenza hemagglutinin fusion pore revealed by the kinetics of individual cell–cell fusion events. *J. Cell. Biol.* 135: 63–72, 1996.

29. Boni, L. T., J. S. Hah, S. W. Hui, P. Mukherjee, J. T. Ho, and C. Y. Jung. Aggregation and fusion of unilamellar vesicles by polyethylene glycol. *Biochim. Biophys. Acta.* 775: 409–418, 1984.

30. Boni, L. T., T. P. Stewart, J. L. Alderfer, and S. W. Hui. Lipid-polyethylene glycol interactions: I. Induction of fusion between liposomes. *J. Membrane Biol.* 62: 65–70, 1981.

31. Boni, L. T., T. P. Stewart, and S. W. Hui. Lipid-polyethylene glycol interactions: II. Formation of defects in bilayers. *J. Membrane Biol.* 62: 71–77, 1981.

32. Boni, L. T., T. P. Stewart, and S. W. Hui. Alterations in phospholipid polymorphism by polyethylene glycol. *J. Membrane Biol.* 80: 91–104, 1984.

33. Bonnett, H. T., and T. Eriksson. Transfer of algal chloroplasts into protoplasts of higher plants. *Planta* 120: 71–79, 1974.

34. Bosch, M. L., P. L. Earl, K. Fargnoli, S. Picciafuoco, F. Giombini, F. Wong-Staal, and G. Franchini. Identification of the fusion peptide of primate immunodeficiency viruses. *Science* 244: 694–697, 1989.

35. Brasseur, R. Differentiation of lipid-associating helices by use of three-dimensional molecular hydrophobicity potential calculations. *J. Biol. Chem.* 266: 16120–16127, 1991.

36. Brasseur, R., M. Vandenbranden, B. Cornet, A. Burny, and J. M. Ruysschaert. Orientation into the lipid bilayer of an asymmetric amphipathic helical peptide located at the N-terminus of viral fusion proteins. *Biochim. Biophys. Acta* 1029: 267–273, 1990.

37. Breckenridge, L. J., and W. Almers. Currents through the fusion

pore that forms during exocytosis of a secretory vesicle. *Nature* 328: 814–817, 1987.

38. Broder, C. C., and E. A. Berger. CD4 molecules with a diversity of mutations encompassing the CDR3 region efficiently support the human immunodeficiency virus type 1 envelope glycoprotein-mediated cell fusion. *J. Virol.* 67: 913–926, 1993.

39. Broder, C. C., D. S. Dimitrov, R. Blumenthal, and E. A. Berger. The block to HIV-1 envelope glycoprotein-mediated membrane fusion in animal cells expressing human CD4 can be overcome by human cell components. *Virology* 193: 483–491, 1993.

39a. Bullough, P. A., F. M. Hughson, J. J. Skehel, and D. C. Wiley. Structure of influenza haemagglutinin at the pH of membrane fusion. *Nature* 371: 37–43, 1994.

40. Burgoyne, R. D. Secretory vesicle-associated proteins and their role in exocytosis. *Annu. Rev. Physiol.* 52: 647–659, 1990.

41. Burkly, L. C., D. Olson, R. Shapiro, G. Winkler, J. J. Rosa, D. W. Thomas, C. Williams, and P. Chisholm. Inhibition of HIV entry by a CD4 domain 2 specific monoclonal antibody: evidence for structural alterations upon CD4/HIV gp120 binding. *J. Immunol.* 149: 1779–1787, 1992.

42. Burns, A. L., K. Magendzo, A. K. Shirvan, A. Srivastava, E. Rojas, M. R. Alijani, and H. B. Pollard. Calcium channel activity of purified human synexin and structure of the human synexin gene. *Proc. Natl. Acad. Sci. U. S. A.* 86: 3798–3802, 1989.

43. Capers, C. R. Multinucleation of skeletal muscle in vitro. *J. Biophys. Biochem. Cytol.* 7: 559–566, 1960.

43a. Carr, C. M., and P. S. Kim. A spring-loaded mechanism for the conformational change of influenza hemagglutinin. *Cell* 73: 823–832, 1993.

44. Celada, F., C. Cambiaggi, J. Maccari, S. Burastero, T. Gregory, E. Patzer, J. Porter, C. McDanal, and T. Matthews. Antibody raised against soluble CD4-rgp120 complex recognizes the CD4 moiety and blocks membrane fusion without inhibiting CD4-gp120 binding. *J. Exp. Med.* 172: 1143–1150, 1990.

45. Chandler, D. E. Exocytosis involves highly localized membrane fusions. *Biochem. Soc. Trans.* 12: 961–963, 1984.

46. Chandler, D. E. Membrane fusion as seen in rapidly frozen secretory cells. *Ann. NY Acad. Sci.* 635: 234–245, 1991.

47. Chandler, D. E., and J. E. Heuser. Membrane fusion during secretion. Cortical granule exocytosis in sea urchin eggs as studied by quick freezing and freeze-fracture. *J. Cell Biol.* 83: 91–108, 1979.

48. Chandler, D. E., and J. E. Heuser. Arrest of membrane fusion events in mast cells by quick freezing. *J. Cell Biol.* 86: 666–674, 1980.

49. Chang, D. C., B. M. Chassy, J. A. Saunders, and A. E. Sowers. *Guide to Electroporation and Electrofusion.* San Diego: Academic Press, 1992.

50. Chen, Y., and R. Blumenthal. On the use of self-quenching fluorophores in the study of membrane fusion kinetics. *Biophys. Chem.* 34: 283–292, 1989.

51. Chen, Y. H., C. Ebenbichler, R. Vornhagen, T. F. Schulz, F. Steindl, G. Bock, H. Katinger, and M. P. Dierich. HIV-1 gp41 contains 2 sites for interaction with several proteins on the helper T-lymphoid cell line, H9. *AIDS* 6: 533–539, 1992.

52. Chernomordik, L., S. S. Vogel, E. Leukina, and J. Zimmerberg. Inhibition of biological membrane fusion by amphipathic compounds. *Biophys. J.* 61: a499, 1992.

53. Chernomordik, L. V., G. B. Melikyan, and Y. A. Chizmadzhev. Biomembrane fusion: a new concept derived from model studies using two interacting planar lipid bilayers. *Biochim. Biophys. Acta* 906: 309–352, 1987.

54. Chernomordik, L. V., and A. E. Sowers. Evidence that the spectrin network and a nonosmotic force control the fusion product morphology in electrofused erythrocyte ghosts. *Biophys. J.* 60: 1026–1037, 1991.

55. Chow, R. H., L. Von Ruden, and E. Neher. Delay in vesicle fusion revealed by electrochemical monitoring of single secretory events in adrenal chromaffin cells. *Nature* 356: 60–63, 1992.

56. Clague, M. J., C. Schoch, and R. Blumenthal. Delay time for influenza hemagglutinin-induced membrane fusion depends on the haemagglutinin surface density. *J. Virol.* 65: 2402–2407, 1991.

57. Clague, M. J., C. Schoch, L. Zech, and R. Blumenthal. Gating kinetics of pH-activated membrane fusion of vesicular stomatitis virus with cells: stopped flow measurements by dequenching of octadecylrhodamine fluorescence. *Biochemistry* 29: 1303–1308, 1990.

58. Clapham, P. R. Human immunodeficiency virus infection of non-haematopoietic cells. The role of CD4-independent entry. *Rev. Med. Virol.* 1: 51–58, 1991.

59. Clary, D. O., I. C. Griff, and J. E. Rothman. SNAPs, a family of NSF attachment proteins involved in intracellular membrane fusion in animals and yeast. *Cell* 61: 709–721, 1990.

60. Coakley, T. W., H. Darmani, and A. J. Baker. Membrane contact induced between erythrocytes by polycations, lectins and dextran. In: *Cell and Model Membrane Interactions,* edited by S. Ohki, New York: Plenum Press, 1991, p. 25–45.

61. Coakley, W. T., and D. Gallez. Membrane-membrane contact: involvement of interfacial instability in the generation of discrete contacts. *Biosc. Rep.* 9: 675–691, 1989.

62. Cohen, F. S., M. H. Akabas, J. Zimmerberg, and A. Finkelstein. Parameters affecting the fusion of unilamellar phospholipid vesicles with planar bilayer membranes. *J. Cell Biol.* 98: 1054–1062, 1984.

63. Colombo, M. I., L. S. Mayorga, P. J. Casey, and P. D. Stahl. Evidence of a role for heterotrimeric GTP-binding proteins in endosome fusion. *Science* 255: 1695–1697, 1992.

64. Colwin, L. H., and A. L. Colwin. Formation of sperm entry holes in the vitelline membrane of Hydroides hexagonus (Annelida) and evidence of their lytic origin. *J. Biophys. Biochem. Cytol.* 7: 315–320, 1960.

65. Compans, R. W., A. Helenius, and M. Oldstone. *Cell Biology of Virus Entry, Replication and Pathogenesis.* New York: Alan R. Liss, 1989, p. 1–449.

66. Creutz, C. Cis-unsaturated fatty acids induce the fusion of chromaffin granules aggregated by synexin. *J. Cell Biol.* 91: 247–256, 1981.

67. Creutz, C. E. The annexins and exocytosis. *Science* 258: 924–931, 1992.

68. Creutz, C. E., L. G. Dowling, J. J. Sando, C. Villar-Palasi, J. H. Whiple, and W. J. Zaks. Characterization of the chromobindins: Soluble proteins that bind to the chromaffin granule membrane in the presence of Ca^{2+} *J. Biol. Chem.* 258: 14664–14674, 1983.

69. Creutz, C. E., C. J. Pazoles, and H. B. Pollard. Identification and purification of an adrenal medullary protein (synexin) that causes calcium dependent aggregation of isolated chromaffin granules. *J. Biol. Chem.* 253: 2858–2866, 1978.

70. Cunningham, J. M. Cellular entry by murine retroviruses. *Seminars in Virology* 3: 85–89, 1992.

71. Curran, M., F. S. Cohen, D. E. Chandler, P. Munson, and J. Zimmerberg. Exocytotic fusion pores exhibit semi-stable states. *J. Membrane Biol.* 133: 61–75, 1993.

72. Curran, M. J., M. S. Brodwick, and C. Edwards. Direct visualization of exocytosis in mast cells. *Biophys. J.* 45: 170a, 1984.

73. Daiz, R., L. Mayorga, P. J. Weidman, J. E. Rothman, and P. D. Stahl. Vesicle fusion following receptor-mediated endocytosis requires a protein active in Golgi transport. *Nature* 339: 398–400, 1989.

74. Dalgleish, A. G., P.C.L. Beverley, P. R. Clapham, D. H. Crawford, M. F. Greaves, and R. A. Weiss. The CD4 (T4) antigen is an essential component of the receptor for the AIDS retrovirus. *Nature* 312: 763–767, 1984.

75. Del Castillo, J., and B. Katz. Quantal components of the endplate potential. *J. Physiol.* 124: 560–573, 1954.

75a. Deng, H., R. Liu, W. Ellmeier, S. Choe, D. Unutmaz, M. Burkhart, P. Di Marzio, S. Marmon, R. E. Sutton, C. M. Hill, C. B. Davis, S. C. Peiper, T. J. Schall, D. R. Littman, and N. R. Landau. Identification of a major co-receptor for primary isolates of HIV-1. *Nature* 381: 661–666, 1996.

76. Dimitrov, D. S. A hydrodynamic theory of the bilayer membrane formation. *Biophys. J.* 36: 21–25, 1981.

77. Dimitrov, D. S. Instability of thin liquid films between membranes. *Coll. Pol. Sci.* 260: 1137–1144, 1982.

78. Dimitrov, D. S. Dynamic interactions between approaching surfaces of biological interest. *Progr. Surface Sci.* 14: 295–424, 1983.

79. Dimitrov, D. S. Approach and instability of membranes: physicochemical mechanisms and applications to adhesion, fusion, dielectrophoresis, electroporation and liposome formation. *Proc. 8th School Bioph. Membrane Transport, Mierki, Poland* 1: 51–82, 1986.

80. Dimitrov, D. S. Membrane fusion kinetics. In: *Guide to Electroporation and Electrofusion,* edited by D. C. Chang, B. M. Chassy, J. A. Saunders, and A. E. Sowers. Orlando: Academic Press, 1992, p. 155–166.

81. Dimitrov, D. S. Electroporation and electrofusion of membranes. In: *Handbook of Physics of Biological Systems, Structure and Dynamics of Membranes,* edited by E. Sackmann and R. Lipowsky, Amsterdam: Elsevier, Vol. 1b Ch.18 852–901, 1995.

81a. Dimitrov, D. S. Fusin—a place for HIV-1 and T4 cells to meet. *Nature Med.* 2: 640–641, 1996.

82. Dimitrov, D. S., and R. Blumenthal. Kinetics of intermembrane interactions leading to fusion. In *Cell and Model Membrane Interactions.* edited by S. Ohki. New York: Plenum Press, 1991, p. 115–133.

82a. Dimitrov, D. S., and R. Blumenthal. Photoinactivation and kinetics of membrane fusion mdiated by the human immunodeficiency virus type I envelope glycoprotein. *J. Virol.* 68: 1956–1961, 1994.

83. Dimitrov, D. S., and R. K. Jain. Membrane stability. *Biochim. Biophys. Acta* 779: 437–468, 1984.

84. Dimitrov, D. S., C. C. Broder, E. A. Berger, and R. Blumenthal. Calcium ions are required for cell fusion mediated by the CD4-HIV-1 envelope glycoprotein interaction. *J. Virol.* 67: 1647–1652, 1993.

85. Dimitrov, D. S., H. Golding, and R. Blumenthal. Initial steps in HIV-1 envelope glycoprotein mediated cell fusion monitored by a new assay based on redistribution of fluorescence markers. *AIDS Res. Human Retroviruses* 7: 799–805, 1991.

86. Dimitrov, D. S., K. Hillman, J. Manischewitz, R. Blumenthal, and H. Golding. Kinetics of interaction of sCD4 with cells expressing HIV-1 envelope and inhibition of syncytia formation. *AIDS* 60: 249–256, 1992.

87. Dimitrov, D. S., K. Hillman, J. Manischewitz, R. Blumenthal, and H. Golding. Kinetics of binding of sCD4 to cells expressing HIV-1 envelope glycoprotein. *J. Virol.* 66: 132–138, 1992.

88. Dimitrov, D. S., I. Panaiotov, P. Richmond, and L. Ter-Minassian-Saraga. Dynamics of insoluble monolayers. I. Dilational or elastic modulus, friction coefficient and Marangoni effect for dipalmitoyllecithin. *J. Colloid Interface Sci.* 65: 483–494, 1978.

89. Dimitrov, D. S., and A. E. Sowers. Membrane electroporation—fast molecular exchange by electroosmosis. *Biochim. Biophys. Acta* 1022: 381–392, 1990.

90. Dimitrov, D. S., and A. E. Sowers. A delay in membrane fusion: lag times observed by fluorescence microscopy of individual fusion events induced by an electric field pulse. *Biochemistry* 29: 8337–8344, 1990.

91. Dimitrov, D. S., R. Willey, M. Martin and R. Blumenthal. Kinetics of HIV-1 interactions with sCD4 and CD4 + cells: implications for inhibition of virus infection and initial steps of virus entry into cells. *Virology* 187: 398–406, 1992.

92. Dimitrov, D. S., R. L. Willey, H. Sato, L.-Ji Chang, R. Blumenthal, and M. A. Martin. Quantitation of HIV-1 infection kinetics. *J. Virol.* 67: 2182–2190, 1993.

93. Doms, R. W., A. Helenius, and J. White. Membrane fusion activity of the influenza virus hemagglutinin: the low pH-induced conformational change. *J. Biol. Chem.* 260: 2973–2981, 1985.

93a. Dragic T., V. Litwin, G. P. Allaway, S. R. Martin, Y. Huang, K. A. Nagashima, C. Cayanan, P. J. Maddon, R. A., Koup, J. P. Moore, and W. A. Paxton. HIV-1 entry into CD4 + cells is mediated by the chemokine receptor CC-CKR-5. *Nature* 381: 667–673, 1996.

94. Dragic, T., P. Charneau, F. Clavel, and M. Alizon. Complementation of murine cells for human immunodeficiency virus envelope/CD4-mediated fusion in human/murine heterokaryons. *J. Virol.* 66: 4794–4802, 1992.

95. Dukhin, S. S., N. N. Rulyov, and D. S. Dimitrov. *Coagulation and Dynamics of Thin Layers.* Kiev: Naukova Dumka, 1986.

96. Duzgunes, N., and J. Bentz. Fluorescence assays for membrane fusion. In: *Spectroscopic Membrane Probes,* edited by L. M. Loew, Boca Raton, Florida: CRC Press, 1988, p. 117–159.

97. Duzgunes, N., and F. Bronner. *Membrane Fusion in Fertilization, Cellular Transport, and Viral Infection: Current Topics in Membranes and Transport,* vol. 32. San Diego: Academic Press, 1988.

98. Earl, P. L., R. W. Doms, and B. Moss. Multimeric CD4 binding exhibited by human and simian immunodeficiency virus envelope protein dimers. *J. Virol.* 66: 5610–5614, 1992.

99. Ebata, S. N., M. J. Cote, C. Y. Kang, and K. Dimock. The fusion and hemagglutinin-neuraminidase glycoproteins of human parainfluenza virus 3 are both required for fusion. *Virology.* 183: 437–441, 1991.

100. Edwards, W., J. H. Phillips, and S. J. Morris. Structural changes in chromaffin granules induced by divalent cations. *Biochim. Biophys. Acta* 356: 164–173, 1974.

101. Ehrenstein, G., E. F. Stanley, S. L. Pocotte, M. Jia, K. H. Iwasa, and K. E. Krebs. Evidence for a model of exocytosis that involves calcium-activated channels. *Ann. NY Acad. Sci.* 635: 297–306, 1991.

102. Ellens, H., J. Bentz, D. Mason, F. Zhang, and J. M. White. Fusion of influenza hemagglutinin-expressing fibroblasts with glycophorin-bearing liposomes: role of hemagglutinin surface density. *Biochemistry.* 29: 9697–9707, 1990.

103. Epel, D., and V. D. Vacquier. Membrane fusion events during invertebrate fertilization. *Cell Surface Rev.* 5: 1–63, 1978.

104. Evered, D., and J. Whelan. *Cell Fusion.* London: Pitman, 1984.

105. Fauci, A. S. The human immunodeficiency virus: infectivity

and mechanisms of pathogenesis. *Science* 239: 617–622, 1988.

105a. Feng, Y., C. C. Broder, P. E., Kennedy, and E. A. Berger. HIV-1 entry cofactor: functional cDNA cloning of a seven-transmembrane, G protein-coupled receptor. *Science* 272: 872–877, 1996.

106. Fernandez, J. M., E. Neher, and B. D. Gomperts. Capacitance measurements reveal stepwise fusion events in degranulating mast cells. *Nature* 312: 453–455, 1984.

107. Fernandez-Larsson, R., K. K. Srivastava, S. Lu, and H. L. Robinson. Replication of patient isolates of human immunodeficiency virus type 1 in T cells: a spectrum of rates and efficiencies of entry. *Proc. Natl. Acad. Sci. U.S.A.* 89: 2223–2226, 1992.

108. Formanowski, F., S. A. Wharton, L. J. Calder, C. Hofbauer, and H. Meier-Ewert. Fusion characteristics of influenza C viruses. *J. Gen. Virol.* 71: 1181–1188, 1990.

109. Freed, E. O., E. L. Delwart, G. L. Jr. Buchschacher, and A. T. Panganiban. A mutation in the human immunodeficiency virus type 1 transmembrane glycoprotein gp41 dominantly interferes with fusion and infectivity. *Proc. Natl. Acad. Sci. U. S. A.* 89: 70–74, 1992.

110. Freed, E. O., D. J. Myers, and R. Risser. Characterization of the fusion domain of the human immunodeficiency virus type 1 envelope glycoprotein gp41. *Proc. Natl. Acad. Sci. U. S. A.* 87: 4650–4654, 1990.

111. Freistadt, M. S., and V. R. Racaniello. Mutational analysis of the cellular receptor for poliovirus. *J. Virol.* 65: 3873–3876, 1991.

112. Fridberger, A., J. Sundeling, V. D. Vacquier, and P. A. Peterson. Amino acid sequence of an egg lysin protein from abalone spermatozoa that solubilizes the vitelline layer. *J. Biol. Chem.* 260: 9092–9099, 1985.

113. Fujioka, A., M. Ohtsuki, M. Nagano, and S. Mori. Membrane fusion events during endocytosis in mouse kidney tubule cells detected by rapid freezing followed by freeze-substitution. *J. Electron Micr.* 39: 356–362, 1990.

114. Fuller, A. O., and W. C. Lee. Herpes simplex virus type 1 entry through a cascade of virus-cell interactions requires different roles of gD and gH in penetration. *J. Virol.* 66: 5002–5012, 1992.

115. Gad, A. E., B. L. Silver, and G. D. Eytan. Polycation-induced fusion of negatively-charged vesicles. *Biochim. Biophys. Acta* 690: 124–132, 1982.

116. Gallo, R. C., S. Z. Salahuddin, M. Popovic, G. M. Shearer, M. Kaplan, B. F. Haynes, T. J. Palker, R. Redfield, J. Oleske, B. Safai, G. White, P. Foster, and P. D. Markham. Frequent detection and isolation of cytopathic retroviruses (HTLV-III) from patients with AIDS and at risk for AIDS. *Science* 224: 500–503, 1984.

117. Gelderblom, H. R. Assembly and morphology of HIV: potential effect of structure on viral function. *AIDS* 5: 617–638, 1991.

118. Gelderblom, H. R., M. Ozel, and G. Pauli. Morphogenesis and morphology of HIV. Structure-function relations. *Arch. Virol.* 106: 1–13, 1989.

119. Gething, M. J., R. W. Doms, D. York, and J. White. Studies on the mechanism of membrane fusion: site-specific mutagenesis of the hemagglutinin of influenza virus. *J. Cell Biol.* 102: 11–23, 1986.

120. Gething, M. J., and J. Sambrook. Cell-surface expression of influenza haemagglutinin from a cloned DNA copy of the RNA gene. *Nature* 293: 620–625, 1981.

121. Glabe, C. C. Interaction of the sperm adhesive protein, bindin, with phospholipid vesicles. II. Bindin induces the fusion of mixed phase vesicles that contain phosphatidylcholine and phosphatidylserine in vitro. *J. Cell Biol.* 100: 800–806, 1985.

122. Godley, L., J. Pfeifer, D. Steinhauer, B. Ely, G. Shaw, R. Kaufmann, E. Suchanek, C. Pabo, J. J. Skehel, D. C. Wiley, and A. L. Et. Introduction of intersubunit disulfide bonds in the membrane-distal region of the influenza hemagglutinin abolishes membrane fusion activity. *Cell* 68: 635–645, 1992.

123. Golding, H., D. S. Dimitrov, R. Blackburn, J. Manischewitz, R. Blumenthal, and B. Golding. Fusion of human B cell lines with HIV-1 envelope-expressing T cells is enhanced by antigen-specific immunoglobulin receptors: a possible mechanism for elimination of gp120-specific B cells in vivo. *J. Immunol.* 150: 2506–2516, 1993.

124. Golding, H., D. S. Dimitrov, and R. Blumenthal. LFA-1 adhesion molecules are not involved in early stages of HIV-1-env-mediated cell membrane fusion. *AIDS Res. Human Retroviruses* 8: 1607–1612, 1992.

125. Grewe, C., A. Beck, and H. R. Gelderblom. HIV: early virus-cell interaction. *J. AIDS* 3: 965–974, 1990.

126. Gruenberg, J., and M. J. Clague. Regulation of intracellular membrane transport. *Curr. Opin. Cell Biol.* 4: 593–599, 1992.

127. Guy, H. R., S. R. Durell, C. Schoch, and R. Blumenthal. Analyzing the fusion process of influenza hemagglutinin by mutagenesis and molecular modeling. *Biophys. J.* 62: 113–116, 1992.

128. Healey, D., L. Dianda, J. P. Moore, J. S. McDougal, M. J. Moore, P. Estess, D. Buck, P. D. Kwong, P.C.L. Beverley, and Q. J. Sattentau. Novel anti-CD4 monoclonal antibodies separate human immunodeficiency virus infection and fusion of CD4 + cells from virus binding. *J. Exp. Med.* 172: 1233–1242, 1990.

129. Herrmann, A., M. J. Clague, and R. Blumenthal. The role of the target membrane structure in fusion with influenza virus: effect of modulating erythrocyte transbilayer phospholipid structure. *Membr. Biochem.* 10: 3–15, 1993.

130. Heuser, J. E., T. S. Reese, M. J. Dennis, Y. Jan, L. Jan and L. Evans. Synaptic vesicle exocytosis captured by quick freezing and correlated with quantal transmitter release. *J. Cell Biol.* 81: 275–300, 1979.

131. Ho, D. D., J. C. Kaplan, I. E. Rackauskas, and M. E. Gurney. Second conserved domain of gp120 is important for HIV infectivity and antibody neutralization. *Science* 239: 1021–1023, 1988.

132. Hoekstra, D., T. De Boer, K. Klappe, and J. Wilschut. Fluorescence method for measuring the kinetics of fusion between biological membranes. *Biochemistry* 23: 5675–5681, 1984.

133. Hoekstra, D., K. Klappe, H. Hoff, and S. Nir. Mechanism of fusion of Sendai virus: role of hydrophobic interactions and mobility constraints of viral membrane proteins: effects of PEG. *J. Biol. Chem.* 264: 6786–6792, 1989.

134. Hoekstra, D., and J. W. Kok. Entry mechanisms of enveloped viruses. Implications for fusion of intracellular membranes. *Biosci. Rep.* 9: 273–305, 1989.

135. Hoggan, M. D., and B. Roizman. The isolation and properties of a variant of herpes simplex virus producing multinucleate giant cells in monolayer cultures in the presence of antibody. *Am. J. Hyg.* 70: 208–219, 1959.

136. Holtzer, H., J. Abbott, and J. Lash. The formation of multinucleate myotubes. *Anat. Rec.* 131: 567, 1958.

137. Hong, K., N. Duzgunes, R. Ekerdt, and D. Papahadjopoulos. Modulation of membrane fusion by calcium-binding proteins. *Biophys. J.* 37: 297–305, 1982.

138. Hong, K., N. Duzgunes, and D. Papahadjopoulos. Role of synexin in membrane fusion. *J. Biol. Chem.* 256: 3641–3644, 1981.

139. Hong, K., N. Duzgunes, and D. Papahadjopoulos. Synexin facilitates fusion of specific phospholipid vesicles at divalent cation concentrations found intracellulary. *Proc. Natl. Acad. Sci. U. S. A.* 79: 4942–4944, 1982.

140. Hong, K., and V. D. Vacquier. Fusion of liposomes induced by a cationic protein from the acrosome granule of abalone spermatozoa. *Biochemistry* 25: 543–549, 1986.

141. Horth, M., B. Lambrecht, M. C. Khim, F. Bex, C. Thiriart, J. M. Ruysschaert, A. Burny, and R. Brasseur. Theoretical and functional analysis of the SIV fusion peptide. *EMBO. J.* 10: 2747–2755, 1991.

142. Horvath, C. M., and R. A. Lamb. Studies on the fusion peptide of a paramyxovirus fusion glycoprotein: roles of conserved residues in cell fusion. *J. Virol.* 66: 2443–2455, 1992.

143. Hsu, M.-C., A. Scheid, and P. Choppin. Reconstruction of membranes with individual paramyxovirus glycoproteins and phospholipid and cholate solution. *Virology* 95: 476–491, 1979.

144. Hu, X. L., R. Ray and R. W. Compans. Functional interactions between the fusion protein and hemagglutinin-neuraminidase of human parainfluenza viruses. *J. Virol.* 66: 1528–1534, 1992.

145. Hui, S. W., and L. T. Boni. Membrane fusion induced by polyethylene glycol. In: *Membrane Fusion,* edited by J. Wilschut and D. Hoekstra, New York: Marcel Dekker, 1991, p. 231–253.

146. Hui, S. W., T. Isac, L. T. Boni, and A. Sen. Action of polyethylene glycol on the fusion of human erythrocyte membranes. *J. Membrane Biol.* 84: 137–146, 1985.

147. Hui, S. W., T. P. Stewart, and L. T. Boni. The nature of lipidic particles and their roles in polymorphic transitions. *Chem. Phys. Lipids* 33: 113–126, 1983.

148. Hui, S. W., T. P. Stewart, L. T. Boni, and P. L. Yeagle. Membrane fusion through point defects in bilayers. *Science* 212: 921–923, 1981.

149. Ivanov, I. B. *Thin Liquid Films.* New York: Marcel Dekker, 1988.

150. Ivanov, I. B., and D. S. Dimitrov. Hydrodynamics of thin liquid films. Effects of surface viscosity on thinning and rupture of foam films. *Coll. Pol. Sci.* 252: 982–990, 1974.

151. Ivanov, I. B., and D. S. Dimitrov. Thin film drainage. In: *Thin Liquid Films,* edited by I. B. Ivanov, New York: Marcel Dekker, 1988.

152. Ivanov, I. B., B. P. Radoev, E. Manev, and A. Scheludko. Dynamics and stability of thin liquid films. *Trans. Faraday Soc.* 66: 1262–1273, 1970.

153. Jaffe, L. A., S. Hagiwara, and R. T. Kado. The time course of cortical vesicle fusion in sea urchin eggs observed as membrane capacitance changes. *Dev. Biol.* 67: 243–248, 1978.

154. Jasin, M., K. A. Page, and D. R. Littman. Glycosyl-phosphatidylinositol-anchored CD4/Thy-1 chimeric molecules serve as human immunodeficiency virus receptors in human, but not mouse, cells and are modulated by gangliosides. *J. Virol.* 65: 440–444, 1991.

155. Johann, S. V., J. J. Gibbons, and B. O'Hara. GLVR1, a receptor for gibbon ape leukemia virus, is homologous to a phosphate permease of Neurospora crassa and is expressed at high levels in the brain and thymus. *J. Virol.* 66: 1635–1640, 1992.

156. Kahn, R. A. Fluoride is not an activator of the smaller (20–25 kDa) GTP-binding proteins. *J. Biol. Chem.* 266: 15595–15597, 1991.

157. Kao, K. N., and M. R. Michayluk. A method for high-frequency intergeneric fusion of plant protoplasts. *Planta* 115: 355–367, 1974.

158. Kaplan, D., J. Zimmerberg, A. Puri, D. P. Sarkar, and R. Blumenthal. Single cell fusion events induced by influenza hemagglutinin: studies with rapid-flow, quantitative fluorescence microscopy. *Exp. Cell. Res.* 195: 137–144, 1991.

159. Katz, B., and R. Miledi. The effect of temperature on the synaptic delay at the neuromuscular junction. *J. Physiol.* 181: 656–670, 1965.

160. Keller, P. M., S. Person, and W. Snipes. A fluorescence enhancement assay of cell fusion. *J. Cell Sci.* 28: 167–177, 1977.

161. Kemble, G. W., D. L. Bodian, J. Rose, I. A. Wilson, and J. M. White. Intermonomer disulfide bonds impair the fusion activity of influenza virus hemagglutinin. *J. Virol.* 66: 4940–4950, 1992.

162. Klatzmann, D., E. Champagne, S. Chamaret, J. Gruest, D. Guetard, T. Hercend, J.-C. Gluckman, and L. Montagnier. T-lymphocyte T4 molecule behaves as the receptor for human retrovirus LAV. *Nature* 312: 767–768, 1984.

163. Knoll, G., and H. Plattner. Ultrastructural analysis of biological membrane fusion and a tentative correlation with biochemical and biophysical aspects. In: *Electron Microscopy of Subcellular Dynamics,* edited by H. Plattner, Boca Raton, FL. CRC Press, 1989, p. 95–117.

164. Knutton, S. Studies of membrane fusion. V. Fusion of erythrocytes with non-haemolytic Sendai virus. *J. Cell Sci.* 36: 85–96, 1979.

165. Knutton, S. Studies of membrane fusion. IV. Fusion of HeLa cells with Sendai virus. *J. Cell Sci.* 36: 73–84, 1979.

166. Knutton, S. Studies of membrane fusion. III. Fusion of erythrocytes with polyethylene glycol. *J. Cell Sci.* 36: 61–72, 1979.

167. Konigsberg, I. R., N. McElvain, M. Tootle, and H. Hermann. The dissociability of deoxyribonucleic acid synthesis from the development of multinuclearity of muscle cells in culture. *J. Biophys. Biochem. Cytol.* 8: 333–343, 1960.

168. Kowalski, M., J. Potz, L. Basiripour, T. Dorfman, W. C. Goh, E. Terwilliger, A. Dayton, C. Rosen, W. Haseltine, and J. Sodroski. Functional regions of the envelope glycoprotein of human immunodeficiency virus type 1. *Science* 237: 1351–1355, 1987.

169. Kozlov, M. M., and V. S. Markin. Possible mechanism of membrane fusion. *Biophysics* 28: 255–261, 1983.

170. Kuhn, T. S. *The Structure of Scientific Revolutions.* Chicago: University of Chicago Press, 1970.

171. Laggner, P. X-Ray studies on biological membranes using synchrotron radiation. *Top. Curr. Chem.* 145: 173–202, 1988.

172. Laggner, P., and M. Kriechbaum. Phospholipid phase transitions: kinetics and structural mechanisms. *Chem. Phys. Lipids* 57: 121–145, 1991.

173. Langhans, T. Uber Resenzellen mit wandstandigen Kernen in Tuberkelen und die fibrose Form des Tuberkels. *Arch. Pathol. Anat. Physiol. Klin. Med.* 42: 382–404, 1868.

174. Lawson, D., M. C. Raff, B. Gomperts, C. Fewtrell, and N. B. Gilula. Molecular events during membrane fusion: a study of exocytosis in rat peritoneal mast cells. *J. Cell Biol.* 72: 242–259, 1977.

175. Layne, S. P., M. J. Merges, M. Dembo, J. L. Spouge, and P. L. Nara, HIV requires multiple gp120 molecules for CD4-mediated infection. *Nature* 346: 277–279, 1990.

176. Levy, J. A., A. D. Hoffman, S. M. Kramer, J. M. Shimabukuro, and L. S. Oshiro. Isolation of lymphocytopathic retroviruses from San Francisco patients with AIDS. *Science* 225: 840–842, 1984.

177. Lewis, W. H. The formation of giant cells in tissue culture and their similarity to those in tuberculous lesions. *Am. Rev. Tuberc.* 15: 616–628, 1927.

178. Liao, M.-J., and J. H. Prestegard. Fusion of phosphatidic acid phosphatidylcholine mixed lipid vesicles. *Biochim. Biophys. Acta* 550: 157–173, 1979.

179. Lifson, J. D., M. B. Feinberg, G. R. Reyes, L. Rabins, B. Banapour, S. Chakrabarti, B. Moss, F. Wong-Staal, K. S. Steimer, and E. G. Engleman. Induction of CD4-dependent cell fusion by the HTLV-III/LAV envelope glycoprotein. *Nature* 323: 725–728, 1986.

180. Lifson, J. D., G. R. Reyes, M. S. McGrath, S. B. Stein, and E. G. Engleman. AIDS retrovirus induced cytopathology: giant cell formation and involvement of CD4 antigen. *Science* 232: 1123–1127, 1986.

181. Lindau, M. Time-resolved capacitance measurements: monitoring exocytosis in single cells. *Q. Rev. Biophys.* 24: 75–101, 1991.

182. Lindau, M., and E. Neher. Patch-clamp techniques for time-resolved capacitance measurements in single cells. *Pflugers Arch.* 411: 137–146, 1988.

183. Lineberger, D. W., D. J. Graham, J. E. Tomassini, and R. J. Colonno. Antibodies that block rhinovirus attachment map to domain 1 of the major group receptor. *J. Virol.* 64: 2582–2587, 1990.

184. Lowy, R. J., D. P. Sarkar, and R. Blumenthal. Influenza virus-cell fusion: Observations of protein and lipid movements using low light-level fluorescent video microscopy. *J. Cell Biochem.* 16: 131–131, 1992.

185. Lowy, R. J., D. P. Sarkar, Y. Chen, and R. Blumenthal. Observation of single influenza virus-cell fusion and measurement by fluorescence video microscopy. *Proc. Natl. Acad. Sci. U.S.A.* 87: 1850–1854, 1990.

186. Loyter, A., V. Citovsky, and R. Blumenthal. The use of fluorescence dequenching methods to follow viral fusion events. *Methods Biochem. Anal.* 33: 128–164, 1988.

187. Loyter, A., M. Tomasi, A. G. Gitman, L. Etinger, and O. Nussbaum. The use of specific antibodies to mediate fusion between Sendai virus envelopes and living cells. *Ciba Found. Symp.* 103: 163–180, 1984.

188. Loyter, A., and D. J. Volsky. Reconstituted Sendai virus envelopes as carriers for the introduction of biological material into animal cells. In: *Cell Surface Reviews,* vol 8, *Membrane Reconstitution,* edited by G. Poste and G. L. Nicolson, Amsterdam: Elsevier/North-Holland, 1982, p. 215–266.

189. Lucy, J. A. Mechanism of chemically induced fusion. *Cell Surface Rev.* 5: 267–304, 1978.

190. Lucy, J. A. Do hydrophobic sequences cleaved from cellular polypeptides induce membrane fusion reaction in vivo? *FEBS Lett.* 166: 223–231, 1984.

191. Maddon, P. J., A. G. Dalgleish, J. S. McDougal, P. R. Clapham, R. A. Weiss, and R. Axel. The T4 gene encodes the AIDS virus receptor and is expressed in the immune system and the brain. *Cell* 47: 333–348, 1986.

192. Malhotra, V., L. Orci, B. S. Glick, M. R. Block, and J. E. Rothman. Role of an N-ethylmaleimide-sensitive transport component in promoting fusion of transport vesicles with cisternae of the Golgi stack. *Cell* 54: 221–227, 1988.

193. Markwell, M. A., A. Portner, and A. L. Schwartz. An alternative route of infection for viruses: entry by means of the asialoglycoprotein receptor of a Sendai virus mutant lacking its attachment protein. *Proc. Natl. Acad. Sci. U.S.A.* 82: 978–982, 1985.

194. Marsh, M., and A. Helenius. Virus entry into animal cells. *Adv. Virus Res.* 36: 107–151, 1989.

195. Maruyama, Y. Ca^{2+}-induced excess capacitance fluctuation studies by phase-sensitive detection method in exocrine pancreatic acinar cells. *Pflugers Arch.* 407: 561–569, 1986.

196. Mayorga, L. S., R. Diaz, and P. D. Stahl. Regulatory role for GTP-binding proteins in endocytosis. *Science* 244: 1475–1477, 1989.

197. McClure, M. O., M. Marsh, and R. Weiss. Human immunodeficiency virus infection of CD4-bearing cells occurs by a pH-independent mechanism. *EMBO J.* 7: 513–518, 1988.

198. McDougal, J. S., M. S. Kennedy, J. N. Sligh, S. P. Cort, A. Mawle, and J.K.A. Nicholson. Binding of HTLV-III/LAV to T4+ cells by a complex of the 110K viral protein and the T4 molecule. *Science* 231: 382–385, 1986.

199. McKeating, J. A., A. McKnight, and J. P. Moore. Differential loss of envelope glycoprotein gp120 from virions of human immunodeficiency virus type 1 isolates: effects on infectivity and neutralization. *J. Virol.* 65: 852–860, 1991.

200. Meers, P., K. Hong, and D. Papahadjopoulos. Free fatty acid enhancement of cation-induced fusion of liposomes: sinergism with synexin and other promoters of vesicle aggregation. *Biochemistry* 27: 6784–6794, 1988.

201. Meers, P., K. L. Hong, and D. Papahadjopoulos. Role of specific lipids and annexins in calcium-dependent membrane fusion. *Ann. NY Acad. Sci.* 635: 259–272, 1991.

202. Melancon, P., B. S. Glick, V. Malhotra, P. J. Weidman, T. Serafini, M. L. Gleason, L. Orci, and J. E. Rothman. Involvement of GTP-binding "G" proteins in transport through the Golgi stack. *Cell* 51: 1053–1062, 1987.

203. Miller, N., and L. M. Hutt-Fletcher. A monoclonal antibody to glycoprotein gp85 inhibits fusion but not attachment of Epstein-Barr virus. *J. Virol.* 62: 2366–2372, 1988.

204. Moore, J. P., B. A. Jameson, R. A. Weiss, and Q. J. Sattentau. The HIV-cell fusion reaction. In: *Viral Fusion Mechanisms,* edited by J. Bentz, Boca Raton, FL: CRC Press, p. 233–289, 1993.

205. Morgan, A., and R. D. Burgoyne. Relationship between arachidonic acid release and Ca^{2+}-dependent exocytosis in digitonin-permeabilized bovine adrenal chromaffin cells. *Biochem. J.* 271: 571–574, 1990.

206. Morgan, A., and R. D. Burgoyne. Exo1 and Exo2 proteins stimulate calcium-dependent exocytosis in permeabilized adrenal chromaffin cells. *Nature* 355: 833–836, 1992.

207. Morris, S. J., D. Bradley, G. C. Gibson, P. D. Smith, and R. Blumenthal. Use of membrane-associated fluorescence probes to monitor fusion of vesicles: rapid kinetics of aggregation and fusion using pyrene excimer/monomer fluorescence. In: *Spectroscopic Membrane Probes,* edited by L. Loew, Boca Raton, FL: CRC Press, 1988, p. 161–191.

208. Morris, S. J., C. C. Gibson, P. D. Smith, P. C. Greif, C. W. Stirk, D. Bradley, D. M. Haynes, and R. Blumenthal. Rapid kinetics of Ca^{2+}-induced fusion of phosphatidylserine-phosphatidylethanol vesicles. *J. Biol. Chem.* 260: 4122–4127, 1985.

209. Morris, S. J., D. P. Sarkar, J. M. White, and R. Blumenthal. Kinetics of pH-dependent fusion between 3T3 fibroblasts expressing influenza hemagglutinin and red blood cells. *J. Biol. Chem.* 264: 3972–3978, 1989.

210. Morrison, T., C. McQuain, and L. McGinnes. Complementation between avirulent Newcastle disease virus and a fusion protein gene expressed from a retrovirus vector: requirements for membrane fusion. *J. Virol.* 65: 813–822, 1991.

211. Moscona, A., and R. W. Peluso. Fusion properties of cells infected with human parainfluenza virus type 3: receptor requirements for viral spread and virus-mediated membrane fusion. *J. Virol.* 66: 6280–6287, 1992.

212. Munson, J. B., and G. W. Sypert. Properties of single fibre excitatory post-synaptic potentials in triceps surae motoneourons. *J. Physiol.* 296: 329–342, 1979.

213. Murata, M., S. Takahashi, Y. Shirai, S. Kagiwada, R. Hishida, and S. Ohnishi. Specificity of amphiphilic anionic peptides for fusion of phospholipid vesicles. *Biophys. J.* 64: 724–34, 1993.

214. Neher, E., and A. Marty. Discrete changes of cell membrane capacitance observed under conditions of enhanced secretion in bovine adrenal chromaffin cells. *Proc. Natl. Acad. Sci. U.S.A.* 79: 6712–6716, 1982.

215. Nemerow, G. R., and N. R. Cooper. CR2 (CD21) mediated infection of B lymphocytes by Epstein-Barr virus. *Semin. Virol.* 3: 117–124, 1992.

216. Neumann, E., and K. Rosenheck. Permeability changes induced by electric pulses in vesicular membranes. *J. Membrane Biol.* 10: 279–290, 1972.

217. Neumann, E., G. Gerisch, and K. Opatz. Cell fusion induced by high electric impulses applied to Dictyostelium. *Naturwissenschaften* 67: 414–415, 1980.

218. Neumann, E., M. Schaefer-Ridder, Y. Wang and P. H. Hofschneider. Gene transfer into mouse lyoma cells by electroporation in high electric fields. *EMBO J.* 1: 841–845, 1982.

219. Neumann, E., A. E. Sowers, and C. A. Jordan. *Electroporation and Electrofusion in Cell Biology.* New York: Plenum Press, 1989.

220. Nusse, O., and M. Lindau. The dynamics of exocytosis in human neutrophils. *J. Cell Biol.* 107: 2117–2126, 1989.

221. Oberhauser, A. F., J. R. Monck, and J. M. Fernandez. Events leading to the opening and closing of the exocytotic fusion pore have markedly different temperature dependencies. Kinetic analysis of single fusion events in patch-clamped mouse mast cells. *Biophys. J.* 61: 800–809, 1992.

222. Ohki, S. Effects of divalent cations, temperature, osmotic pressure gradient, and vesicle curvature on phosphatidylserine vesicle fusion. *J. Membrane Biol.* 77: 265–275, 1984.

223. Ohki, S. *Cell and Model Membrane Interactions.* New York: Plenum Press, 1991.

224. Ohki, S., D. Doyle, T. D. Flanagan, S. W. Hui, and E. Mayhew. *Molecular Mechanisms of Membrane Fusion.* New York: Plenum Press, 1988.

225. Ohnishi, S. Fusion of viral envelopes with cellular membranes. *Curr. Top. Membr. Transp.* 32: 257–296, 1988.

226. Okada, Y. The fusion of Ehrlich's tumor cells caused by HVJ virus in vitro. *Biken's J.* 1: 103–110, 1958.

227. Okada, Y. Analysis of giant polynuclear cell formation caused by HVJ virus from Ehrlich's ascites tumor cells I. Microscopic observation of giant polynuclear cell formation. *Exp. Cell Res.* 26: 98–107, 1962.

228. Okada, Y. Sendai virus-mediated cell fusion. *Curr. Top. Membr. Transp.* 32: 297–336, 1988.

229. Orci, L., and A. Perrelet. Ultrastructural aspects of exocytotic membrane fusion. *Cell Surface Rev.* 5: 629–656, 1978.

230. Orloff, G. M., S. L. Orloff, M. S. Kennedy, P. J. Maddon, and J. S. McDougal. Penetration of CD4 T cells by HIV-1. The CD4 receptor does not internalize with HIV, and CD4-related signal transduction events are not required for entry. *J. Immunol.* 146: 2578–2587, 1991.

231. Ornberg, R. L., and T. S. Reese. Beginning of exocytosis captured by rapid-freezing of Limulus amebocytes. *J. Cell Biol.* 90: 40–54, 1981.

232. Pak, C. C., M. Krumbiegel, and R. Blumenthal. Intermediates in influenza virus PR/8 haemagglutinin-induced membrane fusion. *J. Gen. Virol.* 75: 395–399, 1994.

233. Palade, G. Intracellular aspects of the process of protein synthesis. *Science* 189: 347–358, 1975.

234. Palade, G. E., and R. R. Bruns. Structural modulations of plasmalemmal vesicles. *J. Cell Biol.* 37: 633–649, 1968.

235. Panaiotov, I., D. S. Dimitrov, and M. Ivanova. Influence of bulk-to-surface diffusion interchange on dynamic surface tension excess at uniformly compressed interface. *J. Colloid Interface Sci.* 69: 318–325, 1979.

236. Panaiotov, I., D. S. Dimitrov, and L. Ter-Minassian-Saraga. Dynamics of insoluble monolayers. II. Viscoelastic behavior and Marangoni effect for mixed protein phospholipid films. *J. Colloid Interface Sci.* 72: 49–53, 1979.

237. Pang, D. T., J.K.T. Wang, F. Valtorta, F. Benfenati, and P. Greengard. Protein tyrosine phosphorylation in synaptic vesicles. *Proc. Natl. Acad. Sci. U.S.A.* 85: 762–766, 1988.

238. Papahadjopoulos, D. Calcium-induced phase changes and fusion in natural and model membranes. *Cell Surface Rev.* 5: 765–790, 1978.

239. Papahadjopoulos, D., S. Nir, and N. Duzgunes. Molecular mechanisms of calcium-induced membrane fusion. *J. Bioenerg. Biomembranes* 22: 157–179, 1990.

240. Papahadjopoulos, D., G. Poste, B. E. Schaeffer, and W. J. Vail. Membrane fusion and molecular segregation in phospholipid vesicles. *Biochim. Biophys. Acta* 352: 10–28, 1974.

241. Papahadjopoulos, D., W. J. Vail, K. Jacobson, and G. Poste. Cochleate lipid cylinders: formation by fusion of unilamellar lipid vesicles. *Biochim. Biophys. Acta* 394: 483–491, 1975.

242. Papahadjopoulos, D., W. J. Vail, C. Newton, S. Nir, K. Jacobson, G. Poste, and R. Lazo. Studies on membrane fusion. III. The role of calcium-induced phase changes. *Biochim. Biophys. Acta* 465: 579–598, 1977.

243. Papahadjopoulos, D., W. J. Vail, W. A. Pangborn, and G. Poste. Studies on membrane fusion. II. Induction of fusion in pure phospholipid membranes by calcium ions and other divalent metals. *Biochim. Biophys. Acta* 448: 245–264, 1976.

244. Parsegian, V. A., and R. P. Rand. Membrane interaction and deformation. Surface forces in biological systems. *Ann. NY Acad. Sci.* 416: 1–12, 1983.

245. Parsegian, V. A., and R. P. Rand. Forces governing lipid interaction and rearrangement. In: *Membrane Fusion,* edited by J. Wilschut and D. Hoekstra, New York: Marcel Dekker, 1991, p. 65–85.

246. Paterson, R. G., S. W. Hiebert, and R. A. Lamb. Expression at the cell surface of biologically active fusion and hemagglutinin/neuraminidase proteins of the paramyxovirus simian virus 5 from cloned cDNA. *Proc. Natl. Acad. Sci. U.S.A.* 82: 7520–7524, 1985.

247. Pauza, C. D., and T. M. Price. Human immunodeficiency virus infection of T cells and monocytes proceeds via receptor-mediated endocytosis. *J. Cell Biol.* 107: 959–968, 1988.

248. Pinto Da Silva, P., and M. L. Nogueira. Membrane fusion during secretion. A hypothesis based on electron microscopy observation of Phytophthora palmivora zoospores during encystment. *J. Cell Biol.* 73: 161–181, 1976.

249. Plattner, H. Regulation of membrane fusion during exocytosis. *Int. Review of Cytology* 119: 197–286, 1989.

250. Pollard, H. B., A. L. Burns, and E. Rojas. A molecular basis for synexin driven calcium-dependent membrane fusion. *J. Exp. Biol.* 139: 267–286, 1988.

251. Pollard, H. B., E. Rojas, and A. L. Burns. Synexin and chromaffin granule membrane fusion. A new "hydrophobic bridge" hypothesis for driving and directing of the fusion process. *Ann. NY Acad. Sci.* 493: 524–541, 1987.

252. Pollard, H. B., E. Rojas, R. W. Pastor, E. M. Rojas, H. R. Guy,

and A. L. Burns. Synexin: molecular mechanism of calcium-dependent membrane fusion and voltage-dependent calcium-channel activity. Evidence in support of the "hydrophobic bridge hypothesis" for exocytotic membrane fusion. *Ann. NY Acad. Sci.* 635: 328–351, 1991.

253. Pollard, H. B., O. Zinder, O. G. Hoffman, and O. Nikodejevic. Regulation of the transmembrane potential of isolated chromaffin granules by ATP, ATP analogs and external pH. *J. Biol. Chem.* 251: 4544–4550, 1976.

254. Pontecorvo, G. Production of indefinitely multiplying mammalian somatic cell hybrids by polyethylene glycol (PEG) treatment. *Somatic. Cell Genet.* 1: 397–400, 1975.

255. Poole, A. R., J. I. Howell, and J. A. Lucy. Lysolecithin in cell fusion. *Nature* 227: 810–814, 1970.

256. Popovic, M., M. G. Sarngadharan, E. Read, and R. C. Gallo. Detection, isolation and continuous production of cytopathic retroviruses (HTLV-III) from patients with AIDS and pre-AIDS. *Science* 224: 497–500, 1984.

257. Poste, G., and L. Nicolson. *Membrane Fusion.* Amsterdam: Elsevier North-Holland Biomedical Press, 1978.

258. Poulin, L., L. A. Evans, S. B. Tang, A. Barboza, H. Legg, D. R. Littman, and J. A. Levy. Several CD4 domains can play a role in human immunodeficiency virus infection in cells. *J. Virol.* 65: 4893–4901, 1991.

259. Primakoff, P., H. Hyatt, and J. Tredick-Kline. Identification of a sperm surface protein with a potential role in sperm-egg membrane fusion. *J. Cell Biol.* 104: 141–149, 1987.

260. Puri, A., F. Booy, R. W. Doms, J. M. White, and R. Blumenthal. Conformational changes and fusion activity of influenza hemagglutinin of the H2 and H3 subtypes: effects of acid pretreatment. *J. Virol.* 64: 3824–3832, 1990.

261. Puri, A., M. J. Clague, C. Schoch and R. Blumenthal. Kinetics of fusion of enveloped viruses with cells. *Methods. Enzymol.* 220: 277–287, 1993.

262. Puri, A., S. J. Morris, P. Jones, M. Ryan, and R. Blumenthal. Heat-resistant factors in human erythrocyte membranes mediate CD4-dependent fusion with cells expressing HIV-1 envelope glycoprotein. *Virology* 219: 262–267, 1996.

263. Qureshi, N. M., D. H. Coy, R. F. Garry, and L. A. Henderson. Characterization of a putative cellular receptor for HIV-1 transmembrane glycoprotein using synthetic peptides. *AIDS* 4: 553–558, 1990.

264. Radoev, B. P., D. S. Dimitrov, and I. B. Ivanov. Hydrodynamics of thin liquid films. Effects of surfactants on the rate of thinning. *Colloid Polymer Sci.* 252: 50–55, 1974.

265. Rafalski, M., A. Ortiz, A. Rockwell, L. C. Van Ginkel, J. D. Lear, W. F. Degrado, and J. Wilschut. Membrane fusion activity of the influenza virus hemagglutinin: interaction of HA2 N-terminal peptides with phospholipid vesicles. *Biochemistry.* 30: 10211–10220, 1991.

266. Rand, R. P., and V. A. Parsegian. Hydration forces between phospholipid bilayers. *Biochem. Biophys. Acta* 988: 351–376, 1989.

266a. Rey F. A., F. X. Heinz, C. Mandl, C. Kunz, and S. C. Harrison. The envelope glycoprotein from tick-borne encephalitis virus at 2 Å resolution. *Nature* 375: 291–298, 1995.

267. Ringertz, N. R., and R. E. Savage. *Cell Hybrids.* New York: Academic Press, 1976.

268. Roizman, B. Polykaryocytosis. *Cold Spring Harbor Symp. Quant. Biol.* 27: 327–342, 1962.

269. Roos, D. S., and P. W. Choppin. Biochemical studies on cell fusion. *J. Cell Biol.* 101: 1578–1598, 1985.

270. Roos, D. S., R. Davidson, and P. Choppin. Control of membrane fusion in polyethylene glycol-resistant cell mutants: application to fusion technology. In: *Cell Fusion,* edited by A. E. Sowers, New York: Plenum Press, 1987, p. 123–144.

271. Rothman, J. E., and L. Orci. Molecular dissection of the secretory pathway. *Nature* 355: 409–415, 1992.

272. Rubin, R. J., and Y. Chen. Diffusion and redistribution of lipid-like molecules between membranes in virus-cell and cell-cell fusion systems. *Biophys. J.* 58: 1157–1167, 1990.

273. Ruigrok, R. W., S. R. Martin, S. A. Wharton, J. J. Skehel, and P. M. Bayley. Conformational changes in the hemagglutinin of influenza virus which accompany heat-induced fusion of virus with liposomes. *Virology* 155: 484–497, 1986.

274. Ryu, S.-E., P. D. Kwong, A. Truneh, T. G. Porter, J. Arthos, M. Rosenberg, X. Dai, N.-H. Xuong, R. Axel, R. W. Sweet, and W. A. Hendrickson. Crystal structure of an HIV-binding recombinant fragment of human CD4. *Nature* 348: 419–426, 1990.

275. Sackmann, E. Molecular and global structure and dynamics of membranes and lipid bilayers. *Can. J. Phys.* 68: 999–1012, 1990.

276. Sakai, Y., and H. Shibuta. Syncytium formation by recombinant vaccinia viruses carrying bovine parainfluenza 3 virus envelope protein genes. *J. Virol.* 63: 3661–3668, 1989.

277. Sale, A.J.H., and W. A. Hamilton. Effects of high electric fields on microorganisms. I. Killing of bacteria and yeasts. *Biochim. Biophys. Acta* 148: 781–788, 1967.

278. Sale, A.J.H., and W. A. Hamilton. Effects of high electric fields on microorganisms. III. Lysis of erythrocytes and protoplasts. *Biochim. Biophys. Acta* 163: 37–43, 1968.

279. Saling, P. M., G. Irons, and R. Waibel. Mouse sperm antigens that participate in fertilization. I. Inhibition of sperm fusion with the egg plasma membrane using monoclonal antibodies. *Biol. Reprod.* 33: 515–526, 1985.

280. Sarkar, D. P., S. J. Morris, O. Eidelman, J. Zimmerberg, and R. Blumenthal. Initial stages of influenza hemagglutinin-induced cell fusion monitored simultaneously by two fluorescent events: cytoplasmic continuity and lipid mixing. *J. Cell Biol.* 109: 113–122, 1989.

281. Satir, B. Ultrastructural aspects of membrane fusion. *J. Supramol. Structr.* 2: 529–537, 1974.

282. Sato, H., J. Orstein, D. Dimitrov, and M. Martin. Cell-to-cell spread of HIV-1 occurs within minutes and may not involve the participation of virus particles. *Virology* 186: 712–724, 1992.

283. Sattentau, Q. J. CD4 activation of HIV fusion. *Int. J. Cell Cloning* 10: 323–332, 1992.

284. Sattentau, Q. J., and J. P. Moore. Conformational changes induced in the human immunodeficiency virus envelope glycoproteins by soluble CD4 binding. *J. Exp. Med.* 174: 407–415, 1991.

285. Sattentau, Q. J., and R. A. Weiss. The CD4 antigen: physiological ligand and HIV receptor. *Cell* 60: 697–700, 1988.

286. Sauter, N. K. Hemagglutinins from two influenza virus variants bind to sialic acid derivatives with millimolar dissociation constants: a 500-MHz proton nuclear magnetic resonance study. *Biochemistry.* 28: 8388–8396, 1989.

287. Scheid, A., and P. W. Choppin. Two disulfide-linked polypeptide chains constitute the active F protein of paramyxoviruses. *Virology.* 80: 54–66, 1977.

288. Scheludko, A. *Colloid Chemistry.* Amsterdam: North Holland, 1967.

289. Schierenberg, E. Altered cell-division rates after laser-induced cell fusion in nematode embryos. *Dev. Biol.* 101: 240–245, 1984.

290. Schoch, C., R. Blumenthal, and M. J. Clague. Identification and characterization of a long lived state for influenza virus-erythrocyte complexes committed to fusion a neutral pH. *FEBS Lett.* 311: 221–225, 1992.

291. Scholtissek, C. Stability of infectious influenza A viruses at low pH and at elevated temperature. *Vaccine* 3: 215–218, 1985.

292. Sechoy, O., M. Vidal, J. R. Philippot, and A. Bienvenue. Interactions of human lymphoblasts with targeted vesicles containing Sendai virus envelope proteins. *Exp. Cell Res.* 185: 122–131, 1989.

293. Segawa, A., S. Terakawa, S. Yamashina, and C. R. Hopkins. Exocytosis in living salivary glands: direct visualization by video-enhanced microscopy and confocal laser microscopy. *Eur. J. Cell Biol.* 54: 322–330, 1991.

294. Senda, M., J. Takeda, S. Abe, and T. Nakamura. Induction of cell fusion of plant protoplasts by electrical stimulation. *Plant Cell Physiol.* 20: 1441–1443, 1979.

295. Siegel, D. P. Inverted micellar structures in bilayer membranes: formation rates and half-lives. *Biophys. J.* 45: 399–420, 1984.

296. Siegel, D. P. Modeling protein-induced fusion mechanisms: insights from the relative stability of lipidic structures. In: *Viral Fusion Mechanisms,* edited by J. Bentz, Boca Raton, FL: CRC Press, 1993, p. 475–512.

297. Siegel, D. P., J. Banschbach, and P. L. Yeagle. Stabilization of H2 phases by low levels of diglycerides and alkanes: an NMR, calorimetric, and X-ray diffraction study. *Biochemistry* 28: 5010–5019, 1989.

298. Sinangil, F., A. Loyter, and D. J. Volsky. Quantitative measurement of fusion between human immunodeficiency virus and cultured cells using membrane fluorescence dequenching. *FEBS Lett.* 239: 88–92, 1988.

299. Skehel, J. J., P. M. Bayley, E. B. Brown, S. R. Martin, M. D. Waterfield, J. M. White, I. A. Wilson, and D. C. Wiley. Changes in the conformation of influenza virus hemagglutinin at the pH optimum of virus-mediated membrane fusion. *Proc. Natl. Acad. Sci. U.S.A.* 79: 968–972, 1982.

300. Sodroski, J., W. C. Goh, C. Rosen, K. Campbell, and W. A. Haseltine. Role of the HTLV-III/LAV envelope in syncytium formation and cytopathicity. *Nature* 322: 470–474, 1986.

301. Sorieul, S., and B. Ephrussi. Karyological demonstration of hybridization of mammalian cells in vitro. *Nature* 190: 653–654, 1961.

302. Sowers, A. E. Characterization of electric field-induced fusion in erythrocyte ghost membranes. *J. Cell Biol.* 99: 1989–1996, 1984.

303. Sowers, A. E. A long-lived fusogenic state is induced in erythrocyte ghosts by electric pulses. *J. Cell Biol.* 102: 1358–1362, 1986.

304. Sowers, A. E. *Cell Fusion.* New York: Plenum Press, 1987.

305. Spear, P. G. Membrane fusion induced by herpes simplex virus. In: *Viral Fusion Mechanisms,* edited by J. Bentz, Boca Raton, FL: CRC Press, 1993, p. 201–232.

306. Spruce, A. E., L. J. Breckenridge, A. K. Lee and W. Almers. Properties of the fusion pore that forms during exocytosis of a mast cell secretory vesicle. *Neuron* 4: 643–654, 1990.

307. Spruce, A. E., A. Iwata, and W. Almers. The first milliseconds of the pore formed by a fusogenic viral envelope protein during membrane fusion. *Proc. Natl. Acad. Sci. U. S. A.* 88: 3623–3627, 1991.

308. Spruce, A. E., A. Iwata, J. M. White, and W. Almers. Patch clamp studies of single cell-fusion events mediated by a viral fusion protein. *Nature* 342: 555–558, 1989.

309. Steffy, K. R., G. Kraus, D. J. Looney, and F. Wong-Staal. Role of the fusogenic peptide sequence in syncytium induction and

infectivity of human immunodeficiency virus type 2. *J. Virol.* 66: 4532–4535, 1992.

310. Stegmann, T., J. M. Delfino, F. M. Richards, and A. Helenius. The HA2 subunit of influenza hemagglutinin inserts into the target membrane prior to fusion. *J. Biol. Chem.* 266: 18404–18410, 1991.

311. Stegmann, T., J. M. White, and A. Helenius. Intermediates in influenza-induced membrane fusion. *EMBO J.* 9: 4231–4241, 1990.

312. Stein, B. S., S. D. Gowda, J. D. Lifson, R. C. Penhallow, K. G. Bensch, and E. G. Engleman. pH-independent HIV entry into CD4-positive T cells via virus envelope fusion to the plasma membrane. *Cell* 49: 659–668, 1987.

313. Steinhauer, D. A., N. K. Sauter, J. J. Skehel, and D. C. Wiley. Receptor binding and cell entry by influenza viruses. *Semin. Virol.* 3: 91–100, 1992.

314. Stenger, D. A., and S. W. Hui. Kinetics of ultrastructural changes during electrically-induced fusion of human erythrocytes. *J. Membrane Biol.* 93: 43–53, 1986.

315. Stenger, D. A., and S. W. Hui. Human erythrocyte electrofusion kinetics monitored by aqueous contents mixing. *Biophys. J.* 53: 833–838, 1988.

316. Stenger, D. A., and S. W. Hui. Electrofusion kinetics: studies using electron microscopy and fluorescence contents mixing. In: *Electroporation and Electrofusion in Cell Biology,* edited by E. Neumann, A. E. Sowers, and C. A. Jordan, New York: Plenum Press, 1989, p. 167–180.

317. Stow, J. L., J. B. De Almeida, N. Narula, E. J. Holtzman, L. Ercolani, and D. A. Ausiello. A heterotrimeric G protein, G alpha i-3, on Golgi membranes regulates the secretion of a heparan sulfate proteoglycan in LLC-PK1 epithelial cells. *J. Cell Biol.* 114: 1113–1124, 1991.

318. Sugar, I. P., W. Forster, and E. Neumann. Model of cell electrofusion: membrane electroporation, pore coalescence and percolation. *Biophys. Chem.* 26: 321–335, 1987.

319. Tate, M. W., E. Shyamsunder, and S. M. Gruner. Kinetics of the lamellar-inverse hexagonal phase transition determined by time-resolved X-ray diffraction. *Biochemistry* 31: 1081–1092, 1992.

320. Teissie, I., and M. P. Rols. Fusion of mammalian cells in culture is obtained by creating the contact between cells after their electropermeabilization. *Biochem. Biophys. Res. Commun.* 140: 258–266, 1986.

321. Thali, M., C. Furman, E. Helseth, H. Repke, and J. Sodroski. Lack of correlation between soluble CD4-induced shedding of the human immunodeficiency virus type 1 exterior envelope glycoprotein and subsequent membrane fusion events. *J. Virol.* 66: 5516–5524, 1992.

322. Thomas, L., K. Harting, D. Langosch, H. Rehm, E. Bamberg, W. W. Franke, and H. Betz. Identification of synaptophysin—a hexameric channel protein of the synaptic vesicle membrane. *Science* 2073: 1050–1053, 1988.

323. Tilcock, C.P.S., and D. Fisher. The interaction of phospholipid membranes with polyethylene glycol vesicle aggregation and lipid exchange. *Biochim. Biophys. Acta* 688: 645–652, 1982.

323a. Tse, F. W., A. Iwata, and W. Almers. Membrane flux through the pore formed by a fusogenic viral envelope protein during cell fusion. *J. Cell. Biol.* 121: 543–552, 1993.

324. Vacquier, V. D., and G. W. Moy. Isolation of bindin: the protein responsible for adhesion of sperm to sea urchin eggs. *Proc. Natl. Acad. Sci. U. S. A.* 74: 2456–2460, 1977.

325. Van Meer, G., J. Davoust, and K. Simons. Parameters affecting low-pH-mediated fusion of liposomes with the plasma mem-

brane of cells infected with influenza virus. *Biochemistry.* 24: 3593–3602, 1985.

326. Verkleij, A. J., and P.H.J.Th. Ververgaert. Freeze-fracture morphology of biological membrane. *Biochim. Biophys. Acta* 515: 303–327, 1978.

327. Verwey, E. J. W., and T.Th.G. Overbeek. *Theory of Stability of Lyophobic Colloids.* Amsterdam: Elsevier, 1948.

328. Walter, A., D. Margolis, R. Mohan, and R. Blumenthal. Apocytochrome c induces pH-dependent vesicle fusion. *Membr. Biochem.* 6: 217–237, 1986.

329. Walter, A., C. J. Steer, and R. Blumenthal. Polylysine induces pH-dependent fusion of acidic phospholipid vesicles: a model for polycation-induced fusion. *Biochim. Biophys. Acta* 861: 319–330, 1986.

330. Wang, H., M. P. Kavanaugh, R. A. North, and D. Kabat. Cell-surface receptor for ecotropic murine retroviruses is a basic amino-acid transporter [see comments]. *Nature* 352: 729–731, 1991.

331. Wang, J., Y. Yan, T.P.J. Garrett, J. Liu, D. W. Rodgers, R. L. Garlick, G. E. Tarr, Y. Husain, E. L. Reinherz, and S. C. Harrison. Atomic structure of a fragment of human CD4 containing two immunoglobulin-like domains. *Nature* 348: 411–418, 1990.

332. Wang, K. S., R. J. Kuhn, E. G. Strauss, S. Ou, and J. H. Strauss. High-affinity laminin receptor is a receptor for Sindbis virus in mammalian cells. *J. Virol.* 66: 4992–5001, 1992.

333. Weaver, J. C., and K. T. Powell. Theory of electroporation. In: *Electroporation and Electrofusion in Cell Biology,* edited by E. Neumann, A. E. Sowers, and C. A. Jordan, New York: Plenum Press, 1989, p. 111–126.

334. Weber, H., W. Forster, H.-E. Jacob, and H. Berg. Microbiological implications of electric field effects. *Z. Allg. Microbiol.* 21: 555–562, 1981.

335. Weidman, P. J., P. Melancon, M. R. Block, and J. E. Rothman. Binding of an N-ethylmaleimide-sensitive fusion protein to Golgi membranes requires both a soluble protein(s) and an integral membrane receptor. *J. Cell Biol.* 108: 1589–1596, 1989.

336. Weis, W., J. H. Brown, S. Cusack, J. C. Paulson, J. J. Skehel, and D. C. Wiley. Structure of the influenza virus haemagglutinin complexed with its receptor, sialic acid. *Nature* 333: 426–431, 1988.

337. Weiss, C. D., J. A. Levy, and J. M. White. Oligomeric organization of gp120 on infectious human immunodeficiency virus type 1 particles. *J. Virol.* 64: 5674–5677, 1990.

338. Weiss, R. A. Human immunodeficiency virus receptors. *Semin. Virol.* 3: 79–84, 1992.

339. Wharton, S. A., S. R. Martin, R. W. Ruigrok, J. J. Skehel, and D. C. Wiley. Membrane fusion by peptide analogues of influenza virus haemagglutinin. *J. Gen. Virol.* 69: 1847–1857, 1988.

340. White, J. M. Viral and cellular fusion proteins. *Annu. Rev. Physiol.* 52: 675–697, 1990.

341. White, J. M., and I. A. Wilson. Anti-peptide antibodies detect steps in a protein conformational change: low-pH activation of the influenza virus hemagglutinin. *J. Cell Biol.* 56: 365–394, 1987.

342. White, J., M. Kielian, and A. Helenius. Membrane fusion proteins of enveloped animal viruses. *Q. Rev. Biophys.* 16: 151–195, 1983.

343. Whitt, M. A., P. Zagouras, B. Crise, and J. K. Rose. A fusion-defective mutant of the vesicular stomatitis virus glycoprotein. *J. Virol.* 64: 4907–4913, 1990.

344. Wild, T. F., E. Malvoisin, and R. Buckland. Measles virus: both the haemagglutinin and fusion glycoproteins are required for fusion. *J. Gen. Virol.* 72: 439–442, 1991.

345. Wiley, D. C., and J. J. Skehel. The structure and function of the hemagglutinin membrane glycoprotein of influenza virus. *Annu. Rev. Biochem.* 56: 365–394, 1987.

346. Willey, R., D. H. Smith, L. A. Lasky, T. S. Theodore, P. L. Earl, B. Moss, D. J. Capon, and M. A. Martin. In vitro mutagenesis identifies a region within the envelope gene of the human immunodeficiency virus that is critical for infectivity. *J. Virol.* 62: 139–147, 1988.

347. Wilschut, J. Membrane fusion in lipid vesicle systems: an overview. In: *Membrane Fusion,* edited by J. Wilschut and D. Hoekstra, New York: Marcel Dekker, 1991, p. 89–126.

348. Wilschut, J., N. Duzgunes, R. Fraley, and D. Papahadjopoulos. Studies on the mechanism of membrane fusion: kinetics of calcium ion induced fusion of phosphatidylserine vesicles followed by a new assay for mixing of aqueous vesicle contents. *Biochemistry* 19: 6011–6021, 1980.

349. Wilschut, J., N. Duzgunes, and D. Papahadjopoulos. Calcium/magnesium specificity in membrane fusion: kinetics of aggregation and fusion of phosphatidylserine vesicles and the role of bilayer curvature. *Biochemistry* 20: 3126–3133, 1981.

350. Wilschut, J., and D. Hoekstra. *Membrane Fusion.* New York: Marcel Dekker, 1991.

351. Wilson, D. W., C. A. Wilcox, G. C. Flynn, E. Chen, W.-J. Kuang, et al. A fusion protein required for vesicle-mediated transport in both mammalian cells and yeast. *Nature* 339: 355–359, 1989.

352. Wilson, I. A., J. J. Skehel, and D. C. Wiley. Structure of the hemagglutinin membrane glycoprotein of influenza virus at 3 A resolution. *Nature* 289: 366–373, 1981.

353. Zaks, W. J., and C. E. Creutz. Evaluation of the annexins as potential mediators of membrane fusion in exocytosis. *J. Bioenerg. Biomembr.* 22: 97–120, 1990.

354. Zhelev, D. V., D. S. Dimitrov, and P. Doinov. Correlationbetween physical parameters in electrofusion and electroporation of protoplasts. *Bioelectrochem. Bioenerg.* 20: 155–167, 1988.

355. Zieseniss, E., and H. Plattner. Synchronous exocytosis in Paramecium cells involves very rapid (<1 s), reversible dephosphorylation of a 65 kD-phosphoprotein in exocytosis-competent strains. *J. Cell Biol.* 101: 2028–2035, 1985.

356. Zimmerberg, J., M. Curran, and F. S. Cohen. A lipid/protein hypothesis for exocytotic fusion pore formation. *Ann. NY Acad. Sci.* 635: 307–317, 1991.

357. Zimmerberg, J., M. Curran, F. S. Cohen, and M. Brodwick. Simultaneous electrical and optical measurements show that membrane fusion precedes secretory granule swelling during exocytosis of beige mouse mast cells. *Proc. Natl. Acad. Sci. USA* 84: 1585–1589, 1987.

357a. Zimmerberg, J., R. Blumenthal, D. P. Sarkar, M. Curran, and S. J. Morris. Restricted movement of lipid and aqueous dyes through pores formed by influenza hemagglutinin during cell fusion. *J. Cell. Biol.* 127: 1885–1894, 1994.

358. Zimmermann, U., and P. Scheurich. High frequency fusion of plant protoplasts by electric fields. *Planta* 151: 26–32, 1981.

359. Zimmermann, U., W. M. Arnold, and W. Mehrle. Biophysics of electroinjection and electrofusion. *J. Electrostatics* 21: 309–345, 1988.

360. Zimmermann, U., G. Pilwat, and F. Rienmann. Reversibler dielektrischer Durchbruch von Zellmembranen in electrostatischen Feldern. *Z. Naturforsch.* 29c: 304–305, 1974.

361. Zimmermann, U., J. Schulz, and G. Pilwat. Transcellular ion flow in Escherichia coli B and electrical sizing of bacterias. *Biophys. J.* 13: 1005–1013, 1973.

15. The ER–Golgi membrane system: compartmental organization and protein traffic

HANS-PETER HAURI | *Department of Pharmacology, Biozentrum, University of Basel, Basel, Switzerland*
Hoffmann-La Roche, Basel, Switzerland

ANJA SCHWEIZER | *Department of Medicine, Washington University School of Medicine, St. Louis, Missouri*

THE FUNDAMENTAL RELATIONSHIP BETWEEN THE ENDO-
PLASMIC RETICULUM (ER) AND THE GOLGI APPARATUS
was established in the 1960s by Jamieson and Palade [reviewed in 358] who proposed a role for carrier vesicles in the transport of secretory proteins. This concept was subsequently expanded to membrane proteins and to the endocytotic pathway. Not until 1980 was it clear that proteins pass through the Golgi apparatus from the *cis* to the *trans* face (38, 40, 141). The elucidation of the molecular machinery of vesicular transport began with the reconstitution of intracellular transport in cell-free systems (124) and permeablized cell systems (33, 464) and the isolation of secretory mutants in yeast (334). These approaches have led to a number of key components that control the budding, docking, and fusion of transport vesicles some of which are universal. In parallel the discovery of novel marker proteins (106, 274, 408, 412, 428, 442, 528) and studies on the maturation of glycoproteins (222) have considerably increased our understanding of the compartmental organization of the secretory pathway. Likewise important are the recent findings that newly synthesized proteins are subjected to a sophisticated quality control in the ER before they attain transport competence (190) and the discovery of retrograde transport pathways that are followed by a limited set of important proteins (259, 260, 363, 366). The molecular analysis of the secretory pathway is now well under way and has become a central goal in cell biological research.

In this chapter we attempt to survey major current issues concerning organization, function, and protein traffic of the early secretory pathway. A number of excellent reviews have been published on various aspects of the ER-Golgi system. They will be cited in the corresponding sections of this chapter.

ORGANIZATION OF THE ER–GOLGI MEMBRANE SYSTEM

The ER–Golgi membrane system is composed of three major morphologically distinct membrane compartments, the ER, the ER–Golgi intermediate compartment (ERGIC), and the Golgi apparatus (Fig. 15.1). Depending on the presumed relationship to its neighbor organelles the ERGIC has also been referred to as "cis-Golgi network" (186, 191, 294), "ER exit sites" (182) or "salvage compartment" (540). Because of its distinct morphological appearance and unique characteristics we propose the ERGIC to constitute a separate compartment (see later under ER—Golgi Intermediate Compartment).

Endoplasmic Reticulum

The ER, discovered by Porter and colleagues (388) and characterized in the 1950s, is organized into a netlike labyrinth of branching tubules and flattened sacs throughout the cytoplasm. All the elements of the ER seem to be connected (508) so that the ER forms a continuous albeit dynamic network (245). The visual-

ization of this network by immunofluorescence microscopy requires proper fixation of the cells, preferably including glutaraldehyde, and an antibody against an abundant ER marker protein (Table 15.1). Low abundant ER markers tend to give the impression of a punctate rather than continuous network pattern. Impressive ER networks are apparent in cells in which ER proteins are overproduced by transfection.

Classically the ER is subdivided into two different domains: the rough ER, including the outer nuclear membrane, and the smooth ER. The ribosome-coated rough ER plays a central part in lipid synthesis and is the port of entry for proteins destined for secretion and for all the organelles of the secretory pathway. Additional major functions of the rough ER are listed in Figure 15.1. Regions of the ER that lack bound ribosomes are collectively termed smooth ER. Elements of smooth ER are more difficult to detect because in most cells they are less obvious and less abundant than rough ER and there is only a single accepted marker

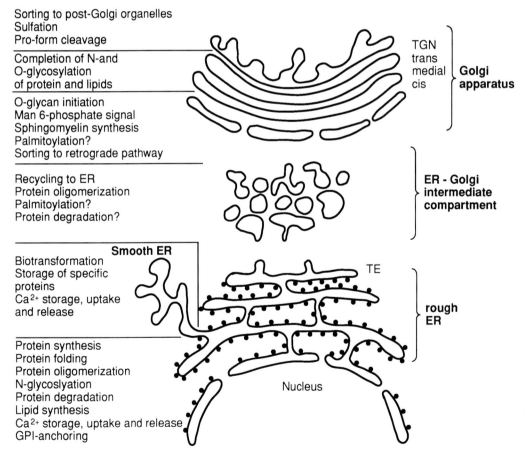

FIG. 15.1. Structural and functional compartmentalization of the early secretory pathway of mammalian cells. *ER*, endoplasmic reticulum; *TE*, transitional element of the rough ER. Note: Unlike as is depicted in this scheme, the ER-Golgi intermediate compartment consists of a multitude of tubulovesicular membrane clusters, some of which are not localized near the Golgi apparatus.

TABLE 15.1. *Popular Organelle Marker Proteins for Which Localization Has Been Firmly Established*

Organelle	Protein	Comment	Reference
ER	Set of at least 6 ER proteins including calnexin	Defined by an antiserum against purified ER, stains RER and SER	271
RER	Glucose-6-phosphatase	Most popular enzyme marker, also present in SER	244
	Ribophorin I and II	Membrane proteins	85, 283
	Docking protein	Membrane protein	183
	PDI	Soluble KDEL-protein, may leak into the secretory pathway in some cells; also present in some SER subregions	132, 467, 495
	BiP	Soluble KDEL-protein, may leak into the secretory pathway in some cells; also present in some SER subregions	46, 132, 467
smooth ER	Epoxide hydrolase	Membrane protein, localization established for hepatocytes	128
ERGIC	ERGIC-53/p58	Recycling membrane protein, high concentration in ERGIC, low concentration in Golgi and RER	196, 428, 430, 442, 443
cis-Golgi	KDEL-receptor	Membrane protein recyling between cis-Golgi and ER, some contribution of medial-Golgi and ERGIC	145, 499
	Peptide-GalNAc-transferase	Membrane protein, first enzyme of O-glycosylation	412, 447
medial-Golgi	Mannosidase I	Membrane protein, also in trans-Golgi	528
	GlcNAc transferae I	Membrane protein, overlap with galactosyl-transferase	106, 327
	Mannosidase II	Membrane protein, also in trans-Golgi	528
trans-Golgi	Galactosyltransferase	Membrane protein, some overlap with GlcNAc-transferase	327, 408, 442
TGN	Sialytransferase	Membrane protein, also in trans-Golgi	394, 409, 410
	TGN38	Membrane protein, recylces between TGN and cell surface	273, 274

protein, epoxide hydrolase, that is specific for smooth ER (128). The validity of this marker has been established for hepatocytes only, although the expression of epoxide hydrolase is quite common in different cell types (Hauri, unpublished). The smooth ER is not generally considered to be an intermediate in protein traffic between rough ER and Golgi in either direction. Based on evidence by immunofluorescence microscopy, however, it was proposed that vesicular stomatitis virus (VSV) G protein has access to and can exit from the smooth ER in virus-infected UT1 cells in which the smooth ER was induced by lovastatin (mevinolin) (41). Whether this pathway is present in normal, noninfected cells remains to be established.

Rough ER Domains. The view of subdividing the ER simply into rough and smooth membranes may require some refinement because both ER domains may have subdomains: the rough ER in terms of protein synthesis or transport and the smooth ER regarding specialized functions. However, there is still limited evidence for a functional subcompartmentalization of the rough ER with the exception of the transitional elements involved in protein exit from the rough ER (358). An attractive hypothesis—that the biosynthesis of the hemoprotein cytochrome P450 preferentially occurs in mitochondria-associated rough ER elements—was not supported by detailed analysis of apoprotein and heme synthesis in mitochondria-associated and non-mitochondria-associated rough ER (291). Several authors, on the other hand, reported on nonuniform distribution of proteins within the rough ER, suggesting subcompartmentalization (94, 110, 530) but the the basis for this apparent regionality is not known.

Smooth ER. The major function of the rough ER is synthesis, translocation, glycosylation, and folding of proteins; whereas the smooth ER has diverse functions including protein storage and biotransformation as well as calcium storage. It is not surprising, therefore, that some of these functions are compartmentalized (467) which is most apparent in a few specialized cell types such as Purkinje neuronal cells of the cerebellum and muscle cells. Most completely characterized is the sarcoplasmic reticulum of skeletal muscle cells that harbors a number of activities determining the relaxation-contraction cycle. Although the protein pattern of the smooth ER is considerably more complex than that of the sarcoplasmic reticulum in which the calcium storage protein calsequestrin and Ca^{2+}-ATPase account for the bulk of the protein, the ER nature of the SR is indicated by the presence of immunoglobulin binding protein (BiP) and protein disulfide isomerase (PDI) (533). Equivalents of sarcoplasmic reticulum may also exist in nonmuscle cells and have been termed calciosomes (532). In Purkinje cells, for example, smooth membrane elements were identified that are enriched in calsequestrin and Ca^{2+}-ATPase. Less clear is the case for most other cells that do not express calsequestrin. In these cells the calcium storage function may be supported by calreticulin (297) but it remains to be shown if calciosome equivalents exist in all cells. It should be noted that the rough ER also stores Ca^{2+} bound to various low-affinity high-capacity luminal proteins (322, 355, 425, 546).

Another kind of subcompartmentalization has been noticed in Purkinje neurons: part of the ER is arranged in smooth-surfaced stacks, most likely due to high expression levels of the inositol–trisphosphate receptor (432, 531) for such stacks can also be induced in COS cells by overexpression of the inositol-trisphosphate receptor (494). A three dimensional morphometric analysis of the Purkinje cell stacks showed that their external cisternae are often in direct contact with rough ER elements whereas the internal cisternae are not, nor do they seem to be connected with each other (420).

The size of the smooth ER drastically increases under conditions that lead to high expression levels of ER proteins. A well-known case is the proliferation of smooth ER in hepatocytes in response to phenobarbital, which leads to the induction of more than 30 different proteins including various cytochromes P450 and epoxide hydrolase (118, 341, 543). The induction process involves phenobarbital responsive elements in the 5′ flanking region of responsive genes (356). In yeast, the membrane anchor of a cytochrome P450 is sufficient to induce ER proliferation (296). Similarly the C-terminal anchor of cytochrome b5 (that can even be replaced by a nonphysiological polyleucine

sequence) mediates ER proliferation (529). Perhaps the most striking example of smooth ER proliferation was observed in UT-1 cells grown in high concentrations of compactin or lovastatin, inhibitors of the 3-hydroxy 3-methylglutaraldehyde coenzyme A (HMG-CoA) reductase (361, 465). Under these conditions the smooth ER forms large cisternal stacks (also termed crystalloid ER) apparently due to the overexpression of the ER protein HMG-CoA reductase. Approximately 24% of the protein in the crystalloid smooth ER corresponds to HGM-CoA reductase (221) whereas phenobarbital-inducible cytochromes P450 and epoxide hydrolase are not induced. The crystalloid ER also contains BiP that is more typically restricted to the rough ER (46). A common prerequisite for ER induction by cytochrome P450, cytochrome b5, and HGM-CoA reductase is membrane insertion pointing to a common mechanism, even though the morphological appearance of ER induced by these three proteins differs somewhat. Obviously, the sensor that mediates ER proliferation responds to a hydrophobic amino acid sequence. Proliferation of smooth ER membranes was also observed in cells transfected with a cDNA encoding rubella virus E1 glycoprotein. Unassembled E1 glycoprotein accumulated in a tubular network of smooth membranes that is in continuity with rough ER but has properties distinct from either the ER, ERGIC or Golgi (181). This network may arise as a consequence of extensive proliferation of transitional elements of the rough ER. It appears, however, that the arrest of the E1 glycoprotein in this subcompartment does not block exocytotic vesicular traffic to the Golgi apparatus.

Smooth ER elements also enlarge as a general response of the cell to the accumulation of abundant, nondegradable secretory proteins that fail to exit the rough ER. Such elements, also termed Russell bodies, have been seen in myeloma cells transfected with a mutant immunoglobulin heavy chain and were proposed to constitute a subcompartment of the smooth ER (522). It is not known if Russell bodies correspond to a subregion of the ER present also in normal cells or if they are formed de novo in response to overexpression on nondegradable proteins.

The physical relationship of smooth ER subdomains with each other is unclear in most cases for various reasons: *(1)* In contrast to rough ER, smooth ER elements of cultured cells often appear vesicular in electron micrographs—which is at least in part due to the difficulty of fixing these fragile membranes properly (226); *(2)* studies on the compartmentalization by immune electron microscopy are still at an early stage and limited to a few cell types (128, 531, 533); and *(3)* domain-specific membrane markers are lacking. The presence or absence of membrane continuities at an ultrastructural level is of limited value because organelle membranes are extremely dynamic. Clearly the establishment of compartmental boundaries will require the isolation of different subdomains, which in turn will allow us to determine whether communication among them occurs by homotypic fusion or vesicular pathways.

Golgi Apparatus

Structural and Functional Compartmentalization. The major functions of the Golgi apparatus (Fig. 15.1) are first, to receive and sort membrane and soluble cargo arriving at the cis-side from the rough ER via the ERGIC; second, to complete the glycosylation of glycoproteins and glycolipids; and third, to sort membrane and soluble cargo at the trans-side to post-Golgi organelles. There is some controversy on the extent of compartmentalization of the Golgi apparatus as summarized in the next section. Three-dimensional reconstruction (398) has shown that the Golgi apparatus consists of a variable number of flattened cisternae arranged in stacks that are laterally connected with each other, at least on the trans-side, by dynamic membrane tubular connections (79, 409). Proteins pass through the Golgi from *cis* to *trans* (38, 40, 141) during which time N-linked oligosaccharide side chains are trimmed and reglycosylated in an ordered fashion (222, 411) and O-linked glycans are attached to selected serine and threonine residues. Protein sorting occurs preferentially on the cis-side as well as on the trans-side. Controversial views concern the number of topologically separate compartments within the Golgi apparatus. A compartment can be defined as a physically and biochemically distinct membrane entity. Two neighbor compartments communicate with each other by shuttle vesicles—much in contrast to subcompartments, which are biochemically distinct regions of the same membrane.

Why the Golgi apparatus has such a remarkable cisternal organization is a mystery. It is generally assumed that the Golgi stack forms an assembly line for glycan biosynthesis, with each cisterna containing a set of enzymes for only one or a few reactions and that the vectorial movement of glycoproteins through the stack is essential to guarantee a sequential order of trimming and reglycosylation (105, 222). However, substrate specificity of the processing enzymes alone should suffice to ensure correct synthesis of a defined oligosaccharide rendering a strict compartmentalization unnecessary. In fact, it has become clear that the Golgi apparatus is functionally much less compartmentalized than initially surmised. Individual cisternae cannot be identified by single enzymes because their *cis-trans* distribution considerably overlaps (see Table 15.1) as most directly shown by immune electron microscopy for the medial-Golgi enzyme GlcNAc-transferase and the trans-Golgi enzyme galactosyltransferase (327). Moreover, there is some variation between cell types (528) the most extreme example being the enterocyte of the small intestine. In these cells trans-Golgi enzymes such as sialytransferase (410) and galactosyltransferase (442) have been localized by immune electron microscopy to all cisternae with the exception of the first fenestrated cis-cisterna. Nevertheless, in most other cells these markers are confined to only one or two cisternae on the trans-side (327, 394, 408, 442, 468, 488). Less well known is the fact that low-temperature incubations frequently used to block protein traffic at various stages of the secretory pathway may alter the compartmentalization of Golgi enzymes. A case in point is galactosyltransferase in HepG2 cells, which at 37°C is confined to one or two trans-cisternae. When the cells are cultured at 15°C for 3 h, however, up to 4 cisternae label positive for galactosyltransferase (Klumperman, J., and H.-P. Hauri, unpublished). The basis for this change is unknown.

A certain Golgi compartmentalization can also be seen in cell fractionation experiments. On density gradients medial Golgi markers, such as mannosidase I, GlcNAc transferase, and mannosidase II can be separated from trans-Golgi markers, such as galactosyltransferase or sialyltransferase (104, 137). On the other hand, it was not possible to separate the cis-Golgi membranes containing KDEL-receptor and GalNAc-transferase from medial-Golgi proteins (196). This may be due to marker overlap or physicochemical similarity of these membranes. The mere fact that cis/medial Golgi membranes can be separated from trans-Golgi membranes does not necessarily indicate that cis- and trans-Golgi are truly separate compartments, nor is the inability to separate neighbor cisternae of cis- and medial-Golgi a valid argument for excluding a vesicular intermediate step.

Cisternal stacking of the Golgi apparatus is perhaps more related to protein sorting than to glycosylation (413). According to this view the stack would provide multiple opportunities to extract escaped ER resident proteins in a multistep process akin to fractional distillation. Although experiments with reporter proteins tagged with the tetrapeptide sequence KDEL at their COOH termini document that the capacity to retrieve ER proteins extends to the trans-side of the Golgi apparatus (280, 298) at present it is unknown if the capacity for multistage retrieval is indeed essential.

Cis-Golgi. To date no marker protein has been localized exclusively to the first fenestrated cis-Golgi cisterna. Unexpectedly, mannosidase I, the first glycosidase to modify N-linked oligosaccharides upon their arrival in the Golgi, was found to be localized to medial/trans Golgi cisternae rather than to the very cis-Golgi (528). Previously, mannosidase I had always been considered to reside in the cis-Golgi (104, 137). The best documented marker for cis-Golgi is peptide-GalNAc-transferase, the enzyme that catalyzes the initial step of O-glycosylation and leads to the formation of GalNAc-O-threonine/serine linkages. Most biochemical studies indicate that O-glycan initiation is a Golgi event (360, 379). O-Glycosylation of the M (formerly E1) glycoprotein of mouse hepatitis virus already starts in the ERGIC (226, 511) but it is unclear if this finding is due to virus infection or cell type variability. Peptide-GalNAc-transferase has now been localized to the cis-Golgi by immune electron microscopy (412) and on density gradients this enzyme cofractionates with cis-medial Golgi membranes (196, Foguet, M. and H.-P. Hauri, unpublished) and not with the ERGIC (447). GlcNAc phosphotransferase, the first enzyme involved in the generation of the lysosomal address mannose-6-phosphate, is also likely to be predominantly associated with the cis-Golgi (363, 444) although in previous fractionation experiments with BW5147.3 cells this enzyme codistributed with the ER enzymes glucosidase I and II (137). Clearly these biochemical findings await confirmation by immune electron microscopy.

Another Golgi protein with major localization in the first cis-cisterna of the Golgi apparatus is the KDEL-receptor (145, 499). The cis-most cisterna also harbors a minor fraction of ERGIC-53/p58 (174, 442) as well as gp74 (5) and a 220 kDa membrane-associated antigen (405). These proteins can recycle to pre-Golgi locations, which supports the notion that protein sorting is a major function of the cis-Golgi.

Medial- and trans-Golgi. Medial- and trans-Golgi cisternae are mainly involved in the terminal processing of N-linked and O-linked oligosaccharide side chains (222, 411). After initial trimming in the ER, further processing of N-linked glycans involves the sequential action of mannosidase I, GlcNAc transferase I, and mannosidase II in the medial-Golgi (106, 528). At this stage the glycans become resistant to endoglycosidase H. In the trans-Golgi and the trans-Golgi network (TGN), N-linked glycans are then converted to mature complex oligosaccharides structures by addition of GlcNAc, fucose, galactose, and sialic acid residues.

Trans-Golgi network. The TGN is thought to be physically connected to the trans-Golgi (142, 143). In the TGN, proteins are segregated into different transport vesicles and dispatched to their final destinations: plasma membrane, endosomes, lysosomes, or secretory vesicles (133, 142). The most popular marker for the TGN is TGN-38/41 (273, 274) (Table 15.1). To what extent this protein is is also present in the trans-Golgi has not been determined by double-labeling experiments at the ultrastructural level. The TGN is thought to be continuous with the trans-Golgi. Nevertheless it remains a separate entity in brefeldin A (BFA)-treated cells, contrary to the cis/medial- and trans-elements of the Golgi apparatus (218, 235, 400). A novel feature of the TGN has recently been established by high-voltage electron microscopy and computed axial tomography (236). There are at least two morphologically distinct coats on vesicles emanating from TGN tubules: clathrin and a novel lace-like coat. Individual tubules do not carry buds of both types and they appear to be consumed in the process of vesicle budding. Assuming that the two types of coated profiles reflect two classes of departing vesicles it follows that sorting of molecules occurs prior to the TGN already in the trans-Golgi (236). This hypothesis is at variance with the widely held view that sorting events take place within the TGN (142, 294).

Alternative Golgi Models. Reflecting the uncertainty of how many compartments constitute the Golgi apparatus different models have been proposed (Fig. 15.2). In the minimal model (Fig. 15.2a) the Golgi consists of three compartments: cis-Golgi network (CGN), medial-Golgi, and TGN (294). The CGN would consist of the cis-most Golgi cisterna and the associated vesicles and tubules that include the intermediate compartment defined by ERGIC-53/p58 (430, 442, 443). Its main function is receipt and sorting of membrane and soluble cargo arriving at the cis-Golgi from the ER. In addition, it would have a limited function in glycosylation that would mostly be restricted to O-glycan initiation by peptide-GalNAc transferase (226, 412, 511). In this model the CGN would also include the "salvage compartment" (364, 540), the major site of protein recycling to the ER (185). Other proposed functions of the CGN are fatty acylation of membrane proteins (48, 526) and phosphotransferase-mediated modification of lysosomal enzymes (363). Independently of the number of cisternae, the medial-Golgi would represent a single continuous compartment and function primarily as glycosylation device for N-linked and O-linked glycosylation reactions as well as glycolipid synthesis. This simplified view of the medial-Golgi is based on the assumption that sequential action of glycosyltransferases does not require separate compartments. The fact that terminal glycosylation steps occur not exclusively in the trans- but also in the medial-Golgi of absorptive enterocytes is supportive evidence for this notion (327, 394, 411). The TGN (142) is viewed as the third compartment. Besides mediating sorted exit

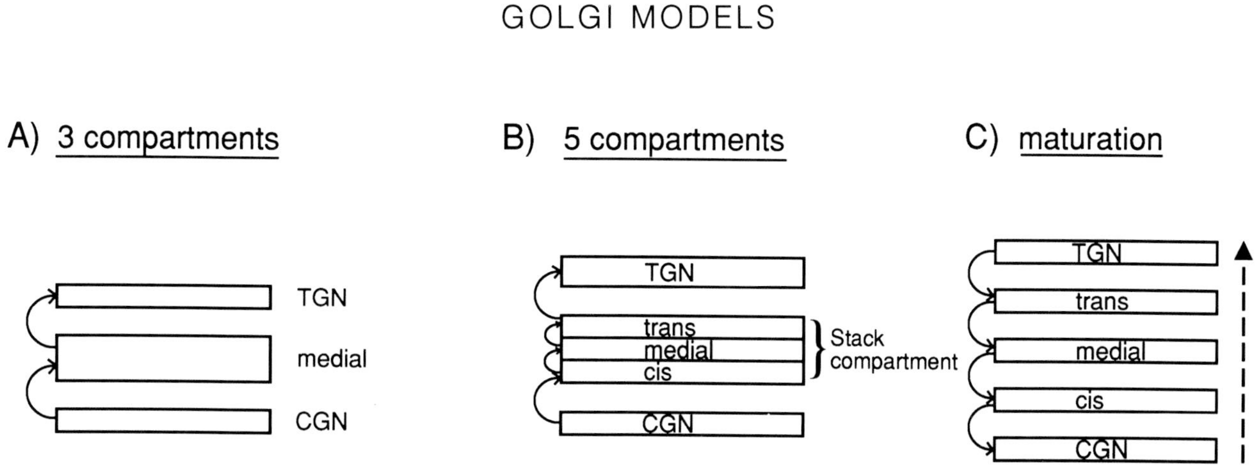

GOLGI MODELS

A) 3 compartments   B) 5 compartments   C) maturation

FIG. 15.2. Alternative Golgi models. *A.* In the minimal compartment model (294) the Golgi apparatus is believed to consist of three compartments. Secretory proteins traverse the Golgi apparatus from cis to trans in two vesicular steps *(arrows)*. *B.* In the five-compartment model (417) the central Golgi stack is further subdivided into three additional compartments named cis, medial, and trans. Transit through the Golgi requires four vesicular steps *(arrows)*. *C.* In the maturation model (258, 307, 429) the cis-Golgi network (CGN) cisterna gradually matures to the trans-Golgi network (TGN) (indicated by dotted arrow) and hence anterograde protein transport through the Golgi would not require the formation of vesicles. Compartmentalization of Golgi enzymes would be maintained by their recycling to earlier cisternae *(arrows)* either by vesicular or tubular transport.

of material from the Golgi apparatus to various post-Golgi organelles it can also receive material from the endocytotic system (111, 223). To consider the TGN as a separate compartment is based mainly on the observation that TGN proteins in general fail to enter the ER upon treatment of cells with BFA (68, 235, 273, 400, 518) suggesting that the TGN stays a separate membrane entity although, at least in some cells (261, 566) the degree of continuity with endosomes and plasma membrane appears to be increased.

The five-compartment model (417) (Fig. 15.2b) is in essence similar to the three-compartment model with the exception of further subdividing the medial Golgi of the first model into cis-, medial-, and trans-compartments. In that model four sequential vesicular steps would operate in transport of proteins from the CGN to the TGN.

A third conceptually different model proposes that the Golgi apparatus is a dynamic maturing structure in which cisternae move and mature in cis to trans direction (73, 258, 307, 429). In the maturation model (Fig. 15.2c) forward transport of proteins through the Golgi apparatus does not require transport vesicles. Functional cis-trans polarity would reflect different degrees of cisternal maturation, which is achieved by progressive recycling of Golgi-specific components from the maturing cisterna to earlier, less mature, cisternae. The cisternal maturation model was originally proposed to explain the movement of algal scales too large to enter forward transport by vesicles (32).

Such a model predicts that Golgi enzymes, not cargo, will be found in the retrograde vesicles. The levels of cargo were indeed lower than expected and Golgi enzymes were found in these (COPI, see below) vesicles (Warren, G., personal communication). However, the average densities of Golgi enzymes in the vesicles were lower than the starting membrane so that budding actually resulted in an increase in the average density of the Golgi enzymes that remained behind. Such a result is incompatible with the cisternal maturation model unless another, yet-to-be identified retrograde vesicle is involved. Nevertheless, there is presently insufficient evidence to discount the maturation model.

ER-Golgi Intermediate Compartment

Structure. Studies on the intracellular transport of Semliki forest virus (427) suggested that protein transport from ER to Golgi involves a complex membrane system that is interposed in between rough ER and cis-Golgi. A recent three-dimensional reconstruction study of the rough ER–Golgi interphase in rat pancreatic acinar cells has readdressed the question of compartment boundaries (458). In these cells the interphase consist of at least four elements: *(1)* transitional elements of the rough ER from which omega-shaped and tubular profiles protrude. Some of these profiles exhibit an electron-dense coat. *(2)* Putative 56 nm transport vesicles. *(3)* Convoluted tubules apparently not, or not permanently, connected with transitional rough ER or cis-Golgi that include profiles of budding (or fusing) coated-vesicles. These tubulo-vesicular elements correspond to the "peripheral elements" (342, 358). (4) Tubular elements in continuity with the first fenestrated cis-cisterna of the Golgi apparatus.

We refer to the tubulo-vesicular ("peripheral") elements as ER-Golgi intermediate compartment (ERGIC). The ERGIC is presently best visualized with antibodies against the human 53 kDa marker protein ERGIC-53 (442) and its rat homologue p58 (238, 239, 428, 430). It consists of a multitude of tubulo-vesicular clusters concentrated mainly near the Golgi apparatus but also present throughout the cytoplasm (Fig. 15.3). Transport studies with the VSV G protein established ERGIC-53 as a marker for the site at which ER-to-Golgi transport is arrested at 15°C (270, 443). The small GTP-binding proteins rab2 (67, 470) and rab1 (145, 431) and the vesicular coat protein beta-COP (147, 342) are also associated with the ERGIC to a great extent.

There are discordant views as to whether the ERGIC is a transiently existing pleiomorphic transport intermediate or a stable compartment (167, 196, 258, 368, 429). The discussion is reminiscent of the issue of whether to consider endosomes permanent or maturing membranes (144, 311). In addition to the general difficulty in determining compartmental boundaries in the secretory pathway, as discussed for the Golgi apparatus, there are two major problems. First, although the concentration of ERGIC-53/p58 is highest in the ERGIC the protein is also present in cis-Golgi and ER (174, 196, 430, 442) reflecting the fact that it recycles in the early secretory pathway. There is at present no membrane marker known that is exclusively and stably associated with the ERGIC. A second complication concerns the extreme dynamics of the ERGIC. With the immunofluorescence microscope, the ERGIC can be seen to undergo drastic morphological changes when transport pathways are blocked by low temperatures, energy depletion, and activators of GTP-binding proteins (196, 430, 443). Therefore, immunofluorescence microscopy alone does not tell us to what extent an observed effect reflects dynamics of the ERGIC or recycling of ERGIC-53.

A number of recent observations, both by light and immune electron microscopy as well as subcellular fractionation, support the notion of a stable albeit highly dynamic intermediate compartment: (1) the ERGIC clusters do not fuse with the ER in BFA-treated cells (260, 430) in which the Golgi apparatus completely disappears (259, 260). Likewise the ERGIC persists in an END4 mutant of Chinese hamster ovary cells under conditions in which the Golgi apparatus

vanishes (209). (2) When ER–Golgi protein transport is impeded by lowering the culture temperature to 15°C, the ERGIC clusters progressively concentrate near the Golgi and upon rewarming to 37°C rapidly reestablish the original pattern of distribution without fusing with the cis-Golgi (196). In fact, the average number of clusters remains constant during temperature manipulations (196, 270). (3) By density gradient centrifugation the ERGIC can be separated from its neighbor organelles, rough ER and cis-Golgi (196, 444). (4) Syntaxin 5 (the mammalian homologue of yeast Sed5p), a putative vesicle targeting molecule that is required for transport between ER and Golgi apparatus in yeast (158) has been localized to the ERGIC in mammalian cells (23). Given the presumed role of Sed5p as a vesicle receptor (471) it follows that the ERGIC is the first post-ER compartment (368). (5) Evidence by subcellular fractionation in conjunction with immunofluorescence and electron microscopy indicates that a major fraction of ERGIC-53 bypasses the cis-Golgi on its recycling pathway to the ER and that the ERGIC receives not only anterograde traffic from the ER but also retrograde traffic from the cis-Golgi (14, 196, 500).

Function. The major function of the ERGIC most likely is to sort proteins in anterograde and retrograde direction. The ERGIC is the first way station from where proteins can recycle to the ER (14, 196). No function has been found yet that is unique for the ERGIC, but the ERGIC contributes to functions that had been ascribed to the ER, including protein assembly such as (higher order) oligomerization of hepatitis B surface antigen (188) and influenza hemagglutinin (503), peptide transport from the cytosol to the secretory pathway by TAP proteins (220), and protein degradation (167, 392, 399). Moreover, mouse corona virus buds into the ERGIC (226, 511) and vaccinia virus assembly also starts at the ERGIC (470). Occasionally mutated proteins appear to accumulate in the ERGIC. Moran and Caras (305) noticed that a hybrid molecule of human growth hormone and the glyophosphatidyl–inositol-anchored decay accelerating factor with a mutated, noncleavable glycophosphatidyl–inositol signal accumulated in an intracellular compartment when overexpressed in COS cells. The compartment partially co-stained for ERGIC-53, and it was speculated that a mechanism exists in the ERGIC to retain proteins containing an uncleaved glycophosphatidyl–inositol signal as part of a system for quality control. However, because the cumulated hybrid behaved in a Golgi-like rather than ERGIC-like manner in BFA-treated cells, the precise site of accumulation remains to be established.

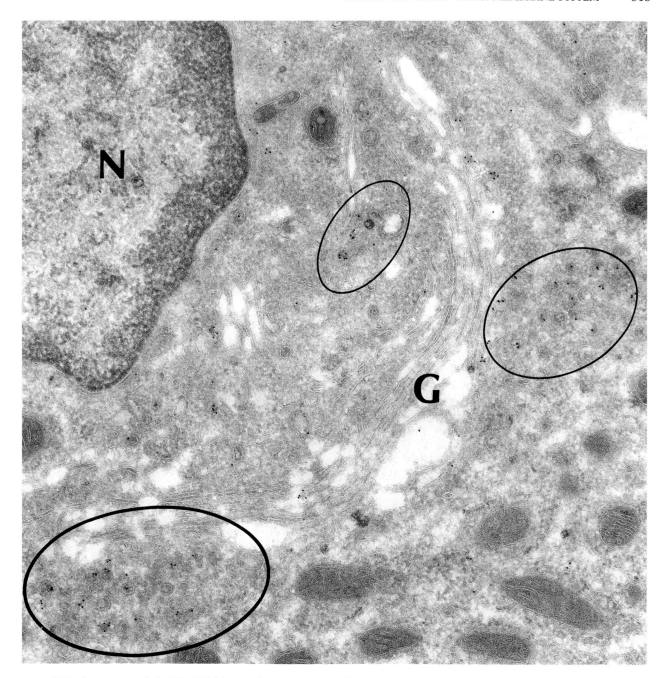

FIG. 15.3. Appearance of the ER–Golgi intermediate compartment (ERGIC) in ultrathin cryosections in enterocytes of a human small intestinal biopsy sample (courtesy of J.A.M. Fransen). The ERGIC was labeled with a mouse mAb against ERGIC-53 followed by incubations with rabbit-anti-mouse and 5 nm colloidal gold-protein A. The ERGIC consists of tubulovesicular membrane clusters *(circled)*. G, Golgi apparatus, N, nucleus.

ANTEROGRADE (EXOCYTOTIC) PROTEIN TRAFFIC FROM ER TO GOLGI

Twenty years ago Palade (358) wrote in his Nobel lecture "The geometry of the connections between the ER and the Golgi complex is still a matter of debate. According to some investigators the two compartments are permanently connected by continuous tubules; according to us, the connection is intermittent and is probably established by shuttling vesicles." It has generally been accepted meanwhile that protein transport within the secretory pathway involves a number of vesicular steps. Although the precise number of such steps from ER to plasma membrane remains to be determined, remarkable progress in understanding vesicle-mediated protein transport has been achieved by the convergence of morphological, biochemical, and genetic approaches (37, 368, 376, 390, 415, 417, 418, 435, 493). The elucidation of the molecular machinery of vesicular transport began with the reconstitution of intracellular transport in cell-free systems (19, 124) and permeabilized cells (33, 464) and the isolation of the secretory mutants in yeast (334, 335).

On the other hand there is also evidence from morphological studies for the existence of tubular connections between organelles including ER and Golgi apparatus (226, 253, 254, 398). Such connections are most convincingly seen by high-voltage electron microscopy of thick sections, but it is still unclear whether they are permanently present and to what extent they contribute to interorganelle transport. Perhaps connections between ER and Golgi cisternae are responsible for retrograde protein transport that is mediated by membrane tubules (196, 260) or they may originate from uncoupled fusion events (108). Alternatively, it was proposed that the observed membrane connections would render vesicular transport unnecessary up to cis-Golgi (226). Tubule formation was found to be temperature dependent. This may explain in part the difficulties encountered in morphological studies which have tried to define whether or not vesicular transport steps occur between the rough ER and the ERGIC (486).

Nevertheless, there is little doubt that anterograde transport between ER and Golgi involves at least one vesicular step. This has most convincingly been demonstrated in yeast, where short-lived transport vesicles can be forced to accumulate as stable intermediates when a component of the vesicle-targeting machinery is inactivated by mutation (390). When isolated, such vesicles can mediate protein transport from ER to Golgi in perforated yeast spheroplasts or microsomal preparations (17, 403).

Vesicular transport from the ER involves the following major steps: *(1)* After translocation the cargo protein undergoes folding and acquires a transport-competent state in the ER. *(2)* The transport competent protein is packaged into a transport vesicle that buds from the ER. *(3)* The vesicle docks to and fuses with the membrane of the first post-ER compartment (that is, the ERGIC in mammalian cells). These processes are discussed in the following sections.

Acquisition of Transport Competence in the ER

During and after translocation newly synthesized secretory proteins undergo a number of folding steps in the ER that ultimately lead to transport competence, that is, proper structural and functional integrity allowing

their export from the ER. This phenomenon is also known as "quality control" (156, 190) or "architectural editing" (217). Folding involves disulfide-bond formation and isomerization as well as prolyl-peptide isomerization and in some cases proper glycosylation. Recent advances have made clear that protein folding and maturation in the cell are complex processes (122, 129, 132, 171, 200). The fact that a protein can direct its own folding in a test tube in vitro (9) does not guarantee its ability to do so in the cell. This has focused attention on chaperone factors necessary for successful folding of proteins within the ER.

Conditions in the ER Lumen. The conditions inside the ER can be considered a compromise to maximize the rate and yield of folding of proteins of vastly different sizes and structures. The Ca^{2+} concentration in the ER is in the millimolar range (approximately 5 mM). It favors the maturation of many proteins that are dependent on Ca^{2+} (266, 268, 490, 556), and the release of ER Ca^{2+} by inositol trisphosphates is a key step in many cellular signaling pathways (42). The redox conditions in the lumen of the ER are approximately 20 to 100 times more oxidizing than those in the cytosol. Glutathion provides the main redox buffer (192). From in vitro experiments (275, 276) the oxidizing conditions of the ER are known to be optimal for spontaneous disulfide formation during protein folding (134, 192). Reducing agents such as dithiothreitol (5 to 10 mM) can unfold partially folded and occasionally fully oxidized proteins in the ER (56). When added to cultured cells in 0.5 to 1 mM concentrations, dithiothreitol prevents folding of newly synthesized proteins without affecting synthesis and translocation (269, 570). Such unfolded proteins either accumulate or are degraded in the ER, demonstrating the importance of proper folding for protein exit from the ER. The transport of proteins that do not have disulfide bonds is unaffected by dithiothreitol (269, 502). Thus, the processes of vesicular budding and fusion function efficiently in the presence of dithiothreitol. Because the dithiothreitol effect is rapidly reversible in cell culture and isolated microsomes, this agent provides a valuable tool to accumulate newly synthesized proteins in the ER and upon removal of dithiothreitol to study their folding and transport to the Golgi (208, 285, 502). It should be noted, however, that some proteins are highly unstable when folding is suppressed by dithiothreitol and are rapidly degraded in the ER. It appears that unfolding by dithiothreitol involves ER-associated components—most likely PDI and other adensoine triphosphate (ATP)-dependent factors (208). In some cases the reducing agent mercaptoethanol promoted rather than inhibited intracellular transport of unassembled proteins (4) suggesting that exposure to free sulfhydryl groups may result in retention.

Protein folding also requires ATP. When cells are depleted of ATP, VSV G protein and influenza hemagglutinin, two well-studied model membrane glycoproteins, form unfolded, disulfide-linked aggregates stably associated with BiP (55). ATP is transported from the cytosol through the ER membrane by an ATP translocator (71, 214) and is required because some of the proteins that facilitate folding such as PiP and PDI (325) depend on ATP for their function (132, 414).

Chaperones. Protein folding in the ER is assisted by molecular chaperones and folding enzymes (129, 132, 162, 200, 234) some of which are present in close to millimolar concentrations. A chaperone is defined as a protein that promotes the folding and assembly of a polypeptide chain into its mature conformation. The most abundant chaperones of the ER are BiP, GRP 94 (293), and the more recently discovered calnexin (39, 153). Under normal growth conditions BiP makes up about 5% of the luminal content of the ER. BiP is also required for translocation of newly synthesized proteins across the ER membrane (426, 434) and for oligomeric protein assembly. Molecular chaperones are thought to recognize hydrophobic sequences or surfaces of proteins that are exposed in the unfolded but not folded state. BiP, the only member of the heat shock protein (hsp) 70 protein family localized in the ER, preferentially binds short peptides containing a subset of aromatic and hydrophobic amino acids in alternating positions (44). Peptide binding to BiP stimulates its ATPase activity (117). BiP either may have a passive role, in which binding merely prevents misfolding and aggregation, or may assist folding by determining a hierarchy of folding of different polypeptide domains.

Calnexin, a membrane-anchored lectin of the ER, binds to newly synthesized soluble and membrane proteins (199, 242, 353, 397, 535) carrying monoglucosylated oligosaccharides (155, 156, 168, 169, 212, 343, 539). Calnexin is believed to contribute to glycoprotein maturation and quality control in conjunction with the ER enzyme UDP-glucose: glycoprotein glucosyltransferase (516) which adds a glucose residue to glycans of still unfolded proteins that have undergone trimming by glucosidase I and II (153, 155, 156). Calnexin and glucosyltransferase would keep glycoproteins in the ER until they are completely folded. Folded proteins are then released from the re- and deglucosylation cycle because they are no longer a substrate for glucosyltransferase and after final deglucosylation by glucosidase II can no longer bind to calnexin. Folding of the VSV G protein involves the sequential interaction of BiP and calnexin (154, 168, 170). When glucose trimming of the VSV G protein is inhibited by castanospermine the association with BiP is prolonged and folding is impaired. It should be noted, however, that folding of only a few proteins depends on calnexin, as indicated by the fact that glucosidase inhibitors do not interfere with folding or transport of most proteins. This indicates that alternative pathways exist for the maturation of most glycoproteins when their binding to calnexin is prevented. An alternative model for the chaperone mechanism of calnexin proposes that interaction with monoglucosylated oligosaccharide chains is just the first step to promote close interaction of the nascent polypeptide with calnexin. In a subsequent step the unfolded glycoprotein binds to calnexin through segments of its polypeptide chain perhaps by hydrophobic interaction (65, 539).

The soluble ER protein calreticulin, which shows marked similarity to calnexin in its amino acid sequence, plays a chaperone-like role in the ER similar to calnexin besides its proposed role as a Ca^{2+}-storage protein (64, 169, 170, 318). Surprisingly, calreticulin was found to copurify with endomannosidase from a Golgi fraction suggesting that the dissociation through glucose removal of calreticulin–glycoprotein complexes may at least in part occur in a post-ER compartment by endomannosidase rather than glucosidase II (476).

Folding enzymes. Folding enzymes are required for catalysis of the necessary covalent changes involved in the folding process of nascent polypeptide chains. The major folding enzymes in the ER are protein disulfide isomerase (PDI) (123, 132, 331) and peptidyl-prolyl cis-trans isomerase (51, 52, 127). PDI (approximately 10% of soluble protein content of the ER lumen) constitutes a family of at least four homologous proteins: PDI, RP60 (277), ERp72 (234, 290, 324) and P5/CaBP1 (66, 125, 419). Besides disulfide bond formation and isomerization, PDI of higher eukaryotes is also a component of the enzyme prolyl-4-hydroxylase, which catalyzes the formation of hydroxyproline in collagens (215, 534) and a subunit of a triacyl glycerol transfer protein that facilitates the incorporation of lipids into newly synthesized core lipoproteins within the ER (549). Studies of coexpression in vivo and of refolding in vitro suggest that the role of PDI in these heterooligomeric enzyme complexes is to maintain the other subunits in a soluble, nonaggregating form. Experiments with yeast have shown that PDI is essential for spore germination and possibly viability. Mammalian PDI and ERp72 can complement the absence of yeast PDI (149). PDI may be both an enzyme and a chaperone (537). This suggestion stems from the observation that in many instances some folding is required so that the respective thiol groups pair correctly to catalyze the formation of the native disulfides. This is unlikely to occur by spontaneous folding, but the evidence for a chaperone function of PDI is still indirect.

Prolyl cis-trans isomerization is a rate-limiting step in the in vitro folding of some proteins. In the ER this process most likely is catalyzed by the immunophilins cyclophilin B and FK506-binding protein (51, 52, 127) although the prolyl cis-trans isomerase function of these proteins in the secretory pathway is less well-documented than that of the related cytosolic immunophilins. When the cis-trans isomerase activities of cyclophilin B and FK506-binding protein are inhibited by cyclosporin A and FK506, respectively, maturation of collagen (482) and transferrin receptor (267) is retarded. On the other hand cyclophilin B is present throughout the secretory pathway (13, 389) and also bound to the plasma membrane. Upon treatment with cyclosporin A cyclophilin B is secreted (389). Immunophilins may therefore have additional functions in the secretory pathway.

Protein Degradation. Misfolded or unassembled proteins either accumulate (140, 512, 522) or more typically are degraded in the ER (216, 477, 554) and ERGIC (167, 392). The degradation is inhibited by membrane-permeable cystein protease inhibitors or sulfhydryl reagents (193, 477, 553, 557). Surprisingly the PDI homologues ERp60 (520) and ERp72 (521) have proteolytic acivity inhibitable by cysteine protease inhibitors, suggesting that these two enzymes may be part of the ER degradation system in addition to their folding function. ER degradation is unaffected by GTP-γ-S, EDTA or depletion of ATP. In some cases ER degradation is slowed in cells depleted of ATP (241, 477, 553) but the ER proteases themselves are not ATP-dependent (570). It is possible that the redox buffer within the ER lumen regulates ER degradation, because treatment of cells with reducing agents selectively disrupts proteolysis. Dithiothreitol induced the rapid degradation of two T-cell receptor subunits that are stable in the ER without dithiothreitol treatment. Reducing conditions did not, however, cause uncontrolled proteolysis of luminal ER resident proteins (570). Protein degradation in the ER does not only serve the purpose of degrading misfolded proteins. This system also regulates the half-life of some proteins including HMG CoA reductase, the rate-limiting enzyme in cholesterol biosynthesis (138, 243). HMG CoA reductase and T-cell alpha-subunit are differentially degraded, indicating that ER degradation may occur by several different

mechanisms but generation of free thiol may be a common requirement (119, 193). Some transmembrane proteins including the cystic fibrosis transmembrane regulator (CFTR) can be degraded in the ER from the cytosolic rather than luminal side by the proteasome, the primary site in mammalian cells for degradation of cytosolic and nuclear proteins (204, 538). It is also possible that this system degrades some secretory proteins at a relatively early stage of their biosynthetic processing in the ER or after reexport from ER to cytosol.

Protein Traffic Defects in Human Genetic Disease. An understanding of the mechanisms leading to transport competence may suggest novel strategies for the treatment of genetic diseases in humans in which physiological important secretory or plasma membrane proteins are synthesized in an active form but fail to be exported from the ER (7, 242, 548). Perhaps the most relevant example is cystic fibrosis (548) which is due to mutations in the CFTR chloride channel. In nearly 70% of the patients, mutant CFTR fails to be transported to the plasma membrane because of a deletion of phenylalanine at residue 508 (deltaF508). Instead mutant CFTR is retained and degraded in the ER (70, 93, 204, 509, 538, 568). Strategies designed to relocate CFTR deltaF508 could be beneficial. Initial experiments have already lead to the observation that in cells grown at 23° to 30°C some of the mutant protein escapes from the ER and is delivered to the cell surface, where it retains at least some function (92). Obviously the mutation is temperature-sensitive, perhaps because folding of CFTR-deltaF508 is improved at reduced temperature. Improper folding at 37°C but not at 28°C has indeed been reported for several cathepsin D/pepsinogen chimeric proteins expressed in B lymphoblasts (136). Not always are improperly folded proteins retained and degraded in the ER, as documented in congenital sucrase-isomaltase deficiency (CSID) in humans. In most patients affected with CSID, the brush border enzyme sucrase-isomaltase is synthesized but its transport to the cell surface is either blocked in the ER or the Golgi apparatus, or the enzyme is mistargeted to the basolateral membrane (121, 165, 314). Epitope mapping with conformation-specific antibodies showed incomplete folding not only of ER-retained but also of Golgi-arrested sucrase-isomaltase (289, 314) suggesting that post-ER compartments can contribute to quality control/transport competence. In one of the CSID cases the transport defect was elucidated and found to be due to a single glutamine to proline substitution. This point mutation leads to arrest and degradation of sucrase-isomaltase in ER, ERGIC and cis-Golgi (354).

In summary, transport competence involves a complex series of folding reactions assisted by folding enzymes and chaperones in a tightly controlled luminal milieu. These factors, in conjunction with a poorly characterized protein degradation system, are responsible for quality control and secure the functional integrity of proteins leaving the ER. In addition to folding, oligomerization may also precede protein exit from the ER (190) although there are notable exceptions to this rule (95, 201, 282, 312). In fact, oligomerization events are now known for most organelles of the secretory pathway. Quality control in the ER also contributes to asynchronous protein transport (166, 190, 264, 265, 407) but exit from the ER is not the only rate-limiting step for slowly transported proteins (485).

There are multiple reasons for quality control in the ER. This mechanism prevents incompletely folded or defective proteins from having access to later compartments of the secretory pathway where they may have harmful effects. Moreover, subunits of heterooligomers are often produced in nonstochiometric proportions creating a need for their retention until correct assembly occurs (217). Quality control also provides a means for posttranslational regulation of protein expression. There may be yet another level of complexity. A transport-competent protein is usually considered to leave the ER passively by bulk flow (see later under PROTEIN RETENTION AND RETRIEVAL). In light of recent findings that some secretory proteins appear to concentrate during exit from the ER (21, 303) transport competence may also involve the acquisition of proper structure to enable interaction with a (putative) sorting machinery.

Precisely how folding enzymes, chaperones, and proteases cooperate in quality control of cargo proteins in the ER is not currently known. There is evidence that this cooperation is facilitated by the assembly of folding factors into a macromolecular complex as a "chaperosome" (84). Observations that some enzymes have dual participation in folding and proteolysis is also intriguing. Similarly some ER chaperones may act as proteases in particular situations, or in conjunction with a protease system, decide about survival or degradation. Examples of the latter possibility are known for the prokaryotic Hsp70 and Hsp100 families and for mitochondrial Hsp70 (84), but it remains unclear what controls partitioning of proteins between systems for folding and proteolysis.

COPI- and COPII-Coated Transport Vesicles

COPI-coated Vesicles in vitro. Transport vesicles were first documented in a cell-free intercisternal Golgi transport assay (19) in which vesicle budding from mammalian Golgi membranes was induced by cytosol and ATP (344, 345). When the nonhydrolyzable nucleotide GTPγS was present in the transport reaction the vesicles accumulated and could therefore be purified (281). These vesicles carry a proteinaceous coat and hence were termed COP (coat protein)-coated vesicles (102, 456, 541). Because there is now a second type of non-clathrin coated vesicles in yeast (28) mammalian Golgi-derived coated vesicles are coined COPI-coated vesicles.

The coat of COPI vesicles consists of the small GTP-binding protein ADP-ribosylation factor (ARF) (457) and a complex of seven proteins termed coatomer (82, 102, 160, 224, 231, 483, 542) whose subunits are α, β, β', γ, δ, ϵ, and ζ (Table 15.2). ARF and the approximately 800 kDa coatomer complex exist separately in the cytosol but coassemble to form coats when the GTP form of ARF binds to a (postulated) receptor on the donor membrane following a brefeldin A-sensitive nucleotide exchange step (100, 172, 173) (Fig. 15.4a). Membrane-bound ARF triggers the binding of coatomer (101, 359) which are transfered *en bloc* (157) and budding occurs (348). The budding process depends on coatomer, ARF, and GTP, whereas vesicle fission requires fatty acyl-coenzyme A (352). Apparently ARF and coatomer are the only cytosolic proteins necessary for budding (349). Hydrolysis of bound GTP by ARF triggers coat disassembly (501) a process inhibitable by GTPγS or aluminum fluoride (AlF$_4^-$).

COPI-coated Vesicles in vivo. Are these vesicles generated in vitro the mediators of interorganelle protein traffic in the secretory pathway in vivo? Correlative evidence (225, 416, 417, 423, 435) from cell-free transport reactions argues in favor of the view that COP-coated vesicles mediate protein transport from ER to Golgi and between successive cisternae of the Golgi apparatus. Additional biochemical and genetic studies appear to support this notion. ER-to-Golgi transport of vesicular stomatitis virus G protein was blocked by microinjection of an anti-peptide antibody against β-COP (369) and a cDNA encoding epsilon-COP was able to correct the mutant phenotype of ldlF CHO cells that exhibit a block of ER-to-Golgi transport at the restrictive temperature (150, 385). In addition, a yeast homologue of γ-COP encoded by SEC21 (483) is essential for secretion (184) and β-COP (encoded by SEC26) and β' COP (encoded by SEC27) also seem to play some role (103, 484). Secretion in yeast is moreover blocked when the two genes encoding ARF are deleted (480) and the expression of a dominant activating mutant of human ARF1 in NRK cells inhibits protein secretion (572). Furthermore, ARF1 and ARF2 overexpression in yeast is synthetically lethal with Sec27-1

TABLE 15.2. *Coat Proteins of COPI- and COPII-Coated Transport Vesicles*

Protein	Mr(kD)	Features/Function	References
COPI			
Arf1	21	Most abundant member of Arf GTPase family, regulates coat binding	97
α-COP (ret 1p)	160 (150)	Contains WD-40 repeats	131, 246, 541
β-COP (sec26p)	110 (109)	Homology to beta-adaptin of clathrin-coated vesicles	102, 103, 456
β'-COP	102 (99)	No homology to beta-COP but to clathrin heavy chain, contains WD-40 repeats	103, 160, 161, 484
γ-COP (sec21p)	98		184,483
δ-COP (ret2p)	61	Homology to adaptin subunites AP47 and AP19	82, 396, 541
ε-COP	36		150, 157
ξ-COP (ret3p)	20	Homologous to AP17 and AP19 adaptor subunits of clathrin-coated vesicles	82, 231
COPII			
Sar1p	21	GTPase required for budding	316
sec23-complex	400		
sec23p	85	Sar1-specific GAP	177, 178
sec24p	105		178
sec13-complex	700		
sec13p	34	Contains WD-40 repeats	391, 460
p150	150	Contains WD-40 repeats	422

Note: COPI coats are most completely known for mammalian cells whereas COPII-coats have been identified in yeast cells. Proteins and Mr values in parentheses refer to yeast homologus. WD-40 repeats are also present in beta-subunits of trimeric GTPases (523). They may mediate protein-protein interaction.

(103). Overall these data demonstrate an important role for COPI-coated vesicles in the early secretory pathway, and they suggest that the regulation of coatomer binding to membranes in yeast and mammalian cells may be similar.

Although COPI-coated vesicles are good candidates for carrier vesicles it remains unclear to what extent they are obligatory transport intermediates in cell-free intercisternal transport assays. Under certain conditions cell-free intra-Golgi transport was observed in the absence of COP I-coated vesicles: in the presence of brefeldin A (347) and in the absence of ARF (504) and/or coatomer (108). The interpretation of these observations is controversial. Whereas Taylor and colleagues (504) conclude that cell-free intra-Golgi transport procedes via a COPI-vesicle-independent mechanism regardless of vesicle formation on Golgi cisternae, Elazar and colleagues (108) consider lowering of ARF and coatomer concentrations nonphysiological. They propose that low cytosol conditions (that is, low ARF and coat protein concentrations) uncouple fusion from budding and result in a switch from vesicle-mediated transport to cisternal fusion. This conclusion is based mainly on the use of a third Golgi population present as a silent acceptor of transport vesicles in the cell-free intercisternal Golgi transport assay. In the presence of ARF and coatomer the transport signal was competed by the silent acceptor, arguing for a vesicular transport under these conditions. In the absence of the two components or in the presence of BFA no competition was observed, indicating direct fusion between Golgi cisternae. Unexplainably, however, these fusion conditions did not increase either the extent or the rate of the in vitro transport reaction. Clearly more experiments are necessary to unravel if indeed coat components not only drive vesicle budding but also couple fusion to budding.

It is also unclear which transport steps are actually mediated by COPI-coated vesicles in vivo. Evidence is accumulating that these vesicles are involved in retrograde rather than anterograde transport. In support of a retrograde involvement is the finding that a coatomer

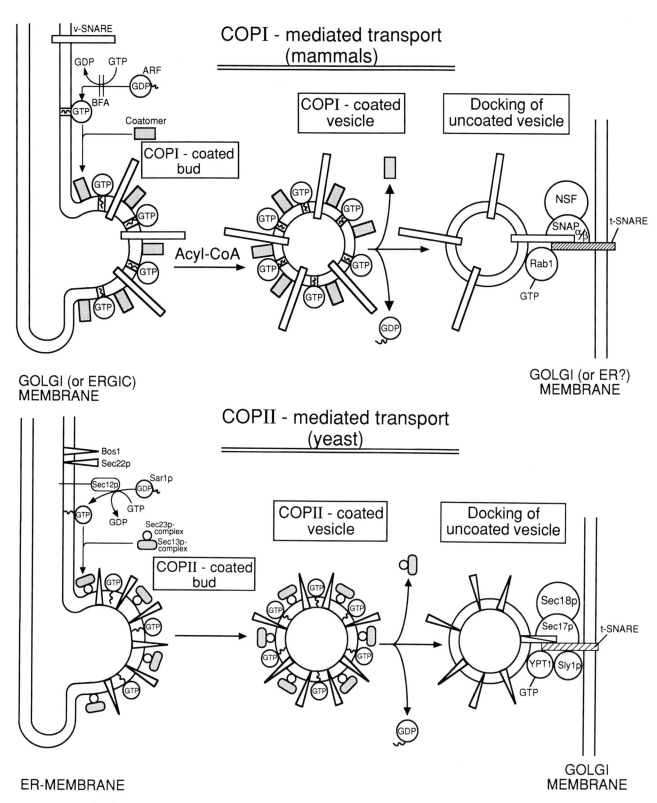

FIG. 15.4. Models for budding and targeting of transport vesicles in the early secretory pathway illustrating similarities between COPI-mediated transport *(A)* and COPII-mediated transport *(B)*. Precisely which pathways depend on the two different coat complexes is currently debated. A speculative view is that COPI-mediated transport (mainly established in a mammalian in vitro system) (415) operates in anterograde direction between ERGIC and cis-Golgi as well as in retrograde direction from cis-Golgi and ERGIC back to the ER. COPII vesicles (mainly established for yeast) (28, 435) may mediate anterograde transport between ER and ERGIC. The v-SNAREs and t-SNAREs of the COPI dependent pathway(s) have not been identified. V-SNARE/t-SNARE interactions may be multimeric.

in vitro binds to the COOH-terminal-KKXX ER retrieval motif (81) present in a number of type 1 ER membrane proteins (197, 198) and that coatomer mutations in α-, δ-, and ζ-COP in yeast lead to loss of pre-Golgi retention of-KKXX-bearing ER proteins, in some cases not affecting transport of secretory proteins (82, 246). Ultrastructural data would be consistent with this notion: β-COP localizes to the ERGIC (147, 196, 342) and ERGIC-53 which recycles between ERGIC and ER was mainly localized to COP-coated budding profiles of the ERGIC (Klumperman, J., and H.-P. Hauri, unpublished). It seems also possible that both ERGIC to cis-Golgi anterograde and ERGIC to ER retrograde transport depend on closely related COPI-coated vesicles (see earlier under Role of the ERGIC in ER-to-Golgi Transport).

COPI-coated vesicles are also involved in mitotic fragmentation of Golgi cisternae. In in vitro studies COPI vesicles were shown to consume up to 55% of Golgi membranes and 45% was consumed by a COP-I-independent pathway (300). Thus vesiculation of the Golgi apparatus during mitosis is caused in part by the continued budding of COPI-coated transport vesicles but an inhibition of their fusion with their target membranes (299). These mitotic COPI-coated vesicles contain Golgi resident enzymes, whereas one would expect interphase COPI-coated vesicles mediating intra-Golgi traffic to exclude such enzymes which has not been formally shown. Provided interphase COPI-coated vesicles do not exclude Golgi resident proteins (352), a role in homotypic (horizontal) interactions between qualitatively identical cisternae of Golgi stacks must also be considered.

COPII-coated vesicles. Through a combined genetic and biochemical approach (390), a set of proteins, Sar1p (27, 316, 340), Sec23p complex (177), and Sec13p complex (391) was identified (Table 15.2) that in the presence of GTP promotes vesicle budding from yeast endoplasmic reticulum in vitro (423) (Fig. 15.4b). While Sar1p dissociates from the vesicles, Sec23p complex and Sec13p complex form a coat around the ER-derived vesicles termed COPII coat (28). Sec12p, an ER nucleotide exchange factor (GEF) for Sar1p is also required for budding but is not included into transport vesicles (26, 315). The Sec23p is a Sar1p-specific GTPase activating protein (569). Just as coatomer is the assembly promotor of the COPI coat, Sec13p and Sec23p complexes are considered assembly promotors of the COPII coat. In addition, two other proteins involved in vesicle targeting, sec22p and Bet1p (319) were associated with these vesicles. Both COPI-vesicles (352) and COPII-vesicles formed in the presence of GTP are fusion competent but are unable to

fuse with an acceptor membrane when the coat is locked by a nonhydrolyzable analogue of GTP. As a further similarity COPI- and COPII formation is initiated by the related GTP-binding proteins Sar1p and ARF, respectively (316, 336). However, the subunits of COPI and COPII are molecularly unrelated (Table 15.2) and palmitoyl coenzyme A is not required in the formation of COPII-coated vesicles. Unexpectedly, although the sec21, sec26, and sec27 mutants (encoding mutant coatomer subunits) block ER-to-Golgi transport (103, 184), COPI proteins are not required for COPII-vesicle formation in yeast (28). This implies either that COPI replaces COPII on ER-derived vesicles or more likely that COPI- and COPII-vesicles are involved in different pathways. In vitro COPI and COPII vesicles can be generated from the same yeast nuclear membrane but only the COPII vesicles carry the major cargo proteins which raises the possibility of two qualitatively different anterograde vesicular pathways from the ER (36, 435).

Vesicle Docking and Fusion

After shedding their coats, transport vesicles dock and fuse with the respective target membrane. Docking must be based on the pairwise interaction of vesicles and target membrane while the fusion itself is triggered by the ATPase N-ethylmaleimide-sensitive fusion protein (NSF encoded by SEC17) and the soluble NSF attachment proteins (SNAPs; alpha-SNAP encoded by SEC18) (72, 550, 551, 562). According to the SNARE-hypothesis (473) docking is believed to result from the specific interaction of a vesicle receptor (v-SNARE = vesicle SNAP receptor) and a t-SNARE (that is, target SNAP receptor). Putative v-SNARE/t-SNARE pairs have first been isolated from brain extracts and appear to promote fusion of synaptic vesicles with the presynaptic membranes (37, 113, 373, 418, 473, 474). A number of candidate v-SNARES and t-SNAREs have now been found (250, 313, 489). The ER–Golgi docking complex isolated from yeast secretion mutants in which transport vesicles are unable to fuse with the Golgi apparatus includes the v-SNAREs, Sec22p, Bos1p (and possibly Ykt6p and Bet1p) and the t-SNARE Sed5p (471). The assembly of this complex requires the functional activity of the small GTP-binding rab protein Ypt1p (471). Previous studies have already suggested a role for Sec22p, Bos1p, Bet1p, Ypt1p (16, 18, 452, 453) and Sed5p (158, 159) in ER–Golgi transport (496). SEC22 and BET1 encode small C-terminally anchored membrane proteins that are structurally similar to the synaptic vesicle protein synaptobrevin (319, 320). BOS1, a gene identified by its ability to suppress the bet1 mutant also encodes a C-terminally anchored protein that is required for ER to Golgi transport (462). Isolated carrier vesicles contain Bos1p, Sec22p, and Ypt1p (252) and depending on the method of isolation also Bet1p (403). Depletion of Bos1 from the ER leads to the formation of transport-incompetent vesicles, and an antibody against Bos1p blocks fusion of carrier vesicles with the Golgi apparatus. Because isolated carrier vesicles do not carry a coat but are nevertheless transport competent (403) it follows that coat proteins need not be maintained on vesicles to retain targeting competence. The number of accessory proteins playing a role in docking and fusion is still increasing (272, 376).

In summary, v- and t-SNAREs form the core of a docked vesicle complex but accessory proteins are required to protect SNAREs from promiscuous binding and to activate SNAREs under conditions in which transport vesicle docking should occur (376). Ultimately, the number of cognate pairs of v- and t-SNAREs should define the number of vesicular steps in the secretory pathway and hence the number of compartments. It is interesting to note that such v-SNARE/t-SNARE pairs have not been identified yet for intercisternal Golgi but two candidate Golgi v-SNAREs are known (24, 313, 489). Likewise, if transport through the Golgi stack indeed occurs by multiple vesicular steps we are also short of rab proteins that were proposed to catalyze the SNARE complex assembly (471). Rab6 is the only rab known to be associated with the stacked Golgi, even though more than 30 rabs have been identified. It has been implicated in vesicle budding from the TGN (205) or transport through the Golgi apparatus (286).

GTPases

By alternating between GDP- and GTP-bound forms, GTP-binding proteins, also referred to as GTPases,

function as molecular switches in diverse cellular functions (54). The role of GTPases in membrane traffic was first demonstrated by studies on Sec4p which encodes a ras-related small molecular weight GTPase (424). Mutations of SEC4 result in accumulation of post-Golgi transport vesicles that fail to fuse with the plasma membrane (335). Following the observation by Melancon and colleagues (292) that the activators of GTPases, GTP-γ-S and AIF_4^-, block vesicular transport in a cell-free intercisternal Golgi transport assay, members of seven groups of GTPases have been identified as regulators of vesicular traffic. They include the Ras-like Arf, Sar1, Rab/Ypt, Rac/CDC42 and Rho families, as well as heterotrimeric G protein and dynamin families (337). Most relevant for the early secretory pathway are the Arf, Sar1, Rab/Ypt, and the heterotrimeric G protein families (Table 15.3).

The roles of Sar 1 and Arf [see (97)] have already been discussed in the previous section. While Arf is clearly essential for the recruitment of coatomer to Golgi (and most likely to ERGIC) membranes (101, 157, 352, 359, 505) the suggestion that Arf is also required for vesicle budding from the ER of mammalian cells (337) is more difficult to reconcile with the finding that COPII but not COPI is responsible for this process in yeast (28). The view that Arf1 promotes budding from the ER is based on the following observations: first, the amino-terminal Arf peptide in semipermeable cells or excess levels of myristylated Arf1 prevent budding of carrier vesicles from the ER (20); second, overexpression in vivo of a dominant inhibitory Arf1 mutant, Arf1 (T31N), with preferential affinity for GDP reduces export from the ER, triggers the release of beta-COP from intracellular membranes, and results in the collaps of the Golgi apparatus into the ER comparable to the effect of BFA (88); third, antibodies against synthetic peptides derived from different regions of β-COP inhibit export from the ER in

TABLE 15.3. *GTPases Involved in Membrane Traffic in the Early Secretory Pathway*

GTPase	Location	Function	Reference
Sar1	ER, TE, vesicles	COPII-coated vesicle budding from ER	27, 232, 316, 340
Arf	Golgi, ERGIC	COPI-coated vesicle budding Clathrin-coated vesicle budding	99, 101, 359, 457, 480, 481 69
Ypt1	ER-Golgi carrier vesicles, Golgi	Targeting/fusion	203, 252, 402, 452, 453, 471
Rab1	ERGIC, cis-Golgi	ER-to-Golgi and intra-Golgi transport	382, 431
Rab2	ERGIC	ER-to-Golgi transport	67, 145, 510
Rab6	Golgi	Vesicle budding from the TGN	10, 139, 205
Heterotrimeric	Golgi, ERGIC	ER-to-Golgi and intra-Golgi, protein exit from TGN	91, 339, 380, 381, 441, 487

permeabilized mammalian cells (371). However, it is possible that the inhibitory effect of the Arf peptide was due to its toxicity rather than to a specific transport block (544). The block in ER-to-Golgi transport by the Arf (T31N) mutant could be an indirect consequence of the disappearance of the Golgi apparatus. The β-COP antibody data (371) are at variance with those of Pepperkok and colleagues (369) who showed that microinjection of a beta-COP anti-peptide antibody blocked protein exit from the ERGIC and not from the ER in living cells. Furthermore, coatomer was unable to promote vesicle budding from the mammalian ER in vitro and the cytosol fraction that promoted efficient export from the ER lacked Arf (372). Collectively all these observations are more compatible with the view that Arf promotes vesicle budding from Golgi and ERGIC membranes but less likely from ER membranes.

Recent studies have raised the possibility that Arf is also a regulator of phospholipid metabolism via phospholipase D (60, 76, 89, 207). Arf may act on coat assembly in a catalytic fashion rather than—as previously proposed (415)—in a stoichiometric fashion. Such a mechanism is supported by the finding that cytosolic Arf is not necessary for initiating coat assembly on Golgi membranes from cell lines with high phospholipase D activity (230). Moreover, formation of COPI-coated vesicles is sensitive to ethanol at concentrations that inhibit the production of phosphatidic acid by phospholipase D. It was therefore proposed that phosphatitic acid may facilitate budding by altering the curvature of the membrane and in cooperation with PIP_2 stimulate coatomer binding (230, 263). If this model is correct, one would predict that the lipid composition of vesicles is different from that of the donor membrane, what remains to be shown.

Rab/Ypt Family. Rab proteins, the mammalian counterparts of yeast sec4p and Ypt1p, represent a family of over 30 proteins that are localized to the cytoplasmic side of distinct membrane compartments (67, 112, 336, 337, 375, 571). Small pools of each Rab are also present in the cytosol where they are bound to Rab GDP-dissociation inhibitor (GDI) (401). The two functionally interchangeable isoforms of Rab1 (337), Rab1a and Rab1b (92% identical), are highly homologous to yeast Ypt1p, which controls vesicle docking/fusion (see above). The homology suggests that Rab1 has an equivalent function in mammalian vesicle docking. Consistent with this are the findings that overexpression of mouse Rab1a can rescue yeast mutants lacking Ypt1 (163), and that Rab1b is required during an initial step in export of protein from the ER in

semi-intact cells (383). Further, Rab1b and β-COP form a functional precoat complex in the cytosol of mammalian cells (371), and the dominant negative Rab1a mutant, rab1a(N124I), defective for guanine nucleotide binding, leads to the accumulation of VSV G protein in pre–cis-Golgi vesicles and tubulovesicular clusters containing Rab1b and beta-COP (382). Unexpectedly, a GDP-bound form of Rab1, the Rab1a (S25N) mutant, was shown to prevent export of proteins from the ER rather than lead to the accumulation of nonfunctional transport vesicles (338). In addition, this mutant leads to the disassembly of the Golgi apparatus (561). These results are believed to indicate a requirement of Rab1 in vesicle budding from ER and Golgi membranes (338), a notion that is difficult to reconcile with the fact that genetic and biochemical studies in yeast have not revealed such a requirement. Moreover, at the ultrastructural level Rab1 was localized to the ERGIC and cis-Golgi (145, 431) not supporting a role for Rab1 in vesicle budding from the rough ER. It is conceivable therefore that the GDP mutant of rab1a exerts an indirect effect by neutralizing a general accessory factor such as GDI, that is required for progression of other GTP-binding proteins including perhaps Sar1 through the GTPase cycle. Besides GDI promoting the dissociation of Rabs from membranes (375) and thereby inhibiting transport (109), the Rab GTPase cycle also involves a guanine nucleotide exchange factor that converts GDP-Rab into GTP-Rab (GEF) (472, 519), a Rab escort protein (REP) that chaperones GTP-Rab to the membrane (6, 59), and a GTPase-activating protein (GAP) catalyzing GTP hydrolysis.

Two other rab proteins, Rab2 associated with the ERGIC (67) and Rab6 associated with medial- and trans-Golgi (10, 139), may also participate in controlling specific trafficking steps in the secretory pathway. GTP-binding mutants of Rab 2 inhibit vesicular transport from ER to Golgi in semi-intact cells (510). Rab6 is associated with the TGN marker TGN38, an integral membrane protein that recycles between the TGN and the plasma membrane (205, 479). Addition of antibodies to Rab6 inhibited the in vitro formation of exocytotic vesicles from rat liver Golgi, implying a role for vesicle budding from the TGN. On the other hand, transient overexpression of wild-type Rab6 or a GTP-bound mutant of Rab6 greatly reduced intra-Golgi transport without affecting transport from ER to Golgi or from TGN to plasma membrane (286). The reason for this discrepancy remains to be elucidated.

Heterotrimeric G Proteins. A role for heterotrimeric G proteins in regulating membrane traffic in the early secretory pathway was suggested by the observation that AlF_4^- inhibits intercisternal Golgi (292) and ER to Golgi (441) transport. AlF_4^- activates heterotrimeric G proteins (43) but not the small-molecular-weight monomeric GTPases (206). Research on the role of the heterotrimeric G proteins in trafficking is still in an early phase and predominantly includes studies on the modulation of protein transport by G protein effectors such as AlF_4^-, pertussis toxin (inactivates G-α-i/G-α-o), cholera toxin (activates G-α-s), mastoparan (activates G-α-i), and excess β/γ dimer (inhibits G-α activity) (29, 47, 337). In permeabilized cells, protein export from the ER is sensitive to mastoparan and excess β/γ dimers but insensitive to pertussis and cholera toxin (441). AlF_4^- promotes the association of coatomer to Golgi and ERGIC (98, 99, 342, 347) and inhibits protein exit from the ERGIC both in anterograde (to Golgi) and retrograde (to ER) direction in vivo (196). Mastoparan also promotes binding of β-COP to Golgi membranes and prevents brefeldin A-induced dissociation of β-COP (229). These effects were diminished in pertussis toxin-treated cells, suggesting the involvement of a G-α-i protein. Moreover, overexpression of G-i-α3 slows protein transit through the Golgi apparatus (487) and antisera to the carboxyterminus of Gi3 in permeabilized erythroleukemia cells inhibits O-linked glycan maturation (560). Exit from the TGN to the apical and basolateral cell surface seems to be regulated by Gs and Gi proteins, respectively (380). The Gs-dependent stimulation of the apical pathway is most likely mediated by adenylate cyclase, the classical downstream effector of Gs, for it was also observed in the presence of activators of protein kinase A including forskolin, dibutyryl-cAMP and isobutylmethylxanthine (381). Likewise, trimeric G proteins appear to regulate the formation of immature secretory granules from the TGN: negatively by pertussis-sensitive G-proteins of the Gi/Go class and positively by cholera toxin-sensitive Gs proteins (30, 251). This regulatory pathway involves cytosolic serine/threonine protein phosphorylation (339).

Clearly, the elucidation of the precise role of heterotrimeric G proteins will require their identification and characterization that is well under way (91). It is conceivable that these G proteins transduce signals originating from the lumen of the secretory pathway via unidentified membrane receptor to various effector systems and thereby regulate protein traffic by modulating coat assembly and release, intraorganellar pH (454), or calcium homeostasis at different way stations of the secretory pathway. An alternative exciting possibility is that heterotrimeric G proteins directly regulate Arf and thereby modulate vesicle budding. Such a mechanism seems plausible in light of the novel finding that heterotrimeric G proteins interact with Arf (74, 120).

Role of the ERGIC in ER-to-Golgi Transport

Morphological and biochemical studies have established the ERGIC as an intermediate in viral and endogenous (Foguet, M., and H.-P. Hauri, unpublished) protein transport from the rough ER to the Golgi stack in mammalian cells (196, 270, 430, 443). Because β-COP is mainly associated with the ERGIC and cis-Golgi, transport from, rather than into, the ERGIC may be dependent on COPI-coated vesicles. Density gradient analysis of the trafficking of ERGIC-53 in HepG2 cells that were exposed to different traffic inhibitors support this notion (196). Most notably, AlF_4^- (an activator of heterotrimeric G proteins that locks COPI coats to membranes) and BFA (which dissociates COPI coats) fail to block transport from ER to ERGIC. Conversely, mammalian sec23p, sec13p, and Sar1p subunits of the COPII protein complex in yeast are associated with transitional elements of the rough ER but absent from the ERGIC in mammalian cells (232, 346, 460) and Sed5p is localized in the ERGIC but absent from Golgi stacks (23). Given the roles of sec23, sec13p, and Sar1p in vesicle budding from the ER in yeast and the presumed role of Sed5p as a t-SNARE vesicle receptor (471) it appears likely that the ER-to-ERGIC transport is mediated by COPII-coated vesicles. Subsequent transport from ERGIC to cis-Golgi membranes (defined by peptide-GalNAc transferase) is inhibitable by AlF_4^- as well as brefeldin A (196) suggesting a trimeric G protein- and Arf-dependent transport step typical for COPI-coated vesicles. This same step is also sensitive to the Ca^{2+} ionophore A23187. Collectively these findings are consistent with the previous ranking in temporal order of the biochemical requirements for ER-to-Golgi transport in permeabilized cells (34, 35, 384): transport arrest at 15°C (now known to occur in the ERGIC) (270, 443) precedes GTP hydrolysis, which in turn precedes a Ca^{2+} sensitive step related to fusion. The biochemical requirements for ER-to-Golgi transport in mammalian cells in vivo are schematically depicted in Figure 15.5.

Recently coatomer subunits were also found to be associated with a subdomain of the ER that is different from the classical transitional elements of the rough ER defined by sec23p (351). Because buds were not evident on coatomer-rich ER cisternae this subcompartment may operate as an acceptor of retrograde transport vesicles from the ERGIC. It is also conceivable that coatomer-rich ER functions in retaining dilysine proteins in the ER.

It is unlcear if there is an ERGIC equivalent in yeast. The yeast *Saccharomyces cerevisiae* has three functionally different post-ER membranes involved in glycosylation. They are considered equivalent to cisternae of a mammalian Golgi apparatus but they do not form stacks. Emp47, a protein distantly related to ERGIC-53 is localized in the Golgi apparatus (439). Perhaps transport from the ER to the early Golgi compartment, where alpha1,6-linked mannose is added onto the core N-linked oligosaccharides in yeast corresponds to ER to ERGIC transport in mammalian cells. Alternatively, simple organisms like yeast may not require an ERGIC.

PROTEIN RETENTION AND RETRIEVAL

Membrane Flow, Selective Transport, and Protein Retention

Transport of proteins from the ER to the plasma membrane is believed not to require transport signals but to occur by default (374). This notion is mainly based on the measurement of transport rates of a tripeptide with a hydrophobically modified amino-terminus, a blocked carboxy-terminus, and the sequence Asn–Tyr–Thr necessary for N-glycosylation. This peptide is taken up by living cells, glycosylated and finally secreted into the culture medium (552). Because transport of this signalless glycotripeptide by bulk flow was as fast as or faster than the fastest secretory protein, it was concluded that protein transport to the plasma membrane would not require a signal. Although it has now been shown that such a tripeptide, at least in part, also acquires complex glycans and hence must have passed through the Golgi apparatus before secretion, it is somewhat worrying that its secretion is not inhibitable by BFA contrary to that of normal proteins (525). This raises the possibility that the tripeptide may not necessarily follow the normal secretory pathway and the application of the glycotripeptide as an accurate measure for vesicular bulk flow is therefore still uncertain.

In another study the rate of bulk flow from cis-Golgi to plasma membrane was determined to be 10 min at 37°C, again as fast as or faster than that of any protein (211). This result is based on measuring the secretion of a water-soluble sphingomyelin analogue. Although synthesized in the cis/medial Golgi (126, 202, 447) sphingomyelin transport to the cell surface occurs independently of protein secretion by a BFA-insensitive, presumably vesicular pathway (461). It remains to be shown, therefore, which pathway is followed by the soluble sphingomyelin probe used by Karrenbauer and colleagues (211).

Although these considerations do not necessarily invalidate the general concept of bulk flow, they raise some doubts on the methods used to measure it. Moreover, the re-evaluation of transport kinetics, taking into account the time required for correct folding and oligomerization in the ER, reveal that the intracellular transport of cargo proteins is considerably faster than previously assumed (25).

The bulk flow concept of protein transport predicts that secretory proteins are not concentrated when packaged into transport vesicles budding from the ER. In contrast, two recent reports claim that vesicular stomatitis virus G protein in semipermeable cells (21) and albumin in vivo (303) are considerably concentrated during this process but not during later steps of secretion supporting a concept of receptor-mediated exit from the ER (264, 265). The validity of these studies has been debated (22, 146, 466). A point not addressed in this discussion is the fact that the budding process itself was poorly documented. It is possible therefore that the observed protein concentration actually occurred in the ERGIC by means of membrane recycling as the ERGIC appears to be a major recycling compartment of the early secretory pathway (see later under Recycling from cis-Golgi). Nevertheless, selective export from the ER would render protein transport considerably more efficient than bulk flow. Selective transport would require an export signal in cargo proteins that can interact with a sorting receptor. Emp24p, a protein of ER-dervided COPII-vesicles in yeast, is a candiate sorting receptor (437). Absence of Emp24 causes a defect in transport of selected proteins to the Golgi apparatus. This protein has numerous homologues in yeast and mammalian cells (478). Another putative sorting receptor is the mannose-selective lectin ERGIC-53/p58 that may facilitate the transport of glycoproteins (196a) and contains and ER exit signal (Kappeler, F., and H.-P. Hauri, unpublished).

Residence of organelle proteins along the secretory pathway occurs by retention, retrieval, or both. Accordingly, a large number of targeting signals has been defined during the past few years (273, 278, 329, 332, 558) (Table 15.4) as discussed in the following sections.

Soluble ER proteins

Most soluble ER proteins possess a carboxyterminal-Lys-Asp-Glu-Leu (KDEL), HDEL, or related tetrapeptide. When attached to reporter proteins it localizes them to the ER demonstrating that it is necessary and sufficient for ER localization (364, 365). KDEL/HDEL is not a retention signal in the true sense but a retrieval determinant that is recognized by a KDEL receptor, termed Erd2 (186, 247, 248, 455) residing in the cis-

TABLE 15.4. *Retention and Retrieval Signals*

Location	Protein	Topology	Signal	Type of signal	Reference
ER	Adenovirus E3/19	I	c: KKMP	Retrieval (KKXX)	197, 198, 463
	Glucuronosyltransferase	I	c: KGKRD	Retrieval (KXKXX)	198
	HMG-CoA reductase	III	c: KKTA	Retrieval (KKXX)	198
	Wbp1p (yeast)	I	c: KKTN	Retrieval	130
	p63	II	l, tm, c (X^+G)	Retention, Aggregation	448
	Invariant chain, alternative spliced form	II	c: MHRRR−	Retrieval?	440
	Cytochrome b5	C	last 10 aa	Retention?	302, 362
ERGIC	ERGIC-53/p58	I	c: KKFF	Retrieval (KKXX)	194, 196
				Endocytosis	195, 210
			c: RSQQE	?	194
	Sed5p/syntaxin 5	C	tm, c	Retention	23
cis-Golgi	Avian coronavirus protein	III	tm (Asn Thr Gln)	Retention	279, 492
	Erd2 (KDEL receptor)	III	tm7 (Asp 193)	Retrieval	513
medial-Golgi	GlcNAc-transferase	II	l, tm, c	Retention	61, 62, 498
trans-Golgi	Galactosyltransferase	II	tm	Retention	11, 287, 326, 421, 506, 567
TGN	Sialyltransferase	II	l (stem), tm	Retention	75, 309, 564
	TGN38	I	c: YQRL	Retrieval	50, 189, 565
			tm	Retention	386, 387
	Furin	I	c	Retrieval	53, 304, 433
	Mouse coronavirus M protein	III	tm, c	Retention	15, 227
	Kex2p (yeast)	I	c (Tyr)	Retrieval	555
	Kex1p (yeast)	I	c	Retrieval	78
	DPAP A (yeast)	II	c (Tyr)	Retrieval	333

Note: Topology refers to the transmembrane orientation of the protein, which can be type I, II, or III, or C-terminally-anchored (C). Signals can be in the cytoplasmic (c), luminal (l), or transmembrane (tm) domain.

Golgi and the ERGIC (145, 499). Upon ligand binding Erd2 recycles KDEL/HDEL-proteins to the ER (249). In general, retrieval of KDEL-proteins does not occur beyond the first cis-cisterna of the stacked Golgi (77) even though the capacity of retrieval extends to the trans-Golgi (280, 298, 370). This is consistent with the fact that most ER glycoproteins are not complex-glycosylated. Mutant ER proteins lacking the C-terminal tetrapeptide are only slowly secreted (164, 475) demonstrating that retention in the ER requires an additional mechanism. This mechanism is poorly characterized but may involve formation of a protein matrix via calcium bridges in the lumen of the rough ER (49, 425, 535). Indeed a mutant of the KDEL-bearing protein calreticulin that lacks the Ca^{2+}-binding domain was less efficiently retained in the ER than wild-type calreticulin (475). In some cell types such as exocrine pancreas, KDEL proteins are distributed throughout the se-

cretory pathway (495) as the result of inefficient retention. In vitro studies have shown that binding of KDEL proteins to Erd2 is stronger at acidic than at neutral pH (563) suggesting a pH-dependent mechanism of ligand binding in the slightly acidic Golgi lumen (8) and release in the (presumed) neutral ER lumen.

ER and ERGIC Membrane Proteins

Only few ER membrane proteins such as the yeast type II transmembrane proteins Sec20p and Sed4p possess a C-terminal KDEL/HDEL determinant in their luminal domain (159, 491). A growing family of type I ER proteins contains a COOH-terminal KKXX or KXKXX signal (where X can be almost any amino amino acid) (197, 463) that is believed to function as a retrieval determinant (130, 198, 514). In vitro this motif specifically binds to coatomer (81) and coatomer mutations in yeast selectively affect the retention of KKXX-proteins (82, 246), suggesting an involvement of coatomer in the retrieval process. In some type II transmembrane proteins, retention is determined by two critical arginines within the first five amino-terminal residues and it has been postulated that this double arginine motif is also a recycling determinant (440).

Surprisingly, ERGIC-53/p58 also carries a COOH-terminal KKXX (that is, KKFF) ER retrieval signal (438) although this protein is most highly concentrated in the ERGIC (442) contrary to other KKXX proteins (197). Intracellular retention of ERGIC-53 is saturable in transfected cells (210) due to the presence of two terminal phenylalanines (194).

Obviously, proper localization of membrane proteins to the ER or ERGIC requires at least a second signal with true retention features as indicated by the observation that inactivation of the KKXX motif in glucurono-syltransferase (198) or ERGIC-53 (194) does not lead to increased cell surface expression. In the case of glucuronosyltransferase additional retention is confined to the transmembrane and/or luminal domain, whereas correct localization of ERGIC-53 requires the presence of a at least two additional determinants.

A mechanistically different way of protein retention was observed for the human type II transmembrane protein p63 (446, 448). Previously p63 (and its animal homologue p62) was believed to be associated with a compartment between ER and Golgi apparatus (308, 445), but has recently been shown to localize to the rough ER largely excluding the outer nuclear membrane (449). All three domains of p63 are required to achieve complete intracellular retention. Retention perfectly correlated with the formation of large detergent-insoluble complexes, suggesting that self-association may be a major mechanism by which p63 is retained within the cells. There are other classes of ER proteins whose mechanism of retention is less clear. They include cytochromes P450 that are anchored via an N-terminal domain and C-terminally anchored proteins that are inserted into the membrane post-translationally (233). The C-terminal anchors may contribute to localization (257, 302).

Golgi Membrane Proteins

There are three classes of Golgi membrane proteins: single-pass transmembrane proteins with a large luminal domain, multispanning proteins, and (discovered more recently) C-terminally anchored proteins. Contrary to the ER, the Golgi apparatus apparently does not harbor soluble proteins in the cisternal lumen with

one exception (436). Most single pass proteins have a type II membrane topology and are involved in glycan trimming and elongation. A number of studies have shown that the membrane-spanning domain and part of its flanking region contain sufficient information for Golgi localization (278, 329) (Table 15.4). A recent study of GlcNAc-transferase suggests, however, that Golgi localization may be mediated by interactions spanning the entire length of the molecule rather than a discrete retention signal (62). Likewise, at least two domains are required for retaining mouse corona virus M protein in the TGN (15, 227).

Based on the importance of the transmembrane domain for retention and the finding that the replacement of the transmembrane domain of VSV G protein with the first of the three transmembrane domains of coronavirus M (formerly E1) glycoprotein results in a chimeric protein that aggregates in the Golgi apparatus (547) it was proposed that each Golgi protein is retained by hetero-oligomer formation with other Golgi proteins of the same cisterna (also termed "kin recognition") (328). Retention would be achieved by attaching the cytoplasmic domains of the proteins in the complex to an intercisternal matrix. Two observations are in accord with the kin recognition model. First, when the medial-Golgi enzymes GlcNAc-transferase was kept in the ER by replacing its cytosolic domain with that containing a double arginine retrieval signal, another medial enzyme, mannosidase II, accumulated in the ER (330). Second, the same two enzymes bind to a Golgi matrix in vitro in an identical salt-dependent manner (469). It should be noted, however, that a direct hetero-oligomeric interaction of Golgi proteins with each other remains to be shown. Inconsistent with the mechanistic explanation of Golgi protein targeting and retention by means of the formation of large, relatively immobile protein complexes is the recent finding of a high diffusional mobility of several Golgi proteins as measured with the fluorescence photobleaching recovery technique (73).

Although kin recognition may function in Golgi retention of some proteins, it cannot be the only mechanism as illustrated by the Golgi protein giantin. The distribution of giantin largely overlaps with that of GlcNAc-transferase and mannosidase II (451), yet it does not form kin-recognition complexes with the two enzymes (Linstedt et al., unpublished). Giantin is a C-terminally anchored transmembrane protein (255, 257, 451) and therefore its retention mechanism may differ from that of conventionally anchored Golgi proteins.

While the retention of medial- and trans-Golgi proteins is not saturable by overexperession (with the exception of galactosyltransferase expressed in COS cells: see Fig. 2 in 219) the retention machinery for retaining the TGN proteins DPAP A and Kex2p in yeast appears to be saturable (332). Retention of these proteins depends on a critical cytoplasmic tyrosine. Both the clathrin heavy chain (450) and the dynamin-like protein Vps1p (559) are required for retaining these two TGN proteins in the Golgi apparatus. In yeast these proteins are believed to shuttle between the TGN and a late endosomal compartment (332). The precise mechanism by which clathrin and Vps1p mediate protein retention in the TGN remains to be elucidated.

The targeting of TGN38 to the TGN of mammalian cells also depends on a critical tyrosine in the cytoplasmic domain (Table 15.4). The tyrosine is part of a retrieval signal related to endocytotic signals (273) that is necessary and sufficient to target integral membrane proteins to the TGN. An additional signal exists in the membrane spanning-domain of TGN38 that is also sufficient for localization to the TGN and most likely constitutes a retention signal comparable to that of medial and trans-Golgi enzymes (387).

A fundamentally different model proposes that membrane thickness rather than protein–protein interaction determines protein retention in the Golgi apparatus (58). It is believed that membrane thickness in the Golgi increases in cis to trans direction as a result of progressively increased cholesterol and sphingolipid content. Because Golgi proteins have consistently shorter membrane-spanning regions (17 amino acids on average) than plasma membrane proteins (21 amino acids on average) they would be trapped during passage through this organelle as soon as membrane thickness matches the length of their membrane-spanning domain. Experiments in support of this view showed that the primary sequence of the transmembrane domain of sialyltransferase was not required for Golgi localization (87, 309) but extension of the domain from 17 to 23 hydrophobic amino acids allowed transport to the cell surface (309). Likewise, when the transmembrane domain of a type I plasma membrane protein was shortened from 23 to 17 leucines, it accumulated in the Golgi apparatus (310). If this model is correct one would predict that cholesterol depletion affects the retention of Golgi proteins. A preliminary report on the localization of several Golgi resident proteins expressed in cholesterol-depleted insect cells suggests that cholesterol is not required for correct targeting of these proteins to the Golgi apparatus (406) contradicting the general validity of the lipid retention model. Nevertheless, it remains possible that lipids contribute to protein retention in the Golgi apparatus in conjunction with kin recognition (367).

RETROGRADE PROTEIN TRAFFIC

Insights from Brefeldin A

The maintenance of anterograde transport from ER to Golgi (or ERGIC) requires retrieval of both membrane lipids and integral membrane proteins involved in vesicle transport such as v-SNAREs. How is the recycling of these components achieved and what are the structures involved? A widely held view assumes that BFA-induced changes of the Golgi apparatus reflect exaggerated normal transport back to the ER (218). The fungal metabolite BFA blocks protein secretion (301) and disrupts the Golgi apparatus. BFA impedes the nucleotide exchange step in the Arf GTPase cycle and thereby prevents COPI binding to membranes (100, 172, 173). In BFA-treated pancreatic acinar (but not other cells) dissociated coatomer forms large aggregates (176, 350). Upon treatment of cells with BFA the Golgi apparatus forms long tubules that fragment into vesicles that rapidly fuse with the ER. This leads to partial consumption of the Golgi apparatus by the ER (90, 175, 179). It has been suggested, therefore, that the effects of BFA are due to the inhibition of forward transport without corresponding inhibition of retrograde flow resulting in a net accumulation of Golgi proteins in the ER (259). Surprisingly, Golgi resident proteins, including mannosidase II and galactosyltransferase, entered these putative retrograde tubules immediately after addition of the drug, even though these proteins do not recycle in untreated cells. Recent studies show that Golgi proteins that follow a recycling pathway in untreated cells accumulate in the ERGIC in BFA-treated cells and not in the ER (5, 196). The difference between noncycling and cycling Golgi resident proteins is best illustrated with GalNAc-transferase (noncycling) and KDEL-receptor (cycling), both of which reside in the same cis-Golgi cisternae in untreated cells: GalNAc-transferase redistributes to the ER upon BFA treatment while KDEL receptor rapidly colocalizes with ERGIC-53 (196). A peripheral cis-Golgi associated protein of unknown function also redistributes to the ERGIC in BFA-treated cells (405). It appears that the redistribution to the ERGIC does not involve prior passage through the ER. Collectively the data show that in BFA-treated cells the noncycling Golgi proteins do not indicate the retrograde pathway followed by known proteins that naturally recycle in nontreated cells. However, these studies suggest a novel application for BFA: discrimination between cycling and noncycling proteins. A temperature-sensitive mutant of CHO cells (DS28-6) expresses key phenotypic changes associated with BFA treatment (573) and may therefore contribute to the elucidation of the mechanism underlying redistribution of Golgi membranes to the ER.

Recycling from the cis-Golgi

The KDEL/HDEL-receptor Erd2 [two genes in humans: (186, 247, 249)] is the most completely studied protein that recycles from the cis-Golgi in a ligand-regulated fashion. In mammalian cells the receptor is localized to the cis-Golgi and in part to the ERGIC (145, 499) but can be induced to move to the ER by high-level expression of an appropriate ligand (249, 515). A mutational analysis of the human KDEL receptor revealed distinct structural requirements for Golgi retention, ligand binding, and retrograde transport (513). Retrograde transport is unaffected by mutations in the cytoplasmic loops but critically dependent upon an aspartic acid residue (D193) in the seventh transmembrane domain. A self-association model was proposed to explain the recycling (513). According to this model the occupied receptors undergo a conformational change that allows them to oligomerize with each other, producing a patch of protein that is incorporated into a retrograde carrier. Oligomerization would be triggered by D193. Such an oligomeric complex would still require a third component that drives the complex into the retrograde carriers, for instance, a transmembrane protein that links cargo to coats.

The KDEL retrieval system also contributes to quality control. Misfolded VSV G protein can leave the ER in a complex with the KDEL-bearing protein BiP. It is subsequently recycled to the ER from the cis-Golgi, most likely by the KDEL receptor (152). Recycling via cis-Golgi was also observed for unassembled MHC class I molecules (185) and may occur by the same mechanism.

Role of the ERGIC and Possible Role of ERGIC-53

ERGIC-53/p58 is the only endogenous protein for which constitutive recycling in the early secretory pathway of higher eukaryotes has been documented. Recent morphological and biochemical evidence suggests that ERGIC-53 recycles to the ER from two sites, from the ERGIC and from the cis-Golgi (196). The recycling of ERGIC-53 can be partially synchronized by incubating cells at 15°C. At this temperature the protein accumulates in the tubulovesicular ERGIC elements that have moved closer to the Golgi apparatus. Upon rewarming, ERGIC-53 follows a tubular retrograde pathway to the ER. Recycling in the ERGIC can be blocked almost completely with AlF_4^- in a reversible manner. Synchronization by AlF_4^- in conjunction with density gradient analysis allows us to directly compare the biochemical

requirements for retrograde and anterograde transport in vivo (Fig. 15.5). ER-to-ERGIC anterograde and ERGIC-to-ER retrograde transport are clearly different concerning sensitivity to low temperature, deoxyglucose/azide (that is, low ATP) and AlF_4^- (that is, most likely activated trimeric G proteins). In contrast, the requirements for ERGIC-to-cis Golgi and ERGIC-to-ER transport are more similar, though not identical. Current evidence suggests that recycling of ERGIC-53 (and KDEL-receptor) from cis-Golgi to ER involves the ERGIC as an intermediated step (Fig. 15.5). It remains to be shown if an additional direct cis-Golgi to ER pathway exists not involving the ERGIC.

The sequence of ERGIC-53 (438) reveals features that suggest this protein connects vesicular cargo with cytoplasmic coats. The luminal domain of ERGIC-53 comprises a lectin-binding domain (114), and it has been suggested that ERGIC-53 and the homologous VIP36, a protein presumably recycling between plasma membrane and Golgi (115), are members of a new class of animal lectins. There is indeed functional evidence that ERGIC-53 is a mannose-selective lectin (12, 196a). By recognizing sugar residues, ERGIC-53 may facilitate anterograde transport of glycoproteins. Alternatively, ERGIC-53 may function in glycoprotein maturation by imposing a post-ER level of quality control downstream of calnexin. The C-terminal KKFF recycling motif of ERGIC-53's cytoplasmic domain, on the other hand, binds coatomer and thereby controls recycling (Kappeler, F., and H.-P. Hauri, unpublished).

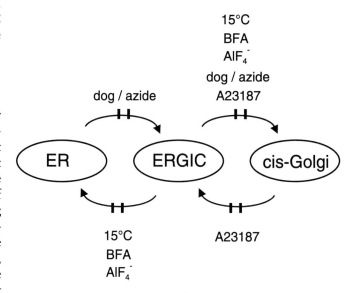

FIG. 15.5. Transport inhibitors for anterograde and retrograde transport in the early secretory pathway of mammalian cells *in vivo* as deduced from morphological and subcellular fractionation experiments (196, 270, 384, 443). *dog,* deoxy-glucose.

REGULATION OF MEMBRANE TRAFFIC IN THE EARLY SECRETORY PATHWAY AND MAINTENANCE OF ORGANELLE STRUCTURE

Recent studies point to the importance of membrane traffic for the structural maintenance of secretory organelles, in particular of the Golgi apparatus. Various general transport inhibitors as well as the permanent activation or inactivation of key components of vesicular traffic lead to vesiculation or disappearance of the Golgi apparatus. In the case of BFA these changes are considered a consequence of an imbalance of anterograde ER-to-Golgi and retrograde Golgi-to-ER traffic due to the loss of COPI coats that results in the disappearance of the Golgi apparatus and its partial fusion with the ER (96, 217, 259). In contrast, locking coatomers to membranes by GTP-γS or AIF$_4^-$ (116, 292) transforms the Golgi apparatus into at least two populations (317) of fusion-incompetent vesicles: COPI-coated and p200-coated vesicles originating from the cis- and trans-Golgi, respectively. A similar effect is observed by overexpressing a dominant allele of ARF1 (572). Moreover, disruptions in Golgi structure and membrane traffic in a conditional lethal mutant of CHO cells are corrected by transfection with ϵ-COP cDNA (150).

Golgi vesiculation also occurs in the presence of the sea sponge metabolite ilimaquinone, which inhibits intra-Golgi protein transport (1, 497). Unlike BFA, ilimaquinone treatment does not cause a rapid release of COPs in vivo (although it prevents association of beta-COP and ARF to Golgi membranes in vitro) and there is no retrograde transport of Golgi enzymes into the ER under these conditions. The relationship between vesicles produced by ilimaquinone and the COPI-coated transport vesicles is unclear at present.

Microinjection of a constitutively GDP-bound mutant of Rab1a (Rab1aS25N) also promotes disassembly and vesiculation of the Golgi apparatus (561). Again, this kind of Golgi disassembly differs from BFA-induced disruption because beta-COP remains membrane associated and Golgi enzymes do not redistribute to the ER. Given the need for the rab1 homologue YPT1 in yeast for ER-Golgi transport it appears likely that the rab1a mutant prevents vesicle docking and thereby leads to the observed Golgi vesiculation.

The above examples clearly indicate that the structure of the Golgi apparatus is tightly coupled to vesicular transport. Obviously the maintenance of secretory organelles requires coordinate regulation of individual transport pathways into and out of the Golgi apparatus both in anterograde and retrograde direction, but what is the basis of this regulation?

There is increasing evidence suggesting that membrane traffic is accelerated by microtubule-dependent transport whereby traffic over long distances is more dependent on a microtubular network than short distance transport (213, 237, 536). This general rule is based on differential sensitivity of individual transport steps to the microtubule poison nocodazole. It is conceivable that trafficking of all vesicles between compartments is powered to some extent by microtubule-dependent motor proteins. Studies with the video-enhanced differential interference contrast microscope showed that in living cells, directed movement of small vesicles ceased after 5 min upon exposure to nocodazole (151). Various pharmacological agents including the serine/threonine phosphatase inhibitor okadaic acid and activators of adenyl cyclase as well as the tyrosine kinase inhibitor genistein and the calcium ionophore A23187 were found to increase the microtubule-dependent vesicle movement. Remarkably, the directionality of vesicle movement was not significantly altered by these agents, supporting a model of coordinate regulation of anterograde (in this case primarily post-Golgi exocytotic) and retrograde (endocytotic) traffic involving regulation of microtubule-dependent motors. This notion is in line with biochemical studies that have demonstrated inefficient protein transport along various post-Golgi pathways in the presence of nocodazole (3, 57, 107, 135, 148, 187, 288).

The role of microtubules in vesicular traffic of the early secretory pathway is less clear. Efficient retrograde transport of Golgi resident proteins to the ER induced by BFA is dependent on an intact microtubular network (260) and morphological data on the temperature-dependent redistribution of p58 were taken as evidence for a role of microtubules in ER-to-Golgi anterograde transport (430). In contrast, recent subcellular fractionation studies with HepG2 cells failed to demonstrate an effect of nocodazole on the bidirectional transport of ERGIC-53 between ER, ERGIC, and cis-Golgi (Foguet, M., and H.-P. Hauri, unpublished). This study clearly shows that an intact microtubular network is not absolutely required for efficient transport, but it does not rule out the possibility that in untreated cells vesicular traffic in the early secretory pathway is mediated by microtubule-dependent motors. Vesicular transport along microtubular tracks may explain the finding that in heterokaryons made between two human cell lines the transfer of a secretory protein occurred more rapidly between homologous ER and Golgi elements than between heterologous ones (524). Selective experimental inactivation or depletion of motor molecules, such as kinesin or dynein, without disrupting the microtubules may provide more direct insight into the role of microtubules in the early secretory pathway (237, 262, 284).

The analysis of the role of microtubules in vesicular transport, in particular by morphological approaches, is complicated because the organelles themselves interact with the microtubular network (517) and thus it is not always clear to what extent microtubule disruption affects organelle structure rather than vesicular traffic. The ER can expand by binding to and moving along microtubules in a kinesin-dependent (that is, plus-end directed) manner, whereas Golgi elements move along microtubules during telophase in the opposite direction (that is, dynein-dependent minus-end direction) (80, 86, 180, 507). In fact, microtubules have a general function in organizing the cytoplasm in concert with actin filaments and intermediate filaments, and in so doing constitute a major determinant of overall cell shape. The mode of interactions of organelles with microtubules is poorly understood. Besides motor proteins, linker proteins may play important roles as exemplified by CLIP-170, a cytoplasmic protein that mediates interactions of endosomes with microtubules (377, 378, 404), and dynactin, the putative receptor for kinesin (63). Similarly, a soluble Golgi-associated 58 kDa protein binds to polymerized tubulin in vitro and it may therefore be involved in the attachment of the Golgi apparatus to microtubules (45). Another potential determinant for Golgi structure and perhaps function is beta-spectrin (31). Disruption of both Golgi structure and function, either in mitotic cells or following addition of BFA, is accompanied by loss of beta-spectrin from Golgi membranes, whereas the protein remains associated with Golgi fragments in nocodazole treated cells. Organelle–microtubule interactions may also be mediated by integral organelle proteins such as giantin (255). Giantin is a large (386 kDa) protein that is C-terminally anchored to the Golgi apparatus (257). Its cytoplasmic domain, comprising over 3000 amino acids, is largely composed of heptad repeats and forms an approximately 250 nm rod. Giantin from interphase cells is resistant to extraction with non-ionic detergents and binds to taxol-polymerized microtubules in vitro (256). During mitosis the protein becomes readily extractable and phosphorylated, suggesting that giantin is involved in cell cycle-dependent changes in Golgi structure and that giantin may link the Golgi apparatus to the cytoskeleton.

In addition to membrane traffic other factors must be involved in organelle reformation and maintenance because the Golgi stack reassembles during telophase before arrival of proteins transported from the ER and Golgi stack formation in ilimaquinone wash-out experiments can occur at 16°C, which blocks ER to Golgi transport in the ERGIC (527). Golgi reassembly can indeed be reconstituted in a cell-free system in the absence of membrane traffic and microtubules (393).

Strikingly, in this in vitro model individual cisternae of a defined length are reformed by vesicle fusion before stacking occurs and further growth of the Golgi stack continues. Obviously, a ruler determines cisternal length and one may speculate that large molecules like giantin may constitute such a ruler. The initial stacking process is independent on microtubules, but can be blocked by the phosphatase inhibitor microcystin (393), the formation of larger Golgi complexes, however, requires an intact microtubular network (2, 527).

A protein clearly required for the reassembly of the Golgi during mitosis is p97 (2, 295, 395). P97 is a homologue of the fusion factor NSF but its mode of action is most likely quite different because in vitro, p97-mediated Golgi assembly did neither require exogenous SNAP nor a Rab protein. Likewise, a homologue of p97 (Cdc48p) participates in the homotypic fusion of ER membranes (240).

The regulation of organelle size and function must also involve signaling to the nucleus because organelles have to grow to their original size after mitosis by de novo synthesis of organelle components. How this is achieved is a mystery. A signaling pathway to the nucleus has been uncovered for the regulation of ER chaperone proteins in yeast (83, 306, 459). The synthesis of these proteins is regulated according demand from the inside of the ER: it is triggered by the accumulation of unfolded proteins. Signal transduction from the ER lumen to the nucleus involves a transmembrane protein kinase encoded by the IRE1 gene (also termed ERN1). Ire1p appears to be the proximal sensor of misfolding events in the ER. It was postulated that Ire1p transduces the information across the ER membrane very much like class I growth factor receptors, leading to activation of a specific set of transcription factors. Higher eukaryotes possess at least two distinct ER-to-nucleus signalling pathways. One responds specifically to the presence of unfolded proteins and activates the glucose-responsive proteins including BiP. A second pathway senses the amount of proteins processed by the ER and signals ER overload. This second pathway is apparently mediated by the transcription factor NF-kappaB which upon induction translocates from the cytoplasm to the nucleus (357).

CONCLUSIONS AND PROSPECTS

Recent progress in understanding the secretory pathway is remarkable. The convergence of morphological, biochemical, and genetic approaches during the past 15 years has led to novel insights into the organization of the secretory pathway and the molecular basis of protein transport. A reasonable view now is that the early secretory pathway consist of at least three highly dynamic compartments—rough ER, ERGIC, and Golgi apparatus—that are connected by vesicular pathways both in anterograde and retrograde direction. Anterograde protein transport from ER to ERGIC is most likely mediated by COP II-coated vesicles whereas COPI coat proteins are involved in ERGIC (and most likely cis-Golgi) to ER retrograde transport. Forward transport from the ERGIC to the Golgi stack may also involve COPI proteins.

Sorting of secretory proteins from ER resident proteins is achieved by a multistep process that starts in the ER, where ER proteins are immobilized by a still poorly characterized retention system and secretory proteins are packed into transport vesicles departing from the transitional elements of the rough ER, either by bulk flow or the aid of (postulated) transport receptors. Before leaving the ER, secretory proteins are subjected to a complex series of folding reactions often termed "quality control." Sorting in the ER is incomplete for some ER resident proteins. ER proteins leaking into the secretory pathway are recycled to the ER from the ERGIC and later points by receptor-mediated and/or coat-mediated mechanisms.

Many details of the formation targeting and fusion of COPI- and COPII-coated vesicles have been elucidated and fitted into a conceptual framework for vesicular transport in which a protein coat drives vesicle budding and the specific pairwise interaction of vesicle and target receptors (that is, v-SNAREs and t-SNAREs) controls targeting and fusion in conjunction with accessory factors, including small GTPases, and an ubiquitous fusion machinery. Work on vesicular transport has undoubtedly laid important groundwork for a molecular description of secretion. Considering the plethora of unanswered questions, however, it would appear that the secretory pathway is far from having revealed its major secrets. We still do not know how many true compartments the secretory pathway is composed of and where the organelle's boundaries are. When vesicular transport is the mechanism by which protein transport occurs, what is the function of tubular membrane continuities observed between organelles? What is the mechanism underlying homotypic membrane interaction (for example, ER with ER or Golgi with Golgi)? How are organelles properly positioned within the cytoplasm? What determines the structure of a compartment—for instance, why is the Golgi apparatus stacked and the ER reticular? How is the size of a compartment regulated? The latter question draws our attention to signaling pathways that must operate between secretory organelles and the nucleus. How many different transport vesicles exist and how can they be defined molecularly in respect to budding targeting and fusion? How is anterograde and retrograde transport coordinated? To what extent is anterograde protein traffic nonselective, that is, occurring by bulk flow versus occurring by means of transport receptors? Answers to these challenging questions will come not only from the continued analysis of transport vesicles but also from many different areas of cell biology, including research on organelle–cytoskeleton interactions, molecular motors, and signal transduction.

We thank Jack Fransen for providing the electron micrograph of Figure 15.3. Work in the author's (H.-P.H.) laboratory was supported by the Swiss National Science Foundation, the Kantons of Basel, Sandoz Pharma Ltd. (Basel), and the Emilia Guggenheim-Schnurr Foundation.

REFERENCES

1. Acharya, U., J. M. McCaffery, R. Jacobs, and V. Malhotra. Reconstitution of vesiculated Golgi membranes into stacks of cisternae: requirement of NSF in stack formation. *J. Cell Biol.* 129: 577–590, 1995.

2. Acharya, U., R. Jacobs, J.-M. Peters, N. Watson, M. G. Farquhar, and V. Malhotra. The formation of Golgi stacks from vesiculated Golgi membranes requires two distinct fusion events. *Cell* 82: 895–904, 1995.

3. Achler, C., D. Filmer, C. Merte, and D. Drenckhahn. Role of microtubules in polarized delivery of apical membrane proteins to the brush border of the intestinal epithelium. *J. Cell Biol.* 109: 179–189, 1989.

4. Alberini, C. M., P. Bet, C. Milstein, and R. Sitia. Secretion of immunoglobulin M assembly intermediates in the presence of reducing agents. *Nature* 347: 485–487, 1990.

5. Alcade J., G. Egea, and I. V. Sandoval. gp74 a membrane glycoprotein of the cis-Golgi network that cycles through the endoplasmic reticulum and intermediate compartment. *J. Cell Biol.* 124: 649–665, 1994.

6. Alexandrov, K., H. Horiuchi, O. Steele-Mortimer, MC. Seabra, and M. Zerial. Rab escort protein-1 is a multifunctional protein that accompanies newly prenylated rab proteins to their target membranes. -*EMBO J.* 13: 5262–5273, 1994.

7. Amara, J. F., S. H. Cheng, and A. E. Smith. Intracellular protein trafficking defects in human disease. *Trends Cell Biol.* 2: 145–149, 1992.

8. Anderson, R.G.W., and R. K. Pathak. Vesicles and cisternae in the trans Golgi apparatus of human fibroblasts are acidic compartments. *Cell* 40: 635–643, 1985.

9. Anfinson, C. B. Principles that govern the folding of protein chains. *Science* 181: 223–230, 1973.

10. Anthony, C., C. Cibert, G. Géraud, A. Santa-Maria, B. Maro, V. Mayau, and B. Goud. The small GTP-binding protein rab6p is distributed from medial Golgi to the trans-Golgi network as determined by a confocal microscopy approach. *J. Cell Sci.* 103: 785–796, 1992.

11. Aoki, D., N. Lee, N. Yamaguchi, C. Dubois, and M. N. Fukuda. Golgi retention of a trans-Golgi membrane protein, galactosyltransferase, requires cysteine and histidine residues within the transmembrane-anchoring domain. *Proc. Natl. Acad. Sci. U.S.A.* 89: 4319–4323, 1992.

12. Arar, C., C. Carpentier, J.-P. le Caer, M. Monsigny, A. Legrand, and A.-C. Roche. ERGIC-53, a membrane protein of the endo-

plasmic reticulum-Golgi intermediate compartment, is identical to MR60, an intracellular mannose-specific lectin of myelomonocytic cells. *J. Biol. Chem.* 270: 3551–3553, 1995.

13. Arber, S., K.-H. Krause, and P. Caroni. S-cyclophilin is retained intracellularly via a unique COOH-terminal sequence and colocalizes with the calcium storage protein calreticulin. *J. Cell Biol.* 116: 113–125, 1992.

14. Aridor, M., S. I. Bannykh, T. Rowe, and W. E. Balch. Sequential coupling between CopII and CopI vesicle coats in endoplasmic reticulum to Golgi transport. *J. Cell Biol.* 131: 875–893, 1995.

15. Armstrong, J., and S. Patel. The Golgi sorting domain of corona virus E1 protein. *J. Cell Sci.* 98: 567–575, 1991.

16. Bacon, R. A., A. Salminen, H. Ruohola, P. Novick, and S. Ferro-Novick. The GTP-binding protein Ypt1 is required for transport in vitro: The Golgi apparatus is defective in ypt1 mutants. *J. Cell Biol.* 109: 1015–1022, 1989.

17. Baker, D., L. Hicke, M. Rexach, M. Schleyer, and R. Schekman. Reconstitution of Sec gene product-dependent intercompartmental protein transport. *Cell* 54: 335–344, 1988.

18. Baker, D., L. Wuestenhube, R. Schekman, D. Botstein, and N. Segev. GTP-binding Ypt1 protein and Ca^{2+} function independently in a cell-free protein transport reaction. *Proc. Natl. Acad. Sci. U.S.A.* 87: 355–359, 1990.

19. Balch, W. E., W. G. Dunphy, W. A. Braell, and J. E. Rothman. Reconstitution of the transport of protein between successive compartments of the Golgi measured by the coupled incorporation of N-acetylglucosamine. *Cell* 39: 405–416, 1984.

20. Balch, W. E., R. A. Kahn, and R. Schwaninger. ADP-ribosylation factor is required for vesicular trafficking between the endoplasmic reticulum and the cis-Golgi compartment. *J. Biol. Chem.* 267: 13053–13061, 1992.

21. Balch, W. E., J. M. McCaffery, H. Plutner, and M. G. Farquhar. Vesicular stomatitis virus glycoprotein is sorted and concentrated during export from the endoplasmic reticulum. *Cell* 76: 841–852, 1994.

22. Balch, W. E. and M. G. Farquhar. Beyond bulk flow. *Trends Cell Biol.* 5: 16–19, 1995.

23. Banfield, D. K., M. J. Lewis, C. Rabouille, G. Warren, and H. R. Pelham. Localization of Sed5, a putative vesicle targeting molecule, to the cis-Golgi network involves both its transmembrane and cytoplasmic domains. *J. Cell Biol.* 127: 357–371, 1994.

24. Banfield, D. K., M. J. Lewis, and H.R.B. Pelham. A SNARE-like protein required for traffic through the Golgi complex. *Nature* 375: 806–809, 1995.

25. Bannykh, S., M. Aridor, H. Plutner, T. Rowe, and W. E. Balch. Regulated export of cargo from the endoplasmic reticulum of mammalian cells. Cold Spring Harbor; *Symp. Quantitat. Biol.* 60: 127–137, 1995.

26. Barlowe, C., and R. Schekman. Sec12 encodes a guanine-nucleotide-exchange factor essential for transport vesicle budding from the ER. *Nature* 365: 347–349, 1993.

27. Barlowe, C., C. d'Enfert, and R. Schekman. Purification and characterization of Sar1p, a small GTP-binding protein required for transport vesicle formation from the endoplasmic reticulum. *J. Biol. Chem.* 268: 873–879, 1993.

28. Barlowe, C., L. Orci, T. Yeung, M. Hosobuchi, S. Hamamoto, N. Salama, M. F. Rexach, M. Ravazzola, M. Amherdt, and R. Schekman. COPII: A membrane coat formed by sec proteins that drive vesicle budding from the endoplasmic reticulum. *Cell* 77: 895–907, 1994.

29. Barr, F. A., A. Leyte, and W. B. Huttner. Trimeric G proteins and vesicle formation. *Trends Cell Biol.* 2: 91–94, 1992.

30. Bauerfeind, R., and W. B. Huttner. Biogenesis of constitutive secretory vesicles, secretory granules and synaptic vesicles. *Curr. Opin. Cell Biol.* 5: 628–635, 1993.

31. Beck, K. A., J. A. Buchanan, V. Malhotra, and W. J. Nelson. Golgi spectrin: identification of an erythroid beta-spectrin homolog associated with the Golgi complex. *J. Cell Biol.* 127: 707–723, 1994.

32. Becker, B., B. Bölinger, and M. Melkonian. Anterograde transport of algal scales through the Golgi complex is not mediated by vesicles. *Trends Cell Biol.* 5: 305–307, 1995.

33. Beckers, C.J.M., D. S. Keller, and W. E. Balch. Semi-intact cells permeable to macromolecules: use in reconstitution of protein transport from the endoplasmic reticulum to the Golgi complex. *Cell* 50: 523–534, 1987.

34. Beckers, C.J.M., and W. E. Balch. Calcium and GTP: essential components in vesicular trafficking between the endoplasmic reticulum and Golgi apparatus. *J. Cell Biol.* 108: 1245–1256, 1989.

35. Beckers, C. J. M., H. Plutner, H. W. Davidson, and W. E. Balch. Sequential intermediates in the transport of protein between the endoplasmic reticulum and the Golgi. *J. Biol. Chem.* 265: 18298–18310, 1990.

36. Bednarek, S. Y., M. Ravazzola, M. Hosobuchi, M. Amherdt, A. Perrelet, R. Schekman, and L. Orci. COPI- and COPII-coated vesicles bud directly from the endoplasmic reticulum in yeast. *Cell* 83: 1183–1196, 1995.

37. Bennet, M. K., and R. H. Scheller. A molecular description of synaptic vesicle membrane trafficking. *Annu. Rev. Biochem.* 63: 63–100, 1994.

38. Bergeron, J.J.M., G. J. Kotwal, G. Levine, P. Bilan, R. Rachubinski, M. Hamilton, G. C. Shore, and H. P. Gosh. Intracellular transport of the transmembrane glycoprotein G of vesicular stomatitis virus through the Golgi apparatus as visualized by electron microscopy radioautography. *J. Cell Biol.* 94: 36–41, 1982.

39. Bergeron, J.J.M., M. B. Brenner, D. Y. Thomas, and D. B. Williams. Calnexin: a membrane-bound chaperone of the endoplasmic reticulum. *Trends Biol. Sci.* 19: 124–128, 1994.

40. Bergmann, J. E., K. T. Tokuyasu, and S. J. Singer. Passage of an integral membrane protein, the vesicular stomatitis virus glycoprotein, through the Golgi apparatus en route to the plasma membrane. *Proc. Natl Acad. Sci. U.S.A.* 78: 1746–1750, 1981.

41. Bergmann, J. E., and P. J. Fusco. The G protein of vesicular stomatitis virus has free access into and egress from the smooth endoplasmic reticulum of UT-1 cells. *J. Cell Biol.* 110: 625–635, 1990.

42. Berridge, M. J. Inositol trisphosphate and calcium signalling. *Nature* 361: 315–325, 1993.

43. Bigay, J., P. Deterre, C. Pfister, and M. Chabre. Fluoride complexes of aluminum and beryllium act on G proteins as reversibly bound analogues of the gamma phosphate of GTP. *EMBO J.* 6: 2907–2913, 1987.

44. Blond-Elguindi, S., S. E. Cwirla, W. J. Dower, R. J. Lipshutz, S. R. Sprang, J. F. Sambrook, and M.-J. Gething. Affinity panning of a library of peptides displayed on bacteriophages reveals the binding specificity of BiP. *Cell* 75: 717–728, 1993.

45. Bloom, G. S., and T. A. Brashear. A novel 58-kDa protein associated with the Golgi apparatus and microtubules. *J. Biol. Chem.* 264: 16083–16092, 1989.

46. Bole, D. G., Dowin, R., Doriaux, M., and J. D. Jamieson. Immunocytochemical localization of BiP to the rough endoplasmic reticulum: Evidence for protein sorting by selective retention. *J. Histochem. Cytochem.* 12: 1817–1823, 1989.

47. Bomsel, M. and K. Mostov. Role of heterotrimeric G proteins in membrane traffic. *Mol. Biol. Cell.* 3: 1317–1328, 1992.
48. Bonatti, S., G. Migliaccio, and K. Simons. Palmitylation of viral membrane glycoproteins takes place after exit from the endoplasmic reticulum. *J. Biol. Chem.* 264: 12590–12595, 1989.
49. Booth, C., and G.L.E. Koch. Perturbation of cellular calcium induces secretion of luminal ER proteins. *Cell* 59: 729–737, 1989.
50. Bos, K., C. Wraight, and K. K. Stanley. TGN38 is maintained in the trans-Golgi network by a tyrosine-containing motif in the cytoplasmic domain. *EMBO J.* 12: 2219–2228, 1993.
51. Bose, S., and R. B. Freedman. Peptidyl prolyl cis-trans-isomerase activity associated with the lumen of the endoplasmic reticulum. *Biochem. J.* 300: 865–870, 1994.
52. Bose, S., M. Mücke, and R. B. Freedman. The characterization of a cyclophilin-type peptidyl prolyl cis-trans-isomerase from the endoplasmic reticulum lumen. *Biochem. J.* 300: 871–875, 1994.
53. Bosshart, H., J. Humphrey, E. Deignan, J. Davidson, J. Drazba, L. Yuan, V. Oorschot, P. J. Peters, and J. S. Bonifacino. The cytoplasmic domain mediates localization of furin to the trans-Golgi network en route to the endosomal/lysosomal system. *J. Cell Biol.* 126: 1157–1172, 1994.
54. Bourne, H. R., D. A. Sanders, and F. McCormick. The GTPase superfamiliy: conserved structure and molecular mechanism. *Nature* 349: 117–127, 1991.
55. Braakman, I, J. Helenius, and A. Helenius. Role of ATP and disulphide bonds during protein folding in the endoplasmic reticulum. *Nature* 356: 260–262, 1992.
56. Braakman, I, J. Helenius, and A. Helenius. Manipulating disulfide bond formation and protein folding in the endoplasmic reticulum. *EMBO J.* 11: 1717–1722, 1992.
57. Breitfeld, P. P., W. McKinnon, and K. E. Mostov. Effect of nocodazole on vesicular traffic to the apical and basolateral surfaces of polarized MDCK cells. *J. Cell Biol.* 111: 2365–2373, 1990.
58. Bretscher, M. S., and S. Munro. Cholesterol and the Golgi apparatus. *Science* 261: 1280–1281, 1993.
59. Brown, M. S., and J. L. Goldstein. Protein prenylation. Mad bet for rab. *Nature* 366: 14–15, 1993.
60. Brown, H. A., S. Gutowski, C. R. Moomaw, C. Slaughter, and P. C. Sternweis. ADP-ribosylation factor, a small GTP-dependent regulatory protein, stimulates phospholipase D activity. *Cell* 75: 1137–1144, 1993.
61. Burke, J., J. M. Pettitt, H. Schachter, M. Sarkar, and P. A. Gleeson. The transmembrane and flanking sequences of beta1,2-N-acetylglucosaminyltransferase I specify medial Golgi localization. *J. Biol. Chem.* 267: 24433–24440, 1992.
62. Burke, J., J. M. Pettitt, D. Humphris, and P. A. Gleeson. Medial-Golgi retention of N-acetylglucosyminyltransferase I. Contribution from all domains of the enzyme. *J. Biol. Chem.* 269: 12049–12059, 1994.
63. Burkhardt, J. K. In search of membrane receptors for microtubule-based motors—Is kinectin a kinesin receptor? *Trends Cell Biol.* 6: 127–130, 1996.
64. Burns, K., and M. Michalak. Interactions of calreticulin with proteins of the endoplasmic and sarcoplasmic reticulum membranes. *FEBS Lett.* 318: 181–185, 1993.
65. Cannon, K. S., D. N. Hebert, and A. Helenius. Glycan-dependent and independent association of vesicular stomatitis virus G protein with calnexin. *J. Biol. Chem.* 271: 1428–14284, 1996.
66. Chaudhuri, M. M., P. N. Tonin, W. H. Lewis, and P. R. Srinivasan. The gene for a novel protein, a member of the protein disulphide isomerase/form I phosphoinositide-specific phospholipase c family, is amplified in hydroxyurea-resistant cells. *Biochem. J* 281: 645–650, 1992.
67. Chavrier, P., R. G. Parton, H.-P. Hauri, K. Simons, and M. Zerial. Localization of low molecular weight GTP binding proteins to exocytic and endocytic compartments. *Cell* 62: 317–329, 1990.
68. Chege, N. W., and S. R. Pfeffer. Compartmentation of the Golgi complex: brefeldin A distinguishes trans-Golgi cisternae from the trans-Golgi network. *J. Cell Biol.* 111: 893–899, 1990.
69. Chen, Y.-G., and D. Shields. ADP-ribosylation factor-1 stimulates formation of nascent secretory vesicles from the trans-Golgi Network of endocrine cells. *J. Biol. Chem.* 271: 5297–5300, 1996.
70. Cheng, S. H., R. J. Gregory, J. Marshall, S. Paul, D. W. Souza, G. A. White, C. R. O'Riordan, and A. E. Smith. Defective intracellular transport and processing of CFTR is the molecular basis of most cystic fibrosis. *Cell* 63: 827–834, 1990.
71. Clairmont, C. A., A. DeMaio, and C. B. Hirschberg. Translocation of ATP into the lumen of rough endoplasmic reticulum-derived vesicles and its binding to luminal proteins including BiP (GRP 78) and GRP 94. *J. Biol. Chem.* 267: 3983–3990, 1992.
72. Clary, D. O., I. C. Griff, and J. E. Rothman. SNAPs, a family of NSF attachment proteins involved in intracellular membrane fusion in animals and yeast. *Cell* 61: 709–721, 1990.
73. Cole, N. B., C. L. Smith, N. Sciaky, M. Terasaki, M. Edidin, J. Lippincott-Schwartz. Diffusional mobility of Golgi proteins in membranes of living cells. *Science* 273: 797–801, 1996.
74. Colombo, M. I., Inglese, J., C. D'Souza-Schorey, W. Beron, and P. D. Stahl. Heterotrimeric G proteins interact with the small GTPase ARF. Possibilities for the regulation of vesicular traffic. *J. Biol. Chem.* 270: 24564–24571, 1995.
75. Colley, K. J., E. V. Lee, and J. C. Paulson. The signal anchor and stem regions of the beta-galactoside alpha2,6-sialytransferase may each act to localize the enzyme to the Golgi apparatus. *J. Biol. Chem.* 267: 7784–7793, 1992.
76. Conckcroft, S., G.M.H. Thomas, A. Fensome, B. Geny, E. Cunningham, I. Gout, I. Hiles, N. F. Totty, O. Truong, and J. J. Hsuan. Phospholipase D: A downstream effector of ARF in granulocytes. *Science* 263: 523–526, 1994.
77. Conolly, C. N., C. E. Futter, A. Gibson, C. R. Hopkins, and D. F. Cutler. Targeting into and out of the Golgi complex studied by transfecting cells with cDNAs encoding horseradish peroxidase. *J. Cell. Biol.* 127: 641–652, 1994.
78. Cooper, A., and H. Bussey. Yeast Kex1p is a Golgi-associated membrane protein: deletions in a cytoplasmic targeting domain result in mislocalization to the vacuolar membrane. *J. Cell Biol.* 119: 1459–1468, 1992.
79. Cooper, M. S., A. H. Cornell-Bell, A. Chernjavsky, J. W. Dani, and S. J. Smith. Tubulovesicular processes emerge from the trans-Golgi cisternae, extend along microtubules, and interlink adjacent trans-Golgi elements into a reticulum. *Cell* 61: 135–145, 1990.
80. Corthésy-Theulaz, I., A. Pauloin, and S. R. Pfeffer. Cytoplasmic dynein participates in the centrosomal localization of the Golgi complex. *J. Cell Biol.* 118: 1333–1345, 1992.
81. Cosson, P., and F. Letourneur. Coatomer interaction with di-lysine endoplasmic reticulum retention motifs. *Science* 263: 1629–1631, 1994.
82. Cosson, P., C. Démollière, S. Hennecke, R. Duden, and F. Letourneur. Delta- and zeta-COP, two coatomer subunits homologous to clathrin-associated proteins, are involved in ER retrieval. *EMBO J.* 15: 1792–1798, 1996.
83. Cox, J. S., C. E. Shamu, and P. Walter. Transcriptional induction

of genes encoding endoplasmic reticulum resident proteins requires a transmembrane protein kinase. *Cell* 73: 1197–1206, 1993.

84. Craig, E. A., J. S. Weissman, and A. L. Horwich. Heat shock proteins and molecular chaperones: mediators of protein conformation and turnover in the cell. *Cell* 78: 365–372, 1994.

85. Crimaudo, C., M. Hortsch, H. Gausepohl, and D. I. Meyer. Human ribophorins I and II: the primary structure and membrane topology of two highly conserved rough endoplasmic reticulum-specific glycoproteins. *EMBO J.* 6: 75–82, 1987.

86. Dabora, S. L., and M. P. Sheetz. The microtubule-dependent formation of a tubulovesicular network with characteristics of the ER from cultured cell extracts. *Cell* 54: 27–35, 1988.

87. Dahdal, R. Y., and K. J. Colley. Specific sequences in the signal anchor of the beta-galactoside alpha-2,6-sialyltransferase are not essential for Golgi localization. *J. Biol. Chem.* 268: 26310–26319, 1993.

88. Dascher, C., and W. E. Balch. Dominant inhibitory mutants of Arf1 block endoplasmic reticulum to Golgi transport and trigger disassembly of the Golgi apparatus. *J. Biol. Chem.* 269: 1437–1448, 1994.

89. De Camilli, P., S. D. Emr, P. S. McPherson, and P. Novick. Phosphoinositides as regulators in membrane traffic. *Science* 271: 1533–1539, 1996.

90. De Lemos-Chiarandini, C., N. E. Ivessa, V. H. Black, Y. S. Tsao, I. Gumper, and G. Kreibich. A Golgi-related structure remains after the brefeldin A-induced fusion of an ER-Golgi hybrid compartment. *Eur. J. Cell Biol.* 58: 187–201, 1992.

91. Denker, S. P., J. M. McCaffery, G. E. Palade, P. A. Insel, and M. G. Farquhar. Differential distribution of alpha and beta/gamma subunits of heterotrimeric G proteins on Golgi membranes of the exocrine pancreas. *J. Cell Biol.* 133: 1027–1040, 1996.

92. Denning, G. M., M. P. Anderson, J. F. Amara, J. Marshall, A. E. Smith, and M. J. Welsh. Processing of mutant cystic fibrosis transmembrane conductance regulator is temperature-sensitive. *Nature* 358: 761–764, 1992.

93. Denning, G. M., L. S. Ostegaard, and M. J. Welsh. Abnormal localization of cystic fibrosis transmembrane conductance regulator in primary cultures of cystic fibrosis airway. *J. Cell Biol.* 118: 551–559 1992.

94. Deschuyteneer, M., A. E. Eckhardt, J. Roth, and R. L. Hill. The subcellular localization of apomucin and non-reducing terminal N-acetylgalactosamine in porcine submaxillary glands. *J. Biol. Chem.* 263: 2452, 1986.

95. DeTomaso, A. W., G. Blanco, and R. W. Mercer. The alpha and beta subunits of the Na,K-ATPase can assemble at the plasma membrane into functional enzyme. *J. Cell Biol.* 127: 55–69, 1994.

96. Doms, R. W., G. Russ, and J. W. Yewdell. Brefeldin A redistributes resident and itinerant Golgi proteins to the endoplasmic reticulum. *J. Cell Biol.* 109: 61–72, 1989.

97. Donaldson, J. G., and R. D. Klausner. ARF: a key regulatory switch in membrane traffic and organelle structure. *Curr. Opin. Cell Biol.* 6: 527–532, 1994.

98. Donaldson, J. G., J. Lippincott-Schwartz, and R. D. Klausner. Guanine nucleotides modulate the effects of brefeldin A in semipermeable cells: regulation of the association of a 110-kD peripheral membrane protein with the Golgi apparatus. *J. Cell Biol.* 112: 579–588, 1991.

99. Donaldson, J. G., R. A. Kahn, J. Lippincott-Schwartz, and R. D. Klausner. Binding of ARF and beta-COP to Golgi membranes: possible regulation by a trimeric G protein. *Science* 254: 1197–1199, 1991.

100. Donaldson, J. G., D. Finazzi, and R. D. Klausner. Brefeldin A inhibits Golgi-membrane-catalized exchange of guanine nucleotide onto ARF protein. *Nature* 360: 350–352, 1992.

101. Donaldson, J. G., D. Cassel, R. A. Kahn, and R. D. Klausner. ADP-ribosylation factor, a small GTP-binding protein, is required for binding of the coatomer protein beta-COP to Golgi membranes. *Proc. Natl. Acad. Sci. U.S.A.* 89: 6408–6412, 1992.

102. Duden, R., G. Griffiths, R. Frank, P. Argos, and T. E. Kreis. Beta-COP, a 110 kD protein associated with non-clathrin-coated vesicles and the Golgi complex, shows homology to beta-adaptin. *Cell* 64: 649–665, 1991.

103. Duden, R., M. Hosobuchi, S. Hamamoto, W. Winey, B. Byers, and R. Schekman. Yeast beta- and beta′-coat proteins (COP). Two coatomer subunits essential for endoplasmic reticulum-to-Golgi protein traffic. *J. Biol. Chem.* 269: 24486–24495, 1994.

104. Dunphy, W. G., and J. E. Rothman. Compartmentation of asparagine-linked oligosaccharide processing in the Golgi apparatus. *J. Cell Biol.* 97: 270–275, 1983.

105. Dunphy, W. G., and J. E. Rothman. Compartmental organization of the Golgi stack. *Cell* 42: 13–21, 1985.

106. Dunphy, W. G., R. Brands, and J. E. Rothman. Attachment of terminal N-acetylglucosamine to asparagine-linked oligosaccharides occurs in central cisternae of the Golgi stack. *Cell* 40: 463–472, 1985.

107. Eilers, U., J. Klumperman, and H.-P. Hauri. Nocodazole, a microtubule-active drug, selectively interferes with apical protein delivery in cultured intestinal epithelial cells (Caco-2). *J. Cell Biol.* 108: 13–22, 1989.

108. Elazar, Z., L. Orci, J. Ostermann, M. Amherdt, G. Tanigawa, and J. E. Rothman. ADP-ribosylation factor and coatomer couple fusion to vesicle budding. *Cell Biol.* 124: 415–424, 1994.

109. Elazar, Z., T. Mayer, and J. E. Rothman. Removal of rab GTP-binding proteins from Golgi membranes by GDP dissociation inhibitor inhibits inter-cisternal transport in the Golgi stack. *J. Biol. Chem.* 269: 794–797, 1994.

110. Ellinger, A., and M. Pavelka. Subdomains of the rough endoplasmic reticulum in colon goblet cells of the rat: lectin-cytochemical characterization. *J. Histochem. Cytochem.* 40: 919–930, 1992.

111. Farquhar, M. G. Progress in unraveling pathways of Golgi traffic. *Annu Rev. Cell Biol.* 1: 447–488, 1985.

112. Ferro-Novick, S., and P. Novick. The role of GTP-binding proteins in transport along the exocytic pathway. *Annu. Rev. Cell Biol.* 9: 575–599, 1993.

113. Ferro-Novick, S., and R. Jahn. Vesicle fusion from yeast to man. *Nature* 370: 191–193, 1994.

114. Fiedler, K., and K. Simons. A putative novel class of animal lectins in the secretory pathway homologous to leguminous lectins. *Cell* 77: 625–626, 1994.

115. Fiedler, K., R. G. Parton, R. Kellner, T. Etzold, and K. Simons. VIP36, a novel component of glycolipid rafts and exocytic carrier vesicles in epithelial cells. *EMBO J.* 13: 1729–1740, 1994.

116. Finazzi, D., D. Cassel, J. G. Donaldson, and R. D. Klausner. Aluminum fluoride acts on the reversibility of ARF-1 dependent coat protein binding to Golgi membranes *J. Biol. Chem.* 269: 13325–13330, 1994.

117. Flynn, G. C., J. Pohl, M. T. Flocco, and J. E. Rothman. Peptide-binding specificity of the molecular chaperone BiP. *Nature* 353: 726–731, 1991.

118. Fournier, T., N. Mejdoubi, C. Lapoumeroulie, J. Hamelin, J. Elion, G. Durand, and D. Porquet. Translational regulation

of rat alpha1-acid glycoprotein gene by phenobarbital. *J. Biol. Chem.* 269: 27175–27178, 1994.

119. Fra, A. M., C. Fagioli, D. Finazzi, R. Sitia, and C. M. Alberini. Quality control of ER synthesized proteins: an exposed thiol group as a three-way switch mediating assembly, retention and degradation. *EMBO J.* 12: 4755–4761, 1993.

120. Franco, M., S. Paris, M. Chabre. The small G-protein ARF1$_{GDP}$ binds to the G$_t$beta/gamma subunit of transducin, but not to G$_t$alpha$_{GDP}$-G$_t$beta/gamma. *FEBS Lett.* 362: 286–290, 1995.

121. Fransen, J.A.M., H.-P. Hauri, L. A. Ginsel, and H. Y. Naim. Naturally occurring mutations in intestinal sucrase-isomaltase provide evidence for the existence of an intracellular sorting signal in the isomaltase subunit. *J. Cell Biol.* 115: 45–57, 1991.

122. Freedman, R. B. Protein folding in the cell. In: Protein Folding, edited by T. E. Creighton. New York: W. H. Freeman and Company, p. 455–539, 1992.

123. Freedman, R. B., T. R. Hirst, and M. F. Tuite. Protein disulphide isomerase: building bridges in protein folding. *Trends Biochem. Sci.* 19: 331–336, 1994.

124. Fries, E., and J. E. Rothman. Transport of vesicular stomatitis virus glycoprotein in a cell-free extract. *Proc. Natl. Acad. Sci. U.S.A.* 77: 3870–3874, 1980.

125. Füllekrug, J., B. Sönnichsen, U. Wünsch, K. Arseven, P. Nguyen Van, H.-D. Söling, and G. Mieskes. CaBP1, a calcium binding protein of the thioredoxin familiy, is a resident KDEL protein of the ER and not of the intermediate compartment. *J. Cell Sci.* 107: 2719–2727, 1994.

126. Futerman, A. H., B. Stieger, A. L. Hubbard, and R. E. Pagano. Sphingomyelin synthesis in rat liver occurs predominantly at the cis and medial cisternae of the Golgi apparatus. *J. Biol. Chem.* 265: 8650–8657, 1990.

127. Galat, A. Peptidylproline cis-trans-isomerases: immunophilins. *Eur. J. Biochem.* 216: 689–707, 1993.

128. Galteau, M.-M., B. Antoine, and H. Reggio. Epoxide hydrolase is a marker for the smooth endoplasmic reticulum in rat liver. *EMBO J.* 4: 2793–2800, 1985.

129. Gaut, J. R., and L. M. Hendershot. The modification and assembly of proteins in the endoplasmic reticulum. *Curr. Opin. Cell Biol.* 5: 589–595, 1993.

130. Gaynor, E. C., S. te Heesen, T. R. Graham, M. Aebi, and S. D. Emr. Signal-mediated retrieval of a membrane protein from the Golgi to the ER in yeast. *J. Cell Biol.* 127: 653–665, 1994.

131. Gerich, B., L. Orci, H. Tschochner, F. Lottspeich, M. Ravazzola, M. Amherdt, F. Wieland, and C. Harter. Non-clathrin-coat protein α is a conserved subunit of coatomer and in Saccharomyces cerevisiae is essential for growth. *Proc. Natl. Acad. Sci. U.S.A.:* 3229–3233, 1995.

132. Gething, M.-J., and J. Sambrook. Protein folding in the cell. *Nature* 355: 33–45, 1992.

133. Geuze, H. J., and D. J. Morré. Trans-Golgi reticulum. *J. Electron Microsc. Tech.* 17: 24–34, 1991.

134. Gilbert, H. F. Molecular and cellular aspects of thiol-disulfied exchange. *Adv. Enzymol.* 63: 69–172, 1990.

135. Gilbert, T., A. LeBivic, A. Quaroni, and E. Rodriguez-Boulan. Microtubular organization and its involvement in the biogenetic pathways of plasma membrane proteins in Caco-2 intestinal epithelial cells. *J. Cell. Biol.* 113: 275–288, 1991.

136. Glickman, J. N., and S. Kornfeld. Mannose 6-phosphate-independent targeting of lysosomal enzymes in I-cell disease B lymphoblasts. *J. Cell Biol.* 123: 99–108, 1993.

137. Goldberg, D. E., and S. Kornfeld. Evidence for extensive subcellular organization and asparagine-linked oligosaccharide processing and lysosomal enzyme phosphorylation. *J. Biol. Chem.* 258: 3159–3165, 1983.

138. Goldstein, J. L., and M. S. Brown. Regulation of the mevalonate pathway. *Nature* 343: 425–430, 1990.

139. Goud, B., A. Zahraoui, A. Tavitian, and J. Saraste. Small GTP-binding protein associated with the Golgi cisternae. *Nature* 345: 553–556, 1990.

140. Graham, K. S., A. Let, and R. N. Sifers. Accumulation of the insoluble PiZ variant of human alpha-1-antitrypsin within the hepatic endoplasmic reticulum does not elevate the steady-state level of BiP. *J. Biol. Chem.* 265: 20463–20468, 1990.

141. Green, J., G. Griffiths, P. Quinn, and G. Warren. Passage of viral membrane proteins through the Golgi complex. *J. Mol. Biol.* 152: 663–698, 1981.

142. Griffiths, G., and K. Simons. The trans Golgi network: sorting at the exit site of the Golgi complex. *Science* 234: 438–443, 1986.

143. Griffiths, G., S. D. Fuller, R. Back, M. Hollinshead, S. Pfeiffer, and K. Simons. The dynamic nature of the Golgi complex. *J. Cell Biol.* 108: 277–297, 1989.

144. Griffiths, G., and J. Gruenberg. The arguments for pre-existing early and late endosomes. *Trends Cell Biol.* 1: 5–9, 1991.

145. Griffiths, G., M. Ericsson, J. Krjinse-Locker, T. Nilsson, B. Gould, H.-D. Söling, B. L. Tang, S. H. Wong, and W. Hong. Localization of the Lys, Asp, Glu, Leu tetrapeptide receptor to the Golgi complex and the intermediate compartment in mammalian cells. *J. Cell Biol.* 127: 1557–1574, 1994.

146. Griffiths, G., R. W. Doms, T. Mayhew, and J. Lucocq. The bulk flow hypothesis: not quite the end. *Trends Cell Biol.* 5: 9–13, 1995.

147. Griffiths, G., R. Pepperkok, J. Krijnse Locker, and T. E. Kreis. Immunocytochemical localization of beta-COP to the ER-Golgi boundary and the TGN. *J. Cell Sci.* 108: 2839–2856, 1995.

148. Gruenberg, J., and K. E. Howell. Membrane traffic in endocytosis: Insights from cell-free assays. *Annu. Rev. Cell Biol.* 5: 453–481, 1989.

149. Günther, R., M. Srinivasan, S. Haugejorden, M. Green, I. M. Ehbrecht, and H. Küntzel. Functional replacement of Sccharomyces cerevisiae Trg1/Pdi1 protein by members of the mammalian protein disulfide isomerase. *J. Biol. Chem.* 268: 7728–7732, 1993.

150. Guo, Q., E. Vasile, and M. Krieger. Disruptions in Golgi structure and membrane traffic in a conditional lethal mammalian cell mutant are corrected by epsilon-COP. *J. Cell Biol.* 125: 1213–1224, 1994.

151. Hamm-Alvarez, S. F., P. Y. Kim, and M. P. Scheetz. Regulation of vesicle transport in CV-1 cells and extracts. *J. Cell Sci.* 106: 955–966, 1993.

152. Hammond, C., and A. Helenius. Quality control in the secretory pathway: retention of a misfolded viral membrane glycoprotein involves recycling between the ER, intermediate compartment, and Golgi apparatus. *J. Cell Biol.* 126: 41–52, 1994.

153. Hammond, C. and A. Helenius. A chaperone with a sweet tooth. *Curr. Biol.* 3: 884–886, 1994.

154. Hammond, C., and A. Helenius. Folding of VSV G protein: Sequential interaction with BiP and calnexin. *Science* 266: 456–458, 1994.

155. Hammond, C., I. Braakman, and A. Helenius. Role of N-linked oligosaccharide recognition, glucose trimming, and calnexin in glycoprotein folding and quality control. *Proc. Natl. Acad. Sci. U.S.A.* 91: 913–917, 1994.

156. Hammond, C. and Helenius, A. Quality control in the secretory pathway. *Curr. Opin. Cell Biol.* 7: 523–529, 1995.

157. Hara-Kuge, S., O. Kuge, L. Orci, M. Amherdt, M. Ravazzola, F. T. Wieland, and J. E. Rothman. En bloc incorporation of

coatomer subunits during the assembly of COP-coated vesicles. *J. Cell. Biol.* 124: 883–892, 1994.

158. Hardwick, K. G., and H.R.B. Pelham. SED5 encodes a 39-kD integral membrane protein required for vesicular transport between the ER and the Golgi complex. *J. Cell Biol.* 119: 513–521, 1992.

159. Hardwick, K. G., J. C. Broothroyd, A. D. Rudner, and H.R.B. Pelham. Genes that allow yeast cells to grow in the absence of the HDEL receptor. *EMBO J.* 11: 4187–4195, 1992.

160. Harrison-Lavoie, K. J., V. A. Lewis, G. M. Hynes, K. S. Collison, N. Nutland, and K. R. Willison. A 102 kDA subunit of a Golgi-associated particle has homology to beta-subunits of trimeric G proteins. *EMBO J.* 12: 2847–2853, 1993.

161. Harter, C., E. Draken, F. Lottspeich, and F. T. Wieland. Yeast coatomer contains a subunit homologous to mammalian beta′-COP. *FEBS Lett.* 332: 71–73, 1993.

162. Hartl, F.-U., R. Hlodan, and T. Langer. Molecular chaperones in protein folding: the art of avoiding sticky situations. *Trends Biochem. Sci.* 19: 20–25, 1994.

163. Haubruck, H. R. Prange, C. Vorgias, and D. Gallwitz. The ras-related mouse ypt1 protein can functionally replace the YPT1 gene product in yeast. *EMBO J.* 8: 1427–1432, 1989.

164. Haugejorden, S. M., M. Srinivasan, and M. Green. Analysis of the retention signals of two resident luminal endoplasmic reticulum proteins by in vitro mutagenesis. *J. Biol. Chem.* 266: 6015–6018, 1991.

165. Hauri, H.-P., J. Roth, E. Sterchi, and M. Lentze. Transport to cell surface of intestinal sucrase-isomaltase is blocked in the Golgi apparatus in a patient with congenital sucrase-isomaltase deficiency. *Proc. Natl. Acad. Sci. U.S.A.* 82: 4423–4427, 1985.

166. Hauri, H.-P., E. E. Sterchi, D. Bienz, J. A. M. Fransen, and A. Marxer. Expression and intracellular transport of microvillus membrane hydrolases in human intestinal epithelial cells. *J. Cell Biol.* 101: 838–851, 1985.

167. Hauri, H.-P., and A. Schweizer. The endoplasmic reticulum—Golgi intermediate compartment. *Curr. Opin. Cell Biol.* 4: 600–608, 1992.

168. Hebert, D. N., B. Foellmer, and A. Helenius. Glucose trimming and reglucosylation determine glycoprotein association with calnexin. *Cell* 81: 425–433, 1995.

169. Hebert, D. H., J. F. Simons, J. R. Peterson, and A. Helenius. Calnexin, calreticulin, and Bip/Kar2p in protein folding. *Cold Spring Harbor Symposium on Quantitative Biology* 60: 405–415, 1995.

170. Hebert, D. N., F. Foellmer, and A. Helenius. Calnexin and calreticulin promote folding, delay oligomerization and suppress degradation of influenza hemagglutinin in microsomes. *EMBO J.* 15: 2961–2968, 1996.

171. Helenius, A., T. Marquardt, and I. Braakman. The endoplasmic reticulum as a protein-folding compartment. *Trends Cell. Biol.* 2: 227–231, 1992.

172. Helms, J. B., and J. E. Rothman. Inhibition by brefeldin A of a Golgi membrane enzyme that catalyzes exchange of guanine nucleotide bound to ARF. *Nature* 360: 352–354, 1992.

173. Helms, J. B., D. J. Palmer, and J. E. Rothman. Two distinct populations of ARF bound to Golgi membranes. *J. Cell Biol.* 121: 751–760, 1993.

174. Hendricks, L. C., C. A. Gabel, K. Suh, and M. G. Farquhar. A 58-kDa resident protein of the cis-Golgi cisterna is not terminally glycosylated. *J. Biol. Chem.* 266: 17559–17565, 1991.

175. Hendricks, L. C., S. L. McClanahan, M. McCaffery, G. E. Palade, and M. G. Farquhar. Golgi proteins persist in the tubulovesicular remnants found in brefeldin A-treated pancreatic acinar cells. *Eur. J. Cell Biol.* 58: 202–213, 1992.

176. Hendricks, L. C., M. McCaffery, G. E. Palade, and M. G. Farquhar. Disruption of endoplasmic reticulum to Golgi transport leads to the accumulation of large aggregates containing beta-COP in pancreatic acinar cells. *Mol. Biol. Cell.* 4: 413–424, 1993.

177. Hicke, L. and R. Schekman. Yeast sec23p acts in the cytoplasm to promote protein transport from the endoplasmic reticulum to the Golgi complex in vivo and in vitro. *EMBO J.* 8: 1677–1684, 1989.

178. Hicke, L., T. Yoshihisa, and R. Schekman. Sec 23p and a novel 105 kD protein function as a multimeric complex to promote vesicle budding and protein transport from the ER. *Mol. Cell Biol.* 3: 667–676, 1992.

179. Hidalgo, J., R. Garcia-Navarro, F. Garcia-Navarro, J. Perez-Vilar, and A. Velasco. Presence of Golgi remnant membranes in the cytoplasm of brefeldin A-treated cells. *Eur. J. Cell Biol.* 58: 214–227, 1992.

180. Ho, W. C., V. J. Allen, G. van Meer, E. G. Berger, and T. E. Kreis. Reclustering of scattered Golgi elements occurs along microtubules. *Eur. J. Cell Biol.* 48: 250–263, 1989.

181. Hobman, T. C., L. Woodward, and M. G. Farquhar. The rubella virus E1 glycoprotein is arrested in a novel post-ER, pre-Golgi compartment. *J. Cell Biol.* 118: 795–811, 1992.

182. Hong, W. and B. L. Tang. Protein trafficking along the exocytic pathway. *BioEssays* 15: 231–238, 1993.

183. Hortsch, M., G. Griffiths, and D. I. Meyer. Restriction of docking protein to the rough endoplasmic reticulum: Immunocytochemical localization in rat liver. *Eur. J. Cell Biol.* 38: 271–279, 1985.

184. Hosobuchi, M., Kreis, T., and R. Schekman. Sec21 is a gene required for ER to Golgi transport that encodes a subunit of the yeast coatomer. *Nature* 360: 603–605, 1992.

185. Hsu, V.W.L., L. C. Yuan, J. G. Nuchtern, J. Lippincott-Schwartz, G. J. Hammerling, and R. D. Klausner. A recycling pathway between the endoplasmic reticulum and the Golgi apparatus for retention of unassembled MHC class I molecules. *Nature* 352: 441–444, 1991.

186. Hsu, V. W., N. Shah, and R. D. Klausner. A brefeldin A-like phenotype is induced by overexpression of a human ERD-2 like protein, ELP1. *Cell* 69: 625–635, 1992.

187. Hunziker, W., P. Male, and I. Mellman. Differential microtubule requirements for transctosis in MDCK cells. *EMBO J.* 9: 3515–3525, 1990.

188. Huovila, A.-P.J, A. M. Eder, and S. D. Fuller. Hepatitis B surface antigen assembles in a post-ER, pre-Golgi compartment. *J. Cell Biol.* 118: 1305–1320, 1992.

189. Humphrey, J. S., P. J. Peters, L. C. Yuan, and J. S. Bonifacino. Localization of TGN38 to the trans-Golgi network: Involvement of a cytoplasmic tyrosine-containing sequence. *J. Cell Biol.* 120: 1123–1135, 1993.

190. Hurtley, S. M, and A. Helenius. Protein oligomerization in the endoplasmic reticulum. *Annu. Rev. Cell Biol.* 5: 277–307, 1989.

191. Huttner, W. B. and S. A. Tooze. Biosynthetic protein transport in the secretory pathway. *Curr. Opin. Cell Biol.* 1: 648–654, 1989.

192. Hwang, C., A. J. Sinskey, and H. F. Lodish. Oxidized redox state of glutathione in the endoplasmic reticulum. *Science* 257: 1496–1502, 1992.

193. Inoue, S., and R. D. Simoni. 3-hydroxy-3-methylglutaryl-coenzyme A reductase and T cell receptor alpha-subunit are differentially degraded in the endoplasmic reticulum. *J. Biol. Chem.* 267: 9080–9086, 1992.

194. Itin, C., R. Schindler, and H.-P. Hauri. Targeting of protein

ERGIC-53 to the ER/ERGIC/cis-Golgi recycling pathway. *J. Cell Biol.* 131: 57–67, 1995.

195. Itin, C., F. Kappeler, A. D. Linstedt, and H.-P. Hauri. A novel endocytosis signal related to the KKXX ER-retrieval signal. *EMBO J.* 10: 2250–2256, 1995.

196. Itin, C., M. Foguet, F. Kappeler, J. Klumperman, and H.-P. Hauri. Recycling of ERGIC-53 in the secretory pathway. *Biochem. Soc. Trans.* 23: 541–543, 1995.

196a. Itin, C., A.-C. Roche, M. Monsigny, and H.-P. Hauri. ERGIC-53 is a functional mannose-selective and calcium-dependent human homologue of leguminous lectins. *Mol. Biol. Cell.* 7: 483–493, 1996.

197. Jackson, M. R., T. Nilsson, and P. A. Peterson. Identification of a consensus motif for retention of transmembrane proteins in the endoplasmic reticulum. *EMBO J.* 9: 3153–3162, 1990.

198. Jackson, M. R., T. Nilsson, and P. A. Peterson. Retrieval of transmembrane proteins to the endoplasmic reticulum. *J. Cell Biol.* 121: 317–333, 1993.

199. Jackson, M. R., M. F. Cohen-Doyle, P. A. Peterson, and D. B. Williams. Regulation of MHC class I transport by the molecular chaperone, calnexin (p88, IP90). *Science* 263: 384–387, 1994.

200. Jaenike, R. Role of accessory proteins in protein folding. *Curr. Opin. Struct. Biol.* 3: 104–112, 1993.

201. Jascur, T., K. Matter, and H.-P. Hauri. Oligomerization and intracellular protein transport: Dimerization of intestinal dipeptidylpeptidase IV occurs in the Golgi apparatus. *Biochemistry* 30: 1908–1915, 1991.

202. Jeckel, D., A. Karrenbauer, R. Birk, R. R. Schmidt, and F. Wieland. Sphingomyelin is synthesized in the cis-Golgi. *FEBS Lett.* 261: 155–157, 1990.

203. Jedd, G., C. Richardson, R. Litt, and N. Segev. The Ypt1 GTPase is essential for the first two steps of the yeast secretory pathway. *J. Cell Biol.* 131: 583–590, 1995.

204. Jensen, T. J., M. A. Loo, S. Pind, D. B. Williams, A. L. Goldberg, and J. R. Riordan. Multiple proteolytic systems, including the proteasome, contribute to CFTR processing. *Cell* 83: 129–135, 1995.

205. Jones, S. M., J. R. Crosby, J. Salamero, and K. E. Howell. A cytosolic complex of p62 and rab6 associates with TGN38/41 and is involved in budding of exocytic vesicles from the trans-Golgi network. *J. Cell Biol.* 122: 775–788, 1993.

206. Kahn, R. A. Fluoride is not an activator of the smaller (20–25 kDa) GTP-binding proteins. *J. Biol. Chem.* 266: 15595–15597, 1991.

207. Kahn, R. A., J. K. Yucel, and V. Malhotra. Arf signalling: a potential role for phospholipase D in membrane traffic. *Cell* 75: 1045–1048, 1993.

208. Kaji, E. H., and H. F. Lodish. In vitro unfolding of retinol-binding protein by dithiothreitol. Endoplasmic reticulum-associated factors. *J. Biol. Chem.* 268: 22195–22202, 1993.

209. Kao, C.-Y., and R. K. Draper. Retention of secretory proteins in an intermediate compartment and disappearance of the Golgi complex in and END4 mutant of Chinese hamster ovary cells. *J. Cell Biol.* 117: 701–715, 1992.

210. Kappeler, F., C. Itin, R. Schindler, and H.-P. Hauri. A dual role for COOH-terminal lysine residues in pre-Golgi retention and endocytosis of ERGIC-53. *J. Biol. Chem.* 269: 6279–6281, 1994.

211. Karrenbauer, A., D. Jeckel, W. Just, R. Birk, R. R. Schmidt, J. E. Rothman, and F. T. Wieland. The rate of bulk flow from the Golgi to the plasma membrane. *Cell* 63: 259–267, 1990.

212. Kearse, K. P., D. B. Williams, and A. Singer. Persistence of glucose residues on core oligosaccharides prevents association

of TCR-alpha and TCR-beta proteins with calnexin and results specifically in accelerated degradation of nascent TCR-alpha proteins within the endoplasmic reticulum. *EMBO J.* 13: 3678–3686, 1994.

213. Kelly, R. B. Microtubules, membrane traffic, and cell organization. *Cell* 61: 5–7, 1990.

214. Kim, S-h., S-j. Shin, and J-s. Park. Identification of the ATP transporter of rat liver rough endoplasmic reticulum via photoaffinity labeling and partial purification. *Biochemistry* 35: 5418–5425, 1996.

215. Kivirikko, K. I., R. Myllyla, and T. Pihlajaniemi. Protein hydroxylation: prolyl 4-hydroxylase, an enzyme with four cosubstrates and a multifunctional subunit. *FASEB J.* 3: 1609–1617, 1989.

216. Klausner, R. D. and R. Sitia. Protein degradation in the endoplasmic reticulum. *Cell* 62: 611–614, 1990.

217. Klausner, R. D., Lippincott-Schwartz, J. S. Bonifacino. Architectural editing: Regulating the surface expression of the multicomponent T-cell antigen receptor. *Curr. Top. Membr. Transport* 36: 31–51, 1990.

218. Klausner, R. D., J. G. Donaldson, and J. Lippincott-Schwartz. Brefeldin A: insights into the control of membrane traffic and organelle structure. *J. Cell Biol.* 116: 1071–1080, 1992.

219. Kleene, R., and E. G. Berger. The molecular and cell biology of glycosyltransferases. *Biochim. Biophys Acta* 1154: 283–325, 1993.

220. Kleijmeer, M, A. Kelly, H. J. Geuze, J. W. Slot, A. Townsend, and J. Trowsdale. Location of MHC-encoded transporters in the endoplasmic reticulum and cis-Golgi. *Nature* 357: 342–344, 1992.

221. Kochevar, D. T., and R. G. W. Anderson. Purified crystalloid endoplasmic reticulum from UT-1 cells contains multiple proteins in addition to 3-hydroxy-3-methylglutaryl coenzyme A reductase. *J. Biol. Chem.* 262: 10321–10326, 1987.

222. Kornfeld, R. and S. Kornfeld. Assembly of asparagine-linked oligosaccharides. *Annu. Rev. Biochem.* 54: 631–664, 1985.

223. Kornfeld, S., and I. Mellman. The biogenesis of lysosomes. *Annu. Rev. Cell Biol.* 5: 483–525, 1991.

224. Kreis, T. E., and R. Pepperkok. Coat proteins in intracellular membrane transport. *Curr. Opin. Cell Biol.* 6: 533–537, 1994.

225. Kreis, T. E., M. Lowe, and R. Pepperkok. COPs regulating membrane traffic. *Annu. Rev. Cell Biol.* 11: 677–706, 1995.

226. Krjinse-Locker, J., M. Ericsson, P.J.M. Rottier, and G. Griffiths. Characterization of the budding compartment of mouse hepatitis virus: evidence that transport from the RER to the Golgi complex requires only one vesicular transport step. *J. Cell Biol.* 124: 55–70, 1994.

227. Krjinse-Locker, J., J. Klumperman, V. Oorschot, M. C. Horzinek, H. J. Geuze, and P.J.M. Rottier. The cytoplasmic tail of mouse hepatitis virus M protein is essential but not sufficient for its retention in the Golgi complex. *J. Biol. Chem.* 269: 28263–28269, 1994.

229. Ktistakis, N. T., M. E. Linder, and M. G. Roth. Action of brefeldin A blocked by activation of a pertussis-toxin-sensitive G protein. *Nature* 356: 344–346, 1992.

230. Ktistakis, N. T., H. A. Brown, G. Waters, P. Sternweis, and M. G. Roth. Evidence that phospholipase D mediates ADP ribosylation factor-dependent formation of Golgi coated vesicles. *J. Cell Biol.* 134: 295–306, 1996.

231. Kuge, O., S. Hara-Kuge, L. Orci, M. Ravazzola, M. Amherdt, G. Tanigawa, F. T. Wieland, and J. E. Rothman. Zeta-COP, a subunit of coatomer is required for COP-coated vesicle assembly. *J. Cell Biol.* 123: 1727–1734, 1991.

232. Kuge, O., C. Dascher, L. Orci, T. Rowe, M. Amherdt,

H. Plutner, M. Ravazzola, G. Tanigawa, J. E. Rothman, and W. E. Balch. Sar1 promotes vesicle budding from the endoplasmic reticulum but not Golgi compartments. *J. Cell Biol.* 125: 51–65, 1994.

233. Kutay, U., E. Hartmann, and T. A. Rapoport. A class of membrane proteins with a C-terminal anchor. *Trends Cell Biol.* 3: 72–75, 1993.

234. Kuznetsov, G., L. B. Chen, and S. K. Nigam. Several endoplasmic reticulum stress proteins, including ERp72, interact with thyroglobulin during its maturation. *J. Biol. Chem.* 269: 22990–22995, 1994.

235. Ladinsky, M. S., and K. E. Howell. The trans-Golgi network can be dissected structurally and functionally from the cisternae of the Golgi complex by brefeldin A. *Eur. J. Cell Biol.* 59: 92–105, 1992.

236. Ladinsky, M. S, J. R. Kremer, P. S. Furcinitti, J. R. McIntosh, and K. E. Howell. HVEM Tomography of the trans-Golgi network: Structural insights and identification of a lace-like vesicle coat. *J. Cell Biol.* 127: 29–38, 1994.

237. Lafont, F., J. K. Burkhardt, and K. Simons. Involvement of microtubule motors in basolateral and apical transport in kidney cells. *Nature* 372: 801–803, 1994.

238. Lahtinen, U., Dahllöf, B., and J. Saraste. Characterization of a 58 kDa cis-Golgi protein in pancreatic exocrine cells. *J. Cell Sci.* 103: 321–333, 1992.

239. Lahtinen, U., U. Hellman, C. Wernstedt, J. Saraste, and R. F. Pettersson. Molecular cloning and expression of a 58-kDa cis-Golgi and intermediate compartment protein. *J. Biol. Chem.* 271: 4031–4037, 1996.

240. Latterich, M., K.-U. Fröhlich, and R. Schekman. Membrane fusion and the cell cycle: Cdc48p participates in the fusion of ER membranes. *Cell* 82: 885–893, 1995.

241. Le, A., K. S. Graham, and R. N. Sifers. Intracellular degradation of the transport-impaired human PiZ alpha1-antitrypsin variant. Biochemical mapping of the degradative event among compartments of the secretory pathway. *J. Biol. Chem.* 265: 14001–14007, 1990.

242. Le, A., J. L. Steiner, G. A. Ferell, J. C. Shaker, and R. N. Sifers. Association between calnexin and secretion-incompetent variant of human alpha1-antitrypsin. *J. Biol. Chem.* 269: 7514–7519, 1994.

243. Lecureux, L. W., and B. M. Wattenberg. The regulated degradation of a 3-hydroxy-3-methylglutaryl-coenzyme A reductase reporter construct occurs in the endoplasmic reticulum. *J. Cell Sci.* 107: 2635–2642, 1994.

244. Leskes, A., P. Siekevitz, and G. E. Palade. Differentiation of endoplasmic reticulum in hepatocytes. I. Glucose-6-phosphatase distribution in situ. *J. Cell Biol.* 49: 264–287, 1971.

245. Lee, C., and L. B. Chen. Dynamic behavior of endoplasmic reticulum in living cells. *Cell* 54: 37–86, 1988.

246. Letourneur, F., E. C. Gaynor, S. Hennecke, C. Démollière, R. Duden, S. D. Emr, H. Riezman, and P. Cosson. Coatomer is essential for retrieval of dilysine-tagged proteins to the ER. *Cell* 79: 1199–1207, 1994.

247. Lewis, M. J., and H.R.B. Pelham. A human homologue of the yeast HDEL receptor. *Nature* 348: 162–163, 1990.

248. Lewis, M. J., D. J. Sweet, and H.R.B. Pelham. The ERD2 gene determines the specificity of the luminal ER protein retention system. *Cell* 61: 1359–1363, 1990.

249. Lewis, M. J., and H.R.B. Pelham. Ligand-induced redistribution of a human KDEL receptor from the Golgi complex to the endoplasmic reticulum. *Cell* 68: 353–363, 1992.

250. Lewis, M. J. and H.R.B. Pelham. SNARE-mediated retrograde traffic from the Golgi complex in the endoplasmic reticulum. *Cell* 85: 205–215, 1996.

251. Leyte, A., F. A. Barr, R. H. Kehlenbach, and W. B. Huttner. Multiple trimeric G-proteins on the trans-Golgi network exert stimulatory and inhibitory effects on secretory vesicles. *EMBO J.* 11: 4795–4804, 1992.

252. Lian, J. P., and S. Ferro-Novick. Bos1p, and integral membrane protein of the endoplasmic reticulum to Golgi transport vesicles, is required for their fusion competence. *Cell* 73: 735–745, 1993.

253. Lindsey, J. D., and M. H. Ellisman. The neuronal endomembrane system. I. Direct links between rough endoplasmic reticulum and cis-element of the Golgi apparatus. *J. Neurosci.* 5: 3111–3123, 1985.

254. Lindsey, J. D., and M. H. Ellisman. The neuronal endomembrane system. II. The multiple forms of the Golgi apparatus cis element. *J. Neurosci.* 5: 3124–3134, 1985.

255. Linstedt, A. D., and Hauri, H.-P. Giantin, a novel conserved Golgi membrane protein containing a cytoplasmic domain of at least 350 kD. *Mol. Biol. Cell* 4: 679–693, 1993.

256. Linstedt, A. D., and H.-P. Hauri. Cell-cycle dependent phosphorylation of giantin, a cytoskeletal-like Golgi membrane protein. *Mol. Biol. Cell* 5: 240a, 1994.

257. Linstedt, A. D., M. Foguet, M. Renz, H. P. Seelig, B. S. Glick, and H.-P. Hauri. A C-terminally-anchored protein is inserted into the endoplasmic reticulum and then transported to the Golgi apparatus. *Proc. Natl. Acad. Sci. U.S.A.*, 1995.

258. Lippincott-Schwartz, J. Bidirectional membrane traffic between the endoplasmic reticulum and Golgi apparatus. *Trends Cell Biol.* 3: 81–88, 1993.

259. Lippincott-Schwartz, J., L. C. Yuan, J. S. Bonifacino, and R. D. Klausner. Rapid redistribution of Golgi proteins into the ER in cells treated with brefeldin A: evidence for membrane cycling from Golgi to ER. *Cell* 56: 801–813, 1989.

260. Lippincott-Schwartz, J., J. G. Donaldson, A. Schweizer, E. G. Berger, H.-P. Hauri, L. C. Yuan, and R. D. Klausner. Microtubule-dependent retrograde transport of proteins into the ER in the presence of brefeldin A suggests an ER recycling pathway. *Cell* 60: 821–836, 1990.

261. Lippincott-Schwartz, J., L. Yuan, C. Tipper, M. Amherdt, L. Orci, and R. D. Klausner. Brefeldin A's effects on endosomes, lysosomes, and the TGN suggests a general mechanism for regulating organelle structure and membrane traffic. *Cell* 67: 601–616, 1991.

262. Lippincott-Schwartz, J., N. B. Cole, A. Marotta, P. A. Conrad, and G. S. Bloom. Kinesin is the motor for microtubule-mediated Golgi-to-ER membrane traffic. *J. Cell Biol.* 128: 293–306, 1995.

263. Liscovitch, M., and L. C. Cantley. Signal transduction and membrane traffic: The PITP/phosphoinositide connection. *Cell* 81: 659–662, 1995.

264. Lodish, H. F. Transport of secretory and membrane proteins from the rough enoplasmic reticulum to the Golgi. *J. Biol. Chem.* 263: 2107–2110, 1988.

265. Lodish, H. F., N. Kong, M. Snider, and G.J.A.M. Strous. Hepatoma secretory proteins migrate from rough endoplasmic reticulum to Golgi at characteristic rates. *Nature* 304: 80–83, 1983.

266. Lodish, H. F., and N. Kong. Perturbation of cellular calcium blocks exit of secretory proteins from the rough endoplasmic reticulum. *J. Biol. Chem.* 265: 10893–10899, 1990.

267. Lodish, H. F., and N. Kong. Cyclosporin A inhibts an initial

step in folding of transferrin within the endoplasmic reticulum. *J. Biol. Chem.* 266: 14835–14838, 1991.

268. Lodish, H. F., N. Kong, and L. Wikström. Calcium is required for folding of newly made subunits of asialoglycoprotein receptor within the endoplasmic reticulum. *J. Biol. Chem.* 267: 12753–12760, 1992.

269. Lodish, H. F., and N. Kong. The secretory pathway is normal in dithiothreitol-treated cells, but disulfide-bonded proteins are reduced and reversibly retained in the endoplasmic reticulum. *J. Biol. Chem.* 268: 20598–20605, 1993.

270. Lotti, L. V., M.-R. Torrisi, M. C. Pascale, and S. Bonatti. Immunocytochemical analysis of the transfer of vesicular stomatitis virus G glycoprotein from the intermediate compartment to the Golgi complex. *J. Cell Biol.* 118: 43–50, 1992.

271. Louvard, D., Reggio, H., and G. Warren. Antibodies to the Golgi complex and the rough endoplasmic reticulum. *J. Cell Biol.* 92: 92–107, 1982.

272. Lupashin, V. V., S. Hamamoto, R. W. Schekman. Biochemical requirements for targeting and fusion of ER-derived transport vesicles with purified yeast Golgi membranes. *J. Cell Biol.* 132: 277–289, 1996.

273. Luzio, J. P., and G. Banting. Eukaryotic membrane traffic: retrieval and retention mechanisms to achieve organelle residence. *TIBS* 18: 395–398, 1993.

274. Luzio, J.-P., B. Brake, G. Banting, K. E. Howell, P. Braghetta, and K. K. Stanley. Identification, sequencing, and expression of an integral membrane protein of the trans Golgi network, TGN 38. *Biochem. J.* 270: 97–102, 1990.

275. Lyles, M. M., and H. F. Gilbert. Catalysis of the oxidative folding of ribonuclease A by protein disulfide isomerase: Dependence of the rate on the composition of the redox buffer. *Biochemistry* 30: 613–619, 1991.

276. Lyles, M. M., and H. F. Gilbert. Catalysis of the oxidative folding of ribonuclease A by protein disulfide: isomerase: Presteady-state kinetics and utilization of the oxidizing equivalents of the isomerase. *Biochemistry* 30: 619–625, 1991.

277. Macer, D. R. J., and G. L. E. Koch. Identification of a set of calcium-binding proteins in reticuloplasm, the luminal content of the endoplasmic reticulum. *J. Cell Sci.* 92: 61–70, 1988.

278. Machamer, C. E. Targeting and retention of Golgi membrane porteins. *Curr. Opin. Cell Biol.* 5: 606–612, 1993.

279. Machamer, C. E., M. G. Grim, A. Esquela, S. W. Chung, M. Rolls, K. Ryan, and R. K. Swift. Retention of a cis-Golgi protein requires polar residues on one face of a predicted alpha-helix in the transmembrane domain. *Mol. Biol. Cell* 5: 695–704, 1993.

280. Majoul, I. V., P.I.H. Bastiaens, and H.-D. Söling. Transport of an external Lys-Asp-Glu-Leu (KDEL) protein from the plasma membrane to the endoplasmic reticulum: studies with cholera toxin in Vero cells. *J. Cell Biol.* 133: 777–789, 1996.

281. Malhotra, V., T. Serafini, L. Orci, J. C. Sheperd, and J. E. Rothman. Purification of a novel class of coated vesicles mediating biosynthetic protein transport through the Golgi stack. *Cell* 58: 329–336, 1989.

282. Mandon, E., E. S. Kempner, M. Ishihara, and C. Hirschberg. A monomeric protein in the Golgi membrane catalyzes both N-deacylation and N-sulfation of heparan sulfate. *J. Biol. Chem.* 269: 11729–11733, 1994.

283. Marcantonio. E. E., Amar-Costesec, and G. Kreibich. Segregation of the polypeptide translocation apparatus to regions of the endoplasmic reticulum containing ribophorins and ribosomes. II Rat liver microsomal subfractions contain equimolar amounts of ribophorins and ribosomes. *J. Cell Biol.* 99: 2254–2259, 1984.

284. Marks, D. L., J. M. Larkin, and M. A. McNiven. Association of kinesin with the Golgi apparatus in rat hepatocytes. *J. Cell Sci.* 107: 2417–2426, 1994.

285. Marquart, T., D. H. Hebert, and A. Helenius. Post-translational folding of influenza hemagglutinin in isolated endoplasmic reticulum-derived microsomes. *J. Biol. Chem.* 268: 19618–19625, 1993.

286. Martinez, O., A. Schmidt, J. Salamero, B. Hoflack, M. Roa, and B. Goud. The small GTP-binding protein rab6 functions in intra-Golgi transport. *J. Cell Biol.* 127: 1575–1588, 1994.

287. Masibay, A. S., P. V. Balaji, E. E. Boeggeman, P. K. Qasba. Mutational analysis of the Golgi retention signal of bovine beta-1,4 galactosyltransferase. *J. Biol. Chem.* 268: 9908–9916, 1993.

288. Matter, K., K. Bucher, and H.-P. Hauri. Microtubule perturbation retards both the direct and the indirect apical pathway but does not affect sorting of plasma membrane proteins in intestinal epithelial cells. *EMBO J.* 9: 3163–3170, 1990.

289. Matter, K., and H.-P. Hauri. Intracellular transport and conformational maturation of intestinal brush border hydrolases. *Biochemistry* 30: 1916–1923, 1991.

290. Mazzarella, R. A., M. Srinivasan, S. M. Haugejordan, and M. Green. ERp72, an abundant luminal endoplasmic reticulum protein contains three copies of the active site sequences of protein disulfide isomerase. *J. Biol. Chem.* 265: 1094–1101, 1990.

291. Meier, P., R. Gasser, H.-P. Hauri, B. Stieger, and U. A. Meyer. Biosynthesis of rat liver cytochrome P-450 in mitochondria associated rough endoplasmic reticulum and in rough microsomes in vivo. *J. Biol. Chem.* 259: 10194–10200, 1984.

292. Melancon, P., B. S. Glick, V. Malhotra, P. J. Weidman, T. Serafini, M. Gleason, L. Orci, and J. E. Rothman. Involvement of GTP-binding "G" proteins in transport through the Golgi stack. *Cell* 51: 1053–1062, 1987.

293. Melnick, J., S. Aviel, and Y. Argon. The endoplasmic reticulum protein GRP94, in addition to BiP, associates with unassembled immunoglobulin chains. *J. Biol. Chem.* 267: 21303–21306, 1992.

294. Mellman, I., and K. Simons. The Golgi complex: In vitro veritas? *Cell* 68: 829–840, 1992.

295. Mellman, I. Enigma variations: Protein mediators of membrane fusion. *Cell* 82: 869–872, 1995.

296. Menzel, R., E. Kärgel, F. Vogel, C. Böttcher, and W.-H. Schunk. Topogenesis of a microsomal cytochrome P450 and induction of endoplasmic reticulum membrane proliferation in Saccharomyces cervisiae. *Arch. Biochem. Biophys.* 330: 97–109, 1996.

297. Michalak, M., R. E. Milner, K. Burns, and M. Opas. Calreticulin. *Biochem. J.* 285: 681–692, 1992.

298. Miesenböck, G., and J. E. Rothman. The capacity to retrieve escaped ER proteins extends to the trans-most cisterna of the Golgi stack. *J. Cell Biol.* 129: 309–319, 1995.

299. Misteli, T., and G. Warren. COP-coated vesicles are involved in the mitotic fragmentation of Golgi stacks in a cell-free system. *J. Cell Biol.* 125:269–282, 1994.

300. Misteli, T. and Warren, G. A role for tubular networks and a COPI-independent pathway in the mitotic fragmentation of Golgi stacks in a cell-free system. *J. Cell Biol.* 130: 1027–1039, 1995.

301. Misumi, I., K. Miki, A. Takatsuki, G. Tamura, and Y. Ikehara. Novel blockade by brefeldin A of intracellular transport of

secretary proteins in cultured rat hepatocytes. *J. Biol. Chem.* 261: 11398–11403, 1986.

302. Mitoma, J., and A. Ito. The carboxy-terminal 10 amino acid residues of cytochrome b5 are necessary for its targeting to the endoplasmic reticulum. *EMBO J.* 11: 4197–4203, 1992.

303. Mizuno, M., and S. J. Singer. A soluble secretory protein is first concentrated in the endoplasmic reticulum before transfer to the Golgi apparatus. *Proc. Natl. Acad. Sci. U.S.A.* 90: 5732–5736, 1993.

304. Molloy, S. S., L. Thomas, J. K. Van Slyke, P. E. Stenberg, and G. Thomas. Intracellular trafficking and activation of the furin proprotein convertase: localization to the TGN and recycling from the cell surface. *EMBO J.* 13: 18–33, 1994.

305. Moran, P., and I. W. Caras. Proteins containing an uncleaved signal glycophosphatidylinositol membrane anchor attachment are retained in a post-ER compartment. *J. Cell Biol.* 119: 763–772, 1992.

306. Mori, K., W. Ma, M.-J. Gething, and J. Sambrook. A transmembrane protein with a cdc+/CDC28-related kinase activity is required for signaling from the ER to the nucleus. *Cell* 74: 743–756, 1993.

307. Morré, D. J., J. Kartenbeck, and W. W. Franke. Membrane flow and interconversions among endomembranes. *Biochim. Biophys. Acta* 559: 71–152, 1979.

308. Mundy, D. I., and G. Warren. Mitosis and inhibition of intracellular transport stimulate palmitoylation of a 62-kD protein. *J. Cell Biol.* 116: 135–146, 1992.

309. Munro, S. Sequences within and adjacent to the transmembrane segment of alpha-2,6-sialyltransferases specify Golgi retention. *EMBO J.* 10: 3577–3588, 1991.

310. Munro, S. An investigation of the role of transmembrane domains in Golgi protein retention. *EMBO J.* 14: 4695–4704, 1995.

311. Murphy, R. F. Maturation models for endosome and lysosome biogenesis. *Trends Cell Biol.* 1: 77–82, 1991.

312. Musil, L. S., and D. A. Goodenough. Multisubunit assembly of an integral plasma membrane channel protein, gap junction connexin 43, occurs after exit from the ER. *Cell* 74: 1065–1077, 1993.

313. Nagahama, M., L. Orci, M. Ravazzola, M. Amherdt, L. Lacomis, P. Tempst, J. Rothman, and T. Söllner. A v-SNARE implicated in intra-Golgi transport. *J. Cell Biol.* 133: 507–516, 1996.

314. Naim, Y., J. Roth, E. E. Sterchi, M. Lentze, P. Milla, J. Schmitz, and H.-P. Hauri. Sucrase-isomaltase deficiency in humans: Different mutations disrupt intracellular transport, processing and function of an intestinal brush border enzyme. *J. Clin. Invest.* 82: 667–6679, 1988.

315. Nakano, A., D. Brada, and R. Schekman. A membrane glycoprotein, Sec12p, required for transport from the endoplasmic reticulum to the Golgi apparatus in yeast. *J. Cell Biol.* 107: 851–863, 1988.

316. Nakano, A., and M. Maramatsu. A novel GTP-binding protein, Sar1, is involved in transport from the endoplasmic reticulum to the Golgi apparatus. *J. Cell Biol.* 109: 2677–2691, 1989.

317. Narula, N., and J. L. Stow. Distinct coated vesicles labeled for p200 bud from trans-Golgi network membranes. *Proc. Natl. Acad. Sci U.S.A.* 92: 2874–2878, 1995.

318. Nauseef, W. M., S. J. McCormick, and R. A. Clark. Calreticulin functions as a molecular chaperone in the biosynthesis of myeloperoxidase. *J. Biol. Chem.* 270: 4741–4747, 1995.

319. Newman, A. P., J. Shim, and S. Ferro-Novick. BET1, BOS1 and SEC22 are members of a group of interacting genes required for

320. Newman, A. P., M. E. Groesch, and S. Ferro-Novick. Bos1p, a membrane protein required for ER to Golgi transport in yeast, co-purifies with the carrier vesicles and with Bet1p and the ER membranes. *EMBO J.* 11: 3609–3617, 1992.

321. Newman, A. P., J. Graf, P. Mancini, G. Rossi, J. P. Lian, and S. Ferro-Novick. Sec22 and SLY2 are identical. *Mol. Cell Biol.* 12: 3663–3664, 1992.

322. Nguyen Van, P., F. Peter, and H.-D. Söling. Four intracisternal calcium-binding glycoproteins from rat liver microsomes with high affinity for calcium. No indication for calsequestrin-like proteins in inositol 1,4,5-trisphosphate-sensitive calcium sequestering rat liver vesicles. *J. Biol. Chem.* 264: 17494–17501, 1989.

324. Nguyen Van, P., K. Rupp, A. Lampen, and H.-D. Söling. CaBP2 is a rat homolog of ERp72 with proteindisulfide isomerase activity. *Eur. J. Biochem.* 213: 789–795, 1993.

325. Nigam, S. K., A. L. Goldberg, S. Ho, M. F. Rohde, K. T. Bush, and M. Yu. Sherman. A set of endoplasmic reticulum proteins possessing properties of molecular chaperones includes Ca^{2+}-binding proteins and members of the thioredoxin superfamily. *J. Biol. Chem.* 269: 1744–1749, 1994.

326. Nilsson, T., J. M. Lucocq, D. Mackay, and G. Warren. The membrane spanning domain of beta-1,4 galactosyltransferase specifies trans-Golgi localization. *EMBO J.* 10: 3567–3575, 1991.

327. Nilsson, T., M. Pypaert, M. H. Hoe, P. Slusarevicz, E. G. Berger, and G. Warren. Overlapping distribution of two glycosyltransferases in the Golgi apparatus of HeLa cells. *J. Cell Biol.* 120: 5–13, 1993.

328. Nilsson, T., Slusarewicz, M. H. Hoe, and G. Warren. Kin recognition. A model for the retention of Golgi enzymes. *FEBS Lett.* 330: 1–4, 1993.

329. Nilsson, T., and G. Warren. Retention and retrieval in the endoplasmic reticulum and the Golgi apparatus. *Curr. Opin. Cell Biol.* 6: 517–521, 1994.

330. Nilsson, T., M. H. Hoe, P. Slusarewicz, C. Rabouille, R. Watson, F. Hunte, G. Watzele, E. G. Berger, and G. Warren. Kin recognition between medial Golgi enzymes in HeLa cells. *EMBO J.* 13: 562–574, 1994.

331. Noiva, R., and W. J. Lennarz. Protein disulfide isomerase. A multifunctional protein resident in the lumen of the endoplasmic reticulum. *J. Biol. Chem.* 267: 3553–3556, 1992.

332. Nothwehr, S. F., and T. H. Stevens. Sorting of membrane proteins in the yeast secretory pathway. *J. Biol. Chem.* 269: 10185–10188, 1994.

333. Nothwehr, S. F. C. J. Roberts, and T. H. Stevens. Membrane protein retention in the yeast Golgi aplparatus: Dipeptidyl aminopeptidase A is retained by a cytoplasmic signal containing aromatic residues. *J. Cell Biol.* 121: 1197–1209, 1993.

334. Novick, P., C. Field, and R. Schekman. Identification of 23 complementation groups required for post-translational events in the yeast secretory pathway. *Cell* 21: 205–215, 1980.

335. Novick, P., S. Ferro, and R. Schekman. Orders of events in the yeast secretory pathway. *Cell* 25: 461–469, 1981.

336. Novick, P., and P. Brennwald. Friends and family: the role of Rab GTPases in vesicular traffic. *Cell* 75: 597–601, 1993.

337. Nuoffer, C., and W. E. Balch. GTPase: multifunctional molecular switches regulating vesicular traffic. *Annu. Rev. Biochem.* 63: 949–90, 1994.

338. Nuoffer, C., H. W. Davidson, J. Matteson, J. Meinkoth, and W. E. Balch. A GDP-bound form of Rab1 inhibits protein

export from the endoplasmic reticulum and transport between Golgi compartments. *J. Cell Biol.* 125: 225–237, 1994.

339. Ohashi, M., and W. B. Huttner. An elevation of cytosolic protein phosphorylation modulates trimeric G-protein regulation of secretory vesicle formation from the trans-Golgi network. *J. Biol. Chem.* 269: 24897–24905, 1994.

340. Oka, T., and A. Nakano. Inhibition of GTP hydrolysis by Sar1p causes accumulation of vesicles that are a functional intermediate of the ER-to-Golgi transport in yeast. *J. Cell Biol.* 124: 425–434, 1994.

341. Okey, A. B. Enzyme induction in the cytochrome P-450 system. In: Pharmacogenetics of Drug Metabolism, edited by W. Kalow. New York: Pergamon Press Inc., p. 549–608, 1992.

342. Oprins, A., R. Duden, T. E. Kreis, H. J. Geuze, and J. W. Slot. Beta-COP localizes mainly to the cis-Golgi side in exocrine pancreas. *J. Cell Biol.* 121: 49–59, 1993.

343. Ora, A. and A. Helenius. Calnexin fails to associate with substrate proteins in glucosidase-deficient cell lines. *J. Biol. Chem.* 270: 2660–26062, 1996.

344. Orci, L., B. S. Glick, and J. E. Rothman. A new type of coated vesicular carrier that appears not to contain clathrin: its possible role in protein transport within the Golgi stack. *Cell* 46: 171–184, 1986.

345. Orci, L., V. Malhotra, M. Amherdt, T. Serafini, and J. E. Rothman. Dissection of a single round of vesicular transport: sequential intermediates for intercisternal movement in the Golgi stack. *Cell* 56: 357–368, 1989.

346. Orci, L., M. Ravazzola, P. Meda, C. Holcomb, H.-P. Moore, L. Hicke, and R. Schekman. Mammalian sec 23p homologue is restricted to the endoplasmic reticulum transitional cytoplasm. *Proc. Natl. Acad. Sci. U.S.A.* 88: 8611–8615, 1991.

347. Orci, L., T. Tagaya, M. Amherdt, A. Perrelet, J. G. Donaldson, J. Lippincott-Schwartz, R. D. Klausner, and J. E. Rothman. Brefeldin A, a drug that blocks secretion, prevents the assembly of non-clathrin coated buds on Golgi cisternae. *Cell* 64: 1183–1195, 1991.

348. Orci, L., D. J. Palmer, M. Ravazzola, A. Perrelet, M. Amherdt, and J. E. Rothman. Budding from Golgi membranes requires coatomer complex of non-clathrin coated proteins. *Nature* 362: 648–652, 1993.

349. Orci, L., D. J. Palmer, M. Amherdt, and J. E. Rothman. Coated vesicles assembly in the Golgi requires only coatomer and ARF proteins from the cytosol. *Nature* 364: 732–734, 1993.

350. Orci, L., A. Perrelet, M. Ravazzola, F. T. Wieland, R. Schekman, and J. E. Rothman. "BFA bodies": A subcompartment of the endoplasmic reticulum. *Proc. Natl. Acad. Sci. U.S.A.* 90: 11089–11093, 1993.

351. Orci, L., A. Perrelet, M. Ravazzola, M. Amherdt, J. E. Rothman, and R. Schekman. Coatomer-rich endoplasmic reticulum. *Proc. Natl. Acad. Sci. U.S.A.* 91: 11924–11928, 1994.

352. Ostermann, J., L. Orci, K. Tani, M. Amherdt, M. Ravazzola, Z. Elazar, and J. E. Rothman. Stepwise assembly of functionally active transport vesicles. *Cell* 75: 1015–1025, 1993.

353. Ou, W.-J., P. H. Cameron, D. Y. Thomas, and J.J.B. Bergeron. Association of folding intermediates of glycoproteins with calnexin during protein maturation. *Nature* 364: 771–776, 1993.

354. Ouwendijk, J., C.E.C. Moolenaar, W. J. Peters, C. P. Hollenberg, L. A. Ginsel, J.A.M. Fransen, and H. Y. Naim. Congenital sucrase-isomaltase deficiency. Identification of a glutamine to proline substitution that leads to a transport block of sucrase-isomaltase in a pre-Golgi compartment. *J. Clin. Invest.* 97: 633–641, 1996.

355. Ozawa, M., and T. Maramatsu. Reticulocalbin, a novel endoplasmic reticulum resident Ca^{2+}-binding protein with multiple EF-hand motifs and a carboxyl-terminal HDEL sequence. *J. Biol. Chem.* 268: 699–705, 1993.

356. Padmanaban, G., C. S. Nirodi. Regulation of phenobarbitone-inducible cytochrome P-450 gene expression. In: Frontiers in Biotransformation, Vol. 9, edited by K. Ruckpaul and H. Rein. Berlin: Akademie Verlag, p. 60–84, 1994.

357. Pahl, H. L., and P. A. Baeuerle. A novel signal transduction pathway from the endoplasmic reticulum to the nucleus is mediated by transcription factor NF-kappaB. *EMBO J.* 14: 2580–2588, 1996.

358. Palade, G. Intracellular aspects of the process of protein secretion. *Science* 189: 347–358, 1975.

359. Palmer, D. J., J. B. Helms, C. J. M. Beckers, L. Orci, and J. E. Rothman. Binding of coatomer to Golgi membranes requires ADP-ribosylation factor. *J. Biol. Chem.* 268: 12083–12089, 1993.

360. Pascale, M. C., N. Malagolini, F. Serafini-Cessi, G. Migliaccio, A. Leone, and S. Bonatti. Biosynthesis and oligosaccharide structure of human CD8 glycoprotein expressed in a rat epithelial cell line. *J. Biol. Chem.* 267: 9940–9947, 1992.

361. Pathak, R. K., K. L. Luskey, and R.G.W. Anderson. Biogenesis of the crystalloid endoplasmic reticulum in UT-1 cells: Evidence that newly formed endoplasmic reticulum emerges from the nuclear envelope. *J. Cell Biol.* 102: 2158–2168, 1986.

362. Pedrazzini, E., A. Villa, and N. Borgese. A mutant cytochrome b5 with a lengthened membrane anchor escapes from the endoplasmic reticulum and reaches the plasma membrane. *Proc. Natl. Acad. Sci U.S.A.* 93: 4207–42012, 1996.

363. Pelham., H.R.B. Evidence that luminal ER proteins are sorted from secreted proteins in post-ER compartment. *EMBO J.* 7: 913–918, 1988.

364. Pelham, H.R.B. Control of protein exit from the endoplasmic reticulum. *Annu. Rev. Cell Biol.* 5: 1–23, 1989.

365. Pelham, H.R.B. The retention signal for soluble proteins of the endoplasmic reticulum. *Trends. Biochem. Sci.* 15: 483–486, 1990.

366. Pelham, H.R.B. Recycling of proteins between the endoplasmic reticulum and Golgi complex. *Curr. Opin. Cell Biol.* 3: 585–591, 1991.

367. Pelham, H.R.B., and S. Munro. Sorting of membrane proteins in the secretory pathway. *Cell* 75: 603–605, 1993.

368. Pelham, H.R.B. Sorting and retrieval between the endoplasmic retculum and Golgi apparatus. *Curr. Opin. Cell Biol.* 7: 530–535, 1995.

369. Pepperkok, R., J. Scheel, H. Horstmann, H.-P. Hauri, G. Griffiths, and T. E. Kreis. Beta-COP is essential for biosynthetic membrane transport from the endoplasmic reticulum to the Golgi complex in vivo. *Cell* 74: 71–82, 1993.

370. Peter, F., P. Nguyen Van, and H.-D. Söling. Different sorting of Lys-Asp-Glu-Leu proteins in rat liver. *J. Biol. Chem.* 267: 10631–10637, 1992.

371. Peter, F., H. Plutner, H. Zhu, T. E. Kreis, and W. E. Balch. Beta-COP is essential for transport of protein from the endoplasmic reticulum to the Golgi in vitro. *J. Cell Biol.* 122: 1155–1167, 1993.

372. Peter, F., C. Nuoffer, S. N. Pind, and W. E. Balch. Guanine nucleotide dissociation inhibitor is essential for Rab1 function in budding from the endoplasmic reticulum and transport through the Golgi stack. *J. Cell Biol.* 126: 1393–1406, 1994.

373. Pevsner, J., and R. H. Scheller. Mechanism of vesicle docking

and fusion: insights from the nervous system. *Curr. Opin. Cell Biol.* 6: 555–560, 1994.

374. Pfeffer, S. R., and J. E. Rothman. Biosynthetic protein transport and sorting by the endoplasmic reticulum and Golgi. *Annu. Rev. Biochem.* 56: 829–852, 1987.

375. Pfeffer, S. R. Rab GTPases: master regulators of membrane trafficking. *Curr. Opin. Cell Biol.* 6: 522–526, 1994.

376. Pfeffer, S. R. Transport vesicle docking: SNAREs and associates. *Annu. Rev. Cell Biol.* 12: 441–461, 1996.

377. Pierre, P., J. Scheel, J. E. Richard, and T. E. Kreis. CLIP-170 links endocytic vesicles to microtubules. *Cell* 70: 887–900, 1992.

378. Pierre, P., R. Pepperkok, and T. E. Kreis. Molecular characterization of two functional domains of CLIP-170 in vivo. *J. Cell Sci.* 107: 1909–1920, 1994.

379. Piller, V., F. Piller, and M. Fukuda. Biosynthesis of truncated O-glycans in the T cell line Jurkat. Localization of O-glycan initiation. *J. Biol. Chem.* 265: 9264–9271, 1990.

380. Pimplikar, S. W., and K. Simons. Regulation of apical transport in epithelial cells by a Gs class of heterotrimeric G protein. *Nature* 362: 456–458, 1993.

381. Pimplikar, S. W., and K. Simons. Activators of protein kinase A stimulate apical but not basolateral transport in epithelial Madine-Darby canine kidney cells. *J. Biol. Chem.* 269: 19054–19059, 1994.

382. Pind, S. N., C. Nuoffer, J. M. Mc Caffery, H. Plutner, H. W. Davidson, M. G. Farquhar, and W. E. Balch. Rab 1 and Ca^{2+} are required for the fusion of carrier vesicles mediating enoplasmic reticulum to Golgi transport. *J. Cell Biol.* 125: 239–252, 1994.

383. Plutner, H., A. D. Cox, S. Pind, R. Khosravi-Far, J. R. Bourne, R. Schwaninger, C. J. Der, and W. E. Balch. Rab1b regulates vesicular transport between the endoplasmic reticulum and successive Golgi compartments. *J. Cell Biol.* 115: 31–43, 1991.

384. Plutner, H., H. W. Davidson, J. Saraste, and W. E. Balch. Morphological analysis of protein transport from the ER to Golgi membranes in digitonin-permeabilized cells: role of the 58 containing compartment. *J. Cell Biol.* 119: 1097–1116, 1992.

385. Podos, S. D., P. Reddy, J. Ashkenas, and M. Krieger. LDLC encodes a brefeldin A-sensitive peripheral Golgi protein required for normal Golgi function. *J. Cell Biol.* 127: 679–691, 1994.

386. Ponnambalam, S., C. Rabouille, J. P. Luzio, T. Nilsson, and G. Warren. The TGN38 glycoprotein contains two non-overlapping signals that mediate localization to the trans-Golgi network. *J. Cell Biol.* 125: 253–268, 1994.

387. Ponnambalan, S., M. Girotti, M.-L. Yaspo, C. E. Owen, A.C.F. Perry, T. Suganuma, T. Nilsson, M. Fried, G. Banting, and G. Warren. Primate homologues of rat TGN38: primary structure, expression and functional implications. *J. Cell Sci.* 109: 675–685, 1996.

388. Porter, K. R., A. Claude, and E. Fullam. A study of tissue culture cells by electron microscopy. *J. Exp. Med.* 81: 233–245, 1945.

389. Price, E. R., M. Jin, D. Lim, S. Pati, C. T. Walsh, and F. D. McKeon. Cyclophilin B trafficking through the secretory pathway is altered by binding of cyclosporin A. *Proc. Natl. Acad. Sci. U.S.A.* 91: 3931–3935, 1994.

390. Pryer, N. K., L. J. Wuestenhube, and R. Schekman. Vesicle-mediated protein sorting. *Annu. Rev. Biochem.* 61: 471–516, 1992.

391. Pryer, N. K., Salama, N. R., R. Schekman, and C. A. Kaiser. Cytosolic Sec13p complex is required for vesicle formation from the endoplasmic reticulum in vitro. *J. Cell Biol.* 120: 865–875, 1993.

392. Rabinovich, E., S. Bar-Nun, R. Amitay, I. Shachar, B. Gur, M. Taya, and J. Haimovich. Different assembly species of IgM are directed to distinct degradation sites along the secretory pathway. *J. Biol. Chem.* 268: 24145–24148, 1993.

393. Rabouille, C., T. Misteli, R. Watson, and G. Warren. Reassembly of Golgi stacks from mitotic Golgi fragments in a cell-free system. *J. Cell Biol.* 129: 605–618, 1995.

394. Rabouille, C., N. Hui, R. Kiekbusch, E. G. Berger, G. Graham, and T. Nilsson. Mapping of the distribution of Golgi enzymes involved in the construction of complex oligosaccharides. *J. Cell Sci.* 108: 1617–1627, 1995.

395. Rabouille, C., T. P. Levine, J.-M. Peters, and G. Warren. An NSF-like ATPase, p97, and NSF mediate cisternal regrowth from mitotic Golgi fragments. *Cell* 82: 905–914, 1995.

396. Radice, P., V. Pensotti, C. Jones, H. Perry, M. A. Perotti, and A. Tunnacliffe. The human archain gene, ARCN1, has high conserved homologs in rice and Drosophila. *Genomics* 26: 101–106, 1995.

397. Rajagopalan, S., Y. Xu, and M. B. Brenner. Retention of unassembled components of integral membrane proteins by calnexin. *Science* 263: 387–390, 1994.

398. Rambourg, A., and Y. Clermont. Three-dimensional electron microscopy: structure of the Golgi apparatus. *Eur. J. Cell Biol.* 51: 189–200, 1990.

399. Raposo, G., H. M. van Santen, R. Leijendekker, H. J. Geuze and H. L. Ploegh. Misfolded major histocompatibility complex class I molecules accumulate in an expanded ER-Golgi intermediate compartment. *J. Cell Biol.* 131: 1403–1419, 1995.

400. Reaves, B., and G. Banting. Perturbation of the morphology of the trans-Golgi network following brefeldin A treatment: redistribution of a TGN-specific integral membrane protein, TGN 38. *J. Cell Biol.* 116: 85–94, 1992.

401. Regazzi, R., A. Kikuchi, Y. Takai, and C. B. Wolheim. The small GTP-binding proteins in the cytosol of insulin-secreting cells are complexed to GDP dissociation inhibitor proteins. *J. Biol Chem.* 267: 17512–17519, 1992.

402. Rexach, M. F., and R. W. Schekman. Distinct biochemical requirements for the budding, targeting, and fusion of ER-derived transport vesicles. *J. Cell Biol.* 114: 219–229, 1991.

403. Rexach, M. F., M. Latterich, and R. W. Schekman. Characteristics of endoplasmic reticulum-derived transport vesicles. *J. Cell Biol.* 126: 1133–1148, 1994.

404. Richard, J. E., and T. E. Kreis. CLIPs for organelle-microtubule interactions. *Trends Cell Biol.* 6: 178–183, 1996.

405. Rios, R. M., A.-M. Tassin, C. Celati, C. Anthony, M.-C. Boissier, J.-C. Homberg, and M. Bornens. A peripheral protein associated with the cis-Golgi network redistributes in the intermediate compartment upon brefeldin A treatment. *J. Cell Biol.* 125: 997–1013, 1994.

406. Rolls, M. M., M. T. Marquart, M. Kielian, and C. Machamer. Cholesterol-independent targeting of Golgi membrane proteins. *Mol. Biol. Cell* 5: 241à, 1994.

407. Rose, J. K., and R. W. Doms. Regulation of protein export from the endoplasmic reticulum. *Annu. Rev. Cell Biol.* 4: 257–288, 1988.

408. Roth, J., and E. G. Berger. Immunocytochemical localization of galactosyltransferase in HeLa cells: codistribution with thiamine pyrophosphatase in trans-Golgi cisternae. *J. Cell Biol.* 93: 223–229, 1982.

409. Roth, J., Taatjes, D. J., J. M. Lucocq, J. Weinstein, and J. C.

Paulson. Demonstration of an extensive trans-tubular network continuous with the Golgi apparatus stack that may function in glycosylation. *Cell* 43: 287–295, 1985.

410. Roth, J., D. J. Taatjes, J. Weinstein, J. C. Paulson, P. Greenwell, and W. M. Watkins. Differential subcompartmentation of terminal glycosylation in the Golgi apparatus of intestinal absorptive and goblet cells. *J. Biol. Chem.* 261: 14307–14312, 1986.

411. Roth, J. Subcellular organization of glycosylation in mammalian cells. *Biochem. Biophys. Acta.* 906: 405–436, 1987.

412. Roth, J., Y. Wang, A. Eckhardt, and R. L. Hill. Subcellular localization of the UDP-N-acetyl-D-galactosamine: polypeptide N-acetylgalactosaminyltransferase-mediated O-glycosylation reaction in the submaxillary gland. *Proc. Natl. Acad. Sci. U.S.A.* 91: 8935–8939, 1994.

413. Rothman, J. E. The Golgi apparatus: two organelles in tandem. *Science* 213: 1212–1219, 1981.

414. Rothman, J. E. Polypeptide chain binding proteins: Catalysts of protein folding and related processes in cells. *Cell* 59: 591–601, 1989.

415. Rothman, J. E. Mechanisms of intracellular protein transport. *Nature* 372: 55–63, 1994.

416. Rothman, J. E., and L. Orci. Movement of proteins through the Golgi stack: a molecular dissection of vesicular transport. *FASEB J.* 4: 1460–1468, 1990.

417. Rothman, J. E., and L. Orci. Molecular dissection of the secretory pathway. *Nature* 355: 409–415, 1992.

418. Rothman, J. E., and G. Warren. Implications of the SNARE hypothesis for intracellular membrane topology and dynamics. *Curr. Biol.* 4: 220–233, 1994.

419. Rupp, K., U. Birnbach, J. Lundström, P. Nguyen Van, and H.-D. Söling. Effects of CaBP2, the rat analog of ERp72, and of CaBP1 on the refolding of denatured reduced proteins. Comparison with protein disulfide isomerase. *J. Biol. Chem.* 229: 2501–2507, 1994.

420. Rusakov, D. A., P. Podini, A. Villa, and J. Meldolesi. Tridimensional organization of Purkinje neuron cisternal stacks, a specialized endoplasmic reticulum subcompartment rich in inositol 1,4,5-trisphosphate receptors. *J. Neurocytol.* 22: 273–282, 1993.

421. Russo, R. N., N. L. Shaper, D. J. Taatjes, and J. H. Shaper. Beta1,4-galactosyltransferase: A short NH$_2$-terminal fragment that includes the cytoplasmic and transmembrane domain is sufficient for Golgi retention. *J. Biol. Chem.* 267: 9241–9247, 1992.

422. Salama, N. R., T. Yeung, and R. W. Schekman. The Sec13p complex and reconstitution of vesicle budding from the ER with purified cytoplasmic proteins. *EMBO J.* 12: 4073–4082, 1993.

423. Salama, N. R. and R. W. Schekman. The role of coat proteins in the biosynthesis of secretory proteins. *Curr. Opin. Cell Biol.* 7: 536–543, 1995.

424. Salminen, A., and P. Novick. A ras-like protein is required for a post-Golgi event in yeast secretion. *Cell* 49: 527–538, 1987.

425. Sambrook, J. F. The involvement of calcium in transport of secretory proteins from the endoplasmic reticulum. *Cell* 61: 197–199, 1990.

426. Sanders, S. L., K. M., Whitefield, J. P. Vogel, M. D. Rose, and R. W. Schekman. Sec 61p and Bip directly facilitate polypeptide translocation into the ER. *Cell* 69: 353–365, 1992.

427. Saraste, J., and E. Kuismanen. Pre- and post-Golgi vacuoles operate in the transport of Semliki Forest virus membrane glycoproteins to the cell surface. *Cell* 38: 535–549, 1984.

428. Saraste, J., G. E. Palade, and M. G. Farquhar. Antibodies to rat pancreas Golgi subfractions: Identification of a 58-kD cis-Golgi protein. *J. Cell Biol.* 105: 2021–2029, 1987.

429. Saraste, J., and E. Kuismanen. Pathways of protein sorting and membrane traffic between the rough endoplasmic reticulum and the Golgi complex. *Sem. Cell Biol.* 3: 343–355, 1992.

430. Saraste, J., and K. Svensson. Distribution of the intermediate elements operating in ER to Golgi transport. *J. Cell Sci.* 100: 415–430, 1991.

431. Saraste, J., V. Lahtinen, and B. Gould. Localization of the small GTP-binding protein rab1p to the early compartments of the secretory pathway. *J. Cell Sci.* 108: 1541–1552, 1995.

432. Satoh, T., C. A. Ross, A. Villa, S. Supattapone, T. Pozzan, S. H. Snyder, and J. Meldolesi. The inositol 1,4,5-trisphosphate receptor in cerebellar Purkinje cells: Quantitative immunogold labeling reveals concentration in an ER subcompartment. *J. Cell Biol.* 111: 615–624, 1990.

433. Schäfer, W., A. Stroh, S. Berghöfer, J. Seiler, M. Vey, M.-L. Kruse, H. F. Kern, H.-D. Klenk and W. Garten. Two independent targeting signal in the cytoplasmic domain determine trans-Golgi network localization and endosomal trafficking of the proprotein convertase furin. *EMBO J.* 14: 2424–2435, 1995.

434. Schekman, R. Translocation gets a push. *Cell* 78: 911–913, 1994.

435. Schekman, R., and L. Orci. Coat proteins and vesicle budding. *Science* 271: 1526–1533, 1996.

436. Scherer, P. E., G. Z. Lederkremer, S. Williams, F. Fogliano, G. Baldini, and H. F. Ladish Cab45, a novel Ca^{2+}-binding protein localized to the Golgi lumen. *J. Cell Biol.* 133: 257–268, 1996.

437. Schimmöller, F., B. Singer-Krüger, S. Schröder, U. Krüger, C. Barlowe, and H. Riezman. The absence of Emp24p, a component of ER-derived COPII-vesicles, causes a defect in transport of selected proteins to the Golgi. *EMBO J.* 7: 1329–1333, 1995.

438. Schindler, R., C. Itin, M. Zerial, F. Lottspeich, and H.-P. Hauri. ERGIC-53, a membrane protein of the ER-Golgi intermediate compartment carries an ER retention motif. *Eur. J. Cell Biol.* 61: 1–9, 1993.

439. Schröder, S., F. Schimmöller, B. Singer-Krüger, and H. Riezman. The Golgi localization of yeast Emp47p depends on its di-lysine motif but is not affected by the ret1–1 mutation in alpha-COP. *J. Cell Biol.* 131: 895–912, 1995.

440. Schutze, M.-P., P. A. Peterson, and M. R. Jackson. An N-terminal double-arginine motif maintains type II membrane proteins in the endoplasmic reticulum. *EMBO J.* 13: 1696–1705, 1994.

441. Schwaninger, R., H. Plutner, G. M. Bokoch, and W. E. Balch. Multiple GTP-binding proteins regulate vesicular transport from the ER to Golgi membranes. *J. Cell Biol.* 119: 1077–1096, 1992.

442. Schweizer, A., J.A.M. Fransen, T. Bächi, L. Ginsel, and H.-P. Hauri. Identification, by a monoclonal antibody, of a 53-kD protein associated with a tubulovesicular compartment at the cis-side of the Golgi apparatus. *J. Cell Biol.* 107: 1643–1653, 1988.

443. Schweizer, A., J. A. Fransen, K. Matter, T. E. Kreis, L. Ginsel, and H.-P. Hauri. Identification of an intermediate compartment involved in protein transport from endoplasmic reticulum to Golgi apparatus. *Eur. J. Cell Biol.* 53: 185–196, 1990.

444. Schweizer, A., K. Matter, C. M. Ketcham, and H.-P. Hauri. The isolated ER-Golgi intermediate compartment exhibits

properties that are different from ER and cis-Golgi. *J. Cell Biol.* 113: 45–54, 1991.

445. Schweizer, A., M. Ericsson, T. Bächi, G. Griffiths, and H.-P. Hauri. Characterization of a novel 63 kDa membrane protein: implications for the organization of the ER-to-Golg pathway. *J. Cell Sci.* 104: 671–683, 1993.

446. Schweizer, A. J. Roher, P. Jenö, A. DeMaio, T. G. Buchman, and H.-P. Hauri. A reversibly palmitoylated resident protein (p63) of an ER-Golgi intermediate compartment is related to a circulatory shock resucitation protein. *J. Cell Sci.* 104: 685–694, 1993.

447. Schweizer, A., H. Clausen, G. van Meer, and H.-P. Hauri. Localization of O-glycan initiation, sphingomyelin synthesis, and glucosylceramide synthesis in Vero cells with respect to the endoplasmic reticulum—Golgi intermediate compartment. *J. Biol. Chem.* 269: 4035–4041, 1994.

448. Schweizer, A., J. Rohrer, H.-P. Hauri, and S. Kornfeld. Retention of p63 in an ER-Golgi intermediate compartment depends on the presence of all three of its domains and on its ability to form oligomers. *J. Cell Biol.* 126: 25–39, 1994.

449. Schweizer, A., J. Rohrer, J. W. Slot, H. J. Geuze, and S. Kornfeld. Reassessment of the subcellular localization of p63. *J. Cell Sci.* 108: 2477–2485, 1995.

450. Seeger, M., and G. S. Payne. Selective and immediate effects of clathrin heavy chain mutations on Golgi membrane protein retention in Saccharomyces cerevisiae. *J. Cell Biol.* 118: 531–540, 1992.

451. Seelig, H. P., P. Schranz, H. Schröter, C. Wiemann, G. Griffiths, and M. Renz. Molecular genetic analyses of a 376-kD Golgi complex membrane protein (Giantin). *Mol. Cell Biol.* 14: 2564–2576, 1994.

452. Segev, N., J. Mulholland, and D. Botstein. The yeast GTP-binding protein YPT1 and a mammalian counterpart are associated with the secretion machinery. *Cell* 52: 915–924, 1988.

453. Segev, N. Mediation of the attachment or fusion step in vesicular transport by the GTP-binding Ypt1 protein. *Science* 252: 1553–1556, 1991.

454. Seksek, O., J. Biwersi, and A. S. Verkan. Direct measurement of trans-Golgi pH in living cells and regulation by second messengers. *J. Biol. Chem.* 270: 4967–4970, 1995.

455. Semenza, J. C., K. G. Hardwick, N. Dean, and H.R.B. Pelham. ERD2, a yeast gene required for receptor-mediated retrieval of luminal ER proteins from the secretory pathway. *Cell* 61: 1349–1357, 1990.

456. Serafini, T., G. Stenbeck, A. Brecht, F. Lottspeich, L. Orci, J. Rothman, and F. T. Wieland. A coat subunit of Golgi-derived non-clathrin coated vesicles with homology to the clathrin coated vesicle coat protein beta-adaptin. *Nature* 349: 215–220, 1991.

457. Serafini, T., L. Orci, M. Amherdt, M. Brunner, R. A. Kahn, and J. E. Rothman. ADP-ribosylation factor is a subunit of the coat of Golgi-derived COP-vesicles: a novel role for a GTP-binding protein. *Cell* 67: 239–253, 1991.

458. Sesso, A., F. Paulo de Faria, E.S.M. Iwamura, and H. Correa. A three-dimensional reconstruction study of the rough ER-Golgi interface in serial thin sections of the pancreatic acinar cell of the rat. *J. Cell Sci.* 107: 517–528, 1994.

459. Shamu, C. E., and P. Walter. Oligomerization and phosphorylation of the Ire1p kinase during intracellular signaling from the endoplasmic reticulum to the nucleus. *EMBO J.* 12: 3028–3039, 1996.

460. Shaywitz, D. A., L. Orci, M, Ravazzola, A. Swaroop, and C. A. Kaiser. Human SEC13Rp functions in yeast and is located on transport vesicles budding from the endoplasmic reticulum. *J. Cell Biol.* 128: 769–777, 1995.

461. Shiao, Y.-J., and J. E. Vance. Sphingomyelin transport to the cell surface occurs independently of protein secretion in rat hepatocytes. *J. Biol. Chem.* 268: 26085–26092, 1993.

462. Shim, J., A. P. Newman, and S. Ferro-Novick. The BOS1 gene encodes an essential 27-kD putative membrane protein that is required for vesicular transport from the ER to the Golgi complex in yeast. *J. Cell Biol.* 112: 55–64, 1991.

463. Shin, J. S., R. L. Dunbrack, S. Lee, and J. L. Strominger. Signals for retention of transmembrane proteins in the endoplasmic reticulum studied with CD4 truncation mutants. *Proc. Natl. Acad. Sci. U.S.A.* 88: 1918–1922, 1991.

464. Simons, K., and H. Virta. Perforated MDCK cells support intracellular transport. *EMBO J.* 6: 2241–2247, 1987.

465. Singer, I. I., S. Scott, D. M. Kazias, and J. W. Huff. Lovastatin, an inhibitor of cholesterol synthesis, induces hydroxy-methylglutaryl-coenzyme A reductase directly on membranes of expanded smooth endoplasmic reticulum. *Proc. Natl. Acad. Sci. U.S.A.* 85: 5264–5268, 1988.

466. Singer, S. J. It's important to concentrate. *Trends Cell Biol.* 5: 14–15, 1995.

467. Sitia, R., and J. Meldolesi. Endoplasmic reticulum: A dynamic patchwork of specialized subregions. *Mol. Biol. Cell* 3: 1067–1072, 1992.

468. Slot, J. W., and H. J. Geuze. Immunoelectron microscopic exploration of the Golgi complex. *J. Histochem. Cytochem.* 31: 1049–1056, 1983.

469. Slusarewicz, P., T. Nilsson, N. Hui, R. Watson, and G. Warren. Isolation of a matrix that binds medial Golgi enzymes. *J. Cell Biol.* 124: 405–413, 1994.

470. Sodeik, B., R. W. Doms, M. Ericsson, G. Hiller, C. E. Machamer, W. van't Hof, G. van Meer, B. Moss, and G. Griffiths. Assembly of vaccinia virus: Role of the intermediate compartment between the endoplasmic reticulum and the Golgi stacks. *J. Cell Biol.* 121: 521–541, 1993.

471. Sogaard, M., K. Tani, R. R. Ye, S. Geromanos, P. Tempst, T. Kirchhausen, J. E. Rothman, and T. Söllner. A rab protein is required for the assembly of SNARE complexes in the docking of transport vesicles. *Cell* 78: 937–948, 1994.

472. Soldati, T., A. D. Shapiro, A.B.D. Svejstrup, and S. R. Pfeffer. Membrane targeting of the small GTPase Rab9 is accompanied by nucleotide exchange. *Nature* 369: 76–78, 1994.

473. Söllner, T., S. W. Whiteheart, M. Brunner, H. Erdjument-Bromage, S. Geromanos, P. Tempst, and J. E. Rothman. SNAP receptors implicated in vesicle targeting and fusion. *Nature* 362: 318–324, 1993.

474. Söllner, T., M. K. Bennett, S. W. Whiteheart, R. H. Scheller, and J. E. Rothman. A protein assembly-disassembly pathway in vitro that may correspond to sequential steps of synaptic vesicle docking, activation and fusion. *Cell* 75: 409–418, 1993.

475. Sönnichsen, B., J. Füllekrug, P. Nguyen Van, W. Diekmann, D. G. Robinson, and G. Mieskes. Retention and retrieval: both mechanisms cooperate to maintain calreticulin in the endoplasmic reticulum. *J. Cell Sci.* 107: 2705–2717, 1994.

476. Spiro, R. G., Q. Zhu, V. Bhoyroo, and H.-D. Söling. Definition of the lectin-like properties of the molecular chaperone, calreticulin, and demonstration of its copurification with endomannosidase from rat liver Golgi. *J. Biol. Chem.* 271: 11588–11594, 1996.

477. Stafford, F. J., and J. S. Bonifacino. A permeabilized cell system identifies the endoplasmic reticulum as a site of protein degradation. *J. Cell Biol.* 115: 1225–1236, 1991.

478. Stammes, M. A., M. W. Craighead, M. H. Hoe, N. Lampen, S. Geromanos, P. Tempst, and J. Rothman. An integral membrane component of coatomer-coated transport vesicles defines a familiy of proteins involved in budding. *Proc. Natl. Acad. Sci. U.S.A.* 92: 8011–8015, 1995.

479. Stanley, K. K., and K. E. Howell. TGN38/41: a molecule on the move. *Trends Cell Biol.* 3: 252–255, 1993.

480. Stearns, T., R. A. Kahn, D. Botstein, and M. A. Hoyt. ADP ribosylation factor is an essential protein in Saccharomyces cerevisiae and is encoded by two genes. *Mol. Cell. Biol.* 10: 6690–6699, 1990.

481. Stearns, T., M. C. Willingham, D. Botstein, and R. A. Kahn. ADP ribosylation factor is functionally and physically associated with the Golgi complex. *Proc. Natl. Acad. Sci. U.S.A.* 87: 1238–1242, 1990.

482. Steinmann, B., P. Bruckner, and A. Superti-Furga. Cyclosporin A slows collagen triple-helix formation in vivo: indirect evidence for a physiologic role of peptidyl-prolyl cis-trans-isomerase. *J. Biol. Chem.* 266: 1299–1303, 1991.

483. Stenbeck, G., R. Schreiner, S. Auerbach, F. Lottspeich, J. E. Rothman, and F. T. Wieland. Gamma-COP, a coat subunit of non-clathrin-coated vesicles with homology to sec21p. *FEBS Lett.* 314: 195–198, 1992.

484. Stenbeck, G., C. Harter, A. Brecht, D. Hermann, F. Lottspeich, L. Orci, and F. T. Wieland. Beta'-COP, a novel subunit of coatomer. *EMBO J.* 12: 2841–2845, 1993.

485. Stieger, B., K. Matter, B. Baur, M. Höchli, and H.-P. Hauri. Dissection of the asynchronous transport of intestinal microvillar hydrolases to the cell surface. *J. Cell Biol.* 106: 1853–1861, 1988.

486. Stinchcombe, J. C., H. Nomoto, D. F. Cutler, and C. R. Hopkins. Anterograde and retrograde traffic between the endoplasmic reticulum and the Golgi complex. *J. Cell Biol.* 131: 1387–1401, 1995.

487. Stow, J. L., J. B. de Almeida, N. Narula, E. J. Holzman, L. Ercolani, and D. A. Ausiello. A heterotrimeric G protein, G-alpha-i-3, on Golgi membranes regulates the secretion of a heparan sulfate proteoglycan in LLC-PK1 epithelial cells. *J. Cell Biol.* 114: 1113–1124, 1991.

488. Strous, G. J., P. van Kerkhof, O. R. Willemsen, H. J. Geuze, and E. G. Berger. Transport and topology of galactosyltransferase in endomembranes of HeLa cells. *J. Cell Biol.* 97: 723–727, 1983.

489. Subramanian, V. N., F. Peter, R. Philip, S. H. Wong, and W. Hong. GS28, a 28-kilodalton Golgi SNARE that participates in ER-Golgi transport. *Science* 272: 1161–1163, 1996.

490. Suzuki, C. K., J. S. Bonifacino, A. Y. Lin, M. M. Davis, and R. D. Klausner. Regulating the retention of T-cell receptor alpha chain variants within the endoplasmic reticulum: Ca-dependent association with BiP. *J. Cell Biol.* 114: 189–205, 1991.

491. Sweet, D. J., and H.R.B. Pelham. The Saccharomyces cerevisiae SEC20 gene is sorted by the HDEL retrieval system. *EMBO J.* 11: 423–432, 1992.

492. Swift, A. M., and C. E. Machamer. A Golgi retention signal in a membrane spanning domain of Coronavirus E1 protein. *J. Cell Biol.* 115: 19–30, 1991.

493. Sztul, E. S., P. Melancon, and K. E. Howell. Targeting and fusion in vesicular transport. *Trends Cell Biol.* 2: 381–386, 1992.

494. Takei, K., A. Mignery, E. Mugnaini, T. C. Südhof, and P. De Camilli. Inositol 1,4,5-trisphosphate receptor causes formation of ER cisternal stacks in transfected fibroblasts and cerebellar Purkinje cells. *Neuron* 12: 327–342, 1994.

495. Takemoto, H., T. Yoshimori, A., Yamamoto, Y. Miyata, I. Yahara, K. Inoue, and Y. Tashiro. Heavy chain binding protein (BiP/GRP78) and endoplasmin are exported from the endoplasmic reticulum in rat exocrine pancreatic cells, similar to protein disulfide-isomerase. *Arch. Biochem. Biophys.* 296: 129–136, 1992.

496. Takizawa, P. A., and V. Malhotra. Coatomers and SNAREs in promoting membrane traffic. *Cell* 75: 593–596, 1993.

497. Takizawa, P. A., J. K. Yucel, B. Veit, D. J. Faulkner, T. Deerinck, G. Soto, M. Ellisman, and V. Malhotra. Complete vesiculation of Golgi membranes and inhibition of protein transport by a novel sea sponge metabolite, ilimaquinone. *Cell* 73: 1079–1090, 1993.

498. Tang, B. L., S. H. Wong, S. H. Low, and W. Hong. The transmembrane domain of N-acetylglucosaminyltransferase I contains a Golgi retention signal. *J. Biol. Chem.* 267: 10122–10126, 1992.

499. Tang, B. L., H. Wong, X. L. Qi, S. H. Low, and W. Hong. Molecular cloning, characterization, subcellular localization and dynamics of p23, the mammalian KDEL receptor. *J. Cell Biol.* 120: 325–338, 1993.

500. Tang, B. L., S. H. Low, H.-P. Hauri, and W. Hong. Segregation of ERGIC-53 and the mammalian KDEL receptor upon exit from the 15°C compartment. *Eur. J. Cell Biol* 68: 398–410, 1995.

501. Tanigawa, G., L. Orci, M. Amherdt, M. Ravazzola, J. B. Helms, and J. E. Rothman. Hydrolysis of bound GTP by ARF proteins triggers uncoating of Golgi-derived COP-coated vesicles. *J. Cell Biol.* 123: 1365–1371, 1993.

502. Tatu, U., I. Braakman, and A. Helenius. Membrane glycoprotein folding, oligomerization and intracellular transport: effects of dithiotreitol in living cells. *EMBO J.* 12: 2151–2157, 1993.

503. Tatu, U., C. Hammond, and A. Helenius. Folding and oligomerization of influenza hemagglutinin in the ER and the intermediate compartment. *EMBO J.* 14: 1340–1348, 1995.

504. Taylor, T. C., M. Kanstein, P. Weidman, and P. Melancon. Cytosolic ARFs are required for vesicle formation but not for cell-free intra-Golgi transport: Evidence for coated vesicle-independent transport. *Mol. Biol. Cell* 5: 237–252, 1994.

505. Teal, S. B., V. W. Hsu, P. J. Peters, R. D. Klausner, and J. G. Donaldson. An activating mutation in ARF1 stabilizes coatomer binding to Golgi membranes. *J. Biol. Chem.* 269: 3135–3138, 1994.

506. Teasdale, R. D., G. D'Agostaro, and P. A. Gleeson. The signal for Golgi retention of bovine beta1,4-galactosyltransferase is in the trans-membrane domain. *J. Biol. Chem.* 267: 4084–4096, 1992.

507. Terasaki, M., L. B. Chen, and K. Fujiwara. Microtubules and the endoplasmic reticulum are highly interdependent structures. *J. Cell Biol.* 103: 1557–1568, 1986.

508. Terasaki, M., and L. A. Jaffe. Organization of the sea urchin egg endoplasmic reticulum and its reorganization at fertilization. *J. Cell Biol.* 114: 929–940, 1991.

509. Thomas, P. J, P, Shenbagamurthi, J. Sondek, J. M. Hullihen, and P. L. Pedersen. The cystic fibrosis transmembrane conductance regulator. Effects of the most common cystic fibrosis-causing mutation on the secondary structure and stability of a synthetic peptide. *Biol. Chem.* 267: 5727–5730, 1992.

510. Tisdale, E. J., J. R. Bourne, R. Khosravi-Far, C. J. Der, and W. E. Balch. GTP-binding mutants of rab1 and rab2 are potent inhibitors of vesicular transport from the endoplasmic reticulum to the Golgi complex. *J. Cell Biol.* 119: 749–761, 1992.

511. Tooze, S. A., J. Tooze, and G. Warren. Site of addition of N-acetyl-galactosamine to the E1 glycoprotein of a mouse hepatitis virus-A59. *J. Cell Biol.* 106: 1475–1487, 1988.

512. Tooze, J., H. F. Kern, S. D. Fuller, and K. E. Howell. Condensation-sorting events in the rough endoplasmic reticulum of exocrine pancreatic cells. *J. Cell. Biol.* 109: 35–50, 1989.

513. Townsley, F. M., D. W. Wilson, and H.R.B. Pelham. Mutational analysis of the human KDEL receptor: distinct structural requirements for Golgi retention, ligand binding and retrograde transport. *EMBO J.* 12: 2821–2829, 1993.

514. Townsley, F., and H.R.B. Pelham. The KKXX signal mediates retrieval of membrane proteins from the Golgi to the ER in yeast. *Eur. J. Cell Biol.* 64: 211–216, 1994.

515. Townsley, F. M., G. Frigerio, and H.R.B. Pelham. Retrieval of HDEL proteins is required for growth of yeast cells. *J. Cell Biol.* 127: 21–28, 1994.

516. Trombetta, S. E., and A. J. Parodi. Purification to apparent homogeneity and partial characterization of rat liver UDP-glucose: glycoprotein glucosyltransferase. *J. Biol. Chem.* 267: 9236–9240, 1992.

517. Turner, J. R., and A. M. Tartakoff. The response of the Golgi complex to microtubule alterations: the roles of metabolic energy and membrane traffic in the Golgi complex organization. *J. Cell Biol.* 109: 2081–2088, 1989.

518. Ulmer, J. B., and G. E. Palade. Effects of brefeldin A on the Golgi complex, endoplasmic reticulum and viral envelope glycoproteins in murine erythroleukemia cells. *Eur. J. Cell Biol.* 54: 38–54, 1991.

519. Ullrich, O., H. Horiuchi, C. Bucci, and M. Zerial. Membrane association of Rab5 mediated by GDP-dissociation inhibitor and accompanied by GDP/GTP exchange. *Nature* 368: 157–160, 1994.

520. Urade, R., M. Nasu, T. Moriyama, K. Wada, and K. Kito. Protein degradation by the phosphoinositide-specific phospholipase C-alpha familiy from rat liver endoplasmic reticulum. *J. Biol. Chem.* 267: 15152–15159, 1992.

521. Urade, R., Y. Takenaka, and M. Kito. Protein degradation by ERp72 from rat and mouse liver endoplasmic reticulum. *J. Biol. Chem.* 268: 22004–22009, 1993.

522. Valetti, C., C. E. Grossi, C. Milstein, and R. Sitia. Russel bodies: A general response of secretory cells to synthesis of a mutant immunolgobulin which can neither exit from, nor be degraded in, the endoplasmic reticulum. *J. Cell Biol.* 115: 983–994, 1991.

523. Van der Voorn, L., and H. L. Ploegh. The WD-40 repeat. *FEBS Lett.* 307: 131–134, 1992.

524. Valtersson, C., A. H. Dutton, and S. J. Singer. Transfer of secretory proteins from the endoplasmic reticulum to the Golgi apparatus: Discrimination between homologous and heterologous transfer in intact heterokaryons. *Proc. Natl. Acad. Sci. U.S.A.* 87: 8175–8179, 1990.

525. Van Leyen, K., and F. Wieland. Complete Golgi passage of glycoptripeptides generated in the endoplasmic reticulum. *FEBS Lett.* 352: 211–215, 1994.

526. Veit, M, and M.F.G. Schmidt. Timing of palmitoylation of influenza hemagglutinin. *FEBS Lett.* 336: 243–247, 1993.

527. Veit, B., J. K. Yucel, and V. Malhotra. Microtubule independent vesiculation of Golgi membranes and reassembly of vesicles into Golgi stacks. *J. Cell Biol.* 122: 1197–1206, 1993.

528. Velasco, A. L., Hendricks, K. W. Moremen, D. R. Tulsiani, O. Touster, and M. G. Farquhar. Cell type-dependent variations in the subcellular distribution of alpha-mannosidase I and II. *J. Cell Biol.* 122: 39–51, 1993.

529. Vergères, G., T.S.B. Yen, J. Aggeler, J. Lausier, and L. Waskell. A model system for studying membrane biogenesis. Overexpression of cytochrome b5 in yeast results in marked proliferation of the intracellular membrane. *J. Cell Sci.* 106: 249–259, 1993.

530. Vertel, B. M., A. Velasco, S. LaFrance, L. Walters, and K. Kaczman-Daniel. Precursors of chondroitin sulfate proteoglycan are segregated within a subcompartment of the chondrocyte endoplasmic reticulum. *J. Cell. Chem.* 109: 1827–1836, 1989.

531. Villa, A., P. Podini, D. O. Clegg, T. Pozzan, and J. Meldolesi. Intracellular Ca^{2+} stores in chicken Purkinje neurons: Differential distribution of the low affinity-high capacity Ca^{2+} binding protein, calsequestrin, of Ca^{2+}-ATPase and of the ER lumenal protein, BiP. *J. Cell Biol.* 113: 779–791, 1991.

532. Volpe, P., K.-H. Krause, S. Hashimoto, F. Zorzato, T. Pozzan, J. Meldolesi, and D. P. Lew. "Calciosome," a cytoplasmic organelle: The inositol 1,4,5-trisphosphate-sensitive Ca^{2+} store of nonmuscle cells? *Proc. Natl. Acad. Sci. U.S.A.* 85: 1091–1095, 1988.

533. Volpe, P., A. Villan, P. Podini, A. Martini, A. Nori, M. C. Panzeri, and J. Meldolesi. The endoplasmic reticulum—sarcoplasmic reticulum connection: Distribution of endoplasmic reticulum markers in the sarcoplasmic reticulum of skeletal muscle fibers. *Proc. Natl. Acad. Sci. U.S.A.* 89: 6142–6146, 1992.

534. Vuori, K., T. Pihlajaniemi, M. Marttila, and K. I. Kivirikko. Characterization of the human prolyl 4-hydroxylase tetramer and its multifunctional protein disulfide isomerase subunit synthesized in a bacculovirus expression system. *Proc. Natl. Acad. Sci. U.S.A.* 89: 7467–7470, 1992.

535. Wada, I., D. Rindress, P. H. Cameron, W.-J. Ou, J. J. Doherty, D. Louvard, A. W. Bell, D. Dignard, D. Y. Thomas, and J. J. M. Bergeron. SSR-alpha and associated calnexin are major calcium binding proteins of the endoplasmic reticulum. *J. Biol. Chem.* 266: 19599–19610, 1991.

536. Walker, R. A., and M. P. Sheetz. Cytoplasmic microtubule-associated motors. *Annu. Rev. Biochem.* 62: 429–451, 1993.

537. Wang, C.-C., and C.-L. Tsou. Protein disulfide isomerase is both an enzyme and a chaperone. *FASEB J.* 7: 1515–1517, 1993.

538. Ward, C. L., S. Omura, and R. R. Kopito. Degradation of CFTR by the ubiquitin-proteasome pathway. *Cell* 83: 121–127, 1995.

539. Ware F., E., A. Vassilakos, P. A. Peterson, M. R. Jackson, M. A. Lehrman, D. B. Williams. The molecular chaperone calnexin binds $Glc_1Man_9GlcNAc_2$ oligosaccharide as an initial step in recognizing unfolded glycoproteins. *J. Biol. Chem.* 270: 4697–4704, 1995.

540. Warren, G. Signals and salvage sequences. *Nature* 227: 17–18, 1987.

541. Waters, M. G., T. Serafini, and J. E. Rothman. Coatomer: a cytosolic protein complex containing subunits of non-clathrin-coated Golgi transport vesicles. *Nature* 349: 248–251, 1991.

542. Waters, M. G., C.J.M. Beckers, and J. E. Rothman. Purification of coat protomers. *Methods Enzymol.* 219: 331–337, 1992.

543. Waxman, D. J., and L. Azaroff. Phenobarbital induction of cytochrome P-450 expression. *Biochem J.* 281: 577–592, 1992.

544. Weidman, P. J., and W. M. Winter. The G protein-activating peptide, mastoparan, and the synthetic NH2-terminal ARF peptide, ARFp13, inhibit in vitro Golgi transport by irreversibly damaging membranes. *J. Cell Biol.* 127: 1815–1827, 1994.

546. Weis, K., G. Griffiths, and A. I. Lamond. The endoplasmic reticulum calcium-binding protein of 55 kDa is a novel EF-

hand protein retained in the endoplasmic reticulum by a carboxyl-terminal His-Asp-Glu-Leu motif. *J. Biol. Chem.* 269: 19142–19150, 1994.

547. Weisz, O. A., A. M. Swift, and C. E. Machamer. Oligomerization of a membrane protein correlates with its retention in the Golgi complex. *J. Cell Biol.* 122: 1185–1196, 1993.

548. Welsh, M. J., and A. S. Smith. Molecular mechanisms of CFTR chloride channel dysfunction in cystic fibrosis. *Cell* 73: 1251–1254, 1993.

549. Wetterau, J. R., K. A. Combs, L. R. McLean., S. N. Spinner, and, L. P. Aggerbeck. Protein disulfide isomerase appears necessary to maintain the catalytically active structure of microsomal triglyceride transfer protein. *Biochemistry* 30: 9728–9735, 1991.

550. Whiteheart, S. W., I. C. Griff, M. Brunner, D. O. Clary, T. Mayer, S. A. Buhrow, and J. E. Rothman. SNAP family of NSF attachment proteins includes a brain-specific isoform. *Nature* 362: 353–355, 1993.

551. Whiteheart, S. W., K. Rossnagel, S. A. Buhrow, M. Brunner, R. Jaenicke, and J. E. Rothman. N-ethylmaleimide-sensitive fusion protein: A trimeric ATPase whose hydrolysis of ATP is required for membrane fusion. *J. Cell Biol.* 126: 945–954, 1994.

552. Wieland, F. T., M. L. Gleason, T. A. Serafini, and J. E. Rothman. The rate of bulk flow from the endoplasmic reticulum to the cell surface. *Cell* 50: 289–300, 1987.

553. Wikström, L., and H. F. Lodish. Nonlysosomal, pre-Golgi degradation of unassembled asialoglycoprotein receptor subunits. A TLCK- and TPCK-sensitive cleavage within the ER. *J. Cell Biol.* 113: 997–1007, 1991.

554. Wikström, L., and H. F. Lodish. Unfolded H2b asialoglycoprotein receptor subunit polypeptides are selectively degraded within the endoplasmic reticulum. *J. Biol. Chem.* 268: 14412–14416, 1993.

555. Wilcox, C. A., K. Redding, R. Wright, and R. S. Fuller. Mutation of a tyrosine localization signal in the cytosolic tail of yeast Kex2 protease disrupts Golgi retention and results in default transport to the vacuole. *Mol. Biol. Cell.* 3: 1353–1371, 1992.

556. Wileman, T., L. P. Kane, G. R. Carson, and C. Terhost. Depletion of cellular calcium accelerates protein degradation in the endoplasmic reticulum. *J. Biol. Chem.* 266: 4500–4507, 1991.

557. Wileman, T., L. P. Kane, and C. Terhost. Degradation of T-cell receptor chains in the endoplasmic reticulum is inhibited by inhibitors of cystein proteinases. *Cell Regul.* 2: 753–765, 1991.

558. Wilsbach, K., and G. S. Payne. Dynamic retention of TGN membrane proteins in Saccharomyces cerevisiae. *Trends Cell Biol.* 3: 426–432, 1993.

559. Wilsbach, K. and G. S. Payne. Vsp1p, a member of the dynamin GTPase familiy, is necessary for Golgi membrane protein retention in Saccharomyces cervisiae. *EMBO J.* 12: 3049–3059, 1993.

560. Wilson, B. S., G. E. Palade, and M. G. Farquhar. Endoplasmic reticulum—through-Golgi transport assay based on O-glycosylation of native glycophorin in permeabilized erythroleukemia cells: Role for Gi3. *Proc. Natl. Acad. Sci. U.S.A.* 90: 1681–1685, 1993.

561. Wilson, B. S., C. Nuoffer, J. L. Meinkoth, M. McCaffery, J. R. Feramisco, W. E. Balch, and M. G. Farquhar. A rab1 mutant affecting guanine nucleotide exchange promotes disassembly of the Golgi apparatus. *J. Cell Biol.* 125: 557–571, 1994.

562. Wilson, D. W., C. A. Wilcox, G. C. Flynn, E. Chen, W. Y. Kuang, M. R. Block, A. Ullrich, and J. E. Rothman. A fusion protein required for vesicle-mediated transport in both mammalian cells and yeast. *Nature* 339: 355–359, 1989.

563. Wilson, D. W., M. J. Lewis, and H.R.B. Pelham. pH-dependent binding of KDEL to its receptor in vitro. *J. Biol. Chem.* 268: 7465–7468, 1993.

564. Wong, S. H., S. H. Low, and W. Hong. The 17 residue transmembrane domain of beta-galactoside alpha2,6-sialyltransferase is sufficient for Golgi retention. *J. Cell Biol.* 117: 245–258, 1992.

565. Wong, S. H., and W. Hong. The SXYQRL sequence in the cytoplasmic domain of TGN38 plays a major role in trans-Golgi network localization. *J. Biol. Chem.* 268: 22853–22862, 1993.

566. Wood, S. A., J. E. Park, and W. J. Brown. Brefeldin A causes a microtubule-mediated fusion of trans-Golgi network and early endosomes. *Cell* 67: 591–600, 1991.

567. Yamaguchi, N., and M. N. Fukuda. Golgi retention mechanism of beta-1,4-galactosyltransferase. Membrane-spanning domain-dependent homodimerization and association with alpha- and beta-tubulins. *J. Biol. Chem.* 270: 12170–12176, 1995.

568. Yang, Y., S. Janich, J. A. Cohn, and J. M. Wilson. The common variant of cystic fibrosis transmembrane conductance regulator is recognized by hsp70 and degraded in a pre-Golgi nonlysosomal compartment. *Proc. Natl. Acad. Sci. U.S.A.* 90: 9480–9484, 1993.

569. Yoshihisa, T., C. Barlowe, and R. Schekman. Requirement for a GTPase-activating protein in vesicle budding from the endoplasmic reticulum. *Science* 259: 1466–1468, 1993.

570. Young, J., L. P. Kane, M. Exley, and T. Wileman. Regulation and selective protein degradation in the endoplasmic reticulum by redox potential. *J. Biol. Chem.* 268: 19810–19818, 1993.

571. Zerial, M., and H. Stenmark. Rab GTPases in vesicular transport. *Curr. Opin. Cell Biol.* 5: 613–620, 1993.

572. Zhang, C.-j., A. G. Rosenwald, M. C. Willingham, S. Skuntz, J. Clark, and R. A. Kahn. Expression of a dominant allele of human ARF1 inhibits membrane traffic in vivo. *J. Cell Biol.* 124: 289–300, 1994.

573. Zuber, C., J. Roth, A. Misteli, A. Nakano, and K. Moremen. DS28-6, a temperature-sensitive mutant of Chinese hamster ovary cells, expresses key phenotypic changes associated with brefeldin A treatment. *Proc. Natl. Acad. Sci. U.S.A.* 88: 9818–9822, 1991.

16. Regulated exocytosis in mammalian secretory cells

DOLA SENGUPTA

JACK A. VALENTIJN

JAMES D. JAMIESON

Department of Cell Biology, Yale University School of Medicine
New Haven, Connecticut

CHAPTER CONTENTS

STUDIES ON THE MECHANISMS INVOLVED IN EXO-CYTOSIS in eukaryotic cells have yielded considerable information in recent years. In most cells, including simpler systems such as yeast, secretion consists of the "constitutive" pathway where proteins are continually released; the regulated pathway (as occurs in pancreatic acinar cells) requires that proteins destined for release be temporarily stored in granules whose contents are discharged in response to secretagogues (22). The molecular mechanisms involved in exocytosis from both secretory pathways are beginning to be elucidated. Most evidence indicates that exocytosis from eukaryotic cells consists of vesicle/vacuole targeting and fusion with the plasmalemma that involves specific sets of membrane interactions governed by domain-specific membrane proteins. *Endocytosis,* the process whereby ligands, receptors, and associated membrane are internalized into cells, appears to utilize at least some of the same mechanisms of membrane targeting and fusion as does exocytosis. In this chapter, we will focus on the mechanisms of regulated exocytosis, and on endocyto-sis as it relates to retrieval of membrane contributed to the cell surface during regulated exocytosis.

Because much of the original work on the processing and secretion of proteins was carried out on the pancreatic acinar cell (56, 114), we will refer extensively to studies on this cell type, using it as a paradigm for the process of regulated exocytosis. Figure 16.1 is a diagrammatic representation of a pancreatic acinar cell. This diagram emphasizes that this and indeed all other regulated secretory cells are polarized in structure and function, "classic" regulated exocytosis occurring at the apical plasmalemma of the cell. The basolateral plasmalemma is the site of receipt of neuroendocrine signals from the extracellular space and is also regarded as the major site of constitutive secretion of proteins from all polarized epithelial cells, the prime secretory products being components of the basement membrane and extracellular matrix. The diagram also indicates the so-called constitutive-like secretory pathway which directs excess proteins and membrane away from the forming secretory granule, presumably to deliver these proteins to the apical pole of the cell (2–4).

ENTRY OF PROTEINS INTO THE SECRETORY PATHWAY

Mechanism of Sorting of Membrane and Secretory Proteins into the Secretory Pathway

A great deal of current interest in cell biology centers on the mechanisms of sorting of soluble and membrane proteins into the secretory pathway. Chapter 15 examines in detail the sorting and trafficking of proteins from the endoplasmic reticulum (ER) to the Golgi complex and the factors controlling this segment of the secretory pathway. It is now clear that all proteins synthesized in the rough endoplasmic reticulum (RER)—membrane, lysosomal, and bonafide secretory—reside together in the trans Golgi network (TGN) prior to being sorted elsewhere in the cell. Perhaps

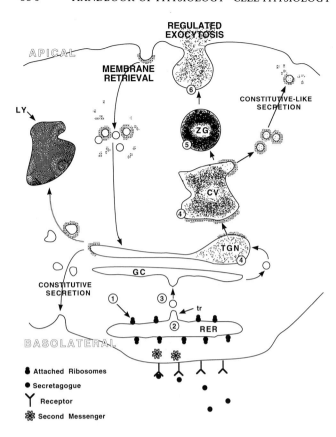

FIG. 16.1. Schematic representation of polarized, regulated secretory cell indicating main steps involved in intracellular transport and exocytosis of secretory proteins. Note existence of at least three pathways of exocytosis: classic regulated exocytosis pathway, constitutive secretion from basolateral pole of cell, and constitutive-like pathway from forming secretory granules to cell surface. *RER*, rough endoplasmic reticulum; *TGN*, trans Golgi network; *ZG*, zymogen granule; *CV*, condensing vacuole; *GC*, Golgi cisterna; *LY*, lysome; *tr*, transitional element.

the best-understood sorting mechanism is that of the mannose-6-P receptor which conveys lysosomal hydrolases to late endosomes and lysosomes (116). Despite a considerable search, similar receptor-mediated sorting for other TGN proteins has not been identified. In general, it appears that proteins which exit the cell via the constitutive secretory pathway do so without further sorting and enter this pathway passively—a process termed "*bulk flow*" (128). This pathway has also been termed the *default secretory pathway* (30) and appears to be shared by constitutively secreted proteins and membrane proteins that are being transferred to the cell surface during plasmalemmal biogenesis. The process of sorting and trafficking of membrane proteins from the TGN to other intracellular compartments and the plasma membrane is dealt with in Chapter 17.

Entry of proteins into the regulated secretory path-

way, on the other hand, is more complex. In many cells, both exocrine and endocrine, a process termed *condensation sorting* (158) takes place in which secretory proteins aggregate together and are effectively removed from the constitutive pathway and de facto have entered the regulated pathway. The mechanism underlying condensation appears to vary considerably depending on the cell type and class of protein being synthesized. For example, in the case of insulin in the pancreatic β cell, disulfide bond formation between monomers of proteolytically processed insulin leads to their aggregation and sorting into the regulated secretory pathway (61, 81).

Irrespective of the mechanisms of sorting of secretory proteins into the regulated secretory pathway, it is clear that the contents of forming secretory granules or condensing vacuoles must undergo considerable (up to 20-fold) concentration. As a consequence of this, excess membrane area accumulates around the forming secretory granule and must be removed during maturation of the secretory granule. It has been shown (3, 5, 80) that the excess membrane is pinched off from the forming secretory granule along with a "sample" of noncondensed secretory protein that is carried along in the vesicle content. The pinched-off vesicle with its content migrates to the cell surface at a rate that is considerably faster than that of proteins stored in regulated secretory granules (5). This form of secretion is termed *constitutive-like* (3) to distinguish it from ordinary constitutive secretion which, while also relatively rapid in comparison to traversal of the regulated pathway, carries a different set of proteins compared to the sample of regulated secretory proteins conveyed in the constitutive-like pathway.

Studies also indicate that formation of constitutive-like secretory vesicles may require the assembly of a clathrin coat on the cytoplasmic surface of the vesicle (3), implying that control mechanisms exist similar to those for the other steps on the secretory pathway (see later).

Mechanism of Movement of Secretory Vesicles to the Plasma Membrane and Endocytic Vesicles from the Plasma Membrane

Following their formation in the TGN, secretory granules/vesicles must be transferred to the cell surface where they await the stimulus that initiates regulated secretion. Despite a great deal of investigation, it is still unclear what the actual propulsion mechanisms for moving secretory granules to the cell surface actually are (see Chapters 13 and 20 for further information). It has been proposed that microtubules and their associated dynein motors may be involved in secretory granule movement although direct evidence for attachment of regulated secretory granules to microtubules and attendant movement is not available. In this regard, studies on MDCK cells in which specific dynein and kinesin motors were depleted from the cytosol showed that while apical transport of vesicles carrying newly synthesized membrane proteins requires both types of motors, basolateral transport depends only on kinesin. These studies also suggest that the requirement for different intracellular motors used in differential transport to the apical or basolateral plasmalemma may be influenced by differential microtubule organization (83). Similarly, movement of endocytic vesicles from their site of formation at the plasmalemma to various sites within the cell also has been thought to require various motors and interactions with the cytoskeleton (1).

Regardless of the propulsive mechanisms required to move secretory granules to the cell surface, in living cells secretory granules remain eerily still until the cell receives the stimulus for regulated exocytosis. At the time of exocytosis, video microscopy indicates a rather abrupt movement of secretory granules to the plasma membrane followed rapidly by fusion and release of granule content (138). At the light and electron microscopic levels, numerous observations have suggested that in the resting cell, an apical actin cytoskeleton, equivalent to the terminal web in intestinal epithelial cells, forms a barrier which is presumed to prevent secretory granules from prematurely fusing with the plasmalemma. This potential barrier has been termed an *actin clamp* (103, 107, 167). According to the actin clamp model, the fusion machinery by itself is constitutive; the regulatory control would be at a step distal to the formation of the docking and fusion complex and involve disassembly of the cytoskeleton. By inference, the clamp is relaxed when regulated exocytosis occur. During exocytosis, local elevations of Ca^{2+} may facilitate dissolution of the cytoskeleton, thus relieving the clamp. In this regard, recent studies (105, 157) have shown that stimulation of exocytosis in polarized regulated secretory cells leads to a wave of Ca^{2+} movement from the secretory granule field in the apex of the cell to the basal pole of the cell, presumably as a result of local release of Ca^{2+} from intracellular stores in response to formation of second messengers at the site of receptor activation in the abluminal pole of the cell (105).

In regard to the apposition of secretory granules with the cell surface, a large family of proteins now termed the *annexins* (23–25, 36, 126) has been implicated in exocytosis for many years. Originally described as synexius (124) for their ability to bind to and aggregate chromaffin granules, the annexins are characterized by their ability to bind phospholipids to become membrane associated in the presence of Ca^{2+}, and to contain a repeat motif that is capable of forming helical structures (126). Annexins have also been found to bind F-actin and possibly other cytoskeletal proteins such as fodrin and are capable of driving the association of actin into bundles (74, 75). These observations further implied a role for the annexins in exocytosis. More recent morphological studies using quick-freeze deep-etch electron microscopy of living cells (104) show that immunocytochemically identifiable annexin forms thread-like bridges between chromaffin granules and the plasma membrane. Interestingly, some of the images (139) suggest that where annexin appears to link secretory granules to the plasma membrane, the peripheral actin network appeared to have parted. These types of observations are intriguing and need to be extended to other regulated secretory systems.

MECHANISMS OF MEMBRANE INTERACTIONS ON THE SECRETORY PATHWAY

The SNARE Hypothesis of Vesicular Targeting and Membrane Fusion

Over the last few years, studies on the control of constitutive exocytosis from yeast, release of neurotransmitters from synaptic vesicles, and studies on membrane fusion among Golgi stacks in vitro have led to a remarkable convergence of information which indicates that the basic mechanisms of membrane interaction along the secretory pathway, including exocytosis, share many common features (Fig. 16.2). These studies indicate that the basic features of intracellular vesicular transport, including exocytosis, have been highly conserved during evolution (46, 69). This has led to the birth to the SNARE (*Soluble NSF Attachment protein REceptor*) hypothesis for membrane fusion (130, 152). According to this hypothesis, each transport vesicle on the secretory pathway has a specific set of proteins (v-SNAREs) that pairs with its cognate receptor on the target membrane (t-SNAREs), providing the basis for specificity of membrane interaction. Given the complexity of membrane interactions along the exocytic and endocytic secretory pathway in polarized regulated secretory cells (see Fig. 16.1), we would also expect that several isoforms of the SNARE proteins will be involved in these cell types.

The pioneering work in regard to specificity of membrane fusion came from the in vitro studies on vesicular transport between the Golgi stacks (122, 129, 169). These studies led to the identification of a cytoplasmic ATPase called NSF (*N*-ethylmaleimide-Sensitive Factor) that is essential for membrane transport (170).

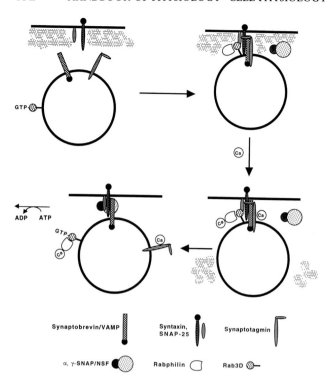

FIG. 16.2. Diagram of early steps of regulated exocytosis prior to actual fusion step and exocytosis. Interaction of v- and t-SNARE proteins in initial recognition and docking of secretory granules with apical plasmalemma is shown, as is parting of actin "clamp" that normally prevents spontaneous exocytosis. *SNAP,* soluble NSF attachment protein; *SNARE,* soluble NSF attachment protein receptor; *v,* transport vesicle; *t,* target membrane; *VAMP,* vesicle-associated membrane proteins.

NSF is a homotetramer, each subunit being 76k, and has two ATP-binding domains. The stability and attachment of NSF to membranes are critically determined by the binding and hydrolysis of ATP. Another soluble factor called SNAP (Soluble NSF Attachment Protein) is required for mediating the binding of NSF with the interacting membranes (168). Three monomeric forms of this protein, namely, α, β, and γ SNAP (M_r = 35k, 36k, and 39k, respectively), have been purified from brain (35). Both NSF and SNAP proteins, which are also involved in fusion of early endosomes (43) and in vesicular traffic between the ER and Golgi stacks (9), thus seem to be the universal components of most intracellular fusion events.

NSF can bind to the membranes only when SNAPs are bound to their receptor proteins on the membranes, called SNAREs (*SNAP REceptor*). The best-characterized SNAREs are those that function in the docking and fusion of the synaptic vesicles with the plasmalemma at the nerve terminals. These are *(1)* the synaptobrevins/VAMP [v-SNAREs; (8, 159)], *(2)* the syntaxins [t-SNAREs; (10, 11)], and *(3)* SNAP 25 [t-SNARE; *Synaptosome-Associated Protein, M_r = 25;* (41, 112); different from the α, β and γ SNAPs]. Two independent approaches provided strong evidence that they are involved in neuronal fusion machinery. First, all three proteins are selectively cleaved by the clostridial neurotoxins, which are highly specific metalloproteases that are also potent inhibitors of neuronal exocytosis (15, 88, 134). Second, these same three proteins copurify during affinity fractionation of brain extracts on immobilized α SNAP bound to NSF (152). These studies suggest that the synaptobrevins, syntaxins, and SNAP 25 constitute the SNAREs that are involved in the fusion of synaptic vesicles with the plasmalemma at the nerve terminals. Based on the observation that these SNAREs can form tight complexes even before the soluble factors can bind to them (151), it has been proposed (152) that specific integral membrane proteins present on the vesicles (v-SNAREs; synaptobrevins in the nerve terminals) initially pair with only specific proteins on the target membranes (t-SNAREs; syntaxin in nerve terminals) to initiate docking. Only correct matching of compartment-specific pairs of v-SNAREs and t-SNAREs will result in productive formation of a complex, thus imparting the selectivity that is essential for intracellular membrane trafficking (13, 46, 154). Subsequent studies have shown that syntaxin and SNAP 25 are present as a tight complex on the plasma membrane. Synaptobrevin itself binds very weakly to either syntaxin or SNAP 25 alone; however, in presence of a syntaxin–SNAP 25 complex it forms a tight 7s complex that is SDS resistant (57). Similarly, α SNAP does not bind to synaptobrevin alone and binds only weakly to isolated syntaxin, SNAP 25, or a syntaxin–SNAP 25 complex. However, after the SDS-resistant 7s complex is formed, α SNAP can bind rapidly to it, followed by the binding of NSF, leading to the formation of a 20s complex (97). Hydrolysis of adenosine triphosphate (ATP) by NSF leads to the disassembly of the 20s particle (152) and is presumed to be directly responsible for membrane fusion. The v-SNARE is probably recycled after fusion as suggested by the absence of accumulation of v-SNAREs on the target membrane (46).

Homologues of each of these SNAREs as well as NSF and SNAP have been identified in such diverse systems as yeast, *Drosophila* (153), and *Caenorhabditis elegans.* Some of the yeast homologues—for example, SEC 18, which is a homologue of mammalian NSF—can replace the mammalian protein in in vitro assays for intra-Golgi transport (35, 170). A list of the yeast homologues with their neuronal counterparts can be found in a recent review (46). Moreover, isoforms of these SNAREs in non-neuronal tissues have also been widely reported (150, 153). Identification of isoforms or functional equivalents of these essential components of the docking and fusion machinery in such diverse tissues and species strongly suggests the universal applicability of the SNARE hypothesis. In the following paragraphs we will describe some of the v- and t-SNAREs in detail.

v-SNARES. Synaptobrevins (SBs) or VAMPs (Vesicle-Associated Membrane Proteins) are integral membrane proteins with a very short cytoplasmic C-terminus, a single transmembrane domain, and an N-terminus exposed on the cytoplasmic face. Originally identified in the rat and human brain, they have since been found in transport vesicles from diverse species including *Torpedo* cholinergic nerve endings, *Drosophila* and yeast (46), as well as in a wide variety of non-neuronal tissues such as the GLUT4 transporters in adipocytes and skeletal muscle (27, 85, 167a), *Drosophila* malphigian tubules and gut, non-neuronal coated vesicles, and endosomes (89, 98). Three isoforms are known: SB 1 and 2 are 18 kDa proteins identified initially in brain. SB 3 or cellubrevin (CB) (89, 98, 167a, 167b) is the ubiquitous smaller isoform (14 kDa) which has been found in all tissues examined. All three isoforms have high degrees of homology, especially in their central domains. However, they diverge considerably toward their unique N-termini and in the transmembrane region. All these isoforms are highly sensitive to the clostridial neurotoxins botulinum D and F (BoNT/D & F) (88, 171) as already noted. Cellubrevin and SB 2 are also sensitive to BoNT/B and tetanus toxin (TeTx). The clostridial neurotoxins, which are potent inhibitors of neurotransmitter release, cleave these proteins at single sites near the centrally conserved region. The pathological effects of these toxins are mediated by cleavage of the SBs, thereby impairing fusion of synaptic vesicles with the plasma membrane. Introduction of these toxins into streptolysin O–permeabilized cells have indicated that the absence of SB 2 or CB leads to only partial loss of exocytosis (18, 49, 50). These studies suggest that there might be toxin-insensitive isoforms of SB 2 and CB.

SB isoforms have been identified in the pancreatic acinar cell. Recent studies (18) have shown by immunocytochemistry and Western blotting that a BoNT-insensitive 18 kDa immunoreactive species is associated with zymogen granules and suggest that it may be the SB 1 isoform or a novel species. It has also been shown (49) that SB 2 is associated with a zymogen granule membrane fraction and proposed that it may be involved in exocytosis since TeTx inhibits by ~30% Ca^{2+}-induced release of amylase from streptolysin O–

permeabilized acinar cells, although a direct correlation with intracellular SB 2 cleavage was not reported. Other studies have examined the effects of clostridial neurotoxins on permeabilized acinar cells and have shown no inhibition of amylase discharge (145); effects of the toxins on SB or other proteins were not examined. Studies from our laboratory (139a) indicate that the rat pancreatic acinar cell expresses a 14 kDa immunoreactive form of cellubrevin. This 14 kDa species is neurotoxin sensitive and is enriched in a smooth microsomal fraction enriched in elements of the Golgi complex. Confocal immunocytochemistry shows that cellubrevin is localized primarily to the Golgi complex with a decreasing gradient of labeling toward the apical zymogen granule field, while electron microscopic immunogold localization of cellubrevin shows labeling of vesicles and tubules in the TGN and of condensing vacuoles. Thus the role of cellubrevin and the SBs in acinar cell exocytosis and membrane trafficking remains complex. We suggest that one role for cellubrevin in the acinar cell may be in the maturation of secretory granules at the TGN and/or constitutive secretion that occurs as the excess membrane is removed from the nascent secretory granule during condensation of the secretory protein content.

t-SNAREs. The best-characterized t-SNARE known to date is syntaxin. Five isoforms of this protein, namely syntaxin 1, 2, 3, 4, and 5, have been cloned and identified (11). These proteins are about 35 kDa in size and some of them are expressed in a highly tissue-specific manner.

For example, syntaxin 1 is expressed only in the neuronal tissues (11) but syntaxins 2, 3, 4, and 5 are expressed in a wide variety of tissues (11, 20a, 56a, 155a). In vitro studies have indicated that different isoforms of syntaxin bind to different isoforms of synaptobrevins (28), thus explaining how the SNAREs can provide the specificty needed for intracellular vesicular trafficking. Syntaxin 4 has been identified in pancreatic acinar cells by Western blot analysis with a syntaxin 4-specific antibody (Sengupta and Jamieson, unpublished observation) as well as by molecular cloning (Richard Scheller, personal communication). Preliminary immunocytochemical localizations of this protein in our laboratory indicated that it localizes to the lateral membranes of the pancreatic acinar cells. This raises the possibility that syntaxin 4 is involved in the secretion of the basolateral membrane components like laminin and integrin. Absence of syntaxin 4 from the apical plasma membrane of the pancreas strongly suggests that this protein is not involved in the regulated secretion of zymogen granules in pancreatic acinar cells. It is not known at this point if the other isoforms of syntaxin such as syntaxin 2 and/or 3 are involved in regulated secretion.

The other well-characterized t-SNARE is SNAP 25, which was thought to be involved in neurite formation (111). Three lines of evidence indicate that SNAP 25 is part of the fusion machinery at the nerve terminal: (1) SNAP 25, together with syntaxin, forms a complex with synaptobrevin 2 (152); (2) SNAP 25 is homologous to the C-terminal 231 amino acid residues of Sec9p which is required for the fusion of post-Golgi transport vesicles with the plasma membrane in yeast (41, 46), (3) SNAP 25 is cleaved by clostridial neurotoxins *botulinum* A and E, which are potent neurotoxins (46). This protein is a membrane-associated cytoplasmic protein which may be anchored to the membranes through one or more of the palmityl side chains attached to the four cysteine residues clustered in the region between amino acids 84 and 92 (112).

Finally, recent studies using the introduction of antibody to NSF into streptolysin O (SLO)-permeabilized MDCK cell lines indicates that while basolateral membrane traffic is inhibited by an antibody against NSF, apical transport is not affected at all (64). It is possible that apical transport in this system involves an as yet unknown isoform of NSF which is insensitive to this antibody. Alternatively, it is possible that there are alternate routes for vesicular transport that do not require some or all of the components of the fusion machinery described above. Future studies along this line should help to clarify this issue.

Control Proteins in Exocytosis

A great deal of interest in the control of exocytosis has focused on the synaptotagmin family of integral membrane proteins (42, 45, 125, 160). To date, nine isoforms of synaptotagmin has been cloned. Synaptotagmins 1–1V are exclusively expressed in the brain while the others have wider tissue distributions (161). All of these isoforms have a single transmembrane region and two cytoplasmic repeats that are homologous to the C2 domain of protein kinase C (118). Interestingly, each of the C2 domains of synaptotagmin binds to the adaptor protein, AP-2 (87, 174), which, as we discuss below, is involved in formation of coated pits and vesicles that may be involved in membrane retrieval following regulated exocytosis. Only some of these isoforms of synaptotagmin undergo Ca^{2+}-dependent phospholipid binding (20, 40) and bind to syntaxin (12, 87). Only the second C2 domain is needed for AP-2 binding (174). The Ca^{2+} and phospholipid binding fold is present in only the first C2 domain (149). Synaptotagmin 1 also binds to the cyto-

plasmic domain of neurexin that includes the α-latrotoxin receptor (119) (Fig. 16.3).

Synaptotagmin is believed to be a negative regulator of exocytosis, acting as a clamp preventing the fusion of even properly docked vesicles in the absence of relevant stimuli. This ensures that secretion does not occur spontaneously when the two SNARE partners find each other, thus delineating the pathway of constitutive secretion from that of regulated secretion. Functional support for a role of synaptotagmin in exocytosis has come from microinjection studies which show that antibodies against synaptotagmin or peptides encompassing its C2 domain inhibit secretion from PC12 cells or the giant squid axon, respectively (16, 45, 125). Genetic studies also implicate synaptotagmin in Ca^{2+}-dependent neurotransmitter release in *Drosophila* (91). However, studies on mice that are homozygous for mutations in the gene for synaptotagmin 1 show that synaptotagmin is selectively required for Ca^{2+} triggering of exocytosis but not for exocytosis per se (54). This finding is in keeping with the findings that synaptotagmin acts as a Ca^{2+} sensor (20, 54) and undergoes conformational changes on binding Ca^{2+}

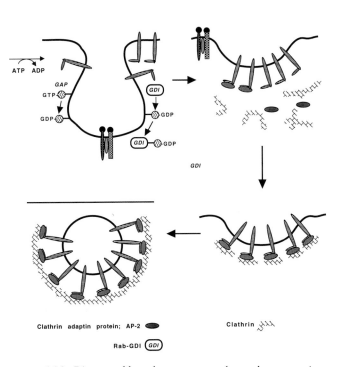

FIG. 16.3. Diagram of fate of secretory granule membrane proteins following insertion into plasmalemma that accompanies regulated exocytosis. Reorientation of synaptotagmin in plasma membrane and subsequent interaction with adaptor proteins and clathrin to form coated vesicles is suggested. *GDI*, guanosine diphosphate dissociation inhibitor.

(40). The four neuronal isoforms exhibit very different distribution pattern in the brain, suggesting that they may serve distinct but overlapping functions in neuronal exocytosis (160). The bindings of synaptotagmin to syntaxin, neurexin, and AP-2 is Ca^{2+} independent. Although this fails to explain the role of synaptotagmin in Ca^{2+}-dependent exocytosis, it suggests a link between synaptotagmin 1 and the fusion apparatus.

Like synaptotagmin, two other proteins, Rabphilin 3A and Doc2, of molecular weight 80 kDa and 44 kDa respectively, also have two repeats of the C2 domains in their C terminal region that can interact with Ca^{2+} and phospholipid. Both of these are expressed in brain and have been implicated in neurotransmitter release (78, 110a, 110b, 140a, 140b). Unlike Doc2, Rabphilin 3A also has a rab3A-binding domain at its N-terminal region. Recent studies have shown that Rabphilin 3A binds to the synaptic vesicles by its interaction with GTP-rab3 via its N-terminal domain comprising a novel Zn^{2+} finger motif (144a).

A 68 kDa protein called rbSec1 or n-Sec1 has been isolated from rat brain. This protein is homologous to the yeast Sec 1 protein that is involved in the constitutive secretory pathway between the Golgi complex and the plasma membrane. It also has homology with unc-18 protein in *C. elegans,* rop protein in *Drosophila,* and the Sly1 and Ssp1 proteins in the yeast. RbSec1 can be detected in both cytosol and membranes. It does not bind to synaptobrevin 2 or SNAP 25 or even to the SNAP 25-syntaxin complex. However, it binds to syntaxin alone and in the presence of the latter becomes membrane-bound. These findings indicate that rbSec1 could be involved in the regulation of synaptic vesicle docking and fusion (51, 52, 120). More recently, two proteins, complexin I (18 kDa) and complexin II (19 kDa), have been identified (96) that are highly enriched in neurons. Complexin II is also found in all tissues at low levels. In neurons they colocalize with syntaxin and SNAP 25. Although complexin does not bind to synaptobrevin 2 or SNAP 25 and binds only weakly to syntaxin alone, the binding is greatly enhanced by formation of a core complex consisting of these v- and t-SNAREs. Also, it competes with α SNAP for binding to the core complex but not with synaptotagmin 1. This suggests that the complexins may be involved in the regulation of sequential interaction of α SNAP and synaptotagmin 1 with the SNAREs during exocytosis (96).

The Role of rab Proteins in Exocytosis and Endocytosis

General Background. Rab proteins are low molecular weight GTPases that belong to the superfamily of small

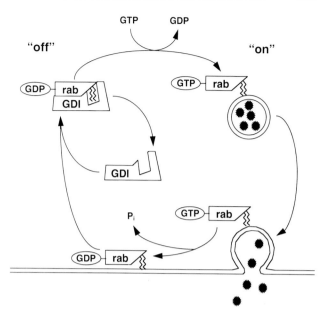

FIG. 16.4. Schematic representation depicting membrane-cytosol cycle of rab proteins. In GTP-bound or "on" state, isoprenylated rab protein is associated with donor membrane compartment, illustrated here as secretory vesicle. Upon fusion of donor membrane with its target—that is, plasma membrane—rab protein hydrolyzes GTP into GDP and becomes cytosolic, forming a complex with rab-GDI; rab protein is now in "off" state until it exchanges its GDP for GTP and becomes newly membrane associated. During cycle, rab proteins probably interact with other regulatory proteins such as rabphilin, GAP and GDS, but precise site of action is yet unclear, and they have therefore not been included in this model.

GTP-binding proteins including ras, rho, and ARF. They also show homology with the yeast GTP-binding proteins Ypt1 and Sec4 (46). To date, over 30 isoforms of rab proteins have been identified in mammalian cells. They have been implicated in regulating critical steps along both the biosynthetic and endocytotic pathways of intracellular membrane trafficking, as well as in cytoskeletal organization (Fig. 16.4).

All rab proteins possess four highly conserved regions which appear to form a single GTP-binding site (162). Despite this complex topological organization of the GTP-recognition site, and in contrast with the GTP-binding α-subunit of heterotrimeric GTPases, rab proteins retain their ability to bind GTP after denaturing gel electrophoresis and electrotransfer to carrier membranes (84). The N- and C-terminal portions of rab proteins are hypervariable and thereby add structural specificity to their targeting to intracellular compartments as well as to their post-translational modifications (32, 106, 146, 147). Nevertheless, all rab proteins contain one or two cysteine residues at their C-termini. These residues are posttranslationally geranylgeranylated and as such contribute to the membrane association of rab proteins. The isoprenylation is carried out by a specific enzyme called rab geranylgeranyl transferase and requires a cofactor termed *rab escort protein* (REP) (55). Geranylgeranylation also seems to be necessary for the association of rabs with the cytosolic protein rab guanosine diphosphate dissociation inhibitor (GDI) (GDP dissociation inhibitor) (121). Several rab proteins are also known to be phosphorylated by specific protein kinases (6, 76).

According to our current understanding, rab proteins oscillate between a membrane-bound and a cytosolic state (Fig. 16.1). In the membrane-bound state the rab proteins bind GTP and interact with effector proteins such as a SNAP or SNARE complex, whereas upon hydrolysis of GTP into GDP the rabs become cytosolic and form a complex with rab-GDI. According to this view rab proteins would act as "molecular on/off switches." Other regulatory proteins interacting with rabs, either in the membrane-bound or the cytosolic state, that have been identified include rabphilin (78, 86), Mss4 (26), GAP (GTPase-activating protein) (17, 106, 121), GDS (GDP dissociation stimulator) (132), rabin (19), and rabaptin (148).

Rab Proteins and Exocytosis. A large body of evidence indicates that rab proteins are implicated in controlling final steps in exocytosis. First, it has been shown that the assembly of SNARE complexes required for vesicle docking is catalyzed by a rab protein (142). Moreover, the triggering of exocytosis has been shown to be concurrent with GTP hydrolysis by rabs, and with the translocation of rabs from the secretory vesicles to the plasma membrane (47, 144). The rab proteins involved in this process belong to the rab3 subfamily, comprising rab3A, rab3B, rab3C, and rab3D, which all display a high degree of sequence homology. To date, rab3A is the best-characterized protein of the rab3 subfamily. Rab3A has long been regarded as a neuron-specific protein, but recent evidence suggests that it is also expressed in other cell types such as adrenal chromaffin cells, adipocytes, and ATt-20 cells (7, 37, 38). Peptides of the rab3A effector domain increase the rate of secretion in various cell types permeabilized with either streptolysin O or digitonin (110, 113, 127, 173). These peptides also increase membrane capacitance, which is a measure of exocytosis (108). (The use of capacitance measurements to monitor exocytosis in regulated secretory cells is covered in Chapter 14). In contrast, C-terminal domain peptides of rab3A and rab3B are potent inhibitors of secretion in neuroendocrine cells, possibly a result of competing with the endogenous functional rab3 proteins for their effector site (92, 117). The nature of this effector site is unknown, which

raises some concern as to the mode of action of the rab3 peptides. In fact, it has been shown in permeabilized pancreatic acinar cells that the increase in secretion induced by the rab3A effector domain peptides is concomitant with an increase in IP$_3$ production (123). In addition, Ca^{2+}-dependent exocytosis in pituitary cells was found to be inhibited by an 18-amino-acid synthetic peptide of the rab effector domain (39). Furthermore, it has been shown that scrambled peptides of the rab3A effector domain induce similar secretory responses in an in vitro zymogen granule–plasmalemma fusion system (94). However, in this system, exocytosis is not dependent on Ca^{2+}, which may indicate that the form of secretion being examined was constitutive rather than regulated (44).

The exact role of rab3A in exocytosis has become even more controversial in light of the recent finding that in genetically manipulated mice which do not express the rab3A protein, most electrophysiological parameters remain normal. Furthermore, it was shown in this study that the expression of over 20 other synaptic proteins was not altered in mice lacking rab3A, and that except for rabphilin no compensatory changes could be detected in the expression of other GTPases or proteins interacting with rab3 (53). Yet another study has demonstrated that the selective inhibition of rab3A expression in adrenal chromaffin cells by means of intracellular injection of specific antisense oligonucleotides increases the cells' responsiveness to repetitive stimulation, suggesting that rab3A controls exocytosis negatively (73). It is thus conceivable that the mode of action of rab3 proteins in exocytosis is tissue specific. Alternatively, the rab3 function may depend on the excitatory state—for example, the intracellular calcium concentration or the membrane potential—of the tissues and cells studied. Also, several investigations have reported that a variety of cell types including neurons and exocrine and endocrine cells, express more than one rab3 isoform, suggesting that the different rab3 proteins, although highly homologous, may control distinct steps in the exocytotic transport machinery. This possibility is corroborated by the fact that in a number of cell types expressing several rab3 proteins, each isoform is confined to a separate membrane compartment (7, 163).

Recent studies from our laboratory suggest that rab3D and as yet unidentified rab3-like proteins are intimately involved in the final stages of regulated exocytosis from regulated secretory cells such as pancreatic acinar cells (163) and gastric parietal cells (156a). First, rab3D is localized to secretory granules in these exocrine cells (163). Second, the rab3-like protein relocates from zymogen granules to the TGN upon induction of exocytosis (72) without an apparent transient association with the plasma membrane or residing as a soluble complex in the cytosol. Third, expression and membrane association of rab3D in the developing pancreas coincides with the onset of postnatal regulated secretion (163a) as we will discuss below.

Rab Proteins and Endocytosis. At least four members of the rab family have been implicated in regulating the endocytotic pathway. Rab7 and rab9 are both localized to late endosomes (93, 143). In a cell-free transport assay, rab9 was found to stimulate the translocation of mannose-6-phosphate receptors from late endosomes to the TGN, whereas functional rab7 was ineffective (93). These observations not only define a role for rab9 but also emphasize the emerging concept that different rab proteins confined to the same membrane compartment may play distinct regulatory roles.

Rab4 and rab5 are both localized on early endosomes, although it remains to be established whether they actually colocalize or are present on different endosomal subpopulations (33, 166). Rab5 is also associated with the plasma membrane and clathrin-coated vesicles (21). In vitro and in vivo studies have demonstrated that rab5 is required for the fusion of endocytotic vesicles with early endosomes as well as for the fusion between early endosomes themselves. For example, one study showed that the overexpression in cells of a mutant rab5 protein defective in GTP binding induced an accumulation of endocytotic vesicles beneath the plasma membrane, whereas the overexpression of wild-type rab5 generated atypically large early endosomes (21). Rab4, on the other hand, appears to control—that is, enhance—the transport from early endosomes to recycling vesicles (165). An additional role for rab4 has been speculated—namely, that of shunting the fusion pathway between endocytotic vesicles and early endosomes so that newly formed coated vesicles will fuse directly with the recycling vesicles (165). A noteworthy feature of rab4 is the fact that it contains a carboxy-terminal phosphorylation site ($S^{196}PRR$) for the cell-cycle-specific $p34^{cdc2}$ kinase (164). This site is phosphorylated by $p34^{cdc2}$ kinase during mitosis and as a consequence rab4 relocates from the membrane to the cytosol. Upon exit of cells from mitosis, the cytosolic rab4 is dephosphorylated and reassociates with membranes. These findings are particularly interesting given that membrane traffic is significantly reduced during the mitotic metaphase.

Rab Proteins and Their Involvement in Secretogenesis. Regulated (that is, agonist-induced) secretion by the acinar cells of the exocrine rat pancreas starts to occur only 1 day after birth, which coincides with the initial activation of the gastrointestinal tract as a result of

suckling by the neonates. In contrast, 1 day before birth the acinar cells do not secrete digestive enzymes in response to stimulation by secretagogues such as cholecystokinin (CCK), although functional CCK receptors coupled to the increase of intracellular Ca^{2+} concentration are already present as well as zymogen granules packed with secretory products (31, 70). It thus appears that the onset of regulated secretion in the pancreatic acinar cell is controlled by additional factors. Rab proteins would be good candidates for such factors. Indeed, recent findings (163a, 163b) indicate that the developmental expression of two small GTP-binding proteins belonging to the rab subfamily, rab3D and rab4, correlates with secretogenesis. Rab3D, which is present in abundance on the zymogen granules of adult acinar cell (163) exists almost exclusively as a cytosolic protein in the fetal pancreas. But at around the time of birth rab3D becomes associated with membranes. High levels of rab4 are detected only during the first 40 h of neonatal life. The transiently expressed rab4 is associated with the cores of microvilli and the actin terminal web in the apical region of acinar cells. The exact roles of rab3D and rab4 in secretogenesis are currently a matter of speculation, although it is tempting to suggest that rab4 is involved in the disinhibition of a cytoskeletal clamp withholding zymogen granules from fusing with the plasma membrane, the latter event being mediated by membrane-bound rab3D. It is clear, however, that the rab4 expressed in the neonatal pancreas is not associated with endosomes and therefore is unlikely to play a role in endocytosis. This suggests that the function of rab proteins is tissue or cell-type specific. In this respect, it is interesting to note that rab13 is present on cytoplasmic vesicles in nonpolarized cells, whereas it colocalizes with tight junctions in polarized epithelial cells (172).

Rab Proteins and Defects in Secretion. Rat pancreatic acinar tumor cells show defects in stimulus-secretion coupling comparable to those in developing acinar cells (67, 68) in that they are relatively insensitive to secretagogues and express a secretory protein pattern similar to day 21 fetal rat pancreas. Acinar cell tumors are unable to secrete basement membrane components and lack polarized organization of their plasma membrane (65, 66). Human pancreatic carcinomas uniformly (>90%) show mutations in the c-K-ras oncogene (140) and are often similarly lacking in epithelial polarity and defective in basement membrane organization or deposition. Whether these abnormalities in the acinar tumor cells are correlated with defects in the expression of small-M_r GTP-binding proteins and their regulators deserves to be studied.

Preliminary data from our laboratory suggest that the acinar cell tumor DL6, from which the AR42J tumor cell line is derived, possesses two distinct populations of secretory granules. Interestingly, only one of these two populations seems to express rab3D (9a).

MEMBRANE RETRIEVAL FOLLOWING REGULATED EXOCYTOSIS

The culmination of regulated exocytosis is the insertion of massive amounts of intracellular membrane from secretory granules into the plasmalemma, which must be compensated for in order to maintain cell-surface area and membrane composition. In several systems, ranging from nerve terminals to cortical granule exocytosis from sea urchin oocytes, it has been observed that coated pits are assembled and likely function in the retrieval of exocytosed membrane (48, 60, 95). The pancreatic acinar cell also is faced with a major problem of membrane retrieval during regulated exocytosis since we estimate that complete discharge of zymogen granules leads to ~30-fold expansion of the apical plasmalemma where granule discharge exclusively occurs (71, 72). We previously reported that secretagogue-induced exocytosis is associated at early times of stimulation with a two to fourfold expansion of the apical plasmalemma, but thereafter, compensatory membrane retrieval keeps pace with continued secretory granule exocytosis, and this is correlated with an approximately twofold expansion of the surface area of the Golgi complex. Expansion of the Golgi complex is also associated with the premature concentration of secretory proteins in the TGN that results in the formation of small secretory granules (71). In addition, we have also observed a large increase in the number of clathrin-coated vesicles under the apical plasmalemma (71) and in the TGN during zymogen granule exocytosis. A number of electron microscopic tracer studies have demonstrated the potential for continuity between the apical plasmalemma and elements of the Golgi complex during regulated exocytosis from glandular cells (58, 59, 109). The properties of the retrieved membrane and its route back to the Golgi complex have not been completely defined, but retrieved membrane may even be reutilized in the formation of new sets of secretory granules. It is not known for certain if the membrane domain retrieved has the same composition as the inserted zymogen granule membrane, if it is derived from nongranule apical plasma membrane, or if it is a mixture of membrane components.

The triggers for membrane internalization during coupled exo/endocytosis are not known, although it

has been proposed that additional members of the rab family, such as rab4 and 5 (34), which are associated with early endosomes, could be involved (see earlier in this chapter). The relationship between exocytosis and biogenesis of the plasmalemma is not clear although it is likely that membrane biogenesis takes place via delivery of membrane to the cell surface by the constitutive secretory pathway (3, 62). In this regard, previous studies (29, 99, 100) have shown that the composition of secretory granule membrane proteins and lipids is markedly different from that of the plasmalemma.

In the context of membrane retrieval related to exocytosis, it is important to point out that two general mechanisms appear to operate during endocytosis. In one form of endocytosis, the cytoplasmic domain of transmembrane receptors and other membrane proteins such as synaptotagmin (see below) interact with adaptor proteins (116), which in turn induce the assembly of clathrin coats on the cytoplasmic surface of the plasmalemma (63, 116, 135). Formation of "coated pits" in turn is followed by invagination of the coated region into the cell as clathrin-coated vesicles. During the invagination process, dynamin, a GTP-binding protein, appears to play a key role in the mechanical retraction of the internalizing membrane (77, 156). Subsequently, the clathrin coat disassembles and the uncoated vesicle moves into the cell, where it enters either the terminal lysosomal degradative pathway, recycles to the cell surface, or possibly returns to the Golgi complex (14, 79, 101, 102, 155). A second type of membrane retrieval involves the formation of caveolae that are characterized by the presence of a coat comprised of the integral membrane protein caveolin (82, 90, 115, 131). The relationship of caveolin-coated vesicles and compensatory membrane retrieval is discussed below.

Recently, an important study has presented evidence that helps to clarify the coupling between exocytosis and coated-vesicle-mediated membrane retrieval. In this study, it was shown that synaptotagmin I serves as a high-affinity receptor for the clathrin AP-2 adaptor that initiates formation of plasmalemmal coated pits (174). This has led to a model in which synaptotagmin in the resting state inhibits membrane fusion by interacting with the SNARE complex, but upon initiation of regulated secretion, synaptotagmin dissociates from the SNARE complex, possibly upon binding Ca^{2+}, allowing fusion to occur. At the same time, synaptotagmin is relocated to the plasma membrane, where its binding sites for AP-2 are accessible. It now is able to initiate clathrin assembly into coated pits. In this way, exocytosis is coupled to membrane retrieval. Subsequent to formation of coated pits and vesicles, other proteins may be involved in regulation of endocytosis/recycling to the Golgi complex. It is unlikely that SB 2 and cellubrevin are involved since fusion of endosomes (89) and a portion of receptor recycling (50) are insensitive to neurotoxins.

As noted above, a second form of endocytosis occurs via "smooth-surfaced" membrane invaginations (141) that have associated with them the glycosyl phosphatidylinositol-linked membrane protein caveolin (90, 115, 131). Caveolin-containing endocytic vesicles are found in the endocytic vesicles involved in internalization and vesicular transport in capillary endothelial cells (136, 137) and adipocytes (133), among other cell types. The connection of caveolin-containing endocytic vesicles with compensatory membrane retrieval following regulated exocytosis is not known.

REFERENCES

1. Allan, V. Membrane traffic motors. *FEBS Lett.* 369: 101–106, 1995.
2. Arvan, P., and J. D. Castle. Phasic release of newly synthesized secretory proteins in the unstimulated rat exocrine pancreas. *J. Cell Biol.* 104: 243–252, 1987.
3. Arvan, P., and J. D. Castle. Protein sorting and secretion granule formation in regulated secretory cells. *Trends Cell Biol.* 2: 327–331, 1992.
4. Arvan, P., and A. Chang. Constitutive protein secretion from the exocrine pancreas of fetal rats. *J. Biol. Chem.* 262: 3886–3890, 1987.
5. Arvan, P., R. Kuliawat, D. Prabakaran, A.-M. Zavacki, D. Elahi, S. Wang, and D. Pilkey. Protein discharge from immature secretory granules displays both regulated and constitutive characteristics. *J. Biol. Chem.* 266: 14171–14174, 1991.
6. Bailly, E., M. McCaffrey, N. Touchot, A. Zahraoui, B. Goud, and M. Bornens. Phosphorylation of two small GTP-binding proteins of the Rab family by p34^{cdc2}. *Nature* 350: 715–718, 1991.
7. Baldini, G., P. E. Scherer, and H. F. Lodish. Nonneuronal expression of Rab3A: induction during adipogenesis and association with different intracellular membranes than Rab3D. *Proc. Natl. Acad. Sci. U.S.A.* 92: 4284–4288, 1995.
8. Baumert, M., P. R. Maycox, F. Navone, P. De Camilli, and R. Jahn. Synaptobrevin: an integral membrane protein of 18,000 daltons present in small synaptic vesicles of rat brain. *EMBO J.* 8: 379–384, 1989.
9. Beckers, C. J. M., M. R. Block, B. S. Glick, J. E. Rothman, and W. E. Balch. Vesicular transport between the endoplasmic reticulum and the Golgi stack requires the NEM-sensitive fusion protein. *Nature* 339: 397–398, 1989.
9a. Bell, R. H. Jr., E. T. Kuhlmann, R. T. Jensen, and D. S. Longnecker. Overexpression of cholecystokinin receptors in azaserine-induced neoplasms of the rat pancreas. *Cancer Res.* 52: 3295–3299, 1992.
10. Bennett, M. K., N. Calakos, and R. H. Scheller. Syntaxin: a synaptic protein implicated in docking of synaptic vesicles at presynaptic active zones. *Science* 257: 255–259, 1992.
11. Bennett, M. K., J. E. García-Arrarás, L. A. Elferink, K. Peterson, A. M. Fleming, C. D. Hazuka, and R. H. Scheller. The syntaxin family of vesicular transport receptors. *Cell* 74: 863–873, 1993.
12. Bennett, M. K., and R. H. Scheller. The molecular machinery for secretion is conserved from yeast to neurons. *Proc. Natl. Acad. Sci. U.S.A.* 90: 2559–2563, 1993.
13. Bennett, M. K., and R. H. Scheller. A molecular description of synaptic vesicle membrane trafficking. *Annu. Rev. Biochem.* 63: 63–100, 1994.
14. Betz, W. J., and L. G. Wu. Synaptic transmission—kinetics of synaptic-vesicle recycling. *Curr. Biol.* 5: 1098–1101, 1995.
15. Blasi, J., E. R. Chapman, S. Yamasaki, T. Binz, H. Niemann, and R. Jahn. Botulinum neurotoxin C1 blocks neurotransmitter release by means of cleaving HPC-1/syntaxin. *EMBO J.* 12: 4821–4828, 1993.
16. Bommert, K., M. P. Charlton, W. M. DeBello, G. J. Chin, H. Betz, and G. J. Augustine. Inhibition of neurotransmitter release by C2-domain peptides implicates synaptotagmin in exocytosis. *Nature* 363: 163–165, 1993.
17. Bourne, H. R. GTPases: a turn-on and a surprise. *Nature* 366: 628–629, 1993.
18. Braun, J.E.A., B. A. Fritz, S.M.E. Wong, and A. W. Lowe. Identification of a vesicle-associated membrane protein (VAMP)-like membrane protein in zymogen granules of the rat exocrine pancreas. *J. Biol. Chem.* 269: 5328–5335, 1994.
19. Brondyk, W. H., C. J. McKiernan, K. A. Fortner, P. Stabila, R. W. Holz, and I. G. Macara. Interaction cloning of Rabin3, a novel protein that associates with the Ras-like GTPase Rab3A. *Mol. Cell. Biol.* 15: 1137–1143, 1995.
20. Brose, N., A. G. Petrenko, T. C. Südhof, and R. Jahn. Synaptotagmin: a calcium sensor on the synaptic vesicle surface. *Science* 256: 1021–1025, 1992.
20a. Brumell, J. H., A. Volchuk, H. Sengelov, N. Borregaard, A. M. Cieutat, D. F. Bainton, S. Grinstein, and A. Klip. Subcellular distribution of docking/fusion proteins in neutrophils, secretory cells with multiple exocytic compartments. *J. Immunol.* 155: 5750–5759, 1995.
21. Bucci, C., R. G. Parton, I. H. Mather, H. Stunnenberg, K. Simons, B. Hoflack, and M. Zerial. The small GTPase rab5 functions as a regulatory factor in the early endocytic pathway. *Cell* 70: 715–728, 1992.
22. Burgess, T. L., and R. B. Kelly. Constitutive and regulated secretion of proteins. *Annu. Rev. Cell Biol.* 3: 243–293, 1987.
23. Burgoyne, R. D., and M. J. Clague. Annexins in the endocytic pathway. *Trends Biochem. Sci.* 19: 231–232, 1994.
24. Burgoyne, R. D., S. E. Handel, A. Morgan, M. E. Rennison, M. D. Turner, and C. J. Wilde. Calcium, the cytoskeleton and calpactin (annexin II) in exocytotic secretion from adrenal chromaffin and mammary epithelial cells. *Biochem. Soc. Trans.* 19: 1085–1090, 1991.
25. Burgoyne, R. D., and A. Morgan. Evidence for a role of calpactin in calcium-dependent exocytosis. *Biochem. Soc. Trans.* 18: 1101–1104, 1990.
26. Burton, J. L., M. E. Burns, E. Gatti, G. J. Augustine, and P. De Camilli. Specific interactions of Mss4 with members of the Rab GTPase subfamily. *EMBO J.* 13: 5547–5558, 1994.
27. Cain, C. C., W. S. Trimble, and G. E. Lienhard. Members of the VAMP family of synaptic vesicle proteins are components of glucose transporter-containing vesicles from rat adipocytes. *J. Biol. Chem.* 267: 11681–11684, 1992.
28. Calakos, N., M. K. Bennett, K. E. Peterson, and R. H. Scheller. Protein-protein interactions contributing to the specificity of intracellular vesicular trafficking. *Science* 263: 1146–1149, 1994.
29. Cameron, R. S., P. L. Cameron, and J. D. Castle. A common spectrum of polypeptides occurs in secretion granule membranes of different exocrine glands. *J. Cell Biol.* 103: 1299–1313, 1986.
30. Caplan, M. J., and K. S. Matlin. Sorting of membrane and

secretory proteins in polarized epithelial cells. In: *Functional Epithelial Cells in Culture,* edited by K. S. Matlin and J. Valentich. New York: Alan R. Liss, 1989, p. 71–127.

31. Chang, A., and J. D. Jamieson. Stimulus-secretion coupling in the developing exocrine pancreas: secretory responsiveness to cholecystokinin. *J. Cell Biol.* 103: 2353–2365, 1986.

32. Chavrier, P., J.-P. Gorvel, E. Stelzer, K. Simons, J. Gruenberg, and M. Zerial. Hypervariable C-terminal domain of rab proteins acts as a targeting signal. *Nature* 353: 769–772, 1991.

33. Chavrier, P., R. G. Parton, H. P. Hauri, K. Simons, and M. Zerial. Localization of low molecular weight GTP binding proteins to exocytic and endocytic compartments. *Cell* 62: 317–329, 1990.

34. Chavrier, P., K. Simons, and M. Zerial. The complexity of the Rab and Rho GTP-binding protein subfamilies revealed by a PCR cloning approach. *Gene* 112: 261–264, 1992.

35. Clary, D. O., I. C. Griff, and J. E. Rothman. SNAPs, a family of NSF attachment proteins involved in intracellular membrane fusion in animals and yeast. *Cell* 61: 709–721, 1990.

36. Creutz, C. E. The annexins and exocytosis. *Science* 258: 924–931, 1992.

37. Darchen, F., J. Senyshyn, W. H. Brondyk, D. J. Taatjes, R. W. Holz, J.-P. Henry, J.-P. Denizot, and I. G. Macara. The GTPase Rab3a is associated with large dense core vesicles in bovine chromaffin cells and rat PC12 cells. *J. Cell Sci.* 108: 1639–1649, 1995.

38. Darchen, F., A. Zahraoui, F. Hammel, M.-P. Monteils, A. Tavitian, and D. Scherman. Association of the GTP-binding protein Rab3A with bovine adrenal chromaffin granules. *Proc. Natl. Acad. Sci. U.S.A.* 87: 5692–5696, 1990.

39. Davidson, J. S., A. Eales, R. W. Roeske, and R. P. Millar. Inhibition of pituitary hormone exocytosis by a synthetic peptide related to the rab effector domain. *FEBS Lett.* 326: 219–221, 1993.

40. Davletov, B. A., and T. C. Südhof. Ca²⁺-dependent conformational change in synaptotagmin I. *J. Biol. Chem.* 269: 28547–28550, 1994.

41. De Camilli, P. Secretion: exocytosis goes with a SNAP. *Nature* 364: 387–388, 1993.

42. DeBello, W. M., H. Betz, and G. J. Augustine. Synaptotagmin and neurotransmitter release. *Cell* 74: 947–950, 1993.

43. Diaz, R., L. S. Mayorga, P. J. Weidman, J. E. Rothman, and P. D. Stahl. Vesicle fusion following receptor-mediated endocytosis requires a protein active in Golgi transport. *Nature* 339: 398–400, 1989.

44. Edwardson, J. M., and P. U. Daniels-Holgate. Reconstitution *in vitro* of a membrane-fusion event involved in constitutive exocytosis. A role for cytosolic proteins and a GTP-binding protein, but not for Ca²⁺. *Biochem. J.* 285: 383–385, 1992.

45. Elferink, L. A., M. R. Peterson, and R. H. Scheller. A role for synaptotagmin (p65) in regulated exocytosis. *Cell* 72: 153–159, 1993.

46. Ferro-Novick, S., and R. Jahn. Vesicle fusion from yeast to man. *Nature* 370: 191–193, 1994.

47. Fischer von Mollard, G., T. C. Südhof, and R. Jahn. A small GTP-binding protein dissociates from synaptic vesicles during exocytosis. *Nature* 349: 79–81, 1991.

48. Fisher, G. W., and L. I. Rebhun. Sea urchin egg cortical granule exocytosis is followed by a burst of membrane retrieval via uptake into coated vesicles. *Dev. Biol.* 99: 456–472, 1983.

49. Gaisano, H. Y., L. Sheu, J. K. Foskett, and W. S. Trimble. Tetanus toxin light chain cleaves a vesicle-associated membrane protein (VAMP) isoform 2 in rat pancreatic zymogen granules

and inhibits enzyme secretion. *J. Biol. Chem.* 269: 17062–17066, 1994.

50. Galli, T., T. Chilcote, O. Mundigl, T. Binz, H. Niemann, and P. De Camilli. Tetanus toxin-mediated cleavage of cellubrevin impairs exocytosis of transferrin receptor-containing vesicles in CHO cells. *J. Cell Biol.* 125: 1015–1024, 1994.

51. Garcia, E. P., E. Gatti, M. Butler, J. Burton, and P. De Camilli. A rat brain Sec1 homologue related to Rop and UNC18 interacts with syntaxin. *Proc. Natl. Acad. Sci. U.S.A.* 91: 2003–2007, 1994.

52. Garcia, E. P., P. S. McPherson, T. J. Chilcote, K. Takei, and P. De Camilli. rbSec1A and B colocalize with syntaxin 1 and SNAP-25 throughout the axon, but are not in a stable complex with syntaxin. *J. Cell Biol.* 129: 105–120, 1995.

53. Geppert, M., V. Y. Bolshakov, S. A. Siegelbaum, K. Takei, P. DeCamilli, R. E. Hammer, and T. C. Südhof. Rab3A function in neurotransmitter release. *Nature* 369: 493–497, 1994.

54. Geppert, M., Y. Goda, R. E. Hammer, C. Li, T. W. Rosahl, C. F. Stevens, and T. C. Südhof. Synaptotagmin I: a major Ca²⁺ sensor for transmitter release at a central synapse. *Cell* 79: 717–727, 1994.

55. Glomset, J. A., and C. C. Farnsworth. Role of protein modification reactions in programming interactions between Ras-related GTPases and cell membranes. *Annu. Rev. Cell Biol.* 10: 181–205, 1994.

56. Gorelick, F. S., and J. D. Jamieson. The pancreatic acinar cell: structure-function relationships. In: *Physiology of the Gastrointestinal Tract,* edited by L. R. Johnson, D. H. Alpers, E. D. Jacobson, J. Christensen, and J. H. Walsh. New York: Raven Press, 1994 p. 1353–1376.

56a. Hackam, D. J., O. D. Rotstein, M. K. Bennett, A. Klip, S. Grinstein, and M. F. Manolson. Characterization and subcellular localization of target membrane soluble NSF attachment protein receptors (t-SNAREs) in macrophages—Syntaxins 2, 3 and 4 are present on phagosomal membranes. *J. Immunol.* 156: 4377–4383, 1996.

57. Hayashi, T., H. McMahon, S. Yamasaki, T. Binz, Y. Hata, T. C. Südhof, and H. Niemann. Synaptic vesicle membrane fusion complex: action of clostridial neurotoxins on assembly. *EMBO J.* 13: 5051–5061, 1994.

58. Herzog, V., and M. G. Farquhar. Luminal membrane retrieved after exocytosis reaches most Golgi cisternae in secretory cells. *Proc. Natl. Acad. Sci. U.S.A.* 74: 5073–5077, 1977.

59. Herzog, V., and H. Reggio. Pathways of endocytosis from luminal plasma in rat exocrine pancreas. *Eur. J. Cell Biol.* 21: 141–150, 1980.

60. Heuser, J. E., and T. S. Reese. Evidence for recycling of synaptic vesicle membrane during transmitter release at the frog neuromuscular junction. *J. Cell Biol.* 57: 315–344, 1973.

61. Huang, X. F., and P. Arvan. Intracellular transport of proinsulin in pancreatic β-cells—structural maturation probed by disulfide accessibility. *J. Biol. Chem.* 270: 20417–20423, 1995.

62. Hubbard, A. L., B. Stieger, and J. R. Bartles. Biogenesis of endogenous plasma membrane proteins in epithelial cells. *Annu. Rev. Physiol.* 51: 755–770, 1989.

63. Hurtley, S. M. Clathrin- and non-clathrin-coated vesicle adaptors. *Trends Biochem. Sci.* 16: 165–166, 1991.

64. Ikonen, E., M. Tagaya, O. Ullrich, C. Montecucco, and K. Simons. Different requirements for NSF, SNAP, and Rab proteins in apical and basolateral transport in MDCK cells. *Cell* 81: 571–580, 1995.

65. Ingber, D. E., J. A. Madri, and J. D. Jamieson. Neoplastic disorganization of pancreatic epithelial cell-cell relations. Role

of basement membrane. *Am. J. Pathol.* 121: 248–260, 1985.

66. Ingber, D. E., J. A. Madri, and J. D. Jamieson. Basement membrane as a spatial organizer of polarized epithelia exogenous basement membrane reorients pancreatic epithelial tumor cells in vitro. *Am. J. Pathol.* 122: 129–139, 1986.

67. Iwanij, V., and J. D. Jamieson. Comparison of secretory protein profiles in developing rat pancreatic rudiments and rat acinar tumor cells. *J. Cell Biol.* 95: 742–746, 1982.

68. Iwanij, V., and J. D. Jamieson. Biochemical analysis of secretory proteins synthesized by normal rat pancreas and by pancreatic acinar tumor cells. *J. Cell Biol.* 95: 734–741, 1982.

69. Jahn, R., and T. C. Südhof. Synaptic vesicles and exocytosis. *Annu. Rev. Neurosci.* 17: 219–246, 1994.

70. Jamieson, J. D., F. S. Gorelick, and A. Chang. Development of secretagogue responsiveness in the pancreas. *Scand. J. Gastroenterol.* 23: 98–103, 1988.

71. Jamieson, J. D., and G. E. Palade. Synthesis, intracellular transport, and discharge of secretory proteins in stimulated pancreatic exocrine cells. *J. Cell Biol.* 50: 135–158, 1971.

72. Jena, B. P., F. D. Gumkowski, E. M. Konieczko, G. Fischer von Mollard, R. Jahn, and J. D. Jamieson. Redistribution of a rab-3 like GTP binding protein from secretory granules to the Golgi complex in pancreatic acinar cells during regulated exocytosis. *J. Cell Biol.* 124: 43–53, 1994.

73. Johannes, L., P.-M. Lledo, M. Roa, J.-D. Vincent, J.-P. Henry, and F. Darchen. The GTPase Rab3a negatively controls calcium-dependent exocytosis in neuroendocrine cells. *EMBO J.* 13: 2029–2037, 1994.

74. Johnstone, S. A., I. Hubaishy, and D. M. Waisman. Regulation of annexin I-dependent aggregation of phospholipid vesicles by protein kinase C. *Biochem. J.* 294: 801–807, 1993.

75. Jones, P. G., S. Fitzpatrick, and D. M. Waisman. Chromaffin granules release calcium on contact with annexin VI: implications for exocytosis. *Biochemistry* 33: 8180–8187, 1994.

76. Karniguian, A., A. Zahraoui, and A. Tavitian. Identification of small GTP-binding rab proteins in human platelets: thrombin-induced phosphorylation of rab3B, rab6, and rab8 proteins. *Proc. Natl. Acad. Sci. U.S.A.* 90: 7647–7651, 1993.

77. Kelly, R. B. Endocytosis: ringing necks with dynamin. *Nature* 374: 116–117, 1995.

78. Kishida, S., H. Shirataki, T. Sasaki, M. Kato, K. Kaibuchi, and Y. Takai. *Rab3*A GTPase-activating protein-inhibiting activity of Rabphilin-3A, a putative *Rab3*A target protein. *J. Biol. Chem.* 268: 22259–22261, 1993.

79. Kornfeld, S., and I. Mellman. The biogenesis of lysosomes. *Annu. Rev. Cell Biol.* 5: 483–525, 1989.

80. Kuliawat, R., and P. Arvan. Protein targeting via the "constitutive-like" secretory pathway in isolated pancreatic islets: Passive sorting in the immature granule compartment. *J. Cell Biol.* 118: 521–529, 1992.

81. Kuliawat, R., and P. Arvan. Distinct molecular mechanisms for protein sorting within immature secretory granules of pancreatic β-cells. *J. Cell Biol.* 126: 77–86, 1994.

82. Kurzchalia, T. V., P. Dupree, and S. Monier. VIP21-caveolin, a protein of the *trans*-Golgi network and caveolae. *FEBS Lett.* 346: 88–91, 1994.

83. Lafont, F., J. K. Burkhardt, and K. Simons. Involvement of microtubule motors in basolateral and apical transport in kidney cells. *Nature* 372: 801–803, 1994.

84. Lapetina, E. G., and B. R. Reep. Specific binding of [α-^{32}P]GTP to cytosolic and membrane-bound proteins of human platelets correlates with the activation of phospholipase C. *Proc. Natl. Acad. Sci. U.S.A.* 84: 2261–2265, 1987.

85. Laurie, S. M., C. C. Cain, G. E. Lienhard, and J. D. Castle. The glucose transporter GluT4 and secretory carrier membrane proteins (SCAMPs) colocalize in rat adipocytes and partially segregate during insulin stimulation. *J. Biol. Chem.* 268: 19110–19117, 1993.

86. Li, C., K. Takei, M. Geppert, L. Daniell, K. Stenius, E. R. Chapman, R. Jahn, P. De Camilli, and T. C. Südhof. Synaptic targeting of rabphilin-3A, a synaptic vesicle Ca^{2+}/phospholipid-binding protein, depends on rab3A/3C. *Neuron* 13: 885–898, 1994.

87. Li, C., B. Ullrich, J. Z. Zhang, R.G.W. Anderson, N. Brose, and T. C. Südhof. Ca^{2+}-dependent and -independent activities of neural and non-neural synaptotagmins. *Nature* 375: 594–599, 1995.

88. Link, E., L. Edelmann, J. H. Chou, T. Binz, S. Yamasaki, U. Eisel, M. Baumert, T. C. Südhof, H. Niemann, and R. Jahn. Tetanus toxin action: inhibition of neurotransmitter release linked to synaptobrevin proteolysis. *Biochem. Biophys. Res. Commun.* 189: 1017–1023, 1992.

89. Link, E., H. McMahon, G. Fischer von Mollard, S. Yamasaki, H. Neimann, T. C. Südhof, and R. Jahn. Cleavage of cellubrevin by tetanus toxin does not affect fusion of early endosomes. *J. Biol. Chem.* 268: 18423–18426, 1993.

90. Lisanti, M. P., P. E. Scherer, J. Vidugiriene, Z. Tang, A. Hermanowksi-Vosatka, Y.-H. Tu, R. F. Cook, and M. Sargiacomo. Characterization of caveolin-rich membrane domains isolated from an endothelial-rich source: implications for human disease. *J. Cell Biol.* 126: 111–126, 1994.

91. Littleton, J. T., M. Stern, K. Schulze, M. Perin, and H. J. Bellen. Mutational analysis of Drosophila *synaptotagmin* demonstrates its essential role in Ca^{2+}-activated neurotransmitter release. *Cell* 74: 1125–1134, 1993.

92. Lledo, P.-M., P. Vernier, J.-D. Vincent, W. T. Mason, and R. Zorec. Inhibition of Rab3B expression attenuates Ca^{2+}-dependent exocytosis in rat anterior pituitary cells. *Nature* 364: 540–544, 1993.

93. Lombardi, D., T. Soldati, M. A. Riederer, Y. Goda, M. Zerial, and S. R. Pfeffer. Rab9 functions in transport between late endosomes and the trans Golgi network. *EMBO J.* 12: 677–682, 1993.

94. MacLean, C. M., G. J. Law, and J. M. Edwardson. Stimulation of exocytotic membrane fusion by modified peptides of the rab3 effector domain: re-evaluation of the role of rab3 in regulated exocytosis. *Biochem. J.* 294: 325–328, 1993.

95. Maycox, P. R., E. Link, A. Reetz, S. A. Morris, and R. Jahn. Clathrin-coated vesicles in nervous tissue are involved primarily in synaptic vesicle recycling. *J. Cell Biol.* 118: 1379–1388, 1992.

96. McMahon, H. T., M. Missler, C. Li, and T. C. Südhof. Complexins: cytosolic proteins that regulate SNA Preceptor function. *Cell* 83: 111–119, 1995.

97. McMahon, H. T., and T. C. Südhof. Synaptic core complex of synaptobrevin, syntaxin, and SNAP25 forms high affinity α-SNAP binding site. *J. Biol. Chem.* 270: 2213–2217, 1995.

98. McMahon, H. T., Y. A. Ushkaryov, L. Edelmann, E. Link, T. Binz, H. Niemann, R. Jahn, and T. C. Südhof. Cellubrevin is a ubiquitous tetanus-toxin substrate homologous to a putative synaptic vesicle fusion protein. *Nature* 364: 346–349, 1993.

99. Meldolesi, J., J. D. Jamieson, and G. E. Palade. Composition of cellular membranes in the pancreas of the guinea pig 3. Enzymatic activities. *J. Cell Biol.* 49: 150–158, 1971.

100. Meldolesi, J., J. D. Jamieson, and G. E. Palade. Composition

of cellular membranes in the pancreas of the guinea pig II. Lipids. *J. Cell Biol.* 49: 130–149, 1971.

101. Mellman, I. Enigma variations: protein mediators of membrane fusion. *Cell* 82: 869–872, 1995.

102. Mellman, I., E. Yamamoto, J. A. Whitney, M. Kim, W. Hunziker, and K. Matter. Molecular sorting in polarized and nonpolarized cells: common problems, common solutions. *J. Cell Sci.* Suppl. 106(17): 1–7, 1993.

103. Muallem, S., K. Kwiatkowska, X. Xu, and H. L. Yin. Actin filament disassembly is a sufficient final trigger for exocytosis in nonexcitable cells. *J. Cell Biol.* 128: 589–598, 1995.

104. Nakata, T., K. Sobue, and N. Hirokawa. Conformational change and localization of calpactin I complex involved in exocytosis as revealed by quick-freeze, deep-etch electron microscopy and immunocytochemistry. *J. Cell Biol.* 110: 13–25, 1990.

105. Nathanson, M. H., P. J. Padfield, A. J. O'Sullivan, A. D. Burgstahler, and J. D. Jamieson. Mechanism of Ca^{2+} wave propagation in pancreatic acinar cells. *J. Biol. Chem.* 267: 18118–18121, 1992.

106. Novick, P., and P. Brennwald. Friends and family: the role of the Rab GTPases in vesicular traffic. *Cell* 75: 597–601, 1993.

107. O'Konski, M. S., and S. J. Pandol. Effects of caerulein on the apical cytoskeleton of the pancreatic acinar cell. *J. Clin. Invest.* 86: 1649–1657, 1990.

108. Oberhauser, A. F., J. R. Monck, W. E. Balch, and J. M. Fernandez. Exocytotic fusion is activated by Rab3a peptides. *Nature* 360: 270–273, 1992.

109. Oliver, C., and A. R. Hand. Membrane retrieval in exocrine acinar cells. *Methods Cell Biol.* 23: 429–444, 1981.

110. Olszewski, S., J. T. Deeney, G. T. Schuppin, K. P. Williams, B. E. Corkey, and C. J. Rhodes. Rab3A effector domain peptides induce insulin exocytosis via a specific interaction with a cytosolic protein doublet. *J. Biol. Chem.* 269: 27987–27991, 1994.

110a. Orita, S., K. Kaibuchi, S. Kuroda, K. Shimizu, H. Nakanishi, and Y. Takai. Comparison of kinetic properties between two mammalian *ras* p21 GDP/GTP exchange proteins, *ras* guanine nucleotide-releasing factor and *smg* GDP dissociation stimulator. *J. Biol. Chem.* 268: 25542–25546, 1993.

110b. Orita, S., T. Sasaki, R. Komuro, G. Sakaguchi, M. Maeda, H. Igarashi, and Y. Takai. Doc2 enhances Ca^{2+}-dependent exocytosis from PC12 cells. *J. Biol. Chem.* 271: 7257–7260, 1944

111. Osen-Sand, A., M. Catsicas, J. K. Staple, K. A. Jones, G. Ayala, J. Knowles, G. Grenningloh, and S. Catsicas. Inhibition of axonal growth by SNAP-25 antisense oligonucleotides *in vitro* and *in vivo*. *Nature* 364: 445–448, 1993.

112. Oyler, G. A., G. A. Higgins, R. A. Hart, E. Battenberg, M. Billingsley, F. E. Bloom, and M. C. Wilson. The identification of a novel synaptosomal-associated protein, SNAP-25, differently expressed by neuronal subpopulations. *J. Cell Biol.* 109: 3039–3052, 1989.

113. Padfield, P. J., W. E. Balch, and J. D. Jamieson. A synthetic peptide of the rab3a effector domain stimulates amylase release from permeabilized pancreatic acini. *Proc. Natl. Acad. Sci. U.S.A.* 89: 1656–1660, 1992.

114. Palade, G. E. Intracellular aspects of the process of protein synthesis. *Science* 189: 347–356, 1975.

115. Parton, R. G., B. Joggerst, and K. Simons. Regulated internalization of caveolae. *J. Cell Biol.* 127: 1199–1215, 1994.

116. Pearse, B.M.F., and M. S. Robinson. Clathrin, adaptors, and sorting. *Annu. Rev. Cell Biol.* 6: 151–171, 1990.

117. Perez, F., P.-M. Lledo, D. Karagogeos, J.-D. Vincent, A. Prochi-

antz, and J. Ayala. Rab3A and Rab3B carboxy-terminal peptides are both potent and specific inhibitors of prolactin release by rat cultured anterior pituitary cells. *Mol. Endocrinol.* 8: 1278–1287, 1994.

118. Perin, M. S., V. A. Fried, G. A. Mignery, R. Jahn, and T. C. Südhof. Phospholipid binding by a synaptic vesicle protein homologous to the regulatory region of protein kinase C. *Nature* 345: 260–263, 1990.

119. Petrenko, A. G., M. S. Perin, B. A. Davletov, Y. A. Ushkaryov, M. Geppert, and T. C. Südhof. Binding of synaptotagmin to the α-latrotoxin receptor implicates both in synaptic vesicle exocytosis. *Nature* 353: 65–68, 1991.

120. Pevsner, J., S.-C. Hsu, and R. H. Scheller. n-Sec1: a neural-specific syntaxin-binding protein. *Proc. Natl. Acad. Sci. U.S.A.* 91: 1445–1449, 1994.

121. Pfeffer, S. R., A. B. Dirac-Svejstrup, and T. Soldati. Rab GDP dissociation inhibitor: putting Rab GTPases in the right place. *J. Biol. Chem.* 270: 17057–17059, 1995.

122. Pfeffer, S. R., and J. E. Rothman. Biosynthetic protein transport and sorting by the endoplasmic reticulum and Golgi. *Annu. Rev. Biochem.* 56: 829–852, 1987.

123. Piiper, A., D. Stryjek-Kaminska, J. Stein, W. F. Caspary, and S. Zeuzem. A synthetic peptide of the effector domain of *rab3A* stimulates inositol 1,4,5-triphosphate production in digitonin-permeabilized pancreatic acini. *Biochem. Biophys. Res. Commun.* 192: 1030–1036, 1993.

124. Pollard, H. B., E. Rojas, R. W. Pastor, E. M. Rojas, H. R. Guy, and A. L. Burns. Synexin: molecular mechanism of calcium-dependent membrane fusion and voltage-dependent calcium-channel activity: evidence in support of the "hydrophobic bridge hypothesis" for exocytotic membrane fusion. *Ann. NY Acad. Sci.* 635: 328–351, 1991.

125. Popov, S. V., and M. Poo. Synaptotagmin: a calcium-sensitive inhibitor of exocytosis? *Cell* 73: 1247–1249, 1993.

126. Raynal, P., and H. B. Pollard. Annexins: the problem of assessing the biological role for a gene family of multifunctional calcium- and phospholipid-binding proteins. *Biochim. Biophys. Acta Rev. Biomembr.* 1197: 63–93, 1994.

127. Richmond, J., and P. G. Haydon. Rab effector domain peptides stimulate the release of neurotransmitter from cell cultured synapses. *FEBS Lett.* 326: 124–130, 1993.

128. Rothman, J. E. Mechanisms of intracellular protein transport. *Nature* 372: 55–63, 1994.

129. Rothman, J. E., and L. Orci. Movement of proteins through the Golgi stack: a molecular dissection of vesicular transport. *FASEB J.* 4: 1460–1468, 1990.

130. Rothman, J. E., and G. Warren. Implications of the SNARE hypothesis for intracellular membrane topology and dynamics. *Curr. Biol.* 4: 220–233, 1994.

131. Sargiacomo, M., P. E. Scherer, Z. L. Tang, E. Kübler, K. S. Song, M. C. Sanders, and M. P. Lisanti. Oligomeric structure of caveolin: implications for caveolae membrane organization. *Proc. Natl. Acad. Sci. U.S.A.* 92: 9407–9411, 1995.

132. Sasaki, T., M. Kato, T. Nishiyama, and Y. Takai. The nucleotide exchange rates of *rho* and *rac* small GTP-binding proteins are enhanced to different extents by their regulatory protein *smg* GDS. *Biochem. Biophys. Res. Commun.* 194: 1188–1193, 1993.

133. Scherer, P. E., M. P. Lisanti, G. Baldini, M. Sargiacomo, C. C. Mastick, and H. F. Lodish. Induction of caveolin during adipogenesis and association of GLUT4 with caveolin-rich vesicles. *J. Cell Biol.* 127: 1233–1243, 1994.

134. Schiavo, G., F. Benfenati, B. Poulain, O. Rossetto, P. Polverino de Laureto, B. R. Dasgupta, and C. Montecucco. Tetanus and

botulinum-B neurotoxins block neurotransmitter release by proteolytic cleavage of synaptobrevin. *Nature* 359: 832–835, 1992.

135. Schmid, S. L., and H. Damke. Coated vesicles: a diversity of form and function. *FASEB J.* 9: 1445–1453, 1995.

136. Schnitzer, J. E., J. Liu, and P. Oh. Endothelial caveolae have the molecular transport machinery for vesicle budding, docking, and fusion including VAMP, NSF, SNAP, annexins, and GTPases. *J. Biol. Chem.* 270: 14399–14404, 1995.

137. Schnitzer, J. E., P. Oh, B. S. Jacobson, and A. M. Dvorak. Caveolae from luminal plasmalemma of rat lung endothelium: microdomains enriched in caveolin, Ca^{2+}-ATPase, and inositol triphosphate receptor. *Proc. Natl. Acad. Sci. U.S.A.* 92: 1759–1763, 1995.

138. Segawa, A., S. Terakawa, S. Yamashina, and C. R. Hopkins. Exocytosis in living salivary glands: direct visualization by video-enhanced microscopy and confocal laser microscopy. *Eur. J. Cell Biol.* 54: 322–330, 1991.

139. Senda, T., T. Okabe, M. Matsuda, and H. Fujita. Quick-freeze, deep-etch visualization of exocytosis in anterior pituitary secretory cells: localization and possible roles of actin and annexin II. *Cell Tissue Res.* 277: 51–60, 1994.

139a. Sengupta, D., F. D. Gumkowski, L. H. Tang, T. J. Chilcote, and J. D. Jamieson. Localization of cellubrevin to the Golgi complex in pancreatic acinar cells. *Eur. J. Cell Biol.* 70: 306–316, 1996.

140a. Shirataki, H., K. Kaibuchi, T. Sakoda, S. Kishida, T. Yamaguchi, K. Wada, M. Miyazaki, and Y. Takai. Rabphilin-3A, a putative target protein for *smg* p25A/*rab*3A p25 small GTP-binding protein related to synaptotagmin. *Mol. Cell. Biol.* 13: 2061–2068, 1993.

140b. Shirataki, H., K. Kaibuchi, T. Yamaguchi, K. Wada, H. Horiuchi, and Y. Takai. A possible target protein for *smg*-25A/*rab*3A small GTP-binding protein. *J. Biol. Chem.* 267: 10946–1992.

141. Smart, E. J., Y. S. Ying, and R.G.W. Anderson. Hormonal regulation of caveolae internalization. *J. Cell Biol.* 131: 929–938, 1995.

142. Sogaard, M., K. Tani, R. R. Ye, S. Geromanos, P. Tempst, T. Kirchhausen, J. E. Rothman, and T. Söllner. A rab protein is required for the assembly of SNARE complexes in the docking of transport vesicles. *Cell* 78: 937–948, 1994.

143. Soldati, T., C. Rancaño, H. Geissler, and S. R. Pfeffer. Rab7 and Rab9 are recruited onto late endosomes by biochemically distinguishable processes. *J. Biol. Chem.* 270: 25541–25548, 1995.

144. Stahl, B., G. Fischer von Mollard, C. Walch-Solimena, and R. Jahn. GTP-cleavage by the small GTP-binding protein Rab3A is associated with exocytosis of synaptic vesicles induced by a-latrotoxin. *J. Biol. Chem.* 269: 24770–24776, 1994.

144a. Stahl, B., J. H. Chou, C. Li, T. C. Südhof, and R. Jahn. Rab3 reversibly recruits rabphilin to synaptic vesicles by a mechanism anlogous to raf recruitment by ras. *EMBO J.* 15: 1799–1996.

145. Stecher, B., G. Ahnert-Hilger, U. Weller, T. P. Kemmer, and M. Gratzl. Amylase release from streptolysin O-permeabilized pancreatic acinar cells. Effects of Ca^{2+}, guanosine 5′-[gamma-thio]triphosphate, cyclic AMP, tetanus toxin and botulinum A toxin. *Biochem. J.* 283: 899–904, 1992.

146. Steele-Mortimer, O., M. J. Clague, L. A. Huber, P. Chavrier, J. Gruenberg, and J.-P. Gorvel. The N-terminal domain of a rab protein is involved in membrane-membrane recognition and/or fusion. *EMBO J.* 13: 34–41, 1994.

147. Stenmark, H., A. Valencia, O. Martinez, O. Ullrich, B. Goud,

and M. Zerial. Distinct structural elements of rab5 define its functional specificity. *EMBO J.* 13: 575–583, 1994.

148. Stenmark, H., G. Vitale, O. Ullrich, and M. Zerial. Rabaptin-5 is a direct effector of the small GTPase Rab5 in endocytic membrane fusion. *Cell* 83: 423–432, 1995.

149. Sutton, R. B., B. A. Davletov, A. M. Berghuis, T. C. Südhof, and S. R. Sprang. Structure of the first C_2 domain of synaptotagmin I: a novel Ca^{2+}/phospholipid-binding fold. *Cell* 80: 929–938, 1995.

150. Söllner, T. SNAREs and targeted membrane fusion. *FEBS Lett.* 369: 80–83, 1995.

151. Söllner, T., M. K. Bennett, S. W. Whiteheart, R. H. Scheller, and J. E. Rothman. A protein assembly-disassembly pathway in vitro that may correspond to sequential steps of synaptic vesicle docking, activation, and fusion. *Cell* 75: 409–418, 1993.

152. Söllner, T., S. W. Whiteheart, M. Brunner, H. Erdjument-Bromage, S. Geromanos, P. Tempst, and J. E. Rothman. SNAP receptors implicated in vesicle targeting and fusion. *Nature* 362: 318–324, 1993.

153. Südhof, T. C., M. Baumert, M. S. Perin, and R. Jahn. A synaptic vesicle membrane protein is conserved from mammals to *Drosophila*. *Neuron* 2: 1475–1481, 1989.

154. Südhof, T. C., P. De Camilli, H. Niemann, and R. Jahn. Membrane fusion machinery: insights from synaptic proteins. *Cell* 75: 1–4, 1993.

155. Südhof, T. C., and R. Jahn. Proteins of synaptic vesicles involved in exocytosis and membrane recycling. *Neuron* 6: 665–677, 1991.

155a. Sumitani, S., T. Ramlal, Z. Liu, and A. Klip. Expression of syntaxin 4 in rat skeletal muscle and rat skeletal muscle cells in culture. *Biochem. Biophys. Res. Commun.* 213: 462–468, 1995

156. Takei, K., P. S. McPherson, S. L. Schmid, and P. De Camilli. Tubular membrane invaginations coated by dynamin rings are induced by GTP-gammaS in nerve terminals. *Nature* 374: 186–190, 1995

156a. Tang, L. H., F. D. Gumkowski, D. Sengupta, I. M. Modlin, and J. D. Jamieson. Rab3D protein is a specific marker for zymogen granules in gastric chief cells of rats and rabbits. *Gastroenterology* 110: 809–820, 1996

157. Toescu, E. C., and O. H. Petersen. Region-specific activity of the plasma membrane Ca^{2+} pump and delayed activation of Ca^{2+} entry characterize the polarized, agonist-evoked Ca^{2+} signals in exocrine cells. *J. Biol. Chem.* 270: 8528–8535, 1995.

158. Tooze, J., H. F. Kern, S. D. Fuller, and K. E. Howell. Condensation-sorting events in the rough endoplasmic reticulum of exocrine pancreatic cells. *J. Cell Biol.* 109: 35–50, 1989.

159. Trimble, W. S., D. M. Cowan, and R. H. Scheller, VAMP-1: a synaptic vesicle-associated integral membrane protein. *Proc. Natl. Acad. Sci. U.S.A.* 85: 4538–4542, 1988.

160. Ullrich, B., C. Li, J. Z. Zhang, H. McMahon, R.G.W. Anderson, M. Geppert, and T. C. Südhof. Functional properties of multiple synaptotagmins in brain. *Neuron* 13: 1281–1291, 1994.

161. Ullrich, B., and T. C. Südhof. Differential distributions of novel synaptotagmins: comparison to synapsins. *Neuropharmacology* 34: 1371–1377, 1995.

162. Valencia, A., P. Chardin, A. Wittinghofer, and C. Sander. The *ras* protein family: evolutionary tree and role of conserved amino acids. *Biochemistry* 30: 4637–4648, 1991.

163. Valentijn, J. A., D. Sengupta, F. D. Gumkowski, L. H. Tang, E. M. Konieczko, and J. D. Jamieson. Rab3D localizes to

secretary granules in rat pancreatic acinar cells. *Eur. J. Cell Biol.* 70: 33–41, 1996.

163a. Valentijn, J. A., F. D. Gumkowski, and J. D. Jamieson. The expression pattern of rab3D in the developing rat exocrine pancreas coincides with the acquisition of regulated exocytosis. *Eur. J. Cell Biol.* 71: 129–136, 1996.

163b. Valentijn, J. A., D. Q. La Civita, F. D. Gumkowski, and J. D. Jamieson. Rab4 associates with the actin terminal web in developing rat pancreatic acinar cells. *J. Cell Biol.* (In press) 1996.

164. Van der Sluijs, P., M. Hull, L. A. Huber, P. Mâle, B. Goud, and I. Mellman. Reversible phosphorylation-dephosphorylation determines the localization of rab4 during the cell cycle. *EMBO J.* 11: 4379–4389, 1992.

165. Van der Sluijs, P., M. Hull, P. Webster, P. Mâle, B. Goud, and I. Mellman. The small GTP-binding protein rab4 controls an early sorting event on the endocytic pathway. *Cell* 70: 729–740, 1992.

166. Van der Sluijs, P., M. Hull, A. Zahraoui, A. Tavitian, B. Goud, and I. Mellman. The small GTP-binding protein rab4 is associated with early endosomes. *Proc. Natl. Acad. Sci. U.S.A.* 88: 6313–6317, 1991.

167. Vitale, M. L., E. P. Seward, and J.-M. Trifaró. Chromaffin cell cortical actin network dynamics control the size of the release-ready vesicle pool and the initial rate of exocytosis. *Neuron* 14: 353–363, 1995.

167a. Volchuk, A., Y. Mitsumoto, L. He, Z. Liu, E. Habermann, W. Trimble, and A. Klip. Expression of vesicle-associated membrane protein 2 (VAMP-2)/synaptobrevin II and cellubrevin in rat skeletal muscle and in a muscle cell line. *Biochem. J.* 304: 139–145, 1996.

167b. Volchuk, A., R. Sargeant, S. Sumitani, Z. Liu, L. He, and A. Klip. Cellubrevin is a resident protein of insulin-sensitive GLUT4 glucose transporter vesicles in 3T3-L1 adipocytes. *J. Biol. Chem.* 270: 8233–8240, 1995.

168. Weidman, P. J., P. Melançon, M. R. Block, and J. E. Rothman. Binding of an *N*-ethylmaleimide-sensitive fusion protein to Golgi membranes requires both a soluble protein(s) and an integral membrane receptor. *J. Cell Biol.* 108: 1589–1596, 1989.

169. Wilson, D. W., S. W. Whiteheart, L. Orci, and J. E. Rothman. Intracellular membrane fusion. *Trends Biochem. Sci.* 16: 334–337, 1991.

170. Wilson, D. W., C. A. Wilcox, G. C. Flynn, E. Chen, W. J. Kuang, W. J. Henzel, M. R. Block, A. Ullrich, and J. E. Rothman. A fusion protein required for vesicle-mediated transport in both mammalian cells and yeast. *Nature* 339: 355–359, 1989.

171. Yamasaki, S., A. Baumeister, T. Binz, J. Blasi, E. Link, F. Cornille, B. Roques, E. M. Fykse, T. C. Südhof, R. Jahn, and H. Niemann. Cleavage of members of the synaptobrevin/VAMP family by types D and F Botulinal neurotoxins and Tetanus toxin. *J. Biol. Chem.* 269: 1–9, 1994.

172. Zahraoui, A., G. Joberty, M. Arpin, J. J. Fontaine, R. Hellio, A. Tavitian, and D. Louvard. A small rab GTPase is distributed in cytoplasmic vesicles in non polarized cells but colocalizes with the tight junction marker ZO-1 in polarized epithelial cells. *J. Cell Biol.* 124: 101–115, 1994.

173. Zeuzem, S., D. Stryjek-Kaminska, W. F. Caspary, J. Stein, and A. Piiper. Effect of a Rab3A effector domain-related peptide, CCK, and EGF in permeabilized pancreatic acini. *Am. J. Physiol.* 267 (*Gastrointest. Liver Physiol.* 30): G350–G356, 1994.

174. Zhang, J. Z., B. A. Davletov, T. C. Südhof, and R.G.W. Anderson. Synaptotagmin I is a high affinity receptor for clathrin AP-2: implications for membrane recycling. *Cell* 78: 751–760, 1994.

17. Epithelial cell polarity: challenges and methodologies

MICHAEL J. CAPLAN | Department of Physiology, Yale University School of Medicine, New Haven, Connecticut

ENRIQUE RODRIGUEZ-BOULAN | Department of Cell Biology and Anatomy, Cornell University Medical College, New York, New York

CHAPTER CONTENTS

POLARIZED EPITHELIAL CELLS form the interface between an organism and its environment. In this capacity, they stand as the sole physical barrier ensuring the separation of the intra- and extracorporeal fluid spaces. The plasma membranes of epithelial cells, in concert with the junctions which interconnect them, prevent the flow of material between these compartments and thus allow the organism to maintain an internal composition which is quite distinct from that of its surroundings. In addition to safeguarding an organism's integrity through their barrier function, epithelial cells also play a dynamic role in determining the organism's chemical makeup. This function is accomplished through the participation of polarized epithelia in the tightly regulated vectorial absorption of nutrients and secretion of waste products.

In order to serve in these roles, the plasma membranes of epithelial cells are divided into two domains whose distinct biochemical compositions reflect their individual roles. The basolateral plasmalemmal domain rests upon the epithelial basement membrane and is bathed in extracellular fluid. In contrast, the apical domain generally faces and forms the limiting surface of a lumen or duct which is topologically continuous with the outer surface of the organism. The boundary between the apical and basolateral portions of the epithelial plasma membrane is defined by the presence of tight or occluding junctions. Different populations of membrane proteins occupy each plasmalemmal domain and endow them with their individual physiologic properties (7, 13c, 124).

The biological utility of epithelial polarity is illustrated by the mechanisms involved in the vectorial transepithelial transport of a solute. Epithelial cells of the mammalian small intestine and renal proximal tubule are capable of transporting glucose from their lumina to the extracellular fluid compartment against steep concentration gradients (135a). This flux is accomplished through the simultaneous participation of two membrane transport systems working in series at opposite poles of the epithelial cells (Fig. 17.1). The Na,K-ATPase which occupies the basolateral plasma membranes of these cells catalyzes the ATP-driven efflux of three sodium ions in exchange for the influx of two potassium ions, resulting in a low intracellular sodium concentration. An apical population of Na-glucose co-transporters couples the movement of sodium across the apical membrane to that of glucose. The sodium gradient created by the Na,K-ATPase favors the entry of sodium across the apical membrane and its energy can be exploited to drive the accumulation of glucose against its concentration gradient. Glucose exits the cell through a passive carrier system present at the basolateral surface. In order for this elegant scheme to succeed in driving the "uphill" transepithelial transport of glucose from the lumen to the extracellular fluid compartment, the transport proteins

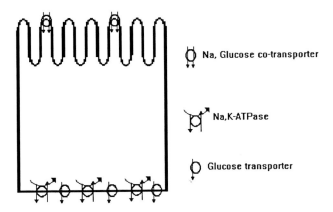

FIG. 17.1. Vectorial transport of fluid and solutes is dependent on polarized distribution of membrane proteins in apical and basolateral plasma membrane domains of epithelial cells. Resorption of glucose by epithelial cells of small intestine and renal proximal tubule involves participation of apical Na,glucose cotransporter, basolateral glucose carrier, and basolateral Na,K-ATPase. Low intracellular Na concentration maintained by the Na,K-ATPase allows apical Na,glucose transporter to mediate uphill glucose transport. Glucose flows down concentration gradient and out of cell through basolateral glucose carrier. Scheme functions productively only if asymmetric distributions of transport proteins are maintained.

involved must be distributed in a polarized fashion. The capacity to mediate vectorial transport, which is one of the most striking physiologic characteristics manifested by epithelia, is in turn entirely dependent upon the cell biologic property of plasma membrane polarity.

The generation and maintenence of this polarity requires that epithelial cells possess a number of specialized skills. In the case of epithelia in a developing organism or in tissue culture, these cells must be able to create domains de novo by using environmental cues to establish a polarized orientation (137, 157). Organization of these domains' individual biochemical identities necessitates the segregation of cell-surface proteins to the appropriate regions as well as stabilizing mechanisms to prevent their redistribution by lateral diffusion (Fig. 17.2). Furthermore, the cells must be able to continually replenish these domains with the appropriate populations of newly synthesized membrane proteins in order to preserve polarity in the face of ongoing membrane turnover.

Since essentially all plasma membrane proteins are synthesized in association with the membranes of the rough endoplasmic reticulum and traverse the same biosynthetic processing pathway en route to the cell surface, mechanisms must exist to sort newly synthesized membrane proteins from one another and to ensure that they become concentrated at the appropriate sites (13c, 124). The cell's capacity to perform

this sorting operation implies the existence of signals embedded within the structure of the plasmalemmal proteins themselves which specify their ultimate localization as well as a sorting machinery which can interpret these signals and act upon them.

This review will focus on what has been discovered about the devices which polarized epithelial cells have devised to solve these multifaceted and interdependent sorting problems. As will be emphasized throughout this review, very little is known about the signals and machinery which polarized epithelials cells rely upon to generate their striking anisotropy. Numerous themes have emerged, however, as well as a variety of fascinating experimental models that illustrate the multiplicity of adaptations which epithelial cells have perfected to meet their architectural needs. We will, therefore, emphasize the open questions in this field as well as the experimental techniques which are being brought to bear in their analysis.

QUESTIONS

Although physiologists and cell biologists have long been aware of the phenomenon of epithelial polarity, investigations into its underlying mechanisms were hampered until two decades ago by a lack of suitable experimental tools. Studies examining the dynamic nature of the sorting process proved difficult to perform on epithelia in situ or acutely isolated from intact animals. The diversity of cell types and complex architectures which endow epithelial tissues with many of their functional and asthetic attributes can serve as rather formidable impediments to those intent on studying subcellular events. Continuous lines of polarized epithelial cells have been propagated in tissue culture for at least 40 yr (51). Their potential utility for research into the biogenesis of polarity, however, required the development of reagents that would permit individual proteins to be followed throughout their postsynthetic processing before it could be fully realized. An observation on the polarity of viral budding reported in 1978 established a biologically interesting and experimentally accessible model system that permitted the steps involved in the generation of polarity to be individually dissected (128).

The genetic material of an enveloped RNA virus is encapsulated in a lipid bilayer membrane which is derived from the plasmalemma or from an organellar membrane of the host cell in which it replicates (140). Embedded within this bilayer are transmembrane spike glycoproteins, which serve as the receptors through which the virus recognizes, attaches to, and fuses with the membranes of host cells. Fusion of the viral envelope with a host cell membrane delivers the viral genome to the cytoplasm and thus sets in motion the program of infection. During the course of their synthesis and processing the viral spike glycoproteins are handled by the host cell as if they were endogenous transmembrane proteins. They are translated on bound polysomes and inserted co-transitionally into the membranes of the rough endoplasmic reticulum (RER). They pursue the standard secretory pathway from the ER, through the Golgi complex to the cell surface, and undergo the stereotypic set of post-translational modifications which accompany this passage (5, 6). When they arrive at their final destination, their cytoplasmic tails participate in interactions with viral nu-

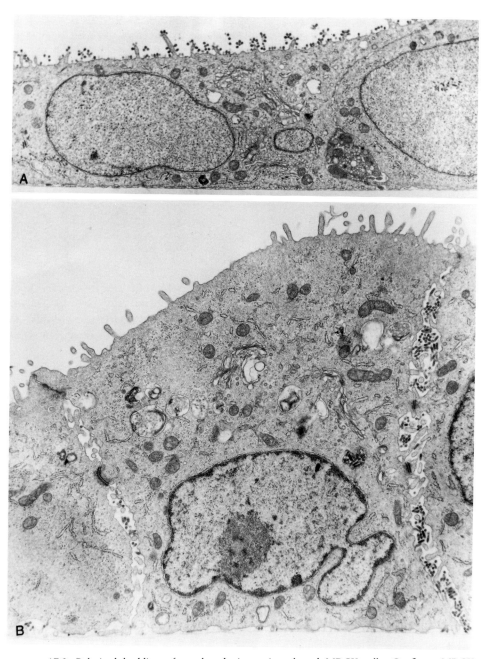

FIG. 17.2. Polarized budding of enveloped viruses in cultured MDCK cells. Confluent MDCK monolayers were infected with either influenza HA *(A)* or Vesicular Stomatitis Virus (VSV) *(B)*. Influenza virus buds exclusively from apical membrane and VSV is assembled at basolateral membrane. Sites of assembly of viruses are determined by prior accumulation of their envelope glycoproteins—namely, influenza hemagglutinin and VSV G protein.

cleocapsid proteins and RNA to generate macromolecular complexes which contain all of the components necessary to generate an infectious viral particle. Assembly of these aggregates leads to evagination of the surrounding host membrane, which subsequently pinches off to create a complete, self-contained virus. In the case of the influenza and vesicular stomatitis viruses, this process of viral budding occurs at the cell surface membrane.

Rodriguez-Boulan and Sabatini (1978) noticed that when cultured canine kidney epithelial cells were infected with the influenza and vesicular stomatitis viruses, the newly produced viral particles were not shed uniformly over the entire surface of the plasmalemma (128). Influenza departed host cells through their apical surfaces, while the bullet-shaped vesicular stomatitis virus (VSV) exited exclusively through the basolateral membrane (Fig. 17.2). Furthermore, it was later found that the spike glycoproteins synthesized by the infected cells obeyed the same rules of asymmetry. The influenza hemagglutinin (HA) protein accumulated at the apical membrane, while the VSV glycoprotein (VSV G) massed at the basolateral surface (126). Since viral assembly and budding is driven by interactions involving the cytoplasmic tails of the spike glycoproteins, these observations suggested that the polarity of virus formation is, in fact, driven by the polarized distribution of the viral spike glycoproteins. Since these spike glycoprotiens are conducted through the host cell's normal secretory pathway, their sorting to the two subdomains of the plasmalemma presumably occurs according to the same algorithms that the host cell applies to the sorting of its own endogenous membrane proteins. Thus, viral budding provided investigators with a readily accessible porthole through which to view the dynamic events associated with membrane protein sorting.

It was subsequently recognized that the viral spike glycoproteins themselves contain all of the information necessary to ensure their own proper targeting. Transfection studies revealed that no other viral components are involved in generating the influenza HA and VSV G proteins' dramatically polarized distributions (129, 141). The mechanisms which underly polarized viral budding are thus the same mechanisms which epithelial cells employ in their day-to-day lives to create and maintain the anisotropic organization of their own plasmalemmal proteins. The sorting of the viral spike glycoproteins must be the product of a dialogue between the proteins themselves and the cellular components involved in their processing. The recognition of this simple truth helped to crystalize the four interrelated and more or less fundamental questions which

have continued to drive polarity research to this day.

Perhaps the two most obvious of these questions relate to the nature of the two partners in this dialogue. First, if viral spike glycoproteins contain all of the information necessary to specify their subcellular destinations, what is the nature of this information? How is it encoded within the structures of these proteins and how does it exert its influence? How general are these sorting signals and how many other aspects of a protein's life cycle do they control? The impressively large body of research which has grown up around these issues has seen a great deal of recent progress and will be discussed in detail later under SORTING SIGNALS AND MECHANISMS. The existence of sorting signals necessarily implies the existence of sorting machinery which is capable of reading these signals and acting upon the instructions they dictate (13c). Presumably, this machinery must interact physically with a protein's sorting signal and, through this interaction, initiate the polypeptide's proper segregation and targeting. What is the nature of the cellular components that recognize sorting signals? Since sorting signals are only just beginning to be characterized, it is not surprising that research into the extremely difficult family of questions relating to the cellular components which interpret them is still fairly embryonic. The directions in which this field of research is moving will also be reviewed later under SORTING SIGNALS AND MECHANISMS and FUTURE PROSPECTS:IN VITRO SYSTEMS, GENETIC MODELS, AND THE SEARCH FOR SORTING MACHINERY.

In its simplest form, the third question of the polarity canon can be reduced to, How do they get there? Newly synthesized apical and basolateral proteins are synthesized in association with the membranes of the same rough endoplasmic reticulum and pass along the same secretory pathway through the stacks of the same Golgi complex (6, 36, 120). Where along this route do they part company? Where within the cell does sorting occur and when do proteins become committed to divergent directions? Are proteins delivered vectorially and exclusively to their sites of ultimate functional residence or are they initially misdelivered and subsequently retrieved and redistributed? Where are the signals read and where does the machinery function? What happens to proteins which do not carry any sorting signal—that is, What is the default pathway? A great deal of research over the past two decades has focused on such questions. This emphasis reflects both the central importance of these issues and their experimental accessibility. The techniques involved in tracing a protein's postsynthetic course through a cell were developed during the great cell biology renaissance of the 1960s. Although the methods needed to be specially adapted to answer the questions posed by polarized epithelia and new tools have had to be invented along the way, the questions relating to sorting pathways were the first to be available to intense scrutiny. The results of these investigations are surprisingly complicated. It appears that different cell types answer these questions with a remarkable number of different solutions. Furthermore, the same cell can apparently answer these questions differently for different proteins. Presumably, the sorting pathways employed by specific epithelial cell types reflect their own individual adaptations to the physiologic functions which they are specialized to perform. A discussion of sorting pathways is presented later under PROTEIN TRAFFICKING PATHWAYS IN EPITHELIAL CELLS.

Finally, a cell's ability to target newly synthesized proteins to distinct regions of its plasmalemma is predicated upon the existence of these stable and defined subdomains within its cell surface membrane, separated by tight junctions (Fig. 17.2). The fourth questions is, What is the nature of the environmental cues that epithelial cells use to generate, organize, and maintain these polarized domains. In developing organisms as well as in tissue culture dishes, epithelial cells display the capacity to orient themselves with respect to their surroundings and thus to create asymmetry. The sensors, signals, and cellular effectors responsible for this ability are the subject of intense research, and their identification will probably go a long way toward answering all four of the questions posed in this section. The section ESTABLISHMENT OF EPITHELIAL POLARITY focuses on the establishment of polarized domains.

TOOLS AND TECHNIQUES WITH WHICH TO STUDY EPITHELIAL POLARITY

Epithelial Monolayers Grown on Permeable Substrata

Epithelial cell lines that reproduce in vitro polarity properties of the parental native epithelium have become essential tools in the study of the establishment of polarity. Epithelial cell lines are derived from many different tissues. Several of them have become extensively utilized. A partial list of model epithelial lines includes kidney cell lines such as MDCK I and II and LLC-PK, human colon carcinoma epithelial lines such as Caco-2, T-84, and HT29 (51), the rat thyroid cell line FRT (168), and the retinal epithelial cell line RPE-J (104). Epithelial cell lines arise spontaneously from primary cultures or after transformation—for example, with SV40 T antigen (22, 104).

Morphological Techniques

Structural techniques, both at the optical and electron microscopic level, usually constitute the first line of analysis of polarity (56). Although immunofluorescence with standard fluorescence microscopes is usually a good initial screening procedure, laser scanning confocal microscopy provides the most reliable method, because it produces a cross-sectional image built by computer reconstruction from a series of successive *en face* images. Confocal microscopy has the potential for quantitation of polarity, although adequate protocols have not been developed yet. Confocal microscopy has all but replaced other more labor-intensive techniques, such as immunofluorescence of semithin (1µm) frozen sections, particularly when analyzing in vitro mono layers. However, immunocytochemical analysis of plastic or frozen sections remains the main approach for the study of polarity in native epithelial tissues. The most precise morphologic quantitation of polarity is provided, at least in theory, by immunogold labeling of ultrathin (0.1 µm) frozen sections. However, mastering this technique is difficult, and appropriate quantitation requires careful consideration of a variety of experimental problems, such as penetration of the antibody and gold reagents, sidedness of access, thickness of the section, etc. (50).

Biochemical Techniques

Ligand Binding. When a ligand is available it is usually easy to measure the number of binding sites on the apical or the basolateral surface. This technique usually requires that the epithelium be placed in a Ussing-like chamber (148). It was initially applied to native epithelia that can be easily isolated as continuous layers of relatively large dimensions, such as frog skin, cornea, retinal pigment epithelium, gut, gallbladder, or urinary bladder. Classical studies of the polarity of Na,K-ATPase and other ion transporters were carried out this way (7). With the advent of tissue culture, ligand-binding studies have been carried out on epithelial monolayers grown on tissue culture chambers consisting of a permeable substratum (nylon mesh coated with collagen, nitrocellulose, or polycarbonate filters) (19, 161). Ligand-binding studies and other polarity-measuring techniques on isolated epithelial monolayers require careful control of equal accessibility of the ligand to both surfaces (25). This is usually not trivial since a variety of factors may selectively reduce accessibility to either the apical or the basolateral surface. On the apical side, accessibility may be reduced by a thick glycocalyx, whereas on the basal side, attachment to other cells or to the substrate, basement membranes, the filter itself may prevent easy access of the ligand to large segments of the membrane (43). An additional factor that needs to be considered in the determination of ligand-binding polarity is whether ligand affinity is the same on both membranes. (Dissociation constants [Kds] should be determined in all cases.) Ligand affinity may be affected by the expression of different isoforms of the receptor, different lipid environment, and differential post-translational processing in either surface.

Antibody Binding. This is a special case of ligand labeling but is treated separately here because of the generality of this approach. All the caveats mentioned above apply to antibody labeling. Antibodies are relatively large molecules and it is clearly preferable for binding studies to utilize, whenever possible, monovalent Fab fragments. These have the advantage of a smaller size, decreased "stickiness" due to removal of the Fc fragment, and the inability to cause alterations in the distribution of the antigen due to crosslinking. This is particularly important since it has been recently become apparent that redistribution of proteins (for example, glycosyl phosphatidylinositol (GPI) anchored proteins) to certain cellular structures (that is, caveolae) may occur even after fixation with paraformaldehyde, unless monovalent Fab fragments or fixatives more effective in cross-linking proteins are used (92). Preparation of Fab fragments from polyclonal antibodies is quite standard but monoclonal antibodies vary widely in their protease sensitivity and require careful experimental adjustment of the cleavage conditions. Antibodies can be used *(1)* coupled to ^{125}I, *(2)* unlabeled followed in a second step by iodinated protein A or protein G, or *(3)* biotinylated followed by iodinated streptavidin.

Cell Fractionation. Various techniques are available for the purification of apical or basolateral plasma membranes. Plasma membranes usually fractionate together with endosomes and Golgi by isopycnic density gradient analysis and therefore cannot be highly purified by this procedure alone, with the exception of the highly organized brush borders of intestinal and proximal tubular cells (97). Domain-selective adsorption and immunoaffinity methods are increasingly used to obtain highly purified apical or basal membrane fractions. The first method can be used almost exclusively on cultured epithelial cells. Basal membranes can be obtained, as for nonpolarized cells, by sonication of epithelial monolayers grown on tissue culture beads or tissue culture plates (21). Apical membranes can be obtained by applying colloidal silica and polyanion layers to the exposed apical surfaces of cells grown on

filters followed by peeling the apical surface with a glass surface (134). Direct peeling of the apical surface by applying a dry nitrocellulose filter is usually inefficient and results only in the production of large holes (138). Immunoaffinity methods with beads coupled to antibodies against the ectoplasmic domain of apical or basolateral antigens may be an excellent choice for both cultured and native epithelial cells, when antibodies are available (66). A possible variation of this procedure may involve the selective biotinylation of apical or basolateral surfaces (135) followed by cell disruption and adsorption to streptavidin-coupled beads. A variety of beads coupled to second antibodies and streptavidin are available commercially. Magnetic beads are now available in a variety of sizes and provide a fast method to separate the immunoadsorbed fractions.

If the researcher is only interested in the analysis of the lipids in the apical or basolateral membranes, a procedure that can result in the recovery of highly purified apical or basolateral membrane fractions involves the use of enveloped viruses that bud asymmetrically from the cell surface (149). Epithelial monolayers are infected with either influenza virus, which buds from the apical cell surface, or with vesicular stomatitis virus, which buds from the basolateral surface of MDCK, Caco-2, retinal pigment epithelium, or other epithelial cells (104, 128). The virus envelope is in fact a patch of the original plasma membrane where the virally coded envelope glycoproteins have replaced the endogenous plasma membrane proteins. The lipid content of this patch is believed to represent faithfully the lipid content of the original plasma membrane, and on the basis of this assumption (backed up by some experimental evidence), differences in the composition of apical and basolateral membrane have been observed (149, 150).

Biotinylation Techniques. Selective biotinylation has become established as a preferred method to quickly identify apical and basolateral proteins, to measure the polarized distribution of endogenous and transfected proteins, and to study the polarized targeting of proteins to the cell surface during biogenesis (80, 135). Confluent monolayers are exposed to water-soluble, membrane-impermeant, and tight-junction-impermeant sulfo-N-hydroxysuccinimide (NHS) biotin added to the apical or to the basolateral side. After extraction with detergent, proteins on the apical or basolateral domain are identified by SDS–polyacrylamide gel electrophoresis and immunoblotting with streptavidin conjugates. Derivatized biotin is also available as a long-chain sulfo-NHS-LC derivative that provides better accessibility of the small biotin molecule to the streptavidin, and as a biotin hydrazide for labeling of sugar residues, very effective for highly sialylated glycoproteins (79). For protein targeting assays, monolayers on filters are metabolically pulse labeled, usually with a ^{35}S-methionine or ^{35}S-cysteine, chased in normal medium containing an excess of methionine and/or cysteine, and biotinylated apically or basolaterally at different times. Proteins targeted apically or basolaterally are then identified by sequential precipitation with specific antibodies and streptavidin-agarose beads (74–76).

Results obtained with biotin-labeling procedures, as with all other procedures utilized to assess polarity, must be interpreted carefully and, if possible, in combination with other methods. It has recently become evident that certain proteins have different accessibility to biotin on the apical or on the basolateral membrane and that labeling at different pHs may result in variations in the protein patterns observed (43). Furthermore, for unknown reasons precipitation with streptavidin is not quantitative and therefore must be compared with the total protein remaining in the supernatant.

Overall, because of their simplicity and versatility, biotin methods provide excellent tools with which to study epithelial polarity.

PROTEIN TRAFFICKING PATHWAYS IN EPITHELIAL CELLS

Biogenetic Pathways

Intracellular sorting of apical and basolateral proteins before delivery to the cell surface provides a major mechanism to generate epithelial membrane asymmetry (85, 96). Studies on the trafficking of the membrane glycoproteins of enveloped RNA viruses provided the initial information on the sites of intracellular sorting of apical and basolateral proteins. Influenza HA and VSV G protein are produced in a common site (the endoplasmic reticulum), transported to the Golgi apparatus, and, at a distal site of this organelle, they are segregated into different populations of vesicles (Fig. 17.3). Transport vesicles bud from a sorting organelle,

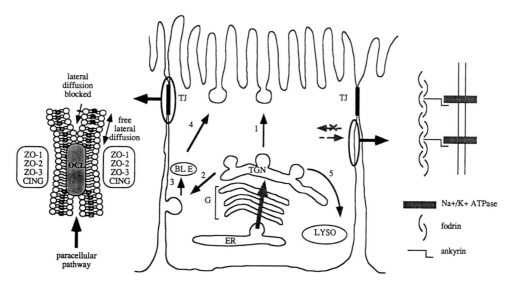

FIG. 17.3. Three mechanisms function in generation and maintenance of epithelial polarity. *Left:* Schematic representation of tight junction. Though several protein components have been identified, precise structural organization within the junction remains to be elucidated. (19). Tight junctions block lateral diffusion of membrane components and regulate passage of molecules between adjacent cells (paracellular pathway) *OCL,* occludin; *CING,* cingulin. *Middle:* Direct and indirect pathways direct polarized targeting of proteins in epithelia. In direct pathway, apically and basolaterally destined proteins pass through secretory pathway together until sorted and segregated at the TGN into apical and basolateral specific transport vesicles. Vesicles are then directly targeted to either apical *(1)* or basolateral surface *(2).* In indirect pathway, all proteins are first shunted to basolateral surface *(2).* Apical proteins are then endocytosed *(3)* and sorted and redirected to apical surface via transcytotic vesicles *(4).* TGN is also site of sorting of lysosomal hydrolases toward the lysosome *(5).* G, Golgi complex; *TGN,* trans-Golgi network; *ER,* endoplasmic reticulum; *BLE,* basolateral endosome; *TJ,* tight junction. *Right:* Simplified model depicting selective retention of Na,K-ATPase at basolateral membrane due to cytoskeletal interactions. Figure presents only subset of proteins making up polarized cytoskeleton. *Gray arrows* demonstrate delivery of Na,K-ATPase to basolateral surface and inhibition of its internalization.

the trans-Golgi network (TGN), comprising a conglomerate of tubules in the distal or trans region of the Golgi apparatus (49). HA and VSV G protein can be accumulated in the TGN by incubation during synthesis at 20°C, which blocks exit from this organelle (84). Upon removal of the temperature block, vesicles are assembled and released into the cytosol, where they can be purified by antibodies against the cytoplasmic domain of the viral glycoproteins (154). Apical and basolateral transport vesicles were shown to have distinct compositions, with some shared and some unique proteins. The fact that both VSV and influenza viruses block the synthesis of all host cell proteins suggests that vesicle proteins identified in infected cells are components of the "sorting machinery."

Vesicles assembled in the TGN are transported across the cytosol along organized microtubule tracks which confer directionality to their movement and increase the fidelity of docking at the respective pole of the cell. In fully confluent monolayers, microtubules are oriented along the major apical-basal axis of the cell, with the minus end located immediately under the apical membrane and the plus end located immediately above the basal membrane (2, 39). At both the apical and the basal poles microtubules end, forming a diffuse network underlying the respective surface. Microtubule inhibitors were shown to slow down the transport of apical proteins but did not seem to affect significantly the transport of basolateral proteins (39, 87, 121). However, more recent experiments performed in MDCK cells in which kinesin and dynein motors in the cytosol were either immunodepleted or inactivated provided evidence for a role of both kinesin and dynein in the transport of apical vesicles, and for the participation of dynein in the transport of basolateral vesicles (70). Microtubules appear not to be indispensable for apical and basolateral transport but, rather, facilitators and enhancers of the fidelity of the delivery process (39, 87, 121, 132). Other mechanisms that contribute to vectorial transport involve molecules that mediate the docking of apical and basolateral vesicles to the respective surfaces. Molecules identified in both vesicles and the target membranes, vesicular-soluble N-ethylmalemide-sensitive-factor attachment protein receptors (v-SNAREs) and target (t) SNAREs, that participate in the docking and fusion of synaptic vesicles and axonal plasma membrane (131) may also play a role in constitutive transport in both neurons and epithelial cells (64). Recent experimental data suggest that docking and fusion of basolateral vesicles is probably directed by v-SNARE and t-SNARE complexes and involves N-ethylmaleimide-sensitive factor (NSF), a ubiquitous factor that mediates fusion between different endocytic or exocytic organelles (131).

Transcytosis

In an alternative pathway present in all epithelial cells, some apical proteins are transported from the TGN to the basolateral surface, endocytosed, and sorted from basolateral proteins in a second sorting organelle, the basolateral endosome [Fig. 17.3; (98)]. Basolateral proteins are recycled to the basolateral surface, whereas apical proteins are transported to the apical surface via distinct *transcytotic vesicles* (indirect or trascytotic pathway). The process of transcytosis has been characterized in detail for the polymeric Ig receptor. This receptor is expressed normally in a variety of mucosal epithelia where it performs the function of binding immune IgA produced by specialized plasma cells in the lamina propria, transporting it across the cell, and releasing it from the apical surface for defense against potentially invasive agents such as parasites, bacteria, etc. (99). The study of transcytosis was facilitated by the demonstration that the transcytotic pathway can be faithfully reconstituted upon transfection of Poly Ig receptor cDNA into MDCK cells (102). In MDCK cells, newly synthesized Poly Ig receptor is transported directly from the TGN to the basolateral surface (101). Here the receptor is endocytosed within minutes into basolateral endosomes, where it is sorted from basolateral proteins which are recycled to the basolateral membrane by incorporation into distinct transcytotic vesicles. The process of incorporation into transcytotic vesicles is not 100% efficient since a fraction of the receptor recycles to the basolateral membrane. This is followed by a second round of endocytosis and sorting at the basolateral endosome. By recycling multiple times to the basolateral endosome, the fidelity of the sorting process is greatly enhanced. From the basolateral endosome, Poly Ig receptor is transported to a juxtanuclear compartment distinct from late endosomes (99). The final stages of apical transport of the Poly IgA receptor appear to involve incorporation into apical endosomes (3) and cycling to the apical surface. At the apical surface the receptor is cleaved by a protease resulting in the release of Ig coupled to a proteolytic segment termed the "secretory component." Transcytosis of Poly Ig receptor is carried out with a half-time of 30 min and is facilitated by microtubules (10, 87). Treatment with nocodazole of MDCK cells transfected with Poly Ig receptor and pre-bound with fluorescent antibody against the receptor results in the accumulation of the receptor in a population of vesicles localized mainly at the bottom half of the cells. Transcytotic vesicles have been isolated from hepatocytes using antibody to the cytoplasmic domain of the Poly IgA receptor (143).

In summary, proteins can be diverted to the apical

route in the TGN or in a post-TGN compartment that generates a transcytotic pathway. Research has identified the post-TGN compartment as a set of early endosomes in functional communication with the basolateral membrane. At these *early endosomes,* sorting decisions are made as to whether a protein will be incorporated in vesicles bound for the basolateral membrane, for the apical membrane, or for late endosomes which results ultimately in delivery to lysosomes and degradation. Note that vesicles leaving the basolateral endosome have destinations similar to those of post-TGN vesicles. The similar fate of the post-TGN and post-BL endosomal vesicles suggests a biochemical resemblance in the sorting mechanisms in the two organelles (91). In fact, recent work suggests that alteration of the signals that direct TGN sorting also modify sorting in the basolateral endosome (90). Elucidation of the degree of resemblance of TGN and BL endosome-sorting mechanisms awaits the identification of molecules that regulate these processes.

Recycling Pathways

Fluid-phase markers added to the apical or the basolateral medium of MDCK cells are incorporated within 5–10 min into distinct populations of endocytic vesicles. Early apical endosomes are preferentially found underneath the apical membrane whereas early basolateral endosomes are localized immediately above the basal membrane and beneath the lateral membrane (9). Simultaneous incorporation of different markers from the apical and basolateral media does not detect any overlap between early apical and early basolateral endosomes. Upon longer incubations of 15–30 min, apical and basolateral markers are transported to deeper late endosomes where they largely colocalize. Transport to late endosomes can be blocked by microtubule inhibitors (8). Longer incubations result in transport to a common set of lysosomes. In vitro experiments have shown that each set of early endosomes can fuse with late endosomes but cannot fuse with the opposite set of early endosomes (8). Fusion of early endosomes with late endosomes is facilitated by the addition of microtubules to the in vitro assay.

Many of the highly endocytic receptors appear to localize to the basolateral surface and endosomes (77). For example, transferrin and asialoglycoprotein receptors cycle between the basolateral surface and the basolateral endosome dozens of times during their lifetime (37, 38). The process of recycling of transferrin receptors has been reconstituted in MDCK cells (117). These experiments show that, whereas endocytosis of the transferrin receptor requires energy, its recycling to the basolateral membrane is energy independent. Recent

studies in which the G-coupled receptor for thyrotropin release hormone (TRH receptor) was introduced via an adenovirus into MDCK cells has shown that the receptor is endocytosed with equal efficiency from the apical membrane and the basolateral membrane upon binding of TRH and activates similar levels of inositol tri-phosphate (IP_3) second messenger (161). Down regulation of receptors has been observed at both the apical and the basolateral membranes.

Tissue-Specific Variation of Protein Polarity and Targeting Pathways in Epithelial Cells

The polarized epithelial phenotype is highly "flexible." The final destination of plasma membrane proteins and the pathways to the cell surface of individual proteins vary widely among different epithelial cells (127). Transgenic mice express human low density lipoprotein (LDL) receptor at the basolateral surface in the hepatocyte and the enterocyte but, strikingly, express it on the apical surface in proximal tubular cells (162). Since there is no evidence that the receptor is subject to different post-translational modifications in the different cell types, these data clearly indicate that the mechanisms that decode basolateral sorting information in kidney proximal tubular cells differ from those operating in intestinal or liver epithelial cells.

Other examples of reversed protein localization exist. Glycosyl-phosphatidylinositol (GPI)–anchored proteins, expressed on the apical surface of kidney (MDCK, LLC-PK) and intestinal (Caco-2) epithelial cells (80, 81), are largely basolateral in the thyroid epithelial cell line FRT (167) or are unpolarized in a mutant MDCK cell line resistant to killing by concanavalin A (55). Interestingly, the distribution of transmembrane apical proteins is not altered in FRT cells, indicating that these two groups of proteins may use different mechanisms for surface targeting. Gp114, an apical glycoprotein in MDCK cells, was found to be basolateral in an RCA-resistant MDCK cell line (73). Na,K-ATPase, basolaterally placed in most transporting epithelial cells, is restricted to the apical surface of the eye's retinal pigment epithelium, as well as in choroid plexus epithelium (53, 83).

The examples mentioned above involve proteins that are identical in the two cell types where they have different distributions (LDL-receptor, N,K-ATPase, GPI-anchored proteins) or have alterations in the carbohydrates in the ectodomain (lectin-resistant mutants). It should be noted that, in some of these examples, an apical protein becomes unpolar or basolateral, suggesting that there are instances in which the basolateral route may behave as a default route for apical proteins. Conversely, alterations in the cytoplasmic domain of some basolateral proteins may result in their unpolar distribution or in targeting to the apical surface. It has been suggested that this is evidence for a default apical route followed by all proteins that have no basolateral determinants in the cytoplasmic domain. However, the examples of mislocalized apical proteins mentioned above suggest that either pathway may act as a default pathway for proteins normally targeted to the opposite domain. The mechanistic bases of the cell-specific variations in pathways and final distributions of proteins in epithelial cells are unknown.

An important aspect of the flexible epithelial phenotype is the wide variation among different epithelial cells in the utilization of direct or transcytotic routes to the cell surface. Whereas the transcytotic pathway is ubiquitous, direct pathways to both apical and basolateral surfaces are expressed together only in some epithelial cells. For example, MDCK cells utilize exclusively the direct route for apical proteins, but hepatocytes appear to transport all their apical proteins via the transcytotic route (59). Native intestinal cells and intestinal cell lines (for example, Caco-2) appear to utilize one or the other route for different apical proteins (74, 86). For instance, the apical hydrolase DPPIV (see below) is directly delivered to the apical surface in MDCK cells but is transcytosed in hepatocytes and in Caco-2 cells (18). Not only cell-specific variations have been detected but, also, time-dependent trafficking divergences. In thyroid (FRT) cells, DPPIV is transcytosed to the apical surface in monolayers confluent for only 1 d, but the enzyme is progressively restricted to the direct route as the monolayer matures (164, 166). In MDCK cells, E-cadherin and desmoglein 1 are randomly delivered to the cell surface during the initial 36 h of monolayer development upon switching from micromolar to millimolar calcium levels (160).

SORTING SIGNALS AND MECHANISMS

In light of the early and prominent role that the enveloped viruses played in catalyzing research into epithelial polarity, it is not surprising that efforts to identify sorting signals began with these virus spike glycoproteins. A number of investigators have prepared mutagenized, chimeric, and deletion constructs of these polypeptides and expressed them by transfection in polarized epithelial cells. The analysis of the resultant proteins' sorting behaviors has produced a large and complicated literature from which a few general themes can be extracted.

When the cytoplasmic and transmembrane domains of the influenza HA protein are deleted, the remaining "anchor minus" ectodomain behaves as a secretory protein which is released at the apical surfaces of both MDCK cells and the MA104.11 line of rhesus monkey kidney epithelial cells (130). Subsequent research has demonstrated that the default pathway for secretory proteins which lack sorting signals leads to equal release at the apical and basolateral poles of MDCK cells (15, 47, 68). Taken together, these observations suggest that the ectodomain of the influenza HA protein contains a sorting signal which is sufficient to ensure its apical targeting. Apical sorting information also appears to reside in the ectodomain of the Poly Ig receptor (100). As was noted above, this protein pursues a complex itinerary through the cell, arriving first at the basolateral cell surface and subsequently undergoing transcytosis and delivery to the apical plasma membrane (101). Deletion of this protein's cytosolic tail produces a membrane protein which is transported directly from the TGN to the apical membrane. Further deletion of the transmembrane domain to produce the "anchor minus" construct results in a secretory protein which is released apically (101). Similarly, an anchor-minus construct of the apical protein aminopeptidase N (152) is also secreted apically in MDCK cells. Since the influenza HA protein (85, 96), the aminopeptidase N (152), the "tail-less" Poly Ig receptor, and other apically secreted proteins (127), travel vectorially from the TGN to the apical surface, the sorting machinery which interprets the apical sorting information present in their ectodomains must be exposed at the Golgi lumen.

The distribution of sorting information within the structure of the VSV G protein presents a marked contrast to the pattern described above for HA. The ectodomain of the VSV G protein does not appear to carry the sorting information required to direct it to the basolateral surface. An "anchor-minus" VSV G construct is secreted equally from both surfaces of MDCK cells, suggesting that this molecule pursues the default pathway for secretion (41). In light of this observation, attention focused on the VSV G protein's cytoplasmic tail as the most likely repository of sorting information. This hypothesis is now widely accepted, but its validation was complicated by the conflicting results produced in a number of studies employing chimeric constructs prepared from complementary portions of the VSV G and influenza HA proteins. Thus, a chimera in which the HA ectodomain was fused to the transmembrane and cytoplasmic domains of the VSV G protein behaved as an apical membrane protein in transfected MDCK cells (94). Similarly, in two separate experiments, a chimera composed of the VSV G ectodomain fused to the influenza HA transmembrane domain and tail appeared to behave as a basolateral protein (24, 95). However, a similar construct prepared from the VSV G ectodomain and transmembrane span and the HA cytoplasmic tail was present at both cell surfaces in MDCK cells (119). More recently, deletion of the VSV G protein cytoplasmic tail was shown to generate a protein which manifests an unpolarized distribution in MDCK cells (146). Furthermore, a chimera consisting of the influenza HA ecto- and transmembrane domains fused to the VSV G cytoplasmic tail has recently been shown to accumulate at the MDCK basolateral cell membrane (147).

In spite of the apparent contradictions presented above, the VSV G/influenza HA chimera literature illustrates several important points which are relevant to all of the studies described in this section. Clearly, homologous chimeras can produce dramatically and unpredictably different sorting patterns. In all likelihood this behavior is attributable to subtle differences in construct design, which may translate into dramatic differences in the tertiary structure of the resultant protein. It is quite likely that the ability of the cellular sorting machinery to recognize a sorting signal depends not only upon the linear amino acid sequence of that signal but also upon the conformation in which it is presented. Although two chimeras may be composed of grossly similar combinations of protein domains, different design strategies may result in small variations at the fusion point, such as the deletion of a few residues or the addition of linker sequences. If these variations differentially affect the chimeras' folding patterns, then they may unequally perturb the recognizability a sorting signal whose amino acid sequence is present in both constructs. In this case, two apparently identical chimeras might pursue very different routes through the cell. If one of the chimeras' sorting signals were obscured by misfolding, this protein would not be diverted to the targetting pathway dictated by the sorting information it was designed to embody. Instead, this functionally "signal-less" polypeptide would pursue whatever default pathway is available to proteins unable to interact with the apical or basolateral sorting machinery (13c).

The specific topography of this default pathway is another variable which may account for some of the heterogeneity in the chimera results presented above. It is natural to anticipate that a protein lacking a sorting signal will be delivered without polarity to both cell surface domains. This expectation receives support from studies on the behavior of "signal-less" secretory proteins, which are released at both the apical and basolateral surfaces of MDCK cells (15, 47, 68). It is quite possible, however, that the default pathway for membrane protein delivery may itself produce polarized distributions (13c). Imagine, for example, that the

process through which proteins are incorporated into basolaterally directed vesicles actively excludes those polypeptides which lack a sorting signal, whereas the apical pathway is less discriminating and permits "signal-less" proteins to enter. In this case, the default pathway for proteins lacking sorting signals would lead to their accumulation at the apical surface. A similar scenario can be envisioned which would lead to the concentration of "signal-less" proteins within the basolateral membrane. Evidence and arguments favoring each of these three alternatives (1: apical and basolateral; 2: apical; 3: basolateral) as the MDCK default pathway have been put forward (13c, 34, 88, 139). Furthermore, studies on secretory protein default sorting suggest that the features of the default pathway can vary dramatically among cell types (61, 122). Since the molecular compositions of sorting signals are only just beginning to be recognized, no reliable model of an "unsorted" membrane protein has yet been identified. Consequently, the precise definition of the default pathway for membrane proteins in MDCK or any other cell type has yet to be formulated. Uncertainty as to the details of default traffic may account in part for some of the conflicting interpretations associated with the discussion of viral protein chimeras presented above.

Recent evidence from site-directed mutagenesis studies demonstrates conclusively that the cytoplasmic tail of the VSV G protein contains a functional basolateral sorting signal. Thomas and Roth generated a construct in which the transmembrane and extracytoplasmic domains of the influenza HA protein were grafted onto the cytoplasmic domain of the VSV G protein (147). The resultant polypeptide was efficiently sorted to the basolateral surfaces of MDCK cells. Mutation of a single tyrosine residue in the cytoplasmic tail produced a protein that accumulated predominantly at the apical surface. Exhaustive mutagenesis of the residues surrounding this tyrosine have revealed the existence of a basolateral sorting motif, whose fundamental features are shared among a wide variety of sorting signals present in a diverse array of proteins.

Thomas and Roth determined that the sequence Y-X-X-aliphatic, where Y is tyrosine and X can be any amino acid, can serve as a powerful basolateral sorting signal in MDCK cells (147). In the case of the VSV G protein, where the aliphatic residue is an isoleucine, this motif is both necessary and sufficient to ensure basolateral targeting. The identification of this sequence is important for a number of reasons. It is now recognized that this sequence or its close homologues are responsible for the basolateral sorting of a number of proteins in MDCK cells (16, 35, 38, 63, 77). Furthermore, mutation of the cytoplasmic tail of the influenza

HA protein to replace cys 543 and cys 546 with tyr and phe, respectively creates a YRIF motif which is interpreted as a basolateral sorting signal by MDCK cells (11). Although the HA Y543,F546 protein presumably retains its ectodomain apical sorting signal, creation of the cytoplasmic YRIF sequence results in the protein's accumulation predominantly at the basolateral surface. Thus, this four-amino-acid motif is an autonomous basolateral signal which is able to overwhelm the HA protein's endogenous sorting information.

The biological utility of the Y-X-X-aliphatic motif appears not to be restricted to basolateral sorting in epithelial cells. Sequences of this form mediate targeting to lysosomes, the trans-Golgi network (62), and endosomes (1, 23, 40, 72). This apparent degeneracy implies that similar or shared molecular machinery must be responsible for interpreting the signals in each of these settings. Certainly the best-studied situation in which these signals operate is that of receptor-mediated endocytosis. It has been demonstrated that cell surface membrane proteins bearing the Y-X-X-aliphatic sequence are aggregated into forming coated pits and consequently into coated vesicles (28). Their association with the developing coated vesicle is mediated through interactions with adaptin proteins (110). Adaptins are multimeric cytoplasmic proteins which bind both to clathrin and to membrane proteins bearing the appropriate recognition signals. In vitro reconstitution studies have demonstrated that Y-X-X-aliphatic and related sequences will support adaptin binding (4, 110, 111). Consequently, proteins bearing these motifs can serve as the membrane anchors on which a clathrin lattice assembles. As a result of this assembly process, the membrane anchors become incorporated into the invaginating patch.

Current concepts regarding the mechanism through which the Y-X-X-aliphatic sequence mediates sorting in epithelial cells are predicated upon the endocytosis model. At least two subclasses of adaptins have been defined, one of which is associated with the plasmalemma while the other appears to be involved in the formation of vesicles at the TGN (110). It is likely that more adaptin isotypes with organelle-specific distributions and functions will be recognized in the future. Adaptins or adaptor-related proteins presumably associate with newly synthesized membrane proteins exhibiting the Y-X-X-aliphatic basolateral targeting signal during their passage through the TGN and thus mediate their segregation into basolaterally directed vesicles. Subtle differences in the amino acid composition of a given Y-X-X-aliphatic sequence, or in the secondary structure it assumes, probably determine

which adaptin isotype the protein bearing that sequence will interact with and, consequently, where within the cell it will come to reside (147).

Sequences unrelated to Y-X-X-aliphatic can also mediate interactions with adaptor proteins and thus function in subcellular sorting. Several proteins which are subject to rapid endocytosis in clathrin-coated pits carry a di-leucine motif which is necessary and sufficient to ensure their internalization (26). In at least one case, this di-leucine motif has also been shown to mediate basolateral sorting. The accumulation of the FcIIB2 receptor at the basolateral surface of MDCK cells is dependent upon an intact di-leucine sorting signal (89).

Although Y-X-X-aliphatic and di-leucine sequences are both capable of mediating basolateral sorting in MDCK cells and both presumably function through interactions with adaptor proteins, these sequences do not always encode targeting to the same subdomain of the epithelial cell plasma membrane. In LLC-PK$_1$ cells, which derive from the pig kidney and resemble the renal proximal tubule epithelium, proteins which carry the Y-X-X-aliphatic motif accumulate at the apical surface, while the FcIIB2 receptor behaves as a basolateral protein (46). These observations demonstrate that the same signal can produce distinct distributions in morphologically and functionally related epithelial cells. Thus, the cellular context in which a particular protein is expressed may determine the manner in which its sorting information is interpreted. It seems likely that the differential sorting behaviors manifest by MDCK and LLC-PK$_1$ cells may be attributable to expression of distinct adaptin isotypes or subclasses. Molecular and biochemical characterization of the adaptor populations in both cell types will be required to test this hypothesis.

Basolateral sorting signals have also been identified which bear no structural or functional resemblance to the sequence determinants involved in endocytosis. The cytoplasmic tail of the LDL receptor, for example, is endowed with both a tyrosine-based internalization signal as well as an autonomous basolateral sorting signal which plays no role in internalization (88, 91). When this primary basolateral signal is deleted the endocytosis signal is capable of functioning as a weak basolateral targeting signal. The proteins with which the autonomous basolateral signal interact have yet to be identified. It is clear, however, that its function is dependent upon its position with respect to both the membrane and the LDL receptor's C-terminus. Deletion constructs in which the signal is moved closer to the transmembrane domain or to the C-terminus are sorted less efficiently. This position dependence of signal function appears to be a fairly general theme and applies as well to signals of the Y-X-X-aliphatic class (147).

The Poly Ig receptor also possesses a 14-amino-acid autonomous basolateral targeting signal which is distinct from the sequences which mediate its endocytosis (16, 108). The Poly Ig receptor basolateral signal exhibits no obvious homology to the LDL receptor signal domain. Following its synthesis, the Poly Ig receptor is delivered directly to the basolateral plasmalemma in polarized epithelial cells. The protein is subsequently endocytosed and either transcytosed to the apical membrane or recycled to the basolateral surface (101). It has been shown that phosphorylation of a serine residue which resides within the basolateral signal markedly increases the fraction of internalized receptor which is transcytosed (17). Mutation of this serine to an alanine results in a receptor which is unable to participate in transcytosis. The mutant protein recycles continually between endosomes and the basolateral surface. It would appear, therefore, that the efficacy of the Poly Ig receptor basolateral sorting signal can be modulated. Phosphorylation appears to inactivate the signal, perhaps allowing the apical sorting information present in the protein's ectodomain to predominate and thus to initiate transcytosis.

So far this discussion of sorting signals has focused upon sequences embedded within the cytoplasmic and extracytoplasmic domains of transmembrane polypeptides. In the case of proteins anchored to the membrane by attachment to phospolipids, however, sorting appears to be mediated by lipid rather than protein determinants. Glycosyl-phosphatidylinositol-linked (GPI-linked) proteins are concentrated at the apical surfaces of most polarized epithelial cells, both **in situ** and in culture (80, 81). A large body of experimental evidence has been gathered which proves that the lipid linkage itself is sufficient to ensure the apical targeting of a GPI-linked protein. Molecular constructs in which the ectodomains of basolateral membrane proteins are fused to GPI attachment sites, for example, have been shown to accumulate at the apical surfaces of MDCK cells (12, 78, 118). The mechanism through which GPI attachment ensures apical sorting remains unclear. It has been demonstrated, however, that GPI-linked proteins self-associate into detergent insoluble "rafts" during their passage through the Golgi apparatus (13). The results of fluorescence photobleaching recovery experiments suggest that the proteins continue to participate in these clusters during and shortly after their delivery to the apical surface (55). It is interesting to speculate that this lipid-mediated self-association in the plane of the bilayer may serve a function similar to

that accomplished by the binding of adaptor proteins with Y-X-X-aliphatic sequences. In both cases these interactions mediate the segregation and immobilazation of membrane proteins and may thus facilitate their incorporation into the carrier vesicles which will participate in their transit to the cell surface. It should also be noted that, as with the Y-X-X-aliphatic sequence, GPI-linked proteins are not identically sorted in all epithelial cell types. In Fisher Rat Thyroid cells (FRT) GPI-linked proteins accumulate at the basolateral surface (165). A partially similar behavior is also observed in a lectin-resistant line of MDCK cells which are deficient in certain forms of glycosylation (168). The cellular components responsible for this differential sorting phenomenon have yet to be identified.

Recent progress has been made in identifying the sorting signals which function in multimeric membrane proteins. The neuronal GABA A receptor is composed of two subunits, each of which exists in multiple isoforms. When the α1 isoform is expressed individually in MDCK cells it behaves as a basolateral protein, whereas the β1 isoform is sorted to the apical surface (112). Coexpression of these proteins leads to the formation of a heterodimer which is targeted apically. Thus, the sorting information present in the β1 subunit appears to be dominant over that embedded within the α1. The asialoglycoprotein receptor (ASGPR) is also composed of two subunits, H1 and H2. The H1 polypeptide encodes a tyrosine-based sequence which mediates the protein's rapid endocytosis and which is sufficient to serve as a basolateral sorting signal when H1 is expressed alone (34). Mutation of the H1 signal produces a protein which is present in equal quantities on both surfaces of MDCK cells. Co-expression of the mutant H1 with H2, however, results in the formation of a dimeric complex which is basolaterally targeted. While it is known that the H2 signal is not tyrosine-based, its precise nature has yet to be identified.

Similar results have been obtained in studies on the sorting of dimeric ion pumps. Both the Na,K-ATPase and the gastric H,K-ATPase are composed of α- and β-subunits (28a). While the Na,K-ATPase is present at the basolateral surfaces of most polarized epithelial cell types (13b), the gastric H,K-ATPase is restricted to the apical membranes and pre-apical storage compartments in acid-secreting gastric parietal cells (57). Recent studies reveal that the β-subunit of the H,K-ATPase contains a Y-X-X-aliphatic sequence which will drive its basolateral sorting in MDCK cells (46). As would be expected from the preceding discussion, when this protein is expressed in LLC-PK$_1$ cells it behaves as a component of the apical membrane (45). Chimera studies demonstrate that the α-subunits also possess sorting information. A construct in which the first ~500 amino acids derive from the H,K-ATPase α-subunit while the C-terminal ~500 residues are taken from the Na,K-ATPase α protein assembles with the Na,K-ATPase β-subunit and the resultant complex is apically sorted in LLC-PK$_1$ cells (45). The complementary chimera assembles with the H,K-ATPase β-subunit and is targeted to the basolateral membrane in this cell type. Thus, sorting information is clearly present in both the α- and β-subunits of the H,K-ATPase. Recent results suggest that the H,K-ATPase α-subunit's apical signal may reside in an ectodomain loop or in a portion of a transmembrane domain. Furthermore, the α signal appears to be dominant over the β signal when the two are in conflict. The biological utility of incorporating multiple sorting signals into protein complexes is currently unclear. Similarly, the nature of the interactions that determine which signal will dominate have yet to be elucidated. Further analysis of the molecular composition of these sorting signals and the proteins which interpret them will be required before these questions can be answered.

ESTABLISHMENT OF EPITHELIAL POLARITY

The establishment of epithelial polarity is strongly dependent on interactions with other cells and with the substrate (124). Epithelial cells dissociated from each other and from the substrate with trypsin-EGTA acquire a round shape in which apical and basolateral domains are largely intermixed, as detected by the overlapping distribution of domain-specific markers (116, 125, 163). Furthermore, when MDCK cells are attached to the substratum, they spread and quickly polarize some of their surface markers. Certain apical proteins are the first to apparently become segregated, within a few hours, to the free surface domain, whereas basolateral markers appear initially to have a relatively unpolar distribution and to acquire polarity relatively slowly as a function of time (24–48 h) upon establishment of cell–cell contacts (151). Infection of MDCK cells attached singly to the substratum results in polarized budding of influenza virus from the free surface and, to a smaller extent, of VSV from the basolateral surface (125). Furthermore, MDCK cells attached to each other but not to the substrate also show polarized budding of viruses, with influenza budding from the free surface and VSV from regions where cell–cell contacts have been established. These experiments suggest that cell–cell and cell–substratum contacts are important in the establishment of polarity.

Important information on the morphogenesis of polarity has been provided by studies on the formation of epithelial cysts in culture. If thyroid cells (20, 106) or MDCK cells (155, 156) are grown on surfaces that do not favor attachment (such as bacteriological petri dishes or petri dishes covered with a thin layer of agar), they form vesicular structures that expose the apical surface while hiding the lateral and basal surfaces. Tight junctions are located in these cysts at their usual place, segregating apical from lateral domains and sealing the paracellular space to the passage of molecules and ions. When these cysts are surrounded by a collagen gel, a substratum on which MDCK cells may attach due to the possession of appropriate receptors, the cysts progressively reverse their polarity, over a period of 24 h. Initially the nuclei are localized close to the inner surface, facing the lumen, with the Golgi apparatus located immediately above them facing the external (apical) surface. The first morphological change detected is the change in the position of the nucleus to a position close to the surface opposite the new collagen matrix surface, which is also accompanied by a shift of the Golgi position to a supranuclear position facing the lumen. At 12 h, microvilli are observed on the luminal facing surfaces, while they have not yet disappeared from the surface facing the collagen gel. At about 16–24 h after transfer to collagen gel, all microvilli have disappeared from the outside surface and are only present in the luminal surface of the cysts. Immunofluorescence analysis demonstrates that an apical marker initially present on the outside surface is internalized into vesicular structures in the cytoplasm during the initial 6 h and appears in large amounts on the luminal surface 16 h after transfer. Tight junctions remain unchanged for about 12 h. At this time new tight junctions appear surrounding the luminal surfaces, as detected by the deposition of the tight junctional marker ZO-1. The old tight junctions surrounding the former apical surfaces facing the collagen gel are still unchanged at this time but start to disintegrate between 16 and 20 h. It is important to note that the apical marker inserted on the new apical (luminal) surface is not the one removed from the old apical (facing the collagen) surface, but is newly synthesized protein, and the same is true for ZO-1. It appears that the epithelial cell remodels its surface not by relocating old components to new locales but by removing and degrading old components from the wrong surface and synthesizing de novo its new surface domains.

The experiments just described suggest that molecules involved in cell–cell and cell–substrate contacts play an important role in the establishment of epithelial cell polarity. In fact, cyst polarity reversal could be blocked by adding a monoclonal antibody against $\beta1$ integrin to the collagen gel; under these conditions, the removal of apical microvilli and biochemical markers was inhibited (107). Furthermore, addition of 1,10-phenantroline, a metalloproteinase inhibitor, blocked the removal of collagen type IV deposited in the lumen of the cysts, which prevented insertion of apical markers in the luminal domain (155). Thus, remodeling of the epithelial cell surface is controlled by the deposition/removal of basement membrane material which interacts with the cell via specific integrins.

The Role of E-Cadherin

E-cadherin is the main adhesive molecule of simple transporting epithelia and appears to play a critical role in the development of epithelial polarity (123). Its appearance very early in mammalian development, at the 8 to 16-cell stage of the blastocyst, results in close apposition of the cells, a process known as *compaction* (67). E-cadherin knock-out in mice blocks development past the blastocyst stage (71). Addition of E-cadherin antibodies to developing MDCK monolayers retards the formation of tight junctions and polarization of the monolayer (52, 65). On the other hand, MDCK monolayers plated in medium containing micromolar levels of Ca, instead of the usual millimolar levels, remain adherent to the substratum but fail to establish cell–cell contacts due to the block in adhesive function of E-cadherin and other cadherins (such as the desmogleins in the desmosome) (42, 151). Cells in monolayers incubated under these conditions are dome shaped. If after several hours of incubation in low-calcium medium they are transferred to normal calcium levels, the cells establish cell–cell contacts and fully functional tight junctions within 3–4 h. The process of tight junction formation is so fast that some proteins are trapped within the wrong surface domain. An example of this is Na,K-ATPase (25), which is then progressively removed from the apical domain and degraded. As in the case of the cysts described above, Na,K-ATPase inserted in the newly formed basolateral surface is newly synthesized and progressively accumulates in the lateral domain of the cell.

How does the establishment of cadherin-mediated cell–cell contact lead to polarization of the cell surface? Initial experiments showed that several apical antigens were less extractable from the apical surface whereas a set of basolateral proteins were less extractable from the basolateral surface (133). This was also shown to be the case for Na,KATPase, which had a longer half-residence time at the basolateral surface (54). In one strain of MDCK cells, Na,K-ATPase was apparently not sorted intracellularly and was delivered to the cell surface without polarity. The steady-state basolateral polarity of the enzyme was proposed to result from stabilization of the enzyme by interactions with elements of the submembrane cytoskeleton that concentrate under the basolateral membrane under the influence of E-cadherin. Support for this hypothesis is provided by three lines of evidence. First, transfection of E-cadherin into fibroblasts or retinal pigment epithelial cells (which do not express E-cadherin, see below) led to a concentration of Na,K-ATPase at regions of cell–cell contact which were also the site of accumulation of E-cadherin (82, 93). Second, Na,K-ATPase was shown to bind with high affinity to ankyrin, which, together with fodrin (the nonerythroid form of spectrin), forms a basolateral submembrane cytoskeleton of MDCK cells (105). Third, the retinal pigment epithelium, which has an important functional apical interaction with photoreceptors, lacks E-cadherin and expresses both Na,K-ATPase and the ankyrin-fodrin cytoskeleton at the apical surface *in situ* (53, 82). Upon culture of RPE in vitro (in the absence of photoreceptors), Na,K-ATPase becomes unpolarized. A similar nonpolar distribution of Na,K-ATPase is observed in a cell line derived from RPE-denominated *RPE-J*. Upon transfection of E-cadherin into RPE-J cells, the pump undergoes a striking polarization to the lateral membrane, where it codistributes with E-cadherin (82). Interestingly, E-cadherin expression causes some other very interesting changes in RPE-J's phenotype, such as induction of desmosomal genes, of a renal ankyrin isoform gene, and of two apical marker proteins originally found in MDCK cells.

Na,K-ATPase is delivered basolaterally from the TGN in other strains of MDCK cells (14, 44) and in the thyroid epithelial cell line FRT (165, 167). Thus it appears that epithelial cells have two complementary mechanisms to polarize Na,K-ATPase (that is, vectorial targeting and cytoskeletal retention) which are expressed to different extents in different cell types.

E-cadherin appears to play an important organizational role in the establishment of the epithelial phenotype. The mechanisms by which E-cadherin interacts with downstream elements of the cytoskeleton or sends signals to the nucleus are starting to be unraveled. Very important interactions are those established between the cytoplasmic domain of E-cadherin and three cytosolic components, α, β, and γ catenins (109), which interact with actin and with some integral plasma membrane proteins, such as EGF receptor (58) and the tumor supressor protein APC. It is likely that some direct specific interaction will be found between the E-cadherin system and tight junctional proteins, which would account for the important role of E-cadherin in the assembly of these structures.

FUTURE PROSPECTS: IN VITRO SYSTEMS, GENETIC MODELS, AND THE SEARCH FOR SORTING MACHINERY

It is no doubt clear from the preceding discussions that major gaps remain in our understanding of the means through which epithelial cells generate and maintain the polarized state. Gains have been made in identifying the extracellular cues that cells use to orient themselves in establishing their polarized domains. The dissection of the cellular apparatus which transduces this information has begun and is yielding insights relevant not only to cell biologists but to developmental biologists as well. Similarly, a handful of sorting signals have been defined, and steady progress toward cracking the cellular code that specifies localization is likely to continue. The nature and composition of the cellular machinery which recognizes proteins endowed with sorting signals and assigns them to vesicular carriers which will transport them to the appropriate surface remain mysterious. What is missing is an understanding of the proteins which connect the intramolecular phenomenon of sorting signals to the macromolecular assemblies that create and support polarized domains.

Perhaps the most important reason that this machinery has proved so elusive relates to the fact that, until quite recently, no good in vitro systems had been developed which would render the intracellular processes of protein sorting amenable to biochemical analysis. Furthermore, genetic models in which mutations which perturb protein sorting could be generated or identified have also been scarce. The utility of such in vitro and genetic systems is elegantly exemplified by the extraordinary progress which has been made both in identifying the cellular machinery mediating the insertion of newly synthesized membrane and secretory proteins into or across membranes (153) and in isolating the components involved in regulating inter- and intraorganellar vesicular traffic (30). A great deal of emphasis is now being placed on developing methodologies through which these powerful tools can be applied to research in epithelial polarity.

Prior to the development of in vitro systems, insight into the sorting machinery was dependent upon the application of pharmacologic interventions. Through the use of the cytoskeletal disruptors colchicine and cytochalasin D, for example, it was determined that microtubules are required for the polarized delivery of apical but not basolateral proteins (121). Disruption of actin filaments does not affect sorting to either

membrane domain in MDCK cells. Further differentiation between the biochemical requirements of the apical and basolateral pathways has been provided by experiments employing membrane permeant agents that affect the function of trimeric G proteins. Treatment of MDCK cells with AlF_4, which activates all classes of G proteins, was found to stimulate apical transport of the influenza HA protein and inhibit basolateral transport of the VSV G protein (114). Addition of pertussis toxin, which ADP-ribosylates and inactivates $G_{\alpha i}$, enhanced basolateral protein secretion, whereas cholera toxin, which ADP-ribosylates and constitutively activates $G_{\alpha s}$, increased secretion from the apical surface. Finally, activators of protein kinases A and C stimulate apical protein delivery. This stimulation is blocked by protein kinase inhibitors (115). Taken together, these observations demonstrate that traffic to the apical and basolateral membranes may be governed by the same systems cells use to generate and respond to second messengers. Furthermore, distinct subsets of these regulatory proteins appear to be involved in transport to the two surfaces of epithelial cells. The mechanism through which these signaling molecules exert their influence remains unclear, although it has been postulated that trimeric G proteins may participate in the budding of vesicles from the TGN or in their subsequent recognition by a target membrane (7, 114, 142).

While pharmacologic experiments of this sort permit inferences to be drawn regarding the nature of the sorting machinery, they do not allow that machinery to be examined directly. Consequently, a great deal of effort has been devoted to the generation of suitable in vitro systems. Toward this end, investigators have taken advantage of ripped or lysed cell preparations to collect the carrier vesicles which transport membrane proteins to either surface domain. In these experiments, newly synthesized proteins are allowed to accumulate in the TGN by incubating MDCK cells at 20°C. Following cell rupture the temperature block is relieved and the vesicles which bud from the TGN are recovered by immunoisolation with antibodies directed against the cytoplasmic tails, of apical or basolateral cargo proteins (154). Analysis of these transport vesicles' membrane proteins has led to the identification of a number of interesting polypeptides. Two proteins, designated VIP-21 and VIP-36, are present in both apically and basolaterally directed carriers (33, 69). Molecular cloning reveals that VIP-21 is identical to caveolin, a component of the caveolar invaginations of the plasma membrane, while VIP-36 manifests structural homology to certain plant lectins. The role of these two proteins in sorting and trafficking remains to be established. This line of research has also led to the recognition of annexin 13b as an enriched component of apical vesicles (32). Annexins are phospholipid-binding proteins (27) whose possible relevance to membrane traffic in polarized epithelial cells will be discussed briefly below.

Several groups have now succeeded in defining experimental parameters according to which the surface membranes of MDCK cells can be permeabilized without destroying their capacity to mediate the polarized delivery of newly synthesized membrane proteins. By incubating MDCK cells with the bacterial toxin streptolysin O under carefully controlled conditions, holes can be introduced into either the apical or basolateral plasma membrane domains, which will permit the passage of exogenously added enzymes, toxins, and antibodies (29, 48). Application of these techniques has confirmed and extended the findings from the pharmacologic studies presented above. Thus, mastoparan, a peptide inhibitor of $G_{\alpha i}^2$ stimulates basolateral protein delivery in permeabilized cells (114). Addition of antibodies specific for $G_{\alpha s}^2$ inhibits apical protein delivery without exerting a comparable effect on basolateral traffic (114). Very recent experiments using in vitro systems to assess the budding of apical and basolateral vesicles from the TGN show that a definite site of action of trimeric G proteins is at the level of vesicle assembly (103).

Monitoring sorting in permeabilized cells has also permitted the roles of other components of the vesicular traffic apparatus to be assessed. In MDCK cells the small GTP-binding protein Rab 8 is distributed along the cytosolic surfaces of the TGN and the basolateral membrane. Introduction of a peptide whose sequence is derived from Rab 8 into permeabilized cells blocks the delivery of basolateral but not apical membrane proteins (60). Similarly, probes specific for components of the recently defined complex of proteins involved in vesicular fusion affect delivery to the basolateral but not the apical surface (64). Antibodies directed against NSF block basolateral delivery, while the addition of α SNAP protein stimulates this process. Clostridial neurotoxins, which block vesicular fusion by enzymatically cleaving SNARE proteins, also specifically inactivate basolateral traffic. It would appear, therefore, that the delivery of proteins from the TGN to the basolateral membrane makes use of vesicular fusion machinery which is common to a number of intracellular transport phenomena. Other molecules appear to mediate the interaction between apically directed carrier vesicles and the apical plasmalemma (64). In this context, it is interesting to recall that apically directed vesicles are generously endowed with annexin 13b (32). Members of the annexin family may be involved in a variety of membrane fusion events (27). Addition of anti–

annexin 13b antibodies to permeabilized MDCK cells blocks apical but not basolateral transport (32). Thus, the proteins which mediate the fusion of apical and basolateral vesicles appear to belong to two quite different mechanistic classes. The biologic function served by this disparity has yet to be established.

Finally, it should be noted that the permeabilized cell techniques have been used to characterize processes associated with the budding as well as the fusion of carrier vesicles. Addition of a peptide whose sequence is derived from that of the cytoplasmic tail of the VSV G protein prevents the basolateral delivery of the VSV G protein in MDCK cells (113). Apical membrane protein traffic is unaffected. The presence of the peptide prevents the departure of the VSV G protein from the TGN. A peptide in which the tyrosine residue critical to VSV G protein sorting is altered to an alanine blocks transport with fivefold-lower affinity. Presumably, the peptide is able to competitively inhibit interactions between the VSV G protein's cytosolic tail and components of the cytosolic sorting machinery. Cross-linking techniques have been used in initial attempts to identify the proteins which interact with this sequence. One candidate protein, which has been designated TIN-2, and has a molecular weight of ~200–300 kDa, interacts with the VSV G protein cytoplasmic tail only during this protein's residence in the TGN (113). It may, therefore, participate in the segregation and sorting of the VSV G protein into the subdomains of the TGN from which basolateral carrier vesicles arise. The molecular structure of TIN-2 has yet to be established, and thus it remains to be determined whether it bears any resemblance to members of the adaptin family.

The generation and characterization of the sec mutants in yeast provided the key to a remarkably rich source of insights into multiple facets of vesicular transport and protein processing (31). While no systems have yet been devised which will allow a similarly powerful description of the genes responsible for the generation and maintenance of epithelial polarity, efforts to establish a genetic approach are underway. One possible model system which is beginning to be exploited for this purpose can be found in the developing embryo of the fruit fly Drosophila melanogaster. The first thirteen mitotic divisions in the Drosophila embryo occur without accompanying cytokinesis, resulting in the formation of a syncytium whose nuclei are dispersed throughout the cytoplasm. After 3 h these nuclei migrate to the margins of the cell and become surrounded by membranes derived both from invaginations of the surface plasmalemma and from cytoplasmic vesicles. When this process of cellularization is complete, the surface of the embryo is composed of a monolayer of polarized epithelial cells whose basolateral membranes face the embryo's central yolk space and whose apical membranes appose the outside world (137). All of the embryo's internal structures, which include a broad array of polarized epithelial tissues, derive from invaginations of this surface epithelium. An understanding of the processes through which these epithelia become polarized and retain their polarity throughout ontogeny may lead to the development of a screening assay that can be used to search for interesting mutations.

Recently, membrane protein sorting has been studied in transgenic *Drosophila* embryos which express polypeptides whose polarized distributions have been established in MDCK cells (136). The GPI-linked human placental alkaline phosphatase (PLAP) protein accumulates at the apical surface when it is expressed in MDCK cells. A construct in which the PLAP ectodo-

main is coupled to the transmembrane and cytoplasmic domains of the VSV G protein behaves as a basolateral polypeptide in MDCK (12). When these two proteins are produced in Drosophila embryos under the control of a heat shock promoter, their subcellular distributions undergo a profound change during the course of development (136). Throughout embryogenesis the cells of the surface epithelium sort both the GPI and VSV-anchored forms of PLAP to their basolateral plasma membranes (Fig. 17.4). In contrast, these proteins exhibit their MDCK sorting patterns in the cells of the internal epithelia such as salivary gland and gut. The transition in sorting behavior is coincident with invagination. PLAP begins to gain access to the apical membrane in epithelial cells undergoing internalization from the embryonic surface. Since the formation of internal structures from the surface epithelium can

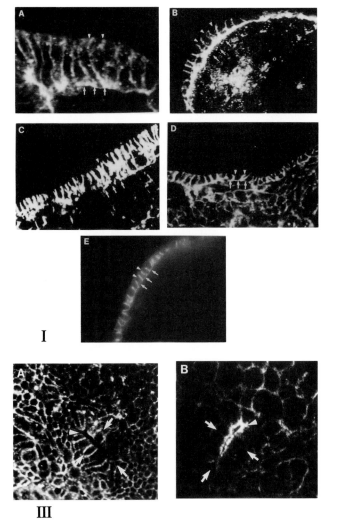

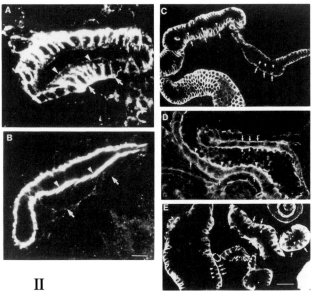

II

FIG. 17.4. Membrane protein sorting is developmentally regulated in *Drosophila* embryos. GPI-linked human placental alkaline phosphatase *(PLAP)* and chimeric version of this protein coupled to transmembrane segment of the VSV G protein *(PLAPG)* were expressed in transgenic *Drosophila* embryos. Distributions of both proteins in embryonic epithelial structures were determined by confocal immunofluorescence microscopy. *Panel I (top left):* In surface ectoderm of early *(A and B)* and late-stage embryos *(C and D)*, both PLAPG *(A and C)* and PLAP *(B and D)* share basolateral localization characteristic of Na,K-ATPase *(E)*. *Panel II (top right):* In salivary gland *(A and B)* and gut *(C, D and E)* PLAP behaves as apical protein *(B and D)* while PLAP "G" *(A and C)* and Na,K-ATPase *(E)* retain their basolateral distributions. *Panel III (bottom left):* Transition in PLAP sorting is coincident with invagination from surface ectoderm. PLAPG is basolateral in cells of a surface-connected tracheal pit *(A)*, whereas PLAPG is present at apical surface in this structure *(B)*. [*Adapted with permission from M. J. Shiel and M. J. Caplan (136, 137)*].

occur without any cell division (13a), it is clear that the same cell can markedly alter its sorting behavior in response to temporal and positional cues. Identification of the genes which are activated during invagination to bring about this transition would doubtless contribute to the characterization of the cellular sorting machinery.

Mutations have been identified which affect the polarity of the embryonic epithelia. The *crumbs* protein is a transmembrane polypeptide which is sorted to the apical plasmalemmas of both surface and internal embryonic epithelial cells (145). Null mutations at the *crumbs* locus are lethal to the embryo and produce a markedly dysmorphic embryonic phenotype. In less severe *crumbs* alleles the protein is made but is missorted to both surfaces of the embryonic epithelial cells. By generating heterozygotes expressing mild and severe *crumbs* alleles it has been determined that functional *crumbs* protein is required for its own apical sorting as well as for the apical sorting of other proteins expressed by the embryonic epithelia (158). Overexpression of the *crumbs* protein appears to be sufficient to induce the formation of "ectopic" apical membrane, suggesting that the *crumbs* protein participates in the specification of membrane identity (159). A second mutation, *stardust*, appears to interact with *crumbs* and is also required for the maintenance of appropriate polarity (144). It remains to be determined how the *crumbs* and *stardust* proteins participate in the formation of epithelial polarity. Since *crumbs* normally resides in the apical membrane, it is conceivable, for example, that this protein serves as a receptor for the docking of apically targeted carrier vesicles. Similarly, it is possible that *crumbs* functions as an anchor for cytoskeletal components that participate in the perpetuation of apical membrane integrity. Obviously, further study, both genetic and biochemical, will be required to determine the *crumbs* protein's precise function. The example of the *crumbs* and *stardust* mutations, however, demonstrates the enormous potential of genetic analysis, which is not predicated upon any mechanistic preconceptions, to reveal previously unpredicted relationships and structures.

Research into epithelial polarity has succeeded both in identifying some of the intramolecular signals which proteins use to specify their localizations and in identifying some of the intercellular signals which epithelial cells use to create their membrane domains. The challenge for the future lies in discovering the machinery which interprets these two classes of signals and utilizes the information inherent in both to generate anisotropy. Since polarity is not unique to epithelia, but is shared at least to some degree by myriad other cell types, the elucidation of this transduction apparatus will almost certainly be aided by and impact upon a broad range of disciplines in cellular physiology.

REFERENCES

1. Alvarez, E., N. Girones, and R. J. Davis. A point mutation in the cytoplasmic domain of the transferrin receptor inhibits endocytosis. *Biochem. J.* 267: 31–35, 1990.
2. Bacallao, R., C. Antony, C. Dotti, E. Karsenti, E.H.K. Stelzer, and K. Simons. The subcellular organization of Madin-Darby Canine Kidney cells during the formation of a polarized epithelium. *J. Cell Biol.* 109: 2817–2832, 1989.
3. Barroso, M., and E. S. Sztul. Basolateral to apical transcytosis in polarized cells is indirect and involves BFA and trimeric G protein sensitive passage through the apical endosome. *J. Cell Biol.* 124: 83–100, 1994.
4. Beltzer, J. P., and M. Spiess. In vitro binding of the asialoglycoprotein receptor to the b adaptin of plasma membrane coated vesicles. *EMBO J* 10: 3735–3742, 1991.
5. Bergmann, J. E., A. Kupfer, and S. J. Singer. Membrane insertion at the leading edge of motile fibroblasts. *Proc. Natl. Acad. Sci. U.S.A.* 80: 1367–1371, 1983.
6. Bergmann, J. E., and S. J. Singer. Immunoelectron microscopic studies of the intracellular transport of the membrane glycoprotein (G) of vesicular stomatitis virus in infected Chinese hamster ovary cells. *J. Cell Biol.* 97: 1777–1787, 1983.
7. Berridge, M. J., and J. L. Oschman. *Transporting Epithelia.* N.Y.: Academic Press, 1972.
8. Bomsel, M., R. Parton, S. A. Kuznetsov, T. A. Schroer, and J. Gruenberg. Microtubule- and motor-dependent fusion in vitro between apical and basolateral endocytic vesicles from MDCK cells. *Cell* 62: 719–731, 1990.
9. Bomsel, M., K. Prydz, R. G. Parton, J. Gruenberg, and K. Simons. Endocytosis in filter grown Madin-Darby canine kidney cells. *J. Cell Biol.* 109: 3243–3258, 1989.
10. Breitfeld, P. P., W. C. Mckinnon, and K. E. Mostov. Effect of nocodazole on vesicular traffic to the apical and basolateral surfaces of polarized Madin-Darby canine kidney cells. *J. Cell Biol.* 111: 2365–2373, 1990.
11. Brewer, C. B., and M. G. Roth. A single amino acid change in the cytoplasmic domain alters the polarized delivery of influenza viral hemagglutin. *J. Cell Biol.* 114: 413–421, 1991.
12. Brown, D. A., B. Crise, and J. K. Rose. Mechanism of membrane anchoring affects polarized expression of two proteins in MDCK cells. *Science* 245: 1499–1501, 1989.
13. Brown, D. A., and J. K. Rose. Sorting of GPI-anchored proteins to glycolipid enriched membrane subdomains during transport to the apical cell surface. *Cell* 68: 533–544, 1992.
13a. Campos-Ortega, J. A., and V. Hartenstein. *The Embryonic Development of Drosophila melanogaster.* Berlin: Springer-Verlag, 1985, 227 pp.
13b. Caplan, M. J. Biosynthesis and sorting of the sodium potassium-ATPase. In: *Regulation of Potassium Transport Across Biological Membranes,* edited by L. Reuss, J. M. Russell, and G. Szabo. Austin: University of Texas Press, 1990, p. 77–101.
13c. Caplan, M. J., and K. S. Matlin. Sorting of membrane and secretory proteins in polarized epithelial cells. In: *Functional Epithelial Cells in Culture,* edited by K. S. Matlin and J. D. Valentich. N.Y.: Alan R. Liss, 1989, p. 71–127.
14. Caplan, M. J., G. E. Palade, and J. D. Jamieson. Newly synthesized Na,K-ATPase alpha-subunit has no cytosolic intermediate in MDCK cells. *J. Biol. Chem.* 261: 2860–2865, 1986.
15. Caplan, M. J., J. L. Stow, A. P. Newman, J. Madri, H. C. Anderson, M. G. Farquhar, G. E. Palade, and J. D. Jamieson.

Dependence on pH of polarized sorting of secreted proteins. *Nature* 329: 632–635, 1987.

16. Casanova, J. E., G. Apodaca, and K. E. Mostov. An autonomous signal for basolateral sorting in the cytoplasmic domain of the polymeric immunoglobulin receptor. *Cell* 66: 65–75, 1991.

17. Casanova, J. E., P. P. Breitfeld, S. A. Ross, and K. E. Mostov. Phosphorylation of the polymeric immunoglobulin receptor required for its efficient transcytosis. *Science* 248: 742–745, 1990.

18. Casanova, J. E., Y. Mishumi, Y. Ikehara, A. L. Hubbard, and K. E. Mostov. Direct apical sorting of rat liver dipeptidylpeptidase IV expressed in Madin-Darby Kidney Cells. *J. Biol. Chem.* 266: 24428–24432, 1991.

19. Cereijido, M., A. Ponce, and L. Gonzalez-Mariscal. Tight junctions and apical/basolateral polarity. *J. Membr. Biol.* 110: 1–9, 1989.

20. Chambard, M., B. Verrier, J. Gabrion, and J. Mauchamp. Polarity reversal of inside-out thyroid follicles cultured within collagen gel: reexpression of specific functions. *Biol. Cell* 51: 315–325, 1984.

21. Chaney, L. K., and B. S. Jacobson. *J. Biol. Chem.* 258: 10062–10072, 1983.

22. Chou, J. Y. Establishment of rat fetal liver lines and characterization of their metabolic and hormonal properties: use of temperature sensitive SV40 mutants. *Methods Enzymol.* 109: 385–396, 1985.

23. Collawn, J. F., M. Stangel, L. A. Kuhn, V. Esekogwu, S. Jing, I. S. Trowbridge, and J. A. Tainer. Transferrin receptor internalization sequence YXRF implicates a tight turn as a structural recognition motif for endocytosis. *Cell* 63: 1061–1072, 1990.

24. Compton, T., I. E. Ivanov, T. Gottlieb, M. Rindler, M. Adesnik, and D. D. Sabatini. A sorting signal for the basolateral delivery of the vesicular stomatitis virus (VSV) G protein lies in its luminal domain: analysis of the targeting of VSV G-influenza hemagglutinin chimeras. *Proc. Natl. Acad. Sci. U.S.A.* 86: 4112–4116, 1989.

25. Contreras, G., G. Avila, C. Gutierrez, J. J. Bolivar, L. Gonzalez-Mariscal, A. Darzon, G. Beaty, E. Rodriguez-Boulan, and M. Cereijido. Repolarization of NaK pumps to the basolateral membrane during establishment of epithelial monolayers. *Am. J. Physiol.* 257 (*Cell Physiol.* 26): C896–C905, 1989.

26. Corvera, S., A. Chawla, R. Chakrabarti, M. Joly, J. Buxton, and M. P. Czech. A double leucine within the Glut4 Glucose transporter COOH terminal domain functions as an endocytosis signal. *J. Cell Biol.* 126: 979–989, 1994.

27. Creutz, C. E. The annexins and exocytosis. *Science* 258: 924–932, 1992.

28. Davis, C. G., M. A. Lehrman, D. W. Russell, R.G.W. Anderson, M. S. Brown, and J. L. Goldstein. The J. D. mutation in familial hypercholesterol amino acid substitution in cytoplasmic domain impedes internalization of LDL receptors. *Cell* 45: 15–24, 1986.

28a. De Pont, J.J.H.H.M., M. L. Helmich-de Jong, A.T.P. Skrabanja, and H.T.W.M. van der Hijden. H,K-ATPase: Na,K-ATPase's Stepsister. In: *The Na,K Pump* edited by J. C. Skou, J. G. Norby, A. V. Maunsbach, and M. Esmann. N.Y.: Alan R. Liss, 1988, p. 585–602.

29. Esparis-Ogando, A., C. Zurzolo, and E. Rodriguez-Boulan. Permeabilization of MDCK cells with cholesterol binding agents: dependence on substratum and confluencty. *Am. J. Physiol.* 267 (*Cell Physiol.* 36): C166–C176, 1994.

30. Ferro-Novick, S., and R. Jahn. Vesicle fusion from yeast to man. *Nature* 370: 191–193, 1994.

31. Ferro-Novick, S., and P. Novick. The role of GTP-binding proteins in transport along the exocytic pathway. *Annu. Rev. Cell Biol.* 9: 575–599, 1993.

32. Fiedler, K., F. LaFont, R. G. Parton, and K. Simons. Annexin XIIIb: a novel epithelial specific annexin is implicated in vesicular traffic to the apical plasma membrane. *J. Cell Biol.* 128: 1043–1053, 1995.

33. Fiedler, K., R. G. Parton, R. Kellner, T. Etzold, and K. Simons. VIP36, a novel component of glycolipid rafts and exocytic carrier vesicles in epithelial cells. *EMBO J.* 13: 1729–1740, 1994.

34. Fuhrer, C., I. Geffen, K. Huggel, and M. Spiess. The two subunits of the asialoglycoprotein receptor contain different sorting information. *J. Biol. Chem.* 269: 3277–3282, 1994.

35. Fuhrer, C., I. Geffen, and M. Speiss. Endocytosis of the ASGP receptor H1 is reduced by mutation of Tyrosine-5 but still occurs via coated pits. *J. Cell Biol.* 114: 423–432, 1991.

36. Fuller, S. D., R. Bravo, and K. Simons. An enzymatic assay reveals that proteins destined for the apical or basolateral domains of an epithelial cell line share the same late Golgi compartments. *EMBO J.* 4: 297–307, 1985.

37. Fuller, S. D., and K. Simons. Transferrin receptor polarity and recycling accuracy in "tight" and "leaky" strains of Madin-Darby canine kidney cells. *J. Cell Biol.* 103: 1767–1779, 1986.

38. Geffen, I., C. Fuhrer, B. Leitinger, M. Weiss, K. Huggel, G. Griffiths, and M. Speiss. Related signals for endocytosis and basolateral sorting of the asialoglycoprotein receptor. *J. Biol. Chem.* 268: 20772–20777, 1993.

39. Gilbert, T., A. Le Bivic, A. Quaroni, and E. Rodriguez-Boulan. Microtubular organization and its involvement in the biogenetic pathways of plasma membrane proteins in Caco-2 intestinal epithelial cells. *J. Cell Biol.* 113: 275–288, 1991.

40. Girones, N., E. Alvarez, A. Seth, I. M. Lin, D. A. Latour, and R. J. Davis. Mutational analysis of the cytoplasmic tail of the human transferrin receptor: identification of a sub-domain that is required for rapid endocytosis. *J. Biol. Chem.* 266: 19006–19012, 1991.

41. Gonzalez, A., L. Rizzolo, M. Rindler, M. Adesnik, D. D. Sabatini, T. Gottlieb. Nonpolarized secretion of truncated forms of the influenza hemagglutinin and the vesicular stomatitis virus G protein from MDCK cells. *Proc. Natl. Acad. Sci. U.S.A.* 84: 3738–3742, 1987.

42. Gonzalez-Mariscal, L., B. Chavez de Ramirez, and M. Cereijido. Tight-junction formation in cultured epithelial cells (MDCK). *J. Membr. Biol.* 86: 113–125, 1985.

43. Gottardi, C., L. A. Dunbar, and M. J. Caplan. Biotinylation and the assessment of epithelial membrane protein polarity. *Am. J. Physiol.* 268: (*Renal Fluid Electrolyte Physiol.* 37): F285–F295, 1995.

44. Gottardi, C. J., and M. J. Caplan. Delivery of Na,K-ATPase in polarized epithelial cells. *Science* 260: 550–552, 1993b.

45. Gottardi, C. J., and M. J. Caplan. An ion-transporting ATPase encodes multiple apical localization signals. *J. Cell Biol.* 121: 283–293, 1993c.

46. Gottardi, C. J., H. Y. Naim, M. G. Roth, and M. J. Caplan. Tyrosine-based membrane protein sorting signals are differentially interpreted by polarized MDCK and LLC-PK1 cells. (manuscript in preparation), 1996.

47. Gottlieb, T. A., G. Baudry, L. Rizzolo, A. Colman, M. Rindler, M. Adesnik, and D. D. Sabatini. Secretion of endogenous and exogenous proteins from polarized MDCK cell monolayers. *Proc. Natl. Acad. Sci. U.S.A.* 83: 2100–2104, 1986.

48. Gravotta, D., M. Adesnik, and D. D. Sabatini. Transport of influenza HA from the trans-Golgi network to the apical surface of MDCK cells permeabilized in their basolateral membranes: energy dependence and involvement of GTP-binding proteins. *J. Cell Biol.* 111: 2893–2908, 1990.

49. Griffiths, G., and K. Simons. The trans Golgi network: sorting

at the exit site of the Golgi complex. *Science* 234: 438–443, 1986.

50. Griffiths, G., K. Simons, G. Warren, and K. T. Tokuyasu. Immunoelectron microscopy using thin, frozen sections: application to studies of the intracellular transport of Semliki Forest virus spike glycoproteins. *Methods Enzymol.* 96: 466–485, 1983.

51. Gstraunthaler, G.J.A. Epithelial cells in tissue culture. *Renal Physiol. Biochem.* 11: 1–42, 1988.

52. Gumbiner, B., and K. Simons. The role of uvomorulin in the formation of epithelial occluding junctions. *Ciba Found. Symp.* 125: 168–186, 1987.

53. Gundersen, D., J. Orlowski, and E. Rodriguez-Boulan. Apical polarity of Na,K-ATPase in retinal pigment epithelium is linked to a reversal of the ankyrin-fodrin submembrane cytoskeleton. *J. Cell Biol.* 112: 863–872, 1991.

54. Hammerton, R. W., K. A. Krzeminski, R. W. Mays, T. A. Ryan, D. A. Wollner, and W. J. Nelson. Mechanism for regulating cell surface distribution of Na + K + -ATPase in polarized epithelial cells. *Science.* 254: 847–850, 1991.

55. Hannan, L. A., M. P. Lisanti, E. Rodriguez-Boulan, and M. Edidin. Correctly sorted molecules of a GPI-anchored protein are clustered and immobile when they arrive at the apical surface of MDCK cells. *J. Cell Biol.* 120: 353–358, 1993.

56. Hanzel, D., I. R., Nabi, C. Zurzolo, S. K. Powell, and E. Rodriguez-Boulan. New techniques lead to advances in epithelial cell polarity. *Semin Cell Biol.* 2: 341–353, 1991.

57. Hirst, B. H., and J. G. Forte. Redistribution and characterization of H, K-ATPase membranes from resting and stimulated gastric parietal cells. *Biochem. J.* 231: 641–649, 1985.

58. Hoschuetzky, H., H. Aberle, R. Kemler. Beta catenin mediates the interaction of the cadherin catenin complex with epidermal growth factor receptor. *J. Cell Biol.* 127:1375–1380, 1994.

59. Hubbard, A. L., and B. Stieger. Biogenesis of endogenous plasma membrane proteins in epithelial cells. *Annu. Rev. Physiol.* 51: 755–770, 1989.

60. Huber, L. A., S. Pimplikar, R. Parton, H. Virta, M. Zerial, and K. Simons. Rab8, a small GTPase involved in vesicular traffic between the TGN and the basolateral plasma membrane. *J. Cell Biol.* 123: 35–46, 1993a.

61. Hughson, E. J., D. F. Cutler, and C. R. Hopkins. Basolateral secretion of kappa light chain in the polarised epithelial cell line, Caco-2. *J. Cell Sci.* 94: 327–332, 1989.

62. Humphrey, J. S., P. J. Peters, L. C. Yuan, J. S. Bonifacino. Localization of TGN38 to the trans-Golgi-network: involvement of a cytoplasmic Tyrosin-containing sequence. *J. Cell Biol.* 120: 1123–1135, 1993.

63. Hunziker, W., and I. Mellman. Expression of macrophage-lymphocyte Fc Receptors in Madin-Darby Canine Kidney cells: polarity and transcytosis differ for isoforms with or without coated pit localization domains. *J. Cell Biol.* 109: 3291–3302, 1989.

64. Ikonen, E. M. Tagaya, O. Ullrich, C. Montecucco, and K. Simons. Different requirements for NSF, SNAP, and Rab Proteins in apical and basolateral transport in MDCK cells. *Cell* 81: 571–580, 1995.

65. Imhof, B. A., H. P. Vollmers, S. L. Goodman, and W. Birchmeier. Cell-cell interaction and polarity of epithelial cells: specific perturbation using a monoclonal antibody. *Cell* 35: 667–675, 1983.

66. Jones, S., J. R. Crosby, J. Salamero, and K. Howell. A cytosolic complex of p62 and rab6 associates with TGN38/41 and is involved in budding of exocytic vesicles from the trans-Golgi Network. *J. Cell Biol.* 122: 775–788, 1993.

67. Kemler, R. Calcium dependent cell-adhesion molecules. *Curr. Opin. Cell Biol.* 1: 892–897, 1989.

68. Kondor-Koch, C., R. Bravo, S. D. Fuller, D. Cutler, and H. Garoff. Exocytic pathways exist to both the apical and the basolateral cell surface of the polarized epithelial cell MDCK. *Cell* 43: 297–306, 1985.

69. Kurzchalia, T. V., P. Dupree, R. G. Parton, R. Kellner, H. Virta, M. Lehnert, and K. Simons. VIP21, a 21-kD membrane protein is an integral component of Trans-Golgi-network-derived transport vesicles. *J. Cell Biol.* 118: 1003–1014, 1992.

70. Lafont, F., J. Burkhardt, and K. Simons. Involvement of microtubule motors in basolateral and apical transport in kidney cells. *Nature* 372: 801–803, 1994.

71. Larue, L., M. Ohsugi, J. Hirchenhain, and R. Kemler. E-cadherin null mutant embryos fail to form a trophectoderm epithelium. *Proc. Natl. Acad. Sci. U.S.A.* 91: 8263–8267, 1994.

72. Lazarovits, J., and M. Roth. A single amino acid change in the cytoplasmic domain allows the influenza virus hemagglutinin to be endocytosed through coated pits. *Cell* 53: 743–752, 1988.

73. Le Bivic, A., M. Garcia, and E. Rodriguez-Boulan. Ricin resistant Madin-Darby Canine Kidney cells missort a major endogenous apical sialoglycoprotein. *J. Biol. Chem.* 10: 6909–6916, 1993.

74. Le Bivic, A., A. Quaroni, B. Nichols, and E. Rodriguez-Boulan. Biogenetic pathways of plasma membrane proteins in Caco-2, a human intestinal epithelial cell line. *J. Cell Biol.* 111: 1351–1361, 1990.

75. Le Bivic, A., F. X. Real, and E. Rodriguez-Boulan. Vectorial targeting of apical and basolateral plasma membrane proteins in a human adenocarcinoma epithelial cell line *Proc. Natl. Acad. Sci. U.S.A.* 86: 9313–9317, 1989.

76. Le Bivic, A., Y. Sambuy, K. Mostov, and E. Rodriguez-Boulan. Vectorial targeting of an endogenous apical membrane sialoglycoprotein and uvomorulin in MDCK cells. *J. Cell Biol.* 110: 1533–1539, 1990.

77. Le Bivic, A., Y. Sambuy, A. Patzak, N. Patil, M. Chao, and E. Rodriguez-Boulan. An internal deletion in the cytoplasmic tail reverses the apical localization of human NGF receptor in transfected MDCK cells. *J. Cell Biol.* 115: 607–618, 1991.

78. Lisanti, M., I. P. Caras, M. A. Davitz, and E. Rodriguez-Boulan. A glycophospholipid membrane anchor acts as an apical targeting signal in polarized epithelial cells. *J. Cell Biol.* 109: 2145–2156, 1989a.

79. Lisanti, M., A. Le Bivic, M. Sargiacomo, and E. Rodriguez-Boulan. Steady state distribution and biogenesis of endogenous MDCK glycoproteins: evidence for intracellular sorting and polarized cell surface delivery. *J. Cell Biol.* 109: 2117–2128, 1989b.

80. Lisanti, M., M. Sargiacomo, L. Graeve, A. Saltiel, and E. Rodriguez-Boulan. Polarized apical distribution of glycosyl phosphatidylinositol anchored proteins in a renal epithelial line. *Proc. Natl. Acad. Sci. U.S.A.* 85: 9557–9561, 1988.

81. Lisanti, M. P., A. Le Bivic, A. Saltiel, and E. Rodriguez-Boulan. Preferred apical distribution of glycosyl-phosphatidylinositol (GPI) anchored proteins: a highly conserved feature of the polarized epithelial cell phenotype. *J. Membr. Biol.* 113: 155–167, 1990.

82. Marrs, J., C. Andersone-Fissone, M. Jeong, L. Cohen-Gould, C. Zurzolo, I. Nabi, E. Rodriguez-Boulan, and W. Nelson. Plasticity in epithelial phenotype: modulation by expression of different cadherin cell adhesion molecules. *J. Cell Biol.* 129: 507–520, 1995.

83. Marrs, J. A., E. W. Napolitano, C. Murphy-Erdosh, R. W. Mays, L. F. Reichardt, and W. J. Nelson. Distinguishing roles of the membrane-cytoskeleton and cadherin mediated cell-cell

adhesion in generating different Na+,K(+)-ATPase distributions in polarized epithelia. *J. Cell Biol.* 123: 149–164, 1993.

84. Matlin, K. S., and K. Simons. Reduced temperature prevents transfer of a membrane glycoprotein to the cell surface but does not prevent terminal glycosylation. *Cell* 34: 233–243, 1983.

85. Matlin, K. S., and K. Simons. Sorting of an apical plasma membrane glycoprotein occurs before it reaches the cell surface in cultured epithelial cells. *J. Cell Biol.* 99: 2131–2139, 1984.

86. Matter, K., M. Brauchbar, K. Bucher, and H. P. Hauri. Sorting of endogenous plasma membrane proteins occurs from two sites in cultured human intestinal epithelial cells (Caco-2). *Cell* 60: 429–437, 1990a.

87. Matter, K., K. Bucher, and H. P. Hauri. Microtubule perturbation retards both the direct and the indirect apical pathway but does not affect sorting of plasma membrane proteins in intestinal epithelial cells (Caco-2). *EMBO J.* 9: 3163–3170, 1990b.

88. Matter, K., W. Hunziker, and I. Mellman. Basolateral sorting of LDL receptor in MDCK cells: the cytoplasmic domain contains two Tyrosine-dependent targeting determinants. *Cell* 71: 741–753, 1992.

89. Matter, K., and I. Mellman. Mechanisms of cell polarity: sorting and transport in epithelial cells. *Curr. Opin. Cell Biol.* 6: 545–554, 1994.

90. Matter, K., J. A. Whitney, E. M. Yamamoto, and I. Mellman. Common signals control low density lipoprotein receptor sorting in endosomes and in the Golgi complex of MDCK cells. *Cell* 74: 1053–1064, 1993.

91. Matter, K., E. M. Yamamoto, and I. Mellman. Structural requirements and sequence motifs for polarized sorting and endocytosis of LDL and Fc receptors in MDCK cells. *J. Cell Biol.* 126: 991–1004, 1994.

92. Mayor, S., J. F. Presley, and F. R. Maxfield. Sorting of membrane components from endosomes and subsequent recycling to the cell surface occurs by a bulk flow process. *J. Cell Biol.* 121: 1257–1269, 1993.

93. McNeill, J., M. Ozawa, R. Kemler, and J. Nelson. Novel function of the cell adhesion molecule uvomorulin as an inducer of cell surface polarity. *Cell* 62: 309–316, 1990.

94. McQueen, N. L., D. P. Nayak, E. B. Stephens, and R. W. Compans. Polarized expression of a chimeric protein in which the transmembrane and cytoplasmic domains of the influenza virus hemagglutinin have been replaced by those of the vesicular stomatitis virus G protein. *Proc. Natl. Acad. Sci. U.S.A.* 83: 9318–9322, 1986.

95. McQueen, N. L., D. P. Nayak, E. B. Stephens, and R. W. Compans. Basolateral expression of a chimeric protein in which the transmembrane and cytoplasmic domains of vesicular stomatitis virus G protein have been replaced by those of the influenza virus hemagglutinin. *J. Biol. Chem.* 262: 16233–16240, 1987.

96. Misek, D. E., E. Bard, and E. Rodriguez-Boulan. Biogenesis of epithelial cell polarity: intracellular sorting and vectorial exocytosis of an apical plasma membrane glycoprotein. *Cell* 39: 537–546, 1984.

97. Mooseker, M. S., and R. E. Stephens. Brush-border alpha-actinin? Comparison of two proteins of the microvillus core with alpha-actinin by two-dimensional peptide mapping. *J. Cell Biol.* 86: 466–474, 1980.

98. Mostov, K., G. Apodaca, B. Aroeti, and C. Okamoto. Plasma membrane protein sorting in polarized epithelial cells. *J. Cell Biol.* 116: 577–583, 1992.

99. Mostov, K. E. Transepithelial transport of immunoglobulins. *Annu. Rev. Immunol.* 12: 63–84, 1994.

100. Mostov, K. E., P. Breitfeld, and J. M. Harris. An anchor-minus form of the polymeric immunoglobulin receptor is secreted predominantly apically in Madin-Darby canine kidney cells. *J. Cell Biol.* 105: 2031–2036, 1987.

101. Mostov, K. E., and D. L. Deitcher. Polymeric immunoglobulin receptor expressed in MDCK cells transcytoses IgA. *Cell* 46: 613–621, 1986.

102. Mostov, K. E., M. Friedlander, and G. Blobel. The receptor for transepithelial transport of IgA and IgM contains multiple immunoglobulin-like domains. *Nature* 308: 37–43, 1984.

103. Muesch, A., H. Xu, D. Shields, and E. Rodriguez-Boulan. Role of basolateral signals and trimeric G proteins in the assembly of apical and basolateral vesicles from the TGN. (Submitted), 1996.

104. Nabi, I. R., A. P. Mathews, L. Cohen-Gould, D. Gundersen, and E. Rodriguez-Boulan. Immortalization of polarized rat retinal pigment epithelium. *J. Cell Science* 104: 37–49, 1993.

105. Nelson, W. J., and P. J. Veshnock. Ankyrin binding to (Na+, K+) ATPase and implications for the organization of membrane domains in polarized cells. *Nature* 328: 533–536, 1987.

106. Nitsch, L., and S. H. Wollman. Ultrastructure of intermediate stages in polarity reversal of thyroid epithelium in follicles in suspension culture. *J. Cell Biol.* 86: 875–880, 1980.

107. Ojakian, G. K., and R. Schwimmer. Regulation of epithelial cell surface polarity reversal by b1 integrins. *J. Cell Sci.* 107: 561–576, 1994.

108. Okamoto, C. T., S. P. Shia, C. Bird, K. E. Mostov, and M. G. Roth. The cytoplasmic domain of the polymeric immunoglobulin receptor contains two internalization signals that are distinct from its basolateral sorting signal. *J. Biol. Chem.* 267: 9925–9932, 1992.

109. Ozawa, M., H. Baribault, and R. Kemler. The cytoplasmic domain of the cell adhesion molecule uvomorulin associates with three independent proteins structurally related in different species. *EMBO J.* 8: 1711–1717, 1989.

110. Pearse, B. M., and M. S. Robinson. Clathrin, adaptors, and sorting. *Annu. Rev. Cell Biol.* 6: 151–171, 1990.

111. Pearse, B.M.F. Receptors compete for adaptors found in plasma membrane coated pits. *EMBO J.* 7: 3331–3336, 1988.

112. Perez-Velazquez, J. L., and K. J. Angelides. Assembly of GABA A receptor subunits determines sorting and localization in polarized cells. *Nature* 361: 457–460, 1993.

113. Pimplikar, S. W., E. Ikonen, and K. Simons. Basolateral protein transport in streptolysin O-permeablized MDCK cells. *J. Cell Biol.* 125: 1025–1035, 1994.

114. Pimplikar, S. W., and K. Simons. Regulation of apical transport in epithelial cells by a Gs class of heterotrimeric G protein. *Nature* 362: 456–458, 1993.

115. Pimplikar, S. W., and K. Simons. Activators of protein kinase A stimulate apical but not basolateral transport in epithelial Madin-Darby canine kidney cells. *J. Biol. Chem.* 269: 19054–19059, 1994.

116. Pisam, M., and P. Ripoche. Redistribution of surface macromolecules in dissociated epithelial cells. *J. Cell Biol.* 71: 907–920, 1976.

117. Podbilewicz, B., and I. Mellman. ATP and cytosol requirements for transferrin recycling in intact and disrupted MDCK cells. *EMBO J.* 9: 3477–3487, 1980.

118. Powell, S. K., B. A. Cunningham, G. M. Edelman, and E. Rodriguez-Boulan. Transmembrane and GPI anchored forms of NCAM are targeted to opposite domains of a polarized epithelial cell. *Nature* 353: 76–77, 1991.

119. Puddington, L., C. Woodgett, and J. K. Rose. Replacement of

the cytoplasmic domain alters sorting of a viral glycoprotein in polarized cells. *Proc. Natl. Acad. Sci. U.S.A.* 84: 2756–2760, 1987.

120. Rindler, M. J., I. E. Ivanov, H. Plesken, E. Rodriguez-Boulan, and D. D. Sabatini. Viral glycoproteins destined for apical or basolateral plasma membrane domains traverse the same Golgi apparatus during their intracellular transport in doubly infected Madin-Darby canine kidney cells. *J. Cell Biol.* 98: 1304–1319, 1984.

121. Rindler, M. J., I. E. Ivanov, and D. D. Sabatini. Microtubule-acting drugs lead to the nonpolarized delivery of the influenza hemagglutinin to the cell surface of polarized Madin-Darby canine kidney cells. *J. Cell Biol.* 104: 231–241, 1987.

122. Rindler, M. J., and M. G. Traber. A specific sorting signal is not required for the polarized secretion of newly synthesized proteins from cultured intestinal epithelial cells. *J. Cell Biol.* 107: 471–479, 1988.

123. Ringwald, M., R. Schuh, D. Vestweber, H. Eistetter, F. Lottspeich, J. Engel, R. Dolz, F. Jahnig, J. Epplen, S. Mayer, C. Muller, and R. Kemler. The structure of cell adhesion molecule uvomorulin. Insights into the molecular mechanism of CA2+-dependent cell adhesion. *EMBO J.* 6: 3647–3653, 1987.

124. Rodriguez-Boulan, E., and W. J. Nelson. Morphogenesis of the polarized epithelial cell phenotype. *Science* 245: 718–725, 1989.

125. Rodriguez-Boulan, E., K. T. Paskiet, and D. D. Sabatini. Assembly of enveloped viruses in Madin-Darby canine kidney cells: polarized budding from single attached cells and from clusters of cells in suspension. *J. Cell Biol.* 96: 866–874, 1983.

126. Rodriguez-Boulan, E., and M. Pendergast. Polarized distribution of viral envelope proteins in the plasma membrane of infected epithelial cells. *Cell* 20: 45–54, 1980.

127. Rodriguez-Boulan, E., and S. K. Powell. Polarity of epithelial and neuronal cells. *Annu. Rev. Cell Biol.* 8: 395–427, 1992.

128. Rodriguez-Boulan, E., and D. D. Sabatini. Asymmetric budding of viruses in epithelial monlayers: a model system for study of epithelial polarity. *Proc. Natl. Acad. Sci. U.S.A.* 75: 5071–5075, 1978.

129. Roth, M. G., R. W. Compans, L. Giusti, A. R. Davis, D. P. Nayak, M. J. Gething, and J. Sambrook. Influenza virus hemagglutinin expression is polarized in cells infected with recombinant SV40 viruses carrying cloned hemagglutinin DNA. *Cell* 33: 435–443, 1983.

130. Roth, M. G., D. Gundersen, N. Patil, and E. Rodriguez-Boulan. The large external domain is sufficient for the correct sorting of secreted or chimeric influenza virus hemagglutinins in polarized monkey kidney cells. *J. Cell Biol.* 104: 769–782, 1987.

131. Rothman, J. E., and G. Warren. Implications of the SNARE hypothesis for intracellular membrane topology and dynamics. *Curr. Biol.* 4: 220–233, 1994.

132. Salas, P.J.I., D. Misek, D. E. Vega-Salas, D. Gundersen, M. Cereijido, and E. Rodriguez-Boulan. Microtubules and actin filaments are not critically involved in the biogenesis of epithelial cell polarity. *J. Cell Biol.* 102: 1853–1857, 1986.

133. Salas, P.J.I., D. Vega-Salas, J. Hochman, E. Rodriguez-Boulan, and M. Edidin. Selective anchoring in the specific plasma membrane domain: a role in epithelial cell polarity. *J. Cell Biol.* 107: 2363–2376, 1988.

134. Sambuy, Y., and E. Rodriguez-Boulan. Isolation and characterization of the apical surface of polarized Madin-Darby canine kidney epithelial cells. *Proc. Natl. Acad. Sci. U.S.A.* 85: 1529–1533, 1988.

135. Sargiacomo, M., M. Lisanti, L. Graeve, A. Le Bivic, and E. Rodriguez-Boulan. Integral and peripheral protein composi-tions of the apical and basolateral membrane domains in MDCK cells. *J. Membr. Biol.* 107: 277–286, 1989.

135a. Schlutz, S. G. Cellular models of epithelial ion transport. In: *Physiology of Membrane Disorders,* edited by T. E. Andreoli, J. F. Hoffman, D. D. Fanestil, and S. G. Schultz.) N.Y.: Plenum, 1986, p. 519–534

136. Shiel, M. J., and M. J. Caplan. Epithelial membrane protein polarity is developmentally regulated during Drosophila embryogenesis. *Am. J. Physiol.* 269 (*Cell Physiol.* 38): C207–C216, 1995a.

137. Shiel, M. J., and M. J. Caplan. The generation of epithelial polarity in mammalian and Drosophila embryos. *Semin. Dev.* 6: 39–46, 1995b.

138. Simons, K., and H. Virta. Perforated MDCK cells support intracellular transport. *EMBO J.* 6: 2241–2247, 1987.

139. Simons, K., and A. Wandinger-Ness. Polarized sorting in epithelia. *Cell* 62: 207–210, 1990.

140. Stephens, E. B., and R. W. Compans. Assembly of animal viruses at cellular membranes. *Annu. Rev. Microbiol.* 42: 489–516, 1988.

141. Stephens, E. B., R. W. Compans, P. Earl, and B. Moss. Surface expression of viral glycoproteins is polarized in epithelial cells infected with recombinant vaccinia viral vectors. *EMBO J.* 5: 237–245, 1986.

142. Stow, J. L., and J. B. de Almeida. Distribution and role of heterotrimeric G proteins in the secretory pathway of polarized epithelial cells. *J. Cell Sci.* 17: 33–39, 1993.

143. Sztul, E., A. Kaplin, L. Saucan, and G. Palade. Protein traffic between distinct plasma membrane domains: isolation and characterization of vesicular carriers involved in transcytosis. *Cell* 64: 81–89, 1991.

144. Tepass, U., and E. Knust. Crumbs and Stardust act in a genetic pathway that controls the organization of epithelia in *Drosophila melanogaster. Dev. Biol.* 159: 311–326, 1993.

145. Tepass, U., C. Theres, and E. Knust. Crumbs encodes an EGF-like protein expressed on apical membranes of *Drosophila* epithelial cells and required for organization of epithelia. *Cell* 61: 787–799, 1990.

146. Thomas, D. C., C. B. Brewer, and M. G. Roth. Vesicular stomatitis virus glycoprotein contains a dominant cytoplasmic basolateral sorting signal dependent upon a tyrosine. *J. Biol. Chem.* 268: 3313–3320, 1993.

147. Thomas, D. C., and M. G. Roth. The basolateral targeting signal in the cytoplasmic domain of glycoprotein G from vesicular stomatitis virus resembles a variety of intracellular targeting motifs related by primary sequence but having diverse targeting activities. *J. Biol. Chem.* 269: 15732–15739, 1994.

148. Ussing, H. H. The use of tracers in the study of active ion transport across animal membranes. *Cold Spring Harbor Symp. Quant. Biol.* 13: 193, 1948.

149. van Meer, G., and K. Simons. Viruses budding from either the apical or the basolateral plasma membrane domain of MDCK cells have unique phospholipid compositions. *EMBO J.* 1: 847–852, 1982.

150. van Meer, G., K. Simons, K.J.A. Op den, and L. M. van Deenen. Phospholipid asymmetry in Semliki Forest virus grown on baby hamster kidney (BHK-21) cells. *Biochemistry* 20: 1974–1981, 1981.

151. Vega Salas, D. E., P. J. Salas, D. Gundersen, and E. Rodriguez-Boulan. Formation of the apical pole of epithelial (Madin-Darby canine kidney) cells: polarity of an apical protein is independent of tight junctions while segregation of a basolateral marker requires cell-cell interactions. *J. Cell Biol.* 104: 905–916, 1987.

152. Vogel, L. K., M. Spiess, H. Sjostrom, and O. Noren. Evidence for an apical sorting signal on the ectodomain of human aminopeptidase N. *J. Biol. Chem.* 267: 2794–2797, 1992.

153. Walter, P., and A. E. Johnson. Signal sequence receptors and protein targeting to the endoplasmic reticulum membrane. *Annu. Rev. Cell Biol.* 10: 87–119, 1994.

154. Wandinger-Ness, A., M. K. Bennett, C. Antony, and K. Simons. Distinct transport vesicles mediate the delivery of plasma membrane proteins to the apical and basolateral domains of MDCK cells. *J. Cell Biol.* 111: 987–1000, 1990.

155. Wang, A., G. K. Ojakian, and W. J. Nelson. Steps in the morphogenesis of a polarized epithelium. 2. Disassembly and assembly of plasma membrane domains during reversal of epithleial polarity in multicellular epithelial (MDCK) cysts. *J. Cell Sci.* 95: 153–165, 1990b.

156. Wang, A. Z., G. K. Ojakian, and W. J. Nelson. Steps in the morphogenesis of a polarized epithelium. I. Uncoupling the roles of cell-cell and cell substratum contact in establishing plasma membrane polarity in multicellular epithelial (MDCK) cysts. *J. Cell Sci.* 95: 137–151, 1990a.

157. Wiley, L. M., G. M. Kidder, and A. J. Watson. Cell polarity and development of the first epithelium. *Bioessays* 12: 67–73, 1990.

158. Wodarz, A., F. Grawe, and E. Kinst. CRUMBS is involved in the control of apical protein targeting during Drosophila epithelial development. *Mech. Dev.* 44: 175–187, 1993.

159. Wodarz, A., U. Hinz, M. Engelbert, E. Knust. Expression of crumbs confers apical character on plasma membrane domains of ectodermal epithelia of Drosophila. *Cell* 82: 67–76, 1995.

160. Wollner, D. A., K. A. Krzeminski, and W. J. Nelson. Remodeling the cell surface distribution of membrane proteins during the development of epithelial cell polarity. *J. Cell Biol.* 116: 889–899, 1992.

161. Yeaman, C., M. Heinflink, E. Falck-Petersen, E. Rodriguez-Boulan, and M. C. Gershengorn. Polarity of thyrotropin releasing hormone receptors following adenovirus-mediated gene transfer into Madin-Darby canine kidney cells 1996 (submitted).

162. Yokode, M., R. K. Pathak, R. E. Hammer, M. S. Brown, J. L. Goldstein, and R. G. W. Anderson. Cytoplasmic sequence required for basolateral targeting of LDL receptor in livers of transgenic mice. *J. Cell Biol.* 117: 39–46, 1992.

163. Ziomek, C. A., S. Schulman, and M. Edidin. Redistribution of membrane proteins in isolated mouse intestinal cells. *J. Cell Biol.* 86: 849–857, 1980.

164. Zurzolo, C., A. Le Bivic, A. Quaroni, L. Nitsch, and E. Rodriguez-Boulan. Modulation of transcytotic and direct targeting pathways in a polarized thyroid cell line. *EMBO J.* 11: 2337–2344, 1992.

165. Zurzolo, C., M. P. Lisanti, I. W. Caras, L. Nitsch, and E. Rodriguez-Boulan. Glycosylphosphatidylinositol-anchored proteins are preferentially targeted to the basolateral surface in Fischer Rat Thyroid epithelial cells. *J. Cell Biol.* 121: 1031–1039, 1993.

166. Zurzolo, C., C. Polistina, M. Saini, R. Gentile, G. Migliaccio, S. Bonatti, and L. Nitsch. Opposite polarity of virus budding and of viral envelope glycoprotein distribution in epithelial cells derived from different tissues. *J. Cell Biol.* 117: 551–564, 1992.

167. Zurzolo, C., and E. Rodriguez-Boulan. Delivery of Na+, K+-ATPase in polarized epithelial cells. *Science* 260: 550–552, 1993.

168. Zurzolo, C., W. van't Hof, G. van Meer, and E. Rodriguez-Boulan. VIP21/caveolin, glycosphingolipid clusters and the sorting of glycosylphosphatidylinositol anchored proteins in epithelial cells. *EMBO J.* 13: 42–53, 1994.

18. Mechanisms of transmembrane signaling

KEVIN WICKMAN

KAREN E. HEDIN

CARMEN M. PEREZ-TERZIC

GRIGORY B. KRAPIVINSKY

LISA STEHNO-BITTEL

BRATISLAV VELIMIROVIC

DAVID E. CLAPHAM

Department of Pharmacology, Mayo Foundation, Rochester, Minnesota

CHAPTER CONTENTS

CELLS ARE INSULATED BY A LIPID LAYER BARRIER, the plasma membrane. This barrier necessitated the evolution of elements capable of sensing and responding to changes in the extracellular environment. While early unicellular sensors simply maintained metabolic homeostasis, more complex organisms required sensing systems that responded to both environmental and intercellular stimuli. To this end, eukaryotic cells were equipped to receive and respond appropriately to an incredible number of stimuli, including ions, amino acids, polypeptides, byproducts of cellular metabolism, and photons. Even more diverse families of receptors and coupling elements generate cellular responses to changes in the extracellular environment.

Research of the last three decades has revealed multiple paradigms of cellular signaling, each with characteristic components, response times, durations, and specificity. Theoretically, the most efficient and specific signaling system employs two components: a ligand and an effector. This type of signaling design is common in the nervous system where the activity of many plasma membrane ion channels is directly related to their occupancy by ligands. The rapid transmission of information critical for mobility, thought processes, and proper execution of reflexes requires minimal delays between the binding of ligand to the ion channel and channel activation. Simplicity in design does not imply limited potential, as ligand-gated ion channels regulate responses dependent on transmembrane potential and intracellular ion composition, responses as diverse as action potential propagation and transcription. Similarly, steroids directly target intracellular receptors, and the resultant complex exerts powerful control over gene transcription. However, most universal transmembrane signaling systems involve multiple components. Positioning elements between receptors and the physiological response increase the potential for regulation and integration of stimuli within a specific cellular context.

Heterotrimeric guanosine triphosphate (GTP)-binding proteins (G proteins) are the common element in one major signaling paradigm. This relatively small family couples an enormous number of transmembrane receptors to a (currently) relatively small family of effectors. Stimulation of particular G protein–coupled signaling systems results in a variety of responses, from adjustments in membrane potential and second-messenger concentrations to mitogenesis. G protein–coupled pathway activity is based on a switch mechanism, "on" and "off" corresponding to GTP-bound and GDP-bound G proteins, respectively. The conversion between GDP-bound and GTP-bound states depends on several factors, but signaling is initiated when a ligand-bound receptor catalyzes the replacement of guanosine diphosphate (GDP) for GTP on the G proteins. Receptors linked directly or indirectly to tyrosine kinase activity are the hallmark of another major signaling paradigm. In contrast to the switch mechanism found in G protein systems, activated tyrosine-kinase-linked receptors assemble signaling complexes. Ligand-bound receptors provide docking sites for a growing family of proteins with important signaling functions. These proteins either have enzymatic activity themselves or link enzymes to the activated receptor. Generally, activation of tyrosine-kinase-linked receptors (RTKs) leads to the differentiation or proliferation of the targeted cell.

Variations on these two paradigms might govern the signaling functions of other membrane-associated receptors, including cell adhesion molecules and the integrins. Initially presumed to subserve functions such as cell–cell or cell–extracellular matrix attachment, it now appears that binding of these proteins by their ligands initiates some of the identical intracellular responses mentioned above [reviewed in (127, 328)]. By itself, this suggests that although the molecular characteristics are unique, Nature sculpted signaling systems from similar material. Because the intricacies of signaling initiated by cell adhesion molecules, integrins, cellular stress, and other membrane proteins are only beginning to be revealed, and since ligand-gated ion channels are the topic of another chapter in this book, this review focuses on the two most well-characterized signaling paradigms, G protein–coupled and tyrosine-kinase-linked signaling.

Both paradigms involve a ligand, an integral membrane receptor, and components designed to amplify and specify the signal initiated by the receptor. Although their multicomponent aspect makes these signaling pathways slow compared to ligand-gated channel responses, stimulation of G protein–coupled or tyrosine-kinase-linked pathways can affect or alter the target cell for extended durations. In both cases, the activated receptor initiates a cascade of events leading to regulation of an effector(s). Insertion of multiple components between the receptor and effector(s) allows the cell to amplify the signal and "fine-tune" the response. Some ligands bind multiple receptor subtypes while some receptors bind multiple ligands. Some coupling elements interact with multiple effectors and effector isoforms. At any of these levels of signaling, the cell can exert positive or negative control over the pathway. Because the signaling components often have multiple targets, activation of one pathway can alter the activity of another pathway. This phenomenon is referred to as *crosstalk*. Crosstalk is critical for an integrated cellular response to a stimulus and probably underlies the subtle differences in the final response of cells to stimuli that elicit similar proximal responses.

Given the existence of isoforms with unique regulatory properties for each position in a signaling pathway, why do some stimuli which could elicit numerous responses initiate only a single response? Compartmentalization, or the active colocalization of system components, is one potential source of specificity. In higher eukaryotes, the existence of multiple organ systems is compartmentalization resulting from a tissue-specific temporal pattern of gene expression. Cells express proteins which allow them to function appropriately in the environment, and this expression is malleable. Analogous to complex organisms, individual cells have developed subcellular specializations. In neurons, the protein composition of dendrites and axons is different and determines their respective ability to integrate passively incoming signals and actively propagate action potentials. Whether this level of compartmentalization is important for G protein–coupled and tyrosine-kinase-linked signaling pathways is unclear. A recurring theme in both types of signaling, however, is the localization of components to the membrane. In addition to receptors, coupling elements and most proximal effectors are either naturally associated with or recruited to the membrane upon pathway stimulation.

Membrane-bound protein "anchors" and post-translational lipid modifications often mediate shifts in the distribution of key signaling components to the membrane. In many cases, translocation of a cytosolic enzyme to the membrane environment provides access to its physiological substrate. The compression of volume due to the shift from three-dimensional to a two-dimensional space can catalyze important reactions (1).

The existence of microdomains for specific pathway components within the cytoplasm or plasma membrane is a matter of speculation. Microdomains would concentrate pathways which might otherwise interact with inappropriate or counterproductive elements. This type of compartmentalization appears critical for signaling involving integrins, where the complex "focal adhesion" specialization organizes components into a signaling complex (328). Similarly, *caveolae,* essential endocytic organelles that concentrate substances before delivery into the cell, have the intriguing potential to form signaling centers (7). The lack of noninvasive, high-resolution assays for analysis of signaling specificity makes it difficult to address the role of subcellular compartmentalization in signal transduction.

Most signaling pathways have negative-feedback mechanisms that prevent constitutive activation of effectors. The importance of these signaling "brakes" is underscored by the variety of pathophysiological states observed when the activity of one or more components in the G protein–coupled or tyrosine-kinase-linked systems goes unchecked. The broad term used to describe the progressive physiological loss of cellular response to a stimulus is *desensitization.* Binding of ligand can desensitize the target receptor to subsequent identical stimuli, termed *homologous* desensitization. In addition, the activity of unrelated signaling pathways can trigger decreased cellular responsiveness to all ligands, termed *heterologous* desensitization. Desensitization can determine the duration of the cellular response or shape the response initiated by receptor stimulation.

Common themes underlie all signaling systems. Currently, the details of specific pathways are under intense investigation. This overview intends to familiarize the reader with G protein–coupled and tyrosine-kinase-linked signaling systems. Several well-studied pathways are delineated, and recently identified intriguing interconnections between these pathways are highlighted.

SIGNAL TRANSDUCTION VIA HETEROTRIMERIC GTP-BINDING PROTEINS

G Protein Coupled Receptors (GCRs)

G protein–coupled receptors (GCRs) funnel a diverse array of ligands into common signaling pathways involving a seemingly small number of heterotrimeric GTP-binding proteins (G proteins) and effectors. Until the mid-1980s, the identity and characteristics of GCRs could be inferred only from ligand-binding assays and successful receptor reconstitutions. These studies, particularly those involving the β_2-adrenergic receptor, advanced our understanding of transmembrane signaling by demonstrating functional coupling between the receptor and a GTP-binding element (G protein), and presaged the existence of receptor isoforms that bind the same ligand yet elicit distinct responses. Recently, an avalanche of GCR clonings has revealed the diversity of this important family and allowed probing of receptor functional domains. Common features among GCR family members suggest structural and functional conservation throughout evolution. Unique features of individual GCRs permit their interactions with ions, enzymes, sensory stimuli, peptides, and biogenic amines (Fig. 18.1). In addition, the existence of multiple receptor isoforms which couple to distinct effector pathways may explain the pleiotropic effects associated with many ligands. Current research is designed to reveal the ligand-binding and G protein–interaction domains within GCRs. Initial results stress the importance of discontinuous receptor domains and caution against the uncritical extrapolation of data obtained from studies on one receptor subfamily to others.

General Structure of GCRs. The overwhelming majority of cloned GCRs contain seven hydrophobic domains,

Biogenic Amines	Peptides
Acetylcholine	Adrenocorticotropin (ACTH)
γ-aminobutyric acid (GABA)	Angiotensin II
Epinephrine	Bombesin
Glutamate	Bradykinin
Histamine	C5A
Norepinephrine	Calcitonin
Serotonin	Cholecystokinin (CCK)
	Chorionic gonadotropin (CG)
	Corticotropin-releasing Hormone (CRF)
	Endothelins
Sensory Stimuli	Follicle-Stimulating Hormone (FSH)
	f-Met-Leu-Phe (chemoattractant)
Light (*cis*-retinal)	Glucagon
Odorants	Gonadotropin-releasing Hormone (GnRH)
Tastants	Growth-hormone-releasing Hormone (GRF)
	Gut endocrine Peptide (PYY)
	Luteinizing Hormone (LH)
Lipid derivatives	Melanocyte-stimulating Hormone (MSH)
	Mating factors (yeast)
Leukotrienes	Neuropeptide Y (NPY)
Platelet activating Factor	NK1 (Substance P)
Prostacyclins	NK2 (Substance K)
Prostaglandins	NK3 (Neurokinin B)
Thromboxanes	Opioids
	Oxytocin
	Pancreatic Polypeptide (PP)
Other	Secretin
	Somatostatin (SST)
Adenosine, AMP, ADP, ATP	Thrombin
cAMP (Dictyostelium)	Thyroid-stimulating Hormone (TSH)
Cannabinoids	Thyrotropin-releasing Hormone (TRH)
Serum Ca^{2+}	Vasoactive Intestinal Peptide (VIP)

FIG. 18.1. Diversity of ligands for G protein–coupled receptors.

each presumed to traverse the membrane as an α-helix. GCRs family members, therefore, are referred to as "heptahelical," "serpentine," or "seven-transmembrane-spaning (7-TM)" receptors. Many GCRs have been cloned based on similarity within these domains. In fact, cloning by homology has revealed a class of "orphan" receptors (clones with seven membrane-spanning domains but without identified ligand) (63, 249, 277). The identification of the IGF-II subfamily of GCRs whose members contain a single putative hydrophobic domain suggests that alternative structures may exist which can couple to G proteins (285). Although we acknowledge exceptions to the classical receptor structure, we will focus on the 7-TM receptors, the most well-characterized and predominant form of GCR.

The generic structural model for GCRs includes an extracellular amino-terminus, seven membrane-spanning domains that form three intra- and three extracellular loops, and a cytoplasmic carboxyl-terminus. GCRs commonly differ in the length and composition of the extracellular amino-terminus, the intracellular

loops, and the carboxyl-terminal tail. Most GCRs contain at least one glycosylation consensus sequence. Certain GCRs are rich in putative glycosylation sites, modifications that may support ligand binding. Ligand-binding domains, however, are not conserved among receptors. In certain receptors (for example, rhodopsin and β₂-adrenergic receptor), post-translational addition of the lipid palmitate probably anchors part of the carboxyl-terminal tail to the membrane and forms a fourth intracellular loop (281, 293). Most GCRs appear to be substrates for serine/threonine kinases. Figure 18.2 illustrates the β₂-adrenergic receptor topology.

Seven membrane-spanning domains are predicted based on hydrophobicity analyses of primary sequences. There are at least three lines of experimental evidence to support this predicted topology. First is the electron diffraction and electron microscopy resolution of bacteriorhodopsin (149, 150). Bacteriorhodopsin is an ion transporter with seven transmembrane α-helical domains that binds the photoactivatable ligand retinal, similar to rhodopsin, the mammalian GCR involved in

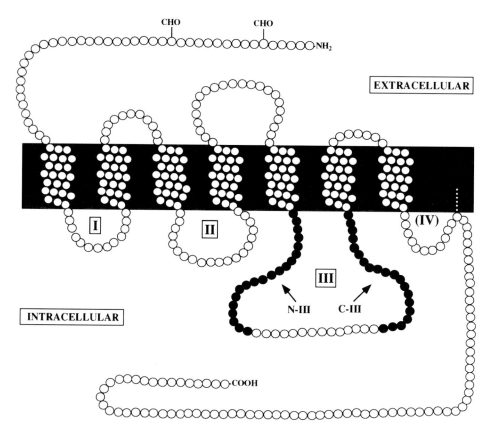

FIG. 18.2. Proposed topology of β₂-adrenergic receptor, model for GCR family. *CHO* indicates N-glycosylation sites of β₂-adrenergic receptor. Domains N-III and C-III have been implicated in specifying G protein interactions in several different GCRs. Membrane-proximal portion of carboxyl-terminus likely forms fourth intracellular loop by virtue of palmitic acid attachment.

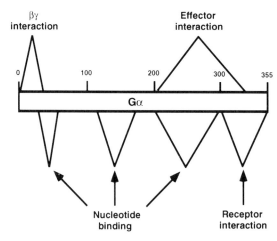

PLATE 18.1. *A*. Linear map of Gα functional domains. *B*. Crystal structure of Gα$_{i1}$ bound to GTPγS; 32 amino-terminal and 11 carboxyl-terminal residues were not resolved. Bound nucleotide and Mg^{2+} (ball-and-stick model) are seen clearly in structure. GTP binding results in aggregation and protrusion of three "Switch" regions: Switch I = Linker 2. Switch II = α2. Switch III = domain linking β4 and α3. Switch regions were identified by comparison of structures of FTP and GDP-bound forms of Gα$_t$. Three-dimensional picture of domain (β1: unresolved amino-terminus) implicated by proteolysis and antibody studies in interaction with Gβγ is not available. Receptor interaction is probably mediated by structures in carboxyl-terminus (β6, α5). Evidence suggests that exposed regions α2/β4, α3/β5, and α4/β6 are involved in effector interactions. [From Coleman et al. (78), Lambright et al. (213), Noel et al. (278), figure used with permission from Dr. A. G. Gilman.]

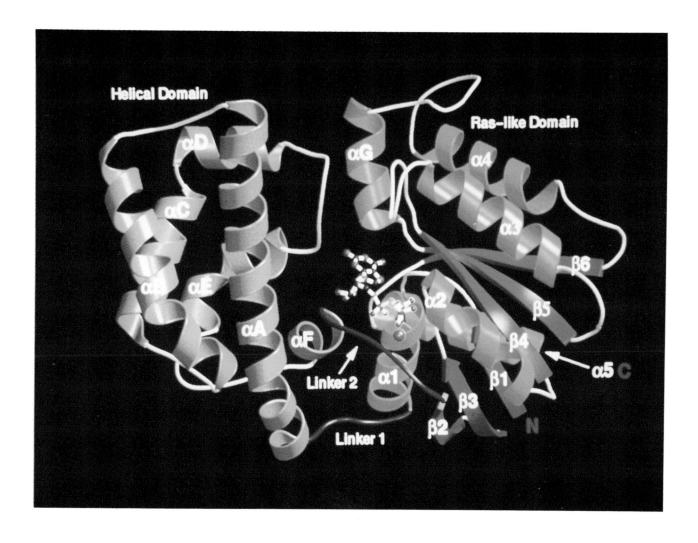

phototransduction. However, bacteriorhodopsin does not couple to G proteins (217). Strict comparisons between bacteriorhodopsin structure and GCR family members should be made cautiously, since the structure of eukaryotic GCRs probably depends on post-translational modifications lacking in bacteria. The binding of antibodies raised against peptides corresponding to predicted extra- and intracellular domains from the β_2-adrenergic receptor confirm the general receptor orientation within the membrane (399). Finally, a low-resolution projection structure of rhodopsin supports the model of seven transmembrane domains mainly oriented perpendicular to the membrane (329).

Mapping Critical Regions of GCRs. The binding of ligands to GCRs initiates a cascade of cellular events by activating heterotrimeric G proteins. The receptor acts as a catalyst, converting multiple inactive G protein heterotrimers into active, dissociated $G\alpha$ and $G\beta\gamma$ subunits. The activated G protein subunits proceed to interact directly with downstream elements (enzymes, ion channels, regulatory proteins) which elicit or shape cellular responses. Inherent in this activation scheme is an amplification potential. Although a single ligand binds a single receptor, the receptor can activate several G proteins. Activation of a single G protein generates two functional signaling molecules ($G\alpha$-GTP and $G\beta\gamma$). Research devoted to understanding specificity in GCR-mediated transmembrane signaling focuses on ligand-binding domains, domains specifying G protein interactions, tissue distribution of receptors, and modes of desensitization utilized by specific receptors. We will summarize these areas of study without delving into extensive detail. Readers interested in detailed accounts of GCR structure/function studies are referred to several recent reviews (147, 162, 197, 354).

Two techniques have been employed extensively in the mapping of important functional domains within GCRs. Site-directed mutagenesis, the precise mutation or deletion of specific amino acids, is a powerful technique for assessing the contribution of single residues to a specific function. A variation on receptor mutagenesis is to exchange domains from receptors with distinct functional characteristics and to assay functional specificity of the resulting receptor "chimeras." Both techniques require expression of the mutated receptors in cells that normally do not express either receptor used to generate the chimera. For receptor mutagenesis studies to be informative, expression levels, membrane incorporation, and general structural characteristics must be similar between the mutant and normal receptors. Therefore, researchers often choose receptors that are evolutionarily related yet possess unique ligand- or G protein–interaction specificities. The families of adren-

ergic and muscarinic receptors have been particularly useful in chimeric studies designed to localize G protein–interaction domains. Although highly homologous to other family members, certain isoforms in these two families utilize different G protein classes to differentially modulate intracellular levels of cyclic adenosine monophosphate (AMP), Ca^{2+}, and inositol-1,4,5-trisphosphate ($InsP_3$). Similarly, chimeric studies of glycoprotein hormone receptors have been useful for localizing the ligand-binding domains of this subclass of GCR.

Ligand-binding domains. One clinical benefit derived from the characterization of GCR ligand-binding domains is the potential for intelligent design of receptor-specific agonists and antagonists. Generalizations regarding GCR ligand-binding domains can be made based on the size and nature of the ligand. Smaller ligands tend to bind receptors with small extracellular domains and to interact with residues within the membrane-spanning domains. For example, muscarinic, adrenergic, serotonergic, and dopaminergic GCRs share a conserved aspartate in the third transmembrane domain that is believed to bind the protonated amines of their endogenous ligands (33, 162). In addition, less well-conserved residues have been shown to alter agonist or antagonist affinities for receptors for biogenic amines (33). In contrast, follicle-stimulating hormone (FSH), lutropin/chorionic gonadotropin (LH/CG), and thyroid-stimulating hormone (TSH), all relatively large glycoproteins, bind to receptors containing large amino-terminal extracellular domains (162). Two approaches demonstrate that the extracellular amino-terminus of glycoprotein hormone receptors forms the high-affinity binding site for their respective hormones. First, the expressed extracellular domain of LH/CG receptor binds human CG with high affinity (422). Second, chimeric studies between these three receptors confirm the role of the extracellular domain in conferring ligand specificity (42, 268). Notable features of the glycoprotein hormone-binding domains are multiple leucine-rich repeats, capable of interacting with both hydrophilic and hydrophobic surfaces, and putative glycosylation signals (33). There is at least one exception to the correlation of ligand size to ligand-binding domain characteristics: The primary sequence of the metabotropic glutamate receptor predicts a surprisingly large extracellular amino terminus, yet glutamate is a small ligand (247).

Generally, the formation of a ligand–receptor complex from free ligand and unbound receptor is a simple diffusion-limited reaction. The affinity of ligand for receptor is relatively low and the release of ligand occurs quickly, allowing for rapid signaling termination upon ligand diffusion (236). However, the glycoprotein hormones (for example, LH/CG) have very high (picomolar) affinities for their receptors (253). Similarly, the ligand (retinal) and receptor (rhodopsin) in phototransduction are linked covalently. In these systems, mechanisms of stimulus adaptation must exist to prevent constitutive signaling from the tight ligand–receptor complex. Another variant of the ligand–receptor interaction involves the thrombin receptor, which is activated upon cleavage by the protease thrombin (396). Thrombin cleaves a portion of the extracellular amino-terminus of the receptor, revealing the "hidden" ligand. Finally, it is important to remember that the ligand-binding domains of many cloned and uncloned GCRs have not been characterized. Each of these receptors may exhibit novel mechanisms of ligand interaction.

G protein–interaction domains. Receptors were first shown to discriminate between classes of G proteins based on the variable sensitivity of the G proteins to certain bacterial toxins. For example, activation of the atrial K^+ channel, I_{KACh}, by acetylcholine can be blocked by pertussis toxin (303), which ADP-ribosylates and inactivates specifically the $G_{i/o}$ class of G proteins. In many cases, native receptor–G protein complexes have been co-isolated following receptor stimulation (138, 218, 219, 342). A newer technique utilizes "antisense" oligonucleotides to specifically reduce the levels of mRNA which code for a particular protein(s). Complementary oligonucleotides, introduced into the cell using various techniques, are used to bind specific mRNA species. The resultant hybrids are subsequently degraded by enzymes that recognize such cytoplasmic anomalies. Correlating the loss of a specific protein with the disruption of a signaling pathway implicates that protein in the natural transduction of the signal. The "antisense" technique has been used to dissect the species of G proteins that interacts with receptors and effectors in vivo. For example, nuclear microinjection of antisense oligonucleotides specific for G protein–subunit isoforms was used to address the physiological mediators of GCR-Ca^{2+} channel coupling (194–196). From these studies, one could infer that receptors and effectors exhibit incredible specificity for particular G protein–subunit isoforms. A more conservative interpretation is that structures in *both* $G\alpha$ and $G\beta\gamma$ determine the specificity of receptor–G protein–effector coupling. Whether this discrimination is determined by compartmentalization and/or functional differences between isoforms is unclear. While recent developments in the isolation of purified receptors, G proteins, and effectors are allowing researchers to address possible functional differences between isoforms, the importance of compartmentalization in certain systems has long been appreciated. The most well-

characterized example is phototransduction. The vision organs in mammals express GCRs and G protein subunits not expressed in any other body region. Expression of these particular components is maximized to achieve efficient phototransduction.

Most cells express several G protein subclasses, yet a specific complement is activated upon stimulation of a single receptor type. For instance, the β_2-adrenergic receptor stimulates adenylyl cyclase by activating G_s, not G_i or G_o. However, reconstitution of purified receptors and G proteins in lipid vesicles and overexpression of receptors in heterologous systems a single receptor can couple to multiple G protein classes in vitro. Heterologously expressed α_2-adrenergic receptors, which normally couple to $G_{i/o}$ classes of G proteins, can couple to multiple G protein types, including G_s and G_q (60). The thyrotropin receptor can stimulate stimulate both G_s and $G_{q/11}$ classes, as shown in *native* membrane preparations (3). Chimeric receptor studies have localized domains in several receptors that determine which class(es) of G proteins will be stimulated by receptor activation. Most studies investigating these domains rely on the heterologous expression of chimeric receptors. Unfortunately, the precise effects of cellular environments and receptor overexpression in heterologous systems on signaling specificity are not known. Also, since some GCRs interact productively with several classes of G proteins, important changes in GCR–G protein coupling in chimeric studies may be overlooked if only the expected endpoints (like the levels of a single second messenger) are measured. However, assuming expression levels, ligand binding, and membrane incorporation are similar between normal receptors and receptor chimeras, chimeric studies can be informative.

Different intracellular receptor regions have been implicated in the coupling of receptors to specific G protein classes. Initial studies on the adrenergic and muscarinic receptor families suggested that the third intracellular loop (Fig. 18.2) is important for specifying the receptor–G protein interaction. For example, exchange of the third intracellular loop in the α_2-adrenergic receptor for the analogous domain from the β_2-adrenergic receptor generates a receptor chimera initiating a response characteristic of a β_2-adrenergic receptor (198). A similar role for the third intracellular loop in the muscarinic receptor family was suggested when receptor chimeras were assayed (208, 221, 408). Although promising, these results may not be generalizable to all GCRs. For example, the first cytoplasmic loop was implicated in determining the particular G protein interaction in the TSH receptor (62). Also, the amino-terminal portion of the carboxyl-terminal tail of the β_2-adrenergic receptor (fourth cytoplasmic loop) has been shown to be important for G protein coupling (43). Although not necessarily involved in specifying G protein interactions, other domains have been shown to be critical for efficient G protein interactions. Targeting of the carboxyl-terminal tail of rhodopsin by antisera (405), and deletions of residues in both the second and third cytoplasmic loops of rhodopsin (111), were both deleterious to G protein coupling. It seems plausible that GCRs utilize their entire cytoplasmic surface for specifying and efficiently stimulating particular classes of G proteins.

Studies involving membrane-proximal regions in the third intracellular loop (N-III and C-III; Fig. 18.2) in muscarinic and β-adrenergic receptors suggest one potential mechanism for G protein activation (54, 282). These membrane-proximal regions form cationic amphipathic α-helices, a structural feature of the wasp venom peptide mastoparan. Mastoparan is a potent receptor-independent stimulator of G protein activity (151). Subsequent studies involving synthetic peptides corresponding to N-III and C-III sequences from several receptors revealed their ability to stimulate G protein activity as well (285). Interestingly, mutations which disrupt the amphipathic, α-helical structure of the N-III in the m3 muscarinic receptor inhibit signaling via G proteins, supporting a role for this structure in receptor-mediated G protein activation (102).

Several problems with peptide and receptor mutagenesis studies counsel cautious interpretation of results. First, some peptides stimulate G proteins by mechanisms distinct from those used by the activated receptor, indicating they do not necessarily mimic the activated receptor (286). Second, the nonspecific effects of large concentrations of peptides in typical assays of G protein activity are poorly characterized, and it is possible that the peptides can disrupt domains within the receptors or G proteins, directly interact with effectors, or activate antagonistic G protein–dependent pathways (289). Receptor mutation has been found to generate unexpected signaling. For example, an m1 muscarinic receptor containing part of β₂-adrenergic receptor N-III was shown to stimulate both adenylyl cyclase and phosphoinositol-phospholipase C (PI-PLC) by distinct G protein classes, indicating the receptor chimera retained the function of both the donor and recipient receptors (87). Unexpected activation of antagonistic or transmodulating pathways is a significant concern when the assay involves quantification of downstream second-messenger concentrations.

GCR Desensitization.

Desensitization refers to the physiological mechanisms employed to minimize cellular responses to subsequent or prolonged stimuli. Although desensitization is achieved through several distinct mechanisms, all can be categorized as either homologous or heterologous depending on the causative stimulus. Desensitization can result from alterations in function or number to any component of a signaling pathway, but the receptor has been the most-studied target in G protein–coupled systems. Repetitive or prolonged exposure to stimuli either functionally "uncouples" the GCR from G proteins or results in "down-regulation" (the loss of receptor numbers). Phosphorylation and sequestration result in the functional uncoupling of many GCRs while down-regulation results from receptor degradation, reduced synthesis, or both. Functional uncoupling is generally fast, occurring in seconds to minutes, while down-regulation is observed only after hours of exposure to ligand.

Whether desensitization is involved in the termination of the physiological response depends upon the desensitization rate relative to the diffusion or removal of ligand. In many cases, response termination depends on diffusion of the ligand from the receptor. In the case of rhodopsin and the LH/CG receptors, which exhibit extremely high affinities for their ligands, desensitization may be the primary mode of signal adaptation and/or termination. In general, desensitization is crucial for decreasing responses to subsequent exposures of most ligands, and it can terminate the cell response to other stimuli. Desensitization has been reported in all G protein–coupled systems, but the precise mechanism varies. Unfortunately, our understanding of GCR desensitization mechanisms comes almost exclusively from studies of the β₂-adrenergic receptor. Since β₁ and β₃-adrenergic receptors show differences in virtually every aspect of desensitization compared to β₂ receptors, the characteristics of desensitization for every GCR are probably unique. However, the well-characterized β₂-adrenergic receptor highlights several key aspects of GCR desensitization. [For recent reviews on GCR desensitization, see (227, 236).]

Functional uncoupling of GCRs. Receptor phosphorylation is a critical step in rapid desensitization. Rapid homologous desensitization results from phosphorylation of receptors by receptor-specific kinase. Receptor-specific kinases preferentially phosphorylate the intracellular domains of agonist-bound receptors. As a result, only activated receptors serve as substrates for phosphorylation. In addition, certain receptor-specific kinases require membrane anchoring by activated G protein subunits to phosphorylate the receptor. Both these characteristics of receptor-specific kinases ensure the selective phosphorylation of those receptors engaged in signaling. The kinases that phosphorylate rhodopsin (rhodopsin kinase, RK) and the β₂-adrenergic receptor (β₂-adrenergic receptor kinase, βARK) are prototypical examples of the growing family of G protein–coupled receptor-specific kinases (GRKs) (144, 227). βARK is a cytosolic kinase while RK associated with the membrane by virtue of a post-translational lipid modification. βARK is recruited to the membrane by the free β-γ complex of the activated G protein ($G_{\beta\gamma}$), where it phosphorylates serine and threonine in the carboxyl-terminal tail of the β_2-adrenergic receptor (164). Although RK and βARK are homologous, RK lacks a domain found in βARK demonstrated to bind Gβγ. The requirement for Gβγ in the rapid homologous desensitization of the β_2-adrenergic receptor may restrict the substrate specificity of βARK, and it likely delays the onset of desensitization. RK apparently recognizes the activated receptor directly, and due to its covalent lipid modification, does not require translocation to the membrane. Each GCR probably does not require a unique GRK since significant overlap in substrate specificity is observed. Diversity and tissue specificity, however, will probably prove important for defining fine details in pathway desensitization. Receptor phosphorylation by GRKs is critical for desensitization, but does not interfere directly with receptor–G protein interactions (19). Phosphorylated rhodopsin binds a molecule called arrestin, and this completely interferes with receptor–G protein interactions (413). Isoforms of arrestin have been identified and shown to function analogously in other systems (237). The presence of multiple members of the arrestin family increases the potential for signaling specificity. However, it is not clear whether they function differently.

GCRs can be phosphorylated by their downstream effectors (for example, protein kinase A [PKA] and protein kinase C [PKC]). In contrast to GRKs, effector kinases such as PKA tend not to discriminate between agonist-bound and unbound β_2-adrenergic receptors (20). Phosphorylation of β_2-adrenergic receptors by these kinases occurs on residues distinct from those phosphorylated by the GRKs, and desensitization initiated by PKA or PKC does not involve the subsequent binding of arrestin-like molecules (306). Therefore, activation of these kinases by any mechanism can result in the heterologous desensitization of multiple GCRs. PKA seems to contribute (together with GRK-mediated phosphorylation) to rapid homologous desensitization in olfaction (34, 94, 331) Homologous desensitization by effector kinases is an example of direct negative feedback, where the effector ultimately regulates its own cellular activity.

Receptors can be physically uncoupled from signaling pathways by sequestration. Sequestration refers to the translocation of GCRs from the membrane to an intracellular compartment not accessible to hydrophilic ligands after several minutes of ligand exposure. The nature of this internal compartment is not clear, but endosomes are likely candidates. Signals initiating sequestration are currently unknown. Interestingly, a tyrosine positioned near the cytoplasmic surface in the seventh membrane-spanning domain is critical for the sequestration of the β_2-adrenergic receptor (14). The high conservation of this residue in the GCR family suggests that sequestration is a common phenomenon among GCRs. Once in the sequestered state, receptors can recycle to the plasma membrane. After ligand exposure, sequestered β_2-adrenergic receptors are less phosphorylated (less desensitized) than receptors in the plasma membrane [reviewed in (236)]. Therefore, sequestration may be a mechanism for receptor resensitization.

Down-regulation. Receptor down-regulation refers to a decrease in the number of receptors due to increased receptor degradation or decreased receptor synthesis. GCR down-regulation occurs after several hours of ligand exposure. The little that is known about mechanisms for receptor down-regulation comes from studies on the β_2-adrenergic receptor [reviewed in (236)]. Phosphorylation seems to play an important role in degradation, as evidenced by the impaired degradation of β_2-adrenergic receptor mutants lacking sites demonstrated to undergo phosphorylation by PKA. However, βARK-mediated phosphorylation does not enhance receptor down-regulation. The mechanisms of down-regulation and sequestration of the β_2-adrenergic receptor appear distinct. For example, many mutations within the third intracellular loop of the β_2-adrenergic receptor result in receptors that sequester normally but show decreased agonist-provoked degradation. Conversely, mutation of a tyrosine in the amino-terminal portion of the carboxyl tail results in a receptor that shows normal degradation, yet sequestration is impaired (14). Little is known about the mechanisms of down-regulation involving decreased receptor synthesis, although reductions in GCR mRNA have been described for several distinct receptors (236).

Heterotrimeric GTP-Binding Proteins (G Proteins)

Heterotrimeric G proteins localize to the cytoplasmic surface of eukaryotic cell membranes where they couple ligand-bound receptors to different effectors. The conservation of G proteins in diverse species such as plant, yeast, and human speaks to their importance in cell function. Three protein products from distinct genes comprise one heterotrimeric G protein. G protein α subunits (Gα) bind guanine nucleotides (GTP, GDP) with high affinity and possess an intrinsic GTPase activity that hydrolyzes GTP to GDP. G protein β (Gβ) and γ (Gγ) subunits are tightly, but not covalently, associated (Gβγ) under normal physiological condi-

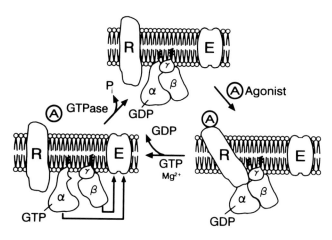

FIG. 18.3. G protein activity cycle. Binding of agonist *(A)* to G protein–coupled receptor induces conformational change in receptor which promotes release of GDP from heterotrimeric G protein $\alpha\beta\gamma$ complex. In presence of Mg^{2+}, GTP can bind $G\alpha$, presumably causing dissociation of G protein from receptor, and $G\alpha$-GTP from $G\beta\gamma$. Intrinsic GTPase activity of $G\alpha$ hydrolyzes GTP to GDP and $G\alpha\beta\gamma$ heterotrimer is reformed.

tions. In the absence of stimulation by ligand-bound receptor, $G\alpha$ and $G\beta\gamma$ exist as a stable complex, $G\alpha\beta\gamma$, with $G\alpha$ bound to GDP.

G proteins function as molecular switches with "on" and "off" states corresponding to GTP and GDP-bound $G\alpha$, respectively. The G protein cycle is depicted in Figure 18.3. Interaction of a G protein with an agonist-bound receptor promotes the release of GDP from $G\alpha$ (in the presence of Mg^{2+}). Because the intracellular GTP concentration exceeds that of GDP by at least one order of magnitude, the nucleotide-binding site on $G\alpha$ becomes occupied by GTP. GTP binding presumably results in the dissociation of the heterotrimer into $G\alpha$-GTP and $G\beta\gamma$ subunits. Dissociated, GTP-bound $G\alpha$ ($G\alpha$-GTP) and/or free $G\beta\gamma$ proceeds to interact with enzymes, ion channels, and regulatory proteins. These interactions result in diverse cellular responses, including the generation or degradation of important second messengers, protein phosphorylation, gene transcription, secretion, cytoskeletal reorganization, and membrane depolarization or hyperpolarization. After hydrolyzing bound GTP to GDP, $G\alpha$ reassociates with $G\beta\gamma$ to form the inactive heterotrimer, the preferential substrate for the ligand-bound receptor (346).

G Protein α subunits ($G\alpha$)

Classification. Heterotrimeric G protein α subunits ($G\alpha$) belong to a superfamily of GTP-binding proteins that includes small GTP-binding proteins like the *ras* proto-oncogene product and the elongation and initia-

tion factor (EF-Tu) of ribosomal protein synthesis (40). Currently, at least 17 distinct genes code for distinct $G\alpha$ polypeptides. Alternative splicing of mRNA generates even more diversity. Comparison of the predicted primary structures of $G\alpha$ from cloned cDNA sequences reveals a consistent pattern of conserved and divergent sequences. The conserved regions probably perform similar or identical functions in the different $G\alpha$, while the nonconserved domains reflect unique characteristics such as specific receptor, effector, and $G\beta\gamma$ interactions (186, 346). The first identified $G\alpha$ were classified based on their ability to regulate adenylyl cyclase. Therefore, the G_s subclass "stimulates" and the G_i subclass "inhibits" adenylyl cyclase (179, 360, 380). The G_o subclass, having no effect on adenylyl cyclase (274, 361), is the "other" subclass. The G_s, G_i, and G_o nomenclature has survived despite the considerable expansion of the family. However, as more novel $G\alpha$ and effectors are identified, it has become apparent that these three subclassifications are insufficient for all $G\alpha$ family members. Today, one can argue conservatively for the existence of at least four subclasses of $G\alpha$ subunits (Fig. 18.4; Table 18.1).

Lipid and toxin modifications. The primary structure of $G\alpha$ indicates no membrane-spanning domains. A reasonable hypothesis is that the hydrophobic $G\beta\gamma$ binds inactive $G\alpha$ to the membrane, while $G\alpha$ acylation, particularly myristoylation and palmitoylation, probably ties active $G\alpha$ ($G\alpha$-GTP) to the membrane (353). $G\alpha$ myristoylation is the stable, co-translational attachment of the 14-carbon fatty acid myristate to a glycine (Gly2) in the $G\alpha$ amino-terminus (124). Many $G\alpha$ including members of the $G_{i/o}$ family, are myristoylated. Mutation of Gly2 in both $G\alpha_o$ and $G\alpha_i$ results in their cytosolic localization (178, 262). In addition,

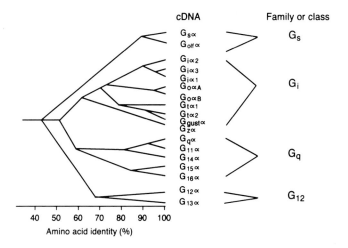

FIG. 18.4. Evolutionary tree of mammalian $G\alpha$. [Modified from Simon et al. (346) and used with permission.]

TABLE 18.1. *Characteristics of Mammalian Gα*

α Subunit	Family	Size, kDa	Expression	Example of Effector	Toxin	Lipid Modification	Reference
$G\alpha_{s_1}$		52					
$G\alpha_{s_2}$		52					
$G\alpha_{s_3}$	G_s	45	Ubiquitous	↑ Adenylyl cyclase	CTX	Palmitoylation	(186)
$G\alpha_{s_4}$		45					
$G\alpha_{olf}$		45	Olfactory neurons				
$G\alpha_{10}$							
$G\alpha_{gust}$		41	Taste buds		PTX CTX (?)	?	(254)
$G\alpha_{t_1}$		39	Retina (rod)	↑ cGMP PDE		Palmitoylation and myristoylation	
$G\alpha_{t_2}$		40	Retina (cone)		PTX & CTX		(368)
$G\alpha_{i_1}$		41	Brain				(43)
$G\alpha_{i_2}$	G_i	40	Ubiquitous	↓ Adenylyl cyclase			(17)
$G\alpha_{i_3}$		41			PTX	Myristoylation	(191)
$G\alpha_{o_1}$		39	Brain	↓ Ca^{2+} channel			(361)
$G\alpha_{o_2}$		39					(274)
$G\alpha_z$		41	Brain, platelets	↓ K^+ channel		—	(287)
$G\alpha_q$		41	Ubiquitous			Palmitoylation	(363)
$G\alpha_{11}$		41					
$G\alpha_{14}$	G_q	42	Stroma cells	↑ PLC-β	—		(365, 414)
$G\alpha_{15}$		43	B lymphocytes				(414)
$G\alpha_{16}$		43	Monocyte and T cells			—	(4)
$G\alpha_{12}$	G_{12}	44	Ubiquitous	?			(364)
$G\alpha_{13}$		44					

nonmyristoylated Gα are inefficient in effector and Gβγ interactions (234). Since other Gα are membrane-associated but not myristoylated, different modifications or protein characteristics must underly their membrane association. Indeed, some Gα (for example G_s, G_q) are palmitoylated (233, 297). Palmitoylation involves the post-translational addition of the 16-carbon fatty acid palmitate to an amino-terminal cysteine (336). Palmitoylation of $G\alpha_s$ and $G\alpha_q$ is necessary for effector stimulation and membrane association (403). Certain Gα (for example, $G_{i/o}$) are both myristoylated and palmitoylated, and studies from these Gα indicate that palmitoylation, not myristoylation, is responsible for strong membrane association. Stability of the lipid modification reveals an important distinction between palmitoylation and myristoylation. Myristoylation is stable while palmitoylation is reversible, permitting the dynamic regulation of palmitoyl content in certain Gα.

In addition to lipid modifications, certain subclasses of Gα are subject to modification by the bacterial toxins from *Bordetella Pertussis* (PTX) and *Vibrion cholerae* (CTX) (Table 18.1). PTX catalyzes the ADP-ribosylation of a cysteine fourth from the Gα carboxyl-terminus (259, 409). CTX catalyzes the ADP-ribosylation of a central arginine conserved among all Gα (161, 265). PTX uncouples G proteins from receptors, locking the G protein in the heterotrimeric state. In contrast, CTX modification inhibits the intrinsic GTPase activity of Gα. Since there is a small basal level of nucleotide exchange in the absence of ligand-bound receptors, CTX modification creates constitutively active Gα, and leads to chronic stimulation of downstream effectors (29). These toxins have preferred substrates: PTX modifies the heterotrimer, while CTX mainly targets free Gα. PTX and CTX are useful tools for dissection of G protein–mediated signaling pathways. Sensitivity of a receptor-mediated cellular response to either toxin suggests a role for G proteins. Also, because only certain Gα are sensitive to either toxin, it is often possible to identify the subclass of Gα involved in a particular signaling pathway.

Gα structure and function. Regions critical to the function of Gα have been mapped using complementary approaches. Comparisons with ras are useful since both Gα and ras are members of the GTP-binding protein superfamily, yet Gα possess a much higher GTPase activity and associate with GCRs and G$\beta\gamma$. Mutagenesis studies have attempted to identify Gα domains involved in G$\beta\gamma$ interaction, receptor recognition, nucleotide binding, GTP hydrolysis, and effector regulation. Crystal structures of Gα in inactive (GDP-bound) and active (GTPγS, AlF$_4^-$-bound) states suggest possible conformational changes underlying the transition from inactive to active Gα (78, 213, 278, 349). The crystal structures of Gα_t and Gα_{i_1} confirm the presence of two distinct domains: *(1)* a six-stranded β sheet surrounded by five α-helices (α1–α5) arranged identically to the ras protein and *(2)* an α-helical domain not present in *ras*, formed by one long central helix (αA) surrounded by five shorter helices (αB–αF) (78, 96, 213, 278, 349). The binding cleft for nucleotide resides between these two domains. Plate 18.1 (facing p. 692) illustrates a linear map of functional domains (mapped by a variety of experimental approaches) in a generic Gα and the corresponding domains in the three-dimensional context of the Gα_{i_1} crystal structure.

Several sequences are conserved among members of the superfamily of GTP-binding proteins. Presumably, these regions are important for the binding and hydrolysis of GTP. Mutations within these discontinuous regions slow the intrinsic GTPase activity of Gα, block transitions between the inactive and active Gα conformations, and alter the affinity of Gα for guanine nucleotides. Mutations of Gln227, which inhibit the intrinsic GTPase activity of Gα_s, are found in certain growth-hormone-secreting human pituitary tumors (214). Gln227 mutations generate GTPase-deficient Gα_s, resulting in the constitutive activation of adenylyl cyclase. Similarly, mutation of the arginine ADP-ribosylated by CTX inhibits the GTPase activity, leading to the constitutive activation of certain Gα (214). A conserved region in the amino-terminus is also involved in nucleotide binding and hydrolysis. Subtle differences in this region of the amino-terminus of Gα_z compared to other Gα might explain its slower rates of GDP release and GTPase activity (59). Gα possesses a greater GTPase activity than either ras or EF-Tu. Ras and EF-Tu in-

crease their GTPase activities by interactions with other proteins (GTPase Activating Proteins, GAP). Because a ~ 120-amino-acid insert in Gα has no counterpart in ras or EF-Tu, it is thought that this region in Gα is a "built-in" GAP. Indeed, expression of this insert as a recombinant protein demonstrated that it can function independently as a GAP (244). The crystal structures of Gα_t bound to GDP and GTPγS (or AlF$_4^-$) (213, 278, 349), show that very subtle structural changes distinguish the active and inactive protein conformations. The structural changes in Gα_t are localized to three adjacent regions in the ras-like domain, namely Switch I (173–183), Switch II (195–215), and Switch III (227–238). In the active state, these regions protrude slightly from the face of Gα (213, 278).

Although Gα probably interacts with G$\beta\gamma$ via multiple contacts, several approaches have implicated the Gα amino-terminus in G$\beta\gamma$ interactions. An antibody (mAB 4A) directed against the amino terminus of Gα_t causes it to dissociate from G$\beta\gamma$ (250). The proteolytic removal of 17–25 residues in the Gα_t amino-terminus renders Gα_t incapable of G$\beta\gamma$ interactions (270) and unrecognizable by mAB 4A (251). Unfortunately, the crystal structures of Gα_t and Gα_{i_1} do not reveal the precise structure of the amino-terminal domain, since crystals were obtained from Gα lacking 25 (Gα_t) and 33 Gα_{i_1} amino-terminal residues. Because G$\beta\gamma$ are promiscuous in their interaction with Gα, a common structural domain in Gα might mediate G$\beta\gamma$ interactions. However, G$\beta\gamma$ do not randomly associate with all Gα (121). It is not clear, however, whether the nonrandom association of Gα and G$\beta\gamma$ is due to selective binding of isoforms or to differential expression of certain subtypes in a particular cell.

Evidence suggests that the Gα carboxyl-terminus forms a contact site with receptors. For example, the site of PTX modification of G$\alpha_{i/o}$ is found within the carboxyl-terminal domain (PTX treatment uncouples G proteins from receptors; (259, 409). Further evidence comes from an S49 lymphoma cell line *unc*, which demonstrates deficient coupling between endogenous β_2-adrenergic receptors and endogenous Gα_s. Subsequently, it was discovered that Gα_s from *unc* contains an arginine-to-proline point mutation in the α6 helix in the carboxyl-terminus (371). Furthermore, independent studies suggest that specific coupling of G$\alpha_{s/i}$ chimeras to different receptors (purinergic, thrombin, and β_2-adrenergic receptors) depends on the nature of the carboxyl-terminal domain of the chimera (although the length of the exchanged domain varies; (132, 246). Finally, synthetic peptides corresponding to regions within the Gα_t carboxyl-terminal domain blocked Gα_t-rhodopsin interactions at low concentrations and mimicked the interactions at higher concentrations (141).

More recently, the extreme carboxyl-terminal amino acids have been implicated in the specification of receptor–G protein interactions. Replacement of only three carboxyl-terminal amino acids of $G\alpha_q$ with corresponding residues form $G\alpha_i$ generated proteins that mediated PI-PLC stimulation (target of G_q) by characteristic G_i linked receptors (79). However, the extreme carboxyl-terminus is probably not the sole determinant of $G\alpha$-receptor interactions. For example, two splice variants of $G\alpha_o$ couple to different receptors despite possessing the identical eight carboxyl terminal residues (193). Therefore, specific interactions between G proteins and receptors must also depend on other domains.

The notable difference between GDP- and GTP-bound forms of $G\alpha_t$ is the protrusion formed by the aggregation of the three Switch regions located between residues 173 and 238 (39). One reasonable hypothesis is that these regions are involved in effector interactions. Indeed, some evidence indicates that the effector-activating region of $G\alpha_s$ includes Switch II (167). However, results from other studies suggest that this hypothesis may oversimplify $G\alpha$/effector interactions. $G\alpha_s/G\alpha_i$ chimeric studies identified a 121-residue stretch (residues 236–356) from $G\alpha_s$ that confers adenylyl cyclase stimulatory ability upon $G\alpha_i$ (21). Within this linear stretch, three additional regions were identified as critical for adenylyl cyclase activation. Similarly, in vitro studies involving $G\alpha_t$ and its effector, cyclic GMP-phosphodiesterase, suggest that the effector interaction domain is contained within a relatively short continuous sequence. A short synthetic peptide corresponding to residues 293–314 of $G\alpha_t$ mimics the activity of $G\alpha_t$-GTP (314) (that is stimulate phosphodiesterase activity).

G Protein β-γ Subunits (Gβγ).

Relatively fewer cDNAs corresponding to distinct Gβ and Gγ have been isolated form mammalian tissues. Only five mammalian Gβ (117, 224, 370, 395, 401) and seven Gγ clones have been identified (51, 346), each family demonstrating limited diversity. Four Gβ are between 79% and 90% identical and over 80% identical with homologs from other species. However, a unique mammalian Gβ ($G\beta_5$) isoform demonstrating only 53% identity with previously characterized Gβ was recently cloned (401). Gβγ was once thought to play a passive role in signaling, orienting $G\alpha$ to the receptor. Given the explosion in identified effectors for Gβγ, it is apparent that Gβγ forms a second arm of G protein–mediated signal transduction, and probably interacts with components from other signaling paradigms [reviewed in (74)]. Characteristics of the cloned mammalian Gβ and Gγ are summarized in Table 18.2.

Structures of Gβ and Gγ. A clear picture is lacking of the domains responsible for Gβγ interactions with receptor, $G\alpha$, and effectors. Predicted primary structures of the cloned Gβ suggest the predominance of two types of structures [reviewed in (271, 272)]. The extreme amino-terminus (30–40 residues) probably forms an amphipathic α-helix and is the most divergent region among Gβ. The remainder of the polypeptide consists of seven homologous cassettes, each containing a signature central tryptophan-aspartate (WD-40) motif surrounded by ~40 residues (110). The WD-40 motif is found in a number of diverse proteins but its significance in function and structure of Gβγ is uncertain (393). Gγ form a slightly larger and more diverse family than the Gβ. Attention to Gγ structure has focused on a conserved CAAX (C = Cys, A = aliphatic residues, X = leucine or serine) motif found in the extreme Gγ carboxyl-terminus (120). This motif signals protein isoprenylation, a modification critical for association of many proteins (including ras) to the membrane. Isoprenylation is only one step of a concerted modification that involves the addition of either a farnesyl or a geranylgeranyl moiety on cysteine, endoproteolytic cleavage of AAX, and the carboxy-methylation of the exposed isoprenylated cysteine (152). All mammalian Gγ are thought to be isoprenylated. Gγ isoprenylation is not required for Gβ and Gγ interactions. However, Gβγ lacking the isoprenylation modification are cytosolic and are inefficient in effector regulation and $G\alpha$ interactions (152, 163). The nature of the isoprenoid modification is likely to be critical for Gβγ function. For example, recombinant $G\beta_1\gamma_1$ is much less potent than $G\beta_1\gamma_2$, $G\beta_1\gamma_5$, and $G\beta_1\gamma_7$ in the stimulation of the muscarinic-gated potassium channel I_{KACh} (235, 412). $G\gamma_1$ is farnesylated while the

TABLE 18.2. *Characteristics of Mammalian Gβ and Gγ Isoforms*

Subunit	Size, kDA	Expression	Reference
β_1	36	Ubiquitous	(370)
β_2	35	Ubiquitous	(117)
β_3	36	Cone photoreceptor	(224)
β_4	37	Ubiquitous	(377)
β_5	39	Brain	(401)
γ_1	8	Photoreceptors	(346)
γ_2	6	Brain, adrenal, other (?)	(287)
γ_3	7	Brain, testes	(121)
γ_4	?	Kidney, retina (?)	(121)
γ_5	7	Ubiquitous	(51)
γ_6	7	Brain, other (?)	(378)
γ_7	7	Ubiquitous	(51)

remaining Gγ have either been shown to be, or are predicted to be, geranylgeranylated (152).

Gβγ interactions. Combination of different Gβ and Gγ isoforms possibly results in signaling molecules that possess unique effector, receptor, or Gα interaction characteristics. However, not every Gγ isoform can combine with every Gβ isoform. For example, Gβ$_2$ associates with Gγ$_2$ but not Gγ$_1$ (163, 311, 333). Restrictions on isoform combinations may minimize the extent of crosstalk in cells that rely on Gβγ to transduce signals to multiple effectors. The amino-terminus of Gβ appears to be involved in Gγ interactions, as determined by chemical cross-linking (47). Consistent with this result, the expression of mutant Gβ lacking amino-terminal residues generates Gβ that are incapable of interacting with Gγ as assayed by the extent of membrane localization of the recombinant Gβγ complexes (119). Deletion of 33 amino-terminal residues of Gβ$_1$, for instance, abolishes Gβ$_1$γ$_2$ formation. Structural models based on amino acid sequence suggest the intriguing possibility that Gγ interacts with Gβ through a coiled-coil interaction (the supercoiling of two or three strongly amphipathic α-helices (119). Coiled-coil interactions are responsible for the dimerization of several DNA-binding proteins, including GCN4 (290), fos, and jun (291). The putative coiled-coil formed by Gβ and Gγ may not be the only structure determining subunit assembly, since putative α-helical domains are found in all mammalian Gβ isoforms, but not all combinations of Gβ and Gγ associate. Indeed, a small (14 amino acid) internal domain in Gγ$_1$ and Gγ$_2$ has been implicated in their discrimination of Gβ$_1$ and Gβ$_2$ (356).

Although the amino-terminal domain of Gα has been shown by several approaches to be involved in interactions with Gβ, little is known about the domains within Gβ important for this interaction. Sequence alignment between the five cloned Gβ indicates that the amino terminal (putative α-helical) domain is the most divergent region of Gβ (401). Experiments employing antipeptide antibodies against defined regions of Gβ$_1$ and Gγ$_1$ (subunit composition of native transducin Gβγ) have implicated regions in both subunits that are potentially involved in the interaction with Gα$_t$. These regions include the extreme amino-terminus in Gβ, the extreme carboxyl-terminus of Gγ, and two internal domains of Gβ predicted to be highly hydrophilic (located on the protein surface) (264). In contrast, antipeptide antibodies directed against the carboxyl-terminus of Gβ$_t$ and the amino-terminus of Gγ$_t$ did not interfere with Gα interactions.

In yeast, several Gβ mutants were generated and characterized which blocked signaling from the pheromone receptor (GCR) to an unidentified effector that stimulates the mitogen-activated protein kinase pathway (see under Ras Stimulates the MAPK Pathway Via Its Effector, Raf-1). These "dominant-negative" Gβ mutations (mutant Gβ that could not regulate effector activity *and* prevented normal Gβ from regulating effector activity) were clustered in two distinct regions of the yeast Gβ, one corresponding to the α-helical domain (extreme amino-terminus) of mammalian Gβ, the other corresponding to a yeast-specific Gβ sequence. Because mutations in both regions generated nonfunctional Gβγ which also blocked the function of normal Gβ the simplest conclusion is that the mutant Gβ can interact with Gγ, but they compete for effector interaction sites (220). This study implicates the amino-terminus of Gβ in mediating effector interactions.

A weakly conserved domain found in several diverse proteins potentially mediates protein interactions with Gβγ. The pleckstrin homology (PH) domain was initially studied as an internal repeat in pleckstrin, the main substrate for receptor-mediated PKC phosphorylation in platelets (388). PH domain–containing proteins include many with important signaling roles in multiple signaling paradigms: the phospholipase PI-PLC-γ, the receptor tyrosine-kinase-binding protein GRB-7, the ras-GTPase enhancer protein p120GAP, and GCR-specific kinase βARK [reviewed in (267)]. The PH domain in βARK is positioned in a carboxyl-terminal region, a region shown to bind Gβγ directly (201). PH domains from other proteins have also been reported to bind Gβγ (383). The numerous, diverse signaling complexes containing PH domains suggest a much wider role for Gβγ than previously appreciated. However, the physiological roles for these interactions have not been determined. Because Gα does not possess a PH domain, multiple Gβγ interaction domains probably exist.

Gβγ has been implicated in the regulation of a wide variety of effectors [reviewed in (74)]. The nature of the Gβγ-dependent βARK regulation suggests that one function of Gβγ is to mediate membrane association of enzymes. Gβγ-dependent interaction possibly recruits cytosolic enzymes to the membrane surface and/or receptor, closer to a substrate. Interestingly, the receptor for activated protein kinase C is a Gβ homolog, suggesting that the Gβ family subserves a general protein-anchoring function (324). However, as several of the reported Gβγ effectors are already membrane-associated, Gβγ must regulate effector activity in other ways as well.

G Protein Subunit Regulation

Covalent modification. As mentioned above, the palmitate modification of certain Gα can be dynamically regulated (95,263). Stimulation of β-adrenergic receptors in S49 *cyc−* cells (cell line with β-adrenergic receptors but lacking functional $G\alpha_s$) resulted in the increased turnover of palmitate in transfected $G\alpha_s$, as well as the cytosolic translocation of $G\alpha_s$ (402). Since mutant $G\alpha_s$ lacking the palmitate modification are incapable of efficient effector (adenylyl cyclase) and membrane interactions (403), regulation of palmitate content may provide subtle control over the localization and activity of $G\alpha$. Some evidence suggests that phosphorylation of G protein subunits occurs in vitro and in vivo and can affect their function. For example, $G\alpha_z$ is reported to be phosphorylated by protein kinase C in vitro, and in vivo following treatment of cells by phorbol 12-myristate 13-acetate (258) (PMA: direct stimulator of PKC). Src, a nonreceptor tyrosine kinase, phosphorylates several classes of purified $G\alpha$ on tyrosine residues (146). Since $G\beta\gamma$ inhibits this phosphorylation, the $G\alpha$ amino-terminus (the region implicated in $G\beta\gamma$ interactions) may be the target of src. Interestingly, overexpression of src in fibroblasts enhanced adenylyl cyclase activity and intracellular cAMP increases in response to β-adrenergic agonists (49, 273). Further evidence that G protein subunit phosphorylation occurs in vivo comes from the cAMP-activated signaling pathway in *Dictyostelium*. Exposure to cAMP results in a transient serine phosphorylation of $G\alpha$ (131). There is currently no strong evidence for a role for phosphorylation in $G\alpha$, $G\beta$, or $G\gamma$ function.

Cell-cycle-dependent subunit expression. Several experimental observations support the hypothesis that G protein expression is both cell- and cell-cycle–dependent, and is subject to change upon differentiation or maturation. For example, the calcitonin receptor couples to increases in cAMP levels (G_s-mediated) or PKC activity (G_i-mediated) depending upon the position of the cell in the cell cycle (61). Also, an increase in $G\alpha_o$ expression in the neuroectoderm is observed immediately prior to the initiation of neural induction in the maturing embryo (307). Similarly, monitoring the expression of $G\alpha_o$, $G\alpha_i$ and $G\alpha_s$, during *Xenopus* embryo development suggests that different classes were expressed almost exclusively in the gastrula ectoderm and neuroectoderm (292). Some reports have demonstrated that mRNA and/or protein content of specific G proteins is altered during differentiation induced by steroid hormones, growth factors, lymphokines, or second messengers such as cAMP. For example, $G\alpha_{i2}$ and $G\alpha_{i3}$ are transcriptionally activated in a coordinated fashion during differentiation, yet the transcription of the two genes differs in response to steroids (154). The mechanisms underlying all of these observations are not understood.

Alteration of G protein cycling. The ratio of GTP to GDP-bound $G\alpha$ can be modified by two classes of regulatory proteins: those that stimulate the dissociation of GDP from $G\alpha$ (GDS: GDP Dissociation Stimulator) and those that catalyze the hydrolysis of GTP (GAP: GTPase Activating Protein). The distinction between a GDS and a GAP is subtle, since an increase in GTP hydrolysis rate can be explained by stimulation of $G\alpha$ GDP release or stimulation of $G\alpha$ GTPase activity. To distinguish between a GDS and a GAP, one needs to determine the rates of GDP dissociation (or GTP uptake) and GTP hydrolysis with purified components. Clear examples of both types of proteins have been shown for the family of small GTP-binding proteins (see under Ras Activation by RTKs). Only the ligand-bound GCR and certain peptides like mastoparan have been shown to possess GDS activity for heterotrimeric G proteins (151). GAP activities in G protein–coupled signaling systems have been reported. For example, the inhibitory subunit of the effector in phototransduction, cGMP phosphodiesterase, exhibits GAP activity toward transducin-GTP (8, 85). Similarly, phospholipase C-β stimulates the GTPase activity of $G\alpha_q$ (22). The acceleration of GTPase activity of $G\alpha$ by downstream effectors may represent a common characteristic of signaling systems.

GAP-43 (<u>G</u>rowth-<u>A</u>ssociated <u>P</u>rotein, not GTPase-Activating Protein), a neuron-specific protein, has been shown to stimulate GTP binding to G_o (367). G_o and GAP-43 are predominant proteins in the neuronal growth cone. Studies of purified GAP-43 indicate that depalmitoylated GAP-43 stimulates G_o nucleotide exchange (369). Palmitoylated GAP-43, however, is ineffective. Thus, membrane bound, palmitoylated GAP-43 might be a reservoir of inactive protein, while depalmitoylation generates the pool of active GAP-43. It is not understood whether G_o/GAP-43 interaction enhances receptor signaling through G_o or disrupts G_o–receptor interactions or G_o–effector interactions. Similarly, it has been reported that an associated ras-GTP/p120-GAP complex can functionally uncouple G proteins from the muscarinic receptor (m2) to the ion channel I_{KACh} by increasing the rate of G protein cycling (427). In light of the new evidence for a direct interaction between $G\beta\gamma$ and p120-GAP, it is possible that the observed uncoupling was due to sequestration of $G\beta\gamma$, the physiological mediator of I_{KACh} activation (see later under Ion Channels).

Phosducin is a retinal and pineal phosphoprotein that copurifies with transducin $G\beta\gamma$ during isolation. Recombinant phosducin has been shown to inhibit the GTPase activity of G_s, G_i, and G_o, probably by acting as a trap for $G\beta\gamma$ ($G\beta\gamma$ stimulates the GTPase activity of Gα by stabilizing the end product, Gα-GDP) (15). Interestingly, PKA-mediated phosphorylation of phosducin prevents its association with Gβγ. Phosducin homologs have been isolated from other tissues, suggesting it serves a general (yet unknown) function in G protein–mediated signaling pathways.

Diseases Linked to GCRs and G Proteins

Given the diversity in GCR ligands, a multitude of pathophysiological states could derive from inherited or acquired alterations in elements of G protein–coupled pathways [reviewed in (72, 316, 354)]. Surprisingly, only a few diseases have been linked *conclusively* to GCRs and G protein subunits (Table 18.3). Disease severities range from color blindness due to mutations in cone-specific opsins (GCR), to potentially fatal disorders such as neonatal severe hyperparathyroidism and familial hypocalciuric hypercalcemia resulting from mutations in the serum calcium-sensing GCR (human homolog of BoPCaR1). Consistent with a role for certain G protein–coupled pathways in mitogenesis, some GCRs and G protein subunits are proto-oncogenes (113). Oncogenic potential of GCRs is supported by the enhanced mitogenesis and tumorigenicity of fibroblasts overexpressing constitutively-active α_{1b}-adrenergic receptor mutants (2). In addition, the *mas* and *gip*2 oncogenes belong to the GCR and Gα families, respectively. Because many GCR ligands serve important roles in the function of the nervous system, the involvement of G protein components in a variety of neurological disorders such as Alzheimer's disease, Parkinson's disease, schizophrenia, manic depression, alchoholism, and obesity is currently being studied. Drugs used to treat some of these afflictions target specific GCRs. For example, clozapine, a common treatment for psychomotor disorders such as Parkinson's disease and schizophrenia, is a dopamine receptor antagonist. Currently, evidence linking specific defects in G protein–coupled pathway elements to the pathogenesis of these diseases is circumstantial. Because of the complexity of G protein–coupled pathways, GCRs or G proteins need not be the primary targets. Indeed, disruption of ligand synthesis or degradation, effector function, or regulatory components could lead to a variety of disease states. Also, similar GCR or G protein isoforms could rescue serious disorders by substituting for improperly functioning elements. Despite the

TABLE 18.3. *GCR and G Protein–Linked Diseases** *

Disease	Defective Component
Familial hypocalciuric hypercalcemia	Human analog of BoPCaR1
Neonatal severe hyperparathyroidism	Human analog of BoPCaR1 (homozygous)
Hyperthyroidism (thyroid adenomas)	Thyrotropin receptor
Precocious puberty	Lutenizing hormone receptor
X-linked nephrogenic diabetes insipidus	V2 vasopressin receptor
Retinitis pigmentosa	Rhodopsin
Color blindness, spectral sensitivity variations	Cone opsin (iodopsin)
Familial glucocorticoid deficiency, isolated glucocorticoid deficiency	Adrenocorticotropic hormone receptor
Albright hereditary osteodystrophy (AHO)	$G\alpha_s$
Cholera	$G\alpha_s$
Whooping cough	$G\alpha_{i/o}$
Somatotroph adenomas	$G\alpha_s$
Pituitary tumors	$G\alpha_s$ (*gsp* oncogene)
Pituitary, thyroid, adrenal cortical, and ovarian tumors	$G\alpha_{i_2}$ (*gip* oncogene)
McCune-Albright syndrome (MAS)	$G\alpha_s$
Dilated cardiomyopathy	$G\alpha_i$
Testotoxicosis and type Ia Pseudohypoparathyroidism	$G\alpha_s$ (160)
Type I lissencephaly with Miller-Dieker syndrome	Gβ-like protein?? (317)

* For reviews see (72, 354, 355).

redundancy in GCR and G protein families, however, the list of diseases associated with G protein–pathway components will continue to grow.

Examples of G Protein–Regulated Effectors and Signaling Pathways

Currently, G protein subunits are known to modulate effector activity by various mechanisms. Some effectors (cyclic GMP-phosphodiesterase and I_{KACh}, discussed below) are regulated exclusively by either $G\alpha$-GTP or $G\beta\gamma$. In contrast, different adenylyl cyclase isoforms demonstrate stimulation solely by $G\alpha_s$, inhibition by either $G\beta\gamma$ or $G\alpha_i$, synergistic interactions between $G\alpha_s$, and $G\beta\gamma$, and indirect G protein stimulation by increased intracellular Ca^{2+} concentration. Both $G\alpha$ and $G\beta\gamma$ can stimulate certain phospholipase C isoforms. Clearly, one of the most exciting recent discoveries is the multiplicity of isoforms within G protein effector families. Multiple, differentially regulated isoforms in effector families potentially link the effector activity to various transmembrane signaling paradigms. Four effectors associated with G protein–regulated systems are described below. Although many other effectors exhibit regulation by G protein subunits, these four effector families (cGMP-PDE, adenylyl cyclases, phospholipases, and ion channels) and their relevant pathways have been relatively well studied. Examination of these systems gives an indication of the unique adaptations utilized by effector families and signaling pathways to respond appropriately to multiple forms of stimuli.

Phototransduction. Phototransduction involves the conversion of photon energy into electrochemical signals that alter the activity of image-processing neurons in the brain. Specialized photosensory neurons (photoreceptors) located in the retina mediate the conversion of light into an electrochemical signal. The phototransduction system is extremely sensitive and responds rapidly to stimuli. These characteristics result from unique signal-transducing components expressed exclusively in the retina. Retinal-specific GCRs and G protein subunits function according to the general G protein signaling scheme, yet they possess unique characteristics allowing for rapid responses to stimuli and enhanced signal amplication. In addition, specific desensitization mechanisms at every level of signaling permit rapid system adaptation in the continued stimulus exposure. Based on morphology and function, two classes of vertebrate photoreceptors coexist in the retina. Photoreceptor cones and photoreceptor rods mediate color and black/white vision, respectively. Since rods are more abundant, the majority of research on

phototransduction has relied on studies of rods. The rod photoreceptor is cylindrical and is divided into two parts: the rod inner segment containing the nucleus and other common cellular structures, and the rod outer segment (ROS), which is comprised of ~ 1000 flat, membranous discs arranged in a stack and surrounded by plasma membrane. The elements that mediate photoreception and signal conversion are located in the easily isolated ROS, permitting the study of this unique signaling organelle in isolation. Enrichment of the components of phototransduction in ROS minimizes signaling delays due to diffusion. Compartmentalization of phototransduction elements also minimizes crosstalk from other pathways, affording clear interpretation of experimental results. For these reasons, the mechanisms of phototransduction are understood better than those of any other signaling pathway.

A simplified model of phototransduction in rods is presented in Figure 18.5. The photoexcited GCR, rhodopsin, undergoes a conformational change and interacts with the photoreceptor-specific G protein, transducin, located in ROS membranes. Formation of the receptor–G protein complex results in transducin activation, which proceeds to stimulate cGMP phosphodiesterase (cGMP-PDE), also located in ROS membranes. cGMP-PDE hydrolyzes cGMP, shown to activate specific cation-selective ion channels in ROS plasma membranes. The decrease in cGMP concentration in the cytoplasm results in closing of cGMP-dependent channels and hyperpolarization of the plasma membrane, which modulates neurotransmitter release at the synapse. Detailed descriptions of pathway components and regulatory mechanisms are discussed below. The mechanism of phototransduction is the subject of several recent reviews (203, 304, 426).

Rhodopsin, the photon-sensing GCR, is found in extremely high density in ROS membranes. Approximately 10^9 rhodopsin molecules are found in a single ROS (107). High rhodopsin density is required for efficient light absorbance. Rhodopsin is linked covalently to the photoisomerizable chromophore, 11-*cis*-retinal. A single photon isomerizes 11-*cis*-retinal very rapidly (<200 femtoseconds) into 11-*trans*-retinal (334). The photoisomerization of *cis*-retinal is sensed by rhodopsin and translated into conformational changes in the receptor. One of the intermediate conformational states of rhodopsin, Meta-II, initiates the phototransduction cascade by stimulating the release of GDP from transducin (158).

Transducin $G\alpha_t\beta\gamma$ is activated by Meta-II rhodopsin. $G\alpha_t\beta\gamma$ is also very abundant in ROS ($\sim 10\%$ of rhodopsin density) and is bound weakly to ROS membranes in the basal (dark) state. Photoactivation of rhodopsin results in the formation of a tight complex

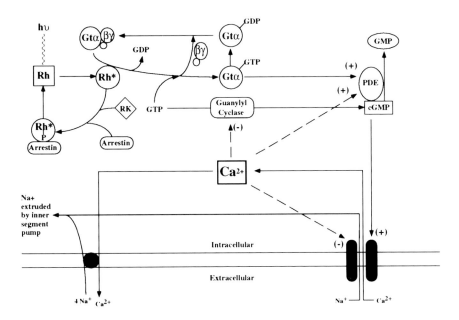

FIG. 18.5. Simplified representation of phototransduction system. A photon (hυ) stimulates rhodopsin *(Rh)*, which activates G protein transducin (Gα$_t$βγ). Gα$_t$-GTP stimulates cyclic GMP phosphodiesterase *(PDE)*, which decreases intracellular cGMP concentrations. Decreased cGMP closes a Ca^{2+}-permeable ion channel, decreasing intracellular Ca^{2+} levels, which inhibits the release of neurotransmitter at the photoreceptor synapse. Regulatory mechanisms are discussed in text. [Modified from Koutalos and Yau (203) and used with permission.]

with Gα$_t$βγ-GDP and the promotion of GDP release. In fact, under these conditions (rhodopsin activation in the absence of GTP), transducin binds rhodopsin so strongly that purification of Gα$_t$βγ was achieved under these conditions. The crystal structure of rod Gα$_t$ shows that bound nucleotide is buried within a deep cleft (278, 349). Meta-II rhodopsin may pry open this cleft and facilitate dissociation of GDP and binding of GTP. In the presence of GTP, the rhodopsin–transducin complex is dissociated. Also, Gα$_t$-GTP dissociates from Gβγ. Changes in the state of transducin lead to changes in the infrared light-scattering profiles of ROS suspensions (390). Monitoring these changes is noninvasive and provides excellent temporal resolution, making infrared scattering an ideal technique for studying transducin behavior in phototransduction. Using this technique, it was shown that the interaction of activated rhodopsin with transducin, as well as GDP-GTP exchange, takes ~1 ms (397). Subsequently, activated rhodopsin is free to catalyze nucleotide exchange on another Gα$_t$βγ. One molecule of activated rhodopsin can activate hundreds of Gα$_t$βγ-GDP. The activation of multiple Gα$_t$βγ-GDPs by one activated rhodopsin is the primary mechanism for achieving signal amplification in phototransduction (397).

Gα$_t$-GTP activates a specific phosphodiesterase (PDE) which hydrolyzes cyclic GMP (cGMP) to GMP. cGMP-PDE is bound peripherally to the cytoplasmic face of disc membranes and consists of four subunits: two distinct catalytic subunits (PDEα and PDEβ: PDEαβ) and two identical inhibitory subunits (PDEγ). Removal of the inhibitory subunits by limited proteolysis results in the impressive, 1000-fold stimulation of enzyme activity in vitro (159, 231) Gα$_t$-GTP stimulates cGMP-PDE to nearly the same extent in vitro. These observations led to the proposal that Gα$_t$-GTP activates cGMP-PDE by removing PDEγ from the enzyme. Support for this proposal came from experiments showing constitutive stimulation of cGMP-PDE and physical release of PDEγ from the membrane by washing ROS with GTP-containing solutions (424). In addition, immunoprecipitation by anti Gα$_t$ monoclonal antibodies demonstrated that stoichiometric complexes of Gα$_t$/PDEγ were formed with Gα$_t$-GTP (114).

The role of cGMP as an intracellular messenger in phototransduction was established by several experimental approaches. First, long and intense exposure of retinas or isolated rods to light resulted in a large (~50%) and rapid decrease in intracellular cGMP concentrations (30, 84). Second, a functional role for cGMP was observed when injection of cGMP into photoreceptors increased the light-sensitive membrane current (184, 256). Similarly, cGMP-PDE inhibitors increased the light-sensitive current as well as intracellular cGMP concentrations. cGMP was identified as a direct regulator of the light-sensitive current by patch-clamp experiments (109, 428). cGMP activates ROS plasma membrane cation-specific channels identical to those activated by light. Half-maximal activation of these channels is achieved by 20–30 μM cGMP. Importantly, channel activation exhibits positive cooperativity with respect to cGMP concentrations, making channel activity responsive to small changes in cGMP levels. cGMP effects on these channels are not mediated by protein kinases or other cellular factors.

The ROS cGMP-dependent channel protein was isolated by monitoring cGMP-dependent channel activity in lipid bilayer channel reconstitutions following various protein purification schemes (80). Incorporation of the purified channel in lipid vesicles or planar lipid membranes resulted in the functional reconstitution of cGMP-dependent ion fluxes with properties similar to those of ROS cGMP-dependent channels. Subsequently, the cDNA coding for the channel was cloned and was shown to generate a 63 kDa product located exclusively in the plasma membrane (181). The heterologous expression of in vitro transcribed channel mRNA in *Xenopus* oocytes resulted in the appearance of a cGMP-dependent conductance. Interestingly, a putative second subunit of the cGMP-gated channel has been cloned recently (66). Co-expression of this subunit with the channel subunit in *Xenopus* oocytes results in a "flickery" appearance of the channel openings characteristic of the native channel.

The activity of the cGMP-dependent channel from ROS is regulated by several mechanisms not involving cGMP. cGMP-dependent channel conductance is regulated by Ca^{2+}/calmodulin which binds to an uncharacterized protein associated with the channel (155). Also, decreased levels of intracellular Ca^{2+} during the photoresponse results in the increase of channel affinity for cGMP, thereby maintaining maximal sensitivity of the system to minor changes of cGMP concentrations. This is important because although the concentration of cGMP in ROS is high, the amount available for hydrolysis after normal exposures to light is small [about 5% of total; (85, 86)]. Finally, phosphorylation of the channel or auxiliary subunits by protein kinase C (PKC) may play a role in photoreceptor light adaptation, modulating the sensitivity of the channel to cGMP (125).

The activity of cGMP-dependent channels returns to basal levels within seconds (depending on stimulus intensity) of a pulse of light (230). Turning off the phototransduction cascade involves the deactivation of three species: activated rhodopsin, $G\alpha_t$-GTP, and PDE$\alpha\beta$, as well as restoration of the basal cGMP concentration. Photoreceptor responses to light terminate within a matter of seconds, due primarily to homologous desensitization at the receptor level. Rhodopsin is inactivated via phosphorylation by rhodopsin kinase, a member of the GRK family located in ROS membranes (294, 295). Rhodopsin kinase specifically phosphorylates photoactivated rhodopsin in multiple points on its carboxyl-terminal tail (413). In vitro experiments approximating native conditions demonstrate that receptor phosphorylation occurs within 1 s of stimulation (330). Phosphorylation, however, only partially reduces the ability of rhodopsin to activate transducin. The receptor–G protein interaction is completely uncoupled by the binding of the soluble ROS protein arrestin (413) to the phosphorylated carboxyl-terminal tail of the receptor.

$G\alpha_t$-GTP is inactivated by GTP hydrolysis. Recent observations show that PDEγ promotes the enhancement of transducin GTPase activity (8, 9). PDEγ remains bound to $G\alpha_t$-GDP even after hydrolysis of GTP, and is released from $G\alpha_t$-GDP only after phosphorylation of PDEγ by a ROS-specific protein kinase (386) and recombination of $G\alpha_t$-GDP with $G\beta\gamma_t$ (424). Subsequently, PDEγ reassociates with and inactivates PDE$\alpha\beta$. One mechanism that potentially accelerates desensitization at the effector level involves intracellular Ca^{2+}. Ca^{2+} enters ROS through the cation-selective cGMP-gated channels. In the dark, Ca^{2+} entry through these channels is exactly compensated by its efflux through $Na^+/Ca^{2+}/K^+$ exchanger (255). Upon illumination, Ca^{2+} influx through cGMP-gated channels decreases, while its extrusion by the $Na^+/Ca^{2+}/K^+$ exchanger remains constant. Exposure to light, then, results in a decrease of intracellular Ca^{2+} levels from ~300 to ~70 nM. It has been shown that PDE light sensitivity and inactivation rate is regulated by the Ca^{2+}-binding protein S-modulin in a Ca^{2+}-dependent manner (182, 183, 185). The decrease in intracellular Ca^{2+} levels results in lower PDE activity and an increased inactivation rate.

Restoration of the dark level of intracellular cGMP requires the stimulation of a cGMP synthesizing enzyme, guanylyl cyclase. Unlike PDE, guanylyl cyclase is not stimulated by light in disrupted photoreceptor membranes. However, light does stimulate cGMP synthesis in intact photoreceptor cells (5). Apparently, the decrease in intracellular Ca^{2+} levels (due to the closing of Ca^{2+}-permeable channels during the photoresponse) stimulates guanylyl cyclase (199). Activation of guanylyl cyclase by low intracellular Ca^{2+} levels is mediated by an ~20 kDa protein purified from bovine rod outer segments (123). In a reconstituted system, this protein restores the Ca^{2+}-sensitive regulation of guanylyl cyclase (97, 123). Following a stimulus, the decrease in intracellular Ca^{2+} level produces the Ca^{2+}-free form of the ~20 kDa protein, which proceeds to stimulate guanylate cyclase. Under basal conditions, the relatively higher level of intracellular Ca^{2+} levels renders this enzyme inactive.

Adenylyl Cyclases. Adenylyl cyclases convert ATP to cAMP. cAMP elicits responses from several targets, including ion channels and receptors *(Dictyostelium)*. However, the pervasive influence of cAMP is mediated by the family of cAMP dependent protein kinases (PKA), which phosphorylate enzymes, transcription factors, ion channels, transmembrane transporters, and cytoskeletal proteins in response to increased cytosolic cAMP levels. The cloning and characterization of several isoforms of adenylyl cyclase has revealed the amazing potential of this family for linking distinct signaling pathways to a common enzymatic activity. Distinct regulatory properties and tissue-specific expression of isoforms, receptors, and G proteins probably shape the particular cAMP response to diverse stimuli in diverse cell types.

Mammalian adenylyl cyclases are integral membrane glycoproteins. Isolation and sequencing of peptide fragments from a purified adenylyl cyclase led to the cloning of the first mammalian adenylyl cyclase gene (206). Subsequently, at least six more distinct isoforms of adenylyl cyclase have been discovered (12, 52, 108, 117, 148, 180, 207, 309, 429). Predictions based on primary structure suggest a common topology consisting of two similar cytoplasmic domains (~350 residues) separated by two hydrophobic, membrane-spanning segments (Fig. 18.6). The similar cytoplasmic domains are highly conserved among all mammalian adenylyl cyclases, and their importance is revealed by the disruption of enzymatic activity due to various point mutations within either domain (378). Interestingly, each subunit of the heterodimeric soluble *guanylyl* cyclases possesses homologous domains, suggesting their importance in the function of both families of enzymes (202). Expression of either the amino- or carboxyl-terminal halves of adenylyl cyclase (type 1) alone gave no functional enzymatic activity, while coexpression restored enzyme function (379).

All mammalian adenylyl cyclases (AC) are stimulated by $G\alpha_s$-GTP and forskolin. The interaction between $G\alpha_s$-GTP and AC isoforms has been demonstrated using native purified proteins, recombinant purified proteins, and heterologous coexpression experiments. Stimulation of AC by $G\alpha_s$ and forskolin in in vitro assays using purified components suggests that the

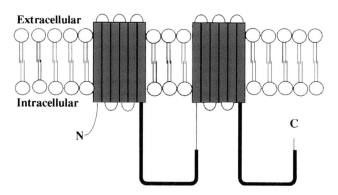

FIG. 18.6. Topology of mammalian adenylyl cyclase. Two homologous cytoplasmic domains *(thick lines)* are flanked by two membrane-spanning domains containing six putative α-helices each.

interaction is direct. However, there is no clear picture of the regions involved in adenylyl cyclase activation. Members of the $G\alpha_i$ family appear capable of inhibiting the activity of most AC isoforms (64, 380, 381). Binding studies suggest independent interaction sites for $G\alpha_s$ and $G\alpha_i$ (381). AC-II and AC-III activities were only partially inhibited when coexpressed with $G\alpha_i$ (168). $G\alpha_o$ was shown to inhibit only AC-I (381). For some isoforms, the inhibition by $G\alpha_i$ is suppressed by PKC activation. Heterologous expression and purification of individual isoforms permitted the characterization of the unique regulatory mechanisms controlling the activity of each of the adenylyl cyclases. The regulatory properties of each isoform are summarized in Table 18.4.

The unique signaling potentials of individual AC isoforms stem from their differential responses to $G\beta\gamma$, intracellular Ca^{2+}, and phosphorylation. $G\beta\gamma$ effects can either be inhibitory [AC-I (379)] or synergistic with $G\alpha_s$ [AC-II and AC-IV (116, 374)]. The inhibitory effect of $G\beta\gamma$ on AC-I is not due to sequestration of $G\alpha_s$, and appears to be mediated by an independent binding site for $G\beta\gamma$ (381). There is currently debate concerning the physiological relevance of these effects (305, 375).

Intracellular Ca^{2+} also regulates adenylyl cyclase in an isoform-specific manner. For example, AC-I, AC-III, and AC-VII are stimulated by Ca^{2+}/calmodulin) (52, 68, 379), while AC-V and AC-VI are inhibited by micromolar concentrations of intracellular Ca^{2+} in vitro (82, 83, 429). Importantly, Ca^{2+}/calmodulin activation of AC-I and AC-III does not require $G\alpha_s$ opening the regulation of these isoforms to non-$G\alpha_s$-coupled receptors, and there is evidence for in vivo regulation of AC-I and AC-VI by Ca^{2+} (83). Because most transmembrane signaling directly or indirectly regulates intracellular Ca^{2+}, these effector isoforms are particu-

TABLE 18.4. *Regulation of the Various Types of Adenylyl Cyclase* *

Regulatory Entity	Mode of Regulation	Adenylyl Cyclase Type
$G\alpha_s$	Activation	1–7
$G\alpha_i$	Inhibition	1,5,6,2(?),3(?)
$G\alpha_o$	Inhibition	1
$G\beta\gamma$	Inhibition	1
	Activation	2,4
Ca^{2+}/CaM	Activation	1,3,7
Free Ca^{2+}	Inhibition	5,6
PKC	Activation	2,1(?),3(?),7(?)

* Modified from Tang and Gilman (376) and used with permission.

larly intriguing as points of convergence for many pathways. Phosphorylation has also been implicated in adenylyl cyclase regulation. For example, indirect studies on intact and permeabilized cells suggest that stimulation of PKC results in the activation of adenylyl cyclase (67, 258). Recently, in vitro PKC-dependent phosphorylation of adenylyl cyclase was shown to stimulate cAMP production by AC-II (170, 430). AC-

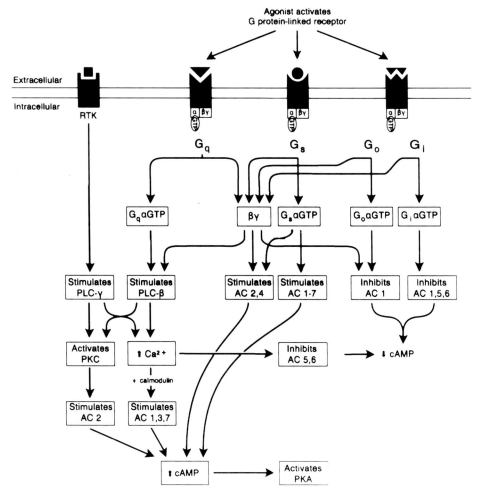

FIG. 18.7. Crosstalk in receptor coupling to adenylyl cyclase isoforms. [Modified from Pieroni et al. (305) and used with permission.]

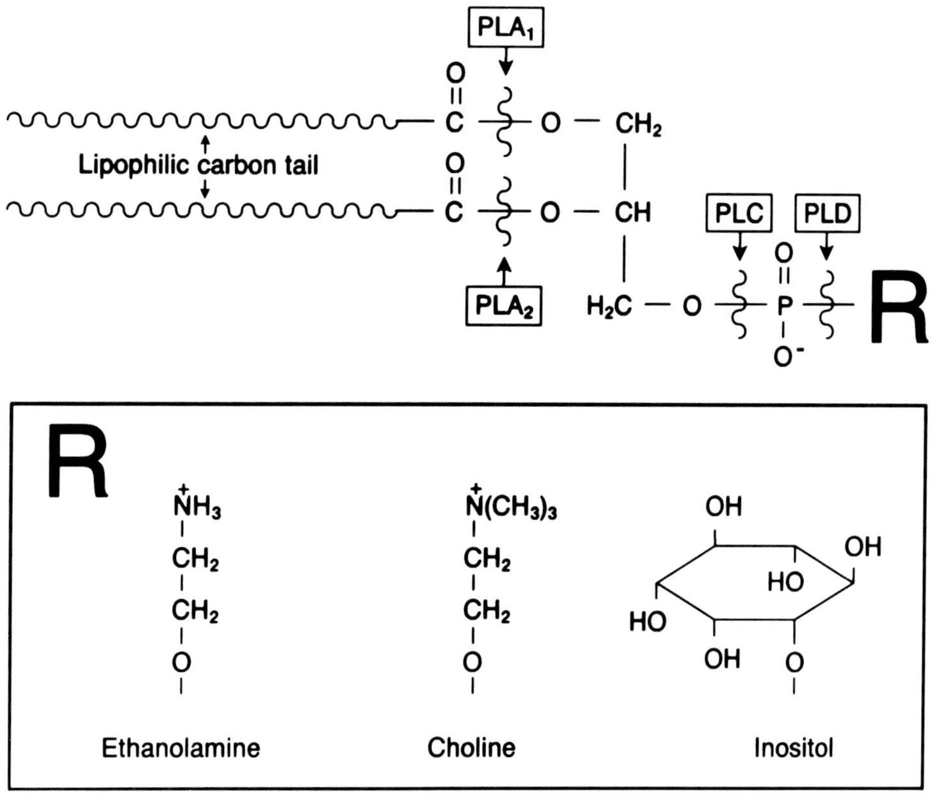

FIG. 18.8. Preferential substrate for family of phospholipases is phosphoglyceride with either ethanolamine, choline, or inositol group in R position.

I and AC-III (67), as well as AC-VII are probably also activated by PKC phosphorylation (169). Phospholipase C activity (see the section on Phospholipases) can have complicated effects on cAMP levels by simultaneously increasing intracellular Ca^{2+} and stimulating PKC (via diacylglycerol). High intracellular Ca^{2+} concentrations can stimulate AC-I and AC-VII, or inhibit AC-V and AC-VI, while stimulation of PKC can activate AC-I, AC-II, AC-III, or AC-VII.

Assuming a cell expresses multiple adenylyl cyclase isoforms, the potential for AC regulation from distinct receptors is impressive (Fig. 18.7). AC-I and AC-II can be stimulated in the absence of activated G_s, AC-II is synergistically activated by $G\beta\gamma$ and stimulated by PKC. In contrast, AC-IV is potentially inhibited by many pathways that increase intracellular Ca^{2+}. The situation in a typical cell is probably not as depicted in Figure 18.7. Although isoform-specific antibodies to adenylyl cyclase are not yet available, analysis of mRNA levels in various tissues suggests that isoform expression is not uniform. Variable cAMP responses in different cells to identical stimuli indicate that cells must be capable of uncoupling adenylyl cyclases from certain stimuli. For example, NGF often increases

cAMP levels in target cells. Because NGF can activate PLC activity which raises intracellular Ca^{2+} and stimulates PKC, stimulation of either AC-I or AC-II can explain this increase in cAMP. However, NGF does not increase cAMP levels in all target cells (305). Whether this discrepancy is due to cell-type-dependent expression of adenylyl cyclase isoforms or receptor to effector coupling components specific to the NGF signaling pathway is not well understood.

Phospholipases. Cellular responses to the activation of many hormone, neurotransmitter, and growth factor receptors are mediated by a family of enzymes, termed *phospholipases*, which hydrolyze phospholipids (phosphoglycerides). Three subclasses of phospholipases (phospholipases A, C, and D) have been characterized based on the specific bond hydrolyzed in the predominant phospholipase substrates (inositol-containing phospholipids, phosphatidylcholine, and phosphatidylethanolamine). Depending upon the substrate and the phospholipase, unique cleavage products are generated (Fig. 18.8). Some of these cleavage products have proven to be important second-messenger molecules (Fig. 18.9). The most ubiquitous and well-characterized

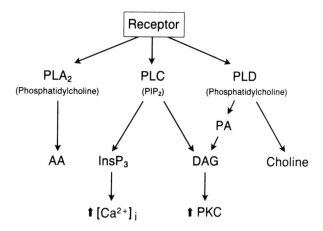

FIG. 18.9. Stimulation of many transmembrane receptors leads to activation of phospholipases. Many important second messengers are generated from action of these phospholipases, including *AA* = arachadonic acid, *InsP₃* = inositol-1,4,5-trisphosphate, and *DAG* = diacylglycerol.

phospholipid-derived second messengers are the soluble inositol trisphosphate (InsP₃), which promotes the release of Ca^{2+} from intracellular stores, and the lipophilic compound diacylglycerol (DAG), which stimulates PKC (26). InsP₃ and DAG are generated by the actions of phospholipase C (PLC) on phosphoinositol-bisphosphate (PIP₂) (25).

All members of the PLC family preferentially hydrolyze inositol-containing phospholipids (phosphoinositides): phosphotidylinositol (PI), phosphatidylinositol-4-monophosphate (PIP), and phosphatidylinositol-4,5-bisphosphate (PIP₂) (321). Therefore, stimulation of PLC activity by any mechanism results in the generation of InsP₃ and DAG. PLC isoforms have been

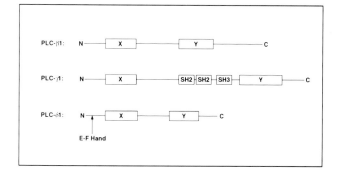

FIG. 18.10. Structural alignment of members of *PLC* (phospholipase C) subfamilies. *X* and *Y* denote critical catalytic domains found within all PLC isoforms. SH2 and SH3 domains, located in PLC-γ isoforms, interact with activated tyrosine-kinase-linked receptors and effector molecules, respectively. EF-hand is a putative Ca^{2+}-binding motif located exclusively in PLC-δ isoforms.

isolated from bovine brain, guinea pig uterus, rat brain and liver, and human platelets and range in size between 85 kDa (PLC-δ) and 150 kDa (PLC-β) (321). The PLC family can be divided into three subclasses based on primary structure comparison of characterized isoforms. The PLC-β, PLC-γ, and PLC-δ subclasses are each comprised of several isoforms: PLC-β₁₋₄, PLC-γ₁₋₂, and PLC-δ₁₋₃ (320). Primary structure similarity between the PLC subclasses is low except in two regions, commonly designated X and Y (Fig. 18.10). Regions X and Y contain ~150 and ~120 amino acids, respectively. Between PLC subclasses, 80 amino acids are either invariant or conservatively substituted within these two domains, while only eight residues are conserved outside these regions (320). Several conserved amino acids in regions X and Y are critical for enzymatic activity. In the PLC-β and PLC-δ families, short segments of 50–70 amino acids separate the X and Y regions. In contrast, the X and Y regions in PLC-γ are separated by ~400 amino acids. Another distinction between the PLC subclasses is the length of the domain following the Y region (carboxyl-terminal tail). This domain is ~450 residues long in PLC-β, while virtually nonexistent in PLC-δ (320).

Regulation of PLC-β and PLC-γ activities is relatively well characterized, while regulation of PLC-δ isoforms is not. PLC-γ isoforms are stimulated by activated growth factor receptors (RTKs). The extended domain separating regions X and Y in PLC-γ isoforms contains src homology domains (SH2 and SH3) which permit the association of PLC-γ with RTKs (320). Association with (and phosphorylation by) RTKs is required for PLC-γ activation. In contrast, PLC-β isoforms are regulated by activated G protein subunits. Initial purification of PLC from the cytosol was puzzling since, given the membrane localization of G proteins and the PLC substrates, it was assumed that PLC was membrane-associated. Successful purification of the first membrane-associated PLC-β isoform required prior stimulation of a PLC-β-linked GCR (321). The membrane-associated PLC-β was identical to a PLC characterized previously from the cytosol, suggesting that PLC-β translocates between the cytosol and membrane upon pathway stimulation. Although PLC translocation has been verified repeatedly, the mechanism is unknown. PLC translocation may regulate PLC-mediated signaling by bringing the enzyme near the substrate and/or ensuring efficient interaction with the activated G protein (425).

All members of the PTX-insensitive $G\alpha_q$ family directly activate PLC-β₁ in vitro (23, 222, 320, 348, 382). $G\alpha_q$ stimulates PLC-β₁ ≈ β₄ > β₃ >> β₂ (173, 298, 320). An in vivo interaction between $G\alpha_q$ and PLC-β₁ has been inferred from transfection studies.

Transfection of $G\alpha_q$ cDNA into COS cells resulted in an increase in $InsP_3$ generation. Furthermore, cotransfection of $G\alpha_q$ and PLC-β_1 cDNAs resulted in even higher levels of $InsP_3$ formation (421). In vitro, PLC-β_1 increases the intrinsic rate GTP hydrolysis of $G\alpha_q$ and $G\alpha_{11}$ by 50-fold over the rate in the absence of effector (22). In this manner, PLC-β_1 acts as a GTPase-activating protein (GAP) for $G\alpha$, rapidly terminating its own activation (22). Mutations and deletions of amino acids in PLC-β_1 suggest that the carboxyl-terminal domain (extended in PLC-β subclass; Fig. 18.8) is required for activation by $G\alpha_q$ but is not involved in the intrinsic enzyme function (419). Also, the carboxyl-terminal domain, specifically residues 903–1030, was shown to be important for PLC-β_1 translocation to the membrane (419).

Both $G\alpha_q$ and $G\beta\gamma$ appear capable of stimulating PLC-β_2-activity (55, 298, 420) $G\beta\gamma$ effectively stimulates PLC-β_2 as demonstrated using isolated membranes from mammalian cells transfected with PLC-β_1 and PLC-β_2 cDNA (55). This role for $G\beta\gamma$ probably explains the isolated reports of pertussis-toxin–sensitive, receptor-mediated activation of PLC. $G\beta\gamma$ stimulates PLC-$\beta_3 > \beta_2 > \beta_1$ and β_4 (75, 320). To test the specificity of PLC-β_2 for distinct $G\beta\gamma$ complexes, recombinant $G\beta_1\gamma_2$, $G\beta_1\gamma_3$, $G\beta_2\gamma_2$ and $G\beta_2\gamma_3$, as well as purified bovine brain $G\beta\gamma$ and turkey erythrocyte $G\beta\gamma$, were tested for their ability to activate turkey erythrocyte PLC-β_2 and brain PLC-β_1. All $G\beta\gamma$ combinations inhibited $G\alpha_{11}$ activation of PLC-β_1, presumably by forming the inactive heterotrimer (41). All $G\beta\gamma$ tested stimulated turkey erythrocyte PLC-β_2, but not bovine brain PLC-β_1. Of the four recombinant $G\beta\gamma$ complexes, three stimulated PLC activity similarly, while $G\beta_2\gamma_3$ was relatively ineffective (41). Coexpression of $G\alpha_{16}$ and $G\beta_1\gamma_1$ in COS-7 cells synergistically activated recombinant PLC-β_2. Based on this synergistic activation of PLC-β_2, the authors concluded that the regions of PLC-β_2 required for activation by $G\beta\gamma$ and $G\alpha$ are physically separate, and the nature of $G\gamma$ protein may determine the relative specificity of the G protein complex (420).

PLC signaling via both PLC-β_1 and PLC-β_2 is attenuated by PKC and PKA (320). Evidence suggests that the GCRs are one target for these kinases. In addition, PKC phosphorylation of the EGF receptor (RTK) prevents the RTK-dependent tyrosine phosphorylation of PLC-γ_1 (320). Since PLC stimulation generates $InsP_3$ and DAG, and DAG (in the presence of Ca^{2+} and lipids) stimulates PKC activity, inhibition of PLC activity induced by various GCRs and RTKs is a potential direct-feedback mechanism.

Despite the wide distribution of PLC-δ isozymes in diverse organisms (yeast to mammals), the receptor which stimulates receptor-linked PLC-δ activation is not known. The PLC-δ family has an EF hand motif (putative Ca^{2+}-binding domain) not found in the other PLC subclasses. Interestingly, despite the lack of an EF hand motif, PLC-β and PLC-γ activities are sensitive to Ca^{2+} (222). The EF-hand in PLC-δ lies ~150 amino acids upstream from region X (11). The function of this motif is uncertain. Surprisingly, in Chinese hamster ovary (CHO) cells overexpressing PLC-δ, thrombin-induced $InsP_3$ formation (presumably mediated by PLC-δ) was dependent on extracellular Ca^{2+} concentration (13).

In contrast to the PLC subfamily of phospholipases, phosphatidylcholine (PC) and phosphatidlyethanolamine (PE) are the preferential substrates for PLA. These PLA isoforms hydrolyze the acyl or alkyl ester bond at the second position of phosphatidylcholine (Fig. 18.8). Arachidonic acid is the preferred fatty acid in this position over other polyunsaturated fatty acids (103). PLA activity generates free fatty acids and lysophospholipids (100). The liberation of arachidonic acid is the rate-limiting step in the formation of prostaglandin, thromboxane, and leukotrienes, all mediators of inflammatory and allergic reactions (204, 232) Lysophospholipids can be converted into platelet-activating factor, another potent mediator of the inflammatory response (232).

Two isoforms of PLA (PLA_1 and PLA_2) have been characterized. The involvement of PLA_1 in signal transduction is uncertain. PLA_2 isoforms are grouped loosely as secretory or cytosolic PLA_2 (204) Secretory PLA_2 ($sPLA_2$) isoforms are convenient for biochemical and biophysical studies since they are enriched in pancreatic secretions and snake and insect venom (337). However, the role of $sPLA_2$ in transmembrane signaling is not as well defined as the role for cytosolic PLA_2. Cytosolic PLA_2 ($cPLA_2$) generates arachidonic acid in response to many stimuli including ATP, thrombin, phorbol esters (which stimulates PKC), and Ca^{2+} ionophores (103, 232). G proteins control $cPLA_2$ activity in some cells [HL60 cells, platelets, Swiss 373 fibroblasts, thyroid cells and osteoblasts (103)]. as shown by generation of free arachidonic acid by nonhydrolyzable GTP analogs and AlF_4^-. In addition, antibodies to $G\alpha_i$ and $G\alpha_o$ inhibit the histamine- and thrombin-stimulated activation of $cPLA_2$ (103). Direct physical and functional coupling of G proteins to $cPLA_2$ has not been reconstituted with purified components. $cPLA_2$ is strongly activated by μM concentrations of Ca^{2+} (a level of Ca^{2+}_i that can be achieved during receptor stimulation). Increasing intracellular Ca^{2+} levels result in the translocation of $cPLA_2$ from the cytosol to the membrane. A putative Ca^{2+}-binding domain is present near in the amino-terminus of $cPLA_2$. Several studies have shown that, in addition to stimulation by Ca^{2+}, $cPLA_2$ is stimulated by protein kinases, including PKC (103, 232). Mutation of the phosphorylated residue decreases agonist-induced arachidonic acid release (232). Interestingly, the mitogen-activated protein kinase (MAPK) also phosphorylates $cPLA_2$, a requirement for full activation of $cPLA_2$ (232). PKC phosphorylation and MAPK phosphorylation sites are different (103). Therefore, any pathway capable of elevating intracellular Ca^{2+} concentrations and/or activating PKC, and/or MAPK could potentially stimulate $cPLA_2$.

Phospholipase D (PLD) removes the polar head group of PC leaving phosphatidic acid and a base such as choline. The enzyme reacts with PC and PE, but only weakly with phosphatidylinositides. DAG is formed by the concerted actions of PLD and phosphatidate phosphohydrolase. Although PLD has not been purified completely or cloned, it has been partially purified from brain, lung, and HL60 cells (103). G proteins have been implicated in PLD activation. PTX inhibits the agonist-stimulated PLD activity in neutrophils, HL-60 cells and rat 1a fibroblasts (103). Following bombesin exposure, PLD activation was detectable in Swiss 3T3 cells. The response was delayed, however, suggesting that PLD is activated by downstream events (398). PDGF, basic FGF, bombesin, and EGF also stimulate PLD activity. In addition, small GTP-binding proteins have been shown to interact functionally with PLD (45). A direct interaction between PLD and G proteins has not been proven.

Ion Channels. Ion channels mediate the passive flux of physiological ions (Na^+, Ca^{2+}, K^+, Cl^-) across hydrophobic membranes. Electrophysiological studies, particularly those involving excitable tissues (brain, heart, skeletal muscle), have shown that ion channel activity can be regulated by direct binding of neurotransmitters, covalent modification, diffusible and membrane-bound second messengers, transmembrane potential, and combinations thereof. Diversity in ion channel selectivity and regulation permits their control over unique cellular activities, from propagating action potentials in excitable tissues to maintaining osmotic homeostasis in epithelial cells. Importantly, a single channel can be regulated by multiple mechanisms. Similarly, stimulation of a single receptor can alter the activity of many different ion channels in a cell. Thus, the particular profile of cellular ion channel activity represents an integrated response of channels to all modes of regulation. Techniques in protein biochemistry and molecular biology have begun to unravel the molecular structures underlying ion channel selectivity and regulation. For recent comprehensive reviews on G protein regulation of ion channels, see (71, 411).

Most channels are modulated in different ways and to varying extents by signaling cascades initiated by receptor stimulation. For example, G protein–stimulated cGMP-PDE controls the activity of the cGMP-gated channel of the phototransduction system (see the section phototransduction). Similarly, G protein–dependent increases in intracellular Ca^{2+} can modulate the activity of several Ca^{2+}-dependent channels. Control over cAMP, DAG, and Ca^{2+} allows GCRs to recruit members of the main classes of serine/threonine kinases into signaling functions. Recently, effects of tyrosine kinases on ion channel activity have suggested a role for these kinases (156). Most channels studied appear to be substrates for kinases.

The role of channel phosphorylation is understood in only a few clear cases. For example, the sympathetic regulation of heart rate initiated by β-adrenergic receptor agonists involves the phosphorylation of several voltage-dependent channel types that are responsible for shaping the action potential. The importance of cAMP and PKA in this response was appreciated before the identification of G proteins in signal transduction (318, 385), and the cloning of the affected channel types. Now, the cloned channel subunits have been shown to be substrates for various kinases in vitro, and mutagenesis of the phosphorylation sites has supported the role for various kinases in channel regulation. By inference, many channel phosphorylations are assumed to be important regulatory mechanisms when the actual physiological role is poorly understood. We acknowledge the important role for G protein–stimulated second-messenger pathways in channel regulation but suggest the interested reader consult recent comprehensive reviews regarding this topic (71, 411). Instead, we focus on two well-studied examples of membrane-delimited ion channel regulation.

Channel regulation by G proteins is often independent of cytosolic elements. In these cases, all elements required for channel regulation are membrane-associated. As such, this type of channel regulation is referred to as "membrane-delimited" (44). Since receptors, G proteins, and channels are all membrane-associated, the simplest explanation for membrane-delimited channel regulation is a direct interaction between the channel and the functional G protein subunit. However, many components of cell membranes remain to be characterized. Since the list of G protein–coupled effectors is growing steadily, and there currently is evidence for channel regulation by lipophilic compounds such as arachidonic acid (a PLA_2 product), alternative mechanisms can explain membrane-delimited channel regulation. Incontrovertible evidence for a direct functional interaction between ion channels and G protein subunits requires satisfying two criteria: *(1)* A direct, reversible, physical interaction between purified ion channel and G protein subunit must be demonstrated. *(2)* A functional interaction between the ion channel and G protein must be reconstituted in a pure system (lipid bilayer) using purified components. Currently, no ion channel regulated by G proteins in a membrane-delimited fashion has satisfied both criteria. Because direct interactions between ion channels and G proteins are probably common, the dearth of supporting evidence is blamed on the extremely low density of ion channels in membranes and on the difficulties associated with functional ion channel reconstitution.

Support for G protein regulation of ion channels can come from several experimental approaches. The requirement for GTP in a GCR-mediated channel response is suggestive of a role for G proteins. In addition, regulation of channel activity by nonhydrolyzable analogs of GTP strongly supports a role for G proteins. Targeting the G protein heterotrimer or activated G protein subunits is also a useful tool for identifying G proteins in channel regulation. For instance, block of GCR-mediated channel regulation by pertussis toxin implicates the $G_{i/o/t}$-subclasses of G proteins. Antibodies or compounds that inhibit the function of activated G protein subunits can be used to identify G proteins responsible for channel regulation. For example, inhibition of $G\beta\gamma$ effects on channel activity by application of $G\alpha$-GDP or by PH domain–containing proteins supports a functional role for $G\beta\gamma$ in channel regulation (200). Alternatively, targeting G protein subunit mRNA to degradation using antisense oligonucleotides has been used to identify specific G proteins involved in channel regulation (194–196). The most direct approach is to apply purified G protein subunits to exposed cytoplasmic faces of patches of membranes. Advances in preparation of recombinant G protein subunits have greatly facilitated this approach for studying G protein–coupled ion channels.

Potassium-selective (K^+) ion channels comprise the largest and most complex class of ion channels and demonstrate virtually the entire spectrum of channel regulation (73, 153). Although the number of unique K^+ channels characterized is large, the opening of all K channels drives the transmembrane potential toward the equilibrium potential specified by the concentration of K^+ on both sides of the membrane. In most mammalian cells, the equilibrium potential for K^+ is between -70 and -90 mV. Because most voltage-gated channels in excitable tissues are inactive in this range, the effect of K^+ channel activation in excitable tissues is to hyperpolarize the membrane, dampening cell electrical activity. Conversely, blocking K^+ channels depolarizes (positive change in membrane potential) the membrane (70).

Stimulation of m2 muscarinic receptors by acetylcholine in pacemaker regions in mammalian atria stimulates a K channel (I_{KACh}) resulting in a slower heart rate. Although classically referred to as the muscarinic (acetylcholine)-gated potassium channel, I_{KACh} activity is also stimulated by a vast number of neurotransmitters in a membrane-delimited manner (279). Considerable controversy revolved around the identity of the activated G protein subunit responsible for I_{KACh} stimulation. The suggested role for $G\beta\gamma$-dependent I_{KACh} activation was met with resistance, grounded in the belief that $G\alpha$ provides the function and specificity to all G protein–linked signaling systems. Reports of extreme $G\alpha$ potency and suggestions that the observed $G\beta\gamma$ activation of I_{KACh} was artifactual supported the case for channel regulation by $G\alpha$ [reviewed in (44)]. However, work from several laboratories has established that $G\beta\gamma$ is the physiological activator of I_{KACh} (211, 319, 412, 423). The G protein regulation of I_{KACh} was examined recently using recombinant subunits (412). Six combinations of $G\beta$ and $G\gamma$ isoforms activated I_{KACh} with minimal differences, while all cases of activation by recombinant $G\alpha$-GTPγS were due to release of GTPγS from $G\alpha$. Indeed, since GTPγS alone activates many channels by stimulation of endogenous G proteins, it is critical to ensure that recombinant activated $G\alpha$ preparations are free of GTPγS (or other nonhydrolyzable GTP analogs). Inhibition of channel activity elicited by GTPγS was observed by $G\alpha$-GDP, indicating that $G\beta\gamma$ is the physiological regulator of I_{KACh} (Fig. 18.11). Evidence from the heterologous expression of a putative I_{KACh} clone [GIRK1 (209) or KGA (92)] supports the primary role for $G\beta\gamma$ in receptor-mediated I_{KACh} activation (319).

Whether $G\beta\gamma$ activation of I_{KACh} involves a direct or indirect interaction is a matter of current intense investigation, and it is likely that initial findings will precede publication of this chapter. It was reported that PLA$_2$ was a $G\beta\gamma$-regulated effector in a different system (172). Since arachidonic acid (product of PLA$_2$) stimulated I_{KACh} and an anti-PLA$_2$ antibody inhibited $G\beta\gamma$-dependent channel activation, it was suggested that $G\beta\gamma$ acts indirectly through PLA$_2$ to stimulate I_{KACh} activity. However, arachidonic acid stimulation of I_{KACh} was always significantly weaker than $G\beta\gamma$ stimulation (190) and PLA$_2$ inhibitors did not disrupt $G\beta\gamma$-mediated I_{KACh} activation [reviewed in (211)]. The physiological role of PLA$_2$-mediated channel activation is not clear, but our current view is that GCR-mediated I_{KACh} activation does not involve an interaction between $G\beta\gamma$ and PLA$_2$. Rather, GCR-mediated stimulation of I_{KACh} activity involves a direct $G\beta\gamma$-channel interaction. Recently, several K^+ channels similar to I_{KACh} in structure and function, yet with unique regulatory properties, have been cloned. Knowledge of the gene sequences of these channels permits intelligent experimentation designed to localize potential G protein (or unknown intermediate) interaction domains.

Voltage-dependent calcium channels (VDCC) are coupled to diverse cellular responses, including neurotransmitter release, hormone secretion, muscle contraction, and the activity of many crucial enzymes. Several different classes of VDCC exist based on pharmacology, localization, and specific voltage-dependent channel characteristics. G proteins mediate VDCC modulation in many cell types. Unfortunately, many studies to not identify the subtype of channels underlying the Ca^{2+} currents, making comparisons difficult. ω-conotoxin-sensitive (N-type) Ca^{2+} channels are found in many brain regions and are common targets of neurotransmitter modulation. Angiotensin II, acetylcholine (via m2 muscarinic receptors), norepinephrine (via α_2 adrenergic receptors), luteinizing hormone-releasing hormone (36) γ-aminobutyric acid (335), and glutamate (372) inhibit N-type Ca^{2+} channels in a membrane-delimited fashion via a pertussis-toxin–sensitive G protein(s) (248, 340, 341). The mechanism of G protein–mediated channel inhibition is not completely understood. Recent evidence suggests that a G protein–mediated conductance decrease through N-type Ca^{2+} channels underlies this inhibition (210).

SIGNAL TRANSDUCTION VIA RECEPTOR TYROSINE KINASES

Introduction to Receptor Tyrosine Kinases (RTKs)

The RTKs form a large class of proteins (>40 members) with related structures, cellular effects, and mechanisms of signal transduction. RTKs are found in both vertebrates and invertebrates but not in unicellular organisms, which led to the suggestion that RTKs have evolved with the necessity for cell-to-cell communication (143). RTKs transmit signals that culminate in cellular proliferation and differentiation. Not surprisingly, RTK mutants are implicated in mammalian cancers, and a number of viral oncogenes encode constitutively active RTK variants (322). For instance, overexpression of the RTK ligand, epidermal growth factor (EGF), or of the EGF receptor itself is a characteristic of 12% of all pancreatic cancers, 86% of adenomas of the salivary gland, and 68% of prostate carcinomas (310). These abundant "natural" RTK mutants have for over 10 yr stimulated research into pathways of cell growth and carcinogenesis. We shall describe here the well-known aspects of the RTK signal transduction paradigm, especially those common elements that permit RTKs to regulate cellular growth, and that also have a role in mitogenic signaling by non-RTK receptors such as G protein–coupled receptors. Several excellent reviews discuss RTK signal transduction in

a. Purified Native G protein subunits

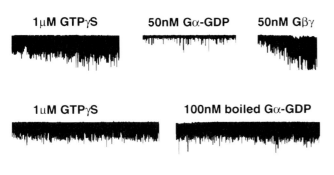

b. Purified recombinant G protein subunits

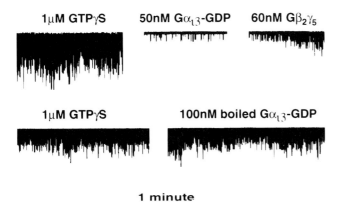

FIG. 18.11. I_{KACh} in inside-out patches from atrial myocyes is activated by $G\beta\gamma$ and GTPγS. GTPγS activation of I_{KACh}, occurring by activation of endogenous G proteins, completely inhibited by purified $G\alpha$-GDP, either bovine brain $G\alpha$ or recombinant $G\alpha_{i3}$. I_{KACh} is restimulated by a molar excess of $G\beta\gamma$ (from bovine brain or recombinant $G\beta_2\gamma_5$).

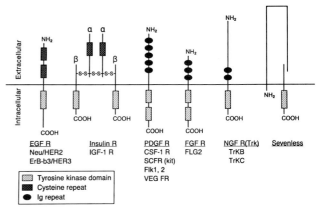

FIG. 18.12. Schematic depiction of families of RTKs.

more detail (28, 112, 157, 266, 296, 301, 332, 391).

In general, RTKs consist of single polypeptides that form an extracellular ligand-binding domain, a single transmembrane-spanning hydrophobic α-helical domain, and an intracellular domain with tyrosine kinase activity (Fig. 18.12). Structural similarities, presumably reflecting evolutionary relatedness, permit grouping of the well-characterized RTKs into the six families shown (157). In addition, three more families of proteins have been identified that possess unmistakable tyrosine kinase domains, typified by EPH (387), ark (283), and c-met (308). Like other RTKs, EPH and ARK can transmit mitogenic signals; however, little is known about their functional mechanisms or ligands. c-met binds hepatocyte growth factor and scatter factor and is a potent oncogene for hepatocytes. The six RTK families we discuss here are exemplified by the receptors for EGF, insulin, platelet-derived growth factor (PDGF), fibroblast growth factor (FGF), nerve growth factor (NGF), and the sevenless RTK of *Drosophila melanogaster.*

All RTKs contain tyrosine kinase domains that are essential for signal transduction. The protein tyrosine kinase domains of RTKs range from 250 to 300 amino acid residues in length and appear in the intracellular regions of RTK polypeptides. Highly conserved amino acids within the catalytic domain play important roles in catalysis and also appear in the catalytic domains of serine/threonine kinases and nonreceptor protein tyrosine kinases. The sequence G-X-G-X-X-G is required for nucleotide binding, and a specific lysine residue is necessary for catalysis. Mutation of the lysine results in a "dead" kinase, that is, a kinase unable to perform catalytically. "Dead" kinase mutants are frequently generated experimentally to determine if the activity of a specific kinase is required in a signaling pathway. While the kinase domain is absolutely required for signal transduction, it is not necessary for RTK expression and targeting to the plasma membrane. Near the center of the kinase domain lies a region that defines it as a member of the tyrosine kinase or serine/threonine kinase families. This region has been targeted for the design of degenerate oligonucleotides for use in screening cDNA libraries to identify novel members of both kinase families (143). A distinctive and important characteristic of RTKs of the PDGF and NGF receptor families is the interruption of the tyrosine kinase domain by an insert region of 70–100 amino acids. Although the kinase insert separates the nucleotide-binding and phosphotransferase regions of the tyrosine kinase, it does not appear to interfere with tyrosine kinase activity. Instead, the kinase insert region forms specific binding sites for proteins involved in signal transduction.

The RTK subfamilies families differ in the extracellular structures used for binding ligands and in the arrangement of intracellular regulatory regions. EGF receptor-like RTKs possess two cysteine-rich regions in their extracellular domains that facilitate ligand binding (57). Similar cysteine-rich motifs are found in the extracellular regions of RTKs in the insulin receptor family. As discussed below, the EGF receptor and most other RTK polypeptides dimerize only upon binding ligand; however, members of the insulin receptor family consist of distinct α and β subunits that are linked by disulfide bonds and even when the ligand is not present. The extracellular ligand-binding domains of the PDGF receptor and FGF receptor families are related and clearly structurally distinct from those of EGF and insulin receptors, consisting of five and three immunoglobulin-like (Ig) repeats, respectively. RTKs of the NGF receptor (trk) family possess two Ig repeats plus a leucine-rich motif at the amino-terminus. The Ig repeats probably fold into the stable scaffold structure of immunoglobulins. In immunoglobulins, this basic structure can be almost infinitely adapted to provide high-affinity binding sites for diverse small molecules including peptides and metabolites. Thus it is not surprising that this structure is also used by receptors to bind ligands.

The unique extracellular region of the sevenless RTK merits its classification as the single member of a distinct RTK family [reviewed in (139)]. Sevenless functions in the eye of the fruit fly, *Drosophila melanogaster,* to determine the fate of the R7 photocell. The extracellular domain of sevenless is synthesized as a 280 kDa precursor that is proteolytically cleaved into 220 kDa and 58 kDa pieces. After cleavage, the two pieces remain noncovalently associated. The smaller subunit forms the transmembrane region and the intracellular tyrosine kinase domain of sevenless and the larger subunit forms the ligand-binding region. In a departure from the structures of other RTK, the amino-terminal region of the larger subunit of sevenless is predicted to encode a second transmembrane-spanning domain (345). The distinctive extracellular domain of sevenless may be required for binding its unusual ligand, called "boss," which is an integral plasma membrane protein on an adjacent photoreceptor cell. Despite its distinctive extracellular domain, the intracellular domain of sevenless and its method of signal transduction closely resemble those of the EGF receptor.

Several RTK domains are capable of functioning independently in the context of any RTK. Although exceptions exist, this remarkable property has been confirmed and exploited in a large number of signal transduction studies. For example, the functions of chimeric receptors that have been created using molecular biology techniques to mix and match RTK domains are predictable: The source of the extracellular domain determines ligand specificity and the source of the intracellular domain determines the signal. This property has been proven by a large number of experiments using chimeric PDGF, EGF, FGF, and insulin receptors, and it is useful because it permits analysis of RTK signaling pathways even when the ligand is not known. For example, the mitogenic signals transduced by the neu RTK were studied by linking the intracellular domain of neu to the extracellular domain of EGF, transfecting the chimeric receptor into cells, and analyzing the effects of EGF application (18). This principle applies even to chimeric receptors constructed entirely from domains of non-RTK receptor polypeptides (166), suggesting that this property might possibly apply to all cell surface receptors consisting of polypeptides that span the membrane once. In another example, RTKs truncated to remove the intracellular tyrosine kinase domain still bind their ligand normally. In fact, some cells normally make such truncated RTKs that may play important physiological roles. Depending on whether the transmembrane domain is also present, these truncated RTKs provide soluble [mek4, EGFR, c-erbB-2; in (387)] or membrane-bound [trkB; (157)] receptors that bind and "mop up" excess ligand without contributing to signal transduction. Similar "extracellular-domain-only" RTK mutants have been generated experimentally to permit in vitro analysis of RTK ligand-binding structures.

Initial Steps of RTK Signal Transduction

The general scheme for signal transduction by most RTKs is diagrammed in Figure 18.13 (112, 332). Most

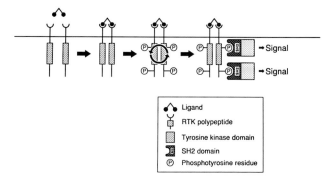

FIG. 18.13. General scheme of signal transduction by RTKs. Ligand binding leads to RTK dimerization, which activates RTK tyrosine kinase and leads to autophosphorylation. Binding of SH2 domains of intracellular proteins to specific receptor phosphotyrosine residues mediates signaling through ligand-bound receptor.

RTKs are monomeric in the absence of ligand, and ligand binding promotes RTK dimerization. Dimerization stimulates the activity of the RTK tyrosine kinase which autophosphorylates various tyrosine residues of the RTK. These phosphotyrosines form binding sites for proteins that mediate RTK signaling, as discussed below.

Ligand Binding Promotes Dimerization. Ligand binding induces dimerization of RTKs. A model for the EGF receptor proposes that two cysteine-rich regions are in contact with each other and are close to the plasma membrane, and that they form the EGF-binding site in a cleft between them (391). In this model, binding of one EGF molecule would bridge two EGF receptor polypeptides and promote their dimerization. It is also very common for the RTK ligand to be itself a dimer, and in such cases, receptor dimerization naturally follows ligand binding. For instance, each monomer of the dimer that forms PDGF binds to one PDGF receptor polypeptide, a process that promotes PDGF receptor dimerization (332). Amino acid residues in the transmembrane domains of some RTKs can also influence dimerization. An oncogenic form of the neu RTK exhibits constitutively active tyrosine kinase activity, apparently because a point mutation in the neu transmembrane domain enhances the tendency of neu polypeptides to dimerize (18). As suggested by this result, mutations that promote RTK dimerization are generally oncogenic (322).

Dimerization Activates the Tyrosine Kinase. Mutations of RTKs that interfere with dimerization generally prevent activation of the tyrosine kinase (332). RTK dimerization stimulates the activity of intracellular tyrosine kinase domains by two mechanisms that are used to different extents by different RTKs. First, dimerization can transmit activating conformational changes to the tyrosine kinase domain. Second, dimerization increases the probability that RTK polypeptides will be tyrosine phosphorylated by other RTK polypeptides (called "trans-autophosphorylation") at sites that stimulate or are required for the activity of their tyrosine kinases. Normal PDGF receptors highly overexpressed in insect cells are constitutively active, supporting the idea that close proximity alone can activate some RTK. The necessity for trans-autophosphorylation has been shown for EGF, PDGF, and FGF receptors using dominant-negative mutant RTKs. "Dominant-negative" is a term for a mutant protein that can interfere with functions of the normal protein when they are expressed together in the same cell. When normal RTKs are expressed together with large amounts of the same RTK containing the dominant-negative "dead" kinase mutation, ligand binding can no longer stimulate the RTK tyrosine kinase because every receptor dimer contains at least one "dead kinase" that is incapable of *trans*-autophosphorylation. Another RTK mutation that has the same effect as the "dead" kinase is an RTK truncated to remove the tyrosine kinase domain altogether, which is still capable of forming dimers with the normal RTK (389). Members of the insulin receptor RTK family do not utilize dimerization as a method for stimulating tyrosine kinase activity, since the insulin receptor polypeptides are linked by disulfide bonds in the absence of ligand. A single insulin molecule binds to both α and β subunits of the insulin receptor (223) and this event is proposed to initiate structural stresses that result in the allosteric activation of the intracellular tyrosine kinase.

RTK Autophosphorylation Leads to Binding of SH2- and SH3- Containing Proteins. Activation of the intrinsic tyrosine kinases of RTKs leads to tyrosine phosphorylation of other proteins and of the RTKs themselves (autophosphorylation). The main purpose of tyrosine phosphorylation of RTKs appears to be the generation of high-affinity binding sites for SH2 domains. SH2 domains, originally identified in the nonreceptor tyrosine kinase src (hence the name, s̲rc h̲omology region 2̲), are found in many proteins with important signaling functions [Fig. 18.14; reviewed in (112, 301)]. Many SH2-containing proteins also contain a distinct src homology region, SH3. SH3 domains are also found in proteins not obviously involved in signal transduction, for instance, cytoskeletal proteins [reviewed in (266, 301)].

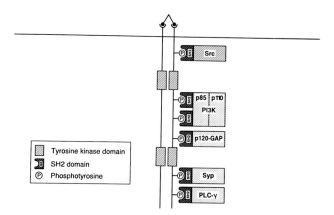

FIG. 18.15. SH2-containing proteins bind to specific phosphotyrosine residues of activated PDGF receptor. [Modified from Pawson and Schlessinger (301) and used with permission.]

SH2 and SH3 domains are formed by approximately 100 and 60 contiguous amino acids, respectively. SH2 and SH3 domains appear singly or in multiples and at various positions in proteins, suggesting that they form independently folded structures capable of functioning either alone or with a variety of other protein structures. This idea was confirmed by x-ray crystallographic studies of SH2 and SH3 domains that show their independent and conserved folding, and by the discovery that both SH2 and SH3 domains can mediate protein–protein interactions either as isolated domains or in their native protein contexts.

Both SH2 and SH3 domains mediate protein–protein interactions. SH2 domains bind with high affinities to short peptides, or to similar regions in proteins, that contain a phosphorylated tyrosine residue. Subclasses of SH2 domains discriminate among phosphotyrosine-containing peptides depending on the amino acid sequence surrounding the phosphotyrosine. This discrimination was shown by elegant experiments in which different SH2 domains were tested for binding to a "library" of >5000 different phosphotyrosine-containing peptides (350, 351). Activated RTKs use a similar mechanism to recruit proteins for signal transduction, but the sequence variations among RTKs generate distinctive ensembles of signaling molecules [see Table 18.5; (112, 187, 350)]. For instance, when the PDGF receptor binds PDGF, several intracellular tyrosine residues become phosphorylated, forming specific binding sites for the SH2 domains of different proteins [Fig. 18.15; (176, 301)] Recent results suggest that SH2 domains can also bind to phosphoserine- and phosphothreonine-containing peptides with low affinity (112). The physiological significance of low-affinity binding sites for SH2 domains is not known, but SH2 domains capable of such interactions are found in PLC-

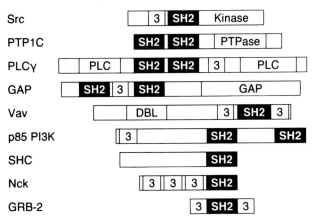

FIG. 18.14 SH2- and SH3-containing proteins. SH2 domains are depicted in linear amino acid sequence as *black box*. SH3 domains are depicted by *boxes labeled "3."* PTPase = protein tyrosine phosphatase, *PLC* = phospholipase C, *GAP* = GTPase-activating protein, *PI3K* = phosphatidylinositol-3-OH kinase, and *DBL* = Dbl homology. [Modified from Pawson and Gish (300) and used with permission.]

γ, p120-GAP, and src. The cognate ligands of SH3 domains are not as well defined, but SH3 domains clearly bind selectively to proline-rich sequences of nine or ten amino acids that appear in several proteins (407).

Ras Activation by RTK

Ras. Ras proteins regulate a mitogenic signaling pathway that is initiated by almost all RTKs and that culminates in activation of p42 and p44 mitogen-activated protein kinases (MAPK) (31, 77, 89, 93, 99, 245). Though this pathway is the most well-characterized outcome of RTK signal transduction, ras and MAPK are also activated by non-RTK receptors (discussed below). Insights from RTK and ras signaling may, therefore, pertain to signal transduction in general. Ras is a central regulator of cell growth, although in some cells it controls differentiation (189). Mutant *ras* genes are found in 30% of all human tumors and are responsible for the potent transforming activity of four oncogenic retroviruses. Ras proteins are members of a much larger class (>50 members) of small GTP-binding proteins, known as the ras superfamily, that are distinct from heterotrimeric G proteins. Other members of the ras superfamily regulate cytoskeletal reorganization (rho, rac) and membrane trafficking (rab).

Four ras polypeptides are expressed by mammalian cells: H-ras, N-ras, and two kinds of K-ras as (35, 37). The four ras isoforms are ubiquitously expressed and exhibit almost identical biochemical properties. All four ras isoforms are farnesylated at a specific cysteine near their carboxyl-termini, and some are also palmitoylated. Such lipid modifications promote membrane association, which is necessary for ras to function as a mitogenic signaling molecule (189). Both ras proteins and heterotrimeric G proteins utilize the GTP binding and hydrolysis "switch" for signal transduction. X-ray crystallographic analysis shows that most of the ras polypeptide folds into a single GTPase domain resembling the guanine nucleotide-binding core of Gα (78, 213, 278, 349) [reviewed in (417)]. Ras and Gα also contain similar blocks of amino acid sequence motifs, evidence of their evolutionary relatedness. However, ras proteins have no additional subunits, are much smaller than Gα (21 kDa vs. 39–52 kDa), and are not substrates for PTX or CTX ADP-ribosyltransferases.

Like Gα, ras proteins are active when bound to GTP and inactive after its hydrolysis to GDP. Regulatory proteins assist this cycle, as shown in Figure 18.16 (35, 189). Activation of ras in vivo is initiated by a GDP-dissociation stimulator, or GDS. Although this process resembles activation of heterotrimeric G proteins by

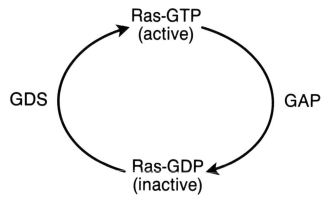

FIG. 18.16. Ras cycle. GAP (GTPase-activating proteins) and GDS (GDP-dissociating stimulators) catalyze activation and inactivation of ras, respectively.

ligand-bound GCRs, GDS proteins and GCRs are otherwise not alike. GCRs cannot directly activate ras, and neither do GDS proteins act on heterotrimeric G proteins. Compared to that of Gα, the intrinsic GTPase of ras is extremely inefficient. One plausible explanation is that the domain which stimulates the intrinsic GTPase of Gα is missing from the ras structure (244, 417). In vivo, hydrolysis of GTP bound to ras is achieved via direct interactions with GTPase-activating proteins (GAPs). Most ras mutations in tumors produce proteins that are defective in hydrolysis of bound GTP even when GAPs proteins are present, and are therefore referred to as "activating" mutations. One class of ras-activating mutation changes the glycine at codon 12 to any other amino acid except proline (338). Ras-V12 is one such activated mutant created for experimental purposes; v-ras is a naturally occurring activated mutant produced by a transforming retrovirus. There appear to be multiple GDS proteins and GAPs capable of interacting with ras. In addition, a protein that inhibits GDS (GDI) has been identified, and inhibitors of GAP activity (anti-GAP) could also exist.

Several approaches have been used to study the functions of normal and oncogenic mutant ras proteins. First, an increased ratio of ras-GTP compared to ras-GDP in a cell indicates that ras has been activated. Second, ras can be experimentally stimulated by GTPγS, or activated ras mutants can be expressed in a cell and the effects observed. AlF_4^- can be used to rule out GTPγS effects on heterotrimeric G proteins, because AlF_4^- activates heterotrimeric G proteins but not ras. Third, specific ras inhibitors can be used to determine if a signaling pathway requires ras. Ras inhibitors include monoclonal antibodies that block ras effects (for instance, Y13-259), a mutant ras protein (ras-N17), and rap-1a, a ras superfamily member capable of antagonizing cellular transformation by v-ras. Ras-N17 is also called "dominant-negative ras" because it can block signaling by normal ras. Fourth, the gene(s) encoding normal ras can be disrupted, an approach feasible in some systems.

The above methods have demonstrated that the effects of many growth factors require ras activation (189). For example, microinjection of monoclonal antibody Y13-259 prevents cells from proliferating in response to growth factors in serum (261). Similarly, ras-N17 expression prevents activation of MAPK in response to EGF, NGF, and FGF (418), and interferes with proliferation in response to PDGF, EGF, and FGF (50). Besides implicating ras in RTK signaling, such approaches have shown that mitogenic signaling by some non-RTK receptors, including GCRs, cytokine receptors (122, 192), and lymphocyte antigen receptors (53, 166), also involves ras. Thus, ras constitutes a basic signaling switch that can be co-opted by diverse signaling pathways in order to regulate mitogenesis.

Another experimental approach was used to address how ras interacts with its downstream targets which, for years, were completely unknown. Many natural and experimentally induced mutations of ras proteins were analyzed. Some of these mutant ras proteins were unable to promote mitogenesis despite binding GTP normally. These mutations clustered in a discrete region of the ras polypeptide (between amino acid residues 32 and 40) that was hypothesized to interact with the unknown ras targets, and which was therefore dubbed the "effector" domain of ras. As discussed below, it is now known that this domain directly binds to at least two important ras effectors, illustrating the accuracy of this area.

The EGF Receptor Activates Ras by Directly Binding to GRB-2/SOS.

The combined effects of GDS and GAP, and potentially GDI and anti-GAP, together regulate the amount of ras in its active, GTP-bound state (Fig. 18.17). Current information suggests that RTKs mainly stimulate ras by localizing GDS proteins into the vicinity of membrane-bound ras, using an indirect mechanism involving SH2- and SH3-containing proteins (112, 187, 189, 299, 301). The left part of Figure 18.17 summarizes this mechanism for the EGF receptor. Signal transduction begins when the EGF receptor binds ligand and dimerizes, stimulating autophosphorylation of the EGF receptor cytoplasmic domain. Phosphorylation of a specific tyrosine forms a high-affinity binding site for the SH2 domain of a protein called GRB-2. GRB-2 has no known intrinsic enzymatic activity. Instead, GRB-2 is composed almost entirely of SH2

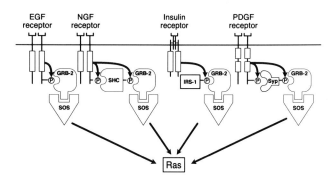

FIG. 18.17. Different RTKs use different methods to stimulate ras. After binding ligand (not shown), RTK phosphorylate specific tyrosines on themselves or other proteins. These phosphotyrosines form binding sites for SH2 domains of proteins that can recruit GRB-2/SOS.

and SH3 domains that permit it to act as a link between other proteins. While the SH2 domain of GRB-2 binds the activated EGF receptor, the SH3 domain of GRB-2 binds tightly to SOS, a GDS protein. Therefore, when GRB-2 binds the activated EGF receptor using its SH2 domain, it brings SOS along. SOS localized to the membrane in this manner has the opportunity to interact with ras, which is constitutively associated with the plasma membrane. SOS then stimulates GDP dissociation and GTP binding by ras, resulting in ras activation.

Several different research groups worked out this mechanism in the years between 1991 and 1993 by integrating RTK signaling research using worms, flies, and mammals [Fig. 18.18; reviewed in (105, 252)]. The genetic approaches permitted by the nonmammalian systems were particularly helpful. Starting with a signaling pathway that has a defined outcome (differentiation of a specific cell, for instance), one mutagenizes large numbers of organisms and then selects for those in which the signaling pathway no longer occurs (for example, those in which the cell no longer differentiates). Even without knowing the molecular nature of

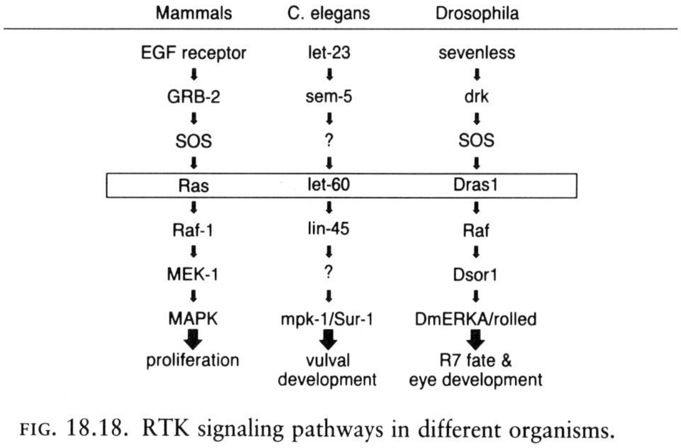

Mammals	C. elegans	Drosophila
EGF receptor	let-23	sevenless
GRB-2	sem-5	drk
SOS	?	SOS
Ras	let-60	Dras1
Raf-1	lin-45	Raf
MEK-1	?	Dsor1
MAPK	mpk-1/Sur-1	DmERKA/rolled
proliferation	vulval development	R7 fate & eye development

FIG. 18.18. RTK signaling pathways in different organisms.

the mutation, each of these mutants can be tested for its position in the signaling pathway using genetics. For instance, each mutation is combined with other mutations as well as with the normal genome to establish its dominance and/or membership in a complementation group. The short generation time, ease of genome manipulation, and haploid forms that are available for some nonmammalian organisms make this process feasible. Finally, a gene with a particularly interesting mutation can be identified and sequenced using the large number of chromosome libaries, genome maps, and other mutants that have previously been characterized for these organisms.

In the worm *Caenorhabditis elegans*, genetic analysis showed that a developmental signaling pathway beginning with an RTK (let-23) required expression of the sem-5 protein. Genetic analysis placed sem-5 in the pathway somewhere downstream of let-23 and upstream of the *C. elegans* ras (let-60; Fig. 18.18). Cloning and sequencing of sem-5 revealed it to have an SH3-SH2-SH3 structure (76). Meanwhile, a novel technique designed to clone proteins that bound tightly to the tyrosine-phosphorylated EGF receptor cytoplasmic domain identified a mammalian protein with the same structure that was named GRB-2 (239). The homology between GRB-2 and sem-5 suggested that GRB-2, by directly binding to the EGF receptor, somehow transmitted signals that activated ras in mammalian cells. This idea was supported by the discovery that injecting both GRB-2 and ras proteins into fibroblasts stimulated DNA synthesis, a known effect of ras activation (239). The missing link between GRB-2 and ras was provided by studies on the sevenless signaling pathway in *Drosophila*. A genetic screen showed that mutations in either a ras protein (Dras1) or a protein called son of sevenless (SOS) decreased the ability of sevenless (an RTK) to specify the fate of R7. When cloned and sequenced, it was discovered that the carboxyl-terminal portion of SOS contained a region of homology with the yeast *(Saccharomyces cerevisiae)* protein CDC25. Since CDC25 had just been discovered to be capable of promoting ras GDP dissociation and GTP binding in vitro, SOS was proposed to function in the sevenless signaling pathway as a direct upstream activator of Dras1 (38, 344). The connections between these results were unclear until two groups cloned a *Drosophila* GRB-2 *(drk)*,showed it was required in the sevenless signaling pathway using genetics, and demonstrated that it could directly interact with both sevenless and SOS in vitro (288, 344). Mammalian homologues of SOS were quickly cloned, and direct functional interactions between the EGF receptor and GRB-2 and SOS proteins from the various organisms were demonstrated in vitro and in vivo [reviewed in] (105, 252).

Other Mechanisms Used by RTK to Activate Ras. Figure 18.18. shows other ways used by RTKs to recruit GRB-2/SOS, in addition to the direct binding of GRB-2/SOS used by the EGF receptor (112, 187, 260, 301). A second pathway utilizes SHC proteins. When activated, EGF, NGF, and PDGF receptors bind a SHC protein and induce its tyrosine phosphorylation. An SHC phosphotyrosine forms a binding site for the SH2 of GRB-2, thereby bringing GRB-2 and SOS into the vicinity of the RTK. This localization of GRB-2/SOS apparently stimulates ras, since SHC binding to the NGF receptor is important for promoting the neuronal differentiation of PC12 cells (280, 359) an event known to require ras activation. Other stimuli also promote tyrosine phosphorylation of SHC, including insulin, cytokines, and oncogenic nonreceptor tyrosine kinases, suggesting that this is a widespread method of ras activation (312). The insulin receptor also employs a third distinctive method of ras activation utilizing a separate "docking protein" called IRS-1 to bind GRB-2/SOS (410). IRS-1 becomes tyrosine phosphorylated at many sites upon insulin stimulation, and one or more of these phosphotyrosines form sites where the SH2 of GRB-2 can bind. A fourth way RTKs can recruit GRB-2/SOS is suggested by research on a cytoplasmic phosphotyrosine phosphatase with two SH2 domains variously known as Syp, Csw, PTP-1D, or SH-PTP2. Growth factor treatment induces binding of syp to EGF and PDGF receptors (58), which apparently results in tyrosine phosphorylation of syp. Syp phosphorylation appears to create a binding site for the SH2 of GRB-2, since syp has been shown to mediate binding of GRB-2/SOS to the activated PDGF receptor (229). Thus, some RTKs may use syp to activate ras, an idea corroborated by genetic evidence from a *Drosophila* signaling pathway involving an RTK called Torso. Tyrosine phosphorylated IRS-1 also binds syp, which potentially provides the insulin receptor with a second way to stimulate ras activity. Interestingly, besides permitting GRB-2/SOS binding, growth-factor-induced tyrosine phosphorylation of syp also stimulates its tyrosine phosphatase activity. A signaling role for the enzymatic activity of syp is not known. Syp does not dephosphorylate the autophosphorylated PDGF receptor (394).

It is not clear why RTKs use so many different mechanisms to activate ras, but experiments using the pheochromocytoma cell line PC12 seem to address this question. The receptors for EGF and NGF both activate ras, but they transduce clearly different signals when expressed in PC12 cells. Application of EGF makes PC12 cells proliferate, while NGF makes them differentiate into neurons that can no longer divide (280, 359). Both RTKs activate ras, but with different kinetics: Compared to EGF, NGF activates ras for a more extended period (280, 359). Both RTKs bind tyrosine phosphorylated SHC, but only EGF receptors bind directly to GRB-2 (Table 18.5), suggesting that the RTKs might utilize different methods of ras activation in order to control the kinetics of ras activation. Alternatively, different substrates targeted by the EGF and NGF receptor tyrosine kinases might promulgate their different effects (313).

Other GDS Proteins Also Activate Ras. Although all RTKs apparently use GRB-2/SOS to activate ras, other GDSs exist which could mediate ras activation by other types of receptors [reviewed in (37, 189)]. Tyrosine-kinase-associated receptors like the T lymphocyte antigen receptor (TCR) may activate ras using a GDS called vav (90, 128, 130). Vav is expressed only in white blood cells and becomes tyrosine phosphorylated upon stimulation of the TCR. In vitro, tyrosine phosphorylation stimulates the GDS activity of vav, providing a means by which the TCR could activate ras. The GDS activity of vav is also stimulated by IL-1 or ceramide via a mechanism that does not require tyrosine kinases, suggesting that vav can mediate ras activation by additional receptor types (129). Other GDSs include ras-GEF from human and CDC25 from mammals and yeasts. Ras-GEF and CDC25 are not activated by known receptors and do not bind GRB-2. They also do not contain SH2 or SH3 domains, and no evidence links them to pathways that involve tyrosine kinases. CDC25 is, however, clearly involved in signal transduction. In the yeast *S. cerevisiae*, CDC25 activates ras (37, 177) in a pathway that leads to increased adenylyl cyclase activity. Mammalian ras proteins can substitute for the yeast ras, and mammalian homologues of

CDC25 have been cloned. The conservation of CDC25 throughout evolution suggests that it functions in an important pathway. The future identification of the receptor that regulates CDC25 will undoubtedly reveal other ways that plasma membrane signal transduction processes can activate ras.

Role of GAP in Signaling by RTKs. Figure 18.16 indicates that cellular levels of GTP-bound ras could also be elevated by decreasing GAP activity, in addition to, or instead of, increasing the GDS activities of proteins like SOS. Mammalian and invertebrate cells express several GAP proteins (189). p120-GAP, a ubiquitous protein, shows potent GAP activity towards ras in vitro, stimulating the ras GTPase 10,000-fold. p120-GAP similarly inhibits ras activation in vivo, since its overexpression in mammalian cells blocks the mitogenic functions of ras. p120-GAP binds to specific tyrosine-phosphorylated sites of many RTKs (Table 18.5), suggesting that in addition to its role as a general negative regulator of ras, p120-GAP might contribute to ras activation by RTKs. This idea is discussed below. It has also been suggested that p120-GAP might be an effector of ras (189), an idea based on the observations that p120-GAP binds to the effector domain of ras and that oncogenic ras mutants all bind to p120-GAP. However, p120-GAP cannot be the ras effector that promulgates mitogenesis since, as mentioned above, overexpression of p120-GAP blocks the mitogenic effects of ras. It is possible, however, that p120-GAP is a ras effector that acts in some unknown pathway. Besides p120-GAP, a GAP called NF-1 is expressed ubiquitously by most tissues (137). Aside from a ho-

TABLE 18.5. *SH2-Containing Proteins that Bind to Activated Tyrosine Kinases*

SH2-Containing Protein	Receptor		Tyrosine	Kinase	
	PDGF R (β)	EGF R	NGF R (trk)	c-kit	IRS-1
PLC-γ	+	+	+	−	−
p120-GAP	+	+	+	−	−
PI3 K	+	+	+	+	+
src	+	?	?	?	−
GRB-2/SOS	−	+	−	?	+
syp	+	+	?	+ *	+
SHC	+	+	+	?	−
nck	+	+	?	?	+

Distinctive ensembles of signaling molecules generated by RTKs. PLC = phospholipase C, GAP = GTPase-activating protein, P13 K = phosphotidylinositol-3-OH kinase, SOS = son of sevenless, PDGF R = platelet-derived growth factor receptor, EGF R = epidermal growth factor receptor, NGF R = nerve growth factor receptor.
    * Predicted in (350).

mologous GAP domain, NF-1 is not similar to p120-GAP. NF-1 does not bind RTKs, and no evidence suggests that it participates in RTK signaling.

Could p120-GAP contribute to ras activation by RTKs? An early proposal, that activated RTKs might activate ras by decreasing the activity of p120-GAP, now seems difficult to reconcile with experimental results. Stimulation of most RTKs produces no changes in GAP activity, and NGF treatment actually increases cellular GAP activity in PC12 cells (228). Despite the lack of direct evidence, there are several other ways p120-GAP could theoretically contribute to RTK activation of ras. First, RTKs binding both GRB-2/SOS and p120-GAP might generate ras-GTP in different quantities or with different kinetics, compared to RTKs that bind only GRB-2/SOS. This idea is supported by experiments on PC12 cells that show a correlation between extended kinetics of ras activation and neuronal differentiation (320, 425). Second, the idea that p120-GAP could contribute to RTK signaling involving ras is suggested by recent discoveries concerning the ras effector, raf-1 (discussed below). Ras-GTP brings raf-1 into a stable complex with other signaling proteins that is located at the membrane, apparently by binding raf-1 in a one-to-one manner. Once in the complex, raf-1 no longer requires ras for localization, but it is not clear if ras-GTP can readily dissociate and mediate the localization of another raf-1 molecule. Ras-GDP does not bind this complex, which suggests a role for p120-GAP in facilitating the dissociation of ras for another round of raf-1 localization. Third, some non-GAP function of p120-GAP might be important for ras activation by RTKs. Since the GAP catalytic domain comprises only the carboxyl-terminal one-third of p120-GAP, the amino-terminal portion might well have another function, also suggested by the presence of SH2, SH3, and PH domains in this portion of p120-GAP. Microinjecting activated ras proteins into *Xenopus* oocytes normally results in their maturation. Maturation was blocked by also microinjecting a monoclonal antibody which bound the SH3 domain of p120-GAP, or peptides from this region of p120-GAP (101). The peptides also blocked oocyte maturation induced by insulin binding to the insulin receptor, suggesting that the SH3 domain of p120-GAP is involved in ras activation by the insulin RTK in these cells. These results should be interpreted cautiously, however, since the p120-GAP peptides used in these experiments might act differently from full-length p120-GAP, and since the SH3 domain of this region of p120-GAP might interact with inappropriate proteins if expressed at a high level. Non-GAP signaling by p120-GAP might employ two proteins which become tightly bound to it upon RTK activation. One is p62, a major tyrosine kinase substrate in transformed cells, and the other is p190 (189). When p190 binds p120-GAP, it inhibits its GAP activity. The sequence of p190 suggests that it may have multiple functions, and p190 acts as a GAP for rho and rac proteins in vitro. A pathway has been proposed by which ras, p120-GAP and p190, might stimulate rac in growth-factor–treated cells (189).

Role of Other Signaling Molecules Bound by RTKs. Some enzymatic activities bound by activated RTKs do not directly regulate ras activity, or have other functions besides activating ras [Table 18.5; (58, 112, 187, 229, 301, 410)]. Phosphatidylinositol-3-OH (PI3) kinase phosphorylates the 3' position of phosphoinositides. PI3 kinase consists of two subunits: p85, which binds activated RTK using two SH2 domains; and p110, which is the catalytic subunit. RTK binding of PI3 kinase potentially promotes PI3 kinase activity in several ways. First, the catalytic functions of the PI3 kinase holoenzyme undergo allosteric activation upon RTK binding of the p85 SH2 domains. Second, localization of PI3 kinase to the membrane brings PI3 kinase near its lipid substrate, PI(3,4) bisphosphate, which would be expected to promote its activity. Third, the plasma membrane is also the site of ras, and recent evidence suggests that the activity of PI3 kinase is stimulated by a direct interaction with ras-GTP (323). PI3 kinase activity has consistently been found to increase in cells after RTK stimulation, and may synergize with other mitogenic signals. However, in one cell line, PI3 kinase binding to the PDGF receptor appears to mainly regulate protein targeting to subcellular membranous organelles (176). The yeast PI3 kinase performs a similar protein targeting function, providing additional evidence that PI3 kinase regulates protein subcellular localization rather than mitogenesis.

PI-PLC-γ also binds activated RTKs using a SH2 domain. This interaction activates the enzymatic activity of PI-PLC-γ because membrane localization brings it near its lipid substrate, PIP2. In addition, PI-PLC-γ becomes tyrosine phosphorylated after RTK stimulation, which substantially stimulates its enzymatic activity. Mitogenic signals from RTKs have been proposed to be mediated by PI-PLC-γ in addition to ras; however, more recent research does not strongly support this idea. Several RTKs have been shown to transmit mitogenic signals despite mutations that abrogate RTK binding and activation of PI-PLC-γ (187). Activated PI-PLC-γ generates both DAG and InsP$_3$, which stimulate protein kinase C and Ca^{2+} release from internal stores, respectively. Second messengers arising from PI-PLC activity are important for nonmitogenic signaling and often synergize with mitogenic signals, but rarely induce proliferation by themselves (187). An exception is the strong proliferation of T lymphocytes in response to protein kinase C stimulation and increased intracellular Ca^{2+} levels.

The roles of other proteins that bind activated RTKs are less clear. Binding of the SH2 domains of src family kinases to RTKs apparently relieves autoinhibition of the src kinase catalytic domains, thereby activating their tyrosine kinase activities. Although mutant src family kinases are well-known oncogenic agents, the purpose of nonmutant src family kinases binding to RTKs is not known. Src kinases might tyrosine phosphorylate docking proteins such as SHC that are localized to the same RTK, and thereby promote activation of ras and/or additional signaling pathways. Tyrosine phosphorylated IRS-1, the docking protein of the insulin receptor, binds several SH2-containing proteins besides GRB-2/SOS and Syp. IRS-1 also binds PI3 kinase, and nck, a protein that has no obvious catalytic domain but which might, like GRB-2, provide a link to some important enzymatic activity. Nck also binds directly to the activated RTK for PDGF and EGF. The SH2 domain of the recently cloned GRB-7 protein binds tightly to an RTK related to the EGF receptor that is overexpressed in breast cancers (357). The function of GRB-7 is unknown, but in addition to the SH2 domain, GRB-7 contains a proline-rich region and a PH domain which could provide interactions with SH3-containing proteins and G$\beta\gamma$, respectively. Activated RTKs also bind multiple, uncharacterized proteins. Thus, it seems likely that future research will greatly expand our understanding of RTK signal transduction.

Ras Stimulates the MAPK Pathway Via Its Effector, Raf-1

The MAPK Pathway. We have delineated above various methods used by growth factors to activate ras. We will describe here a pathway that is important for activated ras to stimulate cell growth. The serine/threonine kinase, raf-1, appears to be a primary effector of ras that is required for the mitogenic actions of RTKs. Raf-1 activation leads to activation of MAPK (mitogen-activated protein kinases), also known as ERK (extracellular signal-regulated kinases), by a pathway utilized by multiple mitogenic and differentiating stimuli [reviewed in (31, 77, 89, 93, 99, 245)]. RTKs stimulate MAPK by utilizing ras, raf-1, and a kinase signaling cascade, shown in Figure 18.19. The first step involves a GDS such as SOS that promotes GTP binding and activation of ras. Ras-GTP activates the serine/theonine kinase activity of raf-1 by a mechanism involving a direct interaction between the two proteins and the membrane localization of raf-1. Raf-1 then phospho-

rylates and activates MEK, the MAP/ERK kinase (also called MAPKK). MEK then phosphorylates and activates the MAPK. The cellular outcomes of MAPK activation include proliferation, differentiation, and metabolic regulation of mammalian cells, cell fate determination in *Drosophila* and *C. elegans,* and cell-cycle arrest necessary for mating of yeasts. It should be noted, however, that MAPK activation is not the only way to induce proliferation of mammalian cells. For example, activin A, a ligand for a receptor serine/threonine kinase, stimulates proliferation of Swiss-3T3 fibroblasts without activating MAPK (327), and IL-4, a ligand for a cytokine-type receptor, promotes proliferation of myeloid cells independently of ras or MAPK activation (406).

MAPK were discovered through studies on proteins that become phosphorylated when cells are stimulated by growth factors, based on the idea that protein phosphorylation states could act as a switch for controlling mitogenesis. Growth factors such as insulin, PDGF, FGF, EGF, and NGF were found to induce

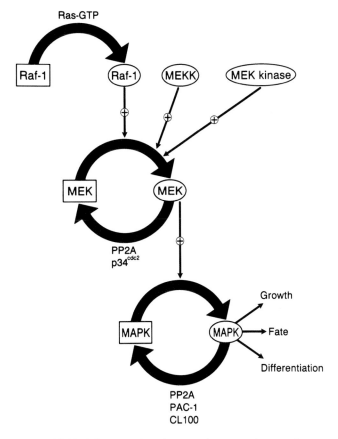

FIG. 18.19. Mitogen-activated protein kinase *(MAPK)* pathway. Catalytically active proteins are shown in *ovals;* inactive forms are shown in *rectangles.* (Described in the section Ras Stimulates the MAPK Pathway Via Its Effector, Raf-1)

phosphorylation of the S6 ribosomal subunit protein. Working backward from this point, S6 kinases were characterized, and subsequent research focused on identifying activators of the S6 kinases. Although it is still unclear how (or if) S6 phosphorylation regulates the cell cycle, these studies led to the discovery of MAPK as the kinase that phosphorylates and activates an S6 kinase, p90$_{rsk}$, in response to growth factors. Hence, MAPK is also known as S6 kinase kinase.

The kinase domains of MEK and MAPK are distinct from those of serine/theonine and tyrosine kinases. MEK exhibit the novel property of dual specificity— that is, they can phosphorylate their substrates both on tyrosine and threonine. MAPK in fact requires both tyrosine and threonine phosphorylation to become active. MEK, MAPK, and other kinases are members of a newly discovered family of homologous signal-transducing kinases (99). There are two MAPK in mammalian cells, p42 and p44, that are encoded by separate genes (77). Other kinases homologous to MAPK have been cloned from mammalian cells (77, 142, 212), but they cannot substitute for the p42 and p44 MAPK in the signaling pathways that we describe here. There are also at least two isoforms of MEK, called MEK1 and MEK2 (325). MEK, MAPK, and other elements of the MAPK pathway shown in Figure 18.19 are highly conserved throughout evolution (see Fig. 18.18).

Activation of Raf-1 by Ras. Raf-1 was initially identified in a truncated, constitutively active form (v-raf) that is the oncogenic agent of a murine sarcoma retrovirus. Subsequently, it was found that all normal cells express full-length raf-1. Cells expressing v-raf become "transformed." This term refers to the unusual growth properties of cancer cells and usually means that the cells can grow in suspension or soft agar and/or form tumors when injected into immunodeficient mice. Cells transformed by v-raf cannot be reverted to their normal phenotype by the microinjection of inhibitory anti-ras antibodies, suggesting that raf-1 functions downstream of ras. Further, anti-ras antibodies inhibit transformation by an activated form of the nonreceptor tyrosine kinase src, suggesting that tyrosine kinases function upstream of ras and raf-1 in a linear signaling pathway to control mitogenesis (347). Similar results were obtained using the dominant-negative mutant, ras-N17, to block ras function (106). A dominant-negative mutant raf-1 protein, mutated at the conserved lysine of the kinase domain to prevent ATP binding, provides additional evidence that raf-1 mediates ras activation of MAPK. When this raf-1 mutant protein was expressed in *Xenopus* embryos, it blocked FGF-stimulated mesoderm induction, a process that involves RTK activation of ras and MAPK (240).

Part of the mechanism by which ras-GTP leads to raf-1 activation has now been worked out [reviewed in (140)]. Raf-1 binds tightly to ras-GTP, but not ras-GDP, in vitro. Although this interaction is not sufficient to stimulate the raf-1 kinase activity, it permits membrane localization of raf-1 which is required for its activation. raf-1 is normally cytoplasmic, but in ras-transformed cells, raf-1 localizes to the plasma membrane in a complex that includes ras, MEK, MAPK, and proteins of unknown function (257). Two groups showed that a mutant raf-1 provided with its own membrane-localization signal no longer required ras for activation (226, 257) confirming that the main role for ras-GTP in this pathway is to localize raf-1 to the membrane. Additional results contribute to the current model of raf-1 activation by ras. In resting cells, raf-1 is inactive and cytoplasmic. When ras binds GTP and becomes active, ras recruits raf-1 to the membrane by directly binding to the amino-terminus of raf-1. This interaction employs the "effector" domain of ras, which was predicted by genetic analysis to link ras to its mitogenic effector. Once at the membrane, raf-1 seems to form independent links to the cytoskeleton because it no longer requires ras for localization. The final step is stimulation of the kinase activity of membrane-associated raf-1 by some as-yet-undetermined mechanism. Raf-1 activation at this step might involve its phosphorylation, based on results in mammalian cells (245), and by analogy with a yeast pathway (below). A group of proteins called 14.3.3 that bind to raf-1 may also be required for ras to activate raf-1 (165).

Raf-1 Activates MEK. Raf-1 activates MEK by phosphorylating two serine residues that are not targets for other serine/theonine kinases (433). Mutating either serine prevented MEK activation in cells following EGF stimulation or after incubation with raf-1 in vitro (433). When activated, MEK phosphorylates MAPK on specific threonine and tyrosine residues, which is required for MAPK activation. Strong evidence confirming that MEK lies in the ras-to-MAPK mitogenic pathway comes from recent reports that expression of mutant MEK with constitutively active kinase activities transform kidney and fibroblast cell lines (243, 359). These activated MEK mutants also induce neuronal differentiation of PC12 cells, another well-known outcome of ras activation (359). The role of MEK was further confirmed by showing that dominant-negative mutant MEK revert fibroblasts transformed by ras or src, and also interfere with NGF-induced differentiation of PC12 cells (359). Interestingly, both the activated and dominant-negative MEK mutants contain point mutations at the serines that are normally phosphorylated by raf-1 to activate MEK;

replacing the serines with glutamic or aspartic acid residues results in constitutively active MEK, while replacing them with alanine created the dominant-negative MEK.

MEK Is a Regulatory Node.

MEK can be regulated by several means besides phosphorylation by raf-1, making MEK a regulatory node for the MAPK pathway (Fig. 18.19). MEK is phosphorylated and activated by a second kinase called MEKK (216). MEKK is a mammalian homologue of the yeast STE-11 protein, which activates yeast MEK in response to signaling by a yeast GCR (6). STE-11 and MEKK, along with two other recently cloned MEKK (215), form a family of MEK kinases that are distinctly different from raf-1 [reviewed in (174)]. Like raf-1, MEKK is stimulated by upstream ras activity (215). The functions of MEKK and STE-11 are not exactly homologous, however, because STE-11 does not function downstream of ras and because MEKK cannot substitute for STE-11 in the yeast pathway (32). Recent observations suggest that besides the p42 and p44 MAPK, MEKK may have other MAPK-like substrates that permit it to promulgate slightly different cellular responses as compared with raf-1 (174, 215). A point mutation of STE-11 constitutively activates its kinase activity; a homologous mutation in MEKK might produce a constitutively active MEKK that could be used to identify the different MAPK substrates of MEK. A third MEK kinase is activated in the mammalian cell line, Raf-1a, following stimulation of GCRs. This mammalian G protein-regulated MEK kinase must be distinct from MEKK and raf-1 because it is activated by a pathway that does not require ras (174) (discussed below). Cloning and characterization of this kinase will very likely reveal another distinct MEK kinase family. The cell-cycle regulatory kinase $p34^{cdc2}$ is another MEK kinase. But instead of activating MEK, $p34^{cdc2}$ phosphorylation of MEK1 (but not MEK2) at two threonines inhibits MEK1 activity (325). This may provide a means for negative feedback regulation by the proliferative processes that are initiated by the MAPK pathway. In addition to kinases, MEK activity can be controlled by phosphatases. MEK (and MAPK) activity can be stimulated by agents, such as the SV40 small tumor antigen protein, which interfere with the function of protein phosphatase 2A because this phosphatase removes the activating phosphates from MEK and MAPK (352). PAC1 and CL100 are related MAPK phosphatases also capable of negatively regulating this pathway (245, 400).

Targets of MAPK.

Following activation, MAPK translocates to the nucleus where it can phosphorylate several targets (93). These include transcription factors such as jun, myc, and NF-IL6; ATF-2, which binds the cAMP response element in multiple promoters; and elk-1, which is part of the serum-response-element–binding complex. MAPK also associates in vivo with AP-1 (24), a complex of fos and jun, and induces transcription of the *fos* gene. Kinases are also targets for MAPK. MAPK phosphorylates the EGF receptor, MAPKAP kinase-2, which may be involved in regulation of glycogen storage, and the S6 kinase, $p90^{rsk}$, which is associated with mitogenic signaling. MAPK phosphorylates phospholipase A_2 in vitro, increasing its activity. MAPK may also regulate the cytoskeleton through phosphorylation of microtubule-associated proteins. Recent description of a constitutively active MAPK mutant (identified in *Drosophila*) should help to define more targets and effects of MAPK activation (46).

RTK Signaling Mechanisms Are Used by Other Receptor Families

Receptor Serine/Theonine Kinases.

Figure 18.20 shows other families of receptors expressed at the plasma membrane that employ variations of the RTK paradigm for signal transduction. The first example is the family of receptor serine/theonine kinases that include the receptors for transforming growth factor-β (TGF-β) and activin (27, 56). Actions of TGF-β require formation of a heterodimer between TGF-β receptor polypeptides type I and II. Type I polypeptides do not bind the ligand; type II polypeptides bind ligand but cannot signal without forming a dimer with type I. The structures of both type I and type II TGF-β receptor polypeptides resemble those of RTKs except that the intracellular tyrosine kinase domain is replaced by a serine/theonine kinase domain (Fig 18.20). Binding of TGFβ is thought to induce formation of the I/II hetero-

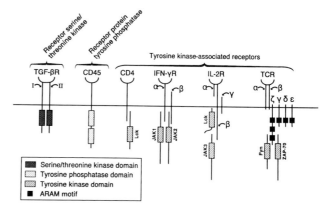

FIG. 18.20. Aspects of RTK signaling paradigm are used by other families of receptors.

dimer and activation of the serine/theonine kinases that is required for signal transduction. The next steps of TGF-β receptor signal transduction are uncertain, but by analogy with the RTKs, might be expected to involve phosphorylation of receptor polypeptides and recruitment of intracellular signaling proteins. Stimulation of TGF-β type II receptors expressed in *Xenopus* oocytes induced mesoderm by a process requiring ras (27), suggesting that TGF-β-like receptors and RTKs couple to similar signaling pathways. However, activin was recently shown to promote cellular proliferation without activating MAPK (327), indicating that differences exist as well.

Transmembrane Protein Tyrosine Phosphatases. A second family of proteins with structures like RTKs are the transmembrane protein tyrosine phosphatases. These proteins resemble RTKs and receptor serine/theonine kinases in structure, except that the intracellular region contains two tyrosine phosphatase catalytic domains instead of a kinase domain [Fig. 18.20(384)]. Little is known about the activation mechanisms of these molecules, including the best-studied member, CD45. CD45 is expressed by lymphocytes and produces signals crucial for signal transduction by the TCR. Cross-linking CD45 extracellular regions using monoclonal antibodies decreases transcription from a specific promoter element (16), suggesting that CD45 also generates relevant signals in the absence of TCR stimulation. Alternate splicing of the CD45 extracellular domain generates multiple possible ligand-binding sites but no physiological ligand has been identified. CD45 can dephosphorylate and activate src family kinases and a distinct tyrosine kinase csk (10). Csk phosphorylates CD45 in resting T lymphocytes, which increases CD45 tyrosine phosphatase activity and also forms a binding site on CD45 for the SH2 domain of the src family kinase, lck (10). In contrast, the behavior of a chimeric receptor constructed from the EGF receptor extracellular domain and the CD45 intracellular domain suggests that when CD45 binds its ligand in the absence of TCR stimulation, CD45 dimerizes and *decreases* its tyrosine phosphatase activity [discussed in (384)]. These studies, together with the recent cloning of a leukocyte-specific 30 kDa protein that binds specifically to CD45 (373), should help to unravel CD45 signaling mechanisms.

Tyrosine-Kinase-Associated Receptors. A third receptor signaling mechanism reminiscent of the RTK paradigm is used by a large number of plasma-membrane receptor proteins that signal though separate tyrosine kinase proteins that associate with their cytoplasmic domains.

Examples of tyrosine-kinase-associated receptors, shown in Figure 18.20, include both relatively simple single-transmembrane-spanning receptors, such as CD4, and the receptors for IL-2 and interferon (IFN)-γ, as well as the complex multisubunit structures exemplified by the T lymphocyte antigen receptor (TCR). None of these receptor polypeptides possesses intrinsic protein tyrosine kinase activity, yet tyrosine kinase activity is required for their signaling. This seeming paradox was resolved by the discovery that the cytoplasmic domains of these receptors associate with nonreceptor tyrosine kinases of the src and JAK families. The T lymphocyte cell-surface glycoprotein CD4 was the first tyrosine-kinase-associated receptor to be discovered. CD4 binds nonvariable regions of class II major histocompatibility (MHC) molecules expressed on the surfaces of other cells. Unlike RTK and TGF-β receptors, CD4 cannot by itself transmit mitogenic signals. Ligation of CD4 by its ligand, MHC class II proteins, enhances and modulates signaling by the TCR, thereby regulating T lymphocyte activation and differentiation. A cysteine motif in the short intracellular portion of CD4 mediates its noncovalent association with an src-family tyrosine kinase, lck [reviewed in (302)]. Cross-linking of CD4 by monoclonal antibodies, thought to mimic binding of the endogenous CD4 ligand, results in increased lck tyrosine kinase activity. Mutating the cysteines in the CD4 cytoplasmic domain inhibits binding of lck. When expressed in T cells of transgenic mice, these mutant CD4 molecules no longer function to regulate T cell development in the way that overexpressed, normal CD4 does. Similar experiments using a T cell line demonstrate that lck association is also important for CD4 to enhance TCR signaling.

Other tyrosine-kinase-associated receptors are the interferon (IFN) receptors (91). The interferons, IFN-α and IFN-γ, are not mitogenic; instead, they inhibit the growth of several types of cells in response to viral infection. IFN-α and IFN-γ bind to specific cell surface receptors composed of polypeptides that span the membrane once, activating intracellular signaling molecules that enhance the transcription of specific genes (91). A recent model for IFN-γ signal transduction shows remarkable similarities to the mechanism of RTK activation and signal transduction (126). IFN-γ is a dimeric ligand that through binding induces dimerization of the IFN-γ receptor α polypeptide. The α chain dimer then apparently complexes with the IFN-γ β chain. Evidence suggests that both the IFN-γ receptor α and β chains associate with JAK family tyrosine kinases, JAK1 and JAK2, and when the receptor chains are brought together by the ligand, the JAK kinases are proposed to phosphorylate and activate each other.

These kinases are also thought to mediate phosphorylation of tyrosines in the α and β chain intracellular regions that occurs after IFN-γ binding. Phosphorylation of tyrosine 440 of the IFN-γ receptor α chain creates a binding site for the SH2 domain of a signaling protein called p91. Bringing p91 into proximity with the receptor promotes its tyrosine phosphorylation, possibly also by the JAK kinases. Tyrosine phosphorylation activates p91, which causes p91 to detach from the receptor, dimerize, migrate to the nucleus, and stimulate transcription of specific genes. p91 or a homologous protein also appears to be involved in signal transduction by IFN-α and other cytokines, and may also mediate gene transcription in responses to RTK stimulation. Members of the cytokine receptor family are defined by a common extracellular ligand-binding motif and homologous intracellular regions. Many (but not all) members of this family transmit mitogenic, and/or differentiative signals that, like signals from RTKs, activate ras, raf-1, PI3 kinase, and MAPK (122, 192).

The family of so-called cytokine receptors also signal by associating with tyrosine kinases (192). The receptor for the mitogenic cytokine, IL-2, is shown schematically in Figure 18.20. IL-2 signal transduction requires formation of a heterodimer consisting of the β and γ chains of the receptor (269, 276); the α chain contributes only to IL-2 binding. Lck interacts with the cytoplasmic domain of the IL-2 receptor β chain. The IL-2 receptor β chain also binds JAK kinases (JAK1 and 3) in an IL-2–dependent manner, and this event is important for IL-2 mitogenic signaling (175, 192, 416). Binding of IL-2 induces tyrosine phosphorylation of the IL-2 receptor β chain, which is thought to create binding sites for SH2 domain–containing signaling proteins. In fact, IL-2 receptors bind and promote tyrosine phosphorylation of SHC (315), suggesting that IL-2 receptors could also use SHC to activate ras. This mechanism may be common among cytokine receptors, since IL-3, IL-5, and GM-CSF also induce tyrosine phosphorylation of SHC (98). In another parallel between cytokine receptor and RTK signal transduction, treatment of some cell types with IL-4 induces tyrosine phosphorylation of IRS-1, which is the mechanism used by the insulin receptor to activate ras [Fig. 18.17; (188)].

A final important family of tyrosine-kinase-associated receptors is represented in Figure 18.20 by the T lymphocyte antigen receptor, or TCR. This family also includes the B lymphocyte antigen receptor and some receptors that bind the Fc portions of antibodies (53, 136, 404). The TCR is a tightly associated complex of six different polypeptides, the α β, γ, δ, ϵ, and ζ subunits, each of which possesses a single-transmembrane-spanning domain. Occasionally a seventh subunit, and isoform of ζ called η, is present. Despite its distinctive structure, the TCR functions by activating many of the same signaling molecules as RTKs, including tyrosine kinases, PLC-γ, PI3 kinase, ras, raf-1, MEK1, and MAPK. TCR signal transduction is crucial for mobilizing T lymphocytes to participate in the immune response because it provokes T cell proliferation, cytokine secretion, and cytotoxic behavior. The α and β subunits bind the TCR ligand, which consists of an antigenic peptide bound to a major histocompatibility (MHC) protein expressed on the surface of another cell. The other TCR subunits participate in TCR signal transduction by using their cytoplasmic domains to interact with src-family kinases (fyn and lck) and a kinase of a distinctive family called ZAP-70. All of these signaling subunits of the TCR contain in their cytoplasmic domains at least one 26-amino-acid motif (called ARAM) that is required for TCR signaling. When placed in the context of chimeric receptors, the cytoplasmic portions of the ϵ and ζ subunits are individually capable of mediating many or all of the signals generated by the entire TCR. For example, a chimeric receptor constructed from the extracellular region of the T lymphocyte cell-surface glycoprotein CD8 and the cytoplasmic portion of ζ is a surprisingly good model for TCR function, producing TCR-like tyrosine kinase activity inside the cell when its extracellular regions are cross-linked by anti-CD8 monoclonal antibodies (166). ARAM motifs are also present in the cytoplasmic portions of proteins in the B lymphocyte antigen receptor complex and these motifs have been demonstrated to be sufficient for signal transduction by this receptor. Upon TCR stimulation, the ARAM motifs of ζ become tyrosine phosphorylated by a src family kinase, forming a binding site for the SH2 domain of ZAP-70. A current model of TCR activation (53) proposes that in response to ligand, fyn and ZAP-70 act sequentially to promote tyrosine phosphorylation of TCR subunits, permitting SH2-containing proteins including ZAP-70 itself, PI-PLC-γ, PI3 kinase p85, and p120-GAP to associate with the receptor to effect signal transduction in a scheme reminiscent of RTK signal transduction. Although useful, this model most likely does not completely reflect the complexity of signaling by the normal, multichain TCR, which also requires the catalytic activities associated with other cell surface proteins like CD4 and CD45 (343). A recent report suggests that the TCR uses GRB-2 to recruit SOS, PI-PLC-γ, and an unidentified tyrosine phosphorylated protein to the plasma membrane (343), suggesting the existence of additional correlations and distinctions among TCR and RTK signaling mechanisms.

MITOGENIC CROSSTALK: GCRs, RTKs, AND THE MAPK PATHWAY

The two major receptor signaling paradigms described above, those involving G protein–coupled receptors (GCR) and receptor tyrosine kinases (RTK), can interact. Such "crosstalk" of signaling pathways that are normally thought of as distinct has the potential to integrate signals from multiple receptors and, since crosstalk effects are often cell-type-specific, for customizing the signaling pathways of each cell. One type of crosstalk concerns the ability of some GCR ligands to provoke mitogenesis, an effect that is classically associated with RTK signaling [reviewed in (133)]. GCR-induced mitogenesis is expected in some cases—for instance, in cells where the activities of adenylyl cyclase or PI-PLC are known to provoke proliferation. Other cases of GCR-induced mitogenesis are independent of PI-PLC or adenylyl cyclase activity (174, 339, 392), and this form of crosstalk is more difficult to explain using the classical receptor signaling paradigms. We concentrate here on one form of crosstalk: the activation by GCR signaling of MAPK, the crucial mitogenic regulator that is stimulated by many RTKs. G protein signaling in mammalian cells can regulate the MAPK pathway at various points, as shown in Figure 18.21.

A GCR Induces Tyrosine Phosphorylation of SHC

First, thyrotropin-releasing hormone (TRH), a GCR ligand, stimulates MAPK activity and promotes tyrosine phosphorylation of SHC (284). It seems likely that TRH-induced tyrosine phosphorylation of SHC activates ras, leading to stimulation of MAPK by the mechanism shown in Figure 18.20. A similar mecha-

nism might explain the mitogenic effects on certain cells of thrombin, endothelin, vasopressin, angiotensin II, bradykinin, bombesin, and carbachol, all of which bind to GCRs and promote increased tyrosine kinase activity (133, 225, 366, 431).

Ras Activation by Gβγ

A second point where GCR signaling can stimulate the MAP kinase pathway was shown by two groups working with a mammalian cell line called COS (88, 104). A common approach used for these and related signaling investigations involves transfecting tissue culture cells with genes encoding GCRs and/or other proteins, and assaying the effects of expression or overexpression of these proteins on cellular biochemistry or behavior a short time later (usually 24 h). Different behavior of transfected cells compared to untransfected (or mock-transfected) control cells is attributed to the presence of the protein(s) introduced by transfection. Although results from this approach may be misinterpreted if the cell makes dramatic adaptations to the transfected proteins during the time needed for their complete expression, such experiments have provided valuable insight into many signaling pathways. Misinterpretation can be minimized by transfecting several different proteins to test the same hypothesis and by combining this approach with others, such as in vitro biochemical experiments, where possible.

Genes for GCRs and/or other proteins were transfected into COS cells to test their effects on MAPK (88, 104). Assays of MAPK activity were simplified by coexpression of a MAPK that was modified slightly to include an epitope tag. The tag did not affect MAPK function and made it easy to purify the MAPK for the assay using a monoclonal antibody. Stimulation of all the GCR provoked MAP kinase activity in COS cells, including those using G_i, G_q, or G_s for signal transduction. MAPK stimulation by G_q or G_s could be explained by the fact that either increased PLC activity or intracellular cAMP stimulates MAPK in these cells (104), but the mechanism for MAPK activation by G_i-coupled GCR was not defined. PTX blocked the MAPK stimulation via G_i-coupled GCR, confirming that G_i proteins mediate this process. Overexpression of activated $G\alpha_i$ subunits did not stimulate MAPK (104). Instead, MAPK was stimulated by overexpression of Gβγ subunits (88, 104), revealing yet another important signaling function for mammalian Gβγ (74). Both Gβ and Gγ were required. Moreover, expression of either a Gβ plus a Gγ that was mutated to prevent membrane association (88), or a Gβγ pair that is naturally unable to form a complex (104), was not effective. Thus, Gβγ

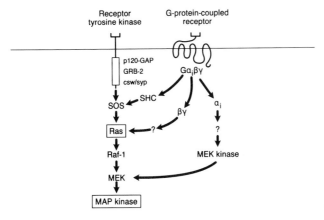

FIG. 18.21. Regulation of MAPK pathway by G protein–coupled receptors.

complexes such as those freed by GCR activation of G proteins appear to strongly activate the MAPK pathway of COS cells. As predicted by this hypothesis, expression of excess free Gα (Gα$_t$) inhibited MAPK activation via GCRs employing G$_i$; presumably because Gα$_t$ binds and sequesters Gβγ.

The ras inhibitor proteins H-ras(N17) and rap-1a were used to identify the step of the MAPK pathway that was regulated by Gβγ. Expressing either of these ras inhibitor proteins nearly abolished the MAPK activity that was induced by Gβγ or carbachol. In contrast, expression of activated raf-1 could override the effects of the ras inhibitors and activate MAPK. These experiments clearly show that Gβγ stimulates MAPK in COS cells by a mechanism involving ras activation. Besides COS cells, Gβγ may also promote ras activation in the many cell types where mitogenic responses to GCR ligands have been found to be PTX-sensitive (133). The PI3 kinase activity observed in neutrophils following the overexpression of Gβγ subunits (358) may be another example of Gβγ activation of ras, since recent results show that the p110 PI3 kinase catalytic subunit can be activated through a direct interaction with GTP-bound ras (323). Gβγ may also be capable of activating ras in starfish oocytes since microinjection of Gβγ subunits promotes oocyte maturation, an effect that is thought to be produced by MAPK activation (171).

What is the mechanism by which Gβγ activates ras in COS cells? The experimental results are consistent with Gβγ stimulating the MAPK pathway at ras, as in Figure 18.21, or at any step upstream of ras—for example, at SOS. Gβγ might regulate ras-GAP or ras-

GDS proteins by binding to their PH domains, an idea suggested by the recent demonstration that Gβγ binds tightly to the PH domain of βARK (383). Gβγ might also stimulate ras activity by a mechanism similar to that used in yeast and mammalian cells for activating CDC42, a rac-like member of the ras superfamily. Mammalian CDC42 is activated through a direct interaction with ACK, a protein that exhibits anti-GAP activity toward CDC42 (241). Activated CDC42 binds and activates a serine/threonine kinase that was unexpectedly found to be homologous to the yeast kinase, STE-20 (242). STE-20 is a member of a well-known yeast MAPK pathway that is initiated by Gβγ (Fig. 18.22, discussed below). Thus, it seems possible that Gβγ stimulates ras activity in mammalian cells by stimulating the anti-ras-GAP activity of an ACK-like protein.

Gβγ Stimulation of MAPK in Yeast

In *S. cerevisiae*, Gβγ activates MAPK by a mechanism apparently independent of ras (6, 69, 81, 205, 275, 432). Both genetic and biochemical experiments demonstrate that a mating factor pheromone binds a classic seven transmembrane-spanning GCR that signals through free Gβγ with Gα acting as a passive negative regulator. Downstream of Gβγ is a protein kinase cascade that culminates in the activation of yeast MAPK (Fig. 18.22) and results in the cell-cycle arrest necessary for mating. This yeast pathway may pertain to some cases in mammalian cells where Gβγ activates MAPK. Yeast mating factor signal transduction requires at least two proteins downstream of Gβγ that have no known mammalian homologues: STE20 and STE5. STE20 is a serine/threonine kinase that functions upstream of STE5, a protein of unclear function. Both STE20 and STE5 are upstream of STE11 (MEK kinase), STE7(MEK), and FUS3/KSS1(MAPK), which are homologues of the mammalian signaling proteins described above.

The role of STE5 has been very unclear because genetic data cannot place STE5 in a simple linear pathway leading to MAPK activation. Figure 18.22 shows three proposed models for STE5 function: *(A)* STE5 acts as a nucleus to form a complex that enhances productive interactions between STE11, STE7, and FUS3; *(B)* STE5 functions in a linear pathway between STE20 and STE11; and *(C)* STE5 functions both upstream of STE11 and downstream of FUS3. Model *A* is strongly supported by recent biochemical and genetic results (69, 145, 205, 275). Different regions of STE5 bind directly to STE11, STE7, and FUS3, and these four proteins co-immunoprecipitate in a large complex from yeast cells. This complex appears to have an

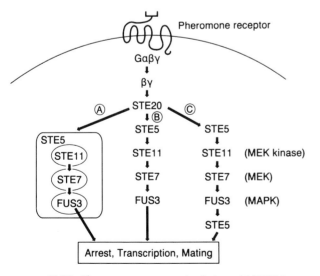

FIG. 18.22. Pheromone receptor stimulation of MAPK in yeast. Three possible pathways are shown, supported by genetic analysis of *S. cerevisiae*.

important signaling function, since the binding of STE5 to FUS3 promotes FUS3 activation by STE11 by STE7. Moreover, the complex is present in resting cells and it falls apart after FUS3 is activated, an event that would allow activated FUS3 to phosphorylate targets in the nucleus. Interestingly, activation of MAPK in mammalian cells also involves a complex containing several proteins in the pathway (362). STE5 may have an additional function downstream of FUS3, as in models *B* and *C,* an idea supported by some genetic data and by a DNA-binding motif present in STE5. Despite these recent advances, it is still unclear how STE20 stimulates formation of this complex, or how G$\beta\gamma$ stimulates the kinase activity of STE20. Genetic analysis indicates that two regions of Gβ, the aminoterminal region and the region between the second and third WD-40 repeats, are important for yeast Gβ to function in this pathway (220). It is curious that this yeast MAPK pathway does not involve ras, especially in light of the recent demonstration that G$\beta\gamma$ activates ras and MAPK in mammalian cells (above). Although the two ras proteins of *S. cerevisiae* (RAS1 and RAS2) participate in adenylyl cyclase activation, another member of the ras superfamily may eventually be found to act in the yeast pheromone MAPK pathway. STE20, STE11, and STE7 function in another yeast pathway (the differentiation of diploid cells in response to nitrogen starvation) that appears to be stimulated by RAS2 (6). Further characterization of this pathway may reveal additional correlations between yeast and mammalian MAPK control systems.

Gα_i Stimulation of a MEK Kinase in Rat-1a Cells

A third point where G protein signaling can control the MAPK pathway is the stimulation of MEK by a mechanism apparently independent of ras and raf-1 (Fig. 18.21). MEK activation by GCRs was discovered when m2 muscarinic receptors, which normally signal via G$_i$ proteins, were expressed in Rat-1a cells [reviewed in (174)]. Addition of the m2 receptor agonist carbachol provoked activation of MAPK by a pathway that was independent of PI-PLC. This pathway was also insensitive to a tyrosine kinase inhibitor, ruling out participation of SHC. Carbachol stimulation of MAPK was inhibited by PTX treatment, suggesting a role for G$_{i/o}$ (415). An additional argument for G$_i$ involvement was provided by results of expressing the constitutively active Gα_{i2} oncogene, *gip2.* Gip2 expression transformed Rat-1a cells and increased the activities of MEK and MAPK (118, 134, 135). This result also indicates that the mitogenic signaling proceeds via the α, not $\beta\gamma$, subunit of G$_i$. The point where Gα_i impinges on the MAPK pathway was addressed by

analyzing the effects of expressing constitutively activated forms of MAPK signaling pathway proteins. Surprisingly, neither activated ras nor activated raf-1 could activate MEK and MAPK in Rat-1a cells to the extent that gip2 did, suggesting that gip2 regulation of MEK bypasses ras and raf-1 (118, 134). This result was confirmed by showing that the dominant-negative ras, H-ras (N17), did not inhibit the abnormal growth of Rat-1a cells caused by *gip2* expression (135). A current model, shown in Figure 18.21, proposes that the MAPK activity stimulated by m2 receptors on Rat-1a cells arises from generation of Gα_i and the activation of a novel, uncharacterized MEK kinase. The recently cloned MEK kinase called MEKK is a candidate for the Gα_i-activated MEK kinase (216). However, MEKK is activated by ras in PC12 cells (215), while the MEK kinase of Rat-1a cells is stimulated only a small amount by an activated ras protein (118). Gα_i stimulation of MEK appears to be highly cell-type-specific; for example NIH-3T3 and Swiss-3T3 fibroblasts are not transformed by *gip2* (135), and gip2 does not stimulate MAPK activity in NIH-3T3 fibroblasts (115). This explains why *gip2* is a tissue-specific oncogene (133), and suggests that the specificity of receptor signaling to MAPK can be regulated in many ways.

GCR-Mediated Inhibition of Mitogenic Signaling

In some cases, GCR signaling actually inhibits RTK signaling and the MAPK cascade. This occurs via several mechanisms. First, PKC and PKA activity induced by the traditional G protein effectors adenylyl cyclase and PLC can phosphorylate, down-modulate, or otherwise inhibit signaling by RTK (391). Second, elevated intracellular Ca^{2+} levels, such as that induced by activity of the classical G protein effector, PI-PLC-β, severely inhibits the activity of yes (432), a member of the src family of nonreceptor tyrosine kinases that presumably mediates signals from tyrosine-kinase-associated receptors. Finally, recent evidence shows that elevated cAMP levels severely inhibit the ability of ras to activate raf-1 (48, 326) or MEKK (215), providing a probable explanation for the deleterious effects of activated Gα on MAPK activation (65) and the antagonistic relationship between high cAMP and proliferation shown by many cells.

CONCLUSION

Cells organize signaling pathways to respond predictably to environmental changes. Identifying the components of these pathways, like fitting the last piece of a jigsaw puzzle into its proper position, satisfies our compulsion to view complete pictures. The current era in signal transduction research is rich due to the advances in molecular biology that have given us the structures of critical "puzzle pieces" and the tools to reveal the rest. At some point in the future, we will have identified all the pieces. Still, questions will remain regarding cellular function because linear representations of signaling pathways are only snapshots of the dynamic network of interactions occurring within the cell, much like pictures are only two-dimensional representations of a three-dimensional subject.

We are extremely grateful to Larry Karnitz and Ray Ghanbari for critical reading of the manuscript, and to Lori Volkman for secretarial assistance.

REFERENCES

1. Adam, G., and M. Delbruck. *Structural Chemistry and Molecular Biology*, edited by A. Rich and M. Delbruck. San Franscisco: Freeman, 1969, p. 198–215.
2. Allen, L. F., R. J. Lefkowitz, M. G. Caron, and S. Cotecchia. G-protein-coupled receptor genes as protooncogenes: constitutively activating mutation of the α_{1B}-adrenergic receptor enhances mitogenesis and tumorigenicity. *Proc. Natl. Acad. Sci. U.S.A.* 88: 11354–11358, 1991.
3. Allgeier, A., S. Offermanns, J. Van Sande, K. Spicher, G. Schultz, and J. E. Dumont. The human thyrotropin receptor activates G-proteins G_s and $G_{q/11}$. *J. Biol. Chem.* 269: 13733–13735, 1994.
4. Amatruda, T.T.I., D. A. Steele, V. Z. Slepak, and M. I. Simon. $G\alpha_{16}$, a G protein α subunit specifically expressed in hematopoietic cells. *Proc. Natl. Acad. Sci. U.S.A.* 88: 5587–5591, 1991.
5. Ames, A., T. F. Walseth, R. A. Heyman, M. Barad, R. M. Graeff, and N. D. Goldberg. Light-induced increases in cGMP metabolic flux correspond with electrical responses of photoreceptors. *J. Biol. Chem.* 261(28): 13034–13042, 1986.
6. Ammerer, G. Sex, stress and integrity: the importance of MAP kinases in yeast. *Curr. Opin. Genet. Dev.* 4: 90–95, 1994.
7. Anderson, R.G.W. Plasmalemmal caveolae and GPI-anchored membrane proteins. *Curr. Opin. Cell Biol.* 5: 647–652, 1993.
8. Angleson, J. K., and T. G. Wensel. Enhancement of rod outer segment GTPase accelerating protein activity by the inhibitory subunit of cGMP phosphodiesterase. *J. Biol. Chem.* 269: 16290–16296, 1994.
9. Arshavsky, V., and M. D. Bownds. Regulation of deactivation of photoreceptor G protein by its target enzyme and cGMP. *Nature* 357(6377): 416–417, 1992.
10. Autero, M., J. Saharinen, T. Pessa-Morikawa, M. Soula-Rothhut, C. Oetken, M. Gassmann, M. Bergman, K. Alitalo, P. Burn, C. G. Gahmberg, and T. Mustelin. Tyrosine phsophorylation of CD45 phosphotyrosine phosphatase by p50csk kinase creates a binding site for p56lck tyrosine kinase and activates the phosphatase. *Mol. Cell. Biol.* 14: 1308–1321, 1994.
11. Bairoch, A., and J. A. Cox. EF-hand motifs in inositol phospholipid-specific phospholipase C. *FEBS Lett.* 269(2): 454–456, 1990.
12. Bakalyar, H. A., and R. R. Reed. Identification of a specialized adenylyl cyclase that may mediate odorant detection. *Science* 250: 1403–1406, 1990.
13. Banno, Y., Y. Okano, and Y. Nazawa. Thrombin-mediated phosphoinositide hydrolysis in Chinese hamster ovary cells over-expressing phospholipase D-1. *J. Biol. Chem.* 269(22): 15846–15852, 1994.
14. Barak, L. S., M. Tiberi, N. J. Freedman, M. M. Kwatra, R. J. Lefkowitz, and M. G. Caron. A highly conserved tyrosine residue in G protein-coupled receptors is required for agonist-mediated β_2-adrenergic receptor sequestration. *J. Biol. Chem.* 269: 2790–2795, 1994.
15. Bauer, P. H., S. Muller, M. Puzicha, S. Pippig, B. Obermaier, E.J.M. Helmreich, and M. J. Lohse. Phosducin is a protein kinase A-regulated G-protein regulator. *Nature* 358: 73–76, 1992.
16. Baur, A., S. Garber, and B. M. Peterlin. *J. Immunol.* 152: 976–983, 1994.
17. Beals, C. R., C. B. Wilson, and R. M. Perlmutter. A small multigene family encodes G_i signal-transduction proteins. *Proc. Natl. Acad. Sci. U.S.A.* 84: 7886–7889, 1987.
18. Ben-Levy, R., E. Peles, R. Goldman-Michael, and Y. Yarden. An oncogenic point mutation confers high affinity ligand binding to the *neu* receptor. *J. Biol. Chem.* 267(24): 17304–17313, 1992.
19. Benovic, J. L., H. Kuhn, I. Weyand, J. Codina, M. G. Caron, and R. J. Lefkowitz. Functional desensitization of the isolated β-adrenergic receptor by the β-adrenergic receptor kinase: potential role of an analog of the retinal protein arrestin (48-kDa protein). *Proc. Natl. Acad. Sci. U.S.A.* 84(24): 8879–8882, 1987.
20. Benovic, J. L., L. J. Pike, R. A. Cerione, C. Staniszewski, T. Yoshimasa, J. Codina, M. G. Caron, and R. J. Lefkowitz. Phosphorylation of the mammalian β-adrenergic receptor by cyclic AMP-dependent protein kinase. Regulation of the rate of receptor phosphorylation and dephosphorylation by agonist occupancy and effects on coupling of the receptor to the stimulatory guanine nucleotide regulatory protein. *J. Biol. Chem.* 260(11): 7094–7101, 1985.
21. Berlot, C. H., and H. R. Bourne. Identification of effector-activating residues of $G\alpha_s$. *Cell* 68: 911–922, 1992.
22. Berstein, G., J. L. Blank, D.-Y. Jhon, J. H. Exton, S. G. Rhee, and E. M. Ross. Phospholipase C-β_1 is a GTPase-activating protein for $G_{q/11}$ its physiological regulator. *Cell* 70: 411–418, 1992.
23. Berstein, G., J. L. Blank, A. V. Smrcka, T. Higashijima, P. C. Sternweis, J. H. Exton, and E. M. Ross. Reconstitution of agonist-stimulated phosphatidylinositol 4,5-bisphosphate hydrolysis using purified m1 muscarinic receptor, $G\alpha_{q/11}$ and phospholipase C-β_1. *J. Biol. Chem.* 267: 8081–8088, 1992.
24. Bernstein, L. R., D. K. Ferris, N. H. Colburn, and M. E. Sobel. A family of mitogen-activated protein kinase-related proteins interacts *in vivo* with activator protein-1 transcription factor. *J. Biol. Chem.* 269(13): 9401–9404, 1994.
25. Berridge, M. J. Inositol trisphosphate and calcium signalling. *Nature* 361: 315–325, 1993.
26. Berridge, M. J., and R. F. Irvine. Inositol trisphosphate, a novel second messenger in cellular signal transduction. *Nature* 312: 315–321, 1984.
27. Bhushan, A., H. Y. Lin, H. F. Lodish, and C. R. Kintner. The transforming growth factor beta type II receptor can replace the activin type II receptor in inducing mesoderm. *Mol. Cell. Biol.* 14: 4280–4285, 1994.
28. Birge, R. B., and H. Hanafusa. Closing in on SH2 specificity. *Science* 262: 1522, 1993.
29. Birnbaumer, L., J. Abramowitz, and A. M. Brown. Receptor-effector coupling by G proteins. *Biochim. Biophys. Acta* 1031: 163–224, 1990.
30. Blazynski, C., and A. I. Cohen. Rapid declines in cyclic GMP

of rod outer segments of intact frog photoreceptors after illumination. *J. Biol. Chem.* 261(30): 14142–14147, 1986.

31. Blenis, J. Signal transduction via the MAP kinases: proceed at your own RSK. *Proc. Natl. Acad. Sci. U.S.A.* 90: 5889–5892, 1993.

32. Blumer, K. L., G. L. Johnson, and C. A. Lange-Carter. Mammalian mitogen-activated protein kinase kinase kinase (MEKK) can function in a yeast mitogen-activated protein kinase pathway downstream of protein kinase C. *Proc. Natl. Acad. Sci. U.S.A.* 91: 4925–4929, 1994.

33. Bockaert, J. G proteins and G-protein-coupled receptors: structure, function and interactions. *Curr. Opin. Neurobiol.* 1: 32–42, 1991.

34. Boekhoff, I., and H. Breer. Termination of second messenger signaling in olfaction. *Proc. Natl. Acad. Sci. U.S.A.* 89: 471–474, 1992.

35. Boguski, M. S., and F. McCormick. Proteins regulating ras and its relatives. *Nature* 366: 643–654, 1993.

36. Boland, L. M., and B. P. Bean. Modulation of N-type calcium channels in bullfrog sympathetic neurons by luteinizing hormone-releasing hormone: kinetics and voltage dependence. *J. Neurosci.* 13(2): 516–533, 1993.

37. Bollag, G., and F. McCormick. Regulators and effectors of RAS proteins. *Annu. Rev. Cell. Biol.* 7: 601–632, 1991.

38. Bonfini, L., C. A. Karlovich, C. Dasgupta, and U. Banerjee. The *Son of sevenless* gene product: a putative activator of *ras. Science* 255: 603–606, 1992.

39. Bourne, H. R. The importance of being GTP. *Nature* 369: 611–612, 1994.

40. Bourne, H. R., D. A. Sanders, and F. McCormick. The GTPase superfamily: conserved structure and molecular mechanism. *Nature* 349: 117–127, 1991.

41. Boyer, J. L., S. G. Graber, G. L. Waldo, T. K. Harden, and J. C. Garrison. Selective activation of phospholipase C by recombinant G-protein α- and βγ-subunits. *J. Biol. Chem.* 269: 2814–2819, 1994.

42. Braun, T., P. R. Schofield, and R. Sprengel. Amino-terminal leucine-rich repeats in gonadotropin receptors determine hormone selectivity. *EMBO J.* 10: 1885–1890, 1991.

43. Bray, P., A. Carter, V. Guo, C. Puckett, J. Kamholz, A. Spiegel, and M. Nirenberg. Human cDNA clones for an α subunit of Gi signal-transduction protein. *Proc. Natl. Acad. Sci. U.S.A.* 84(15): 5115–5119, 1987.

44. Brown, A. M., and L. Birnbaumer. Direct G protein gating of ion channels. *Am. J. Physiol.* 254 (*Heart Circ. Physiol.* 23): H401–H410, 1988.

45. Brown, H. A., S. Gutowski, C. R. Moomaw, C. Slaughter, and P. C. Sternweis. ADP-ribosylation factor, a small GTP-dependent regulatory protein, stimulates phospholipase D activity. *Cell* 75: 1137–1144, 1993.

46. Brunner, D., N. Oellers, J. Szabad, W.H.I. Biggs, S. L. Zipursky, and E. Hafen. A gain-of-function mutation in Drosophila MAP kinase activates multiple receptor tyrosine kinase signalling pathways. *Cell* 76: 875–888, 1994.

47. Bubis, J., and H. G. Khorana. Sites of interaction in the complex between β and γ subunits of transducin. *J. Biol. Chem.* 265: 12995–12999, 1990.

48. Burgering, B.M.T., G. J. Pronk, P. C. Van Weeren, P. Chardin, and J. L. Bos. cAMP antagonizes p21ras-directed activation of extracellular signal-regulated kinase 2 and phosphorylation of mSos nucleotide exchange factor. *EMBO J.* 12: 4211–4220, 1993.

49. Bushman, W. A., L. K. Wilson, D. K. Luttrell, J. S. Moyers, and S. J. Parsons. Overexpression of c-src enhances β-adrenergic-induced cAMP accumulation. *Proc. Natl. Acad. Sci. U.S.A.* 87(19): 7462–7466, 1990.

50. Cai, H., J. Szeberenyi, and G. M. Cooper. Effect of a dominant inhibitory Ha-Ras mutation on mitogenic signal transduction in NIH 3T3 cells. *Mol. Cell. Biol.* 10: 5314–5323, 1990.

51. Cali, J. J., E. A. Balcueva, I. Rybalkin, and J. D. Robishaw. Selective tissue distribution of G protein γ subunits, including a new form of the γ subunits identified by cDNA cloning. *J. Biol. Chem.* 267: 24203–24027, 1992.

52. Cali, J. J., J. C. Zwaagstra, N. Mons, D. M. Cooper, and J. Krupinski. Type VIII adenylyl cyclase. A Ca²⁺/calmodulin-stimulated enzyme expressed in discrete regions of rat brain. *J. Biol. Chem.* 269: 12190–12195, 1994.

53. Cambier, J. C., and W. A. Jensen. The hetero-oligomeric antigen receptor complex and its coupling to cytoplasmic effectors. *Curr. Opin. Genet. Dev.* 4: 55–63, 1994.

54. Campbell, P. T., M. Hnatowich, B. F. O'Dowd, M. G. Caron, R. J. Lefkowitz, and W. P. Hausdorff. Mutations of the human β₂-adrenergic receptor that impair coupling to Gs interfere with receptor down-regulation but not sequestration. *Mol. Pharmacol.* 39: 192–198, 1991.

55. Camps, M., A. Carozzi, P. Schnabel, A. Scheer, P. J. Parker, and P. Gierschik. Isozyme-selective stimulation of phospholipase C-β₂ by G protein βγ-subunits. *Nature* 360: 684–686, 1992.

56. Carcamo, J., F.M.B. Weis, F. Ventura, R. Wieser, J. L. Wrana, L. Attisano, and J. Massague. Type I receptors specify growth-inhibitory and transcriptional responses to transforming growth factor beta and activin. *Mol. Cell. Biol.* 14: 3810–3821, 1994.

57. Carpenter, G., and S. Cohen. Epidermal growth factor. *J. Biol. Chem.* 265: 7709–7712, 1990.

58. Case, R. D., E. Piccione, G. Wolf, A. M. Benett, R. J. Lechleider, B. G. Neel, and S. E. Shoelson. SH-PTP2/Syp SH2 domain binding specificity is defined by direct interactions with platelet-derived growth factor β-receptor, epidermal growth factor receptor, and insulin receptor substrate-1-derived phosphopeptides. *J. Biol. Chem.* 269: 10467–10474, 1994.

59. Casey, P. J., H.K.W. Fong, M. I. Simon, and A. G. Gilman. Gz, a guanine nucleotide-binding protein with unique biochemical properties. *J. Biol. Chem.* 265: 2383–2390, 1990.

60. Chabre, O., B. R. Conklin, S. Brandon, H. R. Bourne, and L. E. Limbird. Coupling of the α₂A-adrenergic receptor to multiple G-proteins. Efficiency in a transient expression system. *J. Biol. Chem.* 269: 5730–5734, 1994.

61. Chakraborty, M., D. Chatterjee, S. Kellokumpu, H. Rasmussen, and R. Baron. Cell-cycle dependent coupling of the calcitonin receptor to different G proteins. *Science* 251: 1078–1082, 1991.

62. Chazenbalk, G. D., Y. Nagayama, D. Russo, H. L. Wadsworth, and B. Rapoport. Functional analysis of the cytoplasmic domains of the human thyrotropin receptor by site-directed mutagenesis. *J. Biol. Chem.* 265(34): 20970–20975, 1990.

63. Chee, M. S., S. C. Satchwell, E. Preddie, K. M. Weston, and B. G. Barrell. Human cytomegalovirus encodes three G protein-coupled receptor homologues. *Nature* 344: 774–777, 1990.

64. Chen, J., and R. Iyengar. Inhibition of cloned adenylyl cyclases by mutant-activated Gi-alpha and specific suppression of type 2 adenylyl cylcase inhibition by phorbol ester treatment. *J. Biol. Chem.* 268(17): 12253–12256, 1993.

65. Chen, J., and R. Iyengar. Suppression of Ras-induced transformation of NIH 3T3 cells by activated Gαs. *Science* 263: 1278–1281, 1994.

66. Chen, T. Y., Y. W. Peng, R. S. Dhallan, B. Ahamed, R. R. Reed, and K. W. Yau. A new subunit of the cyclic nucleotide-gated cation channel in retinal rods. *Nature* 362(6422): 764–747, 1993.

67. Choi, E. J., S. T. Wong, A. H. Dittman, and D. R. Storm. Phorbol ester stimulation of the type I and type III adenylyl cyclases in whole cells. *Biochemistry* 32(8): 1891–1894, 1993.

68. Choi, E. J., Z. Xia, and D. R. Storm. Stimulation of the type III olfactory adenylyl cyclase by calcium and calmodulin. *Biochemistry* 31(28): 6492–6498, 1992.

69. Choi, K.-Y., B. Satterberg, D. M. Lyons, and E. A. Elion. Ste5 tethers multiple protein kinases in the MAP kinase cascade required for mating in S. cerevisiae. *Cell* 78: 499–512, 1994.

70. Clapham, D. E. A mysterious new influx factor? *Nature* 364: 763–764, 1993.

71. Clapham, D. E. Direct G protein gating of ion channels? *Annu. Rev. Neurosci.* 17: 441–64, 1994.

72. Clapham, D. E. Mutations in G protein-linked receptors: new insights for disease. *Cell* 75: 1237–1239, 1993.

73. Not cited.

74. Clapham, D. E., and E. J. Neer. New roles for G protein βγ-dimers in transmembrane signalling. *Nature* 365: 403–406, 1993.

75. Not cited.

76. Clark, S. G., M. J. Stern, and H. R. Horvitz. C. elegans cell-signalling gene *sem-5* encodes a protein with SH2 and SH3 domains. *Nature* 356: 340–344, 1992.

77. Cobb, M. H., T. G. Boulton, and D. J. Robbins. Extracellular signal-regulated kinases: ERKs in progress. *Cell Regul.* 2: 965–978, 1991.

78. Coleman, D. E., A. M. Berghuis, E. Lee, M. E. Linder, A. G. Gilman, and S. R. Sprang. Structures of active conformations of Giα 1 and the mechanism of GTP hydrolysis. *Science* 265: 1405–1412, 1994.

79. Conklin, B. R., Z. Farfel, K. D. Lustig, D. Julius, and H. R. Bourne. Substitution of three amino acids switches receptor specificity of Gα$_q$ to that of Gα$_i$. *Nature* 363: 274–276, 1993.

80. Cook, N. J., W. Hanke, and U. B. Kaupp. Identification, purification, and functional reconstitution of the cyclic GMP-dependent channel from rod photoreceptors. *Proc. Natl. Acad. Sci. U.S.A.* 84(2): 585–589, 1987.

81. Cook, S., and F. McCormick. Ras blooms on sterile ground. *Nature* 369: 361–362, 1994.

82. Cooper, D. M., and G. Brooker. Ca^{2+}-inhibited adenylyl cyclase in cardiac tissue. [Review]. *TIPS* 14(2): 34–36, 1993.

83. Cooper, D.M.F., M. Yoshimura, Y. Zhang, M. Chiono, and R. Mahey. Capacitative Ca^{2+} entry regulates Ca^{2+}-sensitive adenylyl cyclases. *Biochem. J.* 297: 437–440, 1994.

84. Cote, R. H., M. S. Biernbaum, G. D. Nicol, and M. D. Bownds. Light-induced decreases in cGMP concentration precede changes in membrane permeability in frog rod photoreceptors. *J. Biol. Chem.* 259(15): 9635–9641, 1984.

85. Cote, R. H., M. D. Bownds, and V. Y. Arshavsky. Cgmp binding sites on photoreceptor phosphodiesterase—role in feedback regulation of visual transduction. *Proc. Natl. Acad. Sci. U.S.A.* 91(11): 4845–4849, 1994.

86. Cote, R. H., and M. A. Brunnock. Intracellular cGMP concentration in rod photoreceptors is regulated by binding to high and moderate affinity cGMP binding sites. *J. Biol. Chem.* 268(23): 17190–17198, 1993.

87. Cotecchia, S., J. Ostrowski, M. A. Kjelsberg. M. G. Caron, and R. J. Lefkowitz. Discrete amino acid sequences of the α$_1$-adrenergic receptor determine the selectivity of coupling to phosphatidylinositol hydrolysis. *J. Biol. Chem.* 267(3): 1633–1639, 1992.

88. Crespo, P., N. Xu, W. F. Simonds, and J. S. Gutkind. Ras-dependent activation of MAP kinase pathway mediated by G-protein βγ subunits. *Nature* 369: 418–420, 1994.

89. Crews, C. M., and R. L. Erikson. Extracellular signals and reversible protein phosphorylation: what to Mek of it all. *Cell* 74: 215–217, 1993.

90. Darby, C., R. L. Geahlen, and A. D. Schreiber. Stimulation of macrophage FcγRIIIA activates the receptor-associated protein tyrosine kinase Syk and induces phosphorylation of multiple proteins including p95Vav and p62/GAP-associated protein. *J. Immunol.* 152: 5429–5437, 1994.

91. Darnell, J.E.J., I. M. Kerr, and G. R. Stark. Jak-STAT pathways and transcriptional activation in response to IFNs and other extracellular signaling proteins. *Science* 264: 1415–1421, 1994.

92. Dascal, N., W. Schreibmayer, N. F. Lim, W. Wang, C. Chavkin, L. DiMagno, C. Labarca, B. L. Kieffer, R. C. Gaveriaux, D. Trollinger, H. A. Lester, and N. Davidson. Atrial G protein-activated K$^+$ channel: expression cloning and molecular properties. *Proc. Natl. Acad. Sci. U.S.A.* 90(21): 10235–10239, 1993.

93. Davis, R. J. The mitogen-activated protein kinase signal transduction pathway. *J. Biol. Chem.* 268(20): 14553–14556, 1993.

94. Dawson, T. M., J. L. Arriza, D. E. Jaworsky, F. F. Borisy, H. Attramadal, R. J. Lefkowitz, and G. V. Ronett. Beta-adrenergic receptor kinase-2 and beta-arrestin-2 as mediators of odorant-induced desensitization. *Science* 259: 825–829, 1993.

95. Degtyarev, M. Y., A. M. Spiegel, and T.L.Z. Jones. The G protein α$_s$ subunit incorporates [^{3}H]palmitic acid and mutation of cysteine-3 prevents this modification. *Biochemistry* 32: 8057–8061, 1993

96. DeVos, A., L. Tong, M. Milburn, P. Matias, J. Jancerik, K. Miuna, E. Ohitsuka, S. Nogushi, S. Nishimura, and S. Kim. Three dimensional structure of an oncogene protein: catalytic domain of human cH-ras p^{21}. *Science* 239: 888–893, 1988.

97. Dizhoor, A. M., D. G. Lowe, E. V. Olshevskaya, R. P. Laura, and J. B. Hurley. The human photoreceptor membrane guanylylcyclase, RetGC, is present in outer segments and is regulated by calcium and a soluble activator. *Neuron* 12(6): 1345–1352, 1994.

98. Dorsch, M., H. Hock, and T. Diamantstein. Tyrosine phosphorylation of SHC is induced by IL-3, IL-5 and GM-CSF. *Biochem. Biophys. Res. Commun.* 200: 562–568, 1994.

99. Douville, E., P. Duncan, N. Abraham, and J. C. Bell. Dual specificity kinases—a new family of signal transducers. *Cancer Metastasis Rev.* 13: 1–7, 1994.

100. Downes, C. P., P. T. Hawkins, and L. Stephens. Identification of the stimulated reaction in intact cells, its substrate supply and the metabolism of inositol phosphates. In: *Inositol Lipids in Cell Signalling*, edited by R. H. Michell, A. H. Drummond, and C. P. Downes. London: Academic Press, 1989, p. 1–38.

101. Duchesne, M., F. Schweighoffer, F. Parker, F. Clerc, Y. Frobert, M. N. Thang, and B. Tocque. Identification of the SH3 domain of GAP as an essential sequence for Ras-GAP-mediated signaling. *Science* 259: 525–528, 1993.

102. Duerson, K., R. Carroll, and D. E. Clapham. α-helical distorting substitutions disrupt coupling between m3 muscarinic receptors and G proteins. *FEBS Lett.* 324: 103–108, 1993.

103. Exton, J. H. Phosphatidylcholine breakdown and signal transduction. *Biochim. Biophys. Acta* 1212: 26–42, 1994.

104. Faure, M., T. A. Voyno-Yasenetskaya, and H. R. Bourne. cAMP and βγ subunits of heterotrimeric G proteins stimulate the mitogen-activated protein kinase pathway in COS-7 cells. *J. Biol. Chem.* 269: 7851–7854, 1994.

105. Feig, L. A. The many roads that lead to Ras. *Science* 260: 767–768, 1993.

106. Feig, L. A., and G. M. Cooper. Inhibition of NIH 3T3 cell

proliferation by a mutant *ras* protein with preferential affinity for GDP. *Mol. Cell Biol.* 8(8): 3235–3243, 1988.

107. Fein, A. A., and E. Z. Szuts. *Photoreceptors: Their Role in Vision.* Cambridge: Cambridge University Press, 1982.

108. Feinstein, P. G., K. A. Schrader, H. A. Bakalyar, W. J. Tang, J. Krupinski, A. G. Gilman, and R. R. Reed. Molecular cloning and characterization of a Ca^{2+}/calmodulin-insensitive adenylyl cyclase from rat brain. *Proc. Natl. Acad. Sci. U.S.A.* 88(22): 10173:–10177, 1991.

109. Fesenko, E. E., S. S. Kolesnikov, and A. L. Lyubarsky. Induction by cyclic GMP of cationic conductance in plasma membrane of retinal rod outer segment. *Nature* 313(6000): 310–313, 1985.

110. Fong, H.K.W., T. T. Amatruda, B. W. Birren, and M. I. Simon. Distinct forms of the β subunit of GTP-binding regulatory proteins identified by molecular cloning. *Proc. Natl. Acad. Sci. U.S.A.* 84: 3792–3796, 1987.

111. Franke, R. R., T. P. Sakmar, R. M. Graham, and H. G. Khorana. Structure and function in rhodopsin. *J. Biol. Chem.* 267(21): 14767–14774, 1992.

112. Fry, M. J., G. Panayotou, G. W. Booker, and M. D. Waterfield. New insights into protein-tyrosine kinase receptor signaling complexes. *Protein Sci.* 2: 1785–1797, 1993.

113. Frye, R. Involvement of G proteins, cytoplasmic calcium, phospholipases, phospholipid-derived second messengers, and protein kinases in signal transduction from mitogenic cell surface receptors. In: *Oncogenes and Tumor Suppressor Genes in Human Malignancies,* edited by C. Benz, and E. Liu. Boston: Kluwer Academic Publishers, 1993, vol. 14, p. 18.

114. Fung, B. K., and I. Griswold-Prenner. G protein-effector coupling: binding of rod phosphodiesterase inhibitory subunit to transducin. *Biochemistry* 28(8): 3133–3137, 1989.

115. Gallego, C., S. K. Gupta, L. E. Heasley, N.-X. Qian, and G. L. Johnson. Mitogen-activated protein kinase activation resulting from selective oncogene expression in NIH 3T3 and rat 1a cells. *Proc. Natl. Acad. Sci. U.S.A.* 89: 7355–7359, 1992.

116. Gao, B. N., and A. G. Gilman. Cloning and expression of a widely distributed (type IV) adenylyl cyclase. *Proc. Natl. Acad. Sci. U.S.A.* 88(22): 10178–10182, 1991.

117. Gao, B., A. G. Gilman, and J. D. Robishaw. A second form of the β subunit of signal-transducing G proteins. *Proc. Natl. Acad. Sci. U.S.A.* 84(17): 6122–6125, 1987.

118. Gardner. A. M., R. R. Vaillancourt, and G. L. Johnson. Activation of mitogen-activated protein kinase/extracellular signal-regulated kinase kinase by G protein and tyrosine kinase oncoproteins. *J. Biol. Chem.* 268: 17896–17901, 1993.

119. Garritsen, A., P.J.M. van Galen, and W. F. Simonds. The N-terminal coiled-coil domain of β is essential for γ association: a model for G-protein $\beta\gamma$ subunit interaction. *Proc. Natl. Acad. Sci. U.S.A.* 90(16): 7706–7710, 1993.

120. Gautam, N., M. Baetscher, R. Aebersold, and M. I. Simon. A G protein gamma subunit shares homology with ras proteins. *Science* 244: 971–974, 1989.

121. Gautam, N., J. Northrup, H. Tamir, and M. I. Simon. G protein diversity is increased by associations with a variety of γ subunits. *Proc. Natl. Acad. Sci. U.S.A.* 87: 7973–7977, 1990.

122. Gold, M. R., V. Duronio, S. P. Saxena, J. W. Schrader, and R. Aebersold. Multiple cytokines activate phosphatidylinositol 3-kinase in hemopoietic cells: association of the enzyme with various tyrosine-phosphorylated proteins. *J. Biol. Chem.* 269: 5403–5412, 1994.

123. Gorczyca, W. A., K. M. Gray, P. B. Detwiler, and K. Palczewski. Purification and physiological evaluation of a guanylate cyclase activating protein from retinal rods. *Proc. Natl. Acad. Sci. U.S.A.* 91(9): 4014–4018, 1994.

124. Gordon, J. J. Protein N-myristoylation: simple questions unexpected answers. *Clin. Res.* 38: 517–528, 1992.

125. Gordon, S. E., D. L. Brautigan, and A. L. Zimmerman. Protein phosphatases modulate the apparent agonist affinity of the light-regulated ion channel in retinal rods. *Neuron* 9(4): 739–748, 1992.

126. Greenlund, A. C., M. A. Farrar, B. L. Viviano, and R. D. Schreiber. Ligand-induced IFN-gamma receptor tyrosine phosphorylation couples the receptor to its signal transduction system (p91). *EMBO J.* 13: 1591–1600, 1994.

127. Grunwald, G. The structural and functional analysis of cadherin calcium-dependent cell adhesion molecules. *Curr. Opin. Cell Biol.* 5: 797–805, 1993.

128. Gulbins, E., K. M. Coggeshall, G. Baier, S. Katzav, P. Burn, and A. Altman. Tyrosine kinase-stimulated guanine nucleotide exchange activity of Vav in T cell activation. *Science* 260: 822–825, 1993.

129. Gulbins, E., K. M. Coggeshall, G. Baier, D. Telford, C. Langlet, G. Baier-Bitterlich, N. Bonnefoy-Berard, P. Burn, A. Wittinghofer, and A. Altman. Direct stimulation of Vav guanine nucleotide exchange activity for ras by phorbol esters and diglycerides. *Mol. Cell Biol.* 14(7): 4749–4758, 1994.

130. Gulbins, E., C. Langlet, G. Baier, N. Bonnefoy-Berard, E. Herbert, A. Altman, and K. M. Coggeshall. Tyrosine phosphorylation and activation of Vav GTP/GDP exchange activity in antigen receptor-triggered B cells. *J. Immunol.* 152: 2123–2129, 1994.

131. Gunderson, R., and P. Devreotes. *In vivo* receptor-mediated phosphorylation of a G protein in *Dictyostelium. Science* 248: 591–593, 1990.

132. Gupta, S. K., E. Diez, L. E. Heasley, S. Osawa, and G. L. Johnson. A G protein mutant that inhibits thrombin and purinergic receptor activation of phospholipase A_2. *Science* 249: 662–666, 1990.

133. Gupta, S. K., C. Gallego, and G. L. Johnson. Mitogenic pathways regulated by G protein oncogenes. *Mol. Biol. Cell* 3: 123–128, 1992.

134. Gupta, S. K., C. Gallego, G. L. Johnson, and L. E. Heasley. MAP kinase is constitutively activated in *gip2* and *src* transformed rat 1a fibroblasts. *J. Biol. Chem.* 267: 7987–7990, 1992.

135. Gupta, S. K., C. Gallego, J. M. Lowndes, C. M. Pleiman, C. Sable, B. J. Eisfelder, and G. L. Johnson. Analysis of the fibroblast transformation potential of GTPase-deficient *gip2* oncogenes. *Mol. Cell Biol.* 12: 190–197, 1992.

136. Gupta, S., A. Weiss, G. Kumar, S. Wang, and A. Nel. The T-cell antigen receptor utilizes Lck, Raf-1, and MEK-1 for activating mitogen-activated protein kinase. *J. Biol. Chem.* 269: 17349–17357, 1994.

137. Gutmann, D. H., and F. S. Collins. The neurofibromatosis Type 1 gene and its protein product, neurofibromin. *Neuron* 10: 335–343, 1993.

138. Gutowski, S., A. Smrcka, L. Nowak, D. Wu, M. Simon, and P. C. Sternweis. Antibodies to the α_q subfamily of guanine nucleotide-binding regulatory protein α subunits attenuate activation of phosphatidylinositol 4,5-bisphosphate hydrolysis by hormones. *J. Biol. Chem.* 266: 20519–20524, 1991.

139. Hafen, E., B. Dickson, D. Brunner, and T. Raabe. Genetic dissection of signal transduction mediated by the sevenless receptor tyrosine kinase in Drosophila. *Prog. Neurobiol.* 42: 287–292, 1994.

140. Hall, A. A biochemical function for Ras—at last. *Science* 264: 1413–1414, 1994.

141. Hamm, H., A. Deretic, A. Arendt, P. Hargrave, B. Koenig, and K. Hofmann. Site of G protein binding to rhodopsin mapped with synthetic peptides from the α subunit. *Science* 241: 832–835, 1988.

142. Han, J., J.-D. Lee, L. Bibbs, and R. J. Ulevitch. A MAP kinase targeted by endotoxin and hyperosmolarity in mammalian cells. *Science* 265: 808–811, 1994.

143. Hanks, S. K., A. M. Quinn, and T. Hunter. The protein kinase family: conserved features and deduced phylogeny of the catalytic domains. *Science* 241: 42–52, 1988.

144. Haribabu, B., and R. Snyderman. Identification of additional members of human G-protein-coupled receptor kinase multigene family. *Proc. Natl. Acad. Sci. U.S.A.* 90: 9398–9402, 1993.

145. Hasson, M. S., D. Blinder, J. Thorner, and D. D. Jenness. Mutational activation of the *STE5* gene product bypasses the requirement for G protein β and γ subunits in the yeast pheromone response pathway. *Mol. Cell Biol.* 14: 1054–1065, 1994.

146. Hausdorff, W. P., J. A. Pitcher, D. K. Luttrell, M. E. Linder, H. Kurose, S. J. Parsons, M. G. Caron, and R. J. Lefkowitz. Tyrosine phosphorylation of G protein α subunits by pp60^{c-src}. *Proc. Natl. Acad. Sci. U.S.A.* 89: 5720–5724, 1992.

147. Hedin, K., K. Duerson, and D. E. Clapham. Specificity of receptor-G protein interactions: searching for the structure behind the signal. *Cell. Signal.* 5: 505–518, 1993.

148. Hellevuo, K., M. Yoshimura, M. Kao, P. L. Hoffman, D. M. Cooper, and B. Tabakoff. A novel adenylyl cyclase sequence cloned from the human erythroleukemia cell line. *Biochem. Biophys. Res. Commun.* 192(1): 311–318, 1993.

149. Henderson, R., J. Baldwin, T. H. Ceska, F. Zemlin, E. Beckmann, and K. Downing. Model for the structure of bacteriorhodopsin based on high resolution electron microscopy. *J. Mol. Biol.* 213: 899–929, 1990.

150. Henderson, R., and P. N. Unwin. Three-dimensional model of purple membrane obtained by electron microscopy. *Nature* 257: 28–32, 1975.

151. Higashijima, T., J. Burnier, and E. Ross. Regulation of Gi and Go by mastoparan, related amphiphilic peptides and hydrophobic amines. Mechanisms and structural determinants of activity. *J. Biol. Chem.* 265: 14176–14186, 1990.

152. Higgins, J. B., and P. J. Casey. *In vitro* processing of recombinant G protein γ subunits. *J. Biol. Chem.* 209: 9067–9073, 1994.

153. Hille, B. *Ionic Channels of Excitable Membranes.* Sunderland, MA: Sinauer; 1992.

154. Holtzman, E. J., T. B. Kinane, K. West, B. W. Soper, H. Karga, D. A. Ausiello, and L. Ercolani. Transcriptional regulation of G-protein α_i subunit genes in LLC-PK$_1$ renal cells and characterization of the porcine Gα_{i-3} gene promoter. *J. Biol. Chem.* 268(6): 3964–3975, 1993.

155. Hsu, Y. T., and R. S. Molday. Modulation of the cGMP-gated channel of rod photoreceptor cells by calmodulin [see comments]. *Nature* 361(6407): 76–79, 1993.

156. Huang, X.-Y., A. D. Morielli, and E. G. Peralta. Tyrosine kinase-dependent suppression of a potassium channel by the G protein-coupled m1 muscarinic acetylcholine receptor. *Cell* 75: 1145–1156, 1993.

157. Hunter, T., R. A. Lindberg, D. S. Middlemas, S. Tracy, and P. Van der Geer. Receptor protein tyrosine kinases and phosphatases. *Cold Spring Harb. Symp. Quant. Biol.* 57(25): 25–42, 1992.

158. Hurley, J. B. Molecular properties of the cGMP cascade of vertebrate photoreceptors. [Review.] *Annu. Rev. Physiol.* 49: 793–812, 1987.

159. Hurley, J. B., and L. Stryer. Purification and characterization of the gamma regulatory subunit of the cyclic GMP phosphodiesterase from retinal rod outer segments. *J. Biol. Chem.* 257(18): 11094–11099, 1982.

160. Iiri, T., P. Herzmark, J. M. Nakamoto, C. Van Dop, and H. R. Bourne. Rapid GDP release from Gα_s in patients with gain and loss of endocrine function. *Nature* 371: 164–167, 1994.

161. Iiri, T., M. Tohkin, N. Morishima, Y. Ohoka, M. Ui, and T. Katada. Chemotactic peptide receptor-supported ADP-ribosylation of a pertussis toxin substrate GTP-binding protein by cholera toxin in neutrophil-type HL-60 cells. *J. Biol. Chem.* 264: 21394–21400, 1989.

162. Iismaa, T., and J. Shine. G protein coupled receptors. *Curr. Opin. Cell. Biol.* 4: 195–202, 1992.

163. Iniguez-Lluhi, J. A., M. I. Simon, J. D. Robishaw, and A. G. Gilman. G protein $\beta\gamma$ subunits synthesized in Sf9 cells. *J. Biol. Chem.* 267: 23409–23417, 1992.

164. Inglese, J., W. J. Koch, M. G. Caron, and R. J. Lefkowitz. Isoprenylation in regulation of signal transduction by G-protein-coupled receptor kinases. *Nature* 359: 147–150, 1992.

165. Irie, K., Y. Gotoh, B. M. Yashar, B. Errede, E. Nishida, and K. Matsumoto. Stimulatory effects of yeast and mammalian 14-3-3 proteins on the Raf protein kinase. *Science* 265: 1716–1719, 1994.

166. Irving, B. A., and A. Weiss. The cytoplasmic domain of the T cell receptor ζ chain is sufficient to couple to receptor-associated signal transduction pathways. *Cell* 64: 891–901, 1991.

167. Itoh, H., and A. G. Gilman. Expression and analysis of Gα_s mutants with decreased ability to activate adenylylcyclase. *J. Biol. Chem.* 266(24): 16226–16231, 1991.

168. Iyengar, R. Molecular and functional diversity of mammalian G$_s$-stimulated adenylyl cyclases. *FASEB J.* 7(9): 768–775, 1993.

169. Iyengar, R. Multiple families of G$_s$-regulated adenylyl cyclases. *Adv. Second Messenger Phosphoprotein Res.* 28(27): 27–36, 1993.

170. Jacobowitz, O., J. Chen, R. T. Premont, and R. Iyengar. Stimulation of specific types of G$_s$-stimulated adenylyl cyclases by phorbol ester treatment. *J. Biol. Chem.* 268(6): 3829–3832, 1993.

171. Jaffe, L. A., C. J. Gallo, R. H. Lee, Y. K. Ho, and T.L.Z. Jones. Oocyte maturation in starfish is mediated by the $\beta\gamma$-subunit complex of a G-protein. *J. Cell Biol.* 121: 775–783, 1993.

172. Jelsema, C. L., and J. Axelrod. Stimulation of phospholipase A$_2$ activity in bovine rod outer segments by the $\beta\gamma$ subunits of transducin and its inhibition by the α subunit. *Proc. Natl. Acad. Sci. U.S.A.* 84: 3623–3627, 1987.

173. Jiang, H., D. Wu, and M. I. Simon. Activation of phospholipase C β_4 by heterotrimeric GTP-binding proteins. *J. Biol. Chem.* 269: 7593–7596, 1994.

174. Johnson, G. L., A. M. Gardner, C. Lange-Carter, N.-X. Qian, M. Russell, and S. Winitz. How does the G protein, Gi2, transduce mitogenic signals? *J. Cell. Biochem.* 54: 415–422, 1994.

175. Johnston, J. A., M. Kawamura, R. A. Kirken, Y.-Q. Chen, T. B. Blake, K. Shibuya, J. R. Ortaldo, D. W. McVicar, and J. J. O'Shea. Phosphorylation and activation of the Jak-3 janus kinase in response to IL-2. *Nature* 370: 151–153, 1994.

176. Joly, M., A. Kazlauskas, F. S. Fay, and S. Corvera. Disruption of PDGF receptor trafficking by mutation of its PI-3 kinase binding sites. *Science* 263: 684–687, 1994.

177. Jones, S., M.-L. Vignais, and J. R. Broach. The *CDC25* protein of *Saccharomyces cerevisiae* promotes exchange of guanine nucleotides bound to Ras. *Mol. Cell. Biol.* 11: 2641–2646, 1991.

178. Jones, T.L.Z., W. F. Simonds, J. J. Merendino, M. R. Brann, and A. M. Spiegel. Myristoylation of an inhibitory GTP-binding protein α is essential for its membrane attachment. *Proc. Natl. Acad. Sci. U.S.A.* 87: 568–572, 1990.

179. Katada, T., K. Kusakabe, M. Oinuma, and M. Ui. A novel mechanism for the inhibition of adenylate cyclase via inhibitory GTP-binding proteins. *J. Biol. Chem.* 262(25): 11897–11900, 1987.

180. Katsushika, S., L. Chen, J. Kawabe, R. Nilakantan, N. J. Halnon, C. J. Homcy, and Y. Ishikawa. Cloning and characterization of a sixth adenylyl cyclase isoform: types V and VI constitute a subgroup within the mammalian adenylyl cyclase family. *Proc. Natl. Acad. Sci. U.S.A.* 89(18): 8774–8778, 1992.

181. Kaupp, U. B., T. Niidome, T. Tanabe, S. Terada, W. Bonigk, W. Stuhmer, N. J. Cook, K. Kangawa, H. Matsuo, T. Hirose *et al.* Primary structure and functional expression from complementary DNA of the rod photoreceptor cyclic GMP-gated channel. *Nature* 342(6251): 762–766, 1989.

182. Kawamura, S. Light-sensitivity modulating protein in frog rods. *Photochem. Photobiol.* 56(6): 1173–1180, 1992.

183. Kawamura, S., O. Hisatomi, S. Kayada, F. Tokunaga, and C. H. Kuo. Recoverin has S-modulin activity in frog rods. *J. Biol. Chem.* 268(20): 14579–14582, 1993.

184. Kawamura, S., and M. Murakami. Intracellular injection of cyclic-GMP increases sodium conductance in gecko photoreceptors. *Jpn. J. Physiol.* 33(5): 789–800, 1983.

185. Kawamura, S., and M. Murakami. Calcium-dependent regulation of cyclic GMP phosphodiesterase by a protein from frog retinal rods. *Nature* 349(6308): 420–423, 1991.

186. Kaziro, Y., H. Itoh, T. Kozasa, M. Nakafuku, and T. Satoh. Structure and function of signal-transducing GTP-binding proteins. *Annu. Rev. Biochem.* 60: 349–400, 1991.

187. Kazlauskas, A. Receptor tyrosine kinases and their targets. *Curr. Opin. Genet. Dev.* 4: 5–14, 1994.

188. Keegan, A. D., K. Nelms, M. White, L.-M. Wang, J. H. Pierce, and W. E. Paul. An IL-4 receptor region containing an insulin receptor motif is important for IL-4-mediated IRS-1 phosphorylation and cell growth. *Cell* 76: 811–820, 1994.

189. Khosravi-Far, R., and C. J. Der. The Ras signal transduction pathway. *Cancer Metastasis Rev.* 13: 67–89, 1994.

190. Kim, D., D. L. Lewis, L. Graziadei, E. J. Neer, D. Bar-Sagi, and D. E. Clapham. G-protein $\beta\gamma$ subunits activate the cardiac muscarinic K^+-channel via phospholipase A_2. *Nature* 337: 557–560, 1989.

191. Kim, S., S. L. Ang, D. B. Bloch, K. D. Bloch, Y. Kawahara, C. Tolman, R. Lee, J. G. Seidman, and E. J. Neer. Identification of cDNA encoding an additional α subunit of human tissues and cell lines. *Proc. Natl. Acad. Sci. U.S.A.* 85: 4153–4157, 1988.

192. Kishimoto, T., T. Taga, and S. Akira. Cytokine signal transduction. *Cell* 76: 253–262, 1994.

194. Kleuss, C., J. Hescheler, C. Ewel, W. Rosenthal, G. Schultz, and B. Wittig. Assignment of G-protein subtypes to specific receptors inducing inhibition of calcium currents. *Nature* 353: 43–48, 1991.

195. Kleuss, C., H. Scherubl, J. Hescheler, G. Schultz, and B. Wittig. Different β-subunits determine G-protein interaction with transmembrane receptors. *Nature* 358: 424–426, 1992.

196. Kleuss, C., H. Scherubl, J. Hescheler, G. Schultz, and B. Wittig. Selectivity in signal transduction determined by γ subunits of heterotrimeric G proteins. *Science* 259: 832–834, 1993.

197. Kobilka, B. Adrenergic receptors as models for G protein coupled receptors. *Annu. Rev. Neurosci.* 15: 87–114, 1992.

198. Kobilka, B., T. Kobilka, K. Daniel, J. Regan, M. Caron, and R. Lefkowitz. Chimeric α_2, β_2-adrenergic receptors: delineation of domains involved in effector coupling and ligand binding specificity. *Science* 240: 1310–1316, 1988.

199. Koch, K. W., and L. Stryer. Highly cooperative feedback control of retinal rod guanylate cyclase by calcium ions. *Nature* 334(6177): 64–66, 1988.

200. Koch, W. J., B. E. Hawes, J. Inglese, L. M. Luttrell, and R. J. Lefkowitz. Cellular expression of the carboxyl terminus of a G protein-coupled receptor kinase attenuates $G\beta\gamma$-mediated signaling. *J. Biol. Chem.* 269: 6193–6197, 1994.

201. Koch, W. J., J. Inglese, C. W. Stone, and R. J. Lefkowitz. The binding site for the $\beta\gamma$ subunits of hetertrimeric G proteins on the β-adrenergic receptor kinase. *J. Biol. Chem.* 268: 8256–8260, 1993.

202. Koesling, D., G. Schultz, and E. Bohme. Sequence homologies between guanylyl cyclases and structural analogies to other signal-transducing proteins. [Review]. *FEBS Lett.* 280(2): 301–306, 1991.

203. Koutalos, Y., and K.-W. Yau. A rich complexity emerges in phototransduction. *Curr. Opin. Neurobiol.* 3: 513–519, 1993.

204. Kramer, R. M. *Structure, function, and regulation of mammalian phospholipases A2.* In: *Advances in Second Messenger and Phosphoprotein Research,* edited by B. L. Brown, and P. R. Dobson. New York: Raven Press, 1993, p. 81–89.

205. Kranz, J. E., B. Satterberg, and E. A. Elion. The MAP kinase Fus3 associates with and phosphoyrlates the upstream signaling component STE5. *Genes Dev.* 8: 313–327, 1994.

206. Krupinski, J., F. Coussen, H. A. Bakalyar, W. J. Tang, P. G. Feinstein, K. Orth, C. Slaughter, R. R. Reed, and A. G. Gilman. Adenylyl cyclase amino acid sequence: possible channel- or transporter-like structure. *Science* 244(4912): 1558–1564, 1989.

207. Krupinski, J., T. C. Lehman, C. D. Frankenfield, J. C. Zwaagstra, and P. A. Watson. Molecular diversity in the adenylylcyclase family. Evidence for eight forms of the enzyme and cloning of type VI. *J. Biol. Chem.* 267(34): 24858–24862, 1992.

208. Kubo, T., H. Bujo, I. Akiba, J. Nakai, M. Mishina, and S. Numa. Location of a region of the muscarinic acetylcholine receptor involved in selective effector coupling. *FEBS Lett.* 241: 119–125, 1988.

209. Kubo, Y., E. Reuveny, P. A. Slesinger, Y. N. Jan, and L. Y. Jan. Primary structure and functional expression of a rat G-protein-coupled muscarinic potassium channel. *Nature* 364: 802–806, 1993.

210. Kuo, C. C., and B. P. Bean. G-protein modulation of ion permeation through N-type calcium channels. *Nature* 365(6443): 258–262, 1993.

211. Kurachi, Y., R. T. Tung, H. Ito, and T. Nakajima. G protein activation of cardiac muscarinic K^+ channels. *Prog. Neurobiol.* 39: 229–246, 1992.

212. Kyriakis, J. M., P. Banerjee, E. Nikolakaki, T. Dai, E. A. Rubie, M. F. Ahmad, J. Avruch, and J. R. Woodgett. The stress-activated protein kinase subfamily of c-Jun kinases. *Nature* 369: 156–160, 1994.

213. Lambright, D. G., J. P. Noel, H. E. Hamm, and P. B. Sigler. Structural determinants for activation of the α-subunit of a heterotrimeric G protein. *Nature* 369: 621–628, 1994.

214. Landis, C. A., S. B. Masters, A. Spada, A. M. Pace, H. R. Bourne, and L. Vallar. GTPase inhibiting mutations activate

the α chain of G$_s$ and stimulate adenylyl cyclase in human pituitary tumours. *Nature* 340: 692–696, 1989.

215. Lange-Carter, C., and G. L. Johnson. Ras-dependent growth factor regulation of MEK kinase in PC12 cells. *Science* 265: 1458–1461, 1994.

216. Lange-Carter, C., C. M. Pleiman, A. M. Gardner, K. J. Blumer, and G. L. Honson. A divergence in the MAP kinase regulatory network defined by MEK kinase and Raf. *Science* 260: 315–319, 1993.

217. Larhammar, D., A. Blomquist, and C. Wahlestedt. The receptor revolution—multiplicity of G-protein-coupled receptors. *Drug Design Discovery* 9: 179–188, 1993.

218. Law, S. F., D. Manning, and T. Reisine. Identification of the subunits of GTP-binding proteins coupled to somatostatin receptors. *J. Biol. Chem.* 266: 17885–17897, 1991.

219. Law, S. F., K. Yasuda, G. I. Bell, and T. Reisine. Giα$_3$ and Goα selectively associate with the cloned somatostatin receptor subtype SSTR2. *J. Biol. Chem.* 268: 10721–10727, 1993.

220. Leberer, E., D. Dignard, L. Hougan, D. Y. Thomas, and M. Whiteway. Dominant-negative mutants of a yeast G-protein β subunit identify two functional regions involved in pheromone signalling. *EMBO J.* 11(13): 4805–4813, 1992.

221. Lechleiter, J. L., R. Hellmiss, K. Duerson, D. Ennulat, D. E. Clapham, and E. G. Peralta. Distinct sequence elements control the specificity of G protein activation by muscarinic acetylcholine receptor subtypes. *EMBO J.* 9: 4381–4390, 1990.

222. Lee, C. H., D. Park, D. Wu, S. G. Rhee, and M. I. Simon. Members of the G$_q$ α subunit gene family activate phospholipase C β isozymes. *J. Biol. Chem.* 267: 16044–16047, 1992.

223. Lee, J., and P. F. Pilch. The insulin receptor: structure, function, and signaling. *Am. J. Physiol.* 266 (*Cell Physiol.* 35): C319–C334, 1994.

224. Lee, R. H., B. S. Lieberman, H. K. Yamane, D. Bok, and B.-K. Fung. A third form of the G protein β subunit. *J. Biol. Chem.* 267: 24776–24781, 1992.

225. Leeb-Lundberg, L.M.F., and X.-H. Song. Bradykinin and bombesin rapidly stimulate tyrosine phosphorylation of a 120-kDa group of proteins in Swiss 3T3 cells. *J. Biol. Chem.* 266: 7746–7749, 1991.

226. Leevers, S. J., H. F. Paterson, and C. J. Marshall. Requirement for Ras in Raf activation is overcome by targeting Raf to the plasma membrane. *Nature* 369: 411–414, 1994.

227. Lefkowitz, R. J. G protein-coupled receptor kinases. *Cell* 74(3): 409–412, 1993.

228. Li, B.-Q., D. Kaplan, H.-F. Kung, and T. Kamata. Nerve growth factor stimulation of the Ras-guanine nucleotide exchange factor and GAP activities. *Science* 256: 1456–1459, 1992.

229. Li, W., R. Nishimura, A. Kashishian, A. G. Batzer, W.J.H. Kim, J. A. Cooper, and J. Schlessinger. A new function for a phosphotyrosine phosphatase: Linking GRB2-Sos to a receptor tyrosine kinase. *Mol. Cell. Biol.* 14: 509–517, 1994.

230. Liebman, P. A., K. R. Parker, and E. A. Dratz. The molecular mechanism of visual excitation and its relation to the structure and composition of the rod outer segment. [Review.] *Annu. Rev. Physiol.* 49: 765–791, 1987.

231. Liebman, P. A., and E. Pugh Jr. The control of phosphodiesterase in rod disk membranes: kinetics, possible mechanisms and significance for vision. *Vision Res.* 19(4): 375–380, 1979.

232. Lin, L.-L., M. Wartmann, A. Y. Lin, J. L. Knopf, A. Seth, and R. J. Davis. cPLA$_2$ is phosphorylated and activated by MAP kinase. *Cell* 72: 269–278, 1993.

233. Linder, M. E., P. Middleton, J. R. Hepler, R. Taussig, A. G. Gilman, and S. M. Mumby. Lipid modifications of G proteins:

α subunits are palmitoylated. *Proc. Natl. Acad. Sci. U.S.A.* 90: 3675–3679, 1993.

234. Linder, M. E., I.-H. Pang, R. J. Duronio, J. I. Gordon, P. C. Sternweis, and A. G. Gilman, Lipid modifications of G protein subunits: myristoylation of Goα increases its affinity for βγ. *J. Biol. Chem.* 266: 4654–4659, 1991.

235. Logothetis, D. E., D. Kim, J. K. Northup, E. J. Neer, and D. E. Clapham. Specificity of action of guanine nucleotide-binding regulatory protein subunits on the cardiac muscarinic K$^+$ channel. *Proc. Natl. Acad. Sci. U.S.A.* 85: 5814–5818, 1988.

236. Lohse, M. J. Molecular mechanisms of membrane receptor desensitization. *Biochim. Biophys. Acta Mol. Cell. Res.* 1179(2): 171–188, 1993.

237. Lohse, M. J., J. L. Benovic, J. Codina, M. G. Caron, and R. J. Lefkowitz. β-arrestin: a protein that regulates β-adrenergic receptor function. *Science* 248: 1547–1550, 1990.

238. Lounsbury, K. M., B. Schlegel, M. Poncz, L. F. Brass, and D. R. Manning. Analysis of G$_z$ alpha by site-directed mutagenesis. Sites and specificity of protein kinase C-dependent phosphorylation. *J. Biol. Chem.* 268(5): 3494–3498, 1993.

239. Lowenstein, E. J., R. J. Daly, A. G. Batzer, W. Li, B. Margolis, R. Lammers, A. Ullrich, and J. Schlessinger. The SH2 and SH3 domain-containing protein GRB2 links receptor tyrosine kinases to *ras* signaling. *Cell* 70: 431–442, 1992.

240. MacNicol, A. M., A. J. Muslin, and L. T. Williams. Raf-1 kinase is essential for early *Xenopus* development and mediates the induction of mesoderm by FGF. *Cell* 73: 571–583, 1993.

241. Manser, E., T. Leung, H. Salihuddin, L. Tan, and L. Lim. A non-receptor tyrosine kinase that inhibits the GTPase activity of p21cdc42. *Nature* 363: 364–367, 1993.

242. Manser, E., T. Leung, H. Salihuddin, Z.-S. Zhao, and L. Lim. A brain serine/threonine protein kinase activated by CDC42 and Rac1. *Nature* 367: 40–46, 1994.

243. Mansour, S. J., W. T. Matten, A. S. Hermann, J. M. Candia, S. Rong, K. Fukasawa, G. F. Vande Woude, and N. G. Ahn. Transformation of mammalian cells by constitutively active MAP kinase kinase. *Science* 265: 966–970, 1994.

244. Markby, D. W., R. Onrust, and H. R. Bourne. Separate GTP binding and GTPase activating domains of a Gα subunit. *Science* 262: 1895–1901, 1993.

245. Marshall, C. J. MAP kinase kinase kinase, MAP kinase kinase, and MAP kinase. *Curr. Opin. Genet. Dev.* 4: 82–89, 1994.

246. Masters, S., K. Sullivan, T. Miller, B. Beiderman, N. Lopez, J. Ramachandran, H. Bourne. Carboxy-terminal domain of Gsα specifies coupling of receptors to stimulation of adenylyl cyclase. *Science* 241: 448–451, 1988.

247. Masu, M., Y. Tanabe, K. Tsuchida, R. Shigemoto, and S. Nakanishi. Sequence and expression of a metabotropic glutamate receptor. *Science* 349: 760–765, 1991.

248. Mathie, A., L. Bernheim, and B. Hille. Inhibition of N- and L-type calcium channels by muscarinic receptor activation in rat sympathetic neurons. *Neuron* 8: 907, 1992.

249. Matsuda, L. A., S. J. Loliat, M. J. Brownstein, A. C. Young, and T. I. Bonner. Structure of a cannabinoid receptor and functional expression of the cloned cDNA. *Nature* 346: 561–564, 1990.

250. Mazzoni, M. R., and H. Hamm. Effect of monoclonal antibody binding on α-βγ subunit interaction in the rod outer segment G protein, G$_t$. *Biochemistry* 28: 9873–9880, 1989.

251. Mazzoni, M. R., J. A. Malinski, and H. E. Hamm. Structural analysis of rod-GTP binding protein, G$_t$. *J. Biol. Chem.* 266: 14072–14081, 1991.

252. McCormick, F. How receptors turn Ras on. *Nature* 363: 15–16, 1993.

253. McFarland, K. C., R. Sprengel, H. S. Phillips, M. Kohler, N. Rosemblit, K. Nikolics, D. L. Segaloff, and P. H. Seeburg. Lutropin-choriogonadotropin receptor: An unusual member of the G protein-coupled receptor family. *Science* 245: 494–499, 1989.

254. McLaughlin, S. K., P. J. McKinnon, and R. F. Margolskee. Gustducin is a taste-cell-specific G protein closely related to the transducins. *Nature* 357: 563–569, 1992.

255. Miller, J. L., and J. I. Korenbrot. In retinal cones, membrane depolarization in darkness activates the cGMP-dependent conductance. A model of Ca^{2+} homeostasis and the regulation of guanylate cyclase. *J. Gen. Physiol.* 101(6): 933–960, 1993.

256. Miller, W. H. Physiological evidence that light-mediated decrease in cyclic GMP is an intermediary process in retinal rod transduction. *J. Gen. Physiol.* 80(1): 103–123, 1982.

257. Moodie, S. A., B. M. Willumsen, M. J. Weber, and A. Wolfman. Complexes of ras-GTP with raf-1 and mitogen-activated protein kinase kinase. *Science* 260: 1658–1661, 1993.

258. Morimoto, B. H., and D. J. Koshland. Conditional activation of cAMP signal transduction by protein kinase C. The effect of phorbol esters on adenylyl cyclase in permeabilized and intact cells. *J. Biol. Chem.* 269(6): 4065–4069, 1994.

259. Moss, J., and M. Vaughan. ADP ribosylation of guanyl nucleotide binding regulatory proteins by bacterial toxins. *Adv. Enzymol. Relat. Areas Mol. Biol.* 61: 303–379, 1988.

260. Mourey, R. J., and J. E. Dixon. Protein tyrosine phosphatases: characterization of extracellular and intracellular domains. *Curr. Opin. Genet. Dev.* 4: 31–39, 1994.

261. Mulcahy, L. S., M. R. Smith, and D. W. Stacey. Requirement for *ras* proto-oncogene function during serum-stimulated growth of NIH 3T3 cells. *Nature* 313: 241–243, 1985.

262. Mumby, S. M., R. O. Heuckeroth, J. I. Gordon, and A. G. Gilman. G protein α subunit expression, myristoylation, and membrane association in COS cells. *Proc. Natl. Acad. Sci. U.S.A.* 87: 728–732, 1990.

263. Mumby, S. M., C. Kleuss, and A. G. Gilman. Receptor regulation of G-protein palmitoylation. *Proc. Natl. Acad. Sci. U.S.A.* 91: 2800–2804, 1994.

264. Murakami, T., W. F. Simonds, and A. L. Spiegel. Site-specific antibodies directed against G protein β and γ subunits: effects on α and βγ subunit interaction. *Biochemistry* 31: 2905–2911, 1992.

265. Murayama, T., and M. Ui. [³H] GDP release from rat and hamster adipocyte membranes independently linked to receptors involved in activation or inhibition of adenylate cyclase. Differential susceptibility to two bacterial toxins. *J. Biol. Chem.* 259(4): 761–769, 1984.

266. Musacchio, A., T. Gibson, V.-P. Lehto, and M. Saraste. SH3—an abundant protein domain in search of a function. *FEBS Lett.* 307(1): 55–61, 1992.

267. Musacchio, A., T. Gibson, P. Rice, J. Thompson, and M. Saraste. The PH domain: a common piece in the structural patchwork of signalling proteins. *Trends Biochem. Sci.* 18: 343–348, 1993.

268. Nagayama, Y., H. L. Wadsworth, G. D. Chazenbalk, D. Russo, P. Seto, and B. Rapoport. Thyrotropin-luteinizing hormone/chorionic gonadotropin receptor extracellular-domain chimeras as probes for thyrotropin receptor function. *Proc. Natl. Acad. Sci. U.S.A.* 88: 902–905, 1991.

269. Nakamura, Y., S. M. Russell, S. A. Mess, M. Friedmann, M. Erdos, C. Francois, Y. Jacques, S. Adelstein, and W. J. Leonard. Heterodimerization of the IL-2 receptor beta and gamma chain cytoplasmic domains is required for signaling. *Nature* 369: 330–333, 1994.

270. Navon, S. E., and B.K.-K. Fung. Characterization of transducin from bovine retinal rod outer segments. Participation of the amino-terminal region of Tα in subunit interaction. *J. Biol. Chem.* 263: 15476–15751, 1987.

271. Neer, E. J. G proteins: critical control points for transmembrane signals. *Protein Sci.* 3: 3–14, 1994.

272. Neer, E. J., and D. E. Clapham. *Structure and function of the βγ subunit.* In: *The G Proteins,* edited by L. Birnbaumer. Orlando, FL: Academic Press, 1989, p. 41–61.

273. Neer, E., and J. M. Lok. Partial purification and characterization of a pp60^vsrc-related tyrosine kinase from bovine brain. *Proc. Natl. Acad. Sci. U.S.A.* 82: 6025–6029, 1985.

274. Neer, E. J., J. Lok, and L. Wolf. Purification and properties of the inhibitory guanine nucleotide regulatory unit of brain adenylate cyclase. *J. Biol. Chem.* 259: 14222–14229, 1984.

275. Neiman, A. M., and I. Herskowitz. Reconstitution of a yeast protein kinase cascade in vitro: activation of the yeast MEK homologue STE7 by STE11. *Proc. Natl. Acad. Sci. U.S.A.* 91: 3398–3402, 1994.

276. Nelson, B. H., J. D. Lord, and P. D. Greenberg. Cytoplasmic domains of the IL-2 receptor beta and gamma chains mediate the signal for T cell proliferation. *Nature* 369: 333–336, 1994.

277. Nicholas, J., K. R. Cameron. and R. W. Honess. Herpesvirus saimiri encodes homologues of G protein-coupled receptors and cyclins. *Nature* 355: 362–265, 1992.

278. Noel, J. P., H. E. Hamm, and P. B. Sigler. The 2.2 Å crystal structure of transducin-α complexed with GTPγS. *Nature* 366: 654–663, 1993.

279. North, R. A. Drug receptors and the inhibition of nerve cells. *Br. J. Pharmacol.* 98: 13–28, 1989.

280. Obermeier, A., R. A. Bradshaw, K. Seedorf, A. Choidas, J. Schlessinger, and A. Ullrich. Neuronal differentiation signals are controlled by nerve growth factor receptor/Trk binding sites for SHC and PLCγ. *EMBO J.* 13: 1585–1590, 1994.

281. O'Dowd, B., M. Hnatowich, M. Caron, R. Lefkowitz, and M. Bouvier. Palmitoylation of the human β₂-adrenergic receptor. *J. Biol. Chem.* 264(13): 7564–7569, 1989.

282. O'Dowd, B. F., M. Hnatowich, J. W. Regan, W. M. Leader, M. G. Caron, and R. J. Lefkowitz. Site-directed mutagenesis of the cytoplasmic domains of the human β₂-adrenergic receptor. Localizations of regions involved in G protein-receptor coupling. *J. Biol. Chem.* 263(31): 15985–15992, 1988.

283. Ohashi, K., K. Mizuno, K. Kuma, T. Miyata, and T. Nakamura. Cloning of the cDNA for a novel receptor tyrosine kinase, Sky, predominantly expressed in brain. *Oncogene* 9: 699–705, 1994.

284. Ohmichi, M., T. Sawada, Y. Kanda, K. Koike, K. Kirota, A. Miyake, and A. R. Saltiele. Thyrotropin-releasing hormone stimulates MAP kinase activity in GH3 cells by divergent pathways: evidence of role for early tyrosine phosphorylation. *J. Biol. Chem.* 269: 3783–3788, 1994.

285. Okamoto, T., T. Katada, Y. Muriyama, M. Ui, E. Ogata, and I. Nishimoto. A simple structure encodes G protein-activating function of IGF-II mannose 6-phosphate receptor. *Cell* 62: 709–717, 1990.

286. Okamoto, T., and I. Nishimoto. Detection of G protein-activator regions in M₄ subtype muscarinic, cholinergic, and α₂-adrenergic receptor based upon characteristics in primary structure. *J. Biol. Chem.* 267(12): 8342–8346, 1992.

287. Olate, J., and J. E. Allende. Structure and function of G proteins. *Pharmacol. Ther.* 51: 403–419, 1991.

288. Olivier, J. P., T. Raabe, M. Henkemeyer, B. Dickson, G. Mbamula, B. Margonlis, J. Schlessinger, E. Hafen, and T. Pawson. A Drosophila SH2-SH3 adaptor protein implicated

in coupling the sevenless tyrosine kinase to an activator of *Ras* guanine nucleotide exchange, Sos. *Cell* 73: 179–191, 1993.

289. Oppi, C., T. Wagner, A. Crisari, B. Camerini, and G.P.T. Valentini. Attenuation of GTPase activity of recombinant $G\alpha_o$ by peptides representing sequence permutations of mastoparan. *Proc. Natl. Acad. Sci. U.S.A.* 89: 8268–8272, 1992.

290. O'Shea, E. K., J. D. Klemm, P. S. Kim, and T. Alber. X-ray structure of the GCNA leucine zipper, a two-strand, parallel coiled coil. *Science* 254: 539–544, 1991.

291. O'Shea, E. K., R. Rutkowski, and P. S. Kim. Mechanism of specificity in the *Fos-Jun* oncoprotein heterdimer. *Cell* 68: 699–708, 1992.

292. Otte, A. P., L. L. McGrew, J. Olate, N. M. Nathanson, and R. T. Moon. Expression and potential functions of G-protein α subunits in embryos of *Xenopus laevis Development* 116: 141–146, 1992.

293. Ovchinnikov, Y. A., N. G. Abdulaev, and A. S. Bogachuk. Two adjacent cysteine residues in the C-terminal cytoplasmic fragment of bovine rhodopsin are palmitoylated. *FEBS Lett.* 230(1,2): 1–5, 1988.

294. Palczewski, K., and J. L. Benovic. G-protein-coupled receptor kinases. [Review.] *Trends Biochem. Sci.* 16(10): 387–391, 1991.

295. Palczewski, K., G. Rispoli, and P. B. Detwiler. The influence of arrestin (48K protein) and rhodopsin kinase on visual transduction. *Neuron* 8(1): 117–126, 1992.

296. Panayotou, G., and M. D. Waterfield. The assembly of signaling complexes by receptor tyrosine kinases. *Bioessays* 15: 171–177, 1993.

297. Parenti, M., M. A. Vigano, M. H. Newman, G. Milligan, and A. I. Magee. A novel N-terminal motif for palmytoylation of G proteins α subunits. *Biochem. J.* 291: 349–352, 1993.

298. Park, D., D.-Y. Jhon, R. Kriz, J. Knopf, and S. G. Rhee. Cloning, sequencing, expression, and G_q-independent activation of phospholipase C-β_2. *J. Biol. Chem.* 267: 16048, 1992.

299. Pawson, T. Signal transduction—a conserved pathway from the membane to the nucleus. *Dev. Genet.* 14: 333–338, 1993.

300. Pawson, T., and G. D. Gish. SH2 and SH3 domains: from structure to function. *Cell* 71: 359–362, 1992.

301. Pawson, T., and J. Schlessinger. SH2 and SH3 domains. *Curr. Biol.* 3(7): 434–442, 1993.

302. Perlmutter, R. M., S. D. Levin, M. W. Appleby, S. J. Anderson, and I. J. Alberola. Regulation of lymphocyte function by protein phosphorylation. *Annu. Rev. Immunol.* 11: 451–499, 1993.

303. Pfaffinger, P. J., J. M. Martin, D. D. Hunter, N. M. Nathanson, and B. Hille. GTP-binding proteins couple cardiac muscarinic receptors to a K channel. *Nature* 317: 536–538, 1985.

304. Pfister, C., N. Bennett, F. Bruckert, P. Catty, A. Clerc, F. Pages, and P. Deterre. Interactions of a G-protein with its effector—transducin and cGMP phosphodiesterase in retinal rods. *Cell. Signal.* 5(3): 235–251, 1993.

305. Pieroni, J. P., O. Jacobwitz, J. Chen, and R. Iyengar. Signal recognition and integration by G_s-stimulated adenylyl cyclases. *Curr. Opin. Neurobiol.* 3: 345–351, 1993.

306. Pitcher, J., M. J. Lohse, J. Codina, M. G. Caron, and R. J. Lefkowitz. Desensitization of the isolated β_2-adrenergic receptor by β-adrenergic receptor kinase, cAMP-dependent protein kinase, and protein kinase C occurs via distinct molecular mechanisms. *Biochemistry* 31(12): 3193–3197, 1992.

307. Pituello, F., V. Homburger, J.-P. Saint-Jeannet, Y. Audigier, J. Bockaert, and A.-M. Duprat. Expression of the guanine nucleotide-binding protein G_o correlates with the state of neural

308. Ponzetto, C., A. Bardelli, Z. Shen, F. Maina, P. dalla Zonca, S. Giordano, A. Graziani, G. Panayotou, and P. M. Comoglio. A multifunctional docking site mediates signaling and transformation by the hepatocyte growth factor/scatter factor receptor family. *Cell* 77: 261–271, 1994.

309. Premont, R. T., J. Chen, H. W. Ma, M. Ponnapalli, and R. Iyengar. Two members of a widely expressed subfamily of hormone-stimulated adenylyl cyclases. *Proc. Natl. Acad. Sci. U.S.A.* 89(20): 9809–9813, 1992.

310. Prigent, S. A., and N. R. Lemoine. The type I (EGFR-related) family of growth factor receptors and their ligand. *Prog. Growth Factor Res.* 4: 1–24, 1992.

311. Pronin, A. N., and N. Gautam. Interaction between G-protein β and γ subunit types is selective. *Proc. Natl. Acad. Sci. U.S.A.* 89: 6220–6224, 1992.

312. Puil, L., J. Liu, G. Gish, G. Mbamalu, D. Bowtell, P. G. Pelicci, R. Arlinghaus, and T. Pawson. Bcr-Abl oncoproteins bind directly to activators of the Ras signalling pathway. *EMBO J.* 13: 764–773, 1994.

313. Rabin, S. J., V. Cleghon, and D. R. Kaplan. SNT, a differentiation-specific target of neurotrophic factor-induced tyrosine kinase activity in neurons and PC12 cells. *Mol. Cell. Biol.* 13: 2203–2213, 1993.

314. Rarick, H. M., N. O. Artemyev, and H. E. Hamm, A site on rod G protein α subunit that mediates effector activation. *Science* 256: 1031–1033, 1992.

315. Ravichandran, K. S., and S. J. Burakoff. The adaptor protein Shc interacts with the interleukin-2 (IL-2) receptor upon IL-2 stimulation. *J. Biol. Chem.* 269: 1599–1602, 1994.

316. Raymond, J. R. Hereditary and acquired defects in signaling through the hormone-receptor-G protein complex. *Am. J. Phys.* 266 (*Renal Fluid Electrolyte Physiol.* 3): F163–F174, 1994.

317. Reiner, O., R. Carrozo, Y. Shen, M. Wehnert, F. Faustinella, W. B. Dobyns, C. T. Caskey, and D. H. Ledbetter. Isolation of a Miller-Dieker lissencephaly gene cointaining G protein β-subunit like repeats. *Nature* 364: 717–721, 1993.

318. Reuter, H. Properties of two inward membrane currents in the heart. *Annu. Rev. Physiol.* 41: 413–424, 1979.

319. Reuveny, E., P. A. Slesinger, J. Inglese, J. M. Morales, J. A. Iniguez-Lluhi, R. J. Lefkowitz, H. R. Bourne, Y. N. Jan, and L. Y. Jan. Activation of the cloned muscarinic potassium channel by G protein $\beta\gamma$ subunits. *Nature* 370: 143–146, 1994.

320. Rhee, S. G., and K. D. Choi. Regulation of Inositol phospholipid-specific phospholipase C isozymes. *J. Biol. Chem.* 267(18): 12393–12396, 1992.

321. Rhee, S. G., P.-G. Suh, S.-H. Ryu, and S. Y. Lee. Studies of inositol phospholipd-specific phospholipase C. *Science* 244: 546–550, 1989.

322. Rodrigues, G. A., and M. Park. Oncogenic activation of tyrosine kinases. *Curr. Opin. Genet. Dev.* 4: 15–24, 1994.

323. Rodriguez-Viciana, P., P. H. Warne, R. Dhand, B. Vanhaesebroeck, I. Gout, M. J. Fry, M. D. Waterfield, and J. Downward. Phosphatidylinositol-3-OH kinase as a direct target of Ras. *Nature* 370(6490): 527–532, 1994.

324. Ron, D., C. H. Chen, J. Caldwell, L. Jamieson, E. Orr, and D. Mochly-Rosen. Cloning of an intracellular receptor for protein kinase C: a homolog of the β subunit of G proteins. *Proc. Natl. Acad. Sci. U.S.A.* 91: 839–843, 1994.

325. Rossomando, A. J., P. Dent, T. W. Sturgill, and D. R. Marshak. Mitogen-activated protein kinase kinase 1 (MKK1) is negatively regulated by threonine phosphorylation. *Mol. Cell. Biol.* 14(3): 1594–1602, 1994.

326. Russell, M., S. Winitz, and G. L. Johnson. Acetylcholine muscarinic m1 receptor regulation of cyclic AMP synthesis controls growth factor stimulation of Raf activity. *Mol. Cell Biol.* 14: 2343–2351, 1994.

327. Sakurai, T., Y. Abe, Y. Kasuya, N. Takuwa, R. Shiba, T. Yamashita, T. Endo, and K. Goto. Activin A stimulates mitogenesis in Swiss 3T3 fibroblasts without activation of mitogen-activated protein kinase. *J. Biol. Chem.* 269(19): 14118–14122, 1994.

328. Sastry, S., and A. Horwitz. Integrin cytoplasmic domains: mediators of cytoskeletal linkages and extra- and intracellular initiated transmembrane signaling. *Curr. Opin. Cell Biol.* 5: 819–831, 1993.

329. Schertler, G.F.X., C. Villa, and R. Henderson. Projection structure of rhodopsin. *Nature* 362: 770–772, 1993.

330. Schleicher, A., H. Kuhn, and K. P. Hofmann. Kinetics, binding constant, and activation energy of the 48-kDa protein-rhodopsin complex by extra-metarhodopsin II. *Biochemistry* 28(4): 1770–1775, 1989.

331. Schleicher, S., I. Boekhoff, J. Arriza, R. J. Lefkowitz, and H. Breer. A β-adrenergic receptor kinase-like enzyme is involved in olfactory signal termination. *Proc. Natl. Acad. Sci. U.S.A.* 90: 1420–1424, 1993.

332. Schlessinger, J., and A. Ullrich. Growth factor signaling by receptor tyrosine kinases. *Neuron* 9: 383–391, 1992.

333. Schmidt, C. J., T. C. Thomas. M. A. Levine, and E. J. Neer. Specificity of G protein β and γ subunit interactions. *J. Biol. Chem.* 267: 13807–13810, 1992.

334. Schoenlein, R. W., L. A. Peteanu, R. A. Mathies, and C. V. Shank. The first step in vision: femtosecond isomerization of rhodopsin. *Science* 254(5030): 412–415, 1991.

335. Scholz, K. P., and R. J. Miller. GABA-B receptor-mediated inhibition of Ca^{2+} currents and synaptic transmission in cultured rat hippocampal neurons. *J. Physiol. (Lond.)* 444: 669–686, 1991.

336. Schultz, A. M., L. E. Henderson, and S. Oroszlan. Fatty acylation of proteins. *Annu. Rev. Cell Biol.* 4: 611–643, 1988.

337. Scott, D. L., S. P. White, Z. Otwinowski, W. Yuan, M. H. Gelb, and P. B. Sigler. Interfacial catalysis: the mechanism of phospholipase A_2. *Science* 250: 1541–1546, 1990.

338. Seeburg, P. H., W. W. Colby, D. J. Capon, D. V. Goeddel, and A. D. Levinson. Biological properties of human c-Ha-ras1 genes mutated at codon 12. *Nature* 312: 71–75, 1984.

339. Seuwen, K., C. Kahan, T. Hartmann, and J. Pouyssegur. Strong and persistent activation of inositol lipid breakdown induces early mitogenic events but not G_o to S phase progression in hamster fibroblasts. *J. Biol. Chem.* 265: 22292–22299, 1990.

340. Shapiro, M. S., and B. Hille. Substance P and somatostatin inhibit calcium channels in rat sympathetic neurons via different G protein pathways. *Neuron* 10(1): 11–20, 1993.

341. Shapiro, M. S., L. P. Wollmuth, and B. Hille. Angiotensin II inhibits calcium and M current channels in rat sympathetic neurons via G proteins. *Neuron* 12: 1319–1329, 1994.

342. Shenker, A., P. Goldsmith, C. G. Unson, and A. M. Spiegel. The G protein coupled to the thromboxane A_2 receptor in human platelets is a member of the novel G_q family. *J. Biol. Chem.* 266: 9309–9313, 1991.

343. Sieh, M., A. Batzer, J. Schlessinger, and A. Weiss. GRB2 and PLCγ_1 associate with a 36 to 38 kilodalton phosphotyrosine protein after T-cell receptor stimulation. *Mol. Cell Biol.* 14: 4435–4442, 1994.

344. Simon, M. A., D.D.L. Bowtell, G. S. Dodson, T. R. Laverty, and G. M. Rubin. Ras1 and a putative guanine nucleotide exchange factor perform crucial steps in signaling by the sev-

enless protein tyrosine kinase. *Cell* 67: 701–716, 1991.

345. Simon, M. A., R. W. Carthew, M. E. Fortini, U. Gaul, G. Mardon, and G. M. Rubin. Signal transduction pathway initiated by activation of the sevenless tyrosine kinase receptor. *Cold Spring Harb. Symp. Quant. Biol.* 57(375): 375–380, 1992.

346. Simon, M. I., M. P. Strathmann, and N. Gautam. Diversity of G proteins in signal transduction. *Science* 252: 802–808, 1991.

347. Smith, M. R., S. J. DeGudicibus, and D. W. Stacey. Requirement of c-*ras* proteins during viral oncogene transformation. *Nature* 320: 540–543, 1986.

348. Smrcka, A. V., J. R. Hepler, K. O. Brown, and P. C. Sternweis. Regulation of polyphosphoinositide-specific phospholipase C activity by purified G_q. *Science* 251: 804–807, 1991.

349. Sondek, J., D. G. Lambright, J. P. Noel, H. E. Hamm, and P. B. Sigler. GTPase mechanism of G proteins from the 1.7-Å crystal structure of transducin α-GDP-AlF$_4^-$. *Nature* 372: 276–279, 1994.

350. Songyang, Z., *et al.* SH2 domains recognize specific phosphopeptide sequences. *Cell* 72: 767–778, 1993.

351. Songyang, Z., S. E. Shoelson, J. McGlade, P. Olivier, T. Pawson, X. R. Bustelo, M. Barbacid, H. Sabe, H. Hanafusa, T. Yi, R. Ren, D. Baltimore, S. Ratnofsky, R. A. Feldman, and L. C. Cantley. Specific motifs recognized by the SH2 domains of Csk, 3BP2, fps/fes, GRB-2, HCP, SHC, Syk, and Vav. *Mol. Cell Biol.* 14: 2777–2785, 1994.

352. Sontag, E., S. Fedorov, C. Kamibayashi, D. Robbins, M. Cobb, and M. Mumby. The interaction of SV40 small tumor antigen with protein phosphatase 2A stimulates the Map kinase pathway and induces cell proliferation. *Cell* 75: 887–897, 1993.

353. Spiegel, A. M., P.S.J. Backlund, J. E. Butrynski, T.L.Z. Jones, and W. F. Simonds. The G protein connection: molecular basis of membrane association. *Trends Biochem. Sci.* 16: 338–341, 1991.

354. Spiegel, A. M., A. Shenker, and L. S. Weinstein. Receptor-effector coupling by G proteins: implications for normal and abnormal signal transduction. *Endocr. Rev.* 13: 536–565, 1992.

355. Spiegel, A. M., L. S. Weinstein, A. Shenker, S. Hermouet, and J. J. Merendino. G proteins: from basic to clinical studies. *Adv. Second Messenger Phosphoprotein Res.* 28(37): 37–46, 1993.

356. Spring, D. J., and E. J. Neer. A 14-amino acid region of the G protein γ subunit is sufficient to confer selectivity of γ binding to the β subunit. *J. Biol. Chem.* 269(36): 22882–22886, 1994.

357. Stein, D., J. Wu, S.A.W. Fuqua, C. Roonprapunt, V. Yajnik, P. D'Eustachio, J. J. Moskow, A. M. Buchberg, C. K. Osborne, and B. Margolis. The SH2 domain protein GRB-7 is co-amplified, overexpressed and in a tight complex with HER2 in breast cancer. *EMBO J.* 13(6): 1331–1340, 1994.

358. Stephens, L., A. Smrcka, F. T. Cooke, T. R. Jackson, P. C. Sternweis, and P. T. Hawkins. A novel phosphoinositide 3 kinase activity in myeloid-derived cells is activated by G protein βγ subunits. *Cell* 77: 83–93, 1994.

359. Stephens, R. M., D. M. Loeb, T. D. Copeland, T. Pawson, L. A. Greene, and D. R. Kaplan. Trk receptors use redundant signal transduction pathways involving SHC and PLC-γ_1 to mediate NGF responses. *Neuron* 12: 691–705, 1994.

360. Sternweis, P. C., J. K. Norhup, M. D. Smigel, and A. G. Gilman. The regulatory component of adenylate cyclase. *J. Biol. Chem.* 256: 11517–11526, 1981.

361. Sternweis, P. C., and J. Robishaw. Isolation of two proteins with high affinity for guanine nucleotides from membranes of bovine brain. *J. Biol. Chem.* 259: 13806–13813, 1984.

362. Stokoe, D., S. G. Macdonald, K. Cadwallader, M. Symons, and J. F. Hancock. Activation of Raf as a result of recruitment to the plasma membrane. *Science* 264: 1463–1467, 1994.

363. Strathmann, M., and M. I. Simon. G protein diversity: a distinct class of α subunits is present in vertebrates and invertebrates. *Proc. Natl. Acad. Sci. U.S.A.* 87: 9113–9117, 1990.

364. Strathmann, M. P., and M. I. Simon. $G\alpha_{12}$ and $G\alpha_{13}$ subunits define a fourth class of G protein α subunits. *Proc. Natl. Acad. Sci. U.S.A.* 88: 5582–5586, 1991.

365. Strathmann, M., T. M. Wilkie, and M. I. Simon. Diversity of the G-protein family: sequences from five additional α subunits in the mouse. *Proc. Natl. Acad. Sci. U.S.A.* 86: 7407–7409, 1989.

366. Stratton, K. R., P. F. Worley, R. L. Huganir, and J. M. Baraban. Muscarinic agonists and phorbol esters increase tyrosine phosphorylation of a 40-kilodalton protein in hippocampal slices. *Proc. Natl. Acad. Sci. U.S.A.* 86: 2498–2501, 1989.

367. Strittmatter, S. M., D. Valenzuela, T. E. Kennedy, E. J. Neer, and M. C. Fishman. G_o is a major growth cone protein subject to regulation by GAP-43. *Nature* 344: 836–841, 1990.

368. Stryer, L. Cyclic GMP cascade of vision. *Annu. Rev. Neurosci.* 9: 87–119, 1986.

369. Sudo, Y., D. Valenzuela, A. G. Beck-Sickinger, M. C. Fishman, and S. M. Strittmatter. Palmitoylation alters protein activity: blockade of G_o stimulation by GAP-43. *EMBO J.* 11(6): 2095–2102, 1992.

370. Sugimoto, K., T. Nukada, T. Tanabe, H. Takahashi, M. Noda, N. Minamino, K. Kangawa, H. Matsuo, T. Hirose, S. Inayama, and S. Numa. Primary structure of the β subunit of bovine transducin deduced from the cDNA sequence. *FEBS Lett.* 191(2): 235–240, 1985.

371. Sullivan, K., R. T. Miller, S. B. Masters, B. Beiderman, W. Heideman, and H. Bourne, Identification of receptor contact site involved in receptor-G protein coupling. *Nature* 330: 758–760, 1987.

372. Swartz, K. J., and B. P. Bean. Inhibition of calcium channels in rat CA3 pyramidal neurons by a metabotropic glutamate receptor. *J. Neurosci.* 12(11): 4358–4371, 1992.

373. Takeda, A., A. L. Maizel, K. Kitamura, T. Ohta, and S. Kimura. Molecular cloning of the CD45-associated 30-kDa protein. *J. Biol. Chem.* 269: 2357–2360, 1994.

374. Tang, W., and A. G. Gilman. Type-specific regulation of adenylyl cyclase by G protein βγ subunits. *Science* 254: 1500–1503, 1991.

375. Tang, W.-J., and A. G. Gilman. Adenylyl cyclases. *Cell* 70: 869–872, 1992.

377. Tang, W. J., J. A. Iñiguez-Lluhi, S. Mumby, and A. G. Gilman. Regulation of mammalian adenylyl cyclases by G-protein α and βγ subunits. *Cold Spring Harb. Symp. Quant. Biol.* 57: 135–144, 1992.

378. Tang, W. J., L. J. Iniguez, S. Mumby, and A. G. Gilman. Regulation of mammalian adenylyl cyclases by G-protein alpha and beta gamma subunits. [Review.] *Cold Spring Harb. Symp. Quant. Biol.* 57(135): 135–144, 1992.

379. Tang, W. J., J. Krupinski, and A. G. Gilman. Expression and characterization of calmodulin-activated (type I) adenylylcyclase. *J. Biol. Chem.* 266(13): 8595–8603, 1991.

380. Taussig, R., Iñiguez-Lluhi, J. A. and Gilman, A. G. Inhibition of adenylyl cyclase by $G\alpha_i$. *Science* 261: 218–221, 1993.

381. Taussig, R., Tang, W. J., Hepler, J. R. and Gilman, A. G. Distinct patterns of bidirectional regulation of mammalian adenylyl cyclases. *J. Biol. Chem.* 269(8): 6093–6100, 1994.

382. Taylor, S. J., H. Z. Chae, S. G. Rhee, and J. H. Exton. Activation of the β_1 isozyme of phospholipase C by α subunits

383. of the G_q class of G proteins. *Nature* 350: 516–518, 1991.

383. Touhara, K., J. Inglese, J. A. Pitcher, G. Shaw, and R. J. Lefkowitz. Binding of G protein βγ-subunits to pleckstrin homology domains. *J. Biol. Chem.* 269: 10217–10220, 1994.

384. Trowbridge, I. S. CD45: an emerging role as a protein tyrosine phosphatase required for lymphocyte activation and development. *Annu. Rev. Immunol.* 12: 85–116, 1994.

385. Tsien, R. W. Cyclic AMP and contractile activity in heart. *Adv. Cyclic Nucleotide Res.* 8: 363–420, 1977.

386. Tsuboi, S., H. Matsumoto, K. W. Jackson, K. Tsujimoto, T. Williams, and A. Yamazaki. Phosphorylation of an inhibitory subunit of cGMP phosphodiesterase in Rana catesbiana rod photoreceptors: characterization of the phosphorylation. *J. Biol. Chem.* 269(21): 15016–15023, 1994.

387. Tuzi, N. L., and W. J. Gullick. *eph*, the largest known family of putative growth factor receptors. *Br. J. Cancer* 69: 417–421, 1994.

388. Tyers, M., R. A. Rachubinski, M. I. Stewart, A. M. Varrichio, R.G.L. Shorr, R. J. Haslam, and C. B. Harley. Molecular cloning and expression of the major protein kinase C substrate of platelets. *Nature* 333: 470–473, 1988.

389. Ueno, H., J. A. Escobedo, and L. T. Williams. Dominant-negative mutations of platelet-derived growth factor (PDGF) receptors. *J. Biol. Chem.* 268(30): 22814–22819, 1993.

390. Uhl, R., and H. Desel. Optical probes of intradiskal processes in rod photoreceptors. II: light-scattering study of ATP-dependent light reactions. *J. Photochem. Photobiol. B* 3(4): 549–564, 1989.

391. Ullrich, A., and J. Schlessing. Signal transduction by receptors with tyrosine kinase activity. *Cell* 61: 203–212, 1990.

392. van Corven, E. J., P. L. Hordijk, R. H. Medema, J. L. Bos, and W. H. Moolenaar. Pertussin toxin-sensitive activation of p21ras by G protein-coupled receptor agonists in fibroblasts. *Proc. Natl. Acad. Sci. U.S.A.* 90: 1257–1261, 1993.

393. van der Voorn, L., and H. L. Ploegh. The WD-40 repeat. *FEBS Lett.* 307: 131–134, 1992.

394. Vogel, W., R. Lammers, J. Huang, and A. Ullrich. Activation of a phosphotyrosine phosphatase by tyrosine phosphorylation. *Science* 259: 1611–1614, 1993.

395. von Weizsacker, E., M. P. Strathmann, and M. I. Simon. Diversity among the beta subunits of heterotrimeric GTP-binding proteins: characterization of a novel beta subunit cDNA. *Biochem. Biophy. Res.* 183: 350–356, 1992.

396. Vu, T.-K., D. T. Hung, V. I. Wheaton, and S. R. Coughlin. Molecular cloning of a functional thrombin receptor reveals a novel proteolytic mechanism of receptor activation. *Cell* 64: 1057–1068, 1991.

397. Vuong, T. M., M. Chabre, and L. Stryer. Millisecond activation of transducin in the cyclic nucleotide cascade of vision. *Nature* 311(5987): 659–661, 1984.

398. Wakelam, M. J., T. R. Pettitt, P. Kaur, C. P. Briscoe, A. Stewart, A. Paul, A. Paterson, M. J. Cross, S. D. Gardner, S. Currie, E. E. MacNulty, R. Plevin, and S. J. Cook. *Phosphatidylcholine hydrolysis: a multiple messenger generating system.* In: *Advances in Second Messenger and Phosphoprotein Research,* Edited by B. L. Brown, and P. R. Dobson. New York: Raven Press, 1993, p. 73–80.

399. Wang, H., L. Lipfert, C. C. Malbon, and S. Bahouth. Site-directed anti-peptide antibodies define the topography of the β-adrenergic receptor. *J. Biol. Chem.* 264: 14424–14431, 1990.

400. Ward, Y., S. Gupta, P. Jensen, M. Wartmann, R. J. Davis, and K. Kelly. Control of MAP kinase activation by the mitogen-

induced threonine/tyrosine phosphatase PAC1. *Nature* 367: 651–654, 1994.

401. Watson, A. J., A. Katz, and M. I. Simon. A fifth member of the mammalian G-protein β-subunit family. *J. Biol. Chem.* 269: 22150–22156, 1994.

402. Wedegaertner, P. B., and H. R. Bourne. Activation and depalmitoylation of Gα_s. *Cell* 77: 1063–1070, 1994.

403. Wedegaertner, P. B., D. H. Chu, P. T. Wilson, M. J. Levis, and H. R. Bourne. Palmitoylation is required for signaling functions and membrane attachment of Gα_q and Gα_s. *J. Biol. Chem.* 268(33): 25001–25008, 1993.

404. Weiss, A., and D. R. Littman. Signal transduction by lymphocyte antigen receptors. *Cell* 76: 263–274, 1994.

405. Weiss, E. R., D. J. Kelleher, and G. J. Johnson. Mapping sites of interaction between rhodopsin and transducin using rhodopsin antipeptide antibodies. *J. Biol. Chem.* 263(13): 6150–6154, 1988.

406. Welham, M. J., V. Duronio, and J. W. Schrader. Interleukin-4-dependent proliferation dissociates p44^{erk-1}, p42^{erk-2}, and p21^{ras} activation from cell growth. *J. Biol. Chem.* 269(8): 5865–5873, 1994.

407. Weng, A., S. M. Thomas, R. J. Rickles, J. A. Taylor, A. W. Brauer, C. Seidel-Dugan, W. M. Michael, G. Dreyfuss, and J. Brugge. Identification of Src, Fyn, and Lyn SH3-binding proteins: implications for a function of SH3 domains. *Mol. Cell Biol.* 14: 4509–4521, 1994.

408. Wess, J., T. I. Bonner, F. Dorje, and M. R. Brann. Delineation of muscarinic receptor domains conferring selectivity of coupling to guanine nucleotide-binding proteins and second messengers. *Mol. Pharmacol.* 38: 517–523, 1990.

409. West, R., J. Moss, M. Vaughan, and T. Liu. Pertussis toxin catalyzed ADP ribosylation of transducin. Cysteine 347 is the ADP-ribose acceptor site. *J. Biol. Chem.* 260: 14428–14430, 1985.

410. White, M. F. The IRS-1 signaling system. *Curr. Opin. Genet. Dev.* 4: 47–54, 1994.

411. Wickman, K. D., and D. E. Clapham. Ion channel regulation by G proteins *Physiol. Rev.* 75: 865–85, 1995.

412. Wickman, K., J. Iñiguez-Lluhi, P. Davenport, R. A. Taussig, G. B. Krapivinsky, M. E. Linder, A. Gilman, and D. E. Clapham. Recombinant Gβγ Activates the muscarinic-gated atrial potassium channel I_{KACh}. *Nature* 368: 255–257, 1994.

413. Wilden, U., S. W. Hall, and H. Kuhn, Phosphodiesterase activation by photoexcited rhodopsin is quenched when rhodopsin is phosphorylated and binds the intrinsic 48-kDa protein of rod outer segments. *Proc. Natl. Acad. Sci. U.S.A.* 83(5): 1174–1178, 1986.

414. Wilkie, T. M., P. A. Scherle, M. P. Strahmann, V. Z. Slepak, and M. I. Simon. Characterization of G-protein α subunits in the G_q class: expression in murine tissues and in stromal and hematopoietic cell lines. *Proc. Natl. Acad. Sci. U.S.A.* 88(22): 10049–10053, 1991.

415. Winitz, S., M. Russell, N.-X. Qian, A. Gardner, L. Dwyer, and G. L. Johnson. Involvement of Ras and Raf in the G_i-coupled acetylcholine muscarinic m2 receptor activation of mitogen-activated protein (MAP) kinase kinase and MAP kinase. *J. Biol. Chem.* 268: 19196–19199, 1993.

416. Witthuhn, B. A., O. Silvennoinen, O. Miura, K. S. Lai, C. Cwik, E. T. Liu, and J. N. Ihle. Involvement of the Jak-3 janus kinase in signaling by interleukins 2 and 4 in lymphoid and myeloid cells. *Nature* 370: 153–157, 1994.

417. Wittinghofer, A. The structure of transducing Gα_t: more to view than just Ras. *Cell* 76: 201–204, 1994.

418. Wood, K. W., C. Sarnecki, T. M. Roberts, and J. Blenis. Ras mediates nerve growth factor receptor modulation of three signaling-transducting protein kinases: MAP kinase, Raf-1, and RSK. *Cell* 68: 1041–1050, 1992.

419. Wu, D., H. Jiang, A. Katz, and M. I. Simon. Identification of critical regions on phospholipase C-β_1 required for activation by G-proteins. *J. Biol. Chem.* 268(5): 3704–3709, 1993.

420. Wu, D., A. Katz, and M. I. Simon. Activation of phospholipase Cβ_2 by the α and βγ subunits of trimeric GTP-binding protein. *Proc. Natl. Acad. Sci. U.S.A.* 90: 5297–5301, 1993.

421. Wu, D., C. H. Lee, S. G. Rhee, and M. I. Simon. Activation of phospholipase C by the alpha subunits of the Gq and G11 proteins in transfected Cos-7 cells. *J. Biol. Chem.* 267: 1811–1817, 1992.

422. Xie, Y.-B., H. Wang, and D. L. Segaloff. Extracellular domain of lutropin/choriogonadotropin receptor expressed in transfected cells binds choriogonadotropin with high affinity. *J. Biol. Chem.* 265: 21411–21414, 1990.

423. Yamada, M., A. Jahangir, Y. Hosoya, A. Inanobe, T. Katada, and Y. Kurachi. G_K* and brain G_{βγ} activate muscarinic K+ channel through the same mechanism. *J. Biol. Chem.* 268(33): 24551–24554, 1993.

424. Yamazaki, A., F. Hayashi, M. Tatsumi, M. W. Bitensky, and J. S. George. Interactions between the subunits of transducin and cyclic GMP phosphodiesterase in Rana catesbiana rod photoreceptors. *J. Biol. Chem.* 265(20): 11539–11548, 1990.

425. Yang, L. J., S. G. Rhee, and J. R. Williamson. Epidermal growth factor-induced activation and translocation of phospholipase C-γ_1 to the cytoskeleton in rat hepatocytes. *J. Biol. Chem.* 269(10): 7156–7162, 1994.

426. Yarfitz, S., and J. B. Hurley. Transduction mechanisms of vertebrate and invertebrate photoreceptors. *J. Biol. Chem.* 269(20): 14239–14332, 1994.

427. Yatani, A., K. Okabe, P. Polakis, R. Halenbeck, F. McCormick, and A. M. Brown, ras p21 and GAP inhibit coupling of muscarinic receptors to atrial K+ channels. *Cell* 61: 769–776, 1990.

428. Yau, K. W., and K. Nakatani. Light-suppressible, cyclic GMP-sensitive conductance in the plasma membrane of a truncated rod outer segment. *Nature* 317(6034): 252–255, 1985.

429. Yoshimura, M., and D. M. Cooper. Cloning and expression of a Ca^{2+}-inhibitable adenylyl cyclase from NCB-20 cells. *Proc. Natl. Acad. Sci. U.S.A.* 89(15): 6716–6720, 1992.

430. Yoshimura, M., and D. M. Cooper. Type-specific stimulation of adenylylcyclase by protein kinase C. *J. Biol. Chem.* 268(7): 4604–4607, 1993.

431. Zachary, I., J. Sinnett-Smith, and E. Rozengurt. Bombesin, vasopressin, and endothelin stimulation of tyrosine phosphorylation in Swiss 3T3 cells. *J. Biol. Chem.* 267: 19031–19034, 1992.

432. Zhao, Y., H. Uyttendaele, J. G. Krueger, M. Sudol, and H. Hanafusa. Inactivation of c-Yes tyrosine kinase by elevation of intracellular calcium levels. *Mol. Cell. Biol.* 13: 7507–7514, 1993.

433. Zheng, C.-F., and K.-L. Guan. Activation of MEK family kinases requires phosphorylation of two conserved Ser/Thr residues. *EMBO J.* 13: 1123–1131, 1994.

19. Cellular immunology

EPHRAIM FUCHS | *Johns Hopkins Oncology Center, Baltimore, Maryland*

CHAPTER CONTENTS

TO SURVIVE AND REPRODUCE, animals must be able to protect themselves from the possibility of damage inflicted by physical agents or by other living things. For instance, animals must protect themselves from attack by other animals as well as by pathogenic microbes: bacteria, fungi, viruses, and other parasites. Defense against the potentially lethal effects of microbial infection is the responsibility of the innate and adaptive immune systems. The innate immune system comprises physical barriers, such as the skin and mucosa, serum-derived proteins such as complement, and blood-derived, mobile phagocytic cells (Greek *phagein*, to eat + *kytos*, cell), which include polymorphonuclear leukocytes (neutrophils) and macrophages. The adaptive immune system consists of T and B lymphocytes, lymphokines and antibodies, and antigen-presenting cells (APCs). Numerous animal phyla have innate defense mechanisms consisting of phagocytic cells and secreted molecules, but the adaptive immune system is limited to vertebrates and almost certainly evolved later. The distinguishing features of the adaptive immune system are that it responds specifically to the infecting agent, remembers past infections, and responds more quickly and strongly upon reinfection. Recovery from infection leads to immunity from reinfection (Latin *immunitas*, freedom from burden), a fact which has been known since ancient times. Cellular immunology is the study of how lymphocytes mediate their defensive functions, and is the subject of this chapter.

OVERVIEW

The Adaptive Immune System Combats Intracellular and Extracellular Pathogens

Lymphocytes detect infection through cell surface receptors that bind microbial constituents; ligation of these receptors triggers an activity aimed at ridding the body of the microbe. Yet underlying this apparently simple process is an extraordinarily complex network of interacting cells and molecules (Fig. 19.1). This complexity reflects the necessity of dealing with bacteria, fungi, and viruses, any of which can cause tissue damage and avoid host defenses in multifarious ways. For instance, most bacteria live outside host cells and cause disease by secreting toxins or by stealing nutrients present in the extracellular milieu. Although many strains are easily ingested and killed inside macrophages and neutrophils, some have evolved mechanisms to resist phagocytosis or destruction inside phagosomes. Therefore, the immune system must either kill the bacteria in the extracellular fluid or somehow get them into phagocytes where they can be killed inside. Furthermore, toxins must be prevented from binding to and acting on cells. B lymphocytes (named for their origin in the *Bursa of Fabricius* of chickens) participate in these processes by secreting antibodies (or immunoglobulins) that specifically bind microbial constituents. The molecules (proteins, polysaccharides, nucleic acids) that are capable of generating an antibody response are called *antigens.* Certain antibodies serve as the locus of action for the proteins of the complement system, which drill holes in bacterial cell membranes and kill by osmotic rupture of the cell. Other classes of antibodies act as opsonins; that is, they facilitate the ingestion of microbes by phagocytic cells. Finally, antibodies neutralize toxins by binding them and preventing their action on cell surface receptors.

Microbes that live inside host cells pose a more difficult problem for the host. Certain bacteria, such as *Listeria* and *Mycobacterium,* have evolved mechanisms to resist killing once inside macrophages. And viruses actually depend on getting inside host cells, where they subvert the metabolic machinery of the cell for their own purposes. Effective elimination of these microbes largely depends on the action of T lymphocytes (named because they develop in the thymus gland, an organ lying above the heart). T cells bear on their surface heterodimeric receptors for antigen consisting of either paired α and β or paired γ and δ chains. $\alpha\beta^+$ T cells bearing the surface CD4 molecule (CD4$^+$ T cells) are the critical regulatory cells of the immune system, a fact underscored by the consequences of their selective depletion in the acquired immune deficiency syndrome (AIDS), a disease marked by recurrent opportunistic infections by viruses and fungi. CD4$^+$ T lymphocytes are called "helper" T cells because of their role in the induction of antibody synthesis by B cells and in the generation of intracellular killing by phagocytic cells. Their help is mediated by providing activation ligands for B cell surface molecules, or through the secretion of lymphokines, or both. Recently, two

distinct subtypes of CD4$^+$ T cells have been discovered: T$_h$1 cells secrete interleukin 2 (IL-2), interferon gamma (IFN-γ), and lymphotoxin and are responsible for the induction of macrophage microbicidal activity. T$_h$2 cells secrete the lymphokines interleukin-4 (IL-4), IL-5, IL-6, and IL-10 and are important in the induction of synthesis of certain types of antibody. $\alpha\beta$ T cells bearing the CD8 surface molecule (CD8$^+$ T cells) are endowed with cytolytic capabilities; that is, the capacity to recognize and destroy infected cells. This destruction entails the scission of all the nucleic acids within the cell; thus any virus lurking inside is destroyed as well. The function of T cells bearing the $\gamma\delta$ surface receptor is less clear, but these cells are known to seed portals of infection (skin, gastrointestinal tract, female reproductive tract) and may serve as a first line of defense. T cells bearing the $\gamma\delta$ receptor do not express either the CD4 or CD8 molecules on their surface. Finally, antibodies secreted by B cells can limit the spread of viruses by binding those molecules of the virus which are necessary for penetration into cells.

Undoubtedly, the greatest challenge that microbes present to vertebrate defense mechanisms is their ability to mutate rapidly. This is not a problem for the defense against bacteria whose pathogenicity depends upon the presence of certain highly conserved molecules for which cells of the innate immune system have evolved receptors. For instance, gram negative aerobic bacteria have lipopolysaccharides (LPS) in their membranes, and neutrophils have LPS receptors which are used to capture and internalize the bacteria. Once inside the neutrophil, the bacteria are easily destroyed by oxidation. Thus, neutrophils are entirely sufficient for the elimination of these organisms. Viruses are an entirely different story. While certain molecules such as lipopolysaccharides or formylated proteins reliably identify bacterial cells, distinctive surface molecules which can be used as flags for large classes of viruses are absent. This fact, coupled with the ability of viruses to mutate rapidly, means that evolutionarily conserved receptors are useless against viruses. This problem may have been the selective force behind the evolution of the immense diversity of antigen-specific receptors expressed by the pool of B and T lymphocytes. Each developing lymphocyte generates a unique receptor for antigen by the combinatorial rearrangement of diverse minigenes encoding distinct domains of a full receptor (Fig. 19.2; also see section on Organization and Rearrangement of Immunoglobulin Genes). This somatic recombination of minigenes endows humans with the ability to form about 10^9 distinct antibody receptors, each with distinct antigen-binding capacity, and with each individual lymphocyte placing on its surface multiple copies of receptors with identical antigen-binding

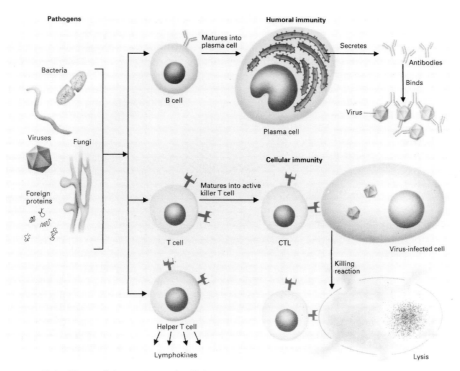

FIG. 19.1. Humoral immunity and cellular immunity. When pathogen invades the body, immune system responds with three types of reaction. Cells of humoral system (B cells) secrete antibodies that can bind to pathogen. Cells of cellular system (T cells) carry out two major types of functions. CD8$^+$ T cells develop ability to kill pathogen-infected cells. CD4$^+$ helper T cells, responding to pathogen, secrete protein factors (lymphokines) that stimulate body's overall response. [Figure and legend adapted from Darnell et al. (28) with permission from the publisher.]

sites. This diverse pool of mature lymphocytes is generated during fetal development, and so at birth there are several lymphocytes capable of binding any microbe. Introduction of antigen stimulates the division of those lymphocytes with receptors capable of binding the antigen, providing an expanded pool of specific lymphocytes to respond more quickly and vigorously upon reinfection. Additionally, appropriately stimulated B lymphocytes mutate their antibody genes, so some daughter cells have receptors with increased affinity for the stimulating antigen. The processes of somatic recombination (in developing B and T lymphocytes) and mutation (in mature B lymphocytes only) endow the immune system with the diversity and plasticity to respond to a rapidly changing universe of microbes.

The Antigen-Specific Receptors on B and T Lymphocytes Bind Distinct Ligands

As was noted above, vertebrates respond to pathogens that are outside or inside host cells through the actions of B or T lymphocytes, respectively. This division of labor occurs because the antigen-specific receptors of

these two cell types recognize distinct classes of ligands. Antibodies and their corresponding receptors on the surface of B lymphocytes bind portions of whole molecules in their native or denatured conformation. The portion of the molecule that is bound by antibody is called the *determinant* or *epitope* (Fig. 19.3). Globular proteins, which contain multiple unique determinants, can be bound by distinct antibodies, whereas polysaccharides containing repeating motifs can be bound by multiple copies of the same antibody. Antibodies bind these molecules whether they are present in extracellular fluid or on the surfaces of microbes or host cells. By contrast, T cells recognize antigens only on the surface of other host cells called *antigen-presenting cells (APCs)*. The ligand for the T cell receptor is generated by the APC through processing and presentation of the antigen (Fig. 19.4): Proteins in intracellular compartments are degraded into peptide fragments, and these peptides then associate with specialized antigen-presenting molecules encoded by the highly polymorphic genes of the major histocompatibility complex (MHC). This MHC–peptide complex is exported to the cell surface, and the T cell receptor is

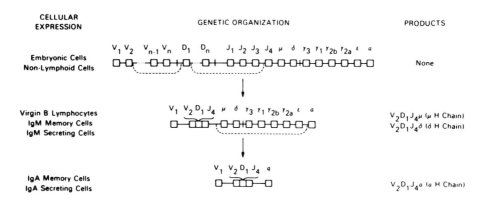

FIG. 19.2. Organization and translocation of IgH genes. Immunoglobulin H chains are encoded by four distinct genetic elements—Igh-V *(V)*, Igh-D *(D)*, Igh-J *(J)*, and Igh-C *(C)* genes. V, D, and J genes together specify variable region of H chain. IgH-C gene specifies C region. Same V region can be found in association with each of C regions (for example, μ, δ, γ_3, γ_1, γ_{2b}, γ_{2a}, ϵ, and α). In germline genome, V, D, and J genes are far apart and there are multiple forms of each of these genes. In course of lymphocyte development, VDJ gene complex is formed by translocation of individual V and D genes so that they lie next to one of the J genes, with excision of intervening genes. This VDJ complex is initially expressed with m and d C genes but may subsequently be translocated so that it lies near one of other C genes (for example, α) and in that case leads to expression of VDJ_α chain. [Figure and legend from Paul (104) with permission from the publisher.]

then ligated by a complex determinant whose configuration is contributed to by both the peptide and portions of the MHC molecule (Fig. 19.5). This "MHC-restricted" T cell recognition of peptide antigen ensures that T lymphocytes only respond to antigens on host cell surfaces.

MHC Class I and II Molecules Present Antigens to CD8[+] and CD4[+] T Cells, Respectively

Two broad classes of MHC molecules participate in antigen presentation, and they are distinguished by their cellular distribution, the origin of bound peptides, and the T cells that recognize these MHC–peptide complexes. MHC class I molecules, which are expressed on the surface of nearly all nucleated cells, mainly present peptides, eight to 11 amino acids in length, that have been generated by the cleavage of proteins synthesized in the cell cytoplasm. The MHC–peptide complex is formed in the endoplasmic reticulum (where peptides derived from cytoplasmic proteins are actively transported), and from there the complex is brought to the cell surface for recognition by CD8[+] T cells. In contrast, class II molecules present primarily 12–24-amino-acid-length fragments of proteins that have been taken in from outside the cell. MHC class II proteins are constitutively expressed on the surface of a limited number of cell types, including thymic epithelium, B lymphocytes, activated macrophages, and dedicated APCs such as dendritic cells of spleen and lymph nodes. The peptides are generated in and associate with class II molecules in endocytic vesicles, and the complexes are then brought to the surface for recognition by CD4[+] T cells.

With greater than 50 alleles at each of the class I and II loci, the MHC is the most highly polymorphic genetic locus in both human and mouse; this polymorphism leads to large amino acid sequence variability in

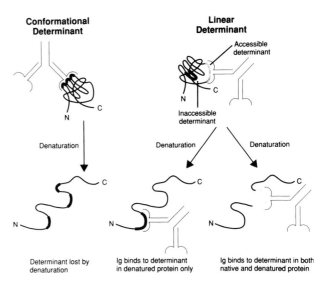

FIG. 19.3. Antigenic determinants (epitopes). Antigenic determinants in proteins (shown as *thick lines*) may depend on protein conformation as well as upon covalent structure. Some linear determinants are accessible in native proteins, whereas others are exposed only upon denaturation. [Figure and text adapted from Abbas et al. (1) with permission from the publisher.]

regions of the molecule bordering a peptide binding groove (see Fig. 19.10). To deposit stably within the groove, the peptide must form hydrogen bonds with specific residues of the MHC molecule; therefore, allelic differences at the MHC have profound effects on the repertoire of peptides that can be bound. A T cell response to a peptide is not possible if this peptide does not associate with an MHC molecule, and for this reason the MHC proteins exert a strong influence on immune responses. T cells are also responsible for rejection of foreign tissue grafts, and a perhaps unintended consequence of the focus of T cell recognition on MHC molecules is that T cells rapidly reject tissue grafts from individuals differing from the recipient in MHC alleles.

The Clonal Selection Theory of Acquired Immunity Provides a Mechanism for Immunologic Specificity and Memory

Although the preceding sections have emphasized the role of the adaptive immune system in the defense

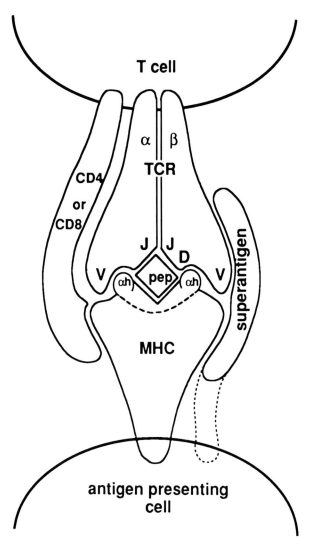

FIG. 19.5. Receptor–ligand interactions involved in T cell recognition of conventional antigens and superantigens. Peptide fragments *(pep)* of protein antigens are held in a groove on membrane-distal aspect of MHC molecules on APCs; peptide-binding groove and adjacent α-helices (αh) form polymorphic region of MHC molecules. During T-APC interaction, MHC-bound peptides and flanking MHC α-helices come into contact with hypervariable region of T-cell receptor (TCR); this region is product of variable (V), diversity (D), and joining (J) gene segments which rearrange during T cell ontogeny. Superantigens, which can be either soluble or cell bound, cause T cells and APCs to stick together by adhering to sides of MHC and TCR molecules; these regions of MHC and TCR molecules are relatively nonpolymorphic, which means superantigens are immunogenic for very high proportion of T cells. For recognition of both conventional antigens and superantigens, T-APC interaction is strengthened by binding of CD4 or CD8 molecules to nonpolymorphic sites on MHC molecules. These "co-receptor" molecules also play a role in T-cell triggering. [Figure and legend reproduced from Sprent (125) with permission.]

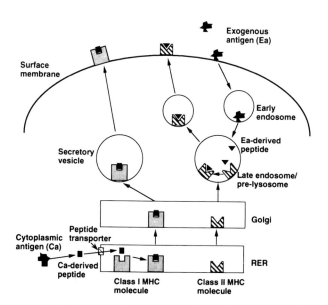

FIG. 19.4. Pathways of antigen processing and presentation. Cytoplasmic antigens *(Ca)* are degraded in cytoplasm and then enter rough endoplasmic reticulum *(RER)* through peptide transporter. In RER, Ca-derived peptides are loaded into class I MHC molecules that move through Golgi apparatus into secretory vesicles and are then expressed on cell surface, where they may be recognized by CD8 + T cells. Exogenous antigen *(Ea)* enters cell through endocytosis and is transported from early endosomes into late endosomes or pre-lysosomes, where it is fragmented and where peptide resulting from it (Ea-derived peptide) is loaded into class II MHC molecules. Latter have been transported from RER through Golgi apparatus to peptide-containing vesicles. Class II MHC molecule–Ea-derived peptide complexes are then transported to cell surface where they may be recognized by CD4 + T cells. [Figure and legend from Paul (103) with permission.]

against complex, mutating pathogens, immunologists have often preferred to study the response to simpler, static antigens. This is often accomplished by injecting animals with a highly purified protein of known sequence and examining the subsequent immune response. This process of immunization results in a state of *active immunity,* reflecting the active participation of the immune system of the recipient. The cells or antibodies that are responsible for this state of protection are carried within the bloodstream, and so an unimmunized individual may be protected by the transfusion ("adoptive transfer") of cells or serum from an immune individual. This state of protection is called *passive immunity,* as the recipient of the transfusion played no role in its development. The adoptive transfer of immunity has allowed experimenters to distinguish protection mediated by serum components, called *humoral immunity,* from that which is mediated by blood cells, called *cellular immunity.*

A simple experiment, as shown in Fig. 19.6, illustrates three of the cardinal features of the adaptive immune system. In this experiment a mouse is injected with albumins of different species, and the amount of specific antibody formed in response is measured at various times after immunization.

1. *Specificity.* The response to the first injection of bovine albumin ("primary response") is marked by the formation of antibodies which bind cow but not horse or mouse albumin.

2. *Specific Memory.* Repeat injection of bovine albumin leads to a secondary response that is both faster and quantitatively larger than that obtained by the first injection of albumin. This heightened response is

specific to bovine albumin as simultaneous injection of horse albumin leads to the generation of specific antibody characteristic of a primary response.

3. *Self-Tolerance.* Injection of a mouse with mouse albumin elicits no antibody response at all. Thus an animal is said to be immunologically tolerant of its own proteins.

A fourth cardinal feature of the adaptive immune system is the enormous *diversity* of the antibody repertoire. Early experiments in the field showed that injection of almost any substance, including proteins, sugars, nucleic acids, even synthetic molecules, induces the formation of specific antibodies. And while some small molecules such as dinitrophenol (DNP) are nonimmunogenic, specific anti-DNP antibodies are formed if DNP is coupled to an immunogenic carrier, such as bovine serum albumin. (A molecule which elicits antibodies only when coupled to an immunogenic carrier is called a *hapten.*) Furthermore, the immune response to DNP is quite specific, as anti-DNP antibodies do not bind trinitrophenol (TNP).

How could the immune system produce uniquely specific antibodies to almost anything that is injected? This question of antibody formation dominated immunological debate for the first half of the 20th century. *Instructive* theories of antibody formation proposed that antigen instructed the shape of the antibody by serving as a template around which the antibody wrapped. However, instructive theories fell out of favor when it was found that antibodies could be denatured and renatured in the absence of antigen while retaining their specificity. Instructive models were eventually superseded by *selective* theories of antibody formation, such as the one initially proposed by Lewis Thomas, modified by David Talmadge and Niels Jerne, and championed, in the final form as the *clonal selection theory of acquired immunity,* by MacFarlane Burnet [Fig. 19.7; (17)]. There are three essential features of the clonal selection theory: *(1)* Prior to the introduction of antigen, a diverse repertoire of antibodies exists as the cell surface receptors for antigen on B lymphocytes; *(2)* the antibody molecules expressed on the surface of an individual B cell are identical, and so each B cell is specific for one antigen only; and *(3)* introduction of antigen *selects* antigen-specific B lymphocytes by binding their cell surface antibody, causing them to proliferate and secrete specific antibody molecules. Each feature of the model has been confirmed, and the clonal selection theory remains the governing paradigm of immunological thinking. Furthermore, the theory provides mechanisms for the cardinal features of the immune system outlined above:

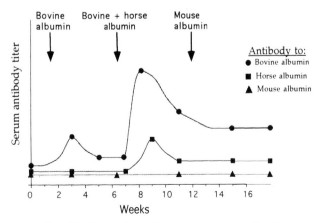

FIG. 19.6. Simple experiment demonstrates three cardinal features of adaptive immune system: specificity, memory, and self-tolerance. See text for details.

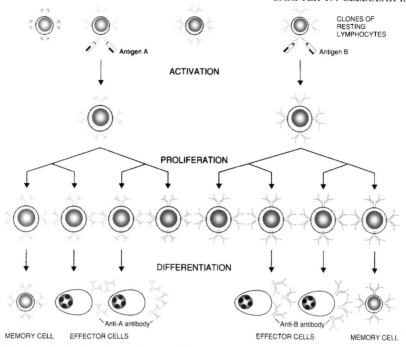

FIG. 19.7. Clonal selection theory of acquired immunity. Each antigen (A or B) selects preexisting clone of specific lymphocytes and stimulates its proliferation and differentiation. Diagram shows only B lymphocytes giving rise to antibody-secreting cells and memory cells, but same principle applies to T lymphocytes as well. [Figure and text adapted from Abbas et al. (1) with permission.]

1. *Specificity.* Only those clones of B lymphocytes bearing antibody receptors specific for the immunogen are triggered to produce antibody.

2. *Specific Memory.* An animal responds more rapidly and vigorously upon reintroduction of a particular antigen (boosting), because the primary response entails the proliferation of antigen-specific B cells. The expanded number of specific clones provides for the more rapid, vigorous response when the animal is boosted with the same antigen.

3. *Self-Tolerance.* Clonal selection theory predicts that self-tolerance is generated by the physical or functional elimination of those clones of lymphocytes bearing receptors specific for self-antigens.

Although the clonal selection theory is quite remarkable in its scope, there are two fundamental questions in immunology that it does not address:

1. Diversity of the antibody repertoire: How is such a diverse repertoire generated from a limited number of antibody genes?

2. Self/nonself discrimination: How does the immune system know which are the antigens that should generate, upon receptor ligation, clonal inactivation or elimination (as for self antigens) vs. those that should cause lymphocyte activation and effector function?

The answer to the question on the origin of antibody diversity will be saved until after a full discussion of the structure of the antibody molecule. The issue of self/nonself discrimination will be briefly touched upon now and at the end of this chapter.

Lymphocytes Require Two Signals for Activation to Effector Function

The enormous array of antibody specificities existing prior to and independent of the introduction of antigen is a powerful tool that the vertebrate organism can use to defend itself against the microbial world. In fact, this feature may be essential for dealing with pathogens capable of rapid mutation. In this case, the immune system is prepared with an antibody capable of binding the mutant molecule, and B cells producing this antibody are triggered to proliferate. Yet, as has just been mentioned, the random generation of antibody specificities prior to the introduction of antigen leads inevitably to the existence of self-reactive B cells that must be eliminated in order to avoid autoimmune reactions. Yet how can antigen receptor ligation lead to the elimination of self-reactive B cells and activation of B cells specific for foreign antigens?

This question, the issue of self/nonself discrimination, has not yet been answered satisfactorily. However, the theories that have been put forth to resolve this issue have shed much light on the normal functioning of the immune system and have been valuable guides to experimentation in immunology. For this reason, two models, the "temporal" model of Joshua Lederberg (71) and the "two-signal model" of Peter Bretscher and Mel Cohn (16), will be discussed in some detail.

Any attempt to provide a mechanism of self/nonself discrimination must begin with a definition of what is "self" vs. "nonself" and then to go on and explore the differences between the two that the immune system might exploit. For Lederberg, the principal difference between self and nonself antigens is that self antigens are present during fetal development, whereas the organism is first exposed to nonself antigens only after birth. This difference led Burnet to predict in 1949 (prior to his clonal selection hypothesis) that any foreign antigen introduced into an organism in its embryonic development would subsequently be treated as self and tolerated (19). This prediction received experimental confirmation in 1953 when Rupert Billingham, Leslie Brent, and Peter Medawar demonstrated that specific tolerance to genetically dissimilar (allogeneic) skin grafts could be obtained by injecting embryonic or neonatal mice with spleen cells from the same strain of the skin graft donor; that is, mice injected at or before birth with allogeneic spleen cells failed to reject skin grafts from mice genetically identical to the spleen cell donor (13). Following the clonal selection theory, the challenge was to convert these observations on whole organisms into models involving lymphocytes. Lederberg proposed that lymphocytes undergo a developmental switch: During embryonic development, antigen receptor ligation leads only to lymphocyte inactivation, whereas in a mature lymphocyte, antigen receptor ligation leads only to activation. As we shall see, there is some truth to this model as applied to T lymphocytes. When developing in the thymus, a thymocyte (the precursor of the mature T lymphocyte) undergoes inactivation upon encountering antigen presented by certain APCs. Yet the exact same encounter for a mature T lymphocyte leads to its activation (84).

There are some obvious flaws to the Lederberg model of self/nonself discrimination. The first is a geographic consideration: How can B lymphocytes, which develop in the bone marrow (the "central" or "primary" organ of B lymphocyte development), encounter self antigens present only outside of the marrow cavity? This problem led some to extend the "inactivation only" state of B cells to include a period during which the cells take a grand tour of peripheral tissues. The second flaw is one that Lederberg could not have been aware

of at the time he formulated his theory. In the 1960s immunologists discovered the process of affinity maturation of an antibody response (37) due to somatic mutation of immunoglobulin genes; Namely, *following* stimulation of a B cell by foreign antigen, the cell's antibody genes mutate at increased frequency (approximately 10^{-3}/base pair per division) and so produce new substrates for clonal selection. Mutant B cells that no longer bound the antigen would fail to be restimulated, whereas other mutant cells whose antibody receptors bound antigen more tightly than its ancestors would be selectively stimulated. The problem envisioned by Bretscher and Cohn was that a mature B cell specific for foreign antigen could mutate following antigen stimulation to a cell whose antibody receptor binds a self antigen. Thus, a situation in which an autoreactive B cell in the inducible-only state would be created. This would lead to a situation of uncontrolled autoimmunity.

To overcome such a problem, Bretscher and Cohn formulated a two-signal model of lymphocyte activation (16). This model contains the following three assumptions: *(1)* Lymphocytes require two signals for activation to effector function; *(2)* antigen receptor ligation (signal 1) in the absence of a second signal results in paralysis of the antigen-specific cell; and *(3)* signal 2 is delivered by another cell specific for a separate portion (determinant) of the antigen. According to this model, a self/nonself discrimination is then achieved because autoreactive cells capable of delivering signal 2 are absent. There is now considerable evidence (to be discussed below) to support the notion that lymphocytes require two signals for activation; for B cells and cytotoxic T cells, signal 2 is delivered by helper T cells, and for helper T cells, signal 2 is delivered by the antigen presenting cell.

Summary

Before proceeding on to a more detailed description of the immune system, the preceding sections are summarized.

1. The immune system evolved to defend vertebrates from infection and disease caused by microorganisms.

2. The immune system of vertebrates comprises the innate and adaptive immune systems. The adaptive immune system is composed of antigen-presenting cells and lymphocytes, which have the unique properties of clonally distributed, antigen-specific receptors, the capacity for memory of past infections, and self-tolerance.

3. B lymphocytes secrete antibodies, which bind whole proteins. T lymphocyte receptors recognize frag-

ments of proteins noncovalently bound to MHC molecules. This MHC-restricted recognition of antigen by T lymphocytes ensures that T cell reactivity is limited to cell surfaces.

4. The specificity of the immune response is achieved by antigen binding and selecting preexisting clones of antigen-specific lymphocytes, thereby stimulating their division. The expanded pool of lymphocytes specific for the antigen provides an explanation for the heightened response to reimmunization.

5. Ligation of the antigen-specific receptor alone is an insufficient signal to activate a lymphocyte and instead results in physical or functional elimination. Full activation to effector function requires a second signal, which for B cells and cytotoxic T cells is delivered by helper T cells.

MEMBERS OF THE IMMUNOGLOBULIN SUPERFAMILY

Immunoglobulins

Antibody Structure. As was noted above, antibodies are the mediators of humoral immunity; that is, they are the molecules in the serum of animals that can mediate specific resistance to infection. Therefore, early studies of antibodies were confined to studying their activity in whole serum (serology). However, a more detailed analysis of their function required their separation from other proteins not involved in immunity. When serum proteins are subjected to an electric field (electropheresis), antibody molecules collectively migrate as a broad band called the *gamma region*. Thus, antibodies are also called *gamma globulins* or *immunoglobulins*. In 1958 Rodney Porter found that treatment of pooled rabbit gamma globulin with the proteolytic enzyme papain generated two fragments, F_{ab} (antigen-binding fraction) and F_c (crystallizable fraction) (105). This led to the concept of functional domains of antibody molecules, but because of the heterogeneity of antibodies in serum, detailed sequence analysis was not yet possible. However, in 1955 Henry Kunkel made the observation that electrophoresis of the serum of patients with multiple myeloma (a bone marrow cancer) revealed a protein spike in the gamma region. These patients suffer from a monoclonal proliferation of plasma cells, descendents of B cells that secrete large amounts of the identical antibody molecule in the serum. The availability of purified myeloma proteins allowed the complete sequence analysis of antibody, as reported by Edelman et al., as well as the first pictures of the crystal structure of the F_{ab} (35).

A typical monomeric immunoglobulin molecule is depicted in Figure 19.8A. The complete structure is

composed of two identical light chains (about 24 kDa) and two identical heavy chains (55 or 70 kDa). Each light chain associates with a single heavy chain, and the heavy chains bind each other. This leads to the characteristic "Y" shape of antibody, with the antigen-binding domain composed of the N-termini of a single heavy–light-chain dimer. A single antibody monomer thus contains two antigen-binding sites; that is—it is divalent.

Closer analysis of antibody structure reveals that both heavy and light chains are organized into separate domains of approximately 110 amino acids, each containing an internal disulfide bond. Each domain as-

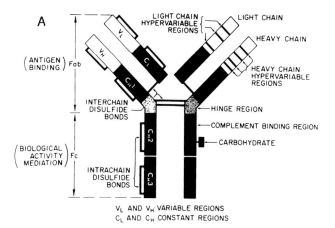

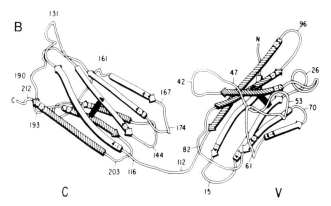

FIG. 19.8. *A.* Schematic representation of IgG molecule, with presence of variable and constant region domains indicated as *light and dark bars*, respectively. Note presence of hypervariable regions within V regions of both heavy and light chains, and division of the Ig molecule into antigen-binding and biologic effector regions. [From Wasserman and Capra (139) with permission.] *B.* Schematic drawing of V- and C-region domains of antibody light chain. Each domain contains two β-pleated sheets containing three and four β strands (indicated by *white and shaded arrows*), and intrachain disulfide bonds linking opposing sheets are presented as *black bars*. Amino acids are *numbered* from amino terminus, with loops containing amino acids 96, 26, and 53 making up three hypervariable regions responsible for antigen binding. [From Edmundson et al. (36) with permission.]

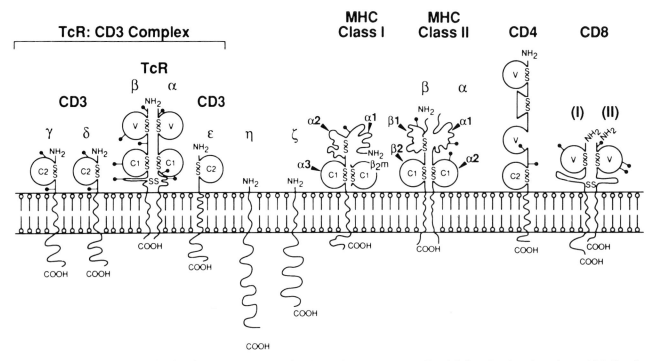

FIG. 19.9. Molecules involved in lymphocyte recognition and activation. With exception of ζ and η, molecules all show sequence homology with Ig molecules and are viewed as part of a single family, the immunoglobulin (Ig) superfamily; these molecules probably arose by gene duplication and divergence from common primordial single domain structure. *Circles* in figure signify domains (predicted from primary sequences); V and C domains show homology with IgV and IgC domains, respectively; based on sequence homology, C domains fall into two broad groups, C1 and C2. [Figure adapted from Hunkapiller and Hood (60) with permission. Text adapted from Paul (104) with permission.]

sumes a characteristic three-dimensional structure, the immunoglobulin fold, which consists of seven β-pleated sheets organized into two opposing layers. Light chains contain two such Ig domains (Fig. 19.8B), whereas heavy chains contain four or five. The immunoglobulin domain structure is not unique to antibody molecules, but in fact is found in several proteins with or without immune function (Fig. 19.9), including the T cell receptor for antigen, MHC proteins, and the CD4 and CD8 co-receptor molecules. Because of the common structure of these proteins, they are grouped as members of an immunoglobulin "superfamily" (60). It is tempting to speculate that the evolution of the Ig domain was critical to the appearance of the adaptive immune system more than 350 million years ago.

Sequence analysis of a number of myeloma proteins done by Kabat and Wu reveals that amino acid variability of the immunoglobulin molecule is largely confined to the N-terminal Ig domain on both the heavy and light chains (142). These domains are therefore called variable regions (V_H and V_L for heavy and light chains, respectively), and C_H and C_L refer to the constant regions of the heavy and light chains. The amino acid variability is not dispersed throughout the variable domain but is instead focused in three (L

chain) or four (H chain) "hypervariable" regions (Fig. 19.8A). Not unexpectedly, the hypervariable regions of the heavy and light chains associate to form the antigen-binding site; therefore, they are also called the complementarity-determining regions (CDRs) of antibody. The framework regions of the variable domain represent the amino acids flanking the hypervariable regions and are largely responsible for the conserved three-dimensional motif of the Ig domain.

In contrast to the amino acid sequences of the variable regions, which differ markedly from antibody to antibody, sequence analysis of the constant regions demonstrates the existence of broad classes (isotypes) of light and heavy chains. Within a given heavy- or light-chain isotype, there is little or no variability in the sequence of the constant regions. Light chains of Ig molecules belong to one of two isotypes, κ and λ. The function of these distinct isotypes is as yet unknown. The five isotypes of antibody, IgM, IgD, IgG, IgA, and IgE, are named after the isotypes of heavy chains, μ, δ, γ, α, and ε, respectively, that they contain. The IgG isotype contains four subclasses in both human and mouse. Each antibody isotype can be secreted, in which case it carries out the effector function of the B cell, or it can exist in the membrane as the B cell surface receptor for antigen. The membrane form is expressed on the surface as a monomer and is distinguished by the presence of a membrane-spanning hydrophobic sequence of 26 amino acids and a short cytoplasmic tail at the carboxy-terminus of the heavy chain. Mature B cells that have not yet been stimulated by antigen express exclusively the IgM and IgD isotypes on the cell surface, whereas previously stimulated (memory) B cells can express any isotype. The isotype expressed on the surface of a memory B cell is identical to the isotype that it is committed to secrete. IgG and IgE are secreted as monomers, but IgM and IgA circulate as pentamers and dimers, respectively. in these cases, individual monomers of IgM or IgA are joined to each other by a 15 kDa joining (J) chain.

The effector function of an antibody molecule is strictly determined by its heavy-chain isotype (Table 19.1). For instance, complement proteins are only activated by antibodies that have polymerized on a microbial cell surface. Since IgM is secreted as a pentamer, a single IgM molecule is sufficient to activate complement. Neutrophils and macrophages bear receptors for the F_c portion of IgG, so the IgG isotype is particularly effective at opsonization. Another killing mechanism, called antibody-dependent cell-mediated cytotoxicity (ADCC), is carried out by natural killer cells which possess distinct F_c receptors for the IgG subtype. Helminths resist killing inside phagocytic cells, but are effectively killed by tissue mast cells. These cells contain Fc receptors for the IgE subtype; IgE is therefore important for the defense against these parasites. Finally, the α heavy chain of the IgA molecule contains a secretory piece which facilitates the accumulation of IgA at mucosal surfaces. The precise function of secreted IgD is unknown because it is present in trace amounts in the serum.

Because antibodies are globular proteins they are themselves antigens containing multiple determinants. As an experiment, a purified myeloma protein from a mouse is injected into an identical twin without myeloma, into a genetically dissimilar mouse, or into a rabbit. This experiment allows one to classify each determinant, or epitope, on the myeloma protein as belonging to one of three types. An *isotope* is defined as a determinant which elicits an antibody response only upon cross-species immunization. Multiple identical isotopes (for instance, the same determinant on multiple copies of the identical antibody) collectively define an *isotype*. Anti-isotypic antibodies bind determinants in the constant regions of the heavy or light chains. Just as mice do not make antibody to their own albumin (Fig. 19.6), they cannot be immunized to mouse isotypes, because the identical determinants are abundantly represented in the serum of all mice and the immune system is tolerant of these determinants. An *allotope* is a determinant on the myeloma protein which also elicits antibodies upon injection into genetically dissimilar mice. Anti-allotypic antibodies recognize allele-specific polymorphisms found in constant or framework regions of antibodies. Finally, an *idiotope* is a determinant on the Ig molecule that can elicit an antibody response even on injection into an identical twin. Anti-idiotypic antibodies bind determinants that are unique to a given antibody molecule, almost always the antigen-binding site. For this reason, each antibody uniquely binds two distinct molecules: the foreign antigen for which it is specific, and its anti-idiotype. To Niels Jerne, this situation presented a formidable problem (62). If a single idiotope on an antibody is immunogenic, then every antibody molecule present in serum would evoke the formation of an anti-idiotype, every anti-idiotype would evoke the formation of an anti-anti-idiotype, and so on, leading to an uncontrolled cascade of antibody production. Just as bad, if every idiotype induces a state of immunologic tolerance (as is the case with isotypes), then the immune system could not respond to foreign antigens, because the universe of idiotypes recapitulates the universe of for-

TABLE 19.1 *Antibody Isotypes* *

	Properties				
	IgM	IgD	IgG	IgE	IgA
Heavy chain	μ	δ	γ	ϵ	α
Mean human adult serum level (mg/ml)	1	0.03	12	0.0003	2
Number of four-chain monomers	5	1	1	1	1, 2, or 3
Special properties	Early appearance; fixes complement; activates macrophages	Found mainly on cell surfaces; traces in serum	Activates complement; crosses placenta; binds to macrophages and granulocytes	Stimulates mast cells to release histamines; found mainly in tissues	Found mainly in secretions

* Adapted from (28) with permission.

TABLE 19.2. *Numbers of Gene Segments in Immunoglobulin and T Cell Receptor Loci* *

	Immunoglobulin			TcR $\alpha\beta$		TcR $\gamma\delta$	
	H chain	κ	λ	α	β	γ	δ
Variable segments	250–1000	250	2	75	25	7	10
Diversity segments	12	0	0	0	2	0	2
Joining segments	4	4	3	50	12	2	2
Potential repertoire (including junctional diversity)	10^{11}			10^{16}		10^{18}	

* [Adapted from (1) with permission of publisher.]

eign antigens. To deal with this problem, Jerne put forth a network theory of the immune response. In it, he proposed that the concentration of any given idiotype in the serum is too low either to immunize or to tolerize, but that the rise in idiotype concentration that occurs with immunization with foreign antigen does evoke the formation of anti-idiotype. Furthermore, he postulated that anti-idiotype inhibits the further formation of idiotype, leading to the regulation of the immune response to the foreign antigen. This "network" theory has had a significant influence on (mostly European) immunological theory since its introduction in the early 1970s.

Organization and Rearrangement of Immunoglobulin Genes. A central prediction of the clonal selection theory of acquired immunity is that the antibody repertoire is generated prior to the introduction of foreign antigen. But how could an organism with 10^5 genes encode 10^9 distinct antibody specificities? This was perhaps the central question in immunology just following the proposal of the clonal selection theory, and its solution represents the greatest contribution of molecular biology to the field of immunology (133).

One clue to the origin of antibody diversity was the finding that among several distinct myeloma proteins of a given H-chain isotype, the amino acid sequence variability is confined exclusively to the N-terminal 116 amino acids. In 1965, Dreyer and Bennett proposed that for each isotype of immunoglobulin there is a single constant (C)-region gene separated in the DNA from multiple variable (V)-region genes, and that during development the gene for the complete polypeptide chain is generated by the physical apposition of a single V-region gene with the C-region gene (34). This prediction violated the dogma of "one gene, one polypeptide," but it was confirmed when in 1976 Hozumi and Tonegawa demonstrated by Southern blot hybridization that the structure of the Ig genes in antibody-producing cells is different from that found in precursors of B cells from the same individual (59).

Thus, immunoglobulin genes were found to undergo a process of somatic recombination during B cell development. Subsequent work has demonstrated that the chains of the antigen-specific receptors on both B and T cells (Ig heavy chain, κ and λ light chains, and T cell receptor [TcR] α, β, γ, and δ chains) are encoded by recombining genes and that all genes employ a common recombination mechanism. The following discussion will focus on the generation of a complete Ig heavy chain, but differences specific to each of the chains will also be mentioned.

A schematic of the mouse immunoglobulin heavy-chain locus is shown in Figure 19.2. The entire H-chain locus spans about 1000–2000 kb of DNA. At the 5′ end of the locus are the V_H-region exons, each about 300 bp long and separated from each other by noncoding DNA of differing lengths. The 3′ end contains the exons encoding the various C-region isotypes, in the order C_μ, C_δ, $C_\gamma 3$, $C_\gamma 1$, $C_\gamma 2b$, $C_\gamma 2a$, C_ϵ, and C_α. Lying in between the V_H- and C_H-region genes are additional exons about 30 bp long each coding the joining (J_H) segments and the diversity (D_H) segments. J-region genes are found in all recombining loci, whereas D-region genes are only found in the IgH, TcR β, and TcR δ loci. Table 19.2 enumerates the number of distinct gene segments encoding the various chains of the immunoglobulin and T cell receptors.

The process by which various gene segments are brought together to form a full variable-region coding sequence is called V(D)J recombination (114). At all recombining loci, this process is dependent upon the presence of recombination signal sequences (RSS) which are found flanking each coding exon. These conserved sequence motifs consist of a dyad symmetric heptamer sequence (CACAGTG) adjacent to the coding exon and an A/T-rich nonamer separated from the heptamer by 12 or 23 nonconserved nucleotides. Efficient joining of two coding regions requires the formation of a loop structure by the complementary pairing of two RSS, one containing the 12-bp spacer and the other containing the 23-bp spacer. Depending on the orientation of the two coding sequences with respect to each other, the coding sequences are brought together either by inversion or through deletion of the intervening noncoding DNA. Deletion of intervening DNA appears to be the predominant mechanism used to join the coding segments and results in the excision of a circle of DNA containing both heptamer components of the RSS next to each other. Interestingly, whereas the formation of signal joints is a precise event, the formation of coding joints is much less precise. This imprecise recombination results in out-of-frame joins, nucleotide deletions, and nucleotide additions, which may be templated ("P" nucleotides) or untemplated ("N" nucleotides). Addition of untemplated nucleotides to coding joints is mediated by the enzyme terminal deoxynucleotide transferase, an enzyme expressed exclusively in developing lymphocytes. N-region nucleotides can result in the addition of one to ten amino acids to the third hypervariable region (CDR3) of the antibody molecule.

V(D)J Recombination in Developing B Cells Results in Heavy-Chain Allelic Exclusion and Light-Chain Isotype Exclusion.

Another central prediction of the clonal selection theory is that each B cell expresses on its surface a population of antibody receptors with identical antigen-binding sites. Given that both H- and L-chain V regions contribute to the binding site for antigen, and knowing that the germline contains two alleles for each of the IgH, Igκ, and Igλ loci, the clonal selection theory predicts that only one heavy-chain V gene and only one light-chain V gene are expressed. This prediction requires that the heavy-chain locus be subject to allelic exclusion (only one allele is expressed) and that the light-chain loci be subject to both isotypic and allelic exclusion. (Only one allele of one isotype is expressed.) The orderly progression of V(D)J recombination assures that this is the case.

The first rearrangement event that occurs in cells committed to the lymphocyte lineage is apposition of one D_H segment to one downstream J_H segment. These $D_H \rightarrow J_H$ joins occur on both alleles of the IgH locus and have been found in both B and T cells; therefore, $D_H \rightarrow J_H$ joins do not signify commitment of a cell to the B lymphocyte lineage. Following this event, one of the numerous V_H genes joins the previously rearranged D–J gene segment. This V $\rightarrow$ D–J join occurs first on only one of the two alleles of the IgH locus and results in a fully rearranged VDJ gene encoding the variable region of the immunoglobulin molecule. The VDJ gene is still separated from the C_μ gene by an intron which is excised from the primary RNA transcript, generating a full-length mRNA encoding the μ heavy chain.

As a result of deletions, mutations, or frameshifts that occur during the V $\rightarrow$ D–J joining process, some pre–B cells generate premature stop codons and nonproductive VDJ rearrangements. If a nonproductive rearrangement takes place, the other IgH allele then attempts a V $\rightarrow$ D–J rearrangement. A second nonproductive rearrangement event leads to the inability to produce surface Ig molecules and cell death, as further development of the pre–B cell requires signals through surface Ig molecules. However, if a productive VDJ join occurs, the μ chain protein that is produced associates with "surrogate" light-chain proteins V pre-B and $\lambda5$, and this μ chain–surrogate light-chain complex is then exported to the cell surface (109). Further development of the cell appears to require signals through this Ig-like receptor. These signals have two effects on the recombination process. First, V $\rightarrow$ D–J rearrangements on the other IgH allele are prevented from occurring, thereby accounting for the phenomenon of H-chain allelic exclusion. Second, the presence of surrogate Ig chain on the cell surface somehow increases the efficiency of V $\rightarrow$ J rearrangement at the κ light-chain locus.

VJ recombination at the light-chain loci also appears to be a regulated process in which productive rearrangement is attempted first at one κ light-chain allele, then at the other κ light-chain allele, then at one λ light-chain allele, and finally at the remaining λ light-chain allele. At any time during this process, a productive (in frame) rearrangement event leads to the synthesis of a functional light-chain protein which then associates with existing heavy chains. Fully assembled, antigen-specific antibody molecules are then exported to the surface where again a signal delivered through the Ig molecule leads to cessation of any further light-chain gene rearrangements. This process leads to the phenomenon of light-chain isotypic and allelic exclusion and to the export of a monospecific population of antibody receptors on the cell surface, thereby fulfilling a central prediction of the clonal selection theory.

The Somatic Generation of Antibody Diversity (133).

The diversity of antigen-binding sites on the population of antibodies reflects the diversity of fully rearranged V-region nucleotide sequences at both the heavy- and light-chain loci. Thus, multiple factors contribute to the diversity of the primary, or unmutated, antibody repertoire. These factors include multiple V, D, and J genes, combinatorial association of the genes, imprecise joining of the coding sequences yielding a variable codon at the site of the join, N and P nucleotide additions, and combinatorial pairing of H and L chains. The true potential for diversity of the antibody repertoire can only be guessed at because the number of functional V_H and V_L genes has not been accurately determined. Also, it is possible that some combinations of heavy and light chains do not dimerize properly. However, as an exercise, we will make a conservative estimate that there are 100 each of V_H and V_κ genes, and that all H–L chain pairs are possible. In this case there are 100 (V_H) × 12 (D_H) × 4 (J_H) × 4 (flexible recombinations on both sides of the D region) = 19,200 possible H chains, and 100 (V_κ) × 4 (J_κ) × 2 (flexible codon for aa 96 resulting from the V–J join) = 800 possible L chains, yielding 19,200 × 800 = $1.54 × 10^7$ possible antibody molecules. This is a vast underestimate in that it does not take into account the additional diversity resulting from P and N nucleotide additions or from alternate reading frames of D regions. Since these factors may not be quantified properly, estimates of the true potential diversity of the primary antibody repertoire are all over the map, but an estimate of 10^9–10^{11} antibodies is not an unreasonable guess.

Components of the V(D)J Recombinase. The presence of the identical recombination signal sequences (RSS) flanking all the rearranging elements of the immunoglobulin and T cell receptor loci strongly suggests that the same enzymes are used for all recombination events. However, no components of the recombinase had been clearly identified until 1988, when Schatz et al. identified two closely linked genes, called recombinase-activating genes 1 and 2 (RAG-1 and RAG-2), whose protein products are clearly involved in the induction of somatic recombination (98, 113). These genes were found to be both necessary and sufficient to activate recombination when introduced into nonlymphoid cells carrying a recombination-dependent selectable gene marker. The names RAG-1 and RAG-2 are intentionally ambiguous as it is not yet known whether they represent lymphoid-specific elements of the recombinase or whether they activate a recombinase which is present in all cells. Nonetheless, mice deficient in RAG-1 or RAG-2 have been created by the technique of targeted gene disruption in embryonic stem cells; in these mice, B and T cell development is arrested at stages prior to the occurrence of any rearrangements (87, 120).

The means by which somatic recombination of immunoglobulin genes proceeds in an orderly fashion, from rearrangements at the heavy-chain allele(s) to rearrangements at the light-chain allele(s), are also unclear. One hypothesis which is gaining support is that the likelihood that a locus will recombine is related to the accessibility of the locus (that is, openness of chromatic structure, access of transcription factors) to the recombinase machinery (3). Transcripts of isolated

rearranging elements, such as a V_H-only transcript, have been found in recombining cells, and mutational analyses have shown that cis-acting nucleotide sequences that function as transcriptional enhancers also greatly enhance the likelihood of recombination at a rearranging locus.

T Cell Receptor Structure, Genes, and the Generation of T Cell Receptor Diversity

Crystallographic data on the structure of the variable domain of the α chain and extracellular portion of the β chain of the T cell antigen receptor have only recently become available (6a, 39a); these structures are homologous to corresponding areas of immunoglobulin but apparently display more restricted conformations. Other details of T cell receptor structure can only be inferred from comparisons to immunoglobulin based upon genetic organization, amino acid, and nucleotide sequence homologies. These comparisons have yielded the following conclusions. First, the α, β, γ, and δ chains each have two immunoglobulin domains consisting of two opposing sheets of four and five β strands held in opposition by an internal disulfide bond. Both chains of the αβ heterodimer are membrane-anchored glycoproteins, and each chain contains a hydrophobic transmembrane region and a cytoplasmic tail of only five amino acids. The transmembrane region of each chain contains a positively charged lysine residue which is thought to associate with acidic amino acids of signal-transducing molecules of the CD3 complex (see section on T Cell Activation). As with Ig, the N-terminal Ig domain of each chain is variable from receptor to receptor, and the site that contacts the complex of peptide + MHC takes contributions from six hypervariable regions, three from each chain. The determinant recognized by these hypervariable regions is thought to be a flat surface, 80% of which is occupied by the MHC molecule and 20% by the peptide. It is possible that the CDR1 and CDR2 are biased toward recognition of the MHC component and that CDR3 (which is the most variable region of both chains) is biased toward recognition of the peptide (32), but there is no firm evidence to support this claim.

The genetic organization of the TcR loci does exhibit some distinctive features. First, the gene for the δ chain of the γδ T cell receptor is nested inside the α locus in between V_α and J_α. Therefore, when a $V_\alpha \rightarrow J_\alpha$ rearrangement occurs, a circle of DNA containing an unrearranged δ locus is deleted. α chain diversity in the mouse is contributed to substantially by the presence of 50 J_α elements, whereas there are only about three to 20 cross-hybridizing V_α gene families and only one C_α gene. The β chain locus consists of 20 V_β families, one

D_β-region gene, seven J_β-region genes, and two C_β-region genes whose functional differences are unknown.

The process of V(D)J recombination in T cells destined to bear $\alpha\beta$ receptors begins with $D_\beta \rightarrow J_\beta$ rearrangements on both alleles followed by a $V_\beta \rightarrow D_\beta J_\beta$ rearrangement on only one allele of the β locus. Both $D \rightarrow J$ and $V \rightarrow DJ$ joins may be accompanied by the addition of N-region nucleotides. Successful rearrangement of the first allele appears to suppress further rearrangements on the second allele, and it has been suggested that this may result from a signal transduced by cell surface β chain homodimers (64). $V_\alpha \rightarrow J_\alpha$ rearrangements on both alleles then follow, but the $\alpha\beta$ heterodimers expressed by mature T cells appear to contain α chains generated by only one of the two alleles. Unlike the L chain of Ig, the $V_\alpha \rightarrow J_\alpha$ join can be accompanied by N nucleotide addition, and because of the greater junctional diversity of T cell receptor chains as compared to Ig H and L chains, the potential repertoire of $\alpha\beta$ T cells is greater. It should be noted that both CD4$^+$ and CD8$^+$ T cells express receptors derived from the same pool of V, D, and J elements.

The generation of T cells bearing the $\gamma\delta$ T cell receptor is interesting in two respects. First, these cells account for 100% of mature T cells produced in the thymus on days 15–17 of fetal development in the mouse (which has a gestation of approximately 20 days), but by birth they only account for 3% of thymic output. Second, cells bearing γ chains encoded by different V_γ regions are not produced randomly and do not migrate randomly; rather, they appear in discrete waves during fetal development and seed distinct areas of the body. The first wave occurs during days 15–17 of fetal development and produces $V_\gamma 3^+$ T cells which localize to the epidermis, the second wave produces $V_\gamma 4^+$ T cells which are destined for the tongue and female reproductive tract, and finally $V_\gamma 1^+$ and $V_\gamma 2^+$ T cells seed the secondary lymphoid organs. $V_\gamma 5^+$ T cells are present in the intestinal epithelium, but these cells appear to develop by a thymus-independent pathway. Finally, despite their potential diversity, many if not most T cells bearing $V_\gamma 3$ and $V_\gamma 4$ have virtually identical antigen-binding sites, a feature which most likely results from precise VJ and VDJ rearrangements and no N nucleotide additions. For instance, 90% of the $\gamma\delta$ T cells found in the skin have identical receptors containing $V_\gamma 3-J_\gamma 2 + V_\delta 1 D_\delta 2 J_\delta 1$ with no junctional diversity. For this reason, it has been speculated that these T cells are specific for conserved determinants, either on common pathogens or on self molecules involved in signaling distress (heat shock proteins).

Major Histocompatibility Complex (MHC) Molecules

Inbred Strains of Mice Facilitate the Analysis of Genetic Influences on Immune Responses. Immunology began largely as a medical discipline concerned with the prevention of infectious diseases, but at the turn of the 20th century both tumor biologists and surgeons—without realizing it—found themselves concerned with problems related to the immune system. Tumor biologists needed to be able to transplant tumors for experimental analysis; the surgeons wished, through improvements in anesthesia and surgical technique, to save lives by performing organ transplants. Both groups were greatly assisted in their efforts by the availability of inbred strains of mice. These strains are produced by brother–sister matings for over 20 successive generations. The end result is a group of mice that is homozygous at all genetic loci and identical to one another (except for males and females, which differ by the Y chromosome). From a group of wild mice, several such inbred strains can be generated. In the early 1900s the cancer biologists found that tumors arising in one strain were easily transplanted to other members of the same strain but were rejected by members of another strain, and these findings were duplicated by surgeons grafting organs within and between mouse strains (123). The surgeons also found that the progeny (F_1) of a mating of different strains accept organ grafts from either parental strain, but the members of either parental strain do not accept a graft from the F_1 offspring. Experiments by Peter Gorer in 1936 then established that a single genetic locus is responsible for the rapid rejection of tissue grafts between strains, a locus which he named H-2 and which is now called the major histocompatibility complex (MHC) (51). Therefore, the discovery of the role of MHC proteins in histocompatibility reactions preceded the discovery of their role in T cell responses by approximately 40 yr. Other cellular proteins not encoded by the MHC but which can trigger a slower graft rejection are termed *minor histocompatibility (H) antigens*. *MHC congenic strains*, which are identical at all minor histocompatibility loci but which differ at MHC alleles, have also been generated.

We now know that the MHC molecules involved in rapid graft rejection and T cell responses are encoded by closely linked, highly polymorphic genes that are members of the immunoglobulin gene superfamily. The structures of the class I and class II molecules are shown in Figures 19.9 and 19.10. The MHC class I molecule (Fig. 19.10A) consists of a 40–45 kDa heavy chain which is noncovalently associated with the 12 kDa, non-MHC encoded β_2 microglobulin (β_2m). The class I heavy chain is a type I transmembrane protein

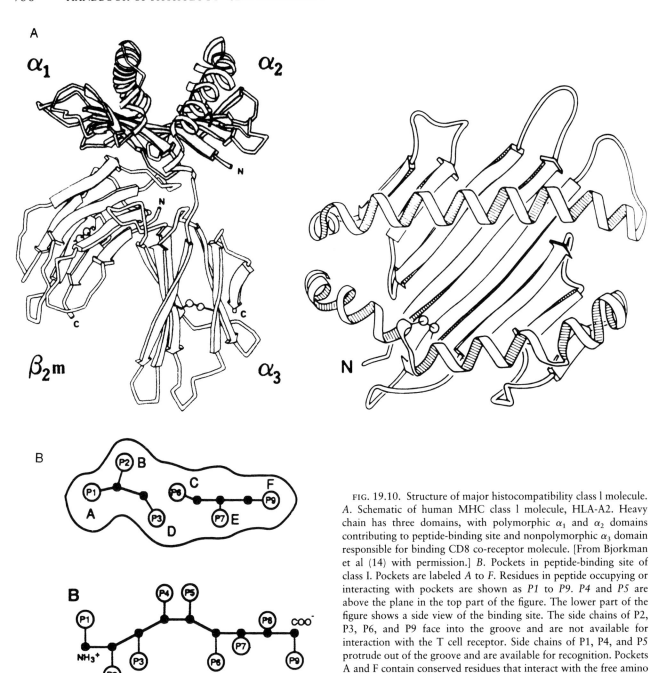

FIG. 19.10. Structure of major histocompatibility class l molecule. A. Schematic of human MHC class l molecule, HLA-A2. Heavy chain has three domains, with polymorphic α_1 and α_2 domains contributing to peptide-binding site and nonpolymorphic α_3 domain responsible for binding CD8 co-receptor molecule. [From Bjorkman et al (14) with permission.] B. Pockets in peptide-binding site of class I. Pockets are labeled A to F. Residues in peptide occupying or interacting with pockets are shown as P1 to P9. P4 and P5 are above the plane in the top part of the figure. The lower part of the figure shows a side view of the binding site. The side chains of P2, P3, P6, and P9 face into the groove and are not available for interaction with the T cell receptor. Side chains of P1, P4, and P5 protrude out of the groove and are available for recognition. Pockets A and F contain conserved residues that interact with the free amino and carboxyl termini of the peptide; pockets B to E are more variable among alleles. [Figure from Matsumura et al (81) with permission. Text from Germain (46) with permission.]

containing three immunoglobulin domains at the amino-terminus and a transmembrane segment and short cytoplasmic tail at the carboxyl-terminus. The α_1 and α_2 domains are the membrane distal regions of the molecule, and the crystal structure of the molecule (14) reveals that these two domains come together to form a peptide-binding groove (Fig. 19.10B). This groove is closed at both ends and so can accommodate short

peptides only eight to 11 amino acids in length. The peptide binds to the MHC molecule via "anchor" residues at its N- and C-termini, and the bound peptide is essentially stretched out in an extended conformation. Peptides greater than 8 amino acids long must accommodate the groove by kinking in the middle. Both α_1 and α_2 domains contribute to the β-pleated sheet which constitutes the floor of the groove, whereas the walls are made up of α helices, one from amino acids of the α_1 domain and one from amino acids of the α_2 domain. The polymorphism of class I molecules is concentrated in areas that form pockets that accommodate amino acid side chains of the bound peptide; for this reason, these polymorphisms have marked effects on the repertoire of peptides that a given class I molecule binds. By contrast, the α_3 domain is highly conserved among distinct class I allelomorphs, and acidic residues at amino acids 223, 227, and 229 form the determinant which is bound by the CD8 molecule expressed by CD8$^+$ T cells.

The class II molecule is also a heterodimer, but it is composed of two type I transmembrane glycoproteins, both of which are encoded by the MHC. Both the α and β chains comprise two immunoglobulin domains, and the peptide-binding groove is generated from the noncovalent association of the α_1 domain of the α chain and the β_1 domain of the β chain. The peptide binding groove of class II molecules has open ends and so accommodates peptides 12–24 amino acids long. Again, the polymorphism of both chains of the class II molecule is thought to be concentrated around those residues which form contact sites for the bound peptide, whereas the β_2 immunoglobulin domain is nonpolymorphic and contains a binding site for the CD4 molecule on CD4$^+$ T cells.

The physical organization of the mouse H-2 complex is illustrated in Fig. 19.11. As can be seen, this complex contains two genes, K and D, encoding class I heavy chains, and a I region locus encoding two sets of class II heterodimers, I-A and I-E. The I-A heterodimer is made up of an A$_\alpha$ and A$_\beta$ chain, and the I-E hetero-

dimer contains an E$_\alpha$ and E$_\beta$ chain. Under normal circumstances there appears to be strict fidelity in pairing of A$_\alpha$ with A$_\beta$ chains and E$_\alpha$ with E$_\beta$ chains. Thus, an animal which is heterozygous at all MHC alleles expresses four distinct class I molecules and eight distinct class II molecules (due to pairing of α and β chains encoded on the same or different chromosomes) on the surface of its B cells. The MHC class III region encodes proteins which are not involved in histocompatibility reactions.

Although the surgeons examining graft rejection in inbred strains worked independently of immunologists studying the body's defense against infection, they did not fail to notice that rejection responses triggered by MHC proteins bore the hallmarks of other immunologic phenomena. That is, a mouse of strain A that has rejected an allogeneic strain B skin graft in 12 days rejects a second strain B skin graft in 9 days, a strain C skin graft in 12 days, and a strain A skin graft not at all. Thus, the strain A recipient displays specificity in the rejection response, specific memory, and a faster response upon regrafting, and self tolerance. Soon after the discovery of T cells in 1961 (48, 85) it was found that these lymphocytes are responsible for the rejection of MHC-disparate grafts, and it was also found that T lymphocyte responses to MHC molecules can be generated and studied in vitro by the co-culture of lymphocytes from strains that differ at the MHC locus. In this procedure, the *mixed lymphocyte culture (MLC)*, lymphocytes from one strain are metabolically inactivated by irradiation and so serve as stimulators for the allogeneic lymphocytes. Proliferation of the responder lymphocytes can be measured by the uptake of radioactive thymidine, and a fraction of the responders develop into cells with cytolytic capacity. In general, CD4$^+$ and CD8$^+$ T cells proliferate in response to MHC class I and II differences, respectively (128), whereas CD8$^+$ T cells develop into cytotoxic T cells that lyse target cells expressing the stimulatory class I molecules. This cytotoxicity is responsible for the destruction of grafted organs in vivo. However, perhaps the most remarkable aspect of alloreactivity is that a huge proportion (from 1 to 10%) of all T cells from unimmunized mice respond to the MHC molecules expressed by a given MHC allogeneic strain (5, 12). This precursor frequency of alloreactive T cells is at least 1000-fold higher than the percentage of T cells in naive mice capable of responding to a given soluble antigen ($\ll$ 0.1%). Why T cells are so interested in allogeneic MHC molecules is a source of considerable debate among immunologists and of hand-wringing among transplantation biologists. Any theory of T cell development and reactivity must provide an explanation for this phenomenon.

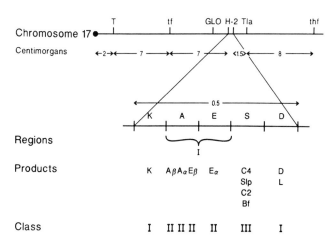

FIG. 19.11. Linear map of H-2 complex. Map units measured in centimorgans were determined by recombination frequencies between indicated loci. *T*, brachyury; *tf*, tufted; *GLO*, glyoxalase; *Tla*, thymus leukemia antigen; *thf*, thin fur. [Text and figure from Hansen et al. (56) with permission from the publisher.]

Although MHC proteins had long been implicated in the phenomenon of graft rejection, their precise role in immune responses was elucidated only after the discovery of T cell–B cell collaboration. This came in 1968, when Miller and Mitchell found that the generation of antibodies to sheep red blood cells requires the participation of two types of lymphocytes—one bone marrow-derived and the actual source of antibody secretion, the other thymus-derived and involved in the induction of antibody synthesis by the bone marrow-derived cells (86). The other functions of T cells were then revealed by the observation that nude mice, which lack a thymus gland and are congenitally deficient in T cells, fail to reject MHC-disparate skin grafts or mount delayed-type hypersensitivity responses. These findings generated intense speculation on two separate but related problems. First, what is the nature of the antigens recognized by T cells? Second, what is the role of the thymus, an organ with intense cell turnover during embryonic development, in T cell development?

Because of the critical role of T cells in alloreactivity and antibody responses to soluble antigens, most immunologists assumed that T cell receptors, like antibody molecules, recognized soluble or cell surface antigens, and this assumption was integral to the first theories on the nature of the T cell repertoire. In 1971, Niels Jerne set forth a theory to account for the phenomenon of alloreactivity and the intense cell turnover in the thymus (61). In it, he proposed that the germline contained approximately 100 T cell receptor V genes, each of which had evolved toward recognition of specific MHC molecules expressed by the species. Since these receptor genes are inherited independently of the genes encoding MHC molecules, an individual would be expected to generate some T cells with receptors reactive to self MHC molecules as well as other T cells reactive to allogeneic MHC molecules. According to Jerne, the role of the thymus is to eliminate the self-reactive T cells and to generate a repertoire of T cells reactive to foreign, non-MHC antigens. This is where the process of intense cell turnover came in. T cells that bound self MHC molecules are stimulated to divide. Those T cells whose receptors failed to mutate supposedly died, but T cells whose receptors mutated away from self-reactivity and toward reactivity to foreign antigens were allowed to emigrate from the thymus as mature functional T cells. Although many features of this theory have been proven to be wrong, Jerne's idea that developing T cells are *positively selected* in the thymus for reactivity to foreign antigens continues to have a profound influence on immunologists studying thymic T cell development.

T Cell Recognition of Antigen is MHC-Restricted. One of the first indications that Jerne's theory was not strictly correct came from the work of Rolf Zinkernagel and Peter Doherty, who were interested in the cytolytic T cell response to viral infections (144). Mice of the BALB/C strain (H-2^d) were infected with lymphocytic choriomeningitis virus (LCMV), and following recovery from infection spleen cells were tested for in vitro killing of infected target cells. The important result was that the T cells killed infected targets of the syngeneic BALB/C but did not kill allogeneic CBA targets (H-2^k) infected with the same virus. Similarly, cytotoxic T cells derived from LCMV-infected CBA mice killed only infected CBA but not infected BALB/C targets. Therefore, T cell recognition of antigen is MHC-restricted.

Two alternative hypotheses were formulated to account for MHC-restricted recognition of antigen. The dual-receptor model proposes that individual T cells bear receptors for the MHC molecule and distinct receptors for antigen, whereas the altered self model proposes that MHC restriction is accomplished by a single T cell receptor that recognizes a determinant formed by the interaction of antigen and the MHC molecule. Both models have fierce advocates, and the arguments between the two camps have not been enlightened by the discovery that T cell receptor genes encode a single heterodimeric structure that recognizes peptides bound inside a groove of the MHC molecule. Rather, a form of two-receptor model has emerged that proposes that CDR1 and CDR2 of the T cell receptor V region domain are responsible for MHC binding and CDR3 is responsible for peptide recognition (32).

Zinkernagel and Doherty's discovery of MHC restriction of T cell responses necessitated several modifications of Jerne's original theory, but a subsequent discovery by Michael Bevan galvanized support for his notion of positive selection in the thymus (11). Bevan was interested in the T cell response to minor histocompatibility (H) antigens which, like responses to viruses, are MHC-restricted. To understand where and how T cells become MHC-restricted, he generated mice with F1 (H-2^a × H-2^b) T cells and APCs inside a parental H-2^a body. This was done by giving the H-2^a mouse a dose of radiation sufficient to kill all blood-derived cells and reconstituting the animal with bone marrow from an (H-2^a × H-2^b) F$_1$ donor. Several weeks after reconstitution, the T cells in these "radiation chimeras" are of (H-2^a × H-2^b) F$_1$ genotype but have matured in an H-2^a thymus. These chimeric animals were then immunized with cells of (H-2^a × H-2^b) F$_1$ MHC type but bearing multiple minor H antigen incompatibilities, and the immunized T cells were then tested for responses to the minor H antigens on cells of either parental MHC type. Although both the T cells and the immunizing cells could present the minor antigens associated with either MHC type, Bevan found that the T cell response to the minor antigens was strongly restricted to the host H-2^a MHC type. Subsequent experiments revealed that the thymus is the site of the development of this MHC restriction; that is, T cells are educated to respond to antigens in association with the MHC molecules expressed by the thymus. The concept of positive selection was resurrected to account for the skewing of the T cell repertoire to reactivity to thymic MHC-associated antigens. Models of positive selection of T cells will be discussed briefly in the section on T Cell Development in the thymus.

LYMPHOCYTE DEVELOPMENT

Monoclonal Antibodies and Flow Cytometry Facilitate the Identification of Distinct Lymphocyte Subpopulations

Detailed analysis of cellular immunity required the separation of lymphocyte subpopulations and subsequent analysis of their functional capabilities. Prior to the 1970s this was primarily achieved by physical methods of separation (for instance, passage of spleen cells through nylon wool enriches for T lymphocytes) or by depletion of subpopulations of cells with polyclonal antibodies and complement. Then, two techniques developed in the 1970s provided immunologists with powerful tools for separation and analysis of lymphocytes.

The first technique is the production of specific monoclonal antibodies to cell surface antigens by hybridoma technology (Fig. 19.12). Hybridomas are the product of the fusion of myeloma cells with antibody-producing cells from an immunized animal (67). As an example, suppose a monoclonal antibody against a mouse B cell surface marker is desired. Rats are immunized and boosted with mouse B cells, and the rat's spleen cells (some of which produce antibody against mouse B cells) are then fused with a myeloma cell line. The myeloma fusion partner that is commonly used is incapable of Ig secretion and also lacks hypoxanthine-guanine phosphoribosyltransferase (HGPRT), an enzyme involved in the salvage of purine nucleotides. Fused cells are then cultured in medium containing aminopterin, which blocks the de novo synthesis of purines. The only cells which continue to grow and secrete antibody in this medium are somatic cell hybrids of the myeloma (which confers immortality) and an antibody-producing cell (which is HGPRT^+ and secretes antibody). These hybridomas are then cloned

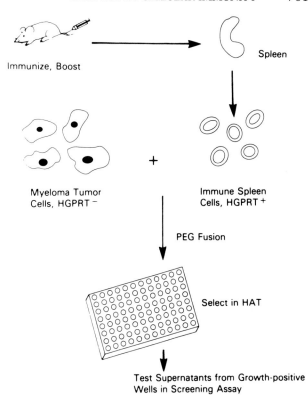

FIG. 19.12. Production of hybridomas. Steps in derivation of hybridomas can be outlined as shown. Spleen cells from immunized donors are fused with myeloma cells bearing selection marker. Fused cells are then cultured in selective medium until visible colonies grow, and their supernatants are then screened for antibody production. [Figure and legend from Berzofsky et al. (9) with permission from publisher.]

and screened for the production of the desired antibody to mouse B cell surface antigens.

The second technique that has facilitated lymphocyte analysis is flow cytometry and fluorescence-activated cell sorting (FACS). The monoclonal antibody against the mouse B cell surface antigen can be chemically coupled to fluorescent molecules (such as fluorescein isothiocyanate or phycoerythrin). Fluorescein-tagged monoclonal antibodies to mouse B cells will bind such cells in heterogeneous populations and allow their identification by fluorescence. The FACS analyzer (Fig. 19.13) contains a fluid flow system in which the cells for analysis are carried single file through a sensing region illuminated by one or two focused laser beams. The laser causes the excitation of fluorescent molecules and light is scattered in all directions. Dichroic filters positioned at a 90° angle to the direction of the laser divide the scattered light into different wavelengths which are then measured by separate detectors. The light is then converted by photomultiplier tubes into an electronic signal whose magnitude is proportional to the intensity of light scattered. In this way, each cell which is bound by antibody molecules will register an electronic signal whose magnitude is proportional to the number of antibodies bound and the number of fluorescent molecules per antibody. Digitized versions of the results are then sent to a computer for purposes of storage and analysis. Furthermore, cells can be sorted based upon fluorescence by applying an electronic charge to those drops containing fluorescing cells and then deflecting these "positive" cells into a container.

Using the hybridoma technique, monoclonal antibodies have been produced against numerous leukocyte surface antigens of the mouse, rat, and human. So many distinct antigens have been identified by monoclonal antibodies that a standard nomenclature has been adopted. Leukocyte antigens now have a CD (for Cluster of Differentiation) designation, and to date over 100 CD antigens have been identified. It should be noted here that all these molecules have specific functions, such as intercellular adhesion and signal transduction, but for immunologists they often serve merely to identify distinct subclasses of lymphocytes

or lymphocytes of one subclass at different stages of differentiation. For instance, it is known that a 220 kDa isoform of the mouse CD45 molecule (B220) has a cytoplasmic domain with phosphatase activity, but B220 is exclusively expressed by cells of the B lymphocyte lineage and can be used to specifically identify B cells from the pro–B cell stage all the way through the plasma cell stage. Likewise, the CD4 and CD8 molecules are critical molecules involved in T cell activation but also identify mutually exclusive subsets of mature, peripheral T cells. In the following discussion on lymphocyte development, then, surface markers will be referred to solely to identify stages of differentiation, but with the knowledge that they may indeed also have critical adhesion or signaling functions.

B Cell Development

Both B and T lymphocytes derive ultimately from the pluripotent hematopoietic stem cell (HSC), which is identified by its capacity for self-renewal and generation of all cells contained within the blood, including neutrophils, basophils, eosinophils, monocytes, red blood cells, and platelets. In mice, γ-irradiation in excess of 1000 rad is sufficient to kill the HSC, and so a lethally irradiated mouse injected with bone marrow from another mouse will eventually generate all blood lineages derived from the marrow donor. Injection of increasing doses of marrow fractions into irradiated recipients permits quantitative analysis of the marrow's ability to protect against the lethal effects of radiation and to generate multilineage hematopoietic reconstitution (132).

Hematopoiesis proceeds through the progressive, irreversible generation of committed progenitor cells that after several stages of differentiation become functional cells that are exported into the bloodstream. B lymphopoiesis in the mouse is carried out first in embryonic blood and placenta and then shifts to fetal liver, omentum, and spleen. However, by the time of birth and thereafter, B cells are produced almost exclusively in the bone marrow, a highly cellular cavity containing blood cell precursors as well as stromal cells that provide critical factors for B lymphopoiesis. It is estimated that a mouse produces 5×10^7 B cells per day in the bone marrow (99), of which only 4%–10% are exported to the periphery. B cell development begins on the inner surface of the bone and then proceeds radially toward the center. During its travels, the developing B cell is essentially tested for the surface expression of a functional receptor; if it passes this test the mature, naive B cell is allowed to exit via a draining sinusoid into the peripheral circulation and secondary lymphoid organs (spleen and lymph nodes).

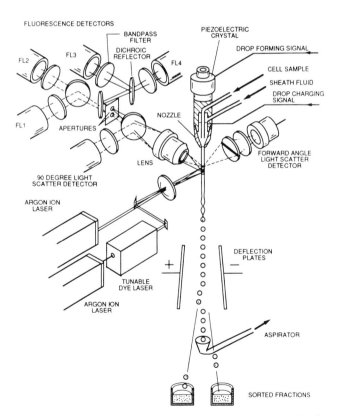

FIG. 19.13. Schematic of a two-laser fluorescence-activated cell sorter, containing fluid, mechanical, and optical elements. See text for details. [Figure and legend adapted from Parks et al. (102a) with permission from the publisher.]

B lymphopoiesis is divided into three stages according to the status of the immunoglobulin receptor genes (109). The Ig genes of the pro–B cell are in germline configuration, but the presence of the $\gamma 5$ and V–pre B surrogate light chains mark this cell as a B lymphocyte precursor. Pro–B cells also express the recombinase activating genes RAG-1 and RAG-2 but fail to express B220. The onset of immunoglobulin heavy-chain rearrangement marks the transition to the pre–B cell stage. Pro–B cells and early pre–B cells appear to depend on stromal cell contact and stromal-cell-derived factors, such as interleukin 7 (IL-7), stem cell factor (SCF), and insulin-like growth factor I (IGF), for their continued survival and/or differentiation. In fact, clones of pro– and pre–B cells have been derived in vitro and found to be separable into sequential differentiation stages that can be maintained for prolonged periods of time via contact with bone marrow stromal cells, with stromal cells + IL-7, or with IL-7 alone. The surface of stromal cells is packed with dividing pre–B cells, and it is possible that daughter cells which have detached from the stromal cell surface due to overcrowding are induced to initiate $V_H \rightarrow D_H J_H$ and $V_L \rightarrow J_L$ rearrangements. Productive rearrangement of both Ig heavy- and light-chain loci leads to the expression surface Ig, the hallmark of the mature B cell stage. The pre–B to B cell transition in the marrow is marked by significant cell loss, either due to nonproductive rearrangements or to the elimination of autoreactive B cells.

Because B-lineage-committed cells with nonproductively rearranged immunoglobulin genes do not emerge from the bone marrow, it has long been suspected that a signal through surface Ig is necessary to complete the maturation process. And teleologically speaking, a mechanism to screen developing B cells for the presence of a productive heavy-chain gene rearrangement prior to the initiation of light-chain rearrangements would also be useful. Evidence for the involvement of surface immunoglobulin or Ig surrogates in B cell development has been obtained from mice rendered deficient (by genetic engineering) of the $\gamma 5$ proteins or surface μ chains (65, 66). Differentiation of pre–B cells is impeded, but not totally blocked, in the $\gamma 5$-deficient mice, and mice which cannot express surface μ heavy chains are totally deficient in mature B cells in the peripheral circulation. Finally, the mature surface Ig-expressing cells must be screened for reactivity to self antigens; autoreactive cells that bind self antigens present in the bone marrow appear to be physically or functionally deleted from the final repertoire of B cells reactive to foreign antigens (50, 93). Whether B cells are also *positively* selected based upon specificity for self antigens expressed within the bone marrow is controver-sial. The differential V gene expression of newly developed marrow B cells vs. mature, naive B cells in the peripheral lymphoid organs has been taken as evidence of positive selection of B cells.

T Cell Development

The process of T cell development commences when hematopoietic progenitors in the bone marrow migrate to the thymus and culminates in the exit of a repertoire of functional, MHC-restricted CD4$^+$ CD8$^-$ T cell receptor (TcR) $\alpha\beta^+$, CD4$^-$ CD8$^+$ TcR $\alpha\beta^+$, and CD4$^-$ CD8$^-$ TcR $\gamma\delta^+$ T cells. These mature T cells are also tolerant of (unresponsive to) the MHC antigens that are expressed in the thymus. The major questions for immunologists studying the process of thymic development therefore have been:

1. How do T cells decide whether to express $\alpha\beta$ or $\gamma\delta$ T cell receptors?
2. How do TcR $\alpha\beta^+$ cells decide whether to express the CD4 or the CD8 co-receptor molecule?
3. How is MHC restriction acquired?
4. How is tolerance to self-MHC acquired?

None of these questions has been fully answered, and all are the subjects of intense investigation. The following discussion will discuss what is known about T cell development but in the process will briefly address all these issues.

The thymic anlage is largely composed of epithelial cells derived from the third and fourth pharyngeal pouches and descends into the area overlying the heart on day 10 of fetal development in the mouse. On day 11, lymphoid stem cells migrate to the thymus, presumably under the influence of chemotactic peptides. By day 14 of gestation nearly all of the developing T cells, or thymocytes, are Thy1$^+$ (a pan–T cell marker) CD4$^-$ CD8$^-$ blast cells. The first two waves of T cell production occur in the next 3 d and generate TcR $\gamma\delta^+$ T cells bearing Vγ3 and Vγ4, respectively (2). These cells do not develop in fetal thymi colonized with adult bone marrow cells or adult thymi colonized with fetal stem cells, and so both the fetal microenvironment and factors pertaining to the thymocyte appear to influence the recombination decisions. (60a) The first T cells bearing the $\alpha\beta$ heterodimeric receptor appear on fetal day 18, and by birth they make up 97% of total thymic output.

Development of TcR $\alpha\beta^+$ T cells is best followed by tracking the expression of Thy1, the T cell receptor, and CD4 and CD8 molecules [Fig. 19.14; (80)]. This approach has led to the identification of three prominent developmental stages. Just after entry into the thymus, T cell progenitors express the Thy1 molecule

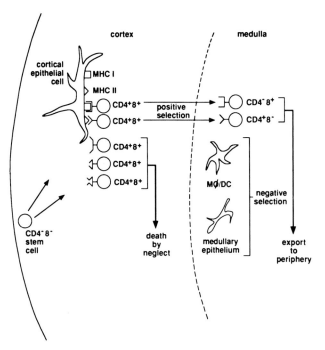

FIG. 19.14. T cell selection in thymus. CD4⁻ 8⁻ stem cells under capsule of thymus spawn huge numbers of CD4⁺ 8⁺ immature T cells expressing low density of ab TcR molecules. Small proportion of these cells are capable of binding MHC class I or II molecules displayed on cortical epithelium. These T cells are positively selected: Selected cells up-regulate TcR expression, down-regulate expression of one of accessory molecules (CD4 or CD8), and move to medulla, where they appear as typical CD4⁺ or CD8⁺ mature T cells. Some of these cells have high affinity for self-MHC molecules and die when T cells pass through dense network of APCs (macrophages and dendritic cells) and epithelial cells in medulla; some deletion of cells can also occur in cortex through contact with cortical epithelium. Only very small proportion (1%–2%) of total thymocytes are exported from thymus; vast majority fail either positive or negative selection steps. [Figure and legend adapted from Sprent (125) with permission.]

but lack expression of T cell receptor, CD4, or CD8. These CD4⁻ CD8⁻ "double negative" (DN) thymocytes proliferate extensively and then, due to expression of RAG-1 and RAG-2, begin to rearrange β chain genes on one chromosome. Productive rearrangement at the β chain locus appears to have two effects on the thymocyte: (1) cessation of rearrangement of the other β chain allele; and (2) up-regulation of expression of both CD4 and CD8 molecules. These effects may be triggered by a signal generated through a β-β homodimer expressed on the cell surface (64). Productive rearrangement at the α chain locus then leads to the expression of the T cell receptor on CD4⁺ CD8⁺ "double positive" T cells, which comprise about 85% of total thymocytes. The evolutionary reason for the existence of a CD4⁺ CD8⁺ intermediate on the way to production of CD4⁺ CD8⁻ and CD4⁻ CD8⁺ single positive T cells is not obvious. However, these are the

first cells which express significant levels of the αβ TcR heterodimer, and as such they may be the targets of selection events based upon TcR specificity for antigen. But unlike events occurring at the β chain locus, productive rearrangement of the α chain and expression of an αβ heterodimer on the cell surface do not lead to cessation of rearrangement at the other α chain. Rather, down-regulation of RAG-1 and RAG-2 expression and cessation of other rearrangements are contingent upon signaling through the T cell receptor (15,136). The general consensus is that this signaling represents the *positive selection* step which leads to self-MHC restriction, and that it occurs by an interaction of the DP thymocyte with epithelial cells in the thymic cortex. The role of MHC-associated self peptides in this selection step is unclear. If a signal through the αβ TcR does not occur at this stage, whether due to nonproductive rearrangements or due to insufficient avidity of the antigen-specific receptor for self-MHC, the thymocyte dies. However, the cells that survive are then subjected to a process of *negative selection* or *clonal deletion* in which cells with overt autoreactivity die by apoptosis (92). Thymocytes that pass the gauntlet of both positive and negative selection are then allowed to emigrate from the thymus as mature CD4⁺ CD8⁻ or CD4⁻ CD8⁺ "single positive" T cells.

What causes a double-positive thymocyte to down-regulate one of its accessory molecules (CD4 or CD8)? This is a matter of considerable debate, and when considering the alternative explanations it is important to remember that the expression of the accessory molecule correlates with restriction specificity—CD4⁺ T cells are MHC class II restricted, and CD8⁺ T cells are MHC class I restricted. The two models that have been put forth to account for co-receptor molecule expression recall the debate on the origin of antibody formation. The *instructive* model of thymic development proposes that the ligand of the newly expressed T cell receptor, MHC class I or MHC class II, instructs the thymocyte to down-regulate CD4 or CD8, respectively (108). The *selective* model states that a double-positive thymocyte randomly commits to the down-regulation of the CD4 or CD8 molecule and that the MHC class I or II subsequently selects those cells whose T cell receptors and accessory molecules both can bind the appropriate MHC molecule (31,137). That is, a cell whose antigen-specific receptor binds MHC class I will only survive if it is committed to the expression of the CD8 molecule.

Finally, the notion of a positive selection step based upon specific T cell receptor recognition of *self* MHC leading to *self* MHC restriction leads to a problem. That is, how can both positive and negative selection be mediated by T cell binding to self MHC molecules? Three models have been proposed to resolve this paradox. The first model states that different cells mediate the two events; that is, positive selection occurs by interactions of thymocytes with epithelial cells of the thymic cortex, and that negative selection occurs by interactions of thymocytes with bone-marrow-derived dendritic cells in the medulla of the thymus. It is possible, then, that a unique set of self peptides expressed by thymic cortical epithelium mediates the positive selection process (78). The second model states that the affinity of the T cell receptor for self MHC determines the outcome of the interaction; a low-affinity interaction generates positive selection, whereas a high-affinity interaction generates negative selection (126). The third model proposes that T cells are not positively selected for preferential binding of self MHC but simply for the presence of a functional receptor on the cell surface (82). In this model, T cell receptors are just like antibodies (that is, they can be ligated by any molecule whose shape is complementary to the antigen-binding site), and MHC restriction is imparted by the requirement of signaling through CD4 or CD8 for T cell activation. This problem is complex and continues to be debated actively.

PERIPHERAL LYMPHOID ORGANS AND LYMPHOCYTE CIRCULATION

Although differentiation of lymphocytes within the primary lymphoid organs requires signals delivered through the antigen-specific receptors, these signals are generated in the absence of those external antigens to which these cells are reactive. However, differentiation of naive lymphocytes in the periphery to memory and effector cells requires stimulation by exogenous antigens. This step occurs in secondary lymphoid organs, the sites of the stimulation and differentiation of both naive and memory lymphocytes. Antigens from the likeliest sites of infection drain into distinct secondary lymphoid organs: the spleen filters blood-borne antigens, lymph nodes drain the skin, and mucosa-associated lymphoid tissues (MALT) collect antigens entering via the lung and the gastrointestinal and reproductive tracts. The lymphocytes reactive to these antigens are exported from the primary lymphoid organs directly into the bloodstream and circulate either directly through the spleen or enter lymph nodes or MALT via high endothelial venules (HEV), specialized postcapillary venules that express adhesion molecules complementary to homing receptors ("addressins") on naive lymphocytes. Once inside a lymph node, naive T and B cells percolate radially inward through a gauntlet of soluble and cell-bound antigens. If a lymphocyte does not encounter its specific antigen during its travel through the node, it exits via efferent lymphatics, reen-

ters the bloodstream, and travels through more secondary lymphoid organs. Naive lymphocytes are long-lived and continue to recirculate from blood to lymph until they encounter their antigen or die of senescence (133a, 138).

The architecture of a lymph node is depicted in Figure 19.15. Soluble or cell-associated antigens are carried into lymph nodes from the skin via afferent lymphatics which open into the subcapsular sinus, a slit-like opening under the capsule of the lymph node and overlying the subjacent cortex. Once inside the lymph node, fluid and cells travel first through the lymphocyte-rich cortex, which contains two distinct zones. The T cell zones of the paracortex contain T cells as well as bone-marrow-derived, interdigitating dendritic cells (to be described below). The B cell zones of the cortex are arranged into spherical structures called *follicles*. Primary follicles consist largely of small resting B cells as well as a loose network of non-bone-marrow-derived follicular dendritic cells (FDCs), which retain antigens in the form of immune complexes on their surface for the continuous restimulation of previously activated lymphocytes. Secondary follicles are distinguished by the presence of germinal centers containing a dense network of FDCs and tangible body macrophages (TBMs) responsible for the disposal of apoptotic B lymphocytes that fail the process of antigen-driven selection. The "dark zone" of the germinal center contains actively dividing, B cells with low levels of membrane Ig (centroblasts) that ultimately slow down and reexpress surface Ig as germinal centrocytes of the "light zone." Fluid and cells that traverse the cortex then pass through the medullary cords, collect into medullary sinuses, and are carried out into the efferent lymph and eventually back into the bloodstream via the thoracic duct.

Protein antigens destined to stimulate T cells in the paracortex are carried to the lymph node by bone-marrow-derived cells of dendritic morphology. Although these cells have different names depending on their location, they are probably directly related to each other. Thus, the epidermal Langerhans cell captures antigens, travels back to the lymph node as the veiled cell of afferent lymph, and finally rests in T cell zones as the interdigitating dendritic cell. On its journey back to the lymph node, the dendritic lineage cell acquires potent T cell–stimulating capacity through up-regulation of MHC class II, adhesion, and costimulatory ("signal 2") molecules. Here they present antigens to naive T cells that have entered the lymph node cortex via the high endothelial venule. The importance of dendritic cells in carrying antigens to lymph nodes via afferent lymphatics is underscored by the finding that ablation of the afferent lymphatics prevents the immunization of naive T cells (8).

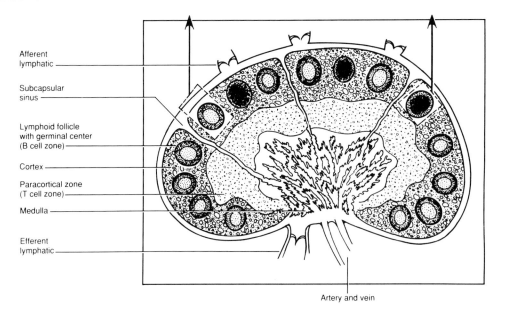

Afferent
lymphatic

Subcapsular
sinus

Lymphoid follicle
with germinal center
(B cell zone)

Cortex

Paracortical zone
(T cell zone)

Medulla

Efferent
lymphatic

Artery and vein

FIG. 19.15. Schematic diagram of lymph node, showing distinct cortex with lymphoid follicles and dense lymphocytes and medulla with lymphatic cords and vessels. [Figure and text from Abbas et al. (1) with permission.]

B cells also enter lymph nodes via HEV and first travel to the T cell zones of the paracortex. In the presence of antigen, the B cell differentiates either into antibody-secreting plasma cells or, in the presence of help from T cells, migrates to primary follicles. The plasma cells settle into the medullary cords of the lymph node and are responsible for the low-affinity, low-titer, predominantly IgM antibody response seen within 3–4 days after immunization. A second phase of the primary immune response is generated by the B cells that have migrated to the primary follicles in response to antigen + T cell help. In the follicle, B cells stimulated by antigens present in immune complexes on the surface of FDCs proliferate rapidly (dividing every 6–7 h) and give rise to the germinal centers of secondary follicles. It is here that centroblasts undergo somatic mutation, affinity maturation, and isotype switching; those cells that lose antigen-binding capacity undergo apoptosis and are ingested by TBMs. Cells retaining antigen-binding capacity are "positively" selected at the germinal centrocyte stage and differentiate either into memory cells or plasma cells responsible for the secretion of high-affinity, high-titer, isotype-switched antibody.

T lymphocytes that have been activated to become effector cells (lymphokine-secreting helper T cells or killer T cells) must then seek out and eliminate their antigens at the site of infection. Indeed, activated effector T cells display altered circulation patterns due to the up-regulation of "homing" receptors (77).

Rather than circulate from blood to lymph and back to blood as naive cells do, effector cells and possibly memory cells have the ability to migrate into and through tissues by transmigrating through capillary endothelium. These activated T cells course through tissues in search of the pathogen; following the performance of their effector function these cells then either die or slow back down to the resting memory cell state. Cells that have settled down presumably return to the hematolymphoid circulation via afferent lymphatics; in this regard, it has been found that the afferent lymphatics contain only cells of the activated or memory type. It is not yet clear whether resting memory cells downregulate their homing receptors and recirculate only through blood and lymph. However, recent evidence does indicate that memory cells are not inherently long-lived but must be periodically restimulated by antigen in secondary lymphoid organs [(53, 97); see section on Immunologic Memory].

LYMPHOCYTE ACTIVATION

Antigen Processing and Presentation: The Generation of Signal 1

As was just mentioned, professional APCs, such as dendritic lineage cells, ferry antigens to secondary lymphoid organs for T cell recognition. However, T cells are not activated by whole proteins but only by fragments of proteins presented in the peptide-binding groove of MHC molecules. Thus, the APC first degrades the antigen and inserts peptide fragments into MHC molecules destined for the cell surface. Sometimes, however, the dendritic cell needs only to pick up and present peptides that have been released from cells killed by viruses or by cytotoxic cells; however, the following discussion is restricted to the consideration of the processing and presentation of whole proteins.

The primary, if not exclusive, function of the MHC glycoproteins is to show to T cells fragments of every protein present within the cell. This is the only way for the immune system to detect the presence of intracellular parasites, and it allows for the recognition of subtle differences in the primary amino acid sequences of proteins, even those found within the interior of globular proteins. The distinction between MHC class I and class II molecules is the source of the peptides presented: MHC class I molecules present fragments of proteins made in the cytoplasm, whereas MHC class II molecules display peptides derived from proteins within endocytic compartments (endosomes, phagosomes, and lysosomes). From an evolutionary perspective, the class I pathway is useful for displaying viral proteins synthesized in the cytoplasm, and the class II pathway is useful for presenting proteins of intracellular bacteria that resist killing in phagosomes. The functional distinction between the two classes of MHC molecules reflects differences in their synthesis, assembly, intracellular trafficking, peptide association, and export to the cell surface, and extraordinary progress has been made within the past few years in understanding these processes.

The Class I Pathway of Antigen Processing and Presentation (Fig. 19.4). MHC class I molecules are synthesized in the endoplasmic reticulum and transported to the ER lumen via a signal sequence. The Class I heavy chain is inserted into the endoplasmic reticulum (ER) membrane and is held there by an 88 kDa protein, calnexin, along with a 35 kDa protein, as yet unnamed. Although subsequent events are unclear, it is likely that release of heavy chain from this complex occurs upon association of heavy chain with β_2 microglobulin (β_2m). Although a small fraction of class I dimers reach the surface in the absence of associated peptide, these "empty" dimers are inherently unstable and their half life on the surface is extremely short in the absence of bound peptides. Association of peptide with the class I dimer within the ER lumen leads to the formation of a stable trimolecular complex that is then exported directly from the Golgi to the cell surface.

What is the source of the peptides that bind MHC class I molecules within the ER lumen? Proteins synthe-

sized in the cytoplasm are degraded enzymatically (perhaps by a multiunit complex called the *proteasome*) and fragments are transported into the ER lumen through an ATP-dependent, peptide-selective heterodimeric transporter protein (119). The two chains of this dimer, TAP1 and TAP2, are encoded by genes located within the class II region of the MHC, and the TAP proteins can be found in the membrane of the ER and cis-Golgi. Both chains are necessary for transport of peptides into the ER lumen, as nonfunction of either chain leads to the inability to deliver peptides to the ER and transport of unstable, empty class I molecules to the surface. In rat and human, the *tap1* and *tap2* genes are highly polymorphic, and these polymorphisms modify the repertoire of peptides that are carried into the ER lumen. Once in the ER lumen, peptides of eight or nine amino acids bind to the groove located between the α_2 and α_3 domains of the class I heavy chain, and the stable heterotrimeric complex of heavy chain, β_2m, and peptide are then released into the Golgi apparatus for eventual export to the cell surface. Recently, immunologists have been able to elute, separate, and sequence naturally processed peptides from purified MHC class I molecules, and have found that the majority of peptides are derived from cytoplasmic and nuclear proteins.

Class I molecules do present peptides from exogenous proteins, and the evidence for this was obtained by Bevan in his studies of the immune response to minor histocompatibility (H) antigens (10). Minor H antigens are peptides, derived from normal intracellular proteins, presented in the groove of the MHC class I molecule, and rejection of a skin graft differing only in minor H antigens is mediated by class I–restricted, CD8$^+$ cytotoxic T cells (CTLs). When Bevan immunized an F_1 (A × B) mouse with spleen cells from an animal of MHC "A" but differing in minor antigens (we shall call this A′), the cytotoxic T cells that developed could kill cells displaying the minor antigens in the context of either MHC A or MHC B; that is, the CTLs killed both A′ and B′ targets. The only way CTLs recognizing B′ targets could have been generated is by presentation of antigens made by the A′ cells by class I molecules of the F_1 (A × B) recipient. This phenomenon is called *cross-priming* and may be physiologically relevant for priming CTLs to antigens made only by cells with poor APC function (that is, cells unable to deliver signal 2).

The Class II Pathway of Antigen Processing and Presentation (Fig. 19.4). The unique trafficking of the MHC class II heterodimer, as well as its presentation of primarily exogenous antigens, can be attributed to its association with a third chain, alternatively called the invariant (Ii) or γ chain (25). Shortly following synthesis of α and β chains in the endoplasmic reticulum, the γ chain associates with the αβ dimer and prevents binding of peptides within the ER. The γ chain itself forms a homotrimer, so MHC class II molecules traverse the ER as a trimer of αβγ heterotrimers. As with the association of the class I heavy chain and $β_2$m in the absence of peptide, an empty αβ heterodimer is also inherently unstable, and association with γ appears to increase the stability of association. Once in the trans-Golgi reticulum, the αβγ heterotrimer is then directed to endocytic compartments by virtue of a targeting signal contained within the cytoplasmic domain of the γ chain. During transport to these compartments the γ chain is degraded by proteolysis, and the αβ dimer is then free to bind peptide fragments of proteins that have been internalized via coated pits, pinocytosis, or receptor-mediated uptake. The same enzymes are probably responsible for the degradation of both the γ chain and the exogenous proteins, and this proteolysis depends upon the low pH found in endosomes and lysosomes. Thus, lysosomotropic agents that increase intracellular pH, such as NH_4Cl or chloroquine, inhibit class II–mediated antigen presentation while leaving the class I pathway intact. It should also be noted that class II molecules can also present endogenous peptides. Whether this occurs by binding of peptides in the ER or through the fusion of autophagosomes with endocytic vesicles is unclear.

Superantigens. There is one exception to the general rule that T cells recognize only peptides in the groove of MHC molecules. The "superantigens" comprise molecules with binding sites for nonpolymorphic regions of both MHC class II molecules and specific T cell receptor β chains (79). For instance, the staphylococcal enterotoxin A directly cross-links MHC class II molecules with T cells expressing β chains encoded by $V_β$s 1, 3, 10, 11, 12, and 17. The term superantigen is therefore based on their ability to stimulate the proliferation of a large percentage of T cells (from 2% to 20%) solely based upon β chain expression and regardless of antigen specificity. There are two classes of superantigens. Exogenous superantigens are toxins derived from bacteria, such as *Staphylococcus, Streptococcus,* and *Mycoplasma;* these toxins impair host defenses by causing massive cytokine release and a shock syndrome. Endogenous superantigens are encoded in the open reading frames of endogenous retroviruses; these have only been observed in mice. Rather than causing massive stimulation of the immune system, these superantigens actually cause massive clonal deletion of thymocytes bearing the reactive $V_β$ chains (141). For instance, in appropriate strains of mice the presence of the *Mtv-7* gene causes deletion of thymocytes bearing the $V_β$ 6, 7, 8.1, and 9 chains. Perhaps this clonal deletion may be evolutionarily advantageous for mice by preventing the potentially lethal effects of ingestion of exogenous superantigens later on in life.

T Cell Activation

Activation of naive $CD4^+$ T lymphocytes to proliferate and differentiate into memory or effector cells occurs in spleen or lymph node through recognition of peptide antigens presented by dendritic cells. The cell conjugates formed between dendritic cells and naive T cells are initially nonantigen-specific and due to the interaction of adhesion molecules: the LFA-1 molecule on the T cell with ICAM-1, which is constitutively highly expressed by the dendritic cell. If its antigen-specific receptor is not ligated, the T cell loses interest in the interaction and continues on its way. However, ligation of this receptor ("signal 1") initiates signals that prolong the contact between the two cells and commits the T cell to a pathway leading either to activation or inactivation. The decision to activate is then contingent upon the delivery of additional nonantigen-specific "costimulatory" signals ("signal 2") to the T cell by the antigen-presenting cell (91). One such costimulatory signal is generated by ligation of the CD28 molecule, expressed by naive T cells, by the B7 family of molecules which is expressed by dendritic cells. The extraordinary potency of dendritic cells in stimulating naive $CD4^+$ T cells is therefore a function of their high expression of MHC class II proteins, adhesion molecules such as ICAM-1, and costimulatory molecules such as B7-1. The activation of the naive $CD4^+$ T cell is a central step in the generation of an immune response to protein antigens and leads to multiple effects. First, the $CD4^+$ T cell is induced to express cell surface receptors for IL-2 and secretes IL-2 as well. IL-2 then binds to its receptor, leading to proliferation and clonal expansion of the antigen-specific $CD4^+$ T cell. Furthermore, the secreted IL-2 may act on other cells in the vicinity which have up-regulated their IL-2 receptors but which may be incapable of IL-2 secretion, such as is often the case with $CD8^+$ cytolytic T cells. The expanded population of antigen-specific T cells is one of the hallmarks of immunological memory. In addition, the activated $CD4^+$ T cell may secrete additional cytokines, such as IFN-γ and TNF-α, which serve to activate natural killer cells and macrophages, respectively. Finally, depending on the milieu, activation of a naive $CD4^+$ T cell induces that cell or its progeny to differentiate into effector cells which are particularly suited to induce inflammation (T_h1 T cells) or B cell activation (T_h2 T cells).

Following the formation of nonantigen-specific conjugates between the T lymphocyte and dendritic cell, ligation of the antigen-specific receptor by peptide-MHC triggers a cascade of intracellular signaling events. It is unlikely that the TcR $\alpha\beta$ heterodimer itself generates these signals, as the cytoplasmic tails of both TcR α and β chains are only five amino acids long. Rather, a complex associated with the TcR at the cell surface is responsible for transducing TcR ligation into intracellular signals. This complex was identified as proteins which co-immunoprecipitate with the TcR in mild detergent extracts of T cells. Two-dimensional electrophoresis of these immunoprecipitates in reducing and nonreducing conditions revealed the presence of five invariant polypeptides: TcR ζ and η, and CD3 γ, δ, and ϵ (112). The majority of the 16 kDa ζ chains of the TcR exist on the surface as ζ-ζ homodimers whereas a minority of ζ chains form heterodimers with TcR η. The 20 kDa ϵ chain, the 25 kDa δ chain, and the 20 kDa γ chain of the CD3 complex are all transmembrane proteins of the Ig gene superfamily and are present on the surface as $\delta\epsilon$ or $\gamma\epsilon$ heterodimers. The precise stoichiometry of a single TcR complex is uncertain but most likely consists of six chains: two from a TcR $\alpha\beta$ heterodimer, two from a CD3 $\delta\epsilon$ or $\gamma\epsilon$ heterodimer, and two from a TcR $\zeta\zeta$ homodimer or TcR $\zeta\eta$ heterodimer (Fig. 19.16). As will be discussed below, each of the invariant chains of the TcR complex

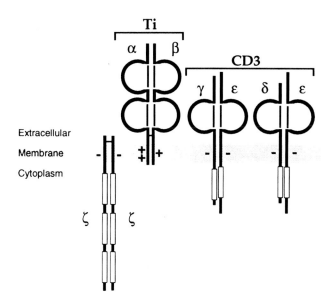

FIG. 19.16. T cell antigen receptor complex. Illustrated schematically is antigen-binding subunit comprising $\alpha\beta$ heterodimer, Ti, and associated invariant CD3 and ζ chains. Acidic (−) and basic (+) residues located within plasma membrane are indicated. *Open rectangular boxes* indicate motifs within cytoplasmic domains that interact with cytoplasmic protein tyrosine kinases. [Figure and legend from Weiss (140) with permission of the publisher.]

contains a shared sequence motif which is critical to the protein's downstream signaling function.

Biochemists studying T cell activation must be able to correlate events at the cell membrane with the intracellular signals that result from these events. To do this, they must be able to distinguish signaling events directly resulting from T cell receptor ligation from nonantigen-specific signals that are a consequence of T cell–APC interactions. For instance, CD2, CD4, CD28, and CD45 generate signals upon ligation, and many of these surface molecules utilize or impinge upon the same signaling pathways as are used by the TcR. To adequately assess the consequences of T cell receptor ligation, then, it is necessary both to obtain a population of T cells rigorously depleted of APCs and to ligate exclusively the T cell receptor for antigen. There are several potential sources of purified T cells, including FACS-sorted splenic, lymph node, or peripheral blood T cells, in vitro propagated T cell clones, T cell hybridomas, and T cell leukemia lines. Purified T cells in the absence of APCs may then be stimulated by monoclonal antibodies to TcR $\alpha\beta$, CD3 ϵ chain, peptides presented by purified MHC molecules incorporated into planar lipid membranes, or plant lectins (such as concanavalin A or phytohemagglutinin) which cross-link the TcR through multivalent binding of carbohydrate moieties.

A schematic of the intracellular events following T cell receptor ligation is shown in Figure 19.17. The earliest intracellular biochemical change that is detected upon T cell receptor ligation, occurring within seconds to minutes, is phosphorylation of several proteins on tyrosine residues. These phosphorylation events are thought to result from the activation of the protein tyrosine kinases (PTks) p59^{fyn} and ZAP-70, which have been isolated in immunoprecipitates of the T cell receptor complex, and P56^{lck}, associated with CD4 or CD8. In contrast, CD45 ligation activates a phosphatase which dephosphorylates and activates p56^{lck}. The involvement of PTks in T cell signal transduction is implicated by the following evidence: *(1)* Inhibitors of tyrosine kinases abolish T cell activation; *(2)* mice rendered deficient of p56^{lck} by gene "knockout" techniques have profound T cell developmental abnormalities and T cell deficiencies; and *(3)* a targeted mutation of the CD4 molecule that prevents its association with p56^{lck} profoundly impairs T cell activation. Several studies have demonstrated that optimal T cell activation requires that the co-receptor molecule, CD4 or CD8, be ligated by the same MHC molecule that is presenting peptide to the antigen-specific T cell receptor. One biochemical consequence of the close association of CD4 with the T cell receptor complex is the phosphorylation of TcR ζ chain, CD3 ϵ, or both on a

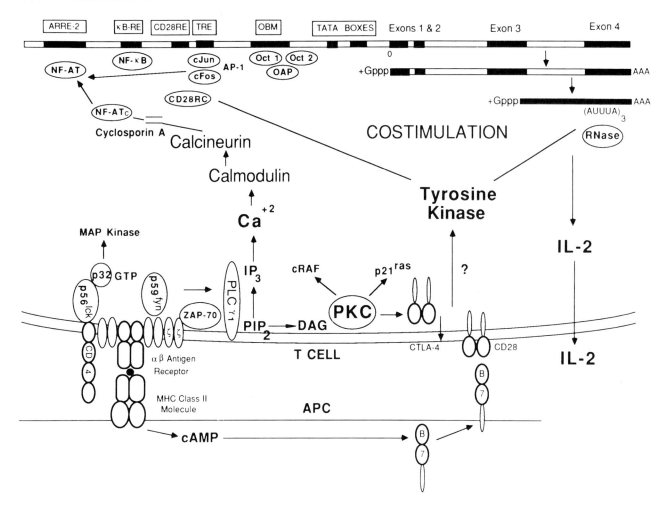

FIG. 19.17. Intracellular events following ligation of T cell receptor for antigen. *APC*, antigen-presenting cell; *IL-2*, interleukin-2; *PLCg1*, phospholipase Cg1; *PIP$_2$*, phosphatidylinositol bisphosphate; *IP$_3$*, inositol 1,4,5-triphosphate; *DAG*, diacylglycerol; *PKC*, protein kinase C; *NF-ATc*, cytoplasmic component of nuclear factor of activated T cells; *NF-κb*, nuclear factor of κ-chain gene enhancer;

ARRE, antigen-receptor response element; *CD28RC*, CD28 response complex; *CD28RE*, CD28 response element; *TRE*, TPA (phorbol ester) response element; *OBM*, octamer-binding motif; *Oct 1*, octamer binding protein 1. See text for details. [Figure from Schwarz (116) with permission.]

tyrosine residue contained within the shared motif. It is currently believed that this phosphotyrosine residue is the site which is bound by the PTK ZAP-70, which then mediates the downstream signals generated by TcR ligation. T cell activation proceeds only when p56^{lck} is brought into proximity with its substrates, CD3ε or TcRζ, which occurs only when the CD4 or CD8 co-receptor molecule is near the T cell receptor complex. MHC molecules are unique among cell surface molecules in that they possess binding sites for both the TcR and CD4 or CD8, and so are uniquely capable of bringing them together. The ability of MHC molecules to bind both TcR and co-receptor molecule may therefore account for the phenomenon of MHC-restriction of T cell activation. p56^{lck} may have other

important effects on T cell activation other than those mediated by its tyrosine kinase activity, because T cells containing mutated p56^{lck} that lack kinase activity can still be activated normally.

Another protein critically involved in T cell signal transduction that is activated by tyrosine phosphorylation is phospholipase Cγ$_1$ (PLCγ$_1$). CD45 signaling is also required for the activation of PLCγ$_1$, implying that the PTK that phosphorylates PLCγ$_1$ is activated by dephosphorylation. Upon activation, PLCγ$_1$ hydrolyzes the rare membrane phospholipid, phosphatidylinositol bisphosphate (PIP$_2$), yielding (1, 4, 5) inositol triphosphate (IP$_3$) and diacylglycerol (DG) (140). These products of PIP$_2$ cleavage have several effects on T cells; as most studies of these effects have focused upon the events leading to IL-2 gene transcription, these events will be described here. IP$_3$ leads to the release of calcium from intracellular stores, elevating intracellular calcium from 100 nM to >1.0 mM. In addition, the plasma membrane may contain receptors for IP$_3$ which, when bound, lead to the influx of extracellular calcium. The sustained rise in intracellular calcium leads to the activation of calcineurin, a calcium- and calmodulin-dependent phosphatase which is thought to have downstream actions on IL-2 gene transcription. Evidence for the involvement of calcineurin in T cell activation has been obtained by the discovery that it is the molecular target of the action of the immunosuppressive drugs cyclosporine A and FK506 (22, 96). The DG that is formed by PIP$_2$ hydrolysis regulates the activity of protein kinase C (PKC), a serine/threonine kinase. Although the effects of PKC activation are multiple and incompletely understood, it is clear that PKC is also involved in generating signals which promote transcription of the IL-2 gene. The rise in intracellular calcium and activation of PKC that result from hydrolysis of PIP$_2$ are events that are likely to be relevant to the T cell activation process, as treatment of T cells with a combination of a calcium ionophore (which raises intracellular calcium) and phorbol ester (which activates PKC) is sufficient to initiate IL-2 gene transcription.

In addition to the biochemical events described above, T cell activation leads to the induction of transcription of multiple genes required for cell division and the expression of effector function (24). The transcription of many of these genes is in turn contingent upon the binding of specific nuclear factors to promoter regions in the 5' region adjacent to the gene. Again, as IL-2 production is central to the process of clonal expansion and the generation of immunologic memory, the control of IL-2 gene transcription in response to cell surface stimuli is perhaps best studied. A 275 bp segment upstream from the transcription initiation site of the IL-2 gene has been found to contain at least six sites known to bind nuclear factors controlling transcriptional activation. Five of these sites are bound by nuclear proteins that are induced by T cell receptor ligation, and binding activity at one site is induced by ligation of the CD28 surface molecule (41). This finding provides a satisfying genetic correlate to the two-signal model of lymphocyte activation; that is, T cells are activated only upon the delivery of two signals that are necessary for the binding of the complement of nuclear factors required for optimal transcription of the interleukin 2 gene. CD28 ligation also increases IL-2 production by CD4$^+$ T cells by a process involving the stabilization of IL-2 mRNA (73).

T cell receptor ligation alone is sufficient to up-regulate the expression of a number of molecules on the cell surface, including the α chain of the IL-2 receptor. This leads to the appearance of a high-affinity IL-2 receptor ($K_d = 10^{-11}$M) consisting of α, β, and γ chains. A low-affinity receptor ($K_d = 10^{-8}$M) consisting of a $\beta\gamma$ heterodimer exists on resting T cells, but this receptor is incapable of binding IL-2 at physiologic concentrations. Because of this, the IL-2 that is secreted by a CD4$^+$ T cell will cause the proliferation of only those T cells which have been activated by seeing antigen.

B Cell Activation

The intracellular signaling events that follow ligation of surface immunoglobulin are remarkably similar to those seen in T cells following TcR ligation. For example, Ig ligation is followed within seconds to minutes by tyrosine phosphorylation of multiple proteins, activation of PLCγ_1 (which is again contingent upon CD45 signaling), generation of IP$_3$ and DG, increased intracellular calcium concentration, and activation of protein kinase C. These second-messenger cascades are not initiated directly by the Ig heavy chains, which have short cytoplasmic tails, but by an associated signaling complex consisting of Igα and Igβ chains. As is the case with the invariant CD3 chains of the T cell receptor complex, Igα and Igβ are both members of the Ig superfamily and possess long cytoplasmic tails containing motifs which may be responsible for the association with protein tyrosine kinases. Four members of the src family of PTKs are activated following mIg cross-linking: p53/56^{lyn}, p59^{fyn}, p56^{lck}, and p55^{blk}. A fifth protein tyrosine kinase that is activated following Ig cross-linking, p72^{syk}, is homologous to ZAP-70 and is likewise responsible for the coupling of receptor ligation to subsequent downstream events. In some cases the differences between T and B cell activation are more apparent than real. For instance, the affinity of the T cell receptor for the MHC–peptide complex is rather low, with a K_d of about 10^{-5} M, and it is believed that one function of the CD4 or CD8 co-receptor molecule (its other function is to focus T cell recognition on MHC molecules) is to increase the avidity of the T cell receptor for its ligand to a level sufficient to allow activation upon recognition of a small number of MHC–peptide complexes on APCs. The affinity of the germline-encoded Ig receptors on naive B cells is also rather low and probably insufficient to generate intracellular signaling upon binding antigens on microbial surfaces. However, a recently discovered complex expressed by B cells binds complement fragments and increases the overall avidity of B cells for antigens by about 100-fold. This complex, called

CR2-CD19-TAPA1, may act to focus naive B cell recognition on pathogens that have been coated by complement components (39).

T Cell–Dependent B Cell Activation. The features of B and T cell activation that are shared between the two cell types are all a consequence of the delivery of signal 1, the ligation of the antigen-specific receptor. Differences in the requirements for activation of the two cell types are revealed when considering the requirements for and the nature of signal 2. For CD4$^+$ T cells, signal 2 is delivered by the antigen-presenting cell. But for B cells responding to typical protein antigens, signal 2 appears to be delivered by activated but not resting CD4$^+$ helper T cells. The process of B cell activation is therefore not a simple matter of binding soluble antigen and ligation of membrane Ig (mIg) but reflects a complex cascade of cellular interactions. For this reason, the process will be described in some detail [reviewed in (102)].

After Claman et al. discovered the participation of at least two distinct cell populations in antibody formation (20), Miller and Mitchell found that T cells must participate in the formation of antibodies to sheep red blood cells (86), and the nature of their participation became a matter of considerable interest (48, 81). In 1971, N. Avrion Mitchison discovered the requirement for "linked recognition" of antigen by B and T cells for effective T–B collaboration (86a). He immunized one mouse with the dinitrophenyl hapten coupled to chicken gamma globulin (DNP-CGG) to obtain DNP-specific B cells, and a second mouse with bovine serum albumin (BSA) to obtain BSA-specific T cells. Spleen cells from both mice were then transferred to lethally irradiated recipients and the requirements for a secondary anti-DNP response were tested. Mitchison found that BSA-specific T cells could provide "help" to DNP-specific B cells in a secondary antibody response only if DNP was chemically coupled to BSA. To account for the phenomenon of linked recognition of T and B cell epitopes, the antigen was envisioned as a bridge, bringing the T and B cell close enough in solution for effective help to be delivered. However, doubt was cast on this model when it became apparent that T and B cells recognize different forms of the antigen. Rather, the discovery of MHC restriction of T–B collaboration led to the correct model—the B cell acting as an antigen-specific APC, picking up the antigen via the DNP hapten and presenting determinants on BSA to T cells. As it turns out, B cell presentation of antigen to the antigen-specific T cell is the first step of T cell–dependent B cell activation.

How do B cells act as antigen-specific antigen-presenting cells? Clearly, many determinants on proteins are monovalent, and B cells specific for these determinants cannot have their immunoglobulin receptors cross-linked by several protein molecules. Therefore, Ig receptor ligation is not necessary for the effective processing and presentation of antigens captured by surface Ig. Rather, B cells normally cycle their Ig molecules from the cell surface into endosomes and back to the cell surface (30). However, if antigen is bound, it and the Ig molecule are cleaved proteolytically in the endosome and fragments of the antigen (as well as fragments of Ig) are presented by MHC class II molecules on the surface of the B cell. At present, it is not known how "naked" Ig molecules are relatively protected from proteolytic degradation, whereas antigen-bound antibody molecules are cleaved. An additional mechanism for antigen capture by APCs, including B cells, is through pinocytosis of antigen; however, Ig-mediated endocytosis of antigen is roughly 10,000-fold more efficient than nonreceptor-mediated pinocytosis by B cells or other APCs (70). For this reason, antigens circulating in extremely low concentrations in vivo are likely to be presented exclusively by B lymphocytes.

Antigens present in multiple copies on the surface of bacteria or viruses are capable of generating Ig cross-linking. If this occurs, the biochemical events described above lead to the increased expression of several molecules involved in antigen presentation. For instance mIg cross-linking leads to the increased expression of MHC class II molecules (29), adhesion molecules such as ICAM-1 (27), and costimulatory molecules such as B7-2 (within 8 h) and B7-1 and BB-1 (by about 48 h) (42, 58, 72). When mIg cross-linking occurs, lymphokines secreted by T cells may be sufficient to induce B cells to proliferate and secrete antibody. For monovalent ligands, which do not cross-link surface Ig, proliferation and antibody production require a signal generated through T cell contact. In this case, conjugation with helper T cells is due to the presentation of specific peptides on the surface of MHC class II molecules. The cross-linking of TcR $\alpha\beta$ on the CD4$^+$ T cell by the MHC–peptide complex induces the expression by the T cell of a 39 kDa glycoprotein called gp39 or CD40 ligand (CD40L), whose receptor, CD40, is present on B cells (4). Once the helper T cell expresses CD40L, it can interact with and activate B cells independently of specific recognition of antigen. That is, activated helper T cells can deliver "noncognate" help to B cells. Ligation of CD40 up-regulates the expression of the B7 costimulatory molecule, which then provides signal 2 to the helper T cell by ligating CD28. Equally importantly, CD40 may function as signal 2 for the B cell whose Ig receptor has been ligated (135), thereby enabling proliferation and antibody secretion. There is some evidence for this notion. Following initial stimulation with antigen, B cells proliferating in germinal centers undergo programmed cell death unless they receive a signal through CD40 (76). The requirement for this signal ensures that B cell activation is tightly controled by CD4$^+$ T cells. As germinal centers are the site of somatic mutation of immunoglobulin genes (7), any autoreactive B cell arising by mutation would die due to the absence of activated T cells reactive to auto-antigens.

Cross-linking of surface Ig and CD40 may be sufficient for the induction of B cell proliferation and antibody secretion. However, surface Ig is not cross-linked on B cells responding to monovalent ligands, and CD40 ligation alone, while sufficient for B cell proliferation, is insufficient for the induction of antibody secretion. In this circumstance, IL-4 secreted by activated cells of the T$_h$2 phenotype may be sufficient to complement CD40 ligation and lead to full activation of the B cell. It is possible that signals delivered by Ig cross-linking and IL-4 secretion are also sufficient for the induction of B cell proliferation and antibody secretion, but antibody secretion is probably not induced by interaction of cells of the T$_h$1 phenotype and B cells whose mIg has not been cross-linked by antigen.

In summary, T cell–dependent B cell activation may be divided into a cognate and a noncognate phase. The cognate phase of activation includes uptake, processing, and presentation of antigen by the B cell, TcR ligation by the complex of peptide + MHC, and upregulation by the T cell of the surface expression of CD40L. The noncognate phase of activation consists of the ligation of CD40 leading to the increased expression of costimulatory molecules (especially B7-2) by the B cell, ligation of CD28 on the T cell surface, secretion of lymphokines by the T cell, and subsequent proliferation and antibody production by the B cell. For B cells responding to multivalent ligands, Ig and CD40 cross-linking may be sufficient to elicit proliferation and antibody secretion. However, B cells responding to monovalent ligands may require CD40 ligation and IL-4 for proper activation, proliferation, and antibody secretion. Whether the latter situation represents an exception to the two-signal model of lymphocyte activation (as signal 1, cross-linking of surface Ig, is not occurring) is an open issue.

Immunoglobulin Isotype Switching. B cell activation by invading microbes leads to the generation of specific memory B cells and the production of specific antibodies designed to neutralize or eliminate the pathogen. Although antigen binding is a property of the sequence of the antibody variable region, the specific effector function mediated by the antibody is determined by its constant-region isotype. In the mouse, IgM, IgG2a, and IgG2b have the greatest ability to fix complement, IgA localizes to mucosal secretions and can resist the action of proteolytic enzymes, and IgE and IgG1 preferentially activate mast cells to resist infection by nematode parasites. It is not surprising that the ability to resist infection is critically dependent upon the expression of an appropriate heavy-chain isotype. Recent evidence indicates that helper T cell–derived cytokines determine the expression of Ig isotypes by proliferating B cells. These cytokines often have mutually antagonistic effects on the isotype of Ig that is secreted. For instance, IL-4 promotes isotype switching to production of IgG1 and IgE and inhibits the production of IgG2a, whereas IFN-γ promotes the production of the IgG2a isotype and inhibits the production of the IgG1 and IgE isotypes (122). The secretion of IgG3 and IgA is generally promoted by transforming growth factor beta (TGF-β). Although many cell types are capable of secreting TGF-β, IFN-γ and IL-4 secretion are the properties of T$_h$1 and T$_h$2 cells, respectively (90).

How is Ig isotype expression controlled at the molecular level? C$_H$ genes are located 3′ of the V$_H$DJ$_H$ genes within a 200 kb region in the order 5′-μ, δ, γ_3, γ_1, γ_{2b}, γ_{2a}, ϵ, and δ-3′, and sequencing of this region reveals the presence of switch recombination sequences 5′ of each C$_H$ gene except for Cδ. Naive (unstimulated) B cells express both surface IgM and IgD. Expression of IgM is achieved by the production of a transcript extending through the membrane exon of C$_\mu$, whereas IgD expression is generated by producing a transcript including C$_\mu$ and C$_\delta$ followed by splicing the C$_\mu$ sequence out of the primary transcript. Activation of B cells is followed by proliferation in germinal centers and, under the influence of T cell–derived cytokines, by isotype switching. Lymphokines appear to influence the switching process by increasing the accessibility of a particular C$_H$ region to both transcription factors and the switch recombination machinery. Evidence for this notion has been obtained by the finding that IL-4 treatment of a switchable B cell line increases the steady-state level of C$_{\gamma 1}$ mRNA prior to the onset of switching to IgG1. The actual switch is then accomplished by the deletion of all C$_H$ genes 5′ to the one targeted for expression and juxtaposition of the rearranged V$_H$DJ$_H$ segment 5′ to the expressed C$_H$ gene.

T CELL EFFECTOR FUNCTIONS

Cytokine Secretion

T cells mediate a variety of effects not directly but through interacting with and altering the behavior of antigen-presenting cells through the action of cytokines. This is a slight variation on the strategy taken by B cells, where antigen specificity and effector function are a property of different regions of the same antibody molecule. Although cytokines are not antigen-specific molecules, they are made only by those T cells that have been activated by antigen, and they are secreted directionally so as to act preferentially on the cell presenting the antigen. Cytokines have multiple sources as well as pleiotropic and oftentimes overlapping or redundant effects, as evidenced by the overall health of mice in which individual cytokine genes have been "knocked out" by homologous recombination of embryonic stem cells. [This has already been observed for mice lacking IL-2 or IL-4 (69, 115)]. Although an exhaustive description of their biologic activities is beyond the scope of this chapter, properties of a few cytokines will be discussed.

Interleukin 2 (IL-2). IL-2 (121) is secreted primarily by activated T cells, particularly $CD4^+8^-$ T cells, and has effects on T cells, B cells, macrophages, and natural killer (NK) cells. Its principal function in vivo is to stimulate the autocrine and paracrine growth of antigen-activated T cells, leading to the clonal expansion of antigen-specific cells that is a hallmark of immunological memory. IL-2 also has an important role in promoting the differentiation of $CD4^-$ $CD8^+$ cytolytic T lymphocyte (CTL) precursors into active CTLs. Furthermore, IL-2 augments IFN-γ production, proliferation, and cytolytic capacity of NK cells. As mentioned previously, IL-2 also appears to costimulate proliferation and antibody secretion by B cells whose surface Ig has been cross-linked.

Interleukin 4 (IL-4). Interleukin 4 (124) is primarily a costimulator of the growth and differentiation of B cells, and is secreted primarily by activated $CD4^+8^-$ T cells, basophils, and mast cells. Interestingly, all these cell types express CD40 ligand and so can provide all of the essential signals to B cells required for growth and differentiation. IL-4 secreted by T cells has been shown to costimulate the proliferation of B cells stimulated by antigen, antibody to IgM, and antibody to CD40. Furthermore, IL-4 induces secretion of the IgG1 and IgE isotypes by promoting Ig switch recombination to Cγ1 and Cϵ and antagonizes the IFN-γ–stimulated switch to IgG2a. IL-4 also promotes the differentiation of naive $CD4^+$ T cells stimulated by antigen to acquire the T_h2 phenotype, to be discussed below.

Interleukin 3 (IL-3), Interleukin 5 (IL-5), and Granulocyte–Macrophage Colony-Stimulating Factor (GM-CSF). These lymphokines are mentioned together because they share several features in common, including the near exclusive secretion by activated $CD4^+$ T cells and stimulation of differentiation of hematopoietic progenitors (21). IL-5 is a potent eosinophil differentiation factor, whereas IL-3 and GM-CSF act on primitive, multipotential hematopoietic progenitor cells. Because of this, T cells play a crucial role in recruiting the production of functional white blood cells in response to infection or marrow damage due to drugs or toxins.

Interleukin 10 (IL-10) and Interleukin 12 (IL-12). These two cytokines are mentioned together because of their mutually antagonistic effects on the differentiation and action of $CD4^+$ T cells. IL-10 is made by B and T cells as well as numerous other cell types within the body. It was originally called "cytokine synthesis inhibitory factor," as it was found to inhibit the production of IFN-γ by activated T cells of the T_h1 phenotype (40). Furthermore, IL-10 has been found to inhibit the APC function of macrophages and dendritic cells by down-regulating expression of MHC class II molecules and possibly costimulatory molecules (88). The effect of IL-10 on T cells is not a direct one but is probably mediated through the inhibition of the secretion of multiple cytokines by macrophages, particularly IL-12.

IL-12 is produced by macrophages in response to infection by intracellular pathogens. In contrast to the effects of interleukins 4 and 10, IL-12 acts on activated naive T cells to promote their differentiation into cells of the T_h1 phenotype (134). IL-12 has also been found to augment the activity of NK cells and CTL and to promote the secretion of IFN-γ by lymphocytes.

Interferon gamma. T cells and NK cells are the only sources of IFN-γ (33), a cytokine whose name derives from its ability to interfere with viral replication. IFN-γ enhances microbicidal activity of macrophages by stimulating the production of hydrogen peroxide and nitrous oxides, and enhances antigen presentation to T cells by the up-regulation of MHC class II expression on monocytes, macrophages, and dendritic cells. As with IL-12, IFN-γ antagonizes the activity of IL-4 and IL-10 by promoting the generation of T_h1 T cells from naive T cell precursors. IFN-γ promotes isotype switching of B cells to the production of IgG2a, which has potent complement-fixing activity. Lastly, IFN-γ is a potent activator of natural killer and cytotoxic T cells.

Delayed-Type Hypersensitivity (DTH)

A DTH response is characterized by warmth, redness, pain, induration, and swelling occurring 2–3 days after the application of immunogenic proteins on or within the skin; this phenomenon is also known as *contact sensitivity*. This reaction is an important mechanism of defense against infections by intracellular bacteria such as *Mycobacterium* sp. and *Listeria monocytogenes*, whose name derives from the mononuclear cell (lymphocytic) infiltrates seen at sites of infection. In a DTH reaction, microbial antigens presented by macrophages activate antigen-specific CD4$^+$ T cells to secrete IFN-γ, which instructs the macrophage to augment its microbicidal activity through the generation of hydrogen peroxide and hydroxyl radicals (the "respiratory burst"). The role of lymphocytes in DTH was revealed long ago by the observation that specific DTH reactivity can be transferred to an unimmunized host by primed, antigen-specific lymphocytes.

Cytolysis

Whereas extracellular and intracellular bacteria are handled primarily by antibodies and DTH reactions, respectively, cell-mediated cytotoxicity is a crucial means of defense against infection by viruses. Although some CD4$^+$ T cells are endowed with the ability to lyse cells, cytotoxicity is primarily the function of MHC class I restricted CD8$^+$ T cells. This is as it should be, as nearly all host cells express MHC class I proteins and nearly all are susceptible to infection by some type of virus. Peptides derived from the degradation of viral proteins synthesized in the cytoplasm are, like peptides of host cell proteins, transported into the endoplasmic reticulum where they associate with MHC class I proteins destined for export to the cell surface. There, the viral peptide-MHC complex is recognized by virus-specific CD8$^+$ T cells and a chain of events is initiated that eventually leads to the destruction of the infected cell. It is also worth noting that foreign tissue grafts within a species are rejected by cell-mediated cytotoxicity.

Because of their circulation through blood and lymph only, priming of naive CD8$^+$ T cell most likely occurs in a secondary lymphoid organ by presentation of antigen on a "professional" antigen-presenting cell that has migrated from a site of infection. Naive CD8$^+$ T cells cannot lyse cells, but their activation leads to their differentiation into memory CD8$^+$ T cells and effector cytotoxic T lymphocytes (CTLs). Activated CTLs express antigens that allow their transmigration through capillaries, thereby allowing them to home to sites of infection. There they recognize and destroy cells bearing the same antigen–MHC complex as was expressed by the professional antigen-presenting cell in the lymphoid organ.

Lysis of infected cells proceeds in four phases. The first phase consists of *recognition* of the target and is mediated by the T cell receptor for antigen. Recognition is soon followed by *attachment*, whereby the activated CTL forms a tight conjugate, in effect a synapse, with the target cell. During the period of attachment, the CTL undergoes a process of polarization in which cytoplasmic lytic granules stream toward the area of contact with the target. The *lethal hit* is then delivered to the target by the exocytosis of the lytic granules across the synapse. Only some of the lytic granules are expended on one target, so killing of the target can be followed by *detachment and recycling* of the CTL.

CD8$^+$ T cells have at least two murder weapons at their disposal. Their lytic granules contain perforin, a 65 kDa protein which inserts into the target cell membrane and polymerizes to form a pore with an internal diameter of 50 Å (143). Multiple such pores in the target membrane lead to loss of membrane integrity, ion fluxes, and cell swelling. Perforin-mediated cytolysis is Ca^{2+}-dependent, yet the second method of target cell destruction can occur in the absence of extracellular calcium or in conditions under which exocytosis of granules does not occur (23). In the 1970s, immunologists observed that cells being killed by CTL underwent characteristic morphologic changes of shrinkage, chromatin condensation, zeiosis (cell boiling or blebbing), and formation of sealed membrane vesicles. This process is termed *apoptosis* and is accompanied by the activation of endogenous endonucleases which cleave the cell's DNA into oligonucleosomal fragments. Cleavage of viral DNA in infected cells has also been observed and may be why cell killing entails destruction of virus. Given that CTLs can induce apoptosis in the absence of granule exocytosis, apoptosis may be triggered by a signaling event at the cell surface. Targets expressing the Fas cell surface molecule undergo apoptosis when Fas is cross-linked by its ligand, FasL, expressed by activated CD4$^+$ or CD8$^+$ CTLs (110). Fas is highly expressed by activated lymphocytes themselves; thus, the Fas-FasL interaction may be crucial in maintaining lymphocyte homeostasis. This hypothesis is supported by the observation that mice deficient in either Fas or FasL expression have swollen lymphoid organs, excessive lymphocyte numbers, and autoimmune disease.

Given the ability of CTL to kill host cells, it is not surprising that their activity is tightly regulated. Effector CTLs are only permitted a few rounds of killing before they either settle down to memory cells or die through exhaustive differentiation. Furthermore, resting memory CTL also require second signals from CD4$^+$ T cells for activation; ligation of the T cell antigen receptor in the absence of help from CD4$^+$ T cells results in antigen-specific CTL tolerance (55). The nature of the second signal for memory CTL remains to be elucidated. It may be a lymphokine, such as IL-2, or it is possible that CD4$^+$ T cells induce the APC to express a costimulatory signal that is required for the activation of memory CD8$^+$ T cells.

Control of T Cell Effector Functions

In 1971, Parish and Liew noted that cellular and humoral immunity to a protein antigen are inversely correlated; that is, doses of antigen that provoke a vigorous antibody response elicit a weak DTH response, and doses of the same antigen that generate a strong DTH reaction elicit meager antibody formation (101). A similar phenomenon occurs in the immune response of different mouse strains to the intracellular parasite *Leishmania major*. In one strain of mouse, infection by *Leishmania* results in an antibody response that is ineffective, and the host suffers progressive skin

ulceration with eventual death of the animal. However, in a second strain infection is followed by healing associated with a vigorous DTH response to *Leishmania promastigotes* (118).

It is now clear that the different types of response to infection can be accounted for by the cytokine secretion pattern of *Leishmania*-specific CD4$^+$ T cells in the different strains. In 1986, Mosmann et al. discovered that CD4$^+$ T cell clones could be divided into the T$_h$1 vs. T$_h$2 phenotype based upon lymphokine secretion profiles (89); the T$_h$1 phenotype is distinguished by the secretion of IL-2, IFN-γ, and TNF-β, whereas T$_h$2 cells are marked by the secretion of IL-4, IL-5, and IL-10. A third phenotype, T$_h$0, is capable of secreting all of these cytokines, but secretes large amounts of IL-2 and is probably an undifferentiated precursor of T$_h$1 and T$_h$2 cells. Based upon lymphokine secretion, the major function of T$_h$1 cells is primarily the activation of

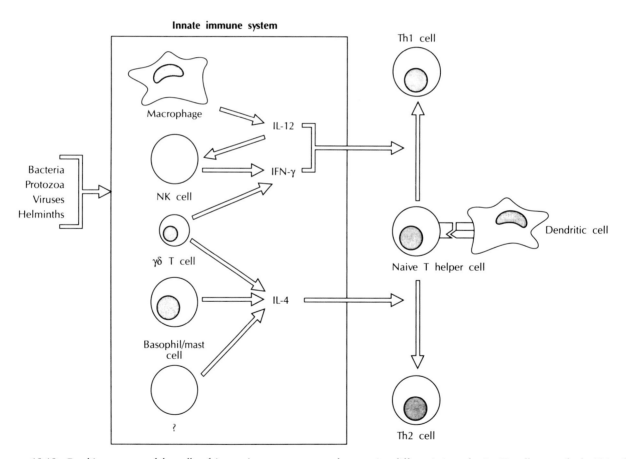

FIG. 19.18. Cytokines generated by cells of innate immune response drive CD4$^+$ T$_h$ cell subset differentiation. Pathogens and their products interact with several components of innate immune system, including macrophages, NK cells, $\gamma\delta$ T cells, and/or basophils/mast cells. Some pathogens (quite frequently bacteria, protozoa, and viruses) interact with macrophages, stimulating production of IL-12, enhancing IFN-γ production by NK cells and/or $\gamma\delta$ T cells, and promoting differentiation of naive T$_h$ cells towards the T$_h$1 cell phenotype. Other pathogens (frequently helminths) stimulate cells to produce IL-4 that drives T$_h$2 cell differentiation. The cells producing IL-4 during early stages of infection have not been defined but could include basophils or mast cells, $\gamma\delta$ T cells, or undefined cell population. All of in vitro models suggest that most efficient APCs are dendritic cells. [Figure and legend from Scott (118) with permission.]

macrophages, DTH, and help for CTL, whereas, T_h2 cells provide effective help for B cells, stimulation of IgE production and allergy, and eosinophil stimulation. It has recently been found that the ability to heal following *Leishmania* infection correlates with the production of a T_h1-type response, whereas strains that produce T_h2 responses develop ulcerative lesions and die. Furthermore, treatment of "healer" mice with antibodies to IFN-γ at the time of primary infection converts them to nonhealers (6), whereas treatment of nonhealers with antibodies to IL-4 at the time of initial infection allows recovery from infection (111). The distinction between T_h1 and T_h2 responses may also account for Parish and Liew's original observation of the reciprocal relationship between cellular and humoral immunity in response to priming.

All fine and good, but what is responsible for the generation of a T_h1 vs. a T_h2 response? This is an area of active interest, and it appears that the CD4$^+$ T cell response is determined by cytokines produced by the innate immune system (Fig. 19.18) Naive T cells primed in the presence of exogenous IL-4 differentiate to become T_h2 cells (129), whereas priming of naive T cells in the presence of IL-12 enhances the generation of T_h1 cells (134). The sources of this nonlymphocyte-derived IL-4 and IL-12 appear to be mast cells and macrophages, respectively. Also, IL-10 produced by macrophages inhibits their production of IL-12 (26). How do helminthic parasites or intracellular bacteria direct the production of IL-4 or IL-12 from mast cells and macrophages, respectively? A complete answer is not yet available.

In summary, the adaptive immune system employs distinct defense mechanisms to handle extracellular bacteria, intracellular bacteria and fungi, helminthic parasites, and viruses. Antibodies and phagocytic cells are important in fighting extracellular bacteria; intracellular bacteria and fungi are handled by DTH responses; mast cells and eosinophils are responsible for killing helminthic parasites; and CD8$^+$ CTL defend against viruses. The activity of both the adaptive and innate immune systems is oftentimes directed by CD4$^+$ T cell–derived lymphokines, whose profile is determined by cytokines produced by the innate immune system in response to initial infection.

IMMUNOLOGICAL MEMORY

One of the most important and distinctive characteristics of the adaptive immune system is its ability to respond more vigorously and effectively upon reencountering a specific pathogen; that is, the immune system displays specific memory of past antigenic in-

sults (52). The overall consequence of immunologic memory is that an individual who has survived an infection is much less likely to get sick or die upon reinfection. As I have mentioned, the clonal selection theory of acquired immunity proposes that specific immunologic memory is a consequence of the selective clonal expansion of antigen-specific antibody-forming cells. Under this model, the long-lived nature of immunologic memory is accounted for by the longevity of the memory cell, which is a direct descendent of the naive, previously unstimulated lymphocyte. As we shall see, although clonal expansion is clearly one characteristic of immunologic memory, the phenomenon of memory in both B and T cells is certainly more multifaceted, and recent developments have vastly enhanced our understanding of this complex phenomenon.

B Cells: Somatic Mutation, Affinity Maturation, and Memory

Introduction of an antigen into an unprimed animal results in the generation of a primary antibody response and the production of memory cells. Antibodies formed in the primary response bind antigen weakly and are predominantly of the IgM isotype. Re-challenge of the animal with antigen generates a secondary antibody response which has the three following characteristics: *(1)* A greater number of B cells are activated and so more antibody molecules are secreted than in the primary response; *(2)* the antibodies that are produced have higher affinity for antigen; and *(3)* the responding B cells have switched to the production of isotypes other than IgM, such as IgG, IgA, or IgE. Although the higher affinity of antibodies in the secondary response may be partly accounted for by the selective proliferation of B cells whose Ig molecules have high affinity, it is now well established that affinity maturation of the antibody response also occurs as a consequence of the somatic mutation of immunoglobulin genes and subsequent selection of high-affinity mutants. This process of mutation and selection requires the participation of helper T cells and takes place only within the germinal centers of spleen and lymph nodes (7). These structures arise as proliferating foci within lymphoid follicles in lymph nodes as early as 3–4 days after immunization, and in the spleen at 1 wk following the introduction of antigen. In addition to B and T cells, germinal centers contain follicular dendritic cells, large cells with arborizing dendrites whose function may be to hold antigens on their surface in the form of antigen–antibody complexes (130). Shortly after immunization, B cells enlarge to form germinal centroblasts, proliferate rapidly with division times of 6–7 h, and then give rise to centrocytes, smaller cells which are the immediate precursors of memory cells. During this time of rapid proliferation, and under the influence of helper T cells, B cells destined to become memory cells undergo mutations in the immunoglobulin genes at a rate of about one in 10^3 bp per division. The vast majority of the mutations that are found in memory B cells are found in the nucleotides encoding the complementarity-determining regions (CDRs), and amino acid and nucleotide sequence comparison indicates a high frequency of mutations leading to amino acid replacements, suggesting that selection by antigen occurs. The mutations occurring in the Ig genes of centroblasts have unpredictable effects and may give rise to cells whose antibody molecules bind antigen with greater or lesser affinity, or may even give rise to cells reactive to self antigens. This is where the FDC comes into play. Cells whose Ig is not ligated by FDC-bound antigen are apparently doomed to death by apoptosis, whereas Ig ligation may rescue the cell to survive on as a memory B cell. It is likely that ligation of CD40, either by its ligand on activated T cells or the FDC itself, is also required for continued survival of the B cell. Nonetheless, only 10% or so of B cells initially activated by antigen to proliferate within germinal centers survive on and join the pool of recirculating cells; these cells contain mutations in their antibody genes and constitute the pool of high-affinity memory B lymphocytes. These cells are not intrinsically long-lived but depend upon periodic restimulation by FDC-bound antigen for their continued survival (54). In this regard, it is remarkable that antigens can be found as part of immune complexes on the surface of FDCs for months to years after primary immunization, thereby providing a depot of antigen for long-term immunologic memory.

Until very recently, immunologists had assumed that immunization of naive B lymphocytes gives rise, by an asymmetric differentiation step, to IgM-secreting plasma cells of the primary response and the somatically mutated, isotype-switched memory B cells. As simple as this model is, it also appears to be wrong. Instead, strong evidence has accumulated that the precursors of memory B cells represent a lineage distinct from the primary antibody-forming cells (74). These two populations can be separated based upon the expression of the heat-stable antigen as identified by the monoclonal antibody J11d.2. Only those cells expressing low levels of the J11d antigen can form germinal centers and undergo somatic mutation, affinity maturation, and isotype switching (75). Therefore, immunization of an animal leads to the stimulation of two types of B cells—J11d^{hi} cells differentiate to form IgM-secreting plasma cells, and J11d^{lo} cells form germinal centers, proliferate rapidly, mutate, switch their isotype, and reenter the circulation as memory cells.

T Cell Memory

The generation of memory T cells is much more poorly understood, in part because at present there is no reliable way to identify them. Although priming of CD4$^+$ and CD8$^+$ T cells leads to their clonal expansion, they do not undergo somatic mutation or isotype switching, and they do not appear to be confined, as are B cells, to a specific location. However, maintenance of T cell memory also depends on continued restimulation by persisting antigen, as memory is lost when primed T cells are transferred in the absence of antigen to unprimed animals (53, 97). Once again, immune complexes on the surface of FDCs probably serve as the depot of antigen responsible for the maintenance of T cell memory. In this scenario, B cells are also critical for T cell memory—not only do they secrete the antibody that is part of the immune complexes on FDCs, but they process these complexes for presentation of antigen in the restimulation of the memory T lymphocytes (68).

IMMUNOLOGICAL TOLERANCE

One of the cardinal features of the immune system is that an animal is immunologically tolerant of its own constituents; for example, injection of a mouse with mouse and human albumin elicits antibodies against the human albumin only, and these antibodies do not cross-react with mouse albumin. Thus the immune system is said to be capable of discriminating self from nonself, ignoring all that is self and reacting to all that is nonself. Just how it accomplishes this feat has been perhaps the most vexing of all questions in immunology.

One conclusion that can be deduced a priori is that immunologic tolerance is learned rather than inherited. Consider two MHC-disparate mouse strains, A and B. As members of one strain can make antibody against the cells of the other strain, it may be concluded that individuals of strain A possess genes for anti–B antibodies, and animals of strain B possess genes for anti–A antibodies. An (A × B) F$_1$ offspring inherits both antibody genes but nonetheless is incapable of making antibody against any of its MHC proteins; therefore the tolerance has been acquired. One clue to how tolerance is generated came from Raymond Owen's discovery that dizygotic twin calves, which share a placental circulation in utero, are hematopoietic chimeras and tolerant of each other's blood cells (100).

Medawar subsequently established that such calves accept each other's skin grafts as adults. Based upon this experiment of nature, Burnet went on to propose that immunologic tolerance is learned during ontogeny and predicted that the introduction of allogeneic blood cells into the fetus would lead to immunologic tolerance and chimerism (19). This prediction was confirmed by Billingham, Brent, and Medawar, who showed that the injection of blood cells into allogeneic fetal or neonatal mice led to chimerism and specific tolerance of skin grafts taken from the blood donor (13). Following the discovery of lymphocytes and the development of the clonal selection theory, Lederberg proposed that developing lymphocytes pass through a phase in which encounter with antigen leads to inactivation only (71). This model involving the *clonal deletion* of autoreactive developing lymphocytes has now received ample experimental confirmation. In 1987, Kappler et al. (63) showed that mature T cells expressing β chains encoded by Vβ17 proliferate strongly to an endogenous superantigen presented by the MHC class II molecule, I-E. However, in strains of mice encoding both I-E and the endogenous superantigen, mature Vβ17$^+$ T cells are nearly absent. When thymocytes of these mice were examined, Kappler et al. found that Vβ17$^+$ thymocytes are present in normal numbers in the CD4$^+$ 8$^+$ double-positive (DP) population but are nearly absent in the mature, CD4$^+$ 8$^-$ and CD4$^-$ 8$^+$ T cell compartment. This provided the first evidence that autoreactive T cells are clonally deleted at the immature DP stage of development. Experiments in 1989 by Nemazee and Burki showed that a similar phenomenon occurs in B cells; in the bone marrow, B cells reactive with molecules expressed on the surface of self cells are also physically eliminated (93).

Clonal deletion of developing lymphocytes is an effective method of self-tolerance induction only for those self antigens that are adequately represented in the primary lymphoid organs. This is an efficient method for inducing tolerance to antigens expressed by cells that are present in the thymus, such as thymic epithelium or developing thymocytes themselves. Interestingly, the cells that are the most potent APCs for the activation of naive T cells, bone-marrow-derived dendritic cells, are also the most potent APCs for the induction of T cell tolerance in the thymus (84); this lends substantiation to the claim that thymocytes are indeed in an inactivatable-only state. However, it is likely that some antigens expressed in or on non-lymphoid tissues (liver, lung, kidney, brain, etc.) are not present in the bone marrow or thymus, so immature lymphocytes specific for these antigens would not be

deleted. For these cells, a second mechanism of immunologic tolerance, *clonal anergy* (*an*- without + *erg*-work), has been invoked. First proposed by Nossal and Pike as a mechanism of tolerance in B cells (95), clonal anergy is said to be induced when the antigen-specific receptors of a mature lymphocyte are cross-linked ("signal 1") in the absence of a required accessory stimulus ("signal 2") (117). This model is of course an adaptation of the two-signal model first proposed by Bretscher and Cohn to deal with autoreactive B lymphocytes that arise as a consequence of the somatic mutation of immunoglobulin genes following stimulation by antigen (16). They proposed that signal 2 is delivered by another cell specific for antigen, and there is now substantial evidence in support of the notion that for B lymphocytes, signal 2 is in fact delivered by the antigen-specific CD4$^+$ helper T cell, perhaps through intercellular contact leading to the ligation of CD40 on the B cell (94). The two-signal model was then adapted by Lafferty and Cunningham to apply to T lymphocytes (69a); in this model signal 1 is still ligation of the antigen-specific receptor, but signal 2 represents a signal delivered by the nonantigen-specific antigen presenting cell (APC). Recent experiments indicate that mature CD4$^+$ T cells are functionally inactivated following ligation of the antigen-specific receptor in the absence of APC-derived costimulatory signals, such as the B7-1 molecule found on dendritic cells (57). One cell type that may induce tolerance in mature T cells in vivo is the B lymphocyte. Resting B cells do not express significant levels of the B7 family of costimulatory molecules, and *naive* T cells specific for soluble protein and histocompatibility antigens are rendered tolerant by these antigens presented by resting B cells in vivo (38, 45). There is also mounting evidence that even activated B cells induce tolerance when presenting antigens to naive T cells. However, memory T cells are able to induce the expression of B7-2 on antigen-presenting B cells (via ligation of the CD40 molecule on the B cell) (107), so they are instead activated by antigens presented by resting or activated B cells. The second signal requirements for the activation of mature CD8$^+$ T cells are slightly less clear, although it appears that memory CD8$^+$ T cells are rendered tolerant in the absence of help provided by antigen-specific CD4$^+$ T cells (55).

The ultimate fate of an anergic lymphocyte is unclear. Theoretically, a cell that receives signal 1 in the absence of signal 2 may simply die or persist in the circulation as a functionless entity. Reversal of clonal anergy in T lymphocytes has been accomplished by cycling "anergic" cells in IL-2 in vitro, or in both B and T lymphocytes by transferring the "anergic" cells into animals in the absence of the tolerizing antigen (49, 106). These results have been cited as evidence that anergic lymphocytes are not committed to die.

Another mechanism for generating unresponsiveness in mature T cells is *T lymphocyte-mediated immunosuppression*. In 1971, Gershon and Kondo showed that mice injected with large doses of sheep red blood cells (SRBCs) were unable to make antibodies to SRBCs when given in doses known to be immunogenic (47). Furthermore, they demonstrated that thymocytes taken from these "tolerant" animals could, upon adoptive transfer, render previously unimmunized recipients unable to respond to the immunogenic dose of SRBCs. This suppression of the immune response was also found to be antigen-specific; that is, T cells from mice tolerant of SRBCs could only suppress the response to SRBCs and not to other antigens. The existence of suppressor cells was also demonstrated in model systems involving soluble and histocompatibility antigens; however, strenuous efforts to clone these cells in vitro or to identify them as a separate lymphocyte subpopulation have been uniformly unsuccessful. Because of these failures, some immunologists deny the existence of phenotypically distinct suppressor T cells and explain previous results as a consequence of altered lymphokine secretion and "immune deviation." As was discussed earlier, cellular and humoral immunity are inversely correlated, perhaps as a consequence of the selective generation of T_h1 vs. T_h2 helper T cells, respectively, but not both. According to the immune deviation model, the inability to form antibodies against SRBCs in animals given high doses is interpreted as being due to the selective generation of SRBC-specific T_h1 cells, which upon transfer to naive recipients would also direct the response of naive T cells along the T_h1 differentiation pathway. This model remains to be proven, and until then the suppressor T cell will continue to be an intriguing and elusive entity.

IMMUNOLOGICAL SELF/NONSELF DISCRIMINATION

Early experiments in serology established that an animal can make antibodies against nearly all foreign proteins while remaining unable to react against self constituents. Naturally, this led immunologists to conclude that the immune system is capable of discriminating self from nonself, ignoring all that is self and reacting to all that is nonself, and the last half century has been devoted to figuring out just how the immune system might accomplish this distinction. To understand just where we stand with respect to answering this question, it is perhaps best at this point to review the historical development of thinking in this area.

The first clue to a solution was Owen's observation that nonidentical twin calves are immunologically tolerant of each other by virtue of the intermingling of blood cells in a shared placenta. Shortly thereafter, Burnet's clonal selection theory of acquired immunity stimulated the formulation of Lederberg's theory of self/nonself discrimination, which states that immature lymphocytes pass from an inactivatable-only state during development to an activatable-only mature state in the peripheral circulation. According to this model, lymphocytes specific for self antigens are purged from the repertoire during development, and any antigen not present during development would be recognized as foreign and eliminated. It should be noted here that the theory of immunological surveillance against cancer, first proposed by Lewis Thomas (131) and championed by MacFarlane Burnet (18), was a direct consequence of this model: Tumor neoantigens expressed on the surface of cancers arising after lymphoid ontogeny would be detected as foreign and the cancer cells would be rejected.

The notion that mature lymphocytes are activatable only was rendered implausible following the discovery that immunoglobulin genes mutate following stimulation by antigen. To Bretscher and Cohn, this finding meant that a B cell initially reactive to foreign antigens could mutate so as to become autoreactive; to deal with this problem they proposed that lymphocytes must be both inducible and paralyzable at all stages of their existence. Their two-signal model of lymphocyte activation proposed that antigen-receptor ligation is an obligatory commitment step and that the decision between activation and tolerance is determined by the presence or absence of a second signal. Although a two-signal model is widely accepted by immunologists as applying to mature lymphocytes, a modern immunological synthesis incorporates features of both models of self/nonself discrimination: Developing lymphocytes pass through an inactivatable-only state on their way to being a mature cell in which they are both inducible and paralyzable.

During the past 5 years the two-signal model has gained wide acceptance and immunologists have made significant inroads into characterizing the second signals required for lymphocyte activation. B lymphocytes and memory $CD8^+$ T cells appear to rely upon $CD4^+$ T cells to deliver the second signal for their activation, whereas the activation/tolerance decision of naive $CD4^+$ T lymphocytes is determined by the presence or absence of a costimulatory signal on the APC. "Professional" APCs, such as dendritic cells, are potent stimulators of naive T cells, whereas resting and activated B cells turn off naive T cells.

Does a complete, coherent model for self/nonself

discrimination emerge from all of these details? I would maintain that, to the contrary, a priori reasoning about the evolutionary driving forces behind the development of the immune system, combined with the available experimental data, suggest that the immune system cannot and should not want to discriminate self from nonself (43, 44). Although the immune system evolved to handle dangerous pathogens, there is more to the external universe (that is, the entire set of nonself molecules) than these microbes. Some nonself entities are either harmless or even beneficial to the vertebrate organism, such as food or the commensal bacteria in the gut that synthesize essential vitamins. We wouldn't want the immune system to eliminate these commensal organisms, nutrients should be allowed to enter cells, and we certainly don't want pregnant mothers to reject the developing fetus. It is therefore reasonable to speculate that the immune system might be less concerned with what is self vs. nonself than with what is dangerous or not. Easily said, but what defines an antigen as dangerous and how does the immune system recognize it? Perhaps the best answer is that the immune system judges danger in retrospect, that is, only after a pathogen has caused tissue damage. In this model, the "professional" APC serves as the sentinel of tissue destruction, picking up the antigens released when a cell is destroyed, traveling back to the lymph node, and presenting these antigens to naive T cells. The reason for the unique activity of "professional" APCs is that they are the only cells capable of delivering signal 2 to a naive T cell. The antigens that are presented to naive T cells include those that belong to the APC and those that belong to the pathogen which initiated the damage. But since T cells specific for APC antigens have been purged from the repertoire in the thymus, the immune response is directed only to the "new" antigens—that is, to those of the agent causing tissue damage. On the other hand, a microbe that infects a nonprofessional APC and does not cause damage does not alert the immune system to its presence, because microbial antigens are not presented by the "professional" APCs required to get the immune response started. In fact, if such a microbe were to reside in exclusively B cells, I would expect tolerance to its antigens to result. Unfortunately, space does not allow a full elaboration of the theory, but the reader is referred to a current review (83).

FUTURE DIRECTIONS

Immunology is now a fully mature science, and it has benefited greatly from the participation of molecular biologists, biochemists, cell biologists, and physicians.

Vaccines have been developed against measles, mumps, rubella, polio, diphtheria, tetanus, hepatitis B, and many other diseases. Successful immune-based therapies for cancer and autoimmune diseases are elusive but always seem to be just around the corner. The clonal selection theory has proved to be a useful paradigm for understanding immunological specificity, and the discovery of somatic recombination of immunoglobulin genes completely explains the origin of the primary antibody repertoire. Biochemists and molecular biologists are just beginning to elucidate the intracellular pathways leading to lymphocyte activation and tolerance, and transgene and gene knockout technologies have proven to be powerful tools for understanding the precise functions of various molecules in vivo. Where does progress need to be made? More work needs to be done to understand the relationships between the innate and adaptive immune system that are important in determining the effector class of the immune response. Precious little is known about the mechanisms of immunologic memory, and the issue of T cell–mediated immunosuppression needs to be explored again. Very little indeed is known about how homeostasis is maintained in the immune system. How are lymphocyte numbers sensed and maintained? How does an ongoing immune response quiet down? Many of these questions can only be answered by examining the rich cellular interactions that occur in vivo. In any case, immunology will provide constant surprises and challenges for biologists and physicians for years to come.

REFERENCES

1. Abbas, A., A. Lichtman, and J. Pober. *Cellular and Molecular Immunology*. Philadelphia, PA: W. B. Saunders, 1991.
2. Allison, J. P., and W. L. Havran. The immunobiology of T cells with invariant gamma delta antigen receptors. *Annu. Rev. Immunol.* 9: 679–705, 1991.
3. Alt, F., T. Blackwell, and G. Yancopoulous. Development of the primary antibody repertoire. *Science* 238: 1079–1087, 1987.
4. Armitage, R. J., W. C. Fanslow, L. Strockbine, T. A. Sato, K. N. Clifford, B. M. Macduff, D. M. Anderson, S. D. Gimpel, T. Davis-Smith, C. R. Maliszewski, et al. Molecular and biological characterization of a murine ligand for CD40. *Nature* 357: 80–82, 1992.
5. Ashwell, J. D., C. Chen, and R. H. Schwartz. High frequency and nonrandom distriction of alloreactivity in T cell clones selected for recognition of foreign antigen in association with self class II molecules. *J. Immunol.* 136 (2): 389–395, 1986.
6. Belosevic, M., D. S. Finbloom, P. H. Van Der Meide, M. V. Slayter, and C. A. Nacy. Administration of monoclonal anti-IFN-gamma antibodies *in vivo* abrogates natural resistance of C3H/HeN mice to infection with *Leishmania major. J. Immunol.* 143: 266–274, 1989.
6a. Bentley, G. A., G. Boulot, K. Karjalainen, and R. A. Mariuzza. Crystal structure of the beta chain of a T cell antigen receptor. *Science* 267: 1984–1987, 1995.

7. Berek, C., A. Berger, and M. Apel. Maturation of the immune response in germinal centers. *Cell* 67: 1121–1129, 1991.

8. Bergstresser, P. R. Sensitization and elicitation of inflammation in contact dermatitis. In: *Immunologic Mechanisms in Cutaneous Disease,* edited by D. Norns. New York: Marcel Dekker, 1988, p. 220.

9. Berzofsky, J. A., I. J. Berkower, and S. L. Epstein. Antigen-antibody interactions and monoclonal antibodies. In: *Fundamental Immunology,* edited by W. E. Paul. New York; NY: Raven Press, 1993, p. 455.

10. Bevan, M. J. Cross-priming for a secondary cytotoxic response to minor H antigens with H-2 congenic cells which do not cross-react in the cytotoxic assay. *J. Exp. Med.* 143: 1283–1288, 1976.

11. Bevan, M. J. In a radiation chimaera, host H-2 antigens determine immune responsiveness of donor cytotoxic cells. *Nature* 269: 417–418, 1977.

12. Bevan, M. J., R. E. Langman, and M. Cohn. H-2 antigen-specific cytotoxic T cells induced by Concanavalin A: estimation of their relative frequency. *Eur. J. Immunol.* 6(3): 150–156, 1976.

13. Billingham, R. E., L. Brent, and P. B. Medawar. Actively acquired tolerance of foreign cells. *Nature* 172: 603–606, 1953.

14. Bjorkman, P. J., M. A. Saper, B. Samraoui, W. S. Bennett, J. L. Strominger, and D. C. Wiley. Structure of the human class I histocompatibility antigen, HLA-A2. *Nature* 329: 506, 1987.

15. Borgulya, P., H. Kishi, Y. Uematsu, and H. von Boehmer. Exclusion and inclusion of alpha and beta T cell receptor alleles. *Cell* 69: 529–537, 1992.

16. Bretscher, P., and M. Cohn. A theory of self-nonself discrimination: paralysis and induction involve the recognition of one and two determinants on an antigen, respectively. *Science* 169: 1042–1049, 1970.

17. Burnet, F. M. *The Clonal Selection Theory of Acquired Immunity.* Cambridge: Cambridge University Press, 1959.

18. Burnet, F. M. The concept of immunological surveillance. *Prog. Exp. Tumor Res.* 13: 1–27, 1970.

19. Burnet, F. M., and F. Fenner. *The Production of Antibodies* (2nd ed.). London: Macmillan, 1949, pp 102–105.

20. Claman, H. N., E. A. Chaperon, and R. F. Triplett. Thymus-marrow cell combination. Synergism in antibody production. *Proc. Soc. Exp. Biol. Med.* 122: 1167, 1966.

21. Clark, S. C., and R. Kamen. The human hematopoietic colony-stimulating factors. *Science* 236: 1229, 1992.

22. Clipstone, N. A., and G. R. Crabtree. Identification of calcineurin as a key signalling enzyme in T-lymphocyte activation. *Nature* 357: 695–697, 1992.

23. Cohen, J. J., R. C. Duke, V. A. Fadok, and K. S. Sellins. Apoptosis and programmed cell death in immunity. *Annu. Rev. Immunol.* 10: 267–293, 1992.

24. Crabtree, G. R. Contingent genetic regulatory events in T lymphocyte activation. *Science* 243: 355–361, 1989.

25. Cresswell, P. Chemistry and functional role of the invariant chain. *Curr. Opin. Immunol.* 4: 87–92, 1992.

26. D'Andrea, A., M. Aste-Amezaga, N. M. Valiante, X. Ma, M. Kubin, and G. Trinchieri. Interleukin-10 (IL-10) inhibits human lymphocyte interferon gamma-production by suppressing natural killer cell stimulatory factor/IL-12 synthesis in accessory cells. *J. Exp. Med.* 178: 1041–1048, 1993.

27. Dang, L. H., and K. L. Rock. Stimulation of B lymphocytes through surface Ig receptors induces LFA-1 and ICAM-1 dependent adhesion. *J. Immunol.* 146: 3273–3279, 1991.

28. Darnell, J., H. Lodish, and D. Baltimore. (Eds) *Molecular Cell Biology.* New York: Scientific American Press, 1990.

29. Davidson, H. W., P. A. Reid, A. Lanzavecchia, and C. Watts. Processed antigen binds to newly synthesized MHC class II molecules in antigen-specific B lymphocytes. *Cell* 67: 105–116, 1991.

30. Davidson, H. W., M. A. West, and C. Watts. Endocytosis, intracellular trafficking, and processing of membrane IgG and monovalent antigen/membrane IgG complexes in B lymphocytes. *J. Immunol.* 144: 4101–4109, 1990.

31. Davis, C. B., N. Killeen, M. E. Crooks, D. Raulet, and D. R. Littman. Evidence for a stochastic mechanism in the differentiation of mature subsets of T lymphocytes. *Cell* 73: 237–247, 1993.

32. Davis, M. M., and P. J. Bjorkman. T-cell antigen receptor genes and T-cell recognition. *Nature* 334(6181): 395–402, 1988.

33. De Maeyer, E., and J. De Maeyer-Guignard. Interferon-gamma. *Curr. Opin. Immunol.* 4: 321–326, 1992.

34. Dreyer, W. J., and J. C. Bennett. The molecular basis of antibody formation. *Proc. Natl. Acad. Sci. U.S.A.* 54: 864–869, 1965.

35. Edelman, G. M., B. A. Cunningham, W. E. Gall, P. D. Gottlieb, U. Rutishauser, and M. J. Waxdal. The covalent structure of an entire γG immunoglobulin molecule. *Proc. Natl. Acad. Sci. U.S.A.* 63: 78–85, 1969.

36. Edmundson, A. B., K. R. Ely, E. E. Abola, M. Schiffer, and N. Panagiotopoulous. Rotational allomorphism and divergent evolution of domains in immunoglobulin light chains. *Biochemistry* 14: 3953–3961, 1975.

37. Eisen, H. N., and G. W. Siskind. Variations in affinities of antibodies during the immune response. *Biochemistry* 3: 996–1008, 1964.

38. Eynon, E. E., and D. C. Parker. Small B cells as antigen-presenting cells in the induction of tolerance to soluble protein antigens. *J. Exp. Med.* 175: 131–138, 1992.

39. Fearon, D. T. The CD19-CR2-TAPA-1 complex, CD45 and signaling by the antigen receptor of B lymphocytes. *Curr. Opin. Immunol.* 5: 341–348, 1993.

39a. Fields, B. A., B. Ober, E. L. Malchiodi, M. I. Lebedeva, B. C. Braden, X. Ysern, J. K. Kim, X. Shao, E. S. Ward, and R. A. Mariuzza. Crystal structure of the V alpha domain of a T cell antigen receptor. *Science* 270: 1821–1824, 1995.

40. Fiorentino, D. F., M. W. Bond, and T. R. Mosmann. Two types of mouse T helper cell. IV. T_h2 clones secrete a factor that inhibits cytokine production by T_h1 clones. *J. Exp. Med.* 170: 2081–2095, 1989.

41. Fraser, J. D., B. A. Irving, G. R. Crabtree, and A. Weiss. Regulation of interleukin-2 gene enhancer activity by the T cell accessory molecule CD28. *Science* 251: 313–316, 1991.

42. Freeman, G. J., F. Borriello, R. J. Hodes, H. Reiser, K. S. Hathcock, G. Laszlo, A. J. McKnight, J. Kim, L. Du, D. B. Lombard, G. S. Gray, L. M. Nadler, and A. H. Sharpe. Uncovering of functional alternative CTLA-4 counterreceptor in B7-deficient mice. *Science* 262: 907–909, 1993.

43. Fuchs E. Two signal model of lymphocyte activation [letter; comment]. *Immunol. Today* 13: 462, 1992.

44. Fuchs, E. Two signal models of lymphocyte activation [reply]. *Immunol. Today* 14(5): 235–237, 1993.

45. Fuchs, E. J., and P. Matzinger. B cells turn off virgin but not memory T cells. *Science* 258: 1156–1159, 1992.

46. Germain, R. N. Antigen processing and presentation. In: *Fundamental Immunology,* edited by W. E. Paul. New York: Raven Press, 1993, p. 639.

47. Gershon, R. K., and K. Kondo. Infectious immunological tolerance. *Immunology* 21: 903–914, 1971.

48. Good, R. A., A. O. Dalmasso, C. Martinez, O. K. Archer, J. C.

Pierce, and B. W. Papermaster. The role of the thymus in development of immunologic capacity in rabbits and mice. *J. Exp. Med.* 116: 773–796, 1962.

49. Goodnow, C. C., R. Brink, and E. Adams. Breakdown of self-tolerance in anergic B lymphocytes. *Nature* 352: 532–536, 1991.

50. Goodnow, C. C., J. Crosbie, S. Adelstein, T. B. Lavoie, S. J. Smith-Gill, R. A. Brink, H. Pritchard-Briscoe, J. S. Wotherspoon, R. H. Loblay, K. Raphael, et al. Altered immunoglobulin expression and functional silencing of self-reactive B lymphocytes in transgenic mice. *Nature* 334(6184): 676–682, 1988.

51. Gorer, P. A. The detection of antigenic differences in mouse erythrocytes by the employment of immune sera. *Br. J. Exp. Pathol.* 17: 42–50, 1936.

52. Gray, D. Immunological memory. *Annu. Rev. Immunol.* 11: 49–77, 1993.

53. Gray, D., and P. Matzinger. T cell memory is short-lived in the absence of antigen. *J. Exp. Med.* 174: 969–974, 1991.

54. Gray, D., and H. Skarvall. B-cell memory is short-lived in the absence of antigen. *Nature* 336: 70–73, 1988.

55. Guerder, S., and P. Matzinger. A fail-safe mechanism for maintaining self-tolerance. *J. Exp. Med.* 176: 553–564, 1992.

56. Hansen, T. H., B. M. Carreno, and D. H. Sachs. The major histocompatibility complex. In: *Fundamental Immunology,* edited by W. E. Paul. New York: Raven Press, 1993, p. 590.

57. Harding, F. A., J. G. McArthur, J. A. Gross, D. H. Raulet, and J. P. Allison. CD28-mediated signalling co-stimulates murine T cells and prevents induction of anergy in T-cell clones. *Nature* 356: 607–609, 1992.

58. Hathcock, K. S., G. Laszlo, H. B. Dickler, J. Bradshaw, P. Linsley, and R. J. Hodes. Identification of a CTLA-4 ligand costimulatory for T cell activation. *Science* 262: 905–907, 1993.

59. Hozumi, N., and S. Tonegawa. Evidence for somatic rearrangement of immunoglobulin genes coding for variable and constant regions. *Proc. Natl. Acad. Sci. U.S.A.* 73(10): 3628–3632, 1976.

60. Hunkapiller, T., and L. Hood. Diversity of the immunoglobulin superfamily. *Adv. Immunol.* 44: 1–63, 1989.

60a. Ikuta, K., T. Kina, I. MacNeil, N. Uchida, B. Peault, Y. H. Chien, and I. L. Weissman. A developmental switch in thymic lymphocyte maturation potential occurs at the level of hematopoietic stem cells. *Cell* 62:863–874, 1990.

61. Jerne, N. K. The somatic generation of immune recognition. *Eur. J. Immunol.* 1: 1, 1971.

62. Jerne, N. K. Towards a network theory of the immune response. *Ann. Immunol. (Paris)* 125C: 373–389, 1974.

63. Kappler, J. W., N. Roehm, and P. Marrack. T cell tolerance by clonal elimination in the thymus. *Cell* 49: 273–280, 1987.

64. Kishi, H., P. Borgulya, B. Scott, K. Karjalainen, A. Traunecker, J. Kaufman, and H. von Boehmer. Surface expression of the beta T cell receptor (TcR) chain in the absence of other TcR or CD3 proteins on immature T cells. *EMBO J.* 10: 93–100, 1991.

65. Kitamura, D., A. Kudo, S. Schaal, W. Muller, F. Melchers, and K. Rajewsky. A critical role of lambda 5 protein in B cell development. *Cell* 69(5): 823–831, 1992.

66. Kitamura, D., J. Roes, R. Kuhn, and K. Rajewsky. A B cell-deficient mouse by targeted disruption of the membrane exon of the immunoglobulin mu chain gene. *Nature* 350(6317): 423–426, 1991.

67. Kohler, G., and C. Milstein. Continuous cultures of fused cells secreting antibody of predefined specificity. *Nature* 256: 495–497, 1975.

68. Kosco, M. H., A. K. Szakal, and J. G. Tew. *In vivo* obtained antigen presented by germinal center B cells to T cells in vitro. *J. Immunol.* 140: 354–360, 1988.

69. Kuhn, R., K. Rajewsky, and W. Muller. Generation and analysis of interleukin-4 deficient mice. *Science* 254: 707–710, 1991.

69a. Lafferty, K. J. and A. J. Cunningham. A new analysis of allogeneic interactions. *Aust. J. Exp. Biol. Med. Sci.* 53: 27–42, 1975.

70. Lanzavecchia, A. Antigen-specific interaction between T and B cells. *Nature* 314: 517–519, 1985.

71. Lederberg, J. Genes and antibodies: do antigens bear instructions for antibody specificity or do they select cell lines that arise by mutation? *Science* 129: 1649–1653, 1959.

72. Lenschow, D. J., G. Huei-Ting Su, L. A. Zuckerman, N. Nabavi, C. L. Jellis, G. S. Gray, J. Miller, and J. A. Bluestone. Expression and functional significance of a second ligand for CTLA-4. *Proc. Natl. Acad. Sci. U.S.A.* 90: 11054–11058, 1993.

73. Lindsten, T., C. H. June, J. A. Ledbetter, G. Stella, and C. B. Thompson. Regulation of lymphokine messenger RNA stability by a surface-mediated T cell activation pathway. *Science* 244: 339–343, 1989.

74. Linton, P. J., D. J. Decker, and N. R. Klinman. Primary antibody-forming cells and secondary B cells are generated from separate precursor cell subpopulations. *Cell* 59: 1049–1059, 1989.

75. Linton P. J., D. Lo, and L. Lai. Among naive precursor cell subpopulations only progenitors of memory B cells originate germinal centers. *Eur. J. Immunol.* 22: 1293–1297, 1992.

76. Liu, Y. J., D. E. Joshua, G. T. Williams, C. A. Smith, J. Gordon, and I. C. MacLennan. Mechanism of antigen-driven selection in germinal centres. *Nature* 342: 929–931, 1989.

77. Mackay, C. R., W. L. Marston, and L. Dudler. Naive and memory T cells show distinct pathways of lymphocyte recirculation. *J. Exp. Med.* 171: 801–817, 1990.

78. Marrack, P., and J. Kappler. The T cell receptor. *Science* 238: 1073–1079, 1987.

79. Marrack, P., G. M. Winslow, Y. Choi, M. Scherer, A. Pullen, J. White, and S. Kappler. The bacterial and mouse mammary tumor virus superantigens: two different families of proteins with the same functions. *Immunol. Rev.* 131: 79–92, 1993.

80. Mathieson, B. J., and B. J. Fowlkes. Cell surface antigen expression on thymocytes: development and phenotypic differentiation of intrathymic subsets. *Immunol. Rev.* 82: 141–173, 1984.

81. Matsumura, M., D. H. Fremont, P. Peterson, and I. A. Wilson. Emerging principles for the recognition of peptide antigens by MHC class I molecules. *Science* 257: 927–934, 1992.

82. Matzinger, P. Why positive selection? *Immunol. Rev.* 135: 81–118, 1993.

83. Matzinger, P. Tolerance, danger, and the extended family. *Annu. Rev. Immunol* 12: 991–1045, 1994.

84. Matzinger, P., and S. Guerder. Does T cell tolerance require a dedicated antigen-presenting cell? *Nature* 338: 74–76, 1989.

85. Miller, J.F.A.P. Immunological function of the thymus. *Lancet* 2: 748–749, 1961.

86. Miller, J.F.A.P., and G. F. Mitchell. Cell to cell interaction in the immune response. I. Hemolysin-forming cells in neonatally thymectomized mice reconstituted with thymus or thoracic duct lymphocytes. *J. Exp. Med.* 128: 801–820, 1968.

86a. Mitchison, N. A. The carrier effect in the secondary response to hapten-protein conjugates. II. Cellular cooperation. *Eur. J. Immunol.* 1: 18–27, 1971.

87. Mombaerts, P., J. Iacomini, R. S. Johnson, K. Herrup, S. Tonegawa, and V. E. Papaioannous. RAG-1-deficient mice have no mature B and T lymphocytes. *Cell* 68: 869–877, 1992.

88. Moore, K. W., A. O'Garra, R. de Waal Malefyt, P. Vieira, and

T. R. Mosmann. Interleukin-10. *Annu. Rev. Immunol.* 11: 165–190, 1993.

89. Mosmann, T. R., H. Cherwinski, M. W. Bond, M. A. Giedlin, and R. L. Coffman. Two types of murine helper T cell clone. I. Definition according to profiles of lymphokine activities and secreted proteins. *J. Immunol.* 136: 2348–2357, 1986.

90. Mosmann, T. R., and R. L. Coffman. T_h1 and T_h2 cells: different patterns of lymphokine secretion lead to different functional properties. *Annu. Rev. Immunol.* 7: 145, 1989.

91. Mueller, D. L., M. K., Jenkins, and R. H. Schwartz. Clonal expansion versus functional clonal inactivation: a costimulatory signaling pathway determines the outcome of T cell antigen receptor occupancy. *Annu. Rev. Immunol.* 7: 445–480, 1989.

92. Murphy, K. M., A. B. Heimberger, and D. Y. Loh. Induction by antigen of intrathymic apoptosis of $CD4^+$ $CD8^+$ TcR^{lo} thymocytes in vivo. *Science* 250: 1720–1723, 1990.

93. Nemazee, D. A., and K. Burki. Clonal deletion of B lymphocytes in a transgenic mouse bearing anti-MHC class I antibody genes. *Nature* 337(6207): 562–566, 1989.

94. Noelle, R. J., J. A. Ledbetter, and A. Aruffo. CD40 and its ligand, an essential ligand-receptor pair for thymus-dependent B-cell activation. *Immunol. Today.* 13: 431–433, 1992.

95. Nossal, G.J.V., and B. L. Pike. Clonal anergy: persistence in tolerant mice of antigen-binding B lymphocytes incapable of responding to antigen or mitogen. *Proc. Natl. Acad. Sci. U.S.A.* 77: 1602–1606, 1980.

96. O'Keefe, S. J., J. Tamura, R. L. Kincaid, M. J. Tocci, and E. A. O'Neill. FK-506- and CsA-sensitive activation of the interleukin-2 promoter by calcineurin. *Nature* 357: 692–694, 1992.

97. Oehen, S., H. Waldner, R. M. Kundig, H. Hengartner, and R. M. Zinkernagel. Antivirally protective cytotoxic T cell memory to lymphocytic choriomeningitis virus is governed by persisting antigen. *J. Exp. Med.* 176: 1273–1281, 1992.

98. Oettinger, M. A., D. G. Schatz, C. Gorka, and D. Baltimore. RAG-1 and RAG-2, adjacent genes that synergistically activate V(D)J recombination. *Science* 248: 1517–1523, 1990.

99. Opstelten, D., and D. G. Osmond. Pre-B cells in the bone marrow: immunofluorescence strathmokinetic studies of the proliferation of cytoplasmic I-chain bearing cells in normal mice. *J. Immunol.* 131: 2635–2640, 1983.

100. Owen, R. D. Immunogenetic consequences of vascular anastomoses between bovine twins. *Science* 102: 400–401, 1945.

101. Parish, C. R., and F. Y. Liew. Immune response to chemically modified flagellin. 3. Enhanced cell-mediated immunity during high and low zone antibody tolerance to flagellin. *J. Exp. Med.* 135: 298–311, 1972.

102. Parker, D. C. T cell-dependent B cell activation. *Annu. Rev. Immunol.* 11: 331–360, 1993.

102a. Parks, D. R., L. A. Herzenberg, and L. A. Herzenberg. Flow cytometry and fluorescence Activated Cell Sorting In: *Fundamental Immunology,* edited by W. E. Paul. New York: Raven Press, 1993, p. 783.

103. Paul, W. E. Development and function of lymphocytes. In: *Inflammation,* edited by J. Gallin, I. Goldstein, and R. Snyderman. New York: Raven Press, 1992, p. 776.

104. Paul, W. E. (Ed) *Fundamental Immunology.* New York, NY: Raven Press, 1993.

105. Porter, R. R. Separation and isolation of fractions of rabbit g-globulin containing the antibody and antigenic combining sites. *Nature* 182: 670–671, 1958.

106. Ramsdell, F., and B. J. Fowlkes. Maintenance of *in vivo* toler-

ance by persistence of antigen. *Science* 257(5073): 1130–1134, 1992.

107. Ranheim, E. A., and T. J. Kipps. Activated T cells induce expression of B7/BB1 on normal or leukemic B cells through a CD40-dependent signal. *J. Exp. Med.* 177: 925–935, 1993.

108. Robey, E. A., B. J. Fowlkes, J. W. Gordon, D. Kioussis, H. von Boehmer, F. Ramsdell, and R. Axel. Thymic selection in CD8 transgenic mice supports an instructive model for commitment to a CD4 or CD8 lineage. *Cell* 64: 99–107, 1991.

109. Rolink, A., and F. Melchers. B lymphopoiesis in the mouse. *Adv. Immunol.* 53: 123–156, 1992.

110. Rouvier, E., M.-F. Luciani, and P. Golstein. Fas involvement in Ca^{2+}-independent T cell-mediated cytotoxicity. *J. Exp. Med.* 177: 195–200, 1993.

111. Sadick, M. D., F. P. Heinzel, B. J. Hoaday, R. T. Pu, R. S. Dawkins, and R. M. Locksley. Cure of murine leishmaniasis with anti-interleukin-4 monoclonal antibody. Evidence for a T cell-dependent, interferon gamma-independent mechanism. *J. Exp. Med.* 171: 115–127, 1990.

112. Samelson, L. E., H. B. Harford, and R. D. Klausner. Identification of the components of the murine T cell antigen receptor complex. *Cell* 43: 223–231, 1985.

113. Schatz, D. G., M. A. Oettinger, and D. Baltimore. The V(D)J recombination activating gene, RAG-1. *Cell* 59(6): 1035–1048, 1989.

114. Schatz, D. G., M. A. Oettinger, and M. S. Schlissel. V(D)J recombination: molecular biology and regulation. *Annu. Rev. Immunol.* 10: 359–383, 1992.

115. Schorle, H., T. Holtschke, T. Hunig, A. Schimpl, and I. Horak. Development and function of T cells in mice rendered interleukin-2 deficient by gene targeting. *Nature* 352: 621–624, 1991.

116. Schwartz, R. H. Costimulation of T lymphocytes: the role of CD28, CTLA-4, and B7/BB1 in interleukin-2 production and immunotherapy. *Cell* 71: 1065–1068, 1992.

117. Schwartz, R. H. T cell anergy. *Sci. Am.* 269: 62–63, 66–71, 1993.

118. Scott, P. T cell subsets and T cell antigens in protective immunity against experimental leishmaniasis. *Curr. Top. Microbiol. Immunol.* 155: 35–52, 1990.

119. Shepherd, J. C., T. N. Schumacher, P. G. Ashton-Rickardt, S. Imaeda, H. L. Ploegh, C. A. Jr. Janeway, and S. Tonegawa. TAP1-dependent peptide translocation in vitro is ATP dependent and peptide selective. *Cell* 74: 577–584, 1993.

120. Shinkai, Y., G. Rathbun, K. P. Lam, et al. RAG-2-deficient mice lack mature lymphocytes owing to inability to initiate V(D)J rearrangement. *Cell* 68: 855–867, 1992.

121. Smith, K. A. Interleukin-2: inception, impact, and implications. *Science* 240: 1169, 1988.

122. Snapper, C. M., and W. E. Paul. Interferon-gamma and B cell stimulatory factor-1 reciprocally regulate Ig isotype production. *Science* 236: 944–947, 1987.

123. Snell, G. D., and J. H. Stimpfling. Genetics of tissue transplantation. In: *Biology of the Laboratory Mouse,* edited by E. Green. New York: McGraw-Hill, 1966, p. 457.

124. Spits, H. (Ed.) *IL-4: Structure and Function.* Boca Raton, FL: CRC Press, 1992.

125. Sprent, J. The thymus and T cell tolerance. *Today's Life Sci.* 3: 14–20, 1991.

126. Sprent, J., D. Lo, E.-K. Gao, and Y. Ron. T cell selection in the thymus. *Immunol. Rev.* 101: 173–190, 1988.

127. Sprent, J., M. Schaefer, M. Hurd, C. D. Surh, and Y. Ron.

Mature murine B and T cells transferred to SCID mice can survive indefinitely and many maintain a virgin phenotype. *J. Exp. Med.* 174: 717–728, 1991.

128. Sprent, J., and S. R. Webb. Function and specificity of T cell subsets in the mouse. *Adv. Immunol.* 41: 39–133, 1987.

129. Swain, S. L., A. D. Weinberg, M. English, and G. Huston. IL-4 directs the development of T_h2-like helper effectors. *J. Immunol.* 145: 3796–3806, 1990.

130. Tew, J. G., M. H. Kosco, G. F. Burton, and A. K. Szakal. Follicular dendritic cells as accessory cells. *Immunol. Rev.* 117: 185–211, 1990.

131. Thomas, L. In: *Discussion of Cellular and Humoral Aspects of the Hypersensitive States,* edited by H. S. Lawrence. New York: Hoeber-Harper, 1959, p. 529–532.

132. Till, J. E. and E. A. McCulloch. A direct measurement of the radiation sensitivity of normal mouse bone marrow cells. *Radiat. Res.* 14: 213–222, 1961.

133. Tonegawa, S. Somatic generation of antibody diversity. *Nature* 302: 575–581, 1983.

133a. Tough, D. F., and J. Sprent. Turnover of naive- and memory-phenotype T cells. *J. Exp. Med.* 179: 1127–1135, 1994.

134. Trinchieri, G. Interleukin-12 and its role in the generation of T_h1 cells. *Immunol. Today* 14: 335–338, 1993.

135. Tsubata, T., J. Wu, and T. Honjo. B-cell apoptosis induced by antigen-receptor crosslinking is blocked by a T-cell signal through CD40. *Nature* 364: 645–648, 1993.

136. Turka, L. A., D. G. Schatz, M. A. Oettinger, J. J. Chun, C. Gorka, K. Lee, W. T. McCormack, and C. B. Thompson. Thymocyte expression of RAG-1 and RAG-2: termina-tion by T cell receptor cross-linking. *Science* 253: 778–781, 1991.

137. van Meerwijk, J. P., and R. N. Germain. Development of mature CD8 + thymocytes: selection rather than instruction? *Science* 261: 911–915, 1993.

138. von Boehmer, H., and K. Hafen. The life span of naive alpha/beta T cells in secondary lymphoid organs. *J. Exp. Med.* 177: 891–896, 1993.

139. Wasserman, R. L., and J. D. Capra. Immunoglobulins. In: *The Glycoconjugates,* edited by M. I. Horowitz, and W. Pigman. Orlando, FL: Academic Press, 1977, p. 323–348.

140. Weiss, A., and J. Imboden. Cell surface molecules and early events involved in human T lymphocyte activation. *Adv. Immunol.* 41: 1–38, 1987.

141. White, J., A. Herman, A. M. Pullen, R. Kubo, J. W. Kappler, and P. Marrack. The V beta-specific superantigen staphylococcal enterotoxin B: stimulation of mature T cells and clonal deletion in neonatal mice. *Cell* 56: 27–35, 1989.

142. Wu, T. T., and E. A. Kabat. An analysis of the sequences of the variable regions of Bence-Jones proteins and myeloma light chains and their implications for antibody complementarity. *J. Exp. Med.* 132: 211–250, 1970.

143. Yagita, H., M. Nakata, A. Kawasaki, Y. Shinkai, and K. Okumura. Role of perforin in lymphocyte-mediated cytolysis. *Adv. Immunol.* 51: 215–242, 1992.

144. Zinkernagel, R. M., and P. C. Doherty. Restriction of *in vitro* T cell-mediated cytotoxicity in lymphocytic choriomeningitis within a syngeneic or semiallogeneic system. *Nature* 248: 701–702, 1974.

20. Cilia and related microtubular arrays in the eukaryotic cell

PETER SATIR | *Department of Anatomy and Structural Biology, Albert Einstein College of Medicine, Bronx, New York*

Cell Biology of Microtubules
   Microtubules and the cytoskeleton
   Structure, assembly, and disassembly
   Stability and pattern formation
   Primary cilia
Evolution of Centrioles, Cilia, and Microtubules
Motor Molecules of Cilia and Microtubule-Based Arrays
   Dynein- and microtubule-based motility
   Axonemal dynein
   Cytoplasmic dynein
   Kinesin
   The kinesin superfamily
   Interactions between dynein and kinesin
   Other putative motors
Motile Cilia and the Mechanism of Motility
   Introduction to ciliary structure
   Experimental analysis of motility
      The mechanochemistry of sliding
      The switch point hypothesis
   Cellular control of ciliary motility

MICROTUBULE-BASED ORGANELLES such as the cilium have played a definitive role in cell biology. The cilium was perhaps the first organelle to be clearly differentiated in a single cell, by Leeuwenhoek in 1665. Ciliary metachronism was seen in the mid-19th century by Purkinje and Valentine, when intensive cytological study of the organelle began. That era culminated in the studies summarized by Gray (67). By that time, Ballowitz (10) and Dellinger (41) had shown that each cilium is composed of a number of "fibrils." Similarly, cilia were among the first organelles carefully delineated by electron microscopy by Manton (117), Fawcett and Porter (45), Afzelius (1), Gibbons and Grimstone (57), and Gibbons (51). The fibrils seen by light microscopy proved to be microtubules arranged in the now-classic 9+2 pattern: nine doublet microtubules surrounding a central pair of simple microtubules. Porter (144) confirmed that the motile cilium was closely related to nonmotile structures such as the outer segment of the rod, and that the ciliary basal body and the centriole were virtually identical, theories that arose during light microscopic observation. The term *microtubule* was introduced by Slautterback (182). When glutaraldehyde (153) and room temperature fixation replaced fixation solely by osmium tetroxide on ice, it soon became apparent that microtubules were normal components of eukaryotic cell cytoplasm (145) and that certain microtubules were easily disassembled, while others, including the ciliary microtubules, were quite stable (15). Tilney et al. (206) showed that even stable arrays such as the heliozoan axopod could be depolymerized, by pressure for example, and that they could reassemble. Among the prominent microtubule arrays with dynamic instability is the mitotic apparatus (88). About this time, Borisy and Taylor (24) showed that a colchicine-binding protein that could be extracted from sea urchin eggs, from isolated mitotic apparatuses, and also from brain was the subunit of the microtubule. Mohri (120) suggested the name *tubulin* for this protein.

An extensive literature exists on the biology of microtubules and microtubule-related organelles, particularly the cilium and the mitotic apparatus, including more detailed reviews of the history of discovery in this field. The reader is referred to Dustin (43), Soifer (188, 189), Roberts and Hyams (151), Sleigh (183, 184), Gibbons (53), and Inoue (87) for further details.

CELL BIOLOGY OF MICROTUBULES

Microtubules and the Cytoskeleton

Porter (144) used the term *cytoskeleton* in connection with microtubule-based arrays in the cell cytoplasm. It was inevitable that the term would become more broadly descriptive of fibrous elements and their connections in the cytoplasmic matrix when operational definitions were provided. The cytoskeleton can be defined as the residue of elements derived from the cytoplasmic matrix remaining in the cell after extensive detergent treatment and washing to remove membranes and soluble components [*cf.* (46)]. Since microtubules are readily depolymerized, care must be taken to preserve their organization upon detergent treatment.

The main structural elements of the cytoskeleton are actin microfilaments, intermediate filaments, and microtubules. They are interconnected or otherwise controlled by a large number of proteins with specific functions [cf. (166)]; a number of such proteins that interact specifically with microtubules are known as *microtubule-associated proteins* or *MAPs*, a term introduced by Sloboda et al. (186). Although many different MAPs have been described, the best known, most abundant, and most readily isolated of the several high molecular weight MAPs associated with brain microtubules or tissue culture cells are MAP1 and MAP2. Even these are heterogeneous categories, MAP1 consisting of at least three proteins (MAP 1A,B,C) and MAP2 of at least 2 proteins (MAP 2A,B) (218). In addition, there is a second class of lower molecular weight (M_r 53–67 kDa) MAPs, first identified by Kirschner's group (227), known as *tau proteins;* this also is a heterogeneous category. Tau proteins are abundant in the tangles that characterize Alzheimer's disease (101). General reviews of MAP characteristics are found in Olmsted (131, 132) and Amos and Amos (5).

The microtubule-associated proteins as a group control microtubule assembly and length, stability and self-association, bundling and/or association with other cellular elements, especially intermediate filaments. In tissue culture cells, intermediate filaments are spun out along microtubules by attachments to MAPs, or by kinesin (70). The intermediate filament network collapses when such cells are treated with colchicine, only to reform as microtubules repolymerize (181). One spectacular example of microtubule displays is found in nerve cells, where microtubules intermingled with intermediate filaments (neurofilaments) form the structural cores of axons and dendrites. The MAP lattice helps determine the spacing between adjacent microtubules or between microtubules and intermediate filaments in such cells. MAP2 and its homologs may act to connect microtubules and intermediate filaments (219).

In vitro, MAPs have binding domains that interact with microtubules at specific points along the tubulin lattice. Microtubules saturated with MAP2, for instance, are heavily decorated with long filamentous projections at an average spacing of 32 nm. The MAP2 molecule is L-shaped; the short foot of the L interacts with the microtubule, while the remainder forms the projection. A MAP2 superlattice has been superimposed on the microtubule (4).

The species of MAPs, taus, and microtubule motors are clearly cell specific and in large measure these determine certain characteristics of the microtubule array in that cell. There are several general patterns of microtubule organization, dependent on cell type and cell cycle stage (Fig. 20.1). In many cells, microtubules originate from an organizing center (29). Characteristically, but not universally, this *microtubule organizing center—or MTOC—*contains a centriole, a pair of centrioles, or the basal body of a primary cilium [see (228)] surrounded by a cloud of amorphous material. This characteristic complex is now commonly called the *centrosome.* The single centrosome of nonconfluent mammalian tissue culture cells in G_1 of interphase, for example, normally lies in the proximity of the nucleus, within the cell cytoplasm, surrounded by Golgi elements. Microtubules radiate out from this structure toward the cell surface in a loose starburst pattern. The microtubules do not grow from the centriole wall itself, but rather from the field of amorphous material. This material contains γ-tubulin (129), intermediate-filament, and centrin-like proteins, as defined by immunofluorescence, as well as other components yet to be elucidated (91, 94). Sometime in late S phase, the centrosomes duplicate, but remain functioning as a single unit. The G_2 cell normally has four centrioles, a pair present in each potential microtubule-organizing center.

Microtubule organization within the cell abruptly changes at mitosis. This is mediated by a cyclin-dependent phosphorylation cascade (128) that affects tubulin and intermediate filament proteins, including the nuclear lamins, to produce a massive reorganization of components. Chromosomes condense and the nuclear membrane dissolves into fragments as the lamin network solubilizes. As part of the reorganization, the duplicated centrosomes separate to opposite sides of the reorganized nucleus, and each centrosome becomes competent to initiate microtubules. The mitotic spindle forms as microtubules assemble from the organizing centers at opposite poles and are captured by chromosome kinetochores (150). The details of centrosome function, mitotic reorganization and microtubule-based chromosome movements, are complex [see (9)]. This chapter, which emphasizes more stable microtubule patterns of the cell cytoplasm, especially cilia, will not deal further with problems associated with these phenomena, except in passing, as they help to illustrate general principles of microtubule function.

Structure, Assembly, and Disassembly

Microtubules that normally assemble on centrosomes or analogous organizing centers elongate by adding 100 kDa heterodimers composed of 1:1 α- and β-tubulin to their growing ends. α- and β-tubulin are especially abundant in brain and are among the best

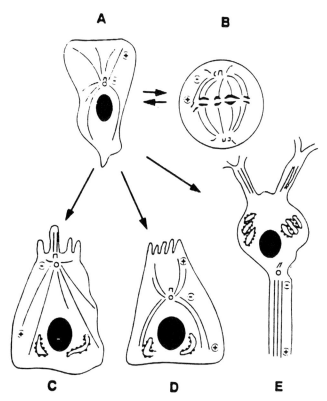

FIG. 20.1. Some representative microtubule patterns in cells with (+) and (−) ends of microtubules indicated. *A:* Cycling, motile interphase cells. *B:* Mitotic cell. *C:* G_0 cell with 1° cilium. *D:* G_0 cell with centrioles. *E:* Neuron. [From Satir et al. (173) with permission.]

protein level is probably relatively unimportant in terms of microtubule function in many situations.

Complete amino acid sequences are known for many tubulins. Structural domains of α- and β-tubulin are summarized in Figure 20.2 (114, 115). There is a tryptic cleavage site at arg 339 in α-tubulin and a chymotryptic cleavage site at tyr 281 in β-tubulin, which define N- and C-terminal domains. Conserved sequences across many species include *(1)* the N-terminal sequence in β-tubulin involved in the autoregulation of tubulin synthesis (230), *(2)* a region around residue 70 in α-tubulin which contains a site for binding the quanine residue of GTP, and *(3)* a glycine-rich region around residue 140 in α-tubulin that contains a loop that binds the phosphate region of GTP. Corresponding nucleotide-binding sites are found on β-tubulin, but the sites differ in that α-tubulin binds GTP only and the bound GTP is not readily exchangeable (N-site), while β-tubulin binds either GTP or GDP and the nucleotides are exchangeable (E-site). X-ray diffraction (14, 222) and antibody binding place the glutamic-acid-rich C-terminal domains of both α- and β-tubulin on the outside of the microtubule, while the N-terminal domains probably mainly lie inside. The molecule is asymmetrical, with the C-terminal domain of each monomer lying on a projection near one side. The intradimer bond is formed by the N-terminal domain of α-tubulin and the C-terminal domain of β-tubulin (Fig. 20.2).

To form a microtubule, the heterodimers probably first associate with one another longitudinally in strings of two or three to form a short protofilament. The interdimer bond along one protofilament involves the C-terminal end of an α-tubulin binding to the N-terminal domain of a β-tubulin. Dimers then assemble alongside the initial short protofilament to give an associated group or nucleus. This stable group of associated short protofilaments nucleates the growing microtubule where dimers are normally added to each end. Protofilaments associate with one another in at least two different ways: a staggered helix—where the α subunit of one protofilament lies next to β subunit of the next protofilament, the A lattice—or an unstaggered helix where the α subunit lies adjacent to a second α subunit, the B lattice. In the ciliary axoneme, the A lattice is the arrangement in subfiber A of each doublet microtubule, the B lattice in subfiber B. The B lattice, if complete, would produce a microtubule with a seam [*cf.* (5)]. Cytoplasmic microtubules usually have 13 protofilaments, but may occasionally have small variations of this number. In the ciliary axoneme, subfiber A is a complete microtubule with 13 protofilaments; subfiber B is incomplete. However, certain portions of the doublet microtubule, possibly specific

studied proteins in terms of their molecular biology. Only a brief review follows here [see also (5)]. The tubulins are highly conserved in evolution; α- and β-tubulin show about 40% homology (111). Co-assembly of dimers between different species is possible. In vertebrates, there are six distinct β-tubulin isotype families with tissue-specific expression, but in experimental tests divergent isotypes almost always freely substitute for one another and then are incorporated into all classes of microtubules in cells, usually with no obvious effects on phenotype [reviewed by Sullivan (198)]. They are, in effect, "isomorphs"—that is, their shape permits them to fit into the microtubule lattice, regardless of minor chemical modifications. A similar family of α-tubulins is present in mammals. The small sequence differences that do occur, mainly at the carboxyl-terminus of the molecule, may influence ligand binding, phosphorylation, or stability, and this may have some physiological significance, particularly during sperm gametogenesis (84). However, yeast *(Saccharomyces cerevisiae)* has only one β-tubulin and two α-tubulin genes; therefore, tubulin diversity at the

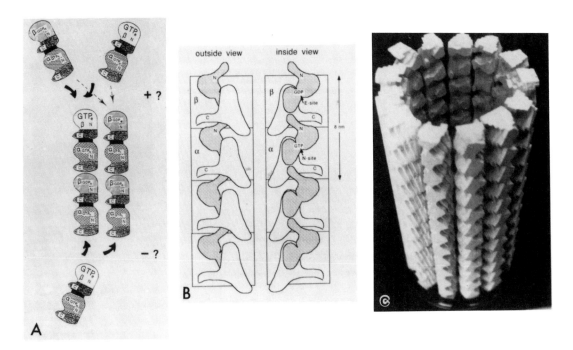

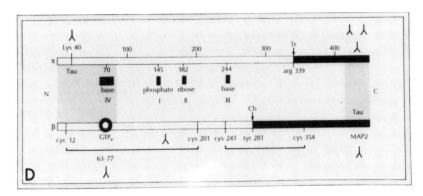

FIG. 20.2. *A:* Domain organization of tubulin dimers and protofilaments. α and β subunits of heterodimers are shown adding to ends of two protofilaments. One end has cap of exposed β subunits; other, a cap of α subunits. [From Mandelkow and Mandelkow (114) with permission.] *B:* Outside and inside views of microtubule protofilament. Speculative model based on x-ray diffraction analysis. [From Amos and Amos (5) with permission.] *C:* Model of microtubule calculated by 3D reconstruction methods. [From Amos and Amos (5) with permission.] *D:* Summary of some structural features of tubulin. Representations of the 1° structure of α and β tubulin. N *(open)* and C *(black)* terminal domains are defined by cleavage sites at arg 339 and cys 354. Regions I–IV are predicted to participate in nucleotide binding; regions of Tau, MAP2, and monoclonal antibody binding are indicated. [From Mandelkow and Mandelkow (115) with permission.]

protofilaments, are not tubulin, but tektin—an intermediate filament protein homolog (191).

Post-translational modifications occur on α- and β-tubulin (69), particularly in polymerized microtubules. These include reversible phosphorylation of β-tubulin, the detyrosination and retyrosination of α-tubulin, and the acetylation of α-tubulin. The significance of the modifications is not yet clear, although stable or longer-lived microtubules tend to be acetylated and detyrosinated. The tubulins of fungi—for example, yeast—lack the amino acid residues where modification usually occurs.

In vitro, where microtubules essentially assemble in the nucleation-elongation reaction described above, a critical concentration of tubulin and specific environmental conditions, such as GTP, low Ca^{2+}, temperatures around 37°C, and normal hydrostatic pressure, are absolutely required to initiate the process. Because one end of each protofilament of the growing microtubule exposes a cap of β-tubulin, while the other exposes α-tubulin (Fig. 20.2), the rates of addition and removal of dimers differ at the two ends, leading to faster elongation at one end (+, the N-terminal domain of β-tubulin) than the other (−, the C-terminal domain of α-tubulin). Under some conditions of steady state, growth at the (+) end may be compensated for by depolymerization at the (−) end. To accomplish this, a dimer with GTP at the exchangeable site on β-tubulin

is added at the (+) end, and then moves through the microtubule. GTP is hydrolyzed (37) on this dimer, as new GTP-tubulin is added behind it. Finally, the dimer—now with GDP at the exchangeable site—is released from the (−) end of microtubule. In solution, GDP is again exchanged for GTP and the process begins again. The microtubule behaves as a treadmill with respect to its subunits (118).

In reality, however, pure treadmilling, if it occurs, is rare. Mitchison and Kirschner (119) observed that individual microtubules grow, but then shorten, even to the point of complete disassembly. This constant

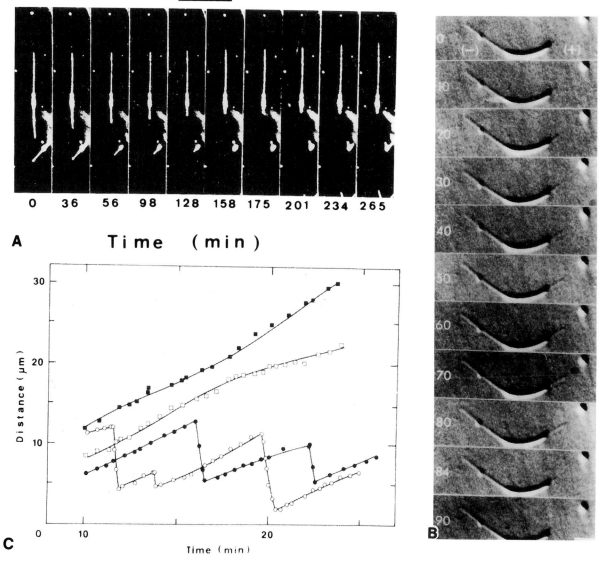

FIG. 20.3. *A:* Treadmilling in microtubule. Thickened region moves through microtubule toward (−) end. [From Hotani and Horio (83) with permission.] Bar, 10 μm. *B:* Dynamic instability of microtubules. At 84 s, (+) end undergoes catastrophic shortening. [From Walker et al. (223) with permission.] Bar, 2 μm *C:* Effect of MAPs on dynamic instability of microtubules. Changes in microtubule length with time at (+) *(black symbols)* and (−) *(open symbols)* ends of microtubule are shown. In absence of MAPs *(circles)* catastrophic depolymerization is seen. This is suppressed in presence of MAPs *(squares).* [From Hotani and Horio (83) with permission.]

change, or "dynamic instability," can be demonstrated by video-enhanced microscopy of single microtubules [Fig. 20.3; cf. (81, 83, 223)]. Growth at both ends of the microtubule is linear and steady vs. an almost instantaneous, catastrophic shortening. After shortening, growth (termed "rescue") resumes. Kinetic constants have been determined by Walker et al. (223) [see also (49)]. The rapid fluctuations in microtubule length predicted by this model have important consequences for physiological processes, where microtubules dramatically change their cellular distribution. Under conditions where microtubule length seems fixed and stable, dynamic instability implies that special stabilizing mechanisms must be present. At the (−) end, the presence of γ-tubulin could represent one such mechanism.

The alternating, catastrophic shortening and rescue phases are hypothesized to be regulated by the presence of a GTP-cap—that is, tubulin with GTP at the exchangeable site which is added to the growing microtubule, but not yet hydrolyzed to tubulin-GDP (75, 119). Since both addition and hydrolysis are stochastic, on occasion hydrolysis will outstrip tubulin-GTP addition and a tubulin-GDP end will be exposed. Rapid depolymerization of the tubulin-GDP end is thought to result in the catastrophic shortening. This model has been subjected to numerous experimental tests, which generally provide support for it, although our understanding of the mechanism is still incomplete (36).

Microtubule growth and depolymerization have been examined by electron microscopy (116). The majority of growing microtubules have blunt ends; some others have sheets of protofilaments protruding. Depolymerizing microtubules, on the other hand, often have coiled protofilaments at their ends. This suggests that depolymerization may not occur via release of individual dimers but rather by fragmentation of short oligomers. It is not generally appreciated exactly how fast depolymerization of microtubules can occur in life. The literature cites as an extreme rate about 6500 subunits per second, based on the the observations by Chen and Schliwa (38). However, axopods of the heliozoan *Heterophyrs marina* are 50 μm microtubule-based organelles that can contract with total microtubule depolymerization within 10 ms of stimulation— that is, at a rate of ca. 5000 μm s^{-1} (40) or about 10^7 subunits s^{-1}.

Stability and Pattern Formation

Particularly interesting and unusual stable patterns of microtubules occur in some cells; these include the cortical ribbons of ciliates and axostyles of termite flagellates. Microtubules in these ribbons slide relative to one another or one sheet of microtubules moves with relation to its neighbor, either passively or with the active participation of a motor molecule, to permit the cell to change shape. In ciliates, the microtubule ribbons actively produce cell elongation; they operate antagonistically to a second system, most likely a fiber system containing the Ca^{2+}-binding protein centrin, that causes cell contraction. No microtubule system, including such ribbons, or even cilia or flagella, is completely and permanently stable; in protozoa all the very stable microtubule arrays are broken down, often completely, and totally reorganized at cell division.

It is not entirely certain why some microtubules are more stable than others, but what can be demonstrated is that MAPs and other molecules will alter the dynamic instability of microtubules in vitro. One interesting molecule that may stabilize certain microtubules, for example, in sperm tails, is histone H1 (123).

The suppression of catastrophes by MAPs can be demonstrated directly in vitro on isolated microtubules [Fig. 20.3; (83)] or for microtubules assembling from centrosomes (28). An interesting speculation is that the cell regulates microtubule stability by MAP phosphorylation. Binding of MAPs to microtubules is reduced by MAP phosphorylation. Other factors influencing microtubule stability are bundling proteins and proteins which bind microtubules to other cellular structures such as membranes or intermediate filaments [cf. (49)].

Microtubule assembly–disassembly is dramatically decreased when cells enter G_0, so the stable, characteristic pattern of the differentiated tissue cell is assumed. In many cell types, and particularly in confluent epithelia, the appearance of the mature microtubule cytoskeleton coincides with the growth of a primary cilium. This also coincides with the specific intermediate filament differentiation in the cell, and it may be that these processes are interdependent. In addition to a stable, characteristic array, G_0 epithelial cells, like protists, probably have a more dynamic subpopulation of microtubules. For example, in MDCK cells, Wadsworth and McGrail (221) have found that many microtubules had a half-life of about 18 min, while a second more stable population had a half-life of 72 min. The percentage of stable microtubules increases with confluency. In confluent cultures, as indicated in Figure 20.1, MDCK cells have primary cilia and an array of microtubules running along the lateral surfaces of the cells in an apical–basal direction with the plus-ends of the microtubules near the base of the cell [(8) and others].

Using confocal microscopy, Novikoff et al. (127) have described the three-dimensional array of microtubules in hepatocytes in short-term culture (Fig. 20.4).

These cells normally have no primary cilia. Microtubules originate from a centrosome near the nucleus, fanning out as they extend toward the top and bottom of the cell, where they flow into an array that cascades along the outer margins of the lateral cell surfaces. Although more extensive, the array is reminiscent of the microtubule ribbons extending from the basal bodies in *Chlamydomonas*. Since microtubules are generally linked to membrane trafficking in cells, this distribution probably defines the paths of exo- and endocytosis in the hepatocyte. For example, an endocytic vesicle transferred to the microtubule array near the cell surface would probably move first along the surface track and then flow into the cell toward the centrosome [*cf.* (170)]. A similar array of microtubules seems responsible for sperm transport across the Sertoli cell (147). With or without cilia, a cage of stable microtubules is probably present in most differentiated cells.

Burnside (35) has studied an interesting array of

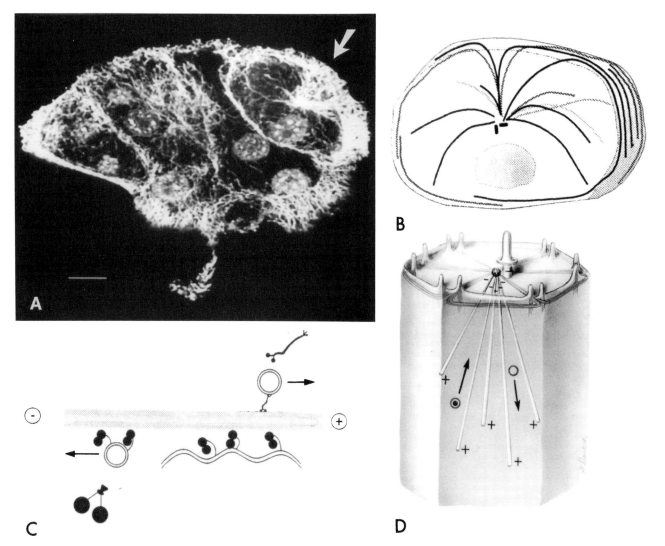

FIG. 20.4. *A:* Microtubule distribution in cultured hepatocytes. Confocal microscopic reconstruction of group of cells showing tubulin immunofluorescent pattern. *Arrow* marks cell where array radiating from centrioles is seen. [From Novikoff et al. (127) with permission.] Bar, 10 μm. *B:* Model of microtubule distribution in cultured hepatocyte. *C:* Vesicular trafficking along microtubule. *Top:* Kinesin-based motility. *Bottom:* Cytoplasmic dynein-based motility. The (+) and (−) ends of the microtubule are indicated. [From Satir et al. (173) with permission.] *D:* Model of microtubule-based vesicular trafficking in a polarized epithelial cell with 1° cilium. Microtubule-based motors could move vesicles in basal to apical direction or apical to basal direction along polarized array of microtubules. This would require, respectively, (−) end–directed motor, such as cytoplasmic dynein and (+) end–directed motor such as kinesin to interact with vesicular membranes. [From Phillips and Satir (140) with permission.]

microtubules in the teleost retinal cone. This cell changes its shape in response to light, elongating in the dark by as much as 100 μm. This elongation occurs by the sliding of uniformly polarized, parallel microtubules that occupy a special zone, the *myoid region*, of the cell, as can be demonstrated directly in reactivation experiments using permeabilized cells. cAMP as well as ATP and low Ca^{2+} is required for elongation. Sliding at slow rates of 2 μm min^{-1} depends on a microtubule motor protein, inhibitable by vanadate in a manner similar to dynein, but with some differences. The microtubules work antagonistically to an actin microfilament system that causes shortening of the cell body when the cone is exposed to light.

Primary Cilia

A primary cilium defines the apical surface of many confluent tissue culture cells in G_0 and of many natural epithelia; it is also found on certain cells in connective tissue, blood, muscle, and nerve. It is a neglected organelle, but possibly a highly important one. As the cells become confluent or differentiate to enter G_0, one centriole interacts with the cell membrane (or a Golgi vesicle and then the cell membrane) to grow a primary cilium (47, 78, 207, 208). Primary cilia have no central microtubules, no dynein-arms, and no active beat. Their axonemes are referred to as $9+0$, although this pattern sometimes is present at only the part of the organelle immediately distal to the basal body. The membrane–axonemal relationships seem unaltered in the $9+0$ segments of primary cilia. All primary cilia, including those which become dendrites of sense cells, have a ciliary necklace (60).

Technical improvements have made it possible to visualize primary cilia in living tissues (52); in some cells they reach lengths of over 17 μm. They are most easily seen and were first described as the receptor derivatives of sense cells (144), especially the outer segment in vertebrate photoreceptors and the insect ear. It seems likely that sensory cilia could have evolved from $9+2$ motile cilia of earlier organisms which invented transduction mechanisms whereby, to regulate ciliary beat, stimuli at the ciliary membrane induced changes in the concentration of second-messenger molecules in the organelle. Signal transduction in the rods and cones, and in olfactory cilia (a transitional case, perhaps), has received much recent attention.

The hypothesis that primary cilia have the capability of signal transduction in cells in general is a plausible one; it is particularly attractive in the case of mechanoreception, where passive bending of the primary cilium might induce a cellular signal (121). Bowser and Leonard (25) have shown that primary cilia of cultured kidney epithelial cells will bend passively in response to fluid motion. Their mechanical properties make them suitable as cellular flowmeters. More generally, primary cilia could integrate spatial or mechanical events around the cell. In migrating 3T3 cells, the primary cilium is always at the leading edge (2). Since the basal body of the primary cilium functions as an MTOC to organize the cytoskeleton and membrane flow pathways, coupled to a signal transduction mechanism, this could be a key element in chemotactic responses of certain moving cells. A detailed review of the possible sensory functions of primary cilia is to be found in Poole et al. (143).

EVOLUTION OF CENTRIOLES, CILIA, AND MICROTUBULES

Like most of the cytoskeletal proteins, tubulin is solely eukaryotic protein; neither tubulin nor tubulin genes have been found in prokaryotes in nature as yet. Similarly, cilia and centrioles do not exist in prokaryotes. Speculations about the origins of these organelles profitably illuminate certain aspects of their organization [*cf.* (185)]. Essentially, three hypotheses have been formulated to account for the origin of centrioles, cilia and microtubules: *(1)* de novo differentiation in a proto-eukaryote cell cytoplasm. This hypothesis provides no rationale for the special replicative properties of centrioles, including the possibility that some centrioles contain DNA [(72) but see (90)], and no rationale for ninefold symmetry. *(2)* endosymbiotic origin via spirochaete invasion of the proto-eukaryote cytoplasm. L. Margulis has long been a key proponent of this hypothesis, which provides a rationale for centriolar DNA (it would come from the spirochaete) and which derives a measure of suggestiveness because special flagellates in the guts of termites attach spirochaetes in rows to their body and use the motile apparatus of these attached organisms in part to swim. But spirochaete motility is prokaryotic motility that depends upon bacterial-type flagella comprised of an extracellular passive protein (flagellin) wire and a rotary motor closely associated with the cell membrane, and not on a cytoplasmically placed tubulin–dynein sliding system at all. Furthermore, this origin would predict that, like mitochondria—which probably do originate from endosymbiotic bacteria—the cilium would be surrounded by two membranes—one derived from the spirochaete and one derived from the host cell—which clearly is not the case. *(3)* endosymbiotic origin via viral invasion. This hypothesis readily accounts for the self-assembly properties of centrioles and their symmetries, since many viruses have complex symmetrical designs; it accounts for cytoplasmic assembly of centrioles from amorphous fibrogranular masses [cf. (42)]. The hypothesis also accounts for nuclear control of synthesis and replication and the positioning of descendent genes on nuclear chromosomes, as the viral genome becomes incorporated into the nucleus. It accounts for the latency of centriolar synthesis seen sometimes, as in Naegleria and Gingko, where the centriole disappears from the cell at some stage only to reappear at another. It accounts for the apparent self-duplication of the centriole in other cases and the possibility that sometimes DNA or perhaps RNA could be present in the organelle. However, it does not account for motor molecules associated with microtubule-based motility. Cytoskeletal motors are solely found in eukaryotes, but perhaps they could be derived secondarily from proto-eukaryotic ATPases with other functions.

The living descendants of the most primitive unicellular eukaryotes, such as chrysomonads and other small flagellates, have motile 9 + 2 cilia. Motility clearly has selective advantages for small microorganisms for dispersion, or for accessibility to new nutrient-rich environments. Further, a cilium maneuvers food onto a receptive cell surface. In a motile cell, there would be a good reason for linking duplication of the cilium to the cell cycle. If the cilium has not duplicated when the cell divides, motility would be lost in the daughter, and this would be selected against. The basal bodies of cilia sometimes act in place of centrioles. For these reasons, it seems likely that, after the invasive period (if that is what occurred), the motile 9 + 2 cilium evolved and became stabilized as the most ancient microtubule-containing organelle. Loss of motility and of the central pair would lead to 9 + 0 primary cilia in more complex metazoan cells, where tissue specialization occurred; loss of the axoneme proper would lead to centrioles; duplication of tubulin genes coupled to assembly mutations would give rise to cytoplasmic microtubules. It is curious that axonemal doublet microtubules assemble directly onto basal body microtubules, but cytoplasmic microtubules, like axonemal central pair microtubules, assemble unconnected to preexisting microtubules. Is the amorphorous material around the centriole a displaced organizing center with γ-tubulin originally developed for the organizing central pair in 9 + 2 axonemes? Although this hypothesis has many attractive features it does not adequately account for the origin of the mitotic apparatus, which is crucial to eukaryotic cell evolution. However, it seems likely that the evolution of microtubule-based anaphase separations of chromosomes occurred in stages, in that in many protistan mitoses the nuclear envelope does not break down and the main microtubule-based motile event seems to be nuclear elongation, comparable to anaphase B of vertebrate cell mitosis.

In epithelial cells of metazoa, the basal body organizes not only the microtubule- and intermediate cell cytoskeletal elements but also the apical actin network (148). The most primitive pattern consists of a motile cilium surrounded by a ring of microvilli. This is found in unicellular choanoflagellates (105) and it persists in many epithelia, in sense organs including the vertebrate eye, and in cells with primary cilia. It is a basic way to organize the apical cytoskeleton of a cell. This organization permits the evolutionarily primitive food-trapping function of the cell surface to be retained. In this way, the apical surface of epithelial cell could evolve into a specialized focus of endocytosis and exocytosis, with microtubule tracts centered on the basal body playing a key role in directed vesicle movement.

MOTOR MOLECULES OF CILIA AND MICROTUBULE-BASED ARRAYS

Dynein- and Microtubule-Based Motility

The striking fact of ciliary motility led people to ask by what mechanism this occurred. In this century, ideas flowed into the field primarily from muscle physiology. By 1950, peripheral fibrillar (microtubule) contraction was a favored model (27, 68), but by the end of the decade, several suggestions favored sliding models (1, 57) taken after the sliding filament model of muscle contraction. The question of whether the cilium was an active organelle (or passively moved by the cell) was settled when laser ablation experiments (63) showed that isolated cilia could bend and swim. Glycerol-treated cilia were reactivated by ATP, the known energy source for ciliary motility, by Hoffmann-Berling (79) and electron microscopy soon showed that membranes in such preparations, like the corresponding muscle preparations, were destroyed [cf. (162)]. This approach was extended by Gibbons and Gibbons (50), who matched the form of living swimming isolated cilia to reactivated axonemes with great fidelity. Therefore, the 9 + 2 microtubular structure or axoneme of the cilium was solely responsible for motility.

While reports of actin and myosin in ciliary axoneme dot the literature, it soon became apparent that the major ciliary ATPase was different from myosin (52). Gibbons and Rowe (59) named this new protein dynein. Gibbons showed that dynein corresponded to the arms, identified by Afzelius (1), projecting from one end of each doublet. That end to which the arms are permanently attached (in the absence of disassembly) could be identified as subfiber A. The arms project from subfiber A toward the back of the adjacent doublet, subfiber B. Afzelius (1) had proposed that the arms of one microtubule (numbered N) might walk upon the next ($N+1$) to cause sliding, but since contractile models were then in fashion, in the absence of further evidence, he abandoned this idea in the early 1960s. However, Satir (163, 164) was able to demonstrate that the ciliary microtubules did not evidently change length as the organelle bent in two different directions, which suggested that some axonemal microtubules did indeed slide past others. Although the amount of sliding (Δln) in micrometers for any doublet was very small, it was related to the bend by simple geometry in that:

$$\Delta \text{ln} = d_n \, \Sigma \alpha \qquad (20\text{--}1)$$

where d_n is a constant related to axonemal diameter and $\Sigma\alpha$ is the amount of bend in radians. For a bend of 100°, the displacement of one doublet vs. its neighbor is about 0.1 μm. In the cilia studied, each microtubule doublet could be numbered, following Afzelius (1). Doublets 5 and 6 at the leading edge of the bend are connected by a bridge and usually do not slide relative to one another. By labeling doublets with gold beads, Brokaw (30, 32) has shown that sliding displacements occur and that equation 20–1 holds, not only after fixation, but as axonemes reactivated by ATP actually form bends (Fig. 20.5).

The sliding microtubule model of ciliary motility attracted the attention of workers on other microtubule-based arrays, and it was tentatively extended to both axonal transport and anaphase of mitosis, with appropriate alterations to fit the physiology, mainly in speculative ways. For cilia, the model was confirmed by further development of in vitro sliding systems, first and most dramatically by Summers and Gibbons (199, 200). Then the search began for a cytoplasmic dynein that could power sliding microtubules in anaphase or cause vesicles to move during axonal transport or other processes. The search was greatly facilitated when first Takahashi and Tonomura (203)—extending Gibbons' (52) observations—and then Haimo et al. (71) and Satir et al. (176), showed that axonemal dynein would bind to and could be released from axonemal and other microtubules. Satir (165) proposed a method by which dyneins could be isolated from cell extracts in general. The addition of microtubules to the extract in the absence of ATP would cause dynein to bind to the microtubules, and the addition of ATP would cause release. Dynein could then be separated from the microtubules and be identified by its ATPase activity, molecular weight, and other characteristics. This approach was fully realized for axonemal dynein by Nasr and Satir (125). This micro-

tubule affinity or "alloaffinity" procedure can be used cyclically to purify dynein considerably.

A second major advance in the identification of cytoplasmic dynein was the development of a drug, taxol, that stabilized polymerized cytoplasmic microtubules in the face of low temperature or high Ca^{2+} (82). Taxol-stabilized microtubules could be added to any cell extract and, in the absence of ATP, they would bind motor molecules such as dynein, which could then be released by ATP. These methods permitted isolation of cytoplasmic dynein in a variety of cells.

It then became important to develop stringent criteria for the identification of dynein, as opposed to other possible microtubule motors. Gibbons's group (55) [as well as (99)] showed that dynein ATPase was particularly sensitive to inhibition by vanadate—in the range of a few μM. Further, they showed that upon irradiation with 350 nm UV light, in the presence of vanadate and ATP or ADP and certain cations, each heavy (H) chain ATPase of dynein was photocleaved into two specific fragments (58, 106). The extraordinary molecular weight of the H chains and their susceptibility to specific vanadate photocleavage were strong presumptive evidence that a given microtubule-associated motor was dynein. This could be further confirmed by in vitro assays of microtubule motility and by electron microscopy of the isolated molecule.

Axonemal Dynein

Axonemal dynein is the grandfather of microtubule motors. In sliding axonemes, doublet N on which the dynein arms are assembled always pushes doublet $N+1$ (to which the arms transiently attach) tipward—that is, toward the (+) end of the axonemal microtubules (159). By Newton's third law, doublet N itself moves in a (−) end direction. Axonemal dynein in general moves the structure to which it is attached in a (−) end direction. Dynein consistently acts as a (−) end–directed microtubule motor. Takahashi et al. (202) carefully measured the speed of sliding produced by axonemal dynein in this assay. They found a $Vmax$ of about 19 μm s^{-1} in a mM ATP concentration range.

It is now possible to study microtubule translocation by isolated axonemal dynein in a completely defined in vitro assay. The dynein to be studied is generally placed on a glass slide and purified brain microtubules, usually with one end labeled in some manner, are added above the substratum (Fig. 20.5). In the presence of Mg^{2+}-ATP, the microtubules move, uniformly with their (+) end leading, at rates that approximate up to half the maximal sliding rate of the sliding axonemal microtubules (138, 215). All axonemal dyneins and dynein fragments that can be shown to move microtubules work in the same way. Force is generated during a mechanochemical cycle, which has been studied extensively for axonemal dynein (89, 146). The cycle is reminiscent of that myosin. It involves a rapid release of the ATP-sensitive tubulin-binding regions of the dynein from the translocating microtubule when ATP is added—a fact used in the affinity procedure for purifying dynein. Release precedes hydrolysis of ATP to bound ADP and Pi, after which the dynein rebinds to the microtubule in a force-generating state. The rate-limiting step appears to be the release of ADP from

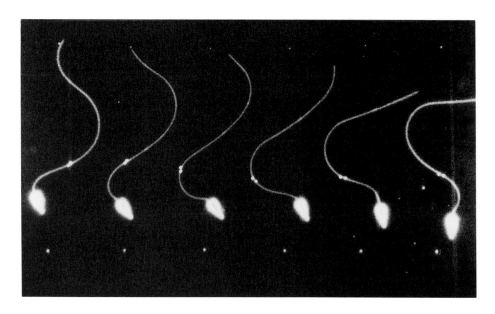

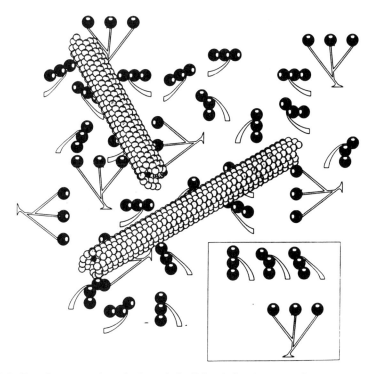

FIG. 20.5. *Top:* Demonstration of microtubule sliding in beating sea urchin sperm axoneme. Gold beads placed on different doublets of axoneme move apart as sliding proceeds. [From Brokaw (32) with permission.] *Bottom:* Diagram of in vitro translocation system at high magnification. Individual three-headed dynein molecules are placed on substratum. Microtubules are introduced. Dynein binds to microtubules, translocating them across field when ATP is added. *Inset:* Two structural forms of dynein. [Courtesy K. Barkalow and T. Hamasaki.]

the hydrolytic site. In the presence of microtubules, release is stimulated and ATPase activity correspondingly activated (133).

Johnson and Wall (93) showed that isolated 22S dynein from *Tetrahymena* ciliary axonemes took the form of a bouquet of three large globular heads attached to stalks or stems that ended in a base (Fig. 20.6). Other axonemal dyneins are two-headed molecules (158) or possibly single-headed molecules, although these are usually taken to be breakdown products of the multiple-headed species. The three globular-heads of 22S dynein correspond to three distinct high molecular weight bands (α, β, γ) in gel electrophoresis, suggesting that each 22S dynein molecule is a heterotrimer composed of three distinct heavy chains. Similarly, the two-headed molecules are comprised of two distinct heavy chains, and these in turn are not the same as those of the three-headed dyneins from the same organism. This is particularly apparent in *Chlamydomonas* (142) where dynein heavy-chain heterogeneity has been well studied and mutants in specific arms are known. In *Chlamydomonas*, there are three types of inner arms, each of which is a two-headed molecule with two distinct heavy chains, subject to

mutation. Some inner arm mutants lack inner arms at specific axonemal locations (187).

Molecular weights of the heavy chains are on the order of about 450–500 kDa, making these chains some of the longest single-protein chains known. Intermediate chains and light chains also form parts of the molecule; in sea urchin dynein a specific intermediate chain is associated with the β heavy chain. In general, the intermediate chains are thought to be located near the base of the molecule (98), which is the site of attachment to the A subfiber of each axonemal doublet.

The bouquet is not the in situ form of the arm, which is more compact and folded (11). Some models of folding attempt to place all three heavy chains away from the doublet (66), so that each chain can interact independently with subfiber B of the adjacent doublet during an arm cycle. However, it can be shown that, even though each heavy chain contains ATPase sites, the physiological role of each chain appears to be different (122), and recent work suggests that the α heavy chain may correspond to a single head that extends closest to interact with the B subfiber (154, 155). An alternate structural model (167) where only

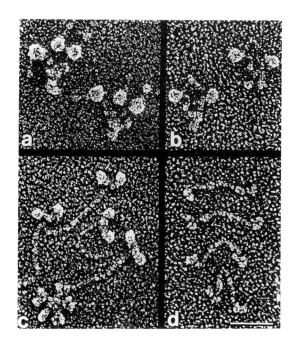

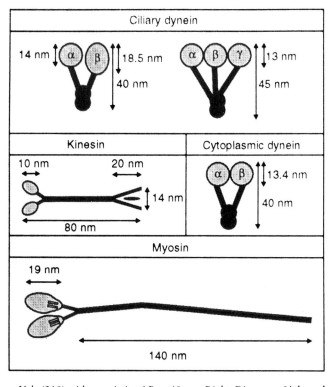

FIG. 20.6. Molecular motors. *Left:* Rotary shadowed images of *(a)* ciliary dynein, *(b)* cytoplasmic dynein, *(c)* myosin II, and *(d)* kinesin. Courtesy of John Heuser, Washington University. [From Vale (210) with permission.] Bar, 40 nm. *Right:* Diagrams. [Adapted from Vale (210).]

one heavy chain—possibly the α chain—is in position to interact with the adjacent doublet, while the other two heavy chains form the body of the arm and influence the interactions more indirectly, has been proposed.

A complete deduced sequence of one of the dynein heavy chains (the β chain of sea urchin dynein) has been obtained using molecular cloning techniques (56, 130). The isolated β heavy chain is of particular interest since alone it is capable of translocating microtubules in an in vitro assay system (157, 214). Mapping and sequencing were made feasible because the molecule can be split into major fragments, either by conventional proteolysis or by vanadate-sensitive photocleavage (Fig. 20.7). The deduced sequence of 511,811 daltons has 4466 residues and contains consensus motifs for five nucleotide-binding sites. The probable site for ATP hydrolysis coupled to microtubule translocation is at residues 1852–1859, a position that corresponds closely to the V1 site of vanadate-mediated photocleavage. Secondary-structure predictions indicate that there are two long regions of unbroken heptad repeats, potential α helices, at residues 3028–3147 (120 residues) and 3255–3304 (50 residues). The dynein heavy chain does not show high homology to other known motor or microtubule-binding proteins.

Gibbons's group has constructed probes around the ATP consensus sequence for use in identifying dynein isoforms by polymerase chain reaction and other molecular genetic procedures and Southern blot analysis. In this way, a large number of dynein isoforms have been identified in a number of organisms including sea urchin, *Drosophila,* and *Paramecium* [*cf.* (6)]. For example, in *Paramecium,* eight unique sequences of dynein heavy chains have been identified, seven of which correspond to the ciliary sequence pattern found in sea urchin, while the eighth corresponds to cytoplasmic dynein from yeast (6a, 13). It is well known that the mRNA concentration for tubulin and other ciliary proteins increases for a time after deciliation, presumably as massive protein synthesis for new cilia begins. Only one mRNA for the heavy-chain isoforms—the one that corresponds to the putative cytoplasmic dynein—fails to increase upon deciliation in sea urchin blastulas and in *Paramecium.*

The question arises as to the significance of so many axonemal isoforms that are expressed. A likely explanation is that each of the three heads of the outer dynein arm corresponds to a different gene product and each of the two heads of each flavor of inner arm corresponds to a still-different gene product. In *Chlamydomonas,* this could account for at least nine different axonemal dynein genes, without considering the potential that some other sites in the axoneme,

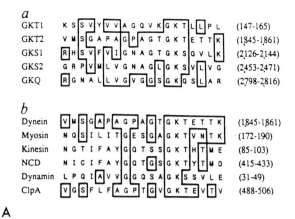

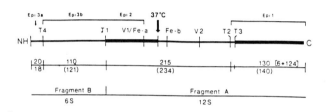

FIG. 20.7. *A: a:* Alignment of putative nucleotide-binding sites in dynein β heavy chain. Residues identical in two or more sites appear in *boxes. b:* Alignment of probable hydrolytic ATP-binding sites of dynein β chain *(GKT2)* with other motors and *E. coli* Clp protease A subunit. Residues identical to dynein GKT2 sequence appear in *boxes.* [From Gibbons et al. (56) with permission.] *B:* Map of principal tryptic and proteolytic cleavage sites on dynein β chain. *Numbers* above middle line give masses of tryptic peptides $(M_r \times 10^{-3})$ for comparison to *numbers in parentheses* calculated from derived sequence. N-termini of the 110k, 215k, 130k, and 124k tryptic peptides are indicated as *T4, T1, T2,* and *T3,* respectively. *V1* and *V2* vanadate-sensitive photocleavage sites are indicated. *Thick regions* of map indicate peptides that are stable to trypsin at 37°C. 6S region may represent tail, 12S region the globular domain of heavy chain. [From Gibbons et al. (56) with permission.]

such as the spoke head, could also contain a dynein heavy chain. This interpretation is consistent with a unique function and a unique position for each head in the axoneme. It will be interesting to know whether the more closely related isoform sequences correspond to different heads of the same arm.

Cytoplasmic Dynein

Cytoplasmic dynein was an ephemeral protein for many years, postulated but not defined. However, Pallini et al. (137) showed that a MAP from bovine brain clearly had dynein-like properties. It differed from axonemal dynein in that it hydrolyzed CTP at a faster rate than ATP. Using microtubule affinity isolation, Vallee et al. (220) identified the brain cytoplasmic dynein definitively as MAP 1C and provided specific evidence that this was a microtubule-translocating motor by using the molecule in in vitro gliding assays. Like axonemal dynein, MAP 1C proved to be a (−) end—directed motor. The rate of movement produced in gliding assays (1–2 μm s^{-1}) is consistent with MAP 1C being the retrograde motor in axonal transport. CTP does not support in vitro motility. A brief review of the properties of cytoplasmic dynein is given in Vallee (216). Other reviews include Porter and Johnson (146), which compares cytoplasmic and axonemal dyneins, and a series of articles in Warner and McIntosh (226).

In addition to two heavy chains, brain cytoplasmic dynein is characterized by intermediate-chain subunits which are distinct from those of axonemal dynein. A polyclonal antibody to the 74 kDa intermediate chain was used in experiments that support localization of cytoplasmic dynein to vesicles along microtubule tracks in axons and in cultured cells (77, 108). Cytoplasmic dynein may also play a role in mitosis, since antibodies to the molecule are also localized at kinetochores (139, 193). Goltz et al. (64) have evidence that receptor–ligand sorting during receptor-mediated endocytosis may involve cytoplasmic dynein binding to the ligand-containing portion of an endocytic vesicle, pulling the vesicle apart and moving that portion to the lysosome.

The deduced sequence for cytoplasmic dynein around the V1-related hydrolysis site is the most different of all the dynein sequences. The presence of a single gene suggests that the two-headed molecule is a homodimer. There is considerable homology between the C-terminal two-thirds of the sequences of sea urchin axonemal vs. *Dictyostelium* cytoplasmic dynein (32% identity) (100). This suggests that this region is the conserved mechanochemical domain. The N-terminal region, which is much less homologous between axonemal and cytoplasmic forms, may specify differences between the molecules, which permit interactions of cytoplasmic dynein with vesicles or chromosome kinetochores, or alternatively of axonemal dynein with axonemal microtubules.

Kinesin

Microtubule affinity methods were turned with considerable success to identifying and isolating other potential microtubule motors, first from nerve axons and then from a variety of cells and organisms. With the use of video-enhanced light microscopy, R.D. Allen and co-workers (3) showed bidirectional motility of small vesicles along single microtubules in nerve axons.

This was exploited by Vale et al. (212, 213), who showed that high-speed supernatants from squid axoplasm would induce ATP-dependent movement of microtubules on glass or movement of beads coated with proteins from the supernatant fractions along microtubules. The protein purified from these axoplasmic extracts, kinesin, could be shown to have properties substantially different from dynein and to move microtubules in the opposite direction.

Kinesin is a (+) end microtubule motor that is probably responsible for vesicular trafficking in an antero-grade direction in nerve axons and possibly more generally in the cell cytoplasm (Fig. 20.4). Movement of taxol-stabilized, MAP-free microtubules along a kinesin-coated glass coverslip is generally smooth and unidirectional. A single kinesin molecule produces sufficient force to move a many-microns-long microtubule at maximum velocity (17). Kinesin-coated beads move linearly along microtubule protofilaments with steps of about 8 nm (201). The velocity of kinesin-induced movement is about 0.5–0.6 μm. s^{-1}, much slower than the microtubule sliding rate produced by dynein in axonemes and somewhat slower than rates produced by outer arm dynein in vitro. Kinesin is separated from dynein by use of the analog 5'-adenyl imidophosphate (AMP-PNP) [*cf.* (104)], which releases dynein from microtubules but retains the rigor state of kinesin. Kinesin is a Mg^{2+}-dependent ATPase whose ATPase activity is stimulated in the presence of microtubules (102), which suggests that ATP hydrolysis is tightly coupled to force production (209). Kinesin is also a GTPase and in vitro motility is supported by GTP as well. When tubulin is polymerized in various ways, it can be demonstrated that kinesin will not bind to the inside surface of the protofilament lattice but that full cylindrical closure of the microtubule is not necessary for motility (95). Single protofilaments are sufficient for kinesin-mediated motility, but smooth gliding requires several parallel protofilaments. A helical path of protofilaments in microtubules with 12 or 14 protofilaments leads to gliding along a helical path. Smooth gliding will also occur even if the microtubular lattice changes from doublet to singlet microtubule as at the tip of an axoneme.

Kinesin has been purified from many cell types [*cf.* (48, 61)]. It consists of two heavy chains, each with a molecular mass of 110–130 kDa, and two light chains of 58–75 kDa [Fig. 20.6; (19, 103)]. The heavy chains are elongate and each ends in a globular head of about 450 amino acids. The heads are the sites of ATP hydrolysis and of microtubule translocation since they bind to microtubules in an ATP-sensitive manner. The heads lead into an elongate α-helical domain, both heavy chains forming a 50–60 nm long coiled coil.

The tail end of the molecule fans out and presumably the light chains are located in this region (76).

The Kinesin Superfamily

A number of other genes produce products that share high sequence homology with the head region of the kinesin heavy chain, especially with the globular motor region that usually is found at the amino terminus of the molecule. These molecules differ significantly in other domains, particularly in their tail ends, and sometimes also in the placement of the putative motor region along the chain (61). Interestingly, one of the molecules (ncd), where the motor domain is not at the N-terminal end, is a (−) end translocator. The variations in the tail domains suggest that the tails could bind to specific organelles to specify the direction of organelle transport and targeting, a hypothesis enhanced by the finding that there may be as many 35 members of the kinesin superfamily in one organism *(Drosophila).* Even in yeast *(S. cerevisiae),* at least five different genes encode kinesin-like proteins. In certain tissue culture cells at least six different members of the superfamily are expressed in a single cell type (194). However, the evidence that all these molecules function primarily in organelle transport is derived almost exclusively from their sequence homology to authentic kinesin heavy chains and from localization studies—and is correspondingly circumstantial. Nonetheless, kinesin-like molecules often are found in places where specific microtubule motors would be helpful in moving things around in the cell cytoplasm—in vesicular transport and particularly in mitosis and meiosis, where the proteins are associated with the mitotic spindle—either at the kinetochore [that is, centrosomal protein-G (CENP-G): (229)] or at the midzone [that is, midzone kinesin related protein (MKLP)-1: (126)]. Mutants in gene coding for kinesin homologs such as ncd and nod in *Drosophila* lead to chromosome loss and nondisjunction during female meiosis, while mutants in authentic kinesin heavy chain have defects only in neuronal function, probably because axonal transport is deficient (177). Some members of the superfamily may act with extreme specificity for certain cellular processes, while others have overlapping functions with other members of the superfamily or with other motors.

Interactions Between Dynein and Kinesin

It seems likely that dynein and kinesin work together, perhaps with other motor molecules, to bring about the variety of chromosome movements associated with mitosis, and also in membrane trafficking in cells. Vale et al. (211) have devised an interesting assay that illustrates bidirectional movement, based on kinesin, working as a (+) end motor, and axonemal or cytoplasmic dynein, working as a (−) end motor. When both proteins are used together in microtubule translocation assays, a given microtubule moves unidirectionally first one way, then the other—switching between the two motors stochastically with one switching event occurring within 10 s, about 40% of the time. This is consistent with a model where translocation in one direction inactivates the motors working in the opposite direction and where reversals require temporary dissociation of the active motor from the microtubule, allowing the inactive motors to become force-generating. However, while kinesin remains strongly attached to the microtubule and force-generating during most of its mechanochemical cycle, dynein, like myosin, has a much shorter duty cycle. Where kinesin and dynein work oppositely on the same microtubule or vesicle, as in the Vale et al. (211) experiments, a single kinesin will overcome many dyneins. Bidirectional movements characterize certain phases of chromosome movement during mitosis and also some vesicular movements.

Other Putative Motors

Affinity assays, in vitro motility assays, and high-resolution video microscopy have made possible a search for putative microtubule motors that do not belong to the kinesin superfamily or to the dyneins. A possible candidate is dynamin, a 100 kDa protein in calf brain and other tissue, whose binding to microtubules is nucleotide sensitive. Dynamin is a GTPase that bundles microtubules under certain circumstances. There have been preliminary reports that dynamin will support microtubule translocation in vitro, as a (+) end–directed motor. However, this process utilizes GTP in preference to ATP. This could be because the dynamin preparations used are contaminated with a low level of kinesin or a kinesin-related molecule, which is the true motor. Brief reviews of the properties of dynamin are given by Collins (39) and by Vallee (217). Dynamin has been cloned and sequenced; aside from its nucleotide-binding site, the molecule has little in common with kinesin, cytoplasmic dynein, or myosin. On the other hand, dynamin is highly homologous to certain gene products in yeast (VSP1, MGM-1) or *Drosophila* (shibire) involved in vesicle recycling. Most of the evidence now suggests that it plays some role in membrane sorting as a GTP-binding proteins, perhaps independently of microtubule-based motility [cf. (113)]. Dynamin is the first, but surely not the last, of the new motor molecules that may not in fact be motors; it is a cautionary tale, suggesting that rigorous criteria for defining new motor classes should be developed and applied before such classes are hyped quite so extensively.

MOTILE CILIA AND THE MECHANISM OF MOTILITY

Introduction to Ciliary Structure

Although microtubules form the major structural element of the ciliary axoneme and axonemal dynein is the axonemal motor, the axoneme is comprised of many proteins other than dynein and tubulin. Over 250 axonemal polypeptides (112) have been found. The best understood are those that form part of the structurally identified components of the axoneme, such as the spokes, since these polypeptides can be identified from spots on gels that fail to appear when the structure itself is missing in mutant axonemes. However, these remain a small subset of the total, perhaps 10%–20%. There is little understanding of the role of the unidentified components of the axoneme, but since they all remain with the axoneme, even after detergent treatment and extensive washing, they can be considered microtubule-associated elements. This suggests that our understanding of proteins that can interact with microtubules in the cytoplasm is correspondingly incomplete.

The structural elements of the axoneme are embedded in a specialized cytosol, sometimes referred to as the *matrix,* and the whole is enclosed by the ciliary membrane, an extension of the cell membrane with specified, localized membrane microdomains.

Ciliary membranes contain transmembrane glycoproteins responsible for serotype or mating responses in certain unicellular organisms, as well as specific receptors and channels in all organisms; these function for the most part in the signal transduction pathway controlling ciliary motility. Other pathways controlled by membrane receptors or channels may include induction of deciliation by Ca^{2+}-mediated contraction of strategically placed centrin fibers above the ciliary necklace (160) or processes related to ciliary growth and assembly or disassembly. An extensive review of the properties of ciliary membranes is found in Bloodgood (18).

The membrane controls ciliary motility by generating appropriate second messengers that interact with the axoneme; presumably, these and other small molecules are part of the cytosol or matrix around the axoneme and are lost upon dissolution of the membrane by detergents. As noted earlier, when the membrane is removed, the cilium stops moving, but motion can be restored by replacing the cytosol with simple buffers containing Mg^{2+}-ATP. Ciliary behavior returns if the correct second messengers are added. Swimming in *Paramecium* provides a good example, since mild detergent treatment need not perturb cortical organization and addition of Mg^{2+}-ATP at pCa7 then restores movement of enough cilia to permit hydrodynamic coupling between them and the consequent induction of metachronism and smooth, helical swimming at speeds of about half the average velocity of living cells (107). If Ca^{2+} is increased toward pCa6, swimming slows and at Ca^{2+} concentrations greater than $10^{-6}M$ the cells swim backwards. If, instead, pCa7 is maintained, but 10 μM cyclic AMP (cAMP) is added to the reactivation buffer, forward swimming speed increases (20). The axoneme is accordingly not only the locus of motility mechanisms of the cilium, but also the locus of mechanisms of response to second messengers.

The major structures comprising the axoneme are for the most part well known, and a complete three-dimensional reconstruction of the structure based on electron microscope analysis has been achieved at a resolution approaching 4 nm [Fig. 20.8; (197)]. In addition to the doublet microtubules and dynein arms, the main features of the 9 + 2 axoneme includes several sets of nonarm linkers between microtubules, the most prominent of which are the radial spokes. The radial spokes extend from a specific protofilament of subfiber A to interact with the projections around the central single microtubules, known as the central sheath. Spokes occur in groups of three, with spacing 32:24:40 nm along the microtubule of many cilia to give a longitudinal repeat of 96 nm; the third spoke is missing in each group in *Chlamydomonas* axonemes. Radial spoke assembly is usually coordinated with the growing axoneme, but where spokes are added to fully assembled spokeless axonemes, assembly surprisingly occurs backward from the tip of the axonemes downward toward the base (92).

In straight regions of the axoneme, spokes generally lie perpendicular to the longitudinal axis of their microtubule of origin and to the central sheath, suggesting that they are unattached to the sheath. However, the heads of spokes in or near the neutral plane in certain bent cilia do firmly attach to the central sheath, and the attachment evidently varies with position of the axoneme (10, 174). Spoke–central-sheath attachment may therefore change during a beat; this could be important in the conversion of sliding to bending. In *Chlamydomonas,* there are 17 radial spoke components and mutants defective in spoke assembly have been identified. Central pair mutants, where the central sheath is correspondingly unassembled, have also been found. In both types of mutants, consistent with the importance of a spoke–central-sheath interaction in

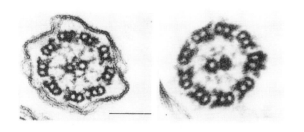

FIG. 20.8. *Above:* Cross sections of an intact cilium *(left)* and axoneme after detergent treatment *(right)*. [From Satir (168) with permission.] Bar = 0.1 μm. *Below:* Three-dimensional computer model of ciliary axoneme at resolution approaching 4 nm, based on electron micrographs of ciliary axoneme. See [Sugrue et al. (197) for details. With permission.]

the bending mechanism, motility is absent or very limited [*cf.* (85)]; however, microtubule sliding disintegration of the axonemes will still take place. The velocity of microtubule sliding is diminished in the spokeless mutants, but it can be restored if dynein from axonemes with wild-type radial spokes is appropriately recombined with the axonemes (187). The radial spokes evidently signal the dynein arms in such a way that some instantaneous activation of the arm occurs. The signal results in a change in the dynein molecule that survives extraction of the arms from the axoneme.

Suppressor mutants have been found where beat is reinitiated without spoke reassembly. These have long been a puzzle, but these new findings suggest that in the suppressor mutants the appropriate arm activation change leading to a bend might occur, despite bypassing of the radial spoke signal. In the evolution of sperm, a corresponding bypass might lead to motile 9 + 0 sperm, which are present in a few organisms.

The central sheath to which the spoke heads may transiently attach generally consists of a series of projections around the central microtubules. These vary somewhat in morphological details from species to species. In *Tetrahymena,* one central microtubule is enclosed in a series of D-shaped rings which repeat every 16 nm. There are other projections around the second microtubule, making the members of the pair readily distinguishable. Individual spoke heads from four to six doublets could interact with the D-rings, while spoke heads from the other doublets could interact with the additional projections. Little is known about the biochemistry of the central sheath.

Near the bases of the spokes, a set of circumferential interdoublet links is seen. These links occur in pairs every 96 nm along the doublet, the basal link in each pair being aligned with spoke 2 in each spoke group, the distance between members of the pair being 16 nm. It seems likely that the links extend from near the midwall of one doublet to the midwall of the next doublet. The links are virtually inextensible, so they detach from doublet $N + 1$ reversibly when sliding between adjacent doublets exceeds about 20 nm (26, 225). They remain attached to the doublet wall after the extraction of the dynein arms and serve to maintain the form of the axoneme. In some organisms the axonemal microtubules splay apart at the tips of the cilia, possibly because the links are not assembled in the distal parts of the axoneme.

A second set of links may be present that repeats with a 24 nm period and connects subfiber A of doublet N to subfiber B of doublet $N + 1$. Where not obscured by the dynein arms, these structures appear as fibers in the interdoublet gap in freeze-etch preparations (197). However, Goodenough and Heuser (65) interpret such fibers (B-links) as integral parts of the arm. These linkages must eventually break in a cooperative manner during active sliding of adjacent microtubules (80). They may be similar to high molecular weight MAPs. Nevertheless, the elastic properties of both sets of links may be important in determining how much displacement is possible in a small, defined region of the axoneme, and thus how sliding is transduced into a growing bend. The radial spokes, central sheath, and linkage molecules work together, primarily with the inner arms.

Experimental Analysis of Motility

The Mechanochemistry of Sliding. In the absence of the constraints such as spoke–central-sheath interactions, circumferential links, or an intact regulatory system, in the presence of Mg^{2+}-ATP the axoneme telescopes apart. In contrast to the situation in the beating cilium, the sliding of axonemal microtubules on one another continues to completion, producing a displacement of 10 μm or more, so adjacent doublets end up with only a small overlap region (0.5 μm or so) connecting them (Fig. 20.9). The velocity of sliding of a given doublet is constant as the axoneme elongates to this point (202). Over a wide range, then, sliding velocity is independent of the number of dyneins present and of the load—that is, the whole microtubule length is being pushed by one dynein. Velocity depends on ATP concentration with a K_m of 3×10^{-4} M Mg^{2+}-ATP.

In this sliding system, the force produced by one arm is about 1 pN (96). This is about the same force as is produced by a single kinesin molecule, which is capable of moving a 10 μm–long microtubule for the time that the molecule remains attached to the microtubule (the duty cycle). As discussed previously, Vale et al. (211) have demonstrated that the duty cycle of dynein is much shorter than that of kinesin. Dynein remains attached for only a small fraction of each arm cycle, perhaps 1–2 ms in a cycle of perhaps 33 ms. For continuous sliding at maximum velocity, enough dyneins need to be working so that one dynein is actively pushing the microtubule at all times. As this number decreases, sliding velocity will fall proportionately.

Sliding requires at least one of the globular domains of the dynein molecule to push the microtubule and in in vitro assays, one head can be shown to be sufficient to generate the needed force (214). In the sliding axoneme, the attachment cycle is accompanied by large conformational changes in the relationships of the multiple heads of each dynein molecule. Presumably, such conformational changes are driven by the release of products after the hydrolysis of ATP. In negative stain, arms can be found extended to their full length and tilted about 40° toward the (−) end of the microtubule, but rigor arms lie perpendicular to the long axes of the doublet microtubules and are attached at both ends (7). A sliding step of 16 nm can be calculated from this geometry (176). The conformational changes in this cycle have been modeled by Sugrue et al. (97).

One probable conclusion from these measurements is that continuous sliding of a microtubule for several micrometers at rates of up to 19 μm s^{-1} requires the action of many different dynein arms, cycling out of phase. At 10 μm s^{-1}, for example, a displacement of

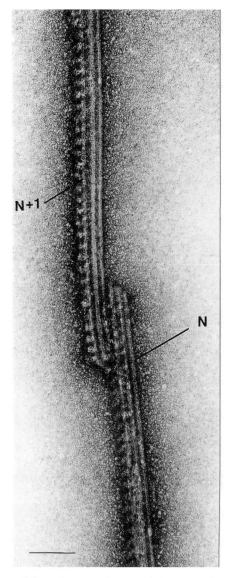

FIG. 20.9. Sliding of axonemal microtubules captured in negative stain electron microscopy; active arms on subfiber A of doublet N push doublet N+1 in a tipward (+) direction, while doublet N moves relatively baseward. [Satir (168) with permission.] Bar, 0.1 μm.

one-third of a μm will take 33 ms, about 1 arm cycle time, but such a displacement would require about 20 arm steps. Twenty asynchronous steps of 1–2 ms would provide the continuous force needed for smooth motion, but would require a minimum of 20 arms or groups of arms cycling with small phase differences. If the arm step size were much larger than 16 nm, fewer steps would be needed, but the microtubule would move in fits and starts in a low Reynolds number regime. The 20 arms could not be localized to one small part of the microtubule (other than the very tip) or motion would cease when sliding passed this point, which infers that hydrolysis is probably stochastic with respect to position along the microtubule.

The Switch Point Hypothesis. Although microtubule sliding in protease-treated axonemes goes to completion and microtubule gliding in translocation assays using dynein as a motor is continuous in one direction, microtubule sliding in an intact axoneme is extremely limited both in time and amount—and it is bidirectional and cyclic. However, sliding in normally bending, intact axonemes occurs at rates that are similar to, or only marginally different from, in vitro translocation rates or rates of unconstrained sliding. In the example given earlier, a bend of 100° is caused by sliding displacements of about 0.1 μm between adjacent doublets, but this phase of bending and sliding lasts perhaps 10 ms in a cilium beating normally at 25 Hz (Table 20.1). The velocity of sliding during this period is 10 μm s^{-1} (the same velocity considered in the last paragraph), but the duration of unidirectional sliding and the distance moved are only 1% of the sliding seen after proteolysis or in an in vitro assay. Furthermore, in a bending axoneme some doublets are moving in the wrong direction, as if they could not be actively translocated by dynein. In these doublets, $N+1$ moves baseward with respect to doublet N. Therefore, during the brief duration of bend formation that is being discussed, some doublets have active arms, while others do not.

Active arms are those that can hydrolyze ATP and complete a mechanochemical cycle; inactive arms are temporarily not cycling. The best explanation for inac-

TABLE 20.1. *Control of Ciliary Beat Frequency* *

1. $F_2 > F_1$		
2. $\Delta l_E = d_n \Sigma \alpha_E = V_E t_E$		
Quantitative Considerations		
	$F_1 = 16.7$ Hz	$F_2 = 25$ Hz
One beat takes	60 ms	40 ms
α_E	100°	100°
t_E	~15 ms	~10 ms
Δl_E N vs. N+1	0.1 μm	0.1 μm
v_E	6.7 μm/s	10 μm/s

* This table calculates the velocity of microtubule sliding (v_E) for the same cilium beating at two different frequencies (F_1 and F_2), assuming that there is no change in beat form. The formation of a bend (α_E) from a straight position during the effective stroke takes about one-quarter of the stroke time (t_E). The amount of sliding (Δl_E) is calculated from equation (1), where d_n is a constant related to the distance between adjacent doublets N and $N+1$. [From Satir et al. (171) with permission.]

tivity is analogous to that proposed by Vale et al. (211) for the bidirectionality of sliding with mixed motor molecules: When a microtubule is moving in the wrong direction with respect to the polarity of doublet $N+1$, arm attachment becomes negligible. Assuming that the step size during one cycle is 16 nm, the number of active steps needed to move adjacent doublets by 0.1 μm is about six. This is consistent with a duty cycle of 1–2 ms. During a complete effective stroke, which would last less than 20 ms at 25 Hz, between 10 and 20 arm cycles would be needed per active doublet as $\Sigma\alpha$ changes from $-80°$ to $+100°$. Why then are there hundreds of outer and inner arms per 10 μm–long doublet in the axoneme? A probable answer is that, in the axoneme, arm activity is stochastic, as in the example in the last section, and cooperative—in that groups of four arms corresponding to a spoke group probably cycle synchronously (190)—that is, one cannot specify which 10%–20% of the groups are hydrolyzed during any given effective stroke, any more than the position of dyneins that hydrolyzed during doublet sliding in vitro can be specified. But something happens at the end of the effective stroke that does not happen in doublet sliding after protease digestion of the spokes and links: The direction of sliding reverses.

The switch point hypothesis states that at the end of the effective stroke, active sliding in one set of doublets is switched off and a new set of doublets becomes activated to move the axoneme in the recovery stroke. The hypothesis identifies the doublets that are active by their direction of sliding at any given stage of the beat. It follows that half of the axoneme has active arms when moving in the effective stroke direction, while the other half has active arms when moving in the recovery stroke direction. In mussel gill ciliary axonemes, for example, during the effective stroke, where doublets 5–6 slide tipward, doublets 1–4 have active arms, but during the recovery stroke, where doublet 1 slides tipward, doublets 6–9 are active (Fig. 20.10). Switching is probably controlled by the regulatory complex of the axoneme consisting of spokes, links, and inner arms, which measures the amount of distortion of the axoneme by some unknown mechanism. The spoke–central-sheath signal seems important in this process.

One line of experimentation that has lent strong support to the hypothesis involves the use of inhibitors to produce ciliary arrest. In a variety of organisms, two different arrest positions have been described; these are often produced in physiological situations such as nerve stimulation. One position is near the beginning of the effective stroke and is labeled "hands up"; the other is near the beginning of the recovery stroke and is labeled "hands down" (Fig. 20.11). Axo-

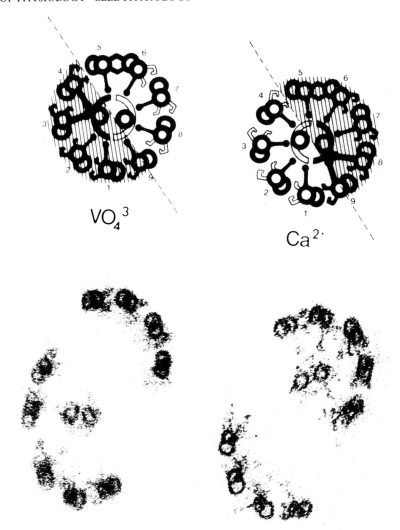

FIG. 20.10. Axonemal splitting in "hands down" vs. "hands up" axonemes supports "switch point hypothesis" of dynein arm activity. *Above:* Predictions of arm activity in axonemes arrested by ions shown. *Shaded areas* and *black arms* indicate predicted active arms. *Below:* Split axonemes *Left:* "hands down." *Right:* "hands up." The central pair splits with active half axoneme. [From Satir (169) with permission.] Bar = 0.1 μm.

nemes can be moved from the "hands up" to the "hands down" position and vice versa without restarting beat by changing inhibitors [*cf.* (149)]. This implies that such inhibitors do not inhibit sliding per se, but rather act on the switching mechanism. There are two switches, one where doublets 1–4 switch on, one where doublets 6–9 switch on, that can be inhibited independently.

To test the hypothesis, Satir and Matsuoka (174) have used known inhibitors to freeze the intact axonemes of mussel gill lateral cilia in either "hands up" or "hands down" positions where one would predict

that half of the doublets possessed active arms and half possessed inactive arms. Then they digested the axonemes briefly with protease and reactivated them in the presence of the inhibitor to prevent switching. The active arms should be capable of producing sliding in their half of the axoneme, but the inactive arms should not, so half of the doublets should potentially slide away from the other, passive half. The doublets that are active actually produce a torque that tears the axoneme apart, with the separation occurring at a border between active and inactive half axonemes. In "hands up" vs. "hands down" axonemes the split

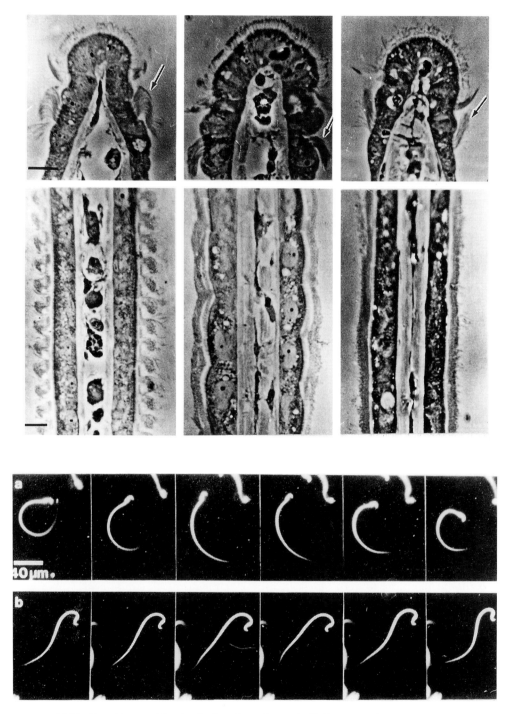

FIG. 20.11. *Top:* Cross and longitudinal sections of mussel gill filaments with active *(left panels)* and arrested *(middle, right panels)* lateral cilia *(arrows)*. Repeating pattern shows metachronal waves of beating cilia captured by fixation. Arrested cilia are in either "hands down" *(middle panels)* or "hands up" *(right panels)* configuration, corrresponding to end of effective and recovery strokes, respectively. [From Satir and Matsuoka (174) with permission.] Bar, 10 μm. *Bottom:* Rat sperm flagellar axonemes reactivated in low ATP concentrations (0.3 mM) under conditions where only half beats are present *(a)* in EGTA with no added Ca^{2+} or *(b)* with Ca^{2+}, balancing chelator concentration. Bends form in one direction only. At higher ATP concentrations, full beating resumes. [From Lindemann and Goltz (109) with permission.]

patterns are reversed (Fig. 20.10), demonstrating that different sets of doublets are active.

According to the switch point hypothesis, the effective stroke in ciliary axonemes corresponds to the formation of a principal bend in sperm motility, while the recovery stroke corresponds to the formation of a reverse bend (172). Lindemann and Goltz (109) produced axonemes where principal bend formation in rat sperm was inhibited, but when ATP was added, the reverse bend still formed, relaxed, and reformed; where the reverse bend was inhibited, principal bends still formed (Fig. 20.11). This demonstrates quite dramatically that opposite halves of the axoneme act independently of one another to form oppositely directed bends. Further, in sliding experiments with similar axonemes (110), two distinct "split" patterns are observed, essentially confirming the Satir and Matsuoka result with a completely different system. Similar results are seen by Sale (156) and Tamm and Tamm (205).

Comparison of the behavior of sperm flagella with cilia in these experiments reveals a similarity of basic mechanism that is one of the attractive features of the switch point hypothesis. One problem that any hypothesis accounting for ciliary motility must address is that beat phenotype varies greatly, but axonemal structure, dynein mechanochemistry, and sliding mechanisms are pretty much the same everywhere. The switch point hypothesis accounts for variations in beat form characteristic of flagella vs. cilia in different eukaryotic organisms, and for variations of beat phenotype during behavioral responses in an organism like *Chlamydomonas*, by assuming only that the timing of switching can be varied in an axoneme. The amplitude of a new bend depends solely on the "on-time" of dynein-arm activity in the half-axoneme after the overall Δln for every doublet returns to zero ($\Sigma\alpha = 0$), a condition that could reset the elastic constraints within the regulatory system that are presumably responsible for the timing mechanism.

$$\alpha_p = v_p\, t_p \text{ and } \alpha_r = v_r\, t_r \qquad (20\text{-}2)$$

where α = amplitude of the new principal (p) or reverse (r) bend, v is the velocity of sliding, and t is the "on-time" of arm activity for the half-axoneme (175). Where the "on-times" of the two half-axonemes are identical, the principal and reverse bends will have the same amplitude and the beat will be more or less undulatory; where the "on-times" differ, the bends will by asymmetrical, the longer "on-time" corresponding to the larger, principal, bend in the effective stroke direction; the shorter to the smaller, reverse, bend in the recovery stroke direction. One particular difficulty for most mechanisms—an explanation of ciliary rever-

sal, where the effective stroke of a cilium suddenly occurs in the opposite direction from its normal orientation—is resolved simply by reversing the "on-times" of the half-axonemes.

The switch point hypothesis is not universally accepted, and several modifications may be necessary. It may be that in some instances, all the arms along a particular doublet may not activate simultaneously, but rather in progression with bend propagation. In a very careful analysis of sea urchin sperm flagellar beat, Brokaw and Gibbons (33) and Gibbons (54) have distinguished two forms of curvature change: "synchronous sliding," which corresponds to what we have been discussing in the switch point mechanism, and "metachronous sliding," where there is no overall change in bend angle, but rather a displacement of the bends along the axoneme. Essentially, metachronous sliding is bend propagation [see also (32)]. The switch point hypothesis does not specify how bend propagation occurs—this could be a mechanism of local dynein-activated sliding (although this would require a different mechanism from the simultaneous arm activation involved in bend formation) or it might involve elastic restoring forces in the spokes and links with no arm activation at all.

There are extensive feedback systems in the axoneme relating bend generation and propagation. As noted above, the nature of the clock mechanism is also unspecified by the switch point hypothesis. Beat form is controlled by the regulatory complex and in particular by the inner arm dyneins, but the way this is translated into switching is unknown. Although the activity of dynein probably depends on the direction of doublet sliding relative to its neighbor, the question of when and why this changes has not been answered and it is not clear whether switching has a mechanical, as well as a biochemical, component. Curvature control models, where physical forces regulate arm activity, are successful to a point (31). An interesting approach to the clock mechanism has been developed by Eshel et al. (44) using a sea urchin flagellum entrained to a vibrating micropipette. They rapidly displaced the pipette, which interrupted bend formation, and stopped the vibration. The flagellum then started to beat spontaneously, initiating the bend that had been delayed because of the displacement, at least for short time periods. This suggests that information regarding which switch is on lasts through the displacement, independent of duration of the displacement.

An additional problem is that bend development can occur as an opposing pair (62) so that, at all times during the growth of the pair, $\Sigma\alpha = 0$. This has been elegantly demonstrated by Shingyoji et al. (180). When both ends of an axoneme are incapable of sliding, development of a matched, opposing pair of bends is the only way in which sliding can be accommodated, but this does not speak to whether all doublets are active or, as predicted by the switch point hypothesis, only a specific subset of active doublets produces both bends. In a variation of the Shingyoji experiment, Yeung and Woolley (231) demonstrated that depending on initial position, upon ATP addition in hamster sperm flagella, new sets of bends would initiate with either the principal bend basally and the reverse bend distally or vice versa, consistent with activation of specific doublet sets 1–4 or 6–9.

A more minor modification of the switch point hypothesis has also been considered, particularly by Sugino and Machemer (195, 196). In this variation, all the doublets in a half-axoneme would not activate simultaneously, but rather with a defined phase relationship, so that there would be nine switches, rather than two. This is attractive especially if the central pair could rotate around the axoneme, like a commutator, activating each doublet in turn during a beat. Evidence that the central pair can sometimes rotate during a beat is adduced (134), but in other instances—for example, in ctenophore (204), sea urchin (156), or mussel gill lateral cilia (164), rotation during normal beat (that is, at say 25 Hz) seems absent or highly unlikely. (Otherwise, why is the relationship between the plane of the central pair and doublets 3–8—the neutral plane—always the same, regardless of beat position?)

Although the sets of doublets in the two half-axonemes are usually fixed physiologically, they can be changed, in some instances experimentally. Shingyoji et al. (179) used a sea urchin flagellum entrained to a vibrating micropipette. They rotated the plane of pipette vibration around the axis, inducing a corresponding rotation in the plane of beating in both live and reactivated sperm. Markers on the doublets were unaffected by the rotation. This would imply that neither the plane of beat nor the direction of the principal bend depends on a specific set of doublets. Both can be forced to change by rotating some internal structure in the axoneme. The logical candidate is the central pair. The central pair is in the perfect position to divide the axoneme into the appropriate halves. Even though it does not revolve with every beat, these experiments imply that the central pair can be moved freely around the axoneme to produce different half-axonemal positions. The axoneme unwinds spontaneously back to its original position, however, if the imposed rotation is released, suggesting that the normal half-axonemes predicted by the switch point hypothesis are the physiologically relevant, stable ones under most conditions. Nonetheless, particularly in protozoa, but possibly also in sperm, there may be biological ways to reset the central pair so that the plane of the beat changes with respect to the cell. This could have important implications for the behavioral responses of such cells.

Cellular Control of Ciliary Motility

Ciliary beat is under cellular control. Either the frequency of beat or the form of beat can be changed. The cell can also arrest and reinitiate motility. Changes generally are brought about by second-messenger molecules that interact with the axoneme, as illustrated briefly earlier for *Paramecium*. In addition to *Paramecium*, *Chlamydomonas* provides one of the best studied and most revealing systems in this regard, but the same principles apply in many details to invertebrate and vertebrate ciliated epithelial cells and to sperm.

Pairs of demembranated *Chlamydomonas* flagella remain attached when isolated appropriately, and they will reactivate in ATP solutions to swim with the usual breast stroke characteristic of flexural ciliary motion in forward-swimming cells. Hyams and Borisy (86) studied the beat form as a function of Ca^{2+} in the reactivation medium. Normal beat of both axonemes is seen at pCa8. Ca^{2+} acts as a second messenger to change the beat form. At pCa's below 6, an undulatory beat more characteristic of eukaryotic flagella or sperm axonemes is seen. Single isolated axonemes treated with Ca^{2+} also undergo this response (16). Kamiya and Witman (97) showed that submicromolar levels of Ca^{2+} affect the two axonemes of *Chlamydomonas*, *cis* and *trans* to the eyespot, differentially; this provides a basis for turning associated with phototactic behavior. This implies that a Ca^{2+}-sensing system is an intrinsic part of the axoneme in these cells. Changes in beat form occur when Ca^{2+} binds to some axonemal sensor in a reversible manner.

Two fundamental questions can be asked about the interaction of Ca^{2+} with the axoneme: *(1)* What is the Ca^{2+} sensor? and *(2)* How does the sensor work to cause the changes in beat when Ca^{2+} concentration changes? The dimensions of these problems have been defined, but the answers remain obscure. Two Ca^{2+}-binding proteins are known to be present as fixed components of the axoneme: calmodulin (CaM) and centrin (caltractin)—but there may be others. The stoichiometry of CaM in the axoneme of many organisms is remarkably constant, at about 1 CaM per 60 tubulin dimers (135). The stoichiometry of centrin along the axoneme is also fixed and centrin may be part of the regulatory complex located at selected sites near the inner arms in *Chlamydomonas* (141). Evidence that CaM is the Ca^{2+} sensor is limited, but suggestive. Anticalmodulin drugs reverse Ca^{2+}-induced behavior in the axoneme in the presence of high Ca^{2+} (pCa < 6) (136). In *Paramecium*, two axonemal polypeptides (95 kDa and 105 kDa) bind CaM at low Ca^{2+} concentrations, but binding is inhibited when Ca^{2+} concentration increases beyond 10^{-6} M. The possible role of CaM as an axonemal sensor has been reviewed by Otter (135).

The switch point hypothesis is based in part on the assumption that Ca^{2+} operates on the switching mechanism of the axoneme. At high Ca^{2+} concentrations, in many cilia, Ca^{2+} inhibits one switch and produces ciliary arrest. The "hands up" position in mussel gill lateral cilia is produced in this way (149, 174); the cilia move from any position through the stroke to be caught and held in the arrest position. In *Chlamydomonas*, these arrests occur as part of the photophobic response, before beat form changes from ciliary to undulatory type or back again. This implies that in this system, as in *Paramecium*, too, the Ca^{2+} switch has more than one setting. Arrest occurs in a single small range of Ca^{2+} concentrations. Above and below this range, the switch point hypothesis predicts that differences in timing of sliding in the Ca^{2+}-sensitive half-axoneme vs. the other half will lead to changes in beat form. Assuming that CaM is part of the Ca^{2+} sensor, the ranges for normal beat, arrest, and altered beat could correspond to Ca^{2+}-free CaM, CaM with 1 Ca^{2+} bound, and CaM with multiple Ca^{2+}'s bound, respectively (166a), but a second possibility is that the multiplicity of responses means that more than one kind of Ca^{2+}-binding protein is present

as part of the sensor. Otter (135) points out that Ca^{2+} might affect the switch mechanism in different ways: either by a structural mechanism, analogous to troponin C regulation in striated muscle, or enzymatically, probably via phosphosphorylation or dephosphorylation of specific axonemal proteins, or by a combination of these methods. As we would predict, the locus of action would seem to be primarily the inner arm regulatory complex (141). When the regulatory system of spokes etc. is destroyed by proteolysis, although switching is destroyed, sliding per se is little affected by Ca^{2+} at concentrations that produce ciliary arrest (224).

More recent studies make it probable that there is a secondary effect of Ca^{2+} on sliding velocity in many systems, which translates into an effect on beat frequency and swimming speed. In mammalian cilia, for example, Ca^{2+} is a major player in the increase of beat frequency (161). The effects of Ca^{2+} on both the switching mechanism controlling doublet activation, which affects beat form, and on the rate of doublet sliding, which affects beat frequency, indicate that there will usually be considerable crosstalk between the independent systems controlling beat form and beat frequency.

Mutants of *Chlamydomonas* missing inner arms can beat at wild-type frequencies but with highly modified beat patterns, while those missing outer arms can beat with normal bend patterns, but at frequencies that are greatly reduced [Fig. 20.12; (34)]. This suggests that, as a first approximation, the inner arms and regulatory

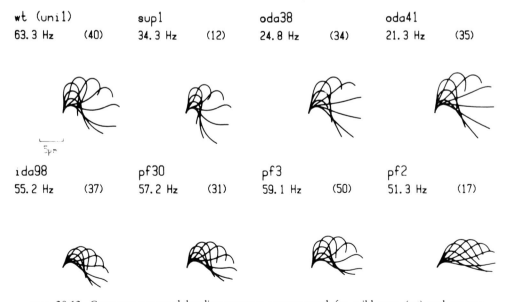

FIG. 20.12. Computer-generated bending patterns reconstructed for wild type *(wt)* and mutant *Chlamydomonas*. Each pattern shows images of 0.125 cycle intervals with cell body fixed in position. *Above:* Mutants missing outer dynein arms compared to wt in form and beat frequency. *Below:* Mutants missing inner dynein arms. Sample size indicated. [From Brokaw and Kamiya (34) with permission.]

complex work together primarily to govern the characteristics of bend generation and beat form, while the outer arms (three-headed 22–23S dyneins) regulate beat frequency. In the absence of changes in beat form, beat-frequency changes depend only upon the velocity of sliding of the active doublets. At a frequency of 16.7 Hz, a bend of 100° will be produced in about 15 ms, but at a frequency of 25 Hz the same amount of sliding, causing the same bend, will take only 10 ms [Table 20.1; (171)]. In general, beat-frequency increases depend on increases of doublet sliding velocity produced by outer arm dynein. A 50% increase in frequency is produced by a 50% increase in sliding velocity. Any agent that changes the properties of the mechanochemical cycle of the outer arm will alter beat frequency.

Many cells control the beat frequency of their cilia by increases in intracellular cAMP. For example, the frequency of beat of mussel gill lateral cilia is under nervous control; the neurotransmitter that produces a beat frequency increase is serotonin, which acts via membrane receptors to increase cAMP [(124), see review by (192)]. In *Paramecium,* a mechanosensitive K^+ channel seems directly coupled to adenylyl cyclase; opening of this channel leads to increased intracellular cAMP and to fast forward swimming by the cell (178). Bonini et al. (21) and Bonini and Nelson (22) showed that monobutyryl-cAMP increases the swimming speed of living *Paramecium.* Recall that cAMP added directly to reactivated permeabilized *Paramecium* increases their swimming speed.

In *Paramecium* a 29 kDa axonemal polypeptide (p29) is phosphorylated in a cAMP-dependent manner in swimming permeabilized cells and in isolated axonemes (23, 74). This cAMP-dependent phosphoryla-

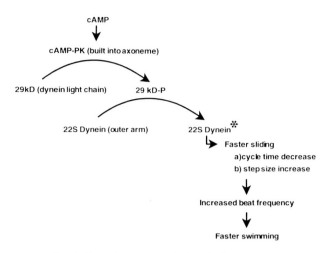

FIG. 20.14. Transduction cascade by which cAMP activates 22S outer arm dynein to increase microtubule sliding velocity, beat frequency, and speed of ciliate swimming. [Adapted from Satir et al. (171) with permission.]

tion is inhibited at high Ca^{2+} concentration. The kinase for this reaction is an integral component of the axoneme. p29 cosediments with 22S dynein, but not with *Paramecium* 14S dynein, and is probably a dynein regulatory light chain.

To test whether cAMP-dependent phosphorylation of this polypeptide regulates microtubule sliding velocity, Hamasaki et al. (73) prepared 22S dynein with and without this phosphorylated species and measured in vitro translocation velocities. If the 29 kDa polypeptide is phosphorylated (actually, thiophosphorylated to prevent the action of phosphatases during the dynein extraction procedure), a 40% increase in velocity is observed (Fig. 20.13), consistent with the direct control of outer arm sliding velocity and hence beat frequency by this phosphorylation. Using chymotryptic digestion, Barkalow et al. (12) have shown that p29 is associated with one specific head of the three-headed 22S molecule; perhaps this is the head that interacts with subfiber B of doublet $N + 1$ in the force-producing step of the mechanochemical cycle.

To test the physiological significance of these findings, swimming speed of permeabilized cells was determined under conditions where p29 was thiophosphorylated, but cAMP itself was removed. A 40%–135% increase in swimming speed was observed (73). The proposed mechanism of cAMP action is diagrammed in Fig. 20.14. Cells evidently control beat frequency by regulating microtubule sliding velocity of outer arm dynein in a manner analogous to the regulation that occurs in some myosins. Sometimes Ca^{2+} is used as the second messenger instead of cAMP. It will be interesting to learn whether cytoplasmic dynein can be controlled in a similar manner. The presence of high

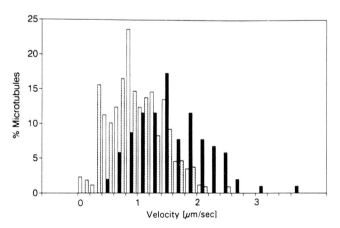

FIG. 20.13. Velocity profiles of microtubule translocation over *Paramecium* 22S dynein comparing controls *(open bars),* dynein pretreated with cAMP to phosphorylate p29 *(black bars),* and dynein pretreated with cAMP and Ca^{2+} *(hatched bars).* [From Hamasaki et al. (73) with permission.]

concentrations of cAMP increases the elongation rate caused by microtubule sliding in permeabilized fish cones (35), which suggests that regulation of cytoplasmic dynein by cAMP-dependent phosphorylation is a possibility.

The cellular control of beat frequency by an increase in microtubule sliding rate mediated by a second-messenger cascade that acts specifically on outer arm dynein is a paradigm for other regulatory controls of microtubule motor molecules, but such controls may be more complicated, involving more components of the system, possibly in a variety of ways, as has been discussed for Ca^{2+} control of beat form. Even for this relatively straightforward cascade, the resultant mechanochemical changes that produce faster sliding are not well understood. Nevertheless, it will be important to sort out the mechanisms of control more completely to comprehend the functions of microtubules and their motors in cytoplasmic processes and in motile cilia.

I thank my colleagues T. Hamasaki, K. Barkalow, N. Isaac, S. Nair, and M. Holwill for helpful discussion on certain aspects of this manuscript. N. Isaac also provided excellent assistance in assembling the photographic and other materials for the figures. Ann Holland provided outstanding secretarial and bibliographic support, for which I am very grateful. Partial support for this work was provided by DK-41918.

REFERENCES

1. Afzelius, B. A. Electron microscopy of the sperm tail: results obtained with a new fixative. *J. Biophys. Biochem. Cytol.* 5: 269–278, 1959.
2. Albrecht-Buehler, G. The orientation of centrioles in migrating 3T3 cells. *Exp. Cell Res.* 120: 111–118, 1979.
3. Allen, R. D., D. G. Weiss, J. H. Hayden, D. T. Brown, H. Fujiwake, and M. Simpson. Gliding movement of and bidirectional organelle transport along single native microtubules from squid axoplasm: evidence for an active role of microtubules in cytoplasmic transport. *J. Cell Biol.* 100: 1736–1752, 1985.
4. Amos, L. A. Arrangement of high molecular weight associated proteins on purified mammalian brain microtubules. *J. Cell Biol.* 72: 642–654, 1977.
5. Amos, L. A., and W. B. Amos. *Molecules of the Cytoskeleton.* New York: Guilford Press, 1991, 253 pp.
6. Asai, D. J., and C. J. Brokaw. Dynein heavy chain isoforms and axonemal motility. *Trends Cell Biol.* 3: 398–402, 1993.
6a. Asai, D. J., S. M. Beckwith, K. A. Kandl, H. H. Keating, H. Tjandra, and J. D. Forny. The dynein genes of *Paramecium tetraurelia. J. Cell Sci.* 107: 839–847.
7. Avolio, J., S. Lebduska, and P. Satir. Dynein arm substructure and the orientation of arm-microtubule attachments. *J. Mol. Biol.* 173: 389–401, 1984.
8. Bacallao, R., C. Anthony, C. Dotti, E. Karsenti, E. H. Stelzer, and K. Simons. The subcellular organization of Madin-Darby canine kidney cells during the formation of a polarized epithelium. *J. Cell Biol.* 109: 2817–2832, 1989.

9. Bailey, E., and N. Bornens. Centrosome and cell division. *Nature* 355: 300–301, 1992.
10. Ballowitz, E. Untersuchungen uber die Structur der Spermatozoen, zugleich ein Beitrag zur Lehre von Feineren bau der contractilen Elemente. *Arch. Mikr. Anat.* 32: 401–473, 1888.
11. Barkalow, K., J. Avolio, M.E.J. Holwill, and P. Satir. Structural and geometrical constraints on the outer dynein arm *in situ. Cell Motil. Cytoskeleton* 27: 299–312, 1994.
12. Barkalow, K., T. Hamasaki, and P. Satir. Regulation of 22S dynein by a 29 kDa light chain. *J. Cell Biol.* 126: 727–735, 1994.
13. Beckwith, S., K. Kandl, J. Forney, and D. Asai. Ciliary and cytoplasmic dynein isoform expression in *Paramecium tetraurelia,* 1993 (personal communication).
14. Beese, L., G. Stubbs, and C. Cohen. Microtubule structure at 18Å resolution. *J. Mol. Biol.* 194: 257–264, 1987.
15. Behnke, O., and A. Forer. Evidence for four classes of microtubules in individual cells. *J. Cell Sci.* 2: 169–192, 1967.
16. Bessen, M., R. B. Fay, and G. B. Witman. Calcium control of waveform in isolated flagellar axonemes of *Chlamydomonas. J. Cell Biol.* 86: 446–455, 1980.
17. Block, S. M., L.S.B. Goldstein, and B. J. Schnapp. Bead movement by single kinesin molecules. Studied with optical tweezers. *Nature* 348: 348–352, 1990.
18. Bloodgood, R. A. *Ciliary and Flagellar Membranes.* New York: Plenum, 1990, 431 pp.
19. Bloom, G. S., M. C. Wagner, K. K. Pfister, and S. T. Brady. Native structure and physical properties of bovine brain kinesin, and identification of the ATP-binding subunit polypeptide. *Biochemistry* 27: 3409–3416, 1988.
20. Bonini, N. M., T. C. Evan, L.A.P. Miglietta, and D. L. Nelson. The regulation of ciliary motility in *Paramecium* by Ca^{2+} and cyclic nucleotides. *Adv. 2nd Messenger Phosphoprotein Res.* 23: 227–272, 1991.
21. Bonini, N. M., M. C. Gustin, and D. L. Nelson. Regulation of ciliary motility by membrane potential in *Paramecium:* a role for cyclic AMP. *Cell Motil. Cytoskeleton* 6: 256–272, 1986.
22. Bonini, N. M., and D. L. Nelson. Differential regulation of *Paramecium* ciliary motility by cAMP and cGMP. *J. Cell Biol.* 106: 1615–1623, 1988.
23. Bonini, N. M., and D. L. Nelson. Phosphoproteins associated with cyclic nucleotide stimulation of ciliary motility in *Paramecium. J. Cell Sci.* 95: 219–230, 1990.
24. Borisy, G. G., and E. W. Taylor. The mechanism of action of colchicine. Colchicine binding to sea urchin eggs and the mitotic apparatus. *J. Cell Biol.* 34: 535–548, 1967.
25. Bowser, S. S., and L. Leonard. Properties of primary (9 + 0) cilia in cultured kidney epithelial cells. *Anat. Rec.* 232: 14A, 1992.
26. Bozkurt, H. H., and D. M. Woolley. Morphology of nexin links in relation to interdoublet sliding in the sperm flagellum. *Cell Motil. Cytoskeleton* 24: 109–118, 1993.
27. Bradfield, J. R. G. Fibre patterns in animal flagella and cilia. *Soc. Exp. Biol. Symp.* 9: 306–334, 1955.
28. Bre, M., and E. Karsenti. Effect of brain microtubule-associated proteins on microtubule dynamics and nucleating activity of centrosomes. *Cell Motil. Cytoskeleton* 15: 88–98, 1990.
29. Brinkley, B. R. Microtubule-organizing centers. *Annu. Rev. Cell Biol.* 1: 145–172, 1985.
30. Brokaw, C. J. Direct measurements of sliding between outer doublet microtubules in swimming sperm flagella. *Science* 243: 1593–1596, 1989a.
31. Brokaw, C. J. Operation and regulation of the flagellar oscilla-

tor. In: *Cell Movement,* edited by F. D. Warner, P. Satir, and I. R. Gibbons. New York: Alan Liss, 1989b, vol. 1, p. 267–278.

32. Brokaw, C. J. Microtubule sliding in swimming sperm flagella: direct and indirect measurements on sea urchin and tunicate spermatozoa. *J. Cell Biol.* 114: 1201–1215, 1991.

33. Brokaw, C. J., and I. R. Gibbons. Mechanisms of movement in flagella and cilia. In: *Swimming and Flying in Nature,* edited by T. Y-T Wu, C. J. Brokaw, and C. Brennan. New York: Plenum, 1975, vol. 1, *p.* 89–126.

34. Brokaw, C. J., and R. Kamiya. Bending patterns of *Chlamydomonas* flagella IV. Mutants with defects in inner and outer dynein arms indicate differences in dynein arm function. *Cell Motil. Cytoskeleton* 8: 68–75, 1987.

35. Burnside, B. Microtubule sliding and the generation of force for cell shape change. In: *Cell Movement,* edited by F. D. Warner and J. R. McIntosh. New York: Alan Liss, 1989, vol. 2, p. 169–189.

36. Caplow, M. Microtubule dynamics. *Curr. Opin. Cell Biol.* 4: 58–65, 1992.

37. Carlier, M. F., D. Didry, and D. Pantaloni. Microtubule elongation and guanosine 5′-triphosphate hydrolysis. Role of guanine nucleotides in microtubule dynamics. *Biochemistry* 26: 4428–4437, 1987.

38. Chen, Y.-T., and M. Schliwa. Direct observation of microtubule dynamics in Reticulomyxa: unusually rapid length changes and microtubule sliding. *Cell Motil. Cytoskeleton* 17: 214–226, 1990.

39. Collins, C. A. Dynamin: a novel microtubule-associated GTPase. *Trends Cell Biol.* 1: 57–60, 1991.

40. Davidson, L. A. Studies of the Actinopods *Heterophyrs marina* and *Ciliophyrs* Marina: Energetic and Structural Analysis of Their Contractile Axopods, General Ultrastructure and Phylogenetic Relationships. Berkeley: University of California, 1975. *PhD thesis.*

41. Dellinger, O. P. The cilium as a key to the structure of contractile protoplasm. *J. Morphol.* 20: 171–210, 1909.

42. Dirksen, E. Centriole and basal body formation during ciliogenesis revisited. *Biol. Cell* 72: 31–38, 1991.

43. Dustin, P. *Microtubules (2nd ed.)* New York: Springer-Verlag, 1984, 482 *pp.*

44. Eshel, D., S. Shingyoji, K. Yoshimura, C. R. Gibbons, and K. Takahashi. The phase of sperm flagellar beating is not conserved over a brief imposed interruption. *Exp. Cell Res.* 202: 552, 1992.

45. Fawcett, D. W., and K. R. Porter. A study of the fine structure of ciliated epithelia. *J. Morphol.* 94: 221–281, 1954.

46. Fey, E. G., D. G. Capeo, G. Krochmalnic, and S. Penman. Epithelial structure revealed by clinical dissection and unembedded electron microscopy. *J. Cell Biol.* 99: 203s–208s, 1984.

47. Fonte, V. G., R. L. Searls, and S. R. Hilfer. The relationship of cilia with cell division and differentiation. *J. Cell Biol.* 49: 226–229, 1971.

48. Gelfand, V. I. Cytoplasmic microtubular motors. *Curr. Opin. Cell Biol.* 1: 63–66, 1989.

49. Gelfand, V. I., and A. D. Bershadsky. Microtubule dynamics: mechanisms, regulation and function. *Annu. Rev. Cell Biol.* 7: 93–116, 1991.

50. Gibbons, B. H., and I. R. Gibbons. Flagellar movement and adenosine triphosphatase activity in sea urchin sperm extracted with Triton X-100. *J. Cell Biol.* 54: 75–97, 1972.

51. Gibbons, I. R. The relationship between fine structure and beat in the gill cilia of a lammelibranch mollusc. *J. Biophys. Biochem. Cytol.* 11: 179–205, 1961.

52. Gibbons, I. R. Studies on the protein components of cilia from Tetrahymena pyriformis. *Proc. Natl. Acad. Sci. USA* 50: 1002–1010, 1963.

53. Gibbons, I. R. Cilia and flagella of eukaryotes. *J. Cell Biol.* 91(3)2: 107s–124s, 1981.

54. Gibbons, I. R. Sliding and bending in sea urchin flagella. *Symp. Soc. Exp. Biol.* 35: 225–287, 1982.

55. Gibbons, I. R., M. P. Cosson, J. A. Evans, B. H. Gibbons, B. Houck, K. H. Martinson, W. S. Sale, and W. J-Y. Tang. Potent inhibition of dynein adenosine triphosphatase and of the motility of cilia and flagella by vanadate. *Proc. Natl. Acad. Sci. U.S.A.* 75: 2220–2224, 1978.

56. Gibbons, I. R., B. H. Gibbons, G. Mocz, and D. J. Asai. Multiple nucleotide-binding sites in the sequence of dynein β-heavy chain. *Nature* 352: 640–643, 1991.

57. Gibbons, I. R., and A. V. Grimstone. On flagellar structure in certain flagellates. *J. Biophys. Biochem. Cytol.* 7: 697–716, 1960.

58. Gibbons, I. R., A. Lee Eiford, G. Mocz, C. A. Phillipson, W.J.Y. Tang, and B. H. Gibbons. Photosensitized cleavage of dynein heavy chains. *J. Biol. Chem.* 262: 2780–2786, 1987.

59. Gibbons, I. R., and A. J. Rowe. Dynein: a protein with adenosine triphosphatase activity from cilia. *Science* 149: 424–425, 1965.

60. Gilula, N. B., and P. Satir. The ciliary necklace: a ciliary membrane specialization. *J. Cell Biol.* 53: 494–509, 1972.

61. Goldstein, L.S.B. The kinesin superfamily: tails of functional redundancy. *Trends Cell Biol.* 1: 93–98, 1991.

62. Goldstein, S. F. Morphology of developing bends in sperm flagella. In: *Swimming and Flying in Nature,* edited by T. Y-T Wu, C. J. Brokaw, and C. Brennan. New York: Plenum, 1975, vol. 1, *p.* 127–132.

63. Goldstein, S. F., M.E.J. Holwill, N. R. Silvester. The effects of laser microbeam irradiation on the flagellum of *Crithidia (Strigomonas) oncopelti. J. Exp. Biol.* 53: 401–409, 1970.

64. Goltz, J. S., A. W. Wolkoff, P. Novikoff, R. J. Stockert, and P. Satir. A role for microtubules in sorting of endocytic vesicles in rat hepatocytes. *Proc. Natl. Acad. Sci.* 89: 7026–7030, 1992.

65. Goodenough, U. W., and J. E. Heuser. Structural comparison of purified dynein proteins with *in situ* dynein arms. *J. Mol. Biol.* 180: 1083–1118, 1984.

66. Goodenough, U. W., and J. E. Heuser. Structure of the soluble and in situ ciliary dyneins visualized by quick-freeze deep etch microscopy In: *Cell Movement,* edited by F. D. Warner, P. Satir, and I. R. Gibbons. New York: Alan Liss, 1989, vol. 1, p. 121–140.

67. Gray, J. *Ciliary Movement.* U.K.: Cambridge University Press, 1928, 162 *pp.*

68. Gray, J. The movement of sea-urchin spermatozoa. *J. Exp. Biol.* 32: 775–801, 1955.

69. Greer, K., and J. L. Rosenbaum. Post-translational modifications of tubulin. In: *Cell Movement,* edited by F. D. Warner and J. R. McIntosh. New York: Alan Liss, 1989, vol. 2, p. 47–66.

70. Gyoeva, F. K., and V. I. Gelfand. Coalignment of vimentin intermediate filaments with microtubules depends on kinesin. *Nature* 353: 445–448, 1991.

71. Haimo, L. T., B. R. Telzer, and J. L. Rosenbaum. Dynein binds to and crossbridges cytoplasmic microtubules. *Proc. Natl. Acad. Sci. U.S.A.* 76: 5759–5763, 1979.

72. Hall, J. L., Z. Ramanis, and D. J. L. Luck. Basal body/centriolar DNA: molecular genetic studies in *Chlamydomonas. Cell* 59: 121–132, 1989.

73. Hamasaki, T., K. Barkalow, J. Richmond, and P. Satir. A cAMP-stimulated phosphorylation of an axonemal polypeptide that copurifies with the 22S dynein arm regulates microtubule trans-

location velocity and swimming speed in *Paramecium*. *Proc. Natl. Acad. Sci. U.S.A.* 88: 7918–7922, 1991.

74. Hamasaki, T., T. Murtaugh, B. H. Satir, and P. Satir. *In vitro* phosphorylation of paramecium axonemes and permeabilized cells. *Cell Motil. Cytoskeleton* 12: 1–11, 1989.

75. Hill, T., and M. F. Carlier. Steady-state theory of the interference of GTP hydrolysis in the microtubule assembly. *Proc. Natl. Acad. Sci. U.S.A.* 80: 7234–7238, 1983.

76. Hirokawa, N., K. K. Pfister, H. Yori Fuji, M. C. Wagner, S. T. Brady, and G. S. Bloom. Submolecular domains of bovine brain kinesin identified by electron microscopy and monoclonal antibody decoration. *Cell* 56: 867–878, 1989.

77. Hirokawa, N., R. Sato-Yoshitake, T. Yoshida, and T. Kawashima. Brain dynein (MAP 1C) localizes on both anterogradely and retrogradely transported membranous organelles *in vivo*. *J. Cell Biol.* 111: 1027–1037, 1990.

78. Ho, P. T-C., and R. M. Tucker. Centriole ciliation and cell cycle variability during G1 phase of BALB/C 3T3 cells. *J. Cell Physiol.* 139: 398–406, 1989.

79. Hoffmann-Berling, H. Geisselmodele und Adenosinetriphosphate. *Biochem. Biophys. Acta* 16: 146–154, 1955.

80. Holwill, M. E. J., and P. Satir. A physical model of microtubule sliding in ciliary axonemes. *Biophys. J.* 58: 905–918, 1990.

81. Horio, T., and H. Hotani. Visualization of the dynamic instability of individual microtubules by dark-field microscopy. *Nature* 321: 605–607, 1986.

82. Horwitz, S. B., J. Parness, P. B. Schiff, and J. J. Manfredi. Taxol: a new probe for studying the structure and function of microtubules. *Cold Spring Harb. Symp. Quant. Biol.* 46: 219–226, 1981.

83. Hotani, H., and T. Horio. Dynamics of microtubules visualized by darkfield microscopy, treadmilling and dynamic instability. *Cell Motil. Cytoskeleton* 10: 229–236, 1988.

84. Hoyle, H. D., and E. C. Raff. Two *Drosophila* beta tubulin isoforms are not functionally equivalent. *J. Cell Biol.* 111: 1009–1026, 1990.

85. Huang, B. P-H. *Chlamydomonas reinhardtii*: a model system for the genetic analysis of flagellar motility. *Int. Rev. Cytol.* 99: 181–215, 1986.

86. Hyams, J., and G. G. Borisy. Isolated flagellar apparatus of *Chlamydomonas*: characterization of forward swimming and alteration of wave form and reversal of swimming by calcium ions *in vitro*. *J. Cell Sci.* 33: 235–253, 1978.

87. Inoue, S. Cell division and the mitotic spindle. *J. Cell Biol.* 91(3)2: 131s–147s, 1981.

88. Inoue, S., and H. Sato. Cell motility by labile association of molecules: the nature of the mitotic spindle fibers and their role in chromosome movement. *J. Gen. Physiol.* 50(6)2: 259–288, 1967.

89. Johnson, K. A. Pathway of the microtubule-dynein ATPase and the structure of dynein: a comparison with actomyosin. *Annu. Rev. Biophys. Chem.* 14: 161–188, 1985.

90. Johnson, K. A., and J. L. Rosenbaum. Basal bodies and DNA. *Trends Cell Biol.* 1: 145–149, 1991.

91. Johnson, K. A., and J. L. Rosenbaum. Replication of basal bodies and centrioles. *Curr. Opin. Cell Biol.* 4: 80–85, 1992a.

92. Johnson, K. A., and J. L. Rosenbaum. Polarity of flagellar assembly in *Chlamydomonas*. *J. Cell Biol.* 119: 1605–1611, 1992b.

93. Johnson, K. A., and J. S. Wall. Structure and molecular weight of the dynein ATPase. *J. Cell Biol.* 96: 669–678, 1983.

94. Kalt, A., and M. Schliwa. Molecular components of the centrosome. *Trends Cell Biol.* 3: 118–128, 1993.

95. Kamimura, S., and E. Mandelkow. Tubulin protofilaments and kinesin-dependent motility. *J. Cell Biol.* 118: 865–875, 1992.

96. Kamimura, S., and K. Takahashi. Direct measurement of the force of microtubule sliding in flagella. *Nature* 293: 566–568, 1981.

97. Kamiya, R., and G. B. Witman. Submicromolar levels of calcium control the balance of beating between the two flagella in demembranated models of *Chlamydomonas*. *J. Cell Biol.* 98: 97–107, 1984.

98. King, S., and G. B. Witman. Localization of an intermediate chain of outer arm dynein by immunoelectron microscopy. *J. Biol. Chem.* 265: 19807–19811, 1990.

99. Kobayashi, T., T. Martensen, J. Nath, and M. Flavin. Inhibition of dynein ATPase by vanadate, and its possible use as a probe for the role of dynein in cytoplasmic motility. *Biochem. Biophy. Res. Commun.* 81: 1313–1318, 1978.

100. Koonce, M. P., P. M. Grissom, and J. R. McIntosh. Dynein from *Dictyostelium*: primary structure comparisons between a cytoplasmic motor enzyme and flagellar dynein. *J. Cell Biol.* 119: 1597–1604, 1992.

101. Kosik, K. S. Tau protein and Alzheimer's disease. *Curr. Opin. Cell Biol.* 2: 101–104, 1990.

102. Kuznetsov, S. A., and V. I. Gelfand. Bovine brain kinesin is a microtubule-activated ATPase. *Proc. Natl. Acad. Sci. USA* 83: 8530–8534, 1986.

103. Kuznetsov, S. A., E. A. Vaisberg, N. A. Sharina, N. N. Magretova, V. Y. Chenryak, and V. I. Gelfand. The quarternary structure of bovine brain kinesin. *EMBO J.* 7: 353–356, 1988.

104. Lasek, R. J., and S. T. Brady. Attachment of transported vesicles to microtubules in axoplasm is facilitated by AMP-PNP. *Nature* 316: 645–647, 1985.

105. Leadbeater, B. S. C. Developmental studies on the loricate choanoflagellate *Stephanoeca diplocosta*. *Protoplasma* 136: 1–15, 1987.

106. Lee Eiford, A., R. A. Ow, and I. R. Gibbons. Specific cleavage of dynein heavy chains by ultraviolet irradiation in the presence of ATP and vanadate. *J. Biol. Chem.* 261: 2337–2342, 1986.

107. Lieberman, S. J., T. Hamasaki, and P. Satir. Ultrastructure and motion analysis of permeabilized Paramecium capable of motility and regulation of motility. *Cell Motil. Cytoskeleton* 9: 73–84, 1988.

108. Lin, S.X.H., and C. A. Collins. Immunolocalization of cytoplasmic dynein to lysosomes in cultured cells. *J. Cell. Sci.* 101: 125–137, 1992.

109. Lindemann, C. B., and J. S. Goltz. Calcium regulation of flagellar curvature and swimming pattern in Triton X-100 extracted rat sperm. *Cell Motil. Cytoskeleton* 10: 420–431, 1988.

110. Lindemann, C. B., A. Orlando, and K. S. Kanous. The flagellar beat of rat sperm is organized by the interaction of two functionally distinct populations of dynein bridges with a stable central axonemal partition. *J. Cell Sci.* 102: 249–260, 1992.

111. Little, M., and T. Seehaus. Comparative analysis of tubulin sequence. *Compar. Biochem. Physiol.* 90B: 655–670, 1988.

112. Luck, D. Genetic and biochemical dissection of the eucaryotic flagellum. *J. Cell Biol.* 98: 789–794, 1984.

113. Maeda, K., T. Nakata, Y. Noda, R. Sato-Yoshitake, and N. Hirokawa. Interaction of dynamin with microtubules: its structure and GTPase activity investigated by using highly purified dynamin. *Mol. Biol. Cell* 3: 1181–1194, 1992.

114. Mandelkow, E., and E.-M. Mandelkow. Microtubular structure and tubulin polymerization. *Curr. Opin. Cell Biol.* 1: 5–9, 1989.

115. Mandelkow, E., and E.-M. Mandelkow. Microtubular structure and tubulin polymerization. *Curr. Opin. Cell Biol.* 2: 3–9, 1990.

116. Mandelkow, E. M., E. Mandelkow, and R. A. Milligan. Micro-

tubule dynamics and microtubule caps: a time-resolved cryo-electron microscopy study. *J. Cell Biol.* 114: 977–991, 1991.

117. Manton, I. The fine structure of plant cilia. *Soc. Exp. Biol. Symp.* 6: 306–319, 1952.

118. Margolis, R. L., and L. Wilson. Opposite end assembly and disassembly of microtubules at steady state *in vitro*. *Cell* 13: 1–8, 1978.

119. Mitchison, T. J., and M. W. Kirschner. Dynamic instability of microtubule growth. *Nature* 312: 237–242, 1984.

120. Mohri, H. Amino acid composition of 'tubulin' constituting microtubules of sperm flagella. *Nature* 217: 1053–1054, 1968.

121. Moran, D. T., F. G. Varela, and J. C. Rowley III. Evidence for an active role of cilia in sensory transduction. *Proc. Natl. Acad. Sci. U.S.A.* 74: 793–797, 1977.

122. Moss, A., W. S. Sale, L. A. Fox, and G. B. Witman. The α subunit of sea urchin outer arm dynein mediates structural and rigor binding to microtubules. *J. Cell Biol.* 118: 1189–200, 1992.

123. Multigner, L., J. Gagnon, A. Dorsselaer, and D. Job. Stabilization of sea urchin flagellar microtubules by histone H1. *Nature* 360: 33–39, 1992.

124. Murakami, A. Control of ciliary beat frequency in the gill of *Mytilus*—I. Activation of the lateral cilia by cyclic AMP. *Compar. Biochem. Physiol.* 86C: 273–279, 1987.

125. Nasr, A., and P. Satir. Alloaffinity filtration: a general approach to the purification of dynein and dynein-like molecules. *Anal. Biochem.* 151: 97–108, 1985.

126. Nislow, C., V. Lombillo, R. Kuriyama, and J. R. McIntosh. A plus-end directed motor enzyme that moves antiparallel microtubules *in vitro* localizes to the interzone of mitotic spindles. *Nature* 359: 543–547, 1992.

127. Novikoff, P., M. Cammer, L. Tao, H. Oda, R. Stockert, A. W. Wolkoff and P. Satir. Three dimensional organization of the rat hepatocyte cytoskeleton: relation to the asialoglycoprotein endocytosis pathway. *J. Cell Sci.* 109: 21–32, 1996.

128. Nurse, P. Universal control mechanism regulating onset of M phase. *Nature* 344: 503–507, 1990.

129. Oakley, B. R. γ-tubulin: the microtubule organizer? *Trends Cell Biol.* 2: 1–5, 1992.

130. Ogawa, K. Four ATP-binding sites in the midregion of the β-heavy chain of dynein. *Nature* 352: 643–645, 1991.

131. Olmsted, J. B. Microtubule-associated proteins. *Annu. Rev. Cell Biol.* 2: 421–457, 1986.

132. Olmsted, J. B. Non-motor microtubule-associated proteins. *Curr. Opin. Cell Biol.* 3: 52–58, 1991.

133. Omoto, C. K., and K. A. Johnson. Activation of the dynein adenosine triphasphatase by microtubules. *Biochemistry* 25: 419–427, 1986.

134. Omoto, C. K., and C. Kung. Rotation and twist of the central-pair microtubules in the cilia of *Paramecium*. *Nature* 279: 532–534, 1980.

135. Otter, T. Calmodulin and control of flagellar movement in: *Cell Movement*, edited by F. D. Warner, P. Satir, and I. R. Gibbons. New York: Alan Liss, 1989, vol. 1, p. 281–289.

136. Otter, T., B. H. Satir, and P. Satir. Trifluoperazine-induced changes in swimming behavior of *Paramecium*: evidence for two sites of drug action. *Cell Motil.* 4: 249–267, 1984.

137. Pallini, V., M. Bugnoli, C. Mencarelli, and G. Scapigliati. Biochemical properties of ciliary, flagellar and cytoplasmic dyneins. *Symp. Soc. Exp. Biol.* 35: 339–352, 1982.

138. Paschal, B. M., S. M. King, A. G. Moss, C. A. Collins, R. B. Vallee, and G. B. Witman. Isolated flagella outer arm dynein translocates brain microtubules *in vitro*. *Nature* 330: 672–674, 1987.

139. Pfarr, C. M., M. Cove, P. M. Grissom, T. S. Hays, M. E. Porter, and J. R. McIntosh. Cytoplasmic dynein is localized to kinetochores during mitosis. *Nature* 345: 263–265, 1990.

140. Phillips, M. J., and P. Satir. The cytoskeleton of the hepatocyte: organization, relationship, and pathology In: *The Liver: Biology and Pathobiology (2nd ed.)*, edited by I. M. Arias, W. B. Jakoby, H. Popper, D. Schachter, and D. A. Shafritz. New York: Raven Press, 1988, *p.* 11–27.

141. Piperno, G., K. Mead, and W. Shestak. The inner dynein arms I2 interact with a "dynein regulatory complex" in *Chlamydomonas* flagella. *J. Cell Biol.* 118: 1455–1463, 1992.

142. Piperno, G., Z. Ramanis, E. F. Smith, and W. S. Sale. Three distinct inner dynein arms in *Chlamydomonas* flagella: molecular composition and location in the axoneme. *J. Cell Biol.* 110: 379–389, 1990.

143. Poole, C. A., M. H. Flint, and B. W. Beaumont. Analysis of the morphology and function of primary cilia in connective tissues: a cellular cybernetic probe? *Cell Motil.* 5: 175–193, 1985.

144. Porter, K. R. The submicroscopic morphology of protoplasm. *Harvey Lect. Ser.* 51: 175–228, 1957.

145. Porter, K. R. Cytoplasmic microtubules and their functions. In: *Principles of Biomolecular Organization*, edited by G. E. W. Wolstenholme, and M. O'Connor. Ciba Foundation Symp., Boston: Little Brown, 1966, p. 308–335.

146. Porter, M. E., and K. A. Johnson. Dynein structure and function. *Annu. Rev. Cell Biol.* 5: 119–151, 1989.

147. Redenbach, D. M., and A. W. Vogl. Microtubule polarity in Sertoli cells: a model for microtubule-based spermatid transport. *Eur. J. Cell Biol.* 54: 277–290, 1991.

148. Reed, W., J. Avolio, and P. Satir. The cytoskeleton of the apical border of the lateral cells of freshwater mussel gill: integration of microtubule- and the microfilament-based organelles. *J. Cell Sci.* 68: 1–33, 1984.

149. Reed, W., and P. Satir. Spreading ciliary arrest in a mussel gill epithelium: characterization by quick fixation. *J. Cell Physiol.* 126: 191–205, 1986.

150. Rieder, C. L. Mitosis: towards a molecular understanding of chromosome behavior. *Curr. Opin. Cell Biol.* 3: 59–66, 1991.

151. Roberts, K., and J. S. Hyams. *Microtubules*. New York: Academic Press, 1979, 595pp.

152. Roth, K. E., C. L. Reider, and S. S. Bowser. Flexible-substratum technique for viewing cells from the side: some *in vivo* properties of primary (9 + 0) cilia in cultured kidney epithelia. *J. Cell Sci.* 89: 457–466, 1988.

153. Sabatini, D. O., K. Bensch, and R. J. Barnett. Cytochemistry and electron microscopy. The preservation of cellular ultrastructure and enzymatic activity by aldehyde fixation. *J. Cell Biol.* 17: 19–58, 1963.

154. Sakakibara, H., D. R. Mitchell, and R. Kamiya. *Chlamydomonas* outer arm dynein mutant missing the α-heavy chain. *J. Cell Biol.* 113: 615–622, 1991.

155. Sakakibara, H., S. Takada, S. M. King, G. B. Witman, and R. Kamiya. A *Chlamydomonas* outer arm dynein mutant lacking the β heavy chain. *J. Cell Biol.* 122: 653–661, 1993.

156. Sale, W. S. The axonemal axis and Ca²⁺-induced asymmetry of active microtubule sliding in sea urchin sperm tails. *J. Cell Biol.* 102: 2042–2052, 1986.

157. Sale, W. S., and L. A. Fox. The isolated β-heavy chain subunit of dynein translocates microtubules *in vitro*. *J. Cell Biol.* 107: 1793–1797, 1988.

158. Sale, W. S., U. W. Goodenough, and J. E. Heuser. The substructure of isolated and *in situ* outer dynein arms of sea urchin sperm flagella. *J. Cell Biol.* 101: 1400–1422, 1985.

159. Sale, W. S. and P. Satir. Direction of active sliding of microtu-

bules in *Tetrahymena* cilia. *Proc. Natl. Acad. Sci. U.S.A.* 74: 2045–2049, 1977.

160. Sanders, M. A., and J. L. Salisbury. Centrin-mediated microtubule severing during flagellar excision in *Chlamydomonas reinhardtii. J. Cell Biol.* 108: 1751–1760, 1989.

161. Sanderson, M., and E. R. Dirksen. Mechanosensitive and beta-adrenergic control of ciliary beat frequencies of mammalian respiratory tract cells in culture. *Am. Rev. Respir. Dis.* 139: 432–440, 1989.

162. Satir, P. Structure and function in cilia and flagella. *Protoplamatologia* III-E: 52, 1965a.

163. Satir, P. Studies on cilia: II. Examination of the distal region of the ciliary shaft and the role of filaments in motility. *J. Cell Biol.* 26: 805–834, 1965b.

164. Satir, P. Studies on cilia. III. Further studies on the cilium tip and a "sliding filament" model of ciliary motility. *J. Cell Biol.* 39: 77–94, 1968.

165. Satir, P. Approaches to potential sliding mechanisms of cytoplasmic microtubules. *Cold Spring Harb. Symp. Quant. Biol.* 46: 285–292, 1982.

166. Satir, P. The cytoplasmic matrix: old and new questions. *J. Cell Biol.* 99: 235s–238s, 1984.

166a. Satir, P. Switching mechanisms in the control of ciliary motility, *Modern Cell Biology* 4: 1–46, 1985.

167. Satir, P. Structural analysis of the dynein cross-bridge cycle. In: *Cell Movement, Vol. 1: The Dynein ATPases,* edited by F. D. Warner *et al.* New York: Alan Liss, 1989a, p. 219–234.

168. Satir, P. The role of axonemal components in ciliary motility. *Compar. Biochem. Physiol.* 94A: 351–357, 1989b.

169. Satir, P. Mechanisms of ciliary movement: contributions from electron microscopy. *Scanning Microsc. Intl.* 6: 573–579, 1992.

170. Satir, P. Motor molecules of the cytoskeleton: possible functions in the hepatocyte. In: *The Liver Biology and Pathology (3rd ed.),* edited by I. Arias *et al.* New York: Raven Press, 1994, p. 45–52.

171. Satir, P., K. Barkalow, and T. Hamasaki. The control of ciliary beat frequency. *Trends Cell Biol.* 3: 409–412, 1993.

172. Satir, P., J. S. Goltz, N. Isaac, and C. B. Lindemann. Switching arrest in cilia and the calcium response of rat sperm: a comparison. In: *Comparative Spermatology 20 Years After,* edited by B. Baccetti. New York: Raven Press, 1991, vol. 75, p. 377–384.

173. Satir, P., J. S. Goltz, and A. W. Wolkoff. Microtubule-based cell motility: the role of microtubules in cell motility and differentiation. *Cancer Invest.* 8: 685–690, 1990.

174. Satir, P., and T. Matsuoka. Splitting the ciliary axoneme: implications for a 'switch point' model of dynein arm activity in ciliary motion. *Cell Motil. Cytoskeleton.* 14: 345–358, 1989.

175. Satir, P., and M. A. Sleigh. The physiology of cilia and mucociliary interactions. *Annu. Rev. Physiol.* 52: 137–155, 1990.

176. Satir, P., J. Wais-Steider, S. Lebduska, A. Nasr, and J. Avolio. The mechanochemical cycle of the dynein arm. *Cell Motil.* 1: 303–327, 1981.

177. Saxton, W. M., J. Hicks, L. S. B. Goldstein, and E. C. Raff. Kinesin heavy chain is essential for viability and neuromuscular function in *Drosophila,* but mutants show no defect in mitosis. *Cell* 64: 1093–1102, 1991.

178. Schultz, J. E., S. Klumpp, R. Benz, W.J.Ch. Schürhoff-Goeters, and A. Schmid. Regulation of adenylyl cyclase from *Paramecium* by an intrinsic potassium conductance. *Science* 255: 600–603, 1992.

179. Shingyoji, C., J. Katada, K. Takahashi, and I. R. Gibbons.

Rotating the plane of imposed vibration can rotate the plane of flagellar bending in sea-urchin sperm without twisting the axoneme. *J. Cell Sci.* 98: 175–182, 1991.

180. Shingyoji, C., A. Murakawi, and K. Takahashi. Local reactivation of Triton-extracted flagella by iontophoretic application of ATP. *Nature* 265: 269–270, 1977.

181. Singer, S. J., E. H. Ball, B. Geiger, and W-T Chen. Immunolabelling studies of cytoskeletal associations in cultured cells. *Cold Spring Harb. Symp. Quant. Biol.* 46: 303–316, 1981.

182. Slautterback, D. B. Cytoplasmic microtubules I. Hydra. *J. Cell Biol.* 18: 367–388, 1963.

183. Sleigh, M. A. *The Biology of Cilia and Flagella.* New York: Macmillan, 1962, 242 *pp.*

184. Sleigh, M. A. *Cilia and Flagella.* New York: Academic Press, 1974, *500 pp.*

185. Sleigh, M. A. The origin of flagella—autogenous or symbiontic? *Cell Motil. Cytoskeleton* 6: 96–98, 1986.

186. Sloboda, R. D., S. A. Rudolph, J. L. Rosenbaum, and P. Greengard. Cyclic AMP-dependent endogenous phosphorylation of a microtubule-associated protein. *Proc. Natl. Acad. Sci. U.S.A.* 72: 177–181, 1975.

187. Smith, E. F., and W. S. Sale. Structural and functional reconstitututon of inner dyneins arms in Chlamydomonas flagellar axonemes. *J. Cell Biol.* 117: 573–581, 1992.

188. Soifer, D. *The Biology of Cytoplasmic Microtubules.* Ann. NY Acad. Sci. 253: 848 pp. 1975.

189. Soifer, D. *Dynamic Aspects of Microtubule Biology.* Ann. NY Acad. Sci. 466: 978 pp. 1986.

190. Spungin, B., J. Avolio, S. Arden, and P. Satir. Dynein arm attachment probed with a non-hydrolyzable ATP analog: Structural evidence of patterns of activity. *J. Mol. Biol.* 197: 671–677, 1987.

191. Steffen, W., and R. W. Linck. Tektins in ciliary and flagellar microtubules and their association with other cytoskeletal systems. In: *Cell Movement,* edited by F. D. Warner and J. R. McIntosh. New York: Alan Liss, 1989, vol. 2, p. 67–81.

192. Stephens, R. E., and E. W. Stommel. Role of cyclic adenosine monophosphate in ciliary and flagellar motility In: *Cell Movement,* edited by F. D. Warner, P. Satir, and I. R. Gibbons. New York: Alan Liss, 1989, vol. 1, p. 299–316.

193. Steuer, E. R., L. Wordeman, T. A. Schroer, and M. P. Sheetz. Localization of cytoplasmic dynein to mitotic spindles and kinetochores. *Nature* 345: 266–268, 1990.

194. Stewart, R. J., P. A. Pesavento, D. N. Woerpel, and L.S.B. Goldstein. Identification and partial characterization of six new members of the kinesin superfamily in *Drosophila. Proc. Natl. Acad. Sci. U.S.A.* 88: 8470–8471, 1991.

195. Sugino, K., and H. Machemer. Axial view recordings: an approach to assess the third dimension of the ciliary cycle. *J. Theor. Biol.* 25: 67–82, 1987.

196. Sugino, K., and H. Machemer. Depolarization-controlled parameters of the ciliary cycle and axonemal function. *Cell Motil. Cytoskeleton* 16: 251–265, 1990.

197. Sugrue, P., J. Avolio, P. Satir, and M.E.J. Holwill. Computer modelling of Tetrahymena axonemes at macromolecular resolution: interpretation of electron micrographs. *J. Cell Sci.* 98: 5–16, 1991.

198. Sullivan, K. F. Structure and utilization of tubulin isotypes. *Annu. Rev. Cell Biol.* 4: 687–716, 1988.

199. Summers, K. E., and I. R. Gibbons. Adenosine triphosphate-induced sliding of tubules in trypsin-treated flagella of sea-urchin sperm. *Proc. Natl. Acad. Sci. U.S.A.* 68: 3092–3096, 1971.

200. Summers, K. E., and I. R. Gibbons. Effects of trypsin digestion

on flagellar structures and their relationship to motility. *J. Cell Biol.* 58: 618–629, 1973.

201. Svoboda, K., C. F. Schmidt, B. J. Schapp, and S. M. Block. Direct observation of kinesin stepping by optical trapping interferometry. *Nature* 365: 721–727, 1993.

202. Takahashi, K., C. Shingyoji, and S. Kamimura. Microtubule sliding in reactivated flagella. *Symp. Soc. Exp. Biol.* 35: 159–177, 1982.

203. Takahashi, M., and Y. Tonomura. Binding of 30S dynein with a B-tubule of the outer doublet of axonemes from *Tetrahymena pyriformis* and adenosine triphosphate-induced dissociation of the complex. *J. Biochem. (Tokyo)* 84: 1339–1355, 1978.

204. Tamm, S. L., and S. Tamm. Ciliary reversal without rotation of axonemal structures in ctenophore comb plates. *J. Cell Biol.* 89: 495–509, 1981.

205. Tamm, S. L., and S. Tamm. Alternate patterns of doublet microtubule sliding in ATP-disintegrated macrocilia of the ctenophore *Beröe. J. Cell Biol.* 99: 1364–1371, 1984.

206. Tilney, L. G., Y. Hiramoto, and D. Marsland. Studies on the microtubules in heliozoa. III. A pressure analysis of the role of these structures in the formation and maintenance of the axopodia of *Actinosphaerium nucleofilum* (Barrett) *J. Cell Biol.* 29: 77–96, 1966.

207. Tucker, R. W., A. B. Pardee, and K. Fujiwara. Centriole ciliation is related to quiescence and DNA synthesis in 3T3 cells. *Cell* 17: 527–535, 1979a.

208. Tucker, R. W., C. D. Scher, and C. D. Stiles. Centriole deciliation associated with the early response of 3T3 cells to growth factors but not to SV40. *Cell* 18: 1065–1072, 1979b.

209. Vale, R. D. Intracellular transport using microtubule-based motors. *Annu. Rev. Cell Biol.* 3: 347–378, 1987.

210. Vale, R. D. Microtubule-based motor proteins. *Curr. Opin. Cell Biol.* 2: 15–22, 1990.

211. Vale, R. D., F. Malik, and D. Brown. Directional instability of microtubule transport in the presence of kinesin and dynein, two opposite polarity motors. *J. Cell Biol.* 119: 1589–1596, 1992.

212. Vale, R. D., T. S. Reese, and M. P. Sheetz. Identification of a novel force-generating protein, kinesin, involved in microtubule-based motility. *Cell* 42: 39–50, 1985a.

213. Vale, R. D., B. J. Schapp, T. J. Mitchison, E. Steuer, T. S. Reese, and M. P. Sheetz. Different axoplasmic proteins generate movement in opposite directions along microtubules *in vitro. Cell* 43: 623–632, 1985b.

214. Vale, R. D., D. M. Soll, and I. R. Gibbons. One dimensional diffusion of microtubules bound to flagellar dynein. *Cell* 59: 915–925, 1989.

215. Vale, R. D., and Y. Y. Toyoshima. Rotation and translocation

216. Vale, R. B. Cytoplasmic dynein: advances in microtubule-based motility. *Trends Cell Biol.* 1: 25–29, 1991.

217. Vallee, R. B. Dynamin: motor protein or regulatory GTPase. *Muscle Res. Cell Motil.* 13: 493–496, 1992.

218. Vallee, R. B., and G. S. Bloom. High molecular weight microtubule-associated proteins (MAPs). *Mod. Cell Biol.* 3: 21–75, 1984.

219. Vallee, R. B., G. S. Bloom, and W. E. Theurkauf. Microtubule-associated proteins: subunits of the cytomatrix. *J. Cell Biol.* 99: 385s–445s, 1984.

220. Vallee, R. B., J. S. Wall, B. M. Paschal, and H. S. Shpetner. Microtubule-associated protein 1C from brain is a two-headed cytosolic dynein. *Nature* 332: 561–563, 1988.

221. Wadsworth, P., and M. McGrail. Interphase microtubule dynamics are cell specific. *J. Cell Sci.* 95: 23–32, 1990.

222. Wais-Steider, C., N. S. White, D. S. Gilbert, and P.A.M. Eagles. X-ray diffraction patterns from microtubules and neurofilaments in axoplasm. *J. Mol. Biol.* 197: 205–218, 1989.

223. Walker, R. A., E. T. O'Brien, N. K. Pryer, and M. F. Soboeiro. Dynamic instability of individual microtubules analyzed by video light microscopy: rate constants and transition frequencies. *J. Cell Biol.* 107: 1437–1448, 1988.

224. Walter, M. F., and P. Satir. Calcium does not inhibit active sliding of microtubules from mussel gill cilia. *Nature* 278: 69–70, 1979.

225. Warner, F. D. Organization of interdoublet links in Tetrahymena cilia. *Cell Motil.* 3: 321–332, 1983.

226. Warner, F. D., and J. R. McIntosh. *Kinesin, Dynein and Microtubule Dynamics, Cell Movement*, New York: Alan Liss, 1989, vol. 2.

227. Weingarten, M. D., A. H. Lockwood, S-Y Hwo, and M. W. Kirschner. A protein factor essential for microtubule assembly. *Proc. Natl. Acad. Sci. U.S.A.* 72: 1858–1862, 1975.

228. Wheatley, D. N. *The Centriole. A Central Enigma of Cell Biology.* New York: Elsevier, 1982, 232 pp.

229. Yen, T. J., G. Li, B. Schaar, I. Szilak, and D. W. Cleveland. CENP-E is a putative kinetochore motor that accumulates just before mitosis. *Nature* 359: 540–543, 1992.

230. Yen, T. J., P. S. Machlin, and D. W. Cleveland. Autoregulated instability of β-tubulin mRNAs by recognition of the nascent amino terminus of β-tubulin. *Nature* 334: 580–584, 1988.

231. Yeung, C. H., and D. M. Woolley. Localized reactivation of the principal piece of demembranated hamster sperm by iontophoretic application of ATP. *J. Submicrosc. Cytol.* 15: 327–331, 1983.

21. Principles of cell cycle control

CHRIS NORBURY | Imperial Cancer Research Fund Molecular Oncology Laboratory, University of Oxford Institute of Molecular Medicine, John Radcliffe Hospital, Oxford, England

CHAPTER CONTENTS

HOW DOES ONE CELL BECOME TWO? The question is perhaps the most fundamental in the whole of biology and underlies all manner of more complex issues of organismal and cellular proliferation. The production of two viable progeny requires that all essential cellular materials are segregated appropriately between the daughter cells. In most cases this will mean the distribution of subcellular components that have been approximately doubled in quantity in the period since the previous cell division. If the components are relatively numerous, as in the case of soluble enzymes or mitochondria, effective segregation between the nascent daughter cells is assured provided there exists some mechanism to divide the total cell mass into two approximately equal parts. Certain low-copy-number organelles, notably the Golgi stacks, can be segregated according to the same principle, but only because their disassembly and fragmentation precedes cell division. Clearly this strategy is of no value in the case of chromosomal DNA, which is by contrast replicated and subsequently segregated with great precision. Progress through this chromosomal cycle is in many cases necessarily linked to cell growth, which can be viewed as replication of the high-copy-number cellular constituents. In some cells, such as those of the fission yeast *Schizosaccharomyces pombe*, critical cell size requirements must be met before DNA replication and mitosis are initiated (62, 193). There is no general requirement for cell growth to drive the chromosomal cycle, however, and in some instances the two processes can be dissociated completely. This is particularly apparent in the case of early embryonic divisions of species such as the intensively studied *Drosophila melanogaster* and *Xenopus laevis*, where the early cell divisions proceed rapidly in the absence of simultaneous cell growth. Conversely, the diversity of cell sizes within multicellular organisms indicates that cell growth can proceed without necessarily triggering chromosomal duplication and cell division. In considering the regulation of the chromosomal cycle it is therefore important to bear in mind its dissociation from the process of growth and yet to make allowance for the fact that these two aspects of cell physiology are frequently coordinated.

Controls over cell growth can be viewed as specialized extensions of a fundamental program for cell proliferation that operates at the chromosomal level. While this view applies equally to prokaryotes, this chapter is concerned exclusively with the fundamental cell cycle program that operates in eukaryotes. For unicellular organisms, proliferation is governed primarily by extrinsic factors, in particular the availability of nutrients; in multicellular organisms the construction of new cells is frequently subject to rigorous controls intrinsic to the organism in question, though such controls are often mediated by extracellular factors, including "juxtacrine" interactions with the surfaces of neighboring cells. The molecular details of these types of control are gradually unfolding as a result of painstaking biological and genetical investigations (43, 110, 157, 213). Nowhere is the importance of such controls more obvious than in the field of cancer research, though normal development also requires the generation of a number of cells appropriate to the structure being made. The high degree of specialization of growth controls reflects the diversity of biological signals that are integrated to promote or inhibit cell proliferation. Extracellular signals that are interpreted either as stimulatory or inhibitory for proliferation clearly do not impinge directly upon the cell cycle regulatory machinery; instead the signals are transduced by means of elaborate networks of intracellular signaling molecules. Such signals can have pleiotropic effects on diverse aspects of cell metabolism and behavior in addition to effects on proliferation, and are largely beyond the scope of this chapter. The complexity of this signaling network is illustrated by extreme cases such as that of transforming growth factor β (TGFβ), which can stimulate proliferation of mesenchymal cells but blocks proliferation of epithelial cells (157); presumably this difference reflects differences between the two cell types in their patterns of expression of signal transduction components, their intracellular "wiring diagrams." In considering cell cycle regulation, then, we are interested not so much in these evolutionarily diverse signaling networks as in those biochemical events that are common to many cell types and which represent the points at which signaling networks converge to influence commitment to, initiation of, or progress through key events of the chromosomal cycle. The two most conspicuous events of the chromosomal cycle—namely, DNA replication in the synthesis (S) phase and chromosome segregation in mitosis or meiosis (M phase)—are effectively irreversible. Segregation is furthermore not generally prone to interruption once initiated, except in special cases such as the meiotic arrest seen in some species during oocyte maturation, or in the case of experimental intervention. Neither is protracted arrest usually seen in cells that have embarked upon DNA replication, though temporary delay can result if DNA damage is sustained during S phase. There are consequently "points of no return" that must be crossed before each successive S and M phase is initiated. These points of commitment are a general feature of cell cycle regulation in diverse eukaryotes, and are becoming increasingly well understood at the molecular level, as discussed in detail below. Perhaps surprisingly, the process of commitment to S phase is typically completed some time before DNA replication itself is first detectable, during the gap phase termed G1 that begins with the completion of the previous M phase. By contrast, the point of commitment to M-phase entry is not found within the G2 gap phase that lies between S and M, but is generally inseparable from the first M phase events themselves. Additional controls exist to ensure that the S and M phases usually alternate. In principle this could be achieved if there were a structural dependence of S-phase initiation on the completion of some M phase event, and vice versa. This is unlikely to be the case, however, as two successive M phases occur during meiosis, and multiple successive rounds of DNA replication are seen in certain specialized somatic cells such as megakaryocytes. An additional control acts to ensure that each sequence in the nuclear genome is replicated once, but only once, during each S phase. Again, this does not seem to reflect an inescapable structural constraint imposed by the biochemistry of replication, as there are examples in biology of both regulated and uncontrolled endoreduplication of limited chromosomal regions. This is not to say that there are no structural dependences within the cell cycle; initiation of DNA replication, for example, requires not only traversal of the S-phase commitment point (a nonstructural control), but also the provision of sufficient deoxynucleotide precursors (a structural requirement), synthesized by means of pathways that are typically activated in parallel with the S-phase commitment pathway. In recent years the term "checkpoint" has come to be applied to a variety of cell cycle regulatory steps, though the term was initially used to describe a control that mediates the dependence of one cell cycle event on the completion of an earlier event (92). As far as it is possible to generalize, the key cell cycle regulatory controls seem to be mediated not through structural dependences (for example, through alterations in the rate of synthesis of deoxynucleotides), but through nonstructural checkpoint dependences. In some circumstances, notably the rapid and synchronous early embryonic cell cycles, there also appear to be roles for autonomous cellular oscillators that provide a time base for cell cycle events without reference to checkpoint controls. Thus inhibition of DNA replication or nuclear division in *Drosophila* early embryos does not appear to perturb the normal periodic duplication of the centrosomes (70, 230).

It is useful to make a conceptual distinction between the enzymatic machinery responsible for cell cycle events, such as the polymerases and ligases required for S phase or the microtubule-associated motors required for M phase, and the regulatory processes that are rate-limiting for progression into and through these phases. In experimental terms these can be distinguished because *advancement of the timing of execution of a rate-limiting process will advance the timing of whichever cell cycle event is controlled by that process;* such advancement cannot be achieved simply by synthesizing the appropriate enzymatic machinery in greater quantities. This criterion of cell cycle advancement has been satisfied in surprisingly few experimental systems; more frequently, strategies involving inhibition of cellular processes have been used. Such strategies, employing, for example, loss-of-function mutations or chemical inhibitors, cannot distinguish mechanical components from regulatory ones, except in those cases where negative regulation is relieved by the inhibition.

The common structural features of the cell cycle in evolutionarily distant eukaryotes suggest common biochemical mechanisms, a suggestion supported in recent years by a variety of biochemical and genetical experiments. Cell-cycle-associated proteins in many cases display conservation of amino acid sequence across large evolutionary distances. This conservation is a feature not only of structural proteins such as tubulin, but also of many of the enzymes involved in DNA metabolism and a number of key regulatory proteins.

In the examination of fundamental cell cycle regulatory elements, simple unicellular eukaryotes—in particular, yeasts—have proven enormously powerful tools. For these rapidly dividing organisms, growth regulation does not extend beyond responses to the availability of nutrients and to the presence of mating pheromones. Importantly, the two commonly used yeasts— namely, the budding yeast *Saccharomyces cerevisiae* and the fission yeast *S. pombe*—are amenable to classical and molecular genetical analyses. Recessive (loss of function) mutants can be isolated in haploid strains and the corresponding genes can be cloned using genetic complementation (that is, correction of the genetic defect by supplying a wild-type copy of the gene in *trans*) after introduction of plasmid gene libraries. More informatively, gain-of-function mutations can identify key regulatory elements. Such approaches have allowed the identification of regulatory gene products conserved throughout the eukaryotes in their overall structure and, in some cases, in their detailed regulation. In the most striking cases, cell cycle regulatory components in yeast can be fully replaced by their human analogs. Studies on unicellular eukaryotes therefore provide a framework within which cell cycle regulation in multicellular organisms is coming to be understood (Fig. 21.1).

MITOSIS

Of all the phases of the cell cycle, mitosis is arguably the best understood in terms of the molecular detail of

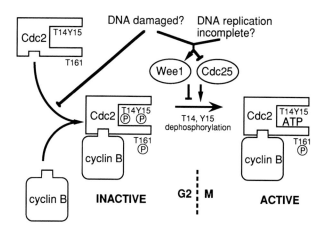

FIG. 21.2. Activation of Cdc2 at onset of mitosis. Assembly of Cdc2/cyclin B heterodimers promotes phosphorylation and inactivation of Cdc2 subunit *(left)*. Dephosphorylation of Cdc2 by Cdc25 phosphatase(s) at positions T14 and Y15 leads to generation of active MPF heterodimers *(right)*. Signals relating to DNA damage and completeness of replication impact on balance of Wee1 (a Cdc2 inhibitory kinase) and Cdc25 activities to delay entry into M phase. Additional DNA damage signal may influence aspect of Cdc2 activation other than Y15 phosphorylation (see text).

its regulation (Fig. 21.2). Furthermore, this understanding may provide a basis for the study of other key cell cycle events, which could be controlled by molecular mechanisms organized along broadly similar lines. Note that much of what is known about the control of mitotic M phase appears also to apply to meiosis, though the latter has been less widely studied.

The cellular reorganization that occurs at the onset of M phase is among the most dramatic events in cell biology, and is brought about by the activation of a number of parallel pathways (185). One such pathway leads in prophase to hypercondensation of the replicated chromosomes into discrete structures that can be segregated into the daughter nuclei (51, 156); chromatin hypercondensation persists until telophase, when the interphase level of condensation is reestablished.

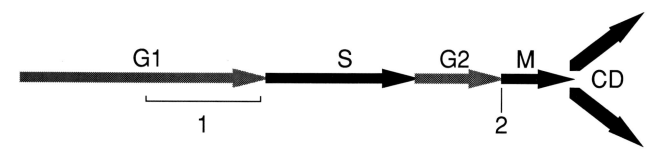

FIG. 21.1. Points of commitment in eukaryotic cell cycle. Schematic representation of cell cycle is shown, with major events *S* (DNA synthesis), *M* (mitosis), gap phases *(G1, G2)*, and cell division *(CD)* indicated. In yeast and mammalian cells commitment to enter S phase occurs during period in late G1 *(1)*, while commitment to enter mitosis *(2)* is indistinguishable from G2/M transition (for discussion see text).

Another pathway promotes the dismantling of the cytoplasmic network of microtubules, followed by the assembly of new tubules to generate a spindle with which the condensed chromosomes will be pulled apart. In many eukaryotes further pathways act simultaneously to cause disassembly of the nuclear lamina, an intermediate filament network that lies beneath the nuclear membrane, as well as vesicularization of the nuclear membrane itself (76, 185). Yet further pathways lead to the cessation of membrane vesicular trafficking, disassembly of the Golgi stacks and the intermediate filament network, repression of protein synthesis, and, in many cases, cell rounding. Remarkably, these diverse effector pathways appear to be triggered, in some cases directly, by the activation of a highly conserved protein phosphokinase, as described in the following section. Activation of this protein kinase both commits a cell to M phase and is responsible for establishing the distinctive M-phase features.

Cdc2: The Archetypal Cyclin-Dependent Kinase

The identification of Cdc2 as a universal eukaryotic M-phase-promoting activity represented a resoundingly successful combination of yeast genetics with studies of biochemistry and cell biology (though, perhaps unsurprisingly, the relative contributions of the different disciplines have subsequently been debated at length). The criterion that a putative cell cycle regulator should be able to advance the cell cycle timing of a particular event, in this case M phase, was satisfied independently by both approaches. Genetic studies by Nurse and colleagues using *S. pombe* identified dominantly acting mutations in a gene called *cdc2* (for *cell division cycle*) that could advance the cell cycle timing of mitosis (195). Other mutations in the same gene were recessive, loss-of-function defects that resulted in cell cycle arrest immediately before M phase (195, 196). Together, these observations suggested that the *cdc2* gene encodes a component that is not only required for the initiation of mitosis but is also responsible for determining the timing of this process. Following the development of plasmid shuttle vectors, genetic complementation experiments led to the discovery that the *cdc2* gene of *S. pombe* is functionally interchangeable with the *CDC28* gene of the distantly related budding yeast *S. cerevisiae* (10, 11). DNA sequence analysis and biochemical assays revealed that both genes encode highly related protein kinases of 34 kDa (98, 147, 233, 256), suggesting that the key rate-limiting function performed by Cdc2 involves protein phosphorylation.

In parallel investigations by Masui and others, a combination of biochemistry and cell biology was used to investigate the nature of a proteinaceous factor that

upon microinjection into immature frog oocytes was able to stimulate them to complete meiosis and arrest in metaphase II, without any requirement for additional protein synthesis (158, 259). This maturation-promoting factor (MPF) was originally isolated from mature oocytes themselves, but was subsequently found to be present in a variety of somatic cells as they traversed mitosis (120, 268), and even in simple eukaryotes such as yeast (271). Cell fusion experiments also indicated the existence of a dominantly acting regulator of the G2–M-phase transition (231). Using biochemical fractionation of *Xenopus laevis* egg extracts, in conjunction with at first the oocyte microinjection assay and later a cell-free assay, in which induction of nuclear envelope breakdown is scored (144), MPF activity was purified several-thousand-fold (145). The purest fractions contained two proteins of about 32 and 45 kDa, which together had protein kinase activity. By this time genetic complementation in *S. pombe* had been used to establish that human cells, too, contain a *cdc2*-related gene, and that this gene is fully able to substitute for its yeast counterpart (135). The availability of the human gene sequence permitted the production of antibodies specific to a perfectly conserved region of the Cdc2 protein, and the circle was finally closed with the demonstration that these antibodies recognized the smaller protein in the purified MPF fractions, identifying it as the *X. laevis* Cdc2 homolog (74).

The second component of MPF was subsequently found to be a cyclin B protein (73, 130). Cyclins A and B had been identified previously as proteins that are synthesized and degraded in a cyclical manner during each of the early cell cycles of cleaving sea urchin eggs (59). Injection of cyclin A mRNA was found to be capable of inducing resumption of meiosis in *X. laevis* oocytes (270), suggesting a general role for cyclins in the induction of meiotic and mitotic M phases. Translation of cyclin B mRNA was found to be essential for the appearance of mitotic events in a *Xenopus* cell-free system (167). This, along with other experiments in which total mRNA was ablated before adding back cyclin mRNA (179), supported the suggestion that cyclin synthesis constitutes the protein synthesis requirement typical of early embryonic cell cycles. Genetical studies in *S. pombe* demonstrated the evolutionary conservation of cyclin B structure and function. The fission yeast *cdc13* gene was first identified by means of temperature-sensitive cell cycle mutants arresting in a mitotic-like state, though without disassembling the interphase microtubular array or forming a spindle (87, 183). This unusual cell cycle arrest was shown to result from a partial retention of Cdc13 function, as deletion of the *cdc13* gene resulted in a premitotic arrest similar to that seen with loss-of-function *cdc2* mutants (87, 172). Allele-specific interactions between *cdc2* and *cdc13* implied a direct interaction between the two gene products (17), and DNA sequence analysis revealed that *cdc13* encodes a protein of the cyclin B family (18, 80, 87, 260). The prediction that Cdc2 and Cdc13 should form a heteromeric protein kinase responsible for driving the onset of mitosis in fission yeast has since been confirmed by immunochemical methods (20, 67). The active form of the MPF protein kinase consists of a 1:1 heterodimer of Cdc2 and cyclin B in all eukaryotic species that have been examined, and the catalytic activity of Cdc2 appears to depend absolutely on its association with the cyclin subunit (130, 262). The conservation of this activity means that observations made concerning the kinase in one organism are in general applicable to other eukaryotes, no matter how distantly related. In comparison with Cdc2, the cyclin B subunit has not been as highly conserved in its amino acid sequence during evolution. Comparison with cyclin A reveals amino acid sequence similarity that is restricted to a region termed the "cyclin box" that is also found in more recently identified and functionally distinct members of the cyclin protein family. This relatively well-conserved cyclin motif includes amino acid residues that are important for interaction with the protein kinase catalytic subunit (121). While cyclin A was originally identified on the basis of its cyclin B–like behavior, and despite the ability of its mRNA to induce meiotic maturation, the precise cell cycle roles of this cyclin are less well defined than that of cyclin B. Cyclin A is able to form complexes with Cdc2 (46), but these complexes do not copurify with MPF activity, and presumably therefore have a role distinct from the establishment of the M-phase state. The situation is complicated by the dimerization of cyclin A with a distinct Cdc2-related kinase subunit termed Cdk2 (for *cyclin-dependent kinase*) (54, 55, 210, 223); this heterodimer is thought to have functions earlier in the cell cycle, as discussed toward the end of this chapter.

Regulation of Cdc2

The protein kinase activity of Cdc2 is markedly periodic in the cell cycle, typically rising gradually as cells move through S and G2 phases, before abruptly increasing to give a sharp spike of activity as cells enter M phase. The level of activity remains high until anaphase, but by telophase has been reduced to a baseline value. This sharply periodic profile ensures a sharp distinction between interphase and M phase, important in preventing both partial M-phase entry and the persistence of M-phase features after nuclear

division. In principle it might be possible to achieve this abrupt activation of Cdc2 by tightly regulating the synthesis or degradation of one or both of the components of the MPF heterodimer. In fact, although the periodic synthesis and destruction of cyclin B plays an important part in determining the overall Cdc2 activity profile, post-translational mechanisms—specifically, phosphorylation of Cdc2—are of overriding importance (82, 83, 126, 190, 191), and changes in the abundance of Cdc2 itself are probably significant only in terms of long-term growth regulation (45, 71, 134).

Cdc2 is regulated by phosphorylation at two conserved amino acid residues in all eukaryotes that have been examined. The first is a tyrosine residue at position 15 (Y15); phosphorylation at this position inhibits Cdc2 activity (83) and is mediated by the Wee1 protein kinase (161, 211). A second protein kinase, Mik1, probably phosphorylates *S. pombe* Y15 to a lesser extent (152). Mutational inactivation of Wee1 (or mutation of Y15 to an unphosphorylatable phenylalanine residue) leads to advancement of the cell cycle timing of mitosis in *S. pombe* (83, 193), while overexpression of Wee1 delays the onset of mitosis (239), indicating that the phosphorylation state of Y15 is responsible for determining the timing of Cdc2 activation in this species. Phosphorylation at a second site, a threonine residue (T167 in *S. pombe* Cdc2; T161 in the mammalian protein), is by contrast positively required for attainment of a biologically significant level of Cdc2 activity (82). This site has been shown to be phosphorylated by MO15/CAK (cyclin-dependent kinase-activating kinase), an evolutionarily distant relative of Cdc2 itself (65, 227, 261, 263). Intriguingly, MO15/CAK is itself a cyclin-dependent kinase (66, 154), demonstrating that there is at least the potential for amplification of biological signals through Cdk "cascades." An inhibitor of MPF activity purified from *Xenopus* eggs and termed INH (34) contains phosphatase 2A activity and appears to act by direct dephosphorylation of T161 (136). In higher eukaryotes, phosphorylation at a second threonine residue 14 (T14) provides an additional level of inhibition (126, 190), and is thought to be carried out by a protein kinase distinct from Wee1 (124). Other sites of phosphorylation on Cdc2 have been reported but appear to be neither widespread nor of clear regulatory significance. Association between Cdc2 and cyclin B promotes phosphorylation at T14, Y15, and T161 (262), providing a means by which the heterodimers can be assembled and yet maintained in an inactive state, sometimes referred to as "pre-MPF" in the literature. Both T14 and Y15 are dephosphorylated by a phosphoprotein phosphatase equivalent to to the *S. pombe* prototype, the product of the *cdc25* gene, to bring about the abrupt activation of Cdc2 at the G2/M boundary (128, 237, 266). This abruptness is probably heightened by positive feedback loops that act to inhibit Wee1 kinase activity and activate Cdc25 phosphatase activity once a small amount of active Cdc2 has been generated. The details of these loops have yet to be fully determined, though a further kinase, Nim1 (63, 238), appears to phosphorylate and inactivate Wee1 (28, 272), while Cdc25 is activated by phosphorylation in vitro (72, 101, 107, 129). Mitosis-specific phosphorylation of Wee1 has been shown to result in a substantial decrease in Wee1 activity (272). Positive-feedback mechanisms of this sort would help to explain experimental observations of "autocatalytic" amplification of MPF in the absence of new protein synthesis after microinjection into immature oocytes (158, 178). Thus, while cyclin B is absolutely required for MPF activity, its availability is not generally the factor that limits Cdc2 activity. Again, early embryonic cell cycles provide the exception to the rule, as the rapid oscillations of MPF activity that are seen in cleaving eggs may well be driven by the corresponding oscillations in cyclin B levels (180). Significantly, cyclical tyrosine phosphorylation/dephosphorylation of Cdc2 was not seen during these early cycles, demonstrating that dephosphorylation of T14/Y15, while frequently rate-limiting, is not an absolute requirement for Cdc2 activation (64).

Additional control over Cdc2 activity is mediated by regulation of the subcellular localization of the Cdc2–cyclin B heterodimers; these are accumulated predominantly as inactive pre-MPF in the cytoplasm of cells that are in S phase or G2, before being suddenly relocated to the nucleus in mitotic prophase (224). There is some evidence that the nuclear Wee1 protein kinase has a role in blocking entry of Cdc2–cyclin B into the nucleus (94); this could in turn deny access of active Cdc2 to key nuclear substrates whose phosphorylation precedes nuclear envelope breakdown, as well as preventing the amplification of Cdc2 activity by means of the putative positive-feedback loops outlined above. There appears to be a further level of subcellular localization of a fraction of Cdc2 to the microtubule-organizing centers in a number of cell types, suggesting a direct role for the kinase in directing microtubular reorganization (6, 20).

Substrates of Cdc2

Ever since it became clear that Cdc2 is a protein kinase responsible for the onset of M phase in diverse eukaryotes, much effort has been directed toward identification of substrates of the kinase in vivo. An early development in this area was the discovery that Cdc2

phosphorylates the spacer histone H1 efficiently in vitro, in comparison with other commonly used assay substrates such as casein (44, 145, 172). In fact, a previously identified growth-associated or M-phase-specific histone H1 kinase and Cdc2 appear to be one and the same (131, 133). Cdc2 phosphorylates H1 during mitosis at multiple serine and threonine residues corresponding to the consensus sequence S/TPXK/R (S = serine, T = threonine, P = proline, X = any amino acid, K = lysine, and R = arginine) (133). More than 20 yr ago it was proposed that H1 phosphorylation might play a role in chromatin condensation (21), and although clear-cut evidence remains elusive, this has formed the basis of a simple and attractive model linking Cdc2 activation directly with the hypercondensation typical of M phase. The results of genetical experiments described below complicate this simple view, however, by implicating additional intermediate elements in this pathway. Demonstrating a physiologically meaningful relationship between a protein kinase and a candidate substrate is indeed far from straightforward (173). For Cdc2 this difficulty is compounded both by the looseness of the substrate site consensus (some known substrate sites contain only the S/TP component of the motif given above) and by the existence of a considerable number of related protein serine/threonine kinases with similar if not identical substrate specificities in vitro. Candidate Cdc2 substrates should be proteins that are phosphorylated by the kinase in vitro at sites known to be maximally phosphorylated in vivo during M phase. In the best cases these phosphorylation events should have implications for the biological activity of the substrate that are compatible with a role for that substrate in one of the known parallel pathways of M phase. Among the most clearly validated substrates are the nuclear lamins A and C, phosphorylation of which by Cdc2 is required for disassembly of the nuclear lamina (93, 153, 218, 285). Conservative replacement of the Cdc2 target sites by nonphosphorylatable residues using site-directed mutagenesis generated lamin protomers that could assemble into filaments but which could not be disassembled at mitosis. Furthermore, lamina disassembly was induced by addition of purified active Cdc2–cyclin B to permeabilized cells in vitro. Lamina disassembly may be involved in precipitating nuclear envelope breakdown, but is unlikely to be sufficient in itself to trigger membrane vesicularization. Reduction in the rate of vesicle fusion can be induced by active Cdc2–cyclin B in a cell-free system (277), indicating that a fundamental change in membrane behavior is induced at M phase, either directly or indirectly by Cdc2; the appropriate target(s) for the kinase in this case is as yet unidentified. A second known Cdc2 substrate is vimentin (4, 26, 58), like the lamins an intermediate filament (IF) protein. Phosphorylation of vimentin filaments by Cdc2 leads to disassembly of these filaments in vitro, and IFAP 300, a 300 kDa IF-crossbridging protein, appears also to be phosphorylated by Cdc2 at a mitosis-specific site (257). Cdc2 may therefore effect reorganization of the vimentin IF network both by disrupting IF crossbridges mediated by IFAP 300 and by disassembling individual IFs into monomeric vimentin. Phosphorylation of the 220 kDa microtubule-associated protein MAP4 may provide the link between Cdc2 activation and the dramatic alteration of microtubule dynamics that occurs at the G2/M transition (255). Cdc2-mediated phosphorylation of the myosin II light chain at two sites is thought to inhibit actin-dependent myosin ATPase activity (243). This provides an attractive mechanism by which premature contraction of the actin ring, required for cytokinesis, could be prevented while Cdc2 activity remains high in metaphase. Nonmuscle caldesmon is also thought to be a significant Cdc2 target, phosphorylation of which could promote microfilament contraction, perhaps also related to cytokinesis or cell rounding (294, 295). Note that the above-mentioned Cdc2 substrates are all relatively abundant components of the cytoarchitecture, indicating that Cdc2 acts directly to trigger some of the physical processes of mitosis. The protein tyrosine kinases Src, Abl, and the serine/threonine-specific casein kinase II constitute a second category of Cdc2 substrates (118, 142, 175, 177, 252). Though the biological consequences of their M-phase-specific phosphorylation are by no means clear, these kinases could provide a basis for diversification of the MPF signal by promoting phosphorylation at amino acid residues that cannot be targeted by Cdc2 directly; alternatively, the Cdc2-mediated phosphorylations could modulate the substrate specificities of the target kinases. Members of the Cdc25 family of phosphatases are phosphorylated by Cdc2 as part of the putative positive feedback loop, discussed earlier, that could ensure the abruptness of MPF activation. By altering the activities of multiple kinases (including, indirectly, itself) and phosphatases, Cdc2 triggers a wide-ranging shift in the pattern of cellular phosphoproteins that has for some time been known as a hallmark of the M-phase state (155). Considering the impact of genetics in other areas of cell cycle regulation, surprisingly few Cdc2 substrates have so far been identified in yeasts. The only good example to date is the S. cerevisiae Swi5 protein (171), which is required for the specialized cell-cycle-related process of mating type switching. Phosphorylation of Swi5 by the S. cerevisiae Cdc2 homolog (termed Cdc28 for reasons of historical precedent) occurs within a region of the protein required for its translocation into

the nucleus and restricts Swi5 to the cytoplasm while the kinase is active. Still other candidate Cdc2 substrates have been described in a variety of species, including the nucleolar proteins NO38 and nucleolin (12, 217), the γ and δ subunits of the EF-1 translational elongation factor (13, 166), glial fibrillary acidic protein (159, 276), and the chromatin-associated high-mobility group (HMG) proteins P1, I, and Y (163). In these cases cell-cycle-associated functions have yet to be clearly defined. For a number of other reported in vitro substrates of Cdc2, phosphorylation in vivo is not maximal in M phase and is likely to be carried out by distinct, Cdc2-related kinases.

In summary, Cdc2 substrates have been identified that provide clues to the mechanisms underlying M-phase-specific disassembly of the nuclear lamina and other intermediate filaments, microtubular reorganization, and microfilament contraction. Other M-phase events could be brought about through less direct pathways involving protein kinases and phosphatases downstream of Cdc2.

Induction of Chromatin Condensation

Although the list of Cdc2 substrates above includes a number of chromatin-associated proteins, a clear understanding of how Cdc2 activation leads to chromatin condensation is still lacking. Mutation of the top2 gene that encodes DNA topoisomerase II in S. pombe causes defective condensation (280), and topoisomerase II is required for chromosome condensation in vitro (1), but the activity or localization of wild-type topoisomerase II has not been shown to change radically at the G2/M boundary, as might be expected for an enzyme with a regulatory role at this stage of the cell cycle. By contrast the NimA protein kinase of Aspergillus nidulans has been shown to be a positive effector of chromatin condensation (151, 203, 205) and seems likely to be the rate-limiting component that links activation of the A. nidulans Cdc2 homolog (termed NimX) to chromatin condensation itself (197, 204). Like Cdc2, NimA satisfies the fundamental criterion used to distinguish cell cycle regulators, as premature activation of NimA leads to advancement of the cell cycle timing of condensation, though in this case without simultaneously triggering the appearance of other features of M phase (197, 205). Another similarity with Cdc2 is the sharply periodic profile of NimA protein kinase activity, but in this case the spike of activity in mitosis appears to be brought about by transient stabilization of the normally unstable NimA protein. It seems likely that there is a role for the A. nidulans Cdc2 homolog in this transient stabilization, though this has yet to be established. When its expression was artificially deregulated, NimA was also able to drive chromatin condensation in *S. pombe* and human cells (197); this process did not require Cdc2 activity and could occur from all points of the cell cycle. The latter observation suggests there is no role for Cdc2-mediated phosphorylation of chromatin-associated proteins in NimA-driven condensation, but does not rule out a role for these phosphorylation events in fine-tuning condensation under physiologically normal conditions. Furthermore, despite its ability to promote condensation in diverse organisms, in some of which protein kinases with sequences similar to that of NimA have been reported (245), the extent to which functional NimA homologs have been conserved in eukaryotic evolution is not yet known. Authentic substrates of NimA must also be identified in order to obtain a coherent view of the pathway linking Cdc2 activation to chromatin condensation.

Controls Operating within M Phase

Under most normal physiological conditions mitosis is completed rapidly in comparison with other cell cycle phases, but despite this rapidity the various events triggered by Cdc2 activation are executed in a highly coordinated manner. A checkpoint control operates to ensure that the paired chromosomes are aligned at the metaphase plate before anaphase is initiated (297). Though the molecular details of this mechanism have yet to be determined, Cdc2 activity remains high and is still required until the metaphase–anaphase transition. This situation is paralleled in unfertilized vertebrate eggs, which are arrested with active Cdc2 in meiotic metaphase II until fertilization. This meiotic arrest is mediated by a cytostatic factor (CSF) that contains the protein kinase product of the c-*mos* proto-oncogene (158, 241). Related activities might act to maintain high Cdc2 activity until the correct alignment of chromosomes at the mitotic metaphase plate is achieved. In murine cells the p53 tumor suppressor protein has been implicated in such a control mechanism, as cells lacking p53 activity exit mitosis without segregating their chromosomes after exposure to drugs that inhibit mitotic spindle formation (33a).

At fertilization a transient increase in cytosolic Ca^{2+} relieves the CSF-mediated stabilization of MPF activity (84, 288); the transducer of this calcium transient signal has been shown to be the calmodulin-dependent protein kinase II (146). This kinase may also be responsible for the signaling of MPF inactivation in mitotic M phase. Cdc2 inactivation at the onset of anaphase does not involve a simple reversal of the activation process. Instead the directionality of mitosis is ensured

by the abrupt proteolytic degradation of the cyclin B subunit (46, 59, 222, 270). A role for proteases in the metaphase–anaphase transition was demonstrated through experiments in which protease inhibitors were injected into oocytes, causing arrest in meiotic metaphase (221). MPF inactivation can also be prevented by the introduction into cells (or cell-free mitotic extracts) of a cyclin B mutant that lacks an N-terminal region of approximately 90 amino acid residues (180). This part of the cyclin protein is required for its proteolytic degradation, which appears to involve cyclin ubiquitination (79). In *Xenopus* egg extracts this process can be triggered by the addition of a constitutively active mutant of the calmodulin-dependent protein kinase II (146), but the steps in the pathway that must link this kinase and the ligation of ubiquitin to cyclin B have yet to be described. Despite early reports in the literature describing metaphase arrest induced by the N-terminally truncated cyclin B, it now appears that Cdc2 inactivation and cyclin B degradation are not required for the initiation of anaphase (102, 269). In extracts of *Xenopus* eggs addition of the nondegradable cyclin B blocks MPF inactivation but does not prevent sister chromatid separation. Inhibition of the ubiquitin-dependent protease responsible for cyclin degradation nevertheless blocks both MPF inactivation and the onset of anaphase events. Together these results suggest that a single protease activity is responsible both for Cdc2 inactivation and, independently, for triggering sister chromatid separation (102). Cyclin B degradation requires the assembly of a microtubular spindle (104, 139, 169), indicating the existence of a further checkpoint mechanism within M phase. A number of proteins involved in the initiation of anaphase have been identified, but it is not yet known whether these include key targets of the ubiquitin-dependent protease. Topoisomerase II, for example, is required for sister chromatid separation, but inhibition of this enzyme using specific catalytic inhibitors or temperature-sensitive mutations does not appear to block the initiation of anaphase (42, 81, 103, 250, 281). This suggests that there is no checkpoint mechanism for monitoring completion of chromosome decatenation before anaphase onset. Another protein required for the metaphase–anaphase transition is the product of the *three rows* gene in *Drosophila*, mutation of which causes mitotic arrest with metaphase-like chromosomes (219). In these arrested cells cyclin B is degraded normally and, without chromosome separation, the nucleus regains an interphase appearance, though no further mitosis is seen. Note that while it is apparently not a prerequisite for the initiation of anaphase, Cdc2 inactivation is ultimately required for reestablishment of the interphase state (77, 180). Activation of specific phosphatase activities may also contribute to reversal of Cdc2-driven M-phase phosphorylation events; in diverse eukaryotes protein phosphatases of the type I class are required both for the completion of anaphase and for the inactivation of Cdc2 (5, 19, 40, 117, 198, 220). This latter function could involve type I phosphatases directly in the activation of the ubiquitin-dependent protease; alternatively, checkpoint mechanisms might exist that link initiation of cyclin B proteolysis to the dephosphorylation of M-phase phosphoproteins. In fission yeast two distinct catalytic subunits with type I phosphatase activity interact with the protein product of the *sds22* gene, which appears to regulate substrate specificity and is essential for the onset of anaphase (199, 265). An explanation of how the modulation of type I phosphatase activities is integrated with cyclin B proteolysis and the initiation of anaphase remains a major challenge for investigators in this area.

CYCLIN-DEPENDENT KINASES OTHER THAN Cdc2

Numerous protein kinases have been described that bear a degree of structural similarity to Cdc2 (88, 165, 186, 202, 210, 292). In some of these cases protein kinase activity has been shown to depend on association with a member of what has turned out to be a large family of cyclin proteins. Some of these cyclin-dependent kinases (Cdks) are able to complement mutations in the budding yeast *CDC28* gene (homologous to *S. pombe cdc2*) (55, 165). Other members of the Cdk family have been identified without reference to biological function, through polymerase chain reaction approaches exploiting sequence similarity with Cdc2/Cdc28. The physiological roles of these Cdc2-related kinases are not currently as well understood as that of Cdc2 itself. One early view was that each Cdk might be responsible for regulation of a discrete cell cycle checkpoint, but subsequent experiments have shown that this is not the case. The kinase designated Cdk5, for example, appears to function mainly in postmitotic neurons as a neurofilament kinase (96, 138, 254), while the budding yeast Cdk encoded by *PHO85* is involved in regulation of phosphate metabolism and is dispensable for normal cell cycle progression (273). Furthermore the Cdk component of CAK, also referred to as Cdk7, has also been identified as a component of the general transcription factor TFIIH (254a). At the same time, at least two of the novel kinases, Cdk2 and Cdk3, are required for cell cycle progression in vertebrates (15, 61, 275, 282), and Cdk2 may well act to control

progression through a rate-limiting step, in this case in the G1 phase (see below).

G1

Gap (G1, G2) phases separating M from S and S from M are a feature of most cell cycles, yet are not an absolute requirement. This is most clearly demonstrated in early embryos, which in many species complete their first rapid cell cycles apparently without G1 or G2 phases. Gap phases first appear at or about the time when zygotic gene expression is activated, between mitoses 13 and 16 in *Drosophila*, for example (52). Up until this period of development, the embryo exploits the pool of maternally derived proteins that exists at fertilization, and the transition from rapid to gapped cell cycles appears to be triggered by the attainment of a critical nuclear/cytoplasmic ratio (119, 184). Thus a requirement for de novo gene expression is at least part of the explanation for the existence of gap phases. Gap phases also provide opportunities for the integration of developmental and proliferative cues, such as the presence of specific mitogens or the availability of nutrients. Checkpoint mechanisms—for example, those monitoring DNA damage—can also act to extend G1 and G2 under appropriate circumstances.

Absence of cell-specific growth factors or nutrients (or in some cases insufficient adherence to a matrix) can result in the establishment of a quiescent state characterized by reduced rates of RNA and protein synthesis. Although this state is frequently referred to as G0, implying the existence of a distinct cell cycle phase, the usefulness of this distinction is not clear. The quiescent state can be entered from either G1 or G2 (29, 75, 214) and resumption of cell cycle progression would appear simply to represent a change in cellular growth status.

Cell Cycle Commitment

Physiological experiments with cultured mammalian cells and genetical experiments with yeasts have identified points in G1 beyond which cells are committed to a new mitotic cell cycle. In budding and fission yeasts this point has been termed "start" and is defined as the period in G1 where haploid cells become refractory to conjugation with cells of the opposite mating type (91, 194, 196). For these G1 yeast cells conjugation represents a developmental alternative to S-phase entry. Temperature-sensitive *cdc* mutants were used to discriminate between gene products required before completion of start (defects in which cause cell cycle arrest but allow conjugation) and those required later in the cell cycle (in which cases no conjugation is seen). Remarkably, the gene products required for completion of start include Cdc2 in *S. pombe* and the homologous Cdc28 in *S. cerevisiae* (194, 232). In these simple eukaryotes a single gene product that triggers entry into M phase is thus also required to perform a distinct function during G1. Note that while dominant *cdc2* alleles can advance the timing of mitosis (195), no activating mutations of *cdc2* or *CDC28* have been described that advance the cell cycle timing of start or S-phase entry. Furthermore, the prestart G1 function of Cdc2 in *S. pombe* is executed at a point when Cdc2 kinase activity as measured *in vitro* is low, in comparison with the level seen in mitosis (172). While the biochemical details of the *S. pombe* Cdc2 function in G1 remain obscure, it appears that Cdc2 has a fundamental role in distinguishing the G1 phase from G2, ensuring that S phase is followed by mitosis, which is in turn followed by another S phase. This sequence can be disrupted experimentally by means of specific *cdc2* mutations, which can direct a G2 cell into an additional round of DNA replication without an intervening mitosis (22). This suggests that Cdc2 exists in two interconvertible forms, one directing entry into S phase from G1 and the other entry into M phase from G2. Investigation of the *S. cerevisiae* cell cycle has led to the conclusion that these forms of Cdc2/Cdc28 are distinguished by means of their association with alternative cyclin subunits (182). Such nonmitotic cyclins were first identified in *S. cerevisiae* through genetical experiments in which mutants were sought that had a shortened G1 phase or were deficient in the cell cycle arrest response to mating pheromone (32, 181, 267). Such mutants satisfy the basic criterion of cell cycle advancement (in this case advancement of the execution of the start control) that has proven useful in the identification of cell cycle regulators. Subsequent molecular genetical analysis has shown that these mutations lie within a cyclin gene termed *WHI1/DAF1/CLN3* and act in a dominant fashion, probably to perturb the normal proteolytic degradation of the cyclin protein (32, 181, 279). Subsequent experiments identified two further G1 cyclin genes through their ability when overexpressed to suppress a *cdc28* mutation in *S. cerevisiae* (86, 235). The three G1 cyclins appear to be functionally redundant, in that disruption of any two of the genes leaves the yeast cell viable, while deletion of all three causes cell cycle arrest in G1. This multiplicity of cyclin genes may be, in part, a peculiarity of the budding yeast, in which many genes (including the mitotic B cyclin genes) appear to be duplicated. While *S. pombe* has nonmitotic cyclin genes, at least one of which is able to suppress a G1

cyclin defect in *S. cerevisiae* (68), the proteins encoded by these genes have not so far been shown to act as G1 cyclins in *S. pombe* (69). This latter observation should be borne in mind when considering the likely functions of genes from multicellular organisms also identified by means of the G1 cyclin complementation approach, as discussed below. Nonetheless, the information on G1 progression that has gradually emerged from the genetical analysis of *S. cerevisiae* has generated the most comprehensive model yet available for the mechanisms that could underlie G1 progression in animal cells. Cdk complexes each consisting of a Cdc28 catalytic subunit in combination with a G1 cyclin appear to be responsible for the activation of transcription factor complexes that control the expression of a variety of G1-specific genes (108, 149), though additional G1 targets for Cdc28 might well exist. According to the model, two transcription factors are targeted, one containing the Swi4 and Swi6 proteins and a second, termed dsc1 (for *DNA synthesis control*), that also contains Swi6, in this case combined with the 120 kDa protein Mbp1 (37, 148, 228). The latter factor directs transcription of a variety of genes encoding proteins such as thymidylate synthase and DNA polymerase I required in the preparation for S phase (108). The Swi4/Swi6 factor acts through a distinct sequence element found in the promoters of genes encoding the Cln1 and Cln2 G1 cyclins themselves, among others. This observation suggests that a positive feedback loop might operate to generate a dramatic increase in G1 cyclin–associated Cdc28 activity once a threshold level of G1 cyclins is attained (38). This would be analogous to the spike of Cdc2 activity seen at the onset of M phase, though achieved in this case by a molecular mechanism based on cyclin accumulation. Attractive as this model may be, it has yet to receive solid support from biochemical investigations of the kind that have confirmed the existence of a universal mechanism for M-phase entry. Further data also suggest that Swi4-mediated cell cycle regulation of Cln2 expression does not involve the putative Swi4/Swi6 responsive (SCB) elements in the *CLN2* gene promoter (33). An additional pair of B-type cyclins also appear to have a role in the G1–S transition (246), probably acting downstream from the Cln cyclins.

In mammals the cell cycle controls that determine progression through G1 have been identified through physiological, rather than genetical, investigations. Points of commitment to S-phase entry have been defined both for cells that have been stimulated to enter the cycle from a quiescent state and for cells already actively cycling [reviewed in (191, 209)]. Commitment has been defined in this case not, as in yeast, with reference to the potential for implementing alternative developmental programs, but in terms of escape from requirements for specific growth factors or high rates of protein synthesis. The general significance of these studies was to some extent limited by the nature of the growth factors and cells (frequently embryonic fibroblasts) used, but their conclusions do provide a general basis for understanding G1 controls in animal cells and make specific predictions about the behavior of regulatory molecules likely to be involved in these controls. Commitment to enter S phase has not generally been distinguished in the literature from commitment to completion of an entire cell cycle. While there is a point in G1 at which fibroblasts become growth factor independent and committed to entering S phase (208, 296), the time interval between this control point and the initiation of DNA replication itself varies from cell to cell over a range of several hours in mammalian fibroblasts. Within this interval the cells enter S phase with approximate first-order kinetics at a rate determined by the concentration of growth factors in the surrounding medium (23, 258). It follows that in animal cells, as in the yeasts, commitment to enter S phase can occur well before DNA replication itself begins. In fibroblasts the requirement for a rapid rate of protein synthesis appears to be confined to the same early part of G1 in which serum growth factors are required; kinetic experiments suggested that fibroblasts escape from their growth factor and essential amino acid requirements at a single "restriction" (R) point (208). On this basis it was proposed that the pre-R-point growth factor requirement reflects the need for synthesis of a critical regulatory labile protein (236). The half-life of this putative protein was estimated at 2–3 h, and a critical threshold level appeared to be required in order for cells to pass the R point. With the benefit of hindsight, it is clear that mammalian equivalents of the budding yeast G1 cyclins would be excellent candidates for such rate-limiting, labile proteins.

Mammalian Nonmitotic Cyclins

In addition to the B cyclins that combine with Cdc2 to form MPF, all vertebrates that have been examined contain a variety of functionally and structurally distinct cyclin proteins (253). As with the Cdks, the "nonmitotic" cyclins have been identified by a variety of means, including complementation of budding yeast mutants and strategies exploiting the regions of amino acid sequence conservation among cyclin proteins. It is reasonable to assume, and is in some cases established, that these proteins combine with cyclin-dependent protein kinases to perform their biological functions, but as pointed out above, the functions of the respective Cdk/cyclin complexes need not necessarily involve cell cycle regulation. At the time of writing, a wealth of molecular biological information has outstripped the capacity of physiological experimentation, such that the functions of many mammalian cyclin/Cdk complexes remain obscure. Two types of mammalian cyclin are nonetheless implicated in rate-limiting processes in G1; these are the D and E cyclins (122, 160, 176), which have been shown to cause advancement into S phase when constitutively overexpressed at a moderate level in fibroblasts (201, 229). The status of D cyclins is complicated by a number of additional experimental observations; when expressed transiently at a high level, cyclin D1 blocked both replicative and repair DNA synthesis (207), and both D and E cyclins are expressed at high levels in senescent fibroblasts, in the absence of associated kinase activity (47). An early assumption, based on the MPF paradigm, that cyclins function simply through interaction with their respective Cdk subunits, has also turned out to be an oversimplification. Cyclin D is able to form heteromeric complexes with the DNA polymerase δ accessory subunit PCNA (proliferating cell nuclear antigen, somewhat confusingly itself termed "cyclin" in the early literature), and with the protein product (pRb) of the retinoblastoma susceptibility gene (41, 60, 112, 292). Similarly, cyclin E has been shown to associate with the transcription factor E2F, as well as with the pRb-related protein p107 (137). These associations could provide mechanisms by which catalytic subunits [principally Cdk4 and Cdk6 in the case of cyclin D1 (9), and Cdk2 in the case of cyclin E (122, 123)] are targeted to specific substrates. Certainly, pRb and E2F appear to be substrates for Cdks in vivo, though the kinases responsible have yet to be unambiguously identified (50, 57, 141). Equally, the stable nature of these associations could reflect structural or inhibitory roles for one or more of the protein subunits. Transcription from promoter elements containing E2F recognition sequences is generally activated by E2F, but is repressed when pRb is also present (287), and appropriate recognition sequences are found upstream of a wide variety of proliferation-associated genes (95). When constitutively overexpressed, pRb accumulated primarily in the hypophosphorylated form and induced an arrest of proliferation. This arrest could be overcome by co-expression of cyclin A or E, in which case the majority of pRb was seen to be hyperphosphorylated (99). These findings, combined with the additional observation that pRb is first phosphorylated (and, it is assumed, inactivated) some hours before replicative DNA synthesis is first detectable (170), provide the basis for a model of G1 progression in animal cells. In this model, sequential activation of different cyclin/Cdk complexes would drive the transition from a G1 state, in which active, hypophosphorylated pRb represses transcription from proliferation-associated gene promoters, to a state in which Rb repression of E2F-mediated transcription is relieved by phosphorylation of Rb. If the targets of E2F include genes encoding proteins that are partly rate-limiting for S-phase entry, the model could help to explain the observed shortening of G1 on overexpression of cyclin D or E. In support of this model, the protein kinase activity associated with cyclin E has been shown to be highly cell cycle phase specific (49, 123), with the timing of maximal activity suggesting a role in G1; by contrast the cell cycle kinetics of cyclin D–associated kinase activity are not currently so well defined, being complicated by the existence of three subtypes whose synthesis may depend on the continued presence of growth factors (160, 249). A major attraction of this model, despite its current preliminary status, is its broad parallelism with the emerging view of G1/S-phase control in yeasts, as described above. A tentative conclusion would be that the G1 role performed in yeasts by Cdc2/Cdc28 is performed in metazoans by distinct Cdks in concert with cyclin D and/or E, and that an important aspect of this role is in each case the activation of specific transcription factors, with the resulting altered pattern of gene expression providing the connection between cell cycle commitment and the initiation of S phase. This view places cyclins D and E and their associated Cdk partners center stage in the cell cycle commitment process but leaves unexplained the interval of variable duration that separates commitment from the G1/S-phase boundary in animal cells (296).

Cell Cycle Commitment and Cancer

Understandably, a great deal of attention is being paid to those molecules potentially involved in cell cycle commitment/R point control in human cells, as loss of this control appears to be a common feature of tumorigenesis. Originally identified as a tumor suppressor, inactivated by mutation or deletion in specific human tumors, pRb is now also widely viewed as a negative regulator of cell cycle progression (27). Importantly, cyclin D1 has been found to be deregulated by gene amplification or translocation in a different subset of tumor types (3, 176, 248, 289). Any view of cancer as a disease of cell cycle disregulation carries with it the awkward implication that cell cycle entry is sufficient to drive unrestrained cell proliferation and that cell growth is therefore a secondary consequence of cell cycle progression. Such a view could not be easily reconciled with the (by now familiar) yeast models, in which cell proliferation and cell cycle progression are both normally dependent on cell growth. Further-

more, animal cells frequently increase in size and total protein content before commitment (8), and the duration of G1 in fibroblasts was shown to display an inverse relationship to cell size at birth (116). It is important to bear in mind that cancer is a multifactorial, polygenic condition. While cell cycle disregulation appears likely to be a contributory factor in tumorigenesis (105), additional genetic alterations affecting general cell growth are likely to be equally important, along with mutations suppressing apoptotic cell death.

A number of diverse gene products have been implicated in G1 regulation on the basis of experiments in which S-phase entry is blocked (rather than advanced) as a result of altered expression of the protein in question. Examples of these include the tumor suppressor p53 (36) and the prohibitin protein, overexpression of which causes cells to accumulate in G1 (192). Conversely Cdk2, Cdk3, s6 kinase, and the Ccg1 component of the TFIID transcription factor all appear to be positively required for S-phase entry (100, 132, 275, 282). This type of approach is analogous in some respects to the isolation of *cdc* mutants in yeast (91, 196), and is in principle a route by which elements required for cell cycle progression could be identified. Indeed, potential cell cycle functions of Cdk2 and p53 are becoming increasingly well defined; the former in association with cyclin E (see above) and the latter as an inducer of the p21 inhibitor of Cdks (see below). However, inhibition of general protein synthesis in G1 also blocks S-phase entry; this simple observation shows that many of the gene products capable of effecting S-phase delay may have functions only distantly related to cell cycle control.

Cdk INHIBITORS

Recent studies have identified a number of proteins able to interact with and directly inhibit the protein kinase activity of mammalian cyclin/Cdk complexes (39, 90, 97, 226, 274, 291). These structurally diverse proteins provide additional possible routes by which cell cycle progression (or, in more general terms, Cdk-dependent events) could be negatively regulated. Examples of this class of Cdk inhibitor in yeast may provide useful insight into the regulation of mammalian Cdk inhibitors. The best understood is Far1 in budding yeast, which was identified through loss-of-function mutations that render cells resistant to pheromone-induced cell cycle arrest, and which normally acts to inhibit Cdc28/Cln protein kinase activity and arrest cells in G1 in preparation for conjugation when cells are exposed to mating pheromone (25, 162, 215, 216, 278). Expression of Far1 is controlled by the Ste12 transcription factor, which is in turn activated by an intracellular signaling pathway that begins with the transmembrane pheromone receptor and includes a heterotrimeric G protein and members of the MAP kinase signaling family. The most obvious analog in mammals of the yeast mating pheromone-induced activation of Far1 involves the Cdk inhibitor p15 (89). The levels of p15 mRNA increase dramatically in epithelial cells exposed to TGFβ, under which circumstances the cells become arrested in late G1. While the details of this signaling pathway involved are not fully understood, the early components clearly include a G protein–linked transmembrane receptor that, as in the yeast system, ultimately signals transcriptional activation (157). Direct binding of p15 to Cdk4 and Cdk6 leads to inhibition of these kinase activities, providing at least part of the explanation for TGFβ-mediated cell cycle arrest. A structurally unrelated Cdk inhibitor termed p27 was independently found to bind Cdk2/cyclin E complexes in TGFβ-arrested cells, but unlike p15, p27 does not appear to be transcriptionally regulated (225, 226). In fission yeast the Rum1 protein, though structurally unrelated to Far1, also appears to delay progression through start (174). Rum1 additionally functions to repress entry into mitosis from prestart G1, probably by direct interaction with Cdc2, and is part of the mechanism that maintains the dependence of S-phase initiation on the completion of mitosis (Fig. 21.3).

The emerging families of Cdk inhibitors also provide further insights into the relationship between tumorigenesis and cell cycle control. The gene encoding the p15-related Cdk inhibitor p16 was found to be deleted in a large number of tumor cell lines (109, 189) [though in a small proportion of some types of primary tumor

(16)], and the p27-related p21 (alternatively known as Cip1, Waf1, Sdi1, and Cap20) is transcriptionally activated by wild-type p53 (48, 53, 140). Induction of cell cycle delay in G1 in response to DNA damage requires wild-type p53 function (111, 127) and is thought to be brought about through inactivation of Cdk/cyclin complexes by p21. A general distinction between normal diploid human fibroblasts and their transformed derivatives has also been described: In the former, cyclins A, B, and D each associate with their respective Cdk partners in higher-order complexes that also include p21 and PCNA (293). Transformation appears to result in the loss of p21 from the complexes containing cyclin A or B. More extensive characterization of the yeast Cdk inhibitors will hopefully help to clarify the modes of action of the analogous mammalian proteins.

LINKING CELL CYCLE COMMITMENT TO S-PHASE ENTRY

What might be the key targets of transcription factor complexes required for progression from G1 into S phase, and how might these events relate to the start control? Studies in fission yeast have perhaps come closest to providing answers to these questions. The fission yeast equivalent of the dsc1 transcription factor contains the product of the *cdc10* gene (150), previously defined genetically as encoding a function required for progression through start (194, 196). A search for cDNAs able to compensate for a partial loss of Cdc10 activity led to the molecular cloning of the *cdc18* gene (114), also previously identified through a *cdc* mutant unable to complete S phase properly at its restrictive temperature (183). Constitutive expression of *cdc18* allowed a *cdc10* mutant to grow at its restrictive temperature, and the normal transcription of *cdc18* was found to depend on *cdc10* activity, with *cdc18* transcripts oscillating in a cell-cycle-dependent manner, peaking in late G1/early S phase (114). Complete deletion of the *cdc18* gene blocked entry into S phase altogether, with a secondary phenotype of lethal attempted entry into mitosis with unreplicated DNA. Taken together, these data suggest that Cdc18 is a key downstream target of the fission yeast dsc1/Cdc10 transcription factor and that Cdc18 furthermore has a direct role in the initiation of DNA replication, as well as a secondary role in the suppression of premature mitosis in cells that have passed start. If correct, this interpretation places Cdc18 in the key position linking prestart activation of dsc1 (directly or indirectly by a G1 form of Cdc2) to the earliest poststart cell cycle event, the initiation of DNA replication (Fig. 21.4).

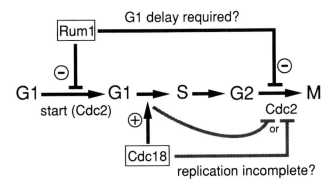

FIG. 21.3. Two checkpoints that prevent premature mitosis in fission yeast. In cells that require delay of G1 progression (for example, cells that are too small to satisfy the minimum size requirement for S-phase entry) Rum1 acts both to delay start and to prevent entry into mitosis. In cells that have passed start, Cdc18 acts to prevent premature mitosis as long as DNA replication is in progress.

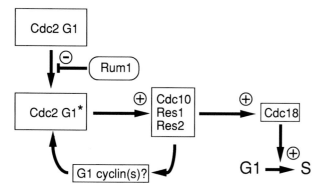

FIG. 21.4. A pathway linking start to the initiation of S phase in fission yeast. Conversion of Cdc2 from "prestart" form to a "poststart" form (asterisk) is inhibited by Rum1. Once activated for its start function, Cdc2 activates transcription factors containing Cdc10 protein and accessory subunits; key target for these transcription factors is *cdc18* gene, which encodes protein required for initiation of DNA replication (see text).

The extent to which the Cdc18 function is conserved in other organisms has yet to be determined, but it is encouraging to note that temperature-sensitive lesions in the Cdc6 protein in budding yeast also cause S-phase defects (24), overexpression of Cdc6 causes delayed M-phase entry, and Cdc6 displays significant overall protein sequence similarity to Cdc18 (114).

S PHASE AND G2

While much progress has been made in unraveling the mechanisms that link cell cycle commitment to the initiation of DNA replication, the key events required for the activation of replication origins themselves remain ill-defined. Much of this uncertainty stems from the elusive nature of chromosomal replication origins in higher eukaryotes, though major contributions [reviewed elsewhere (31, 113, 264)] have been made by researchers using either yeast or viral origins. One intriguing observation is that the PCNA polymerase δ accessory protein is required for DNA replication in the in vitro reconstituted SV40 viral model. The p21 Cdk inhibitor is capable of directly inhibiting DNA replication through interaction with PCNA (283), and it is possible that this represents a physiologically meaningful link between Cdk/cyclin activity in G1 and the activity of replication origins themselves. A further area in need of clarification is the role of cyclin A, which, despite displaying maximal levels of associated kinase activity in G2 (46, 168, 223), has also been reported to be required either for initiation or elongation in DNA replication (78, 206), though in extracts of *Xenopus* eggs, replication appeared not to require

cyclin A (61). Association between cyclin A/Cdk2 and the E2F transcription factor in S phase may inhibit E2F and be partly responsible for the down-regulation of late G1-specific transcription (125), but cyclin A has also been implicated as a positive factor mediating anchorage-dependent growth (85). DNA appears to be "licensed" for a single round of replication by an event occurring in late M phase, and the nuclear membrane plays an important role in preventing multiple rounds of initiation in the following S phase (14, 31). Much effort is currently being directed toward identification of the replication licensing factor (RLF). Reconstitution of RLF activity in vitro required two highly purified but biochemically distinct fractions termed RLF-B and RLF-M (25a, 29a). The RLF-M component is a complex mixture of proteins of the conserved Mcm family, originally identified through budding yeast mutants deficient in the maintenance of mini-chromosomes. It remains to be seen if mammalian functional equivalents of the *S. pombe* Cdc18 protein exist, and if these participate in the generation of active replication complexes. By extension of the yeast model, such complexes could provide a means of signaling ongoing DNA replication and therefore preventing premature mitosis. The guanine nucleotide exchange factor RCC1 may also be involved in this checkpoint control (188, 247). RCC1 appears to be a relatively abundant chromatin protein that interacts with the Ras-related small GTP-binding protein Ran/TC4 to promote GNP exchange (234), and a temperature-sensitive hamster cell mutant defective in RCC1 enters mitosis prematurely when incubated at the restrictive temperature (187, 188, 200). In fission yeast, however, an RCC1 homolog appears to act later in the cell cycle, during the reestablishment of the interphase state at the end of mitosis (244). The precise cell cycle functions of RCC1 and Ran/TC4 are further obscured by reports of their activity in diverse processes including nuclear transport, DNA and RNA synthesis, and RNA processing (30, 164, 234).

Repression of mitosis is a central feature of the cell cycle controls that operate in S phase and G2. In many metazoan cells, the very existence of G2 may reflect the requirement to ensure that DNA replication is indeed complete and that any outstanding DNA damage has been repaired. Note that the G2 phase in actively proliferating cells is defined only in terms of absence of detectable replicative synthesis, and that the final stages of replication could be significantly slower than the bulk replication in S phase. The cell size requirement for M-phase entry in *S. pombe* is largely responsible for the unusually extended G2 phase in this organism (193), but such rigid size controls may not be widespread among the eukaryotes. In general, the cell cycle controls that suppress mitosis during S phase and G2 involve maintenance of Cdc2/cyclin B in an inactive state. The best-understood mechanism for this inactivation involves phosphorylation of Cdc2 on residues T14 and Y15 (83, 126, 190), which is stimulated in response to the signals of DNA damage and incomplete DNA replication (35, 115, 143). In the search for components of the pathways linking these signals to Cdc2 inactivation, genetical strategies in yeasts are once again coming to the fore. These strategies have involved the identification of mutants sensitive to radiation or to inhibition of chromosomal DNA replication (2, 56, 242, 284, 286). Whereas the normal cellular response to these insults is to arrest the cell cycle in G2 with inactive Cdc2, the mutants fail to arrest, and proceed into mitosis with damaged or incompletely replicated DNA. The gene products that are defective in these mutants are currently being subjected to intensive study. In fission yeast the pathway signaling incomplete replication is expected to involve Cdc18 at an early stage and to impact upon the relative activities of Wee1 and Cdc25, thereby modulating the phosphorylation state of Cdc2 at Y15. The sensor components involved in the pathway signaling DNA damage have yet to be identified, but the class of large proteins exemplified by the human ATM (*ataxia telangiectasia mutated*) gene product (243a) appear to act early in this pathway. The 350 kDa ATM protein is required for efficient cell cycle arrest in G1, S, and G2 phases following the induction of double-strand DNA breaks. ATM has similarity in a carboxyl-terminal domain to phosphoinositol 3-kinase, but it seems likely that its true biochemical function is to phosphorylate proteins rather than lipids. ATM function is required for the normal induction of the p53 protein after exposure to ionizing radiation, implicating p53-mediated transcription of the p21 Cdk inhibitor gene, along with other p53 effectors, in the checkpoint pathways downstream from ATM (56a). In other organisms similar checkpoint roles are performed by related members of the ATM family, including Rad3 in fission yeast and Mec1 and Tel1 in budding yeast (84a, 107a, 175a). Experiments in fission yeast suggest that an additional mechanism independent of Y15 phosphorylation acts to inhibit Cdc2 activation in response to DNA damage (7, 251), but the details of this mechanism and its applicability to other systems are not yet clear. The Wee1 protein kinase also appears to have a role in protection against DNA damage that is dissociable from its Cdc2 Y15-phosphorylating activity (2, 7). The conservation of Wee1 and Cdc25 from yeast to humans suggests that the overall structure of G2 checkpoint mechanisms operative in yeast will also be found in human cells (106, 240), but this conjecture awaits confirmation. The nuclear Wee1 protein is clearly capable of inhibiting progression from G2 into mitosis in human cells (161), possibly in part at least by excluding active Cdc2 from the nucleus (94). Wee1 is itself subject to regulation by inhibitory phosphorylation, both in its carboxyl-terminal catalytic domain, where it is phosphorylated by the Nim1 protein kinase, and in its amino-terminal domain, where it is phosphorylated by kinase(s) as yet unidentified (28, 212, 272, 290). Experiments with catalytic inhibitors of DNA topoisomerase II have led to the identification of a further checkpoint mechanism that acts to delay entry into mitosis when topoisomerase II is inhibited (42a). This checkpoint appears to operate in the absence of detectable DNA damage and could serve to monitor decatenation of sister chromatids at the end of S phase.

CONCLUDING REMARKS

Enormous progress is being made toward an integrated understanding of the cell cycle and its regulation. Continued special attention should be given to the comparatively small number of cellular components that are capable of advancing the timing of execution of cell cycle events, as these components are likely to be targets for key regulatory pathways. Ongoing research promises to illuminate the links between molecules involved in oncogenic transformation and those involved in cell cycle progression, as well as the distinct connections between each of these processes and cell growth.

Unfortunately, early optimism that most (or all) of the mechanisms of cell cycle regulation would be conserved from yeast to humans has proved premature. Evolution of multicellular organisms has been characterized not only by the superimposition of ever more elaborate growth control networks on the core cell cycle regulators, but also by diversification of some of these regulators themselves, in particular the components regulating entry into and progression through S phase. For this reason, studies of cell cycle regulation in multicellular organisms will continue to be of fundamental importance, alongside the frequently illuminating studies of simpler eukaryotes.

REFERENCES

1. Adachi, Y., M. Luke, and U. K. Laemmli. Chromosome assembly in vitro: topoisomerase II is required for condensation. *Cell* 64: 137–148, 1991.
2. Al-Khodairy, F., and A. M. Carr. DNA repair mutants defining G2 checkpoint pathways in *Schizosaccharomyces pombe*. *EMBO J.* 11: 1343–1350, 1992.

3. Arnold, A., T. Motokura, T. Bloom, H. Kronenberg, J. Ruderman, H. Juppner, and H. G. Kim. The putative oncogene prad1 encodes a novel cyclin. *Cold Spring Harb. Symp. Quant. Biol.* 56: 93–97, 1991.

4. Aubin, J. E., M. Osborn, W. W. Franke, and K. Weber. Intermediate filaments of the vimentin type and cytokeratin type are distributed differently during mitosis. *Exp. Cell Res.* 129: 149–165, 1980.

5. Axton, J. M., V. Dombradi, P. T. Cohen, and D. M. Glover. One of the protein phosphatase 1 isoenzymes in Drosophila is essential for mitosis. *Cell* 63: 33–46, 1990.

6. Bailly, E., M. Doree, P. Nurse, and M. Bornens. p34cdc2 is located in both nucleus and cytoplasm; part is centrosomally associated at G2/M and enters vesicles at anaphase. *EMBO J.* 8: 3985–3995, 1989.

7. Barbet, N. C., and A. M. Carr. Fission yeast wee1 protein-kinase is not required for DNA damage-dependent mitotic arrest. *Nature* 364: 824–827, 1993.

8. Baserga, R. Growth in size and cell DNA replication. *Exp. Cell Res.* 151: 1–5, 1984.

9. Bates, S., L. Bonetta, D. MacAllan, D. Parry, A. Holder, C. Dickson, and G. Peters. Cdk6 (plstire) and cdk4 (psk-J3) are a distinct subset of the cyclin-dependent kinases that associate with cyclin D1. *Oncogene* 9: 71–79, 1994.

10. Beach, D., B. Durkacz, and P. Nurse. Functionally homologous cell cycle control genes in budding and fission yeast. *Nature* 300: 706–709, 1982.

11. Beach, D., M. Piper, and P. Nurse. Construction of a Schizosaccharomyces pombe gene bank in a yeast bacterial shuttle vector and its use to isolate genes by complementation. *Mol. Gen. Genet.* 187: 326–329, 1982.

12. Belenguer, P., F. M. Caizergues, J. C. Labbé, M. Dorée, and F. Amalric. Mitosis-specific phosphorylation of nucleolin by p34cdc2 protein kinase. *Mol. Cell. Biol.* 10: 3607–3618, 1990.

13. Bellé, R., J. Derancourt, R. Poulhe, J.-P. Capony, R. Ozon, and O. Mulner-Lorillon. A purified complex from Xenopus oocytes contains a p47 protein, an in vivo substrate of MPF, and a p30 protein respectively homologous to elongation factors EF-1γ and EF-1β. *FEBS Lett.* 255: 101–104, 1989.

14. Blow, J. J., and R. A. Laskey. A role for the nuclear envelope in controlling DNA replication within the cell cycle. *Nature* 332: 546–548, 1988.

15. Blow, J. J., and A. M. Sleeman. Replication of purified DNA in Xenopus egg extract is dependent on nuclear assembly. *J. Cell Sci.* 95: 383–391, 1990.

16. Bonetta, L. Open questions on p16. *Nature* 370: 180, 1994.

17. Booher, R., and D. Beach. Interaction between cdc13+ and cdc2+ in the control of mitosis in fission yeast; dissociation of the G1 and G2 roles of the cdc2+ protein kinase. *EMBO J.* 6: 3441–3447, 1987.

18. Booher, R., and D. Beach. Involvement of cdc13+ in mitotic control in Schizosaccharomyces pombe: possible interaction of the gene product with microtubules. *EMBO J.* 7: 2321–2327, 1988.

19. Booher, R., and D. Beach. Involvement of a type 1 protein phosphatase encoded by bws1+ in fission yeast mitotic control. *Cell* 57: 1009–1016, 1989.

20. Booher, R. N., C. E. Alfa, J. S. Hyams, and D. H. Beach. The fission yeast cdc2/cdc13/suc1 protein kinase: regulation of catalytic activity and nuclear localization. *Cell* 58: 485–97, 1989.

21. Bradbury, E. M., R. J. Inglis, and H. R. Matthews. Control of cell division by very lysine rich histone (f1) phosphorylation. *Nature* 247: 257–261, 1974.

22. Broek, D., R. Bartlett, K. Crawford, and P. Nurse. Involvement of p34cdc2 in establishing the dependency of S phase on mitosis. *Nature* 349: 388–394, 1991.

23. Brooks, R. F. Regulation of the fibroblast cell cycle by serum. *Nature* 260: 248–250, 1976.

24. Bueno, A., and P. Russell. Dual functions of Cdc6—a yeast protein required for DNA-replication also inhibits nuclear division. *EMBO J.* 11: 2167–2176, 1992.

25. Chang, F., and I. Herskowitz. Phosphorylation of far1 in response to alpha-factor—a possible requirement for cell-cycle arrest. *Mol. Biol. Cell* 3: 445–450, 1992.

25a. Chong, J. P., H. M. Mahbubani, C. Y. Khoo, and J. J. Blow. Purification of an MCM-containing complex as a component of the DNA replication licensing system. *Nature* 375: 418–421, 1995.

26. Chou, Y.-H., J. R. Bischoff, D. Beach, and R. D. Goldman. Intermediate filament reorganization during mitosis is mediated by p34cdc2 phosphorylation of vimentin. *Cell* 62: 1063–1071, 1990.

27. Cobrinik, D., S. F. Dowdy, P. W. Hinds, S. Mittnacht, and R. A. Weinberg. The retinoblastoma protein and the regulation of cell cycling. *Trends Biochem. Sci.* 17: 312–315, 1992.

28. Coleman, T. R., Z. H. Tang, and W. G. Dunphy. Negative regulation of the wee1 protein-kinase by direct action of the nim1/cdr1 mitotic inducer. *Cell* 72: 919–929, 1993.

29. Costello, G., L. Rogers, and D. Beach. Fission yeast enters the stationary phase G0 state from either mitotic G1 or G2. *Curr. Genet.* 11: 119–125, 1986.

29a. Coue, M., S. E. Kearsey, and M. Mechali. Chromatin binding, nuclear localization and phosphorylation of *Xenopus* cdc21 are cell-cycle dependent and associated with the control of initiation of DNA replication. *EMBO J.* 15: 1085–1097, 1996.

30. Coutavas, E., M. D. Ren, J. D. Oppenheim, P. Deustachio, and M. G. Rush. Characterization of proteins that interact with the cell-cycle regulatory protein Ran/TC4. *Nature* 366: 585–587, 1993.

31. Coverley, D., and R. A. Laskey. Regulation of eukaryotic DNA replication. *Annu. Rev. Biochem.* 63: 745–776, 1994.

32. Cross, F. DAF1, a mutant gene affecting size control, pheromone arrest and cell cycle kinetics of S. cerevisiae. *Mol. Cell. Biol.* 8: 4675–4684, 1988.

33. Cross, F. R., M. Hoek, J. D. McKinney, and A. H. Tinkelenberg. Role of Swi4 in cell cycle regulation of Cln2 expression. *Mol. Cell. Biol.* 14: 4779–4787, 1994.

33a. Cross, S. M., C. A. Sanchez, C. A. Morgan, M. K. Schimke, S. Ramel, R. L. Idzerda, W. H. Raskind, and B. J. Reid. A p53-dependent mouse spindle checkpoint. *Science* 267: 1353–1356, 1995.

34. Cyert, M., and M. Kirschner. Regulation of MPF activity in vitro. *Cell* 53: 185–195, 1988.

35. Dasso, M., and J. Newport. Completion of DNA replication is monitored by a feedback system that controls the initiation of mitosis in vitro: studies in *Xenopus*. *Cell* 61: 811–823, 1990.

36. Diller, L., J. Kassel, C. E. Nelson, M. A. Gryka, G. Litwak, M. Gebhardt, B. Bressac, M. Ozturk, S. J. Baker, B. Vogelstein, and S. H. Friend. p53 functions as a cell cycle control protein in osteosarcomas. *Mol. Cell. Biol.* 10: 5772–5781, 1990.

37. Dirick, L., T. Moll, H. Auer, and K. Nasmyth. A central role for Swi6 in modulating cell cycle start-specific transcription in yeast. *Nature* 357: 508–513, 1992.

38. Dirick, L., and K. Nasmyth. Positive feedback in the activation of G1 cyclins in yeast. *Nature* 351: 754–757, 1991.

39. Donovan, J. D., J. H. Toyn, A. L. Johnson, and L. H. Johnston. P40 (sdb25), a putative cdk inhibitor, has a role in the m/G1

transition in *Saccharomyces cerevisiae*. *Genes Dev.* 8: 1640–1653, 1994.

40. Doonan, J. H., and N. R. Morris. The bimG gene of Aspergillus nidulans, required for completion of anaphase, encodes a homolog of mammalian phosphoprotein phosphatase 1. *Cell* 57: 987–996, 1989.

41. Dowdy, S. F., P. W. Hinds, K. Louie, S. I. Reed, A. Arnold, and R. A. Weinberg. Physical interaction of the retinoblastoma protein with human D cyclins. *Cell* 73: 499–511, 1993.

42. Downes, C. S., A. M. Mullinger, and R. T. Johnson. Inhibitors of DNA topoisomerase II prevent chromatid separation in mammalian cells but do not prevent exit from mitosis. *Proc. Natl. Acad. Sci. U.S.A.* 88: 8895–8899, 1991.

42a. Downes, C. S., D. J. Clarke, A. M. Mullinger, J. F. Gimenez-Abian, A. M. Creighton, and R. T. Johnson. A topoisomerase II-dependent G2 cycle checkpoint in mammalian cells. *Nature* 372: 467–470, 1994.

43. Downward, J. Regulatory mechanisms for ras proteins. *Bioessays* 14: 177–184, 1992.

44. Draetta, G., and D. Beach. Activation of cdc2 protein kinase during mitosis in human cells: cell cycle-dependent phosphorylation and subunit rearrangement. *Cell* 54: 17–26, 1988.

45. Draetta, G., D. Beach, and E. Moran. Synthesis of p34, the mammalian homolog of the yeast cdc2$^+$/CDC28 protein kinase, is stimulated during adenovirus-induced proliferation of primary baby rat kidney cells. *Oncogene* 2: 553–557, 1988.

46. Draetta, G., F. Luca, J. Westendorf, L. Brizuela, J. Ruderman, and D. Beach. cdc2 protein kinase is complexed with both cyclin A and B: evidence for proteolytic inactivation of MPF. *Cell* 56: 829–838, 1989.

47. Dulic, V., L. F. Drullinger, E. Lees, S. I. Reed, and G. H. Stein. Altered regulation of G1 cyclins in senescent human-diploid fibroblasts—accumulation of inactive cyclin E-cdk2 and cyclin D1-cdk2 complexes. *Proc. Natl Acad. Sci. U.S.A.* 90: 11034–11038, 1993.

48. Dulic, V., W. K. Kaufmann, S. J. Wilson, T. D. Tlsty, E. Lees, J. W. Harper, S. J. Elledge, and S. I. Reed. P53-dependent inhibition of cyclin-dependent kinase-activities in human fibroblasts during radiation-induced G1 arrest. *Cell* 76: 1013–1023, 1994.

49. Dulic, V., E. Lees, and S. I. Reed. Association of human cyclin E with a periodic G1-S phase protein-kinase. *Science* 257: 1958–1961, 1992.

50. Dynlacht, B. D., O. Flores, J. A. Lees, and E. Harlow. Differential regulation of e2f transactivation by cyclin cdk2 complexes. *Genes Dev.* 8: 1772–1786, 1994.

51. Earnshaw, W. C. Mitotic chromosome structure. *Bioessays* 9: 147–150, 1988.

52. Edgar, B., and P. O'Farrell. Genetic control of cell division patterns in the Drosophilla embryo. *Cell* 57: 177–187, 1989.

53. Eldeiry, W. S., T. Tokino, V. E. Velculescu, D. B. Levy, R. Parsons, J. M. Trent, D. Lin, W. E. Mercer, K. W. Kinzler, and B. Vogelstein. Waf1, a potential mediator of p53 tumor suppression. *Cell* 75: 817–825, 1993.

54. Elledge, S. J., R. Richman, F. L. Hall, R. T. Williams, N. Lodgson, and J. W. Harper. Cdk2 encodes a 33-kda cyclin-a-associated protein-kinase and is expressed before cdc2 in the cell-cycle. *Proc. Natl. Acad. Sci. U.S.A.* 89: 2907–2911, 1992.

55. Elledge, S. J., and M. R. Spottswood. A new human p34 protein kinase, CDK2, identified by complementation of a cdc28 mutation in Saccharomyces cerevisiae, is a homolog of Xenopus Eg1. *EMBO J.* 10: 2653–2659, 1991.

56. Enoch, T., A. M. Carr, and P. Nurse. Fission yeast genes involved in coupling mitosis to completion of DNA-replication. *Genes Dev.* 6: 2035–2046, 1992.

56a. Enoch, T., and C. Norbury. Cellular responses to DNA damage: cell-cycle checkpoints, apoptosis and the roles of p53 and ATM. *Trends Biochem. Sci.* 20: 426–430, 1995.

57. Evans, G. A., and W. L. Farrar. Retinoblastoma protein-phosphorylation does not require activation of p34^{cdc2} protein-kinase. *Biochem. J.* 287: 965–969, 1992.

58. Evans, R. M., and L. M. Fink. An alteration in the phosphorylation of vimentin-type intermediate filaments is associated with mitosis in cultured mammalian cells. *Cell* 29: 43–52, 1982.

59. Evans, T., E. Rosenthal, J. Youngbloom, D. Distel, and T. Hunt. Cyclin: a protein specified by maternal mRNA in sea urchin eggs that is destroyed at each cleavage division. *Cell* 33: 389–396, 1983.

60. Ewen, M. E., H. K. Sluss, C. J. Sherr, H. Matsushime, J. Y. Kato, and D. M. Livingston. Functional interactions of the retinoblastoma protein with mammalian D-type cyclins. *Cell* 73: 487–497, 1993.

61. Fang, F., and J. W. Newport. Evidence that the G1-S and G2-M transitions are controlled by different cdc2 proteins in higher eukaryotes. *Cell* 66: 731–742, 1991.

62. Fantes, P., and P. Nurse. Control of the timing of the cell division in fission yeast. Cell size mutants reveal a second control pathway. *Exp. Cell. Res.* 115: 317–329, 1978.

63. Feilotter, H., P. Nurse, and P. G. Young. Genetic and molecular analysis of cdr1/nim1 in Schizosaccharomyces pombe. *Genetics* 127: 309–318, 1991.

64. Ferrell, J. J., M. Wu, J. C. Gerhart, and G. S. Martin. Cell cycle tyrosine phosphorylation of p34^{cdc2} and a microtubule-associated protein kinase homolog in Xenopus oocytes and eggs. *Mol. Cell. Biol.* 11: 1965–1971, 1991.

65. Fesquet, D., J. C. Labbe, J. Derancourt, J. P. Capony, S. Galas, F. Girard, T. Lorca, J. Shuttleworth, M. Doree, and J. C. Cavadore. The MO15 gene encodes the catalytic subunit of a protein-kinase that activates cdc2 and other cyclin-dependent kinases (cdks) through phosphorylation of Thr161 and its homologs. *EMBO J.* 12: 3111–3121, 1993.

66. Fisher, R. P., and D. O. Morgan. A novel cyclin associates with mo15/cdk7 to form the cdk-activating kinase. *Cell* 78: 713–724, 1994.

67. Fleig, U. N., K. L. Gould, and P. Nurse. A dominant negative allele of p34^{cdc2} shows altered phosphoamino acid content and sequesters p56^{cdc13} cyclin. *Mol. Cell. Biol.* 12: 2295–2301, 1992.

68. Forsburg, S. L., and P. Nurse. Identification of a G1-type cyclin *puc1*$^+$ in the fission yeast *Schizosaccharomyces pombe*. *Nature* 351: 245–248, 1991.

69. Forsburg, S. L., and P. Nurse. Analysis of the *Schizosaccharomyces pombe* cyclin puc1: evidence for a role in cell cycle exit. *J. Cell Sci.* 107: 601–613, 1994.

70. Freeman, M., C. Nusslein-Volhard and D. M. Glover. The dissociation of nuclear and centrosomal division in gnu, a mutation causing giant nuclei in Drosophila. *Cell* 46: 457–468, 1986.

71. Furukawa, Y., H. Piwnica-Worms, T. J. Ernst, Y. Kanakura, and J. D. Griffin. cdc2 gene expression at the G1 to S transition in human T lymphocytes. *Science* 250: 805–808, 1990.

72. Galaktionov, K., and D. Beach. Specific activation of cdc25 tyrosine phosphatases by b-type cyclins—evidence for multiple roles of mitotic cyclins. *Cell* 67: 1181–1194, 1991.

73. Gautier, J., J. Minshull, M. Lohka, M. Glotzer, T. Hunt, and J. L. Maller. Cyclin is a component of MPF from Xenopus. *Cell* 60: 487–494, 1990.

74. Gautier, J., C. Norbury, M. Lohka, P. Nurse, and J. Maller.

Purified maturation-promoting factor contains the product of a Xenopus homolog of the fission yeast cell cycle control gene cdc2+. *Cell* 54: 433–439, 1988.

75. Gelfant, S. Initiation of mitosis in relation to the cell division cycle. *Exp. Cell Res.* 26: 395–403, 1962.

76. Gerace, L., and G. Blobel. The nuclear lamina is reversibly depolymerised during mitosis. *Cell* 19: 277–287, 1980.

77. Ghiara, J. B., H. E. Richardson, K. Sugimoto, M. Henze, D. J. Lew, C. Wittenberg, and S. I. Reed. A cyclin B homolog in S. cerevisiae: chronic activation of the Cdc28 protein kinase by cyclin prevents exit from mitosis. *Cell* 65: 163–174, 1991.

78. Girard, F., U. Strausfeld, A. Fernandez, and N.J.C. Lamb. Cyclin A is required for the onset of DNA replication in mammalian fibroblasts. *Cell* 67: 1169–1179, 1991.

79. Glotzer, M., A. W. Murray, and M. W. Kirschner. Cyclin is degraded by the ubiquitin pathway. *Nature* 349: 132–138, 1991.

80. Goebl, M., and B. Byers. Cyclin in fission yeast. *Cell* 54: 739–740, 1988.

81. Gorbsky, G. J. Cell-cycle progression and chromosome segregation in mammalian-cells cultured in the presence of the topoisomerase-II inhibitors icrf-187 [(+)-1,2-bis (3,5-dioxopiperazinyl-1-yl)propane; adr-529] and icrf-159 (razoxane). *Cancer Res.* 54: 1042–1048, 1994.

82. Gould, K. L., S. Moreno, D. J. Owen, S. Sazer, and P. Nurse. Phosphorylation at Thr 167 is required for Schizosaccharomyces pombe p34cdc2 function. *EMBO J.* 10: 3297–3309, 1991.

83. Gould, K. L., and P. Nurse. Tyrosine phosphorylation of the fission yeast cdc2+ protein kinase regulates entry into mitosis. *Nature* 342: 39–45, 1989.

84. Grandin, N., and M. Charbonneau. Intracellular free calcium oscillates during cell division of Xenopus embryos. *J. Cell Biol.* 112: 711–718, 1991.

84a. Greenwell, P. W., S. L. Kronmal, S. E. Porter, J. Gassenhuber, B. Obermaier, and T. D. Petes. TEL1, a gene involved in controlling telomere length in S. cerevisiae is homologous to the human ataxia telangiectasia gene. *Cell* 82: 823–829, 1995.

85. Guadagno, T. M., M. Ohtsubo, J. M. Roberts, and R. K. Assoian. A link between cyclin-a expression and adhesion-dependent cell-cycle progression. *Science* 262: 1572–1575, 1993.

86. Hadwiger, J. A., C. Wittenberg, H. E. Richardson, M. De Barros Lopes, and S. I. Reed. A family of cyclin homologs that control the G1 phase in yeast. *Proc. Natl. Acad. Sci. U.S.A.* 86: 6255–6259, 1989.

87. Hagan, I., J. Hayles, and P. Nurse. Cloning and sequencing of the cyclin-related cdc13+ gene and a cytological study of its role in fission yeast mitosis. *J. Cell Sci.* 91: 587–595, 1988.

88. Hanks, S. K., A. M. Quinn, and T. Hunter. The protein kinase family: conserved features and deduced phylogeny of the catalytic domains. *Science* 241: 42–52, 1988.

89. Hannon, G. J., and D. Beach. p15INK4B is a potential effector of TGFβ-induced cell cycle arrest. *Nature* 371: 257–261, 1994.

90. Harper, J. W., G. R. Adami, N. Wei, K. Keyomarsi, and S. J. Elledge. The p21 cdk-interacting protein cip1 is a potent inhibitor of g1 cyclin-dependent kinases. *Cell* 75: 805–816, 1993.

91. Hartwell, L., J. Culotti, and J. Pringle. Genetic control of the cell division cycle in yeast. *Science* 183: 46, 1974.

92. Hartwell, L., and T. Weinert. Checkpoints: controls that ensure the order of cell cycle events. *Science* 246: 629–634, 1989.

93. Heald, R., and F. McKeon. Mutations of phosphorylation sites in lamin A that prevent nuclear lamina disassembly in mitosis. *Cell* 61: 579–589, 1990.

94. Heald, R., M. McLoughlin, and F. McKeon. Human wee-1 maintains mitotic timing by protecting the nucleus from cytoplasmically activated cdc2 kinase. *Cell* 74: 463–474, 1993.

95. Helin, K., and E. Harlow. The retinoblastoma protein as a transcriptional repressor. *Trends Cell Biol.* 3: 43–45, 1993.

96. Hellmich, M. R., H. C. Pant, E. Wada, and J. F. Battey. Neuronal cdc2-like kinase—a cdc2-related protein-kinase with predominantly neuronal expression. *Proc. Natl Acad. Sci. U.S.A.* 89: 10867–10871, 1992.

97. Hengst, L., V. Dulic, J. M. Slingerland, E. Lees, and S. I. Reed. A cell-cycle-regulated inhibitor of cyclin-dependent kinases. *Proc. Natl Acad. Sci. U.S.A.* 91: 5291–5295, 1994.

98. Hindley, J., and G. A. Phear. Sequence of the cell division gene CDC2 from Schizosaccharomyces pombe; patterns of splicing and homology to protein kinases. *Gene* 31: 129–134, 1984.

99. Hinds, P. W., S. Mittnacht, V. Dulic, A. Arnold, S. I. Reed, and R. A. Weinberg. Regulation of retinoblastoma protein functions by ectopic expression of human cyclins. *Cell* 70: 993–1006, 1992.

100. Hisatake, K., S. Hasegawa, R. Takada, Y. Nakatani, M. Horikoshi, and R. G. Roeder. The p250 subunit of native tata box-binding factor-TFIId is the cell-cycle regulatory protein CCG1. *Nature* 362: 179–181, 1993.

101. Hoffmann, I., P. R. Clarke, M. J. Marcote, E. Karsenti, and G. Draetta. Phosphorylation and activation of human cdc25-C by cdc2-cyclin B and its involvement in the self-amplification of mpf at mitosis. *EMBO J.* 12: 53–63, 1993.

102. Holloway, S. L., M. Glotzer, R. W. King, and A. W. Murray. Anaphase is initiated by proteolysis rather than by the inactivation of maturation-promoting factor. *Cell* 73: 1393–1402, 1993.

103. Holm, C., T. Goto, J. Wang, and D. Bostein. DNA topoisomerase II is required at the time of mitosis in yeast. *Cell* 41: 553–563, 1985.

104. Hoyt, M. A., L. Trotis, and B. T. Roberts. S. cerevisiae genes required for cell cycle arrest in response to loss of microtubule function. *Cell* 66: 507–517, 1991.

105. Hunter, T., and J. Pines. Cyclins and cancer. *Cell* 66: 1071–1074, 1991.

106. Igarashi, M., A. Nagata, S. Jinno, K. Suto, and H. Okayama. Wee1+-like gene in human cells. *Nature* 353: 80–83, 1991.

107. Izumi, T., D. H. Walker, and J. L. Maller. Periodic changes in phosphorylation of the *Xenopus* Cdc25 phosphatase regulates its activity. *Mol. Biol. Cell* 3: 927–939, 1992.

107a. Jimenez, G., J. Yucel, R. Rowley, and S. Subramani. The rad3+ gene of *Schizosaccharomyces pombe* is involved in multiple checkpoint functions and in DNA repair. *Proc. Natl. Acad. Sci. U.S.A.* 89: 4952–4956, 1992.

108. Johnston, L. H. Cell cycle control of gene expression in yeast. *Trends Cell Biol.* 2: 353–357, 1992.

109. Kamb, A., N. A. Gruis, J. Weaverfeldhaus, Q. Y. Liu, K. Harshman, S. V. Tavtigian, E. Stockert, R. S. Day, B. E. Johnson, and M. H. Skolnick. A cell-cycle regulator potentially involved in genesis of many tumor types. *Science* 264: 436–440, 1994.

110. Kapeller, R., and L. C. Cantley. Phosphatidylinositol 3-kinase. *Bioessays* 16: 565–576, 1994.

111. Kastan, M. B., Q. M. Zhan, W. S. Eldeiry, F. Carrier, T. Jacks, W. V. Walsh, B. S. Plunkett, B. Vogelstein, and A. J. Fornace. A mammalian-cell cycle checkpoint pathway utilizing p53 and gadd45 is defective in ataxia-telangiectasia. *Cell* 71: 587–597, 1992.

112. Kato, J., H. Matsushime, S. W. Hiebert, M. E. Ewen, and C. J. Sherr. Direct binding of cyclin-d to the retinoblastoma

gene-product (prb) and prb phosphorylation by the cyclin d-dependent kinase cdk4. *Genes Dev.* 7: 331–342, 1993.

113. Kelly, T. J. DNA-replication in mammalian-cells—insights from the SV40 model system. *Harvey Lect.* 85: 1991.

114. Kelly, T. J., G. S. Martin, S. L. Forsburg, R. J. Stephen, A. Russo, and P. Nurse. The fission yeast cdc18 + gene-product couples s-phase to start and mitosis. *Cell* 74: 371–382, 1993.

115. Kharbanda, S., A. Saleem, R. Datta, Z. M. Yuan, R. Weichselbaum, and D. Kufe. Ionizing-radiation induces rapid tyrosine phosphorylation of p34^{cdc2}. *Cancer Res.* 54: 1412–1414, 1994.

116. Killander, D., and A. Zetterberg. A quantitative cytochemical investigation of the relationship between cell mass and initiation of DNA synthesis in mouse fibroblasts. *Exp. Cell Res.* 40: 12–20, 1965.

117. Kinoshita, N., H. Ohkura, and M. Yanagida. Distinct, essential roles of type 1 and 2A protein phosphatases in the control of the fission yeast cell division cycle. *Cell* 63: 405–415, 1990.

118. Kipreos, E. T., and J. Y. J. Wang. Differential phosphorylation of c-Abl in cell cycle determined by cdc2 kinase and phosphatase activity. *Science* 248: 217–220, 1990.

119. Kirschner, M., J. Newport, and J. Gerhart. The timing of early developmental events in Xenopus. *Trends Genet.* 1: 41–47, 1985.

120. Kishimoto, T., R. Kuriyama, H. Kondo, and H. Hanatani. Generality of the action of various maturation-promoting factors. *Exp. Cell Res.* 137: 121–126, 1982.

121. Kobayashi, H., E. Stewart, R. Poon, J. P. Adamczewski, J. Gannon, and T. Hunt. Identification of the domains in cyclin-a required for binding to, and activation of, p34^{cdc2} and p32^{cdk2} protein-kinase subunits. *Mol. Biol. Cell* 3: 1279–1294, 1992.

122. Koff, A., F. Cross, A. Fisher, J. Schumacher, K. Le Guellec, M. Philippe, and J. M. Roberts. Human cyclin E, a new cyclin that interacts with two members of the *CDC2* gene family. *Cell* 66: 1217–1228, 1991.

123. Koff, A., A. Giordano, D. Desai, K. Yamashita, J. W. Harper, S. Elledge, T. Nishimoto, D. O. Morgan, B. R. Franza, and J. M. Roberts. Formation and activation of a cyclin E-Cdk2 complex during the G1 phase of the human cell-cycle. *Science* 257: 1689–1694, 1992.

124. Kornbluth, S., B. Sebastian, T. Hunter, and J. Newport. Membrane localization of the kinase which phosphorylates p34^{cdc2} on threonine 14. *Mol. Biol. Cell* 5: 273–282, 1994.

125. Krek, W., M. E. Ewen, S. Shirodkar, Z. Arany, W. G. Kaelin, and D. M. Livingston. Negative regulation of the growth-promoting transcription factor e2f-1 by a stably bound cyclin a-dependent protein-kinase. *Cell* 78: 161–172, 1994.

126. Krek, W., and E. A. Nigg. Differential phosphorylation of vertebrate p34cdc2 kinase at the G1/S and G2/M transitions of the cell cycle: identification of major phosphorylation sites. *EMBO J.* 10: 305–316, 1991.

127. Kuerbitz, S. J., B. S. Plunkett, W. V. Walsh, and M. B. Kastan. Wild-type p53 is a cell-cycle checkpoint determinant following irradiation. *Proc. Natl. Acad. Sci. U.S.A.* 89: 7491–7495, 1992.

128. Kumagai, A., and W. G. Dunphy. The cdc25 protein controls tyrosine dephosphorylation of the cdc2 protein in a cell-free system. *Cell* 64: 903–914, 1991.

129. Kumagai, A. D., W. G. Dunphy. Regulation of the Cdc25 protein during the cell cycle in *Xenopus* extracts. *Cell* 70: 139–151, 1992.

130. Labbe, J. C., J. P. Capony, D. Caput, J. C. Cavadore, M. K. Derancourt, J. M. Lelias, A. Picard, and M. Dorée. MPF from starfish oocytes at first meiotic metaphase is a heterodimer containing one molecule of cdc2 and one molecule of cyclin B. *EMBO J.* 8: 3053–3058, 1989.

131. Labbe, J. C., A. Picard, E. Karsenti, and M. Doree. An M-phase-specific protein kinase of Xenopus oocytes: partial purification and possible mechanism of its periodic activation. *Dev. Biol.* 127: 157–169, 1988.

132. Lane, H. A., A. Fernandez, N. J. C. Lamb, and G. Thomas. P70^{S6K} function is essential for G1 progression. *Nature* 363: 170–172, 1993.

133. Langan, T. A., J. Gautier, M. Lohka, R. Hollingsworth, S. Moreno, P. Nurse, J. Maller, and R. A. Sclafani. Mammalian growth-associated H1 histone kinase: a homolog of cdc2 + / CDC28 protein kinases controlling mitotic entry in yeast and frog cells. *Mol. Cell Biol.* 9: 3860–3868, 1989.

134. Lee, M. G., C. J. Norbury, N. K. Spurr, and P. Nurse. Regulated expression and phosphorylation of a possible mammalian cell-cycle control protein. *Nature* 333: 676–679, 1988.

135. Lee, M. G., and P. Nurse. Complementation used to clone a human homologue of the fission yeast cell cycle control gene cdc2. *Nature* 327: 31–35, 1987.

136. Lee, T. H., M. J. Solomon, M. C. Mumby, and M. W. Kirschner. INH, a negative regulator of MPF, is a form of protein phosphatase 2A. *Cell* 64: 415–423, 1991.

137. Lees, E., B. Faha, V. Dulic, S. I. Reed, and E. Harlow. Cyclin E-cdk2 and cyclin A-cdk2 kinases associate with p107 and E2F in a temporally distinct manner. *Genes Dev.* 6: 1874–1885, 1992.

138. Lew, J., R. J. Winkfein, H. K. Paudel, and J. H. Wang. Brain proline-directed protein-kinase is a neurofilament kinase which displays high sequence homology to p34^{cdc2}. *J. Biol. Chem.* 267: 25922–25926, 1992.

139. Li, R., and A. W. Murray. Feedback control of mitosis in budding yeast. *Cell* 66: 519–531, 1991.

140. Li, Y., C. W. Jenkins, M. A. Nichols, and Y. Xiong. Cell-cycle expression and p53 regulation of the cyclin-dependent kinase inhibitor p21. *Oncogene* 9: 2261–2268, 1994.

141. Lin, B.T.Y., S. Gruenwald, A. Morla, W.-H. Lee, and J.Y.J. Wang. Retinoblastoma cancer suppressor gene product is a substrate of the cell cycle regulator cdc2. *EMBO J.* 10: 857–864, 1991.

142. Litchfield, D. W., B. Luscher, F. J. Lozeman, R. N. Eisenman, and E. G. Krebs. Phosphorylation of casein kinase-II by p34^{cdc2} in vitro and at mitosis. *J. Biol. Chem.* 267: 13943–13951, 1992.

143. Lock, R. B. Inhibition of p34cdc2 kinase activation, p34cdc2 tyrosine dephosphorylation, and mitotic progression in Chinese hamster ovary cells exposed to etoposide. *Cancer Res.* 52: 1817–1822, 1992.

144. Lohka, M., and J. Maller. Induction of nuclear membrane breakdown, chromosome condensation and spindle formation in cell free extracts. *J. Cell Biol.* 101: 518–523, 1985.

145. Lohka, M. J., M. K. Hayes, and J. L. Maller. Purification of maturation-promoting factor, an intracellular regulator of early mitotic events. *Proc. Natl. Acad. Sci. U.S.A.* 85: 3009–3013, 1988.

146. Lorca, T., F. H. Cruzalegui, D. Fesquet, J. C. Cavadore, J. Mery, A. Means, and M. Doree. Calmodulin-dependent protein kinase-II mediates inactivation of MPF and CSF upon fertilization of *Xenopus* eggs. *Nature* 366: 270–273, 1993.

147. Lorincz, A., and S. Reed. Primary structure homology between the product of the yeast cell division control gene CDC28 and vertebrate oncogenes. *Nature* 307: 183–185, 1984.

148. Lowndes, N. F., A. L. Johnson, L. Breeden, and L. H. Johnston. Swi6 protein is required for transcription of the periodically

expressed DNA-synthesis genes in budding yeast. *Nature* 357: 505–508, 1992.

149. Lowndes, N. F., A. L. Johnson, and L. H. Johnston. Coordination of expression of DNA synthesis genes in budding yeast by a cell-cycle regulated trans factor. *Nature* 350: 247–250, 1991.

150. Lowndes, N. F., C. J. McInerny, A. L. Johnson, P. A. Fantes, and L. H. Johnston. Control of DNA-synthesis genes in fission yeast by the cell-cycle gene cdc10$^+$. *Nature* 355: 449–453, 1992.

151. Lu, K. P., S. A. Osmani, and A. R. Means. Properties and regulation of the cell cycle-specific NIMA protein kinase of *Aspergillus nidulans. J. Biol. Chem.* 268: 8769–8776, 1993.

152. Lundgren, K., N. Walworth, R. Booher, M. Dembski, M. Kirschner, and D. Beach. mik1 and wee1 cooperate in the inhibitory tyrosine phosphorylation of cdc2. *Cell* 64: 1111–1122, 1991.

153. Luscher, B., L. Brizuela, D. Beach, and R. N. Eisenman. A role for p34cdc2 kinase and phosphatases in the regulation of phosphorylation and disassembly of lamin B2 during the cell cycle. *EMBO J.* 10: 865–875, 1991.

154. Makela, T. P., J. P. Tassan, E. A. Nigg, S. Frutiger, G. J. Hughes, and R. A. Weinberg. A cyclin associated with the cdk-activating kinase MO15. *Nature* 371: 254–257, 1994.

155. Maller, J. L., and D. S. Smith. Two-dimensional polyacrylamide gel analysis of changes in protein phosphorylation during maturation of Xenopus oocytes. *Dev. Biol.* 109: 150–156, 1985.

156. Marsden, M., and U. K. Laemmli. Metaphase chromosome structure: evidence for a radial loop model. *Cell* 17: 849–858, 1979.

157. Massague, J. The transforming growth-factor-beta family. *Annu. Rev. Cell Biol.* 6: 597–641, 1990.

158. Masui, Y., and C. Markert. Cytoplasmic control of nuclear behaviour during meiotic maturation of frog oocytes. *J. Exp. Zool.* 177: 129–146, 1971.

159. Matsuoka, Y., K. Nishizawa, T. Yano, M. Shibata, S. Ando, T. Takahashi, and M. Inagaki. 2 different protein-kinases act on a different time schedule as glial filament kinases during mitosis. *EMBO J.* 11: 2895–2902, 1992.

160. Matsushime, H., M. F. Roussel, R. A. Ashmun, and C. J. Sherr. Colony-stimulating factor 1 regulates novel cyclins during the G1 phase of the cell cycle. *Cell* 65: 701–713, 1991.

161. McGowan, C. H., and P. Russell. Human wee1 kinase inhibits cell-division by phosphorylating p34cdc2 exclusively on tyr15. *EMBO J.* 12: 75–85, 1993.

162. McKinney, J. D., F. Chang, N. Heintz, and F. R. Cross. Negative regulation of far1 at the start of the yeast-cell cycle. *Genes Dev.* 7: 833–843, 1993.

163. Meijer, L., A. Ostvold, S. I. Walaas, T. Lund, and S. G. Laland. High-mobility-group proteins P1, I and Y as substrates of the M-phase-specific p34cdc2/cyclincdc13 kinase. *Eur. J. Biochem.* 196: 557–567, 1991.

164. Melchior, F., B. Paschal, J. Evans, and L. Gerace. Inhibition of nuclear-protein import by nonhydrolyzable analogs of GTP and identification of the small GTPase ran/tc4 as an essential transport factor. *J. Cell Biol.* 123: 1649–1659, 1993.

165. Meyerson, M., G. H. Enders, C. L. Wu, L. K. Su, C. Gorka, C. Nelson, E. Harlow, and L. H. Tsai. A family of human cdc2-related protein-kinases. *EMBO J.* 11: 2909–2917, 1992.

166. Minella, O., P. Cormier, J. Morales, R. Poulhe, R. Belle, and O. Mulnerlorillon. Cdc2 kinase sets a memory phosphorylation signal on elongation-factor EF-1δ during meiotic cell-division, which perdures in early development. *Cell. Mol. Biol.* 40: 521–525, 1994.

167. Minshull, J., J. J. Blow, and T. Hunt. Translation of cyclin

168. Minshull, J., R. Golsteyn, C. S. Hill, and T. Hunt. The A- and B-type cyclin associated cdc2 kinases in Xenopus turn on and off at different times in the cell cycle. *EMBO J.* 9: 2865–2875, 1990.

169. Minshull, J., J. Pines, R. Golsteyn, N. Standart, S. Mackie, A. Colman, J. Blow, J. V. Ruderman, M. Wu, and T. Hunt. The role of cyclin synthesis, modification and destruction in the control of cell division. *J. Cell Sci. Suppl.* 12: 77–97, 1989.

170. Mittnacht, S., J. A. Lees, D. Desai, E. Harlow, D. O. Morgan, and R. A. Weinberg. Distinct subpopulations of the retinoblastoma protein show a distinct pattern of phosphorylation. *EMBO J.* 13: 118–127, 1994.

171. Moll, T., G. Tebb, U. Surana, H. Robitsch, and K. Nasmyth. The role of phosphorylation and the Cdc28 protein kinase in cell cycle-regulated nuclear import of the *S. cerevisiae* transcription factor Swi5. *Cell* 66: 743–758, 1991.

172. Moreno, S., J. Hayles, and P. Nurse. Regulation of p34cdc2 protein kinase during mitosis. *Cell* 58: 361–372, 1989.

173. Moreno, S., and P. Nurse. Substrates for p34^{cdc2}: in vivo veritas? *Cell* 61: 549–551, 1990.

174. Moreno, S., and P. Nurse. Regulation of progression through the G1 phase of the cell-cycle by the rum1 + gene. *Nature* 367: 236–242, 1994.

175. Morgan, D. O., J. M. Kaplan, J. M. Bishop, and H. E. Varmus. Mitosis-specific phosphorylation of p60^{c-src} by p34^{cdc2}-associated protein kinase. *Cell* 57: 775–786, 1989.

175a. Morrow, D. M., D. A. Tagle, Y. Shiloh, F. S. Collins, and P. Hieter. TEL1, an *S. cerevisiae* homolog of the human gene mutated in ataxia telangiectasia, is functionally related to the yeast checkpoint gene MEC1. *Cell* 82: 831–840, 1995.

176. Motokura, T., T. Bloom, H. G. Kim, H. Juppner, J. V. Ruderman, H. M. Kronenberg, and A. Arnold. A novel cyclin encoded by a bcl1-linked candidate oncogene. *Nature* 350: 512–515, 1991.

177. Mulner-Lorillon, O., P. Cormier, J. C. Labbe, M. Doree, R. Poulhe, H. Osborne, and R. Belle. M-phase-specific cdc2 protein kinase phosphorylates the beta subunit of casein kinase II and increases casein kinase II activity. *Eur. J. Biochem.* 193: 529–534, 1990.

178. Murray, A. W. Turning on mitosis. *Curr. Biol.* 3: 291–293, 1993.

179. Murray, A. W., and M. W. Kirschner. Cyclin synthesis drives the early embryonic cell cycle. *Nature* 339: 275–280, 1989.

180. Murray, A. W., M. J. Solomon, and M. W. Kirschner. The role of cyclin synthesis and degradation in the control of maturation promoting factor activity. *Nature* 339: 280–286, 1989.

181. Nash, R., G. Tokiwa, S. Anand, K. Erickson, and A. B. Futcher. The WHI1 + gene of *Saccharomyces cerevisiae* tethers cell division to cell size and is a cyclin homolog. *EMBO J.* 7: 4335–4346, 1988.

182. Nasmyth, K. Control of the yeast cell cycle by the Cdc28 protein kinase. *Curr. Opin. Cell Biol.* 5: 166–179, 1993.

183. Nasmyth, K., and P. Nurse. Cell division cycle mutants altered in DNA replication and mitosis in the fission yeast Schizosaccharomyces pombe. *Mol. Gen. Genet.* 182: 119–124, 1981.

184. Newport, J., and M. Kirschner. Regulation of the cell cycle during early Xenopus development. *Cell* 37: 731–742, 1984.

185. Newport, J., and T. Spann. Disassembly of the nucleus in mitotic extracts: membrane vesicularization lamin disassembly and chromosome condensation are independent process. *Cell* 48: 219–230, 1987.

186. Ninomiya-Tsuji, J., S. Nomoto, H. Yasuda, S. I. Reed, and

K. Matsumoto. Cloning of a human cDNA encoding a CDC2-related kinase by complementation of a budding yeast *cdc28* mutation. *Proc. Natl. Acad. Sci. U.S.A.* 88: 9006–9010, 1991.

187. Nishimoto, T., E. Eilen, and C. Basilico. Premature chromosome condensation in a ts DNA-mutant of BHK cells. *Cell* 15: 475–483, 1978.

188. Nishitani, H., M. Ohtsubo, K. Yamashita, H. Iida, J. Pines, H. Yasudo, Y. Shibata, T. Hunter, and T. Nishimoto. Loss of RCC1, a nuclear DNA-binding protein, uncouples the completion of DNA replication from the activation of cdc2 protein kinase and mitosis. *EMBO J.* 10: 1555–1564, 1991.

189. Nobori, T., K. Miura, D. J. Wu, A. Lois, K. Takabayashi, and D. A. Carson. Deletions of the cyclin-dependent kinase-4 inhibitor gene in multiple human cancers. *Nature* 368: 753–756, 1994.

190. Norbury, C., J. Blow, and P. Nurse. Regulatory phosphorylation of the p34^{cdc2} protein kinase in vertebrates. *EMBO J.* 10: 3321–3329, 1991.

191. Norbury, C. J., and P. Nurse. Animal cell cycles and their control. *Annu. Rev. Biochem.* 61: 441–470, 1992.

192. Nuell, M. J., D. A. Stewart, L. Walker, V. Friedman, C. M. Wood, G. A. Owens, J. R. Smith, E. L. Schneider, R. Dell'Orco, C. K. Lumpkin, D. B. Danner, and J. K. McKlung. Prohibitin, an evolutionarily conserved intracellular protein that blocks DNA synthesis in normal fibroblasts and HeLa cells. *Mol. Cell. Biol.* 11: 1372–1381, 1991.

193. Nurse, P. Genetic control of cell size at cell division in yeast. *Nature* 256: 547–551, 1975.

194. Nurse, P., and Y. Bissett. Gene required in G1 for commitment to cell cycle and in G2 for control of mitosis in fission yeast. *Nature* 292: 558–560, 1981.

195. Nurse, P., and P. Thuriaux. Regulatory genes controlling mitosis in the fission yeast Schizosaccharomyces pombe. *Genetics* 96: 627–637, 1980.

196. Nurse, P., P. Thuriaux, and K. Nasmyth. Genetic control of the cell division cycle in the fission yeast *Schizosaccharomyces pombe. Mol. Gen. Genet.* 146: 167–178, 1976.

197. O'Connell, M., C. Norbury, and P. Nurse. Premature chromatin condensation upon accumulation of NimA. *EMBO J.* 13: 4926–4937, 1994.

198. Ohkura, H., N. Kinoshita, S. Miyatani, T. Toda, and M. Yanagida. The fission yeast dis2 + gene required for chromosome disjoining encodes one of two putative type 1 protein phosphatases. *Cell* 57: 997–1007, 1989.

199. Ohkura, H., and M. Yanagida. S. pombe gene sds22 + essential for a midmitotic transition encodes a leucine-rich repeat protein that positively modulates protein phosphatase-1. *Cell* 64: 149–157, 1991.

200. Ohtsubo, M., R. Kai, N. Furuno, T. Sekiguchi, M. Sekiguchi, M. Hayashida, K. Kuma, T. Miyata, S. Fuukushige, T. Murotsu, K. Matsubara, and T. Nishimoto. Isolation and characterisation of the active cDNA of the human cell cycle gene (RCC1) involved in the regulation of onset of chromosome condensation. *Genes Dev.* 1: 585–593, 1987.

201. Ohtsubo, M., and J. M. Roberts. Cyclin-dependent regulation of g (1) in mammalian fibroblasts. *Science* 259: 1908–1912, 1993.

202. Okuda, T., J. L. Cleveland, and J. R. Downing. Pctaire-1 and pctaire-3, two members of a novel cdc2/cdc28-related protein-kinase gene family. *Oncogene* 7: 2249–2258, 1992.

203. Osmani, A. H., S. L. McGuire, K. L. Odonnell, R. T. Pu, and S. A. Osmani. Role of the cell-cycle-regulated NIMA protein-kinase during G2 and mitosis—evidence for two pathways of mitotic regulation. *Cold Spring Harb. Symp. Quant. Biol.* 56: 549–555, 1991.

204. Osmani, A. H., N. Vanpeij, M. Mischke, M. J. O'Connell, and S. A. Osmani. A single p34^{cdc2} protein kinase (encoded by nimx^{cdc2}) is required at G1 and G2 in *Aspergillus nidulans. J. Cell Sci.* 107: 1519–1528, 1994.

205. Osmani, S., R. Pu, and N. Morris. Mitotic induction and maintenance by over-expression of a G2-specific gene that encodes a potential protein kinase. *Cell* 53: 237–244, 1988.

206. Pagano, M., R. Pepperkok, F. Verde, W. Ansorge, and G. Draetta. Cyclin A is required at two points in the human cell cycle. *EMBO J.* 11: 961–971, 1992.

207. Pagano, M., A. M. Theodoras, S. W. Tam, and G. F. Draetta. Cyclin D1-mediated inhibition of repair and replicative DNA synthesis in human fibroblasts. *Genes Dev.* 8: 1627–1639, 1994.

208. Pardee, A. A restriction point for control of normal animal cell proliferation. *Proc. Natl Acad. Sci. U.S.A.* 71: 1286–1290, 1974.

209. Pardee, A., R. Dubrow, J. Hamlin, and R. Kleitzien. Animal cell cycle. *Annu. Rev. Biochem.* 47: 715–178, 1978.

210. Paris, J., R. Le Guellec, A. Couturier, K. Le Guellec, F. Omilli, J. Camonis, S. MacNeill, and M. Philippe. Cloning by differential screening of a *Xenopus* cDNA coding for a protein highly homologous to cdc2. *Proc. Natl Acad. Sci. U.S.A.* 88: 1039–1043, 1991.

211. Parker, L., and H. Piwnica-Worms. Inactivation of the p34cdc2-cyclin B complex by the human WEE1 tyrosine kinase. *Science* 257: 1955–1957, 1992.

212. Parker, L. L., S. A. Walter, P. G. Young, and H. Piwnicaworms. Phosphorylation and inactivation of the mitotic inhibitor wee1 by the nim1/cdr1 kinase. *Nature* 363: 736–738, 1993.

213. Pazin, M. J., and L. T. Williams. Triggering signaling cascades by receptor tyrosine kinases. *Trends Biochem. Sci.* 17: 374–378, 1992.

214. Pedersen, T., and S. Gelfant. G2-population cells in mouse kidney and duodenum and their behaviour during the cell division cycle. *Exp. Cell Res.* 59: 32–36, 1970.

215. Peter, M., A. Gartner, J. Horecka, G. Ammerer, and I. Herskowitz. Far1 links the signal-transduction pathway to the cell-cycle machinery in yeast. *Cell* 73: 747–760, 1993.

216. Peter, M., and I. Herskowitz. Direct inhibition of the yeast cyclin-dependent kinase cdc28-cln by far1. *Science* 265: 1228–1231, 1994.

217. Peter, M., J. Nakagawa, M. Dorée, J. C. Labbé, and E. A. Nigg. Identification of major nucleolar proteins as candidate mitotic substrates of cdc2 kinase. *Cell* 60: 791–801, 1990.

218. Peter, M., J. Nakagawa, M. Dorée, and E. A. Nigg. In vitro disassembly of the nuclear lamina and M phase-specific phosphorylation of lamins by cdc2 kinase. *Cell* 61: 591–602, 1990.

219. Philp, A. V., J. M. Axton, R.D.C. Saunders, and D. M. Glover. Mutations in the *Drosophila melanogaster* gene 3 *rows* permit aspects of mitosis to continue in the absence of chromatid segregation. *J. Cell Sci.* 106: 87–98, 1993.

220. Picard, A., J. P. Capony, D. L. Brautigan, and M. Dorée. Involvement of protein phosphatases 1 and 2A in the control of M phase-promoting factor activity in starfish. *J. Cell Biol.* 109: 3347–3354, 1989.

221. Picard, A., G. Peaucellier, F. LeBouffant, C. LePeuch, and M. Dorée. Role of protein synthesis and proteases in production and inactivation of maturation-promoting activity during meiotic maturation of starfish oocytes. *Dev. Biol.* 109: 311–320, 1985.

222. Pines, J., and T. Hunter. Isolation of a human cyclin cDNA: evidence for cyclin mRNA and protein regulation in the cell cycle and for interaction with p34^{cdc2}. *Cell* 58: 833–846, 1989.

223. Pines, J., and T. Hunter. Human cyclin A is adenovirus E1A-

associated protein p60 and behaves differently from cyclin B. *Nature* 346: 760–763, 1990.

224. Pines, J., and T. Hunter. Human cyclins A and B1 are differentially located in the cell and undergo cell cycle-dependent nuclear transport. *J. Cell Biol.* 115: 1–17, 1991.

225. Polyak, K., J. Y. Kato, M. J. Solomon, C. J. Sherr, J. Massague, J. M. Roberts, and A. Koff. P27 (kip1), a cyclin-cdk inhibitor, links transforming growth-factor-beta and contact inhibition to cell-cycle arrest. *Genes Dev.* 8: 9–22, 1994.

226. Polyak, K., M. H. Lee, H. Erdjumentbromage, A. Koff, J. M. Roberts, P. Tempst, and J. Massague. Cloning of p27 (kip1), a cyclin-dependent kinase inhibitor and a potential mediator of extracellular antimitogenic signals. *Cell* 78: 59–66, 1994.

227. Poon, R.Y.C., K. Yamashita, J. P. Adamczewski, T. Hunt, and J. Shuttleworth. The cdc2-related protein p40 (mo15) is the catalytic subunit of a protein-kinase that can activate p33^{cdk2} and p34^{cdc2}. *EMBO J.* 12: 3123–3132, 1993.

228. Primig, M., S. Sockanathan, H. Auer, and K. Nasmyth. Anatomy of a transcription factor important for the start of the cell cycle in *Saccharomyces cerevisiae*. *Nature* 358: 593–597, 1992.

229. Quelle, D. E., R. A. Ashmun, S. A. Shurtleff, J. Y. Kato, D. Barsagi, M. F. Roussel, and C. J. Sherr. Overexpression of mouse d-type cyclins accelerates g (1) phase in rodent fibroblasts. *Genes Dev.* 7: 1559–1571, 1993.

230. Raff, J. W., and D. M. Glover. Nuclear and cytoplasmic mitotic cycles continue in Drosophila embryos in which DNA synthesis is inhibited with aphidicolin. *J. Cell Biol.* 107: 2009–2019, 1988.

231. Rao, P., and R. Johnson. Mammalian cell fusion studies on the regulation of DNA synthesis and mitosis. *Nature* 225: 159–164, 1970.

232. Reed, S. The selection of *S. cerevisiae* mutants defective in the start event of cell division. *Genetics* 95: 566–577, 1980.

233. Reed, S. I., J. A. Hadwiger, and A. T. Lorincz. Protein kinase activity associated with the product of the yeast cell division cycle gene CDC28. *Proc. Natl. Acad. Sci. U.S.A.* 82: 4055–4059, 1985.

234. Ren, M. D., G. Drivas, P. D'Eustachio, and M. G. Rush. Ran/tc4—a small nuclear GTP-binding protein that regulates DNA-synthesis. *J. Cell Biol.* 120: 313–323, 1993.

235. Richardson, H. E., C. Wittenberg, F. Cross, and S. I. Reed. An essential G1 function for cyclin-like proteins in yeast. *Cell* 59: 1127–1133, 1989.

236. Rossow, P. W., V. G. H. Riddle, and A. B. Pardee. Synthesis of labile, serum-dependent protein in early G1 controls animal cell growth. *Proc. Natl. Acad. Sci. U.S.A.* 76: 4446–4450, 1979.

237. Russell, P., and P. Nurse. cdc25 + functions as an inducer in the mitotic control of fission yeast. *Cell* 45: 145–153, 1986.

238. Russell, P., and P. Nurse. The mitotic inducer nim1 + functions in a regulatory network of protein kinase homologs controlling the initiation of mitosis. *Cell* 49: 569–576, 1987.

239. Russell, P., and P. Nurse. Negative regulation of mitosis by wee1 +, a gene encoding a protein kinase homolog. *Cell* 49: 559–567, 1987.

240. Sadhu, K., S. I. Reed, H. Richardson, and P. Russell. Human homolog of fission yeast cdc25 mitotic inducer is predominantly expressed in G2. *Proc. Natl Acad. Sci. U.S.A.* 87: 5139–5143, 1990.

241. Sagata, N., N. Watanabe, G. F. Vande Woude, and Y. Ikawa. The c-mos proto-oncogene product is a cytostatic factor responsible for meiotic arrest in vertebrate eggs. *Nature* 342: 512–518, 1989.

242. Saka, Y., and M. Yanagida. Fission yeast *cut5*$^+$, Required for S phase onset and M phase restraint, is identical to the radiation-damage repair gene *rad4*$^+$. *Cell* 74: 383–393, 1993.

243. Satterwhite, L. L., M. J. Lohka, K. L. Wilson, T. Y. Scherson, L. J. Cisek, J. L. Corden, and T. D. Pollard. Phosphorylation of myosin-II regulatory light chain by cyclin-p34^{cdc2}—a mechanism for the timing of cytokinesis. *J. Cell Biol.* 118: 595–605, 1992.

243a. Savitsky, K., A. Bar Shira, S. Gilad, G. Rotman, Y. Ziv, L. Vanagaite, D. A. Tagle, S. Smith, T. Uziel, S. Sfez, et al. A single ataxia telangiectasia gene with a product similar to PI-3 kinase. *Science* 268: 1749–1753, 1995.

244. Sazer, S., and P. Nurse. A fission yeast rcc1-related protein is required for the mitosis to interphase transition. *EMBO J.* 13: 606–615, 1994.

245. Schultz, S. J., and E. A. Nigg. Identification of 21 novel human protein kinases, including 3 members of a family related to the cell cycle regulator *nimA* of *Aspergillus nidulans*. *Cell Growth Differ.* 4: 821–830, 1993.

246. Schwob, E., and K. Nasmyth. Clb5 and Clb6, a new pair of B-cyclins involved in DNA replication in *Saccharomyces cerevisiae*. *Genes Dev.* 7: 1160–1175, 1993.

247. Seino, H., H. Nishitani, T. Seki, N. Hisamoto, T. Tazunoki, N. Shiraki, M. Ohtsubo, K. Yamashita, T. Sekiguchi, and T. Nishimoto. Rcc1 is a nuclear-protein required for coupling activation of cdc2 kinase with DNA-synthesis and for start of the cell-cycle. *Cold Spring Harb. Symp. Quant. Biol.* 56: 367–375, 1991.

248. Seto, M., K. Yamamoto, S. Iida, Y. Akao, K. R. Utsumi, I. Kubonishi, I. Miyoshi, T. Ohtsuki, Y. Yawata, M. Namba, T. Motokura, A. Arnold, T. Takahashi, and R. Ueda. Gene rearrangement and overexpression of prad1 in lymphoid malignancy with t (11/14) (q13/q32) translocation. *Oncogene* 7: 1401–1406, 1992.

249. Sewing, A., C. Burger, S. Brusselbach, C. Schalk, F. C. Lucibello, and R. Muller. Human cyclin d1 encodes a labile nuclear-protein whose synthesis is directly induced by growth-factors and suppressed by cyclic-amp. *J. Cell Sci.* 104: 545–554, 1993.

250. Shamu, C. E., and A. W. Murray. Sister chromatid separation in frog egg extracts requires DNA topoisomerase-II activity during anaphase. *J. Cell Biol.* 117: 921–934, 1992.

251. Sheldrick, K. S., and A. M. Carr. Feedback controls and G2 checkpoints: fission yeast as a model system. *Bioessays* 15: 775–782, 1993.

252. Shenoy, S., J. K. Choi, S. Bagrodia, T. D. Copeland, J. L. Maller, and D. Shalloway. Purified maturation promoting factor phosphorylates pp60c-src at the sites phosphorylated during fibroblast mitosis. *Cell* 57: 763–774, 1989.

253. Sherr, C. J. Mammalian G1 cyclins. *Cell* 73: 1059–1065, 1993.

254. Shetty, K. T., W. T. Link, and H. C. Pant. Cdc2-like kinase from rat spinal-cord specifically phosphorylates kspxk motifs in neurofilament proteins—isolation and characterization. *Proc. Natl Acad. Sci. U.S.A.* 90: 6844–6848, 1993.

254a. Shiekhattar, R., F. Mermelstein, R. P. Fisher, R. Drapkin, B. Dynlacht, H. C. Wessling, D. O. Morgan, and D. Reinberg. Cdk-activating kinase complex is a component of human transcription factor TFIIH. *Nature.* 374: 283–287, 1995.

255. Shiina, N., T. Moriguchi, K. Ohta, Y. Gotoh, and E. Nishida. Regulation of a major microtubule-associated protein by mpf and map kinase. *EMBO J.* 11: 3977–3984, 1992.

256. Simanis, V., and P. Nurse. The cell cycle control gene cdc2 + of fission yeast encodes a protein kinase potentially regulated by phosphorylation. *Cell* 45: 261–268, 1986.

257. Skalli, O., Y. H. Chou, and R. D. Goldman. Cell cycle-dependent changes in the organization of an intermediate filament-associated protein—correlation with phosphorylation by p34^{cdc2}. *Proc. Natl. Acad. Sci. U.S.A.* 89: 11959–11963, 1992.

258. Smith, J. A., and L. Martin. Do cells cycle? *Proc. Natl. Acad. Sci. U.S.A.* 70: 1263–1267, 1973.

259. Smith, L. D., and R. E. Ecker. The interaction of steroids with R. pipiens oocytes in the induction of maturation. *Dev. Biol.* 25: 233–247, 1971.

260. Solomon, M., R. Booher, M. Kirschner, and D. Beach. Cyclin in fission yeast. *Cell* 54: 738–739, 1988.

261. Solomon, M. J. Cak, a p34^{cdc2} activating kinase. *Mol. Biol. Cell* 3: 1992.

262. Solomon, M. J., M. Glotzer, T. H. Lee, M. Phillippe, and M. Kirschner. Cyclin activation of p34cdc2. *Cell* 63: 1013–1024, 1990.

263. Solomon, M. J., J. W. Harper, and J. Shuttleworth. Cak, the p34^{cdc2} activating kinase, contains a protein identical or closely-related to p40 (MO15). *EMBO J.* 12: 3133–3142, 1993.

264. Stillman, B., S. P. Bell, A. Dutta, and Y. Marahrens. DNA replication and the cell cycle. *Ciba Found. Symp.* 170: 147–160, 1992.

265. Stone, E. M., H. Yamano, N. Kinoshita, and M. Yanagida. Mitotic regulation of protein phosphatases by the fission yeast sds22-protein. *Curr. Biol.* 3: 13–26, 1993.

266. Strausfeld, U., J.-C. Labbé, D. Fesquet, J. C. Cavadore, A. Picard, K. Sadhu, P. Russell, and M. Dorée. Dephosphorylation and activation of a p34^{cdc2}/cyclin B complex in vitro by human CDC25 protein. *Nature* 351: 242–245, 1991.

267. Sudbery, P. E., A. R. Goodey, and B.L.A. Carter. Genes which control cell proliferation in the yeast Saccharomyces cerevisiae. *Nature* 288: 401–404, 1980.

268. Sunkara, P., D. Wright, and P. Rao. Mitotic factors from mammalian cells induces germinal vesicle breakdown and chromosome condensation in amphibian oocytes. *Proc. Natl. Acad. Sci. U.S.A.* 76: 2799–2802, 1979.

269. Surana, U., A. Amon, C. Dowzer, J. McGrew, B. Byers, and K. Nasmyth. Destruction of the cdc28 clb mitotic kinase is not required for the metaphase to anaphase transition in budding yeast. *EMBO J.* 12: 1969–1978, 1993.

270. Swenson, K. I., K. M. Farrell, and J. V. Ruderman. The clam embryo protein cyclin A induces entry into M phase and the resumption of mitosis in Xenopus oocytes. *Cell* 47: 861–870, 1986.

271. Tachibana, K., N. Yanagishima, and T. Kishimoto. Preliminary characterization of maturation-promoting factor from yeast Saccharomyces cerevisiae. *J. Cell Sci.* 88: 273–281, 1987.

272. Tang, Z. H., T. R. Coleman, and W. G. Dunphy. 2 distinct mechanisms for negative regulation of the wee1 protein-kinase. *EMBO J.* 12: 3427–3436, 1993.

273. Toh-e, A., K. Tanaka, Y. Uesono, and R. B. Wickner. PHO85, a negative regulator of the PHO system, is a homolog of the protein kinase gene, CDC28, of Saccharomyces cerevisiae. *Mol. Gen. Genet.* 214: 162–164, 1988.

274. Toyoshima, H., and T. Hunter. P27, a novel inhibitor of g1 cyclin-cdk protein-kinase activity, is related to p21. *Cell* 78: 67–74, 1994.

275. Tsai, L. H., E. Lees, B. Faha, E. Harlow, and K. Riabowol. The cdk2 kinase is required for the g1-to-s transition in mammalian-cells. *Oncogene* 8: 1593–1602, 1993.

276. Tsujimura, K., J. Tanaka, S. Ando, Y. Matsuoka, M. Kusubata, H. Sugiura, T. Yamauchi, and M. Inagaki. Identification of phosphorylation sites on glial fibrillary acidic protein for cdc2 kinase and ca2 + -calmodulin-dependent protein-kinase-II. *J. Biochem.* 116: 426–434, 1994.

277. Tuomikoski, T., M.-A. Félix, M. Dorée, and J. Gruenberg. Inhibition of endocytic vesicle fusion in vitro by the cell-cycle control protein kinase cdc2. *Nature* 342: 942–944, 1989.

278. Tyers, M., and B. Futcher. Far1 and Fus3 link the mating pheromone signal-transduction pathway to three G1-phase Cdc28 kinase complexes. *Mol. Cell. Biol.* 13: 5659–5669, 1993.

279. Tyers, M., G. Tokiwa, R. Nash, and B. Futcher. The Cln3-Cdc28 kinase complex of *Saccharomyces cerevisiae* is regulated by proteolysis and phosphorylation. *EMBO J.* 11: 1773–1784, 1992.

280. Uemura, T., H. Ohkura, Y. Adachi, K. Morino, K. Shiozaki, and M. Yanagida. DNA topoisomerase II is required for condensation and separation of mitotic chromosomes in S. pombe. *Cell* 50: 917–925, 1987.

281. Uemura, T., and M. Yanagida. Mitotic spindle pulls but fails to separate chromosomes in type II DNA topoisomerase mutants: uncoordinated mitosis. *EMBO J.* 5: 1003–1010, 1986.

282. Vandenheuvel, S., and E. Harlow. Distinct roles for cyclin-dependent kinases in cell-cycle control. *Science* 262: 2050–2054, 1993.

283. Waga, S., G. J. Hannon, D. Beach, and B. Stillman. The p21 inhibitor of cyclin-dependent kinases controls DNA-replication by interaction with pcna. *Nature* 369: 574–578, 1994.

284. Walworth, N., S. Davey, and D. Beach. Fission yeast *chk1* protein kinase links the *rad* checkpoint pathway to *cdc2*. *Nature* 363: 368–371, 1993.

285. Ward, G. E., and M. W. Kirschner. Identification of cell cycle-regulated phosphorylation sites on nuclear lamin C. *Cell* 61: 561–577, 1990.

286. Weinert, T., and L. Hartwell. The RAD9 gene controls the cell cycle response to DNA damage in Saccharomyces cerevisiae. 241: 317–322, 1988.

287. Weintraub, S. J., C. A. Prater, and D. C. Dean. Retinoblastoma protein switches the e2f site from positive to negative element. *Nature* 358: 259–261, 1992.

288. Whitaker, M., and R. Patel. Calcium and cell cycle control. *Development* 108: 525–542, 1990.

289. Williams, M. E., S. H. Swerdlow, C. L. Rosenberg, and A. Arnold. Chromosome-11 translocation breakpoints at the prad1/cyclin d1 gene locus in centrocytic lymphoma. *Leukemia* 7: 241–245, 1993.

290. Wu, L., and P. Russell. Nim1 kinase promotes mitosis by inactivating wee1 tyrosine kinase. *Nature* 363: 738–741, 1993.

291. Xiong, Y., G. J. Hannon, H. Zhang, D. Casso, R. Kobayashi, and D. Beach. P21 is a universal inhibitor of cyclin kinases. *Nature* 366: 701–704, 1993.

292. Xiong, Y., H. Zhang, and D. Beach. D-type cyclins associate with multiple protein-kinases and the DNA-replication and repair factor pcna. *Cell* 71: 505–514, 1992.

293. Xiong, Y., H. Zhang, and D. Beach. Subunit rearrangement of the cyclin-dependent kinases is associated with cellular-transformation. *Genes Dev.* 7: 1572–1583, 1993.

294. Yamashiro, S., Y. Yamakita, H. Hosoya, and F. Matsumura. Phosphorylation of non-muscle caldesmon by p34cdc2 kinase during mitosis. *Nature* 349: 169–172, 1991.

295. Yamashiro, S., Y. Yamakita, R. Ishikawa, and F. Matsumura. Mitosis-specific phosphorylation causes 83k non-muscle caldesmon to dissociate from microfilaments. *Nature* 344: 675–678, 1990.

296. Zetterberg, A., and O. Larsson. Kinetic analysis of regulatory events in G1 leading to proliferation or quiescence of Swiss 3T3 cells. *Proc. Natl. Acad. Sci. U.S.A.* 82: 5365–5369, 1985.

297. Zirkle, R. E. UV-microbeam irradiation of newt cell cytoplasm: spindle destruction, false anaphase, and delay of true anaphase. *Radiat. Res.* 41: 516–537, 1970.

22. Regulation of development: differentiation and morphogenesis

HYNDA K. KLEINMAN | *National Institute of Dental Research, National Institutes of Health, Bethesda, Maryland*

MERTON BERNFIELD | *Department of Pediatrics, Harvard Medical School, Boston, Massachusetts*

CHAPTER CONTENTS

DEVELOPMENTAL BIOLOGY attempts to answer two major questions: How does a single cell, the fertilized egg or zygote, give rise to all of the highly specialized cells of the fully developed organism and how do the cells organize at specific places and times to form tissues of specific shape? The first question asks how cells become functionally distinct, and developmental biologists invoke two processes, known as *cell determination* and *cell differentiation*, for this specialization. The second question asks how the various cellular behaviors are precisely integrated to yield organs with unique forms, and developmental biologists term this process *morphogenesis*.

During embryonic life, cells become specialized while they organize into the tissues that make up the organs. Although each organ has a unique function, the underlying developmental mechanisms apply to the formation of all organs. To acquire specialized functions and to organize, cells undergo selective gene repression and expression, adhesion to other cells and the extracellular matrix, change in shape, and localized proliferation and death. These cellular behaviors are controlled by genetic factors within the cell, and molecular signals from the cellular microenvironment, both soluble factors and insoluble extracellular matrix. These cellular behaviors and their controls are interdependent, and coordinated intricately in time and space.

THE CELLULAR PROCESSES OF DETERMINATION AND DIFFERENTIATION

Cell Determination

The genes responsible for the specialized cellular functions and the cellular organization into tissues are present in all cells. The fertilized egg, or zygote, and its early daughter cells are *totipotent*, meaning that they are capable of giving rise to any type of cell in the body. As development proceeds, these cells give rise to progressively more specialized cells which have a less extensive developmental repertoire but can still develop into multiple different cell types. Finally, the cells become restricted in their ultimate fate, meaning that they become committed to a particular type of differentiation. This commitment or determination involves a gradual reduction of developmental options that ultimately renders the cells capable of becoming only a single cell type. This restriction of developmental fate occurs gradually and some cells in a lineage become committed long after others. For example, most adult tissues contain some cells that can give rise to various differentiated progeny, and thus, by definition, they are not fully committed. The process of determination causes stable, heritable restrictions in a cell's developmental potential. However, this loss of potential is not the result of loss of genes. These stable phenotypes are responsible for maintaining differences between cells in mature organisms. The determination process also means that, following development, cell types that are lost cannot be replaced unless a preexisting parental cell type or an uncommitted stem cell exists. This means that there is no organ regeneration in the mature human.

Determination most likely involves the expression of one or a few regulatory genes that control the subsequent expression of many other genes in the hierarchy. These restrictions of gene expression can be considered to be developmental "decisions." These decisions can occur without any apparent change in the phenotype of the cells. Therefore, one cannot establish when a cell has become committed until after the determination has occurred. Nuclear transplantation experiments have demonstrated that the stability of the determined state depends on the cytoplasm. In these studies, introduction of the nucleus from a determined cell into the cytoplasm of an uncommitted cell "deprograms" the transplanted nucleus in such a way that it expresses genes that had previously been restricted. Indeed, this type of experiment approximates fertilization, in which the nucleus of the sperm cell enters the egg cytoplasm and becomes able to express previously restricted genes. The cytoplasmic factors involved, however, are almost wholly unknown.

Cell Differentiation

In contrast to cell determination, which narrows developmental options, differentiation is the acquisition of specialized functions. Differentiation can be affected by a cell's hormonal milieu and physical environment, but these do not reverse a cell's commitment. Cells usually have some differentiated characteristics even though they have not yet become fully committed. The differentiation of a cell is defined by the proteins that it synthesizes and by its shape, both characteristics that establish its function. Because cells produce a broad range of proteins, not all classes of proteins establish whether a cell is differentiated. Some proteins are for "housekeeping" use and are present in all cells, for example, mitochondrial proteins or glycolytic enzymes, or are common to several cell types, for example, the collagens produced by various fibroblasts or the cytokeratins produced by various epithelia. The proteins that establish differentiation of a cell type are those that enable the cell to carry out a unique function, for example, hemoglobin in red blood cells or IgG in plasma cells. Cell differentiation arises, in part, from controlling the synthesis of these "specialization" proteins.

During differentiation, cells also assume characteristic shapes that enable the cell to function effectively.

The distinct shapes result from different arrangements of the various cytoskeletal components inside the cell, including microfilaments, microtubules, and intermediate filaments. The shape of the cell may even control the production of specific proteins. For example, milk-producing mammary epithelial cells can synthesize milk proteins only when allowed to assume a cuboidal shape, and chondrocytes can produce cartilage only when allowed to round up.

Some differentiated cells may no longer be able to undergo cell division. Tissues that contain cells in such a terminally differentiated state may also contain undifferentiated stem-cell populations that are able to divide. These stem cells will subsequently differentiate, leading to the terminally differentiated cells, for example, red blood cells, plasma cells. Other tissues are not capable of producing new cells because they consist solely of terminally differentiated cells, for example, neurons, cardiac myocytes.

Genes Regulate Determination and Differentiation

During development, some genes appear to function as switches. When activated, they permit or inhibit the expression of other genes, which in turn affect the expression of yet more genes. Thus, there is a regulatory hierarchy in which activation of "master" genes turns on other genes until entire developmental sequences are activated or repressed. The number of master genes required to regulate development is not known, but it is likely that a relatively small number of genes control the many steps of the determination and differentiation processes.

Some of the genes regulating development have been identified from studying organisms with short life cycles in which developmental mutations can be readily defined. Especially useful has been the fruit fly *Drosophila melanogaster*. The early development of *Drosophila* is controlled by genes that are homologs of genes found in vertebrates, and their temporal and spatial patterns of expression suggest that the roles of these genes in vertebrates are similar to their roles in insects (17, 25).

Regulatory genes control the genes that are expressed differentially at precise times during the development of various cell types (28). This regulation can include modifying the synthesis, processing, and degradation of both RNA and protein. RNA synthesis can be regulated by factors that change the conformation (or shape) of DNA, including specific chemical modifications of DNA, such as DNA methylation (11). Such modifications can alter the ability of certain DNA regions, viz. promoters or enhancers, to bind transcription-regulating proteins. Such proteins (*tran-*

scription factors) interact with DNA sequences to increase or decrease the rates of transcription (36). These proteins can activate genes for ubiquitous proteins or they may be more selective. Protein synthesis can be controlled by processes affecting messenger RNA stability and transfer RNA availability, and posttranslational protein modifications can also affect a cell's phenotype (18, 61).

THE PROCESS OF MORPHOGENESIS

As cells specialize they must concomitantly organize into correct arrangements to form functional organs, a process called *morphogenesis.* Morphogenesis transforms the one-dimensional array of genetic information in DNA into the three-dimensional arrangements of cells and tissues in mature organs. Although the cellular arrangements are unique to each tissue and organ, the processes that produce the arrangements are not. Indeed, cells have only a limited repertoire of behaviors that are responsible for all of morphogenesis. These consist of cell adhesion, either to each other or to the extracellular matrix, change in cell shape (which accounts for both the motility of single cells and the movement of cell populations), cell proliferation, and cell death. These behaviors, which involve the coordination of a variety of individual gene products, are integrated by insoluble extracellular matrix and diffusible peptide growth factors. These integrating factors act on cells through receptors at their surfaces which, when occupied, act across the transmembranes to influence intracellular events.

Cell Adhesion

Tissues can be dissociated into single cells that can adhere and recognize one another and reassemble and reform structures resembling the original tissue. Cell adhesion and recognition establish tissue boundaries and determine the relative placement of cells and are accomplished by cell adhesion molecules (CAMs). Subsequently, specialized cell junctions arise that can maintain and stabilize the adhesions and can allow intercellular communication.

A variety of CAMs have been identified and are classified on the basis of whether they require calcium ions to mediate adhesion and whether the molecule they bind to on adjacent cells is identical (homophilic) or nonidentical (heterophilic). For example, calcium-dependent CAMs include E-cadherin, N-cadherin, and P-cadherin (named for their discovery on epithelial, nerve, and placental cells, respectively). Calcium-independent molecules include N-CAM and ICAM-1,

N-CAM and E-cadherin bind in a homopholic manner, but ICAM-1 binds to LFA-1 in an example of heterophilic binding. CAMs can have specific tissue and developmental distribution, established by immunohistochemistry, and their roles have been studied using antibodies that block their adherence function (19, 20, 66).

Change in Cell Shape

When a cell differentiates, it acquires a specific shape that is one of its differentiated characteristics. During development, change in the shape of cells can help determine which proteins are synthesized, and specific cell shapes apparently also help to maintain differentiated functions. Importantly, cells can move when they change shape. If the cells are linked together, as in an epithelial sheet, change in cell shapes results in a change in the form of the entire sheet. If individual cells change shape, they can begin a locomotory sequence in which coordinated changes in adhesion to the substratum result in cell migration.

There are two major determinants of cell shape: interaction with adjacent surfaces, either other cells or the extracellular matrix, and the cytoplasmic framework of fibrillar structures, the cytoskeleton. The cytoskeleton is made up of polymers of fibril-forming proteins and various accessory proteins organized together into microfilaments, intermediate filaments, and microtubules (10). The accessory proteins affect filament assembly or link the filaments to one another or to other cell components. The cytoskeleton is not a permanent structure but changes rapidly in response to cellular events. Indeed, the characteristic shape and intermediate filament proteins of epithelial cells can change to those of mesenchymal cells during normal developmental sequences. Various external factors, transduced by a variety of signaling systems at the surface and within cells, result in changes in cell shape (15).

Localized Cell Proliferation

Localized differences in rates of proliferation change the form of developing organs. The rate, symmetry, and orientation of cell division are all regulated and are linked to specific phases of the cell cycle (50). The cycle begins and ends with mitosis (m), which is followed by the first gap (G1) before DNA synthesis (s), which is followed by the second gap (G2) which precedes mitosis. Noncycling cells are said to be in G0. Two families of intracellular proteins, the H1 kinases and the cyclins, appear to interact to control the transition into and out of mitosis. The rate of cycling can be modified by peptide growth factors on inhibitory factors, including extracellular matrix composites such as the basement membrane (48).

Localized Cell Death

Cell death at specific localized sites is a normal event in morphogenesis, which is illustrated in the interdigital areas of the limb buds, where the death of cells allows separation of the digits. This cell death, called *apoptosis,* is apparently an autonomous phenomenon. In the developing central and peripheral nervous system, cell death is common. Survival or death of a neuron appears to be determined by events in its axonal projection field. Although cell death was presumed to cause loss of epithelia at several sites during organogenesis, such as at the site of palatal fusion, loss of some of these epithelia is now thought to result from epithelial to mesenchymal conversions.

STRATEGIES OF DEVELOPMENT

Determination, differentiation, and morphogenesis can account for the formation of organs, but how are these processes regulated? How is the aggregate behavior of cells coordinated? How do the cells know where they are in a population? How is form controlled so that scale and proportion are achieved? There are two general strategies for this regulation—cell lineage and cell and tissue interactions.

Cell Lineage

In cell lineages the information dictating the developmental behavior of a cell is solely that which is encoded in its genome and is passed directly from parental cell to daughter cell. Events within the cell instruct it when to do something and what to do. In strict examples of this strategy, the developmental behavior of a cell is independent of the behavior or even the existence of neighboring cells. Such regulation leads to development in which the organism is assembled from more or less independent parts. Several invertebrates, for example, *Caenorhabditis elegans,* the well-studied nematode, develop predominantly, but not exclusively, in this manner.

Cell Interactions

In cell interactions the information dictating the development of a cell is, in large part, provided by interactions between it and its neighboring cells. For example, in the early mouse embryo, although the individual blastomeres are totipotent, the actual cell types that each will ultimately form depend on its position within the embryo. This regulation leads to a greater degree of developmental plasticity, and mammalian development involves many such interactions. However, a cell's ability to respond to its neighbors will depend on its lineage, and its response at any time will increasingly depend on how it was previously influenced. These constraints reduce the extent of plasticity with ensuing developmental age.

To produce organs of the correct shape, size, and pattern, cells must behave according to their position in the group. Moreover, the cells must act together in a coordinated way. The process proposed to account for these behaviors is *positional information,* which involves a cell "knowing" its position in a population. Positional information is thought to result from the cell existing within a concentration gradient of some substance, often termed a *morphogen.* Retinoic acid, a vitamin A derivative, applied to developing chick limbs can mimic the behavior of a presumed morphogen. A gradient of retinoic acid presumably causes the limb bud cells to change positional information and to develop normal-appearing but misplaced digits. The evidence for retinoic acid being a morphogen is not unequivocal, and other signals or even mechanisms could specify positional information.

Tissue Interactions

Embryonic inductions are interactions between tissues of dissimilar origin that change the developmental behavior of the interacting tissues. These interactions, which occur during the formation of nearly every organ, influence the determination, differentiation, and morphogenesis of the interacting tissues. Thus, they are an important means of coordinating the developmental behaviors of adjacent cell groups. In this way, inductions also establish proper spatial relationships among developing tissues. For example, an interaction between the optic stalk, an extension of the brain, and the ectodermal cells on the surface of the embryo causes the surface cells to thicken and form the lens placode. The optic stalk gives rise to the retina and the lens placode to the lens; this interaction assures that these tissues are properly aligned. Interactions between embryonic epithelia and mesenchyme are reciprocal, influencing the behavior of both tissues. Embryonic epithelia, destined to become parenchymal tissues, interact with loose embryonic connective tissue, or mesenchyme, to form a variety of organs, including the lung, kidney, liver, and salivary glands.

The molecular mechanisms involved in these inductions are not clear. However, inducing molecules involved in the differentiation of early mesoderm in amphibia are growth factors, including members of the transforming growth factor (TGF)-β and fibroblast growth factor (FGF) families. These same peptides are also involved in later developmental stages and as signals even in the adult. An interaction thought to lead to the formation of branched epithelial organs is remodeling by the mesenchyme of the basement membrane which lies beneath the epithelia. Here, the epithelium produces the basement membrane and stimulates the mesenchyme to produce matrix materials and various matrix-degrading enzymes, which in turn act on the basement membrane. Local alterations in the basement membrane cause the epithelial cells to arrange themselves into the shapes that are characteristic of the tissue.

In both embryonic induction and positional information the signals appear to trigger cellular behaviors that are already within the repertoire of the responding cells. Thus, these signals do not appear to instruct the responding cells; rather they merely accelerate or select a developmental cell behavior.

EFFECT OF THE CELLULAR MICROENVIRONMENT ON MORPHOGENETIC BEHAVIOR

Morphogenesis is wholly dependent on the molecules that organize the behavior of individual cells into morphologically and functionally defined groups. Two classes of molecules can serve this function, the extracellular matrix components and the peptide growth factors. These molecules are the predominant influence on the differentiation, proliferation, shape, and migration of cells. They often act simultaneously to mediate important tissue interactions. The extracellular matrix components are large, multidomain molecules that are extensively cross-linked to each other, predominantly by noncovalent bonds, to form an insoluble matrix that extends over multiple cells. The growth factors are small, soluble peptides that can diffuse locally to affect groups of cells and that may bind with high avidity to components of the extracellular matrix.

The actions of the growth factors appear to be highly dependent on a cellular "context." Thus, a peptide growth factor may at one time stimulate proliferation while later stimulating the same cell type to differentiate. Cell adhesion to the matrix is the major determinant of cell shape, which in turn influences whether a cell responds to the action of a soluble peptide growth factor by differentiating, proliferating, or migrating. Thus, the extracellular matrix is probably the most important factor in generating the cellular context that influences the actions of growth factor peptides.

Extracellular matrix composites and peptide growth factors differ in fundamental ways, yet share biologic effects. Matrix composites are insoluble and constitutive and usually present at a constant concentration, whereas the growth factors are soluble and inducible and vary in concentration. However, they both act *(1)* locally, influencing a particular cohort of cells; *(2)* via autocrine or paracrine mechanisms, influencing their tissue of origin or adjacent tissues; and *(3)* at the cell surface, by binding receptors. These receptors transmit signals into cells either by activating second messengers or by interacting with the actin-containing cytoskeleton. The insoluble matrix can modulate the effect of the otherwise diffusible growth factors. The matrix binds some growth factor peptides and thus can partition their effects, maintain them in a protected state, or serve as a potential reservoir of growth factors.

Extracellular Matrix Components

Cells synthesize macromolecules that are deposited into a meshwork, or "matrix," in the extracellular spaces where they serve several functions, including providing structural support, separating tissues, providing the substrate for cell migrations, and regulating the differentiated phenotype of the resident cells (34). Depending on the cell and tissue types, the extracellular matrix components vary considerably in composition and function. Variations occur during development even within the same tissue. The three major components present in all extracellular matrices are collagens, proteoglycans, and adhesive glycoproteins. Many growth factors/cytokines are also present due to their interactions with these matrix macromolecules. Most of the components present in extracellular matrices are biologically active when tested individually in vitro, viz. they can stimulate cells to behave in a fashion similar to that observed in vivo. These proteins likely function alone and/or in concert with the other factors to influence morphogenetic events in development. The cellular receptors for these components can also regulate tissue form and function and allow a further level of sensitivity to the activity of the extracellular matrices.

Collagens. Collagens are a family of structural glycoproteins which have one or several domains that have a characteristic triple-helical structure due to a high glycine and proline content (71) and can contain hydroxylysine and hydroxyproline. The collagen molecules are designated by Roman numerals which were assigned based on the order in which they were characterized (Table 22.1). Because the collagens are the major structural proteins of almost every organ, they constitute nearly half of the total body protein. Collagens are triple-helical molecules and many genetically distinct chains have been identified which form homo- and heterotrimers. Some of the trimeric forms can self-assemble to form fibrillar structures while others do not and are considered nonfibrillar. The collagen molecules have tissue-specific and developmentally specific localizations suggesting functions other than structural properties (Table 22.1). Several genetic diseases, such as Ehlers-Danlos syndrome and Alport's syndrome, result from defects in the primary structure of collagens or due to a lack of translation of the collagen gene (42).

In addition to their importance as structural macromolecules, collagens have been found to influence cell behavior (2, 39), and several cell surface receptors have been described. For example, the fibrillar collagens induce platelet aggregation and inhibitors of collagen synthesis have been shown to block cell growth. Interestingly, *cis*-hydroxyproline blocks cell growth but growth is not affected if the cells are cultured on collagen. Likewise, salivary gland morphogenesis is blocked by other inhibitors of collagen synthesis, L-azetidine and α-dipyridyl.

Proteoglycans. Proteoglycans are molecules whose protein core is covalently attached to one or more strongly anionic glycosaminoglycan chains and often N- or O-linked oligosaccharides (31). The glycosaminoglycan chains are composed of repeating disaccharide units usually consisting of a uronic acid and a hexosamine with the predominant forms being chondroitin sulfate,

keratan sulfate, heparan sulfate, or heparin (Table 22.2). Many proteoglycans such as aggrecan and syndecan-1 contain two types of glycosaminoglycan chains and the amount and size of the chains can vary during development, aging, and disease. There is enormous variation in the size and structure of proteoglycans.

Proteoglycans have been found to have many functions. In cartilage, the amount of aggrecan exceeds that of collagen. Since aggrecan is 90% carbohydrate, it causes tissue hydration which is necessary for the cushioning properties of cartilage. The high negative charge of perlecan is also responsible for the filtration function of basement membranes. Perlecan regulates the passage of macromolecules in and out of the circulation and kidney.

Several proteoglycans including syndecan-1, CD44 (lymphocyte homing receptor), and thrombomodulin are integral membrane proteins at cell surfaces (79). Syndecan-1 may be important in development since its synthesis is induced in early tissue condensations during kidney, tooth, and limb bud formation. Syndecan-1 has several isoforms which vary in glycosaminoglycan content. Syndecan-1 can bind collagens I, III, and V, fibronectin, and thrombospondin and thus impart cells with a strong interaction with the surrounding matrix (6). CD44 has many variants and was originally described on lymphocytes as an important mediator in homing to lymphoid tissue and in binding to endothelial venules. Because it binds hyaluronan, CD44 func-

TABLE 22.1. *The Collagens: Type, Chain, Locale*

Type	Chains	Localization
I	$\alpha 1$ (I), $\alpha 2$ (I)	Skin, bone, tendon
II	$\alpha 1$ (II)	Cartilage, vitreous
III	$\alpha 1$ (III)	Skin, vessels, organs
IV*	a1 (IV), a2 (IV), $\alpha 3$ (IV), $\alpha 4$ (IV), a5 (IV)	Basement membrane
V	$\alpha 1$ (V), $\alpha 2$ (V), $\alpha 3$ (V)	Skin, muscle, bone, placenta
VI*	$\alpha 1$ (VI), a2 (VI), $\alpha 3$ (VI)	Vessels, skin, intervertebral disc
VII*	α (VII)	Dermoepidermal junction
VIII*	$\alpha 1$ (VIII)	Endothelial cells
IX*	$\alpha 1$ (IX), $\alpha 2$ (IX), $\alpha 3$ (IX)	Cartilage, vitreous tumor
X*	$\alpha 1$ (X)	Hypertrophic cartilage
XI	$\alpha 1$ (XI), $\alpha 2$ (XI), $\alpha 3$ (XI)	Cartilage
XII*	α (XII)	Embryonic tendon, skin periodontal ligament
XIII*	α (XIII)	Endothelial cells
XIV*	α (XIV)	Fetal skin, tendon

*Denotes nonfibrillar collagen.

TABLE 22.2. *Proteoglycans: Content, Locale, Function*

Name	Glycosaminoglycan Content*	Localization	Functions
Aggrecan (cartilage PG)	CS/KS	Cartilage	Compression, water holding
Biglycan	CS	Cartilage, skin, tendon	Binds to matrix?
Decorin (PG 40, PG 11)	CS	Cartilage	Binds collagens I, II, fibronectin, TGFβ
Fibromodulin	KS	Cartilage, sclera Tendon, cornea	Binds collagens I, II
Perlecan (HSPG)	HS	Basement membrane	Filtration; binds colagen IV, laminin, and growth factors
Syndecans	HS/CS	Cell surface	Co-Receptor; binds collagens I, III, V, thrombospondin, bFGF, aFGF
Versican (PG 350)	CS	Fibroblasts	Anticoagulant, spacer
Heparin	H	Mast cells, lung	Binds proteins in secretory vescicles
Hyaluronic acid†	HA	Embryonic tissues, cartilage	Binds many proteoglycans, receptor
CD44 (Hermes antigen)	CS	Cell surface	Binds collagen, fibronectin, hyaluronan acid
Thrombomodulin	CS	Cell surface	Anticoagulant

* CS = chondroitin sulfate (includes dermatan sulfate),   KS = keratan sulfate,   HS = heparan sulfate,
H = heparin,   HA = hyaluronic acid.   †Not linked to protein and, thus, not a true proteoglycan.

tions as a hyaluronan receptor and thus is important in cell adhesion to the matrix. Thrombomodulin inhibits thrombin-induced clotting of fibrinogen, promotes inactivation of thrombin by antithrombin, and activates the zymogen form of the anticoagulant protein C.

A number of growth factors, cytokines, and proteases accumulate in the extracellular matrix due to their interactions with proteoglycans, particularly heparan sulfate proteoglycans (57). This binding is important as a tissue reservoir of bioactive factors which can be released during inflammation or development by proteases (26). Decorin, a chondroitin-sulfate-containing proteoglycan, binds TGF-β, whereas syndecan-1 binds members of the FGF family. Since syndecan-1 is a cell surface receptor, this growth factor binding may be important in the cellular response to these mediators. Likewise, heparan sulfate has been found to be a cell surface receptor for acidic fibroblast growth factor (59). Due to variations in the structure of proteoglycans and their large range of biological properties, their importance in tissue development and structure is certain, but only recently appreciated (6).

Adhesive Glycoproteins. Specialized adhesion glycoproteins are present in the extracellular matrix and in some cases in serum (Table 22.3). They have a wide variety of tissue-specific structures and cell- and protein-binding activities (34). Due to their multiple interactions with other matrix macromolecules, these adhesive glycoproteins organize the collagens, proteo-

glycans, and cells into ordered structures and tissues, and thus are important during development, wound healing, and disease. For example, during neural crest cell migration, matrices enriched in either fibronectin or laminin are encountered.

Fibronectin promotes spreading, migration, and matrix production, whereas laminin promotes neurite outgrowth, cell polarity, and differentiation. Presumably, these activities do not occur simultaneously but, in part, are regulated by the presence or availability of specific active sites on the adhesion molecules as well as by the specific interacting cellular receptors. Matrix molecules generally contain distinct functionally active polypeptide domains specialized for binding other matrix molecules, for self-association, and for binding to cell surfaces. Surprisingly, short amino acid sequences from these proteins have been found to stimulate many of the biological responses observed using the whole molecules and to interact with specific cellular receptors.

Fibronectin (M_r 460,000) is a dimeric, multifunctional glycoprotein found in serum and in many tissues including fibrous connective tissue and in embryonic but not adult basement membrane, the extracellular matrix underlying epithelial cells (78). Although derived from one gene, some 20 forms exist due to alternative splicing. Several active domains have been described and some are encoded in the alternatively spliced domains, giving rise to an additional mechanism for regulating the biological activity and inter-

TABLE 22.3. *Adhesion Matrix Glycoproteins: Locale, Interaction, Function*

Glycoprotein	Tissue Localization	Interaction and Function
Fibronectin *	Plasma, fibrous connective tissue, embryonic basement membrane, many tissues	Binds fibrin, collagens, heparin, itself; promotes adhesions, spreading migration
Laminin *	Basement membrane	Binds collagen IV, heparan sulfate proteoglycan, heparin, entactin, itself; promotes adhesion, polarity, migration, growth, neurite outgrowth, differentiation, tumor growth, and spread
Entactin (nidogen) *	Basement membrane	Binds laminin, collagen IV, fibrinogen, calcium ions; promotes adhesion, migration
Vitronectin * (serum-spreading factor complements protein)	Plasma, many tissues	Binds glass, C5b-9 complex, PAI-1; promotes adhesion spacing, retinal outgrowth
Thrombospondin (TSP), many tissues, platelets	Many tissues, platelets	Promotes adhesion, neurite outgrowth; acts as an antiadhesive; blocks angiogenesis
Tenascin * (hexabranchion) (JI, cytotactin myotendinous antigen)	Edges of wounds, developing nervous system, many tissues	Binds aggrecan, fibronectin; promotes adhesion, antiadhesion, migration
SPARC (BM40, osteonectin), bone, basement membrane, platelets, many tissues	Bone, basement membrane, platelets, many tissues	Binds collagen I, PDGF, promotes adhesion, antiadhesion; regulates growth

* Denotes presence of RGD Sequence.
Abbreviations: SPARC = secreted protein acidic and rich in cepteine;   BM40 = basement membrane 40; PDGF = platelet-derived growth factor.

active cell type. Fibronectin binds to other matrix molecules including collagen, heparin/heparan sulfate, decorin, CD44, and tenascin. It also binds to itself and has a role in matrix assembly. Fibronectin is thought to bind first to an assembly receptor and then organize into extracellular fibrils. Many cellular receptors have been described, including the family of adhesion receptors known as integrins: The $\alpha5\beta1$ integrin is most prominant (56).

Fibronectin is best known for its ability to promote cell adhesion, spreading, and migration. At least six sites which can mediate cell adhesion have been described (77). Most information is known about the arg-gly-asp (or RGD) sequence which promotes cell adhesion, spreading, and migration, and blocks colonization of the lungs by metastatic melanoma cells in mice. Although these short peptide sequences are clearly active, they are much less effective than either the intact molecule or large proteolytic fragments. For example, a second interactive site is required for the full activity of the RGD domain. The RGD sequence is present in many adhesion molecules in addition to fibronectin (77). Additional active sequences for cell binding in fibronectin are in the alternatively spliced

domain and in the heparin-binding domain. Interestingly, some of these sequences show cell type specificity, which is likely important during development.

Laminin, a large (M_r 900,00) basement membrane matrix-derived glycoprotein, is composed of homologous protein chains designated A, B1, and B2 which are held together by disulfide bonds (4, 44). Laminin does not have alternatively spliced forms; rather, several genetically distinct variants of the chains have been identified (21). Laminin is considered an important organizational and regulatory molecule (14), and, indeed, is the first extracellular matrix molecule to appear in the developing embryo. For example, laminin can bind other basement membrane components, including collagen IV, perlecan, entactin, and itself. Since the laminin A chain can regulate cell polarity, laminin likely initiates tissue differentiation in embryos (37).

Laminin has multiple biological activities including promotion of cell adhesion, migration, growth, differentiation, neurite outgrowth, and tumor growth and metastases (44). At least eight biologically active sites on laminin have been defined by synthetic peptides and others possibly exist (4). As in fibronectin, these active sites have been found to promote cell adhesion and some of the sites are specific for different cellular activities. For example, the SIKVAV site on the A chain promotes neurite outgrowth with several types of neural cells whereas other sites are generally inactive. Some of the sites appear to have opposing activities. For example, the YIGSR site on the B1 chain inhibits angiogenesis and tumor growth whereas the SIKVAV site of the A chain stimulates angiogenesis and tumor growth (58, 64). Not all the sites on the laminin molecule may be exposed or available for interactions, due either to conformational changes, or to steric blocking by interacting components, or because variant chains lack the requisite sequence. Interestingly, the RGD sequence in laminin does not appear to be a major site of biological activity, possibly due to lack of availability under some conditions. This diversity in the availability of active sites is likely important in the cellular interactions of cells with multifunctional proteins, resulting in additional regulation. Several different types of cellular receptors for laminin have been identified, including several members of the integrin family, sulfatides, lectins, and cell surface proteins (46).

Vitronectin (M_r 75,000) is a plasma and matrix protein named due to its avidity for glass. It is a single polypeptide chain which is proteolytically cleaved to M_r 65,000 and 10,000 forms. In tissues, it is present in fibrillar form and often colocalizes with fibronectin. Although vitronectin contains the RGD sequence, it is recognized by several receptors including a distinct integrin receptor. Vitronectin can bind to heparin/heparan sulfate and, with a conformational change, to collagens I, II, III, and IV.

Vitronectin was first described as a serum protein involved in cell spreading, that is, serum spreading factor, but it also plays a role in tissue remodeling/repair, and possibly in tumor metastasis. It can regulate protease activity in serum, due to its binding to the C5b-9 complex which protects thrombin from inactivation, and can bind to and stabilize the activity of plasminogen activator inhibitor (PAI-1) (69), and it blocks cell lysis by complement. The role of vitronectin in tissues is less well understood but it can promote retinal neurite outgrowth, an activity shared with thrombospondin (49).

Entactin (M_r 150,000) is a sulfated basement-membrane-specific glycoprotein which binds to laminin, collagen IV, fibrinogen, and itself (12, 76). Due to its strong interaction with laminin, nearly all laminin preparations contain entactin (approximately 5%). These proteins appear to codistribute in basement membranes (60). Entactin promotes the adhesion and migration of many cells, including neutrophils, via its RGD sequence (62), interacting with the integrin receptor $\alpha 3\beta 1$ (16). It is not yet clear if there are

other biologically active sites. Its binding to fibrinogen suggests a potential role in hemostasis and it may have a role in wound healing.

Thrombospondin (M_r 520,000) is a trimeric molecule found primarily in platelet α granules, where it is the most abundant protein (8), but also in various extracellular matrices, particularly embryonic basement membranes and around peripheral nerves. The products of the three thrombospondin genes appear to have distinctive cellular localizations, suggesting specific developmental functions (74). Thrombospondin binds to heparin/heparan sulfate, collagens I and V, fibrinogen, plasminogen, and fibronectin, and regulates cell adhesion, growth, tumor metastases, and retinal neurite outgrowth. Thrombospondin has an important role in coagulation. After release from activated platelets, it self-aggregates, binds to fibrinogen and to fibrin, and is incorporated into the hemostatic plug. Two active sites have been described; one contains an RGD sequence (1), while the other is in the heparin-binding domain. Exogenous heparin blocks its activity, suggesting that where thrombospondin in tissues is bound to matrix via the heparin-binding site, the nearby cell adhesion site is unavailable. The role of thrombospondin in cell adhesion is variable depending on the cell type (70); some cells show direct adhesion whereas thrombospondin can interfere with the adhesion of other cells to fibronectin. Both integrin and nonintegrin cellular receptors have been described.

Tenascin (M_r 1.9×10^6) has six chains (M_r 190,000–230,000) which form a hexabrachion structure (22). Alternative splicing suggests at least three forms. Tenascin binds aggrecan, syndecan-1, and fibronectin. It is expressed at early and specific times in the developing nervous system and during chondrogenesis. Tenascin is present in basement membranes and at the edges of wounds. While it affects cell migration via its antiadhesive activity and ability to block fibronectin-mediated adhesion (13), it also contains an RGD site which can support cell adhesion (24).

Another matrix protein with antiadhesive activity is SPARC (*s*ecreted *p*rotein *a*cid *r*ich in *c*epteine) (M_r 40,000) (45). SPARC, present in a variety of tissues, binds several matrix molecules including collagen I and platelet-derived growth factor (PDGF). SPARC promotes the rounding of confluent endothelial, smooth muscle, and fibroblast cells, is present in areas of active tissue morphogenesis and remodeling, and reduces cell growth possibly due to an interaction with PDGF (52).

Other less well understood matrix adhesion molecules with biological effects have been described. Newly described tissue-specific factors or variants of these molecules will dominate future studies.

Matrix-Associated Effectors of Cell Behavior

Several cytokines, growth factors, proteases, and antiproteases have been described in various extracellular matrices and especially in basement membranes (72). In some cases the factors are produced by the resident cells and secreted into the matrix, whereas in vessels, for example, circulating factors may be sequestered by the matrix components. Binding to matrix components may also protect the factors from proteolysis. The matrix likely functions as a storage depot with factors released by specific signals. For example, following injury, proteolysis could result in release of the factors, and theoretically, because many factors are bound to proteoglycans, factors could be solubilized by heparin released during certain types of inflammation (6, 73).

Members of the fibroblast growth factor (FGF) family and explicitly basic FGF are stored within basement membranes due to interaction with perlecan (23) in the cornea. Various other basement membranes, including those underlying vessels, contain bFGF. Basic FGF has also been found in cartilage, nerve, and bone matrices at specific developmental stages. Basic FGF can bind cell surface proteoglycans such as syndecan-1, which has a highly regulated expression during development. bFGF is involved in development and repair to promote cell migration and growth.

An extract enriched in basement membrane components from the Englebreth-Holm-Swarm (EHS) tumor has been found to contain bFGF as well as epidermal growth factor (EGF), insulin like growth factor (IGF)-1, PDGF, and TGF-β (75a). TGF-β is in authentic basement membranes as well as in bone matrix and binds to both type IV collagen and to decorin. Bone matrices are also enriched in other growth factors, including bFGF, aFGF, EGF, and IGF-1 (32). Thus, multiple growth factors are present in basement membrane and in bone matrices and probably in other extracellular matrices. There they likely function normally during development and in tissue repair and may be involved pathogenically in processes such as atherosclerosis.

Biological Activities of Extracellular Matrix Components

The interactions of cells with the extracellular matrix are important determinants of cell behavior and in particular of cell differentiation. The response of the cells is determined not only by the cell type but also by the type of extracellular matrix. Our understanding of the biological activity of extracellular matrices has been derived mainly from in vitro studies using isolated or reconstituted components (34, 38, 39). Many cell types are routinely cultured on collagen type I, gelatin, fibronectin, or laminin to increase cell adhesion and in some cases proliferation and differentiation (40). For example, striated muscle cells are generally cultured on collagen or gelatin to enhance their differentiation into myotubes (33), whereas neuronal cells are plated on laminin to increase adhesion and promote formation of neurites (3). Many defined media which lack serum contain fibronectin to promote cell adhesion. The routine use of these matrix components has greatly facilitated the ability to culture and study different cell types.

Tissue-derived matrices have also been used in vitro and been found to promote cell differentiation. For example, the matrices of lens and amnion are relatively easy to prepare, but their use in vitro is complicated by the inability to observe cells directly due to their density. Similarly, an insoluble residue of tissues termed *biomatrix* has been found to promote hepatocyte differentiation and survival, but it is difficult to prepare and to spread evenly on the culture dish, and it is not translucent, making viewing of the cells difficult (54). Use of such tissue-derived material has been beneficial in improving cell differentiation in vitro. Certain in vivo uses have been demonstrated; for example, amnion has been used to increase nerve regeneration over long distances (Varon group). The matrix deposited by cells in culture has also been used as a culture substratum. The cells (often corneal endothelial cells) are grown for several days and then removed to leave behind an extracellular matrix rich in growth factors and active in promoting the differentiation of cells.

Reconstituted tissue matrices have been used successfully to study cell differentiation. Collagen gels supplemented with proteoglycans and growth factors have been used to promote cell differentiation in culture and have the potential for repair of injury or burns in vivo (5). A reconstituted basement membrane has also been used in vitro and in vivo to maintain and promote the differentiation of several epithelial and endothelial cell types (38). This widely used basement membrane matrix, termed *matrigel,* is obtained from a tumor, but its components are chemically and antigenically identical to those in authentic basement membrane (41). When cells or tissue rudiments are placed on a substratum of this basement membrane matrix, proliferation is reduced and differentiation specific to the cell type is observed. For example, isolated salivary gland rudiments (65) and an established salivary gland cell line [Fig. 22.1; (35)] both form branching glandular structures when plated on matrigel, whereas no such morphogenesis is observed on plastic. Likewise, endothelial cells form a network of capillary-like structures with a lumen within 18 h of plating (Fig. 22.2). This rapid morphologic differentiation has been used as a prelimi-

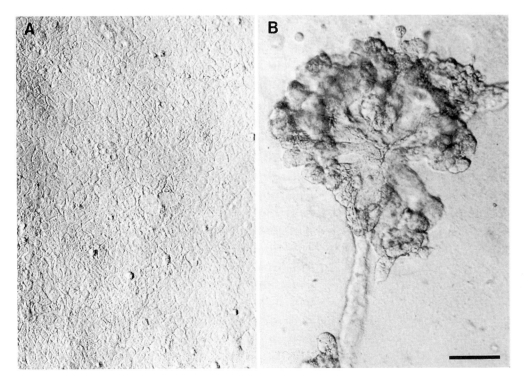

FIG. 22.1. Appearance of A253 salivary gland cells on plastic *(left)* and basement membrane matrigel *(right)* after 3 d.

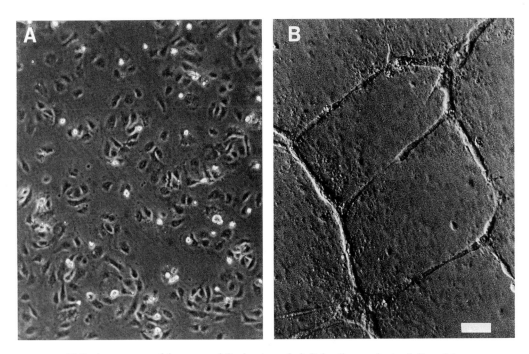

FIG. 22.2. Appearance of human umbilical vein endothelial cells on plastic *(left)* and basement membrane matrigel *(right)* after 18 h in culture.

nary screen for the identification of angiogenic and antiangiogenic compounds (29). Tissue explants can also be placed on this matrix. For example, dorsal root ganglia show extensive neurite outgrowths containing Schwann cells wrapped around neuritic processes and myelin production within 4 d of culture (9). Basement membrane has also been used in vivo to increase survival and repair of intestinal transplants (68). The cells maintain both morphological and functional differentiation. For example, on a basement membrane substratum mammary epithelial cells form cyst-like structures which secrete casein into the center (43) and show an 80-fold increase in casein mRNA levels over that found when the cells are grown on plastic. Likewise, hepatocytes maintain elevated levels of albumin synthesis and cytochrome P450 (7). The use of such extracellular matrices provides important in vitro model systems which are easily manipulated and can be used to define molecular events in cell differentiation, repair, and pathogenesis.

Peptide Growth Factors

These factors were originally identified by their actions on the proliferation of cultured cells and thus are usually referred to as growth factors. However, although some affect cell growth, they have a much wider range of actions, altering responsiveness to other hormones or growth factors, inducing or repressing differentiation, and changing the composition of the extracellular matrix. Most, if not all, are multifunctional molecules with a wide range of effects (47, 63). A combination of growth factor peptides can act differently than each alone, and their effects can differ after changes in several families of growth factor peptides. Each family has several members that share some amino acid sequence and activity, that may be recognized by the same receptor, and that have related forms in a wide variety of animal species. For example, the product of a gene in *Drosophila* is a member of the TGF-β family. Some oncogenes encode family members of growth factors or their receptors. Oncogenes are genes whose products confer on cells properties common to cells of many malignant tumors. For example, the PDGF B chain is the cellular homolog of the *sis* gene of the simian sarcoma virus, a part of the EGF receptor is the product of the erb-B oncogene, and a close homolog of basic FGF is encoded by the int-2 oncogene.

Binding of growth factor peptides to their cell surface receptors activates intracellular signaling mechanisms. The events following receptor binding generally lead to alteration in the transcription of various proteins and to internalization and degradation or recycling of the receptor. Members of the EGF, IGF, FGF, and PDGF families bind to receptors that have tyrosine-specific protein kinase activity.

Localization of growth factors and their messenger RNAs has provided valuable insights into the functions of these factors in development, but it requires cautious interpretation because their actions are so complex. For example, the presence of the messenger RNA for a factor in a developing tissue does not show that the message is transcribed, that the active form of the factor is secreted, that it encounters functional receptors, or what other influences might modify its actions.

Epidermal growth factor (EGF) shares sequence and a receptor with transforming growth factor-α (TGF-α). Both are formed as precursor molecules that are inserted into the plasma membrane, where they may be capable of activating receptors on adjacent cells. Specific proteases can liberate active, diffusible forms, which are each about 6 kDa. EGF/TGF-α stimulates the proliferation of many stem cells and expression of a differentiated phenotype in many mature cell types, principally epithelial cells. In the mouse fetus, functional receptors are found from about day 10, and the major EGF-like activity isolated is TGF-α. However, in the rat, TGF-α is expressed predominantly in decidua, not in fetal tissues, and it is unclear whether it crosses the placental barrier to affect the fetus.

Platelets were the first identified source of platelet-derived growth factors (PDGFs), but PDGFs are produced and secreted by many cell types, for example, vascular endothelial cells, tissue macrophages, fibroblasts, and smooth muscle cells. Native PDGF is a dimer of closely related ca. 15 kDa peptides. PDGFs are potent mitogens for cells of mesenchymal origin types and are chemoactic, especially for vascular smooth muscle cells (55). The PDGF receptor is complex and related structurally to the immunoglobulin family. It consists of two distinct subunits, which are brought together as dimers by PDGF binding. The expression of the different receptor subunits varies from one cell type to another and affects the binding affinity of and response to the PDGFs. Receptors are found on all mesenchymal cells that form connective tissue, but not on epithelial cells.

Most, if not all, cells secrete transforming growth factor-β (TGF-β), which is a "growth factor" that can inhibit the growth of many cells, particularly epithelial cells and cells of the immune system. TGF-β is a 25 kDa homodimer, and there are several closely related molecules (known as TGF-β 1, TGF-β 2, etc.) which have similar if not identical activity (51). Other members of the "family" include müllerian inhibitory substance, the inhibins, and the activins, factors that affect development of the genital tract. TGF-β has wide-ranging effects, including induction of the expression of other mitogens (such as PDGF) and accumulation of matrix components by stimulating their synthesis and reducing their degradation (53).

The prototypes of the fibroblast growth factors (FGFs) are the acidic and basic fibroblast growth factors. These molecules are 16–17 kDa peptides that contain highly basic and hydrophobic domains. They are not secreted in the usual manner, because they lack classic hydrophobic leader sequences and may be released only by leakage from damaged cells. The FGFs are mitogenic for most, if not all, nonterminally differentiated cell types of mesodermal and neuroectodermal origin. These growth factors are also potent promoters of angiogenesis. They are mitogenic and chemotactic for vascular endothelial cells and stimulate secretion of proteolytic enzymes by these cells. These peptides bind to heparin but, most important physiologically, to the heparan sulfate proteoglycans, which are widely distributed on cell surfaces and in basement membranes. This binding can modify the activity of FGF by preventing degradation and can partition its effects by sequestering it near the cells that produce it (67).

The insulin-like growth factors (IGFs), or somatomedins, resemble proinsulin in amino acid sequence, tertiary structure, and receptor-binding characteristics. The IGF-2 and insulin genes are contiguous. Each IGF has its own widely distributed receptor but both cross-react weakly with the insulin receptor and both bind to specific carrier proteins in plasma, so they may have distant effects in addition to local ones. Insulin itself is expressed in extrapancreatic sites early in development and may have paracrine effects in those organs in the embryo. The IGFs promote cellular maturation in some cells and stimulate cell division in others. They are found early in gestation and are predominantly expressed in mesenchyme-derived tissues.

REPARATIVE RESPONSES TO INJURY ANALOGOUS TO DEVELOPMENTAL PROCESSES

When organs are injured, several responses are activated, including processes to inactivate injurious agents, remove debris, and reestablish healthy tissue. The cell behaviors needed in these processes are analogous to those responsible for organogenesis. Determination and differentiation produce the diversity of specialized form and function needed to form organs. The responses to injury also require changes in cell specialization. In other instances, differentiated func-

tions are suppressed, perhaps to redirect cell metabolism to functions necessary for cell survival. Responses to injury also require cell determination events to take place in special subpopulations of cells called *stem cells*. Stem cells persist into adult life but have undergone some, not all, of the steps of determination (30).

The cellular behaviors of morphogenesis, changes in cell adhesion, shape, and number, are essential for regeneration and repair. For example, replacement of injured cells requires migration of cells. These cells use the extracellular matrix as a substratum and are stimulated to change their adhesion and shape, in part by peptide growth factors. For example, aggregation of platelets leads to their degranulation and the release of PDGF, TGF-α, and TGF-β which affect the migration and proliferation of fibroblasts and endothelial cells. Fibroblasts are induced to deposit matrix, which in turn provides the substratum on which macrophages, endothelial cells, and the replacing cells can migrate into the injured tissue. Accumulation of matrix in a form that does not duplicate the original tissue results in a scar. However, if the basement membrane is undamaged, the normal histoarchitecture of a tissue can be restored with minimal scarring (75).

REFERENCES

1. Asch, A. S., J. Tepler, S. Silbyer, and R. L. Nachman. Cellular attachment to thrombospondin. *J. Biol. Chem.* 266: 1740–1747, 1991.
2. Aumailley, M., and R. Timpl. Attachment of cells to basement membrane collagen IV. *J. Cell Biol.* 103: 1569–1575, 1986.
3. Baron van Evercooren, A., H. K. Kleinman, S. Ohno, P. Marangos, J. P. Schwartz, and M. DuBois-Dalcq. Nerve growth factor, laminin and fibronectin promote neurite outgrowth in human fetal sensory ganglia cultures. *J. Neurosci. Res.* 8: 179–193, 1982.
4. Beck, K., I. Hunter, and J. Engel, Structure and function of laminin: anatomy of a multidomain glycoprotein. *FASEB J.* 4: 148–160, 1990.
5. Bell, E., B. Ivarsson, and C. Merrill. Production of a tissue-like structure by construction of collagen lattices by human fibroblasts of different proliferative potential in vitro. *Proc. Natl. Acad. Sci. U.S.A.* 76: 1274–1278, 1979.
6. Bernfield, M., R. Kokenyesi, M. Kato, M. T. Hinkes, J. Spring, R. L. Gallo, and E. J. Lose. Biology of the syndecans: a family of transmembrane heparan sulfate proteoglycans. *Annu. Rev. Cell Biol.* 8: 365–398, 1992.
7. Bissell, J. M., D. M. Arenson, J. J. Maker, and F. J. Roll. Support of cultured hepatocytes by a laminin rich gel. *J. Clin. Invest.* 79: 801–812, 1987.
8. Bornstein, P. Thrombospondins: structure and regulation of expression. *FASEB J.* 6: 3290–3299, 1992.
9. Carey, D. J., M. S. Todd, and C. M. Rafferty. Schwann cell myelination: induction by exogenous basement membrane-like extracellular matrix. *J. Cell Biol.* 102: 2254–2263, 1986.
10. Carraway, K. L., and C.A.C. Carraway. Membrane-cytoskeleton

interactions in animal cells. *Biochim. Biophys. Acta.* 988: 147–171, 1989.

11. Cedar, H. DNA methylation and gene activity. *Cell* 53: 3–4, 1988.

12. Chakravarti, S., M. F. Tam, and A. E. Chung. The basement membrane glycoprotein entactin promotes cell attachment and binds calcium. *J. Biol. Chem.* 265: 10597–10603, 1990.

13. Chiquet-Ehrismann, R., P. Kalla, C. A. Pearson, K. Beck, and M. Chiquet. Tenascin interferes with fibronectin action. *Cell* 53: 383–390, 1988.

14. Cooper, A. R., and H. A. MacQueen. Subunits of laminin are differentially synthesized on mouse eggs and early embryos. *Dev. Biol.* 96: 467–471, 1983.

15. Cunningham, B. A., J. J. Hemperly, B. A. Murray, et al. Neural cell adhesion molecule: structure, immunoglobulin-like domains, cell surface modulation, and alternative RNA splicing. *Science* 236: 799–806, 1987.

16. Dedhar, S., K. Jewell, M. Rojiani, and V. Gray. The receptor for the basement membrane glycoprotein entactin is the integrin α3/β1. *J. Biol. Chem.* 267: 18908–18914, 1992.

17. Dressler, G. R., and P. Gruss. Do multigene families regulate vertebrate development? *Trends Genet.* 4: 214–219, 1988.

18. Dynan, W. S. Modularity in promoters and enhancers. *Cell* 58: 1–4, 1989.

19. Edelman, G. M. Cell adhesion molecules in the regulation of animal form and tissue pattern. *Annu. Rev. Cell Biol.* 2: 81–116, 1986.

20. Ekblom, P., D. Vestweber, and R. Kemler. Cell-matrix interactions and cell adhesion during development. *Annu. Rev. Cell Biol.* 2: 27–47, 1986.

21. Engvall, E., D. Earwicker, T. Haaparanta, E. Ruoslahti, and J. R. Sanes. Distribution and isolation of four laminin variants; tissue restricted distribution of heterotrimers assembled from five different subunits. *Cell Regulation* 1: 731–740, 1990.

22. Erickson, H. P., and M. A. Bourdon. Tenascin: an extracellular matrix protein prominent in specialized embryonic tissues and tumors. *Annu. Rev. Cell Biol.* 5: 71–92, 1989.

23. Folkman, J., M. Klagsbrun, J. Sasse, M. Wadzinski, D. Ingber, and I. Vlodavsky. A heparin-binding angiogenesis protein—basic fibroblast growth factor is stored within basement membranes. *Am. J. Pathol.* 130: 393–400, 1988.

24. Friedlander, D. R., S. Hoffman, and G. M. Edelman. Functional mapping of cytotactin: proteolytic fragments active in cell-substrate adhesion. *J. Cell Biol.* 107: 2329–2340, 1988.

25. Gehring, W. J. Homeo boxes in the study of development. *Science* 236: 1245–1252, 1987.

26. Gonzalez, A. M., M. Buscaglia, M. Org, and A. Bural. Distribution of basic fibroblast growth factor in basement membrane of diverse tissues. *J. Cell Biol.* 110: 753–765, 1990.

27. Gospodarowicz, D., D. Delgado, and I. Vlodavsky. Permissive effect of the extracellular matrix on cell proliferation in vitro. *Proc. Natl. Acad. Sci. U.S.A.* 77: 4094–4098, 1980.

28. Gossler, A., A. L. Joyner, J. Rossant, and W. C. Skarnes. Mouse embryonic stem cells and reporter constructs to detect developmentally regulated genes. *Science* 244: 463–465, 1989.

29. Grant, D. S., K. Tashiro, B. Segui-Real, Y. Yamada, G. R. Martin, and H. K. Kleinman. Two different laminin domains mediate the differentiation of human endothelial cells into capillary-like structures. *Cell* 58: 933–943, 1989.

30. Hall, P. A., and F. M. Watt. Stem cells: the generation and maintenance of cellular diversity. *Development* 106: 619–633, 1989.

31. Hardingham, T. E., and A. J. Fosang. Proteoglycans: many forms and many functions. *FASEB J.* 6: 861–870, 1992.

32. Hauschka, P. V., T. L. Chen, and A. E. Mavrakos. Polypeptide growth factors in bone matrix. *CIBA Found. Symp.* 136: 207–225, 1988.

33. Hauschka, S. D., and I. R. Konigsberg. The influence of collagen on the development of muscle colonies. *Proc. Natl. Acad. Sci. U.S.A.* 55: 119–126, 1966.

34. Hay, E. D. (Ed) *The Biology of Extracellular Matrix* (2nd ed.). New York: Plenum Press, 1991.

35. Kibbey, M. C., L. S. Royce, M. Dym, B. J. Baum, and H. K. Kleinman. Glandular morphogenesis of a human submandibular cell line by basement membrane components in vitro. *Exp. Cell Res.* 198: 343–357, 1992.

36. Klausner, R. D., and J. B. Harford, *Cis-trans* models for post-transcriptional gene regulation. *Science* 246: 870–872, 1989.

37. Klein, G., M. Lagegger, R. Timpl, and P. Ekblom. Role of laminin A chain in development of epithelial cell polarity. *Cell* 55: 331–341, 1988.

38. Kleinman, H. K., J. Graf, Y. Iwamoto, G. T. Kitten, R. C. Ogle, M. Sasaka, Y. Yamada, G. R. Martin, and L. Luckinbill-Edds. Role of basement membranes in cell differentiation. *Ann. N.Y. Acad. Sci.* 513: 134–145, 1987.

39. Kleinman, H. K., R. J. Klebe, and G. R. Martin. Role of collagenous matrices in the adhesion and growth of cells. *J. Cell Biol.* 88: 473–485, 1981.

40. Kleinman, H. K., L. L. Luckenbill-Edds, F. B. Cannon, and G. Sephel. Use of extracellular matrix components for cell culture. *Anal. Biochem.* 166: 1–13, 1987.

41. Kleinman, H. K., M. C. McGarvey, J. R. Hassell, V. L. Star, F. B. Cannon, G. W. Laurie, and G. R. Martin. Basement membrane complexes with biological activity. *Biochemistry* 25: 321–328, 1986.

42. Krieg, T., R. Hein, A. Hamatochi, and M. Aumailley. Molecular and clinical aspects of connective tissue. *Eur. J. Clin. Invest.* 18: 105–123, 1988.

43. Li, L., M. J. Aggeler, C. Halter, J. R. Hassell, and M. J. Bissell. Influence of a reconstituted basement membrane and its components on casein gene expression and secretion in mouse mammary epithelial cells. *Proc. Natl. Acad. Sci. U.S.A.* 84: 136–140, 1986.

44. Martin, G. R., and R. Timpl. Laminin and other basement membrane components. *Annu. Rev. Cell Biol.* 3: 57–85, 1987.

45. Mason, I. J., A. Taylor, J. G. Williams, H. Sage, and B.L.M. Hogan. Evidence from molecular cloning that SPARC, a major product of mouse embryo parietal endoderm, is related to an endothelial cell culture heat shock glycoprotein of M_r43,000. *EMBO J.* 5: 1465–1472, 1986.

46. Mecham, R. P. Receptors for laminin on mammalian cells. *FASEB J.* 5: 2538–2546, 1991.

47. Mercola, M., C. D. Stiles. Growth factor superfamilies and mammalian embryogenesis. *Development* 102: 451–460, 1988.

48. Murray, A. W., and M. W. Kirschner. Dominoes and clocks: the union of two views of the cell cycle. *Science* 246: 614–621, 1989.

49. Neugebauer, K. M., C. J. Emmett, K. A. Venstrom, and L. F. Reichardt. Vitronectin and thrombospondin promote retinal neurite outgrowth: developmental regulation and role of integrins. *Neuron* 6: 345–358, 1991.

50. O'Farrell, P. H., B. A. Edgar, D. Lakich, and C. F. Lehner. Directing cell division during development. *Science* 246: 635–640, 1989.

51. Pelton, R. W., S. Nomura, H. L. Moses, and B.L.M. Hogan. Expression of transforming growth factor beta-2 RNA during murine embryogenesis. *Development* 106: 759–767, 1989.

52. Raines, E. W., T. F. Lane, M. L. Iruela-Arispe, R. Ross, and

E. H. Sage. The extracellular glycoprotein SPARC interacts with platelet-derived growth factor (PDGF)-AB and -BB and inhibits the binding of PDGF to its receptor. *Proc. Natl. Acad. Sci. U.S.A.* 87: 1281–1285, 1992.

53. Rizzino, A. Transforming growth factor-beta: multiple effects on cell differentiation and extracellular matrices. *Dev. Biol.* 130: 411–422, 1988.

54. Rojkind, M., Z. Gatmaitan, S. Mackensen, M. Giambrone, P. Ponce, and L. M. Reid. Connective tissue biomatrix: its isolation and utilization for long term cultures of normal rat hepatocytes. *J. Cell Biol.* 87: 255–263, 1980.

55. Ross, R. Platelet-derived growth factor. *Lancet* 1: 1179–1182, 1989.

56. Ruoslahti, E. Integrins. *J. Clin. Invest.* 87: 1–5, 1991.

57. Ruoslahti, E., and Y. Yamaguchi. Proteoglycans as modulators of growth factor activities. *Cell* 64: 687–689, 1991.

58. Sakamoto, N., M. Iwahana, F. G. Tanaka, and Y. Osada. Inhibition of angiogenesis and tumor growth by a synthetic laminin peptide CDPGYIGSR-NH$_2$. *Cancer Res.* 51: 903–906, 1991.

59. Sakazuchi, K., M. Yanagishita, Y. Takeuchi, and G. D. Aurbach. Identification of heparan sulfate proteoglycan as a high affinity receptor for acidic fibroblast growth factor (aFGF) in a parathyroid cell line. *J. Cell Biol.* 266: 7270–7278, 1991.

60. Schittny, J. C., R. Timpl, and J. Engel. High resolution immunoelectron microscope localization of functional domains of laminin, nidogen, and heparan sulfate proteoglycan in epithelial basement membrane of mouse cornea reveals different topological orientations. *J. Cell Biol.* 107: 1599–1610, 1988.

61. Schule, R., M. Muller, C. Kaltschmidt, R. Renkawitz. Many transcription factors interact synergistically with steroid receptors. *Science* 242: 1418–1420, 1988.

62. Senior, R. M., H. D. Gresham, G. L. Griffin, E. J. Brown, and A. E. Chung. Entactin stimulates neutrophil adhesion and chemotaxis through interactions between its arg-gly-asp (RGD) domain and the leukocyte response integrin. *J. Clin. Invest.* 90: 2251–2257, 1992.

63. Slack, J.M.W. Peptide regulatory factors in embryonic development. *Lancet* 1: 1312–1315, 1989.

64. Sweeney, T. M., M. C. Kibbey. M. Zain, R. Fridman, and H. K. Kleinman. Basement membrane and the laminin peptide SIKVAV promote tumor growth and metastases. *Cancer Metastasis. Rev.* 10: 245–254, 1991.

65. Takahashi, Y., and H. Nogawn. Branching morphogenesis of salivary epithelium in basement membrane-like substratum separated from mesenchyme by the membrane filter. *Development* 111: 327–335, 1991.

66. Takeichi, M. The cadherins: cell-cell adhesion molecules controlling animal morphogenesis. *Development* 102: 639–655, 1988.

67. Thomas, K. A. Fibroblast growth factors. *FASEB J.* 1: 434–440, 1987.

68. Thompson, J. S. Basement membrane components stimulate epithelialization of intestinal defects in vivo. *Cell Tissue Kinet.* 23: 443–451, 1990.

69. Tomasini, B. R., and D. F. Mosher. Thrombospondin *Prog. Hemost. Thromb.* 10: 269–305, 1991.

70. Tuszynski, G. P., V. L. Rothman, M. Papole, B. Hamilton, and J. Eyal. Identification and characterization of a tumor cell receptor for CSVTCG, a thrombospondin adhesive domain. *J. Cell Biol.* 120: 513–521, 1993.

71. Van der Rest, M., and R. Garrone. Collagen family of proteins. *FASEB J.* 5: 2814–2823, 1991.

72. Vlodavsky, I., R. Bar-Shavit, G. Korner, and Z. Tuks. Extracellular matrix bound growth factors, enzymes and plasma proteins. In: *Molecular and Cellular Aspects of Basement Membranes,* edited by D. H. Rohrbach and R. Timpl Orlando, FL: Academic Press, 1993 p 327–343, 1993.

73. Vlodavsky, I., R. Ishai-Michaeli, P. Bashkin, G. Korner, and R. Bar-Shavit. Heparan sulfate sequestration and release of basic fibroblast growth factor. *Trends Biochem. Sci.* 16: 268–271, 1991.

74. Vos, H. L., S. Devarayalu, Y. de Vries, and P. Bornstein. Thrombospondin 3 (THBs 3), a new member of the thrombospondin gene family. *J. Biol. Chem.* 267: 12192–12196, 1992.

75. Vracko, R. Significance of basal lamina for regeneratio of injured lung. *Virchows Arch A Pathol. Anat. Histopathol.* 355: 264–274, 1972.

75a. Vukicevic, S., H. K. Kleinman, F. Luyten, A. Roberts, and A. H. Reddi. Growth factors in reconstituted basement membrane (matrigel) regulate the network formation of MC 3T3-E1 osteoblastic cells. *Exp. Cell. Res.,* 202: 1–8, 1992.

76. Wu, C., and A. E. Chung. Potential role of entactin in hemostasis. *J. Biol. Chem.* 266: 18802–18807, 1991.

77. Yamada, K. M. Adhesive recognition sequences. *J. Biol. Chem.* 266: 128090–12812, 1991.

78. Yamada, K. Fibronectin and other cell interactive glycoproteins. In: *Cell Biology of Extracellular Matrix,* edited by E. D. Hay. New York: Plenum, 1991, p111–146.

79. Yanagishita, M., and V. C. Hascall. Cell surface heparan sulfate proteoglycans. *J. Biol. Chem.* 267: 9451–9454, 1992.

23. Activation of sperm and egg during fertilization

DAVID EPEL | Hopkins Marine Station, Department of Biological Sciences, Stanford University, Pacific Grove, California

CHAPTER CONTENTS

FERTILIZATION encompasses the events surrounding sperm–egg fusion and initiation of the developmental program that leads to a new individual. The phenomenon is of interest to the developmental biologist because the event marks the beginning of embryogenesis. But it is also of interest to the cell biologist and physiologist because in the few seconds and minutes surrounding fertilization there occurs in microcosm, cell biological events that are easily studied. For example, in the course of the sperm finding, recognizing, and fusing with the egg, such cell biological phenomena as chemotaxis, cell recognition, membrane fusion, and exocytosis all occur.

The resultant sperm–egg binding and fusion initiate a cascade of signal transduction events that cause changes in the level of intracellular calcium, and these changes somehow initiate the developmental program. Profound changes in egg structure occur as part of the fertilization program, these result in an actin-mediated incorporation of the sperm into the egg, in a microtubule-mediated movement of the sperm and egg nucleus toward the egg center, and in an accompanying alteration of the cell endoplasmic reticulum. Initiation of DNA synthesis and entry into a rapid series of cell divisions follow. This reorganization of the egg at fertilization results in placement of special cytoplasms in proper positions; this is necessary for later temporal and spatial gene actions that are part of the developmental program.

Excellent model systems are available for use in studying the phenomena surrounding fertilization. The best ones are provided by aquatic organisms, such as sea urchins, mollusks, fish, and frogs. Their gametes are easily fertilized in vitro. Large amounts of material are available, permitting isolation of structures associated with the sperm–egg recognition complexes, of structures associated with exocytosis, and of large amounts of material useful in studying DNA or protein synthesis in vitro. Recent advances in cell culture and reproductive biology have also made available the gametes of mammals as tractable and experimental systems [(201); see Chapter 24, this volume], although the problem of the amount of material will always be a challenge.

Easily manipulated genetic systems useful in studying fertilization are not yet available. One can study part of the phenomenon, such as cell–cell recognition in the mating systems of yeast or protozoa (43, 55, 119). But for the study of "true fertilization" as seen in plants or animals, there is no good genetic system, since fertilization occurs internally in such genetic model organisms as fruit flies or nematodes. Yet, some interesting mutants for later stages have been described in *Caenorhabditis elegans* [for example, (146)], and recent work on *Arabidopsis* suggests that genetic approaches to fertilization might be possible in this convenient plant system (131).

In this review, I will look at fertilization from a temporal vantage point, beginning with the finding of the egg by the sperm (chemotaxis) and ending with the arousal of the egg at fertilization that results in the initiation of DNA synthesis and the rapid cell divisions characteristic of early development. The review will not be exhaustive in terms of covering all organisms or all phenomena associated with the meeting of sperm and egg. Rather, it will emphasize important aspects of fertilization, concentrating on those systems which have provided the best information—in particular, the sea urchin system. I will also focus on areas and questions where the field is lacking information; my intent is to foment new work in this exciting interface of cell, reproductive, and developmental biology.

CHEMOTAXIS (OR, THE DATING GAME)

Historical Background

How does the sperm find the egg? Is it random? Or is there a directed motility, a chemotactic movement of the sperm to the egg? Historically, the existence of chemotaxis has been controversial. Although clearly demonstrated in plants [see recent review (9)], it was denied for many years that the phenomenon was present in the animal kingdom. Was there really a directed motility (that is, chemotaxis)? Or simply activation of motility by egg factors (chemokinesis)? Miller first demonstrated chemotaxis definitively in hydroid gametes and later showed that chemotaxis exists between sperm and egg in gametes from many animal phyla (116). Indeed, there has been recent evidence in mammals that secretions from the follicle might elicit a chemotactic response from the sperm [reviewed (184, 201)] and that such chemotaxis might even exist in humans (133). Part of the problem in studying chemo-

taxis relates to the absence of simple technologies for measuring and documenting this phenomenon (34, 133).

The alteration of sperm behavior by the egg, attributable to products secreted by the egg or egg-associated cells, falls into two patterns. In the first, for which the horseshoe crab (*Limulus*) and herring are models, the sperm is released in an immotile form and motility is only initiated when the sperm comes into the vicinity of the egg (19, 128). This motility initiation is induced by a peptide or protein released by the egg, but the resultant motility is not directed. Rather, the sperm begin to move randomly; those that move in a direction away from the egg cease motility as the concentration of the chemical stimulus is reduced, whereas those that begin to move toward the egg continue in that direction since the concentration of the motility-inducing factor remains high. The consequence is a net accumulation of sperm at the egg surface, with resultant fertilization (19, 128). The mechanism by which sperm motility is turned on in these sperm is not known.

The second and more common situation is for sperm to be released as actively moving cells and to have their motility altered as sperm encounters egg. If this motility alteration results in an accumulation of sperm near the egg, it is considered to be chemotaxis. Chemotaxis has been well described in those plants where fertilization occurs in an aqueous environment, such as in lower algae and mosses, and the identity of a number of the chemotactic molecules is known [see review (9)].

As noted, the identity of chemotactic molecules in animal sperm was historically more controversial, but recent work has resulted in the identification of the active agents and an understanding of their mechanism. In the well-studied sea urchin system, it was known for many years that factors released from the egg jelly, a layer which surrounds the egg and is therefore the first structure that the sperm encounters, had profound affects on sperm behavior (178). However, the early workers failed to see any alteration that could be classified as chemotaxis, and most of the attention was directed to a jelly-induced "agglutination response" and the induction of the acrosome reaction. This agglutination was species-specific and was viewed as a model for the species-specific cell–cell recognition system seen in fertilization, with properties similar in kind to the immune recognition system [see, for example, (178)].

The last major study of this agglutination response was by Collins (21), who reevaluated this phenomenon and showed that the agglutination has two components. First, he found that the major effect is in fact not an agglutination but an aggregation response, in which the sperm reversibly swarm as large clusters

(21). The mechanism of this response is still not understood. One hypothesis is that this aggregation, or swarming behavior, is similar to that seen in slime mold amoeba, in which one amoeboid cell becomes a "founder cell" that releases a potent chemotactic product [see review (109)]. If a similar mechanism applies to the jelly-induced aggregation, one sperm begins to make this putative aggregating factor, which attracts more sperm to it, and then these sperm put out similar chemotactic products until many sperm are drawn to the aggregate.

The second part of this agglutination/aggregation response described by Collins (21), which is irreversible, is the induction of the acrosome reaction of the sperm by some component(s) of the egg jelly (see later) and a resultant adhesion of sperm to each other by the tips of their exposed acrosomal process. This results in the production of rosettes of sperm, stuck head-to-head, which of course are rendered nonfertilizable by this treatment.

Motility and Chemotactic Substances in Sea Urchin Gametes

Subsequent work on sea urchin eggs has revealed that the jelly layer contains a variety of compounds responsible for changes in behavior and morphology of sperm. The acrosome-inducing substance appears to be a glycoconjugate, which binds to a receptor on the sperm and thence induces the sequence of events leading to the acrosome reaction (94).

The motility changes (and chemotaxis, when present) result from peptides also present in the egg jelly. These peptides cause large increases in motility and respiration of sperm (164). and a variety of these peptides have now been isolated and characterized from the jelly layers of a number of sea urchins [see review (163)]. They are small (generally decapeptides) and their effects on sperm are well characterized, including increased motility and respiration along with increases in Na, Ca, pH_i, and cyclic nucleotides. The best characterized of these peptides are resact (seen in *Arbacia punctulata* eggs) and speract (from *Strongylocentrotus purpuratus* eggs).

Using a cross-linkable labeled peptide analog, Garbers and his colleagues have attempted to find the receptor for these peptides. In *Arbacia*, the peptide (resact) bound to a 160,000 kDa protein which was subsequently identified as guanyl cyclase with remarkable homology to the atrial naturitic factor from humans (156). Speract, the similar peptide in *Strongylocentrotus purpuratus* gametes, also bound to a surface protein, but its receptor was a 77,000 kDa protein with homologies to the low density lipoprotein receptor (26), and the receptor had no sequence similarity to the guanyl cyclase seen in *Arbacia*. It is puzzling that peptides that induce similar changes in their respective sperm should bind to such different receptors.

Similar studies on different species of sea urchins suggest that in fact the peptides are binding to a complex of several proteins. Thus, looking at the peptide SAP III in *Clypeaster japonicus*, Yoshino and Suzuki (202) found that an analog of SAP III could be cross-linked to three distinct proteins (126K, 87K, and 64K) [see also review (184)].

What are the consequences of receptor occupancy? In all cases so far examined, the addition of peptide results in elevation of both cyclic AMP (cAMP) and cyclic GMP (cGMP). The alteration of cGMP levels is best understood, and appears to involve a transient activation of a membrane-bound guanyl cyclase, which is associated with a dephosphorylation of the enzyme [see (16, 186)]. Not understood, however, is how the change in cGMP is related to activation of the sperm. The peptides also affect ion conductances and the resultant changes could affect motility (24).

Do these peptide-induced changes relate to chemotaxis? Surprisingly, the answer seems to vary with the species. Resact, the peptide from *Arbacia* egg jelly, induces chemotaxis in *Arbacia* sperm. But speract, a peptide from *S. purpuratus* egg jelly, does not induce chemotaxis in *S. purpuratus* sperm (185). It is puzzling that the chemotactic response is not universal, but perhaps this indicates that the general role of the peptide is not related to chemotaxis but to the general increase in motility mediated by the cyclic nucleotide and ionic changes. Another possibility is that the cGMP elevation is not related just to motility, but to later activation of the egg if the sperm–egg fusion results in increased cGMP levels in the egg (see later).

In cases where chemotaxis is present, how does it work? Presumably there are graded motility differences that are dependent on concentration of the chemotactic substances, resulting in accumulation of the sperm at the egg. One possibility is based on the finding that the peptides alter the H^+ and K^+ conductance of the membrane (101) and that this change occurs very early in the activation process. Cook and Babcock (23) found that the response of ion channels varied with peptide concentration and intracellular cGMP, and Cook et al. (24) found that internal calcium ion and flagellar beat also varied with peptide content. Cook et al. (24) propose that the peptide effect on the K^+ conductance is reflected in a change in calcium ion, and that these differences then account for the differential motility characteristic of chemotactic behavior.

Does Chemotaxis Exist in the Real World?

Finally, one might ask—why is there chemotaxis? The obvious answer is that chemotaxis helps the sperm find the egg. This could certainly apply in situations where fertilization is internal. In the mammal, for example, few sperm find their way into the fallopian tube and fertilization probably occurs with only several sperm around the egg (201). Here one could imagine selection for sperm that can better find the egg, as by a chemotactic attraction to some egg product.

But can this selection for chemotaxis apply to aquatic organisms where fertilization might occur in a turbulent environment? For example, if fertilization occurs in the intertidal or in surge channels, as is believed to be the case for sea urchins, could a chemotactic gradient be established and last sufficiently long to attract the sperm (115)? This seems improbable, but as noted, there are chemotactic substances released from these eggs. This suggests that there may be situations where the water is quiet enough so that a gradient will form and where chemotaxis would therefore "pay off" in terms of better fertilization success. Thus, chemotaxis might be important when fertilization occurs in quiet tide pools or surge channels, but be of no import when fertilization occurs in a wave-swept situation (29).

THE ACROSOME REACTION

Description of the Phenomenon

The sperm has now found the egg, be it by random or directed motion. A new recognition process now begins, which might be referred to as a "rite of passage," related first to the binding of the sperm to the outer egg coat and then to passage of the sperm through these coats.

Passage is required since eggs are typically not "naked" but are enveloped with all sorts of outer membranes, chorions, etc. One role of these coats is to protect the egg from physical damage. Another role may be to prevent accidental fusion of the egg with other cells; as the egg surface is highly fusible, it might fuse with the wrong cells. For example, the hamster egg with outer egg coat present (the *zona pellucida*) cannot be fertilized by a human sperm, but if the *zona* is removed, the human sperm will now fuse with this egg [see review (201)]. These types of observations suggest that another role of the outer egg coat is to act as a recognition barrier to only admit the proper sperm.

The passage problem then is two processes: the species-specific recognition of the outer egg coat and then passage through this coat. These two problems are often interrelated, and they always first involve the

exposure of recognition molecules on the sperm to effect specific sperm–egg binding. Often associated with this is the concomitant exposure of molecules that allow the sperm to pass through these outer egg coats and then reach the egg surface.

Just as the egg plasma membrane is highly fusible, so too is the sperm plasma membrane. And just as the egg plasma membrane is protected from the environment by the chorion and other structures, the sperm's fusible surface is similarly covered/protected until the sperm comes in contact with the egg.

This uncovering of the sperm surface is through the acrosome reaction, an exocytotic event which exposes a new surface for attaching the sperm to the egg, for penetrating the outer egg coat, and later for fusing with the egg [see reviews (177, 184)]. The consequences of the acrosome reaction vary with the organism (Fig. 23.1). In many invertebrates, such as the sea urchin, the

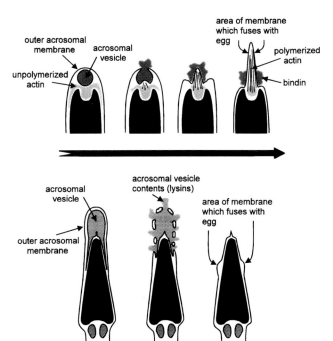

FIG. 23.1. Diagram of acrosome reaction. *Upper part* depicts reaction in invertebrate (sea urchin) and *lower part* comparable reaction in mammal (hamster). In both cases exocytosis of the acrosomal vesicle results in exposure of new surface which will later fuse with egg plasma membrane. In sea urchin case, surface exposed by the acrosome reaction is pushed in front of sperm by polymerization of actin. The new acrosomal process which is formed is also coated with bindin, a protein contained within acrosomal vesicle, which is externalized by acrosomal exocytosis and thence involved in sperm–egg adhesion.

In mammal (*lower part* of figure), there is no comparable acrosomal process formed by actin polymerization. Rather, exocytosis exposes surfaces which later fuse with egg surface, although initial fusion is with subacrosomal sperm surfaces which are already exposed before acrosome reaction.

acrosome reaction uncovers binding proteins (bindin) which coat the surface of the newly exposed sperm plasma membrane and help the sperm to attach to the outer egg coat (the vitelline layer). Actin polymerization is often involved (177), and it results in the presentation of the binding molecules on the tip of a newly formed sperm acrosomal process. The newly exposed plasma membrane underneath the bindin will later fuse with the egg plasma membrane.

In mammals, such as the mouse, the sperm protein that binds to the outer egg coat (the zona) is already exposed and does not require an acrosome reaction to display this recognition protein. Binding to the zona is then followed by induction of the acrosome reaction by other proteins in the zona, with the resultant exposure of a new surface which, as in the invertebrate situation, will later fuse with the egg plasma membrane (see Chapter 24, this volume).

Signal Transduction Molecules

What are the signals involved in the initiation of the acrosome reaction and how are these signals transduced into the exocytosis? Typically, the signaling molecules are associated with the egg structures, usually macromolecules of the outer egg surface such as those of the zona pellucida in the mammal (see Chapter 24, this volume) or the jelly layer in the sea urchin (94, 163). In general, it appears that several factors are involved, a large-molecular-weight component that is part of the outer egg coat, and a smaller component that might be diffusible. The significance of these dual systems is not clear but might relate to fail-safe mechanisms to reduce the possibility of premature acrosome reactions.

These molecules have been identified in a number of systems. In starfish, there are at least three components, of both small and large molecular weight, which must interact together (73, 74). In the sea urchin the major molecule appears to be a glycoprotein of the jelly layer (94), and its action in inducing the acrosome reaction is also facilitated by smaller molecules, such as the sperm-activating peptides of the sea urchin egg jelly (200). The situation may be similarly complex in the mammal, where ZP2, one of the zona proteins, is involved in induction of the acrosome reaction, but this reaction can be also be modulated by a small-molecular-weight cofactor, progesterone (8, 199).

How do these inducers work? One of their major consequences is to increase the intracellular calcium levels, and this seems to be adequate to induce an acrosome reaction [see (184)]. Thus, increasing the calcium level artificially as with ionophores will induce a normal acrosome reaction (22). In sperm of some species the acrosome reaction is also associated with an increase in intracellular pH (137). The physiology of this pH_i change has been well studied in sea urchin sperm, which possess a unique type of Na^+–H^+ exchanger which is stoichiometrically 1:1 but which is not electroneutral, is not inhibited by amiloride or its derivatives, and is modulated by the membrane potential [(27), reviewed (184, 101)].

These ionic changes appear to be the consequence of receptor binding to the sperm plasma membrane. The binding initiates opening of channels, which results in the calcium influx and the pH increase [see reviews (27, 51, 184)]. These ionic fluxes might also be coupled to protein phosphorylation, and in the case of many invertebrate sperm the pH changes are required for actin polymerization and formation of the acrosomal process (177).

SPERM–EGG BINDING

The initial binding of sperm to egg is to the outer egg coat (such as the vitelline layer in the sea urchin, the chorion in abalone or the zona in mammals), as opposed to the egg plasma membrane. In the invertebrates and lower vertebrates, this sperm–egg binding requires the acrosome reaction to expose the egg-binding sites on the sperm. In the mammals, conversely, the sperm first binds to a specific molecule on the zona pellucida—ZP3—and the interaction of the sperm surface with another zona molecule (ZP2) leads to the acrosome reaction (see Chapter 24).

The nature of this sperm–egg binding differs in different organisms. In the sea urchin, the acrosome reaction exposes the adhesive molecule, bindin [recent review (183)], which coats the acrosomal process and attaches to the vitelline layer on the egg surface. The sequence and structure of this protein are well studied (107), and the characteristics of its synthesis during spermiogenesis (12, 120) have been described.

How does this protein bind to the sea urchin egg? Until recently, such questions were difficult to answer since the egg receptor had not been characterized (that is, the receptor for bindin). The recent isolation and cloning of the receptor from Foltz and Lennarz's lab will soon cause this void to fill (47, 105, 124). From its structure, the bindin receptor is seen to be a single transmembrane protein with the external-facing component variable between species, which fits in with the species specificity of fertilization. The transmembrane and cytoplasmic components are conserved, however, suggesting a retention of some essential role, perhaps in signal transduction (46). The external-facing moiety has homology with heat-shock proteins, but the significance of this is unclear (47).

A different mechanism is involved in the mammal. An important difference is that in the sea urchin the acrosome reaction is a prerequisite to binding to the vitelline layer, while in the mammal the sperm first binds to the ZP (via the ZP3 protein), and subsequent binding to another ZP protein (ZP2) then initiates the acrosome reaction. The details of this interaction have been exquisitely worked out and are described in Chapter 24 of this volume.

PASSAGE OF THE SPERM THROUGH THE EGG COAT

As noted above, the sperm attaches to an outer egg envelope which is usually not part of the plasma membrane and indeed is often many microns distant from the egg surface proper. Examples would be the zona in the mammals or the chorion in many mollusks. In other situations, such as the sea urchin, the outer egg coat or vitelline layer is closely apposed to the plasma membrane. Irrespective, the attached sperm must now somehow get to the egg surface and hence must pass through these outer egg coats to abut directly against the egg surface.

There is apparently no single mechanism for this rite of passage, and in any one species several different actions are probably involved. One paradigm suggests that the sperm uses hydrolytic enzymes to pass though the outer egg coat [see review (72)]. There may be problems with this idea, however. For example, in the mammals it was assumed that the sperm acrosomal protease, acrosin, is involved in passage of the sperm through the zona. This has been best studied in the mammals, where a trypsin-like protease (acrosin) is exposed upon the acrosome reaction. A role for this protease action is suggested since fertilization in the presence of protease inhibitors inhibits penetration of the zona (6). However, Baba et al. (4) found that mouse sperm containing a mutated and nonfunctional acrosin were still capable of fertilizing the egg.

An alternative mechanism for passage through the outer egg coat has been discovered in some molluscan gametes. In the turban snail *Tegula* (63) and in the abalone *Haliotis* (180) the sperm acrosome contains a protein, named *sperm lysin,* which seems to "digest" the egg chorion, but the mechanism does not involve any lytic or hydrolytic action. Rather, the mechanism appears to be noncatalytic, involving instead a structural destabilization of the chorion by the lysin molecule. The mechanism is stoichiometric, with one molecule of lysin binding to a complementary protein on the chorion and this causing the destabilization of

structure so that a hole is created through which the sperm can now pass through (148). This mechanism has so far been seen in several mollusks. It will be very interesting to see how widespread—or restricted—this mechanism is in the fertilization process.

SPERM–EGG RECEPTORS AND SPECIATION

Early studies on the specificity of fertilization showed that the process was species-specific (although some hybridization could be achieved under some conditions). The basis of this specificity presumably lies in the species-specific structure of the sperm–egg receptor/ligand complex (and in the case of abalone, in the species-specific nature of the sperm lysin action on the egg chorion). Recent studies on the sequence of these molecules indicates that these molecules are indeed species-specific and thus support this contention [e.g., (117, 180, 181)].

An important question raised by these findings is how any changes can arise in these molecules. If one of these complementary proteins is altered, will they still be functional in sperm–egg binding? One possibility is that there is a looseness in the structure-binding relationship, so an altered molecule can still function. A related question asks whether alterations in these proteins can provide a molecular basis for speciation. Thus, reproductive isolation is often considered a prerequisite for speciation; does variation in the sperm–egg receptors promote this isolation and thus promote speciation? Or do sperm–egg receptors diverge during geographic isolation and preserve reproductive isolation when the geographic barriers are no longer present?

An important insight into these questions comes from comparative intraspecies studies on the structure of the bindin genes/proteins in sea urchins and the lysin genes in abalone. Bindin contains a central core protein which is not involved in sperm–egg adhesion and which varies little between species. There are, however, two tandem repeats of a binding motif, which varies between species. Site-directed mutation studies indicate that this portion is essential for species-specific adhesion [see review (71, 183)].

Hofmann and Glabe (71) suggest a mechanism by which polymorphism in ligands could occur in a population while still allowing normal fertilization. They propose that variation in one of the two tandem repeats would not affect binding greatly and would allow accumulation of variants with different sequences. If these provided some advantage, there could be selection for them. Alternatively, if the changes were neutral but the different populations were geographically isolated, then an accumulation of these variants in the different populations could result in them eventually changing sufficiently so that reproductive isolation would be the case should these populations become contiguous again.

A different insight comes from studies on the lysin of abalone sperm. Here a part of the species-specific block to hybridization comes from the specificity of the lysin—the lysin from one species is not as effective against the chorion of eggs of related species. Studies of the lysin gene reveal that the genes coding for the lysins are highly nonsynonymous—that is, contain triplet codons in which any variation will result in an amino acid variant (104, 171). This phenomenon is rare and suggests that any mutation in a single base will result in a change in amino acid sequence—and that therefore these genes will change extremely rapidly. This suggests a high rate of evolutionary change, implying some selective force driving this change. Vacquier notes that the only other examples of nonsynonymous evolution involve defense molecules typically associated with the immune system or with evading the immune system. Does this mean that the pressure driving this change in the abalone lysin gene relates to predation, as by microbes, on the egg [see (181) for discussion of this intriguing idea]?

A DESCRIPTION OF EGG ACTIVATION

The above discussion on the molecular basis of sperm–egg interaction centers on the sperm and egg molecules involved in sperm recognition and binding to the outer egg coats. These coats are usually quite far removed from the egg surface, and as noted, the sperm must pass through the coats to get to the egg surface proper. Indeed, one might look at these coats as barriers or condoms that protect the fusible egg surface from meeting the sperm of other species (in the case of aquatic organisms) or meeting other cells of the reproductive tract (in the case of animals with internal fertilization).

The following section focuses on the nature of the sperm receptors on the egg surface and their role in signal transduction and activation of the developmental program. But first we must describe the sequelae of fertilization, what happens at fertilization, and then go backward in this cascade to dissect the myriad events that occur and their interconnections.

What Happens When Sperm Meets Egg?

Structural Changes. In dealing with the response of the egg to the sperm, I will primarily concentrate on sea

urchin egg activation, since this is the best-understood case. The large amounts of gametes available from sea urchins allow one to discern by biochemical or physiological approaches what happens when egg and sperm meet. Where appropriate, the state of knowledge in other forms and their similarities and differences are then described.

I first present a brief overview of the microscopically visible changes that accompany sea urchin fertilization and then go into detail on the mechanism of these changes (Fig. 23.2). Upon insemination, numerous sperm attach to the vitelline layer—up to 1800 (182). All sperm that attach continue to move until about 10–15 s after the first sperm attaches. Then the "fertilizing"

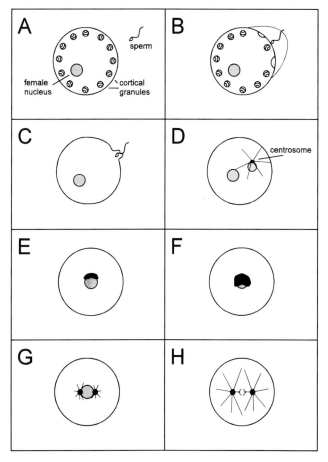

FIG. 23.2. Cartoon illustrating microscopically visible changes of fertilization in sea urchin. Female pronucleus of unfertilized egg (A) is placed eccentrically in cytoplasm, and prominent feature of egg cortex is presence of numerous cortical granules imbedded in cytoplasm. Sperm–egg fusion (B) initiates wave of cortical granule exocytosis and elevation of fertilization membrane. Sperm enters through actin-filled fertilization cone (C). Once inside cytoplasm, sperm centrosome is activated and organizes microtubules around it which are involved in moving sperm and egg pronucleus to egg center (D). Centrosome duplicates and surrounds zygote nucleus (E and F) and then organizes mitotic apparatus (G and H) for first mitosis.

sperm, that single sperm that will fertilize the egg, stops moving (40, 138). {Earlier, and invisible to the eye but associated with the "selection" of the fertilizing sperm, is a block to fusion of any other sperm, referred to as the "block to polyspermy," [reviewed (81)].}

Suddenly the vitelline layer begins to elevate at the site of fusion of this fertilizing sperm, and the remaining sperm, the unsuccessful suitors for the egg, are lifted off with the elevating vitelline layer (182). A translucent cone of actin-filled cytoplasm begins to surround the fertilizing sperm (85) and this sperm then slowly enters into the egg cytoplasm. Upon entry, the sperm turns 180°, and several minutes later a microtubule-containing structure, the sperm aster, appears in the cytoplasm, associated with the sperm centrosome. This microtubule structure is somehow involved in the subsequent movement of both sperm and egg pronucleus toward the egg center [see review (138) of these sperm–egg interactions]. The sperm and egg nucleus meet about 20 min later at the zygote center and fuse with each other; then DNA synthesis begins (70). The egg had earlier lost its centrosome following the meiotic divisions, and the only functional centrosome in the zygote is that brought in by the sperm [(112), reviewed (139)]. The sperm centrosome now replicates, moves around the nucleus, and sets up the organization of the future mitotic apparatus (126). Cell division occurs about 60–90 min later, the exact time depending on the species.

What are the cellular and molecular bases for the above changes? To answer this, we first delve into the changes in the egg upon sperm–egg contact and their relationship to the above fertilization responses.

Electrical Changes and the Block to Polyspermy.

The earliest postfertilization change so far described is an electrical change, seen as a depolarization of the plasma membrane, which goes from −70 mV to +10–20 mV (79). The electrophysiological properties of the unfertilized egg plasma membrane are well described, and the nature of the changes that occur at fertilization is understood [see (108); earlier review (193)]. These studies show that the membrane of the unfertilized egg exhibits low conductance and high resistance. Within 1 s of sperm–egg contact, a small change in egg membrane potential is seen that most likely represents the channels of the fused sperm, and this is followed by a fertilization potential change of 30–90 mV.

How do these changes relate to sperm–egg fusion? Chambers, Longo, and their co-workers have elegantly determined the time of fusion in relation to the electrical changes using a combination of electrophysiological and ultrastructural approaches (106, 113). The time of fusion was determined by capacitance changes, using

an ingenious technique in which the fertilizing sperm is isolated electrically and the change in capacitance associated with the fertilizing sperm is then measured (Fig. 23.3). These measurements are done with simultaneous measurement of ion current changes and indicate that fusion occurs simultaneously with the initial depo-

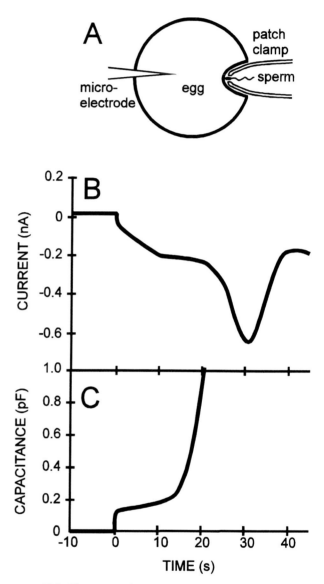

FIG. 23.3. Illustration of experiment in which current changes and capacitance changes associated with fertilization are simultaneously measured. *A* depicts experimental setup, in which microelectrode measures voltage around entire egg and patch clamp measures capacitance and ion currents in region of egg where sperm–egg fusion occurs. Sperm is introduced in pipette, which isolates membrane and allows capacitance measurements to be made in this small area of membrane. As seen in parts *B* and *C*, capacitance and ion changes occur simultaneously, indicating that fusion and current changes are temporally associated. Membrane potential change (not shown) also occurs at same time. [Adapted from study of McCulloh and Chambers (113).]

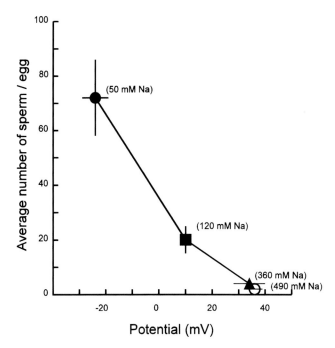

FIG. 23.4. Polyspermy as function of membrane potential in eggs of *Urechis caupo*. Membrane potential is varied by fertilizing eggs in different concentrations of extracellular sodium (concentration noted in *parentheses*). As seen, there is increase in number of sperm that can enter egg in relation to degree of depolarization [Adapted from study of Gould-Somero et al. (59).]

and extensive studies, have examined the role of the various ionic changes as determined by patch clamp and voltage clamp studies. These workers first determined the ionic bases of the potential changes and then carefully studied the consequences for fertilization when the ionic current changes were prevented.

Their work indicates that the egg must undergo voltage changes in order for the sperm to enter during the rapid rise in the membrane potential. Specifically, they found that if the egg were clamped at the unfertilized level—(−70 mV)—the sperm would not be incorporated into the egg. This could mean that the depolarization is necessary for sperm entry. But, as might be expected from Jaffe's findings, they also observed that higher voltages prevented sperm–egg fusion. Specifically, they found that going higher than 0 mV and approaching −20 mV resulted in a progressively decreased probability of sperm incorporation. This implies that the fertilizing sperm has to pass through steps leading to incorporation during the period that the egg membrane potential is between −70 and −20 mV, a period lasting less than 1 s. Their findings agree with the earlier work of Shen and Steinhardt (152) that less than 1 s at a permissive voltage was adequate for sperm–egg fusion. These results then indicate that fusion and incorporation can only occur in a brief voltage window whose time parameter is related to the rapid rise in membrane potential.

THE ROLE OF CALCIUM IN EGG ACTIVATION

Historical Background

Do the initial ionic changes associated with the electrical events have any other role besides steps leading to sperm incorporation and the block to polyspermy? One approach to answering this question is to experimentally induce the voltage change and then assess the consequences. This can be done by voltage clamp, and when assessed morphologically, there are no visible changes in the egg from this simple depolarization (79). The role of the depolarization can also be assessed by artificially depolarizing the membrane with high external potassium (83, 143), and here too there are no overt morphological changes, such as a cortical reaction or nuclear centration (see below for further description of these changes). So, the depolarization per se appears to have no role in egg activation.

The change that appears to be most important for egg activation is a transient rise in intracellular calcium levels. This is not an unexpected finding in the 1990s, but the early studies on egg activation were among the first to document the important role of calcium in

larization changes. These capacitance measurements agree with electron microscope studies which reveal that a prefusion stage can be seen within 4 s of the electrical changes, but fusion, as detected ultrastructurally, does not occur until about 5–8 s later (106).

These changes in electrical properties ensue from associated fluxes of sodium and calcium ions. The subsequent increase in membrane potential also functions to prevent any supernumerary sperm from entering and is the basis for the "block to polyspermy" (81). Thus, the initial establishment of fusion by the first or fertilizing sperm initiates a depolarization, which thence precludes any other sperm from entering.

Part of the evidence that the depolarization is involved in precluding polyspermy is that experimental treatments that slow or prevent the depolarization (such as low Na+ or voltage clamp) increase the incidence of polyspermy [Fig. 23.4; (59)]. As noted above, the sperm–egg fusion is responsible for the initial electrical change. Studies with interspecies hybrids show that the sperm is also the target for increased membrane potential, so by introducing these early ionic changes the fertilizing sperm insures that no other sperm will enter (80, 82, 113).

Chambers and his colleagues (108, 113), in elegant

triggering changes in cell activity. The original indication that this might be the case came from Mazia's work in Heilbrunn's lab in 1937. Mazia (110) found a change in free or "dialyzable" calcium following fertilization of sea urchin eggs. Nevertheless, aside from Heilbrunn's espousal of calcium's role in many areas of cell stimulation in his important physiology text (66), interest in calcium lay dormant until the 1970s when techniques and approaches for altering and measuring this cation became available.

The first indication that calcium was critical came from studies with the calcium ionophore A23187. Steinhardt and Epel (160) and Chambers et al. (15) showed that this ionophore would initiate the early changes of fertilization, such as the cortical reaction, respiratory burst, protein synthesis, and DNA synthesis. This occurred in the absence of extracellular calcium or magnesium, suggesting that the ionophore was releasing a divalent ion from some intracellular store. This ion was most likely calcium, since analysis of cell homogenates indicated that magnesium was largely free whereas calcium was largely bound (160), in agreement with Mazia's earlier 1937 study (110).

The Natural History of Calcium Changes at Fertilization

There is today much evidence to support the view that an increase in calcium is indeed the trigger for egg activation. First there are numerous studies showing that free calcium increases at fertilization. These include measurements with calcium-indicating proteins such as aquoerin (135), and fluorescent calcium-indicating dyes, such as Fura-2 (54). The initial studies on fertilization in fish eggs showed that the calcium passed through the egg as a wave (53). This has since been seen in a variety of eggs, including eggs of sea urchins, tunicates, and mammals (25, 54, 157). In some cases, such as the fish and sea urchins, there is one major pulse of calcium (53, 54). In other cases, such as in eggs of mammals and tunicates, there are subsequent oscillations in calcium level following the initial burst in calcium (25, 157).

An examination of the events that occur upon fertilization in the sea urchin egg indicate that the calcium rise is one of the earliest changes in the sequence of egg activation. As seen in Figure 23.5 and discussed in detail below, there are two phases of calcium release: an early cortical increase involving low levels of calcium and a later massive calcium release that envelops the entire egg. The early change closely follows or is part of the membrane potential changes. Both phases are coincident with a rapid increase in protein tyrosine phosphorylation (18). The later massive change in cal-

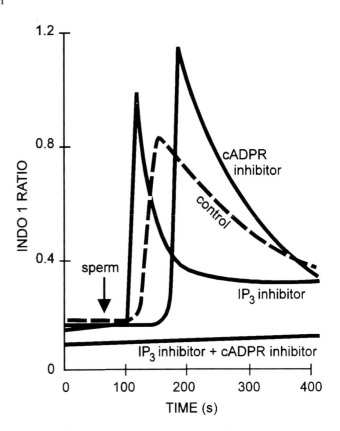

FIG. 23.5. Program of events following fertilization in eggs of sea urchin *S. purpuratus* at 16° C. The timing is in seconds, presented in a logarithmic fashion on the left-hand side of the figure. Period of calcium release is highlighted in the period between 30 and 90 s after fertilization.

cium coincides with the cortical granule exocytosis (125) and the activation of the calmodulin-dependent enzyme nicotinamide-adenine dinucleotide (NAD) kinase (35, 41). The calcium increase is coincident or precedes the wave of cortical actin polymerization. The Ca^{+2} rise precedes a major increase in cytosolic pH, large increases in protein synthesis, and initiation of DNA synthesis [reviewed (39, 193)].

Many studies have been done assessing the consequences of inducing or preventing the calcium rise. As noted earlier, the induction of a calcium rise by ionophore (160) will lead to the cortical reaction, to increased protein synthesis, and to initiation of DNA synthesis and entry into mitosis. [Actual chromosome separation and cleavage do not occur because the centrosome—normally brought in by the sperm—is absent, and this body is essential for organizing a bipolar and functional mitotic apparatus; see for example, (112).] Similarly, injection of calcium buffers to modulate calcium levels will lead to a cortical reaction and nuclear events when the calcium is above a critical level (64).

The opposite side, preventing the calcium release, precludes egg activation. This was initially achieved by injection of calcium chelators; under these conditions there are no overt signs of egg activation, such as the cortical reaction and the migration of the nucleus to the center of the egg (203). In addition, the later consequences of fertilization do not occur, such as nuclear envelope breakdown and entry into mitosis. (Studies on macromolecule synthesis can't be done, since one is injecting single eggs; but the absence of entry into the cell cycle suggests that these synthetic events have not taken place.)

The spatial localization of the calcium increase has now been observed by confocal microscopy. These studies reveal two phases to the calcium release (54). The first phase, probably arising from the membrane depolarization, is a cortical rise in calcium. It is present simultaneously throughout the cortex, and is not propagated. The second phase corresponds to the major wave of calcium that begins at the site of sperm–egg fusion, and this is propagated though the cytoplasm as a wave.

The early increase in cortical calcium appears to have no role in terms of egg activation. As noted earlier, inducing a calcium influx by high potassium (83, 143), or turning on the calcium channels by voltage clamping (79, 108) appears to have no overt affect on the egg—such as inducing cortical granule exocytosis Yet, as the confocal studies now reveal, there is a measurable increase in calcium level which is localized to the cortex. One possibility is that this calcium rise leads to cytoskeletal alterations involved in subsequent sperm incorporation. Another possible role is in the polyspermy block. Or the influx may have no role; it may just reflect the presence of voltage-gated calcium channels in the egg membrane.

A different view is that the cortical calcium rise indeed has a role, but that there is a requirement for a second response in addition to calcium. For example, calcium alone may be inadequate as a stimulus, and there may be some yet-unknown molecule produced at fertilization that functions as a powerful cofactor in addition to calcium; this molecule is not produced when the membrane is artificially depolarized as by a voltage clamp or high potassium.

Another explanation for the apparent lack of effect of the early cortical calcium rise is that the calcium level never gets high enough to trigger a response and the calcium level in the cortex does not reach this level. This view is also consistent with the study of Hamaguchi and Hiramoto (64), who injected calcium buffers into eggs and showed that the free calcium content needed to reach at least 1 μM to initiate a cortical reaction. A third possibility, which could also be related to concentration, is that the calcium is not in the right place—that a cortical locale is inadequate to stimulate enough of the postfertilization events to result in egg activation.

Mechanisms of the Calcium Increase

Sources of Calcium for Activation—Protostomes vs. Deuterostomes. What is the mechanism of the calcium rise? In eggs of the protostome evolutionary line (annelids, mollusks, and arthropods) the comparative evidence available indicates that the source of calcium is external, presumably entering through a voltage-gated channel [see review (86)]. This is best described in the oocytes of the echiuroid worm Urechis, in which fertilization induces a membrane potential change and associated entry of calcium (59, 92). These eggs can also be activated by artificial depolarization of the membrane, as by elevated potassium (a characteristic of a variety of protostome eggs, such as mollusks [see, for example (92)]. A protein extracted from the Urechis sperm acrosome will by itself cause depolarization and egg activation (56, 57). These results suggest that the trigger for calcium entry is the binding of this sperm protein to a receptor on the plasma membrane, which then opens a ligand-activated calcium channel with resultant activation of the oocyte.

The situation is different in the eggs of deuterostomes, such as sea urchins, tunicates, amphibians, and mammals. Here the source of calcium is from internal stores, as indicated by studies that show that eggs can be activated under conditions where external calcium is limiting, such as ionophore activation in calcium-free seawater (15, 160). Also, one can activate sea urchin eggs with acrosome-reacted sperm in a calcium-free environment (143). Although high sperm concentrations are required for this activation, the ability to activate in the calcium-free condition indicates that internal stores are sufficient.

Inositol Phosphate Pathway. The above results then suggest that mechanisms responsible for calcium release from internal stores are operative in these eggs, and now a wide variety of studies indicate that the inositol phosphate pathway is involved [see (194)]. Additionally, it appears that in the sea urchin egg another system is also utilized that uses a newly described second messenger, cyclic ADP ribose derived from NAD (20), and possibly another pyridine nucleotide derivative originating from NADP (98).

Let us first look at the situation in the sea urchin egg. The evidence for inositol phosphate involvement is that increases in turnover of the inositol phospholipids accompany fertilization, that there are increases in the accompanying diacylglycerol, and that these increases precede or are coincident with the calcium rise (17, 194). One can inject IP$_3$ into eggs and this releases calcium and results in egg activation (170). Also, one can isolate microsomal fractions from eggs and these will release calcium in response to IP$_3$ (122); these vesicles contain a calsequestrin-like protein that is most likely involved in binding high levels of calcium in the vesicles (123). These vesicle systems can be visualized by antibody location (68) or by cytochemical procedures to detect calcium stores (130). Such studies reveal an extensive system of calcium-rich trabeculae that are especially concentrated in the cortex of the egg (68, 130).

The above data suggest that sperm binding somehow results in PIP$_2$ hydrolysis, releasing IP$_3$ with resultant calcium increase. The initial calcium release could then induce additional calcium release, perhaps through calcium activation of a PIP kinase (121) and PIP$_2$ phospholipase C (190). However, several discordant findings indicate that the situation is more complex and that multiple systems are involved in calcium release in sea urchin eggs.

The critical observation was that injection of heparin into the egg, at levels that should prevent IP$_3$-induced release of calcium, did in fact not reduce calcium release, and egg activation by sperm still occurred (132). This datum was made more compelling by the observation that injection of IP$_3$ into a heparin-injected egg did not release calcium (so this should have prevented activation by sperm if only IP$_3$ were involved). Yet, if these eggs were now fertilized, there was an apparently normal calcium release and egg activation. This suggested that fertilization utilizes dual systems for releasing calcium, or that it possesses a secondary system that it could call into play if the first system was rendered inoperative (132).

Cyclic ADP Ribose Pathway.

The available evidence now appears to favor the idea that two and maybe three releasing systems are operative in sea urchin eggs. One candidate for the second system is cyclic ADP ribose (cADPR), a potent calcium-releasing compound discovered in Lee's laboratory (20). Clapper and colleagues observed that microsome fractions from sea urchin eggs would slowly release calcium in response to the addition of NAD or NADP to the crude homogenate. Their studies indicated that the pyridine nucleotide was not the inducer itself, but was the parent compound for a calcium-releasing agent that was being made enzymatically in the homogenate. The active factor when NAD was the substrate was isolated and identified as cADPR, produced from NAD by the enzyme cADPR cyclase. It releases calcium from microsome

fractions at nM levels. Importantly, the cADPR-sensitive store is different than the IP$_3$-sensitive store [(20), reviewed (102)]. [The NADP derivative has recently been identified, but its role is not known; see (98).]

Several observations on sea urchin eggs provide additional insights into the potential mode of action of cADPR. One is that a ryanodine receptor (presumably the cADPR site) is particularly enriched in the egg cortex (114). A second is that cGMP can activate the cADPR synthase (50). A related provocative finding is that sea urchin sperm are an extremely rich source of guanyl cyclase (16). One hypothesis, then, is that injection of a small amount of cGMP at the site of sperm–egg fusion would activate the cADPR synthase, which would then release localized calcium which could then propagate through the egg, perhaps in concert with the IP$_3$ system.

An alternate possibility, suggested by the calmodulin sensitivity of the cADPR system (48, 99), is that the IP$_3$ system *is* the primary response, releasing the initial calcium, and that this Ca^{2+} then activates the cADPR system.

Is the cADPR system the second mechanism for calcium release in the sea urchin egg? To answer this, Lee synthesized analogs of cADPR and showed that these would prevent the action of the cADPR in releasing calcium in vitro and also in vivo (100). As shown in Fig. 23.6, injection of the analogs into living eggs prevented any calcium rise in response to injected cADPR, but the analogs did not affect the rise in calcium produced by injected IP$_3$. When these eggs, injected with the cADPR analogs, were fertilized, there was a rise in calcium and egg activation occurred. However, if the cADPR analogs were injected into heparin-treated eggs (which prevents IP$_3$-induced Ca^{2+} release) the rise in calcium and subsequent activation were inhibited. These results complement the heparin result, which as noted above similarly did not prevent the calcium rise and egg activation if only heparin was injected. Therefore, both systems (the IP$_3$ and cADPR) have to be inhibited in order to prevent a calcium rise. These results and those of Galione and his colleagues using different inhibitors (49) indicate that fertilization depends on two systems for egg activation. If one system is inhibited, the other compensates for this—at least in terms of increasing the calcium to normal levels—even in the absence of the alternate system.

These results raise important questions. First, is there an increase in cADPR at fertilization? There is some evidence that cADPR can potentiate calcium release (48), so perhaps the cADPR level is constant and there is a change in cADPR sensitization. Also, it is possible that there is even a third system for releasing calcium,

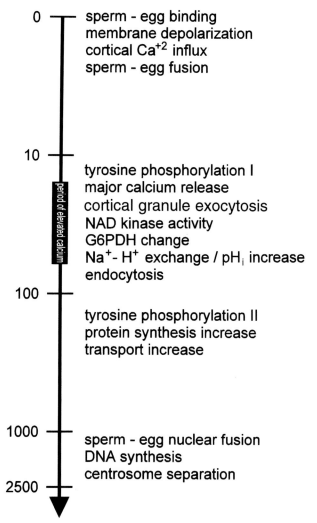

0	sperm - egg binding membrane depolarization cortical Ca^{+2} influx sperm - egg fusion
10	
	tyrosine phosphorylation I major calcium release cortical granule exocytosis NAD kinase activity G6PDH change Na$^+$- H$^+$ exchange / pH$_i$ increase endocytosis
100	
	tyrosine phosphorylation II protein synthesis increase transport increase
1000	sperm - egg nuclear fusion DNA synthesis centrosome separation
2500	

period of elevated calcium

FIG. 23.6. Experiment depicting rise in intracellular calcium following fertilization in sea urchin eggs in which various calcium release mechanisms have been prevented. In the IP$_3$ case, IP$_3$-induced calcium release was prevented by injection of egg with IP$_3$ inhibitor heparin. In cyclic ADP ribose (cADPR) case, cADPR-induced calcium release was prevented by co-injection of cADPR analog. As seen, in either case there was release of calcium following fertilization. However, if egg was injected with both inhibitors (lowest tracing), there was no calcium release in response to fertilization [From study of Lee et al. (100).]

since there is recent evidence for a calcium-releasing factor that is a metabolite of NADP (98).

If there is an increase in cADPR, what is the nature of increased cADPR production by the sperm? cADPR synthase is soluble, so perhaps the coupling is not through some membrane ligand–receptor complex, but through some other mechanism. One possibility is through the cGMP stimulation of the cADPR synthase, with the cGMP brought in by the sperm [but see (188)]. Another possibility is that the nitric oxide (NO) system

is involved. NO applied to eggs will activate guanyl cyclase and increase the calcium levels (99, 195a). If NO increased at fertilization, the activation of NO synthase by the sperm–egg contact would increase cGMP, which could then turn on cADPR synthase.

How Does the Sperm Trigger the Calcium Increase?

Receptor Hypothesis. Although it is clear that IP$_3$ increases at fertilization, how this increase is coupled to the fertilization process is not known. In somatic cells, increases in PIP$_2$ breakdown are often linked to receptor-mediated PIP$_2$ hydrolysis, usually via some G-protein. Such receptors could be operative in protostome eggs, such as *Urechis*, where as noted the postfertilization calcium increase ensues from entry of extracellular calcium in response to binding of a sperm protein (57). Do similar receptors exist in deuterostome eggs, where as noted the calcium release comes from internal stores?

Most of the work on sperm receptors in deuterostome eggs has been on "envelope receptors," that is, those on outer egg coats such as the zona pellucida, which are not involved in egg activation. The bindin receptor isolated from the sea urchin egg vitelline layer (VL) is an exception in that the VL is also part of the plasma membrane (47), and recent work indicates tyrosine phosphorylation of this receptor associated with sperm occupancy (1). The amino acid sequence of this molecule indicates it is a single-spanning membrane protein, and it has no homology to other known receptors (47). The recent work showing associated tyrosine phosphorylation (1) suggests it is coupled to other molecules in the membrane; further description will be awaited eagerly.

An interesting series of studies of Jaffe and her colleagues shows that phospholipase C-receptors can be coupled to egg activation. Jaffe and her colleagues have expressed mammalian hormone receptors in frog (97) and starfish oocytes (154). The receptors expressed were ones that are normally coupled to G-proteins and thence to phospholipase C and hydrolysis of polyphosphoinositides. These workers showed that adding the hormone ligand for the expressed receptor resulted in a normal calcium increase and the activation of such postfertilization events as fertilization membrane elevation. Presumably the receptors were coupled to G-proteins and phospholipase C and the occupancy by hormone then turned on this sequence of events. This observation, however, does not prove that such receptors are involved in egg activation—only that the potential for coupling exists (such as endogenous G proteins and associated phospholipases). For example, these could be there for use in later development;

when the receptors are expressed in the experimental situation, they can become operative.

Injector Hypothesis. Another possibility is that there is no receptor but that the sperm might inject an activating molecule. Whitaker and his colleagues have actively espoused such an idea, initially based on the existence of a 15–20-s "latent" period between sperm–egg fusion and the first indications of the major, propagated increase in calcium (195). They propose that during this latent period the sperm injects some activating molecule that accumulates to a critical level and then induces calcium release.

There is indeed evidence for such a molecule in mammalian sperm. A protein can be extracted from mouse sperm which causes calcium release in the mouse oocyte in a temporal pattern very similar to that following sperm-induced fertilization, including magnitude and oscillation (166–168). This factor was originally thought to release calcium directly, but might function to sensitize the egg to calcium-induced calcium release (167).

Although this is a provocative and potentially important finding, demonstration of activation by injection of a sperm factor is not proof that this is the normal route of egg arousal, as the activation could be an artifact unrelated to the fertilization process. (For example is the factor a phospholipase C involved in the acrosome reaction or some other aspect of sperm function which when injected into the egg is capable of causing calcium release?) Nevertheless, this is an intriguing observation and further work on this will help resolve these issues.

Another candidate that the sperm might inject to activate the egg are small-molecular-weight calcium-releasing agents, such as IP_3 or cGMP. Sperm have large amounts of these compounds—or a high titer of enzymes to produce cGMP or IP_3 (16, 78)—but it is unclear whether enough could be injected, and in enough time, to initiate a self-propagating response.

Another candidate is calcium itself. Following the acrosome reaction, the sperm begins to pump prodigious amounts of calcium from the seawater into the sperm cell (103). Could this calcium then be transferred into the just-fused egg and induce a type of calcium-induced calcium release which could then propagate throughout the egg (87)? Arguing against this idea, however, is that fertilization can occur in calcium-free seawater (143). However, to achieve fertilization in calcium-free seawater, the sperm acrosome reaction must first be induced in calcium-containing seawater, so perhaps there is sufficient calcium introduced during this period to effect activation (87). But another problem with this idea is that most of the calcium goes into the sperm mitochondria, which makes difficult

to imagine a direct transfer of calcium to the egg (103, 130).

Tyrosine Kinase Hypothesis. A final possibility, which could be related to a number of the above alternatives, is that the release of calcium in the egg is tied to tyrosine kinase action. This idea is suggested by recent work of Ciapa and Epel (18) which showed an early increase in tyrosine phosphorylation on several plasma membrane proteins. This phosphorylation is first seen by 15 s after insemination, which precedes the major calcium rise at 30s. Such tyrosine phosphorylations are often related to receptor-mediated events, independent of G-proteins, suggesting a sperm–egg ligand system that directly leads to calcium release. This is consistent with the recent work from Foltz's lab showing that tyrosine phosphorylation is coupled to the bindin receptor (1).

Interestingly, tyrosine phosphorylation also occurred following activation with calcium ionophores, which indicates that this phosphorylation need not be receptor-driven (18). Rather, the ionophore stimulation implies that the phosphorylation can be driven by calcium, which suggests the hypothesis that if a receptor is involved, then the receptor might be coupled to a calcium release mechanism and that the resultant calcium rise then turns on additional calcium-dependent phosphorylation with additional calcium release. If so, this could be a primary mechanism for calcium release, with the initial event being receptor-mediated tyrosine phosphorylation with resultant calcium rise, and the calcium then driving additional tyrosine phosphorylation and propagated calcium release.

WHAT IS CALCIUM DOING TO ACTIVATE THE EGG?

The Timetable of Fertilization Responses in the Sea Urchin Egg

Earlier I presented evidence of the critical nature of the calcium rise. Given that this rise is essential, I now consider what this calcium is doing and start by looking at some of the changes that occur when the calcium level is high. Referring to Fig. 23.6 on the program/timetable of changes at fertilization, we see that a number of events occur at the time the calcium level is elevated. These include cortical granule exocytosis and resultant elevation of the fertilization membrane (125), the activation of NAD kinase (35), changes in location and activity of glucose-6-phosphate dehydrogenase (G6PDH) (172), activation of a Na^+–H^+ exchanger and resultant rise in pH_i (91), and the beginning of a major reorganization of the cortex centering on actin polymerization (159). As the calcium level falls, there is also seen a bout of endocytosis and membrane recycling (44) and the insertion of new transport systems (37). This is followed by a variety of cytoplasmic changes, including fragmentation and then reassociation of the endoplasmic reticulum (84), a large increase in protein synthesis (36), and mobilization of microtubule-containing structures related to movement of the sperm pronuclei (138). Coincident with pronuclear movements is a "swelling" of the nucleus and change in type of nuclear proteins (11). As the sperm and egg nuclei fuse, DNA synthesis begins (70). The sperm centrosome separates and moves the egg nucleus to set up the future mitosis (126). The "fertilization events" are now over as the new embryo enters the first cell division.

The calcium-mediated changes, which occur during the first 2 min when calcium level is elevated, must be critical for subsequent events. The changes that occur during this critical period after fertilization, when the calcium level is elevated, are highlighted in Fig. 23.5, and below I discuss these changes.

Cortical Granule Exocytosis and Fertilization Membrane Formation.

One of the best-studied calcium-dependent events is the cortical reaction, which is an exocytotic event involving some 15,000 cortical granules embedded in the egg cortex. A complex composed of these exocytotic vesicles and plasma membrane can be isolated and the cortical exocytosis can then be induced by addition of micromolar calcium, which is similar to the in vivo level that is attained during fertilization [see, for example, (179)]. In agreement with its calcium requirements, there is a temporal coincidence *in vivo* with the propagation of the calcium wave and the induction of the cortical reaction. This calcium-initiated exocytosis has been extensively studied as a model for exocytosis [see (189) for review].

The cortical granule exocytosis leads to the formation of the fertilization membrane [for reviews, see (93)], which has roles in protecting the embryo and forming a physical "late block" to polyspermy. An adhesive protein, hyaline, is also present in the cortical granules and this is secreted onto the egg surface, and in conjunction with other proteins that are secreted later, is involved in the early adhesion of blastomeres [see (3)]. The exocytosis is followed by an extensive reorganization of the plasma membrane of the egg, but this exocytosis and its consequences are not needed for later development. Thus, one can prevent the exocytosis with high hydrostatic pressure (142), but the eggs will develop normally except for the cell adhesion problems resulting from the absence of the hyaline-layer adhesive proteins. [the embryos will still develop, since there are later bouts of secretion of adhesion proteins (3).]

Redox Changes: NAD Kinase and Glucose-6-Phosphate Dehydrogenase activity.

Another calcium-related event is the activation of NAD kinase, which converts about one-third of the cells NAD into NADP, beginning at about 45 s after fertilization (35). This is a calmodulin-dependent event (41), and in vivo occurs at the time the calcium is elevated (35). The NADP is almost immediately converted into NADPH, presumably by the action of the pentose shunt, of which the limiting enzyme is glucose-6-phosphate dehydrogenase (G6PDH). NADPH is used for the production of hydrogen peroxide via an NADPH oxidase which is also activated after fertilization [see (93)] and this peroxide is then utilized for a peroxidatic cross-linking of tyrosine residues for hardening of the fertilization membrane (45).

The oxidase is controlled by protein kinase C, which is regulated by the postfertilization increases in both calcium and diacylglycerol (67). The massive production of hydrogen peroxide could cause oxidative stress problems in the embryo, but these are mitigated by the presence in oocytes of high levels of a thiol derivative of histidine, called ovothiol, which serves to protect the embryo from free radical damage [see (147) for review].

The increase in NADPH could also have an important role in regulating metabolism. The combined action of NAD kinase and G6PDH activity results in an increase in the ratio of NADPH/NADP, which shifts from a ratio of 1:1 in the unfertilized egg to a ratio of 4:1 after fertilization (144). This means that the newly formed zygote would now have a tendency toward metabolic reduction for those reactions that involve NADPH. As will be described later, one of these is an NADPH-mediated activation of a protein synthesis initiation factor. An intriguing question is, Are there additional redox mediated events that are involved in turning on the egg at fertilization?

The activation of glucose-6-phosphate dehydrogenase, the enzyme most likely responsible for the reduction of the newly formed NADP, is related to this question. This enzyme is apparently activated in vivo at the time that the calcium level is elevated. This was initially suggested by studies showing that this enzyme is bound in the unfertilized egg and is soluble after fertilization (77). Swezey has studied this with many different approaches, including cell-free systems (172), permeabilized cells (173), and in intact eggs (174).

The most reliable assessment of in vivo activity has come from studies with caged substrates that are introduced into the eggs by electroporation (174). These studies reveal that the activity of this enzyme is severely repressed in the unfertilized egg and is turned on after fertilization. This turn-on is only 2%–3% of the potential activity, but is sufficient to account for the observed production of hydrogen peroxide at 2 min after insemination (174). By 20 min after fertilization, the activity of G6PDH has decreased, but the activity is still above the unfertilized rate. The role of the increased shunt metabolism is not known, but could include (1) ribose and deoxyribose production, as well as other reductive events possibly involved in housekeeping chores, and (2) the regulation of synthetic events that are triggered by the shift in NADPH/NADP level to a more reducing level (see below).

The nature of the regulation of G6PDH is still unclear. As noted, the enzyme is bound to yet-unidentified structures in the unfertilized egg and becomes soluble after fertilization. This change is seen in homogenates prepared in low-ionic-strength media, implying a low-affinity binding (172). The binding is associated with decreased activity, so this could be the mechanism of repression of activity in the unfertilized egg. However, there is a complete dissociation of the enzyme after fertilization—as assessed in homogenates—whereas there is only a partial derepression of activity as assessed in vivo with the caged compound approach (174). This suggests there are additional factors controlling the activity, or that the regulation cannot be assessed by the in vitro approach (174). One possible regulator is a small molecule that increases after fertilization that can cause disassociation of the enzyme in vitro (172). This factor is unidentified.

This regulation of G6PDH may be an example of a general phenomenon. In mammalian kidney cells in culture, the addition of growth factors leads to increased G6PDH activity in vivo, and this is associated with a "solubilization" of the enzyme, similar to what happens during sea urchin fertilization (176).

Intracellular pH Changes

A potentially pervasive change that begins at the time of the calcium rise is the activation of a sodium–hydrogen exchange mechanism (91). This is a classic amiloride-sensitive exchanger [indeed, work on sea urchin eggs provided the original description of this exchanger; see review (38)], and it is activated at about 1 min postfertilization, after the major calcium rise passed through the egg (but while the calcium level is still elevated). The activity of the exchanger removes sufficient protons from the egg cytoplasm so that the pH_i is elevated 0.3–0.5 pH units. The egg's buffering capacity is such that to accomplish this about 3 mM of protons is pumped out of the cell.

The Na^+–H^+ exchanger is regulated by both protein kinase C activity (169) and calcium, probably through a calcium-regulated protein kinase (150). These conclusions are based on inhibitor studies which show that

the exchange activity can be turned on by protein kinase C activators, such as phorbol esters, and this activation by the phorbol esters is prevented by protein kinase C (PK-C) inhibitors. However, the activation of the exchanger in vivo is not affected by these same protein kinase C inhibitors, but it is prevented by calmodulin antagonists (150). This suggests that although PK-C can activate the exchanger, the normal route is probably via the calcium rise acting on some protein kinase. The final resolution of this will come from work on the isolated exchanger.

Role of the pH Changes. This pH_i rise is essential for further development. If one prevents the pH_i rise, as by fertilization in regular seawater followed by rapid transfer to sodium-free seawater or by fertilization in the presence of the Na^+–H^+ exchange inhibitor, amiloride, then further development does not occur (14, 91). The early calcium-dependent events take place following activation in sodium-free media, such as the cortical reaction and the turn-on of NAD kinase, but later events such as those associated with the pronuclear migrations and entry into mitosis do not occur [see review (38)]. Similar effects are seen if the pH_i is brought back to the unfertilized pH_i, as by incubation of the eggs in weak acids such as acetate (33, 140). Although protein synthesis appears to be sensitive to pH as measured in cell-free systems (197), this pH sensitivity is not seen in vivo. In vivo studies, where the pH_i is not even allowed to rise after fertilization, reveal that holding the pH_i at the unfertilized level has little effect on protein synthesis (134).

These results indicate that the pH_i increase after fertilization in the sea urchin is critical for subsequent development, since if the pH_i does not rise, postfertilization events related to microtubule polymerization and entry into DNA synthesis do not occur. It is not known, however, why the pH_i has to rise. One possibility is that the low pH_i characteristic of the unfertilized egg limits the activity of the egg, that is, the pH_i of the unfertilized egg is a nonpermissive one for microtubule events. Another possibility is that the pH_i rise is acting as a trigger, to initiate some uniquely pH-sensitive event.

Both mechanisms might hold, as suggested by experiments in which eggs are fertilized and then the intracellular pH is dropped back to the unfertilized level by incubating the eggs in weak acids at a low external pH. When this is done, it is seen that the early phases—from fertilization to the fusion of the two pronuclei—are particularly pH-sensitive, and development arrests when the pH_i is clamped to the unfertilized level (140). This, however, contrasts to the situation when the pH_i is lowered after pronuclear fusion. Here, development will proceed, albeit very slowly and with a reduced-size mitotic apparatus, dysfunctional mitotic progression, and an absence of cytokinesis (33). These results indicate that the pH$_i$ rise is particularly important for initiating early events, especially those related to microtubule function, and then necessary later for proper tempo and coordination of the mitotic events.

What might be so pH-sensitive during the early period after fertilization? Relating to microtubule function, one such event could be a pH-sensitive release or pH-dependent change in microtubule protein, as suggested by Suprenant (162). One possibility raised by Suprenant's work is that there is a store of tubulin that is unavailable for polymerization, and that this is released as a consequence of the pH$_i$ rise.

Another intriguing pH-sensitive target is a pH-sensitive tyrosine kinase. The activity of this kinase is evidenced by a phosphorylation of a 350 kDa protein that begins about 3 min after insemination (89). This timing suggests that it is not an immediate consequence of the calcium rise. It most likely arises from the pH$_i$ increase, since the phosphorylation can be induced by treatments which raise pH$_i$ (such as incubation in weak bases) and is prevented by treatments which prevent the pH$_i$ rise (such as activation in sodium-free seawater).

Additional evidence comes from studies on the properties of the tyrosine kinase. In vitro, activity is highly sensitive to pH and activity increases severalfold over the range of pH changes that occur in vivo (89). This pH sensitivity and the time course of activity in vivo indicate that this may be a major target of the pH$_i$ change. Given the pervasive consequences of tyrosine kinase activity, these findings suggest that the cascade of events linked to the pH$_i$ rise might emanate from this phosphorylation activity.

The nature of the substrates and action of this kinase will be important to discern [see, for example, the recent work of Jiang et al. (89)]. Moore and Kinsey's (118) work with tyrosine kinase inhibitors in fact suggests that the early activity of these kinases may not be critical and that their action comes later in events leading to pronuclear fusion. Ultimately, the pH increase (and tyrosine kinase activity?) must relate to some microtubule-mediated event, since the events associated with pronuclear fusion depend on these structures.

pH Changes in Other Eggs. pH$_i$ increases have also been seen in the oocytes of *Urechis* (58) and the frog (187). Although similar measurements have not been made on mollusk eggs, many of these eggs release large amounts of acid after fertilization, suggesting that a sodium–hydrogen exchange is also occurring here [for example, (76, 127)]. However, similar pH$_i$ changes have not been seen in starfish and mammalian eggs (5, 90), suggesting that pH$_i$ changes are not a universal concomitant of the egg activation at fertilization.

It is puzzling that pH changes are seen sporadically. Perhaps different eggs use different suites of triggering mechanisms. Or perhaps this is related to degree of metabolic repression in these different eggs—related to how long the eggs are stored in the ovary. Thus, sea urchins can hold their eggs for several months, and a low pH$_i$ might be part of the mechanism to hold these eggs in a metabolically repressed state.

ENTRY INTO THE CELL CYCLE

Synthesis of Cyclin as a Prerequisite for Mitosis

One of the best-described changes that follows fertilization of the sea urchin egg is a large increase in protein synthesis (36). This increase results from increased translation from preexisting mRNA stored in the oocyte [reviewed (28)]. This translational control is also seen in other eggs as a consequence of fertilization, and has been intensively studied as a model for translational control (155).

In the context of entry into the cell cycle, this protein synthesis increase is relevant since one of the new proteins made after fertilization, cyclin B, is necessary for initiating mitosis through its control of the cell division protein, cdc2 [see, for example, (136)]. When present in sufficient titer, this protein binds to inactive cdc2 and this binding results in phosphorylation and activation of this cell division protein, thus causing the cell to enter mitosis [for example, (129)].

How is protein synthesis increased, and how is the specific synthesis of cyclin controlled? As noted, the two major ionic changes of fertilization are the transient increase in calcium ion and the permanent change in pH$_i$. The role of these ionic changes in regulating protein synthesis has been approached by experiments in which the ionic changes are induced—or prevented—and the consequences are then noted.

Early experiments indicated that both ionic changes were required (198). Emphasis has especially focused on the pH$_i$ change, since a significant stimulation of protein synthesis could be induced simply by placing unfertilized oocytes into weak bases, such as ammonia, which raises the pH$_i$ (42, 151). Incubation in ammonia also initiated cycles of DNA synthesis along with bouts of nuclear envelope breakdown, chromosome condensation, and formation of monasters [(111); division did not occur since the sperm centrosome necessary for mitosis is not present]. Importantly, a calcium rise does not occur (161) and the normal consequences of the calcium rise—cortical exocytosis and NAD kinase activation—did not take place (41, 42).

Although there was no question about the consequences of ammonia activation, problems soon arose concerning the idea that the pH_i rise induced by ammonia was the critical consequence of ammonia addition. For example, one could raise pH_i directly by raising the outside pH; yet there was not a marked stimulation of protein synthesis nor was there entry into the cell cycle (32). More importantly, recent experiments of Rees et al. show that one can clamp the pH_i of the egg so it does not rise at fertilization; nevertheless, protein synthesis increases to a large extent in response to fertilization, indicating that the increase can occur in the absence of the normal pH_i rise (134).

What then accounts for the increase in protein synthesis, how is calcium related to this, and how does the ammonia stimulation occur? Studies on the protein synthesis machinery indicate that several factors are limiting in the unfertilized egg. One factor is mRNA, which is bound to ribonuclear protein complexes and released from these complexes quickly after fertilization (62, 96). Studies with protein kinase C activators and inhibitors suggest that this kinase activity is involved with the "unmasking" of the mRNA (60).

A second factor is that studies on protein synthesis initiation factors show that the limiting factor in the unfertilized egg is the activity of the initiation factor IEF2B, also known as GEF or the guanine nucleotide exchange factor (2, 30, 69). This factor is somehow activated after fertilization, and the link to the earlier changes seems to be the activation of GEF by some reaction linked to NADPH (2, 69). This factor can be activated by NADPH either allosterically by this cofactor (30), or through an enzymatic reduction mediated by NADPH and thioredoxin (2, 75). If this NADPH-linked change is the only one necessary to increase protein synthesis, then the link to calcium is via the calcium activation of NAD kinase and the activation of G6PDH, which together account for the large increase in NADPH after fertilization. Interestingly, this could also account for the stimulation of protein synthesis by ammonia; ammonia by some not-understood mechanism also increases the NADPH/NADP ratio in the absence of the NAD kinase reaction (192).

The changes in mRNA location and the stimulation of initiation factors do not explain the specific turn-on of cyclin mRNA. This specific turn-on may be related to a brief bout of polyadenylation of cytoplasmic mRNA that occurs after fertilization (196), which could also account for part of the increased turn-on of protein synthesis (149, 155). This, perhaps coupled with specific aspects of mRNA structure and their interactions with initiation factors, could result in the turn-on of specific protein synthesis. Whatever the mechanism, the specific synthesis of cyclins, combined with a cascade of other events, such as the increased availability of tubulins and the cascade of other phosphorylations resulting in active cell division proteins, could account for the turn-on of the egg and its entry into the cell cycle.

Activation of Protein Kinase Activity As a Prerequisite for Mitosis

Another postfertilization change that may also be critical for initiating the cell cycle could be related to the pH_i changes. As noted above, the pH_i change results in activation of a number of protein kinase activities. Protein kinase C is presumably activated very early after sperm–egg contact since it is involved in promoting the pH_i increase and in turning on the NADPH oxidase (67, 169), but it might have other roles in initiating later changes related to entry into the cell cycle (7).

Tyrosine kinase activity changes might be particularly critical. I noted above the early changes in tyrosine phosphorylation that precede the calcium rise (18) and the later pH-sensitive tyrosine protein kinase activity whose action might be similarly important in promoting new cell activity (89). Other phosphorylations on tyrosine have recently been described (1, 89), and these may also have critical roles. Especially intriguing is the activity of tyrosine kinases related to the phosphorylation of cell cycle protein kinases, such as phosphorylation of the cdc2 family [see, for example, (191)], and also changes in MAP kinases (153).

STRUCTURAL CHANGES AS AN IMPORTANT CONCOMITANT OF FERTILIZATION

Besides the above biochemical events, there also occur important structural changes that have critical consequences for subsequent development. Some of these changes are related to changing the egg from a cell set up to receive the sperm and block polyspermy to a cell that will turn into an embryo. Some of these postfertilization changes set up the subsequent cell division. Others result in structural changes essential for setting up the axis of the future embryo as well as further development. Below I discuss some of these alterations and their portent for fertilization and development.

Microfilaments

One of the earliest changes is in cortical and cytoskeletal actin. For example, the unfertilized sea urchin egg has abundant actin in the cortex, but most of this is not filamentous [see review (159)]. By 15 s after sperm–egg binding, there occurs a localized actin filament formation involved in incorporation of the fertilizing sperm [see (85)]. This occurs at the site of sperm–egg fusion where the newly polymerized actin attaches to and surrounds the sperm and is involved in pulling the sperm into the egg. As predicted, incubation of eggs in actin-disrupting drugs, such as cytochalasin, prevents this sperm incorporation (10). The remaining actin in the cortex also enters a more filamentous state, and here the actin filaments form circumferentially around the cortex, which gives the cortex added rigidity (159). There occurs considerable elongation of the microvilli of the egg, which again is related to this actin polymerization (145).

The factors responsible for this abrupt and dramatic change in actin state are not well understood. From studies in which the various postfertilization changes are manipulated, it is known that both the calcium and pH changes are involved, but whether these act directly on the actin, on actin-associated proteins, or through their effects on enzymes is still not clear (13, 159).

Microtubules

Although the egg has abundant tubulin, little of this is present as microtubules. Radical changes in microtubules occur within minutes of fertilization. Part of this is related to the introduction of a functional centrosome into the egg (see below), and part is related to the above-mentioned elevation of the internal pH (140). The result is a localized microtubule formation around the sperm centrosome and the resultant microtubule-mediated motility is then involved in moving the sperm and egg pronuclei so that they meet and fuse as part of the fertilization process [see review (141)]. Introduction of the centrosome may not by itself lead to microtubule formation; Suprenant (162) has shown that alkaline pH favors microtubule assembly in homogenates of eggs and this relates in part to increased solubility of tubulin. There may be other pH-sensitive components involved in increased tubulin availability (140), such as the aforementioned action of the pH-sensitive tyrosine kinase (89).

Finally, the centrosome is involved in setting up the future mitotic apparatus (126). An earlier consequence of this centrosome activity is the establishment of a filamentous array of microtubules that form a cytoskeletal array of microtubules throughout the zygote [see review (141)].

Endoplasmic Reticulum

The advent of the confocal microscope, along with the development of vital dyes that allow the staining of structures in vivo and observation of the living egg, has revolutionized the way we study cells and the mode in which we look at cells. As an example, Terasaki and Jaffe (175) stained the endoplasmic reticulum (ER) with one of these vital dyes and found that the ER undergoes dynamic changes as a consequence of fertilization. There is considerable fragmentation and movement of the ER and then a return to the original structure in a few minutes. Later, during mitosis, the ER changes form again in association with the mitotic apparatus. The functional aspects of these changes are not known.

Introduction of the Centrosome as a Critical Event

One of the most important and, until recently, most unappreciated aspects of egg activation is the introduction of a functional centrosome by the sperm. As first described by Boveri at the turn of the 20th century, the oocyte loses its functional centrosome at the completion of the meiotic divisions [see (112)]. The absence of a normal centrosome means that the egg cannot enter mitosis and the egg is essentially nonfunctional in terms of mitotic activity. For example, one can "activate" eggs by increasing their calcium level, and many of the sequelae of normal fertilization follow, such as DNA synthesis (160). Yet, these cells will form at best a monopolar spindle, but they will not form a normal bipolar mitotic apparatus and enter division (112).

In almost all cases (the exception is the mouse, where the oocyte centrosome remains functional) fertilization reintroduces the centrosome into the egg cytoplasm by providing a functional centrosome from the sperm. Upon entry into the egg cytoplasm, the sperm centrosome forms microtubules around it, which are necessary for the movement of the sperm and egg pronucleus in the cytoplasm as well as to provide the information for the mitotic apparatus later in the cell cycle [see review (139)].

The centrosome's role in organization of microtubules and hence organization of the cell (95) implies that the egg and zygote have radically different organization. The egg loses its functional centrosome at the end of the maturation divisions, and is essentially free of a centrosome until the sperm brings in its own centrosome and restores this structure to the egg. One might therefore look at the egg as a "tabula rasa," a blank slate in terms of cell organization, into which the sperm introduces the organizing factor, the centrosome. This reorganization is also referred to as *ooplasmic segregation,* and is described below.

Ooplasmic Segregation

Another concomitant of fertilization is a reorganization of the egg cytoplasm that has relevance to later developmental events. These were first described in the early 1900s in frog and tunicate eggs as structural changes that can be visualized in the living eggs as changes in the color or localization of pigment in the fertilized zygote. For example, in the frog there are changes in pigment pattern that result in formation of a gray crescent (52); and in tunicate eggs similar changes result in formation of an orange crescent (88).

Today, of course, we know that these changes result from alterations in the cytoskeleton, and it is clear now that they ensue from alterations in both microtubules and microfilaments [see (52, 88)]. This reorganization of the cytoplasm is critical for normal development, and parts of the changes are related to the activity of microtubules, which in turn is coordinated to the activity of the centrosome, brought into the egg with the sperm as a part of the fertilization process (139). In eggs that exhibit this ooplasmic segregation, one might look on the egg structure as a blank slate in which the placement and orientation of the fertilizing sperm (and its centrosome) dictate the organization of the microtubules and the future structure of the egg and specification of parts of the future embryo [see (139, 158)].

A part of this reorganization is also related to the emplacement of specific mRNAs in the cytoplasm (165), and this can be related to specific attachment of ribosomes and associated mRNA. The reorganization of the embryo after fertilization could be related to the movement of these mRNAs with resultant developmental implications [see (65) for role of microtubules in this localization]. This type of localization appears to be on a continuum in terms of timing. In some oocytes this localization occurs during oogenesis, and in others the final emplacement is dictated by the fertilizing sperm [see (28)].

EPILOGUE—IS FERTILIZATION A UNIQUE EVENT IN THE LIFE HISTORY OF THE ORGANISM?

As is clear from this review, fertilization is a singular process in the life history of the organism, insuring that sperm and egg meet and that this interaction transforms the egg from a nondividing cell that is dormant into a dividing cell that is synthetically active and can develop into a new individual. A major question is whether the principles surrounding fertilization are unique to gametes or whether the cell biology is similar or modified from the cell biological principles that are used in the daily activities of cells. As seen in the above review, some of the cell biology is indeed similar, such as the use of calcium as a signaling molecule and the utilization of similar responding pathways. Thus, signaling in gametes or somatic cells can depend on calcium-level increases via the phosphoinositide pathway, and cells can respond to the calcium increase through binding to common calcium-sensing molecules, such as calmodulin.

But there are differences. In somatic cells, the signals leading to calcium release usually center around receptor-mediated signaling, such as through G-proteins. But the situation may be different in gametes. In the mammalian egg, for example, the calcium-signaling mechanism might emanate from a sperm protein that is injected into the egg as part of the fertilization process. So this aspect of fertilization biology—the injection of a calcium-modulating protein—may indeed be unique. It is of interest that this mirrors the cell–cell fusion of sperm to egg, which is another singular aspect of the fertilization process.

To be sure, receptors are involved in the fertilization process, but the receptors so far described also reflect other unique aspects of the fertilization process. The bulk of the described receptors are involved in initial recognition at the outer egg coats, and these are probably not part of the signal transduction process related to egg activation. As we get closer to the egg surface, we see some interesting variations that are in part related to evolutionary origin. Thus, there are receptors coupled to calcium influx in protostome eggs (the *Urechis* situation), and this may be the case for eggs that depend on a calcium source from outside the cell. But there are also receptors for sperm-binding proteins that are membrane spanning (such as the sea urchin bindin receptor), which are phosphorylated at fertilization. Is this involved in egg activation? Or is this involved in the block to polyspermy and/or the sperm incorporation mechanism?

Above I have concentrated on the egg, but there is also the activation of the sperm, involving motility changes often associated with chemotaxis and resultant arrival at the egg surface. This arrival is associated with specific recognition processes, which result in the induction of the acrosome reaction, adhesion to the outer egg coat, and thence passage through these coats to arrive at the egg surface. These changes also have unique aspects, such as the use of stoichiometric mechanisms to destabilize egg coats as seen in some gastropod eggs, as opposed to enzymatic hydrolysis mechanisms to pass through these outer egg coats. Finally, it is remarkable that the sperm carries out all this complex behavior in a cell that contains relatively little cytoplasm; the entire process seems to be programmed into the plasma membrane receiving system of the sperm during spermiogenesis.

Fertilization research has in the past focused—in a highly successful manner—on the description of the process and its regulation. A significant part of this research has not only illuminated the cell biology of fertilization but has also provided unexpected insights into general aspects of cell biology (for example, the early implication of calcium in cell regulation, the role of intracellular pH changes, the discovery of cyclic ADP ribose). Now we are moving from the general aspects to specifics, which are also providing important insights, such as the description of receptors involved in the initial sperm–egg recognition and also the receptors (or injected factors) involved in the initial signaling aspects of fertilization.

Unanswered areas that current work indicates will be profitable and exciting range from general—applicable to many areas of biology—to specific areas—focused more narrowly on fertilization but still of general import. Examples of the general include such cell and developmental questions as the linkage between fertilization and the cell cycle and the linkage between fertilization and the specification of the embryonic axis. These could all be related, and might well emanate from the calcium signals produced at fertilization to the introduction of the centrosome at fertilization.

Finally there are questions related to fertilization at the ecological and the population level. How does fertilization vary in relation to the ecology of the adult organism? Is the biology of fertilization different when the gametes are released into a quiet marine tide pool as opposed to the turbulent surf? or in freshwater as opposed to saltwater? And how does the biology of the fertilization process relate to the isolation of species and the generation of new species? These are all questions that are just beginning to be looked at, and interesting answers will soon be forthcoming.

REFERENCES

1. Abassi, Y. A., and K. R. Foltz. Tyrosine phosphorylation of the egg receptor for sperm at fertilization. *Dev. Biol.* 164: 430–443, 1994.

2. Akkaraju, G. R., L. J. Hansen, and R. Jagus. Increase in eukaryotic initiation factor 2B activity following fertilization reflects changes in redox potential. *J. Biol. Chem.* 266: 24451–24459, 1991.

3. Alliegro, M. C., and D. R. McClay. Storage and mobilization of extracellular matrix proteins during sea urchin development. *Dev. Biol.* 125: 208–216, 1988.

4. Baba, T., S. Azuma, S. Kashiwabara, and Y. Toyoda. Sperm from mice carrying a targeted mutation of the acrosin gene can penetrate the oocyte zona pellucida and effect fertilization. *J. Biol. Chem.* 269: 31845–31849, 1994.

5. Baltz, J. M., C. Lechene, and J. D. Biggers. A novel proton permeability dominating intracellular pH in the early mouse embryo. *Development* 118: 1353–1361, 1993.

6. Barros, C., C. Capote, C. Perez, J. A. Crosby, M. I. Becker, and A. De Ioannes. Immunodetection of acrosin during the acrosome reaction of hamster guinea-pig and human spermatozoa. *Biol. Res.* 25: 31–40, 1992.

7. Bement, WM. Signal transduction by calcium and protein kinase activation. *J Exp. Zool.* 263: 382–397, 1992.

8. Blackmore, P. F. Rapid non-genomic actions of progesterone stimulate Ca-2 + influx and the acrosome reaction in human sperm. *Cell. Signal.* 5: 531–538, 1993.

9. Brownlee, C. Tansley Review No. 70. Signal transduction during fertilization in algae and vascular plants. *New Phytol.* 127: 399–423, 1994.

10. Byrd, W., and G. Perry. Cytochalasin B blocks sperm incorporation but allows activation of the sea urchin egg. *Exp. Cell Res.* 126: 333–342, 1980.

11. Cameron, L. A. and D. L. Poccia. In vitro development of the sea urchin male pronucleus. *Dev. Biol.* 162: 568–578, 1994.

12. Cameron, R. A., J. E. Minor, D. Nishioka, R. J. Britten, and E. H. Davidson. Locale and level of bindin mRNA in maturing testis of the sea urchin, Strongylocentrotus purpuratus. *Dev. Biol.* 142: 44–49, 1990.

13. Carron, C. P., and F. J. Longo. Relation of cytoplasmic alkalinization to microvillar elongation and microfilament formation in the sea urchin egg. *Dev. Biol.* 89: 128–137, 1982.

14. Chambers, E. L. Na is essential for activation of the inseminated sea urchin egg. *J. Exp. Zool.* 197: 149–154, 1976.

15. Chambers, E. L., B. C. Pressman, and R. E. Rose. The activation of sea urchin eggs by the divalent ionophores A23187 and X-537A. *Biochem. Biophys. Res. Commun.* 60: 126–132, 1974.

16. Chinkers, M., and D. L. Garbers. Signal transduction by guanylyl cyclases. *Annu. Rev. Biochem.* 60: 553–576, 1991.

17. Ciapa, B., B. Borg, and M. Whitaker. Polyphosphoinositide metabolism during the fertilization wave in sea urchin eggs. *Development* 115: 187–195, 1992.

18. Ciapa, B., and D. Epel. A rapid change in phosphorylation on tyrosine accompanies fertilization of sea urchin eggs. *FEBS Lett.* 295: 167–170, 1991.

19. Clapper, D. L., and D. Epel. Sperm motility in the horseshoe crab. III. Isolation and characterization of a sperm motility initiating peptide. *Gamete Res.* 6: 315–326, 1982.

20. Clapper D. L., T. F. Walseth, P. J. Dargie, and H. C. Lee. Pyridine nucleotide metabolites stimulate calcium release from sea urchin egg microsomes desensitized to inositol trisphosphate. *J. Biol. Chem.* 262: 9561–9568, 1987.

21. Collins, F. D. A reevaluation of the fertilizin hypothesis of sperm agglutination and the description of a novel form of sperm adhesion. *Dev. Biol.* 49: 381–394, 1976.

22. Collins, F. D., and D. Epel. The role of calcium ions in the acrosome reaction of sea urchin sperm: regulation of exocytosis. *Exp. Cell Res.* 106: 211–222, 1977.

23. Cook, S. P., and D. F. Babcock. Selective modulation by cGMP of the K+ channel activated by speract. *J. Biol. Chem.* 268: 22402–22407, 1993.

24. Cook, S. P., C. J. Brokaw, C. H. Muller, and D. F. Babcock.

Sperm chemotaxis: egg peptides control cytosolic calcium to regulate flagellar responses. *Dev. Biol.* 165: 10–19, 1994.

25. Cuthbertson, K. S., D. G. Whittingham, and P. H. Cobbold. Free Ca2 + increases in exponential phases during mouse oocyte activation. *Nature* 294: 754–757, 1981.

26. Dangott, L. J., and D. L. Garbers. Identification and partial characterization of the receptor for speract. *J. Biol. Chem.* 259: 13712–13716, 1984.

27. Darszon, A., P. Labarca, C. Beltran, J. Garcia-Soto, and A. Lievano. Sea urchin sperm: an ion channel reconstitution study case. *Methods (Orlando)* 6: 37–50, 1994.

28. Davidson, E. H. *Gene activity in early development* (3rd ed.). Orlando: Academic Press, 1986.

29. Denny, M. W., and M. F. Shibata. Consequences of surf-zone turbulence for settlement and external fertilization. *Am. Nat.* 134: 859–889, 1989.

30. Dholakia, J. N., and J. A. Wahba. Phosphorylation of the guanine nucleotide exchange factor from rabbit reticulocytes regulates its activity in polypeptide chain initiation. *Proc. Natl. Acad. Sci. U.S.A.* 85: 51–54, 1988.

31. Dholakia, J. N., Xu, M. B. Hille, and A. J. Wahba. Purification and characterization of sea urchin initiation factor 2: the requirement of guanine nucleotide exchange factor for the release of eukaryotic polypeptide chain initiation factor 2-bound GDP. *J. Biol. Chem.* 265: 19319–19323, 1990.

32. Dube, F., and D. Epel. The relation between intracellular pH and rate of protein synthesis in sea urchin eggs and the existence of a pH-independent event triggered by ammonia. *Exp. Cell Res.* 162: 191–204, 1986.

33. Dube, F., T. Schmidt, C. H. Johnson, and D. Epel. The hierarchy of requirements for an elevated intracellular pH during early development of sea urchin embryos. *Cell* 40: 657–666, 1985.

34. Eisenbach, M., and D. Ralt. Precontact mammalian sperm-egg communication and role in fertilization. *Am. J. Physiol.* 262 (*Cell Physiol.* 31): C1095–C1101, 1992.

35. Epel, D. A primary metabolic change of fertilization: interconversion of pyridine nucleotides. *Biochem. Biophys. Res. Commun.* 17: 62–68, 1964.

36. Epel, D. Protein synthesis in sea urchin eggs: a 'late' response to fertilization. *Proc. Natl. Acad. Sci. U.S.A.* 57: 899–906, 1967.

37. Epel, D. Activation of an Na$^+$-dependent amino acid transport system upon fertilization of sea urchin eggs. *Exp. Cell Res.* 72: 74–89, 1972.

38. Epel, D. Na$^+$–H$^+$ exchange and fertilization. In: *Na$^+$–H$^+$ Exchange,* edited by S. Grinstein. Boca Raton, FL: CRC Press, 1988, p. 209–226.

39. Epel, D. The initiation of development at fertilization. *Cell Diff. Dev.* 29: 1–13, 1990.

40. Epel, D., N. S. Cross, and N. Epel. Flagellar motility is not involved in the incorporation of the sperm into the egg at fertilization. *Dev. Growth Diff.* 19: 15–21, 1977.

41. Epel, D., C. Patton, R. W. Wallace, and W. Y. Cheung. Calmodulin activates NAD kinase of sea urchin eggs: an early event of fertilization. *Cell* 23: 543–549, 1981.

42. Epel, D., R. Steinhardt, T. Humphreys, and D. Mazia. An analysis of the partial metabolic derepression of sea urchin eggs by ammonia. The existence of independent pathways. *Dev. Biol.* 40: 245–255, 1974.

43. Fields, S. Pheromone response in yeast. *Trends Biochem. Sci.* 15: 270–273, 1990.

44. Fisher, G. W., and L. I. Rebhun. Sea urchin egg cortical granule exocytosis is followed by a burst of membrane retrieval via uptake into coated vesicles. *Dev. Biol.* 99: 456–472, 1983.

45. Foerder, C. A., and B. M. Shapiro. Release of ovoperoxidase

from sea urchin eggs hardens the fertilization membrane with tyrosine crosslinks. *Proc. Natl. Acad. Sci. U.S.A.* 75: 3183–3187, 1977.

46. Foltz, K. R. The sea urchin egg receptor for sperm. *Semin. Dev. Biol.* 5: 243–253, 1994.

47. Foltz, K. R., J. S. Partin and W. J. Lennarz. Sea urchin egg receptor for sperm: sequence similarity of binding domain and hsp70. *Science* 259: 1421–1425, 1993.

48. Galione, A., H. C. Lee, and W. B. Busa. Calcium-induced calcium release in sea urchin egg homogenates: modulation by cyclic ADP-ribose. *Science* 253: 1143–1146, 1991.

49. Galione, A., A. Mcdougall, W. B. Busa, N. Willmott, I. Gillot, and M. Whitaker. Redundant mechanisms of calcium-induced calcium release underlying calcium waves during fertilization of sea urchin eggs. *Science* 261: 348–352, 1993.

50. Galione, A., A. White, N. Willmott, M. Turner, B.V.L. Potter, and S. P. Watson, cGMP mobilizes intracellular calcium. *Nature* 365: 456–459, 1993.

51. Garbers, D. L. Molecular basis of fertilization. *Annu. Rev. Biochem.* 58: 719–742, 1989.

52. Gerhart, J., and R. Keller. Region-specific cell activities in amphibian gastrulation. *Annu. Rev. Cell Biol.* 2: 201–292, 1986.

53. Gilkey, J. C., L. F. Jaffe, E. B. Ridgeway, and G. T. Reynolds. A free calcium wave traverses the activating egg of the Medaka, Oryzias latipes. *J. Cell Biol.* 76: 448–466, 1978.

54. Gillot, I., and M. Whitaker. Imaging calcium waves in eggs and embryos. *J. Exp. Biol.* 184: 213–219, 1993.

55. Goodenough, U. W., E V. Armbrust, A. M. Campbell, P. J. Ferris, and R. L. Jones. Molecular genetics of sexuality in Chlamydomonas. *Annu. Rev. Plant Physiol. Plant Mol. Biol.* 46: 21–44, 1995.

56. Gould, M., and J. L. Stephano. Electrical responses of eggs to acrosomal protein similar to those induced by sperm. *Science* 235: 1654–1656, 1987.

57. Gould, M. C., and J. L. Stephano. Peptides from sperm acrosomal protein that initiate egg development. *Dev. Biol.* 146: 509–518, 1991.

58. Gould, M. C., and J. L. Stephano. Nuclear and cytoplasmic pH increase at fertilization in Urechis caupo. *Dev. Biol.* 159: 608–617, 1993.

59. Gould-Somero, M., L. A. Jaffe, and L. Z. Holland. Electrically-mediated fast polyspermy block in eggs of the marine worm *Urechis caupo. J. Cell. Biol.* 82: 426–440, 1979.

60. Grainger, J. L. The unmasking of maternal mRNA during oocyte maturation and fertilization. *Genet. Eng. NY* 16: 229–239, 1994.

61. Grainger, J. L., V. M. Trumpbour, and A. C. Corbett. Mechanism of activation of sea urchin egg messenger rnps. *J. Cell Biol.* III (2): 369A, 1990.

62. Grainger, J. L., and M. M. Winkler. Fertilization triggers unmasking of maternal messenger RNAs in sea urchin eggs. *Mol. Cell. Biol.* 7: 3947–3954, 1987.

63. Haino-Fukushima, K. Studies on the egg-membrane lysin of Tegula pfeifferi. The reaction mechanism of the egg-membrane lysin. *Biochim. Biophys. Acta.* 352: 179–191, 1974.

64. Hamaguchi, Y., and Y. Hiramoto. Activation of sea urchin eggs by microinjection of calcium buffers. *Exp. Cell Res.* 134: 171–179, 1981.

65. Hamill, D., J. Davis, J. Drawbridge, and K. A. Suprenant. Polyribosome targeting to microtubules: enrichment of specific mRNAs in a reconstituted microtubule preparation from sea urchin embryos. *J. Cell Biol.* 127: 973–984, 1994.

66. Heilbrunn, L. V. *An Outline of General Physiology.* Philadelphia: Saunders, 1937.

67. Heinecke, J. W., K. E. Meier, J. A. Lorenzen, and B. M. Shapiro. A specific requirement for protein kinase C in activation of the respiratory burst oxidase of fertilization. *J. Biol. Chem.* 265: 7717–7720, 1990.

68. Henson, J. H., D. A. Begg, S. M. Beaulieu, D. J. Fishkind, E. M. Bonder, M. Teraski, and M. Lebeche. A calsequestrin-like protein in the endoplasmic reticulum of the sea urchin: localization and dynamics in the egg and first cell cycle embryo. *J. Cell Biol.* 109: 149–162, 1989.

69. Hille, M. B., J. N. Dholakia, A. Wahba, E. Fanning, L. Stimler, Z. Xu, and Z. Yablonka-Reuveni. In-vivo and in-vitro evidence supporting co-regulation of translation in sea-urchin eggs by polypeptide initiation factors, pH optimization, and mRNAs. *J. Reprod. Fertil. Suppl.* 42: 235–248, 1990.

70. Hinegardner, R. T., B. Rao, and D. E. Feldman. The DNA synthetic period during early development of the sea urchin egg. *Exp. Cell Res.* 36: 53–59, 1964.

71. Hofmann, A., and C. Glabe. Bindin, a multifunctional sperm ligand and the evolution of new species. *Semin. Dev. Biol.* 5: 233–242, 1994.

72. Hoshi, M. Lysins. In: *Biology of Fertilization,* edited by C. B. Metz, and A. Monroy. San Diego: Academic Press, 1985, vol. 2, p. 431–462.

73. Hoshi, M., T. Nishigaki, A. Ushiyama, T. Okinaga, K. Chiba, and M. Matsumoto. Egg-jelly signal molecules for triggering the acrosome reaction in starfish spermatozoa. *Int. J. Dev. Biol.* 38: 167–174, 1994.

74. Hoshi, M., T. Okinaga, K. Kontani, T. Araki, and K. Chiba. Acrosome reaction-inducing glycoconjugates in the jelly coat of starfish eggs. In: *Comparative Spermatology: 20 Years After,* edited by B. Baccetti. New York: Raven Press, 1991, p. 175–183.

75. Hunt T., P. Herbert E. A. Campbell, C. Delidakis, and R. J. Jackson. The use of affinity chromatography on 2′5′ ADP-sepharose reveals a requirement for NADPH, thioredoxin and thioredoxin reductase for the maintenance of high protein synthesis activity in rabbit reticulocyte lysates. *Eur. J. Biochem.* 131: 303–311, 1983.

76. Ii, I., and L. I. Rebhun. Acid release following activation of surf clam (Spisula solidissima) eggs. *Dev. Biol.* 72: 195–200, 1979.

77. Isono, N. Carbohydrate metabolism in sea urchin eggs. *J. Fac. Sci. Univ. Tokyo* 10: 37–53, 1963.

78. Iwasa, K. H., G. Ehrenstein, L. J. DeFelice, and J. T. Russell. High concentration of inositol 1,4,5-triphosphate in sea urchin sperm. *Biochem. Biophys. Res. Commun.* 172: 932–938, 1990.

79. Jaffe, L. A. Fast block to polyspermy in sea urchin eggs is electrically mediated. *Nature* 261: 68–71, 1976.

80. Jaffe, L. A., N. L. Cross, and B. Picheral. Studies of the voltage-dependent polyspermy block using cross-species fertilization of amphibians. *Dev. Biol.* 98: 319–326, 1983.

81. Jaffe, L. A., and M. Gould. Polyspermy preventing mechanisms. In: *Biology of Fertilization,* edited by C. B. Metz, and A. Monroy. San Diego: Academic Press, 1985, vol. 3, p. 223–251.

82. Jaffe, L. A., M. Gould-Somero, and L. Holland. Studies of the mechanisms of the electrical poslyspermy block using cross-species fertilization. *J. Cell Biol.* 92: 616–621, 1982.

83. Jaffe, L. A., and K. R. Robinson. Membrane potential of the unfertilized sea urchin egg. *Dev. Biol.* 62: 215–282, 1978.

84. Jaffe, L. A., M. Terasaki. Structural changes of the endoplasmic reticulum of sea urchin eggs during fertilization. *Dev. Biol.* 156: 566–573, 1993.

85. Jaffe, L. A., and L. G. Tilney. Actin, microvilli, and the fertilization cone of sea urchin eggs. *J. Cell Biol.* 87: 771–782, 1980.

86. Jaffe, L. F. The role of calcium explosions, waves and pulses in activating eggs. In *Biology of Fertilization,* edited by C. B. Metz and A. Monroy. San Diego: Academic Press, 1985, vol. 3, p. 127–165.

87. Jaffe, L. F. The roles of intermembrane calcium in polarizing and activating eggs. In: *Mechanisms of Fertilization: Plants to Humans,* edited by B. Dale. Berlin: Springer-Verlag, 1990, p. 389–418.

88. Jeffery, W. R., and B. J. Swalla. The myoplasm of ascidian eggs: a localized cytoskeletal domain with multiple roles in embryonic development. *Semin. Cell Biol.* 1: 373–381, 1990.

89. Jiang, W. P., P. A. Veno, R. W. Wood, G. Peaucellier, and W. H. Kinsey. pH regulation of an egg cortex tyrosine kinase. *Dev. Biol.* 146: 81–88, 1991.

90. Johnson, C. H., and D. Epel. Starfish oocyte maturation and fertilization: intracellular pH is not involved in activation. *Dev. Biol.* 92: 461–469, 1982.

91. Johnson, J. D., D. Epel, and M. Paul. $Na^+–H^+$ exchange is required for activation of sea urchin eggs after fertilization. *Nature* 262: 661–664, 1976.

92. Johnston, R. N., and M. Paul. Calcium influx following fertilization of Urechis caupo eggs. *Dev. Biol.* 57: 364–374, 1977.

93. Kay, E. S., and B. M. Shapiro. The formation of the fertilization membrane of the sea urchin egg. In *Biology of Fertilization,* edited by C. B. Metz and A. Monroy. San Diego: Academic Press, 1985, vol. 3, p. 45–81.

94. Keller, S. H., and V. D. Vacquier. The isolation of acrosome-reaction-inducing glycoproteins from sea urchin egg jelly. *Dev. Biol.* 162: 304–312, 1994.

95. Kellogg, D. R., M. Moritz, and B. M. Alberts. The centrosome and cellular organization. *Annu. Rev. Biochem.* 63: 639–674, 1994.

96. Kelso-Winemiller, L. C., and M. M. Winkler. Unmasking of stored maternal messenger RNAs and the activation of protein synthesis at fertilization in sea urchins. *Development* 111: 623–634, 1991.

97. Kline, D., L. Simoncini, G. Mandel, R A. Maue, R. T. Kado, and L. A. Jaffe. Fertilization events induced by neurotransmitters after injection of messenger RNA in Xenopus eggs. *Science* 241: 464–467, 1988.

98. Lee, H. C., and R. Aarhus. A derivative of NADP mobilizes calcium stores insensitive to inositol trisphosphate and cyclic ADP-ribose. *J. Biol. Chem.* 270: 2152–2157, 1995.

99. Lee, H. C., R. Aarhus, and R. M. Graeff. Sensitization of calcium-induced calcium release by cyclic ADP-ribose and calmodulin. *J. Biol. Chem.* 270: 9060–9066, 1995.

100. Lee, H. C., R. Aarhus, and T. F. Walseth. Calcium mobilization by dual receptors during fertilization of sea urchin eggs. *Science* 261: 352–355, 1993.

101. Lee, H. C., and D. L. Garbers. Modulation of the voltage-sensitive Na +/H + exchange in sea urchin. *J. Biol. Chem.* 261: 16026–16032, 1986.

102. Lee, H. C., A. Galione, and T. F. Walseth. Cyclic ADP-ribose: metabolism and calcium mobilizing function. *Vitam. Horm.* 48: 199–257, 1994.

103. Lee, H. C., C. Johnson, and D. Epel. Changes in internal pH associated with initiation of motility and acrosome reaction of sea urchin sperm. *Dev. Biol.* 95: 31–45, 1983.

104. Lee, Y.-H., T. Ota, and V. D. Vacquier. Positive selection is a general phenomenon in the evolution of abalone sperm lysin. *Mol. Biol. Evol.* 12: 231–238, 1995.

105. Lennarz, W. J. Fertilization in sea urchins: how many different molecules are involved in gamete interaction and fusion? *Zygote* 2: 1–4, 1994.

106. Longo, F. J., S. Cook, D. H. McCulloch, P. I. Ivonnet, and E. L. Chambers. Stage leading to and following fusion of sperm and egg plasma membranes. *Zygote* 2: 317–331, 1994.

107. Lopez, A., S. J. Miraglia, and C. G. Glabe. Structure/function analysis of the sea urchin sperm adhesive protein bindin. *Dev. Biol.* 156: 24–33, 1993.

108. Lynn, J. W., D. H. Mcculloh, and E. L. Chambers. Voltage clamp studies of fertilization in sea urchin eggs: II. Current patterns in relation to sperm entry, nonentry, and activation. *Dev. Biol.* 128: 305–323, 1988.

109. Maruta, H. Chemotaxis during the development of cellular slime molds. In: *Biology of Fertilization,* edited by C. B. Metz, and A. Monroy. San Diego: Academic Press, 1985, vol. 2, p. 255–276.

110. Mazia, D. The release of calcium in Arbacia eggs upon fertilization. *J. Cell Comp. Physiol.* 10: 291–304, 1935.

111. Mazia, D. Chromosome cycles turned on in unfertilized sea urchin eggs exposed to NH4OH. *Proc. Natl. Acad. Sci. U.S.A.* 71: 690–693, 1974.

112. Mazia, D. The chromosome cycle and the centrosome cycle in the mitotic cycle. *Int. Rev. Cytol.* 100: 49–92, 1987.

113. Mcculloh, D., and E. L. Chambers. Fusion of membranes during fertilization: increases of the sea urchin egg's membrane capacitance and membrane conductance at the site of contact with the sperm. *J. Gen. Physiol.* 99: 137–175, 1992.

114. Mcpherson, S. M., P. S. Mcpherson, L. Mathews, K. P. Campbell, and F. J. Longo. Cortical localization of a calcium release channel in sea urchin eggs. *J. Cell Biol.* 116: 1111–1121, 1992.

115. Mead, K. S., and D. Epel. Beakers versus breakers: how fertilization in the laboratory differs from fertilization in nature. *Zygote* 3: 95–99, 1995.

116. Miller, R. L. Sperm chemo-orientation in the metazoa. In: *Biology of Fertilization,* edited by C. B. Metz, and A. Monroy. San Diego: Academic Press, 1985, vol. 2, p. 276–340.

117. Minor, J. E., D. R. Fromson, R. J. Britten, and E. H. Davidson. Comparison of the bindin proteins of Strongylocentrotus franciscanus, Strongylocentrotus purpuratus, and Lytechinus variegatus: sequences involved in the species specificity of fertilization. *Mol. Biol. Evol.* 8: 781–795, 1991.

118. Moore, K. L., and W. H. Kinsey. Identification of an abl-related protein tyrosine kinase in the cortex of the sea urchin egg: possible role at fertilization. *Dev. Biol.* 164: 444–455, 1994.

119. Nielsen, O., and J. Davey. Pheromone communication in the fission yeast Schizosaccharomyces pombe. *Semin. Cell Biol.* 6: 95–104, 1995.

120. Nishioka, D., R. D. Ward, D. Poccia, C. Kostacos, and J. E. Minor. Localization of bindin expression during sea urchin spermatogenesis. *Mol. Reprod. Dev.* 27: 181–190, 1990.

121. Oberdorf, J., C. Vilar, and D. Epel. The localization of PI and PIP kinase in the sea urchin egg and their modulation following fertilization. *Dev. Biol.* 131: 236–243, 1989.

122. Oberdorf, J. A., J. F. Head, and B. Kaminer. Calcium uptake and release by isolated cortices and microsomes from the unfertilized egg of the sea urchin Strongylocentrotus droebachiensis. *J. Cell Biol.* 102: 2205–2210, 1986.

123. Oberdorf, J. A., D. Lebeche, J. F. Head, and B. Kaminer. Identification of a calsequestrin-like protein from sea urchin eggs. *J. Biol. Chem.* 263: 6806–6809, 1988.

124. Ohlendieck, K., J. S. Partin, R. L. Stears, and W. J. Lennarz. Developmental expression of the sea urchin egg receptor for sperm. *Dev. Biol.* 165: 53–62, 1994.

125. Paul, M., and D. Epel. Fertilization-associated light scattering changes in eggs of the sea urchin, Strongylocentrotus purpuratus. *Exp. Cell Res.* 65: 281–288, 1971.

126. Paweletz, N., D. Mazia, and E. -M. Finze. Fine structural studies of the bipolarization of the mitotic apparatus in the fertilized sea urchin egg: I. The structure and behavior of centrosomes before fusion of the pronuclei. *Eur. J. Cell Biol.* 44: 195–204, 1987.

127. Peaucellier, G. Acid release at meiotic maturation of oocytes in the polychaete annelid Sabellaria alveolata. *Experientia* 34: 789–90, 1978.

128. Pillai, M. C., T. S. Shields, R. Yanagimachi, and G. N. Cherr. Isolation and partial characterization of the sperm motility initiation factor from eggs of the Pacific herring, Clupea pallasi. *J. Exp. Zool.* 265: 336–342, 1993.

129. Pines, J. Cyclins and cyclin-dependent kinases: a biochemical view. *Biochem. J.* 308: 697–711, 1995.

130. Poenie, M., and D. Epel. Ultrastructural localization of intracellular calcium stores by a new cytochemical method. *J. Hist. Cytochem.* 35: 939–956, 1987.

131. Preuss, D. Being fruitful: genetics of reproduction in Arabidopsis. *Trends Genet.* 11: 147–153, 1995.

132. Rakow, T. L., and S. S. Shen. Multiple stores of calcium are released in the sea urchin egg during fertilization. *Proc. Natl. Acad. Sci. U.S.A.* 87: 9285–9289, 1990.

133. Ralt, D., M. Goldenberg, P. Fetterolf, D. Thompson, J. Dor, S. Mashiach, D. L. Garbers. and M. Eisenbach. Sperm attraction to a follicular factor(s) correlates with human egg fertilizability. *Proc. Natl. Acad. Sci. U.S.A.* 88: 2840–2844, 1991.

134. Rees, B. B., C. Patton, J. L. Grainger, and D. Epel. Protein synthesis increases after fertilization of sea urchin eggs in the absence of an increase in intracellular pH. *Dev. Biol.* 169: 683–698, 1995.

135. Ridgway, E. B., J. C. Gilkey, and L. F. Jafe. Free calcium increases explosively in activating medaka eggs. *Proc. Natl. Acad. Sci. U.S.A.* 74: 623–627, 1977.

136. Ruderman, J., F. Luca, E. Shibuya, K. Gavin, T. Boulton, and M. Cobb. Control of the cell cycle in early embryos. *Cold Spring Harb. Symp. Quant. Biol.* vol. Lvi. The Cell Cycle, p. 495–502, 1991.

137. Schackmann, R. W., R. Christen, and B. M. Shapiro. Membrane potential depolarization and increased intracellular pH accompany the acrosome reaction of sea urchin sperm. *Proc. Natl. Acad. Sci. U.S.A.* 78: 6066–6070, 1981.

138. Schatten G. Motility during fertilization. *Int. Rev. Cytol.* 79: 59–163, 1982.

139. Schatten, G. The centrosome and its mode of inheritance: the reduction of the centrosome during gametogenesis and its restoration during fertilization. *Dev. Biol.* 165: 299–335, 1994.

140. Schatten, G., T. Bestor, R. Balczon, J. Henson, and H. Schatten. Intracellular pH shift leads to microtubule assembly and microtubule-mediated motility during sea urchin fertilization: correlations between elevated intracellular pH and microtubule activity and depressed intracellular pH and microtubule disassembly. *Eur. J. Cell Biol.* 36: 116–127, 1985.

141. Schatten, G., and H. Schatten. Fertilization: motility, the cytoskeleton and the nuclear architecture. *Oxf. Rev. Reprod. Biol.* 9: 322–378, 1987.

142. Schmidt, T., and D. Epel. High hydrostatic pressure and the dissection of fertilization responses I. The relationship between cortical granule exocytosis and proton efflux during fertilization of the sea urchin egg. *Exp. Cell Res.* 146: 235–248, 1983.

143. Schmidt, R., C. Patton, and D. Epel. Is there a role for the Ca2+ influx during fertilization of the sea urchin egg? *Dev. Biol.* 90: 284–290, 1982.

144. Schomer, B. 1995 (unpublished results).

145. Schroeder, T. E. Surface area change at fertilization: resorption of the mosaic membrane. *Dev. Biol.* 70: 306–326, 1979.

146. Shakes, D. C., and S. Ward. Mutations that disrupt the morphogenesis and localization of a sperm-specific organelle in Caenorhabditis elegans. *Dev. Biol.* 134: 307–316, 1989.

147. Shapiro, B. M. The control of oxidant stress at fertilization. *Science* 252: 533–536, 1991.

148. Shaw, A., D. E. McRee, V. D. Vacquier, and C. D. Stout. The crystal structure of lysin, a fertilization protein. *Science* 262: 1864–1867, 1993.

149. Sheets, M. D., C. A. Fox, T. Hunt, G. Vande Woude, and M. Wickens. The 3'-untranslated regions of c-mos and cyclin mRNAs stimulate translation by regulating cytoplasmic polyadenylation. *Genes Dev.* 8: 926–938, 1994.

150. Shen, S. S. Na+-H+ antiport during fertilization of the sea urchin egg is blocked by W-7 but is insensitive to K252a and H-7. *Biochem. Biophys. Res. Commun* 161: 1100–1108, 1989.

151. Shen, S. S., and R. A. Steinhardt. Direct measurement of intracellular pH during metabolic derepression of the sea urchin egg. *Nature* 272: 253–254, 1978.

152. Shen, S. S., and R. A. Steinhardt. Time and voltage windows for reversing the electrical block to fertilization. *Proc. Natl. Acad. Sci. U.S.A.* 81: 1436–1439, 1984.

153. Shibuya, E. K., A. J. Polverino, E. Chang, M. Wigler, and J. V. Ruderman. V. Oncogenic Ras triggers the activation of 42-kDa mitogen-activated protein kinase in extracts of quiescent Xenopus oocytes. *Proc. Natl. Acad. Sci. U.S.A.* 89: 9831–9835, 1992.

154. Shilling, F., G. Mandel, and L. A. Jaffe. Activation by serotonin of starfish eggs expressing the rat serotonin 1c receptor. *Cell Regul.* 1: 465–470, 1990.

155. Simon, R., J. P. Tassan, and J. D. Richter. Translational control by poly(A) elongation during Xenopus development: differential repression and enhancement by a novel cytoplasmic polyadenylation element. *Genes Dev.* 6: 2580–2591, 1992.

156. Singh, S., D. G. Lowe, D. S. Thorpe, H. Rodriguez, W.-J. Kuang, L. J. Dangott, M. Chinkers, D. V. Goeddel, and D. L. Garbers. Membrane guanylate cyclase is a cell-surface receptor with homology to protein kinases. *Nature* 334: 708–712, 1988.

157. Speksnijder, J., D. W. Corson, C. Sardet, and L. F. Jaffe. Free calcium pulses following fertilization in the ascidian egg. *Dev. Biol.* 135: 182–190, 1989.

158. Speksnijder, J., C. Sardet, and L. F. Jaffe. The activation wave of calcium in the ascidian egg and its role in ooplasmic segregation. *J. Cell Biol.* 110: 1589–1598, 1990.

159. Spudich, A. Actin organization in the sea urchin egg cortex. In: *Current Topics in Developmental Biology, Cytoskeleton in Development,* edited by E. L. Bearer. San Diego, CA: Academic Press, 1992, vol. 26, p. 9–21.

160. Steinhardt, R. A., and D. Epel. Activation of sea urchin eggs by a calcium ionophore. *Proc. Natl. Acad. Sci. U.S.A.* 71: 1915–1919, 1974.

161. Steinhardt, R., R. Zucker, and G. Schatten. Intracellular calcium release at fertilization in the sea urchin egg. *Dev. Biol.* 58: 185–196, 1977.

162. Suprenant, K. A. Alkaline pH favors microtubule self-assembly in surf clam, Spisula solidissima, oocyte extracts. *Exp. Cell Res.* 184: 167–180, 1989.

163. Suzuki, N. Structure and function of sea urchin egg jelly molecules. *Zool. Sci.* (Tokyo) 7: 355–370, 1990.

164. Suzuki, N. Structure, function and biosynthesis of sperm-activating peptides and fucose sulfate glycoconjugate in the extracellular coat of sea urchin eggs. *Zool. Sci.* (Tokyo) 12: 13–27, 1995.

165. Swalla, B, and W. R. Jeffery. A maternal RNA localized in the yellow crescent is segregated to the larval muscle cells during ascidian development. *Dev. Biol.* 170: 353–364, 1995.

166. Swann, K. How does a spermatozoon activate an oocyte?: The soluble sperm oscillogen hypothesis. *Zygote* 1: 273–276, 1993.

167. Swann, K. Ca-2+ oscillations and sensitization of Ca-2+ release in unfertilized mouse eggs injected with a sperm factor. *Cell Calcium* 15: 331–339, 1994.

168. Swann, K., and J.-P. Ozil. Dynamics of the calcium signal that triggers mammalian egg activation. *Int. Rev. Cytol.* 152: 183–222, 1994.

169. Swann, K., and M. Whitaker. Stimulation of the Na/H exchanger of sea urchin eggs by phorbol ester. *Nature* 314: 274–277, 1985.

170. Swann, K., and M. Whitaker. The part played by inositol trisphosphate and calcium in the propagation of the fertilization wave in sea urchin eggs. *J. Cell Biol.* 103: 2333–2342, 1986.

171. Swanson, W., and V. D. Vacquier. Extraordinary divergence and positive Darwinian selection in a fusogenic protein coating the acrosomal process of abalone spermatozoa. *Proc. Natl. Acad. Sci. U.S.A.* 92: 4957–4961, 1995.

172. Swezey, R. R., and D. Epel. Regulation of glucose-6-phosphate dehydrogenase activity in sea urchin eggs by reversible association with cell structural elements. *J. Cell. Biol.* 103: 1509–1515, 1986.

173. Swezey, R. R., and D. Epel. Enzyme stimulation upon fertilization is revealed in electrically permeabilized sea urchin eggs. *Proc. Natl. Acad. Sci. U.S.A.* 85: 812–816, 1988.

174. Swezey, R. R., and D. Epel. The in vivo rate of glucose-6-phosphate dehydrogenase activity in sea urchin eggs determined with a photolabile substrate. *Dev. Biol.* 169: 733–744, 1995.

175. Terasaki, M., and L. A. Jaffe. Organization of the sea urchin egg reorganization at fertilization. *J. Cell Biol.* 114: 929–940, 1991.

176. Tian, W.-N., J. N. Pignatare, and R. C. Stanton. Signal transduction proteins that associate with the platelet-derived growth factor (PDGF) receptor mediate the PDGF-induced release of glucose-6-phosphate dehydrogenase from permeabilized cells. *J. Biol. Chem.* 269: 14798–14805, 1994.

177. Tilney, L. The acrosomal reaction. In: *Biology of Fertilization,* edited by C. B. Metz, and A. Monroy. San Diego: Academic Press, 1985, vol. 2, p. 157–215.

178. Tyler, A., and B. S. Tyler. Physiology of fertilization and early development. In: *Physiology of Echinodermata* edited by R. Boolootian. New York: Interscience 1966, p. 683–741.

179. Vacquier, V. D. Dynamic changes of the egg cortex. *Dev. Biol.* 84: 1–26, 1981.

180. Vacquier, V. D., K. R. Carner, and C. D. Stout, Species-specific sequences of abalone lysin, the sperm protein that creates a hole in the egg envelope. *Proc. Natl. Acad. Sci. U.S.A.* 87: 5792–5796, 1990.

181. Vacquier, V. D., and Y.-H. Lee. Abalone sperm lysin: unusual mode of evolution of a gamete recognition protein. *Zygote* 1: 181–196, 1993.

182. Vacquier, V. D., and J. E. Payne. Methods for quantitating sea

urchin sperm-egg binding. *Exp. Cell Res.* 82: 227–235, 1973.

183. Vacquier, V. D., W. J. Swanson, and M. E. Hellberg. What have we learned about sea urchin sperm binding? *Dev. Growth Diff.* 37: 1–10, 1995.

184. Ward, C. R., and G. S. Kopf. Molecular events mediating sperm activation. *Dev. Biol.* 158: 9–34, 1993.

185. Ward, G. E., C. J. Brokaw, D. L. Garbers, and V. D. Vacquier. Chemotaxis of Arbacia punctulata spermatozoa to resact, a peptide from the egg jelly layer. *J. Cell. Biol.* 101: 2324–2329, 1985.

186. Ward, G. E., G. W. Moy, and V. D. Vacquier. Dephosphorylation of sea urchin sperm guanylate cyclase during fertilization. *Adv. Exp. Med. Biol.* 207: 359–382, 1986.

187. Webb, D. J., and R. Nuccitelli. Direct measurement of intracellular pH changes in Xenopus eggs at fertilization and cleavage. *J. Cell Biol.* 91: 562–567, 1981.

188. Whalley, T., A. McDougall, I. Crossley, K. Swann, and M. Whitaker. Internal calcium release and activation of sea urchin eggs by cGMP are independent of the phosphoinositide signaling pathway. *Mol. Biol. Cell* 3: 373–383, 1992.

189. Whitaker, M. Exocytosis in sea urchin eggs. *Ann. NY Acad. Sci.* 710: 248–253, 1994.

190. Whitaker, M., and M. Aitchison. Calcium-dependent polyphosphoinositide hydrolysis is associated with exocytosis in vitro. FEBS. *Lett.* 182: 119–124, 1985.

191. Whitaker, M., and R. Patel. Calcium and cell cycle control. *Development* 108: 525–542, 1990.

192. Whitaker, M. J., and R. A. Steinhardt. The relation between the increase in reduced nicotinamide nucleotides and the initiation of DNA synthesis in sea urchin eggs. *Cell* 25: 95–103, 1981.

193. Whitaker, M. J., and R. A. Steinhardt. Ionic signalling in the sea urchin egg at fertilization. In: *Biology of Fertilization*, edited by C. B. Metz, and A. Monroy. San Diego: Academic Press, 1985, vol. 3, p. 168–222.

194. Whitaker, M., and K. Swann. Lighting the fuse at fertilization. *Development*. 117: 1–12, 1993.

195. Whitaker, M., K. Swann, and I. Crossley. What happens during the latent period at fertilization? In: *Mechanisms of Egg Activation*, edited by R. Nuccitelli, G. N. Cherr, and W. H. Clark, Jr. New York: Plenum Press, 1987, p. 157–172.

195a. Willmott, N., J. K. Sethi, T. F. Walseth, H. C. Lee, A. M. White, and A. Galione. Nitric oxide-induced mobilization of intracellular calcium via the cyclic ADP-ribose signaling pathway. *J. Biol. Chem.* 271: 3699–3705, 1996.

196. Wilt, F. H. Polyadenylation of maternal RNA of sea urchin eggs after fertilization. *Proc. Natl. Acad. Sci. U.S.A.* 70: 2345–2349, 1973.

197. Winkler, M. M., and R. A. Steinhardt. Activation of protein synthesis in a sea urchin cell free system. *Dev. Biol.* 84: 432–439, 1981.

198. Winkler, M. M., R. A. Steinhardt, J. L. Grainger, and L. Minning. Dual ionic controls for the activation of protein synthesis at fertilization. *Nature* 287: 558–560, 1980.

199. Wistrom, C. A., and S. Meizel. Evidence suggesting involvement of a unique human sperm steroid receptor-Cl− channel complex in the progesterone-initiated acrosome reaction. *Dev. Biol.* 159: 679–690, 1993.

200. Yamaguchi, M., M. Kurita, and N. Suzuki. Induction of the acrosome reaction of Hemicentrotus pulcherrimus spermatozoa by the egg jelly molecules, fucose-rich glycoconjugate and sperm-activating peptide I. *Dev. Growth Diff.* 31: 233–240, 1989.

201. Yanagimachi, R. Mammalian fertilization. In: *The Physiology Of Reproduction*, (2nd ed.), edited by E. Knobil, and J. D. Neil. 1994, New York: Raven, vols. 1 and 2, Xxv + 1878 p. (vol. 1); Xxv + 1372 p. (vol. 2).

202. Yoshino, K-I., and N. Suzuki. Two classes of receptor specific for sperm-activating peptide III in sand-dollar spermatozoa. *Eur. J. Biochem.* 206: 887–893, 1992.

203. Zucker, R. S., and R. A. Steinhardt. Prevention of the cortical reaction in fertilized sea urchin eggs by injection of calcium-chelating ligands. *Biochim. Biophys. Acta* 541: 459–466, 1978.

24. Cellular mechanisms during mammalian fertilization

PAUL M. WASSARMAN | *Department of Cell Biology and Anatomy, Mount Sinai School of Medicine, New York, New York*

HARVEY M. FLORMAN | *Department of Anatomy and Cellular Biology, Tufts University School of Medicine, Boston, Massachusetts*

CHAPTER CONTENTS

FERTILIZATION is a ubiquitous process that occurs in plants and animals alike and is essential for maintenance of species. Strictly speaking, the term "fertilization" refers specifically to one event; the fusion of a single sperm and an unfertilized egg to form the zygote or one-cell embryo. However, a much broader definition of fertilization includes other events that precede and follow sperm–egg fusion. The latter definition has been adopted for this chapter.

Fertilization is a subject of considerable inherent interest. This is reflected by an enormous literature that dates back to the turn of the century. For obvious reasons, much of this literature concerns nonmammals and, in particular, marine organisms. Research on fertilization in mammals is a more recent undertaking. For example, it was not until 1968 that in vitro fertilization of mouse eggs was accomplished successfully. (In vitro fertilization of rabbit eggs was reported in 1959, hamster eggs in 1964, and human eggs in 1970.) Despite this, substantial progress has been made in understanding cellular mechanisms that govern mammalian fertilization. To a large degree, the progress is attributable to the fact that such mechanisms have a great deal in common with those that govern the behavior of somatic cells. This chapter focuses on certain of these mechanisms and emphasizes research carried out primarily with mice.

This chapter is not encyclopedic in its coverage of either fertilization, in general, or mammalian fertilization, in particular. Instead it focuses on relatively recent advances on just a few topics in mammalian fertilization research; topics that are particularly relevant for a text on cellular physiology. These topics include the final preparation of gametes for fertilization (that is, capacitation of sperm and meiotic maturation of oocytes), binding of capacitated sperm to ovulated eggs, the acrosome reaction, fusion of sperm and egg, and the cortical reaction. On the other hand, many comprehensive monographs and reviews are available and the reader is encouraged to draw on these, both for an historical perspective of the fertilization field and an introduction to other interesting areas of mammalian fertilization research (21, 22, 26, 148, 149, 178, 256, 267, 294, 319, 391, 392, 469, 504, 523, 548, 549, 565, 638, 657, 660, 715, 717).

GENERAL INTRODUCTION

As indicated above, fertilization is a complex process that includes many events in which eggs and/or sperm participate. Each of these events must take place successfully in order to produce a zygote that, in turn, is able to produce a new individual who exhibits characteristics of the species. Failure of any one of these events to take place successfully can result, at the individual level, in the inability to reproduce and, at the organism level, in the disappearance of the species. Of course, artificial interference with certain of these events could provide a means of fertility control.

Among the critical events that constitute the process of fertilization in mammals are the capacitation of sperm and meiotic maturation of oocytes; species-specific binding of free-swimming sperm to the ovulated egg extracellular coat, or zona pellucida (ZP); the acrosome reaction (AR), a form of exocytosis that sperm must complete in order to be able to penetrate the ZP and reach the egg plasma membrane; fusion of sperm and egg to form a single cell, the zygote; and the cortical reaction (CR), a form of exocytosis that

fertilized eggs must complete in order to be able to establish a block to polyspermy. It must be emphasized that these are only some of the critical events that lead to formation of a viable mammalian zygote. However, each of these events has been the subject of considerable research in recent years and is discussed here.

OOGENESIS AND SPERMATOGENESIS

Oogenesis: Primordial Germ Cells to Eggs

In mammals, oogenesis begins relatively early in fetal development and ends, months to years later, in the sexually mature adult (39, 40, 89–91, 126, 222, 253, 256, 294, 319, 469, 483, 523, 570, 694, 715–717). Oogenesis begins with primordial germ cell (PGC) formation and encompasses a series of cellular transformations, from PGCs to oogonia (fetus), from oogonia to oocytes (fetus), and from oocytes to eggs (adult). This exquisitely orchestrated process results in a cell uniquely able to give rise to a new individual that expresses and maintains characteristics of the species. A description of oogenesis in mammals follows.

Origin and Behavior of Primordial Germ Cells. Eggs originate from a small number of stem cells, the PGCs, that have an extragonadal origin (97, 99, 184, 185, 187, 222, 256, 275, 294, 319, 402, 694). Formation of these cells in presomite embryos signals the beginning of oogenesis. In 8 day mouse embryos (four pairs of somites), as few as 15 and as many as 100 PGCs are recognizable because of their size, distinctive morphology, and characteristic cytochemical staining properties. These large cells (~12 μm in diameter) are found in the yolk sac endoderm and in that region of the allantois arising from the primitive streak. Several lines of evidence suggest that the embryonic rudiment of the allantois and caudal end of the primitive streak may be considered regions of primordial germ cell formation. In this context, it has been found that primitive ectoderm, taken from the caudal end of 7–7.5 day egg cylinders and cultured in vitro, differentiates into cells having properties characteristic of PGCs. Subsequently, PGCs migrate, first by passive transfer into the endodermal epithelium of the hindgut (170–350 PGCs are found in or near the hindgut epithelium in 9 day embryos) and then by ameboid movement along the dorsal mesentery of the genital ridges (present in 10–11 day embryos) found in the roof of the coelom, the site of gonad development. It appears likely that chemotactic mechanisms operate during migration of PGCs from extraembryonic sites to the presumptive gonad. As a result of continuous mitotic activity, the number of PGCs increases to approximately 5000 in 11–12 day embryos and to more than 20,000 by the time genital ridges of 13–14 day embryos are fully colonized. (Note: This represents a doubling of the number of PGCs every 18 h or so, between day 8 and day 14 of embryogenesis; it is estimated that they divide seven to eight times during the 4 day migration period.) PGCs proliferate for only 2–3 days after migrating to the genital ridges; less than 1% of the germ cells exhibit an S phase in 16–17 day embryos. These PGCs are the sole source of adult germ cells. It is noteworthy that the origin and migration of PGCs to the genital ridges is the same in males and females; gonadal sex differentiation, to either testis or ovary, occurs in the 12–13 day embryo. In summary, the germ cell line originates before or during primitive streak formation (~7 day embryo), but precisely where, when, and how are still not clear. Mouse PGCs are found sequentially in four different sites as embryogenesis proceeds: (1) First, they appear in the extraembryonic tissues of the yolk sac and allantois of 8–9 day embryos. (2) Then they appear in the hindgut epithelium of 9–10 day embryos. (3) Then they are found in the dorsal mesentery of the gut of 10–12 day embryos. (4) Finally they are found in the developing gonads of 10.5–11 day embryos. These PGCs are the sole source of adult germ cells.

Primordial Germ Cells to Oogonia. Upon reaching the surface epithelium of the gonad, PGCs move into the cortex and, together with supporting epithelial cells, give rise to the cortical sex cords (222, 253, 319, 716). The somatic components of the ovary arise from coelomic epithelium and mesenchyme of the dorsal body wall, and the mesonephros probably makes a contribution. In 13 day mouse embryos (52–60 pairs of somites), containing a differentiated ovary, migration of PGCs is complete, with virtually all of the cells converted to actively dividing oogonia in the sex cords. Oogonia exhibit a characteristic morphology (including the presence of intercellular bridges connecting adjacent germ cells) and a high frequency of mitotic division.

Oogonia to Nongrowing Oocytes. As early as day 12 of embryogenesis, a few oogonia (~5%) enter the preleptotene, and then leptotene, stage of the first meiotic prophase (39, 40, 99, 222, 253, 319, 482, 687, 716). Once meiotic prophase commences, apparently there is no endocrine requirement for continued meiotic progression. Furthermore, since germ cells, either cultured in vitro or located at ectopic sites, enter and progress through meiosis, a gonadal environment is not required for the progress of meiosis. It is during

preleptotene (interphase following the last mitotic division of oogonia) that the final DNA replication takes place in preparation for meiosis. This synthetic activity signals transformation of oogonia into oocytes. It is possible that a factor originating from the rete ovarii, or simply contact with the rete ovarii, induces oogonia to enter meiosis. In 14 day mouse embryos (61–62 pairs of somites), the germ cell population is about equally divided between oogonia and oocytes, and by day 17 (full quota of 65 pairs of somites) the ovary contains only oocytes at various stages of the first meiotic prophase (38–44, 221, 270, 580). Oocytes progress through leptotene in 3–6 h and then take 12–40 h to complete zygotene. During zygotene, homologous chromosomes pair and synapse to form what appear to be single chromosomes but are actually bivalents composed of four chromatids. In 16 day embryos nearly all oocytes are in pachytene of the first meiotic prophase, a stage that lasts about 60 h and involves genetic crossing-over and recombination. Therefore, it takes approximately 4 d to complete nuclear progression from leptotene through pachytene. By day 18 of embryogenesis, the first oocytes are seen in diplotene of the first meiotic prophase, with their chromosomes exhibiting chiasmata that result from crossing-over. By parturition, a majority of oocytes have entered late diplotene ("diffuse diplotene"), or the so-called *dictyate stage,* and by day 5 postpartum, nearly all oocytes have reached the dictyate stage, where they will remain until stimulated to resume meiosis at the time of ovulation. This pool of small (~12–15 μm in diameter), nongrowing oocytes is the sole source of unfertilized eggs in the sexually mature adult.

Nongrowing to Fully Grown Oocytes. Shortly after birth, the mouse ovary is populated with approximately 8000 nongrowing oocytes arrested in meiosis and enclosed within several squamous follicular cells (39, 40, 91, 319, 482, 716). Approximately 50% of these oocytes are lost during the first 2 wk following birth; this is attributable, in large measure, to oocytes leaving the ovary through the surface epithelium. However, during this same period, more oocytes begin to grow (~5%) than at any other period in the lifetime of the mouse. Commencement of oocyte growth is apparently regulated within the ovary, with the number of oocytes entering the growth phase being a function of the size of the pool of nongrowing oocytes. The oocyte and its surrounding follicle grow coordinately, progressing through a series of definable morphological stages (195, 351, 477–480, 482, 483). In sexually mature mice, oocytes complete growth before formation of a follicular antrum; consequently, the vast majority of follicle growth occurs after the oocyte has stopped growing. Growth is continuous, ending in either ovulation of a matured oocyte (unfertilized egg) or degeneration (atresia) of the oocyte and its follicle. Completion of oocyte growth in the mouse takes 2–3 wk, a relatively short period of time in comparison to the months or years required for completion of oocyte growth in many nonmammalian species. The oocyte grows from a diameter of about 12 μm (volume ~0.9 pl) to a final diameter of about 80 μm (volume ~270 pl), not including the ZP (discussed in the sections on structural and biochemical aspects of oocyte growth). Therefore, during its growth phase, while continually arrested in dictyate of the first meiotic prophase, the mouse oocyte undergoes a ~300-fold increase in volume and becomes one of the largest cells of the body. Each oocyte is contained within a cellular follicle (~17 μm in diameter) that grows concomitantly with the oocyte, from a single layer of a few flattened cells to three layers of cuboidal granulosa cells (~900 cells; ~125 μm–diameter follicle) by the time the oocyte has completed its growth. The theca is first distinguishable, outside of and separated by a basement membrane from the granulosa cells, when the granulosa region is two cell layers thick (~400 cells; ~100 μm–diameter follicle). During a period of several days, while the oocyte remains a constant size, the follicular cells undergo rapid division, increasing to more than 50,000 cells and resulting in a *graafian follicle* greater than 600 μm in diameter. The follicle exhibits an incipient *antrum* when it is several layers thick (~6000 cells; sl250 μm–diameter follicle) and, as the antrum expands, the oocyte takes up an acentric position surrounded by two or more layers of granulosa cells. The innermost layer of granulosa cells becomes columnar in shape and constitutes the *corona radiata;* these cells form specialized intercellular junctions, called *gap junctions,* with the oolemma.

Throughout the reproductive life of mammals, ovaries contain pools of nongrowing and growing oocytes arrested in dictyate of the first meiotic prophase. Only fully grown oocytes resume meiosis and are ovulated during each reproductive (estrous) cycle. Recruitment of oocytes into the growing pool is apparently under control of pituitary gonadotropins, although oocyte growth, without follicle development, does occur in hypophysectomized animals and during culture in vitro in the absence of hormones. The volume of a mouse oocyte increases nearly 300-fold during its 2–3 wk growth phase. This tremendous enlargement of the cell is indicative of a period of intense metabolic activity (37, 45, 281, 340, 653) which, in turn, is reflected in marked changes in oocyte ultrastructure, including the appearance (biogenesis) of some novel organelles (266, 588, 603, 653, 710, 432). For example, cortical granules and the ZP, both involved in regulating fertilization, first appear in oocytes during their growth phase (335–339, 488, 512, 516, 264, 710, 603).

Fully Grown Oocytes to Unfertilized Eggs. In sexually mature mice, fully grown oocytes in graafian follicles resume meiosis and complete the first meiotic reductive division just prior to ovulation. Resumption of meiosis can be mediated by a hormonal stimulus in vivo or simply by the release of oocytes from their ovarian follicles into a suitable culture medium in vitro. Oocytes undergo nuclear progression from dictyate of the first meiotic prophase (four times the haploid DNA complement) to metaphase II (two times the haploid DNA complement). They remain at this stage of meiosis in the oviduct, or in culture, until stimulated to complete meiosis when either fertilization or parthenogenetic activation occurs. Progression from the dictyate stage *(oocyte)* to metaphase II *(egg)* of meiosis is called *meiotic maturation.* Meiotic maturation is characterized by dissolution of the oocyte's nuclear germinal vesicle (GV) membrane, condensation of chromatin into distinct bivalents, separation of homologous chromosomes, emission of the first polar body (PB) and arrest of meiosis with chromosomes aligned on the metaphase II spindle (discussed in the section on structural aspects of oocyte maturation). These ovulated eggs complete meiosis, with separation of chromatids and emission of a second PB (second reductive division), upon fertilization.

Spermatogenesis: Primordial Germ Cells to Spermatozoa

The production of sperm begins with decisions by PGCs to form male gametes. For example, PGCs of the nematode *Caenorhabditis elegans* make sequential decisions, first with respect to meiotic/mitotic divisions, and then with respect to sperm/egg differentiation (25, 293, 332, 333). Similar controls may exist in mammals, where germ cells initiate meiosis during prenatal stages of development in female embryos (see above), but the process is repressed in males until a postnatal stage of development. In the nematode, distal tip cells in the gonad apparently secrete a factor that inhibits expression of a meiotic cell division cycle, although neither this factor nor the PGC receptor that mediates its action has been characterized. Recent studies have shown that the proliferation and long-term survival of PGCs in vitro, as well as their reprogramming into pluripotential stem cells, is stimulated by basic fibroblast growth factor (basic FGF) acting in concert with *c-kit* ligand *(Steel* factor) and leukemia inhibitory factor (LIF) (165, 252, 395, 396, 507). It is tempting to speculate about the role of these factors in the regulation of similar transitions in vivo, although the presence of these agents in the genital ridge has not been demonstrated as yet. Thus, in invertebrates and mammals, germ cell differentiation entails the concerted action of a large number of genes (150, 293, 332), but the mechanisms that commit these cells to the formation of sperm are not well understood.

After colonizing the genital ridge of male embryos, PGCs are incorporated, in association with Sertoli cells, into the seminiferous epithelium. Sertoli cells function as nurse cells, providing nourishment and protection for the germ cells. In addition, specialized tight junctions between adjacent Sertoli cells divide the seminiferous tubule into separate basal and adluminal compartments. The latter is immunologically isolated by this blood–testis barrier and constitutes a unique spermatogenic microenvironment (182). Development of the testis from indifferent gonad already has been considered extensively (48).

Mitotic Stages of Spermatogenesis. The principal cell types of the germ lineage in adult males include spermatogonia, spermatocytes (primary and secondary), spermatids, and spermatozoa (353–358). Spermatogonia are diploid, mitotically dividing cells that reside in the basal compartment of the seminiferous tubules between the basal lamina and Sertoli cells. These cells have been further classified into three subdivisions— A, intermediate (In), and B—based on nuclear and cytoplasmic architecture. In rats and mice, successive mitoses produce several distinct generations of A-spermatogonia (A_1-to-A_2-to-A_3-to-A_4). Subsequent division of A_4 cells gives rise to In-spermatogonia which then divide once to give rise to B-type cells. B-spermatogonia undergo the final mitotic division of spermatogenesis to produce pre-leptotene spermatocytes. In addition, a pool of self-regenerating spermatogonia may be present and sustain gametogenesis; this stem cell pool has not been identified (36, 182, 586).

Spermatogonia undergo incomplete cytyokinesis and so cells develop as syncytial cohorts. Progress through these stages is controlled by paracrine and endocrine pathways, which include *(1)* factors elaborated by Sertoli cells or other seminiferous tubular cells, including transforming growth factor (TGF), activin, and other bioactive agents; *(2)* Leydic cell factors, such as the products of pro-opiomelanocortin (β-endorphin, adrenocorticotrophic hormone, and melanocyte-stimulating hormone) and pro-enkephalin; and *(3)* circulating factors, such as EGF and hypophyseal gonadotropins (60, 581). In response to these cues, spermatogonia exhibit the regulated appearance of nuclear protein such as histones, the high mobility group pro-

teins (HMGs), and proto-oncogene products. Within the latter category are several transcription factors (183, 186, 586) that are likely to account for stage-specific gene expression (60, 183, 186).

Meiotic Stages of Spermatogenesis. Pre-leptotene primary spermatocytes lie in direct contact with the basal lamina. These cells have a relatively small nucleus, but otherwise are similar in appearance to their parent cells, the B-spermatogonia. DNA replication occurs during this period and produces a diploid allotment of chromosomes that consist of bivalents (4C). Following DNA synthesis, pre-leptotene spermatocytes move away from the basal lamina, migrating along Sertoli cells into the adluminal compartment. Tight junctions between adjacent Sertoli cells part to admit spermatocytes and then reclose behind them, therby maintaining the integrity of the blood–testis barrier. Cohorts of germ cells remain associated by cytoplasmic bridges throughout the meiotic stage of spermatogenesis. Thus, migration of spermatocytes toward the seminiferous lumen is spatially and temporally coordinated (182, 391, 392, 586).

The meiotic phase of spermatogenesis is characterized by chromosomal recombination and by stage-specific gene expression. Spermatogenic cells contain a number of unique isozymes or variants of somatic cell enzymes and other proteins. Expression of forms specific for male germ cell lineage has been described for glycolytic enzymes, including lactate dehydrogenase-2 and phosphoglycerate kinase-2, as well as for other enzymes (for example, β-1, 4-galactosyltransferase) and antigens (553, 561, 581).

Entry of germ cells into meiosis has been studied extensively in *Xenopus laevis,* where maturation-promoting factor (MPF) and other protein kinases, acting in concert with *c-mos* protein, initiate meiosis. Presumably, similar events occur during spermatogenesis. Spermatocytes progress through a prolonged first meiotic prophase. The component substages of prophase (that is, leptotene, zygotene, pachytene, diplotene, diakenesis) are distinguished cytochemically and include characteristic alterations in chromosomal morphology, as well as association and separation of chromosomes (182, 586). During pachytene the synaptonemal complex is fully formed and chromosomal recombination can occur. The mechanism of chromosomal homology searching and pairing is not particularly well understood, although specialized synaptonemal proteins have been described in mammals (183, 186, 289, 422). In addition, recent studies in yeast provide insight into the structure and potential function of synaptonemal proteins (283, 602).

At the end of prophase I, the nuclear envelope dissolves and chromatids align themselves at the metaphase I equatorial plate. Anaphase I and telophase I result in production of secondary spermatocytes containing a haploid number of chromosomes and a diploid DNA content. The second meiotic division then takes place without intervening DNA replication, thus establishing an haploid state.

Postmeiotic Stages of Spermatogenesis. Spermiogenesis comprises the morphological transformation of postmeiotic male germ cells, spermatids, through a series of well-characterized intermediate stages into spermatozoa. Major structural features of the sperm cell develop during this period, including accretion of the acrosomal vesicle, nuclear condensation, elongation of the flagellum, elimination of most intracellular membrane systems, and reduction in cellular volume. Following these alterations, spermatozoa are released into the lumen of the seminiferous tubule (36, 182, 186, 586).

Spermiogenesis is associated with, and probably is regulated by, stage-specific gene expression through both transcriptional and translational control mechanisms. Transcription continues in spermatids until nuclear condensation occurs. Compaction of chromatin is accompanied by sequential phases of nuclear protein displacement; histones are first replaced by transitional proteins (TP1, TP2), and these are subsequently replaced by protamines. Several transcripts, including those encoding transitional proteins, protamines, and calspermin, first appear during spermiogenesis. Translation may not occur for several days, as is the case for the transitional proteins and protamines (183, 186, 289, 422, 686).

FINAL PREPARATION OF GAMETES FOR FERTILIZATION

Eggs: "Meiotic Maturation"

Meiotic maturation specifically refers to the conversion of fully grown oocytes (present in antral follicles) into unfertilized eggs just prior to ovulation, following the preovulatory gonadotropin [follicle-stimulating hormone (FSH) and luteinizing hormone (LH)] surge. In particular, meiotic maturation involves nuclear progression, from dictyate of the first meiotic prophase (late G_2) to metaphase II (first meiotic reduction), as well as metabolic changes necessary for activation of the egg at fertilization. Unlike amphibian oocytes (389), isolated, fully grown mammalian oocytes undergo meiotic maturation spontaneously during culture in vitro in the absence of hormones (110, 493, 613),

thus providing an advantageous experimental system for structural and biochemical studies. The timing of cytological events (for example, nuclear membrane breakdown, spindle formation, etc.) observed with maturing oocytes in vitro resembles that observed in vivo. Only oocytes that have undergone successful meiotic maturation are capable of being fertilized and developing normally. Mammalian oocytes matured and fertilized in vitro exhibit normal preimplantation development in vitro (428, 637) and develop into viable fetuses following transplantation to the uteri of foster mothers (135, 434, 546).

Acquisition of Meiotic Competence. Dictyate-stage oocytes acquire the ability to undergo meiotic reduction *(meiotic competence)* during oocyte growth (579, 605, 622, 623). This applies to oocytes in vivo as well as to those grown under in vitro conditions. Only mouse oocytes larger than about 60 μm in diameter mature in vitro; smaller oocytes remain arrested at the dictyate stage. However, small meiotically incompetent oocytes are able to mature in vitro following fusion with large meiotically competent oocytes (46, 47, 232). The acquisition of meiotic competence apparently occurs in two steps; growing oocytes first acquire the ability to undergo nuclear (GV) envelope breakdown with progression to metaphase I, followed by acquisition of the ability to progress from metaphase I to metaphase II (579). Certain evidence suggests that the acquisition of meiotic competence is time dependent, but independent of the presence of follicle cells, heterologous cell contacts, and cell growth (104). In this context, it is evident that mouse oocytes should be included among those meiotic and mitotic cell types for which the transition from G_2 to M phase of the cell cycle is regulated by a cytoplasmic factor, the so-called *maturation-promoting factor (MPF)*. MPF activity appears during meiotic maturation of mouse oocytes (46, 578), reaching highest levels at metaphase of meiosis I and II and diminishing at anaphase–telophase of meiosis I, or after completion of meiosis (231).

Although a thorough analysis of the regulation of MPF activity during meiotic maturation in mammalian oocytes has not been completed, essential structural events associated with the G_2-to-M cell-cycle transition in this meiotic cell type have recently been elucidated (6). During the course of meiotic competence acquisition in the mouse, several important modifications in nuclear and cytoplasmic structure occur indicative of a G_2-to-M cell-cycle transition. These changes include the appearance of a perinuclear shell of heterochromatin, the loss of an interphase array of cytoplasmic microtubules, and the appearance of phosphorylated centrosomes (397, 672, 673). Since these changes coincide in the mouse with the formation of the follicular antrum, a process initiated by FSH, it is likely that FSH plays a role in the acquisition of meiotic competence.

Nuclear (Germinal Vesicle) Breakdown. Within a few hours of culture in vitro, fully grown mouse oocytes undergo complete nuclear breakdown (GVBD) or dissolution (103, 167, 577). This process begins with slight undulation of the nuclear envelope a few minutes after placing oocytes in culture, which continues with increasing intensity during the next 1–2 h. These undulations may be related to chromosome condensation that is initiated during this period. Nuclear pores disappear within 1 h or so of culture; breaks in the nuclear envelope are visible after about 2 h, on average. After 3 h of culture, on average, the nuclear envelope is completely dispersed into membrane doublets that apparently join the endoplasmic reticulum pool, perhaps to play a subsequent role in pronuclear and nuclear envelope formation (604). The mean time for complete breakdown of the nuclear envelope, measured from the first sign of membrane disintegration to disappearance of the nucleolus, is 11 min. Disappearance of the nucleolus occurs shortly after contacting invading cytoplasm. In contrast, the nuclear lamina, a fibrous mesh comprised of lamins A, B, and C that lines the inner nuclear membrane, persists throughout GVBD and appears to disintegrate just prior to metaphase of meiosis I (9). It should also be noted that significant changes in the number and distribution of cortical granules also occur during meiotic maturation of mouse oocytes (175, 176).

Chromosome Condensation. During meiotic maturation, oocyte chromosomes (bivalents) pass through metaphase I, anaphase I, and telophase I, arresting at metaphase II without an intervening prophase II. Diffuse dictyate-stage chromosomes (resembling "lampbrush" chromosomes) undergo significant condensation along the inner margin of the nuclear envelope, concomitant with the envelope's undulatory behavior (103, 663). During this period, chiasmata move to the ends of the chromosomes. The chromatin becomes heterochromatic and contains dense granules that increase in number with increasing chromosome condensation. Chromosome condensation occurs within the agranular vestige of the nucleus within 20 min of nucleolar dissolution. After 2–3 h of culture, on average, when chromosome condensation is nearly complete (with "lampbrush" loops withdrawn), the bivalents are V-shaped and telocentric and are often associated with fragments of nuclear envelope. Shortly thereafter, highly condensed chromosomes become circularly arranged in the center of the egg, lose their contacts with

nuclear envelope fragments, and then (after 6–9 h of culture) line up on the equator of the metaphase I spindle.

Spindle Formation. During the period of nuclear envelope breakdown (GVBD) and chromosome condensation, kinetochores with associated microtubules, as well as other microtubule-organizing centers, appear; microtubules extend from these centers, through breaks in the nuclear envelope, into the nucleoplasm (102, 103). Apparently, there is a kinetochore associated with each chromatid of each chromosomal homolog. A small, but morphologically identifiable meiotic spindle appears after about 6 h of culture and is clearly recognizable by 9 h. Unlike the situation in most cells, the poles of the oocyte spindle lack centrioles and are composed solely of bands of so-called *pericentriolar material*. Pericentriolar material becomes associated with the forming spindle poles during prometaphase and is highly phosphorylated and condensed at metaphase; it disappears from the spindle poles during telophase. In parallel with these progressive changes in the organization of the pericentriolar material at the spindle poles, cytoplasmic centrosomes, also containing this material and nucleating microtubules, undergo alterations in number and location during meiotic maturation. These structures are most numerous at prometaphase and anaphase of meiosis and are located in the oocyte cortex, except at the site of spindle attachment (409). The size of the spindle increases progressively, with a fully formed, barrel-shaped metaphase I spindle present within 9–10 h of culture (167, 577, 661). The pole-to-pole distance of the spindle is about 40 μm, and its width is about 25 μm. One of the spindle's poles is located near the cortex of the cell; the spindle is surrounded by a dense area composed of mitochondria, vacuoles, and granules. Cortical granules are excluded from the site of spindle attachment, where a thick, submembranous band of actin filaments is located. As meiosis proceeds to anaphase, bivalents move toward the opposite ends of the spindle, and the spindle rotates through 90°. These movements apparently involve microfilaments. It is noteworthy that maturing oocytes spend about 6 h in prometaphase I (period from initial chromosome condensation to line up of bivalents on the metaphase I plate) and about 4 h in metaphase I. This long period probably reflects the time necessary to synthesize and assemble the spindle apparatus.

Polar Body Emission. Anaphase and telophase figures predominate by 10–13 h of culture in vitro, and it is during this period that a bulge, destined to become the first PB, appears. During telophase the midbody (a membranous and vacuolar basophilic ring around the central region of the telophase spindle) is formed, and pinching-off of the PB is initiated (167, 577, 661). Separation of homologous chromosomes takes place, together with an asymmetric cleavage of oocyte cytoplasm containing one-half of the original chromosomal complement (late telophase). In addition to chromosomes, the first PB contains a variety of organelles, including mitochondria, ribosomes, and cortical granules, as well as the midbody (709). Although a spindle is occasionally found in the first PB, it is rarely a well-defined structure, and PB chromosomes begin to degenerate in late telophase I. An area deficient in organelles, such as cortical granules, but rich in submembranous microfilamentous actin, overlies the metaphase II spindle of unfertilized eggs. Furthermore, plasma membrane overlying the spindle is relatively smooth, whereas the remainder of the egg surface is highly microvillous. Microtubule-organizing centers, not associated with the spindle, are found in the cytoplasm near the cell's cortex. Progression beyond metaphase II, with separation of chromatids and emission of a second PB, awaits fertilization or parthenogenetic activation of the egg.

Regulatory Aspects of Oocyte Maturation

Mammalian oocytes have the unusual ability to undergo meiotic maturation spontaneously when released from follicles and cultured in vitro (197, 198, 393, 394, 492, 493, 622, 623, 665). Unlike oocytes from nonmammalian species, no hormonal stimulus is needed for mammalian oocytes to reinitiate meiotic progression and reach metaphase II (that is, to undergo the transition from oocyte to egg) in vitro. It is in this context that regulation of meiotic maturation, including prolonged arrest of oocytes at the dictyate stage and reentry of oocytes into meiotic progression just prior to ovulation, has been considered experimentally.

Inhibitors of Meiotic Maturation. A variety of agents prevent spontaneous meiotic maturation of mouse oocytes in vitro at specific stages of nuclear progression. Nuclear (GV) breakdown, the initial morphological feature characteristic of meiotic maturation, is inhibited by dibutyryl cyclic adenosine monophosphate (dbcAMP) and agents that increase intracellular levels of cyclic adenosine monophosphate (cAMP) (197, 622, 623). However, even in the presence of dbcAMP, the nuclear membrane becomes extremely convoluted, and chromosome condensation is initiated, but aborts at a stage short of compact bivalents (663). Complete chromosome condensation requires breakdown of the nuclear envelope, probably attributable to the multiple

associations of dictyate chromosomes with the nuclear envelope (103, 604, 663). In the presence of puromycin (protein synthesis inhibitor), colcemid (microtubule assembly inhibitor), or cytochalasin B (microfilament assembly inhibitor), nuclear membrane breakdown and chromosome condensation occur in an apparently normal manner (663, 665). However, nuclear progression is blocked at the circular bivalent stage when oocytes are cultured continuously in the presence of either puromycin or colcemid, whereas oocytes cultured in the presence of cytochalasin B proceed to metaphase I, form a normal spindle, and arrest (571, 663, 665). Although this suggests that nascent protein synthesis is not required for nuclear (GV) breakdown and chromosome condensation, it is possible that spontaneous resumption of meiosis in vitro is, in fact, dependent on some oocyte proteins with very high turnover rates (189). The effect of cytochalasin on meiotic maturation suggests that microfilaments are involved in events following metaphase I spindle formation (for example, separation of homologous chromosomes). Emission of a PB is inhibited by all of these drugs, suggesting that cytokinesis is blocked when any one of the earlier events of maturation fails to take place (663, 665). In this context, it has been found that activators of protein kinase C (PKC; for example, phorbol esters), but not dbcAMP, inhibit PB emission by oocytes that had already undergone nuclear (GV) breakdown in vitro (84).

Role of cAMP. Spontaneous maturation of mouse oocytes does not occur in vitro when oocytes are cultured in the presence of either membrane-permeable analogs of cAMP, such as dbcAMP and 8-bromo-cAMP, or inhibitors of cyclic nucleotide phosphodiesterase (PDE), such as isobutyl methylxanthine (IBMX) and theophylline (119, 155, 386, 550, 552, 663). Similarly, other agents that increase intracellular cAMP levels by activation of adenyl cyclase, such as forskolin, also prevent spontaneous maturation of mouse oocytes in vitro (86, 154, 464, 500, 539, 550, 636). Furthermore, microinjection of PDE into oocytes cultured in the presence of IBMX overcomes the inhibitory effect of IBMX on meiotic maturation (83). An oocyte PDE activity has been identified and characterized and has been shown to be membrane-bound and modulated by calmodulin; a calmodulin-dependent step occurs subsequent to, or concurrently with, the decrease in oocyte cAMP levels (see below) during spontaneous maturation (86). The behavior of oocytes under these conditions suggests that cAMP is involved in the maintenance of meiotic arrest at the dictyate stage, such that a fall in intracellular cAMP levels could signal reentry of oocytes into meiotic progression. In fact, a significant decrease in cAMP levels of oocytes does occur just prior to nuclear envelope (GV) breakdown, both in vitro and in vivo (156, 171, 550, 663). Therefore, just as cAMP is involved in regulating the G_2-to-M transition during the mitotic cell cycle, it appears to be involved in the reentry of oocytes into meiotic progression. It should be noted that exposure of oocytes to dbcAMP following nuclear envelope (GV) breakdown has no effect on subsequent events of meiotic maturation; PB emission with arrest at metaphase II occurs in a normal manner (663, 665).

Involvement of cAMP in meiotic maturation implies that, as in many other biological situations, it acts via cAMP-dependent protein kinase (PK), an enzyme consisting of catalytic (C) and regulatory (R) subunits (239). While the R–C complex is enzymatically inactive, binding of cAMP to the R (inhibitory) subunit results in dissociation of the complex and activation of the C subunit kinase activity. Assuming that continuous phosphorylation of an oocyte protein(s) by PK is essential for maintenance of meiotic arrest, cAMP levels could thereby regulate meiotic maturation by determining the amount of free C subunit available. Under conditions of low levels of free C subunit (low cAMP), the relevant oocyte phosphoprotein(s) would be dephosphorylated and the meiotic maturation would be initiated. In this context, mouse oocytes microinjected with the C subunit of PK fail to undergo spontaneous meiotic maturation during culture in vitro in the absence of dbcAMP and IBMX (83, 84). Furthermore, changes in the phosphorylation patterns of oocyte proteins, under a variety of experimental conditions, are consistent with this proposed regulatory scheme (83, 550). However, other evidence suggests that cAMP is only one of the components of a complex system employed in the maintenance of oocytes in the dictyate stage of the first meiotic prophase (discussed in the sections on the roles of calcium, intercellular communication, steroids, and oocyte maturation inhibitor). In particular, it is noteworthy that increasing the levels of cAMP in cumulus cells, in the presence of FSH or cholera toxin, results in inhibition of meiotic maturation without a detectable rise in oocyte cAMP levels (199, 551). This suggests the possibility that an inhibitory factor other than cAMP is transferred from cumulus cells to oocytes, resulting in inhibition of meiotic maturation in vitro (discussed in the section on the role of oocyte maturation inhibitor).

Finally, phorbol esters and diacylglycerol, known activators of calcium/phospholipid-dependent protein kinase (PK-C), also inhibit spontaneous maturation of oocytes in vitro without a decrease in oocyte cAMP levels (84). These PK-C activators also inhibit the changes in oocyte phosphoprotein metabolism associated with spontaneous maturation, suggesting that PK-C activators inhibit resumption of meiosis by acting distal to a decrease in cAMP-dependent PK activity, prior to changes in oocyte phosphoprotein metabolism that are necessary for resumption of meiosis.

Role of Calcium. Involvement of calmodulin during meiotic maturation implies that there is a role for calcium (86, 152, 374, 624). Verapamil and tetracaine, two inhibitors of transmembrane calcium transport, transiently prevent breakdown of the nuclear (GV) envelope in vitro (468). Furthermore, increasing extracellular calcium concentrations decreases the effectiveness of dbcAMP as an inhibitor of meiotic maturation, whereas verapamil and tetracaine (which decrease intracellular calcium concentration) increase its effectiveness (495). Therefore, intracellular levels of calcium and cAMP may act synergistically, through a calmodulin-dependent step, to regulate meiotic maturation of mammalian oocytes.

Role of Intercellular Communication. Gap junctions represent regions of physical continuity between cellular membranes. As such, in many biological systems they serve as mediators of intercellular communication and metabolic coupling by permitting passage of small molecules between cells; in this manner, one type of cell can influence the function of another (244, 364). Fully grown mouse oocytes are coupled with surrounding cumulus cells by gap junctions (6–8, 12, 13, 243, 244, 285, 288). They are found at areas of contact between the oocyte's plasma membrane and processes that emanate from cumulus cells and traverse the ZP. Furthermore, an extensive network of gap junctions interconnects all follicle cells with the cumulus and, consequently, with the oocyte.

Throughout oocyte growth and folliculogenesis, both in vivo and in vitro, gap junctions mediate intercellular communication between the oocyte and follicle cells. The cells are both metabolically and ionically coupled, and iontophoretically injected dye is transferred between cumulus cells and oocyte (243, 503). However, just prior to ovulation there is a significant decrease in the number of gap junctions, together with a decrease in the extent of ionic coupling, and following ovulation the oocyte and surrounding cumulus cells are no longer coupled as a result of mucification and cumulus expansion (158, 243, 286, 363, 427). In the latter state, cumulus cell processes are retracted away from the oocyte surface. This behavior suggests that spontaneous meiotic maturation takes place in vitro as a result of removing oocytes from the inhibitory influence of follicle cells. In this context, it has been demonstrated that either grafting oocytes to follicle walls or co-culture of oocytes with granulosa cells prevents spontaneous maturation in vitro.

Between 2 and 3 h following administration of an ovulatory stimulus (hCG), there is about a 15-fold decrease in the net area of cumulus cell-gap junction membrane (that is, about 90% of the total gap junction membrane per cumulus cell is lost within an hour (363). The loss of cumulus cell-gap junctions is temporally correlated with nuclear envelope (GV) breakdown and cumulus expansion (196, 363, 536). In this context, the inhibitory effects of both FSH and suboptimal concentrations of dbcAMP are lost when intercellular communication between cumulus cells and the oocyte is disrupted in vitro. However, although the coupling of cumulus cells and the oocyte terminates before ovulation, it is not completed until after nuclear envelope (GV) breakdown. Therefore, while it is tempting to attribute the onset of meiotic maturation exclusively to termination of intercellular communication between cumulus cells and the oocyte (for example, cessation of transfer of an "inhibitory factor" from cumulus cells to the oocyte via gap junctions), there is evidence to the contrary.

Role of Steroids. Although LH induces both maturation of oocytes and progesterone production in intact follicles in vitro, LH apparently does not induce maturation via progesterone synthesis (157, 159, 622, 623). Furthermore, it remains problematic whether or not other steroid hormones, such as estradiol and testosterone, play a role in regulation of meiotic maturation. Whereas a number of steroid hormones potentiate the FSH-induced inhibition of meiotic maturation of cumulus cell-enclosed oocytes in vitro [without an increase in oocyte cAMP levels; see section on nuclear (GV) breakdown], they only inhibit maturation of denuded oocytes when their cAMP levels are elevated by exposure to dbcAMP or forskolin (197, 199, 499, 551). On the other hand, at very high nonphysiological concentrations, testosterone, progesterone, pregnenolone, as well as other steroids, inhibit meiotic maturation, either alone or in conjunction with dbcAMP or agents that affect intracellular cAMP levels (200, 228, 365, 400, 426, 502, 511, 572).

Role of Oocyte Maturation Inhibitor. It has been suggested that oocyte maturation inhibitor (OMI), a product of granulosa cells, maintains oocytes in the dictyate stage of first meiotic prophase (622, 623, 625, 627, 628). This is consistent with the relatively old observation that follicular fluid inhibits oocyte maturation in vitro (106, 113, 268, 626, 627, 629). OMI is a polypeptide that has a molecular weight of about 1–2 xDa, is found in follicular fluid from ovaries of a variety of

mammals, and prevents oocytes from undergoing spontaneous meiotic maturation when cultured in vitro. However, OMI apparently exerts its inhibitory action via cumulus cells, since it prevents maturation of oocytes cultured with their cumulus cells, but does not interfere with maturation of completely denuded oocytes; in this context, it is likely that OMI is small enough to pass through gap junctions between cumulus cells and the oocyte (622, 623, 625). Preovulatory follicles have relatively low concentrations of OMI, consistent with its proposed role in regulating meiotic maturation. It is possible that the inhibitory action of OMI is potentiated by cAMP, suggesting that OMI could account for some of the effects of cAMP on cumulus-enclosed oocytes observed in vitro (85, 172, 199, 551). However, it should be noted that the existence of OMI has been seriously questioned in some quarters and, in the absence of additional biochemical support, remains a highly controversial subject (205, 375, 376, 501). Furthermore, recent evidence suggests that hypoxanthine and/or adenosine (inhibits cAMP phosphodiesterase) are, in fact, the low-molecular-weight components of follicular fluid that prevent spontaneous meiotic maturation of oocytes in vitro (170, 201).

Sperm: "Capacitation"

Early observations in echinoderms, nematodes, and anurans established that gamete maturity was essential for successful fertilization (381, 677). Yet, those studies concentrated on the ripening of the egg's cytoplasm, and it was generally found that the spermatozoa of those species did not require additional maturation after release from the male reproductive tract. In contrast, mammalian sperm are released from the male in an infertile state. While these cells are motile and are morphologically mature, they require subsequent incubation in either the female reproductive environment or under specialized conditions in vitro in order to manifest the ability to fertilize eggs. This time-dependent acquisition of fertilizing ability was reported independently in 1951 by Austin (18, 19, 24) and by Chang (108, 109) and designated "capacitation."

The model that has emerged is that capacitation entails a functional reprogramming of the sperm cell and is reflective of a cluster of physiological alteration. Passage through this pathway is regulated by instructive signals originating in the male and female reproductive tract environment; in some cases signaling molecules have been identified. In contrast, the intracellular effectors of these signals are less well defined. While capacitation is associated with alterations in sperm surface composition, motility, and metabolism, establishing a causal relationship between any such modifications and the acquisition of a fertile state has proven difficult.

Many alterations that are associated with these transitions, but that apparently are not essential for gamete interaction, will be considered only in passing. Instead, this discussion will focus on the regulation of gamete interaction by capacitation. Readers interested in the general topic of capacitation should also refer to comprehensive reviews (55, 112, 696, 697, 699). Specialized reviews have focused on methodology of capacitation in vitro (50, 51, 513) and on regulators of capacitation (207), as well as on the effects of capacitation on the sperm surface, membranes (360, 449, 463), and metabolism (226, 239).

Sperm Activity States

Capacitation is defined as the ability of sperm populations to fertilize eggs (19). The behavioral repertoire of these cells differs in many regards from sperm that are recently obtained from epididymal or seminal secretions. By extension, sperm acquired from cauda epididymal or seminal secretions and incubated in vitro or in vivo under conditions that do not support the development of a fertile state are considered "uncapacitated." Similarly, the fertile state of sperm populations that were previously incubated under capacitating conditions can be mitigated by seminal secretions; such infertile cells are termed "decapacitated" (111, 194).

This terminology is referenced solely to the fertility of sperm populations and implies certain relationships between states. Thus, the application of the term "uncapacitated" to infertile populations of sperm from cauda epididymal and from seminal fluids suggests that these two populations are commensurate. Yet, mixture of epididymal sperm with seminal secretions during mating alters the requirements for capacitation (559), indicating that these populations are in fact nonequivalent. Comparable arguments can be made regarding the term "decapacitation." Observations that capacitated sperm can be rendered infertile by application of seminal fluid constituents (111) and can be subsequently returned to a fertile state imply that capacitation is a reversible process. In this regard, there is no evidence that capacitation is fully reversible or that the decapacitated state is a counterpart of the uncapacitated states.

The term "capacitation" will be retained here for historical reasons and will be restricted to those alterations that are associated with the development of a fertile state. These alterations will be considered within the context of the physiological transitions that occur during sperm progression from the male reproductive tract to the site of fertilization. Three functional states will be considered: the *resting state* of sperm within the cauda epididymis, the *repressed state* produced by contact of sperm with seminal secretions, and the *"capacitated"* or *fertile state*. These cells are apparently incapable of new gene expression. As a result, activity states are regulated by the creation of unique microenvironments, containing specific proteins as well as a profile of inorganic ions that, acting in concert, impose and maintain these states.

The Resting State: Sperm Within the Cauda Epididymis. Sperm are produced in the seminiferous tubules of the testis and exported into the excurrent duct system. In mammals, this system consists of several ductuli efferentes that collect into the ductus epididymis and terminate in the ductus deferens (392). Sperm entering this system appear structurally mature, yet exhibit only weak, random motility in vitro and fail to fertilize eggs. As sperm transit from the rostral portions of the epididymides (the caput and corpus regions) to the distal (cauda) region, extensive modification occurs in their membrane lipid and protein. Alteration in protein composition apparently is due to the combined effects of protein dissociation from and adsorption to sperm, as well as processing of cell surface components by enzymes present in epididymal secretions. Such biochemical modifications accompany, and possibly produce, a wide range of modifications in sperm. Alterations occur in sperm metabolism (226, 699); in functional characteristics of membrane components, as determined by protein and lipid diffusion rates (440, 681, 682, 684, 685); in motility characteristics, as determined by the development of vigorous, linear motility upon dilution in vitro (52, 697, 699); and in the capacity to fertilize eggs. Sperm arrive in the cauda epididymis in a resting state, characterized by a high cell density ($>10^9$/ml), by immotility, and by infertility when challenged with eggs in vitro. During the interval of storage it is essential to maintain a state of functional inhibition, such that spontaneous events that might compromise fertility (such as the AR) are suppressed.

Some of the alterations associated with transit through the epididymides appear to be intrinsically programmed. For example, the development of vigorous linear motility by rabbit sperm is not dependent upon the unique microenvironment of given regions of the epididymides, but rather reflects cellular aging. Sperm that are retained in any region of the epididymides by restraining ligatures will attain the motility patterns of mature cells (52, 56, 465, 466).

These intrinsic programs of development function in concert with protein and low-molecular-weight elements of the epididymal microenvironment to establish and maintain the resting state. Protein constituents of epididymal secretions adsorb to the sperm surface and may regulate cellular function. For example, rabbit epididymal fluids contain a factor that suppresses spontaneous acrosomal exocytosis. ASF, or acrosome-stabilizing factor, is a 260 kDa tetramer composed of two copies each of a large (92 kDa) and small (38 kDa) subunit. Both subunits contain the high-mannose and complex forms of asparagine-linked (N-linked) oligosaccharides (617, 679). The glycoprotein is synthesized by the principle cells of the corpus epididymides and secreted into the epididymal lumen. ASF reversibly inhibits spontaneous exocytosis when added to capacitated sperm in vitro; presumably it exerts similar effects upon addition to epididymal sperm in vivo and thereby contributes to the suppression of constitutive exocytosis (466, 617, 679). The assumed mechanism of ASF action is by adsorption to sperm during transit through the corpus epididymis. Distal flow of epididymal fluid results in a three to four fold concentration of ASF in cauda epididymal fluids, relative to levels in corpus region secretions. This concentration may help drive the reversible association of ASF with sperm-binding sites during prolonged storage in the cauda epididymis (510). The nature of those binding sites and the mechanism of ASF action remain unknown. Additional components of epididymal secretions bind to sperm, including a high-molecular-weight polylactosaminoglycan (403, 566) that will be considered subsequently.

Protein constituents of epididymal fluids may also contribute to the imposition of the resting state through other mechanisms. Several enzymatic activities are present in epididymal fluids. Enzymes such as LDH-X and certain glycolytic enzymes are derived from sperm, either from shed cytoplasmic droplets or from decaying and dying cells. In addition, the epididymal epithelium secretes a number of enzymes into the lumen, including glycosidases (β-N-acetylglucosaminidase, α/β-fucosidase, β-glucuronidase, β-galactosidase, α-mannosidase) and proteases (52, 392, 697, 699). These enzymes may modify the sperm surface and alter functional states.

Additional control of the resting state may be provided by low-molecular-weight constituents of epididymal secretions. A low-molecular-weight factor(s) that also suppresses constitutive exocytosis has been partially characterized in guinea pig epididymal fluids (302). Ionic constituents of the epididymal microenvironment also may contribute to the resting state. Male and female reproductive tract fluids have been recovered from rodents by micropuncture techniques and their elemental composition has been determined by electron microprobe analysis (81, 82, 316). Testicular fluids contain concentrations of Na^+ approximately equivalent to that of serum (>130 mM) but are enriched in K^+ (>10 mM). As sperm migrate through the epididymides the concentration of Na^+ drops progressively and that of K^+ increases, such that the environment in the distal cauda epididymis includes a Na^+:K^+ ratio of >1 (37 mM Na^+, 40 mM K^+) (316). One role of this relative shift in ionic concentrations may be the immobilization of resting-state sperm. This is suggested by observations that the development of progressive motility by these cells upon dilution from epididymal secretions into culture media is Na^+-dependent (689) and is inhibited by elevated K^+ (688). The pH of the cauda epididymal microenvironment may also reinforce or confer the resting state. These secretions are relatively acidic, with values of 6.7–7.0 prevailing in various mammalian species (392). These conditions are expected to acidify sperm intracellular pH (32, 33) and may limit spontaneous opening of sperm Ca^{2+} channels (29, 207).

In conclusion, it is believed that the resting state is imposed and maintained by the convergent actions of several factors. These include intrinsic factors in sperm that are expressed upon aging, as well as specific proteins and inorganic ions present in epididymal fluids.

The Repressed State and Resting–Repressed-State Transitions. Addition of seminal secretions to cauda epididymal sperm propels these cells from the resting state into a repressed state. It is not known whether this transition is reversible. Entrance into the repressed state is associated with further alterations in motility, metabolism (310, 699), and surface properties (360, 699). In addition, repressed-state sperm are more resistant to capacitation than are resting-state cells (559, 699).

This state is imposed in part by the absorption of seminal fluid constituents onto sperm. Binding of seminal components to sperm has been demonstrated both by direct binding studies and by detection of adsorbed antigens (194, 460, 462, 509, 620, 666). Inhibitory factors that suppress fertilization are known to be present in seminal fluids (111, 297, 324, 559, 563, 632). While the function of these seminal factors is frequently unknown, their reversible association and dissociation is correlated with, and likely causal to, the suppression and expression of fertilizing ability, respectively (15, 194, 460, 462, 509).

The composition of seminal secretions differs in both serum and cauda epididymal fluids with regard to both low- and high-molecular-weight components. Selected ions and reducing agents are concentrated in seminal plasma and may contribute to the establishment of the repressed state. In addition, a number of proteins that typically are associated with signal transduction are present, including β-endorphin, *met* and *leu-*

enkephalin, a variety of hormones and growth factors (LH, FSH, prolactin, insulin-like growth factor I, calcitonin), protein kinase inhibitors, calmodulin, and cyclic nucleotides. In many cases it is not clear whether these are specific secretory products or are released from moribund sperm (391, 392, 564). C-type natriuretic peptides, activating ligands for receptor guanylyl cyclases (238), are also present. Receptors of this family are present in echinoderm sperm and mediate the effects of the egg peptides (237, 238); yet their presence in mammalian sperm has not yet been established. High-molecular-weight components include a wide variety of enzymes and other proteins (391, 392, 564). The function of many of these constituents remains unknown. Nevertheless, in some cases regulatory factors have been identified in seminal fluids.

Associated with transition into the repressed state is a further suppression of spontaneous ARs. The Ca^{2+} content of seminal secretions (1–10 mM) is elevated relative to that experienced by resting-state sperm ($\sim$0.2–0.5 mM) (82, 316, 392). This additional driving force might be sufficient to produce increased Ca^{2+} entry and, consequently, ARs. Seminal constituents minimize this process in two ways. First, Ca^{2+} chelators such as citrate and phosphate are present in high concentrations and reduce the free $[Ca^{2+}]$ (392).

Second, seminal fluid factors regulate Ca^{2+} transport in sperm. Resting-state sperm of the bovine accumulate $^{45}Ca^{2+}$ during incubation in vitro and flux is greatly attenuated in the repressed state (34). Inhibitory activity is due to caltrin, a peptide that is synthesized and secreted by the seminal vesicles (362, 522) and binds to the anterior head and to the distal flagellum of the bovine sperm (538). Caltrin inhibits Ca^{2+} influx into sperm membrane vesicles through a pathway that may be Na^+-dependent, although the relationship of this uptake route to the Na^+/Ca^{2+} exchange mechanism that has been characterized in somatic cells (506) and sperm (88, 454) is uncertain. Inhibition of this uptake route might account for the slightly lower Ca_i of cells in the repressed state, as compared with resting-state sperm (125, 217). Caltrin may abate the increase in spontaneous acrosomal exocytosis that is expected to accompany an elevated driving force for Ca^{2+} entry. Additional roles of caltrin will be considered later.

Other inorganic ions may also contribute to the establishment of the repressed state. Seminal secretions contain $\sim$90 mM each of Na^+ and K^+, whereas epididymal fluids consisted of < 40 mM of these ions (82, 564). The functional significance of these monovalent cations is unknown, although these concentrations of K^+ depolarize the membrane potential of resting-state sperm. $Zn^{2:PL}$ is also present at > 1 mM, due to accumulation and secretion by the prostate (391, 392,

564). This ion will inhibit fertilization of mouse eggs in vitro (16, 17). Moreover, nonspecific Zn^{2+} chelators enhance the capacitation of hamster sperm, as determined in a fertilization assay (14). These observations are consistent with a role for this ion in the establishment of the repressed state.

Transitions to the Capacitated State. Sperm can enter the capacitated state from the repressed state, as occurs during natural mating. Alternatively, resting-state cells can transit into the capacitated state following dilution in vitro. It is inferred, though not demonstrated, that these two populations of fertile sperm are equivalent.

It is generally understood that entrance into the capacitated state does not proceed as a single step, but rather through a series of separate events that regulate stages of gamete interaction (207). During capacitation sperm acquire the ability to penetrate the extracellular matrix of the cumulus oophorus, whereas uncapacitated sperm "stick" to the outer margins of the cumulus layer (141, 606). In addition, the capacity for ZP adhesion is gained during capacitation (267, 278, 309, 535). Development of adhesive activity and complete capacitation (that is, fertility) can be differentiated on the basis of (1) kinetics, where adhesion capability is exhibited more rapidly than fertility (59, 105), and (2) incubation conditions, where adhesion capability is largely (though not specifically) dependent on external Ca^{2+} (535) and fertility requires a more complex environment (513, 699). Thus, adhesive capability is regulated in sperm during capacitation, although it cannot be considered a rate-limiting step of that process. A potential mechanism to account for the regulation of adhesion will be considered below. Finally, sperm acquire the ability to undergo the AR in response to ZP3 as a consequence of capacitation. In this case, time course and inhibitor studies suggest that the development of this secretory response is a late, possibly terminal, event in the capacitation process (211, 646).

Several mechanisms act in concert to establish the capacitated state. First, negative regulatory elements dissociate from the sperm surface or are otherwise neutralized. The presence of negative regulators in epididymal and seminal secretions has been discussed previously. The dissociation of surface components from sperm has been well documented. Moreover, in some cases the reversible association and dissociation of these sperm-coating components is correlated with the loss and acquisition of the fertile state, respectively (15, 194, 224, 451, 452, 460–462, 642). Dissociation may unmask critical sperm surface components. For example, a poly-N-acetyllactosaminoglycan of epididymal origin may associate with carbohydrate-binding proteins of the sperm surface, thereby competing with

ZP3 oligosaccharides for sperm adhesion sites and precluding sperm–ZP association (403, 566).

Second, exogenous positive regulatory elements (or, capacitating factors) also adsorb to the sperm surface and their activities must be expressed. Caltrin, the seminal fluid regulator of sperm Ca^{2+} sequestration, is one example of a positive regulator. The transit of bull sperm from the *repressed* state into the capacitated state is associated with the development of ZP3-activated signal transduction pathways; agonist produces elevation of sperm pH_i and Ca_i and induces ARs only after capacitation (211, 217). In contrast, cauda epididymal sperm are relatively insensitive to agonist and fail to display either ionic or exocytotic responses (125, 212, 217). High-affinity agonist response is acquired during incubation of cauda epididymal sperm with seminal fluids or with either native or synthetic caltrin (125, 212).

ZP3-initiated exocytosis is apparent within 1 h of the treatment of cauda epididymal sperm with highly purified caltrin preparations, whereas 4–5 h of incubation in vitro are required when cells are treated with seminal fluids. This is consistent with models in which both positive and negative modulators in seminal fluids and in which negative modulators must be removed or neutralized before the effects of positive modulators are evident. Moreover, seminal secretions enhance the fertility of bovine sperm in vitro and these effects cannot be ascribed exclusively to caltrin (125). Thus, additional or alternative positive modulators may be present in the male, as well as in the female, reproductive tract. In this regard, fractions of bovine seminal proteins that are retained during heparin affinity chromatography support ZP3-initiated ARs (414), although it is not clear whether this activity is due to caltrin or to other regulatory factors. Similar activities may also be present in epididymal secretions, accounting for the high-affinity response of rodent sperm to ZP3 (73, 117, 216). Within the female, oviductal and follicular fluid glycosaminoglycans may bind to sperm and promote capacitation (273, 274, 412, 472).

A third mechanism of capacitation encompasses the extensive modification of the sperm surface: alterations in lipid composition (151, 251, 359, 360, 384, 574), in lectin-binding sites (4, 345), in diffusion rates of membrane proteins (449, 699) and lipids (680), and in the distribution of intramembranous particles (229, 230, 697, 699). These responses are often highly localized on the sperm surface and they may underlie the altered ability of sperm membranes to engage in fusion reactions following capacitation. In addition, the regulated mobility of proteins in the sperm membrane may provide a mechanism for controlling exocytosis, given

the postulated role of membrane protein cross-linking in the initiation of the AR by selected antibodies (1, 2, 385, 401) or by the ZP (380).

Fourth, alterations of key intracellular mediators may contribute to the development of a capacitated state. Sperm accumulate Ca^{2+} during capacitationy in vitro, as indicated by $^{45}Ca^{2+}$ influx (128, 274, 524, 569), although the contribution of mitochondrial sequestration to this process is unknown. Attempts to determine Ca_i, the internal free ionic concentration, using fluorescent indicator dyes have yielded contradictory results (387, 713). Elevations of pH_i are also associated with development of fertility (471, 693). These ions are important intermediaries of the agonist-initiated AR (see below) and it is plausible that capacitation entails alterations in Ca^{2+} and H^+ homeostasis. Modification of signal-transducing proteins may also occur, either by post-translational modification or by altered localization within the cell. The tyrosine phosphate content of several mouse sperm proteins is elevated during capacitation, as demonstrated by reactivity with antiphosphotyrosine antibodies (379). In contrast, sperm calmodulin is not modified but does redistribute during capacitation (619).

Finally, the inorganic ion composition of the female reproductive tract may provide another means of regulation of capacitation. Na^+ rises progressively toward serum levels, whereas K^+ declines from ~90 mM concentrations in seminal secretion to ~20–30 mM through the uterus and oviduct (81, 82). These ions may control portions of the capacitation process, as suggested by ion depletion and ionophore studies (226, 227, 299, 303, 433). In addition, the external pH influences capacitation, as assayed by fertility and by spontaneous exocytosis. Guinea pig sperm readily undergo ARs in alkaline media (pH ~7.8), whereas specialized conditions are required at a physiological pH (301). The pH of biological systems is buffered in large part by the HCO_3^-/CO_2 system. Bicarbonate anion is required for capacitation (298, 370), although its effects are more complex than pH regulation. The principal uptake route for HCO_3^- in sperm may be the HCO_3^-/Cl^- exchange pathway (459). Intracellular HCO_3^- apparently controls cAMP production through a novel HCO_3^--dependent adenylyl cyclase (455, 457, 458). Whether the regulation of Ca_i by HCO_3^- (524) is a consequence of adenylyl cyclase activation has not been determined.

Thus, sperm capacitation entails a complex series of reactions that are separately regulated. Inhibitory and stimulatory activities control progress through select stages of this pathway. Yet the mechanism by which these external agents affect modulation is not well understood. In some cases, such as caltrin, external regulators have been isolated and characterized (125, 377, 522, 644). In addition, some intracellular mediators have been identified. Yet there is little information regarding the nature of the receptors, or of the downstream signaling pathways, that presumably mediate the actions of external modulators.

BINDING OF SPERM TO EGGS

Of the 50 million or so mouse sperm that begin the journey in vivo (about 250 million for human beings), only about 150 sperm (that is, 0.0003%, or about one out of every 300,000 or so ejaculated sperm) actually reach the site of ovulated eggs in the ampulla of the fallopian tube (a number similar to that for human beings). Sperm are found in this region of the oviduct in as short a time as 15 min after insemination (as short as 5 min for human beings) and these sperm remain capable of fertilizing eggs for up to about 6 h (about 38 h for human beings). Whereas it has been thought that the initial binding of free-swimming sperm to unfertilized eggs in the oviduct occurs as a matter of chance, based on recent observations involving human gametes in vitro, it has been suggested that chemotaxis may play a role in this process. Evidence obtained suggests that sperm are attracted to eggs in the oviduct by a chemotactic factor(s) that originates from follicular fluid at the time of ovulation (188, 505).

Structure of the Zona Pellucida

The first direct contact between sperm and eggs occurs between plasma membrane overlying the sperm head and the ZP surrounding unfertilized eggs (70, 649, 650, 699). While all mammalian eggs possess a ZP, its thickness varies from less than 2 μm to more than 25 μm for eggs from different mammals (179). The ZP is a very loose meshwork of fibrillogranular strands and, consequently, is permeable to relatively large macromolecules, including immunoglobulins and enzymes (162). It has been known for some time that the ZP consists of both protein and carbohydrate (162, 314, 323). In fact, the ZP is constructed of glycoproteins that have a protein:carbohydrate ratio of about 3–5:1 (658). The ZP is quite elastic and noncovalent interactions between ZP glycoproteins are responsible for maintaining the extracellular coat's structure.

The mouse ZP is about 7 μm thick (662) and consists of only three glycoproteins, called mZP1 (~200 kDa), mZP2 (~120 kDa), and mZP3 (~83 kDa), that collectively account for the 3–4 ng of protein per ZP (71,

72, 652, 658, 659). Apparently, a similar low number of glycoproteins constitutes the ZP of eggs from a variety of mammals, including hamsters, pigs, rabbits, cows, monkeys, and human beings. In the presence of detergents, mZP2 and mZP3 are monomers, whereas mZP1 is a dimer of ~75 kDa polypeptides held together by intermolecular disulfide bonds (652, 658, 659). The polypeptides of mZP2 and mZP3 are ~81 kDa and ~44 kDa, respectively. All three mouse ZP glycoproteins are acidic (pI 4.1–5.5) and exhibit considerable M_r heterogeneity on sodium dodecylsulfate-polyacrylamide gel electrophoresis (SDS-PAGE). These two characteristics are attributable to the glycoproteins' oligosaccharides, not to their polypeptides. All three mouse ZP glycoproteins possess complex-type, asparagine-linked (N-linked) oligosaccharides, and at least mZP2 and mZP3 possess serine/threonine-linked (O-linked) oligosaccharides. Certain evidence suggests that the glycoproteins are sulfated, probably on their oligosaccharides, and their oligosaccharides are sialylated, contributing to the glycoproteins' low pIs.

As previously mentioned, the ZP is composed of long, interconnected strands or filaments. The filamentous nature of the ZP is revealed in electron micrographs of samples prepared under a variety of experimental conditions. For example, even after isolated mouse ZPs are solubilized by limited treatment with proteinases (either chymotrypsin or elastase), interconnected filaments as long as 2–3 μm are seen in electron micrographs (260, 652, 658). Furthermore, the filaments exhibit a 140–150 Å structural repeat. The structural features of the ZP can be fully accounted for by mZP1, mZP2, and mZP3; no other as-yet-unidentified components need be implicated. The long filaments are polymers of mZP2–mZP3 dimers (~180,000 apparent M_r), with each dimer present every 140–150 Å or so along the filaments (that is, the structural repeat), and mZP1 serves as a filament crosslinker (652, 658).

Identification of a Mammalian Sperm Receptor

Several criteria can be used to evaluate in vitro whether or not a ZP glycoprotein serves as a primary sperm receptor. In this context, mZP3 has been analyzed most extensively and, thus far, has satisfied all requirements of a bona fide primary sperm receptor. Hamster ZP3 (hZP3), one of three hamster ZP glycoproteins, also has been shown to exhibit many characteristics of a sperm receptor (423). Furthermore, hZP3 (337) and human ZP3 (107) (huZP3) polypeptides are very similar to mZP3 polypeptide, and molecular probes that recognize the mZP3 gene also recognize the hZP3 and huZP3 genes. Therefore, it is likely that mZP3, hZP3, and huZP3 play similar roles during fertilization.

Evidence that mZP3 Is a Sperm Receptor.

The following list (1–7) includes some observations made in vitro with purified egg mZP3 that have led to the conclusion that mZP3 is a primary sperm receptor during fertilization in mice (655): (1) Egg mZP3, at nanomolar concentrations, inhibits binding of free-swimming sperm to ovulated eggs and, consequently, prevents fertilization (72, 659). Apparently, in this "competition assay," mZP3 occupies sites on the sperm head (that is, occupies egg-binding proteins) that normally interact with sperm receptors on ovulated eggs and prevents binding of sperm to eggs. (2) Embryo mZP3, at the same concentrations (as in 1), does not inhibit binding of free-swimming sperm to ovulated eggs or prevent fertilization (72, 308, 659). This is a very significant observation, since free-swimming sperm can bind to egg ZP, but not to embryo ZP. Therefore, it is to be expected that a bona fide sperm receptor would be inactivated following fertilization. (3) Egg mZP3 binds to the sperm head (10^4–10^5 molecules/head) and not to other regions (for example, midpiece and tail) of the sperm (74, 430). (4) Egg mZP3 binds only to heads of acrosome-intact sperm, not to acrosome-reacted sperm (74, 430). This is a very significant observation since only acrosome-intact sperm (that is, those with plasma membrane overlying the anterior portion of the sperm head) can bind to mouse eggs. It is assumed that egg-binding proteins which interact with mZP3 are located in plasma membrane overlying heads of acrosome-intact sperm and are lost as a result of the AR. (5) Egg mZP3 (at concentrations slightly higher than those used in the competition assay) induces acrosome-intact sperm to undergo exocytosis, the AR (73, 655). This is a very significant observation since it confirms that mZP3 binds to sperm and affects sperm physiology. As expected, embryo mZP3, which does not bind to sperm, does not induce sperm to undergo the AR. (6) Free-swimming sperm bind to glass beads to which egg mZP3 is covalently attached and sperm bound to the glass beads undergo the AR (641). (7) mZP3 synthesized and secreted by eukaryotic cells (embryonal carcinoma cells; EC cells) stably transfected with the mZP3 gene (driven by a constitutive promoter, mouse phosphoglycerate kinase-1; pgk-1), inhibits binding of sperm to ovulated eggs and induces sperm to undergo the AR (334). Furthermore, free-swimming sperm bind to aggregates of transfected EC cells, but not untransfected EC cells. This provides an independent line of experimental evidence that mZP3 is a sperm receptor. These and other observations strongly suggest that mZP3, and homologous glycoproteins in other mammalian ZP3, are primary sperm receptors.

Characteristics of Mouse Sperm Receptor Glycoprotein (mZP3).

mZP3 exhibits considerable microheterogeneity on gels following one-dimensional SDS-PAGE and migrates with an apparent M_r of ~83 ± 15 kDa (71). On high-resolution two-dimensional gel electrophoresis mZP3 migrates as a family of charge isomers (537). The glycoprotein is quite acidic (pI range ~4.2–5.2 on isoelectric focusing gels), probably possesses intramolecular disulfides, and probably is sulfated and sialylated on its oligosaccharides. Each mZP3 molecule possesses three or four complex-type N (asparagine)-linked oligosaccharides (349, 517, 537) and an undetermined number of O (serine/threonine)-linked oligosaccharides. The M_r and pI heterogeneity of mZP3 that is observed on gels is attributable to the glycoprotein's oligosaccharides, not to its polypeptide. The mZP3 polypeptide itself migrates as a sharp band on one-dimensional SDS-PAGE, as a discrete "spot" on high-resolution two-dimensional gel electrophoresis, and has a pI ~6.5 (517, 537).

The mature mZP3 polypeptide consists of 402 amino acids (~44 kDa; pI ~6.7) (336, 512), is particularly rich in serine plus threonine (71 residues; 17%) and proline (27 residues; 7%) residues (that is, compared to the average vertebrate protein), possesses six potential N-linked glycosylation sites (that is, consensus sequence -Asn-Xaa-Ser/Thr-) (245, 597) and 13 cysteine residues, contains very little α-helix, and, overall, is neither particularly hydrophobic nor hydrophilic, although there are two small domains near the carboxy-terminus that are probably sequestered from an aqueous environment, as is the amino-terminal signal sequence. The polypeptide apparently consists of stretches of five to 29 residues in an extended chain conformation, interrupted by stretches of four to ten residues in reverse turn or coil conformations (120, 240). All six potential N-linked glycosylation sites are located in relatively hydrophilic stretches of sequence that are predicted to form reverse turns or coils. In this context, it has been suggested that the majority of N-linked oligosaccharide acceptor sequences utilized in glycoproteins are present at β-turns or loops (597). The middle third of mZP3 polypeptide consists of a proline-poor domain (1 proline) sandwiched between two proline-rich domains (16 prolines). The mZP3 polypeptide is not significantly similar to any other entries in either the GenBank or Protein Identification Resources data base. It should be noted that nascent mZP3 polypeptide consists of 424 amino acids, with the amino-terminal 22 amino acids constituting a putative signal sequence that, like many other secreted glycoproteins (643, 678), is removed prior to secretion of the mature (83 kDa) glycoprotein.

Molecular Basis of mZP3 Sperm Receptor Activity. Interestingly, O-linked oligosaccharides, not polypeptide, account for the sperm receptor activity (SRA) of mZP3 (652, 654–656, 658). Thus, gamete adhesion in mice is a carbohydrate-mediated event. Several lines of experimental evidence *(1–6)* support this conclusion: *(1)* mZP3 SRA is resistant to a variety of agents, including temperature (−70 to 100 °C), detergents (.0.1% SDS or 1% Triton X-100), and denaturants (7 M urea). *(2)* Extensive digestion of purified mZP3 by insolubilized pronase produces a mixture of very small glycopeptides (1500–6000 M_r) that retains SRA (209). *(3)* Purified mZP3 treated with trifluoromethanesulfonic acid, a reagent that removes N- and O-linked oligosaccharides from glycoproteins, does not exhibit SRA (218). *(4)* Extensive digestion of purified mZP3 by N-glycanase, an enzyme that removes N-linked oligosaccharides from glycoproteins (123, 610), has no effect on SRA. *(5)* Purified mZP3 subjected to mild alkaline hydrolysis, a procedure that removes O-linked oligosaccharides from glycoproteins (β-elimination), does not exhibit SRA (218). *(6)* Purified mZP3 O-linked oligosaccharides, released from mZP3 by mild alkaline hydrolysis under reducing conditions (alkaline-borohydride), exhibit SRA (218).

These and other observations strongly suggest that mZP3 O-linked oligosaccharides are responsible for sperm recognition of and binding to this particular ZP glycoprotein. It should be noted that several other lines of evidence suggest that binding of sperm to eggs in mammals is a carbohydrate-mediated event. This evidence includes results of experiments designed to test the inhibitory effect of various lectins, sugars, and glycosidases on sperm–egg interaction and/or fertilization in vitro (656). Finally, it is noteworthy that free-swimming mammalian sperm display the same binding characteristics with both live and aldehyde-fixed homologous and heterologous eggs (265, 544); a finding consistent with a role for carbohydrate, rather than polypeptide, in gamete adhesion.

Characterization of mZP3 O-Linked Oligosaccharides. Although the number of O-linked oligosaccharide chains on mZP3 has not been determined accurately, it is clear that there are a relatively large number. For example, mZP3 synthesized and secreted in the presence of tunicamycin, an antibiotic that prevents N-linked glycosylation of nascent glycoproteins, has an apparent M_r of 56 kDa (517). Similarly, mZP3 digested

by N-glycanase is approximately 56 kDa. These results are to be compared with the mZP3 polypeptide M_r of 44 kDa.

Not all mZP3 O-linked oligosaccharides exhibit SRA in vitro. For example, SRA was found associated only with void volume fractions when mZP3 oligosaccharides (released by alkaline-borohydride hydrolysis) were subjected to gel-filtration on Bio-Gel P-2 (~1,800 M_r cutoff) (218). Further analysis of pooled void volume fractions by gel-filtration on Bio-Gel P-6 (~6000 M_r cutoff) revealed that only O-linked oligosaccharides having an apparent M_r between 3.4 and 4.6 kDa exhibited SRA. Apparently, a relatively small percentage of mZP3 O-linked oligosaccharides (~3.9 kDa) is responsible for the glycoprotein's sperm receptor function. In fact, these particular O-linked oligosaccharides selectively adhere to sperm added to a mixture of total O-linked oligosaccharides released from mZP3 and, thereby, are partially purified (218).

Extensive digestion of either purified mZP3 or purified mZP3 O-linked oligosaccharides by α-galactosidase, an enzyme that cleaves terminal galactose residues which are α-linked to oligosaccharides, results in release of galactose from each substrate and destroys SRA (75, 654). Digestion of these substrates by α-fucosidase also destroys SRA, but release of fucose (or any other sugar) was not detected. On the other hand, treatment of either purified mZP3 or purified mZP3 O-linked oligosaccharides with galactose oxidase, an enzyme that converts the C-6 position alcohol of terminal galactose and GalNAc residues of oligosaccharides to an aldehyde, destroys SRA. Reduction of galactose-oxidase-treated substrates with $NaBH_4$ (that is, conversion of the C6 aldehyde to an alcohol (562)) restored SRA to both substrates. These results are consistent with the important role proposed for sugar hydroxyls, which are thought to participate extensively in hydrogen bonding in supporting interactions between carbohydrate-binding proteins and sugar ligands in general (498). In this context, LA4, a monoclonal antibody that recognizes a carbohydrate epitope consisting of N-acetyllactosamine with a terminal α-galactose and a fucose on the penultimate galactose, binds to mZP3, but not to other mouse ZP glycoproteins (558).

Whether or not the relative species-specificity of sperm–egg interaction in mammals is attributable to different O-linked oligosaccharides on ZP3 from different mammals remains to be determined. Certainly the great diversity of known oligosaccharide structures (163, 317, 245, 349, 597) is compatible with generation of species specificity for mammalian sperm receptors. The variety of compositions, sequences, branching patterns, conformations, and other features of oligosaccharides associated with glycoproteins provide for a staggering number of combinatorial possibilities. Furthermore, different degrees of sialylation, sulfation, and/or phosphorylation of sugars add to the complexity.

Location of Functional O-Linked Oligosaccharides. Limited digestion of purified egg mZP3 by papain produces a mixture of distinct glycopeptides, ranging in size from less than 10 to ~55 kDa (519). Only the 55,000 M_r glycopeptide, which is derived from the carboxy-terminal half of the mZP3 polypeptide (based on amino acid sequencing and epitope mapping), has SRA. Similarly, digestion of mZP3 by V8 protease also produces a 55 kDa glycopeptide that is derived from the carboxy-terminal half of the mZP3 polypeptide and has SRA. Results of these and other experiments suggest that at least certain of the mZP3 O-linked oligosaccharides that are essential for SRA are located on the carboxy-terminal half of the polypeptide. It should be noted that, to date, no consensus sequence for O-linked glycosylation of polypeptides has been identified.

Other Mammalian Sperm Receptors

Hamster Sperm Receptor Glycoprotein (hZP3). hZP3 is a glycoprotein that exhibits microheterogeneity on gels following one-dimensional SDS-PAGE and migrates with an apparent M_r of ~56 ± 4 kDa (423). The hZP3 polypeptide consists of 400 amino acids (~44 kDa; pI ~6.5) and its primary structure is very similar (81%) to that of mZP3 (337). Only one small region of the hZP3 polypeptide (residues 318–356) is significantly different (~50%) from mZP3 polypeptide. The primary structures of these two sperm receptors exhibit a normal rate of sequence divergence (0.54% sequence divergence/10^6 yr) in the period since mice and hamsters diverged about 37 million yr ago (153). Like mZP3 polypeptide, the middle third of hZP3 polypeptide is a proline-poor domain sandwiched between two proline-rich domains. All 13 cysteine residues present in mZP3 are conserved in hZP3, but only three of the six potential N-linked glycosylation sites in mZP3 are conserved in hZP3; one additional site is present in hZP3, giving a total of four potential N-linked glycosylation sites. Three of the four potential N-linked glycosylation sites are predicted to be located at reverse turns or coils. Hydrophobicity profiles for hZP3 and mZP3 are virtually identical. Like mZP3, nascent hZP3 polypeptide has a putative signal sequence consisting of 22 amino acids at its amino-terminus. It is of interest to note that mouse gametes interact with hamster gametes, suggesting that mZP3 and hZP3 have certain O-linked oligosaccharide epitopes in common.

Porcine Sperm Receptor Glycoprotein (pZP3α). The porcine sperm receptor, called pZP3α (529, 530, 707, 708), is one of three or four glycoproteins that make up the porcine egg ZP (180, 284). pZP3α has been purified and tested in the in vitro competition assay, and exhibits SRA. pZP3α is a 55 kDa glycoprotein that contains both N- and O-linked oligosaccharides. Like mZP3, on high-resolution two-dimensional gel electrophoresis pZP3α appears to be a family of from 15 to more than 25 charge isomers, and its heterogeneous appearance and relatively acidic pI (average, pI ~5) are due to its oligosaccharides. The sugar composition of pZP3α includes fucose, GlcNAc, GalNAc, mannose, galactose, and sialic acid. Chemically deglycosylated trifluoromethane sulfonate (TFMS) pZP3α is 37 kDa and has a pI ~6.5 (531). pZP3α may be related to, but is structurally and immunologically distinct from, another major porcine ZP glycoprotein, pZP3β (55 kDa), and the two glycoproteins are probably different gene products. pZP3α and pZP3β are reported to have 38 and 35 kDa polypeptides, respectively. It is estimated that pZP3α contains three or four N-linked and three O-linked oligosaccharides, and that certain of the oligosaccharides are acidic (probably sulfated or phosphorylated) lactosaminoglycans (Gal-β1,4-GlcNAc). There is evidence to support the notion that, as in the case of mZP3, pZP3α O-linked oligosaccharides are responsible, at least in part, for SRA.

Human Sperm Receptor Glycoprotein (huZP3). Apparently, the human egg ZP is composed of three glycoproteins, 102, 88, and 73 kDa, all of which exhibit pronounced microheterogeneity (charge isomers) on high-resolution two-dimensional gel electrophoresis (556, 557). It is possible that, like pZP3α and pZP3β, huZP3 (73 kDa) exists as two electrophoretically separable species. Recently, by using molecular probes directed against mZP3, huZP3 has been cloned (cDNA) and the primary structure of its polypeptide determined (107). The nascent huZP3 polypeptide consists of 424 amino acids, just like mZP3, is 67% identical to the mZP3 polypeptide, and contains a 22-amino-acid putative signal sequence at its amino-terminus. huZP3 polypeptide contains four potential N-linked glycosylation sites, three of which are conserved in mZP3, and is particularly rich in serine and threonine residues (66 residues), 71% of which are conserved in mZP3. Interestingly, all 13 cysteine residues in huZP3 are conserved in both mZP3 and hZP3. Overall, the predicted secondary structures of mZP3, hZP3, and huZP3 are very similar to each other.

THE ACROSOME REACTION

Once free-swimming, acrosome-intact sperm bind to the mouse egg ZP, they must penetrate the thick extracellular coat and reach the perivitelline space between the ZP and plasma membrane. To penetrate the ZP, bound sperm must first undergo a kind of exocytosis, called the *acrosome reaction*. Sperm also must complete the AR if they are to be able to fuse with egg plasma membrane; acrosome-intact sperm are unable to fuse with ZP-free eggs in vitro (702).

In 1701 Leeuwenhoek reported a specialized structure in the head of chicken sperm and he designated it the "plaga pellucida rotunda" [cited in (35)]. By the early years of the present century it was established that a granule, or acrosome, was present in the apical head region of sperm from many animal species, including all mammals (381, 508, 677). Yet, appreciation that the acrosome was a secretory vesicle required a greater range of analytical tools. This understanding did not come until the years following the Second World War, when knowledge of cellular structure expanded rapidly as a result of the introduction of the electron and the phase-contrast microscopes.

Exocytosis of the acrosome, by a process known as the AR, is a final structural alteration of sperm that must be completed prior to fertilization. The AR of sea urchin sperm was described in the early 1950s by Dan (143–146), although the production of the actin-based acrosomal filament was probably noted by Popa in 1927 (494). The first indications of a mammalian sperm AR came in 1958, when Austin and Bishop reported alterations in the appearance of the sperm head in motile cells during passage through the extracellular matrix of the cumulus oophorus (23). Subsequent ultrastructural studies revealed that these modifications consisted of exocytosis of the acrosomal granule (49, 491).

This section will review the morphology of the AR, as well as the mechanism for initiation of exocytosis. Readers may wish to consult additional reviews that treat the AR comprehensively (696, 697, 699), or that focus on specialized subjects such as the mechanism of exocytosis in sperm (207, 237, 239, 267, 360, 404, 405, 540, 652) and in somatic cells (134, 484, 599, 671), as well as the evolution, structure, and biogenesis of the acrosome (35, 36, 54, 183, 347).

Anatomy of the Acrosome

The acrosome is a membrane-bound organelle in the apical portion of the sperm head, where it is closely applied to the nuclear membrane and immediately subjacent to the plasma membrane. While the acrosomal membrane is continuous, it has been divided into three zones: the *inner acrosomal membrane* apposes the nuclear membrane, the *outer acrosomal membrane* runs beneath the plasmalemma, and the region where the two domains join is the *equatorial segment*. These membranes differ in their susceptibility to chemical and physical disruption (183, 280, 690), susceptibility to complement-mediated lysis (527), and in the patterns displayed in freeze-fracture and freeze-etch microscopy (229). The inner and outer acrosomal membranes are in turn differentiated from those of the equatorial segment, as the latter displays a unique pentalaminar ultrastructure and exhibits characteristic surface replica and freeze-etch patterns (183, 486). The distinct features of the acrosomal membrane have been interpreted as indicating the presence of a regional domain structure (202, 229), similar to the domains that have been demonstrated in the mammalian sperm plasma membrane (130, 485, 680, 682, 683).

Studies of the composition of the acrosome have been hampered by two general problems. First, isolated intact acrosomal granule preparations are not available. For example, the ready isolation of synaptic vesicles has permitted the acquisition of detailed information on the biochemical composition and has contributed to models of neurosecretion (330, 599). In contrast, the inner acrosomal membrane associates tenaciously with the underlying nuclear membrane. While there have been reports of the isolation of intact acrosomes from ovine sperm (279), it is most commonly found that cellular disruption tends to rupture the acrosome, releasing membrane sheets composed of plasmalemma and outer acrosomal membranes while the inner acrosomal membrane is retained with the cell. Second, subcellular localization in sperm is complicated by the relative paucity of cytoplasm. For example, the total cell water in a bull sperm is ~20 μm^3 (272). While periacrosomal cytoplasmic bands have not been accurately measured, these appear <0.25 μm wide in electron micrographs and likely account for <1 μm^3 of volume. Under these constraints, techniques with higher resolution than the light microscope are required in order to determine whether proteins are localized within the acrosomal matrix, in the acrosomal membrane, or in the periacrosomal cytosol.

Despite these technical limitations, considerable information has accumulated regarding the biochemical composition of the mammalian acrosome. A number of hydrolytic enzymes have been identified and, in some cases, molecular clones have been isolated and characterized. These include acrosin, a serine protease that is stored in the acrosome in a zymogen form. The inactive zymogen is processed into functional acrosin following the AR and is thought to assist sperm penetration of the ZP (635). Hyaluronidase is also present within the acrosome, apparently as a spermatogenic cell-specific isoform (276, 392). Additional hydrolytic enzymes or activities located in the mammalian acrosome include separate collagenase-like, cathepsin D–like, and calpain II–like proteases; neuraminidase, β-N-acetylglucosaminidase, β-galactosidase, and β-glucuronidase; and arylsulfatase A, aspartylaminidase, arylaminidase, phospholipase A, and esterase (183, 392, 442, 699).

In addition, several bioactive peptides or peptide precursors have been localized in the region of the acrosome by immunofluorescence, including gastrin (542), pro-cholecystokinin (481), and pro-enkephalin (331). Additional electron microscopic studies verified an intra-acrosomal localization of pro-cholecystokinin (481). Pro-enkephalin is detected in acrosome-intact sperm and is absent following exocytosis, suggesting release during exocytosis (331). It has not yet been determined whether pro-enkephalin is processed into active opiate peptides. Moreover, the functional significance of the class of acrosomal bioactive peptides has not been established. Finally, several acrosomal proteins (27, 699) and antigens of unknown function have been identified.

It is generally believed that these proteins and peptides are stored in an acidic environment. Permeant weak bases, such as acridine orange and methylamine, have been used to determine acrosomal pH. These compounds selectively accumulate within acidic compartments. If correction factors are applied to account for dye binding to protein and nucleic acids, these probes can be used to determine the pH of cellular compartments (518). Based on dye distributions an acidic intra-acrosomal pH (pH < 5) has been calculated for the hamster sperm (406). The mechanism of vesicular acidification has not been established, although a Mg^{2+}-dependent ATPase activity, that may be the widely distributed vacuolar H^+-ATPase (250), has been detected in the head of hamster sperm (691, 692). While the ionic composition of the acrosome has not otherwise been characterized, it is likely that new tools, such as the widely available ion-selective fluorescent dyes (630), will provide greater definition of vesicular contents.

These ionic and macromolecular constituents are retained within a reticulated meshwork that reveals some substructure and regional organization upon electron microscopic examination (202, 229). These domains may provide the structural basis for the sequential deployment of acrosomal enzymes that has been observed during exocytosis (161, 606, 712). The role of the acrosomal matrix in exocytosis remains uncertain. There has been some indication of acrosomal swelling prior to exocytosis (257, 258, 704), although the causal relationship between these events has been questioned (259). In addition, it is uncertain whether swelling is produced only by water entry and the colloidal properties of the matrix (257, 704), or whether the acrosomal matrix possesses "smart" gel properties and is regulated by additional factors (for example, pH alterations), as has been suggested for the mast cell secretory granule matrix (439).

Stages of Exocytosis

The structure of acrosomal exocytosis has been characterized by electron and fluorescence microscopy. Chlortetracycline, the fluorescent chelate probe of Ca^{2+} (105), has been used to examine the early stages of the AR (373, 534, 640). Three different fluorescent patterns are displayed over the head region by mouse sperm, and the associated acrosomal ultrastructure has been confirmed by electron microscopy (191, 216, 344, 372, 534). The "B" pattern is characteristic of capacitated sperm with intact acrosomes and consists of bright fluorescence over the anterior head and midpiece, with relatively low levels of emission from the postacrosomal head region. The "S" pattern is characterized by bright midpiece fluorescence and spotty, punctate fluorescence over the anterior head. Finally, the "AR" pattern is indicative of cells that have completed the AR and is comprised of bright midpiece fluorescence and a uniform level of low emission from the head. The time courses for the appearance and disappearance of these patterns during sperm association with the ZP are consistent with a sequential order of B-to-S-to-AR, with S showing the characteristics of an intermediate stage. Importantly, the B-to-S transition occurs prior to the initiation of membrane fusion–fission reactions or the loss of the pH gradient between the acrosome and the extracellular milieu, as indicated by application of pharmacological agents that trap sperm in the S stage (344).

The molecular basis of the B-to-S-to-AR transitions are unknown. This is due, in part, to the relatively uncharacterized nature of chlortetracycline. Enhanced fluorescence emission is associated with Ca^{2+} binding by the probe in a hydrophobic environment (105). Yet, it is difficult to establish the location of the reporter group, since the protonated, Ca^{2+}-free form readily penetrates cell membranes. Further problems are posed by the lack of probe selectivity. Transition metal cations other than Ca^{2+} produce fluorescent responses (630), as does membrane potential alterations (608). Nevertheless, the recognition that intermediate stages of the AR exist prior to observable membrane interactions represents an important advance.

Sperm that have entered the S stage then commence with the exocytotic phase of the AR. This consists of multiple membrane fusion–fission events (467) between the plasma membrane and the outer acrosomal membrane. The molecular details of the fusion machinery in sperm are poorly understood. According to the generally accepted model, the initial stages consist of the formation of several pores, or fenestra. The relationship of these structures to the fusion pores that are intermediates in somatic cell exocytosis and viral-mediated fusion (10, 11, 424, 582–584, 671) is unknown.

As exocytosis procedes, more extensive fusion events lead to the formation of small vesicles that decorate the apical sperm head. These vesicles are apparently composed of both outer acrosomal and plasmalemmal domains, as indicated by the distinctive appearances of these membranes (526). It is inferred from this model that the entrapped volume within these hybrid vesicles is derived from periacrosomal cytoplasm, although this has not been demonstrated directly.

It is essential that the integrity of the sperm cytoplasmic compartment be maintained during exocytosis. To this end, a band of fusion between the plasma membrane and the acrosomal membrane of the equatorial segment marks the rostral extent of vesiculation. The establishment of this band provides sperm with a continuous cell surface that is composed of plasmalemma and inner acrosomal membrane. Integral membrane proteins from the posterior head are now able to diffuse laterally into the apical region of the sperm head, demonstrating a coherent membrane system (130, 436). In contrast, moribund sperm exhibit a generalized and random breakdown of the membranes, with no maintenance of cytoplasmic integrity.

The final stage of the AR involves clearing of vesicles from the caudal head. It is generally believed that small, hybrid vesicles release from the sperm surface (49, 223, 697, 699). Alternatively, there have been some reports that the hamster ARs entails restricted acrosomal/plasmalemmal fusion in the region of the equatorial segment, with the subsequent lifting off of an intact acrosomal "cap" (703).

Functions of the Acrosome Reaction during Fertilization

According to our present understanding, the AR plays two crucial and related roles in the fertilization process. First, exocytosis acts as a selection filter. Mammalian fertilization operates under conditions of sperm excess. Typically, the male releases $> 10^7$ sperm during mating and these must compete for < 10 eggs (392). The distribution of sperm within the female reproductive tract has been examined both by electron microscopy *in situ* and by surgical recovery. Large sperm colonies are established in the utero–tubule junction rapidly following mating, yet the sperm:egg ratio in the hamster oviduct is $< 1:1$ until approximately half of the eggs have been fertilized (140). These and other observations (139, 347) suggest that mammalian sperm are subjected to selection pressures that provide only a subpopulation of sperm with access to eggs.

Several factors contribute to and account for the selection process. Sperm exhibit pronounced morphological diversity and in many cases abnormal structure results in infertility. In addition, the post-testicular history of these cells includes prolonged storage in the cauda epididymis, followed by a phase of migration to the site of fertilization. During these intervals sperm must adapt to a wide range of environments (81, 82, 207, 316, 392) and must successfully interact with both cellular and matrix components.

The AR is one focus of these selection pressures. It is now believed that exocytosis must occur either on or in the vacinity of the egg's ZP, and that initiation of the AR prior to reaching the egg renders sperm infertile (see Site of the Acrosome Reaction). In this regard, genetic factors that produce either abnormal acrosomal morphology (96, 296) or premature ARs (94) are associated with compromised fertilization. Conversely, sperm that fail to acrosome react are also excluded from contact with eggs. Thus, the coordination of the AR with the availability of eggs serves as one mechanism for filtering "unsuccessful" sperm.

The second role of exocytosis is that of tossing a behavioral switch. Sperm that retain intact acrosomes are relatively long-lived ($\sim$2 d) and are competent to complete the early stages of gamete interaction, including penetration through the cumulus oophorus extracellular matrix and adhesion to the ZP (see Site of the Acrosome Reaction). Yet, these cells are apparently incapable of the plasmalemmal membrane fusion events that comprise sperm–egg fusion. Light and electron microscopy indicates that acrosome-intact sperm can adhere to the plasma membranes of eggs from which the ZP has been removed, but gamete fusion does not occur until after sperm have completed the AR (487, 697, 699, 701, 702). The molecular basis for the acquisition of fusion competence is not well understood although it may involve diffusion, and potential activation, of putative sperm membrane fusion proteins from the posterior head to the inner acrosomal membrane (77, 78, 496, 671). Alternative mechanisms have been suggested, including processing of sperm membrane components by secreted, acrosomal enzymes (699) and alterations in sperm membrane potential (700). Thus, exocytosis promotes physiological alterations in sperm that are essential for the expression of fertility.

Site of the Acrosome Reaction

It is generally accepted that the AR must be completed prior to the penetration by sperm of the ZP. A determination of the site at which the fertilizing sperm undergoes exocytosis in vivo is central to an understanding of gamete interaction. Yet, identification of that site is particularly difficult among mammals. This is due, in part, to our limited ability to peer into the oviduct at the time of fertilization. A second factor is the occurrence of spontaneous exocytosis in sperm. In many secretory cells (326, 589), including mammalian sperm, secretion continues at a basal rate under resting conditions (agonist-independent exocytosis) and is enhanced by stimulatory signals. The rates of spontaneous AR in sperm populations varies both with species and with physiological state (that is, capacitation), but in vitro studies indicate that this can be as high as 1%–10% h^{-1}. This spontaneous AR provides a significant background that may obscure agonist-initiated events.

The identification of the site of the physiological AR has also been impeded by technical difficulties. The acrosomal morphology of sperm has been examined within the female reproductive tract by electron microscopy (53, 585, 711), as well as by the inspection of sperm that were surgically recovered from the female. These approaches are compromised by an inability to observe gametes in vivo with high temporal resolution. For example, electron micrographs of acrosome-reacted sperm within the cumulus oophorus matrix (53) might represent cells in the process of moving toward the egg or cells that underwent premature exocytosis and subsequently were trapped. Thus, in vivo methods available at present are unable to resolve the site of exocytosis of the fertilizing sperm. The development of methods for direct and continuous observation of sperm in situ (173), coupled with advances in image processing–enhanced light microscopy may permit resolution of this question in the future.

As an alternative, the site of the AR has been examined during fertilization in vitro. Although observations are simplified, care must be taken in extrapolating from in vitro models to in vivo conditions. The model systems often use cauda epididymal sperm, due to their ready availability. While these cells will fertilize eggs, it is now apparent that contact with seminal secretions has a number of functional effects on sperm (see above), including modifying their response to ZP3 (125, 208, 212, 217, 246, 414). In addition, many of the model systems are inefficient and require sperm:egg ratios of 10^3–10^5:1 (513), far in excess of the 10^0–10^1:1 ratios observed about the egg in vivo at the time of fertilization. The relative abundance of sperm complicates identification and analysis of the small number of cells that actually fertilize eggs. Moreover, high concentrations of sperm necessarily include moribund cells. The associated release of acrosomal enzymes can alter the structure of the cumulus oophorus extracellular matrix and further muddle interpretation of results. The recent development of highly efficient in vitro models, permitting monospermic fertilization at physiological sperm:egg ratios (129), has also provided insights into the site of exocytosis (161).

Initial models, formulated immediately following the discovery of the mammalian AR, proposed that exocytosis occurred either in the vicinity of or during the early stages of sperm penetration through the extracellular matrix of cumulus oophorus and that exocytosis was essential for traversing the physical barrier posed by the cumulus (20, 54). Supporting this proposal were observations that hyaluronidase was localized in the sperm head (20, 390); that hyaluronidase effectively dissolved the extracellular matrix of the cumulus oophorus (392, 699); and that motile sperm with disrupted acrosomes were detected moving through the cumulus oophorus (23). More recent studies have disputed this model by the demonstrating that hyaluronidase is not required for sperm penetration of the cumulus oophorus (606, 607). In addition, sperm that have completed the AR prior fail to enter the cumulus, but rather adhere to cells that form the outer margin (141).

An alternative model that is now widely, though not universally accepted (429, 435), identifies the ZP as the site of the AR (267, 591, 651). The chain of evidence supporting this model includes first, the demonstration that sperm with intact acrosomes penetrate through the cumulus oophorus and contact the ZP in vivo (137, 138) or in vitro (129, 593, 606). Most sperm recovered from the uterotubule junction, from the uterus, or from more distal regions of the female reproductive tract retain an intact acrosome. In contrast, sperm in contact with the ZP displayed acrosomes that were either intact, partially vesculated, or had completed exocytosis (137, 138). Second, sperm with intact acrosomes preferentially adhere to the ZP in vitro (267, 533, 534). Third, an AR-inducing agonist activity is present in the ZP of all mammalian species that have been examined (62, 73, 117, 136, 211, 216, 328, 450). Finally, sperm that undergo ZP3-initiated ARs can fertilize eggs. Chlortetracycline fluorescence has been used to monitor the progress of exocytosis in living sperm (533). Single sperm that adhered to the ZP with intact acrosomes were monitored undergoing the AR, penetrating the ZP, and fertilizing eggs. Since there was only one sperm bound to each egg, the identification of the fertilizing sperm was unambiguous. Specific inhibition of the ZP3-initiated AR by muscarinic cholinergic agents (213–216) and by monoclonal antibodies against sperm proteins (532) prevents fertilization of eggs in vitro. Based on these and other evidence it has been proposed that the ZP is the site of the physiological AR in vitro.

While some of the observations of fertilization within the oviduct are consistent with this model, a canonical sequence of sperm–egg interactions in vivo has not yet been established. It should be recalled that the cumulus oophorus experiences time-dependent alterations within the oviduct, including a "loosening" and, in some cases, a dispersal (697). The site at which sperm initiate exocytosis may then vary with the state of the egg and the time following ovulation.

In conclusion, the evidence available at present indicates that sperm undergo the AR at the surface of the ZP. Yet, the supporting evidence was culled largely from in vitro model systems. Unequivocal determination of this central question will require additional examination of fertilization in vivo, with particular emphasis on the development of methods for continuous observation.

Mechanisms of the Acrosome Reaction

The AR is a form of regulated secretion in which exocytotic vesicles accumulate within cells until release is initiated by an appropriate stimulus (329). Understanding of the mechanism of regulated secretion has expanded in recent years with the recognition that stimulatory agents function by binding to and activating cell surface receptors. These receptors either (1) possess intrinsic second-messenger-generating capacity, such as the ligand-activated ion channels of GABA$_A$, glutamate, and nicotinic acetylcholine receptors (290, 634), or the enzymatic activity of tyrosine kinase (101, 633) or guanylyl cyclase receptors (118, 121, 238); or (2) regulate intermediary transducing elements, such as G proteins (67, 287).

The extracellular and intracellular mechanisms controlling acrosomal exocytosis have been extensively and intensively examined during the last two decades, contributing to an emerging picture of sperm signal transduction. These studies have established the presence of signaling elements within sperm and their control by extrinsic agonists.

Initiators of Exocytosis

The mammalian sperm AR was initially regarded as a final phase of capacitation. As such, exocytosis was not thought to be regulated directly, but rather occurred in some probabilistic fashion once sperm had completed all preliminary steps of capacitation. Developments in gamete culture conditions during the 1970s permitted the dissociation of capacitation and exocytosis (see above) and suggested that the AR might be independently controlled.

Identification of AR-initiating agonists is complicated by the nature of spontaneous exocytosis in this system. It is now generally accepted that completion of capacitation is a critical prerequisite for acrosomal exocytosis. Yet, in many mammalian model systems, the rate of agonist-independent exocytosis is enhanced by capacitation (207). Factors or agents that enhance the rate of capacitation will then produce an associated and confounding elevation in the incidence of AR. It is therefore necessary to demonstrate that a putative agonist acts directly at the level of exocytosis.

Agonist identification is further complicated by the embarrassingly large number of compounds that have been reported to trigger sperm exocytosis in vitro. Given the specificity and regulation of sperm–egg interaction, it is unlikely that there will be a large number of common agonists. Some agents may act indirectly, either by effecting the efficiency of capacitation or by modulating an intracellullar effector pathway. It is a particular concern that many of these putative agonists have been described in vitro, where model systems may not be optimized and may be prone to culture epiphenomena. By way of comparison, a similar situation has been described for the amphibian oocyte. There, several dozen agents promote meiotic maturation in vitro although only one, progesterone, is physiologically active. The minimal criteria then for an AR-inducing agonist are: (1) that it promote exocytosis rapidly (that is, within minutes); (2) that agonist-induced exocytosis reflect the anticipated specificity and regulation, including a requirement for prior capacitation of sperm; and (3) that agonist be present at the site of the AR.

Zona Pellucida. The presence of a stimulatory agonist in the ZP was first inferred in the 1970s, following inspection of hamster and mouse sperm during the initial stages of gamete interaction in vitro (267, 269, 533, 534). These studies demonstrated a sequential relationship between adhesion and exocytosis. Parallel investigations, in which an echinoderm AR-initiating agonist was isolated and partially characterized from egg jelly (239, 554, 555), also supported the notion that analagous agonists existed in the ZP.

Direct demonstration of an agonist came in the early 1980s, as extracts of the mouse zona initiated ARs (73, 216). Several lines of evidence indicate that ZP3 is the active constituent. Agonist activity is detected in native

ZP3 isolated from follicular oocytes or from ovulated eggs (73, 423, 450), as well as in recombinant ZP3 that has been overexpressed and secreted by mammalian tissue culture cells (57, 334). ZP3 alone accounts for the agonist activity of ZP extracts, while neither ZP1 nor ZP2 has appreciable stimulatory activities.

ZP3 satisfies the minimal requirements for a sperm agonist. First, it is a potent initiator of the AR. Soluble ZP3 and structurally intact zonae pellucidae are comparable in their agonist activity, promoting exocytosis in $>50\%$ of the sperm population within minutes (73, 213, 216). Comparisons of the initial rate of exocytosis in ZP-bound sperm or in sperm incubated with soluble agonist, as compared to the linear rate of spontaneous exocytosis in untreated, free swimming cells, indicates that soluble agonist and intact zonae pellucidae stimulate exocytosis by 30–50-fold (216). The similarities in time course and potency further support the notion that ZP3 accounts agonist activity in the ZP.

Second, the specificity of ZP3 agonist activity is consistent with the observed specificity of fertilization. Both the AR initiated by structurally intact zonae pellucidae and by ZP3 require prior capacitation of sperm. While sperm acquire the ability to adhere to the ZP rapidly ($t_{1/2}$ ~5–10 min) following release from the cauda epididymis in vitro (309, 535), these cells have not completed capacitation and are unable to undergo ZP3-initiated exocytosis (211, 646).

The availability of selective inhibitors provides an additional comparison between the AR initiated by ZP3 and that of sperm bound to the ZP. In each case, the AR exhibited similar sensitivity to *B-ordetella pertussis* exotoxin (PTx) (190, 191, 217, 369, 641, 647, 674). PTx is an ADP-ribosyltransferase that enzymatically inactivates heterotrimeric, GTP-binding regulatory proteins (G proteins) of the G_i family (67, 431). A second, independent class of inhibitors that effect both the ZP- and ZP3-induced processes include the reversible antagonists of muscarinic cholinergic receptors, such as 3′-quinuclidinyl benzilate and atropine (191, 216, 592, 641, 646). Efficacy and selectivity of antagonists suggests that sperm have a binding site that is similar to, but not identical to, the muscarinic receptor that has been isolated and cloned from somatic tissues (215, 705). Examination of exocytosis in sperm labeled with the fluorescent probe chlortetracycline indicates that both types of antagonists act by inhibiting the B–S state transition that is an intermediate step in the AR (190, 191). Importantly, both inhibitor classes specifically attenuated the AR initiated by soluble ZP3 or in sperm bound to intact zonae pellucidae, but had no effect on spontaneous exocytosis (190, 191, 216).

Additional indication of the specificity of the ZP3-initiated AR is provided by anti-ZP3 antibodies. Passive immunization of female mice with a ZP3 peptide (residues 336–342) had contraceptive effects in vivo (411). These studies support the role of ZP3-mediated sperm–egg interaction during fertilization in vivo.

Another comparison between the activities of the ZP and of ZP3 is afforded by agonist inactivation following fertilization. These alterations are thought to form part of the mechanism that prevents polyspermic fertilization. Thus, agonist activity is not readily detected in zonae pellucidae obtained from two-cell embryos or from parthogenically activated eggs and ZP3 isolated from these sources exhibits a similar attenuation (73). Finally, ZP3 is distributed throughout the ZP and is present at the ZP surface, consistent with its proposed role as agonist following sperm adhesion (664).

Thus, ZP3 has a dual function during fertilization; it is the primary adhesion molecule for sperm in the ZP and it initiates the AR. The notion that ZP3 is a physiologically relevant agonist and accounts for the stimulatory activity in the ZP is supported by this comparable biological regulation as well as by other studies, to be discussed later, that describe the signal transduction pathway engaged by ZP3.

Soluble Proteins of Serum or Female Reproductive Tract Secretions. It has been recognized since the 1960s that serum accelerates the rate of AR in sperm populations in vitro. In addition, the appearance of acrosome-reacted sperm in the female reproductive tract varied with the reproductive cycle (697, 699). Taken together, these observations suggested that sperm exocytosis may be controlled by circulating agents and lead to a detailed examination of serum constituents.

Albumin has been proposed as an AR-inducing agent, based on observations that it promotes exocytosis in vitro (697, 699) and that it is present in female reproductive tract secretions. Yet, it is difficult to reconcile the specificity of gamete interaction with the ubiquitous nature of serum albumin. It is more likely that albumin acts indirectly. The presence of macromolecules, including but not limited to albumin, is known to stabilize sperm during incubation in vitro. In addition, this protein capacitates sperm of several mammals, as discussed previously. By influencing both viability and maturation (that is, capacitation), albumin is expected to enhance the incidence of spontaneous exocytosis.

Other serum proteins have also been implicated in the AR. Complement-initiated membrane damage has been documented in sperm of several species. In addition, it has been suggested that regulated lysis through the alternative pathway may produce acrosomal exocytosis (100, 543). The membrane attack complex that is the end product of C3 convertase activation forms aqueous channels in the plasmalemma. These channels are permeable to small molecules, but not to cellular proteins, and the resultant Donnan forces produce osmotic swelling and lysis of targeted cells. Extension of the channel to intracellular membranes has not been reported in somatic cells. At present, a complement-mediated initiation of ARs is considered speculative.

Other Components of Serum or Female Reproductive Tract Secretions. As indicated, serum has a well-known stimulatory effect on acrosomal exocytosis. In addition to the proteins described previously, several low-molecular-weight constituents of serum or of reproductive tract secretions have been shown to initiate ARs.

Progesterone and several other progestins initiate ARs in human sperm (407, 615), although there have been preliminary reports of stimulation in sperm of other mammalian species (407). Activity was detected first in crude ovarian follicular fluids (598, 706) and subsequently attributed to progesterone (68, 615). The steroid-induced AR of human sperm has several of the characteristics anticipated of a physiological agonist, including: *(1)* a rapid time course, with exocytosis within minutes of progestin addition (407, 706); *(2)* a dependence on sperm capacitation, consistent with the observed specificity of gamete interaction (407, 615); and *(3)* appropriate localization of agonists in the vicinity of eggs, in that progestins are produced by cumulus oophorus cells in vitro (407). These and other studies, in which the structure/activity relationships of stimulatory steroids are determined and intracellular effectors of steroid action are described (see below), suggest that progestins act through a specific signaling cascade rather than by compromising cell viability. Steroids, including progesterone and its natural metabolites, have a wide range of nongenomic effects in somatic cells and in the amphibian oocyte (399). These actions entail secondary modulation of other receptors, such as the $GABA_A$ receptor (160, 568), as well as a potential action at specific plasma membrane steroid receptors (399).

Gastrin-releasing peptide (GRP) initiates ARs in vitro in sperm of several species, including human, monkey, and mouse. GRP, a 29 amino peptide of the bombesin subfamily of bombesin-like peptides, has a primary sequence of:

$$NH_2 - APVSTGAGGGTVLAKMYPRGSHWAVGHLM - COOH$$

The AR initiated by this peptide and by several analogs is dependent on prior sperm capacitation, occurs within minutes in vitro, and acts through a different signaling pathway than ZP3, as demonstrated by additive effects and by specific inhibitors. In addition, Northern hybridization demonstrates the presence of GRP receptors in mouse sperm. While GRP appears to be a potent agonist in vitro, these studies must be expanded to demonstrate the presence of peptide in the vicinity of eggs in vivo (438).

Glycosaminoglycans have also been proposed as AR-initiating agonists. Sulfated and unsulfated glycosaminoglycans are present in follicular and oviductal fluids and [^{3}H]heparin binding sites have been detected on sperm (273). Subsequent reexamination indicated that heparin and closely related glycosaminoglycans capacitate bovine sperm in vitro (274, 412, 414, 471, 472), but do not effect the AR directly.

Finally, several other low-molecular-weight components of serum have agonist activity under some conditions. Biogenic amines, such as catecholamines and seratonin, initiate exocytosis in some in vitro models. Yet the presence of receptors has not been demonstrated by ligand binding or by cDNA hybridization strategies. It has also been reported that the human sperm AR is initiated in vitro in the presence of extracellular ATP (220). Adenosine and related compounds have diverse effects on sperm that might influence the rate of exocytosis, including a regulation of adenylyl cyclase activity (95, 304, 425, 587). In addition, there is physiological evidence that adenosine receptors may be present on sperm (225), although direct binding studies have not been reported. It remains unclear whether these putative agonists act directly at the level of exocytosis or indirectly by modulating capacitation. This is a particular concern in the case of ATP effects, where previous studies suggested that adenosine and related compounds influenced capacitation of mouse sperm (225, 226). In addition, it must be demonstrated that these compounds are present at the site of fertilization.

Conclusions

It is now generally accepted that the AR is regulated in vitro by stimulatory agonists. Available information, culled both in vitro and in vivo and reviewed in this and subsequent sections, strongly suggests that ZP3 is the principal physiological agonist that regulates fertilization in vivo. Yet, the relative importance of ZP3, the progestins, and other putative agonists in controlling exocytosis is still a subject of discussion.

The presence of multiple agonists potentially provides an additional level of sperm selection. Subsequent discussion will establish that at least three classes of agonists (ZP3, progestins, and GRP) all produce elevations of sperm internal Ca^{2+} (Ca_i) that are essential for exocytosis. Such convergence of signaling pathways permits agonist summation and synergistic production of ARs. Thus, robust responses to soluble agonist, such as progestins, may initiate premature exocytosis and the associated failure of that sperm to fertilize. In contrast, a graded response to soluble agonists that elevates Ca_i to subthreshold levels may provide two advantagous. First, there is emerging evidence that the motility pattern of mammalian sperm is regulated by Ca_i (382, 383, 669); such altered motility could enhance sperm migration to the ZP. Second, summation of Ca_i responses could provide a responsive subpopulation of sperm with a kinetic advantage for exocytosis at the ZP. Resolution of this issue depends in turn on the demonstration of the site of the AR of the fertilizing sperm in vivo.

Signaling at the Sperm Plasma Membrane

The current understanding that hormones, neurotransmitters, and drugs produce their effects by associating with receptor molecules derives from the work of Langley and Ehrlich (322, 361). Recent advances in biochemical and molecular genetic techniques have expanded the understanding of the structure and function of these receptors. The paradigm for receptor-mediated signaling events in sperm is the guanylyl cyclase receptor system in echinoderms. Egg peptides, including the tetradecapeptide resact and the decapeptide speract, are secreted by sea urchin eggs and have a variety of effects on sperm, including elevation of cGMP and cAMP; stimulation of motility and respiration; induction of transmembrane Na^+, K^+, Ca^{2+}, and H^+ fluxes; and chemotaxis. The receptor that mediates these effects is the particulate form of guanylyl cyclase (30, 118, 237, 238). Thus, this system provides characterization of a receptor at both the protein and cDNA level, as well as a detailed understanding of extracellular ligand and intracellular effectors. Similar delineation of a receptor system mediating acrosomal exocytosis is not available in any animal species. This section will review our knowledge of receptors for ZP3 and progestins on sperm, as well as their mechanism of activation.

ZP3 Agonist Receptors. ZP3 is the best-characterized mammalian sperm agonist. As discussed previously, this protein functions both as the sperm adhesion molecule in the ZP and as an AR-initiating agonist. These dual biological activities are separable; agonist activity is selectively attenuated by enzymatic and chemical modification of the ZP3 polypeptide chain (75, 209, 218), as well as by in situ alterations of ZP3 that follow application of phorbol esters to eggs (192),

while adhesive interactions are unaltered. These and other studies suggest that adhesion activity is attributable to a 3.9 kDa O (serine/threonine)-linked oligosaccharide chain, whereas the polypeptide backbone of ZP3 is essential for initiation of the AR (209, 380). It is not certain whether these dual activities of ZP3 reflect activation of a single sperm receptor system or whether different receptors function. In either case, a ZP3-activated receptor must initiate a signaling cascade.

Several sperm proteins are candidate receptors for ZP3, based on the disruption of gamete interaction. As discussed previously, three proteins associate directly with ZP3 in in vitro assays, including a 95 kDa protein that is a tyrosine kinase substrate (379) and may possess intrinsic tyrosine kinase activity (378); $\beta 1,4$-galactosyltransferase (385, 413); and a 56 kDa galactose-binding protein (76).

The 95 kDa protein(s), when isolated from mouse sperm membranes by gel electrophoresis and subsequent elution into detergent-free solutions, exhibits ZP3-stimulated autophosphorylation (378). In addition, cross-linking of ZP3 glycopeptide-binding sites on sperm with bivalent anti-ZP3 initiates exocytosis (380). By analogy, many growth-factor-regulated tyrosine kinase receptors, such as the EGF receptor, are activated as a consequence of ligand-induced aggregation, stimulation of tyrosine kinase activity, and autophosphorylation (633). A similar mechanism of activation is inferred for the sperm 95 kDa protein. This protein will be considered in greater detail below.

While two other proteins that have affinity for ZP3 in vitro, there is at present no insight into the molecular mechanism of receptor activation. The $\beta 1,4$-galactosyltransferase has been isolated from mouse sperm (567) and is encoded by a spermatogenic cell-specific transcript (561). These and other observations (528, 560) suggest that the enzyme is a type II transmembrane protein with a short, NH_2-terminal cytoplasmic tail and an extracellular COOH-terminal. Two transcripts have been characterized, encoding 11 and 24 amino acid cytoplasmic domains. The coding sequences contain no apparent signaling motifs. Several classes of signal-transducing receptors on somatic cells (242, 311, 667), as well as the speract receptor of sea urchin sperm (147), also lack apparent intracellular signaling domains. In those cases signal transduction is accomplished by the formation of hetero-oligomeric complexes, including a ligand-binding component and a transduction subunit(s). In this regard, it is unlikely that the $\beta 1,4$-galactosyltransferase can propagate the ZP3 stimulus in the absence of associated subunits. Finally, any consideration of the mechanism of receptor activation involving the 56 kDa galactose-binding protein must await more detailed structural analysis of the protein or cDNA.

Progesterone Receptors. As discussed previously, progestins initiate ARs in sperm of several mammalian species. The likely mechanism of steroid action is through association with a cell surface receptor, as indicated by the potent agonist activity of bovine serum-albumin-conjugated progesterone (69, 408). In addition, the rapid response time (exocytosis within 1 min) and the transcriptionally inert state of sperm chromatin are also inconsistent with a genomic effect of steroids.

It is now widely accepted that steroids, including the progestins, act at several plasma membrane sites instead of, or in addition to, their genomic targets. The $GABA_A$ receptor is a chloride channel that is a member of the nicotinic acetylcholine receptor superfamily of ligand-gated ion channels (596). Progesterone metabolites are positive modulators of GABA-evoked chloride currents through these receptors, acting at a site distinct from the benzodiazapine site (399, 473, 568). In addition, progesterone is believed to promote the meiotic maturation of amphibian oocytes (388).

Structure/activity relationships provide some insight into the nature of the sperm-binding site. Pregnane compounds (steroids with a reduced A ring) are, in general, active at $GABA_A$ receptors (399). The selectivity of the progestin site on human sperm was examined by incorporating fura-2, the Ca^{2+}-chelate probe, into sperm and monitoring steroid-initiated Ca^{2+}_i elevations (see below). Progesterone, a compound that lacks a reduced A ring, is only weakly active at $GABA_A$ receptors but is a potent stimulator of sperm Ca^{2+}_i and exocytosis (68, 615). Many of the key compounds have not been tested. Nevertheless, it appears that the sperm site and the $GABA_A$ receptors have unique specificities. Isolation of these receptors has been hindered by the high levels of nonspecific interaction of steroid with phospholipids and membrane proteins.

Mediators of Receptor-Activated Second-Messenger Production in Sperm

The stimulatory signal provided by agonists is processed by the cell and leads to the production of intracellular second-messenger molecules. This section will consider the links between activated receptor and effectors. Several mechanisms of signal processing have been described and coexist within sperm cells.

G Proteins. Many cells couple activated receptor to effector elements through a class of heterotrimeric, GTP-binding regulatory proteins, or G proteins. G proteins are composed of an α subunit that possesses

intrinsic GTPase activity, as well as β and γ subunits. Extensive heterogeneity is exhibited in these subunit: α, β, and γ proteins are encoded by 17, 4, and 6 different genes, respectively (67, 287). Diversity and specificity of signaling are provided by the association of subunit proteins, producing a large (and possibly combinatorial) number of functional G proteins and these, in turn, may differ in receptor and effector specificity (87, 341, 594, 609).

Activation of G protein occurs as a consequence of the binding of agonist to receptor. This either encourages (287) or alters (203) the association of receptor with G protein, thereby promoting guanine nucleotide exchange on the α subunit and the dissociation of α-GTP from $\beta\gamma$ dimers. The liberated α-GTP and $\beta\gamma$ components then interact with target effectors, discharging their signaling functions (67, 287). G proteins regulate a wide range of effectors, including adenylyl cyclase in hormonally responsive and sensory systems; cGMP phosphodiesterase in the visual system; phospholipase A_2, C_β, and D activities; and Na^+, K^+, and Ca^{2+} channel conductance. These effectors in turn generate or degrade second messengers (287).

The presence of G proteins in cells is often demonstrated by immunochemical methods and by application of bacterial exotoxins. The toxins of *Vibrio cholerae* and *B. pertussis* contain intrinsic ADP-ribosyltransferase activity and use several classes of G proteins as substrates. Pertussis toxin covalently modifies the α subunit of many members of the G_i family of G proteins, including α_{i1-3}, α_{o1-2}, α_{t-rod}, α_{t-cone}, and α_{gust}. ADP-ribosylation prevents the promotion of nucleotide exchange by receptor, and hence inhibits G protein activation. In contrast, cholera toxin catalyzes ADP-ribosylation of G_s, leading to an inhibition of GTP hydrolysis and consequent constitutive activation (431). These exotoxins are only relatively specific and have other cellular substrates, as well as nonenzymatic effects on intact cells (416). Nevertheless, they have proven valuable tools for the study of G proteins when applied in conjunction with other techniques.

Mammalian sperm have several different classes of G proteins. The α subunits of G_{i1-3} have been detected by pertussis-toxin-catalyzed [^{32}P]ADP-ribosylation and by immunochemical methods. These proteins were further localized to the acrosomal region of the head by immunocytochemistry (61, 241, 247, 292, 348). The mature sperm cell also contains the α_z, a pertussis-toxin-insensitive member of the G_i family, in the posterior and lateral regions of the head (567, 598). Additional isoforms of the α subunit are present during spermatogenesis, as demonstrated by both immunochemical methods and Northern hybridization. These include a truncated α_s mRNA, although neither translation products nor cholera toxin substrates have been detected (282, 348); α_o, with protein levels diminishing during spermiogenesis (325, 595); and suggestions of additional α isoforms (292). β subunits have also been detected by immunoprecipitation, but to date isotypes have not been differentiated.

Further support for the role of G proteins in sperm function is derived from studies of appropriate receptors in these cells. G protein–coupled receptors often, but not always (443, 453), are disposed across the cell membrane in a heptahelical structure, consisting of seven transmembrane sequences (164). The transmembrane sequences are among the most highly conserved within this family of receptors and have therefore been used as anchors for the identification of putative G protein linked by polymerase-chain-reaction-based molecular cloning strategies. This approach has identified sperm receptors related to the adrenergic receptor (410) and to the odorant receptors in the olfactory epithelium (470). A second class of G protein–coupled receptors includes the mannose-6-phosphate receptor (453), a single tansmembrane-spanning protein that has been localized on sperm (448). The specific functions of these receptors remain unknown. Nevertheless, G proteins are present in sperm and may regulate cellular function during sperm development, as well as in the mature cell.

Several lines of evidence indicate that G proteins mediate an essential component of the ZP3-initiated AR. First, ZP3 directly activates G proteins in permeabilized sperm preparations, as demonstrated by an enhancement of [^{35}S]GTPγS binding and [^{32}P]GTPase activity (647, 674). The stimulatory activity is attenuated by pretreatment with pertussis toxin. Since G_{i1-3} are the only toxin-sensitive G proteins in sperm, it is presumed that one or all of these isoforms are targets of an activated ZP3 receptor. This effect is specific in that no stimulatory effect was observed with ZP1 or ZP2 obtained from oocyte zonae pellucidae. In addition, ZP3 derived from the zonae pellucidae of two-cell embryos failed to activate sperm G proteins (647), consistent with the absence of AR-initiating activity in embryo ZP3 (73).

The second line of evidence implicating G proteins in the ZP3-induced AR is provided by the effects of pertussis toxin on exocytosis. Treatment of sperm from several mammalian species with toxin inhibits exocytosis of sperm bound to the ZP or incubated with soluble ZP3 (190, 217, 369). Exocytosis was inhibited at an early stage of the signal transduction sequence, such that the obligate elevation of sperm intracellular ionic mediators, such as Ca^{2+} and pH, has not occurred (210, 217), and sperm have not progressed from the B stage into the S stage, as reported by chlortetracycline fluorescence microscopy (177).

Third, the distribution of G proteins in sperm is consistent with a role in the AR. As discussed previously, α_i has been localized to the acrosomal region of the mouse sperm head in immunocytochemical experiments utilizing site-directed antisera (247). Similarly, an uncharacterized a subunit, as well as $\beta\gamma$ subunits, was restricted to the acrosomal region of the bovine sperm (191).

Thus, ZP3 initiates exocytosis through a pertussis-toxin-sensitive G protein. In contrast, the AR initiated by progesterone is not inhibited by pertussis toxin (612). It is uncertain whether progestins act through a pertussis-toxin-insensitive G protein, such as G_z or through an unrelated mechanism. While the downstream mediators of progesterone action have been characterized (see below), little is known about the membrane mechanisms.

Receptor Tyrosine Kinases. A second pathway used in many cells to generate intracellular signals entails the activation of receptor tyrosine kinases. In members of this class, including the receptors for PDGF and EGF, information is relayed by ligand-induced receptor dimerization and consequently to inter-receptor tyrosine phosphorylation (633). Phosphotyrosyl residues, embedded within specific sequences (476, 576), interact directly with and consequently activate a class of target proteins that contain a ≈100 amino acid, Src homology 2 (SH2) domain. SH2 proteins either generate second messengers directly, as in the case of phospholipase-Ct, or they modulate second-messenger generation, as in the case of the GTPase-activating protein (GAP) that regulates Ras function and the phosphatidyl inositol 3-kinase (475, 633).

Protein tyrosine kinase activity is evidently present in mammalian sperm, as demonstrated by immunochemical and immunocytochemical examination using antiphosphotyrosine antibodies (66, 379) and by in vitro tyrosine kinase activity toward synthetic peptide or endogenous protein substrates (66, 181, 378). Attempts at in vivo labeling have been unsuccessful (66). This failure may be due to the apparent compartmentalization of ATP pools in this cell (31, 116), with extracellular P_i entering the kinase-accessible pool at low rates (444). Receptor and nonreceptor tyrosine kinases may be distinguished as only the former will autophosphorylate. In this regard, and as discussed previously, the 95 kDa region of a mouse sperm protein gel contains both tyrosine kinase activity and antiphosphotyrosine antibody immunoreactivity (378). Thus, present evidence indicates that sperm contain at least one receptor tyrosine kinase.

Protein tyrosine kinases have been implicated in the ZP3-initiated AR of mouse sperm. First, 95 kDa protein(s) has been purified by gel electrophoresis and shown to both bind ZP3 and antiphosphotyrosine antibodies, as demonstrated in solid-phase and immunoblot assays (379). Second, protein(s) of this size class was recovered by elution from gel slices and shown to autophosphorylate in solution in a ZP3-stimulated manner (378). Previous studies suggested that these proteins were phosphorylated on tyrosyl residues, based on antiphosphotyrosine immunoreactivity (379). Third, a membrane-permeable inhibitor of tyrosine kinase activity produces a concentration-dependent attenuation of the ZP3-induced AR (378). Finally, immunocytochemical localization of phosphotyrosyl residues demonstrates a concentration in the acrosomal region of the head consistent with a role in exocytosis (379).

The proposed mechanism of tyrosine kinase activation involves ligand-induced aggregation (380), similar to that demonstrated for a variety of growth factor receptors (633). These studies entailed production of ZP3 glycopeptides by extensive proteolysis of the parent glycoprotein with pronase. These glycopeptides were previously shown to retain sperm adhesion receptor activity but to lack the ability to promote ARs (209). Subsequent cross-linking of this glycopeptides on the sperm surface with bivalent antibodies directed against ZP3 initiated exocytosis (380). It is presumed, but not yet demonstrated, that only single class of glycopeptide binds to a single receptor and that antibody treatment then produces cross-linked homodimers. The alternative or additional formation of hetero-oligomers is feasible, given the diverse N- and O-linked oligosaccharide chains on ZP3 and the number of carbohydrate recognition proteins on sperm. Further studies are required in order to determine the molecular events of tyrosine kinase activation.

Second-Messenger Production

The models of exocytosis that emerge from studies in a variety of cellular systems have focused on the essential mediatory roles of intracellular ions and cyclic nucleotides (318). A similar role of these second messengers has been examined in mammalian sperm during the spontaneous AR. Yet, the appreciation that physiological ARs are initiated by specific agonist-activated signal-transducing pathways has been gained only in the last decade. This section will review the central role of Ca^{2+} in the agonist-initiated pathways of mammalian sperm as well as consider other ionic and non-ionic second messengers that may regulate sperm exocytosis.

Ca^{2+} and Ca^{2+} Channels

The essential role of Ca_i in coupling stimulatory signals to exocytosis was first appreciated in the pioneering studies of Douglas (169) and of Katz (326, 327) and subsequently extended to other somatic cells. Several lines of evidence indicate that Ca^{2+} is also an obligate intracellular mediator of the sperm acrosomal exocytosis. The spontaneous AR of mammalian sperm is dependent upon external Ca^{2+} (704) and associated with $^{45}Ca^{2+}$ influx (521). In addition, Ca^{2+}/H^+ ionophores (257, 600, 618) or chemical treatments that otherwise elevate Ca_i (445, 695) promote the AR of mammalian and echinoderm sperm. Similar events occur in sea urchin sperm, where the AR that is initiated by its egg-derived agonist (a fucosyl-sulfated glycoconjugate) entails an elevation of Ca_i (263, 540).

Ca^{2+} is also an essential intracellular mediator of the AR induced by ZP3. Ca_i elevations are observed in sperm treated with soluble agonist (125, 217) or with intact zonae pellucidae (592), as reported by intracellular, ion-selective fluorescent dyes. Examination of single cells responding to ZP3, using fluorescence microscopy enhanced by digital image processing has revealed a number of features of the induced Ca_i elevation. A sustained Ca_i elevation occurs prior to the AR, is restricted to the subpopulation of sperm that go on to acrosome react, and is apparent in both the head and tail regions of sperm (217). ZP3-binding sites are limited to the head region of sperm (74), and some mechanism of signal propogation must account for responses in tail regions (see below).

Mammalian sperm possess voltage-gated Ca^{2+} channels with properties resembling the L-type Ca^{2+} channel of vertebrate cardiac and skeletal muscle, as well as on other somatic cells [for reviews of L-channel structure and function, see (290, 415, 631)]. A prolonged elevation of Ca_i is observed when sperm membrane potential is depolarized with external K^+ (32, 210). Comparisons of the response of sperm and of vertebrate muscle L channels reveal several similarities. In both cases, relatively large depolarizations are required in order to evoke channel opening (32, 210). In addition, the depolarization-induced elevation of sperm Ca_i is attentuated by several classes of L-channel antagonists, including dihydropyridines, arylalkylamines, benzothiazapines, and transition metal cations (290). A Ba^{2+}-conducting channel that is partially blocked by a dihydropyridine and by an arylalkylamine has recently been reconstituted from boar sperm membranes (131, 132). Finally, a saturable class of binding sites for [^3H]PN200-110, a dihydropyridine antagonist of muscle and other somatic cell L channels, is present on sperm membranes. These binding sites and L channels exhibit similar selectivity, as demonstrated in competitive displacement programs, as well as similar ligand binding affinities ($K_d \sim 0.5$ nM) (210).

Despite these extensive similarities, the sperm and somatic cell L channels also exhibit some important functional differences. First, dihydropyridine agonists such as Bay K8644 enhance conductance through the somatic cell L channel by increasing the mean open time (that is, the mean time that a channel is open during each opening event). This compound also modulates the binding of dihydropyridine antagonists to L channels (249, 315, 621). In contrast, Bay K8644 has no effect on sperm Ca_i or on the binding of dihydropyridine antagonists to sperm membranes (210). Second, the sperm and somatic cell L channels differ in the selectivity to blockage by divalent metal cations. The Ca_i elevation that is evoked in mammalian sperm by membrane potential depolarization is relatively more sensitive to blockade by Ni^{2+} than by Co^{2+} (210). This selectivity pattern is inverted from that anticipated at somatic cell L channels (249). These differences may reflect the presence of different isoforms of L-channel subunits in sperm, similar to the case in somatic cells where the α_1 subunit (the pore-forming constituent) is encoded by multiple genes as well as by alternatively spliced mRNAs. Finally, sperm L-channel opening is apparently regulated by internal pH (32, 207, 210). While a similar modulation of somatic cell L channels has been reported (321, 350, 490) there remains a question of its functional significance (497). The significance of pH regulation of sperm L channels will be discussed in a subsequent section.

The sperm L channel plays an essential role in the ZP-induced AR. Sperm exocytosis is inhibited in a concentration-dependent fashion by the same antagonists of somatic cell L channels that were previously shown to attenuate the depolarization-induced elevation of sperm Ca_i, including dihydropyridines, arylalkylamines, benzothiazapines, and transition metal cations. The strong association between the effects of these antagonists on these two processes suggests that an agonist-evoked depolarization underlies the initiation of ARs by ZP3 (210). In this regard, direct evidence of a ZP3-mediated depolarization of sperm membrane potential is not yet available.

The ZP activates sperm L channels through a pertussis-toxin-sensitive pathway (217). As discussed previously, this indicates that these channels are downstream effectors of ZP3-regulated G_i. The L channel of somatic cells is a voltage-gated pathway, although conductance through open channels is furthered by G proteins through indirect and direct mechanisms. During indirect modulation G proteins are linked to channels through the activation of a second-messenger-

dependent pathway such as a protein kinase. One of the most thoroughly studied examples of this regulatory mode is the control of cardiac L channels by β-adrenergic agonists. Agonist-activated receptors engage G_s, leading to the stimulation of adenylyl cyclase and the consequent protein kinase A–mediated phosphorylation of L-channel subunits (290, 291). Alternatively, selected G proteins can act directly on selected ion channels. Direct modulation is restricted to the membrane, does not require soluble cytosolic components, and can be mimicked in by application of activated G_a (α-GTPγS) to the inner face of the membrane. In some cases of direct modulation, such as the cardiac K^+[ACh] channel, the opening event is gated specifically by G_a protein, whereas in other cases, such as the the voltage-gated L-type Ca^{2+} channel and Na^+ channels, opening is evoked by membrane potential and modulated by G proteins (67, 93).

It is likely that sperm L channels are regulated by G proteins through indirect modulation. Channel opening in response to ZP3 is inhibited by pertussis toxin. Yet, this block can be circumvented by depolarizing sperm membrane potential while simultaneously alkalinizing internal pH (see below) (210). Thus, channel activation in response to pharmacological manipulation does not require functional G proteins, consistent with an indirect modulation. In addition, the alternative hypothesis of direct modulation is inconsistent with results from somatic cell systems. ZP3 signal transduction is mediated by G_i (647). To date, direct modulation of L channels has been observed with α_s and α_o, but not with α_i (67, 93). Resolution of this issue must await isolation and reconstitution of sperm L channels.

Ca_i is also an important mediator of the progesterone-initiated AR. Progestins produce a transient elevation of Ca_i in human sperm that is apparent within seconds and subsides within 1–2 min (68, 614, 615). The regulated Ca^{2+} transport pathway is blocked by La^{3+} and is also permeable to Mn^{2+}, but otherwise is uncharacterized (68, 615).

Internal pH

Internal pH (pH_i) also appears to be an important mediator of the ZP3-initiated AR. pH_i has been proposed as a regulator of a variety of cellular functions [(79, 98), but also see (236, 616)], and two types of evidence have suggested a role in sperm–ZP interaction. Incubation in alkaline culture medium accelerates the rate of spontaneous exocytosis in guinea pig sperm (301). In addition, the AR of sea urchin sperm that is induced by egg jelly agonist is associated with elevations of pH_i (122, 367, 618), possibly as a consequence of activation of an Na^+-H^+ antiporter (540, 541).

The role of pH_i in the ZP3-initiated AR was examined directly during interaction of sperm with solubilized or structurally intact zonae pellucidae (217, 371, 372). Mammalian sperm apparently maintain a relatively acidic pH_i, with values of 6.5–6.8 indicated by the intracellular fluorescent indicator carboxyfluorescein (28, 32, 217, 471) or by ^{31}P-NMR (nuclear magnetic resonance) of dense cell suspensions (573). Stimulation with soluble ZP agonist produces a further alkalinization of 0.1–0.2 pH unit that is apparent within ~1 min. The ZP3-initiated alkalinization is produced through a pertussis-toxin-sensitive pathway, indicating a direct or indirect modulatory role of sperm G_i (217).

It is likely that alkalinization is required for the Ca^{2+} influx and exocytosis that are initiated by ZP3. As discussed previously, sperm possess an L-type Ca^{2+} channel that opens in response to ZP3 and that mediates an essential component of the agonist-induced elevation of Ca_i (210). Several lines of evidence indicate that the conductance properties of this channel are modulated by pH_i. First, sperm Ca_i is elevated by the depolarization of membrane potential that results from elevated external K^+. The effect is mediated by L channels and the magnitude of the response is dependent upon external pH (32). Second, gramicidin is a pore-forming ionophore that dissipates Na^+, K^+, and H^+ gradients, thereby depolarizing membrane potential and setting pH_i equal to the prevailing external pH (645). Gramacidin promotes the AR of sperm from several mammalian species at slightly alkaline pH but not at acidic pH. Third, in sperm treated with elevated K^+, exocytosis is only promoted when depolarization is accompanied by either an elevation of external pH or by addition of a permeant weak base. In these cases exocytosis is inhibited by L-channel antagonists, indicating a mediatory role for those channels (210). Permeant weak bases, such as NH_4Cl and CH_3NH_2, produce intracellular alkalinization without altering external pH (518) and hence eliminate the possibility of an extracellular site of action.

The mechanisms that control pH_i in mammalian sperm have not been well characterized. For most somatic cells it is thought that the Na^+/H^+ and Cl^-/HCO_3^- exchange activities represent the principal pathways that maintain resting pH_i in the face of acidic and alkaline challenges, respectively (616). In addition, these pathways in somatic cells are often regulated by G proteins (74), consistent with the observed pertussis toxin sensitivity of ZP3-induced alkalinization in sperm (217).

Many features of pH_i regulation in mammalian sperm are inconsistent with the presence of a Na^+/H^+ antiport (207). Alternatively, sperm may regulate pH_i through an anion exchange mechanism. While HCO_3^- is required for the ZP3-induced AR in mouse (646) and bull (211), and for the spontaneous exocytosis of guinea pig sperm (298), an anion uptake activity with several of the characteristics of the Cl^-/HCO_3^- exchange pathway is present in sperm (459). Yet it remains uncertain whether these effects of HCO_3^- are a consequence of its role in pH_i regulation or due to the direct action of this anion on sperm adenylyl cyclase (455–458). At present, the molecular mechanism of ZP3-initiated alkalinization is unknown.

Although the role of pH_i alterations in exocytosis is not well understood, it is clear that alkalinization alone is not sufficient to promote exocytosis (210). Rather, pH_i appears to be an important regulator of the sperm L channel. It may function as a coincidence detector, acting in concert with other signals provided by ZP3. The importance of such systems can be appreciated by considering sperm during storage in the cauda epididymides. Epididymal plasma is relatively enriched in K^+ (Na^+/K^+ ratio ~1/2) (392) and would be expected to depolarize sperm membrane potential and open voltage-gated ion channels. In this regard, the acidic nature of these secretions (pH > 7) may prevent premature exocytosis.

Downstream Effectors of Receptor Activation

A variety of second-messenger-regulated systems are present in mammalian sperm. Yet there is no well-characterized example in which a physiological agonist modulates the function of a key effector protein (for example, an ion channel) through the activation of a second-messenger pathway. In sea urchins, the egg jelly component that induces the AR activates a number of distal effector systems. There, it has been suggested, though, that Ca_i is the primary mediator of exocytosis and accounts for the stimulation of these other pathways (237, 239). Similarly, the downstream events of the mammalian sperm AR may be regulated by, and be secondary to, Ca_i elevation.

In this regard, calmodulin is present in mammalian sperm in high concentrations and likely mediates many of the consequences of elevated Ca_i. An involvement of calmodulin in exocytosis is suggested by the acceleration of spontaneous ARs in guinea pig sperm induced by W7, a calmodulin antagonist (437). Indirect immunofluorescence revealed that calmodulin is deposited in the acrosomal region of the head and in the flagellum (204, 446). Furthermore, this distribution may be altered during capacitation, as sperm become competent to complete exocytosis (619). Several calmodulin targets have been identified in mammalian sperm including calcineurin, a calmodulin-dependent protein phosphatase (611); cyclic nucleotide phosphodiesterase (114); and several other proteins (5, 648). There is, as yet, no evidence of a calmodulin-regulated ion channel in mammalian sperm, similar to that implicated in the egg-jelly-induced AR of sea urchin sperm (262, 263).

In addition, a Ca^{2+}/calmodulin-dependent adenylyl cyclase has been identified and partially purified from mammalian sperm (261, 300, 304, 437). Preliminary studies suggest that solubilized ZP glycoproteins produce a transient elevation of mouse sperm cAMP and that this effect is dependent upon extracellular Ca^{2+} (347). These cells contain multiple isoforms of cAMP-dependent protein kinase (58, 59, 237, 239). Differences in the sensitivity of these isoforms to cAMP analogs and to inhibitors may explain some of the diverse and contradictory results that have been reported. Yet the role of these kinases in ZP3 signal transduction is not well understood.

Additional or alternative signaling pathways are controlled by the turnover of membrane polyphosphoinositide (63). In somatic cells, receptor activation leads to the stimulation of phosphatidylinositol-4,5-bisphosphate (PIP_2)-specific phospholipase C, with the subsequent hydrolysis of PIP_2 and production of inositol 1,3,5-trisphosphate (IP_3) and sn-1,2-diacylglcyerol (DAG). Multiple isoforms of phospholipase C have been identified and cloned, including β and γ forms that are regulated by G proteins and by receptor tyrosine kinases, respectively. Downstream events are controlled by these second messengers through regulation of Ca_i and of PKC, respectively (64).

Mammalian sperm contain several enzymatic components of this pathway, including a PIP_2-specific phospholipase C (474, 639) and PKC (115). In addition, progesterone activation of human sperm leads to stimulation of PIP_2 turnover, to Ca_i elevation, and to exocytosis (615). Three types of observations indicate that PIP_2 turnover may be downstream of, and regulated by, the primary elevation of sperm Ca_i. First, the progesterone effects are attenuated by extracellular La^{3+} (615), a known antagonist of Ca^{2+} influx. Second, PIP_2 turnover is enhanced in ram sperm by the Ca^{2+}/H^+ ionophore, A23187 (277, 514, 515). Finally, a related accumulation of IP_3 is initiated in sea urchin sperm by the egg jelly agonist that promotes exocytosis and this effect is mitigated by Ca^{2+} channel antagonists (166). It is likely that ZP3-induced Ca_i elevations will also activate this pathway. Yet, the role of PIP_2 hydrolysis in the mammalian AR remains to be determined. These and other signaling pathways (237, 347) are present in mammalian sperm and may particate in the AR.

In conclusion, mammalian sperm contain a variety of signaling pathways. Yet, our understanding of the mechanism of agonist-promoted exocytosis is limited.

The isolation of component elements and subsequent reconstitution of other membrane fusion pathways in vitro has provided great insight into the underlying molecular mechanisms of such complex systems as the Golgi transport system (520, 575, 676) and the neurosecretory system (318, 330). The development of similar cell-free systems would advance our understanding of the molecular and cellular biology of the AR.

Gamete Fusion and Cortical Granule Exocytosis

Sperm that have penetrated the ZP come in direct contact with the oolemma, or the egg's plasma membrane. The fusion of gamete plasma membranes then initiates the late events of the fertilization process, including the cortical reaction (CR) and associated events that are believed to comprise the block to polyspermic fertilization; pronuclear formation and union to form an embryonic genome; and the transcriptional and translational alterations that comprise the activation of development. The biochemical and cellular reactions that underlie these final phases of fertilization have been studied extensively, particularly in marine invertebrates and in anurans (150, 677), and reviewed recently (668). Yet, examination of the gamete fusion events that trigger this process have only become amenable to analysis with recent refinements in biochemical and molecular genetic techniques. This section will focus on the events leading from gamete membrane fusion to cortical granule exocytosis. This discussion focuses on events within mammalian eggs, although comparisons will be drawn to well-studied nonmammalian models when appropriate.

Cortical Granules

Cortical granules are small, spherical, membrane-bound organelles that are found in the cortical region of unfertilized eggs and thought to resemble lysosomes (115–118, 158, 264, 304, 441, 547). These granules fuse with the oolema at fertilization and, by releasing their contents (including proteinases) into the perivitelline space, alter functional properties of the ZP (secondary block to polyspermy) (117, 356–360, 370, 489). Mouse eggs contain approximately 4500 cortical granules (ranging in diameter from 200 to 600 nm) within about 2 μm of the plasma membrane (236). Cortical granules first appear during oocyte growth, associated with an expanding Golgi that has moved to the subcortical region of growing oocytes. Although it is clear that cortical granules are derived from Golgi, certain evidence suggests that there is a contribution from multivesicular bodies as well as from granular endoplasmic reticulum. Morphological studies suggest that the cortical-granule population is heterogeneous with respect to contents, with some granules even containing ordered crystalline arrays. It is unclear whether this heterogeneity of cortical-granule contents reflects functional differences or simply different extents of granule maturation. In any case, it is clear that oocyte growth is accompanied by formation and accumulation of increasing numbers of cortical granules in anticipation of ovulation and fertilization. Recently, a mouse egg cortical-granule protein, 75 kDa (p75), was identified as a product of growing oocytes, and should lead to new insights into cortical-granule biosynthesis during oogenesis (262).

Gamete Membrane Fusion and the Receptor Question

In a trivial sense, the external signal that initiates egg activation and cortical-granule exocytosis is contributed by sperm. Yet, the molecular events of mammalian gamete membrane interaction and the nature of the stimulatory signal that it provides remains unknown. In particular, there is still discussion as to whether specific gamete receptors mediate the activation process or whether membrane fusion per se provides the cue.

Recent observations suggest that the adhesive interactions preceding fusion of gamete plasma membranes are mediated by cell surface proteins. Early studies of mammalian gamete fusion revealed a relatively low selectivity and suggested that specific interactions may be absent. Arguments were made that all selectivity occurred at early stages of the fertilization process or that plasma membrane recognition events were highly conserved. Subsequent reexamination revealed that there are extensive species-to-species variations in gamete membrane interactions. Hamster eggs from which zonae pellucidae have been removed are promiscuous, fusing with a wide range of mammalian sperm. In contrast, eggs of other mammalian species exhibit some degree of discrimination, consistent with the participation of surface receptors (16, 17, 206, 698, 699).

PH-30 is a sperm surface protein that has been implicated in gamete fusion based on the ability of monoclonal antibodies to inhibit fertilization at that stage of the process (496). Isolation and characterization of protein and cDNA has revealed that PH-30 is an oligomer, consisting of α and β subunits in an unknown stochiometry. The α and β subunits are composed of polypeptide chains with deduced molecular masses of 29,700 and 39,000, respectively (78). These polypeptide chains are glycosylated and may also be subject to other post-translational modifications, resulting in mature subunit sizes of PH-30 α and β of 60,000 and 44,000, respectively (77, 496). N-terminal

domains of both subunits are related to the disintegrin family of integrin-binding proteins, as based on similarity of deduced amino acid sequences (78). Integrins are integral membrane proteins that mediate a variety of adhesive interactions between cells and extracellular matrix components, often by binding RGD (aspartyl-glycyl-argininyl) sequences in target proteins (305, 320). Earlier observations are consistent with the presence of integrins on sperm (346) and, further, a role in the prefusion interactions between gametes is suggested by observations that synthetic RGD peptides also inhibit the fertilization of ZP-denuded eggs in vitro (92). A 94 kDa protein on the surface of mouse eggs may also participate in gamete interaction, based on the association of fertility and the presence of this protein following a series of protease treatments (80). It is not known whether this 94 kDa protein is the conjugate integrin for PH-30.

In addition to a potential role in adhesion, PH-30 α also contains a region of its putative extracellular domain that may mediate gamete fusion (78). Similar fusion domains have been identified on influenza virus hemagglutinin and are believed to promote viral fusion with cellular membranes, based on genetic, biochemical, and biophysical studies. Signature features of the hemagglutinin fusion region, as well as putative fusion domains on other viral proteins, include an amphipathic helical structures, often separated by a hinge region, in the extracellular domain of the protein (670, 671). A similar domain has been identified in PH-30 α and, by extension, may participate in the fusion of gamete membranes (78). Fusion domains may also participate in the fertilization of marine invertebrate eggs, as acrosomal proteins from sea urchin and abalone sperm promote phospholipid vesicle fusion in vitro (295, 346). Understanding of the mechanism of membrane fusion has been advanced by the characterization of fusion pores as intermediates in this process in somatic cells (10, 11, 424, 582–584, 714) and in sea urchin gametes (398), as well as by the identification of intracellular proteins that are implicated fusion (318, 520, 575, 676). The precise participation of fusion peptides in this process is not certain. In particular, it is not known whether these peptide domains initiate the formation of fusion pores, as suggested by familiar models (for example, 671), or function at downstream sites such as the expansion of the fusion pore (545).

PH-30 or similar sperm components may participate in adhesion and fusion of gamete plasma membranes. Adhesive interactions, including many that are mediated by integrins, are known to provide instructive signals that regulate the function of the interacting cells (320). Thus, it is inferred that egg activation and cortical-granule exocytosis are initiated by specific sperm ligands acting through specific egg receptor proteins. Yet, the issue of whether the stimulatory signal is provided through a receptor pathway or by fusion has not been resolved.

There are two plausible and nonexclusive mechanisms for fusion-induced activation. Fusion results in the recruitment of sperm membrane channels and other components into the oolemma. This could alter membrane characteristics and directly activate signaling pathways (127). A second mechanism is based on the presence of activating factors in sperm cytoplasm that initiate egg-signaling cascades after they have introduced into the egg cytoplasm at fusion. In marine invertebrates gamete fusion precedes the observed activation of intracellular second messengers, such as Ca_i (398). Moreover, cytosolic factors from sea urchin (142) and mammalian sperm (590, 601) will activate oocytes following microinjection. Candidates for this sperm factor include Ca^{2+} (313), cyclic nucleotides, and IP3, although there is no present consensus (174, 668). Similarly, acrosomal extracts from *Urechis* sperm will activate eggs (254, 255), suggesting the presence of specific ligand–receptor interactions. In this regard, cDNAs have recently been isolated for a 350 kDa glycoprotein from sea urchin egg membranes that are believed to mediate the sperm–egg association. The deduced primary sequence has no signature signal transduction sequences in its putative intracellular domain (219). Thus, while there is an expanding appreciation for the intracellular mediators of egg activation, the initiating events are not yet understood.

From Oolemma to Cortical Granule—Signal Transduction Pathways

Sperm activation of eggs produces several responses, including alterations of membrane potential and second messengers. Yet the sequential relationship between these responses and their role in the initiation of cortical-granule secretion are still subjects of discussion. In particular, activation produces a well-characterized elevation of Ca_i that is a key regulator of cortical-granule exocytosis. The mechanisms that control Ca^{2+} mobilization are the subject of intense scrutiny.

Sperm produce well-characterized alterations in egg membrane potential. In sea urchins and other nonmammalian species this response consists of an Na^+-dependent depolarization and has been linked to the block to polyspermic fertilization (271, 398, 447, 668). In contrast, mammalian oocytes exhibit a recurring hyperpolarization (307, 420). The ionic basis of this hyperpolarization and its role in the mammalian fertilization process are not well understood. In particular,

it is not known whether this membrane potential response reflects local elevations of Ca_i that are not able to trigger Ca^{2+}-induced Ca^{2+} release but yet are sufficient to activated Ca^{2+}-dependent K^+ channels (3). Further studies are required to define the coupling between hyperpolarization and Ca^{2+} mobilization.

Several second-messenger systems are present in oocytes and regulate Ca_i. Cyclic ADP-ribose is a potent Ca_i-mobilizing agent in sea urchin eggs; it promotes release of sequestered Ca^{2+} from internal stores, possibly through regulation of the ryanodine-sensitive Ca^{2+} release channel (124, 233–235, 248, 366, 368, 525). There have, to date, been no reports of either the ADP-ribosyl cyclase enzyme or a cyclic ADP-ribose-sensitive Ca_i store in mammalian eggs. In contrast, other signaling pathways have been implicated in the control of Ca_i and cortical-granule secretion in mammalian eggs. Several lines of evidence suggest that oocyte G proteins mediate an early stage of the sperm-activated signaling pathway, leading to cortical-granule exocytosis. The mammalian oocyte contains at least two different classes of G proteins, the pertussis-toxin-sensitive G_i and the pertussis-toxin-insensitive G_q (675). Microinjection of GTPtS, a high-affinity G protein agonist, produces Ca_i oscillations and cortical-granule exocytosis. Conversely, the G protein antagonist GDPβS blocks the sperm-initiated Ca_i elevation of hamster eggs (133, 418). Finally, in mammalian and nonmammalian oocytes G proteins can be activated by expression of mRNA encoding G protein–linked neurotransmitter receptors. Subsequent activation of these receptor produces secretion of cortical granules, as well as a variety of other responses associated with fertilization (342, 343, 675). The emerging consensus from study of mammalian and nonmammalian eggs is that experimental activation of G proteins is sufficient to promote cortical-granule release. Missing from this analysis is the demonstration of G protein stimulation by sperm and, ultimately, the identification of the inferred sperm-activated receptor that controls oocyte G protein function.

Oocytes also possess components of the phosphatidylinoside-linked second-messenger system. This pathway consists of the hydrolysis of polyphosphoinositides by G protein– or by tyrosine-kinase-regulated phospholipase C, leading to the production of diacylglycerol and soluble inositide phosphates, including IP3; these second messengers in turn activate PKC and elevate Ca_i, respectively (64, 65). Injection of IP3 into mammalian oocytes activates the characteristic early developmental events, such as Ca_i transients and cortical-granule exocytosis (133, 353, 418). Similar effects are observed in the eggs of many nonmammalian species, including anurans, teleosts, and echinoderms (174,

668). IP_3 binds to and activates a Ca^{2+} release channel, leading to elevation of Ca_i (64). Injection of antibodies directed against this channel/receptor inhibits the Ca_i elevation that is produced by fertilization (421). Recently a cDNA encoding the IP_3 receptor has been isolated from *Xenopus laevis* oocytes and reveals extensive sequence similarities to the well-characterized receptors from mammalian brain (352). In addition, the treatment of mouse eggs with diacylglycerol or phorbol ester activators of PKC inhibits fertilization and also produces the proteolytic modifications of ZP2 in the ZP that are observed following fertilization (192, 193). These observations are consistent with a role for both limbs of the bifurcated polyphosphoinositide-linked messenger systems in the activation of mammalian eggs following fertilization. Furthermore, the data support a model in which IP_3 is an upstream regulator of Ca_i elevation. In contrast, the sequential relationship between IP_3 generation and Ca_i mobilization in sea urchin oocytes is not understood (668).

Ca_i in egg cells is elevated in response to these or other second messengers. Mammalian eggs display repetitive Ca_i waves. The initial wave may originate from the site of sperm entry, as has been described in nonmammalian eggs (312, 313), and subsequent waves follow every 1–10 min for several hours (306, 419). The mechanism of Ca_i waves and oscillations in eggs and in somatic cells involves both Ca^{2+}-induced Ca^{2+} release from internal stores and the generation of other Ca_i-mobilizing second messengers (64). Yet, features of the regulatory mechanisms present in mammalian eggs may differ from those in sea urchin eggs and in somatic cells, such as the apparent absence in mammalian eggs of caffeine-sensitive internal stores of Ca^{2+} (417).

It is generally believed that Ca_i elevations are essential for cortical-granule exocytosis, and that they additionally may act at other sites during the zygotic activation. Understanding of the role of Ca_i in exocytosis has been advanced by recent studies in neurosecretory systems, where Ca^{2+} channels are localized in active exocytotic zones and where Ca^{2+}-binding proteins are components of the secretory machinery. In particular, the ability of synapsin I, a secretory vesicle protein, to recruit vesicles into a releasable pool is regulated by phosphorylation by Ca^{2+}/calmodulin-dependent protein kinase II. Ca^{2+} may also potentiate exocytosis by modulating the activity of synaptotagmin/P65; this synaptic vesicle protein has a cytoplasmic C-terminal region that is related to the domains of PKC that bind phospholipids in the presence of Ca^{2+}. It is believed that this protein controls a Ca^{2+}-regulated docking of secretory vesicles to release sites (318, 599). Similar Ca^{2+}-activated proteins associated with cortical-granule and plasma membranes may mediate vesicle exocytosis in response to Ca_i elevations.

We are grateful to members of our laboratories for research contributions and valuable advice.

REFERENCES

1. Aarons, D., T. Battle, H. Boettger-Tong, and G. R. Poirier. Role of monoclonal antibody J-23 in inducing acrosome reactions in capacitated mouse spermatozoa. *J. Reprod. Fertil.* 96: 49–59, 1992.
2. Aarons, D., H. Boettger-Tong, and G. R. Poirier. Acrosome reaction induced by immunoaggregation of a proteinase inhibitor bound to the murine sperm head. *Mol. Reprod. Dev.* 30: 258–264, 1991.
3. Adelmen, J. P., K.-Z. Shen, M. P. Kavanaugh, R. A. Warren, Y.-N. Wu, A. Lagrutta, C. T. Bond, and R. A. North. Calcium-activated potassium channels expressed from cloned complementary DNAs. *Neuron* 9: 209–216, 1992.
4. Ahuja, K. K. Carbohydrate determinants involved in mammalian fertilization. *Am. J. Anat.* 174: 207–223, 1985.
5. Aitken, R. J., J. S. Clarkson, M. J. Hulme, and C. J. Henderson. Analysis of calmodulin acceptor proteins and the influence of calmodulin antagonists on human spermatozoa. *Gamete Res.* 21: 93–111, 1988.
6. Albertini, D. F. Regulation of meiotic maturation in the mammalian oocyte: interplay between exogenous cues and microtubule cytoskeleton. *Bioessays* 5: 100–105, 1992.
7. Albertini, D. F., and E. Anderson. The appearance and structure of the intercellular connections during the ontogeny of the rabbit ovarian follicle with special reference to gap junctions. *J. Cell Biol.* 2: 234–250, 1974.
8. Albertini, D. F., D. Fawcett, and P. Olds. Morphological variations in gap junctions of ovarian granulosa cells. *Tissue Cell* 7: 389–405, 1975.
9. Albertini, D. F., D. Wickramasinghe, S. M. Messinger, B. A. Mattson, and C. E. Plancha. Nuclear and cytoplasmic changes during oocyte maturation. In: *Preimplantation Development of the Mammalian Embryo.* Serono Symposia, New York: Springer-Verlag, 1993, p. 3–21.
10. Almers, W. Exocytosis. *Annu. Rev. Physiol.* 52: 607–624, 1990.
11. Almers, W., and F. W. TSE. Transmitter release from synapses: does a preassembled fusion pore initiate exocytosis? *Neuron* 4: 813–818, 1990.
12. Amsterdam, A., R. Josephs, M. Lieberman, and H. Lindner. Organization of intramembrane particles in freeze-cleaved gap junctions of rat graafian follicles: optical diffraction analysis. *J. Cell Sci.* 21: 93–105, 1976.
13. Anderson, E., and E. Albertini. Gap junctions between the oocyte and companion follicle cells in the mammalian ovary. *J. Cell Biol.* 71: 680–686, 1976.
14. Andrews, J. C., and B. D. Bavister. Capacitation of hamster spermatozoa with the divalent cation chelators D-penicillamine, L-histidine, and L-cysteine in a protein-free culture medium. *Gamete Res.* 23: 159–170, 1989.
15. Aonuma, S., T. Mayumi, K. Suzuki, T. Noguchi, M. Iwai, and M. Okabe. Studies on sperm capacitation. I. The relationship between a guinea pig sperm coating antigen and a sperm capacitation phenomenon. *J. Reprod. Fertil. (Abstr.)* 35: 425–432, 1973.
16. Aonuma, S., M. Okabe, and M. Kawaguchi. The effect of zinc

ions on fertilization of mouse ova in vitro. *J. Reprod. Fertil.* 53: 179–183, 1978.

17. Aonuma, S., M. Okabe, M. Kawaguchi, and Y. Kishi. Zinc effects on mouse spermatozoa and in vitro fertilization. *J. Reprod. Fertil.* 63: 463–466, 1981.

18. Austin, C. R. Observations on the penetration of the sperm into the mammalian egg. *Aust. J. Scient. Res.* 4: 581–596, 1951.

19. Austin, C. R. The 'capacitation' of the mammalian sperm. *Nature* 170: 326, 1952.

20. Austin, C. R. Capacitation and the release of hyaluronidase from spermatozoa. *J. Reprod. Fertil.* 3: 310–311, 1960.

21. Austin, C. R. *The Mammalian Egg.* Oxford: Blackwell Science, 1961.

22. Austin, C. R. The egg. In: *Reproduction in Mammals, Vol. 1, Germ Cells and Fertilization,* edited by C. Austin, and R. Short, Cambridge: Cambridge University, 1982, p. 46–62.

23. Austin, C. R., and M.W.H. Bishop. Role of the rodent acrosome and perforatorium in fertilization. *Proc. R. Soc. Lond. [Biol.]* 149: 241–248, 1958.

24. Austin, C. R., and A.W.H. Braden. Passage of the sperm and the penetration of the egg in mammals. *Nature* 170: 919–921, 1952.

25. Austin, J., and J. Kimble. *glp-1* is required in the germ line for regulation of the decision between mitosis and meiosis in *C. elegans. Cell* 50: 589–599, 1987.

26. Austin, C., and R. Short (Eds). *Reproduction in Mammals, Vol. 1, Germ Cells and Fertilization.* Cambridge: Cambridge University, 1982.

27. Baba, T., H. B. Hoff, H. Nemoto, H. Lee, J. Orth, Y. Arai, and G. L. Gerton. Acrogranin, an acrosomal cysteine-rich glycoprotein, is the precursor of the growth-modulating peptides, granulins, and epithelins, and is expressed in somatic as well as male germ cells. *Mol. Reprod. Dev.* 34: 233–243, 1993.

28. Babcock, D. F. Examination of the intracellular ionic environment and of ionophore action by null point measurements employing the fluorescein chromophore. *J. Biol. Chem.* 258: 6380–6389, 1983.

29. Babcock, D. F. Cross talk between voltage-dependent components of the signal transduction machinery of bovine spermatozoa. *Biol. Reprod. (Abstr.)* 38: 60, 1988.

30. Babcock, D. F., M. M. Bosma, D. E. Battaglia, and A. Darszon. Early persistent activation of sperm K^+ channels by the egg peptide speract. *Proc. Natl. Acad. Sci. U.S.A.* 89: 6001–6005, 1992.

31. Babcock, D. F., N. L. First, and H. A. Lardy. Transport mechanism for succinate and phosphate localized in the plasma membrane of bovine spermatozoa. *J. Biol. Chem.* 250: 6488–6495, 1975.

32. Babcock, D. F., and D. R. Pfeiffer. Independent elevation of cytosolic $[Ca^{2+}]$ and pH of mammalian sperm by voltage-dependent and pH-sensitive mechanisms. *J. Biol. Chem.* 262: 15041–15047, 1987.

33. Babcock, D. F., G. A. Rufo, and H. A. Lardy. Potassium-dependent increases in cytosolic pH stimulate metabolism and motility of mammalian sperm. *Proc. Natl. Acad. Sci. U.S.A.* 80: 1327–1331, 1983.

34. Babcock, D. F., J. P. Singh, and H. A. Lardy. Alteration of membrane permeability to calcium ions during maturation of bovine spermatozoa. *Dev. Biol.* 69: 85–93, 1979.

35. Baccetti, B. The evolution of the acrosomal complex. In: *The Spermatozoon,* edited by D. W. Fawcett, and J. M. Bedford. Baltimore, MD: Urban and Schwarzenberg, 1979, p. 305–329.

36. Baccetti, B., and B. A. Afzelius. The biology of the sperm cell. *Monogr. Dev. Biol.* 10: 1–254, 1976.

37. Bachvarova, R. Gene expression during oogenesis and oocyte development in mammals. In: *Developmental Biology: A Comprehensive Synthesis, Vol. 1, Oogenesis,* edited by L. Browder. New York: Plenum, 1985, p. 453–524.

38. Bachvarova, R., J. Burns, I. Speigelman, J. Choy, and R. Chaganti. Morphology and transcriptional activity of mouse oocyte chromosomes. *Chromosoma* 86: 181–196, 1982.

39. Baker, T. Oogenesis and ovarian development. In: *Reproductive Biology,* edited by H. Balin and S. Glasser. Amsterdam: Excerpta Medica, 1972, p. 398–437.

40. Baker, T. Oogenesis and ovulation. In: *Reproduction in Mammals, Vol. 1, Germ Cells and Fertilization,* edited by C. Austin and R. Short. Cambridge: Cambridge University, 1982, p. 17–45.

41. Baker, T., and L. Franchi. The fine structure of oogonia and oocytes in human ovaries. *J. Cell Sci.* 2: 213–224, 1967.

42. Baker, T., and L. Franchi. The structure of the chromosomes in human primordial oocytes. *Chromosoma* 22: 358–377, 1967.

43. Baker, T., and L. Franchi. The fine structure of chromosomes in bovine primordial oocytes. *J. Reprod. Fertil.* 14: 511–513, 1967.

44. Baker, T., and L. Franchi. The fine structure of oogonia and oocytes in the rhesus monkey *(Macaca mulatta). Z. Zellforsch. Mikrosk. Anat.* 126: 53–74, 1972.

45. Bakken, A., and M. McClanahan. Patterns of RNA synthesis in early meiotic prophase oocytes from fetal mouse ovaries. *Chromosoma* 67: 21–40, 1978.

46. Balakier, H. Induction of maturation in small oocytes from sexually immature mice by fusion with meiotic or mitotic cells. *Exp. Cell Res.* 112: 137–141, 1978.

47. Balakier, H., and R. Czolowska. Cytoplasmic control of nuclear maturation in mouse oocytes. *Exp. Cell Res.* 110: 466–469, 1977.

48. Balinski, B. I. *An Introduction to Embryology.* Philadelphia: W. B. Saunders, 1975.

49. Barros, C., J. M. Bedford, L. E. Franklin, and C. R. Austin. Membrane vesiculation as a feature of the mammalian acrosome reaction. *J. Cell Biol.* 34: C1–C5, 1967.

50. Bavister, B. D. Analysis of culture media for *in vitro* fertilization and criteria for success. In: *Fertilization and Embryonic Development In Vitro,* edited by L. Mastroianni and J. D. Biggers. New York: Plenum, 1981, p. 41–60.

51. Bavister, B. D. Animal *in vitro* fertilization and embryo development. In: *Developmental Biology: A Comprehensive Synthesis, Vol. 4. Manipulation of Mammalian Development,* edited by R.B.L. Gwatkin. New York: Plenum, 1986, p. 81–148.

52. Bedford, J. M. Maturation, transport, and fate of spermatozoa in the epididymis. In: *Handbook of Physiology, Endocrinology, Male Reproductive System,* edited by D. W. Hamilton, and R. O. Greep. Washington, DC: Am. Physiol. Soc., 1975, sect. 7, vol. V, p. 303–317.

53. Bedford, J. M. Ultrastructural changes in the sperm head during fertilization in the rabbit. *Am. J. Anat.* 123: 329–358, 1968.

54. Bedford, J. M. Morphological aspects of sperm capacitation in mammals. *Adv. Biosci.* 4: 35–50, 1970.

55. Bedford, J. M. Significance of the need for sperm capacitation before fertilization in eutherian mammals. *Biol. Reprod.* 28: 108–120, 1983.

56. Bedford, J. M., H. Calvin, and G. W. Cooper. The maturation of spermatozoa in the human epididymis. *J. Reprod. Fertil.* 18: 199–213, 1973.

57. Beebe, S. J., L. Leyton, D. Burks, M. Ishikawa, T. Fuerst, J. Dean, and P. M. Saling. Recombinant mouse ZP3 inhibits sperm binding and induces the acrosome reaction. *Dev. Biol.* 151: 48–54, 1992.

58. Beebe, S. J., O. Oyen, M. Sandberg, A. Froysa, V. Hannson, and T. Jahnsen. Molecular cloning of a tissue-specific protein

kinase C from human testis—representing a third isoform for the catalytic subunit on cAMP-dependent protein kinase. *Mol. Endocrinol.* 4: 465–475, 1990.

59. Beebe, S. J., P. Salomonsky, T. Jahnsen, and Y. LI. The Ct subunit is a unique isozyme of the cAMP-dependent protein kinase. *J. Biol. Chem.* 267: 25505–25512, 1992.

60. Bellve, A. R. and W. Zheng. Growth factors as autocrine and paracrine modulators of male gonadal functions. *J. Reprod. Fertil.* 85: 771–793, 1989.

61. Bentley, J. K., D. L. Garbers, S. E. Domino, T. D. Noland, and C. Van Dop. Spermatozoa contain a guanine nucleotide-binding protein ADP-ribosylated by pertussis toxin. *Biochem. Biophys. Res. Commun.* 138: 728–734, 1986.

62. Berger, T., K. O. Turner, S. Meizel, and J. L. Hedrick. Zona pellucida-induced acrosome reaction in boar sperm. *Biol. Reprod.* 40: 525–530, 1989.

63. Berridge, M. J. Inositol triphosphate and diacylglycerol: two interacting second messengers. *Annu. Rev. Biochem.* 56: 159–194, 1987.

64. Berridge, M. J. Inositol trisphosphate and calcium signalling. *Nature* 361: 315–325, 1993.

65. Berridge, M. J., and D. S. Irvine. Inositol phosphates and cell signalling. *Nature* 341: 197–205, 1989.

66. Berruti, G., and E. Martegani. Identification of proteins cross-reactive to phosphotyrosine antibodies and to a tyrosine kinase activity in boar spermatozoa. *J. Cell Sci.* 93: 667–674, 1989.

67. Birnbaumer, L. Receptor-to-effector signaling through G proteins: roles for βt dimers as well as subunits. *Cell* 71: 1069–1072, 1992.

68. Blackmore, P. F., S. J. Beebe, D. R. Danforth, and N. Alexander. Progesterone and 17a-hydroxyprogesterone. Novel stimulators of calcium influx in human sperm. *J. Biol. Chem.* 265: 1376–1380, 1990.

69. Blackmore, P. F., J. Neulen, F. Lattanzio, and S. J. Beebe. Cell surface-binding sites for progesterone mediate calcium uptake in human sperm. *J. Biol. Chem.* 266: 18655–18659, 1991.

70. Bleil, J. D. Sperm receptors of mammalian eggs. In: *Elements of Mammalian Fertilization, Vol. 1, Basic Concepts*, edited by P. M. Wassarman, Boca Raton, FL: CRC, 1991, p. 133–151.

71. Bleil, J. D., and P. M. Wassarman. Structure and function of the zona pellucida: identification and characterization of the proteins of the mouse oocyte's zona pellucida. *Dev. Biol.* 76: 185–203, 1980.

72. Bleil, J. D., and P. M. Wassarman. Mammalian sperm-egg interaction: identification of a glycoprotein in mouse egg zona pellucidae possessing receptor activity for sperm. *Cell* 20: 873–882, 1980.

73. Bleil, J. D., and P. M. Wassarman. Sperm-egg interactions in the mouse: sequence of events and induction of the acrosome reaction by a zona pellucida glycoprotein. *Dev. Biol.* 95: 317–324, 1983.

74. Bleil, J. D., and P. M. Wassarman. Autoradiographic visualization of the mouse egg's sperm receptor bound to sperm. *J. Cell Biol.* 102: 1363–1371, 1986.

75. Bleil, J. D., and P. M. Wassarman. Galactose at the nonreducing terminus of O-linked oligosaccharides of mouse egg zona pellucida glycoprotein ZP3 is essential for the glycoprotein's sperm receptor activity. *Proc. Natl. Acad. Sci. U.S.A.* 85: 6778–6782, 1988.

76. Bleil, J. D., and P. M. Wassarman. Identification of a ZP3-binding protein on acrosome-intact mouse sperm by photoaffinity crosslinking. *Proc. Natl. Acad. Sci. U.S.A.* 87: 5563–5567, 1990.

77. Blobel, C. P., D. G. Myles, P. Primakoff, and J. M. White. Proteolytic processing of a protein involved in sperm-egg fusion correlates with acquisition of fertilization competence. *J. Cell Biol.* 111: 69–78, 1990.

78. Blobel, C. P., T. G. Wolfsberg, C. W. Turck, D. G. Myles, P. Primakoff, and J. M. White. A potential fusion peptide and an integrin ligand domain in a protein active in sperm-egg fusion. *Nature* 356: 248–252, 1992.

79. Bock, J., and J. Marsh. *Proton Passage Across Cell Membranes.* Chichester, UK: Wiley, 1988.

80. Boldt, J., L. E. Gunter, and A. M. Howe. Characterization of cell surface polypeptides of unfertilized, fertilized, and protease-treated zona-free mouse eggs. *Gamete Res.* 23: 91–101, 1989.

81. Borland, R. M., J. D. Biggers, C. P. Lechene, and M. L. Taymor. Elemental composition of fluid in the human fallopian tube. *J. Reprod. Fertil.* 58: 479–482, 1980.

82. Borland, R. M., S. Hazra, J. D. Biggers, and C. P. Lechene. The elemental composition of the environments of the gametes and preimplantation embryo during the initiation of pregnancy. *Biol. Reprod.* 16: 147–157, 1977.

83. Bornslaeger, E., P. Mattei, and R. M. Schultz. Involvement of cAMP-dependent protein kinase and protein phosphorylation in regulation of mouse oocyte maturation. *Dev. Biol.* 114: 453–462, 1986.

84. Bornslaeger, E., W. Poueymirou, P. Mattei, and R. M. Schultz. Effects of protein kinase C activators on germinal vesicle breakdown and polar body emission of mouse oocytes. *Exp. Cell Res.* 165: 507–517, 1986.

85. Bornslaeger, E., and R. M. Schultz. Regulation of mouse oocyte maturation: Effect of elevating cumulus cell cAMP on oocyte cAMP levels. *Biol. Reprod.* 33: 698–704, 1985.

86. Bornslaeger, E., M. Wilde, and R. M. Schultz. Regulation of mouse oocyte maturation: involvement of cyclic AMP phosphodiesterase and calmodulin. *Dev. Biol.* 105: 488–499, 1984.

87. Bourne, H., and L. Stryer. The target sets the tempo. *Nature* 358: 541–543, 1992.

88. Bradley, M. P., and I. T. Forrester. A sodium-calcium exchange mechanism in plasma membrane vesicles isolated from ram sperm flagella. *FEBS Lett.* 121: 15–18, 1980.

89. Brambell, F. The development and morphology of the gonads of the mouse. Part 1. The morphogenesis of the indifferent gonad and of the ovary. *Proc. R. Soc. Lond. [Biol.]* 101: 391 409, 1927.

90. Brambell, F. The development and morphology of the gonads of the mouse. Part III. The growth of the follicles. *Proc. R. Soc. Lond. [Biol.]* 103: 258–272, 1928.

91. Brambell, F. Ovarian changes. In: *Marshall's Physiology of Reproduction*, edited by A. Parkes. New York: Longmans, Green, 1956, vol. 1, p. 397–542.

92. Bronson, R. A., and F. Fusi. Evidence that an Arg-Gly-Asp adhesion sequence plays a role in mammalian fertilization. *Biol. Reprod.* 43: 1019–1025, 1990.

93. Brown, A. M. Membrane-delimited cell signaling complexes: direct ion channel regulation by G proteins. *J. Membr. Biol.* 131: 93–104, 1993.

94. Brown, J., J. A. Cebra-Thomas, J. D. Bleil, P. M. Wassarman, and L. M. Silver. A premature acrosome reaction is programmed by mouse *t* haplotypes during sperm differentiation and could play a role in transmission ratio distortion. *Development* 106: 769–773, 1989.

95. Brown, M. A., and E. R. Casillas. Bovine sperm adenylate cyclase. Inhibition by adenosine and adenosine analogs. *J. Androl.* 5: 361–368, 1984.

96. Bryan, J.H.D. Spermatogenesis revisited: III. The course of spermatogenesis in a male-sterile pink-eyed mutant type in the mouse. *Cell Tissue Res.* 180: 173–186, 1977.

97. Buccione, R., A. C. Schroeder, and J. J. Eppig. Interactions

between somatic cells and germ cells throughout mammalian oogenesis. *Biol. Reprod.* 43: 543–547, 1990.

98. Busa, W. B., and R. Nuccitelli. Metabolic regulation via intracellular pH. *Am. J. Physiol.* 246 (*Regulatoroy Integrative Comp. Physiol.* 15) R409–R438, 1984.

99. Byskov, A. Primordial germ cells and regulation of meiosis. In: *Reproduction in Mammals*, edited by C. Austin, and R. Short. Cambridge: Cambridge University, 1982, vol. 1.

100. Cabot, C. L., and G. Oliphant. The possible role of immunological complement in induction of the rabbit sperm acrosome reaction. *Biol. Reprod.* 19: 666–672, 1978.

101. Cadena, D. L., and G. N. Gill. Receptor tyrosine kinases. *FASEB J.* 6: 2332–2337, 1992.

102. Calarco, P. The kinetochore in oocyte maturation. In: *Oogenesis*, edited by J. Biggers, and A. Schuetz. Baltimore: University Park, 1972, p. 65–86.

103. Calarco, P., R. Donahue, and D. Szollosi. Germinal vesicle breakdown in the mouse oocyte. *J. Cell Sci.* 10: 369–385, 1972.

104. Canipari, R., F. Palombi, M. Riminucci, and F. Mangia. Early programming and maturation competence in mouse oogenesis. *Dev. Biol.* 102: 519–524, 1984.

105. Caswell, A. H., and J. D. Hutchinson. Selectivity of cation chelation to tetracyclines: evidence for special conformation of calcium chelate. *Biochem. Biophys. Res. Commun.* 43: 525–530, 1971.

106. Centola, G., L. Anderson, and C. Channing. Oocyte maturation inhibition activity in porcine granulosa cells. *Gamete Res.* 4: 451–462, 1981.

107. Chamberlin, M. E., and J. Dean. Human homolog of the mouse sperm receptor. *Proc. Natl. Acad. Sci. U.S.A.* 87: 6014–6018, 1990.

108. Chang, M. C. Fertilizing capacity of spermatozoa deposited into fallopian tubes. *Nature* 168: 697–698, 1951.

109. Chang, M. C. Development of fertilizing capacity of rabbit spermatozoa in the uterus. *Nature* 175: 1036–1037, 1955.

110. Chang, M. C. The maturation of rabbit oocytes in culture and their maturation, activation, fertilization, and subsequent development in the fallopian tubes. *J. Exp. Zool.* 128: 378–405, 1955.

111. Chang, M. C. A detrimental effect of seminal plasma on the fertilizing capacity of sperm. *Nature* 179: 258–259, 1957.

112. Chang, M. C., C. R. Austin, J. M. Bedford, B. G. Brackett, R.H.F. Hunter, and R. Yanagimachi. Capacitation of spermatozoa and fertilization in mammals. In: *Frontiers in Reproduction and Fertility Control: A Review of Reproductive Sciences and Fertility Control*, edited by R. O. Greep and M. A. Koblinsky. Cambridge, MA: MIT, 1977, pp. 434–451.

113. Chari, S., T. Hillensjo, C. Magnusson, G. Sturm, and E. Daume. *In vitro* inhibition of rat oocyte meiosis by human follicular fluid fractions. *Arch. Gynecol.* 233: 155–164, 1983.

114. Chaudhry, P. S., and E. R. Casillas. Calmodulin-stimulated cyclic nucleotide phosphodiesterases in plasma membrane of bovine epididymal spermatozoa. *Arch. Biochem. Biophys.* 262: 439–444, 1988.

115. Chaudhry, P. S., and E. R. Casillas. Isotypes of protein kinase C in bovine sperm. *Arch. Biochem. Biophys.* 295: 268–272, 1992.

116. Cheetham, J. A., and H. A. Lardy. Bovine spermatozoa incorporate $^{32}P_i$ into ADP by an unknown pathway. *J. Biol. Chem.* 267: 5186–5192, 1992.

117. Cherr, G. N., H. Lambert, S. Meizel, and D. F. Katz. In vitro studies of the golden sperm acrosome reaction: completion on the zona pellucida and induction by homologous soluble zonae pellucidae. *Dev. Biol.* 114: 119–131, 1986.

118. Chinkers, M., and D. L. Garbers. Signal transduction by guanylyl cyclases. *Annu. Rev. Biochem.* 60: 553–575, 1991.

119. Cho, W.-K., S. Stern, and J. D. Biggers. Inhibitory effect of dibutyryl cAMP on mouse oocyte maturation *in vitro*. *J. Exp. Zool.* 187: 383–386, 1974.

120. Chou, P. Y., and G. D. Fasman. Prediction of the secondary structure of proteins from their amino acid sequence. *Adv. Enzymol.* 47: 47–148, 1978.

121. Chrisman, T. D., S. Schulz, L. R. Potter, and D. L. Garbers. Seminal plasma factors that cause large elevations in cellular cyclic GMP are C-type natriuretic peptides. *J. Biol. Chem.* 268: 3698–3703, 1993.

122. Christen, R., R. W. Schackmann, and B. M. Shapiro. Interactions between sperm and sea urchin egg jelly. *Dev. Biol.* 98: 1–14, 1983.

123. Chu, F.-K. Requirements of cleavage of high mannose oligosaccharides in glycoproteins by peptide N-glycosidase F. *J. Biol. Chem.* 261: 172–177, 1986.

124. Clapper, D. L., T. F. Walseth, P. J. Daegie, and H. C. Lee. Pyridine nucleotide metabolites stimulate calcium release from sea urchin egg microsomes desensitized to inositol trisphosphate. *J. Biol. Chem.* 262: 9561–9568, 1987.

125. Clark, E. N., M. E. Corron, and H. M. Florman. Caltrin, the calcium transport regulatory peptide of spermatozoa, modulates acrosomal exocytosis in response to the egg's zona pellucida. *J. Biol. Chem.* 268: 5309–5316, 1993.

126. Clark, J., and E. Eddy. Fine structural observations on the origin and association of primordial germ cells of the mouse. *Dev. Biol.* 47: 136–155, 1975.

127. Cohen-Armon, M., and M. Sokolovsky. Evidence for involvement of the voltage-dependent Na^+ channel gating in depolarization-induced activation of G proteins. *J. Biol. Chem.* 268: 9824–9838, 1993.

128. Coronel, C. E., and H. A. Lardy. Characterization of Ca uptake by guinea pig epididymal spermatozoa. *Biol. Reprod.* 37: 1097–1107, 1987.

129. Corselli, J., and P. Talbot. *In vitro* penetration of hamster oocyte-cumulus complexes using physiological numbers of sperm. *Dev. Biol.* 122: 227–242, 1987.

130. Cowan, A. E., D. G. Myles, and D. E. Koppel. Lateral diffusion of the PH-20 protein on guinea pig sperm: evidence that barriers to diffusion maintain plasma membrane domains in mammalian sperm. *J. Cell Biol.* 104: 917–923, 1987.

131. Cox, T., P. Campbell, and R. N. Peterson. Ion channels in boar sperm plasma membranes: characterization of a cation selective channel. *Mol. Reprod. Dev.* 30: 135–147, 1991.

132. Cox, T., and R. N. Peterson. Identification of calcium conducting channels in isolated boar sperm plasma membranes. *Biochem. Biophys. Res. Commun.* 161: 162–168, 1989.

133. Cran, D. B., R. M. Moor, and R. F. Irvine. Initiation of the cortical reaction in hamster and sheep oocytes in response to inositol trisphosphate. *J. Cell Sci.* 91: 139–144, 1988.

134. Creutz, C. E. The annexins and exocytosis. *Science* 258: 924–931, 1992.

135. Cross, P., and R. Brinster. *In vitro* development of mouse oocytes. *Biol. Reprod.* 3: 298–307, 1970.

136. Cross, N. L., P. Morales, J. W. Overstreet, and F. W. Hanson. Induction of acrosome reactions by the human zona pellucida. *Biol. Reprod.* 38: 235–244, 1988.

137. Crozet, N. Ultrastructural aspects of in vivo fertilization in the cow. *Gamete Res.* 10: 241–251, 1984.

138. Crozet, N., and M. Dumont. The site of the acrosome reaction during in vivo penetration of the sheep oocyte. *Gamete Res.* 10: 97–105, 1984.

139. Cummins, J. M. Evolution of sperm form: levels of control

and competition. In: *Fertilization in Mammals,* edited by B. D. Bavister, J. Cummins, and E.R.S. Roldan. Norwell, MA: Serono Symposia, 1990, p. 51–64.

140. Cummins, J. M., and R. Yanagimachi. Sperm-egg ratios and the site of the acrosome reaction during in vivo fertilization in the hamster. *Gamete Res.* 5: 239–256, 1982.

141. Cummins, J. M., and R. Yanagimachi. Development of ability to penetrate the cumulus oophorus by hamster spermatozoa capacitated in vitro, in relation to the timing of the acrosome reaction. *Gamete Res.* 15: 187–212, 1986.

142. Dale, B., L. J. Defelice, and G. Ehrenstein. Injection of a soluble sperm extract into sea urchin eggs triggers the cortical reaction. *Experientia* 41: 1068–1070, 1985.

143. Dan, J. C. Studies on the acrosome. I. Reaction to egg water and other stimuli. *Biol. Bull.* 103: 54–66, 1952.

144. Dan, J. C. Studies on the acrosome. III. Effect of calcium deficiency. *Biol. Bull.* 107: 335–349, 1954.

145. Dan, J. C. The acrosome reaction. *Int. Rev. Cytol.* 5: 365–393, 1956.

146. Dan, J., Y. Ohori, and H. Kushida. Studies on the acrosome VII. Formation of the acrosomal process in sea urchin spermatozoa. *J. Ultrastr. Res.* 11: 508–524, 1964.

147. Dangott, L. J., J. E. Jordan, R. A. Bellet, and D. L. Garbers. Cloning of the mRNA for the protein that crosslinks to the egg peptide speract. *Proc. Natl. Acad. Sci. U.S.A.* 86: 2128–2132, 1989.

148. Daniel, J. (Ed). *Methods in Mammalian Embryology.* San Francisco: W. H. Freeman, 1971.

149. Daniel, J. (Ed). *Methods in Mammalian Reproduction.* New York: Academic, 1978.

150. Davidson, E. H. *Gene Activity in Early Development.* New York: Academic, 1986, p. 670.

151. Davis, B. K. Timing of fertilization in mammals: sperm cholesterol/phospholipid ratio as a determinant of the capacitation interval. *Proc. Natl. Acad. Sci. U.S.A.* 78: 7560–7564, 1981.

152. De Felici, M., and G. Siracusa. Survival of isolated, fully-grown mouse ovarian oocytes is strictly dependent on external Ca^{2+}. *Dev. Biol.* 92: 539–543, 1982.

153. De Jong, W. W. Superordinal affinities of rodentia studied by sequence analysis of eye lens protein. In: *Evolutionary Relationships Among Rodents—a Multidisciplinary Analysis,* edited by W. P. Luckett and J.-L. Hartenberger. NY: Plenum, 1985, p. 211–226.

154. Dekel, N., E. Aberdam, and I. Sherizly. Spontaneous maturation *in vitro* of cumulus-enclosed rat oocytes is inhibited by forskolin. *Biol. Reprod.* 31: 244–250, 1984.

155. Dekel, N., and W. Beers. Rat oocyte maturation *in vitro:* relief of cyclic AMP inhibition with gonadotropins. *Proc. Natl. Acad. Sci. U.S.A.* 75: 4369–4373, 1978.

156. Dekel, N., and W. Beers. Development of the rat oocyte *in vitro:* inhibition and induction of maturation in the presence or absence of the cumulus oophorus. *Dev. Biol.* 75: 247–254, 1980.

157. Dekel, N., T. Hillensjo, and P. Kraicer. Maturational effects of gonadotropins on the cumulus-oocyte complex of the rat. *Biol. Reprod.* 20: 191–197, 1979.

158. Dekel, N., P. Kraicer, D. Phillips, R. Sanchez, and S. Segal. Cellular associations in the rat oocyte-cumulus cell complex: morphology and ovulatory changes. *Gamete Res.* 1: 47–57, 1978.

159. Dekel, N., T. Lawrence, N. Gilula, and W. Beers. Modulation of cell-to-cell communication in the cumulus-oocyte complex and the regulation of oocyte maturation by LH. *Dev. Biol.* 80: 356–362, 1981.

160. Delorey, T. M., and R. W. Olsen. γ-Aminobutyric acid_A receptor structure and function. *J. Biol. Chem.* 267: 16747–16750, 1992.

161. Dicarlantonio, G., and P. Talbot. Evidence for sequential deployment of secretory enzymes during the normal acrosome reaction of guinea pig sperm in vitro. *Gamete Res.* 21: 425–438, 1988.

162. Dietl, J. (Ed). *The Mammalian Egg Coat: Structure and Function.* Berlin: Springer-Verlag KG, 1989, p. 156.

163. Dodd, J., and T. M. Jessell. Lactoseries carbohydrates specify subsets of dorsal root ganglion neurons projecting to the superficial dorsal horn of rat spinal cord. *J. Neurosci.* 5: 3278–3294, 1985.

164. Dohlman, H. G., J. Thorner, M. G. Caron, and R. J. Lefkowitz. Model systems for the study of seven-transmembrane-segment receptors. *Annu. Rev. Biochem.* 60: 653–688, 1991.

165. Dolci, S., D. E. Williams, M. K. Ernst, J. L. Resnick, C. I. Brannan, L. F. Lock, S. D. Lyman, H. S. Boswell, and P. J. Donovan. Requirement for mast cell growth factor for primordial germ cell survival in culture. *Nature* 352: 809–811, 1991.

166. Domino, S. E., and D. L. Garbers. The fucose-sulfate glycoconjugate that induces an acrosome reaction in spermatozoa stimulates inositol 1,4,5-trisphosphate accumulation. *J. Biol. Chem.* 263: 690–695, 1988.

167. Donahue, R. Maturation of the mouse oocyte *in vitro.* 1. Sequence and timing of nuclear progression. *J. Exp. Zool.* 169: 237–250, 1968.

168. Doolittle, R. F. *Of Urfs and Orfs.* Mill Valley, CA: University Science Books, 1986, p. 103.

169. Douglas, W. W. Stimulus-secretion coupling: the concept and clues from chromaffin and other cells. *Br. J. Pharm.* 34: 451–474, 1968.

170. Downs, S., D. Coleman, P. Ward-Bailey, and J. Eppig. Hypoxanthine is the principal inhibitor of murine oocyte maturation in a low molecular weight fraction of porcine follicular fluid. *Proc. Natl. Acad. Sci. U.S.A.* 82: 454–458, 1985.

171. Downs, S. M., S.A.J. Daniel, E. A. Bornslaeger, P. C. Hoppe, and J. J. Eppig. Maintenance of meiotic arrest in mouse oocytes by purines: modulation of cAMP levels and cAMP phosphodiesterase activity. *Gamete Res.* 23: 323–334, 1989.

172. Downs, S., and J. Eppig. Cyclic adenosine monophosphate and ovarian follicular fluid act synergistically to inhibit mouse oocyte maturation. *Endocrinology* 114: 418 427, 1984.

173. Drobnis, E. Z., and D. F. Katz. Videomicroscopy of mammalian fertilization. In: *Elements of Mammalian Fertilization. Vol. I. Basic Concepts,* edited by P. M. Wassarman. Boca Raton, FL: CRC, 1991, p. 269–300.

174. Ducibella, T. Mammalian egg cortical granules and the cortical reaction. In: *Elements of Mammalian Fertilization. Vol. I. Basic Concepts,* edited by P. M. Wassarman. Boca Raton, FL: CRC, 1991, p. 205–231.

175. Ducibella, T., E. Anderson, D. F. Albertini, J. Aalberg, and S. Rangarajan. Quantitative changes in cortical granule number and distribution in the mouse oocyte during meiotic maturation. *Dev. Biol.* 130: 184–197, 1988.

176. Ducibella, T., S. Kurasawa, S. Rangarajan, G. S. Kopf, and R. M. Schultz. Precocious loss of cortical granules during mouse oocyte meiotic maturation and correlation with an egg-induced modification of the zona pellucida. *Dev. Biol.* 137: 46–55, 1990.

177. Ducibella, T., S. Rangarajan, and E. Anderson. The development of mouse oocyte cortical granule competence is accompanied by major changes in cortical vesicles and not cortical granule depth. *Dev. Biol.* 130: 789–792, 1988.

178. Dunbar, B. S., and M. G. O'Rand (Eds). *A Comparative Overview of Mammalian Fertilization*. New York: Plenum, 1991, p. 457.

179. Dunbar, B. S., S. V. Prasad, and T. M. Timmons. Comparative structure and function of mammalian zonae pellucidae. In: *A Comparative Overview of Mammalian Fertilization*, edited by B. S. Dunbar, and M. G. O'Rand, New York: Plenum, 1991, p. 97–114.

180. Dunbar, B., N. Wardrip, and J. Hedrick. Isolation, physicochemical properties, and macromolecular composition of zona pellucida from porcine oocytes. *Biochemistry* 19: 356–365, 1980.

181. Duncan, A. E. and L. R. Fraser. Cyclic AMP-dependent phosphorylation of epididymal mouse sperm proteins during capacitation *in vitro*: identification of an M_r 95 000 phosphotyrosine-containing protein. *J. Reprod. Fertil.* 97: 287–299, 1993.

182. Dym, M. The male reproductive system. In: *Histology. Cell and Tissue Biology*, edited by L. Weiss. New York: Elsevier Biomedical, 1983, p. 1000–1053.

183. Eddy, E. M. The spermatozoon. In: *The Physiology of Reproduction*, edited by E. Knobil, and J. Neill. New York: Raven, 1988, p. 27–68.

184. Eddy, E., J. Clark, D. Gong, and B. Fenderson. Origin and migration of primordial germ cells in mammals. *Gamete Res.* 4: 333–362, 1981.

185. Eddy, E., and A. Hahnel. Establishment of the germ line in mammals. In: *7th Symposium of British Society for Developmental Biology*, edited by A. Mclaren, and C. Wylie, Cambridge: Cambridge University, 1983, p. 41–69.

186. Eddy, E. M., D. A. O'Brien, and J. E. Welch. Mammalian sperm development *in vivo* and *in vitro*. In: *Elements of Mammalian Fertilization. I. Basic Concepts*, edited by P. M. Wassarman. Boca Raton, FL: CRC Press, 1990, p. 1–28.

187. Edge, A.S.B., C. R. Faltynek, L. Hof, L. E. Reichert, and P. Weber. Deglycosylation of glycoproteins by trifluoromethanesulfonic acid. *Anal. Biochem.* 118: 131–137, 1981.

188. Eisenbach, M., and D. Ralt. Precontact mammalian sperm-egg communication and role in fertilization. *Am. J. Physiol.* 262 (*Cell Physiol.* 31): C1095–C1101, 1992.

189. Ekholm, C., and C. Magnusson. Rat oocyte maturation: effects of protein synthesis inhibitors. *Biol. Reprod.* 21: 1287–1293, 1979.

190. Endo, Y., M. A. Lee, and G. S. Kopf. Evidence for a role of a guanine nucleotide-binding regulatory protein in the zona pellucida-induced mouse sperm acrosome reaction. *Dev. Biol.* 119: 210–216, 1987.

191. Endo, Y., M. A. Lee, and G. S. Kopf. Characterization of an islet-activating protein sensitive site in mouse sperm that is involved in the zona pellucida-induced acrosome reaction. *Dev. Biol.* 129: 12–24, 1988.

192. Endo, Y., P. Mattei, G. S. Kopf, and R. M. Schultz. Effects of a phorbol ester on mouse eggs: dissociation of sperm receptor activity from acrosome reaction-inducing activity of the mouse zona pellucida protein, ZP3. *Dev. Biol.* 123: 574–577, 1987.

193. Endo, Y., R. M. Schultz, and G. S. Kopf. Effects of phorbol esters and a diacylglycerol on mouse eggs: inhibition of fertilization and modification of the *zona pellucida*. *Dev. Biol.* 119: 199–209, 1987.

194. Eng, L. A., and G. Oliphant. Rabbit sperm reversible decapacitation by membrane stabilization with a highly purified glycoprotein from seminal plasma. *Biol. Reprod.* 19: 1083–1094, 1978.

195. Eppig, J. A comparison between oocyte growth in coculture

196. with granulosa cells and oocytes with granulosa cell-oocyte junctional contact maintained *in vitro*. *J. Exp. Zool.* 209: 345–353, 1979.

196. Eppig, J. The relationship between cumulus cell-oocyte coupling, oocyte meiotic maturation, and cumulus expansion. *Dev. Biol.* 89: 268–272, 1982.

197. Eppig, J. Oocyte-somatic cell interactions during oocyte growth and maturation in the mammal. In: *Developmental Biology: A Comprehensive Synthesis, Vol. 1, Oogenesis*, edited by L. Browder. New York: Plenum, 1985, p. 313–347.

198. Eppig, J. Intercommunication between mammalian oocytes and companion somatic cells. *Bioessays* 13: 569–574, 1991.

199. Eppig, J., R. Freter, P. Ward-Bailey, and R. M. Schultz. Inhibition of oocyte maturation in the mouse: participation of cAMP, steroid hormones, and a putative maturation-inhibitory factor. *Dev. Biol.* 100: 39–49, 1983.

200. Eppig, J., and S. Koide. Effects of progesterone and oestradiol-173 on the spontaneous meiotic maturation of mouse oocytes. *J. Reprod. Fertil.* 53: 99–101, 1978.

201. Eppig, J., P. Ward-Bailey, and D. Coleman. Hypoxanthine and adenosine in murine ovarian follicular fluid: concentrations and activity in maintaining oocyte meiotic arrest. *Biol. Reprod.* 33: 1041–1049, 1985.

202. Fawcett, D. W. The mammalian spermatozoon. *Dev. Biol.* 44: 394–436, 1975.

203. Fay, S. P., R. G. Posner, W. N. Swann, and L. A. Sklar. Real-time analysis of the assembly of ligand, receptor, and G protein by quantitative fluorescence flow cytometry. *Biochemistry* 30: 5066–5075, 1991.

204. Feinberg, J., J. Weinman, S. Weinman, M. P. Walsh, M. C. Harricane, J. Gabrion, and J. G. Demaille. Immunocytochemical and biochemical evidence for the presence of calmodulin in bull sperm flagellum. *Biochim. Biophys. Acta.* 673: 303–311, 1981.

205. Fleming, A., W. Khalil, and D. Armstrong. Porcine follicular fluid does not inhibit maturation of rat oocytes *in vitro*. *J. Reprod. Fertil.* 69: 665–670, 1983.

206. Fleming, A., R. Yanagimachi, and H. Yanagimachi. Spermatozoa of the Atlantic bottlenose dolphin, *Tursiops truncatus*. *J. Reprod. Fertil.* 63: 509–514, 1981.

207. Florman, H. M., and D. F. Babcock. Progress toward understanding the molecular basis of capacitation. In: *Elements of Mammalian Fertilization. Basic Concepts*, edited by P. M. Wassarman. Boca Raton, FL: CRC, 1991, p. 105–132.

208. Florman, H. M., D. F. Babcock, P. M. Bourissa, N. L. First, D. M. Kerlin, and R. M. Tombes. Egg-activated signal transducing pathways in mammalian sperm: what are they and how are they regulated during development and fertilization? In: *Fertilization in Mammals*, edited by B. D. Bavister, J. Cummins, and E.R.S. Roldan. Norwell, MA: Serono Symposia, USA, 1990, p. 77–88.

209. Florman, H. M., K. B. Bechtol, and P. M. Wassarman. Enzymatic dissection of the functions of the mouse egg's receptor for sperm. *Dev. Biol.* 106: 243–255, 1984.

210. Florman, H. M., M. E. Corron, T.D.-H., Kim, and D. Babcock. Activation of voltage-dependent calcium channels of mammalian sperm is required for zona pellucida-induced acrosomal exocytosis. *Dev. Biol.* 152: 304–314, 1992.

211. Florman, H. M., and N. L. First. The regulation of acrosomal exocytosis. I. Sperm capacitation is required for the induction of acrosome reactions by the bovine zona pellucida in vitro. *Dev. Biol.* 128: 453–463, 1988.

212. Florman, H. M., and N. L. First. Regulation of acrosomal exocytosis. II. The zona pellucida-induced acrosome reaction

of bovine spermatozoa is controlled by extrinsic positive regulatory elements. *Dev. Biol.* 128: 464–473, 1988.

213. Florman, H. M., P. M. Saling, and B. T. Storey. Fertilization of mouse eggs in vitro. Time resolution of the reactions preceding penetration of the zona pellucida. *J. Androl.* 3: 373–381, 1982.

214. Florman, H. M., and B. T. Storey. Inhibition of in vitro fertilization of mouse eggs: 3-quinuclidinyl benzilate specifically blocks penetration of zonae pellucidae by mouse spermatozoa. *J. Exp. Zool.* 216: 159–167, 1981.

215. Florman, H. M., and B. T. Storey. Characterization of cholinomimetic agents that inhibit in vitro fertilization in the mouse. Evidence for a sperm-specific binding site. *J. Androl.* 3: 157–164, 1982.

216. Florman, H. M., and B. T. Storey. Mouse gamete interactions: the zona pellucida is the site of the acrosome reaction leading to fertilization in vitro. *Dev. Biol.* 91: 121–130, 1982.

217. Florman, H. M., R. M. Tombes, N. L. First, and D. F. Babcock. An adhesion-associated agonist from the zona pellucida activates G protein-promoted elevations of internal Ca and pH that mediate mammalian sperm acrosomal exocytosis. *Dev. Biol.* 135: 133–146, 1989.

218. Florman, H. M., and P. M. Wassarman. O-linked oligosaccharides of mouse egg ZP3 account for its sperm receptor activity. *Cell* 41: 313–324, 1985.

219. Foltz, K. R., J. S. Partin, and W. J. Lennarz. Sea urchin egg receptor for sperm: sequence similarity of binding domain and hsp70. *Science* 259: 1421–1425, 1993.

220. Foresta, C., M. Rossato, and F. Di Virgilio. Extracellular ATP is a trigger for the acrosome reaction in human spermatozoa. *J. Biol. Chem.* 267: 19443–19447, 1992.

221. Franchi, L., and A. Mandl. The ultrastructure of oogonia and oocytes in the foetal and neonatal rat. *Proc. R. Soc. Lond. [Biol.]* 157: 99–114, 1962.

222. Franchi, L., A. Mandl, and S. Zuckerman. The development of the ovary and the process of oogenesis. In: *The Ovary*, edited by S. Zuckerman. New York: Academic, 1962, vol. 1, p. 1–88.

223. Franklin, L. E., C. Barros, and E. N. Fussell. The acrosomal region and the acrosome reaction in sperm of the golden hamster. *Biol. Reprod.* 3: 180–200, 1970.

224. Fraser, L. R. Mouse sperm capacitation *in vitro* involves loss of a surface-associated inhibitory component. *J. Reprod. Fertil.* 72: 373–384, 1984.

225. Fraser, L. R. Adenosine and its analogues, possibly acting at A₂ receptors, stimulate mouse sperm fertilizing ability during early stages of capacitation. *J. Reprod. Fertil.* 89: 467–476, 1990.

226. Fraser, L. R., and K. K. Ahuja. Metabolic and surface events in fertilization. *Gamete Res.* 20: 491–519, 1988.

227. Fraser, L. R., G. Umar, and S. Sayed. Na⁺-requiring mechanisms modulate capacitation and acrosomal exocytosis in mouse spermatozoa. *J. Reprod. Fertil.* 97: 539–549, 1993.

228. Freter, R., and R. M. Schultz. Regulation of murine oocyte maturation: Evidence for a gonadotropin-induced, cAMP-dependent reduction in a maturation inhibitor. *J. Cell Biol.* 98: 1119–1128, 1984.

229. Friend, D. S. The organization of the spermatozoal membrane. In: *Immunobiology of Gametes*, edited by M. Edinin and M. H. Johnson. Cambridge: Cambridge University, 1977, p. 5–30.

230. Friend, D. S., L. Orci, A. Perrelet, and R. Yanagimachi. Membrane particle changes attending the acrosome reaction in guinea pig spermatozoa. *J. Cell Biol.* 74: 561–577, 1977.

231. Fulka, J., T. Jung, and R. M. Moor. The fall of biological maturation promoting factor (MPF) and histone H1 kinase activity during anaphase and telophase in mouse oocytes. *Mol. Reprod. Dev.* 32: 378–382, 1992.

232. Fulka, J., J. Motlik, J. Fulfa, and N. Crozet. Inhibition of nuclear maturation in fully-grown porcine and mouse oocytes after their fusion with growing porcine oocytes. *J. Exp. Zool.* 235: 255–259, 1985.

233. Galione, A. Ca²⁺-induced Ca²⁺ release and its modulation by cyclic ADP-ribose. *TIPS* 13: 304–306, 1992.

234. Galione, A. Cyclic ADP-ribose: a new way to control calcium. *Science* 259: 325–326, 1993.

235. Galione, A., H. C. Lee, and W. B. Busa. Ca²⁺-induced Ca²⁺ release in sea urchin egg homogenates: modulation by cyclic ADP-ribose. *Science* 253: 1143–1146, 1991.

236. Ganz, M. B., G. Boyarsky, R. B. Sterzel, and W. F. Boron. Arginine vasopressin enhances pHᵢ regulation in the presence of HCO3⁻ by stimulating three acid-base transport systems. *Nature* 337: 648–651, 1989.

237. Garbers, D. L. Molecular basis of fertilization. *Annu. Rev. Biochem.* 58: 719–742, 1989.

238. Garbers, D. L. Guanylyl cyclase receptors and their endocrine, paracrine, and autocrine ligands. *Cell* 71: 1–4, 1992.

239. Garbers, D. L., and G. S. Kopf. The regulation of spermatozoa by calcium and cyclic nucleotides. In: *Advances in Cyclic Nucleotide Research*, edited by P. Greengard and G. A. Robison, New York: Raven, 1980, p. 251–306.

240. Garnier, J., D. Osguthorpe, and B. Robson. Analysis of the accuracy and implications of simple methods for predicting the secondary structure of globular proteins. *J. Mol. Biol.* 120: 97–120, 1978.

241. Garty, N. B., D. Galiani, A. Aharonheim, Y.-K. Ho, D. M. Phillips, N. Dekel, and Y. Salomon. G-proteins in mammalian gametes: an immunocytochemical study. *J. Cell Sci.* 91: 21–31, 1988.

242. Gearing, D. P., M. R. Comeau, D. J. Friend, S. D. Gimple, C. J. Thut, J. McGourty, K. K. Brasher, J. A. King, S. Gillis, B. Mosley, S. F. Ziegler, and D. Cosman. The IL-6 signal transducer, gp130: an oncostatic receptor and affinity converter for the LIF receptor. *Science* 255: 1434–1437, 1992.

243. Gilula, N., M. Epstein, and W. Beers. Cell-to-cell communication and ovulation: a study of the cumulus cell-oocyte complex. *J. Cell Biol.* 78: 58–75, 1978.

244. Gilula, N., O. Reeves, and A. Steinbach. Metabolic coupling, ionic coupling, and cell contacts. *Nature* 235: 262–265, 1972.

245. Ginsburg, V., and P. W. Robbins. (Eds). *Biology of Carbohydrates.* New York: Wiley, 1984, vol. 2, p. 342.

246. Glabe, C. G. Interaction of the sperm adhesive protein, bindin, with phospholipid vesicles. II. Bindin induces fusion of mixed-phase vesicles that contain phosphatidylcholine and posphatidylserine *in vitro. J. Cell Biol.* 100: 800–806, 1985.

247. Glassner, M., J. Jones, I. Kligman, M. J. Woolkalis, G. L. Gerton, and G. S. Kopf. Immunocytochemical and biochemical characterization of guanine nucleotide-binding regulatory proteins in mammalian spermatozoa. *Dev. Biol.* 148: 438–450, 1991.

248. Glick, D. L., M. R. Hellmich, S. Beushausen, P. Tempst, H. Bayley, and F. Strumwasser. Primary structure of a molluscan egg-specific NADase, a second-messenger enzyme. *Cell Regul.* 2: 211–218, 1991.

249. Glossmann, H., and J. Striessnig. Calcium channels. *Vitam. Horm.* 44: 155–328, 1988.

250. Gluck, S. L. The vacuolar H⁺-ATPases: versatile proton pumps participating in constitutive and specialized functions in eukaryotic cells. *Int. Rev. Cytol.* 137C: 105–137, 1993.

251. Go, K. J., and D. P. Wolf. Albumin-mediated changes in the

sperm sterol contents during capacitation. *Biol. Reprod.* 32: 145–152, 1985.

252. Godin, I., R. Deed, J. Cooke, K. Zsebo, M. Dexter, and C. C. Wylie. Effects of the *steel* gene product on mouse primordial germ cells in culture. *Nature* 352: 807–809, 1991.

253. Gondos, B. Oogonia and oocytes in mammals. In: *The Vertebrate Ovary*, edited by R. Jones. New York: Plenum, 1978, p. 83–120.

254. Gould, M., and J. L. Stephano. Electrical responses of eggs to acrosomal protein similar to those induced by sperm. *Science* 235: 1654–1656, 1987.

255. Gould, M., J. L. Stephano, and L. Z. Holland. Isolation of protein *Urechis* sperm acrosomal granules that binds sperm to eggs and initiates development. *Dev. Biol.* 117: 306–318, 1986.

256. Green, E. (Ed). *Biology of the Laboratory Mouse*, New York: Dover, 1975.

257. Green, D.P.L. The induction of the acrosome reaction in guinea pig sperm by the divalent metal cation ionophore A23187. *J. Cell Sci.* 32: 137–151, 1978.

258. Green, D.P.L. Granule swelling and membrane fusion in exocytosis. *J. Cell Sci.* 88: 547–549, 1987.

259. Green, D.P.L. The effect of permeant and impermeant osmoticants on exocytosis in guinea pig sperm. *J. Cell Sci.* 100: 761–769, 1991.

260. Greve, J., and P. M. Wassarman. Mouse egg extracellular coat is a matrix of interconnected filaments possessing a structural repeat. *J. Mol. Biol.* 181: 253–264, 1985.

261. Gross, M. K., D. G. Toscano, and W. A. Toscano. Calmodulin-mediated adenylate cyclase from mammalian sperm. *J. Biol. Chem.* 262: 8672–8676, 1987.

262. Guerrero, A., and A. Darszon. Egg jelly triggers a calcium influx which inactivates and is inhibited by calmodulin antagonists in the sea urchin sperm. *Biochim. Biophys. Acta* 980: 109–116, 1989.

263. Guerrero, A., and A. Darszon. Evidence for the activation of two different Ca^{2+} channels during the egg jelly-induced acrosome reaction of sea urchin sperm. *J. Biol. Chem.* 264: 19593–19599, 1989.

264. Gulyas, B. Cortical granules of mammalian eggs. *Int. Rev. Cytol.* 63: 357–392, 1980.

265. Gulyas, B. J., and E. D. Schmell. Sperm-egg recognition and binding in mammals. In: *Bioregulators of Reproduction*, edited by G. Jargiello and H. Vogel, New York: Academic, 1981, p. 499–520.

266. Guraya, S. S. Morphology, histochemistry, and biochemistry of human oogenesis and ovulation. *Int. Rev. Cytol.* 37: 121–152, 1974.

267. Gwatkin, R.B.L. *Fertilization Mechanisms in Man and Mammals.* New York: Plenum, 1976.

268. Gwatkin, R.B.L., and O. Andersen. Hamster oocyte maturation *in vitro:* inhibition by follicular components. *Life Sci.* 19: 527–536, 1976.

269. Gwatkin, R.B.L., H. W. Carter, and H. Patterson. Association of mammalian sperm with the cumulus cells and the zona pellucida studied by scanning electron microscopy. In: *Scanning Electron Microscopy*, edited by H. Johari, and R. P. Becker. Chicago: IIT Research Inst., 1976, vol. 2, part 2, p. 379–384.

270. Habibi, B., and L. Franchi. Fine-structural changes in the nucleus of primordial oocytes in immature hamsters. *J. Cell. Sci.* 34: 209–223, 1978.

271. Hagiwara, S. *Membrane Potential-Dependent Ion Channels in Cell Membrane. Phylogenetic and Developmental Approaches.* New York: Raven, 1983.

272. Hammerstedt, R. H., A. D. Keith, W. Snipes, R. P. Amann, D. Arruda, and L. C. Griel. Use of spin labels to evaluate effects of cold shock and osmolality on sperm. *Biol. Reprod.* 18: 686–696, 1978.

273. Handrow, R. R., S. K. Boehm, R. W. Lenz, J. A. Robinson, and R. L. Ax. Specific binding of the glycosaminoglycan ^{3}H-heparin to bull, monkey, and rabbit spermatozoa *in vitro. J. Androl.* 5: 51–63, 1984.

274. Handrow, R. R., N. L. First, and J. J. Parrish. Calcium requirement and increased association with bovine sperm during capacitation by heparin. *J. Exp. Zool.* 252: 174–182, 1989.

275. Hardisty, M. Primordial germ cells and the vertebrate germ line. In: *The Vertebrate Ovary*, edited by R. Jones. New York: Plenum, 1978, p. 1–82.

276. Harrison, R.A.P., and S. J. Gaunt. Multiple forms of ram and bull sperm hyaluronidase revealed by using monoclonal antibodies. *J. Reprod. Fertil.* 82: 777–785, 1988.

277. Harrison, R.A.P., E.R.S. Roldan, D. J. Lander, and R. F. Irvine. Ram spermatozoa produce inositol 1,4,5-trisphosphate but not inositol 1,3,4,5-tetrakisphosphate during the Ca^{2+}/ionophore-induced acrosome reaction. *Cell Signal* 2: 277–284, 1990.

278. Hartmann, J. F., R.B.L. Gwatkin, and C. F. Hutchison. Early contact interactions between mammalian gametes *in vitro:* evidence that the vitellus influences adhesion between sperm and zona pellucida. *Proc. Natl. Acad. Sci. U.S.A.* 69: 2767–2769, 1972.

279. Hartree, E. F. Spermatozoa, eggs, and proteinases. *Biochem. Soc. Trans.* 5: 375–383, 1977.

280. Hartree, E. F., and P. N. Strivastava. Chemical composition of the acrosomes of ram spermatozoa. *J. Reprod. Fertil.* 9: 47–60, 1965.

281. Hartung, M., and A. Stahl. Autoradiographic study of RNA synthesis during meiotic prophase in the human oocyte. *Cytogenet. Cell Genet.* 20: 51–58, 1978.

282. Haugen, T. B., E. J. Paulssen, J. O. Gordeladze, and V. Hansson. A unique mRNA species for a subunit of G_s is present in rat haploid germ cells. *Biochem. Biophys. Res. Commun.* 168: 91–98, 1990.

283. Hawley, R. S. and T. Arbel. Yeast genetics and the fall of the classical view of meiosis. *Cell* 72: 301–303, 1993.

284. Hedrick, J. L., and N. J. Wardrip. Isolation of the zona pellucida and purification of its glycoprotein families from pig oocytes. *Anal. Biochem.* 157: 63–70, 1986.

285. Heller, D., D. Cahill, and R. M. Schultz. Biochemical studies of mammalian oogenesis: metabolic cooperativity between granulosa cells and growing mouse oocytes. *Dev. Biol.* 84: 455 464, 1981.

286. Heller, D., and R. M. Schultz. Ribonucleoside metabolism by mouse oocytes: metabolic cooperativity between the fully-grown oocyte and cumulus cells. *J. Exp. Zool.* 214: 355–364, 1980.

287. Hepler, J. R., and A. G. Gilman. G proteins. *Trends Biochem. Sci.* 17: 383–387, 1992.

288. Herlands, R., and R. M. Schultz. Regulation of mouse oocyte growth: probable nutritional role for intercellular communication between follicle cells and oocytes in oocyte growth. *J. Exp. Zool.* 229: 317–325, 1984.

289. Heyting, C., R. J. Dettmers, A.J.J. Dietrich, E.J.W. Redeker, and A.C.G. Vink. Two major components of synatonemal complexes are specific for meiotic prophase nuclei. *Chromosoma* 96: 325–333, 1988.

290. Hille, B. *Ionic Channels of Excitable Membranes.* Sunderland, MA: Sinauer, 1991.

291. Hille, B. G protein-coupled mechanisms and the nervous system. *Neuron* 9: 187–195, 1992.

292. Hinsch, K.-D., E. Hinsch, G. Aumuller, I. Tychowiecka, G. Schultz, and W.-B. Schill. Immunological identification of G protein α- and β-subunits in tail membranes of bovine spermatozoa. *Biol. Reprod.* 47: 337–346, 1992.

293. Hodgkin, J., T. Domaich, and M. Shen. Sex determination pathway in the nematode *Caenorhabditis elegans:* variations on a theme. *Cold Spring Harb. Symp. Quant. Biol.* 50: 585–593, 1985.

294. Hogan, B., F. Costantini, and E. Lacy (Eds). *Manipulating the Mouse Embryo. A Laboratory Manual.* Cold Spring Harbor: Cold Spring Harbor Laboratory, 1986.

295. Hong, K. and V. D. Vacquier. Fusion of liposomes induced by a cationic protein from the acrosome granule of abalone spermatozoa. *Biochemistry* 25: 543–549, 1986.

296. Hunt, D. M., and D. R. Johnson. Abnormal spermiogenesis in two pink-eyed sterile mutants of the mouse. *J. Embryol. Exp. Morph.* 26: 111–121, 1971.

297. Hunter, A. G., and H. O. Nornes. Characterization and isolation of a sperm coating antigen from rabbit seminal plasma with capacity to block fertilization. *J. Reprod. Fertil.* 20: 419–427, 1969.

298. Hyne, R. V. Bicarbonate- and calcium-dependent induction of rapid guinea pig sperm acrosome reactions by monovalent ionophores. *Biol. Reprod.* 31: 312–323, 1984.

299. Hyne, R. V., K. P. Edwards, A. Lopata, and J. D. Smith. Changes in guinea pig sperm intracellular sodium and potassium content during capacitation and treatment with monovalent ionophores. *Gamete Res.* 12: 65–73, 1985.

300. Hyne, R. V., and D. L. Garbers. Regulation of guinea pig sperm adenylate cyclase by calcium. *Biol. Reprod.* 21: 1135–1142, 1979.

301. Hyne, R. V., and D. L. Garbers. Requirement of serum factors for capacitation and the acrosome reaction of guinea pig spermatozoa in buffered medium below pH 7.8 *Biol. Reprod.* 24: 257–266, 1981.

302. Hyne, R. V., and D. L. Garbers. Inhibition of the guinea pig acrosome reaction by a low molecular weight factor(s) in epididymal fluid and serum. *J. Reprod. Fertil.* 67: 151–156, 1982.

303. Hyne, R. V., R. E. Higginson, D. Kohlman, and A. Lopata. Sodium requirement for capacitation and membrane fusion during the guinea pig sperm acrosome reaction. *J. Reprod. Fertil.* 70: 83–94, 1984.

304. Hyne, R. V., and A. Lopata. Calcium and adenosine affect human sperm adenylate cyclase activity. *Gamete Res.* 6: 81–89, 1982.

305. Hynes, R. O. Integrins: versatility, modulation, and signaling in cell adhesion. *Cell* 69: 11–25, 1992.

306. Igusa, Y., and S. Miyazaki. Periodic increase of cytoplasmic free calcium in fertilized hamster eggs measured with calcium-sensitive electrodes. *J. Physiol.* 377: 193–205, 1986.

307. Igusa, Y., S. Miyazaki, and N. Yamashita. Periodic hyperpolarizing responses in hamster and mouse eggs fertilized with mouse sperm. *J. Physiol.* 340: 643–647, 1983.

308. Inoue, M., and D. P. Wolf. Fertilization-associated changes in the murine zona pellucida: a time sequence study. *Biol. Reprod.* 13: 546–551, 1975.

309. Inoue, M., and D. P. Wolf. Sperm binding characteristics of the murine zona pellucida. *Biol. Reprod.* 13: 340–348, 1975.

310. Inskeep, P. B., S. F. Magargee, and R. H. Hammerstedt. Alterations in motility and metabolism with sperm interaction with accessory sex gland fluids. *Arch. Biochem. Biophys.* 241: 1–9, 1985.

311. Ip, N. Y., S. H. Nye, T. G. Boulton, S. Davis, T. Taga, Y. Li, S. J. Birren, K. Yasukawa, T. Kishimoto, D. J. Anderson, N. Stahl, and G. D. Yancopoulos. CNTF and LIF act on neuronal cells via shared signaling pathways that involve the IL-6 signal transducing receptor component gp130. *Cell* 69: 1121–1132, 1992.

312. Jaffe, L. F. Calcium explosions as triggers of development. *Ann. N. Y. Acad. Sci.* 339: 86–101, 1980.

313. Jaffe, L. F. Sources of calcium in egg activation: a review and hypothesis. *Dev. Biol.* 99: 256–276, 1983.

314. Jakoby, F. Ovarian histochemistry. In: *The Ovary,* edited by S. Zuckerman, New York: Academic, 1962, p. 189–246.

315. Janis, R. A., P. J. Silver, and D. J. Triggle. Drug action and cellular calcium regulation. *Adv. Drug Res.* 116: 309–591, 1987.

316. Jenkins, A. D., C. P. Lechene, and S. S. Howards. Concentrations of seven elements in the intraluminal fluids of the rat seminiferous tubules, rete testis, and epididymis. *Biol. Reprod.* 23: 981–987, 1980.

317. Jessell, T. M., and J. Dodd. Structure and expression of differentiation antigens of functional subclasses of sensory neurons. *Philos. Trans. R. Soc. Lond. Biol.* 308: 271–281, 1985.

318. Jessell, T. M., and E. R. Kandel. Synaptic transmission: a bidirectional and self-modifiable form of cell-cell communication. *Neuron* 10: 1–30, 1993.

319. Jones, R. (Ed). *The Vertebrate Ovary.* New York: Plenum, 1978.

320. Juliano, R. L., and S. Haskill. Signal transduction from the extracellular matrix. *J. Cell Biol.* 120: 577–585, 1993.

321. Kaibara, M., and M. Kameyama. Inhibition of the calcium channel by intracellular protons in single ventricular myocytes of the guinea pig. *J. Physiol.* 403: 621–640, 1988.

322. Kandel, E. R., J. H. Schwartz, and T. M. Jessell. *Principles of Neural Science.* New York: Elsevier, 1991.

323. Kang, Y. Development of the zona pellucida in the rat oocyte. *Am. J. Anat.* 139: 535–566, 1974.

324. Kanwar, K. C., R. Yanagimachi, and A. Lopata. Effects of human seminal plasma on fertilizing capacity of human spermatozoa. *Fertil. Steril.* 31: 321–327, 1979.

325. Karnik, N. S., S. Newman, G. S. Kopf, and G. L. Gerton. Developmental expression of G protein a subunits in mouse spermatogenic cells: evidence that G_{ai} is associated with the developing acrosome. *Dev. Biol.* 152: 393–402, 1992.

326. Katz, B. *Nerve, Muscle, and Synapse.* New York: McGraw-Hill, 1966.

327. Katz, B., and R. Miledi. The timing of calcium action during neuromuscular transmission. *J. Physiol.* 189: 535–544, 1967.

328. Kawakami, E., C. A. Vandevoort, C. A. Mahi-Brown, and J. W. Overstreet. Induction of acrosome reactions of canine sperm by homologous zona pellucida. *Biol. Reprod.* 48: 841–845, 1993.

329. Kelly, R. B. Pathways of protein secretion in eukaryotes. *Science* 230: 25–32, 1985.

330. Kelly, R. B. Storage and release of neurotransmitters. *Neuron* 10: 43–53, 1993.

331. Kew, D., K. E. Muffly, and D. L. Kilpatrick. Proenkephalin products are stored in the sperm acrosome and may function in fertilization. *Proc. Natl. Acad. Sci. USA* 87: 9143–9147, 1990.

332. Kimble, J., M. K. Barton, T. B. Schedl, T. A. Rosenquist, and J. Austin. Controls of postembryonic germ line development

in *Caenorhabditis elegans.* In: *Gametogenesis and the Early Embryo,* edited by J. Gall. New York: Alan Liss, 1986, pp. 97–110.

333. Kimble, J., and J. G. White. Control of germ cell development in *Caenorhabditis elegans. Dev. Biol.* 81: 208–219, 1981.

334. Kinloch, R. A., S. Mortillo, C. L. Stewart, and P. M. Wassarman. Embryonal carcinoma cells transfected with ZP3 genes differentially glycosylate similar polypeptides and secrete active mouse sperm receptor. *J. Cell Biol.* 115: 655–664, 1991.

335. Kinloch, R. A., R. J. Roller, and P. M. Wassarman. Organization and expression of the mouse sperm receptor gene. In: *Developmental Biology,* edited by E. H. Davidson, J. V. Ruderman, and J. W. Posakony. New York: Wiley-Liss, 1990, p. 9–20.

336. Kinloch, R. A., R. J. Roller, C. M. Fimiani, D. A. Wassarman, and P. M. Wassarman. Primary structure of the mouse sperm receptor's polypeptide chain determined by genomic cloning. *Proc. Natl. Acad. Sci. U.S.A.* 85: 6409–6413, 1988.

337. Kinloch, R. A., B. Ruiz-Seiler, and P. M. Wassarman. Genomic organization and polypeptide primary structure of zona pellucida glycoprotein hZP3, the hamster sperm receptor. *Dev. Biol.* 142: 414–421, 1990.

338. Kinloch, R. A., and P. M. Wassarman. Nucleotide sequence of the gene encoding zona pellucida glycoprotein ZP3—the mouse sperm receptor. *Nucleic Acids Res.* 17: 2861–2863, 1989.

339. Kinloch, R. A., and P. M. Wassarman. Profile of a mammalian sperm receptor gene. *New Biol.* 1: 232–238, 1989.

340. Kinloch, R. A., and P. M. Wassarman. Specific gene expression during oogenesis in mice. In: *Genes in Mammalian Reproduction,* edited by R.B.L. Gwatkin. NY: Wiley-Liss, 1993, p. 27–44.

341. Kleuss, C., J. Hescheler, C. Ewel, W. Rosenthal, G. Schultz, and B. Wittig. Assignment of G protein subtypes to specific receptors inducing inhibition of calcium currents. *Nature* 353: 43–48, 1991.

342. Kline, D., G. S. Kopf, L. F. Muncy, and L. A. Jaffe. Evidence for the involvement of a pertussis toxin-insensitive G-protein in egg activation of the frog, Xenopus laevis. *Dev. Biol.* 143: 218–229, 1991.

343. Kline, D., L. Simoncini, G. Mandel, R. A. Maue, R. T. Kado, and L. A. Jaffe. Fertilization events induced by neurotransmitters after injection of mRNA in *Xenopus* eggs. *Science* 241: 464–467, 1988.

344. Klingman, I., M. Glassner, B. T. Storey, and G. S. Kopf. Zona pellucida-mediated acrosomal exocytosis in mouse spermatozoa: characterization of an intermediate stage prior to the completion of the acrosome reaction. *Dev. Biol.* 145: 344–355, 1991.

345. Koehler, J. K. Lectins as probes of the spermatozoon surface. *Arch. Androl.* 6: 197–217, 1981.

346. Koehler, J. K., E. D. Nudelman, and S. Hakamori. A collagen-binding protein on the surface of ejaculated rabbit spermatozoa. *J. Cell Biol.* 86: 529–536, 1980.

347. Kopf, G. S., and G. L. Gerton. The mammalian sperm acrosome and the acrosome reaction. In: *Elements of Mammalian Fertilization. Vol I. Basic Concepts,* edited by P. M. Wassarman. Boca Raton, FL: CRC, 1991, p. 153–203.

348. Kopf, G. S., M. J. Woolkalis, and G. L. Gerton. Evidence for a guanine nucleotide-binding regulatory protein in invertebrate and mammalian sperm. *J. Biol. Chem.* 261: 7327–7331, 1986.

349. Kornfeld, R., and S. Kornfeld. Assembly of asparagine-linked oligosaccharides. *Annu. Rev. Biochem.* 54: 631–664, 1985.

350. Krafte, D. S., and R. S. Kass. Hydrogen ion modulation of Ca channel current in cardiac ventricular cells. Evidence for multiple mechanisms. *J. Gen. Physiol.* 91: 641–657, 1988.

351. Krarup, T., T. Pedersen, and M. Faber. Regulation of oocyte growth in the mouse ovary. *Nature* 224: 187–188, 1969.

352. Kume, S., A. Muto, J. Aruga, T. Nakagawa, T. Michikawa, T. Furuichi, S. Nakade, H. Okano, and K. Mikoshiba. The Xenopus IP₃ receptor: structure, function, and localization in oocytes and eggs. *Cell* 73: 555–570, 1993.

353. Kurasawa, S., R. M. Schultz, and G. S. Kopf. Egg-induced modifications of the zona pellucida of mouse eggs: effects of microinjected inositol 1,4,5-trisphosphate. *Dev. Biol.* 133: 295–304, 1989.

354. Kyte, J., and R. F. Doolittle. A simple method for displaying the hydropathic character of a protein. *J. Mol. Biol.* 157: 105–132, 1982.

355. La Valette St. George, Von Ueber die Genese der Samenkorper. I. *Arch. Mikrosk. Anat.* 1: 404–414, 1865.

356. La Valette St. George, Von Ueber die Genese der Samenkorper. IV. Die Spermatogenese bei den Amphibien. *Arch. Mikrosk. Anat.* 12: 797–825, 1876.

357. La Valette St. George, Von Ueber die Genese der Samenkorper. V. Die Spermatogenese bei den Saugetheiren und dem Menschen. *Arch. Mikrosk. Anat.* 15: 261–314, 1878.

358. La Valette St. George, Von Spermatologische Beitrage. V. Ueber die Bildung der Spermatocysten bei den Lepidopteren. *Arch. Mikrosk. Anat.* 30: 426–434, 1887.

359. Langlais, J., F.W.K. Kan, L. Granger, L. Raymond, G. Bleau, and K. D. Roberts. Identification of sterol acceptors that stimulate cholesterol efflux from human spermatozoa during in vitro capacitation. *Gamete Res.* 20: 185–201, 1988.

360. Langlais, J., and K. D. Roberts. A molecular membrane model of sperm capacitation and the acrosome reaction of mammalian spermatozoa. *Gamete Res.* 12: 183–224, 1985.

361. Langley, J. N. On the contraction of musccle, chiefly in relation to the presence of "receptive" substances. *J. Physiol.* 36: 347–384, 1907.

362. Lardy, H. A., and J. San Agustin. Caltrin and calcium regulation of sperm activity. In: *The Cell Biology of Fertilization,* edited by H. Schatten, and G. Schatten. New York: Academic, 1989, p. 29–39.

363. Larsen, W., S. Wert, and G. Brunner. A dramatic loss of cumulus cell gap junctions is correlated with germinal vesicle breakdown in rat oocytes. *Dev. Biol.* 113: 517–521, 1986.

364. Lawrence, T., W. Beers, and N. Gilula. Transmission of hormonal stimulation by cell-to-cell communication. *Nature* 272: 501–506, 1978.

365. Lawrence, T., R. Ginzberg, N. Gilula, and W. Beers. Hormonally induced cell shape changes in cultured rat ovarian granulosa cells. *J. Cell Biol.* 80: 21–36, 1979.

366. Lee, H. C., and R. Aarhus. ADP-ribosyl cyclase: an enzyme that cyclizes NAD⁺ into a calcium-mobilizing metabolite. *Cell Regul.* 2: 203–209, 1991.

367. Lee, H. C., C. Johnson, and D. Epel. Changes in internal pH associated with initiation of motility and acrosome reaction of sea urchin sperm. *Dev. Biol.* 95: 31–45, 1983.

368. Lee, H. C., T. F. Walseth, G. T. Bratt, R. N. Hayes, and D. L. Clapper. Structural determination of a cyclic metabolite of NAD⁺ with intracellular Ca^{2+}-mobilizing activity. *J. Biol. Chem.* 264: 1608–1615, 1990.

369. Lee, M. A., J. H. Check, and G. S. Kopf. A guanine nucleotide-binding regulatory protein in human sperm mediates acrosomal exocytosis induced by the human zona pellucida. *Mol. Reprod. Dev.* 31: 78–86, 1992.

370. Lee, M. A., and B. T. Storey. Bicarbonate is essential for

fertilization of mouse eggs: mouse sperm require it to undergo the acrosome reaction. *Biol. Reprod.* 34: 349–356, 1986.

371. Lee, M. A., and B. T. Storey. Evidence for plasma membrane impermeability to small ions in acrosome-intact mouse spermatozoa bound to mouse zonae pellucidae, using an aminoacridine fluorescent pH probe: time course of the zona-induced acrosome reaction monitored by both chlortetracycline and pH probe fluorescence. *Biol. Reprod.* 33: 235–246, 1985.

372. Lee, M. A., and B. T. Storey. Endpoint of the first stage of zona pellucida-induced acrosome reaction in mouse spermatozoa characterized by acrosomal H^+ and Ca^{2+} permeability: population and single cell kinetics. *Gamete Res.* 24: 303–326, 1989.

373. Lee, M. A., G. S. Trucco, K. B. Bechtol, N. Wummer, G. S. Kopf, L. Blasco, and B. T. Storey. Capacitation and acrosome reactions in human spermatozoa monitored by a chlortetracycline fluorescence assay. *Fertil. Steril.* 48: 649–658, 1987.

374. Leibfried, L., and N. First. Effects of divalent cations on *in vitro* maturation of bovine oocytes. *J. Exp. Zool.* 210: 575–580, 1979.

375. Leibfried, L., and N. First. Follicular control of meiosis in the porcine oocyte. *Biol. Reprod.* 23: 705–709, 1980.

376. Leibfried, L., and N. First. Effect of bovine and porcine follicular fluid and granulosa cells on maturation of oocytes *in vitro*. *Biol. Reprod.* 23: 699–704, 1980.

377. Lewis, R. V., J. San Agustin, W. Kruggel, and H. A. Lardy. The structure of caltrin, the calcium-transport inhibitor of bovine seminal plasma. *Proc. Natl. Acad. Sci. U.S.A.* 82: 6490–6491, 1985.

378. Leyton, L., P. Leguen, D. Bunch, and P. M. Saling. Regulation of mouse gamete interaction by a sperm tyrosine kinasse. *Proc. Natl. Acad. Sci. U.S.A.* 89: 11692–11695, 1992.

379. Leyton, L., and P. M. Saling. 95 kd sperm proteins bind ZP3 and serve as tyrosine kinase substrates in response to zona binding. *Cell* 57: 1123–1130, 1989.

380. Leyton, L., and P. M. Saling. Evidence that aggregation of mouse sperm receptors by ZP3 triggers the acrosome reaction. *J. Cell Biol.* 108: 2163–2168, 1989.

381. Lillie, F. R. *Problems of Fertilization*. Chicago: University of Chicago, 1919.

382. Lindemann, C. B., T. K. Gardner, E. Westbrook, and K. S. Kanous. The calcium-induced curvature reversal of rat sperm is potentiated by cAMP and inhibited by anti-calmodulin. *Cell Motil. Cytoskeleton* 20: 316–324, 1991.

383. Lindemann, C. B., and J. S. Goltz. Calcium regulation of flagellar curvature and swimming pattern in Triton-X-100-extracted rat sperm. *Cell Motil. Cytoskeleton* 10: 420–431, 1988.

384. Llanos, M. N., and S. Meizel. Phospholipid methylation increases during capacitation of golden hamster sperm in vitro. *Biol. Reprod.* 28: 1043–1050, 1983.

385. Macek, M. B., L. C. Lopez, and B. D. Shur. Aggregation of β-1,4-galactosyltransferase on mouse sperm induces the acrosome reaction. *Dev. Biol.* 147: 440–444, 1991.

386. Magnusson, C., and T. Hillensjo. Inhibition of maturation and metabolism of rat oocytes by cyclic AMP. *J. Exp. Zool.* 201: 138–147, 1977.

387. Mahanes, M. S., D. L. Ochs, and L. A. Eng. Cell calcium of ejaculated rabbit spermatozoa before and following *in vitro* capacitation. *Biochem. Biophys. Res. Commun.* 134: 664–670, 1986.

388. Maller, J. L. Interaction of steroids with the cyclic nucleotide system in amphibian oocytes. *Adv. Cyclic Nucl. Res.* 15: 295–336, 1983.

389. Maller, J. L. Oocyte maturation in amphibians. In: *Develop-mental Biology: A Comprehensive Synthesis, Vol. 1, Oogenesis*, edited by L. Browder, New York: Plenum, 1985, p. 238–311.

390. Mancini, R. E., A. Alanso, J. Barquet, and N. R. Nerimovsky. Histo-immunological localization of hyaluronidase in bull testis. *J. Reprod. Fertil.* 8: 325–330, 1964.

391. Mann, T. *The Biochemistry of Semen and of the Male Reproductive Tract*. New York: Wiley, 1964.

392. Mann, T. and C. Lutwak-Mann. *Male Reproductive Function and Semen. Themes and Trends in Physiology, Biochemistry, and Investigative Andrology*. New York: Springer-Verlag, 1981.

393. Masui, Y. Meiotic arrest in animal oocytes. In: *Biology of Fertilization*, edited by C. Metz, and A. Monroy. New York: Academic, 1985, vol. 1, p. 189–220.

394. Masui, Y., and H. Clark. Regulation of oocyte maturation. *Int. Rev. Cytol.* 57: 185–282, 1979.

395. Matsui, Y., D. Toksoz, S. Nishikawa, S.-I. Nishikawa, D. Williams, K. Zsebo, and B.L.M. Hogan. Effect of *Steel* factor and leukaemia inhibitory factor on murine primordial germ cells in culture. *Nature* 353: 750–752, 1991.

396. Matsui, Y., K. Zsebo, and B.L.M. Hogan. Derivation of pluripotential embryonic stem cells from murine primordial germ cells in culture. *Cell* 70: 841–847, 1992.

397. Mattson, B. A., and D. F. Albertini. Oogenesis: chromatin and microtubule dynamics during meiotic prophase. *Mol. Reprod. Dev.* 25: 374–383, 1990.

398. McCulloh, D. H., and E. L. Chambers. Fusion of membranes during fertilization. Increases of the sea urchin egg's membrane capacitance and membrane conductance at the site of contact with the sperm. *J. Gen. Physiol.* 99: 137–175, 1992.

399. McEwen, B. S. Non-genomic and genomic effects of steroids on neural activity. *TIPS* 12: 141–148, 1991.

400. McGaughey, R. The culture of pig oocytes in minimal medium, and the influence of progesterone and estradiol-17β on meiotic maturation. *Endocrinology* 100: 39–45, 1977.

401. McKinnon, C. A., F. E. Weaver, J. A. Yoder, G. Fairbanks, and D. E. Wolf. Cross-linking of a maturation-dependent ram sperm plasma membrane antigen induces the acrosome reaction. *Mol. Reprod. Dev.* 29: 200–207, 1991.

402. McLaren, A., and C. Wylie (Eds). Current problems in germ cell differentiation. In: *7th Symposium British Society of Developmental Biology*. Cambridge: Cambridge University, 1983.

403. McLaughlin, J. D., and B. D. Shur. Binding of caput epididymal mouse sperm to the zona pellucida. *Dev. Biol.* 124: 557–561, 1987.

404. Meizel, S. The importance of hydrolytic enzymes to an exocytotic event, the mammalian sperm acrosome reaction. *Biol. Rev.* 59: 125–157, 1984.

405. Meizel, S. Molecules that initiate or help stimulate the acrosome reaction by their interaction with the mammalian sperm surface. *Am. J. Anat.* 174: 285–302, 1985.

406. Meizel, S., and D. W. Deamer. The pH of the hamster sperm acrosome. *J. Histochem. Cytochem.* 26: 98–105, 1978.

407. Meizel, P., M. C. Pillai, E. Diaz-Perez, and P. Thomas. Initiation of the human sperm acrosome reaction by components of human follicular fluid and cumulus secretions including steroids. In: *Fertilization in Mammals*, edited by B. D. Bavister, J. Cummins, and E.R.S. Roldan. Norwell, MA: Serono Symposia, 1990, p. 205–222.

408. Meizel, S., and K. O. Turner. Progesterone acts at the plasma membrane of human sperm. *Mol. Cell. Endocrinol.* 11: R1–R5, 1991.

409. Messinger, S. M., and D. F. Albertini. Centrosome and micro-

tubule dynamics during meiotic progression in the mouse oocyte. *J. Cell Sci.* 100: 289–298, 1991.

410. Meyerhof, W., R. Muller-Brechlin, and D. Richter. Molecular cloning of a novel putative G protein coupled receptor expressed during rat spermiogenesis. *FEBS Lett.* 284: 155–160, 1991.

411. Millar, S. E., S. M. Chamow, A. W. Baur, C. Oliver, F. Robey, and J. Dean. Vaccination with a synthetic zona pellucida peptide produces long-term contraception in female mice. *Science* 246: 935–938, 1989.

412. Miller, D. J., and R. L. Ax. Carbohydrates and fertilization in animals. *Mol. Reprod. Dev.* 26: 184–198, 1990.

413. Miller, D. J., M. B. Macek, and B. D. Shur. Complementarity between sperm surface β-1,4-galactosyltransferase and egg-coat ZP3 mediates sperm-egg binding. *Nature* 357: 589–593, 1992.

414. Miller, D. J., M. A. Winer, and R. L. Ax. Heparin-binding proteins from seminal plasma bind to bovine spermatozoa and modulate capacitation by heparin. *Biol. Reprod.* 42: 899–915, 1990.

415. Miller, R. J. Voltage-sensitive Ca^{2+} channels. *J. Biol. Chem.* 267: 1403–1406, 1992.

416. Milligan, G. Techniques used in the identification and analysis of function of pertussis toxin-sensitive guanine nucleotide binding proteins. *Biochem. J.* 255: 1–13, 1988.

417. Miyazaki, S. Cell signalling at fertilization of hamster eggs. *J. Reprod. Fertil. Suppl.* 42: 163–175, 1990.

418. Miyazaki, S. Inositol 1,4,5-trisphosphate-induced calcium release and guanine nucleotide-binding protein-mediated periodic calcium rises in golden hamster eggs. *J. Cell Biol.* 106: 3455–353, 1988.

419. Miyazaki, S., N. Hashimoto, Y. Yoshimoto, T. Kishimoto, Y. Igusa, and Y. Hiramoto. Temporal and spatial dynamics of the periodic increase in intracellular calcium at fertilization of golden hamster eggs. *Dev. Biol.* 118: 259–267, 1986.

420. Miyazaki, S., and Y. Igusa. Fertilization potential in golden hamster eggs consists of recurring hyperpolarizations. *Nature* 290: 706–707, 1981.

421. Miyazaki, S., M. Yuzaki, K. Nakeda, H. Shirakawa, S. Nakanishi, S. Nakade, and K. Mikoshiba. Block of Ca^{2+} wave and Ca^{2+} oscillation by antibody to the inositol 1,4,5-trisphosphate receptor in fertilized hamster eggs. *Science* 257: 251–255, 1992.

422. Moens, P. B., C. Heyting, A.J.J. Dietrich, W. Van Raamsdonk, and Q. Chen. Synaptonemal complex antigen location and conservation. *J. Cell Biol.* 105: 93–101, 1987.

423. Moller, C. C., J. D. Bleil, R. A. Kinloch, and P. M. Wassarman. Structural and functional relationships between mouse and hamster zona pellucida glycoproteins. *Dev. Biol.* 137: 276–286, 1990.

424. Monck, J. R., and J. M. Fernandez. The exocytotic fusion pore. *J. Cell Biol.* 119: 1395–1404, 1992.

425. Monks, N. J., and L. R. Fraser. Inhibition of adenosine-metabolizing enzymes modulates mouse sperm fertilizing ability: a changing role for endogenously generated adenosine during capacitation. *Gamete Res.* 21: 267–276, 1988.

426. Moor, R., J. Osborn, D. Cran, and D. Walters. Selective effect of gonadotrophins on cell coupling, nuclear maturation and protein synthesis in mammalian oocytes. *J. Embryol. Exp. Morphol.* 61: 347–365, 1981.

427. Moor, R., M. Smith, and R. Dawson. Measurement of intercellular coupling between oocytes and cumulus cells using intracellular markers. *Exp. Cell Res.* 126: 15–29, 1980.

428. Moor, R., and A. Trounson. Hormonal and follicular factors affecting maturation of sheep oocytes *in vitro* and their subsequent developmental capacity. *J. Reprod. Fertil.* 49: 101–109, 1977.

429. Morales, P., N. L. Cross, J. W. Overstreet, and F. W. Hanson. Acrosome intact and acrosome-reacted human sperm can initiate binding to the zona pellucida. *Dev. Biol.* 133: 385–392, 1989.

430. Mortillo, S., and P. M. Wassarman. Differential binding of gold-labeled zona pellucida glycoproteins mZP2 and mZP3 to mouse sperm membrane compartments. *Development* 113: 141–150, 1991.

431. Moss, J., and M. Vaughan. ADP-ribosylation of guanyl nucleotide-binding regulatory proteins by bacterial toxins. *Adv. Enzymol.* 61: 303–379, 1988.

432. Mount, S. M. A catalogue of splice junction sequences. *Nucleic Acids Res.* 10: 459–472, 1982.

433. Mrsny, R. J., and S. Meizel. Potassium ion influx and Na^+, K^+-ATPase activity are required for the hamster sperm acrosome reaction. *J. Cell Biol.* 91: 77–82, 1981.

434. Mukherjee, A. Normal progeny from fertilization *in vitro* of mouse oocytes matured in culture and spermatozoa capacitated *in vitro*. *Nature* 237: 397–398, 1972.

435. Myles, D. G., H. Hyatt, and P. Primakoff. Binding of both acrosome-intact and acrosome-reacted guinea pig sperm to the zona pellucida during in vitro fertilization. *Dev. Biol.* 121: 559–567, 1987.

436. Myles, D. G., and P. Primakoff. Localized surface antigens of guinea pig sperm migrate to new regions prior to fertilization. *J. Cell Biol.* 99: 1634–1641, 1984.

437. Nagae, T., and P. N. Srivastava. Induction of the acrosome reaction in guinea pig spermatozoa by calmodulin antagonist W-7. *Gamete Res.* 14: 197–208, 1986.

438. Nagalla, S. R., S. Vijayaraghavan, H. M. Florman, A. Archibong, K. K. Li, M. Kelly, B. Barry, D. H. Coy, K. Hermsmeyer, D. P. Wolf, and E. R. Spindel. The bombesin-like peptides gastrin-releasing peptide (GRP) and ARP, a new splice product of the GRP gene, are endogenous regulators of the acrosome reaction and fertilization (Submitted).

439. Nanavati, C., and J. M. Fernandez. The secretory granule matrix: a fast-acting smart polymer. *Science* 259: 963–965, 1993.

440. Nehme, C. L., M. M. Cesario, D. G. Myles, D. E. Koppel, and J. R. Bartles. Breaching the diffusion barrier that compartmentalizes the transmembrane glycoprotein CE9 to the posterior-tail plasma membrane domain of the rat spermatozoon. *J. Cell Biol.* 120: 687–694, 1993.

441. Nicosia, S., D. Wolf, and M. Inoue. Cortical granule distribution and cell surface characteristics in mouse eggs. *Dev. Biol.* 57: 56–74, 1977.

442. Nikolajczyk, B. S., and M. G. O'Rand. Characterization of rabbit testis β-galactosidase and arylsulfatase A: purification and localization in spermatozoa during the acrosome reaction. *Biol. Reprod.* 46: 366–378, 1992.

443. Nishimoto, I., T. Okamoto, Y. Matsuura, S. Takahashi, Y. Murayama, and E. Ogata. Alzheimer amyloid protein precursor complexes with brain GTP-binding protein Go. *Nature* 362: 75–79, 1993.

444. Noland, T. D., N. A. Abumrad, A. H. Beth, and D. L. Garbers. Protein phosphorylation in intact bovine epididymal spermatozoa: identification of the type II regulatory subunit of cyclic adenosine 3',5'-monophosphate-dependent protein kinase as an endogenous phosphoprotein. *Biol. Reprod.* 37: 171–180, 1987.

445. Noland, T. D., and G. E. Olson. Calcium-induced modification

of the acrosomal matrix in digitonin-permeabilized guinea pig spermatozoa. *Biol. Reprod.* 40: 1057–1066, 1989.

446. Noland, T. D., L. J. Van Eldik, D. L. Garbers, and W. H. Burgess. Distibution of calmodulin and calmodulin-binding proteins in membranes from bovine epididymal spermatozoa. *Gamete Res.* 11: 297–303, 1985.

447. Nuccitelli, R., and R. D. Grey. Controversy over the fast, partial, temporary block to polyspermy in sea urchins: a reevaluation. *Dev. Biol.* 103: 1–17, 1984.

448. O'Brien, D. A., C. A. Gabel, D. L. Rockett, and E. M. Eddy. Receptor-mediated endocytosis and differential synthesis of mannpse 6-phosphate receptors in isolated spermatogenic cells and Sertoli cells. *Endocrinologie* 125: 2973–2984, 1989.

449. O'Rand, M. G. Modification of the sperm membrane during capacitation. *Ann. N. Y. Acad. Sci.* 383: 392–404, 1982.

450. O'Rand, M. G., and S. J. Fisher. Localization of zona pellucida binding sites on rabbit spermatozoa and induction of the acrosome reaction by solubilized zonae. *Dev. Biol.* 119: 551–559, 1987.

451. Okabe, K., K. Takada, T. Adachi, Y. Kohama, and T. Miura. Inconsistent reactivity of an antisperm monoclonal antibody and its relation to sperm capacitation. *J. Reprod. Immunol.* 9: 67–71, 1986.

452. Okabe, M., K. Takada, T. Adachi, Y. Kohama, T. Miura, and S. Aonuma. Studies on sperm capacitation using monoclonal antibody-disappearance of an antigen from the anterior part of the mouse sperm head. *J. Pharmacobiodyn.* 9: 55–60, 1986.

453. Okamoto, T., T. Katada, Y. Murayama, M. Ui, E. Ogata, and I. Nishimoto. A simple structure encodes G protein-activating function of the IGF-II/mannose-6-phosphate receptor. *Cell* 62: 709–717, 1990.

454. Okamura, N., A. Fukuda, M. Tanba, Y. Sugita, and T. Nagai. Changes in the nature of calcium transport systems on the porcine sperm plasma membrane during epididymal maturation. *Biochim. Biophys. Acta* 1108: 110–114, 1992.

455. Okamura, N., S. Onoe, K. Kawakura, Y. Tajima, and Y. Sugita. Effects of a membrane-bound trypsin-like proteinase and seminal proteinase inhibitors on the bicarbonate-sensitive adenylate cyclase in porcine sperm plasma membranes. *Biochim. Biophys. Acta* 1035: 83–89, 1990.

456. Okamura, N., and Y. Sugita. Activation of spermatozoan adenylate cyclase by a low molecular weight factor in porcine seminal plasma. *J. Biol. Chem.* 258: 13056–13062, 1983.

457. Okamura, N., Y. Tajima, S. Onoe, and Y. Sugita. Purification of bicarbonate-sensitive sperm adenylylcyclase by 4-acetamido-4'-isothio-cyano-stilbene-2,2'-disulfonic acid-affinity chromatography. *J. Biol. Chem.* 266: 17754–17759, 1991.

458. Okamura, N., Y. Tajima, A. Soejima, H. Masuda, and Y. Sugita. Sodium bicarbonate in seminal plasma stimulates the motility of mammalian spermatozoa through direct activation of adenylate cyclase. *J. Biol. Chem.* 260: 9699–9705, 1985.

459. Okamura, N., Y. Tajima, and Y. Sugita. Decrease in bicarbonate transport activities during epididymal maturation of porcine sperm. *Biochem. Biophys. Res. Commun.* 157: 1280–1287, 1988.

460. Oliphant, G. Removal of sperm-bound seminal plasma components as a prerequisite to induction of the rabbit acrosome reaction. *Fertil. Steril.* 27: 28–38, 1976.

461. Oliphant, G., and B. G. Brackett. Capacitation of mouse spermatozoa in media with elevated ionic strength and reversible decapacitation with epididymal extracts. *Fertil. Steril.* 24: 948–955, 1973.

462. Oliphant, G., and B. G. Brackett. Immunological assessment

of surface changes of rabbit sperm undergoing capacitation. *Biol. Reprod.* 9: 404–414, 1973.

463. Oliphant, G., A. B. Reynolds, and T. S. Thomas. Sperm surface components involved in the control of the acrosome reaction. *Am. J. Anat.* 174: 269–283, 1985.

464. Olsiewski, P., and W. Beers. cAMP synthesis in the rat oocyte. *Dev. Biol.* 100: 287–293, 1983.

465. Orgebin-Crist, M.-C. Maturation of spermatozoa in the rabbit epididymis. Fertilizing ability and embryonic mortality in does inseminated with epididymal spermatozoa. *Ann. Biol. Anim. Biochim. Biophys.* 7: 373–389, 1967.

466. Orgebin-Crist, M.-C. Sperm maturation in the rabbit epididymis. *Nature* 216: 816–818, 1967.

467. Palade, G. Intracellular aspects of the process of protein synthesis. *Science* 189: 347–358, 1975.

468. Paleos, G., and R. Powers. The effect of calcium on the first meiotic division of the mammalian oocyte. *J. Exp. Zool.* 217: 409–416, 1981.

469. Parkes, A. (Ed) *Marshall's Physiology of Reproduction.* New York: Longmans, Green, 1956, vol. 1, part 1.

470. Parmentier, M., F. Libert, S. Schurmans, S. Schiffmann, A. Lefort, D. Eggerickx, C. Ledent, C. Mollereau, C. Gerard, J. Perret, A. Grootegoed, and G. Vassart. Expression of members of the putative olfactory receptor gene family in mammalian germ cells. *Nature* 355: 453–455, 1992.

471. Parrish, J. J., J. L. Susko-Parrish, and N. L. First. Capacitation of bovine sperm by heparin: inhibitory effect of glucose and role of intracellular pH. *Biol. Reprod.* 41: 683–699, 1989.

472. Parrish, J. J., J. L. Susko-Parrish, M. A. Winer, and N. L. First. Capacitation of bovine sperm by heparin. *Biol. Reprod.* 38: 1171–1180, 1988.

473. Paul, S. M., and R. H. Purdy. Neuroactive steroids. *FASEB J.* 6: 2311–2322, 1992.

474. Paulssen, R. H., E. J. Paulssen, J. O. Gordeladze, V. Hansson, and T. B. Haugen. Cell-specific expression of guanine nucleotide-binding proteins in rat testicular cells. *Biol. Reprod.* 45: 566–571, 1991.

475. Pawson, T., and G. D. Gish. SH2 and SH3 domains: from structure to function. *Cell* 71: 359–362, 1992.

476. Pazin, M. J., and L. T. Williams. Triggering signalling cascades by receptor tyrosine kinases. *Trends Biochem. Sci.* 17: 374–378, 1992.

477. Pedersen, T. Follicle growth in the immature mouse ovary. *Acta Endocrinol. (Copenh.)* 62: 117–132, 1969.

478. Pedersen, T. Follicle kinetics in the ovary of the cyclic mouse. *Acta Endocrinol. (Copenh.)* 64: 304–323, 1970.

479. Pedersen, T. Follicle growth in the mouse ovary. In: *Oogenesis,* edited by J. Biggers, and A. Schuetz. Baltimore: University Park, 1972, p. 361–367.

480. Pedersen, T., and H. Peters. Proposal for the classification of oocytes and follicles in the mouse ovary. *J. Reprod. Fertil.* 17: 555–557, 1968.

481. Persson, H., J. F. Rehfeld, A. Ericsson, M. Schalling, M. Pelto-Huikko, and T. Hokfelt. Transient expression of the cholecystokinin gene in male germ cells and accumulation of the peptide in the acrosomal granule: possible role of cholecystokinin in fertilization. *Proc. Natl. Acad. Sci. U.S.A.* 86: 6166–6170, 1989.

482. Peters, H. The development of the mouse ovary from birth to maturity. *Acta Endocrinol. (Copenh.)* 62: 98–116, 1969.

483. Peters, H. Folliculogenesis in mammals. In: *The Vertebrate Ovary,* edited by R. Jones. New York: Plenum, 1978, p. 121–144.

484. Petersen, O. H. Stimulus-secretion coupling: cytoplasmic cal-

cium signals and the control of ion channels in exocrine acinar cells. *J. Physiol.* 448: 1–52, 1992.

485. Phelps, B. M., D. E. Koppel, P. Primakoff, and D. G. Myles. Evidence that proteolysis of the surface is an initial step in the mechanism of formation of sperm cell surface domains. *J. Cell Biol.* 111: 1839–1847, 1990.

486. Phillips, D. M. Surface of the equatorial segment of the mammalian acrosome. *Biol. Reprod.* 16: 128–137, 1977.

487. Phillips, D. M., and R. Yanagimachi. Difference in the manner of association of acrosome-intact and acrosome-reacted hamster spermatozoa with egg microvilli as revealed by scanning electron microscopy. *Dev. Growth Diff.* 26: 543–551, 1982.

488. Philpott, C. C., M. J. Ringuette, and J. Dean. Oocyte-specific expression and developmental regulation of ZP3, the sperm receptor of the mouse zona pellucida. *Dev. Biol.* 121: 568–575, 1987.

489. Pierce, K. E., M. C. Siebert, G. S. Kopf, R. M. Schultz, and P. G. Calarco. Characterization and localization of a mouse egg cortical granule antigen prior to and following fertilization or egg activation. *Dev. Biol.* 141: 381–392, 1990.

490. Pietrobon, D., B. Prod'Hom, and P. Hess. Interactions of protons with single open L type calcium channels. pH fluctuations with Cs^+, K^+, and Na^+ as permeant ions. *J. Gen. Physiol.* 94: 1–21, 1989.

491. Piko, L., and A. Tyler. Fine structural studies of sperm penetration in the rat. *Proc. Vth Int. Congr. Animal Reprod. A. I. Trento* 2: 372–377, 1964.

492. Pincus, G. *The Eggs of Mammals.* New York: Macmillan, 1936.

493. Pincus, G., and E. Enzmann. The comparative behavior of mammalian eggs *in vivo* and *in vitro*. 1. The activation of ovarian eggs. *J. Exp. Med.* 62: 665–675, 1935.

494. Popa, G. T. The distribution of substances in the spermatozoan (*Arbacia* and *Nereis*). *Biol. Bull.* 52: 238–257, 1927.

495. Powers, R., and G. Paleos. Combined effects of calcium and dibutyryl cAMP on germinal vesicle breakdown in the mouse oocyte. *J. Reprod. Fertil.* 66: 1–8, 1982.

496. Primakoff, P., H. Hyatt, and J. Tredick-Kline. Identification and purification of a sperm surface protein with a potential role in sperm-egg membrane fusion. *J. Cell Biol.* 104: 141–149, 1987.

497. Prod'Hom, B., D. Pietrobon, and P. Hess. Interactions of protons with single open L type calcium channels. Location of protonation site and dependence of proton-induced fluctuations on concentration and species of permeant ion. *J. Gen. Physiol.* 94: 23–42, 1989.

498. Quiocho, F. A. Carbohydrate-binding proteins: tertiary structures and protein-sugar interactions. *Annu. Rev. Biochem.* 55: 287–316, 1986.

499. Racowsky, C. Antagonistic actions of estradiol and tamoxifen upon forskolin-dependent meiotic arrest, intercellular coupling, and the cAMP content of hamster oocyte-cumulus complexes. *J. Exp. Zool.* 234: 251–260, 1985.

500. Racowsky, C. Effect of forskolin on the spontaneous maturation and cyclic AMP content of hamster oocyte-cumulus complexes. *J. Exp. Zool.* 234: 87–96, 1985.

501. Racowsky, C., and R. McGaughey. Further studies of the effects of follicular fluid and membrane granulosa cells on the spontaneous maturation of pig oocytes. *J. Reprod. Fertil.* 66: 505–512, 1982.

502. Racowsky, C., and R. McGaughey. In the absence of protein, estradiol suppresses meiosis of porcine oocytes *in vitro. J. Exp. Zool.* 224: 103–110, 1982.

503. Racowsky, C., and R. Satterlie. Metabolic, fluorescent dye and electrical coupling between hamster oocytes and cumulus cells during meiotic maturation *in vivo* and *in vitro. Dev. Biol.* 108: 191–202, 1985.

504. Rafferty, K. *Methods in Experimental Embryology of the Mouse.* Baltimore: Johns Hopkins University, 1970.

505. Ralt, D., M. Goldenberg, P. Fetterolf, D. Thompson, J. Dor, S. Maschiach, D. L. Garbers, and M. Eissenbach. Sperm attraction to follicular factor(s) correlates with human egg fertilizability. *Proc. Natl. Acad. Sci. U.S.A.* 88: 2840–2844, 1991.

506. Reeves, J. P. Molecular aspects of sodium-calcium exchange. *Arch. Biochem. Biophys.* 292: 329–334, 1992.

507. Resnick, J. L., L. S. Bixter, L. Cheng, and P. J. Donovan. Long-term proliferation of mouse primordial germ cells in culture. *Nature* 359: 550–551, 1992.

508. Retzius, G. *Biologische untersuchungen.* Jena: Gustav Fischer, 1909, vol. 14.

509. Reyes, A., G. Oliphant, and B. G. Brackett. Partial purification and identification of a reversible decapacitation factor from rabbit seminal plasma. *Fertil. Steril.* 26: 148–157, 1975.

510. Reynolds, A. B., T. S. Thomas, W. L. Wilson, and G. Oliphant. Concentration of acrosome stabilizing factor (ASF) in rabbit epididymal fluid and species specificity of anti-ASF antibodies. *Biol. Reprod.* 40: 673–681, 1989.

511. Rice, C., and R. McGaughey. Effect of testosterone and dibutyryl cAMP on the spontaneous maturation of pig oocytes. *J. Reprod. Fertil.* 62: 245–256, 1981.

512. Ringuette, M., M. Chamberlin, A. Baur, D. Sobieski, and J. Dean. Molecular analysis of cDNA coding for ZP3, a sperm binding protein of the mouse zona pellucida. *Dev. Biol.* 127: 287–295, 1988.

513. Rogers, B. J. Mammalian sperm capacitation and fertilization in vitro: a critique of methodology. *Gamete Res.* 1: 165–223, 1978.

514. Roldan, E.R.S., and R.A.P. Harrison. Polyphosphoinositide breakdown and subsequent exocytosis in the Ca^{2+}/ionophore-induced acrosome reaction of mammalian spermatozoa. *Biochem. J.* 259: 397–406, 1989.

515. Roldan, E.R.S., and R.A.P. Harrison. Diacylglycerol and phosphatidate production and the exocytosis of the sperm acrosome. *Biochem. Biophys. Res. Commun.* 172: 8–15, 1990.

516. Roller, R. Kinloch, B. Hiraoka, S.S.-L. Li, and P. M. Wassarman. Gene expression during mammalian oogenesis and early embryogenesis: quantification of three messenger RNAs abundant in fully-grown mouse oocytes. *Development* 106: 251–261, 1989.

517. Roller, R. J., and P. M. Wassarman. Role of asparagine-linked oligosaccharides in secretion of glycoproteins of the mouse egg's extracullar coat. *J. Biol. Chem.* 258: 13243–13249, 1983.

518. Roos, A., and W. F. Boron. Intracellular pH. *Physiol. Rev.* 61: 296–434, 1981.

519. Rosiere, T. K., and P. M. Wassarman. Identification of a region of mouse zona pellucida glycoprotein mZP3 that possesses sperm receptor activity. *Dev. Biol.* 154: 309–317, 1992.

520. Rothman, J. E., and L. Orci. Molecular dissection of the secretory pathway. *Nature* 355: 409–415, 1992.

521. Rufo, G. A., P. K. Schoff, and H. A. Lardy. Regulation of calcium content in bovine spermatozoa. *J. Biol. Chem.* 259: 2547–2552, 1984.

522. Rufo, G. A., J. P. Singh, D. F. Babcock, and H. A. Lardy. Purification and characterization of a calcium transport inhibitor protein from bovine seminal plasma. *J. Biol. Chem.* 257: 4627–4632, 1982.

523. Rugh, R. *The Mouse: Its Reproduction and Development.* Minneapolis: Burgess, 1968.

524. Ruknudin, A., and I. A. Silver. Ca^{2+} uptake during capacitation of mouse spermatozoa and the effect of an anion transport inhibitor on Ca^{2+} uptake. *Mol. Reprod. Dev.* 26: 63–68, 1990.

525. Rusinko, N., and H. C. Lee. Widespread occurrence in animal tissues of an enzyme catalyzing the conversion of NAD^+ into a cyclic metabolite with intracellular Ca^+-mobilizing activity. *J. Biol. Chem.* 264: 11725–11731, 1989.

526. Russell, L., R. Peterson, and M. Freund. Direct evidence for formation of hybrid vesicles by fusion of plasma and outer acrosomal membranes during the acrosome reaction in boar spermatozoa. *J. Exp. Zool.* 208: 41–56, 1979.

527. Russo, J., and C. B. Metz. The ultrastructural lesions produced by antibodies and complement in rabbit spermatozoa. *Biol. Reprod.* 10: 293–308, 1974.

528. Russo, R. N., N. L. Shaper, and J. H. Shaper. Bovine β1,4-galacosyltransferase: two sets of mRNA transcripts encode two forms of the protein with different amino-terminal domains. *In vitro* translation experiments demonstrate that both the short and the long forms of the enzyme are type II membrane-bound glycoproteins. *J. Biol. Chem.* 265: 3324–3331, 1990.

529. Sacco, A. G., E. C. Yurewicz, and M. G. Subramanian. Carbohydrate influences the immunogenic and antigenic characteristics of the ZP3 macromolecule (M_r 55,000) of the pig zona pellucida. *J. Reprod. Fertil.* 76: 575–586, 1986.

530. Sacco, A. G., E. C. Yurewicz, A. G. Subramanian, and F. J. Demayo. Zona pellucida composition: Cross-reactivity and contraceptive potential of antiserum to a pig zona antigen. *Biol. Reprod.* 25: 997–1008, 1981.

531. Sacco, A. G., E. C. Yurewicz, M. G. Subramanian, and P. D. Matzat. Porcine zona pellucida: Association of sperm receptor activity with the a-glycoprotein component of the $M_r = 55,000$ family. *Biol. Reprod.* 41: 523–532, 1989.

532. Saling, P. M. Mouse sperm antigens that participate in fertilization. IV. A monoclonal antibody prevents zona penetration by inhibition of the acrosome reaction. *Dev. Biol.* 117: 511–519, 1986.

533. Saling, P. M., J. Sowinski, and B. T. Storey. An ultrastructural study of epididymal mouse spermatozoa binding to zonae pellucidae in vitro: sequential relationship to the acrosome reaction. *J. Exp. Zool.* 209: 229–238, 1979.

534. Saling, P. M., and B. T. Storey. Mouse gamete interactions during fertilization in vitro. Chlortetracycline as a fluorescent probe for the mouse sperm acrosome reaction. *J. Cell Biol.* 83: 544–555, 1979.

535. Saling, P. M., B. T. Storey, and D. P. Wolf. Calcium-dependent binding of mouse epididymal spermatozoa to the zona pellucida. *Dev. Biol.* 65: 515–525, 1978.

536. Salustri, A., and G. Siracusa. Metabolic coupling cumulus expansion and meiotic resumption in mouse cumuli oophori cultured *in vitro* in the presence of FSH or dbcAMP, or stimulated *in vivo* by hCG. *J. Reprod. Fertil.* 68: 335–341, 1983.

537. Salzmann, G., J. Greve, R. Roller, and P. M. Wassarman. Biosynthesis of the sperm receptor during oogenesis in the mouse. *EMBO J.* 2: 1451–1456, 1983.

538. San Agustin, J. T., P. Hughes, and H. A. Lardy. Properties and function of caltrin, the calcium-transport inhibitor of bull seminal plasma. *FASEB J.* 1: 60–66, 1987.

539. Sato, E., and S. Koide. Forskolin and mouse oocyte maturation *in vitro*. *J. Exp. Zool.* 230: 125–129, 1984.

540. Schackmann, R. W. Ionic regulation of the sea urchin sperm acrosome reaction and stimulation by egg-derived peptides. In: *The Cell Biology of Fertilization,* edited by H. Schatten, and G. Schatten. New York: Academic, 1989, p. 3–28.

541. Schackmann, R. W., and B. M. Shapiro. A partial sequence of ionic changes associated with the acrosome reaction of *Strongylocentrotus purpuratus. Dev. Biol.* 81: 145–154, 1981.

542. Schalling, M., H. Persson, M. Pelto-Huikko, L. Odum, P. Ekman, C. Gottlieb, T. Hokfelt, and J. F. Rehfeld. Expression and localization of gastrin messenger RNA and peptide in spermatogenic cells. *J. Clin. Invest.* 86: 660–669, 1990.

543. Schill, W.-B., N. Heimburger, H. Schliessler, R. Stolla, and H. Fritz. Reversible attachment and localization of the acid-stable seminal plasma acrosin-trypsin inhibitors on boar spermatozoa as revealed by the indirect immunofluorescent staining techniques. *Hoppe Seyler Z. Physiol. Chem.* 356: 1473–1475, 1975.

544. Schmell, E., and B. J. Gulyas. Mammalian sperm-egg recognition and binding in vitro. I. Specificity of sperm interaction with live and fixed eggs in homologous and heterologous inseminations of hamster, mouse, and guinea pig oocytes. *Biol. Reprod.* 23: 1075–1085, 1980.

545. Schoch, C., and R. Blumenthal. Role of the fusion peptide sequence in initial stages of influenza hemagglutinin-induced cell fusion. *J. Biol. Chem.* 268: 9267–9274, 1993.

546. Schroeder, A., and J. Eppig. The developmental capacity of mouse oocytes that matured spontaneously *in vitro* is normal. *Dev. Biol.* 102: 493–497, 1984.

547. Schuel, H. Functions of egg cortical granules. In: *Biology of Fertilization,* edited by C. Metz, and A. Monroy, New York: Academic, 1985, vol. 3, p. 1–44.

548. Schultz, R. M. Molecular aspects of mammalian oocyte growth and maturation. In: *Experimental Approaches to Mammalian Embryonic Development,* edited by J. Rossant, and R. A. Pedersen, Cambridge: Cambridge University, 1986, p. 195–237.

549. Schultz, R. M. Meiotic maturation of mammalian oocytes. In: *Elements of Mammalian Fertilization, Vol. 1, Basic Concepts,* edited by P. M. Wassarman. Boca Raton, FL: CRC, 1991, p. 77–104.

550. Schultz, R. M., R. Montgomery, and J. Belanoff. Regulation of mouse oocyte meiotic maturation: implication of a decrease in oocyte cAMP and protein dephosphorylation in commitment to resume meiosis. *Dev. Biol.* 97: 264–273, 1983.

551. Schultz, R. M., R. Montgomery, P. Ward-Bailey, and J. Eppig. Regulation of oocyte maturation in the mouse: possible roles of intercellular communication, cAMP, and testosterone. *Dev. Biol.* 95: 294–304, 1983.

552. Schultz, R. M., and P. M. Wassarman. Biochemical studies of mammalian oogenesis: protein synthesis during oocyte growth and meiotic maturation in the mouse. *J. Cell Sci.* 24: 167–194, 1977.

553. Scully, N. F., J. H. Shaper, and B. D. Shur. Spatial and temporal expression of cell surface galactosyltrasferase during mouse spermatogenesis and epididymal maturation. *Dev. Biol.* 124: 111–124, 1987.

554. Segall, G. K., and W. J. Lennarz. Chemical characterization of the component of the jelly coat from sea urchin eggs responsible for induction of the acrosome reaction. *Dev. Biol.* 71: 33–48, 1979.

555. Segall, G. K., and W. J. Lennarz. Jelly coat and induction of the acrosome reaction in echinoid sperm. *Dev. Biol.* 86: 87–93, 1981.

556. Shabanowitz, R. B. Mouse antibodies to human zona pellucida: evidence that human ZP3 is strongly immunogenic and con-

tains two distinct isomer chains. *Biol. Reprod.* 43: 260–270, 1990.

557. Shabanowitz, R. B., and M. G. O'Rand. Characterization of the human zona pellucida from fertilized and unfertilized eggs. *J. Reprod. Fertil.* 82: 151–161, 1988.

558. Shalgi, R., J. D. Bleil, and P. M. Wassarman. Carbohydrate-mediated sperm-egg interactions in mammals. In: *Advances in Assisted Reproductive Technologies,* edited by Z. Ben-Rafael. New York: Plenum, 1990, p. 437–441.

559. Shalgi, R., R. Kaplan, L. Nebel, and P. F. Kraicer. The male factor in fertilization of rat eggs in vitro. *J. Exp. Zool.* 217: 399–4022, 1981.

560. Shaper, N. L., G. F. Hollis, J. G. Douglas, I. R. Kirsch, and J. H. Shaper. Characterization of the full length cDNA for murine β-1,4-galactosyltransferase. Novel features at the 5'-end predict start sites at two in-frame AUGs. *J. Biol. Chem.* 263: 10420–10428, 1988.

561. Shaper, N. L., W. W. Wright, and J. H. Shaper. Murine β1,4-galactosyltransferase: both the amounts and structure of the mRNA are regulated during spermatogenesis. *Proc. Natl. Acad. Sci. U.S.A.* 87: 791–795, 1990.

562. Sharon, N. *Complex Carbohydrates: Their Chemistry, Biosynthesis, and Functions.* Reading, MA: Addison-Wesley, 1975, p. 466.

563. Shivaji, S., and P. M. Bhargava. Antifertility factors of mammalian seminal fluids. *Bioessays* 7: 13–17, 1987.

564. Shivaji, S., K. H. Scheit, and P. M. Bhargava. *Proteins of Seminal Plasma.* New York: Wiley, 1990.

565. Short, R. The discovery of the ovaries. In: *The Ovary,* edited by S. Zuckerman, and B. Weir. New York: Academic, 1977, vol. 1, p. 1–41.

566. Shur, B. D., and N. G. Hall. Sperm surface galactosyltransferase activities during in vitro capacitation. *J. Cell Biol.* 95: 567–573, 1982.

567. Shur, B. D. and C. A. Neely. Plasma membrane association, purification, and partial characterization of mouse sperm b1,4-galactosyltransferase. *J. Biol. Chem.* 263: 17706–17714, 1988.

568. Sieghart, W. GABA$_A$ receptors: ligand-gated Cl⁻ ion channels modulated by multiple drug-binding sites. *TIPS* 13: 446–450, 1992.

569. Singh, J. P., D. F. Babcock, and H. A. Lardy. Increased calcium-ion influx is a component of capacitation of spermatozoa. *Biochem. J.* 172: 549–556, 1978.

570. Siracusa, G., M. De Felici, and A. Salustri. The proliferative and meiotic history of mammalian female germ cells. In: *Biology of Fertilization,* edited by C. Metz, and A. Monroy. New York: Academic, 1985, vol. 1, p. 253–298.

571. Siracusa, G., D. Whittingham, M. Molinaro, and E. Vivarelli. Parthenogenetic activation of mouse oocytes induced by inhibitors of protein synthesis. *J. Embryol. Exp. Morphol.* 43: 157–166, 1978.

572. Smith, D., and D. Tenney. Effect of steroids on mouse oocyte maturation *in vitro. J. Reprod. Fertil.* 60: 331–338, 1980.

573. Smith, M. B., D. F. Babcock, and H. A. Lardy. A ³¹P-NMR study of the epididymis and epididymal sperm of the bovine and hamster. *Biol. Reprod.* 33: 1029–1035, 1985.

574. Snider, D. R., and E. D. Clegg. Alterations in phospholipids in porcine spermatozoa during in vivo uterus and oviduct incubation. *J. Anim. Sci.* 40: 269–273, 1975.

575. Sollner, T., S. W. Whiteheart, M. Brunner, H. Erdjument-Bromage, S. Geromanos, P. Tempst, and J. E. Rothman. SNAP receptors implicated in vesicle targeting and fusion. *Nature* 362: 318–324, 1993.

576. Songyang, Z., S. E. Shoelson, M. Chaudhuri, G. Gish, T. Pawson, W. G. Haser, F. King, T. Roberts, S. Ratnofsky, R. J. Lechleider, B. G. Neel, R. B. Birge, J. E. Fajardo, M. M. Chau, H. Hanafusa, B. Schaffhausen, and L. C. Cantley. SH2 domains recognize specific phosphopeptide sequences. *Cell* 72: 767–778, 1993.

577. Sorensen, R. A. Cinemicrography of mouse oocyte maturation utilizing Nomarski differential-interference microscopy. *Am. J. Anat.* 136: 265–276, 1973.

578. Sorensen, R. A., M. Cyert, and R. Pedersen. Active maturation-promoting factor is present in mature mouse oocytes. *J. Cell. Biol.* 100: 1637–1640, 1985.

579. Sorensen, R. A., and P. M. Wassarman. Relationship between growth and meiotic maturation of the mouse oocyte. *Dev. Biol.* 50: 531–536, 1976.

580. Speed, R. Meiosis in the foetal mouse ovary. 1. An analysis at the light microscope level using surface-spreading. *Chromosoma* 85: 427–437, 1982.

581. Spiteri-Grech, J., and E. Nieschlag. Paracrine factors relevant to the regulations of spermatogenesis- a review. *J. Reprod. Fertil.* 98: 1–14, 1993.

582. Spruce, A. E., L. Breckenridge, A. K. Lee, and W. Almers. Properties of the fusion pore that forms during exocytosis of a mast cell secretory vesicle. *Neuron* 4: 643–654, 1990.

583. Spruce, A. E., A. Iwata, and W. Almers. The first milliseconds of the pore formed by a fusogenic viral envelope protein during membrane fusion. *Proc. Natl. Acad. Sci. U.S.A.* 88: 3623–3627, 1991.

584. Spruce, A. E., A. Iwata, J. M. White, and W. Almers. Patch clamp studies of single cell-fusion events mediated by a viral fusion protein. *Nature* 342: 555–558, 1989.

585. Stafanini, M., C. Oura, and L. Zamboni. Ultrastructure of fertilization in the mouse. Penetration of the sperm into the ovum. *J. Submicrosc. Cytol.* 1: 1–23, 1969.

586. Steinberger, E., and A. Steinberger. Spermatogenic function of the testis. In: *Handbook of Physiology, Endocrinology, Male Reproductive System,* edited by D. W. Hamilton, and R. O. Greep. Baltimore: Am. Physiol. Soc., 1975, sect. 7, vol. V, p. 1–19.

587. Stengel, D., D. Henry, S. Tomova, A. Borsodi, and J. Hanoune. Purification of the proteolytically solubilized, active catalytic subunit of adenylate cyclase from ram sperm. Inhibition by adenosine. *Eur. J. Biochem.* 161: 241–247, 1986.

588. Stern, S., J. D. Biggers, and E. Anderson. Mitochondria and early development of the mouse. *J. Exp. Zool.* 176: 179–192, 1971.

589. Stevens, C. F. Quantal release of neurotransmitter and long-term potentiation. *Neuron* 10: 55–63, 1993.

590. Stice, S. L., and J. M. Robl. Activation of mammalian oocytes by a factor obtained from rabbit sperm. *Mol. Endocrinol.* 25: 272–280, 1990.

591. Storey, B. T. Sperm capacitation and the acrosome reaction. *Ann. NY Acad. Sci.* 637: 459–473, 1991.

592. Storey, B. T., C. L. Hourani, and J. B. Kim. A transient rise in intracellular Ca²⁺ is a precursor reaction to the zona pellucida-induced acrosome reaction in mouse sperm and is blocked by the induced acrosome reaction inhibitor 3-quinuclidinyl benzilate. *Mol. Reprod. Dev.* 32: 41–50, 1992.

593. Storey, B. T., M. A. Lee, C. Muller, C. R. Ward, and D. G. Wirtshafter. Binding of mouse spermatozoa to the zona pellucida of mouse eggs in cumulus: evidence that the acrosomes remain substantially intact. *Biol. Reprod.* 31: 1119–1128, 1984.

594. Strathmann, M., T. M. Wilkie, and M. I. Simon. Diversity of the G-protein family: sequences from five additional a subunits

in the mouse. *Proc. Natl. Acad. Sci. U.S.A.* 86: 7407–7409, 1989.

595. Strathmann, M., T. M. Wilkie, and M. I. Simon. Alternative splicing produces transcripts encoding two forms of the a subunit of GTP binding protein G_o. *Proc. Natl. Acad. Sci. U.S.A.* 87: 6477–6481, 1990.

596. Stroud, R. M., M. P. McCarthy, and M. Shuster. Nicotinic aectylcholine receptor superfamily of ligand-gated ion channels. *Biochemistry* 29: 11009–11023, 1990.

597. Struck, D. K., and W. J. Lennarz. The function of saccharide-lipids in synthesis of glycoproteins. In: *The Biochemistry of Glycoproteins and Proteoglycans,* edited by W. J. Lennarz. New York: Plenum, 1980, p. 35–84.

598. Suarez, S. S., D. P. Wolf, and S. Meizel. Induction of the acrosome reaction in human spermatozoa by a fraction of human follicular fluid. *Gamete Res.* 14: 107–121, 1986.

599. Sudhof, T. C., and R. Jahn. Proteins of synaptic vesicles involved in exocytosis and membrane recycling. *Neuron* 6: 665–677, 1991.

600. Summers, R. G., P. Talbot, E. M. Keough, B. L. Hylander, and L. E. Franklin. Ionophore A23187 induces acrosome reactions in sea urchin and guinea pig spermatozoa. *J. Exp. Zool.* 196: 381–385, 1976.

601. Swann, K. A cytosolic sperm factor stimulates repetitive calcium increases and mimics fertilization in hamster. *Development* 110: 1295–1302, 1990.

602. Sym, M., J. Engebrecht, and G. S. Roeder. ZIP1 is a synatonemal complex protein required for meiotic chromosomal synapsis. *Cell* 72: 365–378, 1993.

603. Szollosi, D. Changes of some cell organelles during oogenesis m mammals. In: *Oogenesis,* edited by J. Biggers, and A. Schuetz. Baltimore: University Park, 1972, p. 47–64.

604. Szollosi, D., P. Calarco, and R. Donahue. The nuclear envelope: Its breakdown and fate in mammalian oogonia and oocytes. *Anat. Rec.* 174: 325–340, 1972.

605. Szybek, K. *In vitro* maturation of oocytes from sexually immature mice. *J. Endocrinol.* 54: 527–528, 1972.

606. Talbot, P. Sperm penetration through oocyte investments in mammals. *Am. J. Anat.* 174: 331–346, 1985.

607. Talbot, P., G. Dicarlantonio, P. Zao, J. Penkala, and L. T. Haimo. Motile cells lacking hyaluronidase can penetrate the hamster oocyte cumulus complex. *Dev. Biol.* 108: 387–398, 1985.

608. Tang, S., and T. Beeler. Optical response of the indicator chlortetracycline to membrane potential. *Cell Calcium* 11: 425–429, 1991.

609. Tang, W.-J., and A. G. Gilman. Type-specific regulation of adenylyl cyclase by G protein βt subunits. *Science* 254: 1500–1503, 1991.

610. Tarentino, A. L., C. M. Gomez, and T. H. Plummer, Jr. Deglycosylation of asparagine-linked glycans by peptide: N-glycosidase F. *Biochemistry* 24: 4665–4671, 1985.

611. Tash, J. S., M. Krinks, J. Patel, R. L. Means, C. B. Klee, and A. R. Means. Identification, characterization, and functional correlation of calmodulin-dependent protein phosphatase in sperm. *J. Cell Biol.* 106: 1625–1633, 1988.

612. Tesarik, J., A. Carreras, and C. Mendoza. Differential sensitivity of progesterone- and zona pellucida-induced acrosome reactions to pertussis toxin. *Mol. Reprod. Dev.* 34: 183–189, 1993.

613. Thibault, C. Final stages of mammalian oocyte maturation. In: *Oogenesis,* edited by J. Biggers, and A. Schuetz. Baltimore: University Park, 1972, p. 397–411.

614. Thomas, P., and S. Meizel. An influx of extracellular calcium is required for initiation of the human sperm acrosome reaction induced by human follicular fluid. *Gamete Res.* 20: 397–411, 1988.

615. Thomas, P., and S. Meizel. Phosphatidylinositol 4,5-biphosphate hydrolysis in human sperm stimulated with follicular fluid or progesterone is dependent upon Ca^{2+} influx. *Biochem. J.* 264: 539–546, 1989.

616. Thomas, R. C. Bicarbonate and pH_i response. *Nature* 337: 601, 1989.

617. Thomas, T. S., W. L. Wilson, A. B. Reynolds, and G. Oliphant. Chemical and physical characterization of rabbit sperm acrosomal stabilizing factor. *Biol. Reprod.* 35: 691–700, 1986.

618. Tilney, L. G., D. Kiehart, C. Sardet, and M. Tilney. The polymerization of actin. IV . The role of Ca^{2+} and H^+ in the assembly of actin and in membrane fusion in the acrosomal reaction of echinoderm sperm. *J. Cell Biol.* 77: 536–550, 1978.

619. Trejo, R., and A. Mujica. Changes in calmodulin compartmentalization throughout capacitation and acrosome reaction in guinea pig spermatozoa. *Mol. Reprod. Dev.* 26: 366–376, 1990.

620. Triana, L. R., D. F. Babcock, S. P. Lorton, N. L. First, and H. A. Lardy. Release of acrosomal hyaluronidase follows increased membrane permeability to calcium in the presumptive capacitation sequence for spermatozoa of the bovine and other mammalian species. *Biol. Reprod.* 23: 47–59, 1980.

621. Triggle, D. J., and D. Rampe. 1,4-Dihydropyridine activators and antagonists: structural and functional distinctions. *TIPS* 10: 507–511, 1989.

622. Tsafriri, A. Oocyte maturation in mammals. In: *The Vertebrate Ovary,* edited by R. Jones. NY: Plenum, 1978, p. 409–442.

623. Tsafriri, A. The control of meiotic maturation in mammals. In: *Biology of Fertilization,* edited by C. Metz, and A. Monroy. New York: Academic, 1985, vol. 1, p. 221–252.

624. Tsafriri, A., and S. Bar-Ami. Role of divalent cations in the resumption of meiosis in rat oocytes. *J. Exp. Zool.* 205: 293–300, 1978.

625. Tsafriri, A., S. Bar-Ami, and H. Lindner. Control of the development of meiotic competence and of oocyte maturation in mammals. In: *Fertilization of the Human Egg In Vitro,* edited by H. Beier, and H. Lindner. Berlin: Springer-Verlag, 1983, p. 3–17.

626. Tsafriri, A., and C. Channing. An inhibitory influence of granulosa cells and follicular fluid upon porcine oocyte meiosis *in vitro. Endocrinology* 96: 992, 1975.

627. Tsafriri, A., C. Channing, S. Pomerantz, and H. Lindner. Inhibition of maturation of isolated rat oocytes by porcine follicular fluid. *J. Endocrinol.* 75: 258–291, 1977.

628. Tsafriri, A., N. Dekel, and S. Bar-Ami. The role of oocyte maturation inhibitor in follicular regulation of oocyte maturation. *J. Reprod. Fertil.* 64: 541–551, 1982.

629. Tsafriri, A., S. Pomerantz, and C. Channing. Inhibition of oocyte maturation by porcine follicular fluid: partial characterization of the inhibitor. *Biol. Reprod.* 14: 511–516, 1976.

630. Tsien, R. W. Fluorescent indicators of ion concentration. In: *Fluorescent Microscopy of Living Cells in Culture. Part B: Quantitative Fluorescence Microscopy—Imaging and Spectroscopy,* edited by D. L. Taylor, and Y.-L. Wang. New York: Academic, 1989, p. 127–156.

631. Tsien, R. W., P. T. Ellinor, and W. A. Horne. Molecular diversity of voltage-dependent Ca^{2+} channels. *TIPS* 12: 349–354, 1991.

632. Tsunoda, Y., and M. C. Chang. In vitro fertilization of hamster eggs by ejaculated or epididymal spermatozoa in the presence of male accessory secretions. *J. Exp. Zool.* 201: 445–450, 1977.

633. Ullrich, A., and J. Schlessinger. Signal transduction by receptors with tyrosine kinase activity. *Cell* 61: 203–212, 1990.

634. Unwin, N. Neurotransmitter action: opening of ligand-gated ion channels. *Neuron* 10: 31–41, 1993.

635. Urch, U. A. Biochemistry and function of acrosin. In: *Elements of Mammalian Fertilization. Vol. I. Basic Concepts*, edited by P. M. Wassarman. Boca Raton, FL: CRC, 1991, p. 233–248.

636. Urner, F., W. Herrmann, E. Baulieu, and S. Schorderet-Slatkine. Inhibition of denuded mouse oocyte maturation of forskolin, an activator of adenylate cyclase. *Endocrinology* 113: 1170–1172, 1983.

637. Van Blerkom, J., and R. McGaughey. Molecular differentiation of the rabbit ovum. II. During the preimplantation development of *in vivo* and *in vitro* matured oocytes. *Dev. Biol.* 63: 151–164, 1978.

638. Van Blerkom, J., and P. Motta. *The Cellular Basis of Mammalian Reproduction*. Baltimore: Urban & Schwarzenberg, 1979.

639. Vanha-Perttula, T., and J. Kasurinen. Purification and characterization of phosphatidylinositol-specific phospholipase C from bovine spermatozoa. *Int. J. Biochem.* 21: 997–1007, 1989.

640. Varner, D. D., C. R. Ward, B. T. Storey, and R. M. Kenney. Induction and characterization of acrosome reaction in equine sperm. *Am. J. Vet. Res.* 48: 1383–1389, 1987.

641. Vazquez, M. H., D. M. Phillips, and P. M. Wassarman. Interaction of mouse sperm with purified sperm receptors covalently linked to silica beads. *J. Cell Sci.* 92: 713–722, 1989.

642. Vernon, R. B., M. S. Hamilton, and E. M. Eddy. Effect of in vivo and in vitro environments on the expression of a surface antigen of the mouse sperm tail. *Biol. Reprod.* 99: 404–411, 1973.

643. Von Heijne, G. A new method for predicting signal sequence cleavage sites. *Nucleic Acids Res.* 14: 4683–4690, 1986.

644. Wagner, S., J. Freudenstein, and K. H. Scheit. Characterization by cDNA cloning of the mRNA for seminalplasmin, the major basic protein of bull semen. *DNA Cell Biol.* 9: 437–442, 1990.

645. Wallace, B. A. Gramicidin channels and pores. *Annu. Rev. Biophys. Biophys. Chem.* 19: 127–157, 1990.

646. Ward, C. R., and B. T. Storey. Determination of the time course of capacitation in mouse spermatozoa using a chlortetracycline fluorescence assay. *Dev. Biol.* 104: 287–296, 1984.

647. Ward, C. R., B. T. Storey, and G. S. Kopf. Activation of a G_i protein in mouse sperm membranes by solubilized proteins of the zona pellucida, the egg's extracellular matrix. *J. Biol. Chem.* 267: 14061–14067, 1992.

648. Wasco, W. M., R. L. Kincaid, and G. A. Orr. Identification and characterization of calmodulin-binding proteins in mammalian sperm flagella. *J. Biol. Chem.* 264: 5104–5111, 1989.

649. Wassarman, P. M. Oogenesis: synthetic events in the developing mammalian egg. In: *Mechanism and Control of Animal Fertilization*, edited by J. Hartmann. New York: Academic, 1983, p. 1–54.

650. Wassarman, P. M. Early events in mammalian fertilization. *Annu. Rev. Cell Biol.* 3: 109–142, 1987.

651. Wassarman, P. M. The biology and chemistry of fertilization. *Science* 235: 553–560, 1987.

652. Wassarman, P. M. Zona pellucida glycoproteins. *Annu. Rev. Biochem.* 57: 415–442, 1988.

653. Wassarman, P. M. The mammalian ovum. In: *The Physiology of Reproduction*, edited by E. Knobil, and J. Neill. New York: Raven Press, 1988, p. 69–102.

654. Wassarman, P. M. Role of carbohydrates in receptor-mediated fertilization in mammals. In: *Carbohydrate Recognition in Cellular Function*, edited by G. Bock, and S. Harnett, New York: Wiley, 1989, p. 135–155.

655. Wassarman, P. M. Profile of a mammalian sperm receptor. *Development* 108: 1–17, 1990.

656. Wassarman, P. M. Cell surface carbohydrates and mammalian fertilization. In: *Cell Surface Carbohydrates and Cell Development*, edited by M. Fukuda. Boca Raton, FL: CRC, 1991, p. 215–238.

657. Wassarman, P. M. (Ed). *Elements of Mammalian Fertilization*. Boca Raton, FL: CRC, 1991, vol. 1 p. 310, vol. 2 p. 183.

658. Wassarman, P. M. Mammalian fertilization: sperm receptor genes and glycoproteins. *Adv. Dev. Biochem.* 2: 155–195, 1993.

659. Wassarman, P. M., J. D. Bleil, H. M. Florman, J. M. Greve, R. J. Roller, G. S. Salzmann, and F. G. Samuels. The mouse egg's receptor for sperm: what is it and how does it work? *Cold Spring Harb. Symp. Quant. Biol.* 50: 11–19, 1985.

660. Wassarman, P. M., and M. L. Depamphilis (Eds). *Guide to Techniques in Mouse Development, Methods in Enzymology*. Orlando, FL: Academic, 1993, vol. 225.

661. Wassarman, P. M., and K. Fujiwara. Immunofluorescent anti-tubulin staining of spindles during meiotic maturation of mouse oocytes *in vitro*. *J. Cell Sci.* 29: 171–188, 1978.

662. Wassarman, P. M., and W. Josefowicz. Oocyte development in the mouse: an ultrastructural comparison of oocytes isolated at various stages of growth and meiotic competence. *J. Morphol.* 156: 209–235, 1978.

663. Wassarman, P. M., W. Josefowicz, and G. Letourneau. Meiotic maturation of mouse oocytes *in vitro*: inhibition of maturation at specific stages of nuclear progression. *J. Cell Sci.* 22: 531–545, 1976.

664. Wassarman, P. M., and S. Mortillo. Structure of the mouse egg extracellular coat, the zona pellucida. *Int. Rev. Cytol.* 130: 85–110, 1991.

665. Wassarman, P. M., R. Schultz, G. Letourneau, M. Lamarca, and J. Bleil. Meiotic maturation of mouse oocytes *in vitro*. In: *Ovarian Follicular and Corpus Luteum Function*, edited by C. Channing, J. Marsh, and W. Sadler. New York: Plenum, 1979, p. 251–268.

666. Weil, A. J., and J. M. Rodenburg. The seminal vesicle as the source of the spermatozoa coating antigen of seminal plasma. *Proc. Soc. Exp. Biol. Med.* 109: 567–570, 1962.

667. Weiss, A. T cell antigen receptor signal transduction: a tale of tails and cytoplasmic protein-tyrosine kinases. *Cell* 73: 209–212, 1993.

668. Whitaker, M. and K. Swann. Lighting the fuse at fertilization. *Development* 117: 1–12, 1993.

669. White, D. R., and R. J. Aitken. Relationship between calcium, cyclic AMP, ATP, and intracellular pH and the capacity of hamster spermatozoa to express hyperactivated motility. *Gamete Res.* 22: 163–177, 1989.

670. White, J. M. Viral and cellular membrane fusion proteins. *Annu. Rev. Physiol.* 52: 675–697, 1990.

671. White, J. M. Membrane fusion. *Science* 258: 917–924, 1992.

672. Wickramasinghe, D., and D. F. Albertini. Centrosome phosphorylation and the developmental expression of meiotic competence in mouse oocytes. *Dev. Biol.* 152: 62–74, 1992.

673. Wickramasinghe, D., K. M. Ebert, and D. F. Albertini. Meiotic competence acquisition is associated with the appearance of M-phase characteristics in growing mouse oocytes. *Dev. Biol.* 143: 162–172, 1991.

674. Wilde, M. W., C. R. Ward, and G. S. Kopf. Activation of a G protein in mouse sperm by the zona pellucida, an egg-

associated extracellular matrix. *Mol. Reprod. Dev.* 31: 297–306, 1992.

675. Williams, C. J., R. M. Schultz, and G. S. Kopf. Role of G proteins in mouse egg activation: stimulatory effects of acetylcholine on the ZP2 to ZP2$_f$ conversion and pronuclear formation in eggs expressing a functional m1 muscarinic receptor. *Dev. Biol.* 151: 288–296, 1992.

676. Wilson, D. W., S. W. Whiteheart, L. Orci, and J. E. Rothman. Intracellular membrane fusion. *Trends Biochem. Sci.* 16: 334–337, 1991.

677. Wilson, E. *The Cell in Development and Heredity.* New York: Macmillan, 1925.

678. Wilson, I.B.H., Y. Gavel, and G. Von Heijne. Amino acid distributions around O-linked glycosylation sites. *Biochem. J.* 275: 529–534, 1991.

679. Wilson, W. L., and G. Oliphant. Isolation and biochemical characterization of the subunits of the rabbit sperm acrosome stabilizing factor. *Biol. Reprod.* 37: 159–169, 1987.

680. Wolf, D. E., S. S. Hagopian, and S. Isojima. Changes in sperm plasma membrane lipid diffusibility after hyperactivation during in vitro capacitation in the mouse. *J. Cell Biol.* 102: 1372–1380, 1986.

681. Wolf, D. E., S. S. Hagopian, R. G. Lewis, J. K. Voglmayr, and G. Fairbanks. Lateral regionalization and diffusion of a maturation-dependent antigen in the ram sperm plasma membrane. *J. Cell Biol.* 102: 1826–1831, 1986.

682. Wolf, D. E., A. C. Lipscomb, and V. M. Maynard. Causes of nondiffusing lipid in the plasma membrane of mammalian spermatozoa. *Biochemistry* 27: 860–865, 1988.

683. Wolf, D. E., V. M. Maynard, C. A., McKinnon, and D. L. Melchior. Lipid domains in the ram sperm plasma membrane demonstrated by differential scanning calorimetry. *Proc. Natl. Acad. Sci. U.S.A.* 87: 6893–6896, 1990.

684. Wolf, D. E., B. K. Scott, and C. F. Millette. The development of regionalized lipid diffusibility in the germ cell plasma membrane during spermatogenesis in the mouse. *J. Cell Biol.* 103: 1745–1750, 1986.

685. Wolf, D. E., and J. K. Voglmayr. Diffusion and regionalization in membranes of maturing ram spermatozoa. *J. Cell Biol.* 98: 1678–1684, 1984.

686. Wolfes, H., K. Kogawa, C. F. Millette, and G. M. Cooper. Specific expression of nuclear proto-oncogenes before entry into meiotic prophase of spermatogenesis. *Nature* 245: 740–743, 1989.

687. Wolgemuth, D., G. Jagiello, and A. Henderson. Baboon late diplotene oocytes contain micronucleoli and a low level of extra rDNA templates. *Dev. Biol.* 78: 598–604, 1980.

688. Wong, P.Y.D., and W. M. Lee. Potassium movement during sodium-induced motility initiation in the rat cauda epididymal spermatozoa. *Biol. Reprod.* 28: 206–212, 1983.

689. Wong, P.Y.D., W. M. Lee, and A.Y.S. Tsang. The effects of extracellular sodium on acid release and motility initiation in rat caudal epididymal spermatozoa *in vitro. Exp. Cell Res.* 131: 97–103, 1981.

690. Wooding, F.B.P. Studies on the mechanism of the Hyamine-induced acrosome reaction in ejaculated bovine spermatozoa. *J. Reprod. Fertil.* 44: 185–195, 1975.

691. Working, P. K., and S. Meizel. Evidence that an ATPase functions in the maintenance of the acidic pH of the hamster sperm acrosome. *J. Biol. Chem.* 256: 4708–4711, 1981.

692. Working, P. K., and S. Meizel. Preliminary characterization of a Mg^{2+}-ATPase in hamster sperm head membranes. *Biochem. Biophys. Res. Commun.* 104: 1060–1065, 1982.

693. Working, P. K., and S. Meizel. Correlation of increased intra-acrosomal pH with the hamster sperm acrosome reaction. *J. Exp. Zool.* 227: 97–107, 1983.

694. Wylie, C. (Ed). Germ cell development. *Semin. Dev. Biol.* 4: 1993.

695. Yanagimachi, R. Acceleration of the acrosome reaction and activation of guinea pig spermatozoa by detergents and other reagents. *Biol. Reprod.* 13: 519–526, 1975.

696. Yanagimachi, R. The specificity of sperm-egg interaction. In: *Immunobiology of Gametes,* edited by M. Edinin, and M. H. Johnson. Cambridge: Cambridge University, 1977, p. 255–295.

697. Yanagimachi, R. Mechanisms of fertilization in mammals. In: *Fertilization and Embryonic Development In Vitro,* edited by L. Mastroianni, and J. D. Biggers. New York: Plenum, 1981, p. 81–182.

698. Yanagimachi, R. Zona-free hamster eggs: their use in assessing fertilizing capacity and examining chromosomes of human spermatozoa. *Gamete Res.* 10: 187–232, 1984.

699. Yanagimachi, R. Mammalian fertilization. In: *The Physiology of Reproduction,* edited by E. Knobil, and J. Neill. New York: Raven, 1988, p. 135–185.

700. Yanagimachi, R., and A. Bhattacharyya. Acrosome-reacted guinea pig spermatozoa become fusion competent in the presence of extracellular potassium ions. *J. Exp. Zool.* 248: 354–360, 1988.

701. Yanagimachi, R., and Y. D. Noda. Electron microscopic studies of sperm incorporation into golden hamster eggs. *Am. J. Anat.* 128: 429–462, 1970.

702. Yanagimachi, R., and Y. D. Noda. Physiological changes in the postnuclear cap region of mammalian spermatozoa: a necessary preliminary in the membrane fusion between sperm and egg cells. *J. Ultrastr. Res.* 31: 486–493, 1970.

703. Yanagimachi, R., and D. M. Phillips. The status of acrosomal caps of hamster spermatozoa immediately before fertilization in vivo. *Gamete Res.* 9: 1–19, 1984.

704. Yanagimachi, R., and N. Usui. Calcium dependence of the acrosome reaction and activation of guinea pig spermatozoa. *Exp. Cell Res.* 89: 161–174, 1974.

705. Young, R. J., and J. C. Laing. The binding characteristics of cholinergic sites in rabbit spermatozoa. *Mol. Reprod. Dev.* 28: 55–61, 1991.

706. Yudin, A. I., W. Gottlieb, and S. Meizel. Ultrastructural studies of the early events of the human sperm acrosome reaction as initiated by human follicular fluid. *Gamete Res.* 20: 11–24, 1988.

707. Yurewicz, E. C., B. A. Pack, and A. G. Sacco. Isolation, composition, and biological activity of sugar chains of porcine oocyte zona pellucida 55K glycoproteins. *Mol. Reprod. Dev.* 30: 126–134, 1991.

708. Yurewicz, E. C., A. G. Sacco, and M. G. Subramanian. Structural characterization of the M_r = 55,000 antigen (ZP3) of porcine oocyte zona pellucida. *J. Biol. Chem.* 262: 564–571, 1987.

709. Zamboni, L. Ultrastructure of mammalian oocytes and ova. *Biol. Reprod. (Suppl.)* 2: 44–63, 1970.

710. Zamboni, L. Comparative studies on the ultrastructure of mammalian oocytes. In: *Oogenesis,* edited by J. Biggers, and A. Schuetz, Baltimore: University Park, 1972, p. 5–46.

711. Zamboni, L. Fertilization in the mouse. In: *Biology of fertilization and implantation,* edited by K. S. Moghissi, and E.S.E. Hafez. Springfield, IL: Charles C Thomas, 1972, p. 213–262.

712. Zao, P.Z.R., S. Meizel, and P. Talbot. Release of hyaluronidase

and β-N-acetylhexosaminidase during in vitro incubation of hamster sperm. *J. Exp. Zool.* 234: 63–74, 1985.

713. Zhou, R., B. Shi, K.C.K. Chou, M. D. Oswalt, and A. Haug. Changes in intracellular calcium of porcine sperm during in vitro incubation with seminal plasma and a capacitating medium. *Biochem. Biophys. Res. Commun.* 172: 47–53, 1990.

714. Zimmerberg, J., M. Curran, F. S. Cohen, and M. Brodwick. Simultaneous electrical and optical measurements show that membrane fusion precedes secretory granule swelling during exocytosis of beige mouse mast cells. *Proc. Natl. Acad. Sci. U.S.A.* 84: 1585–1589, 1987.

715. Zuckerman, S. (Ed). *The Ovary.* New York: Academic, 1962, vol. 1.

716. Zuckerman, S., and T. Baker. The development of the ovary and the process of oogenesis. In: *The Ovary,* edited by S. Zuckerman, and B. Weir. New York: Academic, 1977, vol. 1, p. 42–68.

717. Zuckerman, S., and B. Weir (Eds). *The Ovary.* New York: Academic, 1977.

Index